# Table of Atomic Masses

| Element | Symbol | Atomic Number | Atomic Mass | Element | Symbol | Atomic Number | Atomic Mass |
|---|---|---|---|---|---|---|---|
| Actinium | Ac | 89 | 227.0278 | Mercury | Hg | 80 | 200.59(2) |
| Aluminum | Al | 13 | 26.981539(5) | Molybdenum | Mo | 42 | 95.94(1) |
| Americium | Am | 95 | 243.0614 | Neodymium | Nd | 60 | 144.24(3) |
| Antimony | Sb | 51 | 121.757(3) | Neon | Ne | 10 | 20.179(6) |
| Argon | Ar | 18 | 39.948(1) | Neptunium | Np | 93 | 237.0482 |
| Arsenic | As | 33 | 74.92159(2) | Nickel | Ni | 28 | 58.6934(2) |
| Astatine | At | 85 | 209.9871 | Niobium | Nb | 41 | 92.90638(2) |
| Barium | Ba | 56 | 137.327(7) | Nitrogen | N | 7 | 14.00674(7) |
| Berkelium | Bk | 97 | 247.0703 | Nobelium | No | 102 | 259.1009 |
| Beryllium | Be | 4 | 9.012182(3) | Osmium | Os | 76 | 190.2(1) |
| Bismuth | Bi | 83 | 208.98037(3) | Oxygen | O | 8 | 15.9994(3) |
| Bohrium | Bh | 107 | 262.12 | Palladium | Pd | 46 | 106.42(1) |
| Boron | B | 5 | 10.811(5) | Phosphorus | P | 15 | 30.973762(4) |
| Bromine | Br | 35 | 79.904(1) | Platinum | Pt | 78 | 195.08(3) |
| Cadmium | Cd | 48 | 112.411(8) | Plutonium | Pu | 94 | 244.0642 |
| Calcium | Ca | 20 | 40.078(4) | Polonium | Po | 84 | 208.9824 |
| Californium | Cf | 98 | 251.0796 | Potassium | K | 19 | 39.0983(1) |
| Carbon | C | 6 | 12.011(1) | Praseodymium | Pr | 59 | 140.90765(3) |
| Cerium | Ce | 58 | 140.115(4) | Promethium | Pm | 61 | 144.9127 |
| Cesium | Cs | 55 | 132.90543(5) | Protactinium | Pa | 91 | 231.03588(2) |
| Chlorine | Cl | 17 | 35.4527(9) | Radium | Ra | 88 | 226.0254 |
| Chromium | Cr | 24 | 51.9961(6) | Radon | Rn | 86 | 222.0176 |
| Cobalt | Co | 27 | 58.93320(1) | Rhenium | Re | 75 | 186.207(1) |
| Copper | Cu | 29 | 63.546(3) | Rhodium | Rh | 45 | 102.90550(3) |
| Curium | Cm | 96 | 247.0703 | Rubidium | Rb | 37 | 85.4678(3) |
| Dubnium | Db | 105 | 262.114 | Ruthenium | Ru | 44 | 101.07(2) |
| Dysprosium | Dy | 66 | 162.50(3) | Rutherfordium | Rf | 104 | 261.11 |
| Einsteinium | Es | 99 | 252.083 | Samarium | Sm | 62 | 150.36(3) |
| Erbium | Er | 68 | 167.26(3) | Scandium | Sc | 21 | 44.955910(9) |
| Europium | Eu | 63 | 151.965(9) | Seaborgium | Sg | 106 | 263.118 |
| Fermium | Fm | 100 | 257.0951 | Selenium | Se | 34 | 78.96(3) |
| Fluorine | F | 9 | 18.9984032.9 | Silicon | Si | 14 | 28.0855(3) |
| Francium | Fr | 87 | 233.0197 | Silver | Ag | 47 | 107.8682(2) |
| Gadolinium | Gd | 64 | 157.25(3) | Sodium | Na | 11 | 22.989768(6) |
| Gallium | Ga | 31 | 69.723(4) | Strontium | Sr | 38 | 87.62(1) |
| Germanium | Ge | 32 | 72.61(2) | Sulfur | S | 16 | 32.066(6) |
| Gold | Au | 79 | 196.96654(3) | Tantalum | Ta | 73 | 180.9479(1) |
| Hafnium | Hf | 72 | 178.49(2) | Technetium | Tc | 43 | 98.9072 |
| Hassium | Hs | 108 | (265) | Tellurium | Te | 52 | 127.60(3) |
| Helium | He | 2 | 4.002602(2) | Terbium | Tb | 65 | 158.92534(3) |
| Holmium | Ho | 67 | 164.93032(3) | Thallium | Tl | 81 | 204.3833(2) |
| Hydrogen | H | 1 | 1.00794(7) | Thorium | Th | 90 | 232.0381(1) |
| Indium | In | 49 | 114.82(1) | Thulium | Tm | 69 | 168.9342(3) |
| Iodine | I | 53 | 126.90447(3) | Tin | Sn | 50 | 118.710(7) |
| Iridium | Ir | 77 | 192.22(3) | Titanium | Ti | 22 | 47.88(3) |
| Iron | Fe | 26 | 55.847(3) | Tungsten | W | 74 | 183.85(3) |
| Krypton | Kr | 36 | 83.80(1) | Ununbium[a] | Uub | 112 | (277) |
| Lanthanum | La | 57 | 138.9055(2) | Ununnilium[a] | Uun | 110 | (269) |
| Lawrencium | Lr | 103 | 262.11 | Unununium[a] | Uuu | 111 | (272) |
| Lead | Pb | 82 | 207.2(1) | Uranium | U | 92 | 238.0289(1) |
| Lithium | Li | 3 | 6.941(2) | Vanadium | V | 23 | 50.9415(1) |
| Lutetium | Lu | 71 | 174.967(1) | Xenon | Xe | 54 | 131.29(2) |
| Magnesium | Mg | 12 | 24.3050(6) | Ytterbium | Yb | 70 | 173.04(3) |
| Manganese | Mn | 25 | 54.93805(1) | Yttrium | Y | 39 | 88.90585(2) |
| Meitnerium | Mt | 109 | (266) | Zinc | Zn | 30 | 65.39(2) |
| Mendelevium | Md | 101 | 258.10 | Zirconium | Zr | 40 | 91.224(2) |

[a]These unnil-names are temporary. Official names and symbols for these elements have not been agreed upon.

# BASIC CONCEPTS
# OF CHEMISTRY

Seventh Edition

## Leo J. Malone

**Saint Louis University**

WILEY

**John Wiley & Sons, Inc.**

**Acquisitions Editor**   *Deborah Brennan*

**Project Editor**   *Jennifer Yee*

**Marketing Manager**   *Robert Smith*

**Senior Production Editor**   *Valerie A.Vargas*

**Senior Designer**   *Karin Gerdes Kincheloe*

**Photo Editor**   *Hilary Newman*

**Illustration Editor**   *Anna Melhorn*

**Cover Photo**   *David Fleetham/Taxi/Getty Images*

This book was set in 10.5/12.6 Times Roman by Techbooks and printed and bound by Von Hoffmann. The cover was printed by Von Hoffmann.

This book is printed on acid free paper. ∞

To order books or for customer service please, call 1(800)-CALL-WILEY (225-5945).

ISBN 0-471-21522-8

Printed in the United States of America

10 9 8 7 6 5 4

# Preface

*Basic Concepts of Chemistry* began life in 1981 as a text directed at students in a preparatory course. That is, it was intended for those planning to proceed to a main sequence chemistry course but who had little or no background in chemistry. Although this continues to be the mission of this text, the acceptance of previous editions indicates that it has a much broader mission. In addition to preparatory chemistry courses, it has found extensive use in one semester, general-purpose courses. In such courses a mix of students may be present. For some of these students a main sequence in chemistry may follow, but for others the course precedes a semester of organic biochemistry to satisfy allied health professions requirements. Still others are there to satisfy a science requirement. The Seventh Edition has been written with this broader appeal in mind. More topics are included than would normally be covered in a one-semester preparatory course for other applications of this text. Still, this text does not cringe from or apologize for the quantitative nature of real chemistry. Thus, it will continue to emphasize the needs of those intending to continue on in chemistry and other sciences.

The most significant change in recent editions concerns an increased focus on reaction chemistry. The chemical equation and chemical reactions are given a centered and detailed discussion in a separate chapter. Chapter 7 on chemical reactions includes descriptions of the activity series in single-replacement reactions, precipitation reactions including the solubility table, and acid–base neutralization reactions. These are illustrated with net ionic equations as well as molecular equations. The mole and stoichiometry are covered in one chapter (Chapter 8) so that this discussion is not interrupted by discussion of chemical equations and types of reactions. To better emphasize these reactions many more blowout diagrams illustrating the molecular nature of chemistry are included in this edition.

The Seventh Edition puts an increased emphasis on flexibility. In this edition, the chapters have been grouped together in *Units*. The Units can be covered in several different orders. Even within the units, changes in order of presentation are possible. There is little agreement among instructors as to the exact order of presentation on a beginning level. The solution for the textbook author is to make the text as flexible as possible. However, some topics must be covered before others or the result is confusion. It is not practical to try to make each chapter stand alone. Perhaps the biggest division among instructors is whether to cover atomic structure and bonding before the quantitative aspects of the mole and stoichiometry or vice versa. It is important to point out that Unit 3 and Unit 4 were written independently so that either may follow Unit 2. Unit 3 and 4 do not use material discussed in the other Unit.

Although each chapter is continuous, the chapter is broken into two or three parts with a unifying topic labeled as *Sections*. Each *Section* ends with a short summary titled **Looking Back**, and review questions on the Section titled **Checking it Out**.

In preparing the Seventh Edition, the author and publisher obtained considerable input about how chemistry should be presented and what is needed by instructors at this level. As a result of our research, we have built on the strengths of previous editions to present what may be described as a "learning text." It is meant to be of maximum aid to the many dedicated teachers who believe that chemistry is a fascinating subject that serves as the basis of all sciences and technologies as well as to their students who have little or no background in chemistry.

There are four major goals in this text. A discussion of these goals follows.

## Goals and Features of the Seventh Edition

The Seventh Edition continues and expands upon some of the pedagogical elements introduced in the earlier Editions. Called "navigational devices," these learning aids help guide students in understanding basic chemical concepts.

### Goal I. To Introduce Chemistry as a Live, Relevant Science.

Presenting subject matter in a clear and readable style and a logical, smoothly flowing sequence helps achieve this goal.

*A Friendly, Nonintimidating Writing Style.* The discussions in this text take nothing for granted. It is assumed that this is the first contact the students will have had with almost all concepts; thus a careful discussion is required. For example, the distinctions between atoms, molecules, and moles of molecules are reinforced with helpful illustrations. This edition continues and expands on the use of simple, understandable analogies that relate abstract concepts to concrete models. For example, in Chapter 5 the placement of electrons in orbitals is compared to the placement of students in dorm rooms. When possible, current topics of interest are used as examples. A case in point is found in Chapter 14 where chlorine is used as an example of a catalyst in a reaction that destroys ozone.

*A Logical Sequence of Topics.* The sequence of chapters in the text allows a logical step-by-step development of the science. Unit 1 contains The Prologue which covers the scientific method and the study of chemistry. Chapter 1 proceeds directly to the measurements used in chemistry. Unit 2 contains Chapters 2, 3, and 4 which follow each other smoothly. The early introduction to the periodic table and chemical nomenclature in Chapter 4 has been well received. Many feel that students benefit from an early introduction to the language of chemistry. After Unit 2, Unit 3 or Unit 4 follows. The two sequences are equally logical. There is certainly flexibility within Units. Types of chemical reactions, discussed in Chapter 7, may be omitted and a class may proceed from balanced chemical equations directly to Chapter 8. There have been some changes in the order of presentation of topics in several chapters in this edition that are intended for a smoother flow.

In the Seventh Edition, Chapter 16 (Organic Chemistry) is now available at the website. A new chapter on Biochemistry has been added (Chapter 17) and is also included on the Web. These two chapters can be downloaded for those wishing to include these topics.

*Smooth Transitions Between Topics.* Sometimes, one hears the complaint that chemistry is difficult to learn because it is a collection of unrelated facts and concepts. We have addressed that concern in two ways. First, each chapter begins with an introductory section, labeled **Setting the Stage,** that sets up the purpose of the chapter by reference to current topics—such as results from the Galileo probe to Jupiter (Chapter 2), global warming (Chapter 7), or the nature of lightning (Chapter 13). Within the chapters, we have made a major effort to smooth the flow between topics by connecting each section to the next. A brief paragraph begins each topic heading (labeled **Looking Ahead**) which lets the student know what comes next, why, and how it relates to the previous topic.

### Goal II. To Encourage not only Learning but Critical Thinking.

Some effective ways of learning chemistry are by summarizing and recitation, critical questioning, and working problems.

***Frequent Summarizing and Recitation.*** The learning of chemistry requires reflection. Short summaries of two or three closely related topics in each *Section* in a chapter (labeled **Looking Back**) are included. The object is to encourage the student to pause, reflect, and mentally gather in the main points before proceeding to recitation and problem reinforcement. As in previous editions, the chapter concludes with a review that summarizes the whole chapter and is labeled as **Putting It Together.** This summary is often presented in a unique way by use of tables, diagrams, or flow charts. Key terms defined in the chapter are now used in the context of a discussion and are shown in bold type. In addition, a complete Glossary of Terms appears in Appendix F for easy reference. After the paragraph labeled **Looking Back,** a section called *Learning Check* follows—which is referred to as **Checking It Out.** The first part of the *Learning Check* summarizes recent topics with key words left out. By filling in the blanks, the student aids learning by recitation.

***Critical Questioning.*** The introductory section (**Setting the Stage**) ends with a series of questions that are to be addressed in the chapter that follows. These questions are labeled **Formulating Some Questions."**

***Working Problems.*** Real chemistry requires problem-solving skills. This text helps develop these skills in four steps.

1. Example problems (labeled **Working It Out**) are carefully worked out in an easy-to-follow manner. Each step in a solution is explained or diagrammed. Most problems have a section labeled *Procedure* in which a strategy is outlined, and a *Solution* where the strategy is carried out. Example problems have descriptive headings.

2. Shortly after presentation of an example problem, the concept is reinforced with similar problems provided in the *Learning Check*. Worked-out answers to these problems in the *Learning Check* are provided at the end of the chapter for easy reference. A few relevant chapter-end problems are referenced in the *Learning Check* for additional practice.

3. Numerous chapter-end problems are listed by topic and range from the simplest to more complex applications. The hardest are indicated by an asterisk. About 60 percent of the answers are provided in Appendix G. Many of the quantitative problems also include worked-out solutions.

   In addition to the chapter-end problems that support a specific topic or concept, a section of uncategorized problems titled *General Questions* is included in which students must decide for themselves which concepts are being tested. Concepts from previous chapters may also be needed for these more challenging problems. Many of these problems require the application of several concepts. Over 100 new problems have been added to the Seventh Edition.

4. Additional problems are provided with the **Interactive Learning** problems at the end of many chapters. The problems are found on the website and contain an interactive tutorial that clearly shows how to set up and solve problems, and provides

feedback. The interactive problems include topics such as conversion of units (Chapter 1), density (Chapter 2), nomenclature (Chapter 4), Lewis structures (Chapter 6), balancing equations (Chapter 7), and the quantitative problems found in Chapters 8, 9, and 11.

5. *Comprehensive Review Tests* are available that integrate the content of two or three relevant chapters. These tests are found after Chapters 4, 6, 8, 11, and 14. They include multiple-choice questions that help test the material learned in recitation.

### Goal III. To Appreciate the Relevance of Chemistry to Our Everyday Life.

The newest feature of this text is the topics labeled **Making It Real**. There are two short topics in each chapter that relate to the subject at hand. For example, when we introduce matter and energy in Chapter 2, we have a short essay on the existence of the exotic "dark matter" and "dark energy." When we introduce the elements in Chapter 3, we have an essay on how the element iridium was the clue to the demise of the dinosaurs. In Chapter 6 on bonding, there is a discussion on how difficult it is to break the triple bond in nitrogen so as to form proteins. These are intended, not only to demonstrate the applications of what has been learned, but to generate excitement. In addition to these topics, many other applications are discussed within the text.

### Goal IV. To Understand and Provide help with the Math Anxieties That Can Distract Students of Chemistry.

The greatest fear of many of the students taking chemistry for the first time is the math involved. This text has been and remains sympathetic and encouraging to those students. It is understood that many, if not most, students require some preparation or review of mathematical concepts used in introductory chemistry. Extensive end-of-book appendixes (A–E) supplement Chapter 1 on measurements. These appendixes provide not only discussion but also worked-out examples and sample problems (with answers provided) in the areas of basic arithmetic, algebra, and scientific notation (including logarithms). In addition, there are separate appendixes on graphs and calculators. These also provide sample problems to test one's understanding. Students have used these appendixes extensively, and they have helped reduce what can become an exaggerated fear of the math involved in this discipline. An acknowledgment of this difficulty and the extensive help provided are unique to this text.

Major efforts have been made throughout the book to ensure the needed math is meaningful. For example, the use of percent is illustrated in some problems concerning the composition of alloys in Chapter 2. Solubility equilibria in Chapter 14 is illustrated by the formation of kidney stones. There are numerous other examples of making math relevant.

## Supplements

*Study Guide/Solutions Manual* is available to accompany this text. In the Study Guide/Solutions Manual, the same topics in a specific section are also grouped in the same manner for review, discussion, and testing. In this manner, the Study Guide/ Solutions Manual can be put to use before the chapter is completed. The Study Guide/Solutions Manual contains worked-out solutions to appropriate problems in the text. The solutions to the quantitative problems in black lettering are included.

*E-grade* is an on-line quizzing and homework management program that allows students to do practice tests and email homework assignments directly to the professor.

*Instructor's Manual and Test Bank,* by Leo J. Malone, St. Louis University and Kyle Beran, University of Texas-Permian Basin, includes chapter objectives, teaching hints, solutions to all text problems, and a test bank.

*Experiments in Basic Chemistry,* by Steven Murov and Brian Stedjee, Modesto Junior College, contains 26 experiments that parallel text organization and provides learning objectives, discussion sections outlining each experiment, easy-to-follow procedures, post-lab questions, additional exercises, and answers to pre-lab questions.

*Instructor's Manual for Experiments in Basic Chemistry,* written by the lab manual authors, contains answers to post-lab questions, lists of chemicals needed, suggestions for other experiments, as well as suggestions for experiment set-ups.

*CD-ROM Presentation Manager,* containing text illustrations on CD-ROM for lecture presentation.

# Acknowledgments

Writing a chemistry text is certainly not an individual project. In particular, I wish to thank my colleagues in the chemistry department at Saint Louis University for their helpful comments in the various editions of this text. I owe special thanks to Dr. Judith Durham, who was particularly helpful. I also wish to thank my wife Meg who demonstrated understanding and gave me encouragement during the preparation of the Seventh Edition. I also wish to acknowledge the support of my children and their spouses Chris and Lisa, Brian and Mary, Rob and Katie, and Bill. I also wish to thank the many people at John Wiley who helped in this project, especially Debbie Brennan. Finally, the following people reviewed the manuscript and offered many useful comments and suggestions:

| | |
|---|---|
| **Alhajie Dumbuya** | *Austin Community College-Northridge* |
| **Roberta M. Eddy** | *Indiana University of Pennsylvania* |
| **Michael Finnegan** | *Washington State University* |
| **Stan Grenda** | *University of Nevada-Las Vegas* |
| **Mark Morvant** | *Texas A&M University-Corpus Christi* |
| **Richard Musgrave** | *St. Petersburg College* |
| **Robert W. Shaw** | *Texas Tech University* |
| **Susan E. Swope** | *Pennsylvania State University-University Park* |

**Leo J. Malone**
*St. Louis*

# Contents

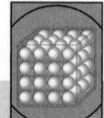

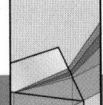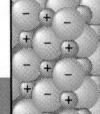

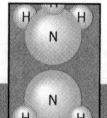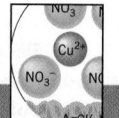

\*Unit 4 may also follow Unit 2.

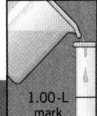 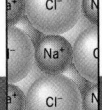 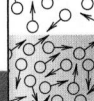

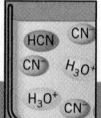  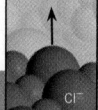

# Introduction to the Study of Chemistry

**Fire is awesome and powerful. Out of control, its destructive force can be disastrous.**

## A. The Mystery of Fire

Who hasn't been mesmerized by the dancing flames in a fireplace or campfire? Fire is such a mysterious and powerful force of nature. Out of control, it causes destruction, pain, and even death. For these reasons, we fear it. Under control, however, it heats our homes, cooks our meals, and transforms ores into metals. For these reasons, we welcome it. We can understand how this awesome force was the source of so much fascination throughout the ages. For example, in Greek mythology there existed a god named Prometheus. He gave animals the special tools they needed to survive in a hostile world. Humans received the gift of fire. Legend has it that Prometheus went too far because this was a tool reserved for the gods themselves. As might be expected, the other gods proceeded to punish him for his unacceptable generosity.

Evidence indicates that fire has been used by humans for at least 400,000 years. It is difficult to imagine how our ancient ancestors could have managed without it. Humans do not have sharp night vision like raccoons, but fire brought light to the long, dark night. We have no protective fur like the deer, but fire lessened the chill of winter. We do not have sharp teeth or powerful jaws like the lion, but fire rendered meat tender. Humans are not as strong or as powerful as the other large animals, but fire repels even the most ferocious of beasts. It seems reasonable to suggest that the taming of fire was one of the most monumental events in the history of the human race. The use of fire made our species dominant over all others. It is truly the tool of the gods.

Let's fast forward in time to near the end of the Stone Age, about 10,000 years ago. In the Stone Age, weapons and utensils were fashioned from rocks, wood, and a few chunks of copper metal that were

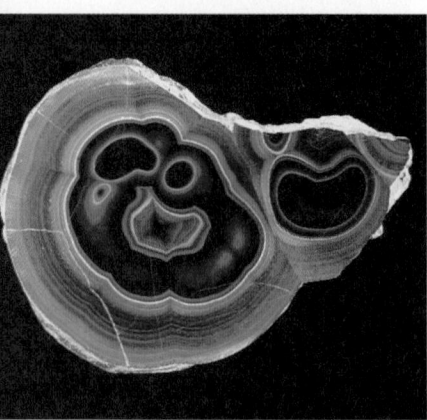

**Figure P-1  Malachite**
**Malachite is a copper ore.**

found in nature. Copper was superior to stone because it could be easily shaped into fine points and sharp blades by pounding. Unfortunately, native copper was quite rare. But about 7000 years ago this changed. Anthropologists speculate that some resident of ancient Persia found copper metal in the ashes of a hot charcoal fire. The free copper had not been there before so it must have come from a green stone called malachite (see Figure P-1), which was present in the fire pit. Imagine the commotion that this discovery must have caused. A stone could be transformed by hot coals into a valuable metal! Fire was the key that brought the human population into the age of metals. The recovery of metals from their ores is now a branch of chemical science called *metallurgy*. The ancient Persians must have considered this discovery a dramatic example of the magic of fire.

Other civilizations used chemistry in various ways. About 3000 B.C., the Egyptians learned how to dye cloth and embalm their dead through the use of certain chemicals found in nature. They were very good at what they did. In fact, we can still determine from ancient mummies the cause of death and even diseases the person may have had. The Egyptians were good chemists, but they had no idea why any of these procedures worked. Every chemical process they used was discovered by accident.

Around 400 B.C., while some Greeks were speculating about their various gods, philosophers were trying to understand and describe nature. These great thinkers argued about why things occurred in the world around them, but they were not inclined (or able) to check out their ideas by experimentation or to put them to practical use. At the time, however, people believed that there were four basic elements of nature—earth, air, water, and fire. Of these, fire was obviously the most mysterious. It was the transforming element; that is, it had the capacity to change one substance into another (e.g., certain rocks into metals). We now call such transformations chemical reactions. Fire itself is simply the hot, glowing gases associated with certain chemical reactions. If fire is a result of an ongoing chemical transformation, then it is reasonable to suggest that chemistry and many significant advances in the human race are very much related.

**Gold has been valued in jewelry since ancient times. This Etruscan jewelry dates from around 400 B.C.**

The early centuries of the Middle Ages (A.D. 500–1600) are usually referred to as the Dark Ages in Europe because of the lack of art and literature and the decline of central governments. In fact, the civilizations that Egypt, Greece, and Rome had previously built disappeared. Chemistry, however, began to grow during this period, especially in the area of experimentation. Chemistry was then considered a combination of magic and art rather than a science. Many of those who practiced chemistry in Europe were known as *alchemists*. Some of these alchemists were simply con artists who tried to convince greedy kings that they could transform cheaper metals such as lead and zinc into gold. Gold was thought to be the perfect metal. Such a task was impossible, of course, so many of these alchemists met a drastic fate for their lack of success. However, all was not lost. Many important laboratory procedures such as distillation and crystallization were discovered. Alchemists also prepared many previously unknown chemicals, which we now know as elements and compounds.

Modern chemistry has its foundation in the late 1700s, when the use of the analytical balance became widespread. Chemistry then became a quantitative science in which theories had to be correlated with the results of direct laboratory experimentation. From these experiments and observations came the modern atomic theory, first proposed by John Dalton around 1803. This theory, in a slightly modified form, is still the basis of our understanding of nature today. Dalton's theory gave chemistry the solid base from which it could serve humanity on an impressive scale. Actually, most of our understanding of chemistry has evolved in the past 100 years. In a way, this makes chemistry a very young science. However, if we mark the beginning of chemistry with the use of fire, it is also the oldest science.

# B. The Scientific Method

**Looking Ahead!** From the ancient Persians five millennia ago, to the Egyptians, to the alchemists of the Middle Ages, various cultures have stumbled on assorted chemical procedures. In many cases, these were used to improve the quality of life. With the exception of the Greek philosophers, there was little attention given to why a certain process worked. The "why" is very important. In fact, the tremendous explosion of scientific knowledge and applications in the past 200 years can be attributed to how science is now approached. This is called the scientific method, which we will discuss in this section.

In ancient history, scientific advances came about by accident. That still happens to some extent, but there have been serious changes in how we approach science so that most modern advances now occur by design. For example, from decades of painstaking and sometimes frustrating scientific research, we are close to actually understanding cancer as a disease and from this a true cure may evolve. If so, it will turn out to be one of the human race's most spectacular achievements. A recent example hints at what may lie ahead. Careful research indicated that most breast cancers are stimulated by a natural hormone (estrogen). This knowledge led a scientist to suggest that an estrogen-like drug may block the harmful effect of estrogen. Such a drug is known as *tamoxifen*. It worked. We now know that it not only can cure some breast cancers but also prevents cancer from appearing in women with a high risk of developing the disease. This is great news since breast cancer is a frightening disease that strikes millions of women annually. In fact, newer chemicals are now being tested that may be even more effective than tamoxifen.

Advancement in cancer research has been achieved by application of the *scientific method*. But the first step in the scientific method is a long way from producing a useful drug. It simply involves *making observations and gathering data*. As an example, imagine that we are the first to make a simple observation about nature—"The sun rises in the east and sets in the west." This never seems to vary and, as far as we can tell from history, it has always been so. In other words, our scientific observation is strictly *reproducible*. So now we ask "Why?" We are ready for a hypothesis. A **hypothesis** *is* *a tentative explanation of observations.* The first plausible hypothesis to explain our observations was advanced by Claudius Ptolemy, a Greek philosopher, in 150 A.D. He suggested that the sun, as well as the rest of the universe, revolves around the Earth from east to west. That made sense. It certainly explained the observation. In fact, this concept became an article of religious faith in much of the western world. However, the hypothesis of Ptolemy did not explain other observations of the time, which included the movement of the planets across the sky and the phases of the moon.

Sometimes new or contradictory evidence means a hypothesis, just like a broken-down old car, must either receive a major overhaul or be discarded entirely. In 1543, a new hypothesis developed by Nicolaus Copernicus explained all of the observations about the sun, moon, and planets by suggesting that the Earth and other planets orbit around the sun. Even though this hypothesis explained the mysteries of the heavenly bodies, it was considered extremely radical and even heretical at the time. (It was believed that God made Earth the center of the universe.) In 1609, a Venetian scientist, Galileo Galilei, built a telescope, a new device recently invented in Holland. His intent was to use this instrument to view ships from the shore that were still far out to sea. But his scientific curiosity quickly got the best of him and he turned his telescope up to the heavens. The magnified view of the planets and stars that he observed produced unmistakable proof that Copernicus was correct. Galileo is sometimes credited with the beginning of the modern scientific method because he provided direct experimental data in support of a concept. The hypothesis had withstood the challenge of experiments and thus could be considered a theory. *A* **theory** *is a well-established hypothesis.* A theory should predict the results of future experiments or observations.

The next part of this story comes in 1684, when an English scientist named Sir Isaac Newton stated a law that governs the motion of planets around the sun. *A* **law** *is a concise scientific statement of fact to which no exceptions are known.* Newton's law of universal gravitation states that planets are held by gravity in stationary orbits around the sun. (See Figure P-2.) Laws usually withstand the test of time, but not always. For example, in the 1930s, it was found that the law of conservation of matter (see Chapter 2) had to be modified to account for nuclear changes such as those that occur in atomic bombs, nuclear power plants, and the interior of the sun.

In summary, these were the steps that led to a law of nature.

1. Reproducible observations (the sun rises in the east).
2. A hypothesis advanced by Ptolemy and then a better one by Copernicus.
3. Experimental data gathered by Galileo in support of the Copernican hypothesis and eventual acceptance of the hypothesis as a theory.
4. The statement by Newton of a universal law based on the theory.

Variations on the scientific method serve us well today as we pursue an urgent search for cures for diseases. An example follows.

## The Scientific Method in Action

The healing power of plants and plant extracts has been known for centuries. For example, it was known that an extract of the willow tree relieved pain. (We now know that it contains a naturally occurring drug very closely related to aspirin.) This is the observation that starts us on our journey to new drugs. An obvious hypothesis comes from this observation, namely that there are many other useful drugs among the plants and soils of the world. We should be able to find them, and we have. In fact, the current top 20 best-selling medicines in the United States originated from plants and other natural sources. These drugs treat conditions such as high blood pressure, cancer, glaucoma, and malaria. The search for effective drugs from natural sources is especially prominent today. One chemical company in the United States randomly tests 3000 plant

Figure P-2 The Solar System The Copernican theory became the basis of a natural law of the universe.

extracts a year for anticancer, anti-arthritic, and anti-AIDS activity. Still, only a small fraction of the 250,000 known species have been tested. Since the greatest variety of plants, molds, and fungi are found in tropical forests, these areas are receiving the most attention. The introduction of a new medicine from a plant involves the following steps.

1. *Collection of materials.* "Chemical prospectors" scour the backwoods of the United States and the tropical forests such as those in Costa Rica collecting and labeling samples of leaves, barks, and roots. Soil samples containing fungi and molds are also collected and carefully labeled.

2. *Testing of activity.* Scientists at several large chemical and pharmaceutical companies make extracts of the sample in the laboratory. These extracts are run through a series of chemical tests to see if there is any antidisease activity among the chemicals in the extract. Recent advances in these laboratory procedures are astounding and allow large-scale testing that was not possible even a few years ago. If there is antidisease activity, it is considered a "hit" and the extract is taken to the next step. (See Figure P-3.)

3. *Isolation and identification of the active ingredient.* The next painstaking task is to separate the one chemical among the "soup" of chemicals present that has the desired activity. Once that's done, the particular structure of the active chemical must be determined. A hypothesis is then advanced about what part of the structure is important and how the chemical works. The hypothesis is tested by attempting to make other more effective drugs (or ones with fewer side effects) based on the chemical's structure. It is then determined whether the chemical or a modified version of it is worth further testing.

4. *Testing on animals.* If the chemical is considered promising, it is now ready to be tested on animals. This is usually done in government and university labs under strictly controlled conditions that minimize the number of animals required and regulate how the animals are used. Scientists study toxicity, side effects, and the chemical's activity against the particular disease for which it is being tested. If, after careful study, the chemical is considered both effective and safe, it is ready for the next step.

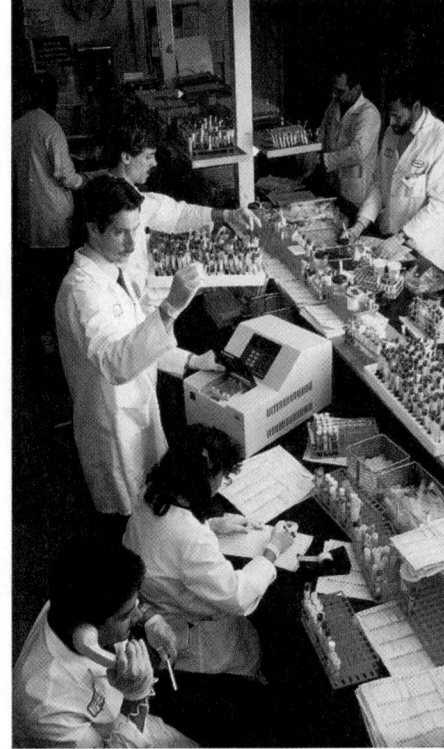

Figure P-3 A Pharmaceutical Laboratory Labs like this screen chemicals for drugs that may cure diseases.

5. *Testing on humans.* The final step is the careful testing on humans in a series of clinical trials carefully monitored by the U.S. Food and Drug Administration. Effectiveness, dosage, and long-term side effects are carefully recorded and evaluated. This process usually takes years, but if all is successful, the drug becomes available to the general public for use.

Only about one in a thousand such new drugs may actually find its way into general use. Still, the process works. A relatively recent success from random testing is an anticancer drug known as *Taxol,* which was originally extracted from the bark of the Pacific Yew tree. (See Figure P-4.) Taxol dramatically increases the life span of some women who suffer from ovarian or metastatic breast cancer. However, the Yew tree is not plentiful so the supply of Taxol would be very limited. Fortunately, chemists have found that the needles of an abundant relative of the Yew tree contain a similar chemical that can be converted into Taxol. It is now readily available to all who need it. By suggesting a hypothesis as to how Taxol attacks cancer cells, scientists hope to develop new drugs that are even more effective but with fewer side effects.

As our previous discussion of the drug tamoxifen indicates, we may be on the verge of even greater discoveries regarding cancer. Because of the discovery of how cancers grow, drugs are now in human testing that may starve the blood supply to tumors and thus shrink or even eliminate them wherever they are in the body. Many scientists express optimism that the disease may yet yield to modern medicine.

Similar work has produced drugs that have allowed thousands of people infected with the AIDS virus to live a relatively normal life. Although the current treatment simply holds the disease in check, many scientists feel that an eventual cure or vaccine will be found.

Besides drugs, our modern scientific method also produces materials for space travel, all sorts of plastics and synthetic fibers, microchips for computers, and processes for genetic engineering. Just a century ago, everything was made out of stone, wood, metal, or natural fibers (wool, cotton, and silk). Our modern society could hardly function on those materials alone.

One good thing about the study of chemistry is that, once we establish a law, we can count on it. Laws of nature never change, but our theories sometimes change or are modified. Social scientists have a somewhat tougher task in this regard. It is difficult to make laws about human nature because there is so much variability among people and people are always changing. If we establish that copper is a brownish red metal, we can be assured that it will always be so, regardless of whether we find the copper in the state of Minnesota or on the planet Mars. That's how sure we are of our facts.

**Figure P-4   The Pacific Yew**
**The bark of this tree is a source of an anticancer drug known as Taxol.**

## C. The Study of Chemistry and Using This Textbook

**Looking Ahead!** **The study of chemistry requires certain skills and study habits. In some ways, it is like the study of math—repetitious problem solving to master a skill. In other ways, it is like the study of a language—basic memorizing and recitation. And it is also like the study of philosophy—the quiet contemplation of concepts. In the next section, we will discuss some specific recommendations about how this science can be mastered and how this text can help.**

Chemistry is *the fundamental natural science.* This is not just an idle boast. Chemistry is concerned with the basic structure and properties of all matter, be it a huge star or a microscopic virus. Biology, physics, and geology, as well as all branches of engineering and medicine, are based on an understanding of the chemical substances of which nature is composed. Chemistry is the beginning point in the course of studies that eventually produces all scientists, engineers, and physicians. But it is also important for all responsible citizens. Our environment is very fragile—more fragile than we realized just a few years ago. Many of the chemicals that make life easier also affect our surroundings. Control of air, water, and land pollution needs as much attention from citizens and scientists as the invention of new materials did in the twentieth century. (See Figure P-5.) We all have a big stake in the future, so it is reasonable that chemistry is a prerequisite not only for courses of study but also for life, especially in these complex times. What follows is a brief list of academic self-disciplines and skills needed in the study of chemistry, and how this text can help.

## 1. Time Management

Chemistry is not unlike basketball or piano—it requires lots of practice. It is not only a question of putting in the time but *how* and *when* you put in the time. One does not wait until the night before the big game to first practice jump shots or the night before the recital to first practice the sonata. A master schedule should be prepared, with chemistry (as well as all other subjects) receiving a regularly scheduled study time. One study period should follow as soon as possible after the lecture. In setting up your schedule, it is wise to select shorter but frequent study periods. Two separate one-hour sessions are more effective than one two-hour period. After a length of time, the mind tends to wander into areas not directly related to chemistry (e.g., baseball—what else?). Reserve the evening before an exam for a comprehensive review followed by a decent night's sleep. The night before the exam is not the time to break new ground.

**Figure P-5  Air Pollution**
Our environment, especially the air we breathe, is very fragile.

### How This Textbook Can Help

The understanding of chemistry is greatly enhanced by proceeding slowly, mastering one concept at a time. In the text, each chapter is divided into two or three sections each followed by a brief summary entitled **Looking Back.** One section is about right for one or two study sessions. The brief summary is followed by a **Learning Check** section, called **Checking It Out,** where you are asked to fill in the blanks in summarizing statements and do a few problems that reinforce the concepts just covered. To rush through the whole chapter and then work a few problems often leads to frustration. The following are examples of **Looking Back** and **Learning Check** sections from Chapter 3.

Looking Back! The matter that we breathe, eat, walk on, or swim in is composed of 88 fundamental substances called elements. These elements are made up of extremely tiny particles called atoms. Molecular elements and compounds are composed of atoms joined together by covalent bonds. A second type of compound is composed of oppositely charged ions. The formulas of these compounds describe either discrete molecular units or the ratios of ions present.

---

### Learning Check A | checking it out

**A-1.**  Fill in the blanks.

The symbol for the element nitrogen is _____ and that of copper is _____ . The symbol Ni represents the element _____ . The fundamental particle of an element is called an _____ . When these fundamental particles are covalently bonded to each other, they form _____ . The designation $C_2H_7N$ for a compound represents its _____ and indicates the presence of _____ atoms of C and _____ atom of N in one molecule. If two compounds have the same formula, they must have different _____ formulas. Besides molecules, compounds can also be composed of _____ . The formula of an ionic compound does not represent discrete molecules but represents the smallest whole number ratio of _____ to _____ .

---

**A-2.**  Acetaminophen (a pain reliever) is a molecular compound composed of eight atoms of carbon, nine atoms of hydrogen, one atom of nitrogen, and two atoms of oxygen. What is its formula?

**A-3.**  Write the formulas of a compound containing three $Na^+$ cations and one $PO_4^{3-}$ anion and a compound containing two $NH_4^+$ cations and one $S^{2-}$ anion.

*Additional Examples: Exercises 3-5, 3-7, 3-17, 3-20, 3-26, and 3-27.*

Many of the problems in the **Learning Check** are preceded by step-by-step examples, which are labeled as **Working It Out.** It is important that you work through these examples with a pencil and not just by reading. Write out the example yourself. There is a strong connection between writing and learning in chemistry. Sometimes learning chemistry requires that you put your feet on the desk, lean back, and think. More often it requires that you put your feet on the floor, lean over, and write.

---

**Calculating the Atomic Mass of an Element from Percent Distribution of Isotopes**

**Example 3-2**

In nature, the element boron occurs as 19.9% $^{10}$B and 80.1% $^{11}$B. If the isotopic mass of $^{10}$B is 10.013 and that of $^{11}$B is 11.009 amu, what is the atomic mass of boron?

**working it out**

**Procedure**

Find the contribution of each isotope toward the atomic mass by multiplying the percent in decimal form by the isotopic mass.

**Solution**

$$^{10}\text{B } 0.199 \times 10.013 \text{ amu} = \underline{\phantom{0}1.99 \text{ amu}}$$
$$^{11}\text{B } 0.801 \times 11.009 \text{ amu} = \underline{\phantom{0}8.82 \text{ amu}}$$
$$\text{atomic mass of boron} = \underline{\underline{10.81 \text{ amu}}}$$

---

After you have completed the entire chapter, you are ready for a comprehensive summary. The final section of each chapter provides such a review. Don't ignore this section. It should help you put the concepts in the chapter into a logical progression. For that reason, we label the review **Putting It Together.** The important terms defined in the chapter are used in a discussion format in this section but are shown in boldface type. Often we can use tables or flowcharts to summarize the whole chapter so that the review is not just a repetitive rehash. The following example is a portion of **Putting It Together** from Chapter 3.

**chapter R eview**                                    **putting it together**

All of the various forms of nature around us, from the simple elements in the air to the complex compounds of living systems, are composed of only a few basic forms of matter called elements. Each element has a name and a unique one- or two-letter **symbol.**

It has been less than 200 years since John Dalton's atomic theory introduced the concept that elements are composed of fundamental particles called **atoms.** It has been about 20 years since we have been able to produce images of these atoms with a special microscope. Most elements are present in nature as aggregates of individual atoms. In some elements, however, two or more atoms are combined by **covalent bonds** to produce basic units called **molecules. Molecular compounds** are also composed of molecules, although, in this case, the atoms of at least two different elements are involved. Each compound has a unique arrangement of atoms in the molecular unit. These are sometimes conveniently represented by **structural formulas.**

## 2. Attendance

The importance of regular class attendance cannot be overemphasized. Any college teacher knows there is a direct correlation between understanding and attendance, whether it is required or not. This text is a great ally, but what material your instructor covers and in what depth can be discovered only in class. Nor can you sense your instructor's emphasis or benefit from his or her problem-solving hints by reading someone else's notes. The key to what will be on the exams is found in the lecture. You need to be there.

### *How This Textbook Can Help*

Before you go to bed the night before a chemistry class, put this book on your dresser so that you will be reminded to go to class. Or perhaps you could put it under your pillow. Whatever it takes, do it.

## 3. Asking Questions

Few students feel confident enough to ask questions in class. This is natural enough—we all tend to think our questions might be "dumb." Nevertheless, you still need to ask the questions. If you hesitate to ask questions in class or the class is too big, take advantage of help sessions or your instructor's office hours. In many cases, instructors are a significantly underutilized resource during their office hours. You may discover that, in the office, the instructor is really human and helpful. The bottom line is that you owe yourself answers and understanding. Do what you have to do.

### *How This Textbook Can Help*

One of the goals of this text is to promote critical thinking by formulating questions. Each chapter begins with an important section labeled **Setting the Stage.** Read this carefully. Our intent is for you to gain a sense of the relevance of the topic that we are about to discuss. Toward the end of this introduction, we state some questions that pertain to the chapter labeled as **Formulating Some Questions.** Our goal is for you also to formulate questions. In other words, the author and your instructor hope to pique your curiosity about nature. Once you get started asking questions, you will get answers. Every time we get an answer, however, it just seems to generate more questions. That is the nature of science.

### Setting the Stage

A few years ago, astronomers in Chile witnessed the catastrophic death of a distant sun. Suddenly one night, a bright point of light appeared in the night sky that was not there the night before. A massive star had literally exploded in an event known as a supernova. It had been over 400 years since a supernova so bright had been witnessed on Earth. The star was dying, but it was also a profound moment of creation. As the star was going through its death throes, elements were being formed. This spectacular event was important to science because it confirmed theories concerning how and where the heavy elements on this planet were originally formed.

For billions of years, stars have been churning, boiling, and eventually undergoing catastrophic explosions such as the one seen recently. Blasted by the force of exploding suns, newly formed elements and some simple compounds of the elements have drifted through cold, dark space as bits of dust and clouds of gases. About four and a half billion years ago, a large cloud containing hydrogen and some heavier elements produced by supernovas condensed to ... net out from this star is ... stems composed of ... e know it—are, in fact,

### Formulating Some Questions

In the past 200 years, the miracle of the human mind, with its boundless curiosity, has unraveled many mysteries about the composition as well as the origin of these elements. For example, it is now universally accepted that elements are composed of very small fundamental units called atoms. The existence of the atom brings up many additional questions that we will discuss in this chapter. How did the concept of the atom come to be accepted? If atoms are the fundamental unit of an element, what is the fundamental unit of a compound? What makes a compound unique? Are the atoms of an element exactly the same? How do the atoms of one element differ from those of another? What determines the mass of an atom? Before we address these questions, however, we have more to say about the number of elements, their distribution, and how we name them.

## 4. Perseverance

Everyone hopes to start off with an A on the first test or quiz. However, few do that, including many who are subsequently very successful in the study of chemistry. If you are disappointed with an exam grade, reanalyze your study habits, make adjustments, and try again. Don't expect better results by doing the same thing. Remember to use your instructor as your primary resource for advice in this matter. Perhaps your problem is not the material but "test anxiety." This is a very real phenomenon and has to be acknowledged. Most colleges have a counseling center that can help you overcome this problem. Sometimes your instructor can give you helpful hints on taking chemistry tests so that you won't have the fear of "freezing up." If you are ultimately not successful in the study of chemistry, at least you can say that you gave it your best try.

### *How This Textbook Can Help*

Your instructor will assign some Exercises provided at the end of each chapter to help check your understanding of basic concepts. Here are some examples of **Exercises** from Chapter 3.

# Exercises

**Names and Symbols of the Elements**

**3-1.** Write the symbols of the following elements. Try to do this without reference to a table of the elements.
(a) bromine   (c) lead   (e) sodium
(b) oxygen   (d) tin   (f) sulfur

**3-2.** The following elements all have symbols that begin with the letter C: cadmium, calcium, californium, carbon, cerium, cesium, chlorine, chromium, cobalt, copper, and curium. The symbols are C, Ca, Cd, Ce, Cf, Cl, Cm, Co, Cr, Cs, and Cu. Match each symbol with an element and then check with the table of elements inside the front cover.

**3-3.** The names of seven elements begin with the letter B. What are their names and symbols?

**3-4.** The names of nine elements begin with the letter S. What are their names and symbols?

**3-5.** Using the table inside the front cover of the text, write the symbols for the following elements.
(a) barium   (c) cesium   (e) manganese
(b) neon   (d) platinum   (f) tungsten

**3-6.** Name the elements corresponding to the following symbols. Try to do this without reference to a table of the elements.
(a) S   (c) Fe   (e) Mg
(b) K   (d) N   (f) Al

**3-7.** Using the table inside the front cover, name the elements corresponding to the following symbols.
(a) B   (c) Ge   (e) Cl   (g) Be
(b) Bi   (d) U   (f) Hg   (h) As

**Molecular Compounds and Formulas**

**3-8.** How do the following concepts relate and differ?
(a) a molecule and an atom
(b) a molecule and a compound
(c) an element and a compound
(d) a molecular element and a monatomic element

**3-9.** Which of the following are formulas of elements rather than compounds?
(a) $P_4O_{10}$   (c) $F_2O$   (e) MgO
(b) $Br_2$   (d) $S_8$   (f) $P_4$

**3-10.** Name the elements in the previous exercise.

**3-11.** What is the difference between Hf and HF?

**3-12.** What is the difference between NO and No?

**3-13.** Which of the following is the formula of a diatomic element? Which is the formula of a diatomic compound?
(a) $NO_2$   (c) $K_2O$   (e) $N_2$
(b) CO   (d) $(NH_4)_2S$   (f) $CO_2$

**3-14.** Name all of the elements in the compounds of the previous exercise.

Other than knowing the material, one of the best ways to prepare for a test is to practice taking exams. A **Cumulative Review** test follows every two or three related chapters. The test includes multiple-choice questions as well as problems. Here is a sample from a Cumulative Review.

# Cumulative Review Chapters 2–4*

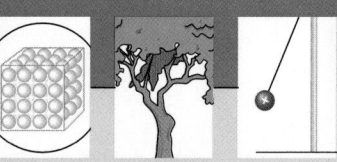

**Multiple Choice**

**The following multiple-choice questions have one correct answer.**

**1.** A substance has a distinct melting point and density. It cannot be decomposed into chemically simpler substances. The substance is a

(a) heterogeneous mixture    (c) compound
(b) solution    (d) element

**2.** A sample of a given liquid has uniform properties, but the temperature at which the liquid boils slowly changes as it boils away. The liquid sample is a

(a) heterogeneous mixture    (c) compound
(b) solution    (d) element

**3.** Which of the following is a chemical property of iodine?

**10.** Which of the following is the formula of a molecular compound with diatomic molecules?

(a) NO    (b) $NH_4Cl$    (c) $H_2$    (d) KBr    (e) $CO_2$

**11.** Which of the following is the correct formula for an element?

(a) $N_4$    (b) $P_2$    (c) $C_2$    (d) $Br_2$

**12.** Which would be the electrical charge on a sulfur atom containing 18 electrons?

(a) $2-$    (b) $1-$    (c) 0    (d) $2+$

**13.** Which would be the formula of an ionic compound made up of $Ba^{2+}$ ions and $Cl^-$ ions?

(a) $Ba_2Cl$    (b) $BaCl_2$    (c) BaCl    (d) $BaCl_3$

**14.** Which of the following elements is a halogen?

(a) H    (b) O    (c) K    (d) Ne    (e) I

The answers to all of these questions and the solutions to the problems are given in the back of the book (Appendix G). Use these review tests as practice for an exam. Take the test within a specific time allotment, as if it were an exam in class. A Study Guide that provides additional worked-out examples and tests is available with the text.

## 5. Study Skills

**a. Organization.** We all wish that we were more organized. It seems that life gets more and more complex. But organization in chemistry is quickly rewarded. We have already spoken to the idea of organizing your time, but there is more. Many of the problems that are required in chemistry require a step-by-step approach. It is a lot like planning an extended trip. Usually, we don't just get in the car and drive. We plan our journey so that we can take the shortest or the easiest route. We may also plan ahead on what we want to accomplish and when we want to do it. The secret is advance planning. So too with chemistry problems. We don't just start doing a calculation—we should take some time to plan the journey through the problem from beginning to end. This requires that we first write down the steps that we will take.

## *How This Textbook Can Help*

We illustrate the textbook's approach to organizing a quantative problem with Example 8-19, which follows. Although this problem may not make sense now, perhaps you can appreciate how it is managed. The ***Procedure*** discusses how we are going to solve the problem in a step-by-step manner (steps *a* through *c*). In the ***Scheme,*** the same procedure is outlined with a color-coded map showing the same three steps along with the proper tools to solve the problem. Finally, in the ***Solution,*** the actual calculation is carried out with each step labeled.

---

**Example 8-19** | **Mass ⟶ Mass Conversions**

What mass of $H_2$ is needed to produce 119 g of $NH_3$?

**Procedure**

As in the last example, we must first convert the mass of what's given (119 g $NH_3$) into moles using the molar mass of $NH_3$ (step *a* below). We then convert moles of $NH_3$ to moles of $H_2$ using mole ratio (5) (step *b* below). Finally, since the mass of $H_2$ is requested, we must convert moles of $H_2$ to mass using the molar mass of $H_2$ as a conversion factor (step *c* below).

**Scheme**

$$\textit{Given (g NH}_3)$$

$$\begin{array}{ccc} (a) & (b) & (c) \\ \times \dfrac{\text{mol NH}_3}{\text{g NH}_3} & \times \dfrac{\text{mol H}_2}{\text{mol NH}_3} & \times \dfrac{\text{g H}_2}{\text{mol H}_2} \end{array} \qquad \textit{Requested (g H}_2)$$

$$\boxed{\text{mass NH}_3} \Longrightarrow \boxed{\text{mol NH}_3} \Longrightarrow \boxed{\text{mol H}_2} \Longrightarrow \boxed{\text{mass H}_2}$$

**Solution**

$$119 \text{ g NH}_3 \times \overset{(a)}{\frac{1 \text{ mol NH}_3}{17.03 \text{ g NH}_3}} \times \overset{(b)}{\frac{3 \text{ mol H}_2}{2 \text{ mol NH}_3}} \times \overset{(c)}{\frac{2.016 \text{ g H}_2}{\text{mol H}_2}} = \underline{\underline{21.1 \text{ g H}_2}}$$

---

**b. Memorization.** Most of us wish that we could just grasp a concept and, with little effort, have it permanently imprinted in our brains. Unfortunately, it rarely works that way. In many cases, we need to memorize a definition or rule so as to be able to categorize facts as fitting the concept or not. We certainly could not begin to play a sport such as soccer without first knowing the object of the game and some basic rules. Likewise, in chemistry we wouldn't be able to write a molecular structure until we first know the steps involved. The use of $3 \times 5$ note cards can help. Carry some cards with you and quiz yourself when you get a chance; it works.

## *How This Textbook Can Help*

After first reading through the chapter-end review, go over each term in boldface type and recite in your own words its meaning or definition. If necessary, return to the chapter text or the glossary (Appendix F) for the definition.

**c. Reading Ahead.** Even in high school, football and basketball opponents are scouted before the game. Every team likes to know what they are up against and what lies ahead. In studying science, the equivalent to scouting is reading ahead. Even if you do not grasp the concepts, reading ahead will give you a feeling for some of the material that you will be discussing in the next lecture or two. If you know something is coming that seems confusing, you will be more alert when the concept is discussed. Reading ahead is also a time saver. When you know that certain definitions or tables are in the book, you can save note-taking time.

## How This Textbook Can Help

A complaint commonly heard about the study of chemistry is that it sometimes seems like an accumulation of isolated facts and concepts. Not so! There are logical progressions of chemical information, with each concept building on the previous one. This text has a unique feature that should help you read ahead. You may have already noticed the connectors that begin sections B and C in the prologue. This also occurs in the text, with a short paragraph at the beginning of each section that explains where we are headed. Consider an example from Chapter 3. The following paragraph (labeled **Looking Ahead**) connects the discussion of two types of chemical substances called compounds.

> **Looking Ahead!** Perhaps one way we could classify the matter around us is as "soft" and "hard." The soft part of nature is the air, the liquids such as water, and some solids that melt at low temperatures. The hard matter is the stuff of minerals, rocks, and even mountains. Although this is a rather crude generalization with many exceptions, the soft part of nature is generally composed of molecular compounds. The hard part of nature—solids with comparatively high melting points—is composed of another type of compound that we will discuss next.

**d. Note Taking.** The most useful approach is a lot of work but it is *extremely* effective. As soon as you can after a lecture, recopy your class notes. As you do this, imagine that you are explaining the material to someone else (out loud, if you can). If you do this, you will become aware of concepts that are still hazy to you. Also, the logical progression of problems may now seem more obvious. As you recopy your notes, leave about one-third of the paper as a blank margin so that you can add thoughts, emphasis, or questions for later review. (See Figure P-6.) Clear, easy-to-read notes correlate with good performance.

Finally, but very important, a fantastic web site is available that directly supports this text. It includes a large amount of additional information

**Figure P-6
Chemistry Notes**
**Good note taking is essential in the study of chemistry.**

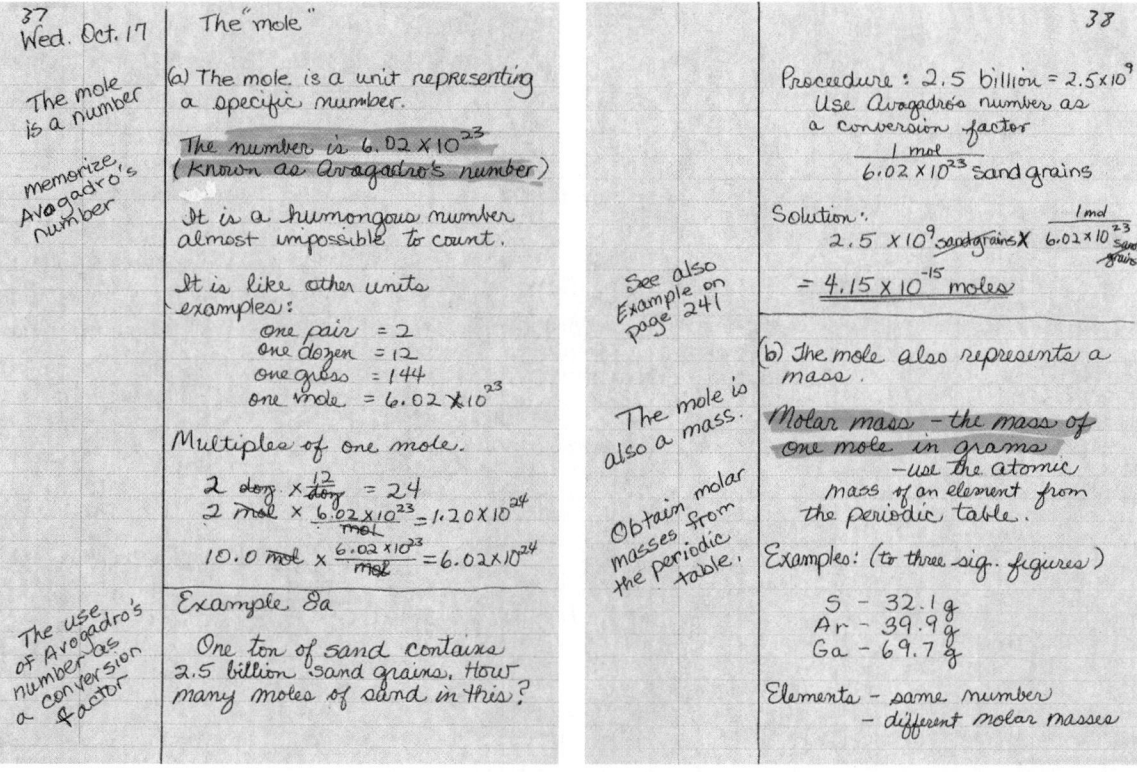

including quizzes with immediate feedback. It also provides many **Interactive Learning** problems where you can find help in a step-by-step manner. This is especially helpful with the quantitative problems. You can find all of this at **http://www.wiley.com/college/malone**. Please log onto this site as soon as possible to take advantage of all that it offers.

## Interactive Learning

**1-118.** A bullet is fired at a speed of 2235 ft/s. What is this speed in kilometers per hour?

**1-119.** A truck weighs $2.07 \times 10^9$ mg. What is its mass in tons?

**1-120.** In 2002, the rate of exchange in China was 8.277 yuan per U.S. dollar. The exchange in Russia was 31.29 rubles per U.S. dollar. If a sandwich costs 22.72 yuan in China, what would it cost in rubles?

**1-121.** Pressure can be measured in pounds per square foot. Convert 36.7 lb/ft$^2$ to g/cm$^2$.

Also at the web site for this text, you will find a link to another great web resource, titled, "Chemistry Webercises". This web site supports the text by Steven Murov and Brian Stedjee, titled, "Experiments and Exercises in Basic Chemistry, Sixth Edition", which may also be a required book for your course. At this dynamic web site, you will find interactive exercises, called Webercises, as well as quizzes, tutorials, and web links that will help enhance and support your study of chemistry. These Webercises have been created and designed to provide you with answers to important and stimulating questions about contemporary chemically related topics that will supplement the material in this text. Also, there is a combination Study Guide/Solutions Manual that accompanies this text that includes brief discussions and many more problems with worked-out solutions. The Study Guide also includes worked-out solutions to all quantitative problems that have answers in the back of the text.

Another new feature of this edition is the **Making It Real** boxes that appear about twice in each chapter. These essays are meant to relate the relevance of chemistry to our everyday world. We know that you will find them interesting. We are now ready to begin the study of chemistry. It is a fascinating subject. Enjoy it!

## MAKING IT REAL

### Roses Are Red; Violets Are Blue—But Why?

From the beauty of flowers and sky to the brightness of rubies and emeralds, we marvel at the colors of light that lie in the visible part of the spectrum (wavelengths 400 nm–800 nm). It was mentioned in this chapter that the sun emits the whole spectrum of colors of the rainbow. Blended together, sunlight appears nearly white. Fireworks emit discrete colors depending on the element that is heated in the explosion. But most of what our eyes receive is a result of reflected light rather than emitted light.

When white light is directed at a surface, the amount of reflection determines its shade. If all of the light is reflected, the surface appears white. If all of the light is absorbed, the surface appears black. (That is why light-colored clothing is preferred in the summer, since it reflects the light energy.) If part of the light is absorbed and part reflected, the surface appears as a shade of gray.

Most of the wonderful colors of our world result from absorption of specific regions of the light spectrum. There are three primary colors of light, as shown in the photograph. Red light is the highest third of the wavelengths of visible light (the lowest energies). Green is the middle third

of the wavelengths, and blue is the lower third of the wavelengths. A blend of all three appears white. It is only when one region of the spectrum is absorbed that we perceive color. A rose appears red because it absorbs the blue-green part of the spectrum and reflects the complement to these colors, which is red. If only the blue part is absorbed and both red and green are reflected, we see a yellow rose. A violet absorbs the red-green part of the spectrum and reflects blue. Leaves appear green because they absorb the red-blue part of the spectrum. Some of that absorbed light energy is converted into chemical energy in the production of sugars by *photosynthesis*.

Color television works on the same principle. Three different signals are shot through different primary-color filters and then transmitted to the TV set. The three signals are then blended back together to provide the proper color. Older sets had red, green, and blue drives that were manually adjusted.

We are lucky. Not all animals see color. Life for us would certainly be dull without the "wonderful world of color."

*The three primary colors.*

## chapter 1

# Measurements in Chemistry

To land this large aircraft safely; the pilot must observe many measurements displayed in the cockpit. Chemistry is a quantitative science that also requires careful measurements. The measurements used in chemistry are the subjects of this chapter.

## Setting the Stage

Before we were even aware of our own existence we were being subjected to measurements. Within the first few minutes of our lives, we were placed on a scale and a tape measure to provide our first weight and length. Years later, as adults, we now find ourselves immersed in a complex world that measures about anything that lends itself to being measured. For example, it is hard to conceive of being able to operate a modern automobile without a speedometer and temperature and fuel gauges. The more expensive the car, the more measurements that are supplied as reported by more gauges and dials. If one pays enough, a global positioning satellite will even measure where you are and tell you how to get where you are going. Pilots, carpenters, artists, teachers, and all other professions and skills are usually centered on some type of measurement. Chemistry is hardly an exception since it is a science based on the natural laws of matter. Many of these laws originate from reproducible quantitative measurements. Since measurements are so important to us, this is where we start even before we get into the science of chemistry itself. We cannot overemphasize the importance of understanding the numbers that are integral to this science.

Measurements, of course, imply the use of mathematics. For many, this "M" word brings on an acute sense of panic. The good news is that most of us can overcome this fear. It's just a matter of getting into shape. If you are not in at least fair *physical* shape, playing a strenuous sport is not fun—it is exhausting and probably frustrating. Likewise, doing chemistry can be frustrating if you are not in at least fair *mathematical* shape. Even though only one year of algebra is required for this course, it is likely that most students need at least a little mathematical workout. For some, a simple review will do, and that is provided in the text (but be prepared to give your pencil a decent physical workout.) At appropriate points in this chapter, you will also be referred to a specific appendix in the back of the book for more extensive review. Review appendixes include basic arithmetical operations (Appendix A), basic algebra operations (Appendix B), and scientific notation (Appendix C). Also, Appendix E can aid you in the use of calculators for the mathematical operations found in the text. *If you are worried about the math, remember that this text was written with your concerns in mind. Just be ready for frequent reviews.*

### Formulating Some Questions

Everyone seems to be busy measuring something, so let's set the tone of this chapter with a formal definition of a measurement. *A* **measurement** *determines the quantity, dimensions, or extent of something, usually in comparison to a specific unit. A* **unit** *is a definite quantity adopted as a standard of measurement.* Thus, a measurement (e.g., 1.23 meters) consists of two parts: a numerical quantity (1.23) followed by a specific unit (meters). In the first three sections of this chapter, we will address questions about the numerical quantity. How reliable are the numbers in a measurement? How reliable is a measurement that results from mathematical operations such as addition or multiplication? Since we are sure to encounter very large and very small numbers, how can such numbers be conveniently expressed and manipulated? In the last three sections, we turn our attention to the units used to express various measurements. What units do we use in chemistry to express dimensions, mass, and volume? How can we convert between related units of measurement (e.g., miles to kilometers) that are directly proportional? Finally, what is temperature and how is it measured? After we prepare ourselves mathematically in this chapter, we will look into what chemistry is all about in the next chapter.

## Section A   THE NUMBERS USED IN CHEMISTRY

### 1-1 The Numerical Value of a Measurement

**Looking Ahead!** Our first goal is to take a close look at some of the numbers that come our way in science. How good are the numbers and what do they mean? We will take a look at some of the tools we have in evaluating the quality of the numbers that are part of the measurement.

A newspaper reports that there were 12,000 people gathered at the rock concert last evening. Did someone actually count them all? Not really; usually, there is someone around who is considered an expert at estimating the size of crowds. Now imagine that eight people leave the concert early. Should the estimate be changed to 11,992 people? Of course not. The original number was not intended to mean exactly 12,000 people. In fact, the three zeros in the number 12,000 were not actually "measured" numbers but simply indicate the magnitude of the number; in other words, they locate the decimal point. In this example, a considerable number of people would have to leave the concert in order for the estimate to be changed to 11,000. In the original number (12,000), only the 1 and 2 are considered significant. *In a measurement, a* **significant figure or digit** *is a digit that is either reliably known or estimated.* In the number 12,000, we can assume the 1 is reliable and reproducible from any number of estimates, but the 2 is estimated. The zeros are not significant since they actually have no numerical meaning. Thus, in our example, there are *two* significant figures: the 1 and the 2. *The number of significant figures or digits in a measurement is simply the number of measured digits and refers to the precision of the measurement.* **Precision** *relates to the degree of reproducibility or uncertainty of the measurement.* For example, two other experts may estimate the crowd at 13,000 and 11,000, respectively. This means that the original estimate had an uncertainty of ±1000. *Indeed, all measured values have an uncertainty that is expressed in the last significant figure to the right.*

Now let us have the same crowd at the concert seated in the bleachers of a stadium instead of milling about. In this case, a more precise estimate is possible since the exact capacity of the stadium is known, which means that the experts can get a better idea of the number of people present. The crowd is now estimated at between 12,400 and 12,600, or an average of 12,500. This is a measurement with three significant figures. The 1 and 2 are now reliable, but the third significant figure, the 5, is estimated. However, the extra significant figure means that the uncertainty is now reduced to ±100. The more significant figures in a measurement, the more precise it is. If the crowd went through a turnstile before entering the stadium, an even more precise number could be given. Notice that the significant figure farthest to the right in a measurement is estimated. (As you may guess, people doing chemistry are far more interested in the degree of precision of a measurement than those interested in the size of crowds at rock concerts.)

Participants in the sport of riflery (target shooting) are judged on two points: how close the bullet holes are to each other (the pattern) and how close the pattern is to the center of the target known as the bull's-eye. The *precision* of the contestant's shooting is measured by the tightness of the pattern. How close the pattern is to the bull's-eye is a measure of the shooter's *accuracy*. **Accuracy** *in a measurement refers to how close the measurement is to the true value.* Usually, the more precise the measurement, the more accurate it is—but not always. In our example, if a certain competitor has a faulty sight on the rifle, the shots may be close together (precise) but offcenter (inaccurate). (See Figure 1-1.)

Accuracy in measurements depends on how carefully the instrument of measurement has been *calibrated* (compared to a reliable standard). For example, what if we attempted to measure length with a plastic ruler that became warped after being left in the hot sun? We obviously would not obtain accurate readings. We would need to recalibrate the ruler by comparing its length divisions to a reliable standard.

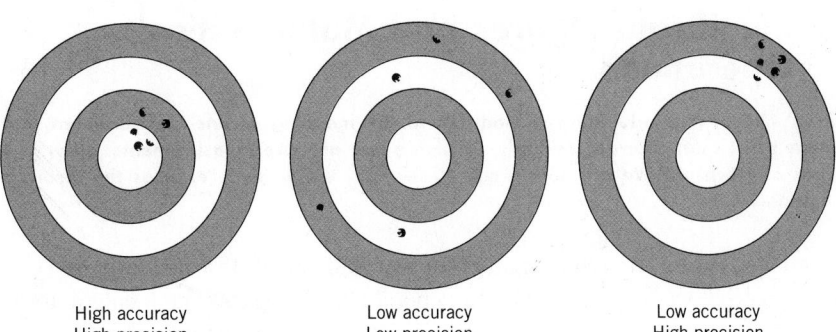

**Figure 1-1 Precision and Accuracy**
**High precision does not necessarily mean high accuracy.**

High accuracy
High precision

Low accuracy
Low precision

Low accuracy
High precision

It would be easy to determine the number of significant figures in a measurement if it were not for the number zero. Unfortunately, zero serves two functions: as a reliable or estimated digit, or simply as a filler to locate the decimal point (such as the zeros in the 12,000 people). Since the zeros look alike in both cases, it is important for us to know whether a zero is significant or is there simply to locate the decimal point. The following rules can be used to tell us about zero. Digits underlined are significant.

1. When a zero is between other digits, it is significant (e.g., 709 has three significant figures).

2. Zeros to the right of a digit and to the right of the decimal point are significant (e.g., 8.0 has two significant figures, as do 7.9 and 8.1; 5.700 has four significant figures).

3. Zeros to the left of the first digit are not significant (e.g., 0.078 and 0.0078 both have two significant figures; 0.04060 has four significant figures).

4. Zeros to the left of an implied decimal point may or may not be significant. In most cases, they are not (e.g., the crowd of 12,000 has two significant figures, and 6600 also has two). What if the zero in a number such as 890 is actually an estimated digit and thus significant? This is a tough question. Some texts use a line over the zero to indicate that it is significant (e.g., 890), and others simply place a decimal point after the zero (e.g., 890.). As we will see in section 1-3, there is a solution to this dilemma. In most problems used in this text, measurements are expressed to three significant figures. Therefore, in calculations where other numbers have three significant figures we will assume that numbers such as 890 also have three significant figures.

| Example 1-1 | Evaluating Zero as a Significant Figure |
|---|---|

**working it out**

a. How many significant figures are in the following measurements?
   **i.** 1508 cm   **ii.** 300.0 ft   **iii.** 20.003 lb   **iv.** 0.00705 gal

b. What is the uncertainty in each of the above measurements?

**Answers**

a. **i.** four (rule 1)   **ii.** four (rule 2) *Note:* Since the zero to the right of the decimal point is significant, the other two zeros are also significant because they lie between significant figures.   **iii.** five (rule 1)   **iv.** three (rules 1 and 3)

b. **i.** $\pm 1$   **ii.** $\pm 0.1$   **iii.** $\pm 0.001$   **iv.** $\pm 0.00001$

# 1-2 Significant Figures and Mathematical Operations

**Looking Ahead!** It is one thing to understand the meaning of one measurement, but what happens if we add, subtract, multiply, or divide two or more measurements, all with various degrees of precision? We are now ready to see how we properly express the results of such calculations.

For under $10 we can own a small computer that can almost instantly carry out any calculation that we may encounter in general chemistry. The hand calculator is truly phenomenal, especially to the "old-timers" who had to carry out these calculations using a slide rule. (This was a primitive but effective device that performed many of the calculations found on a simple electronic calculator. It was about one foot long and could be carried in a case attached to one's belt. However, nonscientific students would sometimes comment on the wearing of such apparel as being "not cool.") The hand calculator does have one serious drawback. It does not realize that many of the numbers that we punch into it are the results of measurements of real things and that each measurement should only be expressed to a certain precision or number of significant figures. *Our calculators assume the numbers are exact.* For example, 7.8 divided by 2.3 reads 3.3913043 and so forth on the display. However, if the numbers represented measurements known to only two significant figures (e.g., 7.8 lb and 2.3 ft), then the calculator has created the illusion that these numbers were known to a much greater precision. Since the calculator can't express the answer to the calculation properly, *we* must know how to prune the answer so that the reported value is honest and appropriate. There are two rules for properly expressing the result of a mathematical operation: one applies to addition and subtraction and the other applies to multiplication and division. We will discuss the rule for addition and subtraction first.

*When numbers are added or subtracted, the answer is expressed to the same number of decimal places as the measurement with the fewest decimal places.* Or, in other words, *the summation must have the same degree of uncertainty as the measurement with the most uncertainty* (e.g., $\pm100$ has more uncertainty than $\pm10$ and $\pm0.1$ more than $\pm0.01$). This is illustrated by the following summation.

The calculator does not give the answer to the proper number of decimal places.

$$
\begin{array}{ll}
10.6871 & \text{(four decimal places or uncertainty of } \pm0.0001) \\
\underline{1.42\phantom{00}} & \text{(two decimal places or uncertainty of } \pm0.01) \\
12.1071 = \underline{12.11} & \text{(two decimal places or uncertainty of } \pm0.01)
\end{array}
$$

Notice 1.42 and the answer have the same uncertainty. Therefore, expressing the "71" in the summation has no meaning and cannot be included except for rounding off purposes.

The rules for rounding off a number are as follows (the examples are all to be rounded off to three significant figures).

1.  If the digit to be dropped is less than 5, simply drop that digit (e.g., 12.44 is rounded off to 12.4).

2.  If the digit to be dropped is 5 or greater, increase the preceding digit by one (e.g., 0.3568 is rounded off to 0.357, and 13.65 is rounded off to 13.7). Note that the number 12.448 is rounded off to 12.45 if it is to be expressed to four significant figures or two decimal places, and to 12.4 if it is to be expressed to three significant figures or one decimal place.

Recall our original example of a crowd estimated at 12,000 people at a rock concert. In cases like this where a decimal point is not involved, we must be careful how we add or subtract numbers from zeros that are not significant. The answer must still show the same precision as before (e.g., $\pm1000$). For example, if we tried to subtract 8 or 80 from 12,000 it would not change our estimate since the resulting numbers must still be rounded off to 12,000. Subtracting 800, however, would affect the result since we can now round off the number to 11,000. These three operations are illustrated as follows.

$$
\begin{array}{ccc}
12{,}000 & 12{,}000 & 12{,}000 \\
-\underline{\phantom{00000}} = 12{,}000 & -\underline{\phantom{00000}} = 12{,}000 & -\underline{\phantom{00000}} \\
11{,}992 = \underline{12{,}000} & 11{,}920 = \underline{12{,}000} & 11{,}200 = \underline{11{,}000}
\end{array}
$$

| Example 1-2 | **Expressing Summations and Subtractions to the Proper Decimal Place** |
| --- | --- |
| working it out | Carry out the following calculations, rounding off the answer to the proper decimal place. |

$$
\begin{array}{rrr}
7.56 & 14{,}000 & \\
+\ 0.375 & +\ 580 & 0.0327 \\
+14.2203 & +\ \ \ 75 & -0.00068 \\
\hline
22.1553 = \underline{\underline{22.16}} & 14{,}655 = \underline{\underline{15{,}000}} & 0.03202 = \underline{\underline{0.0320}}
\end{array}
$$

In multiplication and division, we consider the number of significant figures in the answer rather than the number of decimal places. *The answer is expressed with the same number of significant figures as the multiplier, dividend, or divisor with the least number of significant figures.* In other words, the answer can be only as precise as the least precise part of the calculation (i.e., the chain is only as strong as its weakest link). We must be careful with this rule, however. If one of the multipliers represents a whole-number count, rather than a measurement, then the multiplication is essentially a shortcut to addition. For example, if one cheeseburger weighs 352.4 grams, then three cheeseburgers weigh 1057.2 grams [i.e., 3 (exact) $\times$ 352.4 = 1057.2] grams. In this case we observed the addition and subtraction rules regarding the decimal place since multiplying 352.4 by 3 is the same calculation as 352.4 + 352.4 + 352.4. This is a confusing wrinkle to that rule, but we will remind you of this exception when it shows up in some future exercises. *Just remember that in addition and subtraction we are concerned with the decimal point, but in multiplication and division we are concerned about significant figures.*

| Example 1-3 | **Expressing Multiplications and Divisions to the Proper Number of Significant Figures** |
| --- | --- |
| working it out | Carry out the following calculations. Assume the numbers represent measurements so that the answer should be rounded off to the proper number of significant figures or decimal places. |

**a.** $2.34 \times 3.225$

The answer on the calculator reads 7.5465. Since the first multiplier has three significant figures and the second has four, the answer should be expressed to three significant figures. The answer is rounded off to $\underline{\underline{7.55}}$.

**b.** $\dfrac{11.688}{4.0}$

The answer shown on the calculator is 2.922 but should be rounded off to two significant figures. The answer is $\underline{\underline{2.9}}$.

**c.** $(0.56 \times 11.73) + 22.34$

In cases where we must use both rules, carry out the exercise in parentheses first, round off to the proper number of significant figures, and then add and finally round off to the proper decimal place.

$$(0.56 \times 11.73) = 6.6 \qquad 6.6 + 22.34 = \underline{\underline{28.9}}$$

## Ted Williams and Significant Figures

Ted Williams was possibly the best hitter that baseball has ever known. He played from 1939 to 1960, interrupted by service as a fighter pilot in World War II and again in the Korean War. He was a true American hero. The "Splendid Splinter," as he was known, passed away on July 5, 2002, at age 83.

So what does Ted Williams have to do with chemistry? Not much, actually. But his most important record has a lot to do with significant figures and how numbers are rounded off. Among many other records that he set, Williams was the last major-leaguer to hit the magical "400" in a season. In 1941, Williams, of the Boston Red Sox, had 185 hits in 456 at-bats. Baseball averages are computed by dividing the number of hits by the number of at-bats expressed as a decimal fraction rounded off to three significant figures. Thus his official final average was 406 or 0.406.

But there is more to this story. With one final day of baseball remaining, Williams had 179 hits in 448 at bats for an average of 0.39955, which would officially round off to 0.400. If he had had only one less hit for the whole season his average would have been 0.397—great but still not 400. The Boston manager, Joe Cronin, offered to let Ted sit out the final day's games (a double-header) and preserve his 400 average for the record books. Ted was a true professional, however, so he put his 400 average at risk and played both games. If he had only three hits out of eight at-bats for the day, his average would drop to 0.39912, which now rounds off to 0.399. He would have to go four for eight or at least three for seven to preserve his average. Williams ended the day with a home run, a double, and four singles out of eight at-bats, which brought his average up to 0.4057 for an official average of 0.406.

The great Ted Williams did not want to settle for a record that had to be "rounded off."

*Ted Williams, a baseball legend.*

# 1-3 Expressing Large and Small Numbers: Scientific Notation

**Looking Ahead!** We won't go far into the study of chemistry before we encounter some really large or small numbers. Keeping track of five or more zeros in these numbers is close to impossible, so how can we express the numbers in more compact and readable forms? That is the subject we review in this section, with further information and exercises in Appendix C.

It takes more than a quick glance at a number such as 0.0000078 m to have any idea of its magnitude. However, if the number is written as $7.8 \times 10^{-6}$ m, one can immediately gauge its value. When numbers require the use of so many nonsignificant zeros that one has to carefully count them, we express the numbers in scientific notation. *In scientific notation, a number is expressed with one nonzero digit to the left of the decimal point multiplied by 10 raised to a given power. The* **exponent** *is the appropriate power to which 10 is raised.* Following are some powers of 10 and their equivalent numbers.

$$10^0 = 1 \qquad\qquad 10^{-1} = \frac{1}{10^1} = 0.1$$

$$10^1 = 10 \qquad\qquad 10^{-2} = \frac{1}{10^2} = 0.01$$

$$10^2 = 10 \times 10 = 100 \qquad\qquad 10^{-3} = \frac{1}{10^3} = 0.001$$

$$10^3 = 10 \times 10 \times 10 = 1000 \qquad\qquad 10^{-4} = \frac{1}{10^4} = 0.0001$$

$$10^4 = 10 \times 10 \times 10 \times 10 = 10,000 \qquad \text{etc.}$$
etc.

A huge number that we will soon deal with is 602,200,000,000,000,000,000,000. The way it is expressed, however, is in a much more readable form as follows.

Exponent

$$\underline{6.022} \times 10^{23}$$

Notice that the number (6.022) expresses the proper precision of four significant figures. This expression means that 6.022 is multiplied by $10^{23}$ (which is 1 followed by 23 zeros).

In the following exercises, we will give examples of how numbers are expressed in scientific notation and how scientific notation is handled in multiplication and division. These examples can serve as a brief review, but if further practice is needed, see Appendix C for additional discussion on adding, squaring, and taking square roots of numbers in scientific notation. Appendix E includes discussion of manipulation of scientific notation with calculators. Before we consider the examples, we can see how scientific notation can remove the ambiguity of numbers such as 12,000 where the zeros may or may not be significant. Notice that by expressing the number in scientific notation we can make it clear whether one or more of the zeros are actually significant.

$1.2 \times 10^4$ has two significant figures

$1.20 \times 10^4$ has three significant figures

$1.200 \times 10^4$ has four significant figures

---

| Example 1-4 | **Changing Ordinary Numbers to Scientific Notation** |
| --- | --- |
| **working it out** | |

Express each of the following numbers in scientific notation (one digit to the left of the decimal point).

**a.** 47,500    **b.** 5,030,000    **c.** 0.0023    **d.** 0.0000470

**Solution**

**a.**  The number 47,500 can be factored as $4.75 \times 10,000$. Since $10,000 = 10^4$, the number can be expressed as

$$4.75 \times 10^4$$

A more practical way to transform this number to scientific notation is to count to the left from the old implied decimal point to where you wish to put the new decimal point. The number of places counted to the *left* will be the positive exponent of 10.

$$\overset{4\quad 3\quad 2\quad 1}{4\,7\,5\,0\,0} = 4.75 \times 10^4$$

**b.**
$$\overset{6\quad 5\quad 4\quad 3\quad 2\quad 1}{5,0\,3\,0,0\,0\,0} = 5.03 \times 10^6$$

**c.**  0.0023 can be factored into $2.3 \times 0.001$. Since $0.001 = 10^{-3}$, the number can be expressed as

$$2.3 \times 10^{-3}$$

A more practical way is to count to the right from the old decimal point to where you wish to put the new decimal point. The number of places counted to the *right* will be the negative exponent of 10.

$$\overset{1\quad 2\quad 3}{0.0\,0\,2\,3} = 2.3 \times 10^{-3}$$

**d.**
$$\overset{1\quad 2\quad 3\quad 4\quad 5}{0.0\,0\,0\,0\,4\,7\;0} = 4.70 \times 10^{-5}$$

| Multiplication and Division of Numbers Expressed in Scientific Notation (Consult Appendix E, Section 3, on how to express exponents of 10 on a calculator.) | Example 1-5 |
|---|---|

Carry out the following operations. Express the answer to the proper number of significant figures

**a.** $(8.25 \times 10^{-5}) \times (5.442 \times 10^{-3})$

**b.** $(4.68 \times 10^{16}) \div (9.1 \times 10^{-5})$

### (a) Procedure

Group the digits and the powers of 10. Multiply the digits and add the powers of 10.

### Solution

$$(8.25 \times 5.442) \times (10^{-5} \times 10^{-3}) = 44.9 \times 10^{-8}$$
$$= \underline{4.49 \times 10^{-7}} \text{ (three significant figures)}$$

### (b) Procedure

Group the digits and the powers of 10. Divide the digits and subtract the powers of 10.

### Solution

$$\frac{4.68 \times 10^{16}}{9.1 \times 10^{-5}} = \frac{4.68}{9.1} \times \frac{10^{16}}{10^{-5}}$$
$$= 0.51 \times 10^{16-(-5)}$$
$$= 0.51 \times 10^{21}$$
$$= \underline{5.1 \times 10^{20}} \text{ (two significant figures)}$$

---

**Looking Back!** The evaluation of the numerical part of a measurement has been our first order of business in the study of chemistry. The precision or uncertainty is expressed in the number of significant figures. When we manipulate these numbers in calculations, we turn to the convenient hand calculator. But these devices still need a human brain to express the answer properly since they are clueless in this matter. The very large and small numbers that we will encounter are expressed in a form of exponential notation known as scientific notation. The mathematical skills that we have now introduced can be quickly checked by proceeding to the following Learning Check.

---

## Learning Check A | checking it out

**A-1.** Fill in the blanks.

In a measurement, a significant figure is a number that is either known or _____ . The number of significant figures in a measurement relates to its _____ . How close the number is to the true value is referred to as the _____ of the measurement. In a measurement, zero is a significant figure when it is _____ other digits or to the _____ of a digit and to the _____ of the decimal point. In addition and subtraction of numbers with various degrees of precision, the answer is expressed to the same decimal place as the number with the _____ number of _____ _____ . In multiplication and division, the answer is expressed with the same number of significant figures as the number with the _____ number of _____ _____ . Awkwardly large or small values are expressed in _____ _____ . These numbers consist of a number with _____ digit to the left of the decimal multiplied by 10 raised to a power. The power of 10 is known as an _____ .

**A-2.** How many significant figures are in the following measurements?

**a.** 2.33 ft    **b.** 40.01 lb    **c.** 2.30 L    **d.** 10,200 km    **e.** 0.020 g

**A-3.** Round off the following numbers to three significant figures.

**a.** 23.44    **b.** 483,550    **c.** 0.02203

**A-4.** Carry out the following operations and express the answers properly.

**a.** 19.63 + 0.366    **b.** 0.200 × 12.765    **c.** (12.45 − 11.65) × 2.68

**A-5.** Express the following numbers in scientific notation.

**a.** 456,000,000 to four significant figures    **b.** 0.000340

**A-6.** Carry out the following operations and express the answers properly.

**a.** $(9.41 \times 10^{12}) \times (2.7722 \times 10^{-5})$    **b.** $\dfrac{4.856 \times 10^{10}}{0.020 \times 10^{4}}$

*Additional examples: Exercises 1-4, 1-8, 1-14, 1-18, 1-32, and 1-47.*

---

Section **B**  **THE MEASUREMENTS USED IN CHEMISTRY**

## 1-4 Measurement of Mass, Length, and Volume

**Looking Ahead!** Now that we have examined the numerical part of a measurement, we are ready to examine the other part—the units that are expressed. If one says that "Chicago is 250 away," we have an incomplete statement. Is it 250 miles or kilometers? We need a unit that is a determined quantity compared against a standard. We are now ready to examine some of the units used in chemistry.

In the United States, we use the measurement of length called the *foot*. The standard for this unit was originally the length of certain English king's foot. Fortunately, we have progressed to more reliable and precise standards than some dead king's body part. A disadvantage of the English system is that it can be awkward. For example, a trophy bass out of a midwestern lake tips the scale at 9 lb 6 oz. A highly recruited basketball player for a men's college team tops out at 6 ft 11 in. A major problem with these units is that they lack any systematic relationship between units (see Table 1-1). Thus we often need to use two units (e.g., feet and inches) to report only one measurement. Our monetary system is an exception because it is based on the decimal system. Therefore, only one unit (dollars and decimal fractions of dollars) is needed to show a typical student's dismal financial condition (e.g., $11.98). The old English system required two units (pounds and shillings).

Most of the world and the sciences use the metric system of measurement for length, volume, and mass. In this system, units are conveniently related by multiples of 10. Since 1975, there have been plans to convert to the metric system in the United States, but

**Table 1-1  Relationships Among English Units**

| Length | Volume | Mass |
|--------|--------|------|
| 12 inches (in.) = 1 foot (ft) | 2 pints (pt) = 1 quart (qt) | 16 ounces (oz) = 1 pound (lb) |
| 3 ft = 1 yard (yd) | 4 qt = 1 gallon (gal) | 2000 lb = 1 ton |
| 1760 yd = 1 mile (mi) | 42 gal = 1 barrel (bbl) | |

(a)

(b)

(c)

**Figure 1-2 Length, Volume, and Mass** These properties of a quantity of matter are measured with common laboratory equipment: (*a*) metric ruler; (*b*) graduated cylinders, burets, volumetric flasks, and pipets; (*c*) an electric balance.

action has been implemented in fits and starts. Complete conversion may not occur in our lifetime, but most citizens are becoming more familiar with this system. (See Figure 1-2.) Metric units also form the basis of the SI system, after the French *Système International* (International System). The SI units that will concern us are listed in Table 1-2. (Other SI units designate measurements that are not used in this text.) The basic units of the SI system have very precisely defined standards based on certain precisely known properties of matter and light. For example, the unit of one meter is defined as the distance light travels in a vacuum in 1/299,792,458th of a second. Obviously, this is extremely precise, but when we are aiming a space ship at a planet billions of miles away (e.g., Neptune), we need a great deal of precision in our units. Although some English units were formerly based on a king's anatomy, in modern times many have been redefined more precisely based on a corresponding metric unit. For example, one inch is now defined as exactly equal to 2.54 centimeters. The relationship between the familiar mile and the kilometer originated from a measurement but is now defined as one mile equal to *exactly* 1.609344 km. For simplicity in our calculations, however, the relationship is rounded off to four significant figures. That is, 1.000 mi = 1.609 km.

Use of metric units is simplified by their exact relationships by powers of 10. This is illustrated in Table 1-4 by use of the more common prefixes listed in Table 1-3. In

**Table 1-2 SI Units**

| Measurement | Unit | Symbol |
|---|---|---|
| Mass | kilogram | kg |
| Length | meter | m |
| Time | second | s |
| Temperature | kelvin | K |
| Quantity | mole | mol |
| Energy | joule | J |
| Pressure | pascal | Pa |

**Table 1-3 Prefixes Used in the Metric System**

| Prefix | Symbol | Relation to Basic Unit | Prefix | Symbol | Relation to Basic Unit |
|---|---|---|---|---|---|
| tera- | T | $10^{12}$ | deci- | d | $10^{-1}$ |
| giga- | G | $10^{9}$ | centi- | c | $10^{-2}$ |
| mega- | M | $10^{6}$ | milli- | m | $10^{-3}$ |
| kilo- | k | $10^{3}$ | micro- | $\mu$* | $10^{-6}$ |
| hecto- | h | $10^{2}$ | nano- | n | $10^{-9}$ |
| deca- | da | $10^{1}$ | pico- | p | $10^{-12}$ |

*A Greek letter, mu.

## Worlds from the Small to the Distant—Picometers to Terameters

The meter is just about right for the world of human beings. For example, our height may be a little less than two meters and the room may be about seven square meters. If people were the size of ants, however, the meter would not be a very practical unit of length. Fortunately, the metric system allows us to venture into the far reaches of other worlds both small and distant. All we need is a few prefixes to adjust the unit. First, let's shrink ourselves into the world of the small. Our first stop is the "milli" world ($10^{-3}$ m), where normally small specks now appear large. Those little bugs called mites are of this size and if we lived in this world it would look a lot more hostile than our world of the meter. When we go to the "micro" world ($10^{-6}$ m), we are well into the invisible world of the tiny. Blood cells that circulate in our veins have these dimensions (i.e., 7.5 $\mu$m) Next, we descend to the "nano" world ($10^{-9}$ m), where we encounter one of the tiniest forms of life called viruses. Some of these nasty creatures measure 10 nm across. Our final stop is the "pico" region ($10^{-12}$ m), where we notice that the basic component of matter, atoms, appears as large spheres. The aluminum atom in a can of soda measures 125 pm in diameter.

Now we take ourselves out away from the Earth into the world of the distant. One km ($10^3$ m) from the surface we are not that far up. We can still see cars and buildings. But if we ascend to one megameter ($10^6$ m) we are certainly in outer space. In this region the artificial satellites drift silently by. Our planet still looms large in our vision with a diameter of about 6 Mm. Now let's ascend to one gigameter ($10^9$ m) from the surface. We find ourselves well past the moon, which would lie about halfway between us and Earth. Venus is still 32 Gm away. Finally, at one terameter ($10^{12}$ m) we are nearly to the planet Saturn, which lies about 1.2 Tm from Earth at its closest approach.

The metric system obviously provides convenient units from the smallest to the farthest. Even more units exist that take us to even smaller and farther regions.

Saturn is about 1 Tm from Earth

2 meters tall

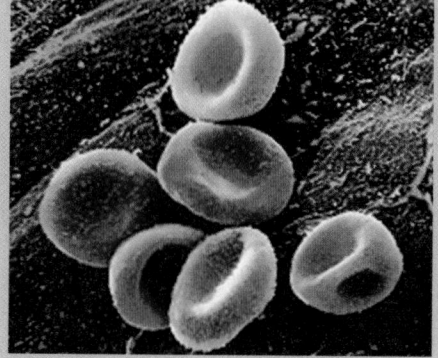

Blood cells are about 1 $\mu$m in diameter

the Making it Real box, we encounter the use of some of these prefixes in smaller and larger dimensions.

There is one other convenient feature of the metric system. There is an exact relationship between length and volume. **Volume** *is the space that a given quantity of*

**Table 1-4  Relationships Among Metric Units Using Common Prefixes**

| Mass Unit | Symbol | Relation to Basic Unit | Volume Unit | Symbol | Relation to Basic Unit | Length Unit | Symbol | Relation to Basic Unit |
|---|---|---|---|---|---|---|---|---|
| kilogram | kg | $10^3$ g | kiloliter | kL | $10^3$ L | kilometer | km | $10^3$ m |
| decigram | dg | $10^{-1}$ g | deciliter | dL | $10^{-1}$ L | decimeter | dm | $10^{-1}$ m |
| centigram | cg | $10^{-2}$ g | centiliter | cL | $10^{-2}$ L | centimeter | cm | $10^{-2}$ m |
| milligram | mg | $10^{-3}$ g | milliliter | mL | $10^{-3}$ L | millimeter | mm | $10^{-3}$ m |
| microgram | $\mu$g | $10^{-6}$ g | | | | nanometer | nm | $10^{-9}$ m |

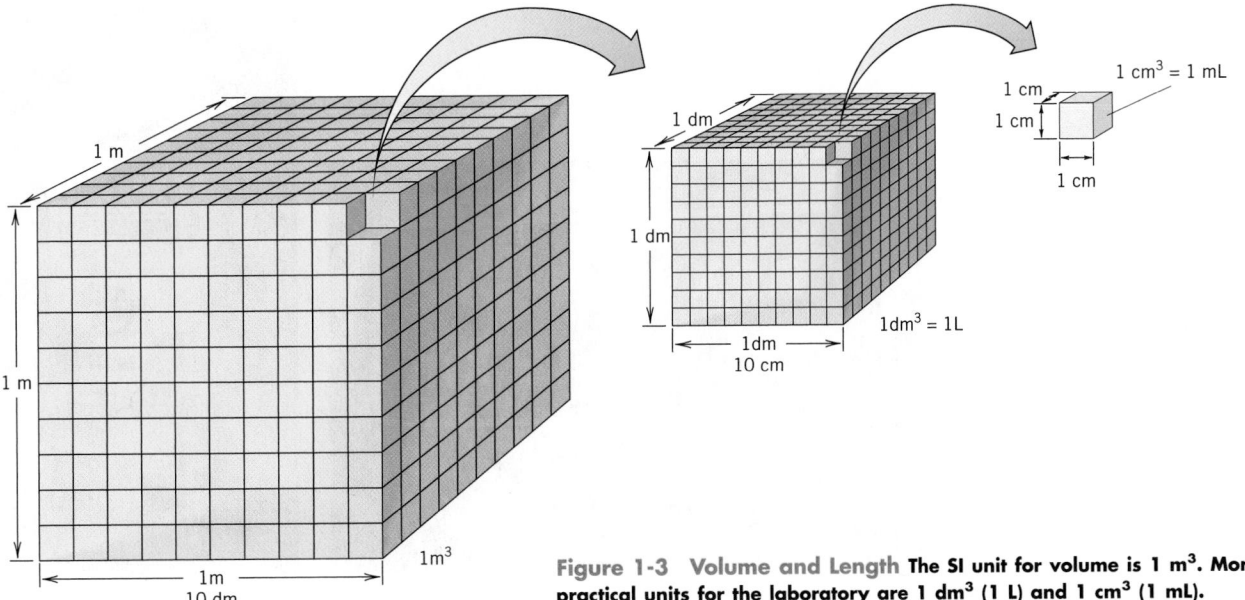

**Figure 1-3 Volume and Length** The SI unit for volume is 1 m³. More practical units for the laboratory are 1 dm³ (1 L) and 1 cm³ (1 mL).

*matter occupies.* The SI unit for volume is the cubic meter ($m^3$). Since this is a rather large volume for typical laboratory situations, the metric unit, known as the liter, is used. One liter is defined as the exact volume of one cubic decimeter (i.e., 1 L = 1 dm³). On a smaller scale, one milliliter is the exact volume of one cubic centimeter (1 mL = 1 cm³ = 1 cc). Thus the units milliliter and cubic centimeter can be used interchangeably when expressing volume. (See Figure 1-3.)

The basic metric unit of mass is the *gram,* but the SI unit is the *kilogram* (kg), which is equal to 1000 grams (g). **Mass** *is the quantity of matter that a sample contains.* The terms "mass" and "weight" are often used interchangeably, but they actually refer to different concepts. **Weight** *is a measure of the attraction of gravity for the sample.* An astronaut has the same mass on the moon as on Earth. Mass is the same anywhere in the universe. An astronaut who weighs 170 lb on Earth, however, weighs only about 29 lb on the moon. In Earth orbit, where the effect of gravity is counteracted, the astronaut is "weightless" and floats free. You can lose all your weight by going into orbit, but obviously your body (mass) is still there. However, since our relevant universe is confined mostly to the surface of the Earth, we often use weight as a measure of mass. In this text, we will use the term "mass," as it is the more scientific term.

Several relationships between the metric and English systems are listed in Table 1-5. (See also Figures 1-4 and 1-5.) The relationships *within* systems (e.g.,

**Table 1-5 The Relationship Between English and Metric Units**

|  | English | Metric Equivalent |
|---|---|---|
| Length | 1.000 in. | 2.540 cm (exact) |
|  | 1.000 mi | 1.609 km |
| Mass | 1.000 lb | 453.6 g |
|  | 2.205 lb | 1.000 kg |
| Volume | 1.057 qt | 1.000 L |
|  | 1.000 gal | 3.785 L |
| Time | 1.000 s | 1.000 s |

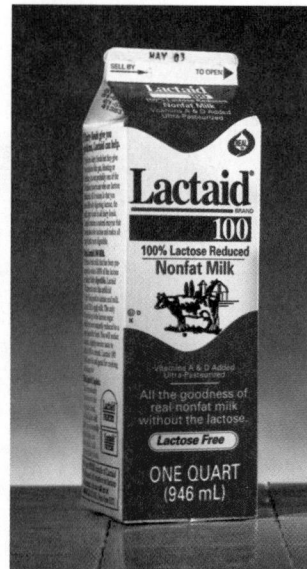

**Figure 1-4   Metric Units** **The use of metric units in this country is becoming increasingly evident.**

Tables 1-1 and 1-4) are always defined and exact numbers. The relationships *between* systems, however, are not necessarily exact, so they can be expressed to various degrees of precision. Most of the relationships shown in Table 1-5 are known to many more than the four significant figures given.

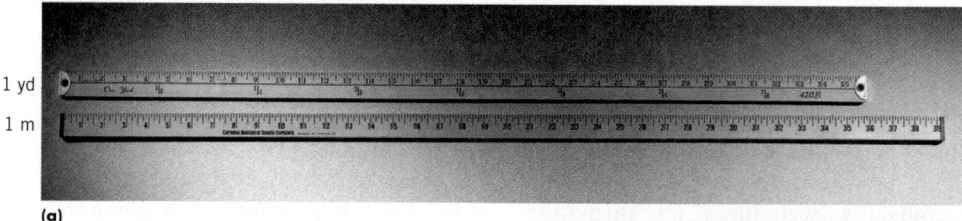

(a)

1 L          1 Qt

(b)

1 kg        1 lb

(c)

**Figure 1-5   Comparison of Metric and English Units** **(a)** **1 meter and 1 yard, (b) 1 quart and 1 liter, and (c) 1 kilogram and 1 pound.**

The astronaut is weightless in Earth orbit but still has mass.

# 1-5 Conversion of Units by the Factor-Label Method

**Looking Ahead!** When one learns a new language, the first goal is to be able to translate sentences from the new to the familiar. Likewise, to become comfortable with a different system of measurement we need to convert from one to the other. Converting between systems can be accomplished with an important tool that we will introduce in this section. This will give us a chance to practice solving many of the types of problems that we will eventually encounter.

In a huge gaffe of NASA in 1999, a rocket destined to land softly on the planet Mars splattered all over the surface instead. The reason is that someone did not convert between English and metric units. We convert between one thing or another all of the time. For example, electrical converters change alternating current (AC) to direct current (DC) for devices such as electrical razors or to charge a cell phone. What we need is a converter to change measurements in one unit to those in another. In chemistry this mathematical converter is known as a conversion factor. *A* **conversion factor** *is a relationship between two units or quantities expressed in fractional form. The* **factor-label method** *(also called* **dimensional analysis***) converts a measurement in one unit to another by the use of conversion factors.* For example, a conversion factor can be constructed from the exact relationship within the metric system.

$$10^3 \, \text{m} = 1 \, \text{km}$$

This equality can be converted to two fractions as shown below.

$$\textbf{(1)} \ \frac{1 \, \text{km}}{10^3 \, \text{m}} \quad \text{and} \quad \textbf{(2)} \ \frac{10^3 \, \text{m}}{1 \, \text{km}} \left( \text{or simply} \ \frac{10^3 \, \text{m}}{\text{km}} \right)$$

These fractional relationships read as "one kilometer per $10^3$ meters" and "$10^3$ meters per one kilometer." The latter fraction is usually simplified to "$10^3$ meters per kilometer." When just a unit (no number) is read or written in the denominator, it is assumed to be 1 of that unit. *Factors that relate a quantity in a certain unit to 1 of another unit are sometimes referred to as* **unit factors.** Unit factors are written without the 1 in the denominator in this text. It should be understood, however, that the 1 in the numerator or implied in the denominator can be considered exact in calculations.

**AC–DC. To recharge this cell phone AC current must be converted into DC current.**

Two other conversion factors can be constructed from an equality between English and metric units. For example, the exact relationship between inches (in.) and centimeters (cm) is

$$1 \text{ in.} = 2.54 \text{ cm}$$

which can be expressed in fractional form as

$$(3) \; \frac{1 \text{ in.}}{2.54 \text{ cm}} \quad \text{and} \quad (4) \; \frac{2.54 \text{ cm}}{\text{in.}}$$

Each of the four fractions can be used to convert one unit into the other. The general procedure for a one-step conversion by the factor-label method is

$$(\text{what's given}) \times (\text{conversion factor}) = (\text{what's requested})$$

In most conversions, "what's given" and "what's requested" each have one unit. The conversion factor has two: the requested unit in the numerator and the given unit in the denominator.

$$\text{A (given unit)} \times \frac{\text{B (requested unit)}}{\text{C (given unit)}} = \text{D (requested unit)}$$

When combined in the calculation, the given units will appear in both the numerator and the denominator and thus cancel just like any numerical quantity. The requested unit survives. This is illustrated as follows, where A, B, C, and D represent the numerical parts of the measurements or definitions.

$$\frac{\text{A (\sout{given unit})} \times \text{B (requested unit)}}{\text{C (\sout{given unit})}} = \text{D (requested unit)}$$

Doing problems by this method is like taking a trip. The two things we know about a trip are where we start and where we want to end up. The conversion factor is "how" we get from here to there. The key to using the factor-label method is to select the right conversion factor that relates the given and requested units. Units are treated like numerical quantities; that is, they are multiplied, divided, or canceled in the course of the calculation.

We will now give some examples of how the factor-label method is used in some one-step conversions. The first problem (Example 1-6) will put to use one of the two conversion factors that we constructed within the metric system, and the next (Example 1-7) will use one of the two conversion factors between the English and metric systems. In the examples, we proceed carefully through four steps.

1. From what is given and what is requested, decide what the conversion factor must do (e.g., convert in. to cm). You can express the procedure in a shorthand method which we will refer to as the *unit map* in this text. (e.g., in. → cm).

2. Find the proper relationship between units from a table (if necessary). Express the relationship as a conversion factor so that the new unit is in the numerator.

3. Put the problem together, making sure that the proper unit cancels and your answer has the requested unit or units.

4. Express your answer to the proper number of significant figures. Be aware of conversion factors that are not the result of measurements but instead are *exact* definitions. The following relationships are examples of exact definitions

$$12 \text{ in.} = 1 \text{ ft} \qquad 4 \text{ qt} = 1 \text{ gal} \qquad 1 \text{ km} = 10^3 \text{ m}$$

That is, there are *exactly* 12 in. (12.0000 etc.) in one foot (1.0000 etc.) and so forth. *Since exact relationships are considered to have unlimited precision in a calculation, they can be ignored in determining the number of significant figures in an answer.* Exact definitions are generally relationships within a measurement system (e.g., in. and ft, km and m). *Measurements are never exact,* so that the number of significant figures shown reflects the precision of the instrument used to make the measurement. The use of exact relationships in calculations can be somewhat confusing at first, so we will remind you when one is being used in calculations in this chapter.

---

**Converting Hours to Days**

**Example 1-6**

working it out

Convert 129 hours to an equivalent number of days.

**Procedure**

1. The unit map for the conversion is

$$hour(hr) \longrightarrow day$$

2. The well-known relationship is that there are exactly 24 hours in one day. This exact relationship can be written as two conversion factors

$$\frac{24 \text{ hr}}{\text{day}} \text{ and } \frac{1 \text{ day}}{24 \text{ hr}}$$

The latter is the appropriate factor since hr (given) is in the denominator and the new requested unit (day) is in the numerator.

**Solution**

$$129 \text{ hr} \times \frac{1 \text{ day*}}{24 \text{ hr}} = \underline{\underline{5.38 \text{ day}}}$$

What would have happened if we had used the first conversion factor (i.e., 24 hr/day) by mistake? In that case, the strange units that result (i.e., $hr^2$/day) would alert us that we goofed. Paying close attention to the units that cancel and those that don't will be a big help in correctly solving equivalent chemistry problems.

---

**Converting Meters to Kilometers**

**Example 1-7**

Convert 0.468 m to (a) km and (b) mm.

**(a) Procedure**

1. The unit map for the conversion is

$$m \longrightarrow km$$

2. From Table 1-4, we find the relationship is $10^3$ m = 1 km. When expressed as the proper conversion factor, *km* is in the numerator (requested) and *m* is in the denominator (given). The conversion factor is

$$\frac{1 \text{ km}}{10^3 \text{ m}}$$

**Solution**

$$0.468 \text{ m} \times \frac{1 \text{ km*}}{10^3 \text{ m}} = 0.468 \times 10^{-3} \text{ km} = \underline{\underline{4.68 \times 10^{-4} \text{ km}}}$$

---

*This is an exact definition.

**(b) Procedure**

1. The unit map for the conversion is

$$m \longrightarrow mm$$

2. From Table 1-4, we find the proper relationship is 1 mm = $10^{-3}$ m. When shown as the proper conversion factor, *mm* is in the numerator (requested) and *m* is in the denominator (given). The conversion factor is

$$\frac{1 \text{ mm}}{10^{-3} \text{ m}}$$

**Solution**

$$0.468 \text{ m} \times \frac{1 \text{ mm}^*}{10^{-3} \text{ m}} = 0.468 \times 10^3 \text{ mm} = \underline{\underline{468 \text{ mm}}}$$

---

**Example 1-8** | **Converting Centimeters to Inches**

Convert 825 cm to in.

**Procedure**

1. The unit map for the conversion is

$$cm \longrightarrow in.$$

2. From Table 1-5, we find the proper relationship is 1 in. = 2.54 cm. Expressed as a conversion factor, *in.* is in the numerator (requested) and *cm* is in the denominator (given). The conversion factor is

$$\frac{1 \text{ in.}}{2.54 \text{ cm}}$$

**Solution**

$$825 \text{ cm} \times \frac{1 \text{ in.}}{2.54 \text{ cm}} = \underline{\underline{325 \text{ in.}}}$$

In conversions between the metric and English systems we will use a conversion factor with four significant figures. (Recall that the relationship between inches and centimeters, used in Example 1-8, is exact, however.) Since most of the measurements are given to three significant figures, this means that the conversion factor does not limit the precision of the answer. The answer should then be expressed to the same number of significant figures as the original measurement.

A one-step conversion like those that have been worked so far is analogous to a direct, nonstop airline flight between your home city and your destination. A multistep conversion is analogous to the situation in which a nonstop flight is not available and you have to make the flight with two or more stops before reaching your destination. Each stop in the flight is a separate journey, but it gets you closer to your ultimate destination. When one plans such a journey, one must carefully plan each step, perhaps with the help of a map. Many of our conversion problems also require more than one step and must be carefully planned. In multistep conversion problems, each step along the way requires a separate conversion factor.

For example, let us consider a problem requiring a conversion between two possible units of quantity—number of apples and boxes of apples. Let us assume we know that exactly six apples can be placed in each sack and exactly four sacks make up one box. The problem is, "How many boxes of apples can be prepared with 254 apples?" Since we don't have a direct relationship between apples and boxes, we need a "game plan." First, we can convert number of apples to number of sacks (conversion **1**) and then number of sacks to number of boxes (conversion **2**). The problem together with the game plan (unit map) are set up and completed as follows.

$$\overset{(1)}{\phantom{x}} \qquad \overset{(2)}{\phantom{x}}$$
$$\text{apples} \longrightarrow \text{sacks} \longrightarrow \text{boxes}$$

$$254 \text{ apples} \times \frac{1 \text{ sack}}{6 \text{ apples}} \times \frac{1 \text{ box}}{4 \text{ sacks}} = \frac{254}{6 \times 4} \text{ boxes} = 10.6 \text{ boxes}$$

Notice that the two conversion factors represent exact relationships, so they do not affect the number of significant figures in the answer.

We are now ready to work through some real multistep conversions between units of measurement.

---

**Converting Dollars to Doughnuts**

**Example 1-9**

**working it out**

How many doughnuts can one purchase for $123 if doughnuts cost $3.25 per dozen?

**Procedure**

1.  A conversion factor between dollars and individual doughnuts is not directly available. We can solve this problem by making a two-step conversion by (a) converting dollars ($) to dozen (doz.) and then (b) converting dozen to number of doughnuts. The unit map for the conversions is shown as

$$\overset{(a)}{\phantom{x}} \qquad \overset{(b)}{\phantom{x}}$$
$$\$ \longrightarrow \text{doz.} \longrightarrow \text{doughnuts}$$

2.  The conversion factors arise from the two relationships: $3.25 per dozen and 12 doughnuts per dozen (an exact relationship). In conversion factor (a) $ must be in the denominator to convert to doz. In conversion factor (b) doz. [the new unit from the conversion in (a)] must be in the denominator to convert to doughnuts. The two conversion factors are

$$\text{(a) } \frac{1 \text{ doz.}}{\$3.25} \quad \text{and} \quad \text{(b) } \frac{12 \text{ doughnuts}}{\text{doz.}}$$

**Solution**

$$\$123 \times \frac{1 \text{ doz.}}{\$3.25} \times \frac{12 \text{ doughnuts*}}{\text{doz.}} = \underline{\underline{454 \text{ doughnuts}}}$$

---

**Converting Liters to Gallons**

**Example 1-10**

Convert 9.85 L to gal.

**Procedure**

1.  A two-step conversion is needed. *L* can be converted into *qt* (a) and *qt* then converted into *gal* (b). The unit map for the conversions is expressed as

$$\overset{(a)}{\phantom{x}} \qquad \overset{(b)}{\phantom{x}}$$
$$L \longrightarrow qt \longrightarrow gal$$

---

*This is an exact definition.

2. The two relationships needed are 1.057 qt = 1.000 L (from Table 1-5) for conversion (a) and 4 qt = 1 gal (from Table 1-1) for conversion (b). The two relationships properly expressed as conversion factors are

$$(a) \frac{1.057 \text{ qt}}{L} \quad \text{and} \quad (b) \frac{1 \text{ gal}}{4 \text{ qt}}$$

**Solution**

$$\quad\quad\quad (a) \quad\quad\quad (b)$$
$$9.85 \, \cancel{L} \times \frac{1.057 \, \cancel{\text{qt}}}{\cancel{L}} \times \frac{1 \text{ gal*}}{4 \, \cancel{\text{qt}}} = \underline{\underline{2.60 \text{ gal}}}$$

---

**Example 1-11** | **Converting Miles per Hour to Meters per Minute**

Convert 55 mi/hr to m/min.

**Procedure**

1. In this case, we can convert *mi/hr* to *km/hr* (a) and then *km/hr* to *m/hr* (b). It is then necessary to change the units of the denominator from *hr* to *min* (c). The unit map for the conversions is expressed as

$$\quad\quad\quad (a) \quad\quad\quad (b) \quad\quad\quad (c)$$
$$\frac{\text{mi}}{\text{hr}} \longrightarrow \frac{\text{km}}{\text{hr}} \longrightarrow \frac{\text{m}}{\text{hr}} \longrightarrow \frac{\text{m}}{\text{min}}$$

2. The needed relationships are 1.000 mi = 1.609 km (from Table 1-5), $10^3$ m = 1 km (from Table 1-4), and 60 min = 1 hr. The relationships expressed as conversion factors are

$$(a) \frac{1.609 \text{ km}}{\text{mi}} \quad (b) \frac{10^3 \text{ m}}{\text{km}} \quad (c) \frac{1 \text{ hr}}{60 \text{ min}}$$

Notice that factor (c) has the requested unit (*min*) in the denominator and the given unit (*hr*) in the numerator. This is because we are changing a denominator in the original unit.

**Solution**

$$\quad\quad (a) \quad\quad\quad (b) \quad\quad\quad (c)$$
$$55 \, \frac{\cancel{\text{mi}}}{\cancel{\text{hr}}} \times \frac{1.609 \, \cancel{\text{km}}}{\cancel{\text{mi}}} \times \frac{10^3 \text{ m*}}{\cancel{\text{km}}} \times \frac{1 \, \cancel{\text{hr}}\text{*}}{60 \text{ min}} = \underline{\underline{1.5 \times 10^3 \text{ m/min}}}$$

---

# 1-6 Measurement of Temperature

**Looking Ahead!** The final measurement that we will consider in this chapter is temperature. Once again, those who grew up in the United States are at a disadvantage. The scale we are most familiar with is not the same as that used by science and most of the rest of the world. We will become more familiar with two other temperature scales in this section.

When someone says an item is "hot," we can interpret this in several ways. It could mean that this item is something we should have, that it looks great, or that it has a somewhat higher temperature than "warm." In this section, we will be concerned with the last concept. "Hot" in science refers to temperature. **Temperature** *is a measure of the intensity of heat of a substance. A* **thermometer** *is a device that* *measures temperature*. The thermometer scale with which we are most familiar in the United States is the **Fahrenheit** scale (°F), but the **Celsius** scale (°C) is commonly used elsewhere and in science. Many U.S. television news shows once broadcast the temperature in both scales, but, unfortunately, this practice is losing popularity.

Thermometer scales are established by reference to the freezing point and boiling point of pure water. These two temperatures are constant and unchanging (under constant air pressure). Also, when pure water is freezing or boiling, the temperature remains constant. We can take advantage of these facts to compare the two temperature scales and establish a relationship between them. In Figure 1-6, the temperature of an ice and water mixture is shown to be exactly 0°C. This temperature was originally established by definition and corresponds to exactly 32°F on the Fahrenheit thermometer. The boiling point of pure water is exactly 100°C, which corresponds to 212°F.

On the Celsius scale, there are 100 equal divisions between these two temperatures, whereas on the Fahrenheit scale, there are $212 - 32 = 180$ equal divisions between the two temperatures. Thus we have the following relationship between the scale divisions:

$$100 \text{ C div.} = 180 \text{ F div.}$$

This relationship can be used to construct conversion factors between an equivalent number of Celsius and Fahrenheit degrees. The relationship results from *exact* definitions, so it does not affect the number of significant figures in the calculation.

$$\frac{100 \text{ C div.}}{180 \text{ F div.}} = \frac{1 \text{ C div.}}{1.8 \text{ F div.}} \quad \text{The inverse relationship is} \quad \frac{1.8 \text{ F div.}}{\text{C div.}}$$

To convert the Celsius temperature [$t$(C)] to Fahrenheit temperature [$t$(F)],

1. Multiply the Celsius temperature by the proper conversion factor (1.8°F/1°C) to get the equivalent number of Fahrenheit degrees.

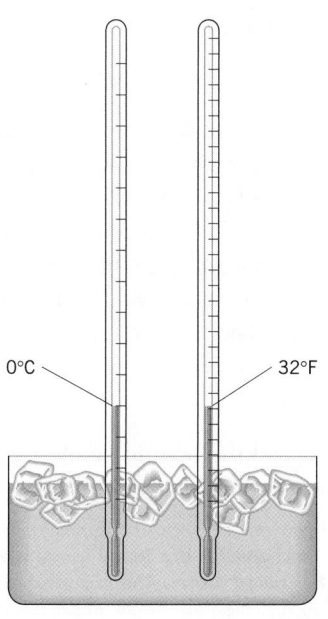

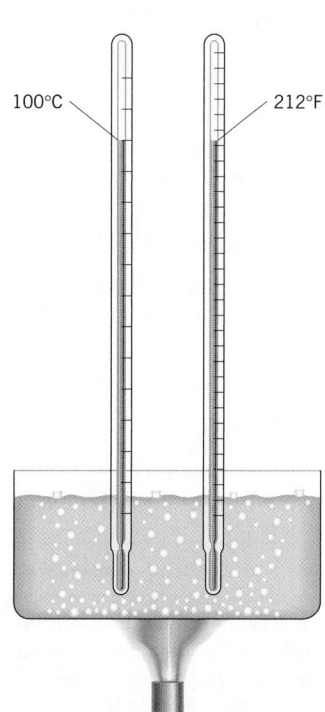

**Figure 1-6  The Temperature Scales**
**The freezing and boiling points of water are used to calibrate the temperature scales.**

2. Add 32°F to this number so that both scales start at the same point (the freezing point of water).

$$t(\text{F}) = \left[ t(\text{C}) \times \frac{1.8°\text{F}}{1°\text{C}} \right] + 32°\text{F} = [t(\text{C}) \times 1.8] + 32$$

To convert the Fahrenheit temperature to the Celsius temperature,

1. Subtract 32°F from the Fahrenheit temperature so that both scales start at the same point.
2. Multiply the number by the proper conversion factor (1°C/1.8°F) to convert to the equivalent number of Celsius degrees.

$$t(\text{C}) = [t(\text{F}) - 32°\text{F}] \times \frac{1°\text{C}}{1.8°\text{F}} = \frac{[t(\text{F}) - 32]}{1.8}$$

| Example 1-12 | Converting Between Fahrenheit and Celsius |
|---|---|
| **working it out** | A person with a cold has a fever of 102°F. What would be the reading on a Celsius thermometer? |

**Solution**

$$t(\text{C}) = \frac{[t(\text{F}) - 32]}{1.8} = \frac{(102 - 32)}{1.8} = \underline{\underline{39°\text{C}}}$$

| Example 1-13 | Converting Between Celsius and Fahrenheit |
|---|---|

On a cold winter day the temperature is −10.0°C. What is the reading on the Fahrenheit scale?

**Solution**

$$t(\text{F}) = [t(\text{C}) \times 1.8] + 32$$
$$t(\text{F}) = (1.8 \times -10.0) + 32.0 = -18.0 + 32.0 = \underline{14.0°\text{F}}$$

The SI temperature unit is called the *kelvin (K)*. Notice that a degree symbol (°) is not shown with a kelvin temperature. The zero on the **Kelvin scale** is theoretically the lowest possible temperature (the temperature at which the heat energy is zero). This corresponds to −273°C (or more precisely −273.15°C). Since the Kelvin scale also has exactly 100 divisions between the freezing point and the boiling point of water, the magnitude of a kelvin and a Celsius degree is the same. Thus we have the following simple relationship between the two scales. The temperature in kelvins is represented by $T(\text{K})$, and $t(\text{C})$ represents the Celsius temperature.

$$T(\text{K}) = t(\text{C}) + 273$$

Thus the freezing point of water is 0°C or 273 K, and the boiling point is 100°C or 373 K. We will use the Kelvin scale more in later chapters.

**Looking Back!** The units of measurement that are used in chemistry are found in the metric system, which is also used in most of the world except the United States. The factor-label method helps us to reliably convert between the two systems. In this method, either exact relationships between units in the same system or equivalent relationships between units in different systems provide the conversion factors needed. The temperature scales used in chemistry and the one used in the United States can also be interconverted.

## Learning Check B | checking it out

**B-1.** Fill in the blanks.

The basic metric unit of length is the _____ , of volume is the _____ , and of mass is the _____ . Other units relate to a basic unit by powers of 10. For example, a quantity of 10 grams is the same as _____ mg and _____ kg. One cubic centimeter in the metric system is the volume of _____ _____ . Changing units by the factor-label method requires the use of _____ factors that are constructed from relationships between units. In a typical problem, the conversion factor has the unit of what's given in the _____ and the units of what's requested in the _____ . The temperature scale used in science is the _____ scale. The _____ scale starts at the lowest possible temperature.

**B-2.** Write a relationship in factor form from Tables 1-1, 1-3, and 1-5 that would make each of the following conversions.

**(a)** feet to yards    **(b)** decimeters to meters    **(c)** gallons to liters    **(d)** pounds to kilograms

**B-3.** How many miles are in 25.0 km?

**B-4.** If the unit price of chili is $0.180/oz, what is the cost of a 12-oz can of chili? How many ounces can you buy at this rate for $1.75?

**B-5.** Convert 12.0 ft to centimeters.

**B-6.** A certain size of nail costs $1.25/lb. What is the cost of 3.25 kg of these nails?

**B-7.** The temperature of cooking oil reaches 248°F. What is this on the Celsius and Kelvin scales?

*Additional examples: Exercises 1-54, 1-61, 1-63, 1-65, 1-67, 1-78, 1-79, and 1-101.*

**chapter** R**eview**                              **putting it together**

Chemistry, probably more than any other science, is a science of **measurements.** Certainly, that is not all there is—many other concepts are also integral to the science—but handling measurements properly is the highest priority. That is why we need to address this subject very early in the text. Our first item of business was to examine the numerical part of a measurement as to the significance of the numbers that are reported. The numerical value has two qualities: **precision** (number of **significant figures**) and **accuracy.**

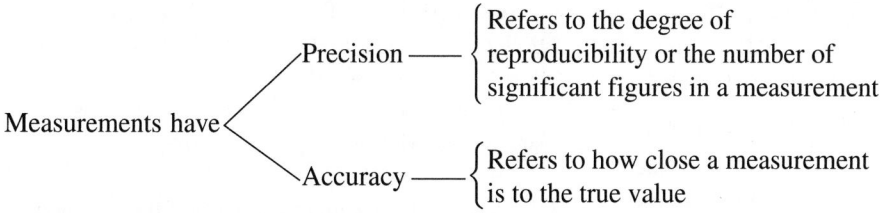

Measurements have
- Precision —— Refers to the degree of reproducibility or the number of significant figures in a measurement
- Accuracy —— Refers to how close a measurement is to the true value

Perhaps a greater challenge than just understanding the precision of one measurement is to handle measurements of varying precision in mathematical operations. The rules are summarized as follows.

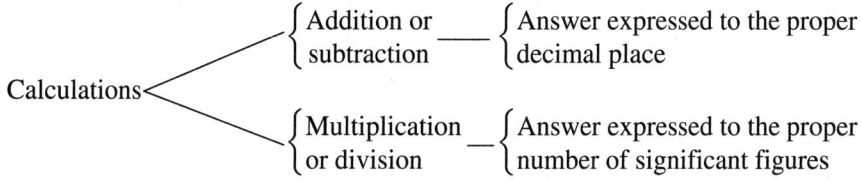

Calculations
- Addition or subtraction —— Answer expressed to the proper decimal place
- Multiplication or division —— Answer expressed to the proper number of significant figures

We won't get too far into the text before we encounter some extremely small and extremely large numbers. Writing large numbers of zeros is cumbersome. **Scientific notation,** however, allows us to write a number in an easy-to-read manner, that is, as a number between one and ten multiplied by ten raised to a power indicated by the **exponent.** Some of the ambiguities of zero as a significant figure are removed when the number is expressed in scientific notation. Remember, help is available in Appendix C if you are uneasy with scientific notation and using it in calculations.

Next, we turned our attention to units of measurements. Most of us have to become familiar with the SI system and the metric system on which it is based, specifically with regard to units of length (meters), **mass** or **weight** (grams), and **volume** (liters). Compared to the rather unsystematic English system, the metric system has the advantage of decimal relationships between units.

Because we need to convert between systems of measurement, we then introduced our main problem-solving technique—the **factor-label method,** also called **dimensional analysis.** In this, we change between units of measurement using **conversion factors** (usually **unit factors**), which are constructed from relationships between units. When a direct relationship is available between what's given and what's requested, conversion can be accomplished in a one-step calculation. If not, the calculation is more involved and requires a game plan that outlines step-by-step conversions from given to requested. This problem-solving method is the language of this text. If you do not follow this method, it is hard to take full advantage of the worked-out examples in this and future chapters.

The final topic of this chapter concerned the measurement of **temperature** using **thermometers.** By comparing the number of divisions between the freezing point and boiling point of water, we can set up conversion factors between the two temperature scales, **Fahrenheit** (°F) and **Celsius** (°C). A third system, closely related to the Celsius scale, is the **Kelvin scale.** Using an algebraic approach, we are able to convert between Fahrenheit and Celsius temperatures, as well as between Celsius and Kelvin.

# Exercises

*Throughout the text, answers to all exercises except those with the exercise number in color are given in Appendix G. The more difficult exercises are marked with an asterisk.*

## Significant Figures

**1-1.** Which of the following measurements is the most precise?
(a) 75.2 gal     (c) 75.22 gal
(b) 74.212 gal     (d) 75 gal

**1-2.** How can a measurement be precise but not accurate?

**1-3.** The actual length of a certain plank is 26.782 in. Which of the following measurements is the most precise and which is the most accurate?
(a) 26.5 in.     (c) 26.202 in.
(b) 26.8 in.     (d) 26.98 in.

**1-4.** How many significant figures are in each of the following measurements?
(a) 7030 g     (e) 4002 m
(b) 4.0 kg     (f) 0.060 hr
(c) 4.01 lb     (g) 8200 km
(d) 0.01 ft     (h) 0.00705 yd

**1-5.** How many significant figures are in each of the following measurements?
(a) 0.045 in.    (d) 21.0 m     (g) 0.0080 in.
(b) 405 ft     (e) 7.060 qt    (h) 2200 lb
(c) 0.340 cm    (f) 2.0010 yd

**1-6.** What is the uncertainty in each of the measurements in Exercise 1-4? Recall that the uncertainty is ±1 expressed in the last significant figure to the right (e.g., 5670 = ±10).

**1-7.** What is the uncertainty in each of the measurements in Exercise 1-5? Recall that the uncertainty is ±1 expressed in the last significant figure to the right.

## Significant Figures in Mathematical Operations

**1-8.** Round off each of the following numbers to three significant figures.
(a) 15.9994    (d) 4885     (g) 301.4
(b) 1.0080     (e) 87,550
(c) 0.6654     (f) 0.027225

**1-9.** Round off each of the following numbers to two significant figures.
**(a)** 115   **(d)** 0.47322   **(g)** 1,557,000
**(b)** 27.678   **(e)** 55.6   **(h)** 321
**(c)** 37,500   **(f)** 0.0396

**1-10.** Express the following fractions in decimal form to three significant figures.
**(a)** $\dfrac{1}{4}$   **(b)** $\dfrac{4}{5}$   **(c)** $\dfrac{5}{3}$   **(d)** $\dfrac{7}{6}$

**1-11.** Express the following fractions in decimal form to three significant figures.
**(a)** $\dfrac{2}{3}$   **(b)** $\dfrac{2}{5}$   **(c)** $\dfrac{5}{8}$   **(d)** $\dfrac{13}{4}$

**1-12.** Without doing the calculation, determine the uncertainty in the answer (e.g., if the answer is to two decimal places, uncertainty = ±0.01).
**(a)** 12.34 + 0.003 + 1.2   **(c)** 45.66 + 12 + 0.002
**(b)** 26,000 + 450 + 132,500   **(d)** 0.055 + 13.43 + 1.202

**1-13.** Without doing the calculation, determine the uncertainty in the answer.
**(a)** 13,330 + 0.8 + 1554
**(b)** 0.0002 + 0.164 + 0.00005
**(c)** 14,230 + 34 + 1932
**(d)** 567 + 7 + 47

**1-14.** Carry out each of the following operations. Assume that the numbers represent measurements and express the answer to the proper decimal place.
**(a)** 14.72 + 0.611 + 173
**(b)** 0.062 + 11.38 + 1.4578
**(c)** 1600 − 4 + 700
**(d)** 47 + 0.91 − 0.286
**(e)** 0.125 + 0.71

**1-15.** Carry out each of the following operations. Assume that the numbers represent measurements and express the answer to the proper decimal place.
**(a)** 0.013 + 0.7217 + 0.04
**(b)** 15.3 + 1.12 − 3.377
**(c)** 35.48 − 4 + 0.04
**(d)** 337 + 0.8 − 12.0

**1-16.** A container holds 32.8 qt of water. The following portions of water are then added to the container: 0.12 qt, 3.7 qt, and 1.266 qt. What is the new volume of water?

**1-17.** A container holds 3760 lb (three significant figures) of sand. The following portions are added to the container: 1.8 lb, 32 lb, and 13.55 lb. What is the final mass of the sand?

**1-18.** Supply the missing measurements.

| **(a)** | 6.03 | **(c)** | (    ) | **(e)** | 0.0468 |
|---|---|---|---|---|---|
| | +(    ) | | + 0.48 | | +(    ) |
| | 13.0 | | 192 | | 3.25 |
| **(b)** | (    ) | **(d)** | 0.5668 | **(f)** | 47.9 |
| | + 0.9 | | −(    ) | | −(    ) |
| | 138 | | 0.122 | | 45.0 |

**1-19.** Supply the missing measurements.

| **(a)** | 98.732 | **(c)** | 98.732 | **(e)** | 9.05 |
|---|---|---|---|---|---|
| | +(    ) | | +(    ) | | +(    ) |
| | 98.780 | | 99.7 | | 11.6 |
| **(b)** | 98.732 | **(d)** | (    ) | **(f)** | 0.0377 |
| | +(    ) | | + 0.0468 | | +(    ) |
| | 98.90 | | 2.227 | | 0.16 |

**1-20.** Without doing the calculation, determine the number of significant figures in the answer from the following.
**(a)** 0.59 × 87 × 23.0   **(c)** $\dfrac{176 \times 0.20}{33.45}$
**(b)** $\dfrac{17.0}{2.334}$   **(d)** $\dfrac{0.30 \times 22.42}{0.03}$

**1-21.** Without doing the calculation, determine the number of significant figures in the answer from the following.
**(a)** 135,200 × 0.330   **(c)** $\dfrac{0.4005 \times 1.2236}{0.0821 \times 298.5}$
**(b)** $\dfrac{0.0303}{6.022}$   **(d)** $\dfrac{23.44 \times 0.050}{0.006 \times 75}$

**1-22.** Supply the missing measurements.
**(a)** 22.4 × (    ) = 136   **(e)** 878.8/(    ) = 221
**(b)** 22.400 × (    ) = 135.5   **(f)** 878.8/(    ) = 221.2
**(c)** 6.482 × (    ) = 55.9   **(g)** 0.7820/(    ) = 1.534
**(d)** 7 × (    ) = 50

**1-23.** Supply the missing measurements.
**(a)** 0.515 × (    ) = 1.65   **(d)** (    )/0.22 = 110
**(b)** 0.517 × (    ) = 0.04   **(e)** 25.1/(    ) = 114
**(c)** 6.482 × (    ) = 55.90   **(f)** 0.782/(    ) = 1.5

**1-24.** Carry out the following calculations. Express your answer to the proper number of significant figures. Specify units.
**(a)** 40.0 cm × 3.0 cm   **(c)** $\dfrac{4.386 \text{ cm}^2}{2 \text{ cm}}$
**(b)** 179 ft × 2.20 ft   **(d)** $\dfrac{14.65 \text{ in.} \times 0.32 \text{ in.}}{2.00 \text{ in.}}$

**1-25.** Carry out the following calculations. Express your answer to the proper number of significant figures. Specify units.
**(a)** $\dfrac{243 \text{ m}^2}{0.05 \text{ m}}$   **(c)** 0.0575 in. × 21.0 in.
**(b)** 3.0 ft × 472 ft   **(d)** $\dfrac{1.84 \text{ yd} \times 42.8 \text{ yd}}{0.8 \text{ yd}}$

**\*1-26.** Carry out the following calculations. Express your answer to the proper number of significant figures or decimal places.
**(a)** $\left(\dfrac{146}{2.3}\right) + 75.0$   **(b)** (157 − 112) × 25.6
**(c)** (12.688 − 10.0) × (7.85 + 2.666)

**1-27.** Carry out the following calculations. Express your answer to the proper number of significant figures or decimal places.
**(a)** (67.43 × 0.44) − 23.456
**(b)** (0.22 + 12.451 + 1.782) × 0.876
**(c)** (1.20 × 0.8842) + (7.332 × 0.0580)

## Scientific Notation

**1-28.** Express the following numbers in scientific notation (one digit to the left of the decimal point).

(a) 157           (e) 0.0349

(b) 0.157         (f) 32,000

(c) 0.0300       (g) 32 billion

(d) 40,000,000 (two    (h) 0.000771

    significant figures)    (i) 2340

**1-29.** Express the following numbers in scientific notation (one digit to the left of the decimal point).

(a) 423,000    (e) 0.00008

(b) 433.8       (f) 82,000,000 (three significant figures)

(c) 0.0020      (g) 75 trillion

(d) 880         (h) 0.00000106

**1-30.** Using scientific notation, express the number 87,000,000 to (a) one significant figure, (b) two significant figures, and (c) three significant figures.

**1-31.** Using scientific notation, express the number 23,600 to (a) one significant figure, (b) two significant figures, (c) three significant figures, and (d) four significant figures.

**1-32.** Express the following as ordinary decimal numbers.

(a) $4.76 \times 10^{-4}$   (c) $788 \times 10^{-5}$   (e) $475 \times 10^{-2}$

(b) $6.55 \times 10^{3}$    (d) $0.489 \times 10^{6}$   (f) $0.0034 \times 10^{-3}$

**1-33.** Express the following as ordinary decimal numbers.

(a) $64 \times 10^{-3}$    (c) $0.022 \times 10^{4}$

(b) $8.34 \times 10^{3}$   (d) $0.342 \times 10^{-2}$

**1-34.** Change the following numbers to scientific notation (one digit to the left of the decimal point).

(a) $489 \times 10^{-6}$    (c) $0.0078 \times 10^{6}$   (e) $4975 \times 10^{5}$

(b) $0.456 \times 10^{-4}$   (d) $571 \times 10^{-4}$    (f) $0.030 \times 10^{-2}$

**1-35.** Change the following numbers to scientific notation.

(a) $0.078 \times 10^{-8}$    (d) $280.0 \times 10^{8}$

(b) $72,000 \times 10^{-5}$   (e) $0.000690 \times 10^{-10}$

(c) $3450 \times 10^{16}$     (f) $0.0023 \times 10^{6}$

**1-36.** Order the following numbers from the smallest to the largest.

(a) 12             (d) $0.084 \times 10^{2}$   (g) 0.0022

(b) $0.042 \times 10^{-3}$   (e) $3.7 \times 10^{6}$

(c) $48 \times 10^{5}$      (f) $8.6 \times 10^{-5}$

**1-37.** Order the following numbers from smallest to largest.

(a) $0.40 \times 10^{2}$   (d) $4.8 \times 10^{5}$   (g) $2.7 \times 10^{-4}$

(b) $40 \times 10^{4}$     (e) $510 \times 10^{2}$

(c) $0.077 \times 10^{-2}$   (f) 8.9

**1-38.** Carry out each of the following operations. Assume that the numbers represent measurements so the answer is expressed to the proper decimal place.

(a) $(1.82 \times 10^{-4}) + (0.037 \times 10^{-4}) + (14.11 \times 10^{-4})$

(b) $(13.7 \times 10^{6}) - (2.31 \times 10^{6}) + (116.28 \times 10^{5})$

(c) $(0.61 \times 10^{-6}) + (0.11 \times 10^{-4}) + (0.0232 \times 10^{-3})$

(d) $(372 \times 10^{12}) + (1200 \times 10^{10}) - (0.18 \times 10^{15})$

**1-39.** Carry out each of the following operations. Assume that the numbers represent measurements so the answer is expressed to the proper decimal place.

(a) $(1.42 \times 10^{-10}) + (0.17 \times 10^{-10}) - (0.009 \times 10^{-10})$

(b) $(146 \times 10^{8}) + (0.723 \times 10^{10}) + (11 \times 10^{8})$

(c) $(1.48 \times 10^{-7}) + (2911 \times 10^{-9}) + (0.6318 \times 10^{-6})$

(d) $(299 \times 10^{10}) + (823 \times 10^{8}) + (0.75 \times 10^{11})$

**1-40.** Carry out the following calculations.

(a) $10^{3} \times 10^{4}$    (c) $\dfrac{10^{26}}{10^{-3}}$

(b) $10^{6} \times 10^{-6}$   (d) $\dfrac{10^{4} \times 10^{-8}}{10^{-13}}$

**1-41.** Carry out the following calculations.

(a) $\dfrac{10^{8}}{10^{-8}}$        (c) $\dfrac{10^{16} \times 10^{-12}}{10^{4}}$

(b) $\dfrac{10^{22} \times 10^{-4}}{10^{17} \times 10^{8}}$   (d) $10^{21} \times 10^{-28}$

**1-42.** Carry out each of the following operations. Assume that the numbers represent measurements so the answer is expressed to the proper number of significant figures.

(a) $(149 \times 10^{6}) \times (0.21 \times 10^{3})$

(b) $\dfrac{0.371 \times 10^{14}}{2 \times 10^{4}}$

(c) $(6 \times 10^{6}) \times (6 \times 10^{6})$

(d) $(0.1186 \times 10^{6}) \times (12 \times 10^{-5})$

(e) $\dfrac{18.21 \times 10^{-10}}{0.0712 \times 10^{6}}$

**1-43.** Carry out each of the following operations. Assume that the numbers represent measurements so the answer is expressed to the proper number of significant figures.

(a) $(76.0 \times 10^{7}) \times (0.6 \times 10^{8})$

(b) $(7 \times 10^{-5}) \times (7.0 \times 10^{-5})$

(c) $\dfrac{0.786 \times 10^{-7}}{0.47 \times 10^{7}}$

(d) $\dfrac{3798 \times 10^{18}}{0.00301 \times 10^{12}}$

(e) $(0.06000 \times 10^{18}) \times (84,921 \times 10^{-9})$

**1-44.** Supply the missing measurement. Express in scientific notation.

(a) $(4.0 \times 10^{12})/(\underline{\hspace{1cm}}) = 2.0$

(b) $(\underline{\hspace{1cm}}) \times (5.18 \times 10^{-8}) = 1.9 \times 10^{9}$

(c) $(6.0 \times 10^{4}) \times (\underline{\hspace{1cm}}) = 3.6 \times 10^{7}$

(d) $(4.0 \times 10^{12})/(\underline{\hspace{1cm}}) = 2 \times 10^{24}$

(e) $(8.520 \times 10^{-8}) \times (\underline{\hspace{1cm}}) = 16$

**1-45.** Supply the missing measurement. Express in scientific notation.

(a) $(\underline{\hspace{1cm}})/(7.890 \times 10^{6}) = 1.552$

(b) $(\underline{\hspace{1cm}})/(7.50 \times 10^{4}) = 1.20 \times 10^{-16}$

(c) $(9.00 \times 10^{-12})/(\underline{\hspace{1cm}}) = 3.0 \times 10^{12}$

(d) $(9.0 \times 10^{4}) \times (\underline{\hspace{1cm}}) = 8 \times 10^{9}$

(e) $(8.002 \times 10^{15})/(\underline{\hspace{1cm}}) = 8 \times 10^{5}$

**1-46.** Supply the missing measurement. Express in scientific notation.

(a) $(\underline{\hspace{1cm}})/(5.32 \times 10^{10}) = 3.4 \times 10^{8}$

(b) $(8.55 \times 10^{12})/(\underline{\hspace{1cm}}) = 2 \times 10^{8}$

(c) $(8.55 \times 10^{12})/(\underline{\hspace{1cm}}) = 4.13$

(d) $(9.00 \times 10^{4}) \times (\underline{\hspace{1cm}}) = 8.1 \times 10^{8}$

## Length, Volume, and Mass in the Metric System

**1-47.** Write the proper prefix, unit, and symbol for the following. Refer to Tables 1-2 and 1-3.
**(a)** $10^{-3}$ L    **(c)** $10^{-9}$ J    **(e)** $10^{-6}$ g
**(b)** $10^2$ g    **(d)** $\dfrac{1}{100}$ m    **(f)** $10^{-1}$ Pa

**1-48.** Write the proper prefix, unit, and symbol for the following. Refer to Tables 1-2 and 1-3.
**(a)** $10^3$ m    **(c)** $\dfrac{1}{1000}$ L    **(e)** $10^{-9}$ m
**(b)** $10^{-3}$ g    **(d)** $10^3$ s    **(f)** $\dfrac{1}{1000}$ mol

**1-49.** Complete the following table.

|  | mm | cm | m | km |
|---|---|---|---|---|
| Example | 108 | 10.8 | 0.108 | $1.08 \times 10^{-4}$ |
| (a) | $7.2 \times 10^3$ | ___ | ___ | ___ |
| (b) | ___ | ___ | 56.4 | ___ |
| (c) | ___ | ___ | ___ | 0.250 |

**1-50.** Complete the following table.

|  | mg | g | kg |
|---|---|---|---|
| (a) | $8.9 \times 10^3$ | ___ | ___ |
| (b) | ___ | 25.7 | ___ |
| (c) | ___ | ___ | 1.25 |

**1-51.** Complete the following table.

|  | mL | L | kL |
|---|---|---|---|
| (a) | ___ | ___ | 6.8 |
| (b) | ___ | 0.786 | ___ |
| (c) | 4452 | ___ | ___ |

## Conversions Between Units of Measurement

**1-52.** Which of the following are "exact" relationships?
**(a)** 12 = 1 doz.    **(d)** 1.06 qt = 1 L
**(b)** 1 gal = 3.78 L    **(e)** $10^3$ m = 1 km
**(c)** 3 ft = 1 yd    **(f)** 454 g = 1 lb

**1-53.** How many significant figures are in each of the following relationships? (If exact, the answer is "infinite.")
**(a)** $10^3$ m = 1 km    **(d)** 1.609 km = 1 mi
**(b)** 4 qt = 1 gal    **(e)** 1 gal = 3.8 L
**(c)** 28.38 g = 1 oz    **(f)** 2 pt = 1 qt

**1-54.** Write a relationship in factor form that would be used in making the following conversions. Refer to Table 1-3.
**(a)** mg to g    **(c)** cL to L
**(b)** m to km    **(d)** mm to km (two factors)

**1-55.** Write a relationship in factor form that would be used in making the following conversions. Refer to Table 1-3.
**(a)** kL to L    **(c)** kg to mg (two factors)
**(b)** mg to g    **(d)** cg to hg (two factors)

**1-56.** Write a relationship in factor form that would be used in making the following conversions. Refer to Tables 1-1 and 1-5.
**(a)** in. to ft    **(d)** L to qt
**(b)** in. to cm    **(e)** pt to L (two steps)
**(c)** mi to ft

**1-57.** Write a relationship in factor form that would be used in making the following conversions. Refer to Tables 1-1 and 1-5.
**(a)** qt to gal    **(c)** gal to L
**(b)** kg to lb    **(d)** ft to km (three steps)

**1-58.** Convert each of these measurement to a unit in the same system that will produce a number between 1 and 100 (e.g., 150 cm converts to 1.5 m).
**(a)** $4.7 \times 10^4$ mL =    **(e)** $9.2 \times 10^{10}$ nm =
**(b)** $98 \times 10^{-5}$ km =    **(f)** 1725 qt =
**(c)** 9780 ft =    **(g)** $32 \times 10^{12}$ mg =
**(d)** 1856 in. =

**1-59.** Convert each of these measurements to a unit in the same system that will produce a number between 1 and 100.
**(a)** $9.5 \times 10^4$ oz =    **(e)** $49 \times 10^5$ cm =
**(b)** 1548 in. =    **(f)** $2.52 \times 10^2$ pt =
**(c)** $8.22 \times 10^6$ mm =    **(g)** $1.5 \times 10^{-12}$ mg =
**(d)** $652 \times 10^{-7}$ kL =

**1-60.** Complete the following table.

|  | mi | ft | m | km |
|---|---|---|---|---|
| (a) | ___ | ___ | $7.8 \times 10^3$ | ___ |
| (b) | 0.450 | ___ | ___ | ___ |
| (c) | ___ | $8.98 \times 10^3$ | ___ | ___ |
| (d) | ___ | ___ | ___ | 6.78 |

**1-61.** Complete the following table.

|  | gal | qt | L |
|---|---|---|---|
| (a) | 6.78 | ___ | ___ |
| (b) | ___ | 670 | ___ |
| (c) | ___ | ___ | $7.68 \times 10^3$ |

**1-62.** Complete the following table.

|  | lb | g | kg |
|---|---|---|---|
| (a) | ___ | ___ | 0.780 |
| (b) | ___ | 985 | ___ |
| (c) | 16.0 | ___ | ___ |

**1-63.** If a person has a mass of 122 lb, what is her mass in kilograms?

**1-64.** The moon is 238,700 miles from the Earth. What is this distance in kilometers?

**1-65.** A punter on a professional football team averaged 28.0 m per kick. What is his average in yards? Should he be kept on the team?

**1-66.** A can of soda has a volume of 355 mL. What is this volume in quarts?

**1-67.** If a student drinks a 12-oz (0.375-qt) can of soda, what volume did she drink in liters?

**1-68.** The meat in a "quarter-pounder" should weigh 4.00 oz. What is its mass in grams?

**1-69.** A prospective basketball player is 6 ft $10\frac{1}{2}$ in. tall and weighs 212 lb. What are his height in meters and his weight in kilograms?

**1-70.** Gasoline is sold by the liter in Europe. How many gallons does a 55.0-L gas tank hold?

**1-71.** If the length of a football field is changed from 100 yd to 100 m, will the field be longer or shorter than the current field? How many yards would a "first and ten" be on the metric field?

**1-72.** Bourbon used to be sold by the "fifth" (one-fifth of a gallon). A bottle now contains 750 mL. Which is greater?

**1-73.** A marathon runner must cover 26 mi 385 yd. How far is this in kilometers?

**1-74.** If the speed limit is 65.0 mi/hr, what is the speed limit in km/hr?

**1-75.** Mount Everest is 29,028 ft in elevation. How high is this in kilometers?

**1-76.** It is 525 mi from St. Louis to Detroit. How far is this in kilometers?

**1-77.** A small pizza has a diameter of 9.00 in. What is this length in millimeters?

**1-78.** Gasoline sold as low as $0.899 per gallon in 2001. What was the cost per liter? What did it cost to fill a 80.0-L tank?

**1-79.** At the price of gas in the preceding exercise, how much does it cost to drive 551 mi if your car averages 21.0 mi/gal? How much does it cost to drive 482 km?

**1-80.** Using the information from the two preceding exercises, how many kilometers can you drive for $45.00?

**1-81.** An aspirin contains 0.324 g (5.00 grains) of aspirin. How many pounds of aspirin are in a 500-aspirin bottle?

**1-82.** A hamburger in Canada sells for $4.55 (Canadian dollars). That seems expensive to a U.S. resident until we realize that the exchange rate in 2002 was $1.56 Canadian per one U.S. dollar. What is the cost in U.S. dollars?

**1-83.** A certain type of nail costs $0.95/lb. If there are 145 nails per pound, how many nails can you purchase for $2.50?

**1-84.** Another type of nail costs $0.92/lb, and there are 185 nails per pound. What is the cost of 5670 nails?

**1-85.** If an automobile gets 24.5 mi/gal of gasoline and gasoline costs $1.22/gal, what would it cost to drive 350 km?

**1-86.** If a train travels at a speed of 85 mi/hr, how many hours does it take to travel 17,000 ft? How many yards can it travel in 37 min?

**1-87.** If grapes sell for $2.15/lb and there are 255 grapes per pound, how many grapes can you buy for $8.15?

**1-88.** A high-speed train in Europe travels 215 km/h. How long would it take for this train to travel nonstop from Boston to Washington, D.C., which is 442 miles?

**1-89.** The exchange rates among currencies vary from day to day and are usually expressed with up to five significant figures. The new currency in Europe since the beginning of 2002 is the Eurodollar (or simply the "euro"). It began trading at about 1.12 euros per U.S. dollar. **(a)** What is the cost in euros of a sandwich that would cost $6.50 in the United States? **(b)** What is the cost in U.S. dollars of one liter of wine in France that sells for 12.65 euros?

**1-90.** The euro is not used in Britain. The British pound traded for 0.695 pound per U.S. dollar in 2002. What is a conversion rate between the pound and the euro using information from the previous exercise? If an automobile sells for 25,500 euros in Germany, how much would it cost in pounds?

**1-91.** In Mexico, the peso exchanged for 9.19 pesos per U.S. dollar in 2002. What is the cost in dollars for a Corona beer that sells for 15 pesos in Cancun? What would it cost in euros and in British pounds? (See the previous two exercises.)

**\*1-92.** At a speed of 35 mi/hr, how many centimeters do you travel per second?

**\*1-93.** The planet Jupiter is about $4.0 \times 10^8$ mi from Earth. If radio signals travel at the speed of light, which is $3.0 \times 10^{10}$ cm/s, how long would it take a radio command from Earth to reach a spacecraft passing Jupiter?

### Temperature

**1-94.** The temperature of the water around a nuclear reactor core is about 300°C. What is this temperature in degrees Fahrenheit?

**1-95.** The temperature on a comfortable day is 76°F. What is this temperature in degrees Celsius?

**1-96.** The lowest possible temperature is −273°C and is referred to as absolute zero. What is this temperature in degrees Fahrenheit?

**1-97.** Mercury thermometers cannot be used in cold arctic climates because mercury freezes at −39°C. What is this temperature in degrees Fahrenheit?

**1-98.** The coldest temperature recorded on Earth was −110°F. What is this temperature in degrees Celsius?

**1-99.** A hot day in the U.S. Midwest is 35.0°C. What is this in degrees Fahrenheit?

**1-100.** Convert the following Kelvin temperatures to degrees Celsius.
**(a)** 175 K    **(d)** 225 K
**(b)** 295 K    **(e)** 873 K
**(c)** 300 K

**1-101.** Convert the following temperatures to the Kelvin scale.
**(a)** 47°C    **(d)** −12°C
**(b)** 23°C    **(e)** 65°F
**(c)** −73°C    **(f)** −20°F

**1-102.** Make the following temperature conversions.
**(a)** 37°C to K    **(d)** 127 K to °F
**(b)** 135°C to K    **(e)** 100°F to K
**(c)** 205 K to °C    **(f)** −25°C to K

**\*1-103.** At what temperature are the Celsius and Fahrenheit scales numerically equal?

### General Problems

**1-104.** Carry out the following calculations. Express the answer to the proper number of significant figures.
**(a)** $\dfrac{12.61 + 0.22 + 0.037}{0.04}$
**(b)** 0.333 g × (23.60 + 1.2) cm
**(c)** $\dfrac{6.286\text{ g}}{(13.68 - 12.48)\text{mL}}$
**(d)** $\dfrac{44.35 + 0.03 + 0.057}{22.35 - 20.018}$

**1-105.** Write in factor form the two relationships needed to convert the following.
**(a)** mg to lb    **(c)** hm to mi
**(b)** L to pt    **(d)** cm to ft

**1-106.** Convert $5.34 \times 10^{10}$ ng to pounds.

**1-107.** Convert $7.88 \times 10^{-4}$ mL to gallons.

**1-108.** If gold costs \$320/oz, what is the cost of 1.00 kg of gold? (Metals are traded as "troy" ounces. There are exactly 12 troy ounces per troy pound, and one troy pound is equal to 373 g, to three significant figures.)

**1-109.** Construct unit factors from the following information: A 82.3-doz. quantity of oranges weighs 247 lb.
**(a)** What is the mass per dozen oranges?
**(b)** How many dozen oranges are there per pound?

**1-110.** A unit of length in horse racing is the "furlong." The height of a horse is measured in "hands." There are exactly 8 furlongs per mile, and one hand is exactly 4 inches. How many hands are there in 12.0 furlongs? (Express the answer in standard scientific notation.)

**1-111.** The unit price of groceries is sometimes listed in cost per ounce. Which has the smaller cost per ounce: 16 oz of baked beans costing \$1.45 or 26 oz costing \$2.10?

**1-112.** A cigarette contains 11.0 mg of tar. How many packages of cigarettes (20 cigarettes per package) would have to be smoked to produce 0.500 lb of tar? If a person smoked two packs per day, how many years would it take to accumulate 0.500 lb of tar?

**1-113.** An automobile engine has a volume of 306 in.$^3$ What is this volume in liters?

**1-114.** A U.S. quarter has a mass of 5.70 g. How many dollars is one pound of quarters worth?

**1-115.** The surface of the sun is at a temperature of about $3.0 \times 10^7$°C. What is this temperature in degrees Fahrenheit? In kelvins?

**1-116.** In Saudi Arabia, gasoline costs 30.0 hillala per liter. If there are exactly 100 hillalas in one ryal and one ryal exchanges for 25.0 cents, what is the cost in cents/gal?

**\*1-117.** If snow were piled up by 1.00 foot on a roof 30.0 ft × 50.0 ft, what would be the mass of snow on the roof in pounds and in tons? Assume that 1 ft$^3$ of snow is equivalent to 0.100 ft$^3$ of water and that 1 ft$^3$ of water has a mass of 62.0 lb.

# Interactive Learning

 **1-118.** A bullet is fired at a speed of 2235 ft/s. What is this speed in kilometers per hour?

 **1-119.** A truck weighs $2.07 \times 10^9$ mg. What is its mass in tons?

 **1-120.** In 2002, the rate of exchange in China was 8.277 yuan per U.S. dollar. The exchange in Russia was 31.29 rubles per U.S. dollar. If a sandwich costs 22.72 yuan in China, what would it cost in rubles?

 **1-121.** Pressure can be measured in pounds per square foot. Convert 36.7 lb/ft$^2$ to g/cm$^2$.

# Solutions to Learning Checks

**A-1.** estimated, precision, accuracy, between, right, left, least, decimal places, least, significant figures, scientific notation, one, exponent

**A-2.** (a) 3 (b) 4 (c) 3 (d) 3 (e) 2

**A-3.** (a) 23.4 (b) 484,000 (c) 0.0220

**A-4.** (a) 20.00 (b) 2.55 (c) 2.1

**A-5.** (a) $4.560 \times 10^8$ (b) $3.40 \times 10^{-4}$

**A-6.** (a) $2.61 \times 10^8$ (b) $2.4 \times 10^8$

**B-1.** meter, liter, gram, $10^4$, $10^{-2}$, 1 mL, conversion, denominator, numerator, Celsius, Kelvin

**B-2.** (a) $\dfrac{1 \text{ yd}}{3 \text{ ft}}$ (b) $\dfrac{10^{-1} \text{ m}}{\text{dm}}$ (c) $\dfrac{3.785 \text{ L}}{\text{gal}}$ (d) $\dfrac{1 \text{ kg}}{2.205 \text{ lb}}$

**B-3.** $25.0 \text{ km} \times \dfrac{1 \text{ mi}}{1.609 \text{ km}} = \underline{\underline{15.5 \text{ mi}}}$

**B-4.** $12 \text{ oz} \times \dfrac{\$0.180}{\text{oz}} = \underline{\underline{\$2.16}}$    $\$1.75 \times \dfrac{1 \text{ oz}}{\$0.180} = \underline{\underline{9.72 \text{ oz}}}$

**B-5.** $12.0 \text{ ft} \times \dfrac{12 \text{ in.}}{\text{ft}} \times \dfrac{2.54 \text{ cm}}{\text{in.}} = \underline{\underline{355 \text{ cm}}}$

**B-6.** $3.25 \text{ kg} \times \dfrac{2.205 \text{ lb}}{\text{kg}} \times \dfrac{\$1.25}{\text{lb}} = \underline{\underline{\$8.96}}$

**B-7.** $^\circ\text{C} = \dfrac{[248^\circ\text{F} - 32^\circ\text{F}]}{1.8^\circ\text{F}} = \dfrac{216^\circ\text{F}}{1.8^\circ\text{F}} = \underline{\underline{120^\circ\text{C}}}$

## chapter 2

# Matter, Changes, and Energy

**In June 1999, the Hubble telescope provided this image of Mars. It is made of the same substances as those found on Earth, including water ice at the poles.**

## Setting the Stage

We are now deep into the exploration of our neighboring planets. Satellites circle Mars, sending back new data. A spacecraft named *Cassini* streaks toward Saturn and its frigid moon, Triton. The spacecraft named *Galileo* has already sent back intriguing data about a moon of Jupiter known as Europa that lies half a billion miles away. What was discovered has excited scientists on Earth. Europa seems to be covered by a thick crust of ordinary ice. But the convoluted and cracked surface indicates that liquid water most certainly lies beneath. It is a long shot, but conditions may be hospitable to a primitive life form in these oceans beneath the ice! The ice that we view on Europa and the other substances identified on these faraway worlds are the same types of stuff that we find on our own Earth. The nature of this stuff (which we call matter) is the domain of chemistry. **Chemistry** *can be defined as the study of matter and the changes it undergoes.* **Matter** *is defined as anything that has mass and occupies space.* We are formed of matter, and more of it surrounds us. Obviously, solid rock is matter, but so is water, and so is the soft wind on our faces. All of the Earth from the center of the planet to the outermost reaches of the atmosphere, the moon, and the flickering stars in the night sky is composed of the same types of matter. The forms of matter that we see on the Earth are often changing. Trees grow, die, and decay. Rocks weather, crumble, and form the fertile soil of the plains or deposits in the oceans. The changes that matter undergoes are another concern of chemistry. But there is even more. *In addition to matter, the universe is also composed of energy.* When a log burns in the fireplace, it is obvious that a change in matter has occurred. The log is transformed into a small pile of ashes and hot gases. But the burning of the log warms us—it has given off heat. The heat and light that were liberated by the burning process are forms of energy. Energy is a more abstract concept than matter. It can't be weighed, and it doesn't have shape, form, or dimensions. But energy can be measured, and it does interact with matter (e.g., it warms us, starts our car, and makes trees grow). Since energy is an integral part of the changes that matter undergoes, it is also important in the study of chemistry. Thus the purpose of this chapter is to examine in detail the two key words in the definition of chemistry: "matter" and "changes."

### Formulating Some Questions

At first glance, the world around us seems so complex. For example, there are billions of people on Earth, so we simplify our understanding of this population by grouping it into general categories such as nationality, gender, race, age, and so forth. Can our understanding of the millions of types of matter be simplified by grouping? We will see that the answer is "yes." But first we may ask, What are the most basic forms of matter? What are the criteria that allow us to distinguish one substance from another? How do we describe mixtures of substances? Finally, how does energy relate to changes in matter? In the first section, we will find that matter can be divided into two general categories.

Section A · **THE TYPES OF MATTER AND ITS PROPERTIES**

## 2-1 Types of Matter: Elements and Compounds

**Looking Ahead!** All that we see around us, from the contents of this room to the farthest stars in space, is composed of fewer than 90 unique types of matter. How these 90 forms of matter exist, either alone or in combinations, is what we will discuss in this section.

Bricks, mortar, metal bars, wood beams—these are some of the basic units of construction of a tall building. Likewise, all of these materials and everything else in the universe are composed of more basic units. These basic forms cannot be broken down by ordinary chemical or physical means into any more fundamental forms of matter, so that is where we begin. *An **element** is the most basic form of matter that exists under ordinary conditions.* Only a few elements are found around us in their free state. The shiny gold in a ring, the life-supporting oxygen in air, the sparkling diamond, and the sturdy but light aluminum in a soda can are all examples of free elements. (See Figure 2-1.) A complete list of elements is found inside the front cover of this text. In the next chapter, we will discuss the names, symbols, and abundances of these elements, and we will then classify the elements into two general groups called metals and nonmetals.

We need iron, sodium, and potassium in our diet for good nutrition. These elements, however, are not ingested as free elements but are components of compounds. *A **compound** is a unique substance that is composed of two or more elements that are chemically combined.* When we say chemically combined we mean not merely a mixture of elements, but that the elements are joined intimately together into a unique form of matter that is distinct from the elements of which the compound is composed. In fact, most of the elements around us are present as components of compounds. Many familiar substances are compounds. For example, water, table salt, and sugar are compounds. Water is composed of the elements oxygen and hydrogen; table salt

Elements

Compounds

**Figure 2-1  Elements and Compounds**
The substances on the left are free elements. Those on the right are common compounds (water, sugar, and salt).

is made up of sodium and chlorine; and sugar consists of carbon, hydrogen, and oxygen. In fact, various chemical combinations of elements make up millions of known compounds. (See Figure 2-1.) Every element and compound has a unique set of properties. Properties *describe the particular characteristics or traits of a substance.* Elements and compounds can be referred to as pure substances. **Pure substances** *have definite compositions and definite, unchanging properties.*

In Figure 2-2, a mixture of two elements, iron and sulfur, is shown. If they are mixed together, we can still separate them with a magnet. If this mixture is ignited with a hot flame, however, the mixture is transformed into a compound. *It is important to note that the compound is no longer considered a mixture even though it was made from a mixture.* After the mixture is ignited, it would take some very involved chemical procedures to change the compound back into its elements. The name of the compound is iron(III) sulfide. The names of many compounds often indicate the presence of two elements. For example, a common gas is known as carbon dioxide. It is not a mixture of carbon and oxygen, however, but a unique substance that just happens to have been made from carbon and oxygen.

Formation of a compound from a mixture is much like baking a cake. We mix eggs, flour, and other ingredients and then heat the mixture. What comes out is a unique substance that no longer looks like, tastes like, or smells like the mixture we put in the oven. The mix of ingredients has been transformed into a new substance (we hope).

Iron and sulfur

Iron(III) sulfide

**Figure 2-2  Formation of Iron(III) Sulfide** In this chemical reaction, a mixture of elements (*left*) is changed into a pure compound (*right*).

# 2-2 Physical Properties of Matter

**Looking Ahead!** When we describe a friend to someone, we usually include something about his or her physical appearance (e.g., tall and thin) and personality (e.g., outgoing and humorous). Likewise, there are two kinds of properties that we use to describe a particular element or compound: physical and chemical. In this section, we will discuss physical properties and then proceed to chemical properties in a later section.

**Figure 2-3 Physical Properties** Potassium nitrate (white) copper(II) chloride (green) and copper(II) sulfate (blue) are compounds with distinct physical properties.

"The suspect in the robbery was a 6-ft 2-in. male with a heavy build, short hair, and a thin mustache." These are a few of the physical properties of a person that we may hear in a news bulletin. These properties can be observed without interacting with this individual or getting too close (fortunately, in this case). As with an individual, the **physical properties** of a substance *are those that can be observed or measured without changing the substance into another substance.* First we will consider the physical properties that we can observe. (See Figure 2-3.) Odor and color are two physical properties that a substance may or may not display. Perhaps the first physical property of a substance that we may report is its physical state under defined conditions such as are found on the surface of the Earth. The three **physical states** of matter are *solid, liquid,* and *gas.* We can appreciate the reasons for the three states of matter if we accept that all forms of matter are composed of very tiny particles. (These particles can be imaged only with the most powerful microscopes known). We will have much more to say about these fundamental particles in the next chapter. When the fundamental particles remain close together in relatively fixed positions, the substance is in the solid state. Movement of the particles is very restricted and confined mostly to vibrations about these positions. Because of the fixed positions of the fundamental particles, **solids** *have a definite shape and a definite volume.* When the fundamental particles remain close together but are able to move past one another, the substance is in the liquid state. Because of the freedom of movement of the fundamental particles, liquids flow and take the shape of the lower part of a container. Thus, **liquids** *have a definite volume but not a definite shape.* If the fundamental particles are not close to each other but move independently in all directions in random motion, the substance is in the gaseous state. The particles in gases fill a container uniformly. Thus, **gases** *have neither a definite volume nor a definite shape.* (See Figure 2-4.)

We are already familiar with many examples of all three physical states. Ice, rock, salt, and steel are substances that exist as solids; water, gasoline, and alcohol exist as liquids; ammonia, natural gas, and the components of air are present in nature as gases. Whether a particular element or compound is a solid, liquid, or gas depends on the nature of the substance and on the temperature. For example, at low temperatures, liquid water freezes to form a solid, and at high temperatures, liquid water boils to form a gas (vapor). If the temperature is very low ($-196°C$), even the gases that form our atmosphere condense to the liquid state. In fact, the temperature at which a pure substance changes to another physical state is a definite, unchanging physical property of that substance. *The temperature at which a particular element or compound changes from the solid state to the liquid state is known as its* **melting point.** (When the reverse change occurs, this same temperature is referred to as the **freezing point.**) At some higher temperature the liquid begins to boil. *The temperature at which bubbles of the vapor begin to form in the liquid and escape to the surroundings is known as the* **boiling point.** (The reverse change occurs at the same temperature, which is then called the **condensation point.**) The boiling point of a liquid is affected by the atmospheric pressure, however. Boiling point temperatures are usually listed as the boiling point of

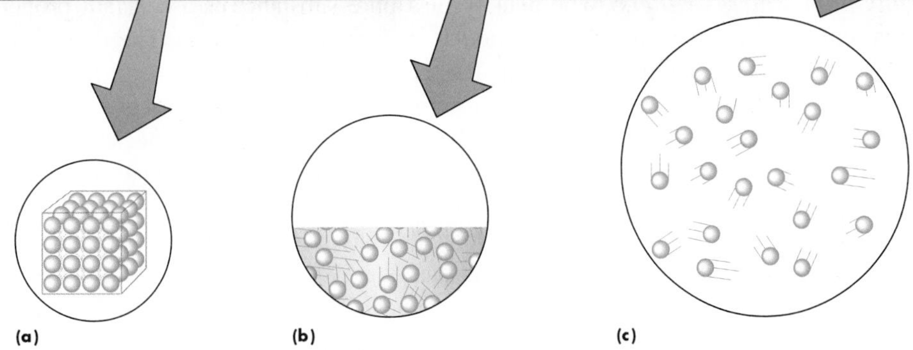

**Figure 2-4 The Three Physical States of Matter** (*a*) Solids have a definite shape and volume. (*b*) Liquids have a definite volume but an indefinite shape. (*c*) Gases have an indefinite shape and volume.

the liquid at average sea-level atmospheric pressure. These phase changes are summarized as follows.

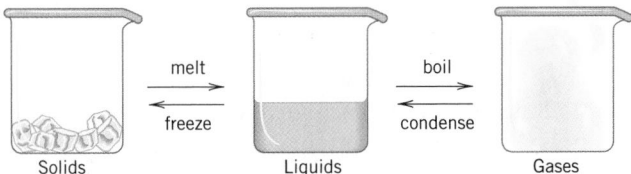

When a liquid freezes or boils, it undergoes a change to another physical state. It is still the same substance, however, so the change is described as physical. *A* **physical change** *in a substance does not involve a change in the composition of the substance but is simply a change in physical state or dimensions.* Liquid water, ice, and steam are all physical states of the same compound.

## 2-3 A Physical Property: Density

**Looking Ahead!** We can describe some physical properties such as color, odor, and physical state by observation. Others such as melting point and boiling point must be measured. Another physical property that can be measured is density. This important physical property is discussed in this section.

A Styrofoam coffee cup is "light," but a lead car battery is "heavy." Actually, by themselves these terms don't mean much because a truckload of Styrofoam would be quite heavy. To give these terms meaning, however, we need to compare equal

**Table 2-1   Density (at 20°C)**

| Substance (Liquid) | Density (g/mL) | Substance (Solid) | Density (g/mL) |
|---|---|---|---|
| Ethyl alcohol | 0.790 | Aluminum | 2.70 |
| Gasoline (a mixture) | ≈0.67 (variable) | Gold | 19.3 |
| Carbon tetrachloride | 1.60 | Ice | 0.92 (0°C) |
| Kerosene | 0.82 | Lead | 11.3 |
| Water | 1.00 | Lithium | 0.53 |
| Mercury | 13.6 | Magnesium | 1.74 |
| | | Table salt | 2.16 |

**Gold is one of the densest substances known.**

volumes. The volume and mass of a substance are variable properties, depending on the size of the sample. A physical property that does not depend on the amount present is density. **Density** *is the ratio of the mass (usually in grams) to the volume (usually in milliliters or liters).* The density of a pure substance is a constant property that can be used to identify the particular element or compound. The densities of several liquids and solids are listed in Table 2-1. (Because the volume of liquids and solids expands slightly as the temperature rises, densities are usually given at a specific temperature. In this case, 20°C is the reference temperature.) Because 1 mL is the same as 1 cm$^3$, density is also expressed as g/cm$^3$. The densities of gases are discussed in Chapter 9.

Density is derived from two measurements, mass and volume, of a particular sample. It makes no difference how large the sample is; the density will be the same. The calculation of density from the two measurements is discussed in the following two examples.

---

**Example 2-1**

**working it out**

**Calculation of Density**

A sample of a pure substance is found to have a mass of 47.5 g. As shown below, a quantity of water has a volume of 12.5 mL. When the substance is added to the water, the volume reads 31.8 mL. The difference in volume is the volume of the substance. What is the density?

**Solution**

The mass is obtained by placing the substance on an electronic balance. The volume is measured by calculating the difference in the volume readings before and after the substance is added. The density is the mass divided by the volume. This is illustrated as follows.

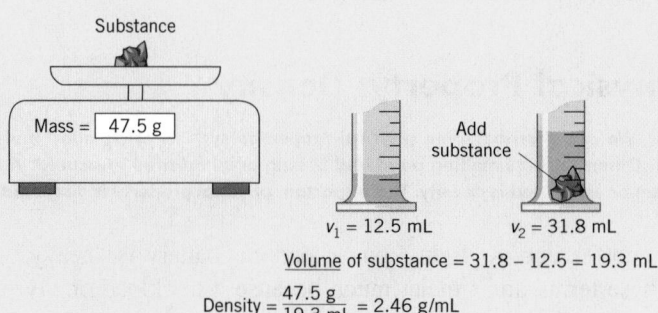

Substance

Mass = 47.5 g

$v_1$ = 12.5 mL       $v_2$ = 31.8 mL

Add substance

Volume of substance = 31.8 − 12.5 = 19.3 mL

$$\text{Density} = \frac{47.5 \text{ g}}{19.3 \text{ mL}} = \underline{2.46 \text{ g/mL}}$$

| Using Density to Identify a Pure Substance | Example 2-2 |

A young woman was interested in purchasing a sample of pure gold having a mass of 89.9 g. Being wise, she wished to confirm that it was actually gold before she paid for it. With a quick test using a graduated cylinder like that shown in the previous example, she found that the "gold" had a volume of 7.96 mL. Was the substance gold?

**Procedure**

From the volume and the mass, the density can be calculated and compared with that of pure gold. (See Table 2-1.)

**Solution**

$$\text{density} = \frac{89.9\ g}{7.96\ mL} = 11.3\ g/mL$$

The sample was *not* gold. It apparently was lead that had been dipped in gold paint.

Density is not only an important physical property but also a handy conversion factor that relates mass to volume of a particular sample. If the density is known, a given mass can be converted to an equivalent volume and vice versa. These two conversions are illustrated in the following examples. Notice that since density originates from two measurements, it is not an exact factor.

| Converting Mass to Volume | Example 2-3 |
| | working it out |

What is the volume in mL occupied by 485 g of table salt?

**Procedure**

1. The unit map for the conversion is

$$g \longrightarrow mL$$

2. The density of table salt is given in Table 2-1 as 2.16 g/mL. Expressed as a conversion factor to convert *g* to *mL*, the relationship is

$$\frac{1\ mL}{2.16\ g}$$

**Solution**

$$485\ \not{g} \times \frac{1\ mL}{2.16\ \not{g}} = \underline{\underline{225\ mL}}$$

| Converting Volume to Mass | Example 2-4 |

What is the mass in grams of 1.52 L of kerosene?

**Procedure**

1. A two-step conversion is necessary. In the first step (a), *L* is converted into *mL*, and in the second step (b), *mL* is converted into *g*. The unit map for the conversion is shown as

$$\begin{array}{ccc} (a) & & (b) \\ L \longrightarrow & mL \longrightarrow & g \end{array}$$

2.  The proper relationships are 1 mL = $10^{-3}$ L (from Table 1-4) and 0.82 g/mL (from Table 2-1). Expressed as proper conversion factors, the relationships are

$$(a) \; \frac{1 \text{ mL}}{10^{-3} \text{ L}} \quad \text{and} \quad (b) \; \frac{0.82 \text{ g}}{\text{mL}}$$

**Solution**

$$1.52 \; \cancel{\text{L}} \times \frac{1 \; \cancel{\text{mL}}}{10^{-3} \; \cancel{\text{L}}} \times \frac{0.82 \text{ g}}{\cancel{\text{mL}}} = \underline{\underline{1.2 \times 10^3 \text{ g}}}$$

Specific gravity is related to density. **Specific gravity** *is the ratio of the mass of a substance to the mass of an equal volume of water under the same conditions.* This definition can be expressed as follows:

$$\text{specific gravity} = \frac{\text{density of a substance}}{\text{density of water}}$$

The density of water at 4°C is 1.00 g/mL. (This was the original definition of the gram in the metric system.) Thus the specific gravity of water is exactly 1 at that temperature. Fortunately, the specific gravity of water changes very little at higher temperatures (e.g., 0.998 at 20°C), so we use a value of 1 in most calculations. The specific gravity of mercury is calculated as follows:

$$\text{density of mercury} = 13.6 \text{ g/mL} \qquad \text{density of water} = 1.00 \text{ g/mL}$$

$$\text{specific gravity} = \frac{13.6 \; \cancel{\text{g/mL}}}{1.00 \; \cancel{\text{g/mL}}} = 13.6$$

Notice that when we use units of g/mL or g/cm$^3$ for density, the specific gravity is numerically the same as the density except that it is expressed without units.

Density or specific gravity is almost always included in a list of distinguishing physical properties of elements and compounds. For example, iron and sulfur are elements and iron(III) sulfide is a compound. (See Figure 2-2.) All three are solids, but they have different and distinct densities, melting points, and colors.

## 2-4 Chemical Properties of Matter

**Looking Ahead!** Besides our physical traits, we also have personality traits. Taken as a whole, our physical and personality traits are unique to each of us. Perhaps the most important aspect of our personality is how we interact with others. Likewise, we can describe how elements and compounds interact with other substances. We will now discuss the properties of matter that are equivalent to a personality.

When water is cooled, it solidifies into ice. When it is allowed to warm, the ice melts back to liquid water. Only the physical state of the substance has changed. On the other hand, when we heat a raw egg, it solidifies. When we cool the egg, however, it stays solid. Obviously, the egg is not the same—it has undergone a profound change. When iron rusts, vegetation decays, and wood burns, the original substances have been transformed into one or more other substances. These processes all describe chemical changes. *When a* **chemical change** *occurs in a substance, the substance is transformed into other substances. The* **chemical properties** *of a pure substance relate to its tendency to undergo chemical changes.* A chemical property

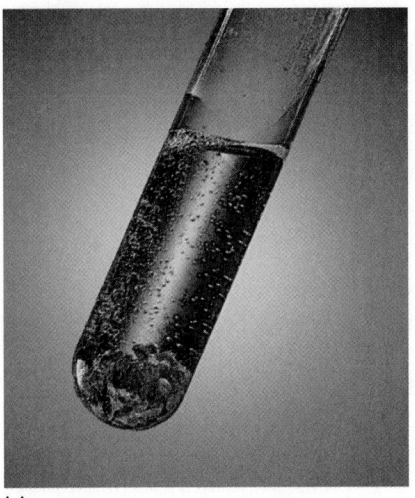
(a)

(b)

(c)

**Figure 2-5  Chemical Properties and Changes** Zinc reacts with hydrochloric acid (*a*), sulfur burns in air (*b*), and iron rusts (*c*).

of the element iron is its ability to react with oxygen from the air to form rust (a compound composed of iron and oxygen). In some cases, chemical properties can be stated in the negative. For example, a chemical property of the element gold is that it resists rusting or tarnishing. In Figure 2-3 we could simply observe the color and physical state of these three compounds. The chemical properties of these three substances can also serve as fingerprints for identification. Three chemical properties are shown in Figure 2-5. In Figure 2-2 we also demonstrated a chemical property of iron and sulfur in that they can be chemically joined to form a compound.

Only two centuries ago, scientists were still puzzled when wood burned, leaving behind only a small portion of the mass in the form of ashes. At that time, the involvement of gases in chemical reactions was not fully understood. The apparent mass loss was explained by a popular theory, but we now know that the mass does not actually disappear. Most of the solid compounds of the wood have been simply transformed in the combustion process into gaseous compounds that drift away in the atmosphere. The mass of the wood plus the mass of the oxygen from the air equals the mass of the ashes plus the mass of the gaseous combustion products. (See Figure 2-6.) Chemical changes illustrate the **law of conservation of mass,** *which states that matter is neither created nor destroyed in a chemical reaction.*

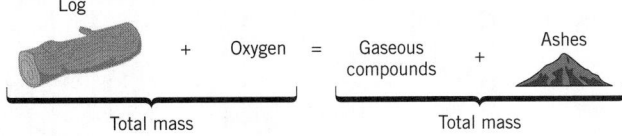

**Figure 2-6  Combustion of Wood**
**In a chemical reaction mass is conserved.**

**Looking Back! Matter is composed of elements that are the most basic forms of ordinary matter. Although a few elements do exist in the free state, most elements combine in countless ways to form millions of compounds. Each element and compound has a unique combination of physical properties and chemical properties. These properties allow further groupings and classifications of pure substances, as we will find throughout the text.**

## Learning Check A | checking it out

**A-1.** Fill in the blanks.

Matter has mass and occupies _____ . Two types of matter are elements and _____ . These are known as _____ substances, and they have unique _____ . The ratio of the mass and the volume of a sample is known as its _____ and is a _____ property. A _____ change transforms one substance into another.

**A-2.** Calcium, an element, is a dull, gray solid that melts at 839°C and has a density of 1.54 g/mL. When it is placed in water, bubbles form, and the calcium slowly dissolves in the water. When the water is evaporated, elemental calcium is not recovered. Which are the physical properties of calcium? Which is a chemical property?

**A-3.** A sample of a given pure liquid has a mass of 254 g and a volume of 159 mL. What might the liquid be? Refer to Table 2-1.

**A-4.** What is the volume of 178 g of aluminum?

*Additional Examples: Exercises 2-3, 2-10, 2-12, 2-17, 2-19, 2-24, and 2-28.*

Section **B** THE COMPOSITION OF MIXTURES

## 2-5 Mixtures of Elements and Compounds

**Looking Ahead!** It is not often in the world around us that we encounter an element or compound in such a highly concentrated state that we may judge it as "pure." Even the water we drink contains other compounds (e.g., salt) and elements (e.g., oxygen). How we describe mixtures is the subject of this discussion. We will also use mixtures to improve our problem-solving skills with the application of percent.

You can't mix oil and water! You've probably heard that often enough. On the other hand, when you place alcohol in water, both liquids disperse into each other and no boundary between the two liquids is apparent. (See Figure 2-7.) Oil and water form a heterogeneous mixture. *A* **heterogeneous mixture** *is a nonuniform mixture containing two or more phases with definite boundaries between the phases. A* **phase** *is one physical state (solid, liquid, or gas) with distinct boundaries and uniform properties.* Alcohol and water form a homogeneous mixture. *A* **homogeneous mixture** *is the same throughout and contains only one phase.* We will discuss heterogeneous mixtures first.

(a)  (b)  (c)  (d)

**Figure 2-7 Mixtures Water and oil (a) form a heterogeneous mixture with two liquid phases (b). Water and an alcohol (c) form a homogeneous mixture or solution with one liquid phase (d).**

Notice in Figure 2-7*b* that the oil floats on top of the water because the density of oil (≈0.90 g/mL) is less than water (1.00 g/mL). Compare this to the homogeneous mixture formed when alcohol and water are mixed, as shown in Figure 2-7*d*.

Besides oil and water (two liquid phases), an obvious heterogeneous mixture is a handful of soil from the backyard. If we look closely, we see bits of sand, some black matter, and perhaps pieces of vegetation. One can easily detect several solid phases with the naked eye. Other examples of heterogeneous mixtures are carbonated beverages (liquid and gas) and dirty water (liquid and solid). Heterogeneous mixtures can often be separated into their homogeneous components by simple laboratory procedures. For example, suspended solid matter can be removed from water by *filtration*. (See Figure 2-8.) When the dirty water is passed through the filter, the suspended matter remains on the filter paper and the liquid phase passes through. In the purification of water for drinking purposes, the first step is the removal of suspended particulate matter.

Sometimes heterogeneous mixtures cannot be detected with the naked eye. For example, creamy salad dressing and smoke both appear uniform at first glance. However, if we were to magnify each, the truth would become apparent. The salad dressing has little droplets of oil suspended in the vinegar (two liquid phases), and the smoke has tiny solid and liquid particles suspended in the air (solid, liquid, and gas phases).

In heterogeneous mixtures, portions of each component are large enough to be detected, although some magnification may be necessary. In homogeneous mixtures, the components disperse uniformly into each other. As mentioned earlier, matter is composed of fundamental particles. In a homogeneous mixture the mixing extends all of the way to the fundamental particle level. Thus, there is no detectable boundary between components. No amount of magnification would reveal pieces of solid salt when it is dissolved in the water. When table salt is added to water, it forms a solution. Although all homogeneous mixtures are technically solutions, *the word* **solution** *usually refers to homogeneous mixtures with one liquid phase.* Thus components of a solution cannot be separated by filtration. However, the two components can be separated by a laboratory procedure called *distillation*. (See Figure 2-9.) In distillation,

**Figure 2-8  Filtration**
**A heterogeneous mixture of a solid and liquid can be separated by filtration.**

**Figure 2-9  Distillation**
**A homogeneous mixture of a solid in liquid or two liquids can be separated by distillation.**

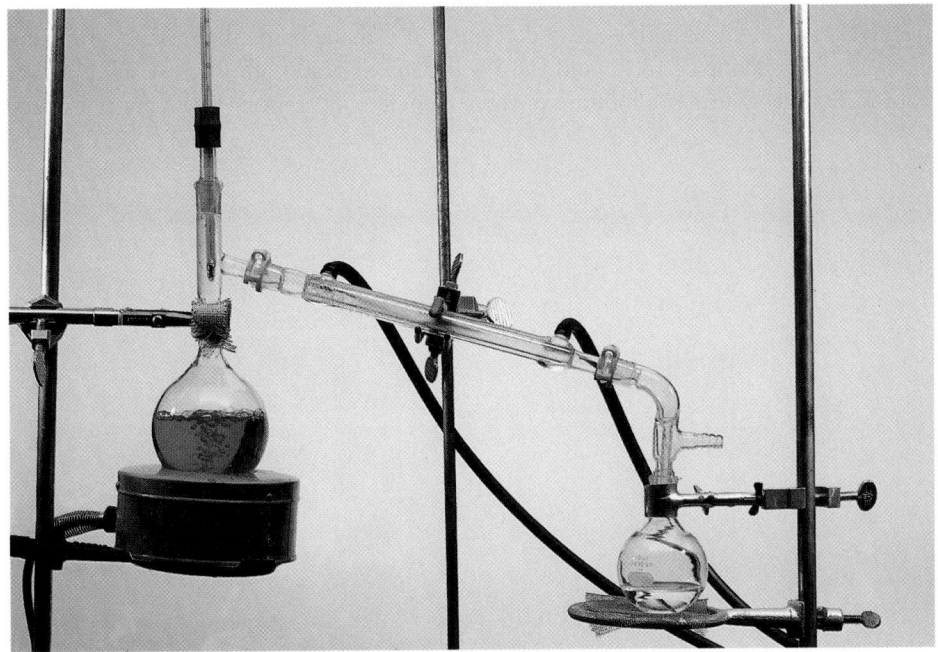

the water is boiled away from the solution and then retrieved by condensation through a water-cooled tube. When all of the water has boiled away, the solid table salt remains behind in the distilling flask.

A glass of salt water and a glass of pure water look exactly the same. They do taste different, however. In fact, *since both solutions and pure substances (elements and compounds) are homogeneous matter, one must examine the physical properties to distinguish between the two.* Mixtures have properties that vary with the proportion of the components. Elements and compounds have definite and unchanging properties. A simple example of a variable property is the taste and color of a cup of coffee. The more coffee that is dissolved in the water, the stronger the taste and the darker the solution.

Now consider the properties of two compounds alone: table salt and water. Solid table salt (sodium chloride) melts at 801°C and water melts at 0°C (32°F). A solution of salt in water begins to freeze anywhere from −18°C to just under 0°C, depending on the amount of salt dissolved. Also, a particular salt water solution does not have a sharp, unchanging boiling point or freezing point as does pure water.

Density is another physical property that is different for a solution compared to the pure liquid component. For example, "battery acid" is a solution of a compound, sulfuric acid, in water. Its density is greater than that of pure water. The more sulfuric acid present, the denser the solution. In a fully charged battery the density is about 1.30 g/mL; if it is mostly discharged, the density is about 1.15 g/mL. *The density of a liquid can be determined with a device called a* **hydrometer.** The hydrometer tube, as shown in Figure 2-10, is exactly balanced with weights in the bottom so that its level in pure water reads exactly 1.00 g/mL. The scale is calibrated in such a manner that the density of a liquid is read directly from the scale at the surface of the liquid. Discharging an automobile battery removes sulfuric acid from the solution, making the solution less dense. When the density drops to about 1.20 g/mL, the battery needs to be recharged.

Many of the metals that we use in our daily lives are actually alloys. *An* **alloy** *is a homogeneous mixture of metallic elements with one solid phase.* Although an alloy is considered a *solid solution,* it is made by mixing the metals in the molten state and then allowing the liquid solution to cool and resolidify. Pure gold (24 K) is a comparatively soft element and is easily deformed. It is made harder by mixing with other elements. For example, 18 K gold is 75% (by mass) gold with the rest silver and/or copper. Stainless steel is a mixture of three elements: 80% iron, 12% chromium, and

The French horn is made of brass, which is a homogeneous solid mixture of copper and zinc.

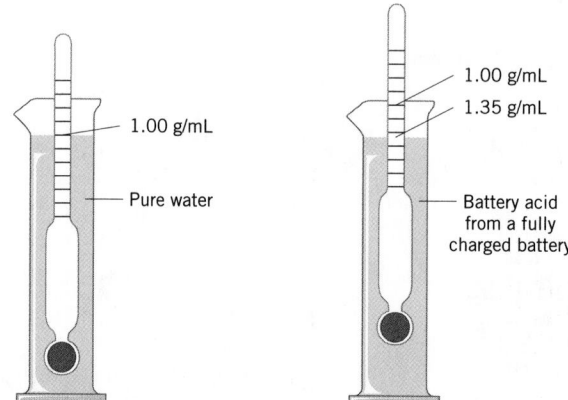

1.00 g/mL — Pure water

1.00 g/mL
1.35 g/mL — Battery acid from a fully charged battery

**Figure 2-10  A Hydrometer**
**This device is used to measure the density of a liquid.**

8% nickel. The percent composition of these alloys can be used to convert between the mass of the alloy and the mass of the component. The use of percent as a conversion factor is illustrated in the following two examples. Further exercises on the use of percent are found in Appendixes A and B.

---

## Conversion of Total Mass to Mass of a Component

**Example 2-5**

working it out

Manganese steel is very strong and finds use as railroad rails. It is composed of 86.0% iron, 13.0% manganese, and 1.0% carbon. What is the mass of each of the three elements in a 254-kg sample of manganese steel?

**Procedure**

Percent means parts per 100. In this case, a conversion factor can be constructed from each percent relating kilograms of a component to kilograms of the steel as follows.

$$86.0\% \text{ iron} = \frac{86.0 \text{ kg iron}}{100 \text{ kg steel}} \qquad 13.0\% \text{ manganese} = \frac{13.0 \text{ kg manganese}}{100 \text{ kg steel}}$$

These conversion factors can now be used as written to convert kilograms of steel to kilograms of a component.

**Solution**

$$254 \text{ kg steel} \times \frac{86.0 \text{ kg iron}}{100 \text{ kg steel}} = \underline{218} \text{ kg iron}$$

$$254 \text{ kg steel} \times \frac{13.0 \text{ kg manganese}}{100 \text{ kg steel}} = \underline{33.0 \text{ kg manganese}}$$

$$254 \text{ kg} - 218 \text{ kg iron} - 33.0 \text{ kg manganese} = \underline{3 \text{ kg carbon}}$$

---

## Conversion of Mass of a Component to Total Mass

**Example 2-6**

A sample of brass is composed of 72% copper and the remainder zinc. What mass of brass can be made from 25 kg of zinc?

**Procedure**

The percent composition of zinc is 100% − 72% = 28% zinc. The conversion factor and its reciprocal are

$$\frac{28 \text{ kg zinc}}{100 \text{ kg brass}} \quad \text{and} \quad \frac{100 \text{ kg brass}}{28 \text{ kg zinc}}$$

The latter factor can now be used to convert kg zinc to kg brass.

**Solution**

$$25 \text{ kg zinc} \times \frac{100 \text{ kg brass}}{28 \text{ kg zinc}} = \underline{89 \text{ kg brass}}$$

---

Actually, everything is a mixture, at least to some extent, since complete purity may be impossible. What we call "fresh drinking water" is hardly pure. It contains dissolved gases and solids. Even rainwater contains some dissolved gases from the

**Figure 2-11 Silicon for Computer Chips**
The silicon used to make computer chips must be "ultrapure."

air. What we consider pure may depend on the application. For example, if a sample of matter is composed of 99% of one element or compound, it may be considered pure for most purposes. On the other hand, the element silicon must be "ultrapure" to be used in computer chips. In this case, it is composed of more than 99.9999% silicon. (See Figure 2-11.)

**Looking Back!** Very little of our surroundings can actually be classified as "pure." Most of what we see is a mixture of the unique substances known as elements and compounds. Sometimes pure substances mix uniformly with no boundaries between components, but sometimes the pure substances form more obvious mixtures with boundaries. Even when the presence of a mixture is not obvious, however, it has properties that are different from the pure component substances. Physical properties distinguish between a mixture and a pure substance.

## Learning Check B | checking it out

**B-1.** Fill in the blanks.

Pure substances can mix in either of two ways. If the mixture is composed of more than one phase, it is _____ . If there are two liquid phases, the liquid on top has the _____ density. If they mix in such a way that only one phase is formed, the mixture is _____ . Mixtures that result in one liquid phase are usually referred to as _____ . These mixtures are distinguished from pure substances by their variable properties such as _____ and _____ points and density. Density of liquids is measured with a device called a _____ . Mixtures of metals that result in one solid phase are referred to as _____ . The percent composition of these mixtures relates the mass of a _____ to the total mass.

**B-2.** Carbon tetrachloride and water form a heterogeneous mixture. Which liquid is on the top? Refer to Table 2-1.

**B-3.** 14 K gold is 58.0% gold. What is the mass of pure gold in 4.00 oz of 14 K gold? What mass of 14 K gold can be made from 4.00 oz of pure gold?

**B-4.** The density of an alloy of aluminum and mercury is 5.40 g/mL. Which element is the principal metal in the alloy? Refer to Table 2-1.

*Additional Examples: Exercises 2-48, 2-55, 2-57, and 2-61.*

 Section **C  ENERGY AND MATTER**

## 2-6 Energy Changes in Chemical and Physical Processes

**Looking Ahead!** The massive stars in distant space are composed of the same basic types of matter as are found on Earth. They announce their existence, however, as tiny points of light in the night sky. The light that they transmit through space is a form of the second component of the universe, and that is energy. How we describe and measure energy is the topic of this section.

Some days we just don't have any energy. It other words, we would prefer not to do any work. Actually, this is almost exactly the definition of energy. **Energy** *is the capacity or the ability to do work.* Work involves the transfer of energy when an object is moved a certain distance. Just as matter has more than one physical state, energy has more than one form. Most of the energy on Earth originates from the sun. Deep in the interior of our ordinary star, transformations of elements occur

## Invisible Matter and Antigravity (Be careful! Don't fall up.)

Invisible people, antigravity drives allowing vehicles to float freely in space—this is the stuff of science fiction movies. Weird stuff—or is it? In recent years, these concepts are not as far out as we once imagined. New discoveries are changing our views of such phenomena.

First consider the possibility of invisible matter. A Swiss astronomer, Fred Zwicky, first proposed its existence in 1933 by observation of the stars. Galaxies are huge groupings of stars rotating around a central core. (Our home galaxy is known as "the Milky Way.") Zwicky noted that the visible matter in the galaxy could not control the motion of the stars around the core. It was like some unseen, exotic part of nature was at work controlling the motions of stars. Most scientists now accept the existence of this invisible stuff, known as *dark matter*. In fact, as much as 95% of the matter in the universe may be classified as dark matter. But it is hard to prove its existence if it is invisible, so we have only indirect evidence of its existence. Weird!

In 1998, more science fiction became reality. While dark matter accounts for the motion of the stars within a galaxy, it now appears that the galaxies of the universe are moving more and more rapidly away from each other. This is a reversal of the thinking of just a few years ago, when we thought the expansion of the universe was slowing. It has been proposed that an exotic form of energy also exists that causes matter to move apart. This "antigravity" energy has been called *dark energy*. It would have an effect the opposite of normal gravity.

Reality and science fiction do not seem to be that far apart. There may be an invisible world around us that leaves only gravity as the marker of its existence. And now we have antigravity—a force that would make us "fall up" rather than "fall down."

The study of dark matter and dark energy is rapidly developing areas of astronomy. We will know more in the next few years as we look for ways to prove their existence.

*The craft from Star Wars used "antigravity."*

that liberate a form of energy called *nuclear energy*. This energy, however, changes in the sun into *light* or *radiant energy* that then travels through space to illuminate Earth. Light energy from the sun bathes our planet and shines on the surface vegetation, where some of it is converted into *chemical energy* by a process called

**Almost all of the energy on Earth originates from the sun.**

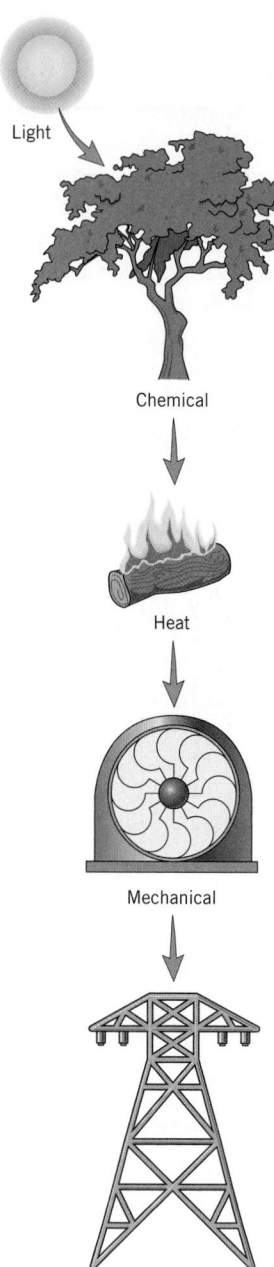

photosynthesis. Chemical energy is stored in the energy-rich compounds that make up the bulk of the vegetation. When logs from a tree are burned, the chemical energy is released in the form of *heat energy*. In a similar process, the metabolism of food in our bodies releases the energy to keep us alive. In the burning or metabolism process, energy-poor compounds are produced, and they recycle to the environment. In the production of electrical power, the heat energy is used to produce steam that turns a turbine. The movement of the turbine is *mechanical energy*. The mechanical energy powers a generator that converts the mechanical energy into *electrical energy*. (See Figure 2-12.) Other conversions between energy forms are possible. For example, in the chemical change that occurs in a car battery, chemical energy is converted directly into electrical energy. In all of these changes, the total energy remains the same. Energy changes are subject to the same law as matter changes in chemical reactions. *The* **law of conservation of energy** *states that energy cannot be created or destroyed but only transformed from one form to another*. In the production of electrical energy in Figure 2-12, only about 35% of the chemical energy is eventually transformed into electrical energy. The rest of the energy is lost as heat energy in the various transformations. The total energy remains constant, however.

Chemical or physical changes may be accompanied by either the release or the absorption of heat energy. *When a change releases heat, it is said to be* **exothermic.** *When a change absorbs heat, it is said to be* **endothermic.** Combustion (burning) is a common example of an exothermic chemical reaction. An example of an endothermic process involves an "instant cold pack." When the compounds ammonium nitrate (a solid) and water are brought together in a plastic bag, a solution is formed. The endothermic solution process causes enough cooling to make an ice pack useful for treating sprains and minor aches. (See Figure 2-13.) The melting of ice is a physical change but is also an endothermic process. Obviously, if we wish to melt ice cubes, we supply heat by taking a tray of cubes out of the freezer and letting it sit at room temperature.

In addition to the *forms* of energy, there are two *types* of energy. These depend on whether the energy is available but not being used or is actually in use. **Potential energy** *is energy that is available because of position or composition*. For example, a weight suspended above the ground has energy available because of its position and the attraction of gravity for the weight. Water stored behind a dam (Figure 2-14), a compressed spring, and a stretched rubber band all have potential energy. The chemical energy stored in the compounds of a tree log is also classified as potential energy. *Energy resulting from motion is known as* **kinetic energy.** A moving baseball, a speeding train, and water flowing down a spillway from a dam (Figure 2-14) all have kinetic energy.

**Figure 2-12  Energy**
**Energy is neither created nor destroyed but can be transformed.**

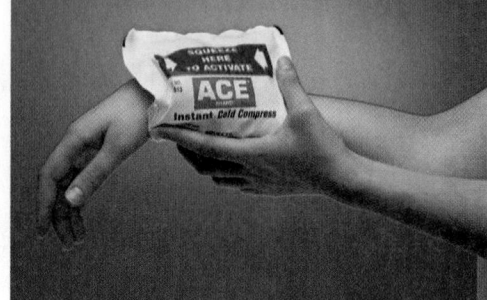

**Figure 2-13  Instant Cold Pack**
**When capsules of ammonium nitrate are broken and mixed with water, a cooling effect results.**

## 2-7 Temperature Change and Specific Heat

**Looking Ahead!** Heat energy interacts with matter in three ways. The application of heat may initiate a chemical change (e.g., the frying of an egg), it may cause a physical change (e.g., the melting of ice), or it may just simply cause the temperature to rise. We will discuss how it initiates chemical changes in Chapter 14 and its relation to phase changes in Chapter 10. Now we are more interested in how heat energy affects the temperature of various substances. Heat energy, just like matter, can be quantified. The quantification of heat energy and how it affects the temperature of various substances is the subject of this section.

It seems to take forever for a pot of water to get hot, but a heavy iron skillet over the same fire heats up quickly. Obviously, the same amount of heat changed the temperature of the skillet more than it did that of the water. A physical property of matter that relates temperature change to the amount of heat is known as specific heat capacity, or simply specific heat. **Specific heat** *is defined as the amount of heat required to raise the temperature of one gram of a substance one degree Celsius (or Kelvin).* Like density, specific heat is the same for any amount of a substance, and it can be used to identify a pure substance.

Units of heat energy are related to the specific heat of water. *The* **calorie** *was originally defined as the amount of heat energy required to raise the temperature of one gram of water from 14.5°C to 15.5°C.* For the most part, the unit of heat energy used in chemistry is the SI unit called the **joule.** *The calorie is now defined in terms of the joule.*

$$1 \; cal = 4.184 \; joule \, (J) \quad (exactly)$$

From the definition of the calorie, it is apparent that the specific heat of water is

$$1.000 \, \frac{cal}{g \cdot °C} = 4.184 \, \frac{J}{g \cdot °C}$$

The units read: calories (or joules) per gram per degree Celsius. The degree Celsius unit represents a *temperature change* and not a specific temperature reading. A change in temperature is represented as $\Delta t$. A formula that can be used to calculate the specific heat is

$$\text{specific heat} = \frac{\text{amount of heat energy} \, (J \text{ or } cal)}{\text{mass (g)} \times \Delta t (°C)}$$

The specific heats of several pure substances are listed in Table 2-2.

The following examples illustrate the calculation of specific heat and the use of specific heat to calculate temperature change.

**Figure 2-14   Potential and Kinetic Energy**
Water stored behind a dam has potential energy. The water flowing over the spillway has kinetic energy.

---

| The Calculation of Specific Heat | Example 2-7 |
|---|---|

**working it out**

It takes 62.8 J to raise the temperature of a 125-g quantity of silver 0.714°C. What is the specific heat of silver in joules and calories?

**Procedure**

The specific heat is calculated by substituting the appropriate quantities in the formula.

**Solution**

$$\text{specific heat} = \frac{62.8 \text{ J}}{125 \text{ g} \times 0.714°C} = \underline{0.704 \, \frac{J}{g \cdot °C}}$$

$$0.704 \, \frac{J}{g \cdot °C} \times \frac{1 \text{ cal}}{4.184 \, J} = \underline{0.168 \, \frac{cal}{g \cdot °C}}$$

| Example 2-8 | The Calculation of Temperature Change |
| --- | --- |

If 1.22 kJ of heat is added to 50.0 g of water at 25.0°C, what is the final temperature of the water? [The specific heat of water is 4.184 J/(g · °C).]

**Solution**

a. $1.22 \text{ kJ} \times 10^3 \dfrac{\text{J}}{\text{kJ}} = 1.22 \times 10^3 \text{ J}$

b. $\text{specific heat} = \dfrac{\text{heat energy}}{\text{mass (g)} \times \Delta t(°\text{C})}$

$$\Delta t(°\text{C}) = \dfrac{\text{heat energy}}{\text{specific heat} \times \text{mass}}$$

$$= \dfrac{1.22 \times 10^3 \text{ J}}{4.184 \dfrac{\text{J}}{\text{g} \cdot °\text{C}} \times 50.0 \text{ g}} = 5.83°\text{C}$$

Since the water has been heated, the final temperature $t$(C) is the initial temperature *plus* $\Delta t$(°C). Thus

$$t(°\text{C}) = 25.0 + 5.83 = \underline{\underline{30.8°\text{C}}}$$

**Nutrition Facts**

Serving Size 1 Tablespoon (14g)
Servings Per Container 96

**Amount Per Serving**

Calories 120      Fat Calories 120

| | % Daily Value* |
| --- | --- |
| **Total Fat** 14g | **22%** |
| Saturated Fat 1g | 6% |
| Polyunsaturated Fat 5g | |
| Monounsaturated Fat 7g | |
| **Sodium** 0mg | **0%** |
| **Total Carbohydrate** 0g | **0%** |
| Protein 0g | **0%** |

**Many people are very conscious of the energy content (Calories) of the food they eat.**

At the beginning of this section, we mentioned that an iron skillet seems to heat faster than water. That is true. The reason it takes so long to heat water is that it has a comparatively high specific heat. Notice in Table 2-2 that it is almost ten times higher than that of the iron in the skillet. Thus, in a calculation similar to the one in Example 2-8, the same amount of heat will raise the temperature of the iron almost 10°C for every 1°C for the same amount of water. (Iron also conducts heat rapidly from the fire to the handle.)

Life in many regions of this planet Earth depends on the high specific heat of water. For example, the sun warms the waters of the Gulf of Mexico. The Gulf Stream is a warm river of water that crosses the Atlantic Ocean all the way to England and the Scandinavian countries. There, the heat absorbed in the warm Gulf Stream is released and warms what would otherwise be uninhabitable countries that lie near the Arctic Circle. The island of Greenland also lies far north, but it is not warmed by the Gulf Stream and is almost entirely covered with ice the year around. That would be the fate of England were it not for the ability of water to absorb and retain vast quantities of heat energy.

**Table 2-2  Specific Heats**

| Substance | Specific Heat [cal/(g · °C)] | Specific Heat [J/(g · °C)] |
| --- | --- | --- |
| Water | 1.000 | 4.184 |
| Ice | 0.492 | 2.06 |
| Aluminum (Al) | 0.214 | 0.895 |
| Gold (Au) | 0.031 | 0.13 |
| Copper (Cu) | 0.092 | 0.38 |
| Zinc (Zn) | 0.093 | 0.39 |
| Iron (Fe) | 0.106 | 0.444 |

The amount of energy that we obtain from food can also be measured and expressed in calories. The nutritional calorie is actually one *kilocalorie* ($10^3$ cal) as defined above. To distinguish between the two calories, the "c" in calorie is capitalized in the nutritional calorie (1 Cal). The Calorie content of a portion of food is determined by burning the dried food in such a way that the heat released is used to heat water. The temperature change of the water is then converted to the amount of heat. This is illustrated in Example 2-9.

---

## The Calories in a Piece of Cake

**Example 2-9**

**working it out**

A piece of cake is dried and burned so that all the heat energy released heats some water. If 3.15 L of water is heated a total of 75.0°C, how many Calories does the cake contain?

**Procedure**

a. Convert the volume of water to the mass of water using the density.

b. Use the formula for specific heat to find the amount of heat energy.

$$\text{amount of heat energy} = \text{mass} \times \Delta t \times \text{specific heat}$$

c. Convert calories to Calories.

**Solution**

a. $3.15 \text{ L} \times \dfrac{10^3 \text{ mL}}{\text{L}} \times 1.00 \dfrac{\text{g}}{\text{mL}} = 3.15 \times 10^3 \text{ g}$

b. $3.15 \times 10^3 \text{ g} \times 75.0°C \times 1.00 \dfrac{\text{cal}}{\text{g} \cdot °C} = 236 \times 10^3 \text{ cal}$

c. $236 \times 10^3 \text{ cal} \times \dfrac{1 \text{ Cal}}{10^3 \text{ cal}} = \underline{\underline{236 \text{ Cal}}}$

---

When two substances at different temperatures are mixed, the hotter item will lose heat energy and the cooler one will gain heat energy. Thus the temperature of the hotter item comes down and the other goes up. Eventually, the two substances come to the same temperature, which is in between the two original temperatures. In other words, "heat lost equals heat gained," assuming no heat is lost to the surroundings. The application of this principle to calculate an unknown specific heat is illustrated in Example 2-10.

---

## The Specific Heat of a Metal by Heat Exchange

**Example 2-10**

**working it out**

A 440-g quantity of a certain metal is heated to 100.0°C. It is immediately thrust into 258 g of water that is initially at 25.0°C. The temperature of the metal-water mixture eventually is 36.5°C. What is the metal? (Refer to Table 2-2.)

**Procedure**

a. The heat gained by the water is given by the equation:

$$\text{heat gained} = \Delta°C(\text{water}) \times \text{g(water)} \times \text{specific heat(water)}$$

b. The heat gained by the water is equal to the heat lost by the metal. Set the heat gained by the water equal to the heat lost by the metal. The heat lost by the metal is:

$$\text{heat lost} = \Delta°C(\text{metal}) \times \text{g(metal)} \times \text{specific heat(metal)}$$

c. Solve for the specific heat of the metal.

**Solution**

$$\text{heat lost} = \text{heat gained}$$

$$\Delta°C(\text{of water}) \times g(\text{water}) \times \text{specific heat(water)}$$
$$= \Delta°C(\text{metal}) \times g(\text{metal}) \times \text{specific heat(metal)}$$

$$(36.5 - 25.0)°C \times 258 \text{ g} \times 4.184 \text{ J/°C g}$$
$$= (100 - 36.6)°C \times 440 \text{ g} \times \text{specific heat(metal)}$$

$$\text{specific heat(metal)} = \underline{0.444 \text{ J/g °C}}$$

From Table 2-2, the metal is probably *iron*.

---

**Looking Back!** An understanding of the nature of energy completes our discussion of the two phenomena of our universe (the other being matter). Energy undergoes numerous transformations as it journeys from the sun to the food we eat or to the electricity we need for our computer. In all of these transformations, energy is conserved. Besides the forms of energy, we have defined the two types of energy (kinetic and potential) and described processes that liberate or absorb heat energy (exothermic or endothermic.) Heat energy interacts with matter in several ways, but in this section we confined our discussion to how it changes the temperature of various substances.

---

## Learning Check C | checking it out

**C-1.** Fill in the blanks.

Energy is defined as the ability to do _____ . Most of the energy used on Earth arrives from the sun in the form of _____ energy. Another form of energy that we rely on to run our appliances is _____ energy. Chemical energy is a type of _____ energy that is stored in compounds. When this energy is released as heat in a chemical change or reaction, the reaction is said to be _____ . When heat is added to matter, it causes an _____ in the temperature. The two units of heat energy that are used are the _____ and the _____ . The specific heat of a substance is a _____ property. It is defined as the amount of heat required to raise one _____ of the substance _____ degree Celsius.

**C-2.** A 10.0-g sample of a certain metal is at 25°C. When 87.4 J of heat energy is added to the sample, the temperature increases to 48°C. Identify the metal from Table 2-2.

**C-3.** A 255-g quantity of gold is heated from 28° to 100°C. How many kilojoules of heat energy were added to the sample? (Refer to Table 2-2.)

*Additional Examples: Exercises 2-63, 2-66, 2-69, 2-70, 2-73, and 2-87.*

---

# chapter Review | putting it together

**Chemistry** is a science concerned with the two components of the universe, **matter** and **energy.** Matter is composed of a few basic substances called **elements.** Actually, we don't find many of the free elements on Earth because they are chemically combined to form **compounds.** Both elements and compounds are known as **pure substances.** Elements and compounds are distinguished by their properties. They display **physical properties** and undergo **physical changes.** An important physical property of a substance is its **physical state: solid, liquid,** or **gas.** A physical change takes place when the substance changes to a different physical state. The temperatures at which an element or compound changes to a different **phase** are known as the **melting point** of a solid (or **freezing point** of a liquid) and **boiling point** of a liquid (or **condensation point** of a gas).

An important physical property of a substance is its **density.** Density relates the mass of a substance to its volume; thus it is independent of the size of the sample. For this reason, it can be used as an identifying property as well as a conversion factor between mass and volume of a sample. When density is expressed in units of g/mL or g/cm$^3$, **specific gravity** has the same numerical quantity but is expressed without units. A pure substance also has **chemical properties** that relate to the **chemical changes** it undergoes. In chemical changes the **law of conservation of mass** is observed.

|  | *Physical* | *Chemical* |
|---|---|---|
| Property | Describes properties that do not result in a change into another substance | Describes the ability of a substance to undergo various types of chemical reactions |
| Change | Changes in dimensions or physical state | Changes into another substance |

Elements and compounds may form either **homogeneous** or **heterogeneous mixtures.** In heterogeneous mixtures of liquids, the less dense liquid floats on the other liquid. Heterogeneous mixtures of solids and liquids can be separated by filtration. Two types of homogeneous mixtures are **solutions,** which are composed of one liquid phase, and **alloys,** which are composed of one solid phase. The components of a solution can be separated by distillation. Aqueous (water) solutions and pure water often look identical. Pure water, however, has definite and unchanging properties. The properties of a solution vary according to the proportions of the mixture. Among those properties that vary for a solution is density. A **hydrometer** can be used to measure the density of a solution as well as that of a pure liquid. In the diagram below, we start with the most complex type of impure matter and proceed to the simplest form of homogeneous matter, an element.

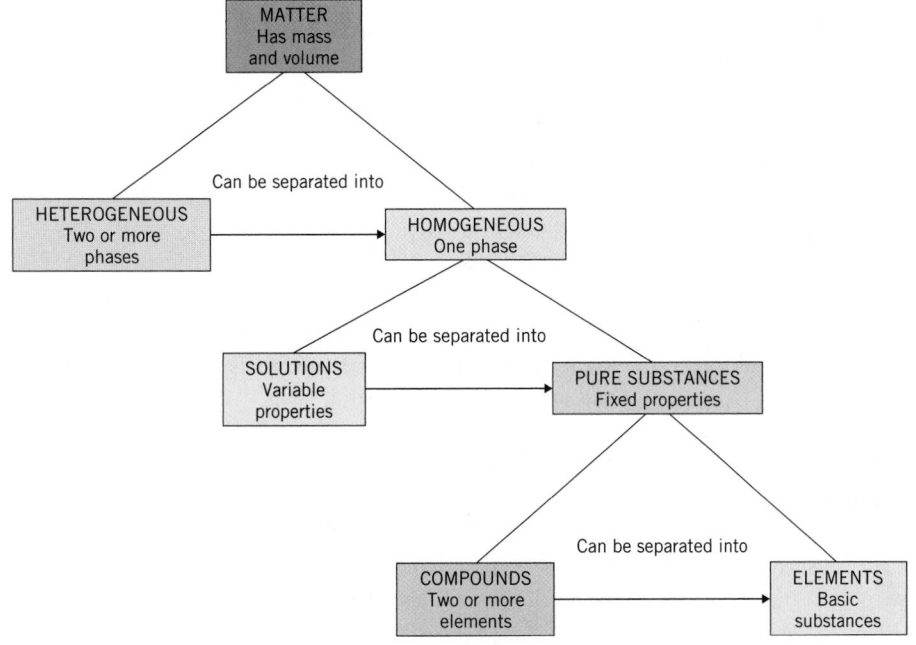

The other component of the universe, energy, is also intimately involved in physical and chemical processes. There are several *forms* of energy such as chemical, heat, and electrical energy. There are also two *types* of energy: **potential energy** and **kinetic energy.** Forms and types of energy can be converted into each other but in all cases the **law of conservation of energy** is observed. When chemical energy is converted into heat energy, the chemical change is said to be **exothermic.** When heat energy is converted into chemical energy, the reaction is said to be **endothermic.**

How much the temperature of a substance is affected by a certain amount of heat is a physical property known as **specific heat.** Specific heat is a measure of the amount of heat in **calories** or **joules** (or kcal or kJ) that will raise the temperature of one gram of the substance one degree Celsius.

# Exercises

## Elements and Compounds

**2-1.** List the elements that you think occur in the free state on the Earth. (Refer to the list of elements inside the front cover.)

**2-2.** What is the difference between an element and a compound?

**2-3.** Identify the following as either elements or compounds. Refer to the list of elements inside the front cover if necessary.
**(a)** carbon monoxide
**(d)** titanium
**(b)** hydrogen
**(e)** potash
**(c)** iron
**(f)** sodium bicarbonate

**2-4.** Identify the following as either elements or compounds. Refer to the list of elements inside the front cover if necessary.
**(a)** ammonia
**(d)** mercury
**(b)** hydrogen peroxide
**(e)** stannous fluoride
**(c)** aspirin
**(f)** uranium

**2-5.** Two centuries ago, water was thought to be an element. Why do you think scientists decided that it is a compound?

**2-6.** A given compound can be decomposed into two other substances. Are the two substances both elements?

## Physical and Chemical Properties and Changes

**2-7.** Which of the following describes the liquid phase?
**(a)** It has a definite shape and a definite volume.
**(b)** It has a definite shape but not a definite volume.
**(c)** It has a definite volume but not a definite shape.
**(d)** It has neither a definite shape nor a definite volume.

**2-8.** Which state of matter is compressible? Why?

**2-9.** A fluid is a substance that flows and can be poured. Which state or states can be classified as fluids?

**2-10.** Identify the following as either a physical or a chemical property.
**(a)** Diamond is one of the hardest known substances.
**(b)** Carbon monoxide is a poisonous gas.
**(c)** Soap is slippery.

**(d)** Silver tarnishes.
**(e)** Gold does not rust.
**(f)** Carbon dioxide freezes at −78°C.
**(g)** Tin is a shiny, gray metal.
**(h)** Sulfur burns in air.
**(i)** Aluminum has a low density.

**2-11.** Identify the following as either a physical or a chemical property.
**(a)** Sodium burns in the presence of chlorine gas.
**(b)** Mercury is a liquid at room temperature.
**(c)** Water boils at 100°C at average sea-level pressure.
**(d)** Limestone gives off carbon dioxide when heated.
**(e)** Hydrogen sulfide has a pungent odor.

**2-12.** Identify the following as a physical or a chemical change.
**(a)** the frying of an egg
**(b)** the vaporization of dry ice
**(c)** the boiling of water
**(d)** the burning of gasoline
**(e)** the breaking of glass

**2-13.** Identify the following as a physical or a chemical change.
**(a)** the souring of milk
**(b)** the fermentation of apple cider
**(c)** the compression of a spring
**(d)** the grinding of a stone

**2-14.** Can an element be distinguished from a compound by examination of its physical properties only? Explain.

**2-15.** When table sugar is heated to its melting point, it bubbles and turns black. When it cools, it remains a black solid. Describe the changes.

**2-16.** Aluminum metal melts at 660°C and burns in oxygen to form aluminum oxide. Identify a physical property and change and a chemical property and change.

**2-17.** A pure substance is a green solid. When heated, it gives off a colorless gas and leaves a brown, shiny solid that melts at 1083°C. The shiny solid cannot be decomposed to simpler substances, but the gas can. List all of the properties given and tell whether they are chemical or physical. Tell whether each substance is a compound or an element.

**2-18.** A pure substance is a greenish yellow, pungent gas that condenses to a liquid at −35°C. It undergoes a chemical reaction with a given substance to form a white solid that melts at 801°C and a brown, corrosive liquid with a specific gravity of 3.12. The white solid can be decomposed to simpler substances, but the gas and the liquid cannot. List all of the properties given and tell whether they are chemical or physical. Tell whether each substance is a compound or an element.

### Density

**2-19.** A handful of sand has a mass of 208 g and displaces a volume of 80.0 mL. What is its density?

**2-20.** A 125-g quantity of iron has a volume of 15.8 mL. What is its density?

**2-21.** A given liquid has a volume of 0.657 L and a mass of 1064 g. What might the liquid be? (Refer to Table 2-1.)

**2-22.** A 25.0-g quantity of magnesium has a volume of 14.4 mL. What is its density? What is the volume of 1.00 kg of magnesium?

**2-23.** What is the volume in milliliters occupied by 285 g of mercury? (See Table 2-1.)

**2-24.** What is the mass of 671 mL of table salt? (See Table 2-1.)

**2-25.** What is the mass of 1.00 L of gasoline? (See Table 2-1.)

**2-26.** What is the mass in pounds of 1.00 gallon of gasoline? (See Tables 2-1 and 1-5.)

**2-27.** What is the mass of 1.50 L of gold? (See Table 2-1.)

**2-28.** What is the volume in milliliters occupied by 1.00 kg of carbon tetrachloride?

**2-29.** The volume of liquid in a graduated cylinder is 14.00 mL. When a certain solid is added, the volume reads 92.45 mL. The mass of the solid is 136.5 g. What might the solid be? (Refer to Table 2-1.)

**2-30.** A 1.05-lb quantity of a solid has a volume of 1.00 L. What is its density? Does this material float on water?

**2-31.** Pumice is a volcanic rock that contains many trapped air bubbles. A 155-g sample is found to have a volume of 163 mL. What is the density of pumice? What is the volume of a 4.56-kg sample? Will pumice float or sink in water? In ethyl alcohol? (Refer to Table 2-1.)

**2-32.** A bottle weighs 44.75 g. When 12.0 mL of a liquid is added to the bottle it weighs 59.20 g. What might the liquid be? (Refer to Table 2-1.) Does it form a layer below water or above water?

**2-33.** The density of diamond is 3.51 g/mL. What is the volume of the Hope diamond if it has a mass of 44.0 carats? (1 carat = 0.200 g.)

**2-34.** A small box is filled with liquid mercury. The dimensions of the box are 3.00 cm wide, 8.50 cm long, and 6.00 cm high. What is the mass of the mercury in the box? (1.00 mL = 1.00 cm³.)

**2-35.** A 125-mL flask has a mass of 32.5 g when empty. When the flask is completely filled with a certain liquid, it weighs 143.5 g. What is the density of the liquid?

**2-36.** Which has a greater volume, 1 kg of lead or 1 kg of gold?

**2-37.** Which has the greater mass, 1 L of gasoline or 1 L of water?

**2-38.** A large nugget of a shiny metal is found in a mountain stream. It weighs 5.65 ounces (16 oz/lb). 25.0 mL of water is placed in a graduated cylinder. When the nugget is placed in the cylinder, the water level reads 33.3 mL. Did we find a nugget of gold? (Refer to Tables 2-1 and 1-5.)

**2-39.** What is the difference between density and specific gravity?

**\*2-40.** What is the mass of 1 gal of carbon tetrachloride in grams? In pounds?

**\*2-41.** Calculate the density of water in pounds per cubic foot (lb/ft³).

**\*2-42.** In certain stars, matter is tremendously compressed. In some cases the density is as high as $2.0 \times 10^7$ g/mL. A tablespoon full of this matter is about 4.5 mL. What is the mass of this tablespoon of star matter in pounds?

### Mixtures and Pure Substances

**2-43.** Carbon dioxide is not a mixture of carbon and oxygen. Explain.

**2-44.** Which of the following is a mixture: water, sulfur dioxide, pewter, or ammonia?

**2-45.** List the following "waters" in order of increasing purity: ocean water, rainwater, and drinking water. Explain.

**2-46.** Iron is attracted to a magnet, but iron compounds are not. How could you use this information to tell whether a mixture of iron and sulfur forms a compound when heated?

**2-47.** When a teaspoon of solid sugar is dissolved in a glass of liquid water, what phase or phases are present after mixing?
**(a)** liquid only    **(b)** still solid and liquid    **(c)** solid only

**2-48.** Identify the following as homogeneous or heterogeneous matter.
**(a)** gasoline    **(d)** alcohol    **(g)** aerosol spray
**(b)** dirt    **(e)** a new nail    **(h)** air
**(c)** smog    **(f)** vinegar

**2-49.** Identify the following as homogeneous or heterogeneous matter.
**(a)** a cloud    **(c)** whipped cream    **(e)** natural gas
**(b)** dry ice    **(d)** bourbon    **(f)** a grapefruit

**2-50.** In which physical state or states does each of the substances listed in exercise 2-48 exist?

**2-51.** In which physical state or states does each of the substances listed in exercise 2-49 exist?

**2-52.** Carbon tetrachloride and kerosene mix, but neither mixes with water. How can water be used to keep the carbon tetrachloride and kerosene apart? Which liquid is on the top? (Refer to Table 2-1.)

**2-53.** How could liquid mercury be used to tell whether a certain sample of metal is lead or gold? (Refer to Table 2-1.)

**2-54.** Why does ice float on water? (Refer to Table 2-1.) Is ice water homogeneous or heterogeneous matter? Pure or a mixture?

**2-55.** Tell whether each of the following properties describes a heterogeneous mixture, a solution (homogeneous mixture), a compound, or an element.
**(a)** a homogeneous liquid that, when boiled away, leaves a solid residue
**(b)** a cloudy liquid that after a time seems more cloudy toward the bottom
**(c)** a uniform red solid that has a definite, sharp melting point and cannot be decomposed into simpler substances
**(d)** a colorless liquid that boils at one unchanging temperature and can be decomposed into simpler substances
**(e)** a liquid that first boils at one temperature but as the heating continues, boils at slowly increasing temperatures (There is only one liquid phase.)

**2-56.** Tell whether each of the following properties describes a heterogeneous mixture, a solution (homogeneous mixture), a compound, or an element.
**(a)** a nonuniform powder that, when heated, first turns mushy and then continues to melt as the temperature rises
**(b)** a colored gas that can be decomposed into a solid and another gas (The entire sample of the new gas seems to have the same chemical properties.)
**(c)** a sample of colorless gas, only part of which reacts with hot copper

**2-57.** A sample of bronze is made by mixing 85 kg of molten tin with 942 kg of molten copper. What is the mass percent tin in this bronze?

**2-58.** A sample of brass weighs 22.8 lb. It contains 14.8 lb of copper and the rest zinc. What is the mass percent zinc in this brass?

**2-59.** A U.S. "nickel" is actually only 25% nickel. What mass of the element nickel is in 255 kg of "nickels"?

**2-60.** 10 K gold is only 42% gold. What mass of gold is in 186 g of 10 K gold?

**2-61.** Duriron is used to make pipes and kettles. It contains 14% silicon, with the remainder iron. What mass of duriron can be made from 122 lb of iron?

**2-62.** A sample of a gold alloy is 7% copper and 10% silver in addition to the gold. What mass of the gold alloy can be made from 175 kg of pure gold?

## Mass and Energy

**2-63.** From your own experiences, tell whether the following processes are exothermic or endothermic.
**(a)** decay of grass clippings
**(b)** melting of ice

**(c)** change in an egg when it is fried
**(d)** condensation of steam
**(e)** curing of freshly poured cement

**2-64.** A car battery can be recharged after the engine starts. Trace the different energy conversions from gasoline to the battery.

**2-65.** Windmills are used to generate electricity. What are all of the different forms of energy involved in generation of electricity by this method?

**2-66.** Identify the principal type of energy (kinetic or potential) exhibited by each of the following.
**(a)** a car parked on a hill       **(d)** an uncoiling spring
**(b)** a train traveling 60 mi/hr    **(e)** a falling brick
**(c)** chemical energy

**2-67.** Identify the following as having either potential or kinetic energy or both.
**(a)** an arrow in a fully extended bow
**(b)** a baseball traveling high in the air
**(c)** two magnets held apart
**(d)** a chair on the fourth floor of a building

**2-68.** When you apply your brakes to a moving car, the car loses kinetic energy. What happens to the lost energy?

**2-69.** When a person plays on a swing, at what point in the movement is kinetic energy the greatest? At what point is potential energy the greatest? Assume that once started, the person will swing to the same height each time without an additional push. At what point is the total of the kinetic energy and the potential energy greatest?

## Specific Heat

**2-70.** It took 73.2 J of heat to raise the temperature of 10.0 g of a substance 8.58°C. What is the specific heat of the substance?

**2-71.** When 365 g of a certain pure metal cooled from 100°C to 95°C, it liberated 56.6 cal. Identify the metal from among those listed in Table 2-2.

**2-72.** A 10.0-g sample of a metal requires 22.4 J of heat to raise the temperature from 37.0°C to 39.5°C. Identify the metal from those listed in Table 2-2.

**2-73.** If 150 cal of heat energy is added to 50.0 g of copper at 25°C, what is the final temperature of the copper? Compare this temperature rise with that of 50.0 g of water initially at 25°C. (Refer to Table 2-2.)

**2-74.** A large cube of ice weighing 558 g is cooled in a freezer to −15.0°C. It is removed and allowed to warm to 0.0°C but does not melt. What is the specific heat of ice if 17.2 kJ of heat is required in the process?

**2-75.** How many joules are evolved if 43.5 g of aluminum is cooled by 13°C? (Refer to Table 2-2.)

**2-76.** What mass of iron is needed to absorb 16.0 cal if the temperature of the sample rises from 25°C to 58°C?

**2-77.** If one has a copper and an iron skillet of the same weight, which would fry an egg the faster? Explain.

**2-78.** When 486 g of zinc absorbs 265 J of heat energy, what is the rise in temperature of the metal?

**2-79.** Given 12.0-g samples of iron, gold, and water, calculate the temperature rise that would occur when 50.0 J of heat is added to each.

**2-80.** A 10.0-g sample of water cools 2.00°C. What mass of aluminum is required to undergo the same temperature change?

**2-81.** A can of diet soda contains 1.00 Cal (1.00 kcal) of heat energy. If this energy were transferred to 50.0 g of water at 25°C, what would be the final temperature?

**2-82.** The specific heat of platinum is one of the lowest of the metals. What is its specific heat if a 5.44-g quantity evolves 7.36 cal when it cools from 55.0°C to 12.3°C?

**2-83.** If 50.0 g of aluminum at 100.0°C is allowed to cool to 35.0°C, how many joules are evolved?

**2-84.** A 22.5-J quantity of heat raises the temperature of a piece of zinc 2.0°C. How many degrees Celsius would the same amount of heat raise the temperature of the same amount of aluminum?

**2-85.** In the preceding exercise, assume that the hot aluminum was added to water originally at 30.0°C. What mass of water was present if the final temperature of the water was 35.0°C?

**2-86.** If 50.0 g of water at 75.0°C is added to 75.0 g of water at 42.0°C, what is the final temperature?

**2-87.** A 100.0-mL volume of water is originally at 25.0°C. When a chunk of lead that is originally at 42.8°C is added to the water, the temperature of the water increases to 28.7°C. If the specific heat of lead is 0.128 J/g · °C, what was the weight of the lead?

**2-88.** If 100.0 g of a metal at 100.0°C is added to 100.0 g of water at 25.0°C, the final temperature is 31.3°C. What is the specific heat of the metal? Identify the metal from Table 2-2.

### General Questions

**2-89.** A 22-mL quantity of liquid A has a mass of 19 g. A 35-mL quantity of liquid B has a mass of 31 g. If they form a heterogeneous mixture, explain what happens.

**2-90.** The volume of water in a graduated cylinder reads 25.5 mL. What does the volume read when a 25.0-g quantity of pure nickel is added to the cylinder? (The density of nickel is 8.91 g/mL.)

**2-91.** A solution of table sugar in water is 14% by mass sugar. The density of the solution is 1.06 g/mL. What is the mass of sugar in 100 mL of the solution?

**2-92.** A solid metal weighing 62.485 g was introduced into a small flask having a total volume of 24.96 mL. To completely fill the flask, 18.22 mL of water was required. What is the density of the metal expressed to the proper number of significant figures?

**2-93.** What is the volume occupied by 22.175 g of the metal in exercise 2-92.

**2-94.** A 10.0-g quantity of table salt was added to 305 mL of water. What is the percent table salt in the solution?

**2-95.** Battery acid is 35% sulfuric acid in water and has a density of 1.29 g/mL. What is the mass of sulfuric acid in 1.00 L of battery acid?

**2-96.** An alloy is prepared by mixing 50.0 mL of gold with 50.0 mL of aluminum. What is the mass percent gold in the alloy?

**2-97.** How many kilocalories are required to raise 1.25 L of ice from −45°C to its melting point? (Refer to Tables 2-1 and 2-2.)

**2-98.** When 215 J of heat is added to a 25.0-g sample of a given substance, the temperature increases from 25°C to 91°C. What is the volume of the sample? (Refer to Tables 2-1 and 2-2.)

**2-99.** A 21.8-mL quantity of magnesium is at an initial temperature of 25.0°C. When 18.4 cal of heat energy is added to the sample, the temperature increases to 27.0°C. What is the specific heat of magnesium? (Refer to Tables 2-1 and 2-2.)

**2-100.** 25.0 mL of a given pure substance has a mass of 67.5 g. How much heat in kilojoules (kJ) is required to heat this sample from 15°C to 88°C? (Refer to Tables 2-1 and 2-2.)

**2-101.** When a log burns, the ashes have less mass than the log. When zinc reacts with sulfur, the zinc sulfide has the same mass as the combined mass of zinc and sulfur. When iron burns in air, the compound formed has more mass than the original iron. Explain each reaction in terms of the law of conservation of mass.

# Interactive Learning

**2-102.** The space shuttle uses liquid hydrogen as a fuel. The external fuel tank used during take-off carries 227,641 lb of hydrogen with a volume of 385,265 gal. Use the concept of specific gravity to calculate the density of liquid hydrogen in units of g/mL. (Express the answer to three significant figures. The density of water is 8.34 lb/gal.)

**2-103.** An alloy is made containing 8.75 kg of magnesium in 148.21 kg of aluminum. What is the percent by mass of magnesium? What mass of magnesium is in 375 g of the alloy? What mass of alloy can be made from 484 kg of aluminum?

**2-104.** Palladium is a metal that is used in catalytic converters in automobiles. If 1.560 kg of Pt is heated to 100.0°C and then placed in 1.000 L of water at 25.0°C, the final volume is 1.150 L and the final temperature of the mixture is 31.3°C. What is the density and specific heat of palladium?

**2-105.** A water-cooled radiator in an automobile engine works by having water remove the excess heat generated in

the engine. If a 250-kg engine made of iron is initially at 125°C, what volume of water originally at 25.0°C is needed to cool the engine to 95.0°C? (Assume that the water and engine end up at 95.0°C.)

# Solutions to Learning Checks

**A-1.** space, compounds, pure, properties, density, physical, chemical

**A-2.** Physical—dull, gray solid, melting point of 839°C, specific gravity of 1.54

Chemical—reacts with water

**A-3.** $\dfrac{254 \text{ g}}{159 \text{ mL}} = 1.60 \text{ g/mL}$ (carbon tetrachloride)

**A-4.** $178 \text{ g aluminum} \times \dfrac{1 \text{ mL}}{2.70 \text{ g aluminum}} = 65.9 \text{ mL}$

**B-1.** heterogeneous, lesser, homogeneous, solutions, melting, boiling, hydrometer, alloys, component

**B-2.** Water floats on top of carbon tetrachloride.

**B-3.** $4.00 \text{ oz 14 K gold} \times \dfrac{58.0 \text{ oz pure gold}}{100 \text{ oz 14 K gold}} = 2.32 \text{ oz pure gold}$

$4.00 \text{ oz pure gold} \times \dfrac{100 \text{ oz 14 K gold}}{58.0 \text{ oz pure gold}} = 6.90 \text{ oz of 14 K gold}$

**B-4.** aluminum (The density of the alloy is closer to that of aluminum.)

**C-1.** work, radiant or light, electrical, potential, exothermic, increase, calorie, joule, physical, gram, one

**C-2.** $\Delta t = 48 - 25 = 23°C$

$\text{specific heat} = \dfrac{87.4 \text{ J}}{10.0 \text{ g} \cdot 23°C} = 0.38 \dfrac{\text{J}}{\text{g} \cdot °\text{C}}$ (copper)

**C-3.** $\Delta t = 100 - 28 = 72°C$

$\text{heat energy} = \text{specific heat} \times \text{mass} \times \Delta t =$

$0.13 \dfrac{\text{J}}{\text{g} \cdot °\text{C}} \times 255 \text{ g} \times 72°C = 2.4 \times 10^3 \text{ J}$

$2.4 \times 10^3 \text{ J} \times \dfrac{1 \text{ kJ}}{10^3 \text{ J}} = 2.4 \text{ kJ}$

## chapter 3

# Elements, Compounds, and Their Composition

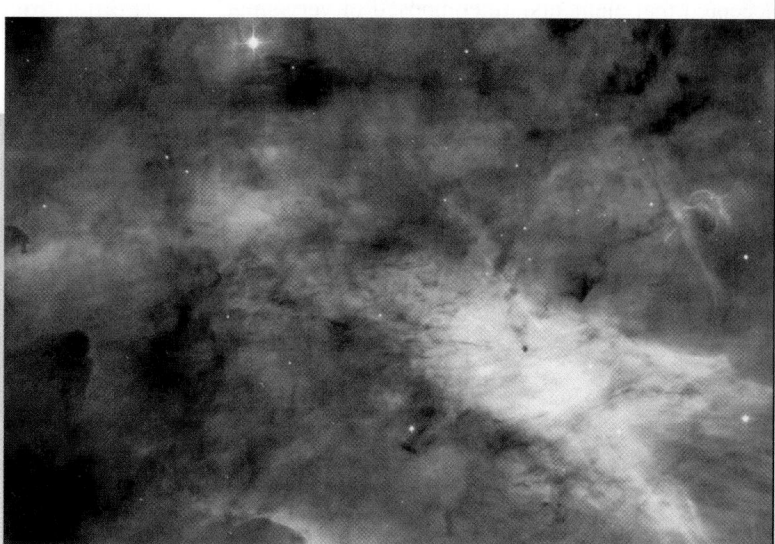

This image of the Omega nebula was taken by the Hubble telescope and released in 2002. The turquoise glow is from oxygen which formed in ancient stars.

### Setting the Stage

A few years ago, astronomers in Chile witnessed the catastrophic death of a distant sun. Suddenly one night, a bright point of light appeared in the night sky that was not there the night before. A massive star had literally exploded in an event known as a supernova. It had been over 400 years since a supernova so bright had been witnessed on Earth. The star was dying, but it was also a profound moment of creation. As the star was going through its death throes, elements were being formed. This spectacular event was important to science because it confirmed theories concerning how and where the heavy elements on this planet were originally formed.

For billions of years, stars have been churning, boiling, and eventually undergoing catastrophic explosions such as the one seen recently. Blasted by the force of exploding suns, newly formed elements and some simple compounds of the elements have drifted through cold, dark space as bits of dust and clouds of gases. About four and a half billion years ago, a large cloud containing hydrogen and some heavier elements produced by supernovas condensed to form a star with nine planets. The third planet out from this star is called Earth. On Earth, there arose living systems composed of these elements. We—and all of nature as we know it—are, in fact, composed of "stardust."

**Formulating Some Questions**

In the past 200 years, the miracle of the human mind, with its boundless curiosity, has unraveled many mysteries about the composition as well as the origin of these elements. For example, it is now universally accepted that elements are composed of very small fundamental units called atoms. The existence of the atom brings up many additional questions that we will discuss in this chapter. How did the concept of the atom come to be accepted? If atoms are the fundamental unit of an element, what is the fundamental unit of a compound? What makes a compound unique? Are the atoms of an element exactly the same? How do the atoms of one element differ from those of another? What determines the mass of an atom? Before we address these questions, however, we have more to say about the number of elements, their distribution, and how we name them.

## Section A  ELEMENTS AND COMPOUNDS AND THEIR COMPOSITION

### 3-1 Names and Symbols of the Elements

**Looking Ahead!** The fundamental building blocks of nature begin with the elements. Elements all have unique names. Many of the names are familiar to all of us, and others are not so familiar. The names, occurrences, and symbolic representation of the elements are the topics to be discussed now.

**Our home planet is composed of less than 90 elements.**

This planet Earth is composed of 88 elements that are present in measurable amounts. The rest of the 114 known elements are products of the nuclear age, as they were synthesized in laboratories or nuclear reactors. The last element was added to the periodic table in 1999. These 26 elements spontaneously change into other elements by radioactive decay. Of the 88 elements found naturally, only 10 constitute over 99% of Earth's crust (the outer portion including the oceans and the atmosphere). About 93% of the mass of our bodies is composed of only three elements—carbon, hydrogen, and oxygen. (See Table 3-1.)

The names of elements come from many sources. Some are derived from Greek, Latin, or German words for a color—for example, bismuth (white mass), iridium (rainbow), rubidium (deep red), and chlorine (greenish yellow). Some relate to the locality where the element was discovered (e.g., germanium, francium, and californium). Four

**Table 3-1   Distribution in the Earth's Crust and Human Body**

| Earth's Crust[a] | | | | Human Body | |
|---|---|---|---|---|---|
| Element | Weight Percent | Element | Weight Percent | Element | Weight Percent |
| Oxygen | 49.1 | Magnesium | 1.9 | Oxygen | 64.6 |
| Silicon | 26.1 | Hydrogen | 0.88 | Carbon | 18.0 |
| Aluminum | 7.5 | Titanium | 0.58 | Hydrogen | 10.0 |
| Iron | 4.7 | Chlorine | 0.19 | Nitrogen | 3.1 |
| Calcium | 3.4 | Carbon | 0.09 | Calcium | 1.9 |
| Sodium | 2.6 | Sulfur | 0.06 | Phosphorus | 1.1 |
| Potassium | 2.4 | All others | 0.50 | Chlorine | 0.9 |
| | | | | All others | 0.9 |

[a] Includes the oceans and the atmosphere.

## The Case of the Element Iridium and the Dead Dinosaurs

"It's elementary, my dear Watson." If Sherlock Holmes had been on the case of the missing dinosaurs, that statement would have been brilliant. The clue to the dead dinosaurs was indeed "elementary," actually "elemental." The first major clue was provided by the element iridium. It is a very rare element on the surface of Earth. When the Earth was molten, four billion years ago, most iridium (which is denser than gold) sank deep into the interior. However, bodies from space including meteors, asteroids, and comets contain comparatively high amounts of this element.

In 1979, American scientists discovered a thin layer of sediment in various locations around the world that was deposited about 65 million years ago. It was interesting that this coincided with the time the dinosaurs became extinct. Indeed, there were dinosaur fossils below the layer but none above. A crucial discovery was that the layer contained comparatively high amounts of iridium. Based on the iridium, it was proposed that this layer is composed of dust and debris from the collision with the Earth of a huge asteroid or comet, about six miles in diameter. A large cloud of dust must have formed, encircling the Earth and completely shutting out the sunlight. A freezing cold followed and most animals and plants quickly died. A hypothesis was proposed that the dinosaurs must have been among the casualties. After many months the dust settled out, forming a thin layer of sediment. Small mammals and some reptiles survived and would inherit the planet.

Since this original hypothesis, more supportive information has been discovered. Evidence of huge tidal waves, over one mile high, have been found in North and Central America. But perhaps the most important was the discovery in 1991 of a huge impact crater near the Yucatan peninsula in Mexico. The 110-mile-wide crater is buried miles into the ground and was formed about 65 million years ago.

Based on all of the evidence, most scientists have now embraced this as a plausible theory. Apparently, the end of the dinosaurs was sudden and dramatic. And now instead of worrying about a nasty Tyrannosaurus Rex lurking in the forest, we can worry about collisions with asteroids.

*Tyrannosaurus Rex.*

elements (yttrium, erbium, terbium, and ytterbium) are named after a town in Sweden (Ytterby). Other elements honor noted scientists (e.g., einsteinium, fermium, and curium) or mythological figures [e.g., plutonium, uranium, titanium, mercury, and promethium (the fire giver)]. Many of the oldest known elements have names with obscure origins. For many years the rights to name elements 104 through 109 were claimed by competing laboratories in the United States and Europe. The issue was finally resolved for good in 1997 by the International Union of Pure and Applied Chemistry (IUPAC), which is the official body that approves the names of new elements. Some of these names (i.e., rutherfordium, seborgium, and meitnerium) honor scientists who were active in research that led to our understanding of the heavy elements. The elements beyond 109 have temporary names that are derived from the Latin words for the number. [For example, the name of element 110 is ununnilium, which is un (1), un (1), and nil (0).]

An element can be designated by a symbol that is shorthand for its full name. *In most cases, the first one or two letters of the name are used as the element's* **symbol.** When an element has a two-letter symbol, the first is capitalized but the second is not. The symbols of the elements are listed along with the names inside the front cover. Some common elements whose symbols are derived from their English names are shown in Table 3-2. The last five elements (110–112,114, and 116) are still officially designated by a three-letter symbol (e.g., ununnilium is Uun).

Among the elements listed there are 11 whose symbols do not seem to relate to their names. These elements have symbols that are derived from their original Latin names (except wolfram, which is German) and are listed in Table 3-3.

The names of compounds are usually based on the elements of which they are composed, and most contain two words. Carbon dioxide, sodium sulfite, and silver nitrate all refer to specific compounds. Notice that the second word ends in *ide, ite,*

**Table 3-2  Some Common Elements**

| Element | Symbol | Element | Symbol |
|---------|--------|---------|--------|
| Aluminum | Al | Iodine | I |
| Bromine | Br | Magnesium | Mg |
| Calcium | Ca | Nickel | Ni |
| Carbon | C | Nitrogen | N |
| Chlorine | Cl | Oxygen | O |
| Chromium | Cr | Phosphorus | P |
| Fluorine | F | Silicon | Si |
| Helium | He | Sulfur | S |
| Hydrogen | H | Zinc | Zn |

**Table 3-3  Elements with Symbols from Earlier Names**

| Element | Symbol | Former Name |
|---------|--------|-------------|
| Antimony | Sb | Stibium |
| Copper | Cu | Cuprum |
| Gold | Au | Aurum |
| Iron | Fe | Ferrum |
| Lead | Pb | Plumbum |
| Mercury | Hg | Hydragyrum |
| Potassium | K | Kalium |
| Silver | Ag | Argentum |
| Sodium | Na | Natrium |
| Tin | Sn | Stannum |
| Tungsten | W | Wolfram |

or *ate*. A few compounds have three words, such as sodium hydrogen carbonate. A number of compounds have common names of one word, such as water, ammonia, lye, and methane. We will talk in more detail about the names of compounds in the next chapter.

# 3-2 Composition of an Element and Atomic Theory

**Looking Ahead!** Our modern understanding of the particulate nature of matter had its beginning over 2000 years ago. A Greek philosopher named Democritus suggested that all matter was like grains of sand on a beach. In other words, he thought that matter was composed of tiny indivisible particles that he called atoms. This view of matter is now the basis of our understanding of the universe around us. In this section, we will come to understand why the atomic theory is now universally accepted.

Just two centuries ago, the view of matter as being composed of atoms was definitely not accepted. Most knowledgeable scientists thought that a sample of an element such as copper could be divided (theoretically) into infinitely smaller pieces without changing its nature. In other words, they believed that matter was continuous. In 1803, however, an English scientist named John Dalton (1766–1844) proposed a different theory of matter based on the original thoughts of Democritus. His ideas are now known as *atomic theory*. The major conclusions of atomic theory are as follows.

1.  Matter is composed of small, indivisible particles called atoms.
2.  Atoms of the same element are identical and have the same properties.
3.  Chemical compounds are composed of atoms of different elements combined in small whole-number ratios.
4.  Chemical reactions are merely the rearrangement of atoms into different combinations.

The atomic theory is now universally accepted as our current view of matter. Thus we may define an **atom** as *the smallest fundamental particle of an element that has the properties of that element.*

The proposals of Democritus 2000 years ago were simply the product of his mind. Dalton's theory was a brilliant and logical explanation of many quantitative experimental

Figure 3-1 Atoms and Marbles On the left is an STM image of gold atoms. The atoms stack together like marbles.

observations and laws that were known at the time but had not been explained. The law of conservation of mass (section 2-4) was among the laws that were explained by atomic theory. Since chemical reactions are simply the shuffling of atoms, none are lost or gained and the mass remains unchanged.

Why are we so sure that Dalton was right? Besides the overwhelming amount of circumstantial evidence, we now have direct proof. In recent years, a highly sophisticated instrument called the scanning tunneling microscope has produced images of atoms of the heavier elements. Although these images are somewhat fuzzy, they indicate that an element such as gold is composed of spherical atoms packed closely together, just as you would find in a container of marbles all of the same size. (See Figure 3-1.)

When we look at a small piece of copper wire, it is hard to imagine that it is not continuous. This is because it is so difficult to comprehend the small size of the atom. Since the diameter of a typical atom is on the order of $10^{-8}$ cm, it would take about $10^{16}$ (ten quadrillion) atoms to appear as a tiny speck. The piece of copper wire is like a brick wall: from a distance it looks completely featureless, but up close we would notice that it is actually composed of closely packed basic units.

## 3-3 Elements and Compounds Composed of Molecules, and Their Formulas

**Looking Ahead!** Very little of the matter in our world is composed of separate atoms. Actually, atoms are usually joined together with other atoms of the same element or other elements into a more complex form of particulate matter. In this section, we will examine one of the principal forms of matter formed by the combined atoms.

About three centuries ago, water was thought to be an element. When a scientist was able to decompose water into hydrogen and oxygen, it became apparent that water is a more complex form of matter we know as a compound. If the basic particle of an element is an atom, the basic particle of one type of compound is known as a molecule. *A* **molecule** *is formed by the chemical combination of two or more atoms. The atoms in a molecule are joined and held together by a force called the* **covalent bond.** The basic particles of **molecular compounds** *are molecules composed of the atoms of two different elements.* If it were possible to magnify a droplet of water so as to visualize its basic particles, we would see that it is an example of a molecular compound. Each molecule of water is composed of two atoms of hydrogen joined by covalent bonds to one atom of oxygen. (See Figure 3-2.) Molecules can contain as few as two atoms or, in the cases of the complex molecules on which life is based, millions of atoms.

**Figure 3-2  A Water Molecule** **A water molecule is composed of two atoms of hydrogen and one atom of oxygen.**

*A compound is represented by using the symbols for the elements of which it is composed. This is called the* **formula** *of the compound.* The familiar formula for water is therefore

$$H_2O$$

This, of course, is pronounced "Aitch-two-Oh." Note that the 2 is written as a subscript, indicating that the molecule has two hydrogen atoms. When there is only one atom of a given element present (e.g., oxygen), a subscript of "1" is assumed but not shown.

What makes one compound different from another? The answer is that each chemical compound has a unique formula or arrangement of atoms in its molecules. For example, there is another compound composed of just hydrogen and oxygen, but it has the formula $H_2O_2$. Its name is hydrogen peroxide, and its properties are distinctly different from those of water ($H_2O$). Figure 3-3 illustrates how the *atoms* of hydrogen and oxygen combine to form the *molecules* of two different compounds. The formulas of other well-known compounds are $C_{12}H_{22}O_{11}$ (table sugar, also known as sucrose), $C_9H_8O_4$ (aspirin), $NH_3$ (ammonia), and $CH_4$ (methane).

Sometimes two or more compounds may share the same chemical formula. What then makes them unique? In this case, their difference stems from the arrangement of the atoms within the molecule. For example, ethyl alcohol and dimethyl ether both have the formula $C_2H_6O$. The difference in the two compounds lies in the order of the bonded atoms.

ethyl alcohol         dimethyl ether

$$
\begin{array}{cc}
\quad\ \text{H}\quad\ \text{H} & \quad\ \text{H}\qquad\ \text{H} \\
\quad\ |\quad\ | & \quad\ |\qquad\ | \\
\text{H}\!-\!\text{C}\!-\!\text{C}\!-\!\text{O}\!-\!\text{H} & \text{H}\!-\!\text{C}\!-\!\text{O}\!-\!\text{C}\!-\!\text{H} \\
\quad\ |\quad\ | & \quad\ |\qquad\ | \\
\quad\ \text{H}\quad\ \text{H} & \quad\ \text{H}\qquad\ \text{H}
\end{array}
$$

Notice that the alcohol has a C—C—O order and the ether a C—O—C order. (The dashes between atoms represent chemical bonds.) The difference in the arrangement

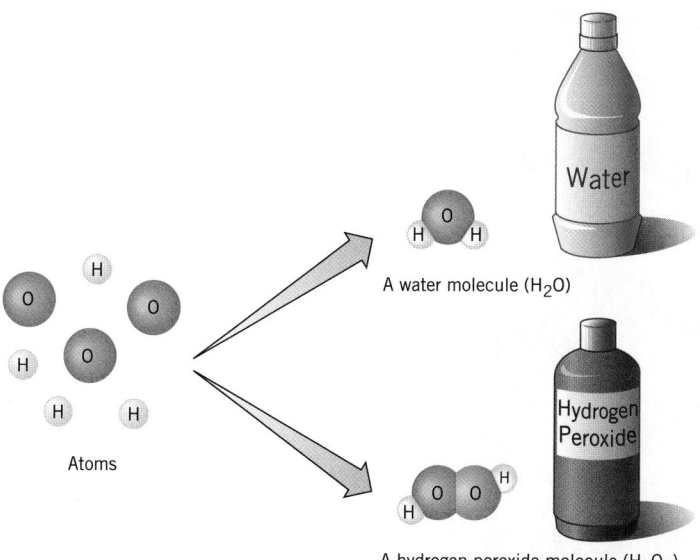

Atoms

A water molecule (H$_2$O)

A hydrogen peroxide molecule (H$_2$O$_2$)

**Figure 3-3 Atoms and Molecules**
**Atoms of hydrogen can combine with atoms of oxygen to form two different compounds.**

has a profound effect on the properties of these two compounds. Ingestion of alcohol causes intoxication, while a similar amount of ether may cause death. *Formulas that show the order and arrangement of specific atoms are known as* **structural formulas.**

Each breath of fresh air that we inhale is primarily just three elements—nitrogen (78%), oxygen (21%), and argon (less than 1%). There are traces of other gases as well. What would we see if we could magnify a sample of air so that the atoms of these three elements could become visible? Perhaps the most noticeable difference among these elements that we would immediately notice is that argon exists as solitary atoms but atoms of nitrogen and oxygen are joined together in pairs to form molecules. (See Figure 3-4.)

Hydrogen, fluorine, chlorine, bromine, and iodine in their elemental form also exist as diatomic (two-atom) molecules under normal temperature conditions. A form of elemental phosphorus consists of molecules composed of four atoms, and a form of sulfur consists of molecules composed of eight atoms. There is also a second form of elemental oxygen, known as ozone, which is composed of three atoms. Molecules composed of two or more atoms of the same element are also referred to by formulas such as I$_2$ (iodine), O$_2$ (oxygen), O$_3$ (ozone), and P$_4$ (phosphorus).

# 3-4 Electrical Nature of Matter: Compounds Composed of Ions

**Looking Ahead!** Perhaps one way we could classify the matter around us is as "soft" and "hard." The soft part of nature is the air, the liquids such as water, and some solids that melt at low temperatures. The hard matter is the stuff of minerals, rocks, and even mountains. Although this is a rather crude generalization with many exceptions, the soft part of nature is generally composed of molecular compounds. The hard part of nature—solids with comparatively high melting points—is composed of another type of compound that we will discuss next.

A streak of lightning shooting through the dark sky is a powerful but frightening force of nature. (See Figure 3-5.) This huge flow of electricity streaking through the air confirms to us that all matter is electrical in nature. On a much smaller scale,

**Figure 3-4  The Composition of the Atmosphere**
**Our atmosphere is composed mostly of nitrogen and oxygen molecules, with a small amount of argon atoms.**

**Figure 3-5  Lightning**
**The phenomenal bolts of electricity known as lightning are a result of electrostatic forces.**

just shuffling across a rug can create a stinging spark of electricity. Sparks and lightning occur because there are two types of electrostatic charge—positive and negative. **Electrostatic forces** *exist between the charges. There is a force of attraction between opposite charges but a force of repulsion between like charges.* (See Figure 3-6.) Coulomb's law states that the forces of attraction become greater as the magnitude of the charges increases and as the charges become closer together. Lightning is a result of the force of attraction. It is a flow of negative charge between the ground that has become negatively charged and a positively charged cloud (or sometimes vice versa). Air normally separates the two charges, but when the charges build to a high level, the positive and negative charges come together through the bolt of lightning.

Atoms can also achieve an electrostatic charge. *When atoms have an electrostatic charge, they are known as* **ions.** *Positively charged ions are known as* **cations,** *and negatively charged ions are called* **anions.** Ordinary table salt is a compound named sodium chloride. In sodium chloride, the sodium is a cation with a single positive charge and the chlorine is an anion with a single negative charge. This is illustrated with a $+$ or $-$ as a superscript to the right of the symbol of the element.

$$Na^+ \quad \text{and} \quad Cl^-$$

Since the sodium cations and the chlorine anions are oppositely charged, the ions are held together by electrostatic forces of attraction. In Figure 3-7, the ions in sodium chloride are shown as they would appear if sufficient magnification were possible. Note that each cation (one of the smaller spheres) is attached not just to one anion. In fact, each ion is surrounded by six oppositely charged ions. This situation is much different from that of the previously discussed molecular compounds, in which atoms bond together to form discrete molecules.

*Compounds consisting of ions are known as* **ionic compounds.** *The electrostatic forces holding the ions together are known as* **ionic bonds.** The ions in ionic compounds are locked tightly in their positions by the strong electrostatic attractions. This results in solid compounds that are almost all hard and rigid, with high melting points. The principal components of many rocks and minerals are ionic compounds.

The formula of sodium chloride is

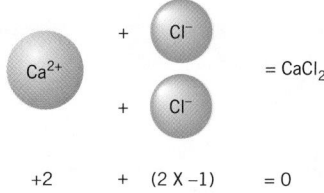

This represents the simplest ratio of cations to anions present (in this case, one to one). This ratio reflects the fact that the two ions have equal and opposite charges and *that in any ionic compound, the anions and cations exist together in a ratio such that the negative charge balances the positive charge* [i.e., $+1 + (-1) = 0$]. Notice that the charges are not displayed in the formula. *The simplest whole-number ratio of ions in an ionic compound is referred to as a* **formula unit.** Ions may also have multiple charges. Calcium chloride is a compound composed of $Ca^{2+}$ cations and $Cl^-$ anions. Two chlorine anions are present to balance the $+2$ charge on the calcium $[+2 + (2 \times -1) = 0]$. Thus the formula is

Ca$^{2+}$ + Cl$^-$ + Cl$^-$ = CaCl$_2$

$+2$ + $(2 \times -1)$ = 0

*Groups of atoms that are covalently bonded to each other may also be cations or anions, and they are known as* **polyatomic ions.** Examples are the nitrate anion ($NO_3^-$), the sulfate anion ($SO_4^{2-}$), and the ammonium cation ($NH_4^+$). When more than one polyatomic ion is in a formula unit, parentheses and a subscript are used. When there is only one polyatomic ion, no parentheses are used. Barium perchlorate (a compound

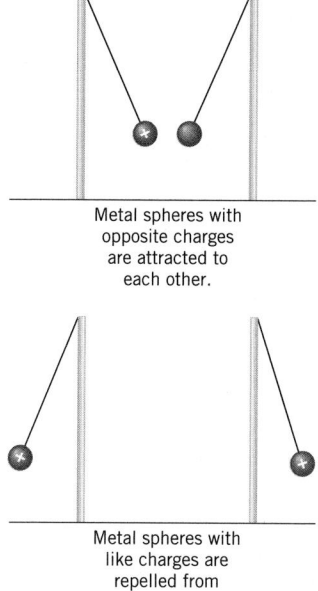

Metal spheres with opposite charges are attracted to each other.

Metal spheres with like charges are repelled from each other.

**Figure 3-6 Electrostatic Forces Opposite charges attract; like charges repel.**

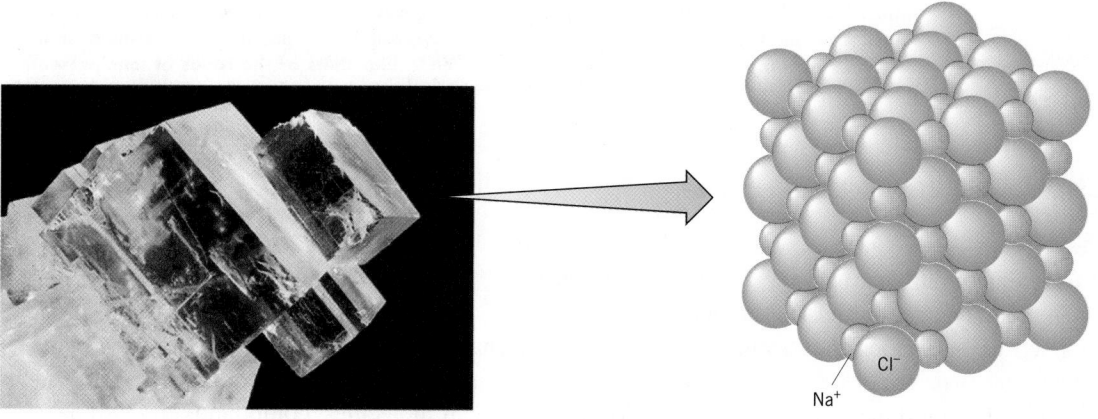

**Figure 3-7 The Ions in Sodium Chloride An ionic compound such as sodium chloride exists as an arrangement of charged particles.**

containing one $Ba^{2+}$ and two $ClO_4^-$) is represented as in (a), and calcium carbonate (a compound containing one $Ca^{2+}$ and one $CO_3^{2-}$) is represented as in (b).

Sometimes, the formulas of compounds are confused with those of ions. For example, $NO_2$ is the formula of a molecular compound known as nitrogen dioxide. It is a brownish gas responsible for some air pollution. (See page xxx.) Notice that no charge is shown by the formula. The $NO_2^-$ species (known as the nitrite ion) shows a negative charge, which means that it is a polyatomic ion. It exists only with a cation as part of an ionic compound (e.g., $NaNO_2$.) Remember that cations and anions do not exist alone but only as parts of an ionic compound. (See Figure 3-8.) We will spend more time on writing and naming ionic compounds in the next chapter.

**Figure 3-8  Molecules and Ions** $NO_2$ **is a gaseous, molecular compound, whereas the** $NO_2^-$ **ion is part of a solid, ionic compound.**

---

**Looking Back!** The matter that we breathe, eat, walk on, or swim in is composed of 88 fundamental substances called elements. These elements are made up of extremely tiny particles called atoms. Molecular elements and compounds are composed of atoms joined together by covalent bonds. A second type of compound is composed of oppositely charged ions. The formulas of these two types of compounds describe either discrete molecular units or the ratios of ions present.

---

# Learning Check A | checking it out

**A-1.** Fill in the blanks.

The symbol for the element nitrogen is _____ and that of copper is _____ . The symbol Ni represents the element _____ . The fundamental particle of an element is called an _____ . When these fundamental particles are covalently bonded to each other, they form _____ . The designation $C_2H_7N$ for a compound represents its _____ and indicates the presence of _____ atoms of C and _____ atom of N in one molecule. If two compounds have the same formula, they must have different _____ formulas. Besides molecules, compounds can also be composed of _____ . The formula of an ionic compound does not represent discrete molecules but represents the smallest whole number ratio of _____ to _____ .

The white cliffs of Dover along the British coast are composed of calcium carbonate, which is an ionic compound.

**A-2.** Acetaminophen (a pain reliever) is a molecular compound composed of eight atoms of carbon, nine atoms of hydrogen, one atom of nitrogen, and two atoms of oxygen. What is its formula?

**A-3.** Write the formulas of a compound containing three $Na^+$ cations and one $PO_4^{3-}$ anion and a compound containing two $NH_4^+$ cations and one $S^{2-}$ anion.

*Additional Examples: Exercises 3-5, 3-7, 3-17, 3-20, 3-26, and 3-27.*

Section  **THE ATOM AND ITS COMPOSITION**

## 3-5 Composition of the Atom

**Looking Ahead! So what is an atom like? Does it have features? What is its composition and how do the atoms of the elements differ? In this section, we will describe the world of the atom as we journey into its tiny confines.**

A bit over 100 years ago, scientists still perceived the atom to be a hard, featureless sphere. However, beginning in the late 1880s and continuing today, the mysteries and complexities of the atom have been slowly discovered and understood. Ingenious experiments of scientists by the names of Thomson, Rutherford, Becquerel, Curie, and Roentgen, among others, contributed to the current model of the atom. Much evidence indicates that, if we could ever peer into the atom through a microscope, what follows is what we would see.

In a journey within the confines of the atom, we would first encounter a very small, almost massless particle called an **electron.** The electron was the first subatomic particle to be identified. In 1897, J. J. Thomson characterized the electron by proving that it had a negative electrical charge (assigned a value of $-1$) and was common to the atoms of all elements. Matter was electrical in nature because atoms themselves contained electrically charged particles. However, since atoms were known to be neutral, it was understood that the atom must also contain a positive charge to counterbalance the negative charge. From this information, Thomson is usually credited with suggesting the "plum pudding" model of the atom, which explained the facts known at the time. (Plum pudding was a popular English dessert.) In this model, the positive charge was thought to be diffuse and evenly distributed throughout the volume of the atom (analogous to pudding). This made sense. The like positive charges would tend to spread out as much as possible. The negative particles (electrons) would be

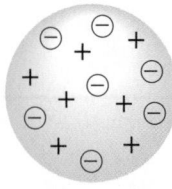

**Figure 3-9 The Plum Pudding Model**
**Electrons were thought to be like tiny particles distributed in a positively charged medium.**

distributed throughout the atom (analogous to raisins in the pudding) with enough electrons to balance the positive charge. (See Figure 3-9.)

The next major development occurred in 1911. Ernest Rutherford in England conducted experiments meant to support the accepted model of Thomson. His results, however, could be explained only by a radically different model. When students in his laboratories bombarded a thin foil of gold with fast-moving alpha particles (positively charged helium ions), they expected these very small particles to pass right through the atoms of the gold with very little effect. The small, hard alpha particles should easily push the tiny electrons aside and pass unaffected through the diffuse positive charge. Instead, a small number of the alpha particles were deflected significantly from their path and a few even came straight back. It would be like shooting a volley of bullets through a bail of cotton. If even one bullet were to ricochet by a large angle the shooter would have to conclude that something dense, such as a rock, was imbedded in the soft cotton. Likewise, Rutherford was forced to conclude that the gold atoms had a small, hard core containing most of the mass of the atom and all of the positive charge. Close encounters with the core caused alpha particles to be deflected because of the repulsion of like positive charges. Occasional "direct hits" would reflect the alpha particle back toward the source. (See Figure 3-10.) This core was called the **nucleus.** Most of the volume of the atom is actually empty space containing the very small electrons. To get an idea about proportions, imagine a nucleus expanded to the size of a softball. In this case, the radius of the atom would extend for about one mile.

Later experiments would show that the nucleus is composed of particles called **nucleons.** There are two types of nucleons: **protons,** which have a positive charge (assigned a value of $+1$, equal and opposite to that of an electron), and **neutrons,** which do not carry a charge. (See Figure 3-11.) The data on the three particles in the atom are summarized in Table 3-4. The proton and neutron have roughly the same mass, which is about "1 amu" ($1.67 \times 10^{-24}$ g). The amu (atomic mass unit) is a convenient unit for the masses of individual atoms and subatomic particles. This unit will be defined more precisely in the next section.

For some time, it was thought that the atom was composed of just these three particles. As experimental procedures became more elaborate and sophisticated, however, the picture became more complicated. It now appears that the three particles are themselves composed of various combinations of even more fundamental particles called quarks. So, of what are atoms composed? The answer is actually quite complex. Fortunately for us, the three-particle model of the atom still meets the needs of the chemist.

**Figure 3-10 Rutherford's Experiment**
**To account for the deflection of some of the positively charged alpha particles, Rutherford proposed that the atom is mostly empty space with the positive charge located in a small, dense core. When an alpha particle encountered a nucleus (a and b), electrostatic repulsions would either deflect it or reflect it back.**

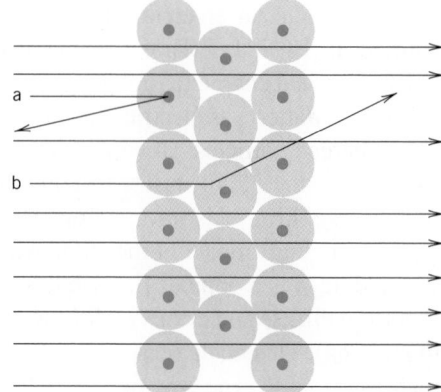

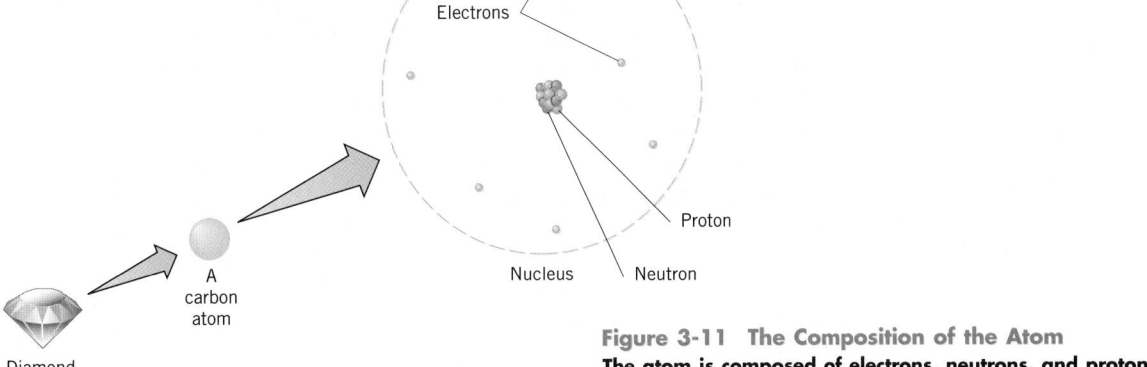

**Figure 3-11   The Composition of the Atom**
**The atom is composed of electrons, neutrons, and protons.**

**Table 3-4   Atomic Particles**

| Name | Symbol | Electrical Charge | Mass (amu) | Mass (g) |
|---|---|---|---|---|
| Electron | $e$ | −1 | 0.000549 | $9.110 \times 10^{-28}$ |
| Proton | $p$ | +1 | 1.00728 | $1.673 \times 10^{-24}$ |
| Neutron | $n$ | 0 | 1.00867 | $1.675 \times 10^{-24}$ |

# 3-6 Atomic Number, Mass Number, and Atomic Mass

**Looking Ahead!** We are now ready to look into how the three particles discussed define the atoms of a particular element. We will see in this section why not all atoms of an element are exactly the same and how the atoms of one element differ from another.

In Dalton's original atomic theory, he suggested that all atoms of an element are the same. But if we look at a number of atoms of most elements, we find that this statement is not exactly true. For example, consider the atoms of the element copper. A typical atom is composed of a nucleus containing a total of 63 nucleons, of which 29 are protons and 34 are neutrons. The atom also contains 29 electrons that exactly balance the positive charge of the protons. Thus the atom is neutral. *The number of protons in the nucleus (which is equal to the total positive charge) is referred to as the atom's* **atomic number.** *The total number of nucleons is called the* **mass number.** Therefore, this particular copper atom has an atomic number of 29 and a mass number of 63. There are other copper atoms that are not exactly the same, however. Some atoms of copper have a mass number of 65 rather than 63. This means that these atoms have 36 neutrons as well as 29 protons. *An atom of a specife element with a specific mass number is known as an* **isotope.** Isotopes of an element have the same atomic number but different mass numbers.

Most elements that are present in nature exist as a mixture of isotopes. *It is the atomic number, however, that distinguishes one element from another.* Any atom with an atomic number of 29, regardless of any other consideration, is an atom of copper. If the atomic number is 28, the element is nickel; if it is 30, the element is zinc.

Specific isotopes are written in a form known as *isotopic notation.* In isotopic notation, the mass number is written as a superscript to the left of the element. Sometimes, the atomic number is written as a subscript, also on the left. The indication of the atomic number is strictly a convenience, however, since the atomic number determines the identity of the element and vice versa. The isotopic notations for the two isotopes of copper are written as follows.

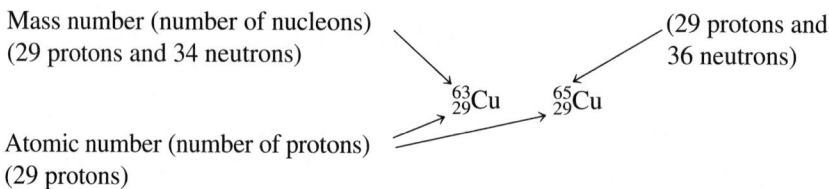

The convention for verbally naming specific isotopes is to give the element followed by the mass number. For example, $^{63}$Cu is called "copper-63" and $^{65}$Cu is called "copper-65." From either the written isotopic notation or the isotope name, we can determine the number of each type of particle in an isotope, as we will see in Example 3-1.

In Chapter 8, an important consideration will be how the mass of one element compares to another. The mass of the electrons is extremely small compared to the masses of the protons and neutrons, so it is not included in the mass of an isotope. Thus the mass number of an isotope is a convenient but rather imprecise measure of its mass. It is imprecise not only because electrons are not included but also because the proton and neutron do not have exactly the same mass. A more precise measure of the mass of one isotope relative to another is known as the isotopic mass. *The* **isotopic mass** *of an isotope is determined by comparison to a standard,* $^{12}C$, *which is defined as having a mass of exactly 12 atomic mass units (amu). Therefore, one* **atomic mass unit** *is a mass of exactly* $\frac{1}{12}$ *of the mass of* $^{12}C$. For example, precise measurements show that the mass of $^{10}$B is 0.83442 times the mass of $^{12}$C, which means it has an isotopic mass of 10.013 amu. From similar calculations we find that the atomic mass of $^{11}$B is 11.009 amu. Since boron, as well as most other naturally occurring elements, is found in nature as a mixture of isotopes, the atomic mass of the element reflects this mixture. *The* **atomic mass** *of an element is obtained from the weighted average of the atomic masses of all isotopes present in nature.* A weighted average relates the isotopic mass of each isotope present to its percent abundance. It can be considered as the isotopic mass of an "average atom," although an *average* atom does not itself exist. Example 3-2 illustrates how atomic mass relates to the distribution of isotopes.

| Example 3-1 | Calculating the Number of Particles in an Isotope |
|---|---|
| **working it out** | How many protons, neutrons, and electrons are present in $^{90}_{38}$Sr? |

**Solution**

$$\text{atomic number} = \text{number of protons} = \underline{\underline{38}}$$
$$\text{number of neutrons} = \text{mass number} - \text{number of protons}$$
$$90 - 38 = \underline{\underline{52}}$$

In a neutral atom, the number of protons and the number of electrons are the same.

$$\text{number of electrons} = \underline{\underline{38}}$$

## Isotopes and the Weather 15,000 Years Ago

It seems like we are always hearing of some great storm or heat record lately. Is something strange going on or are the media just doing a better job of reporting? Actually, we are all anxious to know whether the climate is permanently changing and, if so, whether it is going to get worse. The key to predicting the future, however, may be in understanding the past. An ingenious key to past climates is found with the naturally occurring isotopes of oxygen. Oxygen is composed of 99.76% $^{16}O$, 0.04% $^{17}O$, and 0.20% $^{18}O$, which produces a weighted average of 15.9994. This means that only 24 out of 10,000 oxygen atoms are "heavy" oxygen atoms (i.e., $^{17}O$ or $^{18}O$).

When $^{17}O$ or $^{18}O$ is part of a water molecule, the properties are slightly different from normal water ($H_2^{16}O$). For example, heavy water evaporates just a little slower than normal water. This means that the amount of heavy water is slightly less in fresh water (which comes from precipitation of the evaporated water) than in the ocean. But even this can vary ever so slightly. When the climate is colder than normal, less water evaporates from the ocean so the amount of heavy water in precipitation is less than normal.

Greenland is our natural weather history laboratory. It has been covered with ice for more than 100,000 years. The ice is now two miles deep in places. Year by year, the snow has been accumulating and has been pressed down into layers that look like tree rings. Scientists have bored through the ice and have analyzed the layers for their content of dust, trapped gases, and the ratio of oxygen isotopes. Results of isotope studies indicate that the Northern Hemisphere was a whopping 20°C (or 36°F) colder during the last ice age (about 30,000 years ago) than it is now. More important, variations in the weather from then to now have also been determined. These studies will help us understand and perhaps predict what's going to happen to our weather in the future. That is, is our current warming a natural occurrence or a new phenomenon? We'll see.

*Greenland.*

---

| | |
|---|---|
| **Calculating the Atomic Mass of an Element from Percent Distribution of Isotopes** | **Example 3-2** |

In nature, the element boron occurs as 19.9% $^{10}B$ and 80.1% $^{11}B$. If the isotopic mass of $^{10}B$ is 10.013 and that of $^{11}B$ is 11.009 amu, what is the atomic mass of boron?

*working it out*

**Procedure**

Find the contribution of each isotope toward the atomic mass by multiplying the percent in decimal form by the isotopic mass.

**Solution**

$$^{10}B \quad 0.199 \times 10.013 \text{ amu} = \quad 1.99 \text{ amu}$$
$$^{11}B \quad 0.801 \times 11.009 \text{ amu} = \quad \underline{8.82 \text{ amu}}$$
$$\text{atomic mass of boron} = \quad \underline{10.81 \text{ amu}}$$

---

Before we conclude this section, we need to discuss how ions differ from neutral atoms. A cation such as $Na^+$ has a positive electrical charge because there is an unequal number of electrons and protons. A cation has fewer electrons than protons, and an anion has more electrons than protons. In both cases, it is the electrons that are out of balance, not the protons. A +1 charge on a cation indicates that it has one *less* electron than number of protons (its atomic number), and a +2 charge indicates that the cation has two *fewer* electrons than its atomic number. For example, the $Na^+$ cation has 11 protons (the atomic number of Na) and 10 electrons. The +1 charge arises from this imbalance [i.e., $(11p \times +1) + (10e \times -1) = +1$]. An anion with a −1 charge has one *more* electron than the atomic number of the element, and a −2 charge indicates two *more* electrons than its atomic number. For example, the $S^{2-}$ ion has 16 protons in its nucleus and 18 electrons. The $-2$ charge arises from the two extra electrons [i.e., $(16p \times +1) + (18e \times -1) = -2$]. The charge on polyatomic ions arises from the same considerations. For example, the $CO_3^{2-}$ ion has a total of 30 protons in the four nuclei [i.e., $6(C) + (3 \times 8)(O) = 30$]. The presence of a $-2$ charge indicates that 32 electrons must be present, as shown by the following calculation.

$$(30p \times +1) + (\#e \times -1) = -2$$

Solving for the number of electrons,

$$\#e = 32$$

In a later chapter, we will discuss why specific atoms tend to acquire positive charges while others tend to acquire negative charges.

---

**Looking Back!** Most elements exist in nature as a mixture of isotopes—that is, atoms with the same atomic number but different numbers of neutrons. The mass listed for each element is known as the atomic mass. This is the mass of an "average atom" of the element compared to the mass of the standard ($^{12}C$). When there are more or fewer electrons in an atom or group of atoms than there are protons in the nucleus or nuclei, an ion results.

---

## Learning Check B | checking it out

**B-1.** Fill in the blanks.

An atom is composed of negatively charged _____ , positively charged _____ , and _____ . The nucleus contains _____ and _____ , with both particles having a mass of about _____ amu. The atomic number refers to the number of _____ in the nucleus, and the mass number refers to the number of _____ . Isotopes of an element have the same _____ number but different _____ numbers. The isotopic mass of an isotope is its mass compared to the mass of _____ . The atomic mass is the _____ _____ of all of the naturally occurring isotopes of the element. A cation contains more _____ than _____ .

**B-2.** How many protons and neutrons are in each of the three isotopes of oxygen: $^{16}_{8}O$, $^{17}_{8}O$, $^{18}_{8}O$?

**B-3.** Naturally occurring lead is composed of four isotopes: 1.40% $^{204}Pb$ (203.97 amu), 24.10% $^{206}Pb$ (205.97 amu), 22.10% $^{207}Pb$ (206.98 amu), and 52.40% $^{208}Pb$ (207.98 amu). What is the atomic mass of lead?

**B-4.** How many protons and how many electrons are in the $N^{3-}$ ion?

*Additional Examples: Exercises 3-40, 3-42, 3-46, and 3-60.*

chapter **R**eview                    putting it together

All of the various forms of nature around us, from the simple elements in the air to the complex compounds of living systems, are composed of only a few basic forms of matter called elements. Each element has a name and a unique one- or two-letter **symbol.**

It has been less than 200 years since John Dalton's atomic theory introduced the concept that elements are composed of fundamental particles called **atoms.** It has been about 20 years since we have been able to produce images of these atoms with a special microscope. Most elements are present in nature as aggregates of individual atoms. In some elements, however, two or more atoms are combined by **covalent bonds** to produce basic units called **molecules. Molecular compounds** are also composed of molecules, although, in this case, the atoms of at least two different elements are involved. Each compound has a unique arrangement of atoms in the molecular unit. These are sometimes conveniently represented by **structural formulas.**

There is another type of compound, however. Matter can be positively or negatively charged, with **electrostatic forces** between the charges. Atoms can become electrically charged to form **ions.** Groups of atoms that are covalently bonded together can also have charges and are known as **polyatomic ions.** An **ionic compound** is composed of **cations** (positive ions) and **anions** (negative ions). The interactions of cations and anions are known as **ionic bonds.** The **formula** of a molecular compound represents the actual number of atoms of each element present in a molecular unit. The formula of an ionic compound shows the type and number of ions in a **formula unit.** A formula unit represents the smallest whole-number ratio of cations and anions, which reflects the fact that the positive charge is balanced by the negative charge. The four most common ways that we find atoms in nature are summarized as follows.

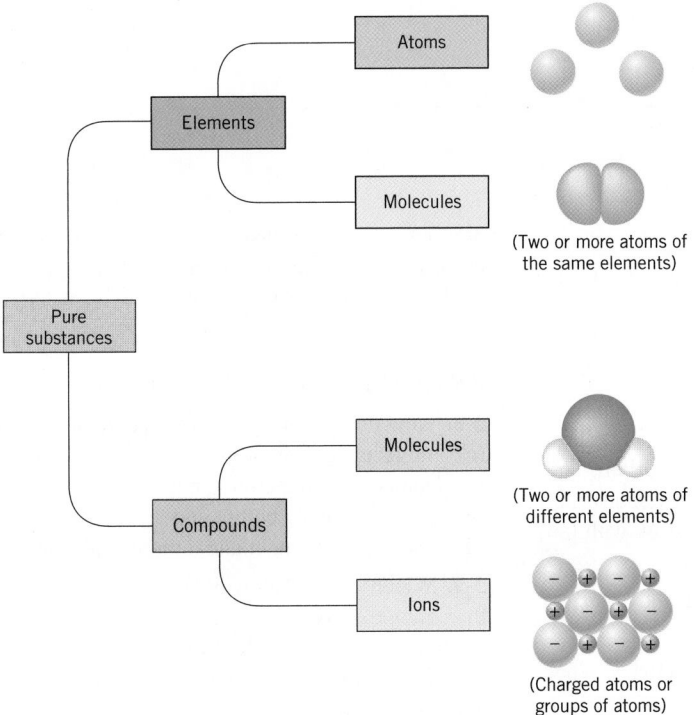

The atom is the unique particle that characterizes an element. It is composed of more basic particles called **electrons** and **nucleons.** There are two types of nucleons, **protons** and **neutrons.** The relative charges and masses of these three particles are summarized in Table 3-4.

Rather than being a hard sphere, the atom is mostly empty space containing the negatively charged electrons. The protons and neutrons are located in a small dense core called the **nucleus.** The number of protons in an atom is known as its **atomic**

**number** (symbolized by $Z$ in the following diagram), which distinguishes the atoms of one element from those of another. The total number of nucleons in an atom is known as its **mass number** (symbolized by $A$ in the diagram). Atoms of the same element may have different mass numbers and are known as **isotopes** of that element. An atom is neutral because it has the same number of electrons as protons. When the number of electrons is greater or fewer than the number of protons, an ion results (the charge, if present, is symbolized by $Q$ in the diagram). This information for a particular isotope is illustrated as follows.

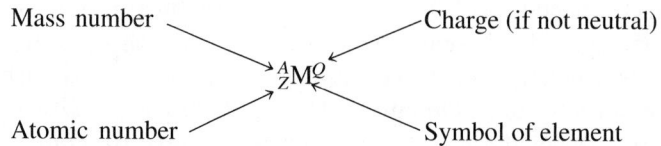

Mass number — Charge (if not neutral)

Atomic number — Symbol of element

Because protons and neutrons do not have exactly the same mass, the mass number is not an exact measure of the comparative masses of isotopes. A more precise measure of mass is the **isotopic mass.** This is obtained by comparing the mass of the particular isotope with the mass of $^{12}C$, which is defined as having a mass of exactly 12 **atomic mass units (amu).** The **atomic mass** of an element is the weighted average of all of the naturally occurring isotopes found in nature.

Finally, an atom can have a net positive charge if there are fewer electrons than protons or a net negative charge when there are more electrons than protons.

# Exercises

## Names and Symbols of the Elements

**3-1.** Write the symbols of the following elements. Try to do this without reference to a table of the elements.
**(a)** bromine  **(c)** lead  **(e)** sodium
**(b)** oxygen  **(d)** tin  **(f)** sulfur

**3-2.** The following elements all have symbols that begin with the letter C: cadmium, calcium, californium, carbon, cerium, cesium, chlorine, chromium, cobalt, copper, and curium. The symbols are C, Ca, Cd, Ce, Cf, Cl, Cm, Co, Cr, Cs, and Cu. Match each symbol with an element and then check with the table of elements inside the front cover.

**3-3.** The names of seven elements begin with the letter B. What are their names and symbols?

**3-4.** The names of nine elements begin with the letter S. What are their names and symbols?

**3-5.** Using the table inside the front cover of the text, write the symbols for the following elements.
**(a)** barium  **(c)** cesium  **(e)** manganese
**(b)** neon  **(d)** platinum  **(f)** tungsten

**3-6.** Name the elements corresponding to the following symbols. Try to do this without reference to a table of the elements.
**(a)** S  **(c)** Fe  **(e)** Mg
**(b)** K  **(d)** N  **(f)** Al

**3-7.** Using the table inside the front cover, name the elements corresponding to the following symbols.
**(a)** B  **(c)** Ge  **(e)** Cl  **(g)** Be
**(b)** Bi  **(d)** U  **(f)** Hg  **(h)** As

## Molecular Compounds and Formulas

**3-8.** How do the following concepts relate and differ?
**(a)** a molecule and an atom
**(b)** a molecule and a compound
**(c)** an element and a compound
**(d)** a molecular element and a monatomic element

**3-9.** Which of the following are formulas of elements rather than compounds?
**(a)** $P_4O_{10}$  **(c)** $F_2O$  **(e)** $MgO$
**(b)** $Br_2$  **(d)** $S_8$  **(f)** $P_4$

**3-10.** Name the elements in the previous exercise.

**3-11.** What is the difference between Hf and HF?

**3-12.** What is the difference between NO and No?

**3-13.** Which of the following is the formula of a diatomic element? Which is the formula of a diatomic compound?
**(a)** $NO_2$  **(c)** $K_2O$  **(e)** $N_2$
**(b)** $CO$  **(d)** $(NH_4)_2S$  **(f)** $CO_2$

**3-14.** Name all of the elements in the compounds of the previous exercise.

**3-15.** Give the name and number of atoms of each element in the formulas of the following compounds.
**(a)** $H_2SeO_3$ **(c)** $NI_3$ **(e)** $Ba(BrO_3)_2$
**(b)** $Na_4SiO_4$ **(d)** $NiI_2$ **(f)** $B_3N_3(CH_3)_6$

**3-16.** What is the total number of atoms in each formula unit for the compounds in exercise 3-15?

**3-17.** Determine the number of atoms of each element in the formulas of the following compounds.
**(a)** $C_6H_4Cl_2$
**(b)** $C_2H_5OH$ (ethyl alcohol)
**(c)** $CuSO_4 \cdot 9H_2O$ ($H_2O$'s are part of a single formula unit)
**(d)** $C_9H_8O_4$ (aspirin)
**(e)** $Al_2(SO_4)_3$
**(f)** $(NH_4)_2CO_3$

**3-18.** What is the total number of atoms in each molecule or formula unit for the compounds listed in exercise 3-17?

**3-19.** How many carbon atoms are in each molecule or formula unit of the following compounds?
**(a)** $C_8H_{18}$ (octane in gasoline) **(c)** $Fe(C_2O_4)_2$
**(b)** $NaC_7H_4O_3NS$ (saccharin) **(d)** $Al_2(CO_3)_3$

**3-20.** Write the formulas of the following molecular compounds.
**(a)** sulfur dioxide (one sulfur and two oxygens)
**(b)** carbon dioxide (one carbon and two oxygens)
**(c)** sulfuric acid (two hydrogens, one sulfur, and four oxygens)
**(d)** acetylene (two carbons and two hydrogens)

**3-21.** Write the formulas of the following molecular compounds.
**(a)** phosphorus trichloride (one phosphorus and three chlorines)
**(b)** naphthalene (ten carbons and eight hydrogens)
**(c)** dibromine trioxide (two bromines and three oxygens)

**3-22.** Neon gas is composed of individual atoms, and chlorine gas is composed of $Cl_2$ molecules. Describe how these two elements appear on the atomic or molecular level.

### Ions and Ionic Compounds

**3-23.** How do the following concepts relate and differ?
**(a)** an atom and an ion
**(b)** a molecule and a polyatomic ion
**(c)** a cation and an anion
**(d)** a molecular and an ionic compound
**(e)** a molecular unit and an ionic formula unit

**3-24.** The gaseous compound HF contains covalent bonds, and the compound KF contains ionic bonds. Describe how the basic particles of these two compounds appear.

**3-25.** Explain the difference between $SO_3$ and $SO_3^{2-}$. Which one would be a gas?

**3-26.** Write the formulas of the following ionic compounds.
**(a)** calcium perchlorate (one $Ca^{2+}$ and two $ClO_4^-$ ions)
**(b)** ammonium phosphate (three $NH_4^+$ ions and one $PO_4^{3-}$ ion)
**(c)** iron(II) sulfate (one $Fe^{2+}$ and one $SO_4^{2-}$ ion)

**3-27.** What is the number of atoms of each element present in the compounds in exercise 3-26?

**3-28.** Write the formulas of the following ionic compounds.
**(a)** calcium hypochlorite (one $Ca^{2+}$ ion and two $ClO^-$ ions)
**(b)** magnesium phosphate (three $Mg^{2+}$ ions and two $PO_4^{3-}$ ions)
**(c)** chromium(III) oxalate (two $Cr^{3+}$ ions and three $C_2O_4^{2-}$ ions)

**3-29.** What is the number of atoms of each element present in the compounds in exercise 3-28?

**3-30.** The formula of an ionic compound indicates one $Fe^{2+}$ ion combined with one anion. Which of the following could be the other ion?
**(a)** $F^-$ **(b)** $Ca^{2+}$ **(c)** $S^{2-}$ **(d)** $N^{3-}$

**3-31.** An ionic compound is composed of two $ClO_4^-$ ions and one cation. Which of the following could be the other ion?
**(a)** $SO_4^{2-}$ **(b)** $Ni^{2+}$ **(c)** $Al^{3+}$ **(d)** $Na^+$

**3-32.** An ionic compound is composed of one $SO_3^{2-}$ ion and two cations. Which of the following could be the cations?
**(a)** $I^-$ **(b)** $Ba^{2+}$ **(c)** $Fe^{3+}$ **(d)** $Li^+$

**3-33.** An ionic compound is composed of two $Al^{3+}$ ions and three anions. Which of the following could be the anions?
**(a)** $S^{2-}$ **(b)** $Cl^-$ **(c)** $Sr^{2+}$ **(d)** $N^{3-}$

**3-34.** Write the formulas of the compounds in exercises 3-30 and 3-32.

**3-35.** Write the formulas of the compounds in exercises 3-31 and 3-33.

### Composition of the Atom

**3-36.** Which of the following were not part of Dalton's atomic theory?
**(a)** Atoms are the basic building blocks of nature.
**(b)** Atoms are composed of electrons, neutrons, and protons.
**(c)** Atoms are reshuffled in chemical reactions.
**(d)** The atoms of an element are identical.
**(e)** Different isotopes can exist for the same element.

**3-37.** Which of the following describes a neutron?
**(a)** +1 charge, mass 1 amu **(c)** 0 charge, mass 1 amu
**(b)** +1 charge, mass 0 amu **(d)** −1 charge, mass 0 amu

**3-38.** Which of the following describes an electron?
**(a)** +1 charge, mass 1 amu **(c)** −1 charge, mass 1 amu
**(b)** +1 charge, mass 0 amu **(d)** −1 charge, mass 0 amu

**3-39.** Give the mass numbers and atomic numbers of the following isotopes. Refer to the table of the elements inside the front cover.
**(a)** $^{193}Au$ **(b)** $^{132}Te$ **(c)** $^{118}I$ **(d)** $^{39}Cl$

**3-40.** Give the numbers of protons, neutrons, and electrons in each of the following isotopes. Refer to the table of the elements inside the front cover.
**(a)** $^{45}Sc$ **(b)** $^{232}Th$ **(c)** $^{223}Fr$ **(d)** $^{90}Sr$

**3-41.** Three isotopes of uranium are $^{234}U$, $^{235}U$, and $^{238}U$. How many protons, neutrons, and electrons are in each isotope?

**3-42.** Using the table of elements inside the front cover, complete the following table for neutral isotopes.

| Isotope | Isotopic Notation | Atomic Number | Mass Number | Subatomic Particles | | |
|---|---|---|---|---|---|---|
| | | | | p | n | e |
| molybdenum-96 | $^{96}_{42}Mo$ | 42 | 96 | 42 | 54 | 42 |
| (a) | $^{?}_{?}Ag$ | | | | | 61 |
| (b) | | 14 | | | | 14 |
| (c) | | | 39 | | | 20 |
| (d) cerium-140 | | | | | | |
| (e) | | | | | 26 | 30 |
| (f) | | 50 | 110 | | | |
| (g) | $^{118}_{?}I$ | | | | | |
| (h) mercury-? | | | | | | 116 |

**3-43.** Using the table of elements inside the front cover, complete the following table for neutral isotopes.

| Isotope | Isotopic Notation | Atomic Number | Mass Number | Subatomic Particles | | |
|---|---|---|---|---|---|---|
| | | | | p | n | e |
| (a) tungsten-? | | | 184 | | | |
| (b) | | | | | 12 | 11 |
| (c) | $^{200}_{?}At$ | | | | | |
| (d) | $^{?}_{?}Pm$ | | | | 87 | |
| (e) | | | 109 | 46 | | |
| (f) | | | 48 | | | 23 |
| (g) | | 21 | | | 29 | |

**3-44.** Write the isotopic notation for an isotope of cobalt that has the same number of neutrons as $^{60}Ni$.

**3-45.** Write the isotopic notation for an isotope of uranium that has the same number of neutrons as $^{240}Pu$.

**3-46.** What are the total number of protons and the total number of electrons in each of the following ions?
(a) $K^+$    (c) $S^{2-}$    (e) $Al^{3+}$
(b) $Br^-$    (d) $NO_2^-$    (f) $NH_4^+$

**3-47.** What are the total number of protons and the total number of electrons in each of the following ions?
(a) $Sr^{2+}$    (c) $V^{3+}$    (e) $SO_3^{2-}$
(b) $P^{3-}$    (d) $NO^+$

**3-48.** Write the element symbol or symbols and the charge for the following ions.
(a) 20 protons and 18 electrons
(b) 52 protons and 54 electrons
(c) one phosphorus and three oxygens with a total of 42 electrons
(d) one nitrogen and two oxygens with a total of 22 electrons

**3-49.** Write the element symbol or symbols and the charge for the following ions.
(a) 50 protons and 48 electrons
(b) 53 protons and 54 electrons
(c) one aluminum and two oxygens with a total of 30 electrons
(d) one chlorine and three fluorines with a total of 43 electrons

**3-50.** A monatomic bromine species has 36 electrons. Does it exist independently?

**3-51.** A species is composed of one chlorine chemically bonded to two oxygens. It has a total of 33 electrons. Is this species most likely a gaseous molecular compound or part of an ionic compound?

## Atomic Mass

**3-52.** How do the following concepts relate and differ?
(a) element and atomic number
(b) atomic mass and atomic number
(c) mass number and atomic mass
(d) isotopes and number of protons
(e) isotopes and number of neutrons

**3-53.** Determine the atomic number and the atomic mass of each of the following elements. Use the table inside the front cover.
(a) Re    (b) Co    (c) Br    (d) Si

**3-54.** About 75% of a U.S. "nickel" is an element with an atomic mass of 63.546 amu. What is the element?

**3-55.** White gold is an alloy of gold containing an element with an atomic mass of 106.4 amu. What is the element?

**3-56.** The elements O, N, Si, and Ca are among several that are composed *primarily* of one isotope. Using the table inside the front cover, write the atomic number and mass number of the principal isotope of each of these elements.

**3-57.** The atomic mass of hydrogen is given inside the front of the book as 1.00794. The three isotopes of hydrogen are $^1H$, $^2H$, and $^3H$. What does the atomic mass tell us about the relative abundances of the three isotopes?

**3-58.** A given element has a mass 5.81 times that of $^{12}C$. What is the atomic mass of the element? What is the element?

**3-59.** The atomic mass of a given element is about $3\frac{1}{3}$ times that of $^{12}C$. Give the atomic mass, the name, and the symbol of the element.

**3-60.** Bromine is composed of 50.5% $^{79}Br$ and 49.5% $^{81}Br$. The isotopic mass of $^{79}Br$ is 78.92 amu and that of $^{81}Br$ is 80.92 amu. What is the atomic mass of the element?

**3-61.** Silicon occurs in nature as a mixture of three isotopes: $^{28}Si$ (27.98 amu), $^{29}Si$ (28.98 amu), and $^{30}Si$ (29.97 amu). The mixture is 92.21% $^{28}Si$, 4.70% $^{29}Si$, and 3.09% $^{30}Si$. Calculate the atomic mass of naturally occurring silicon.

**3-62.** Naturally occurring Cu is 69.09% $^{63}Cu$ (62.96 amu). The only other isotope is $^{65}Cu$ (64.96 amu). What is the atomic mass of copper?

**\*3-63.** Chlorine occurs in nature as a mixture of $^{35}Cl$ and $^{37}Cl$. If the isotopic mass of $^{35}Cl$ is approximately 35.0 amu and that of $^{37}Cl$ is 37.0 amu, and the atomic mass of the mixture as it occurs in nature is 35.5 amu, what is the proportion of the two isotopes?

**\*3-64.** The atomic mass of the element gallium is 69.72 amu. If it is composed of two isotopes, $^{69}Ga$ (68.926 amu) and $^{71}Ga$ (70.925 amu), what is the percent of $^{69}Ga$?

## General Questions

**3-65.** Describe the difference between a molecular and an ionic compound. Of the two types of compounds discussed, is a stone more likely to be a molecular or an ionic compound? Is a liquid more likely to be a molecular or an ionic compound?

**3-66.** Write the symbol, mass number, atomic number, and electrical charge of the element given the following information. Refer to the table of the elements.
**(a)** An ion of Sr contains 36 electrons and 52 neutrons.
**(b)** An ion contains 24 protons, 28 neutrons, and 21 electrons.
**(c)** An ion contains 36 electrons and 45 neutrons and has a −2 charge.
**(d)** An ion of nitrogen contains 7 neutrons and 10 electrons.
**(e)** An ion contains 54 electrons and 139 nucleons and has a +3 charge.

**3-67.** Write the symbol, mass number, atomic number, and electrical charge of the element given the following information. Refer to the table of the elements.
**(a)** An ion of Sn contains 68 neutrons and 48 electrons.
**(b)** An ion contains 204 nucleons and 78 electrons and has a +3 charge.
**(c)** An ion contains 45 neutrons and 36 electrons and has a −1 charge.
**(d)** An ion of aluminum has 14 neutrons and a +3 charge.

**3-68.** Give the number of protons, electrons, and neutrons represented by the following. These elements are composed almost entirely of one isotope which is implied by the atomic mass.
**(a)** Na and Na$^{+}$    **(c)** F and F$^{-}$
**(b)** Ca and Ca$^{2+}$    **(d)** Sc and Sc$^{3+}$

**3-69.** Give the number of protons, electrons, and neutrons represented by the following. These elements are composed

almost entirely of one isotope which is implied by the atomic mass.
**(a)** Cr and Cr$^{2+}$    **(c)** I and I$^{-}$
**(b)** Au and Au$^{3+}$    **(d)** P and P$^{3-}$

**3-70.** An isotope of iodine has a mass number that is 10 amu less than two-thirds the mass number of an isotope of thallium. The total mass number of the two isotopes is 340 amu. What is the mass number of each isotope? (*Hint:* There are two equations and two unknowns.)

**3-71.** An isotope of gallium has a mass number that is 22 amu more than one-fourth the mass number of an isotope of osmium. The osmium isotope is 122 amu heavier than the gallium isotope. What is the mass number of each isotope? (*Hint:* There are two equations and two unknowns.)

**3-72.** A given element is composed of 57.5% of an isotope with an isotopic mass of 120.90 amu and the remainder an isotope with an isotopic mass of 122.90 amu. What is the atomic mass of the element? What is the element? How many electrons are in a cation of this element if it has a charge of "3"? How many neutrons are in each of the two isotopes of this element? What percent of the isotopic mass of each isotope is due to neutrons?

**3-73.** A given isotope has a mass number of 196, and 60.2% of the nucleons are neutrons. How many electrons are in a cation of this element if it has a charge of "2"?

**3-74.** A given isotope has a mass number of 206. The isotope has 51.2% more neutrons than protons. What is the element?

**3-75.** A given molecular compound is composed of one atom of nitrogen and one atom of another element. The mass of nitrogen accounts for 46.7% of the mass of one molecule. What is the other element? What is the formula of the compound? This molecule can lose one electron to form a polyatomic ion. How many electrons are in this ion?

**3-76.** A given molecular compound is composed of one atom of carbon and two atoms of another element. The mass of carbon accounts for 15.8% of the mass of one molecule. What is the other element? What is the formula of the compound?

**3-77.** If the isotopic mass of $^{12}C$ were defined as exactly 8 instead of 12, what would be the atomic mass of the following elements to three significant figures? Assume that the elements have the same masses relative to each other as before; that is, hydrogen still has a mass of one-twelfth that of carbon.
**(a)** H    **(b)** N    **(c)** Na    **(d)** Ca

**3-78.** Assume that the isotopic mass of $^{12}C$ is defined as exactly 10 and that the atomic mass of an element is 43.3 amu on this basis. What is the element?

**3-79.** Assume that the isotopic mass of $^{12}C$ is defined as exactly 20 instead of 12 and that the atomic mass of an element is 212.7 amu on this basis. What is the element?

# Solutions to Learning Checks

**A-1.** N, Cu, nickel, atom, molecules, formula, two, one, structural, ions, cations, anions

**A-2.** $C_8H_9NO_2$

**A-3.** $Na_3PO_4$, $(NH_4)_2S$

**B-1.** electrons, protons, neutrons, protons, neutrons, one, protons, nucleons, atomic, mass, $^{12}C$, weighted average, protons, electrons.

**B-2.** $^{16}O$ (8 p, 8 n) $^{17}O$ (8 p, 9 n) $^{18}O$ (8 p, 10 n)

**B-3.** $0.0140 \times 203.97 = \phantom{00}2.86$ ($^{204}Pb$)
$\phantom{00}0.2410 \times 205.97 = \phantom{0}49.64$ ($^{206}Pb$)
$\phantom{00}0.2210 \times 206.98 = \phantom{0}45.74$ ($^{207}Pb$)
$\phantom{00}0.5240 \times 207.98 = \underline{109.0}$ ($^{208}Pb$)
$\phantom{0000000000000}207.24 = \underline{207.2 \text{ amu}}$

**B-4.** There are seven protons in a nitrogen atom or a nitrogen ion.

$$\text{charge} = \text{number of protons} - \text{number of electrons}$$

$$\text{number of electrons} = \text{number of protons} - \text{charge}$$
$$= 7 - (-3)$$
$$\underline{\text{number of electrons} = 10}$$

## chapter 4

# The Periodic Table and Chemical Nomenclature

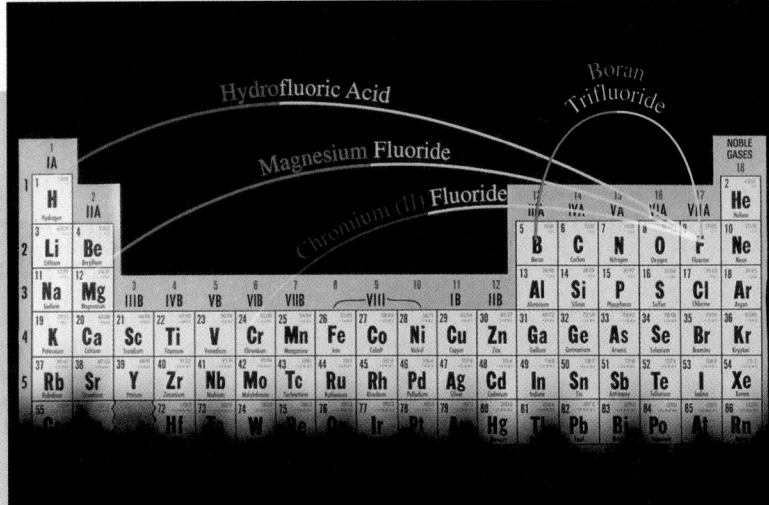

The periodic table helps the chemist systematize the properties of the elements. It is also used for the naming of compounds formed by the elements. The origin of this table and its use in naming compounds are the subjects of this chapter.

## Setting the Stage

Two things may strike the casual and uninformed observer about chemists and students of chemistry: (1) they seem to glance frequently at an ever-present wall poster containing the symbols of the elements, and (2) they seem to have their own language. The observer is correct on both counts. The wall poster is obviously important since it hangs in almost every chemical laboratory and classroom. Known as the *periodic table*, it is as necessary to a chemist as a map is to the world traveler. The location of a city on a map tells us a wealth of information about climate and culture. Likewise, the location of a specific element on the periodic table gives us important information about physical and chemical properties of the element.

Chemists have their own vocabulary for the 10 million or so known compounds. Why is it necessary for us to learn the language of chemistry? The answer is the same as why you learn any language other than your native tongue—you can communicate directly with others who speak the same language. The vocabulary of chemistry is known as **chemical nomenclature.** Like learning any other tongue, learning chemical nomenclature requires some memorization.

These two topics, the periodic table and chemical nomenclature, may at first seem unrelated. However, the periodic table groups the elements into two broad categories, and the naming of compounds relates to these classifications. The periodic table displays this information in an easy-to-read manner that simplifies the rules of chemical nomenclature.

**B** **Formulating Some Questions**

asically, the main function of the periodic table is to provide some order in what otherwise would be total chaos. Many elements have similar properties and can be grouped accordingly. But what kinds of groupings of elements are possible? Why and how did the periodic table originate? What information about the elements does it so conveniently display? How can this information be used in naming compounds? These are some of the questions that we will answer in this chapter.

Section **A** **RELATIONSHIPS AMONG THE ELEMENTS AND THE PERIODIC TABLE**

## 4-1 Grouping the Elements: Metals and Nonmetals, Noble Metals and Active Metals

**Looking Ahead!** Scientists always try to make things easy for themselves by separating types of phenomena into groups or classifications. In biology, living systems can be grouped into flora (trees and plants) and fauna (animals). In geology, the surface of the Earth can be divided into oceans and continents. In chemistry we do the same thing in several ways. These classifications help us simplify some of the information and are the topic of this section.

If we tried to group the human race in two categories we could start with just men and women. We can do a similar classification in chemistry. Elements are either *metals* or *nonmetals*. **Metals** are generally hard, lustrous elements that are *malleable* (can be pounded into thin sheets) and *ductile* (can be drawn into wires). (See Figure 4-1.) We also know they readily conduct electricity and heat. Many metals such as iron, copper, and aluminum form the strong framework on which our modern society is built. The discovery and use of metals over 5000 years ago moved civilization beyond the Stone Age. The second type of element is noted by its *lack* of metallic properties. These are the nonmetals. **Nonmetals** are generally gases or soft solids that do not conduct electricity. (See Figure 4-2.) There are exceptions to these general properties, however. Still, almost everyone has a general idea of what a metal is like. In addition to these physical properties, there are some very important chemical differences between metals and nonmetals, which we will explore in a later chapter. The division between

(a)

(b)

Figure 4-1 **Two Properties of Metals** **Metals can be pressed into thin sheets (malleable) (a) or drawn into wires (ductile) (b).**

**Figure 4-2 Nonmetals**
The bottle on the left contains liquid bromine and its vapor. The flask in the back contains pale green chlorine gas. Solid iodine is in the flask on the right. Powdered red phosphorus is in the dish in the middle, and black powdered graphite (carbon) is in the watch glass in front. Lumps of yellow sulfur are shown in the front.

metallic and nonmetallic properties is not sharp, so some elements have intermediate properties and are sometimes classified as a separate group.

Classifying the elements doesn't stop with the division of elements into these two groups. We find that all metals are not the same, so further classification is possible. It's like classifying the human race into two genders, men and women, but then finding that they can be further subdivided into personality types (e.g., extroverts and introverts). The first thing we note about metals is that some are chemically unreactive. That is, elements such as copper, silver, and gold are very resistant to the chemical reactions of corrosion and rust. These are the metals of coins and jewelry, not only because of their comparative rarity and beauty but also because of this chemical inertness. For this reason, they are known as the *noble metals*. Gold and silver coins on the ocean bottom, deposited from ships that foundered hundreds of years ago, can be easily polished to their original luster. (See Figure 4-3.) Other metals are very much different. They are extremely reactive with air and water. In fact, metals such as lithium, sodium, and potassium must be stored under oil because they react violently (to the point of explosion) with water. These metals are among those known as the *active metals*. Thus, copper, silver, and gold can be placed into one family of metals and lithium, sodium, and potassium into another. Similar relationships among other elements were also noticed and appropriate groupings were made.

**Figure 4-3 One Type of Metal**
With a little polish, these gold and silver coins regained their original luster after three centuries at the bottom of the ocean.

## 4-2 The Periodic Table

**Looking Ahead!** The possibility of relationships and similarities among certain elements was an intriguing mystery for centuries. Was there actually an overall systematic relationship among all of the elements? The answer is "yes," and this is what will be discussed in this section.

Less than 140 years ago, the big picture of the relationships among the elements was presented in a form known as the **periodic table.** (See Figure 4-4.) The earliest version of this table was introduced in 1869 by Dmitri Mendeleev of Russia. Lothar Meyer of Germany independently presented a similar table in 1870. When these two scientists arranged the elements in order of increasing atomic masses (atomic numbers were still unknown), they observed that elements in families appeared at regular

Atomic number: 1 / H / Atomic mass: 1.00794

metal | metalloid | nonmetal | Noble gases

| Periods | IA (1) | IIA (2) | IIIB (3) | IVB (4) | VB (5) | VIB (6) | VIIB (7) | VIIIB (8) | VIIIB (9) | VIIIB (10) | IB (11) | IIB (12) | IIIA (13) | IVA (14) | VA (15) | VIA (16) | VIIA (17) | VIIIA (18) |
|---|---|---|---|---|---|---|---|---|---|---|---|---|---|---|---|---|---|---|
| 1 | 1 H 1.00794 | | | | | | | | | | | | | | | | | 2 He 4.00260 |
| 2 | 3 Li 6.941 | 4 Be 9.01218 | | | | | | | | | | | 5 B 10.811 | 6 C 12.011 | 7 N 14.00674 | 8 O 15.9994 | 9 F 18.99840 | 10 Ne 20.1797 |
| 3 | 11 Na 22.98977 | 12 Mg 24.3050 | | | | | | | | | | | 13 Al 26.98154 | 14 Si 28.0855 | 15 P 30.97376 | 16 S 32.066 | 17 Cl 35.4527 | 18 Ar 39.948 |
| 4 | 19 K 39.0983 | 20 Ca 40.078 | 21 Sc 44.95591 | 22 Ti 47.88 | 23 V 50.9415 | 24 Cr 51.9961 | 25 Mn 54.9380 | 26 Fe 55.847 | 27 Co 58.93320 | 28 Ni 58.69 | 29 Cu 63.546 | 30 Zn 65.39 | 31 Ga 69.723 | 32 Ge 72.61 | 33 As 74.92159 | 34 Se 78.96 | 35 Br 79.904 | 36 Kr 83.80 |
| 5 | 37 Rb 85.4678 | 38 Sr 87.62 | 39 Y 88.90585 | 40 Zr 91.224 | 41 Nb 92.90638 | 42 Mo 95.94 | 43 Tc 98.9072 | 44 Ru 101.07 | 45 Rh 102.90550 | 46 Pd 106.42 | 47 Ag 107.8682 | 48 Cd 112.411 | 49 In 114.82 | 50 Sn 118.710 | 51 Sb 121.75 | 52 Te 127.60 | 53 I 126.90447 | 54 Xe 131.29 |
| 6 | 55 Cs 132.90543 | 56 Ba 137.327 | 57 *La 138.9055 | 72 Hf 178.49 | 73 Ta 180.9479 | 74 W 183.85 | 75 Re 186.207 | 76 Os 190.2 | 77 Ir 192.22 | 78 Pt 195.08 | 79 Au 196.96654 | 80 Hg 200.59 | 81 Tl 204.3833 | 82 Pb 207.2 | 83 Bi 208.98037 | 84 Po 208.9824 | 85 At 209.9871 | 86 Rn 222.0176 |
| 7 | 87 Fr 223.0197 | 88 Ra 226.0254 | 89 †Ac 227.0278 | 104 Rf 261.11 | 105 Db 262.114 | 106 Sg 263.118 | 107 Bh 262.12 | 108 Hs (265) | 109 Mt (266) | 110 Uun (269) | 111 Uuu (272) | 112 Uub (277) | | 114 (285) | | 116 (289) | | |

| | | | | | | | | | | | | | | |
|---|---|---|---|---|---|---|---|---|---|---|---|---|---|---|
| * | 58 Ce 140.115 | 59 Pr 140.90765 | 60 Nd 144.24 | 61 Pm 144.9127 | 62 Sm 150.36 | 63 Eu 151.965 | 64 Gd 157.25 | 65 Tb 158.92534 | 66 Dy 162.50 | 67 Ho 164.93032 | 68 Er 167.26 | 69 Tm 168.93421 | 70 Yb 173.04 | 71 Lu 174.967 |
| † | 90 Th 232.0381 | 91 Pa 231.0359 | 92 U 238.0289 | 93 Np 237.0482 | 94 Pu 244.0642 | 95 Am 243.0614 | 96 Cm 247.0703 | 97 Bk 247.0703 | 98 Cf 242.0587 | 99 Es 252.083 | 100 Fm 257.0951 | 101 Md 258.10 | 102 No 259.1009 | 103 Lr 260.105 |

**Figure 4-4 The Periodic Table**

(periodic) intervals. The periodic table was constructed so that elements in the same family (e.g., Li, Na, and K) fell into vertical columns. At first this did not always happen. Sometimes the next heaviest element did not seem to fit in a certain family. Mendeleev solved this problem by placing the element in a family with similar properties, even if he had to leave a space or two blank. For example, a space was left under silicon and above tin for what Mendeleev suggested was a yet-undiscovered element. Mendeleev called the missing element "eka-silicon." Later, an element was discovered that had properties intermediate between silicon and tin, as predicted by the location of the blank space. The element was later named germanium.

A second problem involved some misfits when the elements were ordered according to atomic mass. For example, notice that tellurium (number 52) is heavier than iodine (number 53). But Mendeleev realized that iodine clearly belonged under bromine and tellurium under selenium, and not vice versa. Mendeleev simply reversed the order, suggesting that perhaps the atomic masses reported were in error. (That was known to happen sometimes.) We now know that this problem does not occur when the elements are listed in order of increasing atomic number instead of atomic mass. This method of ordering conveniently displays the **periodic law,** which states that *the properties of elements are periodic functions of their atomic numbers.*

## 4-3 Periods and Groups in the Periodic Table

**Looking Ahead! We are now ready to look at the periodic table in more detail. Our goal is to become familiar enough with this splendid tool that we can put it to full use. We will examine further classifications of the elements in this section.**

The periodic table allows us to locate families of elements in vertical columns. In fact, there are common characteristics in the horizontal rows as well. *Horizontal rows of elements in the table are called* **periods.** Each period ends with a member of the family of elements called the **noble gases.** These elements, like the noble metals, are chemically unreactive and exist in nature as individual atoms. The first period contains only two elements, hydrogen and helium. The second and third contain 8 each (Figure 4-4), the fourth and fifth contain 18 each, the sixth 32, and the seventh 26. (The seventh would also contain 32 if there were enough elements.)

*Families of elements fall into vertical columns called* **groups.** Each group is designated by a number at the top of the group. The most commonly used label employs Roman numerals followed by an A or a B. Another method, which eventually may be accepted, numbers the groups 1 through 18. It is not clear at this time which method will win out, or if some alternative will yet be proposed and universally accepted. The periodic tables used in the text display both numbering systems. In the discussion, however, we will use the traditional method involving Roman numerals along with the letters A and B.

The groups of elements can be classified even further into four main categories of elements.

### 1. The Main Group or Representative Elements (Groups IA–VIIA)

Most of the familiar elements that we will discuss and use as examples in this text are **main group** or **representative elements.** For example, the three main elements of life—carbon, oxygen, and hydrogen—are in this category. One group of the representative elements includes the highly reactive metals that we discussed earlier. Notice that lithium, sodium, and potassium are found in Group IA. This family of elements (except for hydrogen) is known as the **alkali metals.** Another group of metals that are also chemically reactive is found in Group IIA and are known as the **alkaline earth metals.** Group VIIA are all nonmetals and are known as the **halogens.** Group VIA elements are known as the **chalcogens.** The other representative element groups (IIIA, IVA, and VA) are not generally referred to by a family name. In a later chapter, we will discuss in more detail some of the physical and chemical properties of the representative elements.

### 2. The Noble or Inert Gases (Group VIIIA)

These elements form few chemical compounds. In fact, helium, neon, and argon do not form any compounds. They all exist as individual atoms in nature.

### 3. The Transition Metals (Group B Elements)

**Transition metals** include many of the familiar structural metals such as iron and chromium as well as the noble metals, copper, silver, and gold (Group IB), which we discussed earlier.

### 4. The Inner Transition Metals

The 14 **inner transition metals** between lanthanum (number 57) and hafnium (number 72) are known as the **lanthanides** or *rare earths,* and the 14 metals between actinium (number 89) and rutherfordium (number 104) are known as the **actinides.**

## 4-4 Putting the Periodic Table to Use

**Looking Ahead!** Even more information can be taken from a modern periodic table. The position of an element in the body of the table can tell us a great deal about its chemical and physical properties. We will see what other information can be gleaned from the table in this section.

A quick glance at the periodic table reveals the two broad classifications of the elements that we discussed. The heavy stairstep line in Figure 4-4 separates the

## The Discovery of an Important Nonmetal

It was a cold day in a small seaside town in the Atlantic coast of France in 1811. A few dozen seamen were extracting potassium salts with acid from the sludge of seaweed. Bernard Courtois, the employer of these seamen, was a chemist by training and a graduate of the Polytechnical School in Paris. His factory prepared saltpeter (potassium nitrate) to be used in ammunition for Napoleon's armies. Today, however, the workers' efforts turned fruitless. One of the workers decided to use a more concentrated form of acid. At that point a huge volume of violet fumes rose from the tanks and dark crystals started depositing on every cold surface that was nearby. Their observations would lead to the discovery of a very important element.

Courtois collected those unique crystals for examination and found out they would combine with hydrogen and phosphorus, but not with oxygen. He also discovered they would form an explosive compound with ammonia. He later gave samples to two of his Paris Polytechnical Institute friends, C. Desormes and N. Clement, who published the discoveries two years later. Soon thereafter, Frenchman Joseph Louis Gay-Lussac and Englishman Sir Humphry Davy announced the discovery of a new element, which was first named *iode* (from the Greek word for violet) by Gay-Lussac and finally *iodine*, to give it the same ending as *chlorine*, an element of similar properties. (Iodine would later be classified as a *halogen* in Group VIIA.)

What happened that day in 1811? Seaweed concentrates several salts other than sodium chloride (table salt) but no one had given very much importance to them. That day, however, the iodide salts must have undergone a considerable concentration after extraction of sodium chloride. The concentrated acid converted the iodide salts to elemental iodine. The iodine vaporized (sublimed) but quickly condensed on the cool surfaces.

The practical applications of the new element had an immediate impact on patients of goiter, the enlargement of the thyroid gland. In 1820, Jean Francois Coinder associated the lack of goiter among seamen with the presence of iodide salts in their working environment. The thyroid gland, he concluded, needs iodine to function properly and this may be achieved by adding small amounts of sodium iodide to table salt (i.e., iodized salt).

*Submitted by Paris Sovoronos*
*Queensborough Community College*

*Elemental iodine.*

metals and nonmetals. Metals (about 80% of the elements) are on the left of the line, and nonmetals are on the right. When hydrogen is displayed as a IA element, however, it appears to the left of the line but is definitely a nonmetal. *Many elements on the borderline have properties that are intermediate between metals and nonmetals and are sometimes referred to as either* **metalloids** *or* **semimetals.** These elements are indicated in Figure 4-4. For example, silicon is a brittle solid, typical of nonmetals, but conducts a limited amount of electricity (a semiconductor). Conduction of electricity is a metallic property. In any case, we will use only the two broad classifications (metal and nonmetal) in nomenclature discussed later in the chapter.

Next we will consider what the periodic table can tell us about the physical state of the element. We must be cautious, however, as to the temperature conditions we define. If the temperature is low enough, all elements exist as solids (except helium); if it is high enough, all elements are in the gaseous state. On Triton, a moon of the planet Neptune, the temperature is −236°C (37 K). The atmosphere is very thin because most substances that are gases under Earth conditions are solids or liquids under Triton conditions. At the outer part of the sun, however, the temperature is 50,000°C, so only gases exist. Thus we must come to some agreement as to a reference temperature to define the physical state of an element. **Room temperature,** which is defined as exactly *25°C, is the standard reference temperature* used to describe physical state. At this temperature, all three physical states are found among the elements on the periodic table. Fortunately, except for hydrogen, the gaseous elements are all found at

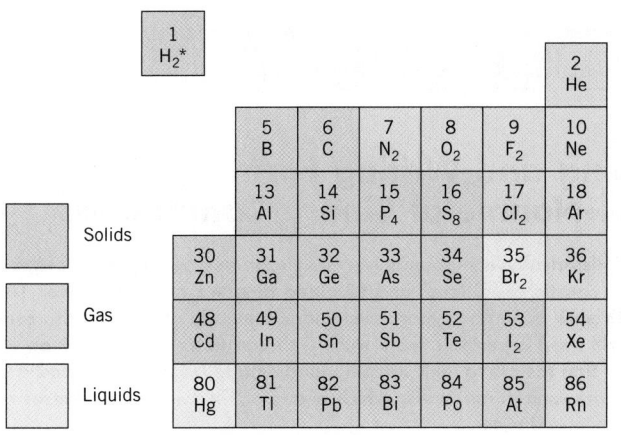

**Figure 4-5 The Physical States of the Elements**

*The subscript 2 indicates diatomic molecules at 25°C

the extreme right top of the table (nitrogen, oxygen, fluorine, and chlorine) and in the right-hand vertical column (the noble gases). There are only two liquids: a metal, mercury, and a nonmetal, bromine. All other elements are solids. Two solids, gallium and cesium, melt to become liquids at slightly above the reference temperature (29°C), however. (See Figure 4-5.)

Many of the nonmetals exist as diatomic molecules rather than individual atoms. The periodic table can help us locate these elements. All of the gaseous elements except for the noble gases, which exist as individual atoms, are composed of diatomic molecules (e.g., $N_2$, $O_2$, and $H_2$). Or one may notice that all of the naturally occurring halogens (Group VIIA) are diatomic. Not all molecular nonmetals are diatomic, however. The formula for the most common form of phosphorus is $P_4$, which indicates the presence of molecules composed of four atoms. The most common form of sulfur is $S_8$, indicating eight-atom molecules. The forms of carbon and less common forms of the other nonmetals will be discussed in Chapter 10.

---

**Looking Back!** It is a great advantage to find that no element is entirely unique. Each element shares certain chemical characteristics with other elements, allowing us to group those elements together. The periodic table is an ingenious display of these groupings of elements. The table makes it possible to quickly identify the elements that belong to a particular group, such as the alkali metals or the halogens. By being able to locate a certain element on the periodic table, we can assume a number of characteristics common to that group and period.

---

## Learning Check A | checking it out

**A-1.** Fill in the blanks.

Of the two general classifications of elements, nickel is a _____ and sulfur is a _____ . Some borderline elements such as germanium are sometimes referred to as _____ . In the periodic table, elements are ordered according to increasing _____ _____ so that _____ of elements fall into vertical columns. Of the four general categories of elements, calcium is a _____ element, nickel is a _____ _____ , and xenon is a _____ _____ . An element in Group IIA, such as calcium, is also known as an _____ _____ metal. A solid nonmetal that is composed of diatomic molecules is in the group known as the _____ . Metals are all solids except for _____ .

*Additional Examples: Exercises 4-5, 4-10, 4-12, and 4-14.*

## 4-5 Naming and Writing Formulas of Metal–Nonmetal Binary Compounds

**Looking Ahead!** Historically, compounds had simple names. Examples of common names include water, ammonia, caustic soda, lime, and laughing gas. We now have over 10 million recorded compounds. With such a number, more systematic methods of naming the compounds must be employed. We will need a periodic table since we need to know whether an element is a metal or nonmetal. Our first task is to look at compounds made from a metal and a nonmetal. For the most part, these are ionic compounds, which is one of the two major classes of compounds described in the previous chapter.

Chemical compounds can be roughly divided into two groups: organic and inorganic. Organic compounds are composed principally of carbon, hydrogen, and oxygen. These are the compounds of life and will not be discussed at this time. All other compounds are called inorganic compounds, and we will focus on these. As mentioned, many inorganic compounds were given common names and many of these names have survived the years. For example, $H_2O$ is certainly known as *water* rather than the more exact systematic name of *dihydrogen oxide.* Ammonia ($NH_3$) and methane ($CH_4$) are two other compounds that are known exclusively by their common names. Other compounds with ancient common names are also described by their more modern systematic names, which we will use here.

One of the most important chemical properties of a metal is its tendency to form positive ions (cations). Most nonmetals, on the other hand, can form negative ions (anions). (The noble gases form neither cations nor anions.) Inorganic compounds composed of just two elements—metal cations and nonmetal anions—will be referred to as *metal–nonmetal binary compounds.* Metals fall into two categories. Some metals form cations with only one charge (e.g., $Ca^{2+}$). Others form cations with two or more charges (e.g., $Fe^{2+}$ and $Fe^{3+}$). The rules for the latter compounds are somewhat different from those for the first. So, we first consider metals with ions of only one charge.

### Metals with Ions of Only One Charge

It is not difficult to recognize metals with only one charge because all except one are in two groups in the periodic table. (There are also several transition metals that form only one cation charge, but they are not included in this discussion.) Representative element metals in Group IA (alkali metals) form +1 cations exclusively. Likewise, the metals in Group IIA (alkaline earth metals) form +2 ions exclusively. Aluminum, in Group IIIA, forms a +3 ion exclusively, but other metals in this group also form a +1 ion. (See Figure 4-6.) When present in metal–nonmetal binary compounds, the nonmetals form one type of anion. Hydrogen and Group VIIA (halogens) form −1 ions, Group VIA elements form −2 ions, and N and P in Group VA form −3 ions. (See Figure 4-6.) *Note that positive and negative ions do not exist separately. They are always found together in ordinary matter.*

In both naming and writing the formula for a binary ionic compound, the metal comes first and the nonmetal second. The unchanged English name of the metal is used. (If a metal cation is named alone, the word "ion" is also included to distinguish it from the free metal.) The name of the anion includes only the English root plus *ide.* For example, *chlorine* as an anion is named *chloride* and *oxygen* as an anion is named *oxide.* So the names for NaCl and CaO are sodium chloride and calcium oxide.

| IA | IIA | IIIA | IVA | VA | VIA | VIIA |
|----|-----|------|-----|-----|-----|------|

| Hydride H⁻ | | | | | | |

Hydride $H^-$

Lithium $Li^+$ | Beryllium $Be^{2+}$ | | Carbide $C^{4-}$ | Nitride $N^{3-}$ | Oxide $O^{2-}$ | Fluoride $F^-$

Sodium $Na^+$ | Magnesium $Mg^{2+}$ | Aluminum $Al^{3+}$ | | Phosphide $P^{3-}$ | Sulfide $S^{2-}$ | Chloride $Cl^-$

Potassium $K^+$ | Calcium $Ca^{2+}$ | | | | Selenide $Se^{2-}$ | Bromide $Br^-$

Rubidium $Rb^+$ | Strontium $Sr^{2+}$ | | | | Telluride $Te^{2-}$ | Iodide $I^-$

Cesium $Cs^+$ | Barium $Ba^{2+}$

**Figure 4-6  Monatomic Ions of the Representative Elements**

Some other examples of writing names from formulas are shown in Example 4-1. Writing formulas from names can be a somewhat more challenging task since we must then determine the number of each element present in the formula. What is important to remember is that the formulas represent neutral compounds where the positive and negative charges add to zero. In other words, the total positive charge is canceled by the total negative charge. Thus, NaCl is neutral because one $Na^+$ is balanced by one $Cl^-$ [i.e., $+1 + (-1) = 0$] and CaO is neutral because one $Ca^{2+}$ is balanced by one $O^{2-}$ [i.e., $+2 + (-2) = 0$]. The formula for magnesium chloride, however, requires two $Cl^-$ ions to balance the one $Mg^{2+}$ ion, so it is written as $MgCl_2$ [i.e., $+2 + (2 \times -1) = 0$]. We will practice writing formulas from names in Example 4-2.

**Naming Metal–Nonmetal Binary Compounds (Metals with Ions of Only One Charge)**

| | | | Example 4-1 |

Name the following binary ionic compounds: $KI$, $Li_2S$, and $Mg_3N_2$.

working it out

**Answers**

| Formula | Metal | Nonmetal | Compound name |
|---------|-------|----------|---------------|
| $KI$ | potassium | iodine | potassium iodide |
| $Li_2S$ | lithium | sulfur | lithium sulfide |
| $Mg_3N_2$ | magnesium | nitrogen | magnesium nitride |

**Writing Formulas of Metal–Nonmetal Binary Compounds (Metals with Ions of Only One Charge)**

Example 4-2

Write the formulas for the following binary metal–nonmetal compounds: aluminum fluoride and calcium selenide.

**Solution**

Aluminum is in Group IIIA, so it forms a cation with a +3 charge exclusively. Fluorine is in Group VIIA, so it forms an anion with a −1 charge. Since the positive charge is balanced by the negative charge, it is obvious that we need three $F^{1-}$

anions to balance one $Al^{3+}$ ion [e.g., $+3 + (\underline{3} \times -1) = 0$]. Therefore, the formula is written as

$$Al^{3+} + 3(F^-) = \underline{\underline{AlF_3}}$$

Another convenient way to establish the formula is to write the ions with their appropriate charges side by side. The numerical value of the charge on the cation becomes the subscript on the anion and vice versa. (The number "1" is understood instead of written in as a subscript.) This is known as the *cross-charge* method.

$$Al^{\textcircled{3}+} \quad F^{\textcircled{1}-} = \underline{\underline{AlF_3}}$$

Calcium is in Group IIA, so it forms a $+2$ cation. Selenium is in Group VIA, so it forms a $-2$ anion. Notice that, by exchanging values of the charge, we first indicate a formula of $Ca_2Se_2$. This is not a correct representation, however. Ionic compounds should be expressed with the simplest whole numbers for subscripts. Therefore, the proper formula is written as CaSe.

$$Ca^{\textcircled{2}+} \quad Se^{\textcircled{2}-} = Ca_2Se_2 = \underline{\underline{CaSe}}$$

## Metals with Ions of More Than One Charge

Except for Groups IA, IIA, and aluminum, other representative metals and most transition metals can form more than one cation. Therefore, a name such as iron chloride would be ambiguous since there are two iron chlorides, $FeCl_2$ and $FeCl_3$. An even more extreme case is that of manganese oxide—there are five different compounds ($MnO$, $Mn_3O_4$, $Mn_2O_3$, $MnO_2$, and $Mn_2O_7$). To distinguish among these compounds, the charge on the metal ion follows the name of the metal in Roman numerals and in parentheses. This is referred to as the **Stock method.** Therefore, the two chlorides of iron are named iron(II) chloride ($FeCl_2$) and iron(III) chloride ($FeCl_3$). There is another method that was widely used at one time. This is known as the *classical method.* In this system, *the name of the metal ion that has the lower charge ends in ous and the higher ends in* ic. If the symbol of the element is derived from a Latin word, the Latin root is generally used rather than an English root. Thus the two chlorides of iron are named ferrous chloride ($FeCl_2$) and ferric chloride ($FeCl_3$). Several common examples of names of ions of metals that form more than one cation are shown in Table 4-1. Since the

### Table 4-1  Metals That Form More Than One Ion

| Ion | Stock Name | Classical Name | Ion | Stock Name | Classical Name |
|-----|-----------|----------------|-----|-----------|----------------|
| $Cr^{2+}$ | chromium(II) | chromous | $Cu^+$ | copper(I) | cuprous |
| $Cr^{3+}$ | chromium(III) | chromic | $Cu^{2+}$ | copper(II) | cupric |
| $Fe^{2+}$ | iron(II) | ferrous | $Mn^{2+}$ | manganese(II) | manganous |
| $Fe^{3+}$ | iron(III) | ferric | $Mn^{3+}$ | manganese(III) | manganic |
| $Pb^{2+}$ | lead(II) | plumbous | $Sn^{2+}$ | tin(II) | stannous |
| $Pb^{4+}$ | lead(IV) | plumbic | $Sn^{4+}$ | tin(IV) | stannic |
| $Au^+$ | gold(I) | aurous | $Co^{2+}$ | cobalt(II) | cobaltous |
| $Au^{3+}$ | gold(III) | auric | $Co^{3+}$ | cobalt(III) | cobaltic |

classical method is rarely used today, it will not be included in examples or problems that follow in this text.

In order to use the Stock method, it is necessary to determine the charge on the metal cation by working backward from the known charge on the anion. For example, in a compound with the formula FeS, we can establish from Figure 4-6 that the charge on the S is $-2$. Therefore, the charge on the one Fe must be $+2$. The compound is named *iron(II) sulfide.* In the following examples, we will use a simple algebra equation that will help us determine the charge on the metal and thus the proper Roman numeral to use for more complex compounds. The equation is

$$[(\text{number of metal cations}) \times (+\text{charge on metal})] +$$
$$[(\text{number of nonmetal anions}) \times (-\text{charge on nonmetal})] = 0$$

---

**Naming Metal–Nonmetal Binary Compounds (Metals with Ions of More Than One Charge)**

Example 4-3

Name the following compounds: $SnO_2$ and $Co_2S_3$.

working it out

**Solution**

For $SnO_2$, the equation is

$$[1 \times (\text{Sn charge})] + [2 \times (\text{O charge})] = 0$$

Substituting the known charge on the oxygen,

$$\text{Sn} + (2 \times -2) = 0$$
$$\text{Sn} - 4 = 0$$
$$\text{Sn} = +4 \,(\text{IV})$$

The name of the compound is, therefore, tin(IV) oxide.

For $Co_2S_3$, we can construct the following equation.

$$[2 \times (\text{Co charge})] + [3 \times (\text{S charge})] = 0$$

The charge on a Group VIA nonmetal is $-2$.

$$2\text{Co} + (3 \times -2) = 0$$
$$2\text{Co} = +6$$
$$\text{Co} = +3\,(\text{III})$$

The name of the compound is cobalt(III) sulfide.

---

**Writing Formulas of Metal–Nonmetal Binary Compounds (Metals with Ions of More Than One Charge)**

Example 4-4

Write the formulas for lead(IV) oxide, iron(II) chloride, and chromium(III) sulfide.

**Solution**

Lead(IV) oxide: If lead has a $+4$ charge, two $O^{2-}$ ions are needed to form a neutral compound.

$$Pb^{4+} \diagdown O^{2-} = Pb_2O_4 = PbO_2$$

Iron(II) chloride: Two chlorides are needed to balance the +2 iron.

$$Fe^{(2)+} \diagdown Cl^{(1)-} = \underline{FeCl_2}$$

Chromium(III) sulfide

$$Cr^{(3)+} \diagdown S^{(2)-} = \underline{Cr_2S_3}$$

# 4-6 Naming and Writing Formulas of Compounds with Polyatomic Ions

**Looking Ahead!** Monatomic ions are the result of an imbalance of the number of electrons and the number of protons in the nucleus of one atom. Cations have fewer electrons than protons, and anions have more electrons than protons. Two or more atoms that are chemically combined with covalent bonds may also have an imbalance of electrons and protons, leading to a charged species called a polyatomic ion. How compounds containing these ions are named is our next topic. Again, you should recognize these as ionic compounds, so we find them as solids under normal conditions.

We use bicarbonates and carbonates for indigestion, as well as sulfites and nitrites to preserve foods. So, we are probably familiar with some of these names, which are commonly used. A list of some common polyatomic ions is given in Table 4-2. Notice that all but one ($NH_4^+$, ammonium) are anions.

Most of the compounds containing polyatomic ions are ionic, as were the compounds discussed in the previous section. Thus we follow essentially the same rules as before. That is, the metal is written and named first. If the metal forms more than one cation, the charge on the cation is shown in parentheses. The polyatomic anion is then named or written.

There is some systematization possible that will help in understanding Table 4-2. In many cases, *the anions are composed of oxygen and one other element.* Thus these anions are called **oxyanions.** When there are two oxyanions of the same element (e.g., $SO_3^{2-}$ and $SO_4^{2-}$), they, of course, have different names. The anion with the smaller

**Table 4-2  Polyatomic Ions**

| Ion | Name | Ion | Name |
|---|---|---|---|
| $C_2H_3O_2^-$ | acetate | $HSO_3^-$ | hydrogen sulfite or bisulfite |
| $NH_4^+$ | ammonium | $OH^-$ | hydroxide |
| $CO_3^{2-}$ | carbonate | $ClO^-$ | hypochlorite |
| $ClO_3^-$ | chlorate | $NO_3^-$ | nitrate |
| $ClO_2^-$ | chlorite | $NO_2^-$ | nitrite |
| $CrO_4^{2-}$ | chromate | $C_2O_4^{2-}$ | oxalate |
| $CN^-$ | cyanide | $ClO_4^-$ | perchlorate |
| $Cr_2O_7^{2-}$ | dichromate | $MnO_4^-$ | permanganate |
| $HCO_3^-$ | hydrogen carbonate or bicarbonate | $PO_4^{3-}$ | phosphate |
| | | $SO_4^{2-}$ | sulfate |
| $HSO_4^-$ | hydrogen sulfate or bisulfate | $SO_3^{2-}$ | sulfite |

number of oxygens uses the root of the element plus *ite*. The one with the higher number uses the root plus *ate*.

$$SO_3^{2-} \qquad \text{sul}fite$$
$$SO_4^{2-} \qquad \text{sul}fate$$

There are four oxyanions containing Cl. The middle two are named as before. The one with one less oxygen than the chlorite has a prefix of *hypo*. The one with one more oxygen than chlorate has a prefix of *per*.

$$ClO^- \qquad \text{hypochlorite}$$
$$ClO_2^- \qquad \text{chlorite}$$
$$ClO_3^- \qquad \text{chlorate}$$
$$ClO_4^- \qquad \text{perchlorate}$$

Certain anions are composed of more than one atom but behave similarly to monatomic anions in many of their chemical reactions. Two such examples in Table 4-2 are the $CN^-$ ion and the $OH^-$ ion. Both of these have *ide* endings similar to the monatomic anions. Thus the $CN^-$ anion is known as the cyanide ion and the $OH^-$ as the hydroxide ion.

Most of the ionic compounds that we have just named are also referred to as salts. *A **salt** is an ionic compound formed by the combination of a cation with an anion.* (Cations combined with hydroxide or oxide form a class of compounds that are not considered salts and are discussed in a later chapter.) For example, potassium nitrate is a salt composed of $K^+$ and $NO_3^-$ ions, and calcium sulfate is a salt composed of $Ca^{2+}$ and $SO_4^{2-}$ ions. Ordinary table salt is NaCl, composed of $Na^+$ and $Cl^-$ ions.

In naming and writing the formulas of compounds with polyatomic ions, as in Example 4-5, we follow the same procedures as with metal–nonmetal compounds. The metal is written first with its charge (if it is not in IA, IIA, or Al) followed by the name of the polyatomic ion. To calculate the charge on the cation, if necessary, we can use the same simple algebra equation as before. For example, consider $Cr_2(SO_4)_3$. Since the sulfate ion has a $-2$ charge, the charge on the chromium is

$$(2 \times Cr) + (3 \times -2) = 0$$
$$2Cr = +6$$
$$Cr = +3$$

The name of the compound is *chromium(III) sulfate*.

When writing formulas from names, as in Example 4-6, we note that when more than one polyatomic ion is present in the compound, parentheses enclose the polyatomic ion. If only one polyatomic ion is present, parentheses are not used (e.g., $CaCO_3$). In Example 4-7 we extend our knowledge of nomenclature to other ions in a specific group.

---

**Naming Compounds with Polyatomic Ions**

Name the following compounds: $K_2CO_3$ and $Fe_2(SO_4)_3$.

**Example 4-5**

working it out

**Solution**

$K_2CO_3$: The cation is $K^+$ (Group IA). The charge is not included in the name because it forms a $+1$ ion only. The anion is $CO_3^{2-}$ (the carbonate ion).

potassium carbonate

$Fe_2(SO_4)_3$: The charge on the Fe cation can be determined from the charge on the $SO_4^{2-}$ (sulfate) ion.

$$2\,Fe + 3SO_4^{2-} = 0$$
$$2Fe + 3(-2) = 0$$
$$2Fe = +6$$
$$Fe = +3$$
$$\underline{\underline{\text{iron(III) sulfate}}}$$

---

**Example 4-6** | **Writing Formulas of Metal–Polyatomic Anion Compounds**

Give the formulas for barium acetate, ammonium sulfate, thallium(III) nitrate, and manganese(III) phosphate.

**Solution**

Barium acetate: Barium is in Group IIA, so it has a $+2$ charge. Acetate is the $C_2H_3O_2^-$ ion.

$$Ba^{2+} \diagdown C_2H_3O_2^{1-} = \underline{\underline{Ba(C_2H_3O_2)_2}}$$

(If more than one polyatomic ion is in the formula, enclose the ion in parentheses.)
Ammonium sulfate: From Table 4-1, ammonium = $NH_4^+$, sulfate = $SO_4^{2-}$.

$$NH_4^{1+} \diagdown SO_4^{2-} = \underline{\underline{(NH_4)_2SO_4}}$$

Thallium(III) nitrate $\qquad Tl^{3+} \diagdown NO_3^{1-} = \underline{\underline{Tl(NO_3)_3}}$

Manganese(III) phosphate $\qquad Mn^{3+} \diagdown PO_4^{3-} = Mn_3(PO_4)_3 = \underline{\underline{MnPO_4}}$

---

**Example 4-7** | **Naming Compounds by Analogy**

Name the following compounds: $NaBrO_4$ and $Cu(IO)_2$.

**Solution**

$NaBrO_4$: The compound is composed of $Na^+$ and $BrO_4^-$ ions. Both Br and Cl are in Group VIIA, so they may form analogous ions. Since the name of the $ClO_4^-$ ion is the perchlorate ion, it is logical to assume that $BrO_4^-$ is named the *perbromate* ion. The name of the compound is

$$\underline{\underline{\text{sodium perbromate}}}$$

$Cu(IO)_2$: The anion is analogous to the $ClO^-$ (hypochlorite) ion since both I and Cl are also in Group VIIA. This is the hypoiodite ion ($IO^-$). The Cu cation must then have a $+2$ charge.

$$\underline{\underline{\text{copper(II) hypoiodite}}}$$

# 4-7 Naming Nonmetal–Nonmetal Binary Compounds

**Looking Ahead!** The compounds that we have named so far are generally ionic compounds. They constitute much of the hard, solid part of nature. When nonmetals bond to other nonmetals, however, molecular compounds are formed. These are very likely to be gases or liquids. How we name molecular compounds is the topic of this section.

When a metal is combined with a nonmetal it is simple to decide which one to name and thus write first. But which do we write first if neither is a metal? In these cases, we generally write the one closer to being a metal first—that is, the nonmetal closer to the metal–nonmetal border in the periodic table (farther down or farther to the left). Thus we write $CO_2$ rather than $O_2C$ but $OF_2$ rather than $F_2O$. In cases where both elements are equidistant from the border, Cl is written before O (e.g., $Cl_2O$) and the others in the order S, N, then Br (e.g., $S_4N_4$ and $NBr_3$). When hydrogen is one of the nonmetals and is combined with nonmetals in Groups VIA and VIIA (e.g., $H_2O$ and HF), hydrogen is written first. When combined with other nonmetals (Groups IIIA, IVA, and VA) and metals, however, it is written second (e.g., $NH_3$ and $CH_4$).*

The nonmetal closer to the metal borderline is also named first using its English name. The less metallic is named second using its English root plus *ide,* as discussed before. If more than one compound of the same two nonmetals exists, the number of atoms of each element present in the compound is indicated by the use of Greek prefixes (see Table 4-3). Table 4-4 illustrates the nomenclature of nonmetal–nonmetal compounds with the six oxides of nitrogen. Notice that if there is only one atom of the nonmetal written first, *mono* is not used. However, if there is only one of the second nonmetal, *mono* is used. (Notice that the *o* in mon*o* is dropped in "monoxide" for ease in pronunciation.) The Stock method is rarely applied to the naming of nonmetal–nonmetal compounds because it can be ambiguous in some cases. For example, both $NO_2$ and $N_2O_4$ could be named nitrogen(IV) oxide.

Chemical names are familiar to us in many common drugs and cleansers.

**Table 4-3  Greek Prefixes**

| Number | Prefix | Number | Prefix | Number | Prefix |
|--------|--------|--------|--------|--------|--------|
| 1 | mono | 5 | penta | 8 | octa |
| 2 | di | 6 | hexa | 9 | nona |
| 3 | tri | 7 | hepta | 10 | deca |
| 4 | tetra | | | | |

**Table 4-4  The Oxides of Nitrogen**

| Formula | Name |
|---------|------|
| $N_2O$ | dinitrogen monoxide (sometimes referred to as nitrous oxide) |
| NO | nitrogen monoxide (sometimes referred to as nitric oxide) |
| $N_2O_3$ | dinitrogen trioxide |
| $NO_2$ | nitrogen dioxide |
| $N_2O_4$ | dinitrogen tetroxide[a] |
| $N_2O_5$ | dinitrogen pentoxide[a] |

[a] The "a" is often omitted from tetra and penta for ease in pronunciation.

Nitrogen dioxide contributes to the brownish haze in polluted air.

*With few exceptions, in organic compounds containing C, H, and other elements, the C is written first followed by H and then other elements that are present.

### Comets—Nature's Huge, Dirty Snowballs

In March 1997, the world was treated to a spectacular celestial sight. The Hale-Bopp comet gave us a brilliant display of a comet with its glowing tail reaching back into space. Comets originate far out in space, beyond the furthest planet, Pluto. Occasionally, one of these hunks of matter is pushed into the inner reaches of the solar system where it begins to glow and produce a characteristic tail. Most comets make one trip around the sun and then disappear back into deep space. Some, like the famous Halley's comet, appear regularly. Halley's comet visits the inner solar system every 76 years.

What are comets and why do they glow? Comets are of interest because they probably formed at the same time and from the same stuff as the Earth about 4.5 billion years ago. Understanding comets will tell us about our own planet's origin and perhaps how life itself began on Earth. Halley's comet last visited in 1986. It was a dud as far as displays go, but it provided a goldmine of scientific information. A total of five spacecraft were sent to close encounters with the comet as it rounded the sun in early 1987. Data returned indicated that the comet is composed mostly of water ice with dust particles containing minerals of silicon, sodium, magnesium, iron, and other elements. The gases escaping were mostly water vapor with some carbon dioxide, ammonia, and methane.

The glow of a comet becomes apparent as it comes close enough to the sun for it to heat the surface. This causes some molecular compounds to vaporize and escape from fissures on the surface. The escaping gases are ionized by the solar wind (high-energy particles from the sun), which then emit a glow. A second tail is from solar radiation interacting with the dust that is ejected along with the gases.

Three spacecraft are on their way or due to be launched soon that will encounter comets. Their names are *Stardust*, *Deep Impact*, and *Rosetta*. *Stardust* was launched by NASA in March 1999. Its mission is to rendezvous with the comet Wild-2 somewhere near Mars in 2004. Scientists hope to capture dust particles given off from the comet and then begin a return journey. An analysis of the compounds in the dust will tell us much about comets and our own origins when *Stardust* returns to Earth in 2006. In 2013, it is hoped that *Rosetta* will actually deploy a lander to study the surface of a comet close up. A fourth expedition named *Contour* was launched in July, 2002 but unfortunately broke apart in space that August.

*The Hale-Bopp comet in 1997.*

Several of these compounds are known only by their common names, such as water ($H_2O$), methane ($CH_4$), and ammonia ($NH_3$).

According to the rules, compounds such as $TiO_2$ and $UF_6$ should be named by the Stock method—for example, titanium(IV) oxide and uranium(VI) fluoride, respectively. Sometimes, however, we hear them named in the same manner as nonmetal–nonmetal binary compounds (i.e., titanium dioxide and uranium hexafluoride.) The rationale for the latter names is that when the charge on the metal exceeds $+3$, the compound has properties more typical of a molecular nonmetal–nonmetal binary compound than an ionic one. For example, $UF_6$ is a liquid at room temperature, whereas true ionic compounds are all solids under these conditions. In any case, in this text, we will identify all metal–nonmetal binary compounds by the Stock method regardless of their properties and confine the use of Greek prefixes to the nonmetal–nonmetal compounds.

## 4-8 Naming Acids

**Looking Ahead! We have one last category of compounds. These involve most of the anions listed in Figure 4-6 and Table 4-2 when combined with hydrogen. Since hydrogen is not a metal, these are molecular compounds in the pure state. However, many of these compounds have an important property when present in water. This special property allows us to give them special names, and we will do so in this section.**

When hydrogen is combined with an anion such as $Cl^-$, the formula of the resulting compound is HCl. The fact that HCl is a gas and not a hard solid at room temperature indicates that HCl is molecular rather than ionic. When dissolved in water,

**Table 4-5  Binary Acids**

| Anion | Formula of Acid | Compound Name | Acid Name |
|-------|-----------------|---------------|-----------|
| $Cl^-$ | HCl | hydrogen chloride | hydrochloric acid |
| $F^-$ | HF | hydrogen fluoride | hydrofluoric acid |
| $I^-$ | HI | hydrogen iodide | hydroiodic acid |
| $S^{2-}$ | $H_2S$ | hydrogen sulfide | hydrosulfuric acid |

however, the HCl is ionized by the water molecules to form $H^+$ ions and $Cl^-$ ions. This ionization is illustrated as

$$HCl \xrightarrow{H_2O} H^+ + Cl^-$$

Most of the hydrogen compounds formed from the anions in Figure 4-6 and Table 4-1 behave in a similar manner, at least to some extent. *This common property of forming $H^+$ in aqueous solution is a property of a class of compounds called* **acids.** Acids are important enough to earn their own nomenclature. The chemical nature of acids is discussed in more detail in Chapters 7 and 12.

The acids formed from the anions listed in Figure 4-6 *are composed of hydrogen plus one other element, so they are called* **binary acids.** These compounds can be named in two ways. In the pure state, the hydrogen is named like a metal with only one charge (+1). That is, HCl is named *hydrogen chloride,* and $H_2S$ is named *hydrogen sulfide.* When dissolved in water, however, these compounds are generally referred to by their acid names. The acid name is obtained by dropping the word "hydrogen," adding the prefix *hydro* to the anion root, and changing the *ide* ending to *ic* followed by the word "acid." Both types of names are illustrated in Table 4-5. Polyatomic anions that have an *ide* ending are also named in the same manner as the binary acids. For example, the acid formed by the cyanide ion ($CN^-$) has the formula HCN and is named hydrocyanic acid.

The following hydrogen compounds of anions listed in Figure 4-6 are not generally considered to be binary acids: $H_2O$, $NH_3$, $CH_4$, and $PH_3$.

*The acids formed by combination of hydrogen with most of the polyatomic anions in Table 4-2 are known as* **oxyacids** *because they are formed from oxyanions.* To name an oxyacid, we use the root of the anion to form the name of the acid. If the name of the oxyanion ends in *ate,* it is changed to *ic* followed by the word "acid." If the name of the anion ends in *ite,* it is changed to *ous* plus the word "acid." Most hydrogen compounds of oxyanions do not exist in the pure state as do the binary acids. Generally, only the acid name is used in the naming of these compounds. For example, $HNO_3$ is called nitric acid and not hydrogen nitrate. Development of the acid name from the anion name is shown for some anions in Table 4-6.

Hydrochloric acid (muriatic acid) is a strong binary acid.

**Table 4-6  Oxyacids**

| Anion | Name of Anion | Formula of Acid | Name of Acid |
|-------|---------------|-----------------|--------------|
| $C_2H_3O_2^-$ | acetate | $HC_2H_3O_2$ | acetic acid |
| $CO_3^{2-}$ | carbonate | $H_2CO_3$ | carbonic acid |
| $NO_3^-$ | nitrate | $HNO_3$ | nitric acid |
| $PO_4^{3-}$ | phosphate | $H_3PO_4$ | phosphoric acid |
| $ClO_2^-$ | chlorite | $HClO_2$ | chlorous acid |
| $ClO_4^-$ | perchlorate | $HClO_4$ | perchloric acid |
| $SO_3^{2-}$ | sulfite | $H_2SO_3$ | sulfurous acid |
| $SO_4^{2-}$ | sulfate | $H_2SO_4$ | sulfuric acid |

In summary, acids are named as follows.

| Ending on anion | Change | Anion example | Acid name |
|---|---|---|---|
| ide | add *hydro* and change ending to *ic* | brom*ide* | *hydro*brom*ic* acid |
| ite | change ending to *ous* | hypochlor*ite* | hypochlor*ous* acid |
| ate | change ending to *ic* | perchlor*ate* | perchlor*ic* acid |

**Example 4-8**

working it out

**Naming Acids**

Name the following acids: $H_2Se$, $H_2C_2O_4$, and $HClO$.

**Answers**

| | |
|---|---|
| $H_2Se$ | hydroselenic acid |
| $H_2C_2O_4$ | oxalic acid |
| $HClO$ | hypochlorous acid |

**Example 4-9**

**Writing Formulas of Acids**

Give formulas for the following: permanganic acid, dichromic acid, and acetic acid.

**Answers**

| | |
|---|---|
| permanganic acid | $HMnO_4$ |
| dichromic acid | $H_2Cr_2O_7$ |
| acetic acid | $HC_2H_3O_2$ |

**Looking Back!** Our first important use of the periodic table is to aid us in the language of chemistry known as chemical nomenclature. The identification of an element as a metal or a nonmetal determines how we name its compounds. There are four major classifications of compounds that we have identified in this chapter: (1) metal–nonmetal binary compounds, (2) metal–polyatomic ion compounds, (3) nonmetal–nonmetal binary compounds, and (4) acids. In addition to these classifications, we also noticed that we represent the charge on a metal in its name if it can form more than one charge; otherwise we do not.

# Learning Check B | checking it out

**B-1.** Fill in the blanks.

Given the following four elements: K, Se, Al, and Br. The two metals are _____ and the two nonmetals are _____ . The charges that these elements have in binary ionic compounds are K = _____ , Se = _____ , Al = _____ , and Br = _____ . The formula of a compound formed between K and Se is _____ , and its name is _____ . The formula of a compound formed between Al and Br is _____ , and its name is _____ . The formula of a compound formed between Se and Br is $SeBr_2$ and its name is _____ . An acid formed from the Se ion has the formula _____ , and its name is _____ .

**B-2.** Fill in the blanks.

Given the following ions: $Ca^{2+}$, $Au^{3+}$, $MnO_4^-$, and $CrO_4^{2-}$. The formula of a compound formed between $Ca^{2+}$ and $MnO_4^-$ is _____ , and its name is _____ . The formula of a compound formed between $Au^{3+}$ and $CrO_4^{2-}$ is _____ , and its name is _____ . An acid formed from the $CrO_4^{2-}$ ion has the formula _____ , and its name is _____ .

**B-3.** Three fluorine compounds formed by Group IIIA elements are $BF_3$, $AlF_3$, and $TlF_3$. All three are named somewhat differently. Name the compounds.

**B-4.** Give the chemical names for $Br_2O$ and $OF_2$. Why is oxygen named first in $OF_2$ and second in $Br_2O$?

*Additional Examples: Exercises 4-18, 4-20, 4-22, 4-34, 4-36, 4-43, 4-45, 4-47, and 4-48.*

**chapter** R**eview**                    **putting it together**

There is no more important time-saving device for the chemist than the **periodic table,** which demonstrates in table form the **periodic law** for the elements. Horizontal rows are known as **periods,** and vertical columns are known as **groups.** Although there are four categories of elements in the table, the category that we will emphasize in the text is the **main group** or **representative elements.** This category includes some named groups known as the **alkali metals,** the **alkaline earth metals,** the **chalcogens,** and the **halogens.** The other three categories are the **noble gases,** the **transition metals,** and the **inner transition metals.** The last category includes the **lanthanides** and the **actinides.** These named groups are summarized in the chart that follows.

One important function of the periodic table is that it shows a clear boundary between elements that are classified as **metals** and **nonmetals.** Some metals and non-metals have intermediate characteristics and may be referred to as **metalloids** or **semimetals.** All three physical states are found among the elements at the reference temperature of 25°C, known as **room temperature.** Some of the nonmetals also exist as molecules rather than individual atoms at the reference temperature.

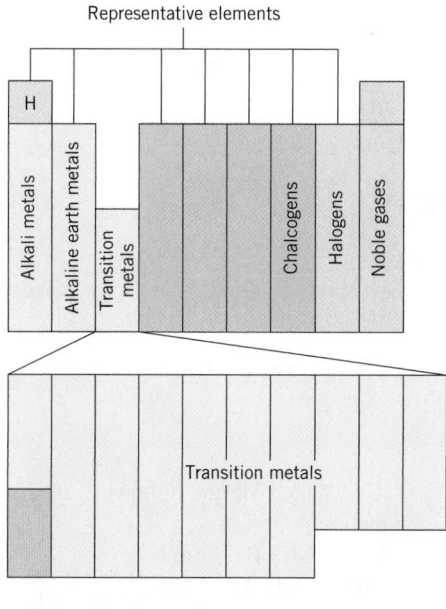

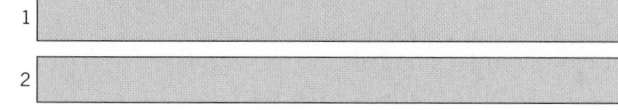

Our first important use of the periodic table is to aid us in one aspect of **chemical nomenclature.** For example, binary compounds containing a metal and a nonmetal are named differently than those composed of two nonmetals. Also, all metals form cations, but some form only one charge and others form more than one. In using the **Stock method** of nomenclature, we need the periodic table to tell us which metals are in the latter group. In addition to the binary compounds, we discussed the naming of **salts** containing **oxyanions.** Finally, a class of hydrogen compounds called **acids** (both **binary** and **oxyacids**) was discussed as a special group of compounds. The naming of compounds is summarized by the chart below.

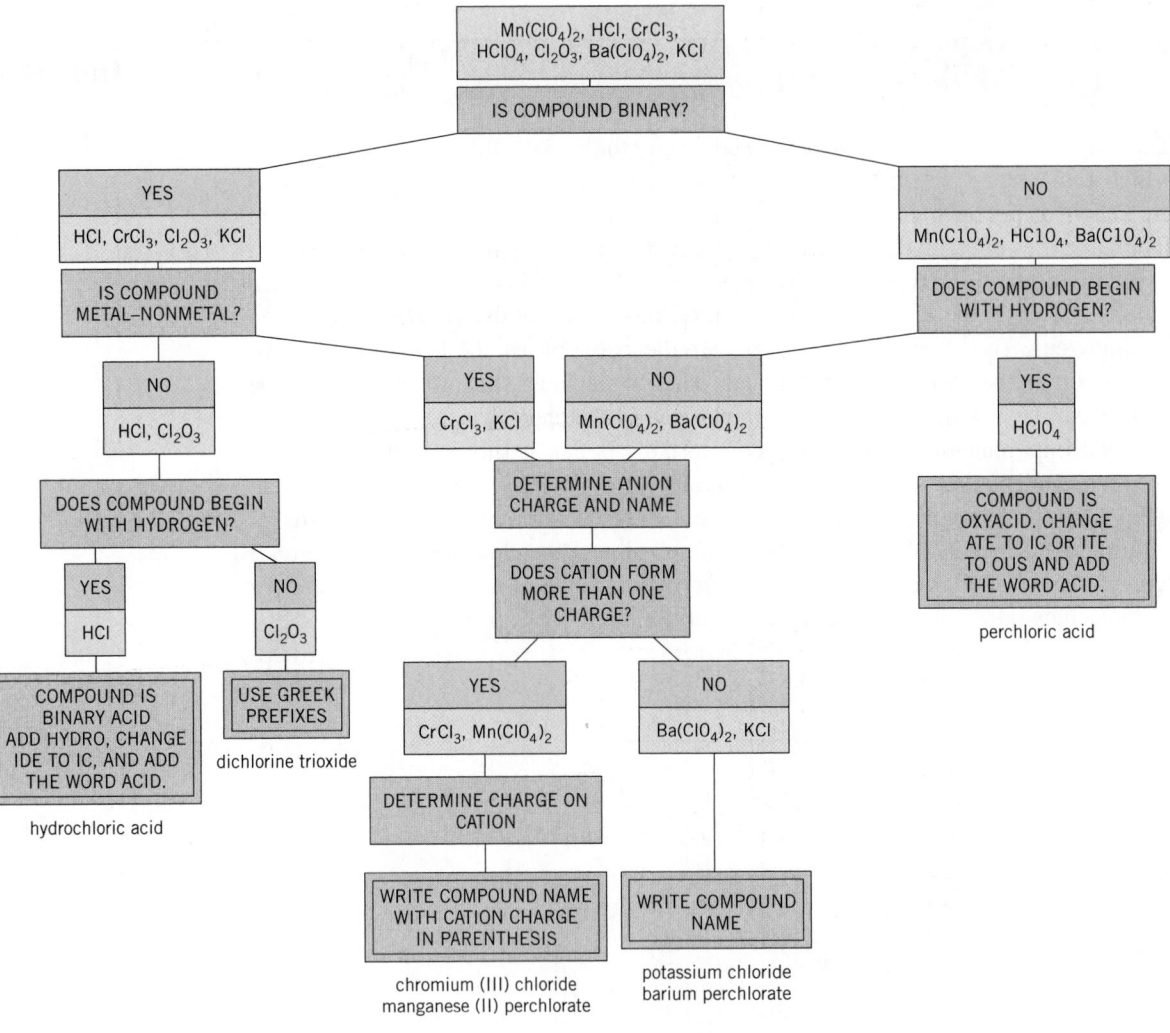

# Exercises

## The Periodic Table

**4-1.** How is an active metal different from a noble metal?

**4-2.** How many elements would be in the seventh period if it were complete?

**4-3.** Which of the following elements are halogens?
(a) $O_2$    (c) $I_2$    (e) Li    (g) $Br_2$
(b) $P_4$    (d) $N_2$    (f) $H_2$

**4-4.** Which of the following elements are alkaline earth metals?
(a) Sr    (c) B    (e) Na
(b) C    (d) Be    (f) K

**4-5.** Classify the following elements into one of the four main categories of elements.
(a) Fe    (c) Pm    (e) Xe    (g) In
(b) Te    (d) La    (f) H

4-6. Classify the following elements into one of the four main categories of elements.
(a) Se   (d) Sr   (g) Er
(b) Ti   (e) Zn
(c) Ni   (f) I

4-7. Which of the following elements are transition metals?
(a) In   (d) Xe   (g) Ag
(b) Ti   (e) Pd
(c) Ca   (f) Tl

## Physical States of the Elements

4-8. What is the most common physical state of the elements at room temperature? Which are more common, metals or nonmetals?

4-9. Which metals, if any, are gases at room temperature? Which metals, if any, are liquids? Which nonmetals, if any, are liquids at room temperature?

4-10. Referring to Figure 4-5, tell which of the following are gases at room temperature.
(a) Ne   (c) B   (e) Br   (g) Na
(b) S   (d) Cl   (f) N

4-11. Referring to Figure 4-5, tell which of the following elements exist as diatomic molecules under normal conditions.
(a) N   (c) Ar   (e) H   (g) Xe
(b) C   (d) F   (f) B   (h) Hg

4-12. Referring to Figure 4-4, tell which of the following elements are metals.
(a) Ru   (c) Hf   (e) Ar   (g) Se
(b) Sn   (d) Te   (f) B   (h) W

4-13. Which, if any, of the elements in exercise 4-12 can be classified as a metalloid?

4-14. Identify the following elements using the information in Figures 4-4 and 4-5.
(a) a nonmetal, monatomic gas in the third period
(b) a transition metal that is a liquid
(c) a diatomic gas in Group VA
(d) the second metal in the second period
(e) the only member of a group that is a metal

4-15. Identify the following elements using the information in Figures 4-4 and 4-5.
(a) a nonmetal, diatomic liquid
(b) the last element in the third period
(c) a nonmetal, diatomic solid
(d) the only member of a group that is a nonmetal

4-16. What would be the atomic number and group number of the next nonmetal after element 112? (Assume that the border between metals and nonmetals continues as before.)

4-17. What is the atomic number of the metal farthest to the right in the periodic table that would appear after element number 112?

## Metal–Nonmetal Binary Compounds

4-18. Name the following compounds.
(a) LiF   (c) $Sr_3N_2$   (e) $AlCl_3$
(b) BaTe   (d) $BaH_2$

4-19. Name the following compounds.
(a) $CaI_2$   (c) BeSe   (e) RaS
(b) FrF   (d) $Mg_3P_2$

4-20. Give formulas for the following compounds.
(a) rubidium selenide   (d) aluminum carbide
(b) strontium hydride   (e) beryllium fluoride
(c) radium oxide

4-21. Give formulas for the following compounds.
(a) potassium hydride   (c) potassium phosphide
(b) cesium sulfide   (d) barium telluride

4-22. Name the following compounds using the Stock method.
(a) $Bi_2O_5$   (c) $SnS_2$   (e) TiC
(b) SnS   (d) $Cu_2Te$

4-23. Name the following compounds using the Stock method.
(a) $CrI_3$   (c) $IrO_4$   (e) $NiCl_2$
(b) $TiCl_4$   (d) $MnH_2$

4-24. Give formulas for the following compounds.
(a) copper(I) sulfide   (d) nickel(II) carbide
(b) vanadium(III) oxide   (e) chromium(VI) oxide
(c) gold(I) bromide

4-25. Give formulas for the following compounds.
(a) yttrium(III) hydride   (c) bismuth(V) fluoride
(b) lead(IV) chloride   (d) palladium(II) selenide

4-26. From the magnitude of the charges on the metals, predict which of the compounds in exercises 4-22 and 4-24 may be molecular compounds.

4-27. From the magnitude of the charges on the metals, predict which of the compounds in exercises 4-23 and 4-25 may be molecular compounds.

## Compounds with Polyatomic Ions

4-28. Which of the following is the chlorate ion?
(a) $ClO_2^-$   (c) $ClO_3^-$   (e) $ClO_3^+$
(b) $ClO_4^-$   (d) $Cl_3O^-$

4-29. Which of the following ions have a −2 charge?
(a) sulfate   (c) chlorite   (e) sulfite
(b) nitrite   (d) carbonate   (f) phosphate

4-30. What are the name and formula of the most common polyatomic cation?

4-31. Which of the following oxyanions contain four oxygen atoms?
(a) nitrate   (d) sulfite   (g) carbonate
(b) permanganate   (e) phosphate
(c) perchlorate   (f) oxalate

4-32. Name the following compounds. Use the Stock method where appropriate.
(a) $CrSO_4$   (d) $RbHCO_3$   (g) $Bi(OH)_3$
(b) $Al_2(SO_3)_3$   (e) $(NH_4)_2CO_3$
(c) $Fe(CN)_2$   (f) $NH_4NO_3$

**4-33.** Name the following compounds. Use the Stock method where appropriate.
**(a)** $Na_2C_2O_4$　**(b)** $CaCrO_4$　**(c)** $Fe_2(CO_3)_3$　**(d)** $Cu(OH)_2$

**4-34.** Give formulas for the following compounds.
**(a)** magnesium permanganate
**(b)** cobalt(II) cyanide
**(c)** strontium hydroxide
**(d)** thallium(I) sulfite
**(e)** iron(III) oxalate
**(f)** ammonium dichromate
**(g)** mercury(I) acetate [The mercury(I) ion exists as $Hg_2^{2+}$.]

**4-35.** Give formulas for the following compounds.
**(a)** zirconium(IV) phosphate
**(b)** sodium cyanide
**(c)** thallium(I) nitrite
**(d)** nickel(II) hydroxide
**(e)** radium hydrogen sulfate
**(f)** beryllium phosphate
**(g)** chromium(III) hypochlorite

**4-36.** Complete the following table. Write the appropriate anion at the top and the appropriate cation to the left. Write the formulas and names in other blanks.

| Cation/Anion | $HSO_3^-$ | _____ | _____ |
|---|---|---|---|
| $NH_4^+$ | $NH_4HSO_3$ ammonium bisulfite | _____ _____ _____ | _____ _____ _____ |
| _____ _____ | _____ _____ | CoTe _____ (name) | _____ _____ |
| _____ _____ | _____ _____ | _____ _____ | ____ (formula) aluminum phosphate |

**4-37.** Complete the following table. Write the appropriate anion at the top and the appropriate cation to the left. Write the formulas and names in the other blanks.

| Cation/Anion | _____ | $C_2O_4^{2-}$ | _____ |
|---|---|---|---|
| _____ | ____ (formula) thallium(I) hydroxide | _____ _____ | _____ _____ |
| $Sr^{2+}$ | _____ _____ | _____ _____ | _____ _____ |
| _____ | _____ _____ | _____ _____ | TiN ─ (name) |

**4-38.** Give the systematic name for each of the following.

| | Common Name | Formula |
|---|---|---|
| **(a)** | table salt | $NaCl$ |
| **(b)** | baking soda | $NaHCO_3$ |
| **(c)** | marble or limestone | $CaCO_3$ |
| **(d)** | lye | $NaOH$ |
| **(e)** | chile saltpeter | $NaNO_3$ |
| **(f)** | sal ammoniac | $NH_4Cl$ |
| **(g)** | alumina | $Al_2O_3$ |
| **(h)** | slaked lime | $Ca(OH)_2$ |
| **(i)** | caustic potash | $KOH$ |

**4-39.** The perzenate ion has the formula $XeO_6^{4-}$. Write formulas of compounds of perzenate with the following.
**(a)** calcium　**(b)** potassium　**(c)** aluminum

**\*4-40.** Name the following compounds. In these compounds, an ion is involved that is not in Table 4-1. However, the name can be determined by reference to other ions of the central element or from ions in Table 4-1 in which the central atom is in the same group.
**(a)** $PH_4F$　**(c)** $Co(IO_3)_3$　**(e)** $AlPO_3$
**(b)** $KBrO$　**(d)** $CaSiO_3$　**(f)** $CrMoO_4$

## Nonmetal–Nonmetal Binary Compounds

**4-41.** The following pairs of elements combine to form binary compounds. Which element should be written and named first?
**(a)** Si and S　**(c)** H and Se　**(e)** H and F
**(b)** F and I　**(d)** Kr and F　**(f)** H and As

**4-42.** The following pairs of elements combine to make binary compounds. Which element should be written and named first?
**(a)** S and P　**(c)** O and Br
**(b)** O and S　**(d)** As and Cl

**4-43.** Name the following.
**(a)** $CS_2$　**(c)** $P_4O_{10}$　**(e)** $SO_3$　**(g)** $PCl_5$
**(b)** $BF_3$　**(d)** $Br_2O_3$　**(f)** $Cl_2O$　**(h)** $SF_6$

**4-44.** Name the following.
**(a)** $PF_3$　**(c)** $ClO_2$　**(e)** $SeCl_4$
**(b)** $I_2O_3$　**(d)** $AsF_5$　**(f)** $SiH_4$

**4-45.** Write the formulas for the following.
**(a)** tetraphosphorus hexoxide
**(b)** carbon tetrachloride
**(c)** iodine trifluoride
**(d)** dichlorine heptoxide
**(e)** sulfur hexafluoride
**(f)** xenon dioxide

**4-46.** Write formulas for the following.
**(a)** xenon trioxide
**(b)** sulfur dichloride
**(c)** dibromine monoxide
**(d)** carbon disulfide
**(e)** diboron hexahydride (also known as diborane)

## Acids

**4-47.** Name the following acids.
**(a)** HCl　**(c)** HClO　**(e)** $HIO_4$
**(b)** $HNO_3$　**(d)** $HMnO_4$　**(f)** HBr

**4-48.** Write formulas for the following acids.
(a) hydrocyanic acid    (d) carbonic acid
(b) hydroselenic acid    (e) hydroiodic acid
(c) chlorous acid    (f) acetic acid

**4-49.** Write formulas for the following acids.
(a) oxalic acid    (c) dichromic acid
(b) nitrous acid    (d) phosphoric acid

**\*4-50.** Refer to the ions in exercises 4-39 and 4-40. Write the acid names for the following.
(a) HBrO    (c) $H_3PO_3$    (e) $H_4XeO_6$
(b) $HIO_3$    (d) $HMoO_4$

**\*4-51.** Write the formulas and the names of the acids formed from the arsenite ($AsO_3^{3-}$) ion and the arsenate ($AsO_4^{3-}$) ion.

## General Problems

**4-52.** A gaseous compound is composed of two oxygens and one chlorine. It has been used to kill anthrax spores in contaminated buildings. Write the formula of the compound and give its name.

**4-53.** The halogen ($A_2$) with the lowest atomic number forms a compound with another halogen ($X_2$) that is a liquid at room temperature. The compound has the formula $A_5X$ or $XA_5$. Write the correct formula with the actual elemental symbols and the name.

**4-54.** A metal that has only a +2 ion and is the third member of the group forms a compound with a nonmetal that has a −2 ion and is in the same period. What are the formula and name of the compound?

**4-55.** The only gas in a certain group forms a compound with a metal that has only a +3 ion. The compound contains one ion of each element. What are the formula and name of the compound? What are the formula and name of the compound the gas forms with a $Ti^{2+}$ ion?

**4-56.** An alkali metal in the fourth period forms a compound with the phosphide ion. What are the formula and name of the compound?

**4-57.** A transition metal ion with a charge of +2 has 25 electrons. It forms a compound with a nonmetal that has only a −1 ion. The anion has 36 electrons. What are the formula and name of the compound?

**4-58.** The lightest element forms a compound with a certain metal in the third period that has a +2 ion and with a nonmetal in the same period that has a −2 ion. What are the formulas and names of the two compounds?

**4-59.** The thiosulfate ion has the formula $S_2O_3^{2-}$. What are the formula and name of the compound formed between the thiosulfate ion and an Rb ion; an Al ion; an $Ni^{2+}$ ion; and a $Ti^{4+}$ ion? What are the formula and name of the acid formed from the thiosulfate ion?

**4-60.** Name the following compounds: $NiI_2$, $H_3PO_4$, $Sr(ClO_3)_2$, $H_2Te$, $As_2O_3$, $Sb_2O_3$, and $SnC_2O_4$.

**4-61.** Name the following compounds: $SiO_2$, $SnO_2$, $MgO$, $Pb_3(PO_4)_2$, $HClO_2$, $BaSO_4$, and $HI$.

**4-62.** Give formulas for the following compounds: tin(II) hypochlorite, chromic acid, xenon hexafluoride, barium nitride, hydrofluoric acid, iron(III) carbide, and lithium phosphate.

**4-63.** Which of the following is composed of a metal that can have one charge and a polyatomic ion?
(a) $H_2CO_3$    (c) $B_2O_3$    (e) $Rb_2C_2O_4$
(b) $Ca_2C$    (d) $V(NO_3)_3$

**4-64.** Which of the following is composed of a metal that can have more than one charge and a monatomic anion?
(a) $Ti(ClO_4)_2$    (c) $Cu_2Se$    (e) $MgCrO_4$
(b) $Mg_2S$    (d) $H_2Se$

**4-65.** The peroxide ion has the formula $O_2^{2-}$. What are the formulas of compounds formed with Rb, Mg, Al, and $Ti^{4+}$? What is the formula of the acid for this anion? What is the name of this compound as a pure compound and as an acid?

**4-66.** The cyanamide ion has the formula $CN_2^{2-}$. What are the formulas of compounds formed with Li, Ba, $Sc^{3+}$, and $Sn^{4+}$? What is its formula as an acid? What is the name of this compound as a pure compound and as an acid?

**4-67.** Give the formulas of the following common compounds.
(a) sodium carbonate    (d) aluminum nitrate
(b) calcium chloride    (e) calcium hydroxide
(c) potassium perchlorate    (f) ammonium chloride.

**4-68.** Give the names of the following common compounds.
(a) $TiCl_4$    (c) LiH    (e) $HNO_3$    (g) $NaClO_4$
(b) $NH_4NO_3$    (d) $Mg(OH)_2$    (f) $H_2SO_4$

**4-69.** Nitrogen is found in five ions mentioned in this chapter. Write the formulas and names of these ions.

**4-70.** Carbon is found in six ions mentioned in this chapter. Write the formulas and names of these ions.

**4-71.** Which of the following is the correct name for $Cr_2(CO_3)_3$?
(a) dichromium tricarbonate    (d) chromium(III) tricarbonate
(b) chromium carbonate    (e) chromium(III) carbonate
(c) chromium(II) carbonate

**4-72.** Which of the following is the correct name for $SiCl_4$?
(a) sulfur tetrachloride    (d) sulfur chloride
(b) silicon tetrachloride    (e) silicon chloride
(c) silicon(IV) chloride

**4-73.** Which of the following is the correct name for $Ba(ClO_2)_2$?
(a) barium dichlorite    (d) barium chlorite(II)
(b) barium(II) chlorite    (e) barium chlorate
(c) barium chlorite

**4-74.** Which of the following is the correct name for $H_2CrO_4$?
(a) hydrogen(I) chromate    (d) dichromic acid
(b) hydrogen chromate    (e) dihydrogen chromate
(c) chromic acid    (f) chromous acid

# Interactive Learning

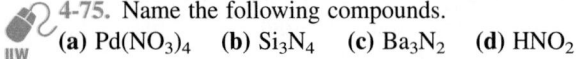

**4-75.** Name the following compounds.
**(a)** $Pd(NO_3)_4$  **(b)** $Si_3N_4$  **(c)** $Ba_3N_2$  **(d)** $HNO_2$

**4-76.** Name the following compounds.
**(a)** $SiS_2$  **(b)** $H_2SO_3$  **(c)** $CsHS$  **(d)** $Au_2(SO_4)_3$

# Solution to Learning Checks

**A-1.** metal, nonmetal, metalloids, atomic number, groups, representative, transition metal, noble gas, alkaline earth, halogens, mercury.

**B-1.** K and Al, Se and Br, $+1$, $-2$, $+3$, $-1$, $K_2Se$, potassium selenide, $AlBr_3$, aluminum bromide, selenium dibromide, $H_2Se$, hydroselenic acid.

**B-2.** $Ca(MnO_4)_2$, calcium permanganate, $Au_2(CrO_4)_3$, gold(III) chromate, $H_2CrO_4$, chromic acid.

**B-3.** $BF_3$—boron trifluoride (molecular compound from two nonmetals)

$AlF_3$—aluminum fluoride (metal has one charge)
$TlF_3$—thallium(III) fluoride (more than one type of metal ion)

**B-4.** $Br_2O$—dibromine monoxide
$OF_2$—oxygen difluoride
Br is more metallic than O, but O is more metallic than F.

# Cumulative Review
## Chapters 2–4*

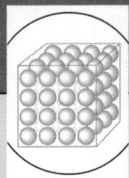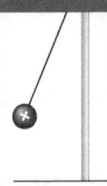

**Multiple Choice**

**The following multiple-choice questions have one correct answer.**

**1.** A substance has a distinct melting point and density. It cannot be decomposed into chemically simpler substances. The substance is a

(a) heterogeneous mixture      (c) compound
(b) solution                    (d) element

**2.** A sample of a given liquid has uniform properties, but the temperature at which the liquid boils slowly changes as it boils away. The liquid sample is a

(a) heterogeneous mixture      (c) compound
(b) solution                    (d) element

**3.** Which of the following is a chemical property of iodine?
(a) It melts at 114°C.
(b) It has a density of 4.94 g/mL.
(c) When heated, it forms a purple vapor.
(d) It forms a compound with sodium.

**4.** A charged car battery has _____ energy that can be converted into _____ energy.
(a) chemical, electrical      (c) light, electrical
(b) heat, mechanical       (d) mechanical, chemical

**5.** Which of the following has the highest density?
(a) a truckload of feathers      (c) a cup of water
(b) an ounce of gold         (d) a gallon of gasoline

**6.** Which of the following has a charge of $+1$ and a mass of 1 amu?

(a) a neutron      (c) a proton
(b) an electron      (d) a helium nucleus

**7.** Isotopes of a specific element have
(a) the same mass number and atomic number
(b) the same mass number but different atomic numbers
(c) different mass numbers and atomic numbers
(d) different mass numbers but the same atomic number

**8.** Which of the following describes an isotope with a mass number of 99 that contains 56 neutrons in its nucleus?
(a) $^{99}_{56}Ba$      (d) $^{56}_{43}Tc$
(b) $^{43}_{56}Ba$      (e) $^{155}_{199}Es$
(c) $^{99}_{43}Tc$

**9.** Which of the following isotopes is used as the standard for atomic mass?
(a) $^{12}C$      (b) $^{16}O$      (c) $^{13}C$      (d) $^{1}H$

**10.** Which of the following is the formula of a molecular compound with diatomic molecules?
(a) NO    (b) $NH_4Cl$    (c) $H_2$    (d) KBr    (e) $CO_2$

**11.** Which of the following is the correct formula for an element?
(a) $N_4$      (b) $P_2$      (c) $C_2$      (d) $Br_2$

**12.** Which would be the electrical charge on a sulfur atom containing 18 electrons?
(a) $2-$      (b) $1-$      (c) 0      (d) $2+$

**13.** Which would be the formula of an ionic compound made up of $Ba^{2+}$ ions and $Cl^-$ ions?
(a) $Ba_2Cl$    (b) $BaCl_2$    (c) BaCl    (d) $BaCl_3$

**14.** Which of the following elements is a halogen?
(a) H    (b) O    (c) K    (d) Ne    (e) I

**15.** Which of the following elements is a transition metal?
(a) tin    (b) nickel    (c) aluminum    (d) krypton

**16.** Which of the following metals forms more than one charge?
(a) Sr    (b) Tl    (c) Al    (d) Li    (e) N

**17.** Which of the following is the correct (Stock) name for $PbO_2$?
(a) lead dioxide      (c) lead oxide
(b) lead(IV) oxide      (d) lead(II) oxide

**18.** The formula of aluminum sulfide is
(a) AlS    (b) $Al_3S_2$    (c) $Al_2S_3$    (d) $AlS_2$    (e) $Al_2S$

**19.** Which of the following is the formula for aluminum hydroxide?
(a) $Al(OH)_3$    (b) $Al(OH)_2$    (c) AlOH    (d) $Al_2O_3$

**20.** Barium superoxide has the formula $Ba(O_2)_2$. What is the charge on the superoxide ion?
(a) $+1$    (b) $+2$    (c) $-2$    (d) $-1$    (e) $-4$

**21.** Which of the following elements is an alkali metal?
(a) Sc      (d) K
(b) Al      (e) H
(c) Cl

**22.** The formula of a compound between a Group IIA metal (M) and a Group VIA nonmetal (X) is
(a) $M_2X_2$    (b) $MX_2$    (c) $M_2X$    (d) MX    (e) $M_2X_3$

**23.** The formula of lithium sulfate is
(a) $LiSO_4$    (b) $(Li)_2SO_3$    (c) $Li_2S$    (d) $Li_2SO_4$

**117**

**24.** Which of the following is the commonly accepted name for $SeO_3$?
**(a)** selenium(VI) oxide
**(b)** selenium oxide
**(c)** selenium dioxide
**(d)** selenium trioxide

**25.** Which of the following compounds or elements is a solid at room temperature?
**(a)** $K_2CO_3$     **(b)** $CO_2$     **(c)** $HBrO$     **(d)** $Cl_2$     **(e)** $Hg$

## Problems

### Carry out the following calculations.

**1.** A cube of metal measures 2.2 cm on a side. It has a mass of 47.68 g. What is its density?

**2.** What is the volume of a 678-g sample of a substance that has a density of 11.3 g/mL?

**3.** The density of liquid iron is 7.05 g/mL. How many milliliters of iron must be mixed with liquid chromium to make 1.00 kg of an alloy that is 72.0% by mass iron?

**4.** The temperature of a given liquid increases by 12.0°C if 69.0 joules of heat is added to a 20.0-g sample. What is the specific heat of the liquid?

**5.** The atomic mass of an element is 10.15 times that of the defined standard. What is the atomic mass of the element? What are the name and symbol of the element?

**6.** The atoms of a given element are distributed between two isotopes: 65.0% of the atoms have a mass number of 116, and the remainder have a mass number of 112. Using mass number as an approximation of isotopic mass, calculate the atomic mass of the element to tenths of an atomic mass unit.

**7.** Answer the following questions about the element aluminum (Al).
**(a)** What are its group number and general classification?
**(b)** Is it a solid, liquid, or gas at room temperature?
**(c)** Is it a metal or nonmetal?
**(d)** What is its electrical charge in an ion?
**(e)** What is the simplest formula when it forms a binary compound with (1) sulfur, (2) bromine, (3) nitrogen?
**(f)** What are the names of the three compounds in question (e)?

*Answers to all questions and problems in Cumulative Review tests are given in Appendix G.

## chapter 5

# Modern Atomic Theory

**Fireworks produce spectacular colors. Each color is caused by the presence of a specific element. The emission of color by the atoms of hot gaseous elements is a topic of this chapter.**

## Setting the Stage

Bursts of bright silver, streaks of green, yellow, red, and blue explode across the sky. How we love it! The Fourth of July and the closing ceremony of the Olympics are highlighted by spectacular displays of fireworks. But what causes such exciting colors? Actually, it has been known for hundreds of years that certain elements in fireworks produce specific colors. Strontium imparts a red color; barium, green; copper, blue; and sodium, yellow. In fact, the specific colors emitted by hot gaseous atoms can be used to identify elements much like fingerprints are used to identify individuals. The study of the colors of these *emission spectra* is what opened the door to a deeper understanding of the nature of the atom.

Often in science, when a theory is developed to explain one phenomenon, other mysteries are explained as well. This happened with the theory that explained the emission spectra of the elements. The theory, which emphasized the electrons in the atom, led to an understanding of the theoretical basis of the periodic table. As we explained in Chapter 4, this marvelous table displays elements that are chemically related in vertical columns called groups. Although the existence of chemically related elements has been known for almost 200 years, only since the 1930s have scientists had a feeling for *why* elements are chemically related.

A theory of the atom that emphasized the electrons was first advanced in 1913 by a student of Lord Rutherford named Niels Bohr. This had the immediate effect of explaining the emission spectrum of hydrogen, the simplest of the elements. More significant to our purposes, however, is that Bohr's theory eventually led to a new, improved theory that explained and predicted chemical similarities among certain elements.

### Formulating Some Questions

In this chapter, we will attempt to answer several important questions. What is the nature of light? How does one color differ from another? How does an atom emit light? What is the relationship between electrons and the periodic table? What trends in atomic properties can be predicted by the periodic table? Our first goal is to examine the nature of light so that we can appreciate how atoms emit specific colors.

## Section A   THE ENERGY OF THE ELECTRON IN THE ATOM

## 5-1 The Emission Spectra of the Elements

**Looking Ahead!** Hot, gaseous atoms in fireworks give off spectacular light, but any substance glows if it is heated to a high enough temperature. We certainly sense the hot sun as it glows in the summer sky. An ordinary study lamp works by the same principle. The tungsten filament in an incandescent lightbulb emits a bright light as a result of heat generated by the flow of electricity. A basic understanding of the nature of light is the object of this section.

Light, which is also known as **electromagnetic radiation,** *is a form of energy like heat or electricity.* Light travels through space in waves much like the waves moving across a pond or a lake. Light waves travel at a given velocity and also have properties called wavelength and amplitude. All light waves travel at the same velocity in a vacuum, which is at the phenomenal rate of $3.0 \times 10^{10}$ cm/s (186,000 mi/s). The amplitude of a wave refers to its height, which in turn relates to the intensity of light. The amplitude or intensity of light is not important to our discussion, so it will not be mentioned further. *The **wavelength** of light is the distance between two adjacent peaks in the wave.* Wavelength is very important for our purposes because it relates to the energy of light. Wavelength is designated by the Greek letter $\lambda$ (lambda).

Let us return to the white light emitted from a tungsten filament in a lightbulb. We can analyze this light by passing it through a glass prism. Since all colors of light do not have the same velocity through a medium such as glass, the white light is separated by the prism into its component colors. (See Figure 5-1.) The white light separates into a continuous range of colors that we associate with a rainbow. *Since one color blends gradually into another, this is known as a **continuous spectrum.*** A rainbow after a rainstorm is a continuous spectrum caused by the separation of sunlight into its component colors by raindrops in the air.

Each color of light in a rainbow has a specific wavelength. Violet light, on one end of the spectrum, has the shortest wavelength; and red, on the other end of the spectrum, has the longest. Important to our consideration of light, however, is that the energy of light is inversely proportional to its wavelength. That is,

$$E \propto \frac{1}{\lambda} \qquad E = \frac{hc}{\lambda}$$

where $h$ is a constant of proportionality known as Planck's constant and $c$ is the velocity of light in a vacuum, which is also a constant.

Because of the inverse relationship, the larger the value of $\lambda$, the lower the energy. Thus red light, with the longest wavelength, has the lowest energy in the visible spectrum, whereas violet, with the smallest wavelength, has the highest energy. The visible part of

**Figure 5-1   The Spectrum of Incandescent Light**
**When a narrow beam of light from an ordinary lightbulb is passed through a glass prism, the white light is found to contain all of the colors of the visible spectrum.**

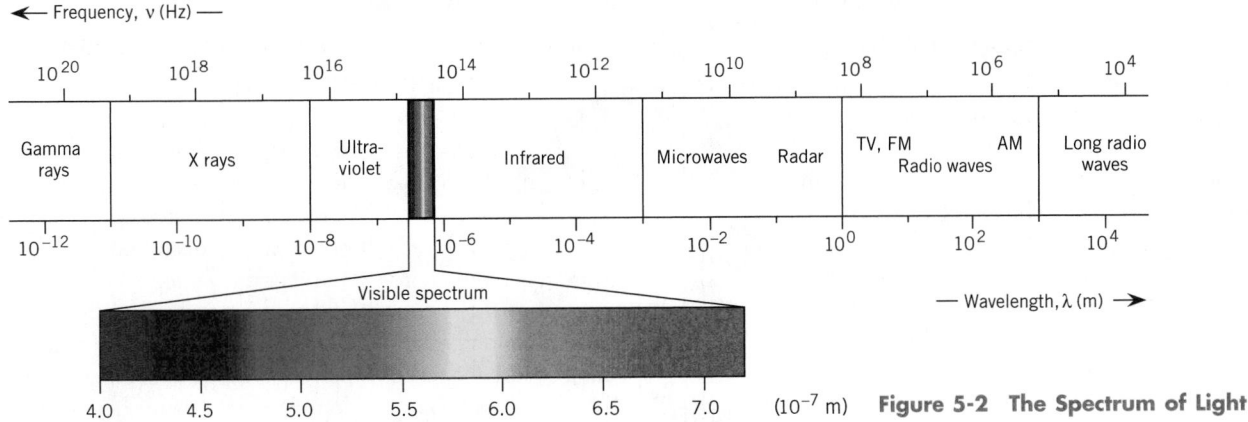

**Figure 5-2  The Spectrum of Light**

the spectrum has wavelengths between 400 nm (1 nm $= 10^{-9}$ m) and about 800 nm. The spectrum of electromagnetic radiation extends in both directions well beyond the wavelengths of visible light. Light that extends beyond violet (*ultraviolet light*) is not visible but is highly energetic. Indeed, it is powerful enough to damage living organisms. Our atmosphere shields us from the more powerful ultraviolet light from the sun. Light with extremely small wavelengths and thus extremely high energies includes X-rays and gamma rays. Only small amounts of this radiation can be tolerated by living systems. On the other side of the visible spectrum is light with wavelengths longer than red (*infrared light*) which extends into the long wavelengths of radio and TV waves. (See Figure 5-2.)

When an element is heated to where the hot gaseous atoms emit light, the results are different from the continuous spectrum emitted by the sun. Only definite, or discrete, colors are produced. Since the colors are discrete, the energies emitted from the atoms are discrete. The spectrum of each element (called its atomic emission spectrum) is unique to that element and is the reason specific elements are used in fireworks. (See Figure 5-3.) *Since only certain colors are produced by the atoms*

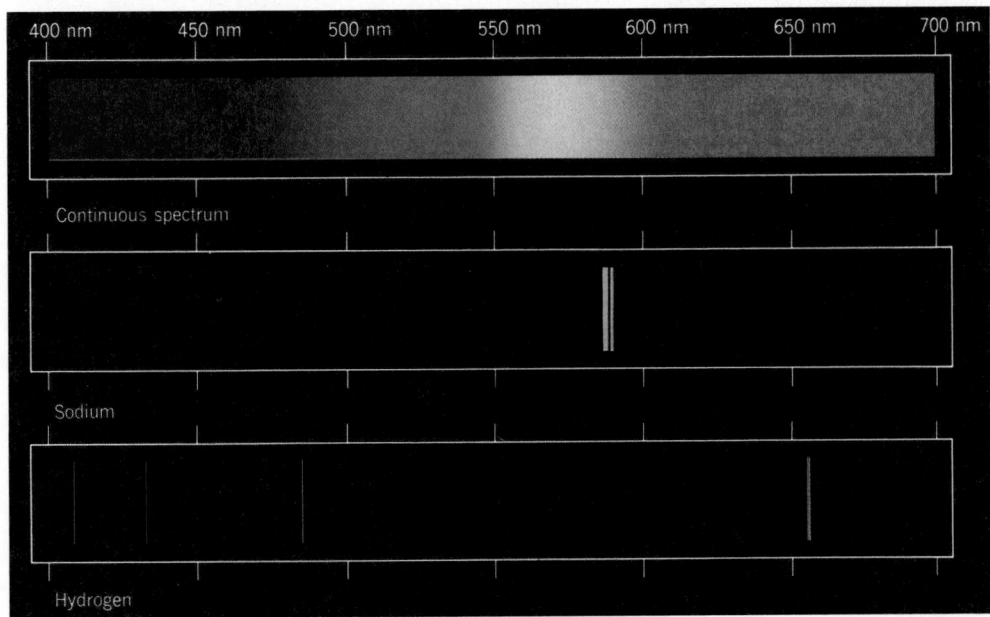

**Figure 5-3  The Continuous Spectrum and the Emission Spectra of Sodium and Hydrogen in the Visible Range** The light from an incandescent light (*top*) is continuous, whereas the atomic spectra of sodium (*middle*) and hydrogen (*bottom*) are composed of discrete colors.

### Roses Are Red; Violets Are Blue—But Why?

From the beauty of flowers and sky to the brightness of rubies and emeralds, we marvel at the colors of light that lie in the visible part of the spectrum (wavelengths 400 nm–800 nm). It was mentioned in this chapter that the sun emits the whole spectrum of colors of the rainbow. Blended together, sunlight appears nearly white. Fireworks emit discrete colors depending on the element that is heated in the explosion. But most of what our eyes receive is a result of reflected light rather than emitted light.

When white light is directed at a surface, the amount of reflection determines its shade. If all of the light is reflected, the surface appears white. If all of the light is absorbed, the surface appears black. (That is why light-colored clothing is preferred in the summer, since it reflects the light energy.) If part of the light is absorbed and part reflected, the surface appears as a shade of gray.

Most of the wonderful colors of our world result from absorption of specific regions of the light spectrum. There are three primary colors of light, as shown in the photograph. Red light is the highest third of the wavelengths of visible light (the lowest energies). Green is the middle third

of the wavelengths, and blue is the lower third of the wavelengths. A blend of all three appears white. It is only when one region of the spectrum is absorbed that we perceive color. A rose appears red because it absorbs the blue-green part of the spectrum and reflects the complement to these colors, which is red. If only the blue part is absorbed and both red and green are reflected, we see a yellow rose. A violet absorbs the red-green part of the spectrum and reflects blue. Leaves appear green because they absorb the red-blue part of the spectrum. Some of that absorbed light energy is converted into chemical energy in the production of sugars by *photosynthesis*.

Color television works on the same principle. Three different signals are shot through different primary-color filters and then transmitted to the TV set. The three signals are then blended back together to provide the proper color. Older sets had red, green, and blue drives that were manually adjusted.

We are lucky. Not all animals see color. Life for us would certainly be dull without the "wonderful world of color."

*The three primary colors.*

of hot gaseous elements, their atomic emission spectra are referred to as **discrete, or line, spectra.** In Figure 5-4, the spectrum of hydrogen in the visible range is shown along with the wavelengths of light of the four lines that appear. Energy is supplied to the hydrogen gas by an electrical discharge similar to that in the familiar neon light.

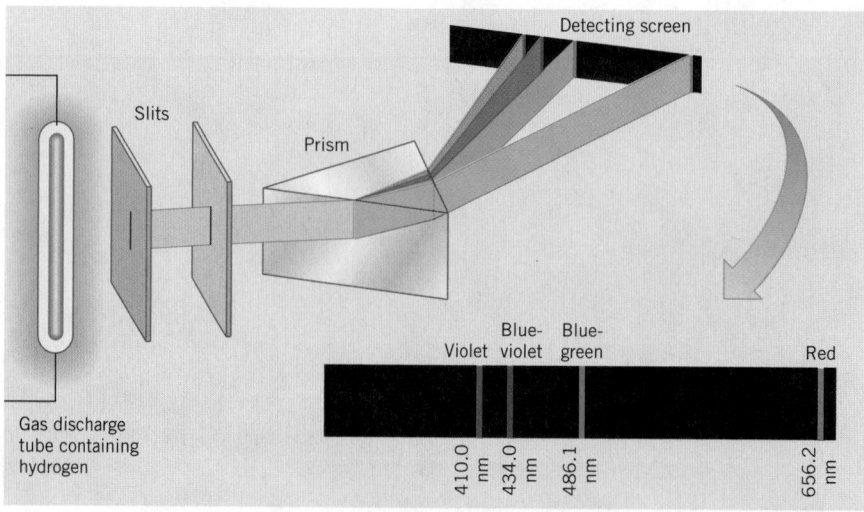

**Figure 5-4  Production of the Line Spectrum of Hydrogen**
The light from excited hydrogen atoms is passed through a glass prism. The prism divides the light into four discrete colors in the visible range.

# 5-2 A Model for the Electrons in the Atom

**Looking Ahead!** Hydrogen, the simplest of elements with one proton and one electron, has the most orderly atomic spectrum (Figure 5-4). The unique symmetry of the converging lines of the hydrogen spectrum made this the most interesting of the elements. We will see in this section how Bohr described the hydrogen atom so as to explain its discrete spectrum.

By 1913, the nuclear theory of the atom proposed by Lord Rutherford was generally accepted by scientists. In that year, Bohr set out to develop this theory further in order to explain the origin of the discrete spectrum of hydrogen. His main goal was to explain how the electron in the hydrogen atom was responsible for the energy emitted in the form of light. Bohr did not realize that his theory would lay the foundation for the explanation of other phenomena such as the periodicity of elements and the way atoms bond to each other.

First, Bohr proposed that the electron in the hydrogen atom revolves around the nucleus in a stable, circular orbit. The electrostatic attraction of the negative electron for the positive proton keeps the electron in a stable orbit. His theory of the hydrogen atom is analogous to the orbiting of the planets around the sun and thus is often referred to as a model. *A* **model** *is a description or analogy used to help visualize a phenomenon.* Bohr's model does not seem so revolutionary at first glance, but classical physics prohibits such a model. Classical physics, which was all that was known at the time, stated that a charged particle (the electron) orbiting around another charged particle (the proton) would lose energy and spin into the nucleus. Bohr side-stepped this problem by postulating (suggesting without proof as a necessary condition) that classical physics did not apply in the small dimensions of the atom. Eventually, he would be proved correct.

Second, Bohr suggested that there are several orbits that the electron may occupy. The orbits available to the electron are said to be **quantized**, *which means that they are at definite, or discrete, distances from the nucleus.* Since the energy of a particular orbits is a function of the distance from the nucleus, the energy of an electron in such an orbit is also quantized. Thus the *discrete orbits available to an electron in a hydrogen atom are referred to as* **energy levels.** (See Figure 5-5.) *Each energy level in the hydrogen atom is designated by an integral number known as the* **principal quantum number.** Thus the first energy level is designated $n = 1$, the second level $n = 2$, etc.

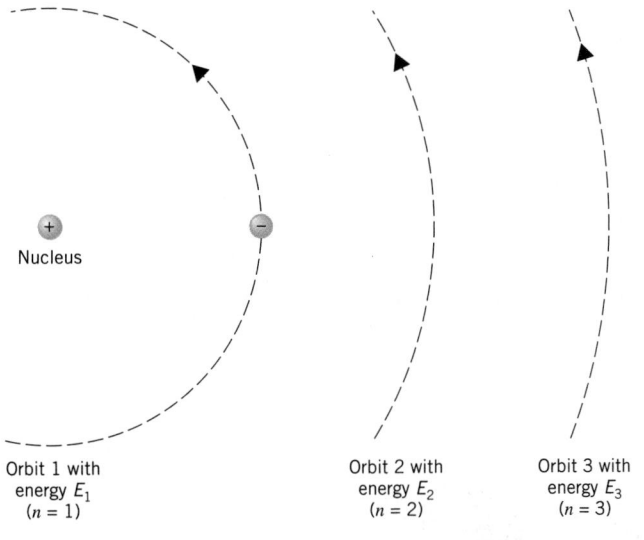

**Figure 5-5  Bohr's Model of the Hydrogen Atom**
**In this model, the electron can exist only in definite, or discrete, energy states.**

Nucleus

Orbit 1 with
energy $E_1$
($n = 1$)

Orbit 2 with
energy $E_2$
($n = 2$)

Orbit 3 with
energy $E_3$
($n = 3$)

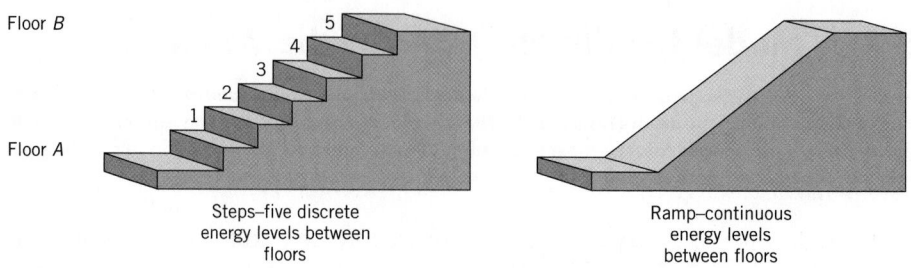

Floor B
Floor A

Steps–five discrete
energy levels between
floors

Ramp–continuous
energy levels
between floors

**Figure 5-6  Discrete Versus Continuous Energy Levels The steps represent discrete
energy levels; the ramp, continuous energy levels.**

The quantized energy levels in the hydrogen atom can be compared with a stairway in your home. In Figure 5-6 you can see that between floor A and floor B there are five steps that are analogous to energy levels in an atom. Since you cannot stand (with both feet together) between steps, we can say that each step represents a discrete, or *quantized,* amount of energy. On the other hand, a ramp between the two floors represents a *continuous,* change in energy. In this case, all energy levels are possible between floors.

Under normal conditions, *the single electron in a hydrogen atom will occupy the lowest energy level, which is the orbit closest to the nucleus. This is called the* **ground state.** When energy is supplied to a hydrogen atom, as when it is heated, the electron can absorb the appropriate amount of this energy to "jump" from the ground state (i.e., the $n = 1$ level) to a higher energy level (i.e., $n = 2, 3, 4$, etc.). Because of its new position in a higher energy level, the electron now has potential energy just like a weight suspended above the ground. *Energy levels higher than the ground state are called* **excited states.** What is most significant according to Bohr was what happened when the electron fell back down to lower excited states or all the way to the ground state. He suggested that the electron gives up this energy in the form of light when it falls back. Since energy levels are quantized, the difference in energy between any two levels is quantized. Thus, when an electron falls back to a lower energy level, it must emit a discrete amount of energy. Since this energy is emitted as light, the light would have a discrete energy, a discrete wavelength, and a discrete color (if the light is in the visible region of the spectrum). (See Figure 5-7).

**Figure 5-7  Light Emitted from a Hydrogen Atom**
**An electron in an excited state emits energy in the
form of light when it drops to a lower energy level.**

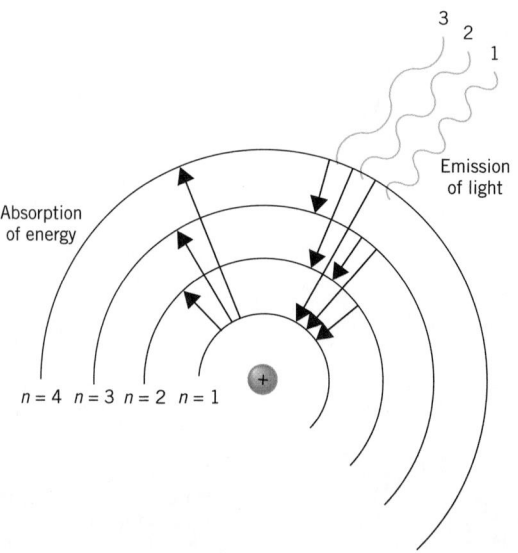

3  2  1

Emission
of light

Absorption
of energy

$n = 4$  $n = 3$  $n = 2$  $n = 1$

This is the qualitative explanation of the discrete spectrum. The real significance of Bohr's model is that he was able to calculate the expected wavelength of light in the hydrogen spectrum from the mathematical relationships. The experimental values listed in Figure 5-4 correspond beautifully with those computed by Bohr. Bohr's model worked specifically for the hydrogen atom. Calculations of the wavelengths of the lines in the discrete spectra of other elements had to await more sophisticated models.

## 5-3 Modern Atomic Theory: A Closer Look at Energy Levels

**Looking Ahead!** Because of the simplicity of Bohr's model of the hydrogen atom, modern scientists still use this picture in certain situations. In the modern view of the atom, however, the nature of the electron had to be modified. Since the characteristics of the electron seemed to be much more complex than Bohr's model could handle, a more sophisticated view of the electron evolved. In this section, we will examine the results of the modern view of the electron in the hydrogen atom.

Soon after Bohr's model was presented, it was realized that the electron in an atom was much more complex than Bohr had suggested. Experiments confirmed that the electron has properties both of a particle (mass) *and* of light (wave nature). Because of its dual nature, the electron could not be viewed as a simple particle circling the nucleus at a definite distance. Also, if the electron moved with a high velocity, as Bohr had claimed, one could not know its location with much certainty. The motion of a rapidly moving electron would be analogous to a fastball thrown by a professional baseball pitcher. The faster the ball is thrown, the less the batter knows about its location and the less likely that the ball will be hit. *The wave nature of the electron and the uncertainty of its location led to a complex mathematical approach to the electron in the hydrogen atom, known as* **wave mechanics.** Using the equations of the wave mechanical model, only the probability of finding the electron in a given region of space at a certain instant could be determined. Bohr calculated the exact radius of the ground state energy level for the electron in hydrogen. In the wave mechanical model,

Electrons have a wave nature, much like waves moving across a pond.

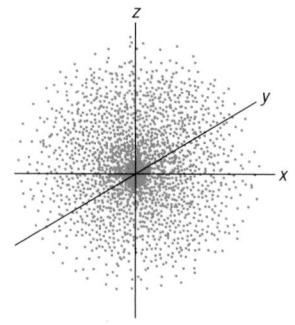

**Figure 5-8 Electron Density**
**The electron density of the ground state of hydrogen is spherical.**

this radius has meaning but represents the highest *probability* of finding the electron. We no longer view the electron as orbiting a certain path around the nucleus. In fact, the motion of an electron in an atom in this model is not understood.

Locating the electron in the vicinity of the nucleus is like trying to locate a fast-moving dancer on a stage with a strobe light. With each flash of the strobe, we find the dancer in a different location. However, if we marked the dancer's position at each flash, eventually we would probably see a pattern and notice that she has a higher probability of being in certain locations on the floor than others. For example, the exact center of the stage has the highest probability. A similar situation would occur if we were to try to measure the position of the electron in a hydrogen atom. That is, each time that we would locate the electron it would most likely be in a different location than before. However, if we took enough measurements, a pattern would eventually develop. For example, in Figure 5-8 a typical pattern of electron probability is shown. The figure includes *x, y,* and *z* coordinates to emphasize the three-dimensional nature of the points. Notice that this pattern suggests that the probability of finding the electron, known as the *electron density,* is in a spherical region of space around the nucleus.

*The region of space where there is the significant probability of finding a particular electron is known as an* **orbital.** (Note that this is a different concept than Bohr's orbits.) An orbital is sometimes referred to as "an electron cloud," as if the electron were spread out in a volume of space like a cloud. An orbital is analogous to a dorm room, which is a region of space with the highest probability of finding a particular student. That is, the student spends more time there than in any other single place. It is difficult to represent an orbital either as a huge number of dots or as a cloud, so, for convenience, we illustrate the shape of this orbital as a solid sphere, which is shown in Figure 5-9*a*. This sphere is understood to represent a volume of space containing about 90% of the electron density. (We do not envision the electron as moving around on the surface of this sphere.) *A spherical volume of probability (the electron cloud) is known as an* **s orbital.** The energy levels or orbits that Bohr described have meaning in our modern approach but are also known as **shells.** We will now refer to the energy levels as shells, which may contain one or more orbitals.

The first shell in hydrogen (*n* = 1), which is known as the ground state of hydrogen because it has the lowest energy, contains only the one spherical *s* orbital. Normally, that's the state occupied by the hydrogen's single electron. Now let's look at the first excited state of hydrogen, which is the second shell (i.e., *n* = 2). The second shell is normally unoccupied, just like a vacant room on the second floor of a home. We find in this shell that there are two kinds of orbitals. First, there is a spherical orbital (an *s* orbital) much like the one in the first level except that its maximum electron density (probability) is farther from the nucleus than the *s* orbital in the first shell. (See Figure 5-9*b*.) The second kind of orbital present

**Figure 5-9 s Orbitals**
**The s orbitals are represented as spheres. The 2s orbital (b) extends farther from the nucleus than the 1s orbital (a).**

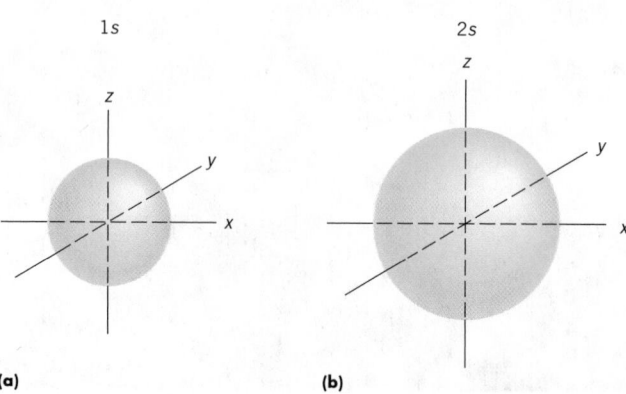

(a)                    (b)

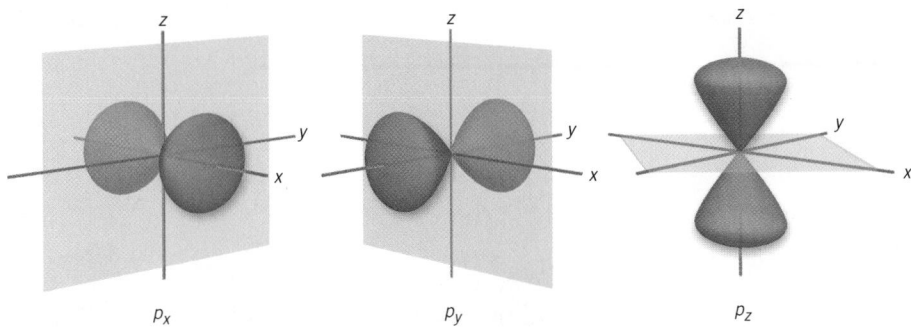

$p_x$　　　　$p_y$　　　　$p_z$

**Figure 5-10**　*p* **Orbitals Each *p* orbital has two lobes.**

in the second shell is known as a ***p* orbital** *and has two regions of high probability called "lobes" on either side of the nucleus.* The electron distribution in a *p* orbital is not spherical but is shaped much like a weird baseball bat with two fat ends. The difference between an *s* and a *p* orbital is analogous to dorm rooms that have different dimensions or shapes. The first-floor rooms may be perfectly square but the second-floor rooms may be rectangular. In fact, there are three distinct *p* orbitals lying at 90° angles to each other. In a three-coordinate graph, if we define *x* as the axis along which the two lobes of a *p* orbital are directed, we can designate that *p* orbital as the $p_x$ orbital. The other two *p* orbitals are each at 90° angles to the $p_x$ orbital and each other. They are referred to as the $p_y$ and the $p_z$ orbitals. The surfaces that represent about 90% of the electron density of the *p* orbitals are shown in Figure 5-10. Again, we do not visualize the electron as a particle moving about the surface of this orbital representation. Thus the second shell has four regions of space that may contain electrons—one *s* orbital and three *p* orbitals. *The orbitals of the same type in each shell make up what is referred to as a* **subshell.** A subshell is labeled with a number corresponding to the shell (1, 2, 3, etc.) and by the type of orbital that makes up that subshell (*s, p,* etc.). Thus the first shell has only a single *s* subshell and is called the 1*s*. The second shell contains an *s* subshell (the 2*s*) and the 2*p* subshell, which is made up of three individual *p* orbitals.

Now let's continue outward in the hydrogen atom and consider the third shell (i.e., *n* = 3). Each successive shell has one additional type of orbital. Thus, the third shell has three types of orbitals, or three subshells. Like the first two, it has a spherical 3*s* orbital and like the second level it has three different 3*p* orbitals. The third type of orbital found in the third shell is known as a ***d* orbital.** There are five *d* orbitals and their shape is more complex. Most of them have four lobes, as illustrated in Figure 5-11. Three of these *d* orbitals have significant electron density between the axes ($d_{xy}$, $d_{xz}$, $d_{yz}$) and two have significant density along the axes

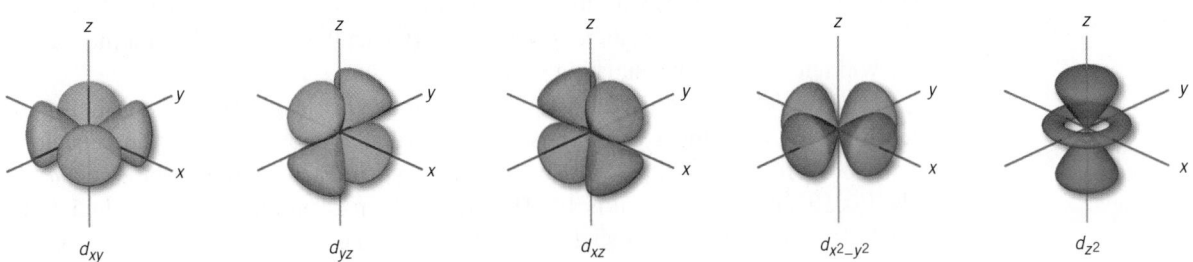

$d_{xy}$　　　　$d_{yz}$　　　　$d_{xz}$　　　　$d_{x2-y2}$　　　　$d_{z2}$

**Figure 5-11**　*d* **Orbitals Except for the *$d_{z^2}$*, the *d* orbitals have four lobes.**

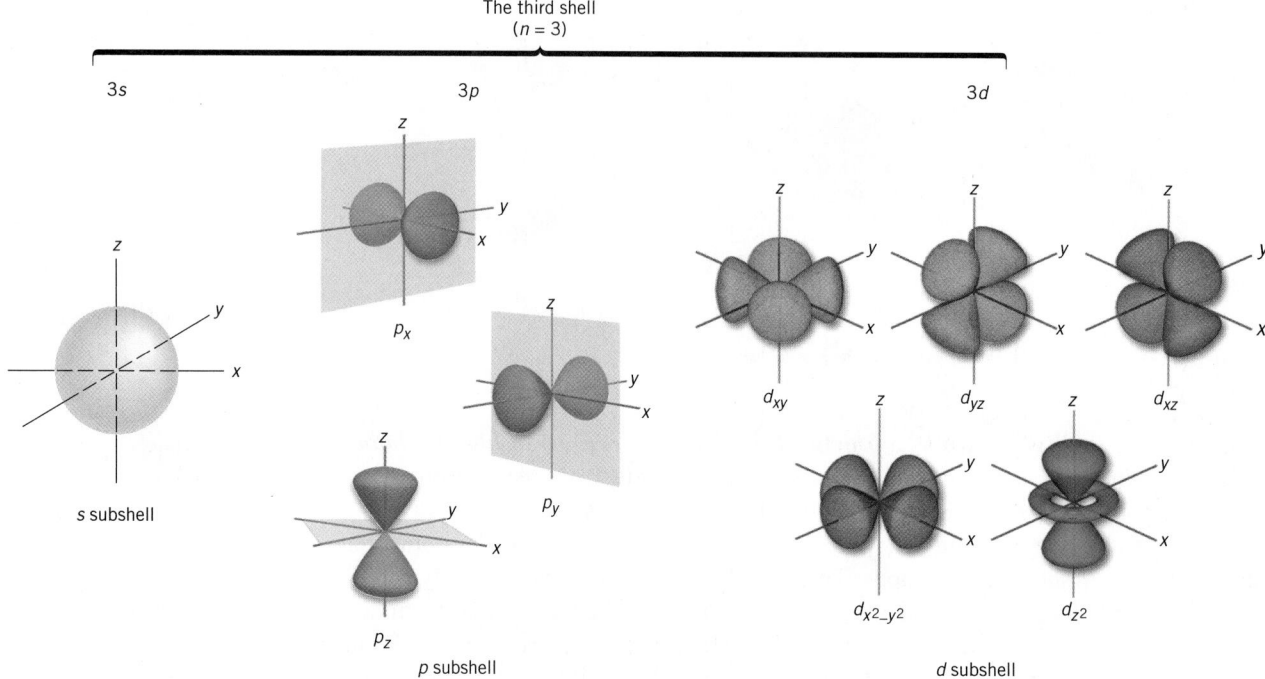

Figure 5-12  **The Third Shell** There are three subshells containing a total of nine orbitals in the third shell.

($d_{x^2-y^2}$, $d_{z^2}$). The third shell with its three subshells and nine orbitals is summarized in Figure 5-12.

The fourth shell ($n = 4$) has all of the same types of orbitals as the third shell (i.e., one 4s, three 4p, and five 4d orbitals) but has one additional type known as an **f orbital.** The shapes of the seven 4f orbitals are even more complex. The shapes of these orbitals are of no consequence to us since electrons occupying these orbitals are not involved in bonding, as we will see in the next chapter. The subshells present in the first four shells are shown in Figure 5-13. Just as successive shells have higher energy, successive subshells also have higher energy. (All of the orbitals in a specific subshell have the same energy.) In atoms with more than one electron, we now find that the energy of the subshells *within a shell* is not the same and increases in the order

$$s < p < d < f$$

increasing energy

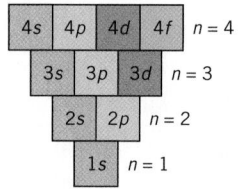

**Figure 5-13  Subshells**
Each successive shell has one additional subshell.

We can now adjust our representation of the shells and subshells shown to the left to account for the difference in energy of the subshells within a shell. In Figure 5-14 we show the first four shells with their subshells in order of increasing energy. Notice that the 4s energy level appears to be lower in energy than the 3d. This is not a mistake, but we will refer back to this situation in the next section.

*Each orbital has a capacity of two electrons.* The capacity of a subshell is determined by the number of orbitals in that particular subshell. Thus the capacity of an s subshell is 2, a p subshell is 6 (two electrons in each of three orbitals), a d subshell is 10, and an f subshell is 14. This information is summarized in Table 5-1. The total electron capacity of a shell, including all of the orbitals in all of the subshells, works out to be conveniently equal to $2n^2$. Thus the first shell ($n = 1$) holds 2 electrons, the second shell ($n = 2$) holds 8, the third 18, and so on.

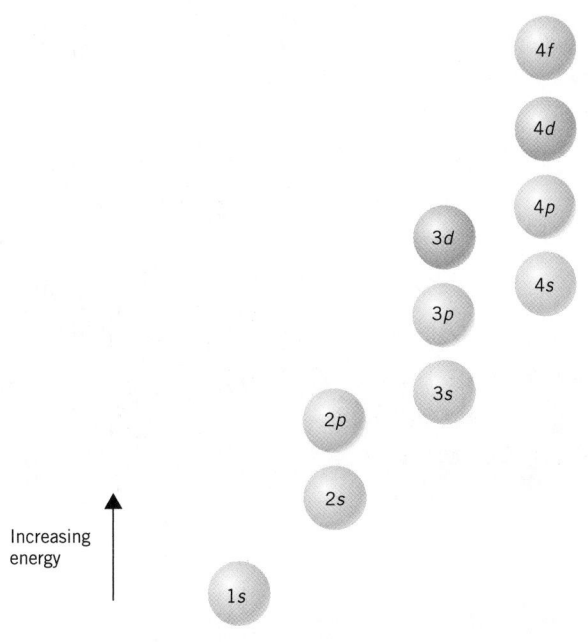

**Figure 5-14  Energy of the Subshells**
**The subshells within a shell have different energy.**

Let's consider an analogy to shells and subshells. Consider a dormitory on campus as analogous to a shell and the floors within that dorm as analogous to subshells. We will assume that there are four dorms, with each dorm located farther and farther up a hill. The dorm lowest on the hill is the lowest in energy since it is the easiest to reach. In Figure 5-15 the dorms are labeled 1 through 4. As we go to higher energy dorms up the hill, they become larger, with each having one additional floor. (Recall that each successive shell has one additional subshell.) Just as the dorms are at different energy levels, so are the floors within a dorm. (In real-world dorms, students are quite aware of the difference in energy of the floors since the elevator is often out of order or wedged open somewhere.) Obviously, the lower the floor in a specific dorm, the lower the energy. In summary, dorms are analogous to shells which represent different energy levels and distances from the nucleus in an atom. Floors within a dorm are analogous to subshells which represent different energy levels within the shell.

---

**Looking Back!** The relationship between the light emitted by hot gaseous atoms and the electrons in the atom was the key that unlocked some hard-core mysteries of the elements. Niels Bohr was responsible for bringing the word "quantum" into frequent usage, at least in the scientific world. After Bohr's model came wave mechanics and the description of the electron as existing in an orbital. Each of the four types of orbitals has a different shape for the region of space of high probability. All of the orbitals of one type in a particular energy level, known as a shell, constitute a subshell. Shells and subshells within a shell can be arranged in order of increasing energy.

---

### Table 5-1  Shells and Subshells

| Shell | 1 | 2 | | 3 | | | 4 | | | |
|---|---|---|---|---|---|---|---|---|---|---|
| Subshell | s | s | p | s | p | d | s | p | d | f |
| Subshell capacity | 2 | 2 | 6 | 2 | 6 | 10 | 2 | 6 | 10 | 14 |
| Shell capacity | 2 | | 8 | | 18 | | | 32 | | |

**Figure 5-15  Floors Within a Dorm**
The floors represent different subdivisions of energy within each dorm, analogous to the subshells within each shell.

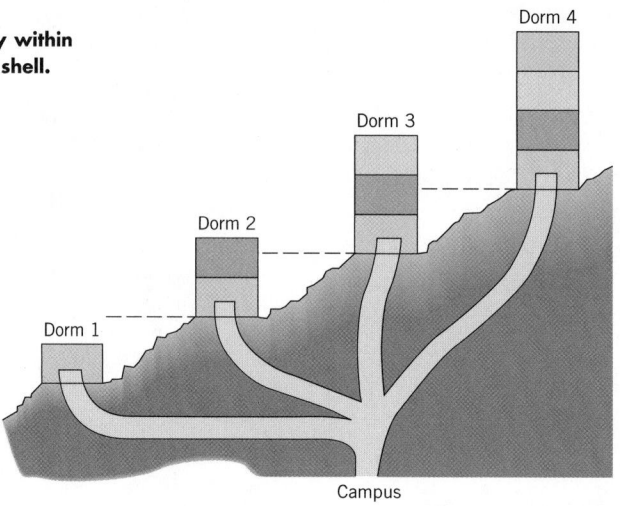

## Learning Check A | checking it out

**A-1.**  Fill in the blanks.

Light has a _____ nature. The wavelength of light is _____ proportional to the energy. Therefore, in the visible spectrum, red light has the _____ wavelength and the _____ energy. When the hot gaseous atoms of an element emit light, it is called a _____ spectrum. Niels Bohr suggested that the orbits of an electron were _____ . The lowest energy state is the _____ state, and the higher states are the _____ states. Wave mechanics tell us that the electron exists in a region of space known as an _____ . The four types of these regions are labeled _____ , _____ , _____ , and _____ . The lowest in energy of these four has a _____ shape. A principal energy level, called a _____ , can be subdivided into _____ . Such a subdivision contains all of the same type of _____ in that particular energy level.

**A-2.**  Consider the $n = 3$ shell.

**a.**  What is its capacity?
**b.**  What subshells are present?
**c.**  What are the capacities of each subshell?
**d.**  How many orbitals are present in each subshell?

*Additional Examples: Exercises 5-3, 5-13, 5-18, 5-19, and 5-20.*

## Section B  THE DISTRIBUTION OF ELECTRONS IN ATOMS

## 5-4 Electron Configurations of the Elements

**Looking Ahead!** We are now ready to apply the information about shells, subshells, and orbitals to the electrons in atoms of all of the elements. By assigning electrons to specific shells and subshells, the basis of the periodic table becomes apparent.

In our analogy to the dorm, an orbital is analogous to a room on a floor. A floor composed of one or more rooms is analogous to a subshell composed of one or more orbitals. Assigning electrons to orbitals in an atom is much like assigning students to rooms in a dormitory. Students will occupy the available rooms on the floors that require the least energy to get to. *In the case of atoms, electrons occupy the available orbitals in the subshells of lowest energy.* This is known as the **Aufbau principle.** *The assignment of all of the electrons in an atom into specific shells and subshells is known as the element's* **electron configuration.** In what follows in this section, we will consider only the electron's shell and subshell designations. In the next section, we will also consider the orbital assignment. In the dorm analogy, it is like considering only the dorm and floor assignment but not the specific room on a floor.

A shell is designated by the principal quantum number ($n$), the subshell by the appropriate letter, and the number of electrons in that subshell by the appropriate superscript number. For example, the existence of three electrons in the $p$ subshell of the fourth shell ($n = 4$) is shown as follows:

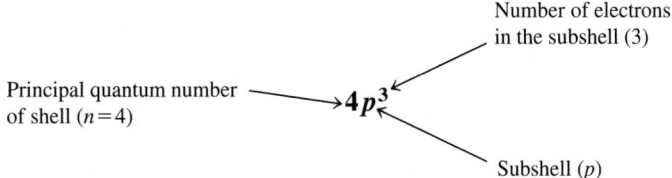

We begin with the ground state of the one electron in the simplest and first element, hydrogen. The electron is in the $n = 1$ shell, which has an $s$ subshell. The electron configuration for H is

$$1s^1$$

The next element, helium, has two electrons. The $1s$ subshell has a capacity of two electrons, so He has the configuration

$$1s^2$$

Next comes Li with three electrons. The first two electrons fill the first shell, but the third is assigned to the second shell ($n = 2$). The second shell has two subshells, the $s$ and the $p$, but the $s$ is lower in energy, so it fills first. The next element is Be. Its four electrons are also assigned to the $1s$ and $2s$ subshells. The electron configurations of Li and Be are

$$\text{Li:} \quad 1s^2 2s^1 \qquad \text{Be:} \quad 1s^2 2s^2$$

The next element is B. Four of its five electrons have the same configuration as Be, but the fifth electron begins the filling of the next subshell, the $2p$. The next five elements, C through Ne, complete the filling of the $2p$ subshell.

$$\text{B:} \quad 1s^2 2s^2 2p^1 \qquad \text{N:} \quad 1s^2 2s^2 2p^3 \qquad \text{F:} \quad 1s^2 2s^2 2p^5$$
$$\text{C:} \quad 1s^2 2s^2 2p^2 \qquad \text{O:} \quad 1s^2 2s^2 2p^4 \qquad \text{Ne:} \quad 1s^2 2s^2 2p^6$$

With the element neon, all of the orbitals in the second shell are full. We now continue with the element sodium, which has 11 electrons. The first 10 electrons fill the first and second shells, so the 11th electron is assigned to the third shell. The third shell has three subshells, the $3s$, $3p$, and $3d$. The lowest in energy is the $3s$ subshell, so the 11th electron is in the $3s$ subshell. At this point, it becomes somewhat tedious to write all of the filled subshells. A shorthand method of writing configurations is to represent all of the filled subshells of a noble gas by the symbol of that noble gas in brackets (e.g., [Ne]). Thus, in the following electron configurations, we will assume

$$[\text{Ne}] = 1s^2 2s^2 2p^6$$

We chose a noble gas to use in our shorthand notation for a good reason. As we will see in the next chapter, electrons in noble gas cores are not involved in bonding, so we can treat them as a group. Using the [Ne] core symbolism, the electron configurations of the next eight elements after neon are shown as follows.

$$Na: [Ne]3s^1 \qquad P: [Ne]3s^23p^3$$
$$Mg: [Ne]3s^2 \qquad S: [Ne]3s^23p^4$$
$$Al: [Ne]3s^23p^1 \qquad Cl: [Ne]3s^23p^5$$
$$Si: [Ne]3s^23p^2 \qquad Ar: [Ne]3s^23p^6$$

Notice that the electron configuration of Li is similar to that of Na (i.e., $[He]2s^1$ and $[Ne]3s^1$), Be to Mg, B to Al, and so forth. Notice also that these pairs of elements are in the same groups in the periodic table. This is very significant. In fact, we can make a general statement for this observation. *The electron configuration of the elements is a periodic property.* Elements in the same group have the same outer subshell electron configurations in successively higher shells. These similar electron configurations are a major reason that the elements in a group have similar properties. The existence of subshells composed of orbitals is a result of our understanding of wave mechanics. By placing electrons in these subshells, we have actually developed the theoretical basis for the existence of the periodic table. Now, however, let us continue with the electron configurations of elements beyond argon. Again, we will use a noble gas core shorthand with [Ar] representing the 18 electrons in argon.

The next element after Ar presents a problem. The third subshell in the third shell ($3d$) is still available, so at first we may be inclined to assign the 19th electron in K to the $3d$ subshell. However, notice that K is under Na and Li in the periodic table. The latter two elements have their last electron in an $s$ subshell. If the basis of the periodic table is correct, then the location of K (under Na) indicates that it should have its last electron in the $4s$ subshell. This is indeed the case. Using noble gas shorthand notation, the configuration of K is

$$[Ar]4s^1$$

To understand how the $4s$ fills before the $3d$, let us return to the analogy of the dorms as shown in Figure 5-15 Although dorm 4 is at a higher level than dorm 3, note that not all floors in dorm 3 are lower than those in dorm 4. In fact, it is easier to proceed farther up the hill to occupy the lowest floor in dorm 4 than to go all the way up to the third floor in dorm 3. In the case of the assignment of the electrons in atoms, a similar phenomenon is true. As shown in Figure 5-14, the $4s$ subshell is lower in energy than the $3d$ subshell; thus the $4s$ fills first. After two electrons are assigned to the $4s$, the $3d$ begins filling with the element Sc. The next nine elements after Sc also involve the filling of the $3d$ subshell. After the $3d$ is filled at Zn, the next higher energy subshell is the $4p$. This subshell is completely filled at the next noble gas, Kr.

$$K: [Ar]4s^1 \qquad Ga: [Ar]4s^23d^{10}4p^1$$
$$Ca: [Ar]4s^2 \qquad \vdots$$
$$Sc: [Ar]4s^23d^1 \qquad Kr: [Ar]4s^23d^{10}4p^6$$
$$\vdots$$
$$Zn: [Ar]4s^23d^{10}$$

At this point, the order of filling of subshells becomes more variable. A scheme may be helpful. In Figure 5-16, the subshells in each shell are written starting with the $1s$ at the top. All of the $s$ subshells are written in a vertical column, as are the other types of subshells. A stairstep line is then drawn on the right. A diagonal arrow is inserted through each corner. The top arrow points to the first subshell filled (the $1s$). The second points to the second subshell filled (the $2s$). The third points to the

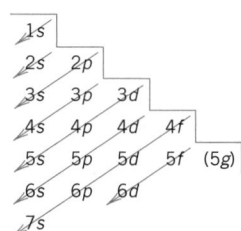

**Figure 5-16  Order of Filling of Subshells**

**The arrows indicate the normal order of filling of subshells.**

next two (the $2p$ followed by the $3s$), The fourth points to the next two (the $3p$ followed by the $4s$). The next arrow tells us that the order of filling of the next three subshells is $3d$ followed by $4p$ followed by $5s$. The scheme continues in a similar manner for subshells of higher shells.

Unfortunately, the electron configurations of many elements are not always exactly as we would predict by following the usual order of filling. For example, there are two notable exceptions among the transition elements when the $d$ subshells are filling. The actual electron configurations of chromium (#24) and copper (#29) are

$$\text{Cr: } [\text{Ar}]4s^13d^5 \qquad \text{Cu: } [\text{Ar}]4s^13d^{10}$$

If the normal order had been followed, we would predict the configuration of Cr to be $[\text{Ar}]4s^23d^4$ and Cu to be $[\text{Ar}]4s^23d^9$. These exceptions are explained on the basis of a particular stability for a half-filled and a completely filled subshell (the $3d$).

There are many other exceptions, especially where $d$ and $f$ subshells are concerned. *We should be aware that such exceptions exist, but it is not important or necessary to know them all.*

Notice that Figure 5-16 predicts the filling of the $4f$ before the $5d$. However, La and Ac do not follow this order. The 57th electron of La and the 89th electron of Ac go into the $5d$ and $6d$ subshells, respectively, rather than the $4f$ and $5f$. Their correct configurations are

$$\text{La: } [\text{Xe}]6s^25d^1 \qquad \text{Ac: } [\text{Rn}]7s^26d^1$$

Thus these two elements are listed in the periodic table under Sc and Y, which also have one electron in a $d$ subshell. The 58th electron of Ce and the 90th electron of Th, however, reside in the $4f$ and $5f$ subshells, respectively. These two electron configurations are

$$\text{Ce: } [\text{Xe}]6s^25d^14f^1 \qquad \text{Th: } [\text{Rn}]7s^26d^15f^1$$

**The Electron Configuration of Elements**

**Example 5-1**

Write the electron configuration of (a) iron and (b) bismuth

working it out

**Solution**

**a.** Iron
From the table inside the front cover, we find that iron has an atomic number of 26, which means that the neutral atom has 26 electrons. Write the subshells in the order of filling plus a running summation of the total number of electrons involved with respect to iron. Use Figure 5-16 to determine the order of filling.

| Subshell | Number of Electrons in Subshell | Total Number of Electrons |
|---|---|---|
| 1s | 2 | 2 |
| 2s | 2 | 4 |
| 2p | 6 | 10 |
| 3s | 2 | 12 |
| 3p | 6 | 18 [Ar] |
| 4s | 2 | 20 |
| 3d | 6 | 26 |

Iron has 6 electrons past the filled $4s$ subshell, but they do not completely fill the $3d$ subshell, which could accommodate 10 electrons. The complete electron configuration of iron is

$$1s^2 2s^2 2p^6 3s^2 3p^6 4s^2 3d^6$$

or, if we use noble gas shorthand,

$$[Ar]4s^2 3d^6$$

**b.** Bismuth

Bismuth is element number 83, which means that we must assign 83 electrons to shells and subshells. We can continue the table from part (a), starting at the $3d$ subshell

| Subshell | Number of Electrons in Subshell | Total Number of Electrons |
|---|---|---|
| $3d$ | 10 | 30 |
| $4p$ | 6 | 36 [Kr] |
| $5s$ | 2 | 38 |
| $4d$ | 10 | 48 |
| $5p$ | 6 | 54 [Xe] |
| $6s$ | 2 | 56 |
| $4f$ | 14 | 70 |
| $5d$ | 10 | 80 |
| $6p$ | 3 | 83 |

The electron configuration of bismuth using the noble gas shorthand that begins with the noble gas xenon (with 54 electrons) is

$$[Xe]6s^2 4f^{14} 5d^{10} 6p^3$$

# 5-5 Orbital Diagrams of the Elements (Optional)

**Looking Ahead!** The energy of an electron in an atom is determined by its shell and subshell. The particular orbitals within a subshell all have the same energy. However, the distribution of electrons in the orbitals of a particular subshell accounts for certain properties of elements. Although these properties are beyond the scope of this text, we can still appreciate how this distribution occurs.

Subshells are composed of one or more orbitals, which are regions of space in which electrons reside. Each orbital can hold a maximum of two electrons. The orbitals in a subshell are analogous to the rooms on a certain floor. The dorm and the floor within the dorm are analogous to the shell and subshell. (See Figure 5-17.) In the following scheme, we will represent an individual orbital by a box. Therefore, an $s$ subshell will have one box, and the three orbitals in a $p$ subshell will be represented by three boxes. Likewise, a $d$ subshell will be represented by five boxes and an $f$ subshell, by seven. Individual electrons will be represented by arrows.

As mentioned earlier, an electron has a dual nature. In some respects it has properties of a wave, and in other respects it has properties of a particle. One

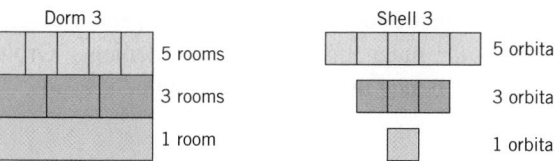

**Figure 5-17 Dorm Rooms and Orbitals**
The rooms represent different regions of space on each floor of a dorm, analogous to the orbitals of each subshell.

particle property is that the electron behaves like a charged particle spinning in either a clockwise or a counterclockwise direction. A spinning charged particle is like a tiny magnet. What if there are two electrons in the same orbital? The **Pauli exclusion principle** *states that no two electrons in the same orbital can have the same spin.*

We will represent the electrons in orbitals by means of orbital diagrams. *The* **orbital diagram** *of an element represents the orbitals in a subshell as boxes and its electrons as arrows. The spin of an electron is indicated by the direction of the arrow pointing either up or down.* Two electrons with opposite spins in the same orbital are said to be *paired*. Thus a doubly occupied orbital is represented as follows:

We will now expand on the electron configuration of the first five elements by including their orbital diagrams.

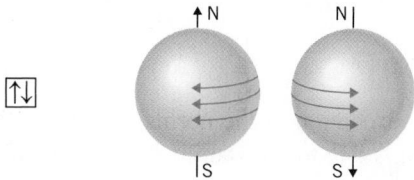

Before going on, we need to consider the placement of the sixth electron in carbon. There are three possibilities. Does it pair with the first $2p$ electron in the same orbital, have opposite spin in a different orbital, or go into a different orbital with the same spin? We have one more rule to guide us. **Hund's rule** *states that electrons occupy separate orbitals in the same subshell with parallel spins.* At least part of this rule is understandable. Since electrons have the same charge, they will repel each other to different regions of space. Electrons "want their space," so they prefer separate orbitals rather than pairing in the same orbital. Pairing occurs when separate empty orbitals in the same subshell are not available. With Hund's rule in mind, we can now write the orbital diagrams of the next five elements.

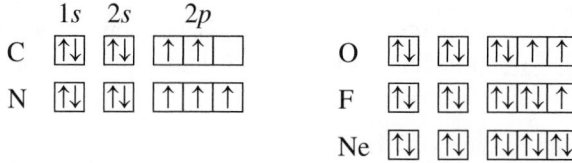

A similar phenomenon occurs with elements that have electrons occupying $d$ or $f$ orbitals. For example, Mn (#25) has the following electron configuration and orbital diagram.

$$\text{Mn: } [\text{Ar}]4s^2 3d^5 \qquad [\text{Ar}] \; \boxed{\uparrow\downarrow} \quad \boxed{\uparrow\,|\,\uparrow\,|\,\uparrow\,|\,\uparrow\,|\,\uparrow}$$

Although we will not pursue the topic in this text, orbital diagrams give us important information about the magnetic properties of elements. Orbital diagrams are also relevant to the types of bonds that a particular element forms.

| Example 5-2 | The Electron Configurations and Orbital Diagrams |
|---|---|
| working it out | Determine the electron configuration and write the orbital diagram of the electrons beyond the noble gas core for the elements S, Cr, and As. |

**Solution**

First, use the table inside the front cover to find the atomic number, which equals the number of electrons in a neutral atom. Then, use Figure 5-16 to determine the electron configuration of the element as in Example 5-1. Finally, show the orbital diagram for the outer orbitals beyond the noble gas core. Apply the Pauli principle and Hund's rule in positioning the arrows.

S: $[Ne]3s^2 3p^4$    [Ne] orbital diagram with $3s$ $[\uparrow\downarrow]$ and $3p$ $[\uparrow\downarrow][\uparrow][\uparrow]$

Cr: $[Ar]4s^1 3d^5$    [Ar] orbital diagram with $4s$ $[\uparrow]$ and $3d$ $[\uparrow][\uparrow][\uparrow][\uparrow][\uparrow]$

Remember that Cr is an exception to the normal order of filling; it has six unpaired electrons. This exception is due to the stability of the half-filled $3d$ subshell.

As: $[Ar]4s^2 3d^{10} 4p^3$    [Ar] orbital diagram with $4s$ $[\uparrow\downarrow]$, $3d$ $[\uparrow\downarrow][\uparrow\downarrow][\uparrow\downarrow][\uparrow\downarrow][\uparrow\downarrow]$, and $4p$ $[\uparrow][\uparrow][\uparrow]$

---

**Looking Back!** The modern view of the hydrogen atom led us to be able to assign electrons in any atom to a specific shell and subshell where the lowest energy level fills first (i.e., the Aufbau principle). Using this method, we see that the electron configurations directly relate to the periodic table. We will eventually see how an understanding of electron configurations leads us to other key periodic trends and properties. Following the Pauli principle and Hund's rule, we can further define the atom by assigning electrons to orbitals in a subshell. Electrons in orbitals are illustrated with orbital diagrams.

---

## Learning Check B | checking it out

**B-1.** Fill in the blanks.

The electron configuration $5d^5$ indicates that there are _____ electrons in the _____ subshell in the _____ shell. "Electrons go into the lowest energy subshell" is a statement of the _____ principle. "Electrons in the same orbital must be paired" is a statement of the _____ principle. "Electrons go into separate orbitals with parallel spins" is a statement of _____ _____ .

**B-2.** What are the electron configurations of the elements Ni (#28), Sr (#38), and At (#85)? Use the noble gas shorthand notations.

**B-3.** Show the electron configuration and orbital diagram of all of the electrons beyond the previous noble gas for Nb (#41).

*Additional Examples: Exercises 5-26, 5-27, 5-31, 5-33, and 5-34.*

## Section C  PERIODIC PROPERTIES

# 5-6 Electron Configuration and the Periodic Table

**Looking Ahead!** In this section, we will classify the elements into various groupings based on common characteristics of their electron configurations. As we come to appreciate the information from such classifications, we will be able to use the table for electron configurations rather than a scheme such as in Figure 5-16.

In Chapter 4, we first presented the periodic table as a valuable tool in locating periods and groups, gases and solids, nonmetals and metals. We have now developed the theoretical basis of the periodic table with the concept of electron configuration. Specifically, we found that groups of elements have the same subshell configuration but consecutive shell assignments. We can now put this understanding to work for us and let the periodic table itself direct us to the electron configuration of any element. In Figure 5-18, we have shown the periodic table with the subshell configuration common to a group at the top of each vertical column. The specific subshell representation is shown within each period. The electron configurations of the subshell that is in the process of filling are shown in more detail for the fourth period from K to Ar. With some practice with a periodic table along with Figure 5-18, we can predict the outer electron configuration of any element.

As we will see in the next chapter, the electrons beyond the previous noble gas core are the ones commonly involved in the formation of chemical bonds. To focus on

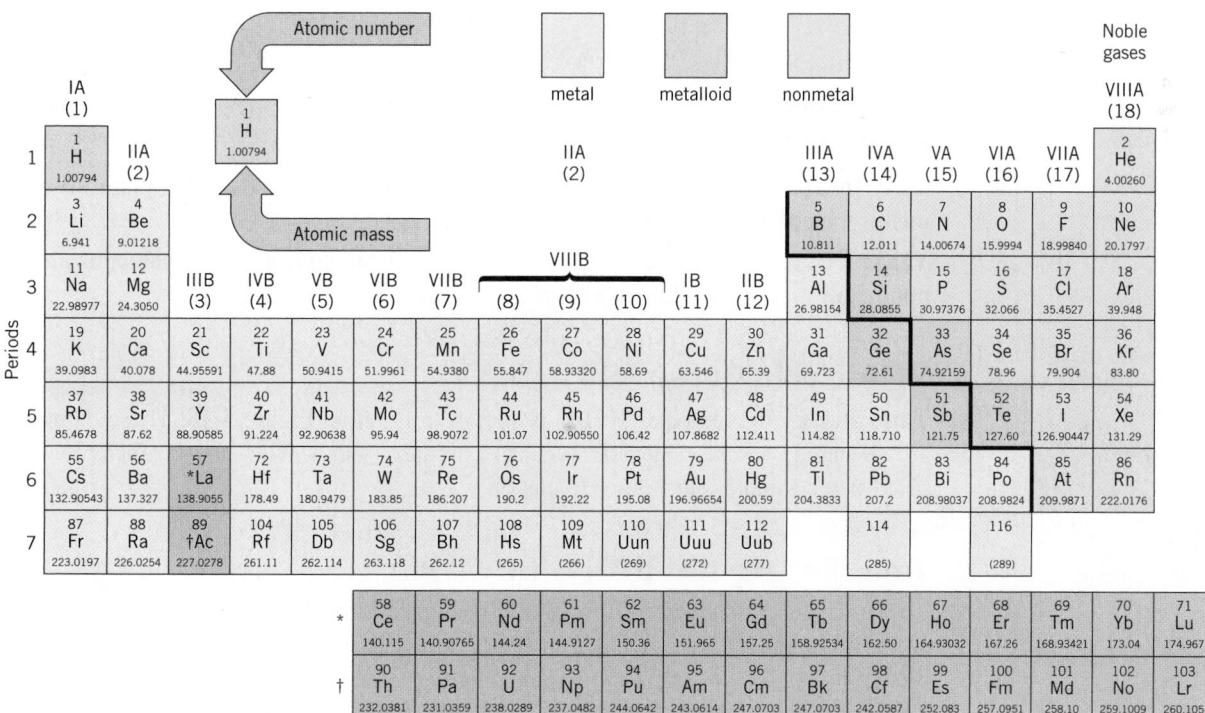

**Figure 5-18  Electron Configuration and the Periodic Table** The electron configuration of an element can be determined from its position in the periodic table. The value of $n$ shown in each box is the shell of the outermost subshell.

these electrons, we use the noble gas shorthand in expressing electron configurations. We will now relate electron configurations with groups and categories of elements discussed in Chapter 4. Once again, the four categories and some of the groups within a category are listed. After the group number, the general electron configuration for that group is noted. For example, [NG]$ns^1$ means a noble gas configuration followed by one electron in the $n$ shell and the $s$ subshell.

## Representative Elements: [NG]$ns^x np^y$

These are the elements shown in blue in Figure 5-18. Four of the groups within this category have family names. The general electron configurations of each group are as follows.

**IA: [NG]$ns^1$.** Except for hydrogen, these are the *alkali metals*. All have one electron beyond a noble gas configuration. Notice that the numbering for the $s$ subshell begins at 1 for hydrogen and is consecutive down the table.

**IIA: [NG]$ns^2$.** These are the *alkaline earth metals*. All have two electrons beyond a noble gas. The numbering begins at 2 in this column with Be. He ($1s^2$) is located to the far right with the noble gases.

**IIIA: [NG]$ns^2 np^1$.** Notice that boron (B) is the first element with an electron in a $p$ subshell. The numbering of the $p$ subshells begins with $2p$ at B and continues consecutively down the table. The elements Ga (#31) and In (#49), as well as all of those to the right of these elements, also have filled $d$ subshells beyond the previous noble gas. Tl (#81) and the elements directly to the right of Tl also have filled $4f$ subshells as well as a filled $6d$ subshell beyond the noble gas. The electron configurations of In (#49) and Tl (#81) are shown below. So as to emphasize the outermost shell, the *filled* inner subshells are sometimes listed first in order of $n$, the shell number.

$$\text{In: } [\text{Kr}]4d^{10}5s^25p^1 \qquad \text{Tl: } [\text{Xe}]4f^{14}5d^{10}6s^26p^1$$

**IVA: [NG]$ns^2 np^2$.**

**VA: [NG]$ns^2 np^3$.**

**VIA: [NG]$ns^2 np^4$.** These elements are known as the *chalcogens*. They are all nonmetals except for Po.

**VIIA: [NG]$ns^2 np^5$.** These elements are known as the *halogens*. The halogens are all nonmetals and one electron short of having a noble gas configuration.

## VIIIA Noble Gases: [NG]$ns^2 np^6$

This category of elements (VIIIA) is shown in green in Figure 5-18. These elements are so named because they rarely form chemical bonds. They are characterized by filled outer $s$ and $p$ subshells. He has only $1s^2$.

## Transition Metals: [NG]$ns^2 (n - 1)d^x$

These 40 elements are shown in yellow in Figure 5-18. The element scandium (#21) is the first element with an electron in a $d$ subshell. Since the $d$ subshells begin with $3d$, Sc has a $4s^23d^1$ configuration and the next element in Group IIIB has a $5s^24d^1$ configuration and so forth down the group. As mentioned previously, there are many exceptions to the expected order of filling in the transition metals, but two important examples involve the VIB and IB elements. For example, Ag (#47) has the configuration [Kr] $5s^14d^{10}$. The $6d$ series of elements are all synthetic and highly unstable.

In most cases, only a few atoms of each element have been produced. Still, some evidence indicates that these elements belong in their appropriate positions in the periodic table.

## Inner Transition Metals: $[\text{NG}]ns^2(n-1)d^1(n-2)f^x$

These elements are shown in orange in Figure 5-18. The orange bar in the transition metals indicates where they would fit in if the periodic table were extended fully to include these elements. There are two series of these elements: the *lanthanides,* which have the general configuration $[\text{Xe}]6s^25d^14f^x$, and the *actinides,* which have the configuration $[\text{Rn}]7s^26d^15f^x$.

---

**The Electron Configuration of Rh from the Periodic Table**

**Example 5-3**

working it out

Write the electron configuration for rhodium.

**Solution**

Locate rhodium (#45) in the periodic table inside the back cover. Now locate Rh in the periodic table shown in Figure 5-18. Rh is the next element below Co, so the highest energy subshell for this element is the $4d$. Counting over from the first $4d$ element (Y), the configuration is $4d^7$. If we write out all subshells preceding the $5s$, the complete electron configuration of Rh is

$$1s^22s^22p^63s^23p^63d^{10}4s^24p^65s^24d^7$$

or, starting with the previous noble gas,

$$[\text{Kr}]5s^24d^7$$

---

**The Identity of an Element from Its Electron Configuration**

**Example 5-4**

What element has the electron configuration $[\text{Kr}]4d^{10}5s^25p^5$?

**Solution**

In Figure 5-18, locate $p^5$ in the upper right of the table. Using the periodic table inside the back cover, locate Kr. The element after Kr in the $p^5$ column in Figure 5-18 has an atomic number 53, which is iodine.

---

## 5-7 Periodic Trends

**Looking Ahead!** Besides electron configuration, there are other characteristics of the atoms of an element that follow trends in the periodic table. These properties will relate directly to the chemical properties of elements that we will pursue in the next chapter. The first characteristic that we will consider in this section is the size of neutral atoms.

We have seen that electron configuration corresponds to the position of an element in the periodic table. But there is much more. The number and types of chemical bonds that an element forms are perhaps the most fundamental property that the periodic table can help us predict and understand. The basis for these trends, however, lies first in the size or radius of the atoms of an element and then in other considerations that relate directly to the atomic size. We will first consider atomic size.

## Atomic Radius

*The **radius** of an atom is the distance from the nucleus to the outermost electrons.* It is not an easy task to measure the radius of an atom. There are both experimental and theoretical problems (e.g., electrons do not have a fixed distance from the nucleus). Despite all of these difficulties, consistent values for the radii of neutral atoms have been compiled. Some generally accepted values are shown in Figure 5-19 for the first three rows of representative elements. Two units of measurement are most often used for such small distances—the *nanometer* ($1 \text{ nm} = 10^{-9}$ m) and the *picometer* ($1 \text{ pm} = 10^{-12}$ m). We will use picometers because the numbers are easier to express and compare (e.g., 37 pm rather than 0.037 nm).

*Notice that there is a general decrease in the radii from left to right across representative element groups, but an increase in radii down a group.* It is understandable that the size of atoms *increases* down a group since it is reasonable to expect that a heavier atom is also a larger atom. In fact, the radii of atoms increase down a group because the outermost electron is in a shell farther from the nucleus. On the other hand, it may seem surprising that the radii *decrease* from Li to F and from Na to Cl, even though the atoms are heavier as we move to the right. Since electrons are being added to the same outer shell as we move from Li to F, one may at first predict that the size of the atoms would not change. However, an additional proton (positive charge) is also being added to each successive nucleus. This increased positive charge increases the attraction between the nucleus and all of the negatively charged electrons. As a result, all of the outer electrons are drawn in tighter around the nucleus and we observe a decrease in radii of the atoms of successive elements as a subshell is filling. This is analogous to a few students sitting at a round table studying by the light of a single lamp in the center of the table. Assume that every time one more student sits down at the table, the lamp is turned up a notch. All students around the table will experience the increased brightness of the lamp. Likewise, all electrons in the same shell feel the increased pull of the positive charge as a subshell is filling.

These same trends are also demonstrated in the transition metals. The radius of Sc (#21) is 144 pm, and the radius of Zn (#30) is 125 pm.

## Ionization Energy

The energy required to form a cation from an atom is an important periodic property. **Ionization energy (I.E.)** *is the energy required to remove an electron from a gaseous*

**Figure 5-19  Atomic Radii**
**Radii of these atoms are given in picometers (pm).**

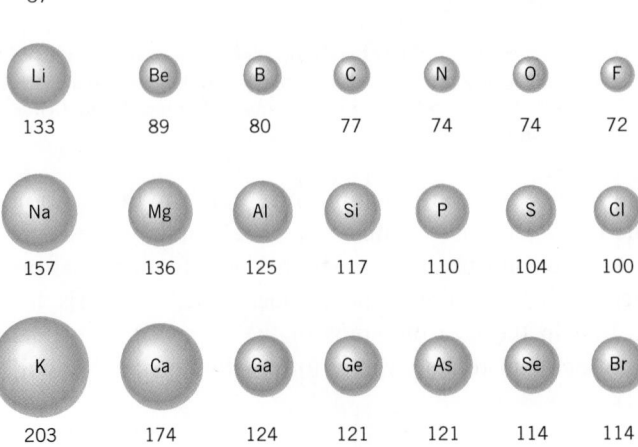

**Table 5-2  Ionization Energy of Some Elements**

| Element | I.E. (kJ/mol) | Element | I.E. (kJ/mol) |
|---------|---------------|---------|---------------|
| Li | 520 | Na | 496 |
| Be | 900 | Mg | 738 |
| B | 801 | Al | 578 |
| C | 1086 | Si | 786 |
| N | 1402 | P | 1102 |
| O | 1314 | S | 1000 |
| F | 1681 | Cl | 1251 |

*atom to form a gaseous ion.* Since the outermost electron is generally the least firmly attached, it will be the first to go.

$$M(g) \longrightarrow M^+(g) + e^-$$

$M(g)$, a gaseous atom, forms a gaseous cation and an electron.

Since all electrons are held by attractive electrostatic forces to the nucleus, it requires energy (an *endothermic* process) to remove an electron. The ionization energy generally increases across a period but decreases down a group. The ionization energies for the second and third periods are shown in Table 5-2. The energy unit abbreviated kJ/mol stands for kilojoules per mole, which is energy per a defined quantity of atoms. Notice that the trends in ionization energy follow from the discussion of atomic radius. That is, the smaller the atom, the harder it is to remove an electron. This inverse relationship between size and ionization energy can be demonstrated graphically by combining the information in Table 5-2 and Figure 5-19 into Figure 5-20. We can

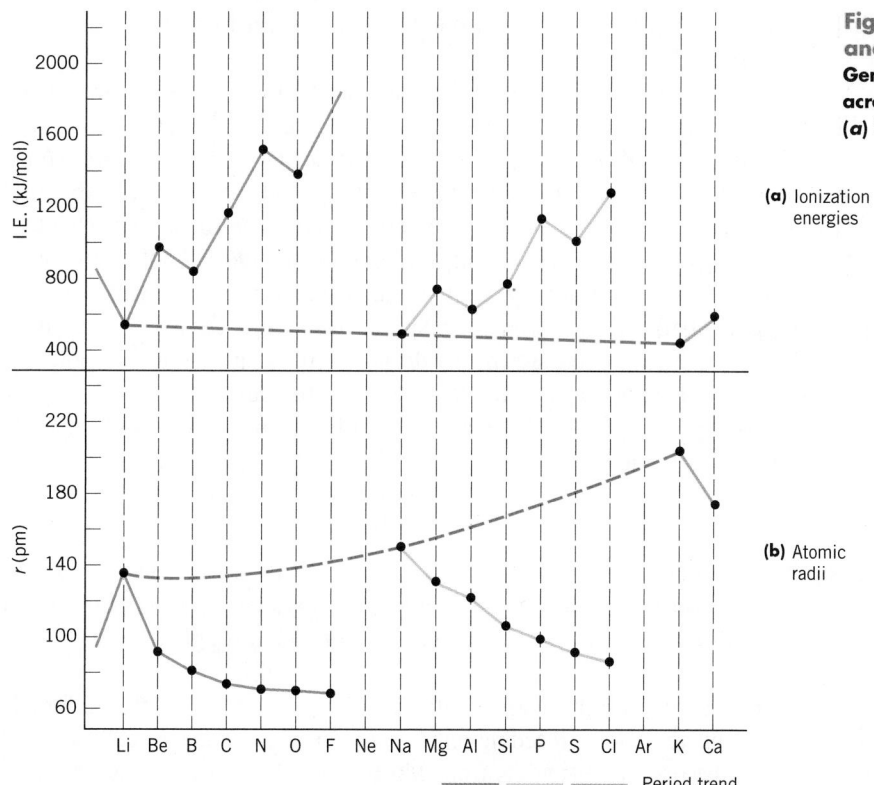

**Figure 5-20  Trends in Atomic Radii and Ionization Energies**
**Generally, as atomic radii decrease across a period (*b*), ionization energies (*a*) increase.**

**(a)** Ionization energies

**(b)** Atomic radii

——— Period trend
- - - Group trend

### Global Warming—It's Getting Hot Out There

Apparently, the Alaskan glaciers are melting at the fastest rate in perhaps the past 2000 years, according to a study released in the summer of 2002. The meltwater is having a measurable effect on rising ocean levels. Pacific islands and low-lying coastal areas could experience increased flooding in a matter of a few years. Is this a normal long-term variation in the weather or is what is known as "global warming," a man-made effect? The answer is still being debated, but most scientists feel that human activities are to blame.

Global warming could be caused by the "greenhouse effect." This is much like what happens on a sunny day inside a closed automobile. Light energy enters through the glass but does not escape at the same rate. As a result, the inside becomes hotter than the outside. Our planet is bathed in life-giving light during the day, which comes to us mostly in the visible region of the spectrum (400–800 nm). At night, the energy radiates back into space in the infrared region of the spectrum. The night side of our planet would look black to us. However, if we could see in the infrared part of the spectrum, we would see the night side of our planet glowing in space.

The major components of our atmosphere, $N_2$, $O_2$, and Ar, are not greenhouse gases. If they were the only components of the atmosphere, all of the heat from the sun that enters during the day would radiate out to space at night. It would get very cold at night even in the summer. Trace gases in our atmosphere such as $CO_2$, $H_2O$, $CH_4$, and $N_2O$ absorb the infrared light radiating from the surface and reflect some of it back instead of letting it all escape into space. We need some of these gases to moderate our climate.

What we do know is that human activity from combustion of fossil fuels and destruction of the forests has increased the $CO_2$ concentration from 300 ppm in the early 1880s to 375 ppm now. It is projected to go higher. Concentrations of methane from cattle and nitrous oxide from automobiles are also increasing.

So, are the six billion humans on the planet causing global warming? We don't know for sure, but a growing number of scientists and concerned citizens want to start doing something about it now.

*Is it hot or what?*

now see how size relates to ionization energy by comparing the same elements in parts (*a*) (ionization energy) and (*b*) (atomic radii). The overall trend is a decrease in radii [red line in part (*b*)] and a corresponding increase in ionization energies [red line in part (*a*)].

Notice that ionization energy generally decreases to the left and down. It is no coincidence that this is the same direction as increasing metallic properties. *In fact, the most significant chemical property of metals is that they lose electrons relatively easily to form cations in compounds.* However, some metals have considerably higher ionization energies than others. How chemically reactive the metal is relates to this energy. For example, the ionization energy of sodium is 496 kJ/mol and of gold is kJ/mol. As a result, sodium is a very reactive metal (i.e., it react explosively with water), yet gold is called the *eternal metal* because of its unreactivity. Nonmetals also have very high ionization energies and so do not form positive ions in compounds. In fact, they form negative ions, as we will notice in the next chapter.

The second ionization energy for an ion involves removal of an electron from a +1 ion to form a +2 ion.

$$M^+(g) \longrightarrow M^{2+}(g) + e^-$$

In a similar manner, the third ionization energy forms a +3 ion and so forth. In all cases it becomes increasingly difficult to remove each succeeding electron. The trends in consecutive ionization energies for Na through Al are shown in Table 5-3.

Note that the second I.E. for Na, the third I.E. for Mg, and the fourth I.E. for Al are all very large compared to the preceding number. For example, it takes 2188 kJ (1450 + 738) to remove the first two electrons from Mg (to form $Mg^{2+}$), but about three times

**Table 5-3  Ionization Energies (kJ/mol)**

| Element | First I.E. | Second I.E. | Third I.E. | Fourth I.E. |
|---------|-----------|-------------|------------|-------------|
| Na | 496 | 4565 | 6912 | 9,540 |
| Mg | 738 | 1450 | 7732 | 10,550 |
| Al | 577 | 1816 | 2744 | 11,580 |

as much energy to remove the third electron (7732 kJ to form $Mg^{3+}$). Why is there such a big jump? The answer lies in the electron configuration of magnesium, which is $[Ne]3s^2$. The first electron removed is from the $3s$ subshell, as is the second. To form $Mg^{3+}$, however, the third electron must be removed from the filled inner shell of the neon configuration (the $2p$). Because the second shell is closer to the nucleus than the third shell, this is very difficult and requires a large amount of energy. The same reasoning holds for the second electron from Na and the fourth from Al. We will refer to this observation in the next chapter. It is important to note here, however, that a filled shell represents a very stable arrangement. Thus, the amount of positive charge that a representative metal can form is limited by the number of electrons beyond a noble gas configuration.

When an electron is removed from an atom, the resulting cation is smaller than the parent atom. The loss of an electron from an outer subshell results in an increased attraction between the nucleus and the remaining electrons. As a result, the loss of electrons results in a contraction of the ion. Also, if the electrons in the outer shell are removed, the remaining inner shells reside closer to the nucleus. If all of the electrons in the outermost shell are removed, the resulting ion will be considerably smaller, as is the case of the two cations shown in Figure 5-21. As we'll see in the next chapter, nonmetals form anions. Just as a cation is smaller than its parent atom, an anion is larger than its neutral parent atom. When an electron is added to an atom, the added repulsions between the electrons cause the radius of the ion to expand. Just as a metal cation becomes smaller as the positive charge increases, a nonmetal anion becomes larger as the negative charge increases. (See Figure 5-21.)

---

**Looking Back!** The periodic relationship of electron configurations is the theoretical basis of the periodic table. The elements in a group have the same subshell configurations but are in successive shells. Thus we can use the periodic table to firmly and easily establish most electron configurations. Further periodic properties among the elements can be generalized. These are the size and ionization energy of the elements. From a look at these trends we discover that compared to nonmetals, metals lose electrons comparatively easily to form cations. We will explore this further in the next chapter.

---

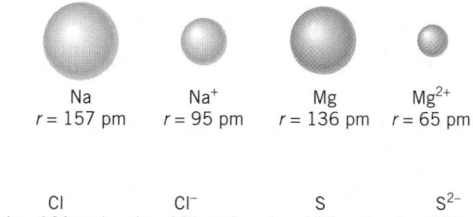

Na
$r = 157$ pm  
Na$^+$
$r = 95$ pm  
Mg
$r = 136$ pm  
Mg$^{2+}$
$r = 65$ pm

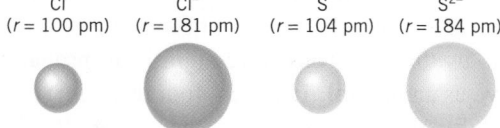

Cl
$(r = 100$ pm$)$  
Cl$^-$
$(r = 181$ pm$)$  
S
$(r = 104$ pm$)$  
S$^{2-}$
$(r = 184$ pm$)$

**Figure 5-21  Ionic Radii**
A positive ion is always smaller than the original neutral atom. A negative ion is always larger than the original atom.

## Learning Check C | checking it out

**C-1.** Fill in the blanks.

An element that has a general configuration beyond a noble gas core of $ns^2np^2$ is in Group _____ , while an element that has a general configuration of $ns^2(n-1)d^2$ is in Group _____ . The radii of neutral atoms generally _____ down a group and _____ to the right across a period. The ionization energies generally _____ down a group and _____ to the right across a period. A metal may form a cation with a charge greater than $+1$ if it has more than one _____ beyond the previous noble gas. The elements with the greatest tendency to gain electrons are the representative element _____ .

**C-2.** Using the periodic table only, give the electron configurations for the elements Zn (#30) and W (#74).

**C-3.** Using the periodic table only, give the symbol of the element with the electron configuration $[Ar]4s^23d^{10}4p^3$.

**C-4.** Give the general electron configurations for the elements in Groups VIA and VIB.

**C-5.** Given the following four elements: As, Se, Sb, and Te.
**a.** Which has the smallest radius?
**b.** Which has the lowest first ionization energy?
**c.** Which is most likely to form a positive ion?

*Additional Examples: Exercises 5-40, 5-42, 5-46, 5-53, 5-67, 5-69, 5-71, and 5-83.*

chapter **R**eview                                          putting it together

This chapter starts us on a journey that ultimately leads to an explanation of why and how atoms of elements bond to each other. The periodic basis of chemical bonding lies in the nature of the electrons in the atoms and the various regions of space in which they exist. Our journey started with a theoretical explanation of the **discrete** or **line spectrum** of **electromagnetic radiation** (light) emitted by hot gaseous hydrogen atoms. This is quite unlike the **continuous spectrum** seen in a rainbow. The **wavelength** of light is inversely related to its energy. In 1913, Niels Bohr proposed a theory, which serves as a **model,** in which the hydrogen electron orbits the nucleus in **quantized energy levels.** He assigned a **principal quantum number (n)** to each energy level, in which the first energy level ($n = 1$) is lowest in energy and known as the **ground state.** In hydrogen, the energy levels higher than $n = 1$ are the **excited states.** Light is emitted from an atom when an electron falls from an excited state to a lower state. The difference in energy between the states becomes the energy of the light wave. If the energy of the light is within the visible part of the spectrum, we see a specific color.

Eventually, Bohr's model had to be adjusted and then mostly discarded as newer, more inclusive theories were advanced. Modern atomic theory, which is known as **wave mechanics,** takes into account the wave nature of the electron. Thus the electron is viewed as having a certain probability of existing in a given region of space. This theory also tells us that electrons have significant probability of existing in a region of space known as an **orbital.** There are four different types of orbitals, known as *s, p, d,* and *f* orbitals. All four have distinctive shapes for the regions of highest probability. The simplest are *s* orbitals, which have a spherical shape. If the principal energy levels are designated as **shells,** then the orbitals of one type within a particular shell make up what is known as a **subshell.** Just as the shells increase in energy from $n = 1$ on, the subshells within a shell increase in energy in the order *s, p, d,* and *f.*

Each shell holds $2n^2$ electrons and has *n* different subshells. The electrons in any atom can be assigned to a given shell and subshell. This is known as the element's **electron configuration.** Electrons fill subshells according to the **Aufbau principle,** which simply means that the lowest energy subshells fill first. As we proceed through the electron configurations of the elements, one fact makes itself apparent. Atoms of elements in vertical columns or groups have the same outermost subshell configuration but successive shells. Thus the basis of the periodic table can be established.

By using **orbital diagrams,** we can expand the representation of electron configuration to include assignment of electrons to orbitals. Two other rules are required to do this successfully. The **Pauli exclusion principle** relates to the spin of electrons in the same orbital, and **Hund's rule** relates to the electron distribution in separate orbitals of the same subshell. The subshells and orbitals in the first four shells are summarized as follows. Each orbital is represented as a box.

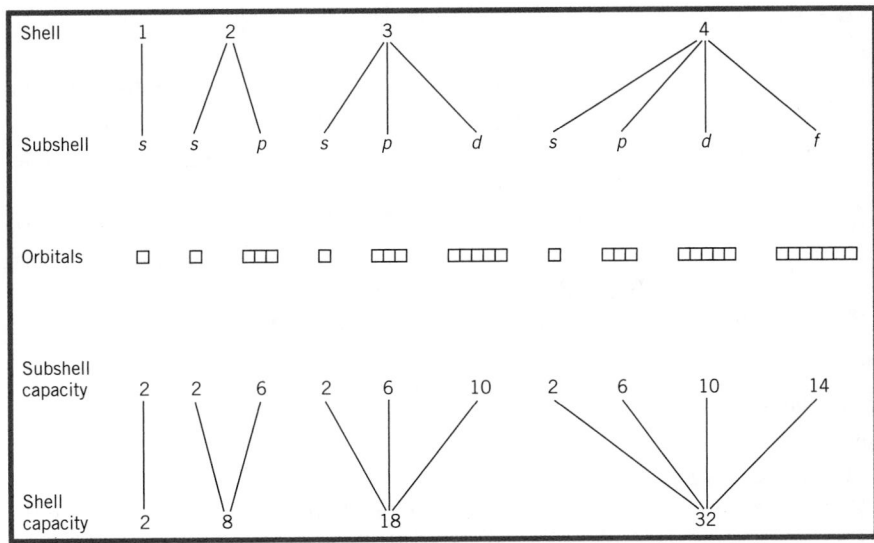

Since the periodic table and electron configuration are interrelated, we now put the periodic table to work in writing electron configurations. We see that various groups can be identified by their specific configurations as well as their properties.

The periodic table tells us even more. There are general trends in the size or **radius** of atoms. In general, atomic radii decrease up and to the right in the table. The radius of an atom is related to the shell of the outermost electrons and to the number of electrons in the outermost subshell. The higher the energy of the shell and the fewer electrons in the outermost subshell, the larger the atom. Thus atoms to the lower left are the largest atoms. The opposite trend occurs if we compare the difficulty in removing the outermost electron, which is known as the element's **ionization energy (I.E.).** The same factors that affect the size of an atom affect the ionization energy. That is, in general, larger atoms have lower ionization energies than smaller atoms. Metals tend to have larger atoms than nonmetals, so they have lower ionization energies. How many electrons can be easily removed from a metal is determined by its outer electron configuration.

Finally, the radius of an ion depends on the position of the element in the periodic table, the sign of the charge, and the magnitude of the charge. Generally, the more positive the ion, the smaller its radius. Thus, if we compare species in the same period, a monatomic anion is larger than a neutral atom, which in turn is larger than a monatomic positive ion.

# Exercises

## Atomic Theory and Orbitals

**5-1.** Ultraviolet light causes sunburns, but visible light does not. Explain.

**5-2.** The $n = 8$ and $n = 9$ energy levels are very close in energy. Using the Bohr model, describe how the wave-lengths of light compare as an electron falls from these two energy levels to the $n = 1$ energy level.

**5-3.** The $n = 3$ energy level is of considerably greater energy than the $n = 2$ energy level. Using the Bohr model, describe how the wavelengths of light compare as an electron falls from these two energy levels to the $n = 1$ energy level.

**5-4.** An electron in the lithium atom is in the third energy level. Is the atom in the ground or excited state? Can the atom emit light? If so, how?

**5-5.** Which of the following types of orbitals do not exist?

$$6s \quad 3p \quad 1p \quad 4d \quad 3f$$

**5-6.** Which of the following types of orbitals do not exist?

$$5f \quad 2d \quad 7p \quad 6d \quad 1s$$

**5-7.** What is meant by the "shape" of an orbital?

**5-8.** Describe the shape of the $4p$ orbitals.

**5-9.** Describe the shape of the $3s$ orbital. How does it differ from that of a $2s$ orbital?

**5-10.** Describe the shape of the $4d$ orbital. How does it differ from a $4p$ orbital?

**5-11.** How many total orbitals are in the third energy level?

**5-12.** How many total orbitals are in the fourth energy level?

## Shells and Subshells

**5-13.** How many orbitals are in the following subshells?
**(a)** $3p$  **(b)** $4d$  **(c)** $6s$

**5-14.** How many orbitals are in the following subshells?
**(a)** $5f$  **(b)** $6p$  **(c)** $2d$

**5-15.** What are the total electron capacities of the $n = 3$ and the $n = 2$ shells?

**5-16.** What is the total electron capacity of the fourth shell?

**5-17.** What is the total electron capacity of each subshell in the fourth shell?

**5-18.** What is the total electron capacity of the fifth shell?

**5-19.** The fifth shell (theoretically) contains a fifth subshell designated the $5g$ subshell. What is the total electron capacity of the $g$ subshell?

**5-20.** How many orbitals are in the $5g$ subshell? (See exercise 5-19.)

## Electron Configuration

**5-21.** Which subshell always fills first?

**5-22.** Which subshell fills second and which subshell fills third?

**5-23.** Explain how a subshell in a particular shell can be lower in energy than a subshell in a lower shell.

**5-24.** Which of the following subshells fills first? (Refer to Figure 5-16 or 5-18.)
**(a)** $6s$ or $6p$  **(b)** $5d$ or $5p$  **(c)** $4p$ or $4f$  **(d)** $4f$ or $4d$

**5-25.** Which of the following subshells is lower in energy? (Refer to Figure 5-16 or 5-18.)
**(a)** $6s$ or $5p$  **(c)** $5s$ or $4d$  **(e)** $4f$ or $6d$
**(b)** $6s$ or $4f$  **(d)** $4d$ or $5p$  **(f)** $3d$ or $4p$

**5-26.** Write the following subshells in order of increasing energy. (Refer to Figure 5-16 or 5-18.)

$$4s, 5p, 4p, 5s, 4f, 4d, 6s$$

**5-27.** Write the total electron configuration for each of the following elements. (Refer to Figure 5-16 or 5-18).
**(a)** Mg  **(b)** Ge  **(c)** Pd  **(d)** Si

**5-28.** Write the total electron configuration for each of the following elements. (Refer to Figure 5-16 or 5-18.)
**(a)** B  **(b)** Ag  **(c)** Se  **(d)** Sr

**5-29.** Write the electron configuration that is implied by [Ar]. (Refer to Figure 5-16 or 5-18.)

**5-30.** Write the electron configuration that is implied by [Xe]. (Refer to Figure 5-16 or 5-18.)

**5-31.** Using the noble gas shorthand, write the electron configurations for the following elements. (Refer to Figure 5-16 or 5-18.)
**(a)** S  **(b)** Zn  **(c)** Pu  **(d)** I

**5-32.** Using the noble gas shorthand, write the electron configurations for the following elements. (Refer to Figure 5-16 or 5-18.)
**(a)** Sn  **(b)** Ni  **(c)** Cl  **(d)** Au

## Orbital Diagrams

**5-33.** Which of the following orbital diagrams is excluded by the Aufbau principle? Which by the Pauli exclusion principle? Which by Hund's rule? Which is correct? Explain how a principle or rule is violated in the others.

$$1s \quad 2s \quad 2p$$

(a) $\uparrow\downarrow$ | $\uparrow\uparrow$ | $\uparrow\downarrow$ $\uparrow$ | |

(b) $\uparrow\downarrow$ | $\uparrow\downarrow$ | $\uparrow$ $\uparrow$ $\uparrow$

(c) $\uparrow\downarrow$ | $\uparrow$ | $\uparrow\downarrow$ $\uparrow$ $\uparrow$

(d) $\uparrow\downarrow$ | $\uparrow\uparrow$ | $\uparrow$ $\uparrow$ $\uparrow$

**5-34.** Write the orbital diagrams for electrons beyond the previous noble gas core for the following elements. (Refer to Figure 5-16 or 5-18.)
**(a)** S **(b)** V **(c)** Br **(d)** Pm

**5-35.** Write the orbital diagrams for electrons beyond the previous noble gas core for the following elements. (Refer to Figure 5-16 or 5-18.)
**(a)** As **(b)** Ar **(c)** Tc **(d)** Tl

**5-36.** How many unpaired electrons are in the atoms of elements in Groups IIB, VB, VIA, VIIA; an atom with electron configuration given by $[ns^1(n-1)d^5]$; and Pm (atomic number 61)?

**5-37.** The atoms of the elements in three groups in the periodic table have all of their electrons paired. What are the groups?

**5-38.** What group or groups have two unpaired $p$ electrons?

**5-39.** What group or groups have five unpaired $d$ electrons?

## Electron Configuration and the Periodic Table

**5-40.** Write the symbol of the first element that has a
**(a)** $5p$ electron **(c)** $4f$ electron
**(b)** $4d$ electron **(d)** filled $3p$ subshell

**5-41.** Write the symbol of the first element that has a
**(a)** filled $4d$ subshell **(c)** $6s$ electron
**(b)** half-filled $4p$ subshell **(d)** filled fourth shell

**5-42.** Which subshell begins to fill for each of the following elements?
**(a)** Al **(b)** In **(c)** La **(d)** Rb

**5-43.** Which subshell begins to fill for each of the following elements?
**(a)** Y **(b)** Th **(c)** Cs **(d)** Ga

**5-44.** What do Groups IIIA and IIIB have in common? How are they different?

**5-45.** What do the elements Al and Ga have in common? How are they different?

**5-46.** Which elements have the following electron configurations? (Use only the periodic table.)
**(a)** $1s^2 2s^2 2p^5$ **(d)** $[Xe]6s^2 5d^1 4f^7$
**(b)** $[Ar]4s^2 3d^{10}4p^1$ **(e)** $[Ar]4s^1 3d^{10}$ (exception to rules)
**(c)** $[Xe]6s^2$

**5-47.** Which elements have the following electron configurations? (Use only the periodic table.)
**(a)** $[He]2s^2 2p^3$ **(d)** $[Rn]7s^2 6d^1$
**(b)** $[Kr]5s^2 4d^2$ **(e)** $[Xe]6s^2 5d^{10}4f^{14}6p^1$
**(c)** $[Ar]4s^2 3d^{10}4p^6$ **(f)** $[Ar]4s^2 3d^{10}$

**5-48.** If any of the elements in exercise 5-46 belong to a numerical group (e.g., Group IA) in the periodic table, indicate the group.

**5-49.** Write the number designation of the two groups that may have five electrons beyond a noble gas configuration.

**5-50.** Write the number designation of a group that has two electrons beyond a noble gas configuration. Write the number designation of a group with 12 electrons beyond a noble gas configuration.

**5-51.** Which two configurations belong to the same group in the periodic table?
**(a)** $[Kr]5s^2$ **(c)** $[Xe]6s^2 5d^1 4f^2$ **(e)** $[Ne]3s^2 3p^2$
**(b)** $[Kr]5s^2 4d^{10}5p^2$ **(d)** $[Ar]4s^2 3d^2$

**5-52.** Which two configurations belong to the same group in the periodic table?
**(a)** $[Kr]5s^2 4d^{10}$ **(c)** $[Ne]3s^1$ **(e)** $[Xe]6s^2 4f^{14}5d^{10}$
**(b)** $[Ar]4s^1 3d^{10}$ **(d)** $[Xe]6s^2$

**5-53.** Which group in the periodic table has the following general electron configuration? ($n$ is the principal quantum number.)
**(a)** $ns^2 np^2$ **(c)** $ns^1(n-1)d^{10}$ (e.g., $4s^1 3d^{10}$)
**(b)** $ns^2 np^6$ **(d)** $ns^2(n-1)d^1(n-2)f^2$ (two elements)

**5-54.** Write the general electron configurations for the following groups.
**(a)** IIA **(b)** IIB **(c)** VIA **(d)** IVB

**5-55.** How does He differ from the other elements in Group VIIIA?

**5-56.** Which of the following elements fits the general electron configuration $ns^2(n-1)d^{10}np^4$?

**(a)** Cr **(b)** Te **(c)** S **(d)** O **(e)** Si

**5-57.** What is the electron configuration of the noble gas at the end of the third period?

**5-58.** What is the electron configuration of the noble gas at the end of the fourth period?

**\*5-59.** What would be the atomic number of an element with one electron in the $5g$ subshell?

**\*5-60.** Element number 114. Scientists believe that element 114 will form a comparatively long-lived isotope. What would be its electron configuration and in what group would it be?

**\*5-61.** How many elements would be in the seventh period if it were complete?

**\*5-62.** How many elements would be in a complete eighth period? (*Hint:* Consider the $5g$ subshell.)

**\*5-63.** Classify the following electron configurations into one of the four main categories of elements.
**(a)** $[Ar]4s^2 3d^2$ **(c)** $[Ar]4s^2 3d^{10}4p^6$
**(b)** $[Kr]5s^2 4d^{10}5p^5$ **(d)** $[Rn]7s^2 6d^1 5f^3$

*5-64. Classify the following electron configurations into one of the four main categories of elements.
(a) $[Ne]3s^2 3p^6$  (c) $[Xe]6s^2 4f^{14} 5d^2$
(b) $[Ne]3s^2 3p^5$  (d) $[Xe]6s^2 5d^1 4f^7$

*5-65. What would be the atomic number and group of the next nonmetal after element 112?

*5-66. What is the atomic number of the heaviest metal that would appear in the periodic table after element number 112?

## Periodic Trends

5-67. Which of the following elements has the larger radius?
(a) As or Se  (b) Ru or Rh  (c) Sr or Ba  (d) F or I

5-68. Which of the following elements has the larger radius?
(a) Tl or Pb  (b) Sc or Y  (c) Pr or Ce  (d) As or P

5-69. Four elements have the following radii (in pm): 117, 122, 129, and 134. The elements, in random order, are V, Cr, Nb, and Mo. Which element has a radius of 117 pm? Which has a radius of 134 pm?

5-70. Four elements have the following radii: 180 pm, 154 pm, 144 pm, and 141 pm. The elements, in random order, are In, Sn; Tl, and Pb. Which element has a radius of 141 pm? Which has a radius of 180 pm?

5-71. Which of the following elements has the higher ionization energy?
(a) Ti or V  (c) Mg or Sr  (e) B or Br
(b) P or Cl  (d) Fe or Os

5-72. Four elements have the following first ionization energies (in kJ/mol): 762, 709, 579, and 558. The elements, in random order, are Ga, Ge, In, and Sn. Which element has an ionization energy of 558 kJ/mol? Which element has an ionization energy of 762 kJ/mol?

5-73. Four elements have the following first ionization energies (in kJ/mol): 869, 941, 1010, and 1140. The elements, in random order, are Se, Br, Te, and I. Which element has an ionization energy of 869 kJ/mol? Which element has an ionization energy of 1140 kJ/mol?

5-74. The first four ionization energies for Ga are 578.8, 1979, 2963, and 6200 kJ/mol. How much energy is required to form each of the following ions: $Ga^+$, $Ga^{2+}$, $Ga^{3+}$, and $Ga^{4+}$? Why does the formation of $Ga^{4+}$ require a comparatively large amount of energy?

5-75. Which of the following monatomic cations is the easiest to form, and which is the hardest to form?
(a) $Cs^+$  (b) $Rb^{2+}$  (c) $Ne^+$  (d) $Sc^{3+}$

5-76. Which of the following atoms would most easily form a cation?
(a) B  (b) Al  (c) Si  (d) C

5-77. Which of the following atoms would most likely form an anion?
(a) Be  (b) Al  (c) Ga  (d) I

5-78. Noble gases form neither anions nor cations. Why?

5-79. Which of the following would have the highest second ionization energy?
(a) Rb  (b) Pb  (c) Ba  (d) Al  (e) Be

5-80. Which of the following would have the lowest third ionization energy?
(a) Na  (b) B  (c) Ga  (d) Mg  (e) N

5-81. Arrange the following ions and atoms in order of increasing radii.
(a) Mg  (c) S  (e) $K^+$
(b) $S^{2-}$  (d) $Mg^{2+}$  (f) $Se^{2-}$

5-82. Arrange the following ions and atoms in order of increasing radii.
(a) Br  (c) K  (e) $Br^-$
(b) $K^+$  (d) $I^-$  (f) $Ca^{2+}$

*5-83. Zirconium and hafnium are in the same group and have almost the same radius despite the general trend down a group. As a result, the two elements have almost identical chemical and physical properties. The fact that these elements have almost the same radius is due to what is referred to as the *lanthanide contraction*. With the knowledge that atoms get progressively smaller as a subshell is being filled, can you explain this phenomenon? (*Hint:* Follow all of the expected trends between the two elements.)

5-84. The first five ionization energies for carbon are 1086, 2353, 4620, 6223, and 37,830 kJ/mol. How much energy is required to form the following ions: $C^+$, $C^{2+}$, $C^{3+}$, $C^{4+}$, and $C^{5+}$? In fact, even $C^+$ does not form in compounds. Compare the energy required to form this ion with that needed to form some metal ions. Explain.

## General Problems

5-85. Write the symbol of the element that corresponds to the following.
(a) the first element with a $p$ electron
(b) the first element with a filled $4p$ subshell
(c) three elements with only one electron in the $4s$ subshell
(d) the first element with one $p$ electron that also has a filled $d$ subshell
(e) the element after Xe that has two electrons in a $d$ subshell

5-86. Write the symbol of the element that corresponds to the following.
(a) the first element with a half-filled $p$ subshell
(b) the element with only three electrons in a $4d$ subshell
(c) the first two elements with a filled $3d$ subshell
(d) the element with only three electrons in a $5p$ subshell

5-87. Write the symbol of the element that corresponds to the following.
(a) a nonmetal with only one electron in a $p$ subshell
(b) an element that is a liquid at room temperature that has five $p$ electrons
(c) the first metal to have three $p$ electrons
(d) a metalloid with two $p$ electrons and no $d$ electrons

**5-88.** Write the symbol of the element that corresponds to the following.
**(a)** a transition metal in the fifth period with three unpaired electrons
**(b)** a representative metal with three unpaired electrons with no $f$ electrons
**(c)** an element that is a liquid at room temperature with no unpaired electrons
**(d)** a metalloid with two unpaired electrons and a filled $d$ subshell

**5-89.** Identify the first element that has a total number of electrons and indicate whether the element is a metal or nonmetal, its category, and group number.
**(a)** 10 $s$ electrons
**(b)** 28 $d$ electrons
**(c)** 15 $p$ electrons

**5-90.** Identify the first element that has a total number of electrons and indicate whether the element is a metal or nonmetal, its category, and group number.
**(a)** 15 $f$ electrons
**(b)** 36 $d$ electrons
**(c)** 24 $p$ electrons

**5-91.** A certain isotope has a mass number of 30. The element has three unpaired electrons. What is the element?

**5-92.** A certain isotope has a mass number of 196. It has two unpaired electrons. What is the element?

**5-93.** Given two elements, X and Z, identify the elements from the following information.
**(a)** They are both metals, but one is a representative element.
**(b)** The first ionization energy of Z is greater than that of X.
**(c)** They are both in the same period that does not have $f$ electrons.
**(d)** They both have two unpaired electrons and neither is used in jewelry.
**(e)** A nonmetal in the same period is a diatomic solid.

**5-94.** Given two elements, Q and R, identify the elements from the following information.
**(a)** One is a gas and one is a solid.
**(b)** One forms a +1 ion and the other does not.
**(c)** Q is larger than R, but both elements are the first elements that have full shells.
**(d)** One is used in coins and the other in fluorescent lights.

**5-95.** Which of the following monatomic cations would require a particularly large amount of energy to form? If a certain ion requires a large amount of energy to form, give the reason.
**(a)** $In^{3+}$    **(c)** $Ca^{2+}$    **(e)** $B^{3+}$
**(b)** $I^+$        **(d)** $K^+$

**5-96.** Which of the following atoms would not be likely to form a +2 cation? Explain.
**(a)** Sr    **(b)** Li    **(c)** B    **(d)** Ba

**5-97.** Chemical reactivity relates to the size of the atom or ion. In the following pairs, which has the larger radius?
**(a)** Be or Ca        **(d)** $S^{2-}$ or $Cl^-$
**(b)** Br or $Br^-$    **(e)** $Na^+$ or $Mg^{2+}$
**(c)** Cl or S

**5-98.** In the following pairs, which has the larger radius?
**(a)** $Na^+$ or $K^+$      **(c)** Ga or Ge    **(e)** $K^+$ or K
**(b)** $O^{2-}$ or $Se^{2-}$  **(d)** Si or Ge

**\*5-99.** On the planet Zerk, the periodic table of elements is slightly different from ours. On Zerk, there are only two $p$ orbitals, so a $p$ subshell holds only four electrons. There are only four $d$ orbitals, so a $d$ subshell holds only eight electrons. Everything else is the same as on Earth, such as the order of filling ($1s$, $2s$, etc.) and the characteristics of noble gases, metals, and nonmetals. Construct a Zerkian periodic table using numbers for elements up to element number 50. Then answer these questions.
**(a)** How many elements are in the second period? In the fourth period?
**(b)** What are the atomic numbers of the noble gases at the ends of the third and fourth periods?
**(c)** What is the atomic number of the first inner transition element?
**(d)** Which element is more likely to be a metal: element number 5 or element number 11; element number 17 or element number 27?
**(e)** Which element has the larger radius: element number 12 or element number 13; element number 6 or element number 12?
**(f)** Which element has a higher ionization energy: element number 7 or element number 13; element number 7 or element number 5; element number 7 or element number 9?
**(g)** Which ions are reasonable?
**(1)** $16^{2+}$    **(5)** $17^{4+}$
**(2)** $9^{2+}$     **(6)** $15^+$
**(3)** $7^+$        **(7)** $1^-$
**(4)** $13^-$

# Interactive Learning

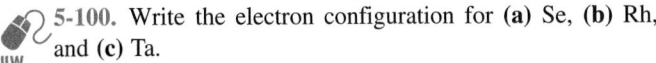 **5-100.** Write the electron configuration for **(a)** Se, **(b)** Rh, and **(c)** Ta.

**5-101.** Write the orbital diagrams for all of the electrons that come after the previous noble gas for **(a)** Se, **(b)** Rh, and **(c)** Ta.

# Solutions to Learning Checks

**A-1.** wave, inversely, longest, lowest, discrete, quantized, ground, excited, orbital, *s, p, d, f,* spherical, shell, subshells, orbitals

**A-2.** For the $n = 3$ shell;

**(a)** shell capacity: $2n^2 = 18$

**(b)** subshells: *s, p,* and *d*

**(c)** subshell capacities: $s = 2$, $p = 6$, $d = 10$

**(d)** orbitals: $s = 1$, $p = 3$, $d = 5$

**B-1.** five, *d,* fifth, Aufbau, Pauli, Hund's rule

**B-2.** Ni [Ar]$4s^2 3d^8$    Sr [Kr]$5s^2$    At [Xe]$6s^2 4f^{14} 5d^{10} 6p^5$

**B-3.** Nb [Kr]$5s^2 4d^3$

$$\begin{array}{cc} 5s & 4d \\ \boxed{\uparrow\downarrow} & \boxed{\uparrow\,|\,\uparrow\,|\,\uparrow\,|\,\phantom{\uparrow}\,|\,\phantom{\uparrow}} \end{array}$$

**C-1.** IVA, IVB, increase, decrease, decrease, increase, electron, nonmetals

**C-2.** Zn [Ar]$4s^2 3d^{10}$    W [Xe]$6s^1 4f^{14} 5d^5$ (an exception to rules)

**C-3.** As

**C-4.** VIA [NG]$ns^2 np^4$    VIB [NG]$ns^1 (n-1)d^5$ (an exception to rules)

**C-5.** **(a)** Se    **(b)** Sb    **(c)** Sb

# chapter 6

# The Chemical Bond

This is our home as seen from far-out space. Its surface and atmosphere are composed of some free elements as well as ionic and molecular compounds. We look deeper into the nature of compounds in this chapter.

## Setting the Stage

This Earth of ours is a fascinating yet complex world of chemicals. First, consider the air. Its major components are free elements: two molecular elements, nitrogen and oxygen, and smaller quantities of noble gases such as argon, helium, and neon that exist as solitary atoms. Other molecular compounds such as carbon dioxide and water are present in small amounts. The surface of the Earth is made up primarily of compounds. Lakes and oceans are composed of water containing dissolved compounds, whereas the solid Earth contains compounds of living things, rocks, and minerals. Perhaps we can begin to bring some order to all of these elements and compounds by dividing all of the substances in, on, and above the Earth into roughly two categories: hard stuff and soft stuff. Under most Earth temperature conditions, the rocks and minerals of the mountains are certainly hard, whereas the air, oceans, and stuff of living things (solid but still flexible) are soft. Let us focus on two compounds that are necessary in life processes and that occur in nature. One is hard—table salt—and one is soft—water. These two compounds represent the two basic types of compounds: ionic and molecular. Although there are several important types of hard substances in nature that are not ionic, such as quartz, diamond, and certain metals, many of Earth's minerals and rocks are ionic compounds that resemble sodium chloride. On the other hand, water is typical of a molecular compound that makes up much of the soft stuff of nature. What can we observe that distinguishes the two

151

compounds? Both are composed of just two elements. But there is an important difference. Sodium chloride is a binary compound formed from a metal, sodium, and a nonmetal, chlorine. Water is also a binary compound, but it is formed from two nonmetal, hydrogen and oxygen. Perhaps we are on to something! Does the combination of a metal and a nonmetal result in ionic compounds, whereas the combination of two nonmetals results in molecular compounds? The answer is "yes" (generally), and we will see why in this chapter.

### Formulating Some Questions

Why do noble gases rarely bond to other elements and therefore exist as solitary atoms in nature? How does this relate to why the atoms of other elements do form bonds? Why do certain elements combine to form ionic and others molecular compounds? Why is the formula of water $H_2O$ and not $H_3O$ or $HO_2$? These are some of the important questions that we will address in this chapter.

These questions all relate to the electron configurations of the elements. In the first section of this chapter, we will take a closer look at why and how representative elements combine.

---

**CHEMICAL BONDS AND THE NATURE OF IONIC COMPOUNDS**

## 6-1 Bond Formation and Representative Elements

**Looking Ahead!** Most elements in the periodic table interact with other elements to form compounds. There are exceptions, however, and they are important. In this section, we will try to answer the question of why some elements react and others do not.

Elemental lithium is a very active element. It easily forms compounds directly with all of the nonmetals except the noble gases. On the other hand, an atom of the noble gas neon forms no chemical bonds. Why does neon so completely resist bond formation? In fact, it has been known for some time that the formation of chemical bonds involves changes in the electron configurations of the atoms involved. Since the atoms of noble gases generally do not bond, they are obviously stable as solitary atoms. In other words, they have no tendency to change their electron configurations. Let us focus more on why this is so. Noble gases (except He) have filled outer $s$ and $p$ subshells (e.g., $ns^2np^6$). Since this is a total of eight electrons, it is referred to as an *octet* of electrons. Eight electrons in the outer $s$ and $p$ subshells form a particularly stable configuration. For example, in the last chapter we found that the energy required to remove an electron from these full subshells was very large. As it turns out, obtaining this stable configuration is the driving force for bond formation for many of the compounds formed by the representative elements. Bonding is correlated by the **octet rule,** *which states that the atoms of the representative elements form bonds so as to have access to eight outer electrons. The outer* s *and* p *electrons in the atoms of a representative element are referred to as the* **valence electrons.** The noble gas helium forms a stable configuration with only two electrons (i.e., a "duet"), however. Thus representative elements that border helium in the periodic table (i.e., H, Li, and Be) follow a duet rule. The octet rule is particularly helpful in describing the bonding of many, but certainly not all, of the compounds of the representative elements. Some elements in the third period (Si through Cl) and in higher periods form compounds that are not explained by the octet rule. In this text, however, we will emphasize the compounds that do follow the octet rule.

**Table 6-1  Lewis Dot Symbols$^a$**

| IA | IIA | IIIA | IVA | VA | VIA | VIIA | VIIIA |
|----|-----|------|-----|-----|-----|------|-------|
| Ḣ |  |  |  |  |  |  | : $\overset{..}{\text{He}}$ |
| L̇i | Be· | B· | ·Ċ· | ·N̈· | ·Ö: | :F̈: | :N̈e: |
| Ṅa | Mg· | Al· | ·Si· | ·P̈· | ·S̈: | :C̈l: | :Är: |
| K̇ | Ca· | Ga· | ·Ge· | ·Äs· | ·S̈e: | :Br: | :Kr: |

$^a$Named after the American chemist G. N. Lewis (1875–1946), who developed this theory of bonding.

How can an atom alter its electron configuration to obtain the octet (or duet) of electrons of a noble gas? There are three ways.

1.  A metal may *lose* one to three electrons to form a cation with the electron configuration of the previous noble gas (i.e., the one with the next lowest atomic number.)

2.  A nonmetal may *gain* one to three electrons to form an anion with the electron configuration of the next noble gas (i.e., the one with the next highest atomic number.)

3.  Atoms (usually two nonmetals) may *share* electrons with other atoms to obtain access to the number of electrons in the next noble gas.

The first two processes complement each other in the formation of ionic compounds. Case 3 produces molecular compounds.

Since the bonding in these three cases involves the loss, gain, or sharing of valence electrons exclusively, we are free to focus on these electrons only. **Lewis dot symbols** *of these elements represent valence electrons as dots around the symbol of the element.* Electrons are represented with up to four pairs on four sides of the element's symbol. Since the elements in each group have the same number of valence electrons (same subshells but different shells), each element in a group has the same number of dots representing electrons. The Lewis dot symbols of the first four periods of representative elements and noble gases are shown in Table 6-1. The dot symbols are usually shown first with one electron on each side of the element (Groups IA through IVA in Table 6-1) and then with paired electrons on each side (Groups VA through VIIIA). Note that the Roman numeral of the group number also represents the number of valence electrons (dots) for a neutral atom. If the groups are labeled 1 through 18 in the table being used, the last digit of the group number represents the number of valence electrons. For example, group 14 has four valence electrons.

# 6-2 Formation of Ions

**Looking Ahead!** In Chapters 4 and 5, we made good use of the fact that we can predict the charges on representative metals and nonmetals with a little help from a periodic table. For example, we noted that calcium forms a +2 cation because it is in Group IIA and that sulfur forms a −2 anion because it is in Group VIA. In this section we will describe why these elements have these particular charges.

First let's focus on metals. Most metals have many familiar *physical* properties such as the ability to conduct heat and electricity and the capacity to be drawn into wires and pounded into sheets. The one *chemical* property of metals that we have established is that it takes a comparatively small amount of energy to remove one or, in some cases, two or even three electrons to form cations. The octet rule tells us how many electrons will be lost and thus the charge. *If a representative metal loses all of its valence electrons, it acquires the octet of the previous noble gas.* The loss of any additional electrons is prohibitively expensive in terms of energy, so it does not occur in compound formation. We can illustrate the octet rule and cation formation using the Lewis dot symbol of sodium.

$$\overset{\displaystyle .}{\text{Na}} \longrightarrow \text{Na}^+ + \text{e}^-$$
$$[\text{Ne}]3s^1 \qquad [\text{Ne}]$$

The Lewis representation of the $\text{Na}^+$ ion does not include any electrons (dots) because the octet of electrons of $\text{Na}^+$ occupy filled inner subshells. All of the metals in Group IA have the same dot symbol, so each can lose one electron to form a $+1$ ion with the electron configuration of the preceding noble gas.

Now consider the alkaline earth metals. Magnesium, for example, can attain the noble gas configuration of the previous noble gas (neon) by losing two electrons.

$$\overset{\displaystyle .}{\text{Mg}}{\cdot} \longrightarrow \text{Mg}^{2+} + 2\text{e}^-$$
$$[\text{Ne}]3s^2 \qquad [\text{Ne}]$$

All other metals in this group form $+2$ ions in the same manner.

Group IIIA metals (Al down) can lose three electrons in order to form an octet of electrons.* Boron is not a metal and does not form a $+3$ ion in its compounds. Boron bonds by electron sharing, which is discussed later in this chapter.

$$\overset{\displaystyle .}{\cdot\text{Al}}{\cdot} \longrightarrow \text{Al}^{3+} + 3\text{e}^-$$
$$[\text{Ne}]3s^23p^1 \qquad [\text{Ne}]$$

Group IVA metals such as tin and lead have four electrons in their outer subshells. Loss of all four of these electrons to produce a $+4$ ion requires a rather large amount of energy. Instead, these metals can lose two of their four outer electrons to form a $+2$ ion that does not follow the octet rule. They do form compounds where all four of their outer electrons are involved, but the bonding in these compounds is best described by electron sharing rather than ion formation. In Group VA, bismuth forms a $+3$ ion that does not follow the octet rule.

Positive ions do not exist alone in compounds. Negative ions must be present to balance the positive charge. In Chapter 5, we found that representative nonmetals complement metals by forming negative ions. First, we will consider ions formed by the VIIA nonmetals and then work our way toward the center of the periodic table.

All of the atoms of the halogens shown in Group VIIA are one electron short of a noble gas configuration. An octet of electrons can be achieved by adding one electron. The result is an anion with a $-1$ charge and the electron configuration of the next noble gas. In this respect, hydrogen is also one electron short of a noble gas

---

*Ions such as $\text{Tl}^{3+}$ and $\text{Ga}^{3+}$ have a filled $d$ subshell in addition to a noble gas configuration. This is sometimes referred to as a pseudo-noble gas configuration. The filled $d$ subshell does not seem to affect the stability of these ions. In this text, we do not distinguish between noble gas and pseudo-noble gas electron configurations. Transition metals also form positive ions, but for the most part, these ions do not relate to a noble gas configuration. Some of these ions were discussed in Chapter 4.

configuration (a duet in this case), so it can also add one electron to form an anion with a $-1$ charge.

$$e^- + H\cdot \longrightarrow H:^- \quad \text{(hydride ion)}$$
$$\underset{1s^1}{} \qquad \underset{1s^2 = [\text{He}]}{}$$

$$e^- + :\overset{\cdot\cdot}{\underset{\cdot\cdot}{Cl}}\cdot \longrightarrow :\overset{\cdot\cdot}{\underset{\cdot\cdot}{Cl}}:^- \quad \text{(chloride ion)}$$
$$\underset{[\text{Ne}]3s^2 3p^5}{} \qquad \underset{[\text{Ne}]3s^2 3p^6 = [\text{Ar}]}{}$$

The atoms of the elements in Group VIA are two electrons short of a noble gas configuration. By gaining two electrons to form a $-2$ ion, they also attain an octet of electrons.

$$2e^- + :\overset{\cdot}{\underset{\cdot\cdot}{O}}\cdot \longrightarrow :\overset{\cdot\cdot}{\underset{\cdot\cdot}{O}}:^{2-} \quad \text{(oxide ion)}$$
$$\underset{[\text{He}]2s^2 2p^4}{} \qquad \underset{[\text{He}]2s^2 2p^6 = [\text{Ne}]}{}$$

Two nonmetals (N and P) in Group VA gain three electrons to form $-3$ ions.

$$3e^- + \cdot\overset{\cdot}{\underset{\cdot}{N}}\cdot \longrightarrow :\overset{\cdot\cdot}{\underset{\cdot\cdot}{N}}:^{3-} \quad \text{(nitride ion)}$$
$$\underset{[\text{He}]2s^2 2p^3}{} \qquad \underset{[\text{He}]2s^2 2p^6 = [\text{Ne}]}{}$$

For the most part, Group IVA nonmetals bond by electron sharing rather than forming monatomic ions. Although there is some evidence for a $C^{4-}$ ion with an octet of electrons, formation of such highly charged ions is an energetically unfavorable process.

# 6-3 Formulas of Binary Ionic Compounds

**Looking Ahead!** If metals easily lose electrons and nonmetals gain electrons, why not put them together so an exchange of electrons can occur? In most cases, that's exactly what happens when we bring these elements together. As we have seen in earlier chapters, the formulas of the resulting ionic compounds are determined by the charges that the elements involved attain. Now we can understand that the magnitude of the charges on the ions is predictable from the octet rule. From this we can acquire a deeper understanding of the formulas of the resulting compounds.

When a small piece of sodium metal is placed in a bottle containing chlorine gas, a chemical change is obvious. (See Figure 6-1.) The sodium ignites, and a white coating of sodium chloride forms on the sides of the bottle. Chemical reactions are discussed in the next chapter, but we can appreciate in this discussion what has happened. Electrons from sodium atoms have been transferred to chlorine atoms to form an ionic compound, sodium chloride.

$$\text{Na} + :\overset{\cdot\cdot}{\underset{\cdot\cdot}{Cl}}: \longrightarrow \text{Na}^+ :\overset{\cdot\cdot}{\underset{\cdot\cdot}{Cl}}:^-$$

$$\text{Formula} = \text{NaCl} \quad \text{(sodium chloride)}$$

As indicated in the previous section, both of the ions formed have octets of electrons. Now let us consider what happens to the electrons when lithium metal comes in contact with the oxygen in the air. The oxygen atom needs two electrons to achieve an octet and form an anion with a $-2$ charge. Since a lithium atom can lose only one electron, two lithium atoms are needed to supply the two electrons.

$$\text{Li}\cdot$$
$$+ \cdot\overset{}{\underset{\cdot\cdot}{O}}: \longrightarrow 2(\text{Li}^+) :\overset{\cdot\cdot}{\underset{\cdot\cdot}{O}}:^{2-}$$
$$\text{Li}\cdot$$
$$\text{Formula} = \text{Li}_2\text{O} \quad \text{(lithium oxide)}$$

**Figure 6-1  Reaction of Sodium with Chlorine** Sodium reacts with chlorine to form sodium chloride.

Notice that the two $+1$ ions balance the charge of the one $-2$ ion [i.e., $2(+1) - 2 = 0$]. The chemical formula of the compound lithium oxide is therefore $Li_2O$.

Now consider the compound formed when calcium combines with bromine. Calcium loses two electrons to form a $+2$ cation, but a bromine can add only one electron to form a $-1$ anion. Two bromine atoms are needed to accept the two electrons lost by one calcium.

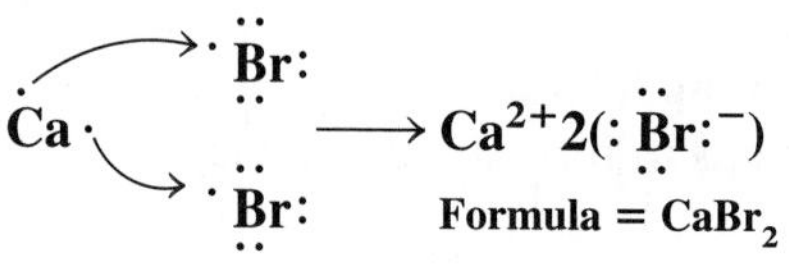

$$Ca \cdot \longrightarrow \quad \ddot{Br}: \quad \longrightarrow Ca^{2+}2(: \ddot{Br}:^-)$$

$$\ddot{Br}: \qquad Formula = CaBr_2 \quad \textbf{(calcium bromide)}$$

When aluminum combines with oxygen, it is somewhat more complex to follow the transfer of three electrons from aluminum to oxygen, which can accept only two. In this case, two Al's give up six electrons, which are then accepted by three O's. The formula is thus $Al_2O_3$, and the charges cancel [$2(+3) + 3(-2) = 0$].

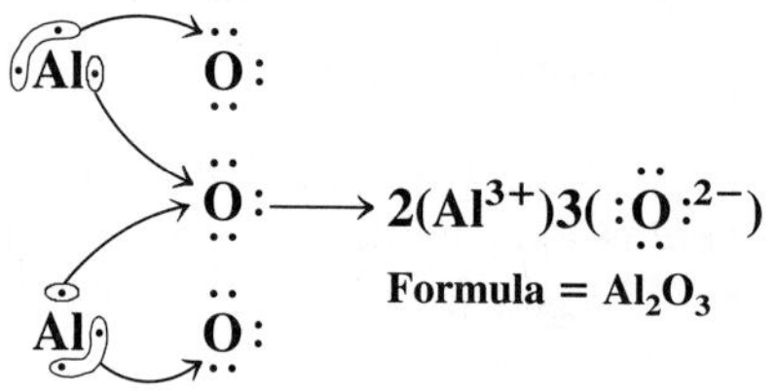

$$2(Al^{3+})3(: \ddot{O}:^{2-})$$

$$Formula = Al_2O_3 \quad \textbf{(aluminum oxide)}$$

---

| Example 6-1 | **The Formulas of Binary Ionic Compounds** |
|---|---|
| **working it out** | What are the formulas of the ionic compounds formed between (a) aluminum and fluorine and (b) barium and sulfur? |

**Solution**

a.  Aluminum is in Group IIIA and fluorine is in Group VIIA. Their dot symbols are

$$\cdot \dot{Al} \cdot \qquad \cdot \ddot{F}:$$

To have a noble gas configuration (an octet), the Al, a metal, must lose all three outer electrons to form a $+3$ ion. Three fluorine atoms are needed to add one electron each to form three $-1$ ions. Note that each fluorine can add only one electron, which gives the $F^-$ ion an octet. The compound formed is

$$Al^{3+} 3(: \ddot{F}:^-) = \underline{\underline{AlF_3}} \quad \textbf{(aluminum fluoride)}$$

b.  Barium is in Group IIA and sulfur is in Group VIA, and they have the dot symbols

$$Ba \cdot \qquad \ddot{S}:$$

One Ba atom gives up two electrons, and one S atom takes up two electrons, forming the compound

$$Ba^{2+} : \ddot{S}:^{2-} = \underline{\underline{BaS}} \quad \textbf{(barium sulfide)}$$

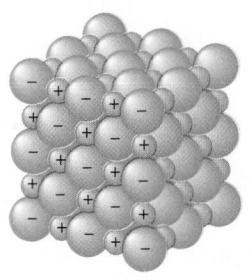
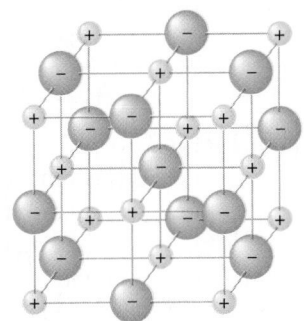

**Figure 6-2  An Ionic Solid**
Each cation (small spheres) is surrounded by six anions. Each anion (large spheres) is surrounded by six cations.

Ionic compounds are solids at room temperature. They tend to have high melting points and are often brittle. As mentioned in the chapter introduction, this type of compound comprises a large portion of the matter that we think of as "hard." If we look into the basic structure of a crystal of table salt, we can see why. Ionic compounds do not exist as discrete molecular units with one $Na^+$ attached to one $Cl^-$. As shown in Figure 6-2, each $Na^+$ is actually surrounded by six $Cl^-$ ions, and each $Cl^-$ ion is surrounded by six $Na^+$ ions in a *three-dimensional array of ions called a* **lattice.** Recall from Chapter 5 that cations are smaller than their parent atoms but anions are larger. Thus, in most cases, we can assume that the anion is larger than the cation. The lattice is held together strongly and rigidly by electrostatic interactions. *These electrostatic attractions are known as* **ionic bonds.** There are several other arrays of ions (lattices). For example, in CsCl both the $Cs^+$ and the $Cl^-$ are surrounded by eight oppositely charged ions.

Besides the monatomic ions, polyatomic ions exist where two or more atoms are bound together by electron sharing, and the total species carries a net charge [e.g., the carbonate ion $(CO_3^{2-})$]. These species exist as ions because they have an imbalance of electrons compared to the total number of protons in their nuclei. We will discuss the bonding within a polyatomic ion later in this chapter, but for now we acknowledge their existence in ionic compounds.

---

**Looking Back!** Metal and nonmetal atoms fit together perfectly. An exchange of electrons between the two leads to charged atoms that satisfy the octet rule. The result is the formation of ionic compounds. Rocks and many other hard substances are composed of ionic compounds. The hard, brittle nature of ionic compounds is explained by the rigid lattices formed by the oppositely charged ions.

---

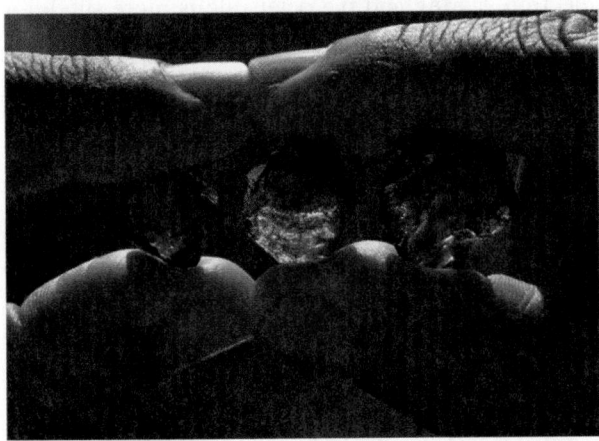

Rubies and sapphires are hard substances found in nature. They are ionic compounds.

## Learning Check A | checking it out

**A-1.** Fill in the blanks.

Elements that have the most stable electron configurations are in Group _____ . The outer electrons are also called the _____ electrons. Since noble gases have _____ valence electrons, other representative elements attain access to this number of electrons by _____ , _____ , or _____ electrons. By following the _____ rule, metals may form _____ ions with a maximum charge of _____ , and nonmetals may form _____ ions. The formula of a binary ionic compound is determined by the _____ on the cation and anion.

**A-2.** What are the Lewis dot symbols for Be and Se?

**A-3.** What are the charges on the ions formed by Be and Se?

**A-4.** What is the formula of a compound formed when Be combines with Se?

*Additional Examples: Exercises 6-1, 6-6, 6-11, 6-14, 6-19, and 6-23.*

Section **B** CHEMICAL BONDS AND THE NATURE OF MOLECULAR COMPOUNDS

## 6-4 The Covalent Bond

**Looking Ahead!** We are now ready to turn our attention to the softer part of nature. Metals are not involved in most of these compounds, so one atom does not give up electrons to another to form ions. In these cases, the octet rule is followed by electron sharing.

Hydrogen and oxygen are both nonmetals. When they are brought together, a dramatic chemical reaction occurs if initiated by a spark and water is the product. In the case of the formation of water, electrons are not exchanged but instead are shared between the hydrogens and the oxygen atom. A neutral molecule results. Compared to the strong interactions between oppositely charged ions, the discrete molecular units of water are only weakly attracted to each other; thus they can move past one another more freely than ions. For this reason, compounds composed of molecules tend to be gases, liquids, or solids with comparatively low melting points.

It is easy to appreciate how a complete exchange of electrons can satisfy the octet rule, but the concept of electron sharing and the octet rule is more subtle. A simple analogy may help. Assume we have a young man (Henry) who has $7. Henry wishes to have access to $8 and, for some strange reason, no more than $8. Now let Henry happen on to an even weirder person who also wishes to have no more than $8 but instead has $9. An exchange of $1 to Henry leaves them both deliriously happy. This is analogous to a metal and a nonmetal forming an ionic bond, with the metal giving its extra electron to the nonmetal. In a second situation, assume that two people have only $7 each and, again, both wish access to $8. There is a solution to the dilemma. If they keep $6 in their own pockets and contribute $1 each (for a total of $2) to a joint checking account, then both can claim access to $8 (but no more than $8). That is, each has $6 plus access to the $2 in the joint account.

A fluorine atom (Group VIIA) has seven valence electrons. Two fluorine atoms form a **covalent bond** by sharing two electrons, one from each fluorine. If each fluorine holds six electrons to itself, each atom achieves an octet of valence electrons by also having access to the two electrons in the bond. The bonding in the $F_2$ molecule, which is like that of all of the other diatomic halogens, is illustrated as follows:

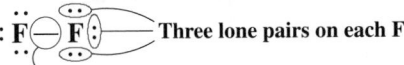

**Shared pair of electrons**
**(one from each F)**

We can now extend the concept of Lewis dot representations to covalent bonds. *A **Lewis structure** for a molecule shows the order and arrangement of atoms in a molecule (the structural formula) as well as all of the valence electrons for the atoms involved.* There are several variations of how Lewis structures represent molecules. A pair of electrons is sometimes shown as a pair of dots (:) or a dash (—). In this text, we use a pair of dots to represent unshared pairs (also called *lone pairs*) of electrons on an atom and a dash to represent a pair of electrons that are shared between atoms. In this way, shared and unshared electrons can be distinguished.

The Lewis structure of $F_2$* is illustrated as follows:

**Total of 14 outer electrons (7 from each F)**

**Three lone pairs on each F**

**Two shared electrons in a covalent bond**

Similarly, other halogens exist as diatomic molecules like $F_2$ and have the same Lewis structures. Hydrogen, which forms the simplest of all molecules, also exists as a diatomic gas with one covalent bond between atoms:

$$H—H$$

Recall that hydrogen follows a duet rule in order to attain the noble gas configuration of He.

Why do two H atoms combine to form an $H_2$ molecule? Again the answer is "because that is a more stable arrangement." Two hydrogen atoms alone have one electron in a $1s$ orbital, which is spherically diffuse. The electron has a probability of existing in any direction from the nucleus in the separate atoms. When two hydrogen atoms come together, however, the two electrons become more localized between the two nuclei in a region where the two $1s$ orbitals overlap. Each positive hydrogen nucleus is attracted to two negative electrons between the atoms rather than just one. Although there are also forces of repulsion between the two electrons, the mutual attraction of two nuclei for two electrons predominates and holds them together. (See Figure 6-3.)

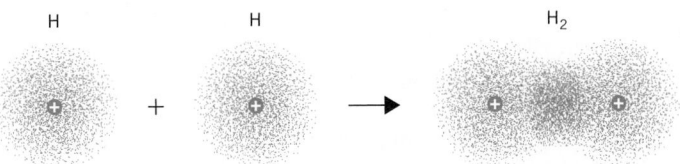

**Figure 6-3   A Covalent Bond** The high electron density between the two nuclei holds the two atoms together.

---

*An F atom has one unpaired electron in a $2p$ orbital. Formation of a covalent bond pairs the electrons in the two F atoms so that the $F_2$ molecule has no unpaired electrons. Although most atoms of the representative elements have unpaired electrons, most molecules or ions formed from these elements do not have unpaired electrons.

Just as we were able to justify the formulas of simple binary ionic compounds by the octet rule, we can do the same with simple binary molecular compounds. In fact, this works so well that we can predict the formulas of compounds based on the octet rule.

First, we will consider the compounds formed by hydrogen with the halogens in Group VIIA. For example, consider the compound formed from hydrogen and fluorine, which has the formula HF. A shared pair of electrons (one from each atom) gives both atoms access to the same number of electrons as a noble gas.

$$\text{H} \bigcirc \ddot{\text{F}}: \longrightarrow \text{H} - \ddot{\text{F}}:$$

Now we will consider the compounds formed between hydrogen and the Group VIA elements. Our primary example, of course, is water. Since an oxygen needs access to two more electrons to have an octet, we will need two hydrogens to form two covalent bonds to one oxygen.

$$\begin{matrix} \text{H} \\ \text{H} \end{matrix} \ddot{\text{O}}: \longrightarrow \text{H} - \ddot{\text{O}}: $$
$$\qquad\qquad\qquad\qquad | $$
$$\qquad\qquad\qquad\quad \text{H}$$

As we move across the periodic table to consider the hydrogen compounds formed between Group VA and Group IVA nonmetals, we see that the octet rule serves us well. Three hydrogens are needed by N (Group VA) and four by C (Group IVA). Recall that hydrogen is written second in binary compounds with Group IVA and VA elements but is written first with Group VIA and VIIA elements. (See also some hydrogen compounds of third-period elements in Figure 6-4.)

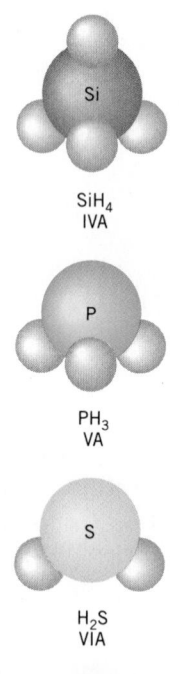

SiH₄
IVA

PH₃
VA

H₂S
VIA

HCl
VIIA

**Figure 6-4 Formulas of Hydrogen Compounds**
**The formulas of some simple hydrogen compounds can be predicted from the octet rule.**

$$\begin{matrix} \text{H} \\ \text{H} \\ \text{H} \end{matrix} \text{N}: \longrightarrow \begin{matrix} \text{H}-\text{N}: \\ | \\ \text{H} \end{matrix}$$

**Ammonia (NH₃)**

$$\begin{matrix} & \text{H} & \\ \text{H} & \text{C} & \text{H} \\ & \text{H} & \end{matrix} \longrightarrow \begin{matrix} \text{H} \\ | \\ \text{H}-\text{C}-\text{H} \\ | \\ \text{H} \end{matrix}$$

**Methane (CH₄)**

Let's try to predict the formula of the simplest compound formed between phosphorus (Group VA) and fluorine (Group VIIA). Since the dot symbol of P indicates that it needs to add three electrons and F needs to add one, the solution points to one P sharing a pair of electrons with each of three different F's.

$$:\ddot{\text{F}} \bigcirc \text{P} \bigcirc \ddot{\text{F}}: \longrightarrow :\ddot{\text{F}}-\text{P}-\ddot{\text{F}}:$$
$$\qquad\quad | \qquad\qquad\qquad |$$
$$\qquad\quad :\ddot{\text{F}}: \qquad\qquad :\ddot{\text{F}}:$$

**Simplest formula = PF₃**

In addition to molecular compounds, atoms within polyatomic ions share electrons in covalent bonds. For example, consider the hypochlorite ion ($ClO^-$). The $-1$ charge on the ion tells us that there is one more electron present in this species than the valence electrons of Cl and O. The total number of electrons is calculated as follows.

$$\begin{aligned} \text{From a neutral Cl} &= \phantom{0}7 \\ \text{From a neutral O} &= \phantom{0}6 \\ \text{Additional electron indicated by charge} &= \underline{\phantom{0}1} \\ \text{Total number of electrons} &= 14 \end{aligned}$$

## From Air to Proteins—The Nitrogen Story

Our bodies are literally held together and run by proteins. Proteins are composed of various amino acids. One of these is glycine ($H_2NCH_2COOH$). Notice that amino acids contain nitrogen. Nitrogen is essential to life, but most of it around us is in a useless form. About 80% of our atmosphere is nitrogen, but getting it from the elemental form and into proteins is not easy.

Elemental nitrogen ($N_2$) contains a triple bond. This is a very strong bond, and strong bonds are naturally hard to break. Before we can get nitrogen from the air and into compounds, the triple bond must be broken. Nature provides two ways (known as *nitrogen fixation*). The energy from a powerful bolt of lightening will do it. When the $N_2$ bond is broken, nitrogen atoms combine with oxygen in the air and form nitric oxide (NO), which comes to Earth in the rain. The second way is more important, and that is through the action of certain plants such as beans and peas (legumes) that have a bacteria attached to their roots. These bacteria contain an enzyme known as *nitrogenase* that has the unusual capacity to break the $N_2$ bond and produce nitrogen compounds to be used by their host plants (See

"Making It Real" on how enzymes work on page 175). This is how nature fertilizes our crops. Nature isn't nearly enough, however, so we must supplement the earth with huge amounts of ammonia ($NH_3$) that is manufactured in an expensive and difficult process. We will discuss this reaction in a later chapter.

Another of science's "Holy Grails" is a cheap and continuous way to fix nitrogen. The goal is to duplicate the action of nitrogenase in plants. The action of the enzyme is very complex and only partially understood. But we haven't given up. There is a lot of research at various laboratories searching for an agent (known as a catalyst) that will absorb atmospheric nitrogen, form bonds between nitrogen and other elements, release the new compound, and then regenerate the original agent. Such a process has so far been elusive, but the search goes on.

One could imagine the formation of a solid compound that could be spread on the surface of a field or garden that would continually generate nitrogen fertilizers. Duplication of nature's genius is proving to be difficult, but it is worth the effort. The payoff for such a substance would be immeasurable.

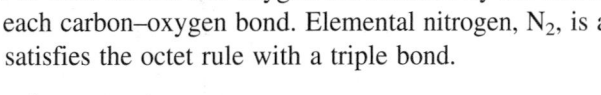

*Bean plants fix nitrogen.*

Two atoms bonded together with 14 electrons have a Lewis structure like $F_2$, which also has 14 electrons.

$$\left[ :\overset{..}{\underset{..}{Cl}}-\overset{..}{\underset{..}{O}}: \right]^-$$

The brackets indicate that the total ion has a $-1$ charge. The extra electron has not been specifically identified because all electrons are identical and belong to the ion as a whole.

In the examples illustrated so far, two atoms share one pair of electrons. There are also examples, especially among the second-period nonmetals (B through O), where two or even three pairs of electrons are shared between two atoms. *The sharing of two pairs of electrons between the same two atoms is known as a* **double bond.** *The sharing of three pairs is known as a* **triple bond.** A double bond is illustrated as ═, and a triple bond is illustrated as ≡. The use of a double bond to satisfy the octet rule is analogous to two people who wish to have access to $8 but only have a total of $12 between them. In this case, each person could have $4 in a private account while sharing $4 in a joint account. A triple bond would be analogous to the two people having only $10 between them. Each could have only $2 in a private account while sharing $6 in the joint account.

The molecules of carbon dioxide have double bonds. In the Lewis structure shown below, notice that the octets of both carbon and oxygen are satisfied by the sharing of two pairs of electrons in each carbon–oxygen bond. Elemental nitrogen, $N_2$, is an example of a molecule that satisfies the octet rule with a triple bond.

$$\overset{..}{\underset{..}{O}}=C=\overset{..}{\underset{..}{O}} \qquad :N\equiv N:$$

**Carbon dioxide is a molecular compound whose molecules contain double bonds. Solid carbon dioxide is known as dry ice.**

# 6-5 Writing Lewis Structures

**Looking Ahead!** The Lewis structures of simple binary compounds with single covalent bonds can be written without too much difficulty by observing the octet rule. The writing of other structures can be more complex, so a set of rules or guidelines is most helpful. In this section, we will see how a sequential approach will make this task quite manageable.

The octet rule is the key that allows us to write the correct Lewis structures for many compounds and polyatomic ions. From this we not only justify the formulas of compounds (e.g., $H_2O$ and not $H_3O$) but can predict other formulas as well. Other features of a compound flow from the correct Lewis structure. For example, later in this chapter we will use the Lewis structure to predict the geometry of some simple molecules. Many physical and chemical properties of a compound are directly related to its geometry. As we progress in the study of chemistry, we will continually refer to the Lewis structures of many compounds. Therefore, writing Lewis structures correctly is considered a fundamental skill to be acquired early in the study of chemistry.

Writing Lewis structures according to the octet rule is quite straightforward when certain guidelines or rules are systematically applied. These rules, of course, require considerable practice in their application. Starting with the formula of a compound (either ionic or molecular), the rules are as follows.

1. Check to see whether any ions are involved in the compounds. Write any ions present.
   a. Metal–nonmetal binary compounds are mostly ionic.
   b. If Group IA or Group IIA metals (except Be) are part of the formula, ions are present. For example, KClO is $K^+ClO^-$ because *K is a Group IA element and forms only a +1 ion*. If K is +1, the ClO must be −1 to have a neutral compound. Likewise, $Ba(NO_3)_2$ contains ions because *Ba is a Group IIA element and forms only a +2 ion*. To maintain neutrality, each $NO_3$ ion must have a −1 charge. The ions are represented as $Ba^{2+}2(NO_3^-)$.
   c. Compounds composed of nonmetals contain only covalent bonds.

2. For a molecule, add all of the outer (valence) electrons of the neutral atoms. For an ion, add (if negative) or subtract (if positive) the number of electrons indicated by the charge.

3. Write the symbols of the atoms of the molecule or ion in a skeletal arrangement.
   a. A hydrogen atom can form only one covalent bond and therefore bonds to only one atom at a time. Hydrogens are situated on the periphery of the molecule.
   b. The atoms in molecules and polyatomic ions tend to be arranged around a central atom. The central atom is generally a nonmetal other than oxygen or hydrogen. Oxygens usually do not bond to each other. Thus $SO_3$ has an S surrounded by three O's.

$$\begin{matrix} & O & \\ & S & \\ O & & O \end{matrix}$$

rather than such structures as

$$\text{S O O O} \qquad \text{O S O O} \qquad \begin{matrix} S & O \\ O & O \end{matrix}$$

In most cases, the first atom in a formula is the central atom, and the other atoms are bound to it.

4. Place a dash representing a shared pair of electrons between adjacent atoms that have covalent bonds (not between ions). Subtract the electrons used for this (two for each bond) from the total calculated in step 2.

5. Distribute the remaining electrons among the atoms (periphery atoms first) so that no atom has more than eight electrons.

6. Check all atoms for an octet (except H). If an atom has access to fewer than eight electrons (usually the central atom), put an electron pair from an adjacent atom into a double bond. Each double bond increases by two the number of electrons available to the atom needing electrons. Remember that you cannot satisfy an octet for an atom by adding any electrons at this point.

An alternative method combines steps 4, 5, and 6. In this method, a count of electrons is used to determine the number of multiple bonds present, and then electrons and bonds are added to the skeletal structure accordingly. This has been conveniently summarized as the "$6N + 2$ rule," where $N$ stands for the number of atoms other than hydrogen in the formula. If the number of valence electrons in the formula equals $6N + 2$, then only single bonds are present. If the number of valence electrons is two less than $6N + 2$, then one double bond is present. (It could also mean that a ring structure is present; these are discussed briefly in Chapter 16.) If the number of valence electrons is four less than $6N + 2$, then one triple bond or two double bonds are present. Consider the following molecules.

**a.** $PF_3$ $\quad N = 4 \quad 6(4) + 2 = 26$
Valence electrons: $5[P] + (3 \times 7)[F] = 26$
Therefore, only single bonds are present.

**b.** $CO_3^{2-}$ $\quad N = 4 \quad 6(4) + 2 = 26$
Valence electrons: $4[C] + (3 \times 6)[O] + 2[\text{charge}] = 24$
$26 - 24 = 2$
Therefore, one double bond is present.

**c.** $C_6H_{10}$ $\quad N = 6 \quad 6(6) + 2 = 38$
Valence electrons: $(6 \times 4)[C] + (10 \times 1)[H] = 34$
$38 - 34 = 4$
Therefore, two double bonds or one triple bond is present in this molecule.

In this text we use the $6N + 2$ rule as a check to confirm the Lewis structure determined by applying steps 1 through 6.

---

**The Lewis Structure of a Molecular Compound**

Write the Lewis structure for $NCl_3$.

**Example 6-2**

**w o r k i n g   i t   o u t**

**Solution**

1. This is a binary compound between two nonmetals. Therefore, it is not ionic.

2. The total number of electrons available for bonding is

$$
\begin{array}{rl}
\text{N } 1 \times 5 = & 5 \\
\text{Cl } 3 \times 7 = & \underline{21} \\
\text{Total} = & 26
\end{array}
$$

3. The skeletal arrangement is

$$\text{Cl} \quad \text{N} \quad \text{Cl}$$
$$\text{Cl}$$

4. Use 6 electrons to form bonds.

$$Cl-N-Cl$$
$$\quad\;\; | \;\;$$
$$\quad\;\; Cl$$

5. Distribute the remaining 20 electrons (26 − 6 = 20).

$$:\ddot{Cl}-N-\ddot{Cl}:$$
$$\qquad | $$
$$\quad\;: \ddot{Cl}:$$

6. Check to make sure that all atoms satisfy the octet rule.

$$:\ddot{Cl} \text{—} N \text{—} \ddot{Cl}:$$
$$\qquad | $$
$$\quad\;: \ddot{Cl}:$$

The $6N + 2$ rule confirms this structure.
For $NCl_3$, $N = 4 \qquad 6(4) + 2 = 26$
From step 2, there are 26 valence electrons, so there are only single bonds.

---

**Example 6-3** | **The Lewis Structure of an Ion**

Write the Lewis structure of the cyanide ion ($CN^-$).

**Solution**

1. This is an ion.

2. The total number of electrons available is

$$
\begin{aligned}
N\; 1 \times 5 &= \;\;5 \\
C\; 1 \times 4 &= \;\;4 \\
\text{From charge} &= \;\underline{\;1} \\
\text{Total} &= \overline{\;10}
\end{aligned}
$$

3. The skeletal arrangement is

$$C \quad N$$

4. Use two electrons to form a bond.

$$C-N$$

5. Distribute the remaining eight electrons (10 − 2 = 8).

$$\left[:\ddot{C}-\ddot{N}:\right]^-$$

6. Notice that both carbon and nitrogen have access to only six electrons each. Use two electrons from the carbon and two electrons from nitrogen to make a triple bond. Now the octet rule is satisfied.

$$\left[:C\equiv N:\right]^-$$

The $6N + 2$ rule confirms the structure.
For $CN^-$, $N = 2 \qquad 6(2) + 2 = 14$
$14 - 10 = 4$   (two double bonds or one triple bond)

---

**Example 6-4** | **The Lewis Structure of an Ionic Compound**

Write the Lewis structure for $CaCO_3$.

**Solution**

1. This is an ionic compound composed of $Ca^{2+}$ and $CO_3^{2-}$ ions. (Since you know that Ca is in Group IIA, it must have a +2 charge; therefore the polyatomic anion must be −2.) A Lewis structure can be written for $CO_3^{2-}$.

2. For the $CO_3^{2-}$ ion the total number of outer electrons available is

$$C\ 1 \times 4 = \ 4$$
$$O\ 3 \times 6 = 18$$
$$\text{From charge} = \underline{\ 2}$$
$$\text{Total} = 24$$

3, 4. The skeletal structure with bonds is

$$\begin{array}{ccc} O & & O \\ & \diagdown \ \diagup & \\ & C & \\ & | & \\ & O & \end{array}$$

5. Add the remaining 18 electrons (24 − 6 = 18).

$$\left[ \ddot{\underset{..}{O}} \diagdown \underset{C}{} \diagup \overset{..}{\underset{..}{O}} \right]^{2-}$$
$$:\!\ddot{O}\!:$$

6. The C needs 2 more electrons, so one double bond is added using one lone pair from one oxygen.

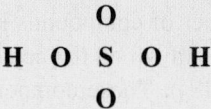

The 6N + 2 rule confirms the structure.
For $CO_3^{2-}$, N = 4      6(4) + 2 = 26
26 − 24 = 2   (one double bond)

**The Lewis Structure of an Acid** | **Example 6-5**

Write the Lewis structure for $H_2SO_4$.

**Solution**

1. All three atoms are nonmetals, which means that all bonds are covalent.

2. The total number of outer electrons available is

$$H\ 2 \times 1 = \ 2$$
$$S\ 1 \times 6 = \ 6$$
$$O\ 4 \times 6 = \underline{24}$$
$$\text{Total} = 32$$

3. In most molecules containing H and O, the H is bound to an O and the O to some other atom, which in this case is S. The skeletal structure is

$$\begin{array}{ccccc} & & O & & \\ H & O & S & O & H \\ & & O & & \end{array}$$

4. Use 12 electrons for the six bonds.

$$
\begin{array}{c}
\text{O} \\
| \\
\text{H}-\text{O}-\text{S}-\text{O}-\text{H} \\
| \\
\text{O}
\end{array}
$$

5. Add the remaining 20 electrons ($32 - 12 = 20$).

$$
\begin{array}{c}
\ddot{\text{O}}\!: \\
| \\
\text{H}-\ddot{\text{O}}-\text{S}-\ddot{\text{O}}-\text{H} \\
| \\
\ddot{\text{O}}\!:
\end{array}
$$

6. All octets are satisfied.
   The $6N + 2$ rule is consistent with this structure.
   For $H_2SO_4$, $N = 5$ (exclude hydrogens)    $6(5) + 2 = 32$
   From step 2, valence electrons = 32
   This indicates that only single bonds are present.

The octet rule is very useful in describing the bonding in most compounds of the representative elements. A significant number of compounds, however, do not follow the octet rule. For example, some molecules have an odd number of valence electrons such as NO (11 valence electrons) and $NO_2$ (17 valence electrons). In both of these cases the nitrogen has access to fewer than 8 electrons. In still others, a Lewis structure may be written that follows the octet rule, but other evidence suggests that the situation is more complex. For example, the ordinary $O_2$ molecule with 12 valence electrons could be written with a double bond as shown below. This representation implies that all electrons are in pairs. However, we know from experiments that $O_2$ has two unpaired electrons. We just need to remember that the Lewis structure is simply the representation of a theory and of the case of $O_2$, the theory doesn't work perfectly. There are other theories of bonding that include the case of $O_2$ quite well. Another example of the conflict between theory and experiments is illustrated by the molecule $BF_3$. Experiments indicate that the B—F bond has little to no double bond character. Thus the correct structure shows the boron with access to only 6 electrons.

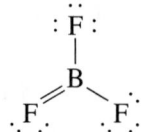

| | | | |
|---|---|---|---|
| : N=O: | :O=O: | | |
| A correct Lewis representation that *does not* follow octet rule | A Lewis representation of $O_2$ that *does not* correspond with experiments | A Lewis representation of $BF_3$ that *does not* correspond with experiments | A Lewis representation of $BF_3$ that *does* correspond with experiments |

There are a significant number of compounds involving representative elements from the third and higher periods in which the central atom has access to more than eight electrons (e.g., $SF_4$ and $ClF_5$). These compounds have not been discussed in this text.

# 6-6 Resonance Structures

**Looking Ahead!** Glance again at the Lewis structure of the carbonate ion shown in Example 6-4. The structure that is displayed implies that the three carbon–oxygen bonds are not all identical in that one bond is double and the other two are single. Is that true? Actually, the answer is "no," but we need to explore this phenomenon in more detail.

If we compare a double bond to a single bond between the same two elements, we find there are significant differences. The sharing of four electrons holds two atoms together more strongly, and thus more closely, than the sharing of two electrons. Likewise, a triple bond is even stronger and shorter than a double bond. The one Lewis structure of the $CO_3^{2-}$ ion shown in Example 6-4 implies that one carbon–oxygen bond is shorter and stronger than the other two. We know from experiment, however, that the ion is perfectly symmetrical, meaning that all three bonds are identical. Experiments also tell us that the lengths of the three identical bonds are somewhere between those expected for a single and a double bond. One Lewis representation of the $CO_3^{2-}$ ion does not convey this information, but three representations (connected by double-headed arrows) indicate that all three bonds are intermediate between a single and a double bond. *The three structures as shown below are known as* **resonance structures.** The actual structure of the molecule can be viewed as a **resonance hybrid** of the three structures.

Resonance structures exist for molecules when equally correct Lewis structures can be written without changing the basic skeletal geometry or the position of any atoms.

Elemental oxygen occurs in nature in two different molecular forms, each with its own properties. *Different forms of the same element are known as* **allotropes.** The most prominent allotrope, the oxygen we breathe, has the formula $O_2$. The other form of oxygen is ozone, $O_3$, which is important in the stratosphere as a shield from ultraviolet light of the sun. Ozone is an example of a molecule whose bonding is described as a resonance hybrid, as shown in the following example.

---

**Resonance Structures of Ozone**

**Example 6-6**

Write a Lewis structure and any equivalent resonance structures for ozone ($O_3$) where one oxygen serves as the central atom.

*working it out*

**Solution**

1. No ions are involved.
2. There are 18 ($3 \times 6$) outer electrons.
3. The skeletal structure is

$$O \quad \overset{O}{\phantom{O}} \quad O$$

4. Use 4 electrons for the two bonds.

$$O \diagdown \overset{O}{\phantom{O}} \diagup O$$

5. Add the remaining 14 electrons.

6. Notice that the central oxygen does not have an octet, so make one double bond to one of the other oxygens.

The $6N + 2$ rule confirms the presence of one double bond. Two resonance structures can be written, which indicates that each oxygen–oxygen bond is a hybrid between a single and a double bond.

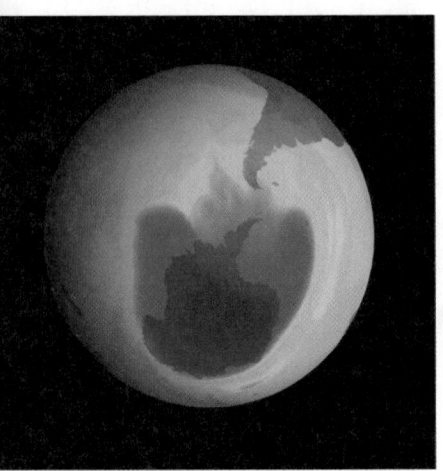

The pale blue over Antarctica (dark blue) in October 1999 shows the area of ozone depletion.

The word *resonance* is associated with vibration or a constantly changing situation. That concept can be misleading in this case. The two oxygen–oxygen bonds in ozone are not changing rapidly back and forth between a single and a double bond. In fact, both bonds exist *at all times* as intermediate between a single and a double bond. It is much like a large, sweet hybrid tomato that you may grow in the garden or buy in the grocery store. This tomato is a hybrid of a large tomato and a small but sweet tomato. It isn't changing rapidly back and forth between these two forms but exists with properties intermediate between the two original species of tomatoes. When we view resonance structures, we try to visualize the molecule as a combination of the two or more structures. This is analogous to trying to taste the hybrid tomato by combining a piece of the large tomato along with a piece of the small, sweet tomato in your mouth at the same time.

**Looking Back!** Binary compounds composed of two nonmetals form molecular compounds because the atoms bond by electron sharing rather than electron exchange. This means that nonmetal–nonmetal binary compounds, as well as molecular elements and the atoms of noble gas elements, exist in discrete units that are not as strongly attracted to each other as are ions. These are the gases, liquids, and soft solids of the world. By applying certain rules, we can easily understand the bonding and formulas of molecular compounds.

## Learning Check B | checking it out

**B-1.** Fill in the blanks.

The sharing of electrons between two atoms is known as a _____ bond. Lewis structures show shared electron pairs as _____ and unshared electrons as _____ _____ _____ . The octet rule helps us understand that the simplest compound between hydrogen and chlorine is _____ , between hydrogen and sulfur is _____ , and between hydrogen and phosphorus is _____ . The Lewis structure of carbon dioxide indicates the sharing of two pairs of electrons between a carbon and oxygen, which is known as a _____ _____ . Elemental nitrogen contains a _______ _______. Three equivalent Lewis structures can be written for the carbonate ion; they are known as _______ structures. The actual structure is a _______ _______ of the three structures. Ozone is an _______ of oxygen.

**B-2.** Write the Lewis structure for (a) $SF_2$, (b) $NO_2Cl$ where N is the central atom, and (c) $Mg(ClO_2)_2$.

**B-3.** In the $NCO^-$ ion, the nitrogen–carbon bond is about midway between a double and a triple bond. Write two resonance structures that are consistent with this conclusion. (The C is the central atom in this ion.)

*Additional Examples: Exercises 6-26, 6-32, 6-38, and 6-44.*

## Section C THE DISTRIBUTION OF CHARGE IN CHEMICAL BONDS

## 6-7 Electronegativity and Polarity of Bonds

**Looking Ahead! In the formation of an ionic bond, electrons are exchanged. In the formation of a covalent bond, electrons are shared. But does this mean that the electrons in the covalent bond are shared equally? As we will see, the situation is not so simple. Atoms of different elements rarely share electrons equally in a bond between them. This unequal sharing leads to a partial charge, as we will see in this section.**

You are probably aware that *sharing* a carton of popcorn at a movie rarely means *equal sharing.* The hungrier, faster popcorn eater usually gets the lion's share. Likewise, in a chemical bond between two different atoms, the pair of electrons is not shared equally, and one atom gets a larger share of the electrons. *The ability of an atom of an element to attract electrons to itself in a covalent bond is known as the element's* **electronegativity.** The value assigned for the electronegativity of each element is shown in Figure 6-5. The most electronegative element is fluorine, which is assigned an electronegativity value of 4.0. Notice that nonmetals tend to have higher values of electronegativity than metals. The values shown in Figure 6-5 were first calculated by Linus Pauling (winner of two Nobel Prizes). Although more refined values are now available, the actual numbers are not as important as how the electronegativity of one element compares with that of another.

Increases →

Decreases ↓

| | | | | | | |
|---|---|---|---|---|---|---|
| 1 H 2.1 | | | | | | |
| 3 Li 1.0 | 4 Be 1.5 | 5 B 2.0 | 6 C 2.5 | 7 N 3.0 | 8 O 3.5 | 9 F 4.0 |
| 11 Na 0.9 | 12 Mg 1.2 | 13 Al 1.5 | 14 Si 1.8 | 15 P 2.1 | 16 S 2.5 | 17 Cl 3.0 |
| 19 K 0.8 | 20 Ca 1.0 | 31 Ga 1.6 | 32 Ge 1.8 | 33 As 2.0 | 34 Se 2.4 | 35 Br 2.8 |
| 37 Rb 0.8 | 38 Sr 1.0 | 49 In 1.7 | 50 Sn 1.8 | 51 Sb 1.9 | 52 Te 2.1 | 53 I 2.5 |
| 55 Cs 0.7 | 56 Ba 0.9 | 81 Tl 1.8 | 82 Pb 1.8 | 83 Bi 1.9 | 84 Po 2.0 | 85 At 2.2 |
| 87 Fr 0.7 | 88 Ra 0.9 | | | | | |

**Figure 6-5 Electronegativity**

Electrons carry a negative charge. When there is a complete exchange of an electron between atoms, as in the formation of an ionic bond, one atom acquires a full negative charge. In a covalent bond between two atoms of different electronegativity, the more electronegative atom attracts the electrons in the bond partially away from the other atom and thus acquires a partial negative charge (symbolized by $\delta^-$). This leaves the less electronegative atom with a partial positive charge (symbolized by $\delta^+$).

*A covalent bond that has a partial separation of charge due to the unequal sharing of electrons is known as a* **polar covalent bond** (or simply, **polar bond**). A polar bond has a negative end and a positive end and is said to contain a **dipole** (two poles). A polar bond is something like Earth itself, which contains a magnetic dipole with a north and south magnetic pole. (The poles in a bond dipole are electrostatic rather than magnetic.) The dipole of a bond is represented by an arrow pointing from the positive to the negative end ($\longmapsto$).

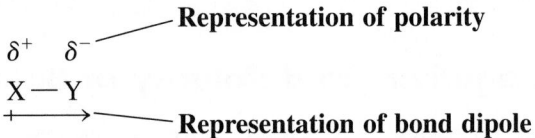

The polarity of bonds has a significant effect on the chemical properties of compounds. For example, the polarity of the H—O bonds in water accounts for many of its familiar properties that we take for granted. We will discuss the chemistry of water in more detail in Chapter 10.

When electrons are shared between atoms of the same element, they are obviously shared equally. *If electrons are shared equally, the bond is known as a* **nonpolar bond.** On the other hand, the greater the difference in electronegativity between two atoms, the more polar the bond. In fact, if the difference is 1.9 or greater, it may indicate that one atom has gained complete control of the pair of electrons. In other words, the bond is ionic.

In summary, when two atoms compete for a pair of electrons in a bond, there are three possibilities for the pair of electrons.

1.  Both atoms share the electrons equally, forming a nonpolar bond (an electronegativity difference between the two atoms of zero or near zero).

2.  The two atoms share electrons unequally, forming a polar bond. This is intermediate between purely ionic and equal sharing (an electronegativity difference of less than 1.9).

3.  The electron pair is not shared since one atom acquires the electrons. This is the ionic bond, in which each atom acquires a complete charge (an electronegativity difference greater than about 1.8).

These three cases are illustrated in Figure 6-6. The bond in $Cl_2$ is nonpolar (case 1) since both atoms are identical. To determine the charge on each Cl, we will assign electrons to the two Cl atoms. In this case, each Cl is assigned the six electrons from its three lone pairs. Since the two atoms share the pair evenly, we can assign exactly one-half of the shared electrons to each Cl for a total of seven each $[6 + (1/2 \times 2) = 7]$. Since Cl is in Group VIIA, seven valence electrons leave each Cl exactly neutral. Thus there is not a positive and a negative end.

The molecule HCl illustrates case 2. There is a significant difference in electronegativity between H and Cl (0.9), indicating a polar bond but not so much as to indicate an ionic bond. Since the Cl is more electronegative than H, it has a

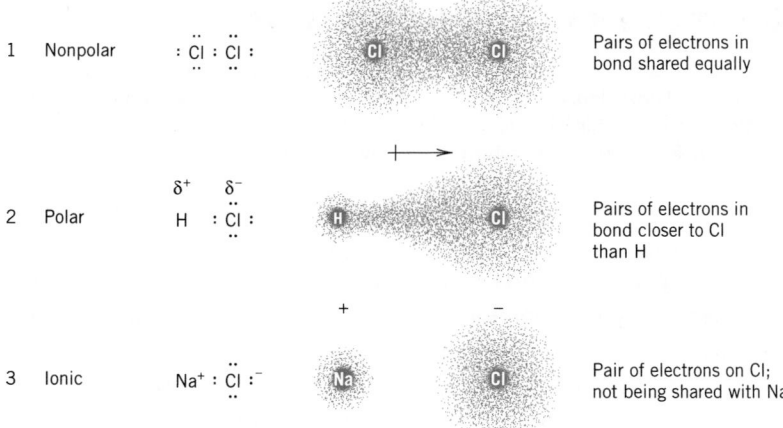

**Figure 6-6  Nonpolar, Polar, and Ionic Bonds**

partial negative charge. In this case, the Cl is still assigned six electrons from its three lone pairs but more than one-half of the pair of electrons in the bond. Since it is assigned more than seven electrons but less than eight, it acquires *a partial negative charge*. The hydrogen, with less than one-half of the electron pair, has fewer than one electron and thus has an equal but opposite *partial positive charge*.

The bond in NaCl illustrates case 3. As mentioned earlier in this chapter, there is a complete exchange of the valence electron from Na to Cl, producing charged ions. Eight electrons on Cl form an anion with a −1 charge. Notice that the difference in electronegativity between Na and Cl is large (2.1), predicting an ionic nature.

---

**The Polarity of Bonds**

Referring to Figure 6-5, rank the following bonds in order of increasing polarity. The positive end of the dipole is written first. On the basis of electronegativity differences, indicate if any of the bonds are predicted to be ionic.

$$Ba—Br \quad C—N \quad Be—F \quad B—H \quad Be—Cl$$

**Example 6-7**

*working it out*

**Solution**

Calculate the difference in electronegativity between the elements.

$$
\begin{array}{ll}
Ba—Br & 2.8 - 0.9 = 1.9 \\
C—N & 3.0 - 2.5 = 0.5 \\
Be—F & 4.0 - 1.5 = 2.5 \\
B—H & 2.1 - 2.0 = 0.1 \\
Be—Cl & 3.0 - 1.5 = 1.5 \\
\end{array}
$$

$$B—H < C—N < Be—Cl < Ba—Br < Be—F$$

The differences in electronegativity suggest that Ba—Br and Be—F have ionic bonds. Notice that the difference in electronegativity suggests that the Be—Cl bond is polar covalent rather than ionic. This is confirmed by a profound difference in the physical properties of Be compounds containing these two bonds.

# 6-8 Geometry of Simple Molecules

**Looking Ahead!** Because the covalent bonds in most molecules are polar, does that mean the molecule itself is polar? Surprisingly, the answer is "not necessarily." The polarity of a molecule depends on its geometry. In this section, we will use Lewis structures to tell us about the geometry of some simple molecules and then return to the subject of molecular polarity in the next section.

Every blink of the eye requires the action of a certain enzyme (a protein) in our body. An enzyme may interact with a specific receptor site in a muscle like a specific key fits into a lock. Like a key, the action of an enzyme is determined by its configuration or geometry. Consider also ordinary water. If water were a linear molecule rather than bent, this would drastically affect its properties. For example, water would be a gas at room temperature rather than a liquid. Life, as we know it, could not exist under those conditions. When properly interpreted, however, the Lewis structure of water indicates its bent nature. The approximate geometry of the atoms around a central atom can be predicted by the **valence shell electron-pair repulsion theory (VSEPR).** *This theory tells us that electron pairs, either unshared pairs or electrons localized in a bond, repel each other to the maximum extent.* In other words, the negatively charged electron groups get as far away from each other as possible (without breaking the bonds) because of their electrostatic repulsion. As a simple example, consider $BeH_2$ (a molecular compound that is an exception to the octet rule), which has two Be—H bonds. To be as far apart as possible, the two electron pairs in the bonds will lie on opposite sides of the beryllium atom at an angle of 180°. The geometry of the molecule is described as *linear.* Other molecules that are linear are $CO_2$ and HCN, as shown below. *Notice that double and triple bonds are considered the same as single bonds.* That is, single, double, and triple bonds are all counted as one group of electrons (even though the latter two constitute larger groups).

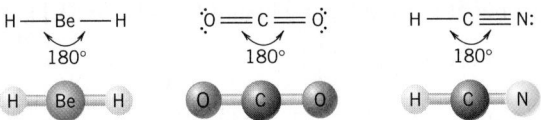

Now consider the $BF_3$ molecule, which has three groups of electrons. In this case, the three groups can get as far away from each other as possible by assuming the geometry of an equilateral triangle with an F—B—F angle of 120°. The geometry of this molecule is described as *trigonal planar.* A similar angle is assumed by the $SO_2$ molecule, which also has three electron groups. The central sulfur atom is bonded to two oxygen atoms, and it has one unshared pair of electrons. (Remember that the two S—O bonds are actually identical since two resonance structures can be written.) The O—S—O angle is approximately that of an equilateral triangle. *The* **molecular geometry** *of a molecule is the geometry described by the bonded atoms and does not include the unshared pairs of electrons.* Thus we describe the molecular geometry of the three atoms in $SO_2$ as *V-shaped or bent.*

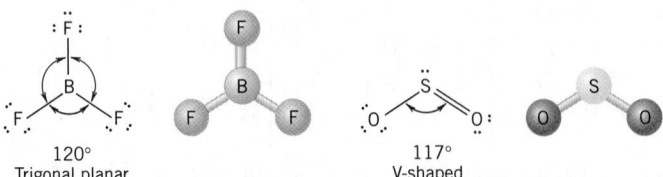

120°
Trigonal planar

117°
V-shaped

In molecules having four groups of bonding electrons, such as $CH_4$, the four hydrogens are found at the corners of a regular tetrahedron. This allows an angle of a little more than 109°. (This is a more open structure than if the hydrogens were located at the corners of a square. In this case, the angle would be only 90°.) Thus the geometry of $CH_4$ is described as *tetrahedral*. Ammonia ($NH_3$) also has four groups of electrons, but one group is an unshared pair of electrons. The H—N—H angle is found to be 107°, which is in good agreement with the angle predicted by this theory. (The angle is somewhat less than 109° because lone pairs of electrons take up more space than bonded pairs.) The molecular geometry of the $NH_3$ molecule is described as *trigonal pyramidal*. Finally, the common $H_2O$ molecule also has four groups but with two unshared pairs of electrons. The H—O—H angle is known to be 105°, which also agrees wth this theory. The molecular geometry of $H_2O$ is described as *V-shaped* or *bent*. (In the V-shaped structure of $SO_2$, the angle of 117° is near the expected trigonal angle of 120°; in the V-shaped structure of $H_2O$, the angle of 105° is near the tetrahedral angle of 109°.)

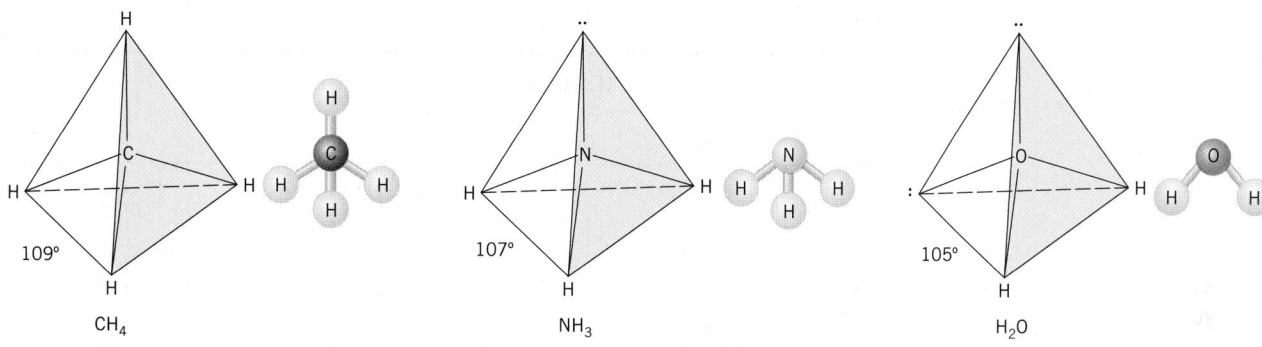

The molecular geometries for these molecules are summarized in Table 6-2.

**Table 6-2  Molecular Geometry**

| Number of Electron Groups on Central Atom (Bonded Atoms + Unshared Pairs) | Number of Atoms Bonded to Central Atom | Number of Unshared Pairs of Electrons on Central Atom | Molecular Geometry | Model |
|:---:|:---:|:---:|---|:---:|
| 2 | 2 | 0 | linear | |
| 3 | 3 | 0 | trigonal planar | |
| 3 | 2 | 1 | V-shaped (near 120°) | |
| 4 | 4 | 0 | tetrahedral | |
| 4 | 3 | 1 | trigonal pyramidal | |
| 4 | 2 | 2 | V-shaped (near 109°) | |

**Example 6-8**

working it out

## The Geometry of Molecules

What is the molecular geometry of each of the following ions and molecules?

**a.** OCN⁻      **b.** H₂CO      **c.** HClO      **d.** SiCl₄

In (a) and (b), the middle atom in the formula is the central atom. In (c), oxygen is the central atom.

**Procedure**

First write the correct Lewis structure. Then count the number of electron groups [bonded atoms (connected to the central atom) + the number of unshared pairs of electrons].

**Solution**

| | Lewis Structure | Number of Electron Groups | Molecular Geometry | |
|---|---|---|---|---|
| (a) OCN⁻ | :O=C=N:⁻ | 2 | linear (180°) | |
| (b) H₂CO | H\C=O: / H | 3 | trigonal planar | |
| (c) HClO | H—O—Cl: | 4 | V-shaped (109°) | |
| (d) SiCl₄ | :Cl: / :Cl—Si—Cl: / :Cl: | 4 | tetrahedral | |

# 6-9 Polarity of Molecules

**Looking Ahead!** A molecule such as CO₂ is composed of two polar bonds, but the molecule itself is not polar. This sounds contradictory, but it is actually predictable. What we need to know to make such a weird prediction, however, is how the geometry of the molecule affects its polarity. We will unite our discussion of bond polarity and geometry in this final section of the chapter.

When a person pulls on a rope attached to a heavy box sitting on the floor, how easily the box moves and in what direction depend on the force that he imparts and the direction in which he pulls. A polar bond in a molecule can also be considered a force with both direction and magnitude. The magnitude of the force of the polar bond depends on the degree of polarity, which relates to the difference in electronegativity of the two atoms. The greater the difference, the greater the partial charges and the larger the magnitude of the dipole. But the force has direction as well. In Figure 6-7, if two

**Figure 6-7   Forces at 180°**
**If forces are equal but opposite (180°), they cancel.**

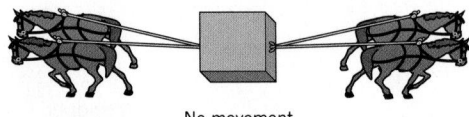

No movement

## Enzymes—The Keys of Life

You can't blink your eye or digest a slice of pizza without the help of enzymes. Enzymes demonstrate quite dramatically the importance of geometry in how they function to do so many jobs. Enzymes are rather complex molecules known as proteins that serve as catalysts. A catalyst is an agent that causes or speeds up a chemical reaction but is not itself consumed in the reaction. Catalysts will be discussed in more detail in later chapters. In effect, an enzyme is much like a specific key that unlocks a specific door. If the geometry of the key is not exact, it does not work. The action of an enzyme in doing a particular job has to do with the geometry of a part of the molecule known as its *active site*.

Glucose, blood sugar, is the only sugar that we can use in our body in metabolism, the combustion reaction that provides the energy to keep us alive. Carbohydrates and other sugars including sucrose (table sugar) must be broken down or converted into glucose before they can be used. Sucrose is actually composed of two simpler sugars, fructose and glucose, chemically bonded together. An enzyme known as *sucrase* is responsible for breaking sucrose

into its components. The sucrase molecule has a specific molecular geometry that is complementary to the sucrose molecule. The polarity of the bonds in the enzyme are also such as to attract the sucrose molecule so as to form an exact fit between the two geometries. The enzyme then causes the sucrose to break into its two components, which are then released. The sucrose molecule can then seek out another sucrose molecule and repeat the process. This is illustrated below. Other enzymes cause the fructose to be converted into glucose so that it too is used in metabolism.

Dairy products contain milk sugar (lactose), which must be broken down by the enzyme, *lactase*. However, some people are "lactose intolerant," which means that they lack the lactase enzyme. To avoid discomfort they must avoid dairy products or take a lactase tablet (i.e., Lactaid) and then they can enjoy a dish of ice cream.

The list of enzymes seems endless. There is a specific enzyme in our body for every biochemical function. In fact, we are just now learning about many of the enzymes that make life possible.

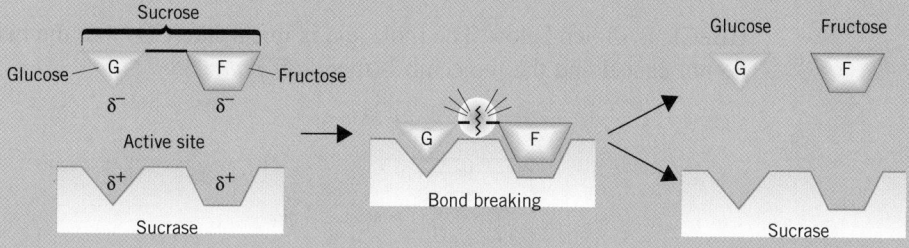

*Enzymes work like a lock and key.*

equal teams of horses are pulling against each other in opposite directions (i.e., an angle of exactly 180°), the forces exactly cancel and as a result there is no movement. In a molecule, *the combined effects of the bond dipoles (the net dipole) are known as the* **molecular dipole.** If the geometry of the molecule is such that equal dipoles cancel, *there is no molecular dipole, which means that the molecule is nonpolar.* For example, in $CO_2$ the bond dipoles are equal and go in exactly opposite directions. $CO_2$ is thus a nonpolar molecule. The same is true for trigonal planar (e.g., $BF_3$) and tetrahedral molecular geometries (e.g., $CCl_4$), where all terminal atoms are the same. (See Figure 6-8.)

In Figure 6.9*a*, if the teams of horses are not equal, the two opposite forces are not equal and there is movement in the direction of the larger team. In the case of molecules, if the terminal atoms connected to the central atom are not identical, the bond dipoles are not all the same and thus do not cancel. Consider the molecule

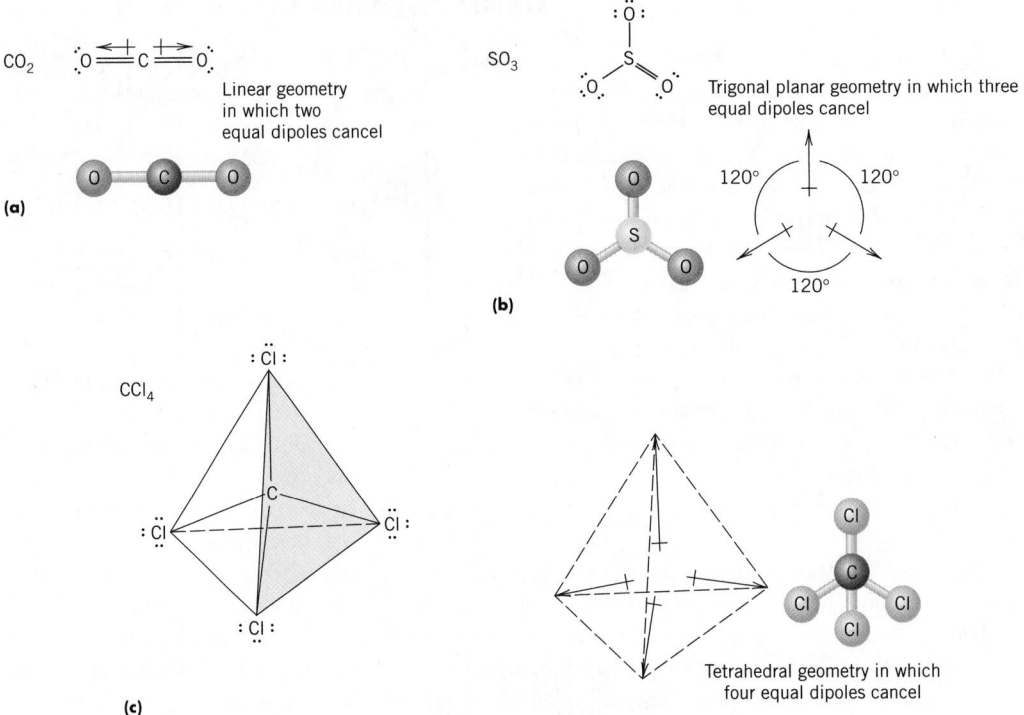

**Figure 6-8  Nonpolar Molecules. If resultant forces exactly cancel in linear (*a*), trigonal planar (*b*), and tetrahedral geometries (*c*), the molecule is nonpolar.**

HBeCl, as shown below. The molecule is linear like $CO_2$, but the two unequal forces do not cancel and the molecule is therefore *polar.*

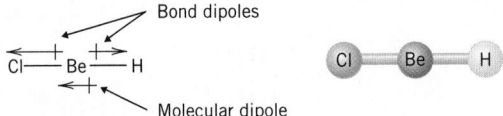

If at least one of the terminal atoms in the examples shown in Figure 6-8 is different from the others, the molecule is polar (e.g., $CCl_4$ is nonpolar but $CHCl_3$ is polar.)

Now let's consider the geometries described as V-shaped and trigonal pyramidal. In these two cases the bond dipoles do not cancel regardless of the nature of the terminal atoms. This is analogous to two equal teams of horses pulling at an angle less than 180°. There is movement along a line between the teams. (See Figure 6-9*b*.) The most important example of a molecule that is polar for this reason is the V-shaped molecule water. The two O—H bonds are at an angle of about 105°, so the bond dipoles do not cancel. As we will see again in later chapters, the fact that water is

**Figure 6-9  Resultant Forces**
**If forces are symmetrical but unequal (*a*), or at an angle such that they do not cancel (*b*), there is a resultant force.**

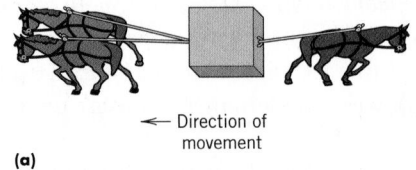

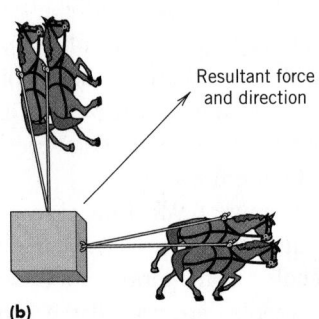

polar (has a molecular dipole) is of critical importance to its role as a room-temperature liquid that can dissolve many ionic compounds.

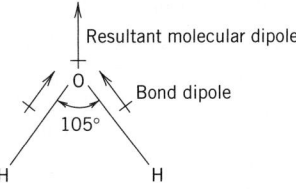

Looking Back! **Bonds between unlike atoms are polar, but the extent of polarity depends on the difference in electronegativity. But whether a molecule is polar or not depends on its molecular geometry. The Lewis structure of simple compounds allows us to predict geometry and, from that, predict whether the molecule itself is polar. In later chapters, the importance of molecular polarity will become apparent, so we will refer back to this discussion at that time.**

## Learning Check C | checking it out

**C-1.** Fill in the blanks.

The most electronegative element is _____ , and the second most electronegative is _____ . In a covalent bond between unlike atoms, the more electronegative atom has a partial _____ charge. The bond is then said to be a _____ covalent bond. If there is a large difference in electronegativity between the two atoms, the bond may be _____ . The geometry of molecules with a central atom is determined by the number of electron _____ . The three symmetrical geometries discussed are _____ , _____ _____ , and _____ . If a molecule has one of these geometries with all the same atoms bonded to the central atom, the molecule is _____ . In other molecules, where the bonded atoms are less than 180° or the atoms bonded to the central atom are not all the same, the molecules is _____ .

**C-2.** If any of the following bonds are polar, indicate with a dipole arrow from the partially positive atom to the partially negative atom.

**a.** Al—Se      **b.** As—S      **c.** S—S      **d.** F—Br

**C-3.** What is the molecular geometry of each of the following molecules or ions?

**a.** $BF_4^-$      **b.** $SO_2$      **c.** $SCl_2$

**C-4.** Discuss the molecular polarity of C-3(b) and C-3(c).

*Additional Examples: Exercises 6-51, 6-54, 6-58, 6-65, and 6-70.*

chapter    eview              **putting it together**

A simple observation followed by a hypothesis leads us to a solid understanding of why representative elements interact with each other as they do. The observation is that noble gas elements exist as solitary atoms, and the hypothesis is that they are that way because their electron configuration (an octet of electrons) is stable. From this we can rationalize why the representative elements adjust their electron configurations to follow the **octet rule.** A metal and a nonmetal combine by an exchange of electrons, thereby achieving an octet for both. The use of **Lewis dot symbols** of the elements aids us in focusing on the **valence electrons.** The ions formed by the electron exchange are arranged in a **lattice** held together by **ionic bonds.** Two nonmetals, on the other hand, follow the octet rule by electron sharing, forming a **covalent bond.** In certain cases, two atoms form **double bonds** or **triple bonds.** Covalent bonds also exist between the atoms comprising polyatomic ions. This is summarized as follows.

| Elements | Type of Bond | Comments |
|---|---|---|
| Metal–nonmetal | Ionic | Nonmetal can form −1, −2, or −3 ion. Metal can form +1, +2, or +3 ion. |
| Nonmetal–nonmetal | Covalent | Single, double, or triple bonds used to form octet. |

By following certain rules, we can become proficient at writing **Lewis structures** of compounds. These rules are summarized for three compounds at the end of this review.

Two or more equally correct Lewis structures can be written for a molecule such as ozone ($O_3$), an **allotrope** of oxygen. These are known as **resonance structures,** and the actual structure is a **resonance hybrid** of all of the Lewis structures.

Atoms of two different nonmetals do not share electrons equally. **Electronegativity** is a measure of the periodic property of the atoms of an element to attract the electrons in the bond. When electrons are not shared equally, the bond is **polar** and contains a **dipole.** Atoms of the same element share electrons equally, so the bond is **nonpolar.**

Whether or not a molecule is polar depends not only on the polarity of the bonds in the molecule but also on its **molecular geometry.** The molecular geometry of simple molecules can be determined from their Lewis structures and the **VSEPR theory.** If equal polar bonds are in a geometric arrangement where their dipoles cancel each other, the compound has no **molecular dipole** and it is nonpolar. If the bond dipoles are not the same in each direction or they do not cancel, the compound is polar.

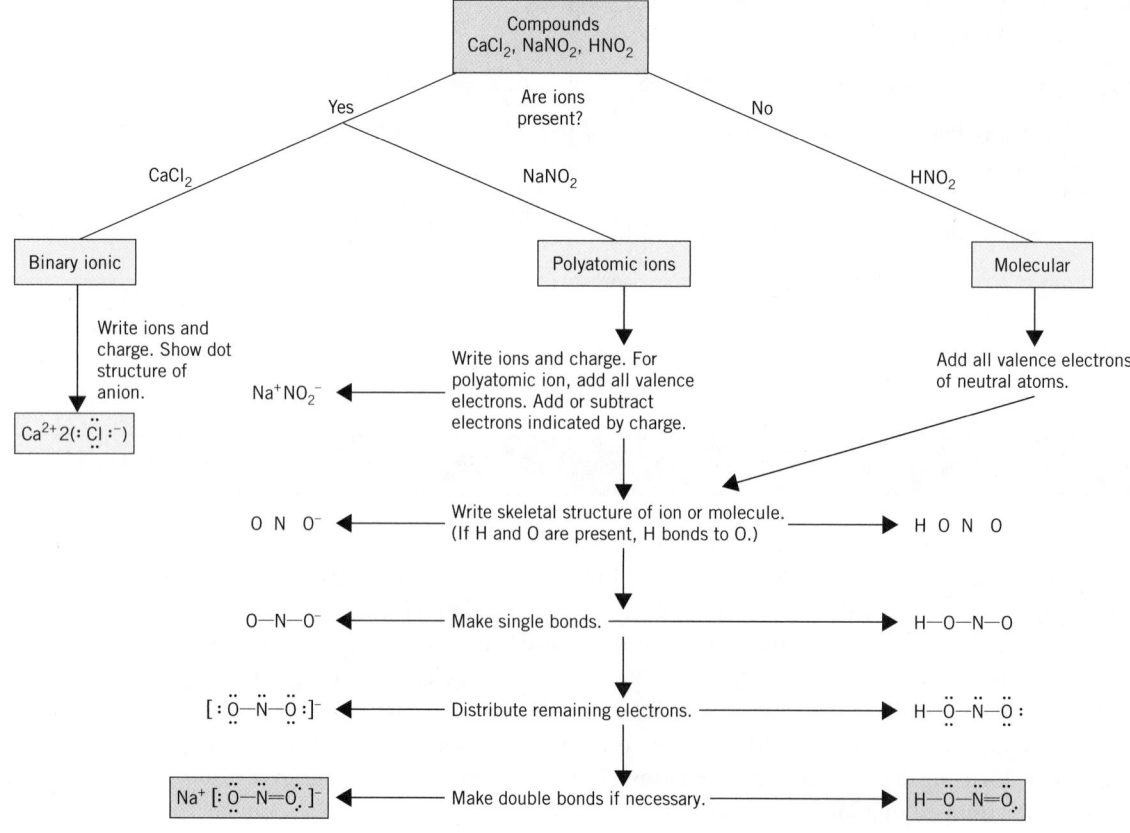

# Exercises

## Dot Symbols of Elements

**6-1.** Write Lewis dot symbols for
(a) Ca (c) Sn (e) Ne
(b) Sb (d) I (f) Bi
(g) all Group VIA elements

**6-2.** Identify the representative element groups from the dot symbols.
(a) $\cdot M \cdot$ (b) $\cdot \ddot{X} \cdot$ (c) $\dot{A} \cdot$

**6-3.** Identify the representative element groups from the dot symbols.
(a) $R \cdot$ (b) $\ddot{:Q} \cdot$ (c) $\cdot \dot{D} \cdot$

**6-4.** Why are only outer electrons represented in dot symbols?

**6-5.** Which of the following dot symbols are incorrect?

(a) $Pb \cdot$ (c) $\ddot{:He:}$ (e) $\cdot \dot{Te} \cdot$

(b) $\cdot \ddot{Bi} \cdot$ (d) $Cs \cdot$ (f) $\cdot Tl \cdot$

## Binary Ionic Compounds

**6-6.** From the periodic table, predict which pairs of elements can combine to form ionic bonds.
(a) H and Cl (c) H and K (e) B and Cl
(b) S and Sr (d) Al and F (f) Xe and F

**6-7.** From the periodic table, predict which pairs of elements can combine to form ionic bonds.
(a) Ba and I (c) C and O (e) P and S
(b) Cs and Se (d) H and Se (f) Cs and P

**6-8.** Which ions would not have a noble gas electron configuration?
(a) $Sr^{2+}$ (c) $Cr^{2+}$ (e) $In^+$ (g) $Ba^{2+}$
(b) $S^-$ (d) $Te^{2-}$ (f) $Pb^{2+}$ (h) $Tl^{3+}$

**6-9.** Which ions would not have a noble gas electron configuration?
(a) $Rb^+$ (c) $Sc^{2+}$ (e) $As^{2-}$
(b) $Si^{4-}$ (d) $I^{2-}$ (f) $Se^{2-}$

**6-10.** The ions $Li^+$, $Be^{2+}$, and $H^-$ do not follow the octet rule. Why?

**6-11.** Write Lewis dot symbols for the following ions.
(a) $K^+$ (c) $I^-$ (e) $Ba^+$ (g) $Sc^{3+}$
(b) $O^-$ (d) $P^{3-}$ (f) $Xe^+$

**6-12.** What is the origin of the octet rule? How does the octet rule relate to $s$ and $p$ subshells and to noble gases?

**6-13.** Which of the ions listed in exercise 6-11 do not follow the octet rule?

**6-14.** For the following atoms, write the charge that would give the element a noble gas configuration.
(a) Mg (d) S
(b) Ga (e) P
(c) Br

**6-15.** For the following atoms, write the charge that would give the element a noble gas configuration.
(a) Rb (b) Ba (c) Te (d) N

**6-16.** Write six ions that have the same electron configuration as Ne.

**6-17.** Write five ions that have the same electron configuration as Kr.

**6-18.** The $Tl^{3+}$ ion does not have the same electron configuration as Xe, even though it lost its three outermost electrons. Explain.

**6-19.** Complete the following table with formulas of the ionic compounds that form between the anions and cations shown.

| Cation/Anion | $Br^-$ | $S^{2-}$ | $N^{3-}$ |
|---|---|---|---|
| $Cs^+$ | CsBr | _____ | _____ |
| $Ba^{2+}$ | _____ | _____ | _____ |
| $In^{3+}$ | _____ | _____ | _____ |

**6-20.** Complete the following table with formulas of the ionic compounds that form between the anions and cations shown.

| Cation/Anion | $F^-$ | $Te^{2-}$ | $P^{3-}$ |
|---|---|---|---|
| $Rb^+$ | _____ | _____ | _____ |
| $Mg^{2+}$ | _____ | _____ | _____ |
| $Cr^{3+}$ | _____ | _____ | _____ |

**6-21.** Write the formulas of the compounds formed between the following nonmetals and the metal calcium.
(a) I (b) O (c) N (d) Te (e) F

**6-22.** Write the formulas of the compounds formed between the following metals and the nonmetal sulfur.
(a) Be (c) Ga
(b) Cs (d) Sr

**6-23.** Most transition metal ions cannot be predicted by reference to the octet rule. Determine the charge on the metal cation from the charge on the anion for each of the following compounds.
(a) $Cr_2O_3$ (c) MnS (e) $NiBr_2$
(b) $FeF_3$ (d) CoO (f) VN

**6-24.** Determine the charge on the metal cation from the charge on the anion for each of the following compounds.
(a) $IrO_4$ (c) $PtF_6$ (e) $Tc_2O_3$
(b) ScN (d) $CoCl_3$ (f) $Ag_2Se$

**6-25.** Why isn't a formula unit of $BaCl_2$ referred to as a molecule?

## Lewis Structures of Compounds

**6-26.** From their Lewis dot symbols, predict the formula of the simplest compound formed by the combination of each pair of elements.
**(a)** H and Se    **(d)** Cl and O
**(b)** H and Ge    **(e)** N and Cl
**(c)** Cl and F    **(f)** C and Br

**6-27.** From their Lewis dot symbols, predict the formula of the simplest compound formed by the combination of each pair of elements.
**(a)** H and I    **(c)** Si and Br
**(b)** Se and Br    **(d)** H and As

**6-28.** From a consideration of the octet rule, which of the following compounds are impossible?
**(a)** $PH_3$    **(c)** $SCl_2$    **(e)** $H_3O$
**(b)** $Cl_3$    **(d)** $NBr_4$

**6-29.** Which of the following binary compounds does not follow the octet rule?
**(a)** $NI_3$    **(c)** $CH_3$    **(e)** $FCl$
**(b)** $F_3O$    **(d)** $HSe$    **(f)** $SCl_2$

**6-30.** Determine the charge on each polyatomic anion from the charge on the cation.
**(a)** $K_2SO_4$    **(c)** $Al_2(SeO_4)_3$    **(e)** $BaC_2$
**(b)** $Ca(IO_3)_2$    **(d)** $Ca(H_2PO_3)_2$

**6-31.** Determine the charge on each polyatomic anion from the charge on the cation.
**(a)** $NaBrO_2$    **(c)** $AlAsO_4$
**(b)** $SrSeO_3$    **(d)** $Mg(H_2PO_4)_2$

**6-32.** Write Lewis structures for the following.
**(a)** $C_2H_6$    **(d)** $SCl_2$
**(b)** $H_2O_2$    **(e)** $C_2H_6O$
**(c)** $NF_3$    (There are two correct answers; both have the lone pairs on the oxygen.)

**6-33.** Write Lewis structures for the following.
**(a)** $N_2H_4$    **(c)** $C_3H_8$
**(b)** $AsH_3$    **(d)** $CH_4O$ (All unshared electrons are on the O.)

**6-34.** Write Lewis structures for the following.
**(a)** $CO$    **(c)** $KCN$
**(b)** $SO_3$    **(d)** $H_2SO_3$ (H's are on different O's.)

**6-35.** In the following four ions, the C is the central atom. Write the Lewis structures.
**(a)** $CN_2^{2-}$        **(c)** $HCO_3^-$ (H on the O)
**(b)** $HCO_2^-$ (H on the C)    **(d)** $ClCO^-$

**6-36.** Use the $6N + 2$ rule to confirm the structures in exercise 6-34.

**6-37.** Use the $6N + 2$ rule to predict the number of multiple bonds in the ions in exercise 6-35.

**6-38.** Use the $6N + 2$ rule to predict the number of multiple bonds (if any) in the following molecules or ions.
**(a)** $N_2O$    **(c)** $AsCl_3$    **(e)** $CH_2Cl_2$
**(b)** $Ca(NO_2)_2$    **(d)** $H_2S$    **(f)** $NH_4^+$

**6-39.** Write Lewis structures for the compounds in exercise 6-38.

**6-40.** Write Lewis structures for the following.
**(a)** $Cl_2O$    **(c)** $C_2H_4$    **(e)** $BF_3$
**(b)** $SO_3^{2-}$    **(d)** $H_2CO$    **(f)** $NO^+$

**6-41.** Use the $6N + 2$ rule to predict the number of multiple bonds (if any) in the following molecules or ions.
**(a)** $CO_2$    **(c)** $BaCl_2$    **(e)** $HOCN$    **(g)** $C_2H_2$
**(b)** $H_2NOH$    **(d)** $NO_3^-$    **(f)** $SiCl_4$    **(h)** $O_3$

**6-42.** Write Lewis structures for the molecules or ions in exercise 6-41.

**6-43.** Write Lewis structures for the following.
**(a)** $Cs_2Se$    **(c)** $LiClO_3$    **(e)** $PBr_3$
**(b)** $CH_3CO_2^-$    **(d)** $N_2O_3$
(All H's are on one C,
and both O's are
bonded to the other C.)

## Resonance Structures

**6-44.** Write all equivalent resonance structures (if any) for the following.
**(a)** $SO_3$    **(b)** $NO_2^-$    **(c)** $SO_3^{2-}$

**6-45.** Write all equivalent resonance structures (if any) for the following.
**(a)** $NO_3^-$
**(b)** $N_2O_4$ (skeletal geometry)

$$
\begin{array}{ccc}
O & & O \\
 & N \quad N & \\
O & & O
\end{array}
$$

**6-46.** Write the equivalent resonance structures for the $H_3BCO_2^{2-}$ anion. The skeletal structure for the ion is

$$
\begin{array}{cccc}
 & H & & O \\
H & B & C & \\
 & H & & O
\end{array}
$$

**6-47.** What is meant by a resonance hybrid? What is implied about the nature of the C—O bonds from the resonance structures in exercise 6-46?

**6-48.** Write any resonance structures possible for the ions in exercise 6-35. What are the implications for any bonds where resonance structures exist?

**\*6-49.** A possible Lewis structure for $CO_2$ involves a triple bond between C and O. Write the two resonance structures involving the triple bond. What is implied about the nature of the C—O bond by these two structures? How does this relate to the common Lewis structure for $CO_2$ involving two double bonds?

## Electronegativity and Polarity

**6-50.** Rank the following elements in order of increasing electronegativity: B, Ba, Be, C, Cl, Cs, F, O.

**6-51.** For bonds between the following elements, indicate the positive end of the dipole by a $\delta^+$ and the negative end by a $\delta^-$. Also indicate with a dipole arrow the direction of the dipole.
(a) N—H    (d) F—O    (g) C—B
(b) B—H    (e) O—Cl   (h) Cs—N
(c) Li—H   (f) S—Se   (i) C—S

**6-52.** Rank the bonds in exercise 6-51 in order of increasing polarity.

**6-53.** Which of the following bonds is ionic, which is polar covalent, and which is nonpolar?
(a) C—F    (b) Al—F    (c) F—F

**6-54.** On the basis of difference in electronegativity, predict whether the following pairs of elements will form an ionic or a polar covalent bond.
(a) Ga, Br   (c) B, Br
(b) H, B     (d) Al, F

**6-55.** On the basis of difference in electronegativity, predict whether the following pairs of elements will form an ionic or a polar covalent bond.
(a) Al, Cl   (c) K, O
(b) Ca, I    (d) Sn, Te

**6-56.** Which of the following bonds is nonpolar?
(a) I—F    (c) C—H    (e) B—N
(b) I—I    (d) N—Br

**6-57.** On the following pairs of bonds, which is the more polar?
(a) N—H or C—H
(b) Si—O or Si—N
(c) S—Cl or S—F
(d) P—Cl or P—I

## Molecular Geometry

**6-58.** From the Lewis structure, determine the molecular geometry of the following molecules.
(a) $SF_2$
(b) $CS_2$
(c) $CCl_2F_2$ (C is the central atom.)
(d) NOCl (N is the central atom.)
(e) $Cl_2O$

**6-59.** From the Lewis structure, determine the molecular geometry of the following molecules or ions.
(a) $BF_2Cl$ (B is the central atom.)
(b) $ClO_3^-$
(c) $N_2O$ (N is the central atom.)
(d) $COCl_2$ (C is the central atom.)
(e) $SO_3$

**6-60.** What are the approximate carbon–oxygen bond angles in the following?
(a) CO    (b) $CO_2$    (c) $CO_3^{2-}$

**6-61.** Which of the following have bond angles of approximately 109°?
(a) $H_2O$    (b) $HCO_2^-$    (c) $NH_4^+$    (d) HCN

**\*6-62.** When a molecule or ion has more than one central atom, the geometry is determined at each central atom site. What are the geometries around the two central atoms in the following?
(a) $H_2NOH$ (N and O)
(b) HOCN (O and C)
(c) $H_3CCN$ (two C's)
(d) $H_3BCO_2^{2-}$ (B and C; see exercise 6-46.)

**\*6-63.** When a molecule or ion has more than one central atom, the geometry is determined at each central atom site. What are the geometries around the two central atoms in the following?
(a) $H_2CCH_2$ (2 C's)    (c) $HClO_2$ (Cl and O)
(b) $Cl_2NNO$ (2 N's)     (d) $H_3COCl$ (C and O)

## Molecular Polarity

**6-64.** How can a molecule be nonpolar if it contains polar bonds?

**6-65.** Discuss the molecular polarities of the molecules in exercise 6-58.

**6-66.** Discuss the molecular polarities of the molecules in exercise 6-59.

**6-67.** Write the Lewis structure of $SO_2Cl_2$ (S is the central atom). Is the molecule polar?

**6-68.** Compare the expected molecular polarities of $H_2O$ and $H_2S$. Assume that both molecules have the same angle.

**6-69.** Compare the expected molecular polarities of $CH_4$ and $CH_2F_2$.

**6-70.** Compare the expected molecular polarities of $CHCl_3$ and $CHF_3$.

**6-71.** $CO_2$ is a nonpolar molecule, but CO is polar. Explain.

**6-72.** $SO_3$ is a nonpolar molecule, but $SO_2$ is polar. Explain.

**6-73.** Propene has the formula $C_3H_6$ ($H_3CCHCH_2$). Write the Lewis structure and determine whether the molecule is polar.

## General Problems

**6-74.** Write the formulas of three binary ionic compounds that contain one cation and one anion. One compound should contain Rb, one Sr, and one N.

**6-75.** There is a noble gas compound formed by xenon, $XeO_3$, that follows the octet rule. Write the Lewis structure of this compound. What is the geometry of the molecule? Is the $XeO_3$ molecule polar?

**6-76.** Which of the following compounds contains both ionic and covalent bonds?
(a) $H_2SO_3$    (d) $H_2S$
(b) $K_2SO_4$    (e) $C_2H_6$
(c) $K_2S$

**6-77.** Write the Lewis structure of $H_3BCO$ (all H's on the B). Can any resonance structures be written? What is the geometry around the B? What is the geometry around the C?

**6-78.** Carbon suboxide has the formula $C_3O_2$. All of the atoms in the molecule are arranged linearly with an oxygen at each end. Write the Lewis structure.

**6-79.** A molecule may exist that has the formula $N_2O_2$. The order of the bonds is O—N—N—O. Write a Lewis structure for the compound that has two N—O double bonds. Write any resonance structures.

**6-80.** Cyanogen has the formula $C_2N_2$. The order of the bonds is N—C—C—N. Write a Lewis structure for cyanogen involving a C—C single bond. What is the geometry around a C atom? Draw dipole arrows for the bonds. Is the molecule polar?

**6-81.** A certain species is composed of one As and four Cl's. It contains 32 valence electrons. Write the Lewis structure of the species. What is the geometry of the species and the approximate Cl—As—Cl bond angle?

**6-82.** A certain species is composed of P and three O's. It contains 26 valence electrons. Write the Lewis structure of the species. What is the geometry of the species and the approximate O—P—O bond angle?

**6-83.** A certain species contains one chlorine and two iodines. It has a total of 20 valence electrons. Write the Lewis structure of the species. What is the geometry of the species and the approximate I—Cl—I bond angle?

**6-84.** A certain species contains one B, three C's, and 9 H's. It has total of 24 valence electrons. Write the Lewis structure if the B is bonded only to the three C's. What is the geometry around the B and the approximate C—B—C bond angle? What is the geometry around the C's and the approximate H—C—H bond angle? Is the species polar?

**\*6-85.** There are two compounds composed of potassium and nitrogen. Write the formula of the compound expected between potassium and nitrogen. Another compound ($KN_3$) is named potassium azide. Write the Lewis structure for the azide ion plus any resonance structures. What is the geometry about the central N atom?

**\*6-86.** Write two resonance structures for hydrogen azide $HN_3$ (HNNN). What is the predicted HNN bond angle in each structure? The actual angle is about 110°. What does this indicate?

**6-87.** Write the Lewis structure for phosphoric acid (hydrogens bonded to oxygens). What is the geometry around the P? What is the geometry around the oxygen in the H—O—P bond? What is the negative end of the dipole in the P—O bond? Is the molecule polar?

**6-88.** Cyanic acid has the formula HOCN (atoms bonded in that order). The H—O—C bond angle is about 105°. Write the Lewis structure that demonstrates this fact. What is the geometry around the C?

**6-89.** Thiocyanic acid has the formula HSCN (atoms bonded in that order). The H—S—C bond angle is about 116°. Write the Lewis structure that demonstrates this fact. What is the geometry around the C?

**6-90.** A compound has the formula $N_2F_2$. Write a Lewis structure that contains a N=N bond.

**6-91.** A second compound of nitrogen and fluorine contains one nitrogen and the expected number of fluorines. A third has the formula $N_2F_4$. Write the Lewis structures of these two compounds. What is the geometry around the N in each compound? Are either or both compounds polar?

**\*6-92.** Two compounds of oxygen are named oxygen difluoride and dioxygen difluoride. Write the Lewis structures of these two compounds. The latter compound contains an O—O bond. Oxygen is usually written and named second in binary compounds. Why not here? Are either or both of these compounds polar?

**\*6-93.** A compound has the formula $Na_2C_2O_4$. Write a Lewis structure for the compound that contains a C—C bond. Write any resonance structures present.

**\*6-94.** If you have already covered Chapter 8, consider the following problem. A compound is 27.4% Na, 14.3% C, 57.1% O, and 1.19% H. What is the formula of the compound? Write the Lewis structure of the anion (H bonded to O). What is the geometry around the C? What is the approximate O—C—O bond angle? What is the geometry around the H—O—C bond and the approximate angle?

**\*6-95.** If you have already covered Chapter 8, consider the following problem. A compound is 21.7% Li, 75.0% C, and 3.16% H. What is the formula of the compound? Write the Lewis structure of the anion. What is the geometry around the central C and the approximate H—C—C bond angle?

**\*6-96.** If you have already covered Chapter 8, consider the following problem. A compound is 19.8% Ca, 1.00% H, 31.7% S, and 47.5% O. What is the formula of the compound? (*Hint:* The formula of the anion should be reduced to its empirical formula.) What is the name of the compound? (Refer to Table 4-2.) What is the geometry around the S in the anion? What is the approximate H—O—S bond angle?

**\*6-97.** Refer to exercise 5-99. Use the periodic table from the planet Zerk to answer the following.

**(1)** What are the simplest formulas of compounds formed between the following elements? (Example: Between 7 and 7 is $7_2$.)

**(a)** 1 and 7    **(d)** 7 and 9    **(g)** 6 and 7
**(b)** 1 and 3    **(e)** 7 and 13   **(h)** 3 and 6
**(c)** 1 and 5    **(f)** 10 and 13

**(2)** Write Lewis structures for all of the above. Indicate which are ionic. (Remember that on Zerk there will be something different from an octet rule.)

**(3)** How would the $6N + 2$ rule be modified on Zerk?

# Interactive Learning

**6-98.** Draw Lewis structures for **(a)** $AsCl_4^+$, **(b)** $ClO_2^-$, and **(c)** $HNO_2$.

**6-99.** Draw resonance structures for the carbonate ion $(CO_3^{2-})$.

**6-100.** Predict the shapes of **(a)** $FCl_2^+$, **(b)** $SbH_3$, and **(c)** $SeO_2$.

# Solutions to Learning Checks

**A-1.** VIIIA, valence, eight, gaining, losing, sharing, octet, positive, +3, negative, charges

**A-2.** Be·   ·Se:

**A-3.** $Be^{2+}$ $Se^{2-}$

**A-4.** BeSe

**B-1.** covalent, dashes, pairs of dots, HCl, $H_2S$, $PH_3$, double bond, triple bond, resonance, resonance hybrid, allotrope

**B-2.** **(a)** $SF_2$   **(b)** $NO_2Cl$

  **(c)** $Mg(ClO_2)_2 = Mg^{2+} + 2ClO_2^-$

:S—F:
:F:

:O:
N—Cl:
:O:

$Mg^{2+}2[:O—Cl—O:]^-$

**B-3.** $[:N \equiv C - \ddot{O}:]^- \leftrightarrow [\ddot{N} = C = \ddot{O}:]^-$

**C-1.** fluorine, oxygen, negative, polar, ionic, groups, linear, trigonal planar, tetrahedral, nonpolar, polar

**C-2.** **(a)** $Al \overset{+}{\longrightarrow} Se$   **(c)** S—S (nonpolar)

  **(b)** $As \overset{+}{\longrightarrow} S$   **(d)** $F \overset{+}{\longleftrightarrow} Br$

**C-3.** **(a)** $BF_4^-$     **(b)** $SO_2$     **(c)** $SCl_2$

$$\left[ \begin{array}{c} :F: \\ | \\ F—B—F \\ | \\ :F: \end{array} \right]^-$$

tetrahedral

:O=S—O:

V-shaped

:Cl—S—Cl:

V-shaped

**C-4.** **(b)** Polar, polar bonds at an angle of about 120°
  **(c)** polar, polar bonds at an angle of about 109°

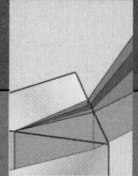

## Multiple Choice

**The following multiple-choice questions have one correct answer.**

**1.** Which of the following types of orbitals are present in the third ($n = 3$) shell?
(a) $s, p$      (d) $s, p, d, f$
(b) $p, d, f$      (e) $s, p, f$
(c) $s, p, d$

**2.** How many orbitals are in a $d$ subshell?
(a) 5      (b) 3      (c) 10      (d) 14      (e) 2

**3.** What is the electron capacity of the $4f$ subshell?
(a) 6      (b) 7      (c) 14      (d) 8      (e) 10

**4.** Which of the following elements has the electron configuration $[Kr]5s^2 4d^2$?
(a) Sn      (b) Zr      (c) Ti      (d) Pr      (e) Sr

**5.** What is the electron configuration of In (atomic number 49)?
(a) $[Kr]4s^2 4p^1$      (d) $[Kr]4s^2 3d^{10} 4p^1$
(b) $[Kr]5s^2 4d^1$      (e) $[Kr]5s^2 4d^6 5p^1$
(c) $[Kr]5s^2 4d^{10} 5p^1$

**6.** Silicon has two unpaired electrons. This is a result of
(a) Hund's rule
(b) the Pauli exclusion principle
(c) the Aufbau principle
(d) Bohr's model

**7.** Which of the following groups of elements have the general electron configuration $ns^2 np^5$?
(a) halogens      (d) Group VA
(b) alkaline earths      (e) transition elements
(c) noble gases

**8.** Which of the following atoms has the smallest radius?
(a) As      (b) Se      (c) Sb      (d) Te

**9.** Which of the following atoms has the highest first ionization energy?
(a) N      (b) O      (c) P      (d) S

**10.** Which of the following ions requires the most energy to form?
(a) $K^+$      (d) $F^-$
(b) $Ca^{3+}$      (e) $O^{2-}$
(c) $Al^{3+}$

**11.** Which of the following ions violates the octet rule?
(a) $I^{2-}$      (d) $N^{3-}$
(b) $S^{2-}$      (e) $Te^{2-}$
(c) $Br^-$

**12.** Which of the following pairs of elements would be expected to form an ionic bond?
(a) Mg, Ca      (d) S, O
(b) B, F      (e) C, O
(c) Mg, Br

**13.** Which of the following is the correct formula for the compound formed between beryllium and bromine?
(a) BeBr      (d) $BeBr_2$
(b) $Br_2Be$      (e) $Be_2Br$
(c) $BBr_2$

**14.** Which of the following pairs of elements would be most likely to form a covalent bond?
(a) Al, F      (c) As, Br
(b) K, Sn      (d) K, H

**15.** Which of the following is the correct formula for the simplest compound formed between As and Cl?
(a) $Cl_2As$      (d) AsCl
(b) $AsCl_2$      (e) $AsCl_4$
(c) $AsCl_3$

**16.** Which of the following compounds contains both ionic and covalent bonds?
(a) $BaCO_3$      (c) $HNO_3$      (e) $CCl_4$
(b) $H_2CO_3$      (d) $K_3N$

**17.** Which of the following is the most electronegative element?
(a) B      (b) Na      (c) O      (d) Cl      (e) N

**18.** Which of the following bonds is the most polar?
(a) C—O      (c) Be—Cl      (e) N—N
(b) B—Cl      (d) Be—F

**19.** What is the approximate F—C—O angle in the ion $FCO^-$?
(a) 180°      (c) 109°
(b) 120°      (d) 90°

**20.** Which of the following is a polar molecule?
(a) $CO_2$      (c) $SO_3$      (e) $BeF_2$
(b) $Cl_2$      (d) $SO_2$

## Problems

Carry out the following calculations.

**1.** Answer the following questions about the element nitrogen (N).

**(a)** What are its group number and general classification?

**(b)** What is its electron configuration?

**(c)** How many unpaired electrons are in an atom of nitrogen?

**(d)** Is it a metal or nonmetal?

**(e)** Does it exist as a solid, liquid, or gas under normal conditions?

**(f)** What is the formula of the element?

**(g)** What is the Lewis structure of the element?

**(h)** What is its electrical charge in a monatomic ion?

**(i)** What noble gas has the same electron configuration as its monatomic ion?

**(j)** What is its polarity (negative or positive) when covalently bonded to **(1)** oxygen, **(2)** boron?

**(k)** What is the simplest formula when it forms a binary compound with **(1)** magnesium, **(2)** lithium, **(3)** fluorine?

**(l)** What are the names of the three compounds in part (k)?

**(m)** What is the Lewis structure of the compound with fluorine in part (k)?

**2.** For each of the following compounds, write **(1)** the formula in ionic form if ions are present, **(2)** the name of the compound, **(3)** the Lewis structure.

| Formula | Ionic Form | Name | Lewis Structure |
|---|---|---|---|
| NaClO | Na⁺ClO⁻ | sodium hypochlorite | Na⁺ [:C̈l—Ö:]⁻ |
| (a) MgSO₄ | | | |
| (b) HNO₂ | | | |
| (c) LiNO₃ | | | |
| (d) Co₂(CO₃)₃ | | | |
| (e) Cl₂O | | | |

**3.** For each of the following compounds, write **(a)** the formula of the compound, **(b)** the formula in ionic form if ions are present, **(c)** the Lewis structure.

| Name | Formula | Ionic Form | Lewis Structure |
|---|---|---|---|
| (a) dinitrogen trioxide | | | |
| (b) chromium(III) sulfite | | | |
| (c) iron(II) hypdroxide | | | |
| (d) strontium oxalate | | | |
| (e) hydroiodic acid | | | |

## chapter 7

# Chemical Reactions

In the rain forest a cycle of life occurs. Green leaves, powered by sunlight, maintain life. The decay of dead trees returns the ingredients for life to the air and earth.

### Setting the Stage

Floods, droughts, hurricanes, and tornadoes—it seems like there is violent, destructive weather somewhere almost every week. Has it always been this bad or is something powerful affecting the weather in recent years? Indeed, there may be. The winter of 2001–2002 was one of the mildest on record. The year 1998 brought us the highest average global temperatures since reliable records were started more than a century ago. In fact, most of the other years in the 1900s rank as among the warmest years recorded. The violent storms are probably spawned by this "global warming." Although there is still debate in the scientific community, the warming may be due to the "greenhouse effect." Activities associated with the human race are likely contributing to this effect and, if so, it may indeed get worse.

Some familiar and common chemical changes play a central role in all of this. The major player in the greenhouse effect is the compound carbon dioxide, which is a normal and necessary part of the atmosphere. Carbon dioxide is a gas that traps heat in the atmosphere much as heat is trapped inside a closed automobile on a sunny day. Carbon dioxide is removed from the atmosphere by the oceans and ocean life and also by trees and other vegetation. In a complex series of chemical reactions called *photosynthesis,* green leaves use energy from the sun, carbon dioxide, and water to produce carbohydrates (carbon, hydrogen, and oxygen compounds) along with elemental oxygen. In a chemical reaction called *combustion,* the carbon compounds from trees or fossil fuels (coal, oil, and natural gas) combine with oxygen in the air to reform carbon dioxide and water and release the energy that originated from the sun as heat. This miraculous cycle of chemical reactions maintains life on this beautiful planet. The problem is that things have gotten out of balance in the past 150 years since the industrial revolution. There is considerably more combustion going on in our world than photosynthesis, so the level of carbon dioxide in the atmosphere is definitely increasing. The existence of six billion people on our globe all burning fuel and needing space at the expense of forests assures that the balance will remain in favor of combustion.

The first chemical reaction probably took place over 10 billion years ago when two hydrogen atoms combined to form a hydrogen ($H_2$) molecule. Today, on Earth, the atoms of the elements combine, separate, and recombine in millions of ways, all indicating unique chemical changes. The rusting of iron, the decay of a fallen tree in the forest, the formation of muscle in our bodies, the cooking of food—all of these everyday, common occurrences are chemical changes. This is the world of chemistry into which we journey in this chapter. But first, we need a simple, symbolic way to represent chemical changes. This is done by means of the chemical equation. Then, with the help of chemical equations, we will group some of the chemical changes into a few simple classifications. This may help us sort through at least a handful of some of the millions of chemical reactions that occur.

### Formulating Some Questions

The important questions that are addressed in this chapter include: How is the chemical equation constructed to represent the chemical changes (which we will refer to as chemical reactions from now on) and what does it tell us? How can we use chemical equations to describe some straightforward classifications of chemical reactions? Since much of our world involves the chemical reactions that occur in water, what is the nature of some of the substances that dissolve in water? Certain metals dissolve in water solutions of acids and others do not. Can we predict which ones dissolve? How do solids form in water and how are acidic solutions neutralized? After studying this chapter, we should have some feeling for the various types of chemical reactions that occur. So, our first order of business is to examine how we represent chemical reactions by means of the chemical equation.

## Section A — THE REPRESENTATION OF CHEMICAL CHANGES AND SOME TYPES OF CHANGES

## 7-1 Chemical Equations

**Looking Ahead!** The main rocket thruster of the space shuttle uses a simple but powerful chemical reaction. Hydrogen combines with oxygen to form water and a lot of heat energy. Is there a simple way to represent this information? The way chemical reactions are symbolized is the topic of this section.

A **chemical equation** *is the representation of a chemical reaction using the symbols of elements and the formulas of compounds.* In the following discussion, we will focus on the matter that undergoes a change in the reaction. In the next chapter, we will

**A powerful chemical reaction blasts the space shuttle into Earth orbit.**

include the heat energy involved. First, let's represent with symbols the basic information about the chemical reaction just mentioned.

$$H + O \longrightarrow H_2O$$

In a chemical equation, *the original reacting species are shown to the left of the arrow and are called the* **reactants.** *The species formed as a result of the reaction are to the right of the arrow and are called the* **products.** In this format, note that the phrase "combines with" (or "reacts with") is represented by a plus sign (+). When there is more than one reactant or product, the symbols or formulas on each side of the equation are separated by a +. The word "produces," also referred to as "yields," may be represented by an arrow (→). Note in Table 7-1 that there are other representations for the yield sign, depending on the situation.

**Table 7-1   Symbols in the Chemical Equation**

| Symbol | Use |
|---|---|
| + | Between the symbols and/or formulas of reactants or products |
| $\longrightarrow$ | Means "yields" or "produces"; separates reactants from products |
| = | Same as arrow |
| $\rightleftharpoons$ | Used for reversible reactions in place of a single arrow (see Chapter 14) |
| (g) | Indicates a gaseous reactant or product |
| ↑ | Sometimes used to indicate a gaseous product |
| (s) | Indicates a solid reactant or product |
| ↓ | Sometimes used to indicate a solid product |
| (l) | Indicates a liquid reactant or product |
| (aq) | Indicates that the reactant or product is in aqueous solution (dissolved in water) |
| $\xrightarrow{\Delta}$ | Indicates that heat must be supplied to reactants before a reaction occurs |
| $\xrightarrow{MnO_2}$ | An element or compound written above the arrow is a *catalyst; a catalyst speeds up a reaction but is not consumed in the reaction.* May also indicate the solvent. |

The chemical equation shown above tells us only about the elements involved. There is much more that the equation can tell us. First, if an element exists as molecules under normal conditions, then the formula of the molecule is shown. Recall from Chapter 4 that both hydrogen and oxygen exist as diatomic molecules under normal conditions. Including this information, the equation is

$$H_2 + O_2 \longrightarrow H_2O$$
$$\text{reactants} \qquad \text{products}$$

An important duty of a chemical equation is to demonstrate faithfully the law of conservation of mass, which states that *mass can be neither created nor destroyed.* In Dalton's atomic theory, this law was explained for chemical reactions. He suggested that reactions were simply rearrangements of the same number of atoms. A close look at the equation above shows that there are two oxygen atoms on the left but only one on the right. To conform to the law of conservation of mass, an equation must be balanced. *A **balanced equation** has the same number and type of atoms on both sides of the equation. An equation is balanced by introducing **coefficients.*** Coefficients are whole numbers in front of the symbols or formulas. The equation in question is balanced in two steps. If we introduce a 2 in front of the $H_2O$, we have equal numbers of oxygen atoms, but the number of hydrogens is now unbalanced.

$$H_2 + O_2 \longrightarrow 2H_2O$$

This problem can be solved rather easily. Simply return to the left and place a coefficient of 2 in front of the $H_2$. The equation is now completely balanced.

$$2H_2 + O_2 \longrightarrow 2H_2O$$

*Note that equations cannot be balanced by changing or adjusting the subscripts of the elements or compounds.* For example, the original equation could seem to be balanced in one step if the $H_2O$ were changed to $H_2O_2$. However, $H_2O_2$ is a compound known as hydrogen peroxide, which is a popular antiseptic but not the same as water.

Finally, the physical states of the reactants and products under the reaction conditions are sometimes added in parentheses after the formula for each substance. Hydrogen and oxygen are gases, and water is a liquid under the conditions of this reaction. Using the proper letters shown in Table 7-1, we have the balanced chemical equation in proper form.

$$2H_2(g) + O_2(g) \longrightarrow 2H_2O(l)$$

Note that if we describe this reaction in words, we have quite a mouthful. "Two molecules of gaseous hydrogen react with one molecule of gaseous oxygen to produce two molecules of liquid water."

Properly balanced equations are a necessity when we consider the quantitative aspects of reactants and products, as we will do in the next chapter. Before we consider some guidelines in balancing equations, there are three points to keep in mind concerning balanced equations.

1. The subscripts of a compound are fixed; they cannot be changed to balance an equation.
2. The coefficients used should be the smallest whole numbers possible.
3. The coefficient multiplies every number in the formula. For example, $2K_2SO_3$ indicates the presence of four atoms of K, two atoms of S, and six atoms of O.

In this chapter, equations will be balanced by *inspection.* Certainly, many complex equations are extremely tedious to balance by this method, but such equations will be reserved for a later chapter in which more systematic methods can be employed. The following rules are helpful in balancing simple equations by inspection.

1. In general, it is easiest to consider balancing elements other than hydrogen or oxygen first. Look to the compound on either side of the equation that contains the most number of atoms of an element other than oxygen or hydrogen. (If poly-atomic ions appear unchanged on both sides of the equation, consider them as single units.) Balance the element in question on the other side of the equation.

2. Balance all other elements except hydrogen and oxygen.

3. Balance hydrogen or oxygen next. Choose the one that is present in the fewer number of compounds first. (Usually, that is hydrogen.)

4. Check to see that the atoms of all elements are balanced. The final balanced equation should have the smallest whole-number ratio of coefficients. If a fractional coefficient has been used in the initial balancing procedure, multiply all of the coefficients by the number in the denominator of the fraction to clear the equation of the fraction. (This is illustrated in Example 7-3.)

| **Example 7-1** | **Balancing a Simple Equation** |
| --- | --- |
| **working it out** | Ammonia is an important industrial commodity that is used mainly as a fertilizer. It is manufactured from its constituent elements. Write a balanced chemical equation from the following word equation: "nitrogen gas reacts with hydrogen gas to produce ammonia gas." |

**Procedure**

The unbalanced chemical equation using the proper formulas of the elements and compound is

$$N_2(g) + H_2(g) \longrightarrow NH_3(g)$$

First consider the $N_2$ molecule since it has the most atoms of an element other than hydrogen or oxygen. Balance the N's on the other side by adding a coefficient of **2** in front of the $NH_3$.

$$N_2(g) + H_2(g) \longrightarrow 2NH_3(g)$$

Now consider the hydrogens. We have "locked in" six hydrogens on the right, so we will need six on the left. By adding a coefficient of **3** in front of the $H_2$, we have completed the balancing of the equation. (See Figure 7-1.)

**Solution**

$$N_2(g) + 3H_2(g) \longrightarrow 2NH_3(g)$$

**Figure 7-1  Nitrogen Plus Hydrogen Yields Ammonia**
**In a chemical reaction, the atoms are simply rearranged into different molecules.**

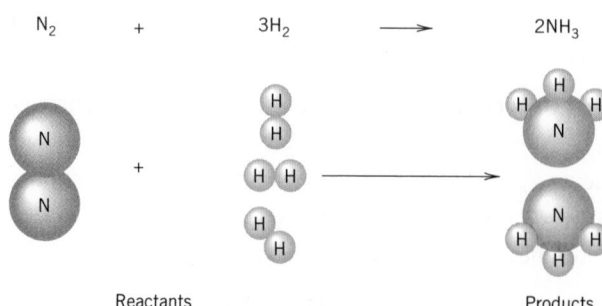

**Balancing an Equation**

**Example 7-2**

Boron hydrogen compounds are being examined as a possible way to produce hydrogen for automobiles that will run on fuel cells. Although $B_2H_6$ is not the compound that will be used (it's explosive), the following equation illustrates how boron–hydrogen compounds react with water to produce hydrogen. Balance the following equation.

$$B_2H_6(g) + H_2O(l) \longrightarrow H_3BO_3(aq) + H_2(g)$$

**Procedure**

First consider the $B_2H_6$ molecule since it has the most atoms of an element other than hydrogen or oxygen. Balance the B by adding a coefficient of **2** in front of the $H_3BO_3$ on the right.

$$B_2H_6(g) + H_2O(l) \longrightarrow 2H_3BO_3(aq) + H_2(g)$$

Next, we notice that oxygen is in the fewer number of compounds, so we balance it next. Since there are 6 oxygens in $2H_3BO_3$, place a **6** before the $H_2O$ on the left. Finally, balance hydrogen. There are 18 on the left that are "locked in." Thus we need a **6** in front of the $H_2$ to have 18 hydrogens on the right. A quick check confirms that we have a balanced equation.

**Solution**

$$B_2H_6(g) + 6H_2O(l) \longrightarrow 2H_3BO_3(aq) + 6H_2(g)$$

**Balancing an Equation**

**Example 7-3**

Most fuels are composed of carbon and hydrogen and some may also contain oxygen. When they react with oxygen gas (burn), they form as products carbon dioxide gas and water. Write a balanced equation showing the burning of liquid "rubbing alcohol" ($C_3H_8O$).

**Procedure**

First, represent the names of the species involved as reactants and products with formulas and indicate their physical states.

$$C_3H_8O(l) + O_2(g) \longrightarrow CO_2(g) + H_2O(l)$$

Next, balance carbon, Place a coefficient of **3** in front of $CO_2$ to balance the carbons in $C_3H_8O$. Balance the 8 hydrogens in $C_3H_8O$ by adding a coefficient of **4** in front of $H_2O$.

$$C_3H_8O(l) + O_2(g) \longrightarrow 3CO_2(g) + 4H_2O(l)$$

Now note that there are 10 oxygens on the right. On the left 1 oxygen is in $C_3H_8O$, so 9 are needed from $O_2$. To get an odd number of oxygens from $O_2$ we need to use a fractional coefficient, in this case $\frac{9}{2}$ (i.e., $\frac{9}{2} \times 2 = 9$)

$$C_3H_8O(l) + \frac{9}{2}O_2(g) \longrightarrow 3CO_2(g) + 4H_2O(l)$$

Finally, we need to clear the fraction so that all coefficients are whole numbers. Multiply the whole equation through by 2 and do a quick check.

**Solution**

$$2C_3H_8O(l) + 9O_2(g) \longrightarrow 6CO_2(g) + 8H_2O(l)$$

# 7-2 Combustion, Combination, and Decomposition Reactions

**Looking Ahead!** All the millions of known chemical changes can be represented by balanced equations. Many of these chemical reactions have aspects in common. So we can group natural phenomena into specific classifications. In the remainder of this chapter, we attempt to do this by considering five basic reactions. We will notice that each type has a characteristic chemical equation. These five types are not the only ways that reactions can be grouped. In later chapters we will find other convenient classifications that suit our purpose at that time. The first three types are the simplest and will be considered in this section.

Fire is certainly dramatic evidence of the occurrence of a chemical reaction. What we see as fire is the hot glowing gases of a combustion reaction. The easiest way to put out a fire is to deprive the burning substance of a reactant (oxygen) by dousing it with water or carbon dioxide from a fire extinguisher. The reaction of elements or compounds with oxygen is the first of three types of reactions that we will discuss in this section.

## Combustion Reactions

One of the most important types of reactions that we may refer to in the future is known as a **combustion reaction.** This type of reaction refers specifically to the reaction of an element or compound with elemental oxygen ($O_2$). Combustion usually liberates a lot of heat energy and is accompanied by a flame. It is often referred to as "burning." When elements undergo combustion, generally only one product is formed. Examples are the combustion of carbon and aluminum shown here.

$$C(s) + O_2(g) \longrightarrow CO_2(g)$$
$$4Al(s) + 3O_2(g) \longrightarrow 2Al_2O_3(s)$$

When compounds undergo combustion, however, two or more combustion products are formed. When carbon–hydrogen compounds undergo combustion in an excess of oxygen, the combustion products are carbon dioxide and water.

$$CH_4(g) + 2O_2(g) \longrightarrow CO_2(g) + 2H_2O(l)$$

The metabolism of glucose ($C_6H_{12}O_6$, blood sugar) that occurs in our bodies to produce the energy to sustain our life is also a combustion reaction that occurs at a steady rate.

$$C_6H_{12}O_6(aq) + 6O_2(g) \longrightarrow 6CO_2(g) + 6H_2O(l)$$

When insufficient oxygen is present (as in the combustion of gasoline, $C_8H_{18}$, in an automobile engine), some carbon monoxide (CO) also forms.

$$2C_8H_{18}(l) + 17O_2(g) \longrightarrow 16CO(g) + 18H_2O(l)$$

## Combination Reactions

All elements and compounds can be characterized by their chemical properties. Elements may be described by whether they do or do not combine with other elements or compounds. This type of reaction concerns the preparation of one compound from two or more elements and/or simpler compounds and is known as a **combination reaction.** For example, an important chemical property of the metal magnesium is that it reacts with elemental oxygen to form magnesium oxide. (See Figure 7-2.) The synthesis (i.e., production of a substance) of MgO is represented at the end of this paragraph by a balanced equation and an illustration of the magnesium and oxygen

**Bursts of energy are obtained by the metabolism of glucose.**

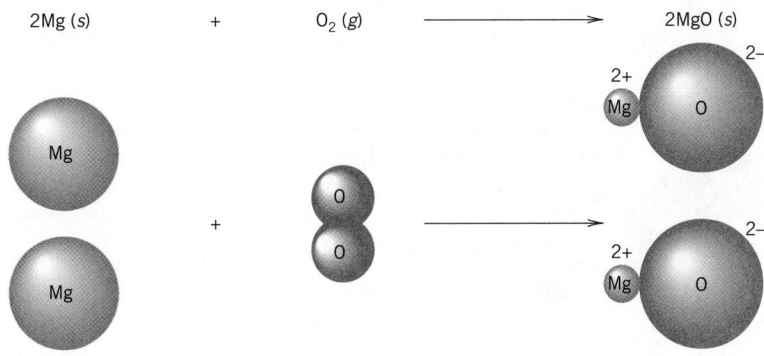

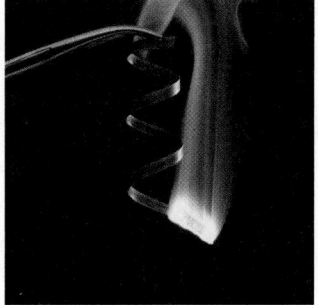

**Figure 7-2  Combination or Combustion Reaction**
When magnesium burns in air, the reaction can be classified as either a combination or a combustion reaction.

atoms in the reaction. Notice in the reaction that an ionic compound is formed from neutral atoms. The $Mg^{2+}$ cation is smaller than the parent atom, while the $O^{2-}$ anion is larger than the parent atom. The reason for this is discussed in Chapter 5. Since one of the reactants is elemental oxygen, this reaction can also be classified as a combustion reaction.

The following equations represent some other important combination reactions.

$$2Na(s) + Cl_2(g) \longrightarrow 2NaCl(s)$$
$$C(s) + O_2(g) \longrightarrow CO_2(g)$$
$$CaO(s) + CO_2(g) \longrightarrow CaCO_3(s)$$

## Decomposition Reactions

A chemical property of a compound may concern its tendency to decompose into simpler substances. This type of reaction is simply the reverse of combination reactions. That is, one compound is decomposed into two or more elements or simpler compounds. Many of these reactions take place only when heat is supplied, which is indicated by a $\Delta$ above the arrow. An example of this type of reaction is the decomposition of carbonic acid ($H_2CO_3$). This **decomposition reaction** causes the fizz in carbonated beverages. The reaction is illustrated here and is followed by other examples. (See also Figure 7-3.)

---

**Looking Back!** The world of chemistry opens before us with the description of the chemical reaction. A reaction is illustrated by a balanced chemical equation that tells us how the atoms have rearranged from reactants to products and, more importantly, it simply tells us "what's happening." If the symbol of an element is the chemist's analog to the alphabet, and the formulas of compounds are equivalent to words, then the chemical equation is like a complete sentence. Three types of these chemical sentences describe combustion, combination, and decomposition reactions.

---

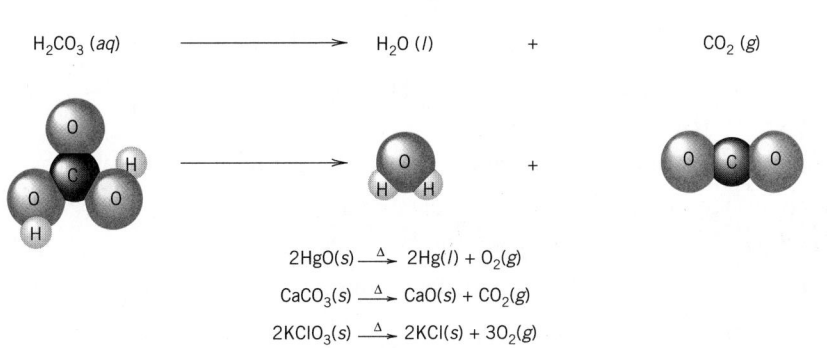

$$2HgO(s) \xrightarrow{\Delta} 2Hg(l) + O_2(g)$$
$$CaCO_3(s) \xrightarrow{\Delta} CaO(s) + CO_2(g)$$
$$2KClO_3(s) \xrightarrow{\Delta} 2KCl(s) + 3O_2(g)$$

**Figure 7-3  Decomposition Reaction**
The fizz of carbonated water is the result of a decomposition reaction.

# MAKING IT REAL

## Life Where the Sun Doesn't Shine—Chemosynthesis

Miles deep in the ocean a phenomenal discovery profoundly changed the way we view the life forms on this planet and perhaps other planets far away. Previously, we understood life as totally dependent on energy from the sun. The chlorophyll in plants produced carbohydrates in a process known as *photosynthesis*, as illustrated below. (Carbohydrates are represented by their empirical formula, $CH_2O$).

$$\text{(solar energy)} + CO_2(g) + H_2O(l) \longrightarrow CH_2O(s) + O_2(g)$$

We use these carbohydrates from plants, or animals that eat these plants, to sustain life. So, our energy is indirectly solar energy.

In 1977, scientists in the research submarine, *Alvin*, were exploring the mile-deep ocean bottom near the Galapagos Islands in the Pacific. Suddenly, they came upon an unbelievable diversity of life that existed in scalding water (350°C). The colony of life extended from bacteria to giant clams, mussels, and tube worms eight feet long. The *Alvin* had come upon ocean vents discharging huge amounts of hot water from deep in the Earth. The water was rich in minerals and chemicals.

One of those chemicals was hydrogen sulfide. This gas is very poisonous to surface life but turned out to be the main source of energy for the bacteria, which served as the bottom of the food chain. These bacteria were able to produce carbohydrates with the chemical energy from hydrogen sulfide rather than the light energy from the sun. This process is called *chemosynthesis*. The equation illustrating this reaction to produce a carbohydrate is

$$2O_2(g) + 8H_2S(aq) + 2CO_2(aq) \longrightarrow$$
$$2CH_2O(aq) + 2S_8(s) + 6H_2O(l)$$

In the late 1990s, even more startling discoveries were reported. Living bacteria have been found 3000 feet deep in the Earth. These life forms seem to use the chemical energy from $H_2$ to maintain life. They do not even need $O_2$, which is produced as a byproduct of photosynthesis, as was found in the chemosynthesis in the deep-sea vents.

Life is much more resilient that we ever thought. Because of these discoveries, many scientists feel much more confident that life could exist on other planets, especially Mars. Currently, we are intensely studying the conditions on that planet. Recent discoveries of the presence of large amounts of water at the poles and buried beneath the surface of Mars has heightened expectations.

*Life in the deep ocean.*

---

## Learning Check A | checking it out

**A-1.** Fill in the blanks.

A chemical reaction is represented with symbols and formulas by means of a chemical _____ . The arrow in an equation separates the _____ on the left from the _____ on the right. To conform to the law of conservation of mass, an equation must be _____ . This is accomplished by introducing _____ in front of formulas rather than changing subscripts in a formula. Reactions involving oxygen as a reactant are known as _____ reactions. Reactions where one product is formed are known as _____ reactions, and those where there is only one reactant are _____ reactions.

**A-2.** Balance the following equations.

**a.** $Cr + O_2 \longrightarrow Cr_2O_3$

**b.** $Co_2S_3 + H_2 \longrightarrow Co + H_2S$

**c.** $C_3H_8 + O_2 \longrightarrow CO_2 + H_2O$

**d.** $H_3BCO + H_2O \longrightarrow B(OH)_3 + CO + H_2$

**A-3.** Represent each of the following word equations with a balanced chemical equation.

**a.** Disilane gas ($Si_2H_6$) undergoes combustion to form solid silicon dioxide and water.

**b.** Solid aluminum hydride is formed by a combination reaction of its two elements.

**c.** When solid calcium bisulfite is heated, it decomposes to solid calcium oxide, sulfur dioxide gas, and water.

*Additional Examples: Exercises 7-1, 7-2, 7-4, 7-16, 7-18, and 7-20.*

## Section B  THE ROLE OF IONS IN WATER AND SOME WAYS THE IONS INTERACT

## 7-3 The Formation of Ions in Water

**Looking Ahead!** A large number of very important chemical reactions take place in water (aqueous) solution, including many that take place in our bodies. In many cases, these reactions involve ions. So, before we discuss two general types of reactions that occur in aqueous solution, we will examine how ions are formed in water.

When one adds a sprinkle of table salt (sodium chloride) to water, the salt soon disappears into the aqueous medium. We observe that the table salt is soluble in water. (See Figure 7-4.) On the other hand, if we added some chalk (calcium carbonate) to water, it slowly settles to the bottom of the container without apparent change. *When an appreciable amount of a substance dissolves in a liquid medium, we say that the substance is* **soluble.** *If very little or none of the substance dissolves, we say that the compound is* **insoluble.** The water and the table salt together make what is called a solution. *A* **solution** *is a homogeneous mixture of a* **solvent** *(usually a liquid such as water) and a* **solute** *(a solid, a liquid, or even a gas).* In Chapters 3 and 4, we described table salt (NaCl) as an ionic compound composed of $Na^+$ cations and $Cl^-$ anions in the solid state. When an ionic compound dissolves in water, in nearly all cases the compound is separated into individual ions. The solution of NaCl in water can be represented by the equation

$$NaCl(s) \xrightarrow{H_2O} Na^+(aq) + Cl^-(aq)$$

The $H_2O$ shown above the arrow indicates the presence of water as a solvent. The solution of another ionic compound, calcium perchlorate, can be represented by the equation

$$Ca(ClO_4)_2(s) \xrightarrow{H_2O} Ca^{2+}(aq) + 2ClO_4^-(aq)$$

Notice that all of the atoms in the perchlorate ion remain together in solution as a complete entity. That is, they do not separate into chlorine and oxygen atoms.

In Chapter 4, a class of compounds known as acids was named. These are molecular compounds, but like ionic compounds, they also produce ions when dissolved in water. In this case, the neutral molecule is "ionized" by the water molecules. **Acids** are so named because *they all form the $H^+(aq)$ ion when dissolved in water.* (The $H^+(aq)$ ion in water is also represented in certain situations as $H_3O^+$, which is known as the hydronium ion.) **Strong acids,** such as hydrochloric acid (HCl), *are completely ionized in water.* We will consider only the action of the six common strong acids (HCl, HBr, HI, $HNO_3$, $HClO_4$, and $H_2SO_4$) in water at this time. The solutions of the strong acids, hydrochloric and nitric acid, are illustrated by the equations

$$HCl(aq) \xrightarrow{H_2O} H^+(aq) + Cl^-(aq)$$
$$HNO_3(aq) \xrightarrow{H_2O} H^+(aq) + NO_3^-(aq)$$

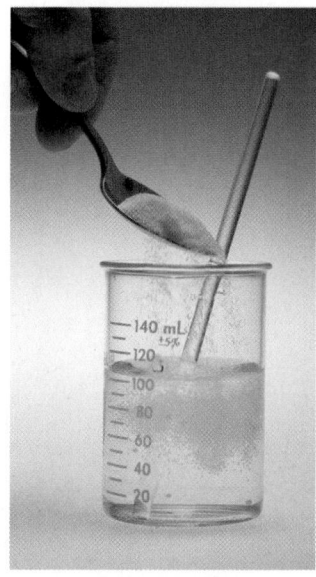

**Figure 7-4  A Soluble Compound**
**Table salt dissolves as it is added to hot water.**

---

**Ions in Solution When Ionic Compounds and Acids Dissolve**

**Example 7-4**

working it out

Write equations illustrating the solution of the following compounds in water: (a) $Na_2CO_3$, (b) $CaCl_2$, (c) $(NH_4)_2Cr_2O_7$, (d) $HClO_4$, and (e) $H_2SO_4$. (Although it is not exactly correct, consider sulfuric acid as completely separated into two $H^+(aq)$ ions in aqueous solution.)

**Procedure**

If necessary, refer to Table 4-1 or the periodic table for the charges on the ions.

**Solution**

a. $Na_2CO_3$ is composed of $Na^+$ and $CO_3^{2-}$ ions.

$$Na_2CO_3 \longrightarrow 2Na^+(aq) + CO_3^{2-}(aq)$$

b. $CaCl_2$ is composed of $Ca^{2+}$ and $Cl^-$ ions.

$$CaCl_2 \longrightarrow Ca^{2+}(aq) + 2Cl^-(aq)$$

c. $(NH_4)_2Cr_2O_7$ is composed of $NH_4^+$ and $Cr_2O_7^{2-}$ ions.

$$(NH_4)_2Cr_2O_7 \longrightarrow 2NH_4^+(aq) + Cr_2O_7^{2-}(aq)$$

d. $HClO_4$ is a strong acid that produces $H^+(aq)$ and $ClO_4^-$ ions.

$$HClO_4 \longrightarrow H^+(aq) + ClO_4^-(aq)$$

e. $H_2SO_4$ is a strong acid that produces $H^+(aq)$ and $SO_4^{2-}$ ions.

$$H_2SO_4 \longrightarrow 2H^+(aq) + SO_4^{2-}(aq)$$

# 7-4 Single-Replacement Reactions

**Looking Ahead!** One of the unique chemical characteristics of water is its ability to hold a variety of ions in solution. Sometimes, however, these ions interact with molecules, atoms, or other ions in chemical reactions. We will examine two basic types of these reactions. In this section, we will see how the ions of one metal may interact with the neutral atoms of another metal.

An interesting thing happens when we immerse a strip of zinc metal in a blue aqueous solution of a copper(II) salt. When we remove the zinc, it now looks as if it has changed into copper. Actually, what happened is that a coating of copper formed on the metal strip. (See Figure 7-5.) This and similar reactions are known as **single-replacement reactions.** In this type of reaction, which most often occurs in aqueous solution, one free element substitutes for another combined in a chemical compound. Silver and gold will also form a coating on a strip of zinc immersed in solutions of compounds containing these metal ions. The replacement of zinc ions for the copper ions and the copper metal for the zinc metal is illustrated here; other examples follow.

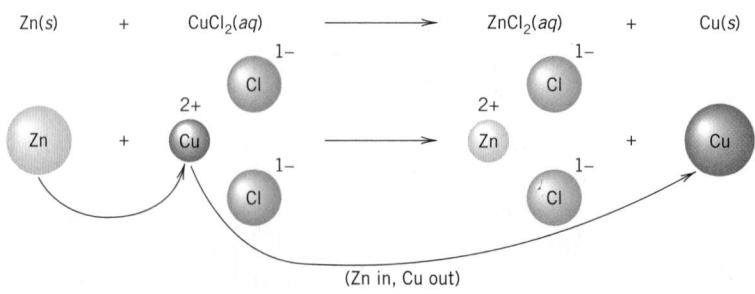

**Figure 7-5 Single-Replacement Reaction**
The formation of a layer of copper on a piece of zinc is a single-replacement reaction.

$Zn(s)$ + $CuCl_2(aq)$ $\longrightarrow$ $ZnCl_2(aq)$ + $Cu(s)$

(Zn in, Cu out)

$$Mg(s) + 2HCl(aq) \longrightarrow MgCl_2(aq) + H_2(g)$$
(Mg in, $H_2$ out)

$$2Cr(s) + 3Pb(NO_3)_2(aq) \longrightarrow 3Pb(s) + 2Cr(NO_3)_3(aq)$$
(Cr in, Pb out)

The equation illustrating the replacement of Zn by Cu is known as the molecular form of the equation. *In a **molecular equation,** all reactants and products are shown as neutral compounds.* To represent the nature of soluble ionic compounds and strong acids in water, it is helpful to show the separate ions as they actually exist in aqueous solution. *When the cations and anions of a compound in solution are shown separately, the resulting equation is known as a **total ionic equation.*** It is illustrated as follows.

$$\text{(molecular equation)} \; Zn(s) + CuCl_2(aq) \longrightarrow ZnCl_2(aq) + Cu(s)$$

$$\text{(total ionic equation)} \; Zn(s) + Cu^{2+}(aq) + 2Cl^-(aq) \longrightarrow Zn^{2+}(aq) + 2Cl^-(aq) + Cu(s)$$

**Figure 7-6 Copper in a Zinc Sulfate Solution**
**When a strip of copper is immersed in a zinc ion solution, no reaction occurs.**

Notice that in this equation, the $Cl^-$ ions appear on both sides of the equation unchanged. Their role is to provide the anions needed to counteract the positive charge of the cations. The presence of the $Cl^-$ ions is certainly necessary because any compound, whether in the pure state or dispersed in solution, must be electrically neutral. However, *ions that are in an identical state on both sides of an equation are called **spectator ions.** If spectator ions are subtracted from both sides of the equation, the remaining equation is known as the **net ionic equation.*** This equation focuses only on the species that have undergone a change in the reaction. By subtracting the two $Cl^-$ spectator ions from both sides of the equation we have the net ionic equation.

$$\text{(net ionic)} \quad Zn(s) + Cu^{2+}(aq) \longrightarrow Zn^{2+}(aq) + Cu(s)$$

A reaction is spontaneous in one direction only so the reverse reaction does not occur. That is, if we were to immerse a copper strip in a $ZnCl_2$ solution, a coating of Zn would not form on the copper. (See Figure 7-6.) This observation allows us to compare the ability of one metal to replace the ions of other metals in solution. *The **activity series** shown in Table 7-2 lists some common metals in decreasing order of their ability to replace metal ions in aqueous solution.* (The metal cations present in solution are shown to the right of the metal.) The metal at the top replaces all of the metal ions below it. The second metal replaces all below it but its ions will be replaced only by Na. In fact, Na and Al are so reactive that they both react with water itself. (Aluminum *appears* unreactive with water because it is coated with $Al_2O_3$, which protects the metal from contact with water.) Thus they replace other metal ions only when the solids are mixed, not when the ions are in water solution. The following balanced equation illustrates such a reaction.

$$2Al(s) + Fe_2O_3(s) \longrightarrow Al_2O_3(s) + 2Fe(l)$$

The above reaction is known as the thermite reaction because it liberates so much heat that the iron formed is molten. As a result, this reaction has application in welding.

In the activity series $H_2$ is treated as a metal and $H^+(aq)$ (from one of the six strong acids) as its metal ion. The activity series can be used to predict which reactions are expected to occur. For example, notice in Table 7-2 that nickel is ranked higher than silver. This allows us to predict that elemental nickel metal replaces the $Ag^+$ ion in solution. Thus, if we immerse a strip of nickel in a solution of aqueous $AgNO_3$, we find that a coating of silver forms on the nickel. The net ionic equation illustrating this reaction is as follows.

$$Ni(s) + 2Ag^+(aq) \longrightarrow 2Ag(s) + Ni^{2+}(aq)$$

**Table 7-2 The Activity Series**

| Metal | Metal Ion |
|---|---|
| Na | $Na^+$ |
| Al | $Al^{3+}$ |
| Zn | $Zn^{2+}$ |
| Cr | $Cr^{3+}$ |
| Fe | $Fe^{2+}$ |
| Ni | $Ni^{2+}$ |
| Sn | $Sn^{2+}$ |
| Pb | $Pb^{2+}$ |
| $H_2$ | $H^+$ |
| Cu | $Cu^{2+}$ |
| Ag | $Ag^+$ |
| Au | $Au^{3+}$ |

Example 7-5

### Predicting Spontaneous Single-Replacement Reactions

Consider the following two possible reactions. If a reaction does occur, write the balanced molecular, total ionic, and net ionic equations illustrating the reactions.

**a.** A strip of tin metal is placed in an aqueous $AgNO_3$ solution.

**b.** A strip of silver metal is placed in an aqueous perchloric acid solution.

**Procedure**

**a.** Notice in the activity series that Sn is higher in the series than Ag. Therefore Sn replaces $Ag^+$ ions from solution.

**b.** In the activity series, $H_2$ is higher than Ag. Therefore, $H_2$ replaces $Ag^+$, but the reverse reaction, the replacement of $H^+$ by Ag, does not occur.

**Solution**

**a.**

$$Sn(s) + 2AgNO_3(aq) \longrightarrow 2Ag(s) + Sn(NO_3)_2(aq) \quad \text{(molecular)}$$

$$Sn(s) + 2Ag^+(aq) + 2NO_3^-(aq) \longrightarrow$$
$$2Ag(s) + Sn^{2+}(aq) + 2NO_3^-(aq) \quad \text{(total ionic)}$$

$$Sn(s) + 2Ag^+(aq) \longrightarrow 2Ag(s) + Sn^{2+}(aq) \quad \text{(net ionic)}$$

**b.** No reaction occurs.

The thermite reaction is a single-replacement reaction and generates a large amount of heat energy.

# 7-5 Double-Replacement Reactions—Precipitation

**Looking Ahead!** In a single-replacement reaction, only ions in one compound are involved. In a double-replacement reaction, both cations and anions are involved. The driving force of these reactions is the formation of a product from the exchange of ions that is insoluble in water, is a molecular compound, or, in a few cases, is both. The first type, formation of a solid, is discussed in this section.

The white cliffs of Dover facing the English Channel have been exposed to the ravages of driving rain for centuries with little effect. The cliffs are composed primarily of chalk (calcium carbonate), which is an ionic compound that is essentially insoluble in water. Of more importance to us in this section is *how* insoluble compounds are formed from soluble compounds. But first, we should bring some order or guidelines to which ionic compounds are soluble and which are not. In Table 7-3, some rules for the solubility of compounds containing common anions are listed. Although this table focuses only on anions, we should note that the compounds formed from cations of Group IA (i.e., $Na^+$, $K^+$) and the $NH_4^+$ ion are water soluble.

Example 7-6

### Predicting Whether an Ionic Compound Is Soluble

Use Table 7-3 to predict whether the following compounds are soluble or insoluble: (a) NaI, (b) CdS, (c) $Ba(NO_3)_2$, and (d) $SrSO_4$

**Solution**

**a.** According to Table 7-3, all alkali metal (Group IA) compounds of the anions listed are soluble. Therefore, <u>NaI is soluble</u>.

## Corrosion of Iron—An Expensive Problem

For five thousand years, iron has been the most important metal on this planet. The Hittites of the ancient Middle East were the first to free iron from its ore, iron(III) oxide. It made them a very powerful people indeed, as their swords were stronger than the bronze weapons used by others. Iron, or its alloy steel, is still the metal that allows us to build huge skyscrapers and expansive bridges. Iron has a long-standing problem, however—it has a strong tendency to corrode. The corrosion takes place in a complex process when iron is in contact with water in which oxygen gas is dissolved. Thus iron in desert climates lasts much longer than in humid areas. When iron corrodes it forms an iron(III) oxide hydrate, which is represented as $Fe_2O_3 \cdot xH_2O$. The x indicates that various amounts of water are associated with the oxide. Unfortunately, as rust forms, it flakes off, exposing more iron underneath. Eventually, the whole piece of iron disintegrates. Other metals such as aluminum, tin, and chromium form oxide coatings but these coatings stick on the metal, protecting the underlying metal from further corrosion.

How can we protect iron from corrosion? The most obvious way is to apply a good coat of paint and avoid scratches. Another way is to coat the iron with a metal such as tin (i.e., *tin cans*) or zinc (*galvanizing*). To preserve many iron objects that are exposed to the weather, a process known as *cathodic protection* is used. For example, iron in buried pipelines and tanks is attached by a wire to a piece of metal such as magnesium or zinc. The higher in Table 7-2 a metal is, the easier a single-replacement reaction such as that shown below occurs.

$$\text{metal}(s) + 2H^+(aq) \longrightarrow \text{metal}^{2+}(aq) + H_2(g)$$

When two metals are in contact, even through an electrical wire, the one that reacts with acid or oxygen easier will do so first. Thus the piece of magnesium or zinc is sacrificed but the iron is protected. Eventually, the magnesium must be replaced to assure continued protection.

Despite all of these measures, bridges and highways (which contain steel rods imbedded in concrete) continue to deteriorate. Research continues on better ways to protect against corrosion, which costs billions of dollars annually.

*Rust is the curse of iron.*

**b.** All $S^{2-}$ compounds are insoluble except those formed with Groups IA and IIA metals and $NH_4^+$. Since Cd is in Group IIB, it is not one of the exceptions. <u>CdS is insoluble.</u>

**c.** All $NO_3^-$ compounds are soluble. Therefore, <u>$Ba(NO_3)_2$ is soluble.</u>

**d.** The $Sr^{2+}$ ion is one of the exceptions to soluble $SO_4^{2-}$ compounds. Therefore, <u>$SrSO_4$ is insoluble.</u>

### Table 7-3 Solubility Rules for Some Ionic Compounds

| Anion | Solubility Rule |
|---|---|
| *Mostly Soluble* | |
| $Cl^-$, $Br^-$, $I^-$ | All cations form *soluble* compounds except $Ag^+$, $Hg_2^{2+}$, and $Pb^{2+}$. ($PbCl_2$ and $PbBr_2$ are slightly soluble.) |
| $NO_3^-$, $ClO_4^-$, $C_2H_3O_2^-$ | All cations form *soluble* compounds. ($KClO_4$ and $AgC_2H_3O_2$ are slightly soluble.) |
| $SO_4^{2-}$ | All cations form *soluble* compounds except $Pb^{2+}$, $Ba^{2+}$, and $Sr^{2+}$. ($Ca^{2+}$ and $Ag^+$ form slightly soluble compounds.) |
| *Mostly Insoluble* | |
| $CO_3^{2-}$, $PO_4^{3-}$ | All cations form *insoluble* compounds except Group IA metals and $NH_4^+$. |
| $S^{2-}$ | All cations form *insoluble* compounds except Groups IA and IIA metals and $NH_4^+$. |
| $OH^-$ | All cations form *insoluble* compounds except Group IA metals, $Ba^{2+}$, $Sr^{2+}$, and $NH_4^+$. [$Ca(OH)_2$ is slightly soluble.] |

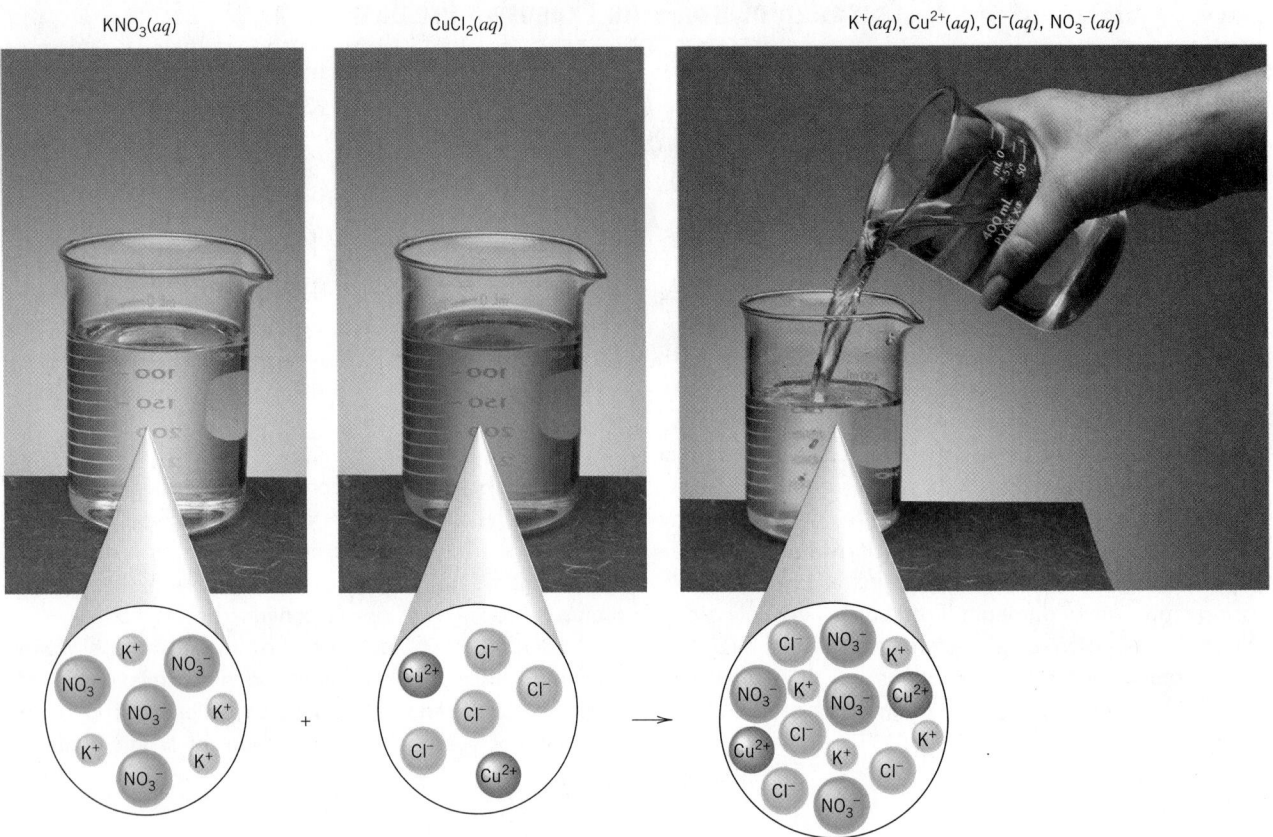

KNO$_3$(aq)      CuCl$_2$(aq)      K$^+$(aq), Cu$^{2+}$(aq), Cl$^-$(aq), NO$_3^-$(aq)

**Figure 7-7  A Mixture of CuCl$_2$ and KNO$_3$ Solutions**
**No reaction occurs when these solutions are mixed.**

What happens when we mix solutions containing soluble ionic compounds? It depends. If we mix a solution of CuCl$_2$ (green) and KNO$_3$, (clear) we simply have a mixture of the four ions in solution, as illustrated in Figure 7-7. No cation is associated with a particular anion.

Now let us consider a case of two solutions, one containing the soluble compound CuCl$_2$ and the other containing the soluble compound AgNO$_3$. When we mix these two solutions, something obviously occurs. The mixture immediately becomes cloudy, and eventually a solid settles to the bottom of the container. *The solid that is formed by the reaction is called a* **precipitate** of the insoluble compound, AgCl. In fact, whenever Ag$^+$ and Cl$^-$ are mixed into the same solution, they come together to form solid AgCl. This leaves the Cu$^{2+}$ and the NO$_3^-$ ions in solution since Cu(NO$_3$)$_2$ is soluble. (See Figure 7-8.)

*The formation of a precipitate by mixing solutions of two soluble compounds is known as a* **precipitation reaction.** This is one of two types of **double-replacement reactions** where the two cations involved exchange anions. The reaction is illustrated by the following molecular equation.

$$2AgNO_3(aq) + CuCl_2(aq) \longrightarrow 2AgCl(s) + Cu(NO_3)_2(aq)$$

In the total ionic equation, the soluble ionic compounds are represented as separate ions on both sides of the equation but the solid precipitate (i.e., AgCl) is shown as a neutral compound since the two ions come together to produce the solid.

$$2Ag^+(aq) + 2NO_3^-(aq) + Cu^{2+}(aq) + 2Cl^-(aq) \longrightarrow$$
$$2AgCl(s) + Cu^{2+}(aq) + 2NO_3^-(aq)$$

**Figure 7-8  A Mixture of AgNO₃ and CuCl₂ Solutions** **When these solutions are mixed, a precipitate forms.**

After spectator ions are subtracted from the equation, we have the net ionic equation for the reaction.

$$Ag^+(aq) + Cl^-(aq) \longrightarrow AgCl(s)$$

As in the single-replacement reactions discussed previously, the net ionic equation focuses on the driving force for the reaction. That is the formation of solid AgCl from two soluble compounds.

There are some very practical applications of precipitation reactions in industry as well as in the laboratory. Our example is of particular value. Silver actually finds more use in development of photographs and film than in jewelry and coins. It is obviously worthwhile to recover this precious metal from film negatives whenever possible. Silver metal in film can be dissolved in aqueous nitric acid to form the water-soluble compound $AgNO_3$. Although the solution contains many other dissolved substances, addition of a soluble compound containing the $Cl^-$ ion (e.g., NaCl) leads to the formation of solid AgCl, as shown in Figure 7-8. As you notice from Table 7-3, very few other cations form precipitates with $Cl^-$, so this is a reaction more or less specific to removing $Ag^+$ from aqueous solution. The AgCl can then be filtered from the solution and silver metal eventually recovered.

In other precipitation reactions, we may wish to recover the soluble compound and discard the insoluble compound. In such a case, we would remove the precipitate by filtration and then recover the soluble compound by boiling away the solvent water.

By careful use of Table 7-3, we can predict the occurrence of many precipitation reactions. To accomplish this, we follow this procedure.

1.  Write the compounds produced in the reaction by "switching partners," making sure that the compounds have the correct formulas based on the ions' charges (which do not change).

2.  Examine Table 7-3 to determine whether one of these compounds is insoluble.

3.  If one of the two new compounds is insoluble in water, a precipitation reaction occurs and we can write the equation illustrating the reaction.

The following examples illustrate the use of Table 7-3 to predict and write precipitation reactions. (In Examples 7-7, 7-8, and 7-9 we will write the balanced molecular equation, the total ionic equation, and the net ionic equation for any reaction that occurs when solutions of the two compounds are mixed.)

---

| **Example 7-7** | **A Possible Precipitation Reaction** |
| --- | --- |
| **w o r k i n g  i t  o u t** | A solution of $Na_2CO_3$ is mixed with a solution of $CaCl_2$. Predict what happens. |

**Solution**

The four ions involved are $Na^+$, $CO_3^{2-}$, $Ca^{2+}$, and $Cl^-$. The combinations of the $Na^+$ and $Cl^-$ and the $Ca^{2+}$ and $CO_3^{2-}$ produce the compounds $NaCl$ and $CaCO_3$. If both of these compounds are soluble, no reaction occurs. In this case, however, reference to Table 7-3 indicates that $CaCO_3$ is insoluble. Thus a reaction occurs that we can illustrate with a balanced reaction written in molecular form.

$$Na_2CO_3(aq) + CaCl_2(aq) \longrightarrow CaCO_3(s) + 2NaCl(aq)$$

The equation written in total ionic form is

$$2Na^+(aq) + CO_3^{2-}(aq) + Ca^{2+}(aq) + 2Cl^-(aq) \longrightarrow$$
$$CaCO_3(s) + 2Na^+(aq) + 2Cl^-(aq)$$

Note that the $Na^+$ and the $Cl^-$ ions are spectator ions. Elimination of the spectator ions on both sides of the equation leaves the net ionic equation.

$$Ca^{2+}(aq) + CO_3^{2-}(aq) \longrightarrow CaCO_3(s)$$

---

| **Example 7-8** | **A Possible Precipitation Reaction** |
| --- | --- |

A solution of KOH is mixed with a solution of $MgI_2$. Predict what happens.

**Solution**

The four ions involved are $K^+$, $OH^-$, $Mg^{2+}$, and $I^-$. An exchange of ions in the reactants produces the compounds KI and $Mg(OH)_2$. Reference to Table 7-3 indicates that $Mg^{2+}$ forms an insoluble compound with $OH^-$. Therefore, a precipitation reaction does occur and is illustrated with the following molecular equation.

$$2KOH(aq) + MgI_2(aq) \longrightarrow Mg(OH)_2(s) + 2KI(aq)$$

The total ionic equation is

$$2K^+(aq) + 2OH^-(aq) + Mg^{2+}(aq) + 2I^-(aq) \longrightarrow$$
$$Mg(OH)_2(s) + 2K^+(aq) + 2I^-(aq)$$

Elimination of spectator ions gives the net ionic equation.

$$Mg^{2+}(aq) + 2OH^-(aq) \longrightarrow Mg(OH)_2(s)$$

**A Possible Precipitation Reaction**

**Example 7-9**

A solution of KBr is mixed with a solution of $Sr(ClO_4)_2$. Predict what happens.

**Solution**

The four ions involved are $K^+$, $Br^-$, $Sr^{2+}$, and $ClO_4^-$. An exchange of ions produces the compounds $KClO_4$ and $SrBr_2$. Both of these compounds are soluble, so no reaction occurs.

## 7-6 Double-Replacement Reactions—Neutralization

**Looking Ahead!** Cations and anions in solution can combine to form a solid ionic compound. In a second type of double-replacement reaction, the ions combine to form a molecular compound. Although there are several examples of this, we will focus on the formation of the simple molecular compound water. This is discussed next.

Strong acids and bases are two compounds that can be difficult, if not dangerous, to handle. They are very corrosive and can cause severe burns. (See Figure 7-9.) When carefully mixed together in the right proportions, however, they become harmless. The corrosive properties of both are neutralized. In section 7-3 we defined a class of compounds known as strong acids. They are characterized by their complete ionization in water to form $H^+(aq)$ and an anion. A second class of compounds, **strong bases,** *dissolve in water to form the hydroxide ion [OH⁻(aq)].* Unlike acids, these compounds are ionic in the solid state and the ions are simply separated in the aqueous solution. The strong bases are hydroxides formed by the IA and IIA metal ions (except for $Be^{2+}$). Examples include sodium hydroxide (lye) and barium hydroxide. Their solution in water is illustrated by the following equations.

$$NaOH(s) \xrightarrow{H_2O} Na^+(aq) + OH^-(aq)$$
$$Ba(OH)_2(s) \xrightarrow{H_2O} Ba^{2+}(aq) + 2OH^-(aq)$$

When solutions of strong acids and bases are mixed, the $H^+(aq)$ from the acid combines with the $OH^-(aq)$ from the base to form the molecular compound water. *The reaction of an acid and a base is known as a* **neutralization reaction.** The neutralization of hydrochloric acid and sodium hydroxide is illustrated below with the molecular, total ionic, and net ionic equations. The ionic compound remaining in solution, NaCl, is known as salt. *A* **salt** *is formed from the cation of the base and the anion of the acid.* If the salt is soluble, its ions are spectator ions and are subtracted from the equation to form the net ionic equation.

$$HCl(aq) + NaOH(aq) \longrightarrow NaCl(aq) + H_2O(l)$$
$$H^+(aq) + Cl^-(aq) + Na^+(aq) + OH^-(aq) \longrightarrow Na^+(aq) + Cl^-(aq) + H_2O(l)$$
$$H^+(aq) + OH^-(aq) \longrightarrow H_2O(l)$$

**Figure 7-9 Strong Acids and Bases**
Containers of these compounds usually include a warning about their corrosive properties.

Unlike precipitation reactions, the net ionic equation for all neutralization reactions between strong acids and strong bases is the same. The driving force for these reactions is the formation of water. The balanced molecular equations for two additional neutralization reactions follow.

$$HBr(aq) + KOH(aq) \longrightarrow KBr(aq) + H_2O(l)$$
$$2HNO_3(aq) + Ca(OH)_2(aq) \longrightarrow Ca(NO_3)_2(aq) + 2H_2O(l)$$

| **Example 7-10** | **Writing Neutralization Reactions** |
| --- | --- |
| **working it out** | Write the molecular, total ionic, and net ionic equations for the neutralization of (a) $HClO_4$ and LiOH and (b) $H_2SO_4$ and KOH. |

**Solution**

To balance the equations, one should make sure that there is one $H^+(aq)$ for each $OH^-(aq)$. Another way is to write the formula of the salt formed and then balance the number of reactant cations (from the base) and reactant anions (from the acid).

**a.**

$$HClO_4(aq) + LiOH(aq) \longrightarrow LiClO_4(aq) + H_2O(l)$$
$$H^+(aq) + ClO_4^-(aq) + Li^+(aq) + OH^-(aq) \longrightarrow Li^+(aq) + ClO_4^-(aq) + H_2O(l)$$
$$H^+(aq) + OH^-(aq) \longrightarrow H_2O(l)$$

**b.**

$$H_2SO_4(aq) + 2KOH(aq) \longrightarrow K_2SO_4(aq) + 2H_2O(l)$$
$$2H^+(aq) + SO_4^{2-}(aq) + 2K^+(aq) + 2OH^-(aq) \longrightarrow$$
$$2K^+(aq) + SO_4^{2-}(aq) + 2H_2O(l)$$
$$H^+(aq) + OH^-(aq) \longrightarrow H_2O(l)$$

The interactions of other types of acids and bases, known as weak acids and bases, are somewhat more involved and will be discussed in more detail in Chapter 12.

**Looking Back!** Water is necessary for life for a number of reasons. One reason, as has been discussed in the last three sections, is that it dissolves a large number of compounds, especially ionic compounds. As a result, chemical reactions occur when certain ions are mixed. These reactions include the formation of a precipitate and the formation of water itself. The formation of a precipitate in our bodies can be harmful, as in the case of kidney stones, but neutralization reactions are necessary to maintain the proper balance of ions in our blood.

# Learning Check B | checking it out

**B-1.** Fill in the blanks.

When a compound dissolves in a liquid, the compound is said to be _____ in that liquid. If it does not dissolve to any appreciable extent, it is _____ in the liquid. A solution is usually composed of two components, a liquid _____ and a _____ that can be any phase.

When an ionic compound dissolves in water, the compound is separated into individual _____ . A class of molecular compounds that form $H^+$ ions in water are known as strong _____ . When one element substitutes for another in a compound, the reaction is known as a _____ -replacement reaction. When the interchange of two ions in a reaction produces a solid compound or a molecular compound, the reaction is known as a _____ -replacement reaction. An equation that illustrates the driving force of these reactions is known as the _____ _____ equation. This equation does not include _____ ions. The order of metals listed according to their ability to replace other metals is known as the _____ series. When a solid forms in a solution, it is known as a _____ , and a reaction forming a solid is known as a _____ reaction. A reaction that produces water from two ions is known as a _____ reaction. In addition to water this reaction forms a _____ .

**B-2.** Write the ions that form when the following compounds dissolve in water: (a) $BaBr_2$, (b) $(NH_4)_2CrO_4$, (c) HBr.

**B-3.** By reference to the activity series, determine whether iron reacts with aqueous solutions of the strong acid $H_2SO_4$. Write the balanced molecular and net ionic equations illustrating this reaction if it occurs.

**B-4.** A solution of $AgNO_3$ is mixed with a solution of $K_2S$. Write the molecular, total ionic, and net ionic equations illustrating the reaction.

**B-5.** Write the balanced molecular, total ionic, and net ionic equations illustrating the neutralization of nitric acid with strontium hydroxide.

*Additional Examples: Exercises 7-24, 7-26, 7-28, 7-34, 7-44, 7-46, 7-55, and 7-57.*

**chapter** **R**eview                                                    **putting it together**

A concise statement of a chemical property is relayed by the **chemical equation.** With symbols, formulas, and other abbreviations, a sizable amount of chemical information can be communicated. This includes the elements or compounds involved as **reactants** and **products,** their physical states, and the number of molecules of each compound involved in the reaction. When the numbers of atoms of each element are made the same on both sides of the equation by use of **coefficients,** the equation is said to be **balanced.**

In this chapter, we considered five different types of reactions that can be conveniently represented by equations. Each type has a general equation that characterizes that kind of reaction. The first three types discussed in this chapter are illustrated below, with letters representing elements, compounds, or ions.

1. **Combustion reactions**

$$A + O_2 \longrightarrow AO$$
$$AB + O_2 \longrightarrow AO + BO$$

2. **Combination reactions**

$$A + B \longrightarrow C$$

3. **Decomposition reactions**

$$C \longrightarrow A + B$$

In addition to these three types of reactions, two other types usually involve reactions that occur in an aqueous solution. When a substance (a **solute**) is dispersed by a liquid (a **solvent**), it forms a homogeneous mixture known as a **solution.** Substances that do dissolve in water are said to be **soluble** and those that do not are **insoluble.** Soluble ionic compounds are separated into their individual ions in aqueous solution.

In **single-replacement reactions,** a metal exchanges places with the cation of a different metal, as illustrated by the following general equation.

$$A + BC \longrightarrow AC + B$$

The ability of metals to replace other metal ions can be compared and ranked in the **activity series.** Hydrogen, although not a metal, is usually included in this series.

**Double-replacement reactions** involve the exchange of ions between two soluble compounds. In a **precipitation reaction,** the two ions combine to form a solid ionic compound known as a **precipitate,** which separates from the solution. In a **neutralization reaction,** the two ions form a molecular compound. A general equation illustrates these two types of reactions.

$$AB + CD \longrightarrow AD + CB$$

In neutralization reactions, **strong acids** [molecular compounds that dissolve in water to form $H^+(aq)$ ions] react with **strong bases** [ionic compounds that dissolve in water to form $OH^-(aq)$ ions]. The reaction produces water and a **salt.**

Single-and double-replacement reactions can be illustrated by three types of equations. In the **molecular equation** all species are represented as neutral compounds. In a **total ionic equation,** soluble ionic compounds and strong acids are represented as separate ions. If the **spectator ions** (those ions that are not directly involved in the reaction) are removed, the result is the **net ionic equation.** The three types of equations for a typical precipitation reaction are shown below.

$$Pb(NO_3)_2(aq) \quad + \quad 2KCl(aq) \longrightarrow PbCl_2(s) \quad + \quad 2KNO_3(aq)$$

$$Pb^{2+}(aq) + 2NO_3^-(aq) + 2K^+(aq) + 2Cl^-(aq) \longrightarrow PbCl_2(s) + 2K^+(aq) + 2NO_3^-(aq)$$

$$Pb^{2+}(aq) + 2Cl^-(aq) \longrightarrow PbCl_2(s)$$

# Exercises

## Chemical Equations

**7-1.** The physical state of an element is included in a chemical equation. Each of the following compounds is a gas, a solid, or a liquid under normal conditions. Indicate the proper physical state by adding $(g)$, $(s)$, or $(l)$ after the formula.

(a) $Cl_2$  (d) $H_2O$  (g) $Br_2$  (j) Na
(b) C  (e) $P_4$  (h) NaBr  (k) Hg
(c) $K_2SO_4$  (f) $H_2$  (i) $S_8$  (l) $CO_2$

**7-2.** Balance the following equations.
(a) $CaCO_3 \xrightarrow{\Delta} CaO + CO_2$
(b) $Na + O_2 \longrightarrow Na_2O$
(c) $H_2SO_4 + NaOH \longrightarrow Na_2SO_4 + H_2O$
(d) $H_2O_2 \longrightarrow H_2O + O_2$

**7-3.** Balance the following equations.
(a) $NaBr + Cl_2 \longrightarrow NaCl + Br_2$
(b) $KOH + H_3AsO_4 \longrightarrow K_2HAsO_4 + H_2O$

(c) $Ti + Cl_2 \longrightarrow TiCl_4$
(d) $Al + H_2SO_4 \longrightarrow Al_2(SO_4)_3 + H_2$

**7-4.** Balance the following equations.
(a) $Al + H_3PO_4 \longrightarrow AlPO_4 + H_2$
(b) $Ca(OH)_2 + HCl \longrightarrow CaCl_2 + H_2O$
(c) $Mg + N_2 \longrightarrow Mg_3N_2$
(d) $C_2H_6 + O_2 \longrightarrow CO_2 + H_2O$

**7-5.** Balance the following equations.
(a) $Ca(CN)_2 + HBr \longrightarrow CaBr_2 + HCN$
(b) $C_3H_6 + O_2 \longrightarrow CO + H_2O$
(c) $P_4 + S_8 \longrightarrow P_4S_3$
(d) $Cr_2O_3 + Si \longrightarrow Cr + SiO_2$

**7-6.** Balance the following equations.
(a) $Mg_3N_2 + H_2O \longrightarrow Mg(OH)_2 + NH_3$
(b) $H_2S + O_2 \longrightarrow S + H_2O$
(c) $Si_2H_6 + H_2O \longrightarrow Si(OH)_4 + H_2$
(d) $C_2H_6 + Cl_2 \longrightarrow C_2HCl_5 + HCl$

**7-7.** Balance the following equations.
(a) $Na_2NH + H_2O \longrightarrow NH_3 + NaOH$
(b) $CaC_2 + H_2O \longrightarrow C_2H_2 + Ca(OH)_2$
(c) $XeF_6 + H_2O \longrightarrow XeO_3 + HF$
(d) $PCl_5 + H_2O \longrightarrow H_3PO_4 + HCl$

**7-8.** Balance the following equations.
(a) $B_4H_{10} + O_2 \longrightarrow B_2O_3 + H_2O$
(b) $SF_6 + SO_3 \longrightarrow O_2SF_2$
(c) $CS_2 + O_2 \longrightarrow CO_2 + SO_2$
(d) $BF_3 + NaH \longrightarrow B_2H_6 + NaF$

**7-9.** Balance the following equations.
(a) $NH_3 + Cl_2 \longrightarrow NHCl_2 + HCl$
(b) $PBr_3 + H_2O \longrightarrow HBr + H_3PO_3$
(c) $Mg + Fe_3O_4 \longrightarrow MgO + Fe$
(d) $Fe_3O_4 + H_2 \longrightarrow Fe + H_2O$

**7-10.** Write balanced chemical equations from the following word equations. Include the physical state of each element or compound.
(a) Sodium metal plus water yields hydrogen gas and an aqueous sodium hydroxide solution.
(b) Potassium chlorate when heated yields potassium chloride plus oxygen gas. (Ionic compounds are solids.)
(c) An aqueous sodium chloride solution plus an aqueous silver nitrate solution yields a silver chloride precipitate (solid) and a sodium nitrate solution.
(d) An aqueous phosphoric acid solution plus an aqueous calcium hydroxide solution yields water and solid calcium phosphate.

**7-11.** Write balanced chemical equations from the following word equations. Include the physical state of each element or compound.
(a) Solid phenol ($C_6H_6O$) reacts with oxygen to form carbon dioxide gas and liquid water.
(b) An aqueous calcium hydroxide solution reacts with gaseous sulfur trioxide to form a precipitate of calcium sulfate and water.
(c) Lithium is the only element that combines with nitrogen at room temperature. The reaction forms lithium nitride.
(d) Magnesium dissolves in an aqueous chromium(III) nitrate solution to form chromium and a magnesium nitrate solution.

**7-12.** Nickel(II) nitrate is prepared by heating nickel metal with liquid dinitrogen tetroxide. In addition to the nitrate, gaseous nitrogen monoxide is formed. Write the balanced equation.

**7-13.** One of the steps in the production of iron involves the reaction of $Fe_3O_4$ with carbon monoxide to produce FeO and carbon dioxide. Write the balanced equation.

### Combustion, Combination, and Decomposition Reactions

**7-14.** Which reactions in exercises 7-2 and 7-4 can be classified as a combustion, a combination, or a decomposition reaction?

**7-15.** Which reactions in exercises 7-3 and 7-5 can be classified as either a combustion, a combination, or a decomposition reaction?

**7-16.** Write combustion reactions when the following compounds react with excess oxygen.
(a) $C_7H_{14}(l)$    (c) $C_4H_{10}O(l)$
(b) $LiCH_3(s)$    (d) $C_2H_5SH(g)$ (a product is $SO_2$)

**7-17.** Write combustion reactions when the following compounds react with excess oxygen.
(a) $C_2H_6O_2(l)$    (c) $C_6H_{12}(l)$
(b) $B_6H_{12}(g)$    (d) $Pb(C_2H_5)_4(s)$
    [a product is    [a product is
    $B_2O_3(s)$]       $PbO_2(s)$]

**7-18.** Write combination reactions that occur when the metal barium reacts with the following nonmetals.
(a) hydrogen    (c) bromine
(b) sulfur    (d) nitrogen

**7-19.** Write combination reactions that occur when the metal aluminum reacts with the following nonmetals.
(a) hydrogen    (c) iodine
(b) oxygen    (d) nitrogen

**7-20.** Write decomposition reactions for the following compounds. Recall that ionic compounds are solids.
(a) $Ca(HCO_3)_2$ into calcium oxide, carbon dioxide, and water
(b) $Ag_2O$ into its elements
(c) $N_2O_3$ gas into nitrogen dioxide gas and nitrogen monoxide gas

**7-21.** Write decomposition reactions for the following compounds. Recall that ionic compounds are solids.
(a) liquid $SbF_5$ into fluorine and solid antimony trifluoride
(b) $PtO_2$ into its elements
(c) gaseous BrF into bromine and gaseous bromine trifluoride

**7-22.** Write balanced equations by predicting the products of the following reactions. Include the physical state of each element or compound.
(a) the combination of potassium and chlorine
(b) the combustion of liquid benzene ($C_6H_6$)
(c) the decomposition of gold(III) oxide into its elements by heating
(d) the combustion of propyl alcohol ($C_3H_8O$).
(e) the combination of phosphorus ($P_4$) and fluorine gas to produce solid phosphorus pentafluoride

**7-23.** Write balanced equations by predicting the products of the following reactions. Include the physical state of each element or compound.
(a) the combustion of liquid butane ($C_4H_{10}$)
(b) the decomposition of aqueous sulfurous acid to produce water and a gas
(c) the combination of sodium and oxygen gas to form sodium peroxide (The peroxide ion is $O_2^{2-}$.)
(d) the decomposition of copper(I) oxide into its elements by heating

## Ions in Aqueous Solution

**7-24.** Write equations illustrating the solution of each of the following compounds in water.
**(a)** $Na_2S$    **(c)** $K_2Cr_2O_7$    **(e)** $(NH_4)_2S$
**(b)** $Li_2SO_4$    **(d)** $CaS$      **(f)** $Ba(OH)_2$

**7-25.** Write equations illustrating the solution of each of the following compounds in water.
**(a)** $Ca(ClO_3)_2$    **(c)** $AlCl_3$
**(b)** $CsBr$        **(d)** $Cs_2SO_3$

**7-26.** Write equations illustrating the solution of the following compounds in water.
**(a)** $HNO_3$    **(b)** $Sr(OH)_2$

**7-27.** Write equations illustrating the solution of the following compounds in water.
**(a)** $LiOH$    **(b)** $HI$

## Single-Replacement Reactions

**7-28.** If any of the following reactions occur spontaneously, write the balanced net ionic equation. If not, write "no reaction." (Refer to Table 7-2.)
**(a)** $Pb + Zn^{2+} \longrightarrow Pb^{2+} + Zn$
**(b)** $Fe + H^+ \longrightarrow Fe^{2+} + H_2$
**(c)** $Cu + Ag^+ \longrightarrow Cu^{2+} + Ag$
**(d)** $Cr + Zn^{2+} \longrightarrow Cr^{3+} + Zn$

**7-29.** If any of the following reactions occur spontaneously, write the balanced net ionic equation. If not, write "no reaction." (Refer to Table 7-2.)
**(a)** $Pb + Sn^{2+} \longrightarrow Pb^{2+} + Sn$
**(b)** $H_2 + Ni^{2+} \longrightarrow H^+ + Ni$
**(c)** $Cr + Ni^{2+} \longrightarrow Cr^{3+} + Ni$
**(d)** $H_2 + Au^{3+} \longrightarrow H^+ + Au$

**7-30.** In the following situations, a reaction may or may not take place. If it does, write the balanced molecular, total ionic, and net ionic equations illustrating the reaction. Assume all involve aqueous solutions.
**(a)** Some iron nails are placed in a $CuCl_2$ solution.
**(b)** Silver coins are dropped in a hydrochloric acid solution.
**(c)** A copper wire is placed in a $Pb(NO_3)_2$ solution.
**(d)** Zinc strips are placed in a $Cr(NO_3)_3$ solution.

**7-31.** In the following situations, a reaction may or may not take place. If it does, write the balanced molecular, total ionic, and net ionic equations illustrating the reaction. Assume all involve aqueous solutions.
**(a)** A solution of nitric acid is placed in a tin can.
**(b)** Iron nails are placed in a $ZnBr_2$ solution.
**(c)** A chromium-plated auto accessory is placed in an $SnCl_2$ solution.
**(d)** A silver bracelet is placed in a $Cu(ClO_4)_2$ solution.

**7-32.** When heated, sodium metal reacts with solid $Cr_2O_3$. Write the balanced molecular and the net ionic equations for this single-replacement reaction.

**7-33.** When heated, aluminum metal reacts with solid $PbO$. Write the balanced molecular and net ionic equations for this single-replacement reaction.

## Solubility and Precipitation Reactions

**7-34.** Referring to Table 7-3, determine which of the following compounds are insoluble in water.
**(a)** $Na_2S$    **(c)** $MgSO_4$    **(e)** $(NH_4)_2S$
**(b)** $PbSO_4$    **(d)** $Ag_2S$     **(f)** $HgI_2$

**7-35.** Referring to Table 7-3, determine which of the following compounds are insoluble in water.
**(a)** $NiS$      **(c)** $Al(OH)_3$    **(e)** $CaS$
**(b)** $Hg_2Br_2$   **(d)** $Rb_2SO_4$    **(f)** $BaCO_3$

**7-36.** Write the formulas of the precipitates formed when $Ag^+$ combines with the following anions.
**(a)** $Br^-$    **(b)** $CO_3^{2-}$    **(c)** $PO_4^{3-}$

**7-37.** Write the formulas of the precipitates formed when $Pb^{2+}$ combines with the following anions.
**(a)** $SO_4^{2-}$    **(b)** $PO_4^{3-}$    **(c)** $I^-$

**7-38.** Write the formulas of the precipitates formed when $CO_3^{2-}$ combines with the following cations.
**(a)** $Cu^{2+}$    **(b)** $Cd^{2+}$    **(c)** $Cr^{3+}$

**7-39.** Write the formulas of the precipitates formed when $OH^-$ combines with the following cations.
**(a)** $Ag^+$    **(b)** $Ni^{2+}$    **(c)** $Co^{3+}$

**7-40.** Which of the following chlorides is insoluble in water? (Refer to Table 7-3.)
**(a)** $NaCl$    **(b)** $Hg_2Cl_2$    **(c)** $AlCl_3$    **(d)** $BaCl_2$

**7-41.** Which of the following sulfates is insoluble in water? (Refer to Table 7-3.)
**(a)** $K_2SO_4$    **(b)** $ZnSO_4$    **(c)** $SrSO_4$    **(d)** $MgSO_4$

**7-42.** Which of the following phosphates is insoluble in water? (Refer to Table 7-3.)
**(a)** $K_3PO_4$      **(c)** $(NH_4)_3PO_4$
**(b)** $Ca_3(PO_4)_2$   **(d)** $Li_3PO_4$

**7-43.** Which of the following hydroxides is insoluble in water? (Refer to Table 7-3.)
**(a)** $Mg(OH)_2$    **(b)** $CsOH$    **(c)** $Ba(OH)_2$    **(d)** $NaOH$

**7-44.** Write the balanced molecular equation for any reaction that occurs when the following solutions are mixed. (Refer to Table 7-3.)
**(a)** $KI$ and $Pb(C_2H_3O_2)_2$
**(b)** $AgClO_4$ and $KNO_3$
**(c)** $Sr(ClO_4)_2$ and $Ba(OH)_2$
**(d)** $BaS$ and $Hg_2(NO_3)_2$
**(e)** $FeCl_3$ and $KOH$

**7-45.** Write the balanced molecular equation for any reaction that occurs when the following solutions are mixed. (Refer to Table 7-3.)
**(a)** $Ba(C_2H_3O_2)_2$ and $Na_2SO_4$
**(b)** $NaClO_4$ and $Pb(NO_3)_2$
**(c)** $Mg(NO_3)_2$ and $Na_3PO_4$
**(d)** $SrS$ and $NiI_2$

**7-46.** Write the total ionic and net ionic equations for any reactions that occurred in exercise 7-44.

**7-47.** Write the total ionic and net ionic equations for any reactions that occurred in exercise 7-45.

**7-48.** Write the total ionic equation and the net ionic equation for each of the following reactions.
**(a)** $K_2S(aq) + Pb(NO_3)_2(aq) \longrightarrow PbS(s) + 2KNO_3(aq)$
**(b)** $(NH_4)_2CO_3(aq) + CaCl_2(aq) \longrightarrow$
$$CaCO_3(s) + 2NH_4Cl(aq)$$
**(c)** $2AgClO_4(aq) + Na_2CrO_4(aq) \longrightarrow$
$$Ag_2CrO_4(s) + 2NaClO_4(aq)$$

**7-49.** Write the total ionic equation and the net ionic equations for each of the following reactions.
**(a)** $Hg_2(ClO_4)_2(aq) + 2HBr(aq) \longrightarrow$
$$Hg_2Br_2(s) + 2HClO_4(aq)$$
**(b)** $2AgNO_3(aq) + (NH_4)_2SO_4(aq) \longrightarrow$
$$Ag_2SO_4(s) + 2NH_4NO_3(aq)$$
**(c)** $CuSO_4(aq) + 2KOH(aq) \longrightarrow Cu(OH)_2(s) + K_2SO_4(aq)$

**\*7-50.** Write the balanced molecular equations indicating how the following ionic compounds can be prepared by a precipitation reaction using any other ionic compounds. In some cases, the equation should reflect the fact that the desired compound is soluble and must be recovered by vaporizing the solvent water after removal of a precipitate.
**(a)** $CuCO_3$   **(c)** $Hg_2I_2$   **(e)** $KC_2H_3O_2$
**(b)** $PbSO_4$   **(d)** $NH_4NO_3$

## Neutralization Reactions

**7-51.** Which of the following is not a strong acid?
**(a)** HBr   **(b)** HF   **(c)** $HNO_3$   **(d)** $HClO_4$

**7-52.** Which of the following is not a strong acid?
**(a)** $HNO_3$   **(b)** HI   **(c)** $H_2SO_4$   **(d)** $HNO_2$

**7-53.** Which of the following is not a strong base?
**(a)** NaOH   **(b)** $Ba(OH)_2$   **(c)** $Al(OH)_3$   **(d)** CsOH

**7-54.** Which of the following is not a strong base?
**(a)** $Be(OH)_2$   **(c)** LiOH
**(b)** $Ba(OH)_2$   **(d)** KOH

**7-55.** Write balanced molecular equations for the neutralization reactions between the following compounds.
**(a)** HI and CsOH          **(c)** $H_2SO_4$ and $Sr(OH)_2$
**(b)** $HNO_3$ and $Ca(OH)_2$

**7-56.** Write balanced molecular equations for the neutralization reactions between the following compounds.
**(a)** $Ca(OH)_2$ and HI      **(c)** $HClO_4$ and $Ba(OH)_2$
**(b)** $H_2SO_4$ and LiOH

**7-57.** Write the total ionic and net ionic equations for the reactions in exercise 7-55.

**7-58.** Write the total ionic and net ionic equations for the reactions in exercise 7-56.

**\*7-59.** Magnesium hydroxide is considered a strong base but has very low solubility in water. It is known as milk of magnesia and is used to neutralize stomach acid (HCl). Write the balanced molecular, total ionic, and net ionic equations illustrating this reaction. (Since magnesium hydroxide is a solid, the total and net ionic equations will be somewhat different.)

**\*7-60.** When calcium hydroxide is neutralized with sulfuric acid, the salt produced is insoluble in water. Write the balanced molecular, total ionic, and net ionic equations illustrating this reaction.

## General Problems

**7-61.** Write the balanced equations representing the combustion of propane gas ($C_3H_8$), butane liquid ($C_4H_{10}$), octane ($C_8H_{18}$) in liquid gasoline, and liquid ethyl alcohol ($C_2H_5OH$) found in alcoholic beverages.

**7-62.** In the combination reaction between sodium and chlorine and in the combustion reaction of magnesium, why could these also be considered electron exchange reactions?

**7-63.** Iron replaces gold ions in solution. Can you think of any practical application of this reaction?

**7-64.** Write balanced equations by predicting the products of the following reactions. Include the physical state of each element or compound.
**(a)** the combination of barium and iodine
**(b)** the neutralization of aqueous rubidium hydroxide with hydrobromic acid
**(c)** a single-replacement reaction of calcium metal with a nitric acid solution
**(d)** the combustion of solid naphthalene ($C_{10}H_8$)
**(e)** a precipitation reaction involving aqueous ammonium chromate and aqueous barium bromide
**(f)** the decomposition of solid aluminum hydroxide into solid aluminum oxide and gaseous water.

**7-65.** Write the total ionic and net ionic equations for parts (b), (c), and (e) of exercise 7-64.

**7-66.** Write balanced equations by predicting the products of the following reactions. Include the physical state of each element or compound.
**(a)** the decomposition of solid sodium azide ($NaN_3$) into solid sodium nitride and nitrogen gas.
**(b)** a precipitation reaction involving aqueous potassium carbonate and aqueous copper(II) sulfate
**(c)** the combustion of solid benzoic acid ($C_7H_6O_2$)
**(d)** a single-replacement reaction of iron metal and an aqueous gold(III) nitrate solution
**(e)** the combination of aluminum and solid sulfur ($S_8$)
**(f)** the neutralization of aqueous sulfuric acid with aqueous barium hydroxide

**7-67.** Write the total ionic and net ionic equations for parts (b), (d), and (f) of exercise 7-66.

**7-68.** Consider a mixture of the following ions in aqueous solution: $Na^+$, $H^+$, $Ba^{2+}$, $ClO_4^-$, $OH^-$, and $SO_4^{2-}$. Write the net ionic equations for any reaction or reactions that occur between a cation and an anion.

**7-69.** Consider a mixture of the following ions in aqueous solution: $NH_4^+$, $Mg^{2+}$, $Ni^{2+}$, $Cl^-$, $S^{2-}$, and $CO_3^{2-}$. Write the net ionic equations for any reaction or reactions that occur between a cation and an anion.

**7-70.** Consider a mixture of the following ions in aqueous solution: $K^+$, $Fe^{3+}$, $Pb^{2+}$, $I^-$, $PO_4^{3-}$ and $S^{2-}$. Write the net ionic equations for any reaction or reactions that occur between a cation and an anion.

# Interactive Learning

**7-71.** Balance the following equations.
**(a)** $Mg(OH)_2 + HBr \longrightarrow MgBr_2 + H_2O$
**(b)** $HCl + Ca(OH)_2 \longrightarrow CaCl_2 + H_2O$
**(c)** $Al_2O_3 + H_2SO_4 \longrightarrow Al_2(SO_4)_3 + H_2O$
**(d)** $KHCO_3 + H_3PO_4 \longrightarrow K_2HPO_4 + H_2O + CO_2$
**(e)** $C_9H_{20} + O_2 \longrightarrow CO_2 + H_2O$

**7-72.** Complete and balance the following equations and then write the ionic and net ionic equations. If all ions cancel, indicate that no reaction (N.R.) takes place.
**(a)** $Na_2SO_3 + Ba(NO_3)_2$
(All sulfites are insoluble except IA and $NH_4^+$ compounds.)
**(b)** $NH_4Br + Pb(C_2H_3O_2)_2$
**(c)** $NH_4ClO_4 + Cu(NO_3)_2$

# Solutions to Learning Checks

**A-1.** equation, reactants, products, balanced, coefficients, combustion, combination, decomposition.

**A-2.** **(a)** $4Cr + 3O_2 \longrightarrow 2Cr_2O_3$
**(b)** $Co_2S_3 + 3H_2 \longrightarrow 2Co + 3H_2S$
**(c)** $C_3H_8 + 5O_2 \longrightarrow 3CO_2 + 4H_2O$
**(d)** $H_3BCO + 3H_2O \longrightarrow B(OH)_3 + CO + 3H_2$

**A-3.** **(a)** $2Si_2H_6(g) + 7O_2(g) \longrightarrow 4SiO_2(s) + 6H_2O(l)$
**(b)** $2Al(s) + 3H_2(g) \longrightarrow 2AlH_3(s)$
**(c)** $Ca(HSO_3)_2(s) \longrightarrow CaO(s) + 2SO_2(g) + H_2O(l)$

**B-1.** soluble, insoluble, solvent, solute, ions, acids, single, double, net ionic, spectator, activity, precipitate, precipitation, neutralization, salt.

**B-2.** **(a)** $BaBr_2 \longrightarrow Ba^{2+}(aq) + 2Br^-(aq)$
**(b)** $(NH_4)_2CrO_4 \longrightarrow 2NH_4^+(aq) + CrO_4^{2-}(aq)$
**(c)** $HBr \longrightarrow H^+(aq) + Br^-(aq)$

**B-3.** Yes, Fe replaces $H^+$ in aqueous solution. (From Table 7-3, notice that $FeSO_4$ is water soluble.)

$$Fe(s) + H_2SO_4(aq) \longrightarrow FeSO_4(aq) + H_2(g)$$
$$Fe(s) + 2H^+(aq) + SO_4^{2-}(aq) \longrightarrow$$
$$Fe^{2+}(aq) + SO_4^{2-}(aq) + H_2(g)$$
$$Fe(s) + 2H^+(aq) \longrightarrow Fe^{2+}(aq) + H_2(g)$$

**B-4.** $2AgNO_3(aq) + K_2S(aq) \longrightarrow Ag_2S(s) + 2KNO_3(aq)$
$$2Ag^+(aq) + 2NO_3^-(aq) + 2K^+(aq) + S^{2-}(aq) \longrightarrow$$
$$Ag_2S(s) + 2K^+(aq) + 2NO_3^-(aq)$$
$$2Ag^+(aq) + S^{2-}(aq) \longrightarrow Ag_2S(s)$$

**B-5.** $2HNO_3(aq) + Sr(OH)_2(aq) \longrightarrow$
$$Sr(NO_3)_2(aq) + 2H_2O(l)$$
$$2H^+(aq) + 2NO_3^-(aq) + Sr^{2+}(aq) + 2OH^-(aq) \longrightarrow$$
$$Sr^{2+}(aq) + 2NO_3^-(aq) + 2H_2O(l)$$
$$H^+(aq) + OH^-(aq) \longrightarrow H_2O(l)$$

## chapter 8

# Quantitative Relationships in Chemistry

**Modern farming techniques require the addition of nitrogen to the soil. The composition of ammonia makes it ideal for delivering a large amount of nitrogen. The composition of compounds is a topic of this chapter.**

## Setting the Stage

There is no way that nearly six billion people could crowd onto this small world unless we could help mother nature. Nature does not supply enough chemically combined nitrogen to the soil so that sufficient crops can be grown to feed this huge throng of humanity. Nitrogen is vital to all growing things and is a constituent of many compounds, such as proteins, that are part of important life cycles. Why do we need to help nature? After all, the air that we breathe contains tremendous amounts of nitrogen (almost 80% of the atmosphere). The trouble is that the nitrogen in the atmosphere is present as a molecular element ($N_2$) and, in this form, does not easily undergo chemical reactions to become part of chemical compounds. The only natural way for nitrogen to get from the huge abundance as a molecule in the air into compounds in living plants and animals is through lightning or by the action of a type of bacteria associated with plants like beans and peas. This is not nearly enough, so *we* must transform nitrogen in the air, by means of chemical reactions, into usable compounds for fertilizers. Two manufactured compounds that find widespread use as a source of nitrogen for the soil are ammonia ($NH_3$) and ammonium nitrate ($NH_4NO_3$). The advantage of ammonia is that 100 lb of ammonia supplies 82 lb of nitrogen to the soil. Its disadvantage is that it is a gas and must be injected into the soil. Ammonium nitrate, an ionic compound, is a solid and can be spread on the surface. Its disadvantage is that it supplies only 35 lb of nitrogen for each

100 lb of fertilizer applied. The decision about which to use depends on cost, availability, and soil conditions. The mass of nitrogen in a given amount of each of these two compounds relates to its formula and to the relative masses of the elements in each compound. Such calculations will be part of the business of this chapter. But a knowledge of the relative masses of elements and compounds has an even greater application.

In the introduction to Chapter 7, we discussed the greenhouse effect and the possibility of global warming. The presence of increasing amounts of carbon dioxide, as well as specific other gases, is what seems to be causing the problem. Besides global warming, the increased amount of the carbon dioxide (a reactant) may cause vegetation (the product) to grow larger and faster. Some early experiments indicate that some vegetation is affected noticeably and others less so. Increasing just one reactant may not lead to more growth unless all other reactants are also increased. It's like building an automobile. We need one engine and four tires per car. Having eight tires does not mean two cars unless we also have two engines as well as other necessary parts. In chemical reactions such as photosynthesis, the mass ratios of reacting compounds as well as the masses of the products that are formed are extremely important. In this chapter, we will employ balanced chemical equations, introduced in Chapter 7, to help us understand the mass relationships between and among reactants and products.

### Formulating Some Questions

There are many important questions that we will deal with in this chapter. How do we measure known amounts of atoms without counting? How does a counting unit, appropriate for our needs, relate to mass? How does the formula relate to the masses of the elements in the compound? How does the chemical equation lead us to the mass relationships between reactants and products? How do chemical equations also tell us about the energy involved in a chemical reaction? But our first order of business is to develop a relationship between the masses of elements and the numbers involved. For this we will need a special counting unit.

## Section A THE MEASUREMENT OF MASSES OF ELEMENTS AND COMPOUNDS

## 8-1 Relative Masses of Elements

**Looking Ahead!** A good place to start is by looking at how the mass of an average atom of one element relates to the average mass of an atom of another element. If we know this, we can see how the masses of a large number of atoms of different elements relate. That is our goal in this section.

The atom is an unbelievably tiny thing. As mentioned in Chapter 3, its smallness is actually quite difficult to comprehend. For example, the period at the end of this sentence contains between $10^{16}$ and $10^{17}$ atoms of carbon. Each of these tiny atoms is a specific isotope of carbon. Most of the atoms are $^{12}C$, which is used as the standard and assigned a mass of exactly 12 amu. The other carbon isotopes and the isotopes of any other element are assigned a mass by comparison to $^{12}C$. Most elements occur in nature as mixtures of isotopes, so the atomic mass in the periodic table represents the weighted average mass of all isotopes of that element. In this chapter, we will treat the atoms of an element as if they were all identical, with the atomic mass representing the mass of this hypothetical "average" atom. The mass of individual atoms has, so far, been expressed in terms of the mass unit *amu* (atomic mass unit). This is valuable when comparing the masses of individual atoms, but it has no practical value in a laboratory situation. For example, the mass of one "average" carbon atom is 12.01 amu. This converts to grams as follows.

$$12.01 \text{ amu} = 1.994 \times 10^{-23} \text{g}$$

Since even the best laboratory balance can detect no less than $10^{-5}$ g, it is obvious that we need many atoms at a time to register on our scales. Because we can't work with individual atoms, we must "scale up" the numbers of atoms so that the amounts are detectable with our laboratory instruments. In order to scale up our measurements, we need an appropriate counting unit for atoms. To use this counting unit, however, we will not be able to actually count the number of atoms. In this case, we will weigh rather than count.

Consider how a grocer may count by weighing. If we wanted to purchase a large number of lemons (e.g., 165) in a grocery store, it would be very tedious to actually count them one by one. It would be far easier and much quicker to use a weight scale and know that we have the desired number. So, if one average lemon has a mass of 145 g and we need 165 lemons, the following calculation tells us how much to weigh.

**Like the chemist, the grocer uses a scale to count.**

$$165 \text{ lemons} \times \frac{145 \text{ g}}{\text{lemon}} \times \frac{1 \text{ kg}}{10^3 \text{ g}} = 23.9 \text{ kg}$$

Now, suppose we want to measure the same number of apples. An average apple has a mass of 282 g, so we need the following mass of apples.

$$165 \text{ apples} \times \frac{282 \text{ g}}{\text{apple}} \times \frac{1 \text{ kg}}{10^3 \text{ g}} = 46.5 \text{ kg}$$

Actually, once we know the mass ratio of apples to lemons we can easily convert an equivalent number of lemons to apples and vice versa. The ratio of the masses can be expressed as follows.

$$\frac{282 \text{ g apple}}{145 \text{ g lemon}} \quad \text{and} \quad \frac{145 \text{ g lemon}}{282 \text{ g apple}}$$

*Whenever the masses of lemons and apples are in the ratio of their individual masses, the same number of each fruit is present.* For example, using the appropriate mass ratio, we can make the previous calculation easier and not even be concerned with the actual number. The following calculation, using the mass ratio expressed in kg, allows us to measure the same number of apples as there are in 23.9 kg of lemons.

$$23.9 \text{ kg lemons} \times \frac{282 \text{ kg apple}}{145 \text{ kg lemon}} = 46.5 \text{ kg apples}$$

We understand from this calculation that 23.9 kg of lemons and 46.5 kg of apples contain the same number of individual fruit.

Now, let us return to the world of atoms and apply the same principles. In many of the calculations that follow in this chapter, we do not need to know the *actual* number of atoms involved in a certain weighed amount, but we do need to know the *relative* numbers of atoms of different elements present. However, if we know the relative masses of the individual atoms, this is no problem. For example, if one helium

**Table 8-1  Mass Relation of C and He**

| C | He | Number of Atoms of Each Element Present |
|---|---|---|
| 12.0 amu | 4.00 amu | 1 |
| 24.0 amu | 8.00 amu | 2 |
| 360 amu | 120 amu | 30 |
| 12.0 g | 4.00 g | $6.02 \times 10^{23}$ |
| 24.0 g | 8.00 g | $1.20 \times 10^{24}$ |
| 12.0 lb | 4.00 lb | $2.73 \times 10^{26}$ |
| 24.0 ton | 8.00 ton | $1.09 \times 10^{30}$ |

atom has a mass of 4.00 amu and one carbon atom has a mass of 12.0 amu, their masses are in the following ratio.

$$\frac{4.00 \text{ amu He}}{12.0 \text{ amu C}}$$

In fact, any time helium and carbon are present in a 4.00:12.0 mass ratio *regardless of the units of mass,* we can conclude that the same number of atoms of each element is present. (See Table 8-1.) We can generalize this statement to all of the elements. *When the masses of samples of any two elements are in the same ratio as that of their atomic masses, the samples have the same number of atoms.* Thus, if we wanted the same number of helium atoms as the number of atoms present in 45.0 g of carbon, we do not have to count or even know what the number is. The following calculation tells us what we want.

$$45.0 \text{ g C} \times \frac{4.00 \text{ g He}}{12.0 \text{ g C}} = 15.0 \text{ g He}$$

---

**Example 8-1**

*working it out*

**The Relative Masses of Elements in a Compound**

The formula of the compound magnesium sulfide (MgS) indicates that there is one atom of Mg for every atom of S. What mass of sulfur is combined with 46.0 lb of magnesium?

**Procedure**

From the atomic masses in the periodic table, note that one atom of magnesium has a mass of 24.31 amu and one atom of sulfur has a mass of 32.07 amu. Likewise 24.31 lb of magnesium and 32.07 lb of sulfur have an equal number of atoms. This statement can be represented by two conversion factors, which we can use to change a mass of one element to an equivalent mass of the other.

$$(1) \quad \frac{24.31 \text{ lb Mg}}{32.07 \text{ lb S}} \qquad (2) \quad \frac{32.07 \text{ lb S}}{24.31 \text{ lb Mg}}$$

Use factor **(2)** to convert the mass of Mg to an equivalent mass of S.

**Solution**

$$46.0 \text{ lb Mg} \times \frac{32.07 \text{ lb S}}{24.31 \text{ lb Mg}} = 60.7 \text{ lb S}$$

Thus 60.7 lb of sulfur has the same number of atoms as 46.0 lb of magnesium.

# 8-2 The Mole and the Molar Mass of Elements

**Looking Ahead!** The atomic mass of an element represents the average mass of one atom of that element expressed in amu. When we express the atomic masses of the elements in grams, we can make the following conclusion: *the atomic mass expressed in grams represents the same number of atoms of each element.* But what is this number and just how big is it? That will be discussed in this section.

We can easily deal with and weigh gram quantities of elements in a laboratory, but such a mass represents an inconceivably large number of atoms. We are now ready for a counting unit that represents this huge number of atoms. *The number of atoms represented by the atomic mass of an element expressed in grams is a unit known as a* **mole.** (The SI symbol is **mol.**) The number that one mole represents is expressed in the following equality.

$$1.000 \text{ mol} = 6.022 \times 10^{23} \text{ objects or particles}$$

This number, $6.022 \times 10^{23}$, is referred to as **Avogadro's number** (named in honor of Amedeo Avogadro, 1776–1856, a pioneer investigator of the quantitative aspects of chemistry). Avogadro's number was determined experimentally by various methods. The formal definition of one mole concerns an isotope of carbon, $^{12}C$. **One mole** *is defined as the number of atoms in exactly 12 grams of* $^{12}C$. This number, of course, is Avogadro's number. *Thus the atomic mass of one mole of any element, expressed in grams, contains the same number of basic particles as there are in exactly 12 grams of* $^{12}C$. Avogadro's number is not an exact, defined number such as 12 in one dozen or 144 in one gross, but it is known to many more significant figures than the four (i.e., 6.022) that are shown and used in this text.

Many common counting units represent a number consistent with their use. For example, a baker sells a dozen doughnuts at a time because 12 is a practical number for that purpose. On the other hand, we buy a ream of typing or computer paper, which is 500 sheets. A ream of doughnuts or a dozen sheets of typing paper are not practical amounts to purchase for most purposes. Counting units of a dozen or a ream are of little use to a chemist because they don't include enough individual objects. For example, grouping $10^{20}$ atoms into about $10^{19}$ dozen atoms does us little good. The counting unit used by chemists includes an extremely large number of individual units in order to be practical. (See Figure 8-1.)

Although Avogadro's number is valuable to a chemist, its size defies the ability of the human mind to comprehend. For example, if an atom were the size of a marble and one mole of marbles were spread over the surface of the Earth, our

**a dozen doughnuts**

**a ream of paper**

**a mole of sand particles**

**Figure 8-1  Counting Units** One dozen, one ream, and one mole all have applications dependent on the amount needed.

**Figure 8-2 Moles of Elements**

**One mole of iron (the paper clips), copper, liquid mercury, and sulfur are shown. Each sample contains Avogadro's number of atoms but has a different mass.**

planet would be covered by a 50-mi-thick layer. Or, if the marbles were laid end to end and extended into outer space, they would reach past the farthest planets almost to the center of the galaxy. It takes light moving at 186,000 miles per second over 30,000 years to travel from Earth to the center of the galaxy. A new supercomputer can count all of the people in the United States in one-quarter of a second, but it would take almost two million years for it to count one mole of people at the same rate. A glass of water, which is about 10 moles of water, contains more water molecules than there are grains of sand in the Sahara desert. That's difficult to imagine.

Note that one mole of a certain element implies two things.

1. *The atomic mass expressed in grams,* which is *different* for each element. This mass is known as the **molar mass of the element.** (Most of the measurements that we will use in the examples and exercises that follow are expressed to three significant figures. We will express the molar mass and Avogadro's number to four significant figures so that these factors will not limit the precision of the answer.) Thus the mass of one mole of oxygen atoms is 16.00 g, the mass of one mole of helium atoms is 4.002 g, and the mass of one mole of uranium atoms is 238.0 g.

2. *Avogadro's number of atoms,* which is the *same* for all elements. (See Figure 8-2.)

At this point, notice that we have established a relationship among a unit (mole), a number ($6.022 \times 10^{23}$), and the mass of an element expressed in grams. It is now necessary for us to be able to convert among these three quantities without too much difficulty. The mechanics of the conversions are not complex, however. So before we do some conversions involving atoms and so on, consider similar conversions involved with one everyday-dozen oranges. In the following two calculations, assume that we are dealing with identical oranges that are weighed and sold by the dozen instead of individually. One dozen of these oranges has a mass of 5.960 lb, so we can consider this to be the "dozen mass" of oranges just as there is a "molar mass" for atoms of elements. The number in one dozen, however, is a much more manageable 12 rather than the $6.022 \times 10^{23}$ in one mole. The relationships between mass of oranges and the number in one dozen can be summarized as follows. Notice that the four ratios that result can be used as conversion factors in calculations using the factor-label method introduced in Chapter 1.

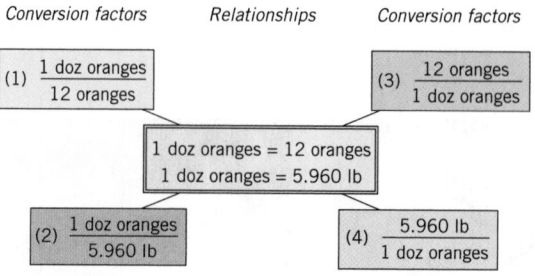

Conversion factors     Relationships     Conversion factors

(1) $\dfrac{1 \text{ doz oranges}}{12 \text{ oranges}}$       (3) $\dfrac{12 \text{ oranges}}{1 \text{ doz oranges}}$

1 doz oranges = 12 oranges
1 doz oranges = 5.960 lb

(2) $\dfrac{1 \text{ doz oranges}}{5.960 \text{ lb}}$       (4) $\dfrac{5.960 \text{ lb}}{1 \text{ doz oranges}}$

**The number of oranges, the mass, and dozens are all related.**

Let's say we are asked to calculate how many dozens of oranges there are in 17.9 lb. In this problem we wish to convert between a mass (lb) and a unit (dozen). Factor (2) can be used for this conversion, as follows.

$$17.9 \cancel{\text{ lb}} \times \frac{1 \text{ doz oranges}}{5.960 \cancel{\text{ lb}}} = \underline{3.00 \text{ doz oranges}}$$

In a second problem, we are asked to calculate the mass of 142 oranges. In this case, a direct conversion between mass in pounds and number of oranges is not given. The conversion can be done in two steps: we first convert the number of oranges to the dozen unit [factor (1)] and in the second step, convert the dozen unit to mass [factor (4)]. The unit map appears as

$$\text{number} \longrightarrow \text{doz} \longrightarrow \text{lb}$$

The calculation is as follows.

$$142 \text{ oranges} \times \frac{1 \text{ doz oranges}}{12 \text{ oranges}} \times \frac{5.960 \text{ lb}}{\text{doz oranges}} = \underline{70.5 \text{ lb}}$$

We are now ready to work very similar problems using moles, grams, and $6.022 \times 10^{23}$ rather than dozens, pounds, and 12. If you're still a little uneasy with scientific notation, see Appendix C for a quick review.

In these problems, we will assume that the atoms are all identical with the same size, shape, and mass. From the periodic table we find that the "molar mass" (the mass of $6.022 \times 10^{23}$) of sodium atoms is 22.99 g. The two relationships just given can be either written as equalities or set up as ratios that can eventually be used as conversion factors.

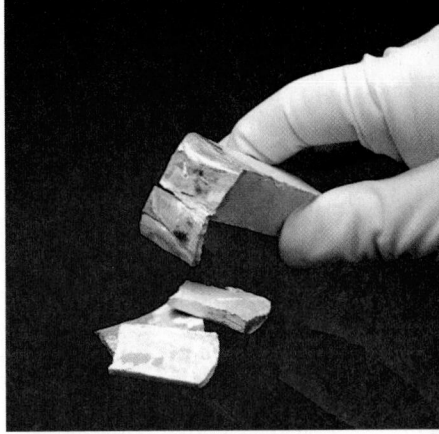

The number of sodium atoms, the mass, and moles are all related.

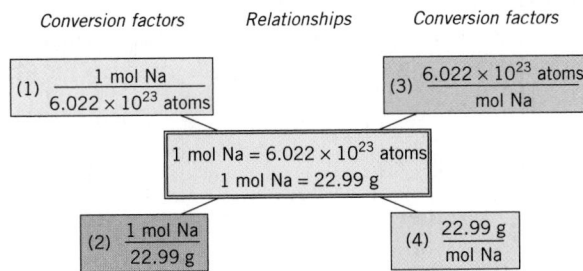

Conversion factors     Relationships     Conversion factors

(1) $\dfrac{1 \text{ mol Na}}{6.022 \times 10^{23} \text{ atoms}}$     (3) $\dfrac{6.022 \times 10^{23} \text{ atoms}}{\text{mol Na}}$

1 mol Na = $6.022 \times 10^{23}$ atoms
1 mol Na = 22.99 g

(2) $\dfrac{1 \text{ mol Na}}{22.99 \text{ g}}$     (4) $\dfrac{22.99 \text{ g}}{\text{mol Na}}$

---

**Moles to Mass**

What is the mass of 3.00 mol of Na?

**Example 8-2**

working it out

**Procedure**

We need a conversion factor that converts units of "moles" to units of mass, or g. Thus the conversion factor would have mole in the denominator and grams in the numerator. This is factor (4).

$$\times \frac{\text{g}}{\text{mol}}$$

$$\text{mol} \Longrightarrow \text{mass}$$

**Solution**

$$3.00 \text{ mol Na} \times \frac{22.99 \text{ g}}{\text{mol Na}} = \underline{\underline{69.0 \text{ g}}}$$

**Example 8-3** | **Mass to Moles**

How many moles are present in 34.5 g of Na?

**Procedure**

This is the reverse of the previous problem, so we need a conversion factor that has grams in the denominator and mole in the numerator. This is factor (2).

$$\times \frac{mol}{g}$$

$$mass \Longrightarrow mol$$

**Solution**

$$34.5 \text{ g} \times \frac{1 \text{ mol Na}}{22.99 \text{ g}} = \underline{\underline{1.50 \text{ mol Na}}}$$

**Example 8-4** | **Number to Mass**

What is the mass of $1.20 \times 10^{24}$ atoms of Na?

**Procedure**

In this conversion, a direct relationship between mass and number is not given. This requires a two-step conversion. In the first step, we convert number of atoms to moles [factor (1)]; in the second step, we convert moles to mass [factor (4)].

$$\times \frac{mol}{6.022 \times 10^{23}} \quad \times \frac{g}{mol}$$

$$number \Longrightarrow mol \Longrightarrow mass$$

**Solution**

$$1.20 \times 10^{24} \text{ atoms Na} \times \frac{1 \text{ mol Na}}{6.022 \times 10^{23} \text{ atoms Na}} \times \frac{22.99 \text{ g}}{\text{mol Na}} = \underline{\underline{45.8 \text{ g}}}$$

**Example 8-5** | **Mass to Number**

How many individual atoms are present in 11.5 g of Na?

**Procedure**

This is the reverse of the previous problem and also requires a two-step conversion. We will convert grams to moles using factor (2) and then moles to number using factor (3).

$$\times \frac{mol}{g} \quad \times \frac{6.022 \times 10^{23}}{mol}$$

$$mass \Longrightarrow mol \Longrightarrow number$$

**Solution**

$$11.5 \text{ g} \times \frac{1 \text{ mol Na}}{22.99 \text{ g}} \times \frac{6.022 \times 10^{23} \text{ atoms Na}}{\text{mol Na}} = \underline{\underline{3.01 \times 10^{23} \text{ atoms Na}}}$$

# 8-3 The Molar Mass of Compounds

**Looking Ahead!** When we express the atomic masses of the elements in grams rather than amu, we successfully scale up our measurements to useful laboratory quantities. Now we are using *moles* of atoms rather than individual atoms. Most of the substances around us, however, are composed of molecules rather than individual atoms. In this section we will extend the concept of the mole from elements to compounds.

Just as the mass of an automobile is the sum of the masses of all its component parts, the mass of a molecule is the sum of the masses of its component atoms. First, we will examine the masses of single molecules. *The **formula weight** of a compound is determined from the number of atoms and the atomic mass (in amu) of each element indicated by a chemical formula.* Recall that chemical formulas represent two types of compounds: molecular and ionic. The formula of a molecular compound represents a discrete molecular unit whereas the formula of an ionic compound represents a formula unit, which is the whole-number ratio of cations to anions (e.g., one formula unit of $K_2SO_4$ contains two $K^+$ ions and one $SO_4^{2-}$ ion). The following examples illustrate the calculation of the formula weight of a molecular compound (Example 8-6) and an ionic compound (Example 8-7).

---

**Calculation of the Formula Weight of a Molecular Compound**

**Example 8-6**

working it out

What is the formula weight of $CO_2$?

**Solution**

| Atom | Number of Atoms in Molecule | | Atomic Mass | | Total Mass of Atoms in Molecule |
|------|------|------|------|------|------|
| C | 1 | × | 12.01 amu | = | 12.01 amu |
| O | 2 | × | 16.00 amu | = | 32.00 amu |
| | | | | | 44.01 amu |

The formula weight of $CO_2$ is

$$\underline{44.01 \text{ amu}}$$

---

**Calculation of the Formula Weight of an Ionic Compound**

**Example 8-7**

What is the formula weight of $Fe_2(SO_4)_3$?

**Solution**

| Atom | Number of Atoms in Formula Unit | | Atomic Mass | | Total Mass of Atoms in Formula Unit |
|------|------|------|------|------|------|
| Fe | 2 | × | 55.85 amu | = | 111.70 amu* |
| S | 3 | × | 32.07 amu | = | 96.21 amu |
| O | 12 | × | 16.00 amu | = | 192.00 amu |
| | | | | | 399.91 amu |

*In these multiplications we observe the rules for addition and subtraction for significant figures because the numbers of atoms are exact and the multiplication could be considered the same as addition.

The formula weight of $Fe_2(SO_4)_3$ is 399.91 amu

$$= \underline{399.9 \text{ amu}}$$

**Figure 8-3 Moles of Compounds**
One mole of white sodium chloride (NaCl), blue copper sulfate hydrate (CuSO$_4$ · 5H$_2$O), yellow sodium chromate (Na$_2$CrO$_4$), and water are shown. Each sample contains Avogadro's number of molecules or formula units.

Certain ionic compounds can have water molecules attached to their cations and/or anions. These compounds are known as *hydrates* and have some distinctive properties compared to their unhydrated (no-water) forms. For example, CuSO$_4$ [copper(II) sulfate] is a pale green solid, whereas CuSO$_4$ · 5H$_2$O [copper(II) sulfate pentahydrate] is a dark blue solid. The waters of hydration can usually be removed by heating the solid. The formula weight of a hydrate includes the mass of the water molecules. For example, the molar mass of CuSO$_4$ · 5H$_2$O is calculated by summing the atomic masses of copper, sulfur, four oxygens, and five waters. That is, 63.55 amu (Cu) + 32.07 amu (S) + (4 × 16.00 amu) (O) + (5 × 18.02 amu) (2H + O) = 249.7 amu.

The formula weights that we have calculated represent one molecule or one formula unit. Once again, we need to scale up our numbers so that we have a workable amount that can be measured on a laboratory balance. Thus we extend the definition of the mole to include compounds. *The mass of one mole (6.022 × 10$^{23}$ molecules or formula units) is referred to as the* **molar mass of the compound** *and is the formula weight expressed in grams.* As was the case with atoms of elements, the molar masses of various compounds differ, but the number of molecules or formula units remains the same. (See Figure 8-3.)

| | |
|---|---|
| **Example 8-8** | **Converting Moles to Mass and Number of Formula Units** |
| *working it out* | **a.** What is the mass of 0.345 mol of Al$_2$(CO$_3$)$_3$? |
| | **b.** How many individual ionic formula units does this amount represent? |

**Procedure**

The problem is worked much like the examples in which we were dealing with moles of atoms rather than compounds. In this case, however, we need to find the molar mass of the compound, which is the formula weight expressed in grams. The units of molar mass are "g/mol." The molar mass equals

$$2Al + 3C + 9O = [(2 \times 26.98) + (3 \times 12.01) + (9 \times 16.00)] = 234.0 \text{ amu}$$

Thus the molar mass is 234.0 g/mol

The conversions are

$$\times \frac{g}{mol}$$

**(a)** mol $\Longrightarrow$ mass

and

$$\times \frac{6.022 \times 10^{23}}{mol}$$

**(b)** mol $\Longrightarrow$ number (formula units)

**Solution**

**a.**
$$0.345 \text{ mol Al}_2(CO_3)_3 \times \frac{234.0 \text{ g}}{\text{mol Al}_2(CO_3)_3} = \underline{\underline{80.7 \text{ g}}}$$

**b.**
$$0.345 \text{ mol Al}_2(CO_3)_3 \times \frac{6.022 \times 10^{23} \text{ formula units}}{\text{mol Al}_2(CO_3)_3}$$
$$= \underline{\underline{2.08 \times 10^{23} \text{ formula units}}}$$

**Converting Mass to Number of Molecules**

**Example 8-9**

How many individual molecules are present in 25.0 g of $N_2O_5$?

**Procedure**

The formula weight of $N_2O_5 = [(2 \times 14.01) + (5 \times 16.00)] = 108.0$ amu. The molar mass is 108.0 g/mol. This is a two-step conversion as follows:

$$\times \frac{mol}{g} \qquad \times \frac{6.022 \times 10^{23}}{mol}$$

mass $\Longrightarrow$ mol $\Longrightarrow$ number (molecules)

**Solution**

$$25.0 \text{ g } N_2O_5 \times \frac{1 \text{ mol}}{108.0 \text{ g } N_2O_5} \times \frac{6.022 \times 10^{23} \text{ molecules}}{\text{mol}} = \underline{\underline{1.39 \times 10^{23} \text{ molecules}}}$$

The term "a mole of molecules" does sound confusing. It is somewhat unfortunate that the counting unit (mole) and the fundamental particle that is being counted (molecule) read so similarly. Try to remember that the *molecule* is the tiny, fundamental

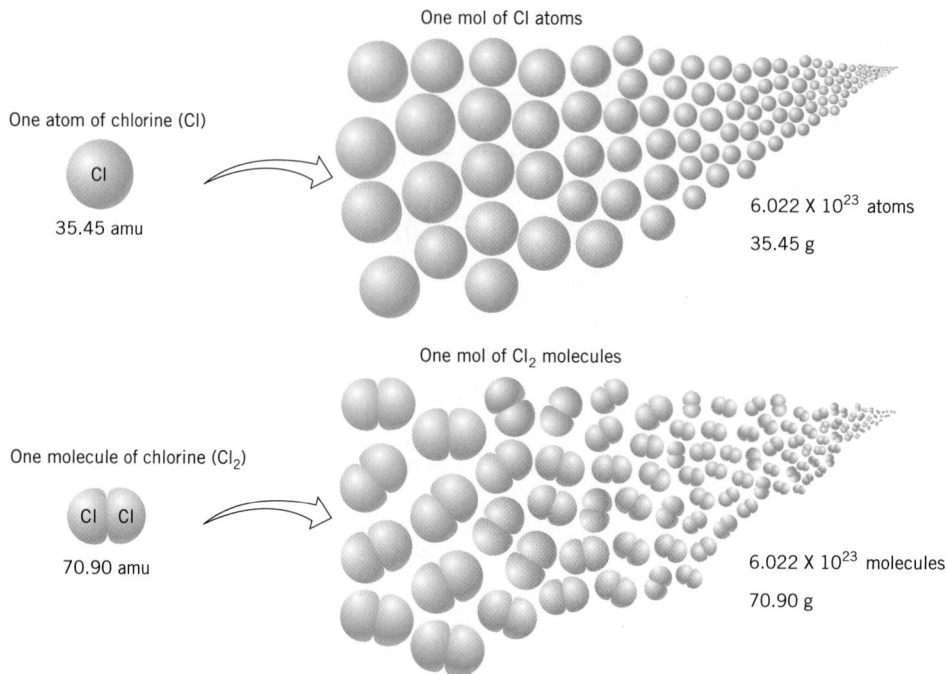

One mol of Cl atoms

One atom of chlorine (Cl)

Cl

35.45 amu

6.022 X $10^{23}$ atoms

35.45 g

One mol of $Cl_2$ molecules

One molecule of chlorine ($Cl_2$)

Cl Cl

70.90 amu

6.022 X $10^{23}$ molecules

70.90 g

**Figure 8-4** **Atoms, Molecules, and Moles**

particle of a compound that is being counted. A *mole* is an incredibly large number of molecules (i.e., $6.022 \times 10^{23}$).

Another point of caution concerns nonmetal elements such as chlorine ($Cl_2$) that exist in nature as molecules rather than solitary atoms. For example, if we report the presence of "one mole of chlorine," it is not always obvious what is meant. Do we have one mole of chlorine atoms ($6.022 \times 10^{23}$ atoms, 35.45 g) or one mole of chlorine molecules ($6.022 \times 10^{23}$ molecules, 70.90 g)? Note that there are *two Cl atoms per molecule* and *two moles of Cl atoms per mole of $Cl_2$ molecules*. (See Figure 8-4.) In future discussions, one mole of chlorine or any of the diatomic elements will refer to the molecules. Thus it is important to be specific in cases where we are using one mole of atoms. This situation is analogous to the difference between a dozen sneakers and a dozen pairs of sneakers. Both the mass and the actual number of the dozen pairs of sneakers are double those of a dozen sneakers.

---

**Looking Back!** Chemistry deals with such unbelievably small atoms that it requires an unbelievably large number of them to even register on a sensitive laboratory balance. However, the mass ratios of individual atoms can be applied to the measurement of large numbers of atoms. Our most useful mass ratios are obtained by expressing the atomic masses of the elements in grams. This is defined as the molar mass of the elements and is the mass of Avogadro's number of atoms. Then, by extending this concept to compounds, we have the molar mass of compounds, which is the mass of Avogadro's number of molecules or formula units.

---

# Learning Check A | checking it out

**A-1.** Fill in the blanks.

The atomic mass of an element expressed in amu represents the mass of _____ atom of the element. The atomic mass expressed in grams represents the mass of _____ atoms of the element and is known as the _____ _____ of the element. This quantity and mass refer to a unit known as the _____ . The mass of one molecule or formula unit is known as the _____ weight. This quantity expressed in grams is known as the _____ _____ of the compound and represents the mass of _____ individual molecules or formula units.

**A-2.** A compound has the formula NO. What mass of oxygen is present in the compound for each 25.0 g of nitrogen?

**A-3.** How many moles of vanadium atoms are represented by $4.82 \times 10^{24}$ atoms? What is the mass of this number of atoms of vanadium?

**A-4.** How many moles of atoms and how many individual atoms are present in 215 g of chlorine?

**A-5.** How many moles of $Cl_2O$ and how many individual molecules are present in 438 g of $Cl_2O$?

*Additional Examples: Exercises 8-1, 8-3, 8-8, 8-14, and 8-29.*

## Section B THE COMPONENT ELEMENTS OF COMPOUNDS

# 8-4 The Composition of Compounds

**Looking Ahead!** When we extended the concept of the mole to compounds, we expanded the scope of the information to a considerable extent. Each compound contains two or more component parts, each present in a given mole and mass ratio. We will now look deeper into the component parts of a compound.

One plus two equals one? Sometimes the math we use doesn't appear to add correctly. However, we do know that *one* frame plus *two* wheels equals *one* bicycle. *Likewise,* one *carbon atom plus* two *oxygen atoms equals* one *molecule of carbon dioxide.* Now consider the compound sulfuric acid ($H_2SO_4$). In Table 8-2, we have illustrated the relation of one mole of compound to all its component parts. All of these relationships can be used to construct conversion factors between the compound and its elements. We will use similar conversion factors in sample problems to separate compounds into their mole components, mass components, and, finally, percent composition.

## The Mole Composition of a Compound

Our first task is to simply separate a quantity of a compound expressed in moles into number of moles of each element that is part of the compound. For this example, let us consider a 0.345-mol quantity of $Al_2(CO_3)_3$.

**Table 8-2  The Composition of One Mole of $H_2SO_4$**

| | One Mole of $H_2SO_4$ | | |
|---|---|---|---|
| | **Number of Atoms** | **Moles of Atoms** | **Mass of Atoms** |
| | $1.204 \times 10^{24}$ H atoms | ⟵ 2 mol H ⟶ | 2.016 g H |
| | $6.022 \times 10^{23}$ S atoms | ⟵ 1 mol S ⟶ | 32.07 g S |
| | $\underline{2.409 \times 10^{24}}$ O atoms | ⟵ $\underline{4\ mol\ O}$ ⟶ | $\underline{64.00\ g\ O}$ |
| Totals | $4.215 \times 10^{24}$ atoms in | 7 mol atoms in | |
| | $\boxed{6.022 \times 10^{23}\ \text{molecules}}$ | ⟵ $\boxed{\text{one mole of molecules}}$ ⟶ | $\boxed{98.09\ g}$ |

| Example 8-10 | **Conversion of Moles of a Compound into Moles of Its Component Elements** |

**working it out**

How many moles of each atom are present in 0.345 mol of $Al_2(CO_3)_3$? What is the total number of moles of atoms present?

**Procedure**

In this problem, note that there are 2 Al atoms, 3 C atoms, and 9 O atoms in each formula unit of compound. Therefore, in 1 mol of compound, there are 2 mol of Al, 3 mol of C, and 9 mol of O atoms. This can be expressed with conversion factors as

$$\frac{2 \text{ mol Al}}{\text{mol Al}_2(\text{CO}_3)_3} \qquad \frac{3 \text{ mol C}}{\text{mol Al}_2(\text{CO}_3)_3} \qquad \frac{9 \text{ mol O}}{\text{mol Al}_2(\text{CO}_3)_3}$$

**Solution**

$$\text{Al: } 0.345 \text{ mol Al}_2(\text{CO}_3)_3 \times \frac{2 \text{ mol Al}}{\text{mol Al}_2(\text{CO}_3)_3} = \underline{\underline{0.690 \text{ mol Al}}}$$

$$\text{C: } 0.345 \text{ mol Al}_2(\text{CO}_3)_3 \times \frac{3 \text{ mol C}}{\text{mol Al}_2(\text{CO}_3)_3} = \underline{\underline{1.04 \text{ mol C}}}$$

$$\text{O: } 0.345 \text{ mol Al}_2(\text{CO}_3)_3 \times \frac{9 \text{ mol O}}{\text{mol Al}_2(\text{CO}_3)_3} = \underline{\underline{3.11 \text{ mol O}}}$$

Total moles of atoms present:

$$0.690 \text{ mol Al} + 1.04 \text{ mol C} + 3.11 \text{ mol O} = \underline{\underline{4.84 \text{ mol atoms}}}$$

## The Mass Composition of a Compound

We can now extend this concept into the component masses of the elements. In the following example, we will calculate the mass of each element present in a specific mole quantity of limestone.

| Example 8-11 | **Conversion of a Mole of a Compound into Masses of Its Component Elements** |

**working it out**

What masses of each element are present in 2.36 mole of limestone ($CaCO_3$).

**Marble is also composed of calcium carbonate.**

**Procedure**

In this problem, convert the moles of each element present in the given quantity of compound to mass of that element. This is a two-step conversion as follows.

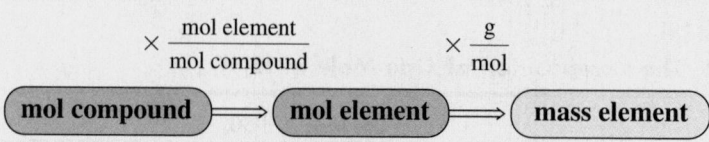

$$\times \frac{\text{mol element}}{\text{mol compound}} \qquad\qquad \times \frac{\text{g}}{\text{mol}}$$

$$\boxed{\text{mol compound}} \Longrightarrow \boxed{\text{mol element}} \Longrightarrow \boxed{\text{mass element}}$$

**Solution**

$$\text{Ca: } 2.36 \text{ mol CaCO}_3 \times \frac{1 \text{ mol Ca}}{\text{mol CaCO}_3} \times \frac{40.08 \text{ g Ca}}{\text{mol Ca}} = \underline{\underline{94.6 \text{ g Ca}}}$$

$$\text{C: } 2.36 \text{ mol CaCO}_3 \times \frac{1 \text{ mol C}}{\text{mol CaCO}_3} \times \frac{12.01 \text{ g C}}{\text{mol C}} = \underline{\underline{28.3 \text{ g C}}}$$

$$\text{O: } 2.36 \text{ mol CaCO}_3 \times \frac{3 \text{ mol O}}{\text{mol CaCO}_3} \times \frac{16.00 \text{ g O}}{\text{mol O}} = \underline{\underline{113 \text{ g O}}}$$

## The Percent Composition of a Compound

Perhaps the most common way of expressing the composition of elements in a compound is percent composition. **Percent composition** *expresses the mass of each element per 100 mass units of compound.* For example, if there is an 82-g quantity of nitrogen present in each 100 g of ammonia ($NH_3$), the percent composition is expressed as 82% nitrogen.

The mass of one mole of carbon dioxide is 44.01 g, and it is composed of one mole of carbon atoms (12.01 g) and two moles of oxygen atoms ($2 \times 16.00 = 32.00$ g). The percent composition is calculated by dividing the total mass of each component element by the total mass (molar mass) of the compound and then multiplying by 100%.

$$\frac{\text{total mass of component element}}{\text{total mass (molar mass)}} \times 100\% = \text{percent composition}$$

In $CO_2$, the percent composition of C is

$$\frac{12.01 \text{ g C}}{44.01 \text{ g } CO_2} \times 100\% = 27.29\% \text{ C}$$

and the percent composition of O is

$$\frac{32.00 \text{ g O}}{44.01 \text{ g } CO_2} \times 100\% = 72.71\%$$

(It is easier to find the percent composition of oxygen by subtracting $100.00\% - 27.29\% = 72.71\%$. In a binary compound, the two percent compositions add to 100%. However, it is sometimes wise to calculate all percentages individually and then check your math by making sure they all add to 100%.)

**Ammonium nitrate, a fertilizer.**

---

### The Percent Composition of Limestone ($CaCO_3$)

What is the percent composition of all of the elements in limestone ($CaCO_3$)?

**Example 8-12**

working it out

**Procedure**

Find the molar mass and convert the *total* mass of each element to percent of molar mass.

**Solution**

One mol of $CaCO_3$ contains

> 1 mol of Ca (40.08 g)
>
> 1 mol of C (12.01 g)
>
> 3 mol of O $\left( 3 \text{ mol O} \times \dfrac{16.00 \text{ g}}{\text{mol O}} = 48.00 \text{ g} \right)$

The molar mass = 40.08 g + 12.01 g + 48.00 g = 100.09 g = 100.1 g.

$$\% \text{ Ca:} \frac{40.08 \text{ g Ca}}{100.1 \text{ g } CaCO_3} \times 100\% = \underline{40.09\%}$$

$$\% \text{ C:} \frac{12.01 \text{ g C}}{100.1 \text{ g } CaCO_3} \times 100\% = \underline{12.00\%}$$

$$\% \text{ O:} \frac{48.00 \text{ g O}}{100.1 \text{ g } CaCO_3} \times 100\% = \underline{47.95\%}$$

| Example 8-13 | The Percent Composition of Borazine |

What is the percent composition of all the elements in borazine ($B_3N_3H_6$)?

**Procedure**

Find the molar mass and convert the *total* mass of each element to percent of molar mass.

**Solution**

One mole of $B_3N_3H_6$ contains

$$3 \text{ mol of B} \left( 3 \text{ mol B} \times \frac{10.81 \text{ g}}{\text{mol B}} = 32.43 \text{ g} \right)$$

$$3 \text{ mol of N} \left( 3 \text{ mol N} \times \frac{14.01 \text{ g}}{\text{mol N}} = 42.03 \text{ g} \right)$$

$$6 \text{ mol of H} \left( 6 \text{ mol H} \times \frac{1.008 \text{ g}}{\text{mol H}} = 6.048 \text{ g} \right)$$

The molar mass $= 32.43 \text{ g} + 42.03 \text{ g} + 6.048 \text{ g} = \underline{80.51 \text{ g}}$

$$\% \text{ B:} \frac{32.43 \text{ g B}}{80.51 \text{ g B}_3\text{N}_3\text{H}_6} \times 100\% = \underline{40.28\%}$$

$$\% \text{ N:} \frac{42.03 \text{ g N}}{80.51 \text{ g B}_3\text{N}_3\text{H}_6} \times 100\% = \underline{52.20\%}$$

$$\% \text{ H:} \frac{6.048 \text{ g H}}{80.51 \text{ g B}_3\text{N}_3\text{H}_6} \times 100\% = \underline{7.51\%}$$

# 8-5 Empirical and Molecular Formulas

**Looking Ahead!** From the formula of a compound, we can calculate either the mass composition of the elements or their percent composition. In this section we will do just the opposite—that is, from the mass composition or the percent composition we will obtain the ratio of the atoms of the elements in a compound.

Two hydrocarbons (carbon–hydrogen compounds) are known as acetylene ($C_2H_2$) and benzene ($C_6H_6$). Actually, these two compounds have very few properties in common. Acetylene is a gas used in welding, and benzene is a liquid solvent used widely in industry. What they do have in common, however, is that they have the same percent composition. This is because they have the same empirical formula (CH). *The **empirical formula** is the simplest whole-number ratio of atoms in the compound.* Most ionic compounds are represented by empirical formulas, but the formula of a molecular compound (the molecular formula) represents the actual number of atoms present in one molecule. One benzene molecule is composed of six carbons and six hydrogens, but the empirical formula is obtained by dividing the subscripts by six. The same empirical formula (CH) is obtained for acetylene by dividing the subscripts by two.

To calculate the empirical formula of a compound from percent composition, we follow three steps.

1. Convert percent composition to an actual mass.
2. Convert mass to moles of each element.
3. Find the whole-number ratio of the moles of different elements.

This procedure is best illustrated by the following examples.

## The Empirical Formula of Laughing Gas

**Example 8-14**

working it out

What is the empirical formula of laughing gas, which is 63.6% nitrogen and 36.4% oxygen?

**Procedure**

Remember that percent means parts per 100. Therefore, if we simply assume that we have a 100-g quantity of compound, the percents convert to specific masses as follows.

$$63.6\% \text{ N} = \frac{63.6 \text{ g N}}{100 \text{ g compound}} \qquad 36.4\% \text{ O} = \frac{36.4 \text{ g O}}{100 \text{ g compound}}$$

We now convert the masses to moles of each element.

$$63.6 \text{ g N} \times \frac{1 \text{ mol N}}{14.01 \text{ g N}} = 4.54 \text{ mol N in 100 g compound}$$

$$36.4 \text{ g O} \times \frac{1 \text{ mol O}}{16.00 \text{ g O}} = 2.28 \text{ mol O in 100 g compound}$$

*The ratio of N to O atoms will be the same as the ratio of N to O moles.* The formula cannot remain fractional, since only whole numbers of atoms are present in a compound. To find the whole-number ratio of moles, divide through by the smaller number of moles, which in this case is 2.28 mol of O.

**Solution**

$$\text{N: } \frac{4.54}{2.28} = 1.99 \approx 2.00 \qquad \text{O: } \frac{2.28}{2.28} = 1.00$$

The empirical formula of the compound is

$$\underline{\underline{N_2O}}$$

## The Empirical Formula of Magnetite

**Example 8-15**

Pure magnetite is composed of an iron–oxygen binary compound. A 3.85-g sample of magnetite is composed of 2.79 g of iron. What is the empirical formula of magnetite?

**Procedure**

a. Find the mass of oxygen by subtracting the mass of the iron from the total mass.

$$3.85 \text{ g} - 2.79 \text{ g (Fe)} = 1.06 \text{ g (O)}$$

b. Convert the masses of iron and oxygen to moles.

$$2.79 \text{ g Fe} \times \frac{1 \text{ mol Fe}}{55.85 \text{ g Fe}} = 0.0500 \text{ mole Fe}$$

$$1.06 \text{ g O} \times \frac{1 \text{ mol O}}{16.00 \text{ g O}} = 0.0663 \text{ mol O}$$

c. Find the whole-number ratio of moles of iron and oxygen. Divide by the smaller number of moles.

$$\text{Fe: } \frac{0.0500}{0.0500} = 1.00 \qquad \text{O: } \frac{0.0663}{0.0500} = 1.33$$

**Magnetite is a mineral composed of iron and oxygen.**

This time we're not quite finished, since $FeO_{1.33}$ still has a fractional number that must be cleared. (You should keep at least two decimal places in these numbers; do not round off a number like 1.33 to 1.3 or 1.) This fractional number can be cleared by multiplying both subscripts by an integer that produces whole numbers. In this case, 1.33 is equivalent to $1\frac{1}{3}$ or $\frac{4}{3}$, so both subscripts (the 1 implied for Fe and the 1.33 for O) can be multiplied by 3 to clear the fraction.

**Solution**

$$Fe_{(1 \times 3)}O_{(1.33 \times 3)} = \underline{\underline{Fe_3O_4}}$$

A decimal value of 0.50 can be multiplied by 2, values of 0.33 and 0.67 can be multiplied by 3, values of 0.25 and 0.75 can be multiplied by 4, and values of 0.20, 0.40, 0.60, and 0.80 can be multiplied by 5. There are few examples that are more complex than these. Since the last decimal place is estimated, values such as 0.49 should be rounded off to 0.50.

To determine the **molecular formula** of molecular compounds, we need to know how many empirical units are present in each molecular unit. Thus it is necessary to know both the molar mass of the compound (g/mol) and the mass of one empirical unit (g/emp. unit). The ratio of these two quantities must be a whole number (represented as $a$ below) and represents the number of empirical units in one mole of compound.

$$a = \frac{\text{molar mass}}{\text{emp. mass}} = \frac{X \text{ g/mol}}{Y \text{ g/emp. unit}} = 1, 2, 3, \text{ etc., emp. unit/mol}$$

The molecular formula is obtained by multiplying the subscripts of the empirical formula by $a$. For example, the empirical formula of borazine (from Example 8-13) is $BNH_2$. The mass of one empirical unit is $[10.81 + 14.01 + (2 \times 1.008)] = 26.84$ g/emp. unit. Its molar mass is 80.5 g/mol.

$$a = \frac{80.5 \text{ g/mol}}{26.84 \text{ g/emp. unit}} = 3 \text{ emp. units/mol}$$

The molecular formula is

$$B_{(1 \times 3)}N_{(1 \times 3)}H_{(2 \times 3)} = B_3N_3H_6$$

| Example 8-16 | The Molecular Formula of a Phosphorus Oxide |
|---|---|
| **working it out** | A phosphorus–oxygen compound has an empirical formula of $P_2O_5$. Its molar mass is 283.9 g/mol. What is its molecular formula? |

**Procedure**

Find the number of empirical units in each mole of compound ($a$). The empirical mass of the compound is found by adding two moles of phosphorus (30.97 g/mol) and five moles of oxygen (16.00 g/mol).

$$(2 \times 30.97 \text{ g}) + (5 \times 16.00 \text{ g}) = 141.9 \text{ g/emp. unit}$$

**Solution**

$$a = \frac{283.9 \text{ g/mol}}{141.9 \text{ g/emp. unit}} = 2 \text{ emp. units/mol}$$

The molecular formula is

$$P_{(2 \times 2)}O_{(2 \times 5)} = \underline{\underline{P_4O_{10}}}$$

The types of calculations performed in this section are of fundamental importance to the advancement of science. Over 10 million unique compounds have been reported and registered by the American Chemical Society, which serves as the world registry of compounds. To establish that a new discovery is a unique compound several steps must be followed. *First,* it must be shown that the compound is indeed a pure substance. This is established by a study of its physical properties. For example, pure substances have definite and unchanging melting and boiling points, as was discussed in Chapter 2. *Second,* the formula of the new compound must be determined. This is mostly done by commercial laboratories that report the percent composition of the elements. In calculations like those we performed here, the empirical formula is obtained from percent composition. *Third,* the molecular formula of the new compound is calculated from the molar mass and the empirical formula. The molar mass may also be obtained commercially or measured experimentally by methods that will be discussed later in this text. *Finally,* the structural formula, which is the order and arrangement of the atoms in the molecule, is determined. Various instruments available in most laboratories can provide information of this nature. The procedure may take anywhere from a few minutes for simple molecules to a few months for very complex molecules such as those associated with life processes.

In many cases, new compounds that are identified by these procedures are tested for anticancer or anti-AIDS activity. Or maybe the new compound has an application as a herbicide, an insecticide, a perfume, a food preservative, or any number of other possibilities. In any case, many of the new compounds that are reported each year have some practical application that improves the quality of our lives.

**A Research Laboratory**
**Preparation and identification of new compounds occur frequently in chemical laboratories.**

---

**Looking Back! A compound is the sum of its component parts. By using the atomic mass of an element along with the number of atoms of that element in a compound, we can establish the number of moles, mass, and percent composition of that element in the compound. In reverse, we can use the moles, mass, or percent composition of all of the elements in a compound to calculate its empirical formula. Ionic compounds are represented by their empirical formulas, but molecular compounds may have a molecular formula that is a multiple of the empirical formula. The molar mass is needed to establish the molecular formula.**

---

## Learning Check B | checking it out

**B-1.** Fill in the blanks.

The mass of an element expressed as the mass per 100 mass units is known as the _____ _____ of the element in the compound. For this to be calculated, one must know two things about the compound: the molar mass and the _____ of the compound. The smallest whole-number ratio of atoms represents the _____ formula, and the actual number of atoms of each element represents the _____ formula. From the percent composition of elements, one can calculate the _____ formula. The _____ _____ of the compound is needed to calculate the molecular formula.

**B-2.** Vitamin C (ascorbic acid) is a compound with the formula $C_6H_8O_6$. The formula weight of the compound is _____ amu. In a 0.650-mol quantity of ascorbic acid, there is _____ mol of carbon, which equals _____ g of carbon. The compound is _____ % carbon, and the empirical formula of the compound is _____ .

**B-3.** A molecular compound is composed of boron and hydrogen. It is 84.4% boron by mass and has a molar mass of 76.96 g/mol. What is its empirical formula? What is its molecular formula?

*Additional Examples: Exercises 8-35, 8-38, 8-45, 8-56, 8-58, and 8-68.*

Section **C**  **MASS AND ENERGY CHANGES IN CHEMICAL REACTIONS**

## 8-6 Stoichiometry

**Looking Ahead!** The balanced chemical equation was introduced in Chapter 7 in terms of individual molecules and formula units. With the help of the mole, we are ready to scale up our measurements so they are appropriate for laboratory situations. We will make important calculations in this section concerning the mass relationships between and among reactants and products.

Any decent cook knows that a good recipe calls for ingredients to be mixed in precise amounts. Likewise, any decent chemical company producing a weed killer or a fertilizer will not waste expensive reactants by using more than the recipe requires. For the chemical company, the recipe is the balanced chemical equation. *The quantitative relationships among reactants and products is known as* **stoichiometry.** One of the most important industrial processes is the production of ammonia ($NH_3$) from hydrogen and nitrogen. We will use the balanced equation representing this reaction as our example. (See Table 8-3.) The most basic relationship of the balanced equation refers to the ratio of molecules and is shown in line 1. But it is important to note that the equation also implies any multiple of the basic molecular ratios, as shown in lines 2, 3, and 4. In lines 5 and 6, the numbers have been changed to mole units and the corresponding mass in grams. We will focus on the mole relationships shown in line 5, as this successfully scales up the stoichiometry of the equation to laboratory situations. The relationships of moles to number (line 4) and to mass (line 6) have been discussed earlier in this chapter.

First we will extract from the balanced equation and express in words the mole relationships between the reactant and product molecules.

*1 mol of $N_2$ produces 2 mol of $NH_3$*

This can be expressed in ratio form as (1) $\dfrac{1 \text{ mol } N_2}{2 \text{ mol } NH_3}$ and (2) $\dfrac{2 \text{ mol } NH_3}{1 \text{ mol } N_2}$.

*1 mol of $N_2$ reacts with 3 mol of $H_2$*

This can be expressed in ratio form as (3) $\dfrac{1 \text{ mol } N_2}{3 \text{ mol } H_2}$ and (4) $\dfrac{3 \text{ mol } H_2}{1 \text{ mol } N_2}$.

*3 mol of $H_2$ produces 2 mol of $NH_3$*

This can be expressed in ratio form as (5) $\dfrac{3 \text{ mol } H_2}{2 \text{ mol } NH_3}$ and (6) $\dfrac{2 \text{ mol } NH_3}{3 \text{ mol } H_2}$.

The mole relation factors generated by the balanced equation will be referred to as the "mole ratios" in this text. *Note that the coefficients in the balanced equation must be*

### Table 8-3  Production of Ammonia

| $N_2(g)$ | + | $3H_2(g)$ | $\longrightarrow$ | $2NH_3(g)$ |
|---|---|---|---|---|
| **1.** 1 molecule | + | 3 molecules | $\longrightarrow$ | 2 molecules |
| **2.** 12 molecules | + | 36 molecules | $\longrightarrow$ | 24 molecules |
| **3.** 1 dozen molecules | + | 3 dozen molecules | $\longrightarrow$ | 2 dozen molecules |
| **4.** $6.022 \times 10^{23}$ molecules | + | $18.1 \times 10^{23}$ molecules | $\longrightarrow$ | $12.0 \times 10^{23}$ molecules |
| **5.** 1 mol | + | 3 mol | $\longrightarrow$ | 2 mol |
| **6.** 28 g | + | 6 g | $\longrightarrow$ | 34 g |

*included in the mole ratios.* The six mole ratios generated by the sample balanced equation will be used in the stoichiometry problems that follow. In these examples, the central conversion is between moles of one reactant or product to moles of another reactant or product. In Examples 8-18, 8-19, and 8-20, additional conversions are necessary.

The examples below illustrate the following conversions.

mole $\longrightarrow$ mole     (Example 8-17)
mass $\longrightarrow$ mole     (Example 8-18)
mass $\longrightarrow$ mass     (Example 8-19)
mass $\longrightarrow$ number     (Example 8-20)

## Mole $\longrightarrow$ Mole Conversions

Example 8-17

working it out

How many moles of $NH_3$ can be produced from 5.00 mol of $H_2$?

**Procedure**

Convert moles of what's given (mol of $H_2$) to moles of what's requested (mol $NH_3$). This requires mole ratio 6, which has *mol $H_2$* in the denominator and *mol $NH_3$* in the numerator.

**Scheme**

*Given* (mol $H_2$)       *Requested* (mol $NH_3$)

$$\times \frac{\text{mol } NH_3}{\text{mol } H_2}$$

$$\boxed{\text{mol } H_2} \Longrightarrow \boxed{\text{mol } NH_3}$$

[mole ratio (6)]

**Solution**

$$5.00 \ \text{mol } H_2 \times \frac{2 \ \text{mol } NH_3}{3 \ \text{mol } H_2} = \underline{\underline{3.33 \ \text{mol } NH_3}}$$

## Mass $\longrightarrow$ Mole Conversions

Example 8-18

How many moles of $NH_3$ can be produced from 33.6 g of $N_2$?

**Procedure**

Before we can convert moles of $N_2$ to moles of $NH_3$, we must first convert the mass of $N_2$ to moles of $N_2$. This means a two-step conversion. In the first step, mass of $N_2$ is converted to moles using the molar mass of $N_2$ as the conversion factor (*a*). In the second step, moles of $N_2$ is converted to moles of $NH_3$ using mole ratio (2) (*b*).

**Scheme**

*Given* (g $N_2$)       *Requested* (mol $NH_3$)

        (*a*)         (*b*)

$$\times \frac{\text{mol } N_2}{\text{g } N_2} \qquad \times \frac{\text{mol } NH_3}{\text{mol } N_2}$$

$$\boxed{\text{mass } N_2} \Longrightarrow \boxed{\text{mol } N_2} \Longrightarrow \boxed{\text{mol } NH_3}$$

**Solution**

         (*a*)          (*b*)

$$33.6 \ \text{g } N_2 \times \frac{1 \ \text{mol } N_2}{28.02 \ \text{g } N_2} \times \frac{2 \ \text{mol } NH_3}{1 \ \text{mol } N_2} = \underline{\underline{2.40 \ \text{mol } NH_3}}$$

**Example 8-19** | Mass $\longrightarrow$ Mass Conversions

What mass of $H_2$ is needed to produce 119 g of $NH_3$?

**Procedure**

As in the last example, we must first convert the mass of what's given (119 g $NH_3$) into moles using the molar mass of $NH_3$ (step *a* below). We then convert moles of $NH_3$ to moles of $H_2$ using mole ratio (5) (step *b* below). Finally, since the mass of $H_2$ is requested, we must convert moles of $H_2$ to mass using the molar mass of $H_2$ as a conversion factor (step *c* below).

**Scheme**

$$\textit{Given } (g\ NH_3) \qquad\qquad\qquad\qquad\qquad \textit{Requested } (g\ H_2)$$

$$\begin{array}{ccc} \textit{(a)} & \textit{(b)} & \textit{(c)} \\ \times \dfrac{mol\ NH_3}{g\ NH_3} & \times \dfrac{mol\ H_2}{mol\ NH_3} & \times \dfrac{g\ H_2}{mol\ H_2} \end{array}$$

$$\boxed{mass\ NH_3} \Longrightarrow \boxed{mol\ NH_3} \Longrightarrow \boxed{mol\ H_2} \Longrightarrow \boxed{mass\ H_2}$$

**Solution**

$$\begin{array}{cccc} & \textit{(a)} & \textit{(b)} & \textit{(c)} \\ 119\ \text{g}\ \cancel{NH_3} \times & \dfrac{1\ \text{mol}\ \cancel{NH_3}}{17.03\ \text{g}\ \cancel{NH_3}} \times & \dfrac{3\ \text{mol}\ \cancel{H_2}}{2\ \text{mol}\ \cancel{NH_3}} \times & \dfrac{2.016\ \text{g}\ H_2}{\cancel{mol\ H_2}} = 21.1\ \text{g}\ H_2 \end{array}$$

---

**Example 8-20** | Mass $\longrightarrow$ Number Conversions

How many molecules of $N_2$ are needed to react with 17.0 g of $H_2$?

**Procedure**

This problem reminds us that the mole relates to a number as well as a mass. In fact, this type of problem generally is not encountered because the actual numbers of molecules involved is not as important as their relative masses. In any case, this problem is much like the previous example except that in the final step moles of $N_2$ is converted to number of molecules (step *c* below).

**Scheme**

$$\textit{Given } (g\ H_2) \qquad\qquad\qquad\qquad \textit{Requested } (\text{molecules } N_2)$$

$$\begin{array}{ccc} \textit{(a)} & \textit{(b)} & \textit{(c)} \\ \times \dfrac{mol\ H_2}{g\ H_2} & \times \dfrac{mol\ N_2}{mol\ H_2} & \times \dfrac{6.022 \times 10^{23}\ N_2}{mol\ N_2} \end{array}$$

$$\boxed{mass\ H_2} \Longrightarrow \boxed{mol\ H_2} \Longrightarrow \boxed{mol\ N_2} \Longrightarrow \boxed{number\ N_2}$$

**Solution**

$$\begin{array}{cccc} & \textit{(a)} & \textit{(b)} & \textit{(c)} \\ 17.0\ \text{g}\ \cancel{H_2} \times & \dfrac{1\ \text{mol}\ \cancel{H_2}}{2.016\ \text{g}\ \cancel{H_2}} \times & \dfrac{1\ \text{mol}\ \cancel{N_2}}{3\ \text{mol}\ \cancel{H_2}} \times & \dfrac{6.022 \times 10^{23}\ \text{molecules}\ N_2}{\cancel{mol\ N_2}} \end{array}$$

$$= 1.69 \times 10^{24}\ \text{molecules}\ N_2$$

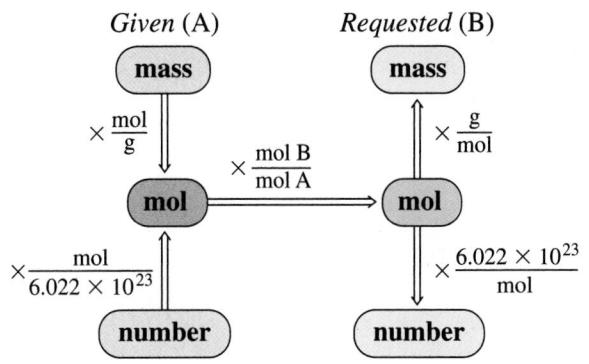

**Figure 8-5 General Procedure for Stoichiometry Problems**
**Any three-step conversion is possible following the proper pathway.**

Before we work through another example of a stoichiometry problem, let us summarize the procedure relating moles, mass, and number of molecules.

1. Write down (A) what is given and (B) what is requested.
2. **a.** If a mass is given, use the molar mass to convert mass to moles of what is given.
   **b.** If a number of molecules is given, use Avogadro's number to convert to moles of what is given.
3. Using the correct mole ratio from the balanced equation, convert moles of what is given to moles of what is requested.
4. **a.** If a mass is requested, convert moles of what is requested to mass of what is requested.
   **b.** If a number of molecules is requested, use Avogadro's number to convert to number of molecules of what is requested.

We represent this information in the general scheme shown in Figure 8-5.

---

**Conversion of Mass of FeS₂ to Mass of SO₂**

**Example 8-21**

working it out

Some sulfur is present in coal in the form of pyrite ($FeS_2$; also known as "fool's gold"). When it burns, it pollutes the air with the combustion product $SO_2$, as shown by the following chemical equation.

$$4FeS_2(s) + 11O_2(g) \longrightarrow 2Fe_2O_3(s) + 8SO_2(s)$$

What mass of $SO_2$ is produced by the combustion of 38.8 g of $FeS_2$?

**Procedure**

Given: 38.8 g of $FeS_2$. Requested: ? g of $SO_2$.
Convert grams of $FeS_2$ to moles of $FeS_2$, moles of $FeS_2$ to moles of $SO_2$, and, finally, moles of $SO_2$ to grams of $SO_2$.

**Scheme**

*Given* (g FeS₂)                    *Requested* (g SO₂)

$\times \dfrac{\text{mol FeS}_2}{\text{g FeS}_2}$  $\times \dfrac{\text{mol SO}_2}{\text{mol FeS}_2}$  $\times \dfrac{\text{g SO}_2}{\text{mol SO}_2}$

$\boxed{\text{mass FeS}_2} \Longrightarrow \boxed{\text{mol FeS}_2} \Longrightarrow \boxed{\text{mol SO}_2} \Longrightarrow \boxed{\text{mass SO}_2}$

**The color of pyrite caused it to be known as "fool's gold."**

The mole ratio that relates moles of given to moles of requested (in the numerator) is

$$\frac{8 \text{ mol SO}_2}{4 \text{ mol FeS}_2}$$

**Solution**

$$38.8 \text{ g FeS}_2 \times \frac{1 \text{ mol FeS}_2}{120.0 \text{ g FeS}_2} \times \frac{8 \text{ mol SO}_2}{4 \text{ mol FeS}_2} \times \frac{64.07 \text{ g SO}_2}{\text{mol SO}_2} = \underline{\underline{41.4 \text{ g SO}_2}}$$

## 8-7 Limiting Reactant

**Looking Ahead! Perhaps the most important stoichiometry problem that we have encountered is the conversion of a given mass of reactant to an equivalent mass of product. So far, we have just assumed that the needed amounts of all other reactants are present. But how do we determine the amount of product formed if there are specific amounts of each reactant present? This is the subject of this section.**

Assume that your instructor is trying to put together a three-page exam for the class. The instructor may have 85 copies of page 1, 85 copies of page 2, but only 80 copies of page 3. Obviously, only 80 copies of the test can be put together even though there is an excess of pages 1 and 2. The page that produces the least number of complete tests can be called "the limiting page." Likewise, *if specific amounts of each reactant are mixed, the reactant that produces the least amount of product is called the* **limiting reactant.** In other words, the amount of product formed is limited by the reactant that is completely consumed. We can illustrate this with the simple example of the production of water from its elements, hydrogen and oxygen.

$$2H_2 + O_2 \longrightarrow 2H_2O$$

The stoichiometry of the reaction tells us that two moles (4.0 g) of hydrogen react with one mole (32.0 g) of oxygen to produce two moles (36.0 g) of water. Thus any time hydrogen and oxygen react in a 4.0 : 32 mass ratio, all reactants are consumed and only product appears. *When reactants are mixed in exactly the mass ratio determined from the balanced equation, the mixture is said to be stoichiometric.*

$$4.0 \text{ g H}_2 + 32.0 \text{ g O}_2 \longrightarrow 36.0 \text{ g H}_2O \text{ (stoichiometric)}$$

What if we mix a 6.0-g quantity of $H_2$ with a 32.0-g quantity of $O_2$? Do we produce 38.0 g water? No, we still produce only 36.0 g of $H_2O$ using only 4.0 g of the $H_2$. Thus $H_2$ is present in excess, and the amount of product is limited by the amount of $O_2$ present. In this case, $O_2$ becomes the limiting reactant.

$$6.0 \text{ g H}_2 + 32.0 \text{ g O}_2 \longrightarrow 36.0 \text{ g H}_2O + 2.0 \text{ g H}_2 \text{ in excess}$$

*($H_2$ in excess, $O_2$ the limiting reactant)*

If we mix a 4.0-g quantity of $H_2$ with a 36.0-g quantity of $O_2$, 36.0 g of $H_2O$ is again produced. In this case, the $H_2$ is completely consumed and limits the amount of water formed. Thus $H_2$ is now the limiting reactant and $O_2$ is present in excess.

$$4.0 \text{ g H}_2 + 36.0 \text{ g O}_2 \longrightarrow 36.0 \text{ g H}_2O + 4.0 \text{ g O}_2 \text{ in excess}$$

*($O_2$ in excess, $H_2$ the limiting reactant)*

## Fizzies and Limiting Reactants

Dehydrated soda–simply add water and have a delicious drink, bubbles and all. Such a product is available but it is not so simple to manufacture. It is made by pressing into tablet form solid citric acid ($H_3C_6H_5O_7$) and solid sodium bicarbonate ($NaHCO_3$). A solid sweetener and some powdered flavoring are added to the tablet to produce a flavored drink.

As dry solids, no reaction takes place. When the tablet is dropped in a glass of water, however, a neutralization reaction occurs forming a solution of sodium citrate and carbon dioxide gas that gives it fizz. This is illustrated by the following equation.

$$H_3C_6H_5O_7(aq) + 3NaHCO_3(aq) \rightarrow$$
$$Na_3C_6H_5O_7(aq) + 3H_2O(l) + 3CO_2(g)$$

The proportions of citric acid and sodium bicarbonate are crucial. Suppose 20.0 g of each reactant were mixed. This represents 0.104 moles of citric acid and 0.238 moles of sodium bicarbonate. From the stoichiometry of the equation we find that sodium bicarbonate would be the limiting reactant and thus be completely consumed. This would not make for a tasty drink as the excess acid would make the drink sour. Let's try it with less citric acid. What if we now mixed 10.0 g of citric acid (0.0521 moles) and 20.0 g of sodium bicarbonate? Now we find that the citric acid is completely consumed so that sodium bicarbonate is in excess. Now the drink tastes too bitter because of the excess base.

The solution is to mix exactly stoichiometric amounts of the two reactants. In this example, 20.0 g of citric acid (0.104 moles) completely reacts with 26.2 g of sodium bicarbonate (0.312 moles). The reaction is now complete and the desired fruity taste is not obscured by either bitterness or sourness from excess base or acid, respectively.

The reactants must be completely dry when mixed. If even a small amount of moisture is present, some carbon dioxide is generated and the package ruptures.

Instant carbonated beverages are a great idea but not that easy to make.

*Submitted by G. Lynn Carlson*
*University of Wisconsin—Parkside*

*Instant carbonated beverages.*

---

When quantities of more than one reactant are given, it is necessary to determine which is the limiting reactant. This is accomplished as follows.

1. Convert the amount of *each reactant* to the number of moles (or mass) of product using the general procedure shown in Figure 8-5.

2. The limiting reactant produces the smallest amount of product.

Notice that each *reactant* must be converted to *product* to solve the problem in the manner outlined. We will illustrate the calculation of limiting reactant with two examples.

| Determination of the Limiting Reactant | Example 8-22 |
|---|---|
| | working it out |

Silver tarnishes (turns black) in homes because of the presence of small amounts of $H_2S$ (a gas that originates from the decay of food and smells like rotten eggs). The reaction is

$$4Ag(s) + 2H_2S(g) + O_2(g) \longrightarrow 2Ag_2S(s) + 2H_2O(l)$$
$$\text{(black)}$$

If 0.145 mol of Ag is present with 0.0872 mol of $H_2S$ and excess $O_2$,

a. What mass of $Ag_2S$ is produced?

b. What mass of the other reactant remains in excess?

**Procedure (a)**

Convert moles of Ag and moles of $H_2S$ to moles of $Ag_2S$ produced. Then convert the number of moles of $Ag_2S$ to mass of $Ag_2S$ based on the limiting reactant.

**Silver tarnishes from the presence of $H_2S$ in the atmosphere.**

**Scheme (a)**

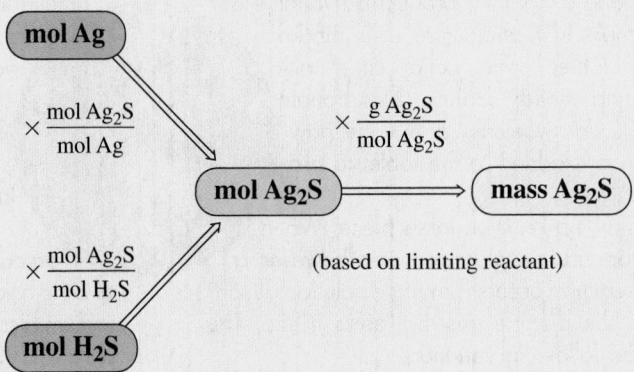

(based on limiting reactant)

**Solution (a)**

$$\text{Ag: } 0.145 \text{ mol Ag} \times \frac{2 \text{ mol Ag}_2\text{S}}{4 \text{ mol Ag}} = 0.0725 \text{ mol Ag}_2\text{S}$$

$$\text{H}_2\text{S: } 0.0872 \text{ mol H}_2\text{S} \times \frac{2 \text{ mol Ag}_2\text{S}}{2 \text{ mol H}_2\text{S}} = 0.0872 \text{ mol Ag}_2\text{S}$$

Since Ag produces the smaller yield of $Ag_2S$, *Ag is the limiting reactant.* To find the mass of $Ag_2S$ formed, convert moles of $Ag_2S$ produced by the Ag to mass of $Ag_2S$.

$$0.0725 \text{ mol Ag}_2\text{S} \times \frac{247.9 \text{ g Ag}_2\text{S}}{\text{mol Ag}_2\text{S}} = \underline{\underline{18.0 \text{ g Ag}_2\text{S}}}$$

**Procedure (b)**

To find the mass of $H_2S$ in excess, we first find the mass of $H_2S$ that was consumed in the reaction along with the Ag. Use the mole ratio relating moles of $H_2S$ to moles of Ag.

**Scheme (b)**

mol Ag ⟹ mol H₂S ⟹ mol H₂S (consumed)

Next, find the moles of $H_2S$ remaining or unreacted by subtracting the moles consumed from the original amount present.

**mol $H_2S$ (original) − mol $H_2S$ (consumed) = mol $H_2S$ (in excess)**

Finally, convert moles of unreacted $H_2S$ to mass.

mol $H_2S$ (in excess) ⟹ mass $H_2S$

**Solution (b)**

$$0.145 \text{ mol Ag} \times \frac{2 \text{ mol H}_2\text{S}}{4 \text{ mol Ag}} = 0.0725 \text{ mol H}_2\text{S consumed}$$

$$0.0872 \text{ mol} - 0.0725 \text{ mol} = 0.0147 \text{ mol H}_2\text{S in excess}$$

$$0.0147 \text{ mol H}_2\text{S} \times \frac{34.09 \text{ g H}_2\text{S}}{\text{mol H}_2\text{S}} = \underline{\underline{0.501 \text{ g H}_2\text{S in excess}}}$$

**Determination of the Limiting Reactant**

Example 8-23

w o r k i n g   i t   o u t

Methanol ($CH_3OH$) is used as a fuel for racing cars. It burns in the engine according to the equation

$$2CH_3OH(l) + 3O_2(g) \longrightarrow 2CO_2(g) + 4H_2O(g)$$

If 40.0 g of methanol is mixed with 46.0 g of $O_2$, what is the mass of $CO_2$ produced?

**Procedure**

1. Convert mass of $CH_3OH$ to moles of $CH_3OH$ and then to moles of $CO_2$, and convert mass of $O_2$ to moles of $O_2$ and then to moles of $CO_2$.

2. Convert the smaller number of moles of $CO_2$ to mass of $CO_2$.

**Scheme**

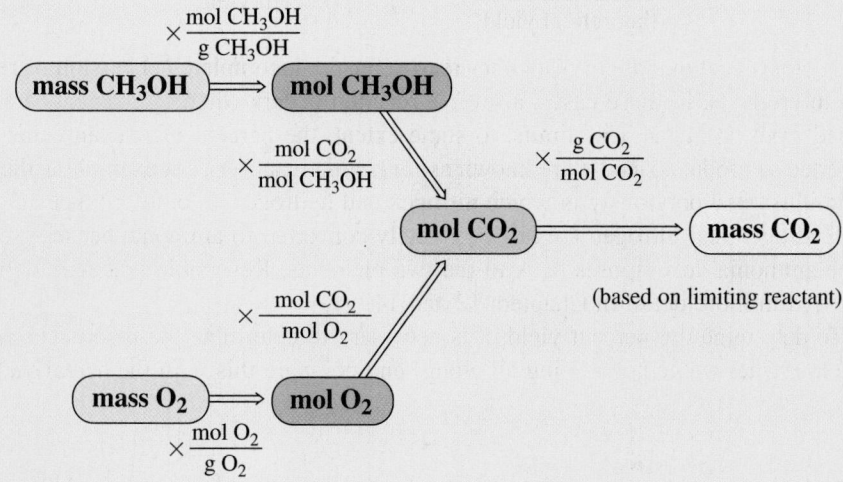

Race cars at the Indianapolis 500 burn methanol as a fuel.

**Solution**

1.

$$CH_3OH: 40.0 \text{ g CH}_3\text{OH} \times \frac{1 \text{ mol CH}_3\text{OH}}{32.04 \text{ g CH}_3\text{OH}} \times \frac{2 \text{ mol CO}_2}{2 \text{ mol CH}_3\text{OH}} = 1.25 \text{ mol CO}_2$$

$$O_2: 46.0 \text{ g O}_2 \times \frac{1 \text{ mol O}_2}{32.00 \text{ g O}_2} \times \frac{2 \text{ mol CO}_2}{3 \text{ mol O}_2} = 0.958 \text{ mol CO}_2$$

Therefore, $O_2$ is the limiting reactant.

2. The yield is determined from the amount of product formed *from the limiting reactant*. Thus we simply convert the 0.958 mol of $CO_2$ produced by the $O_2$ to grams of $CO_2$.

$$0.958 \text{ mol CO}_2 \times \frac{44.01 \text{ g CO}_2}{\text{mol CO}_2} = \underline{\underline{42.2 \text{ g CO}_2}}$$

# 8-8 Percent Yield

**Looking Ahead!** There is something else that we have been taking for granted in the problems worked so far. We have assumed that at least one reactant has been completely converted into products and that the reactants form only the products shown. This is not always the case, however. In this section, we will look into incomplete reactions.

An efficient automobile engine burns gasoline (mainly a hydrocarbon, $C_8H_{18}$) to form carbon dioxide and water. Untuned engines, however, do not burn gasoline efficiently.

They produce carbon monoxide and may even exhaust unburned fuel. The two combustion reactions are shown below.

Complete combustion:   $2C_8H_{18}(g) + 25O_2(g) \longrightarrow 16CO_2(g) + 18H_2O(l)$

Incomplete combustion: $2C_8H_{18}(g) + 17O_2(g) \longrightarrow 16CO(g) + 18H_2O(l)$

Note that if we were asked to calculate the mass of $CO_2$ produced from a given amount of $C_8H_{18}$, our answer would not be correct if all the hydrocarbon were not converted to $CO_2$. *The measured amount of product obtained in any reaction is known as the* **actual yield.** *The* **theoretical yield** *is the calculated amount of product that would be obtained if all of the reactant were converted to a given product. The* **percent yield** *is the actual yield in grams or moles divided by the theoretical yield in grams or moles times 100%.*

$$\frac{\text{actual yield}}{\text{theoretical yield}} \times 100\% = \text{percent yield}$$

In other reactions, there is another reason for the incomplete conversion of reactants to products. In these cases, a reverse reaction occurs whereby reactants are re-formed from products. This limits, to some extent, the percent of reactants that are converted to products. These are known as *reversible* reactions. An example is the reaction illustrated previously in which nitrogen and hydrogen react to produce ammonia. Hydrogen and nitrogen are not completely converted to ammonia because some of the ammonia decomposes back to the two elements. Reversible reactions will be discussed in more detail in Chapters 12 and 14.

To determine the percent yield, it is necessary to determine the theoretical yield (which is what we've been doing all along) and compare this with the actual yield.

| Example 8-24 | **Determination of Percent Yield** |

**working it out**

In a given experiment, a 4.70-g quantity of $H_2$ is allowed to react with $N_2$; a 12.5-g quantity of $NH_3$ is formed. What is the percent yield based on the $H_2$?

**Procedure**

Find the mass of $NH_3$ that would form if all 4.70 g of $H_2$ were converted to $NH_3$ (the theoretical yield). Using the actual yield (12.5 g), find the percent yield.

**Scheme**

*Given* (g $H_2$)                                    *Requested* (g $NH_3$)

$$\times \frac{\text{mol } H_2}{\text{g } H_2} \qquad \times \frac{\text{mol } NH_3}{\text{mol } H_2} \qquad \times \frac{\text{g } NH_3}{\text{mol } NH_3}$$

$$\boxed{\text{mass } H_2} \Longrightarrow \boxed{\text{mol } H_2} \Longrightarrow \boxed{\text{mol } NH_3} \Longrightarrow \boxed{\text{mass } NH_3}$$

The mole ratio needed is

$$\frac{2 \text{ mol } NH_3}{3 \text{ mol } H_2}$$

**Solution**

$$4.70 \text{ g } H_2 \times \frac{1 \text{ mol } H_2}{2.016 \text{ g } H_2} \times \frac{2 \text{ mol } NH_3}{3 \text{ mol } H_2} \times \frac{17.03 \text{ g } NH_3}{\text{mol } NH_3} = 26.5 \text{ g } NH_3$$
$$\text{theoretical yield}$$

$$\frac{12.5 \text{ g}}{26.5 \text{ g}} \times 100\% = \underline{\underline{47.2\% \text{ yield}}}$$

| Determination of Actual Yield from Percent Yield | Example 8-25 |
|---|---|

Zinc and silver undergo a single-replacement reaction according to the equation

$$Zn(s) + 2AgNO_3(aq) \longrightarrow Zn(NO_3)_2(aq) + 2Ag(s)$$

When 25.0 g of zinc is added to the silver nitrate solution, the percent yield of silver is 72.3%. What mass of silver is formed?

**Procedure**

Find the theoretical yield of silver by converting the mass of zinc to the equivalent mass of silver. Use the percent yield to calculate the actual yield. This can be accomplished algebraically as follows:

$$\text{actual yield} = \frac{\text{percent yield}}{100\%} \times \text{theoretical yield}$$

*Given* (g Zn)                                   *Requested* (g Ag)

$$\times \frac{\text{mol Zn}}{\text{g Zn}} \qquad \times \frac{\text{mol Ag}}{\text{mol Zn}} \qquad \times \frac{\text{g Ag}}{\text{mol Ag}}$$

$$\boxed{\text{mass Zn}} \Longrightarrow \boxed{\text{mol Zn}} \Longrightarrow \boxed{\text{mol Ag}} \Longrightarrow \boxed{\text{mass Ag}}$$

**Solution**

$$25.0 \text{ g Zn} \times \frac{1 \text{ mol Zn}}{65.39 \text{ g Zn}} \times \frac{2 \text{ mol Ag}}{1 \text{ mol Zn}} \times \frac{107.9 \text{ g Ag}}{\text{mol Ag}} = 82.5 \text{ g Ag}$$
$$\text{theoretical yield}$$

$$\frac{72.3\%}{100\%} \times 82.5 \text{ g Ag} = \underline{\underline{59.6 \text{ g Ag (actual yield)}}}$$

# 8-9 Heat Energy in Chemical Reactions

**Looking Ahead! Matter is not all that is involved in chemical reactions. The other component of the universe, energy, is also an intimate part of chemical reactions. A definite and measurable amount of energy is involved in any chemical change.**

A chemical reaction involves not only product compounds but also energy. For example, when we barbecue over a charcoal fire, we are putting to delicious use the large amount of heat energy that is liberated by the combustion reaction. In this case, the chemical energy in the charcoal (which is mostly carbon) is being converted into the heat energy of the fire. As mentioned in Chapter 2, this is an *exothermic* reaction. A reaction that absorbs heat energy is an *endothermic* reaction. The amount of heat energy (measured in kilocalories or kilojoules) involved in a reaction is a constant amount that depends on the amount of reactants consumed. For example, if one mole of hydrogen undergoes combustion to form liquid water, 286 kJ of heat is evolved.

*A balanced equation that includes heat energy is referred to as a* **thermochemical equation.** A thermochemical equation can be represented in either of two ways. In the first, the heat is shown separately from the balanced equation using the symbol **ΔH.** This is referred to as "heat of the reaction." The symbol is known technically as the *change in enthalpy.* By convention, *a negative sign for* ΔH *corresponds to an exothermic reaction, and a positive sign corresponds to an endothermic reaction.*

## Hydrogen—The Perfect Fuel

The six billion citizens of this planet all need something to burn. We need fuel to cook our food, to keep our homes warm, and to harvest our crops. Of course many of us drive our cars and other vehicles, which takes even more fuel. Throughout most of history, the main fuel was wood. Now we mainly use fossil fuels such as coal, natural gas, and oil. Many of the forests of the planet have already been lost and there is a limit to the amount of fossil fuels stored in the earth. Nuclear energy can help, but more is needed. Enter the simplest and lightest of the elements—hydrogen.

Why is hydrogen such a good fuel? There are many reasons. First, when it burns it forms only water, which does not accumulate in the atmosphere and cause global warming. Also, it is very light, producing 143 kJ/g compared to 55.6 kJ/g for natural gas and 48.1 kJ/g for gasoline. Hydrogen is already the workhorse fuel for the space program. It is used in the rockets and in the fuel cells that produce electricity in the space shuttles. A fuel cell uses the reaction of hydrogen with oxygen to form water to produce electrical energy rather than heat energy (see section 13-6). Fuel cells are the wave of the future, for both utilities and automobiles.

So, why not use hydrogen more? First, pure hydrogen is expensive. The cheapest way to produce hydrogen is to extract it from fossil fuels such as methane ($CH_4$). That is still comparatively expensive and it produces carbon dioxide, a greenhouse gas. The best hope is in the electrolysis of water. To break water down to its elements (hydrogen and oxygen), however, also requires a large input of energy. However, there is intensive research for an inexpensive way to do this. There is hope that solar energy can be harnessed for this task, but it needs to be made more efficient and less expensive.

Another problem is the storage of hydrogen. Its boiling point is $-253°C$, which means that the liquid state cannot exist at room temperature. Thus it has to be stored as a gas under high pressure, which is inherently dangerous. One way that is being explored is to store the hydrogen in a solid compound. For example, if water is added to sodium borohydride ($NaBH_4$), the reaction generates hydrogen gas as illustrated by the equation below. Still, from the stoichiometry we find that 73.8 g of reactants produces only 8.06 g of $H_2$ (4.00 mol).

$$NaBH_4(s) + 2H_2O(l) \rightarrow NaBO_2(s) + 4H_2(g)$$

The storage of hydrogen is a problem that will be solved, however.

*Fuel cells make efficient use of hydrogen.*

---

Written in this manner, the thermochemical equation for the combustion of two moles of hydrogen is

$$2H_2(g) + O_2(g) \longrightarrow 2H_2O(l) \qquad \Delta H = -572 \text{ kJ}$$

Notice in the preceding equation that the heat evolved is per *two* moles of $H_2$ consumed or per *two* moles of $H_2O$ formed. The second way a thermochemical equation is represented shows the heat energy as if it were a reactant or product. In exothermic reactions heat energy is a product, and in endothermic reactions heat energy is a reactant. A positive sign is used in either case.

$$2H_2(g) + O_2(g) \longrightarrow 2H_2O(l) + 572 \text{ kJ} \quad \text{(exothermic reaction)}$$
$$N_2(g) + O_2(g) + 181 \text{ kJ} \longrightarrow 2NO(g) \qquad \text{(endothermic reaction)}$$

The heat energy involved in a chemical reaction may be treated quantitatively in a manner similar to the amount of a reactant or product. The following example illustrates the calculations implied by a thermochemical equation.

**The heat generated from a campfire is a result of an exothermic reaction.**

## Heat ⟶ Mass Conversion

Acetylene, which is used in welding torches, undergoes combustion according to the following thermochemical equation:

$$2C_2H_2(g) + 5O_2(g) \longrightarrow 4CO_2(g) + 2H_2O(l) \qquad \Delta H = -2602 \text{ kJ}$$

If 550 kJ of heat is evolved in the combustion of a sample of $C_2H_2$, what is the mass of $CO_2$ formed?

**Example 8-26**

working it out

**Procedure**

Two steps are required as illustrated below.

$$\times \frac{\text{mol } CO_2}{\text{kJ}} \qquad \times \frac{\text{g } CO_2}{\text{mol } CO_2}$$

$$\boxed{\text{kJ}} \Longrightarrow \boxed{\text{mol } CO_2} \Longrightarrow \boxed{\text{mass } CO_2}$$

**Solution**

$$550 \text{ kJ} \times \frac{4 \text{ mol } CO_2}{2602 \text{ kJ}} \times \frac{44.01 \text{ g } CO_2}{\text{mol } CO_2} = 37.2 \text{ g } CO_2$$

The combustion of acetylene produces a flame hot enough to melt iron.

---

**Looking Back!** Most of us need a recipe in order to bake a cake from scratch. The recipe tells us the exact proportions of ingredients and how big a cake this will produce. Chemical equations serve as recipes for stoichiometry calculations. They tell us the proper mass proportions of reactants and exactly how much product we may obtain from these reactants. In cases where reactants are not mixed in exactly the stoichiometric proportions, the equation tells which reactant limits the amount of product. Sometimes, even when we mix reactants in stoichiometric amounts, we don't obtain as much product as a calculation would indicate. In these cases we express the actual amount of product as a percent of the calculated amount. The result is the percent yield. Finally, we note that heat energy can be included as part of a balanced equation, which allows us to include heat as part of stoichiometry.

---

## Learning Check C | checking it out

**C-1.** Fill in the blanks.

An important application of the balanced chemical equation is to construct _____ ratios between and among reactants and products. A mass–mass conversion requires a _____ -step calculation. The _____ of the two compounds involved are also needed in the calculation. If a number of molecules is given and we are asked to convert to moles, _____ number is used as a conversion factor. If quantities of more than one reactant are given, the reactant that produces the _____ amount of product is the limiting reactant. The calculated amount of a product that forms if at least one reactant is completely consumed is the _____ yield. The amount of product that is measured is the _____ yield. The ratio of the latter divided by the former times 100% is the _____ yield. When heat energy is included, a balanced equation is called a _____ equation. When the sign of $\Delta H$ is negative or the amount of heat is included as a product, the reaction is _____ .

**C-2.** Consider the following combustion of methylamine ($CH_3NH_2$).

$$4CH_3NH_2(l) + 9O_2(g) \longrightarrow 4CO_2(g) + 2N_2(g) + 10H_2O(l)$$

**a.** How many moles of $O_2$ react with 4.50 moles of methylamine?

**b.** How many moles of $N_2$ are produced from 322 g of methylamine?

**c.** What mass of $H_2O$ is produced along with 6.45 g of $CO_2$?

**d.** How many molecules of $O_2$ are required to produce 0.568 g of $N_2$?

**C-3.** Using the equation from problem C-2, what mass of $H_2O$ is produced when 20.0 g of $CH_3NH_2$ is allowed to react with 40.0 g of $O_2$?

**C-4.** The balanced equation describing the combustion of ethylene ($C_2H_4$) is

$$C_2H_4(g) + 3O_2(g) \longrightarrow 2CO_2(g) + 2H_2O(l)$$

In a given experiment, 25.0 g of $CO_2$ was obtained as a product, which represented a 74.0% yield. What mass of $C_2H_4$ was consumed in this reaction?

**C-5.** In the reaction in problem C-4, $\Delta H = -556$ kJ/mol $CO_2$. Write a thermochemical reaction that includes heat energy as a reactant or product. What mass of $C_2H_4$ is required to produce 250 kJ of heat, assuming 100% yield?

*Additional Examples: Exercises 8-75, 8-78, 8-84, 8-95, 8-109, 8-114, and 8-119.*

## chapter Review

**putting it together**

The atomic masses of the elements can be put to greater use than just comparing the masses of individual atoms. By using the atomic mass as ratios, we can measure equivalent numbers of atoms without knowing the actual value. The most useful measure of a number of atoms is obtained by expressing the atomic mass of an element in grams. This is referred to as the **molar mass of the element,** and it is the mass of a specific number of atoms. This number is known as the **mole (mol).** The mole represents $6.022 \times 10^{23}$ atoms or other individual particles, a quantity known as **Avogadro's number.**

The concept of the mole is then extended to compounds. The **molar mass of a compound** is the **formula weight** of the compound expressed in grams and is the mass of Avogadro's number of molecules or ionic formula units. This information is summarized as follows.

| Unit | Number | Mass |
|---|---|---|
| 1.000 mol | $6.022 \times 10^{23}$ atoms | Atomic mass of element in grams |
| | $6.022 \times 10^{23}$ molecules | Formula or molecular weight in grams |
| | $6.022 \times 10^{23}$ ionic formula units | Formula weight of ionic compound in grams |

One important message of this chapter is for you to become comfortable with the interconversions among moles, mass, and numbers of atoms or molecules. The conversions are summarized as follows.

$$N\text{particles} \times \frac{\text{mol}}{6.022 \times 10^{23} \text{particles}} \qquad \text{mol} \times \frac{\text{g}}{\text{mol}}$$

$\Longrightarrow$      $\Longrightarrow$

**Number $N$**      **Moles (mol)**      **Mass (g)**

$\Longleftarrow$      $\Longleftarrow$

$$\text{mol} \times \frac{6.022 \times 10^{23} \text{particles}}{\text{mol}} \qquad \text{g} \times \frac{\text{mol}}{\text{g}}$$

**Percent composition** can be obtained from the formula of a compound and the atomic masses of the elements. The **empirical formula** of a compound can be determined from its mass composition or percent composition. One needs to know the molar mass of a compound to determine the **molecular formula** from the empirical formula. The molecular formula is determined from the percent composition and the molar mass.

Next we examined how chemical equations tell us about the relationships of masses of reactants to masses of products. These are the calculations of **stoichiometry.** In these problems, the balanced equation provided the mole ratios for conversions between moles of one compound in the reaction to moles of another. All of these conversions are summarized in Figure 8-4.

When reactants are mixed in other than stoichiometric amounts, it is necessary to determine the **limiting reactant.** The limiting reactant is completely consumed; thus it determines the amount of product formed.

Some reactions lead to more than one set of products, and others reach a state of equilibrium before all reactants are converted to products. In either case, the **actual yield** of a product may be less than the **theoretical yield.** The actual yield can be expressed as a percent of the theoretical yield. This is called the **percent yield.**

Finally, we studied the relation of heat energy to chemical reactions. In a **thermochemical equation,** the heat evolved or absorbed is represented as either a reactant or product, or as the heat of the reaction, which is referred to as the change in enthalpy and is symbolized as $\Delta H$.

# Exercises

## Relative Masses of Particles

**8-1.** One penny weighs 2.47 g and one nickel weighs 5.03 g. What mass of pennies has the same number of coins as there are in 1.24 lb of nickels?

**8-2.** A small glass bead weighs 310 mg and a small marble weighs 8.55 g. A quantity of small glass beads weighs 5.05 kg. What does the same number of marbles weigh?

**8-3.** A piece of pure gold has a mass of 145 g. What is the mass of the same number of silver atoms?

**8-4.** A large chunk of pure aluminum has a mass of 212 lb. What is the mass of the same number of carbon atoms?

**8-5.** A piece of copper wire has a mass of 16.0 g; the same number of atoms of a precious metal has a mass of 49.1 g. What is the metal?

**8-6.** In the compound CuO, what mass of copper is present for each 18.0 g of oxygen?

**8-7.** In the compound NaCl, what mass of sodium is present for each 425 g of chlorine?

**8-8.** In a compound containing one atom of carbon and one atom of another element, it is found that 25.0 g of carbon is combined with 33.3 g of the other element. What is the element? What is the formula of the compound?

*__*8-9.__ In the compound $MgBr_2$, what mass of bromine is present for each 46.0 g of magnesium? (Remember, there are two bromines per magnesium.)

*__*8-10.__ In the compound $SO_3$, what mass of sulfur is present for each 60.0 lb of oxygen?

## The Magnitude of the Mole

**8-11.** If you could count two numbers per second, how many years would it take you to count Avogadro's number? If you were helped by the whole human race of 6.0 billion people, how long would it take?

**8-12.** A small can of soda contains 350 mL. Planet Earth contains 326 million cubic miles ($3.5 \times 10^{20}$ gal) of water. How many cans of soda would it take to equal all of the water on Earth? How many moles of cans is this?

**8-13.** What would the number in one mole be if the atomic mass were expressed in kilograms rather than grams? In milligrams?

## Moles of Atoms of Elements

**8-14.** Fill in the blanks. Use the factor-label method to determine the answers.

| Element | Mass in Grams | Number of Moles of Atoms | Number of Atoms |
|---|---|---|---|
| S | 8.00 | 0.250 | $1.50 \times 10^{23}$ |
| (a) P | 14.5 | _____ | _____ |
| (b) Rb | _____ | 1.75 | _____ |
| (c) Al | _____ | _____ | $6.02 \times 10^{23}$ |
| (d) _____ | 363 | _____ | $3.01 \times 10^{24}$ |
| (e) Ti | _____ | _____ | 1 |

**8-15.** Fill in the blanks. Use the factor-label method to determine the answers.

| Element | Mass in Grams | Number of Moles of Atoms | Number of Atoms |
|---|---|---|---|
| (a) Na | 0.390 | _____ | _____ |
| (b) Cr | _____ | $3.01 \times 10^4$ | _____ |
| (c) _____ | 43.2 | _____ | $2.41 \times 10^{24}$ |
| (d) K | $4.25 \times 10^{-6}$ | _____ | _____ |
| (e) Ne | _____ | _____ | $3.66 \times 10^{21}$ |

**8-16.** What is the mass in grams of each of the following?
**(a)** 1.00 mol of Cu **(c)** $6.02 \times 10^{23}$ atoms of Ca
**(b)** 0.50 mol of S

**8-17.** What is the mass in grams of each of the following?
**(a)** 4.55 mol of Be
**(b)** $6.02 \times 10^{24}$ atoms of Ca
**(c)** $3.40 \times 10^5$ mol of B

**8-18.** How many individual atoms are in each of the following?
**(a)** 32.1 mol of sulfur **(c)** 32.0 g of oxygen
**(b)** 32.1 g of sulfur

**8-19.** How many moles of atoms are in each of the following?
**(a)** 281 g of Si
**(b)** $7.34 \times 10^{25}$ atoms of phosphorus
**(c)** 19.0 atoms of fluorine

**8-20.** Which has more atoms, 50.0 g of Al or 50.0 g of Fe?

**8-21.** Which contains the most Ni: 20.0 g, $2.85 \times 10^{23}$ atoms, or 0.450 mol?

**8-22.** Which contains the most Cr: 0.025 mol, $6.0 \times 10^{21}$ atoms, or 1.5 g of Cr?

**8-23.** A 0.251-g sample of a certain element is the mass of $1.40 \times 10^{21}$ atoms. What is the element?

**8-24.** A 0.250-mol sample of a certain element has a mass of 49.3 g. What is the element?

## Formula Weight

**8-25.** What is the formula weight of each of the following? Express your answer to four significant figures.
**(a)** $KClO_2$   **(d)** $H_2SO_4$   **(f)** $CH_3COOH$
**(b)** $SO_3$   **(e)** $Na_2CO_3$   **(g)** $Fe_2(CrO_4)_3$
**(c)** $N_2O_5$

**8-26.** What is the formula weight of each of the following?
**(a)** $CuSO_4 \cdot 6H_2O$ (Include $H_2O$'s; they are included in one formula unit.)
**(b)** $Cl_2O_3$   **(c)** $Al_2(C_2O_4)_3$   **(d)** $Na_2BH_3CO_2$   **(e)** $P_4O_6$

**8-27.** A compound is composed of three sulfate ions and two chromium ions. What is the formula weight of the compound?

**8-28.** What is the formula weight of strontium perchlorate?

## Moles of Compounds

**8-29.** Fill in the blanks. Use the factor-label method to determine the answers.

| Molecules | Mass in Grams | Number of Moles | Number of Molecules or Formula Units |
|---|---|---|---|
| $N_2O$ | 23.8 | 0.542 | $3.26 \times 10^{23}$ |
| (a) $H_2O$ | _____ | 10.5 | _____ |
| (b) $BF_3$ | _____ | _____ | $3.01 \times 10^{21}$ |
| (c) $SO_2$ | 14.0 | _____ | _____ |
| (d) $K_2SO_4$ | _____ | $1.20 \times 10^{-4}$ | _____ |
| (e) $SO_3$ | _____ | _____ | $4.50 \times 10^{24}$ |
| (f) $N(CH_3)_3$ | 0.450 | _____ | _____ |

**8-30.** Fill in the blanks. Use the factor-label method to determine answers.

| Molecules | Mass in Grams | Number of Moles | Number of Molecules or Formula Units |
|---|---|---|---|
| (a) $O_3$ | 176 | _____ | _____ |
| (b) $NO_2$ | _____ | $3.75 \times 10^{-3}$ | _____ |
| (c) $Cl_2O_3$ | _____ | _____ | 150 |
| (d) $UF_6$ | _____ | _____ | $8.50 \times 10^{22}$ |

**8-31.** Tetrahydrocannabinol (THC) is the active ingredient in marijuana. A 0.0684-mol quantity of THC has a mass of 21.5 g. What is the molar mass of THC?

**8-32.** A 0.158-mol quantity of a compound has a mass of 7.28 g. What is the molar mass of the compound?

**8-33.** A 287-g quantity of a compound is the mass of $1.07 \times 10^{24}$ molecules. What is the molar mass of the compound?

**8-34.** A 0.0681-g quantity of ethylene glycol (antifreeze) represents the mass of $7.88 \times 10^{20}$ molecules. What is the molar mass of the compound?

## The Composition of Compounds

**8-35.** How many moles of each type of atom are in 2.55 mol of grain alcohol, $C_2H_6O$? What is the total number of moles of atoms? What is the mass of each element? What is the total mass?

**8-36.** How many moles are in 28.0 g of $Ca(ClO_3)_2$? How many moles of each element are present? How many total moles of atoms are present?

**8-37.** How many moles are in 84.0 g of $K_2Cr_2O_7$? How many moles of each element are present? How many total moles of atoms are present?

**8-38.** What mass of each element is in 1.50 mol of $H_2SO_3$?

**8-39.** What mass of each element is in 2.45 mol of boric acid $(H_3BO_3)$?

**8-40.** How many moles of $O_2$ are in $1.20 \times 10^{22}$ $O_2$ molecules? How many moles of oxygen atoms? What is the mass of oxygen molecules? What is the mass of oxygen atoms?

**8-41.** How many moles of $Cl_2$ molecules are in 985 g of $Cl_2$? How many moles of Cl atoms are present? What is the mass of Cl atoms present?

**8-42.** What is the percent composition of a compound composed of 1.375 g of N and 3.935 g of O?

**8-43.** A sample of a compound has a mass of 4.86 g and is composed of silicon and oxygen. What is the percent composition if 2.27 g of the mass is silicon?

**8-44.** The mass of a sample of a compound is 7.44 g. Of that mass, 2.88 g is potassium, 1.03 g is nitrogen, and the remainder is oxygen. What is the percent composition of the elements?

**8-45.** What is the percent composition of all of the elements in the following compounds?
**(a)** $C_2H_6O$    **(c)** $C_9H_{18}$    **(e)** $(NH_4)_2CO_3$
**(b)** $C_3H_6$    **(d)** $Na_2SO_4$

**8-46.** What is the percent composition of all of the elements in the following compounds?
**(a)** $H_2CO_3$    **(b)** $Cl_2O_7$    **(c)** $Al(NO_3)_3$    **(d)** $NH_4H_2PO_4$

**8-47.** What is the percent composition of all of the elements in borax ($Na_2B_4O_7 \cdot 10H_2O$)?

**8-48.** What is the percent composition of all of the elements in acetaminophen ($C_8H_9O_2N$)? (Acetaminophen is an aspirin substitute.)

**8-49.** What is the percent composition of all of the elements in saccharin ($C_7H_5SNO_3$)?

**8-50.** What is the percent composition of all of the elements in amphetamine ($C_9H_{13}N$)? (Amphetamine is a stimulant.)

**\*8-51.** What mass of carbon is in a 125-g quantity of sodium oxalate?

**\*8-52.** What is the mass of phosphorus in a 25.0-lb quantity of sodium phosphate?

**\*8-53.** What mass of chromium is in a 275-kg quantity of chromium(III) carbonate?

**\*8-54.** Iron is recovered from iron ore, $Fe_2O_3$. How many pounds of iron can be recovered from each ton of iron ore? (1 ton = 2000 lb)

**Empirical Formulas**

**8-55.** Which of the following are not empirical formulas?
**(a)** $N_2O_4$    **(c)** $H_2S_2O_3$    **(e)** $Mn_2O_7$
**(b)** $Cr_2O_3$    **(d)** $H_2C_2O_4$

**8-56.** Convert the following mole ratios of elements to empirical formulas.
**(a)** 0.25 mol of Fe and 0.25 mol of S
**(b)** 1.88 mol of Sr and 3.76 mol of I
**(c)** 0.32 mol of K, 0.32 mol of Cl, and 0.96 mol of O
**(d)** 1.0 mol of I and 2.5 mol of O
**(e)** 2.0 mol of Fe and 2.66 mol of O
**(f)** 4.22 mol of C, 7.03 mol of H, and 4.22 mol of Cl

**8-57.** Convert the following mole ratios of elements to empirical formulas.
**(a)** 1.20 mol of Si and 2.40 mol of O
**(b)** 0.045 mol of C and 0.022 mol of S
**(c)** 1.00 mol of X and 1.20 mol of Y
**(d)** 3.11 mol of Fe, 4.66 mol of C, and 14.0 mol of O

**8-58.** What is the empirical formula of a compound that has the composition 63.2% oxygen and 36.8% nitrogen?

**8-59.** What is the empirical formula of a compound that has the composition 41.0% K, 33.7% S, and 25.3% O?

**8-60.** In an experiment it was found that 8.25 g of potassium combines with 6.75 g of $O_2$. What is the empirical formula of the compound?

**8-61.** Orlon is composed of very long molecules with a composition of 26.4% N, 5.66% H, and 67.9% C. What is the empirical formula of Orlon?

**8-62.** A compound is 21.6% Mg, 21.4% C, and 57.0% O. What is the empirical formula of the compound?

**8-63.** A compound is composed of 9.90 g of carbon, 1.65 g of hydrogen, and 29.3 g of chlorine. What is the empirical formula of the compound?

**8-64.** A compound is composed of 0.46 g of Na, 0.52 g of Cr, and 0.64 g of O. What is the empirical formula of the compound?

**\*8-65.** A compound is composed of 24.1% nitrogen, 6.90% hydrogen, 27.6% sulfur, and the remainder oxygen. What is the empirical formula of the compound?

**\*8-66.** Methyl salicylate is also known as oil of wintergreen. It is composed of 63.2% carbon, 31.6% oxygen, and 5.26% hydrogen. What is its empirical formula?

**\*8-67.** Nitroglycerin is used as an explosive and a heart medicine. It is composed of 15.9% carbon, 18.5% nitrogen, 63.4% oxygen, and 2.20% hydrogen. What is its empirical formula?

**Molecular Formulas**

**8-68.** A compound has the following composition: 20.0% C, 2.2% H, and 77.8% Cl. The molar mass of the compound is 545 g/mol. What is the molecular formula of the compound?

**8-69.** A compound is composed of 1.65 g of nitrogen and 3.78 g of sulfur. If its molar mass is 184 g/mol, what is its molecular formula?

**8-70.** A compound has a composition of 18.7% B, 20.7% C, 5.15% H, and 55.4% O. Its molar mass is about 115 g/mol. What is the molecular formula of the compound?

**8-71.** A compound reported in 1970 has a composition of 34.9% K, 21.4% C, 12.5% N, 2.68% H, and 28.6% O. It has a molar mass of about 224 g/mol. What is the molecular formula of the compound?

**8-72.** Fructose is also known as fruit sugar. It has a molar mass of 180 g/mol and is composed of 40.0% carbon, 53.3% oxygen, and 6.7% hydrogen. What is the molecular formula of the compound?

**8-73.** A compound reported in 1982 has a molar mass of 834 g/mol. A 20.0-g sample of the compound contains 18.3 g of iodine and the remainder carbon. What is the molecular formula of the compound?

**8-74.** Quinine is a compound discovered in the bark of certain trees. It is an effective drug in the treatment of malaria. Its molar mass is 162 g/mol. It is 22.2% carbon, 22.2% hydrogen, 25.9% nitrogen, and 29.6% oxygen. What is the molecular formula of quinine?

## Stoichiometry

**8-75.** Given the balanced equation

$$Mg + 2HCl \longrightarrow MgCl_2 + H_2$$

provide the proper mole ratio for each of the following mole conversions.
(a) Mg to $H_2$    (c) HCl to $H_2$
(b) Mg to HCl    (d) $MgCl_2$ to HCl

**8-76.** Given the balanced equation

$$2C_4H_{10} + 13O_2 \longrightarrow 8CO_2 + 10H_2O$$

provide the proper mole ratio for each of the following mole conversions.
(a) $CO_2$ to $C_4H_{10}$    (c) $CO_2$ to $O_2$
(b) $O_2$ to $C_4H_{10}$    (d) $O_2$ to $H_2O$

**8-77.** Given the balanced equation

$$Cu + 4HNO_3 \longrightarrow Cu(NO_3)_2 + 2NO_2 + 2H_2O$$

provide the proper mole ratio for each of the following mole conversions.
(a) Cu to $NO_2$    (d) $Cu(NO_3)_2$ to $HNO_3$
(b) $HNO_3$ to Cu    (e) $NO_2$ to Cu
(c) $H_2O$ to $NO_2$    (f) Cu to $H_2O$

**8-78.** The reaction that takes place in the reusable solid rocket booster for the space shuttle is shown by the following equation.

$$3Al(s) + 3NH_4ClO_4(s) \longrightarrow$$
$$Al_2O_3(s) + AlCl_3(s) + 3NO(g) + 6H_2O(g)$$

(a) How many moles of each product are formed from 10.0 moles of Al?
(b) How many moles of each product are formed from 3.00 moles of $NH_4ClO_4$?

**8-79.** Phosphine ($PH_3$) is a poisonous gas once used as a fumigant for stored grain. It is prepared according to the following equation.

$$Ca_3P_2(s) + 6H_2O(l) \longrightarrow 3Ca(OH)_2(s) + 2PH_3(g)$$

(a) How many moles of phosphine are prepared from 5.00 moles of $Ca_3P_2$?
(b) How many moles of phosphine are prepared from 5.00 moles of $H_2O$?

**8-80.** Hydrogen cyanide is an important industrial chemical used to make a plastic, acrylonitrile. HCN is prepared according to the following equation.

$$2NH_3(g) + 3O_2(g) + 2CH_4(g) \longrightarrow 2HCN(g) + 6H_2O(l)$$

(a) How many moles of $O_2$ and $CH_4$ react with 10.0 moles of $NH_3$?
(b) How many moles of HCN and $H_2O$ are produced from 10.0 moles of $O_2$?

**8-81.** Iron rusts according to the equation

$$4Fe(s) + 3O_2(g) \longrightarrow 2Fe_2O_3(s)$$

(a) What mass of rust ($Fe_2O_3$) is formed from 0.275 mole of Fe?
(b) What mass of rust is formed from 0.275 mole of $O_2$?
(c) What mass of $O_2$ reacts with 0.275 mole of Fe?

**8-82.** Glass ($SiO_2$) is etched with hydrofluoric acid according to the equation

$$SiO_2(s) + 4HF(aq) \longrightarrow SiF_4(g) + 2H_2O(l)$$

If 4.86 moles of HF reacts with $SiO_2$,
(a) What mass of $SiF_4$ forms?
(b) What mass of $H_2O$ forms?
(c) What mass of $SiO_2$ reacts?

**8-83.** Consider the reaction

$$2H_2 + O_2 \longrightarrow 2H_2O$$

(a) How many moles of $H_2$ are needed to produce 0.400 mole of $H_2O$?
(b) How many moles of $H_2O$ will be produced from 0.640 g of $O_2$?
(c) How many moles of $H_2$ are needed to react with 0.032 g of $O_2$?
(d) What mass of $H_2O$ would be produced from 0.400 g of $H_2$?

**8-84.** Propane burns according to the equation

$$C_3H_8 + 5O_2 \longrightarrow 3CO_2 + 4H_2O$$

(a) How many moles of $CO_2$ are produced from the combustion of 0.450 mole of $C_3H_8$? How many moles of $H_2O$? How many moles of $O_2$ are needed?
(b) What mass of $H_2O$ is produced if 0.200 mol of $CO_2$ is also produced?
(c) What mass of $C_3H_8$ is required to produce 1.80 g of $H_2O$?
(d) What mass of $C_3H_8$ is required to react with 160 g of $O_2$?
(e) What mass of $CO_2$ is produced by the reaction of $1.20 \times 10^{23}$ molecules of $O_2$?
(f) How many moles of $H_2O$ are produced if $4.50 \times 10^{22}$ molecules of $CO_2$ are produced?

**8-85.** The alcohol component of gasohol burns according to the equation

$$C_2H_5OH(l) + 3O_2(g) \longrightarrow 2CO_2(g) + 3H_2O(g)$$

(a) What mass of alcohol is needed to produce 5.45 mol of $H_2O$?
(b) How many moles of $CO_2$ are produced along with 155 g of $H_2O$?
(c) What mass of $CO_2$ is produced from 146 g of $C_2H_5OH$?
(d) What mass of $C_2H_5OH$ reacts with 0.898 g of $O_2$?
(e) What mass of $H_2O$ is produced from $5.85 \times 10^{24}$ molecules of $O_2$?

**8-86.** In the atmosphere, $N_2$ and $O_2$ do not react with each other. In the high temperatures of an automobile engine, however, the following reaction occurs.

$$N_2(g) + O_2(g) \longrightarrow 2NO(g)$$

When the NO reaches the atmosphere through the engine exhaust, a second reaction takes place.

$$2NO(g) + O_2(g) \longrightarrow 2NO_2(g)$$

The $NO_2$ is a brownish gas that contributes to the haze of smog and is irritating to the nasal passages and lungs. What mass of $N_2$ is required to produce 155 g of $NO_2$?

**8-87.** Elemental iron is produced in what is called the *thermite reaction* because it produces enough heat that the iron is initially in the molten state. The liquid iron can then be used to weld iron bars together.

$$2Al(s) + Fe_2O_3(s) \longrightarrow Al_2O_3(s) + 2Fe(l)$$

What mass of Al is needed to produce 750 g of Fe? How many formula units of $Fe_2O_3$ are used in the process?

**8-88.** Antacids, which contain calcium carbonate, react with stomach acid according to the equation

$$CaCO_3(s) + 2HCl(aq) \longrightarrow CaCl_2(aq) + CO_2(g) + H_2O(l)$$

What mass of stomach acid reacts with 1.00 g of $CaCO_3$?

**8-89.** Elemental copper can be recovered from the mineral chalcocite, $Cu_2S$. From the following equation, determine what mass of Cu is formed from $7.82 \times 10^{22}$ molecules of $O_2$.

$$Cu_2S(s) + O_2(g) \longrightarrow 2Cu(s) + SO_2(g)$$

**8-90.** Fool's gold (pyrite) is so named because it looks much like gold. When it is placed in aqueous HCl, however, it dissolves and gives off the pungent gas $H_2S$. Gold itself does not react with aqueous HCl. From the following equation, determine how many individual molecules of $H_2S$ are formed from 0.520 mol of pyrite ($FeS_2$).

$$FeS_2(s) + 2HCl(aq) \longrightarrow FeCl_2(aq) + H_2S(g) + S(s)$$

**8-91.** Nitrogen dioxide may form so-called acid rain by reaction with water in the air according to the equation

$$3NO_2(g) + H_2O(l) \longrightarrow 2HNO_3(aq) + NO(g)$$

What mass of nitric acid is produced from 18.5 kg of $NO_2$?

**8-92.** Elemental chlorine can be generated in the laboratory according to the equation

$$MnO_2(s) + 4HCl(aq) \longrightarrow MnCl_2(aq) + 2H_2O(l) + Cl_2(g)$$

What mass of $Cl_2$ is produced from the reaction of 665 g of HCl?

**8-93.** The fermentation of sugar to produce ethyl alcohol is represented by the equation

$$C_6H_{12}O_6(s) \longrightarrow 2C_2H_5OH(l) + 2CO_2(g)$$

What mass of alcohol is produced from 25.0 mol of sugar?

**8-94.** Methane gas can be made from carbon monoxide gas according to the equation

$$2CO(g) + 2H_2(g) \longrightarrow CH_4(g) + CO_2(g)$$

What mass of CO is required to produce $8.75 \times 10^{25}$ molecules of $CH_4$?

**Limiting Reactant**

**8-95.** Nitrogen gas can be prepared by passing ammonia over hot copper(II) oxide according to the equation

$$3CuO(s) + 2NH_3(g) \longrightarrow N_2(g) + 3Cu(s) + 3H_2O(g)$$

How many moles of $N_2$ are prepared from the following mixtures?
**(a)** 3.00 mol of CuO and 3.00 mol of $NH_3$
**(b)** 3.00 mol of CuO and 2.00 mol of $NH_3$
**(c)** 3.00 mol of CuO and 1.00 mol of $NH_3$
**(d)** 0.628 mol of CuO and 0.430 mol of $NH_3$
**(e)** 5.44 mol of CuO and 3.50 mol of $NH_3$

**8-96.** How many moles remain of the reactant in excess in exercise 8-95 (a) and (c)?

**8-97.** Consider the equation illustrating the combustion of arsenic.

$$4As(s) + 5O_2(g) \longrightarrow 2As_2O_5(s)$$

How many moles of $As_2O_5$ are prepared from the following mixtures?
**(a)** 4.00 mol of As and 4.00 mol of $O_2$
**(b)** 3.00 mol of As and 4.00 mol of $O_2$
**(c)** 5.62 mol of As and 7.50 mol of $O_2$
**(d)** 3.86 mol of As and 4.75 mol of $O_2$

**8-98.** How many moles remain of the reactant in excess in exercise 8-97 (a) and (b)?

**8-99.** Consider the equation

$$2Al + 3H_2SO_4 \longrightarrow Al_2(SO_4)_3 + 3H_2$$

If 0.800 mol of Al is mixed with 1.00 mol of $H_2SO_4$, how many moles of $H_2$ are produced? How many moles of one of the reactants remain?

**8-100.** Consider the equation

$$2C_5H_6 + 13O_2 \longrightarrow 10CO_2 + 6H_2O$$

If 3.44 mol of $C_5H_6$ is mixed with 20.6 mol of $O_2$, what mass of $CO_2$ is formed?

**8-101.** Elemental fluorine is very difficult to prepare by ordinary chemical reactions. In 1986, however, a chemical reaction was reported that produces fluorine.

$$2K_2MnF_6 + 4SbF_5 \longrightarrow 4KSbF_6 + 2MnF_3 + F_2$$

If a 525-g quantity of $K_2MnF_6$ is mixed with a 900-g quantity of $SbF_5$, what mass of $F_2$ is produced?

**8-102.** Consider the equation

$$4HN_3(g) + 3O_2(g) \longrightarrow 2N_2(g) + 6H_2O(l)$$

If a 40.0-g sample of $O_2$ is mixed with 1.50 mol of $NH_3$, which is the limiting reactant? How many moles of $N_2$ form?

**8-103.** Consider the equation

$$2AgNO_3(aq) + CaCl_2(aq) \longrightarrow 2AgCl(s) + Ca(NO_3)_2(aq)$$

If a solution containing 20.0 g of $AgNO_3$ is mixed with a solution containing 10.0 g of $CaCl_2$, which compound is the limiting reactant? What mass of AgCl forms? What mass of one of the reactants remains?

**8-104.** Limestone ($CaCO_3$) dissolves in hydrochloric acid as shown by the equation

$$CaCO_3(s) + 2HCl(aq) \longrightarrow CaCl_2(aq) + CO_2(g) + H_2O(l)$$

If 20.0 g of $CaCO_3$ and 25.0 g of HCl are mixed, what mass of $CO_2$ is produced? What mass of one of the reactants remains?

**8-105.** Consider the balanced equation

$$2HNO_3(aq) + 3H_2S(aq) \longrightarrow 2NO(g) + 4H_2O(l) + 3S(s)$$

If a 10.0-g quantity of $HNO_3$ is mixed with 5.00 g of $H_2S$, what are the masses of each product and of the reactant present in excess after reaction occurs?

### Theoretical and Percent Yield

**8-106.** Sulfur trioxide is prepared from $SO_2$ according to the equation

$$2SO_2(g) + O_2(g) \longrightarrow 2SO_3(g)$$

This is a reversible reaction where not all $SO_2$ is converted to $SO_3$ even with excess $O_2$ present. In a given experiment, 21.2 g of $SO_3$ was produced from 24.0 g of $SO_2$. What is the theoretical yield of $SO_3$? What is the percent yield?

**8-107.** The following is a reversible decomposition reaction.

$$2N_2O_5 \longrightarrow 4NO_2 + O_2$$

When 25.0 g of $N_2O_5$ decomposes, it is found that 10.0 g of $NO_2$ forms. What is the percent yield?

**8-108.** The following equation represents a reversible combination reaction.

$$P_4O_{10} + 6PCl_5 \longrightarrow 10POCl_3$$

If 25.0 g of $PCl_5$ reacts, there is a 45.0% yield of $POCl_3$. What is the actual yield in grams?

**8-109.** Octane in gasoline burns in an automobile engine according to the equation

$$2C_8H_{18}(l) + 25O_2(g) \longrightarrow 16CO_2(g) + 18H_2O(g)$$

If a 57.0-g sample of octane is burned, 152 g of $CO_2$ is formed. What is the percent yield of $CO_2$?

**\*8-110.** In exercise 8-109, the $C_8H_{18}$ that is *not* converted to $CO_2$ forms CO. What is the mass of CO formed? (CO is a poisonous pollutant that is converted to $CO_2$ in a car's catalytic converter.)

**\*8-111.** Given the reversible reaction

$$2NO_2 + 4H_2 \longrightarrow N_2 + 4H_2O$$

what mass of hydrogen is required to produce 250 g of $N_2$ if the yield is 70.0%?

**\*8-112.** When benzene reacts with bromine, the principal reaction is

$$C_6H_6 + Br_2 \longrightarrow C_6H_5Br + HBr$$

If the yield of bromobenzene ($C_6H_5Br$) is 65.2%, what mass of bromobenzene is produced from 12.5 g of $C_6H_6$?

**\*8-113.** A second reaction between $C_6H_6$ and $Br_2$ (see exercise 8-112) produces dibromobenzene ($C_6H_4Br_2$).
**(a)** Write the balanced equation illustrating this reaction.
**(b)** If the remainder of the benzene from exercise 8-112 reacts to form dibromobenzene, what is the mass of $C_6H_4Br_2$ produced?

### Heat in Chemical Reactions

**8-114.** When one mole of magnesium undergoes combustion to form magnesium oxide, 602 kJ of heat energy is evolved. Write the thermochemical equation in both forms.

**8-115.** In exercise 8-86, the reaction between $N_2$ and $O_2$ was discussed. A 90.5-kJ quantity of heat energy must be supplied per mole of NO formed. Write the balanced thermochemical equation in both forms. Is the reaction exothermic or endothermic?

**8-116.** To decompose one mole of $CaCO_3$ to CaO and $CO_2$, 176 kJ must be supplied. Write the balanced thermochemical equation in both forms.

**8-117.** The complete combustion of one mole of octane ($C_8H_{18}$) in gasoline evolves 5480 kJ of heat. The complete combustion of one mole of methane in natural gas ($CH_4$) evolves 890 kJ. How much heat is evolved per 1.00 g by each of these fuels?

**8-118.** Methyl alcohol ($CH_3OH$) is used as a fuel in racing cars. It burns according to the equation

$$2CH_3OH(l) + 3O_2(g) \longrightarrow 2CO_2(g) + 4H_2O(l) + 1750 \text{ kJ}$$

What amount of heat is evolved per 1.00 g of alcohol? How does this compare with the amount of heat per gram of octane in gasoline? (See exercise 8-117.)

**8-119.** The thermite reaction was discussed in exercise 8-87. For the balanced equation, $\Delta H = -850$ kJ. What mass of aluminum is needed to produce 35.8 kJ of heat energy?

**8-120.** Photosynthesis is an endothermic reaction that forms glucose ($C_6H_{12}O_6$) from carbon dioxide, water, and energy from the sun. The balanced equation is

$$6CO_2(g) + 6H_2O(l) \longrightarrow$$
$$C_6H_{12}O_6(aq) + 6O_2(g) \qquad \Delta H = +2519 \text{ kJ}$$

What mass of glucose is formed from 975 kJ of energy?

**8-121.** When butane ($C_4H_{10}$) in a cigarette lighter burns, it evolves 2880 kJ per mole of butane. What is the mass of water formed if 1250 kJ of heat evolves?

### General Problems

**8-122.** The U.S. national debt was about $4.5 trillion in 1994. How many moles of pennies would it take to pay it off?

**8-123.** A compound has the formula MN, where M represents a certain unknown metal. A sample of the compound weighs 1.862 g, and of that, 0.443 g is nitrogen. What is the identity of the metal, M?

**8-124.** What would be the number of particles in one mole if the atomic mass were expressed in ounces rather than grams? (28.375 g = 1 oz.)

**8-125.** A certain alloy of copper has a density of 3.75 g/mL and is 65.0% by mass copper. How many copper atoms are in 16.8 cm$^3$ of this alloy?

**8-126.** The element phosphorus exists as $P_4$. How many moles of molecules are in 0.344 g of phosphorus? How many phosphorus atoms are present in that amount of phosphorus?

**8-127.** Rank the following in order of increasing mass.
**(a)** 100 hydrogen atoms
**(b)** 100 moles of hydrogen molecules
**(c)** 100 grams of hydrogen
**(d)** 100 hydrogen molecules

**8-128.** Rank the following in order of increasing number of atoms.
**(a)** 100 lead atoms
**(b)** 100 moles of helium
**(c)** 100 grams of lead
**(d)** 100 grams of helium

**8-129.** Pyrite, a mineral containing iron, has the formula $FeS_x$. A quantity of $2.84 \times 10^{23}$ formula units of pyrite has a mass of 56.6 g. What is the value of $x$?

**8-130.** A compound has the formula $Na_2S_4O_6$.
**(a)** What ions are present in the compound? (The anion is all one species.)
**(b)** How many grams of sulfur are present in the compound for each 10.0 g of Na?
**(c)** What is the empirical formula of the compound?
**(d)** What is the formula weight of the compound?
**(e)** How many moles and formula units are present in 25.0 g of the compound?
**(f)** What is the percent composition of oxygen in the compound?

**8-131.** Glucose (blood sugar) has the formula $C_6H_{12}O_6$. Calculate how many moles of carbon, individual hydrogen atoms, and grams of oxygen are in a 10.0-g sample of glucose.

**8-132.** A compound has the formula $N_2O_x$. It is 36.8% nitrogen. What is the name of the compound?

**8-133.** A compound has the formula $SF_x$. One sample of the compound contains 0.356 mol of sulfur and $8.57 \times 10^{23}$ atoms of fluorine. What is the name of the compound?

**8-134.** Dioxin is a compound that is known to cause cancer in certain laboratory animals. $4.55 \times 10^{22}$ molecules of this compound have a mass of 24.3 g. Analysis of a sample of dioxin indicated the sample contained 0.456 mol of carbon, 0.152 mol of hydrogen, 0.152 mol of chlorine, and 0.076 mol of oxygen. What is the molecular formula of dioxin?

**8-135.** A certain compound has a molar mass of 166 g/mol. It is composed of 47.1% potassium, 14.4% carbon, and the remainder oxygen. What is the name of the compound?

**8-136.** A compound has the general formula $Cr(ClO_x)_y$ and is 14.9% Cr, 30.4% chlorine, and the remainder oxygen. What is the name of the compound?

**8-137.** Epsom salts has the formula $MgSO_4 \cdot xH_2O$. Calculate the value of $x$ if the compound it 51.1% by mass water.

**8-138.** Potassium carbonate forms a hydrate, $K_2CO_3 \cdot xH_2O$. What is the value of $x$ if the compound is 20.7% by mass water?

**8-139.** Nicotine is a compound containing carbon, hydrogen, and nitrogen. Its molar mass is 162 g/mol. A 1.50-g sample of nicotine is found to contain 1.11 g of carbon. Analysis of another sample indicates that nicotine is 8.70% by mass hydrogen. What is the molecular formula of nicotine?

**8-140.** A hydrocarbon (a compound that contains only carbon and hydrogen) was burned, and the products of the combustion were collected and weighed. All of the carbon in the original compound is now present in 1.20 g of $CO_2$. All of the hydrogen is present in 0.489 g of $H_2O$. What is the empirical formula of the compound? (*Hint:* Remember that all of the moles of C atoms in $CO_2$ and H atoms in $H_2O$ came from the original compound.)

**8-141.** A 0.500-g sample of a compound containing C, H, and O was burned, and the products were collected. The combustion produced 0.733 g of $CO_2$ and 0.302 g of $H_2O$. The molar mass of the compound is 60.0 g/mol. What is the molecular formula of the compound? (*Hint:* Find the mass of C and H in the original compound; the remainder of the 0.500 g is oxygen.)

**8-142.** Calcium cyanamide ($CaCN_2$) is used as a fertilizer. When it reacts with water, it produces ammonia (which fertilizes the soil) and $CaCO_3$ (which counteracts acidity in the soil). Write the balanced equation illustrating the reaction, and calculate the mass of $NH_3$ produced from a 1.00-kg quantity of $CaCN_2$.

**\*8-143.** Liquid iron is made from iron ore ($Fe_2O_3$) in a three-step process in a blast furnace as follows.
**1.** $3Fe_2O_3(s) + CO(g) \longrightarrow 2Fe_3O_4(s) + CO_2(g)$
**2.** $Fe_3O_4(s) + CO(g) \longrightarrow 3FeO(s) + CO_2(g)$
**3.** $FeO(s) + CO(g) \longrightarrow Fe(l) + CO_2(g)$
What mass of iron would eventually be produced from 125 g of $Fe_2O_3$?

**8-144.** A 50.0-g sample of *impure* $KClO_3$ is decomposed to KCl and $O_2$. If a 12.0-g quantity of $O_2$ is produced, what percent of the sample is $KClO_3$? (Assume that all of the $KClO_3$ present decomposes.)

**8-145.** In Example 8-21, $SO_2$ was formed from the burning of pyrite ($FeS_2$) in coal. If a 312-g quantity of $SO_2$ was collected from the burning of 6.50 kg of coal, what percent of the original sample was pyrite?

**\*8-146.** Copper metal can be recovered from an ore, $CuCO_3$, by the decomposition reaction

$$2CuCO_3(s) \longrightarrow 2Cu(s) + 2CO_2(g) + O_2(g)$$

What is the mass of a sample of *impure* ore if it is 47.5% $CuCO_3$ and produces 350 g of Cu? (Assume complete decomposition of $CuCO_3$.)

**\*8-147.** Consider the equation

$$4NH_3(g) + 5O_2(g) \longrightarrow 4NO(g) + 6H_2O(l)$$

When an 80.0-g quantity of $NH_3$ is mixed with 200 g of $O_2$, a 40.0-g quantity of NO is formed. Calculate the percent yield based on the limiting reactant.

**\*8-148.** Consider the equation

$$3K_2MnO_4 + 4CO_2 + 2H_2O \longrightarrow$$
$$2KMnO_4 + 4KHCO_3 + MnO_2$$

How many moles of $MnO_2$ are produced if 9.50 mol of $K_2MnO_4$, $6.02 \times 10^{24}$ molecules of $CO_2$, and 90.0 g of $H_2O$ are mixed?

**\*8-149.** Calcium chloride hydrate ($CaCl_2 \cdot 6H_2O$) is a solid used to melt ice at low temperatures. It is prepared according to the equation

$$CaCO_3(s) + 2HCl(g) + 5H_2O(l) \longrightarrow$$
$$CaCl_2 \cdot 6H_2O(s) + CO_2(g)$$

What mass of the hydrate is prepared from a mixture of 0.250 mole of $H_2O$, $9.50 \times 10^{22}$ molecules of HCl, and 15.0 g of $CaCO_3$?

**\*8-150.** A 2.85-g quantity of gaseous methane is mixed with 15.0 g of chlorine to produce a liquid, compound X, and gaseous hydrogen chloride. Compound X is 14.1% C, 2.35% H, and 83.5% Cl. Its molecular formula is the same as its empirical formula.
**(a)** From the formula of X, write a balanced equation. (*Hint:* Balance hydrogens before chlorines.)
**(b)** What is the theoretical yield of compound X?

**\*8-151.** A 10.00-g sample of gaseous ethane ($C_2H_6$) reacts with chlorine to form gaseous hydrogen chloride and a liquid compound (Y) that has a molar mass of 168 g/mol. Compound

Y is 14.3% carbon, 84.5% chlorine, and the remainder hydrogen. The reaction produces 12.0 g of compound Y, which is 57.0% Yield.
**(a)** Write the balanced equation illustrating the reaction.
**(b)** What mass of ethane was required?

**\*8-152.** The remainder of the ethane from exercise 8-151 reacts to form a liquid compound (Z) that is 18.0% carbon, 79.8% chlorine, and 2.25% hydrogen. The empirical and molecular formulas of this compound are the same.
**(a)** Write the balanced equation illustrating the reaction.
**(b)** What mass of compound Z is formed?

**\*8-153.** If you have already covered Chapter 6, consider the following problem. A compound is 27.4% Na, 14.3% C, 57.1% O, and 1.19% H. What is the formula of the compound? Write the Lewis structure of the anion (H bonded to O). What is the geometry around the C? What is the approximate O—C—O bond angle? What is the geometry around the H—C—O bond and the approximate angle?

**\*8-154.** If you have already covered Chapter 6, consider the following problem. A compound is 21.7% Li, 75.0% C, and 3.16% H. What is the formula of the compound? Write the Lewis structure of the anion. What is the geometry around the central C and the approximate H—C—C bond angle?

**\*8-155.** If you have already covered Chapters 6, consider the following problem. A compound is 19.8% Ca, 1.00% H, 31.7% S, and 47.5% O. What is the formula of the compound? (*Hint:* The formula of the anion should be reduced to its empirical formula.) What is the name of the compound? (Refer to Table 4-2.) What is the geometry around the S in the anion? What is the approximate H—O—S bond angle?

# Interactive Learning

**8-156.** How many moles of $Fe_3O_4$ are required to supply enough iron to prepare 0.260 mol of $Fe_2O_3$?

**8-157.** What mass of oxygen in grams is combined with $7.14 \times 10^{21}$ atoms of nitrogen in the compound $N_2O_5$?

**8-158.** Methyl ethyl ketone (often abbreviated MEK) is a powerful industrial solvent. A sample of this compound (which contains only C, H, and O) weighing 0.822 g was burned in oxygen to give 2.01 g of $CO_2$ and 0.827 g of $H_2O$. What is the empirical formula for MEK? (Remember that all of the C and H in the original compound are now part of the $CO_2$ and $H_2O$, respectively.)

**8-159.** In dilute nitric acid ($HNO_3$) copper metal dissolves according to the equation:

$$3Cu(s) + 8HNO_3(aq) \longrightarrow$$
$$3Cu(NO_3)_2(aq) + 2NO(g) + 4H_2O(l)$$

What mass of $HNO_3$ is needed to dissolve 11.45 g of Cu according to this equation?

**8-160.** An aqueous solution containing 18.0 g of $AgNO_3$ is mixed with a solution containing 32.4 of g $FeCl_3$. A solid precipitate of AgCl forms along with a solution of $Fe(NO_3)_3$. What mass of which reactant remains after the reaction is over?

**8-161.** Aluminum sulfate is used to make the soil in gardens more acidic. It is prepared according to the following equation:

$$2AlCl_3(aq) + 3H_2SO_4(aq) \longrightarrow Al_2(SO_4)_3(aq) + 6HCl(aq)$$

The solid product is obtained by evaporating away the HCl and water and heating the product to 200°C. When 25.0 g of $AlCl_3$ was mixed with 30.0 g of $H_2SO_4$, 28.46 g of pure $Al_2(SO_4)_3$ was eventually obtained. What is the percent yield?

# Solutions to Learning Checks

**A-1.** one, $6.022 \times 10^{23}$, molar mass, mole, formula, molar mass, $6.022 \times 10^{23}$

**A-2.** $25.0 \text{ g N} \times \dfrac{16.00 \text{ g O}}{14.01 \text{ g N}} = \underline{\underline{28.6 \text{ g O}}}$

**A-3.** $4.82 \times 10^{24} \text{ atoms V} \times \dfrac{1 \text{ mol V}}{6.022 \times 10^{23} \text{ atoms V}} =$

$$\underline{\underline{8.00 \text{ mol V}}}$$

$8.00 \text{ mol V} \times \dfrac{50.94 \text{ g V}}{\text{mol V}} = \underline{\underline{408 \text{ g V}}}$

**A-4.** $215 \text{ g Cl}_2 \times \dfrac{1 \text{ mol Cl}_2}{70.90 \text{ g Cl}_2} = \underline{\underline{3.03 \text{ mol Cl}_2}}$

$3.03 \text{ mol Cl}_2 \times \dfrac{2 \text{ mol Cl}}{1 \text{ mol Cl}_2} \times \dfrac{6.022 \times 10^{23} \text{ atoms Cl}}{\text{mol Cl}}$

$$\underline{\underline{3.65 \times 10^{24} \text{ atoms Cl}}}$$

**A-5.** Molar mass of $Cl_2O = (2 \times 35.45) + (16.00) = 86.90 \text{ g/mol}$

$438 \text{ g Cl}_2O \times \dfrac{1 \text{ mol Cl}_2O}{86.90 \text{ g Cl}_2O} = \underline{\underline{5.04 \text{ mol Cl}_2O}}$

$5.04 \text{ mol Cl}_2O \times \dfrac{6.022 \times 10^{23} \text{ molecules}}{\text{mol Cl}_2O} =$

$$\underline{\underline{3.04 \times 10^{24} \text{ moecules}}}$$

**B-1.** percent composition, formula, empirical, molecular, empirical, molar mass

**B-2.** Formula weight $= (6 \times 12.01) + (8 \times 1.008) + (6 \times 16.00) = \underline{176.1 \text{ amu}}$

$0.650 \text{ mol C}_6H_8O_6 \times \dfrac{6 \text{ mol C}}{\text{mol C}_6H_8O_6} = \underline{\underline{3.90 \text{ mol C}}}$

$3.90 \text{ mol C} \times \dfrac{12.01 \text{ g C}}{\text{mol C}} = \underline{\underline{46.8 \text{ g C}}}$

$\dfrac{6 \times 12.01}{176.1} \times 100\% = \underline{\underline{40.92\% C}}$

Empirical formula $= \underline{\underline{C_3H_4O_3}}$

**B-3.** In 100.0 g of compound there are 84.4 g B and $100.0 \text{ g} - 84.4 \text{ g} = 15.6 \text{ g H}.$

$84.4 \text{ g B} \times \dfrac{1 \text{ mol B}}{10.81 \text{ g B}} = \underline{\underline{7.81 \text{ mol B}}}$

$15.6 \text{ g H} \times \dfrac{1 \text{ mol H}}{1.008 \text{ g H}} = \underline{\underline{15.5 \text{ mol H}}}$

$\dfrac{7.81}{7.81} = 1.00 \qquad \dfrac{15.5}{7.81} = 1.98 \approx 2.00$

Empirical formula $= \underline{BH_2}$

$a = \dfrac{76.96 \text{ g/mol}}{10.81 + (2 \times 1.008) \text{ g/emp. unit}} = 6 \text{ emp. units/mol}$

Molecular formula $= B_{(1 \times 6)}H_{(2 \times 6)} = \underline{\underline{B_6H_{12}}}$

**C-1.** mole, three, molar masses, Avogadro's, least, theoretical, actual, percent, thermochemical, exothermic.

**C-2. (a)** $\boxed{\text{mol CH}_3\text{NH}_2} \Longrightarrow \boxed{\text{mol O}_2}$

$4.50 \text{ mol CH}_3NH_2 \times \dfrac{9 \text{ mol O}_2}{4 \text{ mol CH}_3NH_2} = \underline{\underline{10.1 \text{ mol O}_2}}$

**(b)** $\boxed{\text{g CH}_3\text{NH}_2} \Longrightarrow \boxed{\text{mol CH}_3\text{NH}_2} \Longrightarrow \boxed{\text{mol N}_2}$

$CH_3NH_2 [12.01 + (5 \times 1.008) + 14.01] = 31.06 \text{ g/mol}$

$322 \text{ g CH}_3NH_2 \times \dfrac{1 \text{ mol CH}_3NH_2}{31.06 \text{ g CH}_3NH_2} \times \dfrac{2 \text{ mol N}_2}{4 \text{ mol CH}_3NH_2} =$

$$\underline{\underline{5.18 \text{ mol N}_2}}$$

**(c)**

$\boxed{\text{g CO}_2} \Longrightarrow \boxed{\text{mol CO}_2} \Longrightarrow \boxed{\text{mol H}_2\text{O}} \Longrightarrow \boxed{\text{g H}_2\text{O}}$

$6.45 \text{ g CO}_2 \times \dfrac{1 \text{ mol CO}_2}{44.01 \text{ g CO}_2} \times \dfrac{10 \text{ mol H}_2O}{4 \text{ mol CO}_2} \times$

$$\dfrac{18.02 \text{ g H}_2O}{\text{mol H}_2O} = \underline{\underline{6.60 \text{ g H}_2O}}$$

**(d)**

$\boxed{\text{g N}_2} \Longrightarrow \boxed{\text{mol N}_2} \Longrightarrow \boxed{\text{mol O}_2} \Longrightarrow \boxed{\text{molecules O}_2}$

$0.568 \text{ g N}_2 \times \dfrac{1 \text{ mol N}_2}{28.02 \text{ g N}_2} \times \dfrac{9 \text{ mol O}_2}{2 \text{ mol N}_2} \times$

$$\dfrac{6.022 \times 10^{23} \text{ molecules}}{\text{mol O}_2} = \underline{\underline{5.49 \times 10^{22} \text{ molecules}}}$$

**C-3.**

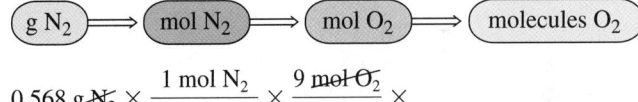

$20.0 \text{ g CH}_3NH_2 \times \dfrac{1 \text{ mol CH}_3NH_2}{31.06 \text{ g CH}_3NH_2} \times \dfrac{10 \text{ mol H}_2O}{4 \text{ mol CH}_3NH_2} =$

$$1.61 \text{ mol H}_2O$$

$40.0 \text{ g O}_2 \times \dfrac{1 \text{ mol O}_2}{32.00 \text{ g O}_2} \times \dfrac{10 \text{ mol H}_2O}{9 \text{ mol O}_2} = 1.39 \text{ mol H}_2O$

Limiting reactant is $O_2$  $1.39 \text{ mol H}_2O \times \dfrac{18.02 \text{ g H}_2O}{\text{mol H}_2O} =$

$$\underline{\underline{25.0 \text{ g H}_2O}}$$

**C-4.** $\dfrac{\text{actual yield}}{\text{theoretical yield}} \times 100\% = \%$ yield

theoretical yield $= \dfrac{\text{actual yield} \times 100\%}{\% \text{ yield}}$   $\dfrac{25.0 \text{ g} \times 100\%}{74.0\%} =$

$33.8 \text{ g CO}_2$

g CO$_2$ $\Longrightarrow$ mol CO$_2$ $\Longrightarrow$ mol C$_2$H$_4$ $\Longrightarrow$ g C$_2$H$_4$

$33.8 \text{ g } CO_2 \times \dfrac{1 \text{ mol } CO_2}{44.01 \text{ g } CO_2} \times \dfrac{1 \text{ mol } C_2H_4}{2 \text{ mol } CO_2} \times$

$\dfrac{28.05 \text{ g C}_2H_4}{\text{mol } C_2H_4} = \underline{\underline{10.8 \text{ g C}_2H_4}}$

**C-5.**

$C_2H_4(g) + 3O_2(g) \longrightarrow 2CO_2(g) + 2H_2O(l) + 1112 \text{ kJ}$

kJ $\Longrightarrow$ mol CO$_2$ $\Longrightarrow$ mol C$_2$H$_4$ $\Longrightarrow$ g C$_2$H$_4$

$250 \text{ kJ} \times \dfrac{1 \text{ mol } CO_2}{556 \text{ kJ}} \times \dfrac{1 \text{ mol } C_2H_4}{2 \text{ mol } CO_2} \times \dfrac{28.05 \text{ g C}_2H_4}{\text{mol } C_2H_4} =$

$\underline{6.31 \text{ g C}_2H_4}$

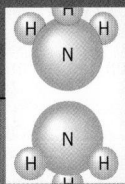
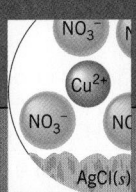

## Multiple Choice

**The following multiple-choice questions have one correct answer.**

**1.** Which of the following elements should be represented in an equation as a diatomic molecule?
**(a)** C    **(b)** He    **(c)** P    **(d)** B    **(e)** Br

**2.** Which of the following compounds or elements is a solid at room temperature?
**(a)** $K_2CO_3$      **(d)** $Cl_2$
**(b)** $CO_2$      **(e)** Hg
**(c)** HBrO

**3.** At high temperatures magnesium combines with nitrogen to form magnesium nitride. What is the coefficient in front of the magnesium in the balanced equation?
**(a)** 5    **(b)** 4    **(c)** 3    **(d)** 2    **(e)** 1

**4.** Which type of reaction does the following equation represent?

$$Ba(s) + 2HNO_3(l) \longrightarrow Ba(NO_3)_2(aq) + H_2(g)$$

**(a)** combustion      **(d)** single-replacement
**(b)** combination      **(e)** double-replacement
**(c)** decomposition

**5.** Which type of reaction does the following equation represent?

$$SiH_4(g) + 2O_2(g) \longrightarrow SiO_2(s) + 2H_2O(l)$$

**(a)** combustion      **(d)** single-replacement
**(b)** combination      **(e)** double-replacement
**(c)** decomposition

**6.** Which of the following completes the equation for the solution of $Na_2SO_4$ in water?

$$Na_2SO_4(s) \xrightarrow{H_2O}$$

**(a)** $Na_2^{2+}(aq) + SO_4^{2-}(aq)$
**(b)** $2Na^+(aq) + SO_4^{2-}(aq)$
**(c)** $Na^{2+}(aq) + SO_4^-(aq)$
**(d)** $2Na^{2+}(aq) + SO_4^-(aq)$

**7.** Which of the following is a net ionic equation?
**(a)** $CaCl_2 + Na_2CO_3 \longrightarrow CaCO_3 + 2NaCl$
**(b)** $Ca^{2+} + 2Cl^- + 2Na^+ + CO_3^{2-} \longrightarrow$
$$CaCO_3 + 2Cl^- + 2Na^+$$
**(c)** $2Cl^- + 2Na^+ \longrightarrow 2Na^+ + 2Cl^-$
**(d)** $Ca^{2+} + CO_3^{2-} \longrightarrow CaCO_3$

**8.** Which of the following equations represents a precipitation reaction?
**(a)** $CaCO_3 \longrightarrow CaO + CO_2$
**(b)** $HCl + KOH \longrightarrow H_2O + KCl$
**(c)** $Na_2S + Ni(NO_3)_2 \longrightarrow NiS + 2NaNO_3$
**(d)** $Zn + CuCl_2 \longrightarrow ZnCl_2 + Cu$
**(e)** $CH_4 + 2O_2 \longrightarrow CO_2 + 2H_2O$

**9.** Which of the equations in question 8 represents a neutralization reaction?

**10.** What is the formula of the salt produced from the neutralization of nitric acid and magnesium hydroxide?
**(a)** $MgNO_3$      **(d)** $Mg(OH)_2$
**(b)** $Mg(NO_2)_2$      **(e)** $Mg(NO_3)_2$
**(c)** $MgH_2$

**11.** When solutions of $K_2CO_3$ and $Cu(ClO_4)_2$ are mixed, a precipitate of $CuCO_3$ forms. The spectator ions are
**(a)** $K^+$ and $Cl^-$      **(d)** $K^{2+}$ and $CO_3^{2-}$
**(b)** $Cu^{2+}$ and $CO_3^{2-}$      **(e)** $Cu^{2+}$ and $K^+$
**(c)** $K^+$ and $ClO_4^-$

**12.** A quantity of aluminum has a mass of 54.0 g. What is the mass of the same number of magnesium atoms?
**(a)** 12.1 g      **(c)** 48.6 g      **(e)** 6.0 g
**(b)** 24.3 g      **(d)** 97.2 g

**13.** To which is the number $6.022 \times 10^{22}$ equivalent?
**(a)** 0.100 mol      **(d)** 0.500 mol
**(b)** 1.00 mol      **(e)** no such number exists
**(c)** 10.0 mol

**14.** What is the mass in grams of 0.250 mol of oxygen atoms?
**(a)** 16.0 g      **(c)** 8.00 g      **(e)** 32.0 g
**(b)** $1.50 \times 10^{23}$ g      **(d)** 4.00 g

**15.** What is the mass of $3.01 \times 10^{24}$ He atoms?
**(a)** 20.0 g      **(c)** 200 g      **(e)** 2.00 g
**(b)** 4.00 g      **(d)** $12.0 \times 10^{24}$ g

**16.** What is the mass of one mole of $H_2C_2O_4$?
**(a)** 90.0 amu      **(c)** 90.0 g      **(e)** 46.0 amu
**(b)** 46.0 g      **(d)** 58.0 g

**17.** How many moles of oxygen atoms are in 0.50 mol of $Ca(ClO_3)_2$?
**(a)** 3.0      **(c)** 0.50      **(e)** 1.50
**(b)** 1.0      **(d)** 6.0

**253**

**18.** How many moles of oxygen atoms are in $3.01 \times 10^{23}$ molecules of $O_2$?
**(a)** 0.50          **(c)** 1.0          **(e)** $1.50 \times 10^{23}$
**(b)** $6.022 \times 10^{23}$          **(d)** 0.25

**19.** Ammonium sulfate is used as a nitrogen fertilizer. What mass of nitrogen is present in a 3.00-mol quantity of the compound?
**(a)** 14.0 g          **(c)** 28.0 g          **(e)** 21.0 g
**(b)** 42.0 g          **(d)** 84.0 g

**20.** A compound is composed of 0.24 mol of Fe, 0.36 mol of S, and 1.44 mol of O. What is its empirical formula?
**(a)** $FeS_3O_6$          **(c)** $Fe_2S_3O_6$          **(e)** $Fe_2S_3O_8$
**(b)** $Fe_4S_6O_{24}$          **(d)** $Fe_2S_3O_{12}$

**21.** A compound has a molar mass of 84.0 g/mol and an empirical formula of $CH_2N$. What is its molecular formula?
**(a)** $C_3H_4N_2O$          **(c)** $C_2H_4N_2$          **(e)** $C_4H_8N_4$
**(b)** $C_3H_6N_3$          **(d)** $C_3H_6N_2$

**22.** Given the equation

$$2C_2H_2 + 5O_2 \longrightarrow 4CO_2 + 2H_2O$$

How many moles of $O_2$ are needed to produce 4 mol of $CO_2$?
**(a)** 4 mol          **(c)** 3 mol          **(e)** 5 mol
**(b)** 2 mol          **(d)** 1 mol

**23.** In the combustion of a certain hydrocarbon 16.0 g of $CO_2$ is produced, which represents a 75% yield. What is the theoretical yield?
**(a)** 12.0 g          **(c)** 21.3 g          **(e)** 44.0 g
**(b)** 8.0 g          **(d)** 32.0 g

**24.** Given the combustion reaction

$$CH_4 + 2O_2 \longrightarrow CO_2 + 2H_2O$$

When 16.0 g of $CH_4$ is burned with 32.0 g of $O_2$, which statement is true?
**(a)** $CH_4$ is the limiting reactant.
**(b)** $CO_2$ is present in excess.
**(c)** $CH_4$ is present in excess
**(d)** $O_2$ is present in excess.
**(e)** $H_2O$ is the limiting reactant.

## Problems

**1.** Fill in the blanks.

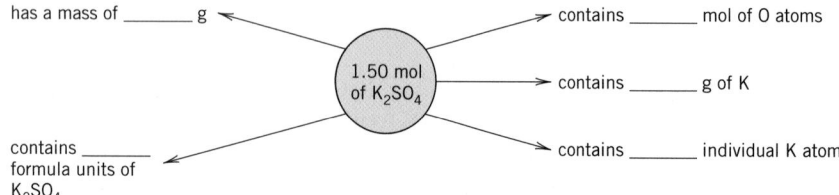

**2.** Fill in the blanks.

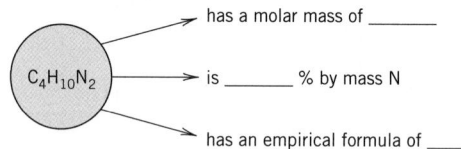

**3.** A compound is composed of 1.75 g of Fe and 0.667 g of O. What is its empirical formula?

**4.** A compound is 9.60% H, 16.4% C, and 74.0% B by mass. Its molar mass is 146 g/mol. What is its molecular formula?

**5.** Complete the following equations. Add subscripts, co-efficients, and physical states where such are missing.
**(a)** $K(\ ) + Al_2Cl_6\ (s) \longrightarrow KCl(\ ) + Al\ (s)$
**(b)** $C_6H_6\ (l) + O\ (g) \longrightarrow CO_2(\ ) + H_2O\ (l)$
**(c)** $Cl_2O_3\ (g) + H_2O\ (l) \longrightarrow HClO_2\ (l)$
**(d)** $K_2S\ (aq) + H_3PO_4\ (aq) \longrightarrow H_2S\ (g) + K_3PO_4\ (aq)$
**(e)** $B_4H_{10}\ (g) + H_2O\ (l) \longrightarrow B(OH)_3\ (aq) + H\ (g)$

**6.** Write balanced molecular and net ionic equations representing the following.
**(a)** a single-replacement reaction involving the reaction of magnesium with an aqueous lead(II) nitrate solution
**(b)** a precipitation reaction involving the reaction of aqueous solutions of lead(II) nitrate and ammonium sulfide
**(c)** a neutralization reaction involving solutions of hydrobromic acid and strontium hydroxide

**7.** Freon-12 ($CCl_2F_2$) is a gas used as a refrigerant. It is prepared according to the equation

$$3CCl_4(l) + 2SbF_3(s) \longrightarrow 3CCl_2F_2(g) + 2SbCl_3(s)$$

**(a)** How many moles of $SbF_3$ are needed to produce 0.0350 mol of Freon?
**(b)** How many moles of $SbCl_3$ are produced if 150 g of Freon is also produced?
**(c)** What mass of $CCl_4$ is needed to react with 850 g of $SbF_3$?
**(d)** How many individual molecules of $CCl_4$ are needed to produce 12.0 kg of $SbCl_3$?

## chapter 9

# The Gaseous State

**Our atmosphere is a protective blanket of gases. The behavior of these gases can be predicted by "gas laws," which are studied in this chapter.**

## Setting the Stage

Our atmosphere of gases is like the combination of a warm blanket and a protective shield. It nurtures life on this planet in numerous ways. Oxygen fuels the metabolism that gives energy to living things. Nitrogen becomes part of proteins and also dilutes the oxygen so that combustion does not occur explosively. Carbon dioxide and water vapor lock in the daytime heat so that the temperature does not become bitterly cold at night. Ozone high in the stratosphere absorbs the harmful portion of the sun's radiation. The thickness of the air burns up most incoming meteorites and other debris from space. Life as we know it requires the constant support and protection of this sea of gases.

Just over 200 years ago there was little understanding of the nature of the gases in which we exist. It wasn't until the late eighteenth century that the experiments of Antoine Lavoisier in France and Joseph Priestley in England gave us some understanding about the gaseous mixture of the atmosphere. These scientists proved that air was not just one substance, as had been previously thought, but was mainly a mixture of two elements: nitrogen and oxygen. Their work also laid the foundation for the development of the law of conservation of mass. This was the beginning of modern quantitative chemistry.

With a few exceptions, gases are invisible. This makes the study of gases somewhat difficult. Except on a windy day, it is easy to forget that this form of matter surrounds us. There is good news when we study the nature of gases, however. In many ways, they all behave similarly, which allows convenient generalizations, including "the gas laws." Solids and liquids are less abstract than gases simply because they are visible. The downside of working with solids and liquids, however, is that few generalizations apply and each substance must be studied individually.

### Formulating Some Questions

Perhaps the first question to ask is why gases are so diffuse compared to the other states of matter. How do the atoms or molecules of gases behave that explains this and other common properties? How is a quantity of gas affected by such variables as pressure, temperature, and the number of molecules? Since gases are part of chemical reactions, how do we include the volume of a gas in stoichiometry calculations? Such questions will be discussed in this chapter. First, we will list some of the unique properties of gases and see how these are explained by an extension of the atomic theory.

## Section A — THE NATURE OF THE GASEOUS STATE AND THE EFFECTS OF CHANGING CONDITIONS

### 9-1 The Nature of Gases and the Kinetic Molecular Theory

**Looking Ahead!** Gases are certainly the softest form of nature. It seems that we move through the air with ease and freedom, especially compared to moving through water. Of course, we have no chance of moving through a solid unless we are in a science fiction movie. Living in a sea of gases, we tend to take the properties of the gaseous state for granted. In this section, we will not only examine these common properties but also provide the accepted theory that explains them.

Five of the common properties of gases are described below.

1. *Gases are compressible.*
   When a glass of water is full, we mean what we say—no more can be added. But when is an automobile tire full of air? Actually, never. We can always add more air (at least until the tire bursts). The nature of the gaseous state allows us to press a volume of gas into a smaller volume. Liquids and solids are not like that. They are essentially incompressible.

2. *Gases have low densities.*
   The density of a typical liquid or solid is about 2 g/mL. The density of a typical gas is about 2 g/L. This means that the density of a solid or liquid is roughly 1000 times greater than that of a gas. This is a big difference.

3. *Gases mix thoroughly.*
   Nothing compares with a good home-cooked meal. Perhaps the best part is the pleasurable fragrances that we detect as the meal is being prepared. It doesn't take long for those familiar odors to drift from the kitchen throughout the entire home. The vapors from cooking mix thoroughly and rapidly with the surrounding air. This is a property of gases that is unique to their state. In contrast, some liquids do not mix at all (e.g., oil and water). If they do mix (e.g., water and alcohol), the mixing process occurs quite slowly.

4. *A gas fills a container uniformly.*

   When we blow air into a round balloon, the balloon becomes spherical and uniform. The air in the balloon obviously pushes out in all directions. If we were to fill the balloon with water instead (for scientific reasons only, of course), we would notice that the balloon sags. The water accumulates in the bottom of the balloon.

5. *A gas exerts pressure uniformly on all sides of a container.*

   This is evident from the preceding property. In the context of a room full of air, it would seem that gases defy gravity. The pressure of the atmosphere is the same on the ceiling, walls, and floor.* If we were to fill the room with water, however, we would notice that the pressure increases markedly the greater the depth. Anyone diving off the high dive at a swimming pool knows the feel of the increase in pressure 10 feet below the water's surface. A balloon "full" of air is spherical because the gas pushes out with the same pressure in all directions.

**The balloon is spherical because a gas exerts equal pressure in all directions.**

Obviously, the gaseous state of matter is quite different from the other two physical states. Gases are composed of either individual atoms (in the case of noble gases) or molecules (for all other gaseous elements and compounds). This fact and others can be summarized in a set of assumptions collectively known as the **kinetic molecular theory,** or simply the **kinetic theory.** The kinetic theory was advanced in the late nineteenth century as a model to explain the common properties of gases. The major points of this theory as applied to gases are as follows.

1. A gas is composed of very small particles called molecules, which are widely spaced and occupy negligible volume. A gas is thus mostly empty space.

2. The molecules of a gas are in rapid, random motion, colliding with each other and the sides of the container. Pressure is a result of these collisions with the sides of the container. (See Figure 9-1.)

3. All collisions involving gas molecules are elastic. (The total energy of two colliding molecules is conserved. A ball bouncing off the pavement undergoes inelastic collisions; it does not bounce as high each time.)

4. Gas molecules have negligible attractive (or repulsive) forces between them.

5. The temperature of a gas is related to the average kinetic energy of the gas molecules. Also, at the same temperature, different gases have the same average kinetic energy (K.E.).

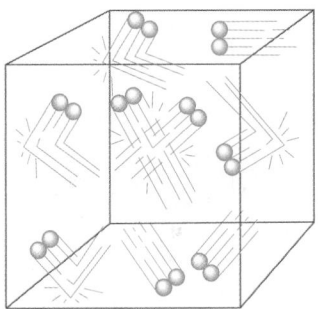

**Figure 9-1 Gases**
**Gas molecules are in rapid, random motion.**

The kinetic theory explains the properties described above as well as the gas laws that follow. For example, since gas molecules have essentially no volume and a gas is mostly empty space, it has a low density and is easily compressed. The rapid motion explains why gases mix rapidly and thoroughly. Because gas molecules are not attracted to each other, they do not "clump together" as do the molecules in liquids and solids. Thus the molecules in gases spread out and fill a container uniformly, exerting pressure equally in all directions.

Another consequence of the kinetic theory concerns the relative velocities of gas molecules at the same temperature. The kinetic energy of a gas is given by the equation

$$\text{K.E.} = \tfrac{1}{2}mv^2 \quad m = \text{mass} \quad \text{and} \quad v = \text{velocity}$$

---

*In fact, gases do not really defy gravity since our atmosphere gets thinner as the altitude increases. However, the thinning of the atmosphere occurs on a scale of several kilometers. It would not be noticed on the scale of a few meters in a room.

**The balloon on the left is filled with air, and the one on the right is filled with helium. One day later, the helium balloon has shrunk because the small atoms of helium have effused out of the balloon.**

If two different gases have the same kinetic energies at the same temperature, then we can derive the following relationship between their average velocities.

$$\text{K.E. (gas 1)} = \tfrac{1}{2}m_1v_1^2 \qquad \text{K.E. (gas 2)} = \tfrac{1}{2}m_2v_2^2$$

Since the two kinetic energies are equal, we have

$$\tfrac{1}{2}m_1v_1^2 = \tfrac{1}{2}m_2v_2^2$$

or, rearranging and simplifying,

$$\frac{v_1}{v_2} = \sqrt{\frac{m_2}{m_1}}$$

The average velocity of a gas molecule relates directly to two other aspects of molecular motion. These are the rates of **diffusion** *(mixing)* and **effusion** *(moving through an opening or hole)*. In fact, the relationship that we have derived from an assumption of the kinetic theory was first proposed in 1832 by Thomas Graham some time before the acceptance of kinetic theory. **Graham's law** *states that the rates of diffusion of gases are inversely proportional to the square roots of their molar masses.*

Consider the comparison of a helium atom to a nitrogen molecule at the same temperature. Since helium is much lighter than nitrogen, it has a higher velocity. (See Figure 9-2.) In Example 9-1, we have calculated the comparative velocities of the two species.

| Example 9-1 | The Relative Velocities of Helium and Nitrogen |
|---|---|
| w o r k i n g   i t   o u t | At the same temperature, how much faster does an He atom travel than an $N_2$ molecule? |

**Solution**

At the same temperature, $(\text{K.E.})_{He} = (\text{K.E.})_{N_2}$. Therefore,

$$m_{He} = 4.003 \text{ g/mol} \qquad m_{N_2} = 28.02 \text{ g/mol}$$

$$\frac{v_{HE}}{v_{N_2}} = \sqrt{\frac{m_{N_2}}{m_{He}}} = \sqrt{\frac{28.02 \text{ g/mol}}{4.003 \text{ g/mol}}} = \sqrt{7.000} = 2.646$$

$$v_{He} = 2.646\, v_{N_2}$$

On the average, He atoms travel <u>2.646 times faster</u> than $N_2$ molecules.

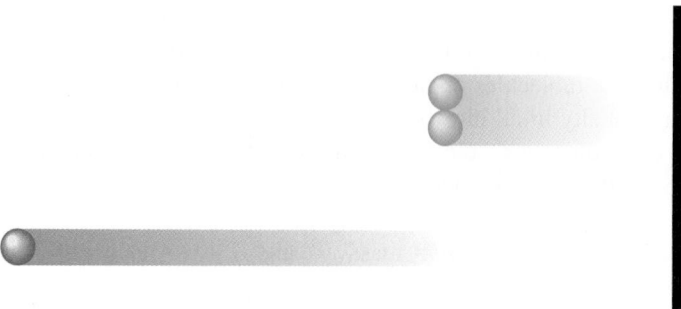

**Figure 9-2   Relative Velocities**
**At the same temperature, a helium atom travels almost three times faster than a nitrogen molecule.**

# 9-2 The Pressure of a Gas

**Looking Ahead!** Since kinetic theory makes the nature of gases seem so simple and reasonable, it is sometimes difficult to appreciate the advances made by those who were not aware of the existence of atoms, molecules, or their motions. Discussing the kinetic theory before the gas laws is somewhat like reading the last chapter of a mystery novel first—we know how it's going to turn out. Before we look at the other gas laws, we will take a closer look at the property known as "pressure."

A newscast on TV would not be complete without the weather forecast. The weather data probably include the atmospheric pressure as read from a **barometer.** Rising pressure usually means improving weather, and a dropping barometer often means that a storm is approaching.

The barometer has been around for quite a while. It was invented by an Italian scientist named Evangelista Torricelli in 1643. Torricelli filled a long glass tube with mercury, a dense liquid metal, and then inverted the tube into a bowl of mercury so that no air would enter the tube. Torricelli found that the mercury in the tube seemed to defy gravity by staying suspended to a height of about 76 cm no matter how long or wide the tube. (See Figure 9-3.)

At the time, many scientists thought that since "nature abhors a vacuum," the vacuum created at the top of the tube suspended the mercury. Torricelli suggested instead that it was the weight of the air on the outside that pushed the mercury up. Otherwise, it was reasoned, the greater vacuum present in tubes 2 and 3 in Figure 9-3 would support a higher level of mercury. Torricelli also suggested that the thinner atmosphere at higher levels in the mountains would support less mercury. He was correct. The higher the elevation, the lower the level of mercury in the barometer.

The weight of a quantity of matter pressing on a surface exerts a **force. Pressure** *is defined as the force applied per unit area.* This can be expressed mathematically as

$$P(\text{pressure}) = \frac{F(\text{force})}{A(\text{area})}$$

In a barometer, the pressure exerted by the atmosphere on the outside is balanced by the pressure exerted by the column of mercury on the inside.

If, at times, it seems like you are under a lot of "pressure," it is not necessarily from the atmosphere. However, you are under a lot of "force" from the atmosphere—

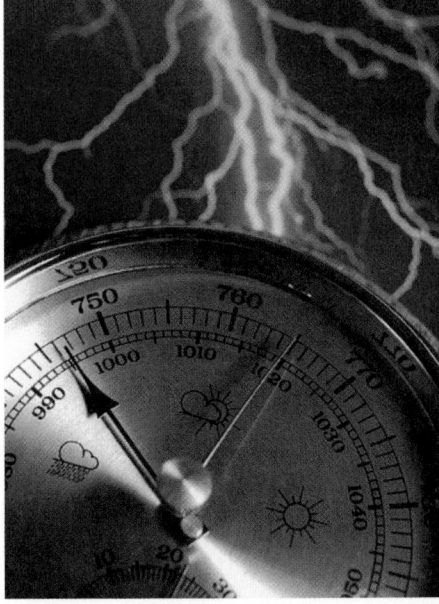

The pressure of the atmosphere is measured by a barometer. The barometric pressure can be used to forecast weather.

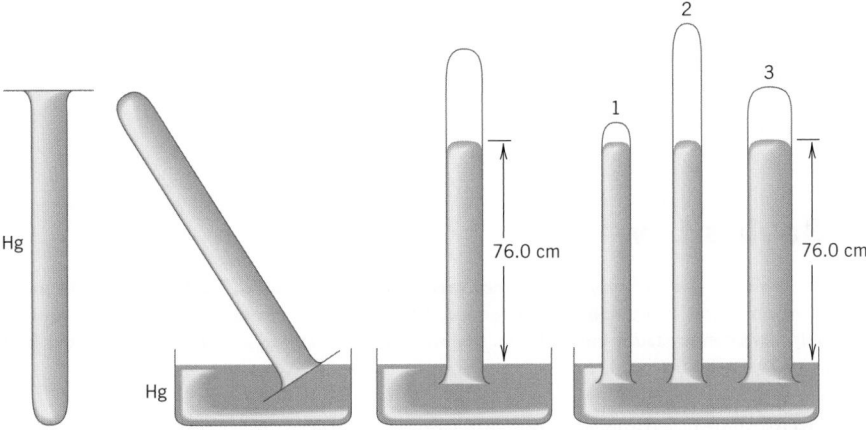

**Figure 9-3 The Barometer** When a long tube is filled with mercury and inverted in a bowl of mercury, the atmosphere supports the column to a height of 76.0 cm.

**Table 9-1   One Atmosphere**

| Unit of Pressure | Special Use |
|---|---|
| 760 mm Hg or 760 torr | Most chemistry laboratory measurements for pressures in the neighborhood of one atmosphere |
| 14.7 lb/in.$^2$ | U.S. pressure gauges |
| 29.9 in. Hg | U.S. weather reports |
| 101.325 kPa (kilopascals) | The metric unit of pressure [1 N (newton)/m$^2$] |
| 1.013 bars | Used in physics and astronomy mainly for very low pressures (millibars) or very high pressures (kilobars) |

over 20,000 lb of it. Fortunately, this force is spread out over your entire body surface (about 1500 in.$^2$, with much individual variation). When we divide the actual force on any one person by that person's body area, the resulting pressure comes out the same for everyone—a reasonable 15 lb/in.$^2$

*One atmosphere (1 atm) is defined as the average pressure of the atmosphere at sea level and is the standard of pressure.* As we have seen, this is equivalent to the pressure exerted by a column of mercury 76.0 cm (760 mm) high.

$$1.00 \text{ atm} = 76.0 \text{ cm Hg} = 760 \text{ mm Hg} = 760 \text{ torr}$$

The unit of mm of Hg is also known as the *torr* in honor of Torricelli. In addition to the torr, there are several other units of pressure (e.g., lb/in.$^2$) that have special uses. The relationships of the units to 1 atm and their applications are listed in Table 9-1. An example of a conversion between units is shown in Example 9-2.

---

**Example 9-2**

**working it out**

**Conversion Between Torr and Atmospheres**

What is 485 torr expressed in atmospheres?

**Procedure**

The conversion factors are

$$\frac{760 \text{ torr}}{\text{atm}} \quad \text{and} \quad \frac{1 \text{ atm}}{760 \text{ torr}}$$

**Solution**

$$485 \text{ torr} \times \frac{1 \text{ atm}}{760 \text{ torr}} = \underline{\underline{0.638 \text{ atm}}}$$

---

## 9-3 Boyle's Law

**Looking Ahead!** We are already aware that gases are compressible. That is, when we increase the pressure on a gas, the volume is decreased. Our next topic examines the actual quantitative relationship between pressure and volume. Historically, this resulted in the first of what are known as the gas laws.

Every breath we take illustrates the interaction between the volume and the pressure of a gas. When the diaphragm under our rib cage relaxes, it moves up, squeezing our lungs and decreasing their volume. The decreased volume increases the air pressure

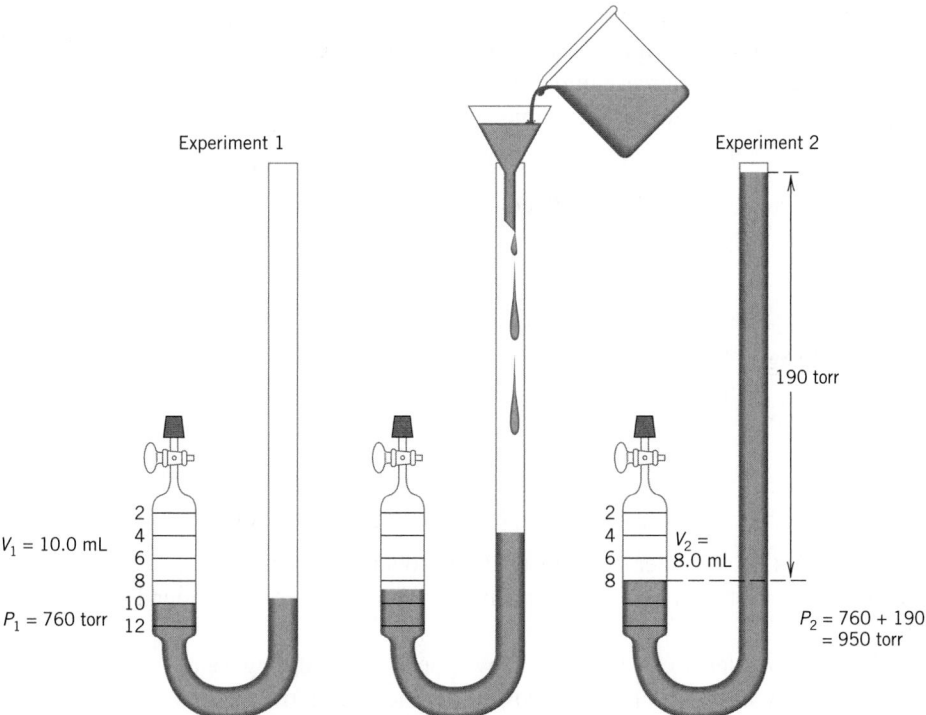

Experiment 1

Experiment 2

190 torr

$V_1 = 10.0$ mL

2
4
6
8
10
12

$P_1 = 760$ torr

2
4
6
8

$V_2 = 8.0$ mL

$P_2 = 760 + 190 = 950$ torr

**Figure 9-4 Boyle's Law Apparatus** **Addition of mercury in the apparatus causes an increase in pressure on the trapped gas. This leads to a reduction in the volume.**

inside the lungs relative to the outside atmosphere, and we expel air. When the diaphragm contracts, it moves down, increasing the volume of the lungs and, as a result, decreasing the pressure. Air rushes in from the atmosphere until the pressures are equal.

The relationship between the volume and pressure of a sample of gas under constant temperature conditions was first advanced in 1660 by the English scientist Robert Boyle. In an apparatus similar to that shown in Figure 9-4, Boyle found that the decrease in volume can be predicted by the amount of pressure increase. This observation is now expressed as **Boyle's law,** *which states that there is an inverse relationship between the pressure exerted on a quantity of gas and its volume if the temperature is held constant.*

The inverse relationship suggested by Boyle is represented as

$$V \propto \frac{1}{P}$$

or as an equality with $k$ the constant of proportionality:

$$V = \frac{k}{P} \quad \text{or} \quad PV = k$$

Boyle's law can be illustrated with the apparatus shown in Figure 9-4. In experiment **1,** a certain quantity of gas ($V_1 = 10.0$ mL) is trapped in a U-shaped tube by some mercury. Since the level of mercury is the same in both sides of the tube and the right side is open to the atmosphere, the pressure on the trapped gas is the same as the atmospheric pressure ($P_1 = 760$ torr). In this experiment,

$$P_1V_1 = 760 \text{ torr} \times 10.0 \text{ mL} = 7600 \text{ torr} \cdot \text{mL} = k$$

At higher altitudes, or in an airplane where the external pressure is lower, the bag of potato chips expands.

When mercury is added to the tube, the pressure on the trapped gas is increased to 950 torr (760 torr originally plus 190 torr from the added mercury). Note in experiment **2** that the increase in pressure has caused a decrease in volume to 8.0 mL. Applying Boyle's law to this experiment, we have

$$P_2V_2 = 950 \text{ torr} \times 8.0 \text{ mL} = 7600 \text{ torr} \cdot \text{mL} = k$$

As predicted by Boyle's law, $PV$ equals the same value in both experiments. Therefore, for a quantity of gas under two sets of conditions at the same temperature,

$$P_1V_1 = P_2V_2 = k$$

We can use this equation to calculate how a volume of gas changes when the pressure changes. For example, if $V_2$ is the new volume that we are to find at a given new pressure, $P_2$, the equation becomes

$$V_2 = V_1 \times \frac{P_1}{P_2}$$

new volume = old volume $\times$ pressure correction factor

If a series of measurements are made of volume and pressure, the results can be graphed. The construction of such a graph illustrating Boyle's law is found in Appendix D on graphs.

Gas law problems can be worked by properly substituting values for the known variables. They can also be worked by logic since we know qualitatively how a gas should react to a change of conditions. For example, let us consider a Boyle's law problem where the pressure on a given volume of gas has been increased ($P_2 > P_1$). We can reason that this would lead to a decrease in the volume of the gas ($V_2 < V_1$). We therefore choose a pressure correction factor that will do the job. This requires the use of a proper fraction (a fraction less than one) with the lower pressure in the numerator of the fraction. The following example emphasizes logic in working Boyle's law problems.

| **Example 9-3** | **Volume and a Change in Pressure** |
| --- | --- |
| w o r k i n g   i t   o u t | Inside a certain automobile engine, the volume of a cylinder is 475 mL when the pressure is 1.05 atm. When the gas is compressed, the pressure increases to 5.65 atm at the same temperature. What is the volume of the compressed gas? |

**Procedure**

First, identify the initial and final conditions. The final volume equals the initial volume times the pressure factor. Since the pressure increases, the volume decreases. Thus the pressure factor must be less than 1.

**Solution**

| Initial Conditions | Final Conditions |
| --- | --- |
| $V_1 = 475$ mL | $V_2 = ?$ |
| $P_1 = 1.05$ atm | $P_2 = 5.65$ atm |

$$V_2 = V_1 \times \frac{P_1}{P_2}$$

$$V_2 = 475 \text{ mL} \times \frac{1.05 \text{ atm}}{5.65 \text{ atm}} = \underline{\underline{88.3 \text{ mL}}}$$

(Note that the units of pressure are the same. *If the initial and final pressures are given in different units, one must be converted to the other.*)

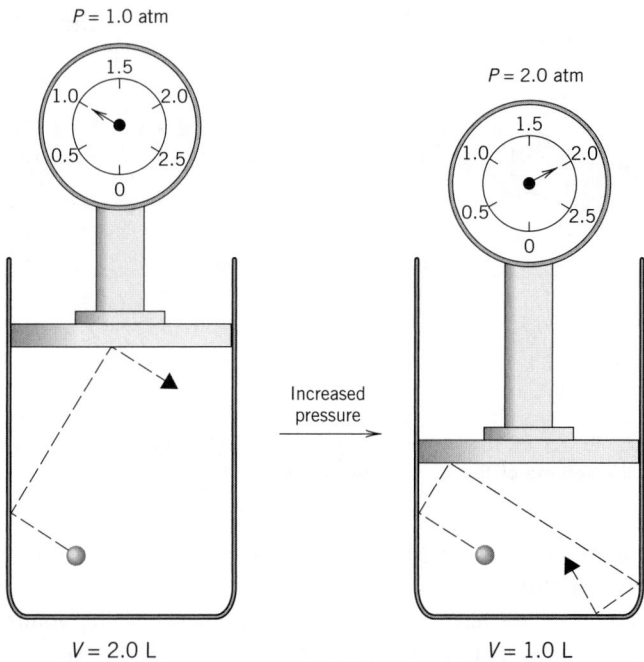

**Figure 9-5  Boyle's Law and Kinetic Molecular Theory**
**When the volume decreases, the pressure increases because of**
**more frequent collisions with the walls of the container.**

Although it certainly wasn't known in the 1660s, Boyle's law is a natural consequence of the kinetic theory of gases. In order to focus on how a change in volume affects the pressure, we will assume that there is an average molecule of a gas
at a given temperature. In Figure 9-5, the path of this average molecule is traced.
The pressure exerted by this molecule is a result of collisions with the sides of the
container. The distance traveled in a given unit of time is represented by the length
of the path. Since this is an average molecule, it travels the same distance per unit
time in both the high-volume situation (on the left) and the low-volume situation
(on the right). Note that, on the right, the lower volume leads to an increased number of collisions with the sides of the container. More frequent collisions mean
higher pressure.

## 9-4 Charles's Law and Gay-Lussac's Law

**Looking Ahead!** We probably sense that all matter expands when it is heated. Gases expand the
most and by the same amount regardless of what the compound is. One hundred years after
Boyle's relationship was advanced, the quantitative relationship between volume and temperature was studied. This and the overall relationships among volume, temperature, and pressure
are discussed in this section.

Very few sights appear more tranquil than brightly colored hot-air balloons drifting
across a blue summer sky. Except for the occasional "swoosh" of gas burners, they
cruise by in majestic silence. With heating, air expands. As it expands, it becomes
less dense than the surrounding cooler air and thus rises. When the hot, expanded air
is trapped, the balloon and attached gondola become "lighter than air" and lift into
the sky. The quantitative effect of temperature on the volume of a sample of gas at
constant pressure was first advanced by a French scientist, Jacques Charles, in 1787.
Charles showed that any gas expands by a definite fraction as the temperature rises.

**Hot-air balloons are practical applications of Charles's law.**

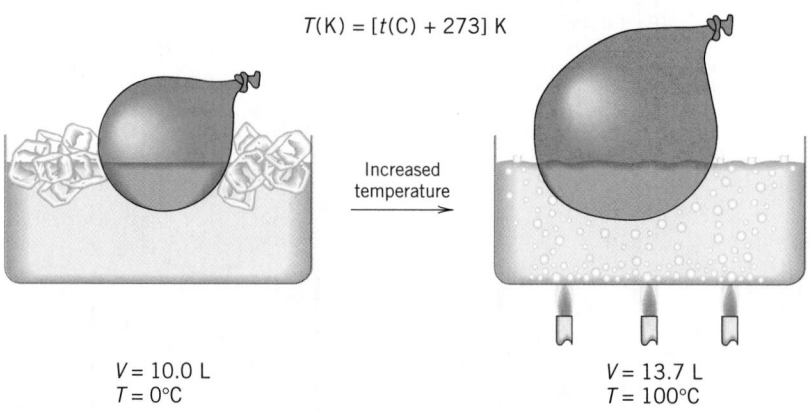

$$T(K) = [t(C) + 273]\ K$$

Increased temperature →

$V = 10.0\ L$
$T = 0°C$

$V = 13.7\ L$
$T = 100°C$

**Figure 9-6   Effect of Temperature on Volume** When the temperature increases, the volume of the balloon increases.

He found that the volume increases by a fraction of 1/273 for each 1°C rise in the temperature. (See Figure 9-6.)

Let us assume that we have made four measurements of the volume of a gas at four temperatures at constant pressure. The results of these experiments are listed in the table on the left in Figure 9-7. The four points are also plotted in a graph on the right. When the four points in the graph are connected, we have a straight line. The relationship between volume and temperature is known as a *linear relationship*. (Refer to Appendix D on graphs.) In the graph, the straight line has been extended to what the volume would be at temperatures lower than those in the experiment. (The procedure of extending data beyond experimental results is known as *extrapolation*.) If extended all the way to where the gas would theoretically have zero volume, the line would intersect the temperature axis at −273.15°C. Certainly, matter could never have zero volume, and it is impossible to cool gases indefinitely. At some point, all gases condense to become liquids or solids. However, this temperature does have significance, because it is the lowest possible temperature. As noted in kinetic molecular theory, the average kinetic energy or velocity of molecules is related to the temperature. *The lowest possible temperature, known as* **absolute zero,** *is the temperature at which translational motion (motion from point to point) ceases. The* **Kelvin scale** *starts at absolute zero.* Thus there are no negative values on the Kelvin scale, just as there are no negative values on any pressure scale. Since the magnitudes of Celsius and Kelvin degrees are the same, we have the following simple relationship between scales, where

| Experiment | $t(C)$ | $V(L)$ |
|------------|--------|--------|
| 1 | 100 | 136.7 |
| 2 | 50 | 118.3 |
| 3 | 0 | 100.0 |
| 4 | −50 | 81.7 |

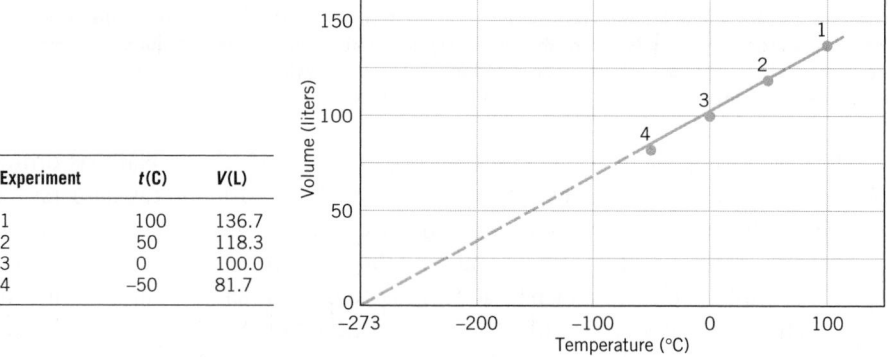

**Figure 9-7   Volume and Temperature**

## Defying Gravity—Hot-Air Balloons

In July 2002, Steve Fossett was the first person to fly solo around the globe in a hot-air balloon. However, hot-air balloons have a long history. Ancient peoples probably realized that hot air rose by observing the ascending embers of a campfire. The first creatures to take advantage of the rising hot air in a balloon were a duck, a chicken, and a sheep on September 19, 1783. The inventors of this device were the Montgolfier brothers. Convinced that the apparatus worked, two men made the first ascent into space later that year. Since the 1960s, hot-air ballooning has become increasingly popular with the use of newer, lightweight materials. Now we have the magnificent displays of balloons of all shapes and sizes floating with the wind over the countryside.

How do they work? They are actually an application of Charles's law, which states that the hotter a gas becomes, the more volume it requires at the same pressure. The expansion of a gas lowers its density, so it becomes buoyant in the surrounding air. Thus it rises just like a cork floats in water because it is less dense. The normal density of air at 25°C is about 1.18 g/L. If the air is heated to 125°C, the expansion causes the density to decrease to 0.88 g/L. The difference is the lifting power of the balloon. Thus a one-liter balloon containing the hot gas could lift 0.30 g. In order to lift the balloon, a gondola, a burner, and a couple of people, the balloon must lift about 1000 lb. That would take a volume of over $1.5 \times 10^6$ L (about 53,000 ft³). A larger balloon or hotter air could lift more weight or go higher.

The other way to have a gas less dense than the surroundings is to use a gas with a low molar mass such as helium (4 g/mol) or hydrogen (2 g/mol). For one liter of gas at 25°C, this means that He can lift 1.0 g and hydrogen 1.1 g. Helium is preferred since it is noncombustible. Hydrogen was used until the 1930s disaster at Lakehurst, New Jersey, when the German airship *Hindenburg* caught fire.

In fact, Fossett's lighter-than-air ship used a combination of hot air and helium. It ascended as high as 35,000 feet to the altitude of jet airliners. It was a marvelous feat that he accomplished on the sixth try. It only took 219 years after the first two men left the bounds of Earth.

*Hot air is less dense than cool air.*

$T$ (K) is the temperature in kelvins and $t$(C) represents the Celsius temperature. In our calculations, we will use the value of the lowest temperature expressed to three significant figures (i.e., −273°C).

The relationship between volume and temperature noted by Charles can now be restated in terms of the Kelvin scale. **Charles's law** *states that the volume of a gas is directly proportional to the Kelvin temperature (T) at constant pressure.* This can be expressed mathematically as a proportion or as an equality with $k$ again serving as the symbol for the constant of proportionality.

$$V \propto T \qquad V = kT \qquad \text{or} \qquad \frac{V}{T} = k$$

For a quantity of gas at two temperatures at the same pressure,

$$\frac{V_1}{T_1} = \frac{V_2}{T_2}$$

This equation can be used to calculate how the volume of gas changes when the temperature changes. For example, if $V_2$ is a new volume that we are to find at a given new temperature $T_2$, the equation becomes

$$V_2 = V_1 \times \frac{T_2}{T_1}$$

final volume = initial volume × temperature correction factor

In this case, if the temperature decreases ($T_2 < T_1$), the volume must also decrease, which means that the temperature factor is less than one. On the other hand, if the temperature increases ($T_2 > T_1$), the volume must increase, which means that the temperature factor is greater than one.

## Example 9-4

**working it out**

### Volume and Change in Temperature

A given quantity of gas in a balloon has a volume of 185 mL at a temperature of 52°C. What is the volume of the balloon if the temperature is lowered to −17°C? Assume that the pressure remains constant.

**Procedure**

As before, identify the initial and final conditions. Note that since the temperature decreases, the volume decreases. The temperature factor must therefore be less than 1. Remember we must use the Kelvin temperature scale in all gas law calculations.

**Solution**   *Initial Conditions*                         *Final Conditions*

$$V_1 = 185 \text{ mL} \qquad\qquad\qquad V_2 = \text{ ?}$$

$$T_1 = (-17 + 273) = 256 \text{ K*} \qquad T_2 = (52 + 273) = 325 \text{ K}$$

$$V_2 = V_1 \times \frac{T_2}{T_1}$$

$$V_2 = 185 \text{ mL} \times \frac{256 \cancel{K}}{325 \cancel{K}} = \underline{\underline{146 \text{ mL}}}$$

*Note that temperature must be expressed in the Kelvin scale. Also, a Celsius reading with two significant figures (e.g., −17°C) becomes a Kelvin reading with there significant figures (e.g., 256 K).

We have now expressed as laws how the volume of a gas is affected by a change in pressure and a change in temperature. Now consider that we have a confined or constant volume of gas that is subjected to a change in temperature. If the volume cannot change, what about the pressure? The answer is found on any pressurized can such as hair spray or deodorant: DO NOT INCINERATE OR STORE NEAR HEAT. This practical warning indicates what will happen. When a confined quantity of gas is heated, the pressure increases. Eventually, the seals on the can break and an explosion follows. The relationship between temperature and pressure was proposed in 1802 by Gay-Lussac. **Gay-Lussac's law** *states that the pressure is directly proportional to the Kelvin temperature at constant volume.* The law can be expressed mathematically as follows.

$$P \propto T \qquad P = kT$$

For a sample of gas under two sets of conditions at the same volume,

$$\frac{P_1}{T_1} = \frac{P_2}{T_2} \qquad P_2 = P_1 \times \frac{T_2}{T_1}$$

**Heat increases the pressure of gas in a confined volume. If heated enough, the can will explode.**

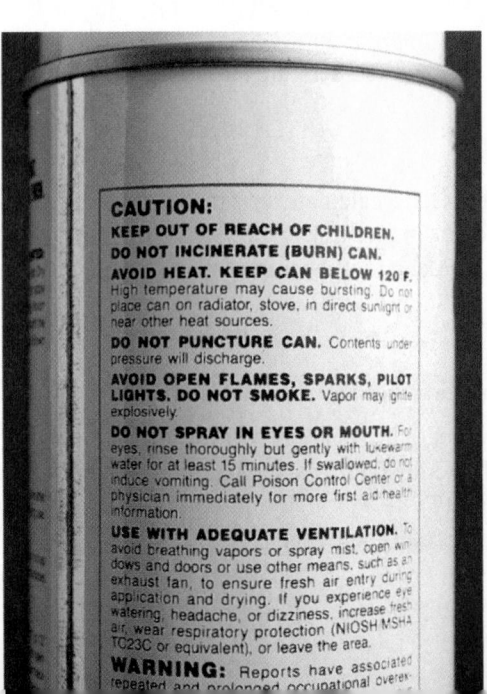

**Pressure and Change in Temperature**

A quantity of gas in a steel container has a pressure of 760 torr at 25°C. What is the pressure in the container if the temperature is increased to 50°C?

**Example 9-5**

working it out

**Procedure**

Since the temperature increases, the pressure increases. The temperature correction factor must therefore be greater than one.

**Solution**

| *Initial Conditions* | *Final Conditions* |
|---|---|
| $P_1 = 760$ torr | $P_2 = ?$ |
| $T_1 = (25 + 273) = 298$ K | $T_2 = (50 + 273) = 323$ K |

$$P_2 = P_1 \times \frac{T_2}{T_1}$$

$$P_2 = 760 \text{ torr} \times \frac{323 \text{ K}}{298 \text{ K}} = \underline{\underline{824 \text{ torr}}}$$

Both Charles's law and Gay-Lussac's law follow naturally from kinetic theory. In Figure 9-8, we again picture one average molecule of gas moving in a container. In the center, we have the situation before any changes are made. To the left, we have raised the temperature at constant pressure. The molecule travels faster on the average at the higher temperature, as shown by its longer path. It also collides with more force on the walls of the container. In order for the pressure to remain constant, the volume must expand correspondingly, which confirms Charles's law. To the right, we have again raised the temperature but this time at constant volume. The more frequent and more forceful collisions mean that the pressure is now higher, which confirms Gay-Lussac's law.

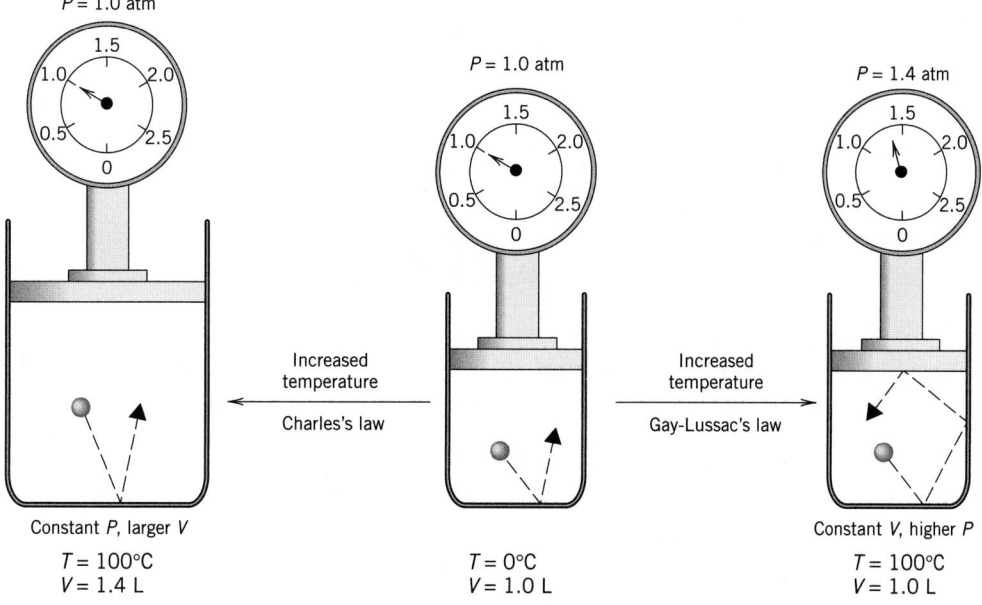

**Figure 9-8  Charles's Law and Gay-Lussac's Law  When temperature increases, molecules travel faster. If the pressure is constant, the volume increases in order to permit the frequency of collisions to remain the same. If the volume is constant, the pressure increases as a result of the more frequent and more forceful collisions.**

What if we have a quantity of gas where both temperature and pressure change? Of course, we could calculate the new volume of the gas by first correcting for the pressure change (Boyle's law) and then correcting for the temperature change (Charles's law). Or we could construct a new relationship that includes both laws and even Gay-Lussac's law as well. This is known, logically, as the **combined gas law** and is expressed as follows, first as a proportionality and second for a quantity of gas under two sets of conditions.

$$\frac{PV}{T} = k \qquad \frac{P_1V_1}{T_1} = \frac{P_2V_2}{T_2}$$

We can now work problems involving simultaneous changes of two conditions. *Notice in the following problems that the calculations require that the units of temperature must be Kelvin and the units of pressure and volume must be the same (i.e., mL or L for volume; torr or atm for pressure).*

| Example 9-6 | **Volume and Change in Pressure and Temperature** |
|---|---|
| working it out | A 25.8-L quantity of gas has a pressure of 690 torr and a temperature of 17°C. What is the volume if the pressure is changed to 1.85 atm and the temperature to 345 K? |

**Procedure**

Identify the two sets of conditions as initial or final. Notice that pressure increases, so its correction factor should be less than 1, but the temperature increases, so its correction factor should be greater than 1.

**Solution**

*Initial Conditions*                      *Final Conditions*

$V_1 = 25.8$ L                         $V_2 = ?$

$$P_1 = 690 \text{ torr} \times \frac{1 \text{ atm}}{760 \text{ torr}} \qquad\qquad P_2 = 1.85 \text{ atm}$$

$$= 0.908 \text{ atm}$$

$$T_1 = (17 + 273) \text{ K} = 290 \text{ K} \qquad T_2 = 345 \text{ K}$$

$$\frac{P_1V_1}{T_1} = \frac{P_2V_2}{T_2}$$

$$V_2 = V_1 \times \frac{P_1}{P_2} \times \frac{T_2}{T_1}$$

                               pressure     temperature

final volume = initial volume × correction × correction

                                      factor         factor

$$V_2 = 25.8 \text{ L} \times \frac{0.908 \text{ atm}}{1.85 \text{ atm}} \times \frac{345 \text{ K}}{290 \text{ K}} = \underline{\underline{15.1 \text{ L}}}$$

Obviously, a major property of a gas is that its volume is very dependent on temperature and pressure. Yet gas is sold by the local gas utility in volume units of cubic feet. Does that mean we get more or less gas in the warm summer than in the winter? Actually, we get the same amount because the volume of the gas is measured under certain universally accepted conditions known as **standard temperature and pressure (STP)**.

*Standard temperature:* 0°C or 273 K

*Standard pressure:* 760 torr or 1 atm

Example 9-7 illustrates the use of STP in the combined gas law.

---

**Temperature and Change in Pressure and Volume**

**Example 9-7**

A 5850-ft$^3$ quantity of natural gas measured at **STP** was purchased from the gas company. Only 5625 ft$^3$ was received at the house. Assuming that all of the gas was delivered, what was the temperature at the house if the delivery pressure was 1.10 atm?

**Procedure**

Notice that the initial conditions are STP and the final conditions are different. In this case, the final temperature is corrected by a pressure and a volume correction factor. Since the final pressure is higher, the pressure factor must be greater than 1 to increase the temperature. The final volume is lower, so the volume correction factor must be less than 1 to decrease the temperature.

**Solution**

| *Initial Conditions* | *Final Conditions* |
|---|---|
| $V_1 = 5850 \text{ ft}^3$ | $V_2 = 5625 \text{ ft}^3$ |
| $P_1 = 1.00 \text{ atm}$ | $P_2 = 1.10 \text{ atm}$ |
| $T_1 = 273 \text{ K}$ | $T_2 = ?$ |

$$\frac{P_1 V_1}{T_1} = \frac{P_2 V_2}{T_2} \qquad T_2 = T_1 \times \frac{P_2}{P_1} \times \frac{V_2}{V_1}$$

$$T_2 = 273 \text{ K} \times \frac{1.10 \text{ atm}}{1.00 \text{ atm}} \times \frac{5625 \text{ ft}^3}{5850 \text{ ft}^3} = 289 \text{ K}$$

$$t(\text{C}) = 289 \text{ K} - 273 = \underline{\underline{16°\text{C}}}$$

---

# 9-5 Avogadro's Law

**Looking Ahead!** It is probably obvious that the volume of a gas is also dependent on the amount of gas present, if all other conditions (pressure and temperature) are constant. We will formalize this quantitative relationship next.

The more air we blow into a balloon, the larger it gets. This very obvious statement is the basis of still another gas law. **Avogadro's law** *states that equal volumes of gases at the same pressure and temperature contain equal numbers of molecules.* Another way of stating this law is that *the volume of a gas is proportional to the number of molecules (moles) of gas present at constant pressure and temperature.* Mathematically, this can be stated in three ways, where *n* represents the number of moles of gas.

$$V \propto n \qquad V = kn \qquad \frac{V_1}{n_1} = \frac{V_2}{n_2}$$

**Figure 9-9 Illustration of Avogadro's Law**
The addition of carbon dioxide to the balloon increases the number of gas molecules, which increases the volume.

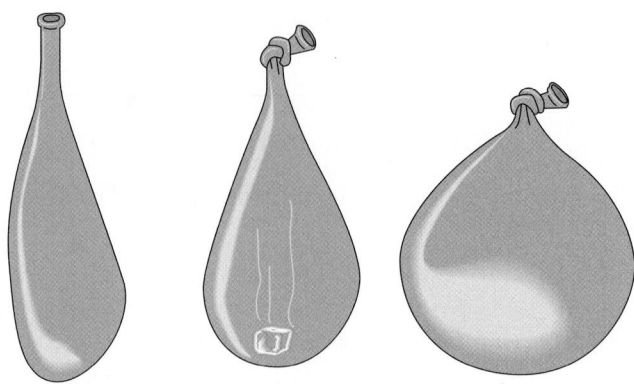

In Figure 9-9, the expansion of a balloon is illustrated by adding carbon dioxide. The following example illustrates the use of Avogadro's law in a sample calculation.

| Example 9-8 | **Volume and Change in Moles of Gas** |
|---|---|
| working it out | |

A balloon that is not inflated but is full of air has a volume of 275 mL and contains 0.0120 mol of air. As shown in Figure 9-9, a piece of dry ice (solid $CO_2$) weighing 1.00 g is placed in the balloon and the neck tied. What is the volume of the balloon after the dry ice has vaporized? (Assume constant $T$ and $P$.)

**Procedure**

Notice that the final condition of moles is the original amount plus the added amount. Since more gas has been added, the volume should increase and the mole correction factor is greater than 1.

**Solution**

*Initial Conditions*

$V_1 = 275$ mL
$n_1 = 0.0120$ mol

*Final Conditions*

$V_2 = ?$
$n_2 = $ mol air + mol $CO_2$

$$= 0.0120 + \left(1.00 \text{ g } CO_2 \times \frac{1 \text{ mol}}{44.01 \text{ g } CO_2}\right)$$

$$= 0.0120 + 0.0227 = 0.0347 \text{ mol}$$

$$\frac{V_1}{n_1} = \frac{V_2}{n_2}$$

$$V_2 = V_1 \times \frac{n_2}{n_1} = 275 \text{ mL} \times \frac{0.0347 \text{ mol}}{0.0120 \text{ mol}}$$

$$V_2 = \underline{795 \text{ mL}}$$

**Looking Back!** The kinetic theory provides a reasonable model for an understanding of the quantitative relationships known as "the gas laws" that had been previously advanced. Historically, the first gas law relates how pressure affects the volume of a sample of gas. The other gas laws that we have studied so far tell us how volume, temperature, and pressure relate to each other and how changing one or two of these conditions affects the others.

## Learning Check A | checking it out

**A-1.** Fill in the blanks.

Unlike solids and liquids, a gas has a _____ density and fills a container _____ . The reason for the unique behavior of gases can be explained by the _____ _____ theory. In this theory, gas molecules are in _____ , _____ motion, exerting pressure by _____ with the walls of the container. Another aspect of this theory is that the temperature relates to the average _____ energy of the gas. By relating the energies of two gases at the same _____ , we can derive a relationship known as _____ law. At the same temperature, the gas with the larger molar mass will have the _____ velocity. Atmospheric pressure is read from a device called a _____ . Pressure is defined as _____ per unit _____ . One atmosphere of pressure is equivalent to _____ torr. An increase in pressure on a volume of gas causes the volume to _____ , whereas an increase in temperature causes the same volume to _____ . Gay-Lussac's law tells us that if the temperature of a confined volume of gas increases, the pressure _____ . To calculate the effect of two changes in conditions, the _____ _____ _____ is used. Increasing the number of moles of gas present will _____ the volume if other conditions are constant.

**A-2.** How does the average velocity of an $SF_6$ molecule compare to an $N_2$ molecule at the same temperature?

**A-3.** What is 0.650 atm expressed in (a) torr, (b) kPa, and (c) $lb/in.^2$?

**A-4.** The volume of 0.0227 mol of gas is 550 mL at a pressure of 1.00 atm and a temperature of 22°C.

**a.** What is the volume at 1.32 atm and 22°C?
**b.** What is the volume at 1.00 atm and 44°C?
**c.** What is the volume if 0.0115 mol of gas is added at the same $T$ and $P$?
**d.** What is the pressure at 122°C if the volume remains 550 mL?
**e.** What is the pressure at 102°C if the volume expands to 825 mL?

**A-5.** Name the law used in each of the calculations in problem A-4.

*Additional Examples: Exercises 9-7, 9-14, 9-20, 9-30, 9-36, 9-46, and 9-55.*

---

Section **B**

# RELATIONSHIPS AMONG QUANTITIES OF GASES, CONDITIONS, AND CHEMICAL REACTIONS

## 9-6 The Ideal Gas Law

**Looking Ahead!** In this section, we will find that we can arrive at the ultimate general gas law by putting the information in the last three sections together in the form of one equation. This very important equation is known as the ideal gas law and can be used to summarize much of our previous discussion and provide new information.

How can we make a balloon larger? In fact, we have discussed three ways. First, we could take it up a mountain to a higher elevation. The lower atmospheric pressure would allow it to expand (Boyle's law). Second, we could raise the temperature. If we placed the balloon in boiling water, it would certainly expand (Charles's law). And finally, we could just put more gas in it (Avogadro's law). We can summarize these three relationships as follows.

$$V \propto \frac{1}{P} \qquad V \propto T \qquad V \propto n$$

We can put all three individual relationships into one general relationship as follows.

$$V \propto \frac{nT}{P}$$

This relationship is an extended version of the combined gas law that now includes Avogadro's law. As before, we can change a proportionality to an equality by introducing a constant of proportionality. The constant used in this case is $R$ and is called the **gas constant.** Traditionally, the constant is placed between the $n$ and $T$.

$$V = \frac{nRT}{P} \qquad \text{or} \qquad PV = nRT$$

This relationship is known as the **ideal gas law.** In fact, all of the laws previously discussed (except for Graham's law) can be derived from this one relationship. For example, consider a confined quantity of gas that is subject to a change of pressure when the temperature is constant. Before the change (conditions designated by "1") we have $P_1V_1 = n_1RT_1$. After the change (conditions designated by "2") we have $P_2V_2 = n_2RT_2$. By solving both equations for the constant $R$ and setting them equal to each other we have

$$R = \frac{P_1V_1}{n_1T_1} = \frac{P_2V_2}{n_2T_2}$$

If we have a confined quantity of gas (i.e., $n_1 = n_2$) and the temperature does not change (i.e., $T_1 = T_2$), we can substitute $n_1$ for $n_2$ and $T_1$ for $T_2$. Notice then that $n$ and $T$ both cancel and we have Boyle's law.

$$\frac{P_1V_1}{n_1T_1} = \frac{P_2V_2}{n_2T_2} = \frac{P_2V_2}{n_1T_1} \qquad P_1V_1 = P_2V_2$$

Similar exercises lead to Charles's, Gay-Lussac's, and the combined gas laws. However, the ideal gas law has a more important application than the exercises we have covered so far, where the effect of changing conditions is calculated. With this law we can calculate one condition (i.e., the volume) of a gas if the other three are known (i.e., $n$, $P$, and $T$). For these calculations, however, we need the value of the gas constant, which is

$$0.082057 \frac{L \cdot atm}{K \cdot mol}$$

For the calculations used in this text, we will use the value of $R$ expressed to three significant figures (i.e., 0.0821). Note that the units of $R$ are a result of specific units of $P$, $V$, and $T$. Thus in all calculations that follow, the pressure must be expressed in *atmospheres,* the temperature in *Kelvin,* and the volume in *liters.* The following three examples illustrate the use of the ideal gas law with a missing variable under one set of conditions.

| Example 9-9 | Calculation of Pressure from *V, T,* and *n* |
|---|---|
| **working it out** | What is the pressure of a 1.45-mol sample of a gas if the volume is 20.0 L and the temperature is 25°C? |

**Procedure**

Write down conditions, making sure the units correspond to $R$. Solve the ideal gas law for $P$ by dividing both sides of the equation by $V$.

**Solution**

$$P = ? \quad V = 20.0\,\text{L} \quad n = 1.45\,\text{mol} \quad T = (25 + 273)\,\text{K} = 298\,\text{K}$$

$$PV = nRT$$

$$P = \frac{nRT}{V} = \frac{1.45\,\cancel{\text{mol}} \times 0.0821\,\dfrac{\cancel{\text{L}} \cdot \text{atm}}{\text{K} \cdot \cancel{\text{mol}}} \times 298\,\cancel{\text{K}}}{20.0\,\cancel{\text{L}}}$$

$$= 1.77\,\text{atm}$$

### Calculation of Volume from *T*, *P*, and *n*

**Example 9-10**

What is the volume of 1.00 mol of gas at STP?

**Solution**

$$P = 1.00\,\text{atm} \quad V = ? \quad T = 273\,\text{K} \quad n = 1.00\,\text{mol}$$

$$PV = nRT$$

$$V = \frac{nRT}{P}$$

$$V = \frac{1.00\,\cancel{\text{mol}} \times 0.0821\,\dfrac{\text{L} \cdot \cancel{\text{atm}}}{\text{K} \cdot \cancel{\text{mol}}} \times 273\,\cancel{\text{K}}}{1.00\,\cancel{\text{atm}}} = \underline{\underline{22.4\,\text{L}}}$$

### Calculation of Temperature from *n*, *P*, and *V*

**Example 9-11**

What is the Celsius temperature of a 1.10-g quantity of oxygen in a 4210-mL container at a pressure of 170 torr?

**Procedure**

Convert mass of oxygen to moles and the other variables to the units required by *R*. Solve for *T* by dividing both sides of the ideal gas equation by *nR*.

**Solution**

$$n = 1.10\,\cancel{\text{g O}_2} \times \frac{1\,\text{mol O}_2}{32.00\,\cancel{\text{g O}_2}} = 0.0344\,\text{mol O}_2$$

$$P = \frac{170\,\cancel{\text{torr}}}{760\,\cancel{\text{torr}}/\text{atm}} = 0.224\,\text{atm} \quad V = 4210\,\cancel{\text{mL}} \times \frac{1\,\text{L}}{10^3\,\cancel{\text{mL}}} = 4.21\,\text{L}$$

$$T = \frac{PV}{nR} = \frac{0.224\,\cancel{\text{atm}} \times 4.21\,\cancel{\text{L}}}{(0.0344\,\cancel{\text{mol}}) \times \left(0.0821\,\dfrac{\cancel{\text{L}} \cdot \cancel{\text{atm}}}{\text{K} \cdot \cancel{\text{mol}}}\right)} = 334\,\text{K}$$

$$^\circ\text{C} = 334 - 273 = \underline{\underline{71^\circ\text{C}}}$$

This law is called "ideal" because it follows from the assumptions of the kinetic theory, which describes an ideal gas. The molecules of an ideal gas have negligible volume and no attraction or repulsion for each other. The molecules of a "real" gas obviously do have a volume, and there is some interaction between molecules, especially at high pressures (i.e., the molecules are pressed close together) and low temperatures (i.e., the molecules move more slowly).

Fortunately, at normal temperatures and pressures found on the surface of Earth, gases have close to ideal behavior. That is, they are far enough apart so that the gas molecules themselves take up no space and they move past one another so fast that interactions between molecules can be ignored. The latter situation is like a young man and young woman rapidly passing each other in jet airliners, where they are unaware of each other's presence. In contrast, assume they pass while jogging in the park, where they may very likely slow down as they exchange attractive glances (or conversely speed up if they are repelled). Therefore, the use of the ideal gas law is justified. On the other hand, regions of the atmosphere of the planet Jupiter are composed of cold gases under very high pressures. Under these conditions, the ideal gas law would not provide acceptable values for a variable (i.e., $PV$ does not exactly equal $nRT$). Other, more complex equations would have to be used that take into account the volume of the molecules and the interactions between molecules.

## 9-7 Dalton's Law of Partial Pressures

**Looking Ahead!** Kinetic theory tells us that gas molecules behave independently. Avogadro's law illustrated this property since we noticed that equal volumes of gases contain equal numbers of molecules regardless of their masses or nature. If the volumes are independent of the nature of the gas, what about the pressure? The nature of the pressure exerted by mixtures of gases was first observed by John Dalton and is discussed in this section.

About 78% of the atmosphere is composed of nitrogen, 21% is oxygen, and a little less than 1% is argon atoms. This means that 21% of the molecules, 21% of the volume of the atmosphere, and 21% of the pressure that we feel from the atmosphere is due to oxygen. Only in terms of the mass of the atmosphere does oxygen not represent 21%, because of the different molar masses of the components. The fact that pressure, like volume, does not depend on the identity of the gas was first suggested by John Dalton, author of modern atomic theory. Another gas law resulted. **Dalton's law** *states that the total pressure of a gas in a system is the sum of the partial pressures of each component gas.* The partial pressure of a specific gaseous element or compound is defined as the pressure due to that substance alone. Mathematically, this can be stated as $P_{tot} = P_1 + P_2 + P_3 + \cdots$ ($P_1$ is the pressure due to gas 1, etc.).

| | |
|---|---|
| **Example 9-12**<br><br>working it out | **Total Pressure from Partial Pressures**<br><br>Three gases, Ar, $N_2$, and $H_2$, are mixed in a 5.00-L container. Ar has a pressure of 255 torr, $N_2$ has a pressure of 228 torr, and $H_2$ has a pressure of 752 torr. What is the total pressure in the container? |

**Solution**

$$P_{Ar} = 255 \text{ torr} \qquad P_{N_2} = 228 \text{ torr} \qquad P_{H_2} = 752 \text{ torr}$$

$$P_{tot} = P_{Ar} + P_{N_2} + P_{H_2} = 255 \text{ torr} + 228 \text{ torr} + 752 \text{ torr}$$

$$= \underline{1235 \text{ torr}}$$

**"Thin air" can be very hard to breathe.**

We can now apply Dalton's law to the composition of our atmosphere. Since 21% of the molecules of the atmosphere are oxygen, 21% of the volume and pressure of the atmosphere is due to oxygen. The partial pressure of oxygen can be calculated by multiplying the percent in decimal fraction form by the total pressure.

$$P_{O_2} = 0.21 \times 760 \text{ torr} = 160 \text{ torr}$$

Most of us function best breathing this partial pressure of oxygen. When we live at higher elevations, the partial pressure of oxygen is lower and our bodies adjust accordingly. On top of the highest mountain, Mt. Everest, the total atmospheric pressure is 270 torr, so the partial pressure of oxygen is only 57 torr, or about one-third of normal. A human cannot survive for long at such a low pressure of oxygen. At that altitude, even conditioned climbers must use an oxygen tank and mask, which give an increased partial pressure of oxygen to the lungs.

---

| Partial Pressures of Atmospheric Gases | Example 9-13 |
|---|---|
| | **working it out** |

On a humid summer day, the partial pressure of water in the atmosphere is 18 torr. If the barometric pressure on this day is a high 766 torr, what are the partial pressures of nitrogen and of oxygen if 78.0% of the dry atmosphere is composed of nitrogen molecules and 21.0% is oxygen? (1% of the dry atmosphere is composed of all other gases.)

**Procedure**

1. Find the pressure of the dry atmosphere by subtracting the pressure of water from the total.
2. Find the partial pressures of $N_2$ and of $O_2$ from the total pressure of the dry atmosphere and the percents.

**Solution**

1. $P_{tot} = P_{N_2} + P_{O_2} + P_{H_2O}$

$$P(\text{dry atmosphere}) = P_{tot} - P_{H_2O} = 766 - 18 = 748 \text{ torr}$$

2. $P_{N_2} = 0.780 \times 748 \text{ torr} = \underline{583 \text{ torr}}$

$P_{O_2} = 0.210 \times 748 \text{ torr} = \underline{157 \text{ torr}}$

---

The total pressure of a mixture of gases can be extended to the ideal gas law. For example, for three gases in one container,

$$P_{tot} = P_1 + P_2 + P_3 = \frac{n_1 RT}{V} + \frac{n_2 RT}{V} + \frac{n_3 RT}{V} = \frac{n_{tot} RT}{V}$$

We see that, at a given temperature, the pressures or volumes of gases depend only on the total number of moles present. This is illustrated in Figure 9-10 and Example 9-14.

---

| Pressure of a Mixture of Gases | Example 9-14 |
|---|---|
| | **working it out** |

What is the pressure (in atmospheres) exerted by a mixture of 12.0 g of $N_2$ and 12.0 g of $O_2$ in a 2.50-L container at 25°C?

**Procedure**

1. Find the number of moles of $N_2$ and $O_2$ present.
2. Use the total number of moles, temperature, and volume in the ideal gas law.

**Solution**

1. $n_{O_2} = 12.0 \text{ g } O_2 \times \dfrac{1 \text{ mol } O_2}{32.00 \text{ g } O_2} = 0.375 \text{ mol } O_2$

$n_{N_2} = 12.0 \text{ g } N_2 \times \dfrac{1 \text{ mol } N_2}{28.02 \text{ g } N_2} = 0.428 \text{ mol } N_2$

$n_{tot} = 0.375 \text{ mol} + 0.428 \text{ mol} = 0.803 \text{ mol of gas}$

$V = 2.50 \text{ L} \quad T = (25 + 273) = 298 \text{ K} \quad R = 0.0821 \dfrac{\text{L} \cdot \text{atm}}{\text{K} \cdot \text{mol}}$

# MAKING IT REAL

## Ozone—Friend and Foe

Oxygen is the element that made animal life possible. Billions of years ago, primitive forms of life on this planet began to use solar energy, carbon dioxide, and water to generate energy-rich compounds with oxygen as a by-product. As the oxygen became more concentrated in the atmosphere, new life forms evolved that used this oxygen for metabolism. But this new life could not have survived on the surface of Earth if it were not for another role that oxygen played in a region around 50 km above the Earth known as the stratosphere. There, solar radiation is powerful enough to break the bond in $O_2$ to form oxygen atoms. (Chemical reactions that are initiated by light energy are known as *photochemical reactions*.) The atoms of oxygen formed another allotrope of oxygen known as ozone. (The solar energy is represented as h$v$ in the equation below.)

$$O_2(g) + hv \longrightarrow O(g) + O(g)$$
$$O(g) + O_2(g) \longrightarrow O_3(g) + \text{heat energy}$$

The solar radiation that breaks the $O_2$ bond lies in the higher-energy ultraviolet region of the spectrum (around 240 nm). However, there are lower-energy wavelengths of ultraviolet light (from 240 nm to 310 nm) that would still come through to the surface if it were not for ozone. Ozone in the stratosphere effectively absorbs these damaging wavelengths of ultraviolet radiation. The equation is represented as

$$O_3(g) + hv \longrightarrow O_2(g) + O(g)$$

Without ozone in the stratosphere, most life on the surface of the Earth would be roasted by ultraviolet radiation. Destruction of ozone by synthetic chemicals is a crucial area of concern and is being addressed. (See Section 14-2.)

On the other hand, in the lower atmosphere, where we live, ozone is a problem. Ozone is associated with what is called *photochemical smog*. Automobiles and daylight are the culprits. $NO_2$ is formed in the atmosphere from NO that is formed in automobiles and other heat sources. In this case, oxygen atoms are produced from visible light from the sun acting on the $NO_2$.

$$NO_2(g) + hv \longrightarrow NO(g) + O(g)$$
$$O(g) + O_2(g) \longrightarrow O_3(g)$$

Ozone is toxic and chemically reactive. It causes trouble for anyone with lung disorders or those exercising outdoors. It is considered a serious pollutant.

So ozone is our essential friend when it is high up in the stratosphere but a serious problem near the surface. "Everything in its place" is a saying that seems appropriate.

*Ozone is associated with air pollution.*

---

2.  $PV = nRT$

$P = \dfrac{nRT}{V}$

$P = \dfrac{0.803 \ \cancel{\text{mol}} \times 0.0821 \ \dfrac{\cancel{L} \cdot \text{atm}}{\cancel{K} \cdot \cancel{\text{mol}}} \times 298 \ \cancel{K}}{2.50 \ \cancel{L}} = \underline{7.86 \text{ atm}}$

---

**Figure 9-10  Pressures of a Pure Gas and a Mixture of Gases**
**Pressure depends only on the number of molecules at a certain temperature, not on their identity.**

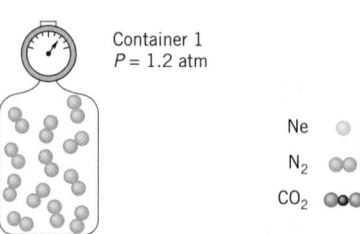

Container 1
$P = 1.2$ atm

Ne
$N_2$
$CO_2$

0.10 mol ($6.0 \times 10^{22}$ molecules) $N_2$

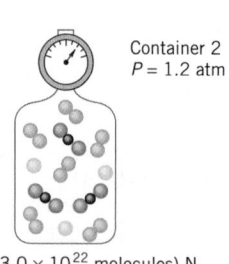

Container 2
$P = 1.2$ atm

0.050 mol ($3.0 \times 10^{22}$ molecules) $N_2$
0.025 mol ($1.5 \times 10^{22}$ molecules) $CO_2$
0.025 mol ($1.5 \times 10^{22}$ atoms) Ne

0.100 mol ($6.0 \times 10^{22}$ particles) total

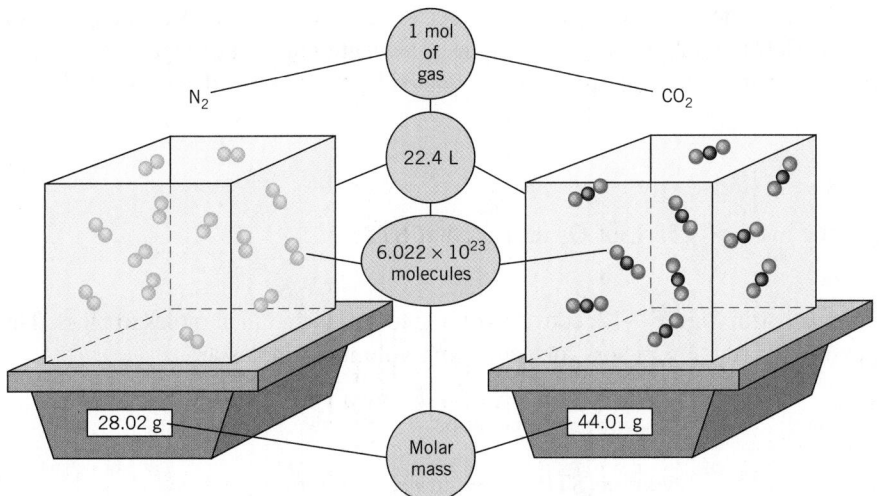

**Figure 9-11  Moles of Two Different Gases**
**One mole of any gas has the same volume at STP, the same number of molecules, but different masses.**

## 9-8 The Molar Volume and Density of a Gas

**Looking Ahead!** In previous chapters, we found that one mole of a compound represents a set number of molecules and a specific mass. If the compound also happens to be a gas, we find that one mole also represents a set volume under specified conditions. We explore this next.

In Example 9-10, we calculated the volume of one mole of a gas at standard temperature and pressure. This result represents the volume of one mole of *any* gas, or even one mole of a mixture of gases, at STP. This phenomenon is certainly not true of equal volumes of liquids and solids. *The volume of one mole of gas at STP, 22.4 L, is known as the* **molar volume.** This corresponds to about six gallons. We can now expand on the significance of the mole that was described in Chapter 8. One mole of a gas refers to three quantities: (1) the molar volume (22.4 L/mol, or 22.414 L/mol to be more precise) and (2) Avogadro's number ($6.022 \times 10^{23}$ molecules/mol), which are independent of the identity of the gas, and (3) the molar mass (g/mol), which depends on the identity of the gaseous compound or element. (See Figure 9-11.)

Before we work two examples of problems relating to the molar volume, it may be helpful to summarize the relationships between moles of gas, volume at STP, and volume at some other $T$ and $P$. In Figure 9-12, we notice that moles of gas can be converted to volume at STP using the molar volume as a conversion factor (path 1). Moles of gas

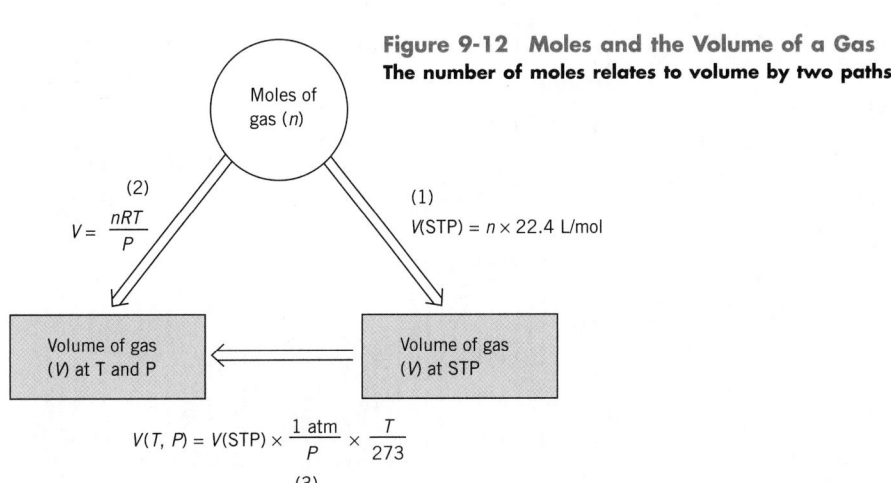

**Figure 9-12  Moles and the Volume of a Gas**
**The number of moles relates to volume by two paths.**

can also be converted to volume at some other temperature ($T$) and pressure ($P$) using the ideal gas law (path 2). Finally, the volume of a quantity of gas at STP and the volume at some other temperature and pressure are related by the combined gas law (path 3).

**Example 9-15** | **Calculation of Mass from Volume at STP**

working it out

What is the mass of 4.55 L of $O_2$ measured at STP?

**Procedure**

Utilizing path 1 in Figure 9-12, convert volume to moles and then moles to mass. The two possible conversion factors for converting volume to moles are

$$\frac{1 \text{ mol}}{22.4 \text{ L}} \quad \text{and} \quad \frac{22.4 \text{ L}}{\text{mol}}$$

$$\text{volume(STP)} \longrightarrow \text{moles} \longrightarrow \text{mass}$$

**Solution**

$$4.55 \text{ L} \times \frac{1 \text{ mol}}{22.4 \text{ L}} \times \frac{32.00 \text{ g}}{\text{mol}} = \underline{\underline{6.50 \text{ g}}}$$

**Example 9-16** | **Calculation of the Molar Mass of a Gas**

A sample of gas has a mass of 3.20 g and occupies 2.00 L at 17°C and 380 torr. What is the molar mass of the gas?

**Procedure**

There are two ways that one can work this problem. The first uses paths 1 and 3 in Figure 9-12.

1.  Convert given volume to volume at STP (reverse of path 3).
2.  Convert volume at STP to moles (reverse of path 1, as in Example 9-15).
3.  Convert moles and mass to molar mass (represented in equations as "MM") as follows.

$$n \text{ (number of moles)} = \text{mass (in g)} \times \frac{1}{\text{molar mass (MM)}}$$

or

$$n = \frac{\text{mass}}{\text{MM}} \qquad \text{MM} = \frac{\text{mass}}{n}$$

**Solution**

| *Initial Conditions* | *Final Conditions* |
|---|---|
| $V_1 = 2.00 \text{ L}$ | $V_2 = ?$ |
| $P_1 = 380 \text{ torr}$ | $P_2 = 760 \text{ torr}$ |
| $T_1 = (17 + 273) \text{ K} = 290 \text{ K}$ | $T_2 = 273 \text{ K}$ |

1.

$$\frac{P_1 V_1}{T_1} = \frac{P_2 V_2}{T_2}$$

$$V_2 = V_1 \times \frac{P_1}{P_2} \times \frac{T_2}{T_1}$$

$$V_2 = 2.00 \text{ L} \times \frac{380 \text{ torr}}{760 \text{ torr}} \times \frac{273 \text{ K}}{290 \text{ K}} = \underline{0.941 \text{ L (STP)}}$$

2.

$$0.941 \text{ L} \times \frac{1 \text{ mol}}{22.4 \text{ L}} = 0.0420 \text{ mol of gas}$$

3.

$$\text{molar mass} = \frac{\text{mass}}{n} = \frac{3.20 \text{ g}}{0.0420 \text{ mol}} = \underline{\underline{76.2 \text{ g/mol}}}$$

**Alternative**

The alternative procedure is more direct, since it uses the ideal gas law as shown in the reverse of path 2 in Figure 9-12. First, use the ideal gas law to find the number of moles of gas. Then use the relationship $n = \text{mass/MM}$ to find the molar mass of the gas.

$$PV = nRT \quad n = PV/RT \qquad T = 17 + 273 = 290 \text{ K}$$

$$P = \frac{380 \text{ torr}}{760 \text{ torr/atm}} = 0.500 \text{ atm}$$

$$n = \frac{0.500 \text{ atm} \times 2.00 \text{ L}}{0.0821 \frac{\text{L} \cdot \text{atm}}{\text{K} \cdot \text{mol}} \times 290 \text{ K}} = 0.0420 \text{ mol}$$

$$n = \frac{\text{mass}}{\text{MM}} \qquad \text{MM} = \frac{\text{mass}}{n} = \frac{3.20 \text{ g}}{0.0420 \text{ mol}} = \underline{\underline{76.2 \text{ g/mol}}}$$

In Chapter 2, the densities of solids and liquids were given in units of g/mL. Since gases are much less dense, units in this case are usually given in g/L (STP). The density of a gas at STP can be calculated by dividing the molar mass by the molar volume.

$$CO_2: 44.01 \text{ g/mol} = 22.4 \text{ L/mol}$$

$$\frac{44.01 \text{ g/mol}}{22.4 \text{ L/mol}} = 1.96 \text{ g/L (STP)}$$

The densities of several gases are given in Table 9-2. The density of air (a mixture) is 1.29 g/L. Gases such as He and $H_2$ are less dense than air. Gases less dense than air rise in the air just as solids or liquids less dense than water float on water. Helium is used as the gas in blimps to make the whole craft "lighter than air." (See Exercise 9-66.)

**Blimps stay suspended because helium is less dense than air.**

# 9-9 Stoichiometry Involving Gases

**Looking Ahead!** **Gases are formed or consumed in many chemical reactions. Since the volumes of gases relate directly to the number of moles, we can use volume as we did mass or number of molecules in previous stoichiometric calculations. For this discussion, we return to the topic of stoichiometry introduced in Chapter 8.**

The beauty of the ideal gas law is that it allows us to convert a given volume of gas at a specified temperature and pressure directly into moles of gas. When

**Table 9-2  Densities of Some Gases**

| Gas | Density [g/L (STP)] | Gas | Density [g/L (STP)] |
| --- | --- | --- | --- |
| $H_2$ | 0.090 | $O_2$ | 1.43 |
| He | 0.179 | $CO_2$ | 1.96 |
| $N_2$ | 1.25 | $CF_2Cl_2$ | 5.40 |
| Air (average) | 1.29 | $SF_6$ | 6.52 |

**Figure 9-13 General Procedure for Stoichiometric Problems**
**Gas volumes relate directly to moles, so they are included in stoichiometric calculations.**

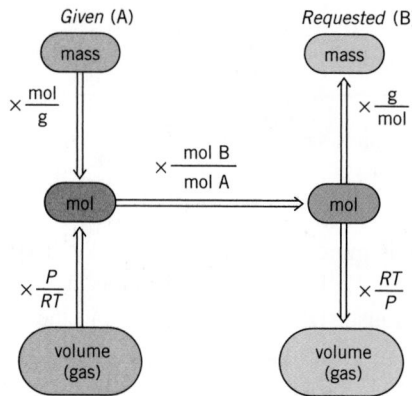

volumes of gases react or are formed as a product in a chemical reaction, we can thus calculate the moles of gases involved. Just as mass and number can be converted to moles of a reactant or product as shown in Figure 8-4, volumes of gases can now be included in the general scheme. (See Figure 9-13.) We have omitted number of molecules from this scheme because problems involving actual numbers of molecules are not usually encountered. The volume and number of moles of a gas are shown to be interrelated by the ideal gas law. In fact, if the volume of the gas is measured at STP, the conversion can be simplified by using the molar volume relationship

$$n = \frac{V(STP)}{22.4 \text{ L/mol}}$$

The following two examples illustrate the relationship of the gas laws to stoichiometry.

**Example 9-17**

**working it out**

**Mass of a Reactant from the Volume of a Product**

Automobiles in the future will very likely run on hydrogen gas in a fuel cell. Storing the hydrogen and providing it conveniently remain problems, however. One possibility is the addition of water to a solid compound such as sodium borohydride. The balanced equation illustrating the generation of hydrogen by this reaction is

$$NaBH_4(s) + 2H_2O(l) \longrightarrow NaBO_2(aq) + 4H_2(g)$$

What mass of $NaBH_4$ is needed to produce 50.0 L of $H_2$ measured at STP?

**Procedure**

The general procedure is shown below. In this case, it is easier to use the molar volume as a conversion factor between volume and moles.

$$Given \ [L(g)H_2] \qquad\qquad Requested \ (g \ NaBH_4)$$

$$\times \frac{\text{mol } H_2}{22.4 \text{ L}} \qquad \times \frac{\text{mol } NaBH_4}{\text{mol } H_2} \qquad \times \frac{\text{g } NaBH_4}{\text{mol } NaBH_4}$$

$$\boxed{\text{volume } H_2} \Longrightarrow \boxed{\text{mol } H_2} \Longrightarrow \boxed{\text{mol } NaBH_4} \Longrightarrow \boxed{\text{mass } NaBH_4}$$

**Solution**

$$50.0 \text{ L(STP)} \times \frac{1 \text{ mol H}_2}{22.4 \text{ L(STP)}} \times \frac{1 \text{ mol NaBH}_4}{4 \text{ mol H}_2} \times \frac{37.83 \text{ g NaBH}_4}{\text{mol NaBH}_4} = \underline{\underline{21.1 \text{ g NaBH}_4}}$$

## Volume of a Product from Mass of a Reactant

Example 9-18

Given the balanced equation

$$4NH_3(g) + 5O_2(g) \longrightarrow 4NO(g) + 6H_2O(l)$$

what volume of NO gas measured at 0.724 atm and 25°C will be produced from 19.5 g of $O_2$?

**Procedure**

$$\textit{Given (g O}_2\textit{)} \qquad\qquad\qquad \textit{Requested [L(g)NO]}$$

$$\times \frac{\text{mol O}_2}{\text{g O}_2} \qquad \times \frac{\text{mol NO}}{\text{mol O}_2} \qquad \times \frac{RT}{P}$$

$$\boxed{\text{mass O}_2} \Longrightarrow \boxed{\text{mol O}_2} \Longrightarrow \boxed{\text{mol NO}} \Longrightarrow \boxed{\text{volume NO}}$$

**Solution**

$$19.5 \text{ g O}_2 \times \frac{1 \text{ mol O}_2}{32.00 \text{ g O}_2} \times \frac{4 \text{ mol NO}}{5 \text{ mol O}_2} = 0.488 \text{ mol NO}$$

1. Using the ideal gas law

$$V = \frac{nRT}{P} = \frac{0.488 \text{ mol} \times 0.0821 \dfrac{\text{L} \cdot \text{atm}}{\text{K} \cdot \text{mol}} \times 298 \text{ K}}{0.724 \text{ atm}} = \underline{\underline{16.5 \text{ L}}}$$

2. Convert moles of NO to volume at STP and then volume at STP to volume at 550 torr and 25°C.

$$0.488 \text{ mol} \times 22.4 \text{ L/mol} = 10.9 \text{ L (STP)}$$

$$V_2 = V_1 \times \frac{P_1}{P_2} \times \frac{T_2}{T_1}$$

*Initial Conditions*      *Final Conditions*

$V_1 = 10.9$ L        $V_2 = ?$

$P_1 = 760$ torr     $P_2 = 550$ torr

$T_1 = 273$ K       $T_2 = (25 + 273)$ K $= 298$ K

$$V_2 = 10.9 \text{ L} \times \frac{760 \text{ torr}}{550 \text{ torr}} \times \frac{298 \text{ K}}{273 \text{ K}} = \underline{\underline{16.4 \text{ L}}}$$

**Looking Back!** Under most Earthly conditions the nature of gases is described accurately by the assumptions of the kinetic theory. Because of this, all of the conditions of a gas (temperature, volume, pressure, and amount) conveniently relate to each other in the ideal gas law. In addition, since gases all behave the same, we find that the volume and the pressure of a mixture depend only on the total number of molecules present. One property that does depend on the identity of the gas is its density, which relates to the molar mass of the gas. Finally, we can use the ideal gas law to relate volumes of gases in stoichiometry problems.

## Learning Check B | checking it out

**B-1.** Fill in the blanks.

The ideal gas law combines most of the gas laws into one relationship, which is _____ . $R$ is known as the _____ _____ and has the value _____ $\dfrac{\text{L} \cdot \text{atm}}{\text{K} \cdot \text{mol}}$. One mole of any gas occupies _____ L at STP, and this is known as the _____ _____ . The density of a gas at STP is obtained by dividing the _____ _____ by the molar volume. In stoichiometry calculations, the volume of a gas can be converted to number of moles by use of the _____ _____ if the volume is measured at STP. If other conditions are present, the volume is related to the number of moles by the _____ _____ _____ .

**B-2.** What is the Celsius temperature of $2.46 \times 10^{-4}$ g of $N_2O$ if it is present in a 2.35-mL container at a pressure of 0.0468 atm?

**B-3.** What is the partial pressure of $O_2$ if 0.0450 mol of $O_2$ is mixed with 0.0328 mol of $N_2$ and the total pressure in the container is 596 torr?

**B-4.** What is the volume occupied by 142 g of $SF_4$ at STP? What is the density of $SF_4$ at STP?

**B-5.** Humans obtain energy from the combustion of blood sugar (glucose) according to the equation

$$C_6H_{12}O_6(aq) + 6O_2(g) \longrightarrow 6CO_2(g) + 6H_2O(l)$$

**a.** What volume of $O_2$ measured at STP is required to react with 0.122 mol of glucose?

**b.** What volume of $CO_2$ measured at 25°C and 1.10 atm pressure is released from the combustion of 228 g of glucose?

**c.** What mass of glucose is required to react with 2.48 L of oxygen measured at 22°C and 655 torr?

*Additional Examples: Exercises 9-59, 9-61, 9-68, 9-73, 9-79, 9-84, 9-92, and 9-96.*

## chapter Review | putting it together

Our modern theories about matter depended on some basic understanding about the most abstract state of matter—gases. Beginning in the 1600s, the quantitative laws that we know as the gas laws began to be advanced. These laws and other observations of the nature of gases led to the accepted model of behavior known as the **kinetic molecular theory.** One important result of **kinetic theory** is the relationship between temperature and the average kinetic energy of molecules. The velocity of the molecules, rate of **effusion,** and rate of **diffusion** all relate to the molar mass of the gas.

An understanding of the gaseous state began in the mid-1600s with Torricelli. His studies of the **barometer** led to a description of **pressure** as a **force** per unit area exerted by gases of the atmosphere. The gas laws now seem quite reasonable and predictable from kinetic theory. Except for average velocity and density, which depend on the mass of the molecules, the other gas laws are independent of the identity of the gas. Avogadro advanced the observation that equal volumes of gases contain equal numbers of molecules under the same conditions, and Dalton observed that pressures depended upon the total amount of gas present.

The gas laws in this chapter and their applications are summarized in the accompanying table.

The gas laws require the use of the **Kelvin** temperature scale, which begins at **absolute zero.** The volumes of gases are often described under **standard temperature**

**and pressure (STP)** conditions. At STP, the volume of one mole of a gas is 22.4 L and is known as the **molar volume.** The molar volume can be used to calculate the

| Gas Law | Relationship | Meaning | Constant Conditions | Application |
|---|---|---|---|---|
| **Graham's** | $v \propto \dfrac{1}{\sqrt{MM}}$ | $v\uparrow MM\downarrow$ | $T$ | Relates MM and $v$ of two different gases at a specific $T$ |
| **Boyle's** | $V \propto \dfrac{1}{P}$ | $V\uparrow P\downarrow$ | $T, n$ | Relates $V$ and $P$ of a gas under two different sets of conditions |
| **Charles's** | $V \propto T$ | $V\uparrow T\uparrow$ | $P, n$ | Relates $V$ and $T$ of a gas under two different sets of conditions |
| **Gay-Lussac's** | $P \propto T$ | $P\uparrow T\uparrow$ | $V, n$ | Relates $P$ and $T$ of a gas under two different sets of conditions |
| **Combined** | $PV \propto T$ | $PV\uparrow T\uparrow$ | $n$ | Relates $P$, $V$, and $T$ of a gas under two different sets of conditions |
| **Avogadro's** | $V \propto n_{tot}$ | $V\uparrow n_{tot}\uparrow$ | $P, T$ | Relates $V$ and $n$ of a gas under two different sets of conditions |
| **Ideal** | $PV \propto n_{tot}T$ | $PV\uparrow n_{tot}T\uparrow$ | — | Relates $P$, $V$, $T$, or $n$ to the other three variables |
| **Dalton's** | $P_{tot} = P_1 + P_2$, etc. | $P_{tot}\uparrow n_{tot}\uparrow$ | $T, V$ | Relates $P_{tot}$ to partial pressures of component gases |

| | | |
|---|---|---|
| $V$ = volume | $P_{tot}$ = total pressure | $v$ = average velocity |
| $T$ = Kelvin temperature | $n$ = moles | $\uparrow$ = quantity increases |
| $P$ = pressure | $n_{tot}$ = total moles | $\downarrow$ = quantity decreases |
| | MM = molar mass | |

density of a gas at STP, or it can be used directly as a conversion factor between moles and volume at STP. To convert between moles and volume under other conditions, the use of the ideal gas law is convenient. In the ideal gas law ($PV = nRT$), $R$ is known as the **gas constant.**

$$\text{volume } (T, P) \xleftarrow{\text{ideal gas law}} \text{moles of gas} \xrightarrow{\text{molar volume}} \text{volume (STP)}$$

In the final section, we related the volume of a gas to stoichiometry as summarized in Figure 9-12.

# Exercises

## The Kinetic Theory of Gases

**9-1.** It is harder to move your arms in water than in air. Explain on the basis of kinetic molecular theory.

**9-2.** A balloon filled with water is pear-shaped, but a balloon filled with air is spherical. Explain.

**9-3.** When a gasoline tank is filled, no more gasoline can be added. When a tire is "filled," however, more air can be added. Explain.

**9-4.** The pressure inside an auto tire is the same regardless of the location of the nozzle (i.e., up, down, or to the side). Explain.

**9-5.** A sunbeam forms when light is reflected from dust suspended in the air. Even if the air is still, the dust particles can be seen to bounce around randomly. Explain.

**9-6.** A bowling ball weighs 6.00 kg and a bullet weighs 1.50 g. If the bowling ball is rolled down an alley at 20.0 mi/hr, what is the velocity of a bullet having the same kinetic energy?

**9-7.** Arrange the following gases in order of increasing average speed (rate of effusion) at the same temperature.
(a) $CO_2$     (c) $N_2$     (e) $N_2O$
(b) $SO_2$     (d) $SF_6$     (f) $H_2$

**9-8.** What is the rate of effusion of $N_2$ molecules compared with Ar atoms at the same temperature?

**9-9.** Compare the rates of effusion of $H_2$ molecules and Kr atoms at the same temperature.

**\*9-10.** A certain gas effuses twice as fast as $SF_6$ molecules. What is the molar mass of the unknown gas?

**\*9-11.** Carbon monoxide effuses 2.13 times faster than an unknown gas. What is the molar mass of the unknown gas?

**\*9-12.** To make enriched uranium for use in nuclear reactors or for weapons, $^{235}U$ must be separated from $^{238}U$. Although $^{235}U$ is the isotope needed for fission, only 0.7% of U atoms are this isotope. Separation is a difficult and expensive process. Since $UF_6$ is a gas, Graham's law can be applied to separate the isotopes. How much faster does a $^{235}UF_6$ molecule travel on the average compared to a $^{238}UF_6$ molecule?

**\*9-13.** The kinetic theory assumes that the volume of molecules and their interactions are negligible for gases. Explain why these assumptions may not be true when the pressure is very high and the temperature is very low.

## Units of Pressure

**9-14.** Make the following conversions.
(a) 1650 torr to atm          (d) 5.65 kPa to atm
(b) $3.50 \times 10^{-5}$ atm to torr   (e) 190 torr to lb/in.$^2$
(c) 185 lb/in.$^2$ to torr       (f) 85 torr to kPa

**9-15.** Make the following conversions.
(a) 30.2 in. of Hg to torr      (c) 57.9 kPa to lb/in.$^2$
(b) 25.7 kilobars to atm        (d) 0.025 atm to torr

**9-16.** Complete the following table.

| torr | lb/in.$^2$ | in. Hg | kPa | atm |
|------|-----------|--------|-----|-----|
| 455  | ——        | ——     | ——  | ——  |
| ——   | 2.45      | ——     | ——  | ——  |
| ——   | ——        | 117    | ——  | ——  |
| ——   | ——        | ——     | 783 | ——  |
| ——   | ——        | ——     | ——  | 0.0768 |

**9-17.** The atmospheric pressure on the planet Mars is 10.3 millibars. What is this pressure in Earth atmospheres?

**9-18.** The atmospheric pressure on the planet Venus is 0.0920 kilobar. What is this pressure in Earth atmospheres?

**9-19.** The density of water is 1.00 g/mL. If water is substituted for mercury in the barometer, how high (in feet) would a column of water be supported by 1 atm? A water well is 40 ft deep. Can suction be used to raise the water to ground level?

## Boyle's Law

**9-20.** A gas has a volume of 6.85 L at a pressure of 0.650 atm. What is the volume of the gas if the pressure is decreased to 0.435 atm?

**9-21.** If a gas has a volume of 1560 mL at a pressure of 81.2 kPa, what is its volume if the pressure is increased to 2.50 atm?

**9-22.** At sea level, a balloon has a volume of 785 mL. What is its volume if it is taken to a place in Colorado where the atmospheric pressure is 610 torr?

**9-23.** A gas has a volume of 125 mL at a pressure of 62.5 torr. What is the pressure if the volume is decreased to 115 mL?

**9-24.** How does the kinetic molecular theory explain Boyle's law?

**9-25.** A few miles above the surface of Earth the pressure drops to $1.00 \times 10^{-5}$ atm. What would be the volume of a 1.00-L sample of gas at sea-level pressure (1.00 atm) if it were taken to that altitude? (Assume constant temperature.)

**\*9-26.** A gas in a piston engine is compressed by a ratio of 15:1. If the pressure before compression is 0.950 atm, what pressure is required to compress the gas? (Assume constant temperature.)

**\*9-27.** The volume of a gas is measured as the pressure is varied. The four measurements are reported as follows.

| Experiment | Volume (mL) | Pressure (torr) |
|------------|-------------|-----------------|
| 1          | 125         | 450             |
| 2          | 145         | 385             |
| 3          | 175         | 323             |
| 4          | 220         | 253             |

Make a graph using volume on the $x$ axis and pressure on the $y$ axis.

## Charles's Law

**9-28.** A balloon has a volume of 1.55 L at 25°C. What would be the volume if the balloon is heated to 100°C? (Assume constant $P$.)

**9-29.** A sample of gas has a volume of 677 mL at 63°C. What is the volume of the gas if the temperature is decreased to 46°C?

**9-30.** A balloon has a volume of 325 mL at 17°C. What is the temperature if the volume increases to 392 mL?

**9-31.** How does the kinetic molecular theory explain Charles's law?

**9-32.** A quantity of gas has a volume of $3.66 \times 10^4$ L. What will be the volume if the temperature is changed from 455 K to 50°C?

**\*9-33.** The temperature of a sample of gas is 0°C. When the temperature is increased, the volume increases by a factor of 1.25 (i.e., $V_2 = 1.25 \, V_1$.) What is the final temperature in degrees Celsius?

**\*9-34.** The volume of a gas is measured as the temperature is varied. The four measurements are reported as follows.

| Experiment | Volume (L) | Temperature (°C) |
|---|---|---|
| 1 | 1.54 | 20 |
| 2 | 1.65 | 40 |
| 3 | 1.95 | 100 |
| 4 | 2.07 | 120 |

Make a graph of the volume on the $x$ axis and the Kelvin temperature on the $y$ axis. What is the average value of the constant of proportionality, $k$?

## Gay-Lussac's Law

**9-35.** A confined quantity of gas is at a pressure of 2.50 atm and a temperature of $-22°C$. What is the pressure if the temperature increases to 22°C?

**9-36.** A quantity of gas has a volume of 3560 mL at a temperature of 55°C and a pressure of 850 torr. What is the temperature if the volume remains unchanged but the pressure is decreased to 0.652 atm?

**9-37.** A metal cylinder contains a quantity of gas at a pressure of 558 torr at 25°C. At what temperature does the pressure inside the cylinder equal 1 atm pressure?

**9-38.** An aerosol spray can has gas under pressure of 1.25 atm at 25°C. The can explodes when the pressure reaches 2.50 atm. At what temperature will this happen? (Do not throw these cans into a fire!)

**9-39.** The pressure in an automobile tire is 28.0 lb/in.$^2$ on a chilly morning of 17°C. After it is driven a while, the temperature of the tire rises to 40°C. What is the pressure in the tire if the volume remains constant?

**9-40.** How does the kinetic molecular theory explain Gay-Lussac's law?

**\*9-41.** The pressure of a confined volume of gas is measured as the temperature is raised. The four measurements are reported as follows:

| Experiment | Pressure (torr) | Temperature (K) |
|---|---|---|
| 1 | 550 | 295 |
| 2 | 685 | 372 |
| 3 | 745 | 400 |
| 4 | 822 | 445 |

Make a graph of the pressure on the $x$ axis and the temperature on the $y$ axis. What is the average value of the constant of proportionality, $k$?

## The Combined Gas Law

**9-42.** Which of the following are legitimate expressions of the combined gas law?
**(a)** $PV = kT$
**(b)** $PT \propto V$
**(c)** $\dfrac{P_1 T_1}{V_1} = \dfrac{P_2 T_2}{V_2}$
**(d)** $\dfrac{P}{T} \propto \dfrac{1}{V}$
**(e)** $VT \propto P$

**9-43.** Which of the following are not STP conditions?
**(a)** $T = 273$ K      **(d)** $P = 1$ atm      **(f)** $P = 760$ torr
**(b)** $P = 760$ atm      **(e)** $t(C) = 273°C$      **(g)** $t(C) = 0°C$
**(c)** $T = 0$ K

**9-44.** In the following table, indicate whether the pressure, volume, or temperature increases or decreases.

| Experiment | P | V | T |
|---|---|---|---|
| 1 | increases | constant | _____ |
| 2 | constant | _____ | decreases |
| 3 | _____ | decreases | constant |
| 4 | increases | increases | _____ |

**9-45.** In the following table, indicate whether the pressure, volume, or temperature increases or decreases.

| Experiment | P | V | T |
|---|---|---|---|
| 1 | decreases | _____ | constant |
| 2 | constant | _____ | T(initial) = 350 K<br>T(final) = 40°C |
| 3 | P(initial) = 1.75 atm<br>P(final) = 2200 torr | constant | _____ |
| 4 | _____ | increases | decreases |

**9-46.** A 5.50-L volume of gas has a pressure of 0.950 atm at 0°C. What is the pressure if the volume decreases to 4.75 L and the temperature increases to 35°C?

**9-47.** A quantity of gas has a volume of 17.5 L at a pressure of 6.00 atm and temperature of 100°C. What is its volume at STP?

**9-48.** A quantity of gas has a volume of 88.7 mL at STP. What is its volume at 0.845 atm and 35°C?

**9-49.** A quantity of gas has a volume of $4.78 \times 10^{-4}$ mL at a temperature of $-50°C$ and a pressure of 78.0 torr. If the volume changes to $9.55 \times 10^{-5}$ mL and the pressure to 155 torr, what is the temperature?

**9-50.** A gas has a volume of 64.2 L at STP. What is the temperature if the volume decreases to 58.5 L and the pressure increases to 834 torr?

**9-51.** A quantity of gas has a volume of $6.55 \times 10^{-5}$ L at 7°C and 0.882 atm. What is the pressure if the volume changes to $4.90 \times 10^{-3}$ L and the temperature to 273 K?

**9-52.** A balloon has a volume of 1.55 L at 25°C and 1.05 atm pressure. If it is cooled in the freezer, the volume shrinks to 1.38 L and the pressure drops to 1.02 atm. What is the temperature in the freezer?

**9-53.** A bubble from a deep-sea diver in the ocean starts with a volume of 35.0 mL at a temperature of 17°C and a pressure of 11.5 atm. What is the volume of the bubble when it reaches the surface? Assume that the pressure at the surface is 1 atm and the temperature is 22°C.

**Avogadro's Law**

**9-54.** A 0.112-mol quantity of gas has a volume of 2.54 L at a certain temperature and pressure. What is the volume of 0.0750 mol of gas under the same conditions?

**9-55.** A balloon has a volume of 188 L and contains 8.40 mol of gas. How many moles of gas would be needed to expand the balloon to 275 L? Assume the same temperature and pressure in the balloon.

**9-56.** A balloon has a volume of 275 mL and contains 0.0212 mol of $CO_2$. What mass of $N_2$ must be added to expand the balloon to 400 mL?

**9-57.** A balloon has a volume of 75.0 mL and contains $2.50 \times 10^{-3}$ mol of gas. What mass of $N_2$ must be added to the balloon for the volume to increase to 164 mL at the same temperature and pressure?

**9-58.** A 48.0-g quantity of $O_2$ in a balloon has a volume of 30.0 L. What is the volume if 48.0 g of $SO_2$ is substituted for $O_2$ in the same balloon?

**Ideal Gas Law**

**9-59.** What is the temperature (in degrees Celsius) of 4.50 L of a 0.332-mol quantity of gas under a pressure of 2.25 atm?

**9-60.** A quantity of gas has a volume of 16.5 L at 32°C and a pressure of 850 torr. How many moles of gas are present?

**9-61.** What mass of $NH_3$ gas has a volume of 16,400 mL, a pressure of 0.955 atm, and a temperature of −23°C?

**9-62.** What is the pressure (in torr) exerted by 0.250 g of $O_2$ in a 250-mL container at 29°C?

**9-63.** A container of $Cl_2$ gas has a volume of 750 mL and is at a temperature of 19°C. If there is 7.88 g of $Cl_2$ in the container, what is the pressure in atmospheres?

**9-64.** What mass of Ne is contained in a large neon light if the volume is 3.50 L, the pressure 1.15 atm, and the temperature 23°C?

**9-65.** A sample of $H_2$ is collected in a bottle over water. The volume of the sample is 185 mL at a temperature of 25°C. The pressure of $H_2$ in the bottle is 736 torr. What is the mass of $H_2$ in the bottle?

**\*9-66.** A blimp has a volume of about $2.5 \times 10^7$ L. What is the mass of He (in lb) in the blimp at 27°C and 780 torr? The average molar mass of air is 29.0 g/mol. What mass of air (in lb) would the blimp contain? The difference between these two values is the lifting power of the blimp. What mass could the blimp lift? If $H_2$ is substituted for He, what is the lifting power? Why isn't $H_2$ used?

**\*9-67.** A good vacuum pump on Earth can produce a vacuum with a pressure as low as $1.00 \times 10^{-8}$ torr. How many molecules are present in each milliliter at a temperature of 27.0°C?

**Dalton's Law**

**9-68.** Three gases are mixed in a 1.00-L container. The partial pressure of $CO_2$ is 250 torr, that of $N_2$ 375 torr, and that of He 137 torr. What is the pressure of the mixture of gases?

**9-69.** The total pressure in a cylinder containing a mixture of two gases is 1.46 atm. If the partial pressure of one gas is 750 torr, what is the partial pressure of the other gas?

**9-70.** Air is about 0.90% Ar. If the barometric pressure is 756 torr, what is the partial pressure of Ar?

**9-71.** A sample of oxygen is collected in a bottle over water. The pressure inside the bottle is made equal to the barometric pressure, which is 752 torr. When collected over water, the gas is a mixture of oxygen and water vapor. The partial pressure of water (known as the vapor pressure) at that temperature is 24 torr. What is the pressure of the pure oxygen?

**9-72.** A container holds two gases, A and B. Gas A has a partial pressure of 325 torr and gas B has a partial pressure of 488 torr. What percent of the molecules in the mixture is gas A?

**9-73.** A volume of gas is composed of $N_2$, $O_2$, and $SO_2$. If the total pressure is 1050 torr, what is the partial pressure of each gas if the gas is 72.0% $N_2$ and 8.00% $O_2$?

**\*9-74.** A mixture of two gases is composed of $CO_2$ and $O_2$. The partial pressure of $O_2$ is 256 torr, and it represents 35.0% of the molecules of the mixture. What is the total pressure of the mixture?

**\*9-75.** A volume of gas has a total pressure of 2.75 atm. If the gas is composed of 0.250 mol of $N_2$ and 0.427 mol of $CO_2$, what is the partial pressure of each gas?

**\*9-76.** The following gases are all combined into a 2.00-L container: a 2.00-L volume of $N_2$ at 300 torr, a 4.00-L volume of $O_2$ at 85 torr, and a 1.00-L volume of $CO_2$ at 450 torr. What is the total pressure?

**\*9-77.** The total pressure of a mixture of two gases is 0.850 atm in a 4.00-L container. Before mixing, gas A was in a 2.50-L container and had a pressure of 0.880 atm. What is the partial pressure of gas B in the 4.00-L container?

**\*9-78.** What is the pressure (in atm) in a 825-mL container at 33°C if it contains 6.25 g of $N_2$ and 12.6 g of $CO_2$?

## Molar Volume and Density

**9-79.** What is the volume of 15.0 g of $CO_2$ measured at STP?

**9-80.** What is the mass (in kilograms) of 850 L of CO measured at STP?

**9-81.** What is the volume of $3.01 \times 10^{24}$ molecules of $N_2$ measured at STP?

**9-82.** A 6.50-L quantity of a gas measured at STP has a mass of 39.8 g. What is the molar mass of the compound?

**9-83.** What is the mass of $6.78 \times 10^{-4}$ L of $NO_2$ measured at STP?

**9-84.** What is the density in g/L (STP) of $B_2H_6$?

**9-85.** What is the density in g/L (STP) of $BF_3$?

**9-86.** A gas has a density of 1.52 g/L (STP). What is the molar mass of the gas?

**9-87.** A gas has a density of 6.14 g/L (STP). What is the molar mass of the gas?

**\*9-88.** A gas has a density of 3.60 g/L at a temperature of 25°C and a pressure of 1.20 atm. What is its density at STP?

**\*9-89.** What is the density (in g/L) of $N_2$ measured at 500 torr and 22°C?

**\*9-90.** What is the density (in g/L) of $SF_6$ measured at 0.370 atm and 37°C?

## Stoichiometry Involving Gases

**9-91.** Limestone is dissolved by $CO_2$ and water according to the equation

$$CaCO_3(s) + H_2O(l) + CO_2(g) \longrightarrow Ca(HCO_3)_2(aq)$$

What volume of $CO_2$ measured at STP would dissolve 115 g of $CaCO_3$?

**9-92.** Magnesium in flashbulbs burns according to the equation

$$2Mg(s) + O_2(g) \longrightarrow 2MgO(s)$$

What mass of Mg combines with 5.80 L of $O_2$ measured at STP?

**9-93.** Oxygen gas can be prepared in the laboratory by decomposition of potassium nitrate according to the equation

$$2KNO_3(s) \xrightarrow{\Delta} 2KNO_2(s) + O_2(g)$$

What mass of $KNO_2$ forms along with 14.5 L of $O_2$ measured at 1 atm and 25°C?

**9-94.** Acetylene ($C_2H_2$) is produced from calcium carbide as shown by the reaction

$$CaC_2(s) + 2H_2O(l) \longrightarrow Ca(OH)_2(s) + C_2H_2(g)$$

What volume of acetylene measured at 25°C and 745 torr would be produced from 5.00 g of $H_2O$?

**9-95.** Nitrogen dioxide is an air pollutant. It is produced from NO (from car exhaust) as follows.

$$2NO(g) + O_2(g) \longrightarrow 2NO_2(g)$$

What volume of NO measured at STP is required to react with 5.00 L of $O_2$ measured at 1.25 atm and 17°C?

**9-96.** Butane ($C_4H_{10}$) burns according to the equation

$$2C_4H_{10}(g) + 13O_2(g) \longrightarrow 8CO_2(g) + 10H_2O(l)$$

**(a)** What volume of $CO_2$ measured at STP would be produced by 85.0 g of $C_4H_{10}$?
**(b)** What volume of $O_2$ measured at 3.25 atm and 127°C would be required to react with 85.0 g of $C_4H_{10}$?
**(c)** What volume of $CO_2$ measured at STP would be produced from 45.0 L of $C_4H_{10}$ measured at 25°C and 0.750 atm?

**9-97.** In March 1979, a nuclear reactor overheated, producing a dangerous hydrogen bubble at the top of the reactor core. The following reaction occurring at the high temperature (about 1500°C) accounted for the hydrogen. (Zr alloys hold the uranium pellets in long rods.)

$$Zr(s) + 2H_2O(g) \longrightarrow ZrO_2(s) + 2H_2(g)$$

If the bubble had a volume of about 28,000 L at 250°C and 70.0 atm, what mass (in kg and tons) of Zr had reacted?

**9-98.** Nitric acid is produced according to the equation

$$3NO_2(g) + H_2O(l) \longrightarrow 2HNO_3(aq) + NO(g)$$

What volume of $NO_2$ measured at −73°C and $1.56 \times 10^{-2}$ atm would be needed to produce $4.55 \times 10^{-3}$ mol of $HNO_3$?

**\*9-99.** Natural gas ($CH_4$) burns according to the equation

$$CH_4(g) + 2O_2(g) \longrightarrow CO_2(g) + 2H_2O(l)$$

What volume of $CO_2$ measured at 27°C and 1.50 atm is produced from 27.5 L of $O_2$ measured at −23°C and 825 torr?

## General Problems

**9-100.** A column of mercury (density 13.6 g/mL) is 15.0 cm high. A cross-section of the column has an area of 12.0 cm$^2$. What is the force (weight) of the mercury at the bottom of the tube? What is the pressure in grams per square centimeter and in atmospheres?

**9-101.** A tube containing an alcohol (density 0.890 g/mL) is 1.00 m high and has a cross-section of 15.0 cm$^2$. What is the total force at the bottom of the tube? What is the pressure? How high would be an equivalent amount of mercury assuming the same cross-section?

**9-102.** A 1.00-L volume of a gas weighs 8.37 g. The gas volume is measured at 1.45 atm pressure and 35°C. What is the molar mass of the gas?

**9-103.** A gaseous compound is 85.7% C and 14.3% H. A 6.58-g quantity of this gas occupies 4500 mL at 77.0°C and a pressure of 1.00 atm. What is the molar mass of the compound? What is its molecular formula?

**9-104.** What is the volume at STP of a mixture of 10.0 g each of Ar, $Cl_2$, and $N_2$? What is the partial pressure of each gas?

**9-105.** What is the volume occupied by a mixture of 0.265 mol of $O_2$, 9.88 g of $N_2$, and $9.65 \times 10^{22}$ molecules of $CO_2$ at a temperature of 37°C and a pressure of 2.86 atmospheres? What is the partial pressure of each gas?

**9-106.** Molecular clouds in space contain 30,000 molecules/mL at a temperature of 10 K. What is the pressure in atmospheres?

**9-107.** Neptune is a planet that orbits the sun about 4.5 billion miles from Earth. It has a moon, Triton, with a thin atmosphere. *Voyager 2* measured a surface temperature of 38 K and a pressure of 10 microbars (1 microbar = $10^{-6}$ bar). What is the density (in g/L) of the atmosphere at the surface? Assume that the atmosphere on Triton is nitrogen. How does this compare with the density of air at STP on Earth?

**9-108.** What is the molar volume at 25°C and 1.25 atm? What is the density of $CO_2$ under these conditions?

**9-109.** A hot-air balloon rises because the heated air trapped in the balloon is less dense than the surrounding air. What is the density of air (assume an average molar mass of 29.0 g/mol) at 400°C and 1 atm pressure? Compare this to the density of air at STP.

**9-110.** A compound is 80.0% carbon and 20.0% hydrogen. Its density at STP is 1.34 g/L. What is its molecular formula?

**9-111.** Given the following *unbalanced* equation

$$H_3BCO(g) + H_2O(l) \longrightarrow B(OH)_3(aq) + CO(g) + H_2(g)$$

A 425-mL quantity of $H_3BCO$ measured at 565 torr and 100°C was allowed to react with excess $H_2O$. What volume of gas was produced measured at 25°C and 0.900 atm?

**9-112.** Given the following *unbalanced* equation

$$C_3H_8O(g) + O_2(g) \longrightarrow CO_2(g) + H_2O(l)$$

What mass of water forms if 6.50 L of $C_3H_8O$ measured at STP is allowed to react with 42.0 L of $O_2$ measured at 27°C and 1.68 atm pressure? Assume that this is the only reaction that occurs.

**9-113.** Given the following *unbalanced* equation

$$Al(s) + F_2(g) \longrightarrow AlF_3(s)$$

An 8.23-L quantity of $F_2$ measured at 35°C and 725 torr was allowed to react with some Al. At the end of the reaction, 3.50 g of $F_2$ remained. What mass of $AlF_3$ formed?

**9-114.** Sulfuric acid is made from $SO_3$, which is obtained from the combustion of sulfur according to the following

*unbalanced* equations.

$$S(s) + O_2(g) \longrightarrow SO_2(g)$$
$$SO_2(g) + O_2(g) \longrightarrow SO_3(g)$$

What volume of $SO_3$ measured at 2.75 atm and 400°C is prepared from 50.0 kg of sulfur?

**9-115.** Liquid $N_2O_3$ decomposes according to the equation

$$N_2O_3(l) \longrightarrow NO_2(g) + NO(g)$$

What is the total volume of gas measured at 1.58 atm and 35°C produced by the decomposition of $2.54 \times 10^{24}$ molecules of $N_2O_3$?

**9-116.** Calcium bicarbonate is formed according to the following equation.

$$CaO(s) + H_2O(l) + 2CO_2(g) \longrightarrow Ca(HCO_3)_2(s)$$

If 80.0 g of CaO is mixed with $7.85 \times 10^{23}$ molecules of $H_2O$ and 30.0 L of $CO_2$ measured at 25°C and 820 torr, what mass of calcium bicarbonate is formed?

**9-117.** What volume of water vapor (gas) measured at 22.0 torr and 25°C contains the same number of molecules as 15.0 mL of ice if the density of ice is 0.917 g/mL?

**9-118.** The following equation represents what happens in swimming pools when ammonia (from people) reacts with sodium hypochlorite (used as a disinfectant). The $N_2H_4$ formed has a serious bad odor.

$$2NH_3(aq) + NaOCl(aq) \longrightarrow N_2H_4(g) + NaCl(aq) + H_2O$$

What volume of $N_2H_4$ (in mL) measured at STP is produced if 20.0 mL of $NH_3$ gas measured at 1.20 atm and 25°C is dissolved in water?

**\*9-119.** A 1.000-g sample of a gaseous compound composed of only nitrogen and fluorine contains 0.269 g of nitrogen. The gas has a density of 4.25 g/L measured at room temperature (25°C) and standard pressure. What is the formula of the gas?

**\*9-120.** A liquid compound composed of only nitrogen and oxygen is 69.6% oxygen. Decomposition of 0.0220 mole of this compound produces 2.03 g of a single gas that has a volume of 1.05 L measured at standard temperature and 715 torr. What is the formula of the orginal compound?

**9-121.** Steering on space vehicles is provided by a propulsion system that produces gaseous products when two liquids are mixed. It can produce short bursts of gases. The reaction is

$$H_2NN(CH_3)_2(l) + 2N_2O_4(l) \longrightarrow$$
$$3N_2(g) + 4H_2O(g) + 2CO_2(g)$$

What volume of gas measured at 1.75 atm and 120°C is produced if 125 g of each of the two reactants are mixed?

**9-122.** Magnesium and lithium both react with elemental nitrogen to form their respective nitrides. Write the balanced equations illustrating these reactions and determine what mass of each metal would react with 256 L of nitrogen measured at 985 torr and 373 K.

# Interactive Learning

**9-123.** A sample of helium at 740 torr and in a volume of 2.58 L was heated from 24.0°C to 75.0°C. The volume of the container expanded to 2.81 L. What was the final pressure (in torr) of helium?

**9-124.** A chemist isolated a gas in a glass bulb with a volume of 255 mL at a temperature of 25°C and a pressure of 10.0 torr. The gas weighed 12.1 mg. What is the molar mass of the gas?

**9-125.** A sample of carbon monoxide was prepared and collected over water at a temperature of 20°C and a total pressure

of 754 torr. It occupied a volume of 268 mL. What mass of the CO was in the sample? (When a gas is collected over water, it is composed of the gas and the water vapor. At 20°C the pressure due to the water vapor is 18 torr.)

**9-126.** What volume (in mL) of $O_2$ measured at 27°C and 654 torr is needed to react completely with 16.8 mL of $CH_4$ measured at 35°C and 725 torr?

# Solutions to Learning Checks

**A-1.** low, uniformly, kinetic molecular, rapid, random, collisions, kinetic, temperature, Graham's, lower, barometer, force, area, 760, decrease, increase, increases, combined gas law, increase

**A-2.** Molar mass of $SF_6 = [32.07 + 6(19.00)] = 146.1$ g/mol, of $N_2 = 2(14.01) = 28.02$ g/mol

$$\frac{v_{SF_6}}{v_{N_2}} = \sqrt{\frac{28.02 \text{ g/mol}}{146.1 \text{ g/mol}}} = 0.438$$

$v_{SF_6} = 0.438\, v_{N_2}$ ($SF_6$ moves less than half as fast as $N_2$.)

**A-3.** $0.650 \text{ atm} \times \dfrac{760 \text{ torr}}{\text{atm}} = \underline{\underline{494 \text{ torr}}}$

$0.650 \text{ atm} \times \dfrac{101.3 \text{ kPa}}{\text{atm}} = \underline{\underline{65.8 \text{ kPa}}}$

$0.650 \text{ atm} \times \dfrac{14.7 \text{ lb/in.}^2}{\text{atm}} = \underline{\underline{9.56 \text{ lb/in.}^2}}$

**A-4. (a)** $550 \text{ mL} \times \dfrac{1.00 \text{ atm}}{1.32 \text{ atm}} = \underline{\underline{417 \text{ mL}}}$

**(b)** $550 \text{ mL} \times \dfrac{(273 + 44) \text{ K}}{(273 + 22) \text{ K}} = \underline{\underline{591 \text{ mL}}}$

**(c)** $550 \text{ mL} \times \dfrac{(0.0227 + 0.0115) \text{ mol}}{0.0227 \text{ mol}} = \underline{\underline{829 \text{ mL}}}$

**(d)** $1.00 \text{ atm} \times \dfrac{(273 + 122) \text{ K}}{(273 + 22) \text{ K}} = \underline{\underline{1.34 \text{ atm}}}$

**(e)** $1.00 \text{ atm} \times \dfrac{550 \text{ mL}}{825 \text{ mL}} \times \dfrac{(273 + 102) \text{ K}}{(273 + 22) \text{ K}} = \underline{\underline{0.847 \text{ atm}}}$

**A-5. (a)** Boyle's law **(d)** Gay-Lussac's law
**(b)** Charles's law **(e)** combined gas law
**(c)** Avogadro's law

**B-1.** $PV = nRT$, gas constant, 0.0821, 22.4, molar volume, molar mass, molar volume, ideal gas law

**B-2.** $T = \dfrac{PV}{nR}$    molar mass

$$= (2 \times 14.01 \text{ g} + 16.00 \text{ g} = 44.01/\text{mol})$$

$$n = \frac{2.46 \times 10^{-4} \text{ g}}{44.01 \text{ g/mol}} = 5.59 \times 10^{-6} \text{ mol}$$

$$T = \frac{0.0468 \text{ atm} \times 2.35 \times 10^{-3} \text{ L}}{5.59 \times 10^{-6} \text{ mol} \times 0.0821 \dfrac{\text{L} \cdot \text{atm}}{\text{K} \cdot \text{mol}}} = 240 \text{ K}$$

$240 \text{ K} - 273 = \underline{\underline{-33°C}}$

**B-3.** $n_{tot} = 0.0450 + 0.0328 = 0.0778$ mol

decimal fraction of $O_2 = \dfrac{0.0450}{0.0778} = 0.578$

$0.578 \times 596 \text{ torr} = \underline{\underline{344 \text{ torr}}}$

**B-4.** $SF_4$: $[32.07 + 4(19.00)] = 108.1$ g/mol

mass $SF_4 \Longrightarrow$ mol $SF_4 \Longrightarrow$ vol $SF_4$

$142 \text{ g SF}_4 \times \dfrac{1 \text{ mol SF}_4}{108.1 \text{ g SF}_4} \times \dfrac{22.4 \text{ L}}{\text{mol SF}_4} = \underline{\underline{29.4 \text{ L (STP)}}}$

$\text{density} = \dfrac{108.1 \text{ g/mol}}{22.4 \text{ L/mol}} = \underline{\underline{4.83 \text{ g/L}}}$

**B-5.**

**(a)** mol $C_6H_{12}O_6 \Longrightarrow$ mol $O_2 \Longrightarrow$ vol $O_2$

$0.122 \text{ mol C}_6\text{H}_{12}\text{O}_6 \times \dfrac{6 \text{ mol O}_2}{1 \text{ mol C}_6\text{H}_{12}\text{O}_6} \times \dfrac{22.4 \text{ L}}{\text{mol O}_2} =$

$\underline{\underline{16.4 \text{ L (STP)}}}$

**(b)** $\boxed{\text{mass } C_6H_{12}O_6} \Longrightarrow \boxed{\text{mol } C_6H_{12}O_6} \Longrightarrow$

$\boxed{\text{mol } CO_2} \Longrightarrow \boxed{\text{vol } CO_2(P, T)}$

$C_6H_{12}O_6 = [(6 \times 12.01) + (12 \times 1.008)$
$+ (6 \times 16.00)] = 180.2 \text{ g/mol}$

$228 \text{ g } C_6H_{12}O_6 \times \dfrac{1 \text{ mol } C_6H_{12}O_6}{180.2 \text{ g } C_6H_{12}O_6} \times \dfrac{6 \text{ mol } CO_2}{1 \text{ mol } C_6H_{12}O_6} =$
$7.59 \text{ mol } CO_2$

$V = \dfrac{nRT}{P} = \dfrac{7.59 \text{ mol} \times 0.0821 \dfrac{\text{L} \cdot \text{atm}}{\text{K} \cdot \text{mol}} \times 298 \text{ K}}{1.10 \text{ atm}} = \underline{\underline{169 \text{ L}}}$

**(c)** $\boxed{\text{vol } O_2} \Longrightarrow \boxed{\text{mol } O_2} \Longrightarrow$

$\boxed{\text{mol } C_6H_{12}O_6} \Longrightarrow \boxed{\text{mass } C_6H_{12}O_6}$

$n = \dfrac{PV}{RT} = \dfrac{\left(\dfrac{655 \text{ torr}}{760 \text{ torr/atm}}\right) \times 2.48 \text{ L}}{0.0821 \dfrac{\text{L} \cdot \text{atm}}{\text{K} \cdot \text{mol}} \times 295 \text{ K}} = 0.0882 \text{ mol } O_2$

$0.0882 \text{ mol } O_2 \times \dfrac{1 \text{ mol } C_6H_{12}O_6}{6 \text{ mol } O_2} \times$

$\dfrac{180.2 \text{ g } C_6H_{12}O_6}{\text{mol } C_6H_{12}O_6} = \underline{\underline{2.65 \text{ g } C_6H_{12}O_6}}$

## chapter 10

# The Solid and Liquid States

Life can flourish on this planet because all three physical states of water can exist. Ice and snow form from the condensation of the vapor or the freezing of the liquid. The solid and liquid states of water are featured in this chapter.

## Setting the Stage

Ice cubes floating in a glass of water—what could be more familiar? What we may not appreciate in this common sight is that it represents very unusual behavior. Most solids are more dense than their liquid states, so the solid sinks to the bottom rather than floating on top. Life on this planet could not occur as we know it if water and ice behaved as most other liquids and their solid forms. Since the ice would sink as it forms, lakes would freeze solid in winter. The hot summer sun would thaw only the top layer of a lake, so very little life could survive. Heat could not be distributed from warm to cold climates by ocean currents. Nothing would be the same on this planet. We wouldn't be here if water were not a very unusual compound.

Water is certainly our most familiar liquid. Water and other liquids and their solid forms are, in a way, easier to study than the gaseous state. These forms of matter are more concrete—we can see them, feel them, and conveniently isolate and measure them. On the other hand, there is a disadvantage in our work with solids and liquids compared to gases. By their very nature, the condensed phases of matter do not lend themselves to such simplifying assumptions as the gas laws.

In this chapter, we will give the solid and liquid states appropriate attention, especially the most common but important compound, water. An understanding of the nature of water on the molecular level prepares us for the discussion of water as a solvent and medium for chemical reactions, which follows in the next chapter. Some of the questions that we will address now include the following: How does the kinetic molecular theory describe solids and liquids? What forces cause molecules to stick together in solids and liquids? How does temperature affect physical state? How does energy cause changes of state? What happens when a liquid evaporates? These and other questions relate to some previously discussed topics such as Lewis structures, electronegativity, molecular geometry, and molecular polarity (Chapter 6). Other topics that we will refer to are kinetic molecular theory, discussed in Chapter 9, and specific heat, discussed in Chapter 2.

## Section A — THE PROPERTIES OF CONDENSED STATES AND THE FORCES INVOLVED

# 10-1 Properties of the Solid and Liquid States

**Looking Ahead!** Our first question concerns the difference between the liquid or solid states and the gaseous state. In this section, we will see how the kinetic theory, introduced in the previous chapter, does and does not apply to liquids and solids.

Most of the properties of the solid and liquid states are obvious to us. Still, it is worthwhile to note these common properties in order to picture the actions and interactions of the ions or molecules that comprise these states.

1. *They have high density.*
   Solids and liquids are about 1000 times denser than a typical gas.

2. *They are essentially incompressible.*
   A tall building can be supported by bricks and other solids because they don't compress as a gas would. Likewise, a hydraulic jack uses a liquid to support weights such as that of a huge truck. Unlike the behavior of a gas, an increase in pressure on a solid or liquid does not result in a significant decrease in volume.

3. *They undergo little thermal expansion.*
   When a bridge is constructed, a small space must be left between sections for expansion on a hot day. Still, this space amounts to only a few inches for a bridge span many yards or meters long. In addition, the degree of expansion varies for different solids and liquids. Gases, on the other hand, expand significantly as the temperature rises, and all gases expand by the same factor.

4. *They have a fixed volume.*
   The volume of a gas is that of the container. Also, the volume is the same for the same number of particles under the same conditions. There is no such convenient relationship for solids and liquids. The same volumes of different solids or liquids have no relationship to the number of molecules present.

In addition to these common characteristics, solids and liquids differ with regard to shape. Solids are rigid and thus have a definite shape. Liquids flow and thus do not have a definite shape. (The shape is determined by the shape of the container.)

The characteristics of gases were adequately explained by the kinetic molecular theory. Two of the basic assumptions of the kinetic theory are also applicable to the other states. That is, solids and liquids are composed of basic particles that have kinetic energy. The average kinetic energy of the particles is related to the temperature. However, to explain the characteristics of the other two states, there are obviously some assumptions related to gases that no longer apply and must be modified. In the solid and liquid states the following circumstances exist:

1. The basic particles have significant attractions for each other and so are held close together.

2. Since the basic particles are close together, the particles occupy a significant portion of the volume of the substance.

3. The basic particles are not in random motion; their motion is restricted by interactions with other, neighboring particles.

The properties of the solid and liquid states are understandable on the basis of these assumptions. Because they are already close together, the basic particles cannot be pressed together easily, so the substances are incompressible and have high densities. In fact, both solids and liquids are referred to as *condensed states*. The attraction of the particles for each other holds them together and essentially counteracts the tendency of heat to move them apart. Thus solids and liquids undergo little thermal expansion.

In Figure 10-1, we illustrate the fundamental differences in behavior of molecules in the three states of matter. We use the water molecule as an example. This simple

**Figure 10-1  Physical States of Water Interactions between water molecules are different in the three physical states.**

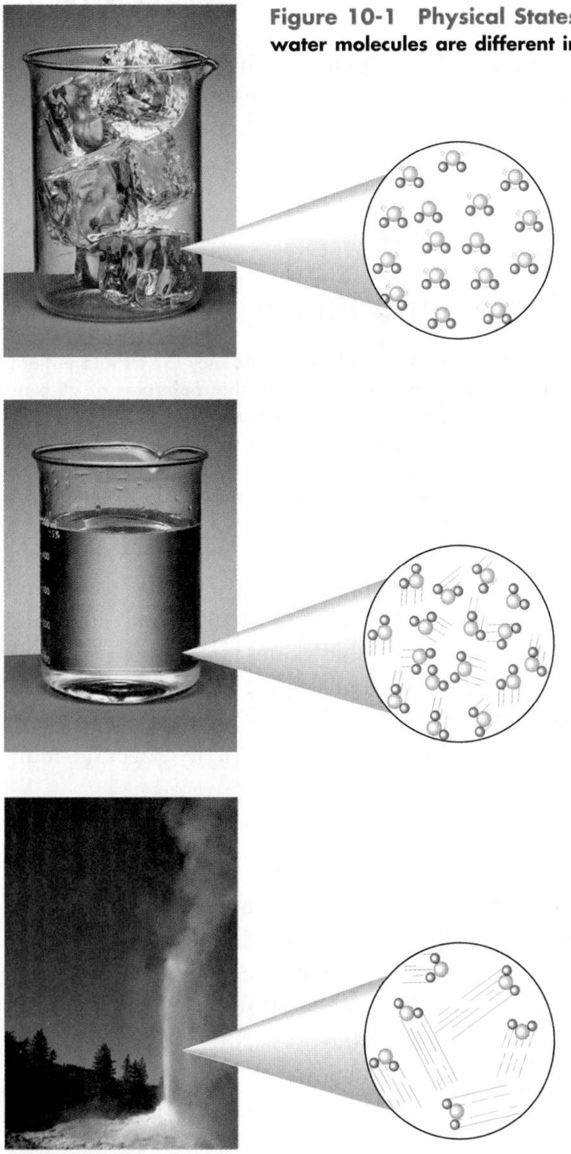

Solid  Molecules in fixed positions, motion within a confined volume highly ordered.

Liquid  Molecules are mobile but attractions hold them together in condensed state. More freedom of movement than solid but less than gas.

Gas  Random motion, very weak interactions, disorder.

but amazing compound exists in all three physical states on Earth: vapor (the gaseous state of a substance normally in the liquid or solid state is sometimes referred to as "vapor"), liquid, and solid (ice). In fact, in a thermos of ice water, all three states exist at once, although the presence of some $H_2O$ molecules in the gaseous state above the ice may not be obvious.

First, let's consider the solid state of water, which we know as ice. In this case, the forces of attraction between molecules hold them in fixed positions relatively close together. In the solid state, the molecules have kinetic energy, meaning that they do have motion. But the motion is restricted to various types of vibrations within a confined space. This is much like a view of a crowded dance floor with all of the dancers shaking and vibrating in one location but very close together. In the liquid state, the water molecules are also held close together by forces of attraction, but the molecules are not held in fixed positions and thus have more freedom of motion than in the solid state. That is, individual molecules or groups of molecules have translational motion as well as vibrational motion. This is like viewing the same dance floor with the dancers not only shaking but also moving around the floor, although staying close together. Since the molecules can move past one another, liquids can flow and take the shape of the bottom of the container. Finally, we are already familiar with the behavior of water molecules in the gaseous state, as discussed in the preceding chapter. In this case, the molecules have so much translational motion that they move freely throughout the whole container and are unaffected by the attractions to other molecules. Their motion and collisions scatter them as far apart as possible. In the dance floor analogy, it's like the dance floor has greatly expanded. The dancers can now move around freely, eventually taking up all of the added space.

# 10-2 Intermolecular Forces and Physical State

**Looking Ahead!** Water molecules in the liquid and solid states have a tendency to "stick together." Obviously, there must be attractive forces between molecules. In fact, the molecules of all compounds have at least some attraction for one another. This attraction varies a great deal for different compounds, however. Before we proceed to further discussions of condensed states, we will explore the forces of attraction that can make molecules (or atoms of noble gases) stick together. This should give us a better appreciation for why some compounds are solids, others are liquids, and still others are gases at the same temperature.

At room temperature ammonia is a gas, water is a liquid, and paraffin is a solid. Yet all three are molecular compounds. The physical state of molecular compounds at a specific temperature depends on how strongly the molecules are attracted to each other. These interactions are known as **intermolecular forces.**

Molecular compounds are held together by attractions called London forces. Some molecules may have additional forces, called dipole–dipole or hydrogen-bonding forces, that may add to the basic London force.

## London Forces

Since atoms and molecules are surrounded by negatively charged electrons, it may seem reasonable that one molecule would *repel* another. In fact, it is just the opposite. There are electrostatic forces of attraction between molecules. These forces are known as **London forces.** London forces also have the rather imposing name of *instantaneous dipole–induced dipole forces*. Let's see if we can make some sense of all of this. In a molecule, positively charged nuclei do exist within negatively charged electron clouds. However, the electrons in these clouds have

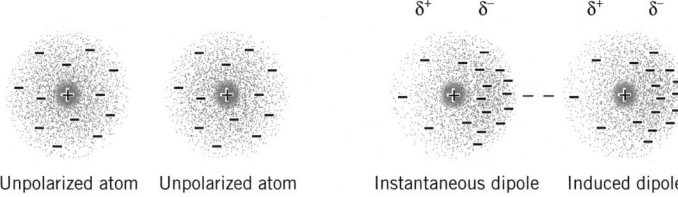

$\delta^+$   $\delta^-$     $\delta^+$   $\delta^-$

Unpolarized atom    Unpolarized atom     Instantaneous dipole    Induced dipole

**Figure 10-2  London Forces An instantaneous dipole in one atom or molecule creates an induced dipole in a neighbor.**

probabilities of being in numerous locations. Because of this, a molecule may have an imbalance of electron charge on one side of the atom or molecule at a given moment. For that instant, the molecule becomes somewhat polar (i.e., forms a dipole with a negative side and an opposite positive side). In other words, the molecule achieves an *instantaneous dipole*. If another molecule happens to be nearby, it is influenced by this instantaneous dipole and that molecule also becomes *polarized*. That is, a dipole is *induced* in the second molecule. Thus there is a force of attraction between the negative side of the instantaneous dipole on one molecule and the positive side of the induced dipole on another. Recall from Chapter 6 (p. 170) that the $\delta^+$ and $\delta^-$ represent partial positive and negative charges respectively. (See Figure 10-2.)

In larger molecules with more electrons, instantaneous dipoles become more likely, so London forces become more significant. Also, larger molecules are generally more *polarizable* than smaller ones because they are surrounded by larger, more diffuse electron clouds. *Since molar mass usually indicates a larger molecule, we can state that the heavier the molecule, the greater the London forces and the more likely we are to find it in the liquid or solid state at a given temperature.* For example, at room temperature, natural gas [$CH_4$ (molar mass = 16 g/mol)] is, of course, a gas. A major component of gasoline, called octane [$C_8H_{18}$ (molar mass = 114 g/mol)], is a liquid, and paraffin [$C_{24}H_{50}$ (molar mass = 338 g/mol)] is a solid. The trend of gas $\longrightarrow$ liquid $\longrightarrow$ solid corresponds to the magnitude of the London forces for the three molecules (i.e., $CH_4 <$ $C_8H_{18} < C_{24}H_{50}$). All molecules are attracted by London forces, but in nonpolar molecules, London forces act exclusively.

## Dipole–Dipole Attractions

Covalent bonds between unlike atoms in a molecule are all polar at least to some extent, but the molecule itself may not be polar. This apparent contradiction was discussed in Chapter 6. In that discussion, with the help of the Lewis structures and VSEPR theory, we were able to predict that certain molecules may be linear (e.g., $CO_2$), trigonal planar (e.g., $BF_3$), or tetrahedral (e.g., $CH_4$). In these highly symmetrical structures the equal bond dipoles cancel, and the molecule as a whole is nonpolar. Consider the case of $CO_2$. Since the two polar carbon–oxygen bonds lie at a 180° angle, the individual bond dipoles cancel and the molecule is nonpolar. On the other hand, carbonyl sulfide (OCS) is also linear, like $CO_2$, but since the bond dipoles are unequal, they do not cancel and the molecule is polar. The $SO_2$ molecule is also polar. In this case, the Lewis structure shows a pair of electrons on the sulfur, so the geometry is V-shaped rather than linear. Thus the two equal bond dipoles are at an angle and do not cancel. The OCS and $SO_2$ molecules each have a permanent dipole, meaning that they have permanent partially negatively and positively charged locations on the molecule. (See p. 175.) *Molecules with a permanent dipole can align*

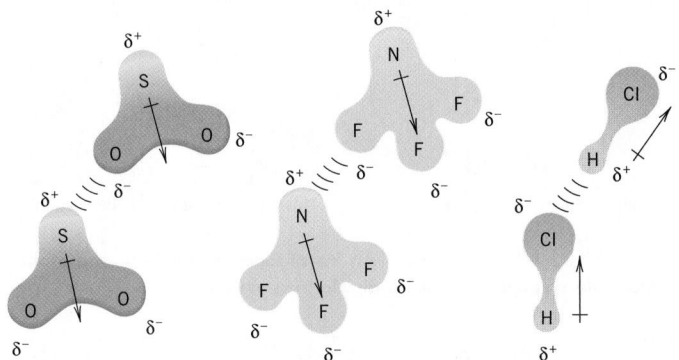

**Figure 10-3 Dipole–Dipole Attractions** $SO_2$, $NF_3$, and HCl are polar molecules and have dipole–dipole attractions. The negative end ($\delta^-$) of one molecule is attracted to the positive end ($\delta^+$) of another.

*themselves so that the negative end of one molecule is attracted to the positive end of another.* (See Figure 10-3.) *These intermolecular interactions are known as* **dipole–dipole attractions** and add to the effect of the London forces. *Thus if we have two compounds of similar molar mass (meaning similar London forces) but one is composed of polar molecules and one nonpolar molecules, the compound with polar molecules is more likely to be in a condensed state at a given temperature.* Having additional dipole–dipole forces of attraction for molecules is like getting 5 extra credit points on a 100-point test. It may not change anything, but it may just be enough to get one a higher grade. In the case of compounds, the added attractions may just be enough to hold the molecules together in a condensed state. For example, at room temperature, $CO_2$ (44 g/mol), a compound with nonpolar molecules, is a gas, whereas $CH_3CN$ (41 g/mol), a compound with polar molecules, is a liquid. It is difficult to compare heavy nonpolar molecules with lighter polar molecules. Generally, the mass of the molecules and the London forces is more important than the additional dipole–dipole forces.

## Hydrogen Bonding

In Chapter 5, we discussed trends in the size of atoms. Atoms of the elements to the upper right in the periodic table were the smallest. In Chapter 6, we also mentioned that these same atoms (fluorine, oxygen, and nitrogen in particular) were the most electronegative elements. This means that these small, highly electronegative atoms tend to attract a significant amount of negative charge to themselves when chemically bonded to other atoms. As an example, consider molecules of water. There is a large difference in electronegativity between oxygen and hydrogen (3.5 − 2.1 = 1.4). This difference is not enough to indicate an ionic bond, but it does point to a highly polar covalent bond. In Chapter 6, we discussed the geometry of water molecules. According to VSEPR theory, the mutually repulsive effect of the two electron pairs on the oxygen atom and the two bonded pairs causes the electron pairs and the hydrogens to be located at the corners of a rough tetrahedron. (See Figure 10-4a.) Since this structure gives $H_2O$ a V-shaped molecular geometry, the bond dipoles do not cancel and the water molecule is significantly polar. (See Figure 10-4b.) The partial positive charge is centered on the hydrogens at two of the corners of the tetrahedron, and the partial negative charge is centered on the electron pairs at the other two corners. As shown in Figure 10-5, in solid ice, hydrogens on two different water molecules interact with the two electron pairs. In liquid water,

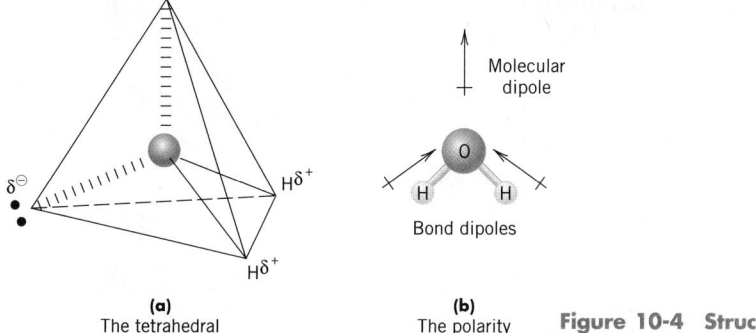

**(a)**
The tetrahedral
representation of water

**(b)**
The polarity
of water

**Figure 10-4   Structure of Water**
**Water molecules are polar because they are V-shaped.**

the structure is less orderly because the interactions between molecules are more random. Individual and groups of water molecules can slide past one another in the liquid state.

In the solid state, the $H_2O$ molecules have a relatively open structure compared to the liquid state. The more compact structure in the liquid state accounts for its higher density and thus why ice cubes float in water. (See Figure 10-5.).

Interactions between a partially positive hydrogen atom on one molecule and the electron pair of the oxygen on another molecule are an example of what is called a hydrogen bond. *A* **hydrogen bond** *is generally restricted to molecules that have an N—H, O—H, or F—H bond, where the hydrogen in these bonds interacts with an unshared electron pair on an F, O, or N atom of another molecule.* Hydrogen bonding usually involves only these three atoms because of their high electronegativity and small size. This interaction at first may seem like a case of an extreme dipole–dipole attraction. In fact, it is more complex than that. A hydrogen bond is not nearly as strong as a regular covalent bond, but it is considerably stronger than typical dipole–dipole attractions. Whereas dipole–dipole attractions usually do not have a large effect on the physical properties of the compound, *hydrogen bonding has a significant effect on the properties of a compound.* (In this case, it is like getting 15–20 extra credit points on a 100-point test. It would most likely make a difference.) Consider the water molecule. It has a very small molar mass of 18 g/mol. If only London forces were present, it would be a gas at temperatures as low as −75°C. Even if regular dipole–dipole attractions were present, it would still boil at a very low temperature. The presence of hydrogen bonding between water molecules provides the "glue" that holds the molecules together so that it exists as a liquid and even a solid under normal conditions on this planet. Other compounds whose properties are altered considerably

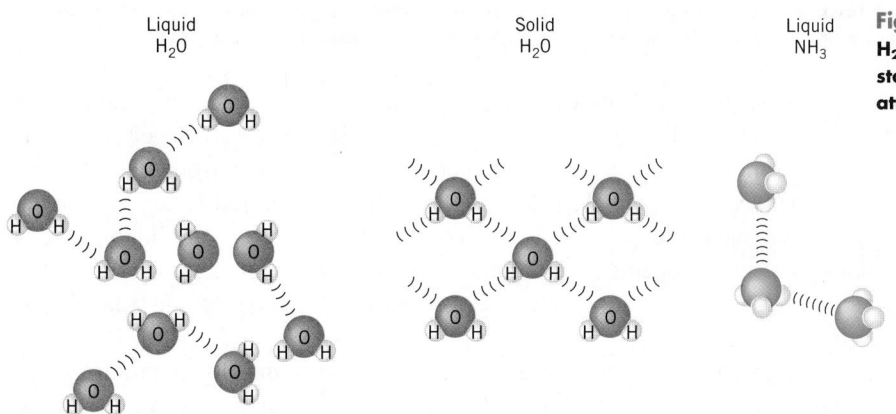

Liquid
$H_2O$

Solid
$H_2O$

Liquid
$NH_3$

**Figure 10-5   Hydrogen Bonding**
**$H_2O$ molecules in both liquid and solid states, as well as $NH_3$ molecules, are attracted by hydrogen bonds.**

**Figure 10-6 Hydrogen Bonding in DNA**
**The double helix structure of DNA consists of two strands twisted about each other. The strands are connected by hydrogen bonds.**

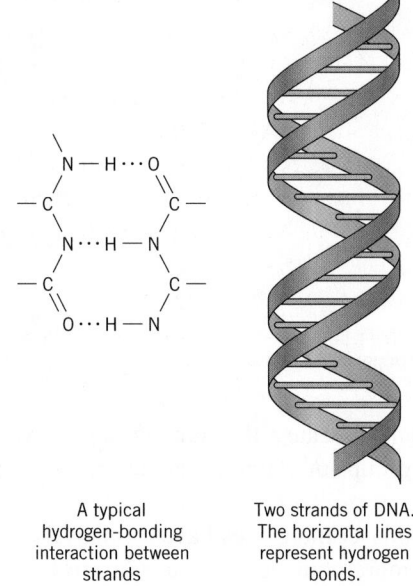

A typical
hydrogen-bonding
interaction between
strands

Two strands of DNA.
The horizontal lines
represent hydrogen
bonds.

by the presence of hydrogen bonding are ammonia ($NH_3$) and hydrogen fluoride (HF). (See Figure 10-5.)

Hydrogen bonding is also important in the huge, complex molecules on which life is based. Consider, for example, a molecule of DNA, the "messenger of life." DNA is an extremely large molecule composed of two strands of covalently bonded atoms offset from each other in a helical arrangement (somewhat like a spiral staircase). One of the miracles of life is that these two strands can separate from each other and replicate themselves from smaller molecules. The two strands can separate because they are held together by hydrogen bonds much like Velcro holds two pieces of clothing together. The hydrogen bonds are not as strong as the covalent bonds within the strands. (See Figure 10-6.)

## 10-3 The Solid State: Melting Point

**Looking Ahead! Now that we know something about the forces holding molecules together, we will look at the solid state and examine how it holds up when heated. Heating a substance causes the molecules to shake and vibrate more violently. This increase in kinetic energy is a disruptive force causing the molecules to separate from one another. The temperature at which the disruptive forces of heat overcome the intermolecular forces that hold the molecules in fixed positions is known as the melting point of the solid. In the following discussion we will examine this temperature not only for molecular solids but for other types of solids as well.**

On a hot summer afternoon, ice cream melts too fast and water evaporates quickly. In fact, we may all feel as if we are "melting." The process of melting is understandable from kinetic theory. As the temperature goes up, the average kinetic energy of molecules increases, which means that they move faster and faster. The increased motion of the molecules eventually overcomes whatever forces are holding the solid or liquid molecules together, and a phase change occurs. It's like a feeling that most of us have experienced. Sometimes we just get too "fidgety" to stay seated and we have to move around.

Why do some solids melt at a certain temperature and others require a higher temperature to melt? The situation is analogous to what happens in an earthquake—

the flimsiest buildings are the first to fall. As the earthquake intensifies, stronger and sturdier buildings may be damaged or even collapse. In order to maintain a rigid structure during a severe shaking, the studs, beams, and walls of a building must be firmly attached. Heating solids is much like subjecting them to an earthquake. A rising temperature causes the molecules or ions of a solid to vibrate more and more vigorously. The flimsiest solids, whose basic particles are not firmly attached to each other, are the first to collapse, so they melt or vaporize at the lower temperatures. At higher temperatures, solids whose molecules have stronger intermolecular forces change to the liquid state.

There are basically two general categories of solids: amorphous and crystalline. **Amorphous solids** *are so named because they have no defined shape.* The basic particles in amorphous solids are not located in any particular positions. Examples of amorphous solids are glass, rubber, and many plastics. *In* **crystalline solids,** *the molecules or ions are arranged in a regular, symmetrical structure called a* **crystal lattice.** A salt crystal, a piece of quartz, and many minerals found on Earth naturally form solids with discrete geometric patterns. These symmetrical shapes reflect the ordered arrangements of the molecules or ions that lie within. (See Figure 10-7.)

In this section, we are interested in how heat causes a change from the solid to the liquid state. *The temperature at which a crystalline solid melts (the melting point) is a definite and constant physical property.* When pure crystalline solids melt, the added heat causes a phase change and the temperature remains constant. When amorphous solids melt, the melting process is a gradual softening that occurs over a temperature range.

Different types of solids have different melting characteristics. We will examine the four types of solids individually.

**When the ice in ice cream melts, the water molecules begin to move past one another.**

## Ionic Solids

**Ionic solids** *are crystalline solids in which ions are the basic particles making up the crystal lattice.* (See Figure 10-8a.) Forces between oppositely charged ions are quite strong, especially when compared to forces between individual molecules. The strong ion–ion forces result in solid compounds that have relatively high melting points, as high as 3000°C, for example, for ZrN. As mentioned in Chapter 6, ionic

Aquamarine

Fluorite

Dolomite

**Figure 10-7  Crystalline Solids** The minerals shown above are crystalline solids that are ionic compounds. Their symmetrical shapes reflect the ordered geometry of the ions within.

**Figure 10-8 Ionic and Molecular Solids**
In ionic compounds, ions occupy lattice points. In molecular solids, molecules occupy lattice points.

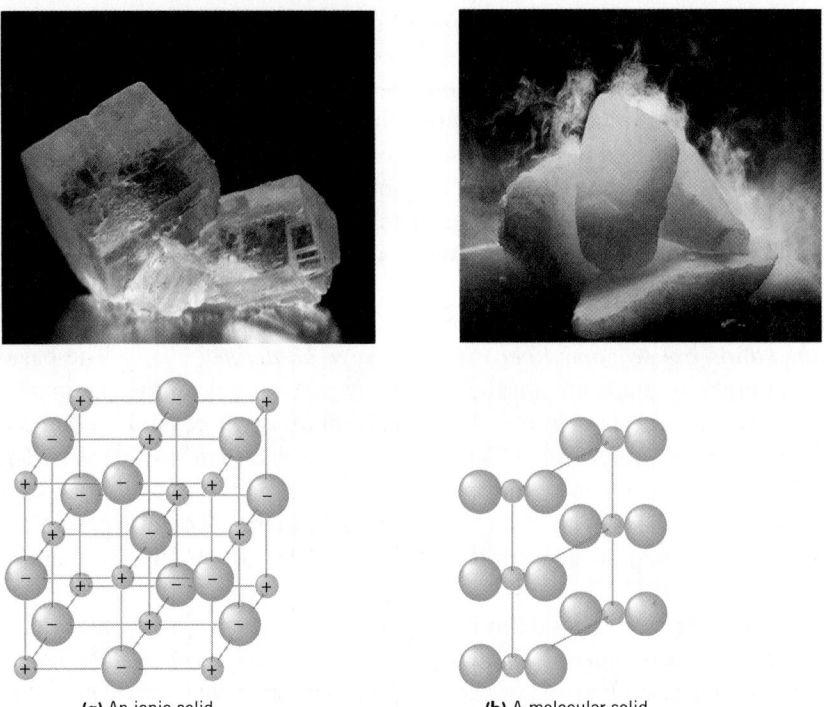

(a) An ionic solid    (b) A molecular solid

The beautiful symmetry of snowflakes reflects the ordering of water molecules within the crystals.

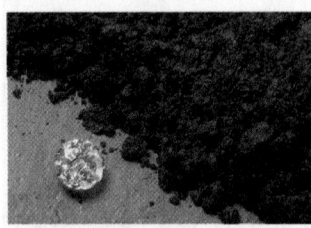

Diamond and graphite are allotropes of carbon.

compounds are the stuff of rocks and minerals. They are always solids at room temperature. Ionic compounds are also very hard and brittle (they shatter into pieces when struck).

## Molecular Solids

*In* **molecular solids,** *the basic particles of the crystal lattice are individual molecules, which are held together by London forces and, in some cases, dipole–dipole attractions or hydrogen bonding.* (See Figure 10-8b.) These attractions are not nearly as strong as the ion–ion attractions found in ionic compounds. As a result, molecular solids have low melting points compared to ionic solids. These melting points range from very low for small, nonpolar molecules such as $N_2$ and $CH_4$ to well above room temperature for large molecules such as table sugar, where molecules are attracted to each other by hydrogen bonds.

## Network Solids

There are a few examples of solids where the atoms are covalently bonded throughout the entire sample of the solid. These are known as **network solids** and generally have very high melting points. Examples of such solids are the three major allotropes of carbon—**diamond, graphite,** and **buckminsterfullerene.** (See Figure 10-9.) In diamond, each carbon is in the center of a tetrahedron and is bonded to four other carbons at the corners of the tetrahedron. These carbons are bonded to four other carbons and so forth throughout the entire crystal. The compact arrangement of atoms makes diamond the hardest material known. To melt diamond, a large number of covalent bonds must be broken, which requires a high temperature to supply the large amount of energy needed. The melting point of diamond is so high (over 4100°C) as to be difficult to establish.

It is hard to believe that an ordinary chunk of charcoal is chemically the same as a precious diamond. The charcoal is mostly composed of graphite, which is

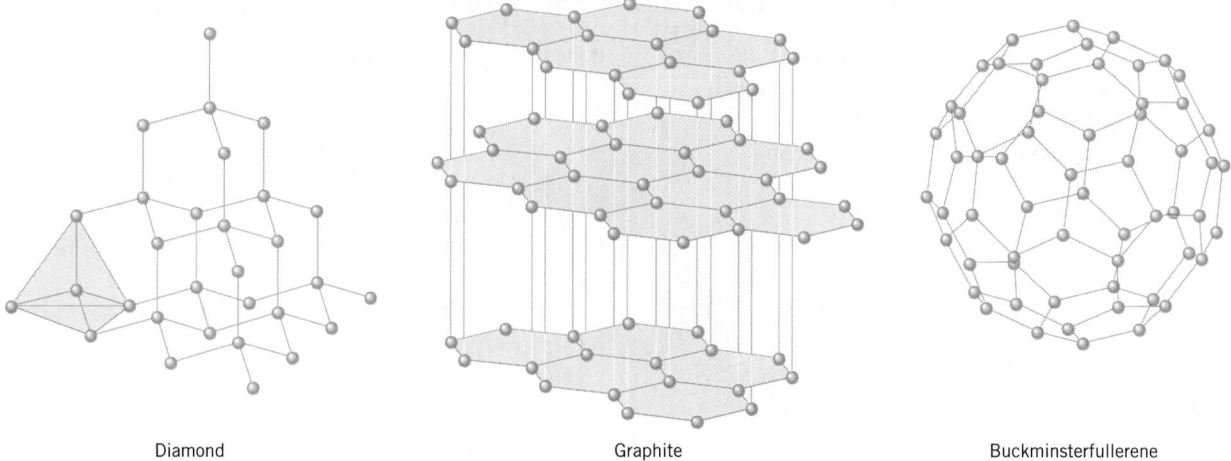

Diamond                     Graphite                  Buckminsterfullerene

**Figure 10-9  The Allotropes of Carbon** **There are three allotropes of carbon.**

another allotrope of carbon. (Diamond is formed deep in the mantle of the Earth from graphite at extremely high temperatures and pressures.) In graphite, the carbons are arranged in parallel planar sheets. Since these sheets can slip past one another, graphite is used as a lubricant and as the "lead" in pencils. The third allotrope of carbon, buckminsterfullerene, was not identified until 1985 and is composed of carbon atoms bonded in spherical shapes somewhat like a soccer ball. It was named after the architect Buckminster Fuller, the designer of the geodesic dome that also reminds one of this allotrope. These "bucky balls," as they are sometimes called, come in a variety of sizes. The two most familiar have the formulas $C_{60}$ and $C_{70}$. Although an important application of this discovery has not yet been forthcoming, we can expect to hear more about this form of carbon in the future. Graphite is also a network solid and has a very high melting point.

**Quartz is a network solid with an empirical formula of $SiO_2$.**

Another example of a network solid is $SiO_2$ (quartz, the major component of ordinary sand). This represents an empirical formula since the oxygens in this compound are each attached to two silicons, forming a network throughout the crystal. Quartz also has a high melting point (1610°C).

## Metallic Solids

Metals are also crystalline solids. As mentioned in Chapter 5, the outer electrons of metals are loosely held. *In* **metallic solids,** *the positive metal ions occupy regular positions in the crystal lattice, with the valence electrons moving freely among these positive ions.* (See Figure 10-10.) This is the reason metals are good conductors

**Figure 10-10  Metallic Solids**
**In metals, positive ions occupy the lattice points.**

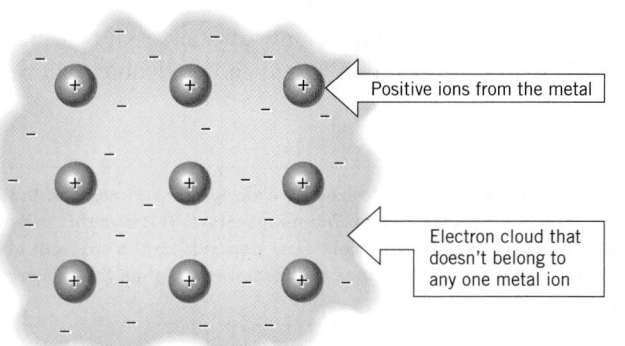

Positive ions from the metal

Electron cloud that doesn't belong to any one metal ion

## The Melting Point of Iron—a Crucial Number

Iron has been the king of metals for thousands of years. Of all of the metals, how did iron gain this exalted position? First, iron ore is plentiful and widely available. Second, the reduction of the ore can be easily accomplished with charcoal, limestone, and a hot fire. But most importantly, it is a strong metal that can support many times its weight in other materials. All large structures like skyscrapers and dams are supported by iron (actually an alloy, known as steel) or concrete reinforced with iron.

But as was tragically illustrated in the destruction of the World Trade Center towers on September 11, 2001, it loses strength at high temperatures. The melting point of iron is 1535°C, which is quite high. However, changes within the internal structure of the metal begin around 900°C and it begins to lose its strength. When the airplanes struck the WTC towers, the strong steel beams, reinforced with concrete, did their job and withstood these powerful collisions. But large amounts of aviation fuel ignited inside the building, raising the temperature to an estimated 1000°C. Eventually, the heat caused the metal to soften and collapse from the weight of the floors.

How can iron be protected from high temperatures? In the past, it was coated with asbestos.* Asbestos is a fibrous form of silica that occurs naturally in the Earth.

It has a high melting point and is a poor conductor of heat. Iron coated with asbestos or similar materials is effectively insulated from heat at least long enough for a typical fire to burn out. In the case of the World Trade Center, the iron was indeed covered by a protective coating, but the force of the collisions blew it away, exposing the metal to the extreme heat.

The space shuttle must also be protected from heat during its reentry into the atmosphere. As it encounters air at 17,000 mph, the friction heats the bottom surface to 1650°C. This temperature would melt most metals, including iron. *Refractory ceramic* tiles are used in this case. A ceramic is a mineral or a mixture of minerals (i.e., quartz and clays) that have been mixed and heated to about 900°C. Dishes and vases are two of many ceramics used in our daily lives. Refractory ceramics have particularly high melting points (above 2000°C) and incorporate metal oxides such as aluminum oxide. Bricks of these ceramics protect the underside of the spacecraft, which is exposed to the heat.

Iron will continue to support our infrastructure. Research continues, however, on various ways to make it stronger and protect it from the effects of extreme heat.

**Submitted by Edward Tokas**
**Carolina Reading and Learning Center**

*Asbestos coatings are no longer applied as they have been found to cause lung disease, including lung cancer, when the small fibers are inhaled over a period of time.

*The space shuttle must be protected against extreme heat.*

Tungsten is used in drill bits because of its strength and high melting point.

of electricity. All metals except mercury (melting point −39°C) are solids at room temperature. Some, like the alkali metals, are soft and melt at comparatively low temperatures. Others such as iron are hard and have high melting points. Tungsten has one of the highest melting points of any substance known (3380°C). Metals are generally ductile (can be drawn into wires) and malleable (can be pounded into sheets).

**Looking Back!** Crystalline solids are characterized by orderly arrangements of the ions, molecules, or atoms in fixed positions in a crystal lattice. The temperature at which this orderly arrangement of basic particles breaks down to form the liquid state depends on the strength of the forces holding them in their fixed positions. The stronger the forces, the higher the temperature needed to break the interactions.

## Learning Check A | checking it out

**A-1.** Fill in the blanks.

Compared to gases, liquids and solids have a _____ density and occupy a _____ volume. The molecules of the condensed states have appreciable _____ attractions for each other. The three types of forces are _____ _____, _____ _____, and _____ _____ . Of these, _____ _____ is the strongest and has a significant effect on physical properties. Nonpolar molecules have only _____ forces. The two types of solids are _____ and _____ . _____ solids have a definite melting point. Of the four types of crystalline solids, _____ solids are composed of ions and are always solids at room temperature. _____ solids have covalent bonds throughout the crystal. An allotrope of carbon that has a very high melting point is _____ . Molecular solids are usually soft and melt at _____ temperatures. _____ solids have positive ions in a sea of _____ .

**A-2.** Write the Lewis structures for $BCl_3$, $NCl_3$, and $HNCl_2$.

**a.** What are the intermolecular forces between molecules in each case?

**b.** Which is more likely to be in a condensed state at a given temperature? Which is least likely?

**A-3.** Which of the following has the higher melting point and why, $CaF_2$ or $SeF_2$?

*Additional Examples: Exercises 10-3, 10-6, 10-8, 10-19, 10-20, and 10-25.*

Section **B** **THE LIQUID STATE AND CHANGES IN STATE**

# 10-4 The Liquid State: Surface Tension and Viscosity

**Looking Ahead!** **All of the elements and compounds, even those that make up our atmosphere, can be found in the solid state if the temperature is low enough. As the temperature rises, these solids, one by one, melt to form the liquid state. In the liquid state, the atoms, molecules, or ions can flow past one another. We now shift our focus to some properties of the liquid state.**

Between the complete disorder of molecules in the gaseous state and the high order of the crystal lattice in the solid state lies the liquid state. Here, the basic particles are still held close together, so liquids remain condensed like in the solid state. On the other hand, the basic particles can move past one another. In this respect, the liquid state is like the gaseous state. At a lower temperature, liquids freeze to the solid state, and at some higher temperature, liquids vaporize to the gaseous state. There are two properties of liquids that we will discuss: surface tension and viscosity.

## Surface Tension

Have you ever noticed that certain insects can walk on water? Also, if one carefully sets a needle or a small metal grate on water, it remains on the surface despite the fact that the metal is much denser than water. (See Figure 10-11.) This tells us that there is some tendency for the surface of the water to stay together. **Surface tension** *is the force that causes the surface of a liquid to contract.* Because of surface tension, drops of liquid are spherical. A molecule within the body of the liquid is equally attracted in all directions by the intermolecular forces. Molecules on the surface are pulled to the side and downward but not upward. This unequal attraction means that a portion of a liquid will tend to have a minimum amount of surface area. A liquid placed on a flat surface draws itself into a "bead," or into a sphere if it is suspended in space. Although raindrops are not completely spherical because of gravity, water released in the space shuttle forms perfectly spherical drops. The force that is required

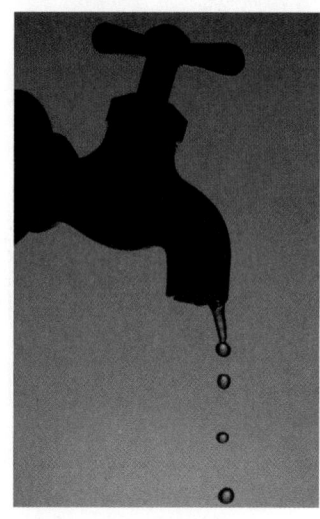

**Water forms spherical drops because of surface tension.**

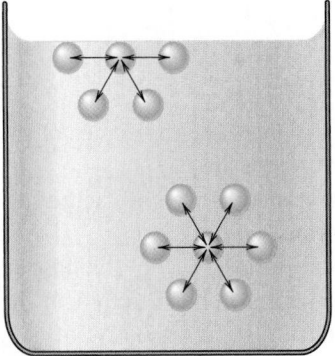

**(a)** Molecules on the surface are pulled down and to the sides.

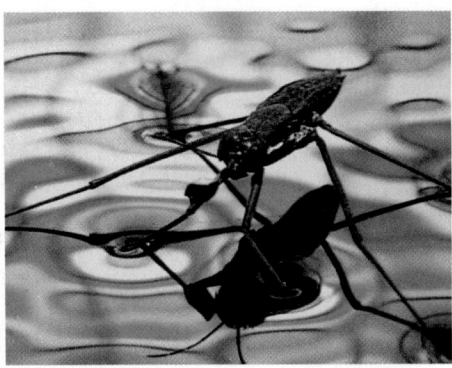

**(b)** An insect can walk on water.

**(c)** The metal grid floats on the surface.

**Figure 10-11  Surface Tension** **The unequal attraction for molecules on the surface accounts for surface tension.**

to break through a surface relates to its surface tension. Water obviously has a high enough surface tension to support small insects. Other liquids, where the intermolecular forces are greater, have higher surface tensions.

Dissolved soaps reduce the surface tension of water, thus allowing the surface to expand and "wet" clothes or skin rather than form beads. An insect would do well to avoid trying to walk on soapy water. Because of the reduced surface tension, the bug would sink.

## Viscosity

**Figure 10-12  Viscosity Syrup is a viscous liquid.**

We all know it seems to take forever to pour ketchup on french fries when we are hungry. Ketchup is more viscous than water. *The **viscosity** of a liquid is a measure of its resistance to flow.* Water and gasoline flow freely because they have low viscosity. Motor oil and syrup flow slowly because they have high viscosity. (See Figure 10-12.) The viscosity of a liquid depends to some extent on the intermolecular forces between molecules. Strong intermolecular attractions usually mean a more viscous liquid. Water is an exception—even though its molecules interact relatively strongly, it has low viscosity. Compounds with long, complex molecules also tend to form viscous liquids because the molecules tangle together.

A breakfast of pancakes with syrup on a cold morning is hard to beat. Unfortunately, the syrup barely moves when it is cold. A little warming of the syrup solves the problem. All liquids become less viscous as the temperature increases. The higher kinetic energy of the molecules counteracts the intermolecular forces, allowing the molecules to move past one another more easily.

## 10-5 Vapor Pressure and Boiling Point

**Looking Ahead!** **A puddle of water after a spring rain soon disappears. Actually, the liquid does not disappear but simply changes into the gaseous or vapor state. The tendency of liquids to vaporize (evaporate) is the subject of this section.**

Perhaps one of the most accepted facts of life is that wet things eventually become dry. *The liquid water changes to the gaseous state in a process known as **vaporization.** When vaporization occurs at temperatures lower than the boiling point, it is known as **evaporation.*** In order for a molecule of a liquid to escape to the vapor state, however, it must overcome the intermolecular forces attracting it to its neighbors in the

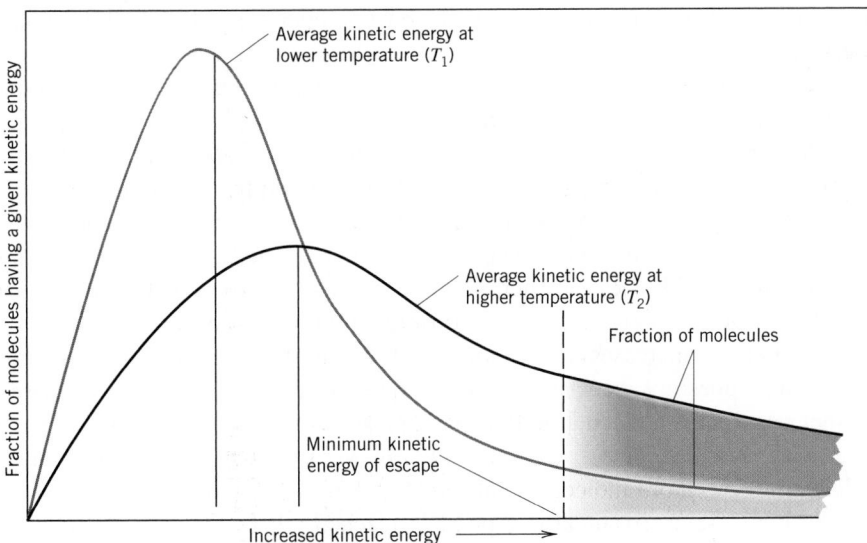

**Figure 10-13  Distribution of Kinetic Energies at Two Temperatures**
**The average kinetic energy is higher at $T_2$ than at $T_1$.**

liquid. Two conditions allow a molecule in a liquid to escape the liquid state to the gaseous state. First, it must be at or near the surface of the liquid. Second, it must have at least the minimum amount of kinetic energy to overcome the intermolecular forces. So, why don't all molecules on the surface escape? To answer this, we must recall that at a given temperature, the molecules have a wide range of kinetic energies. The temperature relates to the *average* kinetic energy. The distribution of kinetic energies of the molecules at two temperatures ($T_1$ and $T_2$) can be represented graphically as in Figure 10-13. The unbroken vertical lines represent the average kinetic energy at each of these two temperatures. $T_2$ represents a higher temperature than $T_1$ because it has a higher average. The broken vertical line represents the minimum kinetic energy that a molecule must have in order to escape from the surface. Notice that a small fraction have the minimum energy at $T_1$, but a larger fraction have the minimum energy at the higher temperature, $T_2$. As a result, a liquid evaporates faster at a higher temperature.

If only the molecules with the highest kinetic energy escape to the vapor state, what effect does that have on the liquid left behind? The effect is similar to what would happen if the top 10% of the grades were left off when the average of the last chemistry test was computed. The remaining average would be lower. When the molecules with the highest kinetic energy escape, the average kinetic energy of the molecules remaining in the liquid state is lowered. This means that the liquid water will be cooled and the gas above the water will be correspondingly heated.

The cooling effect of evaporating water is important to health maintenance in warm climates. Perspiration covers our bodies with a layer of water when it is warm. The evaporation of this liquid cools the water on our bodies and us along with it. The cool feeling after a hot shower is not just a feeling but a reality. Our perspiration cools us but can make life more miserable for the next person in a crowded room. Evaporation cools the liquid but heats the air. If water is allowed to evaporate under a vacuum, the evaporation process occurs faster. In fact, the water cools enough to freeze. Certain food products, such as some instant coffees, advertise that they are "freeze dried." Instant coffee is made by first making a coffee solution and then removing the water, leaving coffee crystals. Boiling the coffee solution to remove the water presumably

**The vapor pressure of water increases as the temperature increases.**

**Mothballs sublime. The vapor repels moths.**

affects the taste. Thus removing the water at lower temperatures (freeze drying) should preserve the flavor.

Now, instead of letting the water vapor escape to the surroundings, we can measure the buildup of pressure by placing a beaker of water in a closed glass container so that the vapor molecules are trapped. Attached to the apparatus is an open-end mercury manometer for measuring the increase in pressure within the container. We will assume that, initially, the air is dry, meaning that any water vapor will come from our beaker of water. Before any water evaporates, the manometer indicates the same pressure inside and outside the container. As the water begins to evaporate, the additional molecules in the gas above the beaker cause the total pressure to increase (Dalton's law). The pressure increases rather rapidly at first, but then increases more and more slowly until it does not increase further. As the number of molecules in the gaseous state increases, some molecules collide with the surface of the liquid and are returned to that state. *The change of state from the gaseous to the liquid state is known as* **condensation.** As more molecules enter the gaseous state, more gaseous molecules enter the liquid state. Eventually, a point is reached where the rate of evaporation equals the rate of condensation and the pressure remains constant. That is, for every molecule that goes from the liquid to the gaseous state, a molecule goes from the gaseous to the liquid state. The system is said to have reached a point of *equilibrium.* (See Figure 10-14.)

The difference in the levels of the manometer when the pressure has reached equilibrium is the pressure exerted by the water vapor. *The pressure exerted by the vapor above its liquid at a given temperature is called its* **equilibrium vapor pressure.** At a higher temperature, more molecules have the minimum energy needed to escape and, as a result, the vapor pressure is higher. (We say the liquid has become more *volatile,* meaning that it has a higher tendency to vaporize.) Solids may also have an equilibrium vapor pressure. *The vaporization of a solid is known as* **sublimation.** Dry ice (solid $CO_2$) has a high vapor pressure and sublimes rapidly. Ordinary ice also has a small but significant vapor pressure. A thin layer of snow will vaporize away even if the temperature remains below the melting point of ice.

In the experiment described above, the vapor pressure at first increases rapidly but then slowly reaches the equilibrium vapor pressure. As the air above the water nears saturation with vapor, more vapor returns to the liquid state and the buildup of pressure slows. We notice this same effect on humid summer days. The relative humidity measures how near the air is to saturation with vapor at a particular temperature. For example, if the relative humidity is 60% and the equilibrium vapor pressure of water at this temperature is 30 torr, the actual vapor pressure of water in the

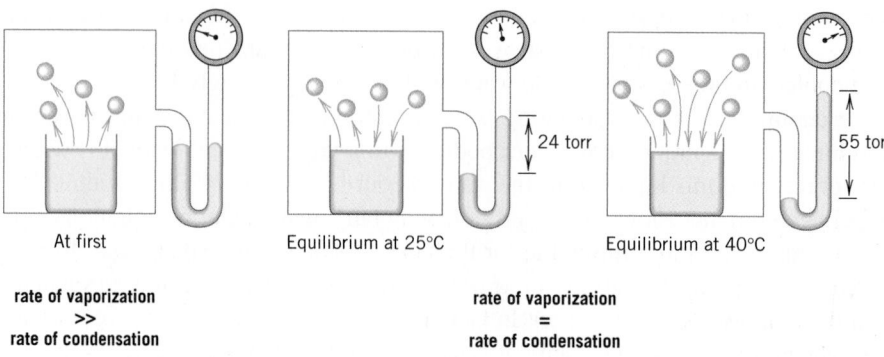

| At first | Equilibrium at 25°C | Equilibrium at 40°C |
|---|---|---|
| | 24 torr | 55 torr |

rate of vaporization
>>
rate of condensation

rate of vaporization
=
rate of condensation

**Figure 10-14 Equilibrium Vapor Pressure A given fraction of molecules escape to the vapor state above a liquid.**

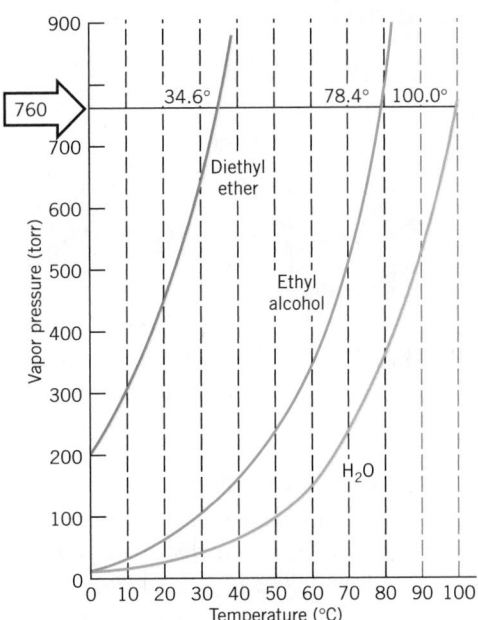

**Figure 10-15 Vapor Pressure and Temperature**
**The boiling points of these liquids are the temperatures at which the vapor pressures equal 760 torr.**

atmosphere is about $0.60 \times 30$ torr $= 18$ torr. The higher the humidity, the slower the evaporation and the less efficient is our personal air-conditioning system known as perspiration. "It's not the heat, it's the humidity." In fact, both factors affect our comfort level. On the weather report, the combined effect of heat and humidity is called the "heat index." This tells us how much hotter it actually feels because of the humidity.

The vapor pressure of liquids varies regularly as a function of temperature. This is represented graphically in Figure 10-15. The vapor pressure curves of water, diethyl ether, and ethyl alcohol are included. Notice that the vapor pressure of water at 100°C is equal to the pressure of the atmosphere (760 torr). *When the vapor pressure*

**At high elevations water boils at a lower temperature.**

### The Oceans of Mars

This magnificent planet where we live exists at a perfect distance from the sun. We receive just the right amount of energy so that water can exist in the liquid state. Of course, it is solid in the colder regions and the discomfort of a humid day tells us it also is present in the gaseous state. The exchange of water among the three states is a continual process powered by the energy from the sun.

Are there oceans on other planets? In the 1970s the scientific world was shocked when spacecraft sent back photos of dry riverbeds on the surface of our neighboring planet, Mars. Scientists have been intrigued and excited ever since. The existence of liquid water implies the possibility that life may have existed on this planet (and maybe still does). More recent visits, including that of the robot *Sojourner* in 1997, confirmed the existence of dry riverbeds. Current conditions on Mars would not allow liquid water. The surface pressure is between 6 and 8 torr, which is about 1% of the pressure on Earth. At this low pressure, water would boil at about 5°C, so the liquid range would only be from 0°C to 5°C. The temperature actually varies from about −75°C at the poles to around 10°C at the equator. Even if the liquid state existed at 3° or 4°C, it would quickly evaporate in the thin atmosphere and freeze from the evaporative process.

Apparently, one to two billion years ago, the temperature and pressure must have been much like Earth's, meaning that liquid water could have existed. What happened to all of the water on Mars? For a long time we thought it all must have escaped into space. Recent data, some reported in 2002, indicate that much of it may still be there. It is either frozen at the surface at its poles or lies in layers beneath the surface. We will know much more in the next few years as more space probes orbit the planet and some make soft landings.

Farther out in space lies a moon of Saturn known as Triton. It has a thick, hazy atmosphere where the cold temperature (−180°C) and pressure of 1.6 atm are just right for oceans of methane ($CH_4$). It is speculated that there may be weather patterns similar to those on Earth, except that the rain in the plain is mostly methane. (Sorry.) The *Cassini* spacecraft is on its way there and is scheduled to arrive in 2004. Many questions about this distant but exotic world should then get some answers.

*Water once flowed on Mars.*

of a liquid equals the restraining pressure, bubbles of vapor form in the liquid and rise to the surface.* This is the **boiling point** of the liquid. *The **normal boiling point** of a liquid is the temperature at which the vapor pressure is equal to 760 torr.* Diethyl ether and ethyl alcohol are more volatile than water, which means that their vapor pressures reach 1 atm at temperatures below that of water. They boil at around 34°C and 78°C, respectively. Unlike the melting point of a solid, the boiling point of a liquid is quite dependent on atmospheric conditions. For example, on Pike's Peak in Colorado, water boils at 86°C because the atmospheric pressure is only 450 torr. The highest point on Earth is the top of Mt. Everest. Water boils at 76°C at this elevation. Cooking takes much longer at high elevations and may not even kill all harmful bacteria because of the lower boiling point of water. Appliances called pressure cookers are sometimes used. They retain some of the water vapor, thus increasing the pressure inside the cooker and increasing the boiling point.

The temperature at which a liquid begins to boil also depends on the attractive forces between the basic particles. Ionic compounds naturally have very high boiling points and molecular compounds have much lower boiling points. Of the molecular compounds, those with large molecules or that have hydrogen bonding have the highest boiling points. The effect of hydrogen bonding has a rather dramatic effect on boiling points. For example, consider the boiling points of the Group VIA hydrogen compounds: $H_2O$, $H_2S$, $H_2Se$, and $H_2Te$. The molar masses of these compounds increase in the order listed. Thus London forces should increase in the same order, and, as a result, so should the intermolecular attractions. From this result alone, we may predict steadily higher boiling points for these compounds. Note in Table 10-1 that this is indeed the case for $H_2S$, $H_2Se$, and $H_2Te$. However, the boiling point of $H_2O$, by far

**Table 10-1  Boiling Points (°C) of Some Binary Hydrogen Compounds**

| Group | 2nd Period | | 3rd Period | | 4th Period | | 5th Period | |
|-------|-----|-----|-----|-----|-----|-----|-----|-----|
| VIIA | HF | 17 | HCl | −84 | HBr | −70 | HI | −37 |
| VIA | $H_2O$ | 100 | $H_2S$ | −61 | $H_2Se$ | −42 | $H_2Te$ | − 2.0 |
| VA | $NH_3$ | −33 | $PH_3$ | −88 | $AsH_3$ | −62 | $SbH_3$ | −18 |

the highest, is way out of line. The explanation for the unusually high boiling point of water is that hydrogen bonding provides significant intermolecular interactions. The other three compounds are polar, but the normal dipole–dipole attractions do not seem to have a major effect on the boiling points compared with the trend in London forces. In Table 10-1, the boiling points of Group VA and Group VIIA hydrogen compounds are also listed. Hydrogen bonding explains the unusually high boiling points of $NH_3$ and HF compared with other hydrogen compounds in their group.

## 10-6 Energy and Changes in State

**Looking Ahead!** The energy required to cause a phase change is specific for a particular substance. The amount of that energy is the subject of this section. We will see that the energy required to melt a solid or vaporize a liquid depends on the nature of the solid or liquid. We will also discuss specific heats, which were examined in Chapter 2.

Large lakes and oceans moderate the climate in a number of ways. When one of the American Great Lakes freezes in winter, heat is released to the surroundings by this process. This heat has the same effect as a giant natural furnace and helps keep the air temperature from falling as much as it otherwise would. Since the Great Lakes do not usually completely freeze over, heat is released all through the winter from the freezing process. On the downside, spring is delayed in this region because the melting ice absorbs heat and keeps the temperature from rising as much as it otherwise would. Regions of Siberia, in Russia, lie about as far north as Minnesota but have few large lakes. As a result, it is much cooler there in the winter but also much hotter in the summer.

An ice cube at its melting point of 0°C remains in the solid state indefinitely unless energy is supplied. The addition of sufficient energy causes the solid to change to the liquid state yet remain at 0°C. The same is true for a liquid at its boiling point. That is, a specific amount of energy must be supplied to change the liquid to vapor at its boiling point. The opposite processes (i.e., fusion and condensation) *release* the same amount of energy as was absorbed in the previous processes. (See Figure 10-16.)

When the Great Lakes freeze in the winter, the heat released warms the surroundings. How can "freezing" water keep the surroundings warm?

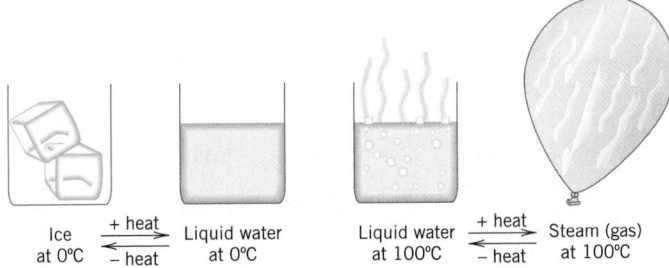

Ice at 0°C  — + heat / − heat →  Liquid water at 0°C          Liquid water at 100°C  — + heat / − heat →  Steam (gas) at 100°C

**Figure 10-16  Heat and Changes in State** Application or removal of heat causes a change in state at the melting point or boiling point of a substance.

**Table 10-2 Heats of Fusion and Melting Points**

| Compound | Type of Compound | Heat of Fusion | | Melting Point (°C) |
|---|---|---|---|---|
| | | (cal/g) | (J/g) | |
| NaCl | ionic | 124 | 519 | 801 |
| H₂O | polar covalent (hydrogen bonding) | 79.8 | 334 | 0 |
| ethyl alcohol | polar covalent (hydrogen bonding) | 24.9 | 104 | −114 |
| ethyl ether | polar covalent | 22.2 | 92.5 | −116 |
| benzene | nonpolar covalent | 30.4 | 127 | 5.5 |
| carbon tetrachloride | nonpolar covalent | 4.2 | 17.6 | −24 |

We understand that the temperature at which a substance melts and boils depends on the forces between the basic particles. The amount of energy it takes to cause changes in physical state also depends on the magnitude of these forces. We will first consider the energy involved in transitions between the solid and the liquid states (melting and freezing), and then transitions between the liquid and the gaseous states (boiling and condensation).

## Melting and Freezing

We are all aware that ice melts in the hot sun and that water changes to ice in the freezer. Melting [i.e., $H_2O(s) \longrightarrow H_2O(l)$] is an endothermic process—so heat must be supplied as in the hot sun. Freezing [i.e., $H_2O(l) \longrightarrow H_2O(s)$] is an exothermic process—so heat must be removed as in a freezer. The same amount of heat energy is released when a given amount of liquid freezes as would be required if the same amount of solid were to melt at the melting point. Each compound requires a specific amount of heat energy to melt a specific mass of sample. *The **heat of fusion** of a substance is the amount of heat in calories or joules required to melt one gram of the substance.* Table 10-2 lists the heats of fusion and the melting points of several substances. Note that sodium chloride, which is ionic, has the strongest attractions between particles of those listed and thus has the highest melting point and the highest heat of fusion.

Nonpolar compounds of low molar mass have small heats of fusion. Water, because of hydrogen bonds, has a rather high heat of fusion for such a light molecule. In fact, it is this relatively high heat of fusion that helps make water so effective at moderating climate such as described for the Great Lakes in the winter.

---

| | |
|---|---|
| **Example 10-1** | **The Heat Released by the Freezing of Water** |
| working it out | How many kilojoules of heat are released when 185 g of water freezes? |

**Procedure**

The heat of fusion can be used as a conversion factor relating mass in grams to joules.

$$\text{mass} \times \text{heat of fusion} = \text{heat}$$

**Solution**

$$185 \ g \times \frac{334 \ J}{g} \times \frac{1 \ kJ}{10^3 \ J} = \underline{\underline{61.8 \ kJ}}$$

**Table 10-3  Heats of Vaporization and Boiling Points**

| Compound | Type of Compound | Heat of Vaporization (cal/g) | Heat of Vaporization (J/g) | Normal Boiling Point (°C) |
|---|---|---|---|---|
| NaCl | ionic | 3130 | 13,100 | 1465 |
| $H_2O$ | polar covalent (hydrogen bonding) | 540 | 2,260 | 100 |
| ethyl alcohol | polar covalent (hydrogen bonding) | 204 | 854 | 78.5 |
| ethyl ether | polar covalent | 89.6 | 375 | 34.6 |
| benzene | nonpolar covalent | 94 | 393 | 80 |
| carbon tetrachloride | nonpolar covalent | 46 | 192 | 76 |

## Boiling and Condensation

If we wish to boil water, we place the container of water on the stove or in the microwave. The vaporization of a liquid [e.g., $H_2O(l) \longrightarrow H_2O(g)$] requires energy and so is also an endothermic process. The opposite process, condensation, is exothermic. As with melting and freezing, the same amount of energy is released when a given amount of vapor condenses [e.g., $H_2O(g) \longrightarrow H_2O(l)$] as would be required if the same amount of liquid were to vaporize at the boiling point. Each compound also requires a specific amount of heat energy to vaporize a specific mass of the sample. *The **heat of vaporization** of a substance is the amount of heat in calories or joules required to vaporize one gram of the substance.* The heats of vaporization and the boiling points of several substances are given in Table 10-3.

Note again that water has an unusually high heat of vaporization compared with other molecular compounds. Once again, the strength and number of hydrogen bonds between $H_2O$ molecules are responsible.

The heats of fusion and vaporization concern how heat can cause phase changes at a constant temperature. In Chapter 2, we discussed how heat also causes the temperature to rise when only one phase is present. All pure solids, liquids, and gases have a physical property that we defined as specific heat. (See Table 2-2.) *Specific heat refers to the amount of heat required to raise one gram of a substance one degree Celsius.* We observed that water also has an unusually high specific heat compared with other compounds or elements. The specific heats for the solid and gaseous states of water are not the same as for the liquid state, however.

The heat required to vaporize water is comparatively large.

---

**The Heat Released by Condensation and Cooling**

Steam causes more severe burns than an equal mass of water at the same temperature. Compare the heat released in calories when 3.00 g of water at 100°C cools to 60°C with the heat released when 3.00 g of steam at 100°C condenses and then cools to 60°C.

**Procedure**

There are two processes to consider. The first is the heat released when the liquid is cooled from 100°C to 60°C, or a change of 40°C. The second is the heat released when the gas condenses at 100°C.

**Example 10-2**

working it out

Use the specific heat of water as a conversion factor to convert mass and temperature change to calories.

$$\text{mass} \times \Delta t \times \text{specific heat} = \text{heat}$$

Use the heat of vaporization as a conversion factor to convert mass to calories.

$$\text{mass} \times \text{heat of vaporization} = \text{heat}$$

**Solution**

Cooling of water

$$3.00 \text{ g} \times (100 - 60)^\circ C \times \frac{1.00 \text{ cal}}{g \cdot {}^\circ C} = \underline{\underline{120 \text{ cal}}}$$

Condensation of steam

$$3.00 \text{ g} \times \frac{540 \text{ cal}}{g} = 1620 \text{ cal}$$

Total heat released from condensation and cooling = 1620 + 120 = $\underline{\underline{1740 \text{ cal.}}}$

Note that almost 15 times more heat energy is released by the steam at 100°C than by the water at the same temperature.

---

**Example 10-3** | **The Heat Required to Melt, Heat, and Vaporize Water**

How many kilojoules of heat are required to convert 250 g of ice at −15°C to steam at 100°C?

**Procedure**

There are four processes to consider. First (1) is the heating of the ice from −15°C to 0°C, or a total of 15°C. Second (2) is the melting of the ice at 0°C. Third (3) is the heating of the liquid from 0°C to 100°C, or a total of 100°C. And finally, fourth (4) is the vaporization of the liquid water to steam at 100°C. In summary, our process is

$$\text{Heat ice (1)} \quad \text{Melt ice (2)} \quad \text{Heat liquid (3)} \quad \text{Vaporize liquid (4)}$$
$$-15^\circ C \longrightarrow 0^\circ C \longrightarrow 0^\circ C \longrightarrow 100^\circ C \longrightarrow 100^\circ C$$

The heats of fusion and vaporization as well as the specific heats of ice and water (Table 2-2) are needed.

1. To calculate the heat required to heat the ice to 0°C, use the specific heat of ice to convert mass and temperature change to heat.

$$\text{mass} \times \Delta t \times \text{sp. heat (ice)} = \text{heat}$$

2. To calculate the heat required to melt the ice, use the heat of fusion to convert mass to heat.

$$\text{mass} \times \text{heat of fusion} = \text{heat}$$

3. To calculate the heat required to heat the water, use the specific heat of water to convert mass and temperature change to heat.

$$\text{mass} \times \Delta t \times \text{sp. heat (water)} = \text{heat}$$

4. To calculate the heat required to vaporize the water, use the heat of vaporization to convert mass to heat.

$$\text{mass} \times \text{heat of vaporization} = \text{heat}$$

**Solution**

To heat ice from $-15°C$ to its melting point ($0°C$), or a total of $15°C$:

$$250 \text{ g} \times 15°C \times 2.06\frac{J}{g \cdot °C} \times \frac{1 \text{ kJ}}{10^3 \text{ J}} = 7.7 \text{ kJ}$$

To melt the ice at $0°C$:

$$250 \text{ g} \times 334\frac{J}{g} \times \frac{1 \text{ kJ}}{10^3 \text{ J}} = 83.5 \text{ kJ}$$

To heat the water from $0°C$ to its boiling point ($100°C$), or a total of $100°C$:

$$250 \text{ g} \times 100°C \times 4.18\frac{J}{g \cdot °C} \times \frac{1 \text{ kJ}}{10^3 \text{ J}} = 105 \text{ kJ}$$

To vaporize the water at $100°C$:

$$250 \text{ g} \times 2260\frac{J}{g} \times \frac{1 \text{ kJ}}{10^3 \text{ J}} = 565 \text{ kJ}$$

$$\text{total} = 7.7 + 83.5 + 105 + 565 = \underline{\underline{761 \text{ kJ}}}$$

# 10-7 Heating Curve of Water

**Looking Ahead!** In the final section of this chapter, we will summarize what happens when heat is applied to a chunk of ice from the freezer so that it changes from a cold solid to a hot vapor. The added heat will cause three temperature changes and two phase changes. As each of these changes occurs, we will try to visualize how the molecules of $H_2O$ are affected.

Assume that we take an ice cube cooled to $-10°C$ from the freezer compartment of the refrigerator. Imagine that the ice crystal could be magnified so that the motions and positions of the individual molecules could be seen. We would notice that all the molecules are in fixed positions and that ice is a crystalline molecular solid. The molecules are certainly in motion, but the motion consists mainly of vibrations about their fixed locations. We are now going to supply heat at a constant rate to the ice cube and observe the changes that occur. *The graphical representation of the temperature as a solid is heated through the two phase changes plotted as a function of the time of heating is called the* **heating curve.** (See Figure 10-17. Note that it is really a series of straight lines despite being called a "curve.") There are five regions of interest in the "curve."

## Heating Ice

From $-10°C$, the added heat causes an increase in the vibrations and rotations of the molecules about the fixed positions of the water molecules in the solid crystal lattice. This is an increase in kinetic energy and only one phase is present, so the temperature rises. The specific heat of a particular phase of a particular compound is a measure of the energy required to cause these kinetic energy changes. The larger the value of the specific heat, the more slowly the temperature rises. Part (*a*) in Figure 10-17 represents the temperature change as the ice is heated.

## Melting Ice

At $0°C$ (and 1 atm pressure), a phase change begins. The vibrations of the molecules become so great that some hydrogen bonds break and the molecules begin to move out of their fixed positions in the crystal lattice. We notice that the solid ice begins

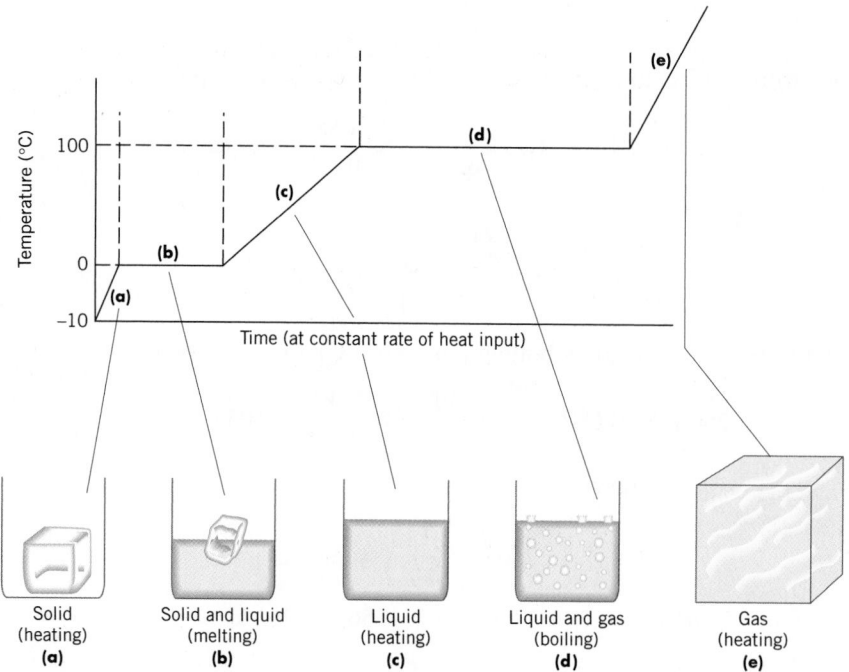

**Figure 10-17 The Heating Curve of Water**

| Solid (heating) **(a)** | Solid and liquid (melting) **(b)** | Liquid (heating) **(c)** | Liquid and gas (boiling) **(d)** | Gas (heating) **(e)** |

to melt. Not all of the hydrogen bonds break, however, so groups of molecules still stick together in the liquid state. The added heat is now causing molecules to move apart rather than move faster. When particles that attract each other move apart, there is an increase in the potential energy. Since changes in temperature are related to changes in kinetic energy only, the temperature remains constant while the ice is melting. A constant temperature as a solid melts is a property of pure matter. (See Figure 10-17*b*.) How long it takes a given solid to melt depends on the heat of fusion.

## Heating Water

Between 0°C and 100°C, only the liquid phase is present, so the added heat causes the temperature to rise. That is because the added heat causes the kinetic energy of the water molecules to increase. Unlike in the solid state, the motion of water molecules in the liquid state includes translational motion (movement from point to point). The rate of rise of temperature of a liquid depends on the specific heat of the liquid. Notice that the temperature does not appear to rise as fast in the liquid as in the solid. This is because the specific heat of water [1.00 cal/(g · °C)] is about twice that of ice [0.490 cal/(g · °C)]. (See Figure 10-17*c*.)

## Vaporizing Water

At 100°C (and 1 atm pressure), a second phase change begins. As the temperature of the liquid increases from 0°C to 100°C, the vapor pressure of the liquid has also been rising. At this temperature, the vapor pressure of water is equal to the atmospheric pressure, and the water begins to boil. The kinetic energy of the water molecules has now become great enough to break the remaining hydrogen bonds holding them in the liquid state. Just as in the melting process, the added heat is now causing molecules to move apart rather than increasing the kinetic energy of the molecules. As with melting, the added heat causes an increase in potential energy, and the temperature remains constant. How long it takes to vaporize a given liquid depends on the heat of vaporization of that liquid. Notice that it takes much longer to vaporize the water than to melt it. This is because the heat of vaporization of water is more than six times that of the heat of fusion. (See Figure 10-17*d*.)

## Heating Vapor

Above 100°C, we again have only one phase, so the added heat causes the temperature of the gas to increase. In the gaseous state, the kinetic energy of the gas molecules is great enough to overcome all of the interactions between molecules. As a result, water vapor acts as any other gas and is subject to the gas laws discussed in the previous chapter. (See Figure 10-17e.)

**Looking Back!** The properties of the liquid state reflect the attractive forces between the molecules. A molecule may escape to the vapor state only if it has enough kinetic energy to overcome these forces. The formation of a liquid from a solid (melting), the formation of a gas from a liquid (vaporization), as well as the heating of a solid, liquid, or gas are all endothermic processes, each requiring a specific amount of heat energy. This is sometimes represented as a heating curve.

## Learning Check B | checking it out

**B-1.** Fill in the blanks.

In the liquid state, _____ _____ causes water to form spherical drops. _____ is a property relating to the rate of flow. When evaporation of a liquid occurs, the remaining liquid becomes _____ . The equilibrium _____ _____ is the pressure exerted by the vapor at a given _____ . The vaporization of a solid is known as _____ . The normal boiling point is the temperature at which the _____ _____ _____ is equal to 760 torr. The energy required to melt one gram of a solid is known as the _____ _____ _____ , and the energy required to vaporize one gram of liquid is known as the _____ _____ . A heating curve represents the time of _____ on one axis and the _____ on the other. When a pure solid is melting, the _____ does not change.

**B-2.** What is the boiling point of water at an altitude where the atmospheric pressure is 550 torr? (See Figure 10-15.) What is the boiling point of diethyl ether at this altitude?

**B-3.** What mass of ethyl ether can be melted with 1.00 kJ of heat energy? What mass of ethyl ether can be vaporized by 1.00 kJ. (Refer to Tables 10-2 and 10-3.)

**B-4.** How many kilojoules of heat energy are released when 10.0 g of ethyl alcohol vapor is condensed at its boiling point, the liquid allowed to cool, and then frozen at its melting point? The specific heat of ethyl alcohol is 2.47 J/g · °C.

*Additional Examples: Exercises 10-38, 10-55, 10-65, 10-70, 10-71, and 10-76.*

**chapter Review**          **putting it together**

This chapter gives the liquid and solid state their proper consideration. The most significant difference between these "condensed" states and the gaseous state is that cohesive forces hold the basic particles together and the basic particles occupy a significant part of the volume. Motions are confined to vibrations about fixed positions in the solid state or vibrational and translational motion in the liquid state. For this reason, solids and liquids are essentially incompressible, undergo little thermal expansion, and have definite volumes and high densities.

The temperature at which a molecular substance undergoes a phase transition (melting or boiling) depends on the **intermolecular forces** between molecules. **London forces** occur between all molecules but act exclusively for nonpolar molecules. These forces depend to a large extent on the size of the molecules. They increase as the size (or molar mass) increases. Compounds composed of heavy molecules generally have higher melting points than lighter ones. If a molecule has a permanent dipole, it has **dipole–dipole attractions** between molecules in addition to London forces. Compounds composed of polar molecules usually have higher melting and boiling points than nonpolar compounds of similar molar mass. **Hydrogen bonding** is a particular type of electrostatic interaction that has a significant effect on physical properties. It involves hydrogen bonded to an F, O, or N interacting with an F, O, or N on another molecule. Hydrogen bonding is responsible for many of the unusual properties of water.

We first examined the solid state of matter. Solids may be either **amorphous** or **crystalline.** Most solids are crystalline, which means they have atoms, ions, or molecules occupying fixed positions in a **crystal lattice.** They display sharp and definite melting points. There are four types of crystalline solids. **Ionic solids** are composed of ions as the basic particles in the crystal lattice. The ion–ion attractions are strong, so ionic compounds are solids at room temperature. **Molecular solids** are composed of molecules in the crystal lattice. The melting points of such solids depend on the intermolecular forces between molecules. Generally, they are soft solids melting at lower temperatures than ionic solids. **Network solids** are somewhat rare but generally melt at high temperatures because covalent bonds must be broken to cause melting. All of the three allotropes of carbon—**diamond, graphite,** and **buckminsterfullerene**—melt at high temperatures, but diamond has the highest melting point. **Metallic solids** have metal cations at the crystal lattice positions surrounded by a sea of electrons. These solids have melting points that range from low to extremely high temperatures.

| Type of Solid | Interaction | Melting points | Examples |
|---|---|---|---|
| ionic | ion–ion | high | $NaCl$, $K_2CO_3$ |
| molecular | London only | low | $O_2$, Ne, $CH_4$ |
| molecular | dipole–dipole | low | $SO_2$, $CH_3F$ |
| molecular | hydrogen bonding | intermediate | $H_2O$, $CH_3OH$ |
| network | covalent | very high | diamond, quartz |
| metallic | metal ion–electron | generally high | Na, Ca, Cr, Fe |

The liquid state is intermediate between the solid and gaseous states. The cohesive forces between the basic particles in the liquid state are evident because of the liquid's **surface tension** and **viscosity.** Some molecules in the liquid state may overcome these cohesive forces and escape to the gaseous state. When this process of **vaporization** is exactly counteracted by the process of **condensation,** an **equilibrium vapor pressure** above the liquid is established. The **boiling point** of the liquid relates to its vapor pressure. The **normal boiling point** is the temperature at which the vapor pressure equals one atm. **Evaporation** occurs when the liquid vaporizes below the boiling point. A solid may also vaporize in a process known as **sublimation.** These phase transitions are summarized as follows.

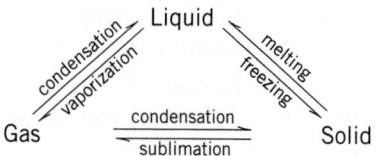

The kinetic molecular theory can be used to explain the behavior of basic particles in these states. This is illustrated with the common example of water.

| Gas | Liquid | Solid |
|---|---|---|
| There are no attractions between $H_2O$ molecules, and motion is random. | Molecules are mobile, but significant hydrogen bonding holds the molecules in a condensed state. | Water molecules are held in fixed positions by additional hydrogen bonding; motion is restricted to vibrations. |

Melting and boiling each take place at a specific temperature for most substances. It also takes a specific amount of energy to change one gram of solid to liquid, which is known as the **heat of fusion,** and a specific amount of energy to change one gram of liquid to vapor, which is known as the **heat of vaporization.** The magnitudes of these quantities depend on the forces between the basic particles of the solid or liquid. A substance with a high melting point usually has a high heat of fusion and a high heat of vaporization. Much of the information on phase transitions can be shown in a **heating curve.** The rate of heating of a single phase is dependent upon the specific heat of that phase. The times required for a solid to melt and a liquid to boil at a given pressure depend on the magnitudes of the heat of fusion and vaporization, respectively. The heating curve of a pure substance shows that, when two phases are present in a phase change, the temperature remains constant. When melting or boiling is occurring, the applied heat energy is increasing the potential energy of the system rather than the kinetic energy.

# Exercises

## Nature of the Solid and Liquid States

**10-1.** Liquids mix more slowly than gases. Why?

**10-2.** Why is a gas compressible, whereas a liquid is not compressible?

**10-3.** Describe why a drop of food coloring in a glass of water slowly becomes evenly distributed without the need for stirring.

**10-4.** What properties do liquids have in common with solids? With gases?

**10-5.** Review the densities of the solids listed in Table 2-1. In general, which have higher densities, solids or liquids? Why is this so?

## Intermolecular Forces

**10-6.** Arrange the following in order of increasing magnitude of the London forces between their molecules.
(a) $CH_4$    (b) $CCl_4$    (c) $GeCl_4$

**10-7.** Arrange the following in order of increasing magnitude of the London forces between their molecules or atoms.
(a) Ne    (b) Xe    (c) $N_2$    (d) $SF_6$

**10-8.** At room temperature, $Cl_2$ is a gas, $Br_2$ is a liquid, and $I_2$ is a solid. Explain the trend.

**10-9.** At room temperature, $CO_2$ is a gas and $CS_2$ is a liquid. Why is this reasonable?

**10-10.** If $H_2O$ were a linear molecule, could it have hydrogen-bonding interactions?

**10-11.** The $H_2S$ molecule is also V-shaped, similar to $H_2O$, but it has a very small molecular dipole. How does one $H_2S$ molecule interact with other $H_2S$ molecules?

**10-12.** Write the Lewis structure of $NH_3$, with the electron pair and the hydrogens in a tetrahedral arrangement. (See Figure 10-4$a$.) How does one $NH_3$ molecule interact with other $NH_3$ molecules?

**10-13.** The $PH_3$ molecule can be represented in a tetrahedral arrangement similar to $NH_3$. How does one $PH_3$ molecule interact with other $PH_3$ molecules?

**10-14.** Which of the following molecules have dipole–dipole attractions in a condensed state?
(a) HBr                     (d) $BF_3$ (trigonal planar)
(b) $SO_2$ (nonlinear)      (e) $N_2$
(c) $CO_2$ (linear)         (f) CO

**10-15.** Which of the following molecules have dipole–dipole interactions in a condensed state?
(a) $SCl_2$ (nonlinear)    (c) $CCl_4$ (tetrahedral)    (e) FCl
(b) $PH_3$                 (d) $O_3$

**10-16.** What is the difference between a covalent bond and hydrogen bonding?

**10-17.** Which of the following molecules can have hydrogen bonding in the liquid state?
(a) HF      (d) $H_2O$      (f)
(b) $NCl_3$      (e) $CH_4$ (tetrahedral)
(c) $H_2NCl$

$$H-C \overset{\displaystyle O}{\underset{\displaystyle O-H}{\diagup}}$$

(g) $CH_3Cl$ (tetrahedral)

**10-18.** Which should have stronger hydrogen bonding, $NH_3$ or $H_2O$?

## The Solid State

**10-19.** Identify the forces that must be overcome to cause melting in the following solids.
(a) KF      (b) HF      (c) HCl      (d) $F_2$

**10-20.** Rank the compounds in exercise 10-19 in order of expected melting points (lowest one first).

**10-21.** Identify the forces that must be overcome to cause melting in the following solids.
(a) diamond      (b) $CF_4$      (c) $CrF_2$      (d) $SCl_2$

**10-22.** Rank the compounds in exercise 10-21 in order of expected melting points (lowest one first).

**10-23.** The two Group IVA oxides $CO_2$ and $SiO_2$ have similar formulas but very different melting points. Why?

**10-24.** The two Group IIA fluorides, $BF_3$ and $AlF_3$, have similar formulas but very different melting points. Why?

**10-25.** Lead forms two compounds with chlorine, $PbCl_2$ and $PbCl_4$. The melting point of $PbCl_2$ is 501°C and that of $PbCl_4$ is −15°C. Interpret in terms of types of solids.

## The Liquid State

**10-26.** Why is motor oil more viscous than water? Does motor oil have a greater surface tension than water?

**10-27.** Explain how molecules of a liquid can go into the vapor state if the temperature is below the boiling point.

**10-28.** Why does a summer rainstorm lower the temperature?

**10-29.** Why does rubbing alcohol feel cool on the skin even if the alcohol is at room temperature when first applied?

**10-30.** Ethyl chloride boils at 12°C. When it is sprayed on the skin, it freezes a small part of the skin and thus serves as a local anesthetic. Explain how it cools the skin.

**10-31.** Given a sample of water at 90°C and a sample of water at 30°C. In which liquid does the temperature change at a faster rate when both are allowed to evaporate?

**10-32.** A beaker of a liquid with a vapor pressure of 350 torr at 25°C is set alongside a beaker of water, and both are allowed to evaporate. In which liquid does the temperature change at a faster rate? Why?

**10-33.** What is implied by the word "equilibrium" in *equilibrium vapor pressure?*

**10-34.** What is the difference between boiling point and normal boiling point?

**10-35.** A liquid has a vapor pressure of 850 torr at 75°C. Is the substance a gas or a liquid at 75°C and 1 atm pressure?

**10-36.** If the atmospheric pressure is 500 torr, what are the approximate boiling points of water, ethyl alcohol, and ethyl ether? (Refer to Figure 10-15.)

**10-37.** How can the boiling point of a pure liquid be raised?

**10-38.** On top of Mt. Everest, the atmospheric pressure is about 260 torr. What is the boiling point of ethyl alcohol at that pressure? If the temperature is 10°C, is ethyl ether a gas or a liquid under conditions on Mt. Everest? (Refer to Figure 10-15.)

**10-39.** The boiling point of water in Death Valley, California, is about 100.2°C. Why is the actual boiling point higher than the normal boiling point?

**10-40.** Propane is used as a fuel to heat rural homes where natural gas pipelines are not available. It is stored as a liquid under normal temperature conditions, although its normal boiling point is −42°C. How can propane remain a liquid in a tank when the temperature is well above its normal boiling point?

**10-41.** The normal boiling point of neon is −246°C and that of argon is −186°C. What accounts for this order of boiling points.

**10-42.** The boiling point of HCl is −84°C and that of HBr is −70°C. Why is this order reasonable on one account but opposite from what one would expect from a consideration of polarity? Which trend is more important in this case?

**10-43.** On the planet Mars the temperature can reach as high as a comfortable 50°F (10°C) at the equator. The atmospheric pressure is about 8 torr on Mars, however. Can liquid water exist on Mars under these conditions? What would the atmospheric pressure have to be before liquid water could exist at this temperature? What would happen to a glass of water set out on the surface of Mars under these conditions? (Refer to Figure 10-15.)

## Energy and Changes of State

**10-44.** At a given temperature, one liquid has a vapor pressure of 240 torr and another measures 420 torr. Which liquid probably has the lower boiling point? Which probably has the lower heat of vaporization?

**10-45.** Ethyl ether ($C_2H_5OC_2H_5$) and ethyl alcohol ($C_2H_5OH$) are both polar covalent molecules. What accounts for the considerably higher heat of vaporization for alcohol?

**10-46.** The heats of vaporization of liquid $O_2$, liquid Ne, and liquid $CH_3OH$ are in the order Ne < $O_2$ < $CH_3OH$. Why?

**10-47.** A given compound has a heat of fusion of about 600 cal/g. Is it likely to have a comparatively high or low melting point?

**10-48.** A given compound has a boiling point of −75°C. Is it likely to have a comparatively high or low heat of vaporization?

**10-49.** A given compound has a boiling point of 845°C. Is it likely to have a comparatively high or low melting point?

**10-50.** Graph the data in Table 10-1 for the Group VIA hydrogen compounds. Plot the boiling points on the $y$ axis and the molar masses of the compounds on the $x$ axis. What would be the expected boiling point of $H_2O$ if only London forces were important, as is the case with the other compounds in this series? (Determine from the graph.)

**10-51.** Graph the data in Table 10-1 for the Group VA hydrogen compounds. Plot the boiling points on the $y$ axis and the molar masses of the compounds on the $x$ axis. What would be the expected boiling point of $NH_3$ if only London forces were important, as is the case with the other compounds in this series?

**10-52.** How many kilojoules are required to vaporize 3.50 kg of $H_2O$ at its boiling point? (Refer to Table 10-3.)

**10-53.** How many joules of heat are released when an 18.0-g sample of benzene condenses at its boiling point? (Refer to Table 10-3.)

**10-54.** Molten ionic compounds are used as a method to store heat. How many kilojoules of heat are released when 8.37 kg of NaCl solidifies at its melting point? (Refer to Table 10-2.)

**10-55.** If 850 J of heat is added to solid $H_2O$, NaCl, and benzene at their respective melting points, what mass of each is changed to a liquid? (Refer to Table 10-2.)

**10-56.** What mass of carbon tetrachloride can be vaporized by addition of 1.00 kJ of heat energy to the liquid at its boiling point? What mass of water would be vaporized by the same amount of heat? (Refer to Table 10-3.)

**10-57.** When a 25.0-g quantity of ethyl ether freezes, how many calories are liberated? When 25.0 g of water freezes, how many calories are liberated? In a large lake, which liquid would be more effective in modifying climate? How many joules of heat are required to melt 125 g of ethyl alcohol at its melting point? (Refer to Table 10-2.)

**10-58.** Air conditioners and refrigerators cool by vaporization of Freon. How many kcal of heat are removed by the vaporization of 1.00 kg of Freon? Will this be enough to freeze one can (325 mL) of soda? Assume the soda is essentially pure water at 0°C. (The heat of vaporization of Freon is 38.5 cal/g).

**10-59.** Refrigerators cool because a liquid extracts heat when it is vaporized. Before the synthesis of Freon ($CF_2Cl_2$), ammonia was used. Freon is nontoxic, and $NH_3$ is a pungent and toxic gas. The heat of vaporization of $NH_3$ is 1.36 kJ/g and that of Freon is 161 J/g. How many joules can be extracted by the vaporization of 450 g of each of these compounds? Which is the better refrigerant on a mass basis?

**10-60.** The heat of vaporization of $BCl_3$ is 4.5 kcal/mol and that of $PCl_3$ is 7.2 kcal/mol. There are two reasons that account for this order. What are they?

**10-61.** How many joules of heat are released when 275 g of steam at 100.0°C is condensed and cooled to room temperature (25.0°C)? (Refer to Table 10-3.)

**10-62.** How many kilocalories of heat are required to melt 0.135 kg of ice and then heat the liquid water to 75.0°C? (Refer to Table 10-2.)

**10-63.** How many calories of heat are needed to heat 120 g of ethyl alcohol from 25.5°C to its boiling point and then vaporize the alcohol? [The specific heat of alcohol is 0.590 cal/(g · °C). Refer to Table 10-3.]

**10-64.** Rubbing alcohol (isopropyl alcohol) helps reduce fevers by cooling the skin when it is rubbed on. How much heat is removed by the vaporization of 25.0 mL of alcohol? (The density of the alcohol is 0.786 g/mL and the heat of vaporization is 705 cal/g.)

**10-65.** How many calories of heat are required to change 132 g of ice at −20.0°C to steam at 100.0°C? [The specific heat of ice is 0.492 cal/(g · °C). Refer to Tables 10-2 and 10-3.]

**10-66.** How many kilojoules of heat are released when 2.66 kg of steam at 100.0°C is condensed, cooled, frozen, and then cooled to −25.0°C? [The specific heat of ice is 2.06 J/(g · °C). Refer to Tables 10-2 and 10-3.]

**\*10-67.** A sample of steam is condensed at 100.0°C and then cooled to 75.0°C. If 28.4 kJ of heat is released, what is the mass of the sample? (Refer to Table 10-3.)

**\*10-68.** What mass of ice at 0°C can be changed into steam at 100°C by 2.00 kJ of heat? (Refer to Tables 10-2 and 10-3.)

**\*10-69.** A 10.0-g sample of benzene is condensed from the vapor at its boiling point, and the liquid is allowed to cool. If 5000 J is released, what is the final temperature of the liquid benzene? [The specific heat of benzene is 1.72 J/(g · °C). Refer to Table 10-3.]

## The Heating Curve

**10-70.** Which of the following processes are endothermic?
**(a)** freezing    **(c)** boiling
**(b)** melting    **(d)** condensation

**10-71.** Which has the higher kinetic energy: $H_2O$ molecules in the form of ice at 0°C, or in the form of liquid water at 0°C?

**10-72.** Which has the higher potential energy: $H_2O$ molecules in the form of ice at 0°C, or in the form of liquid water at 0°C?

**10-73.** Which has the higher potential energy: $H_2O$ molecules in the form of steam at 100°C, or in the form of liquid water at 100°C?

**10-74.** If water is boiling and the flame supplying the heat is turned up, does the water become hotter? What happens?

**10-75.** How many kilocalories are released when 18.0 g of steam at 100°C is condensed and cooled to ice at 0°C?

**\*10-76.** Construct a heating curve for ethyl alcohol. Refer to Tables 10-2 and 10-3. How should the time of constant temperature for melting compare with that for boiling? (Assume that the specific heats of the three phases of water are about twice those of ethyl alcohol.)

**\*10-77.** Construct a heating curve for ethyl ether. Refer to Tables 10-2 and 10-3. How should the time of constant temperature for melting compare with that for boiling? (Assume that the specific heats of the three phases of water are about four times those of ethyl ether.)

## General Problems

**10-78.** The following three compounds have similar molar masses: $C_2H_5NH_2$, $CH_3OCH_3$, and $CO_2$. The temperatures at which these compounds boil are $-78°C$, $-25°C$, and $17°C$. Match the boiling point with the compound and give the respective intermolecular forces that account for this order.

**10-79.** At room temperature, $SF_6$ is a gas and $SnO$ is a solid. Both have similar formula weights. What accounts for the difference in physical states of the two compounds?

**10-80.** At room temperature, $CH_3OH$ is a liquid and $H_2CO$ is a gas. Both are polar and have similar molar masses. What accounts for the difference in physical states of the two compounds?

**10-81.** $CH_3F$ and $CH_3OH$ have almost the same molar mass, and both are polar compounds. Yet $CH_3OH$ boils at $65°C$ and $CH_3F$ at $-78°C$. What accounts for the large difference?

**10-82.** The heat of fusion of gold is 15.3 cal/g and that of silver is 25.0 cal/g, yet it takes more heat to melt the same volume of gold as silver. Calculate the heat required to melt 10.0 mL each of silver and gold. (The density of silver is 10.5 g/mL and that of gold is 19.3 g/mL.)

**10-83.** $SiH_4$, $PH_3$, and $H_2S$ melt at $-185°C$, $-133°C$, and $-85°C$, respectively. Since all have about the same molar mass, what accounts for the order?

**10-84.** The boiling point of $F_2$ is $-188°C$ and that of $Cl_2$ is $-34°C$, yet the boiling point of HF is much higher than that of HCl. Explain.

**10-85.** Nitrogen gas and carbon monoxide have the same molar masses. Carbon monoxide boils at a slightly higher temperature, however ($-191°C$ versus $-196°C$ for $N_2$). Account for the difference.

**10-86.** Liquid sodium metal is used as a coolant in a certain type of experimental nuclear reactor. How many kilojoules does it take to heat 1.00 mol of sodium from a solid at its melting point of $98°C$ to vapor at its boiling point of $883°C$? [For sodium, the heat of fusion is 113 J/g, the heat of vaporization is 3.90 kJ/g, and the specific heat of liquid sodium is 1.18 J/(g · °C).]

**10-87.** The heat of fusion of iron is 266 J/g. Iron is formed in industrial processes in the molten state and is solidified with water. The water vaporizes when it comes in contact with the liquid iron. What mass of water is needed to solidify or freeze 1.00 ton (2000 lb) of iron? (Assume that the water is originally at $25.0°C$ and that the steam remains at $100.0°C$.)

**\*10-88.** On a hot, humid day the relative humidity is 70% of saturation. If the temperature is $34°C$, the vapor pressure of water at 100% of saturation is 39.0 torr. What mass of water is in each 100 L of air under these conditions?

**10-89.** Chromium metal, which is used to make stainless steel, is obtained by reaction of chromium(III) oxide with elemental aluminum. How much heat energy is required to melt the chromium produced from 1.25 kg of chromium(III) oxide? The heat of fusion of chromium is 21.0 kJ/mol.

**10-90.** The heat of fusion of chromium is 21.0 kJ/mol, molybdenum is 28.0 kJ/mol, and tungsten is 35.0 kJ/mol. What mass (in kg) of each of these elements can be melted by the same amount of heat required to change 1.00 kg of liquid water at $25°C$ to steam at $100°C$?

**10-91.** A certain element melts at $-39°C$ and boils at $357°C$. The specific heat of the liquid is 0.139 J/g · °C. The element forms a $+2$ ion. What is the element? What mass of the element can be melted at $-39°C$, heated to its boiling point, and then vaporized at its boiling point by 15.00 kJ of heat? The heat of fusion of the element is 11.5 J/g, and the heat of vaporization is 29.5 J/g.

**10-92.** A common element has a melting point of $114°C$. The element forms a $-1$ ion. What is the element? It has a vapor pressure of 90.5 torr at its melting point. If 50.0 g of the solid element were placed in a 1.00-L container at $114°C$ and 1.00 atm pressure, what percent of the element would eventually be in the vapor state?

**10-93.** Ice has a vapor pressure of 2.50 torr at $-5°C$. Will 50.0 g of ice completely vaporize if placed in a room at $-5°C$ and 1.00 atm pressure with a volume 20.0 kL? Assume that the air in the room was originally dry.

**10-94.** Liquid air is composed mainly of $N_2$ and $O_2$. At $-196°C$, the vapor pressure of $N_2$ is 760 torr. Would you expect the vapor pressure of $O_2$ to be larger or smaller than 760 torr at this temperature? Would liquid $O_2$ boil at a higher or lower temperature than liquid $N_2$? Explain.

# Solutions to Learning Checks

**A-1.** high, definite, intermolecular, London forces, dipole–dipole attractions, hydrogen bonding, hydrogen bonding, London, amorphous, crystalline, crystalline, ionic, network, diamond, low, metallic, electrons

**A-2.**

**(a)** $BCl_3$: London    $NCl_3$: dipole–dipole    $HNCl_2$: hydrogen bonding

**(b)** $HNCl_2$ is most likely to be in the condensed state, and $BCl_3$ is the least likely.

**A-3.** $CaF_2$ has the higher melting point because it is an ionic solid. $SeF_2$ is a molecular solid.

**B-1.** surface tension, viscosity, cooler, vapor pressure, temperature, sublimation, vapor pressure, heat of fusion, heat of vaporization, heating, temperature, temperature

**B-2.** Water boils at about 90°C at this altitude, and diethyl ether at about 25°C.

**B-3.** melting: $1.00 \text{ kJ} \times \dfrac{10^3 \text{ J}}{\text{kJ}} \times \dfrac{1 \text{ g}}{92.5 \text{ J}} = \underline{\underline{10.8 \text{ g}}}$

vaporizing: $1.00 \text{ kJ} \times \dfrac{10^3 \text{ J}}{\text{kJ}} \times \dfrac{1 \text{ g}}{375 \text{ J}} = \underline{\underline{2.67 \text{ g}}}$

**B-4.** condensation at 78.5°C:

$10.0 \text{ g} \times 854 \dfrac{\text{J}}{\text{g}} = \underline{8540 \text{ J}}$

cooling from 78.5°C to −114°C ($\Delta t = 193$°C):

$10.0 \text{ g} \times 193 \text{°C} \times 2.47 \dfrac{\text{J}}{\text{g} \cdot \text{°C}} = \underline{4770 \text{ J}}$

freezing at −114°C:

$10.0 \text{ g} \times 104 \dfrac{\text{J}}{\text{g}} = \underline{1040 \text{ J}}$

$\text{Total} = 14{,}350 \text{ J} = \underline{\underline{14.35 \text{ kJ}}}$

## chapter 11

# Aqueous Solutions

The woman floats like a cork
in the Dead Sea in Israel. The water
in this lake contains high concentrations
of solutes, so is denser than pure water.
The properties of aqueous solutions
are discussed in this chapter.

**Setting the Stage**

The sound and sensation that we feel from the rhythmic pounding of waves against a silvery ocean beach are among nature's most tranquil gifts. This endless body of liquid seems to extend forever past the horizon. Actually, the ocean covers two-thirds of the surface of the Earth and in places it is nearly six miles deep. But we can't drink this water, nor can we use it to irrigate crops. This water is far from pure; it contains a high concentration of dissolved compounds. But the blood coursing in our veins is one example of how our life also depends on this same ability of water to dissolve substances. This red waterway dissolves oxygen and nutrients for life processes and then carries away dissolved waste products for removal by the lungs and kidneys. In fact, water is indispensable for any life form as we know it.

The ability of water to act as a solvent is the emphasis of this chapter. In the last chapter, we examined the unusual properties of this truly unique compound. Other compounds composed of such small molecules are much more volatile than water. But water molecules stick together because of the relatively strong interactions known as hydrogen bonding. Further, the conditions on this planet Earth, the third planet out from a rather ordinary star, are just right for water to be able to exist in all three physical states.

This chapter builds on many topics that have come before. We will note first that water tends to dissolve ionic compounds, so we will need to recall how ionic compounds are identified and how these ions are found in aqueous solution (Chapters 4 and 7). Water is a powerful solvent for polar compounds because water molecules are also polar. Thus we will need to recall how we determine the polarity of a compound (Chapter 6). The hydrogen bonding of water covered in the previous chapter is also important in this discussion. Finally, since many chemical reactions take place in aqueous solution, we will include stoichiometry in these discussions. Stoichiometry, which was discussed in Chapters 8 and 9, will thus be extended to include solutions.

**Formulating Some Questions**

In this chapter, we will address many questions about water as a solvent. How does water interact with a solute to produce a solution? How does temperature affect how much dissolves? How do we indicate the actual amount of a substance that dissolves? And finally, how does the presence of a dissolved substance affect the physical properties of water? Our first order of business is to explore how and why water dissolves so many other compounds.

Section **A**    **SOLUTIONS AND THE QUANTITIES INVOLVED**

## 11-1 The Nature of Aqueous Solutions

**Looking Ahead!** The composition of a solution is familiar territory. In an earlier chapter we defined a solution as a homogeneous mixture of a solute (that which dissolves) in a solvent (the medium that dissolves the solute—usually a liquid). In this section we will look into the process of how and why a solute is dispersed in a solvent.

First, we will briefly consider the mixing of two liquids. *If solution occurs when the solvent and solute are both liquids, we say that the two liquids are* **miscible.** *If two liquids do not mix to form a solution, we say that they are* **immiscible** and remain as a heterogeneous mixture with two liquid phases. Oil and water, for example, are immiscible and thus form a heterogeneous mixture with two obvious phases. Alcohol and water are miscible and form a solution with one phase. This was previously discussed in section 2-5 and illustrated in Figure 2-8. Other types of homogeneous mixtures that qualify as solutions are any mixture of gases, which was discussed in Chapter 9, and metal alloys, discussed in section 2-5.

In this chapter, we will be more concerned with solutions formed by solids and liquids dissolved in water as the solvent. (See Figure 11-1.) The dissolving of table salt, NaCl, is typical of what happens when an ionic compound is placed in water. Recall that this solid compound is composed of alternating $Na^+$ ions and $Cl^-$ ions in a three-dimensional crystal lattice. In Chapters 6 and 10 we discussed how water is a polar compound with a positive dipole located on the hydrogen atoms and a negative dipole located on the oxygen. The water molecules are moderately attracted to each other by hydrogen bonds, but in the presence of the ionic crystal a stronger attraction between the ions in the solid and the dipoles of water exists. (See Figure 11-2.) Certainly, strong forces hold the solid crystal together (ion–ion), but the forces between the water molecules and the ions act to lift the ions on the surface of the solid out of the crystal lattice and into the aqueous medium. *The interactions between the ions and dipoles of solvent molecules are referred to as* **ion–dipole forces.** An individual ion–dipole force is not as strong as one ion–ion force, but there are many ion–dipole forces at work. As a result, a tug-of-war

**Figure 11-1   A Solution**
A solution (the red liquid) is a homogeneous mixture of solute (the solid chromium compound) and solvent (water). The solution is in the same physical state as the solvent.

**Figure 11-2  Interaction of Water and Ionic Compounds**
**There is an electrostatic interaction between the polar water molecules and the ions. This is an ion–dipole force.**

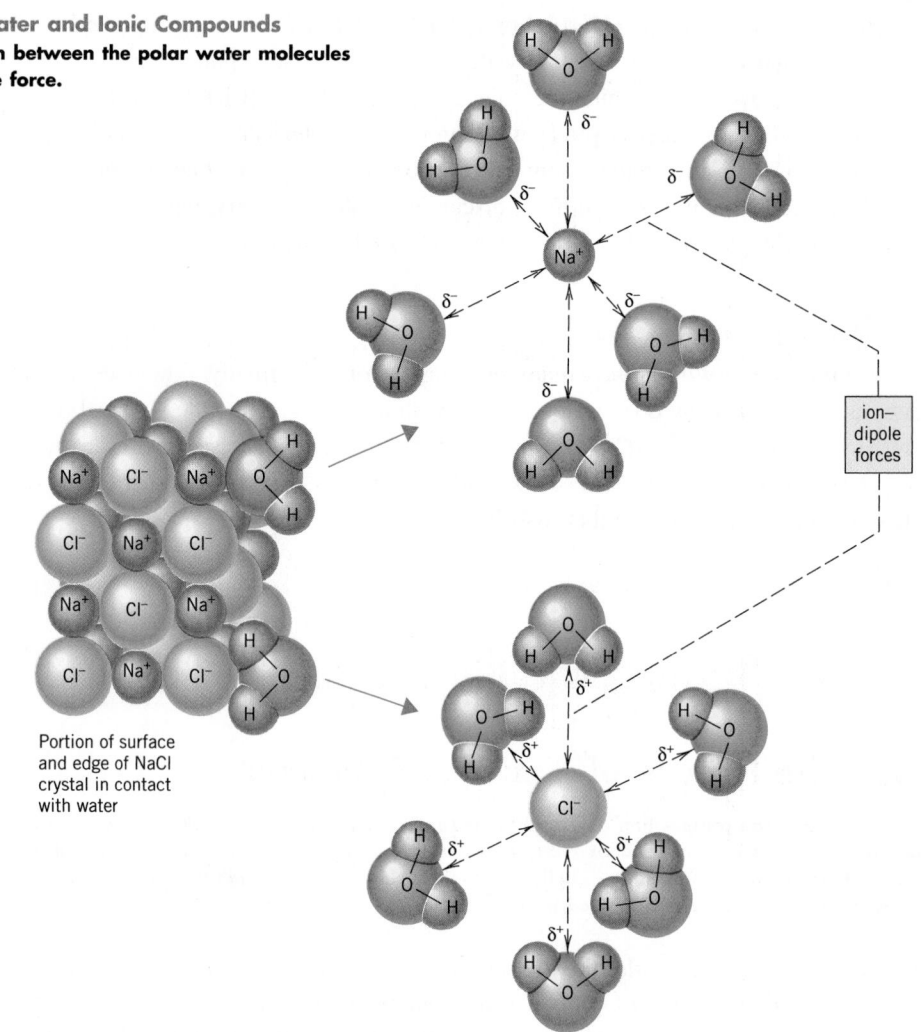

Portion of surface
and edge of NaCl
crystal in contact
with water

ion–dipole forces

develops between the ion–ion forces, holding the crystal together, and the ion–dipole forces, pulling the ions into solution. In the case of NaCl, the solution forces are strong enough to lift the ions, one by one, into the solvent. Since the ions in solution are now surrounded by water molecules, we say that the ions are *hydrated*. In solution, the positive and negative ions are no longer associated with one another. In fact, the charges on the ions are diminished or insulated from each other by the "escort" of hydrating water molecules around each ion. As before, we represent the solution of the ionic compound as going from the ions in the crystal lattice on the left to the hydrated ions on the right. Two examples of the solution of ionic compounds are as follows.

$$\text{NaCl}(s) \xrightarrow{\text{H}_2\text{O}} \text{Na}^+(aq) + \text{Cl}^-(aq)$$
$$\text{K}_2\text{SO}_4(s) \xrightarrow{\text{H}_2\text{O}} 2\text{K}^+(aq) + \text{SO}_4{}^{2-}(aq)$$

The $\text{H}_2\text{O}$ (above the reaction arrow) represents a large, undetermined number of water molecules needed to break up the crystal and hydrate the ions. The *(aq)* after the symbol of each ion indicates that it is hydrated in the solution. Of course, not all ionic compounds dissolve to an appreciable extent in water. In these cases, the ion–ion forces are considerably stronger than the ion–dipole attractions, so the solid crystal remains essentially intact when placed in water. For example, ancient

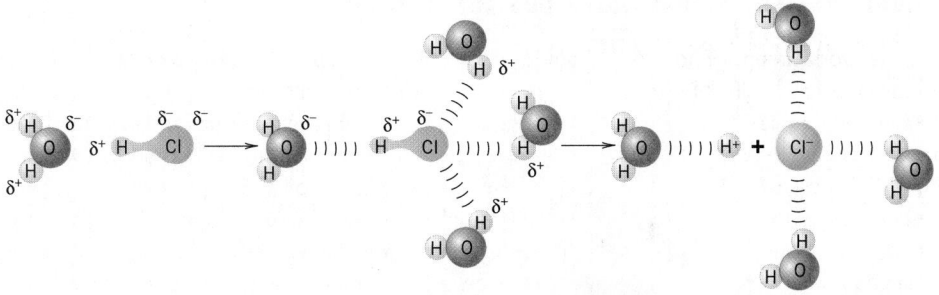

**Figure 11-3 Interaction of HCl and H₂O** The dipole–dipole interaction between H₂O and HCl leads to breaking of the HCl bond.

marble (i.e., CaCO₃) sculptures have suffered the indignities of weather for centuries with little deterioration.

In addition to soluble ionic compounds, certain molecular compounds, known as acids, dissolve in water with the formation of ions in a process known as **ionization.** An example is the HCl molecule, which in the pure gaseous state is held together by a single covalent bond. This is a very polar molecule, however, with the chlorine having a partial negative charge and the hydrogen a partial positive charge. When dissolved in water, a tug-of-war develops. On one side, the H₂O's pull on the HCl molecule because of the dipole–dipole interactions, attempting to break the HCl bond. On the other side, the HCl covalent bond tries to hold the molecule together. For HCl, the H₂O molecules are clearly the winners, since the interaction is strong enough to break the bonds of all dissolved HCl molecules. In this case, however, the pair of electrons in the HCl bond stay with the more electronegative Cl. This means that ions are produced when the bond breaks rather than neutral atoms. In water, the dipole–dipole attractions between water and solute are strong enough to cause the H—Cl bond to break, leading to the formation of hydrated H⁺ and Cl⁻ ions in aqueous solution. The ionization process is illustrated in Figure 11-3. For clarity, we have shown only four water molecules, but actually many more are involved in the interaction. We will have much more to say about the reaction of acids in water in the next chapter. Ionization is represented by the equation

$$HCl(g) \xrightarrow{\text{H}_2\text{O}} H^+(aq) + Cl^-(aq)$$

Other polar molecules dissolve in water without the formation of ions. In these cases, the dipole–dipole interactions between solute and solvent do not lead to ionization, so the solute molecule remains intact in solution. Methyl alcohol, a liquid, is a polar molecular compound that is miscible with water without ionization. (See Figure 11-4.)

If a molecule is nonpolar, there are no centers of positive and negative charge to attract the dipoles of water. Since the water molecules are moderately attracted to one another by hydrogen bonding, there is limited incentive for the water molecules to interact with a solute molecule. *As a result, most nonpolar compounds do not dissolve in polar solvents such as water.* On the other hand, nonpolar compounds do tend to dissolve in nonpolar solvents such as benzene or carbon tetrachloride. This happens because neither solvent nor solute molecules are held together by particularly strong forces, so they are free to mix. For example, if we wish to dissolve a grease stain (i.e., grease is composed of nonpolar molecules), we know that plain water does not work but a nonpolar solvent such as gasoline does. In summary, *like solvents dissolve like solutes (e.g., polar solvents dissolve ionic or polar molecular compounds, whereas nonpolar solvents dissolve nonpolar compounds).*

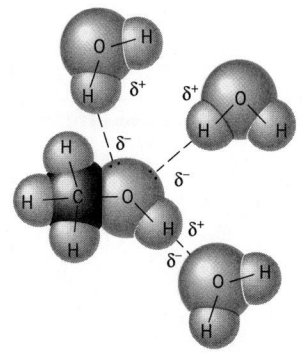

**Figure 11-4 Methyl Alcohol in Water**
**For some polar covalent molecules in water, there are only dipole–dipole attractions between solute and water molecules.**

### Soap—A Compound That Goes Either Way

For how long have the mothers of the world been trying to get their kids to wash their hands and face with soap and water? Actually, we don't know; the history of soap is somewhat obscure. It was apparently used in ancient Babylon in 2800 B.C. to clean garments. However, the use of soap to cleanse the body was not mentioned until about A.D. 200. Now soap and its synthetic cousins, detergents, are a necessary part of our daily hygiene.

Soap is made by mixing animal fat with ashes of plants that were a source of lye (sodium hydroxide). The chemical reaction produces soap (sodium stearate) and glycerin, a smooth, oily compound. Most soap originally made this way probably didn't smell very good and was quite harsh, as it still contained some lye. The use of perfumes, other additives, and controlled conditions now produce a gentler product.

One end of sodium stearate is a long chain of a hydrocarbon group (carbon and hydrogen). This end is nonpolar. The other end of the molecule is highly polar, as it is an ionic group.

$$CH_3CH_2CH_2CH_2CH_2CH_2CH_2CH_2CH_2CH_2 —$$
$$\underbrace{CH_2CH_2CH_2CH_2CH_2CH_2CH_2CO_2^-Na^+}$$

| Nonpolar end | Ionic polar end |

Recall that nonpolar molecules will dissolve in nonpolar liquids (e.g., oils) and ionic compounds dissolve in polar liquids (e.g., $H_2O$). Thus, the nonpolar end of this compound dislikes water (*hydrophobic*, or water-fearing) but the other end is attracted to water (*hydrophilic*, or water-loving). When soapy water is in contact with an oil stain, the hydrophobic end dissolves in the oil and the hydrophilic end stays in contact with the water. Small globs of oil (called *micelles*) then break away and are suspended in the water because of their ionic charge. Since the micelles all have the same charge on the surface, they do not clump together and can be rinsed away.

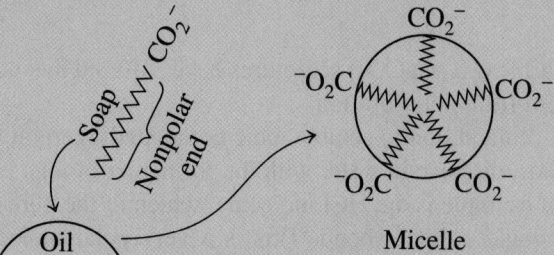

Oil                    Micelle

One of the problems with ordinary soap is that it forms an insoluble compound with $Ca^{2+}$ and $Mg^{2+}$. These ions are present in "hard water," so a precipitate forms when we try to use soap in this water. This precipitate forms a scum that is sometimes referred to as "bathtub ring."

The population of most countries is increasing, and more and more people are crowded closer together. Thank goodness for soap.

*Soap floats away grease.*

## 11-2 Solubility and Temperature

**Looking Ahead!** The terms used to define the relative amount of a substance that dissolves almost constitute a separate vocabulary. Besides the terms that we use, we are also interested in this section in how temperature affects how much of a solute dissolves.

In an earlier chapter we classified an ionic compound as either soluble, meaning that an appreciable amount dissolves, or insoluble, which means that a very small amount dissolves. Whether a substance is labeled soluble or insoluble is determined by the compound's solubility. *The* **solubility** *of a solute is defined as the maximum amount that dissolves in a given amount of solvent at a specified temperature.*

Now let's turn our attention from the solute to the solution. *When a specific amount of solvent contains the maximum amount of dissolved solute, the solution is said to be* **saturated.** *If less than the maximum amount is present, the solution is* **unsaturated.** In certain unusual situations, *an unstable condition may exist in which there is actually more solute present in solution than its solubility would indicate. Such a solution is said to be* **supersaturated.** Supersaturated solutions often shed the excess solute if a tiny "seed" crystal of solute is added or if the solution is shaken. The excess solute solidifies and falls to the bottom of the container as a precipitate.

Given that certain compounds are soluble in water, how do we express the **concentration,** *which is the amount of solute present in a given amount of solvent or*

**Table 11-1  Solubilities of Compounds (at 20°C)**

| Compound | Solubility (g solute/100 g $H_2O$) |
|---|---|
| sucrose (table sugar) | 205 |
| HCl | 63 |
| NaCl | 38 |
| $KNO_3$ | 34 |
| $PbSO_4$ | 0.04 |
| $Mg(OH)_2$ | 0.01 |
| AgCl | $1.9 \times 10^{-4}$ |

*solution?* Several units are available, and the one we choose depends on the calculation involved. Although three additional units are discussed later in the chapter, we will introduce one unit at this time that suits our current needs. The first concentration unit that we will discuss expresses the amount of solute present in a saturated solution as mass of solute present in 100 g of solvent (g solute/100 g $H_2O$ in the case of aqueous solutions). The solubilities of several compounds in water are listed in Table 11-1. For example, note that 205 g of sugar dissolved in 100 g (100 mL) of water produces a saturated solution at 20°C. Although there is no definite dividing point, note that $PbSO_4$, $Mg(OH)_2$, and AgCl have very low solubilities and thus are considered insoluble. In Chapter 7, we indicated that whenever both ions involved in one of these insoluble compounds (e.g., $Mg^{2+}$ and $OH^-$) are mixed from separate solutions, a precipitate of that compound forms [e.g., $Mg(OH)_2$]. The other compounds listed are definitely considered soluble, although to various extents. Table 7-3 can be used as an aid in determining whether compounds of some common anions are considered soluble or insoluble.

It is apparent that the nature of a solute affects its solubility in water. Temperature is also an important factor. From practical experience, most of us know that more sugar or salt dissolves in hot water than in cold. This is generally true. *Most solids and liquids are more soluble in water at higher temperatures.* In Figure 11-5, the

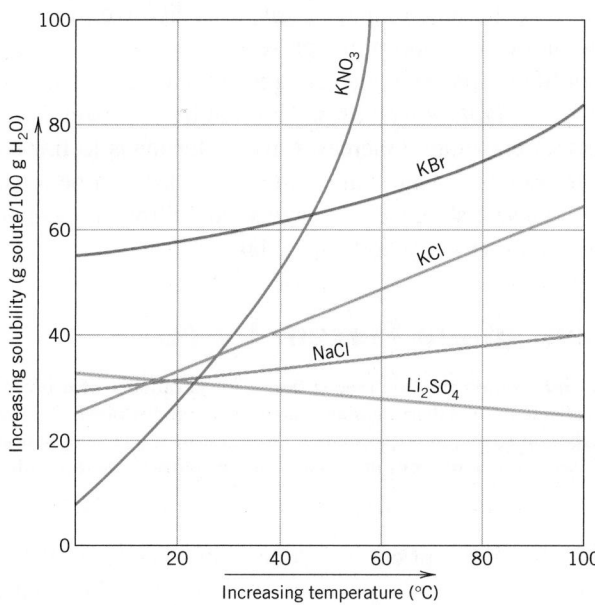

**Figure 11-5  Solubility and Temperature** The solubility of most solids increases as temperature increases.

**Figure 11-6**
**Recrystallization**
**A hot, saturated solution of KBr is being prepared on the left. A precipitate has formed in the cold solution on the right because of its lower solubility.**

solubilities of several ionic compounds are graphed as a function of temperature. Note that all except $Li_2SO_4$ are more soluble as the temperature increases. The information shown in Figure 11-5 has important laboratory implications. An impure solid can be purified by a process called **recrystallization.** In this procedure, a solution is saturated with the solute to be purified at a high temperature such as the boiling point. Insoluble impurities are then filtered from the hot solution. As the solution is allowed to cool, the solvent can hold less and less solute. The excess solute precipitates from the solution as it cools (if it does not become supersaturated.) This solid, now more pure, can then be filtered from the cold solution. The soluble impurities pass through the filter still in the solution. For example, if 100 g of water is saturated with KBr at its boiling point near 100°C, it contains 85 g of dissolved KBr. If the solution is cooled to 0°C in an ice bath, the water can now contain only 55 g of dissolved salt. The difference (i.e., 85 g − 55 g = 30 g) forms a precipitate. (See Figure 11-6.)

Despite their nonpolar nature, many gases also dissolve in water to a small extent (e.g., $O_2$, $N_2$, and $CO_2$). Indeed, the presence of dissolved oxygen in water provides the means of life for fish and other aquatic animals. *Unlike solids, gases become less soluble as the temperature increases.* This can be witnessed by observing water being heated. As the temperature increases, the water tends to fizz somewhat as the dissolved gases are expelled. High temperatures in lakes can be a danger to aquatic animals and may cause fish kills. The lower solubility of oxygen at the higher temperatures can lead to an oxygen-depleted lake.

**Fish maintain life by extracting dissolved oxygen from water.**

## 11-3 Concentration: Percent by Mass

**Looking Ahead!** A *concentrated* solution means that a large amount of a given solute is present in a given amount of solvent. A *dilute* solution means that comparatively little of the same solute is present. Obviously, we need more quantitative methods of expressing concentrations for laboratory situations. Two additional ways of expressing the solute content are discussed in the next two sections.

Basically, there are two types of concentration units. One type relates the amount of solute to the amount of *solvent*. We put such a unit to work in the previous discussion on solubility when we related the grams of solute to 100 grams of solvent. The second type of concentration unit relates the amount of solute to the amount of *solution*. which contains both the solvent and solute. One such unit is **percent by mass,** which expresses the mass of solute per 100 grams of solution. Therefore, in 100 g of a solution that is 25% by mass HCl, there are 25 g of HCl and 75 g of $H_2O$. The formula for percent by mass is

$$\% \text{ by mass (solute)} = \frac{\text{mass of solute}}{\text{mass of solute } + \text{ mass of solvent}} \times 100\%$$

$$= \frac{\text{mass of solute}}{\text{mass of solution}} \times 100\%$$

**Calculation of Mass Percent**

What is the percent by mass of NaCl if 1.75 g of NaCl is dissolved in 5.85 g of $H_2O$?

**Example 11-1**

working it out

**Procedure**

Find the total mass of the solution and then the percent of NaCl.

**Solution**

The total mass of solution is

1.75 g NaCl (solute)

5.85 g $H_2O$ (solvent)

7.60 g solution

$$\frac{1.75 \text{ g NaCl}}{7.60 \text{ g solution}} \times 100\% = \underline{\underline{23.0\% \text{ by mass NaCl}}}$$

**Calculation of Solute Amount from Mass Percent**

A solution is 14.0% by mass $H_2SO_4$. What quantity (in moles) of $H_2SO_4$ is in 155 g of solution?

**Example 11-2**

**Procedure**

1. Find the mass of $H_2SO_4$ in the solution by multiplying the mass of compound by the percent in fraction form.
2. Convert mass to moles using the molar mass of $H_2SO_4$.

**Solution**

1. $155 \text{ g solution} \times \dfrac{14.0 \text{ g } H_2SO_4}{100 \text{ g solution}} = 21.7 \text{ g } H_2SO_4$

2. The molar mass is

$$2.016 \text{ g (H)} + 32.07 \text{ g (S)} + 64.00 \text{ g (O)} = 98.09 \text{ g/mol}$$

$$21.7 \text{ g } H_2SO_4 \times \frac{1 \text{ mol } H_2SO_4}{98.09 \text{ g } H_2SO_4} = \underline{\underline{0.221 \text{ mol } H_2SO_4}}$$

Percent by mass is equivalent to parts per hundred. When concentrations are extremely low, however, two closely related units become more convenient. These units are **parts per million (ppm)** and **parts per billion (ppb),** and they are particularly useful for expressing concentrations of trace amounts of a substance relative to the total amount. For example, one hears of dangerous dioxin levels in the soil in ranges of ppm and even ppb. Parts per million is obtained by multiplying the ratio of the mass of solute to mass of solution by $10^6$ ppm rather than 100%. Parts per billion is obtained by multiplying the same ratio by $10^9$ ppb. For example, if a solution has a mass of 1.00 kg and contains only 3.0 mg of a solute, it has the following concentration in percent by mass, ppm, and ppb.

$$\frac{3.0 \times 10^{-3} \text{ g (solute)}}{1.0 \times 10^3 \text{ g (solution)}} \times 100\% = 3.0 \times 10^{-4}\%$$

$$\frac{3.0 \times 10^{-3} \text{ g}}{1.0 \times 10^3 \text{ g}} \times 10^6 \text{ ppm} = 3.0 \text{ ppm}$$

$$\frac{3.0 \times 10^{-3} \text{ g}}{1.0 \times 10^3 \text{ g}} \times 10^9 \text{ ppb} = 3.0 \times 10^3 \text{ ppb}$$

In this case, the most convenient expression of concentration is in units of ppm.

# 11-4 Concentration: Molarity

**Looking Ahead!** Mass percent relates mass of solute to mass of the solution. In the laboratory it is easier to measure the volume of a solution rather than its mass. So, in this section, we will introduce a concentration unit that relates to the volume of the solution.

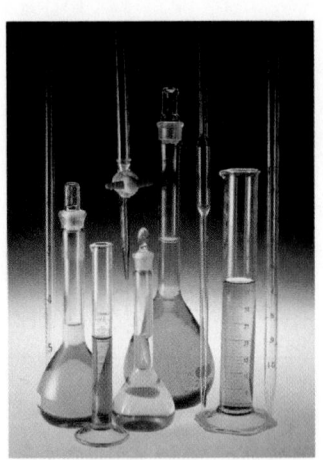

**Solution volume is measured by burets, pipettes, and volumetric flasks.**

Most laboratory procedures require us to measure the amount of a particular substance. If it is a solid or liquid, we can weigh the substance directly on a balance. When it is a gas, we must measure the volume and its temperature and pressure conditions. When the substance is in an aqueous solution, the most convenient way to measure the amount is by measuring the volume of the solution. Graduated cylinders, burets, or other convenient laboratory apparatus allow easy measurement of the volume of a solution. The concentration unit known as molarity is the most useful in this regard because it relates the amount of solute to the volume of the solution. **Molarity (M)** *is defined as the number of moles of solute (n) per volume in liters (V) of solution.* Although molarity may be shown without units it is understood to have units of *mol/L*. This is expressed as follows in equation form.

$$M = \frac{n \text{ (moles of solute)}}{V \text{ (liters of solution)}}$$

The following examples illustrate the calculation of molarity and its use in determining the amount of solute in a specific solution.

| Example 11-3 | **Calculation of Molarity** |
|---|---|
| w o r k i n g   i t   o u t | What is the molarity of $H_2SO_4$ in a solution if 49.0 g of $H_2SO_4$ is present in 250 mL of solution? |

**Procedure**

Write down the formula for molarity and what you have been given, and then solve for what's requested. Recall that the volume is expressed in liters.

**Solution**

$$M = \frac{n}{V}$$

$$n = 49.0 \text{ g } H_2SO_4 \times \frac{1 \text{ mol}}{98.09 \text{ g } H_2SO_4} = 0.500 \text{ mol}$$

$$V = 250 \text{ mL} \times \frac{10^{-3} \text{ L}}{\text{mL}} = 0.250 \text{ L}$$

$$\frac{n}{V} = \frac{0.500 \text{ mol}}{0.250 \text{ L}} = \underline{\underline{2.00 \text{ M}}}$$

**Calculation of Amount of Solute Present**    **Example 11-4**

What mass of HCl is present in 155 mL of a 0.540 M solution?

**Solution**

$$M = \frac{n}{V} \qquad n = M \times V$$

$$M = 0.540 \text{ mol/L}$$
$$V = 155 \text{ mL} = 0.155 \text{ L}$$
$$n = 0.540 \text{ mol/L} \times 0.155 \text{ L} = 0.0837 \text{ mol HCl}$$

$$0.0837 \text{ mol HCl} \times \frac{36.46 \text{ g}}{\text{mol HCl}} = \underline{\underline{3.05 \text{ g HCl}}}$$

**Calculation of Molarity from Percent Composition**    **Example 11-5**

Concentrated laboratory acid is 35.0% by mass HCl and has a density of 1.18 g/mL. What is its molarity?

**Procedure**

Since a volume was not given, you can start with any volume you wish. The molarity will be the *same* for 1 mL as for 25 L. To make the problem as simple as possible, assume that you have exactly 1 L of solution ($V = 1.00$ L) and go from there. The number of moles of HCl ($n$) in 1 L can be obtained as follows.

1. Find the mass of 1 L from the density.
2. Find the mass of HCl in 1 L using the percent by mass and the mass of 1 L.
3. Convert the mass of HCl to moles of HCl.

**Solution**

Assume that $V = 1.00$ L

1. The mass of 1.00 L ($10^3$ mL) is

$$10^3 \text{ mL} \times 1.18 \text{ g/mL} = 1180 \text{ g solution}$$

2. The mass of HCl in 1.00 L is

$$1180 \text{ g solution} \times \frac{35.0 \text{ g HCl}}{100 \text{ g solution}} = 413 \text{ g HCl}$$

3. The number of moles of HCl in 1.00 L is

$$413 \text{ g} \times \frac{1 \text{ mol}}{36.46 \text{ g}} = 11.3 \text{ mol HCl}$$

$$\frac{n}{V} = \frac{11.3 \text{ mol}}{1.00 \text{ L}} = \underline{\underline{11.3 \text{ M}}}$$

## 11-5 Dilution of Concentrated Solutions

**Looking Ahead!** A common laboratory need is to prepare a dilute solution of a specific concentration from a more concentrated solution. This procedure is not complex once we set up a simple relationship between the dilute and concentrated solutions, as we will do in this section.

From soft drinks to medicine, many products are transported as concentrated solutions. Before being sold, these products are carefully diluted to the desired concentration. Dilution is a straightforward procedure requiring a simple calculation. In our laboratory situations, we may be called on to prepare 200 mL of a 2.0 M solution from a large supply of a 6.0 M HCl solution. We simply need to know what volume of the more concentrated solution is needed. (See Figure 11-7.) In the following discussion, the number of moles in the dilute solution is designated $n_d$, and the volume and molarity of that solution are designated $V_d$ and $M_d$, respectively. First we will calculate the number of moles of solute present in the dilute solution we want to prepare.

$$M_d \times V_d = n_d$$

The moles of solute taken from the concentrated solution is designated $n_c$. The moles of solute in the concentrated solution relates to the volume and molarity of the concentrated solution.

$$M_c \times V_c = n_c$$

Since the moles of solute used in the dilute solution is the same as that taken from the concentrated solution, adding more solvent does not change the number of moles.

$$\text{moles solute } = n_d = n_c$$

Because, in algebra, quantities equal to the same quantity are equal to each other, we have the simple relationship between the dilute solution and the volume and molarity of the concentrated solution that is used:

$$M_c \times V_c = M_d \times V_d$$

**Adding water to a concentrate produces a dilute solution.**

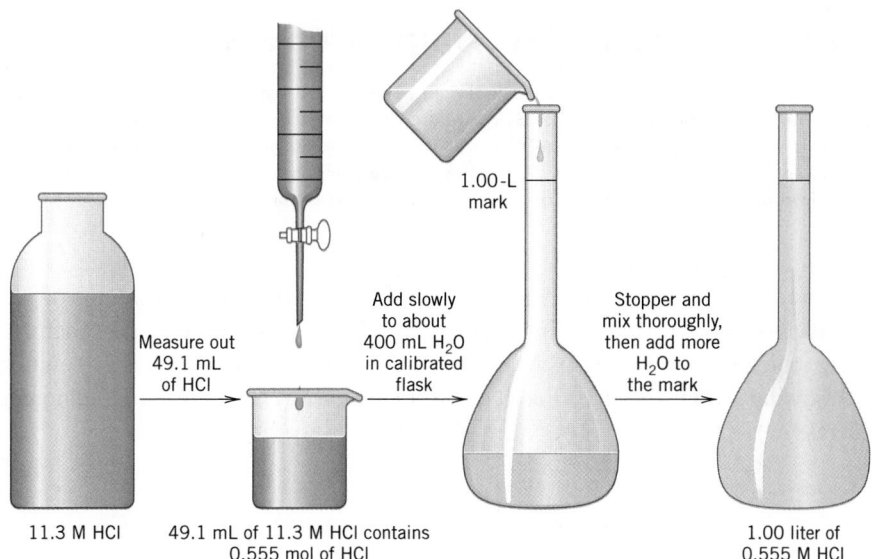

**Figure 11-7   Dilution of Concentrated HCl**
(*Note:* Never add water directly to concentrated acid, because it may splatter and cause severe burns.)

1.00-L mark

Measure out 49.1 mL of HCl

Add slowly to about 400 mL H₂O in calibrated flask

Stopper and mix thoroughly, then add more H₂O to the mark

11.3 M HCl

49.1 mL of 11.3 M HCl contains 0.555 mol of HCl

1.00 liter of 0.555 M HCl

---

**Calculation of Volume of Concentrated Solution**

**Example 11-6**

What volume of 11.3 M HCl must be mixed with water to make 1.00 L of 0.555 M HCl? (See also Figure 11-7.)

*working it out*

**Procedure**

$$M_c \times V_c = M_d \times V_d \qquad V_c = \frac{M_d \times V_d}{M_c}$$

**Solution**

$$M_c = 11.3 \text{ M} \qquad M_d = 0.555 \text{ M}$$
$$V_c = ? \qquad V_d = 1.00 \text{ L}$$

$$V_c = \frac{0.555 \text{ mol/L} \times 1.00 \text{ L}}{11.3 \text{ mol/L}} = 0.0491 \text{ L} = \underline{\underline{49.1 \text{ mL}}}$$

---

**Calculation of Molarity of a Dilute Solution**

**Example 11-7**

What is the molarity of a solution of KCl that is prepared by dilution of 855 mL of a 0.475 M solution to a volume of 1.25 L?

**Procedure**

$$M_c \times V_c = M_d \times V_d \qquad M_d = \frac{M_c \times V_c}{V_d}$$

**Solution**

$$M_c = 0.475 \text{ M} \qquad M_d = ?$$
$$V_c = 855 \text{ mL} = 0.855 \text{ L} \qquad V_d = 1.25 \text{ L}$$

$$M_d = \frac{0.475 \text{ M} \times 0.855 \text{ L}}{1.25 \text{ L}} = \underline{\underline{0.325 \text{ M}}}$$

# 11-6 Stoichiometry Involving Solutions

**Looking Ahead!** Since molarity relates to moles in solution, we can further expand our concept of stoichiometry. Previously, in Chapter 8, we related the moles of a substance to its molar mass and number of molecules. The volume of a gas at a specific temperature and pressure also relates to moles, so gases were included in stoichiometry problems in Chapter 9. The inclusion of solutions into our general stoichiometry scheme is discussed in this section.

It is obvious by now that a large number of chemical reactions take place in aqueous solution. In addition to the balanced equation, we will use the volumes of these solutions as a means of measuring amount. In Figure 11-8, the general scheme for working stoichiometry problems has been extended to include volumes of solutions. The following examples also illustrate the inclusion of solutions in stoichiometry problems.

**Figure 11-8  General Procedure for Stoichiometry Problems**

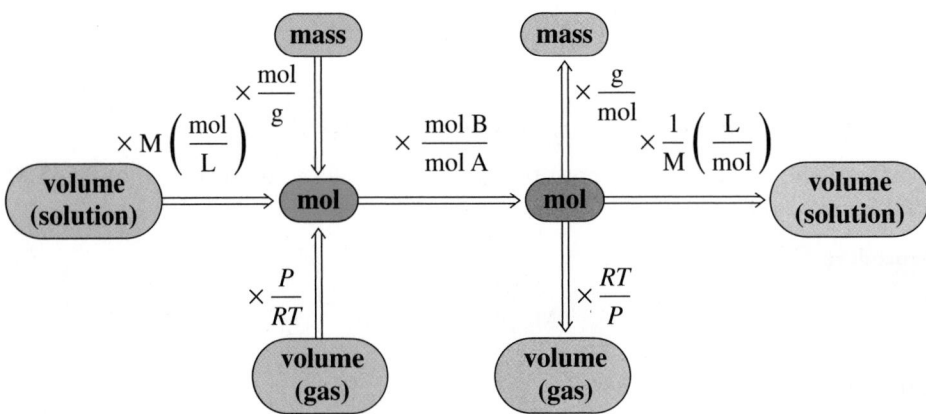

**Example 11-8**

**working it out**

**Calculation of the Volume of a Reactant**

Given the balanced equation

$$3NaOH(aq) + H_3PO_4(aq) \longrightarrow Na_3PO_4(aq) + 3H_2O(l)$$

what volume of 0.250 M NaOH is required to react completely with 4.90 g of $H_3PO_4$?

**Procedure**

$$\text{Given (g } H_3PO_4) \qquad\qquad\qquad\qquad\qquad \text{Requested [L}(aq)\text{ NaOH]}$$

$$\times \frac{\text{mol } H_3PO_4}{\text{g } H_3PO_4} \qquad \times \frac{\text{mol NaOH}}{\text{mol } H_3PO_4} \qquad \times \frac{1}{M}\left(\frac{\text{L NaOH}}{\text{mol NaOH}}\right)$$

$$\boxed{\text{mass } H_3PO_4} \Longrightarrow \boxed{\text{mol } H_3PO_4} \Longrightarrow \boxed{\text{mol NaOH}} \Longrightarrow \boxed{\text{volume NaOH}}$$

**Solution**

$$4.90 \text{ g } H_3PO_4 \times \frac{1 \text{mol } H_3PO_4}{97.99 \text{ g } H_3PO_4} \times \frac{3 \text{ mol NaOH}}{1 \text{ mol } H_3PO_4} = 0.150 \text{ mol NaOH}$$

$$\text{vol NaOH} = \frac{\text{mol NaOH}}{M} = \frac{0.150 \text{ mol}}{0.250 \text{ mol/L}} = \underline{\underline{0.600 \text{ L}}}$$

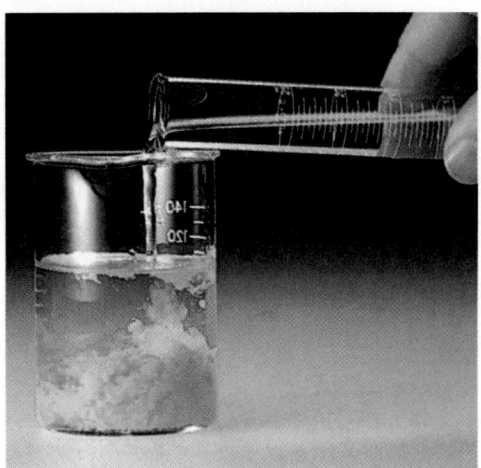

When a $Cd(NO_3)_2$ solution is added to a $Na_2S$ solution, a yellow precipitate of CdS forms.

**Calculation of the Mass of a Product**

**Example 11-9**

Given the following balanced equation

$$Cd(NO_3)_2(aq) + Na_2S(aq) \longrightarrow CdS(s) + 2NaNO_3(aq)$$

what mass of CdS is produced from 158 mL of a 0.122 M $Na_2S$ solution with excess $Cd(NO_3)_2$ present?

**Procedure**

*Given* [mL($aq$)$Na_2S$]                    *Requested* (g CdS)

$$\times\; M\left(\frac{\text{mol Na}_2\text{S}}{\text{L Na}_2\text{S}}\right) \qquad \times\; \frac{\text{mol CdS}}{\text{mol Na}_2\text{S}} \qquad \times\; \frac{\text{g CdS}}{\text{mol CdS}}$$

$$\boxed{\text{volume Na}_2\text{S}} \Longrightarrow \boxed{\text{mol Na}_2\text{S}} \Longrightarrow \boxed{\text{mol CdS}} \Longrightarrow \boxed{\text{mass CdS}}$$

**Solution**

$$\text{mol Na}_2\text{S} = V \times M$$
$$= 0.158\ \cancel{L} \times 0.122\ \text{mol}/\cancel{L} = 0.0193\ \text{mol Na}_2\text{S}$$

$$0.0193\ \cancel{\text{mol Na}_2\text{S}} \times \frac{1\ \cancel{\text{mol CdS}}}{1\ \cancel{\text{mol Na}_2\text{S}}} \times \frac{144.5\ \text{g CdS}}{\cancel{\text{mol CdS}}} = \underline{\underline{2.79\ \text{g Cds}}}$$

**Calculation of the Volume of Gas**

**Example 11-10**

Given the balanced equation

$$2HCl(aq) + K_2S(aq) \longrightarrow H_2S(g) + 2KCl(aq)$$

what volume of $H_2S$ measured at STP would be evolved from 1.65 L of a 0.552 M HCl solution with excess $K_2S$ present?

**Procedure**

*Given* [L($aq$) HCl]                    *Requested* [L($g$) $H_2S$]

$$\times\; M\left(\frac{\text{mol HCl}}{\text{L HCl}}\right) \qquad \times\; \frac{\text{mol H}_2\text{S}}{\text{mol HCl}} \qquad \times\; \frac{22.4\ \text{L}}{\text{mol H}_2\text{S}}$$

$$\boxed{\begin{array}{c}\text{volume HCl}\\\text{(solution)}\end{array}} \Longrightarrow \boxed{\text{mol HCl}} \Longrightarrow \boxed{\text{mol H}_2\text{S}} \Longrightarrow \boxed{\begin{array}{c}\text{volume H}_2\text{S}\\\text{(gas)}\end{array}}$$

**Solution**

$$V\,(\text{solution}) \times M\,(\text{HCl}) = n\,(\text{HCl})$$

$$1.65\ \cancel{L} \times 0.552\ \text{mol}/\cancel{L} = 0.911\ \text{mol HCl}$$

Since the volume of the gas is at STP, the molar volume relationship can be used rather than the ideal gas law.

$$0.911\ \cancel{\text{mol HCl}} \times \frac{1\ \cancel{\text{mol H}_2\text{S}}}{2\ \cancel{\text{mol HCl}}} \times \frac{22.4\ \text{L (STP)}}{\cancel{\text{mol H}_2\text{S}}} = \underline{\underline{10.2\ \text{L (STP)}}}$$

---

**Looking Back!** Water has such an important role in the support of life because of its ability to hydrate ions and polar molecules. Thus many ionic and polar covalent compounds are dissolved by water. Higher temperatures usually lead to a higher solubility for solids and liquids but not for gases. To deal with solutions quantitatively, it is necessary to develop concentration units. Of these, molarity allows us to measure the amount of solute by measuring a volume of solution. With solutions of known molarity we can produce dilute solutions or use these solutions in stoichiometric calculations.

---

# Learning Check A | checking it out

**A-1.** Fill in the blanks.

A solution is composed of a _____ homogeneously dispersed in a _____ . Two liquids that form a solution are _____ . If they do not, they are _____ . When an ionic compound dissolves in water, the _____ – _____ forces between solvent and solute overcome the _____ – _____ forces in the crystal. Polar covalent compounds can also dissolve in water due to the _____ – _____ forces between solute and solvent. If a soluble compound dissolves to the limit of its solubility at a given temperature, the solution is _____ ; if the solute is not dissolved to the limit, the solution is _____ . If more solute is present than indicated by its solubility, the solution is _____ and the excess solute may eventually form a _____ . Solids generally become _____ soluble at higher temperatures, whereas gases become _____ soluble. When the mass of solute is expressed as parts per 100 of the mass of solution, the concentration unit is known as _____ _____ . Molarity is defined as _____ per _____ of solution. The number of moles of a reactant or product is obtained by multiplying the _____ times the _____ .

**A-2.** Write the ions that form when the following ionic compounds dissolve in water:
**a.** $BaBr_2$
**b.** $(NH_4)_2Cr_2O_7$
**c.** $K_2SO_3$

**A-3.** A 5.00-g quantity of $MgCl_2$ is dissolved in 200 mL of water. What are the mass percent and the molarity of the solution? Assume that the volume of the solution is the same as that of the original solvent.

**A-4.** What mass of NaOH is in 15.0 mL of a 0.375 M solution?

**A-5.** What is the molarity of the solution in problem A-4 if the 15.0 mL is diluted to 40.0 mL?

**A-6.** Given the following balanced equation,

$$3HCl(aq) + K_3PO_4(aq) \longrightarrow H_3PO_4(aq) + 3KCl(aq)$$

what mass of $K_3PO_4$ is needed to react with 325 mL of 0.250 M HCl?

*Additional Examples: Exercises 11-1, 11-4, 11-13, 11-24, 11-34, 11-43, and 11-48.*

## 11-7 Physical Properties of Solutions

**Looking Ahead!** Ocean water may look like fresh water, but it certainly doesn't taste like it, freeze like it, or even seem to smell like it. Obviously, ocean water is not pure water since it contains a lot of dissolved solutes. The presence of solutes in the ocean water has changed its properties. We will now examine in more detail how the properties of solutions differ from those of the pure solvents.

We melt ice by spreading salt, we keep electrical appliances away from the bathtub, we put antifreeze in our car's radiator, we cook food faster by putting salt in the boiling water—we take all of these actions because the presence of solutes alters the properties of pure water. The ability of water to conduct electricity, its volatility, and its melting and boiling points are all affected by the presence of dissolved compounds. We will examine these effects in the order mentioned.

### Conductivity

The use of a hair dryer in the bathtub is extremely dangerous. Even though pure water is not a good conductor of electricity, ordinary tap water is. This is because tap water contains a variety of dissolved compounds. It has long been known that the presence of a solute in water may affect its ability to conduct electricity. *Electricity is simply a flow of electrons through a substance called a* **conductor.** Metals are the most familiar conductors and, as such, find use in electrical wires. Because the outer electrons of metals are loosely held, they can be made to flow through a continuous length of wire. *Other substances resist the flow of electricity and are known as* **nonconductors** *or* **insulators.** Glass and wood as well as pure water are examples of nonconductors of electricity. When wires are attached to a charged battery and then to a lightbulb, the light shines brightly. If the wire is cut, the light goes out because the circuit is broken. If the two ends of the cut wire are now immersed in pure water, the light stays out, indicating that water does not conduct electricity. Now let us dissolve certain solutes in water and examine what happens. When compounds such as NaCl and HCl are dissolved in water, the effect is obvious. The light immediately begins to shine, indicating that the solution is a good conductor of electricity. (See Figure 11-9.) *Compounds whose aqueous solutions conduct electricity are known as* **electrolytes.** (Some ionic compounds are not soluble in water, but if their molten state conducts electricity, they are also classified as electrolytes.)

*We now understand that it is the presence of ions in the aqueous solution that allows the solution to conduct electricity.* Almost all soluble ionic compounds form ions in solution, and some polar covalent compounds also dissolve to form ions. For example, both NaCl (ionic) and HCl (polar covalent) are classified as electrolytes because they form ions in aqueous solution. As you will recall, HCl, which in the pure state is a molecular compound, is ionized by the water molecules.

Other compounds such as sucrose (table sugar), glucose (blood sugar), and alcohol dissolve in water, but their solutions do not conduct electricity. *Compounds whose aqueous solutions do not conduct electricity are known as* **nonelectrolytes.** Nonelectrolytes are molecular compounds that dissolve in water without formation of ions. (See Figure 11-9a.)

There are two classes of electrolytes: strong electrolytes and weak electrolytes. *Solutions of* **strong electrolytes** *are good conductors of electricity.* Almost all salts

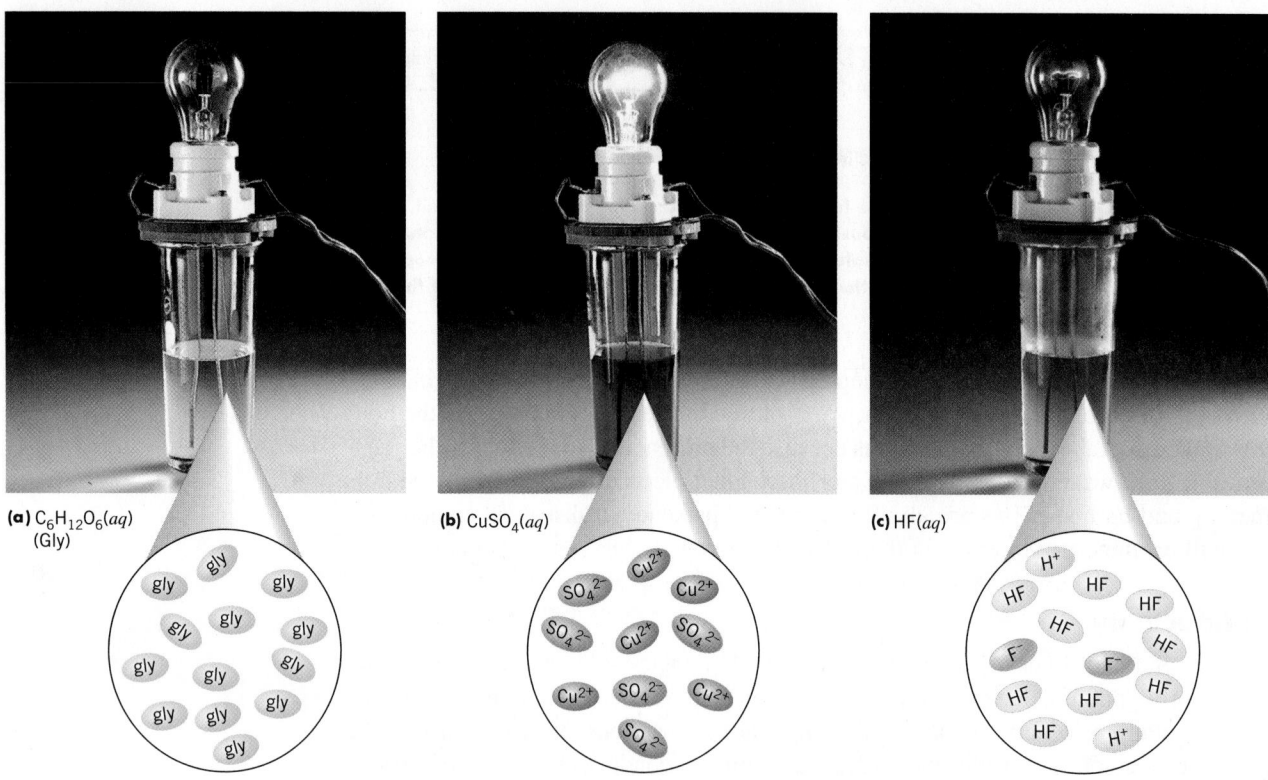

**(a)** $C_6H_{12}O_6(aq)$ (Gly)

**(b)** $CuSO_4(aq)$

**(c)** $HF(aq)$

**Figure 11-9   Nonelectrolytes, Strong Electrolytes, and Weak Electrolytes** **(a) A solution of glucose sugar ($C_6H_{12}O_6$) does not conduct electricity. (b) A $CuSO_4$ solution is a strong electrolyte and conducts electricity. (c) A hydrofluoric acid solution [HF(aq)] is a weak electrolyte and conducts a limited amount of electricity.**

and strong acids are present only as ions in aqueous solution (e.g., NaCl, $CuSO_4$, and HCl) and are thus classified as strong electrolytes. (See Figure 11-9b.) *Solutions of* **weak electrolytes** *allow a limited amount of conduction.* When wires are immersed in solutions of weak electrolytes, the lightbulb glows, but very faintly. (See Figure 11-9c.) Even adding more of the solute does not help. Examples of weak electrolytes are ammonia ($NH_3$) and hydrogen fluoride (HF). Both of these gases are soluble in water but produce only a small concentration of ions at any one time. For example, most of the dissolved HF remains as neutral molecules. Because the number of ions is small compared with the total number of HF molecules dissolved, the solution conducts only a limited amount of electricity. The ionization of HF is an example of a reversible reaction that reaches a point of equilibrium. This type of reaction will be discussed in more detail in the next chapter. The solution and ionization of HF can be represented by the following equations.

$$HF(g) \xrightarrow{H_2O} HF(aq)$$
$$HF(aq) \Longleftrightarrow H^+(aq) + F^-(aq)$$

## Colligative Properties

The presence of a solute in water may or may not affect its conductivity, depending upon whether the solute is an electrolyte or a nonelectrolyte. There are other properties of water, however, that are always affected to some extent by the presence of a solute. Consider again what was mentioned in Chapter 2 as characteristic of a pure substance. Recall that a pure substance has a distinct and unvarying melting point and boiling point. Mixtures, such as aqueous solutions, freeze and boil over a range of temperatures that are lower (for freezing) and higher (for boiling) than those of the pure solvent. The more solute, the more the melting and boiling points are affected. *A property that depends only on the relative amounts of solute and solvent is known as a* **colligative property.**

The conductivity of a solution depends on the nature of the particular solute. In contrast, colligative properties depend only on the amount or number of moles of particles present and not their identity. The particles may be small molecules, large molecules, or ions—only the total number is relevant. This is much like most of the gas laws, which you may recall also depended only on the total number of moles of gas present.

### Vapor Pressure

The Dead Sea in Israel and the Great Salt Lake in the United States contain large concentrations of solutes. These large bodies of water have no outlet to the ocean, so dissolved substances have accumulated, forming saturated solutions. Even though they both exist in semiarid regions with high summer temperatures, both lakes evaporate very slowly compared with a freshwater lake or even the ocean. If water in these lakes evaporated at the same rate as fresh water, both would nearly dry up in a matter of years. Why do they evaporate so slowly?

*The presence of a nonvolatile solute\* in a solvent lowers the equilibrium vapor pressure from that of the pure solvent.* (See section 10-5.) It is not hard to understand why this occurs based on what we have discussed about the interactions between solute and solvent. Because of the attractive forces between solute molecules (or ions) and solvent molecules, the solute particle is surrounded by a sphere of solvent molecules. Solute molecules and ions tend to tie up the solvent molecules, which in effect prevents them from escaping to the vapor. The more solute particles present, the more solvent molecules that are tied up. Thus the solution has a lower equilibrium vapor pressure than the pure solvent. (See Figure 11-10.) As we might predict from this model, the more solute particles present, the lower the vapor pressure of the solution.

### Boiling Point

Cooking food in salt water actually speeds the cooking process. The reason for this is that a salt solution boils at a higher temperature than pure water, so the chemical reactions in the cooking process occur at a faster rate. *A direct effect of the lowered vapor pressure of a solution is a higher boiling point than that of the solvent.* Recall that the normal boiling point of water is 100°C, which is the temperature at which its vapor pressure is equal to 760 torr. Since a solution has a lower vapor pressure at 100°C,

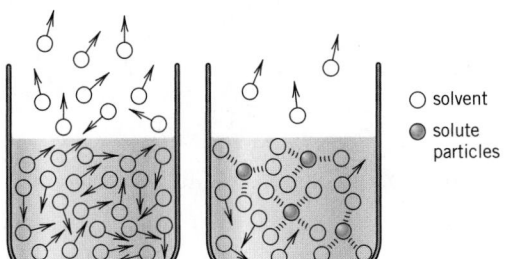

Pure solvent          Solution

**Figure 11-10  Vapor Pressure Lowering**
**A nonvolatile solute reduces the number of solvent molecules escaping to the vapor.**

○ solvent
● solute particles

---

\*A nonvolatile solute is one that has essentially no vapor pressure at the relevant temperatures.

it will not boil under those conditions. The solution must be heated to a temperature above 100°C to cause boiling. Again, the more concentrated the solution, the lower the vapor pressure and the higher the solution must be heated to cause boiling.

The amount of boiling point elevation is given by the equation

$$\Delta T_b = K_b m$$

where

$\Delta T_b$ = the number of Celsius degrees that the boiling point is raised

$K_b$ = a constant characteristic of the solvent [for water $K_b = 0.512$ (°C · kg)/mol. Other values of $K_b$ are given for particular solvents in the exercises.]

$m$ = a concentration unit called molality

**Molality** is a convenient concentration unit for this purpose since it emphasizes the relationship between the relative amounts of solute (expressed in moles) and solvent (expressed in kg) present rather than the volume of the solution, as does molarity. Also, the molality of a solution is not affected by the temperature. (The volume of a solution, which is used in molarity, is affected by the temperature.) The definition of molality is

$$\text{molality } (m) = \frac{\text{moles of solute}}{\text{kg of solvent}}$$

| **Example 11-11** | **Calculation of Molality** |

**working it out**

What is the molality of methyl alcohol in a solution made by dissolving 18.5 g of methyl alcohol ($CH_3OH$) in 850 g of water?

**Procedure**

$$\text{molality} = \frac{\text{mol solute}}{\text{kg solvent}}$$

**Solution**

$$\text{mol solute} = \frac{18.5 \text{ g}}{32.04 \text{ g/mol}} = 0.577 \text{ mol}$$

$$\text{kg solvent} = \frac{850 \text{ g}}{10^3 \text{ g/kg}} = 0.850 \text{ kg}$$

$$\text{molality} = \frac{0.577 \text{ mol}}{0.850 \text{ kg}} = 0.679 \, m$$

**Example 11-12** | **The Boiling Point of a Solution**

What is the boiling point of an aqueous solution containing 468 g of sucrose ($C_{12}H_{22}O_{11}$) in 350 g of water?

**Procedure**

$$\Delta T_b = K_b m = 0.512 \, (\text{°C} \cdot \text{kg})/\text{mol} \times \frac{\text{mol solute}}{\text{kg solvent}}$$

**Solution**

$$\text{mol solute} = \frac{468 \text{ g}}{342.5 \text{ g/mol}} = 1.37 \text{ mol}$$

$$\text{kg solvent} = \frac{350 \text{ g}}{10^3 \text{ g/kg}} = 0.350 \text{ kg}$$

$$\Delta T_b = 0.512 \text{ (°C} \cdot \text{kg)/mol} \times \frac{1.37 \text{ mol}}{0.350 \text{ kg}} = 2.00°C$$

Normal boiling point of water = 100.0°C
Boiling point of the solution = 100.0°C + $\Delta T_b$ = 100.0 + 2.00 = 102.0°C

### Freezing Point

The icy, cold winds of winter can be very hard on automobiles. If there is not enough antifreeze in the radiator, the coolant water may freeze. This could ruin the radiator and even crack the engine block because water expands when it freezes. We illustrate the same principle as antifreeze in radiators when we spread salt on ice-covered streets or sidewalks. In both of these cases we take advantage of the fact that solutions have lower melting points than pure solvents. *Just as the boiling point of a solution is higher than that of the pure solvent, the freezing point is lower.*

The amount of freezing point lowering is given by the equation

$$\Delta T_f = K_f m$$

where

$\Delta T_f$ = the number of Celsius degrees that the freezing point is lowered

$K_f$ = a constant characteristic of the solvent [for water $K_f$ = 1.86 (°C · kg)/mol.]

$m$ = molality of the solution

Antifreeze lowers the freezing point and raises the boiling point of water.

**The Freezing Point of a Solution**

What is the freezing point of the solution in Example 11-12?

**Example 11-13**

working it out

**Procedure**

$$\Delta T_f = K_f m = 1.86 \text{ (°C} \cdot \text{kg)/mol} \times \frac{\text{mol solute}}{\text{kg solvent}}$$

**Solution**

$$\Delta T_f = 1.86 \text{ (°C} \cdot \text{kg)/mol} \times \frac{1.37 \text{ mol}}{0.350 \text{ kg}} = 7.28°C$$

Freezing point of water = 0.0°C
Freezing point of the solution = 0.0°C − $\Delta T_f$ = 0.0°C − 7.3°C = −7.3°C

Because of the **freezing point** *lowering* and **boiling point** *elevation,* the liquid range has been extended by over 9°C (i.e., −7.3°C to 102°C) for the solution compared to pure water.

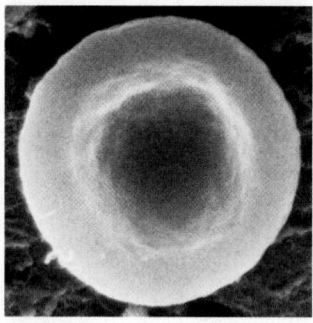

The sizes of these red blood cells is a result of osmosis. The concentrations of solutes in the surrounding liquids determine whether water enters or leaves the cells.

*Osmotic Pressure*

Food is preserved in salt water, drinking ocean water causes dehydration, tree and plant roots absorb water—these are all phenomena related to a colligative effect called osmosis. **Osmosis** *is the tendency for a solvent to move through a thin porous membrane from a dilute solution to a more concentrated solution.* The membrane is said to be *semipermeable,* which means that small solvent molecules can pass through but large hydrated solute species cannot. Figure 11-11 illustrates osmosis. On the right is a pure solvent, and on the left a solution. The two are separated by a semipermeable membrane. Solvent molecules can pass through the membrane in both directions, but the rate at which they diffuse to the right is lower because some of the water molecules on the left side are held back by solute–solvent interactions. As a result of the uneven passage of water molecules, the water level rises on the left and drops on the right. This creates increased pressure on the left, which eventually counteracts the osmosis, and equilibrium is established. *The extra pressure required to establish this equilibrium is known as the* **osmotic pressure.** Like other colligative properties, it depends on the concentration of the solute. In Figure 11-11, the more concentrated the solution on the left (less solvent), the higher the osmotic pressure.

We see an example of the osmosis process whenever we leave our hands in a soapy water or saltwater solution. The movement of water molecules from the cells of our skin to the more concentrated solution causes them to be wrinkled. Pickles are wrinkled because the cells of the cucumber have been dehydrated by the salty brine solution. In fact, brine solutions preserve many foods because the concentrated solution of salt removes water from the cells of bacteria, thus killing the bacteria. Trees and plants obtain water by absorbing water through the semipermeable membranes in their roots into the more concentrated solution inside the root cells. Osmosis has many important applications in addition to life processes. In Figure 11-11, if pressure greater than the osmotic pressure is applied on the left, reverse osmosis takes place and solvent molecules move from the solution to the pure solvent. This process is used in desalination plants that convert seawater (a solution) to drinkable water. This is important in areas of the world such as the Middle East, where there is a shortage of fresh water.

Our final point in this section concerns the difference between electrolytes and nonelectrolytes on colligative properties. Electrolytes have a more pronounced effect on colligative properties than do nonelectrolytes. The reason is that these properties depend only on how many particles are present regardless of whether the particle is

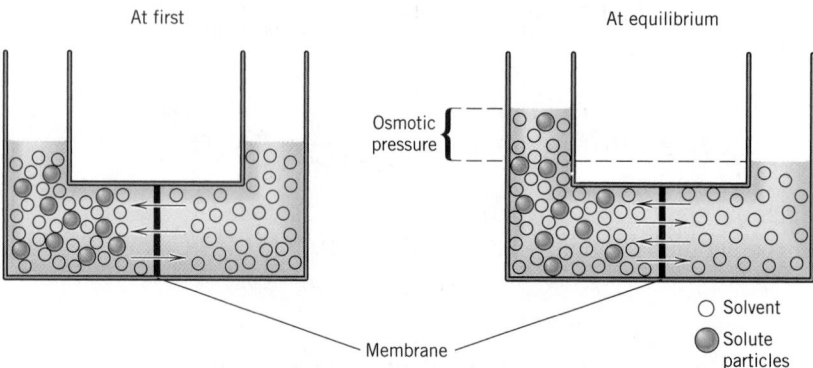

**Figure 11-11  Osmotic Pressure Osmosis causes dilution of the more concentrated solution.**

### Osmosis in a Diaper

Except for the fact that a baby seems to be unusually heavy, one may not realize just how wet a diaper has become. In the old days, a quick touch would tell the condition. Modern superabsorbent diapers make use of a chemical that has a tremendous ability to absorb water. In fact, such diapers contain a white powdery material that can absorb 200 to 500 times its weight in water. Not only is such a diaper convenient to the diaper changer but it is quite a bit more comfortable to the diaper wearer.

The compound used in these diapers is a polymeric material. The polymer used in diapers (sodium polyacrylate) is made of long chains of an ionic unit known as a monomer, sodium acrylate, which is shown below the cute baby photo. Most familiar plastics such as Styrofoam cups, plastic bottles and bags, PVC pipes, and thousands of other everyday materials are made of various types of polymers. More detail about polymers is provided in Chapter 16 on organic chemistry.

The unique property of the diaper polymer is that its surface is semipermeable. Inside the surface, ions are present. The high concentration of ions inside causes water molecules to cross the membrane in the process of osmosis. The water then stays put inside the polymer. In fact, the diaper does not even feel wet, thus protecting the baby from nasty rashes. It is so effective that special "training diapers" can be used when it is time for Junior to move on. These diapers are engineered to allow some wetness and discomfort as an incentive to become "trained."

Consider the absorbent ability of these diapers. A 1-gram quantity of the polyacrylate can absorb up to 500 g of water. If one had an 8-oz glass of water it would take less than 500 mg of superabsorbent to turn the glass of water into a wiggly, gelatinous mass. A 700-lb quantity of superabsorbent would be enough to turn a good-size swimming pool into something like jello.

In 1999, world production of this polymer was 980 million tons. Not all of that went into diapers (fortunately) as it is used in agriculture, crafts, evaporative coolers, fire fighting, toothpaste, and cosmetics.

An unpleasant (odiferous), drippy experience has been changed into something much more tolerable. Baby caregivers of the world are grateful for this chemical advance.

*Submitted by Edward Tokas*
*Carolina Reading and Learning Center*

$$CH_2 = CH$$
$$|$$
$$CO_2^- Na^+$$
***Sodium acrylate***
***(Monomer)***

$$\left[ CH_2 - CH \right]$$
$$|$$
$$CO_2^- Na^+$$
***Sodium polyacrylate***

a molecule, a cation, or an anion. For example, one mole of NaCl dissolves in water to produce two moles of particles, one mole of $Na^+$, and one mole of $Cl^-$.

$$NaCl(s) \xrightarrow{H_2O} Na^+(aq) + Cl^-(aq)$$

Thus one mole of NaCl lowers the freezing point approximately twice as much as one mole of a nonelectrolyte. This effect is put to good use in the U.S. Snow Belt,

**By the process of reverse osmosis, a hand-operated device desalinates seawater for emergency use.**

where sodium chloride is spread on snow and ice to cause melting even though the temperature is below freezing. Even more effective in melting ice is calcium chloride ($CaCl_2$). This compound produces three moles of ions ($Ca^{2+} + 2Cl^-$) per mole of solute and therefore is three times as effective per mole as a nonelectrolyte in lowering the freezing point. Calcium chloride is occasionally used on roads when the temperature is too low for sodium chloride to be effective. Aqueous electrolyte solutions are quite corrosive toward metals because of their electrical conductivity. This is why they are not used as an inexpensive antifreeze in automobile radiators.

Sodium chloride is used to melt ice from sidewalks, streets, and highways.

**Looking Back!** An aqueous solution may have the same appearance as pure water, but its properties can be very different. A solution may (or may not) conduct electricity, but it will always have a lower vapor pressure, a higher boiling point, and a lower melting point than pure water. A solution also has a higher osmotic pressure than pure water.

## Learning Check B | checking it out

**B-1.** Fill in the blanks.

If the presence of a solute in water causes the solution to become a conductor of electricity, the solute is known as an _____ . If it allows only a limited flow of electricity, it is known as a _____ _____ . If the water remains a nonconductor, the solute is a _____ . Properties that depend only on the amount of solute present are known as _____ properties. These are _____ _____ lowering, _____ _____ elevation, and _____ _____ lowering. The movement of solvent through a semipermeable membrane is known as _____ . A solute causes the _____ pressure to _____ .

**B-2.** Sodium chloride is a strong electrolyte. What are the boiling point and the freezing point of a solution made by dissolving 10.0 g of NaCl in 100 g of $H_2O$?

*Additional Examples: Exercises 11-52, 11-55, 11-59, 11-70, and 11-78.*

## chapter Review                                    putting it together

Water acts as an effective solvent, dispersing solutes into a homogeneous mixture known as a solution. When two liquids are **miscible**, they form a solution, but if they are **immiscible**, they remain a heterogeneous mixture. Some ionic compounds may dissolve in water because the ion–water forces (**ion–dipole**) overcome the forces holding the crystal together. Polar covalent molecular compounds may dissolve in water as discrete molecules, or they may undergo **ionization.** Although many ionic compounds are considered soluble, others are said to be insoluble since a very limited amount dissolves. The amount that dissolves—the **solubility** of a compound—is indicated by some convenient unit of **concentration.** How much of a compound can dissolve at a certain temperature to make a **saturated** solution varies from compound to compound. **Unsaturated** solutions contain less than the maximum amount of a compound so that more of the compound may dissolve. **Supersaturated** solutions are unstable solutions containing more of a compound than the solubility would indicate. A precipitate often forms in such a solution. Solid compounds are generally more soluble at higher temperatures, whereas gaseous compounds are less soluble at higher temperatures. This property can be used to purify solids in a process called **recrystallization.**

Besides mass of solute per 100 g of solvent, which was used to illustrate comparative solubilities, other units of concentration are **percent by mass, molarity (M),** and **molality. Parts per million (ppm)** and **parts per billion (ppb)** are used for very small concentrations.

| Percent by Mass | Molarity | Molality |
|---|---|---|
| $\dfrac{\text{mass of solute}}{\text{mass of solution}} \times 100\%$ | $M = \dfrac{\text{mol solute}}{\text{L solution}}$ | $m = \dfrac{\text{mol solute}}{\text{kg solvent}}$ |

Since molarity relates volume of a solution to moles of solute, it can be incorporated into the general scheme for stoichiometry problems along with the mass of a compound (Chapter 8) and the volume of a gas (Chapter 9) (see Figure 11-8).

In the final section, we studied the physical properties of solutions. In the first property mentioned, we found that certain solutes act as **nonelectrolytes** or as either **weak** or **strong electrolytes. Electrolytes** change water from a **nonconductor (insulator)** to a **conductor** of electricity. The information is summarized as follows.

| Type of Solute | Property in Water | Reason | Examples |
|---|---|---|---|
| nonelectrolyte | Solution is a nonconductor of electricity. | Ions are not formed in solution. | $C_{12}H_{22}O_{11}$ (sugar) $CH_3OH$ (methyl alcohol) |
| strong electrolyte | Solution is a good conductor of electricity. | Ions are formed in solution. | $NaCl$ $K_2SO_4$ |
| weak electrolyte | Solution is a weak conductor of electricity. | Limited amounts of ions are formed in solution. | $HF$ $HNO_2$ |

There are also four **colligative properties** of solutions. These are **vapor pressure lowering, boiling point elevation** and **freezing point lowering.** The process of **osmosis** leads to **osmotic pressure elevation,** the fourth colligative property. This is summarized as follows.

| Property | Effect | Result |
|---|---|---|
| vapor pressure | lowered | Solutions evaporate slower than pure solvents. |
| boiling point | raised | Solutions boil at higher temperatures than pure solvents. |
| freezing point | lowered | Solutions freeze at lower temperatures than pure solvents. |
| osmotic pressure | raised | Solvent from dilute solutions diffuses through a semipermeable membrane into concentrated solutions. |

# Exercises

## Aqueous Solutions

**11-1.** When an ionic compound dissolves in water, what forces in the crystal resist the solution process? What forces between water molecules and the crystal remove the ions from the lattice?

**11-2.** When a sample of KOH is placed in water, a homogeneous mixture of KOH is formed. Which is the solute, which is the solvent, and which is the solution?

**11-3.** Calcium bromide readily dissolves in water, but lead(II) bromide does not. Liquid benzene and water form a heterogeneous mixture, but liquid isopropyl alcohol and water mix thoroughly. Which of the above is said to be miscible, which immiscible, which insoluble, and which soluble?

**11-4.** Write equations illustrating the solution of each of the following ionic compounds in water.
(a) LiF      (c) $Na_2CO_3$
(b) $(NH_4)_3PO_4$    (d) $Ca(C_2H_3O_2)_2$

**11-5.** Write equations illustrating the solution of each of the following ionic compounds in water.
(a) $BaCl_2$      (c) $Cr(NO_3)_3$
(b) $Al_2(SO_4)_3$   (d) $Mg(ClO_4)_2$

**11-6.** Formaldehyde ($H_2CO$) dissolves in water without formation of ions. Write the Lewis structure of formaldehyde and show what types of interactions between solute and solvent are involved.

**11-7.** Nitric acid is a covalent compound that dissolves in water to form ions, as does HCl. Write the equation illustrating its solution in water.

### Temperature and Solubility

**11-8.** Referring to Figure 11-5, determine which of the following compounds is most soluble at 10°C: NaCl, KCl, or $Li_2SO_4$. Which is most soluble at 70°C?

**11-9.** Referring to Figure 11-5, determine what mass of each of the following dissolves in 250 g of $H_2O$ at 60°C: KBr, KCl, and $Li_2SO_4$.

**11-10.** Referring to Figure 11-5, determine whether each of the following solutions is saturated, unsaturated, or supersaturated. (All are in 100 g of $H_2O$.)
(a) 40 g of $KNO_3$ at 40°C    (c) 75 g of KBr at 80°C
(b) 40 g of $KNO_3$ at 20°C    (d) 20 g of NaCl at 40°C

**11-11.** A 200-g sample of water is saturated with $KNO_3$ at 50°C. What mass of $KNO_3$ forms as a precipitate if the solution is cooled to the freezing point of water? (Refer to Figure 11-5.)

**11-12.** A 500-mL portion of water is saturated with $Li_2SO_4$ at 0°C. What happens if the solution is heated to 100°C? (Refer to Figure 11-5.)

### Percent by Mass

**11-13.** What is the percent by mass of solute in a solution made by dissolving 9.85 g of $Ca(NO_3)_2$ in 650 g of water?

**11-14.** What is the percent by mass of solute if 14.15 g of NaI is mixed with 75.55 g of water?

**11-15.** A solution is 10.0% by mass NaOH. How many moles of NaOH are dissolved in 150 g of solution?

**11-16.** A solution contains 15.0 g of $NH_4Cl$ in water and is 8.50% $NH_4Cl$. What is the mass of water present?

**11-17.** A solution is 23.2% by mass $KNO_3$. What mass of $KNO_3$ is present in each 100 g of $H_2O$?

**11-18.** A solution contains 1 mol of NaOH dissolved in 9 mol of ethyl alcohol ($C_2H_5OH$). What is the percent by mass NaOH?

**11-19.** Blood contains 10 mg of calcium ions in 100 g of blood serum (solution). What is this concentration in ppm?

**11-20.** A high concentration of mercury in fish is 0.5 ppm. What mass of mercury is present in each kilogram of fish? What is this concentration in ppb?

**11-21.** Seawater contains $1.2 \times 10^{-2}$ ppb of gold ions. If all of the gold could be extracted, what volume in liters of seawater is needed to produce 1.00 g of gold? (Assume the density of seawater is the same as that of pure water.)

**11-22.** The maximum allowable level of lead in drinking water is 50 ppb. What mass of lead in milligrams is contained in a small swimming pool containing 5000 gallons of water? (Assume that the density of the water is the same as that of pure water.)

### Molarity

**11-23.** What is the molarity of a solution made by dissolving 2.44 mol of NaCl in enough water to make 4.50 L of solution?

**11-24.** Fill in the blanks.

| Solute | M | Amount of Solute | Volume of Solution |
|---|---|---|---|
| (a) KI | ——— | 2.40 mol | 2.75 L |
| (b) $C_2H_5OH$ | ——— | 26.5 g | 410 mL |
| (c) $NaC_2H_3O_2$ | 0.255 | 3.15 mol | ——— L |
| (d) $LiNO_2$ | 0.625 | ——— g | 1.25 L |
| (e) $BaCl_2$ | ——— | 0.250 mol | 850 mL |
| (f) $Na_2SO_3$ | 0.054 | ——— mol | 0.45 L |
| (g) $K_2CO_3$ | 0.345 | 14.7 g | ——— mL |
| (h) LiOH | 1.24 | ——— g | 1650 mL |
| (i) $H_2SO_4$ | 0.905 | 0.178 g | ——— mL |

**11-25.** What is the molarity of a solution of 345 g of Epsom salts ($MgSO_4 \cdot 7H_2O$) in 7.50 L of solution?

**11-26.** What mass of $CaCl_2$ is in 2.58 L of a solution with a concentration of 0.0784 M?

**11-27.** What volume in liters of a 0.250 M solution contains 37.5 g of KOH?

**11-28.** What is the molarity of a solution made by dissolving $2.50 \times 10^{-4}$ g of baking soda ($NaHCO_3$) in enough water to make 2.54 mL of solution?

**11-29.** What are the molarities of the hydroxide ion and the barium ion if 13.5 g of $Ba(OH)_2$ is dissolved in enough water to make 475 mL of solution?

**11-30.** What is the molarity of each ion present if 25.0 g of $Al_2(SO_4)_3$ is present in 250 mL of solution?

*11-31. A solution is 25.0% by mass calcium nitrate and has a density of 1.21 g/mL. What is its molarity?

*11-32. A solution of concentrated NaOH is 16.4 M. If the density of the solution is 1.43 g/mL, what is the percent by mass NaOH?

**\*11-33.** Concentrated nitric acid is 70.0% $HNO_3$ and 14.7 M. What is the density of the solution?

### Dilution

**11-34.** What volume of 4.50 M $H_2SO_4$ should be diluted with water to form 2.50 L of 1.50 M acid?

**11-35.** If 450 mL of a certain solution is diluted to 950 mL with water to form a 0.600 M solution, what was the molarity of the original solution?

**11-36.** One liter of a 0.250 M solution of NaOH is needed. The only available solution of NaOH is a 0.800 M solution. Describe how to make the desired solution.

**11-37.** What is the volume in liters of a 0.440 M solution if it was made by dilution of 250 mL of a 1.25 M solution?

**11-38.** What is the molarity of a solution made by diluting 3.50 L of a 0.200 M solution to a volume of 5.00 L?

**11-39.** What volume of water in milliliters should be *added* to 1.25 L of 0.860 M HCl so that its molarity will be 0.545?

**11-40.** What volume of water in milliliters should be *added* to 400 mL of a solution containing 35.0 g of KBr to make a 0.100 M KBr solution?

**\*11-41.** What volume in milliliters of *pure* acetic acid should be used to make 250 mL of 0.200 M $HC_2H_3O_2$? (The density of the pure acid is 1.05 g/mL.)

**\*11-42.** What would be the molarity of a solution made by mixing 150 mL of 0.250 M HCl with 450 mL of 0.375 M HCl?

### Stoichiometry Involving Solutions

**11-43.** Given the reaction

$$3KOH(aq) + CrCl_3(aq) \longrightarrow Cr(OH)_3(s) + 3KCl(aq)$$

what mass of $Cr(OH)_3$ would be produced if 500 mL of 0.250 M KOH were added to a solution containing excess $CrCl_3$?

**11-44.** Given the reaction

$$2KCl(aq) + Pb(NO_3)_2(aq) \longrightarrow PbCl_2(s) + 2KNO_3(aq)$$

what mass of $Pb(NO_3)_2$ is required to react with 1.25 L of 0.550 M KCl?

**11-45.** Given the reaction

$$Al_2(SO_4)_3(aq) + 3BaCl_2(aq) \longrightarrow 3BaSO_4(s) + 2AlCl_3(aq)$$

what mass of $BaSO_4$ is produced from 650 mL of 0.320 M $Al_2(SO_4)_3$?

**11-46.** Given the reaction

$$3Ba(OH)_2(aq) + 2Al(NO_3)_3(aq) \longrightarrow$$
$$2Al(OH)_3(s) + 3Ba(NO_3)_2(aq)$$

what volume of 1.25 M $Ba(OH)_2$ is required to produce 265 g of $Al(OH)_3$?

**11-47.** Given the reaction

$$2AgClO_4(aq) + Na_2CrO_4(aq) \longrightarrow$$
$$Ag_2CrO_4(s) + 2NaClO_4(aq)$$

what volume of a 0.600 M solution of $AgClO_4$ is needed to produce 160 g of $Ag_2CrO_4$?

**11-48.** Given the reaction

$$3Ca(ClO_3)_2(aq) + 2Na_3PO_4(aq) \longrightarrow$$
$$Ca_3(PO_4)_2(s) + 6NaClO_3(aq)$$

what volume of a 2.22 M solution of $Na_3PO_4$ is needed to react with 580 mL of a 3.75 M solution of $Ca(ClO_3)_2$?

**11-49.** Consider the reaction

$$2HNO_3(aq) + 3H_2S(aq) \longrightarrow 2NO(g) + 3S(s) + 4H_2O(l)$$

**(a)** What volume of 0.350 M $HNO_3$ will completely react with 275 mL of 0.100 M $H_2S$?
**(b)** What volume of NO gas measured at 27°C and 720 torr will be produced from 650 mL of 0.100 M $H_2S$ solution?

**\*11-50.** Given the reaction

$$2NaOH(aq) + MgCl_2(aq) \longrightarrow Mg(OH)_2(s) + 2NaCl(aq)$$

what mass of $Mg(OH)_2$ would be produced by mixing 250 mL of 0.240 M NaOH with 400 mL of 0.100 M $MgCl_2$?

**\*11-51.** Given the reaction

$$CO_2(g) + Ca(OH)_2(aq) \longrightarrow CaCO_3(s) + H_2O(l)$$

what is the molarity of a 1.00-L solution of $Ca(OH)_2$ that would completely react with 10.0 L of $CO_2$ measured at 25°C and 0.950 atm?

### Properties of Solutions

**11-52.** Three hypothetical binary compounds dissolve in water. AB is a strong electrolyte, AC is a weak electrolyte, and AD is a nonelectrolyte. Describe the extent to which each of these solutions conducts electricity and how each compound exists in solution.

**11-53.** Chlorous acid ($HClO_2$) is a weak electrolyte, and perchloric acid ($HClO_4$) is a strong electrolyte. Write equations illustrating the different behaviors of these two polar covalent molecules in water.

**11-54.** Explain the difference in the following three terms: 1 mole NaBr, 1 molar NaBr, and 1 molal NaBr.

**11-55.** What is the molality of a solution made by dissolving 25.0 g of NaOH in **(a)** 250 g of water and **(b)** 250 g of alcohol ($C_2H_5OH$)?

**11-56.** What is the molality of a solution made by dissolving 1.50 kg of KCl in 2.85 kg of water?

**11-57.** What mass of NaOH is in 550 g of water if the concentration is 0.720 *m*?

**11-58.** What mass of water is in a 0.430 *m* solution containing 2.58 g of $CH_3OH$?

**11-59.** What is the freezing point of a 0.20 *m* aqueous solution of a nonelectrolyte?

**11-60.** What is the boiling point of a 0.45 *m* aqueous solution of a nonelectrolyte?

**11-61.** When immersed in salty ocean water for an extended period, a person gets very thirsty. Explain.

**11-62.** Dehydrated fruit is wrinkled and shriveled up. When put in water, the fruit expands and becomes smooth again. Explain.

**11-63.** Explain how pure water can be obtained from a solution without boiling.

**11-64.** In industrial processes, it is often necessary to concentrate a dilute solution (much more difficult than diluting a concentrated solution). Explain how the principle of reverse osmosis can be applied.

**\*11-65.** What is the molality of an aqueous solution that is 10.0% by mass $CaCl_2$?

**\*11-66.** A 1.00 $m$ KBr solution has a mass of 1.00 kg. What is the mass of the water?

**\*11-67.** Ethylene glycol ($C_2H_6O_2$) is used as an antifreeze. What mass of ethylene glycol should be added to 5.00 kg of water to lower the freezing point to −5.0°C? (Ethylene glycol is a nonelectrolyte.)

**\*11-68.** What is the boiling point of the solution in exercise 11-67?

**\*11-69.** Methyl alcohol can also be used as an antifreeze. What mass of methyl alcohol ($CH_3OH$) must be added to 5.00 kg of water to lower its freezing point to −5.0°C?

**11-70.** What is the molality of an aqueous solution that boils at 101.5°C?

**11-71.** What is the boiling point of a 0.15 $m$ solution of a solute in liquid benzene? (For benzene, $K_b$ = 2.53, and the boiling point of pure benzene is 80.1°C.)

**11-72.** What is the boiling point of a solution of 75.0 g of naphthalene ($C_{10}H_8$) in 250 g of benzene? (See exercise 11-71.)

**11-73.** What is the freezing point of a solution of 100 g of $CH_3OH$ in 800 g of benzene? (For benzene, $K_f$ = 5.12, and the freezing point of pure benzene is 5.5°C.)

**11-74.** What is the freezing point of a 10.0% by mass solution of $CH_3OH$ in benzene? (See exercise 11-73.)

**11-75.** A 1 $m$ solution of HCl lowers the freezing point of water almost twice as much as a 1 $m$ solution of HF. Explain.

**11-76.** What is the freezing point of automobile antifreeze if it is 40.0% by mass ethylene glycol ($C_2H_6O_2$) in water? (Ethylene glycol is a nonelectrolyte.)

**11-77.** In especially cold climates, methyl alcohol ($CH_3OH$) may be used as an automobile antifreeze. Would a 40.0% by mass of an aqueous solution of methyl alcohol remain a liquid at −40°C? (Methyl alcohol is a nonelectrolyte.)

**\*11-78.** Give the freezing point of each of the following in 100 g of water.
(a) 10.0 g of $CH_3OH$
(b) 10.0 g of NaCl
(c) 10.0 g of $CaCl_2$

## General Problems

**11-79.** A mixture is composed of 10 g of $KNO_3$ and 50 g of KCl. What is the approximate amount of KCl that can be separated using the difference in solubility shown in Figure 11-5.

**11-80.** KBr and $KNO_3$ have equal solubilities at about 42°C. What is the composition of the precipitate if 100 g of $H_2O$ saturated with these two salts at 42°C is then cooled to 0°C? (Refer to Figure 11-6.)

**11-81.** What is the percent composition by mass of a solution made by dissolving 10.0 g of sugar and 5.0 g of table salt in 150 mL of water?

**11-82.** What is the molarity of each ion in a solution that is 0.15 M $CaCl_2$, 0.22 M $Ca(ClO_4)_2$, and 0.18 M NaCl?

**\*11-83.** 500 mL of 0.20 M $AgNO_3$ is mixed with 500 mL of 0.30 M NaCl. What is the concentration of $Cl^-$ ion in the solution? The net ionic equation of the reaction that occurs is

$$Ag^+(aq) + Cl^-(aq) \longrightarrow AgCl(s)$$

**\*11-84.** 400 mL of 0.15 M $Ca(NO_3)_2$ is mixed with 500 mL of 0.20 M $Na_2SO_4$. Write the net ionic equation of the precipitation reaction that occurs. Of the two anions involved, which remains in solution after precipitation? What is its concentration?

**\*11-85.** A certain metal (M) reacts with HCl according to the equation

$$M(s) + 2HCl(aq) \longrightarrow MCl_2(aq) + H_2(g)$$

1.44 g of the metal reacts with 225 mL of 0.196 M HCl. What is the metal?

**\*11-86.** Another metal (Z) also reacts with HCl according to the equation

$$2 Z(s) + 6HCl(aq) \longrightarrow 2 ZCl_3(aq) + 3H_2(g)$$

24.0 g of Z reacts with 0.545 L of 2.54 M HCl. What is the metal? What volume of $H_2$ measured at STP is produced?

**\*11-87.** A certain compound dissolves in a solvent known as nitrobenzene. For nitrobenzene, $K_f$ = 8.10. A solution with 3.07 g of the compound dissolved in 120 g of nitrobenzene freezes at 2.22°C. The freezing point of pure nitrobenzene is 5.67°C. Analysis of the compound shows it to be 40.0% C, 13.3% H, and 46.7% N. What is the formula of the compound?

**11-88.** Given 1.00 $m$ aqueous solutions of (a) $Na_3PO_4$, (b) $CaCl_2$, (c) urea (a nonelectrolyte), (d) $Al_2(SO_4)_3$, and (e) LiBr. Order these solutions from highest to lowest freezing points and explain.

**11-89.** Order the following solutions from lowest to highest boiling points.
(a) 0.30 $m$ sugar (a nonelectrolyte)
(b) pure water
(c) 0.05 $m$ $K_2CO_3$
(d) 0.12 $m$ KCl
(e) 0.05 $m$ $CrCl_3$

*11-90. One mole of an electrolyte dissolves in water to form three moles of ions. A 9.21-g quantity of this compound is dissolved in 175 g of water. The freezing point of this solution is $-1.77°C$. The compound is 47.1% K, 14.5% C, and 38.6% O. What is the formula of this compound?

*11-91. A sample of a metal reacts with water to form 487 mL of a 0.120 M solution of the metal hydroxide along with 753 mL of hydrogen gas measured at 25°C and 0.650 atm. Is the metal Na or Ca?

11-92. What volume of $NH_3$ measured at 0.951 atm and 25°C is needed to form 250 mL of 0.450 M aqueous ammonia?

11-93. A 1.82-L volume of gaseous $H_2S$ measured at 1.08 atm and 20°C is dissolved in water. What is the molarity of the aqueous $H_2S$ if the volume of the solution is 2.00 L?

11-94. Sodium bicarbonate reacts with hydrochloric acid to form water, sodium chloride, and carbon dioxide gas. What volume of $CO_2$ measured at 35°C and 1.00 atm pressure could be released by the reaction of 1.00 L of a 0.340 M solution of sodium bicarbonate with excess hydrochloric acid solution?

11-95. Aqueous calcium hydroxide solutions absorb gaseous carbon dioxide to form calcium bicarbonate solutions. What mass of calcium bicarbonate would be formed by reaction of 25.0 mL of 0.150 M calcium hydroxide with 450 mL of gaseous carbon dioxide measured at STP?

*11-96. The molecules of a compound are composed of one phosphorus and multiple chlorine atoms. A molecule of the compound is described as a trigonal pyramid. This gaseous compound dissolves in water to form a hydrochloric acid solution and phosphoric acid ($H_3PO_3$). What is the molarity of the hydrochloric acid if 750 mL of the gas, measured at STP, dissolves in 250 mL of water?

*11-97. A phosphorus-oxygen compound is 43.7% phosphorus. When the compound dissolves in water, it forms one compound, phosphoric acid. If 0.100 mol of the phosphorus-oxygen compound dissolves in 4.00 L of water to form a 0.100 M solution of phosphoric acid solution, what is the formula of the original compound?

*11-98. A 10.0-g quantity of a compound is dissolved in 100 g of water. The solution formed has a melting point of $-7.14°C$. Is the compound KCl, $Na_2S$, or $CaCl_2$?

11-99. An aqueous solution has a freezing point of $-2.50°C$. What is its boiling point?

*11-100. An aqueous solution of a nonelectrolyte is made by dissolving the solute in 1.00 L of water. The solution has a freezing point of $-1.50°C$. What volume of water must be added to change the freezing point to $-1.15°C$?

*11-101. An aqueous solution of a nonelectrolyte in 500 mL of water has a boiling point of 100.86°C. How many moles of solute must be added so that the solution freezes at $-2.06°C$?

# Interactive Learning

11-102. What mass (in grams) is needed to make 250 mL of 0.100 M $K_2SO_4$?

11-103. What volume (in mL) of water must be added to 150 mL of 2.50 M KOH to give a 1.00 M solution? (Assume volumes are additive.)

11-104. What volume (in mL) of 0.100 M NaOH is needed to completely neutralize 25.0 mL of 0.250 M $H_3PO_4$? The equation is:

$$3NaOH(aq) + H_3PO_4(aq) \longrightarrow Na_3PO_4(aq) + 3H_2O$$

11-105. Consider the reaction of aluminum chloride with silver acetate. How many milliliters of 0.250 M $AlCl_3$ would be needed to completely react with 20.0 mL of 0.500 M $AgC_2H_3O_2$ solution? The net ionic equation for the reaction is:

$$Ag^+(aq) + Cl^-(aq) \longrightarrow AgCl(s)$$

# Solutions to Learning Checks

A-1. solute, solvent, miscible, immiscible, ion–dipole, ion–ion, dipole–dipole, saturated, unsaturated, supersaturated, precipitate, more, less, mass percent, moles, liter, molarity, volume

A-2. (a) $BaBr_2 \longrightarrow Ba^{2+}(aq) + 2Br^-(aq)$
(b) $(NH_4)_2Cr_2O_7 \longrightarrow 2NH_4^+(aq) + Cr_2O_7^{2-}(aq)$
(c) $K_2SO_3 \longrightarrow 2K^+(aq) + SO_3^{2-}(aq)$

**A-3.** 200 m̶L̶ × 1.00 g/m̶L̶ = 200 g $H_2O$

mass of solution = 200 + 5.00 = 205 g

$$\frac{5.00 \text{ g solute}}{205 \text{ g solution}} \times 100\% = \underline{\underline{2.44\%}}$$

$MgCl_2 = [24.30 + 2(35.45)] = 95.21$ g/mol

$$5.00 \not{g} \times \frac{1 \text{ mol}}{95.21 \not{g}} = 0.0525 \text{ mol } MgCl_2$$

$$\frac{n}{V} = \frac{0.0525 \text{ mol}}{0.200 \text{ L}} = \underline{\underline{0.263 \text{ M}}}$$

**A-4.** $n = M \times V = 0.375$ mol/L̶ × 0.015 L̶ = $5.63 \times 10^{-3}$ mol NaOH

$$5.63 \times 10^{-3} \text{ m̶o̶l̶} \times \frac{40.00 \text{ g NaOH}}{\text{m̶o̶l̶}} = \underline{\underline{0.225 \text{ g NaOH}}}$$

**A-5.** $M_d = \dfrac{M_c \times V_c}{V_d} = \dfrac{0.375 \text{ M} \times 15.0 \text{ m̶L̶}}{40.0 \text{ m̶L̶}} = \underline{\underline{0.141 \text{ M}}}$

**A-6.** (volume HCl) ⟹ (mol HCl) ⟹

(mol $K_3PO_4$) ⟹ (mass $K_3PO_4$)

$K_3PO_4 = [(3 \times 39.10) + 30.97 + (4 \times 16.00)]$
= 212.2 g/mol

$$0.325 \not{L} \times \frac{0.250 \text{ m̶o̶l̶ H̶C̶l̶}}{\not{L}} \times \frac{1 \text{ m̶o̶l̶ K̶_3̶P̶O̶_4̶}}{3 \text{ m̶o̶l̶ H̶C̶l̶}} \times$$

$$\frac{212.2 \text{ g } K_3PO_4}{\text{m̶o̶l̶ K̶_3̶P̶O̶_4̶}} = \underline{\underline{5.75 \text{ g } K_3PO_4}}$$

**B-1.** electrolyte, weak electrolyte, nonelectrolyte, colligative, vapor pressure, boiling point, freezing point, osmosis, osmotic, increase

**B-2.** $n(NaCl) = 10.0$ g̶ N̶a̶C̶l̶ $\times \dfrac{1 \text{ mol NaCl}}{58.44 \text{ N̶a̶C̶l̶}}$

= 0.171 mol NaCl

$$m = \frac{\text{mol NaCl}}{\text{kg } H_2O} = \frac{0.171 \text{ mol}}{0.100 \text{ kg}} = 1.71 \text{ } m$$

$\Delta T_b = K_b \times m = 0.512 \times 1.71 = 0.876°C$

Since NaCl is a strong electrolyte producing two ions per mole of NaCl,

$\Delta T_b = m \times 2 = 1.8°C$    Boiling point = 100.0°C +
1.8 = $\underline{\underline{101.8°C}}$

$\Delta T_f = K_f \times m \times 2 = 1.86 \times 1.71 \times 2 = 6.36°C$
Freezing point = 0.0°C − 6.4 = $\underline{\underline{-6.4°C}}$

# Cumulative Review
# Chapters 9–11

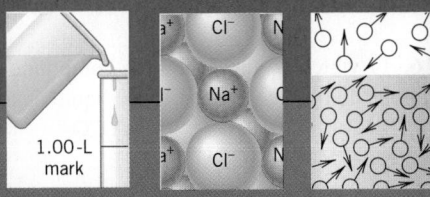

## Multiple Choice
**The following multiple-choice questions have one correct answer.**

**1.** Which of the following is the SI unit of pressure?
**(a)** atm      **(c)** Pa      **(e)** lb/in.$^2$
**(b)** torr      **(d)** in. of Hg

**2.** Which of the following is *not* an assumption of the kinetic molecular theory applied to gases?
**(a)** Molecules have negligible volume.
**(b)** Molecules of all gases have the same average velocity at a given temperature.
**(c)** Gas molecules have negligible interactions.
**(d)** Temperature is related to the average kinetic energy of the system.
**(e)** Molecules are in rapid, random motion.

**3.** Which of the following is a representation of Boyle's law?
**(a)** $P \propto \dfrac{1}{V}$      **(c)** $V \propto T$      **(e)** $V \propto \dfrac{1}{T}$

**(b)** $V \propto P$      **(d)** $P \propto \dfrac{1}{T}$

**4.** The temperature of a volume of gas is increased from 20°C to 40°C at constant pressure. Its volume
**(a)** doubles.
**(b)** decreases by half.
**(c)** increases by a factor of $\frac{313}{293}$.
**(d)** increases by a factor of $\frac{20}{273}$.
**(e)** decreases by a factor of $\frac{293}{313}$.

**5.** Which of the following is a representation of Gay-Lussac's law?
**(a)** $P_1T_1 = P_2T_2$      **(d)** $P_1V_2 = P_2V_1$
**(b)** $P_1V_1 = P_2V_2$      **(e)** $P_1T_2 = P_2T_1$
**(c)** $V_1T_2 = V_2T_1$

**6.** Which of the following is the set of conditions known as standard temperature and pressure (STP)?
**(a)** 0 K and 1 atm      **(d)** 273°C and 1 atm
**(b)** 0°F and 760 torr      **(e)** 273 K and 760 torr
**(c)** 0°C and 760 atm

**7.** Which of the following gases has the highest average velocity at a given temperature?
**(a)** oxygen      **(d)** sulfur dioxide
**(b)** carbon monoxide      **(e)** hydrogen chloride
**(c)** neon

**8.** A mixture of gases has a total pressure of 2.00 atm. If one gas has a partial pressure of 0.50 atm, what part of the mixture is this gas?
**(a)** 50%      **(c)** 25%      **(e)** 1.50 atm
**(b)** 75%      **(d)** 1.00 atm

**9.** A 22.4-L quantity of $O_2$ at STP
**(a)** contains 1 mol of oxygen atoms.
**(b)** has a mass of 16.0 g.
**(c)** contains $1.20 \times 10^{24}$ oxygen atoms.
**(d)** contains 2 mol of $O_2$ molecules.
**(e)** has a mass of 48.0 g.

**10.** A gas has a density of 2.68 g/L (STP). What is the gas?
**(a)** $CO_2$      **(c)** $NO_2$      **(e)** He
**(b)** $SO_2$      **(d)** COS

**11.** Which of the following is a value for the gas constant $R$?
**(a)** $62.4 \dfrac{L \cdot atm}{mol \cdot K}$      **(d)** $0.0821 \dfrac{L \cdot atm}{mol \cdot °C}$

**(b)** $62.4 \dfrac{L \cdot torr}{mol \cdot K}$      **(e)** $0.0821 \dfrac{L \cdot torr}{mol \cdot K}$

**(c)** $82.1 \dfrac{L \cdot atm}{mol \cdot K}$

**12.** Given the equation

$$C(s) + H_2O(l) \longrightarrow CO(g) + H_2(g)$$

what volume of gas measured at STP would be produced from 24.0 g of carbon?
**(a)** 22.4 L      **(c)** 44.8 L      **(e)** 4.0 L
**(b)** 89.6 L      **(d)** 11.2 L

**13.** Which of the following is not a property of the liquid state?
**(a)** surface tension      **(c)** viscosity
**(b)** melting point      **(d)** boiling point

**14.** Which of the following is a property of an ionic compound?
**(a)** high melting point      **(c)** amorphous
**(b)** network bonding      **(d)** soft

**15.** The basic particles (atoms, molecules, or ions) in a liquid
**(a)** have completely random motion.
**(b)** move about fixed points.
**(c)** have translational motion.
**(d)** are not attracted to each other.

**16.** In which of the following compounds would there be dipole–dipole interactions in the liquid state?
(a) NaCl     (c) $CCl_4$     (e) $CO_2$
(b) $H_2$       (d) COS

**17.** Which of the following has hydrogen bonding in the liquid state?
(a) $H_2S$     (c) $CH_4$     (e) $NCl_3$
(b) NaH     (d) $CH_3OH$

**18.** Which of the following nonpolar molecules should have the highest boiling point?
(a) $H_2$      (c) $CO_2$     (e) $N_2$
(b) $SF_6$     (d) $CH_4$

**19.** Based on interactions in the liquid state, which of the following has the highest heat of vaporization?
(a) $H_2O$     (c) $CH_4$     (e) $NH_3$
(b) $H_2S$     (d) $H_2$

**20.** What happens when heat is applied to a liquid at its boiling point?
(a) The heat energy is converted into potential energy.
(b) The heat energy is converted into kinetic energy.
(c) The heat energy is converted into both potential and kinetic energy.
(d) It makes the molecules of the vapor move faster than those in the liquid.

**21.** The normal boiling point of a liquid is defined as the temperature at which
(a) bubbles form in the liquid.
(b) the vapor pressure equals the atmospheric pressure.
(c) the vapor pressure equals 1 atm.
(d) the vapor and the liquid exist in equilibrium.

**22.** A compound has a melting point of 950°C. Which of the following statements is most likely true?
(a) Its molecules are nonpolar.
(b) The compound has a low heat of fusion.
(c) Its molecules are polar covalent.
(d) The compound has a low boiling point.
(e) The compound has a high heat of vaporization.

**23.** When a liquid in an insulated container is allowed to evaporate,
(a) the temperature of the liquid rises.
(b) the temperature of the liquid does not change.
(c) no liquid evaporates unless heat is supplied from the outside.
(d) the vapor that escapes is warmer than the liquid.

**24.** Which of the following represents the solution of $Ca(ClO_4)_2$ in water?
(a) $Ca^+(aq) + (ClO_4)^{2-}(aq)$
(b) $Ca^{2+}(aq) + 2ClO_4^-(aq)$
(c) $Ca^+(aq) + ClO_4^-(aq)$
(d) $Ca^{2+}(aq) + (ClO_4)_2^{2-}(aq)$

**25.** Which of the following forces are the forces between water and the solute when an ionic compound dissolves?
(a) ion–ion           (c) ion–dipole
(b) dipole–dipole    (d) ion–induced dipole

**26.** When the temperature increases, solids generally become _____ soluble and gases become _____ soluble.
(a) more, more      (c) less, less
(b) less, more       (d) more, less

**27.** A 2.0-L quantity of 0.10 M HCl contains
(a) 1.0 mol of HCl.      (d) 0.05 mol of $H_2O$.
(b) 0.20 mol of $H_2O$.    (e) 0.20 mol of HCl.
(c) 20 mol of HCl.

**28.** An experiment calls for 2.00 L of 0.400 M HCl, which must be prepared from 2.00 M HCl. What volume in milliliters of the concentrated acid is needed to be diluted to form the 0.400 M solution?
(a) 400 mL     (c) 800 mL     (e) 500 mL
(b) 200 mL     (d) 250 mL

**29.** Which of the following is not a colligative property?
(a) the boiling point of a solvent
(b) freezing point depression
(c) vapor pressure lowering
(d) osmotic pressure

**30.** What is the freezing point of a 0.100 $m$ aqueous solution of a nonelectrolyte? (For water $K_f = 1.86$.)
(a) $-1.86°C$     (c) $-0.186°C$     (e) $-18.6°C$
(b) $0.186°C$      (d) $18.6°C$

**31.** Compared to pure water, a 1.0 $m$ aqueous sugar solution will have a _____ vapor pressure, a _____ freezing point, and a _____ boiling point.
(a) lower, lower, lower     (d) lower, higher, lower
(b) higher, higher, higher   (e) higher, lower, higher
(c) lower, lower, higher

**32.** If solutions of each of the following have the same concentration, which has the lowest freezing point?
(a) $CH_3OH$ (a nonelectrolyte)    (d) CaS
(b) NaCl                           (e) all are the same
(c) $Na_2SO_4$

## Problems

**1.** Fill in the blanks.

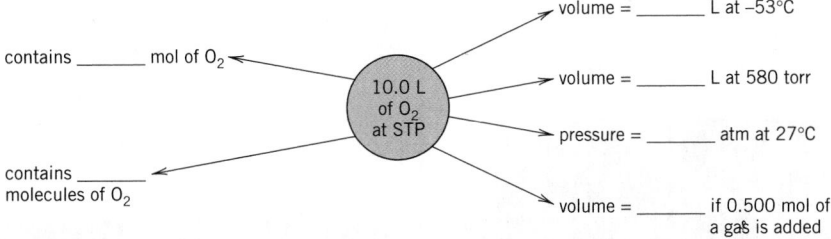

contains _____ mol of $O_2$

10.0 L of $O_2$ at STP

volume = _____ L at –53°C

volume = _____ L at 580 torr

pressure = _____ atm at 27°C

contains _____ molecules of $O_2$

volume = _____ if 0.500 mol of a gas is added

**2.** A compound can be vaporized at 100°C. At that temperature, 1.19 g of the compound occupies 250 mL at 756 torr. Analysis of the compound shows that it is 49.0% C, 48.3% Cl, and 2.72% H. What is the molecular formula of the compound?

**3.** The following compounds have nearly the same formula weights: $(CH_3)_2O$ (polar covalent), $C_2H_5OH$ (polar covalent), $C_3H_8$ (nonpolar covalent), and $BeF_2$ (ionic). The melting points of these compounds are 800°C, −117°C, −138°C, and −189°C. Match the melting point with the compound and explain what forces must be overcome in the solid to allow melting.

**4.** How many calories are required to change 25.0 g of ice at −20.0°C to steam at 100°C? (The specific heat of ice is 0.492 cal/(g · °C), the heat of fusion is 79.8 cal/g, and the heat of vaporization is 540 cal/g.)

**5.** An aqueous solution is made by dissolving 40.0 g of $H_2SO_4$ in enough water to make 850 mL of solution.
**(a)** What is the molarity of the $H_2SO_4$ solution?
**(b)** If 100 mL of this solution is diluted to 500 mL, what is the resulting molarity?
**(c)** What volume of the original solution is needed to form 2.00 L of a 0.150 M solution?

**6.** When aqueous solutions of lead(II) nitrate and sodium iodide are mixed, a precipitate of lead(II) iodide forms.
**(a)** Write the balanced equation representing this reaction.
**(b)** What volume of 0.40 M lead nitrate is needed to react with 2.50 L of 0.55 M sodium iodide?

## chapter 12

# Acids, Bases, and Salts

The lemon is known for its sour taste, which is caused by citric acid. Acids are a unique class of compounds that are discussed in this chapter.

**Setting the Stage**

It can make your mouth pucker, your body shudder, and your eyes water. That's the reaction one gets from taking a bite into a fresh lemon. A taste of vinegar has the same effect. Even carbonated beverages produce a subtle sour taste that peps up the drink. All of these substances have a similar effect because of the presence of a compound that produces a sour taste (but to different degrees). These compounds, known as acids, were characterized over 500 years ago, during the Middle Ages, in the chemical laboratories of alchemists. Substances were classified as acids because of their common properties (such as sourness) rather than a certain chemical composition. Acids are well known to the general population. Some are in foods or drugs, such as citric acid in lemons, acetic acid in vinegar, lactic acid in sour milk, or acetylsalicylic acid in aspirin. Some must be handled with caution, such as sulfuric acid in a car battery or hydrochloric acid used to clean concrete. We also relate the word "acid" with the serious environmental hazard "acid rain." As we will discuss in this chapter, acids undergo many characteristic chemical reactions. When rain is acidic, these reactions lead to the degradation of stone used in buildings and the liberation of poisonous metal ions locked in soil. The control of acid rain has become a subject of international conferences and negotiations.

We briefly introduced acids, bases, and the reaction between them in section 7-6. At that time we restricted our discussion to strong acids and bases. We have covered quite a bit of chemistry since Chapter 7 so we will reexamine this topic in more detail. We can expand on our discussion of these compounds by including weak acids and bases and how we measure acidity.

### Formulating Some Questions

Several important questions will be addressed in this chapter. What active ingredients in acids and bases produce their specific properties? Why are some acids, like citric acid in fruits, comparatively tame and others, like battery acid, so harsh? How do we measure different degrees of acidity? How is acidity controlled? Finally, how does burning coal lead to acid formation in rain? First, we will examine some of the properties of acids and bases.

---

## Section A ACIDS, BASES, AND THE FORMATION OF SALTS

---

# 12-1 Properties of Acids and Bases

**Looking Ahead!** Historically, acids and bases were classified as such as a result of their common properties well before scientists knew much about their compositions. Acids and bases have common properties because they each have a specific "active ingredient." The nature of these specific properties and the active ingredients are the topics of this section.

The sour taste of acids accounts for the origin of the word itself. The word "acid" originates from the Latin *acidus,* meaning "sour," or the closely related Latin *acetum,* meaning "vinegar." This ancient class of compounds has several characteristic chemical properties. Acids are compounds that

1. Taste sour (of course, one *never* tastes laboratory chemicals).
2. React with certain metals (e.g., Zn and Fe), with the liberation of hydrogen gas. (See Figure 12-1.)
3. Cause certain organic dyes to change color (e.g., litmus turns from blue to red in acids).

 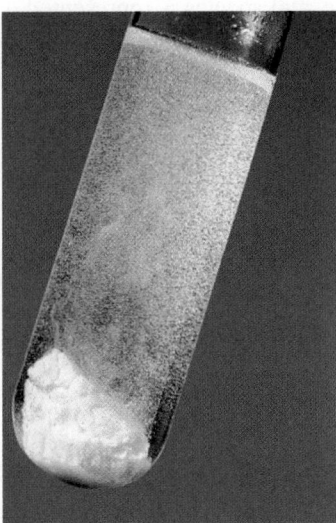

**Figure 12-1 Zinc and Limestone in Acid** Zinc (*left*) reacts with acid to liberate hydrogen; limestone ($CaCO_3$, *right*) reacts with acid to liberate carbon dioxide.

4. React with limestone ($CaCO_3$), with the liberation of carbon dioxide gas. (See Figure 12-1.)

5. React with bases to form salts and water.

The counterparts to acids are bases. Bases are compounds that

1. Taste bitter.
2. Feel slippery or soapy.
3. React with oils and grease.
4. Cause certain organic dyes to change color (e.g., litmus turns from red to blue in bases).
5. React with acids to form salts and water.

The properties listed above relate to what acids and bases do and not to their chemical composition. It was not until 1884 that a Swedish chemist, Svante Arrhenius, suggested that the particular composition of these compounds determined their behavior. *He was the first to suggest that acids produced $H^+$ ions and bases produced $OH^-$ ions in water.* Some familiar acids, their common names, and their formulas are shown below.

| Chemical Name | Common Name | Formula |
|---|---|---|
| hydrochloric acid | muriatic acid | HCl |
| sulfuric acid | oil of vitriol, battery acid | $H_2SO_4$ |
| acetic acid | vinegar | $HC_2H_3O_2$ |
| carbonic acid | carbonated water | $H_2CO_3$ |

The ionization process of acids in water was illustrated in Figure 11-3 in the previous chapter. To illustrate the importance of water in the ionization process, the reaction of an acid with water can be represented as

$$HCl + H_2O \longrightarrow H_3O^+(aq) + Cl^-(aq)$$

*Instead of $H^+$, the acid species is often represented as $H_3O^+$, which is known as the* **hydronium ion.** The hydronium ion is simply a representation of the $H^+$ ion in a hydrated form. The acid species is represented as $H_3O^+$ rather than $H^+$ because it is somewhat closer to what is believed to be the actual species. In fact, the nature of $H^+$ in aqueous solution is even more complex than $H_3O^+$ (i.e., $H_5O_2^+$, $H_7O_3^+$, etc.). In any case, the acid species is represented as $H^+$, $H^+(aq)$, or $H_3O^+(aq)$, depending on the convenience of the particular situation. *Just remember that all refer to the same species in aqueous solution.* If $H^+(aq)$ is used, it should be understood that it is not just a bare proton in aqueous solution but is associated with water molecules. (It is hydrated.)

It is the current practice to list on the label the active ingredient in medicines or drugs. In this regard, the active ingredient of acids is the $H^+(aq)$ ion. We can now see how this ion accounts for some of the behavior of acids listed previously. Equation **1** illustrates the reaction of an acid with a metal. In equation **2,** the reaction of an acid with limestone is illustrated. And, in equation **3,** we have a neutralization reaction. The net ionic equations of these reactions are also shown, which emphasizes the role of the $H^+(aq)$ ion in each case.

1. Acids react with metals (e.g., Zn) and give off hydrogen gas.

$$Zn(s) + 2HCl(aq) \longrightarrow ZnCl_2(aq) + H_2(g)$$
$$Zn(s) + \underline{2H^+(aq)} \longrightarrow Zn^{2+}(aq) + H_2(g)$$

**2.** Acids react with limestone to give off carbon dioxide gas.

$$CaCO_3(s) + 2HNO_3(aq) \longrightarrow Ca(NO_3)_2(aq) + H_2O(l) + CO_2(g)$$
$$CaCO_3(s) + \underline{2H^+(aq)} \longrightarrow Ca^{2+}(aq) + H_2O(l) + CO_2(g)$$

**3.** Acids react with bases.

$$ACID + BASE \longrightarrow SALT + WATER$$
$$HClO_4(aq) + NaOH(aq) \longrightarrow NaClO_4(aq) + H_2O(l)$$
$$\underline{H^+(aq)} + OH^-(aq) \longrightarrow H_2O(l)$$

The last reaction is of prime importance and is discussed in more detail in a later section.

Now let us turn our attention to bases. Bases are compounds that produce $OH^-$ ions in water, forming what are known as basic solutions, sometimes referred to as *alkaline* or *caustic* solutions. Some of the commonly known bases are sodium hydroxide (also known as caustic soda, or lye), potassium hydroxide (caustic potash), calcium hydroxide (slaked lime), and ammonia. Except for ammonia, these compounds are all solid ionic compounds. Solution in water simply releases the $OH^-$ ion into the aqueous medium.

$$NaOH(s) \xrightarrow{H_2O} Na^+(aq) + OH^-(aq)$$
$$Ba(OH)_2(s) \xrightarrow{H_2O} Ba^{2+}(aq) + 2OH^-(aq)$$

The action of ammonia ($NH_3$) as a base is somewhat different from that of the ionic hydroxides and is better described by a more detailed look at acids and bases in the following section.

## 12-2 Brønsted–Lowry Acids and Bases

**Looking Ahead!** So far, we have focused on electrically neutral compounds such as HCl and NaOH that act as acids or bases by formation of $H_3O^+$ or $OH^-$ ions in water. However, to better describe acid strength and the action of ions in water, we need a broader definition of acids and bases. We do this in the following discussion.

Limestone ($CaCO_3$) is quite a versatile compound. We can use it as solid rock in the construction of huge buildings or we can use it as a powder in chalk. It is also the major ingredient of many antacids, which are consumed to neutralize excess stomach acid. In this reaction, the carbonate ion ($CO_3^{2-}$) is the active ingredient and reacts as a base (i.e., an antacid). From our previous definition of acids and bases, it is not immediately obvious how an anion such as $CO_3^{2-}$ behaves as a base. In order to include anions as bases, we need a broader, more inclusive definition than that of Arrhenius, which focused mainly on molecular compounds. We will now focus on the role of the $H^+$ ion in solution. *In the* **Brønsted–Lowry** *definition, an* **acid** *is a proton ($H^+$) donor and a* **base** *is a proton acceptor.* To illustrate this definition, we again look at the reaction of HCl as an acid to form the $H_3O^+$ ion.

$$HCl(aq) + H_2O(l) \longrightarrow H_3O^+(aq) + Cl^-(aq)$$

HCl is an acid by the Arrhenius definition because it produces the $H_3O^+$ ion. It is also an acid by the Brønsted–Lowry definition because *it donates an $H^+$ to $H_2O$.* In this definition, however, the $H_2O$ molecule also takes on the role of a base because it accepts an $H^+$ from the HCl. The reaction of an acid and a base in water can be considered as an exchange of the proton. An acid (HCl) reacts with a base ($H_2O$) to form another acid ($H_3O^+$) and base ($Cl^-$). *The base that remains when an acid donates*

**Tums is composed primarily of calcium carbonate.**

*a proton is known as the* **conjugate base** *of the acid. Likewise, the acid that is formed when the base accepts a proton is known as the* **conjugate acid** *of the base.* Thus $HCl - Cl^-$ and $H_3O^+ - H_2O$ are known as *conjugate acid–base pairs.* The exchange of $H^+$ is illustrated below, where $A_1$ and $B_1$ refer to a specific conjugate acid–base pair and $A_2$ and $B_2$ refer to the other acid–base pair.

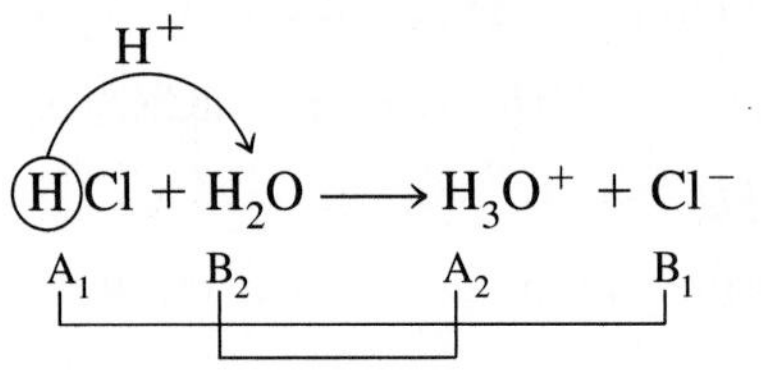

$$\text{(H)Cl} + H_2O \longrightarrow H_3O^+ + Cl^-$$
$$\quad A_1 \qquad B_2 \qquad\quad A_2 \qquad B_1$$

Now let's consider the reaction of ammonia ($NH_3$) in water. Ammonia is a base in the Arrehenius definition because it forms $OH^-$ in aqueous solution even though ammonia itself does not contain the $OH^-$ ion. If we examine its behavior in water as a Brønsted–Lowry base, however, it becomes more obvious how $OH^-$ ions are produced.

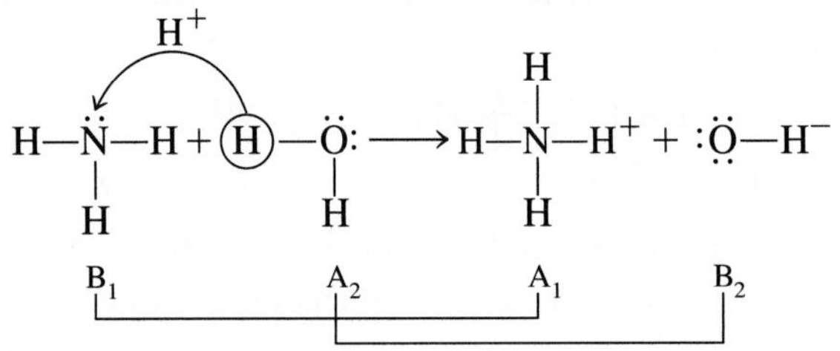

$$H-\overset{\overset{\textstyle H}{|}}{N}-H + \text{(H)}-\overset{\overset{\textstyle }{}}{\underset{\underset{\textstyle H}{|}}{\ddot{O}}}: \longrightarrow H-\overset{\overset{\textstyle H}{|}}{\underset{\underset{\textstyle H}{|}}{N}}-H^+ + :\ddot{O}-H^-$$
$$\quad B_1 \qquad\qquad A_2 \qquad\qquad A_1 \qquad\qquad B_2$$

In the Brønsted–Lowry sense, the reaction can be viewed as simply an exchange of an $H^+$. When the base ($NH_3$) reacts, it adds $H^+$ to form its conjugate acid ($NH_4^+$). When the acid ($H_2O$) reacts, it loses an $H^+$ to form its base ($OH^-$). The $NH_3$, $NH_4^+$ ($B_1$ and $A_1$) and the $H_2O$, $OH^-$ ($A_2$ and $B_2$) pairs are conjugate acid–base pairs.

In this reaction, $H_2O$ is an acid since it donates an $H^+$ to form $NH_4^+$. Recall that $H_2O$ acts as *a base* when HCl is present. *A compound or ion that can both donate and accept $H^+$ ions is called* **amphiprotic.** Water is amphiprotic since it can accept $H^+$ ions when an acid is present or donate $H^+$ when a base is present. An amphiprotic substance has both a conjugate acid and a conjugate base. Examples of other amphiprotic substances include $HS^-$ and $H_2PO_4^-$.

Before we look at other examples of Brønsted–Lowry acid–base reactions, we should emphasize the meaning of conjugate acids and bases. *The conjugate base of a compound or ion results from removal of an $H^+$. The conjugate acid of a compound or ion results from the addition of an $H^+$.*

$$\text{conjugate acid} \underset{+H^+}{\overset{-H^+}{\rightleftharpoons}} \text{conjugate base}$$

For example,

$$H_3PO_4 \underset{+H^+}{\overset{-H^+}{\rightleftharpoons}} H_2PO_4^-$$
$$\quad\text{acid} \qquad\qquad \text{base}$$

Notice that in the formation of a conjugate base, the base species ($H_2PO_4^-$) has one less hydrogen and the charge decreases by one from the acid ($H_3PO_4$). The reverse is true for formation of a conjugate acid from a base. That is, the acid has one more hydrogen and the charge increases by one.

We are now ready to examine how the carbonate ion in calcium carbonate behaves as a base in antacid tablets. The $CO_3^{2-}$ ion relieves acidic stomachs (containing excess $H_3O^+$) as illustrated by the following proton exchange reaction.

$$\overset{\overset{\displaystyle H^+}{\curvearrowleft}}{CO_3^{2-}} + H_3O^+ \longrightarrow HCO_3^- + H_2O$$

Notice that the carbonate ion acts as a base by accepting the proton from $H_3O^+$ to form its conjugate acid ($HCO_3^-$), while the $H_3O^+$ ion forms its conjugate base ($H_2O$). Decreasing the hydronium ion concentration in the stomach is what is meant by "relief of stomach distress."

---

**Determining Conjugate Bases of Compounds**

**Example 12-1**

working it out

What are the conjugate bases of (a) $H_2SO_3$ and (b) $H_2PO_4^-$?

**Procedure**

$$acid - H^+ = conjugate\ base$$

**Solution**

a.  $H_2SO_3 - H^+ = \underline{\underline{HSO_3^-}}$        b.  $H_2PO_4^- - H^+ = \underline{\underline{HPO_4^{2-}}}$

---

**Determining Conjugate Acids of Compounds**

**Example 12-2**

What are the conjugate acids of (a) $CN^-$ and (b) $H_2PO_4^-$?

**Procedure**

$$base + H^+ = conjugate\ acid$$

**Solution**

a.  $CN^- + H^+ = \underline{\underline{HCN}}$        b.  $H_2PO_4^- + H^+ = \underline{\underline{H_3PO_4}}$

---

**Writing Acid–Base Reactions**

**Example 12-3**

Write the equations illustrating the following Brønsted–Lowry acid–base reactions.
a.  $H_2S$ as an acid with $H_2O$
b.  $H_2PO_4^-$ as an acid with $OH^-$
c.  $H_2PO_4^-$ as a base with $H_3O^+$
d.  $CN^-$ as a base with $H_2O$

**Procedure**

A Brønsted–Lowry acid–base reaction produces a conjugate acid and base.

**Solution**

|      | Acid |   | Base |   $\longrightarrow$ | Acid |   | Base |
|------|------|---|------|-----|------|---|------|
| (a)  | $H_2S$ | + | $H_2O$ | $\longrightarrow$ | $H_3O^+$ | + | $HS^-$ |
| (b)  | $H_2PO_4^-$ | + | $OH^-$ | $\longrightarrow$ | $H_2O$ | + | $HPO_4^{2-}$ |
| (c)  | $H_3O^+$ | + | $H_2PO_4^-$ | $\longrightarrow$ | $H_3PO_4$ | + | $H_2O$ |
| (d)  | $H_2O$ | + | $CN^-$ | $\longrightarrow$ | $HCN$ | + | $OH^-$ |

Note that equations (b) and (c) indicate that the $H_2PO_4^-$ is amphiprotic.

# 12-3 Strengths of Acids and Bases

**Looking Ahead!** The properties of acids and bases described in section 12-1 are mostly associated with strong acids and bases. Other substances display these properties but in less dramatic fashion. For example, dilute acetic acid is sour but tame enough to use on a salad, and ammonia dissolves grease but is mild enough to clean oil stains from floors. There is a wide range of behavior that we regard as acidic or basic. This is the subject of this section.

Ammonia as a base would do a poor job of unclogging a stopped-up drain, yet we certainly wouldn't want to use lye to clean an oil spot from a carpet. In the former case, the base is too weak; in the latter, it is much too strong. Strong acids and bases are difficult and dangerous to handle and store. Weak acids and bases are quite easy and safe to have around. The large difference in acid or base behavior relates to the concentration of the active ingredient ($H^+$ or $OH^-$) produced by the acid or base in water. This depends on its strength. First, we will consider the strength of acids.

In Chapter 7, we indicated that strong acids were 100% dissociated into ions in solution. Actually, there are only six common strong acids. In addition to hydrochloric (HCl), which we have already discussed, there are two other binary acids, hydrobromic acid (HBr) and hydroiodic acid (HI). The other strong acids are sulfuric acid ($H_2SO_4$), nitric acid ($HNO_3$), and perchloric acid ($HClO_4$). Sulfuric acid is a somewhat more complex case but will be considered shortly.

We have been using the 100% ionization criteria to describe a strong acid, but we still haven't formally answered the question "strong compared to what?" In the Brønsted–Lowry definition, the reaction of a molecular acid with water is considered a proton ($H^+$) exchange reaction. In fact, there is a competition between the proton-donating abilities of two acids and, like other competitions, the stronger prevails. *The stronger acid reacts with the stronger base to produce a weaker conjugate acid and conjugate base. In the case of a strong acid in water, the molecular acid (e.g., HCl) is a stronger proton donor than $H_3O^+$.* Therefore, the reaction proceeds essentially 100% to the right.

$$\underset{\substack{\text{stronger} \\ \text{acid}}}{HCl(aq)} + \underset{\substack{\text{stronger} \\ \text{base}}}{H_2O(l)} \longrightarrow \underset{\substack{\text{weaker} \\ \text{acid}}}{H_3O^+(aq)} + \underset{\substack{\text{weaker} \\ \text{base}}}{Cl^-(aq)}$$

Most acids that we may be familiar with, such as acetic acid, ascorbic acid (vitamin C), and citric acid, are all considered weak acids. The weak acids that we will discuss in this section are also neutral molecular acids. *A **weak molecular acid** is partially ionized (usually less than 5% at typical molar concentrations).*

The ionization of a weak acid appears to be limited because it is a reversible reaction. Such a reaction was briefly mentioned in section 11-7, since weak acids are examples of weak electrolytes. The extent of ionization of weak acids will be discussed in Chapter 14. However, we need an understanding of the concept of equilibrium at this time. A reaction in which the reverse reaction also occurs to an appreciable extent is illustrated by double arrows ($\rightleftharpoons$) rather than the single arrow that implies an essentially complete reaction, such as was previously shown for HCl. Ionization of the weak acid HF is illustrated as follows.

$$HF(aq) + H_2O(l) \rightleftharpoons H_3O^+(aq) + F^-(aq)$$

The partial ionization of a weak acid is one example of a chemical reaction that reaches a state of equilibrium. *In a reaction at equilibrium, two reactions are occurring simultaneously.* In the ionization of HF, for example, a forward reaction occurs to the right, producing ions ($H_3O^+$ and $F^-$), and a reverse reaction occurs to the left, producing molecular compounds (HF and $H_2O$). In a reaction at equilibrium it is

**Figure 12-2 Strong Acids and Weak Acids**
Strong acids are completely ionized in water; weak acids are only partially ionized.

HF, a weak acid

HCl, a strong acid

apparent that the identity of the reactants as opposed to the products becomes obscure. However, by convention, we still refer to the species on the left of the double arrows as the reactants and on the right as the products.

*Forward:* $\quad HF(aq) + H_2O(l) \longrightarrow H_3O^+(aq) + F^-(aq)$

*Reverse:* $\quad H_3O^+(aq) + F^-(aq) \longrightarrow HF(aq) + H_2O(l)$

*At equilibrium, the forward and reverse reactions occur at the same rate.* For weak acids, the point of equilibrium lies far to the left side of the original ionization equation. This is because the $H_3O^+$ ion is a stronger proton donor than the HF molecular acid and the $F^-$ is a stronger proton acceptor than $H_2O$, opposite the case for strong acids. Thus, in the case of weak acids, reactants are favored over products. Even though the reverse reaction is favored, the forward reaction does occur to an extent, so there is a small but important concentration of $H_3O^+$ present in the solution. In any case, however, most of the fluorine is present in the form of molecular HF rather than fluoride ions. (See Figure 12-2.)

When a system is at equilibrium, the concentrations of all species (reactants and products) remain the same, but the identities of the individual molecules change. The reaction thus *appears* to have gone to a certain extent and then stopped. In fact, at equilibrium, a *dynamic (constantly changing)* situation exists in which two reactions going in opposite directions at the same rate keep the concentrations of all species constant.

In the following discussions and examples we will refer to the percent ionization of the weak acid. Like all percent problems, the actual amount or concentration of ions present is found by the multiplication of the total amount of acid initially present by the percent expressed in decimal form, which is obtained by dividing the percent ionization by 100%. Thus if the original concentration of an acid is 0.20 M and it is 5.0% ionized, the concentration of each ion (the $H_3O^+$ cation and the specific anion) is

$$0.20 \text{ M} \times \frac{5.0\%}{100\%} = 0.20 \text{ M} \times 0.050 = 0.010 \text{ M}$$

The following examples illustrate the difference in acidity (the difference in $H_3O^+$ concentration) between a strong acid and a weak acid. In these examples, *the appearance of a species in brackets (e.g., [$H_3O^+$]) represents the numerical value of the concentration of that species in moles per liter* (M).

| Example 12-4 | **The $H_3O^+$ Concentration in a Strong Acid Solution** |
|---|---|
| working it out | What is $[H_3O^+]$ in a 0.100 M $HNO_3$ solution? |

**Solution**

$HNO_3$ is a strong acid, so the following reaction goes 100% to the right.

$$HNO_3(aq) + H_2O(l) \longrightarrow H_3O^+(aq) + NO_3^-(aq)$$

As in other stoichiometry problems involving complete reactions, the amount (or concentration) of a product is found from the amount (or concentration) of a reactant using a mole ratio conversion factor from the balanced equation.

$$0.100 \; \cancel{\text{mol/L HNO}_3} \times \frac{1 \text{ mol/L } H_3O^+}{1 \; \cancel{\text{mol/L HNO}_3}} = 0.100 \text{ mol/L } H_3O^+$$

$$[H_3O^+] = \underline{\underline{0.100 \text{ M}}}$$

| Example 12-5 | **The $H_3O^+$ Concentration in a Weak Acid Solution** |
|---|---|
| | What is $[H_3O^+]$ in a 0.100 M $HC_2H_3O_2$ solution that is 1.34% ionized? |

**Solution**

Since $HC_2H_3O_2$ is a weak acid, the following ionization reaches equilibrium when 1.34% of the initial $HC_2H_3O_2$ is ionized.

$$HC_2H_3O_2(aq) + H_2O(l) \rightleftharpoons H_3O^+(aq) + C_2H_3O_2^-(aq)$$

The $[H_3O^+]$ is calculated by multiplying the original concentration of acid by the percent *expressed in fraction form*.

$$[H_3O^+] = [\text{original concentration of acid}] \times \frac{\% \text{ ionization}}{100\%}$$

In this case,

$$[H_3O^+] = [0.100] \times \frac{1.34\%}{100\%} = \underline{\underline{1.34 \times 10^{-3} \text{ M}}}$$

In the two preceding examples, we found that the concentration of $H_3O^+$ is about 100 times greater in the strong acid solution, although both were at the same original concentration. Only in the case of strong acids does the original concentration of the acid equal the concentration of $H_3O^+$ ions.

In Example 12-5 we used acetic acid ($HC_2H_3O_2$) as an example of a weak acid. Perhaps one wonders why it is written that way and not as $H_4C_2O_3$. However, if we look at the Lewis representation of acetic acid, we notice that there are two types of hydrogens. The one attached to the oxygen is polar and is potentially acidic. It is ionized (to a limited extent) by the water molecules, as illustrated in Figure 11-3 in the previous chapter. The three attached to the carbon are essentially nonpolar and do not

Three H's on the C in acetic acid do not ionize. (The C—H bond is essentially nonpolar).

The O—H bond is polar, so the H can be ionized.

interact with polar water molecules when placed in aqueous solution. Thus the three hydrogens attached to carbon are not affected by proton exchange and remain as part of the acetate ion.

Now we consider the case of bases. They also exhibit a range of strengths depending on the concentration of $OH^-$ produced by the base. Strong bases are ionic compounds that dissolve in water to form $OH^-$ ions. All alkali metal hydroxides are strong bases and are quite soluble in water. The alkaline earth hydroxides [except $Be(OH)_2$] also completely dissociate into ions in solution. However, $Mg(OH)_2$ has a very low solubility in water and so produces a very small concentration of aqueous $OH^-$. Because of its low solubility, it can be taken internally to combat excess stomach acid (milk of magnesia).

The most familiar example of a weak molecular base and the one we will emphasize is ammonia ($NH_3$), whose reaction as a base was discussed in section 12-2. *A **weak molecular base** is a base that is only partially converted into ions in solution.* The reaction of ammonia as a base is shown by the equation

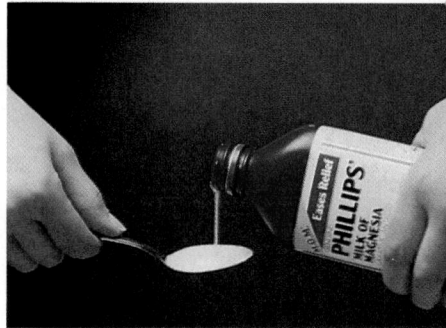

**Milk of magnesia is a base taken for indigestion.**

$$NH_3(aq) + H_2O(l) \rightleftharpoons NH_4^+(aq) + OH^-(aq)$$

The tip-off that ammonia is a weak base is found in the equilibrium arrows rather than the single arrow that implies a complete reaction. As in the case of weak acids, the position of equilibrium lies far to the left. The vast majority of dissolved $NH_3$ molecules remain in the molecular form shown on the left of the double arrows rather than as ions, shown on the right. In the Brønsted–Lowry sense we note that $NH_4^+$ is a stronger proton donor than $H_2O$ and $OH^-$ is a stronger proton acceptor than $NH_3$. Many other neutral nitrogen compounds, such as methylamine ($CH_3NH_2$) and pyridine ($C_5H_5N$), also react with water in a similar manner to produce weakly basic solutions.

## 12-4 Neutralization and Salts

**Looking Ahead!** Compounds or ions that act as acids in water have characteristic properties, as do compounds or ions that act as bases. When solutions of these acids and bases are mixed in the proper amounts, the characteristic properties are destroyed or neutralized. The products of such a reaction are a salt and water. We will look at the interactions of solutions of acids with solutions of bases next.

In Chapter 7 we described one type of a double-displacement reaction between acids and bases known as *neutralization*. If we mix the acid and base in stoichiometric amounts, the products are simply water and a salt. We will begin our discussion of neturalization reactions with a review of the reaction between a strong acid (hydrochloric acid) and a strong base (sodium hydroxide) as described in Chapter 7 and then move on to more complex cases. The molecular, total ionic, and net ionic equations are shown below. In this case, it is more convenient to represent the acid species as simply $H^+(aq)$ rather than $H_3O^+$.

$$\text{ACID} + \text{BASE} \longrightarrow \text{SALT} + \text{WATER}$$

*Molecular:* $HCl(aq) + NaOH(aq) \longrightarrow NaCl(aq) + H_2O(l)$

*Total ionic:* $H^+(aq) + \cancel{Cl^-(aq)} + \cancel{Na^+(aq)} + OH^-(aq) \longrightarrow$
$$\cancel{Na^+(aq)} + \cancel{Cl^-(aq)} + H_2O(l)$$

*Net ionic:* $H^+(aq) + OH^-(aq) \longrightarrow H_2O(l)$

The key to what drives neutralization reactions is found in the net ionic equation. The active ingredient from the acid [$H^+(aq)$] reacts with the active ingredient from the base [$OH^-(aq)$] to form the molecular compound water. A salt is what is left over—usually present as spectator ions if the salt is soluble.

As a vital mineral needed to maintain good health, "salt" refers to just one substance, sodium chloride, as formed in the preceding reaction. Actually, salts can result from many different combinations of anions and cations from a variety of neutralizations. The following neutralization reactions, written in molecular form, illustrate the formation of some other salts.

ACID + BASE $\longrightarrow$ SALT + WATER

1. $2HNO_3(aq) + Ca(OH)_2(aq) \longrightarrow Ca(NO_3)_2(aq) + 2H_2O(l)$
2. $HClO(aq) + LiOH(aq) \longrightarrow LiClO(aq) + H_2O(l)$
3. $H_2SO_4(aq) + 2NaOH(aq) \longrightarrow Na_2SO_4(aq) + 2H_2O(l)$

Each of these three neutralization reactions represents somewhat different situations, so we will look at these reactions one at a time in ionic form. Reaction **1** again represents the neutralization of a strong acid with a strong base. In this case, however, the base, $Ca(OH)_2$, dissolves in water to produce two $OH^-$ ions per formula unit. Thus two moles of acid are needed per mole of base for complete neutralization. The total ionic and net ionic equations for reaction **1** are as follows.

$$2H^+(aq) + 2NO_3^-(aq) + Ca^{2+}(aq) + 2OH^-(aq) \longrightarrow$$
$$Ca^{2+}(aq) + 2NO_3^-(aq) + 2H_2O(l)$$
$$H^+(aq) + OH^-(aq) \longrightarrow H_2O(l)$$

Notice that the net ionic equation is identical to the reaction illustrated in the beginning of this section.

Reaction **2** illustrates a neutralization of a weak acid (HClO) with a strong base. Recall that in the case of most weak acids, the overwhelming majority of molecules are present in solution in the molecular form [i.e., $HClO(aq)$] rather than as ions [i.e., $H^+(aq)$ and $ClO^-(aq)$]. Thus when we write the total ionic and net ionic equations, the acid is displayed in molecular form. These two equations for reaction **2** are shown as follows.

*Total ionic:* $HClO(aq) + Li^+(aq) + OH^-(aq) \longrightarrow$
$$Li^+(aq) + ClO^-(aq) + H_2O(l)$$

*Net ionic:* $HClO(aq) + OH^-(aq) \longrightarrow ClO^-(aq) + H_2O(l)$

Acids are sometimes designated by the number of $H^+$ ions that are available from each molecule. Thus HCl, $HNO_3$, and HClO are known as **monoprotic acids,** *since only one $H^+$ is produced per molecule. Those acids that can produce more than one $H^+$ are known as* **polyprotic acids.** More specifically, polyprotic acids may be **diprotic** (*two $H^+$'s*) such as $H_2SO_4$ or **triprotic** (*three $H^+$'s*) in the case of $H_3PO_4$.

Reaction **3** represents the neutralization of the strong diprotic acid, $H_2SO_4$. In Chapter 7 we represented the ionization of $H_2SO_4$ as being complete, but that is not exactly the case. The first ionization is indeed complete (100%), typical of a strong acid, but the second is not. This ionization reaches an equilibrium typical of weak acids. This reaction is thus represented by a double arrow.

$$H_2SO_4 \longrightarrow H^+(aq) + HSO_4^-(aq) \quad \text{(first ionization of } H_2SO_4\text{)}$$
$$HSO_4^- \rightleftharpoons H^+(aq) + SO_4^{2-}(aq) \quad \text{(second ionization of } H_2SO_4\text{)}$$

Addition of one mole of NaOH to $H_2SO_4$ results in a partial neutralization, forming water and $NaHSO_4$ (sodium bisulfate). Sodium bisulfate (or sodium hydrogen sulfate) is an example of an acid salt. *An acid salt is an ionic compound containing an anion with one or more acidic hydrogens that can be neutralized by a base.*

*Molecular equation:* $H_2SO_4(aq) + NaOH(aq) \longrightarrow NaHSO_4(aq) + H_2O(l)$

*Net ionic equation:* $H^+(aq) + OH^-(aq) \longrightarrow H_2O(l)$

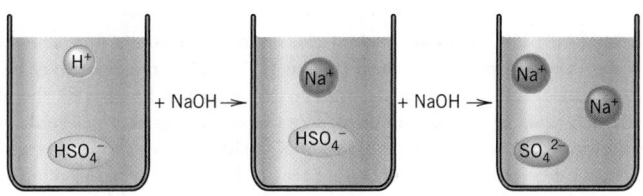

**Figure 12-3  Neutralization of $H_2SO_4$**
The hydrogens of $H_2SO_4$ can be neutralized one at a time.

This reaction is typical of strong acid–strong base neutralizations where the $HSO_4^-$ and the $Na^+$ ions are subtracted out to produce the net ionic equation. A second mole of NaOH added to the $NaHSO_4$ solution completes the neutralization. The net ionic equation for this reaction is typical of a weak acid–strong base neutralization where the anion acts as the acid.

*Molecular equation:*  $NaHSO_4(aq) + NaOH(aq) \longrightarrow Na_2SO_4(aq) + H_2O(l)$

*Net ionic equation:*  $HSO_4^-(aq) + OH^-(aq) \longrightarrow SO_4^{2-}(aq) + H_2O(l)$

The complete neutralization (**3**) can now be represented by the molecular, total ionic, and net ionic equations as follows. (See Figure 12-3.)

$$H_2SO_4(aq) + 2NaOH(aq) \longrightarrow Na_2SO_4(aq) + 2H_2O(l)$$

$$H^+(aq) + HSO_4^-(aq) + 2Na^+(aq) + 2OH^-(aq) \longrightarrow$$
$$2Na^+(aq) + SO_4^{2-}(aq) + 2H_2O(l)$$

$$H^+(aq) + HSO_4^-(aq) + 2OH^- \longrightarrow SO_4^{2-}(aq) + 2H_2O(l)$$

---

**Writing a Complete Neutralization Reaction**

**Example 12-6**

*working it out*

Write the balanced equation in molecular form illustrating the complete neutralization of $Al(OH)_3$ with $H_2SO_4$.

**Procedure**

Complete neutralization requires one $H^+$ for each $OH^-$. Since $Al(OH)_3$ has three available $OH^-$ ions and $H_2SO_4$ can only provide two $H^+$ ions, the reaction requires two moles of $Al(OH)_3$ for three moles of $H_2SO_4$.

**Solution**

$$2Al(OH)_3 + 3H_2SO_4 \longrightarrow Al_2(SO_4)_3 + 6H_2O$$

---

**Writing a Partial Neutralization Reaction**

**Example 12-7**

Write the balanced molecular, total ionic, and net ionic equations illustrating the reaction of 1 mol of $H_3PO_4$ with 1 mol of $Ca(OH)_2$.

**Procedure**

Although 1 mol of $H_3PO_4$ has three available $H^+$ ions to neutralize, 1 mol of $Ca(OH)_2$ can react with only two of them. This would leave the $HPO_4^{2-}$ ion in solution.

**Solution**

*Molecular:*  $H_3PO_4(aq) + Ca(OH)_2(aq) \longrightarrow CaHPO_4(aq) + 2H_2O(l)$

*Total ionic:*  $H_3PO_4(aq) + Ca^{2+}(aq) + 2OH^-(aq) \longrightarrow$
$$Ca^{2+}(aq) + HPO_4^{2-}(aq) + 2H_2O(l)$$

*Net ionic:*  $H_3PO_4(aq) + 2OH^- \longrightarrow HPO_4^{2-}(aq) + 2H_2O(l)$

**Looking Back!** Acids and bases are two ancient classes of compounds that are classified as such because of (a) some common properties and (b) the H$^+$ or OH$^-$ ions that they produce in aqueous solution. In a broader concept of acids and bases, which is known as the Brønsted–Lowry definition, the role of water in the ionization process becomes apparent. Weak acids and bases produce a limited amount of H$^+$ or OH$^-$ ions, so their acid or base properties are more modest. When acids and bases are mixed together, they form a third class of compounds called salts (or, in some cases, acid salts) and water. The formation of water is the driving force for the neutralization reaction between acids and bases.

## Learning Check A | checking it out

**A-1.** Fill in the blanks.

An acid is a compound that produces the _____ ion in solution, which is also written as the hydronium ion ( _____ ). A base is a compound that produces an ion with the formula _____ and the name _____ ion. In the Brønsted–Lowry definition, acids are _____ donors and bases are _____ acceptors. A conjugate acid of a compound or ion results from the _____ of an _____ ion. A substance that has both a conjugate acid and a conjugate base is said to be _____ . Strong acids are essentially _____ % ionized in aqueous solution, whereas weak acids are _____ ionized. Sodium hydroxide is a strong base, but ammonia is a _____ _____ . The reaction between an acid and a base is known as a _____ reaction. The net ionic equation of this reaction always has _____ as a product. The spectator ions of the reaction form a _____ . The partial neutralization of polyprotic acids produces an _____ _____ and water.

**A-2.** Write the formula and name of the acid or base formed from the following ions.

**a.** $ClO_4^-$    **b.** $Fe^{2+}$    **c.** $S^{2-}$    **d.** $Li^+$

**A-3.** Write the conjugate acid and the conjugate base of the $HC_2O_4^-$ ion.

**A-4.** Write the reaction of the $HC_2O_4^-$ ion with **(a)** water acting as a base and **(b)** water acting as an acid.

**A-5.** A certain 0.10 M solution of an acid is 2.50% ionized. Is this a strong or a weak acid? What is the $[H_3O^+]$ in this solution?

**A-6.** Write the balanced molecular, total ionic, and net ionic equations illustrating the neutralization of $HNO_3$ with $Sr(OH)_2$.

**A-7.** Write the balanced molecular, total ionic, and net ionic equations illustrating the reaction of 1 mol of $H_2C_2O_4$ (oxalic acid) and 1 mol of CsOH.

*Additional Examples: Exercises 12-1, 12-3, 12-9, 12-20, 12-24, 12-33, 12-35, and 12-41.*

## Section B   THE MEASUREMENT OF ACID STRENGTH

## 12-5 Equilibrium of Water

**Looking Ahead!** The actual acidity of an aqueous solution is measured by the concentration of H$_3$O$^+$ ions. This depends on the original concentration and the strength of the acid present. We need a convenient way to express the acidity of a solution, but to do this we need to take a closer look at water itself. As we will see in this section, there is more going on in pure water than we originally indicated.

In the previous chapter, pure water was classified as a nonconductor of electricity. This implied that no ions were present. This isn't entirely true, however. With more sensitive instruments, we find that there actually is a very small concentration of ions in pure water. Since there is no solute, the ions must come from the water itself. The presence of ions in water is due to a process known as autoionization. **Autoionization** *produces positive and negative ions from the dissociation of the molecules of the liquid.* For water this is represented as follows, the double arrow again indicating that the reaction is reversible and reaches a state of equilibrium.

$$H_2O + H_2O \rightleftharpoons H_3O^+ + OH^-$$

Although this equilibrium lies very far to the left, there is a small but important amount of $H_3O^+$ ions and $OH^-$ ions coexisting in pure water. It is this small concentration that we will focus on as a means of expressing acid or base behavior and their relative strengths.

The concentration of each ion at 25°C has been found by experiment to be $1.0 \times 10^{-7}$ mol/L. This means that only about one out of every 500 million water molecules is actually ionized at any one time. Other experimental results tell us that the product of the ion concentrations is a constant. This phenomenon will be explained in more detail in Chapter 14, but for now we accept it as fact. Therefore, at 25°C

$$[H_3O^+][OH^-] = K_w \quad \text{(a constant)}$$

Substituting the actual concentrations of the ions, we can now find the numerical value of the constant.

$$[1.0 \times 10^{-7}][1.0 \times 10^{-7}] = 1.0 \times 10^{-14}$$

$K_w(1.0 \times 10^{-14})$ *is known as the* **ion product** *of water* at 25°C. The importance of this constant is that it tells us the concentrations of $H_3O^+$ and $OH^-$ not only in pure water but also in acidic and basic solutions. The following example illustrates this relationship.

---

**Calculation of [OH⁻] from [H₃O⁺]**

**Example 12-8**

**working it out**

In a certain solution, $[H_3O^+] = 1.5 \times 10^{-2}$ M. What is $[OH^-]$ in this solution?

**Procedure**

Use the relationship for $K_w$, $[H_3O][OH^-] = 1.0 \times 10^{-14}$, and solve for $[OH^-]$.

**Solution**

$$[H_3O^+][OH^-] = 1.0 \times 10^{-14}$$

$$[OH^-] = \frac{1.0 \times 10^{-14}}{[H_3O^+]} = \frac{1.0 \times 10^{-14}}{1.5 \times 10^{-2}}$$

$$= \underline{\underline{6.7 \times 10^{-13} \text{ M}}}$$

---

In Figure 12-4, the ion product of water is illustrated. Notice that there is an inverse relationship between the two ion concentrations. That is, the larger the concentration of $H_3O^+$, the smaller the concentration of $OH^-$. In pure water or a neutral solution, the concentrations are both equal to $10^{-7}$ M. If an acid is added to pure water, the balance is tipped toward the $H_3O^+$ side and the solution becomes acidic to some degree. This means that the concentration of $H_3O^+$ rises above $10^{-7}$ M, while the concentration of $OH^-$ drops below $10^{-7}$ M.

**Figure 12-4 The Relationship Between H₃O⁺ and OH⁻ in Water** A large concentration of H₃O⁺ corresponds to a low concentration of OH⁻ in a solution, and vice versa.

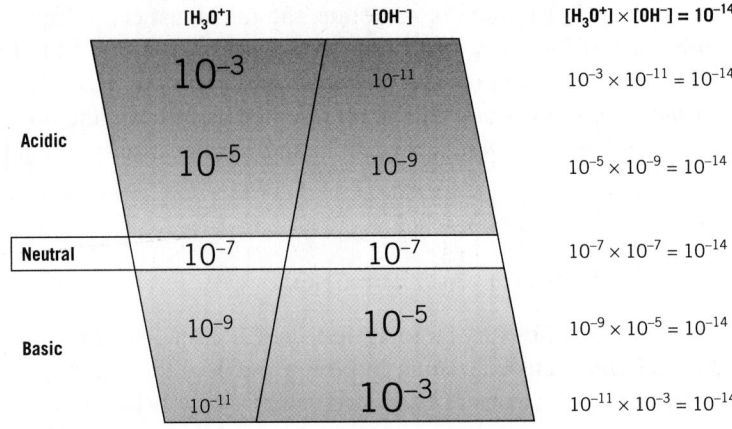

| [H₃O⁺] | [OH⁻] | $[H_3O^+] \times [OH^-] = 10^{-14}$ |
|---|---|---|
| $10^{-3}$ | $10^{-11}$ | $10^{-3} \times 10^{-11} = 10^{-14}$ |
| $10^{-5}$ | $10^{-9}$ | $10^{-5} \times 10^{-9} = 10^{-14}$ |
| $10^{-7}$ | $10^{-7}$ | $10^{-7} \times 10^{-7} = 10^{-14}$ |
| $10^{-9}$ | $10^{-5}$ | $10^{-9} \times 10^{-5} = 10^{-14}$ |
| $10^{-11}$ | $10^{-3}$ | $10^{-11} \times 10^{-3} = 10^{-14}$ |

In summary, an acidic, basic, or neutral solution can now be defined in terms of concentrations of ions.

*Neutral:* $[H_3O^+] = [OH^-] = 1.0 \times 10^{-7}$

*Acidic:* $[H_3O^+] > 1.0 \times 10^{-7}$ and $[OH^-] < 1.0 \times 10^{-7}$

*Basic:* $[H_3O^+] < 1.0 \times 10^{-7}$ and $[OH^-] > 1.0 \times 10^{-7}$

We should now incorporate this information into our understanding of acids and bases. A modern version of the Arrhenius definition of an acid is a substance that produces $H_3O^+$ ions in aqueous solution. But now we see that $H_3O^+$ is present in neutral and even basic solutions as well. A slight modification of the definition solves this problem. *An acid is any substance that increases [H₃O⁺] in water, and a base is any substance that increases [OH⁻] in water.* With our new understanding of the equilibrium, we can see that a substance can be an acid by directly donating $H^+(aq)$ to the solution (e.g., HCl, H₂S), or a substance can be an acid by reacting with OH⁻ ions, thus removing them from the solution.

## 12-6 The pH Scale

**Looking Ahead!** The concentration of hydronium or hydroxide ions in aqueous solution is usually quite small. While scientific notation is a great help in expressing these very small numbers, it is still awkward. There is another way. This involves expressing the numbers as logarithms, which then gives us a three- or four-digit number that tells us the same thing. The expression of these numbers in this manner is discussed in this section.

The producers of commercial television advertising assume that the general population is aware not only of the importance of acidity but also of how it is scientifically expressed. We often hear references to controlled pH in hair shampoo commercials. (See Figure 12-5.) In fact, pH is an important and convenient method for expressing $[H_3O^+]$ in aqueous solution. For example, in one acidic solution $[H_3O^+]$ is equal to $10^{-5}$ M. Scientific notation is certainly better than using a string of nonsignificant zeros (i.e., 0.00001 M), but expressed as pH, the number is simply 5.0. The pH scale represents the negative exponent of 10 as a positive number. The exponent of 10 in a number is the number's common logarithm or, simply, log. **pH** is a logarithmic expression of $[H_3O^+]$.

$$pH = -\log[H_3O^+]$$

Therefore, a solution of pH = 1.00 has $[H_3O^+]$ equal to $1.0 \times 10^{-1}$ M, and pure water has pH = 7.00 ($[H_3O^+] = 1.0 \times 10^{-7}$ M). In expressing pH, the number to the right of the decimal place should have the same number of significant figures as the original number. That is,

$$\text{if } [H_3O^+] = \underline{1.00} \times 10^{-4} \text{ M}, \quad pH = 4.\underline{000}$$

$$\uparrow \qquad\qquad\qquad \uparrow$$

3 significant figures   3 places to the right
of the decimal

A much less popular but valid way of expressing $[OH^-]$ is **pOH.**

$$pOH = -\log [OH^-]$$

A simple relationship between pH and pOH can be derived from the ion product of water.

$$[H_3O^+][OH^-] = 1.0 \times 10^{-14}$$

If we now take $-\log$ of both sides of the equation, we have

$$-\log[H_3O^+][OH^-] = -\log(1.0 \times 10^{-14})$$

Since $\log(A \times B) = \log A + \log B$, the equation can be written as

$$-\log[H_3O^+] - \log[OH^-] = -\log 1.0 - \log 10^{-14}$$

Since $\log 1.0 = 0.00$ and $\log 10^{-14} = -14$, the equation is

$$pH + pOH = 14.00$$

Generally, pOH is not used extensively since pH relates to the $OH^-$ concentration as well as the $H_3O^+$ concentration. However, the relationships among $[H_3O^+]$, $[OH^-]$, pH, and pOH can be summarized as follows.

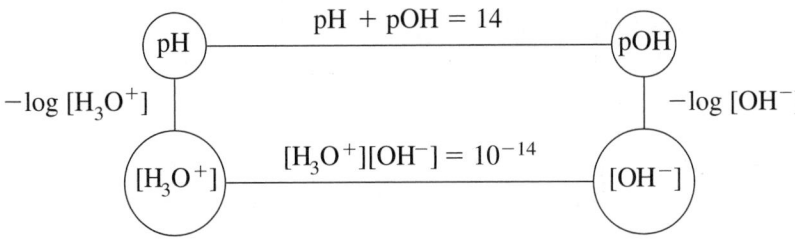

Although most of us are tempted to go straight to our calculators to change from scientific notation to logarithms, it is helpful to review the *meaning* of common logs and some of the rules of their use. You are encouraged to read Appendix C, which contains a brief discussion of common logarithms.

Figure 12-5   pH and Commercial Products **This shampoo is supposedly desirable because it is "pH balanced."**

---

**Expressing the pH and pOH of a Solution**

What is the pH of a solution with $[H_3O^+] = 1.0 \times 10^{-8}$ M? What is the pOH?

**Example 12-9**

working it out

**Procedure**

A calculator is not needed for calculations of pH where the coefficient is exactly 1. Understanding the definition is sufficient. The pH is simply the negative of the exponent of 10.

**Solution**

$$[H_3O^+] = 1.0 \times 10^{-8} \text{ M}$$
$$pH = -\log (1.0 \times 10^{-8}) = \underline{8.00}$$

*(Since there are two significant figures, there should be two places to the right of the decimal.)*

$$pH + pOH = 14.00$$
$$pOH = 14.00 - pH$$
$$pOH = 14.00 - 8.00 = \underline{\underline{6.00}}$$

**Example 12-10** | **Expressing the $[H_3O^+]$ from the pH**

What is the $[H_3O^+]$ of a solution with a pH = 3.00? What is the pOH? What is $[OH^-]$?

**Procedure**

Again, a calculator is not needed for calculations if the pH is a whole number. The $[H_3O^+]$ is simply the antilog (or inverse log) of the pH.

**Solution**

$$pH = 3.00$$
$$3.00 = -\log[H_3O^+]$$
$$-3.00 = \log[H_3O^+]$$

This means that the exponent of 10 is −3 and the coefficient should be expressed to two significant figures.

$$[H_3O^+] = \underline{\underline{1.00 \times 10^{-3}}}$$
$$pOH = 14.00 - 3.00 = \underline{\underline{11.00}}$$
$$[OH^-] = -\text{antilog}[11.00] = \underline{\underline{1.00 \times 10^{-11}}}$$

**pH is read directly with this common laboratory instrument.**

**Example 12-11** | **Expressing the pH and pOH of a Strong Acid Solution**

What is the pH of a $1.5 \times 10^{-2}$ M solution of $HClO_4$? What is the pOH of this solution?

**Procedure**

$HClO_4$ is a strong acid, which means that it is 100% ionized in solution. Therefore, $[H_3O^+]$ is equal to the original $HClO_4$ concentration. In cases where the coefficient is not exactly 1, a calculator is needed.

1. Enter the number. First enter "1.5," then push the exponent key and enter "2". Push the $\boxed{+/-}$ key to change the exponent to −2.
2. Push the $\boxed{\log}$ key.
3. The display reads −1.82, which is the log of the number you entered.
4. Change the reading to 1.82 since the pH is the negative of the log.

**Solution**

$$[H_3O^+] = 1.5 \times 10^{-2} \, M$$
$$\log(1.5 \times 10^{-2}) = -1.82$$
$$pH = -\log(1.5 \times 10^{-2}) = \underline{\underline{1.82}}$$
$$pH + pOH = 14.00$$
$$pOH = 14.00 - 1.82 = \underline{\underline{12.18}}$$

| Conversion of pH to $[H_3O^+]$ | Example 12-12 |

In a given weakly basic solution, pH = 9.62. What is $[H_3O^+]$?

**Procedure**

In this case, it is necessary to take the antilog (or the inverse log on a calculator).

1. Enter the 9.62 on the calculator.
2. Change the sign ($\boxed{+/-}$ key).
3. Press the $\boxed{inv}$, $\boxed{shift}$, or $\boxed{2^{nd}}$ key, then the $\boxed{log}$ key. Consult the instructions for your calculator if these keys are different or not available.
4. The number displayed should be rounded off to two significant figures since there are two numbers to the right of the decimal place in the original number.

**Solution**

$$-\log [H_3O^+] = 9.62 \quad \log [H_3O^+] = -9.62$$
$$[H_3O^+] = \underline{2.4 \times 10^{-10} M}$$

In Figure 12-6 we have included the pH and pOH in addition to the ion concentrations shown in Figure 12-4. Acidic, basic, and neutral solutions can now be defined in terms of pH and pOH.

*Neutral:* pH = pOH = 7.00

*Acidic:* pH < 7.00    and    pOH > 7.00

*Basic:* pH > 7.00    and    pOH < 7.00

In the use of pH, one must remember that a *low* value for pH corresponds to a *high* concentration of $H_3O^+$. Also, *a change in one unit in the pH (e.g., from 4 to 3) corresponds to a tenfold change in concentration (e.g., from $10^{-4}$ to $10^{-3}$ M)*. Another scientific scale that is logarithmic is the Richter scale for measuring earthquakes. This scale measures the amplitude of seismic waves set off by the tremor. An earthquake with a reading of 7.0 on this scale indicates the waves are 10 times larger than an earthquake of magnitude 6.0. In astronomy, the brightness of stars is also measured in units of magnitude where each unit represents a tenfold increase in brightness. In science, a change of one magnitude generally indicates a tenfold change.

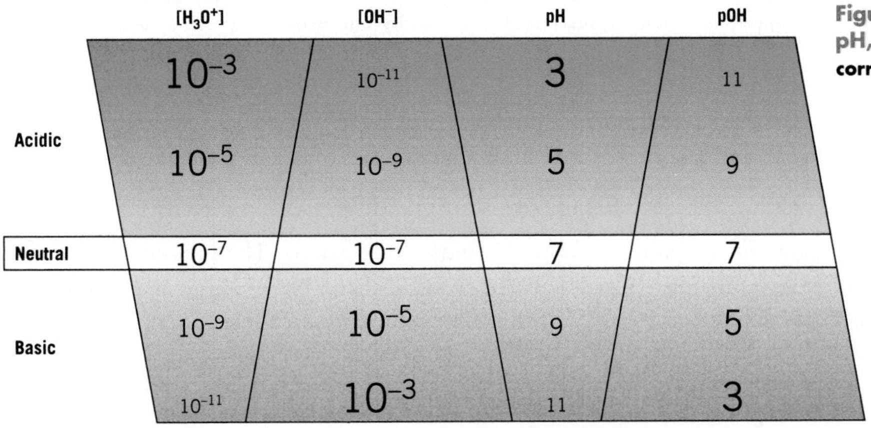

**Figure 12-6 Concentrations of Ions, pH, and pOH A low pH in a solution corresponds to a high pOH.**

| | $[H_3O^+]$ | $[OH^-]$ | pH | pOH |
|---|---|---|---|---|
| Acidic | $10^{-3}$ | $10^{-11}$ | 3 | 11 |
| | $10^{-5}$ | $10^{-9}$ | 5 | 9 |
| Neutral | $10^{-7}$ | $10^{-7}$ | 7 | 7 |
| Basic | $10^{-9}$ | $10^{-5}$ | 9 | 5 |
| | $10^{-11}$ | $10^{-3}$ | 11 | 3 |

**Table 12-1 The pH Scale**

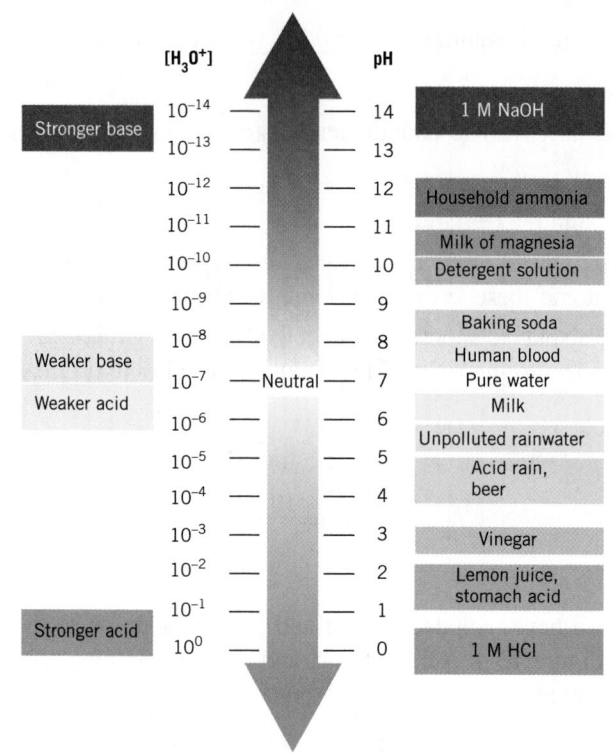

In Table 12-1, we have listed the pH of some common chemicals, foods, or products. A 0.10 M HCl solution (pH = 1) is the most acidic solution listed. Solutions of pH 1 or less are considered *strongly acidic*. Solutions with pH less than 7 but greater than 1 are *weakly acidic*. Even then what we refer to as a weakly acidic solution still covers quite a range. A solution of pH 6 is 100,000 times less acidic than a solution of pH 1. On the other side of the scale, we have a 0.10 M solution of KOH, which has a pH of 13. Solutions of pH 13 or greater are considered *strongly basic*. Solutions greater than 7 but less than 13 are *weakly basic*.

---

**Looking Back!** The situation in water is somewhat more complex then we originally implied. Since water autoionizes, there is an equilibrium concentration of both the active ingredient of acids $[H_3O^+]$ and that of bases $[OH^-]$ in the same solution. Whether a solution is acidic, neutral, or basic depends on which ion predominates. One convenient thing about this, however, is that since both ions are present, we can focus on just one to express either acidity or basicity. This fact is even more apparent in the expression of both acidity and basicity by use of one unit, pH.

---

## Learning Check B | checking it out

**B-1.** Fill in the blanks.

Pure water contains a small but important concentration of _____ and _____ ions. The product of the concentrations of these two ions, known as the _____ _____ of water, is symbolized by _____ and has a value of _____ at 25°C. The pH of a solution is defined as _____ . An acidic solution is one that has an $[H_3O^+]$ greater than _____ and a pH _____ than 7. A basic solution is one that has an $[OH^-]$ greater than _____ and a pH _____ than 7.

**B-2.** What is the $[H_3O^+]$ in a solution that has $[OH^-] = 7.2 \times 10^{-5}$? Is the solution acidic or basic? Is it strongly or weakly acidic or basic?

**B-3.** What is the pH of the solution in problem B-2? What is the pOH?

*Additional Examples: Exercises 12-51, 12-53, 12-57, 12-59, 12-63, and 12-74.*

Section **SALTS AND OXIDES AS ACIDS AND BASES**

## 12-7 The Effect of Salts on pH—Hydrolysis

**Looking Ahead!** Solutions of two cleaning solutions both have a slippery feeling typical of weak bases. They are ammonia and bleach. We have discussed how ammonia acts as a base, but the active ingredient of bleach solutions is a salt, sodium hypochlorite. How salts exhibit acidic or basic characteristics is the subject of this section.

When a salt dissolves in water, both a cation and an anion are produced. The cation has the potential to act as an acid and the anion as a base. *The reaction of a cation as an acid or an anion as a base is known as a* **hydrolysis reaction.** To understand this, we will first consider ions that do not undergo hydrolysis, which means that they have no acid or base character in water and will not change the pH of pure water from 7.0. First, let's consider the $Cl^-$ ion, which is the anion or conjugate base of the strong acid HCl. Recall that in the ionization process, the forward reaction is essentially 100% complete so the reverse reaction does not occur in water. Thus we notice that the $Cl^-$ ion, although formally known as a conjugate base, in fact does not exhibit proton-accepting ability in water to form an HCl molecule. *Therefore, the $Cl^-$ ion does not undergo hydrolysis. In fact, other anions of strong acids (i.e., $NO_3^-$, $ClO_4^-$, $Br^-$, and $I^-$) do not undergo hydrolysis reactions.* The $HSO_4^-$ ion is an exception and is weakly acidic in water, as we indicated previously. The same is true for the cations of strong bases (e.g., $Na^+$, $K^+$, $Ca^{2+}$, and $Mg^{2+}$ and other alkali metal and alkaline earth metal ions). They all exist as independent, hydrated ions in aqueous solution and so do not undergo hydrolysis. Now let's consider the ionization equilibrium for the weak acid HClO.

$$HClO(aq) + H_2O(l) \rightleftharpoons H_3O^+(aq) + ClO^-(aq)$$

Recall that the reverse reaction in this case predominates, indicating that the $ClO^-$ ion demonstrates proton-accepting ability in water. In fact, when this ion is added to water as part of a salt, the following equilibrium occurs to a limited extent.

$$ClO^-(aq) + H_2O(l) \rightleftharpoons HClO(aq) + OH^-(aq)$$

The amount of hydrolysis of anions is generally quite small, however, so the equilibrium lies far to the left. We can now make a general statement. *The anions of weak acids (i.e., their conjugate bases) undergo hydrolysis in aqueous solution, producing a weakly basic solution.* We can make another generalization: the weaker the acid, the stronger is its conjugate base. In fact, the anions (conjugate bases) of strong acids are so weak they don't act as bases at all in water. The anions (conjugate bases) of weak acids, however, are strong enough to exhibit proton-accepting ability in water.

Solutions of ordinary bleach (NaClO) are weakly basic.

Now let's consider the weak base ammonia. As indicated in section 12-3, its reaction with water produces a limited amount of $NH_4^+$ and $OH^-$ ions. This again indicates that the reverse reaction predominates and that the $NH_4^+$ ion demonstrates proton-donating properties in water. Thus, *cations (conjugate acids) of weak bases undergo hydrolysis in aqueous solution, producing a weakly acidic solution.* This is illustrated for the $NH_4^+$ ion as follows.

$$NH_4^+(aq) + H_2O(l) \rightleftharpoons H_3O^+(aq) + NH_3(aq)$$

We are now ready to predict how the solutions of various salts in water affect the pH. To do this, we must look at the cation and the anion individually and decide whether neither, one, or both undergo hydrolysis. We will consider four representative cases.

1. *Neutral solutions of salts.* When neither the cation nor the anion undergoes a hydrolysis reaction, solutions of these salts do not affect the pH. They remain the same as that of pure water, which is 7.0. In such salts, the anions are those of the strong acids and the cations are those of the strong bases [i.e., the alkali and alkaline earth metals (except $Be^{2+}$)]. Examples of neutral salts are NaCl, $Ba(ClO_4)_2$, and $CsNO_3$.

2. *Basic solutions of salts.* When the salt forms from the cation of a strong base and the anion of a weak acid, the cation does not affect the pH but the anion does. In the anion hydrolysis reaction, a small equilibrium concentration of $OH^-$ makes the solution basic. Examples of basic salts are NaOCl, $Ca(NO_2)_2$, and NaF. The pH of solutions of these salts will be greater than 7.0.

3. *Acidic solutions of salts.* When a salt forms from the cation of a weak base and the anion of a strong acid, the anion does not affect the pH but the cation does. The primary example of a cation that undergoes hydrolysis is $NH_4^+$. Since the anions do not undergo hydrolysis, solutions of salts such as $NH_4Br$ and $NH_4NO_3$ are acidic. The pH of such solutions will be less than 7.0.

4. *Complex cases.* There are other salt solutions that are not easy to predict without quantitative data. For example, in a solution of the salt $NH_4NO_2$ both anion and cation undergo hydrolysis, but not to the same extent. In other cases, it is not immediately obvious whether solutions of acid salts are acidic or basic. For instance, the $HCO_3^-$ ion is potentially an acid since it has a hydrogen that can react with a base. However, it is also the conjugate base of the weak acid $H_2CO_3$, so it can act as a base by undergoing hydrolysis. Sodium bicarbonate solutions are actually slightly basic and for that reason are sometimes used as an "antacid" to treat upset stomachs. On the other hand, solutions of the acid salts $NaHSO_3$ and $NaHSO_4$ are acidic. Quantitative information, which is discussed in Chapter 14, would be needed, however, for us to have predicted these facts.

The discussion of the effect on pH when salts dissolve in water is summarized as follows.

| Origin of Cation | Origin of Anion | pH of Solution | Examples |
|---|---|---|---|
| strong acid | strong base | 7.0 | KI, $Ca(NO_3)_2$ |
| strong acid | weak base | $<7.0$ | $KNO_2$, BaS |
| weak acid | strong base | $>7.0$ | $NH_4ClO_4$, $NH_4I$ |
| weak acid | weak base | more information needed | $NH_4ClO$, $N_2H_5NO_2$ |
| strong acid | acid salt | more information needed | $KH_2PO_4$, $Ca(HSO_3)_2$ |

Example 12-13

working it out

**Predicting the Acidity of Salt Solutions**

Indicate whether the following solutions are acidic, basic, or neutral. If the solution is acidic or basic, write the equation illustrating the appropriate reaction.

a. KCN

b. $Ca(NO_3)_2$

c. $(CH_3)_2NH_2^+Br^-$ [$(CH_3)_2NH$ is a weak base like $NH_3$.]

**Solution**

a. KCN: $K^+$ is the cation of the strong base KOH and does not hydrolyze. The $CN^-$ ion, however, is the conjugate base of the weak acid HCN and hydrolyzes as follows.

$$CN^- + H_2O \rightleftharpoons HCN + OH^-$$

Since $OH^-$ is formed in this solution, the solution is <u>basic</u>.

b. $Ca(NO_3)_2$: $Ca^{2+}$ is the cation of the strong base $Ca(OH)_2$ and does not hydrolyze. $NO_3^-$ is the conjugate base of the strong acid $HNO_3$ and also does not hydrolyze. Since neither ion hydrolyzes, the solution is <u>neutral</u>.

c. $(CH_3)_2NH_2^+Br^-$: $(CH_3)_2NH_2^+$ is the conjugate acid of the weak base $(CH_3)_2NH$. It undergoes hydrolysis according to the equation

$$(CH_3)_2NH_2^+ + H_2O \rightleftharpoons (CH_3)_2NH + H_3O^+$$

The $Br^-$ ion is the conjugate base of the strong acid HBr and does not hydrolyze. Since only the cation undergoes hydrolysis, the solution is <u>acidic</u>.

# 12-8 Control of pH—Buffer Solutions

**Looking Ahead!** Perhaps the most important chemical system involves the blood coursing through our veins. Our blood has a pH of about 7.4, but a variation of as little as 0.2 pH unit causes coma or even death. How is the pH of the blood so rigidly controlled despite all of the acidic and basic substances we ingest? The control of pH is the job of buffers, which are discussed next.

The word "buffer" is usually used in the context of "absorbing a shock." For example, a car bumper serves as a buffer for the passengers by absorbing the energy of an impact. A buffer solution has a similar effect on the pH of a solution. *A **buffer solution** resists changes in pH caused by the addition of limited amounts of a strong acid or a strong base.* Commercial products tell us that buffered solutions are important in such items as hair shampoo and aspirin. But to us it is most important because many chemical reactions in our bodies, including those that give us life's energy, must take place in a controlled pH environment.

A buffer works by having a substance in solution that is available to react with any added $H_3O^+$ or $OH^-$. The most logical candidates for this duty are weak bases and weak acids, respectively. Consider how a hypothetical monoprotic weak acid (HX) reacts with added $OH^-$.

$$HX(aq) + OH^-(aq) \longrightarrow X^-(aq) + H_2O(l)$$

In the previous section we found that the conjugate base of the weak acid ($X^-$) can act as a base in water. In fact, it reacts with $H_3O^+$ as follows

$$X^-(aq) + H_3O^+(aq) \longrightarrow HX(aq) + H_2O(l)$$

The buffer in this product controls the pH.

Both of these reactions are essentially complete. So, if we have a solution that contains a significant concentration of both the weak acid and its conjugate weak base, we have a solution that reacts with either added $H_3O^+$ or $OH^-$. One may think that just the weak acid alone would be a buffer because the ionization produces some of its anion. However, the amount from ionization is very small so we need additional anion concentration from another source such as a salt.

*Buffer solutions are made from (a) a solution of a weak acid that also contains a salt of its conjugate base or (b) a solution of a weak base that also contains a salt of its conjugate acid.*

A typical buffer solution would occur in a solution of HCN that also contains its salt, NaCN. Since HCN is a weak acid, it ionizes to a limited extent to form small concentrations of two ions. The NaCN is soluble and a strong electrolyte, so it is completely dissociated into ions in solution. These two reactions are as follows.

$$HCN(aq) + H_2O(l) \rightleftharpoons H_3O^+(aq) + CN^-(aq)$$
$$NaCN(s) \longrightarrow Na^+(aq) + CN^-(aq)$$

The solution of these two compounds together produces three ions, $H_3O^+$, $Na^+$, and $CN^-$, and one molecular compound, HCN. The $Na^+$ is the cation of a strong base. Thus it is a spectator ion and does not affect the pH. It will not be considered further. Our main focus then is on the comparatively large concentration of nonionized HCN and $CN^-$ along with the small concentration of $H_3O^+$. This is illustrated as

$$HCN_{(aq)} + H_2O(l) \rightleftharpoons H_3O^+(aq) + CN^-_{(aq)}$$

large concentration

small concentration due to ionization

large concentration due to NaCN, with a small concentration due to ionization

The job of the HCN molecule and the $CN^-$ ion in this solution is to protect the small concentration of $H_3O^+$. The shock imposed on this fragile system comes from the addition of limited amounts of $H_3O^+$ from a strong acid or $OH^-$ from a strong base. When $OH^-$ is added, the HCN absorbs the shock by reacting with that ion as follows.

$$HCN(aq) + OH^-(aq) \longrightarrow H_2O(l) + CN^-(aq)$$

When $H_3O^+$ is added, the $CN^-$ absorbs the shock by reacting with that ion as follows.

$$H_3O^+(aq) + CN^-(aq) \longrightarrow HCN(aq) + H_2O(l)$$

Thus addition of a limited amount of a strong acid or base is counteracted by a species present in the buffer solution, and the pH changes very little. (See Figure 12-7.)

A strong acid cannot act as a buffer because it is not present in solution in molecular form. That is, it is completely ionized in water. Therefore, there is no reservoir of molecules that can react with added $OH^-$ ions, nor can the conjugate base react with $H_3O^+$ ions.

**Figure 12-7  A Buffer Solution**
**A buffer can react with added $H_3O^+$ or $OH^-$.**

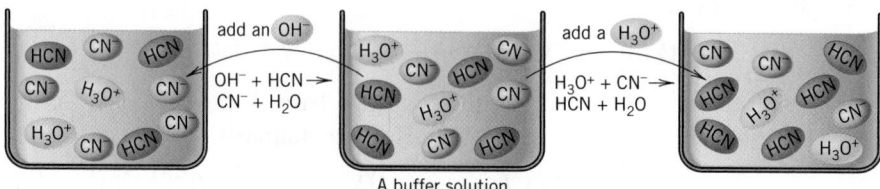

A buffer solution

## The pH Balance in the Blood

The chemistry of the blood is a science unto itself. It is our river of life, delivering fuel and oxygen for the muscles, removing waste products, and keeping our body temperature steady. The pH of the blood is 7.40, just barely alkaline. As mentioned in an earlier Making It Real, enzymes run our body but they function only in a very narrow pH range. Deviations of only 0.2 pH unit either way can cause coma or even be fatal. If the pH drops below 7.20, a condition known as *acidosis* results. If the pH rises to above 7.60, the opposite condition is called *alkalosis*. Either situation seriously upsets our body chemistry.

Perhaps one of the most amazing properties of blood is its ability to maintain a nearly constant pH despite the fact that acids and bases find their way into the bloodstream. The pH of our blood is held constant by two buffering systems. We will look closely at the more important system, which is the carbonic acid–bicarbonate buffer. This system is represented by the equation

$$H_2CO_3(aq) + H_2O \rightleftharpoons HCO_3^-(aq) + H_3O^+(aq)$$

The carbonic acid ($H_2CO_3$) part of the buffer reacts with any extra $OH^-$ ion that enters our blood, and the $HCO_3^-$ ion reacts with any added $H_3O^+$. For example, exercise produces extra $H_3O^+$ from the lactic acid formed in the muscles. The $H_3O^+$ reacts with $HCO_3^-$ to form carbonic acid. This is transported to the lungs where it decomposes to $CO_2$ gas and $H_2O$, which is then exhaled. Our bodies respond to exercise by faster breathing and faster release of carbon dioxide.

If extra $OH^-$ enters the blood, it reacts with $H_2CO_3$ and more bicarbonate is formed, which is eventually excreted by the kidneys. Sometimes when people get nervous they breathe too fast, which is called *hyperventilation*. This causes too much carbon dioxide to be exhaled and the pH of the blood rises above 7.4. If the person faints, the breathing rate decreases and the pH quickly comes back into balance. Actually, the concentration of bicarbonate is about 10 times the concentration of carbonic acid, so we are more protected against extra acid than extra base, which works out well.

Health problems such as diabetes and kidney disease may disturb the buffers of the blood. Generally, however, we couldn't have a better system to maintain pH and with it the smooth functioning of our body.

*The pH of the blood is an important measurement.*

Consider the following analogy. One person has $20 in his pocket with no savings in the bank; another person has $20 in her pocket and $100 in the bank. A $10 expense will change the first person's pocket money drastically. The second person is "buffered" from this expense and can cover it with the bank savings; this person can then maintain the same amount of pocket money. The nonionized HCN is like "money in the bank." It is available to react with added $OH^-$, leaving the concentration of $H_3O^+$ essentially unchanged. The person who keeps all of his money in his pocket is analogous to a strong acid solution in which there is no nonionized acid in reserve. In this case, since all of the acid ionizes to $H_3O^+$, any added $OH^-$ decreases the $H_3O^+$ concentration and thus increases the pH.

There is a limit to how much a buffer system can resist change. If the added amount of $OH^-$ exceeds the reserve of HCN (referred to as the *buffer capacity*), then the pH will rise. This is analogous to a $110 expense for the person with the savings. It is more than she can cover with bank savings, so the amount in her pocket decreases.

Solutions of weak bases and salts containing their conjugate acids (e.g., $NH_3$ and $NH_4Cl$) also serve as buffers. The two relevant reactions in this case are

*Added acid:* $NH_3 + H_3O^+ \longrightarrow NH_4^+ + H_2O$

*Added base:* $NH_4^+ + OH^- \longrightarrow H_2O + NH_3$

The pH of the buffer solution depends on the strength of the acid or base chosen. In Chapter 14, actual pH values of buffer solutions are calculated from quantitative information related to acid strength.

We began this section by mentioning the importance of the buffer system of our blood. The carbonic acid–bicarbonate buffer system is our main protection and is discussed in Making It Real. However, another buffer system is at work in the cells of our bodies. This is the $H_2PO_4^-$–$HPO_4^{2-}$ buffer system. In this system the dihydrogen phosphate ion ($H_2PO_4^-$) acts as the acid species and the hydrogen phosphate ion ($HPO_4^{2-}$) acts as the base species. These two ions can remove $H_3O^+$ or $OH^-$ ions produced in the cells as follows.

$$\textit{Added acid:} \quad HPO_4^{2-} + H_3O^+ \longrightarrow H_2PO_4^- + H_2O$$
$$\textit{Added base:} \quad H_2PO_4^- + OH^- \longrightarrow HPO_4^{2-} + H_2O$$

## 12-9 Oxides as Acids and Bases

**Looking Ahead!** We have one more important question concerning acids. How do they get into the atmosphere so as to make rain acidic? The question of acid rain is a matter of international consequence.

One of the more important consequences of the human race's progress is acid rain. Tall industrial smokestacks disperse exhaust gases high into the atmosphere, where prevailing winds may carry pollutants hundreds of miles before they return to the Earth in the form of acid rain. Probably no other issue is currently more touchy between the United States and Canada and among countries of northern Europe than control of acid rain. Its effect on lakes and forests can be devastating, and there is little doubt that the problem must be faced and solved.

Most acid rain originates from the combustion of coal or other fossil fuels that contain sulfur as an impurity. Combustion of sulfur or sulfur compounds produces sulfur dioxide ($SO_2$). In the atmosphere, sulfur dioxide reacts with oxygen to form sulfur trioxide ($SO_3$).

$$2SO_2(g) + O_2(g) \longrightarrow 2SO_3(g)$$

The sulfur trioxide reacts with water in the atmosphere according to the equation

$$SO_3(g) + H_2O(l) \longrightarrow H_2SO_4(aq)$$

The product is a sulfuric acid solution, which is one of the strong acids—corrosive and destructive. As mentioned earlier, acids react with metals and limestone, which are both used externally in buildings. (See Figure 12-8.) In the above reaction, sulfur trioxide can be considered as simply the dehydrated form of sulfuric acid. It is thus known as an **acid anhydride,** *which means acid without water.* Many nonmetal oxides are acid anhydrides. When dissolved in water, they form acids. Three other reactions of nonmetal oxides to form acids follow.

$$CO_2(g) + H_2O(l) \rightleftharpoons H_2CO_3(aq)$$
carbonic acid
$$SO_2(g) + H_2O(l) \rightleftharpoons H_2SO_3(aq)$$
sulfurous acid
$$N_2O_5(l) + H_2O(l) \longrightarrow 2HNO_3(aq)$$
nitric acid

**Figure 12-8  Statue of George Washington in 1935 (above) and 1994 (below).**
**The deterioration is due to acid rain.**

We have previously mentioned that carbon dioxide, the acid anhydride of carbonic acid, is responsible for the fizz and the tangy taste in carbonated soft drinks and beer. When all of the carbon dioxide escapes, the beverage goes flat. Carbon dioxide is also present in the atmosphere and dissolves in rainwater to make rain naturally acidic. Carbon dioxide by itself lowers the pH to about 5.7. The presence of oxides of sulfur

### Acid Rain—The Price of Progress?

A century ago, it was inconceivable that we could actually change our planet. The human population was less than two billion and the world seemed vast and the environment forgiving. Now we know better. Greenhouse gases, ozone holes, and polluted water are serious problems created by people. One problem that we are doing something about is *acid rain*. Acid rain results from the presence of nonmetal oxides in the atmosphere as industrial or automobile by-products.

In Canada and Scandinavia some lakes have been seriously affected. The pH of these lakes has become progressively lower as a result of acid rain. Fish and any other aquatic creatures have completely disappeared. In these areas of the world, the soil and rocks contain little of the types of minerals that neutralize acids, such as limestone ($CaCO_3$). Limestone protects lakes from excess acid. (See reaction **1**.)

In other situations, the vulnerability of $CaCO_3$ to acid rain causes problems. Marble statues and mortar that have survived for centuries are rapidly deteriorating in the industrialized world. Various attempts are being made to protect these treasures with coatings, but so far nothing completely effective has been found.

The fertility of the soil is being affected by acid rain. Metal oxides are an integral part of the soil. These oxides are insoluble in water but become more soluble at low pH. (See reaction **2**.) Some forests in the world seem to be seriously stressed. Acid rain is the suspected culprit.

Many metals are not affected by ordinary rain, but acid rain is more corrosive. (See reaction **3**.)

1. $CaCO_3(s) + 2H_3O^+(aq) \longrightarrow$
$$Ca^{2+}(aq) + CO_2(g) + H_2O$$
2. $Al_2O_3(s) + 6H_3O^+(aq) \longrightarrow 2Al^{3+}(aq) + 6H_2O$
3. $Zn(s) + 2H_3O^+(aq) \longrightarrow Zn^{2+}(aq) + 2H_2$

There is hope, however. There are strict limits on sulfur emissions from power plants. Catalytic converters on automobiles are reducing the nitrogen emissions. Laws are on the books and more are coming as we attempt to reverse this serious problem. These solutions do not come cheaply, however. We are all paying the billions of dollars necessary to prevent these oxides from entering the atmosphere.

*Acid rain has had a harmful effect on the environment.*

and nitrogen, however, lowers the pH to 4.0 or even lower.* The oxides of nitrogen originate mainly from engines in automobiles. The high temperature in the engine causes the two elements of air, nitrogen and oxygen, to combine to form nitrogen oxides. In eastern North America and western Europe, the acidity is about two-thirds due to sulfuric acid and the remainder due to nitric acid. There is also a small amount of hydrochloric acid in acid rain.

*Ionic metal oxides dissolve in water to form bases and thus are known as* **base anhydrides.** Some examples of these reactions are

$$Na_2O(s) + H_2O(l) \longrightarrow 2NaOH(aq)$$
$$CaO(s) + H_2O(l) \longrightarrow Ca(OH)_2(aq)$$

Salt is formed by the reaction between an acid anhydride and a base anhydride. For example, the following reaction and the neutralization of $H_2SO_3$ with $Ca(OH)_2$ in aqueous solution produce the same salt.

$$SO_2(g) + CaO(s) \longrightarrow CaSO_3(s)$$
$$H_2SO_3(aq) + Ca(OH)_2(aq) \longrightarrow CaSO_3(s) + 2H_2O(l)$$

The first reaction represents a way of removing $SO_2$ from the combustion products of an industrial plant so that some of our abundant high-sulfur coal can be used without harming the environment.

---

*On April 10, 1974, a rain fell on Pilochry, Scotland, that had a pH of 2.4, which is about the same as that of vinegar. This is the most acidic rain ever recorded. Nitrogen oxides formed in automobile engines also contribute to acid rain by forming nitric acid.

**Looking Back!** Besides neutral compounds that act as acids and bases, certain ions have acidic or basic properties. Such a reaction is known as hydrolysis. Thus aqueous salt solutions may affect the pH. That is, they can be neutral, acidic or basic. When we combine weak molecular acids and salts containing their conjugate base, we form a buffer solution. Such a solution resists change in pH. Finally, oxides can dissolve in water to form acidic or basic solutions.

## Learning Check C | checking it out

**C-1.** Fill in the blanks.

The reaction of an ion with water is known as a _____ reaction. Salts composed of cations of strong bases and anions of weak acids are _____ in water. A buffer solution _____ change in pH. It is usually made up of a weak acid and a salt containing its _____ _____ . Most nonmetal oxides react with water to form _____ solutions, and ionic metal oxides react with water to form _____ solutions.

**C-2.** Given the salts $Na_2CO_3$, $BaBr_2$, and $N_2H_5NO_3$, how will the pH be affected by dissolving each of these salts in water? If a hydrolysis reaction occurs, write the reaction. ($N_2H_4$ is a weak base like ammonia.)

**C-3.** A mixture of $H_2C_2O_4$ and $NaHC_2O_4$ can act as a buffer. **(a)** Write the reaction that occurs when $H_3O^+$ is added to the solution. **(b)** Write the reaction when $OH^-$ is added to the solution.

*Additional Examples: Exercises 12-78, 12-81, 12-87, 12-92, 12-96, and 12-99.*

**chapter** R**eview**                                   **putting it together**

Compounds have been classified as acids or bases for hundreds of years on the basis of common sets of chemical characteristics. In the twentieth century, however, acid character was attributed to formation of $H^+$ [also represented as the **hydronium ion** $(H_3O^+)$] in aqueous solution. Base character is due to the formation of $OH^-$ in solution.

Our understanding of acid–base behavior can be broadened somewhat by use of the **Brønsted–Lowry** definition, which defines **acids** as proton $(H^+)$ donors and **bases** as proton acceptors. **Amphiprotic** substances can either donate or accept a proton. An acid–base reaction constitutes an $H^+$ exchange between an acid and a base to form a **conjugate base** and **conjugate acid,** respectively.

Acids and bases can also be classified as to strength. Strong acids and strong bases are 100% ionized in water, whereas **weak molecular acids** and **weak molecular bases** are only partially ionized. Partial ionization of a molecular acid results when a reaction reaches a point of equilibrium in which both molecules (on the left of the equation) and ions (on the right) are present. For weak molecular acids and bases, the point of equilibrium favors the left, or molecular, side of the equation. Therefore, the $H_3O^+$ concentration in a weak acid solution is considerably lower than in a strong acid solution at the same initial concentration of acid.

When acidic and basic solutions are mixed, the two active ions combine in a neutralization reaction to form water. Complete neutralization of a **monoprotic acid** results in a salt and water (reaction **1** below for the hypothetical acid, HX). Incomplete neutralization of a **polyprotic acid** (either **diprotic** or **triprotic**) produces an **acid salt** and water (reaction **2** below for the hypothetical diprotic acid $H_2Y$).

Acid + Base $\longrightarrow$ Salt + Water

1. $HX(aq) + M^+OH^-(aq) \longrightarrow M^+X^-(aq) + H_2O(l)$
2. $H_2Y(aq) + M^+OH^-(aq) \longrightarrow M^+HY^-(aq) + H_2O(l)$

Even in pure water, there is a very small equilibrium concentration of $H_3O^+$ and $OH^-$ ($1.0 \times 10^{-7}$ M) due to the **autoionization** of water. The product of these concentrations is a constant known as the **ion product** of water. The ion product can be used to calculate the concentration of one ion from that of the other in any aqueous solution.

A convenient method to express the $H_3O^+$ or $OH^-$ concentrations of solutions involves the use of the logarithmic expressions **pH** and **pOH**. The following are examples of solutions of various acidities in terms of $[H_3O^+]$ and pH.

| Solution | $[H_3O^+]$ | $[OH^-]$ | pH | pOH |
|---|---|---|---|---|
| strongly acidic | $>10^{-1}$ | $<10^{-13}$ | $< 1.0$ | $>13.0$ |
| weakly acidic | $10^{-4}$ | $10^{-10}$ | 4.0 | 10.0 |
| neutral | $10^{-7}$ | $10^{-7}$ | 7.0 | 7.0 |
| weakly basic | $10^{-10}$ | $10^{-4}$ | 10.0 | 4.0 |
| strongly basic | $<10^{-13}$ | $>10^{-1}$ | $>13.0$ | $< 1.0$ |

The ions of a salt may or may not interact with water as acids or bases. If such a reaction of an ion does occur, it is known as **hydrolysis.** To predict the effect of the solution of a salt on the pH, possible hydrolysis reactions of both the cation and the anion must be examined. Five possible combinations follow.

| Salt | Solution | Comments |
|---|---|---|
| 1. $NaClO_4$ | neutral | Neither cation nor anion hydrolyzes. |
| 2. $K_2S$ | basic | Anion ($S^{2-}$) undergoes hydrolysis. $S^{2-} + H_2O \rightleftharpoons HS^- + OH^-$ |
| 3. $NH_4Br$ | acidic | Cation ($NH_4^+$) undergoes hydrolysis. $NH_4^+ + H_2O \rightleftharpoons NH_3 + H_3O^+$ |
| 4. $(NH_4)_2S$ | unable to predict | Both ions undergo hydrolysis, but more information is needed as to the extent of hydrolysis of each ion. |
| 5. $NaHSO_3$ | unable to predict | This is an acid salt. The anion can ionize further (be acidic) or hydrolyze (be basic). More information is needed concerning the extent of these two reactions. |

When a solution of a weak acid is mixed with a solution of a salt providing its conjugate base, a **buffer solution** is formed. Buffer solutions resist changes in pH from addition of limited amounts of a strong acid or base. In a buffer, the reservoir of non-ionized acid (e.g., HCN) reacts with added $OH^-$, while the reservoir of the conjugate base (e.g., $CN^-$) reacts with added $H_3O^+$. Weak bases and salts providing their conjugate acids also act as buffers (e.g., $NH_3$ and $NH_4^+Cl^-$).

Finally, the list of acids and bases was expanded to include oxides, which can be classified as **acid anhydrides** or **base anhydrides.**

The acids and bases studied in this chapter are summarized as follows.

| Type | Example | Reaction |
|---|---|---|
| **ACIDS** | | |
| 1. molecular hydrogen compounds ($H^+$ + ion) | $HClO_4$ | $HClO_4 + H_2O \longrightarrow H_3O^+ + ClO_4^-$ |
| 2. cations (conjugate acids of weak bases) | $NH_4^+$ | $NH_4^+ + H_2O \rightleftharpoons H_3O^+ + NH_3$ |
| 3. nonmetal oxides | $SO_3$ | $SO_3 + H_2O \longrightarrow H_2SO_4$<br>$H_2SO_4 + H_2O \longrightarrow H_3O^+ + HSO_4^-$ |
| **BASES** | | |
| 1. ionic hydroxides | $Ca(OH)_2$ | $Ca(OH)_2 \xrightarrow{H_2O} Ca^{2+} + 2OH^-$ |
| 2. molecular nitrogen compounds | $(CH_3)_2NH$ | $(CH_3)_2NH + H_2O \rightleftharpoons (CH_3)_2NH_2^+ + OH^-$ |
| 3. anions (conjugate bases of weak acids) | $CN^-$ | $CN^- + H_2O \rightleftharpoons HCN + OH^-$ |
| 4. metal oxides | $K_2O$ | $K_2O + H_2O \longrightarrow 2KOH$<br>$KOH \xrightarrow{H_2O} K^+ + OH^-$ |

# Exercises

## Acids and Bases

**12-1.** Give the formulas and names of the acid compounds derived from the following anions.
(a) $NO_3^-$ (b) $NO_2^-$ (c) $ClO_3^-$ (d) $SO_3^{2-}$

**12-2.** Give the formulas and names of the acid compounds derived from the following anions.
(a) $CN^-$ (b) $CrO_4^{2-}$ (c) $ClO_4^-$ (d) $Br^-$

**12-3.** Give the formulas and names of the base compounds derived from the following cations.
(a) $Cs^+$ (b) $Sr^{2+}$ (c) $Al^{3+}$ (d) $Mn^{3+}$

**12-4.** Give the formulas and names of the acid or base compounds derived from the following anions.
(a) $Ba^{2+}$ (c) $Pb^{2+}$ (e) $H_2PO_4^-$
(b) $Se^{2-}$ (d) $ClO^-$ (f) $Fe^{3+}$

**12-5.** Write reactions illustrating the acid or base behavior in water for the following.
(a) $HNO_3$ (b) $CsOH$ (c) $Ba(OH)_2$ (d) $HBr$

**12-6.** Write reactions illustrating the acid or base behavior in water for the following.
(a) $HI$ (b) $Sr(OH)_2$ (c) $RbOH$ (d) $HClO_4$

## Bronsted–Lowry Acids and Bases

**12-7.** What is the conjugate base of each of the following?
(a) $HNO_3$ (c) $HPO_4^{2-}$ (e) $H_2O$
(b) $H_2SO_4$ (d) $CH_4$ (f) $NH_3$

**12-8.** What is the conjugate acid of each of the following?
(a) $CH_3NH_2$ (c) $NO_3^-$ (e) $HCN$
(b) $HPO_4^{2-}$ (d) $O^{2-}$ (f) $H_2O$

**12-9.** Identify conjugate acid–base pairs in the following reactions.
(a) $HClO_4 + OH^- \longrightarrow H_2O + ClO_4^-$
(b) $HSO_4^- + ClO^- \longrightarrow HClO + SO_4^{2-}$
(c) $H_2O + NH_2^- \longrightarrow NH_3 + OH^-$
(d) $NH_4^+ + H_2O \longrightarrow NH_3 + H_3O^+$

**12-10.** Identify conjugate acid–base pairs in the following reactions.
(a) $HCN + H_2O \longrightarrow H_3O^+ + CN^-$
(b) $HClO_4 + NO_3^- \longrightarrow HNO_3 + ClO_4^-$
(c) $H_2S + NH_3 \longrightarrow NH_4^+ + HS^-$
(d) $H_3O^+ + HCO_3^- \longrightarrow H_2CO_3 + H_2O$

**12-11.** Write reactions indicating Brønsted–Lowry acid behavior with $H_2O$ for the following. Indicate conjugate acid–base pairs.
(a) $H_2SO_3$ (c) $HBr$ (e) $H_2S$
(b) $HClO$ (d) $HSO_3^-$ (f) $NH_4^+$

**12-12.** Write reactions indicating Brønsted–Lowry base behavior with $H_2O$ for the following. Indicate conjugate acid–base pairs.
(a) $NH_3$ (b) $N_2H_4$ (c) $HS^-$ (d) $H^-$ (e) $F^-$

**12-13.** Write equations showing how $HS^-$ can act as a Brønsted–Lowry base with $H_3O^+$ and as a Brønsted–Lowry acid with $OH^-$.

**12-14.** Bicarbonate of soda ($NaHCO_3$) acts as an antacid (base) in water. Write an equation illustrating how the $HCO_3^-$ ion reacts with $H_3O^+$. Bicarbonate is amphiprotic. Write the reaction illustrating its behavior as an acid in water.

## Strengths of Acids and Bases

**12-15.** Describe how a strong acid and a weak acid relate and differ.

**12-16.** When HBr ionizes in water, the reaction is 100% complete. Write the equation illustrating how HBr behaves as an acid. What does this tell us about the strength of the acid? An accepted observation is that *the stronger the acid, the weaker its conjugate base*. Compare the strength of HBr as a proton donor with $Br^-$ as a proton acceptor.

**12-17.** Solutions of $HClO_2$ indicate that the ionization is very limited. Write the reaction illustrating how $HClO_2$ behaves as an acid. What does this tell us about the strength of the acid and the strength of its conjugate base?

**12-18.** Write equations illustrating the reactions with water of the acids formed in exercise 12-1. Indicate strong acids with a single arrow and weak acids with equilibrium arrows.

**12-19.** Write equations illustrating the reactions with water of the acids formed in exercise 12-2. Indicate strong acids with a single arrow and weak acids with equilibrium arrows.

**12-20.** Dimethylamine $[(CH_3)_2NH]$ is a weak base that reacts in water like ammonia $(NH_3)$. Write the equilibrium illustrating this reaction.

**12-21.** Pyridine $(C_5H_5N)$ behaves as a weak base in water like ammonia. Write the equilibrium illustrating this reaction.

**12-22.** The concentration of a monoprotic acid (HX) in water is 0.10 M. The concentration of $H_3O^+$ ion in this solution is 0.010 M. Is HX a weak or a strong acid? What percent of the acid is ionized?

**12-23.** A 0.50-mol quantity of an acid is dissolved in 2.0 L of water. In the solution, $[H_3O^+] = 0.25$ M. Is this a strong or a weak acid? Explain.

**12-24.** What is $[H_3O^+]$ in a 0.55 M $HClO_4$ solution?

**12-25.** What is $[H_3O^+]$ in a 0.55 M solution of a weak acid, HX, that is 3.0% ionized?

**12-26.** What is $[OH^-]$ in a 1.45 M solution of $NH_3$ if the $NH_3$ is 0.95% ionized?

**\*12-27.** What is $[H_3O^+]$ in a 0.354 M solution of $H_2SO_4$? Assume that the first ionization is complete but that the second is only 25% complete.

**12-28.** A 1.0 M solution of HF has $[H_3O^+] = 0.050$. What is the percent ionization of the acid?

**12-29.** A 0.10 M solution of pyridine (a weak base in water) has $[OH^-] = 4.4 \times 10^{-5}$. What is the percent ionization of the base?

**12-30.** The $HSO_4^-$ ion is not amphiprotic in water. Which species cannot exist in water—its conjugate base or conjugate acid? Why not?

## Neutralization and Salts

**12-31.** Identify each of the following as an acid, base, normal salt, or acid salt.
**(a)** $H_2S$    **(c)** $H_3AsO_4$    **(e)** $K_2SO_4$
**(b)** $BaCl_2$    **(d)** $Ba(HSO_4)_2$    **(f)** $LiOH$

**12-32.** Explain why a strong acid is represented in aqueous solution as two ions but a weak acid is represented as one molecule.

**12-33.** Write the balanced molecular equations showing the complete neutralizations of the following.
**(a)** $HNO_3$ by NaOH    **(c)** $HClO_2$ by KOH
**(b)** $Ca(OH)_2$ by HI

**12-34.** Write the balanced molecular equations showing the complete neutralizations of the following.
**(a)** $HNO_2$ by NaOH    **(c)** $H_2S$ by $Ba(OH)_2$
**(b)** $H_2CO_3$ by CsOH

**12-35.** Write the total ionic and net ionic equations for the reactions in exercise 12-33.

**12-36.** Write the total ionic and net ionic equations for the reactions in exercise 12-34.

**12-37.** Write the molecular, total ionic, and net ionic equations of the complete neutralization of $H_2C_2O_4$ with $NH_3$.

**12-38.** Write the molecular, total ionic, and net ionic equations for the complete neutralization of $HC_2H_3O_2$ with $NH_3$.

**12-39.** Write the formulas of the acid and the base that formed the following salts.
**(a)** $KClO_3$   **(b)** $Al_2(SO_3)_3$   **(c)** $Ba(NO_2)_2$   **(d)** $NH_4NO_3$

**12-40.** Write the formulas of the acid and the base that formed the following salts.
**(a)** $Li_2CrO_4$   **(b)** NaCN   **(c)** $Fe(ClO_4)_3$   **(d)** $Mg(HCO_3)_2$

**12-41.** Write balanced acid–base neutralization reactions that would lead to formation of the following salts or acid salts.
**(a)** $CaBr_2$   **(b)** $Sr(ClO_2)_2$   **(c)** $Ba(HS)_2$   **(d)** $Li_2S$

**12-42.** Write balanced acid–base neutralization reactions that would lead to formation of the following salts or acid salts.
**(a)** $Na_2SO_3$   **(b)** $AlI_3$   **(c)** $Mg_3(PO_4)_2$   **(d)** $NaHCO_3$

**12-43.** Write two equations illustrating the stepwise neutralization of $H_2S$ with LiOH.

**12-44.** Write the two net ionic equations illustrating the two reactions in exercise 12-43.

**12-45.** Write three equations illustrating the stepwise neutralization of $H_3AsO_4$ with LiOH. Write the total reaction.

**12-46.** Write the net ionic equations for the three reactions in exercise 12-45.

**12-47.** Write the equation illustrating the reaction of 1 mol of $H_2S$ with 1 mol of NaOH.

**\*12-48.** Write the equation illustrating the reaction between 1 mol of $Ca(OH)_2$ and 2 mol of $H_3PO_4$.

## Equilibrium of Water and $K_w$

**12-49.** If some ions are present in pure water, why isn't pure water considered to be an electrolyte?

**12-50.** Why can't $[H_3O^+] = [OH^-] = 1.00 \times 10^{-2}$ M in water? What would happen if we tried to make such a solution by mixing $10^{-2}$ mol/L of KOH with $10^{-2}$ mol/L of HCl?

**12-51.** (a) What is $[H_3O^+]$ when $[OH^-] = 10^{-12}$ M?
(b) What is $[H_3O^+]$ when $[OH^-] = 10$ M?
(c) What is $[OH^-]$ when $[H_3O^+] = 2.0 \times 10^{-5}$ M?

**12-52.** (a) What is $[OH^-]$ when $[H_3O^+] = 1.50 \times 10^{-3}$ M?
(b) What is $[H_3O^+]$ when $[OH^-] = 2.58 \times 10^{-7}$ M?
(c) What is $[H_3O^+]$ when $[OH^-] = 56.9 \times 10^{-9}$ M?

**12-53.** When 0.250 mol of the strong acid $HClO_4$ is dissolved in 10.0 L of water, what is $[H_3O^+]$? What is $[OH^-]$?

**12-54.** Lye is a very strong base. What is $[H_3O^+]$ in a 2.55 M solution of NaOH? In the weakly basic household ammonia, $[OH^-] = 4.0 \times 10^{-3}$ M. What is $[H_3O^+]$?

**12-55.** Identify the solutions in exercise 12-51 as acidic, basic, or neutral.

**12-56.** Identify the solutions in exercise 12-52 as acidic, basic, or neutral.

**12-57.** Identify each of the following as an acidic, basic, or neutral solution.
(a) $[H_3O^+] = 6.5 \times 10^{-3}$ M
(b) $[H_3O^+] = 5.5 \times 10^{-10}$ M
(c) $[OH^-] = 4.5 \times 10^{-8}$ M
(d) $[OH^-] = 50 \times 10^{-8}$ M

**12-58.** Identify each of the following as an acidic, basic, or neutral solution.
(a) $[OH^-] = 8.1 \times 10^{-8}$ M
(b) $[H_3O^+] = 10.0 \times 10^{-8}$ M
(c) $[H_3O^+] = 4.0 \times 10^{-3}$ M
(d) $[OH^-] = 55 \times 10^{-8}$ M

## pH and pOH

**12-59.** What is the pH of the following solutions?
(a) $[H_3O^+] = 1.0 \times 10^{-6}$ M
(b) $[H_3O^+] = 1.0 \times 10^{-9}$ M
(c) $[OH^-] = 1.0 \times 10^{-2}$ M
(d) $[OH^-] = 2.5 \times 10^{-5}$ M
(e) $[H_3O^+] = 6.5 \times 10^{-11}$ M

**12-60.** What is the pH of the following solutions?
(a) $[H_3O^+] = 1.0 \times 10^{-2}$ M
(b) $[OH^-] = 1.0 \times 10^{-4}$ M
(c) $[H_3O^+] = 1.0$ M
(d) $[OH^-] = 3.6 \times 10^{-9}$ M
(e) $[OH^-] = 7.8 \times 10^{-4}$ M
(f) $[H_3O^+] = 42.2 \times 10^{-5}$ M

**12-61.** What are the pH and pOH of the following?
(a) $[H_3O^+] = 0.0001$ (c) $[H_3O^+] = 0.020$
(b) $[OH^-] = 0.00001$ (d) $[OH^-] = 0.000320$

**12-62.** What are the pH and pOH of the following?
(a) $[H_3O^+] = 0.0000001$ (c) $[OH^-] = 0.0568$
(b) $[OH^-] = 0.0001$ (d) $[H_3O^+] = 0.00082$

**12-63.** What is $[H_3O^+]$ of the following?
(a) pH = 3.00 (c) pOH = 8.00 (e) pH = 12.70
(b) pH = 3.54 (d) pOH = 6.38

**12-64.** What is $[H_3O^+]$ of the following?
(a) pH = 9.0 (c) pH = 2.30
(b) pOH = 9.0 (d) pH = 8.90

**12-65.** Identify each of the solutions in exercises 12-59 and 12-63 as acidic, basic, or neutral.

**12-66.** Identify each of the solutions in exercises 12-60 and 12-64 as acidic, basic, or neutral.

**12-67.** A solution has pH = 3.0. What is the pH of a solution that is 100 times less acidic? What is the pH of a solution that is 10 times more acidic?

**12-68.** A solution has pOH = 4. What is the pOH of a solution that is 1000 times more acidic? What is the pOH of a solution that is 100 times more basic?

**12-69.** What is the pH of a 0.075 M solution of the strong acid $HNO_3$?

**12-70.** What is the pH of a 0.0034 M solution of the strong base KOH?

**12-71.** What is the pH of a 0.018 M solution of the strong base $Ca(OH)_2$?

**12-72.** A weak monoprotic acid is 10.0% ionized in solution. What is the pH of a 0.10 M solution of this acid?

**12-73.** A weak base is 5.0% ionized in solution. What is the pH of a 0.25 M solution of this base? (Assume one $OH^-$ per formula unit.)

**12-74.** Identify each of the following solutions as strongly basic, weakly basic, neutral, weakly acidic, or strongly acidic.
(a) pH = 1.5 (d) pH = 13.0 (g) pOH = 7.5
(b) pOH = 13.0 (e) pOH = 7.0 (h) pH = −1.0
(c) pH = 5.8 (f) pH = 8.5

**12-75.** Arrange the following substances in order of increasing acidity.
(a) household ammonia, pH = 11.4
(b) vinegar, $[H_3O^+] = 2.5 \times 10^{-3}$ M
(c) grape juice, $[OH^-] = 1.0 \times 10^{-10}$ M
(d) sulfuric acid, pOH = 13.6
(e) eggs, pH = 7.8
(f) rainwater, $[H_3O^+] = 2.0 \times 10^{-6}$ M

**12-76.** Arrange the following substances in order of increasing acidity.
(a) lime juice, $[H_3O^+] = 6.0 \times 10^{-2}$ M
(b) antacid tablet in water, $[OH^-] = 2.5 \times 10^{-6}$ M
(c) coffee, pOH = 8.50
(d) stomach acid, pH = 1.8
(e) saliva, $[H_3O^+] = 2.2 \times 10^{-7}$ M
(f) a soap solution, pH = 8.3
(g) a solution of lye, pOH = 1.2
(h) a banana, $[OH^-] = 4.0 \times 10^{-10}$ M

*__12-77.__ What is the pH of a 0.0010 M solution of $H_2SO_4$? (Assume that the first ionization is complete but the second is only 25% complete.)

## Hydrolysis of Salts

**12-78.** Two of the following act as weak bases in water. Write the appropriate reactions illustrating the weak base behavior.
(a) $ClO_4^-$ (b) $C_2H_3O_2^-$ (c) $NH_3$ (d) HF

**12-79.** Two of the following act as weak acids in water. Write the appropriate reactions illustrating the weak acid behavior.
**(a)** $H_2CrO_4$ **(b)** $NH_4^+$ **(c)** $NH_2CH_3$ **(d)** $CrO_4^{2-}$

**12-80.** Three of the following molecules or ions do not affect the pH of water. Which are they and why do they not affect the pH?
**(a)** $K^+$ **(c)** $HCO_3^-$ **(e)** $O_2$
**(b)** $NH_3$ **(d)** $NO_3^-$ **(f)** $N_2H_5^+$

**12-81.** Complete the following hydrolysis equilibria.
**(a)** $S^{2-} + H_2O \rightleftharpoons \underline{\hspace{1cm}} + OH^-$
**(b)** $N_2H_5^+ + H_2O \rightleftharpoons N_2H_4 + \underline{\hspace{1cm}}$
**(c)** $HPO_4^{2-} + H_2O \rightleftharpoons H_2PO_4^- + \underline{\hspace{1cm}}$
**(d)** $(CH_3)_2NH_2^+ + H_2O \rightleftharpoons \underline{\hspace{1cm}} + H_3O^+$

**12-82.** Complete the following hydrolysis equilibria.
**(a)** $CN^- + H_2O \rightleftharpoons HCN + \underline{\hspace{1cm}}$
**(b)** $NH_4^+ + H_2O \rightleftharpoons \underline{\hspace{1cm}} + H_3O^+$
**(c)** $B(OH)_4^- + H_2O \rightleftharpoons H_3BO_3 + \underline{\hspace{1cm}}$
**(d)** $Al(H_2O)_6^{3+} + H_2O \rightleftharpoons Al(H_2O)_5(OH)^{2+} + \underline{\hspace{1cm}}$

**12-83.** Write the hydrolysis equilibria (if any) for the following ions.
**(a)** $F^-$ **(d)** $HPO_4^{2-}$
**(b)** $SO_3^{2-}$ **(e)** $CN^-$
**(c)** $(CH_3)_2NH_2^+$ [$(CH_3)_2NH$ is a weak base like ammonia.] **(f)** $Li^+$

**12-84.** Write the hydrolysis equilibria (if any) for the following ions.
**(a)** $Br^-$ **(b)** $HS^-$ **(c)** $ClO_4^-$ **(d)** $H^-$ **(e)** $Ca^{2+}$

**12-85.** Calcium hypochlorite is used to purify water. When dissolved, it produces a slightly basic solution. Write the equation illustrating the solution of calcium hypochlorite in water and the equation illustrating its basic behavior.

**12-86.** Aqueous NaF solutions are slightly basic, whereas aqueous NaCl solutions are neutral. Write the appropriate equation that illustrates this. Why aren't NaCl solutions also basic?

**12-87.** Predict whether aqueous solutions of the following salts are acidic, neutral, or basic.
**(a)** $Ba(ClO_4)_2$ **(d)** $KBr$
**(b)** $N_2H_5^+NO_3^-$ **(e)** $NH_4Cl$
($N_2H_4$ is a weak base.) **(f)** $BaF_2$
**(c)** $LiC_2H_3O_2$

**12-88.** Predict whether aqueous solutions of the following salts are acidic, neutral, or basic.
**(a)** $Na_2CO_3$ **(b)** $K_3PO_4$ **(c)** $NH_4ClO_4$ **(d)** $SrI_2$

**\*12-89.** Both $C_2^{2-}$ and its conjugate acid $HC_2^-$ hydrolyze 100% in water. From this information, complete the following equation.

$CaC_2(s) + 2H_2O(l) \longrightarrow$
$\underline{\hspace{1cm}} (g) + Ca^{2+}(aq) + 2\underline{\hspace{1cm}} (aq)$

(The gas formed—acetylene—can be burned as it is produced. This reaction was once important for this purpose as a source of light in old miners' lamps.)

**\*12-90.** Aqueous solutions of $NH_4CN$ are basic. Write the two hydrolysis reactions and indicate which takes place to the greater extent.

**12-91.** Aqueous solutions of $NaHSO_3$ are acidic. Write the two equations (one hydrolysis and one ionization) and indicate which takes place to the greater extent.

## Buffers

**12-92.** Identify which of the following form buffer solutions when 0.50 mol of each compound is dissolved in 1 L of water.
**(a)** $HNO_2$ and $KNO_2$ **(f)** $HCN$ and $KClO$
**(b)** $NH_4Cl$ and $NH_3$ **(g)** $NH_3$ and $BaBr_2$
**(c)** $HNO_3$ and $KNO_2$ **(h)** $H_2S$ and $LiHS$
**(d)** $HNO_3$ and $KNO_3$ **(i)** $KH_2PO_4$ and $K_2HPO_4$
**(e)** $HClO$ and $Ca(ClO)_2$

**12-93.** A certain solution contains dissolved HCl and NaCl. Why can't this solution act as a buffer?

**12-94.** Write the equilibrium involved in the $N_2H_4$, $N_2H_5Cl$ buffer system. ($N_2H_4$ is a weak base.) Write equations illustrating how this system reacts with added $H_3O^+$ and added $OH^-$.

**12-95.** Write the equilibrium involved in the $HCO_3^-$, $CO_3^{2-}$ buffer system. Write equations illustrating how this system reacts with added $H_3O^+$ and added $OH^-$.

**12-96.** Write the equilibrium involved in the $HPO_4^{2-}$, $PO_4^{3-}$ buffer system. Write equations illustrating how this system reacts with added $H_3O^+$ and added $OH^-$.

**12-97.** If 0.5 mol of KOH is added to a solution containing 1.0 mol of $HC_2H_3O_2$, the resulting solution is a buffer. Explain.

**12-98.** A solution contains 0.50 mol each of HClO and NaClO. If 0.60 mol of KOH is added, will the buffer prevent a significant change in pH? Explain.

## Oxides as Acids and Bases

**12-99.** Write the formula of the acid or base formed when each of the following anhydrides is dissolved in water.
**(a)** $SrO$ **(c)** $P_4O_{10}$ **(e)** $N_2O_3$
**(b)** $SeO_3$ **(d)** $Cs_2O$ **(f)** $Cl_2O_5$

**12-100.** Write the formula of the acid or base formed when each of the following anhydrides is dissolved in water.
**(a)** $BaO$ **(c)** $Cl_2O$ **(e)** $K_2O$
**(b)** $SeO_2$ **(d)** $Br_2O$

**12-101.** Carbon dioxide is removed from the space shuttle by bubbling the air through an LiOH solution. Show the reaction and the product formed.

**\*12-102.** Complete the following equation.

$$Li_2O(s) + N_2O_5(g) \longrightarrow \underline{\hspace{1cm}} (s)$$

## General Problems

**12-103.** Iron reacts with an acid, forming an aqueous solution of iron(II) iodide and a gas. Write the equation illustrating the reaction.

**12-104.** Aluminum reacts with perchloric acid. Write the equation illustrating this reaction.

**12-105.** Nitric acid reacts with sodium sulfite, forming sulfur dioxide gas and a salt. Write the equation illustrating the reaction.

**12-106.** Perbromic acid reacts with sodium sulfide to form a pungent gas, hydrogen sulfide. Write the equation illustrating the reaction.

**12-107.** There are acid–base systems based on solvents other than $H_2O$. One is ammonia ($NH_3$), which is also amphiprotic. Write equations illustrating each of the following.
**(a)** the reaction of HCN with $NH_3$ acting as a base
**(b)** the reaction of $H^-$ with $NH_3$ acting as an acid
**(c)** the reaction of $HCO_3^-$ with $NH_3$ acting as a base
**(d)** the reaction between $NH_4Cl$ and $NaNH_2$ in ammonia

**12-108.** The conjugate base of methyl alcohol ($CH_3OH$) is $CH_3O^-$. Its conjugate acid is $CH_3OH_2^+$. Write equations illustrating each of the following.
**(a)** the reaction of HCl with methyl alcohol acting as a base
**(b)** the reaction of $NH_2^-$ with methyl alcohol acting as an acid

**12-109.** Sulfite ion ($SO_3^{2-}$) and sulfur trioxide ($SO_3$) look similar at first glance, but one forms a strongly acidic solution whereas the other is weakly basic. Write equations illustrating this behavior.

**12-110.** Tell whether each of the following compounds forms an acidic, basic, or neutral solution when added to pure water. Write the equation illustrating the acidic or basic behavior where appropriate.
**(a)** $H_2S$   **(d)** $NH_3$   **(g)** $Sr(NO_3)_2$   **(i)** $H_2SO_3$
**(b)** KClO   **(e)** $N_2H_5^+Br^-$   **(h)** $LiNO_2$   **(j)** $Cl_2O_3$
**(c)** NaI   **(f)** $Ba(OH)_2$

**12-111.** Tell whether each of the following compounds forms an acidic, basic, or neutral solution when added to pure water. Write the equation illustrating the acidic or basic behavior where appropriate.
**(a)** HBrO   **(d)** $N_2H_4$   **(f)** $Ba(C_2H_3O_2)_2$
**(b)** CaO   **(e)** $SO_2$   **(g)** RbBr
**(c)** $NH_4ClO_4$

**12-112.** In a lab there are five different solutions with pH's of 1.0, 5.2, 7.0, 10.2, and 13.0. The solutions are LiOH, $SrBr_2$, KClO, $NH_4Cl$, and HI, all at the same concentration. Which pH corresponds to which compound? What must be the concentration of all of these solutions?

**12-113.** When one mixes a solution of baking soda ($NaHCO_3$) with vinegar ($HC_2H_3O_2$), bubbles of gas appear. Write equations for two reactions that indicate the identity of the gas.

**\*12-114.** High-sulfur coal contains 5.0% iron pyrite ($FeS_2$). When the coal is burned, the iron pyrite also burns according to the equation

$$4FeS_2(s) + 11O_2(g) \longrightarrow 2Fe_2O_3(s) + 8SO_2(g)$$

What mass of sulfuric acid can eventually form from the combustion of 100 kg of coal? Sulfuric acid is formed according to the following equations.

$$2SO_2(g) + O_2(g) \longrightarrow 2SO_3(g)$$
$$SO_3(g) + H_2O(l) \longrightarrow H_2SO_4(aq)$$

**12-115.** A 2.50-g quantity of HCl is dissolved in 245 mL of water and then diluted to 890 mL. What is the pH of the concentrated and the dilute solution?

**12-116.** A 0.150-mole quantity of NaOH is dissolved in 2.50 L of water. In a separate container, 0.150 mole of HCl is present in 2.50 L of water. What is the pH of each solution? What is the pH of a solution made by mixing the two?

**\*12-117.** A solution is prepared by mixing 10.0 g of HCl with 10.0 g of NaOH. What is the pH of the solution if the volume is 1.00 L?

**\*12-118.** A solution is prepared by mixing 25.0 g of $H_2SO_4$ with 50.0 g of KOH. What is the pH of the solution if the volume is 500 mL?

**\*12-119.** A solution is prepared by mixing 500 mL of 0.10 M $HNO_3$ with 500 mL of 0.10 M $Ca(OH)_2$. What is the pH of the solution after mixing?

**\*12-120.** A solution is prepared by mixing 250 mL of 0.250 M $HClO_4$ with 500 mL of 0.150 M KOH. What is the pH of the solution after mixing?

# Solutions to Learning Checks

**A-1.** $H^+$, $H_3O^+$, $OH^-$, hydroxide, proton ($H^+$), proton, gain, $H^+$, amphiprotic, 100, partially, weak base, neutralization, water, salt, acid salt

**A-2.** **(a)** $HClO_4$, perchloric acid
**(b)** $Fe(OH)_2$, iron(II) hydroxide
**(c)** $H_2S$, hydrosulfuric acid
**(d)** LiOH, lithium hydroxide

**A-3.** $H_2C_2O_4 \longleftarrow H_2CO_4^- \longrightarrow C_2O_4^{2-}$
    conjugate acid          conjugate base

**A-4.** **(a)** $HC_2O_4^- + H_2O \rightleftharpoons C_2O_4^{2-} + H_3O^+$
         base
**(b)** $HC_2O_4^- + H_2O \rightleftharpoons H_2C_2O_4 + OH^-$
         acid

**A-5.** This is a weak acid solution.

$$0.10 \text{ M} \times 0.0250 = 2.50 \times 10^{-3} \text{ M} = [H_3O^+]$$

**A-6.** *Molecular:*   $2HNO_3(aq) + Sr(OH)_2(aq) \longrightarrow$
$$Sr(NO_3)_2(aq) + 2H_2O(l)$$

*Total ionic:* $2H^+(aq) + 2NO_3^-(aq) + Sr^{2+}(aq) + 2OH^-(aq) \longrightarrow$
$$Sr^{2+}(aq) + 2NO_3^-(aq) + 2H_2O(l)$$

*Net ionic:* $H^+(aq) + OH^-(aq) \longrightarrow H_2O(l)$

**A-7.** *Molecular:* $H_2C_2O_4(aq) + CsOH(aq) \longrightarrow$
$$CsHC_2O_4(aq) + H_2O(l)$$

*Total ionic:* $H_2C_2O_4(aq) + Cs^+(aq) + OH^-(aq) \longrightarrow$
$$Cs^+(aq) + HC_2O_4^-(aq) + H_2O(l)$$

*Net ionic:* $H_2C_2O_4(aq) + OH^-(aq) \longrightarrow$
$$HC_2O_4^-(aq) + H_2O(l)$$

**B-1.** $H_3O^+$, $OH^-$, ion product, $K_w$, $1.0 \times 10^{-14}$, $-\log[H_3O^+]$, $10^{-7}$, less, $10^{-7}$, greater

**B-2.** $[H_3O^+] = K_w/[OH^-] = (1.0 \times 10^{-14})/(7.2 \times 10^{-5})$
$$= \underline{\underline{1.4 \times 10^{-10}}}$$

The solution is weakly basic.

**B-3.** pH $= -\log[1.4 \times 10^{-10}] = 9.85$
pOH $= 14.00 - 9.85 = \underline{\underline{4.15}}$

**C-1.** hydrolysis, basic, resists, conjugate base, acidic, basic

**C-2.** $Na_2CO_3$ forms a basic solution because the $CO_3^{2-}$ ion undergoes basic hydrolysis.

$$CO_3^{2-}(aq) + H_2O(l) \rightleftharpoons HCO_3^-(aq) + OH^-(aq)$$

Solutions of $BaBr_2$ are neutral because neither $Ba^{2+}$ nor $Br^-$ hydrolyzes.

Solutions of $N_2H_5NO_3$ are acidic because the $N_2H_5^+$ ion undergoes acidic hydrolysis.

$$N_2H_5^+(aq) + H_2O(l) \rightleftharpoons N_2H_4(aq) + H_3O^+(aq)$$

**C-3.** (a) $HC_2O_4^- + H_3O^+ \longrightarrow H_2C_2O_4 + H_2O$
(b) $H_2C_2O_4 + OH^- \longrightarrow HC_2O_4^- + H_2O$

## chapter 13

# Oxidation–Reduction Reactions

**Electricity has been essential to our lives for over a century. The first electricity was generated by chemical reactions.**

### Setting the Stage

A distant rumble signals the ominous gathering of thunderstorms. We may cast a cautious eye toward the sky and think of shelter. The roll of thunder warns us about one force of nature for which we have great respect, so we try to get out of its way. That, of course, is lightning. Lightning has no doubt caused fear as well as amazement in the human race since people first looked to the sky and wondered about its nature. But this force was not harnessed until modern times. The use of electricity (the same force as lightning) is so common to us now that it is taken for granted. Huge generating plants dot the rural landscape with towering smokestacks discharging smoke and steam. Not many decades ago, however, electricity was mainly a laboratory curiosity, until the experiments of inventors such as James Watt, Alexander Graham Bell, and Thomas Edison tapped its limitless potential. The electricity used by these early investigators originated from chemical reactions. Even now, when we turn on a calculator or a flashlight or start a car, a specific chemical reaction in a battery causes a flow of electrons that is the source of electricity for these uses. Because one reactant has a greater affinity for electrons than the other, the exchange of electrons takes place spontaneously. This type of reaction is much like the acid–base reactions discussed in the preceding chapter that are favorable because one reactant has a greater affinity for an $H^+$ ion than the other. Electron exchanges have not previously been defined as a specific type of reactions, but many of the classifications discussed in Chapter 7 fit into this category. Most combination, all combustion, and all single-replacement reactions can also be categorized as electron exchange reactions.

### Formulating Some Questions

There are many practical questions that will be discussed in this chapter. First of all, what is involved in the electron exchange process and what are the terms that are used? How do we keep track of electrons in these special reactions? Can we predict favorable and unfavorable electron exchange reactions? How do we put favorable reactions to use in batteries? And, finally, how can we make unfavorable reactions occur?

---

**Section A  REDOX REACTIONS—THE EXCHANGE OF ELECTRONS**

## 13-1 The Nature of Oxidation and Reduction

**Looking Ahead!** The driving force of two general types of reactions involves an exchange of parts of a hydrogen atom. In Chapter 12, we described the actions of acids and bases in water as an exchange of a proton. In this chapter, we describe reactions involving the other part of a hydrogen atom—the electron. In this section, we examine this exchange and the terms used to describe it.

Sodium metal and chlorine gas react in a spectacular demonstration of chemical power. A small chunk of sodium placed in a flask filled with chlorine gas immediately glows white hot as the elements combine to form ordinary table salt. (See Figure 13-1.) We will examine the reaction of sodium and chlorine to illustrate the process of an electron exchange reaction. The equation for this reaction is as follows.

$$2Na(s) + Cl_2(g) \longrightarrow 2NaCl(s)$$

In section 6-3, we briefly explained what was happening in the reaction. A sodium atom loses one electron, which is gained by a chlorine atom, and ions form. The gain and loss of an electron was predicted as a logical consequence of the octet rule.

$$Na\odot + \ddot{Cl}: \longrightarrow Na^+ :\ddot{Cl}:^-$$

In the acid–base reactions discussed in the previous chapter, the atoms in the reactants keep their quota of electrons in changing to products. Such is not the case in the reaction shown above, however. This is best illustrated by taking the reaction apart and examining the change that each reactant undergoes as the product is formed. An electron exchange reaction can be viewed as the sum of two half-reactions. *A half-reaction represents either the loss of electrons or the gain of electrons as a separate balanced equation.* Thus the half-reaction involving only sodium is

$$Na \longrightarrow Na^+ + e^-$$

Notice that the neutral sodium atom has lost an electron to form a sodium ion in this half-reaction. *A substance that loses electrons in a chemical reaction is said to be* **oxidized.**

Now let's consider what happens to the chlorine in going from reactant to product.

$$2e^- + Cl_2 \longrightarrow 2Cl^-$$

In this half-reaction, the neutral chlorine molecule has gained two electrons to form two chloride ions. *A substance that gains electrons in a chemical reaction is said to be* **reduced.**

**Figure 13-1  Formation of NaCl**

**An active metal reacts with a poisonous gas to form ordinary table salt (NaCl).**

Obviously, the two processes (oxidation and reduction) complement each other, giving us the basis of this classification of chemical reactions. *Reactions involving an exchange of electrons are known as* **oxidation–reduction** *or, simply,* **redox reactions.**

Instead of identifying a reactant by what happened to it (i.e., it was oxidized or reduced), in this type of reaction we may emphasize what it *does. The substance that causes the oxidation (i.e., accepts electrons) is called the* **oxidizing agent,** *and the substance that causes the reduction (i.e., provides the electrons) is called the* **reducing agent.** Thus the substance reduced is the oxidizing agent and the substance oxidized is the reducing agent. This is summarized in the following diagram.

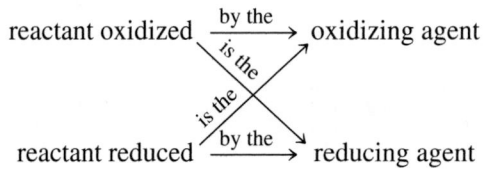

In our example, Na is the reducing agent and is oxidized to $Na^+$. $Cl_2$ is the oxidizing agent and is reduced to $Cl^-$.

Let's now see how the two half-reactions add together to make a complete, balanced equation. *An important principle of a redox reaction is that the electrons gained in the reduction process must equal the electrons lost in the oxidation process.* Note in our sample reaction that the reduction process involving $Cl_2$ requires two electrons. Therefore, the oxidation process must involve two Na's to provide these two electrons. The electrons on both sides of the equation must be equal so that they can be eliminated by subtraction when the two half-reactions are added.

*Oxidation half-reaction:*  $2Na \longrightarrow 2Na^+ + 2e^-$

*Reduction half-reaction:*  $2e^- + Cl_2 \longrightarrow 2Cl^-$

*Total reaction:*  $2Na + Cl_2 \longrightarrow 2Na^+ + 2Cl^-$ (or 2NaCl)

Many other familiar chemical changes are redox reactions. For example, the corrosion of iron to form rust involves an exchange of electrons. The formation of rust [iron(III) oxide] is illustrated below.

$$4Fe(s) + 3O_2(g) \longrightarrow 2Fe_2O_3(s)$$

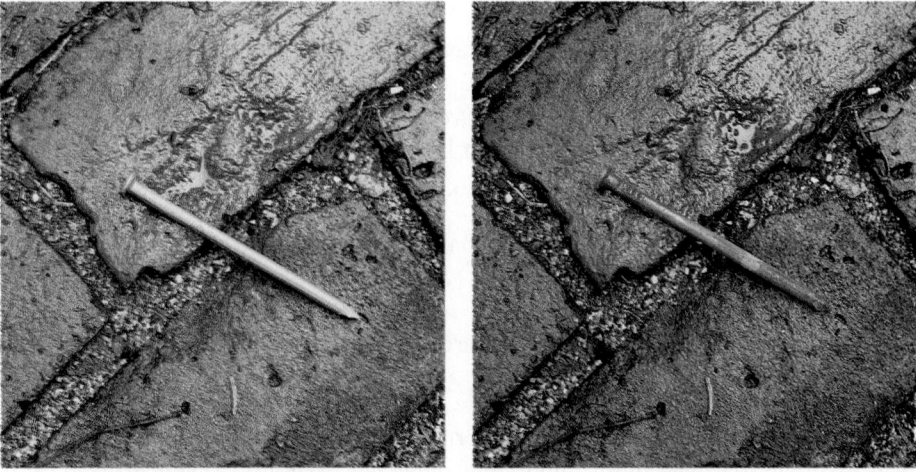

**An iron nail soon forms a coating of rust when exposed to moisture and air.**

### Lightning Bugs (Fireflies)—Nature's Little Night-Lights

Tiny little flashes of light on a summer night—most of us know the sources as "lightning bugs" or "fireflies." They have been a source of curiosity since ancient times, but only recently have we known how these tiny creatures can light up.

In photochemical reactions, such as photosynthesis, light energy *initiates* a chemical reaction. In the firefly we have just the opposite situation. That is, light energy is *produced* by a chemical reaction. The production of light energy by a chemical reaction is known as *chemiluminescence*. If it is produced by living organisms, such as the firefly, it is known as *bioluminescence*.

We now have some understanding of what goes on in the firefly. The chemical reaction is an oxidation–reduction reaction with $O_2$ from the air serving as the oxidizing agent. The other chemicals are complex molecules, so we will simply refer to them by name (luciferin) or initials (ATP). Also involved is an enzyme (luciferase), which acts as a catalyst. This reaction is represented below.

luciferin + ATP + $O_2(g)$ $\xrightarrow{\text{luciferase}}$ products + $h\nu$ (cold light)

By studying the bioluminescent chemistry of the firefly, we also discovered that all fireflies have the same luciferin, but each species produces a different color of light. The color was found to be determined by the luciferase that is unique to each species of firefly. A really amazing discovery about this reaction is that the production of light is incredibly efficient. Eighty out of 100 molecules that react produce light, thus giving it an 80% efficiency.

It wasn't until 1928 that a scientist, H. O. Albrecht, first described a nonbiological chemical reaction that could be conducted in a laboratory to generate chemiluminescence, but with only 0.1% efficiency. The reaction is similar to the one shown above except that nonbiological chemicals available to the scientist were used along with hydrogen peroxide ($H_2O_2$) as the oxidizing agent. In the early 1960s, other chemiluminescent reactions were discovered and patented. Today a substance called Cyalume™, a trademark product of American Cyanamid, is used to produce light with about 5% efficiency.

When you see children with decorative jewelry that glows in the dark, thank the firefly for its amazing contribution to our world of chemistry.

*Submitted by Edward Tokas*
*Carolina Reading and Learning*

*Fireflies use a redox reaction to produce light.*

Rust is the unrelenting enemy of iron in ships or other structures. The corrosion of iron to form rust costs billions of dollars yearly for replacement of damaged structures and parts or protection of exposed metal with paints and coatings. In the reaction, iron metal gives up electrons to oxygen from the air and forms this brown flaky substance.

Originally, the term "oxidation" referred specifically to such a reaction where a substance like iron adds oxygen. In the reverse reaction, oxygen is removed from iron(III) oxide, which reduces its mass. Thus the removal of oxygen was known as "reduction." Now we define the process in terms of the exchange of electrons, so the terms are used regardless of whether oxygen is involved. In many redox reactions, however, the species undergoing electron exchange are not as obvious as the two examples that we have used in this section. So, we need a tool to help us.

## 13-2 Oxidation States

**Looking Ahead!** To appreciate the extent of redox reactions, we need a form of "electron bookkeeping." In other words, in order to identify the species that has given up or gained electrons, we need to know how many electrons the species had in the first place. In this section, we will introduce an important tool that will help us identify electronic changes.

In Chapter 6 we presented Lewis structures for molecules and ions where electrons were shared between two atoms. We are now going to assign electrons that are shared to one of the two bonded atoms. By doing this we will be able to detect changes in assigned electrons when reactants change into products. *The **oxidation state** (or **oxidation number**)* *of an atom in a molecule or ion is the charge that atom would have if all atoms were present as monatomic ions.* This is accomplished by assigning the electrons in each chemical bond to the more electronegative atom. Recall that electronegativity (see section 6-7) is a measure of the atom's ability to attract electrons in a chemical bond. The most electronegative element is fluorine, oxygen is second, and the electronegativity of the elements then decreases in the direction of the metal–nonmetal borderline in the periodic table. It seems like we need to know all of the electronegativities of the elements but, in fact, a few general rules will suffice. These rules and some practice exercises follow.

1. The oxidation state of an element in its free (natural) state is zero [e.g., Cu(*s*), B(*s*)]. This includes polyatomic elements [e.g., $Cl_2(g)$, $P_4(s)$].

2. The oxidation state of a monatomic ion is the same as the charge on that ion (e.g., $Na^+ = +1$ oxidation state, $O^{2-} = -2$, $Al^{3+} = +3$).
   **a.** Alkali metal ions are always $+1$ (same as group number).
   **b.** Alkaline earth metal ions are always $+2$ (same as group number).

3. The halogens are in a $-1$ oxidation state in binary (two-element) compounds, whether ionic or covalent, *when bound to a less electronegative element.*

4. Oxygen in a compound is usually $-2$. Certain compounds (which are rare) called peroxides or superoxides contain O in a lower negative oxidation state. Oxygen is positive when bound to F.

5. Hydrogen in a compound is usually $+1$. When combined with a less electronegative element, usually a metal, H has a $-1$ oxidation state (e.g., LiH).

6. The sum of the oxidation states of all of the atoms in a neutral compound is zero. For a polyatomic ion, the sum of the oxidation states equals the charge on the ion.

The following examples illustrate the assignment of oxidation states.

| Example 13-1 | Calculation of Oxidation States |
|---|---|

**working it out**

What is the oxidation state of the following?

**a.** Fe in FeO          **c.** S in $H_2SO_3$
**b.** N in $N_2O_5$          **d.** As in $AsO_4{}^{3-}$

**Procedure**

An algebraic equation can be constructed from rule 6. For example, assume that we have a hypothetical compound $M_2A_3$. Then from rule 6

$$[2 \times (\text{oxidation state of M})] + [3 \times (\text{oxidation state of A})] = 0$$

or

$$2(\text{ox. state M}) + 3(\text{ox. state A}) = 0$$

If the formula represents a polyatomic ion, the quantity on the left is equal to the charge.

**Solution**

**a.** FeO: The oxidation states of the two elements add to zero (rule 6).

$$(\text{ox. state Fe}) + (\text{ox. state O}) = 0$$

Since the oxidation of state of oxygen is $-2$ (rule 4),

$$\text{ox. state Fe} + (-2) = 0$$

$$\text{ox. state Fe} = \underline{+2}$$

**b.** $N_2O_5$: The oxidation states add to zero (rule 6), as shown by the equation

$$2(\text{ox. state N}) + 5(\text{ox. state O}) = 0$$

Since ox. state O is $-2$ (rule 4),

$$2(\text{ox. state N}) + 5(-2) = 0$$
$$2(\text{ox. state N}) = +10$$
$$\text{ox. state N} = \underline{\underline{+5}}$$

**c.** $H_2SO_3$: The oxidation states add to zero (rule 6).

$$2(\text{ox. state H}) + (\text{ox. state S}) + 3(\text{ox. state O}) = 0$$

H is usually $+1$ and O is usually $-2$ (rules 4 and 5).

$$2(+1) + \text{ox. state S} + 3(-2) = 0$$
$$\text{ox. state S} = \underline{\underline{+4}}$$

**d.** $AsO_4^{3-}$: The sum of the oxidation states of the atoms equals the charge on the ion (rule 6).

$$(\text{ox. state As}) + 4(\text{ox. state O}) = -3$$

Since O is $-2$ (rule 4),

$$\text{ox. state As} + 4(-2) = -3$$
$$\text{ox. state As} = \underline{\underline{+5}}$$

By noting the change of oxidation state of the same atom in going from a reactant to a product, we can trace the exchange of electrons in the reaction. We can now add to our definition of oxidation and reduction in terms of oxidation state. *Oxidation is a loss of electrons as indicated by an increase in the oxidation state. Reduction is a gain of electrons as indicated by a decrease in the oxidation state.* In the reaction of sodium with chlorine discussed in section 13-1, notice that the oxidation state of the sodium increased from zero to $+1$, indicating oxidation, and that of chlorine decreased from zero to $-1$, indicating reduction.

In most compounds, usually only one of the elements undergoes a change in redox reactions. Thus it is often necessary to calculate the oxidation states of all of the elements in all compounds so that we can see which ones have undergone the change. With experience, however, the oxidized and reduced species are more easily recognized. In the following examples, we will find all the oxidation states so that we can identify the changes and label them appropriately.

---

**Identification of Oxidized and Reduced Species**

Example 13-2

In the following unbalanced equations, indicate the reactant oxidized, the reactant reduced, the oxidizing agent, and the reducing agent. Indicate the products that contain the elements that were oxidized or reduced.

working it out

**a.** $Al + HCl \longrightarrow AlCl_3 + H_2$     **b.** $CH_4 + O_2 \longrightarrow CO_2 + H_2O$

**c.** $MnO_2 + HCl \longrightarrow MnCl_2 + Cl_2 + H_2O$

**d.** $K_2Cr_2O_7 + SnCl_2 + HCl \longrightarrow CrCl_3 + SnCl_4 + KCl + H_2O$

**Procedure**

In the equations, we wish to identify the species that contain atoms of an element undergoing a change in oxidation state. At first, it may be necessary to calculate the oxidation state of every atom in the equation until you can recognize the changes by inspection. You will notice that any substance present as a free element is involved in either the oxidation or the reduction process, and that hydrogen and oxygen are generally not oxidized or reduced unless they are present as free elements.

**Solution**

a. Oxidation states of elements:

$$\overset{0}{Al} + \overset{+1-1}{HCl} \longrightarrow \overset{+3-1}{AlCl_3} + \overset{0}{H_2}$$

| Reactant | Change | Agent | Product |
|---|---|---|---|
| Al | oxidized | reducing | AlCl$_3$ |
| HCl | reduced | oxidizing | H$_2$ |

b. Oxidation states of elements:

$$\overset{-4+1}{CH_4} + \overset{0}{O_2} \longrightarrow \overset{+4-2}{CO_2} + \overset{+1-2}{H_2O}$$

| Reactant | Change | Agent | Product |
|---|---|---|---|
| CH$_4$ | oxidized | reducing | CO$_2$ |
| O$_2$ | reduced | oxidizing | CO$_2$, H$_2$O |

c. Oxidation states of elements:

$$\overset{+4-2}{MnO_2} + \overset{+1-1}{HCl} \longrightarrow \overset{+2-1}{MnCl_2} + \overset{0}{Cl_2} + \overset{+1-2}{H_2O}$$

| Reactant | Change | Agent | Product |
|---|---|---|---|
| HCl | oxidized | reducing | Cl$_2$ |
| MnO$_2$ | reduced | oxidizing | MnCl$_2$ |

d. Oxidation states of elements:

$$\overset{+1+6-2}{K_2Cr_2O_7} + \overset{+2-1}{SnCl_2} + \overset{+1-1}{HCl} \longrightarrow \overset{+3-1}{CrCl_3} + \overset{+1-1}{KCl} + \overset{+4-1}{SnCl_4} + \overset{+1-2}{H_2O}$$

| Reactant | Change | Agent | Product |
|---|---|---|---|
| SnCl$_2$ | oxidized | reducing | SnCl$_4$ |
| K$_2$Cr$_2$O$_7$ | reduced | oxidizing | CrCl$_3$ |

# 13-3 Balancing Redox Equations: Oxidation State Method

**Looking Ahead!** An important principle of redox reactions is that "electrons lost equal electrons gained." In the next two sections, we will put this concept to use as the key to balancing some rather complex reactions that would be quite difficult to balance by inspection as we did in Chapter 7. There are two procedures for this endeavor. The oxidation state method, discussed in this section, is useful for balancing molecular equations.

In a typical redox reaction, generally there are only two atoms involved that undergo changes in oxidation states. By identifying these two atoms and calculating the change we can arrive at a balanced equation. *The **oxidation state** or **bridge method** focuses on the atoms of the elements undergoing a change in oxidation state.*

The following reaction will be used to illustrate the procedures for balancing equations by the oxidation state method.

$$HNO_3(aq) + H_2S(aq) \longrightarrow NO(g) + S(s) + H_2O(l)$$

1. Identify the atoms whose oxidation states have changed.

$$
\begin{array}{cccc}
+5 & -2 & +2 & 0 \\
H\underline{N}O_3 + H_2\underline{S} & \longrightarrow & \underline{N}O + \underline{S} + H_2O
\end{array}
$$

2. Draw a bridge between the same atoms whose oxidation states have changed, indicating the electrons gained or lost. This is the change in oxidation state.

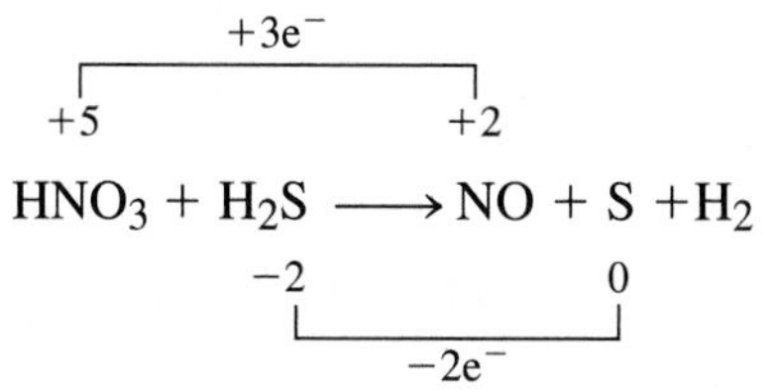

3. Multiply the two numbers ($+3$ and $-2$) by whole numbers that produce a common number. For 3 and 2 the common number is 6. (For example, $+3 \times \underline{2} = +6$; $-2 \times \underline{3} = -6$.) Use these multipliers as coefficients of the respective compounds or elements.

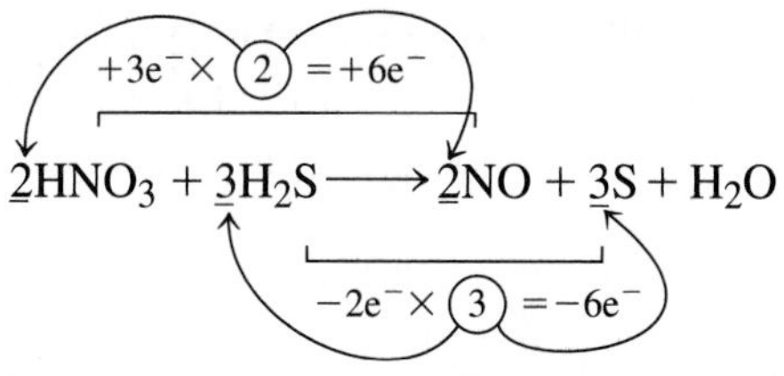

Note that six electrons are lost (bottom) and six are gained (top).

4. Balance the rest of the equation by inspection. Note that there are eight H's on the left, so *four* $H_2O$'s are needed on the right. If the equation has been balanced correctly, the O's should balance. Note that they do.

$$2HNO_3 + 3H_2S \longrightarrow 2NO + 3S + 4H_2O$$

| Example 13-3 | **Balancing Equations by the Oxidation State Method** |
| --- | --- |
| w o r k i n g   i t   o u t | Balance the following equations by the oxidation state method. |

**a.** $Zn + AgNO_3 \longrightarrow Zn(NO_3)_2 + Ag$

$$\overset{\overset{\displaystyle -2e^-}{\overbrace{\hspace{3cm}}}}{\underset{\underset{\displaystyle +1e^-}{\underbrace{\hspace{3cm}}}}{\overset{0 \qquad\qquad\qquad +2}{\underset{+1 \qquad\qquad\qquad\quad 0}{Zn + AgNO_3 \longrightarrow Zn(NO_3)_2 + Ag}}}}$$

The oxidation (top) should be multiplied by 1, and the reduction process (bottom) should be multiplied by 2.

$$\overset{\overset{\displaystyle -2e^- \times 1 = -2e^-}{\overbrace{\hspace{3cm}}}}{\underset{\underset{\displaystyle +1e^- \times 2 = +2e^-}{\underbrace{\hspace{3cm}}}}{Zn + 2AgNO_3 \longrightarrow Zn(NO_3)_2 + 2Ag}}$$

The final balanced equation is

$$Zn + 2AgNO_3 \longrightarrow Zn(NO_3)_2 + 2Ag$$

**b.** $Cu + HNO_3 \longrightarrow Cu(NO_3)_2 + H_2O + NO_2$

$$\overset{\overset{\displaystyle -2e^- \times 1 = -2e^-}{\overbrace{\hspace{3cm}}}}{\underset{\underset{\displaystyle +1e^- \times 2 = +2e^-}{\underbrace{\hspace{3cm}}}}{\overset{0 \qquad\qquad +2}{\underset{+5 \qquad\qquad\qquad\qquad +4}{Cu + HNO_3 \longrightarrow Cu(NO_3)_2 + H_2O + NO_2}}}}$$

The equation, so far, is

$$Cu + 2HNO_3 \longrightarrow Cu(NO_3)_2 + H_2O + 2NO_2$$

Note, however, that four N's are present on the right, but only two are on the left. The addition of two more $HNO_3$'s balances the N's, and the equation is completely balanced with two $H_2O$'s on the right.

$$Cu + 4HNO_3 \longrightarrow Cu(NO_3)_2 + 2H_2O + 2NO_2$$

(In this aqueous reaction, $HNO_3$ serves two functions. Two $HNO_3$'s are reduced to two $NO_2$'s, and the other two $HNO_3$'s provide anions for the $Cu^{2+}$ ion. These latter $NO_3^-$ ions are present in the solution as spectator ions. Spectator ions are not oxidized, reduced, or otherwise changed during the reaction.)

**When a copper penny reacts with nitric acid, nitrogen dioxide gas (brown) is formed.**

**c.** $O_2 + HI \longrightarrow H_2O + I_2$

The elements undergoing a change in oxidation state are oxygen and iodine. *If an atom that has changed is in a compound where it has a subscript other than one, first balance these atoms by adding a temporary coefficient.*

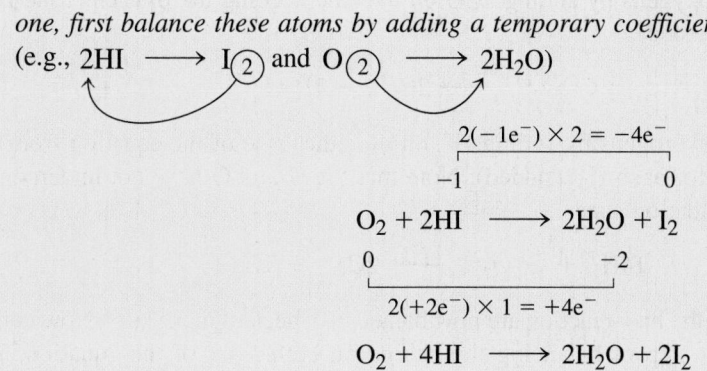

(e.g., $2HI \longrightarrow I_2$ and $O_2 \longrightarrow 2H_2O$)

$$2(-1e^-) \times 2 = -4e^-$$

$$\overset{-1}{\phantom{O_2 + 2}} \overset{\phantom{0}}{\phantom{2H_2O}} \overset{0}{}$$

$$O_2 + 2HI \longrightarrow 2H_2O + I_2$$

$$\overset{0}{\phantom{O_2 +}} \overset{-2}{\phantom{2H_2O}}$$

$$2(+2e^-) \times 1 = +4e^-$$

$$O_2 + 4HI \longrightarrow 2H_2O + 2I_2$$

# 13-4 Balancing Redox Equations: Ion-Electron Method

**Looking Ahead!** **One problem with the oxidation state method is that it may give the impression that some atoms exist as ions when they are actually *part* of a compound or ion (e.g., $N^{5+}$). The second method that we will discuss focuses on the entire ion or molecule containing the atom undergoing a change. In a way, this is a more realistic approach.**

While it is generally true that only two atoms actually undergo oxidation state changes, we can also consider the entire species that changes (e.g., $NO_3^-$ rather than $N^{5+}$). *In the* **ion-electron method** *(also known as the* **half-reaction method**), *the total reaction is separated into half-reactions, which are balanced individually and then added*. The ion-electron method recognizes the complete change of an ion or molecule as it goes from reactant to products. As we will see later in this chapter, a balanced half-reaction is how we represent a specific change that occurs in a battery.

The rules for balancing equations are somewhat different in acidic solution [containing $H^+(aq)$ ion] than in basic solution [containing $OH^-(aq)$ ion]. The two solutions are considered separately, with acid solution reactions discussed first. To simplify the equations, only the net ionic equations are balanced.

The balancing of an equation in aqueous acid solution is illustrated with the following unbalanced equation.

$$H^+(aq) + Cl^-(aq) + Cr_2O_7^{2-}(aq) \longrightarrow Cr^{3+}(aq) + Cl_2(g) + H_2O(l)$$

1. Separate the molecule or ion that contains atoms of an element that has been oxidized or reduced and the product containing atoms of that element. If necessary, calculate the oxidation states of individual atoms until you are able to recognize the species that changes. In this method, it is actually not necessary to know the oxidation state. The reduction process is

$$Cr_2O_7^{2-} \longrightarrow Cr^{3+}$$

2. If a subscript of the atoms of the element undergoing a change in oxidation state is more than one, balance those atoms with a temporary coefficient. In this case, it is the Cr.

$$Cr_2O_7^{2-} \longrightarrow 2Cr^{3+}$$

3. Balance the oxygens by adding $H_2O$ on the side needing the oxygens (one $H_2O$ for each O needed).

$$Cr_2O_7^{2-} \longrightarrow 2Cr^{3+} + 7H_2O$$

4. Balance the hydrogens by adding $H^+$ on the other side of the equation from the $H_2O$'s ($2H^+$ for each $H_2O$ added). Note that the H and O have not undergone a change in oxidation state.

$$14H^+ + Cr_2O_7^{2-} \longrightarrow 2Cr^{3+} + 7H_2O$$

5. The atoms in the half-reaction are now balanced. Check to make sure. Now comes the important step of balancing the charge on both sides of the equation. The charge is determined separately on each side of the equation. This is accomplished by multiplying the coefficient of any ion present times the charge on that ion. Neutral molecules are thus excluded from charge determination. Notice that there are 14 $H^+$ ions. The charge due to $H^+$ is thus 14 times the charge on the proton $(+1)$, or $+14$. This is then added to the charge on the dichromate $(-2)$, which makes a total charge on the left of $+12$ [i.e., $(14 \times +1) + (-2) = +12$]. On the right the charge on the Cr $(+3)$ is multiplied by its coefficient (2), making a charge of $+6$ on the right. The $H_2O$ is a neutral molecule, so it is not included. The total charge on the left is $+12$ and on the right it is $+6$. The charges on both sides of the reaction must now be balanced. To do this, add the appropriate number of *negative* electrons ($-1$ each) to the more *positive* side. Adding $6e^-$ on the left (the more positive side) balances the charges on both sides, and the half-reaction is balanced (i.e., $+12 - 6 = +6$).

$$6e^- + 14H^+ + Cr_2O_7^{2-} \longrightarrow 2Cr^{3+} + 7H_2O$$

6. Repeat the same procedure for the other half-reaction.

$$Cl^- \longrightarrow Cl_2$$
$$2Cl^- \longrightarrow Cl_2$$
$$2Cl^- \longrightarrow Cl_2 + 2e^-$$

7. Before the two half-reactions are added, we must make sure that electrons gained equal electrons lost. Sometimes, the half-reactions must be multiplied by factors that give the same number of electrons. In this case, if the oxidation process is multiplied by 3 (and the reduction process by 1), there will be an exchange of $6e^-$. When these two half-reactions are added, the $6e^-$ can be subtracted from both sides of the equation.

$$3(2Cl^- \longrightarrow Cl_2 + 2e^-)$$
$$6Cl^- \longrightarrow 3Cl_2 + 6e^-$$

Addition produces the balanced net ionic equation.

$$\cancel{6e^-} + 14H^+ + Cr_2O_7^{2-} \longrightarrow 2Cr^{3+} + 7H_2O$$
$$6Cl^- \longrightarrow 3Cl_2 + \cancel{6e^-}$$

$$\overline{14H^+(aq) + 6Cl^-(aq) + Cr_2O_7^{2-}(aq) \longrightarrow 2Cr^{3+}(aq) + 3Cl_2(g) + 7H_2O(l)}$$

An excellent way to check our answer is to make sure the net charge on both sides of the equation is the same. The net charge on the right is $(14 \times +1) + (6 \times -1) + (-2) = \underline{+6}$. On the left the charge is $(2 \times +3) = \underline{+6}$.

**Balancing Redox Equations in Acid Solution**

Balance the following equations for reactions occurring in acid solution by the ion-electron method.

**a.** $MnO_4^-(aq) + SO_2(g) + H_2O(l) \longrightarrow Mn^{2+}(aq) + SO_4^{2-}(aq) + H^+(aq)$

**Example 13-4**

w o r k i n g  i t  o u t

$$\text{Reduction:} \qquad MnO_4^- \longrightarrow Mn^{2+}$$

$$H_2O: \qquad MnO_4^- \longrightarrow Mn^{2+} + 4H_2O$$

$$H^+: \qquad 8H^+ + MnO_4^- \longrightarrow Mn^{2+} + 4H_2O$$

$$e^-: \quad 5e^- + 8H^+ + MnO_4^- \longrightarrow Mn^{2+} + 4H_2O$$

$$\text{Oxidation:} \qquad SO_2 \longrightarrow SO_4^{2-}$$

$$H_2O: \qquad 2H_2O + SO_2 \longrightarrow SO_4^{2-}$$

$$H^+: \qquad 2H_2O + SO_2 \longrightarrow SO_4^{2-} + 4H^+$$

$$e^-: \qquad 2H_2O + SO_2 \longrightarrow SO_4^{2-} + 4H^+ + 2e^-$$

The reduction reaction is multiplied by 2 and the oxidation by 5 to produce 10 electrons for each process, as shown below.

$$2(5e^- + 8H^+ + MnO_4^- \longrightarrow Mn^{2+} + 4H_2O)$$
$$5(2H_2O + SO_2 \longrightarrow SO_4^{2-} + 4H^+ + 2e^-)$$
$$\overline{\cancel{10e^-} + 16H^+ + 2MnO_4^- \longrightarrow 2Mn^{2+} + 8H_2O}$$
$$\underline{10H_2O + 5SO_2 \longrightarrow 5SO_4^{2-} + 20H^+ + \cancel{10e^-}}$$
$$10H_2O + 16H^+ + 5SO_2 + 2MnO_4^- \longrightarrow 5SO_4^{2-} + 2Mn^{2+} + 8H_2O + 20H^+$$

Note that $H_2O$ and $H^+$ are present on both sides of the equation. Therefore, $8H_2O$ and $16H^+$ can be subtracted from *both sides,* leaving the final balanced net ionic equation as

$$2MnO_4^-(aq) + 5SO_2(g) + 2H_2O(l) \longrightarrow 2Mn^{2+}(aq) + 5SO_4^{2-}(aq) + 4H^+(aq)$$

In checking our answer, we notice that the net charge on each side of the equation is $\underline{-2}$.

**b.** $Cu(s) + NO_3^-(aq) \longrightarrow Cu^{2+}(aq) + H_2O + NO(g)$

$$\text{Reduction:} \qquad NO_3^- \longrightarrow NO$$

$$H_2O: \qquad NO_3^- \longrightarrow NO + 2H_2O$$

$$H^+: \qquad 4H^+ + NO_3^- \longrightarrow NO + 2H_2O$$

$$e^-: \quad 3e^- + 4H^+ + NO_3^- \longrightarrow NO + 2H_2O$$

$$\text{Oxidation:} \qquad Cu \longrightarrow Cu^{2+}$$

$$e^-: \qquad Cu \longrightarrow Cu^{2+} + 2e^-$$

Multiply the reduction half-reaction by 2 and the oxidation half-reaction by 3, and then add the two half-reactions.

$$6e^- + 8H^+ + 2NO_3^- \longrightarrow 2NO + 4H_2O$$
$$\underline{3Cu \longrightarrow 3Cu^{2+} + 6e^-}$$
$$8H^+(aq) + 2NO_3^-(aq) + 3Cu(s) \longrightarrow 3Cu^{2+}(aq) + 2NO(g) + 4H_2O(l)$$

Our answer seems correct since the net charge on each side of the equation is $\underline{+6}$.

In a basic solution, $OH^-$ is predominant rather than $H^+$. Perhaps the simplest way to adjust to this condition is to follow the same procedure as in acid solution but neutralize any $H^+$ ions remaining with $OH^-$ ions in an additional, final step. We can do this by adding one $OH^-$ to *both sides* of the equation for each $H^+$ present in the equation. As we learned in the previous chapter, the $H^+$ ion combines with an $OH^-$ ion to form $H_2O$ [i.e., $H^+(aq) + OH^-(aq) \longrightarrow H_2O(l)$]. In effect, this procedure converts an $H^+$ ion to an $H_2O$ on one side, leaving one $OH^-$ ion on the opposite side of the equation. For example, consider the half-reaction shown below, which is balanced in acid solution. We will adjust the equation to a basic solution as follows.

1. Reduction half-reaction balanced in acid solution:

$$2e^- + 2H^+ + ClO^- \longrightarrow Cl^- + H_2O$$

2. Add $2OH^-$ to both sides:

$$2e^- + (2H^+ + \mathbf{2OH^-}) + ClO^- \longrightarrow Cl^- + H_2O + \mathbf{2OH^-}$$

3. Convert $H^+ + OH^-$ to $H_2O$:

$$2e^- + \mathbf{2H_2O} + ClO^- \longrightarrow Cl^- + H_2O + \mathbf{2OH^-}$$

4. Simplify by subtracting one $H_2O$ from both sides of the equation:

$$2e^- + H_2O + ClO^- \longrightarrow Cl^- + 2OH^-$$

Now consider a typical unbalanced equation representing a total redox reaction. The presence of $OH^-$ ion in the equation tells us that this reaction occurs in basic solution.

$$MnO_4^-(aq) + C_2O_4^{2-}(aq) + OH^-(aq) \longrightarrow MnO_2(s) + CO_3^{2-}(aq) + H_2O(l)$$

Because tables of half-reactions are represented as either acidic (with $H^+$ ions) or basic (with $OH^-$ ions), we will change each half-reaction to basic solution. (Alternatively, we could balance the total equation as if in acid solution and then change the final balanced equation to basic solution.)

1. Balance the reduction half-reaction involving $MnO_4^-$ and $MnO_2$ in acid solution.

$$3e^- + 4H^+ + MnO_4^- \longrightarrow MnO_2 + 2H_2O$$

2. Change to basic solution by adding $4OH^-$ to each side.

$$3e^- + (4H^+ + 4OH^-) + MnO_4^- \longrightarrow MnO_2 + 2H_2O + 4OH^-$$

3. Combine $4H^+$ and $4OH^-$ to form $4H_2O$ and then subtract out $2H_2O$.

$$3e^- + 2H_2O + MnO_4^- \longrightarrow MnO_2 + 4OH^-$$

4. Balance the oxidation half-reaction involving $C_2O_4^{2-}$ and $CO_3^{2-}$ in acid solution.

$$2H_2O + C_2O_4^{2-} \longrightarrow 2CO_3^{2-} + 4H^+ + 2e^-$$

5. Change to basic solution as before by adding $4OH^-$ to each side and simplifying by subtracting out $2H_2O$.

$$4OH^- + C_2O_4^{2-} \longrightarrow 2CO_3^{2-} + 2H_2O + 2e^-$$

6. Multiply the reduction reaction by 2 and the oxidation half-reaction by 3 and add equations.

$$6e^- + \overset{4}{\cancel{4H_2O}} + \cancel{12}OH^- + 3C_2O_4^{2-} + 2MnO_4^- \longrightarrow$$

$$6CO_3^{2-} + \overset{2}{\cancel{6}}H_2O + 2MnO_2 + \cancel{8OH^-} + 6e^-$$

7. Simplify by subtracting out electrons, $4H_2O$, and $8OH^-$.

$$4OH^- + 3C_2O_4^{2-} + 2MnO_4^- \longrightarrow 6CO_3^{2-} + 2H_2O + 2MnO_2$$

---

**Balancing Redox Equations in Basic Solution**

**Example 13-5**

Balance the following equation in basic solution by the ion-electron method.

working it out

$$Bi_2O_3(s) + NO_3^-(aq) + OH^-(aq) \longrightarrow BiO_3^-(aq) + NO_2^-(aq) + H_2O(l)$$

Reduction reaction: $\qquad\qquad NO_3^- \longrightarrow NO_2^-$

Balance in acidic: $\qquad 2e^- + 2H^+ + NO_3^- \longrightarrow NO_2^- + H_2O$

Change to basic: $\quad 2e^- + 2H^+ + 2OH^- + NO_3^- \longrightarrow NO_2^- + H_2O + 2OH^-$

Simplify $H_2O$: $\qquad 2e^- + H_2O + NO_3^- \longrightarrow NO_2^- + 2OH^-$

Oxidation reaction: $\qquad\qquad Bi_2O_3 \longrightarrow 2BiO_3^-$

Balance in acidic: $\qquad 3H_2O + Bi_2O_3 \longrightarrow 2BiO_3^- + 6H^+ + 4e^-$

Change to basic: $\quad 6OH^- + 3H_2O + Bi_2O_3 \longrightarrow$

$$2BiO_3^- + (6H^+ + 6OH^-) + 4e^-$$

Simplify $H_2O$: $\qquad 6OH^- + Bi_2O_3 \longrightarrow 2BiO_3^- + 3H_2O + 4e^-$

Multiply the reduction reaction by 2 and add to the oxidation half-reaction.

$$4e^- + Bi_2O_3 + 2H_2O + 2NO_3^- + 6OH^- \longrightarrow$$
$$2BiO_3^- + 2NO_2^- + 4OH^- + 3H_2O + 4e^-$$

Simplify by subtracting out $2H_2O$, $4OH^-$, and $4e^-$:

$$Bi_2O_3(s) + 2NO_3^-(aq) + 2OH^-(aq) \longrightarrow 2BiO_3^-(aq) + 2NO_2^-(aq) + H_2O(l)$$

---

**Looking Back!** Just as acid–base reactions involve an exchange of an $H^+$ ion between reactants, redox reactions involve an exchange of electrons. One substance is oxidized and one is reduced. An important point is that the same number of electrons are gained by the substance reduced as are lost by the substance oxidized. Thus a redox reaction can be balanced by emphasizing the equal exchange of electrons.

---

## Learning Check A | checking it out

**A-1.** Fill in the blanks.

If all of the atoms in a compound were ions, the charge on the ion would be the same as its _____ _____ . For oxygen in compounds, this is usually _____ . Hydrogen is usually _____ in compounds. A substance oxidized undergoes a _____ of electrons and an _____ in oxidation state. This substance is also known as a _____ agent. The principle used to aid us in balancing equations is that electrons _____ equal electrons _____ .

**A-2.** What is the oxidation state of (a) B in $H_3BO_3$ and (b) S in $S_2O_3^{2-}$?

**A-3.** In the following reaction, indicate the substance oxidized, the substance reduced, the oxidizing agent, and the reducing agent.

$$ClO_2 + H_2O_2 \longrightarrow O_2 + Cl^-$$

**A-4.** Balance the following equations by the ion-electron method.

**a.** $NO_3^- + H_2SO_3 \longrightarrow SO_4^{2-} + NO$ (acid solution)

**b.** The reaction in problem A-3, which occurs in basic solution.

*Additional Examples: Exercises 13-1, 13-6, 13-10, 13-12, 13-16, 13-18, 13-20, 13-22, and 13-24.*

Section  **SPONTANEOUS AND NONSPONTANEOUS REDOX REACTIONS**

## 13-5 Predicting Spontaneous Redox Reactions

**Looking Ahead!** In the previous chapter, we found that favorable reactions occur between a stronger acid and a stronger base, forming a weaker acid and a weaker base. We say that such a reaction is spontaneous. Electron exchange reactions may also be favorable and spontaneous. By observing the direction of electron flow in a limited number of reactions, we can establish a table of oxidizing and reducing agents. We can use this table to predict the direction of a large number of reactions.

When zinc metal is immersed in an aqueous solution containing $Cu^{2+}$ ions, an oxidation–reduction reaction occurs spontaneously. (In Chapter 7, we referred to this same reaction as a single-replacement reaction.) The zinc metal is oxidized to $Zn^{2+}$ and the $Cu^{2+}$ ion is reduced to copper metal with the transfer of two electrons from the zinc metal to the copper ion. (See Figure 7-5.) The reverse reaction does not occur spontaneously since reactions are spontaneous (favorable) in one direction only. (See Figure 7-6.) The net ionic equation for the spontaneous reaction is

$$Zn(s) + Cu^{2+}(aq) \longrightarrow Zn^{2+}(aq) + Cu(s)$$

*A spontaneous reaction occurs between the stronger oxidizing agent and the stronger reducing agent to form weaker oxidizing and reducing agents.* From the direction of this reaction from left to right, we can thus draw two conclusions.

1. $Cu^{2+}$ is a better oxidizing agent than $Zn^{2+}$.
2. Zn metal is a better reducing agent than Cu metal.

With simple experiments like the one described involving Cu and Zn, we can make further observations. For example, we find that nickel metal also replaces aqueous $Cu^{2+}$ ions, as illustrated by the following equation.

$$Ni(s) + Cu^{2+}(aq) \longrightarrow Ni^{2+}(aq) + Cu(s)$$

Two more statements can now be made.

3. $Cu^{2+}$ is a better oxidizing agent than $Ni^{2+}$.
4. Ni metal is a better reducing agent than Cu metal.

Now let's compare zinc and nickel since both are better reducing agents than copper. In a third experiment like we have described, we would find that zinc metal forms a coating of nickel when immersed in a $Ni^{2+}$ solution but not the opposite. This spontaneous reaction is represented below.

$$Zn(s) + Ni^{2+}(aq) \longrightarrow Zn^{2+}(aq) + Ni(s)$$

From this we conclude

5. $Ni^{2+}$ is a better oxidizing agent than $Zn^{2+}$.
6. Zn metal is a better reducing agent than Ni metal.

**Table 13-1  Oxidizing Agents and Reducing Agents**

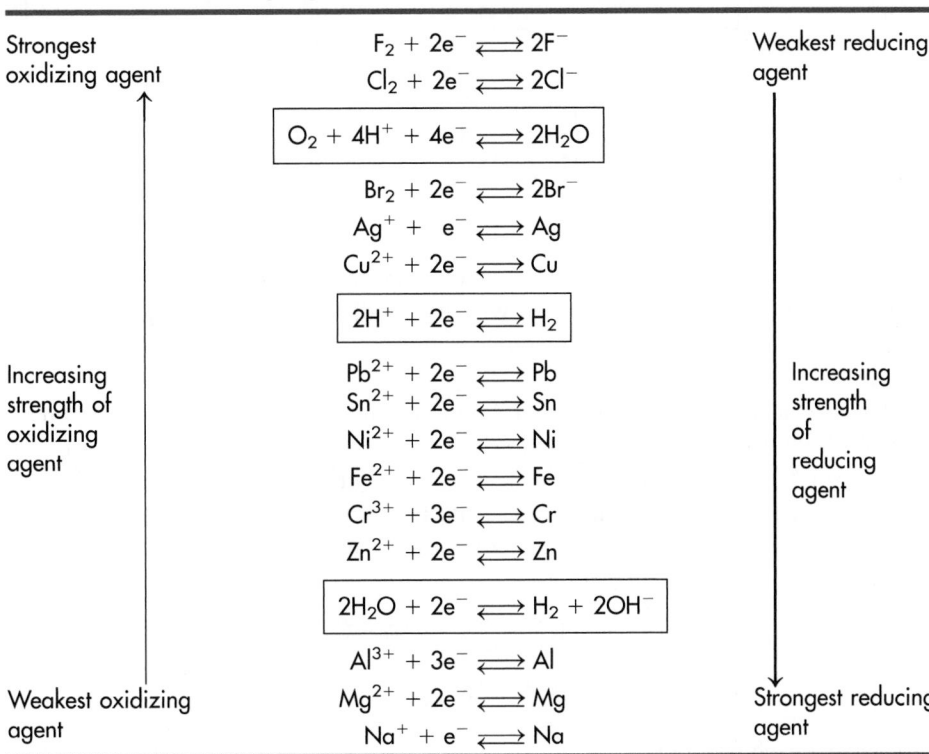

Strongest oxidizing agent

$F_2 + 2e^- \rightleftharpoons 2F^-$

$Cl_2 + 2e^- \rightleftharpoons 2Cl^-$

$O_2 + 4H^+ + 4e^- \rightleftharpoons 2H_2O$

$Br_2 + 2e^- \rightleftharpoons 2Br^-$

$Ag^+ + e^- \rightleftharpoons Ag$

$Cu^{2+} + 2e^- \rightleftharpoons Cu$

$2H^+ + 2e^- \rightleftharpoons H_2$

Increasing strength of oxidizing agent

$Pb^{2+} + 2e^- \rightleftharpoons Pb$

$Sn^{2+} + 2e^- \rightleftharpoons Sn$

$Ni^{2+} + 2e^- \rightleftharpoons Ni$

$Fe^{2+} + 2e^- \rightleftharpoons Fe$

$Cr^{3+} + 3e^- \rightleftharpoons Cr$

$Zn^{2+} + 2e^- \rightleftharpoons Zn$

$2H_2O + 2e^- \rightleftharpoons H_2 + 2OH^-$

$Al^{3+} + 3e^- \rightleftharpoons Al$

Weakest oxidizing agent

$Mg^{2+} + 2e^- \rightleftharpoons Mg$

$Na^+ + e^- \rightleftharpoons Na$

Weakest reducing agent

Increasing strength of reducing agent

Strongest reducing agent

As a result of our observations, we can rank these three ions as oxidizing agents in order of decreasing strength.

$$Cu^{2+} > Ni^{2+} > Zn^{2+}$$

*Notice that the strength of the metals as reducing agents is inversely related to the strength of their ions as oxidizing agents.* Because of the inverse relationship, the ranking of the three metals as reducing agents is in the reverse order.

$$Zn > Ni > Cu$$

These three metals were included in the activity series that was discussed in Chapter 7. In this chapter, we now explain the activity series by comparing the strengths of the metals as reducing agents and of their ions as oxidizing agents.

More experiments can provide additional ions and molecular species for our ranking. Eventually, we can construct a table of oxidizing agents ordered according to strength. Such a ranking is given in Table 13-1 and is an extension of the activity series of metals introduced in Chapter 7. In some cases, instruments are required to give quantitative values, known as *reduction potentials,* which indicate the comparative strength of a given oxidizing agent in relation to a defined standard. The strengths of the oxidizing agents are compared in these measurements at the same concentration for all ions involved (1.00 M) and at the same partial pressure of all gases involved (1.00 atm).

The strongest oxidizing agent ($F_2$) is at the top of Table 13-1, on the left. On the other hand, the reducing agents, shown on the right, become stronger *down* the table. A redox reaction takes place between an oxidizing agent on the left and a reducing agent on the right. *A favorable or spontaneous reaction occurs between an element, ion, or compound on the left (an oxidizing agent) with a species on the right (a reducing*

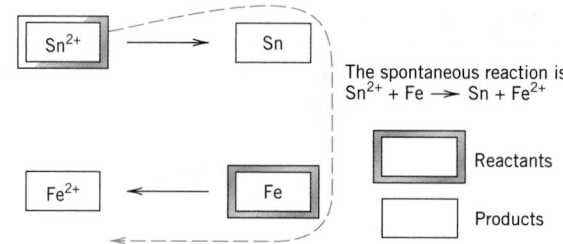

**Figure 13-2  Spontaneous Reaction**
**The stronger oxidizing agent reacts with the stronger reducing agent.**

*agent) that lies below it in the table.* The spontaneous reaction can be visualized as taking place in a clockwise direction, with the oxidizing agent reacting to the right and the reducing agent below it reacting in the opposite direction to the left. (See Figure 13-2.)

Of particular interest is the use of Table 13-1 to predict the reactions of certain elements with water. (All of the reactions shown in the table are assumed to occur in aqueous solution.) The boxed reaction near the top of the table represents the oxidation of water when read from right to left. Note that the gaseous elements $F_2$ and $Cl_2$ spontaneously oxidize water to produce oxygen gas and an acid. (The reaction of $Cl_2$ with water is quite slow, however.)

$$Cl_2 + 2e^- \rightleftharpoons 2Cl^-$$
$$O_2 + 4H^+ + 4e^- \rightleftharpoons 2H_2O$$

The spontaneous reaction is

$$2Cl_2 + 2H_2O \longrightarrow O_2 + 4H^+ + 4Cl^-$$

The boxed reaction near the bottom of the table represents the reduction of water when read from left to right. Note that the metals Al, Mg, and Na spontaneously reduce water to produce hydrogen gas and a base. *(We can say that Na reduces water, or, conversely, we can say that water oxidizes Na.)*

$$2H_2O + 2e^- \rightleftharpoons H_2 + 2OH^-$$
$$Na^+ + e^- \rightleftharpoons Na$$

The spontaneous reaction is

$$2H_2O + 2Na \longrightarrow H_2 + 2Na^+ + 2OH^-$$

These metals are known as *active* metals because of their chemical reactivity with water.

The third boxed reaction, near the middle of the table, represents the reduction of aqueous acid solutions (1.00 M $H^+$) to form hydrogen gas. Note that metals such as Ni and Fe are not oxidized by water but are oxidized by strong acid solutions.

$$2H^+ + 2e^- \rightleftharpoons H_2$$
$$Fe^{2+} + 2e^- \rightleftharpoons Fe$$

The spontaneous reaction is

$$2H^+ + Fe \longrightarrow Fe^{2+} + H_2$$

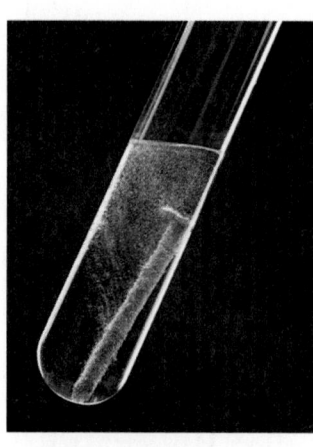

**Iron is oxidized by a strong acid solution to form iron(II) ions and hydrogen gas.**

Acid rain contains a considerably higher $H^+$ concentration than ordinary rain. From this discussion, we can understand why metals such as iron and nickel are more likely to be corroded by acid rain.

Two of the metals shown in the table—Cu and Ag—are not oxidized by either water or acid solutions.* Thus these metals are relatively unreactive and find use in jewelry and coins.

---

### The Direction of a Spontaneous Reaction

**Example 13-6**

*working it out*

A strip of tin is placed in an aqueous solution of $Cr(NO_3)_3$ in one experiment. In another, a strip of chromium is placed in an aqueous solution of $Sn(NO_3)_2$. Write the equation illustrating the spontaneous reaction that occurs.

**Procedure**

In Table 13-1, note that $Sn^{2+}$ is a stronger oxidizing agent than $Cr^{3+}$ and Cr is a stronger reducing agent than Sn. The reactants are, therefore, $Sn^{2+}$ and Cr. A balanced equation is obtained by multiplying the $Sn^{2+}$ half-reaction by 3 and the Cr half-reaction by 2 for an exchange of six electrons.

**Solution**

$$3Sn^{2+}(aq) + 2Cr(s) \longrightarrow 3Sn(s) + 2Cr^{3+}(aq)$$

---

### A Spontaneous Single-Replacement Reaction

**Example 13-7**

A strip of tin metal is placed in an $AgNO_3$ solution. If a reaction takes place, write the equation illustrating the spontaneous reaction.

**Procedure**

In Table 13-1, note that the oxidizing agent $Ag^+$ is above the reducing agent Sn. Therefore, a spontaneous reaction does occur.

**Solution**

$$2Ag^+(aq) + Sn(s) \longrightarrow Sn^{2+}(aq) + 2Ag(s)$$

---

### The Spontaneous Reaction of a Metal with Water

**Example 13-8**

A length of aluminum wire is placed in water. Does the aluminum react with water?

**Procedure**

In Table 13-1, note that aluminum is an active metal and should react with water (as an oxidizing agent).

**Solution**

$$6H_2O(l) + 2Al(s) \longrightarrow 2Al^{3+}(aq) + 6OH^-(aq) + 3H_2(g)$$

Since $Al(OH)_3$ is insoluble in water, however, the equation should be written as

$$6H_2O(l) + 2Al(s) \longrightarrow 2Al(OH)_3(s) + 3H_2(g)$$

Theoretically, aluminum should dissolve in water. Metallic aluminum is actually coated with $Al_2O_3$, which protects the metal from coming into contact with water. Thus it is a useful metal even for the hulls of boats, despite its high chemical reactivity.

---

*Copper is dissolved by nitric acid solutions but the copper is oxidized by the nitrate ion, not the hydronium ion. See Example 13-3(b).

# 13-6 Voltaic Cells

**Looking Ahead!** Spontaneous oxidation–reduction reactions can be put to unlimited important uses. The electrons that are exchanged can be detoured through a wire and put to work. This is the principle used in all the different types of batteries that we depend on to run our calculators, your uncle's pacemaker, and most importantly, the Walk-man. How this is accomplished is the topic of this section.

Releasing the brake on an automobile parked on the side of a hill causes no surprise. The automobile spontaneously rolls down the hill because the bottom of the hill represents a lower potential energy than the top. Chemical reactions occur spontaneously for the same reason. The products represent a position of lower potential energy than the reactants. In the case of the car, the difference in potential energy is due to position and the attraction of gravity; in the case of a chemical reaction, the difference in potential energy is due to the composition of the reactants and products. (See Figure 13-3.) When the car rolls down the hill, the difference in energy is transformed into the kinetic energy of the moving car and heat from friction. When a chemical reaction proceeds spontaneously from reactants to products, the difference is given off as heat, light, or electrical energy.

A **voltaic cell** *(also called a galvanic cell) uses a favorable or spontaneous redox reaction to generate electrical energy through an external circuit.* One of the earliest voltaic cells put to use the spontaneous reaction discussed in the previous section, which is

$$Zn(s) + Cu^{2+}(aq) \longrightarrow Zn^{2+}(aq) + Cu(s)$$

This voltaic cell is known as the Daniell cell and was used to generate electrical current for the new telegraph and doorbells in homes.

To generate electricity, however, the oxidation reaction must be separated from the reduction reaction so that the electrons exchanged can flow in an external wire where they can be put to use. The Daniell cell is illustrated in Figure 13-4. A zinc strip is immersed in a $Zn^{2+}$ solution, and, in a separate compartment, a copper strip is immersed in a $Cu^{2+}$ solution. A wire connects the two metal strips. The two metal strips are called electrodes. *The **electrodes** are the surfaces in a cell at which the reactions take place. The electrode at which oxidation takes place is called the **anode.** Reduction takes place at the **cathode.***

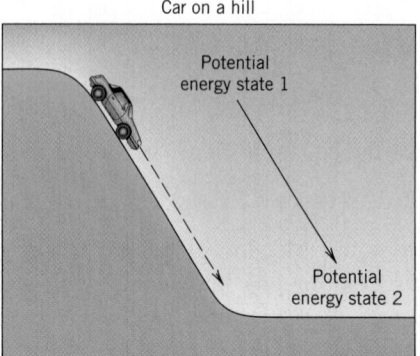
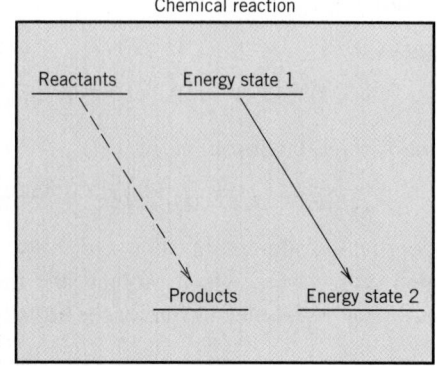

**Figure 13-3 Energy States A chemical reaction proceeds in a certain direction for the same reason that a car rolls down a hill.**

In the compartment on the left, the strip of Zn serves as the anode, since the following reaction occurs when the circuit is connected.

$$Zn \longrightarrow Zn^{2+} + 2e^-$$

When the circuit is complete, the two electrons travel in the external wire to the Cu electrode, which serves as the cathode. The reduction reaction occurs at the cathode.

$$2e^- + Cu^{2+} \longrightarrow Cu$$

To maintain neutrality in the solution, some means must be provided for the movement of a $SO_4^{2-}$ ion (or some other negative ion) from the right compartment where a $Cu^{2+}$ has been removed to the left compartment where a $Zn^{2+}$ has been produced. The *salt bridge* is an aqueous gel that allows ions to migrate between compartments but does not allow the mixing of solutions. (If $Cu^{2+}$ ions wandered into the left compartment, they would form a coating of Cu on the Zn electrode, thus short-circuiting the cell.)

As the cell discharges (generates electrical energy), the Zn electrode becomes smaller but the Cu electrode becomes larger. An important feature of this cell is that the reaction can be stopped by interrupting the external circuit with a switch. If the circuit is open, the electrons can no longer flow, so no further reaction occurs until the switch is again closed.

Two of the most common voltaic cells in use today are the dry cell (flashlight battery) and the lead–acid cell (car battery). *(The word* **battery** *means a collection of one or more separate cells joined together in one unit.)*

A three-cell lead–acid storage battery is illustrated in Figure 13-5. Each cell is composed of two grids separated by an inert spacer. One grid of a fully charged battery contains metallic lead. The other contains $PbO_2$, which is insoluble in $H_2O$. Both grids are immersed in a sulfuric acid solution (battery acid). When the battery is discharged by connecting the electrodes, the following half-reactions take place spontaneously.

*Anode:* $\qquad\qquad\qquad Pb(s) + H_2SO_4(aq) \longrightarrow PbSO_4(s) + 2H^+(aq) + 2e^-$

*Cathode:* $\quad 2e^- + 2H^+(aq) + PbO_2(s) + H_2SO_4(aq) \longrightarrow PbSO_4(s) + 2H_2O$

*Total*
*reaction:* $\qquad\quad Pb(s) + PbO_2(s) + 2H_2SO_4(aq) \longrightarrow 2PbSO_4(s) + 2H_2O$

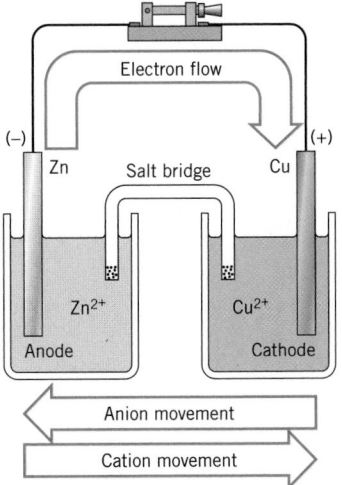

**Figure 13-4 The Daniell Cell** This chemical reaction produced electricity for the first telegraphs.

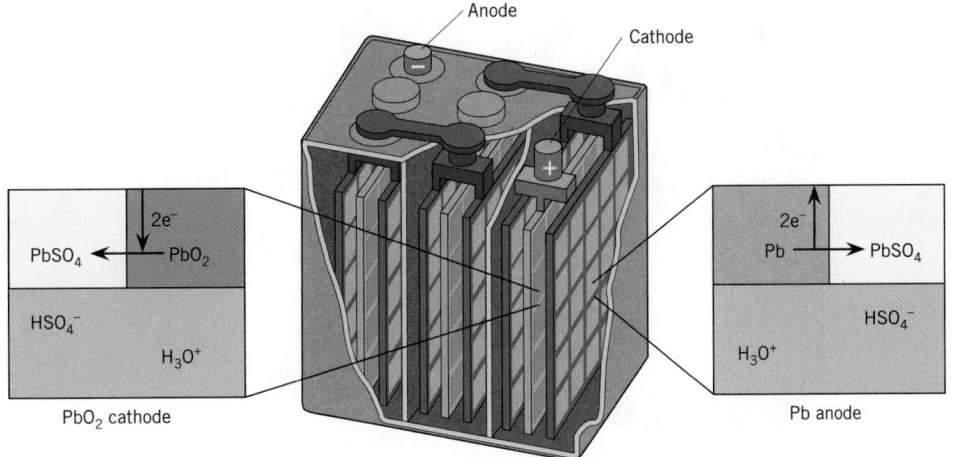
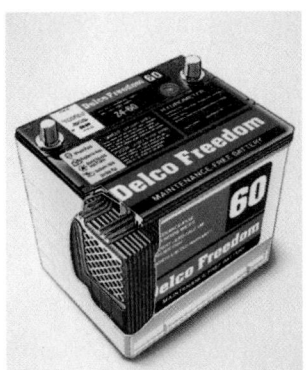

**Figure 13-5 Lead Storage Battery** The lead storage battery is rechargeable.

The electrons released at the Pb anode travel through the external circuit to run lights, starters, radios, or whatever is needed. The electrons return to the $PbO_2$ cathode to complete the circuit. As the reaction proceeds, both electrodes are converted to $PbSO_4$, and the $H_2SO_4$ is depleted. Since $PbSO_4$ is also insoluble, it remains attached to the grids as it forms. The degree of discharge of a battery can be determined by the density of the battery acid. Since the density of a fully discharged battery is 1.05 g/mL, the difference in density between this value and the density of a fully charged battery (1.35 g/mL) gives the amount of charge remaining in the battery. The hydrometer discussed in Chapter 2 is used to determine the density of the acid. As the electrodes convert to $PbSO_4$, the battery loses power and eventually becomes "dead."

The convenience of a car battery is that it can be recharged. After the engine starts, an alternator or generator is engaged to push electrons back into the cell in the opposite direction from which they came during discharge. This forces the reverse, nonspontaneous reaction to proceed.

$$2PbSO_4(s) + 2H_2O \longrightarrow Pb(s) + PbO_2(s) + 2H_2SO_4(aq)$$

When the battery is fully recharged, the alternator shuts off, the circuit is open, and the battery is ready for the next start.

The dry cell (invented by Leclanché in 1866) is not rechargeable to any extent but is comparatively inexpensive and easily portable. (In contrast, the lead–acid battery is heavy and expensive and must be kept upright.) The dry cell illustrated in Figure 13-6 consists of a zinc anode, which is the outer rim, and an inert graphite electrode. (An inert electrode provides a reaction surface but does not itself react.) In between is an aqueous paste containing $NH_4Cl$, $MnO_2$, and carbon. The reactions are as follows.

*Anode:* $\qquad\qquad\qquad Zn(s) \longrightarrow Zn^{2+}(aq) + 2e^-$

*Cathode:* $\quad 2NH_4^+(aq) + 2MnO_2(s) + 2e^- \longrightarrow Mn_2O_3(s) + 2NH_3(aq) + H_2O$

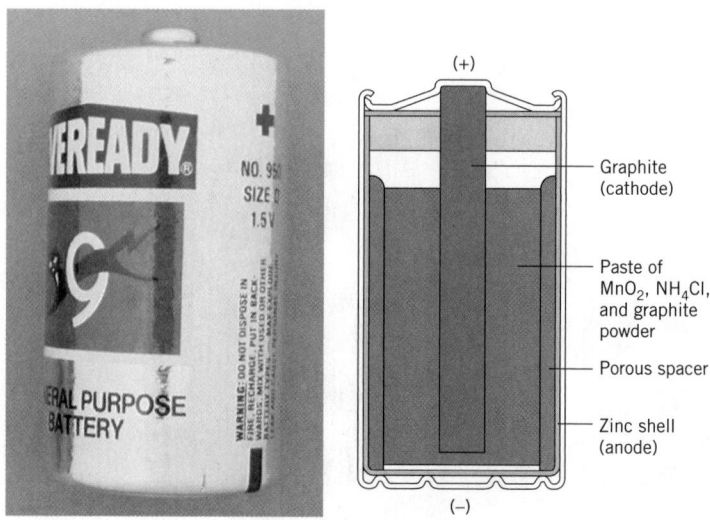

**Figure 13-6  Dry Cell** **The dry cell is comparatively inexpensive, light, and portable.**

A disadvantage of the dry cell is that the $NH_4^+$ ion creates an acidic solution. (In the previous chapter we discussed how this cation undergoes hydrolysis to form a weakly acidic solution.) The zinc electrode slowly reacts with the weakly acidic solution, so the shelf-life of these batteries is a matter of only a few months. In an alkaline battery, NaOH or KOH is substituted for the $NH_4Cl$, so the solution is basic and zinc reacts much more slowly. Alkaline batteries are more expensive but they have much longer shelf-lives. The anode reaction in the alkaline battery is

$$Zn(s) + 2OH^-(aq) \longrightarrow ZnO(s) + H_2O(l) + 2e^-$$

Other types of batteries that have been developed in modern times are more useful for calculators and wristwatches. These batteries are very small and deliver a small amount of current for a long time. Some batteries can produce current for three to five years. One is the *silver battery,* which uses a zinc anode and a silver(I) oxide cathode. Another, known as the *mercury battery,* also uses a zinc anode but a mercury(II) oxide cathode. (See Figure 13-7.) The two reactions in these batteries are

*Silver battery:* $\quad Zn(s) + Ag_2O(s) \longrightarrow ZnO(s) + 2Ag(s)$

*Mercury battery:* $\quad Zn(s) + HgO(s) \longrightarrow ZnO(s) + Hg(l)$

A battery that substitutes for dry cells is known as a *nickel–cadmium battery.* The advantage of this battery is that it is rechargeable. However, it is considerably more expensive initially. The reversible reaction that takes place in this cell is

$$Cd(s) + NiO_2(s) + 2H_2O(l) \longrightarrow Ni(OH)_2(s) + Cd(OH)_2(s)$$

Space travel requires a tremendous source of electrical energy. The requirements are that the source be continuous (no recharging necessary), lightweight, and dependable. Solar energy directly converts rays from the sun into electricity but is not practical for shorter runs such as the space shuttle. Although expensive, a source of power that fills the bill nicely is the fuel cell. A **fuel cell** *uses the direct reaction of hydrogen and oxygen to produce electrical energy.* Figure 13-8 is an illustration of a fuel cell. Hydrogen and oxygen gases are fed into the cell where they form water. As long as the gases enter the cell, power is generated. The water that is formed can be removed and used for other purposes in the spacecraft. Since reactants are supplied

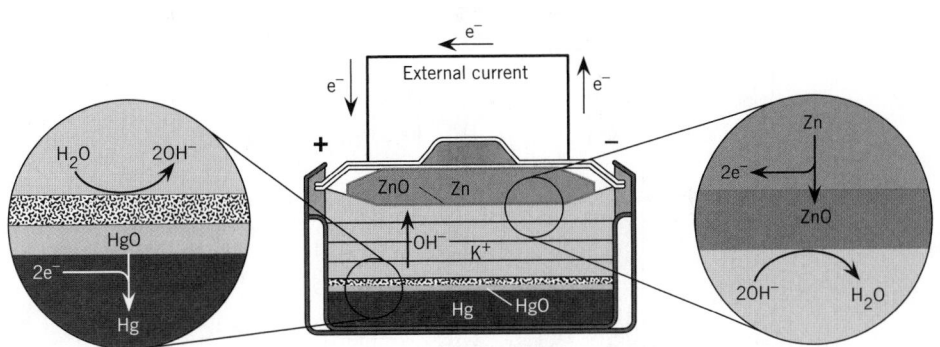

**Figure 13-7  Mercury Battery These small batteries deliver electrical current for long periods.**

## Fuel Cells—The Future Is (Almost) Here

The simplest combustion reaction, hydrogen plus oxygen to form water, is the key to the future. On a mass of fuel basis, it is the most energetic reaction. The combustion of two moles (4 g) of hydrogen produces 570 kJ of energy, much more than any other fuel. Since the combustion of hydrogen is a redox reaction, the energy can be released as electrical energy. This is the principle of the fuel cell, first demonstrated in the nineteenth century. It has long been used in the space program by NASA to continuously generate electricity without releasing toxic products. The two reactions that occur in the fuel cell and a diagram are illustrated in the text.

The use of the fuel cell has been limited by the expense of inert electrodes that have traditionally contained rare platinum and/or palladium, which are more expensive than gold. Also, although hydrogen was previously discussed as the perfect fuel, it is hazardous to handle, it is not cheap in the pure state, nor is it easy to store in large quantities. Another problem involves the high operating temperatures necessary in a fuel cell.

All of these problems, however, are being addressed with huge amounts of investment and research. Much of the hope of making fuel cells practical in the short run rests in using other fuels such as methane (natural gas), butane, methanol, or even gasoline rather than pure hydrogen. These compounds are readily available and easy to transport. The problem with these fuels is that they tend to gunk up the electrodes and they produce carbon dioxide, which is a greenhouse gas. Still, it would be much easier to dispense natural gas or even butane at a service station rather than pure hydrogen. Hydrogen is ideal for the fuel cell, but its large-scale application may have to wait.

If you have not yet heard much about fuel cells, you will. Experimental cars and buses using fuel cells are already on the road. Reports come in almost daily about advancements in the research on fuel cells. The operating temperatures keep coming down, new and cheaper materials for the electrodes are being announced, and more efficient use of fuels other than hydrogen is being developed. Eventually, we will even have fuel cells generating electrical power scattered around a city, eliminating the need for overhead transmission lines. Research is in progress to miniaturize the fuel cell so it can be used in cell phones and calculators. It would be smaller, cheaper (eventually), and easily refueled. Fuel cells are the power source of the future.

*Special vehicles are currently testing fuel cells as a power source.*

from external sources, the electrodes are not consumed, and the cell does not have to be shut down to be regenerated as a car battery does. Fuel cells have had some large-scale application in power generation for commercial purposes, but they are quite expensive. The best inert electrode surfaces at which the gases react are made of the

### Figure 13-8 Fuel Cell
**The fuel cell can generate power without interruption for recharging.**

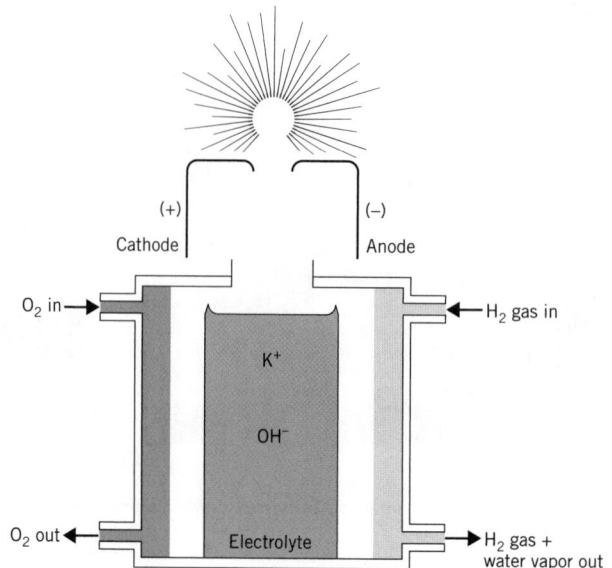

extremely expensive metal platinum. The reactions that take place in a basic solution in the fuel cell are:

$$\text{Anode:} \quad H_2(g) + 2OH^-(aq) \longrightarrow 2H_2O + 2e^-$$

$$\text{Cathode:} \quad O_2(g) + 2H_2O + 4e^- \longrightarrow 4OH^-(aq)$$

$$\text{Overall:} \quad 2H_2(g) + O_2(g) \longrightarrow 2H_2O$$

More efficient and durable batteries are the subject of much research at the current time. For example, an efficient and relatively inexpensive electric automobile is the object of intensive efforts. A small electric car using lead–acid batteries requires at least 18 batteries. These have to be replaced every year or so, depending on use. Also, much of the power of these heavy batteries must be used just to move the batteries around, not including the car and passengers. There have been some encouraging possibilities for lighter, more durable batteries, and several types of electric automobiles are in the testing phase.

## 13-7 Electrolytic Cells

**Looking Ahead!** Spontaneous reactions are like cars rolling down a hill. They go to lower energy states. Can a car or a reaction go up a hill? Of course, but in order for this to happen, energy in the form of a push for a car or energy for a chemical reaction must be supplied. Nonspontaneous redox reactions can occur if sufficient electrical energy is supplied from an outside source. This is the final topic of this chapter.

Back in the good old days before inexpensive but strong plastics, automobiles were equipped with beautiful chrome bumpers and other chromium accessories. Chromium not only looks great, but it does not rust, so it protects the underlying iron (which does rust) from exposure to air and water. How is iron metal coated with a layer of chromium? According to Table 13-1, we would *not* predict that a coating of chromium would spontaneously form on iron immersed in a $Cr^{3+}$ solution. But that doesn't mean we can't *make* it happen. Nonspontaneous redox reactions occur if enough electrical energy is supplied from an outside source. *Cells that convert electrical energy into chemical energy are called* **electrolytic cells.** They involve nonspontaneous redox reactions.

An example of an electrolytic cell is shown in Figure 13-9. When sufficient electrical energy is supplied to the electrodes from an outside source, the following nonspontaneous reaction occurs.

$$2H_2O(l) \longrightarrow 2H_2(g) + O_2(g)$$

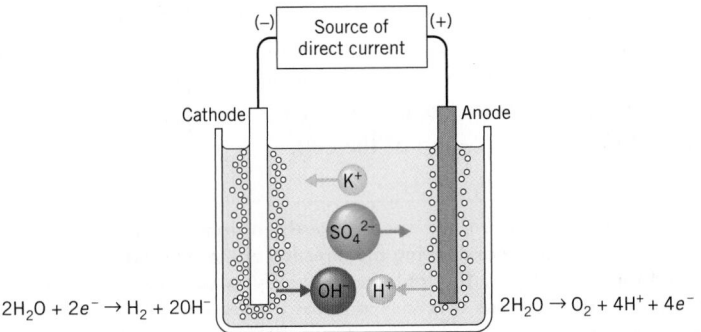

**Figure 13-9  An Electrolytic Cell  Electrolysis of a solution of potassium sulfate gives hydrogen gas and oxygen gas as products.**

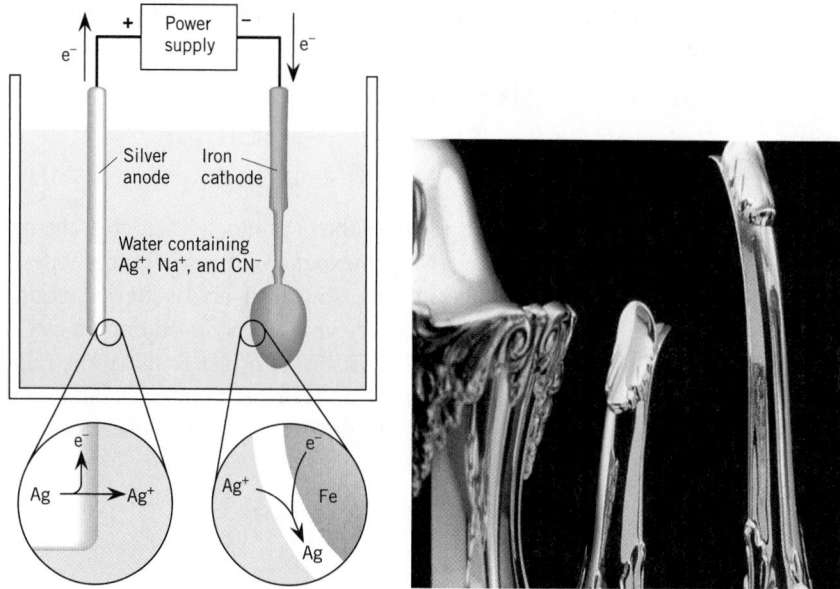

**Figure 13-10 Electroplating With an input of energy, a spoon can be coated with silver. The service has been electroplated with silver.**

For this electrolysis to occur, an electrolyte such as $K_2SO_4$ must be present in solution. Pure water alone does not have a sufficient concentration of ions to allow conduction of electricity.

Another example of an electrolytic cell is the recharge cycle of the lead–acid battery described in an earlier section. When energy from the engine activates the alternator, electrical energy is supplied to the battery, and the nonspontaneous reaction occurs as an electrolysis reaction. This reaction re-forms the original reactants.

Electrolysis has many useful applications. For example, in addition to chromium, silver and gold can be electroplated onto cheaper metals. In Figure 13-10 the metal spoon is the cathode and the silver bar serves as the anode. When electricity is supplied, the Ag anode produces $Ag^+$ ions, and the spoon cathode reduces $Ag^+$ ions to form a layer of Ag. The silver-plated spoon can be polished and made to look as good as sterling silver, which is more expensive.

Electrolytic cells are used to free elements from their compounds. Such cells are especially useful where metals are held in their compounds by strong chemical bonds. Examples are the metals aluminum, sodium, and magnesium. All aluminum is produced by the electrolysis of molten aluminum salts. Commercial quantities of sodium and chlorine are also produced by electrolysis of molten sodium chloride. An apparatus used for the electrolysis of molten sodium chloride is illustrated in Figure 13-11. At the high temperature required to keep the NaCl in the liquid state, sodium forms as a liquid and is drained from the top of the cell.

**Looking Back!** Redox reactions proceed spontaneously in only one direction. By matching oxidizing agents against each other and noting the direction of the reaction, we can construct a table of oxidizing and reducing agents. The table is a powerful predictive tool for many redox reactions. The direction of spontaneous reactions is toward lower energy. The energy difference can be harnessed in voltaic cells. Each common battery type uses a given spontaneous chemical reaction as a source of electricity. Reactions that proceed to a higher energy state can occur with an input of energy. These reactions have important applications.

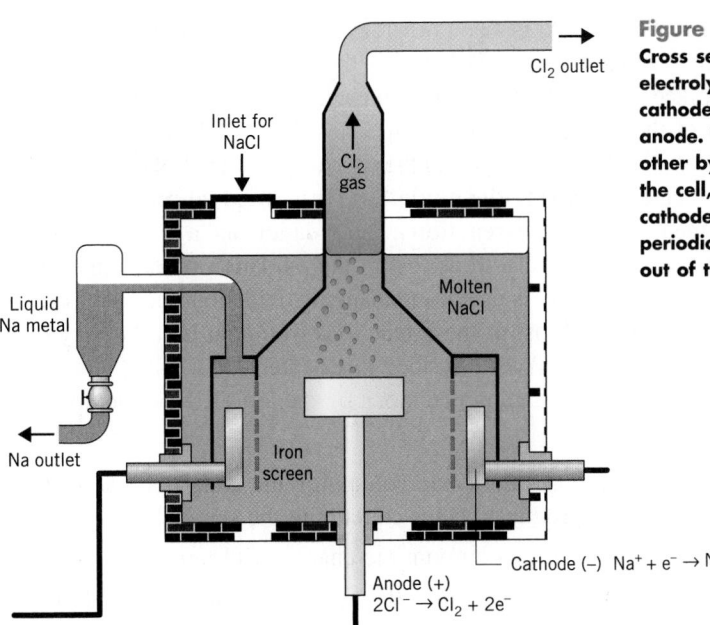

**Figure 13-11  Electrolysis of NaCl**
**Cross section of the Downs cell used for the electrolysis of molten sodium chloride. The cathode is a circular ring that surrounds the anode. The electrodes are separated from each other by an iron screen. During the operation of the cell, molten sodium collects at the top of the cathode compartment, from which it is periodically drained. The chlorine gas bubbles out of the anode compartment and is collected.**

## Learning Check B | checking it out

**B-1.** Fill in the blanks.

Spontaneous redox reactions occur between the _____ oxidizing and reducing agents to produce _____ oxidizing and reducing agents. In a table the stronger oxidizing agent is located _____ in the table than the reducing agent. A spontaneous chemical reaction is used in a _____ cell. An example is the Daniell cell, where _____ is oxidized at the _____ and _____ is reduced at the _____ . An electrolytic balance is maintained by means of a _____ bridge. In the lead–acid battery, _____ is _____ at the anode and _____ is reduced at the _____ . Nonspontaneous reactions occur in _____ cells.

**B-2.** In which of the following would a spontaneous reaction occur? Write the balanced equation for that reaction.

**a.** A strip of Pb is placed in a $Cr^{3+}$ solution.
**b.** A strip of Cr is placed in a $Pb^{2+}$ solution.

**B-3.** A cell is constructed of a $Br_2$, $Br^-$ half-cell connected to an $Fe^{2+}$, Fe half-cell. Write a balanced equation representing the spontaneous reaction that occurs.

**B-4.** Write the reaction that occurs when molten $MgCl_2$ is electrolyzed.

*Additional Examples: Exercises 13-27, 13-29, 13-36, 13-38, 13-40, and 13-43.*

chapter **R**eview                                    putting it together

A common characteristic of a large number of chemical reactions is an exchange of electrons between reactants. These reactions are known as **oxidation–reduction reactions,** or simply **redox reactions.** In such reactions, the reactant that gives up or loses the electrons is **oxidized,** and the reactant that gains the electrons is **reduced.**

The reactant oxidized is also known as a **reducing agent,** and the reactant reduced is known as an **oxidizing agent.**

To keep track of the electron exchange, we can follow the change in **oxidation states** of the elements in the compounds. The oxidation state increases in the substance oxidized and decreases in the substance reduced. We can use this understanding to balance equations by the **oxidation state (bridge) method.** A redox reaction can be divided into two **half-reactions:** an oxidation and a reduction. These two processes take place so that all electrons lost in the oxidation process are gained in the reduction process. This fact is useful in balancing oxidation–reduction reactions by the **ion-electron method.** Most of these reactions would, at best, be very difficult to balance by inspection methods as described in Chapter 7.

The rules for balancing equations by the ion-electron method are summarized as follows.

1. Identify the ion or molecule containing the atom that has a change in oxidation state and the product species containing the same atom.

2. Balance the atom undergoing the change with temporary coefficients.

3. Balance oxygens by adding the appropriate number of $H_2O$'s to the oxygen-deficient side.

4. Balance hydrogens by adding the appropriate number of $H^+$'s to the hydrogen-deficient side.

5. Add the appropriate number of electrons to the more positive side to balance the charge.

6. If the reaction takes place in basic solution, add one $OH^-$ to each side for each $H^+$ remaining in the equation. Combine the $H^+$ and $OH^-$ to form $H_2O$ and simplify the equation if necessary.

7. Multiply each half-reaction by a whole number so that electrons cancel when the half-reactions are added.

8. Subtract any species common to both sides of the equation.

Each substance has its own inherent strength as either an oxidizing agent or a reducing agent. A table can be constructed in which oxidizing and reducing agents are ranked by strength as determined by observation or measurements with electrical instruments. From this table, a great many other spontaneous reactions can be predicted. In Table 13-1, stronger oxidizing agents are ranked higher on the left and stronger reducing agents are ranked lower on the right. Reactions can thus be predicted as shown in Figure 13-2.

The reactions (or lack of reactions) of certain elements with water can be predicted from Table 13-1. Four groups of elements can be classified.

1. Nonmetals (e.g., $F_2$ and $Cl_2$) that are very strong oxidizing agents. These elements *oxidize* water to produce oxygen gas.

2. Metals (e.g., Na and Mg) that are very strong reducing agents. These elements *reduce* water to produce hydrogen gas.

3. Metals (e.g., Cu and Ag) that are very weak reducing agents. These elements are not oxidized by water or a strong acid (1.00 M) solution.

4. Metals (e.g., Ni and Fe) that are moderate reducing agents. These elements are not oxidized by water but are oxidized by aqueous acid.

Spontaneous chemical reactions occur because reactants are higher in chemical (potential) energy than products. We are familiar with many exothermic reactions where the difference in energy is released as heat. The energy can also be released as electrical energy in a **voltaic cell.** In a voltaic cell, the two half-reactions are physically separated so that electrons travel in an external circuit or wire between **electrodes.** The **anode** is the electrode at which oxidation takes place, and the **cathode** is the electrode at which reduction takes place. The Daniell cell, the car **battery,** the dry cell, and the **fuel cell** all involve spontaneous chemical reactions in which the chemical energy is converted directly into electrical energy.

Many reactions that are predicted to be unfavorable or nonspontaneous can be made to occur if electrical energy is supplied from an outside source. These are referred to as **electrolytic cells** and are useful in the commercial production of certain elements and in electroplating.

# Exercises

## Oxidation States

**13-1.** Give the oxidation states of the elements in the following compounds.
(a) $PbO_2$    (c) $C_2H_2$    (e) LiH    (g) $Rb_2Se$
(b) $P_4O_{10}$    (d) $N_2H_4$    (f) $BCl_3$    (h) $Bi_2S_3$

**13-2.** Give the oxidation states of the elements in the following compounds.
(a) $ClO_2$    (c) CO    (e) $Mn_2O_3$
(b) $XeF_2$    (d) $O_2F_2$    (f) $Bi_2O_5$

**13-3.** Which of the following elements form *only* the +1 oxidation state in compounds?
(a) Li    (c) Ca    (e) K    (g) Rb
(b) H    (d) Cl    (f) Al

**13-4.** Which of the following elements form *only* the +2 oxidation state in compounds?
(a) O    (c) Be    (e) Sc    (g) Ca
(b) B    (d) Sr    (f) Hg

**13-5.** What is the only oxidation state of Al in its compounds?

**13-6.** What is the oxidation state of each of the following?
(a) P in $H_3PO_4$    (d) Cr in $CaCr_2O_7$    (f) N in $CsNO_3$
(b) C in $H_2C_2O_4$    (e) S in $SF_6$    (g) Mn in $KMnO_4$
(c) Cl in $ClO_4^-$

**13-7.** What is the oxidation state of the specified atom?
(a) S in $SO_3$    (c) U in $UF_6$    (e) Cr in $K_2CrO_4$
(b) Co in $Co_2O_3$    (d) N in $HNO_3$    (f) Mn in $CaMnO_4$

**13-8.** What is the oxidation state of each of the following?
(a) Se in $SeO_3^{2-}$    (c) S in $Al_2(SO_3)_3$    (e) N in $(NH_4)_2S$
(b) I in $H_5IO_6$    (d) Cl in $HClO_2$

## Oxidation–Reduction

**13-9.** Which of the following reactions are oxidation–reduction reactions?
(a) $2H_2 + O_2 \longrightarrow 2H_2O$
(b) $CaCO_3 \longrightarrow CaO + CO_2$
(c) $2Na + 2H_2O \longrightarrow 2NaOH + H_2$
(d) $2HNO_3 + Ca(OH)_2 \longrightarrow Ca(NO_3)_2 + 2H_2O$
(e) $AgNO_3 + KCl \longrightarrow AgCl + KNO_3$
(f) $Zn + CuCl_2 \longrightarrow ZnCl_2 + Cu$

**13-10.** Identify each of the following half-reactions as either oxidation or reduction.
(a) $Na \longrightarrow Na^+ + e^-$
(b) $Zn^{2+} + 2e^- \longrightarrow Zn$
(c) $Fe^{2+} \longrightarrow Fe^{3+} + e^-$
(d) $O_2 + 4H^+ + 4e^- \longrightarrow 2H_2O$
(e) $S_2O_8^{2-} + 2e^- \longrightarrow 2SO_4^{2-}$

**13-11.** Identify each of the following changes as either oxidation or reduction.
(a) $P_4 \longrightarrow H_3PO_4$    (d) $Al \longrightarrow Al(OH)_4^-$
(b) $NO_3^- \longrightarrow NH_4^+$    (e) $S^{2-} \longrightarrow SO_4^{2-}$
(c) $Fe_2O_3 \longrightarrow Fe^{2+}$

**13-12.** For each of the following unbalanced equations, complete the table below.
(a) $MnO_2 + H^+ + Br^- \longrightarrow Mn^{2+} + Br_2 + H_2O$
(b) $CH_4 + O_2 \longrightarrow CO_2 + H_2O$
(c) $Fe^{2+} + MnO_4^- + H^+ \longrightarrow Fe^{3+} + Mn^{2+} + H_2O$

| Reaction | Reactant Oxidized* | Product of Oxidation | Reactant Reduced | Product of Reduction | Oxidizing Agent | Reducing Agent |
|---|---|---|---|---|---|---|
| (a) | | | | | | |
| (b) | | | | | | |
| (c) | | | | | | |

*Element, molecule, or ion.

**13-13.** For the following two unbalanced equations, construct a table like that in exercise 13-12.
(a) $Al + H_2O \longrightarrow AlO_2^- + H_2 + H^+$
(b) $Mn^{2+} + Cr_2O_7^{2-} + H^+ \longrightarrow MnO_4^- + Cr^{3+} + H_2O$

**13-14.** For the following equations, identify the reactant oxidized, the reactant reduced, the oxidizing agent, and the reducing agent.
(a) $Sn + HNO_3 \longrightarrow SnO_2 + NO_2 + H_2O$
(b) $IO_3^- + SO_2 + H_2O \longrightarrow I_2 + SO_4^{2-} + H^+$
(c) $CrI_3 + OH^- + Cl_2 \longrightarrow CrO_4^{2-} + IO_4^- + Cl^- + H_2O$
(d) $I^- + H_2O_2 \longrightarrow I_2 + H_2O + OH^-$

**13-15.** Identify the product or products containing the elements oxidized and reduced in exercise 13-14.

## Balancing Equations by the Oxidation State Method

**13-16.** Balance each of the following equations by the oxidation state method.

(a) $NH_3 + O_2 \longrightarrow NO + H_2O$

(b) $Sn + HNO_3 \longrightarrow SnO_2 + NO_2 + H_2O$

(c) $Cr_2O_3 + Na_2CO_3 + KNO_3 \longrightarrow$
$$CO_2 + Na_2CrO_4 + KNO_2$$

(d) $Se + BrO_3^- + H_2O \longrightarrow H_2SeO_3 + Br^-$

**13-17.** Balance the following equations by the oxidation state method.

(a) $I_2O_5 + CO \longrightarrow I_2 + CO_2$

(b) $Al + H_2O \longrightarrow AlO_2^- + H_2 + H^+$

(c) $HNO_3 + HCl \longrightarrow NO + Cl_2 + H_2O$

(d) $I_2 + Cl_2 + H_2O \longrightarrow HIO_3 + HCl$

## Balancing Equations by the Ion-Electron Method

**13-18.** Balance the following half-reactions in acidic solution.

(a) $Sn^{2+} \longrightarrow SnO_2$     (d) $I_2 \longrightarrow IO_3^-$

(b) $CH_4 \longrightarrow CO_2$     (e) $NO_3^- \longrightarrow NO_2$

(c) $Fe^{3+} \longrightarrow Fe^{2+}$

**13-19.** Balance the following half-reactions in acidic solution.

(a) $P_4 \longrightarrow H_3PO_4$     (d) $NO_3^- \longrightarrow NH_4^+$

(b) $ClO_3^- \longrightarrow Cl^-$     (e) $H_2O_2 \longrightarrow H_2O$

(c) $S_2O_3^{2-} \longrightarrow SO_4^{2-}$

**13-20.** Balance each of the following by the ion-electron method. All are in acidic solution.

(a) $S^{2-} + NO_3^- + H^+ \longrightarrow S + NO + H_2O$

(b) $I_2 + S_2O_3^{2-} \longrightarrow S_4O_6^{2-} + I^-$

(c) $SO_3^{2-} + ClO_3^- \longrightarrow Cl^- + SO_4^{2-}$

(d) $Fe^{2+} + H_2O_2 + H^+ \longrightarrow Fe^{3+} + H_2O$

(e) $AsO_4^{3-} + I^- + H^+ \longrightarrow I_2 + AsO_3^{3-} + H_2O$

(f) $Zn + H^+ + NO_3^- \longrightarrow Zn^{2+} + NH_4^+ + H_2O$

**13-21.** Balance each of the following by the ion-electron method. All are in acidic solution.

(a) $Mn^{2+} + BiO_3^- + H^+ \longrightarrow MnO_4^- + Bi^{3+} + H_2O$

(b) $IO_3^- + SO_2 + H_2O \longrightarrow I_2 + SO_4^{2-} + H^+$

(c) $Se + BrO_3^- + H_2O \longrightarrow H_2SeO_3 + Br^-$

(d) $P_4 + HClO + H_2O \longrightarrow H_3PO_4 + Cl^- + H^+$

(e) $Al + Cr_2O_7^{2-} + H^+ \longrightarrow Al^{3+} + Cr^{3+} + H_2O$

(f) $ClO_3^- + I^- + H^+ \longrightarrow Cl^- + I_2 + H_2O$

(g) $As_2O_3 + NO_3^- + H_2O \longrightarrow AsO_4^{3-} + NO + H^+$

**13-22.** Balance the following half-reactions in basic solution.

(a) $SnO_2^{2-} \longrightarrow SnO_3^{2-}$

(b) $ClO_2^- \longrightarrow Cl_2$

(c) $Si \longrightarrow SiO_3^{2-}$

(d) $NO_3^- \longrightarrow NH_3$

**13-23.** Balance the following half-reactions in basic solution.

(a) $Al \longrightarrow Al(OH)_4^-$

(b) $S^{2-} \longrightarrow SO_4^{2-}$

(c) $N_2H_4 \longrightarrow NO_3^-$

**13-24.** Balance each of the following by the ion-electron method. All are in basic solution.

(a) $S^{2-} + OH^- + I_2 \longrightarrow SO_4^{2-} + I^- + H_2O$

(b) $MnO_4^- + OH^- + I^- \longrightarrow MnO_4^{2-} + IO_4^- + H_2O$

(c) $BiO_3^- + SnO_2^{2-} + H_2O \longrightarrow$
$$SnO_3^{2-} + OH^- + Bi(OH)_3$$

(d) $CrI_3 + OH^- + Cl_2 \longrightarrow CrO_4^{2-} + IO_4^- + Cl^- + H_2O$

[*Hint:* In (d), two ions are oxidized; include both in one half-reaction.]

**13-25.** Balance each of the following by the ion-electron method. All are in basic solution.

(a) $ClO_2 + OH^- \longrightarrow ClO_2^- + ClO_3^- + H_2O$

(b) $OH^- + Cr_2O_3 + NO_3^- \longrightarrow CrO_4^{2-} + NO_2^- + H_2O$

(c) $Cr(OH)_4^- + BrO^- + OH^- \longrightarrow Br^- + CrO_4^{2-} + H_2O$

(d) $Mn^{2+} + H_2O_2 + OH^- \longrightarrow H_2O + MnO_2$

(e) $Ag_2O + Zn + H_2O \longrightarrow Zn(OH)_2 + Ag$

**13-26.** Balance the following two equations by the ion-electron method, first in acidic solution and then in basic solution.

(a) $H_2 + O_2 \longrightarrow H_2O$     (b) $H_2O_2 \longrightarrow O_2 + H_2O$

## Predicting Redox Reactions

**13-27.** Using Table 13-1, predict whether the following reactions occur in aqueous solution. If not, write N.R. (no reaction).

(a) $2Na + 2H_2O \longrightarrow H_2 + 2NaOH$

(b) $Pb + Zn^{2+} \longrightarrow Pb^{2+} + Zn$

(c) $Fe + 2H^+ \longrightarrow Fe^{2+} + H_2$

(d) $Fe + 2H_2O \longrightarrow Fe^{2+} + 2OH^- + H_2$

(e) $Cu + 2Ag^+ \longrightarrow 2Ag + Cu^{2+}$

(f) $2Cl_2 + 2H_2O \longrightarrow 4Cl^- + O_2 + 4H^+$

(g) $3Zn^{2+} + 2Cr \longrightarrow 2Cr^{3+} + 3Zn$

**13-28.** Using Table 13-1, predict whether the following reactions occur in aqueous solution. If not, write N.R.

(a) $Sn^{2+} + Pb \longrightarrow Pb^{2+} + Sn$

(b) $Ni^{2+} + H_2 \longrightarrow 2H^+ + Ni$

(c) $Cu + F_2 \longrightarrow CuF_2$

(d) $Ni^{2+} + 2Br^- \longrightarrow Ni + Br_2$

(e) $3Ni^{2+} + 2Cr \longrightarrow 2Cr^{3+} + 3Ni$

(f) $2Br_2 + 2H_2O \longrightarrow 4Br^- + O_2 + 4H^+$

**13-29.** If a reaction occurs, write the balanced molecular equation.

(a) Nickel metal is placed in water.

(b) Bromine is dissolved in water that is in contact with tin metal.

(c) Silver metal is placed in an $HClO_4$ solution.

(d) Oxygen gas is bubbled into an HBr solution.

(e) Liquid bromine is placed on a sheet of aluminum.

**13-30.** If a reaction occurs, write the balanced molecular equation.

(a) Bromine is added to an HCl solution.

(b) Sodium metal is heated with solid aluminum chloride.

(c) Iron is placed in a $Pb(ClO_4)_2$ solution.

(d) Oxygen gas is bubbled into an HCl solution.

(e) Fluorine gas is added to water.

**13-31.** Which of the following elements react with water: **(a)** Pb, **(b)** Ag, **(c)** $F_2$, **(d)** $Br_2$, **(e)** Mg? Write the balanced equation for any reaction that occurs.

**13-32.** Which of the following species will be reduced by hydrogen gas in aqueous solution: **(a)** $Br_2$, **(b)** Cr, **(c)** $Ag^+$, **(d)** $Ni^{2+}$? Write the balanced equation for any reaction that occurs.

**13-33.** In Chapter 12, we mentioned the corrosiveness of acid rain. Why does rain containing a higher $H^+(aq)$ concentration cause more damage to iron exposed in bridges and buildings than pure $H_2O$? Write the reaction between Fe and $H^+(aq)$.

**13-34.** $Br_2$ can be prepared from the reaction of $Cl_2$ with NaBr dissolved in seawater. Explain. Write the reaction. Can $Cl_2$ be used to prepare $F_2$ from NaF solutions?

## Voltaic Cells

**13-35.** What is the function of the salt bridge in the voltaic cell?

**13-36.** In an alkaline battery, the following two half-reactions occur.

$$Zn(s) + 2OH^-(aq) \longrightarrow Zn(OH)_2(s) + 2e^-$$

$$2MnO_2(s) + 2H_2O(l) + 2e^- \longrightarrow$$
$$2MnO(OH)(s) + 2OH^-(aq)$$

Which reaction takes place at the anode and which at the cathode? What is the total reaction?

**13-37.** The following overall reaction takes place in a silver oxide battery.

$$Ag_2O(s) + H_2O(l) + Zn(s) \longrightarrow Zn(OH)_2(s) + 2Ag(s)$$

The reaction takes place in basic solution. Write the half-reaction that takes place at the anode and the half-reaction that takes place at the cathode.

**13-38.** The nickel–cadmium (nicad) battery is used as a replacement for a dry cell because it is rechargeable. The overall reaction that takes place is

$$NiO_2(s) + Cd(s) + 2H_2O(l) \longrightarrow$$
$$Ni(OH)_2(s) + Cd(OH)_2(s)$$

Write the half-reactions that take place at the anode and the cathode.

**13-39.** Sketch a galvanic cell in which the following overall reaction occurs.

$$Ni^{2+}(aq) + Fe(s) \longrightarrow Fe^{2+}(aq) + Ni(s)$$

**(a)** What reactions take place at the anode and the cathode? **(b)** In what direction do the electrons flow in the wire? **(c)** In what direction do the anions flow in the salt bridge?

**13-40.** Describe how a voltaic cell could be constructed from a strip of iron, a strip of lead, an $Fe(NO_3)_2$ solution, and a $Pb(NO_3)_2$ solution. Write the anode reaction, the cathode reaction, and the total reaction.

**13-41.** Judging from the relative difference in the strengths of the oxidizing agents ($Fe^{2+}$ vs. $Pb^{2+}$) and ($Zn^{2+}$ vs. $Cu^{2+}$), which do you think would be the more powerful cell, the one in exercise 13-40 or the Daniell cell? Why?

\*13-42. The power of a cell depends on the strength of both the oxidizing and the reducing agents. Write the equation illustrating the most powerful redox reaction possible between an oxidizing agent and a reducing agent *in aqueous solution*. Consider only the species shown in Table 13-1.

## Electrolytic Cells

**13-43.** Chrome plating is an electrolytic process. Write the reaction that occurs when an iron bumper is electroplated using a $CrCl_3$ solution. Are there any metals shown in Table 13-1 on which a chromium layer would spontaneously form?

**13-44.** Why can't elemental sodium be formed in the electrolysis of an aqueous NaCl solution? Write the reaction that does occur at the cathode. How is elemental sodium produced by electrolysis?

**13-45.** Why can't elemental fluorine be formed by electrolysis of an aqueous NaF solution? Write the reaction that does occur at the anode. How is elemental fluorine produced?

**13-46.** A "tin can" is made by forming a layer of tin on a sheet of iron. Is electrolysis necessary for such a process or does it occur spontaneously? Write the equation for this reaction. Is electrolysis necessary to form a layer of tin on a sheet of lead? Write the relevant equation.

**13-47.** Certain metals can be purified by electrolysis. For example, a mixture of Ag, Zn, and Fe can be dissolved so that their metal ions are present in aqueous solution. If a solution containing these ions is electrolyzed, which metal ion would be reduced to the metal first?

## General Problems

**13-48.** Nitrogen exists in nine oxidation states. Arrange the following compounds in order of increasing oxidation state of N: $K_3N$, $N_2O_4$, $N_2$, $NH_2OH$, $N_2O$, $Ca(NO_3)_2$, $N_2H_4$, $N_2O_3$, NO.

**13-49.** Given the following information concerning metal strips immersed in certain solutions, write the net ionic equations representing the reactions that occur.

| Metal Strip | Solution | Reaction |
|---|---|---|
| Cd | $NiCl_2$ | Ni coating formed |
| Cd | $FeCl_2$ | no reaction |
| Zn | $CdCl_2$ | Cd coating formed |
| Fe | $CdCl_2$ | no reaction |

Where does $Cd^{2+}$ rank as an oxidizing agent in Table 13-1?

**13-50.** A hypothetical metal (M) forms a coating of Sn when placed in an $SnCl_2$ solution. However, when a strip of Ni is placed in an $MCl_2$ solution, a coating of the metal M forms on the nickel. Write the net ionic equations representing the reactions that occur. Where does $M^{2+}$ rank as an oxidizing agent in Table 13-1?

**13-51.** A solution of gold ions ($Au^{3+}$) reacts spontaneously with water to form metallic gold. Metallic gold does not react with chlorine but does react with fluorine. Write the equations representing the two spontaneous reactions, and locate $Au^{3+}$ in Table 13-1 as an oxidizing agent.

**13-52.** Given the following *unbalanced* equation,

$$H^+(aq) + Zn(s) + NO_3^-(aq) \longrightarrow$$
$$Zn^{2+}(aq) + N_2(g) + H_2O(l)$$

what mass of Zn is required to produce 0.658 g of $N_2$?

**13-53.** Given the following *unbalanced* equation,

$$MnO_2(s) + HBr(aq) \longrightarrow MnBr_2(aq) + Br_2(l) + H_2O(l)$$

what mass of $MnO_2$ reacts with 228 mL of 0.560 M HBr?

**13-54.** Given the following *unbalanced* equation,

$$H^+(aq) + NO_3^-(aq) + Cu_2O(s) \longrightarrow$$
$$Cu^{2+}(aq) + NO(g) + H_2O(l)$$

what volume of NO gas measured at STP is produced by the complete reaction of 10.0 g of $Cu_2O$?

**\*13-55.** Given the following *unbalanced* equation in acid solution,

$$H_2O(l) + HClO_3(aq) + As(s) \longrightarrow$$
$$H_3AsO_3(aq) + HClO(aq)$$

If 200 g of As reacts with 200 g of $HClO_3$, what mass of $H_3AsO_3$ is produced? (*Hint:* Calculate the limiting reactant.)

**\*13-56.** Given the following *unbalanced* equation in basic solution,

$$Zn(s) + NO_3^-(aq) \longrightarrow NH_3(g) + Zn(OH)_4^{2-}(aq)$$

what volume of $NH_3$ is produced by 6.54 g of Zn? The $NH_3$ is measured at 27.0°C and 1.25 atm pressure.

**13-57.** Solutions of potassium permanganate are a deep-purple color. Permanganate is a strong oxidizing agent that forms the $Mn^{2+}$ ion in acid solution when it is reduced. When these purple solutions are added to a reducing agent, the purple color disappears until all of the reducing agent reacts. So, we have a very convenient way to know when the reducing agent is used up—the solution suddenly turns purple. Balance the following net ionic reactions involving permanganate in acidic solution.

(a) $MnO_4^-(aq) + Fe^{2+}(aq) \longrightarrow Fe^{3+}(aq) + Mn^{2+}(aq)$
(b) $MnO_4^-(aq) + Br^- \longrightarrow Br_2(l) + Mn^{2+}(aq)$
(c) $MnO_4^-(aq) + C_2O_4^{2-}(aq) \longrightarrow CO_2(g) + Mn^{2+}(aq)$

**13-58.** Using the balanced equation from exercise 13-57, determine the volume (in mL) of 0.220 M $KMnO_4$ needed to completely react with 25.0 g of $FeCl_2$ dissolved in water.

**13-59.** Using the balanced equation from exercise 13-57, determine the volume (in mL) of 0.450 M KBr needed to completely react with 125 mL of 0.220 M $KMnO_4$.

**13-60.** The active ingredient in household bleach is sodium hypochlorite, which is a strong oxidizing agent. We can tell how much of the active ingredient is present in a two-step analysis. First the hypochlorite oxidizes excess iodide to elemental iodine according to the unbalanced equation that occurs in basic solution.

$$ClO^-(aq) + I^-(aq) \longrightarrow I_2(s) + Cl^-(aq)$$

Next, a solution of sodium thiosulfate ($Na_2S_2O_3$) is added to react with the iodine, as illustrated by the equation

$$I_2(s) + S_2O_3^{2-}(aq) \longrightarrow I^-(aq) + S_4O_6^{2-}(aq)$$

Balance the two equations.

**\*13-61.** Household bleach is 5.00% by weight sodium hypochlorite. Using the balanced equations from exercise 13-60, determine the volume (in mL) of 0.358 M $Na_2S_2O_3$ solution needed to react with all of the sodium hypochlorite in 100 mL of household bleach. Consider the density of bleach to be the same as water.

# Interactive Learning

**13-62.** Balance the following equations in acidic solution.
(a) $S_2O_3^{2-} + OCl^- \longrightarrow Cl^- + S_4O_6^{2-}$
(b) $IO_3^- + AsO_3^{3-} \longrightarrow I^- + AsO_4^{3-}$

**\*13-63.** A sample of a copper ore with a mass of 0.4225 g was dissolved in acid. The copper in the ore was in the form of $CuCO_3$; so after treatment with acid, it is present in solution as the $Cu^{2+}$ ion. A solution of potassium iodide was then added that caused the reaction

$$2Cu^{2+}(aq) + 5I^-(aq) \longrightarrow I_3^-(aq) + 2CuI(s)$$

The $I_3^-$ that formed reacted quantitatively with exactly 29.96 mL of 0.02100 M $Na_2S_2O_3$ according to the following equation.

$$I_3^-(aq) + 2S_2O_3^{2-}(aq) \longrightarrow 3I^-(aq) + S_4O_6^{2-}(aq)$$

(a) What was the percentage by mass of copper in the ore?
(b) What was the percentage by mass of $CuCO_3$ in the ore?

# Solutions to Learning Checks

**A-1.** oxidation state, $-2$, $+1$, loss, increase, reducing, gained, lost

**A-2.** **(a)** $H_3BO_3$:  $\quad 3H + B + 3O = 0$ $\qquad\qquad$ **(b)** $S_2O_3^{2-}$:  $\quad 2S + 3O = -2$

$\qquad\qquad\qquad 3(+1) + B + 3(-2) = 0$ $\qquad\qquad\qquad\qquad\qquad 2S + 3(-2) = -2$

$\qquad\qquad\qquad\qquad\qquad \underline{\underline{B = +3}}$ $\qquad\qquad\qquad\qquad\qquad\qquad\quad \underline{\underline{S = +2}}$

**A-3.** $ClO_2 \longrightarrow Cl^-$  $\quad ClO_2$ is reduced and is the oxidizing agent.

$\qquad H_2O_2 \longrightarrow O_2$  $\quad H_2O_2$ is oxidized and is the reducing agent.

**A-4.** **(a)** $\qquad\qquad\qquad\qquad 3e^- + 4H^+ + NO_3^- \longrightarrow NO + 2H_2O \qquad\qquad\qquad \times 2$

$\qquad\qquad\qquad\qquad\quad H_2O + H_2SO_3 \longrightarrow SO_4^{2-} + 4H^+ + 2e^- \qquad\qquad\quad \times 3$

$$\overline{\quad 6e^- + 8H^+ + 2NO_3^- + 3H_2O + 3H_2SO_3 \longrightarrow 2NO + \overset{1}{\cancel{4}}H_2O + 3SO_4^{2-} + \overset{4}{\cancel{12}}H^+ + 6e^- \quad}$$

$$\underline{\underline{2NO_3^- + 3H_2SO_3 \longrightarrow 2NO + 3SO_4^{2-} + H_2O + 4H^+}}$$

**(b)** $\qquad\qquad\qquad\qquad\qquad 5e^- + 2H_2O + ClO_2 \longrightarrow Cl^- + 4OH^- \qquad\qquad\qquad \times 2$

$\qquad\qquad\qquad\qquad 2OH^- + H_2O_2 \longrightarrow O_2 + 2H_2O + 2e^- \qquad\qquad\qquad \times 5$

$$\overline{\quad \cancel{10}e^- + \cancel{4}H_2O + 2ClO_2 + 5H_2O_2 + \overset{2}{\cancel{10}}OH^- \longrightarrow 2Cl^- + \cancel{8OH}^- + 5O_2 + \overset{6}{\cancel{10}}H_2O + \cancel{10}e^- \quad}$$

$$\underline{\underline{2ClO_2 + 5H_2O_2 + 2OH^- \longrightarrow 2Cl^- + 5O_2 + 6H_2O}}$$

**B-1.** stronger, weaker, higher, voltaic, Zn, anode, $Cu^{2+}$, cathode, salt, Pb, oxidized, $PbO_2$, cathode, electrolytic

**B-2.** A spontaneous reaction would occur in (b).

$\qquad 3Pb^{2+}(aq) + 2Cr(s) \longrightarrow 3Pb(s) + 2Cr^{3+}(aq)$

**B-3.** $Br_2(l) + Fe(s) \longrightarrow 2Br^-(aq) + Fe^{2+}(aq)$

**B-4.** $MgCl_2(l) \longrightarrow Mg(l) + Cl_2(g)$

## chapter 14

# Reaction Rates and Equilibrium

Because water enters this small pond at the same rate that it leaves, the level of the pond stays the same. This is similar to the equilibrium that occurs in certain chemical reactions.

## Setting the Stage

Who among us has not discovered what was first thought to be a new and possibly alien life form in the refrigerator? Usually it is in the back, perhaps on a long-forgotten slice of bologna. A ghastly, green mold of some sort slowly devours the meat and even spreads to neighboring foods. What we are witnessing is a chemical reaction connected with the decay of foods and the growth of molds. But the situation could be much worse. Without the cool temperature in the refrigerator, the strange life form that took weeks to develop would have evolved in a matter of a few days. At room temperature, meat and dairy products immediately begin to deteriorate as a result of chemical reactions. Decay, growth of alien life forms, and all other chemical reactions slow down as the temperature decreases. How chemical reactions occur and why temperature affects the rate at which they occur are topics of this chapter.

A related topic concerns reactions that attain a "balance." Many of us strive for a healthy balance between the demands of school or job and the need to relax and have fun. Or perhaps we seek a balance between having money coming in as fast as it goes out. In a chemical sense, balance means an equilibrium between two competing reactions, one forming products and the other, in the reverse direction, re-forming the original reactants.

**Formulating Some Questions**

Some of the questions that we will be addressing in this chapter include: How do reactants become transformed into products? What factors besides temperature affect how fast a reaction occurs? Why don't some reactions go entirely to the right? What factors affect the distribution of reactants and products if the reaction does not go entirely to the right? How can we predict the amounts of reactants and products present at equilibrium?

## Section A | COLLISIONS OF MOLECULES AND REACTIONS AT EQUILIBRIUM

## 14-1 How Reactions Take Place

**Looking Ahead!** Chemical reactions don't just happen. Somehow, reactant molecules or ions must find a way to come together so that the atoms can be reshuffled into product molecules or ions. The concept of how this happens is surprisingly straightforward and is the topic of this section.

Getting from Los Angeles to Chicago always raises a question. How are you going to get there? One could drive, fly, or take a bus. In any case, we can't get from point A to point B without some means of transportation. Likewise, chemical reactions must have some way for reactant molecules to be transformed into product molecules. Our modern view of how chemical reactions occur is actually quite simple. For example, we know that because molecules have kinetic energy, they are all in constant motion in one way or another. In the gaseous and liquid states, this motion leads to frequent collisions between molecules. These collisions are responsible for chemical changes. *The assumption that chemical reactions are due to the collisions of molecules is known as the* **collision theory.** In some cases, collisions are such that chemical bonds are broken in the colliding molecules. If the molecular or atomic fragments from the collision re-form in a different arrangement, a chemical reaction results. We will illustrate the collision theory by means of a hypothetical chemical reaction that can be "custom designed" to clearly illustrate the basic principles involved. Real-life reactions are generally more complicated, and explanations are required for each complication or exception.

Our reaction involves the combination of two hypothetical diatomic gaseous elements, $A_2$ and $B_2$, to form two molecules of a gaseous product, AB. The reaction is illustrated by the equation

$$A_2(g) + B_2(g) \longrightarrow 2AB(g)$$

We will assume that products form from the collision of an $A_2$ molecule with a $B_2$ molecule. If all collisions led to products, however, this and all reactions would be essentially instantaneous because of the large number of collisions per second. There are two conditions for the collision to lead to formation of products. First, *the collision between two reactant molecules must take place in the right geometric orientation.* As illustrated in Figure 14-1, for new bonds to form, the two reactant molecules must meet in a side-to-side manner rather than end-to-end or side-to-end. This condition, by itself, severely limits the number of collisions that can lead to a reaction. A second condition is that *the collision must occur with enough energy to break the bonds in the reactants so that new bonds can form in the products.* The second condition even more severely limits the number of collisions that lead to product formation. If the two conditions are not met, the colliding molecules simply recoil from each other unchanged. (See Figure 14-1.)

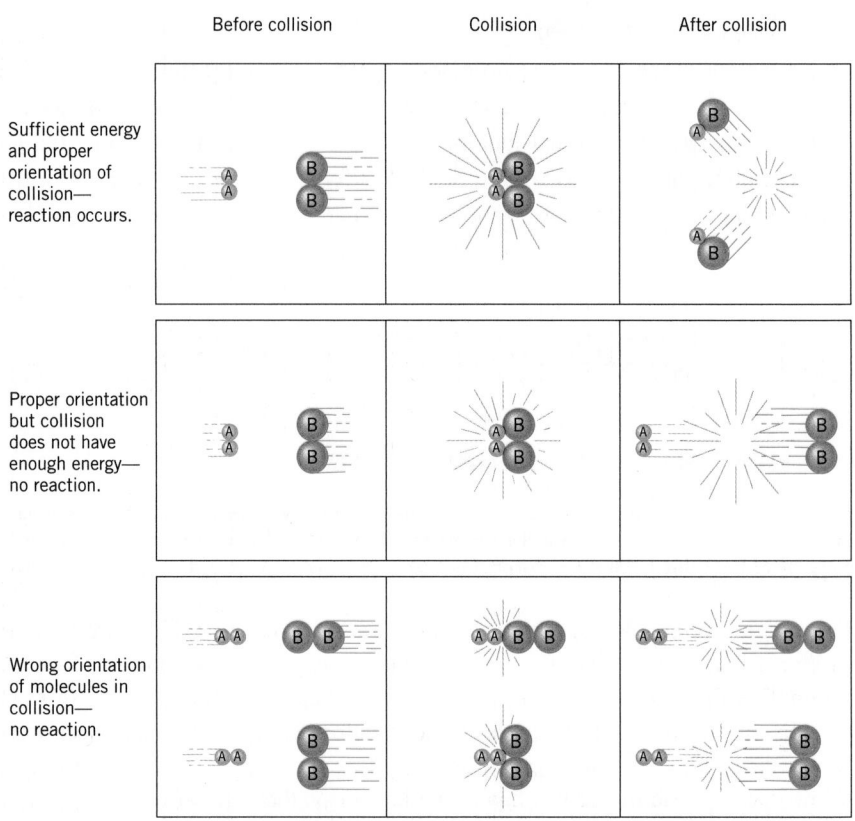

**Figure 14-1   Reaction of A₂ and B₂** **Reactions occur only when colliding molecules have the minimum amount of energy and the right orientation.**

Analogous to this, imagine that you are trying to roll a bowling ball over a small incline. When you swing the ball behind you and then release it, you are imparting motion (kinetic energy) to the ball. As it starts up the hill, however, it begins to slow. It slows because the kinetic energy is being converted into potential energy because of the increased elevation of the ball. If you swing the ball hard enough, it may have enough kinetic energy to overcome the potential energy barrier of the hill. In that case it will make it to the top of the hill and go down the other side. If not, the ball will stop before reaching the top when all of the kinetic energy has been converted into potential energy and then roll back at you as the potential energy converts back to kinetic energy of the moving ball. (See Figure 14-2.)

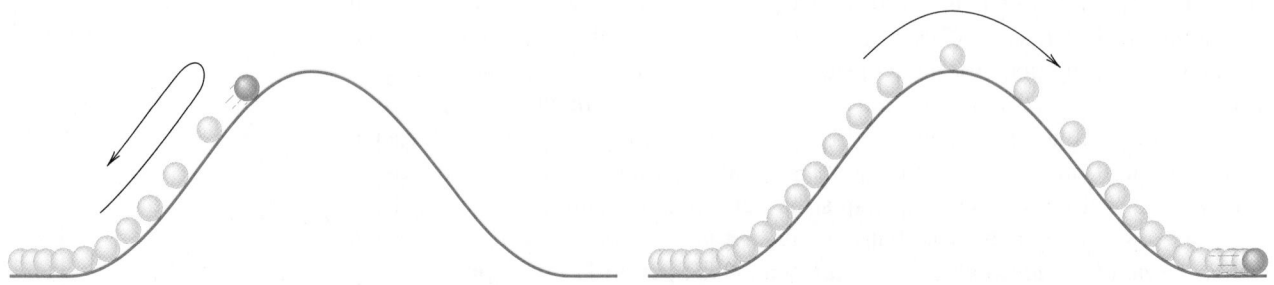

**Figure 14-2   The Potential Energy Barrier** **On the left, the ball does not have enough kinetic energy to overcome the barrier. On the right, the ball has enough kinetic energy to make it over.**

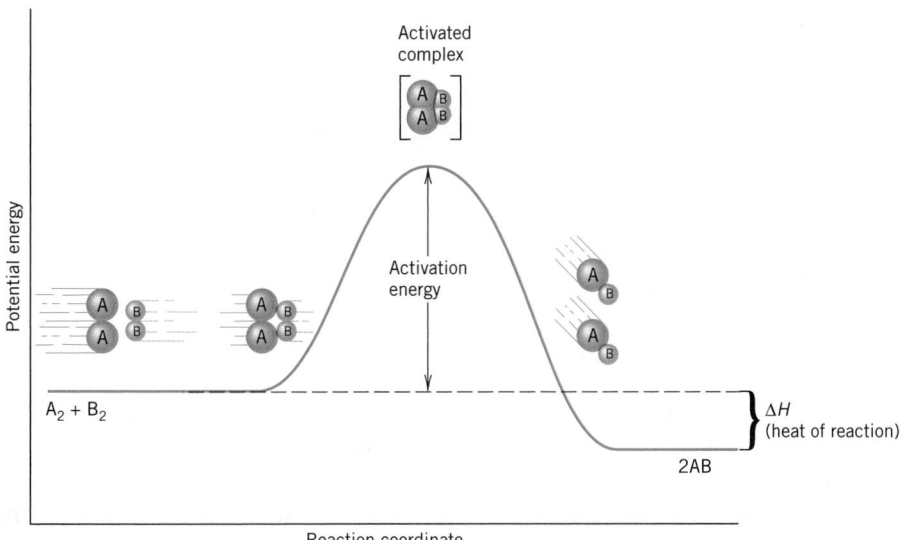

**Figure 14-3 Activation Energy** **The activation energy is the potential energy difference between the activated complex and the reactants.**

Chemical reactions are analogous. *The minimum kinetic energy needed for the reaction to occur is known as the* **activation energy.** If properly colliding molecules have kinetic energy equal to or greater than the activation energy (which represents the potential energy barrier), a reaction can occur; if not, they recoil unchanged. To illustrate, we will follow the fate of two hypothetical diatomic elements $A_2$ and $B_2$ as they move from far apart, to a collision with the proper orientation that has enough energy for reaction, and finally to the formation of products. Recall that energy is conserved, meaning that the sum of the kinetic energy and potential energy of the molecules remains the same as they collide. In Figure 14-3, we attempt to illustrate the progress of the reaction as the molecules collide. The curve represents the potential energy (on the vertical axis) of the species involved in the reaction. Notice that the separate molecules have a certain amount of potential energy before they collide. This energy is present in their chemical bonds. By moving to the right along the horizontal axis in the graph, we can follow the change in potential energy of the reactant molecules as they approach each other, collide, and eventually form products. This horizontal axis is referred to as the *reaction coordinate* (also called *path of the reaction*) and can be viewed as a time axis. The potential energy of the molecules does not change until they make contact and begin to compress together. As they collide, the molecules lose velocity, which means they lose kinetic energy. This kinetic energy changes to potential energy of the compressing molecules, so the potential energy of the compressing molecules rises. At maximum impact, the motion stops for an instant and the potential energy is at a maximum. This is analogous to hitting a tennis ball with a racquet. As the ball makes contact with the racquet, it slows down and compresses. For just an instant, the ball has no motion but has maximum potential energy from compression. As the ball recoils, the potential energy of the ball reconverts to kinetic energy of motion. Likewise, *at maximum impact, both molecules are compressed together, forming what is known as an* **activated complex.** Since we are assuming a reaction takes place, the activated complex represents the state that is intermediate between reactants and products. That is, old bonds are partially broken and new ones are partially formed. The activation energy for the forward reaction is represented by the difference in potential energy of the activated complex (at the peak

**At the instant the tennis ball is not moving, all of the energy is in the form of potential energy.**

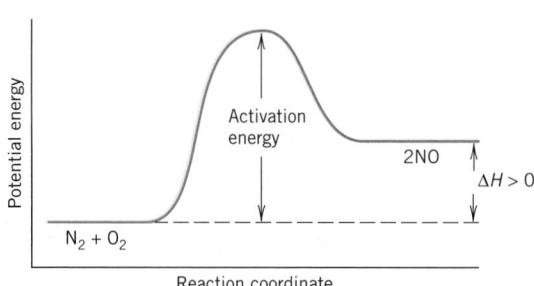

**Figure 14-4 An Endothermic Reaction**
In an endothermic reaction, the potential energy of the reactants is less than that of the products.

of the curve) and the reactants. As product molecules recoil, potential energy decreases as it is converted into the kinetic energy of the recoiling molecules. Notice in Figure 14-3 that the product molecules (AB) eventually have lower potential energy than the original reactants. This is because some of the potential energy in the original chemical bonds (i.e., in $A_2$ and $B_2$) has been released as kinetic energy to the product molecules. Since the product molecules have more kinetic energy than the original reactants, the product molecules move faster and are "hotter."

The difference between the potential energy of the products and that of the reactants represents the *heat of the reaction* and is given the symbol $\Delta H$. *In an exothermic process, the potential energy of the products is lower than the potential energy of the reactants, so heat is released.* An analogy to an exothermic reaction is also found in the game of tennis. The ball should recoil at a faster rate from the racquet than when it first made contact. The player tries to give the ball some "heat."

When the potential energy of the products is higher than that of the reactants, the reaction is endothermic and the product molecules are "cooler." Heat is required for the reaction to occur and is extracted from the environment. An actual endothermic reaction between $N_2$ and $O_2$ to form NO is represented in Figure 14-4.

# 14-2 Rates of Chemical Reactions

**Looking Ahead! We have found that the rate of a reaction is governed by the frequency of collisions between molecules and whether these collisions have the proper orientation and enough energy. Some reactions are essentially instantaneous, such as in an explosion, whereas others may take months, such as the rusting of a nail. Why the big range? We are now ready to examine the factors that affect the rate of a reaction.**

One important question to be answered in deciding how one gets from Los Angeles to Chicago involves how fast you want to get there. If you fly, you may average 500 mi/hr, but if you drive you may average only 50 mi/hr. We measure the rate of travel by a distance per a unit of time, which in this case is hours. Chemical reactions also take place at specific rates. As any chemical reaction occurs, reactants disappear and products appear. *The **rate of a reaction** measures the increase in concentration of a product or the decrease in concentration of a reactant per unit of time.* There are several factors that affect the rate of a reaction. These include the magnitude of the activation energy, the temperature at which the reaction takes place, the concentration of reactants, the size of the particles of solid reactants, and the presence of a catalyst. We will discuss these factors individually.

## Activation Energy

Perhaps one of the most significant characteristics of any chemical reaction is its activation energy, which is determined by the nature of the reactants. Consider, for example, the reactions of two nonmetals with oxygen. An allotrope of phosphorus known as white phosphorus reacts almost instantly with the oxygen in the air, even at room

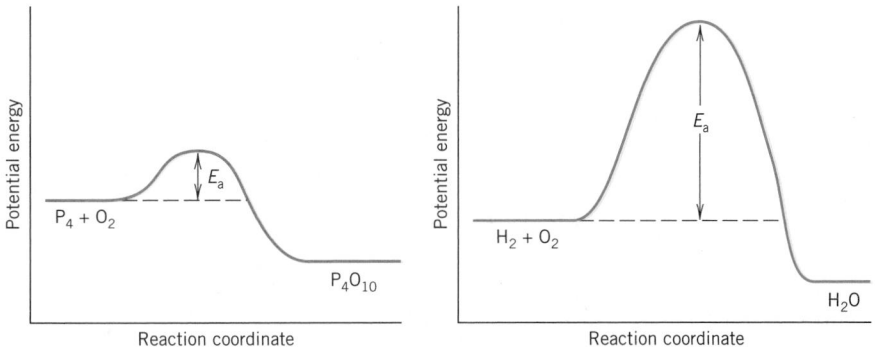

**Figure 14-5  Combustion of P₄ and H₂** **P₄ undergoes combustion at room temperature because of its lower activation energy.**

temperature, producing an extremely hot flame. As a result, this form of phosphorus must be stored under water to prevent exposure to the air. Hydrogen also reacts with oxygen to form water. This is also a highly exothermic reaction that is used in space rockets. A mixture of hydrogen and oxygen, however, does not react at room temperature. In fact, no appreciable reaction occurs unless the temperature is raised to at least 400°C or the mixture is ignited with a spark. The difference in the rates of combustion of these two elements at room temperature lies in the activation energies of the two reactions. Figure 14-5 illustrates the activation energies (labeled $E_a$) for the combustion reactions of hydrogen and phosphorus. Note that the phosphorus reaction has a much lower activation energy than the hydrogen reaction. This explains why collisions between phosphorus molecules and oxygen have enough energy for reaction at room temperature, whereas collisions of the same energy between hydrogen and oxygen do not have the required energy to overcome the activation energy barrier.

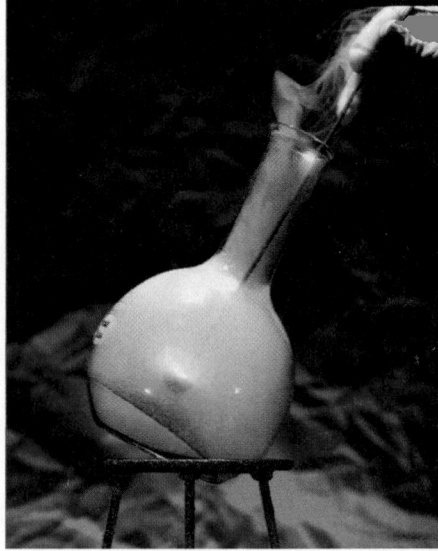

**White phosphorus ignites instantly in air because the reaction has a low activation energy.**

## Temperature

Do not forget to remove a sack of hamburgers from a hot automobile on a summer day! By the next day the car may have to be abandoned because it smells so bad. Food spoilage occurs amazingly fast on hot summer days. In fact, all chemical reactions occur faster on hot summer days. Previously, we established that the rate of a reaction depends on the kinetic energy of the reactant molecules.

$$r(\text{rate}) \propto \text{energy of colliding molecules}$$

In Chapter 10 we discussed the distribution of kinetic energies of molecules at a given temperature. This is again illustrated in Figure 14-6. Note that, since temperature

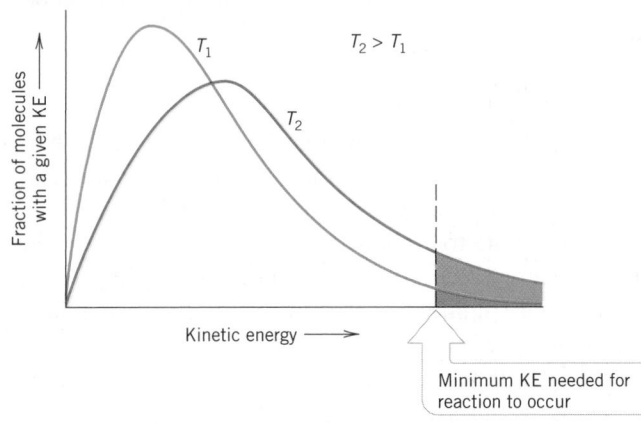

**Figure 14-6  Kinetic Energy Distributions for a Reaction Mixture at Two Different Temperatures** **The sizes of the shaded areas under the curves are proportional to the total fractions of the molecules that possess the minimum activation energy.**

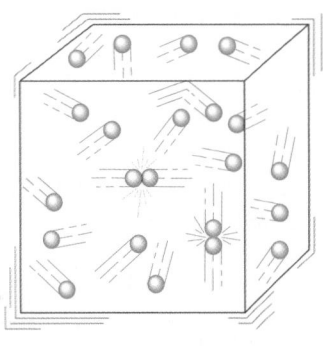 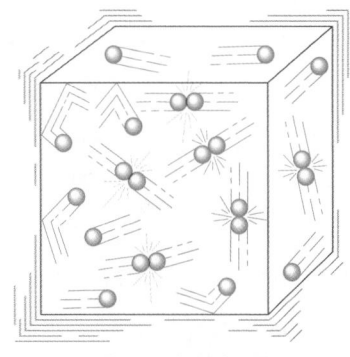

Slow jiggling (low $T$)   Fast jiggling (high $T$)

**Figure 14-7** **Effect of Velocity on Collisions** **Collisions occur more frequently and with more force as the velocity of the balls increases.**

is related to average kinetic energy, the higher the temperature, the higher the average kinetic energy of the molecules. In the figure, the dashed line represents the activation energy ($E_a$). As mentioned, this is the minimum amount of kinetic energy that colliding molecules must have for products to form. Note that at the higher temperature ($T_2$) the curve intersects the dashed line at a higher point than does $T_1$, meaning that a greater fraction of molecules can overcome the activation energy in a collision. As a result, the rates of chemical reactions increase as the temperature increases.

There is another reason why an increase in temperature increases the rate of a reaction. In addition to the energy of the collisions, the rate of a reaction depends on the frequency of collisions (the number of collisions per second).

$$r(\text{rate}) \propto \text{frequency of collisions}$$

Recall that the kinetic energy (K.E.) of a moving object is given by the relationship

$$\text{K.E.} = \tfrac{1}{2}mv^2 \ (m = \text{mass}, v = \text{velocity})$$

This indicates that the more kinetic energy a moving object has, the faster it is moving.

As the temperature increases, the average velocity of the molecules increases. This means that the frequency of collisions also increases. As an analogy, imagine a box containing red and blue Ping-Pong balls as shown in Figure 14-7. If we jiggle the box, the balls move around and collide. If we jiggle the balls faster (analogous to a higher temperature), the balls move around faster and there are more frequent collisions. The increased noise we hear tells us that, indeed, collisions are not only more energetic but more frequent. Most of the increase in the rate of a chemical reaction is due to the increased energy of the collisions, with a lesser contribution from the increased rate of collisions. The rates of many chemical reactions approximately double for each 10°C rise in temperature.

## Concentration of Reactants

Increasing the rate at which we shake a box of Ping-Pong balls obviously increases the rate of collisions between the balls. Another way of increasing the rate of collisions is to increase the number of balls in the box. In Figure 14-8, we have two situations. On the left, there are four red and four blue balls. On the right, there are four red balls, but the number of blue balls has been increased to eight. The concentration (number of balls per unit volume) has been increased. Increasing the number of balls of either color increases the number of blue–red collisions, which is signaled by the more intense noise even though the box is jiggled at the same rate (analogous to

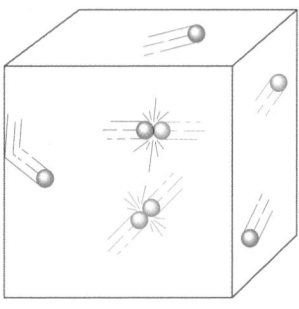

Four red and four blue balls
(low concentration)

Four red and eight blue balls
(high concentration)

**Figure 14-8 Effect of Concentration on Collisions Collisions occur more frequently when there are more balls in the box.**

the same temperature). In the reaction of $A_2$ and $B_2$, the same phenomenon applies. The greater the concentration of either or both of the reactants, the more frequent the collisions and the greater the rate of the reaction.

## Particle Size

If we wish to burn an old dead tree, there is a slow way and there is a quick way. Trying to burn it as one big log could take days. If we cut the tree up into smaller logs, we could burn it in hours. In fact, if we changed the tree into a pile of sawdust and spread it around, we might burn it in a matter of minutes. Related to this is the inherent danger in the storage of grain in large silos. Normally, a pile of grain burns slowly. However, small particles of grain dust can become suspended in air, forming a dangerous mixture that can actually detonate. Such explosions have occurred. In the reaction between a solid and a gas, a solid and a liquid, or a liquid and a gas, the surface

Grain dust and air make an explosive mixture because of the small particle size of the dust. This grain elevator was destroyed by such an explosion.

area of the solid or liquid obviously affects the rate of the reaction. The more area that is exposed, the faster the rate of the reaction.

## Catalysts

One of the principles of road building in the mountains is that if it's too difficult to go over the mountain, go through it. The mountain is like the activation energy of a reaction and the tunnel represents a pathway or mechanism with a lower activation energy. *A* **catalyst** *is a substance that provides an alternate mechanism with a lower activation energy.* (See Figure 14-9.) As a result, *a catalyst increases the rate of a reaction but is itself not consumed in the reaction.*

An excellent example of how catalysts work is provided by the catalytic destruction of ozone ($O_3$) in the stratosphere, which is a topic of serious environmental concern. The following reaction is very slow in the absence of catalysts.

$$O_3(g) + O(g) \longrightarrow 2O_2(g)$$

A tunnel provides an easier, faster route than a high mountain pass.

The atoms of oxygen (O) originate from the dissociation of normal oxygen molecules ($O_2$) by solar radiation. Chlorofluorocarbons (e.g., $CF_2Cl_2$), used primarily as refrigerants, diffuse into the stratosphere and also are dissociated by solar radiation into Cl atoms (not $Cl_2$ molecules). When an atom of chlorine is present, two reactions take place that lead to the rapid conversion of ozone to normal oxygen ($O_2$).

$$\begin{array}{c} Cl + O_3 \longrightarrow ClO + O_2 \\ ClO + O \longrightarrow Cl + O_2 \\ \hline Cl + ClO + O_3 + O \longrightarrow Cl + ClO + 2O_2 \end{array}$$

*Net reaction:* $\quad O_3 + O \longrightarrow 2O_2$

There are two significant facts about these two reactions. First, the presence of chlorine atoms drastically increases the rate of conversion of ozone to oxygen. Of more importance is the fact that a chlorine atom is a reactant in the first reaction but is regenerated as a product in the second. This is how it acts as a catalyst. Because it is not consumed in the reaction, one chlorine atom can destroy thousands of ozone molecules. In fact, the series of reactions leading to the destruction of ozone in the stratosphere appears to be more complex than that shown above. Intense studies are currently under way to help us understand exactly how chlorine atoms interact with ozone. In the meantime, the use of chlorofluorocarbons is being phased out by international agreement.

In the reactions just described, the catalyst is intimately mixed with reactants and is actually involved in the reaction. Catalysts of this nature are referred to as *homogeneous catalysts*. In other cases, a catalyst may simply provide a surface on which reactions take place (*heterogeneous catalyst*).

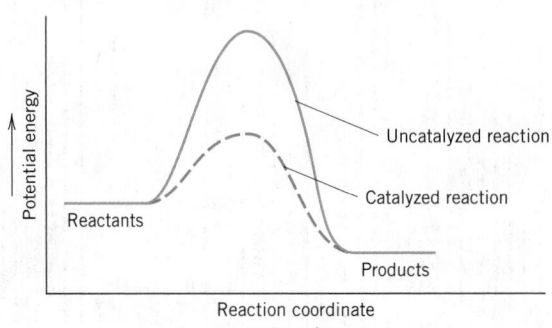

**Figure 14-9 Activation Energy and Catalysis**
**A catalyst speeds a reaction by lowering the activation energy.**

Perhaps the most familiar application of heterogeneous catalysts is in the catalytic converter of an automobile (see Figure 14-10). The exhaust from the engine contains poisonous carbon monoxide, unburned fuel (mainly $C_8H_{18}$), and NO. All contribute to air pollution, which is still a serious environmental concern in many localities. The catalytic converter, which is attached to the exhaust pipe, contains finely divided platinum and/or palladium. These metals provide a surface for the following reactions, which occur only at a very high temperature in the absence of a catalyst.

$$2CO(g) + O_2(g) \longrightarrow 2CO_2(g)$$
$$2C_8H_{18}(g) + 25O_2(g) \longrightarrow 16CO_2(g) + 18H_2O(g)$$
$$2NO(g) \longrightarrow N_2(g) + O_2(g)$$

All three reactions are exothermic, which explains why the catalytic converter becomes quite hot when the engine is running.

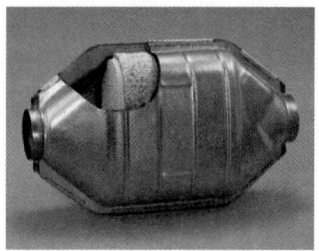

**Figure 14-10 The Catalytic Converter**
**This device on the automobile helps to reduce air pollution.**

## 14-3 Reactions at Equilibrium

**Looking Ahead!** In some chemical reactions, collisions between product molecules may re-form the original reactants. These reactions appear to go to the right so far and then stop. These reactions have reached a point of equilibrium because the reaction is reversible. We consider reversible reactions that reach equilibrium next.

If we were sitting in a leaky boat on a large lake, we would certainly hope to maintain a point of equilibrium. That is, we would need to bail the water out of the boat as fast as it leaks in. If not, we might get very wet. Chemical reactions may also reach a point of equilibrium where product molecules are formed at the same rate as they re-form reactants. To illustrate chemical equilibrium, let's return to the hypothetical reaction discussed in section 14-1 and illustrated in Figure 14-1. We will assume that this reaction is reversible. *A* **reversible reaction** *is one where both a forward reaction (forming products) and a reverse reaction (re-forming reactants) can occur.* Reversible reactions where both reactions occur simultaneously reach a point of equilibrium. *The* **point of equilibrium** *in a reversible process is when both the forward and reverse processes proceed at the same rate, so the concentrations of reactants and products remain constant.*

In our hypothetical reaction, the reaction mixture initially contains only $A_2$ and $B_2$. As the reaction proceeds, the concentrations of these two molecules begin to decrease as the concentration of the product, AB, increases. The rate of buildup of AB begins to decrease until, eventually, the concentration of AB no longer changes despite the presence of excess reactant molecules. To understand this, we turn our attention to the product molecule AB. If we had started with pure AB, we would find that AB slowly decomposes to form $A_2$ and $B_2$, just the reverse of the original reaction. This reaction occurs when two AB molecules collide with the proper orientation and sufficient energy to form $A_2$ and $B_2$. (See Figure 14-11.)

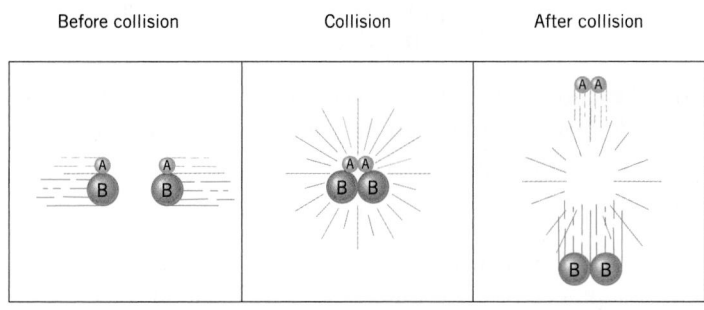

Before collision   Collision   After collision

**Figure 14-11 Reaction of 2AB**
**Collisions between AB molecules having the minimum energy and correct orientation lead to the formation of reactants.**

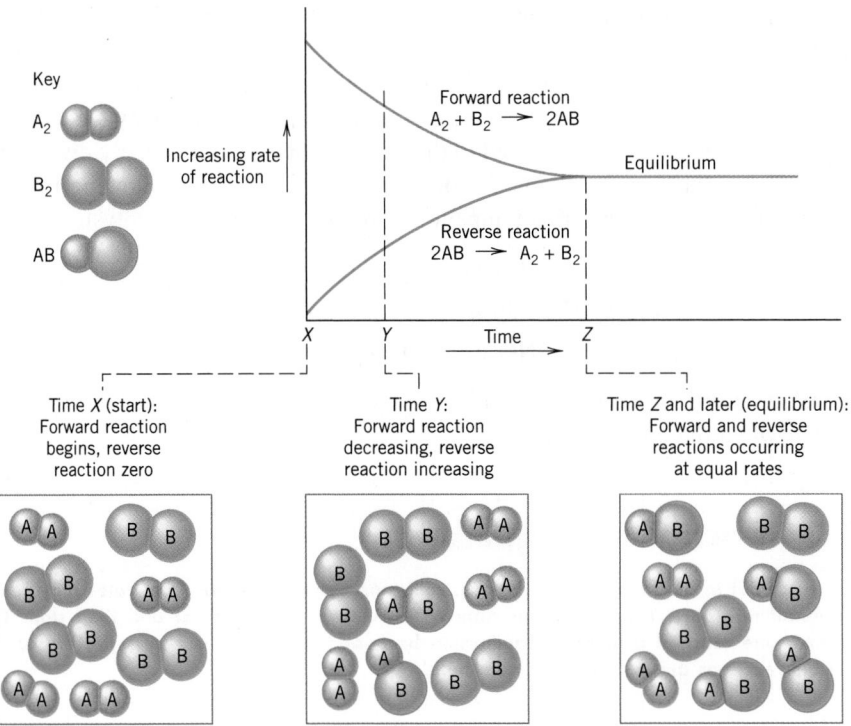

**Figure 14-12  $A_2$, $B_2$, AB Equilibrium** Equilibrium is achieved when the rates of the forward and reverse reactions are equal.

Now we put these two reactions together. If we again start with pure $A_2$ and $B_2$, at first only the forward reaction occurs, leading to the formation of AB. As the concentration of AB increases, however, the reverse reaction begins to occur. Eventually, the rate of formation of products (the forward reaction) is exactly offset by the rate of formation of reactants (the reverse reaction). That is, for every $A_2$ and $B_2$ that react to form two AB's, two AB molecules react to re-form one $A_2$ and $B_2$. At this point the concentrations of all species ($A_2$, $B_2$, and AB) remain constant. This phenomenon is referred to as a **dynamic equilibrium,** *which emphasizes the changing identities of reactants and products despite the fact that the total amounts of each do not change.* In Figure 14-12, the reaction and the point of equilibrium are graphically illustrated.

If we could shrink ourselves to the molecular level, we would immediately become aware of this dynamic (changing) situation. If one A atom in an $A_2$ molecule was marked so that it could be traced, we would notice that at one moment it is present in an $A_2$ molecule, later it is part of an AB molecule, and still later it is part of another $A_2$ molecule.

$$\begin{array}{ccccccc} {}^*A & & {}^*A & & {}^*A & & \\ | & & | & & | & & \\ A & \xrightarrow{+B_2} & B & \xrightarrow{+AB} & A & \xrightarrow{+B_2} & \text{etc.} \end{array}$$

A reversible reaction that reaches a point of equilibrium is indicated by the use of double arrows instead of a single arrow, which implies a complete reaction. Thus our hypothetical reaction is represented as

$$A_2(g) + B_2(g) \rightleftharpoons 2AB(g)$$

In earlier parts of this text, we have mentioned chemical and physical systems that reach a point of equilibrium. Our first mention of this phenomenon was in

Chapter 7, where we discussed the formation of ammonia from its elements. This is a classic example of a system at equilibrium, since ammonia decomposes to its elements, which is the reverse reaction of its formation.

$$N_2(g) + 3H_2(g) \rightleftharpoons 2NH_3(g)$$

We encountered a system in Chapter 10 involving equilibrium between two physical states of water. When water is placed in a closed container at a certain temperature, it establishes an equilibrium vapor pressure. At the point of equilibrium, the vapor pressure is constant, which means that the rate of condensation equals the rate of evaporation. The equation for this equilibrium is

$$H_2O(l) \rightleftharpoons H_2O(g)$$

The equilibrium reactions discussed most recently were in Chapter 12 and involved the weak acids and bases. Although the ionization of a weak acid in water is unfavorable, it does occur to a small but measurable extent. At equilibrium, most molecules are present in water in the molecular rather than the ionized state, as represented by the following equilibrium equation.

$$HF(aq) + H_2O(l) \rightleftharpoons H_3O^+(aq) + F^-(aq)$$

The quantitative aspects of the reactions of weak acids and bases in water are discussed later in this chapter.

# 14-4 Equilibrium and Le Châtelier's Principle

**Looking Ahead!** A system at equilibrium has a remarkable ability to adapt to external changes. We can use this information to influence the point of equilibrium. How and why this happens is the topic of this section.

Life, at times, seems to be a continual attempt to maintain equilibrium despite all the stresses that come our way. If we spend too much money on clothes, we have to counteract that loss by cutting back on expenses somewhere else. If we stay up too late one night (studying, naturally), we must compensate the next night by getting to bed early or, better yet, sleeping in the next morning. Chemical reactions are analogous. There are several external changes of conditions (stresses) that may affect the point of equilibrium and cause the system to compensate.

How a system at equilibrium reacts to a change in conditions is summarized by **Le Châtelier's principle,** which states: *When stress is applied to a system at equilibrium, the system reacts in such a way to counteract the stress.* The changes in a chemical system that affect the point of equilibrium include the following:

1. A change in concentration of a reactant or product
2. A change in the pressure on a gaseous system
3. A change in the temperature

We will discuss these three changes individually. To illustrate how these conditions affect the point of equilibrium, we will use a very important industrial process involving the following reaction.

$$2SO_2(g) + O_2(g) \rightleftharpoons 2SO_3(g)$$

The $SO_2$ is formed by the combustion of sulfur or sulfur compounds. The $SO_3$ produced by the reaction is then allowed to come in contact with water to form sulfuric acid.

$$SO_3(g) + H_2O(l) \longrightarrow H_2SO_4(aq)$$

In 2001, 48 million tons of $H_2SO_4$, with a value of about \$5 billion, were produced by this process in the United States alone.

In such an important industrial process, it is important to convert the maximum amount of $SO_2$ to $SO_3$. In other words, the goal is to shift the equilibrium to the right as much as possible in favor of the product. There are several conditions that can be adjusted in order to do this.

1. If the system is at equilibrium and then an additional portion of a reactant is introduced, the system shifts to produce more products. This happens because the rate of the forward reaction increases by the increased number of collisions that can occur to form products. The concentration of products will increase, which then increases the rate of the reverse reaction. Eventually, equilibrium is reestablished. In our system, addition of $O_2$ will lead to formation of more $SO_3$. The additional $O_2$ can be considered as a stress on a system at equilibrium. Formation of more product removes some of the additional $O_2$ and thus relieves the stress on the system. We can generalize this observation. *An increase in the concentration of a reactant or product ultimately leads to an increase in the concentration of the species on the other side of the equation.* Conversely, a decrease in the concentration of a reactant or product ultimately leads to formation of more of that reactant or product. In reactions where a precipitate forms or a gas is evolved, that product is, in effect, removed from the system. The system shifts to replace this loss, but if the loss is complete, the reaction in the direction of that loss is complete. This is known as an *irreversible reaction.*

2. In a gaseous system, a change in the volume of the reaction container or the pressure also affects the point of equilibrium. Boyle's law tells us that *a decrease in volume corresponds to an increase in pressure.* Increasing the pressure on a gaseous mixture of reactants and products at equilibrium (or decreasing the volume of the container) is another example of stress on a system. The system can counteract the stress of increased pressure by shifting to a lower volume. In our sample system, there are three moles of gaseous reactants ($2SO_2 + O_2$) but only two moles of the gaseous product ($2SO_3$). Therefore, an increase in pressure on this system shifts the point of equilibrium to the right, increasing the concentration of $SO_3$ until equilibrium is eventually reestablished. *In general, an increase in pressure on a gaseous system shifts the point of equilibrium in the direction of the smaller number of moles of gas.* When gases are not involved in a reaction at equilibrium, such as one in aqueous solution, pressure changes have negligible effect since solids and liquids are virtually incompressible.

3. The reaction of $SO_2$ with $O_2$ is an exothermic reaction, which means that heat energy is produced along with $SO_3$. The equation showing heat as a product can be written as follows.

$$2SO_2(g) + O_2(g) \rightleftharpoons 2SO_3(g) + \text{heat}$$

Since heat can be considered a component of the reaction, it can be treated in the same way as any other component, as discussed in **1** above. That is, if we add heat (increase the temperature), the reaction shifts in a direction to remove that heat. If we remove heat by cooling the reaction mixture, the equilibrium shifts in a direction to replace that heat loss, which increases the concentration of species on the other side. *For an exothermic reaction, cooling the reaction mixture increases the concentration of products when equilibrium is reestablished. For an endothermic reaction, where heat can be considered as a reactant, heating the reaction mixture increases the concentration of products at equilibrium.* In our system, the formation of $SO_3$ is favored at low temperatures. Cooling the reaction

## The Lake That Exploded

In 1986, Lake Nyos in Cameroon, a country in Africa, literally blew its top. What happened in this huge lake is the same thing that happens when one opens a well-shaken can of soda. As we all know, gas (carbon dioxide) comes out of solution and suddenly propels the liquid at whatever or whomever the opening of the can is aimed at. In the case of Lake Nyos, tons of carbon dioxide gas suddenly erupted from deep in the lake, forming a thick cloud. Since carbon dioxide is denser than air, the cloud hugged the ground and flowed over the rim of the crater and crept into the surrounding valleys. As the cloud moved, it pushed the air with its life-sustaining oxygen out of the way. Normal air is 0.03% $CO_2$. If the concentration rises to 10%, it can be fatal. Eventually, 1700 people and many thousands of cattle were asphyxiated. Two years earlier it happened at nearby Lake Monoun, causing 37 casualties.

What happened to cause such a calamity? As it turns out, the lake is what is known as a volcanic crater lake. From the hot magma, 50 miles down, carbon dioxide rises through vents underneath the lake and enters the cold water near the bottom. In the equilibrium between the gaseous state and dissolved state, the high pressure near the bottom of the lake holds the gas in solution (i.e., Le Châtelier's principle).

$$CO_2(g) \rightleftharpoons CO_2(aq)$$

As long as it is not disturbed, the carbon dioxide stays in solution in the cold lower layers of the lake. A small earthquake, an underwater landslide, or even a violent storm may have upset the layers. Once the layer of water containing the carbon dioxide came closer to the surface, the reduced pressure allowed the gas to escape. Once it started, it continued violently, like opening a bottle of soda. The cloud of gas expanded for over 20 miles from the lake until it dissipated.

Efforts are underway by scientists and engineers to prevent this from happening again in Lake Nyos and two other lakes like it. In one attempt, pumps extend to the deep parts of the lakes, bringing some of the lower layer to the surface. In this way, the carbon dioxide can be released gradually before it again builds to dangerous levels. Still, much needs to be done or this tragedy could happen again.

*The explosive evolution of carbon dioxide from this lake created a natural disaster.*

mixture removes heat. The system attempts to compensate for this loss by formation of more $SO_3$ and thus more heat.

The formation of products in the endothermic reaction illustrated in Figure 14-4 is favored at high temperatures. As a result, this reaction between the two major components of our atmosphere does not take place under normal atmospheric conditions but does occur to some extent in the hot combustion chambers of automobile engines.

In our example system, we have a potential problem as we lower the temperature to increase the concentration of $SO_3$. The rate of the reaction decreases as the temperature decreases. Since time is also important in industry, we must consider a compromise. The reaction must be run at a high enough temperature that it proceeds to equilibrium in a reasonable amount of time. The discovery of a catalyst for the reaction allows the reaction to be run at a low enough temperature that a reasonable yield of $SO_3$ is possible. That temperature is around 400°C.

| | Effect at Equilibrium | | |
| --- | --- | --- | --- |
| **Change** | **$SO_2$** | **$O_2$** | **$SO_3$** |
| add $O_2$ | decrease | increase | increase |
| increase $P$ | decrease | decrease | increase |
| decrease $V$ | decrease | decrease | increase |
| lower $T$ | decrease | decrease | increase |
| add catalyst | no effect | no effect | no effect |

**433**

Catalysts play a very significant role in many chemical reactions with industrial significance. *A catalyst in a system that reaches a point of equilibrium increases the rate of both the forward and reverse reactions but does not change the point of equilibrium.* The function of a catalyst is simply to reach the point of equilibrium in less time. In many cases, the use of a catalyst allows the reaction to be run at a considerably lower temperature than would otherwise be possible. In the reaction discussed, heterogeneous catalysts such as $V_2O_5$ are effective. In other industrial processes, the exact formulation of the catalyst for a specific reaction may be a closely guarded secret.

In summary, the production of $SO_3$ from $SO_2$ is affected by the changes shown in the table on the previous page.

---

**Looking Back!** Chemical reactions don't just happen. The molecules that react must make contact in order for old bonds to break and new bonds to form. Even then, a reaction does not occur every time two reactant molecules collide. There are important conditions involving the way the molecules interact with each other that make formation of products possible. How fast these reactions occur (the reaction rate) depends on several conditions such as the temperature and the concentration of reactants. Some reactions, however, are reversible and do not proceed completely to the right. Still, we can force the point of equilibrium to the right or left by manipulating conditions appropriately.

---

# Learning Check A | checking it out

**A-1.** Fill in the blanks.

Chemical reactions occur because of _____ between reactant molecules. In order for a reaction to take place, molecules must _____ in the proper orientation and with a _____ amount of energy known as the _____ energy. The rate of appearance of a product as a function of _____ is a measure of the _____ of the reaction. An _____ forms at the point of maximum potential energy. The difference in potential energy between the activated complex and the reactants is the _____ _____ . At _____ temperatures, more molecules can overcome this energy barrier so that the rate of the reaction _____ . Increasing the concentration of colliding molecules _____ the rate of the reaction. The presence of a _____ does not change the point of equilibrium, but it _____ the rate of the reaction. Reactions that are reversible may reach a _____ _____ _____ where the concentrations of reactants and products remain _____ . The point of equilibrium may be affected by a change of _____ of reactants, a change of _____ , and a change of _____ (in a gaseous system).

**A-2.** Given the equilibrium

$$2H_2O(g) \rightleftharpoons 2H_2(g) + O_2(g) \qquad \Delta H = +484 \text{ kJ}$$

will the concentration of $H_2$ at equilibrium be increased, decreased, or not affected by the following?

**a.** an increase in concentration of $H_2O$
**b.** an increase in concentration of $O_2$
**c.** an increase in temperature
**d.** an increase in pressure
**e.** addition of a catalyst

**A-3.** Sketch a reaction profile for the reaction in problem A-2. Sketch a reaction profile when a catalyst is present.

*Additional Examples: Exercises 14-1, 14-3, 14-6, 14-8, 14-10, and 14-12.*

Section B | **THE QUANTITATIVE ASPECTS OF REACTIONS AT EQUILIBRIUM**

## 14-5 Point of Equilibrium and Law of Mass Action

**Looking Ahead!** Long before chemists knew anything about collision theory, reaction rates, or activation energy, they realized that the concentrations of reactants and products were distributed in a predictable manner in a reaction at equilibrium. Equilibrium distributions are determined by a law, just as the conditions of gases are determined by gas laws. The quantitative aspects of the point of equilibrium are our next subject.

To illustrate how the concentrations of reactants and products relate, we will consider a hypothetical reversible reaction involving *a* moles of reactant A combining with *b* moles of reactant B to form *c* moles of product C and *d* moles of product D.

$$aA + bB \rightleftharpoons cC + dD$$

For a reaction at equilibrium, the distribution of reactants and products is given by the following relationship, which is known as the **law of mass action.**

$$K_{eq} = \frac{[C]^c[D]^d}{[A]^a[B]^b}$$

Note that the coefficients (*a, b, c,* and *d*) of the substances (A, B, C, and D) become the exponents of the molar concentrations (mol/L) of the same substances. The products are written in the numerator, and the reactants are written in the denominator. The law of mass action for a particular reaction consists of two parts. $K_{eq}$ *is called the* **equilibrium constant** *and has a definite numerical value at a given temperature. The ratio to the right of the equality is called the* **equilibrium constant expression.** We will apply the law of mass action to gas phase reactions in this section and to certain reactions that occur in aqueous solution in the next section.

The significance of the law of mass action is that, even with a variety of initial concentrations of reactants and products, the distribution of reactants and products is predictable when the reaction reaches the point of equilibrium. This is illustrated in Table 14-1. In a series of three experiments, we start with an initial concentration of reactants (expt. 1), an initial concentration of products (expt. 2), and an initial concentration of reactants and products (expt. 3). In all cases, the eventual distribution follows the law of mass action. That is, by substituting the actual equilibrium concentrations of all species in the equilibrium constant expression, a consistent value for the equilibrium constant is obtained.

**Table 14-1  The Value of $K_{eq}$**

For the reaction $H_2(g) + I_2(g) \rightleftharpoons 2HI(g)$ at 450°C,

$$K_{eq} = \frac{[HI]^2}{[H_2][I_2]} \quad [X] = \text{concentration in mol/L}$$

| | Initial Concentration | | | Equilibrium Concentration | | | |
|---|---|---|---|---|---|---|---|
| **Expt.** | **[H$_2$]** | **[I$_2$]** | **[HI]** | **[H$_2$]** | **[I$_2$]** | **[HI]** | **$K_{eq}$** |
| 1 | 2.000 | 2.000 | 0 | 0.428 | 0.428 | 3.144 | 54.0 |
| 2 | 0 | 0 | 2.000 | 0.214 | 0.214 | 1.572 | 54.0 |
| 3 | 1.000 | 1.000 | 1.000 | 0.321 | 0.321 | 2.358 | 54.0 |

In our first example, we will write the law of mass action for three reversible reactions in the gaseous phase.

| Example 14-1 | The Law of Mass Action |

**working it out**

Write the law of mass action for each of the following reactions.

a. $N_2(g) + 3H_2(g) \rightleftharpoons 2NH_3(g)$

b. $4NH_3(g) + 5O_2(g) \rightleftharpoons 4NO(g) + 6H_2O(g)$

c. $PCl_3(g) + Cl_2(g) \rightleftharpoons PCl_5(g)$

**Solutions**

a. $K_{eq} = \dfrac{[NH_3]^2}{[N_2][H_2]^3}$    b. $K_{eq} = \dfrac{[NO]^4[H_2O]^6}{[NH_3]^4[O_2]^5}$    c. $K_{eq} = \dfrac{[PCl_5]}{[PCl_3][Cl_2]}$

The numerical value of $K_{eq}$ is a measure of the extent of the reaction, so it tells us about the *position* of equilibrium. A large value for $K_{eq}$ indicates that the numerator is much larger than the denominator in the equilibrium constant expression. This means that the concentrations of products are larger than those of reactants. For example, consider the simple equilibrium

$$A \rightleftharpoons B \qquad K_{eq} = 10^2$$

$$K_{eq} = \frac{[B]}{[A]} = 10^2 \quad \text{thus } [B] = 10^2[A] = 100[A]$$

The large value of the equilibrium constant signifies that, for this reaction at the point of equilibrium, the concentration of the product, [B], is 100 times the concentration of the reactant, [A]. That is, the equilibrium lies far to the right. On the other hand, if the value of $K_{eq}$ were small (e.g., $10^{-2}$), then the concentration of the reactant, [A], would be 100 times the concentration of the product, [B], and this equilibrium would lie far to the left.

*The equilibrium constant is calculated from the experimental determination of the distribution of reactants and products at equilibrium.* The following two examples illustrate such calculations.

| Example 14-2 | Calculation of $K_{eq}$ |

**working it out**

What is the value of $K_{eq}$ for the following system at equilibrium?

$$2NO(g) + O_2(g) \rightleftharpoons 2NO_2(g)$$

At a given temperature, the equilibrium concentrations of the gases are [NO] = 0.890, $[O_2]$ = 0.250, and $[NO_2]$ = 0.0320.

**Solution**

For this reaction,

$$K_{eq} = \frac{[NO_2]^2}{[NO]^2[O_2]}$$

$$= \frac{[0.0320]^2}{[0.890]^2[0.250]} = \frac{1.024 \times 10^{-3}}{0.198} = \underline{5.17 \times 10^{-3}}$$

| Calculation of Equilibrium Concentrations and $K_{eq}$ | Example 14-3 |

For the equilibrium

$$N_2(g) + 3H_2(g) \rightleftharpoons 2NH_3(g)$$

Complete the table and compute the value of the equilibrium constant.

| Initial Concentration | | | Equilibrium Concentration | | |
|---|---|---|---|---|---|
| [H$_2$] | [N$_2$] | [NH$_3$] | [H$_2$] | [N$_2$] | [NH$_3$] |
| 0.200 | 0.200 | 0 | ? | ? | 0.0450 |

### Procedure

1. As in other stoichiometry problems, find the [H$_2$] and [N$_2$] that reacted to form the 0.0450 mol/L of NH$_3$ (section 8-6).

2. Find the [H$_2$] and [N$_2$] remaining at equilibrium by subtracting the concentration that reacted from the initial concentration; that is,

$$[N_2]_{eq} = [N_2]_{initial} - [N_2]_{reacted}$$

3. Substitute the concentrations of all compounds present at equilibrium into the equilibrium constant expression and solve to find the value of $K_{eq}$.

### Solution

1. If 0.0450 mol of NH$_3$ is formed, calculate the number of moles of N$_2$ that reacted (per liter).

$$0.0450 \; \text{mol NH}_3 \times \frac{1 \; \text{mol N}_2}{2 \; \text{mol NH}_3} = 0.0225 \; \text{mol N}_2 \; \text{reacted (per liter)}$$

The number of moles of H$_2$ that reacted is

$$0.0450 \; \text{mol NH}_3 \times \frac{3 \; \text{mol H}_2}{2 \; \text{mol NH}_3} = 0.0675 \; \text{mol H}_2 \; \text{reacted (per liter)}$$

2. At equilibrium

$$[N_2]_{eq} = 0.200 - 0.0225 = 0.178$$
$$[H_2]_{eq} = 0.200 - 0.0675 = 0.132$$

3.
$$K_{eq} = \frac{[NH_3]^2}{[N_2][H_2]^3} = \frac{(0.0450)^2}{(0.178)(0.132)^3} = \underline{\underline{4.95}}$$

| Calculation of Equilibrium Concentrations from $K_{eq}$ | Example 14-4 |

In the preceding equilibrium, what is the concentration of NH$_3$ at equilibrium if the equilibrium concentrations of N$_2$ and H$_2$ are 0.22 and 0.14 mol/L, respectively?

### Procedure

In this example, use the value of $K_{eq}$ found in the previous example. The concentration of NH$_3$ can be found by substituting the concentrations of the species given and solving for the one unknown.

**Solution**

$$K_{eq} = \frac{[NH_3]^2}{[N_2][H_2]^3} \qquad \begin{array}{l} [N_2] = 0.22 \\ [H_2] = 0.14 \end{array} \qquad K_{eq} = 4.95$$

$$K_{eq} = \frac{[NH_3]^2}{[0.22][0.14]^3} = 4.95$$

$$[NH_3]^2 = 2.99 \times 10^{-3}$$

$$[NH_3] = 5.5 \times 10^{-2} = 0.055$$

Thus the concentration of $NH_3$ = <u>0.055 mol/L.</u>

# 14-6 Equilibria of Weak Acids and Weak Bases in Water

**Looking Ahead!** We now return to weak acids, weak bases, and buffer solutions, which we first discussed in Chapter 12. Solutions of these substances are in equilibria, so they are also subject to the law of mass action. How this law applies to these particular situations is the subject of this chapter.

If one is in a leaky boat and no one bails—down you go. No equilibrium is established between input and output and the boat fills with water. The reaction of a strong acid in water is like this; that is, the reverse reaction does not occur. As a result, acids like $HNO_3$ are completely ionized in water. In this case, the ionization is represented by a single arrow, meaning that the reaction goes essentially to completion.

$$HNO_3(aq) + H_2O(l) \longrightarrow H_3O^+(aq) + NO_3^-(aq)$$

A weak acid such as hypochlorous acid (HClO), however, produces only a small concentration of ions because the ionization is a reversible reaction in which the equilibrium lies far to the left. The ionization of HClO is represented by the following equation.

$$HClO(aq) + H_2O(l) \rightleftharpoons H_3O^+(aq) + ClO^-(aq)$$

The law of mass action can be written as

$$K_{eq} = \frac{[H_3O^+][ClO^-]}{[H_2O][HClO]}$$

In this equilibrium, the concentrations of $H_3O^+$, $ClO^-$, and HClO in aqueous solution can all be varied. $H_2O$ is the solvent, however, and is present in a very large excess compared with the other species. Since the amount of $H_2O$ actually reacting is very small compared with the total amount present, the concentration of $H_2O$ is essentially a constant. The $[H_2O]$ can therefore be included with the other constant, $K_{eq}$, to produce another constant labeled $K_a$, *which is known as an* **acid ionization constant.**

$$K_{eq}[H_2O] = K_a = \frac{[H_3O^+][ClO^-]}{[HClO]}$$

Ionization of the weak base $NH_3$ is represented as

$$NH_3(aq) + H_2O(l) \rightleftharpoons NH_4^+(aq) + OH^-(aq)$$

Vinegar contains a weak acid and ammonia, a weak base.

For this reaction, *the equilibrium constant is labeled* $K_b$, *which is the* **base ionization constant.**

$$K_b = \frac{[NH_4^+][OH^-]}{[NH_3]}$$

The values of $K_a$ and $K_b$ for weak acids and bases are determined experimentally. To calculate $K_a$ or $K_b$, we need to know, or be able to determine from other data, the concentrations of both ions and the nonionized acid or base. In the following three examples, the needed concentrations are not directly given but must be inferred from the amount that ionizes (Example 14-5), the pH of the solution (Example 14-6), or the percent ionization (Example 14-7). After we illustrate how these constants are calculated, we will discuss their significance and how they are used.

---

| **Calculation of $K_a$ from Equilibrium Concentrations** | **Example 14-5** |
|---|---|

In a 0.20 M solution of $HNO_2$, it is found that 0.009 mol/L of the $HNO_2$ ionizes. What are the concentrations of $H_3O^+$, $NO_2^-$, and $HNO_2$ at equilibrium, and what is the value of $K_a$?

**working it out**

**Procedure**

1. Write the equilibrium equation.
2. Calculate the concentration of nonionized $HNO_2$ present at equilibrium. Remember that the initial concentration given (0.20 M) represents the concentration of nonionized $HNO_2$ present *plus* the concentration that ionizes. Therefore, $[HNO_2]$ at equilibrium is the initial concentration minus the concentration that ionizes.

$$[HNO_2]_{eq} = [HNO_2]_{initial} - [HNO_2]_{ionized}$$

3. Note that one $H_3O^+$ and one $NO_2^-$ are formed for each $HNO_2$ that ionizes (from the equation stoichiometry).

$$[H_3O^+]_{eq} = [NO_2^-]_{eq} = [HNO_2]_{ionized}$$

4. Calculate $K_a$

**Solution**

1. 
$$HNO_2(aq) + H_2O(l) \rightleftharpoons H_3O^+(aq) + NO_2^-(aq)$$

2. 
$$[HNO_2]_{initial} = 0.20 \qquad [HNO_2]_{ionized} = 0.009$$
$$[HNO_2]_{eq} = 0.20 - 0.009 = \underline{0.19}$$

3. 
$$[H_3O^+]_{eq} = [NO_2^-]_{eq} = \underline{0.009}$$

4. 
$$K_a = \frac{[H_3O^+][NO_2^-]}{[HNO_2]} = \frac{(0.009)(0.009)}{0.19}$$
$$= \underline{\underline{4 \times 10^{-4}}}$$

---

| **Calculation of $K_a$ from pH** | **Example 14-6** |
|---|---|

A 0.25 M solution of HCN has a pH of 5.00. What is $K_a$?

**Procedure**

1. Write the equilibrium reaction.
2. Convert pH to $[H_3O^+]$.
3. Note that $[CN^-] = [H_3O^+]$ at equilibrium.
4. Calculate $[HCN]$ at equilibrium, which is

$$[HCN]_{eq} = [HCN]_{initial} - [HCN]_{ionized}$$

$[HCN]_{ionized} = [H_3O^+]_{eq}$, since one $H_3O^+$ is produced at equilibrium for every HCN that ionizes.

5. Use these values to calculate $K_a$.

**Solution**

$$HCN(aq) + H_2O(l) \rightleftharpoons H_3O^+(aq) + CN^-(aq)$$
$$[H_3O^+] = \text{antilog} -5.00$$
$$[H_3O^+] = 1.0 \times 10^{-5} = [CN^-]$$
$$[HCN]_{eq} = 0.25 - (1.0 \times 10^{-5}) \approx 0.25$$

($\approx$ means approximately equal.) Note that the amount of HCN that ionizes ($10^{-5}$ M) is negligible compared with the initial concentration of HCN (0.25 M).* Therefore, $[HCN]_{eq} = [HCN]_{initial}$.

$$K_a = \frac{[H_3O^+][CN^-]}{[HCN]}$$
$$K_a = \frac{(1.0 \times 10^{-5})(1.0 \times 10^{-5})}{0.25}$$
$$= \underline{\underline{4.0 \times 10^{-10}}}$$

---

**Example 14-7** | **Calculation of $K_b$ from Percent Ionization**

A 0.10 M solution of $NH_3$ is 1.34% ionized. What is the value of $K_b$ (the base ionization constant)?

**Procedure**

Find the concentrations of all species in the mass action expression at equilibrium. Substitute these values into the expression and solve for $K_b$.

**Solution**

$$NH_3(aq) + H_2O(l) \rightleftharpoons NH_4^+(aq) + OH^-(aq)$$
$$K_b = \frac{[NH_4^+][OH^-]}{[NH_3]}$$

At equilibrium, 1.34% of the $NH_3$ is ionized, or

$$0.0134 \times 0.10 = 1.3 \times 10^{-3} \text{ mol/L}$$

---

*For the purposes of these calculations (two significant figures), a number is considered negligible compared with another if it is less than 10% of the larger number.

According to the equation, for every $NH_3$ ionized, one $NH_4^+$ and one $OH^-$ form. Therefore, if $1.3 \times 10^{-3}$ mol/L ionizes, at equilibrium

$$[NH_4^+] = [OH^-] = 1.3 \times 10^{-3}$$

The concentration of $NH_3$ at equilibrium is the initial concentration (0.10) minus the concentration that ionizes.

$$[NH_3] = 0.10 - (1.3 \times 10^{-3}) \approx 0.10$$

Substituting these values into the expression,

$$K_b = \frac{(1.3 \times 10^{-3})(1.3 \times 10^{-3})}{0.10} = \underline{\underline{1.7 \times 10^{-5}}}$$

In Chapter 12, we mentioned that weak acids have a range of acid strengths depending on the degree of ionization. The percent ionization of a particular weak acid and the pH of the resulting solution depend on the original concentration of the acid, however, whereas the value for $K_a$ does not. Therefore, the value of $K_a$ is the preferred way of assigning a quantitative value to the acid strength. Also, note that the equilibrium constant expression has the same basic form for all acids. The larger the value of $K_a$, the stronger the acid. In Table 14-2 we have listed values of $K_a$ for some weak acids. $K_b$ values for some weak bases are listed in Table 14-3.

Once a value for $K_a$ or $K_b$ for a specific weak acid or base is known, we have the ability to calculate the pH of a solution of known concentration of a weak acid (Example 14-8) or a weak base (Example 14-9).

## Table 14-2  $K_a$ for Some Weak Acids

$K_a = \dfrac{[H_3O^+][X^-]}{[HX]}$   (HX symbolizes a weak acid, $X^-$ its conjugate base)

| Acid | Formula | $K_a$ | |
|------|---------|-------|--|
| hydrofluoric | HF | $6.7 \times 10^{-4}$ | |
| nitrous | $HNO_2$ | $4.5 \times 10^{-4}$ | |
| formic | $HCHO_2$ | $1.8 \times 10^{-4}$ | Decreasing |
| acetic | $HC_2H_3O_2$ | $1.8 \times 10^{-5}$ | acid |
| hypochlorous | HClO | $3.2 \times 10^{-8}$ | strength |
| hypobromous | HBrO | $2.1 \times 10^{-9}$ | |
| hydrocyanic | HCN | $4.0 \times 10^{-10}$ | |

## Table 14-3  $K_b$ for Some Weak Bases

$K_b = \dfrac{[HB^+][OH^-]}{[B]}$   (B symbolizes a weak base, $HB^+$ its conjugate acid)

| Base | Formula | $K_b$ | |
|------|---------|-------|--|
| dimethylamine | $(CH_3)_2NH$ | $7.4 \times 10^{-4}$ | Decreasing |
| ammonia | $NH_3$ | $1.8 \times 10^{-5}$ | base |
| hydrazine | $N_2H_4$ | $9.8 \times 10^{-7}$ | strength |

| **Example 14-8** | **Calculation of the pH of a Solution of a Weak Acid** |
| :--- | :--- |
| **working it out** | What is the pH of a 0.155 M solution of HClO? |

**Procedure**

1. Write the equilibrium involved.
2. Write the appropriate equilibrium constant expression.
3. Let $x = [H_3O^+]$ at equilibrium; since $[H_3O^+] = [ClO^-]$, $x = [ClO^-]$.
4. At equilibrium, $[HClO]_{eq} = [HClO]_{initial} - [HClO]_{ionized}$
5. Using the value for $K_a$ in Table 14-2, solve for $x$.
6. Convert $x$ to pH.

In summary:

|  | **HClO** | **[H$_3$O$^+$]** | **[ClO$^-$]** |
| :--- | :--- | :--- | :--- |
| initial | 0.155 | 0 | 0 |
| equilibrium | 0.155 − x | x | x |

**Solution**

1.
$$HClO(aq) + H_2O(l) \rightleftharpoons H_3O^+(aq) + ClO^-(aq)$$

2.
$$K_a = \frac{[H_3O^+][ClO^-]}{[HClO]} = 3.2 \times 10^{-8}$$

3. At equilibrium, $[H_3O^+] = [ClO^-] = x$.

4. At equilibrium, $[HClO] = 0.155 - x$.

5.
$$\frac{(x)(x)}{0.155 - x} = 3.2 \times 10^{-8}$$

The solution of this equation appears to require the quadratic equation. However, a simplification of this calculation is possible. Note that $K_a$ is a small number, indicating that the degree of ionization is small (the equilibrium lies far to the left). This means that $x$ is a very small number. Since very small numbers make little or no difference when added to or subtracted from large numbers, they can be ignored with little or no error. (Refer to the example of the large crowd in section 1-1.)

$$0.155 - x \approx 0.155$$

Therefore the expression can now be simplified.

$$\frac{(x)(x)}{0.155 - x} = \frac{x^2}{0.155} = 3.2 \times 10^{-8}$$
$$x^2 = 5.0 \times 10^{-9}$$

To solve for $x$, take the square root of each side of the equation.

$$\sqrt{x^2} = \sqrt{5.0 \times 10^{-9}}$$
$$x = [H_3O^+] = 7.1 \times 10^{-5}$$

(Note that the $x$ is indeed much smaller than 0.155.)

6.
$$pH = -\log[H_3O^+] = -\log(7.1 \times 10^{-5}) = \underline{\underline{4.15}}$$

**Calculation of the pH of a Solution of a Weak Base**

Example 14-9

What is the pH of a 0.245 M solution of $N_2H_4$?

**Solution**

1.
$$N_2H_4(aq) + H_2O(l) \rightleftharpoons N_2H_5^+(aq) + OH^-(aq)$$

2.
$$K_b = \frac{[N_2H_5^+][OH^-]}{[N_2H_4]} = 9.8 \times 10^{-7} \text{ (from Table 14-3)}$$

3. Let $x = [OH^-] = [N_2H_5^+]$ (at equilibrium).

4. $[N_2H_4] = 0.245 - x$ (at equilibrium). Since $K_b$ is very small, $x$ is very small. Therefore,
$$0.245 - x \approx 0.245$$

5.
$$\frac{(x)(x)}{0.245} = 9.8 \times 10^{-7}$$
$$x^2 = 2.4 \times 10^{-7}$$
$$x = 4.9 \times 10^{-4} = [OH^-]$$

6.
$$pOH = -\log[OH^-] = -\log(4.9 \times 10^{-4}) = 3.31$$
$$pH = 14.00 - pOH = 14.00 - 3.31 = \underline{10.69}$$

As mentioned in Chapter 12, buffer solutions are those that resist changes in pH and are made from mixtures of a weak acid (or base) and a salt containing the conjugate base (or acid). They have many important applications, especially in life processes such as maintaining a nearly constant pH of the blood. We can also calculate the pH of buffers using a form of the equation for $K_a$ or $K_b$ outlined previously. In fact, calculations of the pH of buffers are comparatively simple if we rearrange the equilibrium constant expression, as illustrated here with HClO.

$$[H_3O^+] = K_a \times \frac{[HClO]}{[ClO^-]}$$

By taking $-\log$ of both sides of the equation, we now have

$$-\log[H_3O^+] = -\log K_a - \log \frac{[HClO]}{[ClO^-]}$$

Now we substitute pH for $-\log[H_3O^+]$ and $pK_a$ for $-\log K_a$ and invert the ratio of the concentrations.

$$pH = pK_a + \log \frac{[ClO^-]}{[HClO]}$$

Notice that the $ClO^-$ is the base species in the equilibrium and HClO is the acid species. We can thus make this a general equation for calculating the pH of buffers by substituting [acid] for [HClO] and [base] for [ClO^-]. *This general equation is known as the* **Henderson–Hasselbalch equation.** The complementary equation for the pH of buffers made from weak bases and their conjugate acids is also given.

$$pH = pK_a + \log \frac{[base]}{[acid]} \qquad pOH = pK_b + \log \frac{[acid]}{[base]}$$

Since weak acids and bases are used to make buffers, we are quite safe in assuming that concentrations of the acid and base species present in solution are essentially the same as the originally measured amounts. Also, the conjugate acid or base is added in the

## Swimming Pool Chemistry

When we take a refreshing dive into a swimming pool, we may not appreciate the fact that maintenance of this pleasure requires application of some serious chemistry. The balance of chemicals in a pool must be closely monitored on a daily basis by some lucky person.

First, the pool must have the proper content of chlorine, which is usually in the form of hypochlorous acid (HOCl). This is most important as HOCl destroys bacteria and keeps the pool safe. The most common source of chlorine is sodium hypochlorite (bleach) added as a 10% solution. Sodium hypochlorite dissolves to produce the hypochlorite ion, which undergoes basic hydrolysis (reaction **1**). By itself this solution would produce a swimming pool with a pH of around 10, which is way too alkaline. The pH is lowered to about 7.5 by adding sodium bisulfate (reaction **2**). At pH = 7.5, pH = $pK_a$ for this system, which means that equimolar amounts of HOCl and $OCl^-$ are present (reaction **3**). Recall that this is a buffer solution, so the pH is stabilized around 7.5 even if some acid or base enters the pool. However, if the pH does drop below 7.2, the water can sting the eyes and cause corrosion of metal pipes. Sodium carbonate is added until the pH returns to 7.5 (reaction **4**). If it rises above 7.8, the pool may become cloudy and scaling may occur. The pool manager adds more sodium bisulfate to bring the pH down. The reactions that we have discussed are illustrated

by the equations below.

**1.** Add NaOCl:
$$NaOCl(aq) \xrightarrow{H_2O} Na^+(aq) + OCl^-$$
$$OCl^- + H_2O \rightleftharpoons HOCl + OH^- \quad (pH \ 10)$$

**2.** Lower pH:
$$NaHSO_4(s) \xrightarrow{H_2O} Na^+(aq) + HSO_4^-(aq)$$
$$HSO_4^-(aq) + OH^- \longrightarrow SO_4^{2-}(aq) + H_2O$$

**3.** Buffer equilibrium (pH = $pK_a$ = 7.5)
$$HOCl(aq) + H_2O \rightleftharpoons OCl^- + H_3O^+(aq)$$

**4.** Raise pH: $Na_2CO_3(s) \xrightarrow{H_2O} 2Na^+(aq) + CO_3^{2-}(aq)$
$$CO_3^{2-}(aq) + H_3O^+(aq) \longrightarrow HCO_3^-(aq) + H_2O$$

A good clean pool needs even more help. The presence of humans in the pool and their fluids (perspiration, etc.) produces ammonia. Ammonia reacts with HOCl to produce chloramines (i.e., $NH_2Cl$). If "the pool smells like chlorine," the smell is actually due to chloramines. It is removed by "shocking" the pool by addition of another special chemical. If algae form, the pool becomes green and slimy but another chemical clears it up.

Next time you are invited to take a dip in a nice swimming pool, remember how chemistry, work, and lots of money make it all possible as well as pleasant.

**Submitted by G. Lynn Carlson**
**University of Wisconsin—Parkside**

*Control of pH in a swimming pool is essential.*

form of a salt where one of the ions is not involved in the equilibrium (i.e., the $Na^+$ in $NaClO$). The most effective buffers are made from solutions containing equal molar amounts of acid and base species. In such cases, the ratio of base to acid equals unity. Since $\log 1 = 0$, pH equals $pK_a$. Thus, $pK_a$ is sometimes used as a measure of the acid strength and as the pH of an effective buffer formed from this weak acid. The pHs of other ratios of acid and base species can be calculated by simply substituting the amounts in moles or the concentrations into the appropriate Henderson–Hasselbalch equation. Note that this equation is useful only for buffers and not for solutions of weak acids or weak bases alone.

| Example 14-10 | Calculation of the pH of a Buffer Solution |
|---|---|
| working it out | What is the pH of a buffer solution that is made by dissolving 0.30 mol of $HC_2H_3O_2$ and 0.50 mol of $NaC_2H_3O_2$ in enough water to make 1.00 L of solution? |

**Solution**

In this solution, the concentration of the acid species is $[HC_2H_3O_2] = 0.30$ mol/L, and the concentration of the base species is $[C_2H_3O_2^-] = 0.50$ mol/L.

$$pK_a = -\log K_a = -\log (1.8 \times 10^{-5}) = 4.74$$

$$pH = 4.74 + \log \frac{[0.50]}{[0.30]} = 4.74 + 0.22 = \underline{4.96}$$

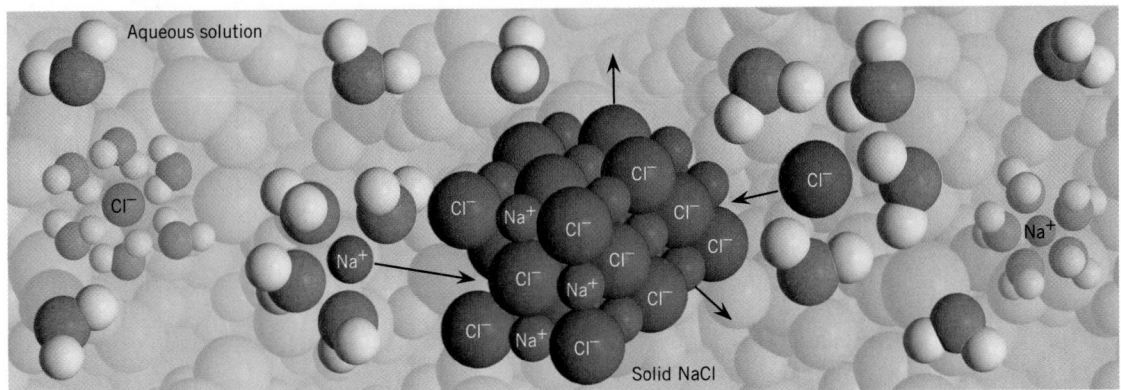

**Figure 14-13 Solubility and Equilibrium** At equilibrium, the rate at which ions are dissolving equals the rate at which ions are forming the solid.

# 14-7 Solubility Equilibria

**Looking Ahead!** We mentioned in Chapters 7 and 11 that salts all have a certain solubility in water. In fact, saturated solutions of salts demonstrate a dynamic equilibrium between the solid salt and the ions in solution. This is reexamined in this section.

One of the more painful maladies that strikes older adults is kidney stones. The most common type of kidney stone is composed of calcium oxalate ($CaC_2O_4$), which is deposited as a solid in the kidney. One can arrest the development of these stones by cutting back on foods that contain either calcium or oxalate ions. By the end of this section, we should have an understanding of why this is so.

First let's review what happens when a salt such as NaCl is added to water. As discussed in Chapter 11, if enough solid is added to the water, the solution eventually becomes saturated. At this point it appears that the solution process has stopped. Instead, we find that the rate of solution of the salt is equal to the rate of deposition of the solid from the ions in solution and we have a dynamic equilibrium situation. (See Figure 14-13.) This is illustrated by the equilibrium shown below.

$$NaCl(s) \rightleftharpoons Na^+(aq) + Cl^-(aq)$$

A common but interesting chemical phenomenon that is often displayed at science fairs is the science (and art) of crystal growing. Crystal growing demonstrates the principle of equilibrium. If a comparatively large crystal of a compound such as $CuSO_4$ is suspended in a saturated solution of that compound, the small crystals at the bottom get smaller and the large one grows larger. The concentrations of $Cu^{2+}$ and $SO_4^{2-}$ in solution remain constant, as does the total mass of solid crystals. The changing size of the crystals, however, indicates that an equilibrium exists in which the identities of ions in the solid and dissolved states are changing. (See Figure 14-14.) This phenomenon occurs because small crystals, with their greater surface area, dissolve faster than large crystals. Thus, although the total mass of the crystals remains constant, the unequal rate of solution favors solution of small crystals and precipitation on the large crystal.

Now consider the equilibrium between solid calcium oxalate and its ions in solution.

$$CaC_2O_4(s) \rightleftharpoons Ca^{2+}(aq) + C_2O_4^{2-}(aq)$$

The extent of the solubility of this compound can be represented by an equilibrium constant known as the **solubility product constant ($K_{sp}$)** or simply the **solubility**

**Figure 14-14 Crystal Growth**
After the CuSO₄ crystal on the left was suspended in the solution for a few days, it grew larger.

**product.** In this case the constant expression is

$$K_{sp} = [Ca^{2+}][C_2O_4^{2-}]$$

It is set up like any other equilibrium constant expression except that the solid (on the left) is not included. The reason for this is that the concentration of a solid is a constant. *So, as long as at least some solid is present, how much solid is present is not relevant to the concentration of ions in solution.* Another example of a solubility product involves the solution of lead(II) chloride.

$$PbCl_2(s) \rightleftharpoons Pb^{2+}(aq) + 2Cl^-(aq) \qquad K_{sp} = [Pb^{2+}][Cl^-]^2$$

Notice that the [Cl⁻] concentration is squared because there are two chlorides in the equilibrium equation.

The equilibrium constant expressions are shown for other salts in Example 14-11. The values for $K_{sp}$ are determined from molar solubility data, as illustrated in Examples 14-12 and 14-13. The use of $K_{sp}$ can in turn be used to determine molar solubility, as illustrated in Example 14-14.

---

**Example 14-11**

**working it out**

**The Expressions for $K_{sp}$**

Write the equilibrium constant expressions ($K_{sp}$) for solution of (a) $Ag_2CrO_4$, (b) $Fe(OH)_3$, and (c) $Ca_3(PO_4)_2$.

**Procedure**

Write the equations for the solution of the salts in water. It may be necessary to review Chapter 4 and specifically Table 4-2 to determine the identity of the ions formed from the compounds.

a. $Ag_2CrO_4(s) \rightleftharpoons 2Ag^+(aq) + CrO_4^{2-}(aq)$
b. $Fe(OH)_3(s) \rightleftharpoons Fe^{3+}(aq) + 3OH^-(aq)$
c. $Ca_3(PO_4)_2(s) \rightleftharpoons 3Ca^{2+}(aq) + 2PO_4^{3-}(aq)$

**Solution**

a. $K_{sp} = [Ag^+]^2[CrO_4^{2-}]$     c. $K_{sp} = [Ca^{2+}]^3[PO_4^{3-}]^2$

b. $K_{sp} = [Fe^{3+}][OH^-]^3$

---

**Example 14-12**

**Determination of $K_{sp}$ from Solubility Data**

The molar solubility of AgCl is $1.3 \times 10^{-5}$ mol/L. What is the value of $K_{sp}$ for AgCl?

**Procedure**

Write the equation for the solution of AgCl and the expression for $K_{sp}$.

$$AgCl(s) \rightleftharpoons Ag^+(aq) + Cl^-(aq) \qquad K_{sp} = [Ag^+][Cl^-]$$

Notice from the stoichiometry of the equilibrium that for every mole of AgCl that dissolves, one mole of $Ag^+$ and one mole of $Cl^-$ are formed. Therefore, the concentration of each ion is equal to the amount that dissolves, or the molar solubility.

**Solution**

$$\text{solubility} = 1.3 \times 10^{-5} \text{ mol/L} = [Ag^+] = [Cl^-]$$
$$[1.3 \times 10^{-5}][1.3 \times 10^{-5}] = 1.7 \times 10^{-10}$$

| Determination of $K_{sp}$ from Solubility Data | Example 14-13 |
|---|---|

What is the value of $K_{sp}$ for $PbCl_2$ if the molar solubility of this compound is $3.9 \times 10^{-2}$ mol/L?

**Procedure**

Write the solution equilibrium and the expression for $K_{sp}$.

$$PbCl_2 \rightleftharpoons Pb^{2+}(aq) + 2Cl^{-}(aq) \qquad K_{sp} = [Pb^{2+}][Cl^{-}]^2$$

Notice from the stoichiometry of the solution that for each mole of $PbCl_2$ that dissolves, it produces one mole of $Pb^{2+}$ and *two* moles of $Cl^-$. Therefore, the concentration of $Pb^{2+}$ is equal to the actual moles/L that dissolve (i.e., the solubility) but the concentration of $Cl^-$ is twice the solubility.

**Solution**

$$[Pb^{2+}] = 3.9 \times 10^{-2} \qquad [Cl^-] = 2 \times 3.9 \times 10^{-2} = 7.8 \times 10^{-2}$$
$$K_{sp} = [3.9 \times 10^{-2}][7.8 \times 10^{-2}]^2 = (3.9 \times 10^{-2})(6.1 \times 10^{-3}) = \underline{\underline{2.4 \times 10^{-4}}}$$

| Determination of Molar Solubility from $K_{sp}$ | Example 14-14 |
|---|---|

What is the molar solubility of $CaC_2O_4$? For $CaC_2O_4$, $K_{sp} = 2.1 \times 10^{-9}$.

**Procedure**

Write the solution equilibrium for $CaC_2O_4$ and the expression for $K_{sp}$.

$$CaC_2O_4(s) \rightleftharpoons Ca^{2+}(aq) + C_2O_4^{2-}(aq) \qquad K_{sp} = [Ca^{2+}][C_2O_4^{2-}]$$

Notice that the number of moles of $Ca^{2+}$ in solution is equal to the number of moles of $C_2O_4^{2-}$ and both are equal to the molar solubility. We will let $x$ = the molar solubility.

**Solution**

If $x$ = the molar solubility, then at equilibrium $[Ca^{2+}] = [C_2O_4^{2-}] = x$.
Thus $[x][x] = K_{sp} = 2.1 \times 10^{-9}$.

Taking the square root of $K_{sp}$, the molar solubility is $\underline{\underline{4.6 \times 10^{-5}}}$ mol/L.

Let's now return to the question of why calcium oxalate (kidney stones) precipitates from aqueous solution. In a complex system such as in our blood, $Ca^{2+}$ ion and $C_2O_4^{2-}$ ion originate from many sources, so they are not exactly equal. However, whenever the product of the concentrations of the two ions is equal to $K_{sp}$, the solution is saturated and at equilibrium. In Example 14-15 we have calculated the concentration of oxalate that is present in a saturated solution with a specific calcium ion concentration. The answer represents the maximum amount of that ion that can be present without formation of solid calcium oxalate. If the concentration of $Ca^{2+}$ and $C_2O_4^{2-}$ is such that their product is greater than the value for $K_{sp}$, the system is not at equilibrium. *To reestablish equilibrium, the concentration of the ions must decrease by coming together to form a precipitate.* (See Figure 14-15.) So how does one avoid calcium oxalate kidney stones? A stone (precipitate) will not form if the product of the ion concentrations is equal to or less than $K_{sp}$. There are three ways to accomplish this. (1) Lower the concentration of both ions by dilution. In other words, drink plenty of

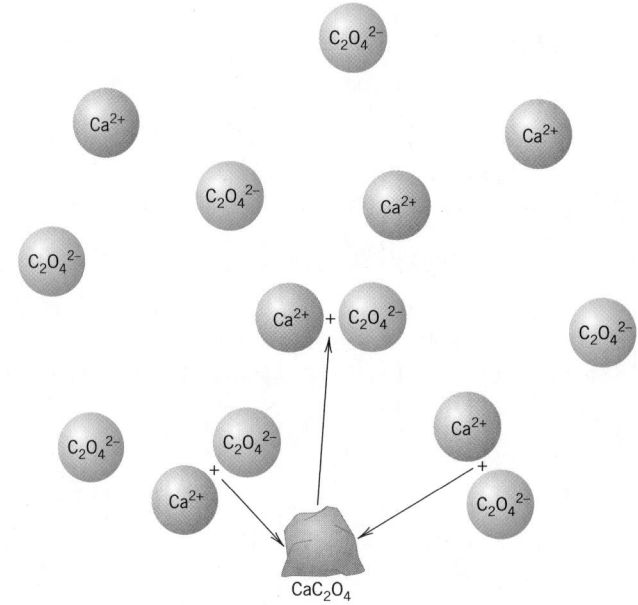

**Figure 14-15 The Formation of Solid Calcium Oxalate**
When the molar concentration of $Ca^{2+}$ times the molar concentration of $C_2O_4^{2-}$ is greater than $K_{sp}$, a solid forms.

water. This is always a good idea. (2) Lower the concentration of $Ca^{2+}$ ion. This is not a good idea. $Ca^{2+}$ is also incorporated into teeth and bones. If we lower the $Ca^{2+}$ ion concentration too much, the solubility equilibrium involving bones may be affected and cause bone tissue to dissolve. We don't want to avoid kidney stones by having brittle bones. (3) Lower the concentration of oxalate ion. Many foods such as brussels sprouts have relatively high concentrations of oxalate. Avoiding foods that have a high amount of oxalate salts should help. If one is susceptible to these types of kidney stones, the best advice is to drink plenty of water and avoid brussels sprouts.

| **Example 14-15** | **The Concentrations of Ions in a Saturated Solution** |
|---|---|
| **working it out** | What is the maximum oxalate ion concentration in a solution where $[Ca^{2+}] = 4.4 \times 10^{-4}$ mol/L? |

**Procedure**

Solve for $[C_2O_4^{2-}]$ in the expression for $K_{sp}$.

**Solution**

$$K_{sp} = [Ca^{2+}][C_2O_4^{2-}]$$
$$[C_2O_4^{2-}] = K_{sp}/[Ca^{2+}] = 2.1 \times 10^{-9}/4.4 \times 10^{-4} = \underline{\underline{4.8 \times 10^{-6} \text{ M}}}$$

If the concentration of oxalate were any higher than the calculated value, a precipitate of $CaC_2O_4$ would form.

---

**Looking Back!** Perhaps one of the most important tools for dealing with reactions that reach a point of equilibrium is provided by the law of mass action. This law permits many calculations that can tell us about the actual distribution of reactants and products present at equilibrium. Once we know the value of the equilibrium constant, we can determine how any mixture of reactants and products will redistribute at equilibrium. A practical application of these calculations allows us to calculate the pH of weak acid and weak base solutions.

## Learning Check B | checking it out

**B-1.** Fill in the blanks.

The law of mass action is composed of a numerical quantity known as an _____ and a ratio known as the _____ . A large value for $K_{eq}$ tells us that the equilibrium lies in favor of _____ . In the calculation of $K_{eq}$, the coefficient of a compound in the balanced equation becomes an _____ of the _____ of the compound in the ratio. For weak acids, the smaller the value of _____ , the _____ the acid in water. One can calculate the pH of a weak base solution from a knowledge of the original _____ of the weak base and the value of _____ .

**B-2.** Write the law of mass action for the following equilibria.

**a.** $CO(g) + 3H_2(g) \rightleftharpoons CH_4(g) + H_2O(g)$

**b.** $HCHO_2(aq) + H_2O(l) \rightleftharpoons H_3O^+(aq) + CHO_2^-(aq)$

**B-3.** What is the value of $K_{eq}$ in problem B-2(a) if there are $2.0 \times 10^{-3}$ mol of CO, $1.0 \times 10^{-2}$ mol of $H_2$, $3.5 \times 10^{-4}$ mol of $CH_4$, and $5.0 \times 10^{-5}$ mol of $H_2O$ present in a 10.0-L container at equilibrium?

**B-4.** A 0.25 M solution of a hypothetical weak base ($NX_3$) is 1.0% ionized. What is $K_b$?

**B-5.** What is the pH of a 0.10 M solution of $HCHO_2$? (Refer to Table 14-2.)

**B-6.** What is the pH of a buffer made by mixing 0.068 mol of $NH_3$ and 0.049 mol of $NH_4Cl$ in 2.0 L of solution? (Refer to Table 14-3.)

**B-7.** What is the molar solubility of barium carbonate? For $BaCO_3$, $K_{sp} = 8.1 \times 10^{-9}$.

*Additional Examples: Exercises 14-15, 14-20, 14-25, 14-30, 14-35, 14-39, 14-45, 14-49, 14-51, 14-56, 14-62, and 14-66.*

**chapter** **R** **eview**                                    **putting it together**

This chapter concerns several important chemistry topics such as how reactions take place, the rates at which they take place, and the point of equilibrium in reversible reactions. **Collision theory** tells us that reactions take place through collisions of molecules. The **rate of a reaction** is determined by the frequency of these collisions and the **activation energy** of the reaction. There are several things that influence these two factors and thus affect the rate of the reaction.

1. *Activation energy*
   Each set of potential reactants requires a different amount of energy to form the **activated complex** that leads to the formation of products. If collisions do not have the necessary kinetic energy, no products form.

2. *Temperature*
   Higher temperatures increase both the energy of colliding molecules and the frequency of collisions. Thus the rate increases.

3. *Concentration of reactants*
   The concentration of reactants relates directly to the frequency of collisions. The higher the concentration of colliding molecules, the faster the reaction.

4. *Particle size*

   Finely divided solids allow more frequent collisions because of increased surface area. Thus reaction rates are faster.

5. *Catalyst*

   The presence of a **catalyst** lowers the activation energy, so the rate of the reaction increases.

**Reversible reactions** often reach a **point of equilibrium** where significant concentrations of reactants and products coexist. In fact, a **dynamic equilibrium** exists where both the forward and reverse reactions are occurring at the same rate. The point of equilibrium can be influenced by reaction conditions. However, changes in these reaction conditions have a predictable effect on the point of equilibrium, which is correlated by **Le Châtelier's principle.** The factors that can shift the point of equilibrium are summarized as follows.

1. *Concentration*

   An increase in concentration on one side of the equation results in an increase in concentration on the other side. A decrease on one side leads to a shift to that side.

2. *Pressure*

   For gas phase reactions, compression of the mixture leads to a lower volume and a higher pressure. The equilibrium shifts to the side with the smaller total number of molecules of gas, thus reducing the pressure.

3. *Temperature*

   For exothermic reactions, an increase in temperature leads to a decrease in the proportion of products. Just the opposite occurs for an endothermic reaction.

Many important industrial reactions do not take place readily, or even at all, without a catalyst present. A catalyst brings a reaction mixture to the point of equilibrium faster but does not change the actual point of equilibrium.

The distribution of concentrations of reactants and products can be predicted from the **law of mass action.** This law is composed of a numerical value ($K_{eq}$) called the **equilibrium constant,** which is equal to a ratio of reactants and products constructed from the balanced equation and known as the **equilibrium constant expression.** Examples of the two types of equilibria discussed are

$$\textit{Gaseous:} \quad A_2(g) + B_2(g) \rightleftharpoons 2AB(g) \qquad K_{eq} = \frac{[AB]^2}{[A_2][B_2]}$$

$$\textit{Ionic:} \quad HX(aq) + H_2O(l) \rightleftharpoons H_3O^+(aq) + X^-(aq) \qquad K_a = \frac{[H_3O^+][X^-]}{[HX]}$$

The value of the equilibrium constant at a given temperature is obtained by experimental measurements of the distribution of reactants and products at equilibrium. The constant and the initial concentrations can then be used to calculate the concentration of a given gas or ion at equilibrium. The **acid ionization constants** for weak acids ($K_a$) and the **base ionization constants** for weak bases ($K_b$) are useful for calculation of the pH of these solutions. We are also able to use equilibrium constants to calculate the pH of buffers. The **Henderson–Hasselbalch equation** can be used to calculate the pH of buffer solutions. The solubility of salts in water is determined by the equilibrium constant $K_{sp}$, which is known as the **solubility product constant** or simply the **solubility product.** $K_{sp}$ can be calculated from solubility data and vice versa.

# Exercises

## Collision Theory and Reaction Rates

**14-1.** Explain why all collisions between reactant molecules do not lead to products.

**14-2.** The equation

$$NO_2(g) + CO(g) \rightleftharpoons NO(g) + CO_2(g)$$

represents a reaction in which the mechanism of the forward reaction involves a collision between the two reactant molecules. Describe the probable orientation of the molecules in this collision.

**14-3.** Explain the following facts from your knowledge of collision theory and the factors that affect the rates of reactions.
**(a)** The rates of chemical reactions approximately double for each 10°C rise in temperature.
**(b)** Eggs cook more slowly in boiling water at higher altitudes where water boils at temperatures less than 100°C.
**(c)** $H_2$ and $O_2$ do not start to react to form $H_2O$ except at very high temperatures (unless initiated by a spark).
**(d)** Wood burns explosively in pure $O_2$ but slowly in air, which is about 20% $O_2$.
**(e)** Coal dust burns faster than a single lump of coal.
**(f)** Milk sours if left out for a day or two but will last two weeks in the refrigerator.
**(g)** $H_2$ and $O_2$ react smoothly at room temperature in the presence of finely divided platinum.

**14-4.** Explain the following facts from your knowledge of collision theory and the factors that affect the rates of reactions.
**(a)** A wasp is lethargic at temperatures below 65°F.
**(b)** Charcoal burns faster if you blow on the coals.
**(c)** A 0.10 M boric acid solution ($[H_3O^+] = 7.8 \times 10^{-6}$) can be used for eyewash, but a 0.10 M hydrochloric acid solution ($[H_3O^+] = 0.10$) would cause severe damage.
**(d)** To keep apple juice from fermenting to apple cider, the apple juice must be kept cold.
**(e)** Coal does not spontaneously ignite at room temperature, but white phosphorus does.
**(f)** Potassium chlorate decomposes at a much lower temperature when mixed with $MnO_2$.

## Equilibrium

**14-5.** Explain the point of equilibrium in terms of reaction rates.

**14-6.** In the hypothetical reaction between $A_2$ and $B_2$, when was the rate of the forward reaction at a maximum? In the same reaction, when was the rate of the reverse reaction at a maximum? (Refer to Figure 14-12.)

**14-7.** Give a reason, besides reaching a point of equilibrium, why certain reactions do not go to completion (100% to the right).

**14-8.** Compare the activation energy for the reverse reaction with that for the forward reaction in Figure 14-3. Which are easier to form, reactants from products or products from reactants? Which system would come to equilibrium faster, a reaction starting with pure products or one starting with pure reactants?

**14-9.** Consider an endothermic reaction that comes to a point of equilibrium such as shown in Figure 14-4. Which is greater, the activation energy for the forward or for the reverse reaction? Which system would come to equilibrium faster, a reaction starting with pure products or one starting with pure reactants?

## Le Châtelier's Principle and Equilibrium

**14-10.** The following equilibrium represents an important industrial process, called the Haber process, used to convert $N_2$ to $NH_3$. The ammonia is used mainly for fertilizer. It requires the use of a catalyst.

$$N_2(g) + 3H_2(g) \underset{}{\overset{500°C}{\rightleftharpoons}} 2NH_3(g) + heat$$

Determine the direction in which the equilibrium will be shifted by the following changes.
**(a)** increasing the concentration of $N_2$
**(b)** increasing the concentration of $NH_3$
**(c)** decreasing the concentration of $H_2$
**(d)** decreasing the concentration of $NH_3$
**(e)** compressing the reaction mixture
**(f)** removing the catalyst
**(g)** How will the yield of $NH_3$ be affected by raising the temperature to 750°C?
**(h)** How will the yield of $NH_3$ be affected by lowering the temperature to 0°C? How will this affect the rate of formation of $NH_3$?

**14-11.** Consider the equilibrium

$$4NH_3(g) + 5O_2(g) \rightleftharpoons 4NO(g) + 6H_2O(g) + heat\ energy$$

How will this system at equilibrium be affected by the following changes?
**(a)** increasing the concentration of $O_2$
**(b)** removing all of the $H_2O$ as it is formed
**(c)** increasing the concentration of NO
**(d)** compressing the reaction mixture
**(e)** increasing the volume of the reaction container
**(f)** increasing the temperature
**(g)** decreasing the concentration of $O_2$
**(h)** adding a catalyst

**14-12.** The following equilibrium takes place at high temperatures.

$$N_2(g) + 2H_2O(g) + heat\ energy \rightleftharpoons 2NO(g) + 2H_2(g)$$

How will the concentration of NO at equilibrium be affected by these changes?
**(a)** increasing $[N_2]$
**(b)** decreasing $[H_2]$
**(c)** compressing the reaction mixture

**(d)** decreasing the volume of the reaction container
**(e)** decreasing the temperature
**(f)** adding a catalyst

**14-13.** Consider the equilibrium

$$2SO_3(g) + CO_2(g) + \text{heat energy} \rightleftharpoons CS_2(g) + 4O_2(g)$$

How will the concentration of $CS_2$ at equilibrium be affected by these changes?
**(a)** decreasing the volume of the reaction vessel
**(b)** adding a catalyst
**(c)** increasing the temperature
**(d)** increasing the original concentration of $CO_2$
**(e)** removing some $O_2$ as it forms

**14-14.** Fortunately for us, the major components of air, $N_2$ and $O_2$, do not react under ordinary conditions. The reaction shown is endothermic.

$$N_2(g) + O_2(g) \rightleftharpoons 2NO(g)$$

Would the formation of NO in an automobile be affected by these changes?
**(a)** a lower pressure          **(c)** a lower concentration of $O_2$
**(b)** a lower temperature

## Law of Mass Action

**14-15.** Write the law of mass action for each of the following equilibria.
**(a)** $CO(g) + Cl_2(g) \rightleftharpoons COCl_2(g)$
**(b)** $CH_4(g) + 2H_2O(g) \rightleftharpoons CO_2(g) + 4H_2(g)$
**(c)** $4HCl(g) + O_2(g) \rightleftharpoons 2Cl_2(g) + 2H_2O(g)$
**(d)** $CH_4(g) + Cl_2(g) \rightleftharpoons CH_3Cl(g) + HCl(g)$

**14-16.** Write the law of mass action for each of the following equilibria.
**(a)** $3O_2(g) \rightleftharpoons 2O_3(g)$
**(b)** $N_2(g) + 2O_2(g) \rightleftharpoons 2NO_2(g)$
**(c)** $C_2H_2(g) + 2H_2(g) \rightleftharpoons C_2H_6(g)$
**(d)** $4H_2(g) + CS_2(g) \rightleftharpoons CH_4(g) + 2H_2S(g)$

## Value of $K_{eq}$

**14-17.** For the hypothetical reaction $A_2 + B_2 \rightleftharpoons 2AB$, $K_{eq} = 1.0 \times 10^8$, are reactants or products favored at equilibrium?

**14-18.** For the hypothetical reaction $2C + 3B \rightleftharpoons 2D + F$, $K_{eq} = 5 \times 10^{-7}$, are reactants or products favored at equilibrium?

**14-19.** For the reaction $H_2(g) + I_2(g) \rightleftharpoons 2HI(g)$, $K_{eq} = 45$ at a given temperature, are reactants or products favored at equilibrium?

**14-20.** Given the system

$$3O_2(g) \rightleftharpoons 2O_3(g)$$

at equilibrium, $[O_2] = 0.35$ and $[O_3] = 0.12$. What is $K_{eq}$ for the reaction under these conditions?

**14-21.** Given the system

$$N_2(g) + 2O_2(g) \rightleftharpoons 2NO_2(g)$$

at a given temperature, there are $1.25 \times 10^{-3}$ mol of $N_2$, $2.50 \times 10^{-3}$ mol of $O_2$, and $6.20 \times 10^{-4}$ mol of $NO_2$ in a 1.00-L container. What is $K_{eq}$ for this reaction at this temperature?

**14-22.** Given the system

$$2SO_3(g) + CO_2(g) \rightleftharpoons CS_2(g) + 4O_2(g)$$

what is $K_{eq}$ if, at equilibrium, $[SO_3] = 2.0 \times 10^{-2}$ mol/L, $[CO_2] = 4.5 \times 10^{-3}$ mol/L, $[CS_2] = 6.2 \times 10^{-4}$ mol/L, and $[O_2] = 1.0 \times 10^{-4}$ mol/L?

**14-23.** Given the system

$$CH_4(g) + 2H_2O(g) \rightleftharpoons CO_2(g) + 4H_2(g)$$

at equilibrium, we find 2.20 mol of $CO_2$, 4.00 mol of $H_2$, 6.20 mol of $CH_4$, and 3.00 mol of $H_2O$ in a 30.0-L container. What is $K_{eq}$ for the reaction? (*Hint:* Convert amount and volume to concentration.)

**14-24.** Given the system

$$C_2H_2(g) + 2H_2(g) \rightleftharpoons C_2H_6(g)$$

at equilibrium, we find 296 g of $C_2H_6$ present along with 3.50 g of $H_2$ and 21.0 g of $C_2H_2$ in a 400-mL container. What is $K_{eq}$? (*Hint:* Convert amount and volume to concentration.)

**\*14-25.** Consider the following system:

$$2HI(g) \rightleftharpoons H_2(g) + I_2(g)$$

**(a)** If we start with $[HI] = 0.60$, what would be $[H_2]$ and $[I_2]$ if all of the HI reacts?
**(b)** If we start with $[HI] = 0.60$ and $[I_2] = 0.20$, what would be $[H_2]$ and $[I_2]$ if all of the HI reacts?
**(c)** If we start with only $[HI] = 0.60$ and 0.20 mol/L of HI reacts, what are $[HI]$, $[I_2]$, and $[H_2]$ at equilibrium?
**(d)** From the information in (c), calculate $K_{eq}$.
**(e)** What is the $K_{eq}$ for the reverse reaction? How does this value differ from the value given in Table 14-1?

**\*14-26.** In the reaction

$$N_2(g) + 3H_2(g) \rightleftharpoons 2NH_3(g)$$

initially 1.00 mol of $N_2$ and 1.00 mol of $H_2$ are mixed in a 1.00-L container. At equilibrium, it is found that $[NH_3] = 0.20$.
**(a)** What are the concentrations of $N_2$ and $H_2$ at equilibrium?
**(b)** What is the $K_{eq}$ for the system at this temperature?

**\*14-27.** At a given temperature, $N_2$, $H_2$, and $NH_3$ are mixed. The initial concentration of each is 0.50 mol/L. At equilibrium, the concentration of $N_2$ is 0.40 mol/L.
**(a)** Calculate the concentrations of $H_2$ and $NH_3$ at equilibrium.
**(b)** What is the $K_{eq}$ at this temperature?

**\*14-28.** At the start of the reaction

$$4NH_3(g) + 5O_2 \rightleftharpoons 4NO(g) + 6H_2O(g)$$

$[NH_3] = [O_2] = 1.00$. At equilibrium, it is found that 0.25 mol/L of $NH_3$ has reacted.
**(a)** What concentration of $O_2$ reacts?
**(b)** What are the concentrations of all species at equilibrium?

(c) Write the equilibrium constant expression and substitute the proper values for the concentrations of reactants and products.

*14-29. At the start of the reaction

$$CO(g) + Cl_2(g) \rightleftharpoons COCl_2(g)$$

[CO] = 0.650 and [Cl$_2$] = 0.435. At equilibrium, it is found that 10.0% of the CO has reacted. What are [CO], [Cl$_2$], and [COCl$_2$] at equilibrium? What is $K_{eq}$?

## Calculations Involving $K_{eq}$

14-30. For the reaction

$$PCl_3(g) + Cl_2(g) \rightleftharpoons PCl_5(g)$$

$K_{eq} = 0.95$ at a given temperature. If [PCl$_3$] = 0.75 and [Cl$_2$] = 0.40 at equilibrium, what is the concentration of PCl$_5$ at equilibrium?

14-31. At a given temperature, $K_{eq} = 46.0$ for the reaction

$$4HCl(g) + O_2(g) \rightleftharpoons 2Cl_2(g) + 2H_2O(g)$$

At equilibrium, [HCl] = 0.100, [O$_2$] = 0.455, and [H$_2$O] = 0.675. What is [Cl$_2$] at equilibrium?

14-32. Using the value for $K_{eq}$ calculated in exercise 14-23, find the concentration of H$_2$O at equilibrium if, at equilibrium, [CH$_4$] = 0.50, [CO$_2$] = 0.24, and [H$_2$] = 0.20.

*14-33. For the following equilibrium, $K_{eq} = 56$ at a given temperature.

$$CH_4(g) + Cl_2(g) \rightleftharpoons CH_3Cl(g) + HCl(g)$$

At equilibrium, [CH$_4$] = 0.20 and [Cl$_2$] = 0.40. What are the equilibrium concentrations of CH$_3$Cl and HCl if they are equal?

*14-34. Using the equilibrium in exercise 14-20, calculate the equilibrium concentration of O$_3$ if the equilibrium concentration of O$_2$ is 0.80 mol/L.

## Equilibria of Weak Acids and Weak Bases

14-35. Write the expression for $K_a$ or $K_b$ for each of the following equilibria. Where necessary, complete the equilibrium.
(a) HBrO + H$_2$O $\rightleftharpoons$ H$_3$O$^+$ + BrO$^-$
(b) NH$_3$ + H$_2$O $\rightleftharpoons$ NH$_4^+$ + OH$^-$
(c) H$_2$SO$_3$ + H$_2$O $\rightleftharpoons$ H$_3$O$^+$ + HSO$_3^-$
(d) HSO$_3^-$ + H$_2$O $\rightleftharpoons$ _____ + SO$_3^{2-}$
(e) H$_3$PO$_4$ + H$_2$O $\rightleftharpoons$ H$_3$O$^+$ + _____
(f) (CH$_3$)$_2$NH + H$_2$O $\rightleftharpoons$ (CH$_3$)$_2$NH$_2^+$ + _____

14-36. Write the expression for $K_a$ or $K_b$ for each of the following equilibria. Where necessary, complete the equilibrium.
(a) N$_2$H$_4$ + H$_2$O $\rightleftharpoons$ N$_2$H$_5^+$ + OH$^-$
(b) HCN + H$_2$O $\rightleftharpoons$ H$_3$O$^+$ + CN$^-$
(c) H$_2$C$_2$O$_4$ + H$_2$O $\rightleftharpoons$ H$_3$O$^+$ + _____
(d) H$_2$PO$_4^-$ + H$_2$O $\rightleftharpoons$ _____ + HPO$_4^{2-}$

14-37. A 0.10 M solution of a weak acid HX has a pH of 5.0. A 0.10 M solution of another weak acid HB has a pH of 5.8. Which is the weaker acid? Which has the larger value for $K_a$?

14-38. A hypothetical weak acid HZ has a $K_a$ of 4.5 × 10$^{-4}$. Rank the following 0.10 M solutions in order of increasing pH: HZ, HC$_2$H$_3$O$_2$, and HClO.

14-39. In a 0.20 M solution of cyanic acid, HOCN, [H$_3$O$^+$] = [OCN$^-$] = 6.2 × 10$^{-3}$.
(a) What is [HOCN] at equilibrium?
(b) What is $K_a$?
(c) What is the pH?

14-40. A 0.58 M solution of a weak acid (HX) is 10.0% dissociated.
(a) Write the equilibrium equation and the equilibrium constant expression.
(b) What are [H$_3$O$^+$], [X$^-$], and [HX] at equilibrium?
(c) What is $K_a$?
(d) What is the pH?

14-41. Nicotine (Nc) is a nitrogen base in water (similar to ammonia). Write the equilibrium equation illustrating this base behavior. In a 0.44 M solution of nicotine, [OH$^-$] = [NcH$^+$] = 5.5 × 10$^{-4}$. What is $K_b$ for nicotine? What is the pH of the solution?

14-42. In a 0.085 M solution of carbolic acid, HC$_6$H$_5$O, the pH = 5.48. What is $K_a$?

14-43. Novocaine (Nv) is a nitrogen base in water (similar to ammonia). Write the equilibrium equation. In a 1.25 M solution of novocaine, pH = 11.46. What is $K_b$?

14-44. In a 0.300 M solution of chloroacetic acid (HC$_2$Cl$_3$O$_2$), [HC$_2$Cl$_3$O$_2$] = 0.277 M at equilibrium. What is the pH? What is $K_a$?

14-45. What is the pH of a 0.65 M solution of HBrO?

14-46. What is the pH of a 1.50 M solution of HNO$_2$?

14-47. What is [OH$^-$] in a 0.55 M solution of NH$_3$?

14-48. What is [H$_3$O$^+$] in a 0.25 M solution of HC$_2$H$_3$O$_2$?

14-49. What is the pH of a 1.00 M solution of (CH$_3$)$_2$NH?

14-50. What is the pH of a 0.567 M solution of N$_2$H$_4$?

14-51. What is the pH of a buffer made by mixing 0.45 mol of NaCN and 0.45 mol of HCN in 2.50 L of solution?

14-52. What is the pH of a buffer made by dissolving 1.20 mol of NH$_3$ and 1.20 mol of NH$_4$ClO$_4$ in 13.5 L of solution?

14-53. What is the pH of a buffer made by dissolving 0.20 mol of KBrO and 0.60 mol of HBrO in 850 mL of solution? What is the pH if the solution is diluted to 2.00 L?

14-54. What is the pH of a buffer solution made by mixing 0.044 mol of HCHO$_2$ with 0.064 mol of KCHO$_2$ in 1.00 L of solution?

14-55. What is the pH of a buffer solution made by mixing 0.058 mol of HClO and 5.50 g of Ca(ClO)$_2$ in a certain amount of water?

14-56. What is the pH of a buffer made by mixing 1.50 g of N$_2$H$_4$ with 1.97 g of N$_2$H$_5^+$Cl$^-$ in 2.00 L of solution?

14-57. What is the pH of a buffer that contains 150 g of HNO$_2$ and 150 g of LiNO$_2$?

*14-58. A buffer of pH = 7.50 is desired. Which two of the following compounds should be dissolved in water in equimolar amounts to provide this buffer solution: $HNO_2$, $HClO$, $HCN$, $NaNO_2$, $NH_3$, $KCN$, $KClO$, $NH_4Cl$?

*14-59. A buffer of pH = 9.25 is desired. Which two of the following compounds should be dissolved in water in equimolar amounts to provide this buffer solution: $HC_2H_3O_2$, $HClO$, $NH_3$, $N_2H_4$, $KC_2H_3O_2$, $NH_4Cl$, $NaClO$, $N_2H_5Br$?

### Solubility Equilbiria

14-60. Write the expression of $K_{sp}$ for the following salts: (a) FeS, (b) $Ag_2S$, (c) $Zn(OH)_2$.

14-61. Write the expression for $K_{sp}$ for the following salts: (a) $PbCO_3$, (b) $Bi_2S_3$ (c) $PbI_2$.

14-62. What is the value of $K_{sp}$ for copper(I) iodide if the molar solubility is $2.2 \times 10^{-6}$ mol/L?

14-63. What is the value of $K_{sp}$ for magnesium carbonate ($MgCO_3$) if a saturated solution contains 0.287 g/L?

14-64. What is the value of $K_{sp}$ for $Ca(OH)_2$ if the molar solubility is $1.3 \times 10^{-2}$ mol/L?

*14-65. What is the value of $K_{sp}$ for $Cr(OH)_3$ if the molar solubility is $3 \times 10^{-8}$ mol/L?

14-66. What is the molar solubility of AgI? For AgI, $K_{sp} = 8.3 \times 10^{-17}$.

*14-67. Lead(II) fluoride is present in the Earth as the mineral fluorite. It is considered insoluble in water. What is the molar solubility of lead(II) fluoride? For $PbF_2$, $K_{sp} = 4.1 \times 10^{-8}$.

14-68. The concentration of $Ag^+$ ion in a solution is $7 \times 10^{-6}$ M and the concentration of $Br^-$ is $8 \times 10^{-7}$ M. Will a precipitate of AgBr form? For AgBr, $K_{sp} = 7.7 \times 10^{-13}$.

14-69. The concentration of $Ag^+$ in a solution is 0.012 M and the concentration of $SO_4^{2-}$ is 0.050 M. Will a precipitate of $Ag_2SO_4$ form? For $Ag_2SO_4$, $K_{sp} = 1.4 \times 10^{-5}$.

### General Problems

14-70. In a given gaseous equilibrium, chlorine reacts with ammonia to produce nitrogen trichloride and hydrogen chloride.
(a) Write the balanced equation illustrating this equilibrium.
(b) Write the expression for $K_{eq}$.
(c) How is the equilibrium concentration of ammonia affected by an increase in pressure?
(d) How is the equilibrium concentration of ammonia affected by the addition of some chlorine gas?
(e) The value of $K_{eq}$ for this reaction is $2.4 \times 10^{-9}$. Are there more products or reactants present at equilibrium?
(f) If the equilibrium concentrations of chlorine = 0.10 M, nitrogen trichloride = $2.0 \times 10^{-4}$ M, and hydrogen chloride = $1.0 \times 10^{-3}$ M, what is the equilibrium concentration of ammonia?

14-71. In a given gaseous equilibrium, acetylene ($C_2H_2$) reacts with HCl to form $C_2H_4Cl_2$.
(a) Write the balanced equation illustrating this equilibrium.
(b) Write the expression for $K_{eq}$.

(c) How does an increase in pressure affect the equilibrium concentration of $C_2H_4Cl_2$?
(d) The reaction is exothermic. How does an increase in temperature affect the value of $K_{eq}$? How does it affect the equilibrium concentration of HCl?
(e) At a given temperature at equilibrium, $[C_2H_2] = 0.030$, $[HCl] = 0.010$, and $[C_2H_4Cl_2] = 0.60$. What is the value of $K_{eq}$?
(f) What does the value of $K_{eq}$ tell us about the point of equilibrium at this temperature?

14-72. Dinitrogen oxide (nitrous oxide) decomposes to its elements in a gaseous equilibrium reaction.
(a) Write the balanced equation illustrating this equilibrium.
(b) Write the expression for $K_{eq}$.
(c) How does an increase in pressure affect the equilibrium concentration of nitrogen?
(d) If the reaction is exothermic, how does a decrease in temperature affect the equilibrium concentration of dinitrogen oxide?
(e) At a given temperature, 0.10 mol of dinitrogen oxide is placed in a 1.00-L container. At equilibrium, it is found that 1.5% of the dinitrogen oxide has decomposed. What is the value of $K_{eq}$?

14-73. Nitrogen dioxide is in a gaseous equilibrium with dinitrogen tetroxide.
(a) Write the balanced equation illustrating this reaction. Show nitrogen dioxide as the reactant and dinitrogen tetroxide as the product.
(b) Write the expression for $K_{eq}$.
(c) How does an increase in pressure affect the concentration of nitrogen dioxide at equilibrium?
(d) At equilibrium at a given temperature, it is found that there are 10.0 g of nitrogen dioxide and 12.0 g of dinitrogen tetroxide in a 2.50-L container. What is the value of $K_{eq}$?

14-74. The bicarbonate ion ($HCO_3^-$) can act as either an acid or a base in water. For the acid reaction

$$HCO_3^- + H_2O \rightleftharpoons H_3O^+ + CO_3^{2-} \qquad K_a = 4.7 \times 10^{-11}$$

For the base reaction

$$HCO_3^- + H_2O \rightleftharpoons OH^- + H_2CO_3 \qquad K_b = 2.2 \times 10^{-8}$$

(a) Based on the values of $K_a$ and $K_b$, is a solution of $NaHCO_3$ acidic or basic?
(b) How is bicarbonate of soda used medically?
(c) Write the reaction that occurs when sodium bicarbonate reacts with stomach acid (HCl).
(d) Can sodium bisulfate act in the same way toward acids as sodium bicarbonate does? (Refer to Table 12-2.)

*14-75. Given 0.10 M solutions of the compounds KOH, NaCl, $HNO_2$, $NH_3$, $HNO_3$, $HCHO_2$, and $N_2H_4$, rank them in order of increasing pH.

14-76. What is the pH of a buffer that is made by mixing 265 mL of 0.22 M $HC_2H_3O_2$ with 375 mL of 0.12 M $Ba(C_2H_3O_2)_2$?

*14-77. What is the pH of a buffer made by adding 0.20 mol of NaOH to 1.00 L of a 1.00 M solution of $HC_2H_3O_2$? Assume no volume change. (*Hint:* First consider the partial neutralization of $HC_2H_3O_2$ by NaOH.)

**\*14-78.** What is the value of $K_{sp}$ for $Mg(OH)_2$ if the pH of a saturated solution is equal to 10.50?

**\*14-79.** What is the pH of a saturated solution of $Fe(OH)_2$? For $Fe(OH)_2$, $K_{sp} = 1.6 \times 10^{-14}$.

**14-80.** Aqueous solutions of barium salts are very poisonous, yet we ingest barium sulfate as a contrasting agent for X-rays of the intestinal tract. What is the $Ba^{2+}$ concentration in a saturated solution of $BaSO_4$? What is the mass of $BaSO_4$ present in one liter of solution? For $BaSO_4$, $K_{sp} = 1.1 \times 10^{-10}$.

# Interactive Learning

**14-81.** At a high temperature, 2.00 mol of HBr is put in a 4.00-L container, where it decomposes according to the equation:

$$2HBr(g) \rightleftharpoons Br_2(g) + H_2(g)$$

At equilibrium the concentration of $Br_2$ is 0.0955 M. What is $K_{eq}(K_c)$ for this reaction?

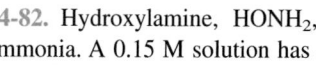

**14-82.** Hydroxylamine, $HONH_2$, is a base in water like ammonia. A 0.15 M solution has a pH of 10.12. What is the value of $K_b$ for hydroxylamine?

# Solutions to Learning Checks

**A-1.** collisions, collide, minimum, activation, time, rate, activated complex, activation energy, higher, increases, increases, catalyst, increases, point of equilibrium, constant, concentration, temperature, pressure

**A-2.** (a) increased  (c) increased  (e) not affected
     (b) decreased  (d) decreased

**A-3.**

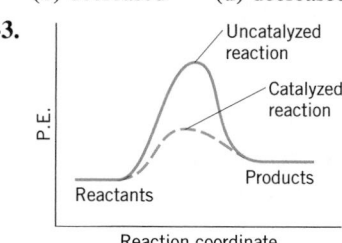

**B-1.** equilibrium constant, equilibrium constant expression, products, exponent, concentration, $K_a$, weaker, concentration, $K_b$

**B-2.** (a) $K_{eq} = \dfrac{[CH_4][H_2O]}{[CO][H_2]^3}$  (b) $K_a = \dfrac{[H_3O^+][CHO_2^-]}{[HCHO_2]}$

**B-3.** $K_{eq} = \dfrac{\left(\dfrac{3.5 \times 10^{-4}\,mol}{10.0\,L}\right)\left(\dfrac{5.0 \times 10^{-5}\,mol}{10.0\,L}\right)}{\left(\dfrac{2.0 \times 10^{-3}\,mol}{10.0\,L}\right)\left(\dfrac{1.0 \times 10^{-2}\,mol}{10.0\,L}\right)^3}$

$= \dfrac{17.5 \times 10^{-11}}{2.0 \times 10^{-13}} = \underline{8.80}$

**B-4.** $NX_3(aq) + H_2O(l) \rightleftharpoons HNX_3^+(aq) + OH^-(aq)$
$[HNX_3^+] = [OH^-] = 0.010 \times 0.25\,M = 2.5 \times 10^{-3}\,M$ (at equilibrium)

$[NX_3] = 0.25 - (2.5 \times 10^{-3}) = 0.25\,M$

$K_b = \dfrac{(2.5 \times 10^{-3})^2}{0.25} = \dfrac{6.25 \times 10^{-6}}{0.25} = \underline{2.5 \times 10^{-5}}$

**B-5.** From Table 14-2 for $HCHO_2$, $K_a = 1.8 \times 10^{-4}$.
$HCHO_2(aq) + H_2O(l) \rightleftharpoons H_3O^+(aq) + CHO_2^-(aq)$
Let $x = [H_3O^+] = [CHO_2^-]$ at equilibrium. Then $[HCHO_2] = 0.10 - x \approx 0.10$ at equilibrium.

$\dfrac{x^2}{0.10} = 1.8 \times 10^{-4}$   $x^2 = 1.8 \times 10^{-5}$   $x = 4.2 \times 10^{-3}$

$pH = -\log(4.2 \times 10^{-3}) = \underline{2.38}$

**B-6.** For a base buffer, $pOH = pK_b + \log\dfrac{[acid]}{[base]}$

$pK_b = -\log K_b = 4.74$

The base species is $NH_3$: $[base] = 0.068\,mol/2.0\,L$
The acid species is $NH_4^+$: $[acid] = 0.049\,mol/2.0\,L$
(The $Cl^-$ is a spectator ion and is not part of the equilibrium.)

$pOH = 4.74 + \log\dfrac{(0.049\,mol/2.0\,L)}{(0.068\,mol/2.0\,L)} = 4.74 - 0.14 = 4.60$

$pH = 14.00 - 4.60 = \underline{9.40}$

**B-7.** $BaCO_3(s) \rightleftharpoons Ba^{2+}(aq) + CO_3^{2-}(aq)$
$K_{sp} = [Ba^{2+}][CO_3^{2-}]$

Let $x =$ the molar solubility of $BaCO_3$. Then $x = [Ba^{2+}] = [CO_3^{2-}]$.

$[x][x] = 8.1 \times 10^{-9}$   $x = \underline{9.0 \times 10^{-5}\,mol/L}$

# Cumulative Review
# Chapters 12–14

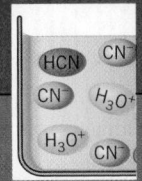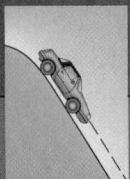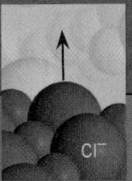

## Multiple Choice

**The following multiple-choice questions have one correct answer.**

**1.** Which of the following is the formula for perchloric acid?
**(a)** HCl    **(c)** $HClO_2$    **(e)** $HClO_4$
**(b)** HClO    **(d)** $HClO_3$

**2.** Which of the following is a balanced neutralization reaction that leads to the formation of an acid salt?

**(a)** $HCl + NaOH \longrightarrow NaCl + H_2O$
**(b)** $H_3PO_4 + KOH \longrightarrow KHPO_4 + H_2O$
**(c)** $H_2SO_4 + 2KOH \longrightarrow K_2SO_4 + H_2O$
**(d)** $NaOH + H_2SO_3 \longrightarrow NaHSO_3 + H_2O$

**3.** Which of the following is a strong acid?
**(a)** $HNO_2$   **(b)** $HNO_3$   **(c)** $H_2SO_3$   **(d)** HClO   **(e)** KOH

**4.** The concentration of an aqueous HCl solution is 0.01 M. The pH is
**(a)** 1    **(b)** 2    **(c)** 3    **(d)** 12    **(e)** 7

**5.** A solution has a pH of 8.5. The solution is
**(a)** strongly basic    **(c)** neutral    **(e)** strongly acidic
**(b)** weakly basic    **(d)** weakly acidic

**6.** In pure water at room temperature, the concentration of $OH^-$ is always equal to
**(a)** 7 mol/L    **(c)** $K_w[H_3O^+]$    **(e)** $10^7$ mol/L
**(b)** $1.0 \times 10^{-14}$ mol/L    **(d)** $[H_3O^+]$

**7.** A sample of milk has pH = 6.0. What is the hydroxide ion concentration?
**(a)** $1.0 \times 10^{-6}$    **(c)** $1.0 \times 10^{-8}$    **(e)** $5.0 \times 10^{-7}$
**(b)** $6.0 \times 10^{-1}$    **(d)** $1.0 \times 10^{-14}$

**8.** Which of the following is the conjugate acid of the $HS^-$ ion?
**(a)** $H_2S^-$    **(c)** $S^-$    **(e)** $H_2S$
**(b)** $S^{2-}$    **(d)** $H_3O^+$

**9.** For the following reaction, identify the reactant base and its conjugate acid.

$$NH_2^-(aq) + H_2O(l) \longrightarrow NH_3(aq) + OH^-(aq)$$

**(a)** $NH_2^-$, $NH_3$    **(c)** $H_2O$, $OH^-$    **(e)** $NH_3$, $NH_4^+$
**(b)** $NH_3$, $OH^-$    **(d)** $NH_2^-$, $OH^-$

**10.** Which of the following conjugate bases does not exhibit basic nature in water?
**(a)** $S^{2-}$   **(b)** $NO_2^-$   **(c)** $ClO_4^-$   **(d)** $PO_4^{3-}$   **(e)** $HS^-$

**11.** A 0.10 M solution of a compound has a pH of 5.4. The compound is
**(a)** HCl    **(c)** $NH_4Cl$    **(e)** NaCl
**(b)** KOH    **(d)** $K_2S$

**12.** Which of the following pairs of compounds does not form a buffer when aqueous solutions of the two are mixed?
**(a)** $NaC_2H_3O_2$ and $HC_2H_3O_2$    **(c)** $LiNO_2$ and $HNO_2$
**(b)** KBr and HBr    **(d)** $NH_3$ and $NH_4Cl$

**13.** Which of the following acids form when $Cl_2O$ dissolves in water?
**(a)** HCl   **(b)** $HClO_4$   **(c)** HClO   **(d)** $HClO_2$   **(e)** $HClO_3$

**14.** Which of the following is an oxidation process?
**(a)** $H_2CO_3 \longrightarrow H_2O + CO_2$    **(d)** $H^+ + OH^- \longrightarrow H_2O$
**(b)** $2Br^- \longrightarrow Br_2$    **(e)** $SO_4^{2-} \longrightarrow S^{2-}$
**(c)** $Ca^{2+} + SO_4^{2-} \longrightarrow CaSO_4$

**15.** What is the oxidation state of the Br in $Ca(BrO_3)_2$?
**(a)** +10    **(b)** +6    **(c)** +5    **(d)** +12    **(e)** −1

**16.** In the following equation, how many electrons have been gained ($+e^-$) or lost ($-e^-$) by one molecule of $O_3$?

$$6H^+ + O_3 \longrightarrow 3H_2O$$

**(a)** $+2e^-$   **(b)** $-2e^-$   **(c)** $+6e^-$   **(d)** $+4e^-$   **(e)** $-6e^-$

**17.** In the Daniell cell, what allows for the migration of anions between compartments?
**(a)** a salt bridge    **(c)** the external wire
**(b)** the electrodes    **(d)** the electrons

**18.** In the following electrolytic cell, _____ is _____ at the anode.

$$2Cl^-(aq) + Fe^{2+}(aq) \longrightarrow Fe(s) + Cl_2(g)$$

**(a)** $Fe^{2+}$, reduced    **(c)** $Cl_2$, reduced    **(e)** Fe, oxidized
**(b)** $Cl^-$, reduced    **(d)** $Cl^-$, oxidized

**19.** The following equation represents a spontaneous reaction.

$$Br_2(aq) + Ni(s) \longrightarrow Ni^{2+}(aq) + 2Br^-(aq)$$

Which of the following statements is correct?
**(a)** $Br_2$ is a stronger reducing agent than $Ni^{2+}$.
**(b)** $Br^-$ is a stronger reducing agent than Ni.
**(c)** Ni is an oxidizing agent.
**(d)** $Br_2$ is oxidized.
**(e)** $Br_2$ is a stronger oxidizing agent than $Ni^{2+}$.

**20.** Which of the following does not affect the rate of a reaction?
(a) the volume of the reaction vessel
(b) the temperature in the reaction vessel
(c) the concentration of a reactant
(d) the shape of the reaction vessel

**21.** Which of the following is not true about a catalyst?
(a) can be homogeneous or heterogeneous
(b) allows more products to form
(c) is not consumed in the reaction
(d) lowers the activation energy

**22.** Given the following equilibrium,

$$2CO(g) + O_2(g) \rightleftharpoons 2CO_2(g) + heat$$

the concentration of $CO_2$ is increased at equilibrium by
(a) increasing the temperature.
(b) decreasing the concentration of CO.
(c) increasing the pressure.
(d) adding a catalyst.

**23.** Given the following equilibrium,

$$A + B \rightleftharpoons C + D \quad K_{eq} = 10^{-6}$$

which of the following is correct?
(a) Products are favored at equilibrium.
(b) Reactants are favored at equilibrium.
(c) There are significant amounts of both products and reactants present at equilibrium.
(d) No conclusions can be made about the point of equilibrium.

**24.** A 0.10 M solution of a weak acid is 1.0% ionized. Its $K_a$ is
(a) $10^{-3}$   (b) $10^{-4}$   (c) $10^{-5}$   (d) $10^{-6}$   (e) $10^{-7}$

**25.** A buffer is made by dissolving an equal number of moles of a weak base $(CH_3)_2NH$ and $(CH_3)_2NH_2{}^+Br^-$. If $K_b = 7.4 \times 10^{-4}$, what is the pH of the solution?
(a) 10.87   (c) 1.57   (e) more information needed
(b) 3.13   (d) $-3.13$

## Problems

**1.** (a) Write the balanced equation representing the complete neutralization of sulfuric acid with lithium hydroxide. (b) What volume (in mL) of 0.25 M sulfuric acid is needed to completely react with 1.80 g of lithium hydroxide?

**2.** Complete the following equations.
(a) $HClO + H_2O \rightleftharpoons$   (c) $KOH + HNO_2 \longrightarrow$
(b) $NH_3 + H_2O \rightleftharpoons$   (d) $H_3AsO_3 + 2NaOH \longrightarrow$

**3.** Write equations representing reactions for the following. If no reaction occurs, write N.R.
(a) $K^+ + H_2O$   (c) $NH_4{}^+ + H_2O$   (e) $ClO^- + H_2O$
(b) $CaO + H_2O$   (d) $CO_2 + H_2O$   (f) $Br^- + H_2O$

**4.** In a 0.25 M solution of a weak base, $[OH^-] = 6.5 \times 10^{-4}$. What is the value of
(a) $[H_3O^+]$   (b) pH   (c) pOH

**5.** (a) When elemental tin is placed in a nitric acid solution, a spontaneous redox reaction occurs, producing $SnO_2$ and $NO_2$. Write the balanced equation representing this reaction.
(b) What mass of $NO_2$ is produced from the complete reaction of 350 mL of 0.20 M nitric acid?

**6.** From the equation in problem 5, answer the following:
(a) What is the reactant oxidized?
(b) What is the reactant reduced?
(c) What is the product of oxidation?
(d) What is the product of reduction?
(e) What is the oxidizing agent?
(f) What is the reducing agent?

**7.** Given the following table of oxidizing and reducing agents, answer the questions below.

strongest $\longrightarrow$ $\quad I_2 + 2e^- \rightleftharpoons 2I^-$
oxidizing $\quad\quad Cr^{3+} + 3e^- \rightleftharpoons Cr$
agent $\quad\quad\quad Mn^{2+} + 2e^- \rightleftharpoons Mn \quad$ strongest
$\quad\quad\quad\quad\quad Ca^{2+} + 2e^- \rightleftharpoons Ca \longleftarrow$ reducing
$\quad\quad\quad\quad\quad\quad\quad\quad\quad\quad\quad\quad\quad\quad$ agent

(a) Is the following reaction spontaneous or nonspontaneous?

$$Mn + Ca^{2+} \longrightarrow Ca + Mn^{2+}$$

(b) Write a balanced equation representing a spontaneous reaction involving the Cr, $Cr^{3+}$ and the Mn, $Mn^{2+}$ half-reactions.
(c) If an aqueous solution of $I_2$ (an antiseptic) is spilled on a chromium-coated bumper of an automobile, will a reaction occur? If so, write the balanced equation for the reaction.

**8.** Given the equilibrium

$$2NO(g) + Br_2(g) \rightleftharpoons 2NOBr(g)$$

(a) Write the law of mass action.
(b) What is the value of $K_{eq}$ if we start with 4.00 mol of NOBr only in a 20.0-L container, and, at equilibrium, it is found that [NOBr] = 0.10?
(c) What is the concentration of NOBr if [NO] = 0.45 and $[Br_2] = 0.22$ at equilibrium?

**9.** What is the value of $K_b$ for the weak base in problem 4?

**10.** What is the pH of a 0.64 M solution of boric acid? Boric acid $(H_3BO_3)$ is a weak monoprotic acid used as an eyewash. $K_a$ for boric acid is $6.0 \times 10^{-10}$.

## chapter 15

# Nuclear Chemistry

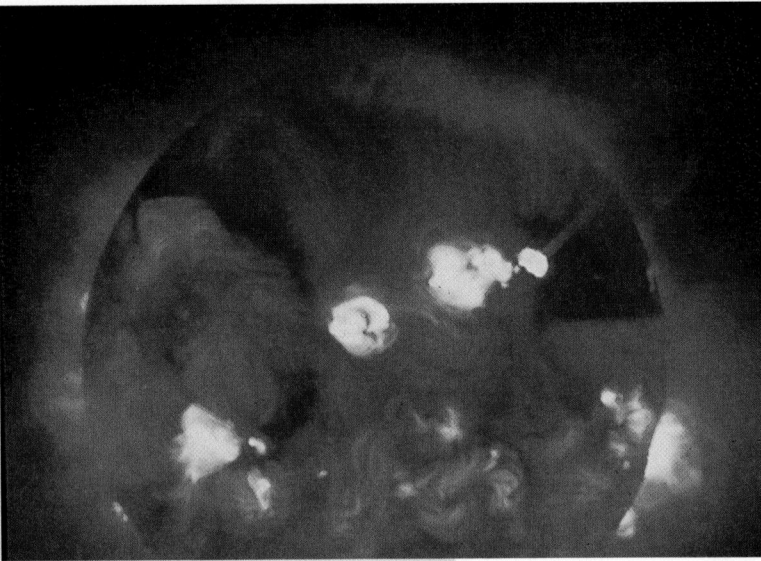

**The sun is the source of life. Deep in its interior, nuclear reactions are occurring that liberate life-giving energy.**

## Setting the Stage

In July 1945, the most powerful explosion yet produced by the human race shook the New Mexico desert. The first nuclear bomb had been detonated, and our world would never be the same—the nuclear age was now upon us. But the dawn of this age actually began quite innocently 49 years earlier. At that time, there was certainly no intent to launch the human race in a radical new direction. In 1896, a French scientist named Henri Becquerel was investigating how sunlight interacted with a uranium mineral called pitchblende. He suspected that sunlight caused pitchblende to give off the mysterious "X-rays" that had been discovered the previous year by Wilhelm Roentgen. X-rays, a high-energy form of light, were known to penetrate the covering of photographic plates, thus exposing the film. One cloudy day, Becquerel was unable to do an experiment, but he left the uranium on top of unexposed film in a photographic plate. Several days later, he developed the film anyway and discovered that the part underneath the sample of pitchblende was exposed. Obviously, the uranium spontaneously emitted some form of radiation. It was this discovery that marked the true beginning of the nuclear age.

Since we first discussed the atom in Chapter 3, we have addressed the results of interactions of the electrons that reside outside a tiny but comparatively massive nucleus. Interactions of electrons account for the way atoms of an element combine chemically with other atoms. Now we turn our attention to the nucleus, which is composed of two types of particles, neutrons and protons, collectively called nucleons. Also recall that elements are composed of isotopes. Isotopes of an element are those atoms that have the same number of protons but different numbers of neutrons. The changes that can occur in the nucleus of a particular isotope account for the phenomenon that Becquerel first observed. We will discover the awesome implications, starting from that first observation, that range from the fear of nuclear bombs to the hope of nuclear medicine.

### Formulating Some Questions

Questions that we will address in this chapter include: What is the nature of the radiation observed by Becquerel and other early investigators? What does this radiation imply for the nucleus? How does radiation cause damage, and how can it be detected? What good can it do? How can we change one element into another? What goes on in the atomic and hydrogen bombs, and how can we tame these reactions for energy?

## Section A  NATURALLY OCCURRING RADIOACTIVITY

# 15-1 Radioactivity

**Looking Ahead!** It was a monumental discovery that one element could change into another through natural processes. We will look closely at the nature of those changes in this section.

The radiation process discovered by Becquerel was more complex than originally thought. The term *radioactive* was first suggested by Marie Curie to describe elements that spontaneously emit **radiation** (*particles or energy*). Marie Curie and her husband, Pierre Curie, pursued the observation of Becquerel and discovered several more naturally **radioactive isotopes** of elements besides uranium (thorium, polonium, and radium). Lord Rutherford later identified three types of natural radiation. We will look at these three processes individually, as well as two additional processes since discovered.

### Alpha (α) Particles

The first nuclear change that was identified involved the emission of a helium nucleus from the nucleus of a heavy isotope. *The emitted helium nucleus is referred to as an* **alpha (α) particle.** This process is conveniently illustrated by a nuclear equation. *A* **nuclear equation** *shows the initial nucleus to the left of the reaction arrow and the product nuclei or particles to the right of the arrow.* Just as chemical equations are balanced, so are nuclear equations. *Nuclear equations are balanced by having the same totals of positive charges and mass numbers on both sides of the equation.* The emission of an alpha particle by $^{238}$U is illustrated in Figure 15-1 and represented by the following nuclear equation.

$$^{238}_{92}\text{U} \longrightarrow {}^{234}_{90}\text{Th} + {}^{4}_{2}\text{He}$$

Note that the loss of four nucleons from the original (parent) nucleus leaves the remaining (daughter) nucleus with 234 nucleons ($238 - 4 = 234$); the loss of two protons leaves 90 protons in the daughter nucleus ($92 - 2 = 90$). Because a nucleus with 90 protons is an isotope of thorium, one element has indeed changed into another.

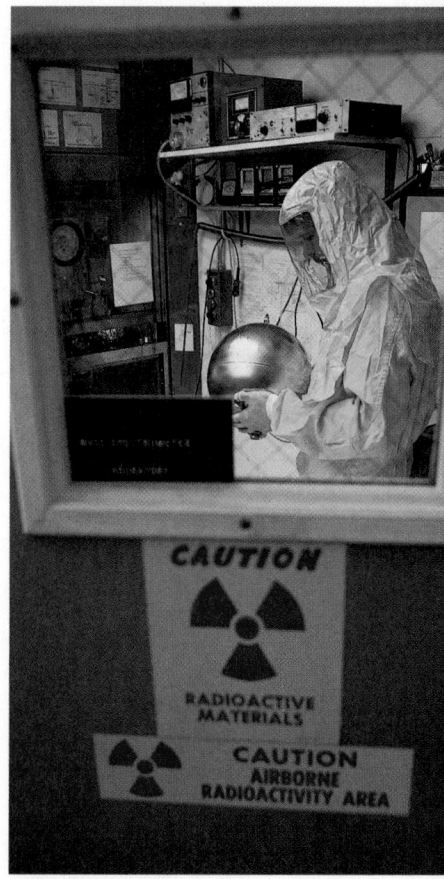

**Labels clearly indicate the danger of radiation.**

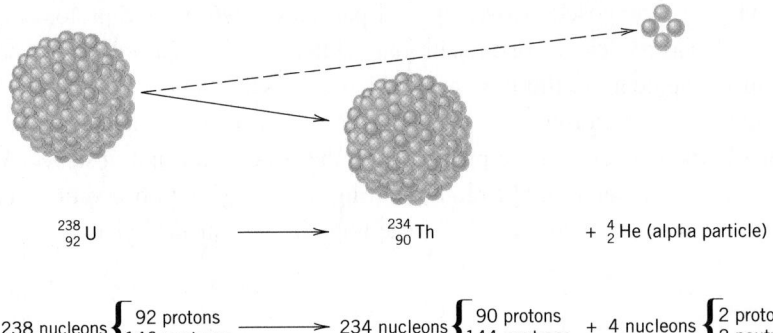

**Figure 15-1  Alpha Radiation**
**An alpha particle is a helium nucleus emitted from a larger nucleus.**

$^{238}_{92}$U $\longrightarrow$ $^{234}_{90}$Th + $^{4}_{2}$He (alpha particle)

238 nucleons $\begin{cases} \text{92 protons} \\ \text{146 neutrons} \end{cases}$ $\longrightarrow$ 234 nucleons $\begin{cases} \text{90 protons} \\ \text{144 neutrons} \end{cases}$ + 4 nucleons $\begin{cases} \text{2 protons} \\ \text{2 neutrons} \end{cases}$

## Beta (β) Particles

A second form of radiation involves the emission of an electron from the nucleus. *An electron emitted from the nucleus is known as a **beta (β) particle.** In effect, beta particle emission changes a neutron in a nucleus into a proton.* This causes the atomic number of the isotope to increase by one while the mass number remains constant. This type of radiation is illustrated in Figure 15-2 and represented by the following nuclear equation.

$$^{131}_{53}\text{I} \longrightarrow ^{131}_{54}\text{Xe} + ^{0}_{-1}\text{e}$$

## Gamma (γ) Rays

*A third type of radioactive decay involves the emission of a high-energy form of light called a **gamma (γ) ray.*** Like all light, gamma rays travel at $3.0 \times 10^{10}$ cm/sec (186,000 mi/sec). This type of radiation may occur alone or in combination with alpha or beta radiation. The following equation illustrates gamma ray emission.

$$^{60m}_{27}\text{Co} \longrightarrow ^{60}_{27}\text{Co} + \gamma$$

Gamma radiation alone does not result in a change in either the mass number or the atomic number of the isotope. In the equation shown above, the *m* stands for *metastable,* which means that the isotope is in a high-energy state. A nucleus in a metastable, or high-energy, state emits energy in the form of gamma radiation. The product nucleus has the same identity but is then in a lower energy state.

## Positron Particles

This type of radiation is not among the three types of naturally occurring radiation but is found in a few artificially produced isotopes. *A **positron ($^{0}_{+1}$e) particle** is an electron with a positive charge rather than a negative charge. In effect, positron emission changes a proton into a neutron.* As a result, the product nucleus has the same mass number but a lower atomic number. A positron is known as a form of *antimatter* because, when it comes in contact with normal matter (e.g., a negatively charged

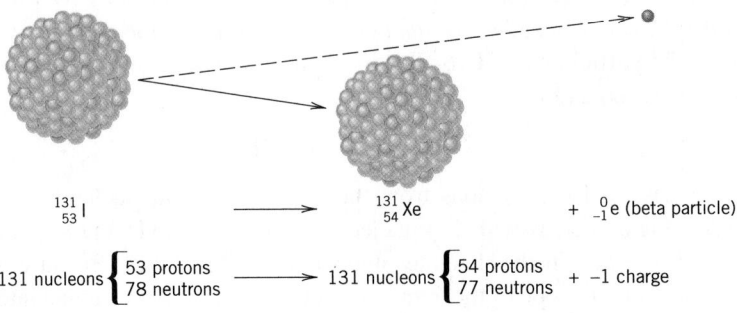

**Figure 15-2  Beta Radiation**
**A beta particle is an electron emitted from a nucleus.**

$^{131}_{53}$I $\longrightarrow$ $^{131}_{54}$Xe + $^{0}_{-1}$e (beta particle)

131 nucleons $\begin{cases} \text{53 protons} \\ \text{78 neutrons} \end{cases}$ $\longrightarrow$ 131 nucleons $\begin{cases} \text{54 protons} \\ \text{77 neutrons} \end{cases}$ + $-1$ charge

**Table 15-1  Types of Radiation**

| Radiation | Mass Number | Charge | Identity |
|---|---|---|---|
| alpha ($\alpha$) | 4 | +2 | helium nucleus |
| beta ($\beta$) | 0 | −1 | electron (out of nucleus) |
| gamma ($\gamma$) | 0 | 0 | light |
| positron | 0 | +1 | positive electron |
| electron capture | 0 | −1 | electron (into nucleus) |

electron), the two particles annihilate each other, producing two gamma rays that travel in exactly opposite directions.

$$^{13}_{7}\text{N} \longrightarrow {}^{13}_{6}\text{C} + {}^{0}_{+1}\text{e}$$

$$^{0}_{+1}\text{e} + {}^{0}_{-1}\text{e} \longrightarrow 2\gamma$$

## Electron Capture

Another nuclear process, known as **electron capture,** involves *the capture of an inner-shell electron by the nucleus.* This process is very rare in nature but is more common in artificially produced elements. When an isotope captures an orbital electron, the effect is the same as positron emission. *That is, a proton changes into a neutron.* Electron capture is detected by the X-rays that are emitted as outer electrons fall into the inner shell, where there is a vacancy due to the captured electron. In most cases, gamma rays are also emitted by the product nuclei. Electron capture is illustrated by the following nuclear equation.

$$^{50}_{23}\text{V} + {}^{0}_{-1}\text{e} \longrightarrow {}^{50}_{22}\text{Ti} + \text{X-rays} + \gamma$$

Although why one isotope is stable and another is not is a somewhat involved topic, we can make some observations. Notice that the atomic mass of some of the lighter elements in the periodic table is about double the atomic number. This is because the most common isotope of that element has an equal number of protons and neutrons, or nearly so (e.g., $^4$He, $^{16}$O, $^{40}$Ca, $^{32}$S). Heavier elements, however, tend to have more neutrons than protons in their most common isotopes. Isotopes that have excess neutrons (e.g., $^{14}_{6}$C) tend to decay by beta particle emission. The result of this radiation is a change of a neutron into a proton and a more stable neutron/proton ratio (e.g., $^{14}_{7}$N + $^{0}_{-1}$e). Isotopes that have excess protons (e.g., $^{10}_{6}$C) tend to decay by positron emission (or electron capture). The result of this radiation is a change of a proton into a neutron and a more stable nucleus (e.g., $^{10}_{5}$B + $^{0}_{+1}$e). Very heavy isotopes decay in such a way that they lose both mass and charge. In this case, alpha particle emission carries away both mass and charge.

The five types of radiation that have been discussed are summarized in Table 15-1.

---

**Nuclear Equations and Radioactivity**

**Example 15-1**

w o r k i n g   i t   o u t

Complete the following nuclear equations

a.  $^{218}_{84}\text{Po} \longrightarrow \underline{\hspace{1.5cm}} + {}^{4}_{2}\text{He}$

b.  $^{210}_{81}\text{Tl} \longrightarrow {}^{210}_{82}\text{Pb} + \underline{\hspace{1.5cm}}$

c.  $\underline{\hspace{1.5cm}} \longrightarrow {}^{8}_{4}\text{Be} + {}^{0}_{+1}\text{e}$

**Procedure**

For the missing isotope or particle:

1. Find the total number of nucleons; it is the same on both sides of the equation.

2. Find the total charge or atomic number; it is also the same on both sides of the equation.

3. If what's missing is the isotope of an element, find the symbol of the element that matches the atomic number from the list of elements inside the front cover.

**Solution**

a. Nucleons: $218 = x + 4$, so $x = 214$.
   Atomic number: $84 = y + 2$, so $y = 82$.

From the inside cover, the element having an atomic number of 82 is Pb. The isotope is

$$^{214}_{82}\text{Pb}$$

b. Nucleons: $210 = 210 + x$, so $x = 0$
   Atomic number: $81 = 82 + y$, so $y = -1$.

An electron or beta particle has negligible mass (compared with a nucleon) and a charge of $-1$.

$$^{0}_{-1}\text{e}$$

c. Nucleons: $x = 8 + 0$, so $x = 8$.
   Atomic number: $y = 4 + 1$, so $y = 5$.

From the inside cover, the element with an atomic number of 5 is B. The isotope is

$$^{8}_{5}\text{B}$$

## 15-2 Rates of Decay of Radioactive Isotopes

**Looking Ahead!** Uranium is a naturally occurring element, but it is radioactive. Why hasn't it all decayed by now? Why some radioactive elements are still present on this ancient Earth is the subject of this section.

The heaviest element in the periodic table with at least one stable isotope is bismuth (#83). All elements heavier than bismuth have no stable isotopes (i.e., all isotopes are radioactive). Some heavy elements, such as uranium, still exist in the Earth in significant amounts, however, because they have very long "half-lives." *The* **half-life** **($t_{1/2}$)** *of a radioactive isotope is the time required for one-half of a given sample to decay.* Each radioactive isotope has a specific and constant half-life. Unlike the rate of chemical reactions, the rate of decay of an isotope is independent of conditions such as temperature, pressure, and whether the element is in the free state or part of a compound. For example, $^{238}_{92}\text{U}$ has a half-life of $4.5 \times 10^9$ years. Since that is roughly the age of Earth, about one-half of the $^{238}_{92}\text{U}$ originally present when Earth formed from hot gases and dust is still present. Half of what is now present will be gone in another 4.5 billion years. The half-lives and modes of decay of some radioactive isotopes are listed in Table 15-2.

**Table 15-2  Half-Lives**

| Isotope | $t_{1/2}$ | Mode of Decay | Product |
|---|---|---|---|
| $^{238}_{92}U$ | $4.5 \times 10^9$ years | $\alpha, \gamma$ | $^{234}_{90}Th$ |
| $^{234}_{90}Th$ | 24.1 days | $\beta$ | $^{234}_{91}Pa$ |
| $^{226}_{88}Ra$ | 1620 years | $\alpha$ | $^{222}_{86}Rn$ |
| $^{14}_{6}C$ | 5760 years | $\beta$ | $^{14}_{7}N$ |
| $^{131}_{53}I$ | 8.0 days | $\beta, \gamma$ | $^{131}_{54}Xe$ |
| $^{218}_{85}At$ | 1.3 sec | $\alpha$ | $^{214}_{83}Bi$ |

Note in Table 15-2 that when $^{238}_{92}U$ decays, it forms an isotope ($^{234}_{90}Th$) that also decays (very rapidly compared with $^{238}_{92}U$). The $^{234}_{91}Pa$ formed from the decay of $^{234}_{90}Th$ also decays and so forth, until finally the stable isotope $^{206}_{82}Pb$ is formed. This is known as the $^{238}_{92}U$ radioactive decay series. *A radioactive decay series* starts with a naturally occurring radioactive isotope with a half-life near the age of Earth. (If it was very much shorter, there wouldn't be any left.) *The series ends with a stable isotope.* There are at least two other naturally occurring decay series: the $^{235}_{92}U$ series, which ends with $^{207}_{82}Pb$, and the $^{232}_{90}Th$ series, which ends with $^{208}_{82}Pb$. As a result of the $^{238}_{92}U$ decay series, where uranium is found in rocks, we also find other radioactive isotopes as well as lead. In fact, by examining the ratio of $^{238}_{92}U/^{206}_{82}Pb$ in a rock, its age can be determined. For example, a rock from the moon showed that about half of the original $^{238}_{92}U$ had decayed to $^{206}_{82}Pb$. This meant that the rock was about $4.5 \times 10^9$ years old.

---

**Half-Life**

**Example 15-2**

*working it out*

If we started with 4.0 mg of $^{14}C$, how long would it take until only 0.50 mg remained?

**Solution**

After each half-life the amount is reduced by one-half.

$$\text{After 5760 years, } \tfrac{1}{2} \times 4.0 \text{ mg} = 2.0 \text{ mg remaining}$$
$$\text{After another 5760 years, } \tfrac{1}{2} \times 2.0 \text{ mg} = 1.0 \text{ mg remaining}$$
$$\text{After another 5760 years, } \tfrac{1}{2} \times 1.0 \text{ mg} = 0.50 \text{ mg remaining}$$
$$\text{Total time} = 17,280 \text{ years}$$

Therefore, in 17,280 years, 0.50 mg remains.

---

# 15-3 Effects of Radiation

**Looking Ahead!** We tend to fear the effects of radiation. Certainly, it may cause cancer and sickness, but this same radiation can save lives when used in medical diagnosis or cancer treatment. How radiation interacts with matter is our next subject.

Anyone born after World War II has grown up with the fear of a nuclear holocaust and the resulting death of millions by the effects of radiation from nuclear explosions. It has been a frightful nightmare. If one wasn't killed by the immediate blast, then death could come later from the high doses of radiation received. Or, if one survived all of that, long-term damage may have caused genetic changes and even one's offspring may have health problems. But this same radiation that we fear so much may also be used to shrink cancerous tumors or diagnose illnesses and diseases that would otherwise go undetected.

Nuclear decay is a process in which a nucleus spontaneously gives off high-energy particles (alpha or beta) or high-energy light (gamma). The interaction of these fast-moving particles or light waves with surrounding matter eventually results in increased velocity for all molecules and a higher temperature. Thus the first effect that we note is that radioactivity causes heating. In fact, much of the heat generated in the interior of the Earth and other planets is a direct result of natural radiation from decaying isotopes. The presence of radioactive isotopes in wastes from nuclear power plants makes these materials extremely hot; in most cases they must be stored in water continuously so that they do not melt.

As human beings we may be more concerned with how radiation interacts with matter to cause *ionization*. Beside producing heat, collisions of radioactive particles or rays leave a trail of ions along their path. The formation of ions is caused by the removal of an electron from a neutral molecule. For example, if high-energy particles collide with an $H_2O$ molecule, an electron may be ejected from the molecule, leaving an $H_2O^+$ ion behind. The properties of the ion are distinctly different from those of the neutral molecule and can be very destructive to life processes. If the molecule in question is a large complex molecule that is part of a cell in a living system, the ionization causes damage, mutation, or even eventual destruction of the cell. As shown in Figure 15-3, an alpha particle causes the most ionization and is the most destructive along its path. However, these particles do not penetrate matter to any extent and can be stopped, even by a piece of paper. The danger of alpha emitters (such as uranium and plutonium) is that they can be ingested through food or inhaled into the lungs. These heavy elements tend to accumulate in bones. There, in intimate contact with the blood-producing cells of the bone marrow, they slowly do damage. Ultimately, the radiation can cause certain cells to change into abnormal cells that reproduce rapidly. These are the dreaded cancer cells such as those found in leukemia.

Beta radiation is less ionizing along its path than alpha radiation but is more penetrating. Still, a thin sheet of metal such as aluminum will stop most beta radiation. This type of radiation can cause damage to surface tissue such as skin and eyes but does not reach internal organs unless ingested. When countries used to test nuclear devices in the atmosphere, large quantities of $^{90}Sr$ (a beta emitter with a half-life of 28 years) were produced. Strontium is an alkaline earth metal like calcium, a major component of bones. So strontium can substitute for calcium, accumulate in the bones, and eventually cause leukemia and bone cancer. Good sense finally prevailed on the major atomic powers, and atmospheric testing of nuclear weapons has not occurred to any extent for many years.

Gamma radiation in comparison causes few ions to form along its path. That is the good news. But the bad news is really bad. Gamma radiation has tremendous penetrating power. Several feet of concrete or thick blocks of lead are needed for protection from gamma radiation. Without such protection, gamma radiation can cause damage from far away.

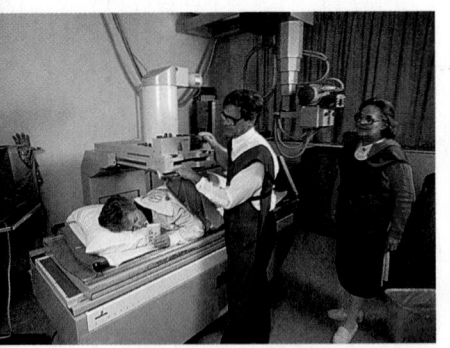

Lead aprons are worn to protect workers from X-rays and gamma rays.

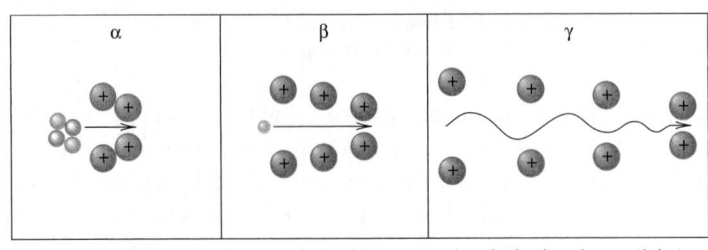

**Figure 15-3 Ionization and Penetration**
**Gamma radiation has the most penetrating power of the three types of radiation.**

High ionization along path but low penetration | Moderate ionization along path but low penetration | Low ionization along path but very high penetration

The damage done to living cells by radiation can be immediate or long term. Small doses may simply cause a temporary decrease in the white blood cell count. Higher exposures may cause nausea, vomiting, and diarrhea. Death may follow in a few days for very high exposures. But radiation also has a rather insidious long-term effect as well. Long exposure to low levels of radiation can be cumulative, leading to a weakening of the system and the formation of malignant tumors even after many years have elapsed. For this reason, the dosage of radiation to which employees are exposed in radioactive environments such as a nuclear research laboratory or nuclear power plant must be carefully and continually monitored. If a worker receives too much radiation over a specified period, that person must be protected from additional exposure for a length of time until the body has had time to repair some of the damage.

# 15-4 Detecting Radiation

**Looking Ahead!** How do we know how much radiation we are absorbing? Since we can't see it or feel it, we must have some method of measuring radiation. The detection and measurement of radiation are obviously important in understanding the danger or the benefit.

Becquerel first discovered radiation by noticing that it exposed photographic film. This is still an important method of both detecting and measuring radiation. Film badges are worn by workers in the nuclear industry or anyone else who works near a source of radiation. (See Figure 15-4.) When the film is developed, the degree of darkening is proportional to the amount of radiation absorbed. Different filters are used so that the amount of each type of radiation received (alpha, beta, or gamma) can be measured.

Radon-222 is a radioactive isotope formed as part of the natural decay series starting with $^{238}$U. Because radon is a noble gas, it can escape from the ground into the air. In certain areas of the United States, radon may accumulate in basements or other enclosed areas through foundation cracks. When inhaled, radon is usually exhaled unchanged, causing no harm. Its half-life, however, is only 3.82 days, which means that some actually does undergo radioactive decay when it is inhaled. If so, it decays to $^{218}$Po, which stays in the lungs because polonium is a solid. This is also a highly radioactive isotope that then decays in the lungs and may cause damage. It is estimated that between 10,000 and 20,000 lung cancer deaths may be caused annually by radon in the United States alone. There is still controversy, however, about these numbers and whether radon is actually that much of a risk. Nonetheless, radon detectors are now sold commercially. When

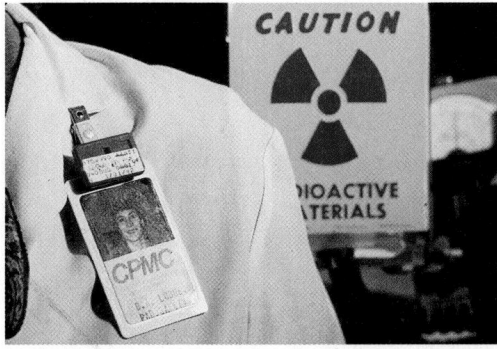

**Figure 15-4 Film Badge and Radon Detector**
**These devices measure radiation by the amount that the film is exposed.**

left in a basement, the detector measures the number of radon disintegrations. After a specified time, the detector is sent to a laboratory where the film is developed and the amount of radiation from radon is determined.

As mentioned in the previous section, radiation causes ionization. Ionization is also used to measure radiation in the *Geiger counter*. The Geiger counter is composed of a long metal tube with a thin film that allows all types of radiation to penetrate at one end. The metal tube is filled with a gas such as argon at low pressure. A metal wire in the center of the tube is positively charged. When a radioactive particle enters the tube, it causes ionization. The electrons from this ionization move quickly to the positive electrode, causing a pulse of electricity to flow. This pulse is amplified and causes a light to flash, produces an audible "beep," or registers on a meter in some unit such as "counts per minute." (See Figure 15-5.)

A third type of radiation detector is a *scintillation counter*. Certain compounds such as zinc sulfide are known as *phosphors*. When radiation strikes a phosphor, a tiny flash of light is emitted. The flash of light is converted into a pulse of electricity that can be counted in the same manner as in the Geiger counter.

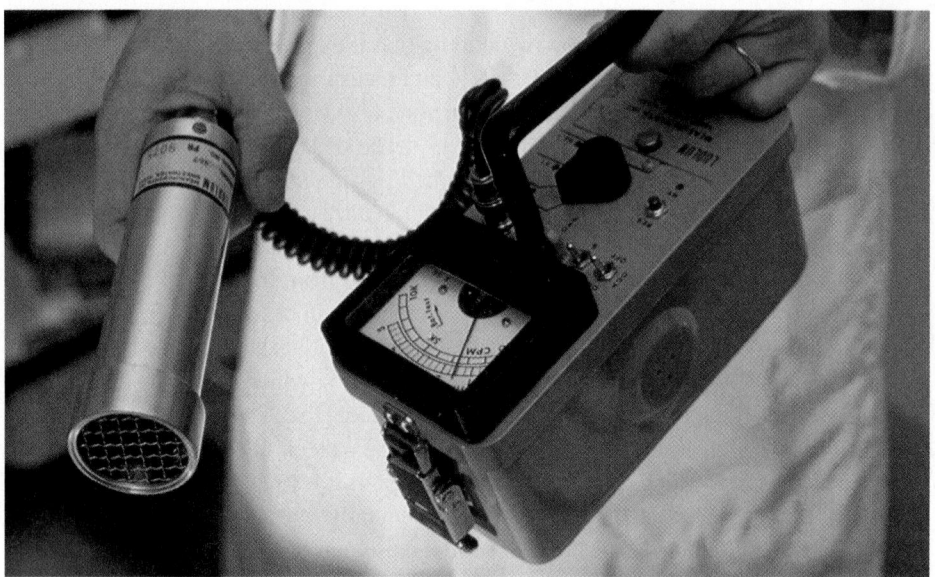

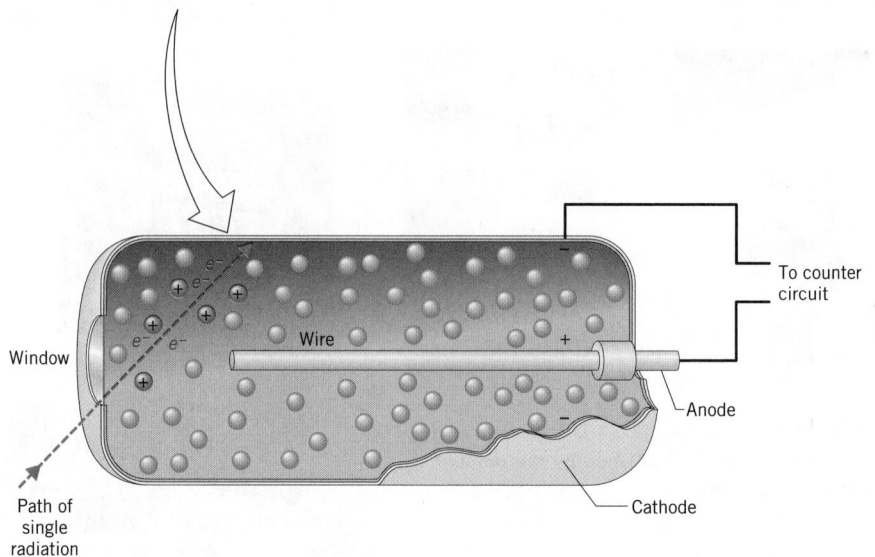

**Figure 15-5  Geiger Counter**
**This device measures radiation by the ionization it causes.**

**Looking Back!** The discovery of natural radioactivity was a key that unlocked many of the mysteries of the atom. Except for gamma rays, the emission of particles from a radioactive nucleus of one element leads to the formation of the nucleus of another element. Each radioactive isotope has a specific rate of decay that is indicated by its half-life. Since radiation causes destruction of living cells, it must be monitored in many locations or situations. Radiation is detected by exposure of film or from the resulting ionization.

## Learning Check A | checking it out

**A-1.** Fill in the blanks.

There are five types of _____ .

1. An alpha particle is a _____ nucleus.
2. A beta particle is an _____ .
3. A gamma ray is a high-energy form of _____ .
4. A position is a _____ _____ .
5. Electron capture has the same effect on a nucleus as _____ emission.

   The rate of decay of an isotope is indicated by its _____ . The most ionizing types of radiation are _____ _____ , whereas the most penetrating are _____ _____ . Radiation is detected by _____ badges, and by _____ and _____ counters.

**A-2.** Write the radioactive isotope that leads to the indicated nucleus and particle or ray.

a. _____ $\longrightarrow$ $^{87}_{38}Sr$ + beta particle

b. _____ $\longrightarrow$ $^{223}_{87}Fr$ + alpha particle

c. _____ $\longrightarrow$ $^{54}_{26}Fe$ + positron

d. _____ + electron capture $\longrightarrow$ $^{123}_{51}Sb$

e. _____ $\longrightarrow$ $^{137}_{55}Cs$ + gamma ray

**A-3.** After 36.0 minutes, it was found that 4.0 $\mu$g remains of a sample of a radioactive isotope that originally weighed 64 $\mu$g. What is the half-life of the isotope?

*Additional Examples: Exercises 15-1, 15-16, 15-18, 15-20, 15-22, and 15-27.*

Section **B** | **INDUCED NUCLEAR CHANGES AND THEIR USES**

## 15-5 Nuclear Reactions

**Looking Ahead!** In the Middle Ages, alchemists tried in vain to change one element into another. Much later we found that it is a spontaneous process in nature. Our next topic involves the actual accomplishment of changing one element into another.

Deep in castle dungeons around 600 years ago, people labored in vain to change one element into another. Actually, it wasn't just any element they wanted—it was gold. It never happened then, but 100 years ago we found that elements do change into others naturally by spontaneous emission of alpha or beta particles. But could we deliberately cause such a change? It wasn't long into the nuclear age before transmutation

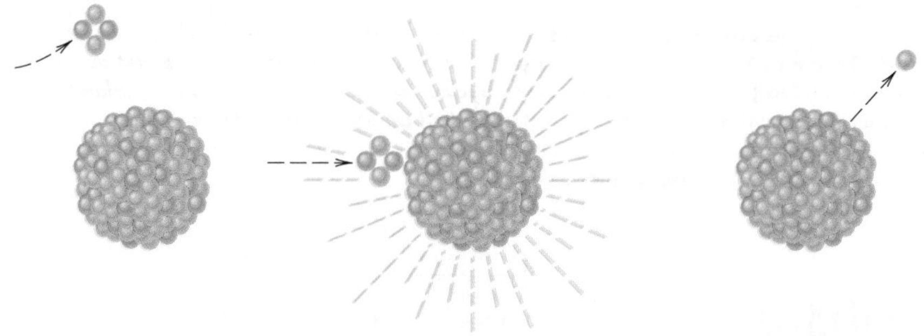

**Figure 15-6 Collision of an Alpha Particle with a Nucleus** Alpha particles must have sufficient energy to overcome the repulsion of the target nucleus for a collision to take place.

Alpha particle does not have enough energy, thus no collision

Collision occurs

High-energy nucleus gives off particle

happened. **Transmutation** *of an element is the conversion of that element into another.* The first example of transmutation was discovered by Ernest Rutherford in 1919; earlier, he had proposed the nuclear model of the atom. Rutherford found that bombarding $^{14}$N atoms with alpha particles caused a nuclear reaction that produced $^{17}$O and a proton. This nuclear reaction is illustrated by the nuclear equation

$$^{14}_{7}\text{N} + {}^{4}_{2}\text{He} \longrightarrow {}^{17}_{8}\text{O} + {}^{1}_{1}\text{H}$$

Note that the total number of nucleons is conserved by the reaction ($14 + 4$ on the left $= 17 + 1$ on the right). The total charge is also conserved during the reaction ($7 + 2$ on the left $= 8 + 1$ on the right). Both the number of nucleons and the total charge must be balanced (the same on both sides of the equation) in a nuclear reaction.

Over a decade later, Irène Curie, daughter of Pierre and Marie Curie, and her husband Frédéric Joliot, produced an isotope of phosphorus that was radioactive. This was the first artificially induced radioactivity, for which Curie and Joliot received the 1935 Nobel Prize in chemistry. The radioactive isotope formed was $^{30}$P, which decays to $^{30}$Si as illustrated by the following nuclear equations.

$$^{27}_{13}\text{Al} + {}^{4}_{2}\text{He} \longrightarrow {}^{30}_{15}\text{P} + {}^{1}_{0}\text{n}$$
$$^{30}_{15}\text{P} \longrightarrow {}^{30}_{14}\text{Si} + {}^{0}_{+1}\text{e}$$

Because the nucleus has a positive charge and many of the bombarding nuclei (except neutrons) also have a positive charge (e.g., $^{4}_{2}\text{He}$, $^{1}_{1}\text{H}$, $^{2}_{1}\text{H}$), the particles must have a high energy (velocity) to overcome the natural repulsion between these two like charges. From the collision of the nucleus and the particle, a high-energy nucleus is formed. As a result, another particle (usually a neutron) is then emitted by the nucleus to carry away excess energy from the collision. This is analogous to a head-on collision of two cars, where fenders, doors, and bumpers come flying off after impact. The nucleus–particle collision is illustrated in Figure 15-6.

The invention of particle accelerators, which speed nuclei and subatomic particles to high velocities and energies, opened the door to vast possibilities for artificial nuclear reactions. The larger accelerators also allow studies of collisions so powerful that the colliding nuclei are blasted apart. From these experiments, scientists have glimpsed the most basic composition of matter. These accelerators are large and expensive. Completed in the early 1970s, the Fermi National Accelerator in Batavia, Illinois, has a circumference of about 6 miles and cost $245 million. A huge accelerator, known as the *Superconducting Super Collider* (SSC), was about one-third completed in Texas and would have had a circumference of 52 miles. (See Figure 15-7.) The cost of about $12 billion was considered too much, however, so it was canceled by the U.S. Congress.

The use of particle accelerators has made possible the extension of the periodic table past element number 94, plutonium. Elements number 95 through the last element reported, number 116, have been synthesized by means of particle accelerators.

**Figure 15-7 The Superconducting Super Collider** Construction of this particle accelerator was suspended in late 1993 because of its high cost.

In 1995 elements number 110 and number 111 and in 1996 number 112 were reported as being synthesized by German scientists in Darmstadt, Germany. The following nuclear reactions illustrate the formation of two heavy elements.

$$^{238}_{92}U + ^{12}_{6}C \longrightarrow ^{244}_{98}Cf + 6^{1}_{0}n$$

$$^{238}_{92}U + ^{16}_{8}O \longrightarrow ^{250}_{100}Fm + 4^{1}_{0}n$$

---

### Nuclear Reactions and Equations

**Example 15-3**

working it out

Complete the following nuclear equations:

a. $^{9}_{4}Be + ^{2}_{1}H \longrightarrow$ _____ $+ ^{1}_{0}n$

b. $^{252}_{98}Cf + ^{10}_{5}B \longrightarrow$ _____ $+ 5^{1}_{0}n$

**Solution**

Same as Example 15-1.

a. Nucleons: $9 + 2 = x + 1$, so $x = 10$
   Atomic number: $4 + 1 = y + 0$, so $y = 5$

From the list of elements, we find that the element is boron (atomic number 5). The isotope is

$$^{10}_{5}B$$

b. Nucleons: $252 + 10 = x + (5 \times 1)$, so $x = 257$.
   Atomic number: $98 + 5 = y + (5 \times 0)$, so $y = 103$.

From the list of elements, the isotope is

$$^{257}_{103}Lr$$

---

# 15-6 Uses of Radioactivity

**Looking Ahead!** Despite their horrible reputation, radioactive isotopes have a huge number of beneficial applications. How we use radioactivity to date artifacts, make our food healthier, and help diagnose and even cure our illnesses is the subject of this section.

The benefits of the nuclear age are awesome. Heart scans tell us whether bypass surgery is indicated. Radiation treatments have saved thousands of lives and prolonged many thousands of others. The shelf-life of perishable food has been extended by weeks. These and other uses have far outweighed the dangers from the misuse of radioactivity. We will begin our discussion of how radioactivity improves our lives with how we use a naturally produced radioactive isotope to date artifacts.

## Carbon Dating

Perhaps one of the most useful applications of radioactive isotopes is the dating of wood or other carbon-containing substances that were once alive. Carbon-14 dating is effective in dating artifacts that are from about 2000 to about 50,000 years old. Carbon-14 is a radioactive isotope with a half-life of 5760 years. It is produced in the stratosphere by a **nuclear reaction** involving *cosmic rays* (a variety of radiation and particles) from the sun.

$$^{14}_{7}N + ^{1}_{0}n \longrightarrow ^{14}_{6}C + ^{1}_{1}H$$

The $^{14}C$ then mixes with the normal and stable isotopes of carbon in the form of carbon dioxide. Carbon dioxide is taken up by living systems and, through photosynthesis,

## Radioactivity and Smoke Detectors

We don't usually think of the presence in our home of radioactive isotopes as a good thing. In fact, the presence of an isotope of Americium ($^{241}$Am) in smoke detectors has saved many lives. Americium is a rather new element on the periodic table. $^{243}$Am ($t_{1/2} = 7370$ yrs) was first discovered in 1945 in airborne dust particles after the first atomic bombs. The isotope used in smoke detectors is a by-product of nuclear reactors. It is formed from the decay of $^{241}$Pu, which is extracted from spent reactor elements. It all starts in the reactor with $^{238}$U, the most abundant isotope of uranium. The neutrons produced in the reactor form $^{241}$Pu in the following steps.

$$^{238}U + {}^{1}_{0}n \longrightarrow {}^{239}U \xrightarrow{\beta^-} {}^{239}Np \xrightarrow{\beta^-}$$

$$^{239}Pu + {}^{1}_{0}n \longrightarrow {}^{240}Pu + {}^{1}_{0}n \longrightarrow {}^{241}Pu$$

Most of the $^{241}$Pu ($t_{1/2} = 14$ days) has decayed to $^{241}$Am ($t_{1/2} = 432$ years) by the time the extraction occurs. It is quite expensive, though. It is produced by the U.S. Atomic Energy Commission, where it sells for about $1500 per gram. However, since only 0.2 mg of $^{241}$Am (as the compound $AmO_2$) is used in each unit, 5000 smoke detectors can be made from one gram of the isotope. That amount of Americium produces about $3 \times 10^4$ decays/sec.

Smoke detectors work on simple principles. The $^{241}$Am decays by alpha particle emission. The alpha particles ($^4$He) interact with the major molecules of air ($N_2$ and $O_2$) to form ions ($N_2^+$ and $O_2^+$). These ions form between plates of two electrodes that are connected to a battery or house current. The ions serve as a conduction pathway between the two electrodes and complete a circuit. When smoke drifts in between the two electrodes, it absorbs the alpha particles, so fewer ions are formed. When the detector senses the circuit is disrupted, the alarm sounds.

Are smoke detectors safe? The answer is yes. Alpha particles have a short range of a few centimeters at most, so they stay within the confines of the detector. The detectors themselves do not have enough radioactivity to pose a problem to the environment. The current necessary for continual operation is very small, so a 9-volt battery should last at least six months.

The modern smoke detector is a life-saving device that is the direct product of the nuclear age. We are glad to have them.

*Submitted by G. Lynn Carlson*
*University of Wisconsin—Parkside*

*Smoke detectors make use of a synthetic element.*

---

becomes part of the carbon structure of the organism. As long as the carbon-based system is alive, the ratio of $^{14}$C to normal carbon is the same as in the atmosphere. When the system dies, the amount of $^{14}$C in the organism begins to decrease. By comparing the amount of radiation from $^{14}$C in the artifact with that in living systems, the age can be determined. (See Example 15-2.) Besides $^{14}$C dating, the decay of other isotopes (e.g., $^{40}$K) can be studied to determine the age of rocks that were formed from molten material millions of years ago.

### Neutron Activation Analysis

Most naturally occurring isotopes are not radioactive. When these isotopes are bombarded by neutrons from a nuclear reactor, however, they often absorb a neutron, which makes them radioactive. An example is the production of $^{60}$Co, which is used in cancer therapy, from a stable cobalt isotope.

$$^{59}_{27}Co + {}^{1}_{0}n \longrightarrow {}^{60m}_{27}Co$$

Another application of neutron activation concerns the analysis of arsenic in human hair. Arsenic compounds can be used as a slow poison, but arsenic accumulates in human hair in minute amounts. By subjecting the human hair to neutron bombardment, the stable isotope of arsenic is changed to a metastable nucleus.

$$^{75}_{33}As + {}^{1}_{0}n \longrightarrow {}^{76m}_{33}As \longrightarrow {}^{76}_{33}As + \gamma$$

The amount of gamma radiation from the metastable arsenic can be measured and is proportional to the amount of arsenic present. The method is very sensitive to even trace amounts of arsenic.

**Figure 15-8 Irradiated Strawberries** The strawberries on the left are moldy after a few days. Those on the right have been irradiated and are still fresh after two weeks.

## Food Preservation

In a very simple procedure, many types of food can be irradiated with gamma radiation, which kills bacteria and other microorganisms that cause food spoilage without changing the taste or appearance of the food. This increases the shelf-life of food before decay sets in. For example, mold begins to form on strawberries, even when refrigerated, in just a few days. After irradiation, the same strawberries can be stored for two weeks without decay. (See Figure 15-8.) There has been some buyer resistance to irradiated foods. The public worries that irradiation somehow may make the food radioactive. The procedure is perfectly safe, however, and as public apprehension declines, we will probably see more and more irradiated foods at the grocery store.

## Medical Therapy

Cobalt-60 gamma radiation can be focused into a narrow beam. Although gamma radiation destroys healthy as well as malignant cells, healthy cells recover faster. By focusing the beam at different angles on tumors located within the body, the gamma rays can be concentrated on the tumor. (See Figure 15-9.) This treatment has many unpleasant side effects because of the destruction of normal, healthy cells. A procedure known as *brachytherapy* is used increasingly in the treatment of prostate cancer. Tiny "seeds" of $^{103}$Pd ($t_{1/2} = 17$ days) or $^{125}$I ($t_{1/2} = 8$ days) inside a titanium capsule are implanted in the cancerous gland. As these isotopes decay by electron capture, they release low-energy X-rays, which destroy prostate tissue but penetrate only about 1 centimeter into the surrounding tissue. Since the radiation has a localized effect, nearby organs are not affected. The radiation lasts only a few months because of the short half-lives, so the seeds are left permanently in place. Although there are several other radioactive isotopes used in therapies, the greatest application of these isotopes is in diagnosis.

## Medical Diagnosis

Today, hardly any large hospital could be without its nuclear medicine division. We have come to rely on the use of radioactive isotopes to help diagnose many diseases and conditions. Various radioactive isotopes can be injected into the body, and their movement through the body or where they accumulate can be detected outside the body with radiation detectors. Iron-59 is used to measure the rate of formation and

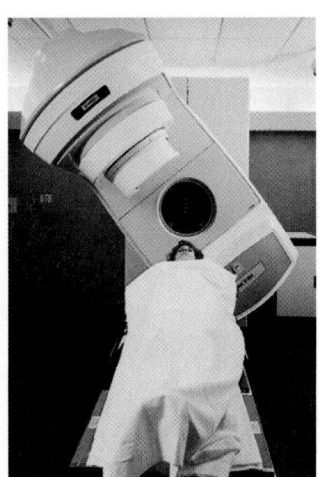

**Figure 15-9** $^{60}$Co Radiation Treatment
**A beam of gamma rays can be used to destroy cancerous tissue.**

lifetime of red blood cells. Technetium-99m is used to image the brain, heart, and other organs. Iodine-131 is used to detect thyroid malfunction and can also be used for the treatment of thyroid tumors. Sodium-24 is used to study the circulatory system.

One of the more useful (and expensive) tests is known as a *PET (positron emission tomography)* scan, which produces an image of a two-dimensional slice through a portion of the body. The body is injected with a compound (e.g., glucose) containing a radioactive isotope such as $^{11}$C. The glucose containing this isotope of carbon is metabolized along with glucose produced by the body containing the stable isotope $^{12}$C. The parts of the brain that are particularly active in metabolism of glucose will display increased radioactivity. Abnormal glucose metabolism in the brain can then be detected, which may indicate a tumor or Alzheimer's disease. This diagnostic procedure is considered noninvasive compared to surgery. (See Figure 15-10.)

The $^{11}$C decays by positron emission.

$$^{11}_{6}C \longrightarrow {}^{11}_{5}B + {}^{0}_{+1}e$$
$$^{0}_{+1}e + {}^{0}_{-1}e \longrightarrow 2\gamma$$

Almost immediate annihilation of the positron by an electron leads to two gamma rays that exit the body in exactly opposite directions. Scintillation counters are positioned around the body so as to detect these two gamma rays and ignore others from background radiation. Computers are then used to translate the density and location of the gamma rays into two-dimensional images.

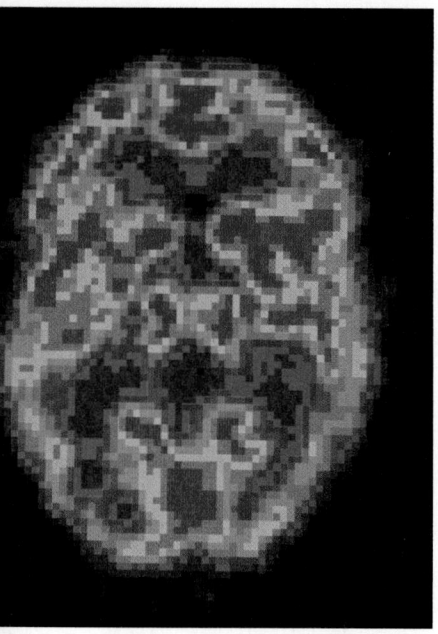

**Figure 15-10  PET Scan**
**This PET scan of the brain reveals areas of activity during sleep.**

# 15-7 Nuclear Fission

**Looking Ahead!** In 1938, a discovery was made that at first suggested massive destruction and, later, on the positive side, inexpensive energy. The discovery was the atomic bomb and then the control of its process in nuclear power plants. The nuclear process that produces this power is the next topic.

In 1934, two Italian physicists, Enrico Fermi and Emilio Segrè, attempted to expand the periodic table by bombarding $^{238}$U with neutrons to produce isotopes that would decay by beta particle emission to form elements with atomic numbers greater than 92 (the last element on the periodic table at the time).

$$^{238}_{92}U + {}^{1}_{0}n \longrightarrow {}^{239}_{92}U \longrightarrow {}^{239}_{93}X + {}^{0}_{-1}e$$

The experiment seemed to work, but they were perplexed by the presence of several radioactive isotopes produced in addition to the presumed element number 93. In 1938, these experiments were repeated by two German scientists, Otto Hahn and Fritz Strassman. They were able to identify the excess radioactivity as coming from isotopes such as barium, lanthanum, and cerium, which had about half the mass of the uranium isotope. Two other German physicists, Lise Meitner and Otto Frisch, were able to show that the rare isotope of uranium ($^{235}$U, 0.7% of naturally occurring uranium) was undergoing fission into roughly two equal parts after absorbing a neutron. **Fission** *is the splitting of a large nucleus into two smaller nuclei of similar size.* It should be noted that there are a variety of products from the fission of $^{235}$U in addition to those represented by the following equation. (See Figure 15-11.)

$$^{235}_{92}U + {}^{1}_{0}n \longrightarrow {}^{139}_{56}Ba + {}^{94}_{36}Kr + 3{}^{1}_{0}n$$

Two points about this reaction had monumental consequences for the world, and scientists in Europe and America were quick to grasp their meaning in a world about to go to war.

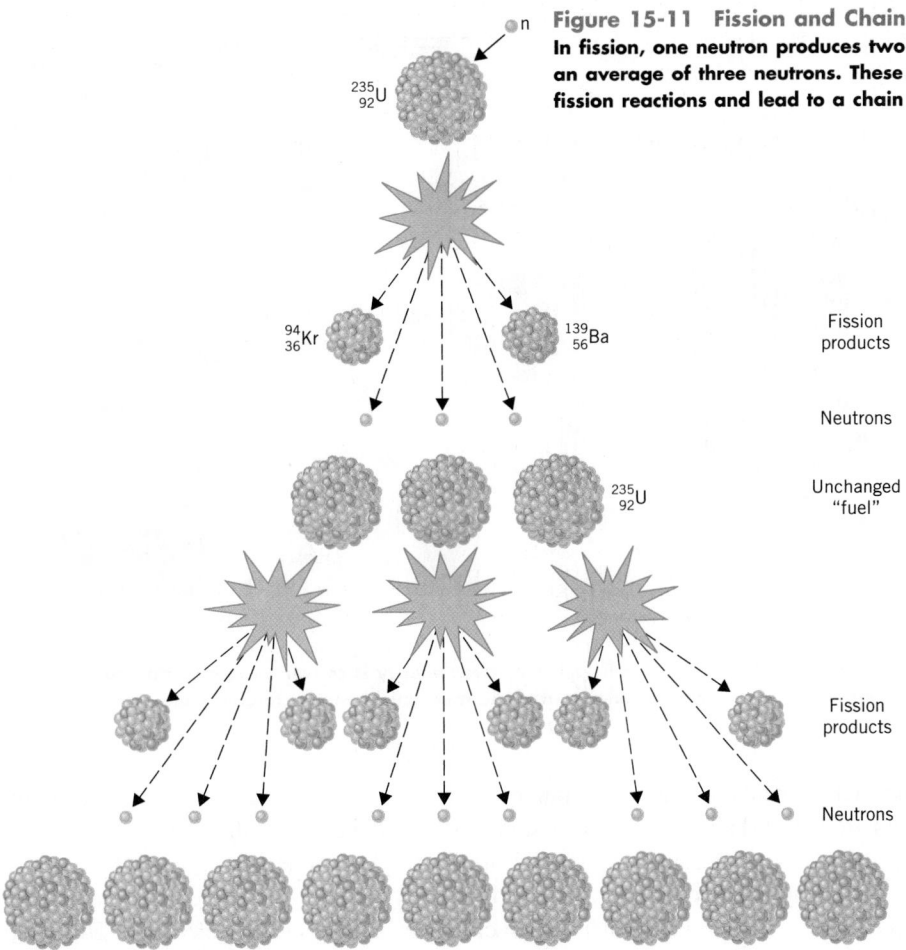

**Figure 15-11 Fission and Chain Reactions**

**In fission, one neutron produces two fragments of the original atom and an average of three neutrons. These three neutrons can cause additional fission reactions and lead to a chain reaction.**

Fission products

Neutrons

Unchanged "fuel"

Fission products

Neutrons

1. Comparison of the masses of the product nuclei with that of the original nucleus indicated that a significant amount of mass was lost in the reaction. According to Einstein's equation ($E = mc^2$), the lost mass must be converted to a tremendous amount of energy. Fission of a few kilograms of $^{235}U$ could produce energy equivalent to tens of thousands of tons of the conventional explosive TNT.

2. What made the rapid fission of a large sample of $^{235}U$ feasible was the potential for a chain reaction. *A nuclear chain reaction is a reaction that is self-sustaining.* The reaction generates the means to trigger additional reactions. Note in Figure 15-11 that the reaction of one original neutron caused three to be released. These three neutrons cause the release of nine neutrons and so forth. If a given densely packed "critical mass" of $^{235}U$ is present, the whole mass of uranium can undergo fission in an instant with a quick release of energy in the form of radiation.

Unfortunately, the world thus entered the nuclear age in pursuit of a bomb. After a massive secret effort, the first nuclear bomb was exploded over Alamogordo Flats in New Mexico on July 16, 1945. This bomb and the bomb exploded over Nagasaki, Japan, were made of $^{239}Pu$, which is a synthetic fissionable isotope. The bomb exploded over Hiroshima, Japan, was made of $^{235}U$.

This enormously destructive device, however, can be tamed. The chain reaction can be controlled by absorbing excess neutrons with cadmium bars. A typical nuclear reactor is illustrated in Figure 15-12. In a reactor core, uranium in the form of pellets is encased in long rods called fuel elements. Cadmium bars are raised and lowered among

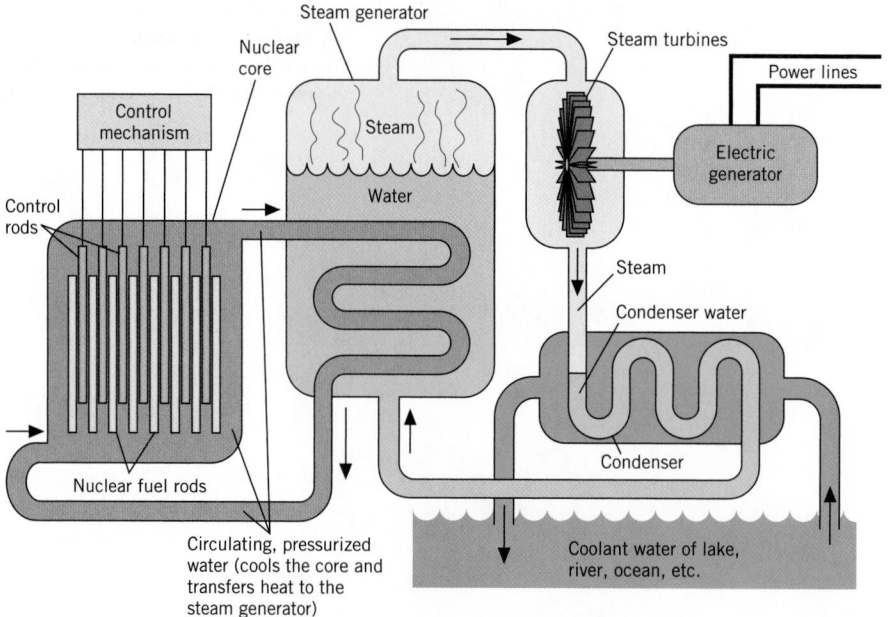

**Figure 15-12  A Nuclear Reactor** Nuclear energy is converted to heat energy, the heat energy to mechanical energy, and the mechanical energy to electrical energy.

the fuel elements to control the rate of fission by absorbing neutrons. If the cadmium bars are lowered all the way, the fission process can be halted altogether. In normal operation, the bars are raised just enough so that the fission reaction occurs at the desired level. Energy released by the fission and the decay of radioactive products formed from the fission is used to heat water, which circulates among the fuel elements. This water, which is called the primary coolant, becomes very hot (about 300°C) without boiling because it is under high pressure. The water heats a second source of water, changing it to steam. The steam from the secondary coolant is cycled outside of the containment building, where it is used to turn a turbine that generates electricity.

Perhaps the greatest advantage of nuclear energy is that it does not pollute the air with oxides of sulfur and carbon as conventional power plants do. Sulfur oxides have been implicated as a major cause of acid rain. It also seems likely that large amounts of carbon dioxide in the atmosphere are leading to significant changes in the weather as a result of the greenhouse effect. Another advantage is that nuclear power does not deplete the limited supply of fossil fuels (oil, coal, and natural gas), which are also used to make plastics and fertilizers. Until the tragedy in Chernobyl, Ukraine, in April 1986, there had been no loss of human life from the use of nuclear energy to generate electricity. At the current time, about 28 countries produce electricity through nuclear power. The United States currently generates about 20% and France about 80% of their electricity from nuclear power. (See Figure 15-13.) Proponents of nuclear power feel that if the proper systems were used with adequate safeguards and backups, catastrophic accidents would be avoided. In fact, new designs of nuclear power plants solve many of the legitimate concerns of citizens. A safe reactor would be one that shuts itself down in case of an emergency and would not overheat or explode. Experts from several countries are also currently planning a relatively safe reactor that will use as a fuel the $^{239}$Pu salvaged from dismantled nuclear bombs of the United States and of the former USSR. This reactor would be highly efficient, converting about 50% of the heat generated into electricity, compared with an efficiency of about 33% for a typical commercial nuclear power plant.

**Figure 15-13  A Nuclear Power Plant**
In such plants, the nuclear energy of uranium is converted into electrical energy.

The disadvantages have been made painfully obvious. The catastrophic explosion and subsequent fire of reactor 4 in Chernobyl, Ukraine, in 1986 dispersed more dangerous radiation into the environment than all of the atmospheric bomb tests by all nations put together (see Figure 15-14.) The surroundings for at least 30 mi around the plant, and even farther downwind, are still uninhabitable. The reactor itself has been entombed in concrete and must remain so for hundreds of years. The extent of the damage is still being assessed. Several thousand people lost their lives and thousands more will have their lives shortened by radiation sickness. Many brave people died as a result of the radiation received while fighting the fire or attempting to contain and secure the building. In the United States, an accident at Three Mile Island, Pennsylvania, in March 1979 took more than 14 years to clean up at a cost greater than $1 billion. Fortunately, there was very little release of radioactivity and no loss of life in this accident.

In both of these accidents, a core "meltdown" did not occur. A meltdown would occur if the temperature of the reactor core exceeded 1130°C, the melting point of uranium. Theoretically, the mass of molten uranium, together with all of the highly radioactive decay products, could accumulate on the floor of the reactor, melt through the many feet of protective concrete, and eventually reach groundwater. If this happened, vast amounts of deadly radioactive wastes could be spread through the groundwater into streams and lakes.

The type of accident that occurred in Ukraine is unlikely in most of the world's commercial reactors. Unlike reactors in the former Soviet Union, these reactors are

**Figure 15-14  The Chernobyl Reactor**
The explosion of this reactor spread radioactive debris for miles around.

surrounded by an extremely strong structure, called a containment building, made of reinforced concrete. These buildings should contain an explosion such as occurred at Chernobyl. Also, most reactors do not contain graphite, which is highly combustible and led to the prolonged fire in Chernobyl that spread so much radioactivity into the atmosphere. There is one reactor similar to the Ukrainian reactors, in Hanford, Washington, but it has been shut down indefinitely.

A growing problem in the use of nuclear energy involves used, or spent, fuel. Originally, used fuel was to be processed at designated centers where the unused fuel could be separated and reused and the radioactive wastes disposed. However, problems in transportation of this radioactive material and the danger of theft of the $^{239}$Pu that is produced in a reactor have hindered solution of the problem. ($^{239}$Pu is also fissionable and could be used in a nuclear bomb.) Currently, spent fuel elements, which must be replaced every four years, are being stored at the reactor sites. These fuel rods are not only highly radioactive but remain extremely hot and must be continually cooled. It is estimated that they will remain dangerously radioactive for at least 1000 years. This is a problem that must soon be resolved.

Because of the accidents, new regulations regarding safeguards have made the cost of a new nuclear power plant considerably higher than that of a conventional power plant using fossil fuels. As a result, no new nuclear plants are planned in the United States in the immediate future.

# 15-8 Nuclear Fusion

**Looking Ahead!** In the late 1930s, another source of nuclear energy was proposed by Hans Bethe. This involved fusing small nuclei together rather than splitting large ones. It was even suggested that this was the major source of energy in the universe because it powered the sun and the stars. We now have a good (but not complete) understanding of the nuclear reactions that eventually made life on Earth possible. This is the subject of the final section of this chapter.

Soon after the process of nuclear fission was demonstrated, an even more powerful type of nuclear reaction was shown to be theoretically possible. *This reaction involved the* **fusion,** *or bringing together, of two small nuclei.* An example of a fusion reaction is

$$^3_1\text{H} + {}^2_1\text{H} \longrightarrow {}^4_2\text{He} + {}^1_0\text{n}$$

($^3_1$H is called tritium; it is a radioactive isotope of hydrogen.)

As in the fission process, a significant amount of mass is converted into energy. Fusion energy is the origin of almost all of our energy, since it powers the sun. Millions of tons of matter are converted to energy in the sun every second. Because of its large mass, however, the sun contains enough hydrogen to "burn" for additional billions of years.

The principle of fusion was first demonstrated on this planet with a tremendously destructive device called the hydrogen bomb. This bomb can be more than 1000 times as powerful as the atomic bomb, which uses the fission process.

Fission is controlled in nuclear power plants. What about fusion? Controlling fusion is technically extremely difficult. Temperatures on the order of 50 million degrees Celsius are needed so that the colliding nuclei have enough kinetic energy to overcome their mutual repulsions and cause fusion. The necessary temperature is hotter than the interior of the sun. No known materials can withstand these temperatures, so alternate containment procedures must be used. Producing electricity from a fusion reactor is perhaps the greatest technical challenge yet faced by the human race. However, the effort is well under way. The design and construction of a large fusion reactor (International Thermonuclear Experimental Reactor, or ITER) is being financed and staffed by an international consortium of nations including the United States, Japan, Russia, and

## The Origin of the Elements

We are all children of the universe since we are literally composed of stardust. This sounds rather melodramatic but it is indeed true. Other than hydrogen, all other atoms in our bodies originated in the violent interiors of ancient stars or from their catastrophic explosion.

The universe itself began violently about 15 billion years ago. All of the matter and energy in the universe originated from a single point called the "singularity." This point exploded in what is called "the big bang." The fundamental particles—electrons, neutrons, and protons—were formed within minutes. Hydrogen was the primeval element from which other elements would form.

Sometime after the big bang, clouds of hydrogen gas began to clump together and contract because of the attraction of gravity. As contraction continued, the hydrogen cloud became denser and hotter (i.e., Charles's law). Eventually, the nuclei of hydrogen had enough energy to fuse together to produce helium. This process liberates huge amounts of energy, so the first stars began to shine brightly and the universe began to light up. After a few million years, the early stars had used up their supply of hydrogen and the helium began to fuse to form even heavier elements like carbon and oxygen, as shown here.

$$\frac{4}{2}He + \frac{4}{2}He \longrightarrow \frac{8}{4}Be + \frac{4}{2}He \longrightarrow \frac{12}{6}C + \frac{4}{2}He \longrightarrow \frac{16}{8}O$$

The fusion reactions continued in stages up to the formation of elements such as iron. At this point, further fusion reactions forming heavier nuclei do not liberate energy and thus do not occur.

So, how did heavier elements like lead and gold form? When the large stars exhausted their supply of elements that can fuse, the nuclear reactions suddenly stopped. The star immediately collapsed and then rebounded, ejecting its elements into outer space along with a huge cloud of neutrons. This violent event is called a "supernova." The elements from the star absorbed the neutrons singly or in batches. This made the nuclei "neutron rich," which means that they would decay by beta emission, forming elements higher in the periodic table. An example is shown here.

$$\frac{99}{39}Y + \frac{1}{0}n \longrightarrow \frac{91}{39}Y \longrightarrow \frac{91}{40}Zr + \frac{0}{-1}e$$

These elements drifted through space as dust and gases. About 4.5 billion years ago a cloud of hydrogen containing the dust from previous supernovas contracted and formed our sun and its planets. We are the inhabitants of the third planet out from this ordinary star.

*Elements are created in the interiors of stars.*

the European Union. The cost is estimated at about $8 billion over a 13-year period. It is hoped that eventually this reactor will exceed the *break-even point* where the same amount of energy is produced by the fusion as is needed to bring about the fusion. In

**This small reactor is being used to study the feasibility of the generation of fusion power.**

about 20 years, if things go well, the first fusion reactor designed to produce electricity could be built. With the amount of fossil fuel on the decline, there are not many alternatives for energy later in the twenty-first century.

The advantages of controlled fusion power are impressive.

1. It should be relatively clean. Few radioactive products are formed.

2. Fuel is inexhaustible. The oceans of the world contain enough deuterium, one of the reactants, to provide the world's energy needs for a trillion years. On the other hand, there is a very limited supply of fossil fuels and uranium.

3. There is no possibility of the reaction going out of control and causing a meltdown. Fusion will occur in power plants in short bursts of energy that can be stopped easily in case of mechanical problems.

**Looking Back!** Nuclear reactions occur when stable nuclei are bombarded with various charged particles or light nuclei. These reactions have led to an expansion of the periodic table from element 93, the heaviest naturally occurring element, to element number 112. These reactions have also produced radioactive isotopes that have broad application in the diagnosis and treatment of diseases. A nuclear reaction involving a neutron and $^{235}U$ led to the discovery of nuclear fission and the production of vast amounts of energy. Nuclear fusion is also a nuclear reaction that may someday provide a major source of energy.

## Learning Check B | checking it out

**B-1.** Fill in the blanks.

Nuclear reactions may cause one element to be transformed into another by a process known as _____ . The energy of the bombarding particle is increased by use of particle _____ . Both naturally occurring and synthetic radioactive isotopes find many uses. The decay of carbon- _____ can be used to date many carbon-containing fossils. Medically useful radioactive isotopes can be prepared by _____ activation. Food can be preserved by radiation with _____ _____ . PET scans make use of isotopes that emit _____ . A nuclear reaction known as fission can occur as a nuclear _____ reaction. The process whereby small nuclei combine to form a larger nucleus and a small particle is known as _____ .

**B-2.** Complete the following nuclear equations.

a. $^{51}_{23}V + ^{2}_{1}H \longrightarrow \underline{\hspace{1cm}} + ^{1}_{1}H$

b. $^{235}_{92}U + ^{1}_{0}n \longrightarrow ^{90}_{38}Sr + \underline{\hspace{1cm}} + 2^{1}_{0}n$

c. $^{246}_{96}Cm + \underline{\hspace{1cm}} \longrightarrow ^{254}_{102}No + 4^{1}_{0}n$

d. $^{2}_{1}H + ^{2}_{1}H \longrightarrow ^{3}_{1}H + \underline{\hspace{1cm}}$

**B-3.** Which of the nuclear reactions in problem B-2 represents fission? Which represents fusion?

*Additional Examples: Exercises 15-30, 15-32, and 15-36.*

The nuclei of **radioactive isotopes** are unstable and emit **radiation.** Although radioactive isotopes exist for each element, all isotopes with an atomic number greater than 83 are unstable. Originally, three types of radiation were discovered, **alpha ($\alpha$) particles, beta ($\beta$) particles,** and **gamma ($\gamma$) rays.** Since then, two other types of radiation have been characterized that occur rarely in nature but more commonly in artificially produced radioactive isotopes. These are **positron ($_{+}^{0}$e) particles** and **electron capture.** These five modes of decay are illustrated by the following **nuclear equations.**

*Alpha particle:* $\qquad \qquad _{96}^{240}\text{Cm} \longrightarrow _{94}^{236}\text{Pu} + _{2}^{4}\text{He}$

*Beta particle:* $\qquad \qquad _{30}^{71}\text{Zn} \longrightarrow _{31}^{71}\text{Ga} + _{-1}^{0}\text{e}$

*Gamma radiation:* $\qquad \quad _{43}^{99m}\text{Tc} \longrightarrow _{43}^{99}\text{Tc} + \gamma$

*Positron particle:* $\qquad \; _{10}^{19}\text{Ne} \longrightarrow _{9}^{19}\text{F} + _{+1}^{0}\text{e}$

*Electron capture:* $\qquad \; _{4}^{7}\text{Be} + _{-1}^{0}\text{e} \longrightarrow _{3}^{7}\text{Li}$

Radioactive isotopes decay at widely different rates. A measure of the rate of decay is called the **half-life ($t_{1/2}$)** of the isotope. Half-lives vary from billions of years to fractions of a second. A long-lived isotope is $^{238}\text{U}$, which begins a **radioactive decay series** and ends with a stable isotope of lead after a series of alpha and beta decays.

Radiation affects the molecules of surrounding matter by causing ionization. In living matter, this ionization may lead to destruction of the cells. Alpha and beta emitters are most dangerous when ingested since they cause a high degree of ionization in close proximity to cells. Gamma rays are less ionizing but very penetrating; thus they are also very dangerous. Radiation is detected and measured by means of film badges, Geiger counters, and scintillation counters.

**Nuclear reactions** occur when nuclei are bombarded by particles or other nuclei. This leads to the artificial **transmutation** of one element into another. **Fission** and **fusion** are also nuclear reactions. Fission is the splitting of one heavy nucleus into two more-or-less equal fragments and neutrons. Fusion is the joining of two light nuclei to form a heavier nucleus and a particle. Examples of these three processes are illustrated with the following nuclear equations.

*Nuclear reaction:* $\qquad _{99}^{253}\text{Es} + _{2}^{4}\text{He} \longrightarrow _{101}^{256}\text{Md} + _{0}^{1}\text{n}$

*Fission:* $\qquad \qquad \quad _{92}^{235}\text{U} + _{0}^{1}\text{n} \longrightarrow _{38}^{94}\text{Sr} + _{54}^{139}\text{Xe} + 3_{0}^{1}\text{n}$

*Fusion:* $\qquad \qquad \quad \; _{1}^{2}\text{H} + _{1}^{2}\text{H} \longrightarrow _{2}^{3}\text{He} + _{0}^{1}\text{n}$

# Exercises

## Nuclear Radiation

**15-1.** Write isotope symbols in the form $_{Z}^{A}\text{M}$ for isotopes of the following compositions.
**(a)** 84 protons and 126 neutrons
**(b)** 46 neutrons and a mass number of 84
**(c)** 100 protons and a mass number of 257
**(d)** lead-206
**(e)** uranium-233

**15-2.** Write isotope symbols in the form $_{Z}^{A}\text{M}$ for isotopes of the following compositions.
**(a)** 86 protons and 134 neutrons
**(b)** 6 protons and a mass number of 13
**(c)** 22 neutrons and a mass number of 40
**(d)** potassium-41
**(e)** americium-243

**15-3.** Give the symbols, including mass number and charge, for
**(a)** an alpha particle     **(c)** a neutron
**(b)** a beta particle

**15-4.** A deuteron has a mass number of two and a positive charge of one. Write its isotope symbol.

**15-5.** A triton has a mass number of three and a positive charge of one. Write its isotope symbol.

**15-6.** Write the isotope symbol, including mass number, for the isotope that results when each of the following emits an alpha particle.
**(a)** $_{84}^{210}\text{Po}$   **(b)** $^{152}\text{Gd}$   **(c)** fermium-252   **(d)** Mt-266

**15-7.** Write the isotope symbol, including mass number, for the isotope that results when each of the following emits an alpha particle.
**(a)** $_{92}^{234}\text{U}$   **(b)** $_{88}^{222}\text{Ra}$   **(c)** $^{210}\text{Bi}$   **(d)** thorium-229

**15-8.** Write the isotope symbol, including mass number, for the isotope that results when each of the following emits a beta particle.
(a) $^3_1H$   (b) $^{153}Gd$   (c) iron-59   (d) sodium-24

**15-9.** Write the isotope symbol, including mass number, for the isotope that results when each of the following emits a beta particle.
(a) $^{131}_{53}I$   (b) $^{234}Pa$   (c) lead-210   (d) nitrogen-16

**15-10.** Give the symbol of a positron. What effect does the emission of a positron have on an isotope? How are positrons detected?

**15-11.** Manganese-51 undergoes positron emission. Write the nuclear equation illustrating this reaction.

**15-12.** A certain isotope undergoes positron emission to form $^{23}Na$. Write the nuclear equation illustrating this reaction.

**15-13.** What happens to an isotope that undergoes electron capture? How is electron capture detected?

**15-14.** Germanium-68 undergoes electron capture. Write the nuclear equation illustrating this reaction.

**15-15.** A certain isotope undergoes electron capture to form $^{55}Mn$. Write the nuclear equation illustrating this reaction.

**15-16.** Complete the following nuclear equations.
(a) $^{214}_{83}Bi \longrightarrow ^{214}_{84}Po + \underline{\quad}$
(b) $^{90}_{37}Rb \longrightarrow \underline{\quad} + ^{0}_{-1}e$
(c) $^{26}_{14}Si \longrightarrow \underline{\quad} + ^{0}_{+1}e$
(d) $^{235}_{92}U \longrightarrow \underline{\quad} + ^{4}_{2}He$
(e) $^{179}_{73}Ta + ^{0}_{-1}e \longrightarrow \underline{\quad}$
(f) $\underline{\quad} \longrightarrow ^{41}_{21}Sc + ^{0}_{-1}e$
(g) $\underline{\quad} \longrightarrow ^{210}_{82}Pb + ^{0}_{-1}e$

**15-17.** Complete the following nuclear equations.
(a) $^{239}_{93}Np \longrightarrow \underline{\quad} + ^{0}_{-1}e$
(b) $\underline{\quad} \longrightarrow ^{93}_{44}Ru + ^{0}_{+1}e$
(c) $^{226}_{86}Ra \longrightarrow ^{222}_{86}Rn + \underline{\quad}$
(d) $\underline{\quad} \longrightarrow ^{235}_{92}U + ^{4}_{2}He$
(e) $^{80}_{37}Rb + \underline{\quad} \longrightarrow ^{80}_{36}Kr$
(f) $^{32}_{15}P \longrightarrow ^{32}_{16}S + \underline{\quad}$

**15-18.** From the following information, write nuclear equations that include all isotopes and particles.
(a) $^{230}_{90}Th$ decays to $^{226}_{88}Ra$.
(b) $^{214}_{84}Po$ emits an alpha particle.
(c) $^{210}_{84}Po$ emits a beta particle.
(d) An isotope emits an alpha particle and forms lead-214.
(e) $^{14}_{6}C$ decays to form $^{14}_{7}N$.
(f) Chromium-50 is formed by positron emission.
(g) Argon-37 captures an electron.

**\*15-19.** The decay series of $^{238}_{92}U$ to $^{206}_{82}Pb$ involves alpha and beta emissions in the following sequence: $\alpha$, $\beta$, $\beta$, $\alpha$, $\alpha$, $\alpha$, $\alpha$, $\alpha$, $\beta$, $\alpha$, $\beta$, $\beta$, $\beta$, $\alpha$. Identify all isotopes formed in the series.

### Nuclear Decay and Half-Life

**15-20.** What fraction of a radioactive isotope remains after four half-lives?

**15-21.** What percent of a radioactive isotope remains after five half-lives?

**15-22.** If one starts with 20 mg of a radioactive isotope with a half-life of 2.0 days, how much remains after each interval?
(a) four days   (b) eight days   (c) four half-lives

**15-23.** The half-life of a given isotope is 10 years. If we start with a 10.0-g sample of the isotope, how much is left after 20 years?

**15-24.** Start with 12.0 g of a given radioactive isotope. After 11 years, only 3.0 g is left. What is the half-life of the isotope?

**15-25.** The isotope $^{14}_{6}C$ is used to date fossils of formerly living systems such as prehistoric animal bones. If the radioactivity of the carbon in a sample of bone from a mammoth is one-fourth of the radioactivity of the current level, how old is the fossil? ($t_{1/2} = 5760$ years)

### Effects of Radiation and Its Detection

**15-26.** Of the three types of radiation, which is the most ionizing? Which is the most penetrating? How does each type of radiation cause damage to cells?

**15-27.** How does radiation cause ionization? How does ionization cause damage to living tissues?

**15-28.** How does a film badge work? How can a film badge tell which type of radiation is being absorbed?

**15-29.** How does a Geiger counter work? How does a scintillation counter work?

### Nuclear Reactions

**15-30.** Complete the following nuclear equations.
(a) $^{35}_{17}Cl + ^{1}_{0}n \longrightarrow \underline{\quad} + ^{1}_{1}H$
(b) $^{27}_{13}Al + \underline{\quad} \longrightarrow ^{25}_{12}Mg + ^{4}_{2}He$
(c) $^{27}_{13}Al + ^{4}_{2}He \longrightarrow \underline{\quad} + ^{1}_{0}n$
(d) $^{238}_{92}U + 15^{1}_{0}n \longrightarrow ^{253}_{100}Fm + \underline{\quad}$
(e) $^{244}_{96}Cm + ^{12}_{6}C \longrightarrow \underline{\quad} + 2^{1}_{0}n$
(f) $\underline{\quad} + ^{2}_{1}H \longrightarrow ^{238}_{93}Np + ^{1}_{0}n$
(g) $^{242}_{94}Pu + ^{22}_{10}Ne \longrightarrow ^{260}_{104}\underline{\quad} + \underline{\quad}$

**15-31.** Complete the following nuclear equations.
(a) $^{249}_{96}Cm + \underline{\quad} \longrightarrow ^{260}_{103}Lr + 4^{1}_{0}n$
(b) $^{15}_{7}N + ^{1}_{1}H \longrightarrow \underline{\quad} + ^{4}_{2}He$
(c) $^{10}_{5}B + ^{4}_{2}He \longrightarrow \underline{\quad} + ^{1}_{1}H$
(d) $^{249}_{98}Cf + \underline{\quad} \longrightarrow ^{263}_{106}\underline{\quad} + 4^{1}_{0}n$

**15-32.** An element on the periodic table was prepared by the following nuclear reaction.

$$^{209}_{83}Bi + ^{58}_{26}Fe \longrightarrow ^{266}_{109}Mt$$

How many neutrons were emitted in the reaction?

**15-33.** Db-260 plus four neutrons is prepared by bombarding Cf-249 with what isotope?

**15-34.** Hs-265 plus one neutron is prepared by bombarding an isotope of lead with iron-58. What is the isotope of lead?

**15-35.** Bismuth-209 can be bombarded with chromium-54, producing one neutron and what heavy element isotope?

## Fission and Fusion

**15-36.** When $^{235}_{92}U$ and $^{239}_{94}Pu$ undergo fission, a variety of reactions take place. Complete the following.
(a) $^{235}_{92}U + ^{1}_{0}n \longrightarrow$ _____ $+ ^{146}_{58}Ce + 3^{1}_{0}n$
(b) $^{239}_{94}Pu + ^{1}_{0}n \longrightarrow ^{141}_{56}Ba +$ _____ $+ 2^{1}_{0}n$

**15-37.** What is the difference between fission and fusion? What is the source of energy for these two processes?

**\*15-38.** In 1989, some scientists reported that they had achieved "cold fusion." This was supposedly accomplished by electrolyzing heavy water (deuterium oxide) with palladium electrodes. It was suggested that the deuterium was absorbed into the electrodes and somehow two deuterium nuclei could overcome the strong repulsions at a low temperature and fuse. Evidence was presented to indicate that more energy came out of the reaction than could be accounted for by a chemical process. There has been little support for these initial experiments, but, for a while, they generated nightly reports on the national news. The fusion of two deuterium nuclei should produce helium-4, which, according to theory, would have too much energy to be stable and so would decompose to either helium-3 or tritium. What other two particles would be produced by this decomposition? Write the appropriate nuclear equations.

## General Problems

**15-39.** Write equations illustrating each of the following nuclear processes.
(a) Palladium-106 absorbs an alpha particle to produce an isotope and a proton.
(b) Mt-266 emits an alpha particle to form an isotope that also emits an alpha particle.
(c) Bismuth-212 emits a beta particle.
(d) An isotope emits a positron to form copper-60.
(e) Plutonium-239 absorbs a neutron to produce cesium-140, another isotope, and three neutrons.
(f) Lead-206 is bombarded by an isotope to produce Sg-257 and three neutrons.
(g) An isotope captures an electron to form niobium-93.

**15-40.** Write equations illustrating each of the following nuclear processes.
(a) An isotope of a heavy element is bombarded with boron-11 to form lawrencium-257 and four neutrons.
(b) An isotope emits an alpha particle to form actinium-231.
(c) Oxygen-14 emits a particle to form nitrogen-14.
(d) Sulfur-31 captures an electron to form an isotope.
(e) Uranium-235 absorbs a neutron to form rubidium-90, cesium-142, and several neutrons.
(f) Two helium nuclei fuse (in the sun) to form lithium-7 and a particle.
(g) An isotope emits a beta particle to form lead-208.

(h) Plutonium-239 is bombarded with an isotope to form curium-242 and a neutron.

**15-41.** The radioactive isotope $^{90}_{38}Sr$ can accumulate in bones, where it replaces calcium. It emits a high-energy beta particle, which eventually can cause cancer.
(a) What is the product of the decay of $^{90}_{38}Sr$?
(b) How long would it take for a 0.10-mg sample of $^{90}_{38}Sr$ to decay to where only $2.5 \times 10^{-2}$ mg was left? (The half-life of $^{90}_{38}Sr$ is 25 years.)

**15-42.** The radioactive isotope $^{131}_{53}I$ accumulates in the thyroid gland. On the one hand, this can be useful in detecting diseases of the thyroid and even in treating cancer at that location. On the other hand, exposure to excessive amounts of this isotope, such as from a nuclear power plant, can cause cancer of the thyroid. $^{131}_{53}I$ emits a beta particle with a half-life of 8.0 days. What is the product of the decay of $^{131}_{53}I$? If one started with $8.0 \times 10^{-6}$ g of $^{131}_{53}I$, how much would be left after 32 days?

**15-43.** The fissionable isotope $^{239}_{94}Pu$ is made from the abundant isotope of uranium, $^{238}_{92}U$, in nuclear reactors. When $^{238}_{92}U$ absorbs a neutron from the fission process, $^{239}_{94}Pu$ eventually forms. This is the principle of the breeder reactor, although $^{239}_{94}Pu$ is formed in all reactors. Complete the following reaction.

$$^{238}_{92}U + ^{1}_{0}n \longrightarrow \underline{\hspace{1cm}} + ^{0}_{-1}e$$
$$\longrightarrow ^{239}_{94}Pu + \underline{\hspace{1cm}}$$

**15-44.** An isotope of hydrogen, known as tritium ($^{3}H$), has a half-life of 12 years. If a sample of tritium was prepared 60 years ago, what was its original mass if its current mass is 0.42 μg?

**15-45.** A particular isotope has a half-life of 10.0 days. What percent of the original sample is left after 30.0 days?

**15-46.** A new element was created in 1994. What is the atomic number and mass number of the isotope if the nuclear reaction involved bombardment of $^{209}Bi$ with $^{64}Ni$ to form the new element plus one neutron?

**15-47.** The isotope $^{269}110$ was also created in 1994. The nuclear reaction involved bombardment of an isotope with $^{62}Ni$ and produced the new element and one neutron. What is the isotope involved in the reaction?

**15-48.** A team in Germany reported the formation of one atom of $^{277}112$ in 1996 by bombarding $^{208}Pb$ with a lighter nucleus. If one neutron is also produced, what is the isotopic notation for the lighter nucleus?

**15-49.** A team in Germany recently reported element $^{282}114$, which, according to theory, should be more stable than other superheavy elements. The reaction involves bombarding $^{208}Pb$ with a lighter isotope. If two neutrons are also produced, what is the isotopic notation of the lighter nucleus?

# Solutions to Learning Checks

**A-1.** radiation, helium, electron, light, positive electron, positron, half-life, alpha particles, gamma rays, film, Geiger, scintillation

**A-2.** (a) $^{87}_{37}\text{Rb}$ (b) $^{227}_{89}\text{Ac}$ (c) $^{54}_{27}\text{Co}$ (d) $^{123}_{52}\text{Te}$ (e) $^{137m}_{55}\text{Cs}$

**A-3.** After four half-lives, 4.0 μg of the original sample is left. $t_{1/2} = 36/4 = 9.0$ min (i.e., after 9.0 min, 32 μg remains; after 18 min, 16 μg; after 27 min, 8.0 μg; and after 36 min, 4.0 μg.)

**B-1.** transmutation, accelerators, 14, neutron, gamma rays, positrons, chain, fusion

**B-2.** (a) $^{52}_{23}\text{V}$ (b) $^{144}_{54}\text{Xe}$ (c) $^{12}_{6}\text{C}$ (d) $^{1}_{1}\text{H}$

**B-3.** Reaction (b) is fission and reaction (d) is fusion.

**Chapters 16 and 17 can be located on the Wiley website: www.wiley.com/college/malone**

# Foreword to the Appendixes

The successful athlete must be "in shape" physically. The successful chemistry student must be "in shape" mathematically.

Why do some (one or two, anyway) students seem so self-assured in the study of chemistry and yet others (all the rest) seem so worried? Most likely, it has a lot to do with preparation. Preparation in this case probably does not mean a prior course in chemistry, but it does mean having a solid mathematical background. Most of the students using this text probably are a little rusty on at least some aspects of basic arithmetic, algebra, and scientific notation. There are several reasons for this. Some have not had a good secondary school background in math courses, and others have been away from their high school or college math courses for a number of years. It makes no difference—most students need access to a few reminders, hints, and review exercises to get in shape. The sooner students admit that they have forgotten some math, the faster they do something about it and start to enjoy chemistry. It is really difficult to appreciate the study of this science if math deficiencies get in the way.

**Appendix A** reviews some of the basic arithmetic concepts such as manipulation of fractions, expressing decimal fractions, and, very importantly, the expression and use of percent. **Appendix B** reviews the manipulation and solution of simple algebra equations, which are so important in the quantitative aspects of chemistry. **Appendix C** supplements the discussion of scientific notation in Chapter 1 with more examples and exercises. Also included in this appendix is a discussion of the concept of logarithms, which is a convenient way to express exponential numbers in certain situations. **Appendix D** is a brief discussion of graphing, an important tool of the social sciences as well as the natural sciences. **Appendix E** contains a discussion of the function of the more common type of calculator used by students in chemistry classes. **Appendix F** is a glossary of terms, and **Appendix G** contains answers to more than half of the exercises at the ends of the chapters.

# Appendix A
# Basic Mathematics

The following is a quick (very quick) refresher of fundamentals of math. This may be sufficient to aid you if you are just a little rusty on some of the basic concepts. For more thorough explanations and practice, however, you are urged to use a more comprehensive math review workbook or consult with your instructor.

One may ask "Why not just use a calculator?" The answer is that serious science students need a "feeling" for the numbers they use. This can only be accomplished by understanding the calculations involved. Therefore, it is well worth the time to go through this appendix *without a calculator*. Being able to do these calculations on your own will certainly pay off.

## A-1 Addition and Subtraction

Since most calculations in this text use numbers expressed in decimal form, we will emphasize the manipulation of this type of number. In addition and subtraction, it is important to line up the decimal point carefully before doing the math.

Subtraction is simply the addition of a negative number. Remember that subtraction of a negative number changes the sign to a plus (two negatives make a positive). For example, $4 - 7 = -3$, but $4 - (-7) = 4 + 7 = 11$.

| Example A-1 | **Addition and Subtraction** |
|---|---|
| working it out | Carry out the following calculations. |

**a.** $16.75 + 13.31 + 175.67$

$$\begin{array}{r} \downarrow \\ 16.75 \\ 13.31 \\ \underline{175.67} \\ \underline{205.73} \end{array}$$

**b.** $11.8 + 13.1 - 6.1$

$$\begin{array}{r} 11.8 \\ \underline{+13.1} \\ 24.9 \end{array} \quad \nearrow \quad \begin{array}{r} 24.9 \\ \underline{-6.1} \\ \underline{18.8} \end{array}$$

**c.** $47.82 - 111.18 - (-12.17)$
This is the same as $47.82 - 111.18 + 12.17$.

$$
\begin{array}{r}
47.82 \\
+12.17 \\
\hline
59.99
\end{array}
\qquad \longrightarrow \qquad
\begin{array}{r}
-111.18 \\
+59.99 \\
\hline
-51.19
\end{array}
$$

## Exercises

**A-1.** Carry out the following calculations.

(a) $47 + 1672$  (e) $0.897 + 1.310 - 0.063$
(b) $11.15 + 190.25$  (f) $-0.377 - (-0.101) + 0.975$
(c) $114 + 26 - 37$  (g) $17.489 - 318.112 - (-0.315) + (-3.330)$
(d) $-97 + 16 - 118$

*Answers:* (a) 1719  (b) 201.40  (c) 103  (d) $-199$  (e) 2.144  (f) 0.699
(g) $-303.638$

## A-2 Multiplication

Multiplication is expressed in various ways as follows:

$$13.7 \times 115.35 = 13.7 \cdot 115.35 = (13.7)(115.35) = 13.7(115.35)$$

If it is necessary to carry out the multiplication in longhand, you must be careful to place the decimal point correctly in the answer. Count the *total* number of digits to the right of the decimal point in both multipliers (three in this example). The answer has that number of digits to the right of the decimal point in the answer. Finally, round off to the proper number of significant figures.

$$
\begin{array}{l}
13.7 \times \underbrace{2.15}_{} = \\
\;\;1 \quad + 2 = \;③ \\
\text{(decimal places)}
\end{array}
\qquad
\begin{array}{r}
13.7 \\
\times\; 2.15 \\
\hline
685 \\
137 \\
274 \\
\hline
29455 \\
③
\end{array}
\; = 29.455 = 29.5*
$$

When a number (called a *base*) is multiplied by itself one or more times, it is said to be raised to a *power*. The power (called the *exponent*) indicates the number of bases multiplied. For example, the exact values of the following numbers raised to a power are

$$
\begin{aligned}
4^2 &= 4 \times 4 = 16 && \text{(``four squared'')} \\
2^4 &= 2 \times 2 \times 2 \times 2 = 16 && \text{(``two to the fourth power'')} \\
4^3 &= 4 \times 4 \times 4 = 64 && \text{(``four cubed'')} \\
(14.1)^2 &= 14.1 \times 14.1 = 198.81 = 199*
\end{aligned}
$$

In the calculations used in this book, most numbers have specific units. In multiplication, the units as well as the numbers are multiplied. For example,

$$3.7 \text{ cm} \times 4.61 \text{ cm} = 17 \,(\text{cm} \times \text{cm}) = 17 \text{ cm}^2$$

$$(4.5 \text{ in.})^3 = 91 \text{ in.}^3$$

---

*Rounded off to three significant figures. See section 1-2.

In the multiplication of a series of numbers, grouping is possible.

$$(a \times b) \times c = a \times (b \times c)$$
$$3.0 \text{ cm} \times 148 \text{ cm} \times 3.0 \text{ cm} = (3.0 \times 3.0) \times 148 \times (\text{cm} \times \text{cm} \times \text{cm})$$
$$= \underline{\underline{1300 \text{ cm}^3}}$$

When multiplying signs, remember:

$$(+) \times (-) = - \qquad (+) \times (+) = + \qquad (-) \times (-) = +$$

For example, $(-3) \times 2 = -6$; $(-9) \times (-8) = +72$.

## Exercises

**A-2.** Carry out the following calculations. For (a) through (d) carry out the multiplications completely. For (e) through (h) round off the answer to the proper number of significant figures and include units.

**(a)** $16.2 \times (-118)$        **(d)** $(-47.8) \times (-9.6)$
**(b)** $(4 \times 2) \times 879$        **(e)** $3.0 \text{ ft} \times 18 \text{ lb}$
**(c)** $(-8) \times (-2) \times (-37)$    **(f)** $17.7 \text{ in.} \times (13.2 \text{ in.} \times 25.0 \text{ in.})$
**(g)** What is the area of a circle where the radius is 2.2 cm? (Area $= \pi r^2$, $\pi = 3.14$.)
**(h)** What is the volume of a cylinder 5.0 in. high with a cross-sectional radius of 0.82 in.? (Volume $=$ area of cross section $\times$ height.)

*Answers:* **(a)** $-1911.6$   **(b)** 7032   **(c)** $-592$   **(d)** 458.88   **(e)** 54 ft · lb
**(f)** 5840 in.$^3$   **(g)** 15 cm$^2$   **(h)** 11 in.$^3$

## A-3 Roots of Numbers

A root of a number is a fractional exponent. It is expressed as

$$\sqrt[x]{a} = a^{1/x}$$

If $x$ is not shown (on the left), it is assumed to be 2 and is known as the *square root*. The square root is the number that when multiplied by itself gives the base $a$. For example,

$$\sqrt{4} = 2 \qquad (2 \times 2 = 4)$$
$$\sqrt{9} = 3 \qquad (3 \times 3 = 9)$$

The square root of a number may have either a positive or a negative sign. Generally, however, we are interested only in the positive root in chemistry calculations.

If the square root of a number is not a whole number, it may be computed on a calculator or found in a table. Without these tools available, an educated approximation can come close to the answer. For example, the square root of 54 lies between 7 ($7^2 = 49$) and 8 ($8^2 = 64$) but closer to 7. An educated guess of 7.3 would be excellent.

The cube root of a number is expressed as

$$\sqrt[3]{b} = b^{1/3}$$

It is the number multiplied by itself two times that gives $b$. For example,

$$\sqrt[3]{27} = 3.0 \qquad (3 \times 3 \times 3 = 27)$$
$$\sqrt[3]{64} = 4.0 \qquad (4 \times 4 \times 4 = 64)$$

A hand calculator (see Appendix E) is the most convenient source of roots of numbers.

# Exercises

**A-3.** Find the following roots. If necessary, first approximate the answer, then check with a calculator.

**(a)** $\sqrt{25}$   **(b)** $\sqrt{36 \text{ cm}^2}$   **(c)** $\sqrt{144 \text{ ft}^4}$   **(d)** $\sqrt{40}$

**(e)** $\sqrt{7.0}$   **(f)** $110^{1/2}$   **(g)** $100^{1/3}$   **(h)** $\sqrt[3]{50}$

**(i)** What is the radius of a circle that has an area of 150 ft$^2$? (Area $= \pi r^2$)

**(j)** What is the radius of the cross section of a cylinder that has a volume of 320 m$^3$ and a height of 6.0 m? (Volume $= \pi r^2 \times$ height)

*Answers:*   **(a)** 5.0   **(b)** 6.0 cm   **(c)** 12.0 ft$^2$   **(d)** 6.3   **(e)** 2.6   **(f)** 10.5   **(g)** 4.64   **(h)** 3.7   **(i)** 6.91 ft   **(j)** 4.1 m

## A-4 Division, Fractions, and Decimal Numbers

Common fractions express ratios or portions of numbers. A *proper fraction* is less than one so it has a larger denominator than numerator. An *improper fraction* is greater than one so it has a smaller denominator than numerator. In chemistry most fractions are expressed as decimal numbers which show proper fractions less than 1.0 and improper fractions greater than 1.0. To convert a common fraction to a decimal number, the numerator is divided by the denominator. Consider the two fractions shown below where the decimal is shown to three digits.

| proper fraction | improper fraction |
|---|---|
| $7/8 = 7 \div 8 = 0.675$ | $11/9 = 11 \div 9 = 1.22$ |
| common            decimal | common            decimal |
| fraction          fraction | fraction          fraction |

Heaven forbid, but someday we may find ourselves in a test without a calculator. In that case, we may be well served by a review of long division. For example, consider the ratio 88.8/2.44. It isn't necessary, but perhaps it is easier to remove the decimals in this ratio so that we are dealing only with whole numbers. As shown below, by multiplying both numerator and denominator by 100 we can move the decimal. Multiplying both numerator and denominator by the same number does not change the value of the fraction.

$$\frac{a}{b} = \frac{a \times c}{b \times c}$$

$$\frac{88.8}{2.44} = \frac{88.8 \times 100}{2.44 \times 100} = \frac{8880}{244} = \underline{36.4}$$

$$2.44\overline{)88.800} = 244\overline{)8880.00} = 36.4 \text{ (three significant figures)}$$

$$
\begin{array}{r}
36.39 \\
\underline{732} \\
1560 \\
\underline{1464} \\
960 \\
\underline{732} \\
2280 \\
2196
\end{array}
$$

Many divisions can be simplified by cancellation, which is the elimination of common factors in the numerator and denominator. This is possible because a number divided by itself is equal to unity (e.g., $25/25 = 1$). As in multiplication, all units also must be divided. If identical units appear in both numerator and denominator, they also can be canceled.

$$\frac{a \times c}{b \times c} = \frac{a}{b}$$

$$\frac{\overset{1}{\cancel{190}} \times 4 \text{ torr}}{\underset{1}{\cancel{190}} \text{ torr}} = \frac{4}{-}$$

$$\frac{2500 \text{ cm}^3}{150 \text{ cm}} = \frac{\overset{1}{\cancel{50}} \times 50 \text{ cm} \times \text{cm} \times \text{cm}}{\underset{1}{\cancel{50}} \times 3 \text{ cm}} = \frac{50 \text{ cm}^2}{3} = 17 \text{ cm}^2$$

$$\frac{2800 \text{ mi}}{45 \text{ hr}} = \frac{\overset{1}{\cancel{5}} \times 560 \text{ mi}}{\cancel{5} \times 9 \text{ hr}} = \frac{62 \text{ mi}}{1 \text{ hr}} = \underline{62 \text{ mi/hr}}$$

This is read as 62 miles "per" one hour or simply 62 miles per hour. The word "per" implies a fraction or a ratio with the unit after "per" in the denominator. If a number is not written or read in the denominator with a unit, it is assumed that the number is unity and is known to as many significant figures as the number in the numerator (i.e., 62 miles per 1.0 hr).

## Exercises

**A-4.** Express the following in decimal form. Express (a)–(c) to three digits.

(a) 3/7     (d) 892 mi ÷ 41 hr     (g) $\dfrac{67.5 \text{ g}}{15.2 \text{ mL}}$     (j) $\dfrac{0.8772 \text{ ft}^3}{0.0023 \text{ ft}^2}$

(b) 14/19     (e) 982.6 ÷ 0.250     (h) $\dfrac{1890 \text{ cm}^3}{66 \text{ cm}}$     (k) $\dfrac{37.50 \text{ ft}}{0.455 \text{ sec}}$

(c) 19/14     (f) 195 ÷ 2650     (i) $\dfrac{146 \text{ ft} \cdot \text{hr}}{0.68 \text{ ft}}$

*Answers:*   (a) 0.429   (b) 0.737   (c) 1.36   (d) 22 mi/hr   (e) 3930   (f) 0.0736
(g) 4.44 g/mL   (h) 29 cm²   (i) 210 hr   (j) 380 ft   (k) 82.4 ft/sec

## A-5 Multiplication and Division of Fractions

When two or more fractions are multiplied, all numbers *and units* in both numerator and denominator can be combined into one fraction.

The division of one fraction by another is the same as the multiplication of the numerator by the *reciprocal* of the denominator. The reciprocal of a fraction is simply the fraction in an inverted form (e.g., $\frac{3}{5}$ is the reciprocal of $\frac{5}{3}$).

$$\frac{a}{b/c} = a \times \frac{c}{b} \qquad \frac{a/b}{c/d} = \frac{a}{b} \times \frac{d}{c} = \frac{a \times d}{b \times c}$$

---

**Example A-2**

**working it out**

**Multiplication and Division**

Carry out the following calculations. Round off the answer to two digits.

**a.** $\dfrac{3}{5} \times \dfrac{75}{4} \times \dfrac{16}{7} = \dfrac{3 \times 75 \times 16}{5 \times 4 \times 7} = \dfrac{3 \times \overset{15}{\cancel{75}} \times \overset{4}{\cancel{16}}}{\underset{1}{\cancel{5}} \times \underset{1}{\cancel{4}} \times 7} = \dfrac{180}{7} = \underline{\underline{26}}$

**b.** $\dfrac{42 \text{ mi}}{\text{hr}} \times \dfrac{3}{7}\text{hr} \times \dfrac{5280 \text{ ft}}{\text{mi}} = \dfrac{\overset{6}{\cancel{42}} \times 3 \times 5280 \text{ mi} \times \text{hr} \times \text{ft}}{\underset{1}{\cancel{7}} \text{ hr} \times \text{mi}} = \underline{\underline{95{,}000 \text{ ft}}}$

**c.** $\dfrac{3}{4}\text{mol} \times \dfrac{0.75 \text{ g}}{\text{mol}} \times \dfrac{1 \text{ mL}}{19.3 \text{ g}} = \dfrac{3 \times 0.75 \times 1 \times \cancel{\text{mol}} \times \cancel{\text{g}} \times \text{mL}}{4 \times 1 \times 19.3 \times \cancel{\text{mol}} \times \cancel{\text{g}}} = \underline{\underline{0.029 \text{ mL}}}$

**d.** $\dfrac{1650}{3/5} = 1650 \times \dfrac{5}{3} = \underline{\underline{2800}}$

**e.** $\dfrac{145 \text{ g}}{7.5 \text{ g/mL}} = 145 \cancel{\text{ g}} \times \dfrac{1 \text{ mL}}{7.5 \cancel{\text{ g}}} = \underline{\underline{19 \text{ mL}}}$

# Exercises

**A-5.** Express the following answers in decimal form. If units are not used, round off the answer to three digits. If units are included, round off to the proper number of significant figures and include units in the answer.

**(a)** $\frac{3}{8} \times \frac{4}{7} \times \frac{21}{20}$

**(b)** $\frac{250}{273} \times \frac{175}{300} \times (-6)$

**(c)** $\frac{4}{9} \times \left(-\frac{5}{8}\right) \times \left(-\frac{3}{4}\right)$

**(d)** 195 g/mL $\times$ 47.5 mL

**(e)** 0.75 mol $\times$ 17.3 g/mol

**(f)** $(3.57 \text{ in.})^2 \times 0.85 \text{ in.} \times \dfrac{16.4 \text{ cm}^3}{\text{in.}^3}$

**(g)** $\dfrac{\frac{150}{350}}{\frac{25}{42}}$

**(h)** $\dfrac{\left(-\frac{3}{7}\right)}{\left(-\frac{4}{9}\right)}$

**(i)** $\dfrac{\left(-\frac{17}{3}\right)}{\frac{8}{9}}$

**(j)** $\dfrac{\frac{16}{9} \times \frac{10}{14}}{\frac{5}{6}}$

**(k)** $\dfrac{75.2 \text{ torr}}{760 \text{ torr/atm}}$

**(l)** $\dfrac{(55.0 \text{ mi/hr}) \times (5280 \text{ ft/mi}) \times (1 \text{ hr/60 min})}{60 \text{ sec/min}}$

**(m)** $\dfrac{305 \text{ K} \times 62.4 \dfrac{\text{L} \cdot \text{torr}}{\text{K} \cdot \text{mol}} \times 0.25 \text{ mol}}{650 \text{ torr}}$

*Answers:* **(a)** 0.225 **(b)** −3.21 **(c)** 0.208 **(d)** 9260 g **(e)** 13 g **(f)** 180 cm³
**(g)** 0.720 **(h)** 0.964 **(i)** −6.38 **(j)** 1.52 **(k)** 0.0989 atm **(l)** 80.7 ft/sec
**(m)** 7.3 L

## A-6 Decimal Numbers and Percent

In the examples of fractions thus far, we have seen that the units of the numerator can be profoundly different from those of the denominator (e.g., miles/hr, g/mL, etc.). In other problems in chemistry we use fractions to express a component part in the numerator to the total in the denominator. In most cases such fractions are expressed without units and in decimal form.

**Decimal Numbers**

**a.** A box of nails contains 985 nails; 415 of these are 6-in. nails, 375 are 3-in. nails, and the rest are roofing nails. What is the fraction of roofing nails in decimal form?

**Example A-3**

*working it out*

**Solution**

$$\text{Roofing nails} = \text{total} - \text{others} = 985 - (415 + 375) = 195$$

$$\frac{\text{component}}{\text{total}} = \frac{195}{375 + 415 + 195} = \underline{\underline{0.198}}$$

**b.** A mixture contains 4.25 mol of $N_2$, 2.76 mol of $O_2$, and 1.75 mol of $CO_2$. What is the fraction of moles of $O_2$ present in the mixture? (This fraction is known, not surprisingly, as "the mole fraction." The mole is a unit of quantity, like dozen.)

**Solution**

$$\frac{\text{component}}{\text{total}} = \frac{2.76}{4.25 + 2.76 + 1.75} = \underline{\underline{0.315}}$$

# Exercises

**A-6.** A grocer has 195 dozen boxes of fruit; 74 dozen boxes are apples, 62 dozen boxes are peaches, and the rest are oranges. What is the fraction of the boxes that are oranges?

**A-7.** A mixture contains 9.85 mol of gas. A 3.18-mol quantity of the gas is $N_2$, 4.69 mol is $O_2$, and the rest is He. What is the mole fraction of He in the mixture?

**A-8.** The total pressure of a mixture of two gases, $N_2$ and $O_2$, is 0.72 atm. The pressure due to $O_2$ is 0.41 atm. What is the fraction of the pressure due to $O_2$?

*Answers:* **A-6** 0.30   **A-7** 0.201   **A-8** 0.57

The decimal numbers that have just been discussed are frequently expressed as percentages. Percent simply means parts per 100. Percent is obtained by multiplying a fraction in decimal form by 100%.

| Example A-4 | **Expressing Percent** |
|---|---|
| working it out | If 57 out of 180 people at a party are women, what is the percent women? |

**Solution**

The fraction of women in decimal form is

$$\frac{57}{180} = 0.317$$

The percent women is

$$0.317 \times 100\% = \underline{\underline{31.7\% \text{ women}}}$$

The general method used to obtain percent is

$$\frac{\text{component}}{\text{total}} \times 100\% = \underline{\hspace{1cm}}\% \text{ of component}$$

To change from percent back to a decimal number, divide the percent by 100%, which moves the decimal to the left two places.

$$86.2\% = \frac{86.2\%}{100\%} = 0.862 \text{ (fraction in decimal form)}$$

# Exercises

**A-9.** Express the following fractions or decimal numbers as percents: $\frac{1}{4}, \frac{3}{8}, \frac{9}{8}, \frac{55}{25}$, 0.67, 0.13, 1.75, 0.098.

**A-10.** A bushel holds 198 apples, 27 of which are green. What is the percent of green apples?

**A-11.** A basket contains 75 pears, 8 apples, 15 oranges, and 51 grapefruit. What is the percent of each?

*Answers:* **A-9** $\frac{1}{4} = 25\%, \frac{3}{8} = 37.5\%, \frac{9}{8} = 112.5\%, \frac{55}{25} = 220\%, 0.67 = 67\%$, $0.13 = 13\%, 1.75 = 175\%, 0.098 = 9.8\%$ **A-10** 13.6% **A-11** 50.3% pears, 5.4% apples, 10.1% oranges, 34.2% grapefruit

We have seen how the percent is calculated from the total and the component part. We now consider problems where percent is given and we calculate either the component part as in Example A-5(a) or the total as in Example A-5(b). Such problems can be solved in two ways. The method we employ here uses the percent as a conversion factor, and the problems are solved by the factor-label method. (See section 1-5.) They also can be solved algebraically as is done in Appendix B.

---

**Using Percent in Calculations**

**Example A-5**

*working it out*

**a.** A crowd at a rock concert was composed of about 87% teenagers. If the crowd totaled 586 people, how many were teenagers?

**Procedure**

Remember that percent means "per 100." In this case it means 87 teenagers per 100 people or, in fraction form,

$$\frac{87 \text{ teenagers}}{100 \text{ people}}$$

If this fraction is then multiplied by the number of people, the result is the component part or the number of teenagers.

$$586 \text{ people} \times \frac{87 \text{ teenagers}}{100 \text{ people}} = (586 \times 0.87) = \underline{\underline{510 \text{ teenagers}}}$$

**b.** A professional baseball player got a hit 28.7% of the times he batted. If he got 246 hits, how many times did he bat?

**Procedure**

The percent can be written in fraction form and then inverted. It thus relates the total at bats to the number of hits:

$$28.7\% = \frac{28.7 \text{ hits}}{100 \text{ at bats}} \quad \text{or} \quad \frac{100 \text{ at bats}}{28.7 \text{ hits}}$$

If this is now multiplied by the number of hits, the result is the total number of at bats.

$$246 \text{ hits} \times \frac{100 \text{ at bats}}{28.7 \text{ hits}} = \frac{246}{0.287} = \underline{\underline{857 \text{ at bats}}}$$

# Exercises

**A-12.** In a certain audience, 45.9% were men. If there were 196 people in the audience, how many women were present?

**A-13.** In the alcohol molecule, 34.8% of the mass is due to oxygen. What is the mass of oxygen in 497 g of alcohol?

**A-14.** The cost of a hamburger in 2003 is 216% of the cost in 1970. If hamburgers cost $0.75 each in 1970, what do they cost in 2003?

**A-15.** In a certain audience, 46.0% are men. If there are 195 men in the audience, how large is the audience?

**A-16.** If a solution is 23.3% by mass HCl and it contains 14.8 g of HCl, what is the total mass of the solution?

**A-17.** An unstable isotope has a mass of 131 amu. This is 104% of the mass of a stable isotope. What is the mass of the stable isotope?

*Answers:*   **A-12** 106 women   **A-13** 173 g   **A-14** $1.62   **A-15** 424 people
**A-16** 63.5 g   **A-17** 126 amu

# Appendix B
# Basic Algebra

There are two aspects to the use of algebra that affect chemistry students. First is the actual skill and application of basic concepts and second is the ability to translate words or quantitative concepts into a proper algebraic relationship. In the first section, we will concentrate on the algebra equation itself and how it can be manipulated. In the two sections following this we will concentrate on how we can express quantitative concepts as equations.

## B-1 Operations of Basic Equations

Many of the quantitative problems of chemistry require the use of basic algebra. As an example of a simple algebra equation we use

$$x = y + 8$$

In any algebraic equation the equality remains valid when identical operations are performed on both sides of the equation. The following operations illustrate this principle.

1.  A quantity may be added to or subtracted from both sides of the equation.

    (add 8)　　$x + 8 = y + 8 + 8$　　$x + 8 = y + 16$

    (subtract 8)　$x - 8 = y + 8 - 8$　　$x - 8 = y$

2.  Both sides of the equation may be multiplied or divided by the same quantity.

    (multiply by 4)　$4x = 4(y + 8) = 4y + 32$

    (divide by 2)　$\dfrac{x}{2} = \dfrac{(y + 8)}{2}$　　$\dfrac{x}{2} = \dfrac{y}{2} + 4$

3.  Both sides of the equation may be raised to a power, or a root of both sides of an equation may be taken.

    (equation squared)　$x^2 = (y + 8)^2$

    (square root taken)　$\sqrt{x} = \sqrt{y + 8}$

4.  Both sides of an equation may be inverted.

    $$\frac{1}{x} = \frac{1}{y + 8}$$

In addition to operation on both sides of an equation, two other points must be recalled.

1. As in any fraction, identical factors in the numerator and the denominator in an algebraic equation may be canceled.

$$\frac{\cancel{4}x}{\cancel{4}} = x = y + 8 \qquad \text{or} \qquad x = \frac{\cancel{z}(y + 8)}{\cancel{z}} = y + 8$$

2. Quantities equal to the same quantity are equal to each other. Thus substitutions for equalities may be made in algebraic equations.

$$x = y + 8$$
$$x = 27$$

Therefore, since $x = x$,

$$y + 8 = 27$$

We can use these basic rules to solve algebraic equations. Usually, we need to isolate one variable on the left-hand side of the equation with all other numbers and variables on the right-hand side of the equations. The operations previously listed can be simplified for this purpose in two ways.

In practice, a number or a variable may be moved to the other side of an equation with a change of sign. For example, if

$$x + z = y$$

then subtracting $z$ from both sides, in effect, gives us

$$x = y - z$$

Also, the numerator of a fraction on the left becomes the denominator on the right. The denominator of a fraction on the left becomes the numerator on the right. For example, consider the following two cases.

If $xz = y$                                    If $\dfrac{x}{k + 5} = B,$

then by dividing both                         then multiplying both
sides by $z$ gives us, in effect,             sides by $k + 5$
gives us

$$x\cancel{z} = y$$                            $$\frac{x}{\cancel{k + 5}} = B$$

$$x = \frac{y}{z}$$                            $$x = B(k + 5)$$

The following examples illustrate the isolation of one variable ($x$) on the left-hand side of the equation.

---

| Example B-1 | **Solving Algebra Equations for a Variable** |
|---|---|
| **working it out** | **a.** Solve for $x$ in $x + y + 8 = z + 6$. |

**Solution**

Move $+y$ and $+8$ to the right by changing signs.

$$x = z + 6 - y - 8$$
$$= z - y + 6 - 8 = \underline{\underline{z - y - 2}}$$

**b.** Solve for $x$ in

$$\frac{x + 8}{y} = z$$

**Solution**

First, move $y$ to the right by multiplying both sides by $y$.

$$\cancel{y} \cdot \frac{x + 8}{\cancel{y}} = z \cdot y$$

This leaves

$$x + 8 = zy$$

Subtract 8 from both sides to obtain the final answer.

$$\underline{\underline{x = zy - 8}}$$

**c.** Solve for $x$ in

$$\frac{4x + 2}{3 + x} = 7$$

**Solution**

First, multiply both sides by $(3 + x)$ to clear the fraction.

$$\cancel{(3 + x)} \cdot \frac{4x + 2}{\cancel{(3 + x)}} = 7(3 + x)$$

This leaves

$$4x + 2 = 21 + 7x$$

To move integers to the right and the $x$ variable to the left, subtract $7x$ and 2 from both sides of the equation. This leaves

$$-3x = 19$$

Finally, divide both sides by $-3$ to move the $-3$ to the right.

$$x = -\frac{19}{3} = \underline{\underline{-6.33}}$$

**d.** Solve for $T_2$ in

$$\frac{P_1 V_1}{T_1} = \frac{P_2 V_2}{T_2}$$

**Solution**

To move $T_2$ to the left, multiply both sides by $T_2$.

$$T_2 \cdot \frac{P_1 V_1}{T_1} = \cancel{T_2} \cdot \frac{P_2 V_2}{\cancel{T_2}} = P_2 V_2$$

Move $P_1 V_1$ to the right by dividing by $P_1 V_1$.

$$\frac{T_2}{\cancel{P_1 V_1}} \cdot \frac{\cancel{P_1 V_1}}{T_1} = \frac{P_2 V_2}{P_1 V_1}$$

Finally, move $T_1$ to the right by multiplying both sides by $T_1$.

$$T_2 = \underline{\underline{\frac{T_1 P_2 V_2}{P_1 V_1}}}$$

**e.** Solve the following equation for $y$.

$$\frac{2y}{3} + x = 9z + 4$$

**Solution**

First, to clear the fraction, multiply both sides by 3. This leaves

$$2y + 3x = 27z + 12$$

Subtract $3x$ from both sides, which leaves

$$2y = 27z + 12 - 3x$$

Finally, divide both sides by 2, which leaves

$$y = \frac{27z + 12 - 3x}{2}$$

# Exercises

**B-1.** Solve for $x$ in $17x = y - 87$.

**B-2.** Solve for $x$ in

$$\frac{y}{x} + 8 = z + 16$$

**B-3.** Solve for $T$ in $PV = (\text{mass}/MM)RT$.

**B-4.** Solve for $x$ in

$$\frac{7x - 3}{6 + 2x} = 3r$$

**B-5.** Solve for $x$ in $18x - 27 = 2x + 4y - 35$. If $y = 3x$, what is the value of $x$?

**B-6.** Solve for $x$ in

$$\frac{x}{4y} + 18 = y + 2$$

**B-7.** Solve for $x$ in $5x^2 + 12 = x^2 + 37$.

**B-8.** Solve for $r$ in

$$\frac{80}{2r} + \frac{y}{r} = 11$$

What is the value of $r$ if $y = 14$?

*Answers:* **B-1** $x = (y - 87)/17$   **B-2** $x = y/(8 + z)$   **B-3** $T = PV \cdot MM/$ mass $\cdot R$   **B-4** $x = 3(6r + 1)/(7 - 6r)$   **B-5** $x = (y - 2)/4$. When $y = 3x$, $x = -2$.   **B-6** $x = 4y(y - 16)$   **B-7** $x = \pm 2.5$   **B-8** $r = (40 + y)/11$. When $y = 14$, $r = \frac{54}{11}$

## B-2 Word Problems and Algebra Equations

Eventually, a necessary skill in chemistry is the ability to translate word problems into algebra equations and then solve. The key is to assign a variable (usually $x$) to be equal to a certain quantity and then to treat the variable consistently throughout the equation. Again, examples are the best way to illustrate the problems.

---

| | |
|---|---|
| **Solving Abstract Word Equations** | **Example B-2** |
| Translate each of the following to an equation. | working it out |

**a.** A number $x$ is equal to a number that is 4 larger than $y$.

$$x = y + 4$$

**b.** A number $z$ is equal to three-fourths of $u$.

$$z = \tfrac{3}{4}u$$

**c.** The square of a number $r$ is equal to 16.9% of the value of $w$.

$$r^2 = 0.169w \quad \text{(change percent to a decimal number)}$$

**d.** A number $t$ is equal to 12 plus the square root of $q$.

$$t = 12 + \sqrt{q}$$

---

# Exercises

**B-9.** Write algebraic equations for the following:
**(a)** A number $n$ is equal to a number that is 85 smaller than $m$.
**(b)** A number $y$ is equal to one-fourth of $z$.
**(c)** Fifteen percent of a number $k$ is equal to the square of another number $d$.
**(d)** A number $x$ is equal to 14 more than the square root of $v$.
**(e)** Four times the sum of two numbers, $q$ and $w$, is equal to 68.
**(f)** Five times the product of two variables, $s$ and $t$, is equal to 16 less than the square of $s$.
**(g)** Five-ninths of a number $C$ is equal to 32 less than a number $F$.

*Answers:* **(a)** $n = m - 85$   **(b)** $y = z/4$   **(c)** $0.15k = d^2$   **(d)** $x = \sqrt{v} + 14$
**(e)** $4(q + w) = 68$   **(f)** $5st = s^2 - 16$   **(g)** $\tfrac{5}{9}C = F - 32$

We now move from the abstract to the real. In the following examples it is necessary to translate the problem into an algebraic expression, as in the previous examples. There are two types of examples that we will use. The first you will certainly recognize, but the second type may be unfamiliar, especially if you have just begun the study of chemistry. However, it is *not* important that you understand the units of chemistry problems at this time. What *is* important is for you to notice that the problems are worked in the same manner regardless of the units.

---

| | |
|---|---|
| **Solving Concrete Word Equations** | **Example B-3** |
| **a.** John is 2 years more than twice as old as Mary. The sum of their ages is 86. How old is each? | working it out |

**Solution**

Let $x$ = age of Mary. Then $2x + 2$ = age of John.

$$x + (2x + 2) = 86$$
$$3x = 84$$
$$x = \underline{\underline{28}} \quad \text{(age of Mary)}$$
$$2(28) + 2 = \underline{\underline{58}} \quad \text{(age of John)}$$

**b.** One mole of $SF_6$ has a mass 30.0 g less than four times the mass of 1 mol of $CO_2$. The mass of 1 mol of $SF_6$ plus the mass of 1 mol of $CO_2$ is equal to 190 g. What is the mass of 1 mol of each?

**Solution**

Let $x$ = mass of 1 mol of $CO_2$. Then $4x - 30$ = mass of 1 mol of $SF_6$.

$$x + (4x - 30) = 190$$
$$x = \underline{\underline{44 \text{ g}}} \quad \text{(mass of 1 mol of } CO_2)$$
$$4(44) - 30 = \underline{\underline{146 \text{ g}}} \quad \text{(mass of 1 mol of } SF_6)$$

**c.** Two students took the same test, and their percent scores differed by 10%. If there were 200 points on the test and the total of their point scores was 260 points, what was each student's percent score?

**Procedure**

Set up an equation relating each person's percent scores to their total points (260).

Let $x$ = percent score of higher test.
Then $x - 10$ = percent score of lower test.

The points that each person scores is the percent in fraction form multiplied by the points on the test.

$$\frac{\% \text{ grade}}{100\%} \times (\text{points on test}) = \text{points scored}$$

**Solution**

$$\left[\frac{x}{100}(200 \text{ points})\right] + \left[\frac{x - 10}{100}(200 \text{ points})\right] = 260 \text{ points}$$
$$200x + 200x - 2000 = 26{,}000$$
$$400x = 28{,}000$$
$$x = 70$$

higher score = $\underline{\underline{70\%}}$    lower score = $70 - 10 = \underline{\underline{60\%}}$

**d.** If an 8.75-g quantity of sugar represents 65.7% of a mixture, what is the mass of the mixture?

**Solution**

Let $x$ = mass of the mixture. Then

$$\frac{65.7}{100}x = 0.657x = 8.75$$

$$x = \frac{8.75}{0.657} = \underline{\underline{13.3 \text{ g}}}$$

**e.** A used car dealer has Fords, Chevrolets, and Plymouths. There are 120 Fords, 152 Chevrolets, and the rest are Plymouths. If the fraction of Fords is 0.310, how many Plymouths are on the lot?

**Solution**

Let $x$ = number of Plymouths.

$$\text{fraction of Fords} = \frac{\text{number of Fords}}{\text{total number of cars}} = 0.310$$

$$\frac{120}{120 + 152 + x} = 0.310$$

$$120 = 0.310(272 + x)$$
$$120 = 84.3 + 0.310x$$
$$x = \underline{115 \text{ Plymouths}}$$

**f.** There is a 0.605-mol quantity of $N_2$ present in a mixture of $N_2$ and $O_2$. If the mole fraction of $N_2$ is 0.251, how many moles of $O_2$ are present?

**Solution**

Let $x$ = number of moles of $O_2$. Then

$$\text{mole fraction } N_2 = \frac{\text{mol of } N_2}{\text{total mol present}} = 0.251$$

$$\frac{0.605}{0.605 + x} = 0.251$$

$$0.605 = 0.251(0.605 + x)$$
$$x = \underline{1.80 \text{ mol}}$$

In the following exercises, a problem concerning an everyday situation is followed by one or more closely analogous problems concerning a chemistry situation. In both cases the mechanics of the solution are similar. Only the units differ.

# Exercises

**B-10.** The total length of two boards is 18.4 ft. If one board is 4.0 ft longer than the other, what is the length of each board?

**B-11.** An isotope of iodine has a mass 10 amu less than two-thirds the mass of an isotope of thallium. The total mass of the two isotopes is 340 amu. What is the mass of each isotope?

**B-12.** An isotope of gallium has a mass 22 amu more than one-fourth the mass of an isotope of osmium. The difference in the two masses is 122 amu. What is the mass of each?

**B-13.** An oil refinery held 175 barrels of oil. When refined, each barrel yields 24 gallons of gasoline. If 3120 gallons of gasoline were produced, what percentage of the original barrels of oil was refined?

**B-14.** A solution contained 0.856 mol of a substance $A_2X$. In solution some of the $A_2X$s break up into As and Xs. (Note that each mole of $A_2X$ yields 2 mol of A.) If 0.224 mol of A is present in the solution, what percentage of the moles of $A_2X$ dissociated (broke apart)?

**B-15.** In Las Vegas, a dealer starts with 264 decks of cards. If 42.8% of the decks were used in an evening, how many jacks (four per deck) were used?

**B-16.** A solution originally contains a 1.45-mol quantity of a compound $A_3X_2$. If 31.5% of the $A_3X_2$ dissociates (three As and two Xs per $A_3X_2$), how many moles of A are formed? How many moles of X? How many moles of undissociated $A_3X_2$ remain? How many moles of particles (As, Xs, and $A_3X_2$s) are present in the solution?

**B-17.** The fraction of kerosene that can be recovered from a barrel of crude oil is 0.200. After a certain amount of oil was refined, 8.90 gal of kerosene, some gasoline, and 18.6 gal of other products were produced. How many gallons of gasoline were produced?

**B-18.** The fraction of moles (mole fraction) of gas A in a mixture is 0.261. If the mixture contains 0.375 mol of gas B and 0.175 mol of gas C as well as gas A, how many moles of gas A are present?

*Answers:*   **B-10** 7.2 ft, 11.2 ft   **B-11** thallium, 210 amu; iodine, 130 amu   **B-12** gallium, 70 amu; osmium, 192 amu   **B-13** 74.3%   **B-14** 13.1%   **B-15** 452 jacks   **B-16** 1.37 mol of A, 0.914 mol of X, 0.99 mol of $A_3X_2$, 3.27 mol total   **B-17** 17.0 gal   **B-18** 0.195 mol

## B-3 Direct and Inverse Proportionalities

There is one other point that should be included in a review on algebra—direct and inverse proportionalities. We use these often in chemistry.

When a quantity is directly proportional to another, it means that an increase in one variable will cause a corresponding increase of the same percent in the other variable. A direct proportionality is shown as

$$A \propto B \quad (\propto \text{ is the proportionality symbol})$$

which is read "$A$ is directly proportional to $B$." A proportionality can be easily converted to an algebraic equation by the introduction of a constant (in our examples designated $k$), called a constant of proportionality. Thus the proportion becomes

$$A = kB$$

or, rearranging,

$$\frac{A}{B} = k$$

Note that $k$ is not a variable but has a certain numerical value that does not change as do $A$ and $B$ under experimental conditions.

A common, direct proportionality that we will study relates Kelvin temperature $T$ and volume $V$ of a gas at constant pressure. This is written as

$$V \propto T \qquad V = kT \qquad \frac{V}{T} = k$$

(This is known as Charles's law.)

In a hypothetical case, $V = 100$ L and $T = 200$ K. From this information we can calculate the value of the constant $k$.

$$\frac{V}{T} = \frac{100 \text{ L}}{200 \text{ K}} = 0.50 \text{ L/K}$$

A change in volume or temperature requires a corresponding change in the other *in the same direction.* For example, if the temperature of the gas is changed to 300 K, we can see that a corresponding change in volume is required from the following calculation.

$$V = kT = 0.50 \text{ L/K} \times 300 \text{ K} = \underline{150 \text{ L}}$$

When a quantity is inversely proportional to another quantity, an increase in one brings about a corresponding *decrease* in the other. An inverse proportionality between $A$ and $B$ is written as

$$A \propto \frac{1}{B}$$

As before, the proportionality can be written as an equality by the introduction of a constant (which has a value different from the example above).

$$A = \frac{k}{B} \quad \text{or} \quad AB = k$$

A common inverse proportionality that we use relates the volume $V$ of a gas to the pressure $P$ at a constant temperature. This is written as

$$V \propto \frac{1}{P} \quad V = \frac{k}{P} \quad PV = k$$

(This is known as Boyle's law.)

In a hypothetical case, $V = 100 \text{ L}$ and $P = 1.50 \text{ atm}$. From this information we can calculate the value of the constant $k$.

$$PV = k = 1.50 \text{ atm} \times 100 \text{ L} = 150 \text{ atm} \cdot \text{L}$$

A change in volume or pressure requires a corresponding change in the other *in the opposite direction*. For example, if the pressure on the gas is changed to 3.00 atm, we can see that a corresponding change in volume is required.

$$V = \frac{k}{P} = \frac{150 \text{ atm} \cdot \text{L}}{3.00 \text{ atm}} = \underline{50.0 \text{ L}}$$

When one variable (e.g., $x$) is directly proportional to two other variables (e.g., $y$ and $z$), the proportionality can be written as the product of the two.

$$\text{If} \quad x \propto y \quad \text{and} \quad x \propto z, \quad \text{then}$$
$$x \propto yz$$

When one variable, (e.g., $a$) is directly proportional to one variable, (e.g., $b$) and inversely proportional to another (e.g., $c$), the proportionality can be written as the ratio of the two.

$$\text{If} \quad a \propto b \quad \text{and} \quad a \propto \frac{1}{c}, \quad \text{then}$$
$$a \propto \frac{b}{c}$$

Quantities can be directly or inversely proportional to the square, square root, or any other function of another variable or number, as illustrated by the examples that follow.

| Solving Equations with Proportionalities | Example B-4 |
|---|---|

**a.** A quantity $C$ is directly proportional to the square of $D$. Write an equality for this statement and explain how a change in $D$ affects the value of $C$.

**working it out**

**Solution**

The equation is

$$C = kD^2$$

Note that a change in $D$ will have a significant effect on the value of $C$. For example,

$$\text{If } D = 1, \text{ then } C = k$$
$$\text{If } D = 2, \text{ then } C = 4k$$
$$\text{If } D = 3, \text{ then } C = 9k$$

Note that when the value of $D$ is doubled, the value of $C$ is increased *fourfold*.

**b.**   A variable $X$ is directly proportional to the square of the variable $Y$ and inversely proportional to the square of another variable $Z$. Write an equality for this statement.

**Solution**

This can be written as two separate equations if it is assumed that $Y$ is constant when $Z$ varies and vice versa.

$$X = k_1 Y^2 \quad (Z \text{ constant})$$
$$X = k_2 / Z^2 \quad (Y \text{ constant})$$

$k_1$ and $k_2$ are different constants. This relationship can be combined into one equation when both $Y$ and $Z$ are variables.

$$X = \frac{k_3 Y^2}{Z^2}$$

$k_3$ is a third constant that is a combination of $k_1$ and $k_2$.

## Exercises

**B-19.**  Write equalities for the following relations.
**(a)** $X$ is inversely proportional to $Y + Z$.
**(b)** $[H_3O^+]$ is inversely proportional to $[OH^-]$.
**(c)** $[H_2]$ is directly proportional to the square root of $r$.
**(d)** $B$ is directly proportional to the square of $y$ and the cube of $z$.
**(e)** The pressure $P$ of a gas is directly proportional to the number of moles $n$ and the temperature $T$, and inversely proportional to the volume $V$.

*Answers:*   **(a)** $X = k/(Y + Z)$   **(b)** $[H_3O^+] = k/[OH^-]$   **(c)** $[H_2] = k\sqrt{r}$   **(d)** $B = ky^2z^3$   **(e)** $P = knT/V$

# Appendix C
# Scientific Notation

Although this topic was first introduced in section 1-3 in this text, we will focus on a review of the mathematical manipulation of numbers expressed in scientific notation in this appendix. Specifically, addition, multiplication, division, and taking the roots of numbers expressed in scientific notation are covered. We conclude this section with how we can simplify the expression of numbers in scientific notation with the use of logarithms. As mentioned in Chapter 1, scientific notation makes use of powers of 10 to express awkward numbers that employ more than two or three zeros that are not significant figures. The exponent of 10 simply indicates how many times we should multiply or divide a number (called the coefficient) by 10 to produce the actual number. For example, $8.9 \times 10^3 = 8.9$ (the coefficient) multiplied by 10 *three* times, or

$$8.9 \times 10 \times 10 \times 10 = 8900$$

Also, $4.7 \times 10^{-3} = 4.7$ (the coefficient) divided by 10 *three* times, or

$$\frac{4.7}{10 \times 10 \times 10} = 0.0047$$

## C-1 Review of Scientific Notation

The method for expressing numbers in scientific notation was explained in section 1-3. However, to simplify a number or to express it in the standard form with one digit to the left of the decimal point in the coefficient, it is often necessary to change a number already expressed in scientific notation. If this is done in a hurry, errors may result. Thus it is worthwhile to practice moving the decimal point of numbers expressed in scientific notation.

**Changing Normal Numbers to Scientific Notation**

Change the following numbers to the standard form in scientific notation.

**a.** $489 \times 10^4$      **b.** $0.00489 \times 10^8$

**Example C-1**

working it out

**Procedure**

All you need to remember is to raise the power of 10 one unit for each place the decimal point is moved to the left, and lower the power of 10 one unit for each place that the decimal point is moved to the right in the coefficient.

**Solution**

a. $489 \times 10^4 = (4\,8\,9) \times 10^4 = 4.89 \times 10^{4+2} = \underline{4.89 \times 10^6}$

b. $0.00489 \times 10^8 = (0.0\,0\,4\,89) \times 10^8 = 4.89 \times 10^{8-3} = 4.89 \times 10^5$

As an aid to remembering whether you should raise or lower the exponent as you move the decimal point, it is suggested that you write (or at least imagine) the coefficient on a slant. For each place that you move the decimal point *up*, add one to the exponent. For each place that you move the decimal point *down*, subtract one from the exponent. Note that the exponent moves up or down with the decimal point. It may be easier to recall "up or down" rather than "right or left."

| Example C-2 | **Changing the Decimal Point in the Coefficient** |

**working it out**

Change the following numbers to the standard form in scientific notation.

a. $4223 \times 10^{-7}$   b. $0.00076 \times 10^{18}$

**Solution**

a. $4223 \times 10^{-7} = \begin{bmatrix} 4 \\ 2 \\ 2 \\ 3 \end{bmatrix} \begin{matrix} \\ +3 \\ +2 \\ +1 \end{matrix} \times 10^{-7} = 4.223 \times 10^{-7+3} = \underline{4.223 \times 10^{-4}}$

b. $0.00076 \times 10^{18} = \begin{bmatrix} 0 \\ 0 \\ 0 \\ 0 \\ 7 \\ 6 \end{bmatrix} \begin{matrix} \\ -1 \\ -2 \\ -3 \\ -4 \end{matrix} \times 10^{18} = 7.6 \times 10^{18-4} = \underline{7.6 \times 10^{14}}$

# Exercises

**C-1.** Change the following numbers to standard scientific notation with one digit to the left of the decimal point in the coefficient.

(a) $787 \times 10^{-6}$   (c) $0.015 \times 10^{-16}$   (e) $49.3 \times 10^{15}$
(b) $43.8 \times 10^{-1}$   (d) $0.0037 \times 10^9$   (f) $6678 \times 10^{-16}$

**C-2.** Change the following numbers to a number with two digits to the left of the decimal point in the coefficient.

(a) $9554 \times 10^4$   (c) $1 \times 10^6$   (e) $0.023 \times 10^{-1}$
(b) $1.6 \times 10^{-5}$   (d) $116.5 \times 10^4$   (f) $0.005 \times 10^{23}$

*Answers:* **C-1:** **(a)** $7.87 \times 10^{-4}$ **(b)** 4.38 **(c)** $1.5 \times 10^{-18}$ **(d)** $3.7 \times 10^{6}$
**(e)** $4.93 \times 10^{16}$ **(f)** $6.678 \times 10^{-13}$ **C-2:** **(a)** $95.54 \times 10^{6}$ **(b)** $16 \times 10^{-6}$
**(c)** $10 \times 10^{5}$ **(d)** $11.65 \times 10^{5}$ **(e)** $23 \times 10^{-4}$ **(f)** $50 \times 10^{19}$

## C-2 Addition and Subtraction

Addition or subtraction of numbers in scientific notation can be accomplished only when all coefficients have the same exponent of 10. When all the exponents are the same, the coefficients are added and then multiplied by the power of 10. The correct number of places to the right of the decimal point, as discussed in section 1-2, must be shown.

| Addition of Numbers in Scientific Notation | Example C-3 |
| --- | --- |
| **a.** Add the following numbers: $3.67 \times 10^{-4}$, $4.879 \times 10^{-4}$, and $18.2 \times 10^{-4}$. | working it out |

**Solution**

$$
\begin{array}{r}
3.67 \ \times 10^{-4} \\
4.879 \times 10^{-4} \\
\underline{18.2 \ \ \times 10^{-4}} \\
26.749 \times 10^{-4} = \underline{26.7 \times 10^{-4}} = \underline{2.67 \times 10^{-3}}
\end{array}
$$

**b.** Add the following numbers: $320.4 \times 10^{3}$, $1.2 \times 10^{5}$, and $0.0615 \times 10^{7}$.

**Solution**

Before adding, change all three numbers to the same exponent of 10.

$$
\begin{array}{r}
320.4 \times 10^{3} = 3.204 \times 10^{5} \\
1.2 \times 10^{5} = 1.2 \ \ \times 10^{5} \\
\underline{0.0615 \times 10^{7} = 6.15 \ \ \times 10^{5}} \\
10.554 \times 10^{5} = \underline{10.6 \times 10^{5}} = \underline{1.06 \times 10^{6}}
\end{array}
$$

## Exercises

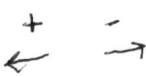

**C-3.** Add the following numbers. Express the answer to the proper decimal place.
**(a)** $152 + (8.635 \times 10^{2}) + (0.021 \times 10^{3})$
**(b)** $(10.32 \times 10^{5}) + (1.1 \times 10^{5}) + (0.4 \times 10^{5})$
**(c)** $(1.007 \times 10^{-8}) + (118 \times 10^{-11}) + (0.1141 \times 10^{-6})$
**(d)** $(0.0082) + (2.6 \times 10^{-4}) + (159 \times 10^{-4})$

**C-4.** Carry out the following calculations. Express your answer to the proper decimal place.
**(a)** $(18.75 \times 10^{-6}) - (13.8 \times 10^{-8}) + (1.0 \times 10^{-5})$
**(b)** $(1.52 \times 10^{-11}) + (17.7 \times 10^{-12}) - (7.5 \times 10^{-15})$
**(c)** $(481 \times 10^{6}) - (0.113 \times 10^{9}) + (8.5 \times 10^{5})$
**(d)** $(0.363 \times 10^{-6}) + (71.2 \times 10^{-9}) + (519 \times 10^{-12})$

*Answers:* **C-3:** **(a)** $1.037 \times 10^{3}$ **(b)** $1.18 \times 10^{6}$ **(c)** $1.254 \times 10^{-7}$
**(d)** $2.44 \times 10^{-2}$ **C-4:** **(a)** $2.9 \times 10^{-5}$ **(b)** $3.29 \times 10^{-11}$ **(c)** $3.69 \times 10^{8}$
**(d)** $4.35 \times 10^{-7}$

## C-3 Multiplication and Division

When numbers expressed in scientific notation are multiplied, the exponents of 10 are *added*. When the numbers are divided, the exponent of 10 in the denominator (the divisor) is subtracted from the exponent of 10 in the numerator (the dividend).

| **Example C-4** | **Multiplication and Division** |
|---|---|
| working it out | **a.** Carry out the following calculation. |

$$(4.75 \times 10^6) \times (3.2 \times 10^5)$$

**Solution**

In the first step, group the coefficients and the powers of 10. Carry out each step separately.

$$(4.75 \times 3.2) \times (10^6 \times 10^5) = 15.200 \times 10^{6+5}$$
$$= 15 \times 10^{11} = \underline{\underline{1.5 \times 10^{12}}}$$

**b.** Carry out the following calculation.

$$(1.62 \times 10^{-8}) \div (8.55 \times 10^{-3})$$

**Solution**

$$\frac{1.62 \times 10^{-8}}{8.55 \times 10^{-3}} = \frac{1.62}{8.55} \times \frac{10^{-8}}{10^{-3}} = 0.189 \times 10^{-8-(-3)}$$
$$= 0.189 \times 10^{-5} = \underline{\underline{1.89 \times 10^{-6}}}$$

# Exercises

**C-5.** Carry out the following calculations. Express your answer to the proper number of significant figures with one digit to the left of the decimal point.
(a) $(7.8 \times 10^{-6}) \times (1.12 \times 10^{-2})$
(b) $(0.511 \times 10^{-3}) \times (891 \times 10^{-8})$
(c) $(156 \times 10^{-12}) \times (0.010 \times 10^4)$
(d) $(16 \times 10^9) \times (0.112 \times 10^{-3})$
(e) $(2.35 \times 10^3) \times (0.3 \times 10^5) \times (3.75 \times 10^2)$
(f) $(6.02 \times 10^{23}) \times (0.0100)$

**C-6.** Follow the instructions in Problem C-5.
(a) $(14.6 \times 10^8) \div (2.2 \times 10^8)$      (d) $(0.0221 \times 10^3) \div (0.57 \times 10^{18})$
(b) $(6.02 \times 10^{23}) \div (3.01 \times 10^{20})$      (e) $238 \div (6.02 \times 10^{23})$
(c) $(0.885 \times 10^{-7}) \div (16.5 \times 10^3)$

**C-7.** Follow the instructions in Problem C-5.
(a) $[(8.70 \times 10^6) \times (3.1 \times 10^8)] \div (5 \times 10^{-3})$
(b) $(47.9 \times 10^{-6}) \div [(0.87 \times 10^6) \times (1.4 \times 10^2)]$
(c) $1 \div [(3 \times 10^6) \times (4 \times 10^{10})]$
(d) $1.00 \times 10^{-14} \div [(6.5 \times 10^5) \times (0.32 \times 10^{-5})]$
(e) $[(147 \times 10^{-6}) \div (154 \times 10^{-6})] \div (3.0 \times 10^{12})$

*Answers:* **C-5: (a)** $8.7 \times 10^{-8}$   **(b)** $4.55 \times 10^{-9}$   **(c)** $1.6 \times 10^{-8}$
**(d)** $1.8 \times 10^6$   **(e)** $3 \times 10^{10}$   **(f)** $6.02 \times 10^{21}$   **C-6: (a)** $6.6$   **(b)** $2.00 \times 10^3$
**(c)** $5.36 \times 10^{-12}$   **(d)** $3.9 \times 10^{-17}$   **(e)** $3.95 \times 10^{-22}$   **C-7: (a)** $5 \times 10^{17}$
**(b)** $3.9 \times 10^{-13}$   **(c)** $8 \times 10^{-18}$   **(d)** $4.8 \times 10^{-15}$   **(e)** $3.2 \times 10^{-13}$

# C-4 Powers and Roots

When a number expressed in scientific notation is raised to a power, the coefficient is raised to the power and the exponent of 10 is *multiplied* by the power.

For a number expressed in scientific notation, we take the root of the coefficient and *divide* the exponent by the root. (A square root is the same as raising the number to the 1/2 power, a cube root to the 1/3 power, etc.) In the interest of easy viewing in the exercises, we will adjust the number so that division of the exponent by the root produces a whole number. It is not necessary to adjust the number when using a calculator.

**Powers and Roots**

Example C-5

working it out

a. Carry out the following calculation.

$$(3.2 \times 10^3)^2 = (3.2)^2 \times 10^{3 \times 2}$$
$$= 10.24 \times 10^6 = \underline{\underline{1.0 \times 10^7}}$$

$$[(10^3)^2 = 10^3 \times 10^3 = 10 \times 10 \times 10 \times 10 \times 10 \times 10 = 10^6]$$

b. Carry out the following calculation.

$$(1.5 \times 10^{-3})^3 = (1.5)^3 \times 10^{-3 \times 3}$$
$$= \underline{\underline{3.4 \times 10^{-9}}}$$

c. Carry out the following calculation.

$$\sqrt{2.9 \times 10^5}$$

**Solution**

First adjust the number so that the exponent of 10 is divisible by 2.

$$\sqrt{2.9 \times 10^5} = \sqrt{29 \times 10^4} = \sqrt{29} \times \sqrt{10^4} = \sqrt{29} \times 10^{4/2}$$
$$= \underline{\underline{5.4 \times 10^2}}$$

d. Carry out the following calculation.

$$\sqrt[3]{6.9 \times 10^{-8}}$$

**Solution**

Adjust the number so that the exponent of 10 is divisible by 3.

$$\sqrt[3]{6.9 \times 10^{-8}} = \sqrt[3]{69 \times 10^{-9}} = \sqrt[3]{69} \times \sqrt[3]{10^{-9}}$$
$$= 4.1 \times 10^{-9/3} = \underline{\underline{4.1 \times 10^{-3}}}$$

# Exercises

**C-8.** Carry out the following operations.
(a) $(6.6 \times 10^4)^2$      (d) $(0.035 \times 10^{-3})^3$
(b) $(0.7 \times 10^6)^3$      (e) $(0.7 \times 10^7)^4$
(c) $(1200 \times 10^{-5})^2$ (It will be easier to square if you change the number to $1.2 \times 10^?$ first.)

**C-9.** Take the following roots. Approximate the answer if necessary.
(a) $\sqrt{36 \times 10^4}$      (c) $\sqrt{64 \times 10^9}$      (e) $\sqrt{81 \times 10^{-7}}$
(b) $\sqrt[3]{27 \times 10^{12}}$      (d) $\sqrt[3]{1.6 \times 10^5}$      (f) $\sqrt{180 \times 10^{10}}$

*Answers:* **C-8:** (a) $4.4 \times 10^9$ (b) $3 \times 10^{17}$ (c) $1.4 \times 10^{-4}$ (d) $4.3 \times 10^{-14}$ (e) $2 \times 10^{27}$ **C-9:** (a) $6.0 \times 10^2$ (b) $3.0 \times 10^4$ (c) $2.5 \times 10^5$ (d) 54 (e) $2.8 \times 10^{-3}$ (f) $1.3 \times 10^6$

## C-5 Logarithms

Scientific notation is particularly useful in expressing very large or very small numbers. In certain areas of chemistry, however, such as in the expression of $H_3O^+$ concentration, even the repeated use of scientific notation becomes tedious. In this situation, it is convenient to express the concentration as simply the *exponent of 10. The exponent to which 10 must be raised to give a certain number is called its* **common logarithm.** With common logarithms (or just logs) it is possible to express both the coefficient and the exponent of 10 as one number.

Since logarithms are simply exponents of 10, logs of exact multiples of 10 such as 100 can be easily determined. Note that 100 can be expressed as $10^2$, so that the log of 100 is exactly 2. Other examples of simple logs of numbers (that we assume to be exact) are

| | | | |
|---|---|---|---|
| $1 = 10^0$ | $\log 1 = 0$ | $0.1 = 10^{-1}$ | $\log 0.1 = -1$ |
| $10 = 10^1$ | $\log 10 = 1$ | $0.01 = 10^{-2}$ | $\log 0.01 = -2$ |
| $100 = 10^2$ | $\log 100 = 2$ | $0.001 = 10^{-3}$ | $\log 0.001 = -3$ |
| $1000 = 10^3$ | $\log 1000 = 3$ | $0.0001 = 10^{-4}$ | $\log 0.0001 = -4$ |

There are two general rules regarding logarithms.

**1.** $\log(A \times B) = \log A + \log B$

**2.** $\log(A/B) = \log A - \log B$

We can see how these rules apply when we multiply and divide multiples of ten.

| | Exponents | Logarithms |
|---|---|---|
| multiplication | $10^4 \times 10^3 = 10^{4+3}$ <br> $= 10^7$ | $\log(10^4 \times 10^3) = \log 10^4 + \log 10^3$ <br> $= 4 + 3 = 7$ |
| division | $\dfrac{10^{10}}{10^4} = 10^{10-4} = 10^6$ | $\log\dfrac{10^{10}}{10^4} = \log 10^{10} - \log 10^4$ <br> $= 10 - 4 = 6$ |

Although we will use the calculator (see Appendix E) to determine the logs of numbers that are not simple multiples of ten, such as those above, it is helpful to have a sense of how the log of a specific number will appear. The log of a number has two parts. *The number to the left of the decimal is known as the* **characteristic** *and represents the exact exponent of 10 in the exponential number. The number to the right of the decimal is known as the* **mantissa** *and is the log of the coefficient in the* exponential number. For example, consider the log of the following exponential number.

Exponential number $\quad 5.7 \times 10^6$

$$(\log 10^6 = 6) \quad + \quad (\log 5.7 = 0.76)$$

$$\log(5.7 \times 10^6) \quad = \quad \boxed{6.76}$$

$$\underbrace{\qquad}_{\text{characteristic}} \qquad \underbrace{\qquad}_{\text{mantissa}}$$

Notice that the mantissa should be expressed with the same number of significant figures as the coefficient of the original exponential number. In this case, that is two significant figures.

Logs of numbers between one and ten are positive numbers where the characteristic is a zero (i.e., $5.8 = 5.8 \times 10^0$). For example,

$$\log 5.8 = \underline{0.76}$$

Logs of numbers greater than ten have a characteristic greater than zero. For example,

$$\log 4.7 \times 10^3 = \underline{3.67}$$

Logs of numbers that are less than one have a negative value since the exponent is a negative value. For example,

number $\quad 0.66 = 6.6 \times 10^{-1}$ $\qquad\qquad 7.3 \times 10^{-4}$
$\qquad\qquad \log 6.6 + \log 10^{-1}$ $\qquad\qquad \log 7.3 + \log 10^{-4}$
$\qquad\qquad \log 0.66 = 0.82 + (-1) = \underline{-0.18} \quad \log(7.3 \times 10^{-4}) = 0.86 - 4 = \underline{-3.14}$

The *antilog* or *inverse log* is the opposite of a logarithm. It is the number whose log has a certain value. For example, consider the antilog ($x$) of 2.

$$\log x = 2 \qquad x = \underline{10^2} \quad \text{since } \log 10^2 = 2$$

Since the log of a number between one and ten is a positive number with zero as the characteristic, then the antilog of a number with zero as a characteristic is between one and ten. If the characteristic of a positive number is greater than zero, that number is the exponent of ten and the antilog of the mantissa is the coefficient. For example,

$$\text{antilog } 0.62 = \underline{4.17} \qquad \text{antilog } 6.62 = \underline{4.17 \times 10^6}$$

The antilog of a negative number is somewhat different. The calculator can only take the antilog of a mantissa with a positive value. To change a mantissa with a negative value to one with a positive value requires a mathematical manipulation. The negative number is separated into its two parts, the mantissa and the characteristic. A value of one is added to the mantissa and a value of one is subtracted from the characteristic. For example, consider the antilogs of $-3.28$ and $-0.18$.

$$
\begin{array}{ccccc}
 & & \text{Add 1 to} & & \text{Subtract} \\
 & & \text{mantissa} & & \text{1 from} \\
 & \text{Separate} & & & \text{characteristic} \\
-3.28 = & \overbrace{-0.28 - 3} = & \overbrace{(-0.28 + 1)} & - & \overbrace{3 - 1} = +0.72 - 4 \\
-0.18 = & -0.18 - 0 = & (-0.18 + 1) & - & 0 - 1 = +0.82 - 1
\end{array}
$$

The antilog of the positive number becomes the coefficient of the exponential number, and the antilog of the negative integer becomes the negative exponent of ten. Thus,

$$\text{antilog}(-3.28) = \text{antilog}(0.72) + \text{antilog}(-4) = \underline{5.2 \times 10^{-4}}$$
$$\text{antilog}(-0.18) = \text{antilog}(0.82) + \text{antilog}(-1) = 6.6 \times 10^{-1} = \underline{0.66}$$

In the following examples and exercises, we will test our understanding of logs by concentrating on the characteristic of the log or the exponent of ten for an antilog. It is suggested that you do this without the benefit of a calculator. You can then plug the number or the log into your calculator to get the actual answer.

| Example C-6 | Logs |
| --- | --- |

**working it out**

Give the value of the characteristic in the logarithm for each of the following numbers. Use xx as the mantissa.

**a.** 5.8         number between one and ten                    $\log = \underline{\underline{0.xx}}$

**b.** $4.7 \times 10^3$     number greater than ten                      $\log = \underline{\underline{3.xx}}$

**c.** 0.085       number less than one expressed in scientific
                notation as $8.5 \times 10^{-2}$
                $\log 8.5 + \log 10^{-2} = +0.yy - 2 = -1.xx$     $\log = \underline{\underline{-1.xx}}$

**d.** $8.7 \times 10^{-7}$   number less than one                       $\log = \underline{\underline{-6.xx}}$

| Example C-7 | Antilogs |
| --- | --- |

Give the value of the exponent of ten for the number (expressed in scientific notation) whose log is the following. Use z.z as the value of the coefficient.

**a.** 0.84       antilog between zero and one
                                 exponent = 0     $z.z \times 10^0 = \underline{\underline{z.z}}$

**b.** 4.65       antilog greater than one            exponent = 4     $z.z \times 10^4$

**c.** $-0.020$    antilog less than zero (antilog $0.080 - 1$)
                                exponent = $-1$    $z.z \times 10^{-1} = \underline{\underline{0.zz}}$

**d.** $-4.54$    antilog less than zero (antilog $0.46 - 5$)
                                exponent = $-5$     $\underline{\underline{z.z \times 10^{-5}}}$

# Exercises

**C-10.** Give the value of the characteristic in the logarithm for each of the following numbers. Use xx as the mantissa.

**(a)** 7.4     **(b)** 0.087     **(c)** 1700     **(d)** $7.3 \times 10^4$     **(e)** $32 \times 10^{-5}$     **(f)** $32 \times 10^5$

**C-11.** Give the value of the exponent of ten for the number (expressed in scientific notation) whose log is the following. Use z.z for the value of the coefficient.

**(a)** 0.34     **(b)** $-5.48$     **(c)** $-0.070$     **(d)** 8.40     **(e)** 10.94     **(f)** $-2.60$

*Answers:* **C-10: (a)** 0.xx     **(b)** $-1.xx$     **(c)** 3.xx     **(d)** 4.xx     **(e)** $-3.xx$
**(f)** 6.xx     **C-11: (a)** z.z     **(b)** $z.z \times 10^{-6}$     **(c)** 0.zz     **(d)** $z.z \times 10^8$
**(e)** $z.z \times 10^{10}$     **(f)** $z.z \times 10^{-3}$

# Appendix D
# Graphs

The graphical representation of data is important in all branches of science, social as well as natural. Trends and cycles in data may be obscure when presented in a table but become immediately apparent in graphical form. For example, consider a graph in which the average monthly temperature (represented on the vertical axis) is plotted as a function of the month (represented on the horizontal axis). The graph is constructed by marking the temperature for each month and drawing a smooth curve through the points. A quick look at the graph tells us what we already knew— the temperature cycles up and down throughout the seasons.

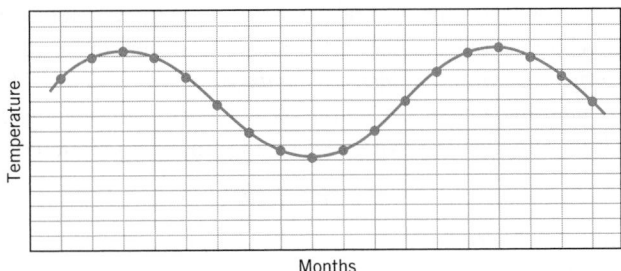

A useful graphs of this nature is shown in Figure 5-20. The two graphs illustrate the periodic nature of atomic radius and ionization energy as atomic number increases.

## D-1 Direct Linear Relationships

Other types of graphs can be useful in predicting data as well as simply illustrating measurements. For example, in Appendix B we discussed proportionalities. In a proportionality a variable ($x$) is related to another variable ($y$) by some mathematical operation (a power, a root, a reciprocal, etc.). A graph translates a proportionality into a line that illustrates the relationship of $x$ to $y$. There are two types of graphs: linear and nonlinear. A linear graph is represented by a straight line and is a result of a direct, first-power proportionality between the variables $x$ and $y$. This is expressed as

$$x \propto y$$

Such a graph is illustrated in Figure 9-7, which relates the volume of a gas to the Celsius temperature. Since the data points fall in a straight line, the line can be extended to provide data points beyond the limits of the experiments. (We say the line has been extrapolated.) Using a linear graph, as few as two experiments can be used to predict the results of other measurements.

The relationship between the volume (*V*) of a gas at constant pressure and the Kelvin temperature (*T*) can be expressed as

$$V \propto T$$

We will use this direct relationship to construct a linear graph. The following two experiments provide the information necessary to construct such a graph.

**Experiment 1**   At a temperature of 350°C, the volume was 12.5 L.

**Experiment 2**   At a temperature of 250°C, the volume was 10.5 L.

Actually, it is always risky to assume that only two experiments provide an accurate graph or give all of the needed information, but for simplicity we will assume that it is valid in this case. We will check the graph with an additional experiment later.

To record this information on a graph, we first need some graph paper with regularly spaced divisions. On the vertical axis (called the ordinate or *y* axis) we put temperature. Pick divisions on the graph paper that spread out the range of temperatures as much as possible. In our example, the temperatures range between 250°C and 350°C, or 100 degrees. Note that the two axes do not need to intersect at *x* = *y* = 0. We then plot the volume on the horizontal axis (called the abscissa or *x* axis). The volumes range between 12.5 L and 10.5 L, or 2.0 L.

The graph is shown in Figure D-1. Dots mark the locations of the two points from the experiments, and a straight line has been drawn between them.

We can now check the accuracy of our graph. A third experiment tells us that at *T* = 300°C, *V* = 11.5 L. Refer back to the graph. Locate 300°C on the ordinate and trace that line to where it intersects the vertical line representing the volume of 11.5 L. Since the point of intersection is right on the line we have confirmation of the accuracy of the original plot. *Normally, four or five measurements are required to provide enough points on the graph so that a line can be drawn through or as near as many points as possible.*

A straight line graph can be expressed by the linear algebraic equation

$$y = mx + b$$

where *y* = a value on the ordinate

   *x* = the corresponding value on the abscissa

   *b* = the point on the *y* axis or ordinate where the straight line intersects the axis

   *m* = the slope of the line

The slope of a line is the ratio of the change on the ordinate to the change on the abscissa. It is determined by choosing two widely spaced points on the line ($x_1$, $y_1$ and $x_2$, $y_2$). The slope is the difference in *y* divided by the difference in *x*.

$$m = \frac{y_2 - y_1}{x_2 - x_1}$$

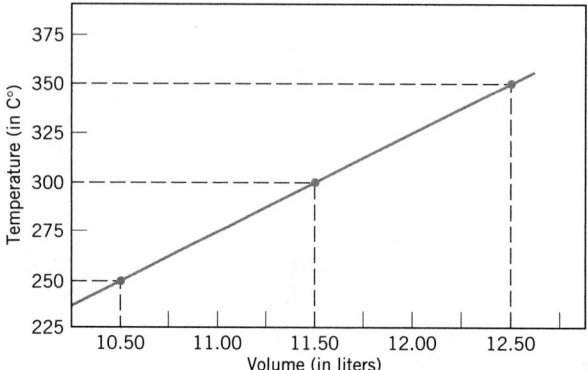

**Figure D-1**
**The Volume of a Gas as a Function of Temperature.**

The larger the slope *for a particular graph,* the steeper the straight line. For the graph shown in Figure D-1, if we pick point 1 to be $x_1 = 10.50$, $y_1 = 250$ and point 2 to be $x_2 = 12.50$, $y_2 = 350$, then

$$m = \frac{350 - 250}{12.50 - 10.50} = \frac{100}{2.00} = 50.0$$

The value for $b$ cannot be determined directly from Figure D-1. The value for $b$ is found by determining the point on the $y$ axis where the line intersects when $x = 0$. If we plotted the data on such a graph, we would find that $b = -273°C$. Therefore, the linear equation for this line is

$$y = 50.0x - 273$$

From this equation we can determine the value for $y$ (the temperature) by substitution of any value for $x$ (the volume).

## Exercises

**D-1.** A sample of water was heated at a constant rate, and its temperature was recorded. The following results were obtained.

| Experiment | Temp (°C) | Time (min) |
|---|---|---|
| 1 | 10.0 | 0 |
| 2 | 20.0 | 4.5 |
| 3 | 30.0 | 7.5 |
| 4 | 55.0 | 17.8 |
| 5 | 85.0 | 30.0 |

(a) Construct a graph that includes all of the information above with temperature on the ordinate and time on the abscissa.
(b) Calculate the slope of the line and the linear equation $y = mx + b$ for the line.
(c) From the graph find the temperature at 11.0 min and compare it with the value calculated from the equation.
(d) From the graph find the time when the temperature is 65.0°C and compare it with the value calculated from the equation.

*Answers:* (a) Draw the line touching or coming close to as many points as possible. Note that the straight line does not go through all points. (b) $m = 2.5$, $y = 2.5x + 10$ (c) From the equation, when $x = 11.0$, $y = 2.5 \cdot 11.0 + 10$, so $y = 37.5°C$. The graph agrees. (d) From the equation, when $y = 65$, $65 = 2.5x + 10$, and $x = 22.0$ min. The graph agrees.

## D-2 Nonlinear Relationships

Nonlinear graphs are represented by a curved line and result from an inverse proportionality, or any type of proportionality other than direct. Examples of such relationships can be expressed as

$$x \propto y^2 \quad \text{and} \quad x \propto \frac{1}{y}$$

An example of an inverse relationship that produces a curve when plotted is the relationship between volume of a gas ($V$) at constant temperature and the pressure ($P$). The relationship is

$$V \propto \frac{1}{P}$$

In the following experiments the pressure on a volume of gas was varied and the volume was measured at constant temperature.

| Experiment | Pressure (torr) | Volume (L) |
|---|---|---|
| 1 | 500 | 15.2 |
| 2 | 600 | 12.6 |
| 3 | 760 | 10.0 |
| 4 | 900 | 8.44 |
| 5 | 1200 | 6.40 |
| 6 | 1600 | 4.80 |

This information has been graphed in Figure D-2. Note that the line has a gradually changing slope, starting with a very steep slope on the left and going to a small slope on the right of the graph. We can interpret this. In the region of the steep slope, the increase in pressure causes little decrease in volume compared with the part of the curve with the small slope to the right. Note that this is a plot of an inverse relationship. That is, the higher the pressure, the lower the volume. If we regraph the information and include much higher pressures and the corresponding volumes, as shown in Figure D-3, we see that *the curve approaches the y axis but never actually touches it,* no matter how high the pressure. Such a curve is said to approach the axis **asymptotically.** In our example, the curve would eventually appear to change to a vertical line parallel to the *y* axis.

## Exercises

**D-2.** Construct a graph from the following experimental information. Plot the time on the abscissa and the concentration on the ordinate. Describe what this graph tells you about the change in concentration of HI as a function of (how it varies with) the time.

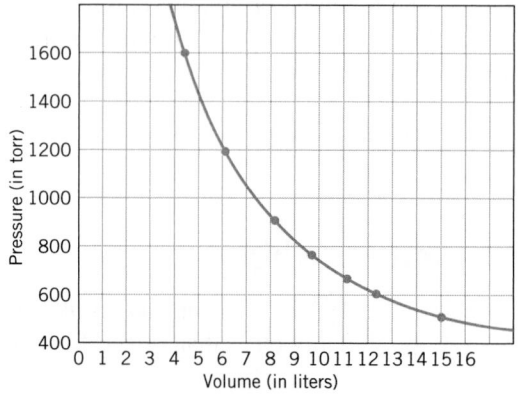

**Figure D-2 An Inverse Relationship—The Volume of a Gas as a Function of Pressure**

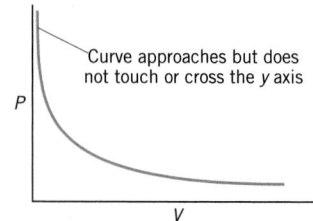

**Figure D-3**

| Experiment | Time (min) | Concentration of HI (mol/L) |
|---|---|---|
| 1 | 0.0 | $2.50 \times 10^{-2}$ |
| 2 | 5.0 | $1.50 \times 10^{-2}$ |
| 3 | 10.0 | $0.90 \times 10^{-2}$ |
| 4 | 15.0 | $0.65 \times 10^{-2}$ |
| 5 | 20.0 | $0.55 \times 10^{-2}$ |
| 6 | 35.0 | $0.50 \times 10^{-2}$ |

(a) From the graph, find the concentration of HI at 12.0 min and at 25 min.

(b) From the graph, find the times at which the concentration of HI is $2.00 \times 10^{-2}$ and $0.75 \times 10^{-2}$.

*Answers:*   The concentration decreases rapidly at the start (the curve has a large slope) but then changes little after about 25 min. At that point the curve appears to change to a straight line parallel to the $x$ axis.   (a) At 12.0 min, the concentration is $0.80 \times 10^{-2}$ mol/L; at 25 min, the concentration is about $0.52 \times 10^{-2}$ mol/L.   (b) When the concentration is $2.00 \times 10^{-2}$ mol/L, the time is 2.4 min; when the concentration is $0.75 \times 10^{-2}$ mol/L, the time is 13.0 min.

# Appendix E
# Calculators

An inexpensive hand calculator has become a necessity for almost all chemistry students. The convenience of this handy instrument is undeniable. Calculations are easy, quick, and accurate. In a way, however, use of calculators creates the risk that we may lose the ability to add, multiply, and divide simple numbers on our own. It is sad for an instructor to see students pull out their calculators when the only exercise is to multiply a number by 10. This appendix is included last because it is hoped that students will first review math operations, scientific notation, and logs before using the calculator. Obviously, it is easy to become overly dependent on the calculator, but it is here to stay.

There are two basic types of hand calculators. Almost all inexpensive hand calculators now in use have a logic system known as Algebraic Operating System (AOS). A few students may have calculators that use a system called Reverse Polish Notation (RPN). Calculators using this system have a key labeled "ENTER." In this appendix, we will describe use of the more common AOS system. If your calculator uses the RPN system, consult your instruction book.

Different brands of calculators vary as to the number of digits shown in the answer, the labeling of some functions, and, of course, the number of functions available. You should consult the instruction booklet for more details. In this appendix, we concentrate on the types of calculations that you will encounter in general chemistry courses. This should save time and confusion when reading through an often complicated instruction booklet. Calculations encountered in chemistry include powers, logs, and scientific notation. Recall that the calculator usually gives at least six digits in an answer. This is almost always far too many for the result of a chemical calculation. You must know how to express the answer to the proper number of significant figures (see section 1-2). A chemistry teacher is not impressed with a student who reports a pH of 8.678346 for a calculation made from a two-significant-figure number. Reporting more numbers than necessary makes an answer look worse, not better.

## E-1 Addition, Subtraction, Multiplication, and Division

In one-step calculations, you simply enter one number and then tell the calculator what you want it to do with the next number (i.e., $+$, $-$, $\times$, or $\div$). You then enter the next number and ask for the answer (i.e., press the $=$ key). The following examples illustrate simple one-step calculations.

**One-Step Calculations**

a.  $6.368 + 43.852$    b.  $0.0835 \div 37.2$

| Enter | Display | Enter | Display |
|-------|---------|-------|---------|
| 6.368 | 6.368 | 0.0835 | 0.0835 |
| + | 6.368 | ÷ | 0.0835 |
| 43.852 | 43.852 | 37.2 | 37.2 |
| = | 50.22 | = | $2.2446237^{-03}$ |

The answer in (a) should be reported as 50.220 to indicate the proper number of decimal places. A calculator does not include zeros that are to the right of the last digit. Our calculator did not realize that the last zero is significant.

The answer in (b) includes a superscript −03, which is the power of 10. We should report this answer to three significant figures in the form $2.24 \times 10^{-3}$.

## E-2 Chain Calculations

One of the most convenient features of the modern hand calculator allows us to do a series of calculations without pausing. For example, many gas law problems require successive calculations as illustrated in the following example.

**Chain Calculations**

a.  Solve the following problem.

$$n = \frac{PV}{RT} = \frac{0.749 \times 7.22}{0.0821 \times 312} = \underline{\qquad} \text{ mol}$$

| Enter | Display |
|-------|---------|
| 0.749 | 0.749 |
| × | 0.749 |
| 7.22 | 7.22 |
| ÷ | 5.40778 |
| 0.0821 | 0.0821 |
| ÷ | 65.86821 |
| 312 | 0.211116 |
| Report as 0.211 | |

You can do this problem in other sequences. For example, you could just as easily carry out the calculation in the following sequence: $(0.749 \div 0.0821) \times 7.22 \div 312 = 0.211$. One sequence that you *cannot do* is to multiply the numerator $(0.74 \times 7.22)$ and then divide by 0.0821 *times* 312. The final calculation must again be a division by 312. To calculate the formula weight of a compound, the atomic masses are multiplied by the number of atoms and the masses totaled. This is illustrated as follows.

**b.** Find the formula weight of $B_2O_3$.

$$\text{Formula weight} = (2 \times 10.8) + (3 \times 16.0) = \underline{\hspace{1cm}} \text{amu.}$$

| Enter | Display |
|-------|---------|
| 2 | 2 |
| × | 2 |
| 10.8 | 10.8 |
| + | 21.6 |
| 3 | 3 |
| × | 3 |
| 16.0 | 16.0 |
| = | 69.6 |

When a negative number is part of a calculation (i.e., $2 \times -6$), the $+/-$ key or equivalent must be pressed after the 6 is entered. The display should now read $-6$, and you can proceed with the calculation.

Some calculations require the use of the parentheses functions which are included on most calculators. These usually involve summations of other operations or summations in a numerator or denominator. For example, consider the following calculation.

$$\frac{7 + 5 + 6}{8}$$

First, press "(" or "[(—" then $7 + 5 + 6$ followed by ")" or "—)]". Next press "÷" then 8 then "=" or "exe" to obtain the answer of 2.25.

Next, consider the calculation

$$(37 \times 56) + \frac{150 \times 10}{2}$$

Each operation must be included within a parenthesis. Carry out the calculation in the order "("$37 \times 56$")" + "$\left(\dfrac{\text{"}150 \times 10\text{"}}{2}\right)$" $= 2822$.

## Exercises

**E-1.** Perform the following calculations.

**(a)** $0.662 \times 7.99$

**(b)** $73.8 + 0.27 + 45.8$

**(c)** $43.8 - 178.34$

**(d)** $7.3 + 8.5 - (-3.4)$

**(e)** $\dfrac{46.8 \times 23.4}{0.00786}$

**(f)** $\dfrac{172 + 67 - 89}{55}$

**(g)** $(4 \times 14.006) + (3 \times 15.998) + (6 \times 1.008)$

**(h)** $(0.810 \times 102.99) + (0.190 \times 104.88)$

*Answers:* **(a)** 5.29 **(b)** 119.9 **(c)** −134.5 **(d)** 19.2 **(e)** 139000 **(f)** 2.7 **(g)** 110.066 **(h)** 103

## E-3 Scientific Notation

Most calculators have the capacity to easily handle scientific notation. This is done with a key marked EE, EXP, or EEX. The coefficient is entered first; then the exponent key is pressed and the exponent entered. If it is a negative exponent, one must press the +/− key to change the sign. The following examples illustrate the use of scientific notation in typical calculations.

**Scientific Notation**

Carry out the following calculations.

**Example E-3**

working it out

**a.** $\dfrac{0.250 \times (6.02 \times 10^{23})}{1.55}$    **b.** $\dfrac{3.76 \times 10^{-18}}{0.862 \times 10^{6}}$

| Enter | Display | |
|-------|---------|----|
| 0.250 | 0.250 | |
| × | 0.25 | |
| 6.02 | 6.02 | |
| EXP | 6.02 | 00 |
| 23 | 6.02 | 23 |
| ÷ | 1.505 | 23 |
| 1.55 | 1.55 | |
| = | 9.71 | 22 |
| | $(9.71 \times 10^{22})$ | |

| Enter | Display | |
|-------|---------|----|
| 3.76 | 3.76 | |
| EXP | 3.76 | 00 |
| 18 | 3.76 | 18 |
| +/− | 3.76 | −18 |
| ÷ | 3.76 | −18 |
| 0.862 | 0.862 | |
| EXP | 0.862 | 00 |
| 6 | 0.862 | 06 |
| = | 4.36 | −24 |
| | $(4.36 \times 10^{-24})$ | |

## E-4 Powers, Roots, Logs, and Antilogs

These are all one-step calculations that require a special function on the calculator. On some calculators a direct key function is not available. However, notice that there are other functions listed above the key that are accessed by pressing the inverse (INV) or second (2nd) keys. For example, an $x^2$ key is not available on some calculators, but the function is obtained by pressing the INV key first and then the $\sqrt{x}$ key. The square is the inverse of the square root. On one calculator, the cube root of a number $\sqrt[3]{x}$ is the inverse of the +/− key. On this calculator, the cube of the number must be obtained by first squaring the number, then multiplying the number again. On another calculator, taking the cube root is the same as raising the number to the 1/3 power. There are several variations of how cube functions are presented, and you should experiment with your own calculator in this regard.

There is also variation in the order of operations. For example, some calculators require that you press the "$\sqrt{\phantom{x}}$" (square root) or the "log" signs first, then the number, followed by "=" or "exe."

---

**Example E-4**

working it out

**Roots and Powers**

Carry out the following calculations.

**a.** $(6.2)^2$

**b.** $(6.2)^3$

| Enter | Display |
|-------|---------|
| 6.2 | 6.2 |
| INV | 6.2 |
| $\sqrt{x}$ | 38.44 |

| Enter | Display |
|-------|---------|
| 6.2 | 6.2 |
| $x^y$ | 6.2 |
| 3 | 3 |
| = | 238.328 |

**c.** $\sqrt{6.2}$

**d.** $\sqrt[3]{6.2}$

| Enter | Display |
|-------|---------|
| 6.2 | 6.2 |
| $\sqrt{x}$ | 2.49 |

| Enter | Display |
|-------|---------|
| 6.2 | 6.2 |
| INV* | 6.2 |
| $x^y$ | 6.2 |
| 3 | 3 |
| = | 1.837 |

*INV $x^y$ is the same as $x^{1/y}$

---

**Example E-5**

**Logs and Antilogs**

Find the following logs (common) and antilogs.

**a.** log 35.6

**b.** log $4.88 \times 10^{-9}$

**c.** the antilog of $-3.876$

| Enter | Display |
|-------|---------|
| 35.6 | 35.6 |
| log | 1.55 |

| Enter | Display | |
|-------|---------|---|
| 4.88 | 4.88 | |
| EXP | 4.88 | 00 |
| 9 | 4.88 | 09 |
| +/− | 4.88 | −09 |
| log | −8.312 | |

| Enter | Display |
|-------|---------|
| 3.876 | 3.876 |
| +/− | −3.876 |
| INV | −3.876 |
| log | 0.000133 or |
| | 1.3304 −04 |

# Exercises

**E-2.** Perform the following calculations on your calculator.

**(a)** $0.230^2$          **(g)** the log of 0.0257

**(b)** $\sqrt{6.80 \times 10^{-3}}$     **(h)** the log of 567

**(c)** $\sqrt[3]{0.986}$         **(i)** the antilog of 0.665

**(d)** $(3.60)^4$          **(j)** the antilog of 2.977

**(e)** $\sqrt[5]{4.32 \times 10^6}$    **(k)** the antilog of $-4.005$

**(f)** the log of 8.33

*Answers* (given to three significant figures):   **(a)** 0.0529    **(b)** 0.0825    **(c)** 0.995    **(d)** 168    **(e)** 21.2    **(f)** 0.921    **(g)** $-1.590$    **(h)** 2.754    **(i)** 4.62    **(j)** 948    **(k)** $9.89 \times 10^{-5}$

# Appendix F
# Glossary

The numbers in parentheses at the end of each entry refer to the chapter in which the entry is first discussed plus any chapter in which the topic is discussed in detail.

## A

**absolute zero.** The lowest possible temperature. The temperature at which translational motion ceases. Defined as zero on the Kelvin scale. (1, 9)

**accuracy.** How close a measurement is to the true value. (1)

**acid.** A compound that increases the $H^+$ concentration in aqueous solution. In the Brønsted definition, an acid is a proton donor. (4, 7, 12, 14)

**acid anhydride.** A molecular oxide that dissolves in water to form an oxyacid. (12)

**acid ionization constant ($K_a$).** The equilibrium constant specifically for the ionization of a weak acid. The magnitude of the constant relates to the strength of the acid. (14)

**acid salt.** An ionic compound containing one or more acidic hydrogens on the anion. (12)

**acidic solution.** A solution with pH $< 7$ or $[H_3O^+] > 10^{-7}$ M. (12)

**actinide.** One of the 14 elements between Ac and Rf. An element whose $5f$ subshell is filling. (4, 5)

**activated complex.** The transition state formed at the instant of maximum impact between reacting molecules. (14)

**activation energy.** The minimum energy needed by reactant molecules to form an activated complex so that a reaction may occur. It is the difference in potential energy between the reactants and the activated complex. (14)

**activity series.** The ability of metals to replace other metal ions can be compared and ranked in an activity series. (7)

**actual yield.** The measured amount of product in a chemical reaction. (8)

**alcohol.** An organic compound containing at least one hydroxyl group (i.e., R—OH). (16)

**aldehyde.** An organic compound containing a carbonyl group (i.e., C=O) bound to at least one hydrogen (i.e., RCHO). (16)

**alkali metal.** An element in Group IA in the periodic table (except H). Elements with the electron configuration $[NG]ns^1$. (4, 5)

**alkaline earth metal.** An element in Group IIA in the periodic table. Elements with the electron configuration $[NG]ns^2$. (4, 5)

**alkane.** A hydrocarbon containing only single bonds (a saturated hydrocarbon). (16)

**alkene.** A hydrocarbon containing at least one double bond (an unsaturated hydrocarbon). (16)

**alkylation.** The formation of large hydrocarbon molecules from small ones in the refining process. (16)

**alkyl group.** A substituent that is an alkane minus a hydrogen atom [e.g., $C_2H_6$ (ethane) $- H = C_2H_5$ (ethyl)]. (16)

**alkyne.** A hydrocarbon containing a triple bond. (16)

**allotropes.** Different forms of the same element. (6)

**alloy.** A homogeneous mixture of metallic elements in one solid phase. (2)

**alpha ($\alpha$) particle.** A helium nucleus ($_2^4He^{2+}$) emitted from a radioactive nucleus. (3, 15)

**amide.** A derivative of a carboxylic acid in which an amine group replaces the hydroxyl group of the acid (i.e., $RCONH_2$). (16)

**amine.** An organic compound containing a nitrogen with a single bond to a hydrocarbon group and a total of two other bonds to hydrocarbon groups or hydrogens (i.e., $RNH_2$, $R_2NH$, $R_3N$). (16)

**amorphous solid.** A solid without a defined shape. The molecules in such a solid do not occupy regular, symmetrical positions. (10)

**amphiprotic.** Refers to a molecule or ion that has both a conjugate acid and a conjugate base. (12)

**anion.** A negatively charged ion. (3)

**anode.** The electrode at which oxidation takes place. (13)

**aromatic compound.** A cyclic hydrocarbon (usually six-membered) that can be written with alternating single and double bonds. (16)

**atmosphere.** The sea of gases above the surface of Earth. Also, the average pressure of the atmosphere at sea level, which is defined as the standard pressure and is abbreviated atm. (9)

**atom.** The smallest fundamental particle of an element that has the properties of that element. (3)

**atomic mass.** The weighted average of the isotopic masses of all of the naturally occurring isotopes of an element. The mass of an "average" atom of a naturally occurring element compared to $^{12}C$. (3)

**atomic mass unit (amu).** A mass that is exactly one-twelfth of the mass of an atom of $^{12}C$, which is defined as exactly 12 amu. (3)

**atomic number.** The number of protons (positive charge) in the nuclei of the isotopes of a particular element. (3)

**atomic radius.** The distance from the nucleus of an atom to the outermost electron. (5)

**atomic theory.** A theory first proposed by John Dalton in 1803 that holds that the basic components of matter are atoms. (3)

**Aufbau principle.** A rule that states that electrons fill the lowest available energy level first. (5)

**autoionization.** The dissociation of a solvent to produce positive and negative ions. (12)

**Avogadro's law.** A law that states that the volume of a gas is directly proportional to the number of moles of gas present at constant temperature and pressure (i.e., $V = kn$). (9)

**Avogadro's number.** The number of objects or particles in one mole, which is $6.02 \times 10^{23}$. (8)

## B

**balanced equation.** A chemical equation that has the same number and type of atoms on both sides of the equation. (7)

**barometer.** A device that measures the atmospheric pressure. (9)

**base.** A compound that increases the $OH^-$ concentration in aqueous solution. In the Brønsted definition, a base is a proton acceptor. (7, 12)

**base anhydride.** An ionic oxide that dissolves in water to form an ionic hydroxide. (12)

**base ionization constant ($K_b$).** An equilibrium constant specifically for the ionization of a weak base. The magnitude of the constant relates to the strength of the base. (14)

**basic solution.** A solution with pH > 7 or $[OH^-] > 10^{-7}$ M. (12)

**battery.** One or more voltaic cells joined together as a single unit. (13)

**beta ($\beta$) particle.** A high-energy electron emitted from a radioactive nucleus. (15)

**binary acid.** An acid composed of hydrogen and one other element. (4)

**Bohr model.** A model of the atom in which electrons orbit around a central nucleus in discrete energy levels. (5)

**boiling point.** The temperature at which a pure substance changes from the liquid to the gaseous state. (2) It is the temperature at which the vapor pressure equals the restraining pressure. (10, 11)

**box diagram.** See *orbital diagram.*

**Boyle's law.** A law that states that the volume of a gas is inversely proportional to the pressure at constant temperature (i.e., $PV = k$). (9)

**Brønsted–Lowry definition.** Acids are proton donors and bases are proton acceptors. (12)

**buffer solution.** A solution that resists changes in pH from the addition of limited amounts of strong acid or base. Made from a solution of a weak acid and a salt containing its conjugate base, or a weak base and a salt containing its conjugate acid. (12)

## C

**calorie.** The amount of heat required to raise one gram of water one degree Celsius. Equal to exactly 4.184 joules. (2)

**carbonyl group.** The functional group of ketones and aldehydes, made up of a carbon with a double bond to an oxygen (i.e., C=O). (16)

**carboxylic acid.** An organic compound containing a carbonyl group attached to a hydroxide group and a hydrocarbon group (i.e., RCOOH). (16)

**catalyst.** A substance that is not consumed in a reaction but whose presence increases the rate of the reaction. A catalyst lowers the activation energy. (14)

**cathode.** The electrode at which reduction takes place. (13)

**cation.** A positively charged ion. (3)

**Celsius.** A temperature scale with 100 equal divisions between the freezing and boiling points of water at average sea level pressure, with exactly zero assigned to the freezing point of water. (1)

**chain reaction.** A self-sustaining reaction. In a nuclear chain reaction, one reacting neutron produces between two and three other neutrons that in turn cause reactions. (15)

**Charles's law.** A law that states that the volume of a gas is directly proportional to the Kelvin temperature at constant pressure (i.e., $V = kT$). (9)

**chemical change.** A change in a substance to another substance or substances. (2)

**chemical equation.** The representation of a chemical reaction using the symbols of the elements and the formulas of compounds. (7)

**chemical property.** The property of a substance relating to its tendency to undergo chemical changes. (2)

**chemical thermodynamics.** The study of heat and its relationship to chemical changes. (8)

**chemistry.** The branch of science dealing with the nature, composition, and structure of matter and the changes it undergoes. (2)

**coefficient.** The number before an element or compound in a balanced chemical equation indicating the number of molecules or moles of that substance. (7)

**colligative property.** A property that depends only on the relative amounts of solute and solvent present, not on their identities. (11)

**collision theory.** A theory that states that chemical reactions are brought about by collisions of molecules. (14)

**combination reaction.** A chemical reaction whereby one compound is formed from two elements and/or compounds. (7)

**combined gas law.** A law that relates the pressure, volume, and temperature of a gas (i.e., $PV = kT$). (9)

**combustion reaction.** A chemical reaction whereby an element or compound reacts with elemental oxygen. (7)

**compound.** A pure substance composed of two or more elements that are chemically combined in fixed proportions. (2)

**concentrated solution.** A solution containing a relatively large amount of a specified solute per unit volume. (11)

**concentration.** The amount of solute present in a specified amount of solvent or solution. (11)

**condensation point.** The temperature at which a pure substance changes from the gas to the liquid or solid state. (2, 10)

**conductor.** A substance that allows a flow of electricity. (11)

**conjugate acid.** An acid formed by addition of one $H^+$ to a base. (12)

**conjugate base.** A base formed by removal of one $H^+$ from an acid. (12)

**conservation of energy.** A law that states that energy is neither created nor destroyed but can be transformed from one form to another. (2)

**conservation of mass.** A law that states that matter is neither created nor destroyed in a chemical reaction. (2)

**continuous spectrum.** The visible spectrum where one color blends gradually into another. A spectrum containing all wavelengths of visible light. (5)

**conversion factor.** A relationship between two units or quantities expressed in fractional form. (1)

**covalent bond.** The force that bonds two atoms together by a shared pair or pairs of electrons. (3, 6)

**cracking.** The formation of small hydrocarbon molecules from large ones in the refining process. (16)

**crystal lattice.** See *lattice*.

**crystalline solid.** A solid with a regular, symmetrical shape where the molecules or ions occupy set positions in the crystal lattice. (10)

**D**

**Dalton's law.** A law that states that the total pressure of a gas in a system is the sum of the partial pressures of all component gases. (9)

**Daniell cell.** A voltaic cell made up of zinc and copper electrodes immersed in solutions of their respective ions that are connected by a salt bridge. (13)

**decomposition reaction.** A chemical reaction whereby one compound decomposes to two or more elements and/or compounds. (7)

**density.** The ratio of the mass (usually in grams) to the volume (usually in milliliters or liters) of a substance. (2)

**derivative.** A compound produced by the substitution of a group or hetero atom on the molecules of the original or parent compound. (16)

**diffusion.** The mixing of one gas or liquid into others. (9)

**dilute solution.** A solution containing a relatively small amount of a specified solute per unit volume. (11)

**dilution.** The preparation of a dilute solution from a concentrated solution by the addition of more solvent. (11)

**dimensional analysis.** See *factor-label method*.

**dipole.** Two poles—one positive and one negative—that may exist in a bond or molecule. (6)

**dipole–dipole attractions.** The force of attraction between a dipole on one polar molecule and a dipole on another polar molecule. (10)

**diprotic acid.** An acid that can produce two $H^+$ ions per molecule. (12)

**discrete spectrum.** A spectrum containing specific wavelengths of light originating from the hot gaseous atoms of an element. (5)

**distillation.** A laboratory procedure where a solution is separated into its components by boiling the mixture and condensing the vapor to form a liquid. (2)

**double bond.** The sharing of two pairs of electrons between two atoms. (6)

**double-replacement reaction.** A chemical reaction whereby the cations and anions in two compounds exchange, leading to formation of either a precipitate or a molecular compound such as water. (7)

**dry cell.** A voltaic cell composed of zinc and graphite electrodes immersed in an aqueous paste of $NH_4Cl$ and $MnO_2$. (13)

**dynamic equilibrium.** See *point of equilibrium*.

**E**

**effusion.** The movement of gases through an opening or hole. (9)

**electricity.** A flow of electrons through a conductor. (11)

**electrode.** A surface in a cell where the oxidation or reduction reaction takes place. (13)

**electrolysis.** The process of forcing electrical energy through an electrolytic cell, thereby causing a chemical reaction. (13)

**electrolyte.** A solute whose aqueous solution or molten state conducts electricity. (11)

**electrolytic cell.** A voltaic cell that converts electrical energy into chemical energy. It involves a nonspontaneous reaction. (13)

**electron.** A negatively charged particle in an atom, with a comparatively very small mass. (3)

**electron capture.** The capture of an orbital electron by a radioactive isotope. (15)

**electron configuration.** The designation of all of the electrons in an atom into specific shells and subshells. (5)

**electronegativity.** The ability of an atom of an element to attract electrons to itself in a covalent bond. (6)

**electrostatic forces.** The forces of attraction between unlike charges and of repulsion by like charges. (3)

**element.** A pure substance that cannot be broken down into simpler substances. The most basic form of matter existing under ordinary conditions. (2)

**empirical formula.** The simplest whole-number ratio of atoms in a compound. (8)

**endothermic reaction.** A chemical reaction that absorbs heat from the surroundings. (2, 8, 14)

**energy.** The capacity or the ability to do work. It comes in several different forms (e.g., light and heat) and two types (kinetic and potential). (2)

**equilibrium.** See *point of equilibrium*.

**equilibrium constant.** A number that defines the position of equilibrium for a particular reaction at a specified temperature. (14)

**equilibrium constant expression.** The ratio of the concentrations of products to reactants, each raised to the power corresponding to its coefficient in the balanced equation. (14)

**equilibrium vapor pressure.** The pressure exerted by a vapor above a liquid at a certain temperature. (10)

**ester.** A derivative of a carboxylic acid in which a hydrocarbon group (R') replaces the hydrogen on the oxygen in the parent acid (i.e., RCOOR'). (16)

**ether.** An organic compound containing an oxygen bonded to two hydrocarbon groups (i.e., ROR'). (16)

**evaporation.** The vaporization of a liquid below its boiling point. (10)

**exact number.** A number that results from a definition or an actual count. (1)

**excited state.** An energy level higher than the lowest available energy level. (5)

**exothermic reaction.** A chemical reaction that occurs with the evolution of heat to the surroundings. (2, 8, 14)

**exponent.** In scientific notation, it is the power to which ten is raised. (1)

## F

**factor-label method.** A problem-solving technique that converts from one unit to another by use of conversion factors. (1)

**Fahrenheit.** A temperature scale with 180 divisions between the freezing and boiling points of water, with exactly 32 assigned to the freezing point of water. (1)

**family.** See *group*.

**filtration.** A laboratory procedure where solids are removed from liquids by passing the heterogeneous mixture through a filter. (2)

**fission.** The splitting of a large, unstable nucleus into two smaller nuclei of similar size, resulting in the production of energy. (15)

**formula.** The symbols of the elements and the number of atoms of each element that make up a compound. (3)

**formula unit.** The simplest whole-number ratio of ions in an ionic compound. (3)

**formula weight.** The mass of a compound (in amu), which is determined from the number of atoms and the atomic mass of each element indicated by the formula. (8)

**freezing point.** The temperature at which a pure substance changes from the liquid to the solid state. (2, 11)

**fuel cell.** A voltaic cell that can generate a continuous flow of electricity from the reaction of hydrogen and oxygen to produce water. (13)

**functional group.** The atom or group of atoms that determines the chemical nature of the molecules of an organic compound. (16)

**fusion (nuclear).** The combination of two small nuclei to form a larger nucleus, resulting in the production of energy. (15)

## G

**gamma (γ) ray.** A high-energy form of light emitted from a radioactive nucleus. (15)

**gas.** A physical state that has neither a definite volume nor a definite shape and fills a container uniformly. (2, 9)

**gas constant.** The constant of proportionality ($R$) in the ideal gas law. (9)

**gas law.** A law governing the behavior of gases that is consistent with the kinetic molecular theory as applied to gases. (9)

**Gay-Lussac's law.** A law that states that the pressure of a gas is directly proportional to the Kelvin temperature at constant volume (i.e., $P = kT$). (9)

**Graham's law.** A law that states that the rates of diffusion of gases are inversely proportional to the square root of their molar masses. (9)

**ground state.** The lowest available energy level in an atom. (5)

**group.** A vertical column of elements in the periodic table. (4)

## H

**half-life.** The time required for one-half of a given sample of an isotope to undergo radioactive decay. (15)

**half-reaction.** The oxidation or reduction process in a redox reaction written separately. (13)

**halogen.** An element in Group VIIA in the periodic table. Elements with the electron configuration [NG] $ns^2np^5$. (4, 5)

**heat of fusion.** The amount of heat in calories or joules required to melt one gram of a substance. (10)

**heat of reaction.** The amount of heat energy absorbed or evolved in a specified chemical reaction. (8)

**heat of vaporization.** The amount of heat in calories or joules required to vaporize one gram of the substance. (10)

**heating curve.** The graphical representation of the temperature as a solid is heated through two phase changes, plotted as a function of the time of heating. (10)

**Henderson–Hasselbalch equation.** Equations that are used to calculate the pH of buffer solutions $\left( \text{i.e., pH} = pK_a + \log \frac{[\text{base}]}{[\text{acid}]}; \text{pOH} = pK_b + \log \frac{[\text{acid}]}{[\text{base}]} \right)$. (14)

**heterogeneous mixture.** A nonuniform mixture containing two or more phases with definite boundaries between phases. (2)

**homogeneous mixture.** A mixture that is the same throughout and contains only one phase. (2)

**homologous series.** A series of related compounds in which each member differs from the next by a constant number of atoms. (16)

**Hund's rule.** A rule that states that electrons occupy separate orbitals with parallel spins if possible. (5)

**hydrocarbon.** An organic compound containing only carbon and hydrogen. (16)

**hydrogen bonding.** A force of attraction between a N, O, or F on one molecule and a hydrogen bonded to a N, O, or F on another. (10)

**hydrolysis reaction.** The reaction of an anion as a base with water or a cation as an acid. (12)

**hydrometer.** A device that measures the density or specific gravity of a liquid. (2)

**hydronium ion.** A representation of the hydrogen ion in aqueous solution ($H_3O^+$). (12)

**hydroxyl group.** The —OH group as found in an organic compound. It is the functional group of alcohols. (16)

**hypothesis.** A tentative explanation of related data. It can be used to predict results of more experiments. (Prologue)

**I**

**ideal gas.** A hypothetical gas whose molecules are considered to have no volume or interactions with each other. An ideal gas would obey the ideal gas law under all conditions. (9)

**ideal gas law.** A relationship between the pressure, volume, temperature, and number of moles of gas (i.e., $PV = nRT$). (9)

**immiscible liquids.** Two liquids that do not mix and thus form a heterogeneous mixture. (11)

**improper fraction.** A fraction whose numerator is larger than the denominator and thus has a value greater than one. (Appendix A)

**infrared light.** Light with wavelengths somewhat longer than those of red light in the visible spectrum. (5)

**inner transition element.** Either a lanthanide, where the $4f$ subshell is filling, or an actinide, where the $5f$ subshell is filling. (4, 5)

**insoluble compound.** A compound that does not dissolve to any appreciable amount in a solvent. (7)

**insulator.** See *nonconductor*.

**intermolecular forces.** The attractive forces between molecules. (10)

**ion.** An atom or group of covalently bonded atoms that has a net electrical charge. (3)

**ion product ($K_w$).** The equilibrium expression of the anion and the cation of water (i.e., $[H_3O^+][OH^-] = K_w$). (12)

**ion–dipole force.** The force between an ion and the dipole of a polar molecule. (11)

**ion-electron method.** A method of balancing oxidation-reduction reactions where two half-reactions are balanced separately and then added. (13)

**ionic bond.** The electrostatic force holding the positive and negative ions together in an ionic compound. (3, 6)

**ionic compound.** Compounds containing positive and negative ions. (3)

**ionic solid.** A solid where the crystal lattice positions are occupied by ions. (10)

**ionization.** The process of forming an ion or ions from a molecule or atom. (5, 11, 12, 15)

**ionization energy.** The energy required to remove an electron from a gaseous atom or ion. (5)

**isomer.** Compounds with the same formula but different structure. (16)

**isotopes.** Atoms having the same atomic number but different mass numbers. (3)

**isotopic mass.** The mass of an isotope compared to $^{12}C$, which is defined as having a mass of exactly 12 amu. (3)

**J**

**joule.** The SI unit for measurement of heat energy. (2)

**K**

**Kelvin scale.** A temperature scale in which 0 K is the lowest possible temperature. $T$ (K) = $t$ (°C) + 273. (1, 9)

**ketone.** An organic compound containing a carbonyl group bonded to two hydrocarbon groups (i.e., RCOR). (16)

**kinetic energy.** Energy as a result of motion; equal to $\frac{1}{2}mv^2$. (2, 9)

**kinetic molecular theory.** A theory advanced in the late 1800s to explain the nature of gases. (9)

**L**

**lanthanide.** One of fourteen elements between La and Hf. Elements whose $4f$ subshell is filling. (4, 5)

**lattice.** A three-dimensional array of ions or molecules in a solid crystal. (6)

**law.** A concise statement or mathematical relationship that describes some behavior of matter. (Prologue)

**Law of conservation of energy.** Energy cannot be created or destroyed but only transformed from one form to another. (2)

**Law of conservation of mass.** Matter is neither created nor destroyed in a chemical reaction. (2)

**law of mass action.** The relationship that describes the relative distribution of the concentrations of reactants and products at equilibrium. It is composed of a constant and a ratio called the equilibrium constant expression. (14)

**lead–acid battery.** A rechargeable voltaic battery composed of lead and lead dioxide electrodes in a sulfuric acid solution. (13)

**Le Châtelier's principle.** A principle that states that when stress is applied to a system at equilibrium, the system reacts in such a way as to counteract the stress. (14)

**Lewis dot symbols.** The representation of an element by its symbol with its valence electrons as dots. (6)

**Lewis structure.** The representation of a molecule or ion showing the order and arrangement of the atoms as well as the bonded pairs and unshared electrons of all of the atoms. (6)

**limiting reactant.** The reactant that produces the least amount of product when that reactant is completely consumed. (8)

**liquid.** A physical state that has a definite volume but not a definite shape. Liquids take the shape of the bottom of the container. (2, 10)

**London forces.** The instantaneous dipole-induced dipole forces between molecules caused by an instantaneous imbalance of electrical charge in a molecule. The force is dependent on the size of the molecule. (10)

**M**

**main group element.** See *representative element*.

**mass.** The quantity of matter (usually in grams or kilograms) in a sample. (1)

**mass number.** The number of nucleons (neutrons and protons) in a nucleus. (3)

**matter.** Anything that has mass and occupies space. (2)

**measurement.** The quantity, dimensions, or extent of something, usually in comparison to a specific unit. (1)

**melting point.** The temperature at which a pure substance changes from the solid to the liquid state. (2)

**metal.** An element with a comparatively low ionization energy that forms positive ions in compounds. Generally, metals are hard, lustrous elements that are ductile and malleable. (4, 5)

**metallic solid.** A solid made of metals where positive metal ions occupy regular positions in the crystal lattice with the valence electrons moving freely among these positive ions. (10)

**metalloid.** Elements with properties intermediate between metals and nonmetals. Many of the elements on the metal–nonmetal borderline in the periodic table. (4)

**metallurgy.** The conversion of metal ores into metals. (Prologue)

**metric system.** A system of measurement based on multiples of 10. (1)

**miscible liquids.** Two liquids that mix or dissolve in each other to form a solution. (11)

**model.** A description or analogy used to help visualize a phenomenon. (5)

**molality.** A temperature independent unit of concentration that relates the moles of solute to the mass (kg) of solvent. (11)

**molar mass.** The atomic mass of an element or the formula weight of a compound expressed in grams. (8)

**molar volume.** The volume of one mole of a gas at STP, which is 22.4 L. (9)

**molarity.** A unit of concentration that relates moles of solute to volume (in liters) of solution. (11)

**mole.** A unit of $6.022 \times 10^{23}$ atoms, molecules, or formula units. It is the same number of particles as there are atoms in exactly 12 grams of $^{12}C$. It also represents the atomic mass of an element or the formula weight of a compound expressed in grams. (8)

**mole ratio.** The ratios of moles from a balanced equation that serve as conversion factors in stoichiometry calculations. (8)

**molecular compound.** A compound composed of discrete molecules. (3)

**molecular dipole.** The combined or net effect of all of the bond dipoles in a molecule as determined by the molecular geometry. (6)

**molecular equation.** A chemical equation showing all reactants and products as neutral compounds. (7)

**molecular formula.** See *formula*.

**molecular geometry.** The geometry of a molecule or ion described by the bonded atoms. It does not include the unshared pairs of electrons. (6)

**molecular solid.** A solid where the individual molecules in the crystal lattice are held together by London forces, dipole–dipole attractions, or hydrogen bonding. (10)

**molecular weight.** The formula weight of a molecular compound. (8)

**molecule.** The basic unit of a molecular compound, which is two or more atoms held together by covalent bonds. (3)

**monoprotic acid.** An acid that can produce only one $H^+$ ion per molecule. (12)

## N

**net ionic equation.** A chemical equation shown in ionic form with spectator ions eliminated. (7)

**network solid.** A solid where the atoms are covalently bonded to each other throughout the entire crystal. (10)

**neutral.** Pure water or a solution with pH = 7. (12)

**neutralization reaction.** A reaction whereby an acid reacts with a base to form a salt and water. The reaction of $H^+$ (*aq*) with $OH^-$ (*aq*). (7, 12)

**neutron.** A particle in the nucleus with a mass of about 1 amu and no charge. (3)

**noble gas.** An element with a full outer *s* and *p* subshell. Group VIIIA in the periodic table. (4, 5)

**nonconductor.** A substance that does not conduct electricity. (11)

**nonelectrolyte.** A solute whose aqueous solution or molten state does not conduct electricity. (11)

**nonmetal.** Elements to the right in the periodic table. These elements generally lack metallic properties. They have relatively high ionization energies. (4, 5)

**nonpolar bond.** A covalent bond in which electrons are shared equally. (6)

**normal boiling point.** The temperature at which the vapor pressure of a liquid is equal to exactly one atmosphere pressure. (10)

**nuclear equation.** A symbolic representation of the changes of a nucleus or nuclei into other nuclei and particles. (15)

**nuclear reactor.** A device that can maintain a controlled nuclear fission reaction. Used either for research or generation of electrical power. (15)

**nucleons.** The protons and neutrons that make up the nucleus of the atom. (3, 15)

**nucleus.** The core of the atom containing neutrons, protons, and most of the mass. (3, 15)

## O

**octet rule.** A rule that states that atoms of representative elements form bonds so as to have access to eight electrons. (6)

**orbital.** A region of space where there is the highest probability of finding a particular electron. There are four types of orbitals; each has a characteristic shape. (5)

**orbital diagram.** The representation of specific orbitals of a subshell as boxes and the electrons as arrows in the boxes. (5)

**organic chemistry.** The branch of chemistry that deals with most of the compounds of carbon. (16)

**osmosis.** The tendency of a solvent to move through a semipermeable membrane from a region of low concentration to a region of high concentration of solute. (11)

**osmotic pressure.** The pressure needed to counteract the movement of solvent through a semipermeable membrane from a region of low concentration of solute to a region of high concentration. (11)

**oxidation.** The loss of electrons as indicated by an increase in oxidation state. (13)

**oxidation–reduction reaction.** A chemical reaction involving an exchange of electrons. (13)

**oxidation state (number).** The charge on an atom in a compound if all atoms were present as monatomic ions. The electrons in bonds are assigned to the more electronegative atom. (13)

**oxidation state method.** A method of balancing oxidation–reduction reactions that focuses on the atoms of the elements undergoing a change in oxidation state. (13)

**oxidizing agent.** The element, compound, or ion that oxidizes another reactant. It is reduced. (13)

**oxyacid.** An acid composed of hydrogen and an oxyanion. (4)

**oxyanion.** An anion composed of oxygen and one other element. (4)

**P**

**partial pressure.** The pressure of one component in a mixture of gases. (9)

**parts per billion (ppb).** A unit of concentration obtained by multiplying the ratio of the mass of solute to the mass of solution by $10^9$ ppb. (11)

**parts per million (ppm).** A unit of concentration obtained by multiplying the ratio of the mass of solute to the mass of solution by $10^6$ ppm. (11)

**Pauli exclusion principle.** A rule that states that no two electrons can have the same spin in the same orbital. (5)

**percent by mass.** The mass of solute expressed as a percent of the mass of solution. (11)

**percent composition.** The mass of each element expressed per 100 mass units of the compound. (8)

**percent yield.** The actual yield in grams or moles divided by the theoretical yield in grams or moles and multiplied by 100%. (8)

**period.** A horizontal row of elements between noble gases in the periodic table. (4)

**periodic law.** A law that states that the properties of elements are periodic functions of their atomic numbers. (4)

**periodic table.** An arrangement of elements in order of increasing atomic number. Elements with the same number of outer electrons are arranged in vertical columns. (4, 5)

**pH.** The negative of the common logarithm of the $H_3O^+$ concentration. (12)

**phase.** A homogeneous state (solid, liquid, or gas) with distinct boundaries and uniform properties. (2)

**physical change.** A change in physical state or dimensions of a substance that does not involve a change in composition. (2)

**physical properties.** Properties that can be observed without changing the composition of a substance. (2)

**physical states.** The physical condition of matter—solid, liquid, or gas. (2)

**pOH.** The negative of the common logarithm of the $OH^-$ concentration. (12)

**point of equilibrium.** The point at which the forward and reverse reactions in a reversible reaction occur at the same rate so that the concentrations of both reactants and products remain constant. (14)

**polar covalent bond.** A covalent bond that has a partial separation of charge due to the unequal sharing of electrons. (6)

**polyatomic ion.** A group of atoms covalently bonded to each other that have a net electrical charge. (3)

**polymer.** A high-molar-mass substance made from repeating units of an alkene (addition) or from combinations of other molecules (condensation). (16)

**polyprotic acid.** An acid that can produce more than one $H^+$ ion per molecule. (12)

**positron.** A positively charged electron emitted from a radioactive isotope. (15)

**potential energy.** Energy as a result of position or composition. (2, 14)

**precipitate.** A solid compound formed in a solution. (7)

**precipitation reaction.** A type of double-replacement reaction in which an insoluble ionic compound is formed by an exchange of ions in the reactants. (7)

**precision.** The reproducibility of a measurement as indicated by the number of significant figures expressed. (1)

**pressure.** The force per unit area. (9)

**principal quantum number ($n$).** A number that corresponds to a particular shell occupied by the electrons in an atom. (5)

**product.** An element or compound in an equation that is formed as a result of a chemical reaction. (7)

**proper fraction.** A fraction whose numerator is smaller than the denominator and thus has a value less than one. (Appendix A)

**property.** A particular characteristic or trait of a substance. (2)

**proton.** A particle in the nucleus with a mass of about 1 amu and a charge of +1. (3)

**pure substance.** A substance that has a definite composition with definite and unchanging properties (i.e., elements and compounds). (2)

**Q**

**quantized energy level.** An energy level with a definite and measurable energy. (5)

**R**

**radiation.** Particles or high-energy light rays that are emitted by an atom or a nucleus of an atom. (15)

**radioactive decay series.** A series of elements formed from the successive emission of alpha and beta particles starting from a long-lived isotope and ending with a stable isotope. (15)

**radioactivity.** The emission of energy or particles from an unstable nucleus. (15)

**rate of reaction.** A measure of the increase in concentration of a product or the decrease in concentration of a reactant per unit time. (14)

**reactant.** An element of compound in an equation that undergoes a chemical reaction. (7)

**recrystallization.** A laboratory procedure whereby a solid compound is purified by saturating a solution at a high temperature and then forming a precipitate at a lower temperature. (11)

**redox reaction.** See *oxidation–reduction reaction.*

**reducing agent.** An element, compound, or ion that reduces another reactant. It is oxidized. (13)

**reduction.** The gain of electrons as indicated by a decrease in the oxidation state. (13)

**reforming.** The rearrangement of long-chain hydrocarbons to branched hydrocarbons and/or removal of hydrogen from hydrocarbons. (16)

**representative element.** Elements whose outer *s* and *p* subshells are filling. The A Group elements in the periodic table (except for VIIIA). (4, 5)

**resonance hybrid.** The actual structure of the molecule as implied by the resonance structures. (6)

**resonance structure.** A Lewis structure showing one of two or more possible Lewis structures. (6)

**reversible reaction.** A reaction where both a forward reaction (forming products) and a reverse reaction (reforming reactants) can occur. (14)

**room temperature.** The standard reference temperature for physical state, which is usually defined as 25°C. (4)

**S**

**salt.** An ionic compound formed by the combination of most cations and anions. Also, the compound that forms from the cation of a base and the anion of an acid. (4, 7, 12)

**salt bridge.** An aqueous gel that allows anions to migrate between compartments in a voltaic cell. (13)

**saturated hydrocarbon.** A hydrocarbon containing only single covalent bonds (an alkane). (16)

**saturated solution.** A solution containing the maximum amount of dissolved solute at a specific temperature. (11)

**scientific method.** The method whereby modern scientists explain the behavior of nature with hypotheses and theories, or describe the behavior of nature with laws. (Prologue)

**scientific notation.** A number expressed with one nonzero digit to the left of the decimal point multiplied by 10 raised to a given power. (1, Appendix C)

**semimetal.** See *metalloid.*

**shell.** The principal energy level which contains one or more subshells. (5)

**SI units.** An international system of units of measurement. (1)

**significant figure.** A digit or number in a measurement that is either reliably known or is estimated. (1)

**single-replacement reaction.** A chemical reaction whereby a free element substitutes for another element in a compound. (7)

**solid.** A physical state with both a definite shape and a definite volume. (2, 10)

**solubility.** The maximum amount of a solute that dissolves in a specific amount of solvent at a certain temperature. (11)

**solubility product constant ($K_{sp}$).** The equilibrium constant associated with the solution of ionic compounds. (14)

**soluble compound.** A compound that dissolves to an appreciable extent in a solvent. (7)

**solute.** A substance that dissolves in a solvent. (7, 11)

**solution.** A homogeneous mixture with one phase. It is composed of a solute dissolved in a solvent. (2, 7, 11)

**solvent.** A medium, usually a liquid, that disperses a solute to form a solution. (7, 11)

**specific gravity.** The ratio of the mass of a substance to the mass of an equal volume of water under the same conditions. (2)

**specific heat.** The amount of heat required to raise the temperature of one gram of a substance one degree Celsius. (2)

**spectator ion.** An ion that is in an identical state on both sides of an equation. (7)

**spectrum.** The separate color components of a beam of light. (5)

**standard temperature and pressure (STP).** The defined standard conditions for a gas, which are exactly 0°C and one atmosphere pressure. (9)

**Stock method.** A method used to name metal–nonmetal or metal–polyatomic ion compounds where the charge on the metal is indicated by Roman numerals enclosed in parentheses. (4)

**stoichiometry.** The quantitative relationship among reactants and products. (8)

**strong acid (base).** An acid (or base) that is completely ionized in aqueous solution. (7, 12)

**structural formula.** Formulas written so that the order and arrangement of specific atoms are shown. (3, 6, 16)

**sublimation.** The vaporization of a solid. (10)

**subshell.** The orbitals of the same type within a shell. The subshells are named for the types of orbitals, that is, *s, p, d,* or *f.* (5)

**substance.** A form of matter. Usually thought of as either an element or a compound. (2)

**substituent.** An atom or group of atoms that can substitute for a hydrogen atom in an organic compound. (16)

**supersaturated solution.** A solution containing more than the maximum amount of solute indicated by the compound's solubility at that temperature. (11)

**surface tension.** The forces of attraction between molecules that cause a liquid surface to contract. (10)

**symbol.** One or two letters from an element's English or, in some cases, Latin name. (3)

**T**

**temperature.** A measure of the intensity of heat of a substance. It relates to the average kinetic energy of the substance. (1, 9)

**theoretical yield.** The calculated amount of product that would be obtained if all of a reactant were converted to a certain product. (8)

**theory.** A hypothesis that withstands the test of time and experiments designed to test the hypothesis. (Prologue)

**thermochemical equation.** A balanced equation that includes the amount of heat energy. (8)

**thermometer.** A device that measures temperature. (1)

**torr.** A unit of gas pressure equivalent to the height of one millimeter of mercury. (9)

**total ionic equation.** A chemical equation showing all soluble compounds that exist primarily as ions in aqueous solution as separate ions. (7)

**transition element.** Elements whose outer $s$ and $d$ subshells are filling. The B Group elements in the periodic table. (4, 5)

**transmutation.** The changing of one element into another by a nuclear reaction. (15)

**triple bond.** The sharing of three pairs of electrons in a bond between two atoms. (6)

**triprotic acid.** An acid that can produce three $H^+$ ions per molecule. (12)

## U

**ultraviolet light.** Light with wavelengths somewhat shorter than those of violet light. (5)

**unit.** A definite quantity adapted as a standard of measurement. (1)

**unit factor.** A fractional expression that relates a quantity in a certain unit to "one" of another unit. (1)

**unit map.** A shorthand representation of the procedure for solving a problem that indicates the conversion of units in one or more steps. (1, 8)

**unsaturated hydrocarbon.** A hydrocarbon that contains at least one double or triple bond. (16)

**unsaturated solution.** A solution that contains less than the maximum amount of solute indicated by the compound's solubility at that temperature. (11)

## V

**valence electron.** An outer $s$ or $p$ electron in the atom of a representative element. (6)

**valence shell electron-pair repulsion theory (VSEPR).** A theory that predicts that electron pairs either unshared or in a bond repel each other to the maximum extent. (6)

**vapor pressure.** See *equilibrium vapor pressure.*

**viscosity.** A measure of the resistance of a liquid to flow. (10)

**volatile.** Refers to a liquid or solid with a significant vapor pressure. (11, 16)

**voltaic cell.** A spontaneous oxidation–reduction reaction that can be used to produce electrical energy. (13)

**volume.** The space that a certain quantity of matter occupies. (1)

## W

**wavelength ($\lambda$).** The distance between two adjacent peaks in a wave. (5)

**wave mechanics.** A complex mathematical approach to the electrons in an atom that considers the electron as having both a particle and a wave nature. (5)

**weak acid (base).** An acid (or base) that is only partially ionized in aqueous solution. (12)

**weak electrolyte.** A solute whose aqueous solution allows only a limited amount of electrical conduction. (11)

**weight.** A measure of the attraction of gravity for a sample of matter. (1)

# Appendix G
# Answers to Exercises

## Chapter 1

**1-1. (b)** 74.212 gal (the most significant figures)

**1-2.** A device used to produce a measurement may provide a reproducible answer to several significant figures, but if the device itself is inaccurate (such as a ruler with the tip broken off) the measurement is inaccurate.

**1-4. (a)** three **(b)** two **(c)** three **(d)** one
**(e)** four **(f)** two **(g)** two **(h)** three

**1-6. (a)** $\pm 10$ **(b)** $\pm 0.1$ **(c)** $\pm 0.01$ **(d)** $\pm 0.01$
**(e)** $\pm 1$ **(f)** $\pm 0.001$ **(g)** $\pm 100$ **(h)** $\pm 0.00001$

**1-8. (a)** 16.0 **(b)** 1.01 **(c)** 0.665 **(d)** 4890
**(e)** 87,600 **(f)** 0.0272 **(g)** 301

**1-10. (a)** 0.250 **(b)** 0.800 **(c)** 1.67 **(d)** 1.17

**1-12. (a)** $\pm 0.1$ **(b)** $\pm 1000$ **(c)** $\pm 1$ **(d)** $\pm 0.01$

**1-14. (a)** 188 **(b)** 12.90 **(c)** 2300 **(d)** 48
**(e)** 0.84

**1-16.** 37.9 qt

**1-18. (a)** 7.0 **(b)** 137 **(c)** 192 **(d)** 0.445
**(e)** 3.20 **(f)** 2.9

**1-20. (a)** two **(b)** three **(c)** two **(d)** one

**1-23. (a)** 6.07 **(b)** 6.049 **(c)** 8.62 **(d)** 7
**(e)** 3.98 **(f)** 3.973 **(g)** 0.5098

**1-26. (a)** $(63) + 75.0 = \underline{138}$ **(b)** $(45) \times 25.6 = \underline{1200}$
**(c)** $(2.7) \times (10.52) = \underline{28}$

**1-28. (a)** $1.57 \times 10^2$ **(b)** $1.57 \times 10^{-1}$
**(c)** $3.00 \times 10^{-2}$ **(d)** $4.0 \times 10^7$ **(e)** $3.49 \times 10^{-2}$
**(f)** $3.2 \times 10^4$ **(g)** $3.2 \times 10^{10}$ **(h)** $7.71 \times 10^{-4}$
**(i)** $2.34 \times 10^3$

**1-30. (a)** $9 \times 10^7$ **(b)** $8.7 \times 10^7$ **(c)** $8.70 \times 10^7$

**1-32. (a)** 0.000476 **(b)** 6550 **(c)** 0.00788
**(d)** 48,900 **(e)** 4.75 **(f)** 0.0000034

**1-34. (a)** $4.89 \times 10^{-4}$ **(b)** $4.56 \times 10^{-5}$ **(c)** $7.8 \times 10^3$ **(d)** $5.71 \times 10^{-2}$ **(e)** $4.975 \times 10^8$
**(f)** $3.0 \times 10^{-4}$

**1-36. (b)** < **(f)** < **(g)** < **(d)** < **(a)** < **(e)** < **(c)**

**1-38. (a)** $1.597 \times 10^{-3}$ **(b)** $2.30 \times 10^7$
**(c)** $3.5 \times 10^{-5}$ **(d)** $2.0 \times 10^{14}$

**1-40. (a)** $10^7$ **(b)** $10^0 = 1$ **(c)** $10^{29}$ **(d)** $10^9$

**1-42. (a)** $3.1 \times 10^{10}$ **(b)** $2 \times 10^9$ **(c)** $4 \times 10^{13}$
**(d)** 14 **(e)** $2.56 \times 10^{-14}$

**1-44. (a)** $2.0 \times 10^{12}$ **(b)** $3.7 \times 10^{16}$ **(c)** $6.0 \times 10^2$
**(d)** $2 \times 10^{12}$ **(e)** $1.9 \times 10^8$

**1-45. (a)** $1.225 \times 10^7$ **(b)** $9.00 \times 10^{-12}$
**(c)** $3.0 \times 10^{-24}$ **(d)** $9 \times 10^4$ **(e)** $1 \times 10^{10}$

**1-47. (a)** milliliter (mL) **(b)** hectogram (hg)
**(c)** nanojoule (nJ) **(d)** centimeter (cm)
**(e)** microgram ($\mu$g) **(f)** decipascal (dPa)

**1-49. (a)** 720 cm, 7.2 m, $7.2 \times 10^{-3}$ km
**(b)** $5.64 \times 10^4$ mm, 5640 cm, 0.0564 km
**(c)** $2.50 \times 10^5$ mm, $2.50 \times 10^4$ cm, 250 m

**1-50. (a)** 8.9 g, $8.9 \times 10^{-3}$ kg **(b)** $2.57 \times 10^4$ mg, 0.0257 kg **(c)** $1.25 \times 10^6$ mg, 1250 g

**1-52. (a)** 12 = 1 doz **(c)** 3 ft = 1 yd
**(e)** $10^3$ m = 1 km

**1-54. (a)** $\dfrac{1\ \text{g}}{10^3\ \text{mg}}$ **(b)** $\dfrac{1\ \text{km}}{10^3\ \text{m}}$ **(c)** $\dfrac{1\ \text{L}}{100\ \text{cL}}$
**(d)** $\dfrac{1\ \text{m}}{10^3\ \text{mm}}, \dfrac{1\ \text{km}}{10^3\ \text{m}}$

**1-56. (a)** $\dfrac{1\ \text{ft}}{12\ \text{in.}}$ **(b)** $\dfrac{2.54\ \text{cm}}{\text{in.}}$ **(c)** $\dfrac{5280\ \text{ft}}{\text{mi}}$
**(d)** $\dfrac{1.057\ \text{qt}}{\text{L}}$ **(e)** $\dfrac{1\ \text{qt}}{2\ \text{pt}}, \dfrac{1\ \text{L}}{1.057\ \text{qt}}$

**1-58. (a)** 47 L **(b)** 98 cm **(c)** 1.85 mi
**(d)** 51.56 yd **(e)** 92 m **(f)** 10.27 bbl
**(g)** 32 Gg

**1-60. (a)** 7.8 km, 4.8 mi, $6 \times 10^4$ ft
**(b)** 2380 ft, 0.724 km, 724 m
**(c)** 1.70 mi, 2.74 km, 2740 m
**(d)** 4.21 mi, $2.22 \times 10^4$ ft, 6780 m

**1-61. (a)** 25.7 L, 27.2 qt
**(b)** 630 L, 170 gal
**(c)** $8.12 \times 10^3$ qt, $2.03 \times 10^3$ gal

**1-63.** 55.3 kg

**1-65.** $28.0\ \text{m} \times \dfrac{10^2\ \text{cm}}{\text{m}} \times \dfrac{1\ \text{in.}}{2.54\ \text{cm}} \times \dfrac{1\ \text{ft}}{12\ \text{in.}} \times \dfrac{1\ \text{yd}}{3\ \text{ft}}$
$= \underline{30.6\ \text{yd}}$ (New punter is needed.)

**1-67.** 0.355 L

**1-69.** 6 ft 10 1/2 in. = 82.5 in.
$82.5\ \text{in.} \times \dfrac{2.54\ \text{cm}}{\text{in.}} \times \dfrac{1\ \text{m}}{10^2\ \text{cm}} = \underline{2.10\ \text{m}}$
$212\ \text{lb} \times \dfrac{1\ \text{kg}}{2.205\ \text{lb}} = \underline{96.1\ \text{kg}}$

**1-70.** 14.5 gal

**1-72.** $0.200 \text{ gal} \times \dfrac{4 \text{ qt}}{\text{gal}} = 0.800 \text{ qt}$

$0.800 \text{ qt} \times \dfrac{1 \text{ L}}{1.057 \text{ qt}} \times \dfrac{1 \text{ mL}}{10^{-3} \text{ L}} = \underline{757 \text{ mL}}$

There is slightly more in a "fifth" than in 750 mL.

**1-74.** 105 km/hr

**1-78.** $\dfrac{\$0.899}{\text{gal}} \times \dfrac{1 \text{ gal}}{4 \text{ qt}} \times \dfrac{1.057 \text{ qt}}{\text{L}} = \$0.238/\text{L}$

$80.0 \text{ L} \times \dfrac{\$0.238}{\text{L}} = \underline{\$19.04}$

**1-79.** $23.59 (551 mi), $12.82 (482 km)

**1-80.** 1690 km

**1-83.** 382 nails

**1-84.** $28.20

**1-85.** $350 \text{ km} \times \dfrac{1 \text{ mi}}{1.609 \text{ km}} \times \dfrac{1 \text{ gal}}{24.5 \text{ mi}} \times \dfrac{\$1.22}{\text{gal}} = \underline{\$10.83}$

**1-88.** 3.31 hr

**1-89.** **(a)** 7.28 euro, **(b)** $11.29

**1-90.** $\dfrac{0.621 \text{ pds}}{\text{euro}}$, 15,800 pds

**1-93.** 2100 s, 0.58 hr

**1-94.** 572°F

**1-95.** 24°C

**1-97.** −38°F

**1-99.** 95.0°F

**1-100.** **(a)** −98°C **(b)** 22°C **(c)** 27°C **(d)** −48°C **(e)** 600°C

**1-101.** **(a)** 320 K **(b)** 296 K **(c)** 200 K **(d)** 261 K **(e)** 291 K **(f)** 244 K

**1-103.** Since $t \, (°C) = t \, (°F)$ substitute $t \, (°C)$ for $t \, (°F)$ and set the two equations equal.

$[t \, (°C) \times 1.8] + 32 = \dfrac{t \, (°C) - 32}{1.8}$

$(1.8)^2 \, t \, (°C) - t \, (°C) = -32 - 32(1.8) \quad t \, (°C) = -40°C$

**1-104.** **(a)** $3 \times 10^2$ **(b)** $8.26 \text{ g} \cdot \text{cm}$ **(c)** 5.24 g/mL **(d)** 19.1

**1-105.** **(a)** $\dfrac{1 \text{g}}{10^3 \text{ mg}}$, $\dfrac{1 \text{ lb}}{453.6 \text{ g}}$ **(b)** $\dfrac{1.057 \text{ qt}}{\text{L}}$, $\dfrac{2 \text{ pt}}{\text{qt}}$ **(c)** $\dfrac{1 \text{ km}}{10 \text{ hm}}$, $\dfrac{1 \text{ mi}}{1.609 \text{ km}}$ **(d)** $\dfrac{1 \text{ in.}}{2.54 \text{ cm}}$, $\dfrac{1 \text{ ft}}{12 \text{ in.}}$

**1-106.** $5.34 \times 10^{10} \text{ ng} \times \dfrac{10^{-9} \text{ g}}{\text{ng}} \times \dfrac{1 \text{ lb}}{453.6 \text{ g}} = \underline{0.118 \text{ lb}}$

**1-108.** $10,300

**1-109.** $\dfrac{247 \text{ lb}}{82.3 \text{ doz}} = \underline{3.00 \text{ lb/doz}}$ $\quad \dfrac{82.3 \text{ doz}}{247 \text{ lb}} = \underline{0.333 \text{ doz/lb}}$

**1-110.** $12.0 \text{ fur} \times \dfrac{1 \text{ mi}}{8 \text{ fur}} \times \dfrac{5280 \text{ ft}}{\text{mi}} \times \dfrac{12 \text{ in.}}{\text{ft}} \times \dfrac{1 \text{ hand}}{4 \text{ in.}}$
$= \underline{2.38 \times 10^4 \text{ hands}}$

**1-112.** 1030 pkgs, 1.41 years

**1-113.** 5.02 L

**1-115.** $5.4 \times 10^7$°F $\quad 3.0 \times 10^7$°C $+ 273 = 3.0 \times 10^7$ K

**1-117.** 4.65 ton

## Chapter 2

**2-1.** Gases: nitrogen, oxygen, helium, neon, argon, krypton, xenon. Solids: gold, silver, copper, platinum, palladium, carbon (diamond and graphite), sulfur (in subsurface deposits).

**2-3.** **(a)** compound **(b)** element **(c)** element **(d)** element **(e)** compound **(f)** compound

**2-5.** It was found that water could be decomposed into simpler substances (hydrogen and oxygen).

**2-6.** Not necessarily. For example, the compound calcium carbonate can be decomposed into two other compounds, calcium oxide and carbon dioxide.

**2-7.** **(c)** It has a definite volume but not a definite shape.

**2-8.** The gaseous state is compressible because the basic particles are very far apart and thus the volume of a gas is mostly empty space.

**2-10.** **(a)** physical **(b)** chemical **(c)** physical **(d)** chemical **(e)** chemical **(f)** physical **(g)** physical **(h)** chemical **(i)** physical

**2-12.** **(a)** chemical **(b)** physical **(c)** physical **(d)** chemical **(e)** physical

**2-14.** No. Both compounds and elements have definite and unchanging properties. A compound can be chemically decomposed, however, but an element cannot.

**2-16.** Physical property: melts at 660°C Physical change: melting Chemical property: burns in oxygen Chemical change: formation of aluminum oxide

**2-17.** Original substance: green, solid (physical); can be decomposed (chemical). Substance is a compound. Gas: gas, colorless (physical); can be decomposed (chemical). Substance is a compound since it can be decomposed. Solid: shiny, solid (physical); cannot be decomposed (chemical). Substance is an element since it cannot be decomposed.

**2-19.** 2.60 g/mL

**2-21.** 1064 g/657 mL = $\underline{1.62 \text{ g/mL}}$ (carbon tetrachloride)

**2-23.** 1450 g

**2-24.** 670 g

**2-26.** 5.6 lb

**2-28.** 625 mL

**2-29.** 1.74 g/mL (magnesium)

**2-30.** 0.476 g/mL Yes, it floats.

**2-31.** 0.951 g/mL, 4790 mL
Pumice floats in water but sinks in alcohol.

**2-34.** 2080 g

**2-35.** 111.0 g/125 mL = 0.888 g/mL

**2-37.** Water: 1000 g Gasoline: 670 g
One liter of water has a greater mass.

**2-38.** 160 g/8.3 mL = 19 g/mL It's gold.

**2-41.** One needs a conversion factor between mL ($cm^3$) and $ft^3$.

$$\left(\frac{2.54 \text{ cm}}{\text{in.}}\right)^3 = \frac{16.4 \text{ cm}^3}{\text{in.}^3} = \frac{16.4 \text{ mL}}{\text{in.}^3}\left(\frac{12 \text{ in.}}{\text{ft}}\right)^3 = \frac{1728 \text{ in.}^3}{\text{ft}^3}$$

$$\frac{1.00 \text{ g}}{\text{mL}} \times \frac{1 \text{ lb}}{453.6 \text{ g}} \times \frac{16.4 \text{ mL}}{\text{in.}^3} \times \frac{1728 \text{ in.}^3}{\text{ft}^3} = \underline{62.5 \text{ lb/ft}^3}$$

**2-42.** $2.0 \times 10^5$ lb (100 tons)

**2-43.** Carbon dioxide is a compound composed of carbon and oxygen. It can be prepared from a mixture of carbon and oxygen, but the compound is no longer a mixture of the two elements.

**2-45.** Ocean water is the least pure because it contains a large amount of dissolved compounds. That is why it is not drinkable and cannot be used for crop irrigation. Drinking water also contains chlorine and some dissolved compounds but not nearly as much as ocean water. Rainwater is most pure but still contains some dissolved gases from the air.

**2-47.** (a) liquid only

**2-48.** (a) homogeneous  (b) heterogeneous
(c) heterogeneous  (d) homogeneous
(e) homogeneous  (f) homogeneous solution
(g) heterogeneous  (h) homogeneous

**2-50.** (a) liquid  (b) various solid phases
(c) gas and liquid  (d) liquid  (e) solid
(f) liquid  (g) solid and liquid  (h) liquid and gas
(i) gas

**2-52.** A mixture of all three would have carbon tetrachloride on the bottom, water in the middle, and kerosene on top. Water and kerosene float on carbon tetrachloride; kerosene floats on water.

**2-54.** Ice is less dense than water. An ice–water mixture is pure but heterogeneous.

**2-55.** (a) solution (a solid dissolved in a liquid)
(b) heterogeneous mixture (probably a solid suspended in a liquid such as dirty water)  (c) element
(d) compound  (e) solution (two liquids)

**2-57.** Mass of mixture = 85 + 942 = 1027 kg

$$\frac{85 \text{ kg}}{1027 \text{ kg}} \times 100\% = \underline{8.3\% \text{ tin}}$$

**2-59.** 64 kg nickel

**2-61.** $122 \text{ lb iron} \times \dfrac{100 \text{ lb duriron}}{86 \text{ lb iron}} = \underline{140 \text{ lb duriron}}$

**2-63.** (a) exothermic  (b) endothermic
(c) endothermic  (d) exothermic  (e) exothermic

**2-64.** Gasoline is converted into heat energy when it burns. The heat energy causes the pistons to move which is mechanical energy. The mechanical energy turns the

alternator, which generates electrical energy. The electrical energy is converted into chemical energy in the battery.

**2-66.** (a) potential  (b) kinetic  (c) potential (It is stored because of its composition.)  (d) kinetic  (e) kinetic

**2-69.** Kinetic energy is at a maximum nearest the ground when the swing is moving the fastest. Potential energy is at a maximum when the swing has momentarily stopped at the highest point. Assuming no gain or loss of energy, the total of the two energies is constant.

**2-70.** 0.853 J/(g · °C)

**2-71.** $\dfrac{56.6 \text{ cal}}{365 \text{ g} \cdot 5.0°\text{C}} = 0.031 \text{ cal/(g} \cdot °\text{C})$  (gold)

**2-73.** $°\text{C} = \dfrac{\text{cal}}{\text{sp. heat} \times \text{g}} = \dfrac{150 \text{ cal}}{0.092\dfrac{\text{cal}}{\text{g} \cdot °\text{C}} \times 50.0 \text{ g}}$
$= 33°\text{C rise}$  $t\ (°\text{C}) = 25 + 33 = \underline{58°\text{C}}$

This compares to a 3.0°C rise in temperature for 50.0 g of water.

**2-75.** 506 J

**2-76.** 58 − 25 = 33°C rise in temperature

$$g = \frac{\text{cal}}{\text{sp. heat} \cdot °\text{C}} = \frac{16.0 \text{ cal}}{0.106\dfrac{\text{cal}}{\text{g} \cdot °\text{C}} \cdot 33°\text{C}} = \underline{4.6 \text{ g}}$$

**2-77.** The copper skillet, because it has a lower specific heat. The same amount of applied heat will heat the copper skillet more than the iron.

**2-79.** Iron: 9.38°C rise  Gold: 32°C rise
Water: 0.997°C rise

**2-81.** 45°C

**2-83.** 2910 J

**2-87.** 860 g

**2-88.** heat lost by metal = heat gained by water
$100.0 \text{ g} \times 68.7°\text{C} \times \text{specific heat} = 100.0 \text{ g} \times 6.3°\text{C} \times \dfrac{1.00 \text{ cal}}{\text{g} \cdot °\text{C}}$  specific heat = $\underline{0.092 \text{ cal/g} \cdot °\text{C}}$
The metal is copper.

**2-89.** Density of A = 0.86 g/mL; density of B = 0.89 g/mL Liquid A floats on liquid B.

**2-91.** 15 g of sugar

**2-92.** 9.27 g/mL

**2-94.** 3.17% salt

**2-96.** $50.0 \text{ mL gold} \times \dfrac{19.3 \text{ g}}{\text{mL gold}} = 965 \text{ g gold}$

$50.0 \text{ mL alum.} \times \dfrac{2.70 \text{ g}}{\text{mL alum.}} = 135 \text{ g alum.}$

$\dfrac{965 \text{ g gold}}{(965 + 135) \text{ g alloy}} \times 100\% = \underline{87.7\% \text{ gold}}$

**2-98.** specific heat = $0.13 \dfrac{\text{J}}{\text{g} \cdot °\text{C}}$ (gold)

$25.0 \text{ g gold} \times \dfrac{1 \text{ mL}}{19.3 \text{ g gold}} = \underline{1.30 \text{ mL}}$

**2-101.** When a log burns, most of the compounds formed in the combustion are gases and dissipate into the atmosphere. Only some solid residue (ashes) is left. When zinc and sulfur (both solids) combine, the only product is a solid so there is no weight change. When iron burns, however, its only product is a solid. It weighs more than the original iron because the iron has combined with the oxygen gas from the air.

# Chapter 3

**3-2.** cadmium, Cd  calcium, Ca  californium, Cf  carbon, C  cerium, Ce  cesium, Cs  chlorine, Cl  chromium, Cr  cobalt, Co  copper, Cu  curium, Cm

**3-5.** (a) Ba  (b) Ne  (c) Cs  (d) Pt  (e) Mn  (f) W

**3-7.** (a) boron  (b) bismuth  (c) germanium  (d) uranium  (e) cobalt  (f) mercury  (g) beryllium  (h) arsenic

**3-8.** (a) They are both basic units of matter. Most elements are composed of individual atoms and many compounds are composed of individual molecules. Molecules are composed of atoms chemically bonded together.
(b) A compound is a pure form of matter. It is composed of individual units called molecules.
(c) They are both pure forms of matter. Compounds, however, are composed of two or more elements chemically combined.
(d) Most elements are composed of individual atoms. Some elements, however, are composed of molecules, which in most cases, contain two atoms.

**3-9.** (b) $Br_2$  (d) $S_8$  (f) $P_4$

**3-10.** P, phosphorus  O, oxygen  Br, bromine  F, fluorine  S, sulfur  Mg, magnesium

**3-11.** Hf is the symbol of the element hafnium. HF is the formula of a compound composed of one atom of hydrogen and one atom of fluorine.

**3-13.** (e) $N_2$, diatomic element  (b) CO, diatomic compound

**3-14.** N, nitrogen  O, oxygen  C, carbon  K, potassium  H, hydrogen  S, sulfur

**3-17.** (a) six carbons, four hydrogens, two chlorines
(b) two carbons, six hydrogens, one oxygen
(c) one copper, one sulfur, 18 hydrogens, 13 oxygens
(d) nine carbons, eight hydrogens, four oxygens
(e) two aluminums, three sulfurs, 12 oxygens
(f) two nitrogens, eight hydrogens, one carbon, three oxygens

**3-18.** (a) 12  (b) 9  (c) 33  (d) 21  (e) 17  (f) 14

**3-19.** (a) 8  (b) 7  (c) 4  (d) 3

**3-20.** (a) $SO_2$  (b) $CO_2$  (c) $H_2SO_4$  (d) $C_2H_2$

**3-22.** Neon would appear as individual spheres that are comparatively widely spaced. Chlorine would appear as molecules of two spheres joined together.

**3-23.** (a) They are both basic forms of matter. Atoms are neutral but ions are atoms that have acquired an electrical charge. Positive and negative ions are always found together.

(b) They are both basic forms of matter containing more than one atom. Molecules are neutral but polyatomic ions have acquired an electrical charge.
(c) Both have an electrical charge. Cations have a positive charge and anions have a negative charge.
(d) Both are classified as compounds, which are composed of the atoms of two or more elements. The basic unit of a molecular compound is a neutral molecule but the basic units of ionic compounds are cations and anions.
(e) Both are the basic units of compounds. Molecules are the basic entities of molecular compounds, and an ionic formula unit represents the smallest whole number of cations and anions representing a net charge of zero.

**3-25.** $SO_3$ represents the formula of a compound. It could be a gas. $SO_3^{2-}$ is an anion and does not exist independently. It is part of an ionic compound with the other part being a cation.

**3-26.** (a) $Ca(ClO_4)_2$  (b) $(NH_4)_3PO_4$  (c) $FeSO_4$

**3-27.** (a) one calcium, two chlorines, and eight oxygens
(b) three nitrogens, 12 hydrogens, one phosphorus, and four oxygens  (c) one iron, one sulfur, and four oxygens

**3-30.** (c) $S^{2-}$

**3-32.** (d) $Li^+$

**3-34.** FeS, $Li_2SO_3$

**3-36.** (b) and (e)

**3-37.** (c)

**3-40.** (a) 21 p, 21 e, 24 n  (b) 90 p, 90 e, 142 n
(c) 87 p, 87 e, 136 n  (d) 38 p, 38 e, 52 n

**3-42.**

| Isotope Name | Isotope Symbol | At. no. | Mass no. | p | n | e |
|---|---|---|---|---|---|---|
| (a) silver-108 | $^{108}_{47}Ag$ | 47 | 108 | 47 | 61 | 47 |
| (b) silicon-28 | $^{28}_{14}Si$ | 14 | 28 | 14 | 14 | 14 |
| (c) potassium-39 | $^{39}_{19}K$ | 19 | 39 | 19 | 20 | 19 |
| (d) cerium-140 | $^{140}_{58}Ce$ | 58 | 140 | 58 | 82 | 58 |
| (e) iron-56 | $^{56}_{26}Fe$ | 26 | 56 | 26 | 30 | 26 |
| (f) tin-110 | $^{110}_{50}Sn$ | 50 | 110 | 50 | 60 | 50 |
| (g) iodine-118 | $^{118}_{53}I$ | 53 | 118 | 53 | 65 | 53 |
| (h) mercury-196 | $^{196}_{80}Hg$ | 80 | 196 | 80 | 116 | 80 |

**3-44.** $^{59}Co$

**3-46.** (a) $K^+$: 19 p, 18 e  (b) $Br^-$: 35 p, 36 e
(c) $S^{2-}$: 16 p, 18 e  (d) $NO_2^-$: 7 + 16 = 23 p, 24 e
(e) $Al^{3+}$: 13 p, 10 e  (f) $NH_4^+$: 7 + 4 = 11 p, 10 e

**3-48.** (a) $Ca^{2+}$  (b) $Te^{2-}$  (c) $PO_3^{3-}$
(d) $NO_2^+$

**3-50.** This is the $Br^-$ ion. It is part of an ionic compound.

**3-52.** (a) The identity of a specific element is determined by its atomic number. An element is a basic form of matter and its atomic number relates to the number of protons in the nuclei of its atoms.
(b) Both relate to the particles in the nucleus of the atoms of an element. The atomic mass is the mass of an average atom

since elements are usually composed of more than one isotope. The atomic number is the number of protons.

(c) Both relate to the number of particles in a nucleus. The mass number relates to the total number of protons and neutrons in an isotope of an element while the atomic number is the number of protons in the atoms of a specific element.

(d) The isotopes of a specific element are determined by the atomic number. Every element is composed of at least one isotope.

(e) Different isotopes of a specific element have the same number of protons but different numbers of neutrons or mass numbers.

**3-53.** (a) Re: at. no. 75, at. wt. 186.2　　(b) Co: at. no. 27, at. wt. 58.9332　　(c) Br: at. no. 35, at. wt. 79.904
(d) Si: at. no. 14, at. wt. 28.086

**3-54.** copper (Cu)

**3-56.** O: at. no. 8, mass no. 16
N: at. no. 7, mass no. 14
Si: at. no. 14, mass no. 28
Ca: at. no. 20, mass no. 40

**3-58.** $5.81 \times 12.00 = 69.7$ amu. The element is Ga.

**3-60.** 79.9 amu

**3-61.** $^{28}$Si: $0.9221 \times 27.98 = 25.80$
$^{29}$Si: $0.0470 \times 28.98 = 1.362$
$^{30}$Si: $0.0309 \times 29.97 = \underline{0.926}$
$28.088 = \underline{28.09 \text{ amu}}$

**3-63.** Let $x$ = decimal fraction of $^{35}$Cl and $y$ = decimal fraction of $^{37}$Cl. Since there are two isotopes present, $x + y = 1$, $y = 1 - x$.
$(x \times 35) + (y \times 37) = 35.5$
$(x \times 35) + [(1 - x) \times 37] = 35.5$
$x = 0.75\ (\underline{75\%\ ^{35}\text{Cl}})$　　$y = 0.25\ (\underline{25\%\ ^{37}\text{Cl}})$

**3-66.** (a) $^{90}_{38}\text{Sr}^{2+}$　　(b) $^{52}_{24}\text{Cr}^{3+}$　　(c) $^{79}_{34}\text{Se}^{2-}$
(d) $^{14}_{7}\text{N}^{3-}$　　(e) $^{139}_{57}\text{La}^{3+}$

**3-68.** (a) Na: 11 p, 12 n, and 11 e; $\text{Na}^+$ has 10 electrons.
(b) Ca: 20 p, 20 n, and 20 e; $\text{Ca}^{2+}$ has 18 electrons.
(c) F: 9 p, 10 n, and 9 e; $\text{F}^-$ has 10 electrons.
(d) Sc: 21 p, 24 n, and 21 e; $\text{Sc}^{3+}$ has 18 electrons.

**3-70.** Let $x$ = mass no. of I and $y$ = mass no. of Tl
Then (1) $x + y = 340$　or　$x = 340 - y$

(2) $x = \frac{2}{3}y - 10$

Substituting for $x$ from (1) and solving for $y$
$y = 210$ amu (Tl)　and　$x = 340 - 210 = 130$ amu (I)

**3-72.** 121.8 (Sb)　$\text{Sb}^{3+}$ has $51 - 3 = 48$ electrons.
$^{121}$Sb has $121 - 51 = 70$ neutrons. $^{123}$Sb has $123 - 51 = 72$ neutrons. $^{121}$Sb, 57.9% due to neutrons; $^{123}$Sb, 58.5% due to neutrons

**3-73.** 118 neutrons and 78 protons [platinum (Pt)] $78 - 2 = 76$ electrons for $\text{Pt}^{2+}$

**3-75.** Mass of other atom = 16 (oxygen)　$\text{NO}^+ = (7 + 8) - 1 = 14$ electrons

**3-77.** (a) H: $\dfrac{1.008}{12.00} \times 8.000 = 0.672$

(b) N: $\dfrac{14.01}{12.00} \times 8.000 = 9.34$

(c) Na: $\dfrac{22.99}{12.00} \times 8.000 = 15.3$

(d) Ca: $\dfrac{40.08}{12.00} \times 8.000 = 26.72$

**3-78.** $\dfrac{43.3}{10.0} = 4.33$ times as heavy as $^{12}$C
$4.33 \times 12.0$ amu = 52.0 amu　The element is Cr.

# Chapter 4

**4-1.** An active metal reacts with water and air. A noble metal is not affected by air, water, or most acids.

**4-2.** 32

**4-3.** (c) $\text{I}_2$, and (g) $\text{Br}_2$

**4-5.** (a) Fe, and (d) La—transition elements
(b) Te, (f) H, and (g) In—representative elements
(e) Xe—noble gas
(c) Pm—inner transition element

**4-7.** (b) Ti, (e) Pd, and (g) Ag

**4-8.** The most common physical state is a solid, and metals are more common than nonmetals.

**4-10.** (a) Ne, (d) Cl, and (f) N

**4-12.** (a) Ru, (b) Sn, (c) Hf, and (h) W

**4-13.** (d) Te and (f) B

**4-14.** (a) Ar　　(b) Hg　　(c) $\text{N}_2$　　(d) Be　　(e) Po

**4-16.** Element 118 is in Group VIIIA (noble gas).

**4-18.** (a) lithium fluoride　　(b) barium telluride
(c) strontium nitride　　(d) barium hydride
(e) aluminum chloride

**4-20.** (a) $\text{Rb}_2\text{Se}$　　(b) $\text{SrH}_2$　　(c) RaO　　(d) $\text{Al}_4\text{C}_3$
(e) $\text{BeF}_2$

**4-22.** (a) bismuth(V) oxide　　(b) tin(II) sulfide
(c) tin(IV) sulfide　　(d) copper(I) telluride
(e) titanium(IV) carbide

**4-24.** (a) $\text{Cu}_2\text{S}$　　(b) $\text{V}_2\text{O}_3$　　(c) AuBr　　(d) $\text{Ni}_2\text{C}$
(e) $\text{CrO}_3$

**4-26.** In 4-22: (a) $\text{Bi}_2\text{O}_5$, (c) $\text{SnS}_2$, and (e) TiC; in 4-24: (e) $\text{CrO}_3$

**4-28.** (c) $\text{ClO}_3^-$

**4-30.** ammonium, $\text{NH}_4^+$

**4-31.** (b) permanganate ($\text{MnO}_4^-$), (c) perchlorate ($\text{ClO}_4^-$), (e) phosphate ($\text{PO}_4^{3-}$), and (f) oxalate ($\text{C}_2\text{O}_4^{2-}$)

**4-32.** (a) chromium(II) sulfate　　(b) aluminum sulfite
(c) iron(II) cyanide　　(d) rubidium hydrogen carbonate
(e) ammonium carbonate　　(f) ammonium nitrate
(g) bismuth(III) hydroxide

**4-34.** (a) $\text{Mg(MnO}_4)_2$　　(b) $\text{Co(CN)}_2$　　(c) $\text{Sr(OH)}_2$
(d) $\text{Tl}_2\text{SO}_3$　　(e) $\text{In(HSO}_4)_3$　　(f) $\text{Fe}_2(\text{C}_2\text{O}_4)_3$
(g) $(\text{NH}_4)_2\text{Cr}_2\text{O}_7$　　(h) $\text{Hg}_2(\text{C}_2\text{H}_3\text{O}_2)_2$

**4-36.**

|        | $HSO_3^-$                          | $Te^{2-}$                          | $PO_4^{3-}$                        |
|--------|------------------------------------|------------------------------------|------------------------------------|
| $NH_4^+$ | $NH_4HSO_3$ <br> ammonium bisulfite | $(NH_4)_2Te$ <br> ammonium telluride | $(NH_4)_3PO_4$ <br> ammonium phosphate |
| $Co^{2+}$ | $Co(HSO_3)_2$ <br> cobalt(II) bisulfite | $CoTe$ <br> cobalt(II) telluride | $Co_3(PO_4)_2$ <br> cobalt(II) phosphate |
| $Al^{3+}$ | $Al(HSO_3)_3$ <br> aluminum bisulfite | $Al_2Te_3$ <br> aluminum telluride | $AlPO_4$ <br> aluminum phosphate |

**4-38.** (a) sodium chloride (b) sodium hydrogen carbonate (c) calcium carbonate (d) sodium hydroxide (e) sodium nitrate (f) ammonium chloride (g) aluminum oxide (h) calcium hydroxide (i) potassium hydroxide

**4-39.** (a) $Ca_2XeO_6$ (b) $K_4XeO_6$ (c) $Al_4(XeO_6)_3$

**4-40.** (a) phosphonium fluoride (b) potassium hypobromite (c) cobalt(III) iodate (d) calcium silicate (actual name is calcium metasilicate) (e) aluminum phosphite (f) chromium(II) molybdate

**4-41.** (a) Si (b) I (c) H (d) Kr (e) H (f) As

**4-43.** (a) carbon disulfide (b) boron trifluoride (c) tetraphosphorus decoxide (d) dibromine trioxide (e) sulfur trioxide (f) dichlorine oxide or dichlorine monoxide (g) phosphorus pentachloride (h) sulfur hexafluoride

**4-45.** (a) $P_4O_6$ (b) $CCl_4$ (c) $IF_3$ (d) $Cl_2O_7$ (e) $SF_6$ (f) $XeO_2$

**4-47.** (a) hydrochloric acid (b) nitric acid (c) hypochlorous acid (d) permanganic acid (e) periodic acid (f) hydrobromic acid

**4-48.** (a) HCN (b) $H_2Se$ (c) $HClO_2$ (d) $H_2CO_3$ (e) HI (f) $HC_2H_3O_2$

**4-50.** (a) hypobromous acid (b) iodic acid (c) phosphorous acid (d) molybdic acid (e) perxenic acid

**4-52.** $ClO_2$ chlorine dioxide

**4-53.** A = F  X = Br  $BrF_5$ (Br is more metallic.) bromine pentafluoride

**4-55.** Gas = $N_2$; Al forms only +3. Thus the formula is AlN ($N^{3-}$), for aluminum nitride.  $Ti_3N_2$, titanium(II) nitride

**4-57.** $Co^{2+}$ and $Br^-$: $CoBr_2$ cobalt(II) bromide

**4-58.** Metal = Mg, nonmetal = S: $MgH_2$, magnesium hydride; $H_2S$, hydrogen sulfide or hydrosulfuric acid

**4-60.** $NiI_2$, nickel(II) iodide  $H_3PO_4$, phosphoric acid  $Sr(ClO_3)_2$, strontium chlorate  $H_2Te$, hydrogen telluride or hydrotelluric acid  $As_2O_3$, diarsenic trioxide  $Sb_2O_3$, antimony(III) oxide  $SnC_2O_4$, tin(II) oxalate

**4-62.** tin(II) hypochlorite, $Sn(ClO)_2$  chromic acid, $H_2CrO_4$  xenon hexafluoride, $XeF_6$  barium nitride, $Ba_3N_2$  hydrofluoric acid, HF  iron(III) carbide, $Fe_4C_3$, lithium phosphate, $Li_3PO_4$

**4-63.** (e) $Rb_2C_2O_4$

**4-65.** $Rb_2O_2$, $MgO_2$, $Al_2(O_2)_3$, $Ti(O_2)_2$  $H_2O_2$, hydrogen peroxide, hydroperoxic acid

**4-66.** (a) $Na_2CO_3$ (b) $CaCl_2$ (c) $KClO_4$ (d) $Al(NO_3)_3$ (e) $Ca(OH)_2$ (f) $NH_4Cl$

**4-69.** $N^{3-}$, nitride  $NO_2^-$, nitrite  $NO_3^-$, nitrate  $NH_4^+$, ammonium  $CN^-$, cyanide

**4-71.** (e) chromium(III) carbonate

**4-73.** (c) barium chlorite

## Cumulative Review Test on Chapters 2–4

### Multiple Choice

| 1. (d) | 2. (b) | 3. (d) | 4. (a) | 5. (b) |
|--------|--------|--------|--------|--------|
| 6. (c) | 7. (d) | 8. (c) | 9. (a) | 10. (a) |
| 11. (d) | 12. (a) | 13. (b) | 14. (e) | 15. (b) |
| 16. (b) | 17. (b) | 18. (c) | 19. (a) | 20. (d) |
| 21. (d) | 22. (d) | 23. (d) | 24. (d) | 25. (a) |

### Problems

1. 4.3 g/mL    2. 60.0 mL    3. 102 mL of iron
4. 0.288 J/g · °C    5. 121.8 amu Antimony, Sb
6. 114.6 amu    7. Aluminum (Al) (a) Group IIIA and a representative element (b) solid (c) metal (d) +3 (e) $Al_2S_3$, $AlBr_3$, AlN (f) aluminum sulfide, aluminum bromide, aluminum nitride

## Chapter 5

**5-1.** Ultraviolet light has shorter wavelengths but higher energy than visible light. Because of this high energy, ultraviolet light can damage living cells in tissues, thus causing a burn.

**5-2.** Since these two shells are close in energy, transitions of electrons from these two levels to the $n = 1$ shell have similar energy. Thus, the wavelengths of light from the two transitions are very close together.

**5-3.** Since these two shells are comparatively far apart in energy, transitions from these two levels to the $n = 1$ shell have comparatively different energies. Thus, the wavelengths of light from the two transitions are quite different. (The $n = 3$ to $n = 1$ transition has a shorter wavelength than the $n = 2$ to the $n = 1$ transition.)

**5-5.** $1p$ and $3f$

**5-8.** A $4p$ orbital is shaped roughly like a two-sided baseball bat with two "lobes" lying along one of the three axes. This shape represents the region of highest probability of finding the $4p$ electrons.

**5-9.** The $3s$ orbital is spherical in shape. There is an equal probability of finding the electron regardless of the orientation from the nucleus. (In fact, the probability lies in three concentric spheres with the highest probability in the sphere farthest from the nucleus.) The highest probability of finding the electron lies farther from the nucleus in the $3s$ than in the $2s$.

**5-11.** One $3s$, three $3p$, and five $3d$, for a total of nine.

**5-13.** (a) $3p$, three    (b) $4d$, five    (c) $6s$, one

**5-15.** $2n^2$: $2(3)^2 = 18$ and $2(2)^2 = 8$

**5-18.** $2(5)^2 = 50$

**5-19.** The first four subshells in the fifth shell ($s$, $p$, $d$, and $f$) hold 32 electrons. The $g$ subshell holds $50 - 32 = 18$ electrons.

**5-20.** Since each orbital holds two electrons, there are nine orbitals in this subshell.

**5-21.** The $1s$ subshell always fills first.

**5-24.** (a) $6s$    (b) $5p$    (c) $4p$    (d) $4d$

**5-26.** $4s$, $4p$, $5s$, $4d$, $5p$, $6s$, $4f$

**5-27.** (a) Mg: $1s^22s^22p^63s^2$
(b) Ge: $1s^22s^22p^63s^23p^64s^23d^{10}4p^2$
(c) Pd: $1s^22s^22p^63s^23p^64s^23d^{10}4p^65s^24d^8$
(d) Si: $1s^22s^22p^63s^23p^2$

**5-29.** $1s^22s^22p^63s^23p^6$

**5-31.** (a) S: $[Ne]3s^23p^4$    (b) Zn: $[Ar]4s^23d^{10}$
(c) Pu: $[Rn]7s^26d^15f^5$    (d) I: $[Kr]5s^24d^{10}5p^5$

**5-33.** (a) This is excluded by Hund's rule since electrons are not shown in separate orbitals of the same subshell with parallel spins.    (b) This is correct.    (c) This is excluded by the Aufbau principle because the $2s$ subshell fills before the $2p$.    (d) This is excluded by the Pauli exclusion principle since the two electrons in the $2s$ orbital cannot have the same spin.

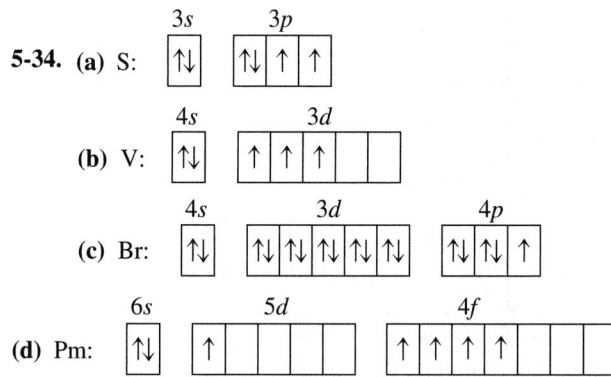

**5-34.** (a) S:

(b) V:

(c) Br:

(d) Pm:

**5-36.** IIB, none; VB, three; VIA, two; VIIA, one; $[ns^1(n-1)d^5]$, six; Pm, five

**5-38.** IVA and VIA

**5-40.** (a) In    (b) Y    (c) Ce    (d) Ar

**5-42.** (a) $3p$    (b) $5p$    (c) $5d$    (d) $5s$

**5-44.** Both have three valence electrons. The outer electron in IIIA is in a $p$ subshell; in IIIB it is in a $d$ subshell.

**5-46.** (a) F    (b) Ga    (c) Ba    (d) Gd    (e) Cu

**5-48.** (a) VIIA    (b) IIIA    (c) IIA    (e) IB

**5-49.** VA: $[NG]ns^2np^3$  VB: $[NG]ns^2(n-1)d^3$

**5-51.** (b) and (e) belong to Group IVA

**5-53.** (a) IVA    (b) VIIIA    (c) IB    (d) Pr and Pa

**5-54.** (a) $[NG]ns^2$    (b) $[NG]ns^2(n-1)d^{10}$
(c) $[NG]ns^2np^4$  or  $[NG]ns^2(n-1)d^{10}np^4$
(d) $[NG]ns^2(n-1)d^2$

**5-55.** Helium does not have a filled $p$ subshell. (There is no $1p$ subshell.)

**5-57.** $[Ne]3s^23p^6$

**5-59.** The theoretical order of filling is $6d$, $7p$, $8s$, $5g$. The $6d$ is completed at element number 112. The $7p$ and $8s$ fill at element number 120. Thus element number 121 would theoretically begin the filling of the $5g$ subshell. This assumes filling in the order indicated by Figure 5-14.

**5-61.** $7s^26d^{10}5f^{14}7p^6$    Total 32

**5-62.** $8s^25g^{18}7d^{10}6f^{14}8p^6$    Total 50

**5-63.** (a) transition    (b) representative    (c) noble gas
(d) inner transition

**5-65.** element number 118 (under Rn)

**5-67.** (a) As    (b) Ru    (c) Ba    (d) I

**5-69.** Cr, 117 pm Nb, 134 pm

**5-71.** (a) V    (b) Cl    (c) Mg    (d) Fe    (e) B

**5-72.** In, 558 kJ/mol   Ge, 762 kJ/mol

**5-73.** Te, 869 kJ/mol   Br, 1140 kJ/mol

**5-75.** (a) $Cs^+$ easiest,    (b) $Rb^{2+}$ hardest

**5-77.** (d) I

**5-79.** (a) Rb

**5-81.** (a) $Mg^{2+}$    (b) $K^+$    (c) S    (d) Mg
(e) $S^{2-}$    (f) $Se^{2-}$

**5-83.** The outer electron in Hf is in a shell higher in energy than that in Zr. This alone would make Hf a larger atom. However, in between Zr and Hf lie several subshells including the long $4f$ subshell (Ce through Lu). The filling of these subshells, especially the $4f$, causes a gradual contraction that offsets the higher shell for Hf.

**5-84.** $C^+$ (1086 kJ/mol), $C^{2+}$ (3439 kJ/mol), $C^{3+}$ (8059 kJ/mol), $C^{4+}$ (14,282 kJ/mol), $C^{5+}$ (52,112 kJ/mol). Notice that the energy required to form $C^+$ (1086 kJ) is about twice the energy required to form $Ga^+$ (579 kJ), which is a metal. Thus it is apparent that metal cations form more easily than nonmetal cations.

**5-85.** (a) B    (b) Ar    (c) K, Cr, and Cu    (d) Ga
(e) Hf

**5-87.** (a) B    (b) Br    (c) Sb    (d) Si

**5-89.** (a) Sr, metal, representative element, IIA
(b) Pt, metal, transition metal, VIIIB
(c) Br, nonmetal, representative element, VIIA

**5-91.** P

**5-93.** Z = Sn, X = Zr

**5-97.** (a) Ca   (b) $Br^-$   (c) S   (d) $S^{2-}$
(e) $Na^+$

**5-99.**

| $s^1$ | | | | | | | | | | | | | | $p^4$ |
|---|---|---|---|---|---|---|---|---|---|---|---|---|---|---|
| 1 | $s^2$ | | | | | | | | | $p^1$ | $p^2$ | $p^3$ | | 2 |
| 3 | 4 | | | | | | | | | 5 | 6 | 7 | | 8 |
| 9 | 10 | $d^1$ | $d^2$ | $d^3$ | $d^4$ | $d^5$ | $d^6$ | $d^7$ | $d^8$ | 11 | 12 | 13 | | 14 |
| 15 | 16 | 17 | 18 | 19 | 20 | 21 | 22 | 23 | 24 | 25 | 26 | 27 | | 28 |
| 29 | 30 | 31 | 32 | 33 | 34 | 35 | 36 | 37 | 38 | 39 | 40 | 41 | | 42 |
| 43 | 44 | 45* | | | | | | | | | | | | |

*46–50 would be in a $4f$ subshell

(a) six in second period, 14 in fourth
(b) third period, #14; fourth period, #28
(c) first inner transition element is #46 (assuming an order of filling like that on Earth)
(d) Elements #11 and #17 are most likely to be metals.
(e) Element #12 would have the larger radius in both cases.
(f) Element #7 would have the higher ionization energy in all three cases.
(g) The ions that would be reasonable are $16^{2+}$ (metal cation), $13^-$ (nonmetal anion), $15^+$ (metal cation), and $1^-$ (nonmetal anion). The $9^{2+}$ ion is not likely because the second electron would come from a filled inner subshell. The $7^+$ ion is not likely because it would be a nonmetal cation. The $17^{4+}$ ion is not likely because #17 has only three electrons in the outer shell.

# Chapter 6

**6-1.** (a) $Ca\cdot$   (b) $\cdot Sb\cdot$   (c) $\cdot Sn\cdot$   (d) $\cdot\ddot{I}:$
(e) $:\ddot{Ne}:$   (f) $\cdot Bi\cdot$   (g) $\cdot VIA:$

**6-2.** (a) Group IIIA   (b) Group VA   (c) Group IIA

**6-4.** The electrons from filled inner subshells are not involved in bonding.

**6-6.** (b) Sr and S   (c) H and K   (d) Al and F

**6-8.** (b) $S^-$   (c) $Cr^{2+}$   (e) $In^+$   (f) $Pb^{2+}$
(h) $Tl^{3+}$

**6-10.** They have the noble gas configuration of He, which requires only two electrons.

**6-11.** (a) $K^+$   (b) $\cdot\ddot{O}:^-$   (c) $:\ddot{I}:^-$   (d) $:\ddot{P}:^{3-}$   (e) $Ba^+$
(f) $\cdot\ddot{Xe}:^+$   (g) $Sc^{3+}$

**6-13.** (b) $O^-$   (e) $Ba^+$   (f) $Xe^+$

**6-14.** (a) $Mg^{2+}$   (b) $Ga^{3+}$ (pseudo-noble gas configuration.)   (c) $Br^-$   (d) $S^{2-}$   (e) $P^{3-}$

**6-17.** $Se^{2-}$, $Br^-$, $Rb^+$, $Sr^{2+}$, $Y^{3+}$

**6-18.** $Tl^{3+}$ has a full $5d$ subshell. This is a pseudo-noble gas configuration.

**6-19.** $Cs_2S$, $Cs_3N$, $BaBr_2$, $BaS$, $Ba_3N_2$, $InBr_3$, $In_2S_3$, $InN$

**6-21.** (a) $CaI_2$   (b) CaO   (c) $Ca_3N_2$   (d) CaTe
(e) $CaF_2$

**6-23.** (a) $Cr^{3+}$   (b) $Fe^{3+}$   (c) $Mn^{2+}$   (d) $Co^{2+}$
(e) $Ni^{2+}$   (f) $V^{3+}$

**6-26.** (a) $H_2S$   (b) $GeH_4$   (c) ClF   (d) $Cl_2O$
(e) $NCl_3$   (f) $CBr_4$

**6-28.** (b) $Cl_3$   (d) $NBr_4$   (e) $H_3O$

**6-30.** (a) $SO_4^{2-}$   (b) $IO_3^-$   (c) $SeO_4^{2-}$
(d) $H_2PO_4^-$   (e) $C_2^{2-}$

**6-32.** (a) 

```
    H  H
    |  |
H—C—C—H
    |  |
    H  H
```

(b) $H-\ddot{O}-\ddot{O}-H$

(c) 

```
:F—N—F:
    |
   :F:
```

(d) $:\ddot{Cl}-\ddot{S}-\ddot{Cl}:$

(e) 

```
    H        H            H  H
    |        |            |  |
H—C—O—C—H      H—C—C—O—H
    |        |            |  |
    H        H            H  H
```

**6-34.** (a) $:C\equiv O:$   (b) $.\ddot{O}\quad\ddot{O}.$ over S over $:\ddot{O}:$   (c) $K^+[:C\equiv N:]^-$

(d) 

```
H—Ö:
    |
  :S—Ö:
    |
 :Ö—H
```

**6-36.** (a) valence electrons = 4 + 6 = 10, N = 2
$6N + 2 = 14$   $14 - 10 = 4$ (two double bonds or one triple bond)
(b) valence electrons = $(3 \times 6) + 6 = 24$, N = 4
$6N + 2 = 26$   $26 - 24 = 2$ (one double bond)
(c) valence electrons for $CN^-$: 4 + 5 + 1 = 10, N = 2
$6N + 2 = 14$   $14 - 10 = 4$ (two double bonds or one triple bond)

**(d)** valence electrons = $(3 \times 6) + 6 + (2 \times 1) = 26, N = 4$
$6N + 2 = 26$    $26 - 26 = 0$ (no multiple bonds)
**6-38. (a)** $N_2O$: valence electrons = $(2 \times 5) + 6 = 16, N = 3$
$6N + 2 = 20$    $20 - 16 = 4$ (two double bonds or one triple bond)
**(b)** $Ca(NO_2)_2$: For the $NO_2^-$ ion, valence electrons = $5 + (2 \times 6) + 1 = 18, N = 3$
$6N + 2 = 20$    $20 - 18 = 2$ (one double bond)
**(c)** $AsCl_3$: valence electrons = $5 + (3 \times 7) = 26, N = 4$
$6N + 2 = 26$    $26 - 26 = 0$ (no multiple bonds)
**(d)** $H_2S$: valence electrons = $(2 \times 1) + 6 = 8, N = 1$
$6N + 2 = 8$    $8 - 8 = 0$ (no multiple bonds)
**(e)** $CH_2Cl_2$: valence electrons = $4 + (2 \times 1) + (2 \times 7) = 20, N = 3$
$6N + 2 = 20$    $20 - 20 = 0$ (no multiple bonds)
**(f)** $NH_4^+$: valence electrons = $(4 \times 1) + 5 - 1 = 8, N = 1$
$6N + 2 = 8$    $8 - 8 = 0$ (no multiple bonds)

**6-39.** **(a)** $\ddot{N}{=}N{=}\ddot{O}$    **(b)** $Ca^{2+}$ [ structure ]

**(c)** $:\!\ddot{Cl}\!-\!\ddot{A}s\!-\!\ddot{Cl}\!:$    **(d)** [ S with H H ]
          $:\!\ddot{Cl}\!:$

**(e)** $:\!\ddot{Cl}\!-\!\overset{H}{\underset{H}{C}}\!-\!\ddot{Cl}\!:$    **(f)** [ H—N—H structure ]$^+$

**6-40.** **(a)** $:\!\ddot{Cl}\!-\!\ddot{O}\!-\!\ddot{Cl}\!:$    **(b)** $[:\!\ddot{O}\!-\!\overset{\cdot\cdot}{\underset{\ddot{O}:}{S}}\!-\!\ddot{O}\!:]^{2-}$

**(c)** [ H₂C=CH₂ structure ]    **(d)** [ H₂C=O structure ]

**(e)** [ BF₃ structure ]    **(f)** $[:\!N{\equiv}O\!:]^+$

**6-44.** **(a)** [ three SO₃ resonance structures ]

**(b)** [ two NO₂⁻ resonance structures ]    **(c)** $[:\!\ddot{O}\!-\!\overset{\cdot\cdot}{\underset{\ddot{O}:}{S}}\!-\!\ddot{O}\!:]^{2-}$
(only one structure)

**6-46.** [ two resonance structures of H—B—C with O's, $^{2-}$ ]

**6-47.** A resonance hybrid is the actual structure of a molecule or ion that is implied by the various resonance structures. Each resonance structure contributes a portion of the actual structure. For example, the two structures shown in exercise 6-46 imply that both carbon–oxygen bonds have properties that are halfway between those of a single and a double bond.

**6-49.** $:\!O{\equiv}C\!-\!\ddot{O}\!: \longleftrightarrow :\!\ddot{O}\!-\!C{\equiv}O\!:$

The two resonance structures imply that both carbon–oxygen bonds are halfway between a single and a triple bond, which is a double bond. This is the same as the one structure shown below.

$$:\!\ddot{O}{=}C{=}\ddot{O}\!:$$

**6-50.** Cs, Ba, Be, B, C, Cl, O, F

**6-51.** **(a)** $\overset{\delta-}{N}\!-\!\overset{\delta+}{H}$    **(b)** $\overset{\delta+}{B}\!-\!\overset{\delta-}{H}$    **(c)** $\overset{\delta+}{Li}\!-\!\overset{\delta-}{H}$
**(d)** $\overset{\delta-}{F}\!-\!\overset{\delta+}{O}$    **(e)** $\overset{\delta-}{O}\!-\!\overset{\delta+}{Cl}$    **(f)** $\overset{\delta-}{S}\!-\!\overset{\delta+}{Se}$
**(g)** $\overset{\delta-}{C}\!-\!\overset{\delta+}{B}$    **(h)** $\overset{\delta+}{Cs}\!-\!\overset{\delta-}{N}$
**(i)** C—S (very low polarity)

**6-52.** **(i)** nonpolar, **(b)** = **(f)**, **(d)** = **(e)** = **(g)**, **(a)**, **(c)**, **(h)**

**6-54.** Only **(d)** Al—F is predicted to be ionic on this basis since the electronegativity difference is 2.5.

**6-56.** **(b)** I—I

**6-58.** **(a)** [ SF₂ structure ] is angular    **(b)** $:\!\ddot{S}{=}C{=}\ddot{S}\!:$ is linear

**(c)** [ CCl₃F structure ] is tetrahedral    **(d)** [ ONCl structure ] is angular

**(e)** [ Cl₂O structure ] is angular

**6-60.** **(a)** No angle between two points    **(b)** 180°
**(c)** 120°

**6-62.**
**(a)** [ H₂N—O—H structure ]  N–trigonal pyramid, O–V-shaped

**(b)** $H\!-\!\ddot{O}\!-\!C{\equiv}N\!:$   O–V-shaped, C–linear

**(c)** $H\!-\!\overset{H}{\underset{H}{C}}\!-\!C{\equiv}N\!:$   C–tetrahedral, C–linear

**(d)** See exercise 6-46. B– tetrahedral, C– trigonal planar

**6-64.** If the molecule has a symmetrical geometry and all bonds are the same, the bond dipoles cancel and the molecule is nonpolar.

**6-65.** **(a)** polar    **(b)** nonpolar    **(c)** polar
**(d)** polar    **(e)** polar

**6-67.** :Ö:
S
:Ö. | .Cl:
.. :Cl: ..

The molecule is polar.

**6-68.** Since the H—S bond is much less polar than the H—O bond, the resultant molecular dipole is much less. The $H_2S$ molecule is less polar than the $H_2O$ molecule.

**6-70.** The $CHF_3$ molecule is more polar than the $CHCl_3$ molecule. The C—F bond is more polar than the C—Cl bond, which means that the resultant molecular dipole is larger for $CHF_3$.

**6-72.**
Ö
‖
S
:Ö. .Ö:

All bond dipoles in $SO_3$ cancel.

Ö.
‖
S
.Ö. .Ö:

Bond dipoles in $SO_2$ do not cancel.

**6-74. (a)** e.g., RbCl    **(b)** e.g., SrO    **(c)** e.g., AlN

**6-75.**
Xe
.Ö. .Ö.
:Ö:

is trigonal pyramid; the molecule is polar.

**6-77.**
H
|
H—B—C≡O:
|
H

No resonance structures.
Geometry around B is tetrahedral.
Geometry around C is linear.

**6-79.**
N—N
.Ö. .Ö.

Geometry is angular for this structure; the molecule is polar.

Other resonance structures:

:Ö—N≡N—Ö: ⟷ :O≡N—N̈—Ö: ⟷ :Ö—N̈—N≡O:

**6-82.** Valence electrons for neutral $PO_3$ = 5(P) + (3 × 6) (O's) = 23. Since the species has 26 electrons it must be an anion with a −3 charge (i.e., $PO_3^{3-}$).

:Ö—P̈—Ö: ³⁻
|
:Ö:

The geometry of the anion is trigonal pyramid with a O—P—O angle of about 109°.

**6-83.** Valence electrons for neutral $ClI_2$ = 3 × 7 = 21. Since the species has 20 valence electrons it must be a cation with a +1 charge (i.e., $ClI_2^+$).

.Cl. ⁺
.Ï. .Ï.

The geometry of the cation is V-shaped with an angle of about 109°.

**6-85.** $K_3N$, potassium nitride $KN_3 = K^+N_3^-$ Resonance structures:

:N̈=N=N̈: ⟷ :N̈—N≡N: ⟷ :N≡N—N̈:

In all resonance structures, the geometry around N is linear.

**6-88.**
H
\
Ö—C≡N:

The angle of 105° indicates that the H—O—C angle is V-shaped with an angle of about 109°. This would involve two bonds and two pairs of electrons on the O. The geometry around the C is linear.

**6-89.**
H
\
S̈=C=N:

The angle of 116° indicates that the H—S—C angle is V-shaped with an angle of about 120°. This would involve three bonds and one pair of electrons on the S. The geometry around the C is linear.

**6-90.** :F̈—N̈=N̈—F̈:

**6-92.** Oxygen difluoride, $OF_2$. Lewis structure:
Ö.
:F̈: :F̈:

Dioxygen difluoride, $O_2F_2$. Lewis structure:
Ö—Ö
:F̈: :F̈:

Oxygen is less electronegative (more metallic) than fluorine and so is named first. Both molecules are angular; thus they are polar.

**6-94.** Formula = $NaHCO_3$ = $Na^+HCO_3^-$

O: ⁻
//
H—Ö—C
\
Ö:

The geometry around the C is trigonal planar with the approximate H—O—C angle of 120°.

**6-96.** Formula = $CaH_2S_2O_6$ = $Ca(HSO_3)_2$ calcium bisulfite or calcium hydrogen sulfite

H—Ö—S̈—Ö: ⁻
|
:Ö:

The geometry around the S is trigonal pyramid. The H—O—S angle is about 109°.

**6-97. (1) (a)** 17    **(b)** 31    **(c)** $51_3$    **(d)** 97
**(e)** 7(13)    **(f)** $10(13)_2$    **(g)** $67_2$    **(h)** $3_26$

**(2)** On Zerk, six electrons fill the outer s and p orbitals to make a noble gas configuration. Therefore, we have a "sextet" rule on Zerk.

**(a)** 1—7̈:    **(b)** $3^+1^-$ (ionic)    **(c)** 1—5—1
|
1

**(d)** $9^+$ :7̈:⁻(ionic)    **(e)** :7̈—1̈3:    **(f)** $10^{2+}$(:1̈3:⁻)₂ (ionic)

**(g)** :7̈—6̈—7̈:    **(h)** $(3^+)_2$:6̈:²⁻ (ionic)

**(3)** $4N + 2$

## Cumulative Review Test on Chapters 4–6

### Multiple Choice

**1. (c)**    **2. (a)**    **3. (c)**    **4. (b)**    **5. (c)**
**6. (a)**    **7. (a)**    **8. (b)**    **9. (b)**    **10. (b)**

| | | | | |
|---|---|---|---|---|
| **11.** (a) | **12.** (c) | **13.** (d) | **14.** (c) | **15.** (c) |
| **16.** (a) | **17.** (c) | **18.** (d) | **19.** (b) | **20.** (d) |

## Problems

**1.** Nitrogen (N) **(a)** Group VA and a representative element **(b)** $1s^2 2s^2 2p^3$ **(c)** three **(d)** nonmetal **(e)** gas **(f)** $N_2$ **(g)** $:N\equiv N:$ **(h)** $-3$ **(i)** neon **(j)** oxygen, positive; boron, negative **(k)** $Mg_3N_2$, $Li_3N$, $NF_3$ **(l)** magnesium nitride, lithium nitride, nitrogen trifluoride

**(m)** $:\ddot{F}-\overset{\displaystyle \underset{|}{}}{N}-\ddot{F}:$

$:\ddot{F}:$

**2.**

| Ionic Form | Name | Lewis Structure |
|---|---|---|
| **(a)** $Mg^{2+}SO_4^{2-}$ | magnesium sulfate | $Mg^{2+}$ |
| **(b)** no ions | nitrous acid | $H-\ddot{O}-N=\ddot{O}:$ |
| **(c)** $Li^+NO_3^-$ | lithium nitrate | $Li^+$ |
| **(d)** $2(Co^{3+})3(CO_3^{2-})$ | cobalt(III) carbonate | $2Co^{3+}\ 3$ |
| **(e)** no ions | dichlorine monoxide | $:\ddot{C}l-\ddot{O}-\ddot{C}l:$ |

**3.**

| Formula | Ionic Form | Lewis Structure |
|---|---|---|
| **(a)** $N_2O_3$ | no ions | |
| **(b)** $Cr_2(SO_3)_3$ | $2Cr^{3+}3(SO_3^{2-})$ | $2Cr^{3+}\ 3$ |
| **(c)** $Fe(OH)_2$ | $Fe^{2+}2OH^-$ | $Fe^{2+}\ 2\left[:\ddot{O}-H\right]^-$ |
| **(d)** $SrC_2O_4$ | $Sr^{2+}C_2O_4^{2-}$ | $Sr^{2+}$ |
| **(e)** $HI$ | no ions | $H-\ddot{I}:$ |

# Chapter 7

**7-1. (a)** $Cl_2(g)$ **(b)** $C(s)$ **(c)** $K_2SO_4(s)$
**(d)** $H_2O(l)$ **(e)** $P_4(s)$ **(f)** $H_2(g)$
**(g)** $Br_2(l)$ **(h)** $NaBr(s)$ **(i)** $S_8(s)$
**(j)** $Na(s)$ **(k)** $Hg(l)$ **(l)** $CO_2(g)$

**7-2. (a)** $CaCO_3 \longrightarrow CaO + CO_2$
**(b)** $4Na + O_2 \longrightarrow 2Na_2O$
**(c)** $H_2SO_4 + 2NaOH \longrightarrow Na_2SO_4 + 2H_2O$
**(d)** $2H_2O_2 \longrightarrow 2H_2O + O_2$

**7-4. (a)** $2Al + 2H_3PO_4 \longrightarrow 2AlPO_4 + 3H_2$
**(b)** $Ca(OH)_2 + 2HCl \longrightarrow CaCl_2 + 2H_2O$
**(c)** $3Mg + N_2 \longrightarrow Mg_3N_2$
**(d)** $2C_2H_6 + 7O_2 \longrightarrow 4CO_2 + 6H_2O$

**7-6. (a)** $Mg_3N_2 + 6H_2O \longrightarrow 3Mg(OH)_2 + 2NH_3$
**(b)** $2H_2S + O_2 \longrightarrow 2S + 2H_2O$
**(c)** $Si_2H_6 + 8H_2O \longrightarrow 2Si(OH)_4 + 7H_2$
**(d)** $C_2H_6 + 5Cl_2 \longrightarrow C_2HCl_5 + 5HCl$

**7-8. (a)** $2B_4H_{10} + 11O_2 \longrightarrow 4B_2O_3 + 10H_2O$
**(b)** $SF_6 + 2SO_3 \longrightarrow 3O_2SF_2$
**(c)** $CS_2 + 3O_2 \longrightarrow CO_2 + 2SO_2$
**(d)** $2BF_3 + 6NaH \longrightarrow B_2H_6 + 6NaF$

**7-10. (a)** $2Na(s) + 2H_2O(l) \longrightarrow H_2(g) + 2NaOH(aq)$
**(b)** $2KClO_3(s) \longrightarrow 2KCl(s) + 3O_2(g)$
**(c)** $NaCl(aq) + AgNO_3(aq) \longrightarrow$
$$AgCl(s) + NaNO_3(aq)$$
**(d)** $2H_3PO_4(aq) + 3Ca(OH)_2(aq) \longrightarrow$
$$Ca_3(PO_4)_2(s) + 6H_2O(l)$$

**7-12.** $Ni(s) + 2N_2O_4(l) \longrightarrow Ni(NO_3)_2(s) + 2NO(g)$

**7-14.** In exercise 7-2, **(a)** and **(d)** are decomposition reactions and **(b)** is a combination and combustion reaction. In exercise 7-4, **(c)** is a combination reaction and **(d)** is a combustion reaction.

**7-16. (a)** $2C_7H_{14}(l) + 21O_2(g) \longrightarrow 14CO_2(g) + 14H_2O(l)$
**(b)** $2LiCH_3(s) + 4O_2(g) \longrightarrow$
$$Li_2O(s) + 2CO_2(g) + 3H_2O(l)$$
**(c)** $C_4H_{10}O(l) + 6O_2(g) \longrightarrow 4CO_2(g) + 5H_2O(l)$
**(d)** $2C_2H_5SH(g) + 9O_2(g) \longrightarrow$
$$2SO_2(g) + 4CO_2(g) + 6H_2O(l)$$

**7-18. (a)** $Ba(s) + H_2(g) \longrightarrow BaH_2(s)$
**(b)** $8Ba(s) + S_8(s) \longrightarrow 8BaS(s)$
**(c)** $Ba(s) + Br_2(l) \longrightarrow BaBr_2(s)$
**(d)** $3Ba(s) + N_2(g) \longrightarrow Ba_3N_2(s)$

**7-20. (a)** $Ca(HCO_3)_2(s) \longrightarrow CaO(s) + 2CO_2(g) + H_2O(l)$
**(b)** $2Ag_2O(s) \longrightarrow 4Ag(s) + O_2(g)$
**(c)** $N_2O_3(g) \longrightarrow NO_2(g) + NO(g)$

**7-22. (a)** $2K(s) + Cl_2(g) \longrightarrow 2KCl(s)$
**(b)** $2C_6H_6(l) + 15O_2(g) \longrightarrow 12CO_2(g) + 6H_2O(l)$
**(c)** $2Au_2O_3(s) \longrightarrow 4Au(s) + 3O_2(g)$
**(d)** $2C_3H_8O(l) + 9O_2(g) \longrightarrow 6CO_2(g) + 8H_2O(l)$
**(e)** $P_4(s) + 10F_2(g) \longrightarrow 4PF_5(s)$

**7-24. (a)** $Na_2S \longrightarrow 2Na^+(aq) + S^{2-}(aq)$
**(b)** $Li_2SO_4 \longrightarrow 2Li^+(aq) + SO_4^{2-}(aq)$
**(c)** $K_2Cr_2O_7 \longrightarrow 2K^+(aq) + Cr_2O_7^{2-}(aq)$

**(d)** $CaS \longrightarrow Ca^{2+}(aq) + S^{2-}(aq)$
**(e)** $2(NH_4)_2S \longrightarrow 4NH_4^+(aq) + 2S^{2-}(aq)$
**(f)** $4Ba(OH)_2 \longrightarrow 4Ba^{2+}(aq) + 8OH^-(aq)$

**7-26. (a)** $HNO_3(aq) \longrightarrow H^+(aq) + NO_3^-(aq)$
**(b)** $Sr(OH)_2(s) \longrightarrow Sr^{2+}(aq) + 2OH^-(aq)$

**7-28. (a)** no reaction
**(b)** $Fe + 2H^+ \longrightarrow Fe^{2+} + H_2$
**(c)** $Cu + 2Ag^+ \longrightarrow Cu^{2+} + 2Ag$
**(d)** no reaction

**7-30. (a)** $CuCl_2(aq) + Fe(s) \longrightarrow FeCl_2(aq) + Cu(s)$
$Cu^{2+}(aq) + 2Cl^-(aq) + Fe(s) \longrightarrow$
$$Fe^{2+}(aq) + 2Cl^-(aq) + Cu(s)$$
$Cu^{2+}(aq) + Fe(s) \longrightarrow Fe^{2+}(aq) + Cu(s)$
**(b)** and **(c)** no reaction

**(d)** $3Zn(s) + 2Cr(NO_3)_3(aq) \longrightarrow$
$$3Zn(NO_3)_2(aq) + 2Cr(s)$$
$3Zn(s) + 2Cr^{3+}(aq) + 6NO_3^-(aq) \longrightarrow$
$$3Zn^{2+}(aq) + 6NO_3^-(aq) + 2Cr(s)$$
$3Zn(s) + 2Cr^{3+}(aq) \longrightarrow 3Zn^{2+}(aq) + 2Cr(s)$

**7-32.** $6Na(l) + Cr_2O_3(s) \longrightarrow 2Cr(s) + 3Na_2O(s)$
$6Na + 2Cr^{3+} \longrightarrow 2Cr + 6Na^+$

**7-34.** Insoluble compounds are **(b)** $PbSO_4$, and
**(d)** $Ag_2S$

**7-36. (a)** $AgBr$ **(b)** $Ag_2CO_3$ **(c)** $Ag_3PO_4$

**7-38. (a)** $CuCO_3$ **(b)** $CdCO_3$ **(c)** $Cr_2(CO_3)_3$

**7-40.** $Hg_2Cl_2$

**7-42. (b)** $Ca_3(PO_4)_2$

**7-44. (a)** $2KI(aq) + Pb(C_2H_3O_2)_2(aq) \longrightarrow$
$$PbI_2(s) + 2KC_2H_3O_2(aq)$$
**(b)** and **(c)** no reaction occurs
**(d)** $BaS(aq) + Hg_2(NO_3)_2(aq) \longrightarrow$
$$Hg_2S(s) + Ba(NO_3)_2(aq)$$
**(e)** $FeCl_3(aq) + 3KOH(aq) \longrightarrow$
$$Fe(OH)_3(s) + 3KCl(aq)$$

**7-46. (a)** $2K^+(aq) + 2I^-(aq) + Pb^{2+}(aq) + 2C_2H_3O_2^-(aq)$
$$\longrightarrow PbI_2(s) + 2K^+(aq) + 2C_2H_3O_2^-(aq)$$
$Pb^{2+}(aq) + 2I^-(aq) \longrightarrow PbI_2(s)$
**(d)** $Ba^{2+}(aq) + S^{2-}(aq) + Hg_2^{2+}(aq) + 2NO_3^-(aq)$
$$\longrightarrow Hg_2S(s) + Ba^{2+}(aq) + 2NO_3^-(aq)$$
$Hg_2^{2+}(aq) + S^{2-}(aq) \longrightarrow Hg_2S(s)$
**(e)** $Fe^{3+}(aq) + 3Cl^-(aq) + 3K^+(aq) + 3OH^-(aq)$
$$\longrightarrow Fe(OH)_3(s) + 3K^+(aq) + 3Cl^-(aq)$$
$Fe^{3+}(aq) + 3OH^-(aq) \longrightarrow Fe(OH)_3(s)$

**7-48. (a)** $2K^+(aq) + S^{2-}(aq) + Pb^{2+}(aq) + 2NO_3^-(aq)$
$$\longrightarrow PbS(s) + 2K^+(aq) + 2NO_3^-(aq)$$
$S^{2-}(aq) + Pb^{2+}(aq) \longrightarrow PbS(s)$
**(b)** $2NH_4^+(aq) + CO_3^{2-}(aq) + Ca^{2+}(aq) + 2Cl^-(aq)$
$$\longrightarrow CaCO_3(s) + 2NH_4^+(aq) + 2Cl^-(aq)$$
$CO_3^{2-}(aq) + Ca^{2+}(aq) \longrightarrow CaCO_3(s)$

**(c)** $2Ag^+(aq) + 2ClO_4^-(aq) + 2Na^+(aq) +$
$\quad CrO_4^{2-}(aq) \longrightarrow Ag_2CrO_4(s) + 2Na^+(aq) +$
$\quad\quad 2ClO_4^-(aq)$

$\quad 2Ag^+(aq) + CrO_4^{2-}(aq) \longrightarrow Ag_2CrO_4(s)$

**7-50.** **(a)** $CuCl_2(aq) + Na_2CO_3(aq) \longrightarrow$
$\quad\quad CuCO_3(s) + 2NaCl(aq)$

Filter the solid $CuCO_3$.

**(b)** $(NH_4)_2SO_4(aq) + Pb(NO_3)_2(aq) \longrightarrow$
$\quad\quad PbSO_4(s) + 2NH_4NO_3(aq)$

Filter the solid $PbSO_4$.

**(c)** $2KI(aq) + Hg_2(NO_3)_2(aq) \longrightarrow$
$\quad\quad Hg_2I_2(s) + 2KNO_3(aq)$

Filter the solid $Hg_2I_2$.

**(d)** $NH_4Cl(aq) + AgNO_3(aq) \longrightarrow$
$\quad\quad AgCl(s) + NH_4NO_3(aq)$

Filter the solid AgCl; the desired product remains after water is removed by boiling.

**(e)** $Ca(C_2H_3O_2)_2(aq) + K_2CO_3(aq) \longrightarrow$
$\quad\quad CaCO_3(s) + 2KC_2H_3O_2(aq)$

Filter the solid $CaCO_3$; the desired product remains after the water is removed by boiling.

**7-51.** **(b)** HF

**7-53.** **(c)** $Al(OH)_3$

**7-55.** **(a)** $HI(aq) + CsOH(aq) \longrightarrow CsI(aq) + H_2O(l)$

**(b)** $2HNO_3(aq) + Ca(OH)_2(aq) \longrightarrow$
$\quad\quad Ca(NO_3)_2(aq) + 2H_2O(l)$

**(c)** $H_2SO_4(aq) + Sr(OH)_2(aq) \longrightarrow$
$\quad\quad SrSO_4(s) + 2H_2O(l)$

**7-57.** **(a)** $H^+(aq) + I^-(aq) + Cs^+(aq) + OH^-(aq) \longrightarrow$
$\quad\quad Cs^+(aq) + I^-(aq) + H_2O(l)$

$\quad H^+(aq) + OH^-(aq) \longrightarrow H_2O(l)$

**(b)** $2H^+(aq) + 2NO_3^-(aq) + Ca^{2+}(aq) + 2OH^-(aq)$
$\quad \longrightarrow Ca^{2+}(aq) + 2NO_3^-(aq) + 2H_2O(l)$

$\quad H^+(aq) + OH^-(aq) \longrightarrow H_2O(l)$

**(c)** $2H^+(aq) + SO_4^{2-}(aq) + Sr^{2+}(aq) + 2OH^-(aq)$
$\quad \longrightarrow SrSO_4(s) + 2H_2O(l)$

Net ionic equation is the same as the total ionic equation since $SrSO_4$ precipitates.

**7-59.** $Mg(OH)_2(s) + 2HCl(aq) \longrightarrow MgCl_2(aq) + 2H_2O(l)$
$Mg(OH)_2(s) + 2H^+(aq) + 2Cl^-(aq) \longrightarrow$
$\quad\quad Mg^{2+}(aq) + 2Cl^-(aq) + 2H_2O(l)$
$Mg(OH)_2(s) + 2H^+(aq) \longrightarrow Mg^{2+}(aq) + 2H_2O(l)$

**7-61.** $C_3H_8(g) + 5O_2(g) \longrightarrow 3CO_2(g) + 4H_2O(l)$

$2C_4H_{10}(g) + 13O_2(g) \longrightarrow 8CO_2(g) + 10H_2O(l)$

$2C_8H_{18}(l) + 25O_2(g) \longrightarrow 16CO_2(g) + 18H_2O(l)$

$C_2H_5OH(l) + 2O_2(g) \longrightarrow 2CO_2(g) + 3H_2O(l)$

**7-62.** In both of these reactions, the reactants change from neutral atoms to cations and anions in the products. Ions are formed from neutral atoms from the loss or gain of electrons.

**7-64.** **(a)** $Ba(s) + I_2(s) \longrightarrow BaI_2(s)$

**(b)** $HBr(aq) + RbOH(aq) \longrightarrow RbBr(aq) + H_2O(l)$

**(c)** $Ca(s) + 2HNO_3(aq) \longrightarrow Ca(NO_3)_2(aq) + H_2(g)$

**(d)** $C_{10}H_8(s) + 12O_2(g) \longrightarrow 10CO_2(g) + 4H_2O(l)$

**(e)** $(NH_4)_2CrO_4(aq) + BaBr_2(aq) \longrightarrow$
$\quad\quad BaCrO_4(s) + 2NH_4Br(aq)$

**(f)** $2Al(OH)_3(s) \longrightarrow Al_2O_3(s) + 3H_2O(g)$

**7-65.** **(b)** $H^+(aq) + Br^-(aq) + Rb^+(aq) + OH^-(aq) \longrightarrow$
$\quad\quad Rb^+(aq) + Br^-(aq) + H_2O(l)$

$\quad H^+(aq) + OH^-(aq) \longrightarrow H_2O(l)$

**(c)** $Ca(s) + 2H^+(aq) + 2NO_3^+(aq) \longrightarrow$
$\quad\quad Ca^{2+}(aq) + 2NO_3^-(aq) + H_2(g)$

$\quad Ca(s) + 2H^+(aq) \longrightarrow Ca^{2+}(aq) + H_2(g)$

**(e)** $2NH_4^+(aq) + CrO_4^{2-}(aq) + Ba^{2+}(aq) + 2Br^-(aq)$
$\quad \longrightarrow BaCrO_4(s) + 2NH_4^+(aq) + 2Br^-(aq)$

$\quad CrO_4^{2-}(aq) + Ba^{2+}(aq) \longrightarrow BaCrO_4(s)$

**7-68.** $H^+(aq) + OH^-(aq) \longrightarrow H_2O(l)$

$Ba^{2+}(aq) + SO_4^{2-}(aq) \longrightarrow BaSO_4(s)$

**7-70.** $Fe^{3+}(aq) + PO_4^{3-}(aq) \longrightarrow FePO_4(s)$

$2Fe^{3+}(aq) + 3S^{2-}(aq) \longrightarrow Fe_2S_3(s)$

$Pb^{2+}(aq) + 2I^-(aq) \longrightarrow PbI_2(s)$

$3Pb^{2+}(aq) + 2PO_4^{3-}(aq) \longrightarrow Pb_3(PO_4)_2(s)$

$Pb^{2+}(aq) + S^{2-}(aq) \longrightarrow PbS(s)$

# Chapter 8

**8-1.** 0.609 lb of pennies

**8-3.** $145 \text{ g Au} \times \dfrac{108 \text{ g Ag}}{197.0 \text{ g Au}} = \underline{\underline{79.5 \text{ g Ag}}}$

**8-4.** 94.4 lb C

**8-6.** 71.5 g Cu

**8-8.** $25.0 \text{ g C} \times \dfrac{x \text{ g}}{12.01 \text{ g C}} = 33.3 \text{ g} \quad x = 16.0 \text{ g (O)}$

The compound is CO.

**8-10.** 40.1 lb S

**8-11.** $6.022 \times 10^{23} \text{ units} \times \dfrac{1 \text{ sec}}{2 \text{ units}} \times \dfrac{1 \text{ min}}{60 \text{ sec}}$
$\quad\quad \times \dfrac{1 \text{ hr}}{60 \text{ min}} \times \dfrac{1 \text{ day}}{24 \text{ hr}} \times \dfrac{1 \text{ year}}{365 \text{ day}}$
$\quad\quad = \underline{\underline{9.548 \times 10^{15} \text{ years (9.548 quadrillion)}}}$

$\dfrac{9.5 \times 10^{15} \text{ years}}{5.5 \times 10^9} = 1.7 \times 10^6 \text{ years (1.7 million)}$

**8-13.** $6.022 \times 10^{26}$ (if mass in kg); $6.022 \times 10^{20}$ (if mass in mg)

**8-14.** **(a)** 0.468 mol P   $2.82 \times 10^{23}$ atoms P

**(b)** 150 g Rb   $1.05 \times 10^{24}$ atoms

**(c)** Al: 27.0 g, 1.00 mol,

**(d)** 5.00 mol X   element is Ge

**(e)** $1.66 \times 10^{-24}$ mol   $7.95 \times 10^{-23}$ g

**8-16.** **(a)** 63.5 g Cu   **(b)** 16 g S   **(c)** 40.1 g Ca

**8-18.** **(a)** $1.93 \times 10^{25}$ atoms   **(b)** $6.03 \times 10^{23}$ atoms
**(c)** $1.20 \times 10^{24}$ atoms

**8-20.** $50.0 \text{ g Al} \times \dfrac{1 \text{ mol Al}}{26.98 \text{ g Al}} = 1.85 \text{ mol Al}$

$50.0 \text{ g Fe} \times \dfrac{1 \text{ mol Fe}}{55.85 \text{ g Fe}} = 0.895 \text{ mol Fe}$

There are more moles of atoms (more atoms) in 50.0 g of Al.

**8-21.** $20.0 \text{ g Ni} \times \dfrac{1 \text{ mol Ni}}{58.69 \text{ g Ni}} = 0.341 \text{ mol Ni}$

$2.85 \times 10^{23} \text{ atoms} \times \dfrac{1 \text{ mol Ni}}{6.022 \times 10^{23} \text{ atoms}}$
$= 0.473 \text{ mol Ni}$

The $2.85 \times 10^{23}$ atoms of Ni contain more atoms than 20.0 g.

**8-23.** $1.40 \times 10^{21}$ atoms $= 2.32 \times 10^{-3}$ mol
$0.251 \text{ g}/(2.32 \times 10^{-3} \text{ mol}) = 108 \text{ g/mol}$ (silver)

**8-25.** **(a)** 106.6 amu  **(b)** 80.07 amu  **(c)** 108.0 amu
**(d)** 98.09 amu  **(e)** 106.0 amu  **(f)** 60.05 amu
**(g)** 459.7 amu

**8-27.** $Cr_2(SO_4)_3$  $(2 \times 52.00) + (3 \times 32.07) +$
$(12 \times 16.00) = \underline{392.2 \text{ amu}}$

**8-29.** **(a)** 189 g $H_2O$, $6.32 \times 10^{24}$ molecules
**(b)** $5.00 \times 10^{-3}$ mol $BF_3$, 0.339 g $BF_3$
**(c)** 0.219 mol $SO_2$, $1.32 \times 10^{23}$ molecules
**(d)** 0.209 g $K_2SO_4$, $7.23 \times 10^{19}$ formula units
**(e)** 7.47 mol $SO_3$, 598 g $SO_3$
**(f)** $7.61 \times 10^{-3}$ mol, $4.58 \times 10^{21}$ molecules

**8-31.** $21.5 \text{ g}/0.0684 \text{ mol} = \underline{314 \text{ g/mol}}$

**8.33.** 161 g/mol

**8-35.** 5.10 mol C, 15.3 mol H, 2.55 mol O  Total =
23.0 mol of atoms 61.3 g C, 15.4 g H, 40.8 g O  Total
mass $= \underline{117.5 \text{ g}}$

**8-36.** 0.135 mol $Ca(ClO_3)_2$, 0.135 mol Ca, 0.270 mol Cl,
0.810 mol O  Total = 1.215 mol of atoms

**8-38.** $1.50 \text{ mol } H_2SO_4 \times \dfrac{2 \text{ mol H}}{\text{mol } H_2SO_4} \times \dfrac{1.008 \text{ g H}}{\text{mol H}}$
$= \underline{3.02 \text{ g H}}$

48.1 g S, 72.0 g O

**8-40.** $1.20 \times 10^{22} \text{ molecules} \times \dfrac{1 \text{ mol } O_2}{6.022 \times 10^{23} \text{ molecules}}$
$= \underline{0.0199 \text{ mol } O_2}$

$0.0199 \text{ mol } O_2 \times \dfrac{2 \text{ mol O atoms}}{\text{mol } O_2} = 0.0398 \text{ mol O atoms}$

$0.0199 \text{ mol } O_2 \times \dfrac{32.00 \text{ g } O_2}{\text{mol } O_2}$
$= \underline{0.637 \text{ g } O_2 \text{ The mass is the same.}}$

**8-42.** Total mass of compound = 1.375 + 3.935 = 5.310 g,
25.89% N, 74.11% O

**8-43.** 46.7% Si, 53.3% O

**8-45.** **(a)** $C_2H_6O$ 52.14% C, 13.13% H, 34.73% O
**(b)** $C_3H_6$ 85.62% C, 14.38% H
**(c)** $C_9H_{18}$ 85.66% C, 14.34% H

**(b)** and **(c)** are actually the same. The difference comes from rounding off.
**(d)** $Na_2SO_4$ 32.36% Na, 22.57% S, 45.07% O
**(e)** $(NH_4)_2CO_3$ 29.16% N, 8.392% H, 12.50% C, 49.95% O

**8-47.** 12.06% Na, 11.34% B, 71.31% O, 5.286% H

**8-49.** $C_7H_5SNO_3$ Formula weight = $(7 \times 12.01) + (5 \times 1.008) + 32.07 + 14.01 + (3 \times 16.00) = 183.2$ amu

C: $\dfrac{84.07 \text{ amu}}{183.2 \text{ amu}} \times 100\% = \underline{45.89\% \text{ C}}$

H: $\dfrac{5.040 \text{ amu}}{183.2 \text{ amu}} \times 100\% = \underline{2.751\% \text{ H}}$

S: $\dfrac{32.07 \text{ amu}}{183.2 \text{ amu}} \times 100\% = \underline{17.51\% \text{ S}}$

N: $\dfrac{14.01 \text{ amu}}{183.2 \text{ amu}} \times 100\% = \underline{7.647\% \text{ N}}$

O: $100\% - (45.89 + 2.751 + 17.51 + 7.647)$
$= \underline{26.20\% \text{ O}}$

**8-51.** $Na_2C_2O_4$  Formula weight = $(2 \times 22.99) + (2 \times 12.01) + (4 \times 16.00) = 134.0$ amu
There is 24.02 g $(2 \times 12.01)$ of C in 134.0 g of compound.

$125 \text{ g } Na_2C_2O_4 \times \dfrac{24.02 \text{ g C}}{134.0 \text{ g } Na_2C_2O_4} = \underline{22.4 \text{ g C}}$

**8-52.** 4.72 lb P

**8-54.** $1.40 \times 10^3$ lb Fe

**8-55.** **(a)** $N_2O_4$ and **(d)** $H_2C_2O_4$

**8-56.** **(a)** FeS  **(b)** $SrI_2$  **(c)** $KClO_3$  **(d)** $I_2O_5$
**(e)** $Fe_2O_{2.66} = Fe_3O_4$  **(f)** $C_3H_5Cl_3$

**8-58.** $N_2O_3$

**8-60.** $KO_2$

**8-62.** $MgC_2O_4$

**8-63.** $CH_2Cl$

**8-65.** $N_2H_8SO_3$

**8-66.** $C_{8/3}H_{8/3}O = C_8H_8O_3$

**8-68.** $C_3H_4Cl_4$ (empirical formula)  $C_9H_{12}Cl_{12}$ (molecular formula)

**8-70.** $B_2C_2H_6O_4$ (molecular formula)

**8-71.** Empirical formula = $KC_2NH_3O_2$  Empirical mass = 112.2 g/ emp. unit

$\dfrac{224 \text{ g/mol}}{112.2 \text{ g/emp. unit}} = 2$ emp. units/mol

$K_2C_4N_2H_6O_4$ (molecular formula)

**8-73.** $I_6C_6$

**8-75.** **(a)** $\dfrac{1 \text{ mol } H_2}{1 \text{ mol Mg}}$  **(b)** $\dfrac{2 \text{ mol HCl}}{1 \text{ mol Mg}}$  **(c)** $\dfrac{1 \text{ mol } H_2}{2 \text{ mol HCl}}$

**(d)** $\dfrac{2 \text{ mol HCl}}{1 \text{ mol } MgCl_2}$

**8-76.** **(a)** $\dfrac{2 \text{ mol } C_4H_{10}}{8 \text{ mol } CO_2}$  **(b)** $\dfrac{2 \text{ mol } C_4H_{10}}{13 \text{ mol } O_2}$

**(c)** $\dfrac{13 \text{ mol } O_2}{8 \text{ mol } CO_2}$  **(d)** $\dfrac{10 \text{ mol } H_2O}{13 \text{ mol } O_2}$

**8-78.** (a) 3.33 mol $Al_2O$, 3.33 mol $AlCl_3$, 10.0 mol NO, 20.0 mol $H_2O$
(b) 1.00 mol $Al_2O_3$, 1.00 mol $AlCl_3$, 3.00 mol NO, 6.00 mol $H_2O$

**8-80.** (a) 15.0 mol $O_2$ and 10.0 mol $CH_4$ react
(b) 6.67 mol HCN and 20.0 mol $H_2O$ produced

**8-82.** (a) 126 g $SiF_4$ and (b) 43.8 g $H_2O$ produced
(b) 73.0 g $SiO_2$ reacts

**8-83.** (a) mol $H_2O \longrightarrow$ mol $H_2$

$$0.400 \ \overline{mol \ H_2O} \times \frac{2 \ mol \ H_2}{2 \ \overline{mol \ H_2O}} = \underline{\underline{0.400 \ mol \ H_2}}$$

(b) g $O_2 \longrightarrow$ mol $O_2 \longrightarrow$ mol $H_2O$

$$0.640 \ \overline{g \ O_2} \times \frac{1 \ \overline{mol \ O_2}}{32.00 \ \overline{g \ O_2}} \times \frac{2 \ mol \ H_2O}{1 \ \overline{mol \ O_2}}$$
$$= \underline{\underline{0.0400 \ mol \ H_2O}}$$

(c) g $O_2 \longrightarrow$ mol $O_2 \longrightarrow$ mol $H_2$

$$0.032 \ \overline{g \ O_2} \times \frac{1 \ \overline{mol \ O_2}}{32.00 \ \overline{g \ O_2}} \times \frac{2 \ mol \ H_2}{1 \ \overline{mol \ O_2}}$$
$$= \underline{\underline{0.0020 \ mol \ H_2}}$$

(d) g $H_2 \longrightarrow$ mol $H_2 \longrightarrow$ mol $H_2O \longrightarrow$ g $H_2O$

$$0.400 \ \overline{g \ H_2} \times \frac{1 \ \overline{mol \ H_2}}{2.016 \ \overline{g \ H_2}} \times \frac{2 \ \overline{mol \ H_2O}}{2 \ \overline{mol \ H_2}}$$
$$\times \frac{18.02 \ g \ H_2O}{\overline{mol \ H_2O}} = \underline{\underline{3.58 \ g \ H_2O}}$$

**8-84.** (a) 1.35 mol $CO_2$, 1.80 mol $H_2O$, 2.25 mol $O_2$
(b) 4.81 g $H_2O$     (c) 1.10 g $C_3H_8$
(d) 44.1 g $C_3H_8$
(e) molecules $O_2 \longrightarrow$ mol $O_2 \longrightarrow$ mol $CO_2$
$$\longrightarrow g \ CO_2$$

$$1.20 \times 10^{23} \ \overline{molecules} \times \frac{1 \ \overline{mol \ O_2}}{6.022 \times 10^{23} \ \overline{molecules}}$$
$$\times \frac{3 \ \overline{mol \ CO_2}}{5 \ \overline{mol \ O_2}} \times \frac{44.09 \ g \ CO_2}{\overline{mol \ CO_2}} = \underline{\underline{5.27 \ g \ CO_2}}$$

(f) 0.0996 mol $H_2O$

**8-86.** 47.2 g $N_2$

**8-88.** 0.728 g HCl

**8-90.** mol $FeS_2 \longrightarrow$ mol $H_2S \longrightarrow$ molecules $H_2S$

$$0.520 \ \overline{mol \ FeS_2} \times \frac{1 \ \overline{mol \ H_2S}}{1 \ \overline{mol \ FeS_2}}$$

$$\times \frac{6.022 \times 10^{23} \ molecules}{\overline{mol \ H_2S}} = \underline{\underline{3.13 \times 10^{23} \ molecules}}$$

**8-91.** 16,900 g (16.9 kg) $HNO_3$

**8-93.** $2.30 \times 10^3$ g (2.30 kg) $C_2H_5OH$

**8-94.** 8140 g (8.14 kg) CO

**8-95.** (a) CuO limiting reactant producing 1.00 mol $N_2$
(b) stoichiometric mixture producing 1.00 mol $N_2$
(c) $NH_3$ limiting reactant producing 0.500 mol $N_2$
(d) CuO limiting reactant producing 0.209 mol $N_2$
(e) $NH_3$ limiting reactant producing 1.75 mol $N_2$

**8-96.** (a)
$$3.00 \ \overline{mol \ CuO} \times \frac{2 \ mol \ NH_3}{3 \ \overline{mol \ CuO}} = 2.00 \ mol \ NH_3 \ used$$
$$3.00 - 2.00 = 1.00 \ mol \ NH_3 \ in \ excess$$
(c) 1.50 mol CuO in excess

**8-99.** $H_2SO_4$ is the limiting reactant and the yield of $H_2$ is $\underline{1.00 \ mole.}$

$$1.00 \ \overline{mol \ H_2SO_4} \times \frac{2 \ mol \ Al}{3 \ \overline{mol \ H_2SO_4}}$$
$$= 0.667 \ mol \ of \ Al \ used$$
$$0.800 - 0.667 = \underline{\underline{0.133 \ mol \ Al \ remaining}}$$

**8-100.** $3.44 \ \overline{mol \ C_5H_6} \times \dfrac{10 \ mol \ CO_2}{2 \ \overline{mol \ C_5H_6}} = 17.2 \ mol \ CO_2$

$$20.6 \ \overline{mol \ O_2} \times \frac{10 \ mol \ CO_2}{13 \ \overline{mol \ O_2}} = 15.8 \ mol \ CO_2$$

Since $O_2$ is the limiting reactant:

$$15.8 \ \overline{mol \ CO_2} \times \frac{44.01 \ g \ CO_2}{\overline{mol \ CO_2}} = \underline{\underline{695 \ g \ CO_2}}$$

**8-102.** Since $NH_3$ produces the least $N_2$, it is the limiting reactant and the yield of $N_2$ is $\underline{0.750 \ mol.}$

**8-103.** $20.0 \ \overline{g \ AgNO_3} \times \dfrac{1 \ \overline{mol \ AgNO_3}}{169.9 \ \overline{g \ AgNO_3}} \times \dfrac{2 \ mol \ AgCl}{2 \ \overline{mol \ AgNO_3}}$
$$= 0.118 \ mol \ AgCl$$

$$10.0 \ \overline{g \ CaCl_2} \times \frac{1 \ \overline{mol \ CaCl_2}}{111.0 \ \overline{g \ CaCl_2}} \times \frac{2 \ mol \ AgCl}{1 \ \overline{mol \ CaCl_2}}$$
$$= 0.180 \ mol \ AgCl$$

Since $AgNO_3$ produces the least AgCl, it is the limiting reactant.

$$0.118 \ \overline{mol \ AgCl} \times \frac{143.4 \ g \ AgCl}{\overline{mol \ AgCl}} = \underline{\underline{16.9 \ g \ AgCl}}$$

Convert moles of AgCl (the limiting reactant) to grams of $CaCl_2$ used.

$$0.118 \ \overline{mol \ AgCl} \times \frac{1 \ \overline{mol \ CaCl_2}}{2 \ \overline{mol \ AgCl}} \times \frac{111.0 \ g \ CaCl_2}{\overline{mol \ CaCl_2}}$$
$$= 6.55 \ g \ CaCl_2 \ used$$

$$10.0 \ g - 6.55 \ g = \underline{\underline{3.5 \ g \ CaCl_2 \ remaining}}$$

**8-105.** Products: 3.53 g $H_2O$, $\underline{4.71 \ g \ S}$, $\underline{2.94 \ g \ NO}$
Reactants remaining: 3.8 g $HNO_3$ remaining

**8-106.** 30.0 g $SO_3$ (theoretical yield) and 70.7% yield

**8-109.** $\underline{86.4\%}$

**8-110.** If 86.4% is converted to $CO_2$, the remainder (13.6%) is converted to CO. Thus, $0.136 \times 57.0 \ g = 7.75$ g of $C_8H_{18}$ is converted to CO. Notice that 1 mole of $C_8H_{18}$ forms 8 moles of CO (because of the eight carbons in $C_8H_{18}$). Thus

$$7.75 \ \overline{g \ C_8H_{18}} \times \frac{1 \ \overline{mol \ C_8H_{18}}}{114.2 \ \overline{g \ C_8H_{18}}} \times \frac{8 \ \overline{mol \ CO}}{1 \ \overline{mol \ C_8H_{18}}}$$
$$\times \frac{28.01 \ g \ CO}{\overline{mol \ CO}} = \underline{\underline{15.2 \ g \ CO}}$$

**8-111.** Theoretical yield $\times$ 0.700 = 250 g (actual yield)

Theoretical yield = 250 g/0.700 = 357 g $N_2$

g $N_2 \longrightarrow$ mol $N_2 \longrightarrow$ mol $H_2 \longrightarrow$ g $H_2$

$$357 \text{ g } N_2 \times \frac{1 \text{ mol } N_2}{28.02 \text{ g } N_2} \times \frac{4 \text{ mol } H_2}{1 \text{ mol } N_2} \times \frac{2.016 \text{ g } H_2}{\text{mol } H_2}$$
$$= 103 \text{ g } H_2$$

**8-114.** $2Mg(s) + O_2 \longrightarrow 2MgO(s) + 1204$ kJ

$2Mg(s) + O_2(g) \longrightarrow 2MgO(s) \quad \Delta H = -1204$ kJ

**8-116.** $CaCO_3(s) + 176$ kJ $\longrightarrow CaO(s) + CO_2(g)$

$CaCO_3(s) \longrightarrow CaO(s) + CO_2(g) \quad \Delta H = 176$ kJ

**8-117.** $1.00 \text{ g } C_8H_{18} \times \dfrac{1 \text{ mol } C_8H_{18}}{114.2 \text{ g } C_8H_{18}} \times \dfrac{5480 \text{ kJ}}{\text{mol } C_8H_{18}}$
$$= 48.0 \text{ kJ}$$

$1.00 \text{ g } CH_4 \times \dfrac{1 \text{ mol } CH_4}{16.04 \text{ g } CH_4} \times \dfrac{890 \text{ kJ}}{\text{mol } CH_4} = 55.6 \text{ kJ}$

**8-119.** kJ $\longrightarrow$ mol Al $\longrightarrow$ g Al

$35.8 \text{ kJ} \times \dfrac{2 \text{ mol Al}}{850 \text{ kJ}} \times \dfrac{26.99 \text{ g Al}}{\text{mol Al}} = 2.27 \text{ g Al}$

**8-120.** 69.7 g $C_6H_{12}O_6$

**8-122.** $7.5 \times 10^{-10}$ mol pennies

**8-123.** $0.443 \text{ g } N \times \dfrac{1 \text{ mol } N}{14.01 \text{ g } N} = 0.0316 \text{ mol } N$

Thus 1.420 g of M also equals 0.0316 mol M since M and N are present in equimolar amounts.

$1.420 \text{ g}/0.0316 \text{ mol} = 44.9 \text{ g/mol [scandium (Sc)]}$

**8-126.** $2.78 \times 10^{-3}$ mol $P_4$

$2.78 \times 10^{-3} \text{ mol } P_4 \times \dfrac{4 \text{ mol } P}{\text{mol } P_4}$

$\times \dfrac{6.022 \times 10^{23} \text{ atoms } P}{\text{mol } P} = 6.70 \times 10^{21} \text{ atoms } P$

**8-127.** 100 mol $H_2$ = 202 g $H_2$ therefore
100 H atoms < 100 $H_2$ molecules < 100 g $H_2$ < 100 mol $H_2$

**8-129.** 120 g/mol of compound 120 − 55.8 = 64 g of S

$\dfrac{64 \text{ g } S}{32.07 \text{ g } S/\text{mol}} = 2 \text{ mol } S$

Formula = $FeS_2$

**8-130.** (a) $2Na^+$ and $S_4O_6^{2-}$    (b) 27.9 g S
(c) $NaS_2O_3$    (d) 270.3 g/mol
(e) 0.0925 mol $Na_2S_4O_6$, $5.57 \times 10^{22}$ formula units
(f) 35.5% oxygen

**8-132.** $\dfrac{2N}{2N + xO} = 0.368 \quad \dfrac{28.02}{28.02 + 16.00x} = 0.368$

$x = 3 \quad N_2O_3$ dinitrogen trioxide

**8-134.** $C_{12}H_4Cl_4O_2$ (molecular formula)

**8-136.** Empirical formula $CrCl_3O_{12}$ Actual formula = $Cr(ClO_4)_3$ chromium(III) perchlorate

**8-137.** Assume exactly 100 g of compound. There is then 51.1 g $H_2O$ and 48.9 g $MgSO_4$.

$MgSO_4$: $48.9 \text{ g } MgSO_4 \times \dfrac{1 \text{ mol}}{120.4 \text{ g } MgSO_4}$
$$= 0.406 \text{ mol } MgSO_4$$

2.94 mol $H_2O$/0.406 mol $MgSO_4$
$$= 7.0 \text{ mol } H_2O/\text{mol } MgSO_4$$

The formula is $MgSO_4 \cdot 7H_2O$

**8-140.** $1.20 \text{ g } CO_2 \times \dfrac{1 \text{ mol } CO_2}{44.01 \text{ g } CO_2} \times \dfrac{1 \text{ mol } C}{\text{mol } CO_2}$
$$= 0.0273 \text{ mol } C$$

$0.489 \text{ g } H_2O \times \dfrac{1 \text{ mol } H_2O}{18.02 \text{ g } H_2O} \times \dfrac{2 \text{ mol } H}{\text{mol } H_2O}$
$$= 0.0543 \text{ mol } H$$

C: $\dfrac{0.0273}{0.0273} = 1.0$   H: $\dfrac{0.0543}{0.0273} = 2.0$   $CH_2$

**8-143.** $125 \text{ g } Fe_2O_3 \times \dfrac{1 \text{ mol } Fe_2O_3}{159.7 \text{ g } Fe_2O_3} \times \dfrac{2 \text{ mol } Fe_3O_4}{3 \text{ mol } Fe_2O_3}$

$\times \dfrac{3 \text{ mol } FeO}{1 \text{ mol } Fe_3O_4} \times \dfrac{1 \text{ mol } Fe}{1 \text{ mol } FeO} \times \dfrac{55.85 \text{ g } Fe}{\text{mol } Fe}$
$$= 87.4 \text{ g } Fe$$

**8-144.** $2KClO_3 \longrightarrow 2 KCl + 3O_2$

Find the mass of $KClO_3$ needed to produce 12.0 g $O_2$.

g $O_2 \longrightarrow$ mol $O_2 \longrightarrow$ mol $KClO_3$
$$\longrightarrow \text{ g } KClO_3$$

$12.0 \text{ g } O_2 \times \dfrac{1 \text{ mol } O_2}{32.00 \text{ g } O_2} \times \dfrac{2 \text{ mol } KClO_3}{3 \text{ mol } O_2}$

$\times \dfrac{122.6 \text{ g } KClO_3}{\text{mol } KClO_3} = 30.7 \text{ g } KClO_3$

percent purity = $\dfrac{30.7 \text{ g}}{50.0 \text{ g}} \times 100\% = 61.4\%$

**8-145.** 4.49% $FeS_2$

**8-147.** (1) $NH_3$ is the limiting reactant.
(2) 141 g NO (theoretical yield), 28.4% yield

**8-149.** $H_2O$ is the limiting reactant producing 11.0 g $CaCl_2 \cdot 6H_2O$

**8-150.** Molecular formula = $CH_2Cl_2$
$CH_4(g) + 2Cl_2(g) \longrightarrow CH_2Cl_2(l) + 2HCl(g)$
$Cl_2$ is the limiting reactant producing 9.00 g $CH_2Cl_2$

**8-153.** Formula = $NaHCO_3 = Na^+ \ HCO_3^-$

The geometry around the C is trigonal planar with the approximate H—O—C angle of 120°.

**8-155.** Formula = $CaH_2S_2O_6 = Ca(HSO_3)_2$ calcium bisulfite or calcium hydrogen sulfite

The geometry around the S is trigonal pyramid. The H—O—S angle is about 109°.

## Cumulative Review Test for Chapters 7 and 8

### Multiple Choice

| | | | | |
|---|---|---|---|---|
| **1.** (e) | **2.** (a) | **3.** (c) | **4.** (d) | **5.** (a) |
| **6.** (b) | **7.** (d) | **8.** (c) | **9.** (b) | **10.** (e) |
| **11.** (c) | **12.** (c) | **13.** (a) | **14.** (d) | **15.** (a) |
| **16.** (c) | **17.** (a) | **18.** (c) | **19.** (d) | **20.** (d) |
| **21.** (b) | **22.** (e) | **23.** (c) | **24.** (c) | |

### Problems

**1.** 1.50 mol $K_2SO_4$ has a mass of 261 g and contains $9.03 \times 10^{23}$ formula units. It contains 6.00 mol O atoms, 117 g K, and $1.81 \times 10^{24}$ K atoms

**2.** $C_4H_{10}N_2$ has a molar mass of 86.14 g/mol, is 32.53% N, and has an empirical formula of $C_2H_5N$.

**3.** $Fe_3O_4$

**4.** $C_2H_{14}B_{10}$

**5.** (a) $6K(s) + Al_2Cl_6(s) \longrightarrow 6KCl(s) + 2Al(s)$
(b) $2C_6H_6(l) + 15O_2(g) \longrightarrow 12CO_2(g) + 6H_2O(l)$
(c) $Cl_2O_3(g) + H_2O(l) \longrightarrow 2HClO_2(aq)$
(d) $3K_2S(aq) + 2H_3PO_4(aq) \longrightarrow 3H_2S(g) + 2K_3PO_4(aq)$
(e) $B_4H_{10}(g) + 12H_2O(l) \longrightarrow 4B(OH)_3(aq) + 11H_2(g)$

**6.** (a) $Mg(s) + Pb(NO_3)_2(aq) \longrightarrow Mg(NO_3)_2(aq) + Pb(s)$
$Mg(s) + Pb^{2+}(aq) + 2NO_3^-(aq) \longrightarrow$
$\qquad\qquad Mg^{2+}(aq) + 2NO_3^-(aq) + Pb(s)$
$Mg(s) + Pb^{2+}(aq) \longrightarrow Mg^{2+}(aq) + Pb(s)$
(b) $Pb(NO_3)_2(aq) + (NH_4)_2S(aq) \longrightarrow$
$\qquad\qquad PbS(s) + 2NH_4NO_3(aq)$
$Pb^{2+}(aq) + 2NO_3^-(aq) + 2NH_4^+(aq) + S^{2-}(aq) \longrightarrow$
$\qquad\qquad PbS(s) + 2NH_4^+(aq) + 2NO_3^-(aq)$
$Pb^{2+}(aq) + S^{2-}(aq) \longrightarrow PbS(s)$
(c) $2HBr(aq) + Sr(OH)_2(aq) \longrightarrow SrBr_2(aq) + 2H_2O(l)$
$2H^+(aq) + 2Br^-(aq) + Sr^{2+}(aq) + 2OH^-(aq) \longrightarrow$
$\qquad\qquad Sr^{2+}(aq) + 2Br^-(aq) + 2H_2O(l)$
$H^+(aq) + OH^-(aq) \longrightarrow H_2O(l)$

**7.** (a) 0.0233 mol $SbF_3$ (b) 0.827 mol $SbCl_3$
(c) $1.10 \times 10^3$ g $CCl_4$ (d) $4.75 \times 10^{25}$ molecules of $CCl_4$

## Chapter 9

**9-1.** The molecules of water are closely packed together and thus offer much more resistance. The molecules in a gas are dispersed into what is mostly empty space.

**9-3.** Since a gas is mostly empty space, more molecules can be added. In a liquid, the space is mostly occupied by the molecules so no more can be added.

**9-5.** Gas molecules are in rapid but random motion. When gas molecules collide with a light dust particle suspended in the air, they impart a random motion to the particle.

**9-6.** 1260 mi/hr

**9-7.** The molecule with the largest formula weight travels the slowest. $SF_6$(146.1 amu) $<$ $SO_2$(64.07 amu) $<$ $N_2O$(44.02 amu) $<$ $CO_2$(44.01 amu) $<$ $N_2$(28.02 amu) $<$ $H_2$(2.016 amu)

**9-8.** $rate_{N_2} = 1.19\ rate_{Ar}$

**9-11.** 127 g/mol

**9-12.** $\dfrac{r_{(235)}}{r_{(238)}} = \sqrt{\dfrac{352\ \text{g/mol}}{349\ \text{g/mol}}}$    $r_{(235)} = 1.004\ r_{(238)}$
(about 0.4% faster)

**9-13.** When the pressure is high, the gas molecules are forced close together. In a highly compressed gas the molecules can occupy an appreciable part of the total volume. When the temperature is low, molecules have a lower average velocity. If there is some attraction, they can momentarily stick together when moving slowly.

**9-14.** (a) 2.17 atm    (b) 0.0266 torr    (c) 9560 torr
(d) 0.0558 atm    (e) 3.68 lb/in.$^2$    (f) 11 kPa

**9-15.** (a) 768 torr    (b) $2.54 \times 10^4$ atm
(c) 8.40 lb/in.$^2$    (d) 19 torr

**9-17.** 0.0102 atm

**9-19.** Assume a column of Hg has a cross-section of 1 cm$^2$ and is 76.0 cm high. Weight of Hg = 76.0 cm $\times$ 1 cm$^2$ $\times$ 13.6 g/cm$^3$ = 1030 g. If water is substituted, 1030 g of water in the column is required. height $\times$ 1 cm$^2$ $\times$ 1.00 g/cm$^3$ = 1030 g   height = 1030 cm

$$1030\ \text{cm} \times \frac{1\ \text{in.}}{2.54\ \text{cm}} \times \frac{1\ \text{ft}}{12\ \text{in.}} = \underline{33.8\ \text{ft}}$$

If a well is 40 ft deep, the water cannot be raised in one stage by suction since 33.8 ft is the theoretical maximum height that is supported by the atmosphere.

**9-20.** 10.2 L

**9-22.** 978 mL

**9-23.** 67.9 torr

**9-26.** $V_{final}\ (V_f) = 15\ V_{initial}\ (V_i)$    $\dfrac{V_f}{V_i} = \dfrac{P_i}{P_f};$

$\dfrac{15\ V_i}{V_i} = \dfrac{0.950\ \text{atm}}{P_f}$

$P_f = 0.950\ \text{atm} \times 15 = \underline{14.3\ \text{atm}}$

**9-27.**

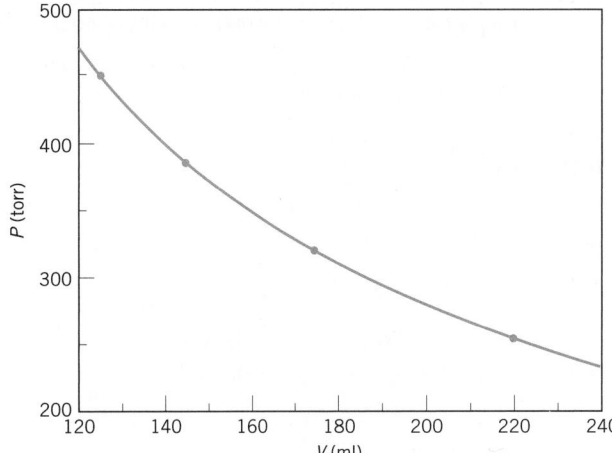

**9-28.** 1.94 L

**9-30.** 77°C

**9-32.** $2.60 \times 10^4$ L

**9-33.** 341 K (68°C)

**9-35.** 2.94 atm

**9-36.** 191 K − 273 = −82°C

**9-38.** 596 K (323°C)

**9-39.** 30.2 lb/in.$^2$

**9-42.** (a) and (d)

**9-44.** Exp. 1, $T$ increases   Exp. 2, $V$ decreases   Exp. 3, $P$ increases   Exp. 4, $T$ increases

**9-46.** 1.24 atm

**9-47.** 76.8 L

**9-49.** 88.5 K = −185°C

**9-52.** 258 K (−15°C)

**9-53.** 409 mL

**9-54.** 1.70 L

**9-55.** Let $x$ = total moles needed in the expanded baloon.

$$188 \text{ L} \times \frac{x}{8.40 \text{ mol}} = 275 \text{ L} \qquad x = 12.3 \text{ mol}$$

$$12.3 - 8.4 = \underline{\underline{3.9 \text{ mol must be added.}}}$$

**9-57.** $n_2 = 2.50 \times 10^{-3} \text{ mol} \times \dfrac{164 \text{ mL}}{75.0 \text{ mL}} = 5.47 \times 10^{-3} \text{ mol}$

$(5.47 \times 10^{-3}) - (2.50 \times 10^{-3}) = 2.97 \times 10^{-3}$ mol added

$$2.97 \times 10^{-3} \text{ mol N}_2 \times \frac{28.02 \text{ g N}_2}{\text{mol N}_2} = \underline{\underline{0.0832 \text{ g N}_2}}$$

**9-59.** 98°C

**9-61.** 13.0 g

**9-62.** 589 torr

**9-64.** 3.35 g

**9-66.** $n = 1.0 \times 10^6$ mol of gas in the balloon. 8800 lb He or 64,000 lb air

Lifting power with He = 64,000 − 8800 = $\underline{55,000 \text{ lb}}$

Lifting power with $H_2$ = 64,000 − 4000 = $\underline{60,000 \text{ lb}}$

Helium is a noncombustible gas whereas hydrogen forms an explosive mixture with $O_2$.

**9-68.** 762 torr

**9-70.** 6.8 torr

**9-73.** $P_{N_2} = 756$ torr; $P_{O_2} = 84.0$ torr; $P_{SO_2} = 210$ torr

**9-74.** 730 torr

**9-76.** $P_{N_2} = 300$ torr    $P_{O_2} = 85 \text{ torr} \times \dfrac{4.00 \text{ L}}{2.00 \text{ L}} = 170$ torr

$P_{CO_2} = 225$ torr    $P_{tot} = 695$ torr

**9-77.** $P_A = 0.550$ atm    $P_B = 0.300$ atm

**9-79.** 7.63 L

**9-81.** 112 L

**9-83.** $1.39 \times 10^{-3}$ g

**9-84.** 1.24 g/L

**9-86.** 34.0 g/mol

**9-88.** $1.00 \text{ L} \times \dfrac{273 \text{ K}}{298 \text{ K}} \times \dfrac{1.20 \text{ atm}}{1.00 \text{ atm}} = 1.10$ L (STP)

$$\frac{3.60 \text{ g}}{1.10 \text{ L}} = \underline{3.27 \text{ g/L (STP)}}$$

**9-89.** Find moles of $N_2$ in 1 L at 500 torr and 22°C using the ideal gas law.

$n = 0.272$ mol $N_2$    mass = 0.762 g $N_2$

Density = $\underline{0.762 \text{ g/L}}$ (500 torr and 22°C)

**9-91.** 25.8 L $CO_2$ (STP)

**9-92.** Vol. $O_2 \longrightarrow$ mol $O_2 \longrightarrow$ mol Mg $\longrightarrow$ g Mg

$5.80 \text{ L } O_2 \times \dfrac{1 \text{ mol } O_2}{22.4 \text{ L } O_2} \times \dfrac{2 \text{ mol Mg}}{1 \text{ mol } O_2} \times \dfrac{24.31 \text{ g Mg}}{\text{mol Mg}}$
$$= \underline{12.6 \text{ g Mg}}$$

**9-94.** 3.47 L

**9-96.** (a) g $C_4H_{10} \longrightarrow$ mol $C_4H_{10} \longrightarrow$ mol $CO_2$
$$\longrightarrow \text{Vol } CO_2$$

$85.0 \text{ g } C_4H_{10} \times \dfrac{1 \text{ mol } C_4H_{10}}{58.12 \text{ g } C_4H_{10}} \times \dfrac{8 \text{ mol } CO_2}{2 \text{ mol } C_4H_{10}} \times$
$$\dfrac{22.4 \text{ L}}{\text{mol } CO_2} = \underline{\underline{131 \text{ L}}}$$

(b) 96.1 L $O_2$    (c) 124 L $CO_2$

**9-97.** 2080 kg Zr (2.29 tons)

**9-99.** $n_{O_2} = 1.46$ mol $O_2$    $1.46 \text{ mol } O_2 \times \dfrac{1 \text{ mol } CO_2}{2 \text{ mol } O_2} =$

0.730 mol $CO_2$ $V = \underline{12.0 \text{ L } CO_2}$

**9-100.** Force = $12.0 \text{ cm}^2 \times 15.0 \text{ cm} \times \dfrac{13.6 \text{ g}}{\text{cm}^3} = 2450$ g

$$P = \frac{2450 \text{ g}}{12.0 \text{ cm}^2} = 204 \text{ g/cm}^2$$

$1 \text{ atm} = 76.0 \text{ cm} \times \dfrac{13.6 \text{ g}}{\text{cm}^3} = 1030$ g/cm$^2$

$204 \text{ g/cm}^2 \times \dfrac{1 \text{ atm}}{1030 \text{ g/cm}^2} = \underline{0.198 \text{ atm}}$

**9-102.** $n = 0.0573$ mol    $\dfrac{8.37 \text{ g}}{0.0573 \text{ mol}} = \underline{146 \text{ g/mol}}$

**9-103.** Empirical formula = $CH_2$    molar mass = 41.9 g/mol    molecular formula = $C_3H_6$

**9-105.** $n_{tot}$ = 0.265 mol $O_2$ + 0.353 mol $N_2$ + 0.160 mol $CO_2$ = 0.778 mol of gas

$V = 6.92$ L

$$P_{O_2} = \frac{0.265}{0.778} \times 2.86 \text{ atm} = 0.974 \text{ atm}$$

$P_{N_2} = 1.30$ atm    $P_{CO_2} = 0.59$ atm

**9-106.** $n = 5 \times 10^{-20}$ mol    $P = 4 \times 10^{-17}$ atm

**9-108.** 19.6 L/mol    density of $CO_2$ = 2.25 g/L

**9-109.** Density $= \dfrac{\text{mass}}{V} = \dfrac{P \times MM}{RT} = 0.525$ g/L (hot air)

density at STP $= 1.29$ g/L

$0.525/1.29 = 0.41$ (Hot air is less than half as dense as air at STP.)

**9-111.** 0.0103 mol of $H_3BCO$ produces 0.0412 mol of gaseous products; $V = 1.12$ L (products at 25°C and 0.900 atm)

**9-113.** $2Al(s) + 3F_2(g) \longrightarrow 2AlF_3(s)$

      original $F_2 = 0.310$ mol $F_2$

leftover $F_2 = 0.0921$ mol

      $0.310 - 0.092 = 0.218$ mol $F_2$ reacts forming

<u>12.2 g $AlF_3$</u>

**9-115.** 135 L

**9-117.** 0.763 mol $H_2O$ in the ice.      V (of vapor) $= \underline{645\ L}$

**9-118.** $9.81 \times 10^{-4}$ mol $NH_3$ produces 11.0 mL $N_2H_4$

**9-120.** Empirical formula $= NO_2$

    **(2)** Find molar mass of product compound.

$$n = \frac{PV}{RT} = \frac{\dfrac{715\ \text{torr}}{760\ \text{torr/atm}} \times 1.05\ L}{0.0821\ \dfrac{L \cdot atm}{K \cdot mol} \times 273\ K} = 0.0441\ \text{mol}$$

$$\text{Molar mass} = \frac{2.03\ g}{0.0441\ \text{mol}} = 46.0\ \text{g/mol}$$

    **(3)** Since one compound decomposes to one other compound, the reactant compound must have the same empirical formula as the product compound. Since the empirical mass of $NO_2 = 46.01$ g/emp unit, then the product must be $NO_2$(MM $= 46.0$ g/mol). Since 0.0220 mol of reactant form 0.0441 mol of product (1:2 ratio), the reaction must be

$$N_2O_4(l) \longrightarrow 2NO_2(g)$$

**9-122.** $6Li(s) + N_2(g) \longrightarrow 2Li_3N(s)$

$3Mg(s) + N_2(g) \longrightarrow Mg_3N_2(s)$

    10.8 mol $N_2$ reacts with <u>450 g Li</u> or <u>788 g Mg</u>

## Chapter 10

**10-1.** Since gas molecules are far apart, they travel a comparatively large distance between collisions. Liquid molecules, on the other hand, are close together so do not travel far between collisions. The farther molecules travel, the faster they mix.

**10-3.** Both the liquid molecules and food coloring molecules are in motion. Through constant motion and collisions the food coloring molecules eventually become dispersed.

**10-5.** Generally, solids have greater densities than liquids. (Ice and water are notable exceptions.) Since the molecules of a solid are held in fixed positions, more of them usually fit into the same volume compared to the liquid state. This is similar to being able to get more people into a room if they are standing still than if they are moving around.

**10-6.** $CH_4 < CCl_4 < GeCl_4$

**10-8.** All are nonpolar molecules with only London forces between molecules. The higher the molar mass, the greater the London forces and the more likely that the compound is a solid. $I_2$ is the heaviest and is a solid; $Cl_2$ is the lightest and is a gas.

**10-10.** If $H_2O$ were linear, the two equal bond dipoles would be exactly opposite and would therefore cancel. Hydrogen bonding can occur only when the molecule is polar.

**10-12.**

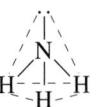

    $NH_3$ molecules interact by hydrogen bonding.

**10-14.** **(a)** HBr     **(b)** $SO_2$     **(f)** CO

**10-17.** **(a)** HF     **(c)** $H_2NCl$     **(d)** $H_2O$     **(f)** HCOOH

**10-19.** **(a)** ion–ion     **(b)** hydrogen bonding plus London **(c)** dipole–dipole plus London     **(d)** London only

**10-20.** $F_2 < HCl < HF < KF$

**10-23.** $CO_2$ is a nonpolar molecular compound. $SiO_2$ is a network solid.

**10-25.** $PbCl_2$ is most likely ionic ($Pb^{2+}, 2Cl^-$), while the melting point of $PbCl_4$ indicates that it is a molecular compound.

**10-26.** Motor oil is composed of large molecules that increase viscosity. Motor oil also has a higher surface tension.

**10-28.** Water evaporates quickly on a hot day and thus lowers the air temperature.

**10-30.** The comparatively low boiling point of ethyl chloride indicates that it has a high vapor pressure at room temperature. In fact, it boils at 25°C, rapidly cooling the skin to below the freezing point of water.

**10-32.** The liquid other than water. The higher the vapor pressure, the faster the liquid evaporates and the liquid cools.

**10-33.** Equilibrium refers to a state where opposing forces are balanced. In the case of a liquid in equilibrium with its vapor, it means that a molecule escaping to the vapor is replaced by one condensing to the liquid.

**10-35.** The substance is a gas at 1 atm and 75°C. It would boil at a temperature below 75°C.

**10-38.** Ethyl alcohol boils at about 52°C at that altitude. At 10°C ethyl ether is a gas at that altitude.

**10-39.** Death Valley is below sea level. The atmospheric pressure is more than 1 atm so water boils above its normal boiling point.

**10-41.** Both exist as atoms with only London forces, but the higher atomic mass of argon accounts for its higher boiling point.

**10-43.** Figure 10-15 is not very precise; at 10°C the actual vapor pressure of water is 9.2 torr. Liquid water could theoretically exist but it would rapidly evaporate and be changed to ice by the cooling effect.

**10-44.** The liquid with the higher vapor pressure (420 torr) has the lower boiling point. It also probably has the lower heat of vaporization.

**10-46.** Molecular $O_2$ is heavier than Ne atoms and so has higher intermolecular forces (London). $CH_3OH$ has hydrogen bonding, which would indicate a much higher heat of vaporization than the other two.

**10-47.** This is a comparatively high heat of fusion, so the melting point is probably also comparatively high.

**10-49.** This is a comparatively high boiling point, so the compound probably also has a high melting point.

**10.50.**

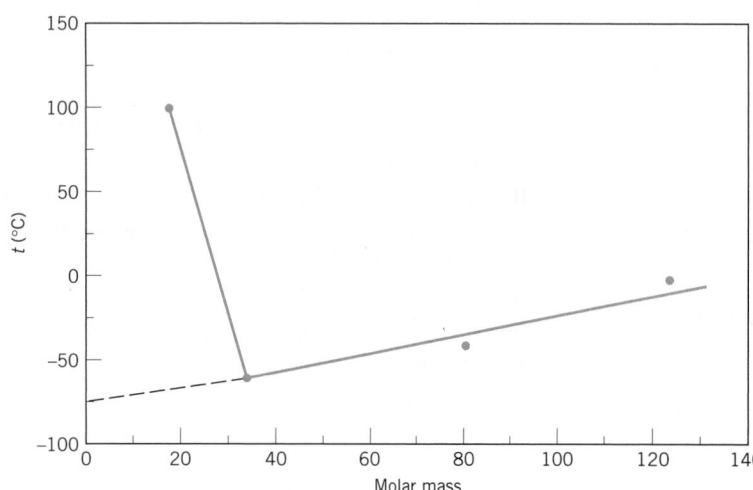

The boiling point of $H_2O$ would be about $-75°C$ without hydrogen bonding.

**10-53.** $7.07 \times 10^3$ J

**10-55.** 2.54 g $H_2O$, 1.64 g NaCl, 6.69 g benzene

**10-57.** 555 cal (ether), 2000 cal ($H_2O$)
   Water would be more effective.
   $1.30 \times 10^4$ J (13.0 kJ)

**10-59.** $NH_3$: $6.12 \times 10^5$ J    Freon: $7.25 \times 10^4$ J
   On the basis of mass, ammonia is the more effective refrigerant.

**10-61.** Condensation: $62.2 \times 10^4$ J    Cooling: $8.6 \times 10^4$ J
   Total $= \underline{7.08 \times 10^5\,J}$ (708 kJ)

**10-63.** $2.83 \times 10^4$ cal

**10-65.** heat ice: $132\ \cancel{g} \times 20°\cancel{C} \times \dfrac{0.492\ \text{cal}}{\cancel{g} \cdot °\cancel{C}} = 1300$ cal

melt ice: $132\ \cancel{g} \times \dfrac{79.8\ \text{cal}}{\cancel{g}} = 10{,}500$ cal

heat $H_2O$: $132\ \cancel{g} \times 100°\cancel{C} \times \dfrac{1.00\ \text{cal}}{\cancel{g} \cdot °\cancel{C}} = 13{,}200$ cal

vap. $H_2O$: $132\ \cancel{g} \times \dfrac{540\ \text{cal}}{\cancel{g}} = 71{,}300$ cal

Total $= \underline{\underline{96{,}300\ \text{cal}\ (96.3\ \text{kcal})}}$

**10-67.** Let $Y =$ the mass of the sample in grams. Then
$$\left(\dfrac{2260\ \text{J}}{\text{g}} \times Y\right) + \left(25.0°C \times Y \times \dfrac{4.18\ \text{J}}{\text{g} \cdot °C}\right)$$
$$= 28{,}400\ \text{J}$$
$Y = \underline{12.0\ \text{g}}$

**10-69.** 18°C

**10-70.** (b) melting    (c) boiling

**10-71.** The average kinetic energy of all molecules of water at the same temperature is the same regardless of the physical state.

**10-73.** Because $H_2O$ molecules have an attraction for each other, moving them apart increases the potential energy. Since the molecules in a gas at 100°C are farther apart than in a liquid at the same temperature, the potential energy of the gas molecules is greater.

**10-74.** The water does not become hotter, but it will boil faster.

**10-76.**

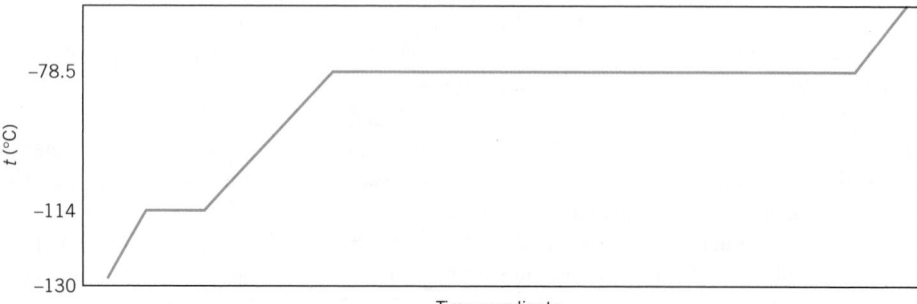

**10-78.** $C_2H_5NH_2$: 17°C, hydrogen bonding

$CH_3OCH_3$: −25°C, dipole–dipole (polar)

$CO_2$: −78°C, London forces only (nonpolar)

**10-79.** $SF_6$ is a molecular compound and SnO is an ionic compound (i.e., $Sn^{2+}$, $O^{2-}$). All ionic compounds are found as solids at room temperature because of the strong ion–ion forces.

**10-81.** There is hydrogen bonding in $CH_3OH$. Hydrogen bonding is a considerably stronger attraction than ordinary dipole–dipole forces. The stronger the attractions the higher the boiling point.

**10-83.** $SiH_4$ is nonpolar. $PH_3$ and $H_2S$ are both polar but $H_2S$ is more polar with more and stronger dipole–dipole attractions.

**10-85.** Carbon monoxide is polar, whereas nitrogen is not. The added dipole–dipole attraction in CO may account for the slightly higher boiling point.

**10-87.** $2000 \text{ lb} \times \dfrac{453.6 \text{ g}}{\text{lb}} \times \dfrac{266 \text{ J}}{\text{g}} = 2.41 \times 10^8 \text{ J}$

Heating 1 g of $H_2O$ from 25.0°C to 100.0°C and the vaporizing the water requires

$(75.0°C \times 4.184 \text{ J/°C}) + 2260 \text{ J} = 2.57 \times 10^3 \text{ J/g } H_2O$

$\dfrac{2.41 \times 10^8 \text{ J}}{2.57 \times 10^3 \text{ J/g } H_2O} = 9.38 \times 10^4 \text{ g } H_2O = \underline{93.8 \text{ kg } H_2O}$

**10-88.** V = 100 L, T = 34 + 273 = 307 K, $P(H_2O) =$ 0.700 × 39.0 torr = 27.3 torr

Use the ideal gas law to find moles of water.

$n = \dfrac{PV}{RT} = \dfrac{\dfrac{27.3 \text{ torr}}{760 \text{ torr/atm}} \times 100 \text{ L}}{0.0821 \dfrac{\text{L} \cdot \text{atm}}{\text{K} \cdot \text{mol}} \times 307 \text{ K}} = 0.143 \text{ mol}$

$0.143 \text{ mol } H_2O \times \dfrac{18.02 \text{ g } H_2O}{\text{mol } H_2O} = \underline{2.58 \text{ g } H_2O}$

**10-90.** Calculate the heat required first to heat the water from 25°C to 100°C then to vaporize the water.

$3.10 \times 10^5$ J (to heat water), $2.260 \times 10^6$ J (to vaporize water) = 2570 kJ total

$Cr: \dfrac{2570 \text{ kJ}}{21.0 \text{ kJ/mol}} \times \dfrac{52.00 \text{ g } Cr}{\text{mol } Cr} \times \dfrac{1 \text{ kg}}{10^3 \text{ g}}$
$= \underline{6.36 \text{ kg } Cr,}$

$\underline{8.81 \text{ kg } Mo}, \underline{13.5 \text{ kg } W}$

**10-91.** The element must be mercury (Hg) since it is a liquid at room temperature and must be a metal since it forms a +2 ion.

The 15.0 kJ represents the heat required to melt $X$ g of Hg, heat $X$ g from −39°C to 357°C (i.e., 396°C), and vaporize $X$ g of Hg.

$(11.5 \text{ J/g} \times X) + \left[ 396°C \times 0.139 \dfrac{\text{J}}{\text{g °C}} \times X \right]$
$+ (29.5 \text{ J/g} \times X) = 15.00 \text{ kJ} \times \dfrac{10^3 \text{ k}}{\text{kJ}}$

$11.5 \text{ J/g } X + 55.0 \text{ J/g } X + 29.5 \text{ J/g } X = 15,000 \text{ J}$

$X = \underline{156 \text{ g Hg}}$

**10-93.** Calculate the mass of water in the vapor state in the room at −5°C (268 K) using the ideal gas law. 2.99 mol $H_2O$ (in vapor) = 53.9 g $H_2O$ (capacity of the room)

Eventually, all of the 50.0 g should sublime since it is less than the room could contain.

# Chapter 11

**11-1.** The ion–ion forces in the crystal hold the crystal together and resist the ion–dipole forces between water and the ions in the crystal. The ion–dipole forces remove the ions from the crystal.

**11-3.** Calcium bromide is soluble, lead(II) bromide is insoluble, benzene and water are immiscible, and alcohol and water are miscible.

**11-4.** (a) $LiF \longrightarrow Li^+(aq) + F^-(aq)$
(b) $(NH_4)_3PO_4 \longrightarrow 3NH_4^+(aq) + PO_4^{3-}(aq)$
(c) $Na_2CO_3 \longrightarrow 2Na^+(aq) + CO_3^{2-}(aq)$
(d) $Ca(C_2H_3O_2)_2 \longrightarrow Ca^{2+}(aq) + 2C_2H_3O_2^-(aq)$

**11-6.**

There is hydrogen bonding between solute and solvent.

**11-8.** At 10°C, $Li_2SO_4$ is the most soluble; at 70°C, KCl is the most soluble.

**11-10.** (a) unsaturated **(b)** supersaturated
(c) saturated **(d)** unsaturated

**11-12.** A specific amount of $Li_2SO_4$ will precipitate unless the solution becomes supersaturated.

$\left( 500 \text{ g } H_2O \times \dfrac{35 \text{ g}}{100 \text{ g } H_2O} \right) - \left( 500 \text{ g } H_2O \times \dfrac{28 \text{ g}}{100 \text{ g } H_2O} \right)$
$= \underline{35 \text{ g } Li_2SO_4 \text{ (precipitate)}}$

**11-13.** 1.49%

**11-15.** 0.375 mol NaOH

**11-17.** 30 g $KNO_3$

**11-18.** 8.81% NaOH

**11-19.** 100 ppm

**11-21.** $8.3 \times 10^7$ L

**11-23.** 0.542 M

**11-24.** (a) 0.873 M **(b)** 1.40 M **(c)** 12.4 L
(d) 41.4 g **(e)** 0.294 M **(f)** 0.024 mol
(g) 307 mL **(h)** 49.1 g **(i)** 2.00 mL

**11-28.** $1.17 \times 10^{-3}$ M

**11-29.** $Ba^{2+}$: 0.166 M  $OH^-$: 0.332 M

**11-31.** 1.84 M

**11-33.** 1.32 g/mL

**11-34.** 0.833 L

**11-36.** Slowly add <u>313 mL</u> of the 0.800 M NaOH to about 500 mL of water in a 1-L volumetric flask. Dilute to the 1-L mark with water.

**11-38.** 0.140 M

**11-39.** 720 mL

**11-41.** 2.86 mL

**11-42.** Find the total moles and the total volume.

$$n = V \times M = 0.150 \, \cancel{L} \times 0.250 \, mol/\cancel{L} = 0.0375 \, mol$$
(solution 1)
$$n = V \times M = 0.450 \, \cancel{L} \times 0.375 \, mol/\cancel{L} = 0.169 \, mol$$
(solution 2)

$$n_{tot} = 0.206 \, mol \qquad V_{tot} = 0.600 \, L$$

$$\frac{n}{V} = \frac{0.206 \, mol}{0.600 \, L} = \underline{0.343 \, M}$$

**11-43.** $0.500 \, \cancel{L} \times \dfrac{0.250 \, \cancel{mol \, KOH}}{\cancel{L}} \times \dfrac{1 \, mol \, Cr(OH)_3}{3 \, \cancel{mol \, KOH}} \times$

$\dfrac{103.0 \, g \, Cr(OH)_3}{\cancel{mol \, Cr(OH)_3}} = \underline{\underline{4.29 \, g \, Cr(OH)_3}}$

**11-45.** 146 g BaSO$_4$

**11-46.** 4.08 L

**11-48.** 0.653 L

**11-50.** NaOH is the limiting reactant, producing 1.75 g of Mg(OH)$_2$.

**11-52.** An aqueous solution of AB is a good conductor of electricity. AB is dissociated into ions such as A$^+$ and B$^-$. A solution of AC is a weak conductor of electricity, which means that AC is only partially dissociated into ions: i.e., $AC \rightleftharpoons A^+ + C^-$. A solution of AD is a nonconductor of electricity because it is present as undissociated molecules in solution.

**11-54.** 1 mole of NaBr has a mass of 102.9 g, 1 molar NaBr is a solution containing 102.9 g of NaBr per liter of solution, and 1 molal NaBr is a solution containing 102.9 g of NaBr per kilogram of solvent.

**11-55.** 2.50 *m* (The molality is the same in both solvents since the mass of solvent is the same.)

**11-57.** 15.8 g NaOH

**11-59.** −0.37°C

**11-61.** The salty water removes water from the cells of the skin by osmosis. After a prolonged period, a person would dehydrate and become thirsty.

**11-64.** We can concentrate a dilute solution by boiling away some solvent if the solute is not volatile. Reverse osmosis can also be used to concentrate a solution if pressure greater than the osmotic pressure is applied on a concentrated solution separated from the solvent by a semipermeable membrane. As the solution becomes more concentrated, the osmotic pressure becomes greater, and the corresponding pressure that is applied must be increased.

**11-65.** 1.00 *m*

**11-67.** 807 g glycol

**11-68.** 101.4°C

**11-70.** 2.9 *m*

**11-73.** −14.5°C

**11-75.** HCl is a strong electrolyte that produces two particles (ions) for each mole of HCl that dissolves. HF is a weak electrolyte that is essentially present as nonionized molecules in solution.

**11.78.** **(a)** −5.8°C   **(b)** −6.4°C   **(c)** −5.0°C

**11-79.** Dissolve the mixture in 100 g of H$_2$O and heat to over 45°C. Cool to 0°C where the solution is saturated with about 10 g of KNO$_3$ and about 25 g of KCl. 50 − 25 = 25 g of KCl precipitates.

**11-81.** 90.9% H$_2$O   3.03% salt   6.06% sugar

**11-83.** 0.10 mol of Ag$^+$ reacts with 0.10 mol of Cl$^-$, leaving 0.05 mol of Cl$^-$ in 1.0 L of solution. [Cl$^-$] = 0.05 M

**11-85.** $0.225 \, \cancel{L} \times \dfrac{0.196 \, \cancel{mol \, HCl}}{\cancel{L}} \times \dfrac{1 \, mol \, M}{2 \, \cancel{mol \, HCl}} =$
$$0.0221 \, mol \, M$$

$$\frac{1.44 \, g}{0.0221 \, mol} = \underline{65.2 \, g/mol} \, (Zn)$$

**11-87.** molality = 0.426 *m*   molar mass = 60.1 g/mol
empirical formula = CH$_4$N   molecular formula = C$_2$H$_8$N$_2$

**11-89.** 0.30 *m* sugar < 0.12 *m* KCl (0.24 *m* ions) < 0.05 *m* CrCl$_3$ (0.20 *m* ions) < 0.05 *m* K$_2$CO$_3$ (0.15 *m* ions) < pure water.

**11-91.** n(hydroxide) = 0.0584 mol   n(H$_2$) = 0.0292 mol H$_2$
$$\frac{0.0584}{0.0292} = 2 \, (2 \, mol \, hydroxide : 1 \, mol \, H_2)$$
Balanced Equations:
$$2Na(s) + 2H_2O(l) \longrightarrow 2NaOH(aq) + H_2(g)$$
$$(2 \, mol \, hydroxide : 1 \, mol \, H_2)$$
$$Ca(s) + 2H_2O(l) \longrightarrow Ca(OH)_2(aq) + H_2(g)$$
$$(1 \, mol \, hydroxide : 1 \, mol \, H_2)$$
The answer is sodium since it produces the correct ratio of hydrogen.

**11-92.** 2.91 L

**11-94.** 8.60 L

**11-96.** The formula must be PCl$_3$ with the Lewis structure

which would be trigonal pyramidal.

The reaction is PCl$_3$(g) + 3H$_2$O $\longrightarrow$ H$_3$PO$_3$(aq) + 3HCl(aq)   <u>0.404 M HCl</u>

**11-98.** (2Na$^+$S$^{2-}$) $10.0 \, \cancel{g \, Na_2S} \times \dfrac{1 \, mol \, Na_2S}{78.05 \, \cancel{g \, Na_2S}}$
$$= 0.128 \, mol \, Na_2S$$

$$\Delta T = 1.86 \times \frac{\dfrac{0.128 \, mol}{100 \, g}}{1000 \, g/kg} \times 3 \, mol \, ions/mol \, Na_2S$$
$$= 7.14°C \, change$$

The answer is Na$_2$S since it would have a freezing point of −7.14°C.

**11-100.** $\Delta T = K_f m$   $1.15°C = 1.86 \times \dfrac{0.806}{1.00 + X}$

Solving for $X$, $X = 0.30$ kg   0.30 kg = 300 g = 300 mL of water added.

## Cumulative Review Test for Chapters 9–11

### Multiple Choice

| | | | |
|---|---|---|---|
| 1. (c) | 2. (b) | 3. (a) | 4. (c) |
| 5. (e) | 6. (e) | 7. (c) | 8. (c) |
| 9. (c) | 10. (d) | 11. (b) | 12. (b) |
| 13. (b) | 14. (a) | 15. (c) | 16. (d) |
| 17. (d) | 18. (b) | 19. (a) | 20. (a) |
| 21. (c) | 22. (e) | 23. (d) | 24. (b) |
| 25. (c) | 26. (d) | 27. (e) | 28. (a) |
| 29. (a) | 30. (c) | 31. (c) | 32. (c) |

### Problems

**1.** 0.446 mol of $O_2$, $2.69 \times 10^{23}$ molecules of $O_2$, 8.06 L at $-53°C$, 13.1 L at 580 torr, 1.10 atm at 27°C, 21.2 L if 0.500 mol of a gas is added

**2.** molar mass = 147 g/mol   empirical formula = $C_3H_2Cl$ molecular formula = $C_6H_4Cl_2$

**3.** $BeF_2$: 800°C   Ion–ion forces, the strongest
$C_2H_5OH$: $-117°C$   A molecular compound with hydrogen bonding
$(CH_3)_2O$: $-138°C$   A polar covalent compound with ordinary dipole–dipole forces
$C_3H_8$: $-189°C$   A nonpolar covalent compound with only London forces between molecules

**4.** $1.82 \times 10^4$ cal

**5. (a)** 0.408 M   **(b)** 0.0873 M   **(c)** 625 mL

**6. (a)** $Pb(NO_3)_2(aq) + 2NaI(aq) \longrightarrow$
$\qquad\qquad PbI_2(s) + 2NaNO_3(aq)$

**(b)** 1.7 L

## Chapter 12

**12-1. (a)** $HNO_3$, nitric acid   **(b)** $HNO_2$, nitrous acid
**(c)** $HClO_3$, chloric acid   **(d)** $H_2SO_3$, sulfurous acid

**12-3. (a)** CsOH, cesium hydroxide   **(b)** $Sr(OH)_2$, strontium hydroxide   **(c)** $Al(OH)_3$, aluminum hydroxide
**(d)** $Mn(OH)_3$, manganese(III) hydroxide

**12-5. (a)** $HNO_3 + H_2O \longrightarrow H_3O^+ + NO_3^-$
**(b)** $CsOH \longrightarrow Cs^+ + OH^-$
**(c)** $Ba(OH)_2 \longrightarrow Ba^{2+} + 2OH^-$
**(d)** $HBr + H_2O \longrightarrow H_3O^+ + Br^-$

**12-7.**

| Acid–Conjugate base | | Acid–Conjugate base | |
|---|---|---|---|
| **(a)** $HNO_3$ | $NO_3^-$ | **(d)** $CH_4$ | $CH_3^-$ |
| **(b)** $H_2SO_4$ | $HSO_4^-$ | **(e)** $H_2O$ | $OH^-$ |
| **(c)** $HPO_4^{2-}$ | $PO_4^{3-}$ | **(f)** $NH_3$ | $NH_2^-$ |

**12-9. (a)** $HClO_4$, $ClO_4^-$ and $H_2O$, $OH^-$
**(b)** $HSO_4^-$, $SO_4^{2-}$ and $HClO$ and $ClO^-$
**(c)** $H_2O$, $OH^-$ and $NH_3$, $NH_2^-$
**(d)** $NH_4^+$, $NH_3$ and $H_3O^+$, $H_2O$

**12-12. (a)** $NH_3 + H_2O \longrightarrow NH_4^+ + OH^-$
$\quad B_1 \qquad\qquad A_1$
$\qquad A_2 \qquad\qquad\qquad B_2$
**(b)** $N_2H_4 + H_2O \longrightarrow N_2H_5^+ + OH^-$
$\quad B_1 \qquad\qquad A_1$
$\qquad A_2 \qquad\qquad\qquad B_2$
**(c)** $HS^- + H_2O \longrightarrow H_2S + OH^-$
$\quad B_1 \qquad\qquad A_1$
$\qquad A_2 \qquad\qquad\qquad B_2$
**(d)** $H^- + H_2O \longrightarrow H_2 + OH^-$
$\quad B_1 \qquad\qquad A_1$
$\qquad A_2 \qquad\qquad\qquad B_2$
**(e)** $F^- + H_2O \longrightarrow HF + OH^-$
$\quad B_1 \qquad\qquad A_1$
$\qquad A_2 \qquad\qquad\qquad B_2$

**12-13.** base: $HS^- + H_3O^+ \rightleftharpoons H_2S + H_2O$
acid: $HS^- + OH^- \rightleftharpoons S^{2-} + H_2O$

**12-16.** The information indicates that HBr is a strong acid. The $Br^-$ ion, however, would be a very weak base in water.

**12-18. (a)** $HNO_3 + H_2O \longrightarrow H_3O^+ + NO_3^-$
**(b)** $HNO_2 + H_2O \rightleftharpoons H_3O^+ + NO_2^-$
**(c)** $HClO_3 + H_2O \rightleftharpoons H_3O^+ + ClO_3^-$
**(d)** $H_2SO_3 + H_2O \rightleftharpoons H_3O^+ + HSO_3^-$

**12-20.** $(CH_3)_2NH + H_2O \rightleftharpoons (CH_3)_2NH_2^+ + OH^-$

**12-22.** HX is a weak acid. The concentration of $H_3O^+$ must equal the concentration of the HX that ionized.

$\dfrac{0.010}{0.100} \times 100\% = \underline{10\% \text{ ionized}}$

**12-24.** $[H_3O^+] = 0.55$ M

**12-25.** $[H_3O^+] = 0.030 \times 0.55 = 0.017$ M

**12-27.** From the first ionization: $[H_3O^+] = 0.354$ M
Of that, 25% undergoes further ionization.
$0.25 \times 0.354 = 0.089$ M $[H_3O^+]$ from the second ionization.
The total $[H_3O^+] = 0.354 + 0.089 = \underline{0.443 \text{ M}}$

**12-28.** $\dfrac{0.050}{1.0} \times 100\% = \underline{5.0\% \text{ ionized}}$

**12-30.** The conjugate acid of $HSO_4^-$ is $H_2SO_4$. This is a strong acid so it completely ionizes in water. Thus it cannot exist in the molecular form in water.

**12-31. (a)** acid   **(b)** normal salt   **(c)** acid
**(d)** acid salt   **(e)** normal salt   **(f)** base

**12-33. (a)** $HNO_3 + NaOH \longrightarrow NaNO_3 + H_2O$
**(b)** $2HI + Ca(OH)_2 \longrightarrow CaI_2 + 2H_2O$
**(c)** $HClO_2 + KOH \longrightarrow KClO_2 + H_2O$

**12-35. (a)** $H^+(aq) + NO_3^-(aq) + Na^+(aq) + OH^-(aq) \longrightarrow$
$\qquad\qquad Na^+(aq) + NO_3^-(aq) + H_2O(l)$
$H^+(aq) + OH^-(aq) \longrightarrow H_2O(l)$   (net ionic)
**(b)** $2H^+(aq) + 2I^-(aq) + Ca^{2+}(aq) + 2OH^-(aq)$
$\longrightarrow Ca^{2+}(aq) + 2I^-(aq) + 2H_2O(l)$
$H^+(aq) + OH^-(aq) \longrightarrow H_2O(l)$   (net ionic)

**(c)** $HClO_2(aq) + K^+(aq) + OH^-(aq) \longrightarrow$
$$K^+(aq) + ClO_2^-(aq) + H_2O(l)$$
$HClO_2(aq) + OH^-(aq) \longrightarrow$
$$ClO_2^-(aq) + H_2O(l) \quad \text{(net ionic)}$$

**12-37.** $H_2C_2O_4(aq) + 2NH_3(aq) \longrightarrow (NH_4)_2C_2O_4(aq)$
$H_2C_2O_4(aq) + 2NH_3(aq) \longrightarrow$
$$2NH_4^+(aq) + C_2O_4^{2-}(aq)$$

Net ionic equation is the same as the total ionic equation.

**12-39. (a)** KOH and $HClO_3$    **(b)** $Al(OH)_3$ and $H_2SO_3$
**(c)** $Ba(OH)_2$ and $HNO_2$    **(d)** $NH_3$ and $HNO_3$

**12-41. (a)** $2HBr + Ca(OH)_2 \longrightarrow CaBr_2 + 2H_2O$
**(b)** $2HClO_2 + Sr(OH)_2 \longrightarrow Sr(ClO_2)_2 + 2H_2O$
**(c)** $2H_2S + Ba(OH)_2 \longrightarrow Ba(HS)_2 + 2H_2O$
**(d)** $H_2S + 2LiOH \longrightarrow Li_2S + 2H_2O$

**12-43.** $LiOH + H_2S \longrightarrow LiHS + H_2O$
$LiOH + LiHS \longrightarrow Li_2S + H_2O$

**12-44.** $H_2S(aq) + OH^-(aq) \longrightarrow HS^-(aq) + H_2O(l)$
$HS^-(aq) + OH^-(aq) \longrightarrow S^{2-}(aq) + H_2O(l)$

**12-47.** $H_2S + NaOH \longrightarrow NaHS + H_2O$

**12-50.** The system would not be at equilibrium if $[H_3O^+]$ = $[OH^-] = 10^{-2}$ M. Therefore, $H_3O^+$ reacts with $OH^-$ until the concentration of each is reduced to $10^{-7}$ M. This is a neutralization reaction,
i.e., $H_3O^+ + OH^- \longrightarrow 2H_2O$

**12-51. (a)** $[H_3O^+] = \dfrac{K_w}{[OH^-]} = \dfrac{10^{-14}}{10^{-12}} = 10^{-2}$ M

**(b)** $[H_3O^+] = 10^{-15}$ M    **(c)** $[OH^-] = 5.0 \times 10^{-10}$ M

**12-53.** $[H_3O^+] = 0.0250$ M    $[OH^-] = 4.00 \times 10^{-13}$ M

**12-55. (a)** acidic    **(b)** basic    **(c)** acidic

**12-57. (a)** acidic    **(b)** basic    **(c)** acidic
**(d)** basic

**12-59. (a)** 6.00    **(b)** 9.00    **(c)** 12.00
**(d)** 9.40    **(e)** 10.19

**12-61. (a)** pH = 4.0, pOH = 10.0    **(b)** pOH = 5.0, pH = 9.0    **(c)** pH = 1.70, pOH = 12.30
**(d)** pOH = 3.495, pH = 10.505

**12-63. (a)** $1.0 \times 10^{-3}$ M    **(b)** $2.9 \times 10^{-4}$ M
**(c)** $1.0 \times 10^{-6}$ M    **(d)** $2.4 \times 10^{-8}$ M
**(e)** $2.0 \times 10^{-13}$ M

**12-65.** for exercise **12-59:** **(a)** acidic    **(b)** basic
**(c)** basic    **(d)** basic    **(e)** basic
     for exercise **12-63:** **(a)** acidic    **(b)** acidic
**(c)** acidic    **(d)** basic    **(e)** basic

**12-67.** pH = 5.0 (less acidic), pH = 2.0 (more acidic)

**12-69.** pH = 1.12

**12-71.** pH = 12.56

**12-72.** $[H_3O^+] = 0.100 \times 0.10 = 0.010 = 1.0 \times 10^{-2}$ M
pH = 2.00

**12-74. (a)** strongly acidic    **(b)** strongly acidic
**(c)** weakly acidic    **(d)** strongly basic    **(e)** neutral
**(f)** weakly basic    **(g)** weakly acidic    **(h)** strongly acidic

**12-75.** Ammonia (pH = 11.4), eggs (pH = 7.8), rainwater ($[H_3O^+] = 2.0 \times 10^{-6}$ M, pH = 5.70), grape juice ($[OH^-]$ = $1.0 \times 10^{-10}$ M, pH = 4.0), vinegar ($[H_3O^+] = 2.5 \times 10^{-3}$ M, pH = 2.60), sulfuric acid (pOH = 13.6, pH = 0.4)

**12-78. (b)** $C_2H_3O_2^- + H_2O \rightleftharpoons HC_2H_3O_2 + OH^-$
**(c)** $NH_3 + H_2O \rightleftharpoons NH_4^+ + OH^-$

**12-80. (a)** $K^+$ The cation of the strong base KOH.
**(d)** $NO_3^-$ The conjugate base of the strong acid $HNO_3$.
**(e)** $O_2$ Dissolves in water without formation of ions.

**12-81. (a)** $HS^-$    **(b)** $H_3O^+$    **(c)** $OH^-$    **(d)** $(CH_3)_2NH$

**12-83. (a)** $F^- + H_2O \rightleftharpoons HF + OH^-$
**(b)** $SO_3^{2-} + H_2O \rightleftharpoons HSO_3^- + OH^-$
**(c)** $(CH_3)_2NH_2^+ + H_2O \rightleftharpoons (CH_3)_2NH + H_3O^+$
**(d)** $HPO_4^{2-} + H_2O \rightleftharpoons H_2PO_4^- + OH^-$
**(e)** $CN^- + H_2O \rightleftharpoons HCN + OH^-$
**(f)** no hydrolysis (cation of a strong base)

**12-85.** $Ca(ClO)_2 \longrightarrow Ca^{2+}(aq) + 2ClO^-(aq)$
$ClO^-(aq) + H_2O \rightleftharpoons HClO(aq) + OH^-(aq)$

**12-87. (a)** neutral (neither ion hydrolyzes)    **(b)** acidic (cation hydrolysis)    **(c)** basic (anion hydrolysis)
**(d)** neutral (neither ion hydrolyzes)    **(e)** acidic (cation hydrolysis)    **(f)** basic (anion hydrolysis)

**12-89.** $CaC_2(s) + 2H_2O(l) \longrightarrow$
$$C_2H_2(g) + Ca^{2+}(aq) + 2OH^-(aq)$$

**12-90.** cation: $NH_4^+ + H_2O \rightleftharpoons NH_3 + H_3O^+$
anion: $CN^- + H_2O \rightleftharpoons HCN + OH^-$
     Since the solution is basic, the anion hydrolysis reaction must take place to a greater extent than the cation hydrolysis.

**12-92. (a), (b), (e), (h),** and **(i)** are buffer solutions.

**12-93.** There is no equilibrium when HCl dissolves in water. A reservoir of nonionized acid must be present to react with any strong base that is added. Likewise, the $Cl^-$ ion does not exhibit base behavior in water, so it cannot react with any $H_3O^+$ added to the solution.

**12-94.** $N_2H_4(aq) + H_2O \rightleftharpoons N_2H_5^+(aq) + OH^-(aq)$
added $H_3O^+$: $H_3O^+ + N_2H_4 \longrightarrow N_2H_5^+ + H_2O$
added $OH^-$: $OH^- + N_2H_5^+ \longrightarrow N_2H_4 + H_2O$

**12-96.** $HPO_4^{2-}(aq) + H_2O(l) \rightleftharpoons$
$$PO_4^{3-}(aq) + H_3O^+(aq)$$
added $H_3O^+$: $H_3O^+ + PO_4^{3-} \longrightarrow$
$$HPO_4^{2-} + H_2O$$
added $OH^-$: $OH^- + HPO_4^{2-} \longrightarrow PO_4^{3-} + H_2O$

**12-99. (a)** $Sr(OH)_2$    **(b)** $H_2SeO_4$    **(c)** $H_3PO_4$
**(d)** CsOH    **(e)** $HNO_2$    **(f)** $HClO_3$

**12-101.** $CO_2(g) + LiOH(aq) \longrightarrow LiHCO_3(aq)$

**12-102.** $2LiNO_3(s)$

**12-103.** $Fe(s) + 2HI \longrightarrow FeI_2(aq) + H_2(g)$

**12-105.** $2HNO_3(aq) + Na_2SO_3(aq) \longrightarrow$
$\qquad\qquad 2NaNO_3(aq) + SO_2(g) + H_2O(l)$

**12-107.** **(a)** $HCN + NH_3 \longrightarrow NH_4^+ + CN^-$
**(b)** $NH_3 + H^- \longrightarrow H_2 + NH_2^-$
**(c)** $HCO_3^- + NH_3 \longrightarrow NH_4^+ + CO_3^{2-}$
**(d)** $NH_4^+Cl^- + Na^+NH_2^- \longrightarrow Na^+Cl^- + 2NH_3$

**12-108.** **(a)** $HCl + CH_3OH \longrightarrow CH_3OH_2^+ + Cl^-$
**(b)** $CH_3OH + NH_2^- \longrightarrow NH_3 + CH_3O^-$

**12-110.** **(a)** acidic: $H_2S + H_2O \rightleftharpoons H_3O^+ + HS^-$
**(b)** basic: $ClO^- + H_2O \rightleftharpoons HClO + OH^-$
**(c)** neutral
**(d)** basic: $NH_3 + H_2O \rightleftharpoons NH_4^+ + OH^-$
**(e)** acidic: $N_2H_5^+ + H_2O \rightleftharpoons N_2H_4 + H_3O^+$
**(f)** basic: $Ba(OH)_2 \longrightarrow Ba^{2+} + 2OH^-$
**(g)** neutral
**(h)** basic: $NO_2^- + H_2O \rightleftharpoons HNO_2 + OH^-$
**(i)** acidic: $H_2SO_3 + H_2O \rightleftharpoons H_3O^+ + HSO_3^-$
**(j)** acidic: $Cl_2O_3 + H_2O \longrightarrow 2HClO_2$;
$\qquad HClO_2 + H_2O \rightleftharpoons H_3O^+ + ClO_2^-$

## Chapter 13

**13-1.** **(a)** Pb +4, O −2    **(b)** P +5, O −2
**(c)** C −1, H +1    **(d)** N −2, H +1    **(e)** Li +1,
H −1
**(f)** B +3, Cl −1    **(g)** Rb +1, Se −2
**(h)** Bi +3, S −2

**13-3.** **(a)** Li, **(e)** K, and **(g)** Rb

**13-5.** +3

**13-6.** **(a)** P = +5    **(b)** C = +3    **(c)** Cl = +7
**(d)** Cr = +6    **(e)** S = +6    **(f)** N = +5
**(g)** Mn = +7

**13-7.** **(a)** S = +6    **(b)** Co = +3    **(c)** U = +6
**(d)** N = +5    **(e)** Cr = +6    **(f)** Mn = +6

**13-9.** **(a)**, **(c)**, and **(f)**

**13-10.** **(a)** oxidation    **(b)** reduction    **(c)** oxidation
**(d)** reduction    **(e)** reduction

**13-12.**

| Reactant Oxidized | Product of Oxidation | Reactant Reduced | Product of Reduction | Oxidizing Agent | Reducing Agent |
|---|---|---|---|---|---|
| $Br^-$ | $Br_2$ | $MnO_2$ | $Mn^{2+}$ | $MnO_2$ | $Br^-$ |
| $CH_4$ | $CO_2$ | $O_2$ | $CO_2, H_2O$ | $O_2$ | $CH_4$ |
| $Fe^{2+}$ | $Fe^{3+}$ | $MnO_4^-$ | $Mn^{2+}$ | $MnO_4^-$ | $Fe^{2+}$ |

**13-13.**

| Reactant Oxidized | Product of Oxidation | Reactant Reduced | Product of Reduction | Oxidizing Agent | Reducing Agent |
|---|---|---|---|---|---|
| $Al$ | $AlO_2^-$ | $H_2O$ | $H_2$ | $H_2O$ | $Al$ |
| $Mn^{2+}$ | $MnO_4^-$ | $Cr_2O_7^{2-}$ | $Cr^{3+}$ | $Cr_2O_7^{2-}$ | $Mn^{2+}$ |

**12-112.** LiOH, strongly basic, pH = 13.0; $SrBr_2$, neutral, pH = 7.0; KClO, weakly basic, pH = 10.2; $NH_4Cl$, weakly acidic, pH = 5.2; HI, strongly acidic, pH = 1.0

    When pH = 1.0, $[H_3O^+] = 0.10$ M. If HI is completely ionized, its initial concentration must be 0.10 M.

**12-114.** 8.2 kg of $H_2SO_4$

**12-115.** pH = 0.553 (con)    pH = 1.113 (dilute)

**12-117.** 0.024 mol of HCl remains after neutralization
pH = 1.62

**12-119.** 0.025 mol $Ca(OH)_2$ remaining in 1.00 L

$$[OH^-] = 0.025 \; \overline{\text{mol } Ca(OH)_2} \times \frac{2 \; \text{mol } OH^-}{\overline{\text{mol } Ca(OH)_2}}$$
$$= 0.050 \; \text{mol/L} \quad pOH = 1.30 \quad \underline{pH = 12.70}$$

**13-16.**

$$\overbrace{\phantom{xxxxxxxxxx}}^{-5e^- \times 4 = -20e^-}$$
$$-3 \qquad\qquad +2$$

**(a)** $4NH_3 + 5O_2 \longrightarrow 4NO + 6H_2O$

$$0 \qquad\qquad\qquad -2$$
$$\underbrace{\phantom{xxxxxxxxxx}}_{+4e^- \times 5 = +20e^-}$$

$$\overbrace{\phantom{xxxxxxx}}^{-4e^- \times 1 = -4e^-}$$
$$0 \qquad\qquad +4$$

**(b)** $Sn + 4HNO_3 \longrightarrow SnO_2 + 4NO_2 + 2H_2O$

$$+5 \qquad\qquad +4$$
$$\underbrace{\phantom{xxxxxxx}}_{+1e^- \times 4 = +4e^-}$$

**(c)** Before the number of electrons lost is calculated, notice that a temporary coefficient of "2" is needed for the

$Na_2CrO_4$ in the products since there are 2 Cr's in $Cr_2O_3$ in the reactants.

$$\overset{\overbrace{\hspace{3cm}}^{+2e^- \times 3 = +6e^-}}{}$$

$$\underset{+6}{Cr_2O_3} + 2Na_2CO_3 + 3KNO_3 \longrightarrow 2CO_2 + \underset{+3}{2Na_2CrO_4} + 3KNO_2$$

$$\underbrace{\hspace{5cm}}_{-6e^- \times 1 = -6e^-}$$

**(d)** 
$$\underbrace{\hspace{4cm}}_{-4e^- \times 3 = -12e^-}$$
$$\underset{0}{3Se} + 2BrO_3^- + 3H_2O \longrightarrow 3H_2SeO_3 + \underset{+4}{2Br^-}$$
$$\underbrace{\hspace{5cm}}_{+6e^- \times 2 = +12e^-}$$

**13-18. (a)** $2H_2O + Sn^{2+} \longrightarrow SnO_2 + 4H^+ + 2e^-$
**(b)** $2H_2O + CH_4 \longrightarrow CO_2 + 8H^+ + 8e^-$
**(c)** $e^- + Fe^{3+} \longrightarrow Fe^{2+}$
**(d)** $6H_2O + I_2 \longrightarrow 2IO_3^- + 12H^+ + 10e^-$
**(e)** $e^- + 2H^+ + NO_3^- \longrightarrow NO_2 + H_2O$

**13-20.**
$$\begin{array}{ll} \text{(a)} \quad S^{2-} \longrightarrow S + 2e^- & \times 3 \\ \underline{3e^- + 4H^+ + NO_3^- \longrightarrow NO + 2H_2O} & \times 2 \\ 3S^{2-} + 8H^+ + 2NO_3^- \longrightarrow 3S + 2NO + 4H_2O & \end{array}$$
**(b)** $2S_2O_3^{2-} + I_2 \longrightarrow S_4O_6^{2-} + 2I^-$
$$\begin{array}{ll} \text{(c)} \quad H_2O + SO_3^{2-} \longrightarrow SO_4^{2-} + 2H^+ + 2e^- & \times 3 \\ \underline{6e^- + 6H^+ + ClO_3^- \longrightarrow Cl^- + 3H_2O} & \times 1 \\ 3SO_3^{2-} + ClO_3^- \longrightarrow Cl^- + 3SO_4^{2-} & \end{array}$$
**(d)** $2H^+ + 2Fe^{3+} + H_2O_2 \longrightarrow 2Fe^{3+} + 2H_2O$
**(e)** $AsO_4^{3-} + 2I^- + 2H^+ \longrightarrow I_2 + AsO_3^{3-} + H_2O$
**(f)** $4Zn + NO_3^- + 10H^+ \longrightarrow$
$$4Zn^{2+} + NH_4^+ + 3H_2O$$

**13-22. (a)** $2OH^- + SnO_2^{2-} \longrightarrow SnO_3^{2-} + H_2O + 2e^-$
**(b)** $6e^- + 4H_2O + 2ClO_2^- \longrightarrow Cl_2 + 8OH^-$
**(c)** $6OH^- + Si \longrightarrow SiO_3^{2-} + 3H_2O + 4e^-$
**(d)** $8e^- + 6H_2O + NO_3^- \longrightarrow NH_3 + 9OH^-$

**13-24. (a)** In acid solution:
$4H_2O + S^{2-} + 4I_2 \longrightarrow SO_4^{2-} + 8I^- + 8H^+$
Add $8OH^-$ to both sides and simplify.
$8OH^- + S^{2-} + 4I_2 \longrightarrow SO_4^{2-} + 8I^- + 4H_2O$
**(b)** $8OH^- + I^- + 8MnO_4^- \longrightarrow$
$$8MnO_4^{2-} + IO_4^- + 4H_2O$$
**(c)** $2H_2O + SnO_2^{2-} + BiO_3^- \longrightarrow$
$$SnO_3^{2-} + Bi(OH)_3 + OH^-$$
**(d)** 
$$\begin{array}{ll} 32OH^- + CrI_3 \longrightarrow & \\ \quad CrO_4^{2-} + 3IO_4^- + 16H_2O + 27e^- & \times 2 \\ \underline{2e^- + Cl_2 \longrightarrow 2Cl^-} & \times 27 \\ 2CrI_3 + 64OH^- + 27Cl_2 \longrightarrow & \\ \quad 2CrO_4^{2-} + 6IO_4^- + 32H_2O + 54Cl^- & \end{array}$$

**13-26. (a)** $2H_2 + O_2 \longrightarrow 2H_2O$
$$\begin{array}{l} \textbf{(b)} \quad H_2O_2 \longrightarrow O_2 + 2H^+ + 2e^- \\ \underline{2e^- + 2H^+ + H_2O_2 \longrightarrow 2H_2O} \\ 2H_2O_2 \longrightarrow O_2 + 2H_2O \\ 2OH^- + H_2O_2 \longrightarrow O_2 + 2H_2O + 2e^- \\ \underline{2e^- + H_2O_2 \longrightarrow 2OH^-} \\ 2H_2O_2 \longrightarrow O_2 + 2H_2O \end{array}$$

**13-27.** Reactions **(a)**, **(c)**, **(e)**, and **(f)** are predicted to be favorable.

**13-29. (a)** no reaction
**(b)** $Br_2(aq) + Sn(s) \longrightarrow SnBr_2(aq)$
**(c)** no reaction
**(d)** $O_2(g) + 4H^+(aq) + 4Br^-(aq) \longrightarrow$
$$2Br_2(l) + 2H_2O(l)$$
**(e)** $3Br_2(l) + 2Al(s) \longrightarrow 2AlBr_3(s)$

**13-31. (c)** $2F_2(g) + 2H_2O(l) \longrightarrow 4HF(aq) + O_2(g)$
**(e)** $Mg(s) + 2H_2O(l) \longrightarrow Mg(OH)_2(s) + H_2(g)$

**13-33.** From Table 13-1:
$Fe + 2H_2O \longrightarrow$ no reaction
$Fe + 2H^+ \longrightarrow Fe^{2+} + H_2$
Acid rain has a higher $H^+(aq)$ concentration, thus making the second reaction more likely.

**13-36.** $Zn(s)$ reacts at the anode and $MnO_2(s)$ reacts at the cathode. The total reaction is

$$Zn(s) + 2MnO_2(s) + 2H_2O(l) \longrightarrow$$
$$Zn(OH)_2(s) + 2MnO(OH)(s)$$

**13-38.** anode: $Cd(s) + 2OH^-(aq) \longrightarrow Cd(OH)_2(s) + 2e^-$
cathode: $NiO_2(s) + 2H_2O(l) + 2e^- \longrightarrow$
$$Ni(OH)_2(s) + 2OH^-(aq)$$

**13-40.** The spontaneous reaction is
$Pb(NO_3)_2(aq) + Fe(s) \longrightarrow Fe(NO_3)_2(aq) + Pb(s)$
anode: $Fe \longrightarrow Fe^{2+} + 2e^-$
cathode: $Pb^{2+} + 2e^- \longrightarrow Pb$

**13-41.** The Daniell cell is more powerful. The greater the separation between oxidizing and reducing agents as shown in Table 13-1, the more powerful is the cell.

**13-43.** $3Fe(s) + 2Cr^{3+}(aq) \longrightarrow 3Fe^{2+}(aq) + 2Cr(s)$
Zinc would spontaneously form a chromium coating as illustrated by the equation:
$$3Zn(s) + 2Cr^{3+}(aq) \longrightarrow 3Zn^{2+}(aq) + 2Cr(s)$$

**13-44.** Sodium reacts spontaneously with water since it is an active metal. The actual cathode reaction is
$$2H_2O + 2e^- \longrightarrow H_2 + 2OH^-$$
Elemental sodium is produced by electrolysis of the molten salt, NaCl.

**13-47.** The strongest oxidizing agent is reduced the easiest. Thus the reduction of $Ag^+$ to Ag occurs first. This procedure can be used to purify silver.

**13-48.** $K_3N$ (−3), $N_2H_4$ (−2), $NH_2OH$, (−1), $N_2$ (0), $N_2O$ (+1), NO (+2), $N_2O_3$ (+3), $N_2O_4$ (+4), $Ca(NO_3)_2$ (+5)

**13-49.** $Cd(s) + NiCl_2(aq) \longrightarrow Ni(s) + CdCl_2(aq)$
$Zn(s) + CdCl_2(aq) \longrightarrow Cd(s) + ZnCl_2(aq)$
These reactions indicate that $Cd^{2+}$ is a stronger oxidizing agent than $Zn^{2+}$ but weaker than $Ni^{2+}$. It appears about the same as $Fe^{2+}$.

**13-51.** The reactions that occur are
$$4Au^{3+} + 6H_2O \longrightarrow 4Au + 12H^+ + 3O_2$$
$$2Au + 3F_2 \longrightarrow 2AuF_3$$
These reactions and the fact that Au does not react with $Cl_2$ rank $Au^{3+}$ above $H^+$ and $Cl_2$ but below $F_2$.

**13-52.** $12H^+(aq) + 5Zn(s) + 2NO_3^-(aq) \longrightarrow$
$$5Zn^{2+}(aq) + N_2(g) + 6H_2O(l)$$
0.658 g $N_2$ requires <u>7.68 g Zn.</u>

**13-54.** $14H^+(aq) + 2NO_3^-(aq) + 3Cu_2O(s) \longrightarrow$
$$6Cu^{2+}(aq) + 2NO(g) + 7H_2O(l)$$
10.0 g of $Cu_2O$ produces <u>1.04 L of NO</u> measured at STP.

**13-56.** $7OH^-(aq) + 4Zn(s) + 6H_2O(l) + NO_3^-(aq) \longrightarrow$
$$NH_3(g) + 4Zn(OH)_4^{2-}(aq)$$
6.54 g of Zn produces <u>0.493 L of $NH_3$</u> at 27.0°C and 1.25 atm pressure.

**13-57.** **(a)** $MnO_4^- + 8H^+ + 5Fe^{2+} \longrightarrow$
$$5Fe^{3+} + Mn^{2+} + 4H_2O$$
**(b)** $2MnO_4^- + 16H^+ + 10Br^- \longrightarrow$
$$5Br_2 + 2Mn^{2+} + 8H_2O$$
**(c)** $2MnO_4^- + 16H^+ + 5C_2O_4^{2-} \longrightarrow$
$$10CO_2 + 2Mn^{2+} + 8H_2O$$

**13-58.** 179 mL

**13-59.** 306 mL

# Chapter 14

**14-1.** Colliding molecules must have the proper orientation relative to each other at the time of the collision, and the colliding molecules must have the minimum kinetic energy for the particular reaction.

**14-3.** **(a)** As the temperature increases, the frequency of collisions between molecules increases, as does the average energy of the collisions. Both contribute to the increased rate of reaction.

    **(b)** The cooking of eggs initiates a chemical reaction that occurs more slowly at lower temperatures.

    **(c)** The average energy of colliding molecules at room temperature is not sufficient to initiate a reaction between $H_2$ and $O_2$.

    **(d)** A higher concentration of oxygen increases the rate of combustion.

    **(e)** When a solid is finely divided, a greater surface area is available for collisions with oxygen molecules. Thus it burns faster.

    **(f)** The souring of milk is a chemical reaction that slows as the temperature drops. It takes several days in a refrigerator.

    **(g)** The platinum is a catalyst. Since the activation energy in the presence of a catalyst is lower, the reaction can occur at a lower temperature.

**14-6.** The rate of the forward reaction was at a maximum at the beginning of the reaction; the rate of the reverse reaction was at a maximum at the point of equilibrium.

**14-7.** It many cases, reactions do not proceed directly to the right because other products are formed between the same reactants. For example, combustion may produce carbon monoxide as well as carbon dioxide.

**14-8.** Products are easier to from because the activation energy for the forward reaction is less than for the reverse

reaction. This is true of all exothermic reactions. The system should come to equilibrium faster starting with pure reactants.

**14-10.** **(a)** right    **(b)** left    **(c)** left    **(d)** right
**(e)** right    **(f)** has no effect    **(g)** yield decreases
**(h)** yield increases but rate of formation decreases

**14-12.** **(a)** increase    **(b)** increase    **(c)** decrease
**(d)** decrease    **(e)** decrease    **(f)** no effect

**14-14.** **(a)** Since there are the same number of moles of gas on both sides of the equation, pressure (or volume) has no effect on the point of equilibrium.

    **(b)** decrease the amount of NO

    **(c)** decrease the amount of NO

**14-15.** **(a)** $K_{eq} = \dfrac{[COCl_2]}{[CO][Cl_2]}$    **(b)** $K_{eq} = \dfrac{[CO_2][H_2]^4}{[CH_4][H_2O]^2}$

**(c)** $K_{eq} = \dfrac{[Cl_2]^2[H_2O]^2}{[HCl]^4[O_2]}$    **(d)** $K_{eq} = \dfrac{[CH_3Cl][HCl]}{[CH_4][Cl_2]}$

**14-17.** products

**14-19.** There will be an appreciable concentration of both reactants and products at equilibrium.

**14-20.** $K_{eq} = 0.34$

**14-21.** $K_{eq} = 49.2$

**14-23.** $K_{eq} = \dfrac{[CO_2][H_2]^4}{[CH_4][H_2O]^2} = \dfrac{(2.20/30.0)(4.00/30.0)^4}{(6.20/30.0)(3.00/30.0)^2} =$
0.0112

**14-25.** **(a)** $[H_2] = [I_2] = 0.30$ mol/L    **(b)** $[H_2] = 0.30$ mol/L; $[I_2] = 0.50$ mol/L    **(c)** $[HI] = 0.40$ mol/L; $[I_2] = [H_2] = 0.10$ mol/L    **(d)** $K_{eq} = 0.063$    **(e)** $K_r = 16$. This is a smaller value than that used in Table 14-1. This indicates that the equilibrium in this problem was established at a different temperature than that of Table 14-1.

**14-27.** **(a)** $[N_2]_{reacts} = 0.50 - 0.40 = 0.10$ mol/L

$$0.10 \text{ mol } N_2 \times \frac{3 \text{ mol } H_2}{1 \text{ mol } N_2} = 0.30 \text{ mol/L } H_2 \text{ (reacts)}$$
$$[H_2]_{eq} = 0.50 - 0.30 = \underline{0.20 \text{ mol/L}}$$
$$0.10 \text{ mol } N_2 \times \frac{2 \text{ mol } NH_3}{1 \text{ mol } N_2} = 0.20 \text{ mol/L } NH_3 \text{ formed}$$
$$[NH_3]_{eq} = 0.50 + 0.20 = \underline{0.70 \text{ mol/L}}$$

**(b)** $K_{eq} = \dfrac{(0.70)^2}{(0.20)^3(0.40)} = 150$

**14-28.** **(a)** The concentration of $O_2$ that reacts is

$$0.25 \text{ mol } NH_3 \times \frac{5 \text{ mol } O_2}{4 \text{ mol } NH_3} = 0.31 \text{ mol/L } O_2$$

**(b)** $[NH_3] = 0.75$ mol/L; $[O_2] = 0.69$ mol/L $O_2$;
$[NO] = 0.25$ mol/L NO (formed)
$$[H_2O] = 0.38 \text{ mol/L } H_2O$$

**(c)** $K_{eq} = \dfrac{[NO]^4[H_2O]^6}{[NH_3]^4[O_2]^5} = \dfrac{(0.25)^4(0.38)^6}{(0.75)^4(0.69)^5}$

**14-30.** $[PCl_5] = 0.28$ mol/L

**14-32.** $[H_2O] = 0.26$ mol/L

**14-33.** Let $x = [HCl] = [CH_3Cl]$

$$K_{eq} = \frac{[HCl][CH_3Cl]}{[CH_4][Cl_2]} = \frac{x^2}{(0.20)(0.40)} = 56$$
$$x^2 = 4.5 \qquad x = \underline{2.1 \text{ mol/L}}$$

**14-35.** (a) $K_a = \dfrac{[H_3O^+][BrO^-]}{[HBrO]}$ (d) $K_a = \dfrac{[H_3O^+][SO_3^{2-}]}{[HSO_3^-]}$

(b) $K_b = \dfrac{[NH_4^+][OH^-]}{[NH_3]}$ (e) $K_a = \dfrac{[H_3O^+][H_2PO_4^-]}{[H_3PO_4]}$

(c) $K_a = \dfrac{[H_3O^+][HSO_3^-]}{[H_2SO_3]}$ (f) $K_b = \dfrac{[(CH_3)_2NH_2^+][OH^-]}{[(CH_3)_2NH]}$

**14-37.** The acid HB is weaker because it produces a smaller hydronium ion concentration (higher pH) at the same initial concentration of acid. The stronger acid HX has the larger value of $K_a$.

**14-39.** (a) $[HOCN]_{eq} = 0.20 - 0.0062 = 0.19$

(b) $K_a = \dfrac{[H_3O^+][OCN^-]}{[HOCN]} = \dfrac{(6.2 \times 10^{-3})(6.2 \times 10^{-3})}{0.19} = 2.0 \times 10^{-4}$

(c) $pH = 2.21$

**14-40.** (a) $HX + H_2O \rightleftharpoons H_3O^+ + X^- \quad K_a = \dfrac{[H_3O^+][X^-]}{[HX]}$

(b) From the equation $[H_3O^+] = [X^-] = 0.100 \times 0.58$, $[HX] = 0.58 - 0.058 = 0.52$
(c) $K_a = 6.5 \times 10^{-3}$ (d) $pH = 1.24$

**14-43.** $Nv + H_2O \rightleftharpoons NvH^+ + OH^- \quad K_b = 6.6 \times 10^{-6}$

**14-44.** $[H_3O^+] = 0.300 - 0.277 = 0.023 \qquad pH = 1.64$
$K_a = 1.9 \times 10^{-3}$

**14-45.** $pH = 4.43$

**14-47.** $[OH^-] = 3.2 \times 10^{-3}$

**14-49.** $pH = 12.43$

**14-51.** $pH = 9.40$

**14-53.** $pK_a = 8.68 \qquad pH = 8.68 + \log \dfrac{0.20/0.850}{0.60/0.850} = 8.68 - 0.48 = \underline{8.20}$

**14-56.** $pK_b = 6.01 \qquad pOH = 6.01 + \log \dfrac{0.0288 \text{ mol acid}}{0.0469 \text{ mol base}}$
$pOH = 6.01 - 0.21 = 5.80 \qquad pH = \underline{8.20}$

**14-58.** When the concentrations of acid and base species are equal, $pH = pK_a$ and $pOH = pK_b$. If a buffer of $pH = 7.50$ is required, then we look for an acid with $K_a = [H_3O^+] = 3.2 \times 10^{-8}$ or a base with $K_b = [OH^-] = 3.2 \times 10^{-7}$. Since $K_a = 3.2 \times 10^{-8}$ for HClO, an equimolar mixture of HClO and KClO produces the required buffer.

**14-60.** (a) $FeS(s) \rightleftharpoons Fe^{2+}(aq) + S^{2-}(aq)$
$K_{sp} = [Fe^{2+}][S^{2-}]$
(b) $Ag_2S(s) \rightleftharpoons 2Ag^+(aq) + S^{2-}(aq)$
$K_{sp} = [Ag^+]^2[S^{2-}]$

(c) $Zn(OH)_2(s) \rightleftharpoons Zn^{2+}(aq) + 2OH^-(aq)$
$K_{sp} = [Zn^{2+}][OH^-]^2$

**14-62.** $K_{sp} = 4.8 \times 10^{-12}$

**14-64.** $Ca(OH)_2(s) \rightleftharpoons Ca^{2+}(aq) + 2OH^-(aq)$
$K_{sp} = [Ca^{2+}][OH^-]^2$
At equilibrium $[Ca^{2+}] = 1.3 \times 10^{-2}$ mol/L $[OH^-]$ $= 2 \times (1.3 \times 10^{-2})$mol/L $= 2.6 \times 10^{-2}$ mol/L $K_{sp}$ $= [1.3 \times 10^{-2}][2.6 \times 10^{-2}]^2 = 8.8 \times 10^{-6}$

**14-66.** $9.1 \times 10^{-9}$ mol/L

**14-68.** $AgBr(s) \rightleftharpoons Ag^+(aq) + Br^-(aq) \qquad K_{sp} = [Ag^+][Br^-] = 7.7 \times 10^{-13}$
$[7 \times 10^{-6}][8 \times 10^{-7}] = 6 \times 10^{-12}$
This is a larger number than $K_{sp}$, so a precipitate of AgBr does form.

**14-70.** (a) $3Cl_2(g) + NH_3(g) \rightleftharpoons NCl_3(g) + 3HCl(g)$
(b) $K_{eq} = \dfrac{[NCl_3][HCl]^3}{[NH_3][Cl_2]^3}$ (c) no effect
(d) decreases $[NH_3]$ (e) reactants
(f) $[NH_3] = 0.083$ mol/L

**14-72.** (a) $2N_2O(g) \rightleftharpoons 2N_2(g) + O_2(g)$
(b) $K_{eq} = \dfrac{[N_2]^2[O_2]}{[N_2O]^2}$ (c) decreases $[N_2]$
(d) decreases $[N_2O]$
(e) $[N_2O]_{eq} = 0.10 - (0.015 \times 0.10) = 0.10$ mol/L;
$[N_2] = 0.015 \times 0.10 = 1.5 \times 10^{-3}$ mol/L;
$[O_2] = \dfrac{1.5 \times 10^{-3}}{2} = 7.5 \times 10^{-4}$ mol/L
$K_{eq} = \underline{1.7 \times 10^{-7}}$

**14-74.** (a) A solution of $NaHCO_3$ is slightly basic because the hydrolysis reaction occurs to a greater extent than the acid ionization reaction. That is, $K_b$ is larger than $K_a$.
(b) It is used as an antacid to counteract excess stomach acidity.
(c) $HCl + H_2O \longrightarrow H_3O^+ + Cl^-$ (HCl is a strong acid.)
$H_3O^+(aq) + HCO_3^-(aq) \longrightarrow H_2CO_3(aq) + H_2O(l)$
$H_2CO_3(aq) \longrightarrow H_2O(l) + CO_2(g)$
(d) No. The $HSO_4^-$ does not react with $H_3O^+$ because it would form $H_2SO_4$, which is a strong acid. The molecular form of a strong acid does not exist in water.

**14-76.** $0.265 \cancel{L} \times 0.22$ mol/$\cancel{L} = 0.058$ mol $HC_2H_3O_2$
$0.375 \cancel{L} \times \dfrac{0.12 \cancel{\text{mol Ba(C}_2\text{H}_3\text{O}_2)_2}}{\cancel{L}} \times$
$\dfrac{2 \text{ mol } C_2H_3O_2^-}{\cancel{\text{mol Ba(C}_2\text{H}_3\text{O}_2)_2}} = 0.090$ mol $C_2H_3O_2^-$ $pH = 4.93$

**14-77.** The addition of the NaOH neutralizes part of the acetic acid to produce sodium acetate.
$HC_2H_3O_2 + NaOH \longrightarrow NaC_2H_3O_2 + H_2O$
$1.00 - 0.20 = 0.80$ mol of $HC_2H_3O_2$ remains, and $0.20$ mol of $C_2H_3O_2^-$ is formed after $0.20$ mol of $OH^-$ is added. $pH = \underline{4.14}$

**14-78.** $[OH^-] = 3.2 \times 10^{-4}$   $[Mg^{2+}] = [OH^-]/2 = 1.6 \times 10^{-4}$

$K_{sp} = [1.6 \times 10^{-4}][3.2 \times 10^{-4}]^2 = 1.6 \times 10^{-11}$

**14-80.** $1.0 \times 10^{-5}$ mol/L, 0.023 g $BaSO_4$/L

## Cumulative Review Test for Chapters 12–14

### Multiple Choice

| | | | | |
|---|---|---|---|---|
| **1. (e)** | **2. (d)** | **3. (b)** | **4. (b)** | **5. (b)** |
| **6. (d)** | **7. (c)** | **8. (e)** | **9. (a)** | **10. (c)** |
| **11. (c)** | **12. (b)** | **13. (c)** | **14. (b)** | **15. (c)** |
| **16. (c)** | **17. (a)** | **18. (d)** | **19. (e)** | **20. (d)** |
| **21. (b)** | **22. (c)** | **23. (b)** | **24. (c)** | **25. (a)** |

### Problems

**1.** (a) $H_2SO_4(aq) + 2LiOH(aq) \longrightarrow Li_2SO_4(aq) + 2H_2O$
(b) 150 mL

**2.** (a) $HClO + H_2O \rightleftharpoons H_3O^+ + ClO^-$
(b) $NH_3 + H_2O \rightleftharpoons NH_4^+ + OH^-$
(c) $KOH + HNO_2 \longrightarrow KNO_2 + H_2O$
(d) $H_3AsO_4 + 2NaOH \longrightarrow Na_2HAsO_4 + 2H_2O$

**3.** (a) N.R.     (b) $CaO + H_2O \longrightarrow Ca(OH)_2$
(c) $NH_4^+ + H_2O \rightleftharpoons NH_3 + H_3O^+$
(d) $CO_2 + H_2O \longrightarrow H_2CO_3$
(e) $ClO^- + H_2O \rightleftharpoons HClO + OH^-$     (f) N.R.

**4.** (a) $[H_3O^+] = 1.5 \times 10^{-11}$     (b) pH = 10.82
(c) pOH = 3.18

**5.** (a) $Sn(s) + 4HNO_3(aq) \longrightarrow$
$$SnO_2(s) + 4NO_2(g) + 2H_2O(l)$$
(b) 3.2 g $NO_2$

**6.** (a) Sn     (b) $HNO_3$     (c) $SnO_2$     (d) $NO_2$
(e) $HNO_3$     (f) Sn

**7.** (a) nonspontaneous
(b) $2Cr^{3+} + 3Mn \longrightarrow 2Cr + 3Mn^{2+}$
(c) Yes. $3I_2 + 2Cr \longrightarrow 2Cr^{3+} + 6I^-$

**8.** (a) $K_{eq} = \dfrac{[NOBr]^2}{[NO]^2[Br_2]}$     (b) $K_{eq} = 20$
(c) [NOBr] = 0.94 mol/L

**9.** $B + H_2O \rightleftharpoons BH^+ + OH^-$     $K_b = 1.7 \times 10^{-6}$

**10.** $H_3BO_3 + H_2O \rightleftharpoons H_3O^+ + H_2BO_3^-$
pH = 4.72

## Chapter 15

**15-1.** (a) $^{210}_{84}Po$     (b) $^{84}_{38}Sr$     (c) $^{257}_{100}Fm$
(d) $^{206}_{82}Pb$     (e) $^{233}_{92}U$

**15-3.** (a) $^4_2He$     (b) $^{0}_{-1}e$     (c) $^1_0n$

**15-4.** $^2_1H$

**15-6.** (a) $^{206}_{82}Pb$     (b) $^{148}_{62}Sm$     (c) $^{248}_{98}Cf$
(d) $^{262}_{107}Bh$

**15-8.** (a) $^3_2He$     (b) $^{153}_{65}Tb$     (c) $^{59}_{27}Co$
(d) $^{24}_{12}Mg$

**15-11.** $^{51}_{25}Mn \longrightarrow ^{51}_{24}Cr + ^{0}_{+1}e$

**15-14.** $^{68}_{32}Ge + ^{0}_{-1}e \longrightarrow ^{68}_{31}Ga$

**15-16.** (a) $^{0}_{-1}e$     (b) $^{90}_{38}Sr$     (c) $^{26}_{13}Al$
(d) $^{231}_{90}Th$     (e) $^{179}_{72}Hf$     (f) $^{41}_{20}Ca$
(g) $^{210}_{81}Tl$

**15-18.** (a) $^{230}_{90}Th \longrightarrow ^{226}_{88}Ra + ^4_2He$
(b) $^{214}_{84}Po \longrightarrow ^{210}_{82}Pb + ^4_2He$
(c) $^{210}_{84}Po \longrightarrow ^{210}_{85}At + ^{0}_{-1}e$
(d) $^{218}_{84}Po \longrightarrow ^{214}_{82}Pb + ^4_2He$
(e) $^{14}_6C \longrightarrow ^{14}_7N + ^{0}_{-1}e$
(f) $^{50}_{25}Mn \longrightarrow ^{50}_{24}Cr + ^{0}_{+1}e$
(g) $^{37}_{18}Ar + ^{0}_{-1}e \longrightarrow ^{37}_{17}Cl$

**15-19.** $^{234}_{90}Th$, $^{234}_{91}Pa$, $^{234}_{92}U$, $^{230}_{90}Th$, $^{226}_{88}Ra$, $^{222}_{86}Rn$, $^{218}_{84}Po$, $^{214}_{82}Pb$, $^{214}_{83}Bi$, $^{210}_{81}Tl$, $^{210}_{82}Pb$, $^{210}_{83}Bi$, $^{210}_{84}Po$, $^{206}_{82}Pb$

**15-20.** $\dfrac{1}{16}$

**15-22.** (a) 5 mg     (b) 1.25 mg     (c) 1.25 mg

**15-23.** 2.50 g

**15-25.** about 11,500 years old

**15-27.** The energy from the radiation causes an electron in an atom or a molecule to be expelled, leaving behind a positive ion. When a molecule in a cell is ionized, it is damaged and may die or mutate.

**15-29.** Radiation entering a chamber causes ionization. The electrons formed migrate to the central electrode (positive) and cause a burst of current that can be detected and amplified. In a scintillation counter, the radiation is detected by phosphors that glow when radiation is absorbed.

**15-30.** (a) $^{35}_{16}S$     (b) $^2_1H$     (c) $^{30}_{15}P$     (d) $8\ ^{0}_{-1}e$
(e) $^{254}_{102}No$     (f) $^{237}_{92}U$     (g) $4\ ^1_0n$

**15-32.** one neutron

**15-34.** $^{206}_{82}Pb$

**15-35.** $^{262}_{107}Bh$

**15-36.** (a) $^{87}_{34}Se$     (b) $^{97}_{38}Sr$

**15-38.**
$$^2_1H + ^2_1H \longrightarrow ^4_2He \nearrow\ ^3_2He + ^1_0n \searrow\ ^3_1H + ^1_1H$$

**15-39.** (a) $^{106}_{46}Pd + ^4_2He \longrightarrow ^{109}_{47}Ag + ^1_1H$
(b) $^{266}_{109}Mt \longrightarrow ^{262}_{107}Bh + ^4_2He$
$^{262}_{107}Bh \longrightarrow ^{258}_{105}Db + ^4_2He$
(c) $^{212}_{83}Bi \longrightarrow ^{212}_{84}Po + ^{0}_{-1}e$
(d) $^{60}_{30}Zn \longrightarrow ^{60}_{29}Cu + ^{0}_{+1}e$
(e) $^{239}_{94}Pu + ^1_0n \longrightarrow ^{140}_{55}Cs + ^{97}_{39}Y + 3^1_0n$
(f) $^{206}_{82}Pb + ^{54}_{24}Cr \longrightarrow ^{257}_{106}Sg + 3^1_0n$
(g) $^{93}_{42}Mo + ^{0}_{-1}e \longrightarrow ^{93}_{41}Nb$

**15-41.** (a) $^{90}_{39}Y$     (b) 50 years

**15-44.** 60 years is five half-lives. $0.42\ \mu g \times 2^5 = \underline{13.4\ \mu g}$

**15-46.** atomic number = 111     mass number = 272

**15-48.** $^{70}Zn$

# Chapter 16

**16-1. (a)**

$$H-\overset{\overset{\displaystyle H}{|}}{\underset{\underset{\displaystyle H}{|}}{C}}-\overset{\cdot\cdot}{\underset{\cdot\cdot}{Br}}:$$

**(b)**

$$\overset{H}{\underset{H}{>}}C=C=C\overset{H}{\underset{H}{<}}$$

**(c)**

$$\overset{H}{\underset{H}{>}}C=C-\overset{H}{\underset{H}{C}}-\overset{H}{\underset{H}{C}}-H \qquad H-\overset{H}{\underset{H}{C}}-C=C-\overset{H}{\underset{H}{C}}-H$$

structures (cyclic)

**(d)**

$$H-\overset{\overset{\displaystyle H}{|}}{\underset{\underset{\displaystyle H}{|}}{C}}-\overset{\cdot\cdot}{\underset{\underset{\displaystyle H}{|}}{N}}-H$$

**(e)**

$$H-\overset{H}{\underset{H}{C}}-\overset{H}{\underset{H}{C}}-\overset{\cdot\cdot}{\underset{H}{N}}-H \qquad H-\overset{H}{\underset{H}{C}}-\overset{\cdot\cdot}{\underset{H}{N}}-\overset{H}{\underset{H}{C}}-H$$

**(f)**

$$H-\overset{H}{\underset{H}{C}}-\overset{H}{\underset{H}{C}}-\overset{H}{\underset{H}{C}}-\overset{\cdot\cdot}{\underset{\cdot\cdot}{O}}-H \qquad H-\overset{H}{\underset{H}{C}}-\overset{H}{\underset{H}{C}}-\overset{H:\overset{\displaystyle H}{\overset{\displaystyle O}{:}}H}{\underset{H}{C}}-H$$

$$H-\overset{H}{\underset{H}{C}}-\overset{H}{\underset{H}{C}}-\overset{\cdot\cdot}{\underset{\cdot\cdot}{O}}-\overset{H}{\underset{H}{C}}-H$$

**16-3.** $CH_3-CH_2-CH_2-CH_2-CH_2-CH_3$

$$CH_3-CH_2-CH_2-\underset{\underset{\displaystyle CH_3}{|}}{CH}-CH_3$$

$$CH_3-CH_2-\underset{\underset{\displaystyle CH_3}{|}}{CH}-CH_2-CH_3 \qquad CH_3-CH_2-\underset{\underset{\displaystyle CH_3}{\overset{\overset{\displaystyle CH_3}{|}}{C}}}{}-CH_3$$

$$CH_3-\underset{\underset{\displaystyle CH_3}{|}}{CH}-\underset{\underset{\displaystyle CH_3}{|}}{CH}-CH_3$$

**16-5. (a)** No. The second compound does not have oxygen.
**(b)** No. The compounds have unequal numbers of carbon atoms.
**(c)** Yes. Both compounds have the formula $C_2H_7N$.
**(d)** No. The compounds have unequal numbers of carbon atoms.
**(e)** Yes. The compounds are identical.

**16-7. (a)** $C_4H_{10}$,   **(d)** $C_{10}H_{22}$, and   **(g)** $C_{18}H_{38}$

**16-8. (a)** $CH_3-CH_2-CH_3$

**(b)** $CH_3-\underset{\underset{\displaystyle CH_3}{|}}{CH}-CH_2-\underset{\underset{\displaystyle CH_2-CH_2-CH_3}{|}}{CH}-CH_2-CH_3$

**(c)**

$$CH_3-CH-\underset{\underset{\displaystyle CH_3}{\overset{\overset{\displaystyle CH_3}{|}}{C}}}{}-CH_3$$

**(d)**

$$CH_3-CH-CH_2-\underset{\underset{\displaystyle CH_2}{|}}{\overset{\overset{\displaystyle CH_3}{|}}{C}}-CH_3$$

with $CH_2$ ring closure

**16-11. (f)** $C_7H_{14}$

**16-12. (b)** $C_3H_7$ and **(e)** $C_7H_{15}$

**16-14. (a)** $CH_3-CH_2-\underset{\underset{\displaystyle CH_3}{|}}{CH}-CH_2-CH_3$

**(b)** $CH_3CH_2CH_2CH_2CH_2CH_3$

**(c)** $CH_3-\underset{\underset{\displaystyle CH_3}{|}}{CH}-CH_2-\underset{\underset{\displaystyle CH_3}{|}}{CH}-\underset{\underset{\displaystyle CH_3}{|}}{CH}-CH_2CH_2CH_3$

**(d)** $CH_3-\underset{\underset{\displaystyle CH_3}{|}}{CH}-CH_2-\underset{\underset{\displaystyle CH_2CH_3}{|}}{CH}-CH_2CH_2CH_3$

**(e)** $CH_3-CH_2-\underset{\underset{\displaystyle CH(CH_3)_2}{|}}{CH}-CH_2CH_2CH_3$

**(f)** $CH_3\underset{\underset{\displaystyle CH_3}{\overset{\overset{\displaystyle CH_3}{|}}{C}}}{}-CH_2-\underset{\underset{\displaystyle C(CH_3)_3}{|}}{CH}-CH_2CH_2CH_2CH_3$

**16-16. (a)** $CH_3\underset{\underset{\displaystyle CH_3}{|}}{\overset{\overset{\displaystyle CH_2CH_3}{|}}{CH}}CHCH_2CH_3$

3,4-dimethylhexane

**(b)** $CH_3\underset{\underset{\displaystyle CH(CH_3)_2}{|}}{\overset{\overset{\displaystyle CH_2CH_2CH_3}{|}}{C}}CH_2CH_2CH_3$

4-isopropyl-4-methyloctane

**16-18. (a)** six (hexane)   **(b)** eight (octane)
**(c)** nine (nonane)   **(d)** seven (heptane)

**16-19. (a)** three methyl   2,3,5-trimethylhexane
**(b)** methyl and ethyl   5-ethyl-2-methyloctane
**(c)** two methyl   3,3-dimethylnonane
**(d)** one methyl   4-methylheptane

**16-21. (a)** propane   **(b)** 4-ethyl-2-methylheptane
**(c)** trimethylbutane

**16-23.** This general formula is that of a straight-chain alkene with one double bond or of a cyclic alkane.

**16-25. (a)** $CH_3CH_2CH=CH_2$   **(b)** $CH_3C\equiv CCH_3$
**(c)** $CH_3CH_2\underset{\underset{\displaystyle CH_3}{|}}{C}=CHCH_3$   **(d)** $CH_3CH_2C\equiv C\underset{\underset{\displaystyle CH_2CH_3}{|}}{CH}CH_3$

**16-27. (a)** $CH_3C{\equiv}CCH_3$    **(b)**  $CH_2{=}CHCHCH_3$
$$\underset{\displaystyle CH_3}{|}$$

**(c)** $CH_3CH{=}CCH_2CH_2CH_3$
$$\underset{\displaystyle CH_2CH_2CH_3}{|}$$

**(d)** $CH_3CH_2C{\equiv}CCHCH_2CH_2CH_3$
$$\underset{\displaystyle CH_3}{|}$$

**16-29.** $CH_3{-}CH_2{-}\underset{\displaystyle \underset{CH_3}{|}}{CH}{-}C{\equiv}CH$   3-methyl-1-pentyne

**16-31. (a)** 1-butene    **(b)** 2-butyne
**(c)** 3-methyl-2-pentene    **(d)** 5-methyl-3-heptyne

**16-33. (a)** $\text{+}CH_2{-}\underset{\displaystyle \underset{CH_3}{|}}{CH}\text{+}_n$    polypropylene

**(b)** $\text{+}CH_2{-}\underset{\displaystyle \underset{F}{|}}{CH}\text{+}_n$    polyvinylfluoride

**(c)** $\text{+}CH_2{-}\underset{\displaystyle \underset{CO_2CH_3}{|}}{CH}\text{+}_n$    polymethyacrylate

**16-34.** $CHF{=}CHCH_3$
**16-35.**

**16-37.** Carboxylic acids contain a carboxyl group, and alcohols contain only a hydroxyl group.

**16-38. (a)** In an ether, the oxygen is between two carbons; in an alcohol the oxygen is between a carbon and a hydrogen.

**(b)** In an aldehyde, a carbonyl (i.e., C=O) is between a carbon and a hydrogen; in a ketone the carbonyl is between two carbons.

**(c)** In an amine, the $NH_2$ group is attached to a hydrocarbon group; in an amide, the $NH_2$ group is attached to a carbonyl group.

**(d)** Carboxylic acids have a hydrogen attached to an oxygen; in esters the hydrogen is replaced by a hydrocarbon group.

**16-40. (a)** ketone    **(b)** alkene    **(c)** alkane
**(d)** aldehyde    **(e)** aromatic    **(f)** alcohol

**16-42. (a)** alkene and aldehyde    **(b)** ketone and amine
**(c)** ether and amide    **(d)** alkyne and alcohol
**(e)** ester    **(f)** carboxylic acid

**16-44. (a)** $CH_3CH_2CH_2CH_2OH$
**(b)** $CH_3CH_2CH_2OCH_2CH_2CH_3$    **(c)** $(CH_3)_3N$
**(d)** $CH_3CH_2CHO$    **(f)** $CH_3CH_2COOH$
**(e)** $CH_3CH_2\underset{\displaystyle \underset{O}{\|}}{C}CH_2CH_3$    **(g)** $CH_3\underset{\displaystyle \underset{O}{\|}}{C}OCH_3$

**(h)** $CH_3CH_2\underset{\displaystyle \underset{O}{\|}}{C}NH_2$

**16-45. (a)** 2-pentanone    **(b)** butanoic acid
**(c)** acetamide    **(d)** *t*-butyl alcohol    **(e)** pentanal

**16-47. (a)** alkene    **(b)** alkane    **(c)** alkene
**(d)** alkyne    **(e)** alkane    **(f)** alkene
**(g)** alkane

**16-49.**

**16-50.**

**(a)** alkane    $CH_3{-}\underset{\displaystyle \underset{CH_3}{|}}{\overset{\displaystyle \overset{CH_3}{|}}{C}}{-}CH_2{-}CH_2{-}CH_3$

**(b)** alkene    $CH_3{-}\underset{\displaystyle \underset{CH_3}{|}}{\overset{\displaystyle \overset{CH_3}{|}}{C}}{-}CH_2{-}CH{=}CH_2$

**(c)** alkyne    $CH_3{-}\underset{\displaystyle \underset{CH_3}{|}}{\overset{\displaystyle \overset{CH_3}{|}}{C}}{-}CH_2{-}C{\equiv}CH$

**(d)** aromatic    $CH_3$

**16-52. (a)** $C_3H_8 + 5O_2 \longrightarrow 3CO_2 + 4H_2O$
**(b)** $CH_2{=}CHCH_3 + H_2O \longrightarrow CH_3CH(OH)CH_3$
**(c)** $C_2H_5Cl + 2NH_3 \longrightarrow C_2H_5NH_2 + NH_4{}^+Cl^-$
**(d)** $(CH_3)_2NH(aq) + HCl(aq) \longrightarrow$
$$(CH_3)_2NH_2{}^+(aq) + Cl^-(aq)$$
**(e)** $CH_3CH_2COOH + (CH_3)_2NH \longrightarrow$
$$CH_3CH_2CON(CH_3)_2 + H_2O$$
**(f)** $HC{\equiv}CH + HBr \longrightarrow CH_2{=}CHBr$
**(g)** $CH_3CH_2COOH + H_2O \rightleftharpoons$
$$CH_3CH_2COO^- + H_3O^+$$

**16-57.**

**(a)** $C_3H_8O$:    $CH_3{-}CH_2{-}CH_2{-}OH$    *n*-propyl alcohol

$\underset{\displaystyle \underset{CH_3}{}}{\overset{\displaystyle \overset{CH_3}{}}{CH{-}OH}}$    isopropyl alcohol

$CH_3{-}CH_2{-}O{-}CH_3$    ethyl methyl ether

**(b)** $C_3H_6O_2$:    $CH_3{-}\overset{\displaystyle \overset{O}{\|}}{C}{-}O{-}CH_3$    methyl acetate (ester)

$H{-}\overset{\displaystyle \overset{O}{\|}}{C}{-}O{-}CH_2{-}CH_3$    ethyl formate (ester)

$CH_3{-}CH_2{-}COOH$    propionic acid

# Photo Credits

*Page 259:* Walter Geirsperger/Corbis Stock Market. *Page 261:* Tom Pantages. *Page 264:* R. Rowan/Photo Researchers. *Page 265:* Jacques Cochin/Agence Vandystadt/Photo Researchers. *Page 266:* Michael Dalton/Fundamental Photographs. *Page 274:* Patrick Morrow. *Page 276:* Sylvain Grandadam/Photo Researchers. *Page 279:* Corbis Images.

## Chapter 10

*Page 291:* Art Wolfe/The Image Bank/Getty Images. *Page 293 (top and center):* Andy Washnik. *Page 293 (bottom):* ImageState. *Page 299 (top):* Bob Jacobson/Corbis Images. *Page 299 (bottom left):* Fred Ward/Black Star. *Page 299 (bottom & center):* Herve Berthoule/Photo Researchers. *Page 299 (bottom right):* M. Claye/Photo Researchers. *Page 300 (top):* Kristian Hilsen/Stone/Getty Images. *Page 300 (center):* Paul Silverman/Fundamental Photographs. *Page 300 (bottom left):* C.D. Winters/Photo Researchers. *Page 300 (bottom right):* Ken Karp. *Page 301 (center):* Paul Silverman/ Fundamental Photographs. *Page 301 (bottom):* Richard Megna/Fundamental Photographs. *Page 302:* Courtesy NASA. *Page 303:* Yoav Levy/Phototake. *Page 304 (top left):* Courtesy R. Stimson Wilcox, SUNY Binghamton. *Page 304 (top right):* Ken Karp/Fundamental Photographs. *Page 304 (center):* Pat LaCroix/The Image Bank/Getty Images. *Page 306 (top):* Michael Dalton/Fundamental Photographs. *Page 306 (center):* Fundamental Photographs. *Page 307:* Tom Stock/Stone/Getty Images. *Page 308:* Chaix/BSIP/Science Source/Photo Researchers. *Page 309:* Barbara Gerlach/Visuals Unlimited. *Page 311:* Charles D. Winters/Photo Researchers.

## Chapter 11

*Page 322:* Alan Puzey/Stone/Getty Images. *Page 323:* Andy Washnik. *Page 326:* PhotoDisc, Inc./Getty Images. *Page 328 (top):* Andy Washnik. *Page 328 (bottom):* Art Wolfe/ Stone/Getty Images. *Page 330:* Michael Watson. *Page 332:* Paul Silverman/Fundamental Photographs. *Pages 335 and 338:* Michael Watson. *Page 341:* Dennis MacDonald/PhotoEdit. *Page 342:* David Phillips/Photo Researchers. *Page 343 (top):* Courtesy Jack Malone. *Page 343 (center):* Richard Hutchings/PhotoEdit. *Page 343 (bottom):* Courtesy Recovery Engineering.

## Chapter 12

*Page 354:* Richard Hamilton Smith/Corbis Images. *Page 355 (left):* Richard Megna/Fundamental Photographs. *Page 355 (right):* Andy Washnik. *Page 357:* Ken Karp. *Page 361:* Andy Washnik. *Page 363:* Robert Capece. *Page 369:* Leonard Lessin/Peter Arnold, Inc. *Page 370:* Courtesy Fisher Scientific. *Page 373:* OPC, Inc. *Page 375:* Andy Washnik. *Page 377:* Lester Lefkowitz/Corbis Stock Market. *Page 378 (top):* NYC Parks Photo Archive/Fundamental Photographs. *Page 378 (bottom):* Kristen Brochmann/Fundamental Photographs. *Page 379:* Will & Deni McIntyre/Stone/Getty Images.

## Chapter 13

*Page 388:* James P. Blair/National Geographic/Getty Images, Inc. *Page 389:* Michael Watson. *Page 390:* Richard Megna/Fundamental Photographs. *Page 391:* Darwin Dale/ Photo Researchers. *Page 396:* Richard Megna/Fundamental Photographs. *Page 404:* Michael Watson. *Page 407:* Courtesy of Delco Remy. *Page 408:* Courtesy of Eveready. *Page 409:* Peter Arnold, Inc. *Page 410:* ©AP/Wide World Photos. *Page 412:* Bill Gallery/Viesti Associates, Inc.

## Chapter 14

*Page 420:* Alex Stewart/The Image Bank/Getty Images. *Page 423:* Lucidio Studio, Inc/Corbis Images. *Page 425:* Andy Washnik. *Page 427:* Courtesy USDA. *Page 428:* David Sailors/Corbis Stock Market. *Page 429:* Courtesy AC Spark Plug. *Page 433:* Peter Turnley/Corbis Images. *Page 438:* Andy Washnik. *Page 444:* PhotoDisc, Inc./Getty Images. *Page 445:* Andy Washnik.

## Chapter 15

*Page 458:* Jisas/Lockheed/Science Photo Library/Photo Researchers. *Page 459:* Dan McCoy/Rainbow. *Page 464:* Larry Mulvehill/Photo Researchers. *Page 465 (left):* Yoav Levy/Phototake. *Page 465 (right):* Richard Megna/Fundamental Photographs. *Page 466:* Hank Morgan/Photo Researchers. *Page 468:* Courtesy of Superconducting Super Collider Laboratory. *Page 470:* Charles D. Winters/Photo Researchers. *Page 471 (top):* Courtesy of Council for Energy Awareness, Washington, D.C. *Page 471 (bottom):* Kelly Culpepper/Transparencies, Inc. *Page 472:* H. Morgan/Photo Researchers. *Page 475 (top):* IFA/Bruce Coleman, Inc. *Page 475 (bottom):* Tass/Sipa Press. *Page 477 (top):* Digital Art/Corbis Images. *Page 477 (bottom):* Courtesy Princeton Plasma Physics Laboratory.

## Chapter 16

*Page 483:* David Fleetham/Taxi/Getty Images. *Page 487:* Andy Washnik. *Page 492 (top):* Brett Froomer/The Image Bank/Getty Images. *Page 492 (bottom):* Ray Ellis/Photo Researchers. *Page 494:* Elie Bernager/Stone/Getty Images. *Page 495 (left):* Peter Lerman. *Page 495 (center):* Robert Tringall//Sports Chrome Inc. *Page 495 (right):* Michael Ventura//Bruce Coleman, Inc. *Page 496:* Charles Peterson/The Image Bank/Getty Images/Getty Images. *Page 500:* Andy Washnik. *Page 503:* Richard Megna/Fundamental Photographs. *Page 504 (top):* Andy Washnik. *Page 504 (bottom):* Courtesy Heinz. *Page 506:* Andy Washnik. *Page 507 (top):* Ken Karp. *Page 507 (bottom left):* Ulli Seer/Stone/Getty Images. *Page 507 (bottom right):* Charles D. Winters/Photo Researchers.

## Chapter 17

Opener: Dr. Tim Evans/SPL/Photo Researchers. Making It Real: Chris Butler/Photo Researchers.

## Appendix A

Opener: Vision/Photo Researchers.

# Index

Page numbers in italics indicate illustrations.

# Health Professions

## Career and Education Directory
## 2007-2008

## 35th Edition

Indispensable information on
6,873 programs in 71 professions

AMA
AMERICAN
MEDICAL
ASSOCIATION

# Health Professions
# Career and Education Directory 2007-2008

Copyright 2007 by the American Medical Association
All rights reserved
Printed in the United States of America

Internet address: www.ama-assn.org

## Order Information

To order additional copies of the *Health Professions Career and Education Directory*, call the American Medical Association toll free at 800 621-8335
Mention product number OP417507

## For other correspondence address inquiries to:

Fred Donini-Lenhoff
Medical Education Products
American Medical Association
515 N State St
Chicago, IL 60610
312 464-5333
312 464-5830 Fax
E-mail: fred.lenhoff@ama-assn.org

ISBN 978-1-57947-814-8
BP19:07-P-017:2/07

# Contents

# Contents

## 465   Section II—Institutions Sponsoring Accredited Programs

## 585   Section III—Accrediting Agencies

# Preface

The *2007-2008 Health Professions Career and Education Directory* now encompasses

- 6,873 programs enrolling more than 200,000 students
- 2,468 sponsoring institutions
- 71 occupations
- 26 accrediting agencies

## Organization of the Directory

The *Health Professions Career and Education Directory* contains information on 6,873 educational programs in 71 health professions.

*Section I: Occupations and Educational Programs* includes occupational descriptions, job descriptions, employment characteristics, and information about educational programs, such as length, curriculum, and educational standards, for the majority of the 71 occupations listed. Section I also provides sources for information on careers, certification, licensure, and registration.

This information is followed by a complete list of accredited/approved educational programs in alphabetical order by state, for each occupation. Most program listings include the educational program name and address; program director name, telephone, fax, and e-mail address; and name of the medical director, if applicable. In addition, Web site addresses are displayed for those programs that completed the 2006 AMA Survey of Health Professions Programs (see below).

This edition features a data chart showing class capacity (per start date or session), month(s) classes begin, program length(s), yearly tuition cost (resident and nonresident), yearly stipend (if any), academic award(s) granted, availability of evening/weekend courses, whether a program offers education/courses in medical/health care terms in non-English languages, and whether the program offers education in cultural competence or in patient communication. These data are shown only those programs that completed the 2006 AMA survey.

*Section II: Institutions Sponsoring Accredited Programs* lists 2,468 institutions sponsoring 6,873 health professions education programs. Sponsoring institutions are listed by state and city in alphabetical order. The majority of entries for each institution include the name, address, and telephone number of the chief executive officer and a list of the accredited health education programs for which the institution is a sponsor.

*Section III: Accrediting Agencies* provides information on the accreditation of health professions education programs by the agencies listed in Table 1 on page 586.

*Section IV: Health Professions Education Data* offers data tables and sources for additional information on the health professions.

## Annual Survey of Health Professions Education Programs

In 2006, the AMA used an online survey instrument to collect data on health professions education programs for the 2005-2006 academic year. The Internet-based survey, available at www.ama-assn.org/go/hpsurvey, helped reduce survey mailing costs and allowed for the collection of racial/ethnic program data, which had been discontinued in 1996 due to the costs and difficulty of obtaining accurate data. Aggregate survey data are published in Section IV and in the *Health Professions Education Data Book* (see below).

## Health Professions e-Letter

This monthly electronic newsletter covers educational trends and career-related issues in the professions included in the *Directory*. The *e-Letter* is distributed free of charge via e-mail to all program

and institution personnel listed in the *Directory* who opt in to receive it. To view the current issue, click on www.ama-assn.org/go/hpe-letter; to subscribe, see www.ama-assn.org/go/enews or e-mail dorothy.grant-bryant@ama-assn.org.

## Health Professions Education Data Book

Data collected on the annual survey are available in the *Health Professions Education Data Book*. For more information or to order, call 312 464-5333 or e-mail enza.perrone@ama-assn.org, or access www.ama-assn.org/go/hpdatabook.

## Health Professions Data Service/Mailing Labels

The AMA's Health Professions Data Service offers a wide range of custom data and mailing labels for the health professions. These data, both current and historical, are collected on the Annual Survey of Health Professions Education Programs. Available data variables include:

*Student data*
- data on program enrollments, attrition, and graduates by gender

*Program data*
- tuition cost(s) (total for first-year student)
- class capacity (per start date or session)
- availability of evening/weekend classes
- program length(s), in months
- program start date(s)
- credential(s) awarded
- data on percentages of recent graduates (within last 6 months) finding employment or seeking additional education
- name, address, telephone/fax numbers, and e-mail addresses of program officials.

For more information, contact Medical Education Products at 312 464-5333 or enza.perrone@ama-assn.org.

## Disclaimer

The AMA does not certify or register health personnel and does not recognize or approve certifying agencies or the professions listed in the *Directory*. The certifying agencies listed in Section I do not represent an inclusive listing. For more information on certifying agencies, contact the National Organization for Competency Assurance at www.noca.org.

## Acknowledgments

The AMA gratefully acknowledges the cooperative relationships with the participating health professions accrediting agencies, listed on the inside front cover; these relationships and the hard work of accrediting agency staff throughout the year are essential for establishing the survey population and for providing information about and updates to accredited programs. The AMA also expresses its deep appreciation for the continued cooperation of program directors, nearly all of whom respond to the Annual Survey of Health Professions Education Programs.

Acknowledgments are also due to Enza Perrone and Arecia Washington for survey and data assistance; Rod Hill for production support; and AMA customer service staff for handling the questions generated by this product. In addition, thanks are due to the following AMA Press staff who worked on this project: Suzanne Fraker, Nancy Baker, John Kinney, Amy Burgess, Ronnie Summers, Boon-Ai Tan, Jean Roberts, and Mark Daniels.

Fred Donini-Lenhoff, MA, Editor
Dorothy Grant-Bryant, Survey and Database Coordinator
Paul H Rockey, MD, MPH, Director, Division of Undergraduate and
   Graduate Medical Education

# Section I

# Occupations and Educational Programs

Section I: *Occupations and Educational Programs* offers descriptions and general information for 71 health-related professions and includes listings for 6,873 accredited educational programs in all 50 states, Puerto Rico, Canadian provinces, and some other countries.

Each profession description contains many if not all of the following elements, identified by an icon:

 **History** describes the development of the profession over time, from the origin and evolution of each profession's key duties and responsibilities, and the establishment of the first sponsoring institutions and programs, to important convocations of decision-makers, accrediting bodies, and market forces that have shaped and defined the current state of the profession. (Additional history for given professions is also available in Section III: Accrediting Agencies.)

 **Occupational Description** details the general duties of the profession within the context of the health care environment in which it exists. Background information on the area of medical care that the profession supports (eg, laboratory testing, physical rehabilitation, surgery, administration, diet and nutrition) is also included.

 **Job Description** gives a more in-depth depiction of the day-to-day activities of each profession and may list a variety of duties and responsibilities that could be expected or assigned, depending on the facility, physical location, staff needs, and work environment.

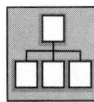 **Employment Characteristics** describes the workplace, facility, or physical location where the profession's duties are typically performed. Lengths of shifts, typical work weeks, and on-call status are also included, if available.

 **Salary ranges,** from a variety of government, professional association, and industry survey sources, are provided if available. (*Note:* Salary data are also shown, by profession, in Table 5, Section IV.)

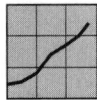 **Employment Outlook,** if available, gives statistics or trend summaries from government, professional associations, and industry surveys about overall projections for demand, growth, and staffing needs within each profession's field of activity.

 **Educational Programs** offers general information regarding the length, prerequisites, typical coursework, and specific subjects of study for accredited educational programs in the chosen profession. This section also describes the various levels of educational attainment (eg, certificate, associate's degree, bachelor's degree, master's level) that are offered, as well as those that are required, to work in the selected profession.

 **Licensure, Certification, and Registration** specifies the legal and/or professional requirements, if any, for practicing in a chosen profession. Renewal cycles for certification and continuing education standards, as well as information on national and local accrediting bodies, may also be included.

 **Inquiries** lists names, addresses, and other contact information for national professional associations, accrediting bodies, and related information resources.

The lists of accredited educational programs are in alphabetical order by state, then city, for each occupation. Most program listings include the name and address of the educational program and name and telephone number of the program director, as well as fax number and e-mail address and name of the medical director, if applicable. In addition, Web site addresses are displayed for those programs that completed the 2006 AMA Survey of Health Professions Programs (see below).

At the end of each profession's section is a data chart showing class capacity (per start date or session), month(s) classes begin, program length(s), yearly tuition cost (resident and nonresident), yearly stipend (if any), academic award(s) granted, availability of evening/weekend courses, whether a program offers education/courses in medical/health care terms in non-English languages, and whether the program offers education in cultural competence or in patient communication. These data are shown only for those programs that completed the 2006 AMA survey.

For more information about a specific educational program, contact the program officials directly.

## Data Sources; Updates to Listings

The data in each of the 6,873 program listings (Section I) and 2,468 institution listings (Section II) are the product of two processes: The cooperative relationships between the AMA and the participating health professions accrediting agencies, and the cooperation of program directors and institutional officials in responding to the Annual Survey of Health Professions Education Programs.

Program directors or institutional officials wishing to update the listings in Sections I or II should make the required changes on the annual survey, available online at www.ama-assn.org/go/hpsurvey. Updates can also be sent to the AMA at the following address:

Medical Education Products
American Medical Association
attn: Dorothy Grant-Bryant
515 N State Street
Chicago, IL 60610
312 464-4936
312 464-5830 Fax
E-mail: dorothy.grant-bryant@ama-assn.org

# Anesthesiologist Assistant

## Occupational Description

The anesthesiologist assistant (AA) functions as a specialty physician assistant under the direction of a licensed and qualified anesthesiologist, principally in medical centers. The AA assists the anesthesiologist in developing and implementing the anesthesia care plan. This may include collecting preoperative data, such as taking an appropriate health history; performing various preoperative tasks, such as the insertion of intravenous and arterial catheters and special catheters for central venous pressure monitoring, if necessary; performing airway management and drug administration for induction and maintenance of anesthesia; assisting in the administering and monitoring of regional and peripheral nerve blockade; administering supportive therapy, for example, with intravenous fluids and cardiovascular drugs; adjusting anesthetic levels on a minute-to-minute basis; performing intraoperative monitoring; providing recovery room care; and functioning in the intensive care unit. The AA may also be used in pain clinics or may participate in administrative and educational activities.

## Job Description

In addition to the duties described in the occupational description above, anesthesiologist assistants provide other support according to established protocols. Such activities may include pretesting anesthesia delivery systems and patient monitors and operating special monitors and support devices for critical cardiac, pulmonary, and neurological systems. AAs may be involved in the operation of bedside electronic computer-based monitors and have supervisory responsibilities for laboratory functions associated with anesthesia and operating room care. They provide cardiopulmonary resuscitation in association with other anesthesia care team members and in accordance with approved emergency protocols.

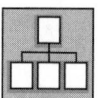

## Employment Characteristics

Anesthesiologist assistants work as members of the anesthesia care team in any locale where they may be appropriately directed by legally responsible anesthesiologists. The AAs most often work within organizations that also employ nurse anesthetists, and their responsibilities are identical. Experience to date has been that AAs are most commonly employed in larger facilities that perform procedures such as cardiac surgery, neurosurgery, transplant surgery, and trauma care, given the training in extensive patient monitoring devices and complex patients and procedures emphasized in AA educational programs. However, AAs are used in hospitals of all sizes and assist anesthesiologists in a variety of settings and for a wide range of procedures.

## Salary

Starting salaries for 2006 graduates are $95,000 up to $120,000 for the 40-hour work week plus benefits and consideration of on-call activity. The high end of the salary range is around $160,000 to $180,000 for experienced anesthetists (including overtime). Refer to Section IV, Table 5 of this *Directory* for more information, or see www.ama-assn.org/go/ salary.

## Educational Programs

**Length.** These postbaccalaureate programs are essentially 24 to 28 months.

**Prerequisites.** The programs require an undergraduate premedical background (premedical courses in biology, chemistry, physics, and math) and a baccalaureate degree. Although any baccalaureate major is acceptable (if premedical requirements are met), majors typically are biology, chemistry, physics, mathematics, computer science, or one of the allied health professions, such as respiratory therapy, medical technology, or nursing.

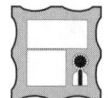

## Licensure, Certification, and Registration

The National Commission for Certification of Anesthesiologist Assistants (NCCAA), which was founded in 1989, provides the certification process for AAs in the United States. The Commission includes anesthesiologists and anesthesiologist assistants. NCCAA contracts with the National Board of Medical Examiners to assist with the certification process, including task analyses, development of content grids, item writing and editing, and administration of examinations. The certification process incorporates a 6-year cycle of certifying examination, examination for continued demonstration of qualifications, and registration of continuing medical education.

## Inquiries

**Careers.** Curriculum inquiries should be directed to the individual programs. Inquiries regarding AA careers should be directed to

American Academy of Anesthesiologist Assistants
PO Box 13978
Tallahassee, FL 32317
866 328-5858
850 656-3038 Fax
www.anesthetist.org

**Certification**
National Commission for Certification of Anesthesiologist
    Assistants
PO Box 15519
Atlanta, GA 30333-0519
404 687-9978 Fax

**Program Accreditation**
Commission on Accreditation of Allied Health Education Programs
    (CAAHEP) in collaboration with:
Accreditation Review Committee for the Anesthesiologist Assistant
Brad Maxwell
c/o CAAHEP
1361 Park Street
Clearwater, FL 33756
727 210-2350

## Anesthesiologist Assistant

### Florida

**Nova Southeastern University**
Anesthesiologist Asst Prgm
3200 S University Dr
Ft Lauderdale, FL 33328-2018
*Prgm Dir:* Robert Wagner, MMSc AA-C
*Tel:* 954 262-1166   *Fax:* 954 262-1181
*E-mail:* rwagner@nsu.nova.edu

### Georgia

**Emory University**
Anesthesiologist Asst Prgm
101 Woodruff Circle Rm 617
Atlanta, GA 30322
www.emoryaaprogram.org
*Prgm Dir:* Richard Brouillard, ScD
*Tel:* 404 727-3188   *Fax:* 404 727-3021
*E-mail:* rick_brouillard@emoryhealthcare.org
*Med Dir:* James R Hall, MD

**South University**
*Cosponsor:* Mercer University School of Medicine
Anesthesiologist Asst Prgm
709 Mall Blvd
Savannah, GA 31406
www.southuniversity.edu
*Prgm Dir:* A William Paulsen, MMSc PhD CCE AAC
*Tel:* 912 201-8000   *Fax:* 912 201-8070
*E-mail:* bpaulsen@southuniversity.edu
*Med Dir:* Stacie Wong, MD, Board Certification in
   Anesthesiology

### Ohio

**Case Western Reserve University**
Anesthesiologist Asst Prgm
11100 Euclid Ave
Lakeside Hospital Rm 2532
Cleveland, OH 44106-5007
www.anesthesiaprogram.com
*Prgm Dir:* Joseph Rifici, AA-C MEd
*Tel:* 216 844-3161   *Fax:* 216 844-7349
*E-mail:* joseph.rifici@uhhs.com
*Med Dir:* Donald Voltz, MD

## Anesthesiologist Assistant

| Programs* | Class Capacity | Begins | Length (months) | Award | Res Tuition | Non-res Tuition | Stipend | Offers:‡ 1 | 2 | 3 |
|---|---|---|---|---|---|---|---|---|---|---|
| **Georgia** | | | | | | | | | | |
| Emory University (Atlanta) | 30 | Jun | 27 | MMSc | $26,400 | $26,400 | | | | |
| South University (Savannah) | 25 | Jun | 28 | MMSc | $24,380 | $24,380 | | | | • |
| **Ohio** | | | | | | | | | | |
| Case Western Reserve University (Cleveland) | 16 | Jun | 24 | MS | $33,000 | $33,000 | | | | • |

*Data are shown only for programs that completed the 2006 AMA Survey of Health Professions Education Programs
‡Key to Offers: 1: Evening or weekend classes; 2: Non-English instruction; 3: Cultural competence instruction

# Art Therapist

### History

Art therapy emerged as a distinct profession in the 1940s when hospitals and rehabilitation facilities increasingly began to include art therapy programs along with traditional "talk therapies," underscoring the recognition that art making enhanced recovery, health, and wellness. Since that time, the profession of art therapy has grown into an effective and important method of treatment and assessment with children, adults, families, and groups in a variety of settings. Currently, the field of art therapy has gained attention in health care facilities throughout the United States and within psychiatry, medicine, psychology, counseling, education, and the arts.

The American Art Therapy Association (AATA), the official member organization for professionals and students, was founded in 1969 to develop and promote educational, professional, and ethical standards for the field of art therapy. The AATA sponsors annual conferences, approves educational programs, and publishes *Art Therapy: Journal of the American Art Therapy Association* (first published in 1983), the quarterly *AATA Newsletter,* the *AATA E-Newsletter,* books, and monographs. In the 1970s, the first master's degrees in art therapy were awarded. Today, there are 34 AATA-approved master's degree programs, and university and college curricula across the US include undergraduate introductory courses and preparatory programs in art therapy.

### Occupational Description

Art therapy is a mental health profession that uses the creative process of art making to improve and enhance the physical, mental, and emotional well being of individuals of all ages. It is based on the belief that the creative process of artistic self-expression helps people to resolve conflicts and problems, develop interpersonal skills, manage behavior, reduce stress, increase self-esteem and self-awareness, and achieve insight.

Art therapy integrates the fields of human development, visual art (drawing, painting, sculpture, and other art forms), and the creative process with models of counseling and psychotherapy. Art therapy is used with children, adolescents, adults, older adults, groups, and families to assess and treat the following: anxiety, depression, and other mental and emotional problems and disorders; substance abuse and other addictions; family and relationship issues; abuse and domestic violence; social and emotional difficulties related to disability and illness; trauma and loss; physical, cognitive, and neurological problems; and psychosocial difficulties related to medical illness. Art therapy programs and art therapists are found in a variety of settings including hospitals, clinics, public, social service, and community agencies, wellness centers, educational institutions, businesses, and private practices.

### Job Description

Art therapists use drawing, painting, and other art processes to assess and treat clients with emotional, cognitive, physical, and/or developmental needs and disorders. Using their skills in evaluation and psychotherapy, they choose materials and interventions appropriate to their clients' needs and design sessions to achieve therapeutic goals and objectives. Art therapists also maintain appropriate charts, records, and periodic reports on client progress as required by agency guidelines and professional standards; participate in professional staff meetings and conferences; and provide information and consultation regarding the client's clinical progress. They also may function as supervisors, administrators, consultants, and expert witnesses.

With the growing acceptance of complementary therapies and recent research findings on art therapy with medical populations, there is an increase in the application of art therapy to a variety of patient groups. For example, art therapists work with cancer, burn, pain, post-surgery, HIV-positive, asthma, and substance abuse patients, among others, and with pediatric, geriatric, and other medical populations. In hospitals, art therapists may be part of psychiatric departments, child life programs, arts in hospital programs, or creative arts therapies or activity therapies departments. Many art therapists also hold credentials in mental health counseling or marriage and family therapy because of their training and experience.

An understanding of the application of various art media and art processes to treatment is central to the practice of art therapy. In general, an art therapist must be sensitive to a variety of human needs and possess emotional stability, patience, interpersonal skills, and a capacity for insight into psychological processes. An art therapist also must be an attentive listener and keen observer and be able to develop a rapport with people. Flexibility and a sense of humor are important in adapting to work with people with a wide range of mental health and healthcare needs and in a variety of settings.

### Employment Characteristics

Art therapists work in a many different healthcare environments, including, but not limited to, the following: hospitals and clinics (medical and psychiatric); outpatient mental health agencies and day treatment facilities; residential treatment centers; domestic violence and homeless shelters; community agencies and nonprofit settings; sheltered workshops; correctional facilities; elder care facilities; art studios; private practice; and schools, colleges, and universities.

An art therapist may work as part of a team that includes physicians, psychologists, nurses, mental health counselors, marriage and family therapists, rehabilitation counselors, social workers, and/or teachers. Together, they determine each client's therapeutic goals and objectives and implement a treatment plan. Other art therapists work independently and maintain private practices with children, adolescents, adults, groups, and/or families.

### Salary

Earnings for art therapists vary depending on type of practice, job responsibilities, and practice location. The average entry-level income is approximately $32,000, median income is between $38,000 and $48,000, and top earning potential for salaried administrators is between $50,000 and $80,000. Art therapists who possess doctoral degrees and/or licensure or who qualify in their state to conduct a private practice can earn an average of $85 to $120 per hour as an independent practitioner. Refer to Section IV, Table 5 of this *Directory* for more information, or see www.ama-assn.org/go/salary.

### Educational Programs

Length. Art therapy master's degree programs are no less than 2 years and must include a minimum of 24 graduate credit hours in the art therapy core curriculum.

**Prerequisites.** Applicants to master's degree programs must hold a baccalaureate degree from an accredited US institution or have equivalent academic preparation from an institution outside the United States. In addition, prospective students must submit a portfolio of original artwork and must document 15 semester hours in studio art and 12 semester hours in psychology.

**Curriculum.** Educational requirements include theories of art therapy, counseling, and psychotherapy; psychopathology; ethics and standards of practice; assessment and evaluation; individual, group, and family techniques; human and creative development; multicultural issues; research methods; and practicum experiences in clinical, community, and/or other settings.

### Licensure, Certification, and Registration

The Art Therapy Credentials Board (ATCB) Inc, an independent organization, grants a credential, Art Therapist Registered (ATR), after reviewing documentation of completion of graduate education and supervised post-graduate experience. Registered art therapists who successfully complete a written examination administered by the ATCB are qualified as Board Certified (ATR-BC). Recertification is required every 5 years by documentation of continuing education credits (CECs).

### Inquiries

**Education, Program Approval, Careers, Resources**
American Art Therapy Association (AATA), Inc
5999 Stevenson Avenue
Alexandria, VA 22304
888 290-0878
703 212-2238
E-mail: info@arttherapy.org
www.arttherapy.org

---

## Art Therapist*

## California

**Notre Dame de Namur University**
Art Therapy Prgm
1500 Ralston Ave
Belmont, CA 94002
*Prgm Dir:* Richard Carolan, EdD
*Tel:* 650 508-3784   *Fax:* 650 508-3426
*E-mail:* arttherapy@ndnu.edu

**Phillips Graduate Institute**
Art Therapy Prgm
5445 Balboa Blvd
Encino, CA 91316-1509
www.pgi.edu
*Prgm Dir:* Noah Hass-Cohen, MA ATR-BC LMFT
*Tel:* 818 386-5611
*E-mail:* noah@pgi.edu

**Loyola Marymount University**
Art Therapy Prgm
Grad Dept of Marital and Family Therapy
1 LMU Dr
Los Angeles, CA 90045-2659
*Prgm Dir:* Debra Linesch, PhD MFT ATR-BC
*Tel:* 310 338-4562   *Fax:* 310 338-4518
*E-mail:* dlinesch@lmu.edu

## Colorado

**Naropa University**
Art Therapy Prgm
2130 Arapahoe Ave
Boulder, CO 80302
*Prgm Dir:* Michael Franklin, MA ATR-BC
*Tel:* 303 546-3545
*E-mail:* michaelf@naropa.edu

---

*Additional information about programs that returned the AMA's survey is available in a table beginning on page 7.

## Connecticut

**Albertus Magnus College**
Art Therapy Prgm
700 Prospect St
New Haven, CT 06511
www.albertus.edu
*Prgm Dir:* William More, MA ATR-BC
*Tel:* 203 773-8903   *Fax:* 203 773-3117
*E-mail:* wmore@albertus.edu

## District of Columbia

**George Washington University**
Art Therapy Prgm
2129 G St NW, Bldg L, Rear
Washington, DC 20052
www.gwu.edu/~artx
*Prgm Dir:* Heidi Bardot, MA ATR-BC
*Tel:* 202 994-6285   *Fax:* 202 994-1404
*E-mail:* hbardot@gwu.edu

## Florida

**Florida State University**
Art Therapy Prgm
126 Milton Carothers Hall
Tallahassee, FL 32306-4480
*Prgm Dir:* Marcia Rosal, PhD
*Tel:* 850 644-5473
*E-mail:* mrosal@mailer.scns.fsu.edu

## Illinois

**Adler School of Professional Psychology**
Art Therapy Prgm
65 E Wacker Pl, Ste 2100
Chicago, IL 60601-7298
www.adler.edu
*Prgm Dir:* Judy Sutherland, PhD ATR-BC LCPC
*Tel:* 312 201-5900, Ext 220   *Fax:* 312 201-5917
*E-mail:* jhs@adler.edu

**School of the Art Institute of Chicago**
Art Therapy Prgm
Dept of Art Therapy
37 S Wabash
Chicago, IL 60603-3103
*Prgm Dir:* Randy M Vick, MS ATR-BC LCPC
*Tel:* 312 899-7481   *Fax:* 312 899-1477
*E-mail:* rvick@artic.edu

## Southern Illinois Univ at Edwardsville
Art Therapy Prgm
Dept of Art and Design
Box 1764
Edwardsville, IL 62026-1764
www.siue.edu/ART/AREAS/art_therapy
*Prgm Dir:* P Gussie Klorer, PhD ATR-BC LCSW LCPC
*Tel:* 618 650-3183   *Fax:* 618 650-3106
*E-mail:* pklorer@siue.edu

## Kansas

**Emporia State University**
Art Therapy Prgm
1200 Commercial
Campus Box 4031
Emporia, KS 66801
*Prgm Dir:* Nancy Slater, PhD ATR
*Tel:* 620 341-5814
*E-mail:* slaterna@emporia.edu

## Kentucky

**University of Louisville**
Art Therapy Prgm
Coll of Education & Human Dev Expressive Therapies
Louisville, KY 40292
*Prgm Dir:* Laura Cherry, PhD ATR
*Tel:* 502 852-8796   *Fax:* 502 852-4598
*E-mail:* etemail@louisville.edu

## Massachusetts

**Lesley University**
Art Therapy Prgm
Expressive Therapies
29 Everett St
Cambridge, MA 02138
*Prgm Dir:* Susan Spaniol, ATR-BC
*Tel:* 617 349-8436, Ext 8432   *Fax:* 617 349-8431
*E-mail:* sspaniol@mail.lesley.edu

**Springfield College**
Art Therapy Prgm
Springfield Coll Visual & Performing Arts
263 Alden St
Springfield, MA 01109-3797
*Prgm Dir:* Simone Alter-Muri, EdD
*Tel:* 413 748-3752   *Fax:* 413 748-3580
*E-mail:* saltermu@spfldcol.edu

## Michigan

**Wayne State University**
Art Therapy Prgm
163 Community Arts Bldg
Detroit, MI 48202
*Prgm Dir:* Holly Feen, PhD ATR-BC
*Tel:* 313 577-1823   *Fax:* 313 993-7558
*E-mail:* hfeen@wayne.edu

## New Mexico

**Southwestern College**
Art Therapy Prgm
PO BOX 4788
Santa Fe, NM 87502-4788
www.swc.edu
*Prgm Dir:* Deborah Schroder, MS ATR-BC LPAT LPAT
*Tel:* 505 471-5756, Ext 22   *Fax:* 505 471-4071
*E-mail:* info@swc.edu

## New York

**Pratt Institute**
Art Therapy Prgm
200 Willoughby Ave
East 3
Brooklyn, NY 11205
www.pratt.edu
*Prgm Dir:* Laurel Thompson, MPS ADTR ATR-BC LCAT
*Tel:* 718 636-3428   *Fax:* 718 636-3597
*E-mail:* lthompso@pratt.edu

**Long Island University - C W Post Campus**
Art Therapy Prgm
720 Northern Blvd
Brookville, NY 11548
*Prgm Dir:* Christine Kerr, PhD ATR-BC CGP
*Tel:* 516 299-2935, Ext 2944   *Fax:* 516 299-2858
*E-mail:* christine.kerr@liu.edu

**Hofstra University**
Art Therapy Prgm
124 Hofstra University
Hempstead, NY 11549-1240
*Prgm Dir:* Joan Bloomgarden, PhD ATR-BC CGP
*Tel:* 516 463-5752   *Fax:* 516 463-6184
*E-mail:* Joan.S.Bloomgarden@Hofstra.edu

**College of New Rochelle**
Art Therapy Prgm
29 Castle Pl
Chapel Hall G 15
New Rochelle, NY 10805
www.cnr.edu
*Prgm Dir:* Patricia St John, EdD ATR-BC
*Tel:* 914 654-5280   *Fax:* 914 654-5593
*E-mail:* pstjohn@cnr.edu

**New York University**
Art Therapy Prgm
34 Stuyvesant St
New York, NY 10003
*Prgm Dir:* Ikuko Acosta, MA ATR-BC
*Tel:* 212 998-5726   *Fax:* 212 995-4320
*E-mail:* ia4@nyu.edu

**Nazareth College of Rochester**
Art Therapy Prgm
4245 East Ave
Rochester, NY 14618-3790
www.naz.edu/dept/art_therapy
*Prgm Dir:* Ellen G Horovitz, PhD ATR-BC
*Tel:* 585 389-2535   *Fax:* 585 586-2452
*E-mail:* ehorovi4@naz.edu

## Ohio

**Ursuline College**
Art Therapy Prgm
2550 Lander Rd
Pepper Pike, OH 44124
*Prgm Dir:* Gail Rule-Hoffman, MEd ATR-BC
*Tel:* 440 646-8139   *Fax:* 440 684-6135
*E-mail:* grulehof@ursuline.edu

## Oregon

**Marylhurst University**
Art Therapy Prgm
Dept of Graduate Studies in Art Therapy Counseling
17600 Pacific Hwy - PO Box 261
Marylhurst, OR 97036
www.marylhurst.edu
*Prgm Dir:* Christine Turner, ATR-BC LPC NCC ACS
*Tel:* 503 699-6244   *Fax:* 503 534-4062
*E-mail:* cturner@marylhurst.edu

## Pennsylvania

**Seton Hill University**
Art Therapy Prgm
PO Box 467F Seton Hill Dr
Greensburg, PA 15601
*Prgm Dir:* Nina Denninger, MA ATR-BC
*Tel:* 724 830-1047, Ext 4368   *Fax:* 724 830-1295
*E-mail:* denninger@setonhill.edu

**Drexel University**
Art Therapy Prgm
245 N 15th St
Philadelphia, PA 19102-1192
www.drexel.edu/cnhp/creative_arts/about.asp
*Prgm Dir:* Nancy Gerber, PhD ATR-BC LPC
*Tel:* 215 762-6928   *Fax:* 215 762-6933
*E-mail:* nancy.gerber@drexel.edu

**Marywood University**
Art Therapy Prgm
Art Dept
2300 Adams Ave
Scranton, PA 18509
www.marywood.edu
*Prgm Dir:* Barbara Parker-Bell, ATR-BC LPC
*Tel:* 570 348-6278, Ext 2525   *Fax:* 570 340-6023
*E-mail:* parkerbell@marywood.edu

## Quebec, Canada

**Concordia University**
Art Therapy Prgm
1455 de Maisonneuve Blvd W
Montreal, QC H3G 1M8
*Prgm Dir:* Josee Leclerc, PhD ATR ATPQ
*Tel:* 514 848-4683   *Fax:* 514 848-4790
*E-mail:* cats@vax2.concordia.ca

## Virginia

**Eastern Virginia Medical School**
Art Therapy Prgm
PO Box 1980
Norfolk, VA 23501
*Prgm Dir:* Kay Stovall, ATR-BC MFT
*Tel:* 757 446-5895   *Fax:* 757 446-6179
*E-mail:* ARTthrpy@evms.edu

## Washington

**Antioch University - Seattle**
Art Therapy Prgm
2326 Sixth Ave
Seattle, WA 98121-1814
*Prgm Dir:* Janice Hoshino, PhD
*Tel:* 206 268-4818
*E-mail:* jhoshino@antiochsea.edu

## Wisconsin

**Mount Mary College**
Art Therapy Prgm
2900 N Menomonee River Pkwy
Milwaukee, WI 53222-4545
www.mtmary.edu
*Prgm Dir:* Bruce Moon, PhD ATR-BC
*Tel:* 414 256-1215, Ext 302   *Fax:* 414 256-1205
*E-mail:* moonb@mtmary.edu

## Art Therapist

| Programs* | Class Capacity | Begins | Length (months) | Award | Res Tuition | Non-res Tuition | Stipend | Offers:‡ 1 | 2 | 3 |
|---|---|---|---|---|---|---|---|---|---|---|
| **California** | | | | | | | | | | |
| Notre Dame de Namur University (Belmont) | 25 | Sep Jan May | 24 | MAAT, MAMFT | $20,000 | $0 | | • | | • |
| Phillips Graduate Institute (Encino) | 50 | Sep Jan | 24, 36 | MA | $675 | $675 | | • | | • |
| **Connecticut** | | | | | | | | | | |
| Albertus Magnus College (New Haven) | 16 | Aug Jan | 28 | Cert, MA | $12,500 | $12,500 | | • | | • |
| **District of Columbia** | | | | | | | | | | |
| George Washington University | 20 | Aug | 24, 48 | Dipl, Cert, MA | $25,000 | $25,000 | $17,000 | • | | • |
| **Illinois** | | | | | | | | | | |
| Adler School of Professional Psychology (Chicago) | 15 | Fall Winter | 24 | Dipl, Cert, MA, PM, CAT | | $0 | | • | | • |
| Southern Illinois Univ at Edwardsville | 10 | Aug | 24 | Dipl, Cert, MA | $3,216 | $6,432 | | | | • |
| **Massachusetts** | | | | | | | | | | |
| Springfield College | 20 | Sep | | MS | $16,296 | $0 | | • | | • |
| **New Mexico** | | | | | | | | | | |
| Southwestern College (Santa Fe) | 65 | Fall (Fulltime) Wint. | 24, 30 | Dipl, Cert, MA | $16,677 | $16,677 | | • | | • |
| **New York** | | | | | | | | | | |
| College of New Rochelle | 25 | Sep Jan May Jul | 24, 36 | Dipl, MS | $14,662 | $14,662 | | • | | |
| Nazareth College of Rochester | 25 | Sep May | 24 | Dipl, MS, MCAT | $17,820 | $17,820 | | • | | • |
| Pratt Institute (Brooklyn) | 36 | Sep Mar Jun | 24, 30 | MPS | $1,010 | $1,010 | | | | • |
| **Oregon** | | | | | | | | | | |
| Marylhurst University | 20 | Sep | 21 | Cert, MA | $14,500 | $14,500 | | • | | • |
| **Pennsylvania** | | | | | | | | | | |
| Drexel University (Philadelphia) | 21 | Aug | 24 | MA | $15,000 | $0 | | | | • |
| Marywood University (Scranton) | 12 | Sep Jan | 27 | Cert, MA | $20,160 | $20,160 | | • | | • |
| **Wisconsin** | | | | | | | | | | |
| Mount Mary College (Milwaukee) | 30 | Sep | 48, 24 | BA, MS | $11,040 | $11,040 | | • | | • |

*Data are shown only for programs that completed the 2006 AMA Survey of Health Professions Education Programs

‡Key to Offers: 1: Evening or weekend classes; 2: Non-English instruction; 3: Cultural competence instruction

PROGRAMS

## Occupational Description

The athletic trainer, with the consultation and direction of attending and/or consulting physicians, is an integral part of the health care system associated with physical activity and sports. Through preparation in both academic education and clinical experience, the athletic trainer provides a variety of services, including injury prevention, assessment, immediate care, treatment, and rehabilitation after physical injury or trauma.

## Job Description

Past role delineation studies have concluded that the role of an athletic trainer includes, but may not be limited to:

- Prevention
- Recognition, evaluation, and assessment
- Immediate care
- Treatment, rehabilitation, and reconditioning
- Organization and administration
- Professional development

## Employment Characteristics

Athletic trainers typically provide their services in one or more of the following settings: secondary schools, colleges and universities, professional athletic organizations, physician offices, hospital-based clinics, private sports medicine and rehabilitation clinics, industrial/occupational, military, and performing arts.

## Salary

According to the National Athletic Trainers' Association, entry-level salaries in 2002 averaged $35,000. The average salary with less than 5 years experience is $40,500. Refer to Section IV, Table 5 of this *Directory* for more information, or see www.ama-assn.org/go/salary.

## Educational Programs

**Length.** Baccalaureate degree programs require 4 years of study. Postbaccalaureate programs are generally 2 years.

**Prerequisites.** Applicants for the 4-year baccalaureate degree programs must have a high school diploma or equivalent and meet institutional entrance requirements. Applicants for postbaccalaureate programs should have a baccalaureate degree that includes appropriate course work and clinical experience, as specified by the institution.

**Curriculum.** The professional curriculum includes formal instruction in risk management and injury/illness prevention, pathology of injury/illness, assessment of injury/illness, general medical conditions and disabilities, therapeutic modalities, therapeutic exercise, rehabilitative techniques, health care administration, weight management and body composition, psychosocial intervention and referral, medical ethics and legal issues, pharmacology, and professional development and responsibilities. The didactic curriculum is augmented by a series of structured laboratory and clinical experiences.

## Professional Certification

Almost all states require that athletic trainers hold the ATC® (Athletic Trainer, Certified) credential, which is issued by the Board of Certification Inc. The ATC credential is supported by three pillars: the BOC certification examination, BOC Standards of Practice and Disciplinary Process, and continuing competence requirements. The 1-day, three-part examination verifies that the knowledge, skills, and abilities required for competent performance as an entry-level athletic trainer have been met.

## Inquiries

**Careers**
National Athletic Trainers' Association, Inc
2952 Stemmons Freeway, Suite 200
Dallas, TX 75247
214 637-6282
800 TRY-NATA
214 637-2206 Fax
www.NATA.org

**Certification**
Board of Certification Inc
BOC Administrative Offices
4223 S 143rd Circle
Omaha, NE 68137
402 559-0091
402 561-0598 Fax
www.BOCATC.org

**Program Accreditation**
Commission on Accreditation of Athletic Training Education (CAATE)
2201 Double Creek Drive, Suite 5006
Round Rock, TX 78664
512 733-9700
512 733-9700 Fax
E-mail: caate@sbcglobal.net
www.caate.net

## Athletic Trainer*

## Alabama

**Samford University**
Athletic Training Prgm
800 Lakeshore Dr
PO Box 292448
Birmingham, AL 35229
*Prgm Dir:* Chris A Gillespie, MEd ATC LAT
*Tel:* 205 726-2379   *Fax:* 205 726-2607
*E-mail:* cagilles@samford.edu
*Med Dir:* Lawrence J Lemak, MD

**University of West Alabama**
Athletic Training Prgm
Station 14
Livingston, AL 35470
http://at.uwa.edu
*Prgm Dir:* R T Floyd, EdD ATC LAT CSCS
*Tel:* 205 652-3714   *Fax:* 205 652-3799
*E-mail:* rtf@uwa.edu
*Med Dir:* E Lyle Cain, MD

**University of Mobile**
Athletic Training Prgm
5735 College Pkwy
Mobile, AL 36663-0220
*Prgm Dir:* William Carroll, EdD JD ATC
*Tel:* 251 442-2324   *Fax:* 251 442-2519
*E-mail:* bcarroll@mail.umobile.edu
*Med Dir:* Anthony Tropeano, MD

**Huntingdon College**
Athletic Training Prgm
1500 E Fairview Ave
Montgomery, AL 36106
www.huntingdon.edu
*Prgm Dir:* Wanda Swiger, EdD
*Tel:* 334 833-4455   *Fax:* 334 833-4206
*E-mail:* wswiger@huntingdon.edu
*Med Dir:* Charles Hartzog, MD

**Troy University**
Athletic Training Prgm
3212 Veterans Stadium Dr, 2nd Fl
Dept of Athletic Training Education
Troy, AL 36082
*Prgm Dir:* John Anderson, MS BS ATC
*Tel:* 334 670-3722, Ext 3722   *Fax:* 334 670-3870
*E-mail:* athtrain@troy.edu
*Med Dir:* James Whiteside, MD

**University of Alabama**
Athletic Training Prgm
Dept of Health Science
Box 870311
Tuscaloosa, AL 35487-0311
*Prgm Dir:* Deidre Leaver-Dunn, PhD ATC
*Tel:* 205 348-9176   *Fax:* 205 348-7568
*E-mail:* dleaver@bama.ua.edu
*Med Dir:* James B Robinson, MD

## Arizona

**Northern Arizona University**
Athletic Training Prgm
PO Box 15094
Flagstaff, AZ 86011
www.nau.edu/hp/at
*Prgm Dir:* Debbie Craig, PhD ATC LAT
*Tel:* 928 523-0704   *Fax:* 928 523-4315
*E-mail:* debbie.craig@nau.edu
*Med Dir:* L George Hershey, DO

**Grand Canyon University**
Athletic Training Prgm
3300 W Camelback Rd
Phoenix, AZ 85017
*Prgm Dir:* Cindy Seminoff, MS ATC CSCS
*Tel:* 602 589-2741   *Fax:* 602 589-2716
*E-mail:* cseminoff@gcu.edu
*Med Dir:* Craig Phelps, DO FAOASM

## Arkansas

**Henderson State University**
Athletic Training Prgm
HSU Box 7552
Arkadelphia, AR 71999-0001
www.hsu.edu/atep
*Prgm Dir:* John K Miller, MSE ATC
*Tel:* 870 230-5360   *Fax:* 870 230-5073
*E-mail:* millerj@hsu.edu
*Med Dir:* James Bryan, MD

**Ouachita Baptist University**
Athletic Training Prgm
Box 3700
Arkadelphia, AR 71998
www.obu.edu/atep
*Prgm Dir:* Terry DeWitt, PhD ATC
*Tel:* 870 245-5264   *Fax:* 870 245-5241
*E-mail:* dewittt@obu.edu
*Med Dir:* Wesley Kluck, MD

**University of Central Arkansas**
Athletic Training Prgm
Prince Center 133E
201 Donaghey Ave
Conway, AR 72035-0001
www.uca.edu/divisions/academic/kped/welcome.html
*Prgm Dir:* Ellen Epping
*Tel:* 501 450-5112   *Fax:* 501 450-5087
*E-mail:* eepping@uca.edu
*Med Dir:* John D Smith, MD

**University of Arkansas**
Athletic Training Prgm
HPER 308h
Fayetteville, AR 72701
*Prgm Dir:* Jeffrey A Bonacci, DA ATC
*Tel:* 479 575-4112   *Fax:* 479 575-3119
*E-mail:* bonacci@uark.edu
*Med Dir:* Al Gordonm, MD

**Southern Arkansas University**
Athletic Training Prgm
100 E University
PO Box 9299
Magnolia, AR 71754
*Prgm Dir:* Jan Kiilsgaard, MS LAT ATC
*Tel:* 870 235-4161   *Fax:* 870 235-4988
*E-mail:* jmkiilsgaard@saumag.edu
*Med Dir:* Gregg Massanelli, MD

**Arkansas State University**
Athletic Training Prgm
PO Box 240
State University, AR 72467
www.clt.astate.edu/hpess/AthleticTrainingProgram/athome.html
*Prgm Dir:* Matthew Comeau, PhD LAT ATC CSCS
*Tel:* 870 972-3066   *Fax:* 870 972-3096
*E-mail:* mcomeau@astate.edu
*Med Dir:* Brian Dixon MD, Spencer Guinn, MD

## California

**Humboldt State University**
Athletic Training Prgm
1 Harpst St
Arcata, CA 95521
www.humboldt.edu
*Prgm Dir:* Thomas "TK" Koesterer, PhD ATC
*Tel:* 707 826-5967   *Fax:* 707 826-5480
*E-mail:* tk@humboldt.edu
*Med Dir:* Raymond Koch, MD

**Azusa Pacific University**
Athletic Training Prgm
Dept of Exercise and Sport Science
701 E Foothill Blvd
Azusa, CA 91702
www.apu.edu/bas/exercisesport/atep
*Prgm Dir:* Christopher Schmidt, PhD ATC
*Tel:* 626 815-6000, Ext 5194   *Fax:* 626 815-5084
*E-mail:* cschmidt@apu.edu
*Med Dir:* Wayne Bowden, MD

**Vanguard University**
Athletic Training Prgm
55 Fair Dr
Costa Mesa, CA 92626
*Prgm Dir:* Terry Zeigler, EdD ATC
*Tel:* 714 556-3610, Ext 280   *Fax:* 714 668-6144
*E-mail:* tzeigler@vanguard.edu
*Med Dir:* Peter Kim, MD

**California State University - Fresno**
Athletic Training Prgm
5275 N Campus Dr
Fresno, CA 93740-8018
*Prgm Dir:* Robert Pettitt, PhD
*Tel:* 209 278-7389   *Fax:* 209 278-7010
*E-mail:* rpettitt@csufresno.edu

**California State University - Fullerton**
Athletic Training Prgm
PO Box 6870
Fullerton, CA 92834-6870
http://hdcs.fullerton.edu/at
*Prgm Dir:* Robert Kersey, PhD ATC CSCS
*Tel:* 714 278-3430   *Fax:* 714 278-5317
*E-mail:* rkersey@fullerton.edu
*Med Dir:* Chris Koutures, MD

**Concordia Univesity - Irvine**
Athletic Training Prgm
1530 Concordia West
Irvine, CA 92612
www.cui.edu
*Prgm Dir:* Christopher S Teich, MS ATC
*Tel:* 949 854-8002, Ext 1445   *Fax:* 949 854-6771
*E-mail:* chris.teich@cui.edu
*Med Dir:* Scott Graham, MD

**University of La Verne**
Athletic Training Prgm
1950 Third St
La Verne, CA 91750
www.ulv.edu/athletictraining
*Prgm Dir:* Marilyn Oliver, MS ATC
*Tel:* 909 593-3511, Ext 4270   *Fax:* 909 392-2760
*E-mail:* oliverm@ulv.edu
*Med Dir:* Christopher Chalian, MD

**California State University - Long Beach**
Athletic Training Prgm
Department of Kinesiology
1250 Bellflower Blvd
Long Beach, CA 90840
*Prgm Dir:* Keith Freesemann, EdD
*Tel:* 562 985-4669
*E-mail:* kfreesemn@sculb.edu

*Additional information about programs that
returned the AMA's survey is available in a table
beginning on page 24.

**California State University - Northridge**
Athletic Training Prgm
18111 Nordhoff St
Department of Kinesiology
Northridge, CA 91330-8287
www.csun.edu
*Prgm Dir:* Shane Stecyk, PhD ATC CSCS
*Tel:* 818 677-3205, Ext 4738   *Fax:* 818 677-3207
*E-mail:* shane.stecyk@csun.edu
*Med Dir:* Eric Sletten, MD

**Chapman University**
Athletic Training Prgm
One University Dr
Orange, CA 92866
www.chapman.edu/educ/atep
*Prgm Dir:* Ky Kugler, EdD ATC
*Tel:* 714 997-6642   *Fax:* 714 997-6991
*E-mail:* kekugler@chapman.edu
*Med Dir:* Miguel Prietto, MD

**California State University - Sacramento**
Athletic Training Prgm
CSUS 6000 J St
Sacramento, CA 95819
www.hhs.csus.edu/KHS/AthleticTraining
*Prgm Dir:* Doris E Fennessy Flores, MS
*Tel:* 916 278-6401   *Fax:* 916 278-7664
*E-mail:* floresde@csus.edu
*Med Dir:* Alan Hirahara, MD

**Point Loma Nazarene University**
Athletic Training Prgm
3900 Lomaland Dr
San Diego, CA 92106
*Prgm Dir:* Leon Kugler, PhD
*Tel:* 619 849-2376   *Fax:* 619 849-2553
*E-mail:* lkugler@ptloma.edu

**San Diego State University**
Athletic Training Prgm
5500 Campanile Dr
Exercise and Nutrition Science Department
San Diego, CA 92182
www.rohan.sdsu.edu/dept/athletic/athletictraining
*Prgm Dir:* Marcia Klaiber, MA ATC
*Tel:* 619 594-4094, Ext n/a   *Fax:* 619 594-6553
*E-mail:* atadvise@mail.sdsu.edu
*Med Dir:* Gregory Gastaldo, MD

**San Jose State University**
Athletic Training Prgm
One Washington Square
Department of Kinesiology
San Jose, CA 95192-0054
www.sjsu.edu/at
*Prgm Dir:* KyungMo Han, PhD ATC CSCS
*Tel:* 408 924-3041   *Fax:* 408 924-3053
*E-mail:* han@kin.sjsu.edu
*Med Dir:* Dan Haber, MD

**University of the Pacific**
Athletic Training Prgm
Dept of Sport Sciences
3601 Pacific Ave
Stockton, CA 95211
*Prgm Dir:* Sharon West, PhD ATC
*Tel:* 209 946-3182   *Fax:* 209 946-3225
*E-mail:* swest@pacific.edu
*Med Dir:* Carlos Meza, MD

**California Lutheran University**
Athletic Training Prgm
60 W Olsen Rd, MC 3400
Thousand Oaks, CA 91360
*Prgm Dir:* James Hand, MS ATC
*Tel:* 805 493-3288   *Fax:* 805 493-3860
*E-mail:* jwhand@clunet.edu
*Med Dir:* Melvin Hayashi, MD

# Colorado

**Metropolitan State College of Denver**
Athletic Training Prgm
Campus Box 25
PO Box 173362
Denver, CO 80217-3362
*Prgm Dir:* Genia Lemonedes, MS ATC
*Tel:* 303 556-3228   *Fax:* 303 556-8301
*E-mail:* lemonede@mscd.edu
*Med Dir:* Douglas Foulk, MD

**Ft Lewis College**
Athletic Training Prgm
1000 Rim Dr
Durango, CO 81301-3999
*Prgm Dir:* Carrie Meyer, EdD
*Tel:* 970 247-7581   *Fax:* 970 247-7568
*E-mail:* meyer_c@fortlewis.edu
*Med Dir:* Joseph Murphy, MD

**Mesa State College**
Athletic Training Prgm
1100 North Ave
Grand Junction, CO 81501
*Prgm Dir:* Robert Ryan, MA
*Tel:* 970 248-1374
*E-mail:* rryan@mesastate.edu
*Med Dir:* Michael Reeder, DO

**University of Northern Colorado**
Athletic Training Prgm
Sch of Kinesiology and Phys Ed
Butler-Hancock 124
Greeley, CO 80639
*Prgm Dir:* Shannon Courtney, ATC
*Tel:* 970 351-2822   *Fax:* 970 351-2018
*E-mail:* shannon.courtney@unco.edu
*Med Dir:* Tom Pazik, MD

**Colorado State University - Pueblo**
Athletic Training Prgm
T250 - 2200 Bonforte Blvd
Pueblo, CO 81001
www.colostate-pueblo.edu
*Prgm Dir:* Roger Clark, PhD ATC
*Tel:* 719 549-2530   *Fax:* 719 549-2519
*E-mail:* roger.clark@colostate-pueblo.edu
*Med Dir:* Roger Davis, MD

# Connecticut

**Sacred Heart University**
Athletic Training Prgm
5151 Park Ave
Fairfield, CT 06432
*Prgm Dir:* Gail Samdperil, MAT ATC
*Tel:* 203 396-8033
*E-mail:* SamdperilG@sacredheart.edu

**Quinnipiac University**
Athletic Training Prgm
275 Mount Carmel Ave
Hamden, CT 06518
www.quinnipiac.edu
*Prgm Dir:* Lennart D Johns, PhD ATC
*Tel:* 203 582-8557
*E-mail:* lenn.johns@quinnipiac.edu
*Med Dir:* Robert Henry, MD

**Central Connecticut State University**
Athletic Training Prgm
1615 Stanley St
New Britain, CT 06050-4010
*Prgm Dir:* Peter Morano, BS
*Tel:* 860 832-2609   *Fax:* 860 832-2159
*E-mail:* moranop@ccsu.edu
*Med Dir:* William Waskowitz, MD

**Southern Connecticut State University**
Athletic Training Prgm
Pelz Gymnasium
501 Crescent St
New Haven, CT 06515
*Prgm Dir:* Gary E Morin, PhD ATC
*Tel:* 203 392-6089   *Fax:* 203 392-6093
*E-mail:* moring1@southernct.edu
*Med Dir:* Arthur Geiger, MD

**University of Connecticut**
Athletic Training Prgm
Dept of Kinesiology
2095 Hillside Rd, U-1110
Storrs, CT 06269-1110
www.uconn.edu
*Prgm Dir:* Douglas Casa, PhD ATC
*Tel:* 860 486-3624   *Fax:* 860 486-1123
*E-mail:* douglas.casa@uconn.edu
*Med Dir:* Jeffrey Anderson, MD

# Delaware

**University of Delaware**
Athletic Training Prgm
Human Performance Laboratory, Fred Rust Ice Arena
541 South College Ave
Newark, DE 19716
www.udel.edu/HNES/AT/Site
*Prgm Dir:* Thomas Kaminski, PHD ATC FACSM
*Tel:* 302 831-6402   *Fax:* 302 831-3693
*E-mail:* kaminski@udel.edu
*Med Dir:* Andrew Reisman, MD

# District of Columbia

**George Washington University**
Athletic Training Prgm
Department of Exercise Science
817 23rd St NW
Washington, DC 20052
*Prgm Dir:* Beverly J Westerman, EdD ATC
*Tel:* 202 994-3862   *Fax:* 202 994-1420
*E-mail:* bev@gwu.edu
*Med Dir:* Kenneth M Fine, MD

# Florida

**University of Miami**
Athletic Training Prgm
PO Box 248065
Coral Gables, FL 33124
www.education.miami.edu/AthleticTraining
*Prgm Dir:* Kysha Harriell, MS
*Tel:* 305 284-3201   *Fax:* 305 284-5168
*E-mail:* kharriell@miami.edu
*Med Dir:* John Uribe, MD

**Florida Gulf Coast University**
Athletic Training Prgm
10501 FGCU Blvd South
Fort Myers, FL 33965
*Prgm Dir:* Jason Craddock, MS
*Tel:* 239 482-5399   *Fax:* 239 482-4353
*E-mail:* jcraddoc@fgcu.edu

**University of Florida**
Athletic Training Prgm
Dept of Applied Physiology and Kinesiology
160 Florida Gym, Box 118205
Gainesville, FL 32611-8205
www.hhp.ufl.edu/apk/undergrad
*Prgm Dir:* Erik Wikstrom, MS ATC LAT
*Tel:* 352 392-0584, Ext 1402   *Fax:* 352 392-5262
*E-mail:* ewikstrom@hhp.ufl.edu
*Med Dir:* Douglas McDonald, MD

**University of North Florida**
Athletic Training Prgm
4567 St Johns Bluff Rd S
Brooks College of Health
Jacksonville, FL 32224
www.unf.edu/brooks/cohatweb.htm
*Prgm Dir:* Joel Beam, EdD ATC LAT
*Tel:* 904 620-1424   *Fax:* 904 620-2848
*E-mail:* jbeam@unf.edu
*Med Dir:* Frederick Beck, MD

**Florida Southern College**
Athletic Training Prgm
111 Lake Hollingsworth Dr
Lakeland, FL 33801-5698
www.flsouthern.edu
*Prgm Dir:* Sue Stanley-Green, MS ATC LAT
*Tel:* 863 680-4262   *Fax:* 863 680-4251
*E-mail:* sstanleygreen@flsouthern.edu
*Med Dir:* Larry Padgett, MD

**Barry University**
Athletic Training Prgm
11300 NE 2nd Ave
Miami Shores, FL 33161-6695
www.barry.edu/hpls
*Prgm Dir:* Carl R Cramer, EdD RKT ATC LAT
*Tel:* 305 899-3497   *Fax:* 305 899-4809
*E-mail:* ccramer@mail.barry.edu
*Med Dir:* Philip R Lozman, MD

**University of Central Florida**
Athletic Training Prgm
4000 Central Florida Blvd, HPA II Rm 121
Orlando, FL 32816-2200
*Prgm Dir:* Kristen Schellhase, MEd ATC CSCS
*Tel:* 407 823-3463
*E-mail:* kschellh@mail.ucf.edu
*Med Dir:* Douglas Meuser, MD

**University of West Florida**
Athletic Training Prgm
11000 University Pkwy
Pensacola, FL 32524
www.uwf.edu
*Prgm Dir:* Richard Frazee
*Tel:* 850 473-7245   *Fax:* 850 474-2106
*E-mail:* rfrazee@uwf.edu
*Med Dir:* Joshua Hackel, MD

**Florida State University**
Athletic Training Prgm
422 Sandels Bldg - NFES
Tallahassee, FL 320306
*Prgm Dir:* Angela Sehgal, EdD
*Tel:* 850 644-1899   *Fax:* 850 645-5000
*E-mail:* asehgal@fsu.edu

**University of South Florida**
Athletic Training Prgm
4202 E Fowler Ave, PED 214
Tampa, FL 33620
*Prgm Dir:* Pradeep Vanguri, PhD ATC
*Tel:* 813 974-4766   *Fax:* 813 974-2885
*E-mail:* vanguri@coedu.usf.edu

**University of Tampa**
Athletic Training Prgm
Box 35F, 401 W Kennedy Blvd
Tampa, FL 33606
www.ut.edu
*Prgm Dir:* J C Andersen, PhD ATC PT SCS
*Tel:* 813 253-3333, Ext 3127   *Fax:* 813 258-7497
*E-mail:* jcandersen@ut.edu
*Med Dir:* Seth Gasser, MD

**Palm Beach Atlantic University**
Athletic Training Prgm
PO Box 24708
West Palm Beach, FL 33416-4708
*Prgm Dir:* Sharon Bohres
*Tel:* 561 803-2342   *Fax:* 561 803-2370
*E-mail:* sharon_bohres@pba.edu
*Med Dir:* Michael Leighton, MD

# Georgia

**University of Georgia**
Athletic Training Prgm
Ramsey Center, 300 River Rd
Department of Kinesiology
Athens, GA 30602
www.coe.uga.edu/kinesiology/exs/athletictraining
*Prgm Dir:* Michael S Ferrara, PhD ATC
*Tel:* 706 542-4801   *Fax:* 706 542-3148
*E-mail:* mferrara@uga.edu
*Med Dir:* Ronald Elliott, MD

**North Georgia College & State University**
Athletic Training Prgm
Memorial Hall
Dahlonega, GA 30597
*Prgm Dir:* Derek Suranie
*Tel:* 706 867-2770   *Fax:* 706 867-2865
*E-mail:* djsuranie@ngcsu.edu
*Med Dir:* Eric Powell, MD

**Georgia College & State University**
Athletic Training Prgm
Campus Box 65
Milledgeville, GA 31061
*Prgm Dir:* Earl R Cooper, Jr, EdD ATC CSCS
*Tel:* 478 445-4072, Ext 1786   *Fax:* 478 445-1790
*E-mail:* bud.cooper@gcsu.edu
*Med Dir:* Steven Niergarth, DO

**Georgia Southern University**
Athletic Training Prgm
PO Box 8076
Hollis Bldg
Statesboro, GA 30460-8076
www.georgiasouthern.edu
*Prgm Dir:* Mary Barron, PhD
*Tel:* 912 681-5264   *Fax:* 912 681-0381
*E-mail:* mjbarron@georgiasouthern.edu
*Med Dir:* John Hodges, MD

**Valdosta State University**
Athletic Training Prgm
Dept KSPE
Valdosta, GA 31698
http://coefaculty.valdosta.edu/spm
*Prgm Dir:* Chuck Conner, MA LAT ATC
*Tel:* 229 333-5354   *Fax:* 229 333-5972
*E-mail:* cconner@valdosta.edu
*Med Dir:* Ben Hogan, MD

# Hawaii

**University of Hawaii at Manoa**
Athletic Training Prgm
Dept of KLS
1337 Lower Campus Rd PE, A Rm 231
Honolulu, HI 96822
*Prgm Dir:* Rebecca Cheema
*Tel:* 808 956-7606   *Fax:* 808 956-3106
*E-mail:* cheema@hawaii.edu
*Med Dir:* Andrew Williams Nichols, MD

# Idaho

**Boise State University**
Athletic Training Prgm
Dept of Kinesiology, K209
1910 University Dr
Boise, ID 83725
*Prgm Dir:* John W McChesney, PhD ATC
*Tel:* 208 426-1481   *Fax:* 208 426-1894
*E-mail:* atep@boisestate.edu

**University of Idaho**
Athletic Training Prgm
Department of HPERD
PO Box 442401
Moscow, ID 83844-2401
www.uidaho.edu
*Prgm Dir:* Jacqueline M Williams, MS LAT ATC
*Tel:* 208 885-2185   *Fax:* 208 885-5929
*E-mail:* jwilli@uidaho.edu
*Med Dir:* G David Rych

# Illinois

**Aurora University**
Athletic Training Prgm
347 S Gladstone Ave
Aurora, IL 60506
www.aurora.edu/shape/at
*Prgm Dir:* Oscar H Krieger, MS ATC CIE
*Tel:* 630 844-4224   *Fax:* 630 844-7809
*E-mail:* okrieger@aurora.edu
*Med Dir:* Robert Paras, MD

**Olivet Nazarene University**
Athletic Training Prgm
1 University Ave
Bourbonnais, IL 60914-2345
*Prgm Dir:* Brian Hyma, MA
*Tel:* 815 939-5115   *Fax:* 815 939-7933
*E-mail:* bhyma@olivet.edu
*Med Dir:* Michael Corcoran, MD

**Southern Illinois Univ at Carbondale**
Athletic Training Prgm
Dept of Kinesiology
1075 S Normal Ave
Carbondale, IL 62901
*Prgm Dir:* Toby Brooks, PhD
*Tel:* 618 453-3116   *Fax:* 618 453-3329
*E-mail:* tbrooks@siu.edu
*Med Dir:* Rollin Perkins, MD

**Eastern Illinois University**
Athletic Training Prgm
Department of Physical Education
600 Lincoln Ave 2220 Lantz Gymnasium
Charleston, IL 61920
www.eiu.edu/~athtrain
*Prgm Dir:* Lee Ann Price, MS LAT ATC
*Tel:* 217 581-7615   *Fax:* 217 581-7973
*E-mail:* lprice@eiu.edu
*Med Dir:* Richard Larson, MD

**North Park University**
Athletic Training Prgm
3225 W Foster Ave
Chicago, IL 60625
*Prgm Dir:* Andrew Lundgren, MA Ed
*Tel:* 773 244-6293   *Fax:* 773 244-4952
*E-mail:* alundgren@northpark.edu

**Millikin University**
Athletic Training Prgm
1184 W Main
Decatur, IL 62522
*Prgm Dir:* Tisha Hess, MS ATC/L
*Tel:* 217 420-6624   *Fax:* 217 363-6414
*E-mail:* thess@mail.millikin.edu
*Med Dir:* Lawrence Li, MD

**Trinity International University**
Athletic Training Prgm
2065 Half Day Rd
Deerfield, IL 60015
www.tiu.edu
*Prgm Dir:* Karl Glass, MS
*Tel:* 847 317-7066   *Fax:* 847 317-8056
*E-mail:* kglass@tiu.edu
*Med Dir:* Jeffrey Mjaanes, MD

**Northern Illinois University**
Athletic Training Prgm
Dept of Kinesiology & Physical Education
DeKalb, IL 60115
*Prgm Dir:* Gretchen A Schlabach, PhD ATC/L
*Tel:* 815 753-1424   *Fax:* 815 753-1413
*E-mail:* gas@niu.edu

**McKendree College**
Athletic Training Prgm
701 College Rd
Lebanon, IL 62254-1299
*Prgm Dir:* Dawn M Hankins, PhD ATC/L
*Tel:* 618 537-6917   *Fax:* 618 537-6876
*E-mail:* dhankins@mckendree.edu
*Med Dir:* Jay E Noffsinger, MD

**Western Illinois University**
Athletic Training Prgm
Brophy Hall 220B
One University Circle
Macomb, IL 61455
www.wiu.edu/kinesiology/atraining.html
*Prgm Dir:* Renee Polubinsky, MEd ATC CSCS
*Tel:* 309 298-2050   *Fax:* 309 298-2981
*E-mail:* RL-Polubinsky@wiu.edu
*Med Dir:* Richard Iverson, MD

**North Central College**
Athletic Training Prgm
30 N Brainard
Naperville, IL 60540
*Prgm Dir:* Heidi M Matthews, MS ATC/L
*Tel:* 630 637-5511   *Fax:* 630 637-5360
*E-mail:* hmmatthews@noctrl.edu
*Med Dir:* Kenneth Sanders, MD

**Illinois State University**
Athletic Training Prgm
208 Horton
Normal, IL 61790-5120
www.cast.ilstu.edu/isuatep
*Prgm Dir:* Todd A McLoda, PhD ATC LAT
*Tel:* 309 438-2605   *Fax:* 309 438-5559
*E-mail:* tamcloda@ilstu.edu
*Med Dir:* Robert Seidl, MD

**Univ of Illinois at Urbana - Champaign**
Athletic Training Prgm
331 Freer Hall
906 S Goodwin Ave
Urbana, IL 61801-3841
*Prgm Dir:* Steven Broglio, PhD
*Tel:* 217 244-7322   *Fax:* 217 244-7322
*E-mail:* broglio@uiuc.edu
*Med Dir:* Jennifer Anderson, MS Ed

# Indiana

**Anderson University**
Athletic Training Prgm
1100 E 5th St
Anderson, IN 46012-1362
*Prgm Dir:* Steven D Risinger, MA ATC LAT
*Tel:* 765 641-4491   *Fax:* 765 641-3841
*E-mail:* sdrisinger@anderson.edu
*Med Dir:* Stephen Hampton, MD

**Indiana University - Bloomington**
Athletic Training Prgm
Sportsmedicine Dept, Assembly Hall
1001 E 17th St
Bloomington, IN 47408-1590
*Prgm Dir:* Katie Grove, PhD LAT ATC
*Tel:* 812 855-3640   *Fax:* 812 855-1810
*E-mail:* kagrove@indiana.edu
*Med Dir:* Andrew Hipskind, MD

**University of Evansville**
Athletic Training Prgm
1800 Lincoln Ave, Wallace Graves Hall
Room 217
Evansville, IN 47722
*Prgm Dir:* Jeffrey Tilly, MS
*Tel:* 812 488-1054   *Fax:* 812 479-2717
*E-mail:* jt3@evansville.edu

**Franklin College**
Athletic Training Prgm
101 Brainigin Blvd
Franklin, IN 46131
www.franklincollege.edu/athweb/training
*Prgm Dir:* Kathy Remsburg, MS
*Tel:* 317 738-8135   *Fax:* 317 738-8248
*E-mail:* kremsburg@franklincollege.edu

**DePauw University**
Athletic Training Prgm
Kinesiology Department
1 Lilly Center
Greencastle, IN 46135
*Prgm Dir:* Marie Pickerill, PhD
*Tel:* 765 658-6689   *Fax:* 765 658-4964
*E-mail:* mariapickerill@depauw.edu
*Med Dir:* Scott Ripple, MD

**University of Indianapolis**
Athletic Training Prgm
1400 E Hanna Ave
Indianapolis, IN 46227
www.athtrg.uindy.edu
*Prgm Dir:* Connie Pumpelly, MS LAT ATC
*Tel:* 317 788-6143   *Fax:* 317 788-3472
*E-mail:* cpumpelly@uindy.edu
*Med Dir:* George DeSilvester, MD

**Indiana Wesleyan University**
Athletic Training Prgm
4201 S Washington St
Marion, IN 46953
http://cas.indwes.edu/hkrss/athletictraining
*Prgm Dir:* Adam Thompson, PhD ATC LAT
*Tel:* 765 677-2335, Ext 2335   *Fax:* 765 677-2328
*E-mail:* adam.thompson@indwes.edu
*Med Dir:* Jeremy Hunt, MD

**Ball State University**
Athletic Training Prgm
HP 209
Muncie, IN 47306
*Prgm Dir:* Thomas Weidner, PhD
*Tel:* 765 285-5039   *Fax:* 765 285-8254
*E-mail:* tweidner@bsu.edu
*Med Dir:* Kent Bullis, MD

**Manchester College**
Athletic Training Prgm
Box PERC
North Manchester, IN 46962
*Prgm Dir:* Mark W Huntington, PED ATC PT
*Tel:* 260 982-5033   *Fax:* 260 982-5405
*E-mail:* mwhuntington@manchester.edu
*Med Dir:* Eric Jenkinson, MD

**Indiana State University**
Athletic Training Prgm
Arena Rm C-09
Terre Haute, IN 47809
www.indstate.edu/athtrn
*Prgm Dir:* Catherine L Stemmans, PhD ATC/L
*Tel:* 812 237-8336   *Fax:* 812 237-4368
*E-mail:* cstemmmans@isugw.indstate.edu
*Med Dir:* Dorene Hojinicki, DO, FACEP

**Purdue University**
Athletic Training Prgm
Dept of Health & Kinesiology
800 W Stadium Ave
West Lafayette, IN 47907
www.purdue.edu/sportsmed
*Prgm Dir:* Larry J Leverenz, PhD ATC
*Tel:* 765 494-3167   *Fax:* 765 496-1239
*E-mail:* llevere@purdue.edu
*Med Dir:* Gregory A Rowdon, MD

# Iowa

**Iowa State University**
Athletic Training Prgm
225 Forker Bldg
Ames, IA 50011
*Prgm Dir:* Mary Meier, MS ATC LAT
*Tel:* 515 294-3587   *Fax:* 515 294-8740
*E-mail:* mkmeier@iastate.edu
*Med Dir:* Marc Shulman, MD

**University of Northern Iowa**
Athletic Training Prgm
203 Wellness Recreation Center
Cedar Falls, IA 50614-0241
www.uni.edu/athtrn
*Prgm Dir:* Thomas Dompier, PhD
*Tel:* 319 273-2180   *Fax:* 319 273-5958
*E-mail:* thomas.dompier@uni.edu
*Med Dir:* Jeff Clark, DO

**Coe College**
Athletic Training Prgm
1220 First Ave NE
Cedar Rapids, IA 52402
*Prgm Dir:* Mitch Doyle, MA LAT ATC
*Tel:* 319 399-8653   *Fax:* 319 399-8721
*E-mail:* mdoyle@coe.edu
*Med Dir:* Mark J Goedken, MD

**Luther College**
Athletic Training Prgm
700 College Dr
Decorah, IA 52101
*Prgm Dir:* Brian Solberg
*Tel:* 563 387-1246
*E-mail:* solberbr@luther.edu
*Med Dir:* Drew Pellett, MD

**Clarke College**
Athletic Training Prgm
1550 Clarke Dr, MS 1757
Dubuque, IA 52001
*Prgm Dir:* Melody "Dee" Higgins
*Tel:* 319 588-6549
*E-mail:* dee.higgins@clarke.edu
*Med Dir:* Scott Schemmel, MD

**Loras College**
Athletic Training Prgm
1450 Alta Vista
Dubuque, IA 52004-0178
*Prgm Dir:* Susan P Wehring
*Tel:* 563 588-7020
*E-mail:* susie.wehring@loras.edu

**Upper Iowa University**
Athletic Training Prgm
605 Washington St
PO Box 1857
Fayette, IA 52142
www.uiu.edu
*Prgm Dir:* Angela Leete, MS ATC LAT
*Tel:* 563 425-5782   *Fax:* 563 425-5379
*E-mail:* leetea@uiu.edu
*Med Dir:* David Tinker, DO

**Simpson College**
Athletic Training Prgm
701 North C St
Indianola, IA 50125
*Prgm Dir:* Mike Hadden, MS
*Tel:* 515 961-1751   *Fax:* 515 961-1279
*E-mail:* hadden@simpson.edu

**University of Iowa**
Athletic Training Prgm
414 FH
Iowa City, IA 52242-1020
*Prgm Dir:* Danny T Foster, PhD ATC LAT
*Tel:* 319 335-9393   *Fax:* 319 335-9398
*E-mail:* danny-foster@uiowa.edu
*Med Dir:* John P Albright, MD

**Graceland University**
Athletic Training Prgm
1 University Place
Lamoni, IA 50140
*Prgm Dir:* Diane Bartholomew, DHSc ATC
*Tel:* 641 784-5043   *Fax:* 641 784-5472
*E-mail:* batholo@graceland.edu
*Med Dir:* Mark Easter, DO

**Northwestern College**
Athletic Training Prgm
101 7th St SW
Orange City, IA 51041-1998
www.nwciowa.edu
*Prgm Dir:* Jennifer J Rogers, MEd ATC
*Tel:* 712 707-7307   *Fax:* 712 707-7280
*E-mail:* jjaacks@nwciowa.edu
*Med Dir:* Mark Muilenburg, MD

**Central College**
Athletic Training Prgm
812 University Box 6600
Pella, IA 50219
www.central.edu
*Prgm Dir:* John Roslien, MS
*Tel:* 641 628-5132   *Fax:* 641 628-5356
*E-mail:* roslienj@central.edu
*Med Dir:* Kurt Vander Ploeg, MD

**Buena Vista University**
Athletic Training Prgm
610 W Fourth St
Storm Lake, IA 50588
www.bvu.edu
*Prgm Dir:* Christopher Todden, MS ATC
*Tel:* 712 749-2022   *Fax:* 712 749-1460
*E-mail:* todden@bvu.edu
*Med Dir:* Rick Wilkerson, DO

# Kansas

**Benedictine College**
Athletic Training Prgm
1020 N Second St
Atchison, KS 66002
*Prgm Dir:* Lanny Leroy
*Tel:* 913 367-5340, Ext 2541
*E-mail:* lannyl@benedictine.edu
*Med Dir:* Tom L Shriwise

**Emporia State University**
Athletic Training Prgm
1200 Commerical
Campus Box 4013
Emporia, KS 66801
www.emporia.edu/hper
*Prgm Dir:* Leslie Kenney, MS ATC
*Tel:* 620 341-5653   *Fax:* 620 341-6400
*E-mail:* lkenney@emporia.edu
*Med Dir:* Marshall A Havenhill II, MD

**Ft Hays State University**
Athletic Training Prgm
600 Park St
Hays, KS 67601
*Prgm Dir:* Mark Stutz, PhD
*Tel:* 785 628-4354   *Fax:* 785 628-4126
*E-mail:* mstutz@fhsu.edu

**Tabor College**
Athletic Training Prgm
400 S Jefferson
Hillsboro, KS 67063
www.tabor.edu
*Prgm Dir:* James Moore, MS ATC CSCS
*Tel:* 620 947-3121, Ext 1306   *Fax:* 620 947-2607
*E-mail:* jimm@tabor.edu
*Med Dir:* Chris Miller, MD

**University of Kansas**
Athletic Training Prgm
1301 Sunnyside Ave
161-C
Lawrence, KS 66045
*Prgm Dir:* W David Carr, PhD ATC
*Tel:* 785 864-0799   *Fax:* 785 864-3343
*E-mail:* wdcarr@ku.edu
*Med Dir:* Lawrence Magee, MD

**Kansas State University**
Athletic Training Prgm
241 Justin Hall
Department of Human Nutrition
Manhattan, KS 66506-1407
https://www.k-state.edu/humec/hn/athleticcover.htm
*Prgm Dir:* Shawna Jordan, MSEd ATC LAT
*Tel:* 785 532-0151   *Fax:* 785 532-3132
*E-mail:* jordan@ksu.edu
*Med Dir:* Keith Wright, MD

**Bethel College**
Athletic Training Prgm
300 E 27th St
North Newton, KS 67117
*Prgm Dir:* Russell P Graber
*Tel:* 316 284-5383   *Fax:* 316 284-5286
*E-mail:* rpgraber@bethelks.edu
*Med Dir:* R M Glover, II

**Mid-America Nazarene University**
Athletic Training Prgm
2030 E College Way
Olathe, KS 66062-1899
*Prgm Dir:* Jennifer Amborn, MEd
*Tel:* 913 791-3753   *Fax:* 913 791-3404
*E-mail:* jamborn@mnu.edu

**Kansas Wesleyan University**
Athletic Training Prgm
100 E Claflin Ave
Salina, KS 67401-6196
*Prgm Dir:* Matthew Williams, ATC DSM
*Tel:* 785 827-5541, Ext 3194   *Fax:* 785 827-0927
*E-mail:* mjw@kwu.edu
*Med Dir:* Gary Harbin, MD

**Sterling College**
Athletic Training Prgm
125 W Cooper
Sterling, KS 67579
*Prgm Dir:* Robert Bradley, MS
*Tel:* 620 278-4393   *Fax:* 620 278-4319
*E-mail:* rbradley@sterling.edu

**Washburn University**
Athletic Training Prgm
1700 SW College
Topeka, KS 66621
www.washburn.edu
*Prgm Dir:* John Burns, MS
*Tel:* 785 670-2155   *Fax:* 785 670-1059
*E-mail:* john.burns@washburn.edu
*Med Dir:* Iris Gonzalez, MD

**Southwestern College**
Athletic Training Prgm
100 College St
Winfield, KS 67156
*Prgm Dir:* Lisa Braun, MEd LAT ATC
*Tel:* 620 229-6226   *Fax:* 620 229-6112
*E-mail:* lbraun@sckans.edu
*Med Dir:* Bryan Dennett, MD

# Kentucky

**Northern Kentucky University**
Athletic Training Prgm
109 Albright Health Center
Highland Heights, KY 41099
*Prgm Dir:* Trey Morgan, MS ATC
*Tel:* 859 572-1399   *Fax:* 859 572-6090
*E-mail:* morgant@nku.edu
*Med Dir:* James Bilbo, MD

**Murray State University**
Athletic Training Prgm
Dept of Wellness & Therapeutic Sciences
102A Carr Health Bldg
Murray, KY 42071
www.murraystate.edu/academics/hshs/wts/
athletictraining.htm
*Prgm Dir:* Jeremy Erdmann, MA ATC
*Tel:* 270 809-4517   *Fax:* 270 809-6803
*E-mail:* jeremy.erdmann@murraystate.edu
*Med Dir:* Hal E Houston, MD

**Eastern Kentucky University**
Athletic Training Prgm
Moberly 224
521 Lancaster Ave
Richmond, KY 40475
*Prgm Dir:* Joseph Beckett, EdD ATC
*Tel:* 859 622-2134   *Fax:* 859 622-1254
*E-mail:* Joe.Beckett@eku.edu
*Med Dir:* Mary Lloyd Ireland, MD

# Louisiana

**Louisiana State University**
Athletic Training Prgm
Department of Kinesiology
112 HP Long Fieldhouse
Baton Rouge, LA 70803
www.lsu.edu/kinesiology
*Prgm Dir:* Ray Castle, PhD ATC LAT
*Tel:* 225 578-7175   *Fax:* 225 578-3680
*E-mail:* rcastll@lsu.edu
*Med Dir:* Jeffery Burnham, MD

**Southeastern Louisiana University**
Athletic Training Prgm
SLU 10845
Hammond, LA 70402
www.selu.edu/atep
*Prgm Dir:* Karen Lew, MEd
*Tel:* 985 549-2350   *Fax:* 985 549-5119
*E-mail:* klew@selu.edu
*Med Dir:* H Reiss Plauche, MD

**University of Louisiana at Lafayette**
Athletic Training Prgm
225 Cajundome Blvd
Lafayette, LA 70506
*Prgm Dir:* Toby Dore, PhD ATC LAT
*Tel:* 337 482-6283  *Fax:* 337 482-6278
*E-mail:* tldore@louisiana.edu
*Med Dir:* Glen Mire, MD

**Louisiana College**
Athletic Training Prgm
1140 College Dr, Box 563
Pineville, LA 71359
*Prgm Dir:* Mike Brunet, PhD
*Tel:* 318 487-7519  *Fax:* 318 487-7174
*E-mail:* mbrunet@lacollege.edu
*Med Dir:* Chris Rich, MD

**Nicholls State University**
Athletic Training Prgm
PO Box 2090
Thibodaux, LA 70310
www.nicholls.edu/ahs/health/athletic
*Prgm Dir:* Gerard White, ATC LAT MEd
*Tel:* 985 493-2612  *Fax:* 985 493-2614
*E-mail:* gerard.white@nicholls.edu
*Med Dir:* Neil J Maki, MD

# Maine

**University of New England**
Athletic Training Prgm
11 Hills Beach Rd
Biddeford, ME 04005
*Prgm Dir:* Wayne Lamarre, MEd
*Tel:* 207 602-2412  *Fax:* 207 282-6379
*E-mail:* wlamarre@une.edu

**University of Southern Maine**
Athletic Training Prgm
37 College Ave
Gorham, ME 04038
*Prgm Dir:* Tina Claiborne, PhD
*Tel:* 207 780-4698  *Fax:* 207 780-4745
*E-mail:* tclaibor@usm.maine.edu
*Med Dir:* William Dexter, MD

**University of Maine - Orono**
Athletic Training Prgm
108 Lengyel Hall
Orono, ME 04469
*Prgm Dir:* Sherrie L Weeks
*Tel:* 207 581-2466
*E-mail:* sherrie_weeks@umit.maine.edu
*Med Dir:* Mark Jackson, MD

**University of Maine - Presque Isle**
Athletic Training Prgm
181 Main St, 209 South Hall
Persque Isle, ME 04769
*Prgm Dir:* Barbara Blackstone, MSS
*Tel:* 207 768-9415  *Fax:* 207 768-9553
*E-mail:* blackstb@umpi.maine.edu

# Maryland

**Frostburg State University**
Athletic Training Prgm
Cordts Physical Education Center
Frostburg, MD 21532
*Prgm Dir:* James Wright, MS
*Tel:* 301 687-4477  *Fax:* 301 687-7961
*E-mail:* jwright@frostburg.edu
*Med Dir:* David Tuel, MD

**Salisbury University**
Athletic Training Prgm
1101 Camden Ave
Salisbury, MD 21801
www.salisbury.edu
*Prgm Dir:* Donna Ritenour, EdD
*Tel:* 410 543-6348  *Fax:* 410 543-6434
*E-mail:* dmritenour@salisbury.edu
*Med Dir:* Christopher Snyder, DO

**Towson University**
Athletic Training Prgm
8000 York Rd
Towson, MD 21252
www.towson.edu
*Prgm Dir:* Michael Higgins, PHD ATC/PT CSCS
*Tel:* 410 704-3166  *Fax:* 410 704-3912
*E-mail:* mhiggins@towson.edu
*Med Dir:* Teri McCambridge, MD

# Massachusetts

**Endicott College**
Athletic Training Prgm
376 Hale St
Beverly, MA 01915
*Prgm Dir:* Deborah Swanton, EdD LATC
*Tel:* 978 232-2433  *Fax:* 978 232-2600
*E-mail:* dswanton@endicott.edu
*Med Dir:* David Fehnel, MD

**Boston University**
Athletic Training Prgm
635 Commonwealth Ave
Boston, MA 02215
*Prgm Dir:* Mark Laursen, MS
*Tel:* 617 353-7570  *Fax:* 617 353-9463
*E-mail:* rmarkl@bu.edu
*Med Dir:* Matthew Pecci, MD

**Northeastern University**
Athletic Training Prgm
304 Dockser Hall
Boston, MA 02115
www.neu.edu
*Prgm Dir:* Jamie L Musler, MS Ed ATC
*Tel:* 617 373-5355  *Fax:* 617 373-5920
*E-mail:* j.musler@neu.edu

**Bridgewater State College**
Athletic Training Prgm
Dept of MAHPLS
Tinsley Center
Bridgewater, MA 02325
*Prgm Dir:* Marcia Anderson, PhD
*Tel:* 508 531-2072  *Fax:* 508 531-1717
*E-mail:* mkanderson@bridgew.edu
*Med Dir:* Stephen C McNeil, MD

**Lasell College**
Athletic Training Prgm
1844 Commonwealth Ave
Newton, MA 02466
www.lasell.edu
*Prgm Dir:* Bill Nowlan, MEd ATC LAT CSCS
*Tel:* 617 243-2262  *Fax:* 617 243-2309
*E-mail:* bnowlan@lasell.edu
*Med Dir:* Mininder Kocher, MD

**Merrimack College**
Athletic Training Prgm
315 Turnpike St
North Andover, MA 01845
*Prgm Dir:* Birgid Hopkins, MS LAT ATC
*Tel:* 978 837-5332  *Fax:* 978 837-5032
*E-mail:* birgid.hopkins@merrimack.edu
*Med Dir:* Thomas Hoerner, MD

**Salem State College**
Athletic Training Prgm
352 Lafayette St
Salem, MA 01970-5353
*Prgm Dir:* Joe Gallo, DSc
*Tel:* 978 542-6585  *Fax:* 978 542-6554
*E-mail:* jgallo@salemstate.edu
*Med Dir:* David St Pierre, MD

**Springfield College**
Athletic Training Prgm
Physical Education Complex
Springfield, MA 01109
http://tinyurl.com/yjm58y
*Prgm Dir:* Mary G Barnum, EdD ATC LAT
*Tel:* 413 748-3763  *Fax:* 413 748-3052
*E-mail:* mary_barnum@spfldcol.edu
*Med Dir:* Mark Kenton, DO

**Westfield State College**
Athletic Training Prgm
Department of Movement Science
577 Western Ave
Westfield, MA 01086-1630
*Prgm Dir:* William N Miller, DPE ATC
*Tel:* 413 572-5450  *Fax:* 413 572-8214
*E-mail:* wmiller@wsc.ma.edu
*Med Dir:* Ronald Agnes, MD

# Michigan

**Albion College**
Athletic Training Prgm
Box 4830
Albion, MI 49224
www.albion.edu
*Prgm Dir:* Robert Moss, PhD ATC
*Tel:* 517 629-0548  *Fax:* 517 629-0619
*E-mail:* rmoss@albion.edu
*Med Dir:* Martin Holmes, MD

**Grand Valley State University**
Athletic Training Prgm
Movement Science Dept
192 Field House
Allendale, MI 49401
www.gvsu.edu/move-sci
*Prgm Dir:* Shari Bartz-Smith, PhD ATC CSCS
*Tel:* 616 331-3044  *Fax:* 616 331-3232
*E-mail:* bartzs@gvsu.edu
*Med Dir:* Steven VanNoord, MD

**Alma College**
Athletic Training Prgm
614 W Superior St
Alma, MI 48801
*Prgm Dir:* Denny Griffin
*Tel:* 989 463-7988
*E-mail:* griffin@alma.edu
*Med Dir:* Bill Thiemkey, MD

**University of Michigan**
Athletic Training Prgm
401 Washtenaw Ave
Kinesiology Bldg
Ann Arbor, MI 48109-2214
*Prgm Dir:* Brian Czajika, MS
*Tel:* 734 647-2702  *Fax:* 734 936-1925
*E-mail:* baczaika@umich.edu
*Med Dir:* C Daniel Hendrickson, MD

**Michigan State University**
Athletic Training Prgm
105 IM Circle
East Lansing, MI 48824
*Prgm Dir:* Tracey Covassin, PhD
*Tel:* 517 353-2010  *Fax:* 517 432-2944
*E-mail:* covassin@msu.edu
*Med Dir:* Jeff Kovan, DO

**Aquinas College**
Athletic Training Prgm
1607 Robinson Rd SE
Hruby Hall 42
Grand Rapids, MI 49506-1799
*Prgm Dir:* Deborah J Springer
*Tel:* 616 632-2897   *Fax:* 616 732-4487
*E-mail:* sprindeb@aquinas.edu
*Med Dir:* Edwin T Kornoelje, DO

**Hope College**
Athletic Training Prgm
PO Box 9000
Holland, MI 49423
www.hope.edu/academic/kinesiology/athtrain
*Prgm Dir:* Richard Ray, EdD ATC
*Tel:* 616 395-7708   *Fax:* 616 395-7175
*E-mail:* ray@hope.edu
*Med Dir:* Patrick Hulst, MD

**Western Michigan University**
Athletic Training Prgm
Health, Physical Education and Recreation
1903 W Michigan Ave #5426
Kalamazoo, MI 49008
www.wmich.edu
*Prgm Dir:* Jennifer O'Donoghue, MA ATC CSCS
*Tel:* 269 387-2703
*E-mail:* ODonoghueJ@groupwise.wmich.edu
*Med Dir:* Robert Baker, MD, PhD, ATC

**Northern Michigan University**
Athletic Training Prgm
Dept of HPER
1401 Presque Isle Ave
Marquette, MI 49855
*Prgm Dir:* Julie A Rochester, MS ATC
*Tel:* 906 227-2026   *Fax:* 906 227-2181
*E-mail:* jrochest@nmu.edu
*Med Dir:* John L Lehtinen, MD

**Central Michigan University**
Athletic Training Prgm
HPB 1173
HPB 1171
Mt Pleasant, MI 48859
www.chp.cmich.edu/atep
*Prgm Dir:* Rene Shingles, PhD ATC
*Tel:* 989 774-2378   *Fax:* 989 774-2923
*E-mail:* shing1rr@cmich.edu

**Lake Superior State University**
Athletic Training Prgm
650 W Easterday Ave
Sault Ste Marie, MI 49783
*Prgm Dir:* Joseph D Susi II, MS ATC
*Tel:* 906 635-2161   *Fax:* 906 635-2753
*E-mail:* jsusi@lssu.edu
*Med Dir:* Robert Mackie, MD

**Saginaw Valley State University**
Athletic Training Prgm
7400 Bay Rd
University Center, MI 48710
www.svsu.edu
*Prgm Dir:* Paul Ballard, EdD
*Tel:* 989 964-7269   *Fax:* 989 964-4564
*E-mail:* pballard@svsu.edu
*Med Dir:* Blake Bergeon, MD

**Eastern Michigan University**
Athletic Training Prgm
310 Porter Bldg
Ypsilanti, MI 48197
*Prgm Dir:* Jodi Johnson, MS ATC
*Tel:* 734 487-7120, Ext 2722   *Fax:* 734 487-2024
*E-mail:* jjohnson@emich.edu
*Med Dir:* Waldemar Roeser, MD

# Minnesota

**University of Minnesota - Duluth**
Athletic Training Prgm
110 Sports and Health Center
1216 Ordean Court
Duluth, MN 55812
www.d.umn.edu/hper/ATEP
*Prgm Dir:* William Gear, PhD
*Tel:* 218 726-8654   *Fax:* 218 726-6243
*E-mail:* wgear@d.umn.edu
*Med Dir:* Stephen Harrington, MD

**Minnesota State University - Mankato**
Athletic Training Prgm
1400 Highland Center
Mankato, MN 56001
www.mnsu.edu/athletictraining
*Prgm Dir:* Patrick Sexton, EdD ATC/R CSCS
*Tel:* 507 389-2092   *Fax:* 507 389-5618
*E-mail:* patrick.sexton@mnsu.edu
*Med Dir:* Todd Kanzenbach, MD

**Minnesota State University - Moorhead**
Athletic Training Prgm
106 D Alex Namzek Hall
Moorhead, MN 56563
*Prgm Dir:* Dawn Hammerschmidt, MEd
*Tel:* 218 477-2318   *Fax:* 218 477-2363
*E-mail:* hammerda@mnstate.edu

**St Cloud State University**
Athletic Training Prgm
720 4th Ave South, Halenbeck Hall 339
St Cloud, MN 56301
*Prgm Dir:* William Picconatto, PhD
*Tel:* 320 308-3079   *Fax:* 320 308-5399
*E-mail:* wjpicconatto@stcloudstate.edu
*Med Dir:* MR Hwang, MD(Orthopedics)

**Bethel University**
Athletic Training Prgm
3900 Bethel Dr
St Paul, MN 55112
*Prgm Dir:* Neal S Dutton
*Tel:* 651 638-6255
*E-mail:* dutnea@bethel.edu
*Med Dir:* Robert Johnson, MD

**Gustavus Adolphus College**
Athletic Training Prgm
800 W College Ave
St Peter, MN 56082
*Prgm Dir:* John Mattson, PhD ATC
*Tel:* 507 933-7674   *Fax:* 507 933-8412
*E-mail:* jmattson@gac.edu
*Med Dir:* Alan Markman, MD

**Winona State University**
Athletic Training Prgm
Memorial Hall 117
Winona, MN 55987-5838
*Prgm Dir:* Shellie F Nelson, EdD ATC
*Tel:* 507 457-5214   *Fax:* 507 457-5606
*E-mail:* snelson@winona.edu
*Med Dir:* Richard Romeyn, MD

# Mississippi

**Delta State University**
Athletic Training Prgm
PO Box B-2
Cleveland, MS 38733
*Prgm Dir:* Timothy Colbert, MS
*Tel:* 662 846-4562   *Fax:* 662 846-4571
*E-mail:* tcolbert@deltastate.edu
*Med Dir:* Scott Nelson, MD

**University of Southern Mississippi**
Athletic Training Prgm
Dept of Human Performance & Recreation
118 College Dr, Box 5142
Hattiesburg, MS 39405-0001
*Prgm Dir:* Benito Velasquez, DA
*Tel:* 601 266-6058   *Fax:* 601 266-4445
*E-mail:* benito.velasquez@usm.edu
*Med Dir:* Suzie Folse, MD

# Missouri

**Southwest Baptist University**
Athletic Training Prgm
1600 University Ave
Bolivar, MO 65613
*Prgm Dir:* Todd A John
*Tel:* 417 328-1988
*E-mail:* tjohn@sbuniv.edu
*Med Dir:* Louis Harris, MD

**Culver-Stockton College**
Athletic Training Prgm
1 College Hill
Canton, MO 63435
*Prgm Dir:* Robert W Carmichael
*Tel:* 573 288-6304   *Fax:* 573 288-6442
*E-mail:* rcarmichael@culver.edu
*Med Dir:* James M Daniels, II, MD

**Southeast Missouri State University**
Athletic Training Prgm
One University Plaza MS7650
Cape Girardeau, MO 63701
*Prgm Dir:* Craig Elder, PhD ATC CSCS
*Tel:* 573 651-5193   *Fax:* 573 651-5150
*E-mail:* celder@semo.edu
*Med Dir:* Charley Pancoast, MD

**Central Methodist University**
Athletic Training Prgm
411 CMC Square
Fayette, MO 65248
www.centralmethodist.edu/athtr
*Prgm Dir:* Wade Welton, MS ATC/L
*Tel:* 660 248-6217   *Fax:* 660 248-6381
*E-mail:* wwelton@centralmethodist.edu
*Med Dir:* Kristin Malaker, MD

**William Woods University**
Athletic Training Prgm
One University Ave
Fulton, MO 65251
*Prgm Dir:* Cynthia Robb, MA
*Tel:* 573 592-1689   *Fax:* 573 592-1641
*E-mail:* crobb@williamwoods.edu
*Med Dir:* Matt Thornburg, MD

**Truman State University**
Athletic Training Prgm
Pershing Bldg
Kirksville, MO 63501
http://hes.truman.edu/atmaj
*Prgm Dir:* Michelle Boyd, MS
*Tel:* 660 785-7364   *Fax:* 660 785-4166
*E-mail:* mboyd@truman.edu
*Med Dir:* Bailey John, DO

**Missouri Valley College**
Athletic Training Prgm
500 E College St
Marshall, MO 65340
www.moval.edu
*Prgm Dir:* Karla Bruntzel, PhD ATC
*Tel:* 660 831-4103   *Fax:* 660 831-4030
*E-mail:* bruntzelk@moval.edu
*Med Dir:* Kelly Ross, DO

**Park University**
Athletic Training Prgm
8700 NW Riverpark Dr
Parkville, MO 64152
www.park.edu
*Prgm Dir:* Thomas Bertoncino, MS ATC
*Tel:* 816 584-6864   *Fax:* 816 587-8907
*E-mail:* thomas.bertoncino@park.edu
*Med Dir:* Steve Smith, MD

**Missouri State University**
Athletic Training Prgm
Professional Bldg 160
901 S National Ave
Springfield, MO 65897
*Prgm Dir:* Gary Ward, MS ATC LAT PT
*Tel:* 417 836-8553   *Fax:* 417 836-8554
*E-mail:* GaryWard@MissouriState.edu
*Med Dir:* Richard A Seagrave, MD

**Lindenwood University**
Athletic Training Prgm
209 S Kings Hwy
St Charles, MO 63301
*Prgm Dir:* Randy Biggerstaff, MS
*Tel:* 636 949-4683   *Fax:* 636 949-4636
*E-mail:* rbiggerstaff@lindenwood.edu
*Med Dir:* Katherine Burns, MD

## Montana

**Montana State University - Billings**
Athletic Training Prgm
PE 119
1500 University Dr
Billings, MT 59101
*Prgm Dir:* Mike Diede, PhD ATC
*Tel:* 406 657-2351
*E-mail:* mdiede@msubillings.edu
*Med Dir:* Jim Elliot, MD

**University of Montana**
Athletic Training Prgm
Health & Human Performance Dept
McGill Hall 101
Missoula, MT 59812-1055
*Prgm Dir:* Scott T Richter, EdM
*Tel:* 406 243-5246   *Fax:* 406 243-6252
*E-mail:* scott.richter@mso.umt.edu
*Med Dir:* Timothy McCue, MD

## Nebraska

**University of Nebraska - Kearney**
Athletic Training Prgm
905 W 25th St
Cushing Bldg Rm 158
Kearney, NE 68849
*Prgm Dir:* Scott Unruh
*Tel:* 308 865-8627
*E-mail:* unruhsa@unk.edu
*Med Dir:* Brad Rogers, MD

**Nebraska Wesleyan University**
Athletic Training Prgm
5000 St Paul Ave
Lincoln, NE 68504
*Prgm Dir:* Laura Steele, MS ATC
*Tel:* 402 465-2277   *Fax:* 402 465-2170
*E-mail:* lsteele@nebrwesleyan.edu
*Med Dir:* Matthew C Reckmeyer, MD

**University of Nebraska - Lincoln**
Athletic Training Prgm
231 Mabel Lee Hall
Lincoln, NE 68588-0234
*Prgm Dir:* Jeffrey P Rudy
*Tel:* 402 472-5978
*E-mail:* jrudy2@unl.edu
*Med Dir:* Lonnie Albers

**Creighton University**
Athletic Training Prgm
2500 California Plaza, KFC 234
Omaha, NE 68178
*Prgm Dir:* Charles Pfeifer, MS
*Tel:* 402 280-2402   *Fax:* 402 280-4732
*E-mail:* pcp@creighton.edu

**University of Nebraska - Omaha**
Athletic Training Prgm
HPER 207
6001 Dodge St
Omaha, NE 68182-0216
www.unocoe.unomaha.edu/hper/athletictraining
*Prgm Dir:* Joshua Nichter, MS ATC CSCS
*Tel:* 402 554-3224   *Fax:* 402 554-3693
*E-mail:* jnichter@mail.unomaha.edu
*Med Dir:* Michael Walsh, MD

## Nevada

**University of Nevada - Las Vegas**
Athletic Training Prgm
4505 Maryland Pkwy
PO Box 453034
Las Vegas, NV 89154-3034
www.unlv.edu/athletics/training
*Prgm Dir:* Mack Rubley, PhD LAT ATC CSCS
*Tel:* 702 895-2457   *Fax:* 702 895-1500
*E-mail:* mack.rubley@unlv.edu
*Med Dir:* Jason Tarno, DO

## New Hampshire

**University of New Hampshire**
Athletic Training Prgm
Dept of Kinesiology
145 Main St, Field House
Durham, NH 03824
www.unh.edu/athletic-training
*Prgm Dir:* Daniel R Sedory, MS ATC
*Tel:* 603 862-1831   *Fax:* 603 862-4198
*E-mail:* dan.sedory@unh.edu
*Med Dir:* Joe Bernard, MD

**Keene State College**
Athletic Training Prgm
229 Main St
Keene, NH 03435-2301
www.keene.edu
*Prgm Dir:* Sherry L Bovinet, PhD LATC
*Tel:* 603 358-2804   *Fax:* 603 358-2075
*E-mail:* sbovinet@keene.edu
*Med Dir:* Leslie Pitts, MD

**Colby-Sawyer College**
Athletic Training Prgm
541 Main St
New London, NH 03257
*Prgm Dir:* John Culp, MS ATC
*Tel:* 603 526-3403   *Fax:* 603 526-3872
*E-mail:* jculp@colby-sawyer.edu
*Med Dir:* Douglas Moran, MD

**Plymouth State University**
Athletic Training Prgm
MSC 22
Plymouth, NH 03264
www.plymouth.edu
*Prgm Dir:* Linda S Levy, EdD ATC
*Tel:* 603 535-2577   *Fax:* 603 535-2395
*E-mail:* levy@plymouth.edu
*Med Dir:* Victor Gennaro, DO

## New Jersey

**Rowan University**
Athletic Training Prgm
201 Mullica Hill Rd
Glassboro, NJ 08028
*Prgm Dir:* Robert Sterner, PhD
*Tel:* 856 256-4500, Ext 376   *Fax:* 856 256-5613
*E-mail:* sterner@rowan.edu

**Seton Hall University**
Athletic Training Prgm
400 S Orange Ave
South Orange, NJ 07079
*Prgm Dir:* Carolyn Goeckel, MA ATC
*Tel:* 973 275-2826
*E-mail:* goeckeca@shu.edu

**Kean University**
Athletic Training Prgm
D'Angola Gym, Morris Ave
Union, NJ 07083
www.kean.edu/~gball
*Prgm Dir:* Gary Ball, EdD MEd BS ATC
*Tel:* 908 737-5437   *Fax:* 908 353-7199
*E-mail:* gbkuatc@verizon.net
*Med Dir:* Michele Gilsenan, DO

**Montclair State University**
Athletic Training Prgm
1 Normal Ave
Upper Montclair, NJ 07043
*Prgm Dir:* David Middlemas, EdD
*Tel:* 973 655-7090   *Fax:* 973 655-5461
*E-mail:* middlemasd@mail.montclair.edu

**William Paterson Univ of New Jersey**
Athletic Training Prgm
Dept of Exercise & Movement Science
300 Pompton Rd
Wayne, NJ 07470
www.wpunj.edu/cos/ex-movsci/atep.htm
*Prgm Dir:* Linda Gazzillo Diaz, EdD ATC
*Tel:* 973 720-2364   *Fax:* 973 720-2034
*E-mail:* gazzillol@wpunj.edu
*Med Dir:* Gary Drillings, MD PA

## New Mexico

**University of New Mexico**
Athletic Training Prgm
South Athletic Complex
Albuquerque, NM 87131
*Prgm Dir:* Susan McGowen
*Tel:* 505 277-5151   *Fax:* 505 277-6227
*E-mail:* yorex@unm.edu
*Med Dir:* Robert Schenck, MD

**New Mexico State University**
Athletic Training Prgm
Box 30001, MSC 3FAC
Las Cruces, NM 88003-0001
www.nmsu.edu
*Prgm Dir:* Leah Putman, MS ATC
*Tel:* 505 646-5038   *Fax:* 505 646-3564
*E-mail:* lputman@nmsu.edu
*Med Dir:* William Baker, DO

## New York

**Alfred University**
Athletic Training Prgm
1 Saxon Dr
Alfred, NY 14802
www.alfred.edu
*Prgm Dir:* Timothy Howell, PhD ATC CSCS
*Tel:* 607 871-2784   *Fax:* 607 871-2712
*E-mail:* howellt@alfred.edu
*Med Dir:* Daniel Curtin, MD

**SUNY at Brockport**
Athletic Training Prgm
260 Tuttle South
Brockport, NY 14420
*Prgm Dir:* Timothy J Henry, PhD ATC
*Tel:* 585 395-5357    *Fax:* 585 395-2771
*E-mail:* thenry@brockport.edu
*Med Dir:* E James Swenson, MD

**Long Island University - Brooklyn Campus**
Athletic Training Prgm
1 University Plaza
HS 312
Brooklyn, NY 11201-8423
www.brooklyn.liu.edu/athletictraining
*Prgm Dir:* Tracye Rawls-Martin, MS ATC
*Tel:* 718 780-4081    *Fax:* 718 488-1432
*E-mail:* tmartin@liu.edu
*Med Dir:* Howard Levy, MD

**Canisius College**
Athletic Training Prgm
2001 Main St
Buffalo, NY 14208-1098
www.canisius.edu/sportsmed
*Prgm Dir:* Peter Koehneke, MS ATC
*Tel:* 716 888-2954    *Fax:* 716 888-3216
*E-mail:* koehneke@canisius.edu
*Med Dir:* Rajiv Jain, MD

**University at Buffalo - SUNY**
Athletic Training Prgm
405 Kimball Tower
3435 Main St
Buffalo, NY 14214
*Prgm Dir:* Paula J Maxwell, PhD ATC
*Tel:* 716 829-2941, Ext 405    *Fax:* 716 829-2428
*E-mail:* pmaxwell@buffalo.edu
*Med Dir:* Robert J Smolinski, MD

**SUNY College at Cortland**
Athletic Training Prgm
PO Box 2000
Cortland, NY 13045
*Prgm Dir:* John Cottone, EdD ATC
*Tel:* 607 753-4962    *Fax:* 607 753-5975
*E-mail:* cottoneJ@cortland.edu
*Med Dir:* Nancy Sternfeld, MD

**Hofstra University**
Athletic Training Prgm
220 Hofstra University
The Dome
Hempstead, NY 11550
*Prgm Dir:* Jayne Kitsos-Ciarlante, MA ATC
*Tel:* 516 463-6952    *Fax:* 516 463-6275
*E-mail:* hprjmk@Hofstra.edu
*Med Dir:* Damion Martins

**Ithaca College**
Athletic Training Prgm
1100 Hill Center
Ithaca, NY 14850
*Prgm Dir:* Paul Geisler, EdD ATC
*Tel:* 607 274-3006    *Fax:* 607 274-1943
*E-mail:* pgeisler@Ithaca.edu
*Med Dir:* Andrew Getzin, MD

**Dominican College**
Athletic Training Prgm
470 Western Hwy
Orangeburg, NY 10962
www.dc.edu
*Prgm Dir:* James T Crawley, MEd ATC PT
*Tel:* 845 848-6004, Ext 6004    *Fax:* 845 398-4895
*E-mail:* jim.crawley@dc.edu
*Med Dir:* Barry Kraushaar, MD

**Marist College**
Athletic Training Prgm
Marist College 3399 North Rd
McCann Center Rm 212
Poughkeepsie, NY 12601
www.marist.edu
*Prgm Dir:* Sally A Perkins, ATC
*Tel:* 845 575-3912    *Fax:* 845 575-3282
*E-mail:* sally.perkins@marist.edu
*Med Dir:* Lawrence Kusior, MD

**Stony Brook University**
Athletic Training Prgm
Sch of Hlth Tech and Management
Health Science Center
Stony Brook, NY 11794-3500
*Prgm Dir:* Kathryn A Koshansky, MS ATC
*Tel:* 631 632-2837    *Fax:* 631 632-7210
*E-mail:* kathryn.koshansky@stonybrook.edu
*Med Dir:* Stuart B Cherney, MD

# North Carolina

**Lees - McRae College**
Athletic Training Prgm
PO Box 128
375 College Dr
Banner Elk, NC 28604
*Prgm Dir:* Rita A Smith
*Tel:* 828 898-8768    *Fax:* 828 898-8814
*E-mail:* smithr@lmc.edu
*Med Dir:* Marion M Herring, MD

**Gardner - Webb University**
Athletic Training Prgm
Department of Physical Education, Wellness & Sport
  Studies
Campus 7257
Boiling Springs, NC 28017
www.gardner-webb.edu
*Prgm Dir:* Ashley White, MS LAT ATC
*Tel:* 704 406-3810    *Fax:* 704 406-3503
*E-mail:* awhite@gardner-webb.edu
*Med Dir:* Kevin James, MD

**Appalachian State University**
Athletic Training Prgm
Hlth Leisure and Exercise Science
Boone, NC 28608
*Prgm Dir:* Jamie Moul, EdD ATC
*Tel:* 828 262-7557    *Fax:* 828 262-7680
*E-mail:* mouljl@appstate.edu
*Med Dir:* Patricia Geiger, MD

**Campbell University**
Athletic Training Prgm
PO Box 10
89 Pope St
Buies Creek, NC 27506
*Prgm Dir:* Mary Jones, PhD
*Tel:* 910 893-1362    *Fax:* 910 893-1424
*E-mail:* jonesm@campbell.edu
*Med Dir:* Kurt Ehlert, MD

**Univ of North Carolina at Chapel Hill**
Athletic Training Prgm
211 Fetzer, CB 8700
Chapel Hill, NC 27599-8700
*Prgm Dir:* Darin Padua
*Tel:* 919 962-5175    *Fax:* 919 962-0489
*E-mail:* dpadua@email.unc.edu
*Med Dir:* Tim Taft, MD

**Univ of North Carolina at Charlotte**
Athletic Training Prgm
226 Belk Gymnasium
Dept of Kinesiology
Charlotte, NC 28223
www.health.uncc.edu/academic_programs.cfm?pname=
  bsat
*Prgm Dir:* Tricia J Hubbard, PhD ATC
*Tel:* 704 687-6202    *Fax:* 704 687-3350
*E-mail:* thubbar1@uncc.edu
*Med Dir:* Robert L Jones, MD

**North Carolina Central University**
Athletic Training Prgm
1801 Fayetteville St
PO Box 19542
Durham, NC 27707
www.nccu.edu
*Prgm Dir:* Dawn Maffucci, MA ATC LAT
*Tel:* 919 530-7239    *Fax:* 919 530-6156
*E-mail:* dmaffucci@nccu.edu
*Med Dir:* Alison Toth, MD

**Elon University**
Athletic Training Prgm
Campus Box 2525
Elon, NC 27244
*Prgm Dir:* Gregory Calone, MS
*Tel:* 336 278-6715    *Fax:* 336 278-5918
*E-mail:* gcalone@elon.edu
*Med Dir:* Ted Armour, MD

**Methodist College**
Athletic Training Prgm
5400 Ramsey St
Fayetteville, NC 28311
*Prgm Dir:* Hugh Harling, EdD
*Tel:* 910 630-7418    *Fax:* 910 630-7676
*E-mail:* hharling@methodist.edu

**Greensboro College**
Athletic Training Prgm
815 W Market St
Greensboro, NC 27401
*Prgm Dir:* Michelle Lesperance, MS
*Tel:* 336 272-7102, Ext 629    *Fax:* 336 217-7237
*E-mail:* mlesperance@gborocollege.edu

**Univ of North Carolina at Greensboro**
Athletic Training Prgm
250 HHP Dept of Exercise & Sport Science
Greensboro, NC 27402
www.uncg.edu/ess/atep
*Prgm Dir:* Jolene Henning, EdD ATC
*Tel:* 336 334-3694    *Fax:* 336 334-3238
*E-mail:* jmhenni2@uncg.edu
*Med Dir:* John Lalonde, MD

**East Carolina University**
Athletic Training Prgm
245 Ward Sports Medicine Bldg
Greenville, NC 27858
*Prgm Dir:* Katie Walsh, EdD ATC-L
*Tel:* 252 328-4560    *Fax:* 252 328-4565
*E-mail:* walshk@ecu.edu
*Med Dir:* John Siegel, MD

**Lenoir - Rhyne College**
Athletic Training Prgm
PO Box 7356
Hickory, NC 28603
*Prgm Dir:* Michael R McGee, EdD LAT ATC
*Tel:* 828 328-7127    *Fax:* 828 267-3445
*E-mail:* mcgee@lrc.edu
*Med Dir:* Robert Liljeberg, MD

### High Point University
Athletic Training Prgm
833 Montlieu Ave
High Point, NC 27262
*Prgm Dir:* Rick Proctor, EdD LAT ATC
*Tel:* 336 841-9267    *Fax:* 336 841-9182
*E-mail:* rproctor@highpoint.edu
*Med Dir:* Edward Weller, MD

### Mars Hill College
Athletic Training Prgm
PO Box 6668
Mars Hill, NC 28754
*Prgm Dir:* Robin Kennel, MS
*Tel:* 828 689-1108    *Fax:* 828 689-1313
*E-mail:* rkennel@mhc.edu
*Med Dir:* Jay Jansen, MD

### Catawba College
Athletic Training Prgm
2300 W Innes St
Salisbury, NC 28144
www.catawba.edu
*Prgm Dir:* Robert Dingle, HSD LAT ATC
*Tel:* 704 637-4455    *Fax:* 704 637-5705
*E-mail:* rdingle@catawba.edu
*Med Dir:* James Comadoll, MD

### Univ of North Carolina - Wilmington
Athletic Training Prgm
601 S College Rd
Wilmington, NC 28403-5956
http://people.uncw.edu/brownk/UNCAThomepage.htm
*Prgm Dir:* Kirk Brown, PhD LAT ATC
*Tel:* 910 962-7184    *Fax:* 910 962-7073
*E-mail:* brownk@uncw.edu
*Med Dir:* William Sutton, MD

### Barton College
Athletic Training Prgm
PO Box 5000
Wilson, NC 27893
www.barton.edu
*Prgm Dir:* Carla Stoddard, MS ATC LAT
*Tel:* 252 399-6377    *Fax:* 252 399-6516
*E-mail:* cstoddard@barton.edu
*Med Dir:* Robert Satterfield, MD

### Wingate University
Athletic Training Prgm
Campus Box 3079
Wingate, NC 28174
www.wingate.edu/academics/sport_sciences/
    athletic_training
*Prgm Dir:* Traci N Gearhart, PhD
*Tel:* 704 233-8179    *Fax:* 704 233-8365
*E-mail:* tgearhar@wingate.edu
*Med Dir:* David Kingery, MD

# North Dakota

### University of Mary
Athletic Training Prgm
7500 University Dr
Bismarck, ND 58504-9652
www.umary.edu/UM/AcademicInformation/
    Undergraduate/hps/AthleticTraining
*Prgm Dir:* Blaine Steiner, MS ATC LAT
*Tel:* 701 355-8188    *Fax:* 701 255-7687
*E-mail:* bsteiner@umary.edu
*Med Dir:* Raymond Gruby, MD

### North Dakota State University
Athletic Training Prgm
Bentson Bunker Fieldhouse 9C
PO Box 5576
Fargo, ND 58105-5576
*Prgm Dir:* Pamela Hansen, EdD ATC
*Tel:* 701 231-8093    *Fax:* 701 231-8872
*E-mail:* pamela.j.hansen@ndsu.edu
*Med Dir:* Bruce Piatt, MD

### University of North Dakota
Athletic Training Prgm
Div of Sports Medicine
2751 2nd Ave N Stop 9013
Grand Forks, ND 58202-9013
*Prgm Dir:* Steven Westereng, LATC MA ATC
*Tel:* 701 777-3886    *Fax:* 701 777-4846
*E-mail:* Wester@medicine.nodak.edu
*Med Dir:* William Mann, MD

# Ohio

### Ohio Northern University
Athletic Training Prgm
Dept of HPSS
#243 Sports Center
Ada, OH 45810
www.onu.edu/a+s/hpss/at
*Prgm Dir:* Michelle Glon, MS ATC
*Tel:* 419 772-2443    *Fax:* 419 772-2437
*E-mail:* m-glon@onu.edu
*Med Dir:* Todd Ignarski, MD

### University of Akron
Athletic Training Prgm
Memorial Hall 77D
Akron, OH 44325-5103
*Prgm Dir:* Stacey Buser, MS ATC LAT
*Tel:* 330 972-7475    *Fax:* 330 972-5293
*E-mail:* buser@uakron.edu
*Med Dir:* Julie Kerr, MD

### Mount Union College
Athletic Training Prgm
1972 Clark Ave
Alliance, OH 44601
*Prgm Dir:* Daniel M Gorman, MS ATC/L
*Tel:* 330 823-4882    *Fax:* 330 823-2399
*E-mail:* gormandm@muc.edu
*Med Dir:* Michael McGrady, MD

### Ashland University
Athletic Training Prgm
916 King Rd
Arthur L & Maxine Sheets Rybolt Sport Sciences Center
Ashland, OH 44805
www.ashland.edu
*Prgm Dir:* Dennis M Gruber, EdD ATC
*Tel:* 419 289-5446    *Fax:* 419 289-5460
*E-mail:* dgruber@ashland.edu
*Med Dir:* Mark Elderbrock, MD

### Ohio University
Athletic Training Prgm
Grover Center E-188
Athens, OH 45701-2979
*Prgm Dir:* Kristi White, PhD LAT
*Tel:* 740 597-1876    *Fax:* 740 593-0284
*E-mail:* whitek2@ohio.edu
*Med Dir:* Rodney Comisar, MD

### Baldwin-Wallace College
Athletic Training Prgm
275 Eastland Rd
Berea, OH 44017
*Prgm Dir:* Karyn A Gentile, ATC/L
*Tel:* 440 826-3463    *Fax:* 440 826-3577
*E-mail:* kgentile@bw.edu
*Med Dir:* James Krcik, MD

### Bowling Green State University
Athletic Training Prgm
Eppler Complex N217
Bowling Green, OH 43403
www.bgsu.edu
*Prgm Dir:* Chris Schommer, MEd ATC
*Tel:* 419 372-6810    *Fax:* 419 372-0383
*E-mail:* schomme@bgsu.edu
*Med Dir:* Richard Barker, MD

### Cedarville University
Athletic Training Prgm
251 N Main St
Cedarville, OH 45314
www.cedarville.edu
*Prgm Dir:* Evan Hellwig, PhD PT ATC
*Tel:* 937 766-7691    *Fax:* 937 766-2795
*E-mail:* hellwige@cedarville.edu
*Med Dir:* Bruce Binder, MD

### College of Mt St Joseph
Athletic Training Prgm
Dept of Health Sciences
5701 Delhi Rd
Cincinnati, OH 45233-1870
www.msj.edu
*Prgm Dir:* Malissa Martin, EdD ATC CSCS
*Tel:* 513 244-4542    *Fax:* 513 244-4654
*E-mail:* Malissa_martin@mail.msj.edu
*Med Dir:* Howard Schertzinger, JD

### University of Cincinnati
Athletic Training Prgm
PO Box 210002
Cincinnati, OH 45221-0002
www.uc.edu/athletictraining
*Prgm Dir:* Pat Graman, MA ATC
*Tel:* 513 556-0576    *Fax:* 513 556-3898
*E-mail:* pat.graman@uc.edu
*Med Dir:* W Kenneth Stephens, MD

### Xavier University
Athletic Training Prgm
3800 Victory Pkwy
Cincinnati, OH 45207-6312
*Prgm Dir:* Tina Davlin, PhD
*Tel:* 513 745-3430    *Fax:* 513 745-4291
*E-mail:* davlin@xavier.edu
*Med Dir:* Henry Stiene, MD

### Capital University
Athletic Training Prgm
Capital Center
1 College and Main
Columbus, OH 43209
*Prgm Dir:* Bonnie Goodwin, MESS ATC
*Tel:* 614 236-6667    *Fax:* 614 236-6178
*E-mail:* bgoodwin@capital.edu
*Med Dir:* Thomas Pommerang, DO

### Ohio State University
Athletic Training Prgm
Athletic Training Division
453 W 10th Ave
Columbus, OH 43210
http://amp.osu.edu/at
*Prgm Dir:* Mark Merrick, PhD ATC
*Tel:* 614 247-6231    *Fax:* 614 292-0210
*E-mail:* merrick.29@osu.edu
*Med Dir:* Tom Best, MD, PhD

### Wright State University
Athletic Training Prgm
Health & Physical Education Dept
Rm 303 Nutter Center
Dayton, OH 45435
*Prgm Dir:* L Tony Ortiz, ATC
*Tel:* 937 775-3259
*E-mail:* tony.ortiz@wright.edu
*Med Dir:* Barry Fisher, MD

### Defiance College
Athletic Training Prgm
701 N Clinton St
Defiance, OH 43512
www.defiance.edu
*Prgm Dir:* Cynthia M Studrawa, MA/ATC
*Tel:* 419 783-2393    *Fax:* 419 783-2369
*E-mail:* cstudrawa@defiance.edu
*Med Dir:* Nathan Fogt, MD

**University of Findlay**
Athletic Training Prgm
1000 N Main St
Findlay, OH 45840
*Prgm Dir:* Donald Fuller, PhD ATC
*Tel:* 419 434-6739
*E-mail:* dfuller@findlay.edu
*Med Dir:* Michael Stump, MD

**Denison University**
Athletic Training Prgm
Dept of Physical Education
Granville, OH 43023
www.denison.edu/phed/ATindex.html
*Prgm Dir:* Eric R Winters, PhD ATC
*Tel:* 740 587-6311   *Fax:* 740 587-5742
*E-mail:* winterse@denison.edu
*Med Dir:* Jim Sturmi, MD

**Kent State University**
Athletic Training Prgm
School of Exercise Leisure and Sport
Rm 161D MACC Gym Annex
Kent, OH 44242
www.kent.edu/athletictraining
*Prgm Dir:* Kimberly S Peer, EdD ATC/L
*Tel:* 330 672-0231   *Fax:* 330 672-4106
*E-mail:* kpeer@kent.edu
*Med Dir:* Mark J Hudak, MD

**Marietta College**
Athletic Training Prgm
Dept of Sports Medicine
215 Fifth St
Marietta, OH 45750-4031
*Prgm Dir:* Richard Crowther, MS ATC
*Tel:* 740 376-4774   *Fax:* 740 376-4405
*E-mail:* crowthers@marietta.edu
*Med Dir:* George Tokodi, DO

**Miami University**
Athletic Training Prgm
PHS Dept
Oxford, OH 45056
www.ohiou.edu
*Prgm Dir:* J Brett Massie, EdD ATC
*Tel:* 513 529-8105   *Fax:* 513 529-5006
*E-mail:* massiejb@muohio.edu
*Med Dir:* Stephen W Dailey, MD

**Shawnee State University**
Athletic Training Prgm
940 Second St
Portsmouth, OH 45662
www.shawnee.edu
*Prgm Dir:* Tony Ward, MS ATC LAT
*Tel:* 740 351-3348   *Fax:* 740 351-3172
*E-mail:* tward@shawnee.edu
*Med Dir:* George Pettit, MD

**Heidelberg College**
Athletic Training Prgm
310 E Market St
Tiffin, OH 44883
*Prgm Dir:* Margo Greicar
*Tel:* 419 448-2290   *Fax:* 419 448-2126
*E-mail:* mgreicar@heidelberg.edu
*Med Dir:* James Anthony, MD

**University of Toledo**
Athletic Training Prgm
College of Health & Human Services
Dept of Kinesiology, 2801 W Bancroft St
Toledo, OH 43606
http://homepages.utoledo.edu/sportsmed
*Prgm Dir:* James M Rankin, PhD ATC
*Tel:* 419 530-2752   *Fax:* 419 530-2477
*E-mail:* james.rankin@utoledo.edu
*Med Dir:* Roger J Kruse, MD

**Urbana University**
Athletic Training Prgm
579 College Way
Urbana, OH 43078
www.urbana.edu
*Prgm Dir:* Bradley K Adams, MEd
*Tel:* 937 652-1225   *Fax:* 937 484-1223
*E-mail:* badams@urbana.edu
*Med Dir:* Boyd C Hoddinott, MD

**Otterbein College**
Athletic Training Prgm
160 Center St, Rike Center
Westerville, OH 43081
*Prgm Dir:* Joan E Rocks, MS LATC
*Tel:* 614 823-3505   *Fax:* 614 823-1966
*E-mail:* jrocks@otterbein.edu

**Wilmington College**
Athletic Training Prgm
251 Ludovic St
Pyle Center Box 1246
Wilmington, OH 45177
*Prgm Dir:* Larry Howard, MA ATC/L
*Tel:* 937 382-6661, Ext 257   *Fax:* 937 383-8527
*E-mail:* larry_howard@wilmington.edu
*Med Dir:* John Turba, MD

# Oklahoma

**East Central University**
Athletic Training Prgm
1000 East 14th
Ada, OK 74820
*Prgm Dir:* Jeff Williams, MHR
*Tel:* 580 310-5357   *Fax:* 580 332-8361
*E-mail:* jwilliams@mailclerk.ecok.edu

**Southern Nazarene University**
Athletic Training Prgm
6729 NW 39th Expressway
Bethany, OK 73008
www.snu.edu
*Prgm Dir:* Sylvia M Goodman, EdD LAT C
*Tel:* 405 717-6263   *Fax:* 405 717-6257
*E-mail:* sgoodman@snu.edu
*Med Dir:* Ami Seims, MD

**Oklahoma State University**
Athletic Training Prgm
191 Colvin Center
Stillwater, OK 74078
http://fp.okstate.edu/ptona/atep
*Prgm Dir:* Tona Palmer Hetzler, EdD
*Tel:* 405 744-9437   *Fax:* 405 744-6507
*E-mail:* tona.palmer@okstate.edu
*Med Dir:* Ken Smith, DO

**University of Tulsa**
Athletic Training Prgm
Chapman Hall 355
600 S College Ave
Tulsa, OK 74104-3189
*Prgm Dir:* Robin Ploeger, EdD ATC/L
*Tel:* 918 631-3170   *Fax:* 918 631-2068
*E-mail:* robin-ploeger@utulsa.edu
*Med Dir:* T Jeffery Emel, MD

**Southwestern Oklahoma State University**
Athletic Training Prgm
100 Campus Dr
Weatherford, OK 73096
*Prgm Dir:* Michael Catterson, MS
*Tel:* 580 774-3073   *Fax:* 580 774-3749
*E-mail:* michael.catterson@swosu.edu
*Med Dir:* Blake Badgett, MD

# Oregon

**Oregon State University**
Athletic Training Prgm
226 Langton Hall
Corvallis, OR 97331-3302
*Prgm Dir:* Mark Hoffman
*Tel:* 541 737-6801   *Fax:* 541 737-2788
*E-mail:* Mark.hoffman@oregonstate.edu
*Med Dir:* Craig Graham, MD

**Linfield College**
Athletic Training Prgm
900 SE Baker St
McMinnville, OR 97128
*Prgm Dir:* Laura Kenow, MS
*Tel:* 503 883-2580   *Fax:* 503 883-2453
*E-mail:* lkenow@linfield.edu
*Med Dir:* Stephen W Teal, MD

**George Fox University**
Athletic Training Prgm
414 N Meridian St #6188
Newberg, OR 97132
*Prgm Dir:* Karen Hostetter, ATC MS
*Tel:* 503 554-2922   *Fax:* 503 554-3864
*E-mail:* khostetter@georgefox.edu
*Med Dir:* Tom Croy, MD, Orthopedic Surgeon

# Pennsylvania

**Neumann College**
Athletic Training Prgm
One Neumann Dr
Aston, PA 19014
www.neumann.edu
*Prgm Dir:* Hurbert Lee, MA ATC CSCS
*Tel:* 610 361-2499   *Fax:* 610 361-2494
*E-mail:* leeh@neumann.edu
*Med Dir:* Chris Mehallo, MD

**University of Pittsburgh - Bradford**
Athletic Training Prgm
300 Campus Dr
Bradford, PA 16701
www.upb.pitt.edu
*Prgm Dir:* Jason Honeck, MSEd ATC
*Tel:* 814 362-7536   *Fax:* 814 362-7503
*E-mail:* honeck@exchange.upb.pitt.edu
*Med Dir:* Jill Owens, MD

**California University of Pennsylvania**
Athletic Training Prgm
250 University Ave
California, PA 15419
*Prgm Dir:* Bruce Barnhart, EdD ATC PTA
*Tel:* 724 938-4562   *Fax:* 724 938-4342
*E-mail:* barnhart@cup.edu
*Med Dir:* Greg Christiansen, MD

**East Stroudsburg University**
Athletic Training Prgm
200 Prospect St
East Stroudsburg, PA 18301
*Prgm Dir:* John R Hauth, EdD
*Tel:* 570 422-3006   *Fax:* 570 422-3616
*E-mail:* jhauth@po-box.esu.edu
*Med Dir:* George A Primiano, MD

**Mercyhurst College**
Athletic Training Prgm
501 E 38th St
Erie, PA 16546-0001
*Prgm Dir:* Suzanne Gusuie, MA ATC
*Tel:* 814 824-2526, Ext 3012   *Fax:* 814 824-2204
*E-mail:* sgushie@mercyhurst.edu
*Med Dir:* Laura McIntosh, MD

**Messiah College**
Athletic Training Prgm
One College Ave
Grantham, PA 17027
www.messiah.edu
*Prgm Dir:* Edwin A Bush, MS ATC
*Tel:* 717 691-6037, Ext 6037    *Fax:* 717 691-6044
*E-mail:* sbush@messiah.edu
*Med Dir:* Kenneth Graf, MD

**Indiana University of Pennsylvania**
Athletic Training Prgm
228 Zink Hall
1190 Maple St
Indiana, PA 15705
www.hhs.iup.edu/hped
*Prgm Dir:* Jose E Rivera, MS Ed ATC
*Tel:* 724 357-5507    *Fax:* 724 357-3777
*E-mail:* jose.rivera@iup.edu

**Lock Haven University**
Athletic Training Prgm
104 Himes Hall
Lock Haven, PA 17745
*Prgm Dir:* Eric L Lippincott, PT ATC
*Tel:* 570 893-2781    *Fax:* 570 893-2220
*E-mail:* elippinc@lhup.edu
*Med Dir:* Michael Greenberg, MD

**Temple University**
Athletic Training Prgm
Dept of Kinesiology
129 Pearson Hall
Philadelphia, PA 19122
*Prgm Dir:* Kathleen Swanik, PhD
*Tel:* 215 204-8836    *Fax:* 215 214-4414
*E-mail:* kswanik@temple.edu
*Med Dir:* Ray Moyer, MD

**Duquesne University**
Athletic Training Prgm
John G Rangos, Sr School of Health Sciences
122 Health Sciences Building
Pittsburgh, PA 15282
www.healthsciences.duq.edu/at/athome.html
*Prgm Dir:* Paula S Turocy, EdD ATC
*Tel:* 412 396-5695    *Fax:* 412 396-4160
*E-mail:* turocyp@duq.edu
*Med Dir:* Graham Johnstone, MD, and Edward Snell, MD

**University of Pittsburgh**
Athletic Training Prgm
4049 Forbes Tower
Pittsburgh, PA 15260
*Prgm Dir:* Kevin M Conley, PhD ATC
*Tel:* 412 383-6737    *Fax:* 412 383-6636
*E-mail:* kconley@shrs.pitt.edu
*Med Dir:* Freddie H Fu, MD

**Alvernia College**
Athletic Training Prgm
400 St Bernardine St
Reading, PA 19607
*Prgm Dir:* Thomas Porrazzo, PhD
*Tel:* 610 796-8311    *Fax:* 610 796-8349
*E-mail:* tom.porrazzo@alvernia.edu
*Med Dir:* John Martin, Jr, MD

**Marywood University**
Athletic Training Prgm
2300 Adams Ave
Health and Physical Education Center
Scranton, PA 18509-1598
www.marywood.edu/departments/health_pe/
    athletictraining
*Prgm Dir:* Christopher W O'Brien, MS ATC
*Tel:* 570 348-6259, Ext 2692   *Fax:* 570 961-4743
*E-mail:* cobrien@marywood.edu
*Med Dir:* P Christopher Metzger, MD

**Slippery Rock University**
Athletic Training Prgm
Department of Exercise & Rehabilitative Sciences
114 West Gym
Slippery Rock, PA 16057
www.sru.edu/ers
*Prgm Dir:* Bonnie Siple, MS ATC
*Tel:* 724 738-2930    *Fax:* 724 738-4890
*E-mail:* bonnie.siple@sru.edu
*Med Dir:* Daniel Ferguson, MD

**Eastern University**
Athletic Training Prgm
1300 Eagle Rd
St David, PA 19087
*Prgm Dir:* Michele Monaco, MS
*Tel:* 610 225-5169    *Fax:* 610 341-1460
*E-mail:* mmonaco@eastern.edu

**Penn State University**
Athletic Training Prgm
275 Recreation Bldg
University Park, PA 16802
*Prgm Dir:* Lauren Kramer, PhD ATC
*Tel:* 814 863-1758    *Fax:* 814 865-1275
*E-mail:* lco100@psu.edu
*Med Dir:* Wayne Sebastianelli, MD

**Waynesburg College**
Athletic Training Prgm
51 W College St
Waynesburg, PA 15370
*Prgm Dir:* Michele Kabay, MEd
*Tel:* 724 852-3309, Ext 309    *Fax:* 724 852-3295
*E-mail:* mkabay@waynesburg.edu
*Med Dir:* Joseph Stracci, MD

**West Chester University**
Athletic Training Prgm
Department of Sports Medicine
West Chester, PA 19383
http://health-sciences.wcupa.edu/sportsmed
*Prgm Dir:* Neil Curtis, EdD ATC
*Tel:* 610 436-2969    *Fax:* 610 436-2803
*E-mail:* ncurtis@wcupa.edu
*Med Dir:* Arthur Bartolozzi, MD

**King's College**
Athletic Training Prgm
133 N River St
Wilkes-Barre, PA 18711
www.kings.edu
*Prgm Dir:* Jeremy Simington, MS ATC
*Tel:* 570 208-5900, Ext 5636   *Fax:* 570 208-5988
*E-mail:* jeremysimington@kings.edu
*Med Dir:* William Charlton, MD

# South Carolina

**Charleston Southern University**
Athletic Training Prgm
9200 University Blvd
Charleston, SC 29406
www.csuniv.edu
*Prgm Dir:* Kelly Harkins, MS
*Tel:* 843 863-7399    *Fax:* 843 863-7675
*E-mail:* kharkins@csuniv.edu
*Med Dir:* Seth Kupferman, MD

**College of Charleston**
Athletic Training Prgm
66 George St
Charleston, SC 29424
*Prgm Dir:* Susan L Rozzi, PhD ATC
*Tel:* 843 953-7163    *Fax:* 843 953-6757
*E-mail:* rozzis@cofc.edu
*Med Dir:* Carek Peter, MD

**University of South Carolina**
Athletic Training Prgm
Blatt Center
Dept of PE
Columbia, SC 29208
*Prgm Dir:* James M Mensch, PhD
*Tel:* 803 777-3846    *Fax:* 803 777-6250
*E-mail:* jmensch@gwm.sc.edu
*Med Dir:* Robert Peele, MD

**Erskine College**
Athletic Training Prgm
PO Box 338
Due West, SC 29639
www.erskine.edu
*Prgm Dir:* Jim Riser, MS ATC
*Tel:* 864 379-8899    *Fax:* 864 379-2197
*E-mail:* riser@erskine.edu
*Med Dir:* Richard Taylor, Jr, MD

**Limestone College**
Athletic Training Prgm
1115 College Dr
Gaffney, SC 29340
www.limestone.edu
*Prgm Dir:* Vanessa B Fulbright, MA ATC
*Tel:* 864 488-8243    *Fax:* 864 488-8211
*E-mail:* vfulbright@llimestone.edu
*Med Dir:* Todd Morgan, MD

**Lander University**
Athletic Training Prgm
PO Box 6026
Greenwood, SC 29649
*Prgm Dir:* Jerald Hawkins, EdD
*Tel:* 864 388-8280    *Fax:* 864 388-8660
*E-mail:* jhawkins@lander.edu

**Winthrop University**
Athletic Training Prgm
117 Peabody Gymnasium
Rock Hill, SC 29733
www.winthrop.edu
*Prgm Dir:* Alice J McLaine, PhD ATC
*Tel:* 803 323-2123    *Fax:* 803 323-2124
*E-mail:* mclainea@winthrop.edu
*Med Dir:* Robert Lesslie, MD

# South Dakota

**South Dakota State University**
Athletic Training Prgm
Health PE & Recreation
Box 2820
Brookings, SD 57007-1497
www.sdstate.edu
*Prgm Dir:* Jim Booher, PhD PT ATC
*Tel:* 605 688-5824    *Fax:* 605 688-5999
*E-mail:* jim.booher@sdstate.edu
*Med Dir:* John Ramsay, MD

**Dakota Wesleyan University**
Athletic Training Prgm
1200 W University, Box 912
Mitchell, SD 57301
www.dwu.edu/catalog/courses/athletic_training.htm
*Prgm Dir:* Dan Wagner, EdD ATC CSCS
*Tel:* 605 995-2145    *Fax:* 605 995-2143
*E-mail:* dnwagner@dwu.edu
*Med Dir:* Robert McWhirter, MD

**National American University**
Athletic Training Prgm
321 Kansas City St
Rapid City, SD 57701
www.national.edu
*Prgm Dir:* Robert D Steele Jr, MS ATC LAT
*Tel:* 605 394-5085    *Fax:* 605 394-4871
*E-mail:* rsteele@national.edu
*Med Dir:* Lew Papendick, MD

**Augustana College**
Athletic Training Prgm
2001 S Summit Ave
Sioux Falls, SD 57197
www.augie.edu
*Prgm Dir:* Brian Gerry, MS ATC
*Tel:* 605 274-5534   *Fax:* 605 274-5298
*E-mail:* brian.gerry@augie.edu
*Med Dir:* Scott Boyens, MD

# Tennessee

**University of Tennessee - Chattanooga**
Athletic Training Prgm
615 McCallie Ave, Dept 6606
Chattanooga, TN 37403-2598
*Prgm Dir:* Marisa A Colston, PhD
*Tel:* 423 425-4743   *Fax:* 423 425-5395
*E-mail:* marisa-colston@utc.edu
*Med Dir:* David Jenkinson, DO

**Lee University**
Athletic Training Prgm
1120 N Ocoee St
Cleveland, TN 37320-3450
*Prgm Dir:* Kelly Lumpkin, MS
*Tel:* 423 614-8474   *Fax:* 423 614-8438
*E-mail:* klumpkin@leeuniversity.edu
*Med Dir:* DeWayne Knight, MD

**Bryan College**
Athletic Training Prgm
721 Bryan Dr, Box 7814
Dayton, TN 37321
www.bryan.edu/atc
*Prgm Dir:* Christy Rodenbeck, MS
*Tel:* 423 775-7254   *Fax:* 423 775-7330
*E-mail:* athletictraining@bryan.edu
*Med Dir:* David Jenkinson, DO

**Tusculum College**
Athletic Training Prgm
60 Old Shiloh Rd
Greeneville, TN 37743
www.tusculum.edu
*Prgm Dir:* Jane C Sandusky, MS ATC/L
*Tel:* 800 729-0256, Ext 5731   *Fax:* 423 636-7404
*E-mail:* jsandusk@tusculum.edu
*Med Dir:* Richard W Pectol, MD

**Lincoln Memorial University**
Athletic Training Prgm
6965 Cumberland Gap Pkwy
Harrogate, TN 37752
www.lmunet.edu
*Prgm Dir:* Jack Mansfield, EdD
*Tel:* 423 869-6322   *Fax:* 423 869-6382
*E-mail:* jack.mansfield@lmunet.edu
*Med Dir:* Ron Dubin, MD

**Union University**
Athletic Training Prgm
1050 Union University Dr
PO Box 1824
Jackson, TN 38305
www.uu.edu
*Prgm Dir:* Cliff Pawley, MEd ATC
*Tel:* 731 661-5529   *Fax:* 731 661-5182
*E-mail:* cpawley@uu.edu
*Med Dir:* Chris Lewis, MD

**Carson - Newman College**
Athletic Training Prgm
Box 71897
MSAC Rm 1013
Jefferson City, TN 37760
*Prgm Dir:* Gary Noble, EdD
*Tel:* 865 471-4650   *Fax:* 865 471-2037
*E-mail:* gnoble@cn.edu
*Med Dir:* Edward Phillips, MD

**Cumberland University**
Athletic Training Prgm
One Cumberland Square
Lebanon, TN 37087
www.cumberland.edu
*Prgm Dir:* Daniel W Rogers, MS ATC/L
*Tel:* 615 444-2562, Ext 1135   *Fax:* 615 444-2569
*E-mail:* drogers@cumberland.edu
*Med Dir:* Charles R Kaelin, MD

**University of Tennessee - Martin**
Athletic Training Prgm
3006 Elam Center
Martin, TN 38238
www.utm.edu/departments/cebs/hhp/
    AthleticTraining.php
*Prgm Dir:* Janet Wilbert, MA ATC
*Tel:* 731 881-7310   *Fax:* 731 881-7319
*E-mail:* jwilbert@utm.edu
*Med Dir:* G Bradford Wright, MD

**Middle Tennessee State University**
Athletic Training Prgm
PO Box 96
Murfreesboro, TN 37132
*Prgm Dir:* Helen Binkley, PhD
*Tel:* 615 904-8129   *Fax:* 615 904-8469
*E-mail:* hbinkley@mtsu.edu
*Med Dir:* Tom Johns, MD

**Lipscomb University**
Athletic Training Prgm
3901 Granny White Pike
Nashville, TN 37204-3951
*Prgm Dir:* Williams Ness, MS
*Tel:* 615 279-7042   *Fax:* 615 279-5852
*E-mail:* nesswr@lipscomb.edu
*Med Dir:* Burton Elrod, MD

# Texas

**Hardin-Simmons University**
Athletic Training Prgm
2200 Hickory St
Box 16180
Abilene, TX 79698
www.hsutx.edu
*Prgm Dir:* David Stuckey, MS ATC LAT EMT-I
*Tel:* 325 670-1378   *Fax:* 325 670-5852
*E-mail:* dstuckey@hsutx.edu
*Med Dir:* Dale Funk, MD

**University of Texas at Arlington**
Athletic Training Prgm
Box 19259
Arlington, TX 76019
www.uta.edu/coed/kinesiology/atep
*Prgm Dir:* Louise Fincher, EdD ATC LAT
*Tel:* 817 272-3107   *Fax:* 817 272-3233
*E-mail:* lfincher@uta.edu
*Med Dir:* Chris Witherspoon, MD

**University of Texas at Austin**
Athletic Training Prgm
1 University Station Shop D3700
Austin, TX 78712
www.edb.utexas.edu/atep
*Prgm Dir:* Brian Farr, MA ATC LAT CSCS
*Tel:* 512 471-9885   *Fax:* 512 471-0946
*E-mail:* bfarr@mail.utexas.edu
*Med Dir:* Robyn McCarty, MD

**University of Mary Hardin-Baylor**
Athletic Training Prgm
900 College St, Box 8010
Belton, TX 76502
*Prgm Dir:* Courtney Burken
*Tel:* 254 295-5514
*E-mail:* cburken@umhb.edu
*Med Dir:* George Burken, MD

**West Texas A&M University**
Athletic Training Prgm
WTAMU Box 60216
Canyon, TX 79016
*Prgm Dir:* Lorna Strong, MS ATC LAT
*Tel:* 806 651-2370
*E-mail:* lstrong@mail.wtamu.edu

**Texas A&M University - Commerce**
Athletic Training Prgm
ATEP Dept of Health
Kinesiology and Sports Studies
Commerce, TX 75429-3011
*Prgm Dir:* Ed Sunderland
*Tel:* 903 886-5553   *Fax:* 903 886-5365
*E-mail:* ed_sunderland@tamu-commerce.edu
*Med Dir:* Pat Prapan, MD

**Texas Christian University**
Athletic Training Prgm
PO Box 297730
3005 Stadium Dr
Ft Worth, TX 76129
www.tcu.edu
*Prgm Dir:* Sean Willeford, MS ATC LAT
*Tel:* 817 257-6737   *Fax:* 817 257-7702
*E-mail:* s.willeford@tcu.edu
*Med Dir:* Sam Haraldson, MD

**Southwestern University**
Athletic Training Prgm
PO Box 770
Georgetown, TX 78627-0770
www.southwestern.edu/athletics/ath_training.html
*Prgm Dir:* Miguel A Benavides, MEd ATC LAT
*Tel:* 512 863-1385   *Fax:* 512 863-1393
*E-mail:* benavidm@southwestern.edu
*Med Dir:* James Bray, MD

**Texas Tech Univ Health Sciences Center**
Athletic Training Prgm
3601 4th St, Stop 6226
Lubbock, TX 79430-6226
www.ttuhsc.edu
*Prgm Dir:* LesLee Taylor, PhD ATC LAT
*Tel:* 806 743-1032   *Fax:* 806 743-3518
*E-mail:* leslee.taylor@ttuhsc.edu
*Med Dir:* Kevin Crawford, MD

**East Texas Baptist University**
Athletic Training Prgm
1209 N Grove
Marshall, TX 75670
*Prgm Dir:* Dawn Johnston, MSE ATC LAT
*Tel:* 903 923-2237   *Fax:* 903 935-0162
*E-mail:* djohnston@etbu.edu
*Med Dir:* Douglas Waldman, MD

**Stephen F Austin State University**
Athletic Training Prgm
PO Box 13015-SFA Station
Nacogdoches, TX 75962
www.kin.sfasu.edu
*Prgm Dir:* Linda Stark-Bobo, PhD ATC LAT
*Tel:* 936 468-3503
*E-mail:* lbobo@sfasu.edu
*Med Dir:* John Miller, MD

**Angelo State University**
Athletic Training Prgm
ASU Station #10899
2601 W Ave N
San Angelo, TX 76909
www.angelo.edu/org/sportmed
*Prgm Dir:* Troy Hill, MS ATC LAT
*Tel:* 325 942-2264, Ext 247   *Fax:* 325 942-2272
*E-mail:* troy.hill@angelo.edu
*Med Dir:* Joe Wilkinson, MD

**University of the Incarnate Word**
Athletic Training Prgm
4301 Broadway, CPO 472
San Antonio, TX 78209
www.uiw.edu/athp
*Prgm Dir:* Brad Robinson, MS ATC LAT
*Tel:* 210 829-2787   *Fax:* 210 829-3174
*E-mail:* athp@uiwtx.edu
*Med Dir:* Ralph Curtis, MD

**Texas State University-San Marcos**
Athletic Training Prgm
Dept of Health, Physical Education & Recreation
Jowers Center 601 University Dr
San Marcos, TX 78666
http://uweb.txstate.edu/~jr41
*Prgm Dir:* Jack Ransone, PhD ATC FACSM
*Tel:* 512 245-8176, Ext na   *Fax:* 512 245-8678
*E-mail:* ransone@txstate.edu
*Med Dir:* Wes Wallis, MD

**Texas Lutheran University**
Athletic Training Prgm
1000 W Court St
Seguin, TX 78155
www.tlu.edu
*Prgm Dir:* Rick Roswell, MS ATC LAT
*Tel:* 830 372-8133   *Fax:* 830 372-8135
*E-mail:* rroswell@tlu.edu
*Med Dir:* David Starch, MD

**Baylor University**
Athletic Training Prgm
One Bear Place #97313
HHPR Dept
Waco, TX 76798-7313
www3.baylor.edu/HHPR/Undergraduate/Programs/atsm
*Prgm Dir:* Michael Chandler, EdD LAT ATC
*Tel:* 254 710-3505   *Fax:* 254 710-3527
*E-mail:* m_chandler@baylor.edu
*Med Dir:* R W Covington, MD

**Midwestern State University**
Athletic Training Prgm
3410 Taft Blvd
Wichita Falls, TX 76308
*Prgm Dir:* Jennifer Lancaster, MS ATC LAT
*Tel:* 940 397-6229
*E-mail:* jennifer.lancaster@mwsu.edu
*Med Dir:* G A Pino, MD

# Utah

**Southern Utah University**
Athletic Training Prgm
351 W University Blvd
Cedar City, UT 84720
www.suu.edu
*Prgm Dir:* Ben Davidson, MS
*Tel:* 435 586-7823   *Fax:* 435 865-8057
*E-mail:* davidson@suu.edu
*Med Dir:* Robert E Nakken, MD

**Weber State University**
Athletic Training Prgm
2801 University Circle
Ogden, UT 84408
http://programs.weber.edu/athletictraining
*Prgm Dir:* Valerie Herzog, EdD
*Tel:* 801 626-7656   *Fax:* 801 626-6228
*E-mail:* valerieherzog@weber.edu
*Med Dir:* Jeffrey Harrison, MD

**Brigham Young University**
Athletic Training Prgm
Dept of Exercise Sciences
267 SFH
Provo, UT 84602-2111
http://exsc.byu.edu/programs/atrain.html
*Prgm Dir:* David Kaiser, EdD ATC
*Tel:* 801 422-1627   *Fax:* 801 422-0555
*E-mail:* david_kaiser@byu.edu
*Med Dir:* Mitchell Pratte, DO

**University of Utah**
Athletic Training Prgm
1850 East 250 South, Rm 241
Salt Lake City, UT 84112-0920
www.health.utah.edu/ess/athletic-trng
*Prgm Dir:* Bradley Hayes, PhD
*Tel:* 801 581-7362   *Fax:* 801 585-3992
*E-mail:* bradley.hayes@hsc.utah.edu
*Med Dir:* Robert T Burks, MD

# Vermont

**University of Vermont**
Athletic Training Prgm
213 A Patrick Gymnasium
97 Spear St
Burlington, VT 05405
www.uvm.edu
*Prgm Dir:* Alan Maynard, MEd LAT ATC
*Tel:* 802 656-7678   *Fax:* 802 656-3492
*E-mail:* alan.maynard@uvm.edu
*Med Dir:* John Porter, MD

**Castleton State College**
Athletic Training Prgm
Glenbrook Gym
Castleton, VT 05735
*Prgm Dir:* Reese Barber, MS
*Tel:* 802 468-1435   *Fax:* 802 468-2189
*E-mail:* reese.barber@castleton.edu
*Med Dir:* Matt Gammons, MD

**Norwich University**
Athletic Training Prgm
158 Harmon Dr
Northfield, VT 05663
*Prgm Dir:* Todd Neuharth, ATC LAT MA
*Tel:* 802 485-2231   *Fax:* 802 485-2333
*E-mail:* tneuhart@norwich.edu
*Med Dir:* Kevin Crowley, MD

# Virginia

**Bridgewater College**
Athletic Training Prgm
402 E College St, Box 66
Bridgewater, VA 22812
*Prgm Dir:* Barbara Long, MS
*Tel:* 540 828-5771   *Fax:* 540 828-5733
*E-mail:* bhlong@bridgewater.edu
*Med Dir:* Irvin Hess, MD

**Averett University**
Athletic Training Prgm
420 W Main St
Danville, VA 24541
*Prgm Dir:* Lee Burton
*Tel:* 434 791-5821   *Fax:* 434 791-5740
*E-mail:* lee.burton@averett.edu
*Med Dir:* Carl Winfield, MD

**Emory & Henry College**
Athletic Training Prgm
PO Box 947
Emory, VA 24327-0947
www.ehc.edu
*Prgm Dir:* Dennis Cobler, MA ATC CSCS
*Tel:* 276 944-6589   *Fax:* 276 944-6738
*E-mail:* dcobler@ehc.edu
*Med Dir:* Timothy G McGarry, MD FACS

**Longwood University**
Athletic Training Prgm
201 High St
Farmville, VA 23909
www.longwood.edu/hrk/athletic_training
*Prgm Dir:* Sharon Menegoni, MS ATC
*Tel:* 434 395-2845   *Fax:* 434 395-2380
*E-mail:* menegonism@longwood.edu
*Med Dir:* Doug Cutter, MD

**James Madison University**
Athletic Training Prgm
Department of Health Sciences
MSC 4301
Harrisonburg, VA 22807
*Prgm Dir:* Herbert K Amato, DA ATC
*Tel:* 540 568-3576   *Fax:* 540 568-3336
*E-mail:* amatohk@jmu.edu
*Med Dir:* David Knitter, MD

**Liberty University**
Athletic Training Prgm
1971 University Blvd
Lynchburg, VA 24502
www.liberty.edu/academics/education/sport/index.cfm?
  PID=85
*Prgm Dir:* Jerry Pickard, EdD
*Tel:* 434 592-3762   *Fax:* 434 582-2412
*E-mail:* vpickard@liberty.edu
*Med Dir:* Richard Lane, MD

**Lynchburg College**
Athletic Training Prgm
1501 Lakeside Dr
Lynchburg, VA 24501
*Prgm Dir:* Timothy Laurent, EdD
*Tel:* 434 544-8726   *Fax:* 434 544-8365
*E-mail:* laurent@lynchburg.edu

**George Mason University**
Athletic Training Prgm
10900 University Blvd, MSN 4E5
Manassas, VA 20110-2203
*Prgm Dir:* Shane Caswell, PhD
*Tel:* 703 993-4638   *Fax:* 703 993-2025
*E-mail:* scaswell@gmu.edu
*Med Dir:* Frank A Pettrone, MD

**Radford University**
Athletic Training Prgm
Dept of Education, Sport & Health Education
PO Box 6957
Radford, VA 24142
*Prgm Dir:* Angela Mickle
*Tel:* 540 831-5330   *Fax:* 540 831-6053
*E-mail:* ammickle@radford.edu
*Med Dir:* Scott Kincaid

**Virginia Commonwealth University**
Athletic Training Prgm
PO Box 842020
Richmond, VA 23284-2020
www.soe.vcu.edu
*Prgm Dir:* Scott E Ross, PhD
*Tel:* 804 828-1948   *Fax:* 804 828-1946
*E-mail:* seross@vcu.edu
*Med Dir:* Michael DePalma, MD

**Roanoke College**
Athletic Training Prgm
221 College Ln
Salem, VA 24153
*Prgm Dir:* James Buriak, MS ATC
*Tel:* 540 375-2343   *Fax:* 540 375-2031
*E-mail:* buriak@roanoke.edu
*Med Dir:* Bertram Spetzler, MD

**Shenandoah University**
Athletic Training Prgm
1460 University Dr
Winchester, VA 22601
*Prgm Dir:* Rose Schmieg, MSPT OCS ATC CSCS
*Tel:* 540 545-7385   *Fax:* 540 545-7387
*E-mail:* rschmieg@su.edu
*Med Dir:* Winston Cameron, MD

# Washington

**Eastern Washington University**
Athletic Training Prgm
200 Physical Education Building, PEHR Dept
Cheney, WA 99004-2476
www.ewu.edu/x3356.xml
*Prgm Dir:* Garth Babcock, PhD ATC
*Tel:* 509 359-2427   *Fax:* 509 359-4833
*E-mail:* gbabcock@mail.ewu.edu
*Med Dir:* Thomas Halvorson, MD

**Washington State University**
Athletic Training Prgm
Dept of Educational Leadership & Counseling Psycholgy
PEB 122
Pullman, WA 99164-1410
http://academics.wsu.edu/fields/study.asp?ID=ATH_T
*Prgm Dir:* Carol Zweifel, MS PE
*Tel:* 509 335-0307   *Fax:* 509 335-6961
*E-mail:* carolz@wsu.edu
*Med Dir:* Dennis Garcia, MD

**Whitworth College**
Athletic Training Prgm
300 W Hawthorne Rd
Spokane, WA 99251-2501
www.whitworth.edu
*Prgm Dir:* Russell J Richardson, MA ATC
*Tel:* 509 777-3244   *Fax:* 509 777-3720
*E-mail:* rrichardson@whitworth.edu
*Med Dir:* Edward J Reisman, MD

# West Virginia

**Concord University**
Athletic Training Prgm
PO Box 1000
Campus Box 77
Athens, WV 24712
*Prgm Dir:* Jennifer LH Koerber, MSEd ATC CIE
*Tel:* 304 384-6063   *Fax:* 304 384-5117
*E-mail:* hvozdovicj@concord.edu
*Med Dir:* Byron Smith, DO

**West Virginia Wesleyan College**
Athletic Training Prgm
59 College Ave
Buckhannon, WV 26201-2995
*Prgm Dir:* Jeremy Sibold, EdD ATC
*Tel:* 304 473-8002   *Fax:* 304 473-8349
*E-mail:* sibold@wvwc.edu
*Med Dir:* Clyde Moxley, DO

**University of Charleston**
Athletic Training Prgm
2300 MacCorkle Ave SE
Charleston, WV 25304
*Prgm Dir:* Ericka Zimmerman, MS ATC
*Tel:* 304 357-4828   *Fax:* 304 357-4965
*E-mail:* erickazimmerman@ucwv.edu
*Med Dir:* David Santrock, MD

**Marshall University**
Athletic Training Prgm
College of Education & Human Services
400 Hal Greer Blvd
Huntington, WV 25755
www.marshall.edu/coehs/essr/athletic.training
*Prgm Dir:* R Daniel Martin, EdD ATC
*Tel:* 304 696-2412   *Fax:* 304 696-2928
*E-mail:* martind@marshall.edu
*Med Dir:* Charles Giangarra, MD

**West Virginia University**
Athletic Training Prgm
PO Box 6116 Coliseum
Morgantown, WV 26506-6116
www.wvu.edu/~physed/attrain/wvattr-1.htm
*Prgm Dir:* Vincent Stilger, HSD ATC
*Tel:* 304 293-3295, Ext 5148   *Fax:* 304 293-4641
*E-mail:* vstilger@mail.wvu.edu
*Med Dir:* Dana Brooks, EdD

**Alderson-Broaddus College**
Athletic Training Prgm
500 College Hill Rd
Box 2062
Philippi, WV 26416
*Prgm Dir:* Eric M Shor, MS ATC
*Tel:* 304 457-6276   *Fax:* 304 457-6291
*E-mail:* shorem@mail.ab.edu
*Med Dir:* Richard Topping MD, Luke Pavlovich MD

# Wisconsin

**University of Wisconsin - Eau Claire**
Athletic Training Prgm
Dept of Kinesiology
213 McPhee Center
Eau Claire, WI 54702
www.uwec.edu/kin
*Prgm Dir:* Robert Stow, PhD
*Tel:* 715 836-2022   *Fax:* 715 836-4074
*E-mail:* stowrc@uwec.edu
*Med Dir:* John Drawbert, MD

**Carthage College**
Athletic Training Prgm
2001 Alford Park Dr
Kenosha, WI 53140
*Prgm Dir:* Daniel Ruffner, MS
*Tel:* 262 551-5741   *Fax:* 262 551-5951
*E-mail:* druffner@carthage.edu

**University of Wisconsin - La Crosse**
Athletic Training Prgm
135 Mitchell Hall
La Crosse, WI 54601
*Prgm Dir:* Mark H Gibson, MS LAT PT
*Tel:* 608 785-8190   *Fax:* 608 785-8172
*E-mail:* gibson.mark@uwlax.edu
*Med Dir:* Scott Escher, MD

**University of Wisconsin - Madison**
Athletic Training Prgm
2000 Observatory Dr, Rm 1037
Madison, WI 53706
*Prgm Dir:* Andrew Winterstein, PhD
*Tel:* 608 265-2503   *Fax:* 608 262-1656
*E-mail:* winterstein@education.wisc.edu
*Med Dir:* Greg Landry, MD

**Concordia University Wisconsin**
Athletic Training Prgm
12800 N Lake Shore Dr
Mequon, WI 53097
www.cuw.edu
*Prgm Dir:* Russell DeLap, MBA ATC CSCS
*Tel:* 262 243-4323, Ext 4323   *Fax:* 262 243-2969
*E-mail:* russell.delap@cuw.edu
*Med Dir:* William B Smith, MD

**Marquette University**
Athletic Training Prgm
PO Box 1881
Milwaukee, WI 53201-1881
*Prgm Dir:* Christopher F Geiser
*Tel:* 414 288-5069   *Fax:* 414 288-6079
*E-mail:* christopher.geiser@mu.edu
*Med Dir:* Carolyn Smith

**University of Wisconsin - Milwaukee**
Athletic Training Prgm
Dept of Human Movement Sciences
Enderis 413
Milwaukee, WI 53201-0413
*Prgm Dir:* Jennifer Earl, PhD
*Tel:* 414 229-3227   *Fax:* 414 229-2619
*E-mail:* jearl@uwm.edu
*Med Dir:* H Richard Weiner, MD

**University of Wisconsin - Oshkosh**
Athletic Training Prgm
169H Kolf Center
Oshkosh, WI 54901
*Prgm Dir:* Hal Strough, PhD LAT/ATC
*Tel:* 920 424-1298   *Fax:* 920 424-1068
*E-mail:* strough@uwosh.edu
*Med Dir:* John Swanson, MD

**University of Wisconsin - Stevens Point**
Athletic Training Prgm
129 HEC
Stevens Point, WI 54481
*Prgm Dir:* Holly Herrmann, PhD
*Tel:* 715 346-2922   *Fax:* 715 346-4655
*E-mail:* hherrman@uwsp.edu
*Med Dir:* James Banovetz, MD

**Carroll College**
Athletic Training Prgm
100 N East Ave, Health Sciences
Waukesha, WI 53186
*Prgm Dir:* Kristopher Hartz, MS
*Tel:* 262 524-7701   *Fax:* 262 524-7690
*E-mail:* khartz@cc.edu

# Wyoming

**University of Wyoming**
Athletic Training Prgm
Div of Kinesiology and Health Dept 3196
1000 E University Ave
Laramie, WY 82071
www.uwyo.edu
*Prgm Dir:* William T Lyons
*Tel:* 307 766-5285, Ext None   *Fax:* 307 766-4090
*E-mail:* iceit@uwyo.edu
*Med Dir:* Robert Curnow, MD

## Athletic Trainer

| Programs* | Class Capacity | Begins | Length (months) | Award | Res Tuition | Non-res Tuition | Stipend | Offers:‡ 1 | 2 | 3 |
|---|---|---|---|---|---|---|---|---|---|---|
| **Alabama** | | | | | | | | | | |
| Huntingdon College (Montgomery) | 32 | Aug | 15 | Dipl, BA | $14,560 | $14,560 | | | | |
| University of West Alabama (Livingston) | 40 | Aug | 48 | Dipl, BS | $3,838 | $7,676 | | | | |
| **Arizona** | | | | | | | | | | |
| Northern Arizona University (Flagstaff) | 45 | Aug | 24 | BS | $4,076 | $12,596 | | | | |
| **Arkansas** | | | | | | | | | | |
| Arkansas State University | 12 | Jan | 36 | Dipl, BS | $2,607 | $5,787 | | | | |
| Henderson State University (Arkadelphia) | 30 | Aug | 36 | BS | $1,820 | $3,640 | | | | • |
| Ouachita Baptist University (Arkadelphia) | 12 | Aug | 18 | BA | $18,900 | $18,900 | | | | |
| Southern Arkansas University (Magnolia) | 8 | Aug | 36 | BS | $1,950 | $2,955 | | • | | |
| University of Central Arkansas (Conway) | 10 | Jan | 25 | BS | $4,500 | $5,700 | | • | | |
| **California** | | | | | | | | | | |
| Azusa Pacific University | 15 | Jan sophomore yr | 20 | BA | $28,000 | $28,000 | | | | |
| California State University - Fullerton | 10 | Aug | 30 | Cert, BS | $1,572 | $6,768 | | | | |
| California State University - Northridge | 15 | Jun | 36 | BS | $1,389 | $5,643 | | | | |
| California State University - Sacramento | 40 | Sep | 48 | BS | $3,280 | $11,416 | | • | | |
| Chapman University (Orange) | 50 | Aug | 36 | BS | $35,000 | $35,000 | | | | |
| Concordia Univesity - Irvine | 30 | Aug | 48 | BA | $28,000 | $28,000 | | | | |
| Humboldt State University (Arcata) | | Aug | 36 | BS | $1,034 | $1,156 | | | | |
| San Diego State University | 70 | Aug | 18 | Dipl, BS | $1,854 | $4,230 | | • | | |
| San Jose State University | 40 | Aug Jan | 48 | BS | $2,500 | $9,700 | | | | |
| University of La Verne | 30 | Fall | 27 | Dipl, BS | $24,260 | $24,260 | | | | • |
| **Colorado** | | | | | | | | | | |
| Colorado State University - Pueblo | 32 | Aug | 20 | Dipl, BS | $1,310 | $5,268 | | | | |
| Ft Lewis College (Durango) | 24 | Aug | 24 | BA | $2,551 | $6,595 | | | | |
| Metropolitan State College of Denver | 20 | Aug | 48 | BS | $2,564 | $9,235 | | | | |
| **Connecticut** | | | | | | | | | | |
| Quinnipiac University (Hamden) | 22 | Sep | 48 | Dipl, BS | $26,280 | $26,280 | | | | |
| University of Connecticut (Storrs) | 12 | Jan | 28 | BS | $9,000 | $17,000 | | | | |
| **Delaware** | | | | | | | | | | |
| University of Delaware (Newark) | 45 | Aug | 36 | BS | $6,980 | $17,690 | | | | |
| **Florida** | | | | | | | | | | |
| Barry University (Miami Shores) | 20 | Sep | 48 | BS | $21,631 | $21,631 | | • | | • |
| Florida Southern College (Lakeland) | 48 | Aug | | Dipl, BS | $18,500 | $19,700 | | | | |
| University of Florida (Gainesville) | 20 | Aug | 24 | BS | $2,955 | $15,827 | | • | | |
| University of Miami (Coral Gables) | 60 | Aug | | Dipl, BS Ed | $31,288 | $31,288 | | | | |
| University of North Florida (Jacksonville) | 55 | Aug | 22 | BS | $3,101 | $6,959 | | | | |
| University of South Florida (Tampa) | 30 | Summer | 24 | Dipl, AT | | | | | | • |
| University of Tampa | 24 | Aug | 36 | Dipl, BS | $23,162 | $0 | | | | |
| University of West Florida (Pensacola) | 24 | Aug | 24 | BS/HLE | $3,000 | $9,500 | | • | | |
| **Georgia** | | | | | | | | | | |
| Georgia Southern University (Statesboro) | 45 | Aug | 36 | Dipl, BSK | $2,814 | $3,525 | | | | • |
| North Georgia College & State University (Dahlonega) | 12 | Aug | 20 | Dipl, BS | $2,560 | $10,242 | | | | |
| University of Georgia (Athens) | 45 | Aug | 24 | BSEd | $4,628 | $16,848 | | • | | |
| Valdosta State University | 16 | Aug | 36 | BS | $1,639 | $5,297 | | • | | |
| **Idaho** | | | | | | | | | | |
| University of Idaho (Moscow) | 30 | Aug | 48 | BS | $4,200 | $9,600 | | | | |
| **Illinois** | | | | | | | | | | |
| Aurora University | 30 | Aug | 27 | BS | $16,090 | $16,090 | | • | | |
| Eastern Illinois University (Charleston) | 18 | Aug | 28 | Dipl, BS | $3,996 | $9,202 | | • | | |
| Illinois State University (Normal) | 16 | Jan | 36 | BS | $6,734 | $12,302 | | | | |
| Olivet Nazarene University (Bourbonnais) | 30 | Aug | 48 | Dipl, BS | $22,590 | $22,590 | | | | |
| Southern Illinois Univ at Carbondale | 40 | Aug | 30 | Dipl, BS | $3,102 | $6,204 | | | | |
| Trinity International University (Deerfield) | 40 | Aug | 36 | BA | $19,000 | $19,000 | | | | |
| Western Illinois University (Macomb) | 33 | Aug Jan | 35, 28 | BS | $5,538 | $7,714 | | • | | |
| **Indiana** | | | | | | | | | | |
| DePauw University (Greencastle) | 8 | Jan | 30 | BA | $27,000 | $0 | | • | | |
| Franklin College | 26 | Aug | 27 | Dipl, BA | | | | | | |
| Indiana State University (Terre Haute) | 16 | Aug | 48, 12 | Dipl, BS, MS | $6,102 | $13,518 | | | | • |
| Indiana Wesleyan University (Marion) | 32 | Sep | 32 | BS | $20,500 | $20,500 | | | | |
| Purdue University (West Lafayette) | 15 | Aug | 27 | BS | $7,096 | $21,266 | | | | |
| University of Indianapolis | 14 | Aug | 48 | Dipl, BS | $19,000 | $19,000 | | | | • |
| **Iowa** | | | | | | | | | | |
| Buena Vista University (Storm Lake) | 27 | Aug | 36 | Dipl, BA | $22,556 | $22,556 | | • | | |
| Central College (Pella) | 32 | Aug | 30 | BA | $23,898 | $23,898 | | • | | |
| Northwestern College (Orange City) | 28 | Aug | 32 | BS | $25,866 | $26,166 | | | | |

*Data are shown only for programs that completed the 2006 AMA Survey of Health Professions Education Programs
‡Key to Offers: 1: Evening or weekend classes; 2: Non-English instruction; 3: Cultural competence instruction

**Athletic Trainer**

| Programs* | Class Capacity | Begins | Length (months) | Award | Res Tuition | Non-res Tuition | Stipend | Offers:‡ 1 | 2 | 3 |
|---|---|---|---|---|---|---|---|---|---|---|
| University of Northern Iowa (Cedar Falls) | 72 | Aug | 36 | BA | $5,086 | $13,002 | | | | |
| Upper Iowa University (Fayette) | 12 | Aug | 36 | BS | $18,778 | $18,778 | | | | |
| **Kansas** | | | | | | | | | | |
| Emporia State University | 24 | Aug Jan | 36 | Dipl, BS | $1,388 | $4,457 | | | | |
| Kansas State University (Manhattan) | 20 | Aug | 30 | Dipl, BS | $1,572 | $6,543 | | • | | |
| Tabor College (Hillsboro) | 16 | Sep | 27 | BA | $15,574 | $15,574 | | | | |
| University of Kansas (Lawrence) | 20 | Fall semester | 18 | Dipl, BSE | $2,266 | $9,034 | | • | | • |
| Washburn University (Topeka) | 12 | Aug | 36 | BS | $3,936 | $8,904 | | • | | |
| **Kentucky** | | | | | | | | | | |
| Eastern Kentucky University (Richmond) | 20 | Aug | 48 | BS | $4,660 | $13,070 | | • | | |
| Murray State University | 36 | Aug | 30 | BS | $4,428 | $12,036 | | | | |
| Northern Kentucky University (Highland Heights) | 30 | Spring Sophomore yr | 30 | Dipl, BS | $4,968 | $9,696 | | • | | |
| **Louisiana** | | | | | | | | | | |
| Louisiana College (Pineville) | 36 | | 36 | Dipl, BS | | | | • | | |
| Louisiana State University (Baton Rouge) | 24 | Aug | 28 | BS | $4,286 | $11,086 | | | | |
| Nicholls State University (Thibodaux) | 15 | Aug | 48 | BS | $3,240 | $8,680 | | | | |
| Southeastern Louisiana University (Hammond) | 48 | Spring Fall | 48 | BS | $1,178 | $2,664 | | | | |
| **Maine** | | | | | | | | | | |
| University of Southern Maine (Gorham) | 36 | Sep | 36 | BS | $4,478 | $11,006 | | | | |
| **Maryland** | | | | | | | | | | |
| Salisbury University | 32 | Aug | 36 | Dipl, BS | $4,814 | $12,708 | | | | |
| Towson University | 60 | Aug | 36 | BS | $7,100 | $16,000 | | | | |
| **Massachusetts** | | | | | | | | | | |
| Boston University | 80 | Sep | 36 | BS | $31,530 | $31,530 | | | | • |
| Bridgewater State College | 20 | Sep | 24 | BS, MS | $1,675 | $3,525 | | | | |
| Lasell College (Newton) | 22 | Sep | 48 | Dipl, BS | $18,700 | $18,700 | | | | • |
| Merrimack College (North Andover) | 12 | Sep | 48 | BS | $26,620 | $26,620 | | • | | |
| Northeastern University (Boston) | 28 | Sep Jan | 54 | Dipl, BS | $13,375 | $13,375 | | | | • |
| Springfield College | 30 | Sep | 48 | BS | $22,715 | $22,715 | | | | |
| **Michigan** | | | | | | | | | | |
| Albion College | | Aug | 36 | BA | $24,012 | $25,012 | | | | |
| Central Michigan University (Mt Pleasant) | 14 | Aug Jan | 36 | BS | $6,024 | $14,016 | | | | • |
| Grand Valley State University (Allendale) | 35 | Aug | 36 | BS | $6,588 | $12,510 | | • | | • |
| Hope College (Holland) | 24 | Aug | 36 | BA | $22,430 | $22,430 | | | | • |
| Lake Superior State University (Sault Ste Marie) | 16 | Junior Year | 24 | BS | $6,558 | $13,116 | | | | • |
| Michigan State University (East Lansing) | 55 | Aug | 36 | BS | $5,666 | $14,014 | | | | |
| Saginaw Valley State University (University Center) | 10 | Fall Winter | 26 | Dipl, BS | $171 | $404 | | • | | |
| University of Michigan (Ann Arbor) | 40 | Sep | 36 | BS | $8,464 | $27,482 | | | | |
| Western Michigan University (Kalamazoo) | 45 | Aug | 24 | BA | $6,478 | $9,378 | | | | • |
| **Minnesota** | | | | | | | | | | |
| Minnesota State University - Mankato | 18 | Aug | 17 | BAT | $2,983 | $6,170 | | • | | |
| St Cloud State University | | | | | | | | | | |
| University of Minnesota - Duluth | 16 | Sep | 36 | Dipl, BASc | $7,605 | $18,712 | | | | |
| **Mississippi** | | | | | | | | | | |
| Delta State University (Cleveland) | 12 | Fall | 5 | BS | $3,762 | $8,950 | | • | | |
| University of Southern Mississippi (Hattiesburg) | 25 | Jun 29 | 20 | Dipl, BS | $2,053 | $2,585 | | • | | |
| **Missouri** | | | | | | | | | | |
| Central Methodist University (Fayette) | 24 | Aug | 36 | BS | $14,590 | $14,590 | | | | |
| Lindenwood University (St Charles) | 32 | Aug | 36 | BS | $12,000 | $12,000 | | | | |
| Missouri State University (Springfield) | 60 | Aug | 46 | BS | $5,100 | $10,200 | | | | |
| Missouri Valley College (Marshall) | 16 | Aug | 27, 9 | BS | $13,900 | $19,100 | 1900 | | | |
| Park University (Parkville) | 40 | Aug | 36 | Dipl, BA | $6,000 | $6,000 | | • | | • |
| Truman State University (Kirksville) | 20 | Aug | 24 | BS | $5,970 | $10,400 | | | | |
| William Woods University (Fulton) | 48 | Aug | 48 | BS | $15,150 | $15,150 | | | | |
| **Montana** | | | | | | | | | | |
| Montana State University - Billings | 10 | Jul 24 | 21 | Dipl, MS | $4,300 | $10,400 | | • | | |
| **Nebraska** | | | | | | | | | | |
| University of Nebraska - Omaha | 40 | Jul Aug | 24, 36 | BS Ed, MA | $3,937 | $11,602 | | • | | |
| **Nevada** | | | | | | | | | | |
| University of Nevada - Las Vegas | 15 | Jan | 29 | BS | $3,660 | $6,563 | | | | |
| **New Hampshire** | | | | | | | | | | |
| Keene State College | 12 | Aug | 32 | BS | $5,410 | $12,250 | | | | |
| Plymouth State University | 32 | Aug | 48 | BS, MEd | $4,750 | $10,800 | | • | | |
| University of New Hampshire (Durham) | 20 | Sep | 36 | BS | $9,778 | $21,498 | | | | |

*Data are shown only for programs that completed the 2006 AMA Survey of Health Professions Education Programs
‡Key to Offers: 1: Evening or weekend classes; 2: Non-English instruction; 3: Cultural competence instruction

## Athletic Trainer

| Programs* | Class Capacity | Begins | Length (months) | Award | Res Tuition | Non-res Tuition | Stipend | Offers:‡ 1 | 2 | 3 |
|---|---|---|---|---|---|---|---|---|---|---|
| **New Jersey** | | | | | | | | | | |
| Kean University (Union) | 24 | Sep | 16 | Dipl, BS | $1,771 | $2,662 | | | | |
| Rowan University (Glassboro) | 24 | Sep | 48 | BA | $9,566 | $12,696 | | | | |
| William Paterson Univ of New Jersey (Wayne) | 36 | Jan | 20 | BS | $9,422 | $15,370 | | • | | |
| **New Mexico** | | | | | | | | | | |
| New Mexico State University (Las Cruces) | 40 | Aug | 32 | BS | $3,936 | $13,206 | | • | | |
| **New York** | | | | | | | | | | |
| Alfred University | 40 | Sep | 48 | BS | $18,498 | $18,498 | | | | |
| Canisius College (Buffalo) | 16 | Sep | 45 | Dipl, BS | $23,000 | $23,000 | | | | |
| Dominican College (Orangeburg) | 36 | Aug | 48 | BS | $23,095 | $0 | | • | | |
| Ithaca College | 36 | Aug | 48 | Dipl, BS | $25,194 | $0 | | | | • |
| Long Island University - Brooklyn Campus | 20 | Sep | 36 | BS/MS | $24,000 | $24,000 | | • | | |
| Marist College (Poughkeepsie) | 60 | Sep | 36, 18 | Dipl, BS | $22,066 | $22,066 | | • | | |
| SUNY College at Cortland | 40 | Sep | 30 | BS | $4,350 | $10,610 | | | | |
| **North Carolina** | | | | | | | | | | |
| Appalachian State University (Boone) | 42 | Feb | 30 | BS | $3,917 | $13,659 | | | | |
| Barton College (Wilson) | 10 | Third semester | 20 | Dipl, BS | $16,200 | $16,200 | | | | |
| Catawba College (Salisbury) | 12 | Aug | 27 | BS | $26,000 | $26,000 | | | | |
| Elon University | 36 | Aug | 36 | Dipl, BS | $11,328 | $12,928 | | | | |
| Gardner - Webb University (Boiling Springs) | 10 | Jan | 35 | Dipl, BS | $14,960 | $14,960 | | | | • |
| High Point University | 15 | Aug | 24 | BS | $14,290 | $14,290 | | | | |
| Mars Hill College | 36 | Aug | 24 | Dipl, Cert, BS | $21,000 | $21,000 | | • | | |
| North Carolina Central University (Durham) | 36 | Jan | 20 | BS | $8,152 | $17,896 | | | | • |
| Univ of North Carolina - Wilmington | 50 | Aug | 48 | Dipl, BA | $1,812 | $6,667 | | | | |
| Univ of North Carolina at Charlotte | 16 | Junior year | 24 | Dipl, BS | | | | | | |
| Univ of North Carolina at Greensboro | 10 | Jun | 24 | Dipl, MS | $3,119 | $13,683 | | | | |
| Wingate University | 45 | Aug | 36 | BS | $24,500 | $24,500 | | | | |
| **North Dakota** | | | | | | | | | | |
| North Dakota State University (Fargo) | 20 | Aug | 36 | Dipl, BS | $2,800 | $7,000 | | | | |
| University of Mary (Bismarck) | 30 | Aug | 30 | BS, BA | $9,990 | $9,990 | | • | | |
| University of North Dakota (Grand Forks) | 15 | Aug | 34 | BS | $4,828 | $5,952 | | | | |
| **Ohio** | | | | | | | | | | |
| Ashland University | 40 | Aug | 36 | BS | $19,134 | $19,134 | | • | | |
| Bowling Green State University | 20 | Aug | 25 | BS | $15,819 | $23,127 | | | | |
| Capital University (Columbus) | 30 | Aug | 18, 27 | BA | $24 | $24,100 | | | | |
| Cedarville University | 12 | Aug | 36 | BA | $17,120 | $17,120 | | | | |
| College of Mt St Joseph (Cincinnati) | 60 | Aug | 36 | BS | $18,500 | $0 | | | | |
| Defiance College (Definace) | 36 | Aug | 36 | BS | $17,780 | $17,780 | | | | |
| Denison University (Granville) | 5 | Aug | 27 | Dipl, Cert, BA | $29,860 | $29,860 | | | | • |
| Kent State University | 60 | Aug | 20 | BS | $7,504 | $14,516 | | • | | • |
| Marietta College | | Aug | 48 | Dipl, BS | $23,200 | $23,200 | | • | | |
| Miami University (Oxford) | 60 | Aug | 24 | BS | $11,538 | $22,618 | | | | |
| Ohio Northern University (Ada) | 18 | Sep | 27 | BA, BS | $28,050 | $28,050 | | | | |
| Ohio State University (Columbus) | 66 | Sep | 27 | BS | $8,667 | $20,562 | | | | |
| Shawnee State University (Portsmouth) | | | | BS | $5,508 | $9,396 | | • | | |
| University of Akron | 40 | Aug | 48 | BS | $2,398 | $2,769 | | | | |
| University of Cincinnati | 35 | Sep | 27 | BS | $8,329 | $21,351 | | | | |
| University of Toledo | 32 | Aug | 33 | BS | $6,429 | $18,206 | | | | |
| Urbana University | 36 | Aug | 36 | BS | $15,050 | $15,050 | | | | |
| **Oklahoma** | | | | | | | | | | |
| Oklahoma State University (Stillwater) | | Aug | | BS | | | | • | | |
| Southern Nazarene University (Bethany) | 40 | Sep | 36 | BS | $14,750 | $14,750 | | | • | |
| Southwestern Oklahoma State University (Weatherford) | 16 | Aug | 24 | Dipl, BS | $3,300 | $7,800 | | • | | |
| **Oregon** | | | | | | | | | | |
| George Fox University (Newberg) | 30 | Aug | 36 | Dipl, BS | $21,710 | $21,720 | | | | |
| Linfield College (McMinnville) | 18 | Aug | 27 | BS, BA | $21,800 | $0 | | | | |
| **Pennsylvania** | | | | | | | | | | |
| Alvernia College (Reading) | 10 | Fall semester | 36 | BS | $20,000 | $20,000 | | | | |
| Duquesne University (Pittsburgh) | 30 | Sep | 32 | BS | $23,132 | $23,132 | | | | • |
| Indiana University of Pennsylvania | 28 | Aug | 18 | Dipl, BS | $3,043 | $6,676 | | • | | |
| King's College (Wilkes-Barre) | 15 | Aug | 36 | Dipl, BS | $22,280 | $22,280 | | | | |
| Marywood University (Scranton) | 30 | Aug | 48 | BS | $21,840 | $21,840 | | • | | • |
| Messiah College (Grantham) | 11 | Feb | 20 | BA | $21,420 | $21,420 | | | | |
| Neumann College (Aston) | 44 | Aug | 48 | Dipl, BS | $17,992 | $17,992 | | • | | |
| Slippery Rock University | 20 | Sep | 40 | Dipl, BS | $5,053 | $12,633 | | | | |
| University of Pittsburgh | 20 | Aug | 24 | BS | $13,512 | $24,284 | | | | |

*Data are shown only for programs that completed the 2006 AMA Survey of Health Professions Education Programs

‡Key to Offers: 1: Evening or weekend classes; 2: Non-English instruction; 3: Cultural competence instruction

## Athletic Trainer

| Programs* | Class Capacity | Begins | Length (months) | Award | Res Tuition | Non-res Tuition | Stipend | Offers:‡ 1 | 2 | 3 |
|---|---|---|---|---|---|---|---|---|---|---|
| University of Pittsburgh - Bradford | 24 | Aug | 48 | BS | $8,614 | $17,926 | | | | |
| Waynesburg College | 32 | Sep | 16 | BS | $14,810 | $14,810 | | | | |
| West Chester University | 24 | Aug | 48 | BS | $4,810 | $12,026 | | | | |
| **South Carolina** | | | | | | | | | | |
| Charleston Southern University | 20 | Aug | 24 | Dipl, BS | $23,230 | $23,230 | | • | | |
| College of Charleston | 24 | Aug | 48 | Dipl, BS | | | | | | |
| Erskine College (Due West) | 24 | Sep Feb | 36 | BS | $23,954 | $23,954 | | | | |
| Limestone College (Gaffney) | 40 | Aug | 48 | BS | $15,000 | $15,000 | | | | |
| Winthrop University (Rock Hill) | 26 | Aug | 28 | BS | $7,816 | $14,410 | | | | |
| **South Dakota** | | | | | | | | | | |
| Augustana College (Sioux Falls) | 24 | Sep | 27 | Dipl, BA | $19,750 | $19,750 | | | | |
| Dakota Wesleyan University (Mitchell) | 26 | Aug | 54 | BA | $16,650 | $16,650 | | • | | |
| National American University (Rapid City) | 16 | Sep | 27 | BA/BS | $9,840 | $9,840 | | | | |
| South Dakota State University (Brookings) | 36 | Aug | 24 | BS | $1,221 | $3,881 | | | | |
| **Tennessee** | | | | | | | | | | |
| Bryan College (Dayton) | 32 | Aug | 48 | Dipl, BS | $19,990 | $19,990 | | | | |
| Carson - Newman College (Jefferson City) | 36 | Aug | 36 | BS | $12,900 | $12,900 | | | | |
| Cumberland University (Lebanon) | 24 | Aug | 30 | BA | $13,344 | $13,344 | | | | |
| Lincoln Memorial University (Harrogate) | 16 | Aug | 48 | BS | $16,400 | $16,400 | | | | |
| Tusculum College (Greeneville) | 32 | Aug | 36 | BA | $15,000 | $0 | | | | |
| Union University (Jackson) | 24 | Aug | 48 | BS | $20,000 | $20,000 | | | | |
| University of Tennessee - Chattanooga | 16 | Jul | 23 | Dipl, MS | $5,434 | $14,830 | | | | • |
| University of Tennessee - Martin | 20 | Aug | 48 | Dipl, BSHHP | $2,075 | $6,202 | | | | |
| **Texas** | | | | | | | | | | |
| Angelo State University (San Angelo) | 32 | Aug | 48 | Dipl, BS | $3,780 | $11,520 | | | | |
| Baylor University (Waco) | 48 | Aug | 36 | BS Ed | $17,900 | $17,900 | | | | |
| Hardin-Simmons University (Abilene) | 20 | Aug | 45 | Dipl, BBS | $495 | $495 | $500 | | | |
| Southwestern University (Georgetown) | 10 | Aug | 31 | BA | $21,900 | $21,900 | $1,800 | | | |
| Stephen F Austin State University (Nacogdoches) | 10 | Summer II semester | 22 | Dipl, Cert, MS | $48 | $282 | | | | • |
| Texas Christian University (Ft Worth) | 12 | Aug | 48 | BS | $22,980 | $22,980 | | • | | |
| Texas Lutheran University (Seguin) | 16 | Aug | 36 | BS | $18,720 | $18,720 | | | | |
| Texas State University-San Marcos | 16 | Aug | 36 | Dipl, BS | $4,016 | $12,026 | | | | |
| Texas Tech Univ Health Sciences Center (Lubbock) | 28 | Late May Early Jun | 24 | MAT | $7,500 | $17,500 | | | | |
| University of Texas at Arlington | 45 | Fall | 48 | Dipl, BS | | | | | | |
| University of Texas at Austin | 45 | Fall semester | 36 | BS | $4,000 | $8,500 | | • | | |
| University of the Incarnate Word (San Antonio) | 45 | Aug | 36 | Dipl, BS | $17,400 | $17,400 | | | | |
| **Utah** | | | | | | | | | | |
| Brigham Young University (Provo) | 20 | Sep Jan | 32 | BS | $3,620 | $7,240 | | • | | |
| Southern Utah University (Cedar City) | 24 | Aug | 48 | BS, BA | $3,054 | $9,008 | | | | • |
| University of Utah (Salt Lake City) | 45 | Aug | 21 | BS | $2,790 | $8,495 | | | | |
| Weber State University (Ogden) | 18 | Fall freshman year | 30, 42 | BS | $2,793 | $9,776 | | • | | • |
| **Vermont** | | | | | | | | | | |
| University of Vermont (Burlington) | 60 | Sep | 36 | Dipl, BS | $9,832 | $24,816 | | | | • |
| **Virginia** | | | | | | | | | | |
| Bridgewater College | 40 | Jan | 22 | BS, BA | $20,190 | $20,190 | | | | • |
| Emory & Henry College | 36 | Aug | 32 | BS, BA | $28,220 | $28,220 | | | | |
| Liberty University (Lynchburg) | 40 | Aug | 48 | BS | $12,020 | $12,020 | | | | |
| Longwood University (Farmville) | 32 | Aug | 30 | BS | $9,000 | $15,000 | | | | • |
| Roanoke College (Salem) | 24 | Jan | 30 | BS | $30,143 | $30,143 | | | | |
| Virginia Commonwealth University (Richmond) | 15 | Aug | 36 | BS | $5,385 | $17,440 | | | | |
| **Washington** | | | | | | | | | | |
| Eastern Washington University (Cheney) | 15 | Sep | 27 | BS | $4,822 | $14,299 | | | | • |
| Washington State University (Pullman) | 15 | Aug | 48 | BS | $5,506 | $14,514 | | | | |
| Whitworth College (Spokane) | 36 | Aug | 36 | BA | $23,850 | $23,850 | | | | |
| **West Virginia** | | | | | | | | | | |
| Marshall University (Huntington) | 24 | Sep Jan | 48 | BA | $2,075 | $5,076 | | • | | • |
| University of Charleston | 35 | Aug | 30 | BS | $20,200 | $20,200 | | | | |
| West Virginia University (Morgantown) | 18 | Aug | 36 | BS | $3,938 | $12,060 | | | | |
| West Virginia Wesleyan College (Buckhannon) | 24 | Aug | 36 | Dipl, BS | $25,875 | $25,875 | | | | |
| **Wisconsin** | | | | | | | | | | |
| Concordia University Wisconsin (Mequon) | 45 | Aug | 48 | BS | $9,300 | $9,300 | | | | |
| University of Wisconsin - Eau Claire | 45 | Sep | 27 | BS | $4,864 | $19,040 | | • | | • |
| University of Wisconsin - Stevens Point | 35 | Sep | 36 | BS | $3,932 | $14,782 | | | | |
| **Wyoming** | | | | | | | | | | |
| University of Wyoming (Laramie) | 64 | Aug | 48 | BS | $3,425 | $9,815 | | | | |

*Data are shown only for programs that completed the 2006 AMA Survey of Health Professions Education Programs
‡Key to Offers: 1: Evening or weekend classes; 2: Non-English instruction; 3: Cultural competence instruction

# Audiologist

## Job Description

Audiologists are professionals who work with people that exhibit hearing, balance, and related ear problems. They examine individuals of all ages and identify those with the symptoms of hearing loss and other auditory, balance, and related neural problems. They then assess the nature and extent of the problems and help the individuals manage them. Using audiometers, computers, and other testing devices, they measure the loudness at which a person begins to hear sounds, the ability to distinguish between sounds, and the impact of hearing loss or balance problems on an individual's daily life. Audiologists interpret these results and may coordinate them with medical, educational, and psychological information to make a diagnosis and determine a course of treatment. Audiologists must effectively communicate diagnostic test results, interpretation, and proposed treatment in a manner easily understood to patients/clients and their families/care givers.

Hearing disorders can result from a variety of causes, including trauma at birth, viral infections, genetic disorders, exposure to loud noise, certain medications, or aging. Treatment may include examining and cleaning the ear canal, fitting and dispensing hearing aids, fitting and tuning cochlear implants, and providing aural rehabilitation. Aural rehabilitation emphasizes counseling on adjusting to hearing loss, training on the use of hearing instruments, and teaching communication strategies for use in a variety of listening environments. Audiologists also may recommend, fit, and dispense personal or large area amplification systems and alerting devices.

Some audiologists specialize in work with the elderly, children, or hearing-impaired individuals who need special therapy programs. Others develop and implement ways to protect workers' ears from on-the-job injuries. They measure noise levels in workplaces and conduct hearing protection programs in factories, as well as in schools and communities.

A graduate degree is necessary to practice as an audiologist in most states. A doctoral degree (PhD) is required for work in some areas.

Audiologists specialize in the study of:
- Normal and impaired hearing
- Prevention of hearing loss
- Identification and assessment of hearing and balance problems
- Rehabilitation of persons with hearing and balance disorders
 In addition, audiologists may:
- Prepare future professionals in colleges and universities
- Manage agencies, clinics, or private practices
- Engage in research to enhance knowledge about normal hearing, and the evaluation and treatment of hearing disorders
- Design hearing instruments and testing equipment

## Employment Characteristics

Audiologists may work in a wide range of settings, including schools, hospitals, rehabilitation centers, skilled nursing facilities, government health facilities, community clinics, geriatric facilities, health maintenance organizations (HMOs), public health departments, research laboratories, private practice, or industrial corporations.

## Salary

Salaries of audiologists depend on educational background, specialty, and experience, along with the geographical location and type of setting in which they work. The median salary range in 2004 for audiologists certified by the American Speech-Language-Hearing Association (ASHA) was $62,000. Persons in supervisory positions—for example, in administration and management—may earn well over $78,000 per year. While the 2004 median starting salary for certified audiologists with 1 to 3 years' experience was $45,000, the median salary for certified audiologists with clinical doctorate degrees was $65,000 and $82,606 for those with research doctorate degrees. Good benefits packages, such as insurance programs and leave, are usually available to audiologists. Refer to Section IV, Table 5 of this *Directory* for more information, or see www.ama-assn.org/go/salary.

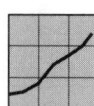

## Employment Outlook

Employment of audiologists is expected to grow about as fast as the average for all occupations through the year 2014 (www.bls.gov/oco/ocos085.htm#outlook). Because hearing loss is strongly associated with aging, rapid growth in the population age 55 and older will cause the number of persons with hearing impairment to increase markedly. Members of the baby boom generation are now entering middle age, when the possibility of neurological disorders and associated hearing impairments increases. Medical advances are also improving the survival rate of premature infants and trauma and stroke victims, who then need assessment and possible treatment. Many states now require that all newborns be screened for hearing loss and receive appropriate early intervention services. Opportunities in research and higher education are expected to increase as baby boomers currently in these positions retire and a clinical doctoral degree is required as the minimum educational requirement for new clinicians. Greater awareness of the importance of early identification and diagnosis of speech, language, and hearing disorders also will increase employment.

## Educational Programs

Approximately 80 universities in the United States offer graduate education programs in audiology at the clinical doctoral level that prepare students for entry into practice. Although master's degree programs had been the prominent degree programs for preparation for professional practice since 1965, clinical doctoral degree programs have replaced these master's programs in the US. Most programs can be identified as Doctor of Audiology programs or AuD programs; some clinical doctoral programs have chosen to use other degree designations such as Clinical PhD or Doctor of Science (ScD). Some advanced degree programs are available for further study with an emphasis on clinical issues or in research. *Note:* As of January 1, 2007, the Council on Academic Accreditation in Audiology and Speech-Language Pathology and Audiology (CAA) no longer accredits master's level programs in audiology. Few master's programs in audiology will be available after that date.
**Length.** Most clinical doctoral programs are designed to be completed in 4 years, including summers, for full-time study; most master's degree programs in audiology usually take at least 2 years, including summers.

**Prerequisites.** Coursework in the biological sciences, physical sciences, mathematics, and behavioral or social sciences is required for graduate study. Undergraduate programs in communication sciences and disorders will provide a background in linguistics, phonetics, psychology, normal speech, and language development, and introductory course work in audiology. Excellent oral and written communication skills are expected.

**Curriculum.** Graduate programs should offer a curriculum to allow a student to meet the knowledge and skills necessary to enter independent practice in audiology. A typical graduate program of study includes courses in genetics; normal and abnormal communication development; auditory, balance, and neural systems assessment and treatment; diagnosis and treatment; pharmacology; and ethics. Opportunities to work in a variety of different clinical settings and with a broad range of clients should be provided during the graduate program of study. Most clinical doctoral programs will expect the completion of a 9- to 12-month clinical externship as part of the degree program.

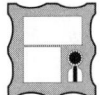

### Licensure and Certification
In most states, audiologists must comply with state regulatory (licensure) standards and/or have state teacher certification to practice in specific settings. Earning a graduate degree, completion of a 9- to 12-month clinical experience, and passage of a national examination are typically required to achieve the credentials. Individuals should contact the appropriate state licensure board or teacher certification agency for more information about requirements. ASHA offers the Certificate of Clinical Competence in Audiology (CCC-A), a nationally recognized credential that offers certificate holders ease in qualifying for state credentials because those requirements are similar or identical to ASHA's CCC requirements, recognition as a "highest qualified provider" of audiology services for reimbursement, and increased opportunities for employment or promotion, as certain positions in hospitals, educational programs, or private practices may require ASHA certification. In 2012, a doctoral degree will be required by ASHA to award certification in audiology; further, some states also will require a doctoral degree for licensure.

### Inquiries
For information about a specific program, write to the director of the audiology program in care of the institution listed.

For additional information about the professions or program accreditation, contact:

American Speech-Language-Hearing Association (ASHA)
10801 Rockville Pike
Rockville, MD 20852
800 498-2071
301 571-0457 Fax
www.asha.org

---

## Audiologist*

## Alabama

**Auburn University**
Audiologist Prgm
Dept of Comm Disorders
1199 Haley Center
Auburn University, AL 36849-5232
*Prgm Dir:* Lawrence F Molt, PhD
*Tel:* 334 844-9613   *Fax:* 334 844-4585
*E-mail:* moltlaw@auburn.edu

**University of South Alabama**
Audiologist Prgm
Dept of Speech Pathology and Audiology
2000 University Commons
Mobile, AL 36688-0002
www.southalabama.edu/alliedhealth/speechandhearing
*Prgm Dir:* Paul Dagenais, PhD CCC-SLP
*Tel:* 334 380-2600, Ext 4-2608   *Fax:* 334 380-2699
*E-mail:* pdagenais@usouthal.edu

## Arizona

**Arizona State University**
Audiologist Prgm
Dept of Speech & Hearing
PO Box 870102
Tempe, AZ 85287-0102
www.asu.edu/clas/shs
*Prgm Dir:* Sid P Bacon, PhD
*Tel:* 480 965-2905   *Fax:* 480 965-8516
*E-mail:* spb@asu.edu

*Additional information about programs that returned the AMA's survey is available in a table beginning on page 33.

**University of Arizona**
Audiologist Prgm
Speech, Language, and Hearing Sciences
PO Box 210071
Tucson, AZ 85721
www.slhs.arizona.edu
*Tel:* 520 621-1644   *Fax:* 520 651-9901
*E-mail:* adamsp@u.arizona.edu

## Arkansas

**University of Arkansas for Medical Sciences**
*Cosponsor:* University of Arkansas at Little Rock
Audiologist Prgm
Dept of Audiology & Speech Pathology
2801 S University Ave
Little Rock, AR 72204-1099
www.uams.edu/chrp
*Prgm Dir:* Laura Smith-Olinde, PhD
*Tel:* 501 569-3155   *Fax:* 501 569-3157
*E-mail:* lso@uams.edu

## California

**California State University - Los Angeles**
Audiologist Prgm
Dept of Communication Disorders
5151 State University Dr
Los Angeles, CA 90032
*Prgm Dir:* Edward S Klein, PhD
*Tel:* 323 343-4754   *Fax:* 323 343-4698
*E-mail:* eklein@cslanet.calstatela.edu

**California State University - Northridge**
Audiologist Prgm
Dept of Communicative Disorders
18111 Nordhoff St
Northridge, CA 91330-8279
*Prgm Dir:* J Stephen Sinclair, PhD
*Tel:* 818 677-2852   *Fax:* 818 677-2632
*E-mail:* steve.sinclair@csun.edu

**California State University - Sacramento**
Audiologist Prgm
Dept of Speech Pathology & Audiology
6000 J St
Sacramento, CA 95819-6071
*Prgm Dir:* Laureen O'Hanlon, PhD
*Tel:* 916 278-7341   *Fax:* 916 278-7730
*E-mail:* ohanlon@csus.edu

**San Diego State University**
*Cosponsor:* University of California - San Diego
Audiologist Prgm
Sch of Speech, Language, & Hearing Sciences
5500 Campanile Dr MC-1518
San Diego, CA 92182-1518
http://chhs.sdsu.edu/slhs/audmain.php
*Prgm Dir:* Steven Kramer, PhD
*Tel:* 619 594-6140   *Fax:* 619 594-7109
*E-mail:* skramer@mail.sdsu.edu

**San Francisco State University**
Audiologist Prgm
College of Health and Human Services
1600 Holloway Ave, Gym 105
San Francisco, CA 94132-4158
www.sfsu.edu
*Prgm Dir:* Marcia Raggio, PhD
*Tel:* 415 338-7653   *Fax:* 415 338-0907
*E-mail:* mraggio@sfsu.edu

## Colorado

**University of Colorado at Boulder**
Audiologist Prgm
2501 Kittredge Loop Rd
UCB 409
Boulder, CO 80309-0409
www.colorado.edu/slhs
*Prgm Dir:* Susan M Moore
*Tel:* 303 492-5284   *Fax:* 303 492-3274
*E-mail:* susan.moore@colorado.edu
*Med Dir:* Kathryn Arehart, PhD

**University of Northern Colorado**
Audiologist Prgm
Audiology & Speech Language Sciences
Gunter 1400, Box 140
Greeley, CO 80639-0030
www.unco.edu/nhs/asls
*Prgm Dir:* Ellen Meyer Gregg, PhD CCC-SLP
*Tel:* 970 351-1597   *Fax:* 970 351-2974
*E-mail:* ellen.gregg@unco.edu

## Connecticut

**Southern Connecticut State University**
Audiologist Prgm
Communication Disorders
501 Crescent St
New Haven, CT 06515
*Prgm Dir:* James Dempsey, PhD
*Tel:* 203 392-5954   *Fax:* 203 392-5968
*E-mail:* dempseyj1@southernct.edu

**University of Connecticut**
Audiologist Prgm
Communication Sciences
850 Bolton Rd, Unit 1085
Storrs, CT 06268-1085
*Prgm Dir:* Harvey R Gilbert
*Tel:* 860 486-2817   *Fax:* 860 486-5422
*E-mail:* harvey.gilbert@uconn.edu

## District of Columbia

**Gallaudet University**
Audiologist Prgm
Audiology & Speech-Lang Path
800 Florida Ave NE
Washington, DC 20002-3695
*Prgm Dir:* James J Mahshie, PhD
*Tel:* 202 651-5329   *Fax:* 202 651-5324
*E-mail:* james.mahshie@gallaudet.edu

**Howard University**
Audiologist Prgm
Comm Sciences and Disorders
525 Bryant St NW
Washington, DC 20059
*Prgm Dir:* Ovetta Harris
*Tel:* 202 806-6990   *Fax:* 202 806-4046
*E-mail:* oharris@howard.edu

## Florida

**Nova Southeastern University**
Audiologist Prgm
3200 S University Dr
Ft Lauderdale, FL 33328
www.nova.edu/aud
*Prgm Dir:* Barry A Freeman, PhD
*Tel:* 954 262-7745   *Fax:* 954 262-1181
*E-mail:* freemanb@nova.edu

**University of Florida**
Audiologist Prgm
Comm Processes and Disorders
335 Dauer Hall, PO Box 117420
Gainesville, FL 32611-7420
*Prgm Dir:* Kenneth J Logan, PhD
*Tel:* 352 392-2113   *Fax:* 352 846-0243
*E-mail:* logan@csd.ufl.edu

**University of South Florida**
Audiologist Prgm
Communication Sciences and Disorders
4202 E Fowler Ave PCD 1017
Tampa, FL 33620-8150
www.cas.usf.edu/csd
*Prgm Dir:* Theresa Chisolm, PhD
*Tel:* 813 974-2006   *Fax:* 813 974-0822
*E-mail:* chisolm@cas.usf.edu

## Idaho

**Idaho State University**
Audiologist Prgm
Communication Sciences & Disorders and Education of
    the Deaf
Box 8116
Pocatello, ID 83209-0009
www.isu.edu/csed
*Prgm Dir:* Jeff Brockett, EdD CCC-A
*Tel:* 208 282-2556   *Fax:* 208 236-4571
*E-mail:* brocjeff@isu.edu

## Illinois

**Univ of Illinois at Urbana - Champaign**
Audiologist Prgm
220 Speech & Hearing Sci Bldg
901 S 6th St
Champaign, IL 61820
*Prgm Dir:* Ron D Chambers, PhD
*Tel:* 217 333-2230   *Fax:* 217 244-2235
*E-mail:* rdc@uiuc.edu

**Rush University Medical Center**
Audiologist Prgm
Comm Disorders and Sciences
1653 W Congress Pkwy
Chicago, IL 60612
www.rushu.rush.edu/cds/Communications.htm
*Prgm Dir:* Dianne Meyer, PhD
*Tel:* 312 942-3289   *Fax:* 312 942-7211
*E-mail:* dianne_h_meyer@rush.edu

**Northern Illinois University**
Audiologist Prgm
Communicative Disorders
DeKalb, IL 60115-2899
*Prgm Dir:* Pamela Jackson, PhD
*Tel:* 815 753-6510   *Fax:* 815 753-9123
*E-mail:* plj@niu.edu

**Northwestern University**
Audiologist Prgm
Comm Sciences and Disorders
2240 Campus Dr
Evanston, IL 60208-3540
*Prgm Dir:* Dean C Garstecki, PhD
*Tel:* 847 491-3066   *Fax:* 847 491-4975
*E-mail:* d-garstecki@northwestern.edu

**Illinois State University**
Audiologist Prgm
Speech Pathology & Audiology
Fairchild Hall #204
Normal, IL 61790
www.ilstu.edu
*Prgm Dir:* Walter Smoski, PhD
*Tel:* 309 438-8643   *Fax:* 309 438-5221
*E-mail:* wsmoski@ilstu.edu
*Med Dir:* Verticchio Heidi, MS

## Indiana

**Indiana University - Bloomington**
Audiologist Prgm
Speech and Hearing Sciences
200 S Jordan Ave
Bloomington, IN 47405-7002
*Prgm Dir:* Phil Connell, PhD
*Tel:* 812 855-4156   *Fax:* 812 855-5531
*E-mail:* pconnell@indiana.edu

**Ball State University**
Audiologist Prgm
Speech Pathology & Audiology
2000 University Ave
Muncie, IN 47306
www.bsu.edu/csh/spa
*Prgm Dir:* David Coffin, AuD
*Tel:* 765 285-8160   *Fax:* 765 285-5623
*E-mail:* dcoffin@bsu.edu

**Purdue University**
Audiologist Prgm
Speech, Language, and Hearing Sciences
500 Oval Dr
West Lafayette, IN 47907-2038
*Prgm Dir:* Robert Novak, PhD CCC-A
*Tel:* 765 494-3788   *Fax:* 765 494-9771
*E-mail:* novakr@purdue.edu

## Iowa

**University of Iowa**
Audiologist Prgm
Speech Pathology & Audiology
119 SHC
Iowa City, IA 52242
www.shc.uiowa.edu
*Prgm Dir:* Paul Abbas
*Tel:* 319 335-8718   *Fax:* 319 335-8851
*E-mail:* paul-abbas@uiowa.edu

## Kansas

**University of Kansas**
Audiologist Prgm
Intercampus Prgm
3901 Rainbow Rd
Kansas City, KS 66160-7605
www.ukans.edu/~splh
*Prgm Dir:* John Ferraro, PhD
*Tel:* 913 588-5937   *Fax:* 913 588-5923
*E-mail:* jferraro@kumc.edu

**Wichita State University**
Audiologist Prgm
Communication Sciences and Disorders
1845 N Fairmount
Wichita, KS 67260-0075
www.wichita.edu/csd
*Prgm Dir:* Kathy Coufal, PhD
*Tel:* 316 978-3171   *Fax:* 316 978-3291
*E-mail:* kathy.coufal@wichita.edu

## Kentucky

**University of Louisville**
Audiologist Prgm
Surgery/Graduate Program in Communicative Disorders
Health Sciences Center, Myers Hall
Louisville, KY 40292
*Prgm Dir:* Barbara M Baker, PhD
*Tel:* 502 852-5274   *Fax:* 502 852-0865
*E-mail:* barbara.baker@louisville.edu

## Louisiana

**Louisiana State Univ and A&M College**
Audiologist Prgm
Communication Disorders
Music & Dramatic Arts Bldg, Rm 163
Baton Rouge, LA 70803-2606
*Prgm Dir:* Paul R Hoffman
*Tel:* 225 578-2545   *Fax:* 225 578-2528
*E-mail:* cdhoff@lsu.edu

**Louisiana State Univ Health Sciences Center**
Audiologist Prgm
Communication Disorders
1900 Gravier St
New Orleans, LA 70112
*Prgm Dir:* Jerry L Cranford, PhD
*Tel:* 504 568-4348    *Fax:* 504 568-4352
*E-mail:* jcranf@lsuhsc.edu

**Louisiana Tech University**
Audiologist Prgm
Dept of Speech
PO Box 3165
Ruston, LA 71272
www.latech.edu/slp-aud
*Prgm Dir:* J Clarice Dans, PhD
*Tel:* 318 257-4764    *Fax:* 318 257-4492
*E-mail:* cdans@latech.edu

# Maryland

**Univ of Maryland at College Park**
Audiologist Prgm
Hearing and Speech Science
Le Frak Hall
College Park, MD 20742
www.bsos.umd.edu/hesp
*Prgm Dir:* Nan Ratner
*Tel:* 301 405-4214    *Fax:* 301 314-2023
*E-mail:* Nratner@hesp.umd.edu
*Med Dir:* Sandra Gordon-Salant, PhD

**Towson University**
Audiologist Prgm
Audiology, Speech Language Pathology & Deaf Studies
8000 York Rd
Towson, MD 21252
www.towson.edu/asld
*Prgm Dir:* Sharon Glennen, PhD
*Tel:* 410 704-4153    *Fax:* 410 704-4131
*E-mail:* sglennen@towson.edu

# Massachusetts

**University of Massachusetts - Amherst**
Audiologist Prgm
Communication Disorders
715 N Pleasant St
Amherst, MA 01003-9304
*Prgm Dir:* Jane Baran, PhD
*Tel:* 413 545-0131    *Fax:* 413 545-0803
*E-mail:* baran@comdis.umass.edu

**Northeastern University**
Audiologist Prgm
Speech-Language Pathology & Audiology
106 Forsyth Bldg, 360 Huntington Ave
Boston, MA 02115
*Prgm Dir:* Linda J Ferrier, PhD
*Tel:* 617 373-3698    *Fax:* 617 373-2239
*E-mail:* l.ferrier@neu.edu

# Michigan

**Wayne State University**
Audiologist Prgm
Communication Sciences and Disorders
207 Rackham Building
Detroit, MI 48202
www.clas.wayne.edu/CSD
*Prgm Dir:* Alex Johnson, PhD
*Tel:* 313 577-3339    *Fax:* 313 577-8885
*E-mail:* ajohnson@wayne.edu

**Michigan State University**
Audiologist Prgm
378 Comm Arts & Sciences Bldg
East Lansing, MI 48824-1220
*Prgm Dir:* Michael W Casby, PhD
*Tel:* 517 353-8780    *Fax:* 517 353-3176
*E-mail:* casby@msu.edu

**Western Michigan University**
Audiologist Prgm
Speech Pathology & Audiology
Kalamazoo, MI 49008-5355
*Prgm Dir:* John M Hanley, PhD
*Tel:* 616 387-8045    *Fax:* 616 381-8044
*E-mail:* john.hanley@wmich.edu

**Central Michigan University**
Audiologist Prgm
Communication Disorders
Health Professions Building 2186
Mt Pleasant, MI 48859
www.chp.cmich.edu/cdo
*Prgm Dir:* Renny H Tatchell, PhD
*Tel:* 989 774-1323    *Fax:* 989 774-2799
*E-mail:* tatch1rh@cmich.edu

# Minnesota

**University of Minnesota - Minneapolis**
Audiologist Prgm
115 Shevlin Hall
164 Pillsbury Dr SE
Minneapolis, MN 55455
www.slhs.umn.edu
*Prgm Dir:* Jennifer Windsor, PhD
*Tel:* 612 624-3322    *Fax:* 612 624-7586
*E-mail:* windsor@umn.edu

# Mississippi

**University of Southern Mississippi**
Audiologist Prgm
Dept of Speech & Hearing Sciences
PO Box 5092
Hattiesburg, MS 39406-5092
*Prgm Dir:* Stephen E Oshrin, PhD
*Tel:* 601 266-5216    *Fax:* 601 266-5224
*E-mail:* steve.oshrin@usm.edu

# Missouri

**Missouri State University**
Audiologist Prgm
Communication Disorders
901 S National Ave
Springfield, MO 65897-0095
www.missouristate.edu/csd
*Prgm Dir:* Neil J DiSarno, PhD
*Tel:* 417 836-5368, Ext 66511    *Fax:* 417 836-4242
*E-mail:* neildisarno@missouristate.edu

**Washington University**
Audiologist Prgm
Program in Audiology & Communication Sciences
660 S Euclid Ave, Campus Box 8042
St Louis, MO 63110
http://pacs.wustl.edu
*Prgm Dir:* William W Clark, PhD
*Tel:* 314 747-0103    *Fax:* 314 747-0105
*E-mail:* clarkw@msnotes.wustl.edu

**Central Missouri State University**
Audiologist Prgm
Speech Pathology & Audiology
Martin Bldg 41
Warrensburg, MO 64093
www.cmsu.edu
*Prgm Dir:* Carl Harlan, PhD CCC-SP
*Tel:* 660 543-4606, Ext 660    *Fax:* 660 543-8234
*E-mail:* tibbits@cmsu1.cmsu.edu

# Nebraska

**University of Nebraska - Lincoln**
Audiologist Prgm
Dept of Special Education & Communication Disorders
301 Barkley Center
Lincoln, NE 68583-0234
*Prgm Dir:* John Bernthal, PhD
*Tel:* 402 472-2145    *Fax:* 402 472-7697
*E-mail:* jbernthal1@unl.edu
*Med Dir:* T Decker, PhD

# New York

**University at Buffalo - SUNY**
Audiologist Prgm
Dept of Communicative Disorders & Sciences
3435 Main St, 122 Cary Hall
Buffalo, NY 14222-3005
*Prgm Dir:* Elaine Stathopoulos, PhD
*Tel:* 716 829-2797    *Fax:* 716 829-3979
*E-mail:* stathop@buffalo.edu

**Adelphi University**
*Cosponsor:* Hofstra University, St. John's University
Audiologist Prgm
Comm Sciences and Disorders
Hy Weinberg Center - Rm 001
Garden City, NY 11530
www.adelphi.edu
*Prgm Dir:* Janet Schoepflin, PhD
*Tel:* 516 877-4770    *Fax:* 516 877-4783
*E-mail:* schoepfl@adelphi.edu

**Hofstra University**
Audiologist Prgm
Dept of Speech - Language - Hearing Sciences
110 Hofstra University
Hempstead, NY 11549
*Prgm Dir:* Carole Ferrand, PhD
*Tel:* 516 463-5511    *Fax:* 516 463-5260
*E-mail:* Carole.T.Ferrand@hofstra.edu

**St John's University**
Audiologist Prgm
Speech and Hearing Ctr
8000 Utopia Pkwy
Jamaica, NY 11439
http://new.stjohns.edu
*Prgm Dir:* Donna Geffner, PhD
*Tel:* 718 990-6480    *Fax:* 718 990-1917
*E-mail:* geffnerd@stjohns.edu

**CUNY Hunter College**
*Cosponsor:* Brooklyn College
Audiologist Prgm
Clinical Doctoral Programs
365 Fifth Ave
New York, NY 10016
www.gc.cuny.edu
*Prgm Dir:* Barabara Weinstein, PhD
*Tel:* 212 817-7981    *Fax:* 212 817-1680
*E-mail:* bweinstein@gc.cuny.edu

**Syracuse University**
Audiologist Prgm
Dept of Communication Sciences & Disorders
805 S Crouse Ave
Syracuse, NY 13244-2280
*Prgm Dir:* Raymond H Colton, PhD
*Tel:* 315 443-9637    *Fax:* 315 443-1113
*E-mail:* rhcolton@syr.edu

# North Carolina

**Univ of North Carolina at Chapel Hill**
Audiologist Prgm
Division of Speech & Hearing Sciences
CB 7190 Wing D Medical School
Chapel Hill, NC 27599-7190
*Prgm Dir:* Jackson Roush, PhD
*Tel:* 919 966-1006    *Fax:* 919 966-0100
*E-mail:* jroush@med.unc.edu

**East Carolina University**
Audiologist Prgm
Dept of Communication Sciences & Disorders
Greenville, NC 27858-4353
www.ecu.edu/csd
*Prgm Dir:* Gregg D Givens, PhD
*Tel:* 252 744-6080    *Fax:* 252 744-6081
*E-mail:* givensg@ecu.edu

# Ohio

**University of Akron**
*Cosponsor:* Northeast Ohio Audiology Consortium
Audiologist Prgm
School of Speech-Language & Audiology
Polsky Building, Rm 184B
Akron, OH 44325-3001
www.uakron.edu/sslpa
*Prgm Dir:* Sharon Lesner, PhD
*Tel:* 330 972-6118    *Fax:* 330 972-7884
*E-mail:* lesner@uakron.edu

**Ohio University**
Audiologist Prgm
School of Hearing, Speech and Language Sciences
Grover Center W218
Athens, OH 45701-2979
*Prgm Dir:* Brooke Hallowell, PhD
*Tel:* 740 593-1356    *Fax:* 740 593-0287
*E-mail:* hallowel@ohio.edu

**University of Cincinnati**
Audiologist Prgm
Communication Sciences & Disorders
Mail Station 379
Cincinnati, OH 45221-0379
*Prgm Dir:* Nancy Creaghead, PhD
*Tel:* 513 558-8501    *Fax:* 513 558-8500
*E-mail:* nancy.creaghead@uc.edu

**Ohio State University**
Audiologist Prgm
Dept of Speech and Hearing Sciences
110 Pressey Hall, 1070 Carmack Rd
Columbus, OH 43210-1002
*Prgm Dir:* Robert Allen Fox, PhD
*Tel:* 614 292-8207    *Fax:* 614 292-7504
*E-mail:* fox.2@osu.edu

# Oklahoma

**Univ of Oklahoma Health Sciences Center**
Audiologist Prgm
Communication Sciences and Disorders
825 NE 14th, PO Box 26901
Oklahoma City, OK 73190
www.ah.ouhsc.edu/csd
*Prgm Dir:* Stephen Painton, PhD
*Tel:* 405 271-4214, Ext 46054    *Fax:* 405 271-3360
*E-mail:* stephen-painton@ouhsc.edu

# Oregon

**Portland State University**
Audiologist Prgm
Speech & Hearing Sciences
PO Box 751
Portland, OR 97207-0751
*Prgm Dir:* Thomas Dolan, PhD
*Tel:* 503 725-3533    *Fax:* 503 725-9171
*E-mail:* dolant@pdx.edu

# Pennsylvania

**Bloomsburg University**
Audiologist Prgm
Dept of Audiology & Speech Pathology
400 E 2nd St, Centennial Hall
Bloomsburg, PA 17815-1301
*Prgm Dir:* Dainne Angelo, PhD
*Tel:* 570 389-4436    *Fax:* 570 389-5022
*E-mail:* dangelo@bloomu.edu

**University of Pittsburgh**
Audiologist Prgm
Communication Science and Disorders
4033 Forbes Tower - Atwood St
Pittsburgh, PA 15260
www.shrs.pitt.edu/csd
*Prgm Dir:* Malcom McNeil, PhD
*Tel:* 412 383-6540    *Fax:* 412 383-6555
*E-mail:* mcneil@pitt.edu

**Penn State University**
Audiologist Prgm
Dept of Communication Sciences and Disorders
110 Moore Bldg
University Park, PA 16802-3100
*Prgm Dir:* Gordon Blood, PhD
*Tel:* 814 865-3177    *Fax:* 814 863-3759
*E-mail:* F2X@psu.edu

# Rhode Island

**University of Rhode Island**
Audiologist Prgm
Dept of Communicative Disorders
Independence Sq, Ste 1
Kingston, RI 02881-0821
*Prgm Dir:* Jay Singer, PhD
*Tel:* 401 874-5969    *Fax:* 401 874-4404
*E-mail:* DrJay@URI.edu

# South Dakota

**University of South Dakota**
Audiologist Prgm
Dept of Communication Disorders
414 E Clark St
Vermillion, SD 57069-2390
www.usd.edu/dcom
*Prgm Dir:* Teri J Bellis, PhD
*Tel:* 605 677-5474    *Fax:* 605 677-5767
*E-mail:* tbellis@usd.edu

# Tennessee

**East Tennessee State University**
Audiologist Prgm
Dept of Communicative Disorders
PO Box 70643
Johnson City, TN 37614-0643
www.etsu.edu/cpah/commdis
*Prgm Dir:* Marc Fagelson, PhD
*Tel:* 423 439-4583    *Fax:* 423 439-4607
*E-mail:* fagelson@etsu.edu

**University of Tennessee - Knoxville**
Audiologist Prgm
Audiology & Speech Pathology
578 S Stadium Hall
Knoxville, TN 37996-0740
*Prgm Dir:* Patricia M Visser
*Tel:* 865 974-5019    *Fax:* 865 974-1539
*E-mail:* pvisser@utk.edu

**University of Memphis**
Audiologist Prgm
School of Audiology & Speech-Language Pathology
807 Jefferson Ave
Memphis, TN 38105
www.ausp.memphis.edu
*Prgm Dir:* Maurice I Mendel, PhD
*Tel:* 901 678-5800    *Fax:* 901 525-1282
*E-mail:* mmendel@memphis.edu

**Vanderbilt University Medical Center**
Audiologist Prgm
Hearing and Speech Sciences
1215 21st Ave South, Rm 8310
Nashville, TN 37232-8718
http://vanderbiltbillwilkersoncenter.com/dhss.html
*Prgm Dir:* Fred H Bess, PhD
*Tel:* 615 936-5000    *Fax:* 615 936-5014
*E-mail:* fred.h.bess@vanderbilt.edu
*Med Dir:* Edward G Conture, PhD

# Texas

**University of Texas at Austin**
Audiologist Prgm
Communication Sciences & Disorders
1 University Station, A1100
Austin, TX 78712-1089
http://csd.utexas.edu/graduate.html
*Prgm Dir:* Craig Champlin, PhD
*Tel:* 512 471-4119    *Fax:* 512 471-2957
*E-mail:* champlin@austin.utexas.edu

**Lamar Institute of Technology**
Audiologist Prgm
Dept of Communication
Box 10076, Lamar Station
Beaumont, TX 77710
*Prgm Dir:* Sumalai Maroonroge, PhD
*Tel:* 409 880-8177    *Fax:* 409 880-2265
*E-mail:* maroonroge@hal.lamar.edu

**University of Texas at Dallas**
Audiologist Prgm
Program in Communication Disorders
1966 Inwood Rd
Dallas, TX 75235-7298
*Prgm Dir:* Robert D Stillman, PhD
*Tel:* 214 905-3060    *Fax:* 214 905-3006
*E-mail:* stillman@utdallas.edu

**University of North Texas**
Audiologist Prgm
Dept of Speech & Hearing Sciences
PO Box 305010
Denton, TX 76203-5010
*Prgm Dir:* Jeffrey Cokely
*Tel:* 940 565-2481    *Fax:* 940 565-4058
*E-mail:* cokely@unt.edu

**Texas Tech Univ Health Sciences Center**
Audiologist Prgm
Dept of Speech, Language, and Hearing Sciences
STOP 6073, 3601 4th St
Lubbock, TX 79430
www.ttuhsc.edu/SAH
*Prgm Dir:* Candace Hicks, PhD
*Tel:* 806 743-5660, Ext 230    *Fax:* 806 743-5670
*E-mail:* candace.hicks@ttuhsc.edu

# Utah

**Utah State University**
Audiologist Prgm
Comm Disorders & Deaf Educ
1000 Old Main Hill
Logan, UT 84322-1000
*Prgm Dir:* James C Blair, PhD
*Tel:* 435 797-1375   *Fax:* 435 797-0221
*E-mail:* jimb@cc.usu.edu

**University of Utah**
Audiologist Prgm
390 South 1530 East
Room #1201 BEH S
Salt Lake City, UT 84322-1000
www.health.utah.edu/cmdis
*Prgm Dir:* Bruce L Smith, PhD
*Tel:* 801 581-6725   *Fax:* 801 581-7955
*E-mail:* angelina.harder@hsc.utah.edu

**University of Utah Health Science Center**
Audiologist Prgm
Communication Sciences and Disorders
1201 Behavioral Science Bldg
Salt Lake City, UT 84112
www.health.utah.edu/cmdis
*Prgm Dir:* Lisa Hunter, PhD
*Tel:* 801 581-6725   *Fax:* 801 571-7955
*E-mail:* lisa.hunter@hsc.utah.edu

# Virginia

**James Madison University**
Audiologist Prgm
Communication Sciences & Disorders
MSC 4304
Harrisonburg, VA 22807
*Prgm Dir:* Vicki Reed, EdD
*Tel:* 540 568-6440   *Fax:* 540 568-8077
*E-mail:* reedva@jmu.edu

# Washington

**Western Washington University**
Audiologist Prgm
Dept of Communication Sciences and Disorders
Parks Hall 17, MS 9078
Bellingham, WA 98225-9078
*Prgm Dir:* Michael T Seilo, PhD
*Tel:* 360 650-3855   *Fax:* 360 650-2843
*E-mail:* mseilo@cc.wwu.edu

**Washington State University**
Audiologist Prgm
Speech and Hearing Sciences
201 Daggy Hall, PO Box 642420
Pullman, WA 99164-2420
*Prgm Dir:* Gail Chermak
*Tel:* 509 335-4526   *Fax:* 509 335-8357
*E-mail:* chermak@wsu.edu

**University of Washington**
Audiologist Prgm
Speech and Hearing Sciences
1417 NE 42nd St
Seattle, WA 98105-6246
*Prgm Dir:* Christopher Moore
*Tel:* 206 685-7402   *Fax:* 206 543-1093
*E-mail:* camoore@u.washington.edu
*Med Dir:* Pamela Souza, Graduate Program Coordinator

# West Virginia

**West Virginia University**
Audiologist Prgm
Speech Pathology & Audiology
805 Allen Hall, Box 6122
Morgantown, WV 26506-6122
*Prgm Dir:* Lynn R Cartwright, EdD
*Tel:* 304 293-4241   *Fax:* 304 293-7565
*E-mail:* lcartwri@wvu.edu

# Wisconsin

**University of Wisconsin - Madison**
*Cosponsor:* University of Wisconsin - Stevens Point
Audiologist Prgm
Communicative Disorders
1975 Willow Dr - Goodnight Hall
Madison, WI 53706
www.comdis.wisc.edu
*Prgm Dir:* Cynthia Fowler, PhD
*Tel:* 608 262-3951   *Fax:* 608 262-6466
*E-mail:* cgfowler@wisc.edu

**University of Wisconsin - Stevens Point**
*Cosponsor:* University of Wisconsin - Madison
Audiologist Prgm
Communicative Disorders
1901 4th Ave
Stevens Point, WI 54481-3897
http://www.uwsp.edu/commD
*Prgm Dir:* Gary Cumley, PhD
*Tel:* 715 346-4699   *Fax:* 715 346-2157
*E-mail:* gcumley@uwsp.edu

# Wyoming

**University of Wyoming**
Audiologist Prgm
Neuroscience Program
Division of Communication Disorders
Laramie, WY 82071
*Prgm Dir:* Mary Hardin-Jones, PhD
*Tel:* 307 766-6427   *Fax:* 307 766-6829
*E-mail:* mhardinj@uwyo.edu

## Audiologist

| Programs* | Class Capacity | Begins | Length (months) | Award | Res Tuition | Non-res Tuition | Stipend | Offers:‡ 1 | 2 | 3 |
|---|---|---|---|---|---|---|---|---|---|---|
| **Alabama** | | | | | | | | | | |
| University of South Alabama (Mobile) | 8 | Aug | 45 | Dipl, AuD | $6,012 | $12,024 | $7,500 | | | • |
| **Arizona** | | | | | | | | | | |
| Arizona State University (Tempe) | 12 | Aug | 48 | Dipl, AuD | | | | | | • |
| University of Arizona (Tucson) | 15 | Fall semester | | Dipl, AuD | | | | | | • |
| **Arkansas** | | | | | | | | | | |
| University of Arkansas for Medical Sciences (Little Rock) | 9 | Fall semester | 48, 21 | Dipl, AuD | $257 | $555 | | • | | • |
| **California** | | | | | | | | | | |
| San Diego State University | 10 | Fall | 48 | Dipl, AuD | | | | | | • |
| San Francisco State University | | | | | | | | | | |
| **Colorado** | | | | | | | | | | |
| University of Colorado at Boulder | 8 | Fall | 48 | Dipl, AUD, PhD | | | | | | • |
| University of Northern Colorado (Greeley) | 8 | Aug | 48 | AuD | $5,118 | $14,832 | | | | • |
| **Florida** | | | | | | | | | | |
| Nova Southeastern University (Ft Lauderdale) | 20 | Sep | 48 | Dipl, AuD | $16,745 | $16,745 | | • | | • |
| University of South Florida (Tampa) | 13 | Aug | 48 | AuD | $7,000 | $27,000 | $10,000 | | | |
| **Idaho** | | | | | | | | | | |
| Idaho State University (Pocatello) | 8 | Aug | 48 | AuD | $4,740 | $11,700 | | | | • |
| **Illinois** | | | | | | | | | | |
| Illinois State University (Normal) | 10 | | 48 | AuD | $3,000 | $6,000 | | • | | |
| Rush University Medical Center (Chicago) | 12 | Fall | 45 | AuD | $5,540 | $5,540 | | | | • |
| **Indiana** | | | | | | | | | | |
| Ball State University (Muncie) | 10 | Aug | 48 | Dipl, AuD | $12,416 | $27,983 | | | | • |

*Data are shown only for programs that completed the 2006 AMA Survey of Health Professions Education Programs
‡Key to Offers: 1: Evening or weekend classes; 2: Non-English instruction; 3: Cultural competence instruction

# Audiologist

## Audiologist

| Programs* | Class Capacity | Begins | Length (months) | Award | Res Tuition | Non-res Tuition | Stipend | Offers:‡ 1 | 2 | 3 |
|---|---|---|---|---|---|---|---|---|---|---|
| **Iowa** | | | | | | | | | | |
| University of Iowa (Iowa City) | 8 | Fall semester | 48 | MA | $6,959 | $18,353 | | | | |
| **Kansas** | | | | | | | | | | |
| University of Kansas (Kansas City) | 10 | Jun Aug Jan | 48 | Dipl, AuD, PhD | $6,000 | $12,000 | | | | • |
| Wichita State University | 8 | Aug | 48 | BA, AuD | $5,949 | $16,446 | | | | |
| **Louisiana** | | | | | | | | | | |
| Louisiana Tech University (Ruston) | 16 | Fall | 48 | AuD | $3,600 | $12,000 | $8,000 | • | | |
| **Maryland** | | | | | | | | | | |
| Towson University | 12 | Fall | 48 | Dipl, AuD | $4,020 | $6,720 | | | | |
| Univ of Maryland at College Park | 15 | Aug | 48 | AuD, PhD | $10,250 | $17,500 | $8,000 | | | |
| **Michigan** | | | | | | | | | | |
| Central Michigan University (Mt Pleasant) | 40 | Aug | 48 | Dipl, AuD | $366 | $366 | | | | • |
| Wayne State University (Detroit) | 12 | Fall term | 45 | AuD | $14,000 | $29,675 | | | | |
| **Minnesota** | | | | | | | | | | |
| University of Minnesota - Minneapolis | 20 | Fall | 44 | Dipl | $8,748 | $15,848 | | | | • |
| **Missouri** | | | | | | | | | | |
| Central Missouri State University (Warrensburg) | 45 | Fall Spring Summer | 24 | Dipl, BS, MS | | | | | | |
| Missouri State University (Springfield) | 60 | Aug | 44 | AuD | $5,860 | $11,718 | | | | • |
| Washington University (St Louis) | 12 | Aug | 48 | Dipl, Cert, AuD | $21,100 | $21,100 | | | | |
| **New York** | | | | | | | | | | |
| Adelphi University (Garden City) | 25 | Sep Jan | | AuD | $22,000 | $22,000 | | • | | • |
| CUNY Hunter College (New York) | 12 | Fall | 4 | Dipl, AuD | | | | • | | • |
| St John's University (Jamaica) | | Sep 1 | 24 | MA | $805 | $0 | | • | | • |
| **North Carolina** | | | | | | | | | | |
| East Carolina University (Greenville) | 10 | Fall | 60 | Dipl, PhD | | | $41,200 | | | • |
| **Ohio** | | | | | | | | | | |
| University of Akron | 20 | Aug | 48 | Dipl, AuD | $11,132 | $20,892 | | | | • |
| **Oklahoma** | | | | | | | | | | |
| Univ of Oklahoma Health Sciences Center (Oklahoma City) | 10 | Aug | 46 | Dipl, AuD | $7,539 | $21,039 | | | | |
| **Pennsylvania** | | | | | | | | | | |
| University of Pittsburgh | 41 | Aug | 45, 48 | Dipl, AuD, PhD | $23,208 | $30,993 | $18 | • | | • |
| **South Dakota** | | | | | | | | | | |
| University of South Dakota (Vermillion) | 10 | Aug Sep | 48 | AuD | $3,006 | $5,448 | | | | • |
| **Tennessee** | | | | | | | | | | |
| East Tennessee State University (Johnson City) | 9 | Fall | 48 | AuD | $4,366 | $12,183 | | • | | • |
| University of Memphis | 10 | Fall semester | 48 | AuD | $9,567 | $25,266 | $4,784 | | | |
| Vanderbilt University Medical Center (Nashville) | 15 | Aug | 48 | Dipl, Cert, AuD, PhD | $25,200 | $32,400 | | • | | • |
| **Texas** | | | | | | | | | | |
| Texas Tech Univ Health Sciences Center (Lubbock) | 10 | Fall | 45 | Dipl, AuD | $7,360 | $19,638 | | | | • |
| University of Texas at Austin | 8 | Fall semester | 45 | Dipl, AuD | $8,520 | $17,270 | $8,520 | | | • |
| **Utah** | | | | | | | | | | |
| University of Utah (Salt Lake City) | | Fall | 48 | AuD | $8,056 | $17,286 | | | | • |
| University of Utah Health Science Center (Salt Lake City) | 8 | | 45 | Dipl, AuD, PhD | $9,000 | $20,000 | | • | | • |
| Utah State University (Logan) | 6 | Aug | 36 | AuD | $2,141 | $7,261 | $5,000 | | | |
| **Washington** | | | | | | | | | | |
| University of Washington (Seattle) | 12 | Sep 15 | 45 | AuD | $20,000 | $38,000 | | | | • |
| **Wisconsin** | | | | | | | | | | |
| University of Wisconsin - Madison | 15 | Fall | 48 | Dipl, AuD, PhD | $12,000 | $36,000 | | | | • |
| University of Wisconsin - Stevens Point | 5 | Fall | | Dipl, Cert, AuD | $528/ credit | $1,530/ credit | $4,000 | • | | |

*Data are shown only for programs that completed the 2006 AMA Survey of Health Professions Education Programs
‡Key to Offers: 1: Evening or weekend classes; 2: Non-English instruction; 3: Cultural competence instruction

# Blindness and Visual Impairment Professions

**Includes:**
- Low vision therapist
- Orientation and mobility specialist
- Rehabilitation teacher

## Low Vision Therapist

### Job Description
Low vision therapists help people learn to use their vision more efficiently, both with and without optical devices. They may work with optometrists or ophthalmologists in helping persons with low vision to do near tasks like reading, writing, crafts, or using a computer, and in doing distance tasks like watching TV or reading signs. They often provide follow-up training to people in their homes, schools, or work sites. Low vision therapists may also offer recommendations for improved lighting, enhanced contrast, reduced glare, and improved organization to assist those with low vision to function more successfully in their environments.

### Employment Characteristics
At least 70% of the visually impaired population is over age 65. Thus, many low vision therapists work with this age group, either in low vision clinics or in outreach settings. Others may work with school-age children in a variety of settings, such as in residential schools for the blind and visually impaired or with school districts.

### Salary
According to the 2002 Salary Survey of the Association for Education and Rehabilitation of the Blind and Visually Impaired (AER), the average full-time salary for low vision therapists is $44,777. Refer to Section IV, Table 5 of this *Directory* for more information, or see www.ama-assn.org/go/salary.

### Educational Programs
Low vision therapists must have at least a bachelor's degree and must have passed a national certification examination administered by the Academy for Certification of Vision Rehabilitation and Education Professionals (ACVREP). To be eligible for certification, applicants must meet the following minimum criteria:
- Possess at least a bachelor's degree in education, rehabilitation, or health care
- Sign a written statement agreeing to uphold a high ethical and professional standard
- Pass a written examination to demonstrate knowledge about low vision rehabilitation

## Orientation and Mobility Specialist

### Occupational Description
Orientation and mobility specialists teach people who are blind or visually impaired the skills and concepts they need to travel independently and safely at home, in the classroom, in their communities, and wherever they may want to go.

Orientation and Mobility Instruction is a sequential process in which visually impaired individuals are taught to utilize their remaining senses to determine their position within their environment and to negotiate safe movement from one place to another. The skills involved in this teaching include but are not limited to:
- Concept development, which includes body image, spatial, temporal, positional, directional. and environmental concepts
- Motor development, including motor skills needed for balance, posture, and gait, as well as the use of adaptive devices and techniques to assist those with multiple disabilities
- Sensory development, which includes visual, auditory, vestibular, kinesthetic, tactile, olfactory, and proprioceptive senses, and the interrelationships of these systems
- Residual vision stimulation and training
- Human guide technique
- Upper and lower protective techniques
- Locating dropped objects
- Trailing
- Squaring-off
- Cane techniques
- Soliciting/declining assistance
- Following directions
- Utilizing landmarks
- Search patterns
- Compass directions
- Route planning
- Analysis and identification of intersections and traffic patterns
- Use of traffic control devices
- Techniques for crossing streets
- Techniques for travel in indoor environments, outdoor residential, small and large business districts, mall travel, and rural areas
- Problem solving
- Use of public transportation
- Evaluation with sun filters for the reduction of glare
- Instructional use of low vision devices

According the 2002 Salary Survey of the AER, the average full-time salary for orientation and mobility specialists is $46,564.

## Rehabilitation Teacher

### Job Description
Rehabilitation teachers instruct persons with vision impairment in the use of compensatory skills and assistive technology that will enable them to live safe, productive, and independent lives. Specific rehabilitation teacher responsibilities include
- Assessing and evaluating learners' needs in home, work, and community environments
- Developing and implementing instructional programs, case management, and record keeping
- Helping persons with visual impairment identify and use local and national resources
- Facilitate psychosocial adjustment to blindness and vision loss

### Employment Characteristics

Rehabilitation teachers work in organizations that enhance vocational opportunities, independent living, and educational development of persons with vision loss. This may include working in center-based or itinerant settings, including clients' homes and workplaces. Rehabilitation teachers provide individualized programs of instruction that accommodate the unique needs of specialized groups, including persons who are aging, deaf-blind, or disabled.

### Salary

According to the 2002 Salary Survey of the AER, the average full-time salary for rehabilitation teachers is $37,055. Refer to Section IV, Table 5 of this *Directory* for more information, or see www.ama-assn.org/go/salary.

### Inquiries

#### Careers/Curriculum

Association for Education and Rehabilitation of the Blind and Visually Impaired
1703 N Beauregard Street, Suite 440
Alexandria, VA 22311-1744

703 671-4500
703 671-6391 Fax
www.aerbvi.org

#### Certification

Academy for Certification of Vision Rehabilitation and Education Professionals
300 N Commerce Park Loop, #200
Tucson, AZ 85745
520 887-6816
520 887-6826 Fax
E-mail: info@acvrep.org
www.acvrep.org

#### Program Accreditation

Association for Education and Rehabilitation of the Blind and Visually Impaired
1703 N Beauregard Street, Suite 440
Alexandria, VA 22311-1744
703 671-4500
703 671-6391 Fax
www.aerbvi.org

## Low Vision Therapist

## Pennsylvania

**Pennsylvania College of Optometry**
*Cosponsor:* Vision Rehabilitation Therapy & O&M Therapy
Low Vision Therapy Prgm
Dept of Graduate Studies in Vision Impairment
8360 Old York Rd
Elkins Park, PA 19027-1598
www.pco.edu
*Prgm Dir:* Maureen A Duffy, CVRT
*Tel:* 215 780-1362   *Fax:* 215 780-1357
*E-mail:* mduffy@pco.edu

## Orientation and Mobility Specialist

## Arizona

**University of Arizona**
Orientation and Mobility Specialist Prgm
PO Box 210069
Tucson, AZ 85721
*Prgm Dir:* Ian Stewart, PhD
*Tel:* 520 621-3887
*E-mail:* istewart@u.arizona.edu

## Arkansas

**University of Arkansas at Little Rock**
Orientation and Mobility Specialist Prgm
2801 S University Ave
Little Rock, AR 72204
www.ualr.edu/orientationandmobility
*Prgm Dir:* William H Jacobson, EdD
*Tel:* 501 569-8505   *Fax:* 501 224-3170
*E-mail:* whjacobson@ualr.edu

## California

**California State University - Los Angeles**
Orientation and Mobility Specialist Prgm
5151 State University Dr
Los Angeles, CA 90032
*Prgm Dir:* Diane Fazzi, PhD COMS
*Tel:* 323 343-4411   *Fax:* 323 343-5605
*E-mail:* dfazzi@calstatela.edu

**San Francisco State University**
Orientation and Mobility Specialist Prgm
1600 Holloway Ave
San Francisco, CA 94132
www.sfsu.edu/~mobility
*Prgm Dir:* Sandra Rosen, PhD COMS
*Tel:* 415 338-1245   *Fax:* 415 338-0566
*E-mail:* srosen@sfsu.edu

## Colorado

**University of Northern Colorado**
Orientation and Mobility Specialist Prgm
NCLID, McKee Hall Campus Box 146
Greeley, CO 80639
*Prgm Dir:* Paula Conroy, EdD TVI COMS
*Tel:* 970 351-1651   *Fax:* 970 351-1061
*E-mail:* paula.conroy@unco.edu

### Low Vision Therapist

| Programs* | Class Capacity | Begins | Length (months) | Award | Res Tuition | Non-res Tuition | Stipend | Offers:‡ 1 | 2 | 3 |
|---|---|---|---|---|---|---|---|---|---|---|
| **Pennsylvania** | | | | | | | | | | |
| Pennsylvania College of Optometry (Elkins Park) | | | | Cert | | | | • | | |

*Data are shown only for programs that completed the 2006 AMA Survey of Health Professions Education Programs
‡Key to Offers: 1: Evening or weekend classes; 2: Non-English instruction; 3: Cultural competence instruction

# Florida

**Florida State University**
Orientation and Mobility Specialist Prgm
205 Stone Bldg
Tallahassee, FL 32306-4459
*Prgm Dir:* Silvia Correa, COMS
*Tel:* 850 644-8413    *Fax:* 850 644-8715
*E-mail:* correa@coe.fsu.edu

# Illinois

**Northern Illinois University**
Orientation and Mobility Specialist Prgm
Dept of Teaching and Learning
DeKalb, IL 60115
www.cedu.niu.edu/tlrn/visualdisabilities
*Prgm Dir:* Jodi Sticken, MSEd COMS
*Tel:* 815 753-8456    *Fax:* 815 753-8594
*E-mail:* jsticken@niu.edu

# Michigan

**Michigan State University**
Orientation and Mobility Specialist Prgm
College of Education
332 Erickson Hall
East Lansing, MI 48824-1034
*Prgm Dir:* Tom T Hwang
*Tel:* 517 355-1871    *Fax:* 517 353-6393
*E-mail:* tomhwang@pilot.msu.edu

**Western Michigan University**
Orientation and Mobility Specialist Prgm
Dept of Blindness and Low Vision Studies
Mail Stop 5218
Kalamazoo, MI 49008
*Prgm Dir:* Paul Ponchillia, PhD RTC
*Tel:* 269 387-3455
*E-mail:* marvin.weessies@wmich.edu

# New York

**CUNY Hunter College**
*Cosponsor:* The Lavelle Fund for the Blind, Inc.
Orientation and Mobility Specialist Prgm
695 Park Ave, 909 W
New York, NY 10021
www.hunter.cuny.edu/education/deptsprograms/
chart.shtml
*Prgm Dir:* Rosanne K Silberman, EdD
*Tel:* 212 772-4740    *Fax:* 212 650-3542
*E-mail:* rsilberm@hunter.cuny.edu

# New Zealand

**Massey University**
Orientation and Mobility Specialist Prgm
School of Health Sciences
Private Bag 11222
Palmerston North, NZ
*Prgm Dir:* Steven La Grow, BS MA EdD COMS
*Tel:* 646 350-5799, Ext 2248    *Fax:* 646 350-5688
*E-mail:* S.J.Lagrow@massey.ac.nz

# North Carolina

**North Carolina Central University**
Orientation and Mobility Specialist Prgm
712 Cecil St
Durham, NC 27707
*Prgm Dir:* Brad R Walker, PhD
*Tel:* 919 715-6342    *Fax:* 919 530-7291
*E-mail:* brad.walker@ncmail.net

# Ontario, Canada

**Mohawk College**
Orientation and Mobility Specialist Prgm
411 Elgin St
Brantford, ON N3T 5V2
*Prgm Dir:* Mary-Maureen Snook-Hill, EdD
*Tel:* 519 758-6029
*E-mail:* mary-maureen.atkin@mohawkcollege.ca

# Pennsylvania

**Pennsylvania College of Optometry**
Orientation and Mobility Specialist Prgm
Dept of Graduate Studies
8360 Old York Rd
Elkins Park, PA 19027
www.pco.edu/acad_progs/grad/om/gs_om.htm
*Prgm Dir:* Laurel Leigh
*Tel:* 215 780-1449    *Fax:* 215 780-1357
*E-mail:* lleigh@pco.edu

**University of Pittsburgh**
Orientation and Mobility Specialist Prgm
Vision Studies Specialization
5316 WWPH
Pittsburgh, PA 15260
*Prgm Dir:* George J Zimmerman, PhD
*Tel:* 412 624-7247    *Fax:* 412 648-7081
*E-mail:* gjz@pitt.edu

# South Carolina

**South Carolina State University**
Orientation and Mobility Specialist Prgm
Dept of Human Services
PO Box 7356, 300 College St
Orangeburg, SC 29117
*Prgm Dir:* Shirley Madison, MA MACOMS CRC
*Tel:* 803 533-3956    *Fax:* 803 533-3636
*E-mail:* smadison@scsu.edu

# Texas

**Texas Tech University**
Orientation and Mobility Specialist Prgm
Personnel Preparation/Visual Impairment
Box 41071
Lubbock, TX 79409-1071
*Prgm Dir:* Nora Griffin-Shirley
*Tel:* 806 742-2345    *Fax:* 806 742-2326
*E-mail:* n.griffin-shirley@ttu.edu

**Stephen F Austin State University**
Orientation and Mobility Specialist Prgm
13019 SFA Station
Nacogdoches, TX 75962
*Prgm Dir:* Bob Bryant
*Tel:* 936 468-1145
*E-mail:* bbryant@sfasu.edu

## Orientation and Mobility Specialist

| Programs* | Class Capacity | Begins | Length (months) | Award | Res Tuition | Non-res Tuition | Stipend | Offers:‡ 1 | 2 | 3 |
|---|---|---|---|---|---|---|---|---|---|---|
| **Arkansas** | | | | | | | | | | |
| University of Arkansas at Little Rock | 25 | Every semester | 30 | Cert, MA, Grad C | $280 | $0 | | | | |
| **California** | | | | | | | | | | |
| San Francisco State University | 12 | Fall Spring Summer | 16 | Cert, MA | $3,674 | $10,170 | $6,000 | • | | • |
| **Illinois** | | | | | | | | | | |
| Northern Illinois University (DeKalb) | 20 | Aug | 16 | MSEd | $6,000 | $12,000 | $10,000 | • | | • |
| **New York** | | | | | | | | | | |
| CUNY Hunter College (New York) | 12 | Fall semester | 12 | Dipl, Cert, MSEd | $1,135 | $1,635 | $2,500 | • | | • |
| **Pennsylvania** | | | | | | | | | | |
| Pennsylvania College of Optometry (Elkins Park) | | | 14, 24 | Cert, MS | $18,632 | $20,916 | | • | | • |
| University of Pittsburgh | 25 | May | 15 | Dipl, Cert, MEd | $17,000 | $34,000 | | • | | |
| **Texas** | | | | | | | | | | |
| Texas Tech University (Lubbock) | | | 18 | Cert, MSEd | $8,000 | $12,000 | $5,000 | | | • |

*Data are shown only for programs that completed the 2006 AMA Survey of Health Professions Education Programs
‡Key to Offers: 1: Evening or weekend classes; 2: Non-English instruction; 3: Cultural competence instruction

## Arkansas

**University of Arkansas at Little Rock**
Rehabilitation Teacher Prgm
Dept of CARE
2801 S University
Little Rock, AR 72204-1099
www.ualr.edu/rehdept
*Prgm Dir:* Patricia Smith, EdD CVRT
*Tel:* 501 569-3169   *Fax:* 501 569-8129
*E-mail:* pbsmith@ualr.edu

## Florida

**Florida State University**
Rehabilitation Teacher Prgm
205 Stone Bldg
Tallahassee, FL 32306-4489
*Prgm Dir:* Lynda Jones, RTC
*Tel:* 850 644-5610   *Fax:* 850 644-8715
*E-mail:* jonesl@coe.fsu.edu

## Illinois

**Northern Illinois University**
Rehabilitation Teacher Prgm
Dept of Teaching and Learning
DeKalb, IL 60115
www.cedu.niu.edu/tlrn/visualdisabilities
*Prgm Dir:* Jodi Sticken, MSEd COMS
*Tel:* 815 753-8456   *Fax:* 815 753-8594
*E-mail:* jsticken@niu.edu

## Michigan

**Western Michigan University**
Rehabilitation Teacher Prgm
Dept of Blindness and Low Vision Studies
3413 Sangren Hall
Kalamazoo, MI 49008-5218
*Prgm Dir:* Susan Ponchillia, EdD MA CVRT
*Tel:* 269 387-3450   *Fax:* 269 387-3567
*E-mail:* susan.ponchillia@wmich.edu

## New York

**CUNY Hunter College**
Rehabilitation Teacher Prgm
Dept of Special Education
695 Park Ave
New York, NY 10021
*Prgm Dir:* Rosanne Silberman, EdD
*Tel:* 212 772-4740   *Fax:* 212 650-3542

## New Zealand

**Massey University**
Rehabilitation Teacher Prgm
School of Health Sciences
Private Bag 11222
Palmerston North, NZ
*Prgm Dir:* Gretchen A Good, BA MA RTC COMS
*Tel:* 646 356-9099, Ext 2245
*E-mail:* G.A.Good@massey.ac.nz

## Ontario, Canada

**Mohawk College**
Rehabilitation Teacher Prgm
11 Elgin St
Brantford, ON N3T 5V2
*Prgm Dir:* Mary-Maureen Snook-Hill
*Tel:* 519 758-6029   *Fax:* 519 758-6043
*E-mail:* mary-maureen.atkin@mohawkcollege.ca

## Pennsylvania

**Pennsylvania College of Optometry**
Rehabilitation Teacher Prgm
Dept of Graduate Studies in Vision Impairment
8360 Old York Rd
Elkins Park, PA 19027-1598
*Prgm Dir:* Maureen A Duffy, RTC
*Tel:* 215 780-1362   *Fax:* 215 780-1357
*E-mail:* mduffy@pco.edu

## Rehabilitation Teacher

| Programs* | Class Capacity | Begins | Length (months) | Award | Res Tuition | Non-res Tuition | Stipend | Offers:‡ 1 | 2 | 3 |
|---|---|---|---|---|---|---|---|---|---|---|
| **Arkansas** | | | | | | | | | | |
| University of Arkansas at Little Rock | Var | Any semester | | MA | $280 | $280 | 5700 | • | | • |
| **Illinois** | | | | | | | | | | |
| Northern Illinois University (DeKalb) | 10 | Aug | 16 | MSEd | $6,000 | $12,000 | 10000 | • | | • |
| **New York** | | | | | | | | | | |
| CUNY Hunter College (New York) | 12 | Fall semester | 30, 12 | Dipl, Cert, MSEd, COMS | $1,135 | $1,535 | 2500 | • | | • |

*Data are shown only for programs that completed the 2006 AMA Survey of Health Professions Education Programs
‡Key to Offers: 1: Evening or weekend classes; 2: Non-English instruction; 3: Cultural competence instruction

# Blood Bank Technology-Specialist

## Occupational Description

Specialists in blood bank technology perform both routine and specialized tests in blood center and transfusion services, using methodology that conforms to the *Standards for Blood Centers and Transfusion Services* of the American Association of Blood Banks (AABB).

## Job Description

Specialists in blood bank technology demonstrate a superior level of technical proficiency and problem-solving ability in such areas as (1) testing for blood group antigens, compatibility, and antibody identification; (2) investigating abnormalities such as hemolytic diseases of the newborn, hemolytic anemias, and adverse reactions to transfusion; (3) supporting physicians in transfusion therapy for patients with coagulopathies (diseases affecting blood clotting), for example, or candidates for organ and cellular transplantation/therapy; and (4) performing blood collection and processing, including selecting donors, collecting blood, typing blood, and performing viral marker testing to ensure the safety of the patient. Accordingly, supervision, management, and/or teaching make up a considerable part of the responsibilities of the specialist in blood bank technology educational program.

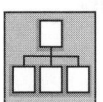

## Employment Characteristics

Specialists in blood bank technology work in many types of facilities, including community blood centers, private hospital blood banks, university-affiliated blood banks, transfusion services, and independent laboratories; they also may be part of a university faculty. Specialists may have some weekend and night duty, including emergency calls. Qualified specialists may advance to supervisory or administrative positions or move into teaching or research activities. The criteria for advancement in this field are experience, technical expertise, and completion of advanced education courses.

## Salary

Salaries for specialists in blood banking range from $45,000 for bench techs and $54,000 for supervisors to $66,000 for managers. Refer to Section IV, Table 5 of this *Directory* for more information, or see www.ama-assn.org/go/salary.

## Educational Programs

**Length.** Most of the educational programs are approximately 12 months. Some programs offer a master's degree and are approximately 24 months.

**Prerequisites.** Applicants must be certified in medical technology by the Board of Registry and possess a baccalaureate degree from a regionally accredited college or university. If applicants are not certified in medical technology by the Board of Registry, they must possess both a baccalaureate degree from a regionally accredited college or university with a major in any of the biological or physical sciences and have work experience in a blood bank.

**Curriculum.** Each specific educational program defines its own criteria for measurement of student achievement, and the sequence of instruction is at the discretion of the medical director and the program director and/or educational coordinator of the program. The clinical material available in the educational program provides the student with a full range of experiences. The educational design and environment are conducive to the development of competence in all technical areas of the modern blood bank and transfusion services. The didactic experience covers all theoretical concepts of blood bank immunohematology and transfusion medicine.

## Inquiries

### Careers/Curriculum

American Association of Blood Banks
8101 Glenbrook Road
Bethesda, MD 20814-2749
301 215-6589                301 951-3729 Fax

### Certification/Registration

Board of Registry
PO Box 12277
Chicago, IL 60612-0277
312 738-1336
E-mail: bor@ascp.org

### Program Accreditation

Commission on Accreditation of Allied Health Education Programs (CAAHEP) in collaboration with:
Committee on Accreditation of SBB Schools
American Association of Blood Banks
8101 Glenbrook Road
Bethesda, MD 20814-2749
301 215-6482 Education Department
301 907-6895 Fax
E-mail: education@aabb.org
www.aabb.org

---

## Specialist in Blood Bank Technology*

### District of Columbia

**Walter Reed Army Medical Center**
Specialist in BB Tech Prgm
US Army Blood Bank Fellowship
Dept of Pathology, 6900 Georgia Ave NW
Washington, DC 20307-5001
*Prgm Dir:* William Turcan, MT (ASCP)SBB
*Tel:* 202 782-6210   *Fax:* 202 782-4502
*E-mail:* william.turcan@na.amedd.army.mil
*Med Dir:* LTC Francis Chiricosta, MD

## Florida

**Transfusion Med Acad Ctr FL Blood Svcs**
Specialist in BB Tech Prgm
10100 Dr Martin Luther King Jr St N
St Petersburg, FL 33716-3806
*Prgm Dir:* Marjorie Doty, MT(ASCP)SBB
*Tel:* 727 568-5433, Ext 1514   *Fax:* 727 568-1177
*E-mail:* mdoty@fbsblood.org
*Med Dir:* German F Leparc, MD

---

*Additional information about programs that returned the AMA's survey is available in a table beginning on page 40.

## Illinois

**University of Illinois at Chicago**
Specialist in BB Tech Prgm
College of Applied Hlth Sciences M/C 530
1919 W Taylor St, Rm 250 AHSB
Chicago, IL 60612-7249
www.bhis.uic.edu
*Prgm Dir:* Veronica N Lewis, MS MT(ASCP)SBB
*Tel:* 312 996-6721   *Fax:* 312 996-8342
*E-mail:* veronica@uic.edu
*Med Dir:* Lou Anne Maes, MD

# Indiana

## Indiana Blood Center
Specialist in BB Tech Prgm
3450 N Meridian St
Indianapolis, IN 46208
*Prgm Dir:* Jayanna Slayten, MS MT(ASCP)SBB
*Tel:* 317 916-5186
*E-mail:* jslayten@cirbc.org
*Med Dir:* Dan Waxman, MD

# Louisiana

## Medical Center of Louisiana
Specialist in BB Tech Prgm
Blood Bank
1541 Tulane Ave Rm 205
New Orleans, LA 70112
*Prgm Dir:* Cindy A Eicher, MHS MT(ASCP) SBB
*Tel:* 504 903-2265   *Fax:* 504 903-2270
*E-mail:* ceiche@lsuhsc.edu
*Med Dir:* Yuan S Kao, MD

# Maryland

## Johns Hopkins Hospital
Specialist in BB Tech Prgm
Carnegie Bldg #667
600 N Wolfe St
Baltimore, MD 21287-6667
*Prgm Dir:* Jan Light, MAS MT(ASCP)SBB
*Tel:* 410 955-6580   *Fax:* 410 955-0618
*E-mail:* jlight5@jhmi.edu
*Med Dir:* Paul M Ness, MD

## NIH Clinical Center Blood Bank
Specialist in BB Tech Prgm
NIH/CC/DTM Bldg 10 Rm 1C 711
10 Center Dr MSC 1184
Bethesda, MD 20892
www.cc.nih.gov/dtm/dtm_training_&_education_sbb_program.htm
*Prgm Dir:* Sherry Sheldon, MT(ASCP)SBB
*Tel:* 301 451-8654   *Fax:* 301 496-9990
*E-mail:* SSheldon@cc.nih.gov
*Med Dir:* David F Stroncek, MD

# Ohio

## University of Cincinnati Medical Center
*Cosponsor:* Hoxworth Blood Center
Specialist in BB Tech Prgm
3130 Highland Ave, PO Box 670055
Cincinnati, OH 45267-0055
*Prgm Dir:* Susan L Wilkinson, EdD MT(ASCP)SBB
*Tel:* 513 558-1271   *Fax:* 513 558-1279
*E-mail:* catherine.beiting@uc.edu
*Med Dir:* Ronald A Sacher, MD

## ARC Blood Svcs - Central Ohio Region
*Cosponsor:* OSU Transfusion Service
Specialist in BB Tech Prgm
995 E Broad
Columbus, OH 43205
*Prgm Dir:* Joanne Kosanke, MT(ASCP)SBB
*Tel:* 614 253-2740, Ext 2270   *Fax:* 614 253-2487
*E-mail:* kosankej@usa.redcross.org
*Med Dir:* Arwa Shana'ah, MD

# Texas

## Univ of Texas Southwestern Med Ctr at Dallas
Specialist in BB Tech Prgm
5323 Harry Hines Blvd
Dallas, TX 75390-8878
*Prgm Dir:* Lynn M Little, PhD MT(ASCP)
*Tel:* 214 648-1780   *Fax:* 214 648-1029
*E-mail:* barbara.fryer@utsouthwestern.edu
*Med Dir:* Laurie Sutor, MD

## University of Texas Medical Branch/SAHS
Specialist in BB Tech Prgm
301 University Blvd
Galveston, TX 77555-1140
www.utmb.edu/sbb
*Prgm Dir:* Janet L Vincent, MS SBB(ASCP)
*Tel:* 409 772-9476   *Fax:* 409 772-9470
*E-mail:* jvincent@utmb.edu
*Med Dir:* Alexander Indrikovs, MD

## Gulf Coast School of Blood Bank Technology
Specialist in BB Tech Prgm
1400 La Concha Ln
Houston, TX 77054-1802
*Prgm Dir:* Clare Wong, MT(ASCP)SBB SLS
*Tel:* 713 791-6201   *Fax:* 713 791-6610
*E-mail:* cwong@giveblood.org
*Med Dir:* Susan Rossmann, MD, PhD

## Univ of Texas Hlth Sci Ctr at San Antonio
*Cosponsor:* University Hospital Health System
Specialist in BB Tech Prgm
Dept of Clinical Laboratory Sci - 6246
7703 Floyd Curl Dr
San Antonio, TX 78229-3900
*Prgm Dir:* Linda A Smith, PhD Bonnie Fodermaier, SBB(ASCP)
*Tel:* 210 567-8869   *Fax:* 210 567-8875
*E-mail:* smithla@uthscsa.edu
*Med Dir:* Chantal R Harrison, MD

# Wisconsin

## Blood Center of Southeast Wisconsin
Specialist in BB Tech Prgm
638 N 18th St
Milwaukee, WI 53233
*Prgm Dir:* Susan T Johnson, MT(ASCP) SBB
*Tel:* 414 937-6403   *Fax:* 414 937-6202
*E-mail:* stjohnson@bcsew.edu
*Med Dir:* Kenneth D Friedman, MD

## Specialist in Blood Bank Technology

| Programs* | Class Capacity | Begins | Length (months) | Award | Res Tuition | Non-res Tuition | Stipend | Offers:‡ 1 | 2 | 3 |
|---|---|---|---|---|---|---|---|---|---|---|
| **District of Columbia** | | | | | | | | | | |
| Walter Reed Army Medical Center (Washington) | 6 | Jul | 18 | Dipl, Cert, SBB, MSHS | | | | | | |
| **Illinois** | | | | | | | | | | |
| University of Illinois at Chicago | 10 | Aug | 12, 16 | Cert | $9,520 | $9,520 | | | | |
| **Louisiana** | | | | | | | | | | |
| Medical Center of Louisiana (New Orleans) | 4 | Jan | 12, 24 | Cert | | | | | | |
| **Maryland** | | | | | | | | | | |
| Johns Hopkins Hospital (Baltimore) | 6 | Jul | 12 | Cert, SBB | | | | | | |
| NIH Clinical Center Blood Bank (Bethesda) | 5 | Jul | 12 | Cert | | | | | | |
| **Ohio** | | | | | | | | | | |
| ARC Blood Svcs - Central Ohio Region (Columbus) | 4 | Jan | 24 | Cert | $1,000 | $0 | | | | |
| University of Cincinnati Medical Center | 2 | Sep | 15 | MS | $10,158 | $19,992 | $15,000 | | | |
| **Texas** | | | | | | | | | | |
| Gulf Coast School of Blood Bank Technology (Houston) | 6 | Jun | 12 | Cert | $2,400 | $2,400 | | | | |
| Univ of Texas Hlth Sci Ctr at San Antonio | 2 | Sep | 12 | Dipl, Cert, MS Opt | | | | | | |
| University of Texas Medical Branch/SAHS (Galveston) | 30 | Mar | 12 | Cert | $2,400 | $2,400 | | | | |

*Data are shown only for programs that completed the 2006 AMA Survey of Health Professions Education Programs
‡Key to Offers: 1: Evening or weekend classes; 2: Non-English instruction; 3: Cultural competence instruction

# Cardiovascular Technologist

## Occupational Description

The cardiovascular technologist performs diagnostic examinations and therapeutic interventions of the heart and/or blood vessels at the request or direction of a physician in one or more of the following:

- Invasive cardiology—Cardiac catheterization
- Noninvasive cardiology—Echocardiography
- Noninvasive peripheral vascular study—Vascular ultrasound

Through subjective sampling and/or recording, the technologist creates an easily definable foundation of data from which a correct anatomic and physiologic diagnosis may be established for each patient.

## Job Description

The cardiovascular technologist is qualified by specific didactic, laboratory, and clinical technological education to perform various cardiovascular/peripheral vascular diagnostic and therapeutic procedures. The role of the cardiovascular technologist may include but is not limited to (1) reviewing and/or recording pertinent patient history and supporting clinical data; (2) performing appropriate clinical procedures and obtaining a record of anatomical, pathological, and/or physiological data for interpretation by a physician; and (3) exercising discretion and judgment in the performance of cardiovascular diagnostic and therapeutic services.

## Employment Characteristics

Cardiovascular technologists may provide their services to patients in any medical setting under the supervision of a doctor of medicine or osteopathy (MD or DO). The procedures performed by the cardiovascular technologist may be found in, but are not limited to, one of the following general settings: (1) invasive cardiovascular laboratories, including cardiac catheterization, blood gas, and electrophysiology laboratories; (2) noninvasive cardiovascular laboratories, including echocardiography, exercise stress test, and electrocardiography laboratories; and (3) noninvasive peripheral vascular studies laboratories, including Doppler ultrasound, thermography, and plethysmography laboratories.

## Salary

In 2006, entry-level salaries ranged from $36,000 to $45,000. The overall average was $50,000 to $65,000, with the upper ranges at $75,000 plus. Refer to Section IV, Table 5 of this *Directory* for more information, or see www.ama-assn.org/go/salary.

## Educational Programs

**Length.** Programs may be from 1 to 4 years, depending on student qualifications and number of areas of diagnostic evaluation selected: invasive cardiology, noninvasive cardiology, or noninvasive peripheral vascular study.
**Prerequisites.** High school diploma or equivalent or qualifications in a clinically related allied health profession.
**Curriculum.** Curricula of accredited programs include didactic instruction, formal laboratory experiences, and patient-based clinical instruction. Suggested areas of instruction in the core curriculum include an introduction to the field of cardiovascular technology, general and/or applied sciences, human anatomy and physiology, basic pharmacology, and basic medical electronics and medical instrumentation. Emphasis, following the core curriculum, is given in the specialty area(s) selected: invasive cardiology, noninvasive cardiology, and noninvasive peripheral vascular study. Both didactic instruction and clinical experiences are provided in these areas.

## Inquiries

### Careers/Curriculum

Society for Vascular Ultrasound
4601 Presidents Drive, Suite 260
Lanham, MD 20706-4365
301 459-7550 or 800 SVT-VEIN
301 459-5651 Fax

American Society of Echocardiography
1500 Sunday Drive, Suite 102
Raleigh, NC 27607
919 861-5574
919 787-4916 Fax
www.asecho.org

Society of Invasive Cardiovascular Professionals
1500 Sunday Drive, Suite 102
Raleigh, NC 27607
919 861-4546
www.sicp.com

### Certification/Registration

Cardiovascular Credentialing International
1500 Sunday Drive, Suite 102
Raleigh, NC 27607
800 326-0268
919 787-4916 Fax
www.cci-online.org

American Registry of Diagnostic Medical Sonographers
51 Monroe Street, Plaza East One
Rockville, MD 20850-2400
301 738-8401
301 738-0312 Fax
www.ardms.org

### Program Accreditation

Commission on Accreditation of Allied Health Education Programs (CAAHEP) in collaboration with:
Joint Review Committee on Education in Cardiovascular Technology
1248 Harwood Road
Bedford, TX 76021-4244
817 283-2835
817 354-8519
E-mail: richwalker@coarc.com

## Cardiovascular Technologist*

## California

### Orange Coast College
Cardiovascular Technology Prgm
2701 Fairview Rd
Costa Mesa, CA 92628-5005
*Prgm Dir:* Darryl Isaac
*Tel:* 714 432-5549   *Fax:* 714 432-5535
*E-mail:* disaac@mail.occ.cccd.edu
*Med Dir:* Robin Day Shaughnessy, MD

### Grossmont College
Cardiovascular Technology Prgm
Invasive/Noninvasive/Vascular
8800 Grossmont College Dr
El Cajon, CA 92020-1799
www.grossmont.edu/healthprofessions
*Prgm Dir:* Rick D Kirby, RICS RCS MA
*Tel:* 619 644-7302   *Fax:* 619 644-7961
*E-mail:* rick.kirby@gcccd.net
*Med Dir:* William Ceretto, MD

### Naval School of Health Sciences, San Diego
Cardiovascular Technology Prgm
Invasive Track
34101 Farenholt Ave
San Diego, CA 92134
http://nshssd.med.navy.mil
*Prgm Dir:* HM1 James M Glasgow, RCIS RN
*Tel:* 619 532-5124   *Fax:* 619 532-6731
*E-mail:* jmglasgow@nshs-sd.med.navy.mil
*Med Dir:* Peter E Linz, MD

## Florida

### Edison College
Cardiovascular Technology Prgm
Invasive Track
8099 College Pkwy SW
Ft Myers, FL 33906
*Prgm Dir:* Robert Jeff Davis, RRT RCIS
*Tel:* 239 489-9430   *Fax:* 239 985-8352
*E-mail:* jdavis@edison.edu
*Med Dir:* Robert Grohowski, MD

### Santa Fe Community College
Cardiovascular Technology Prgm
Invasive/Noninvasive/Vascular Tracks
3000 NW 83rd St
Gainesville, FL 32653
*Prgm Dir:* Reeda Fullington
*Tel:* 352 395-5707   *Fax:* 352 395-5711
*E-mail:* reeda.fullington@sfcc.edu
*Med Dir:* Edward A Geiser, MD

### Central Florida Institute
Cardiovascular Technology Prgm
Invasive/Noninvasive/Cardiovascular Ultrasound
30522 US Hwy 19 N
Palm Harbor, FL 34684
www.cfinstitute.com
*Prgm Dir:* Thomas O'Brien, MBA BBA AS CCT
*Tel:* 727 786-4707   *Fax:* 727 781-9421
*E-mail:* tobrien@cfinstitute.com
*Med Dir:* Patrick Cambier, MD

## Georgia

### Harry T Harper Jr Sch of Cardiac & Vas Tech
*Cosponsor:* Augusta Technical College & USC Columbia
Cardiovascular Technology Prgm
University Hospital
1350 Walton Way
Augusta, GA 30901
*Prgm Dir:* Patricia L Thomas, MBA RCIS BSRT
*Tel:* 706 774-5044   *Fax:* 706 774-8644
*E-mail:* pthomas@uh.org
*Med Dir:* John W Kelly, MD

## Kentucky

### Spencerian College
Cardiovascular Technology Prgm
4627 Dixie Hwy
Louisville, KY 40216
www.spencerian.edu
*Prgm Dir:* Vicki Lemaster
*Tel:* 502 447-1000   *Fax:* 502 449-7866
*E-mail:* vlemaster@spencerian.edu
*Med Dir:* Vincent Degeare, MD

## Louisiana

### Louisiana State Univ Health Sciences Center
Cardiovascular Technology Prgm
1900 Gravier St
New Orleans, LA 70112
*Prgm Dir:* Andy Pellett, PhD
*Tel:* 504 568-4230   *Fax:* 504 568-4249
*E-mail:* apelle@lsuhsc.edu
*Med Dir:* Sudhanva Wadgaonkar, MD

## Maryland

### Howard Community College
Cardiovascular Technology Prgm
Invasive Track
10901 Little Patuxent Pkwy
Columbia, MD 21044
*Prgm Dir:* Mary Patricia English, RTR(CV) MS RLIS
*Tel:* 410 772-4466   *Fax:* 410 772-4494
*E-mail:* penglish@howardcc.edu
*Med Dir:* Jon Rodney Resar, MD

## Michigan

### Carnegie Institute
Cardiovascular Technology Prgm
Noninvasive Peripheral Vascular
550 Stephenson Hwy Ste 100
Troy, MI 48083
www.carnegie-institute.com
*Prgm Dir:* Bonnie M Normile, RN CCVT CMA
*Tel:* 248 589-1078   *Fax:* 248 589-1631
*E-mail:* info@Carnegie-Institute.edu
*Med Dir:* Joel Kahn, MD

## Minnesota

### Northland Community & Technical College
Cardiovascular Technology Prgm
Invasive Track
2022 Central Ave NE
East Grand Forks, MN 56721
*Prgm Dir:* Connie Schimke, BS RCIS
*Tel:* 218 773-3441, Ext 4797   *Fax:* 218 773-4502
*E-mail:* connie.schimke@northlandcollege.edu
*Med Dir:* Noah Chelliah, MD

### St Cloud Technical College
Cardiovascular Technology Prgm
Invasive and Echocardiography
1540 Northway Dr
St Cloud, MN 56303
*Prgm Dir:* Patrick McGuire
*Tel:* 320 308-6010
*E-mail:* pmcguire@sctc.edu

## Nebraska

### BryanLGH College of Health Sciences
*Cosponsor:* BryanLGH Medical Center
Cardiovascular Technology Prgm
5035 Everett St
Lincoln, NE 68506-1398
www.bryanlghcollege.org
*Prgm Dir:* Diane Kathol, MSN MEd
*Tel:* 402 481-8847   *Fax:* 402 481-8421
*E-mail:* dkathol@bryanlgh.org
*Med Dir:* Charles Wilson, MD

## New Jersey

### Morristown Memorial Hospital
Cardiovascular Technology Prgm
Invasive/Noninvasive/Vascular Tracks
100 Madison Ave
Morristown, NJ 07962-1956
www.atlantichealth.org
*Prgm Dir:* Susan Smith, BA RDCS RVS
*Tel:* 973 971-6336   *Fax:* 973 290-7310
*E-mail:* susan.smith@atlantichealth.org
*Med Dir:* Grant Parr, MD

### Univ of Med & Dent of New Jersey
Cardiovascular Technology Prgm
Vascular Technology Program
1776 Raritan Rd
Scotch Plains, NJ 07076
*Prgm Dir:* Clifford T Araki, PhD RVT
*Tel:* 908 889-2468   *Fax:* 973 972-0433
*E-mail:* arakict@umdnj.edu
*Med Dir:* Robert W Hobson II, MD

## New York

### Molloy College
Cardiovascular Technology Prgm
Invasive/Noninvasive/Vascular Tracks
PO Box 5002
Rockville Centre, NY 11571
www.molloy.edu
*Prgm Dir:* Sister Marie Buckley, MS RCPT RCIS RCVT
*Tel:* 516 678-5000, Ext 6749   *Fax:* 516 594-0446
*E-mail:* MBUCKLEY@MOLLOY.EDU
*Med Dir:* Thierry Duchatellier, MD

## Ohio

### University of Toledo
Cardiovascular Technology Prgm
Noninvasive Cardiovascular (Echo) Track
2801 W Bancroft St, Mail Stop 119
Toledo, OH 43606-3390
www.utoledo.edu
*Prgm Dir:* Suzanne Wambold, PhD RN RDCS FASE
*Tel:* 419 530-4688   *Fax:* 419 530-4780
*E-mail:* suzanne.wambold@utoledo.edu
*Med Dir:* Mohammed M Maaieh, MD

---

*Additional information about programs that
returned the AMA's survey is available in a table
beginning on page 44.

# Pennsylvania

**Geisinger Medical Center**
Cardiovascular Technology Prgm
Invasive Track
100 N Academy Ave
Danville, PA 17822-2011
www.geisinger.org
*Prgm Dir:* Stephanie Ranck, BA RCIS
*Tel:* 570 271-6638   *Fax:* 570 271-5962
*E-mail:* sranck@geisinger.edu
*Med Dir:* William Kimber, MD

**Gwynedd-Mercy College**
Cardiovascular Technology Prgm
Invasive/Noninvasive Tracks
1325 Sumneytown Pike - PO Box 901
Gwynedd Valley, PA 19437
*Prgm Dir:* Andrea Reiley-Helzner, MS RCIS
*Tel:* 215 646-7300, Ext 476   *Fax:* 215 641-5559
*E-mail:* reiley.a@gmc.edu
*Med Dir:* Paul Coady, MD, FACC

**Lancaster Gen Coll of Nursing & Hlth Sciences**
Cardiovascular Technology Prgm
410 N Lime St
Lancaster, PA 17602
www.LancasterGeneralCollege.org
*Prgm Dir:* William L Fisher, RCIS MEd
*Tel:* 717 290-5511, Ext 44700   *Fax:* 717 290-5970
*E-mail:* wlfisher@LancasterGeneral.org
*Med Dir:* John P Slovak, MD

# South Carolina

**Sister of Charity Providence Hospital**
Cardiovascular Technology Prgm
School of Cardiovascular Diagnostics
Department 7195
Columbia, SC 29204
*Prgm Dir:* Eric Walker, RDCS RVT
*Tel:* 803 256-5636   *Fax:* 803 256-5913
*E-mail:* eric.walker@providencehospitals.com
*Med Dir:* Claude Smith, MD

# South Dakota

**Southeast Technical Institute**
Cardiovascular Technology Prgm
Invasive/Noninvasive/Vascular Tracks
2320 N Career Ave
Sioux Falls, SD 57107
*Prgm Dir:* Catherine J Miller, BS LPN RVT
*Tel:* 605 367-4634   *Fax:* 605 367-6108
*E-mail:* Cathy.Miller@southeasttech.com
*Med Dir:* Greg Schultz, MD

# Tennessee

**Northeast State Technical Comm College**
Cardiovascular Technology Prgm
Invasive Track
PO Box 246, 2425 Hwy 75
Blountville, TN 37617-0246
*Prgm Dir:* Connie Marshall, RT(R) RCIS
*Tel:* 423 279-3680   *Fax:* 423 477-4859
*E-mail:* cmarshall@NortheastState.edu
*Med Dir:* Herbert Ladley, MD FACC

# Texas

**El Centro College**
Cardiovascular Technology Prgm
Echocardiology Technology Program
Main and Lamar Sts
Dallas, TX 75202
*Prgm Dir:* Catherine Carolan, RDMS RDCS RVT
*Tel:* 214 860-2310   *Fax:* 214 860-2268
*E-mail:* cmc5540@dcccd.edu
*Med Dir:* Beth Brickner, MD

**US Army Medical Dept Center & School**
Cardiovascular Technology Prgm
Invasive/Noninvasive Tracks
MCCS-HMD, 2250 Stanley Rd Bldg 2398
FT Sam Houston, TX 78234-6170
*Prgm Dir:* Harrell Carmichael
*Tel:* 210 221-3472   *Fax:* 210 221-1197
*E-mail:* harrell.carmichael@cen.amedd.army.mil

# Virginia

**Sentara Norfolk General Hospital**
Cardiovascular Technology Prgm
Invasive/Noninvasive/Vascular Tracks
600 Gresham Dr
Norfolk, VA 23507
*Prgm Dir:* Kathy Butterbaugh
*Tel:* 757 668-2900   *Fax:* 757 668-2905
*E-mail:* klbutter@sentara.com
*Med Dir:* John P Parker, MD

# Washington

**Spokane Community College**
Cardiovascular Technology Prgm
Invasive/Noninvasive Tracks
N 1810 Greene St MS 2090
Spokane, WA 99217-5499
*Prgm Dir:* Darren Powell, RCIS FSICP
*Tel:* 509 533-7306   *Fax:* 509 533-8621
*E-mail:* dpowell@scc.spokane.edu
*Med Dir:* Pierre Leimgruber, MD

# Wisconsin

**Milwaukee Area Technical College**
Cardiovascular Technology Prgm
Invasive Track
700 W State St
Milwaukee, WI 53233-1443
*Prgm Dir:* Erwin Wuehr, CP
*Tel:* 414 297-8517   *Fax:* 414 297-8955
*E-mail:* wuehre@matc.edu
*Med Dir:* Thomas Mahn, MD

## Cardiovascular Technologist

| Programs* | Class Capacity | Begins | Length (months) | Award | Res Tuition | Non-res Tuition | Stipend | Offers:‡ 1 | 2 | 3 |
|---|---|---|---|---|---|---|---|---|---|---|
| **California** | | | | | | | | | | |
| Grossmont College (El Cajon) | 54 | Aug | 24 | Cert, AS | $1,430 | $9,735 | | | | • |
| Naval School of Health Sciences, San Diego | 16 | Feb | 12 | Cert | | | | | | • |
| **Florida** | | | | | | | | | | |
| Central Florida Institute (Palm Harbor) | 15 | Varies | 21 | AAS | $22,958 | $22,958 | | | | • |
| Edison College (Ft Myers) | 20 | Aug | 22 | AS | $2,843 | $11,094 | | | | |
| **Kentucky** | | | | | | | | | | |
| Spencerian College (Louisville) | 10 | Jan Mar Jun Sep | 21 | AASD | $12,120 | $12,120 | | | | • |
| **Michigan** | | | | | | | | | | |
| Carnegie Institute (Troy) | 14 | Mar Sep | 18 | Dipl | $11,535 | $11,535 | | • | | |
| **Nebraska** | | | | | | | | | | |
| BryanLGH College of Health Sciences (Lincoln) | 24 | Aug | 21 | AS | $9,600 | $9,600 | | | | • |
| **New Jersey** | | | | | | | | | | |
| Morristown Memorial Hospital | 15 | Sep | 18 | Dipl | $18,000 | $18,000 | | | | |
| Univ of Med & Dent of New Jersey (Scotch Plains) | 12 | Sep | 15 | Cert, BS | $9,000 | $13,500 | | | | |
| **New York** | | | | | | | | | | |
| Molloy College (Rockville Centre) | 12 | Sep | 24 | AAS | $16,280 | $0 | | • | | |
| **Ohio** | | | | | | | | | | |
| University of Toledo | 10 | Aug | 24 | AAS, BS | $12,000 | $25,000 | | | | |
| **Pennsylvania** | | | | | | | | | | |
| Geisinger Medical Center (Danville) | 6 | Aug | 12 | Dipl | $8,500 | $8,500 | | | | |
| Gwynedd-Mercy College (Gwynedd Valley) | 16 | Aug | 24 | Cert, AS | $18,580 | $18,580 | | | | • |
| Lancaster Gen Coll of Nursing & Hlth Sciences | 11 | Aug | 12, 24 | Dipl, AS | $9,400 | $0 | | | | |
| **Texas** | | | | | | | | | | |
| El Centro College (Dallas) | 15 | Jun | 12, 24 | Cert, AAS | $2,100 | $5,600 | | | | • |
| US Army Medical Dept Center & School (FT Sam Houston) | 15 | Apr | 21, 36 | Dipl | | | | | | • |
| **Washington** | | | | | | | | | | |
| Spokane Community College | 30 | Sep | 20 | AAS | $2,864 | $5,348 | | | | • |
| **Wisconsin** | | | | | | | | | | |
| Milwaukee Area Technical College | 12 | Aug | 19 | AAS | $3,100 | $17,500 | | • | | |

*Data are shown only for programs that completed the 2006 AMA Survey of Health Professions Education Programs
‡Key to Offers: 1: Evening or weekend classes; 2: Non-English instruction; 3: Cultural competence instruction

## Includes:

- Clinical assistant
- Clinical laboratory scientist/medical technologist
- Clinical laboratory technician/medical laboratory technician
- Cytogenetic technologist
- Diagnostic molecular scientist
- Histologic technician/histotechnologist
- Pathologists' assistant
- Phlebotomist

## Clinical Assistant

### Occupational Description

Laboratory tests play an important role in the detection, diagnosis, and treatment of diseases. Clinical assistants are formally prepared multiskilled health care providers with a laboratory focus. Clinical assistants are trained to work under the supervision of an appropriately qualified person in chemistry, donor room collection screening, component processing, hematology, immunology, microbiology, phlebotomy, and/or urinalysis.

### Job Description

Clinical assistants follow standard operating procedures to collect specimens; prepare blood and body fluid specimens for analysis according to standard operating procedures; prepare/reconstitute reagents, standards, and controls according to standard operating procedure; perform appropriate tests at the clinical assistant level; perform and record vital sign measurements; and follow established quality control protocols. Clinical assistants follow infection control and safety practices and use information systems necessary to accomplish job functions.

### Employment Characteristics

Many clinical assistants are employed in hospital laboratories. Others are employed in physicians' private laboratories and clinics; by the armed forces; or by city, state, and federal health agencies.

### Salary

Salaries vary depending on the employer and geographic location. Refer to Section IV, Table 5 of this *Directory* for more information, or see www.ama-assn.org/go/salary.

### Educational Programs

**Length.** Approved programs culminate in a postsecondary certificate.

**Prerequisites.** High school diploma or equivalent. The applicant must also meet the admission requirements of the sponsoring institution.

**Curriculum.** Clinical assistant programs are conducted in junior or community colleges, hospitals, medical laboratories, proprietary schools, and other equivalent postsecondary educational institutions.

## Clinical Laboratory Scientist/ Medical Technologist

### Occupational Description

Laboratory tests play an important role in the detection, diagnosis, and treatment of many diseases. Clinical laboratory scientists/medical technologists perform these tests in conjunction with pathologists (physicians who diagnose the causes and nature of disease) and other physicians or scientists who specialize in clinical chemistry, microbiology, or the other biological sciences. Clinical laboratory scientists/medical technologists develop data on the blood, tissues, and fluids of the human body by using a variety of precise methodologies and technologies.

### Job Description

In addition to possessing the skills of clinical laboratory technicians/medical laboratory technicians, clinical laboratory scientists/medical technologists perform complex analyses, fine-line discrimination, and error correction. They are able to recognize the interdependency of tests and have knowledge of physiological conditions affecting test results so that they can confirm these results and develop data that may be used by a physician in determining the presence, extent, and, as far as possible, cause of a disease.

Clinical laboratory scientists/medical technologists assume responsibility and are held accountable for accurate results. They establish and monitor quality assurance and quality improvement programs and design or modify procedures as necessary. Tests and procedures performed or supervised by clinical laboratory scientists/medical technologists in the clinical laboratory focus on major areas of hematology, microbiology, immunohematology, immunology, clinical chemistry, and urinalysis.

### Employment Characteristics

Most clinical laboratory scientists/medical technologists are employed in hospital laboratories. Others are employed in physicians' private laboratories and clinics; by the armed forces; by city, state, and federal health agencies; in industrial medical laboratories; in pharmaceutical houses; in numerous public and private research programs dedicated to the study of specific diseases; and as faculty of accredited programs preparing medical laboratory personnel. While many graduates are employed in the clinical laboratory setting, career options are expanding, with opportunities in all areas of health care. As a clinical laboratory scientist/medical technologist, one may decide to specialize in:

- Biomedical research and development
- Andrology and assisted reproductive technology laboratories
- Organ transplantation
- Genetic testing
- Infection control
- Health information management
- Health care industry
- Consultative and entrepreneurial opportunities
- Forensic testing

### Salary

Salaries vary depending on the employer and geographic location. Based on a 2005 survey published in *Laboratory Medicine*, median salaries ranged from $44,500 to $52,000, and median manager salaries ranged from $69,500 to $72,000. Refer to Section IV, Table 5 of this *Directory* for more information, or see www.ama-assn.org/go/salary.

### Educational Programs

**Length.** Programs are at least 1 year of professional/clinical education in conjunction with either a baccalaureate or a master's degree.

**Prerequisites.** College courses and number of required credits are those necessary to ensure admission of a student who is prepared for the clinical educational program. Content areas should include general chemistry, general biological sciences, organic and/or biochemistry, microbiology, immunology, and mathematics. Survey courses do not qualify as fulfillment of chemistry and biological science prerequisites, and remedial mathematics courses will not satisfy the mathematics requirement.

College/university programs that integrate preprofessional and professional coursework are structured with professional courses in the junior and senior years or at the graduate level.

**Curriculum.** There must be a structured laboratory program, including instruction pertaining to theory and practice in hematology, clinical chemistry, microbiology, immunology, and immunohematology. The program must culminate in a baccalaureate degree for those students not already possessing the degree but may also culminate in a master's degree.

# Clinical Laboratory Technician/ Medical Laboratory Technician

### Occupational Description

Laboratory tests play an important role in the detection, diagnosis, and treatment of many diseases and in the promotion of health. Clinical laboratory technicians/medical laboratory technicians perform these tests under the supervision or direction of pathologists (physicians who diagnose the causes and nature of disease) and other physicians, clinical laboratory scientists/medical technologists, or other scientists who specialize in clinical chemistry, microbiology, or the other biological sciences. Clinical laboratory technicians/medical laboratory technicians develop data on the blood, tissues, and fluids of the human body by using a variety of precise methodologies and technologies.

### Job Description

Associate degree clinical laboratory technicians/medical laboratory technicians perform all the routine tests in an up-to-date medical laboratory and can demonstrate discrimination between closely similar items and correction of errors by the use of preset strategies. The technician has knowledge of specific techniques and instruments and is able to recognize factors that directly affect procedures and results. The technician also monitors quality assurance procedures.

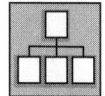

### Employment Characteristics

Most clinical laboratory technicians/medical laboratory technicians work in hospital laboratories, averaging a 40-hour week.

### Salary

Salaries vary, depending upon the employer and geographic location. Based on a 2005 survey published in *Laboratory Medicine,* median salaries ranged from $37,100 to $41,000. Refer to Section IV, Table 5 of this *Directory* for more information, or see www.ama-assn.org/go/salary.

### Educational Programs

**Length.** The period of education is usually 2 academic years for graduates receiving an associate degree; a period of clinical education is usually 12 months for graduates receiving a certificate.

**Prerequisites.** High school diploma or equivalent. The applicant also must meet the admission requirements of the sponsoring educational institution.

**Curriculum.** Associate degree programs are conducted in junior or community colleges, in 2-year divisions of universities and colleges, or in other recognized institutions granting associate degrees. Courses are taught on campus and usually in affiliated hospitals. Certificate programs are conducted in junior and community colleges or in other recognized instructional institutions. The curriculum focuses on knowledge and basic skills and on understanding principles and mastering procedures of laboratory testing, as well as on basic laboratory mathematics, computer technology, communication skills and interpersonal relationships, and social responsibilities. Classroom and laboratory classes focus on general knowledge and basic skills; understanding principles and master procedures of laboratory testing; and basic laboratory mathematics, computer technology, communication skills, and interpersonal relationships and responsibilities. The clinical courses include application of basic principles commonly used in the diagnostic laboratory. Technical instruction includes procedures in hematology, microbiology, immunohematology, immunology, clinical chemistry, and urinalysis.

# Cytogenetic Technologist

### Occupational Description

Laboratory tests play an important role in the detection, diagnosis, and treatment of diseases. Cytogenetic technologists study the morphology of chromosomes and their relationship to disease. Cytogenetic analysis provides important data for the diagnosis, prognosis, and treatment of genetic disorders and malignant diseases.

### Job Description

Cytogenetic technologists evaluate the correct method of collection, transport, and handling of various specimen types for cytogenetic analysis; identify culture techniques based on tissue type and reason for referral; and perform chromosomal staining, microscopic analysis, and karyotyping (organizing chromosomes according to a standardized ideogram). In addition to practicing good general laboratory skills, quality assurance principles, and safety protocols, cytogenetic technologists understand the legal implications of their work environment and exhibit appropriate ethical and professional health care standards while demonstrating professional conduct, stress management, and interpersonal and communication skills with patients, peers, other health care personnel, and the public.

### Employment Characteristics

Cytogenetic technologists are employed in hospital laboratories, private medical laboratories, and research facilities. They may also serve as faculty in cytogenetic education programs.

### Salary

Salaries vary depending upon the employer and geographic location. Refer to Section IV, Table 5 of this *Directory* for more information, or see www.ama-assn.org/go/salary.

### Educational Programs

**Length.** Programs are at least 1 year. Cytogenetic technologists attend a baccalaureate or postbaccalaureate program that includes professional and clinical education. Certification is desired by many employers.

**Prerequisites.** College courses and a number of required credits necessary to ensure admission of a student who is prepared for the clinical education program. College and university programs that integrate preprofessional and professional coursework are structured with professional courses in the junior and senior years or at the graduate level.

**Curriculum.** Cytogenetic technology programs are conducted in colleges and universities, hospitals, private medical laboratories, and other equivalent postsecondary educational institutions. The areas of study that must be included in either the professional program or as prerequisites are biology, chemistry, biochemistry or cellular biology, genetics, cytogenetics, hematology, microbiology, immunology, laboratory information systems, laboratory safety, and quality control.

## Diagnostic Molecular Scientist

### Occupational Description

The diagnostic molecular scientist performs diagnostic assay or testing using a variety of manual techniques and precision instruments. The results of these tests are used to detect and diagnose disease and other abnormalities. The main responsibilities of the diagnostic molecular scientist are all aspects of genetic testing, including DNA and RNA isolation, amplification and detection, infectious disease testing, and viral load analysis.

### Job Description

Diagnostic molecular scientists provide service in the molecular diagnosis of acquired, inherited, and infectious diseases. This includes researching, evaluating, implementing, and monitoring methods of collection, transport, and handling of various specimen types for molecular analysis; researching and developing principles, practices, and applications of molecular-based testing for laboratory utilization and clinical decisions for client outcomes; and performing appropriate techniques utilizing instrumentation and information management systems for molecular analysis and correlating results with acquired, inherited, and infectious diseases. Finally, diagnostic molecular scientists apply the principles of management and supervision when they function as section supervisors and of educational methodology when they teach students.

### Employment Characteristics

Most diagnostic molecular scientists work in hospital laboratories.

### Educational Programs

**Length.** Programs for the diagnostic molecular scientist usually lead to a master's degree.

**Prerequisites.** College level courses as required by the sponsoring institution.

**Curriculum.** The curriculum includes both didactic instruction and practical demonstration in the areas of organic and/or biochemistry, genetics, cell biology, microbiology, immunology, diagnostic molecular biology, principles and methodologies for all major areas commonly practiced by a modern diagnostic molecular laboratory, and clinical significance of laboratory procedures in diagnosis and treatment. It also includes principles and practices of

- Laboratory administration, supervision, safety, and problem solving
- Quality management
- Computer science (including acquisition and evaluation of laboratory information systems)
- Professional conduct

## Histologic Technician

### Occupational Description

Physicians (usually pathologists) and other scientists specializing in biological sciences or related clinical areas such as chemistry work in partnership with medical laboratory workers to analyze blood, tissues, and fluids from humans (and sometimes animals), using a variety of precision instruments. The results of these tests are used to detect and diagnose disease and other abnormalities.

The main responsibility of the histologic technician in the clinical laboratory is preparing sections of body tissue for examination by a pathologist. This includes the preparation of tissue specimens of human and animal origin for diagnostic, research, or teaching purposes. Tissue sections prepared by the histologic technician for a variety of disease entities enable the pathologist to diagnose body dysfunction and malignancy.

### Job Description

Histologic technicians process sections of body tissue by fixation, dehydration, embedding, sectioning, decalcification, microincineration, mounting, and routine and special staining. They identify tissue structures, cell components, and their staining characteristics and relate them to physiological functions; and institute proper procedures to maintain accuracy and precision.

### Employment Characteristics

Most histologic technicians work in hospital laboratories, averaging a 40-hour week.

### Salary

Based on a 2005 survey published in *Laboratory Medicine,* median salaries ranged from $40,250 to $40,580. Refer to Section IV, Table 5 of this *Directory* for more information, or see www.ama-assn.org/go/salary.

### Educational Programs

**Length.** Programs for the histologic technician are 12 months, unless the curriculum is an integral part of a college program.

**Prerequisites.** High school diploma or equivalent.

**Curriculum.** The curriculum includes both didactic instruction and practical demonstration in the areas of medical ethics, medical terminology, chemistry, laboratory mathematics, computer technology, organic and/or biochemistry, immunology, electron microscopy, management, anatomy, histology, histochemistry, quality control, instrumentation, microscopy, processing techniques, preparation of museum specimens, and record and administration procedures. It is recommended that the curriculum be an integral part of a junior or community college program culminating in an associate degree and that the course of study include chemistry, biology, and mathematics.

# Histotechnologist

### Occupational Description

Physicians (usually pathologists) and other scientists specializing in biological sciences or related clinical areas such as chemistry work in partnership with medical laboratory workers to analyze blood, tissues, and fluids from humans (and sometimes animals), using a variety of precision instruments. The results of these tests are used to detect and diagnose disease and other abnormalities.

The main responsibility of the histotechnologist in the clinical laboratory is preparing sections of body tissue for examination by a pathologist. This includes the preparation of tissue specimens of human and animal origin for diagnostic, research, or teaching purposes. Tissue sections prepared by the histotechnologist for a variety of disease entities enable the pathologist to diagnose body dysfunction and malignancy.

### Job Description

Histotechnologists process sections of body tissue by fixation, dehydration, embedding, sectioning, decalcification, microincineration, mounting, and routine and special staining. In addition, histotechnologists perform the more complex procedures for processing tissues. They identify tissue structures, cell components, and their staining characteristics and relate them to physiological functions; implement and test new techniques and procedures; make judgments concerning the results of quality control measures; and institute proper procedures to maintain accuracy and precision. Histotechnologists apply the principles of management and supervision when they function as section supervisors and of educational methodology when they teach students.

### Employment Characteristics

Most histotechnologists work in hospital laboratories, averaging a 40-hour week.

### Salary

Based on a 2005 survey published in *Laboratory Medicine,* median salaries ranged from $44,970 to $49,360. Refer to Section IV, Table 5 of this *Directory* for more information, or see www.ama-assn.org/go/salary.

### Educational Programs

**Length.** Programs for the histotechnologist are 4 years and lead to a baccalaureate degree.

**Prerequisites.** High school diploma or equivalent.

**Curriculum.** The curriculum includes both didactic instruction and practical demonstration in the areas of medical ethics, medical terminology, chemistry, laboratory mathematics, computer technology, organic and/or biochemistry, immunology, electron microscopy, management, anatomy, histology, histochemistry, quality control, instrumentation, microscopy, processing techniques, preparation of museum specimens, and record and administration procedures. The baccalaureate-level program includes coursework designed to provide supervisors and teachers with advanced capabilities.

# Pathologists' Assistant

### Occupational Description

Anatomic pathologists are physicians who examine tissue specimens from patients and perform autopsies to diagnose the disease processes involved. Pathologists' assistants participate in autopsies and in the examination, dissection, and processing of tissue specimens. They function as physician extenders.

### Job Description

The following services are provided under the direct supervision of a licensed and board-certified pathologist and should include, but not be limited to, the following:

*Surgical pathology.* Assisting in the preparation and performance of surgical specimen dissection by ensuring appropriate specimen accessioning, obtaining pertinent clinical information and studies, describing gross anatomic features, dissecting surgical specimens, preparing and submitting tissue for histologic processing, obtaining and submitting specimens for additional analytic procedures (immunostaining, flow cytometry, image analysis, bacterial and viral cultures, toxicology, etc), and assisting in photographing gross and microscopic specimens.

*Autopsy pathology.* Assisting in the performance of postmortem examination by ascertaining proper legal authorization; obtaining and reviewing the patient's chart and other pertinent clinical data and studies; notifying involved personnel of all special procedures and techniques required; coordinating special requests for specimens; notifying involved clinicians and appropriate authorities and individuals; assisting in the postmortem examination; selecting and preparing tissue for histologic processing and special studies; obtaining specimens for biological and toxicologic analysis; assisting in photographing gross and microscopic specimens and photomicrography; and participating in the completion of the autopsy report.

*Additional duties.* Assuming duties as may be assigned relative to teaching, administrative, supervisory, and budgetary functions in anatomic pathology.

### Employment Characteristics

Pathologists' assistants are employed in a variety of settings, including community and regional hospitals, university medical centers, private pathology laboratories, and medical examiners/coroners' offices. Most work 40-55 hours per week.

### Salary

Salaries vary with geographic location and type of employing institution. Entry-level salaries average $55,000-$75,000. Refer to Section IV, Table 5 of this *Directory* for more information, or see www.ama-assn.org/go/salary.

### Educational Programs

**Length.** Minimum of 22 months.

**Degree.** Most programs are at the master's level.

**Prerequisites.** Variable among programs and dependent on the degree offered. Baccalaureate programs require a minimum of 60 hours of acceptable credits, with variable specific requirements.

**Curriculum.** The curriculum includes both didactic and practical training to provide a sound background in the basic medical sciences and the necessary skills to work in an anatomic pathology laboratory. Coursework includes anatomy, physiology, and medical terminology, as well as general, systemic, pediatric, and forensic pathology. Practical training includes autopsy pathology, surgical pathology, forensic pathology, and medical photography.

# Phlebotomist

### Occupational Description

Laboratory tests play an important role in the detection, diagnosis, and treatment of diseases. Phlebotomists collect blood specimens for many of these tests. Phlebotomists practice safe blood collection and handling techniques that protect patients from injury, safeguard themselves from accidents, and produce high-quality specimens while demonstrating compassion for the patient.

### Job Description

Phlebotomists collect, transport, handle, and process blood specimens for analysis; identify and select equipment, supplies, and additives used in blood collection; and understand factors that affect specimen collection procedures and test results. Recognizing the importance of specimen collection in the overall patient care system, phlebotomists adhere to infection control and safety policies and procedures. They monitor quality control within predetermined limits while demonstrating professional conduct, stress management, and communication skills with patients, peers, and other health care personnel as well as with the public.

### Employment Characteristics

Many phlebotomists are employed in hospital laboratories. Others are employed in physicians' private laboratories and clinics; by the armed forces; by city, state, and federal health agencies; in industrial medical laboratories; in pharmaceutical houses; in numerous public and private research programs; and as faculty of approved programs preparing medical laboratory personnel.

### Salary

Salaries vary depending on the employer and geographic location. Based on a 2005 survey published in *Laboratory Medicine,* median salaries ranged from $24,315 to $29,120. Refer to Section IV, Table 5 of this *Directory* for more information, or see www.ama-assn.org/go/salary.

### Educational Programs

**Length.** Approved programs contain at least 100 hours of clinical practicum and culminate in a postsecondary certificate.

**Prerequisites.** High school diploma or equivalent. The applicant must also meet the admission requirements of the sponsoring institution.

**Curriculum.** Phlebotomy programs are conducted in junior or community colleges, hospitals, medical laboratories, proprietary schools, and other equivalent postsecondary educational institutions. The curriculum includes didactic instruction and 100 hours of applied experiences, performance of a minimum of 100 successful unaided collections, and instruction in a variety of collection techniques, including vacuum collection devices, syringe, and capillary/skin-puncture methods.

### Inquiries

#### Careers/Curriculum

American Society for Clinical Laboratory Science
7910 Woodmont Avenue, Suite 530
Bethesda, MD 20814
301 657-2768
301 657-2909 Fax
E-mail: ascls@ascls.org
www.ascls.org

American Society for Clinical Pathology
33 West Monroe, Suite 1600
Chicago, IL 60603
312 541-4999
E-mail: info@ascp.org
www.ascp.org

Association of Genetic Technologists
Executive Office
PO Box 15945-288
Lenexa, KS 66285
913 541-9077

National Society for Histotechnology
4201 Northview Drive, Suite 502
Bowie, MD 20716-2604
301 262-6221
301 262-9188 Fax
E-mail: histo@nsh.org
www.nsh.org

American Association of Pathologists' Assistants
8030 Old Cedar Avenue, Suite 225
Bloomington, MN 55425-1215
800 532-AAPA
E-mail: mspindler@mn.state.net
www.pathologistsassistants.org

#### Certification/Registration

American Society for Clinical Pathology
Board of Registry
PO Box 12270
Chicago, IL 60612
312 738-1336, Ext 1341
E-mail: bor@ascp.org
www.ascp.org

National Certification Agency for Medical Laboratory Personnel
PO Box 15945-289
Lenexa, KS 66285
913 438-5110
E-mail: nca-info@goamp.com
www.nca-info.org

American Medical Technologists
710 Higgins Road
Park Ridge, IL 60068-5765
847 823-5169
E-mail: amtmail@aol.com
www.amtl.org

**Program Accreditation**
Dianne M Cearlock, PhD, MT(ASCP), CLS(NCA), Chief Executive
Officer (CEO)
8410 W Bryn Mawr Avenue, Suite 670
Chicago, IL 60631-3415
773 714-8880
773 714-8886 Fax
E-mail: dcearlock@naacls.org
www.naacls.org

## Clinical Assistant

# Massachusetts

**Springfield Technical Community College**
Clinical Assisting Prgm
One Amory Square
PO Box 9000
Springfield, MA 01105-1204
*Prgm Dir:* Susan A Schneider, MEd MT(ASCP)
*Tel:* 413 755-4846   *Fax:* 413 755-6312
*E-mail:* SASchneider@stcc.edu
*Med Dir:* Krystyna Sikorska, MD

# Michigan

**Northern Michigan University**
Clinical Assisting Prgm
3513 West Science Bldg
1401 Presque Isle Ave
Marquette, MI 49855-5346
*Prgm Dir:* Lucille A Contois, MS MT(ASCP)
*Tel:* 906 227-1660   *Fax:* 906 227-1309
*E-mail:* lcontois@nmu.edu
*Med Dir:* John Weiss, MD

# Oregon

**Clackamas Community College**
Clinical Assisting Prgm
19600 S Molalla Ave
Oregon City, OR 97045-7998
www.clackamas.edu
*Prgm Dir:* Helen Wand, MT(ASCP)
*Tel:* 503 657-6958, Ext 5505   *Fax:* 503 722-5878
*E-mail:* helenw@clackamas.edu
*Med Dir:* David Mitchell, MD

## Clinical Laboratory Scientist/Medical Technologist*

# Alabama

**University of Alabama at Birmingham**
Clin Lab Scientist/Med Technologist Prgm
RMSB 433
1705 University Blvd
Birmingham, AL 35294-1212
www.uab.edu/mt
*Prgm Dir:* Janelle M Chiasera, PhD MT(ASCP)
*Tel:* 205 975-7302   *Fax:* 205 975-7302
*E-mail:* chiaser@uab.edu

**University of South Alabama**
Clin Lab Scientist/Med Technologist Prgm
Dept of Clinical Laboratory Sciences
1504 Springhill Ave, Rm 2309
Mobile, AL 36604
www.usouthal.edu
*Prgm Dir:* George A Harwell, EdD MT(ASCP)SC
CLS(NCA)
*Tel:* 251 434-3468   *Fax:* 251 434-3403
*E-mail:* gharwell@usouthal.edu
*Med Dir:* J Allan Tucker, MD

**Auburn University Montgomery**
Clin Lab Scientist/Med Technologist Prgm
PO Box 244023
Montgomery, AL 36124-4023
www.aum.edu/cls
*Prgm Dir:* Jeffrey Barksdale, PhD MT(ASCP)
*Tel:* 334 244-3606   *Fax:* 334 244-3146
*E-mail:* jbarksda@mail.aum.edu
*Med Dir:* Robert Adams, MD, FCAP

**Baptist Medical Center South**
Clin Lab Scientist/Med Technologist Prgm
2105 E South Blvd
PO Box 11010
Montgomery, AL 36111-0100
www.baptistfirst.org/educationopportunities.htm
*Prgm Dir:* Jeanne Whitney, MS MT(ASCP)CM
*Tel:* 334 286-2899   *Fax:* 334 286-3351
*E-mail:* jwhitney@baptistfirst.org
*Med Dir:* Glen Pinkston, MD

**Tuskegee University**
Clin Lab Scientist/Med Technologist Prgm
BIOE 71-271
Booker T Washington Blvd
Tuskegee, AL 36088
www.tuskegee.edu
*Prgm Dir:* Cheryl G Davis, DHA CLS(NCA)
*Tel:* 334 727-8335   *Fax:* 334 727-8259
*E-mail:* davis@tuskegee.edu

# Alaska

**University of Alaska Anchorage**
Clin Lab Scientist/Med Technologist Prgm
3211 Providence Dr
Anchorage, AK 99508
*Prgm Dir:* Heidi Mannion, MS MT(ASCP)
*Tel:* 907 786-6923   *Fax:* 907 786-6938
*E-mail:* afham@uaa.alaska.edu

# Arizona

**Arizona State University**
Clin Lab Scientist/Med Technologist Prgm
School of Life Sciences
PO Box 874501
Tempe, AZ 85287-4501
*Prgm Dir:* Diana Mass, MA MT(ASCP)CLS
*Tel:* 480 965-7090   *Fax:* 480 965-8126
*E-mail:* Diana.Mass@asu.edu

---

*Additional information about programs that
returned the AMA's survey is available in a table
beginning on page 60.

## Clinical Assistant

| Programs* | Class Capacity | Begins | Length (months) | Award | Res Tuition | Non-res Tuition | Stipend | Offers:‡ 1 | 2 | 3 |
|---|---|---|---|---|---|---|---|---|---|---|
| **Massachusetts** | | | | | | | | | | |
| Springfield Technical Community College | 18 | Sep May | 9 | Cert | $1,408 | $6,016 | | | | • |
| **Oregon** | | | | | | | | | | |
| Clackamas Community College (Oregon City) | 24 | Sep 25 | 9 | Cert | $2,700 | $8,910 | | | | |

*Data are shown only for programs that completed the 2006 AMA Survey of Health Professions Education Programs
‡Key to Offers: 1: Evening or weekend classes; 2: Non-English instruction; 3: Cultural competence instruction

# Arkansas

**Baptist Health**
Clin Lab Scientist/Med Technologist Prgm
11900 Colonel Glenn Rd, Ste 1000
Little Rock, AR 72210-2820
www.baptist-health.org
*Prgm Dir:* Sandra G Ackerman, MT (ASCP)SH
*Tel:* 501 202-7409   *Fax:* 501 202-6220
*E-mail:* sgackerm@baptist-health.org
*Med Dir:* John Slaven, MD

**University of Arkansas for Medical Sciences**
Clin Lab Scientist/Med Technologist Prgm
4301 W Markham St, Slot 597
Little Rock, AR 72205
www.uams.edu/chrp/medtech.htm
*Prgm Dir:* Kathleen M Mugan, MEd MT(ASCP)SH
*Tel:* 501 686-5776   *Fax:* 501 526-6563
*E-mail:* mugankathleenm@uams.edu

**Arkansas State University**
Clin Lab Scientist/Med Technologist Prgm
PO Box 910
State University, AR 72467
www.astate.edu
*Prgm Dir:* R Whitney Williams, PhD MT(ASCP)SC
*Tel:* 870 972-3073   *Fax:* 870 972-2004
*E-mail:* wwilliam@astate.edu
*Med Dir:* Terri Y Green, MD

# California

**California State University - Dominguez Hills**
Clin Lab Scientist/Med Technologist Prgm
1000 E Victoria Blvd
Carson, CA 90747-9960
www.csudh.edu
*Prgm Dir:* Cheryl Jackson-Harris, MS MT(ASCP)SH
*Tel:* 310 243-3748, Ext 3899   *Fax:* 310 516-3865
*E-mail:* charris@csudh.edu
*Med Dir:* Najeeb S Alshak, MD

**Loma Linda University**
Clin Lab Scientist/Med Technologist Prgm
School of Allied Health Professions, Dept of CLS
Nichol Hall, Rm A918
Loma Linda, CA 92350
www.llu.edu/llu/sahp/clinlab/mthome.html
*Prgm Dir:* Rodney Roath, MBA MT(ASCP)
*Tel:* 909 558-4966   *Fax:* 909 558-0458
*E-mail:* rroath@llu.edu
*Med Dir:* James Pappas, MD

**Univ of California Irvine Med Ctr**
Clin Lab Scientist/Med Technologist Prgm
101 City Dr S, Rte 38
Orange, CA 92868
www.ucihs.uci.edu/com/pathology/medtechprog.html
*Prgm Dir:* Laura S Ogata, MA MT(ASCP)
*Tel:* 714 456-6305   *Fax:* 714 456-6090
*E-mail:* logata@uci.edu
*Med Dir:* Luis M De La Maza, MD, PhD

**Eisenhower Memorial Hospital**
Clin Lab Scientist/Med Technologist Prgm
39000 Bob Hope Dr
Rancho Mirage, CA 92270
www.emc.org
*Prgm Dir:* Joan M Steiner-Adler, EdD MA MS MT(ASCP)
*Tel:* 760 773-4525   *Fax:* 760 773-4363
*E-mail:* jsteiner@emc.org
*Med Dir:* Barbara Comess, MD

**Univ of California Davis Health System**
Clin Lab Scientist/Med Technologist Prgm
Specialty Testing Center
3740 Business Dr
Sacramento, CA 95820
http://som.ucdavis.edu/departments/pathology/
    education/clinical_lab
*Prgm Dir:* Jiunn Huang, PhD MT(ASCP)
*Tel:* 916 734-0231   *Fax:* 916 734-0320
*E-mail:* jiunn.huang@ucdmc.ucdavis.edu
*Med Dir:* Hanna M Jensen, MD

**San Francisco State University**
Clin Lab Scientist/Med Technologist Prgm
Ctr for Biomedical Laboratory Sci 211
San Francisco, CA 94132
*Prgm Dir:* Geraldine M Albee, MA CLS(NCA) CNMT
*Tel:* 415 405-3779   *Fax:* 415 338-7747
*E-mail:* galbee@sfsu.edu
*Med Dir:* Paul Ortega, MD

**San Jose State University**
Clin Lab Scientist/Med Technologist Prgm
One Washington Square
San Jose, CA 95192-0100
www.sjsu.edu/cls
*Prgm Dir:* Mara Williams, MS MT(ASCP)
*Tel:* 408 924-4851   *Fax:* 408 924-4840
*E-mail:* cls@science.sjsu.edu

**Santa Barbara Cottage Hospital**
Clin Lab Scientist/Med Technologist Prgm
Pueblo at Bath Sts
PO Box 689
Santa Barbara, CA 93102
www.cottagehealthsystem.org
*Prgm Dir:* Lynette Hansen, MHA MT(ASCP)EdD CLS
*Tel:* 805 569-7378   *Fax:* 805 569-8223
*E-mail:* lhansen@cottagehealthsystem.org
*Med Dir:* Peter Lorenz Morris, MD

# Colorado

**Centura Health/Penrose-St Francis Hlth Serv**
Clin Lab Scientist/Med Technologist Prgm
2222 N Nevada Ave
Colorado Springs, CO 80907-7021
www.penrosestfrancis.org
*Prgm Dir:* Sr Rose V Brown, MEd MT(ASCP) CLS(NCA)
*Tel:* 719 776-5221   *Fax:* 719 776-5584
*E-mail:* rosebrown@centura.org
*Med Dir:* Sig Torgerson,MD, MD

**Colorado Health Foundation**
Clin Lab Scientist/Med Technologist Prgm
1719 E 19th Ave
Denver, CO 80218
*Prgm Dir:* Karen E Myers, MA MT(ASCP)SC
*Tel:* 303 839-6485   *Fax:* 303 869-1720
*E-mail:* kmyers@health1.org
*Med Dir:* Thomas A Merrick, MD

**Parkview Medical Center**
Clin Lab Scientist/Med Technologist Prgm
400 W 16th St
Pueblo, CO 81003
www.parkviewmc.com
*Prgm Dir:* Mary Chasteen, MA MT(ASCP)SBB
    CLS(NCA)
*Tel:* 719 584-4429   *Fax:* 719 584-4658
*E-mail:* mary_chasteen@parkviewmc.com
*Med Dir:* Stuart Marsh, MD

# Connecticut

**Danbury Hospital**
Clin Lab Scientist/Med Technologist Prgm
24 Hospital Ave
Danbury, CT 06810
www.danburyhospital.org/educationandresidencies/
    medical_technology.htm
*Prgm Dir:* Ana H Vicente, MHA MT(ASCP)
*Tel:* 203 797-7804   *Fax:* 203 739-8900
*E-mail:* Ana.Vicente@danhosp.org
*Med Dir:* Ramon N Kranwinkel, MD

**Hartford Hospital**
Clin Lab Scientist/Med Technologist Prgm
School of Allied Health
80 Seymour St
Hartford, CT 06115
www.harthosp.org
*Prgm Dir:* Wayne A Aguiar, MS MT(ASCP)
*Tel:* 860 545-5632   *Fax:* 860 545-5066
*E-mail:* waguiar@harthosp.org
*Med Dir:* Jaber Aslanzadeh, PhD, Chairman advisory
    Committee

**University of Hartford**
Clin Lab Scientist/Med Technologist Prgm
Dana Hall, Rm 429
200 Bloomfield Ave
West Hartford, CT 06117-1599
www.hartford.edu
*Prgm Dir:* Karen Barrett, MS MT(ASCP)
*Tel:* 860 768-4489   *Fax:* 860 768-4558
*E-mail:* barrett@hartford.edu
*Med Dir:* George Barrows, MD

# Delaware

**University of Delaware**
Clin Lab Scientist/Med Technologist Prgm
Dept of Medical Technology
305 Willard Hall Education Building
Newark, DE 19716
www.udel.edu/medtech
*Prgm Dir:* Anna P Ciulla, MCC MT(ASCP)SC CC(NRCC)
    CLS(NCA)
*Tel:* 302 831-2849   *Fax:* 302 831-4180
*E-mail:* aciulla@udel.edu
*Med Dir:* Mark L Mitchell, MD

# District of Columbia

**George Washington University**
Clin Lab Scientist/Med Technologist Prgm
2300 I St NW, Ste 503
Washington, DC 20037
www.gwumc.edu/healthsci/Programs/CLS/cls.htm
*Prgm Dir:* Sylvia Silver, DA MT(ASCP)
*Tel:* 202 994-2945   *Fax:* 202 994-5056
*E-mail:* ssilver@gwu.edu
*Med Dir:* Donald S Karcher, MD

**Howard University**
Clin Lab Scientist/Med Technologist Prgm
6th & Bryant Sts NW
Coll of Phar, Nursing & Allied Hlth Sciences
Washington, DC 20059
www.howard.edu
*Prgm Dir:* Marguerite E Neita, PhD MT(ASCP)
*Tel:* 202 806-5632   *Fax:* 202 806-7918
*E-mail:* mneita@howard.edu
*Med Dir:* Josephine J Marshalleck, MD

**Walter Reed Army Medical Center**
Clin Lab Scientist/Med Technologist Prgm
Clinical Lab Officer Course
6900 Georgia Ave NW Bldg 1 Rm 121C
Washington, DC 20307-5001
*Prgm Dir:* Patricia Fishback, MA MT(ASCP)
*Tel:* 202 782-8353    *Fax:* 202 782-4502
*E-mail:* patricia.fishback@us.army.mil
*Med Dir:* Mark D Brissette, Col MD

**Washington Hospital Center**
Clin Lab Scientist/Med Technologist Prgm
110 Irving St NW
Department of Pathology and Laboratory Medicine
Washington, DC 20010
www.whcenter.org/departments/pathology
*Prgm Dir:* John C Rees, PhD MS MT(ASCP)
*Tel:* 202 877-3346    *Fax:* 202 877-5263
*E-mail:* john.c.rees@medstar.net
*Med Dir:* Thomas Godwin, MD

# Florida

**Shands Jacksonville**
Clin Lab Scientist/Med Technologist Prgm
School of Medical Technology/CLS
655 W Eighth St
Jacksonville, FL 32209
www.shandsjacksonville.com/schools/medtech
*Prgm Dir:* James D Sigler, MT(ASCP) CLSup(NCA) CLS
*Tel:* 904 244-9520    *Fax:* 904 244-2899
*E-mail:* jim.sigler@jax.ufl.edu
*Med Dir:* Roger Bertholf, PhD

**St Vincent's Medical Center**
Clin Lab Scientist/Med Technologist Prgm
1800 Barrs St
Jacksonville, FL 32204
www.jaxhealth.com/medicalscience
*Prgm Dir:* Lynnette Chakkaphak, MS MT(ASCP)
*Tel:* 904 308-3817    *Fax:* 904 308-2970
*E-mail:* Lchak001@stvincentshealth.com
*Med Dir:* Ricardo Ramos, MD

**University of Central Florida**
Clin Lab Scientist/Med Technologist Prgm
4000 Central Florida Blvd, HPA II 335A
Orlando, FL 32816-2360
www.bmsc.ucf.edu
*Prgm Dir:* Dorilyn J Hitchcock, MS MT(ASCP)
*Tel:* 407 823-5220    *Fax:* 407 823-3095
*E-mail:* hitchcod@mail.ucf.edu
*Med Dir:* Raymond B Franklin Jr, MD PhD

**University of West Florida**
Clin Lab Scientist/Med Technologist Prgm
11000 University Pkwy
Pensacola, FL 32514-5751
www.uwf.edu/clinicallabsciences
*Prgm Dir:* Swarna Krothapalli, MS MT(ASCP)
*Tel:* 850 474-2988    *Fax:* 850 474-2749
*E-mail:* skrothap@uwf.edu

**Bayfront Medical Center**
Clin Lab Scientist/Med Technologist Prgm
701 Sixth St S
St Petersburg, FL 33701
*Prgm Dir:* June P Schurig, MT(ASCP)
*Tel:* 727 893-6604    *Fax:* 727 893-6977
*E-mail:* june.schurig@bayfront.org
*Med Dir:* Larry Davis, MD

**Tampa General Hospital**
Clin Lab Scientist/Med Technologist Prgm
PO Box 1289
Tampa, FL 33601
*Prgm Dir:* Donald W Vendrone, MPH MT(ASCP)
*Tel:* 813 844-7284    *Fax:* 813 844-4160
*E-mail:* dvendron@tgh.org
*Med Dir:* Irwin L Browarsky, MD

# Georgia

**Medical College of Georgia**
Clin Lab Scientist/Med Technologist Prgm
EC-3422
Augusta, GA 30912-0500
www.mcg.edu/SAH/brt/MT.htm
*Prgm Dir:* Barbara L Russell, MHE MT(ASCP)SH
*Tel:* 706 721-3046    *Fax:* 706 721-7631
*E-mail:* brussell@mcg.edu

**Armstrong Atlantic State University**
Clin Lab Scientist/Med Technologist Prgm
11935 Abercorn St
Savannah, GA 31419-1997
www.medtech.armstrong.edu
*Prgm Dir:* Hassan Aziz, PhD CLS/NCA
*Tel:* 912 927-5204    *Fax:* 912 921-5585
*E-mail:* azizhass@mail.armstrong.edu
*Med Dir:* John Ralph Edgar, MD

**Thomas University**
Clin Lab Scientist/Med Technologist Prgm
1501 Millpond Rd
Thomasville, GA 31792
www.thomasu.edu
*Prgm Dir:* Jill Dennis, MEd MT(ASCP) CLS(NCA)
*Tel:* 229 226-1621, Ext 146    *Fax:* 229 227-6967
*E-mail:* jdennis@thomasu.edu
*Med Dir:* Jeff Byrd, MD

# Hawaii

**University of Hawaii at Manoa**
Clin Lab Scientist/Med Technologist Prgm
1960 East-West Rd, Bio C206
Honolulu, HI 96822
www.hawaii.edu/medtech/Medtech.html
*Prgm Dir:* Andre Theriault, PhD CLS(C)
*Tel:* 808 956-8557    *Fax:* 808 956-5457
*E-mail:* andret@hawaii.edu

# Idaho

**Idaho State University**
Clin Lab Scientist/Med Technologist Prgm
Dept of Biological Sciences
741 S Seventh, Box 8007
Pocatello, ID 83209
*Prgm Dir:* Kathleen Spiegel, PhD MT(ASCP)
*Tel:* 208 236-4378
*E-mail:* spiekath@isu.edu
*Med Dir:* Charles O Garrison, MD

# Illinois

**St Elizabeth Hospital**
Clin Lab Scientist/Med Technologist Prgm
211 S Third St
Belleville, IL 62222
*Prgm Dir:* JoAnn Denaro, MT(ASCP)
*Tel:* 618 343-0639, Ext 105    *Fax:* 618 343-0687
*E-mail:* jdenaro@tri-lab.com

**Rush University**
Clin Lab Scientist/Med Technologist Prgm
600 S Paulina St Ste 1019 AAC
Chicago, IL 60612
www.rushu.rush.edu/cls
*Prgm Dir:* Herbert J Miller, PhD MT(ASCP) CLS(NCA)
*Tel:* 312 942-7251    *Fax:* 312 942-6464
*E-mail:* herb_j_miller@rush.edu
*Med Dir:* Robert DeCresce, MD

**Northern Illinois University**
Clin Lab Scientist/Med Technologist Prgm
1425 W Lincoln Hwy
DeKalb, IL 60115-2825
www.chhs.niu.edu/ahp/cls
*Prgm Dir:* Jeanne M Isabel, MSEd
*Tel:* 815 753-1382    *Fax:* 815 753-1653
*E-mail:* jisabel@niu.edu
*Med Dir:* David S Laib, MD Board certified in anatomic
    & clinical path

**Evanston Northwestern Healthcare - Evanston**
Clin Lab Scientist/Med Technologist Prgm
2650 Ridge Ave, Rm 1927D
Department of Pathology and Laboratory Medicine
Evanston, IL 60201
www.enh.org/academicprograms
*Prgm Dir:* Marcia Hicks, MAdEd MT(ASCP)
*Tel:* 847 570-2737    *Fax:* 847 570-1938
*E-mail:* mhicks@enh.org
*Med Dir:* James Perkins, MD

**Edward Hines Jr VA Hospital**
Clin Lab Scientist/Med Technologist Prgm
PO Box 5000/113/School
Hines, IL 60141-5113
*Prgm Dir:* Donna M Wray, MT(ASCP)
*Tel:* 708 202-8387, Ext 23212    *Fax:* 708 202-2611
*E-mail:* Donna.Wray@med.va.gov
*Med Dir:* Myron E Rubnitz, MD

**Illinois State University**
Clin Lab Scientist/Med Technologist Prgm
5220 Health Sciences
Normal, IL 61790-5220
www.healthsciences.ilstu.edu
*Prgm Dir:* Meridee Van Draska, MS MT(ASCP) CLS
    (NCA)
*Tel:* 309 438-8269    *Fax:* 309 438-2450
*E-mail:* mrust@ilstu.edu

**Rosalind Franklin Univ of Medicine & Science**
Clin Lab Scientist/Med Technologist Prgm
Dept of Clinical Laboratory Sciences
3333 Green Bay Rd
North Chicago, IL 60064
www.rosalindfranklin.edu
*Prgm Dir:* Janet M Vanik, MS MT(ASCP)
*Tel:* 847 578-3418    *Fax:* 847 578-8651
*E-mail:* Janet.Vanik@rosalindfranklin.edu
*Med Dir:* Nancy L Jones, MD

**OSF St Francis Medical Center**
Clin Lab Scientist/Med Technologist Prgm
530 NE Glen Oak Ave
Peoria, IL 61637
www.osfhealthcare.org
*Prgm Dir:* Carol E Becker, MS MT(ASCP)CLS(NCA)
*Tel:* 309 624-9021    *Fax:* 309 624-9150
*E-mail:* carol.e.becker@osfhealthcare.org
*Med Dir:* Michael P Hayes, MD

**OSF St Anthony Medical Center**
Clin Lab Scientist/Med Technologist Prgm
5666 E State St
Rockford, IL 61108
*Prgm Dir:* James F Beam, MT(ASCP) CLS(NCA)
*Tel:* 815 395-5119    *Fax:* 815 395-5364
*E-mail:* jim.beam@osfhealthcare.org
*Med Dir:* David Laib, MD FASCP FCAP FABP

**St John's Hospital**
Clin Lab Scientist/Med Technologist Prgm
800 E Carpenter St
Springfield, IL 62769
www.st-johns.org
*Prgm Dir:* Gilma Roncancio-Weemer, MS CLS(NCA)
    MT(ASCP)
*Tel:* 217 757-6788    *Fax:* 217 535-3775
*E-mail:* gilma.roncancio-weemer@st-johns.org

**University of Illinois at Springfield**
Clin Lab Scientist/Med Technologist Prgm
One University Plaza
MS HSB 314
Springfield, IL 62703
*Prgm Dir:* Linda J McCown, MS CLS(NCA) MT(ASCP)
*Tel:* 217 206-7550   *Fax:* 217 206-6162
*E-mail:* lmcco2@uis.edu
*Med Dir:* Fritz Lower, MD

# Indiana

**St Francis Hospital & Health Centers**
Clin Lab Scientist/Med Technologist Prgm
1600 Albany St
Beech Grove, IN 46107
*Prgm Dir:* DeAnne S Maxwell, MT(ASCP)
*Tel:* 317 783-8195   *Fax:* 317 783-8801
*E-mail:* deanne.maxwell@ssfhs.org
*Med Dir:* Janice K Fitzharris, MD, PhD

**Parkview Hospital**
Clin Lab Scientist/Med Technologist Prgm
328 Ley Rd
Ft Wayne, IN 46825
*Prgm Dir:* Brian Goff, MA MT(ASCP)
*Tel:* 219 373-9406   *Fax:* 219 373-9418
*E-mail:* brian.goff@parkview.com
*Med Dir:* Darryl R Smith, MD

**St Margaret Mercy Healthcare Centers**
Clin Lab Scientist/Med Technologist Prgm
5454 Hohman Ave
Hammond, IN 46320
www.ssfhs.org
*Prgm Dir:* Rosemary Ann Duda, MS MT(ASCP)I SM
*Tel:* 219 932-2300, Ext 34199   *Fax:* 219 852-2508
*E-mail:* rose.butkiewicz@ssfhs.org
*Med Dir:* Gitie Jaffe, MD

**Clarian Health Partners Inc**
*Cosponsor:* Methodist Hospital
Clin Lab Scientist/Med Technologist Prgm
I-65 at 21st St, PO Box 1367
Indianapolis, IN 46206
*Prgm Dir:* Cheryl Oliver, MS MT(ASCP)
*Tel:* 317 962-8280   *Fax:* 317 962-2102
*E-mail:* coliver@clarian.org
*Med Dir:* Caroline An, MD

**Indiana University**
Clin Lab Scientist/Med Technologist Prgm
School of Medicine
Dept of Pathology & Laboratory Medicine
Indianapolis, IN 46202-5113
www.pathology.iupui.edu/clin_sci.htm
*Prgm Dir:* Linda M Kasper, EdD MS CLS(NCA)
  MT(ASCP)SC
*Tel:* 317 274-1270   *Fax:* 317 278-0643
*E-mail:* lmkasper@iupui.edu
*Med Dir:* John N Eble, MD MBA

**Ball Memorial Hospital**
*Cosponsor:* Pathologists Associated Laboratory
Clin Lab Scientist/Med Technologist Prgm
2401 University Ave
Muncie, IN 47303
*Prgm Dir:* Shirley A Replogle, MT(ASCP)
*Tel:* 765 254-6289, Ext 2289   *Fax:* 765 747-3133
*E-mail:* replogles@palab.com
*Med Dir:* Richard W Pearson, MD

**Good Samaritan Hospital**
Clin Lab Scientist/Med Technologist Prgm
520 S Seventh St
Vincennes, IN 47591
*Prgm Dir:* Michaele R McDonald, MS MT(ASCP)
*Tel:* 812 885-3361   *Fax:* 812 885-3135
*E-mail:* mmcdonald@gshvin.org
*Med Dir:* Kathy Freeman, MD

# Iowa

**Mercy/St Luke's Hospital**
Clin Lab Scientist/Med Technologist Prgm
1026 A Ave NE
Cedar Rapids, IA 52402
www.stlukescr.org
*Prgm Dir:* Nadine Sojka, MS MT(ASCP)SH
*Tel:* 319 369-7309   *Fax:* 319 369-8095
*E-mail:* sojkan@crstlukes.com
*Med Dir:* Dorryl Buck, MD

**Mercy Medical Center**
Clin Lab Scientist/Med Technologist Prgm
1111 6th Ave
Des Moines, IA 50314
*Prgm Dir:* Kyla Deibler, MS MT(ASCP) CLS(NCA)
*Tel:* 515 247-4469   *Fax:* 515 643-8810
*E-mail:* kdeibler@mercydesmoines.org
*Med Dir:* Vijaya Dhannavada, MD

**University of Iowa Hospitals & Clinics**
Clin Lab Scientist/Med Technologist Prgm
1000 Medical Laboratories
Iowa City, IA 52242
www.medicine.uiowa.edu/CLSP
*Prgm Dir:* Barbara R Swanson, MEd MT(ASCP)
*Tel:* 319 384-9380   *Fax:* 319 384-4557
*E-mail:* barbara-swanson@uiowa.edu
*Med Dir:* Fred R Dee, MD

**Mercy Medical Center-Sioux City**
Clin Lab Scientist/Med Technologist Prgm
801 5th St
Sioux City, IA 51101
www.mercysiouxcity.com/services/clinical
*Prgm Dir:* Mary K Smith, MS BS MT(ASCP)
*Tel:* 712 279-2371   *Fax:* 712 279-2372
*E-mail:* smithmk@mercyhealth.com
*Med Dir:* Askar A Qalbani, MD

**St Luke's College**
Clin Lab Scientist/Med Technologist Prgm
2720 Stone Park Blvd
Sioux City, IA 51104
www.stlukes.org
*Prgm Dir:* Pamela G Briese, MS MT(ASCP)SC
*Tel:* 712 279-3967   *Fax:* 712 233-8017
*E-mail:* briesepg@stlukes.org
*Med Dir:* James Quesenberry, MD

# Kansas

**University of Kansas Medical Center**
Clin Lab Scientist/Med Technologist Prgm
3901 Rainbow Blvd, MS 4048G-Eaton
G014 Eaton
Kansas City, KS 66160
www.cls.kumc.edu
*Prgm Dir:* Venus J Ward, PhD MT(ASCP) CLS(NCA)
*Tel:* 913 588-0154   *Fax:* 913 588-5222
*E-mail:* vward@kumc.edu
*Med Dir:* Lowell Tilzer, MD

**Wichita State University**
Clin Lab Scientist/Med Technologist Prgm
1845 Fairmount Box 43
Wichita, KS 67260
*Prgm Dir:* Mary E Conrad, PhD MT(ASCP)
*Tel:* 316 978-5655, Ext 5655   *Fax:* 316 978-3025
*E-mail:* mary.conrad@wichita.edu
*Med Dir:* Mary Conrad, MT(ASCP) PhD

# Kentucky

**St Elizabeth Medical Center**
Clin Lab Scientist/Med Technologist Prgm
One Medical Village Dr
Edgewood, KY 41017
www.stelizabeth.com
*Prgm Dir:* Beth Warning, MS MT(ASCP) CLS (NCA)
*Tel:* 859 301-2417   *Fax:* 859 301-5560
*E-mail:* bwarning@stelizabeth.com
*Med Dir:* Jackson Pemberton, MD

**University of Kentucky**
Clin Lab Scientist/Med Technologist Prgm
Div of Clinical Lab Sci
CHS Bldg 900 S Limestone Rm 209
Lexington, KY 40536-0200
www.mc.uky.edu/CLS
*Prgm Dir:* Linda Gorman, PhD MT(ASCP)
*Tel:* 859 323-1100, Ext 80855   *Fax:* 859 257-2454
*E-mail:* Linda.Gorman@uky.edu
*Med Dir:* Julie Ribes, MD

**Bellarmine University**
Clin Lab Scientist/Med Technologist Prgm
Lansing Sch of Nursing & Hlth Sci, 108 Pasteur Hall
2001 Newburg Rd
Louisville, KY 40205-0671
www.bellarmine.edu/lansing/cls
*Prgm Dir:* Karen Golemboski, MT (ASCP) PhD
*Tel:* 502 452-8387   *Fax:* 502 452-8389
*E-mail:* kgolemboski@bellarmine.edu
*Med Dir:* Carolyn Burns, MD

**Owensboro Medical Health System**
Clin Lab Scientist/Med Technologist Prgm
PO Box 20007
Owensboro, KY 42304-0007
*Prgm Dir:* Lisa Sellars Cecil, MT(ASCP)
*Tel:* 502 688-2934   *Fax:* 502 688-2938
*E-mail:* Lcecil@omhs.org
*Med Dir:* Brian E Ward, MD

**Eastern Kentucky University**
Clin Lab Scientist/Med Technologist Prgm
Dizney 220
521 Lancaster Ave
Richmond, KY 40475-3102
*Prgm Dir:* David C Hufford, PhD MT(ASCP)
*Tel:* 859 622-3078   *Fax:* 859 622-1139
*E-mail:* david.hufford@eku.edu

# Louisiana

**Rapides Regional Medical Center**
Clin Lab Scientist/Med Technologist Prgm
211 Fourth St, PO Box 30101
Alexandria, LA 71301
*Prgm Dir:* Laine O Reeder, MT(ASCP) SC
*Tel:* 318 473-3175, Ext 3809   *Fax:* 318 473-3079
*E-mail:* laine.poe@hcahealthcare.com
*Med Dir:* Sheldon Deshotels, MD

**Our Lady of the Lake College**
Clin Lab Scientist/Med Technologist Prgm
7434 Perkins Rd
Baton Rouge, LA 70808
www.ololcollege.edu
*Prgm Dir:* Todd Casanova, PhD MT(ASCP)
*Tel:* 225 768-1745   *Fax:* 225 768-0819
*E-mail:* tcasanov@ololcollege.edu

**Lake Charles Mem Hosp Sch of Med Tech**
Clin Lab Scientist/Med Technologist Prgm
1701 Oak Park Blvd
Lake Charles, LA 70601
*Prgm Dir:* Dianne D Malveaux, MEd MT(ASCP)
*Tel:* 337 494-3000, Ext 3366   *Fax:* 337 494-2464
*E-mail:* DMalveaux@LCMH.com
*Med Dir:* Carl G Bowling, MD

**McNeese State University**
Clin Lab Scientist/Med Technologist Prgm
4205 Ryan St
Lake Charles, LA 70609-2000
*Prgm Dir:* Donna Calvert, MEd MT(ASCP)
*Tel:* 337 475-5655  *Fax:* 337 475-5677
*E-mail:* dcalvert@mail.mcneese.edu
*Med Dir:* John VanHoose, MD

**St Francis Medical Center**
Clin Lab Scientist/Med Technologist Prgm
309 Jackson St
PO Box 1901
Monroe, LA 71201-1901
*Prgm Dir:* Melanie S Chapman, MEd MT(ASCP)
*Tel:* 318 327-4359  *Fax:* 318 327-4857
*E-mail:* chapmam@stfran.com
*Med Dir:* Richard J Blanchard Jr, MD

**Louisiana State Univ Health Sciences Center**
Clin Lab Scientist/Med Technologist Prgm
1900 Gravier St
New Orleans, LA 70112
www.alliedhealth.lsuhsc.edu/ClinicalLaboratory
*Prgm Dir:* Louann Lawrence, DrPH CLS(NCA)
    MT(ASCP)SH
*Tel:* 504 568-4276  *Fax:* 504 568-6761
*E-mail:* llawre@lsuhsc.edu
*Med Dir:* Jack P Strong, MD

**Louisiana State Univ Hlth Sci Ctr-Shreveport**
Clin Lab Scientist/Med Technologist Prgm
PO Box 33932
1501 Kings Hwy
Shreveport, LA 71130-3932
*Prgm Dir:* John S Davis, MBA MT(ASCP)SC DLM
*Tel:* 318 675-6806  *Fax:* 318 675-6937
*E-mail:* jdavis@lsuhsc.edu

**Overton Brooks VA Medical Center**
Clin Lab Scientist/Med Technologist Prgm
510 E Stoner Ave
Shreveport, LA 71101-4295
*Prgm Dir:* Vanessa Jones Johnson, MBA MT(ASC)
    CLS(NCA)
*Tel:* 318 424-6169  *Fax:* 318 424-6093
*E-mail:* vanessa.johnson@med.va.gov
*Med Dir:* Aubrey A Lurie, MD

# Maine

**Eastern Maine Medical Center**
Clin Lab Scientist/Med Technologist Prgm
489 State St
Bangor, ME 04401
www.medtech.emmc.org
*Prgm Dir:* Ellen M Libby, MS MT(ASCP) CLS(NCA))
*Tel:* 207 973-7616  *Fax:* 207 973-5823
*E-mail:* elibby@emh.org
*Med Dir:* Ann Marie Blenc, MD

# Maryland

**Morgan State University**
Clin Lab Scientist/Med Technologist Prgm
Cold Spring Ln & Hillen Rd
Baltimore, MD 21251
*Prgm Dir:* Diane Wilson, PhD MT(ASCP)
*Tel:* 443 885-3611  *Fax:* 443 885-8285
*E-mail:* dwilson1@jewel.morgan.edu

**University of Maryland**
Clin Lab Scientist/Med Technologist Prgm
100 Penn St, AHB
Baltimore, MD 21201
www.medschool.umaryland.edu/dmrt
*Prgm Dir:* Deirdre D Parsons, MS MT(ASCP)SBB
*Tel:* 410 706-1829  *Fax:* 410 706-0073
*E-mail:* dparsons@som.umaryland.edu
*Med Dir:* Sanford A Stass, MD

**Salisbury University**
Clin Lab Scientist/Med Technologist Prgm
1101 Camden Ave
Salisbury, MD 21801
www.salisbury.edu/healthsci/MDTC1.htm
*Prgm Dir:* Johanna W Laird, MS MT(ASCP) CLS(NCA)
*Tel:* 410 543-6364  *Fax:* 410 548-9185
*E-mail:* jwlaird@salisbury.edu
*Med Dir:* Eric Weaver, MD

# Massachusetts

**Northeastern University**
Clin Lab Scientist/Med Technologist Prgm
360 Huntington Ave
206 Mugar
Boston, MA 02215
www.neu.edu
*Prgm Dir:* Mary Louise Turgeon, EdD CLS(NCA)
*Tel:* 617 373-4192  *Fax:* 617 373-3665
*E-mail:* m.turgeon@neu.edu
*Med Dir:* Sanford Kurtz, MD

**University of Massachusetts - Lowell**
Clin Lab Scientist/Med Technologist Prgm
3 Solomont Way, Ste 4, Weed Hall
Lowell, MA 01854
www.uml.edu/college/she
*Prgm Dir:* Kathleen (Kay) Doyle, PhD MT(ASCP)
*Tel:* 978 934-4425, Ext 4425  *Fax:* 978 934-3006
*E-mail:* Kathleen_Doyle@uml.edu
*Med Dir:* Manilo Lo Conte, MD

**University of Massachusetts - Dartmouth**
Clin Lab Scientist/Med Technologist Prgm
Dept of Med Lab Science
285 Old Westport Rd
North Dartmouth, MA 02747-2300
www.umassd.edu/cas/medlabscience
*Prgm Dir:* Dorothy A Bergeron, MS CLS(NCA)
*Tel:* 508 999-8584  *Fax:* 508 999-8418
*E-mail:* dbergeron@umassd.edu
*Med Dir:* David K Rubin, MD

**Berkshire Medical Center**
Clin Lab Scientist/Med Technologist Prgm
725 North St
Pittsfield, MA 01201
*Prgm Dir:* Lori Moore, MEd MT(ASCP)
*Tel:* 413 447-2580  *Fax:* 413 447-2097
*E-mail:* Lmoore@bhs1.org
*Med Dir:* Rebecca L Johnson, MD

# Michigan

**Andrews University**
Clin Lab Scientist/Med Technologist Prgm
Department of Clinical and Laboratory Sciences
Berrien Springs, MI 49104
www.andrews.edu/cls
*Prgm Dir:* Marcia Kilsby, PhD MT(ASCP) SBB
    CLS(NCA)
*Tel:* 269 471-6294  *Fax:* 269 471-6218
*E-mail:* kilsby@andrews.edu

**Ferris State University**
Clin Lab Scientist/Med Technologist Prgm
VFS210A
200 Ferris Dr
Big Rapids, MI 49307-2740
www.ferris.edu/cls
*Prgm Dir:* Barbara Ross, MS MT(ASCP)
*Tel:* 231 591-2317  *Fax:* 231 591-3788
*E-mail:* rossb@ferris.edu
*Med Dir:* Nicholas J Hruby, MD

**DMC University Laboratories**
Clin Lab Scientist/Med Technologist Prgm
4201 St Antoine
Detroit, MI 48201
www.dmc.org/univ
*Prgm Dir:* Joyce Salancy, MS MT(ASCP)
*Tel:* 313 993-0482  *Fax:* 313 993-2900
*E-mail:* jsalancy@dmc.org
*Med Dir:* Barbara J Anderson, MD

**Wayne State University**
Clin Lab Scientist/Med Technologist Prgm
Eugene Applebaum College of Pharmacy and Health
    Sciences
Ste 4690
Detroit, MI 48201
*Prgm Dir:* Janet M Brown, MS MT(ASCP) CLS(NCA)
*Tel:* 313 577-1384  *Fax:* 313 577-5497
*E-mail:* aa3663@wayne.edu

**Michigan State University**
Clin Lab Scientist/Med Technologist Prgm
322 N Kedzie Hall
East Lansing, MI 48824
www.medtech.cls.msu.edu
*Prgm Dir:* Kathryn M Doig, PhD CLS(NCA)
*Tel:* 517 353-7800, Ext 8  *Fax:* 517 432-2006
*E-mail:* doig@msu.edu

**Hurley Medical Center**
Clin Lab Scientist/Med Technologist Prgm
One Hurley Plaza
Flint, MI 48503
*Prgm Dir:* Sheila Moore, MSA MT(ASCP)SH
*Tel:* 810 762-7092  *Fax:* 810 762-7082
*E-mail:* smoore3@hurleymc.com
*Med Dir:* Cathy O Blight, MD

**Grand Valley State University**
Clin Lab Scientist/Med Technologist Prgm
301 Michigan St NE, Ste 200
Grand Rapids, MI 49503-3314
www.gvsu.edu/cls
*Prgm Dir:* Linda Goossen, PhD MT(ASCP)
*Tel:* 616 331-8540  *Fax:* 616 331-5999
*E-mail:* goossenl@gvsu.edu

**St John Health**
Clin Lab Scientist/Med Technologist Prgm
St John Health Laboratories
19251 Mack Ave, Ste 460
Grosse Pointe Woods, MI 48236
www.stjohn.org/labcareers
*Prgm Dir:* Marilyn Held, SpM MT(ASCP)DLM
*Tel:* 313 343-3508  *Fax:* 313 881-4727
*E-mail:* marilyn.held@stjohn.org
*Med Dir:* Martha Higgins, MD

**Northern Michigan University**
Clin Lab Scientist/Med Technologist Prgm
1401 Presque Isle Ave
3513 West Science
Marquette, MI 49855
www.nmu.edu/cls/cls.htm
*Prgm Dir:* Lucille A Contois, MA MT(ASCP)
*Tel:* 906 227-1660  *Fax:* 906 227-1309
*E-mail:* lcontois@nmu.edu
*Med Dir:* John Weiss, MD

**William Beaumont Hospital**
Clin Lab Scientist/Med Technologist Prgm
3601 W 13 Mile Rd
Department of Clinical Pathology - RI 105
Royal Oak, MI 48073-6769
www.beaumont.edu/alliedhealth
*Prgm Dir:* Nancy E Ramirez, MS MT(ASCP) SH
    CLS(NCA)
*Tel:* 248 551-8023, Ext 15135  *Fax:* 248 551-3694
*E-mail:* nramirez@beaumont.edu
*Med Dir:* Joan C Mattson, MD

**Eastern Michigan University**
Clin Lab Scientist/Med Technologist Prgm
342 Marshall Bldg
Ypsilanti, MI 48197
*Prgm Dir:* Gary Hammerberg, EdD MT(ASCP)
*Tel:* 734 487-3223   *Fax:* 734 487-4095
*E-mail:* gary.hammerberg@emich.edu
*Med Dir:* Jeffrey Warren, MD

# Minnesota

**Fairview Health Services**
Clin Lab Scientist/Med Technologist Prgm
2450 Riverside Ave, F180-56
Minneapolis, MN 55454
www.fairview.org/cls
*Prgm Dir:* Carol T McCoy, PhD MT(ASCP) CLS(NCA)
*Tel:* 612 672-4298   *Fax:* 612 672-4280
*E-mail:* cmccoy1@fairview.org
*Med Dir:* David Dexter, MD

**Hennepin County Medical Center**
Clin Lab Scientist/Med Technologist Prgm
701 Park Ave S, Lab P4
Minneapolis, MN 55415
www.co.hennepin.mn.us
*Prgm Dir:* Robert E Klicker, MS MT(ASCP)
*Tel:* 612 873-3032   *Fax:* 612 904-4229
*E-mail:* robert.klicker@co.hennepin.mn.us
*Med Dir:* James Fink, MD

**University of Minnesota - Minneapolis**
Clin Lab Scientist/Med Technologist Prgm
420 Delaware St SE, MMC 711
Minneapolis, MN 55455
http://medtech.umn.edu
*Prgm Dir:* Donna J Spannaus-Martin, PhD CLS(NCA)
*Tel:* 612 626-6579   *Fax:* 612 625-5901
*E-mail:* spann003@umn.edu

# Mississippi

**University of Southern Mississippi**
Clin Lab Scientist/Med Technologist Prgm
118 College Dr, #5134
Hattiesburg, MS 39406-0001
www.usm.edu/medtech
*Prgm Dir:* M Jane Hudson, PhD MT(ASCP)SM
    CLS(NCA)
*Tel:* 601 266-4908   *Fax:* 601 266-4913
*E-mail:* jane.hudson@usm.edu
*Med Dir:* Levesque Karin, MD

**Mississippi Baptist Medical Center**
Clin Lab Scientist/Med Technologist Prgm
1225 N State St
Jackson, MS 39202
www.mbhs.org
*Prgm Dir:* Jennifer Knight, MHS CLS (NCA) MT(ASCP)
*Tel:* 601 968-3070, Ext 7376   *Fax:* 601 974-6286
*E-mail:* jknight@mbhs.org
*Med Dir:* Gary W Benson, MD

**University of Mississippi Medical Center**
Clin Lab Scientist/Med Technologist Prgm
2500 N State St
Jackson, MS 39216
*Prgm Dir:* Libby M Spence, PhD CLS(NCA)
*Tel:* 601 984-6329   *Fax:* 601 815-1717
*E-mail:* lspence@shrp.umsmed.edu
*Med Dir:* William A Rock, Jr, MD

**North Mississippi Medical Center**
Clin Lab Scientist/Med Technologist Prgm
830 S Gloster
Tupelo, MS 38801
*Prgm Dir:* Lee H Montgomery, MT(ASCP)
*Tel:* 662 377-3066   *Fax:* 662 377-3109
*E-mail:* Lmontgomery@NMHS.net
*Med Dir:* Ishak L Enggano, MD

# Missouri

**Southeast Missouri Hospital**
Clin Lab Scientist/Med Technologist Prgm
1701 Lacey St
Cape Girardeau, MO 63701
www.southeastmissourihospital.com/labschool
*Prgm Dir:* Ann Green, MS MAEd MT(ASCP)
*Tel:* 573 334-6825, Ext 31   *Fax:* 573 339-7805
*E-mail:* agreen@sehosp.org
*Med Dir:* Mark Hosler, MD

**St John's Regional Medical Center**
Clin Lab Scientist/Med Technologist Prgm
2727 McClelland Blvd
Joplin, MO 64804-1695
*Prgm Dir:* Connie Wilkins, MS MT(ASCP)
*Tel:* 417 625-2135   *Fax:* 417 659-6429
*E-mail:* cjwilkin@stj.com

**Saint Luke's Hospital**
Clin Lab Scientist/Med Technologist Prgm
4401 Wornall Rd
Kansas City, MO 64111
www.saint-lukes.org
*Prgm Dir:* Jane M Rachel, MA MT(ASCP)
*Tel:* 816 932-2413   *Fax:* 816 932-2259
*E-mail:* jrachel@saint-lukes.org
*Med Dir:* Marjorie L Zucker, MD

**North Kansas City Hospital**
Clin Lab Scientist/Med Technologist Prgm
2800 Clay Edwards Dr
North Kansas City, MO 64116-3220
www.nkch.org/careers/Lab.pdf
*Prgm Dir:* Marisa James, MA MT(ASCP)
*Tel:* 816 691-1321   *Fax:* 816 346-7304
*E-mail:* marisa.james@nkch.org
*Med Dir:* Mark Stivers, MD

**CoxHealth**
Clin Lab Scientist/Med Technologist Prgm
3801 S National Ave
Springfield, MO 65807
www.coxhealth.com
*Prgm Dir:* Douglas D Hubbard, MT(ASCP)
*Tel:* 417 269-6633   *Fax:* 417 269-4600
*E-mail:* doug.hubbard@coxhealth.com
*Med Dir:* David M Smid, MD

**Saint Louis University**
Clin Lab Scientist/Med Technologist Prgm
3437 Caroline St
St Louis, MO 63104-1111
www.slu.edu
*Prgm Dir:* Mary Lou Vehige, MA MT(ASCP) CLS (NCA)
*Tel:* 314 977-8518   *Fax:* 314 977-8503
*E-mail:* vehigeml@slu.edu
*Med Dir:* Michael H Creer, MD

**St John's Mercy Medical Center**
Clin Lab Scientist/Med Technologist Prgm
615 S New Ballas Rd
St Louis, MO 63141
*Prgm Dir:* Teresa A Taff, MA MT(ASCP)SM
*Tel:* 314 251-6855   *Fax:* 314 251-4580
*E-mail:* taffta@stlo.mercy.net
*Med Dir:* Beverly B Kraemer, MD

# Montana

**Benefis Healthcare**
Clin Lab Scientist/Med Technologist Prgm
1101 26th St S
Great Falls, MT 59405
www.benefis.org/pages/labschool.htm
*Prgm Dir:* Susan Keeney, MA MT(ASCP)
*Tel:* 406 455-5439   *Fax:* 406 455-4970
*E-mail:* keensusc@benefis.org
*Med Dir:* Thomas Carey, MD

# Nebraska

**Nebraska Methodist Hospital**
Clin Lab Scientist/Med Technologist Prgm
8303 Dodge St
Omaha, NE 68114
www.unmc.edu/AlliedHealth/medtech
*Prgm Dir:* Julie Richards, MPA MT(ASCP) BB
*Tel:* 402 354-4563   *Fax:* 402 354-4535
*E-mail:* Julie.Richards@nmhs.org
*Med Dir:* Christine Reyes, MD

**University of Nebraska Medical Center**
Clin Lab Scientist/Med Technologist Prgm
983135 Nebraska Medical Center
Omaha, NE 68198-3135
www.unmc.edu/AlliedHealth/medtech
*Prgm Dir:* Linda L Fell, MS MT(ASCP)SH
*Tel:* 402 559-7739   *Fax:* 402 559-9044
*E-mail:* lfell@unmc.edu
*Med Dir:* James Wisecarver, MD PhD

# Nevada

**University of Nevada - Las Vegas**
Clin Lab Scientist/Med Technologist Prgm
4505 Maryland Pkwy
Mail Box 453021
Las Vegas, NV 89154-3021
www.unlv.edu/Colleges/Health_Sciences/cls
*Prgm Dir:* Janice M Conway-Klaassen, MT(ASCP)SM
    CLS(NCA)
*Tel:* 702 895-3788   *Fax:* 702 895-3872
*E-mail:* janice.klaassen@unlv.edu
*Med Dir:* Linda Veneman, MD

# New Hampshire

**University of New Hampshire**
Clin Lab Scientist/Med Technologist Prgm
210 Kendall Hall
129 Main St
Durham, NH 03824
www.mls.unh.edu
*Prgm Dir:* Adele Marone, MT(ASCP)
*Tel:* 603 862-1376   *Fax:* 603 862-3758
*E-mail:* amarone@cisunix.unh.edu

# New Jersey

**Monmouth Medical Center**
Clin Lab Scientist/Med Technologist Prgm
300 2nd Ave
Long Branch, NJ 07740
*Prgm Dir:* John A Mihok, MS Ed MT SM(ASCP) CLS
*Tel:* 732 923-7367   *Fax:* 732 923-7355
*E-mail:* jmihok@sbhcs.com
*Med Dir:* Louis J Zinterhofer, MD

**Morristown Memorial Hospital**
Clin Lab Scientist/Med Technologist Prgm
100 Madison Ave
Morristown, NJ 07960
*Prgm Dir:* Phyllis R Vail, EdM MT(ASCP)
*Tel:* 973 971-5282   *Fax:* 973 290-7370
*E-mail:* phyllis.vail@ahsys.org
*Med Dir:* Craig Dise, MD, PhD

**Jersey Shore University Medical Center**
Clin Lab Scientist/Med Technologist Prgm
3rd Floor Mehandru Pavilion
1945 Corlies Ave
Neptune, NJ 07753
www.meridianhealth.com
*Prgm Dir:* Perla L Simmons, MPA MT(ASCP)SH
*Tel:* 732 776-4603   *Fax:* 732 776-4592
*E-mail:* psimmons@meridianhealth.com
*Med Dir:* Brian S Erler, MD, PhD

**PROGRAMS**

**Univ of Med & Dent of New Jersey**
Clin Lab Scientist/Med Technologist Prgm
School of Health Related Professions
65 Bergen St
Newark, NJ 07107-3006
www.shrp.umdnj.edu/programs/cls/medic.htm
*Prgm Dir:* Elaine M Keohane, PhD CLS(NCA)
*Tel:* 973 972-5510   *Fax:* 973 972-8527
*E-mail:* keohanem@umdnj.edu
*Med Dir:* Bonita L Bachl, MD

**The Valley Hospital**
Clin Lab Scientist/Med Technologist Prgm
223 N Van Dien Ave
Ridgewood, NJ 07450
*Prgm Dir:* Jacqueline M Opera, MT(ASCP)BB
*Tel:* 201 447-8234   *Fax:* 201 447-8657
*E-mail:* jopera@valleyhealth.com
*Med Dir:* Arthur Christiano, MD

# New Mexico

**University of New Mexico**
Clin Lab Scientist/Med Technologist Prgm
MSC09 5250
1 University of New Mexico
Albuquerque, NM 87131-0001
http://hsc.unm.edu/pathology/medlab
*Prgm Dir:* Leslie Danielson, PhD MT(ASCP)
*Tel:* 505 272-5434   *Fax:* 505 272-8079
*E-mail:* ldanielson@salud.unm.edu
*Med Dir:* Mary Lipscomb, MD

# New York

**New York Methodist Hospital**
Clin Lab Scientist/Med Technologist Prgm
506 Sixth St
Brooklyn, NY 11215
*Prgm Dir:* Lori A Burkard, MS MT(ASCP)
*Tel:* 718 780-3706   *Fax:* 718 780-3673
*E-mail:* lab9026@nyp.org
*Med Dir:* Larry H Bernstein, MD

**Long Island University - C W Post Campus**
Clin Lab Scientist/Med Technologist Prgm
720 Northern Blvd
Brookville, NY 11548
www.liu.edu/health
*Prgm Dir:* Angela L Meisse, MPA MT(ASCP) SBB
   CLS(NCA)
*Tel:* 516 299-3039   *Fax:* 516 299-3106
*E-mail:* angela.meisse@liu.edu
*Med Dir:* Virginia Donovan, MD

**University at Buffalo - SUNY**
Clin Lab Scientist/Med Technologist Prgm
26 Cary Hall, 3435 Main St
Buffalo, NY 14214-3005
www.smbs.buffalo.edu/cls
*Prgm Dir:* Robert Klick, MS MT(ASCP)
*Tel:* 716 829-3630, Ext 109   *Fax:* 716 829-3601
*E-mail:* klick@buffalo.edu
*Med Dir:* Francisco Velazquez, MD

**St Vincent Catholic Med Ctr Brklyn-Queens Reg**
Clin Lab Scientist/Med Technologist Prgm
St Anthony's Health Prof & Nursing Inst
175-07 Horace Harding Exp
Fresh Meadows, NY 11365
*Prgm Dir:* Ann Paula Zero, MS MT(ASCP)
*Tel:* 718 357-0500, Ext 110   *Fax:* 718 357-4588
*E-mail:* azero@svcmcny.org
*Med Dir:* Usha Ruder, MD

**Woman's Christian Association Hospital**
Clin Lab Scientist/Med Technologist Prgm
PO Box 840
207 Foote Ave
Jamestown, NY 14702
www.wcahospital.org/mtschool
*Prgm Dir:* Michele G Harms, MS MT(ASCP)
*Tel:* 716 664-8484   *Fax:* 716 664-8306
*E-mail:* michele.harms@wcahospital.org
*Med Dir:* William A Geary, MD, PhD

**St Vincent's Hospital Manhattan SVCMC**
Clin Lab Scientist/Med Technologist Prgm
153 W 11th St
New York, NY 10011
*Prgm Dir:* Sr Catherine Sherry, MS MT(ASCP)
*Tel:* 212 604-8385   *Fax:* 212 604-8426
*E-mail:* csherry@svcmcny.org
*Med Dir:* John F Gillooley, MD

**Marist College**
Clin Lab Scientist/Med Technologist Prgm
Dept of Medical Laboratory Sciences, DN 228
3399 North Rd
Donnelly Hall, Rm 109
Poughkeepsie, NY 12601-1387
*Prgm Dir:* Robert Sullivan, PhD MT(ASCP)
*Tel:* 845 575-3000, Ext 2505   *Fax:* 845 575-3184
*E-mail:* Robert.Sullivan@marist.edu
*Med Dir:* James Sharp, MD

**Rochester General Hospital**
Clin Lab Scientist/Med Technologist Prgm
1425 Portland Ave
Rochester, NY 14621
*Prgm Dir:* Nancy C Mitchell, MS MT(ASCP)DLM
*Tel:* 585 922-4274   *Fax:* 585 922-2088
*E-mail:* nancy.mitchell@viahealth.org
*Med Dir:* Theodor K Mayer, MD PhD

**Stony Brook University**
Clin Lab Scientist/Med Technologist Prgm
Sch of Hlth Tech and Management
Stony Brook, NY 11794-8205
www.hsc.sunysb.edu/shtm
*Prgm Dir:* Kathleen Finnegan, MS MT(ASCP)SH
*Tel:* 631 444-3224   *Fax:* 631 444-8821
*E-mail:* kathleen.finnegan@stonybrook.edu
*Med Dir:* Jay L Bock, MD

**SUNY Upstate Medical University**
Clin Lab Scientist/Med Technologist Prgm
750 E Adams St
Syracuse, NY 13210
*Prgm Dir:* Susan Graham, MS MT(ASCP)SH
*Tel:* 315 464-4608   *Fax:* 315 464-4609
*E-mail:* grahams@upstate.edu
*Med Dir:* Robert E Hutchinson, MD

# North Carolina

**Univ of North Carolina at Chapel Hill**
Clin Lab Scientist/Med Technologist Prgm
Dept of Allied Hlth Sciences, CB 7145
Chapel Hill, NC 27599-7145
www.med.unc.edu/ahs/clinical
*Prgm Dir:* Susan J Beck, PhD CLS(NCA)
*Tel:* 919 966-3033   *Fax:* 919 966-8384
*E-mail:* sbeck@med.unc.edu

**Carolinas College of Health Sciences**
Clin Lab Scientist/Med Technologist Prgm
PO Box 32861
Charlotte, NC 28232
www.carolinascollege.edu
*Prgm Dir:* Elizabeth T Anderson, MHDL MT(ASCP)
   CLS(NCA)
*Tel:* 704 355-4275   *Fax:* 704 355-5967
*E-mail:* banderson@carolinas.org
*Med Dir:* Edward H Lipford, MD

**Western Carolina University**
Clin Lab Scientist/Med Technologist Prgm
120 Moore Hall
Cullowhee, NC 28723
www.wcu.edu/aps/healths/HS_UG-CLS.htm
*Prgm Dir:* Russell Cheadle, MS MT(ASCP)
*Tel:* 704 227-3515   *Fax:* 704 227-7446
*E-mail:* cheadle@email.wcu.edu
*Med Dir:* Larry Selby, MD

**East Carolina University**
Clin Lab Scientist/Med Technologist Prgm
Department of Clinical Laboratory Science
School of Allied Hlth Sciences, East Carolina University
Greenville, NC 27858-4353
www.ecu.edu/clsc
*Prgm Dir:* Richard Bamberg, PhD MT(ASCP)SH
   CLDir(NCA) CHES
*Tel:* 252 744-6064   *Fax:* 252 744-6018
*E-mail:* bambergw@ecu.edu
*Med Dir:* Peter J Kragel, MD

**Wake Forest University Baptist Medical Center**
Clin Lab Scientist/Med Technologist Prgm
Medical Center Blvd
Winston-Salem, NC 27157
www1.wfubmc.edu/pathology/medtech
*Prgm Dir:* Julie H Simmons, MPH MT(ASCP)SBB
*Tel:* 336 716-4727   *Fax:* 336 716-6359
*E-mail:* jusimmon@wfubmc.edu
*Med Dir:* Abbott Julian Garvin, MD

**Winston-Salem State University**
Clin Lab Scientist/Med Technologist Prgm
CLS Dept FL Atkins Ste 401
601 Martin Luther King Jr Dr
Winston-Salem, NC 27110
*Prgm Dir:* Karen K Hunter, MS MT(ASCP)
*Tel:* 336 750-2513   *Fax:* 336 750-2517
*E-mail:* hunterk@wssu.edu
*Med Dir:* Mabrey Hopkins, MD

# North Dakota

**MeritCare Medical Center**
Clin Lab Scientist/Med Technologist Prgm
737 Broadway
PO Box MC
Fargo, ND 58122-0041
https://www.meritcare.com/jobs
*Prgm Dir:* Sandra Matthey, MS BS MT(ASCP)
*Tel:* 701 234-2481   *Fax:* 701 234-2403
*E-mail:* sandra.matthey@meritcare.com

**University of North Dakota**
Clin Lab Scientist/Med Technologist Prgm
Dept of Pathology/CLS, PO Box 9037
Grand Forks, ND 58202-9037
http://pathology.med.und.nodak.edu/cls
*Prgm Dir:* Ruth Paur, MS CLS(NCA) MT(ASCP)
*Tel:* 701 777-2651   *Fax:* 701 777-2404
*E-mail:* ruthpaur@medicine.nodak.edu
*Med Dir:* Leslie Torgerson, MD

# Ohio

**Ohio Northern University**
Clin Lab Scientist/Med Technologist Prgm
525 S Main St
Ada, OH 45810
www.onu.edu
*Prgm Dir:* Cara L Calvo, MS MT(ASCP)SH CLS(NCA)
*Tel:* 419 772-3084   *Fax:* 419 772-2330
*E-mail:* c-calvo@onu.edu
*Med Dir:* Rogelio Decanio, MD, FASCP, MBA

**Children's Hospital Medical Ctr of Akron**
Clin Lab Scientist/Med Technologist Prgm
One Perkins Square
Akron, OH 44308
www.akronchildrens.org
*Prgm Dir:* Sharon K Shriber, MBA MT(ASCP)SH
*Tel:* 330 543-8676    *Fax:* 330 543-6303
*E-mail:* sshriber@chmca.org
*Med Dir:* Robert Novak, MD

**Bowling Green State University**
Clin Lab Scientist/Med Technologist Prgm
504 Life Science Bldg
Bowling Green, OH 43403
*Prgm Dir:* Robert Harr, MS MT(ASCP)
*Tel:* 419 372-8109    *Fax:* 419 372-0332
*E-mail:* rharr@bgnet.bgsu.edu
*Med Dir:* Fleming Fallon, MD

**University of Cincinnati Medical Center**
Clin Lab Scientist/Med Technologist Prgm
3202 Eden Ave
Room 371 French East, ML 0394
Cincinnati, OH 45267-0394
www.cahs.uc.edu
*Prgm Dir:* Linda J Graeter, PhD MT(ASCP)
*Tel:* 513 558-2018    *Fax:* 513 558-6002
*E-mail:* graetel@email.uc.edu

**Ohio State University**
Clin Lab Scientist/Med Technologist Prgm
453 W 10th Ave
535 Atwell Hall
Columbus, OH 43210-1234
*Prgm Dir:* Sally V Rudmann, PhD MT(ASCP)
*Tel:* 614 292-7303, Ext 5    *Fax:* 614 292-0210
*E-mail:* rudmann.1@osu.edu

**Wright State University**
Clin Lab Scientist/Med Technologist Prgm
235A Biological Sciences Bldg
3640 Colonel Glenn Hwy
Dayton, OH 45435
www.wright.edu
*Prgm Dir:* Denene Lofland, PhD MT(ASCP)
*Tel:* 937 775-2306    *Fax:* 937 775-3320
*E-mail:* denene.lofland@wright.edu
*Med Dir:* James W Funkhouser, MD

**Southwest General Health Center**
Clin Lab Scientist/Med Technologist Prgm
18697 E Bagley Rd
Middleburg Heights, OH 44130
www.swgeneral.com
*Prgm Dir:* Corinne Hartwell, MS MT(ASCP)
*Tel:* 440 816-8879, Ext NA    *Fax:* 440 816-8690
*E-mail:* chartwell@swgeneral.com
*Med Dir:* Karen Gerken, MD

**St Vincent Mercy Medical Center**
Clin Lab Scientist/Med Technologist Prgm
2222 Cherry St
Toledo, OH 43608
www.ehealthconnection.com/regions/toledo/content/
    Integrated_CLS_Mercy_Program.asp
*Prgm Dir:* Karlyn Lange
*Tel:* 419 251-8252    *Fax:* 419 251-7816
*E-mail:* Karlyn_Lange@mhsnr.org

## Oklahoma

**Valley View Regional Hospital**
Clin Lab Scientist/Med Technologist Prgm
430 N Monta Vista
Ada, OK 74820
*Prgm Dir:* Leah Babcock, MSHR MT(ASCP)
*Tel:* 405 421-1596    *Fax:* 405 421-1525
*E-mail:* lbabcock@vvrh.com
*Med Dir:* Larry W Cartmell, MD

**Comanche County Memorial Hospital**
Clin Lab Scientist/Med Technologist Prgm
PO Box 129
Lawton, OK 73502
*Prgm Dir:* Gary Jackson, SH(ASCP)
*Tel:* 405 355-8699, Ext 3362    *Fax:* 405 585-5462
*E-mail:* jacksong@memorialhealthsource.com
*Med Dir:* R J Boatsman, MD

**Muskogee Regional Medical Center**
Clin Lab Scientist/Med Technologist Prgm
300 Rockefeller Dr
Muskogee, OK 74401
www.muskogeehealth.com
*Prgm Dir:* Robert King, MBA MT(ASCP)
*Tel:* 918 684-3453    *Fax:* 918 684-2223
*E-mail:* rking@muskogeehealth.com
*Med Dir:* Brian J Bock, MD

**Saint Francis Hospital**
Clin Lab Scientist/Med Technologist Prgm
6161 S Yale Ave
Tulsa, OK 74136
www.saintfrancislab.com
*Prgm Dir:* Theresa D Foster, MPH MT(ASCP)SH
*Tel:* 918 494-6342    *Fax:* 918 494-1497
*E-mail:* tdfoster@saintfrancis.com
*Med Dir:* Robert S White, MD

## Oregon

**Oregon Health & Science University**
*Cosponsor:* Oregon Insitute of Technology
Clin Lab Scientist/Med Technologist Prgm
3181 SW Sam Jackson Park Rd, MTGH
Portland, OR 97239-3098
www.oit.edu/cls
*Prgm Dir:* Marian Ewell, MT(ASCP)SBB
*Tel:* 503 494-8589    *Fax:* 503 494-2730
*E-mail:* ewellm@ohsu.edu

## Pennsylvania

**Altoona Regional Health System**
Clin Lab Scientist/Med Technologist Prgm
620 Howard Ave
Altoona, PA 16601-4899
www.altoonaregional.org/medtech
*Prgm Dir:* Joseph R Noel, CLS(NCA) MT(ASCP)
*Tel:* 814 889-2835    *Fax:* 814 889-6001
*E-mail:* jnoel@altoonaregional.org
*Med Dir:* Americo B Anton, MD

**Neumann College**
Clin Lab Scientist/Med Technologist Prgm
One Neumann Dr
Aston, PA 19014
www.neumann.edu
*Prgm Dir:* Sandra M Weiss, EdD MT (ASCP) CLS (NCA)
*Tel:* 610 558-5607    *Fax:* 610 361-5314
*E-mail:* sweiss@neumann.edu
*Med Dir:* Harvey Spector, MD

**Saint Vincent Health Center**
Clin Lab Scientist/Med Technologist Prgm
232 W 25th St
Erie, PA 16544-0002
www.saintvincenthealth.com/medicalprofessionals/
    medical_technology.htm
*Prgm Dir:* Stephen M Johnson, MS MT(ASCP)
*Tel:* 814 452-5365    *Fax:* 814 456-4784
*E-mail:* sjohnson@svhs.org
*Med Dir:* Mary E Reitz, MD

**Conemaugh Memorial Medical Center**
Clin Lab Scientist/Med Technologist Prgm
1086 Franklin St
Johnstown, PA 15905-4398
*Prgm Dir:* Abra Elkins, MA MT(ASCP)
*Tel:* 814 534-5578    *Fax:* 814 534-3253
*E-mail:* aelkins@conemaugh.org
*Med Dir:* Sidney A Goldblatt, MD

**Lancaster Gen Coll of Nursing & Hlth Sciences**
Clin Lab Scientist/Med Technologist Prgm
410 N Lime St
Lancaster, PA 17602
www.lancastergeneral.org
*Prgm Dir:* Wendy S Gayle, MT(ASCP)
*Tel:* 717 544-7208    *Fax:* 717 544-5970
*E-mail:* wsgayle@lancastergeneral.org
*Med Dir:* James T Eastman III, MD

**Graduate Hospital**
Clin Lab Scientist/Med Technologist Prgm
1800 Lombard St
Philadelphia, PA 19146
www.graduatehospital.com
*Prgm Dir:* Jean Buchenhorst, MS MT (ASCP)
*Tel:* 215 893-2242    *Fax:* 215 893-4169
*E-mail:* Jean.Buchenhorst@tenethealth.com
*Med Dir:* Manjula Balasubramanian, MD

**Pennsylvania Hospital**
Clin Lab Scientist/Med Technologist Prgm
800 Spruce St
Room 756 Preston Building
Philadelphia, PA 19107
*Prgm Dir:* William H Hunt, MBA CLS(NCA) CLSp(H)
    MT(ASCP)
*Tel:* 215 829-6911    *Fax:* 215 829-7304
*E-mail:* william.hunt@uphs.upenn.edu
*Med Dir:* John S J Brooks, MD

**Thomas Jefferson University**
Clin Lab Scientist/Med Technologist Prgm
130 S 9th St, Ste 1924
Philadelphia, PA 19107
https://www.jefferson.edu/jchp/ls
*Prgm Dir:* Janet Devine, EdD MT(ASCP)
*Tel:* 215 503-8187    *Fax:* 215 503-2189
*E-mail:* Janet.Devine@jefferson.edu
*Med Dir:* Albert A Keshgegian, MD, PhD

**Reading Hospital & Medical Center**
Clin Lab Scientist/Med Technologist Prgm
PO Box 16052
Reading, PA 19612-6052
www.readinghospital.org
*Prgm Dir:* Sharon Kay Strauss, MS CLS(NCA)
    MT(ASCP)SM CLSp(NCA)
*Tel:* 610 988-8273    *Fax:* 610 988-5964
*E-mail:* strausss@readinghospital.org
*Med Dir:* William K Natale, MD JD

**Robert Packer Hospital**
*Cosponsor:* Guthrie Health Systems
Clin Lab Scientist/Med Technologist Prgm
Guthrie Square
Sayre, PA 18840
www.guthrie.org/medtech
*Prgm Dir:* Brian D Spezialetti, MS MT(ASCP)
*Tel:* 570 882-4736    *Fax:* 570 882-6509
*E-mail:* bspezial@ghs.guthrie.org
*Med Dir:* Joseph T King, MD

**Mount Nittany Medical Center**
Clin Lab Scientist/Med Technologist Prgm
1800 E Park Ave
State College, PA 16803
*Prgm Dir:* Michael W Archer, MEd MT(ASCP)
*Tel:* 814 231-7279
*E-mail:* marcher@mountnittany.org

**Williamsport Hosp & Medical Center**
Clin Lab Scientist/Med Technologist Prgm
777 Rural Ave
Williamsport, PA 17701
www.shscares/education
*Prgm Dir:* Loretta A Moffatt, MHA MT(ASCP)
*Tel:* 570 321-2326    *Fax:* 570 321-2727
*E-mail:* lmoffatt@shscares.org
*Med Dir:* Willem Lubbe, MD

**York Hospital**
Clin Lab Scientist/Med Technologist Prgm
1001 S George St
York, PA 17405-7198
www.wellspan.org/EducationResearch/
    AlliedHealthEducation_YHLabScience.htm
*Prgm Dir:* Carolyn Darr, MA MT(ASCP)
*Tel:* 717 851-2473    *Fax:* 717 851-2934
*E-mail:* cdarr@wellspan.org
*Med Dir:* J David Owens, MD

# Puerto Rico

**Pontifical Catholic University**
Clin Lab Scientist/Med Technologist Prgm
2250 Las Americas Ave, Ste 588
Ponce, PR 00717-9997
*Prgm Dir:* Maribel Figueroa Pena, MS MT(ASCP)
*Tel:* 787 841-2000, Ext 1585    *Fax:* 787 651-2015
*E-mail:* tecmed@email.pucpr.edu
*Med Dir:* Adalberto Mendoza, MD

**Inter American University of Puerto Rico**
Clin Lab Scientist/Med Technologist Prgm
Call Box 5100
San German, PR 00683-9801
*Prgm Dir:* Irma Mendez, MS MT(ASCP)
*Tel:* 787 264-1912, Ext 7711    *Fax:* 787 892-6350
*E-mail:* imendez@sg.inter.edu

**Inter - American University - Metro Campus**
Clin Lab Scientist/Med Technologist Prgm
Sein St, Road 1
Francisco Sein State Rd 1
San Juan, PR 00926
www.metro.inter.edu
*Prgm Dir:* Ida Mejias, PhD MT(ASCP)
*Tel:* 787 250-1912, Ext 2406    *Fax:* 787 767-5081
*E-mail:* iamejias@metro.inter.edu

**University of Puerto Rico**
Clin Lab Scientist/Med Technologist Prgm
Med Sciences Campus, College of Health-Related Profs
PO Box 365067
San Juan, PR 00936-5067
http://cprsweb.rcm.upr.edu
*Prgm Dir:* Carmen Ofelia Melendez, EdD MT(ASCP)
*Tel:* 787 758-2525, Ext 2106    *Fax:* 787 756-7220
*E-mail:* ofeliamelendez@cprs.rcm.upr.edu
*Med Dir:* Consuelo Climent-Peris, MD

# Rhode Island

**Our Lady of Fatima Hospital**
Clin Lab Scientist/Med Technologist Prgm
200 High Service Ave
North Providence, RI 02904
*Prgm Dir:* Frances Ingersoll, MS MT(ASCP) CLS(NCA)
*Tel:* 401 456-3416    *Fax:* 401 456-3695
*E-mail:* fingersoll@saintjosephri.com
*Med Dir:* Cecilia Gmuer, MD

**Rhode Island Hospital**
Clin Lab Scientist/Med Technologist Prgm
593 Eddy St
PO Bldg RM #034
Providence, RI 02903
*Prgm Dir:* David J Mello, MS MT(ASCP)CLS
*Tel:* 401 444-5724    *Fax:* 401 444-0151
*E-mail:* djmello@lifespan.org
*Med Dir:* Lewis Glasser, MD

# South Carolina

**Palmetto Health Baptist**
Clin Lab Scientist/Med Technologist Prgm
Taylor at Marion Sts
Columbia, SC 29220
*Prgm Dir:* Jennifer R Gray, MPH MT(ASCP) HT(ASCP)
*Tel:* 803 296-5014
*E-mail:* jennifer.gray@palmettohealth.org
*Med Dir:* A Atwell Coleman III, MD

**McLeod Regional Medical Center**
Clin Lab Scientist/Med Technologist Prgm
555 E Cheves St
PO Box 100551 (mailing address zip 29501-0551)
Florence, SC 29506
www.peedeeahec.net
*Prgm Dir:* Vicki T Anderson, MT(ASCP)
*Tel:* 843 777-2497    *Fax:* 843 777-2071
*E-mail:* vanderson@mcleodhealth.org
*Med Dir:* Sharon Mitchell, MD

**Lexington Medical Center**
Clin Lab Scientist/Med Technologist Prgm
2720 Sunset Blvd
West Columbia, SC 29169
www.lexmed.com
*Prgm Dir:* Ann Beaman, MSW BS MT(ASCP)
*Tel:* 803 936-8126    *Fax:* 803 739-3882
*E-mail:* ajbeaman@lexhealth.org
*Med Dir:* John Carter

# South Dakota

**Rapid City Regional Hospital**
Clin Lab Scientist/Med Technologist Prgm
353 Fairmont Blvd
Rapid City, SD 57701
www.rcrh.org/clsp
*Prgm Dir:* Pam Kieffer, MS CLS(NCA) MT(ASCP)
*Tel:* 605 719-8092    *Fax:* 605 719-2205
*E-mail:* pkieffer@rcrh.org
*Med Dir:* Susan L Eliason, MD

**Sioux Valley Hospital**
Clin Lab Scientist/Med Technologist Prgm
1305 W 18th St
Sioux Falls, SD 57117-5039
www.siouxvalley.org/Employment/
    EducationalOpportunities
*Prgm Dir:* Renee Rydell, MBA MS MT(ASCP)
*Tel:* 605 333-7104    *Fax:* 605 333-1532
*E-mail:* rydellr@siouxvalley.org
*Med Dir:* David W Ohrt, MD PhD

# Tennessee

**Austin Peay State University**
Clin Lab Scientist/Med Technologist Prgm
PO Box 4718
Clarksville, TN 37044
www.apsu.edu/medtech
*Prgm Dir:* Robert D Robison, PhD MT(ASCP)
*Tel:* 931 221-7018    *Fax:* 931 221-6323
*E-mail:* RobisonR@apsu.edu
*Med Dir:* Randall R Haase, DO

**Lincoln Memorial University**
Clin Lab Scientist/Med Technologist Prgm
6965 Cumberland Gap Pkwy
Harrogate, TN 37752
www.lmunet.edu
*Prgm Dir:* Bill Engle, DD ThD MS(CLS) MT(ASCP)
*Tel:* 423 869-6471    *Fax:* 423 869-6244
*E-mail:* bengle@lmunet.edu
*Med Dir:* Lynn F Blake, MD

**Univ of Tennessee Medical Ctr at Knoxville**
Clin Lab Scientist/Med Technologist Prgm
1924 Alcoa Hwy SW
Knoxville, TN 37920
www.utmedicalcenter.org/health_professionals/
    educational_services/medical_technology_program
*Prgm Dir:* Betty M White, MAT MT(ASCP)
*Tel:* 865 544-9087    *Fax:* 865 525-5762
*E-mail:* BMWhite@mc.utmck.edu
*Med Dir:* John Carl Neff, MD

**University of Tennessee Health Science Ctr**
Clin Lab Scientist/Med Technologist Prgm
930 Madison Ave, #664
Memphis, TN 38163
www.utmem.edu/allied/mthome.html
*Prgm Dir:* Linda Ross, MS BS MT(ASCP) CLS(NCA)
*Tel:* 901 448-6304    *Fax:* 901 448-7545
*E-mail:* lross@utmem.edu
*Med Dir:* Sherri Dawn Flax, MD

**Tennessee State University**
Clin Lab Scientist/Med Technologist Prgm
3500 John A Merritt Blvd
Nashville, TN 37203
www.tnstate.edu
*Prgm Dir:* Theola Copeland, MS MT(ASCP)
*Tel:* 615 963-5062    *Fax:* 615 963-2538
*E-mail:* tcopeland@tnstate.edu
*Med Dir:* Digna Saunders Forbes, MD

**Vanderbilt University Medical Center**
Clin Lab Scientist/Med Technologist Prgm
4605D The Vanderbilt Clinic
1165 21st Ave S
Nashville, TN 37232-5310
https://www.mc.vanderbilt.edu/medtech
*Prgm Dir:* Maralie Gaffron Exton, MT(ASCP)SH
*Tel:* 615 322-6940    *Fax:* 615 343-8420
*E-mail:* maralie.exton@vanderbilt.edu
*Med Dir:* David R Head, MD

# Texas

**Austin State Hospital**
Clin Lab Scientist/Med Technologist Prgm
4110 Guadalupe St
Austin, TX 78751-4296
*Prgm Dir:* Judith D Larsen, MT(ASCP)SH
*Tel:* 512 419-2038    *Fax:* 512 419-2039
*E-mail:* judy.larsen@mhmr.state.tx.us
*Med Dir:* Susan J Pacinda, MD

**Christus Hospital - St Elizabeth**
Clin Lab Scientist/Med Technologist Prgm
2830 Calder St, PO Box 5405
Beaumont, TX 77726-5405
www.christusste.org
*Prgm Dir:* Deborah R Zink, MBA MT(ASCP)
*Tel:* 409 899-7150, Ext 4002    *Fax:* 409 899-7991
*E-mail:* debbi.zink@christushealth.org
*Med Dir:* Kathryn Bommer, MD

**Texas A&M University - Corpus Christi**
Clin Lab Scientist/Med Technologist Prgm
College of Science and Technology
6300 Ocean Dr
Corpus Christi, TX 78412
*Prgm Dir:* Christina Thompson, EdD MT(ASCP)SBB
*Tel:* 361 825-2473    *Fax:* 361 825-3719
*E-mail:* cthomp@falcon.tamucc.edu
*Med Dir:* Joe Lewis, MD

**Univ of Texas Southwestern Med Ctr at Dallas**
Clin Lab Scientist/Med Technologist Prgm
5323 Harry Hines Blvd
Dallas, TX 75390-8878
www.utsouthwestern.edu/utsw/medlabsci
*Prgm Dir:* Lynn M Little, PhD MBA CLS(NCA)
   MT(ASCP)M
*Tel:* 214 648-1780   *Fax:* 214 648-1029
*E-mail:* lynn.little@utsouthwestern.edu
*Med Dir:* Robert McKenna, MD

**Univ of Texas - Pan American**
Clin Lab Scientist/Med Technologist Prgm
1201 W University Dr
Edinburg, TX 78541
www.panam.edu/dept/clinlab
*Prgm Dir:* Karen Chandler, MA MT(ASCP) CLS (NCA)
*Tel:* 956 381-2296   *Fax:* 956 318-5253
*E-mail:* kchandler@utpa.edu
*Med Dir:* Domingo Useda, MD

**University of Texas at El Paso**
Clin Lab Scientist/Med Technologist Prgm
1101 N Campbell, Rm 714
El Paso, TX 79902
*Prgm Dir:* M Lorraine Torres, MS MT(ASCP) ABD
*Tel:* 915 747-7282   *Fax:* 915 747-7207
*E-mail:* lorit@utep.edu
*Med Dir:* Arturo Vargas, MD

**Tarleton State University**
Clin Lab Scientist/Med Technologist Prgm
1501 Enderly Place
Ft Worth, TX 76104
*Prgm Dir:* Sally S Lewis, MS MT(ASCP) HTL (ASCP)
*Tel:* 817 926-1101   *Fax:* 817 922-8103
*E-mail:* slewis@tarleton.edu
*Med Dir:* Clifton Ray Daniel, MD

**University of Texas Medical Branch/SAHS**
Clin Lab Scientist/Med Technologist Prgm
Sch of Allied Hlth Sciences
301 University Blvd
Galveston, TX 77555-1140
www.sahs.utmb.edu/cls
*Prgm Dir:* Vicki S Freeman, PhD MT(ASCP)SC
*Tel:* 409 772-3056   *Fax:* 409 772-9470
*E-mail:* vfreeman@utmb.edu
*Med Dir:* Alexander J Indrikovs, MD

**Methodist Hospital**
Clin Lab Scientist/Med Technologist Prgm
6565 Fannin, B154
Houston, TX 77030
*Prgm Dir:* Judy Jobe, MT(ASCP)
*Tel:* 713 441-2599   *Fax:* 713 793-7408
*E-mail:* jjobe@tmh.tmc.edu
*Med Dir:* Christopher Leveque, MD

**Texas Southern University**
Clin Lab Scientist/Med Technologist Prgm
3100 Cleburne
Houston, TX 77004
www.tsu.edu
*Prgm Dir:* Dorothy J Quiller, MEd BS MT(ASCP)
*Tel:* 713 313-7248   *Fax:* 713 313-1094
*E-mail:* quiller_dj@tsu.edu

**Univ of Texas M D Anderson Cancer Ctr**
Clin Lab Scientist/Med Technologist Prgm
1100 Holcombe Blvd, Unit 0204
Houston, TX 77030
www.mdanderson.org/healthsciences
*Prgm Dir:* Karen McClure, MS CLS(NCA)
   MT(ASCP)SBB
*Tel:* 713 745-1688   *Fax:* 713 745-3337
*E-mail:* kmcclure@mdanderson.org
*Med Dir:* Jeffrey J Tarrand, MD

**Texas Tech Univ Health Sciences Center**
Clin Lab Scientist/Med Technologist Prgm
School of Allied Health Sciences
3601 4th St
Lubbock, TX 79430
www.ttuhsc.edu/pages/alh
*Prgm Dir:* Lori Rice-Spearman, MS MT(ASCP)
*Tel:* 806 743-3252   *Fax:* 806 743-3249
*E-mail:* lori.ricespearman@ttuhsc.edu
*Med Dir:* David L Morgan, MD

**Univ of Texas Hlth Sci Ctr at San Antonio**
Clin Lab Scientist/Med Technologist Prgm
7703 Floyd Curl Dr
San Antonio, TX 78229-3900
www.uthscsa.edu/sah/cls/cls.html
*Prgm Dir:* Shirlyn B McKenzie, MS PhD
*Tel:* 210 567-8860, Ext 7-8868   *Fax:* 210 567-8875
*E-mail:* mckenzie@uthscsa.edu
*Med Dir:* John Olson, MD

**Texas State University-San Marcos**
Clin Lab Scientist/Med Technologist Prgm
601 University Dr
San Marcos, TX 78666-4616
www.txstate.edu/cls
*Prgm Dir:* David M Falleur, MEd MT(ASCP) CLS(NCA)
*Tel:* 512 245-3500   *Fax:* 512 245-7860
*E-mail:* dfalleur@txstate.edu
*Med Dir:* Margaret C Young, MD, PhD

**Scott & White Memorial Hospital**
Clin Lab Scientist/Med Technologist Prgm
2401 S 31st St
Temple, TX 76508
www.sw.org
*Prgm Dir:* Mary Ruth Beckham, MEd MT(ASCP
*Tel:* 254 724-5970   *Fax:* 254 724-0819
*E-mail:* mbeckham@swmail.sw.org
*Med Dir:* Kathleen Jones, MD

**Hillcrest Baptist Medical Center**
Clin Lab Scientist/Med Technologist Prgm
3000 Herring Ave
Waco, TX 76708
*Prgm Dir:* Kele D Johnson, MHSM MT(ASCP)
*Tel:* 254 202-8133   *Fax:* 254 202-5675
*E-mail:* kjohnson@hillcrest.net
*Med Dir:* Edwin B Morrison, MD

**United Regional Health Care Systems**
Clin Lab Scientist/Med Technologist Prgm
1600 Eleventh St
Wichita Falls, TX 76301
*Prgm Dir:* Gwendolyn Morman, MS MA MT(ASCP)
*Tel:* 940 764-3187   *Fax:* 940 764-3328
*E-mail:* gmorman@urhcs.org
*Med Dir:* David Flack, MD, Anatomic and Clinical
   Pathologist

# Utah

**Weber State University**
Clin Lab Scientist/Med Technologist Prgm
3905 University Circle
Ogden, UT 84408-3905
http://weber.edu/cls
*Prgm Dir:* Yasmen Simonian, PhD MT(ASCP) CLS
   (NCA)
*Tel:* 801 626-7080   *Fax:* 801 626-7508
*E-mail:* ysimonian@weber.edu
*Med Dir:* Val B Johnson, MD

**Brigham Young University**
Clin Lab Scientist/Med Technologist Prgm
375 WIDB
Provo, UT 84602
http://mmbio.byu.edu/home/page/
   About_the_Program.aspx
*Prgm Dir:* Shauna C Anderson, PhD CLS (NCA)
   MT(ASCP)
*Tel:* 801 422-8757   *Fax:* 801 422-0014
*E-mail:* shauna_anderson@byu.edu

**University of Utah Health Science Center**
Clin Lab Scientist/Med Technologist Prgm
Dept of Pathology
30 North, 1900 East
Salt Lake City, UT 84132
www.path.utah.edu/mls
*Prgm Dir:* Larry Schoeff, MS MT(ASCP)
*Tel:* 801 585-6989   *Fax:* 801 585-2463
*E-mail:* lschoeff@path.utah.edu
*Med Dir:* Joseph A Knight, MD

# Vermont

**University of Vermont**
Clin Lab Scientist/Med Technologist Prgm
Medical Laboratory and Radiation Sciences
302 Rowell Hall
Burlington, VT 05405
www.uvm.edu/mlrs
*Prgm Dir:* J Patrick Reed, MS MT(ASCP)SH
*Tel:* 802 656-3811   *Fax:* 802 656-2191
*E-mail:* pat.reed@uvm.edu
*Med Dir:* Edwin G Bovill, MD

# Virginia

**Inova Fairfax Hospital**
Clin Lab Scientist/Med Technologist Prgm
3300 Gallows Rd
Falls Church, VA 22046
*Prgm Dir:* Amy Shoemaker, MT(ASCP)DLM
*Tel:* 703 698-2891   *Fax:* 703 698-2407
*E-mail:* amy.shoemaker@inova.com
*Med Dir:* C Barrie Cook, MD

**Augusta Medical Center**
Clin Lab Scientist/Med Technologist Prgm
PO Box 1000
78 Medical Center Dr
Fishersville, VA 22939
www.augustamed.com/cls
*Prgm Dir:* Bernadette Bekken, CLS(NCA)MT(ASCP)BB
*Tel:* 540 332-4539   *Fax:* 540 332-4543
*E-mail:* bbekken@augustamed.com
*Med Dir:* Wayne P Jessee, MD

**Rockingham Memorial Hospital**
Clin Lab Scientist/Med Technologist Prgm
235 Cantrell Ave
Harrisonburg, VA 22801
*Prgm Dir:* Sue W Lawton, MS MA MT(ASCP)
*Tel:* 540 564-5407   *Fax:* 540 433-4284
*E-mail:* slawton@rhcc.com
*Med Dir:* Warren D Bannister, MD

**Norfolk State University**
Clin Lab Scientist/Med Technologist Prgm
700 Park Ave
Norfolk, VA 23504
www.nsu.edu
*Prgm Dir:* Mildred Fuller, PhD MT(ASCP) CLS(NCA)
*Tel:* 757 823-2366   *Fax:* 757 823-2909
*E-mail:* mkfuller@nsu.edu
*Med Dir:* Robert L Dillard, MD

## Old Dominion University

Clin Lab Scientist/Med Technologist Prgm
School of Med Laboratory & Radiation Sciences
Spong Hall, 209
Norfolk, VA 23529
*Prgm Dir:* Faye E Coleman, MS MT(ASCP) CLS
*Tel:* 757 683-3588   *Fax:* 757 683-5028
*E-mail:* fcoleman@odu.edu

## Virginia Commonwealth University

*Cosponsor:* VCU Medical Center
Clin Lab Scientist/Med Technologist Prgm
PO Box 980583, Medical Center Campus
301 College St
Richmond, VA 23298-0583
*Prgm Dir:* Teresa Nadder, PhD CLS(NCA) MT(ASCP)
*Tel:* 804 828-9469   *Fax:* 804 828-1911
*E-mail:* tsnadder@vcu.edu
*Med Dir:* Richard A McPherson, MD

## Carilion Medical Center

Clin Lab Scientist/Med Technologist Prgm
Carilion Roanoke Community Hospital
101 Elm Ave SW
Roanoke, VA 24013
*Prgm Dir:* Maribeth Greenway, MEd MT (ASCP)SH
*Tel:* 540 985-8109   *Fax:* 540 224-4457
*E-mail:* mgreenway@carilion.com
*Med Dir:* Robert White, MD

# Washington

## University of Washington

Clin Lab Scientist/Med Technologist Prgm
School of Medicine
Dept of Laboratory Medicine, Box 357110
Seattle, WA 98195-7110
*Prgm Dir:* Mary F Lampe, PhD
*Tel:* 206 598-2135   *Fax:* 206 598-6189
*E-mail:* lampe@u.washington.edu
*Med Dir:* James S Fine, MD

## Sacred Heart Medical Center

Clin Lab Scientist/Med Technologist Prgm
101 W Eighth Ave
Spokane, WA 99220-2555
www.shmclab.org
*Prgm Dir:* Cynthia Hamby, MEd MT(ASCP)
*Tel:* 509 474-3339   *Fax:* 509 474-2052
*E-mail:* hambyc@shmc.org
*Med Dir:* Thomas Allerding, MD

## Yakima Regional CLS Program

Clin Lab Scientist/Med Technologist Prgm
1120 W Spruce St
Yakima, WA 98902
*Prgm Dir:* Claudia R Steen, MS MT(ASCP)
*Tel:* 509 454-6100   *Fax:* 509 454-6117
*E-mail:* yakimacls@yakima.hma-corp.com
*Med Dir:* Stephen Muehleck, MD

# West Virginia

## Marshall University

Clin Lab Scientist/Med Technologist Prgm
One John Marshall Dr
Huntington, WV 25755
*Prgm Dir:* Dorothy J Fike, MS MT(ASCP)
*Tel:* 304 969-3165   *Fax:* 304 696-3243
*E-mail:* fike@marshall.edu

## West Virginia University

Clin Lab Scientist/Med Technologist Prgm
2163E Health Sciences Ctr N, PO Box 9211
Morgantown, WV 26506-9211
www.hsc.wvu.edu/som/medtech
*Prgm Dir:* Martha J Lake, EdD CLS(NCA) MT(ASCP)
*Tel:* 304 293-2069   *Fax:* 304 293-6249
*E-mail:* mlake@hsc.wvu.edu
*Med Dir:* Patricia Canfield, MD

## West Liberty State College

Clin Lab Scientist/Med Technologist Prgm
Dept of Health Sciences
West Liberty, WV 26074
www.wlsc.edu
*Prgm Dir:* William C Wagener, PhD MT(ASCP)
*Tel:* 304 336-8177   *Fax:* 304 336-8266
*E-mail:* wagenerw@wlsc.edu
*Med Dir:* Jaywant Philip Parmar, MD

# Wisconsin

## Affinity Health System

Clin Lab Scientist/Med Technologist Prgm
St Elizabeth Hospital
1506 S Oneida St
Appleton, WI 54915
*Prgm Dir:* Cecelia W Landin, MS MT(ASCP)
*Tel:* 920 738-2132   *Fax:* 920 831-8518
*E-mail:* clandin@affinityhealth.org
*Med Dir:* Peter V Podlusky, MD

## Sacred Heart Hospital

Clin Lab Scientist/Med Technologist Prgm
900 W Clairemont Ave
Eau Claire, WI 54701
*Prgm Dir:* Linda Stai, BS MT(ASCP) MS
*Tel:* 715 839-4232   *Fax:* 715 833-4941
*E-mail:* lstai@shec.hshs.org
*Med Dir:* Thomas W Hadley, MD

## University of Wisconsin - Madison

Clin Lab Scientist/Med Technologist Prgm
1300 University Ave
6175 MSC
Madison, WI 53706
www.clsmedtech.wisc.edu
*Prgm Dir:* Sharon Ehrmeyer, PhD MT(ASCP)
*Tel:* 608 262-2085   *Fax:* 608 262-9520
*E-mail:* ehrmeyer@wisc.edu
*Med Dir:* Teresa Darcy, MD

## St Joseph's Hospital

Clin Lab Scientist/Med Technologist Prgm
Marshfield Laboratories
611 St Joseph Ave
Marshfield, WI 54449
www.marshfieldlaboratories.org/career.asp
*Prgm Dir:* Julie J Seehafer, MS MT(ASCP)SH
*Tel:* 715 387-7202   *Fax:* 715 387-7121
*E-mail:* seehafer.julie@marshfieldclinic.org
*Med Dir:* Gene Shaw, MD

## Clement J Zablocki VA Medical Center

Clin Lab Scientist/Med Technologist Prgm
5000 W National Ave
Milwaukee, WI 53295
*Prgm Dir:* Mark J Maticek, MT(ASCP)
*Tel:* 414 384-2000, Ext 41332   *Fax:* 414 389-4187
*E-mail:* mark.maticek@med.va.gov
*Med Dir:* Bruce Edward Dunn, MD

## Marquette University

Clin Lab Scientist/Med Technologist Prgm
PO Box 1881
Milwaukee, WI 53201-1881
www.marquette.edu/chs/clls
*Prgm Dir:* Linda Milson, MS MT(ASCP)
*Tel:* 414 288-7566   *Fax:* 414 288-5847
*E-mail:* linda.milson@marquette.edu
*Med Dir:* Bruce Dunn, MD

## University of Wisconsin - Milwaukee

Clin Lab Scientist/Med Technologist Prgm
Department of Health Sciences
PO Box 413
Milwaukee, WI 53201
http://cfprod.imt.uwm.edu/chs/academics/
    undergraduate/healthsciences/medicaltechnology
*Prgm Dir:* Cindy Brown, MA(Ed) CLS(NCA) MT(ASCP)
*Tel:* 414 229-5299   *Fax:* 414 229-2619
*E-mail:* cbrown@uwm.edu

## University of Wisconsin - Stevens Point

Clin Lab Scientist/Med Technologist Prgm
D127 Science Bldg
Stevens Point, WI 54481-3897
www.uwsp.edu/hlthsci
*Prgm Dir:* Susan L Raab, EdD MT(ASCP)
*Tel:* 715 346-3777   *Fax:* 715 346-2640
*E-mail:* sraab@uwsp.edu
*Med Dir:* Lloyd Arnold, MD

## Aspirus Wausau Hospital

Clin Lab Scientist/Med Technologist Prgm
333 Pine Ridge Blvd
Wausau, WI 54401
www.aspirus.org/wausau_regional_lab/edu-main.php
*Prgm Dir:* Susan Flaker Johnson, MEd MT(ASCP)
*Tel:* 715 847-2000, Ext 53210   *Fax:* 715 847-2930
*E-mail:* susanj@aspirus.org
*Med Dir:* Edgar Betancourt, MD

## Clinical Laboratory Scientist/Medical Technologist

| Programs* | Class Capacity | Begins | Length (months) | Award | Res Tuition | Non-res Tuition | Stipend | Offers:‡ 1 | 2 | 3 |
|---|---|---|---|---|---|---|---|---|---|---|
| **Alabama** | | | | | | | | | | |
| Auburn University Montgomery | 50 | Aug | 24 | Cert | $6,132 | $18,096 | | | | |
| Baptist Medical Center South (Montgomery) | 8 | Jan Jul | 12 | Cert | | | | | | |
| Tuskegee University | 10 | Aug | 24 | BSCLS | $11,290 | $11,290 | | | | |
| University of Alabama at Birmingham | 24 | Aug | 21 | Cert, BS, MS | $5,457 | $10,914 | | | | |
| University of South Alabama (Mobile) | 14 | Aug | 21 | Dipl, BSCLS | $4,064 | $8,128 | | | | |
| **Arkansas** | | | | | | | | | | |
| Arkansas State University | 20 | Jun | 24 | BS | $142 | $365 | | | | |
| Baptist Health (Little Rock) | 8 | Jul | 12 | Cert | $3,800 | $7,600 | | | | |
| University of Arkansas for Medical Sciences (Little Rock) | 23 | Aug | 18, 12 | Dipl, BSMT, BSMT | $4,785 | $10,923 | | | | |

*Data are shown only for programs that completed the 2006 AMA Survey of Health Professions Education Programs
‡Key to Offers: 1: Evening or weekend classes; 2: Non-English instruction; 3: Cultural competence instruction

## Clinical Laboratory Scientist/Medical Technologist

| Programs* | Class Capacity | Begins | Length (months) | Award | Res Tuition | Non-res Tuition | Stipend | Offers:‡ 1 | 2 | 3 |
|---|---|---|---|---|---|---|---|---|---|---|
| **California** | | | | | | | | | | |
| California State University - Dominguez Hills (Carson) | 24 | Jul | 12 | Cert, BS | $2,986 | $9,715 | $6,000 | | | • |
| Eisenhower Memorial Hospital (Rancho Mirage) | 4 | Sep | 12 | Cert, MT/CLS | | | $12,000 | | | |
| Loma Linda University | 22 | Mid-Aug | 20 | Dipl, BS | $23,000 | $23,000 | | | | • |
| San Jose State University | 20 | Mar Sep | 12 | Cert | $5,000 | $12,000 | $12,000 | | | |
| Santa Barbara Cottage Hospital | 4 | Sep | 12 | Cert, MT | | | $12,000 | | | |
| Univ of California Davis Health System (Sacramento) | 12 | Sep | 12 | Cert | | | $1,000 | | | |
| Univ of California Irvine Med Ctr (Orange) | 6 | Sep | 12 | Cert | | | $6,000 | | | |
| **Colorado** | | | | | | | | | | |
| Centura Health/Penrose-St Francis Hlth Serv (Colorado Springs) | 4 | Aug | 12 | Cert | $1,000 | $1,000 | $900 | | | |
| Colorado Health Foundation (Denver) | 21 | Aug | 12 | Cert, BS | $6,840 | $6,840 | | | | |
| Parkview Medical Center (Pueblo) | 4 | Jul | 12 | Cert | | | | | | • |
| **Connecticut** | | | | | | | | | | |
| Danbury Hospital | 6 | Mid-Jun | 12 | Cert | | | | | | • |
| Hartford Hospital | 24 | Jul Jan | 12 | Cert | $5,000 | $5,000 | | | • | • |
| University of Hartford (West Hartford) | 16 | Jan Jun Sep | 12, 48 | Cert, MT, BS | $24,576 | $24,576 | | | | • |
| **Delaware** | | | | | | | | | | |
| University of Delaware (Newark) | 26 | Sep | 18 | BS | $6,614 | $16,770 | | | | |
| **District of Columbia** | | | | | | | | | | |
| George Washington University | 30 | Aug | 24, 12 | Cert, BS | $14,245 | $14,285 | | • | | • |
| Howard University (Washington) | 16 | Aug | 18 | BS | $10,160 | $10,160 | | | | |
| Walter Reed Army Medical Center (Washington) | 6 | Aug | 12 | Cert | | | | | | • |
| Washington Hospital Center | 14 | Aug | 12 | Cert, BS | $5,000 | $5,000 | $9,000 | | | |
| **Florida** | | | | | | | | | | |
| Bayfront Medical Center (St Petersburg) | 6 | Aug | 12 | Cert | $1,000 | $2,000 | | | | |
| Shands Jacksonville | 6 | Jul Jan | 12 | Cert | | | | | | |
| St Vincent's Medical Center (Jacksonville) | 10 | Jan Jul | 12 | Cert, BS | | | | | | |
| Tampa General Hospital | 6 | Aug | 12 | Cert | $1,500 | $1,500 | | | | |
| University of Central Florida (Orlando) | 20 | Aug | 24 | BS | $2,982 | $15,488 | | | | |
| University of West Florida (Pensacola) | 20 | Jan | 15 | BS | $2,782 | $13,520 | | | | |
| **Georgia** | | | | | | | | | | |
| Armstrong Atlantic State University (Savannah) | 25 | Aug | 16 | Cert, BS | $2,864 | $10,180 | | | | • |
| Medical College of Georgia (Augusta) | 50 | Aug May | 12, 24 | Cert, BS, BS Web | $5,457 | $21,000 | | • | | • |
| Thomas University (Thomasville) | 12 | Jan | 24 | BS | $12,000 | $12,000 | | • | | |
| **Hawaii** | | | | | | | | | | |
| University of Hawaii at Manoa (Honolulu) | 20 | Aug | 30 | Cert, BS | $3,408 | $9,888 | | | | |
| **Illinois** | | | | | | | | | | |
| Edward Hines Jr VA Hospital | 6 | Aug | 11 | Cert | | | | | | |
| Evanston Northwestern Healthcare - Evanston | 8 | Sep | 10 | Cert | $8,000 | $8,000 | | | | |
| Illinois State University (Normal) | 24 | Fall | 24 | BS | $6,734 | $12,302 | | | | |
| Northern Illinois University (DeKalb) | 32 | Aug | 18 | BS | $6,652 | $11,713 | | | | • |
| OSF St Anthony Medical Center (Rockford) | 4 | Aug | 9 | Cert, CLS/MT | | | | | | |
| OSF St Francis Medical Center (Peoria) | 6 | Aug | 11 | BS | $1,000 | $2,000 | | | | • |
| Rosalind Franklin Univ of Medicine & Science (North Chicago) | 20 | May Aug Mar | 24, 15 | BSMT, MSCLS | $14,098 | $14,098 | | | | |
| Rush University (Chicago) | 30 | Sep Jan | 20 | BS, MS | $15,600 | $0 | | | | • |
| St Elizabeth Hospital (Belleville) | 4 | Jul | 10 | Cert | $1,750 | $1,750 | | | | • |
| St John's Hospital (Springfield) | 6 | Jun | 11 | Cert | $800 | $800 | | | | |
| University of Illinois at Springfield | 20 | Aug | 21 | BS | $4,575 | $13,725 | | | | • |
| **Indiana** | | | | | | | | | | |
| Ball Memorial Hospital (Muncie) | 6-8 | Jul | 12 | BS | $3,000 | $3,000 | $1,000 | | | |
| Clarian Health Partners Inc (Indianapolis) | 24 | Jul | 11 | Cert, BS | $3,000 | $3,000 | | | | |
| Good Samaritan Hospital (Vincennes) | 6 | Aug | 12 | Cert, BS | | | $3,000 | | | |
| Indiana University (Indianapolis) | 12 | Aug | 11 | BS | $7,500 | $19,150 | | | | |
| Parkview Hospital (Ft Wayne) | 10 | Jul | 11 | Cert | $2,500 | $2,500 | | | | |
| St Francis Hospital & Health Centers (Beech Grove) | 8 | Jul | 11 | Cert, BA/BS | $3,000 | $3,000 | | | | • |
| St Margaret Mercy Healthcare Centers (Hammond) | 9 | Aug | 12 | Cert | $1,500 | $1,500 | | | | |
| **Iowa** | | | | | | | | | | |
| Mercy Medical Center (Des Moines) | 6 | Aug | 12 | Cert, BS | $2,500 | $2,500 | | | | |
| Mercy Medical Center-Sioux City | 4 | Aug | 12 | Cert, BS | $3,000 | $3,000 | | | | • |
| Mercy/St Luke's Hospital (Cedar Rapids) | 5 | Jul | 12 | Cert | $3,000 | $3,000 | | | | |
| St Luke's College (Sioux City) | 6 | Aug | 12 | Cert | $3,000 | $3,000 | | | | |
| University of Iowa Hospitals & Clinics (Iowa City) | 16 | May | 12 | Cert, BS | $7,512 | $23,817 | | | | |
| **Kansas** | | | | | | | | | | |
| University of Kansas Medical Center (Kansas City) | 30 | Aug | 24 | BS | $6,522 | $18,984 | | | | |
| **Kentucky** | | | | | | | | | | |
| Bellarmine University (Louisville) | 16 | Fall Spring | 20, 16 | Cert, BHS, BHS | $22,600 | $22,600 | | | | • |

*Data are shown only for programs that completed the 2006 AMA Survey of Health Professions Education Programs
‡Key to Offers: 1: Evening or weekend classes; 2: Non-English instruction; 3: Cultural competence instruction

### Clinical Laboratory Scientist/Medical Technologist

| Programs* | Class Capacity | Begins | Length (months) | Award | Res Tuition | Non-res Tuition | Stipend | Offers:‡ 1 | 2 | 3 |
|---|---|---|---|---|---|---|---|---|---|---|
| Eastern Kentucky University (Richmond) | 24 | Jan Aug | 17 | BS | $2,597 | $7,269 | | | | |
| Owensboro Medical Health System | 6 | Jul | 12 | Cert, BS | $1,200 | $1,200 | | | | |
| St Elizabeth Medical Center (Edgewood) | 4 | Jun | 12 | Cert | $3,500 | $3,500 | | | | • |
| University of Kentucky (Lexington) | 12 | Jul | 10 | Cert, BHS | $8,029 | $17,042 | | • | | |
| **Louisiana** | | | | | | | | | | |
| Lake Charles Mem Hosp Sch of Med Tech | 10 | Jan Aug | 12 | Cert | $50 | $0 | | | | |
| Louisiana State Univ Health Sciences Center (New Orleans) | 28 | Jan | 16 | Dipl, BS | $5,289 | $8,414 | | | | • |
| McNeese State University (Lake Charles) | 8 | Aug Jan | 12 | BS | | | | | | |
| Our Lady of the Lake College (Baton Rouge) | 10 | Jun | 18 | BS | $226 | $226 | | | | • |
| Overton Brooks VA Medical Center (Shreveport) | 8 | Jan Jul | 12 | Cert | $1,500 | $1,500 | | | | |
| Rapides Regional Medical Center (Alexandria) | 8 | Jan Jul | 12 | Cert, BS, BA | | | | | | |
| St Francis Medical Center (Monroe) | 8 | Jul Jan | 12 | Cert, BS | | | | | | |
| **Maine** | | | | | | | | | | |
| Eastern Maine Medical Center (Bangor) | 6 | Aug | 11 | Cert | $6,300 | $6,300 | | | | • |
| **Maryland** | | | | | | | | | | |
| Morgan State University (Baltimore) | 10 | Sep | 24 | Dipl, BS | $2,120 | $5,750 | | | | |
| Salisbury University | 15 | Sep | 18 | BS | $4,814 | $12,492 | | | | |
| University of Maryland (Baltimore) | 60 | Aug Jan | 18 | Cert, BSMRT, MSMRT | $6,981 | $13,962 | | | | |
| **Massachusetts** | | | | | | | | | | |
| Berkshire Medical Center (Pittsfield) | 6 | Jul | 12 | Cert, BS | $2,000 | $2,000 | | | | |
| Northeastern University (Boston) | 32 | Sep | 60, 24 | BS, MS | $29,910 | $29,910 | | | | |
| University of Massachusetts - Dartmouth (North Dartmouth) | 24 | Sep | 48 | BS | $8,036 | $17,536 | | | | |
| University of Massachusetts - Lowell | 24 | Sep | 36 | BS | $1,454 | $8,567 | | | | |
| **Michigan** | | | | | | | | | | |
| Andrews University (Berrien Springs) | 24 | Aug | 11 | Cert, BSCLS, MSCLS | | | | | | • |
| DMC University Laboratories (Detroit) | 10 | Aug Jan | 10 | Cert | | | $3,600 | | | |
| Ferris State University (Big Rapids) | 30 | May | 48 | Dipl, BS | $6,740 | $13,480 | | | | • |
| Grand Valley State University (Grand Rapids) | 20 | Jan | 17 | BS | $5,782 | $12,510 | | | | |
| Hurley Medical Center (Flint) | 10 | Sep | 10 | Cert | | | | | | |
| Michigan State University (East Lansing) | 17 | Aug | 30 | Cert, BS | $8,475 | $21,015 | | | | |
| Northern Michigan University (Marquette) | 10 | Aug Jan | 35 | BS | $5,328 | $9,072 | | • | | |
| St John Health (Grosse Pointe Woods) | 12 | Jul Aug | 11 | Cert | | | $5,500 | | | • |
| William Beaumont Hospital (Royal Oak) | 12 | Jul Jan | 11 | Cert, MT | | | | | | • |
| **Minnesota** | | | | | | | | | | |
| Fairview Health Services (Minneapolis) | 10 | Jun 1 | 12 | Cert | $5,000 | $0 | | | | • |
| Hennepin County Medical Center (Minneapolis) | 12 | Sep | 9 | Cert | $2,400 | $2,400 | | | | • |
| University of Minnesota - Minneapolis | 42 | Sep | 15 | BS, MS | $8,900 | $22,900 | | | | • |
| **Mississippi** | | | | | | | | | | |
| Mississippi Baptist Medical Center (Jackson) | 12 | Jun Sep | 12 | Cert | $1,000 | $1,000 | | | | |
| North Mississippi Medical Center (Tupelo) | 12 | Aug | 12 | Cert, BS | | | | | | |
| University of Mississippi Medical Center (Jackson) | 25 | Aug | 18 | BS | $3,232 | $6,904 | | | | |
| University of Southern Mississippi (Hattiesburg) | 15 | Aug Jan | 15 | Dipl, Cert, BS, MS | $4,312 | $9,742 | | | | |
| **Missouri** | | | | | | | | | | |
| CoxHealth (Springfield) | 6 | Jan Jun | 12 | Cert, CLS | $3,000 | $3,000 | | | | |
| North Kansas City Hospital | 5 | Jun | 11 | Cert | $1,000 | $1,000 | | | | |
| Saint Louis University (St Louis) | 25 | Aug Jan | 36 | BS | $26,250 | $26,250 | | | | |
| Saint Luke's Hospital (Kansas City) | 8 | Jul | 11 | Cert | $3,000 | $3,000 | | | | • |
| Southeast Missouri Hospital (Cape Girardeau) | 8 | Jul | 11 | Cert | $7,776 | $7,776 | | | | • |
| St John's Mercy Medical Center (St Louis) | 8 | Last week of Jun | 12 | Cert, BSMT, MT/CLS | $4,800 | $4,800 | | | | |
| St John's Regional Medical Center (Joplin) | 8 | Jan Jun | 12 | Cert, BS | $2,000 | $2,000 | | | | |
| **Montana** | | | | | | | | | | |
| Benefis Healthcare (Great Falls) | 4 | Jul | 12 | Cert | | | | | | |
| **Nebraska** | | | | | | | | | | |
| Nebraska Methodist Hospital (Omaha) | 10 | Jun | 11 | Cert, BS | $6,240 | $18,525 | | | | • |
| University of Nebraska Medical Center (Omaha) | 36 | Jun | 12 | BS | $6,181 | $19,219 | | | | • |
| **Nevada** | | | | | | | | | | |
| University of Nevada - Las Vegas | 14 | Aug | 24 | Dipl, BS | $2,868 | $12,334 | | | | • |
| **New Hampshire** | | | | | | | | | | |
| University of New Hampshire (Durham) | 25 | Sep | 49 | BS | $10,401 | $22,851 | | | | |
| **New Jersey** | | | | | | | | | | |
| Jersey Shore University Medical Center (Neptune) | 10 | Aug | 11 | Cert, BS | $3,000 | $3,000 | | | | |
| Monmouth Medical Center (Long Branch) | 8 | Aug | 11 | Cert, BS | $2,500 | $2,500 | | | | • |
| Morristown Memorial Hospital | 8 | Sep | 11 | Cert, BS | $2,000 | $2,000 | | | | |
| The Valley Hospital (Ridgewood) | 4 | Jun | 12 | Cert, BS | $2,000 | $2,000 | | | | |
| Univ of Med & Dent of New Jersey (Newark) | 24 | Jun | 15 | Cert, BS | $9,768 | $14,652 | | | | • |

*Data are shown only for programs that completed the 2006 AMA Survey of Health Professions Education Programs

‡Key to Offers: 1: Evening or weekend classes; 2: Non-English instruction; 3: Cultural competence instruction

## Clinical Laboratory Scientist/Medical Technologist

| Programs* | Class Capacity | Begins | Length (months) | Award | Res Tuition | Non-res Tuition | Stipend | Offers:‡ 1 | 2 | 3 |
|---|---|---|---|---|---|---|---|---|---|---|
| **New Mexico** | | | | | | | | | | |
| University of New Mexico (Albuquerque) | 25 | Jun Jan Sep | 18 | Cert, BS MLS | $4,200 | $15,000 | | | | • |
| **New York** | | | | | | | | | | |
| Long Island University - C W Post Campus (Brookville) | 20 | Sep | 24 | Dipl, Cert, BS | $22,100 | $22,100 | | • | | |
| Marist College (Poughkeepsie) | 30 | Sep Jan | 35 | BS | $22,066 | $22,066 | | | | |
| New York Methodist Hospital (Brooklyn) | 10 | Aug | 12 | Cert, BS | $6,200 | $6,200 | | | | |
| Rochester General Hospital | 18 | Aug | 12 | Cert, BS | $5,800 | $5,800 | | | | |
| St Vincent Catholic Med Ctr Brklyn-Queens Reg (Fresh Meadows) | 14 | Jun | 12 | Cert, BSMT | $7,000 | $7,000 | | | | |
| St Vincent's Hospital Manhattan SVCMC (New York) | 10 | Jun | 12 | Cert, BS | $6,800 | $6,800 | | | | |
| Stony Brook University | 25 | Aug | 20 | BS | $4,350 | $10,610 | | | | |
| SUNY Upstate Medical University (Syracuse) | 40 | Aug | 21 | BS | $4,350 | $10,610 | | | | |
| University at Buffalo - SUNY | 35 | Aug | 21 | BS, MS | $4,655 | $9,555 | | | | |
| Woman's Christian Association Hospital (Jamestown) | 10 | Aug | 11 | Cert | $5,800 | $5,800 | | | | • |
| **North Carolina** | | | | | | | | | | |
| Carolinas College of Health Sciences (Charlotte) | 12 | Aug Jan | 12 | Cert | $4,500 | $4,500 | | | | |
| East Carolina University (Greenville) | 16 | Aug | 21 | BS | $2,979 | $12,635 | | | | |
| Univ of North Carolina at Chapel Hill | 24 | Aug | 16, 20 | Cert, BS | $4,613 | $18,411 | | | | |
| Wake Forest University Baptist Medical Center (Winston-Salem) | 12 | Jul Jan | 12 | Cert, BS MT | $3,000 | $3,000 | | | | |
| Western Carolina University (Cullowhee) | 14 | Aug | 18 | BS | $3,945 | $13,528 | | | | |
| **North Dakota** | | | | | | | | | | |
| MeritCare Medical Center (Fargo) | 10 | Jun | 12 | Cert, BS | $2,468 | $2,886 | | | | • |
| University of North Dakota (Grand Forks) | 75 | May | 24, 12 | Dipl, Cert, BS | $6,400 | $6,400 | | • | | • |
| **Ohio** | | | | | | | | | | |
| Bowling Green State University | 14 | May | 14 | Cert, BS | $13,590 | $24,552 | | | | |
| Children's Hospital Medical Ctr of Akron | 9 | Jul | 12 | Cert | $4,000 | $4,000 | | | | |
| Ohio Northern University (Ada) | 8 | Jun | 12 | Cert, BS | $8,400 | $8,400 | | | | |
| Ohio State University (Columbus) | 70 | Sep | 22 | Cert, BS | $8,667 | $20,562 | | | | |
| Southwest General Health Center (Middleburg Heights) | 6 | Jul | 12 | Cert | $2,500 | $2,500 | | | | |
| St Vincent Mercy Medical Center (Toledo) | 6 | Jul | 12 | Cert | $500 | $0 | | | | |
| University of Cincinnati Medical Center | 20* | Sep rolling | 12, 26 | Dipl, Cert, BS | $12,508 | $31,872 | | • | | • |
| Wright State University (Dayton) | 12 | Jun | 12 | Cert, BSCLS | $7,278 | $14,004 | | | | • |
| **Oklahoma** | | | | | | | | | | |
| Comanche County Memorial Hospital (Lawton) | 6 | Aug | 12 | Cert | $1,250 | $0 | | | | |
| Muskogee Regional Medical Center | 7 | Jun | 12 | Cert | | | | | | |
| Saint Francis Hospital (Tulsa) | 8 | Jun | 12 | Cert | $2,400 | $2,400 | | | | |
| Valley View Regional Hospital (Ada) | 6 | Late May | 12 | Cert, BS | | | | | | • |
| **Oregon** | | | | | | | | | | |
| Oregon Health & Science University (Portland) | 26 | Sep | 15 | BS | $14,728 | $23,872 | | | | |
| **Pennsylvania** | | | | | | | | | | |
| Altoona Regional Health System | 8 | Jul | 12 | Dipl, BS | $1,000 | $1,000 | | | | |
| Conemaugh Memorial Medical Center (Johnstown) | 8 | Jul | 12 | Cert | $5,900 | $5,900 | | | | |
| Graduate Hospital (Philadelphia) | 8 | Aug | 11 | Cert, MT-BS | $10,000 | $10,000 | | | | |
| Lancaster Gen Coll of Nursing & Hlth Sciences | 12 | Aug | 12 | Cert | $10,500 | $10,500 | | | | • |
| Neumann College (Aston) | 15 | Sep | 24 | BS | $17,992 | $17,992 | | | | |
| Pennsylvania Hospital (Philadelphia) | 6 | Sep | 11 | Cert | $7,000 | $7,000 | | | | • |
| Reading Hospital & Medical Center | 4 | Jul | 11 | Cert | $4,000 | $4,000 | | | | • |
| Robert Packer Hospital (Sayre) | 10 | Aug | 11 | Cert | $5,400 | $5,400 | | | | |
| Saint Vincent Health Center (Erie) | 12 | Aug | 12 | Cert, BS | $6,885 | $6,885 | | | | |
| Thomas Jefferson University (Philadelphia) | 35 | Sep | 12, 22 | BS | $24,000 | $24,000 | | | | |
| Williamsport Hosp & Medical Center | 4 | Aug | 12 | Cert | $3,500 | $3,500 | | | | |
| York Hospital | 6 | Jul | 12 | Cert | $6,500 | $6,500 | | | | • |
| **Puerto Rico** | | | | | | | | | | |
| Inter - American University - Metro Campus (San Juan) | 25 | Feb Aug | 12, 36 | Dipl, Cert, BS, MS | $5,750 | $5,750 | | • | • | |
| Inter American University of Puerto Rico (San German) | 18 | Aug Feb | 12 | Dipl, CLS, BSMT | $5,750 | $5,750 | | | | |
| University of Puerto Rico (San Juan) | 40 | Aug | 17 | Cert, BSMT | $2,550 | $0 | | | • | • |
| **Rhode Island** | | | | | | | | | | |
| Our Lady of Fatima Hospital (North Providence) | 6 | Jul | 11 | Cert | $4,500 | $4,500 | | | | |
| **South Carolina** | | | | | | | | | | |
| Lexington Medical Center (West Columbia) | 4 | Aug | 12 | Cert | | | 2000 | | | |
| McLeod Regional Medical Center (Florence) | 8 | Aug | 12 | Cert | $2,000 | $2,000 | | | | |
| Palmetto Health Baptist (Columbia) | 4 | Jan Jul | 12 | Cert | | | | | | |
| **South Dakota** | | | | | | | | | | |
| Rapid City Regional Hospital | 10 | Jun | 12 | Cert, BS | $3,000 | $3,000 | | | | |
| Sioux Valley Hospital (Sioux Falls) | 10 | Jun | 11 | Cert, BA/BS | $4,000 | $4,000 | | | | • |
| **Tennessee** | | | | | | | | | | |
| Austin Peay State University (Clarksville) | 18 | Jun May | 12 | Cert, BS | $4,932 | $17,316 | | | | |

*Data are shown only for programs that completed the 2006 AMA Survey of Health Professions Education Programs

‡Key to Offers: 1: Evening or weekend classes; 2: Non-English instruction; 3: Cultural competence instruction

## Clinical Laboratory Scientist/Medical Technologist

| Programs* | Class Capacity | Begins | Length (months) | Award | Res Tuition | Non-res Tuition | Stipend | Offers:‡ 1 | 2 | 3 |
|---|---|---|---|---|---|---|---|---|---|---|
| Lincoln Memorial University (Harrogate) | 12 | Aug | 24 | BS | $13,104 | $13,104 | | • | | |
| Tennessee State University (Nashville) | 14 | Aug | 12 | Cert, BS | $6,621 | $20,589 | | | | |
| Univ of Tennessee Medical Ctr at Knoxville | 8 | Jan | 12 | Cert, BS | $8,000 | $23,000 | | | | |
| University of Tennessee Health Science Ctr (Memphis) | 20 | Sep | 21, 48 | BS, MS | $6,384 | $20,952 | | | | |
| Vanderbilt University Medical Center (Nashville) | 10 | Jun | 12 | Cert | $5,000 | $5,000 | | | | |
| **Texas** | | | | | | | | | | |
| Austin State Hospital | 4 | Sep | 12 | Cert | $300 | $0 | $600 | | | |
| Christus Hospital - St Elizabeth (Beaumont) | 6 | Aug | 12 | Cert | $300 | $300 | | | | |
| Hillcrest Baptist Medical Center (Waco) | 5 | Jan Jul | 12 | Cert | | | | | | |
| Methodist Hospital (Houston) | 4 | Aug | 12 | Cert, BSMT | | | $2,400 | | | |
| Scott & White Memorial Hospital (Temple) | 8 | Aug | 12 | Cert | $500 | $500 | | | | |
| Texas A&M University - Corpus Christi | 12 | Aug | 12 | Cert, BS | $6,950 | $18,550 | | | | |
| Texas Southern University (Houston) | 18 | Aug | 24 | BS | $3,098 | $8,760 | | | | • |
| Texas State University-San Marcos | 20 | Sep | 24 | BS | $6,586 | $16,471 | | | | |
| Texas Tech Univ Health Sciences Center (Lubbock) | 38 | Aug | 21 | BS | $4,500 | $9,104 | | | | • |
| United Regional Health Care Systems (Wichita Falls) | 6 | Jul | 12 | Cert, BS | | | $1,500 | | | |
| Univ of Texas - Pan American (Edinburg) | 18 | Sep | 15 | Cert, BS | $116 | $391 | | | | • |
| Univ of Texas Hlth Sci Ctr at San Antonio | 30 | Aug Jan | 19 | Cert, BS, MS | $3,230 | $11,610 | | | | • |
| Univ of Texas M D Anderson Cancer Ctr (Houston) | 20 | Aug | 12 | Dipl, Cert, BS, CLS | $3,750 | $15,500 | | | | • |
| Univ of Texas Southwestern Med Ctr at Dallas | 24 | Aug | 16, 28 | Cert, BSMT | $1,200 | $7,824 | | | | |
| University of Texas at El Paso | 20 | May | 22 | BS | $6,600 | $13,050 | | | | |
| University of Texas Medical Branch/SAHS (Galveston) | 40 | Aug | 24 | Cert, BS | $3,000 | $12,000 | | • | • | • |
| **Utah** | | | | | | | | | | |
| Brigham Young University (Provo) | 14 | Sep Jan | 12 | Dipl, BS | $3,620 | $7,240 | | | | |
| University of Utah Health Science Center (Salt Lake City) | 30 | Aug | 21 | BS | $1,800 | $4,000 | | | | |
| Weber State University (Ogden) | 40 | Aug | 36 | Cert, BS | $2,772 | $8,612 | | | | • |
| **Vermont** | | | | | | | | | | |
| University of Vermont (Burlington) | 24 | Sep | 48 | Cert, BS | $9,452 | $23,638 | | | | |
| **Virginia** | | | | | | | | | | |
| Augusta Medical Center (Fishersville) | 5 | Jul | 12 | Cert | $900 | $900 | | | | |
| Carilion Medical Center (Roanoke) | 12 | Jul | 12 | Cert | $6,500 | $6,500 | | | | |
| Norfolk State University | 11 | Aug | 18 | BS | $5,572 | $17,956 | | | | • |
| Rockingham Memorial Hospital (Harrisonburg) | 8 | Jun | 12 | Cert, BS | $635 | $635 | | | | |
| Virginia Commonwealth University (Richmond) | 35 | Aug | 20 | BS | $5,765 | $17,562 | | | | • |
| **Washington** | | | | | | | | | | |
| Sacred Heart Medical Center (Spokane) | 12 | Jul | 12 | Cert | | | | | | • |
| **West Virginia** | | | | | | | | | | |
| West Liberty State College | 16 | Aug | 24 | BS | $3,686 | $9,054 | | | | |
| West Virginia University (Morgantown) | 29 | Aug | 21 | BS | $4,982 | $15,628 | | | | |
| **Wisconsin** | | | | | | | | | | |
| Affinity Health System (Appleton) | 6 | Sep | 9 | Cert | $1,500 | $1,500 | | | | • |
| Aspirus Wausau Hospital | 7 | Aug | 9 | Cert | $1,000 | $1,000 | | | | • |
| Clement J Zablocki VA Medical Center (Milwaukee) | 10 | Jul | 10 | Cert | | | | | | |
| Marquette University (Milwaukee) | 96 | Fall semester | 37 | BS | $22,950 | $22,950 | | | | |
| Sacred Heart Hospital (Eau Claire) | 4 | Aug | 9 | Cert | $850 | $1,200 | | | | |
| St Joseph's Hospital (Marshfield) | 12 | Aug | 9 | Cert, BS/MT | $850 | $850 | | | | • |
| University of Wisconsin - Madison | 24 | Sep | 18 | BS | $6,730 | $20,730 | | | | |
| University of Wisconsin - Milwaukee | 20 | Sep | 35 | BS, MS | $6,224 | $18,976 | | | | • |
| University of Wisconsin - Stevens Point | 24 | Fall | 6 | BSCLS | | | | | | |

*Data are shown only for programs that completed the 2006 AMA Survey of Health Professions Education Programs
‡Key to Offers: 1: Evening or weekend classes; 2: Non-English instruction; 3: Cultural competence instruction

## Clinical Laboratory Technician/Medical Laboratory Technician*

## Alabama

**Jefferson State Community College**
Clin Lab Technician/Med Lab Technician
2601 Carson Rd
Birmingham, AL 35215
www.jeffstateonline.com
*Prgm Dir:* Candy Hill, MA Ed MT(ASCP) CLS(NCA)
*Tel:* 205 856-6031   *Fax:* 205 856-7725
*E-mail:* chill@jeffstateonline.com
*Med Dir:* Donald R Cantley, MD

**Gadsden State Community College**
Clin Lab Technician/Med Lab Technician
PO Box 227
1001 George Wallace Dr
Gadsden, AL 35902-0227
www.gadsdenstate.edu
*Prgm Dir:* Sunita M Graves, MS MT(ASCP)
*Tel:* 256 549-8470   *Fax:* 256 549-8272
*E-mail:* sgraves@gadsdenstate.edu
*Med Dir:* John B Priest, MD

**Wallace State Community College**
Clin Lab Technician/Med Lab Technician
801 Main St
PO Box 2000
Hanceville, AL 35077-2000
*Prgm Dir:* Julie Welch, MS MT(ASCP)
*Tel:* 256 352-8327   *Fax:* 256 352-8320
*Med Dir:* James Lester Newsome, Jr, MD

## Alaska

**University of Alaska Anchorage**
Clin Lab Technician/Med Lab Technician
3211 Providence Dr
Anchorage, AK 99508-4610
http://alliedhealth.uaa.alaska.edu
*Prgm Dir:* Heidi Mannion, PhD MT(ASCP)
*Tel:* 907 786-6924, Ext 102   *Fax:* 907 786-6938
*E-mail:* afham@uaa.alaska.edu

## Arkansas

**Arkansas State University - Beebe**
Clin Lab Technician/Med Lab Technician
PO Box 1000
Beebe, AR 72012-1000
www.asub.edu
*Prgm Dir:* Jimmy L Boyd, MS MHS CLS(NCA)
*Tel:* 501 882-8214   *Fax:* 501 882-8387
*E-mail:* jlboyd@asub.edu
*Med Dir:* John R Brineman, MD

**South Arkansas Community College**
Clin Lab Technician/Med Lab Technician
300 S West Ave
PO Box 7010
El Dorado, AR 71730
www.southark.edu
*Prgm Dir:* Oliver Borden, MS MT(ASCP)
*Tel:* 870 862-8131, Ext 102   *Fax:* 870 864-7122
*E-mail:* oborden@southark.edu

**North Arkansas College**
Clin Lab Technician/Med Lab Technician
1515 Pioneer Dr
Harrison, AR 72601-5599
www.northark.edu
*Prgm Dir:* Sherry Gibbany, MA MT(ASCP)
*Tel:* 870 391-3288   *Fax:* 870 743-5326
*E-mail:* sgibbany@northark.edu
*Med Dir:* Robert L Miller, MD PhD

**Phillips Community College/U of Arkansas**
Clin Lab Technician/Med Lab Technician
PO Box 785
1000 Campus Dr
Helena, AR 72342
www.pccua.edu
*Prgm Dir:* Julie Byrd, MS MT(ASCP)
*Tel:* 870 338-6474, Ext 1109   *Fax:* 870 338-7542
*E-mail:* jbyrd@pccua.edu
*Med Dir:* John R Brineman, MD

**National Park Community College**
Clin Lab Technician/Med Lab Technician
101 College Dr
Hot Springs, AR 71913-9174
www.npcc.edu
*Prgm Dir:* Carol A Spargo, BS MEd MT(ASCP)
*Tel:* 501 760-4130   *Fax:* 501 760-4141
*E-mail:* cspargo@npcc.edu
*Med Dir:* Jorge F Jimenez, MD

**Arkansas State University**
Clin Lab Technician/Med Lab Technician
PO Box 910
State University, AR 72467
www.clt.astate.edu/cls
*Prgm Dir:* R Whitney Williams, PhD MT(ASCP)
*Tel:* 870 972-3073   *Fax:* 870 972-2004
*E-mail:* wwilliam@astate.edu
*Med Dir:* Terri Green, MD

## California

**Hartnell College**
Clin Lab Technician/Med Lab Technician
156 Homestead Ave
Salinas, CA 93901-1697
*Prgm Dir:* Susan A McQuiston, BS MS JD MT(ASCP)
*Tel:* 831 770-6152   *Fax:* 831 770-6144
*E-mail:* smcquist@hartnell.edu
*Med Dir:* Andrew Wilson, MD

**Naval School of Health Sciences, San Diego**
Clin Lab Technician/Med Lab Technician
34101 Farenholt Ave
San Diego, CA 92134-5291
http://nshssd.med.navy.mil
*Prgm Dir:* Luis A Nunez Jr, LT MSC USN MA MT(ASCP)
*Tel:* 619 532-7330   *Fax:* 619 532-7145
*E-mail:* lanunez@nshs-sd.med.navy.mil
*Med Dir:* Michael Quigley, MD CDR MC USN

## Colorado

**Arapahoe Community College**
Clin Lab Technician/Med Lab Technician
5900 S Santa Fe Dr, PO Box 9002
Littleton, CO 80160-9002
www.arapahoe.edu
*Prgm Dir:* Jennifer Wolff, MT (ASCP)
*Tel:* 303 797-5796   *Fax:* 303 797-5935
*E-mail:* jennifer.wolff@arapahoe.edu

## Connecticut

**Housatonic Community College**
Clin Lab Technician/Med Lab Technician
900 Lafayette Blvd
Bridgeport, CT 06604-4704
www.hcc.commnet.edu
*Prgm Dir:* Phyllis J Gutowski, MS MT(ASCP)
*Tel:* 203 332-5106   *Fax:* 203 332-5123
*E-mail:* pgutowski@hcc.commnet.edu
*Med Dir:* Angelique Levi, MD

## Delaware

**Delaware Tech & Comm Coll - Owens Campus**
Clin Lab Technician/Med Lab Technician
PO Box 610
Georgetown, DE 19947
*Prgm Dir:* Sheridan A Shupe, MEd MT(ASCP)
*Tel:* 302 855-5974   *Fax:* 302 858-5470
*E-mail:* sshupe@college.dtcc.edu
*Med Dir:* William A Diedrich, Jr, MD

## Florida

**Brevard Community College**
Clin Lab Technician/Med Lab Technician
1519 Clearlake Rd
Cocoa, FL 32922
www.brevardcc.edu
*Prgm Dir:* Celine Marilyn Hulme, MEd MT(ASCP)
*Tel:* 321 433-7543   *Fax:* 321 433-7599
*E-mail:* hulmem@brevardcc.edu
*Med Dir:* Carl Smedberg, MD

**Keiser University**
Clin Lab Technician/Med Lab Technician
1500 NW 49th St
Ft Lauderdale, FL 33309
www.keisercollege.edu
*Prgm Dir:* Lureen Samuel, MS BS BS MT(ASCP)
*Tel:* 954 776-4456, Ext 431   *Fax:* 954 776-5157
*E-mail:* lureens@keisercollege.edu

**Indian River Community College**
Clin Lab Technician/Med Lab Technician
3209 Virginia Ave
Ft Pierce, FL 34981
*Prgm Dir:* Marilyn Barbour, MEd MT(ASCP) CLS(NCA)
*Tel:* 772 462-7534   *Fax:* 772 462-7816
*E-mail:* mbarbour@ircc.edu

**Florida Community College - Jacksonville**
Clin Lab Technician/Med Lab Technician
North Campus, 4501 Capper Rd
Jacksonville, FL 32218
*Prgm Dir:* Rhoda S Jost, MS MT(ASCP)
*Tel:* 904 766-6580   *Fax:* 904 766-6654
*E-mail:* rjost@fccj.edu

**Miami Dade College**
Clin Lab Technician/Med Lab Technician
Medical Center Campus
950 NW 20th St
Miami, FL 33127
www.mdc.edu
*Prgm Dir:* Nilia Madan, MBA MT(ASCP)SH
*Tel:* 305 237-4041   *Fax:* 305 237-4278
*E-mail:* nilia.madan@mdc.edu
*Med Dir:* Susan R Baker, MD PhD

**St Petersburg College**
Clin Lab Technician/Med Lab Technician
PO Box 13489
St Petersburg, FL 33733-3489
www.spcollege.edu/hec/medlab
*Prgm Dir:* Valerie Polansky, MEd MT(ASCP)
*Tel:* 813 341-3714   *Fax:* 813 341-3740
*E-mail:* polansky.valerie@spcollege.edu
*Med Dir:* Rehana Nawab, MD

*Additional information about programs that
returned the AMA's survey is available in a table
beginning on page 74.

**Erwin Technical Center**
Clin Lab Technician/Med Lab Technician
2010 E Hillsborough Ave
Tampa, FL 33610-8299
www.erwin.edu
*Prgm Dir:* Linda J Dickson, MT(ASCP) MSPH
*Tel:* 813 231-1800, Ext 1326   *Fax:* 813 231-1820
*E-mail:* dickson_l@firn.edu
*Med Dir:* John Shively, MD

# Georgia

**Darton College**
Clin Lab Technician/Med Lab Technician
2400 Gillionville Rd
Albany, GA 31707
www.darton.edu/programs/AlliedHealth/mlt
*Prgm Dir:* Nancy T Beamon, MS MT(ASCP)
*Tel:* 912 430-6846   *Fax:* 912 430-6910
*E-mail:* beamonn@darton.edu
*Med Dir:* Frank Isele, MD

**Coastal Georgia Community College**
Clin Lab Technician/Med Lab Technician
3700 Altama Ave
Brunswick, GA 31520
*Prgm Dir:* Katherine Nisi Zell, EdS MT(ASCP) SH
   CLS/NCA
*Tel:* 912 264-7382   *Fax:* 912 262-3283
*E-mail:* nzell@cgcc.edu
*Med Dir:* Patrick E T Godbey, MD

**North Georgia Technical College**
Clin Lab Technician/Med Lab Technician
Hwy 197 N, PO Box 65
1500 Highway 197 N
Clarkesville, GA 30523
www.northgatech.edu
*Prgm Dir:* Lauren Strader, MEd MT(ASCP)
*Tel:* 706 754-7757   *Fax:* 706 754-7777
*E-mail:* lstrader@northgatech.edu
*Med Dir:* Susan T Martin, MD

**DeKalb Technical College**
Clin Lab Technician/Med Lab Technician
495 N Indian Creek Dr
Clarkston, GA 30021-2397
www.dekalbtech.org
*Prgm Dir:* Virginia D Roberts, MEd MT(AMT)
*Tel:* 404 297-9522, Ext 1160   *Fax:* 404 294-4234
*E-mail:* robertsg@dekalbtech.org
*Med Dir:* William F McNeill, MD

**Dalton State College**
Clin Lab Technician/Med Lab Technician
650 College Dr
Dalton, GA 30720-3778
http://www.daltonstate.edu
*Prgm Dir:* Tyra D Stalling, BS MSHS CLS(NCA)
   MT(ASCP)
*Tel:* 706 272-2508   *Fax:* 706 272-2517
*Med Dir:* Eugene Fong, MD

**Central Georgia Technical College**
Clin Lab Technician/Med Lab Technician
3300 Macon Tech Dr
Macon, GA 31206
www.centralgatech.edu
*Prgm Dir:* Tony Dugan, MS MT(ASCP)DLM
*Tel:* 478 757-3571   *Fax:* 478 757-3575
*E-mail:* tdugan@centralgatech.edu
*Med Dir:* Robert S Donner, MD

**Lanier Technical College**
Clin Lab Technician/Med Lab Technician
2990 Landrum Education Dr
Oakwood, GA 30566
*Prgm Dir:* Kimberly A Randolph, MT(ASCP) MS
*Tel:* 770 531-6367   *Fax:* 770 531-6306
*E-mail:* krandolph@laniertech.edu
*Med Dir:* Earl Joseph Conway, MD

**Southwest Georgia Technical College**
Clin Lab Technician/Med Lab Technician
15689 US Hwy 19 N
Thomasville, GA 31792
*Prgm Dir:* Richard Miller, PhD MBA MT(ASCP)CM
   CLS(NCA)
*Tel:* 229 225-5203   *Fax:* 229 225-5289
*E-mail:* rmiller@southwestgatech.edu
*Med Dir:* Jeff W Byrd, MD

**Valdosta Technical College**
Clin Lab Technician/Med Lab Technician
4089 Val Tech Rd
Valdosta, GA 31602-0929
www.valdostatech.edu
*Prgm Dir:* Angela Robinson, MEd MT(AMT) MLT(ASCP)
*Tel:* 229 249-4855
*E-mail:* arobinson@valdostatech.edu

**Southeastern Technical College**
Clin Lab Technician/Med Lab Technician
3001 E First St
Vidalia, GA 30474
*Prgm Dir:* Charlotte Bates, MEd MT(ASCP)
*Tel:* 912 538-3183
*E-mail:* cbates@southeasterntech.edu

**West Central Technical College**
Clin Lab Technician/Med Lab Technician
178 Murphy Campus Blvd
Waco, GA 30182
www.westcentraltech.edu
*Prgm Dir:* John W Ridley, PhD MT(ASCP)
*Tel:* 770 537-6043   *Fax:* 770 537-7992
*E-mail:* jridley@westcentraltech.edu

**Okefenokee Technical College**
Clin Lab Technician/Med Lab Technician
1701 Carswell Ave
Waycross, GA 31501
www.okefenokeetech.edu
*Prgm Dir:* Amber G Tuten, MEd MT(ASCP) CLDir(NCA)
*Tel:* 912 287-5838   *Fax:* 912 287-4865
*E-mail:* atuten@okefenokeetech.edu
*Med Dir:* Dwight Mirmow, MD

# Hawaii

**Kapi'olani Community College**
Clin Lab Technician/Med Lab Technician
4303 Diamond Head Rd
Honolulu, HI 96816
*Prgm Dir:* Marcia A Armstrong, MS CLS(NCA)
   MT(ASCP)
*Tel:* 808 734-9231   *Fax:* 808 734-9126
*E-mail:* marciaa@hawaii.edu

# Illinois

**Southwestern Illinois College**
Clin Lab Technician/Med Lab Technician
2500 Carlyle Rd
Belleville, IL 62221-5899
www.southwestern.cc.il.us
*Prgm Dir:* Jean M Deitz, MS MT(ASCP)
*Tel:* 618 235-2700, Ext 5386   *Fax:* 618 235-2052
*E-mail:* Jean.Deitz@swic.edu

**Oakton Community College**
Clin Lab Technician/Med Lab Technician
1600 E Golf Rd
Des Plaines, IL 60016
*Prgm Dir:* Lynne Steele, MS MT(ASCP)
*Tel:* 847 635-1889   *Fax:* 847 635-1987
*E-mail:* lynne@oakton.edu
*Med Dir:* Ebrahim Amir-Mokri, MD

**Elgin Community College**
Clin Lab Technician/Med Lab Technician
1700 Spartan Dr
Elgin, IL 60123-7193
www.elgin.edu
*Prgm Dir:* Wendy Miller, MS CLS(NCA) MT(ASCP)SI
*Tel:* 847 214-7308   *Fax:* 847 214-7527
*E-mail:* wmiller@elgin.edu

**Southern Illinois Collegiate Common Market**
Clin Lab Technician/Med Lab Technician
3213 S Park Ave
Herrin, IL 62948
www.siccm.com
*Prgm Dir:* Paula Ann Berry, MS Ed MT(ASCP)
*Tel:* 618 942-6902   *Fax:* 618 942-6658
*E-mail:* pberry@siccm.com

**Kankakee Community College**
Clin Lab Technician/Med Lab Technician
River Rd, PO Box 888
Kankakee, IL 60901
www.kcc.edu
*Prgm Dir:* Manuela Sawalha, MHS MT(ASCP)
*Tel:* 815 802-8835   *Fax:* 815 802-8101
*E-mail:* nsawalha@kankakee.edu

**Illinois Central College**
Clin Lab Technician/Med Lab Technician
Health and Public Svcs Bldg
201 SW Adams St
Peoria, IL 61635-0001
www.icc.edu
*Prgm Dir:* Anh Strow, MPH MT(ASCP) CLS(NCA)
*Tel:* 309 999-4661   *Fax:* 309 673-9626
*E-mail:* astrow@icc.edu
*Med Dir:* Marvin Schmidt, MD

**Blessing Hospital**
Clin Lab Technician/Med Lab Technician
Broadway at 11th St
Quincy, IL 62305
www.blessinghospital.com
*Prgm Dir:* Heather Ator, MT(ASCP)SM
*Tel:* 217 223-8400, Ext 6205   *Fax:* 217 223-7032
*E-mail:* hator@blessinghospital.org
*Med Dir:* Robert Merrick, MD

# Indiana

**Indiana University Northwest**
Clin Lab Technician/Med Lab Technician
3400 Broadway
Gary, IN 46408
www.iun.edu
*Prgm Dir:* Susan A K Higgins, MS MT(ASCP)SC
*Tel:* 219 980-6923   *Fax:* 219 980-6649
*E-mail:* sahiggin@iun.edu

**Ivy Tech Community College - South Bend**
Clin Lab Technician/Med Lab Technician
220 Dean Johnson Blvd
South Bend, IN 46601-3415
*Prgm Dir:* Pamela B Primrose, MLFSC MT(ASCP)
*Tel:* 574 289-7001, Ext 5401   *Fax:* 574 236-7166
*E-mail:* pprimros@ivytech.edu
*Med Dir:* Robert J Tomec, MD

**Ivy Tech Community College - Terre Haute**
Clin Lab Technician/Med Lab Technician
7999 US Hwy 41 S
Terre Haute, IN 47802-4898
www.goivytech.net
*Prgm Dir:* Janee Gambill, MS MT(ASCP)
*Tel:* 812 298-2243   *Fax:* 812 298-2392
*E-mail:* Jgambill@ivytech.edu
*Med Dir:* M Bashar Kashlar, MD

# Iowa

**Des Moines Area Community College**
Clin Lab Technician/Med Lab Technician
2006 Ankeny Blvd, Bldg 9
Ankeny, IA 50023
www.dmacc.cc.ia.us/programs/medlabtech
*Prgm Dir:* Karen Campbell, MAT MT(ASCP) CLS(NCA)
*Tel:* 515 964-6296   *Fax:* 515 964-6440
*E-mail:* kjcampbell@dmacc.edu

**Iowa Central Community College**
Clin Lab Technician/Med Lab Technician
330 Ave M
Ft Dodge, IA 50501
www.iccc.cc.ia.us/health/files/labtech.htm
*Prgm Dir:* Diane C Edwards, BS MT(ASCP)
*Tel:* 515 576-0099, Ext 2394   *Fax:* 515 576-5656
*E-mail:* EDWARDS@triton.iccc.cc.ia.us
*Med Dir:* Richard Wyatt, MD

**Hawkeye Community College**
Clin Lab Technician/Med Lab Technician
1501 E Orange Rd
PO Box 8015
Waterloo, IA 50704
*Prgm Dir:* Amy R Kapanka, MS MT(ASCP)
*Tel:* 319 296-2329, Ext 1357   *Fax:* 319 296-2874
*E-mail:* akapanka@hawkeyecollege.edu
*Med Dir:* Alan K Brown, MD

# Kansas

**Barton County Community College**
Clin Lab Technician/Med Lab Technician
245 NE 30th Rd
Great Bend, KS 67530-9283
www.bartonccc.edu/mlt/home
*Prgm Dir:* Leonard Bunselmeyer Jr, MS MT(ASCP)
*Tel:* 316 792-2701, Ext 393   *Fax:* 316 786-1164
*E-mail:* bunselmeyerl@bartonccc.edu

**Seward County Community College**
Clin Lab Technician/Med Lab Technician
PO Box 1137
Liberal, KS 67901
www.sccc.edu
*Prgm Dir:* Suzanne Campbell, MS BS MT(ASCP)
*Tel:* 620 626-3077   *Fax:* 620 626-3040
*E-mail:* scampbel@sccc.edu

**Wichita Area Technical College**
Clin Lab Technician/Med Lab Technician
324 N Emporia St
Wichita, KS 67202-2512
www.watc.edu
*Prgm Dir:* Barbara C Wenger, MS MT(ASCP)
*Tel:* 316 677-1378   *Fax:* 316 677-1310
*E-mail:* bwenger@watc.edu

# Kentucky

**HCC/MCC Consortium**
*Cosponsor:* Madisonville Community College
Clin Lab Technician/Med Lab Technician
2000 College Dr
Madisonville, KY 42431
www.madisonville.kctcs.edu
*Prgm Dir:* Karol Conrad, MS MT(ASCP)
*Tel:* 270 824-1741   *Fax:* 270 824-1879
*E-mail:* karol.conrad@kctcs.edu
*Med Dir:* Dr Justin Sedlak, MD

**Southeast Kentucky Comm & Tech College**
Clin Lab Technician/Med Lab Technician
3300 US Highway 25E South
Pineville, KY 40977
*Prgm Dir:* Sheila Gibbs Miracle, MEd CLS(NCA)
H(ASCP)
*Tel:* 606 248-0946   *Fax:* 606 248-3361
*E-mail:* sheila.miracle@kctcs.edu
*Med Dir:* Ann Marshall, MD

**Eastern Kentucky University**
Clin Lab Technician/Med Lab Technician
Dizney 220
521 Lancaster Ave
Richmond, KY 40475-3102
*Prgm Dir:* David C Hufford, PhD MT(ASCP)
*Tel:* 859 622-3078   *Fax:* 859 622-1140
*E-mail:* david.hufford@eku.edu

**Somerset Community College**
Clin Lab Technician/Med Lab Technician
808 Monticello St
Somerset, KY 42501
*Prgm Dir:* Nancy W Powell, MAEd MT(ASCP)
*Tel:* 606 451-6842, Ext 16842   *Fax:* 606 679-3684
*E-mail:* nancy.powell@kctcs.edu
*Med Dir:* Marilyn M McMillen, MD

# Louisiana

**Louisiana State University - Alexandria**
Clin Lab Technician/Med Lab Technician
8100 Hwy 71 S
Alexandria, LA 71302-9121
*Prgm Dir:* Sheryl B Herring, MSA MT(ASCP)
*Tel:* 318 473-6464   *Fax:* 318 473-6588
*E-mail:* slienhop@lsua.edu
*Med Dir:* William E Roberts, MD

**MedVance Institute**
Clin Lab Technician/Med Lab Technician
9255 Interline Ave
Baton Rouge, LA 70809
*Prgm Dir:* Laverne Floyd, MA MT(ASCP)
*Tel:* 225 248-1015   *Fax:* 225 248-9517
*E-mail:* laverne@multipro.com
*Med Dir:* Edgar Shannon Cooper, MD JD

**Our Lady of the Lake College**
Clin Lab Technician/Med Lab Technician
7434 Perkins Rd
Baton Rouge, LA 70808
www.ololcollege.edu
*Prgm Dir:* Todd Casanova, PhD MT(ASCP)
*Tel:* 225 768-1745   *Fax:* 225 768-0819
*E-mail:* tcasanov@ololcollege.edu

**Louisiana Tech College - Lafayette Campus**
Clin Lab Technician/Med Lab Technician
PO Box 4909
Lafayette, LA 70502
www.lafayette.tec.la.us
*Prgm Dir:* Margaret B Boone, BS MT(ASCP) SBB
*Tel:* 337 262-5962, Ext 249   *Fax:* 337 262-5122
*E-mail:* mboone@theltc.net

**Delgado Community College**
Clin Lab Technician/Med Lab Technician
615 City Park Ave
New Orleans, LA 70119
www.dcc.edu
*Prgm Dir:* Sheila M Hickman, MEd MT(ASCP)SBB
*Tel:* 504 483-4189   *Fax:* 504 483-4363
*E-mail:* shickm@dcc.edu
*Med Dir:* Francis Rodwig, MD

**Southern Univ at Shreveport**
Clin Lab Technician/Med Lab Technician
3050 Martin Luther King Jr Dr
Shreveport, LA 71107
www.susla.edu
*Prgm Dir:* Patricia Raphiel-Brown, MA CLS(AMT)
*Tel:* 318 674-3350   *Fax:* 318 676-5495
*E-mail:* pbrown@susla.edu
*Med Dir:* Warren D Grafton, MD

# Maine

**Central Maine Community College**
Clin Lab Technician/Med Lab Technician
1250 Turner St
Auburn, ME 04210
*Prgm Dir:* Valerie V Ferrante, MS MT(ASCP)
*Tel:* 207 755-5420   *Fax:* 207 755-5496
*E-mail:* vferrante@cmcc.edu
*Med Dir:* Douglas Pohl, MD PhD

**University of Maine - Presque Isle**
*Cosponsor:* University of Maine - Augusta
Clin Lab Technician/Med Lab Technician
181 Main St
Presque Isle, ME 04769
www.uma.edu
*Prgm Dir:* Linda Graves, EdD MT(ASCP)
*Tel:* 207 768-9451   *Fax:* 207 768-9553
*E-mail:* graves@umpi.maine.edu
*Med Dir:* John Benzinger, MD

# Maryland

**Allegany College of Maryland**
Clin Lab Technician/Med Lab Technician
12401 Willowbrook Rd SE
Cumberland, MD 21502-2596
*Prgm Dir:* Mary H Saunders-Bloom, MEd MT(ASCP)
*Tel:* 301 784-5548   *Fax:* 301 784-5015
*E-mail:* msaunders@allegany.edu
*Med Dir:* Jonathan R Walburn, MD

# Massachusetts

**Bristol Community College**
Clin Lab Technician/Med Lab Technician
777 Elsbree St
Fall River, MA 02720
www.bristol.mass.edu
*Prgm Dir:* Sandra G Campos, MS CLS
*Tel:* 508 678-2811, Ext 2148   *Fax:* 508 730-3281
*E-mail:* scampos@bristol.mass.edu
*Med Dir:* David Ziemba, MD

**Springfield Technical Community College**
Clin Lab Technician/Med Lab Technician
One Armory Sq
PO Box 9000
Springfield, MA 01105-1204
*Prgm Dir:* Susan Schneider, MEd MT(ASCP)
*Tel:* 413 755-4846   *Fax:* 413 755-6312
*E-mail:* saschneider@stcc.edu
*Med Dir:* Krystyna Sikorska, MD

# Michigan

**Kellogg Community College**
Clin Lab Technician/Med Lab Technician
450 North Ave
Battle Creek, MI 49017
www.kellogg.edu/alliedhealth/mlt/mltmain.html
*Prgm Dir:* Kathleen T Paff, MA CLS(NCA) MT(ASCP)
*Tel:* 269 965-3931, Ext 2316   *Fax:* 269 565-2059
*E-mail:* paffk@kellogg.edu
*Med Dir:* Christopher Flynn, MD

**Ferris State University**
Clin Lab Technician/Med Lab Technician
VFS210A
200 Ferris Dr
Big Rapids, MI 49307-2740
www.ferris.edu/cls
*Prgm Dir:* Barbara A Ross, MS MT(ASCP)
*Tel:* 231 591-2317   *Fax:* 231 591-3788
*E-mail:* rossb@ferris.edu
*Med Dir:* Nicholas J Hruby, MD

**Northern Michigan University**
Clin Lab Technician/Med Lab Technician
3513 West Science Bldg
Marquette, MI 49855
*Prgm Dir:* Lucille A Contois, MA MT(ASCP)
*Tel:* 906 227-1660   *Fax:* 906 227-1309
*E-mail:* lcontois@nmu.edu
*Med Dir:* John D Weiss, MD

**Baker College of Owosso**
Clin Lab Technician/Med Lab Technician
1020 S Washington St
Owosso, MI 48867
www.baker.edu
*Prgm Dir:* Mary Vuckovich, MBA MT(ASCP)
*Tel:* 789 729-3416   *Fax:* 789 729-3411
*E-mail:* mary.vuckovich@baker.edu
*Med Dir:* Qazi Azher, MD

# Minnesota

**Alexandria Technical College**
Clin Lab Technician/Med Lab Technician
1601 Jefferson St
Alexandria, MN 56308
www.alextech.org/MedLabTech&Phlebotomy
*Prgm Dir:* Wanda Haberer, MS MT(ASCP) CLS(NCA)
*Tel:* 320 762-0221   *Fax:* 320 762-4501
*E-mail:* wandah@alextech.edu
*Med Dir:* Sharon Banister, MD

**North Hennepin Community College**
Clin Lab Technician/Med Lab Technician
7411 85th Ave N
Brooklyn Park, MN 55445
www.nhcc.edu
*Prgm Dir:* Nancy C Denny, MAEd MT(ASCP)
*Tel:* 763 424-0768   *Fax:* 763 493-0571
*E-mail:* nancy.denny@nhcc.edu
*Med Dir:* Cynthia J Lais, MD

**Lake Superior College**
Clin Lab Technician/Med Lab Technician
2101 Trinity Rd
Duluth, MN 55811
*Prgm Dir:* Mary Grace Werner, MSEd MT(ASCP)
   CLS(NCA)
*Tel:* 218 733-7679   *Fax:* 218 723-4921
*E-mail:* m.werner@lsc.mnscu.edu
*Med Dir:* Geoffrey A Witrak, MD

**Argosy University/Twin Cities**
Clin Lab Technician/Med Lab Technician
Twin Cities Campus
1515 Central Parkway
Eagan, MN 55121
www.argosyu.edu
*Prgm Dir:* Kevin Swanson, DC MT(ASCP)
*Tel:* 651 846-3555   *Fax:* 651 994-0144
*E-mail:* kswanson@argosyu.edu

**Northland Community & Technical College**
Clin Lab Technician/Med Lab Technician
2022 Central Ave NE
East Grand Forks, MN 56721
www.northlandcollege.edu
*Prgm Dir:* Paula Bowman, MS MT(ASCP)
*Tel:* 218 773-2149   *Fax:* 218 773-4502
*E-mail:* paula.bowman@northlandcollege.edu
*Med Dir:* Albert Marvin Cooley, MD

**South Central College**
Clin Lab Technician/Med Lab Technician
1225 3rd St SW
Faribault, MN 55021
www.southcentral.edu
*Prgm Dir:* Darla Petersen, MA BS
*Tel:* 507 332-5852   *Fax:* 507 332-5888
*E-mail:* darla.petersen@southcentral.edu

**Minnesota State Community and Technical Coll**
Clin Lab Technician/Med Lab Technician
1414 College Way
Fergus Falls, MN 56537
www.minnesota.edu
*Prgm Dir:* Patricia Sjolie, MS MT(ASCP)
*Tel:* 218 736-1594   *Fax:* 218 736-1510
*E-mail:* pat.sjolie@minnesota.edu
*Med Dir:* Gregory M Smith, MD

**Hibbing Community College**
Clin Lab Technician/Med Lab Technician
1515 E 25th St
Hibbing, MN 55746
www.hcc.mnscu.edu
*Prgm Dir:* Mitzi Morris, MT(ASCP)
*Tel:* 218 262-7254   *Fax:* 218 262-6717
*E-mail:* mitzimorris@hibbing.edu

**St Paul College**
Clin Lab Technician/Med Lab Technician
235 Marshall Ave
St Paul, MN 55102-1807
www.saintpaul.edu
*Prgm Dir:* Michelle Briski, MEd MT(ASCP) CLS(NCA)
*Tel:* 651 846-1421   *Fax:* 651 221-1416
*E-mail:* michelle.briski@saintpaul.edu
*Med Dir:* Virginia Dale, MD

**Minnesota West Comm & Tech College**
Clin Lab Technician/Med Lab Technician
1450 Collegeway Dr
Worthington, MN 56187
*Prgm Dir:* Rita MIller, MS (CLS) MT(ASCP)
*Tel:* 507 372-3422   *Fax:* 507 825-4656
*E-mail:* rmiller@wr.mnwest.mnscu.edu

# Mississippi

**Northeast Mississippi Community College**
Clin Lab Technician/Med Lab Technician
101 Cunningham Blvd
Booneville, MS 38829
www.nemcc.edu
*Prgm Dir:* Rilla Jones, MEd MT(ASCP) SM
*Tel:* 601 720-7388   *Fax:* 601 728-1165
*E-mail:* rcjones@nemcc.edu
*Med Dir:* Michael Todd, MD

**Mississippi Gulf Coast Community College**
Clin Lab Technician/Med Lab Technician
2300 Hwy 90
PO Box 100
Gautier, MS 39553
*Prgm Dir:* Peggy Caldwell, MA MT(ASCP)
*Tel:* 228 497-7846   *Fax:* 228 497-7676
*E-mail:* peggy.caldwell@mgccc.edu
*Med Dir:* Lyman J Scripter, MD

**Pearl River Community College**
Clin Lab Technician/Med Lab Technician
5448 US Hwy 49 S
Hattiesburg, MS 39401
www.prcc.edu
*Prgm Dir:* Evelyn Wallace, BS MT(ASCP)
*Tel:* 601 554-5523   *Fax:* 601 554-5511
*E-mail:* ewallace@prcc.edu
*Med Dir:* Timothy L Cole, MD

**Hinds Community College**
Clin Lab Technician/Med Lab Technician
Nursing/Allied Hlth Ctr, 1750 Chadwick Dr
Jackson, MS 39204-3402
*Prgm Dir:* Timothy Henry, BS MT(ASCP)
*Tel:* 601 371-3515   *Fax:* 601 371-3529
*E-mail:* Tghenry@hindscc.edu
*Med Dir:* Barbara Proctor, MD

**Meridian Community College**
Clin Lab Technician/Med Lab Technician
910 Hwy 19 N
Meridian, MS 39307
www.meridiancc.edu/mlt
*Prgm Dir:* Cassandra Johnson, MHS MT(ASCP)
*Tel:* 601 484-8754   *Fax:* 601 484-8743
*E-mail:* cjohnson@meridiancc.edu
*Med Dir:* F M Phillippi, MD

**Mississippi Delta Community College**
Clin Lab Technician/Med Lab Technician
PO Box 668
Moorhead, MS 38761
www.msdelta.edu
*Prgm Dir:* Patricia Kelly, MT(AMT)(ASCP)BB MBA
*Tel:* 662 246-6500   *Fax:* 662 246-6507
*E-mail:* pkelly@msdelta.edu
*Med Dir:* Joyce J Bradshaw, MD

**Copiah-Lincoln Community College**
Clin Lab Technician/Med Lab Technician
PO Box 457
Wesson, MS 39191-0457
www.colin.edu
*Prgm Dir:* Mary E Shivers, MT(ASCP)
*Tel:* 601 643-8391   *Fax:* 601 643-8214
*E-mail:* mary.shivers@colin.edu
*Med Dir:* Robert B Britt, MD

# Missouri

**Three Rivers Community College**
Clin Lab Technician/Med Lab Technician
2080 Three Rivers Blvd
Poplar Bluff, MO 63901
www.trcc.edu
*Prgm Dir:* Dionne M Thompson, MSE MT(ASCP)
*Tel:* 573 840-9677, Ext 677   *Fax:* 573 840-9657
*E-mail:* deethomp@trcc.edu
*Med Dir:* Robert J Cacchione, MD

**St Louis Community College - Forest Park**
Clin Lab Technician/Med Lab Technician
5600 Oakland Ave
St Louis, MO 63110
www.stlcc.edu
*Prgm Dir:* Karen M Kiser, MA MT(ASCP) PBT
*Tel:* 314 644-9645   *Fax:* 314 644-9752
*E-mail:* kkiser@stlcc.edu

# Nebraska

**Central Community College**
Clin Lab Technician/Med Lab Technician
PO Box 1024
E Hwy 6
Hastings, NE 68902-1024
www.cccneb.edu
*Prgm Dir:* Lori Van Boening, MT(ASCP)
*Tel:* 402 461-2451
*E-mail:* lvanboening@cccneb.edu

**Southeast Community College**
Clin Lab Technician/Med Lab Technician
8800 O St
Lincoln, NE 68520
www.southeast.edu
*Prgm Dir:* Janis K Bible, BA MT(ASCP)
*Tel:* 402 437-2760   *Fax:* 402 437-2404
*E-mail:* jbible@southeast.edu
*Med Dir:* Aina I Silenieks, MD

**Mid-Plains Community College**
Clin Lab Technician/Med Lab Technician
North Campus
1101 Halligan Dr
North Platte, NE 69101
www.mpcc.edu/page.cfm?action=show_section&
pg_id=73#
*Prgm Dir:* Martin D Steinbeck, MT(ASCP)
*Tel:* 308 535-3754   *Fax:* 308 535-3794
*E-mail:* steinbeckm@mpcc.edu
*Med Dir:* Byron Barksdale, MD

# Nevada

**Community College of Southern Nevada**
Clin Lab Technician/Med Lab Technician
6375 W Charleston Blvd
Las Vegas, NV 89146
www.ccsn.nevada.edu
*Prgm Dir:* Patricia R Castro, EdD MS MT(ASCP)BB
*Tel:* 702 651-5819   *Fax:* 702 651-5501
*E-mail:* patricia_castro@ccsn.edu

# New Hampshire

**New Hampshire Comm Tech Coll - Claremont**
Clin Lab Technician/Med Lab Technician
1 College Dr
Claremont, NH 03743
www.claremont.nhctc.edu
*Prgm Dir:* Andrea Gordon, MEd MT(ASCP)SH
*Tel:* 603 542-7744, Ext 2536   *Fax:* 603 543-1844
*E-mail:* agordon@nhctc.edu
*Med Dir:* Suellen Balestra, MD

# New Jersey

**Camden County College**
Clin Lab Technician/Med Lab Technician
PO Box 200
College Dr
Blackwood, NJ 08012
*Prgm Dir:* Patricia A Chappell, MA MT(ASCP)
*Tel:* 856 227-7200, Ext 4330   *Fax:* 856 374-4981
*E-mail:* pchappell@camdencc.edu
*Med Dir:* William Harrer, MD

**Middlesex County College**
Clin Lab Technician/Med Lab Technician
2600 Woodbridge Ave
Edison, NJ 08818-3050
www.middlesexcc.edu
*Prgm Dir:* Stephen P Larkin III, MHSA MT(ASCP)SH
*Tel:* 732 906-2581
*E-mail:* slarkin@middlesexcc.edu
*Med Dir:* Frederick C Skvara, MD

**Mercer County Community College**
Clin Lab Technician/Med Lab Technician
PO Box B
1200 Old Trenton Rd
Trenton, NJ 08690
www.mccc.edu
*Prgm Dir:* Jane O'Reilly, MEd MT(ASCP)
*Tel:* 609 586-4800, Ext 3387   *Fax:* 609 689-0762
*E-mail:* oreilly@mccc.edu
*Med Dir:* Richard H Siderits, MD

# New Mexico

**New Mexico State U at Alamogordo**
Clin Lab Technician/Med Lab Technician
2400 N Scenic Dr
Alamogordo, NM 88310
www.nmsua.nmsu.edu
*Prgm Dir:* Judith J O'Brien, BS MT(ASCP) MM
*Tel:* 505 439-3761   *Fax:* 505 439-3759
*E-mail:* jjobrien@nmsua.nmsu.edu
*Med Dir:* William Gordon McGee, MD

**Central New Mexico Community College**
Clin Lab Technician/Med Lab Technician
525 Buena Vista SE
Albuquerque, NM 87106
www.cnm.edu
*Prgm Dir:* Monya Kmetz, MA MT(ASCP)
*Tel:* 505 224-5021   *Fax:* 505 224-5033
*E-mail:* monya@cnm.edu
*Med Dir:* Michael J Crossey, MD PhD

**University of New Mexico - Gallup**
Clin Lab Technician/Med Lab Technician
1901 Mariyana Ave
Gallup, NM 87301
*Prgm Dir:* Michael Nye, MT(ASCP) MSM
*Tel:* 505 863-7598   *Fax:* 505 863-7513
*E-mail:* mnye@gallup.unm.edu
*Med Dir:* James Hathaway, MD

# New York

**Broome Community College**
Clin Lab Technician/Med Lab Technician
PO Box 1017
Binghamton, NY 13902
*Prgm Dir:* Andrea C Wade, PhD MT(ASCP)
*Tel:* 607 778-5211
*E-mail:* wade_a@sunybroome.edu
*Med Dir:* Loren Wolsh, MD

**Farmingdale State University of New York**
Clin Lab Technician/Med Lab Technician
2350 Broadhollow Rd, Gleeson Hall, Rm 304
Farmingdale, NY 11735
*Prgm Dir:* Karen M Escolas, EdD MT(ASCP)
*Tel:* 631 420-2257   *Fax:* 631 420-2784
*E-mail:* Karen.Escolas@Farmingdale.edu
*Med Dir:* Virginia Donovan, MD

**Orange County Community College**
Clin Lab Technician/Med Lab Technician
115 South St
Middletown, NY 10940
www.sunyorange.edu
*Prgm Dir:* Rosamaria B Contarino, MS MT(ASCP)
CLS(NCA)
*Tel:* 845 341-4136   *Fax:* 845 341-4122
*E-mail:* rcontari@sunyorange.edu
*Med Dir:* Joseph B Naplitano, MD

**Clinton Community College**
Clin Lab Technician/Med Lab Technician
Lake Shore Rd, Rte 9 South
136 Clinton Point Dr
Plattsburgh, NY 12901
www.clinton.edu
*Prgm Dir:* Sharon T Columbus, MEd MT(ASCP)
*Tel:* 518 562-4273   *Fax:* 518 562-4158
*E-mail:* Sharon.Columbus@clinton.edu
*Med Dir:* Mike Ladwig, MD

**Dutchess Community College**
Clin Lab Technician/Med Lab Technician
53 Pendell Rd
Poughkeepsie, NY 12601
www.sunydutchess.edu
*Prgm Dir:* Karen Ann Ingham, MT(ASCP)
*Tel:* 845 431-8321   *Fax:* 845 431-8329
*E-mail:* ingham@sunydutchess.edu
*Med Dir:* Neela Pushparaj, MD

**Erie Community College - North Campus**
Clin Lab Technician/Med Lab Technician
6205 Main St
Williamsville, NY 14221
www.ecc.edu
*Prgm Dir:* Marcia T Bermel, MS MT(ASCP)
*Tel:* 716 851-1553   *Fax:* 716 851-1550
*E-mail:* bermel@ecc.edu
*Med Dir:* Adrian Vladutiu, MD PhD

# North Carolina

**Asheville-Buncombe Technical Comm College**
Clin Lab Technician/Med Lab Technician
340 Victoria Rd
Asheville, NC 28801
www.abtech.edu
*Prgm Dir:* Melissa Hyatt, BS MT(ASCP) MHS
*Tel:* 828 254-1921, Ext 266   *Fax:* 828 281-9734
*E-mail:* Mhyatt@abtech.edu
*Med Dir:* Joseph Patrick Sleater, MD

**Central Piedmont Community College**
Clin Lab Technician/Med Lab Technician
PO Box 35009
Charlotte, NC 28235-5009
*Prgm Dir:* Rebecca C Sanders, MSA MT(ASCP)SM
*Tel:* 704 330-5028   *Fax:* 704 330-6131
*E-mail:* becky.sanders@cpcc.edu
*Med Dir:* William Karnes Poston Jr, MD

**Alamance Community College**
Clin Lab Technician/Med Lab Technician
PO Box 8000
Graham, NC 27253-8000
www.alamance.cc.nc.us
*Prgm Dir:* Pamela E Hall, MA MT(ASCP)SBB
*Tel:* 336 506-4196   *Fax:* 336 578-1987
*E-mail:* pam.hall@alamance.cc.nc.us
*Med Dir:* Mary S Olney, MD

**Coastal Carolina Community College**
Clin Lab Technician/Med Lab Technician
444 Western Blvd
Jacksonville, NC 28546-6877
*Prgm Dir:* Christine N Weaver, MS MT(ASCP)
CLS(NCA)
*Tel:* 910 938-6275   *Fax:* 910 938-6806
*E-mail:* weaverc@coastal.cc.nc.us
*Med Dir:* Charles L Garrett Jr, MD

**Davidson County Community College**
Clin Lab Technician/Med Lab Technician
PO Box 1287
Lexington, NC 27293-1287
www.davidsonccc.edu
*Prgm Dir:* Suzanne Rohrbaugh, MA MT(ASCP)
*Tel:* 336 249-8186, Ext 6721   *Fax:* 336 249-9060
*E-mail:* srohr@davidsonccc.edu
*Med Dir:* Guillermo L Restrepo, MD

**Western Piedmont Community College**
Clin Lab Technician/Med Lab Technician
1001 Burkemont Ave
Morganton, NC 28655-0680
*Prgm Dir:* Nancy Shoaf, MA MT(ASCP)
*Tel:* 828 438-6128   *Fax:* 828 430-7183
*E-mail:* nshoaf@wpcc.edu
*Med Dir:* James Parker, MD

**Sandhills Community College**
Clin Lab Technician/Med Lab Technician
3395 Airport Rd
Pinehurst, NC 28374
www.sandhills.edu
*Prgm Dir:* Cheryl P McCormick, PhD MT(ASCP)
*Tel:* 910 695-3839   *Fax:* 910 692-6918
*E-mail:* mccormickc@sandhills.edu
*Med Dir:* Jerry Huey, MD

**Wake Technical Community College**
Clin Lab Technician/Med Lab Technician
9101 Fayetteville Rd
Raleigh, NC 27603
*Prgm Dir:* Pamela Horton, MEd BS MT(ASCP)
*Tel:* 919 212-3818   *Fax:* 919 250-4329
*E-mail:* pbhorton@waketech.edu

**Southwestern Community College**
Clin Lab Technician/Med Lab Technician
447 College Dr
Sylva, NC 28779
www.southwesterncc.edu/mlt
*Prgm Dir:* Andrea Rowland, MBA MT(ASCP)
*Tel:* 828 586-4091, Ext 312    *Fax:* 828 586-3129
*E-mail:* andrea@southwesterncc.edu
*Med Dir:* Michael B Rohlfing, MD

**Beaufort County Community College**
Clin Lab Technician/Med Lab Technician
PO Box 1069
Highway 264 East
Washington, NC 27889
www.beaufort.cc.nc.us
*Prgm Dir:* Arthur S Keehnle, MS MT(ASCP)
*Tel:* 252 946-6194    *Fax:* 252 946-0271
*E-mail:* artk@email.beaufort.cc.nc.us
*Med Dir:* Allan R Smith, MD

**Halifax Community College**
Clin Lab Technician/Med Lab Technician
PO Box 809
Weldon, NC 27890
*Prgm Dir:* Randy C Harris, PhD MT(ASCP)
*Tel:* 804 786-4370    *Fax:* 804 536-4144
*E-mail:* rcharris@hanover.k12.va.us
*Med Dir:* Kamlesh Gupta, MD

**Southeastern Community College**
Clin Lab Technician/Med Lab Technician
PO Box 151
Whiteville, NC 28472
www.southeastern.cc.nc.us
*Prgm Dir:* Patricia Wright, MT(ASCP)
*Tel:* 910 642-7141, Ext 312    *Fax:* 910 642-3257
*E-mail:* pwright@sccnc.edu
*Med Dir:* Mett Ausley, MD

# North Dakota

**Bismarck State College**
Clin Lab Technician/Med Lab Technician
1500 Edwards Ave
Bismarck, ND 58501
www.bismarckstate.edu
*Prgm Dir:* Angela Uhlich, MS MT(ASCP)SBB
*Tel:* 701 323-5482    *Fax:* 701 323-5831
*E-mail:* Angela.Uhlich@bsc.nodak.edu
*Med Dir:* Dwight Hertz, MD

# Ohio

**Cincinnati State Tech & Comm College**
Clin Lab Technician/Med Lab Technician
3520 Central Pkwy
Cincinnati, OH 45223
www.cincinnatistate.edu
*Prgm Dir:* A Janelle Gohn, PhD MT(ASCP)SM
*Tel:* 513 569-1688    *Fax:* 513 487-1688
*E-mail:* janelle.gohn@cincinnatistate.edu

**Cuyahoga Community College**
Clin Lab Technician/Med Lab Technician
Metro Campus Health Careers and Science Bldg 126I
2900 Community College Ave
Cleveland, OH 44115-3196
www.tri-c.edu/MLT/docs/contact.htm
*Prgm Dir:* Amy Gatautis, MT(ASCP)SC MBA
*Tel:* 216 987-4438, Ext 4438    *Fax:* 216 987-4386
*E-mail:* amy.gatautis@tri-c.edu

**Columbus State Community College**
Clin Lab Technician/Med Lab Technician
550 E Spring St
Columbus, OH 43216-1609
www.cscc.edu
*Prgm Dir:* Leslie King, PhD MT(ASCP)
*Tel:* 614 287-2597    *Fax:* 614 287-5144
*E-mail:* lking01@cscc.edu
*Med Dir:* Rose Goodwin, MD

**Lorain County Community College**
Clin Lab Technician/Med Lab Technician
1005 N Abbe Rd
Elyria, OH 44035
*Prgm Dir:* James E Daly, MEd MT(ASCP)
*Tel:* 440 366-7194, Ext 7194    *Fax:* 440 366-4116
*E-mail:* jdaly@loraincc.edu
*Med Dir:* Priscilla Heimann, MD

**Lakeland Community College**
Clin Lab Technician/Med Lab Technician
7700 Clocktower Dr
Kirtland, OH 44094-5198
www.lakelandcc.edu
*Prgm Dir:* Kathryn G Ertter, MSHS MT(ASCP)
*Tel:* 440 525-7169    *Fax:* 440 975-4733
*E-mail:* kertter@lakelandcc.edu
*Med Dir:* David J Keep, MD

**Washington State Community College**
Clin Lab Technician/Med Lab Technician
710 Colegate Dr
Marietta, OH 45750
www.wscc.edu
*Prgm Dir:* Heather Kincaid, MEd MT(ASCP)
*Tel:* 740 374-8716, Ext 1674    *Fax:* 740 373-7496
*E-mail:* hkincaid@wscc.edu
*Med Dir:* F R Macatol, MD

**Marion Technical College**
Clin Lab Technician/Med Lab Technician
1467 Mt Vernon Ave
Marion, OH 43302
http://mtc.edu
*Prgm Dir:* Deborah L Bates, MBA MT(ASCP)SBB
*Tel:* 740 389-4636, Ext 254    *Fax:* 740 725-4018
*E-mail:* batesd@mtc.edu

**Stark State College of Technology**
Clin Lab Technician/Med Lab Technician
6200 Frank Ave NW
North Canton, OH 44720
www.starkstate.edu
*Prgm Dir:* Kozy Corsaut, MS Ed MT(ASCP)
*Tel:* 330 494-6170, Ext 4221    *Fax:* 330 966-6586
*E-mail:* kcorsaut@starkstate.edu
*Med Dir:* G Feszcko, MD

**Shawnee State University**
Clin Lab Technician/Med Lab Technician
940 Second St
Portsmouth, OH 45662
*Prgm Dir:* Marla H Thoroughman, MS MA MT(ASCP)
*Tel:* 740 351-3388    *Fax:* 740 351-3354
*E-mail:* mthoroughman@shawnee.edu

**University of Rio Grande**
Clin Lab Technician/Med Lab Technician
218 N College Ave, F-39
Rio Grande, OH 45674-0500
*Prgm Dir:* Keith Searls, BS MT(ASCP)
*Tel:* 740 245-7319    *Fax:* 740 245-7440
*E-mail:* keith.searls@med.va.gov
*Med Dir:* Frederic La Carbonara, MD

**Clark State Community College**
Clin Lab Technician/Med Lab Technician
570 E Leffel Ln
Springfield, OH 45501-0570
www.clarkstate.edu
*Prgm Dir:* Sandra Jean Horn, MS MT(ASCP)
*Tel:* 937 328-8077    *Fax:* 937 328-6138
*E-mail:* htttp://www.horns@clarkstate.edu
*Med Dir:* Robert V Stewart, MD

**Jefferson Community College**
Clin Lab Technician/Med Lab Technician
4000 Sunset Blvd
Steubenville, OH 43952
www.jcc.edu
*Prgm Dir:* Sondra Sutherland, MEd MT(ASCP)
  CLS(NCA)
*Tel:* 740 264-5591, Ext 165    *Fax:* 740 264-1338
*E-mail:* ssutherlan@jcc.edu
*Med Dir:* Souheil J Nassar, MD

**Youngstown State University**
Clin Lab Technician/Med Lab Technician
Dept of Health Professions
One University Plaza
Youngstown, OH 44555
www.ysu.edu
*Prgm Dir:* Maria E Delost, PhD MT(ASCP) CLS(NCA)
*Tel:* 330 941-1761    *Fax:* 330 941-2921
*E-mail:* medelost@ysu.edu
*Med Dir:* Norton German, MD

**Zane State College**
Clin Lab Technician/Med Lab Technician
1555 Newark Rd
Zanesville, OH 43701
www.zanestate.edu
*Prgm Dir:* Vicki Huntsman, MS MT(ASCP)
*Tel:* 740 588-1311, Ext 1311    *Fax:* 740 454-0035
*E-mail:* vhunstman@zanestate.edu

# Oklahoma

**Northeastern Oklahoma A&M College**
Clin Lab Technician/Med Lab Technician
Second and I Sts NE
Miami, OK 74354
www.neoam.edu
*Prgm Dir:* Rita Kay Harris, MS MT(ASCP)NM
*Tel:* 918 540-6315    *Fax:* 918 540-6471
*E-mail:* kharris@neoam.edu
*Med Dir:* Tammy M Battaglia, MD

**Rose State College**
Clin Lab Technician/Med Lab Technician
6420 SE 15th St
Midwest City, OK 73110
*Prgm Dir:* Evelyn Paxton, MS MT(ASCP)
*Tel:* 405 733-7577    *Fax:* 405 736-0338
*E-mail:* epaxton@rose.edu
*Med Dir:* Kenneth Blick, PhD

**Seminole State College**
Clin Lab Technician/Med Lab Technician
PO Box 351
2701 Boren Blvd
Seminole, OK 74818-0351
*Prgm Dir:* Perthena Latchaw, MS MT(ASCP)
*Tel:* 405 382-9214    *Fax:* 405 382-3122
*E-mail:* latchaw_p@sscok.edu
*Med Dir:* Levi Jones, MD

**Tulsa Community College**
Clin Lab Technician/Med Lab Technician
909 S Boston Ave
Tulsa, OK 74119
*Prgm Dir:* Karen L Holmes, MA MT(ASCP)
*Tel:* 918 595-7008    *Fax:* 918 595-7091
*E-mail:* kholmes@tulsacc.edu
*Med Dir:* Melvin J VanBoven, DO

# Oregon

**Portland Community College**
Clin Lab Technician/Med Lab Technician
PO Box 19000
Portland, OR 97280-0990
www.pcc.edu
*Prgm Dir:* Carol A Enyart, MPA MT(ASCP) CLS(NCA)
*Tel:* 503 978-5195    *Fax:* 503 413-2767
*E-mail:* cenyart@pcc.edu
*Med Dir:* Juan Millan, MD

# Pennsylvania

**Montgomery County Community College**
Clin Lab Technician/Med Lab Technician
340 DeKalb Pike
PO Box 400
Blue Bell, PA 19422-1412
www.mc3.edu
*Prgm Dir:* Debra Lynn Eckman, MS MT(ASCP)
*Tel:* 215 641-6487   *Fax:* 215 619-7178
*E-mail:* deckman@mc3.edu
*Med Dir:* Irwin J Hollander, MD

**Harcum College**
Clin Lab Technician/Med Lab Technician
750 Montgomery Ave
Bryn Mawr, PA 19010
www.harcum.edu
*Prgm Dir:* Donna M Broderick, MS MT(ASCP)CLS(NCA)
*Tel:* 610 526-6662   *Fax:* 610 526-6031
*E-mail:* dbroderick@harcum.edu
*Med Dir:* Albert A Keshgegian, MD PhD

**Harrisburg Area Community College**
Clin Lab Technician/Med Lab Technician
One HACC Dr
Harrisburg, PA 17110-2999
*Prgm Dir:* Ruth A Negley, MEd MT(ASCP)SM CLS(NCA)
*Tel:* 717 337-3855   *Fax:* 717 780-2551
*E-mail:* ranegley@hacc.edu

**Penn State University - Hazleton**
Clin Lab Technician/Med Lab Technician
76 University Dr
Hazleton, PA 18202
www.hn.psu.edu/Academics/MLTassoc.htm?cn21
*Prgm Dir:* Patricia D Ferry, MS MT(ASCP)
*Tel:* 570 450-3090   *Fax:* 570 450-3182
*E-mail:* pdf1@psu.edu

**Community College of Philadelphia**
Clin Lab Technician/Med Lab Technician
1700 Spring Garden St
Philadelphia, PA 19130
*Prgm Dir:* Robin G Krefetz, MEd MT(ASCP) CLS(NCA)
*Tel:* 215 751-8511   *Fax:* 215 751-8937
*E-mail:* rkrefetz@ccp.edu
*Med Dir:* Behnaz C Toorkey, MD

**Reading Area Community College**
Clin Lab Technician/Med Lab Technician
Ten S Second St, PO Box 1706
Reading, PA 19603
www.racc.edu
*Prgm Dir:* Alayne H Fessler, BA CLS(NCA)
*Tel:* 610 372-4721, Ext 5428   *Fax:* 610 607-6254
*E-mail:* afessler@racc.edu
*Med Dir:* Jerome I Marcus, MD

**Comm Coll of Allegheny County**
Clin Lab Technician/Med Lab Technician
1750 Clairton Rd, Rte 885
West Mifflin, PA 15122
www.ccac.edu
*Prgm Dir:* Jane Coughanour, MEd MT(ASCP)
*Tel:* 412 469-6280, Ext 6280   *Fax:* 412 469-6371
*E-mail:* jcoughanour@ccac.edu
*Med Dir:* Nirmal Kotwal, MD

# Rhode Island

**Community College of Rhode Island**
Clin Lab Technician/Med Lab Technician
1762 Louisquisset Pike
Lincoln, RI 02865
*Prgm Dir:* Maggie Joseph
*Tel:* 401 333-7144   *Fax:* 401 333-7441
*E-mail:* mjoseph@ccri.edu
*Med Dir:* Upendra Shah, MD

# South Carolina

**Trident Technical College**
Clin Lab Technician/Med Lab Technician
PO Box 118067
7000 Rivers Ave
Charleston, SC 29423-8067
*Prgm Dir:* Donna J Donaldson, MCLT MT(ASCP)
*Tel:* 843 574-6476   *Fax:* 843 574-6585
*E-mail:* donna.donaldson@tridenttech.edu

**Midlands Technical College**
Clin Lab Technician/Med Lab Technician
PO Box 2408
Columbia, SC 29202
www.midlandstech.edu
*Prgm Dir:* Mary Breci, MAT MT(ASCP)
*Tel:* 803 822-3557   *Fax:* 803 822-3417
*E-mail:* brecim@midlandstech.edu
*Med Dir:* Ron G Burns, MD

**Florence-Darlington Technical College**
Clin Lab Technician/Med Lab Technician
PO Box 100548
Florence, SC 29501-0548
*Prgm Dir:* Kathleen Hanrahan, MHA MT PBT (ASCP)
*Tel:* 843 661-8105   *Fax:* 843 292-0851
*E-mail:* hanrahank@fdtc.edu

**Greenville Technical College**
Clin Lab Technician/Med Lab Technician
PO Box 5616, Station B
Greenville, SC 29606
www.greenvilletech.com
*Prgm Dir:* Tommie H Whitt, MHSA MT(ASCP)
*Tel:* 864 250-8292   *Fax:* 864 250-8462
*E-mail:* Tommie.Whitt@gvltec.edu
*Med Dir:* Michael Wolff, MD

**Orangeburg Calhoun Technical College**
Clin Lab Technician/Med Lab Technician
3250 St Matthews Rd
Orangeburg, SC 29118
*Prgm Dir:* Bonnie D Fanning, MS MT(ASCP)
*Tel:* 803 535-1349   *Fax:* 803 535-1350
*E-mail:* fanningb@octech.edu

**Tri-County Technical College**
Clin Lab Technician/Med Lab Technician
PO Box 587
Pendleton, SC 29670-0587
*Prgm Dir:* Polly Kay, MHS MT(ASCP)
*Tel:* 864 646-1349   *Fax:* 864 646-1892
*E-mail:* pkay@tctc.edu
*Med Dir:* Albert A Hollingsworth, MD

**York Technical College**
Clin Lab Technician/Med Lab Technician
452 S Anderson Rd
Rock Hill, SC 29730
www.yorktech.com
*Prgm Dir:* R Lynne Fantry, MLA MT(ASCP)
*Tel:* 803 981-7082   *Fax:* 803 981-7216
*E-mail:* lfantry@yorktech.com

**Spartanburg Community College**
Clin Lab Technician/Med Lab Technician
PO Drawer 4386
Spartanburg, SC 29305
*Prgm Dir:* Ellen F Romani, MHSA MT(ASCP) BBDLM
*Tel:* 864 592-4866   *Fax:* 864 592-4881
*E-mail:* romanie@stcsc.edu
*Med Dir:* Robert Rainer, MD

# South Dakota

**Presentation College**
Clin Lab Technician/Med Lab Technician
1500 N Main St
Aberdeen, SD 57401
*Prgm Dir:* Terry Piatz, MS MT(ASCP)
*Tel:* 605 229-8526   *Fax:* 605 229-8518
*E-mail:* Terry.Piatz@presentation.edu
*Med Dir:* Roy Burt, MD

**Mitchell Technical Institute**
Clin Lab Technician/Med Lab Technician
821 N Capital
Mitchell, SD 57301
www.mitchelltech.edu
*Prgm Dir:* Lynne M Smith, MT(ASCP) MEd
*Tel:* 605 995-3032   *Fax:* 605 996-3299
*E-mail:* lynne.smith@mitchelltech.edu
*Med Dir:* Kim Lorenzen, MD

**Lake Area Technical Institute**
Clin Lab Technician/Med Lab Technician
200 NE 9th St
Watertown, SD 57201
*Prgm Dir:* Mona Gleysteen, MS CLS(NCA)
*Tel:* 605 882-5284, Ext 324   *Fax:* 605 882-6347
*E-mail:* mgleyste@lakeareatech.edu

# Tennessee

**MedVance Institute**
Clin Lab Technician/Med Lab Technician
1025 Highway 111
Cookeville, TN 38501
www.medvance.org
*Prgm Dir:* LaVerne Floyd, MA MT(ASCP)
*Tel:* 931 528-8589   *Fax:* 931 528-5327
*E-mail:* laverne@multipro.com

**Northeast State Technical Comm College**
Clin Lab Technician/Med Lab Technician
Nave Center, 1000 Jason Witten Way
Elizabethton, TN 37643
*Prgm Dir:* Linda S Lahr, MS MT(ASCP)
*Tel:* 423 547-4907   *Fax:* 423 543-2266
*E-mail:* lslahr@northeaststate.edu
*Med Dir:* David Anthony Sibley, MD

**Volunteer State Community College**
Clin Lab Technician/Med Lab Technician
1480 Nashville Pike
Gallatin, TN 37066
*Prgm Dir:* Katherine Karas, MS MT(ASCP)
*Tel:* 615 452-8600, Ext 3363
*E-mail:* Katherine.Karas@volstate.edu

**Jackson State Community College**
Clin Lab Technician/Med Lab Technician
Allied Health Department
2046 N Parkway
Jackson, TN 38301-3797
www.jscc.edu
*Prgm Dir:* Peter P O'Brien, MBA MT(ASCP) CLS(NCA)
*Tel:* 731 425-2612, Ext 226   *Fax:* 731 425-9551
*E-mail:* pobrien@jscc.edu
*Med Dir:* Christopher S Giampapa, MD

**Southwest Tennessee Community College**
Clin Lab Technician/Med Lab Technician
PO Box 780
761 Linden Ave
Memphis, TN 38101
*Prgm Dir:* Darius Y Wilson, MAT MT(ASCP)
*Tel:* 901 333-5407   *Fax:* 901 333-5391
*E-mail:* dwilson@southwest.tn.edu
*Med Dir:* Michael F Bugg, MD

# Texas

### Amarillo College
Clin Lab Technician/Med Lab Technician
PO Box 447
Amarillo, TX 79178
www.actx.edu/medical_lab
*Prgm Dir:* Jan Martin, AAS BA MEd-MLT/MT(ASCP)
CLT/CLS(NCA)
*Tel:* 806 354-6059   *Fax:* 806 354-6076
*E-mail:* martin-jm@actx.edu

### Austin Community College
Clin Lab Technician/Med Lab Technician
Eastview Campus
3401 Webberville Rd
Austin, TX 78702
www.austincc.edu/health/mlt
*Prgm Dir:* Terry Kotrla, MS MT(ASCP)BB
*Tel:* 512 223-5932   *Fax:* 512 223-5898
*E-mail:* kotrla@austincc.edu
*Med Dir:* Paul LeBourgeosis, MD

### Univ Tx at Brownsville/Tx Southmost Coll
Clin Lab Technician/Med Lab Technician
80 Ft Brown
Brownsville, TX 78520
*Prgm Dir:* Matilde Perez Lozano, MS BS MT(ASCP)
*Tel:* 956 882-5010   *Fax:* 956 554-5012
*E-mail:* mlozano@utb.edu
*Med Dir:* Lawrence Dahm, MD

### Del Mar College
Clin Lab Technician/Med Lab Technician
101 Baldwin Blvd
Corpus Christi, TX 78404
*Prgm Dir:* Duncan F Samo, MEd MT(ASCP) CLS(NCA)
*Tel:* 361 698-1107   *Fax:* 361 698-1598
*E-mail:* dsamo@delmar.edu
*Med Dir:* James M Scherer, MD

### Navarro College
Clin Lab Technician/Med Lab Technician
3200 W 7th Ave
Corsicana, TX 75110-4899
www.navarrocollege.edu
*Prgm Dir:* Evelyn Glass, MS MT(ASCP)
*Tel:* 903 875-7516   *Fax:* 903 875-7525
*E-mail:* evelyn.glass@navarrocollege.edu

### El Centro College
Clin Lab Technician/Med Lab Technician
801 Main St
Dallas, TX 75202-3604
www.elcentrocollege.edu/MedicalLabTech
*Prgm Dir:* Lisa Lock, MBA BSMT MT(ASCP)BB
CLS(NCA)
*Tel:* 214 860-2304, Ext 2304   *Fax:* 214 860-2268
*E-mail:* lal5630@dcccd.edu
*Med Dir:* Tak-Shun Choi, MD

### Grayson County College
Clin Lab Technician/Med Lab Technician
6101 Grayson Dr
Denison, TX 75020
*Prgm Dir:* Alan Jackson, MS MT(ASCP)
*Tel:* 903 463-8779   *Fax:* 903 463-5284
*E-mail:* jacksona@grayson.edu

### El Paso Community College
Clin Lab Technician/Med Lab Technician
PO Box 20500
El Paso, TX 79998
www.epcc.edu
*Prgm Dir:* Veronica Dominguez, MEd CLS(NCA)
*Tel:* 915 831-4085   *Fax:* 915 831-4114
*E-mail:* VeronicaD@epcc.edu
*Med Dir:* Ellen Francis Dudrey, MD

### US Army Medical Dept Center & School
Clin Lab Technician/Med Lab Technician
Academy of Health Sciences
3151 Scott Rd
Ft Sam Houston, TX 78234-6137
www.cs.amedd.army.mil/dcss/lab/mltphas2.html
*Prgm Dir:* Donna S Whitlaker, PhD MT(ASCP)SBB
*Tel:* 210 221-7709   *Fax:* 210 221-7679
*E-mail:* donna.whitlaker@amedd.army.mil
*Med Dir:* LTC Kerry M Brady, MC USA

### Houston Community College
Clin Lab Technician/Med Lab Technician
1900 Pressler
Houston, TX 77030
www.hccs.edu
*Prgm Dir:* Theresa Spain, MEd MT(ASCP) CLS(NCA)
*Tel:* 713 718-5518   *Fax:* 713 718-7653
*E-mail:* theresa.spain@hccs.edu
*Med Dir:* Oscar R Mangini, MD

### Central Texas College
Clin Lab Technician/Med Lab Technician
6200 W Central TX Expressway, PO Box 1800
Killeen, TX 76541-9990
www.ctcd.edu
*Prgm Dir:* Donna E Poteet, MA CLS(NCA)
*Tel:* 254 526-1187   *Fax:* 254 526-1765
*E-mail:* donna.poteet@ctcd.edu
*Med Dir:* Carlton E Hardin, MD

### Laredo Community College
Clin Lab Technician/Med Lab Technician
West End Washington St
Laredo, TX 78040
www.laredo.edu
*Prgm Dir:* Bill Branim, MS MPA MT(ASCP) DLM
*Tel:* 956 796-2155   *Fax:* 956 721-5421
*E-mail:* bill_branim@chs.net

### Lamar State College - Orange
Clin Lab Technician/Med Lab Technician
410 Front St
Orange, TX 77630
www.lsco.edu
*Prgm Dir:* Kathleen Ann Park, MEd MT(ASCP)
*Tel:* 409 882-3914   *Fax:* 409 882-3374
*E-mail:* Kathy.Park@lsco.edu

### San Jacinto College Central
Clin Lab Technician/Med Lab Technician
8060 Spencer Hwy, PO Box 2007
Pasadena, TX 77501-2007
*Prgm Dir:* Terri C Simon, MS
*Tel:* 281 478-2730   *Fax:* 281 478-2754
*E-mail:* terri.simon@sjcd.edu
*Med Dir:* Dorothy Willis, MD

### St Philip's College
Clin Lab Technician/Med Lab Technician
1801 Martin Luther King Dr
San Antonio, TX 78203-2098
www.accd.edu/spc
*Prgm Dir:* Jerri Lee Reynolds, PhD MT(ASCP)
*Tel:* 210 531-3449   *Fax:* 210 531-3459
*E-mail:* jreynold@accd.edu
*Med Dir:* Desiree E D'Orsogna, MD

### 882d Training Group
Clin Lab Technician/Med Lab Technician
382 Training Squadron, XYAC
917 Missile Rd, Ste 3
Sheppard AFB, TX 76311-2263
*Prgm Dir:* Barry T White, MS MT(ASCP)
*Tel:* 940 676-3869   *Fax:* 940 676-3850
*E-mail:* Barry.White@sheppard.af.mil
*Med Dir:* Mark P Burton, MD FCAP FASCP

### Tyler Junior College
Clin Lab Technician/Med Lab Technician
PO Box 9020
Tyler, TX 75711-9020
*Prgm Dir:* Daniel L Spencer, EdD MT(ASCP) CLS(NCA)
*Tel:* 903 510-2367   *Fax:* 903 510-2880
*E-mail:* dspe@tjc.edu
*Med Dir:* Marian Fagan, MD

### Victoria College
Clin Lab Technician/Med Lab Technician
2200 E Red River
Victoria, TX 77901
www.victoriacollege.edu/dept/mlt
*Prgm Dir:* Larry S Dunn, MS MT(ASCP)
*Tel:* 361 572-6455   *Fax:* 361 572-6441
*E-mail:* larry.dunn@victoriacollege.edu
*Med Dir:* Joe David Ibanez, MD

### McLennan Community College
Clin Lab Technician/Med Lab Technician
1400 College Dr
Waco, TX 76708
www.mclennan.edu
*Prgm Dir:* Diane L Schmaus, MA MT(ASCP)
*Tel:* 254 299-8417   *Fax:* 254 299-8397
*E-mail:* dschmaus@mclennan.edu
*Med Dir:* Daniel R Samples, MD PhD

# Utah

### Weber State University
Clin Lab Technician/Med Lab Technician
3905 University Cirlce
Ogden, UT 84408-3905
*Prgm Dir:* Yasmen Simonian, PhD MT(ASCP) CLS(NCA)
*Tel:* 801 626-6509   *Fax:* 801 626-7508
*E-mail:* ysimonian@weber.edu
*Med Dir:* Val B Johnson, MD

### Salt Lake Community College
Clin Lab Technician/Med Lab Technician
PO Box 30808
Salt Lake City, UT 84130
*Prgm Dir:* Karen A Brown, MS MT(ASCP) CLS
*Tel:* 801 581-3544   *Fax:* 801 585-2463
*E-mail:* karen.brown@path.utah.edu
*Med Dir:* Joseph A Knight, MD

# Virginia

### Thomas Nelson Community College
Clin Lab Technician/Med Lab Technician
PO Box 9407
99 Thomas Nelson Dr
Hampton, VA 23670
http://tncc.edu
*Prgm Dir:* Linda A Dezern, MS MT(ASCP)
*Tel:* 757 825-2783   *Fax:* 757 825-3831
*E-mail:* dezernl@tncc.edu

### Centra Health Systems of Lynchburg
Clin Lab Technician/Med Lab Technician
3300 Rivermont Ave
Lynchburg, VA 24503
*Prgm Dir:* Robin L Levandoski, MEd BS MT(ASCP) SC
*Tel:* 434 947-4551   *Fax:* 434 947-4035
*E-mail:* levandoskir@cvcc.vccs.edu
*Med Dir:* David H Cresson, MD

### J Sargeant Reynolds Community College
Clin Lab Technician/Med Lab Technician
PO Box 85622
Richmond, VA 23285-5622
www.jsr.vccs.edu
*Prgm Dir:* Becky Clark, MEd MT(ASCP)
*Tel:* 804 523-5772   *Fax:* 804 786-5298
*E-mail:* bclark@reynolds.edu
*Med Dir:* Brad Siegmund, MD

**PROGRAMS**

**Northern Virginia Community College**
Clin Lab Technician/Med Lab Technician
6699 Springfield Center Dr
Springfield, VA 22150
www.nvcc.edu
*Prgm Dir:* Glenn Flodstrom, MS MT(ASCP)
*Tel:* 703 822-6567  *Fax:* 703 822-6619
*E-mail:* gflodstrom@nvcc.edu

**Wytheville Community College**
Clin Lab Technician/Med Lab Technician
1000 E Main St
Wytheville, VA 24382
*Prgm Dir:* Lorri Huffard, MS MT(ASCP)SBB
*Tel:* 276 223-4828  *Fax:* 276 223-4778
*E-mail:* wchuffl@wcc.vccs.edu
*Med Dir:* Andrew Williams, MD

# Washington

**Clover Park Technical College**
Clin Lab Technician/Med Lab Technician
4500 Steilacoom Blvd SW
Lakewood, WA 98499-4098
www.cptc.ctc.edu
*Prgm Dir:* Anne G O'Neil, MT(ASCP) MEd
*Tel:* 253 589-5625  *Fax:* 253 589-5866
*E-mail:* anne.oneil@cptc.edu

**Shoreline Community College**
Clin Lab Technician/Med Lab Technician
16101 Greenwood Ave N
Seattle, WA 98133
www.shoreline.edu/shoreline.medlablocal.html
*Prgm Dir:* Molly Morse, MS CLS(NCA) MT(ASCP)
*Tel:* 206 546-6947  *Fax:* 206 533-5103
*E-mail:* mmorse@shoreline.edu
*Med Dir:* Richard Patton, MD

**Wenatchee Valley College**
Clin Lab Technician/Med Lab Technician
1300 Fifth St
Wenatchee, WA 98801
www.wvc.edu/go/mlt
*Prgm Dir:* David C Abbott, MSA MT(ASCP)
*Tel:* 509 682-6668  *Fax:* 509 682-6541
*E-mail:* dabbott@wvc.edu
*Med Dir:* Michael Daines, MD

# West Virginia

**Bluefield Regional Medical Center**
Clin Lab Technician/Med Lab Technician
500 Cherry St
Bluefield, WV 24701-3306
*Prgm Dir:* James Gibberson, MS MT(ASCP)
*Tel:* 304 327-1596  *Fax:* 304 327-1591
*E-mail:* jgibberson@brmcwv.org
*Med Dir:* Dennis I Pullins, MD

**Fairmont State Univ**
*Cosponsor:* Pierpont Comm & Tech College
Clin Lab Technician/Med Lab Technician
1201 Locust Ave
Fairmont, WV 26554
www.fairmontstate.edu
*Prgm Dir:* Rosemarie R Romesburg, PhD MT(ASCP)
*Tel:* 304 367-4284  *Fax:* 304 367-4268
*E-mail:* rromesburg@fairmontstate.edu
*Med Dir:* Chinmay Datta, MD

**Marshall University**
Clin Lab Technician/Med Lab Technician
Clinical Lab Science Dept
One John Marshall Dr
Huntington, WV 25755
*Prgm Dir:* Dorothy J Fike, MS MT(ASCP)
*Tel:* 304 696-3165  *Fax:* 304 696-3243
*E-mail:* fike@marshall.edu

**Southern West Virginia Comm & Tech College**
Clin Lab Technician/Med Lab Technician
Logan Campus, PO Box 2900
Mt Gay, WV 25637
www.southern.wvnet.edu
*Prgm Dir:* Vernon R Elkins, MA BS MT(ASCP)
*Tel:* 304 792-7098, Ext 243  *Fax:* 304 792-7053
*E-mail:* vernone@southern.wvnet.edu
*Med Dir:* Alex Racadag, MD

# Wisconsin

**Chippewa Valley Technical College**
Clin Lab Technician/Med Lab Technician
620 W Clairemont Ave
Eau Claire, WI 54701
www.cvtc.edu/Programs/DeptPages/CLT/MLTHomePage.
  html
*Prgm Dir:* Patricia Griffin, MS MT(ASCP)
*Tel:* 715 833-6420  *Fax:* 715 833-6470
*E-mail:* pgriffin@cvtc.edu
*Med Dir:* Thomas W Hadley, MD

**Moraine Park Technical College**
Clin Lab Technician/Med Lab Technician
235 N National Ave
PO Box 1940
Fond du Lac, WI 54936-1940
*Prgm Dir:* Linda Bau, MPA BS MT(ASCP)
*Tel:* 920 924-6373
*E-mail:* lbau@morainepark.edu

**Northeast Wisconsin Technical College**
Clin Lab Technician/Med Lab Technician
2740 W Mason St, PO Box 19042
Green Bay, WI 54307-9042
www.nwtc.edu
*Prgm Dir:* Patricia Moore-Cribb, MT(ASCP) MS
*Tel:* 920 498-6374  *Fax:* 920 498-2660
*E-mail:* patricia.cribb@nwtc.edu

**Western Technical College**
Clin Lab Technician/Med Lab Technician
304 N Sixth St
La Crosse, WI 54601
www.wwtc.edu
*Prgm Dir:* Carolyn Byom, MS MT(ASCP)
*Tel:* 608 789-6284  *Fax:* 608 785-9299
*E-mail:* byomc@wwtc.edu
*Med Dir:* Jeffrey A Degenhardt, MD

**Madison Area Technical College**
Clin Lab Technician/Med Lab Technician
3550 Anderson St
Madison, WI 53704-2599
www.matcmadison.edu
*Prgm Dir:* Sue Ellen S Beglinger, MS MT(ASCP)
  CLS(NCA)
*Tel:* 608 246-6459  *Fax:* 608 246-6013
*E-mail:* beglinger@matcmadison.edu
*Med Dir:* Deborah Turski, MD

**Milwaukee Area Technical College**
Clin Lab Technician/Med Lab Technician
700 W State St
Milwaukee, WI 53233
www.matc.edu
*Prgm Dir:* Dennis Schmidt, MS MT(ASCP)
*Tel:* 414 297-7142  *Fax:* 414 297-6851
*E-mail:* SchmidtD@matc.edu

## Clinical Laboratory Technician/Medical Laboratory Technician

| Programs* | Class Capacity | Begins | Length (months) | Award | Res Tuition | Non-res Tuition | Stipend | Offers:‡ 1 | 2 | 3 |
|---|---|---|---|---|---|---|---|---|---|---|
| **Alabama** | | | | | | | | | | |
| Gadsden State Community College | 17 | Aug Jan | 24 | AAS | $3,780 | $7,560 | | | | |
| Jefferson State Community College (Birmingham) | 18 | May Aug | 24 | Dipl, AAS | $6,528 | $12,288 | | | | |
| **Alaska** | | | | | | | | | | |
| University of Alaska Anchorage | 13 | Open enrollment | 20 | AAS | $2,574 | $8,580 | | | | • |
| **Arkansas** | | | | | | | | | | |
| Arkansas State University | 20 | Aug | 24 | AAS | $142 | $365 | | | | |
| Arkansas State University - Beebe | 15 | Jul | 12 | AAS | $1,080 | $1,824 | | | | • |
| National Park Community College (Hot Springs) | 10 | Jan | 18 | AS | $1,600 | $3,896 | | • | | • |
| North Arkansas College (Harrison) | 12 | Jun | 24 | AAS | $2,040 | $4,366 | | | | |
| Phillips Community College/U of Arkansas (Helena) | 8 | Aug | 22 | Dipl, AAS | $1,500 | $2,910 | | | | |
| South Arkansas Community College (El Dorado) | 10 | Aug | 24 | AAS | $3,500 | $6,498 | | • | | |
| **California** | | | | | | | | | | |
| Naval School of Health Sciences, San Diego | 55 | Sep Jan May | 12 | Cert | | | | | | • |
| **Colorado** | | | | | | | | | | |
| Arapahoe Community College (Littleton) | 30 | Aug | 24 | AAS | $2,500 | $10,400 | | • | | |

*Data are shown only for programs that completed the 2006 AMA Survey of Health Professions Education Programs
‡Key to Offers: 1: Evening or weekend classes; 2: Non-English instruction; 3: Cultural competence instruction

| Programs* | Class Capacity | Begins | Length (months) | Award | Res Tuition | Non-res Tuition | Stipend | Offers:‡ 1 | 2 | 3 |
|---|---|---|---|---|---|---|---|---|---|---|
| **Connecticut** | | | | | | | | | | |
| Housatonic Community College (Bridgeport) | 20 | Sep | 21 | AS | $2,672 | $7,976 | | • | | |
| **Florida** | | | | | | | | | | |
| Brevard Community College (Cocoa) | 16 | Open entry | 22 | Dipl, AS, ATD | $5,149 | $18,772 | | | | |
| Erwin Technical Center (Tampa) | 18 | Jan | 15 | Dipl | $2,002 | $11,067 | | | | • |
| Indian River Community College (Ft Pierce) | 17 | Aug | 23 | AS | $2,246 | $8,500 | | • | | |
| Keiser University (Ft Lauderdale) | 20 | Jan Apr Aug | 22 | AS | $11,408 | $11,408 | | | | |
| Miami Dade College | 45 | Aug Jan | 22 | AS | $4,009 | $14,006 | | • | | |
| St Petersburg College | 40 | Aug | 24 | AS | $2,648 | $9,952 | | | | |
| **Georgia** | | | | | | | | | | |
| Central Georgia Technical College (Macon) | 12 | Oct | 21 | AAT | $1,812 | $1,812 | | | | |
| Coastal Georgia Community College (Brunswick) | 15 | Aug | 21 | AS | $2,631 | $9,567 | | | | |
| Dalton State College | 12 | Aug Jan | 24 | AAS | $1,772 | $6,951 | | | | • |
| Darton College (Albany) | 36 | Aug | 17 | AS | $2,202 | $8,808 | | • | | |
| DeKalb Technical College (Clarkston) | 24 | Oct | 16, 24 | Dipl, MLT, AAT | $2,316 | $5,432 | | | | • |
| Lanier Technical College (Oakwood) | 12 | Oct | 24 | Dipl | $1,860 | $3,720 | | | | • |
| North Georgia Technical College (Clarkesville) | 15 | Oct | 21 | AAT | $1,832 | $3,320 | | • | | |
| Okefenokee Technical College (Waycross) | 14 | Apr | 24 | AAT | $1,750 | $0 | | | | • |
| Southeastern Technical College (Vidalia) | 5 | Winter quarter (Jan) | 8 | AS | | | | | | |
| Southwest Georgia Technical College (Thomasville) | 12 | Apr | 24 | AAT | $1,674 | $2,178 | | | | |
| Valdosta Technical College | 13 | Jan | 20 | AAT | | | | | | |
| West Central Technical College (Waco) | 12 | Oct | 24 | Assoc | $1,600 | $3,200 | | | | • |
| **Hawaii** | | | | | | | | | | |
| Kapi'olani Community College (Honolulu) | 16 | Jan | 17 | AS | $2,218 | $9,070 | | | | |
| **Illinois** | | | | | | | | | | |
| Blessing Hospital (Quincy) | 8 | Aug | 12 | Cert | $1,000 | $1,000 | | | | • |
| Elgin Community College | 15 | Aug | 22 | AAS | $3,300 | $9,300 | | • | | • |
| Illinois Central College (Peoria) | 16 | Aug | 20 | AAS | $2,415 | $5,348 | | • | | |
| Kankakee Community College | 17 | Aug | 18, 12 | AAS | $2,483 | $6,484 | | | | |
| Oakton Community College (Des Plaines) | 20 | Aug | 21 | Dipl, AAS | $3,500 | $7,500 | | | | • |
| Southern Illinois Collegiate Common Market (Herrin) | 24 | Aug | 22 | Cert, AAS | $4,919 | $10,310 | | | | |
| Southwestern Illinois College (Belleville) | 14 | Aug | 20 | AAS | $3,500 | $5,000 | | | | |
| **Indiana** | | | | | | | | | | |
| Indiana University Northwest (Gary) | 18 | Aug | 21 | AS | $5,320 | $12,920 | | | | • |
| Ivy Tech Community College - South Bend | 15 | Aug | 22 | AAS | $3,000 | $6,000 | | | | |
| Ivy Tech Community College - Terre Haute | 24 | Aug | 21 | AAS | $6,143 | $12,495 | | | | |
| **Iowa** | | | | | | | | | | |
| Des Moines Area Community College (Ankeny) | 24 | Aug | 22 | AAS | $3,492 | $6,984 | | | | |
| Iowa Central Community College (Ft Dodge) | 20 | Sep 6 | 23 | AAS | $7,725 | $0 | | | | |
| **Kansas** | | | | | | | | | | |
| Barton County Community College (Great Bend) | 25 | Aug | 24 | Dipl, AAS | $2,700 | $3,330 | | | | |
| Seward County Community College (Liberal) | 15 | Aug Jan | 24 | AAS | $3,300 | $3,300 | | • | | |
| Wichita Area Technical College | 18 | Aug | 24 | AAS | $7,016 | $22,541 | | | | • |
| **Kentucky** | | | | | | | | | | |
| Eastern Kentucky University (Richmond) | 36 | Aug Jan | 22 | AS | $2,597 | $7,269 | | | | |
| HCC/MCC Consortium (Madisonville) | 14 | Aug | 24 | AAS | $3,706 | $11,118 | | | | |
| Somerset Community College | 15 | Aug | 21 | AAS | $3,727 | $9,450 | | | | |
| Southeast Kentucky Comm & Tech College (Pineville) | 14 | Aug | 24 | AAS | $109 | $327 | | | | |
| **Louisiana** | | | | | | | | | | |
| Delgado Community College (New Orleans) | 12 | Jan | 24 | AAS | $768 | $2,258 | | | | |
| Louisiana Tech College - Lafayette Campus | 20 | Aug | 24 | AAS | $1,200 | $2,400 | | • | | |
| MedVance Institute (Baton Rouge) | 30 | Jan Apr Jul Oct | 18 | AOS | $16,663 | $16,663 | | • | | |
| Our Lady of the Lake College (Baton Rouge) | 14 | Jun | 12 | AS CLT | $7,700 | $7,700 | | | | • |
| Southern Univ at Shreveport | 24 | Aug | 24 | AS | $2,000 | $3,700 | | | | |
| **Maine** | | | | | | | | | | |
| University of Maine - Presque Isle | 30 | Sep | 20 | AS | $5,005 | $12,460 | | | | |
| **Maryland** | | | | | | | | | | |
| Allegany College of Maryland (Cumberland) | 18 | Sep | 20 | AAS | $3,060 | $6,868 | | | | |
| **Massachusetts** | | | | | | | | | | |
| Bristol Community College (Fall River) | 17 | Sep | 18 | AS | $3,885 | $7,500 | | • | | • |
| Springfield Technical Community College | 18 | Sep May | 10 | Dipl, AS | $1,408 | $6,016 | | | | • |
| **Michigan** | | | | | | | | | | |
| Baker College of Owosso | 20 | Mar | 13 | AS | $13,680 | $13,680 | | | | • |
| Ferris State University (Big Rapids) | 30 | May | 24 | Dipl, AAS | $6,740 | $13,480 | | | | • |
| Kellogg Community College (Battle Creek) | 16 | Aug | 22 | AAS | $2,624 | $4,131 | | | | |

*Data are shown only for programs that completed the 2006 AMA Survey of Health Professions Education Programs

‡Key to Offers: 1: Evening or weekend classes; 2: Non-English instruction; 3: Cultural competence instruction

## Clinical Laboratory Technician/Medical Laboratory Technician

| Programs* | Class Capacity | Begins | Length (months) | Award | Res Tuition | Non-res Tuition | Stipend | Offers:‡ 1 | 2 | 3 |
|---|---|---|---|---|---|---|---|---|---|---|
| **Minnesota** | | | | | | | | | | |
| Alexandria Technical College | 30 | Aug | 18 | AAS | $4,719 | $9,438 | | • | | |
| Argosy University/Twin Cities (Eagan) | 12 | Oct Apr | 24 | AS | $30,000 | $0 | | • | | |
| Hibbing Community College | 32 | Sep | 24 | AAS | $2,000 | $0 | | | | |
| Lake Superior College (Duluth) | 26 | Sep | 20 | AAS | $3,927 | $7,377 | | • | | • |
| Minnesota State Community and Technical Coll (Fergus Falls) | 15 | Aug | 19 | AS | $4,425 | $8,288 | | • | | • |
| North Hennepin Community College (Brooklyn Park) | 30 | Aug | 24 | AAS | $135 | $242 | | • | | |
| Northland Community & Technical College (East Grand Forks) | 20 | Aug Jan May | 25 | AAS | $4,096 | $4,096 | | | | • |
| South Central College (Faribault) | 50 | Aug Jan | 22 | AAS | $4,680 | $7,560 | | | | |
| St Paul College | 24 | Aug | 20 | AAS | $137 | $265 | | | | • |
| **Mississippi** | | | | | | | | | | |
| Copiah-Lincoln Community College (Wesson) | 30 | Aug Jan | 24 | AAS | $1,800 | $3,500 | | | | |
| Meridian Community College | 16 | Aug | 24 | Dipl, AAS | $3,603 | $5,378 | | | | |
| Mississippi Delta Community College (Moorhead) | 18 | Jun | 24 | AAS | $1,800 | $3,408 | | | | • |
| Northeast Mississippi Community College (Booneville) | 15 | Aug | 22 | AAS | $850 | $1,710 | | | | |
| Pearl River Community College (Hattiesburg) | 20 | Aug | 24 | Dipl, AS | $1,620 | $2,398 | | • | | |
| **Missouri** | | | | | | | | | | |
| St Louis Community College - Forest Park | 20 | Aug | 20 | AAS | $78 | $103,138 | | | | • |
| Three Rivers Community College (Poplar Bluff) | 20 | Aug | 22 | Dipl, AAS | $2,100 | $4,100 | | | | |
| **Nebraska** | | | | | | | | | | |
| Central Community College (Hastings) | 16 | Aug | 24 | AD | $1,798 | $2,639 | | | | |
| Mid-Plains Community College (North Platte) | 20 | Aug | 24 | AAS | $2,470 | $3,078 | | • | | |
| Southeast Community College (Lincoln) | 26 | Jul | 24 | AAS | $2,688 | $3,250 | | • | | • |
| **Nevada** | | | | | | | | | | |
| Community College of Southern Nevada (Las Vegas) | 12 | Jan Aug | 24 | AAS | $1,696 | $4,042 | | • | | |
| **New Hampshire** | | | | | | | | | | |
| New Hampshire Comm Tech Coll - Claremont | 12 | Sep | 20 | AS | $6,400 | $14,664 | | • | | • |
| **New Jersey** | | | | | | | | | | |
| Camden County College (Blackwood) | 16 | May | 24 | AAS | $2,580 | $3,311 | | • | | • |
| Mercer County Community College (Trenton) | 12 | May | 22 | AAS | $2,639 | $4,374 | | | | • |
| Middlesex County College (Edison) | 21 | Sep | 21 | AAS | $4,000 | $8,000 | | • | | |
| **New Mexico** | | | | | | | | | | |
| Central New Mexico Community College (Albuquerque) | 20 | Aug | 20 | AS | $1,501 | $1,747 | | | | • |
| New Mexico State U at Alamogordo | 10 | Aug | 20 | Assoc | $1,600 | $5,310 | | | | |
| University of New Mexico - Gallup | 12 | Aug Jan | 24, 30 | AS | $1,344 | $3,096 | | | | • |
| **New York** | | | | | | | | | | |
| Broome Community College (Binghamton) | 24 | Aug Jan | 20 | AAS | $2,914 | $5,828 | | | | • |
| Clinton Community College (Plattsburgh) | 15 | Aug | 22 | AAS | $2,500 | $5,000 | | | | |
| Dutchess Community College (Poughkeepsie) | 30 | Aug | 20 | AAS | $2,600 | $5,200 | | • | | |
| Erie Community College - North Campus (Williamsville) | 45 | Sep | 21 | AAS | $2,987 | $5,974 | | | | |
| Farmingdale State University of New York | 40 | Sep | 24 | AS | $4,350 | $10,610 | | | | |
| Orange County Community College (Middletown) | 25 | Sep | 18 | AAS | $3,000 | $6,000 | | | | |
| **North Carolina** | | | | | | | | | | |
| Alamance Community College (Graham) | 40 | Aug Jan May | 24 | AAS | $1,500 | $5,600 | | | • | • |
| Asheville-Buncombe Technical Comm College | 16 | Aug | 21 | AAS | $1,748 | $9,706 | | | | |
| Beaufort County Community College (Washington) | 13 | Aug | 22 | AAS | $1,824 | $10,128 | | • | | |
| Central Piedmont Community College (Charlotte) | 20 | Aug | 21 | AAS | $1,500 | $7,500 | | | | |
| Coastal Carolina Community College (Jacksonville) | 20 | Aug | 22 | AAS | $1,126 | $6,134 | | | | |
| Davidson County Community College (Lexington) | 24 | Aug | 21 | AAS | $1,000 | $5,705 | | | | |
| Halifax Community College (Weldon) | 10 | Aug | 21 | AAS | $1,136 | $6,304 | | | | |
| Sandhills Community College (Pinehurst) | 18 | Aug | 21 | AAS | $836 | $4,642 | | | | • |
| Southeastern Community College (Whiteville) | 20 | Aug | 22 | AAS | $1,323 | $7,353 | | • | | |
| Southwestern Community College (Sylva) | 20 | Aug | 22 | AAS | $1,264 | $7,024 | | • | | |
| Wake Technical Community College (Raleigh) | 24 | Aug | 21 | AAS | $1,896 | $10,536 | | | | • |
| Western Piedmont Community College (Morganton) | 16 | Aug | 21 | AAS | $1,824 | $10,128 | | | | |
| **North Dakota** | | | | | | | | | | |
| Bismarck State College | 10 | Aug | 22 | AS | $2,947 | $7,868 | | • | • | • |
| **Ohio** | | | | | | | | | | |
| Cincinnati State Tech & Comm College | 30 | Sep | 24 | AAS | $5,100 | $10,200 | | | | • |
| Clark State Community College (Springfield) | 28+ | Sep Jan | 18 | AAS | $4,599 | $8,687 | | | | |
| Columbus State Community College | 25 | Mar | 21 | AAS | $3,504 | $7,728 | | | | |
| Cuyahoga Community College (Cleveland) | 24 | Aug | 21 | AAS | $2,275 | $5,975 | | | | |
| Jefferson Community College (Steubenville) | 16 | Aug | 24 | AAS | $3,017 | $4,082 | | | | |
| Lakeland Community College (Kirtland) | 24 | Aug | 22 | AAS | $3,500 | $4,289 | | | | • |
| Lorain County Community College (Elyria) | 20 | Aug | 20 | AAS | $2,769 | $3,334 | | | | |
| Marion Technical College | 15 | Sep | 21 | AAS | $6,037 | $8,263 | | | | |

*Data are shown only for programs that completed the 2006 AMA Survey of Health Professions Education Programs
‡Key to Offers: 1: Evening or weekend classes; 2: Non-English instruction; 3: Cultural competence instruction

## Clinical Laboratory Technician/Medical Laboratory Technician

| Programs* | Class Capacity | Begins | Length (months) | Award | Res Tuition | Non-res Tuition | Stipend | 1 | 2 | 3 |
|---|---|---|---|---|---|---|---|---|---|---|
| Shawnee State University (Portsmouth) | 21 | Sep | 21 | AAS | $5,796 | $9,924 | | • | | |
| Stark State College of Technology (North Canton) | 20 | Aug | 21 | AAS | $4,500 | $6,545 | | | | |
| Washington State Community College (Marietta) | 16 | Sep | 21 | AAS | $3,420 | $6,840 | | | | • |
| Youngstown State University | 15 | Aug | 24 | AAS | $6,104 | $11,312 | | • | | |
| Zane State College (Zanesville) | 24 | Sep | 22 | AAS | $4,347 | $8,694 | | • | | • |
| **Oklahoma** | | | | | | | | | | |
| Northeastern Oklahoma A&M College (Miami) | 12 | Aug | 18 | AAS | $1,890 | $4,690 | | | | |
| Rose State College (Midwest City) | 20 | Aug | 24 | AAS | $1,943 | $6,438 | | • | | |
| Seminole State College | 20 | Aug Jan | 12, 24 | AAS | $2,323 | $5,415 | | | | • |
| Tulsa Community College | 12 | Aug | 24 | AAS | $3,000 | $6,000 | | • | | |
| **Oregon** | | | | | | | | | | |
| Portland Community College | 78 | Sep | 21 | Dipl, AAS | $3,552 | $10,545 | | | | |
| **Pennsylvania** | | | | | | | | | | |
| Comm Coll of Allegheny County (West Mifflin) | 20 | Aug | 20 | AS | $96 | $110 | | | | • |
| Harcum College (Bryn Mawr) | 10 | Sep Jan | 24 | MLT | $15,250 | $15,250 | | • | • | |
| Montgomery County Community College (Blue Bell) | 16 | Sep | 21 | AAS | $3,600 | $10,800 | | | | |
| Penn State University - Hazleton | 10 | Aug | 24 | AS | $13,248 | $20,380 | | • | | |
| Reading Area Community College | 24 | Sep | 20 | AAS | $2,436 | $4,872 | | • | | • |
| **South Carolina** | | | | | | | | | | |
| Greenville Technical College | 22 | Aug | 24 | AAS | $4,275 | $4,650 | | | | • |
| Midlands Technical College (Columbia) | 18 | May | 24 | AS | $2,904 | $8,712 | | | | |
| Orangeburg Calhoun Technical College | 18 | Aug | 21 | AAS | $3,816 | $4,788 | | | | • |
| Tri-County Technical College (Pendleton) | 20 | Aug | 21 | AS | $2,618 | $5,964 | | | | • |
| York Technical College (Rock Hill) | 16 | Aug | 21 | AHS | $2,988 | $6,864 | | | | • |
| **South Dakota** | | | | | | | | | | |
| Mitchell Technical Institute | 18 | Aug | 18 | AAS | $2,340 | $2,340 | | | | |
| Presentation College (Aberdeen) | 10 | Aug | 19 | MLT-AD | $12,000 | $12,000 | | • | | |
| **Tennessee** | | | | | | | | | | |
| Jackson State Community College | 20 | Aug | 21 | AAS | $3,116 | $11,311 | | | | |
| MedVance Institute (Cookeville) | 20 | Jan Apr Jul Oct | 18, 21 | AAS | $10,033 | $10,033 | | • | | |
| Northeast State Technical Comm College (Elizabethton) | 15 | Aug | 24 | AAS | $2,418 | $8,860 | | | | |
| Southwest Tennessee Community College (Memphis) | 20 | Aug Jan | 24 | AAS | $1,198 | $4,405 | | • | | |
| **Texas** | | | | | | | | | | |
| 882d Training Group (Sheppard AFB) | 30 | 6-7x/yr | 13 | Cert | | | | | | • |
| Amarillo College | 20 | Aug | 23 | AAS | $1,961 | $2,506 | | • | | • |
| Austin Community College | 17 | Aug | 21 | AAS | $4,155 | $6,356 | | | | |
| Central Texas College (Killeen) | 20 | Aug Jan | 24 | AAS | $1,200 | $2,800 | | | | |
| Del Mar College (Corpus Christi) | 24 | Sep | 21 | AAS | $1,000 | $1,900 | | | | |
| El Centro College (Dallas) | 24 | Aug | 24 | AAS | $950 | $1,655 | | | | • |
| El Paso Community College | 12 | Aug | 24 | AAS | $2,576 | $3,468 | | | | |
| Grayson County College (Denison) | 24 | Aug | 24 | AAS | $1,640 | $1,886 | | | | |
| Houston Community College | 24 | Aug | 24 | AAS | $2,203 | $4,417 | | | | • |
| Lamar State College - Orange | 18 | Aug | 24 | AAS | $3,700 | $14,000 | | | | • |
| Laredo Community College | 12 | Aug | 21 | Dipl, AAS | $1,520 | $2,454 | | • | | |
| McLennan Community College (Waco) | 24 | Aug | 24 | AAS | $1,488 | $2,928 | | | | • |
| Navarro College (Corsicana) | 20 | Aug | 20 | Dipl, AAS | $1,954 | $3,300 | | | | |
| San Jacinto College Central (Pasadena) | 25 | Aug | 23 | AAS | $2,942 | $4,598 | | | | |
| St Philip's College (San Antonio) | 15 | Aug | 24 | AAS | $1,100 | $1,900 | | | | |
| Tyler Junior College | 20 | Aug | 22 | AAS | $4,849 | $6,617 | | • | | • |
| Univ Tx at Brownsville/Tx Southmost Coll | 20 | Aug | 21 | AAS | $2,950 | $12,364 | | | | • |
| US Army Medical Dept Center & School (Ft Sam Houston) | 116 | Every 8 wks | 12 | Cert, ASHS | | | | | | |
| Victoria College | 14 | Aug | 20 | AAS | $1,702 | $2,462 | | | | |
| **Utah** | | | | | | | | | | |
| Salt Lake Community College (Salt Lake City) | 15 | Aug | 24 | AAS | $2,767 | $8,626 | | | | |
| Weber State University (Ogden) | 40 | Aug Jan | 18 | AAS | $2,772 | $8,612 | | | | • |
| **Virginia** | | | | | | | | | | |
| Centra Health Systems of Lynchburg | 10 | Aug | 12 | Cert | $2,400 | $7,300 | | | | |
| J Sargeant Reynolds Community College (Richmond) | 60 | Aug Jan | 22 | AAS | $2,136 | $6,947 | | • | | |
| Northern Virginia Community College (Springfield) | 25 | Aug | 21 | AAS | $2,459 | $7,411 | | • | | • |
| Thomas Nelson Community College (Hampton) | 63 | Jan Jul | 12, 24 | Cert, AAS | $2,751 | $8,614 | | | | |
| **Washington** | | | | | | | | | | |
| Clover Park Technical College (Lakewood) | 13 | Mar | 12 | AAT | $3,362 | $3,362 | | | | |
| Shoreline Community College (Seattle) | 20 | Sep | 24 | Cert, AAAS, MLT | $3,120 | $10,076 | | | | • |
| Wenatchee Valley College | 24 | Jun | 24 | ATS | $3,800 | $4,400 | | | | |

*Data are shown only for programs that completed the 2006 AMA Survey of Health Professions Education Programs
‡Key to Offers: 1: Evening or weekend classes; 2: Non-English instruction; 3: Cultural competence instruction

## Clinical Laboratory Technician/Medical Laboratory Technician

| Programs* | Class Capacity | Begins | Length (months) | Award | Res Tuition | Non-res Tuition | Stipend | Offers:‡ 1 | 2 | 3 |
|---|---|---|---|---|---|---|---|---|---|---|
| **West Virginia** | | | | | | | | | | |
| Fairmont State Univ | 13 | Aug | 22 | AAS | $3,558 | $7,914 | | | | |
| Southern West Virginia Comm & Tech College (Mt Gay) | 20 | Aug | 21 | AAS | $1,704 | $6,626 | | | | |
| **Wisconsin** | | | | | | | | | | |
| Chippewa Valley Technical College (Eau Claire) | 28 | Sep | 21 | AS | $3,200 | $17,372 | | | | |
| Madison Area Technical College | 14 | Aug | 18 | AAS | $4,800 | $150,000 | | | | |
| Milwaukee Area Technical College | 16 | Aug Jan | 19 | AAS | $2,790 | $19,100 | | • | • | • |
| Northeast Wisconsin Technical College (Green Bay) | 16 | Aug | 21 | AS | $2,589 | $14,275 | | • | | • |
| Western Technical College (La Crosse) | 20 | Aug | 20 | AAS | $2,700 | $16,917 | | | | |

*Data are shown only for programs that completed the 2006 AMA Survey of Health Professions Education Programs
‡Key to Offers: 1: Evening or weekend classes; 2: Non-English instruction; 3: Cultural competence instruction

## Cytogenetic Technologist

### Connecticut

**University of Connecticut**
Cytogenetic Technology Prgm
School of Allied Health, 222 Koons Halls
358 Mansfield Rd Unit 2101
Storrs, CT 06269
www.uconn.edu
*Prgm Dir:* Martha Keagle, MEd CT(ASCP) CLSp(CG)
*Tel:* 860 486-0036    *Fax:* 860 486-5375
*E-mail:* martha.keagle@uconn.edu

### Georgia

**Kennesaw State University**
Cytogenetic Technology Prgm
1000 Chastain Rd
Bldg 12
Kennesaw, GA 30144-5591
*Prgm Dir:* Xueya Hauge, PhD FABMG
*Tel:* 770 423-6163    *Fax:* 770 423-6625
*E-mail:* xhauge@kennesaw.edu

### Michigan

**Northern Michigan University**
Cytogenetic Technology Prgm
1401 Presque Isle Ave
3513 W Science Bldg
Marquette, MI 49855-5346
*Prgm Dir:* Lucille A Contois, MA MT(ASCP)
*Tel:* 906 227-1660    *Fax:* 906 227-1309
*E-mail:* lcontois@nmu.edu

### Texas

**Univ of Texas M D Anderson Cancer Ctr**
Cytogenetic Technology Prgm
Unit 146, 1515 Holcombe Blvd
Houston, TX 77030-4095
www.mdanderson.org/healthsciences
*Prgm Dir:* Vicki Hopwood, MS CLSp(CG)
*Tel:* 713 745-1688    *Fax:* 713 745-3337
*E-mail:* vhopwood@mdanderson.org
*Med Dir:* Lynne Abruzzo, MD

**Univ of Texas Hlth Sci Ctr at San Antonio**
Cytogenetic Technology Prgm
Dept of Clin Lab Sci
7703 Floyd Curl Dr
San Antonio, TX 78284-7772
www.uthscsa.edu/sah/cls/cls.html
*Prgm Dir:* Betty Dunn, MS CLSp(CG)
*Tel:* 210 567-8865    *Fax:* 210 567-8875
*E-mail:* dunnb0@uthscsa.edu

## Cytogenetic Technologist

| Programs* | Class Capacity | Begins | Length (months) | Award | Res Tuition | Non-res Tuition | Stipend | Offers:‡ 1 | 2 | 3 |
|---|---|---|---|---|---|---|---|---|---|---|
| **Connecticut** | | | | | | | | | | |
| University of Connecticut (Storrs) | 18 | Aug Jan | 24, 15 | Cert, BS | $6,096 | $18,600 | | | | |
| **Texas** | | | | | | | | | | |
| Univ of Texas Hlth Sci Ctr at San Antonio | 14 | Fall semester | 11 | Dipl, Cert, BS | $4,500 | $24,125 | | | | • |
| Univ of Texas M D Anderson Cancer Ctr (Houston) | 15 | Aug | 12 | Dipl, Cert, BS | $3,312 | $15,870 | | | | • |

*Data are shown only for programs that completed the 2006 AMA Survey of Health Professions Education Programs
‡Key to Offers: 1: Evening or weekend classes; 2: Non-English instruction; 3: Cultural competence instruction

## Diagnostic Molecular Scientist

### Kansas

**University of Kansas Medical Center**
Diagnostic Molecular Scientist Prgm
G014 Eaton
3901 Rainbow Blvd - MS 4048G-Eaton
Kansas City, KS 66160
www.cls.kumc.edu
*Prgm Dir:* Venus J Ward, PhD MT(ASCP) CLS(NCA)
*Tel:* 913 588-0154    *Fax:* 913 588-5222
*E-mail:* vward@kumc.edu
*Med Dir:* Lowell Tilzer, MD

### Michigan

**Northern Michigan University**
Diagnostic Molecular Scientist Prgm
1401 Presque Isle Ave
3513 W Science Bldg
Marquette, MI 49855-5346
*Prgm Dir:* Lucille A Contois, MA MT(ASCP)
*Tel:* 906 227-1660    *Fax:* 906 227-1309
*E-mail:* lcontois@nmu.edu

### Texas

**Univ of Texas M D Anderson Cancer Ctr**
Diagnostic Molecular Scientist Prgm
Program in Molecular Genetic Technology
1515 Holcombe Blvd, Unit 240
Houston, TX 77030
*Prgm Dir:* Peter Hu, MS CLS(NCA) CLSp(CG)
*Tel:* 713 563-3095
*E-mail:* pchu@mdanderson.org

**Texas Tech Univ Health Sciences Center**
Diagnostic Molecular Scientist Prgm
Dept of Laboratory Sciences & Primary Care
3601 Fourth St, 2-B-181, MS 6281
Lubbock, TX 79430
www.ttuhsc.edu
*Prgm Dir:* Lori Rice-Spearman, MS MT(ASCP)
*Tel:* 806 743-3255    *Fax:* 806 743-3249
*E-mail:* lori.ricespearman@ttuhsc.edu

## Histotechnician

### Arizona

**Pima Community College**
Histotechnician Prgm
2202 W Anklam Rd
Tucson, AZ 85709
http://wc.pima.edu/~jlindeberg/histotech
*Prgm Dir:* Jean Lindeberg, MS
*Tel:* 520 206-6031    *Fax:* 520 206-6902
*E-mail:* jean.lindeberg@pima.edu

### Arkansas

**Baptist Health**
Histotechnician Prgm
11900 Colonel Glenn Rd
Little Rock, AR 72210-2820
www.baptist-health.org
*Prgm Dir:* S Shane Jones, BS HT(ASCP)
*Tel:* 501 202-6700    *Fax:* 501 202-7712
*E-mail:* ssjones@baptist-health.org
*Med Dir:* Michelle Riddick, MD

### California

**Mt San Antonio College**
Histotechnician Prgm
1100 N Grand Ave
Walnut, CA 91789-1399
*Prgm Dir:* Virginia Pascoe, MS
*Tel:* 909 594-5611, Ext 4218    *Fax:* 909 468-4170
*E-mail:* vpascoe@mtsac.edu

### Connecticut

**Goodwin College**
Histotechnician Prgm
745 Burnside Ave
East Hartford, CT 06108
*Prgm Dir:* Zoe Ann Durkin, MEd HT(ASCP)
*Tel:* 860 727-6917
*E-mail:* ZDurkin@goodwincollege.org

### Delaware

**Delaware Tech & Comm Coll - Wilmington**
*Cosponsor:* Christiana Care Health Services
Histotechnician Prgm
333 North Shipley St
WSE 308k
Wilmington, DE 19801
www.dtcc.edu/wilmington/ah/htt.html
*Prgm Dir:* Ray Lynch, MD
*Tel:* 302 571-5320    *Fax:* 302 577-6431
*E-mail:* wlynch@dtcc.edu
*Med Dir:* Mary Virginia Iacocca, MD

### District of Columbia

**Armed Forces Institute of Pathology**
Histotechnician Prgm
6825 16th St, NW, Bldg 54
Washington, DC 20306-6000
*Prgm Dir:* Julia Wilson, BS HT(ASCP)
*Tel:* 202 782-2194    *Fax:* 202 782-8150
*E-mail:* wilsonj@afip.osd.mil

### Florida

**Florida Community College - Jacksonville**
Histotechnician Prgm
North Campus, 4501 Capper Rd
Jacksonville, FL 32218
*Prgm Dir:* Merry A Carter, MEd MT(ASCP) SC
*Tel:* 904 766-6511    *Fax:* 904 766-6654
*E-mail:* mcarter@fccj.edu
*Med Dir:* Nancy Lammert, MD

**Miami Dade College**
Histotechnician Prgm
950 NW 20th St
Miami, FL 33127-4693
www.mdc.edu
*Prgm Dir:* Caridad Ivis Gutierrez, BS HTL(ASCP)
*Tel:* 305 237-4231    *Fax:* 305 237-4278
*E-mail:* caridad.gutierrez@mdc.edu
*Med Dir:* Susan Baker, MD

### Georgia

**Darton College**
Histotechnician Prgm
2400 Gillionville Rd
Albany, GA 31707
www.darton.edu
*Prgm Dir:* Nancy T Beamon, MS MT(ASCP)
*Tel:* 229 317-6846
*E-mail:* nancybeamon@darton.edu
*Med Dir:* Frank Isele, MD

### Illinois

**OSF St Francis Medical Center**
Histotechnician Prgm
530 NE Glen Oak Ave
Peoria, IL 61637-0001
www.osfsaintfrancis.org
*Prgm Dir:* Carol E Becker, MS MT(ASCP) CLS(NCA)
*Tel:* 309 624-9021    *Fax:* 309 624-9150
*E-mail:* carol.e.becker@osfhealthcare.org
*Med Dir:* Michael P Hayes, MD

### Diagnostic Molecular Scientist

| Programs* | Class Capacity | Begins | Length (months) | Award | Res Tuition | Non-res Tuition | Stipend | Offers:‡ 1 | 2 | 3 |
|---|---|---|---|---|---|---|---|---|---|---|
| **Kansas** | | | | | | | | | | |
| University of Kansas Medical Center (Kansas City) | 12 | Aug | 24 | BS | $6,522 | $18,984 | | | | |
| **Texas** | | | | | | | | | | |
| Texas Tech Univ Health Sciences Center (Lubbock) | 16 | Jun 1 | 12 | Masters | $6,000 | $16,000 | | | | • |

*Data are shown only for programs that completed the 2006 AMA Survey of Health Professions Education Programs
‡Key to Offers: 1: Evening or weekend classes; 2: Non-English instruction; 3: Cultural competence instruction

PROGRAMS

# Indiana

**Indiana University**
Histotechnician Prgm
635 Barnhill Dr
MS-A128
Indianapolis, IN 46202-5120
http://msa.iusm.iu.edu/hpp/admissions/histo
*Prgm Dir:* Debra Wood, BS HT(ASCP)
*Tel:* 317 278-1690   *Fax:* 317 278-1820
*E-mail:* demwood@iupui.edu
*Med Dir:* Thomas M Ulbright, MD

# Maryland

**Harford Community College**
Histotechnician Prgm
401 Thomas Run Rd
Bel Air, MD 21014
*Prgm Dir:* Floyd M Grimm III, MEd
*Tel:* 410 836-4372   *Fax:* 410 836-4485
*E-mail:* fgrimm@harford.edu
*Med Dir:* Ramiro Lindado, MD

# Michigan

**Lansing Community College**
Histotechnician Prgm
Ingham Intermediate School District
611 Hagadorn Rd
Mason, MI 48854-9592
*Prgm Dir:* Elizabeth Toy-Krummrey, BS MA HT(ASCP)
*Tel:* 517 244-1357   *Fax:* 517 676-3602
*E-mail:* bkrummre@inghamisd.org
*Med Dir:* Patricia K Senagaore, MD

**William Beaumont Hospital**
Histotechnician Prgm
3601 W 13 Mile Rd
Royal Oak, MI 48079-6769
www.beaumont.edu/alliedhealth
*Prgm Dir:* Peggy A Wenk, BA BS HTL(ASCP)SLS
*Tel:* 248 898-9079   *Fax:* 248 898-9054
*E-mail:* pwenk@beaumont.edu
*Med Dir:* Jacqueline Trupiano, MD

# Minnesota

**Argosy University/Twin Cities**
Histotechnician Prgm
Twin Cities Campus
1515 Central Parkway
Eagan, MN 55121
*Prgm Dir:* Joyce Sohrabian, BA HT(ASCP)
*Tel:* 952 844-0064   *Fax:* 952 844-0472
*E-mail:* jsohrabian@argosyu.edu
*Med Dir:* Matt D McCoy, MD

# New York

**SUNY Agric & Tech College at Cobleskill**
Histotechnician Prgm
111 Schenectady Ave
Cobleskill, NY 12043
www.cobleskill.edu
*Prgm Dir:* Pamela Colony, PhD HT(ASCP)
*Tel:* 518 255-5417   *Fax:* 518 255-5113
*E-mail:* colonyp@cobleskill.edu
*Med Dir:* Russell E Newkirk, MD

# Ohio

**Columbus State Community College**
Histotechnician Prgm
550 E Springs St
Columbus, OH 43215
www.cscc.edu
*Prgm Dir:* Peggy Mayo, MEd MLT(ASCP)
*Tel:* 614 287-2608   *Fax:* 614 287-3854
*E-mail:* pmayo@cscc.edu

**Youngstown State University**
Histotechnician Prgm
Dept of Health Professions
One University Plaza
Youngstown, OH 44555
www.ysu.edu
*Prgm Dir:* Maria E Delost, PhD MT(ASCP) CLS (NCA)
*Tel:* 330 941-1761   *Fax:* 330 941-2921
*E-mail:* medelost@ysu.edu
*Med Dir:* Norton I German, MD

# Pennsylvania

**Conemaugh Memorial Medical Center**
Histotechnician Prgm
1086 Franklin St
Johnstown, PA 15905-4305
*Prgm Dir:* Gerard Campagna, PA (ASCP) BS HT (ASCP)
*Tel:* 814 534-9831   *Fax:* 814 534-9372
*E-mail:* gcampag@conemaugh.org
*Med Dir:* Curtis Steven Goldblatt, MD

# Texas

**Houston Community College**
Histotechnician Prgm
1900 Pressler
Houston, TX 77030
www.hccs.edu
*Prgm Dir:* Theresa Spain, MEd MT(ASCP) CLS(NCA)
*Tel:* 713 718-5518   *Fax:* 713 718-7653
*E-mail:* theresa.spain@hccs.edu
*Med Dir:* John Hicks, BS, DDS, MS, PhD, MD

**Univ of Texas M D Anderson Cancer Ctr**
Histotechnician Prgm
1515 Holcombe Blvd, Box 206
Houston, TX 77030
www.mdanderson.org/healthsciences
*Prgm Dir:* Hazel V Dalton, MS HT(ASCP) OIHC
*Tel:* 713 794-5877   *Fax:* 713 745-0172
*E-mail:* hdalton@mdanderson.org
*Med Dir:* Stanley R Hamilton, MD

**Univ of Texas Hlth Sci Ctr at San Antonio**
Histotechnician Prgm
7703 Floyd Curl Dr
San Antonio, TX 78284
*Prgm Dir:* Cynthia Morris, BS HTL/HT(ASCP) QIHC
*Tel:* 210 567-4059   *Fax:* 210 567-2478
*E-mail:* morris@pathology.uthscsa.edu
*Med Dir:* Robert Reddick, MD

# Wisconsin

**St Joseph's Hospital**
*Cosponsor:* Marshfield Clinic
Histotechnician Prgm
611 N St Joseph Ave
1000 N Oak
Marshfield, WI 54449-1898
www.marshfieldlaboratories.org/career.asp
*Prgm Dir:* Katherine Gorman, BS HT/HTL(ASCP)
*Tel:* 715 387-7790   *Fax:* 715 389-5353
*E-mail:* gorman.katherine@marshfieldclinic.org
*Med Dir:* Kathryn Kolquist, MD

## Histotechnician

| Programs* | Class Capacity | Begins | Length (months) | Award | Res Tuition | Non-res Tuition | Stipend | Offers:‡ 1 | 2 | 3 |
|---|---|---|---|---|---|---|---|---|---|---|
| **Arizona** | | | | | | | | | | |
| Pima Community College (Tucson) | 24 | Fall | 20 | Dipl, Cert, Assoc | $44 | $75 | | | | |
| **Arkansas** | | | | | | | | | | |
| Baptist Health (Little Rock) | 5 | Jul | 12 | Cert | $4,500 | $8,145 | | | | |
| **Delaware** | | | | | | | | | | |
| Delaware Tech & Comm Coll - Wilmington | 8 | May | 24 | AAS | $1,956 | $4,890 | | | | |
| **Florida** | | | | | | | | | | |
| Miami Dade College | 20 | Aug | 27 | Dipl, AS | $60 | $211 | | • | | |
| **Georgia** | | | | | | | | | | |
| Darton College (Albany) | 12 | Fall | 8, 24 | Cert, AS | $1,178 | $4,655 | | | | |
| **Illinois** | | | | | | | | | | |
| OSF St Francis Medical Center (Peoria) | 2-3 | Sep | 7 | Cert | $1,000 | $2,000 | | | | |
| **Indiana** | | | | | | | | | | |
| Indiana University (Indianapolis) | 50 | Aug | 10 | Cert | $5,000 | $5,000 | | | | |

*Data are shown only for programs that completed the 2006 AMA Survey of Health Professions Education Programs
‡Key to Offers: 1: Evening or weekend classes; 2: Non-English instruction; 3: Cultural competence instruction

# Histotechnologist

## Histotechnician

| Programs* | Class Capacity | Begins | Length (months) | Award | Res Tuition | Non-res Tuition | Stipend | Offers:‡ 1 | 2 | 3 |
|---|---|---|---|---|---|---|---|---|---|---|
| **Maryland** | | | | | | | | | | |
| Harford Community College (Bel Air) | 6 | Sep | 24, 10 | Cert, AAS | $2,682 | $8,044 | | • | | |
| **Michigan** | | | | | | | | | | |
| Lansing Community College (Mason) | 12 | Aug | 20 | Cert | $780 | $1,032 | | | | |
| William Beaumont Hospital (Royal Oak) | 2-3 | Sep | 7 | Cert | | | | | | • |
| **New York** | | | | | | | | | | |
| SUNY Agric & Tech College at Cobleskill | 16 | Aug Jan | 24, 48 | Dipl, AAS, BT | $4,350 | $7,210 | | | | • |
| **Ohio** | | | | | | | | | | |
| Columbus State Community College | 12 | Every other year | 9 | Cert | $1,343 | $2,737 | | | | |
| Youngstown State University | 5 | Aug | 24 | AAS | $6,104 | $11,312 | | • | | |
| **Pennsylvania** | | | | | | | | | | |
| Conemaugh Memorial Medical Center (Johnstown) | | Jul Jan | 12 | Cert | $5,750 | $5,750 | | | | |
| **Texas** | | | | | | | | | | |
| Houston Community College | 15 | Fall | 21 | AAS | $2,000 | $4,000 | | | | • |
| Univ of Texas M D Anderson Cancer Ctr (Houston) | 4 | Sep | 12 | Cert | $3,330 | $14,818 | | | | • |
| **Wisconsin** | | | | | | | | | | |
| St Joseph's Hospital (Marshfield) | 2 | Mar Sep | 12 | Cert | | | | | | |

*Data are shown only for programs that completed the 2006 AMA Survey of Health Professions Education Programs
‡Key to Offers: 1: Evening or weekend classes; 2: Non-English instruction; 3: Cultural competence instruction

## Histotechnologist

### Florida

**Barry University**
Histotechnology Prgm
11300 Northeast Second Ave
Miami Shores, FL 33161-6695
www.barry.edu
*Prgm Dir:* Alicia A Zuniga, PhD CLSup HTL(ASCP) AHI(AMT)
*Tel:* 305 899-3220   *Fax:* 305 899-3183
*E-mail:* azuniga@mail.barry.edu

### Michigan

**William Beaumont Hospital**
Histotechnology Prgm
3601 W 13 Mile Rd
Royal Oak, MI 48073-6769
www.beaumont.edu/alliedhealth
*Prgm Dir:* Peggy Wenk, BA BS HTL(ASCP)SLS
*Tel:* 248 898-9079   *Fax:* 248 898-9054
*E-mail:* pwenk@beaumont.edu
*Med Dir:* Jacqueline Trupiano, MD

## South Carolina

**Medical University of South Carolina**
Histotechnology Prgm
165 Ashley Ave, Ste 205
PO Box 250332
Charleston, SC 29425
*Prgm Dir:* Nina Epps, MS/HPE MT(ASCP)
*Tel:* 843 792-1906   *Fax:* 843 792-3327
*E-mail:* eppsn@musc.edu

## Texas

**Tarleton State University**
Histotechnology Prgm
1501 Enderly Place
Fort Worth, TX 76104
*Prgm Dir:* Glenda F Hoye, BS HT(ASCP)
*Tel:* 817 926-1101, Ext 234
*E-mail:* hoye@tarleton.edu

## Pathologists' Assistant

### Connecticut

**Quinnipiac University**
Pathologists' Assistant Prgm
275 Mt Carmel Ave
Hamden, CT 06518
www.quinnipiac.edu
*Prgm Dir:* Kenneth V Kaloustian, PhD
*Tel:* 203 582-8676   *Fax:* 203 582-8706
*E-mail:* kenneth.kaloustian@quinnipiac.edu
*Med Dir:* Nelson Gelfman, MD

### Illinois

**Rosalind Franklin Univ of Medicine & Science**
Pathologists' Assistant Prgm
3333 Green Bay Rd
North Chicago, IL 60064
www.rosalindfranklin.edu
*Prgm Dir:* John E Vitale, MHS PA(ASCP)
*Tel:* 847 578-8638
*E-mail:* john.vitale@rosalindfranklin.edu
*Med Dir:* Osvaldo Rubinstein, MD

## Histotechnologist

| Programs* | Class Capacity | Begins | Length (months) | Award | Res Tuition | Non-res Tuition | Stipend | Offers:‡ 1 | 2 | 3 |
|---|---|---|---|---|---|---|---|---|---|---|
| **Florida** | | | | | | | | | | |
| Barry University (Miami Shores) | 14 | Fall | 30, 12 | Dipl, Cert, BS | $20,000 | $0 | $5,000 | • | | |
| **Michigan** | | | | | | | | | | |
| William Beaumont Hospital (Royal Oak) Kinesiotherapist | 4 | Sep | 12 | Cert | | | | | | • |
| **California** | | | | | | | | | | |
| San Diego State University | 20 | Sep | 48 | BS | $2,488 | $16,920 | | • | | |
| **Mississippi** | | | | | | | | | | |
| University of Southern Mississippi (Hattiesburg) | 125 | Aug | 24 | BS | $4,106 | $5,170 | | | | |
| **North Carolina** | | | | | | | | | | |
| Shaw University (Raleigh) | 80 | Aug | 45 | BS | $6,272 | $11,196 | | | | |

*Data are shown only for programs that completed the 2006 AMA Survey of Health Professions Education Programs
‡Key to Offers: 1: Evening or weekend classes; 2: Non-English instruction; 3: Cultural competence instruction

# Indiana

**Indiana University**
Pathologists' Assistant Prgm
School of Medicine
635 Barnhill Dr
Indianapolis, IN 46202
*Prgm Dir:* Randy Stine, PA
*Tel:* 317 274-5781
*E-mail:* crstine@iupui.edu

# Maryland

**University of Maryland**
Pathologists' Assistant Prgm
Dept of Pathology-UMMC
10 S Pine St
Baltimore, MD 21201
www.medschool.umaryland.edu/pathology
*Prgm Dir:* Raymond T Jones, PhD
*Tel:* 410 328-1221, Ext 1   *Fax:* 410 328-5508
*E-mail:* rjones@som.umaryland.edu
*Med Dir:* John C Papadimitriou, MD PhD

# Michigan

**Wayne State University**
*Cosponsor:* Eugene Applebaum College of Pharmacy +
   Health Sci
Pathologists' Assistant Prgm
Department of Fundamental & Appl Sci, Mort Sci
5439 Woodward Ave
Detroit, MI 48202
www.wayne.edu
*Prgm Dir:* Peter D Frade, PhD
*Tel:* 313 577-7874   *Fax:* 313 577-4456
*E-mail:* ab8123@wayne.edu
*Med Dir:* Gail Bentley, MD

# North Carolina

**Duke University Medical Center**
Pathologists' Assistant Prgm
Box 3712
Durham, NC 27710
*Prgm Dir:* Kenneth R Broda, MA PhD
*Tel:* 919 681-4847   *Fax:* 919 668-2767
*E-mail:* broda001@mc.duke.edu
*Med Dir:* Roger McLendon, MD

# Ohio

**Ohio State University**
Pathologists' Assistant Prgm
410 W 10th Ave, N-308 Doan Hall
Columbus, OH 43210
www.pathology.medctr.ohio-state.edu/ext
*Prgm Dir:* Charles Hitchcock, MD PhD
*Tel:* 614 247-7469   *Fax:* 614 293-7273
*E-mail:* charles.hitchcock@osumc.edu
*Med Dir:* Nilsa Ramirez, MD

## Phlebotomist*

# Arkansas

**Phillips Community College/U of Arkansas**
Phlebotomist Prgm
1000 Campus Dr / PO Box 785
Helena-West Helena, AR 72342
www.pccua.edu
*Prgm Dir:* Julie Byrd, MT(ASCP) MS
*Tel:* 870 338-6474, Ext 1109   *Fax:* 870 338-7542
*E-mail:* jbyrd@pccua.edu

# California

**Loma Linda University**
Phlebotomist Prgm
School of Allied Health Professions
Dept of Clinical Lab Sci, NH A918
Loma Linda, CA 92350
*Prgm Dir:* Monique Gilbert, BS MT(ASCP)
*Tel:* 909 558-4966
*E-mail:* mgilbert@sahp.llu.edu
*Med Dir:* James Pappas, MD

**California State University - Long Beach**
Phlebotomist Prgm
Dept Biological Sciences
1250 Bellflower Blvd
Long Beach, CA 90840
www.csulb.edu
*Prgm Dir:* Carol Ann Itatani, PhD MS MT(ASCP)
*Tel:* 562 985-4825   *Fax:* 562 985-8878
*E-mail:* citatani@csulb.edu
*Med Dir:* Natalie Harris

# Florida

**Florida Hospital**
Phlebotomist Prgm
601 E Rollins St
Orlando, FL 32803
www.flhosp.org
*Prgm Dir:* Patricia L Rogers, BS MT(ASCP) SBB
*Tel:* 407 303-1855   *Fax:* 407 893-6943
*E-mail:* pat.rogers@flhosp.org
*Med Dir:* Luis Guarda, MD

*Additional information about programs that
returned the AMA's survey is available in a table
beginning on page 84.

| Pathologists' Assistant | | | | | | | | | | |
|---|---|---|---|---|---|---|---|---|---|---|
| Programs* | Class Capacity | Begins | Length (months) | Award | Res Tuition | Non-res Tuition | Stipend | Offers:‡ 1 | 2 | 3 |
| **Connecticut** | | | | | | | | | | |
| Quinnipiac University (Hamden) | 18 | Jun | 24 | MHS | $25,000 | $0 | | • | | |
| **Illinois** | | | | | | | | | | |
| Rosalind Franklin Univ of Medicine & Science (North Chicago) | 22 | Jun 1 | 22 | MS | $17,210 | $17,210 | | | | |
| **Maryland** | | | | | | | | | | |
| University of Maryland (Baltimore) | 10 | Aug | 22 | Dipl, Cert, MS | $17,000 | $27,000 | | | | |
| **Michigan** | | | | | | | | | | |
| Wayne State University (Detroit) | 10 | Sep | 24 | BS | $16,305 | $37,488 | | | | |
| **North Carolina** | | | | | | | | | | |
| Duke University Medical Center (Durham) | 6 | Aug | 24 | Dipl, Cert, MHS | $21,000 | $21,000 | | | | |
| **Ohio** | | | | | | | | | | |
| Ohio State University (Columbus) | 2-3 | Jul 1 | 24 | Dipl, MS | $12,584 | $30,388 | | | | |

*Data are shown only for programs that completed the 2006 AMA Survey of Health Professions Education Programs
‡Key to Offers: 1: Evening or weekend classes; 2: Non-English instruction; 3: Cultural competence instruction

# Georgia

**Dalton State College**
Phlebotomist Prgm
650 College Dr
Dalton, GA 30720-3778
*Prgm Dir:* Doris M Shoemaker, EdS MLM MT(ASCP)
*Tel:* 706 272-4512    *Fax:* 706 272-2517
*E-mail:* dshoemaker@em.daltonstate.edu
*Med Dir:* Eugene Fong, MD

# Hawaii

**Kapi'olani Community College**
Phlebotomist Prgm
4303 Diamond Head Rd
Honolulu, HI 96816
*Prgm Dir:* Marcia A Armstrong, MS CLS(NCA) MT(ASCP)
*Tel:* 808 734-9231    *Fax:* 808 734-9126
*E-mail:* marciaa@hawaii.edu

# Illinois

**College of Lake County**
Phlebotomist Prgm
19351 W Washington St
Grayslake, IL 60030-1198
www.clcillinois.edu
*Prgm Dir:* Remedios H Tesch, MSEd MT(ASCP)
*Tel:* 847 543-2878    *Fax:* 847 223-1357
*E-mail:* rtesch@clcillinois.edu

**Moraine Valley Community College**
Phlebotomist Prgm
9000 College Parkway
Palos Hills, IL 60465
www.morainevalley.edu/healthsciences
*Prgm Dir:* Susan E Phelan, MHS MT(ASCP)
*Tel:* 708 974-5743    *Fax:* 708 974-0185
*E-mail:* phelan@morainevalley.edu

**South Suburban College**
Phlebotomist Prgm
15800 S State St
South Holland, IL 60473
*Prgm Dir:* Andrea Stone, MPA MT(ASCP)
*Tel:* 708 596-2000, Ext 2260
*E-mail:* astone@ingalls.org

# Indiana

**Indiana University Northwest**
Phlebotomist Prgm
School of Nursing and Health Professions
3400 Broadway
Gary, IN 46408
*Prgm Dir:* Susan A K Higgins, MS MT(ASCP)SC
*Tel:* 219 980-6923    *Fax:* 219 980-6649
*E-mail:* sahiggin@iun.edu

**Ivy Tech Community College - South Bend**
Phlebotomist Prgm
220 Dean Johnson Blvd
South Bend, IN 46619-3829
*Prgm Dir:* Pamela B Primrose, MLFSC MT(ASCP)
*Tel:* 574 289-7001, Ext 5401    *Fax:* 574 236-7166
*E-mail:* pprimros@ivytech.edu
*Med Dir:* Robert Tomec MD, MD

# Louisiana

**Rapides Regional Medical Center**
Phlebotomist Prgm
211 Fourth St
Alexandria, LA 71301-8421
*Prgm Dir:* E Ann Faircloth Dauzart, RN PBT(ASCP)
*Tel:* 318 473-3207    *Fax:* 318 473-3079
*E-mail:* Elizabeth.Faircloth@HCAHealthcare.com
*Med Dir:* Seldon J Deshotels, MD

**Bossier Parish Community College**
Phlebotomist Prgm
3220 East Texas St
Bossier City, LA 71111
www.bpcc.edu
*Prgm Dir:* Pam M Tully, MHS MT(ASCP) PBT(ASCP)
*Tel:* 318 678-6355    *Fax:* 318 678-6199
*E-mail:* ptully@bpcc.edu

**L E Fletcher Technical Community College**
Phlebotomist Prgm
310 St Charles St
PO Box 5033
Houma, LA 70360
*Prgm Dir:* Marcie Guidry, RN MSN
*Tel:* 985 876-8900
*E-mail:* mguidry@lefletcher.edu

**Delgado Community College**
Phlebotomist Prgm
615 City Park Ave
New Orleans, LA 70119
*Prgm Dir:* Erica Cassimere
*Tel:* 504 483-4328    *Fax:* 504 483-4198
*E-mail:* ecassi@dcc.edu
*Med Dir:* Bobby Rodwig, MD

# Maryland

**Hagerstown Business College**
Phlebotomist Prgm
18618 Crestwood Dr
Hagerstown, MD 21742
*Prgm Dir:* Michele Buzard, MS MT (ASCP) BA
*Tel:* 301 739-2680, Ext 201    *Fax:* 301 791-7661
*E-mail:* mbuzard@hagerstownbusinesscol.edu
*Med Dir:* Ernst Uzicanin, MD

# Michigan

**Medright, Inc**
Phlebotomist Prgm
427 Allen Rd
Ferndale, MI 48220
*Prgm Dir:* Gail Lucas, EMT-P BS
*Tel:* 248 547-0834    *Fax:* 248 543-8675
*E-mail:* medright1@aol.com

**Baker College of Owosso**
Phlebotomist Prgm
1020 S Washington St
Owosso, MI 48867
www.baker.edu
*Prgm Dir:* Diane Denard, BHSA MT(HHS) MLT( ASCP)
*Tel:* 989 729-3383    *Fax:* 989 729-3411
*E-mail:* diane.denard@baker.edu
*Med Dir:* Qazi Azher, MD

# Minnesota

**College of St Catherine - Minneapolis**
Phlebotomist Prgm
601 25th Ave S
Minneapolis, MN 55454
http://minerva.stkate.edu/offices/academic/phlebotomist.nsf
*Prgm Dir:* Julie Mumm, MT(ASCP)
*Tel:* 651 690-7764
*E-mail:* jrmumm@stkate.edu

# Missouri

**Saint Luke's Hospital**
Phlebotomist Prgm
4401 Wornall Rd
Kansas City, MO 64111
www.saint-lukes.org
*Prgm Dir:* Brenna Ildza, MT(ASCP)SH PBT(ASCP)
*Tel:* 816 932-2074    *Fax:* 816 932-2259
*E-mail:* bildza@saint-lukes.org
*Med Dir:* Marge Zucker, MD

# New Jersey

**Univ of Med & Dent of New Jersey**
Phlebotomist Prgm
65 Bergen St
Bergen Bldg GB-159
Newark, NJ 07107
www.shrp.umdnj.edu/programs/cls/phlebotomy.htm
*Prgm Dir:* H Jesse Guiles, EdD MT(ASCP)SC
*Tel:* 973 972-6863    *Fax:* 973 972-8527
*E-mail:* guiles@umdnj.edu

# New York

**Trocaire College**
Phlebotomist Prgm
360 Choate Ave
Buffalo, NY 14220
www.trocaire.edu
*Prgm Dir:* Kathy L Smith, BS MT MS-ASCP NCA
*Tel:* 716 826-1200    *Fax:* 716 828-3595
*E-mail:* chernowskim@trocaire.edu

**Nicholas H Noyes Memorial Hospital**
Phlebotomist Prgm
111 Clara Barton St
Dansville, NY 14437
www.noyes-health.org
*Prgm Dir:* Dawn Kennedy
*Tel:* 585 335-4225    *Fax:* 585 335-4370
*E-mail:* dkennedy@noyes-hospital.org
*Med Dir:* CHI KIM, MD

**Orange County Community College**
Phlebotomist Prgm
115 South St
Middletown, NY 10940
www.sunyorange.edu
*Prgm Dir:* Rosamaria B Contarino, MS MT(ASCP) CLS(NCA)
*Tel:* 845 341-4136    *Fax:* 845 341-4122
*E-mail:* rcontari@sunyorange.edu
*Med Dir:* Joseph Napolitano, MD

**Rochester General Hospital**
Phlebotomist Prgm
1425 Portland Ave
Rochester, NY 14621
*Prgm Dir:* Nancy C Mitchell, MS MT(ASCP)DLM
*Tel:* 585 922-4274    *Fax:* 585 922-2088
*E-mail:* nancy.mitchell@viahealth.org
*Med Dir:* Theodor K Mayer, MD, PhD

# North Carolina

**Asheville-Buncombe Technical Comm College**
Phlebotomist Prgm
340 Victoria Rd
Asheville, NC 28801
www.abtech.edu
*Prgm Dir:* Melissa Hyatt, BS MT(ASCP) MHS
*Tel:* 828 254-1921, Ext 266    *Fax:* 828 281-9734
*E-mail:* MHyatt@abtech.edu
*Med Dir:* Joseph Sleater, MD

**Carolinas College of Health Sciences**
Phlebotomist Prgm
PO Box 32861
Charlotte, NC 28232
www.carolinascollege.edu
*Prgm Dir:* Elizabeth T Anderson, MHDL MT(ASCP)
*Tel:* 704 355-4275   *Fax:* 704 355-5967
*E-mail:* banderson@carolinas.org
*Med Dir:* Edward Lipford, MD

**Fayetteville Technical Community College**
Phlebotomist Prgm
2201 Hull Rd, PO Box 35236
Fayetteville, NC 28303
www.faytechcc.edu
*Prgm Dir:* Linda Starling, MT (ASCP) PBT (ASCP)
*Tel:* 910 678-8538   *Fax:* 910 678-8500
*E-mail:* starlinl@faytechcc.edu

**James Sprunt Community College**
Phlebotomist Prgm
PO Box 398, James Sprunt Dr
Kenansville, NC 28349
*Prgm Dir:* Rhonda Ferrell, RN MSN
*Tel:* 910 296-2450   *Fax:* 910 296-1636
*E-mail:* rferrell@jscc.cc.nc.us
*Med Dir:* Roland Draughn, MD

**Wake Technical Community College**
Phlebotomist Prgm
Consortium of Wake Tech & Durham Tech CC
9101 Fayetteville Rd
Raleigh, NC 27603-5696
*Prgm Dir:* Pamela B Horton, MEd MT(ASCP)
*Tel:* 919 212-3818   *Fax:* 919 250-4329
*E-mail:* pbhorton@waketech.edu

**Nash Community College**
Phlebotomist Prgm
522 N Old Carriage Rd, PO Box 7480
Rocky Mount, NC 27804
www.nashcc.edu
*Prgm Dir:* TaMeika Dickens, MT(ASCP)DLM MPH
*Tel:* 252 451-8336   *Fax:* 252 451-8201
*E-mail:* debg@nash.cc.nc.us

**Brunswick Community College**
Phlebotomist Prgm
PO Box 30
50 College Rd
Supply, NC 28462
www.brunswickcc.edu
*Prgm Dir:* Julianna Olsen, PBT(ASCP)
*Tel:* 910 755-7324   *Fax:* 910 754-9609
*E-mail:* olsenj@brunswickcc.edu

**Southwestern Community College**
Phlebotomist Prgm
447 College Dr
Sylva, NC 28779
*Prgm Dir:* Andrea L Rowland, MBA MT(ASCP)
*Tel:* 828 586-4091, Ext 312   *Fax:* 828 586-3129
*E-mail:* andrea@southwesterncc.edu

**Halifax Community College**
Phlebotomist Prgm
PO Drawer 809
Hwy 158
Weldon, NC 27890
www.hcc.cc.nc.us
*Prgm Dir:* Lori M Howard, BSMT(ASCP)SM
*Tel:* 252 536-7284
*E-mail:* howardl@halifaxcc.edu

**Rockingham Community College**
Phlebotomist Prgm
PO Box 38
Wentworth, NC 27375-0038
www.rockinghamcc.edu
*Prgm Dir:* Darlene Vestal, MT ASCP PBT ASCP
*Tel:* 336 342-4261, Ext 2207   *Fax:* 336 634-1253
*E-mail:* guyn@rockinghamcc.edu

**Southeastern Community College**
Phlebotomist Prgm
PO Box 151
Whiteville, NC 28472
www.sccnc.edu
*Prgm Dir:* Patricia Wright, MT(ASCP)
*Tel:* 910 642-7141, Ext 312   *Fax:* 910 642-3257
*E-mail:* pwright@sccnc.edu

**Cape Fear Community College**
Phlebotomist Prgm
411 N Front St
Wilmington, NC 28401-3993
http://cfcc.edu
*Prgm Dir:* Vickie Pridgen, MT(ASCP) AMT
*Tel:* 910 362-7492
*E-mail:* vpridgen@cfcc.edu

# North Dakota

**Bismarck State College**
Phlebotomist Prgm
PO Box 5505
1500 Edwards Ave
Bismarck, ND 58501
www.bismarckstate.edu
*Prgm Dir:* Angela Uhlich, MS MT(ASCP)SBB
*Tel:* 701 323-5482   *Fax:* 701 323-5831
*E-mail:* Angela.Uhlich@bsc.nodak.edu
*Med Dir:* Dwight Hertz, MD (CP/AP)

# Ohio

**Collins Career Center**
Phlebotomist Prgm
11627 State Route 243
Chesapeake, OH 45619
www.collins-cc.k12.oh.us
*Prgm Dir:* Shawn P Boggs, MT(ASCP)
*Tel:* 740 867-6641
*E-mail:* shawn.boggs@kdmc.net

**Cuyahoga Community College**
Phlebotomist Prgm
Metropolitan Campus
2900 Community College Ave
Cleveland, OH 44115-3196
*Prgm Dir:* Barbara Freeman, MEd MT(ASCP)
*Tel:* 216 987-4438   *Fax:* 216 987-4386
*E-mail:* barbara.freeman@tri-c.edu
*Med Dir:* John Anhalt, MD

**Columbus State Community College**
Phlebotomist Prgm
550 E Spring St
Columbus, OH 43215
www.cscc.edu/nursing/phoverview.htm
*Prgm Dir:* Peggy Mayo, MEd MLT (ASCP)
*Tel:* 614 287-2608   *Fax:* 614 287-3854
*E-mail:* pmayo@cscc.edu

**Lorain County Community College**
Phlebotomist Prgm
1005 N Abbe Rd
Elyria, OH 44035
*Prgm Dir:* James E Daly, MEd MT(ASCP)
*Tel:* 800 995-5222, Ext 7194   *Fax:* 440 366-4116
*E-mail:* jdaly@lorainccc.edu
*Med Dir:* Priscilla Heimann, MD

**Trumbull Memorial Hospital**
Phlebotomist Prgm
1350 E Market St
Warren, OH 44482-1269
*Prgm Dir:* James Obermiyer, BA MT(ASCP)
*Tel:* 330 841-9128   *Fax:* 330 841-9110
*E-mail:* jobermiyer@forumhealth.org

**Zane State College**
Phlebotomist Prgm
1555 Newark Rd
Zanesville, OH 43701
www.zanestate.edu
*Prgm Dir:* Jane Poulson-Dunlap, MS MT(ASCP)
*Tel:* 740 588-1228
*E-mail:* jdunlap@zanestate.edu

# Oklahoma

**Tulsa Community College**
Phlebotomist Prgm
909 S Boston Ave
Tulsa, OK 74119
*Prgm Dir:* Karen L Holmes, MT(ASCP)
*Tel:* 918 595-7008   *Fax:* 918 595-7091
*E-mail:* kholmes@tulsacc.edu
*Med Dir:* Melvin VanBoven, DO

# Pennsylvania

**Montgomery County Community College**
Phlebotomist Prgm
340 DeKalb Pike
PO Box 400
Blue Bell, PA 19422-0796
www.mc3.edu
*Prgm Dir:* Debra Eckman, MS MT(ASCP)
*Tel:* 215 641-6487   *Fax:* 215 619-7178
*E-mail:* deckman@mc3.edu
*Med Dir:* Irwin Hollander, MD

**Community College of Beaver County**
Phlebotomist Prgm
1 Campus Dr
Monaca, PA 15061
www.ccbc.edu
*Prgm Dir:* Estelle Del Principe, BA MEd MT(ASCP)
*Tel:* 724 775-8561, Ext 189   *Fax:* 724 775-3003
*E-mail:* estelle.delprincipe@ccbc.edu

**Community College of Philadelphia**
Phlebotomist Prgm
1700 Spring Garden St
Philadelphia, PA 19130
*Prgm Dir:* Robin Gaynor Krefetz, MEd MT(ASCP) CLS
   (NCA)
*Tel:* 215 751-8511   *Fax:* 215 751-8937
*E-mail:* rkrefetz@ccp.edu
*Med Dir:* Benaz Torkkey, MD

# South Dakota

**Western Dakota Technical Institute**
Phlebotomist Prgm
800 Mickelson Dr
Rapid City, SD 57703
www.westerndakotatech.org
*Prgm Dir:* Lynn E Seime, MS MT(ASCP)
*Tel:* 605 394-4034, Ext 303   *Fax:* 605 394-1789
*E-mail:* lseime@wdti.tec.sd.us

# Tennessee

**Southwest Tennessee Community College**
Phlebotomist Prgm
PO Box 780
761 Linden Ave
Memphis, TN 38101-0780
www.southwest.tn.edu
*Prgm Dir:* Darius Y Wilson, MAT BS MT(ASCP)
*Tel:* 901 333-5407   *Fax:* 901 333-5391
*E-mail:* dwilson@southwest.tn.edu
*Med Dir:* Michael F Bugg, Medical Doctor

# Texas

### Austin Community College
Phlebotomist Prgm
Eastview Campus
3401 Webberville Rd
Austin, TX 78702
www.austincc.edu/health/phb
*Prgm Dir:* Terry Kotrla, MS MT(ASCP)BB
*Tel:* 512 223-5932   *Fax:* 512 223-5898
*E-mail:* kotrla@austincc.edu

### Christus Hospital - St Elizabeth
Phlebotomist Prgm
2830 Calder Ave
PO Box 5405
Beaumont, TX 77726-5405
www.christusste.org
*Prgm Dir:* Deborah R Zink, MBA MT(ASCP)
*Tel:* 409 899-7150, Ext 4002   *Fax:* 409 899-7991
*E-mail:* debbi.zink@christushealth.org
*Med Dir:* Kathryn Bommer, MD

# Wisconsin

### Milwaukee Area Technical College
Phlebotomist Prgm
Health Occupations Division
700 W State St
Milwaukee, WI 53233
www.matc.edu
*Prgm Dir:* Dennis Schmidt, MS MT(ASCP)
*Tel:* 414 297-7142   *Fax:* 414 297-6851
*E-mail:* SchmidtD@matc.edu

### Mid-State Technical College
Phlebotomist Prgm
933 Michigan Ave
Stevens Point, WI 54481-3141
*Prgm Dir:* Lori Jasper, MT(ASCP) CMA
*Tel:* 715 342-3106   *Fax:* 715 342-3134
*E-mail:* lori.jasper@mstc.edu
*Med Dir:* Lloyd Arnold, MD

### Northcentral Technical College
Phlebotomist Prgm
1100 W Campus Dr
Wausau, WI 54401
*Prgm Dir:* Janice Miller, BS MS MT(ASCP)
*Tel:* 715 675-3331, Ext 1371
*E-mail:* millerja@ntc.edu

## Phlebotomist

| Programs* | Class Capacity | Begins | Length (months) | Award | Res Tuition | Non-res Tuition | Stipend | Offers:‡ 1 | 2 | 3 |
|---|---|---|---|---|---|---|---|---|---|---|
| **Arkansas** | | | | | | | | | | |
| Phillips Community College/U of Arkansas (Helena-West Helena) | 8 | Aug | 9 | Cert | $1,500 | $2,910 | | | | |
| **California** | | | | | | | | | | |
| California State University - Long Beach | 12 | Jun 5 | 2 | Cert | $1,400 | $0 | | | | • |
| Loma Linda University | 12 | Jun | 1 | Cert | $1,788 | $1,788 | | • | | • |
| **Florida** | | | | | | | | | | |
| Florida Hospital (Orlando) | 10 | Jan Apr Jul Oct | 2 | Cert | $300 | $300 | | | | • |
| **Georgia** | | | | | | | | | | |
| Dalton State College | 16 | Aug Jan May | 6 | Cert | $1,656 | $6,280 | | | | • |
| **Hawaii** | | | | | | | | | | |
| Kapi'olani Community College (Honolulu) | 12 | Jan Apr Jun Sep | 2 | Cert, CC | $950 | $950 | | • | | |
| **Illinois** | | | | | | | | | | |
| College of Lake County (Grayslake) | 20 | Aug | 3 | Dipl, Cert | $70 | $194 | | | | |
| Moraine Valley Community College (Palos Hills) | 60 | Aug Jan May | 8 | Cert | $530 | $0 | | • | | • |
| South Suburban College (South Holland) | 15 | | 4, 1 | Cert | $1,067 | $3,025 | | • | | |
| **Indiana** | | | | | | | | | | |
| Indiana University Northwest (Gary) | 18 | Jan | 5 | Cert | $2,100 | $5,100 | | | | • |
| Ivy Tech Community College - South Bend | 15 | Fall Spring | 4 | | $1,600 | $3,200 | | • | | |
| **Louisiana** | | | | | | | | | | |
| Bossier Parish Community College (Bossier City) | 10 | Jan Aug | 4 | Cert | $1,720 | $3,860 | | • | | • |
| Rapides Regional Medical Center (Alexandria) | 9 | Apr Oct | 6 | Cert | | | | | | • |
| **Maryland** | | | | | | | | | | |
| Hagerstown Business College | 100 | Quarterly | 9 | Cert | $13,323 | $13,323 | | • | | |
| **Michigan** | | | | | | | | | | |
| Baker College of Owosso | 15 | Sep | 8 | Cert | $6,825 | $6,825 | | | | • |
| Medright, Inc (Ferndale) | 42 | Multiple start dates | | Cert | $590 | $590 | | • | | |
| **Minnesota** | | | | | | | | | | |
| College of St Catherine - Minneapolis | 12 | Fall Winter | 4 | Cert | $2,370 | $2,370 | | | | • |
| **Missouri** | | | | | | | | | | |
| Saint Luke's Hospital (Kansas City) | 8 | Jan Apr Jul Oct | 1 | Cert | $700 | $700 | | | | • |
| **New Jersey** | | | | | | | | | | |
| Univ of Med & Dent of New Jersey (Newark) | 27 | Sep | 6 | Cert | $890 | $890 | | • | | • |
| **New York** | | | | | | | | | | |
| Nicholas H Noyes Memorial Hospital (Dansville) | 6 | Fall spring | 2 | Cert | $550 | $550 | | | | • |
| Orange County Community College (Middletown) | 18 | Fall Spring | 4 | Cert | $3,000 | $6,000 | | | | • |
| Rochester General Hospital | 8 | Jan Sep | 3 | Cert | $350 | $0 | | | | |
| Trocaire College (Buffalo) | 16 | Mar Jun Oct | 2 | Cert | $0 | $899 | | • | | |
| **North Carolina** | | | | | | | | | | |
| Asheville-Buncombe Technical Comm College | 32 | Aug | 4 | Cert | $608 | $2,532 | | | | |
| Brunswick Community College (Supply) | 10 | Aug | 4 | Cert | $720 | $2,890 | | | | |
| Cape Fear Community College (Wilmington) | 13 | Fall Spring | 5 | Cert | $503 | $2,663 | | | | |
| Carolinas College of Health Sciences (Charlotte) | 10 | Jan May Sep | 2 | Cert | $505 | $505 | | | | |
| Fayetteville Technical Community College | 15 | Aug Jan Mar | 4 | Cert | $543 | $2,883 | | | | |

*Data are shown only for programs that completed the 2006 AMA Survey of Health Professions Education Programs
‡Key to Offers: 1: Evening or weekend classes; 2: Non-English instruction; 3: Cultural competence instruction

## Phlebotomist

| Programs* | Class Capacity | Begins | Length (months) | Award | Res Tuition | Non-res Tuition | Stipend | Offers:‡ 1 | 2 | 3 |
|-----------|---------------|--------|-----------------|-------|-------------|-----------------|---------|---|---|---|
| Halifax Community College (Weldon) | 8 | Fall Spring | 5 | Cert | | | | | | |
| Nash Community College (Rocky Mount) | 15 | Aug Jan | 3 | Cert | $498 | $2,652 | | | | |
| Rockingham Community College (Wentworth) | 18 | Fall semester | 5 | Cert | $491 | $2,526 | | • | | • |
| Southeastern Community College (Whiteville) | 20 | Jan 4 | 5 | Cert | $632 | $3,512 | | | | |
| Southwestern Community College (Sylva) | 12 | Jan | 4 | Cert | $474 | $2,634 | | • | | |
| Wake Technical Community College (Raleigh) | 15 | Jan Aug | 4 | Cert | $474 | $2,634 | | | | • |
| **North Dakota** | | | | | | | | | | |
| Bismarck State College | 10 | Aug Jan | 5 | Cert | $3,240 | $7,868 | | • | | • |
| **Ohio** | | | | | | | | | | |
| Collins Career Center (Chesapeake) | 30 | Oct | 11 | Cert | | | | • | | • |
| Columbus State Community College | 15 | Jun Jan | 6 | Cert | $558 | $1,077 | | | | • |
| Lorain County Community College (Elyria) | 30 | Aug | 9 | Cert | $1,753 | $2,112 | | | | |
| Zane State College (Zanesville) | 30 | Fall Spring | 5 | Cert | $76 | $152 | | | | • |
| **Pennsylvania** | | | | | | | | | | |
| Community College of Beaver County (Monaca) | 20 | Sep Jan | 4 | Cert | $621 | $1,206 | | | | • |
| Montgomery County Community College (Blue Bell) | 12 | Sep Jan | 4 | Cert | $410 | $820 | | | | |
| **South Dakota** | | | | | | | | | | |
| Western Dakota Technical Institute (Rapid City) | 16 | Aug | 9 | Dipl | $2,112 | $0 | | | | • |
| **Tennessee** | | | | | | | | | | |
| Southwest Tennessee Community College (Memphis) | 15 | Sep 1 Jan 15 | 6 | Cert | $1,198 | $4,405 | | | | |
| **Texas** | | | | | | | | | | |
| Austin Community College | 14 | Jan May Aug | 3 | Cert | $250 | $609 | | | | • |
| Christus Hospital - St Elizabeth (Beaumont) | 10 | Depends on need | 2 | Cert | $250 | $250 | | | | • |
| **Wisconsin** | | | | | | | | | | |
| Milwaukee Area Technical College | 35 | Aug Jan | 6 | Dipl | $2,790 | $19,100 | | • | • | • |

*Data are shown only for programs that completed the 2006 AMA Survey of Health Professions Education Programs
‡Key to Offers: 1: Evening or weekend classes; 2: Non-English instruction; 3: Cultural competence instruction

**Includes:**

- Career Counseling (CrC)
- College Counseling (ClC)
- Community Counseling (CC)
- Gerontological Counseling (GC)
- Marital, Couple and Family Counseling/Therapy (MFC/T)
- Mental Health Counselor (MHC)
- School Counseling (SC)
- Student Affairs (SA)
- Student Affairs-College Counseling (SACC)
- Student Affairs-Professional Practice (SAPP)
- Counselor Education and Supervision (CE)

*Note: Career information and educational programs for rehabilitation counseling and genetic counseling are listed separately.*

## Job Description

The counseling profession differs from other human service professions in its developmental approach to problem solving. Counselors deal with human development concerns through support, therapeutic approaches, consultation, evaluation, teaching, and research. Simply stated, counseling is the art of helping people grow.

## Employment Characteristics

Professional counselors can be found in a variety of settings, including:

- private practice
- elementary, middle, and secondary schools
- colleges/universities
- hospitals
- health maintenance organizations
- insurance firms
- drug and alcohol abuse rehabilitation agencies
- mental health agencies
- correctional institutions
- career development and vocational training facilities

Success in the counseling field requires motivation, a commitment to service, and skills in communication. Counselors will be faced with numerous challenges and opportunities in the future, including drug abuse, homelessness, disaster recovery, and the "graying of America" with an increasing percentage of Americans who are senior citizens. The foundations of the counseling profession have been intertwined with both social and educational reform movements in this century and will continue to be so in the future.

## Salary

Median earnings for full-time educational and vocational counselors were about $42,110 a year in 2004. The bottom 10% earned less than $23,560 a year; the top 10% earned over $67,170 a year. As with most professions, pay scales differ based on education, prior experience, and geographical location. Refer to Section IV, Table 5 of this *Directory* for more information, or see www.ama-assn.org/go/ salary.

## Educational Programs

**Length.** Career Counseling, College Counseling, Community Counseling, Gerontological Counseling, School Counseling, and Student Affairs programs are a minimum of 48 semester hours or 72 quarter hours. Mental Health Counseling and Marital, Couple and Family Counseling/Therapy programs are a minimum of 60 semester hours or 90 quarter hours.

**Curriculum.** Curricular experiences and demonstrated knowledge in each of the eight common core areas are required of all counseling students:

1. Professional Identity
2. Social and Cultural Diversity
3. Human Growth and Development
4. Career Development
5. Helping Relationships
6. Group Work
7. Assessment
8. Research and Program Evaluation

**Clinical Requirements.** Students must complete supervised practicum experiences that total a minimum of 100 clock hours. The practicum provides for the development of counseling skills under supervision. The student's practicum includes all of the following:

1. Forty hours of direct service with clients, including experience in individual counseling and group work.
2. Weekly interaction with an average of 1 hour per week of individual and/or triadic supervision, which occurs regularly over the course of one academic term by a programs faculty member or a supervisor working under the supervision of a program faculty member.
3. An average of 1 ½ hours per week of group supervision, provided on a regular schedule over the course of the student's practicum by a program faculty member or a supervisor under the supervision of a program faculty member.
4. Evaluation of the student's performance throughout the practicum, including a formal evaluation after the student completes the practicum.

The program requires students to complete a supervised internship of 600 clock hours after successful completion of the student's practicum. The internship provides an opportunity for the student to perform, under supervision, a variety of counseling activities that a professional counselor is expected to perform. The student's internship includes all of the following:

1. Two hundred and forty hours of direct service with clients appropriate to the programs of study.
2. Weekly interaction with an average of 1 hour per week of individual and/or triadic supervision, throughout the internship, usually performed by the on-site supervisor.
3. An average of 1 ½ hours per week of group supervision provided on a regular schedule throughout the internship, usually performed by a program faculty member.
4. The opportunity for the student to become familiar with a variety of professional activities in addition to direct service (eg, recordkeeping, supervision, information and referral, in-service and staff meetings).
5. The opportunity for the student to develop program-appropriate audio and/or videotapes of the student's interactions with clients for use in supervision.
6. The opportunity for the student to gain supervised experience in the use of a variety of professional resources, such as assessment instruments, technologies, print and nonprint media, professional literature, and research.

7. A formal evaluation of the student's performance during the internship by a program faculty member, in consultation with the site supervisor.

 **Inquiries**

**Careers**
American Counseling Association
5999 Stevenson Avenue
Alexandria, VA 22304
703 823-9800
www.counseling.org

**Certification**
National Board for Certified Counselors
3 Terrace Way, Suite D
Greensboro, NC 27403
336 547-0607
www.nbcc.org

**Program Accreditation**
Council for Accreditation of Counseling and Related Educational Programs
5999 Stevenson Avenue
Alexandria, VA 22304
703 823-9800, x301
www.cacrep.org

## Counselor

# Alabama

**Auburn University**
Counseling Prgm
2084 Haley Center
Auburn University, AL 36849-5222
www.auburn.edu/coun
*Prgm Dir:* Chippewa Thomas, PhD
*Tel:* 334 844-5160   *Fax:* 334 844-2860
*E-mail:* thoma07@auburn.edu

**University of Montevallo**
Counseling Prgm
CC, SC Programs
College of Education, Station 6380
Montevallo, AL 35115
*Prgm Dir:* Stephanie Puleo, PhD
*Tel:* 205 665-6371
*E-mail:* puleos@montevallo.edu

**Troy University**
Counseling Prgm
CC, MHC, SC Programs
One University Place
Phenix City, AL 36869
*Prgm Dir:* M Kathryn Ness
*Tel:* 334 448-5146   *Fax:* 334 448-5203
*E-mail:* kness@troy.edu

**Troy University**
Counseling Prgm
CC, SC Programs
10 McCartha Hall
Troy, AL 36082
*Prgm Dir:* Dianne Gossett, PhD
*Tel:* 334 670-3350
*E-mail:* dgossett@troy.edu

**University of Alabama**
Counseling Prgm
CC, SC, CE Programs
PO Box 870231
Tuscaloosa, AL 35487-0231
*Prgm Dir:* S Allen Wilcoxon
*Tel:* 205 348-7579   *Fax:* 205 348-7584
*E-mail:* awilcoxo@bamaed.ua.edu

# Arizona

**Northern Arizona University**
Counseling Prgm
CC, SC Programs
Educational Psychology
Flagstaff, AZ 86011-5774
http://coe.nau.edu/academics/EPS
*Prgm Dir:* Ramona Mellott, PhD
*Tel:* 928 523-6534   *Fax:* 928 523-9284
*E-mail:* ramona.mellott@nau.edu

**University of Phoenix**
Counseling Prgm
CC Program
4605 E Elwood St (Mailstop Code AAC-704)
Phoenix, AZ 85040
www.phoenix.edu
*Prgm Dir:* Patricia L Kerstner, PhD
*Tel:* 480 537-2179   *Fax:* 480 929-7164
*E-mail:* patricia.kerstner@phoenix.edu

**Arizona State University**
Counseling Prgm
CC Program
PO Box 870611
Tempe, AZ 85287-0611
*Prgm Dir:* Sharon Kurpius
*Tel:* 480 965-6104   *Fax:* 480 965-8057
*E-mail:* sharon.kurpius@asu.edu

# Arkansas

**University of Arkansas**
Counseling Prgm
CC, SC, CE Programs
136 Graduate Education Bldg
Fayetteville, AR 72701
*Prgm Dir:* Rebecca A Newgent
*Tel:* 479 575-1617
*E-mail:* newgent@uakron.alumlink.com

**Arkansas State University**
Counseling Prgm
SC Program
PO Box 1560
State University, AR 72467-1560
*Prgm Dir:* Nola Christenberry, PhD
*Tel:* 870 972-3064
*E-mail:* nchriste@astate.edu

# British Columbia, Canada

**Trinity Western University**
Counseling Prgm
CC Program
7600 Glover Rd
Langley, BC V2Y 1Y1
*Prgm Dir:* Marvin McDonald, PhD RPsych
*Tel:* 604 513-2034, Ext 3223   *Fax:* 604 513-2150
*E-mail:* mcdonald@twu.ca

**University of British Columbia**
Counseling Prgm
CC, SAP-CC, SC Programs
Faculty of Education
Vancouver, BC V6T 1Z4
*Prgm Dir:* William Borgen
*Tel:* 604 822-5261   *Fax:* 604 822-2328
*E-mail:* william.a.borgen@ubc.ca

# California

**California State University - Fresno**
Counseling Prgm
MFC/T Program
5005 N Maple Ave, M/S3
Fresno, CA 93740-8025
*Prgm Dir:* Sari H Dworkin
*Tel:* 559 278-0328   *Fax:* 559 278-0404
*E-mail:* sarid@csufresno.edu
*Med Dir:* H Dan Smith, MD

**California State University - Los Angeles**
Counseling Prgm
MFC/T, SC Programs
5151 State University Dr
Los Angeles, CA 90032
*Prgm Dir:* Randy Campbell, PhD
*Tel:* 323 343-4250   *Fax:* 323 343-4252
*E-mail:* rcampbe@calstatela.edu

**California State University - Northridge**
Counseling Prgm
CrC, MFC/T, SACC, SC Programs
18111 Nordhoff St
Northridge, CA 91330-8265
*Prgm Dir:* Rie Rogers Mitchell, EdD
*Tel:* 818 677-2558   *Fax:* 818 677-2544
*E-mail:* rie.mitchell@csun.edu

---

*Additional information about programs that returned the AMA's survey is available in a table beginning on page 95.

## Sonoma State University
Counseling Prgm
CC, SC Programs
1801 E Cotati Ave, Rm N220
Rohnert Park, CA 94928
*Prgm Dir:* Adam Hill, PhD
*Tel:* 707 664-2340   *Fax:* 707 664-2038
*E-mail:* adam.hill@sonoma.edu

## San Francisco State University
Counseling Prgm
Dept of Counseling
1600 Holloway Ave
San Francisco, CA 94132
www.sfsu.edu
*Prgm Dir:* Wanda Lee, PhD
*Tel:* 415 338-1398   *Fax:* 415 338-0594
*E-mail:* counsel@sfsu.edu
*Med Dir:* John Blando, PhD

## California Polytechnic State University
Counseling Prgm
MFC/T Program
Psychology & Human Dev Dept
San Luis Obispo, CA 93407
*Prgm Dir:* Michael Selby
*Tel:* 805 756-1617   *Fax:* 805 756-1134
*E-mail:* mselby@calpoly.edu

# Colorado

## Adams State College
Counseling Prgm
CC, SC Programs
Dept of Psychology
Alamosa, CO 81102
*Prgm Dir:* Mark Manzanares, PhD
*Tel:* 719 587-7224
*E-mail:* mgmanzan@adams.edu

## University of Colorado at Colorado Springs
Counseling Prgm
4120 Austin Bluffs Pkwy
PO Box 7150
Colorado Springs, CO 80933-7150
www.uccs.edu/coegen
*Prgm Dir:* Beverly Snyder, EdD LPC NCC
*Tel:* 719 262-4095
*E-mail:* bsnyder@uccs.edu

## U of Colorado at Denver and Health Scis Ctr
Counseling Prgm
CC, MFC/T, SC Programs
Counseling Psychology & Counselor Ed
Denver, CO 80217-3364
*Prgm Dir:* Marsha Wiggins-Frame
*Tel:* 303 556-6032   *Fax:* 303 556-4479
*E-mail:* marsha.frame@cudenver.edu

## Colorado State University
Counseling Prgm
CrC, CC, SC Programs
Education 215
Ft Collins, CO 80523
*Prgm Dir:* Rich Feller
*Tel:* 970 491-6879
*E-mail:* feller@cahs.colostate.edu

## University of Northern Colorado
Counseling Prgm
CC, MFC/T, SC, CE Programs
Division of Professional Psychology
Greeley, CO 80639
*Prgm Dir:* David Gonzalez, PhD
*Tel:* 970 351-1639   *Fax:* 970 351-2625
*E-mail:* david.gonzalez@unco.edu

## Denver Seminary
Counseling Prgm
CC Program
Counseling Division
Littleton, CO 80120
www.denverseminary.edu
*Prgm Dir:* Monte Hasz, PsyD
*Tel:* 303 762-6896, Ext 1292   *Fax:* 303 762-6976
*E-mail:* monte.hasz@denverseminary.edu
*Med Dir:* Zandy Winnerstrom

# Connecticut

## Western Connecticut State University
Counseling Prgm
CC, SC Programs
Education Dept
Danbury, CT 06810
*Prgm Dir:* Michael Gilles, PhD
*Tel:* 203 837-8513
*E-mail:* gillesm@wcsu.edu

## Fairfield University
Counseling Prgm
CC, SC Programs
Graduate School of Ed & Allied Prof
Fairfield, CT 06430-7524
*Prgm Dir:* Virginia Kelly, PhD
*Tel:* 203 254-4000, Ext 3228   *Fax:* 203 254-4047
*E-mail:* Vkelly@mail.fairfield.edu

## Southern Connecticut State University
Counseling Prgm
CC, SC Programs
Counseling & School Psychology Dept
New Haven, CT 06515
*Prgm Dir:* Patricia W De Barbieri
*Tel:* 203 392-5483   *Fax:* 203 392-5917
*E-mail:* debarbierip1@southernct.edu

# Delaware

## Wilmington College
Counseling Prgm
31 Read's Way
Wilson Graduate Center
New Castle, DE 19720
www.wilmcoll.edu/behavioralscience/mcc-prog.html
*Prgm Dir:* Richard C Williams, PhD
*Tel:* 302 295-1150
*E-mail:* r.craig.williams@wilmcoll.edu

# District of Columbia

## Gallaudet University
Counseling Prgm
MHC, SC Programs
800 Florida Ave NE
Washington, DC 20002
*Prgm Dir:* Roger Beach
*Tel:* 202 651-4511   *Fax:* 202 651-5657
*E-mail:* roger.beach@gallaudet.edu

## George Washington University
Counseling Prgm
CC, SC, CE Programs
Graduate School of Ed & Human Dev
Washington, DC 20052
*Prgm Dir:* Pat Schwallie-Giddis
*Tel:* 202 994-6856   *Fax:* 202 994-3436
*E-mail:* drpat@gwu.edu

# Florida

## Florida Atlantic University
Counseling Prgm
MHC, SC Programs
777 Glades Rd
Boca Raton, FL 33431
*Prgm Dir:* Larry Kontosh, PhD
*Tel:* 561 297-2806
*E-mail:* kontosh@fau.edu

## Stetson University
Counseling Prgm
MFC/T, MHC, SC Programs
421 N Woodland Blvd, Unit 8389
Deland, FL 32720
www.stetson.edu/artsci/counselor
*Prgm Dir:* Brigid Noonan, PhD LMHC
*Tel:* 386 822-7238   *Fax:* 386 740-3664
*E-mail:* bnoonan@stetson.edu

## Florida Gulf Coast University
Counseling Prgm
MHC, SC Programs
10501 FGCU Blvd S
Ft Myers, FL 33965-6565
*Prgm Dir:* Madeyln Issacs, PhD
*Tel:* 239 590-7785
*E-mail:* misaacs@fgcu.edu

## University of Florida
Counseling Prgm
MHC Program
1215 Norman Hall, Box 117046
Gainesville, FL 32611
www.ufl.edu
*Prgm Dir:* Sondra Smith-Adcock, PhD
*Tel:* 352 392-0731, Ext 247   *Fax:* 352 846-2697
*E-mail:* ssmith@coe.ufl.edu
*Med Dir:* M Harry Daniels, PhD

## University of Florida
Counseling Prgm
MFC/T program
1215 Norman Hall, Box 117046
Gainesville, FL 32611
www.ufl.edu
*Prgm Dir:* Silvia Echevarria-Doan, PhD
*Tel:* 352 392-0731, Ext 237   *Fax:* 352 846-2697
*E-mail:* silvia@coe.ufl.edu
*Med Dir:* M Harry Daniels, PhD

## University of Florida
Counseling Prgm
Counselor Education Dept
1215 Norman Hall, POB 117046
Gainesville, FL 32611
www.education.ufl.edu/counselor
*Prgm Dir:* Mary Ann Clark, PhD
*Tel:* 352 392-0731, Ext 229   *Fax:* 352 846-2697
*E-mail:* maclark@coe.ufl.edu
*Med Dir:* M Harry Daniels, PhD, Department Chairman

## University of North Florida
Counseling Prgm
MHC, SC Programs
4567 St Johns Bluff Rd S
Jacksonville, FL 32224-2645
*Prgm Dir:* David Whittinghill
*Tel:* 904 620-1749   *Fax:* 904 620-2982
*E-mail:* dwhittin@unf.edu

## Florida International University
Counseling Prgm
MHC, SC Programs
FIU - University Park Campus
Miami, FL 33119
*Prgm Dir:* Adriana McEachern, PhD
*Tel:* 305 348-3202   *Fax:* 305 348-3205
*E-mail:* Adriana.Mceachern@fiu.edu

**Barry University**
Counseling Prgm
MHC, SC, MFC/T Programs
11300 NE 2nd Ave
Miami Shores, FL 33161-6695
www.barry.edu
*Prgm Dir:* Maureen Duffy, PhD
*Tel:* 305 899-3711
*E-mail:* mduffy@mail.barry.edu
*Med Dir:* Scott Gillig, PhD, Professor

**University of Central Florida**
Counseling Prgm
MHC, SC, CE Programs
400 Central Blvd
Orlando, FL 32816
www.ucfcounselored.org
*Prgm Dir:* Mark Young, PhD
*Tel:* 407 823-6314    *Fax:* 407 823-3859
*E-mail:* meyoung@mail.ucf.edu

**Florida State University**
Counseling Prgm
CrC, MHC, SC Programs
215 Stone Bldg
Tallahassee, FL 32306
*Prgm Dir:* F Donald Kelly
*Tel:* 850 644-9439    *Fax:* 850 644-8776
*E-mail:* Railey@coe.fsu.edu

**University of South Florida**
Counseling Prgm
CrC, MHC, SC Programs
4202 E Fowler Ave EDU 162
Tampa, FL 33620-5650
www.coedu.usf.edu/deptpsysoc/ce/eds.htm
*Prgm Dir:* Herbert Exum, PhD
*Tel:* 813 974-8395    *Fax:* 813 974-5814
*E-mail:* Exum@tempest.coedu.usf.edu

**Rollins College**
Counseling Prgm
MHC, SC Programs
1000 Holt Ave, PO Box 2726
Winter Park, FL 32789-4499
*Prgm Dir:* Alicia Homrich
*Tel:* 407 646-2307
*E-mail:* jprovost@rollins.edu

# Georgia

**University of Georgia**
Counseling Prgm
CC, SC Programs
402 Aderhold Hall
Athens, GA 30602-7142
*Prgm Dir:* Georgia Calhoun
*Tel:* 706 542-4130
*E-mail:* gcalhoun@uga.edu

**Georgia State University**
Counseling Prgm
CC, SC, CE programs
Professional Counseling Prgm
Atlanta, GA 30303-3083
www.gsu.edu
*Prgm Dir:* JoAnna White, PhD
*Tel:* 404 651-3427    *Fax:* 404 651-1160
*E-mail:* jwhite@gsu.edu

**State University of West Georgia**
Counseling Prgm
CC, SC Programs
Dept of Counseling & Educational Psychology
Carrollton, GA 30118-5170
*Prgm Dir:* Brent M Snow
*Tel:* 770 836-6554
*E-mail:* bsnow@westga.edu

**Columbus State University**
Counseling Prgm
CC, SC Programs
4225 University Ave
Columbus, GA 31907-5645
www.colstate.edu
*Prgm Dir:* Michael L Baltmore
*Tel:* 706 568-2222    *Fax:* 706 569-3434
*E-mail:* baltimore_michael@colstate.edu

# Hawaii

**University of Hawaii at Manoa**
Counseling Prgm
CC, SC Programs
Coll of Education
Honolulu, HI 96822
*Prgm Dir:* Marta Garrett
*Tel:* 808 956-9820
*E-mail:* martag@hawaii.edu

# Idaho

**Boise State University**
Counseling Prgm
Counseling Dept Ed Bldg, Rm 612
1910 University Dr
Boise, ID 83725-1710
http://education.boisestate.edu/counselored
*Prgm Dir:* Ken Coll
*Tel:* 208 426-1821    *Fax:* 208 426-2046
*E-mail:* kcoll@boisestate.edu

**University of Idaho**
Counseling Prgm
SC, CE programs
College of Education, Rm 206
Moscow, ID 83844-3083
*Prgm Dir:* Jerry Fischer
*Tel:* 208 885-6556    *Fax:* 208 885-6869
*E-mail:* jfischer@uidaho.edu

**Northwest Nazarene University**
Counseling Prgm
CC, SC, MFC Programs
623 Holly St
Nampa, ID 83686
www.nnu.edu/counseloreducation
*Prgm Dir:* Brenda Freeman, PhD LCPC NCC
*Tel:* 208 467-8428    *Fax:* 208 467-8339
*E-mail:* BJFreeman@NNU.Edu

**Idaho State University**
Counseling Prgm
MHC, SC, SACC, CE, MFC/T Programs
PO Box 8120
Pocatello, ID 83209-8120
*Prgm Dir:* David M Kleist
*Tel:* 208 282-3156    *Fax:* 208 282-2583
*E-mail:* kleidavi@isu.edu

# Illinois

**Southern Illinois Univ at Carbondale**
Counseling Prgm
CC, MFC/T, SC, CE Programs
Wham Bldg, Rm 223
Carbondale, IL 62901-4618
www.siu.edu/departments/coe/epse
*Prgm Dir:* Lyle White, PhD LMFT LCPC AAMFT
*Tel:* 618 536-7763    *Fax:* 618 453-7110
*E-mail:* lwhite@siu.edu
*Med Dir:* Karen Prichard

**Eastern Illinois University**
Counseling Prgm
CC, SC Programs
2102 Buzzard Hall
Charleston, IL 61920
www.eiu.edu/~csd
*Prgm Dir:* Richard L Roberts, PhD
*Tel:* 217 581-2400    *Fax:* 217 581-7800
*E-mail:* rlroberts@eiu.edu

**Chicago State University**
Counseling Prgm
CC, SC Programs
9501 S King Dr
Chicago, IL 60629
*Prgm Dir:* Lindsay Bicknell-Hentges, PhD
*Tel:* 773 995-2359
*E-mail:* lbicknel@csu.edu

**Northeastern Illinois University**
Counseling Prgm
CC, SC, MFC/T Programs
5500 N St Louis Ave
Chicago, IL 60625-4699
*Prgm Dir:* Nan Giblin, PhD
*Tel:* 773 442-5552    *Fax:* 773 442-5559
*E-mail:* n-giblin@neiu.edu

**Roosevelt University**
Counseling Prgm
CC, MHC Programs
Department of Counseling and Human Services
Chicago, IL 60605
*Prgm Dir:* Renate Rohde, PhD
*Tel:* 847 619-8827    *Fax:* 847 619-8830
*E-mail:* rrohde@roosevelt.edu

**Northern Illinois University**
Counseling Prgm
CC, SC, CE Programs
Graham Hall 223
DeKalb, IL 60115-2854
*Prgm Dir:* Francesca Giordano
*Tel:* 815 753-8462
*E-mail:* fgiordano@niu.edu

**Western Illinois University**
Counseling Prgm
CC, SC Programs
Quad Cities Counselor Education Department
Moline, IL 61265-5881
*Prgm Dir:* Rori Carson, PhD
*Tel:* 309 762-1876, Ext 273    *Fax:* 309 762-6989
*E-mail:* RR-Carson@win.edu

**Bradley University**
Counseling Prgm
1501 W Bradley Ave
Westlake Hall
Peoria, IL 61625
*Prgm Dir:* Jobie L Skaggs, PhD
*Tel:* 309 677-3193
*E-mail:* jskaggs@bradley.edu

**Concordia University**
Counseling Prgm
CC, SC Programs
Psychology Dept
River Forest, IL 60305-1499
*Prgm Dir:* Dale J Septeowski
*Tel:* 708 209-3059    *Fax:* 708 209-3176
*E-mail:* crfsepteodj@curf.edu

**University of Illinois at Springfield**
Counseling Prgm
CC, SC Programs
One University Plaza
Springfield, IL 62703
*Prgm Dir:* Bill Abler, PhD
*Tel:* 217 206-6504
*E-mail:* abler.bill@uis.edu

**Governors State University**
Counseling Prgm
CC, MFC/T, SC Programs
College of Education
University Park, IL 60466
*Prgm Dir:* Cyrus Ellis
*Tel:* 708 534-4393
*E-mail:* j-yang@govst.edu

# Indiana

**Indiana University - Bloomington**
Counseling Prgm
School of Education 4003
201 N Rose Ave
Bloomington, IN 47405-1006
www.indiana.edu/~counsel/ccepro.html
*Prgm Dir:* Jeff Daniels, PhD
*Tel:* 812 856-8304    *Fax:* 812 856-8333
*E-mail:* jedaniel@indiana.edu

**Butler University/Clarian Health**
Counseling Prgm
SC Program
4600 Sunset Dr
Indianapolis, IN 46208
*Prgm Dir:* John Bloom
*Tel:* 317 940-9490    *Fax:* 317 940-6481
*E-mail:* jbloom@butler.edu

**Indiana Wesleyan University**
Counseling Prgm
CC, MFC/T Programs
4201 S Washington St
Marion, IN 46953
*Prgm Dir:* Jerry E Davis
*Tel:* 765 677-2995    *Fax:* 765 677-2504
*E-mail:* jdavis@indwes.edu

**Ball State University**
Counseling Prgm
CC, SC Programs
Teachers College, Rm 622
Muncie, IN 47306-0585
*Prgm Dir:* Phyllis Gordon
*Tel:* 765 285-8040    *Fax:* 765 285-2067
*E-mail:* pgordon@bsu.edu

**Indiana University South Bend**
Counseling Prgm
CC, SC Programs
1700 Mishawaka Ave
South Bend, IN 46634
*Prgm Dir:* Jannette Alexander, EdD
*Tel:* 574 237-4127    *Fax:* 574 520-5550
*E-mail:* jshaw@iusb.edu

**Indiana State University**
Counseling Prgm
MHC, SC Programs
School of Education, Rm 1517
Terre Haute, IN 47809
http://counseling.indstate.edu
*Prgm Dir:* James L Campbell, PhD
*Tel:* 812 237-4389    *Fax:* 812 237-2729
*E-mail:* jcampbell2@indstate.edu

**Purdue University**
Counseling Prgm
MHC, SC programs
Dept of Educational Studies
West Lafayette, IN 47907-2098
www.edst.purdue.edu/cd/development/cdschool.html
*Prgm Dir:* Jean Peterson, PhD LMHC NCC
*Tel:* 765 494-9742    *Fax:* 765 496-1228
*E-mail:* jeanp@purdue.edu

# Iowa

**University of Northern Iowa**
Counseling Prgm
MHC, SC Programs
508 Schindler Education Center
Cedar Falls, IA 50614-0604
*Prgm Dir:* Ann Vernon
*Tel:* 319 273-2605
*E-mail:* Ann.Vernon@uni.edu

**University of Iowa**
Counseling Prgm
N338 Lindquist Center N
Iowa City, IA 52242-1529
*Prgm Dir:* Debora Liddell
*Tel:* 319 335-5188    *Fax:* 319 335-5291
*E-mail:* debora-liddell@uiowa.edu

# Kansas

**Emporia State University**
Counseling Prgm
MHC, SC, SA Programs
Campus Box 4036
Emporia, KS 66801-5087
*Prgm Dir:* Patricia J Neufeld
*Tel:* 620 341-5794    *Fax:* 620 341-6200
*E-mail:* neufeldp@emporia.edu

**Pittsburg State University**
Counseling Prgm
CC Program
Dept of Psychology & Counseling
Pittsburg, KS 66762-7551
www.pittstate.edu/psych/comcouns.htm
*Prgm Dir:* Donald E Ward, PhD LCPC NCC ACS
*Tel:* 316 235-4530    *Fax:* 316 235-6102
*E-mail:* dward@pittstate.edu

# Kentucky

**Western Kentucky University**
Counseling Prgm
MFC/T, MHC Programs
1906 College Heights Blvd
Bowling Green, KY 42101
*Prgm Dir:* Bill Greenwalt, PhD
*Tel:* 270 745-6003
*E-mail:* bill.greenwalt@wku.edu

**Lindsey Wilson College**
Counseling Prgm
MHC Program
210 Lindsey Wilson St
Columbia, KY 42728
*Prgm Dir:* John R Rigney, EdD NCC LPCC
*Tel:* 502 384-8121    *Fax:* 502 384-8152
*E-mail:* rigneyj@lindsey.edu

**Murray State University**
Counseling Prgm
CC Program
PO Box 9
Murray, KY 42071
*Prgm Dir:* Thomas F Holcomb
*Tel:* 502 762-2795
*E-mail:* tom.holcomb@coe.murraystate.edu

**Eastern Kentucky University**
Counseling Prgm
MHC, SC Programs
521 Lancaster Ave
Richmond, KY 40475
www.education.eku.edu/cel
*Prgm Dir:* Kim A Naugle, PhD
*Tel:* 859 622-1863
*E-mail:* kim.naugle@eku.edu

# Louisiana

**Louisiana State University**
Counseling Prgm
CC, SC Programs
122 Peabody Hall
Baton Rouge, LA 70803
*Prgm Dir:* Gary Gintner, PhD
*Tel:* 225 388-2199    *Fax:* 225 578-6918
*E-mail:* gintner@lsu.edu

**Southeastern Louisiana University**
Counseling Prgm
CC, SACC, SC, MFC/T Programs
SLU Box 10863
Hammond, LA 70402
*Prgm Dir:* Mary Ballard
*Tel:* 504 549-2155
*E-mail:* mballard2@selu.edu

**University of Louisiana at Monroe**
Counseling Prgm
CC, MFC/T, SC Programs
700 University Ave
Monroe, LA 71209-0230
*Prgm Dir:* Charles Pryor, EdD
*Tel:* 318 342-1246    *Fax:* 318 342-1213
*E-mail:* cpryor@ulm.edu

**Northwestern State University**
Counseling Prgm
CIC, SA Programs
College of Education
Natchitoches, LA 71497
*Prgm Dir:* Frances Pearson
*Tel:* 318 357-6915
*E-mail:* pearson@nsula.edu

**Loyola University New Orleans**
Counseling Prgm
CC Program
6363 St Charles Ave, Camp Box 66
New Orleans, LA 70118
*Prgm Dir:* Kevin A Fall, PhD
*Tel:* 504 864-7859
*E-mail:* kfall@loyno.edu

**Our Lady of Holy Cross College**
Counseling Prgm
MFC/T, CC Programs
4123 Woodland Dr
New Orleans, LA 70131
*Prgm Dir:* Timothy F Dwyer, PhD
*Tel:* 504 398-2169    *Fax:* 504 391-2421
*E-mail:* tdwyer@olhcc.edu

**University of New Orleans**
Counseling Prgm
CC, CE, SC Programs
348 Education Bldg
New Orleans, LA 70148-2515
www.uno.edu
*Prgm Dir:* Diana Hulse-Killacky
*Tel:* 504 280-6662    *Fax:* 504 280-6453
*E-mail:* dhulseki@uno.edu

# Maine

**University of Southern Maine**
Counseling Prgm
MHC, SC Programs
400 Bailey Hall
Gorham, ME 04038-1083
*Prgm Dir:* Marijane Fall, EdD
*Tel:* 207 780-5472    *Fax:* 207 780-5043
*E-mail:* mjfall@usm.maine.edu

# Maryland

**Loyola College of Maryland**
Counseling Prgm
4501 N Charles St
Baltimore, MD 21210
https://loyola.edu/education/counseling
*Prgm Dir:* Bradley T Erford, PhD NCC LCPC
*Tel:* 410 617-1509   *Fax:* 410 617-1708
*E-mail:* berford@Loyola.edu

**Univ of Maryland at College Park**
Counseling Prgm
CE, SC Programs
College of Education
College Park, MD 20742
*Prgm Dir:* Courtland Lee, PhD
*Tel:* 301 405-8904   *Fax:* 301 405-9995
*E-mail:* clee5@umd.edu

**Loyola College of Maryland**
Counseling Prgm
CC Program
Pastoral Counseling Department
Columbia, MD 21045
www.loyola.edu/pastoralcounseling
*Prgm Dir:* David Newton
*Tel:* 410 617-7620   *Fax:* 410 617-7644
*E-mail:* dcnewton@Loyola.edu

# Michigan

**Andrews University**
Counseling Prgm
Bell Hall 159
Berrien Springs, MI 49104-0104
www.educ.andrews.edu
*Prgm Dir:* Elvin Gabriel
*Tel:* 269 471-6223   *Fax:* 269 471-6374
*E-mail:* gabriel@andrews.edu

**University of Detroit Mercy**
Counseling Prgm
CC, SC Programs
444 Manning Hall
Detroit, MI 48219-0900
*Prgm Dir:* Nancy Calley
*Tel:* 313 578-0436
*E-mail:* calleyng@udmercy.edu

**Wayne State University**
Counseling Prgm
CC, SC, CE Programs
311 Education Bldg
Detroit, MI 48202
*Prgm Dir:* Daisy B Ellington
*Tel:* 313 577-2435   *Fax:* 313 577-5235
*E-mail:* delling@coe.wayne.edu

**Western Michigan University**
Counseling Prgm
CC, CE, CIC, SC Programs
3102 Sangren Hall
Kalamazoo, MI 49008-5195
*Prgm Dir:* Carla Adkison-Bradley, PhD
*Tel:* 616 387-3504
*E-mail:* carla.bradley@wmich.edu

**Oakland University**
Counseling Prgm
CC, SC Programs
Dept of Counseling, 491B Pawley Hall
Rochester, MI 48309-4494
*Prgm Dir:* Luellen Ramey, PhD
*Tel:* 248 370-4169   *Fax:* 248 370-4141
*E-mail:* ramey@oakland.edu

**Eastern Michigan University**
Counseling Prgm
CC, SC, CIC Programs
304 Porter Bldg
Ypsilanti, MI 48197
*Prgm Dir:* Irene Mass Ametrano
*Tel:* 734 487-0255   *Fax:* 734 487-4608
*E-mail:* irene.ametrano@emich.edu

# Minnesota

**University of Minnesota - Duluth**
Counseling Prgm
CC, SC Programs
320 Bohannon Hall
Duluth, MN 55812
*Prgm Dir:* Bud McClure, PhD
*Tel:* 218 726-8196
*E-mail:* bmcclure@d.umn.edu

**Mankato State University**
Counseling Prgm
CC, SC Programs
PO Box 52
Mankato, MN 56002-8400
*Prgm Dir:* Anne Blackhurst
*Tel:* 507 389-2423
*E-mail:* anne.blackhurst@mnsu.edu

**Minnesota State University - Mankato**
Counseling Prgm
CC, SAPP, SC Programs
107 Armstrong Hall
Mankato, MN 56002-2423
*Prgm Dir:* Diane Coursol
*Tel:* 507 389-5656
*E-mail:* diane.coursol@mnsu.edu

**Minnesota State University - Moorhead**
Counseling Prgm
CC, SACC, SAPP Programs
Counseling & Student Affairs
Moorhead, MN 56563
*Prgm Dir:* Wes Erwin, PhD NCC
*Tel:* 218 477-2044   *Fax:* 218 477-2547
*E-mail:* erwin@mnstate.edu

**Winona State University**
Counseling Prgm
CC, SC Programs
859 30th Ave SE
Rochester, MN 55904
*Prgm Dir:* Tim Hatfield
*Tel:* 507 457-5337
*E-mail:* THatfield@winona.edu

**St Cloud State University**
Counseling Prgm
SC Program
720 4th Ave S
St Cloud, MN 56301
*Prgm Dir:* Terrance Peterson
*Tel:* 320 308-2992
*E-mail:* tpeterson@stcloudstate.edu

# Mississippi

**Delta State University**
Counseling Prgm
CC, SC Programs
Ewing 335
Cleveland, MS 38733
*Prgm Dir:* Matthew Buckley
*Tel:* 662 846-4357   *Fax:* 662 846-4402
*E-mail:* mbuckley@dsu.deltastate.edu

**Mississippi College**
Counseling Prgm
PO Box 4013
Clinton, MS 39058
*Prgm Dir:* Buddy Wagner, PhD
*Tel:* 601 925-3354   *Fax:* 601 925-3951
*E-mail:* bwagner@mc.edu

**University of Southern Mississippi**
Counseling Prgm
CC Program
Southern Station Box 5025
Hattiesburg, MS 39406-5025
*Prgm Dir:* Bonnie Nicholson
*Tel:* 601 266-4598   *Fax:* 601 266-5580
*E-mail:* bonnie.nicholson@usm.edu

**Mississippi State University**
Counseling Prgm
CC, SACC, SC, CE Programs
PO Box 9727
Mississippi State, MS 39762
*Prgm Dir:* E Joan Looby, PhD LPC NCC
*Tel:* 601 325-3426   *Fax:* 601 325-3263
*E-mail:* jlooby@colled.msstate.edu

**University of Mississippi**
Counseling Prgm
CC, SC, CE Programs
Ste 200, School of Education
University, MS 38677-1848
*Prgm Dir:* Sonja Burnham
*Tel:* 662 915-7069
*E-mail:* sburnham@olemiss.edu

# Missouri

**Southeast Missouri State University**
Counseling Prgm
CC, SC Programs
Mail Stop 5550
Cape Girardeau, MO 63701-4799
*Prgm Dir:* Verl Pope, EdD
*Tel:* 573 651-2399   *Fax:* 573 986-6812
*E-mail:* vpope@semo.edu

**Truman State University**
Counseling Prgm
CC, SC, SAPP Programs
Div of Education
Kirksville, MO 63501
*Prgm Dir:* Tricia Brown, PhD
*Tel:* 660 785-4399
*E-mail:* tbrown@truman.edu

**University of Missouri - St Louis**
Counseling Prgm
CC, CrC, SC Programs
469 Marillac Hall
St Louis, MO 63121-4499
*Prgm Dir:* Mark Pope, EdD MAC LPC NCC
*Tel:* 314 516-7121   *Fax:* 314 516-5784
*E-mail:* PopeML@msx.umsl.edu

# Montana

**Montana State University**
Counseling Prgm
MFC/T, MHC, SC Programs
218 Herrick Hall
Bozeman, MT 59717
*Prgm Dir:* Mark Nelson, EdD
*Tel:* 406 994-3810   *Fax:* 406 994-2013
*E-mail:* markn@montana.edu

**University of Montana**
Counseling Prgm
MHC, SC Programs
32 Campus Dr
Missoula, MT 59812-6356
*Prgm Dir:* Catherine Jenni, PhD
*Tel:* 406 243-2608
*E-mail:* cathy.jenni@mso.umt.edu

# Nebraska

**University of Nebraska - Kearney**
Counseling Prgm
Dept of Counseling & School Psychology
College of Education Bldg
Kearney, NE 68849
www.unk.edu/acad.csp
*Prgm Dir:* James Slate Fleming, EdD NCC CPC LPC
*Tel:* 308 865-8358    *Fax:* 308 865-8097
*E-mail:* flemingj@unk.edu

**University of Nebraska - Omaha**
Counseling Prgm
CC, SC Programs
Kayser Hall 421
Omaha, NE 68182-0167
*Prgm Dir:* Jeannette Seaberry, PhD
*Tel:* 402 554-2728    *Fax:* 402 554-3684
*E-mail:* jseaberry@mail.unomaha.edu

# Nevada

**University of Nevada - Las Vegas**
Counseling Prgm
CC, MFC/T, SC Programs
4505 S Maryland Pkwy
Las Vegas, NV 89154-3003
*Prgm Dir:* Randall Astramovich, PhD
*Tel:* 702 895-2948    *Fax:* 702 895-1658
*E-mail:* Randy.Astramovich@ccmail.nevada.edu

**University of Nevada - Reno**
Counseling Prgm
CC, SACC, SC, CE Programs
Psych Dept, 281
Reno, NV 89557-0213
*Prgm Dir:* Marlowe Smaby, PhD
*Tel:* 775 784-1772    *Fax:* 775 784-1990
*E-mail:* smaby@unr.edu

# New Jersey

**College of New Jersey**
Counseling Prgm
Forcina Hall 337
PO Box 7718
Ewing, NJ 08620-0718
*Prgm Dir:* Mark S Kiselica
*Tel:* 609 771-2119
*E-mail:* kiselica@tcnj.edu

**Rider University**
Counseling Prgm
2083 Lawrenceville Rd
Lawrenceville, NJ 08648-3099
*Prgm Dir:* Jesse B Deesch
*Tel:* 609 895-5487    *Fax:* 609 896-5362
*E-mail:* deesch@rider.edu

**Kean University**
Counseling Prgm
CC, SC Programs
1000 Morris Ave
Union, NJ 07083
*Prgm Dir:* Juneau Gary, PhD
*Tel:* 908 737-3842
*E-mail:* jgary@kean.edu

**William Paterson Univ of New Jersey**
Counseling Prgm
CC, SC Programs
Dept of Special Education & Counseling
Wayne, NJ 07170
*Prgm Dir:* Paula R Danzinger
*E-mail:* danzingerp@wpunj.edu

# New Mexico

**University of New Mexico**
Counseling Prgm
CC, SC, CE Programs
College of Education, Simpson Hall
Albuquerque, NM 87131-1246
*Prgm Dir:* Deborah Rifenbary
*Tel:* 505 277-8933
*E-mail:* riffer@unm.edu

**New Mexico State University**
Counseling Prgm
MHC Program
PO Box 30001
Las Cruces, NM 88003
*Prgm Dir:* Michael Nystul
*Tel:* 505 646-4092
*E-mail:* michaelnystul@hotmail.com

# New York

**SUNY at Brockport**
Counseling Prgm
350 New Campus Dr
Brockport, NY 14420-2953
http://brockport.edu
*Prgm Dir:* Susan R Seem, PhD LMHC NCC ACS
*Tel:* 585 395-2258    *Fax:* 585 395-2366
*E-mail:* sseem@brockport.edu
*Med Dir:* Elizabeth Caruso, MS

**Long Island University - C W Post Campus**
Counseling Prgm
MHC, SC Programs
Library Rm 320
Brookville, NY 11548-2814
*Prgm Dir:* Joseph Despres, EdD
*Tel:* 516 299-2814
*E-mail:* joseph.despres@liu.edu

**St John's University**
Counseling Prgm
SC Program
Jamaica & Staten Island Campus
Jamaica, NY 11439
*Prgm Dir:* Carolina K Wilde, PhD
*Tel:* 718 990-1567

**SUNY College at Plattsburgh**
Counseling Prgm
CC, SA, SC Programs
101 Broad St
Plattsburgh, NY 12901
*Prgm Dir:* Beverly A Burnell, PhD
*Tel:* 518 564-4178    *Fax:* 518 564-4161
*E-mail:* beverly.burnell@plattsburgh.edu

**University of Rochester**
Counseling Prgm
CC, SC, CE Programs
Dewey Hall 1-314
Rochester, NY 14627
*Prgm Dir:* Deborah Erickson, PhD
*Tel:* 585 275-1007
*E-mail:* derickson@warner.rochester.edu

**Syracuse University**
Counseling Prgm
SACC, SC, CE, CC Programs
Counseling and Human Services
Syracuse, NY 13244-3240
*Prgm Dir:* Janine Bernard, PhD NCC ACS
*Tel:* 315 443-2266    *Fax:* 315 443-5732
*E-mail:* bernard@syr.edu

# North Carolina

**Appalachian State University**
Counseling Prgm
CC, SC, CIC Programs
Boone, NC 28608
*Prgm Dir:* Lee Baruth
*Tel:* 704 262-2055
*E-mail:* baruthlg@appstate.edu

**Univ of North Carolina at Chapel Hill**
Counseling Prgm
SC Program
CB #3500 Peabody Hall
Chapel Hill, NC 27599-3500
*Prgm Dir:* John P Galassi, PhD
*Tel:* 919 962-9196
*E-mail:* jgalassi@email.unc.edu

**Univ of North Carolina at Charlotte**
Counseling Prgm
CC, CE, SC Programs
9201 University City Blvd
Charlotte, NC 28223-0001
http://education.uncc.edu/counseling
*Prgm Dir:* Susan Furr, PhD
*Tel:* 704 687-8967    *Fax:* 704 687-2916
*E-mail:* SRFurr@email.uncc.edu

**Western Carolina University**
Counseling Prgm
CC, SC Programs
204 Killiam Bldg
Cullowhee, NC 28723
*Prgm Dir:* Mary Deck, PhD NCC LPC
*Tel:* 828 227-7207    *Fax:* 828 227-7388
*E-mail:* deck@wpoff.wcu.edu

**North Carolina A&T State University**
Counseling Prgm
CC, SC Programs
Human Development & Services
Greensboro, NC 27411-1066
*Prgm Dir:* David Lundberg
*Tel:* 336 334-7916
*E-mail:* lundberg@ncat.edu

**Univ of North Carolina at Greensboro**
Counseling Prgm
CC, CE, GC, MFC/T, SACC, SC Programs
PO Box 26171
Greensboro, NC 27402-6170
*Prgm Dir:* Craig S Cashwell, PhD
*Tel:* 336 334-3425    *Fax:* 336 334-3433
*E-mail:* cscashwe@uncg.edu

**North Carolina State University**
Counseling Prgm
CC, SC, CIC, CE Programs
520 Poe Hall
Raleigh, NC 27695-7801
*Prgm Dir:* S Raymond Ting
*Tel:* 919 515-6362    *Fax:* 919 515-6891
*E-mail:* raymond_ting@ncsu.edu

**Wake Forest University**
Counseling Prgm
CC, SC Programs
PO Box 7406
Winston-Salem, NC 27109
http://wfu.edu/counseling
*Prgm Dir:* Sam Gladding, PhD LPC NCC CCMHC
*Tel:* 336 758-4882    *Fax:* 336 758-3129
*E-mail:* stg@wfu.edu

# North Dakota

**North Dakota State University**
Counseling Prgm
CC, SC Programs
School of Education, 210 FLC
Fargo, ND 58105-5057
www.ndsu.nodak.edu/ndsu/counsed
*Prgm Dir:* J Wade Hannon, EdD
*Tel:* 701 231-7204   *Fax:* 701 231-7416
*E-mail:* wade.hannon@ndsu.edu

# Ohio

**University of Akron**
Counseling Prgm
CC, MFC/T, SC, CE Programs
127 Carroll Hall
Akron, OH 44325-5007
*Prgm Dir:* Patricia Parr, PhD
*Tel:* 330 972-8151   *Fax:* 330 972-5292
*E-mail:* pparr@uakron.edu

**Ohio University**
Counseling Prgm
CC, SC, CE Programs
201 McCracken Hall
Athens, OH 45701
www.coe.ohiou.edu
*Prgm Dir:* Dana Levitt, PhD
*Tel:* 740 593-4163   *Fax:* 740 593-0569
*E-mail:* Levitt@ohio.edu

**University of Cincinnati**
Counseling Prgm
PO Box 210002
Cincinnati, OH 45221
www.uc.edu/counselingprogram
*Prgm Dir:* Geoffrey G Yager, PhD LPC
*Tel:* 513 556-3347   *Fax:* 513 556-3898
*E-mail:* geof.yager@uc.edu

**Cleveland State University**
Counseling Prgm
CC, SC Programs
1983 E 24th St
Cleveland, OH 44115
*Prgm Dir:* Kathryn C MacCluskie, EdD
*Tel:* 216 523-7147
*E-mail:* k.maccluskie@csuohio.edu

**John Carroll University**
Counseling Prgm
20700 N Park Blvd
University Heights
Cleveland, OH 44118-4581
www.jcu.edu/graduate
*Prgm Dir:* Christopher Faiver, PhD
*Tel:* 216 397-3001   *Fax:* 216 397-3045
*E-mail:* faiver@jcu.edu

**Wright State University**
Counseling Prgm
CC, SC, MHC Programs
M052 Creative Arts Center
Dayton, OH 45435-0001
*Prgm Dir:* Stephen B Fortson
*Tel:* 937 775-2075
*E-mail:* stephen.fortson@wright.edu

**Kent State University**
Counseling Prgm
CC, SC, CE Programs
310 White Hall
Kent, OH 44242-0001
*Prgm Dir:* Jason McGlothlin, PhD
*Tel:* 330 672-0716   *Fax:* 330 672-3063
*E-mail:* jmcgloth@kent.edu

**University of Toledo**
Counseling Prgm
Counselor Education & School Psychology
Mail Stop 119
Toledo, OH 43606-3390
http://hhs.utoledo.edu/cesp
*Prgm Dir:* Martin Ritchie, EdD LPC NCC
*Tel:* 419 530-4775   *Fax:* 419 530-7879
*E-mail:* martin.ritchie@utoledo.edu

**Youngstown State University**
Counseling Prgm
CC, SC Programs
Department of Counseling, BCOE Rm 3305
Youngstown, OH 44555-0001
*Prgm Dir:* Robert Beebe, EdD
*Tel:* 330 941-3257   *Fax:* 330 941-2369
*E-mail:* rjbeebe@ysu.edu

# Oklahoma

**Oklahoma State University**
Counseling Prgm
CC, SC Programs
421 Willard Hall
Stillwater, OK 74078
*Prgm Dir:* Camille S DeBell, PhD
*Tel:* 405 744-9442
*E-mail:* dcamill@okstate.edu

# Oregon

**Oregon State University**
Counseling Prgm
CC, SC, CE Programs
New School of Education
Corvallis, OR 97331
*Prgm Dir:* Gene Eakin, PhD
*Tel:* 541 737-9215
*E-mail:* gene.eakin@orst.edu

**Portland State University**
Counseling Prgm
CC, SC Programs
PO Box 751
Portland, OR 97207-0751
www.ed.pdx.edu/spedcoun/counpgrm.html
*Prgm Dir:* Rick Johnson, PhD
*Tel:* 503 725-9764   *Fax:* 503 725-5599
*E-mail:* johnsonp@pdx.edu

# Pennsylvania

**Edinboro University of Pennsylvania**
Counseling Prgm
SACC, SAPP, SC, CC Programs
128 Butterfield Hall
Edinboro, PA 16444
*Prgm Dir:* Salene Cowher
*Tel:* 814 732-1116   *Fax:* 814 732-2268
*E-mail:* scowher@edinboro.edu

**Duquesne University**
Counseling Prgm
CC, MFC/T, SC Programs
School of Education
Pittsburgh, PA 15282
*Prgm Dir:* Maura Krushinski, EdD
*Tel:* 412 396-4026   *Fax:* 412 396-1340
*E-mail:* krushinski@duq.edu

**Marywood University**
Counseling Prgm
MHC, SC Programs
2300 Adams Ave
Scanton, PA 18509
*Prgm Dir:* John J Lemoncelli, EdD
*Tel:* 570 348-6211, Ext 2317
*E-mail:* lemoncelli@es.marywood.edu

**University of Scranton**
Counseling Prgm
CC, SC Programs
Dept of Counseling and Human Services
Scranton, PA 18510-4523
*Prgm Dir:* Lee Ann Eschbach
*Tel:* 570 941-6299   *Fax:* 570 941-5882
*E-mail:* eschbach@uofs.edu

**Shippensburg University**
Counseling Prgm
CC, MHC, SC, CIC, SA Programs
1871 Old Main Dr
Shippensburg, PA 17257
*Prgm Dir:* Beverly L Mustaine
*Tel:* 717 477-1658   *Fax:* 717 530-4056
*E-mail:* blmust@ship.edu

**Slippery Rock University**
Counseling Prgm
CC Program
006 McKay Education Bldg
Slippery Rock, PA 16057
*Prgm Dir:* Donald Strano, PhD
*Tel:* 724 738-2274
*E-mail:* donald.strano@sru.edu

**Penn State University**
Counseling Prgm
CE, SC Programs
331 Cedar Bldg
University Park, PA 16828
*Prgm Dir:* Richard Hazler, PhD
*Tel:* 814 863-2415
*E-mail:* hazler@psu.edu

# South Carolina

**Clemson University**
Counseling Prgm
CC, SACC, SAPP, SC Programs
Education and Human Development
Clemson, SC 29634-0710
*Prgm Dir:* Robert Urofsky, PhD
*Tel:* 864 656-0927
*E-mail:* rurofsk@clemson.edu

**University of South Carolina**
Counseling Prgm
MFC/T, SC, CE Programs
Family Counseling Program
Columbia, SC 29208
*Prgm Dir:* Joshua Gold, PhD NCC
*Tel:* 803 777-1936   *Fax:* 803 777-3045
*E-mail:* josgold@sc.edu

**South Carolina State University**
Counseling Prgm
SC Program
300 College St NE
Orangeburg, SC 29177
*Prgm Dir:* Philip M Scriven, PhD
*Tel:* 803 536-7147   *Fax:* 803 536-8841
*E-mail:* pscriven@scsu.edu

**Winthrop University**
Counseling Prgm
CC, SC Programs
143 Withers Bldg
Rock Hill, SC 29733
*Prgm Dir:* M Jane Rankin, PhD
*Tel:* 803 323-2233   *Fax:* 803 323-4755
*E-mail:* rankinj@winthrop.edu

# South Dakota

**South Dakota State University**
Counseling Prgm
CC, SACC, SC Programs
PO Box 507
Brookings, SD 57007-0095
www.sdstate.edu
*Prgm Dir:* Jay Trenhaile, EdD
*Tel:* 605 688-4367    *Fax:* 605 688-5929
*E-mail:* jay_trenhaile@sdstate.edu

**University of South Dakota**
Counseling Prgm
Delzell School of Education
414 E Clark St
Vermillion, SD 57069
*Prgm Dir:* Frank Main
*Tel:* 605 677-5257    *Fax:* 605 677-5438
*E-mail:* fmain@usd.edu

# Tennessee

**University of Tennessee - Chattanooga**
Counseling Prgm
CC, SC Programs
615 McCallie Ave
Chattanooga, TN 37403
*Prgm Dir:* Robin Lee, PhD
*Tel:* 423 425-4544
*E-mail:* Robin-Lee@utc.edu

**East Tennessee State University**
Counseling Prgm
CC, SC Programs
PO Box 70548
Johnson City, TN 37614-0548
*Prgm Dir:* Patricia Robertson, EdD
*Tel:* 423 439-7693    *Fax:* 423 439-7790
*E-mail:* robertsonp@etsu.edu

**University of Tennessee - Knoxville**
Counseling Prgm
MHC, SC, CE Programs
College of Education
Knoxville, TN 37996-3400
*Prgm Dir:* Jeannine Studer, EdD
*Tel:* 865 974-0693    *Fax:* 865 974-0135
*E-mail:* jstuder@utk.edu

**University of Memphis**
Counseling Prgm
Ball Hall, Rm 100
Memphis, TN 38152-0001
*Prgm Dir:* N Dewaine Rice, EdD
*Tel:* 901 678-2841    *Fax:* 901 678-5114
*E-mail:* ndrice@memphis.edu

**Middle Tennessee State University**
Counseling Prgm
SC Program
PO Box 87
Murfreesboro, TN 37132
*Prgm Dir:* Virginia Dansby
*Tel:* 615 898-2559
*E-mail:* vsdansby@mtsu.edu

**Vanderbilt University**
Counseling Prgm
CC, SC Programs
PO Box 322-GPC
Nashville, TN 37203
*Prgm Dir:* Gina Frieden
*Tel:* 615 322-8484
*E-mail:* gina.frieden@vanderbilt.edu

# Texas

**Texas A&M University - Commerce**
Counseling Prgm
202 Education North
Commerce, TX 75429-3011
*Prgm Dir:* Stephen Freeman, LPC
*Tel:* 903 886-5631
*E-mail:* Sfreeman@TAMU-Commerce.edu

**Texas A&M University - Corpus Christi**
Counseling Prgm
CC, CE, MFC/T, SC Programs
6300 Ocean Dr
Corpus Christi, TX 78412
*Prgm Dir:* Robert Smith, PhD
*Tel:* 361 825-2307
*E-mail:* Robert.Smith@mail.tamucc.edu

**Texas Woman's University**
Counseling Prgm
CC, SC Programs
Dept of Family Sciences
Denton, TX 76204-5769
*Prgm Dir:* Susan Adams
*Tel:* 940 898-2692
*E-mail:* SAdams1@mail.twu.edu

**University of North Texas**
Counseling Prgm
CC, SC, CIC, CE Programs
PO Box 310829
Denton, TX 76203-0829
*Prgm Dir:* Janice Holden, EdD LPC-S LMFT NCC
*Tel:* 940 565-2910    *Fax:* 940 565-2905
*E-mail:* holden@coe.unt.edu

**Texas Tech University**
Counseling Prgm
CC, SC, CE Programs
PO Box 41071
Lubbock, TX 79409-1071
*Prgm Dir:* Loretta J Bradley
*Tel:* 806 742-1997, Ext 263    *Fax:* 806 742-2179
*E-mail:* loretta.bradley@ttu.edu

**Stephen F Austin State University**
Counseling Prgm
CC, SC Programs
PO Box 13019
Nacogdoches, TX 75962-3019
*Prgm Dir:* David Lawson, PhD
*Tel:* 936 468-1079
*E-mail:* lawsondm@sfasu.edu

**St Mary's University**
Counseling Prgm
CC, CE, MHC Programs
One Camino Santa Maria
San Antonio, TX 78228-8527
www.stmarytx.edu
*Prgm Dir:* Dana Comstock, PhD
*Tel:* 210 436-3226    *Fax:* 210 431-6886
*E-mail:* dcomstock@stmarytx.edu

**Texas State University-San Marcos**
Counseling Prgm
CC, MFC/T, SACC, SC Programs
601 University Dr
San Marcos, TX 78666-4615
*Prgm Dir:* Thelma Duffey
*Tel:* 512 245-2217    *Fax:* 512 245-8872
*E-mail:* tduffey@satx.rr.com

# Utah

**University of Phoenix - Utah**
Counseling Prgm
MHC Program
5373 S Green St
Salt Lake City, UT 84123
*Prgm Dir:* Donald E Beck, PhD
*Tel:* 801 263-1444, Ext 54222
*E-mail:* don.beck@phoenix.edu

# Vermont

**University of Vermont**
Counseling Prgm
MHC, SC Programs
Mann Hall
Burlington, VT 05405-1757
*Prgm Dir:* Eric Nichols, PhD
*Tel:* 802 656-3888    *Fax:* 802 656-3173
*E-mail:* Eric.Nichols@uvm.edu

# Virginia

**Marymount University**
Counseling Prgm
CC, SC Programs
2807 N Glebe Rd
Arlington, VA 22207
*Prgm Dir:* Lisa Jackson-Cherry
*Tel:* 703 284-1633
*E-mail:* lisa.jackson-cherry@marymount.edu

**Virginia Polytechnic Inst & State Univ**
Counseling Prgm
CC, SC, CE Programs
Counselor Education
Blacksburg, VA 24061-0302
*Prgm Dir:* Gerard Lawson, Assistant Professor
*Tel:* 540 231-9703
*E-mail:* glawson@vt.edu

**University of Virginia**
Counseling Prgm
Counselor Ed-Ruffner Hall #160
405 Emmet St South, PO Box 400269
Charlottesville, VA 22904-2495
http://curry.edschool.virginia.edu/curry/dept/edhs
*Prgm Dir:* Sandra I Lopez-Baez, PhD LPC
*Tel:* 434 243-8716    *Fax:* 434 924-1433
*E-mail:* slopez-baez@virginia.edu

**James Madison University**
Counseling Prgm
Department of Graduate Psychology
MSC 7401
Harrisonburg, VA 22807
www.psyc.jmu.edu
*Prgm Dir:* Lennis Echterling, PhD
*Tel:* 540 568-6522    *Fax:* 540 568-3322
*E-mail:* echterlg@jmu.edu

**Lynchburg College**
Counseling Prgm
CC, SC Programs
1501 Lakeside Dr
Lynchburg, VA 24501-3199
*Prgm Dir:* Jeanne Booth, PhD
*Tel:* 434 544-8551
*E-mail:* nielsen@lynchburg.edu

**Old Dominion University**
Counseling Prgm
CC, SC, SA Programs
College of Education
Norfolk, VA 23529
*Prgm Dir:* Nina Brown
*Tel:* 757 783-5308    *Fax:* 757 683-5756
*E-mail:* nbrow026@odu.edu

**Radford University**
Counseling Prgm
PO Box 6994
Radford, VA 24142
*Prgm Dir:* Donald Anderson, EdD
*Tel:* 540 831-5214   *Fax:* 540 831-6755
*E-mail:* danderso@radford.edu

**Regent University**
Counseling Prgm
1000 Regent University Dr
Virginia Beach, VA 23464-9800
www.regent.edu/psychology
*Prgm Dir:* Eric Scalise, LPC LMFT
*Tel:* 757 226-4868   *Fax:* 757 226-4263
*E-mail:* ericsca@regent.edu

**College of William & Mary**
Counseling Prgm
CC, SC, CE Programs
PO Box 8795
Williamsburg, VA 23187-8795
*Prgm Dir:* Victoria Foster, EdD LPC LMFT NCC
*Tel:* 757 221-2321   *Fax:* 757 221-2988
*E-mail:* vafost@wm.edu

# Washington

**Western Washington University**
Counseling Prgm
Dept of Psychology
516 High St
Bellingham, WA 98225-9089
www.ac.wwu.edu/~psych
*Prgm Dir:* Arleen C Lewis, PhD
*Tel:* 360 650-3523   *Fax:* 360 650-7305
*E-mail:* arleen.lewis@wwu.edu

**Eastern Washington University**
Counseling Prgm
MHC, SC Programs
705 W First Ave
Spokane, WA 99201
*Prgm Dir:* Mark Young, PhD
*Tel:* 509 623-4225   *Fax:* 509 359-4366
*E-mail:* myoung@mail.ewu.edu

# West Virginia

**West Virginia University**
Counseling Prgm
CC, SC Programs
3040 University Ave
Morgantown, WV 26506-6122
*Prgm Dir:* Ed Jacobs, PhD
*Tel:* 304 293-2177   *Fax:* 304 293-4082
*E-mail:* Ed.Jacobs@mail.wvu.edu

# Wisconsin

**University of Wisconsin - Oshkosh**
Counseling Prgm
CC, SACC, SC Programs
800 Algoma Blvd
Oshkosh, WI 54901
*Prgm Dir:* Peg Olson
*Tel:* 920 424-1475   *Fax:* 920 424-0858
*E-mail:* olsonmj@uwosh.edu

**University of Wisconsin - Superior**
Counseling Prgm
CC, SC Programs
McCaskill Hall, Rm 111
Superior, WI 54880
*Prgm Dir:* Clyde W Ekbow, EdD
*Tel:* 715 394-8151   *Fax:* 715 394-8146
*E-mail:* jholter@staff.uwsuper.edu

**University of Wisconsin - Whitewater**
Counseling Prgm
CC, SACC, SC, SAPP Programs
Counselor Educ Dept
Whitewater, WI 53190
*Prgm Dir:* Brenda O'Beirne, PhD LP
*Tel:* 262 472-1452   *Fax:* 262 472-2841
*E-mail:* obeirneb@uww.edu

# Wyoming

**University of Wyoming**
Counseling Prgm
CC, SC, SA, CE Programs
PO Box 3374
Laramie, WY 82071
*Prgm Dir:* Michael Loos
*Tel:* 307 766-4002   *Fax:* 307 766-2366
*E-mail:* mdloos@uwyo.edu

## Counselor

| Programs* | Class Capacity | Begins | Length (months) | Award | Res Tuition | Non-res Tuition | Stipend | Offers:‡ 1 | 2 | 3 |
|---|---|---|---|---|---|---|---|---|---|---|
| **Alabama** | | | | | | | | | | |
| Auburn University | 20 | Fall semester | 24 | MS, MEd | $1,630 | $4,890 | | • | | • |
| **Arizona** | | | | | | | | | | |
| Northern Arizona University (Flagstaff) | 60 | Sep Jan | 24, 18 | MA | $4,902 | $13,862 | $8,885 | • | | • |
| University of Phoenix | 25 | Varies | 30, 36 | Dipl, BSHS, MSC | $13,000 | $0 | | • | | • |
| **California** | | | | | | | | | | |
| San Francisco State University | 130 | Fall | 24 | MS | $3,000 | $10,000 | | • | | • |
| **Colorado** | | | | | | | | | | |
| Denver Seminary (Littleton) | 85 | Aug Jan | 48 | MA | | | | • | | • |
| University of Colorado at Colorado Springs | 60 | Summer | 24 | Dipl, MA | $333 | $897 | | • | | • |
| **Delaware** | | | | | | | | | | |
| Wilmington College (New Castle) | 35 | Fall term | 36 | MS | | | | • | | • |
| **Florida** | | | | | | | | | | |
| Barry University (Miami Shores) | 35 | Fall Spring Winter | 36 | MS | $10,980 | $10,980 | | • | | • |
| Stetson University (Deland) | 20 | Fall Spring Summer | 36 | Dipl, MS | $500 | $500 | | • | | |
| University of Central Florida (Orlando) | 30 | Aug Jan | 30 | MA PhD, MEd | $5,788 | $21,927 | | • | | • |
| University of Florida (Gainesville) | Var | Fall Spring | 48 | MED/ED, MAE/ED | $284/ credit | $915/ credit | | | | • |
| University of Florida (Gainesville) | Var | Fall Spring | 48 | MED/ED, MAE/ED | $284/ credit | $915/ credit | | | | • |
| University of Florida (Gainesville) | Var | Fall Spring | 36, 60 | MEDEDS, PhD | $6,827 | $21,951 | | | | • |
| University of South Florida (Tampa) | 25 | Fall semester | 24, 48 | MA | $5,590 | $21,470 | | | | • |
| **Georgia** | | | | | | | | | | |
| Columbus State University | | Fall | 24 | MS, MEd | $865 | $0 | | • | | • |
| Georgia State University (Atlanta) | 75 | Aug | 24 | Dipl, MS | $4,000 | $0 | | • | | • |
| **Idaho** | | | | | | | | | | |
| Boise State University | 18 | Fall | 54 | MA | $8,500 | $8,500 | | • | | |
| Northwest Nazarene University (Nampa) | | | | MS | $5,850 | $5,850 | | • | | • |
| University of Idaho (Moscow) | 25 | Fall semester | 24 | MEd, MS | $6,800 | $9,200 | | • | | |
| **Illinois** | | | | | | | | | | |
| Bradley University (Peoria) | 150 | Open | 36 | Cert, Masters | $5,085 | $0 | | • | | • |
| Eastern Illinois University (Charleston) | 20 | Aug | 48 | MS | $3,000 | $0 | | • | | • |

*Data are shown only for programs that completed the 2006 AMA Survey of Health Professions Education Programs
‡Key to Offers: 1: Evening or weekend classes; 2: Non-English instruction; 3: Cultural competence instruction

## Counselor

| Programs* | Class Capacity | Begins | Length (months) | Award | Res Tuition | Non-res Tuition | Stipend | Offers:‡ 1 | 2 | 3 |
|---|---|---|---|---|---|---|---|---|---|---|
| Southern Illinois Univ at Carbondale | 10 | Open | 30, 48 | Dipl, Cert, MSEd, PhD CE | $192 | $480 | $11,718 | • | | • |
| Western Illinois University (Moline) | 50 | Spring Fall | 48 | MSEd | $182/ credit | $365/ credit | | • | | |
| **Indiana** | | | | | | | | | | |
| Indiana State University (Terre Haute) | 60 | Jun Aug | 24 | MS, MEd | $5,086 | $10,018 | $4,725 | • | | • |
| Indiana University - Bloomington | 50 | Fall Spring Summer | 24 | MS | $241 | $702 | | • | | • |
| Purdue University (West Lafayette) | 10 | Fall | 24 | MSEd | $6,460 | $19,824 | $12,855 | • | | • |
| **Iowa** | | | | | | | | | | |
| University of Iowa (Iowa City) | 30 | Fall semester | 24 | MA | $6,759 | $18,153 | | | • | • |
| **Kansas** | | | | | | | | | | |
| Pittsburg State University | 20 | Each semester | 24 | MS | $2,500 | $6,300 | $4,410 | • | | • |
| **Kentucky** | | | | | | | | | | |
| Eastern Kentucky University (Richmond) | | Aug Jan May | 60, 48 | MA MHC, MAED | $5,030 | $14,174 | | • | | • |
| **Louisiana** | | | | | | | | | | |
| University of New Orleans | 140 | Aug | 24 | MEd, PhD | $2,612 | $0 | | • | | |
| **Maryland** | | | | | | | | | | |
| Loyola College of Maryland (Baltimore) | 25 | Every semester | | MEd, MA | $9,000 | $9,000 | | • | | • |
| Loyola College of Maryland (Columbia) | 400 | Sep Jan May | 24, 48 | MS, PhD | $10,128 | $10,128 | | • | | |
| **Michigan** | | | | | | | | | | |
| Andrews University (Berrien Springs) | 15 | Aug | 18, 24 | MA | $16,800 | $16,800 | $4,000 | • | | |
| Oakland University (Rochester) | 30 | Sep Jan | 24, 36 | MA | $414 | $717 | | • | | • |
| **Mississippi** | | | | | | | | | | |
| Mississippi College (Clinton) | 20 | Fall semester | 24 | MS | $4,000 | $4,000 | | | • | • |
| **Nebraska** | | | | | | | | | | |
| University of Nebraska - Kearney | 250 | Aug Jan | 36 | Dipl, MS | $3,014 | $5,402 | | • | | • |
| **New Jersey** | | | | | | | | | | |
| College of New Jersey (Ewing) | | | | | | | | | | |
| Rider University (Lawrenceville) | 25 | Fall Spring | 48 | MA, EdS | $14,403 | $0 | | • | | |
| **New York** | | | | | | | | | | |
| SUNY at Brockport | | Fall Spring Summer | | MSed | $6,900 | $10,920 | | • | | • |
| University of Rochester | 25 | Rolling admissions | 24, 48 | MS, PhD | $40,000 | $40,000 | | | | |
| **North Carolina** | | | | | | | | | | |
| Univ of North Carolina at Charlotte | 50 | Summer Fall | 24, 36 | Cert, MA, PhD | $5,328 | $20,640 | | • | | • |
| Wake Forest University (Winston-Salem) | 15 | Aug | 24 | MA | $25,000 | $25,000 | $25,000 | | | • |
| **North Dakota** | | | | | | | | | | |
| North Dakota State University (Fargo) | 24 | Jun | 24 | MS, MEd | $3,500 | $0 | | • | | |
| **Ohio** | | | | | | | | | | |
| John Carroll University (Cleveland) | 30 | Rolling admissions | 36 | MA | | | | • | | • |
| Ohio University (Athens) | 90 | Sep | 24, 30 | Dipl, MEd | $7,500 | $0 | | • | | |
| University of Cincinnati | 20 | Jun | 21 | MEd | $11,661 | $21,495 | $11,000 | • | | • |
| University of Toledo | 20 | Summer Fall Spring | 24, 36 | MA | $8,720 | $17,358 | $10,200 | • | | • |
| **Oregon** | | | | | | | | | | |
| Portland State University | | | | | | | | | | |
| **Pennsylvania** | | | | | | | | | | |
| Duquesne University (Pittsburgh) | 65 | Every semester | 29 | MSEd | $13,000 | $13,000 | | • | | • |
| **South Dakota** | | | | | | | | | | |
| South Dakota State University (Brookings) | 40 | Sep Jan Jun | 24, 36 | MS | $115 | $341 | | • | | |
| University of South Dakota (Vermillion) | 40 | Aug | 30 | MA | $214 | $449 | | • | | • |
| **Tennessee** | | | | | | | | | | |
| University of Memphis | 35 | Each semester | 24 | Dipl, MS, EdD | | | | • | | |
| **Texas** | | | | | | | | | | |
| St Mary's University (San Antonio) | | Every semester | 24, 36 | MA | $14,040 | $14,040 | | • | • | |
| Texas A&M University - Commerce | | | 48 | Masters, BS | | | | • | | |
| University of North Texas (Denton) | 23 | Fall Spring Summer | 24, 36 | MEd, MS | | | | • | | • |
| **Utah** | | | | | | | | | | |
| University of Phoenix - Utah (Salt Lake City) | 14 | Continuous | 36 | Dipl, MS | | | | • | | • |
| **Virginia** | | | | | | | | | | |
| James Madison University (Harrisonburg) | 10 | Sep Aug | 36 | Dipl, MEdEdS | $4,752 | $13,374 | | • | | |
| Radford University | 20 | Aug | 24, 54 | MS | $6,230 | $11,482 | | • | | • |
| Regent University (Virginia Beach) | 60 | Fall | 36, 30 | Dipl, MA, PsyD | $9,600 | $9,600 | | • | | • |
| University of Virginia (Charlottesville) | 10 | Aug | 24 | MEdEdS, EdS | $0 | $25,000 | | | | • |
| **Washington** | | | | | | | | | | |
| Western Washington University (Bellingham) | 6 | Sep | 24 | Cert, MEd | $5,694 | $16,221 | | | | • |

*Data are shown only for programs that completed the 2006 AMA Survey of Health Professions Education Programs

‡Key to Offers: 1: Evening or weekend classes; 2: Non-English instruction; 3: Cultural competence instruction

## Occupational Description

Cytology is the study of the structure and the function of cells. Cytotechnologists are specially trained technologists who work with pathologists to evaluate cellular material from virtually all body sites primarily utilizing the microscope. Paramount to cytotechnologists is the microscopic recognition of normal and abnormal cytologic changes, including, but not limited to, malignant neoplasms, precancerous lesions, infectious agents, and inflammatory processes in gynecologic, non-gynecologic, and fine needle aspiration specimens. Cytotechnologists possess the technical skills for a wide variety of cytologic laboratory specimen preparations and a basic knowledge of contemporary procedures and technologies.

## Job Description

Cell specimens may be obtained from various body sites, such as the female reproductive tract, the lung, or any body cavity shedding cells. Using special techniques, slides are first prepared from these specimens. Cytotechnologists then examine the slides microscopically, mark cellular changes that are most representative of a disease process, and submit to a pathologist for final evaluation. Using the findings of cytotechnologists, the pathologist is then able, in many instances, to diagnose cancer and other diseases long before they can be detected by other methods. In recent years, fine needles have been used to aspirate lesions, often deeply seated in the body, thus greatly enhancing the ability to diagnose tumors located in otherwise inaccessible sites.

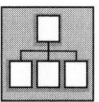

## Employment Characteristics

Most cytotechnologists work in hospitals or in commercial laboratories. With experience, cytotechnologists may also work in private industry or in supervisory, research, and teaching capacities.

## Salary

Employment opportunities and salaries vary depending on geographic location, experience, and ability. According to the ASCT, the average hourly pay for cytotechnologists was $28.99 in 2005. Refer to Section IV, Table 5 of this *Directory* for more information, or see www.ama-assn.org/go/ salary.

## Educational Programs

**Length.** The length of the program depends significantly on its organizational structure. In general, after completion of the prerequisite course work, at least one calendar year of structured professional instruction in cytotechnology is necessary to achieve program objectives and to establish entry-level competencies.

**Prerequisites.** Applicants should be well grounded in the biological sciences and in basic chemistry. This entails that students have a minimum of 28 semester hours of biological sciences and chemistry upon completion of a cytotechnology program, and 3 semester hours of mathematics and/or statistics. In addition, applicants are also required to have a baccalaureate degree in order to qualify for the national certification exam.

**Curriculum.** The curriculum includes the principles of cytopreparation of cell samples, cytologic evaluation of cell samples from all body sites, introduction to principles of management, research, and education as they apply to the cytology laboratory, and cytology as applied in clinical medicine. Also, as molecular diagnostics becomes increasingly important in the field of pathology, programs are incorporating instruction in immunohistochemistry, cytogenetics, in situ hybridization, polymerase chain reaction, and flow cytometry. Upon completion of a cytotechnology program, graduates will possess the technical skills to evaluate a wide variety of cytologic preparations and have a basic knowledge of contemporary procedures and technologies used in cytopathology.

## Inquiries

**Careers/Curriculum**
American Society of Cytopathology
400 West 9th Street, Suite 201
Wilmington, DE 19801
302 429-8802
302 429-8807 Fax
E-mail: asc@cytopathology.org

**Certification/Registration**
ASCP Board of Registry
PO Box 12270
Chicago, IL 60612
312 738-1336
E-mail: bor@ascp.org

**Program Accreditation**
Commission on Accreditation of Allied Health Education Programs (CAAHEP) in collaboration with:
Cytotechnology Programs Review Committee
American Society of Cytopathology
400 West 9th Street, Suite 201
Wilmington, DE 19801
302 429-8802
302 429-8807 Fax
E-mail: dmacintyre@cytopathology.org

## Cytotechnologist*

## Alabama

**University of Alabama at Birmingham**
Cytotechnology Prgm
Sch of Hlth Related Professions
UAB Station-RMSB 440
Birmingham, AL 35294-1212
www.uab.edu/ct
*Prgm Dir:* Vivian Pijuan-Thompson, PhD
*Tel:* 205 934-4863   *Fax:* 205 975-7302
*E-mail:* pijuan@uab.edu
*Med Dir:* Isam-Eldin A Eltoum, MD MRC Path

**Auburn University Montgomery**
Cytotechnology Prgm
204B Moore Hall
PO Box 244023
Montgomery, AL 36124-4023
www.aum.edu/cls
*Prgm Dir:* Sonya Griffin, MS SCT(ASCP) CT(IAC)
*Tel:* 334 244-3480   *Fax:* 334 244-3146
*E-mail:* sgriffi2@mail.aum.edu
*Med Dir:* Darshana Jhala, MD

## Arkansas

**University of Arkansas for Medical Sciences**
Cytotechnology Prgm
College of Health Related Professions
4301 W Markham St, # 517
Little Rock, AR 72205-9985
www.uams.edu/chrp/cytotechnology
*Prgm Dir:* Donald Simpson, MPH CT(ASCP)
*Tel:* 501 686-8448   *Fax:* 501 686-8519
*E-mail:* simpsondonald@uams.edu
*Med Dir:* Michael Johnson, MD

## California

**Loma Linda University**
Cytotechnology Prgm
Dept of Clinical Laboratory Science
Anderson/Barton Rd
Loma Linda, CA 92350
www.llu.edu/llu/sahp
*Prgm Dir:* Marlene Ota, BS SCT(ASCP)
*Tel:* 909 558-4000, Ext 45386   *Fax:* 909 558-0458
*E-mail:* mota@sahp.llu.edu
*Med Dir:* Darryl Heustis, MD

**UCLA Center for Health Sciences**
Cytotechnology Prgm
Department of Pathology
UCLA Medical Center, A3-231S
Los Angeles, CA 90095-1732
*Prgm Dir:* Mary Levin, SCT(ASCP)
*Tel:* 310 794-7135   *Fax:* 310 206-8108
*E-mail:* mlevin@mednet.ucla.edu
*Med Dir:* Sharon Hirschowitz, MD

## Connecticut

**University of Connecticut Health Center**
Cytotechnology Prgm
263 Farmington Ave
Department of Pathology MC-3985
Farmington, CT 06030
www.alliedhealth.uconn.edu
*Prgm Dir:* Nancy J Smith, MS SCT(ASCP)
*Tel:* 860 679-4215   *Fax:* 860 679-4334
*E-mail:* nsmith@nso1.uchc.edu
*Med Dir:* M Melinda Sanders, MD

*Additional information about programs that
returned the AMA's survey is available in a table
beginning on page 100.

## Georgia

**Grady Health System**
Cytotechnology Prgm
80 Jessie Hill Jr Dr, Box 056
Atlanta, GA 30303
*Prgm Dir:* Dean Willis, SCT(ASCP)
*Tel:* 404 616-3650   *Fax:* 404 616-9913
*E-mail:* dwillis@gmh.edu
*Med Dir:* Talaat Tadros, MD

## Indiana

**Indiana University**
Cytotechnology Prgm
Clarian Pathology Laboratory
350 W 11th St, Rm 6002J
Indianapolis, IN 46202
www.pathology.iupui.edu
*Prgm Dir:* William N Crabtree, PhD SCT(ASCP)
*Tel:* 317 491-6221   *Fax:* 317 491-6212
*E-mail:* wcrabtre@iupui.edu
*Med Dir:* Harvey Cramer, MD

## Iowa

**Mercy Medical Center**
Cytotechnology Prgm
1111 6th Ave
Des Moines, IA 50314
www.mercydesmoines.org
*Prgm Dir:* Marty Boesenberg, BS SCT(ASCP)
*Tel:* 515 247-4466   *Fax:* 515 643-8810
*E-mail:* mboesenberg@mercydesmoines.org
*Med Dir:* Joy E Trueblood, MD

## Kansas

**University of Kansas Medical Center**
Cytotechnology Prgm
1600 Bell, Cytology Laboratory
3901 Rainbow Blvd
Kansas City, KS 66160-7281
www.alliedhealth.kumc.edu/programs/cytotech.htm
*Prgm Dir:* Marilee Means, PhD SCT(ASCP)
*Tel:* 913 588-1179   *Fax:* 913 588-1160
*E-mail:* mmeans@kumc.edu
*Med Dir:* Ossama Tawfik, PhD, MD

## Kentucky

**Pathology & Cytology Laboratories**
Cytotechnology Prgm
290 Big Run Rd
Lexington, KY 40503
*Prgm Dir:* Ivan Eads, MS SCT(ASCP) CMIAC
*Tel:* 859 278-9513, Ext 235   *Fax:* 859 277-6063
*E-mail:* pclsct@yahoo.com
*Med Dir:* Richard Lozano, MD

**Bellarmine University**
Cytotechnology Prgm
Lansing Sch of Nursing & Health Sciences
2001 Newburg Rd
Louisville, KY 40205
www.bellarmine.edu/lansing
*Prgm Dir:* Mary Beth Adams, MEd SCT(ASCP)
*Tel:* 502 452-8391   *Fax:* 502 452-8389
*E-mail:* madams@bellarmine.edu
*Med Dir:* Sidney Murphree, MD

## Louisiana

**Nicholls State University**
Cytotechnology Prgm
Dept of Allied Health Sciences
235 Civic Center Blvd
Houma, LA 70360
www.nicholls.edu
*Prgm Dir:* Celdrick Course, MS CT(ASCP)
*Tel:* 985 876-8850   *Fax:* 985 876-8856
*E-mail:* celdrick.course@nicholls.edu
*Med Dir:* David Gyer, MD

## Maryland

**Johns Hopkins Hospital**
Cytotechnology Prgm
600 N Wolfe St, 406 Pathology Bldg
Baltimore, MD 21287-6940
*Prgm Dir:* Frances H Burroughs, SCT(ASCP)
*Tel:* 410 955-1180   *Fax:* 410 614-9556
*E-mail:* fburroug@jhmi.edu
*Med Dir:* Yener S Erozan, MD

## Massachusetts

**Berkshire Medical Center**
Cytotechnology Prgm
725 North St
Pittsfield, MA 01201-4124
www.berkshirehealthsystems.org
*Prgm Dir:* Judith R Shaffer, SCT(ASCP) BS
*Tel:* 413 447-2590   *Fax:* 413 447-2097
*E-mail:* jshaffer@bhs1.org
*Med Dir:* Teri Cooper, MD

## Michigan

**DMC University Laboratories**
Cytotechnology Prgm
4707 St Antoine Blvd
Detroit, MI 48201
www.dmc.org/univlab/ct-clnprog.htm
*Prgm Dir:* Kalyani Naik, MS SCT(ASCP)
*Tel:* 313 745-7122   *Fax:* 313 993-8894
*E-mail:* knaik@dmc.org
*Med Dir:* Mujtaba Husain, MD

## Minnesota

**Mayo School of Health Sciences**
Cytotechnology Prgm
200 First St SW
Rochester, MN 55905
www.mayo.edu/education
*Prgm Dir:* Jill L Caudill, MEd SCT(ASCP)
*Tel:* 507 284-1142   *Fax:* 507 266-1578
*E-mail:* caudill.jill@mayo.edu
*Med Dir:* Diva Salomao, MD

## Mississippi

**University of Mississippi Medical Center**
Cytotechnology Prgm
Department of Diagnostic and Clinical Health Sciences
2500 N State St
Jackson, MS 39216
http://shrp.umc.edu
*Prgm Dir:* Zelma Cason, MS SCT(ASCP)
*Tel:* 601 984-6358   *Fax:* 601 815-1717
*E-mail:* zcason@shrp.umsmed.edu
*Med Dir:* Mithra Baliga, MD

# Missouri

**Barnes-Jewish Coll of Nursing/Allied Health**
Cytotechnology Prgm
306 S Kingshighway Blvd
MS: 90-30-625
St Louis, MO 63110
www.barnesjewishcollege.edu
*Prgm Dir:* Linda Hoechst, MA SCT(ASCP)(IAC)
*Tel:* 314 454-8655  *Fax:* 314 454-5239
*E-mail:* lhoechst@bjc.org
*Med Dir:* Rosa Davila, MD

# Nebraska

**University of Nebraska Medical Center**
Cytotechnology Prgm
987549 Nebraska Medical Center
Omaha, NE 68198-7549
*Prgm Dir:* Amber Donnelly, SCT(ASCP)
*Tel:* 402 552-2043  *Fax:* 402 559-9044
*E-mail:* addonnelly@unmc.edu
*Med Dir:* Stanley J Radio, MD

# Nevada

**Quest Diagnostics - Las Vegas**
Cytotechnology Prgm
4230 S Burnham Ave, Ste 250
Las Vegas, NV 89119
www.questdiagnostics.com
*Prgm Dir:* Robert Gay, CT(ASCP)
*Tel:* 702 733-7866, Ext 3200  *Fax:* 702 733-8941
*E-mail:* robert.w.gay@questdiagnostics.com
*Med Dir:* Jerry J Marty, MD

# New Jersey

**Univ of Med & Dent of New Jersey**
Cytotechnology Prgm
School of Health Related Professions
1776 Raritan Rd
Scotch Plains, NJ 07076-2997
www.shrp.umdnj.edu/programs
*Prgm Dir:* Cecilia B Vallejo, MD
*Tel:* 908 889-2424  *Fax:* 908 889-2487
*E-mail:* vallejcb@umdnj.edu
*Med Dir:* Andrey Gritsman, MD

# New York

**Albany College of Pharmacy**
Cytotechnology Prgm
106 New Scotland Ave
Albany, NY 12208-3492
www.acp.edu
*Prgm Dir:* Indra Balachandran, PhD SCT(ASCP) CFIAC
*Tel:* 518 694-7390  *Fax:* 518 694-7037
*E-mail:* balachai@acp.edu
*Med Dir:* Pratima Kunchala, MD

**Memorial Sloan - Kettering Cancer Ctr**
Cytotechnology Prgm
1275 York Ave C596
New York, NY 10021
www.mskcc.org/mskcc/html
*Prgm Dir:* Rose Marie Gatscha, SCT(ASCP)
*Tel:* 212 639-5902  *Fax:* 212 639-6318
*E-mail:* gatschar@mskcc.org
*Med Dir:* Oscar Lin, MD

**Stony Brook University**
Cytotechnology Prgm
Sch of Health Tech and Management
HSC L-2, Rm 415
Stony Brook, NY 11794-8205
*Prgm Dir:* Catherine M Vetter, MS SCT(ASCP) MT
*Tel:* 631 444-2403  *Fax:* 631 444-7621
*E-mail:* cvetter@notes.cc.sunysb.edu
*Med Dir:* Alan Heimann, MD

**SUNY Upstate Medical University**
Cytotechnology Prgm
750 E Adams St, Silverman Hall
Syracuse, NY 13210
www.upstate.edu/chp/cyto
*Prgm Dir:* Susan B Stowell, MA SCT(ASCP)
*Tel:* 315 464-6900  *Fax:* 315 464-6914
*E-mail:* stowells@upstate.edu
*Med Dir:* Kamal Khurana, MD

**St John's Riverside Hospital**
Cytotechnology Prgm
967 N Broaway
Yonkers, NY 10701
*Prgm Dir:* Galina Filonenko, MPH
*Tel:* 914 345-9395  *Fax:* 914 594-4163
*E-mail:* galina_filonenko@nymc.edu
*Med Dir:* Myron Melamed, MD

# North Carolina

**Univ of North Carolina at Chapel Hill**
Cytotechnology Prgm
Department of Allied Health Sciences, Medical School
Wing E
CB 7120
Chapel Hill, NC 27599-7120
*Prgm Dir:* Allen C Rinas, MS SCT(ASCP) CM(IAC)
*Tel:* 919 843-1065  *Fax:* 919 966-8384
*E-mail:* arinas@med.unc.edu
*Med Dir:* Harsharan Singh, MD

**Central Piedmont Community College**
Cytotechnology Prgm
PO Box 35009
Charlotte, NC 28235
http://cpcc.edu/Health_Sciences/Cytotechnology
*Prgm Dir:* M Arlene Parrish, MS SCT(ASCP) IAC
*Tel:* 704 330-6383  *Fax:* 704 330-6151
*E-mail:* arlene.parrish@cpcc.edu
*Med Dir:* James E McDermott, MD

# North Dakota

**University of North Dakota**
Cytotechnology Prgm
School of Medicine & Health Sciences Rm 3124
501 North Columbia Rd, Stop 9037
Grand Forks, ND 58202
*Prgm Dir:* Kathy Hoffman, MM SCT(ASCP)
*Tel:* 701 777-4466  *Fax:* 701 777-2404
*E-mail:* khoffman@medicine.nodak.edu
*Med Dir:* Timothy L Weiland, MD

# Ohio

**Akron General Medical Center**
Cytotechnology Prgm
400 Wabash Ave
Akron, OH 44307
*Prgm Dir:* Angela T Powell, MD
*Tel:* 330 344-1885  *Fax:* 330 344-6418
*E-mail:* jjohnson@agmc.org
*Med Dir:* Angela T Powell, MD

# Pennsylvania

**Thomas Jefferson University**
Cytotechnology Prgm
Jefferson Coll of Hlth Prof, Dept of Bioscience Tech
Edison Bldg, Ste 1924, 130 S Ninth St, Rm 1924
Philadelphia, PA 19107-5233
*Prgm Dir:* Shirley E Greening, MS JD CT(ASCP) CFIAC
*Tel:* 215 503-8561  *Fax:* 215 503-2189
*E-mail:* shirley.greening@jefferson.edu
*Med Dir:* Ronald D Luff, MD MPH

**Univ Hlth Ctr of PA/Anisa I Kanbour Sch -Cyto**
Cytotechnology Prgm
300 Halket St
Pittsburgh, PA 15213-3180
http://path.upmc.edu/cytotech
*Prgm Dir:* Anisa I Kanbour, MD
*Tel:* 412 641-4664  *Fax:* 412 641-6263
*E-mail:* akanbour@magee.edu
*Med Dir:* Anisa I Kanbour, MD

# Puerto Rico

**University of Puerto Rico**
Cytotechnology Prgm
College of Allied Health Professions
PO Box 365067 Ofc 701
San Juan, PR 00936-5067
www.cprs.rcm.upr.edu
*Prgm Dir:* Alma J Camacho, CT (ASCP ASC) MBA
*Tel:* 787 758-2525, Ext 4604  *Fax:* 787 759-5095
*E-mail:* almacamacho@cprs.rcm.upr.edu
*Med Dir:* Maria Marcos, MD

# Rhode Island

**Rhode Island School of Cytotechnology**
*Cosponsor:* University of Rhode Island
Cytotechnology Prgm
Shepard Bldg, Rm 302
80 Washington St
Providence, RI 02903-1803
www.rischoolofcytotechnology.com and
  www.uri.edu/cels/cmb
*Prgm Dir:* Barbara G Klitz, MSCT(ASCP)
*Tel:* 401 277-5199  *Fax:* 401 277-5060
*E-mail:* bgk@etal.uri.edu
*Med Dir:* Latha Pisharodi, MD

# South Carolina

**Medical University of South Carolina**
Cytotechnology Prgm
Health Prof, Department of Clinical Services
151B Rutledge Ave, Rm B102
Charleston, SC 29425
www.musc.edu/chp/cb
*Prgm Dir:* Karen B Geils, MS CT(ASCP) HT(ASCP)
*Tel:* 843 792-1913  *Fax:* 843 792-0506
*E-mail:* brinkerk@musc.edu
*Med Dir:* Rana Hoda, MD

# Tennessee

**University of Tennessee Health Science Ctr**
Cytotechnology Prgm
Coll of Allied Hlth Sciences
930 Madison Ave, Ste 674
Memphis, TN 38163
www.utmen.edu/allied/cytotechnology_home.html
*Prgm Dir:* Barbara Benstein, PhD SCT(ASCP)
*Tel:* 901 448-6304  *Fax:* 901 448-7545
*E-mail:* bbenstein@utmem.edu
*Med Dir:* Nadeem Zafar, MD

# Texas

**Brooke Army Medical Center**
Cytotechnology Prgm
Interservice Cytotechnology Program
Dept of Pathology and Area Laboratory Services
Ft Sam Houston, TX 78234-6200
www.bamc.amedd.army.mil/dept.htm
*Prgm Dir:* HMC Jorge A Franco, MA Ed SCT(ASCP)IAC
*Tel:* 210 916-0499   *Fax:* 210 916-0877
*E-mail:* jorge.franco@cen.amedd.army.mil
*Med Dir:* Karen K Nauschuetz, MD COL

**Univ of Texas M D Anderson Cancer Ctr**
Cytotechnology Prgm
1515 Holcombe Blvd, Unit 206
Houston, TX 77030-4009
www.mdanderson.org/healthsciences
*Prgm Dir:* Christina M Alapat, MS SCT MP(ASCP) IAC
*Tel:* 713 794-5877   *Fax:* 713 745-0172
*E-mail:* calapat@mdanderson.org
*Med Dir:* Gregg A Staerkel, MD

# Utah

**University of Utah Health Science Center**
Cytotechnology Prgm
ARUP Laboratory
500 Chipeta Way
Salt Lake City, UT 84108
www.path.utah.edu/mls
*Prgm Dir:* Michael C Berry, SCT(ASCP)
*Tel:* 801 583-2787, Ext 2327   *Fax:* 801 584-5213
*E-mail:* berrymc@aruplab.com
*Med Dir:* Evelyn Gopez, MD

# Vermont

**Fletcher Allen Health Care**
Cytotechnology Prgm
111 Colchester Ave
Burlington, VT 05401
https://www.fahc.org/cytoschool
*Prgm Dir:* Sandra Giroux, MS SCT(ASCP) CFIAC
*Tel:* 802 847-5133   *Fax:* 802 847-3632
*E-mail:* Sandra.Giroux@vtmednet.org
*Med Dir:* Pamela Gibson, MD

# Virginia

**Old Dominion University**
Cytotechnology Prgm
209 Spong Hall
Norfolk, VA 23529
*Prgm Dir:* Sophie K Thompson, MHS CT(ASCP)
*Tel:* 757 683-3016   *Fax:* 757 683-5028
*E-mail:* sxthomps@odu.edu
*Med Dir:* Anna B Schiller, MD

# West Virginia

**CAMC Health Education & Research Institute**
Cytotechnology Prgm
3200 MacCorkle Ave SE
Charleston, WV 25304
www.cytologyschool.com
*Prgm Dir:* Carolyn H Stevens, CT(ASCP)
*Tel:* 304 388-5570   *Fax:* 304 388-9833
*E-mail:* carolyn.stevens@camc.org
*Med Dir:* William Mangano, MD

**Cabell Huntington Hospital**
Cytotechnology Prgm
1340 Hal Greer Blvd
Huntington, WV 25701
www.cabellhuntington.org
*Prgm Dir:* Margene Smith, SCT(ASCP)
*Tel:* 304 526-2155   *Fax:* 304 526-2187
*E-mail:* mjsmith@chhi.org
*Med Dir:* Linda G Brown, MD

# Wisconsin

**State Laboratory of Hygiene**
Cytotechnology Prgm
465 Henry Mall
Madison, WI 53706
www.slh.wisc.edu/cytology
*Prgm Dir:* John E Shalkham, MA SCT(ASCP)
*Tel:* 608 265-9191   *Fax:* 608 265-6294
*E-mail:* shalkham@wisc.edu
*Med Dir:* Daniel Kurtycz, MD

**Marshfield Clinic**
*Cosponsor:* St. Joseph's Hospital
Cytotechnology Prgm
1000 N Oak Ave
Marshfield, WI 54449-5795
www.marshfieldlaboratories.org
*Prgm Dir:* Donald Schnitzler, BS CT(ASCP)
*Tel:* 715 387-7440   *Fax:* 715 389-5353
*E-mail:* schnitzler.donald@marshfieldclinic.org
*Med Dir:* George M Rupp, MD

**University of Wisconsin - Milwaukee**
Cytotechnology Prgm
Dept of Health Sciences
PO Box 413
Milwaukee, WI 53201
www.uwm.edu
*Prgm Dir:* Evelyn Dell, BS CT HT(ASCP)
*Tel:* 414 328-7914   *Fax:* 414 328-8505
*E-mail:* evelyn.dell@aurora.org
*Med Dir:* Janet Durham, MD

## Cytotechnologist

| Programs* | Class Capacity | Begins | Length (months) | Award | Res Tuition | Non-res Tuition | Stipend | Offers:‡ 1 | 2 | 3 |
|---|---|---|---|---|---|---|---|---|---|---|
| **Alabama** | | | | | | | | | | |
| Auburn University Montgomery | 8 | Jan | 12 | Cert | $6,259 | $17,959 | | | | |
| University of Alabama at Birmingham | 12 | Aug | 12 | Cert, BS | $8,320 | $20,800 | | | | |
| **Arkansas** | | | | | | | | | | |
| University of Arkansas for Medical Sciences (Little Rock) | 8 | Aug | 12 | BS | $6,925 | $15,165 | | | | |
| **California** | | | | | | | | | | |
| Loma Linda University | 12 | Aug | 12, 18 | Cert, BS | $23,000 | $23,000 | | | | |
| **Connecticut** | | | | | | | | | | |
| University of Connecticut Health Center (Farmington) | 6 | Jun | 12, 24 | Cert, BS, MLS | $5,824 | $14,942 | | | | |
| **Indiana** | | | | | | | | | | |
| Indiana University (Indianapolis) | 8 | Aug | 12 | BS | $5,860 | $18,574 | | | | • |
| **Iowa** | | | | | | | | | | |
| Mercy Medical Center (Des Moines) | 4 | Jul | 12 | Cert | $3,500 | $3,500 | | | | |
| **Kansas** | | | | | | | | | | |
| University of Kansas Medical Center (Kansas City) | 4 | Aug | 12 | BS | $6,271 | $17,259 | | | | |
| **Kentucky** | | | | | | | | | | |
| Bellarmine University (Louisville) | 8 | Jan Aug | 16, 21 | BHS | $10,850 | $10,850 | | | | |
| **Louisiana** | | | | | | | | | | |
| Nicholls State University (Houma) | 4 | Jun | 12 | BS | $4,253 | $12,425 | | | | |
| **Maryland** | | | | | | | | | | |
| Johns Hopkins Hospital (Baltimore) | 5 | Aug | 12 | Cert | $8,650 | $8,650 | | | | • |
| **Massachusetts** | | | | | | | | | | |
| Berkshire Medical Center (Pittsfield) | 4 | Sep | 12 | Cert, BS | $3,500 | $3,500 | $175 | | | |

*Data are shown only for programs that completed the 2006 AMA Survey of Health Professions Education Programs
‡Key to Offers: 1: Evening or weekend classes; 2: Non-English instruction; 3: Cultural competence instruction

## Cytotechnologist

| Programs* | Class Capacity | Begins | Length (months) | Award | Res Tuition | Non-res Tuition | Stipend | Offers:‡ 1 | 2 | 3 |
|---|---|---|---|---|---|---|---|---|---|---|
| **Michigan** | | | | | | | | | | |
| DMC University Laboratories (Detroit) | 5 | Aug | 12 | Cert | | | | | | |
| **Minnesota** | | | | | | | | | | |
| Mayo School of Health Sciences (Rochester) | 8 | Jul | 12 | Cert | $3,500 | $3,500 | | | | • |
| **Mississippi** | | | | | | | | | | |
| University of Mississippi Medical Center (Jackson) | 12 | Aug | 18 | BS | $2,301 | $5,283 | | | | |
| **Missouri** | | | | | | | | | | |
| Barnes-Jewish Coll of Nursing/Allied Health (St Louis) | 12 | Jun | 12 | Cert, BSCT | $15,200 | $15,200 | | | | • |
| **Nevada** | | | | | | | | | | |
| Quest Diagnostics - Las Vegas | 8 | Aug | 12 | Cert | $6,500 | $6,500 | | | | |
| **New Jersey** | | | | | | | | | | |
| Univ of Med & Dent of New Jersey (Scotch Plains) | 10 | Late Aug | 12 | Cert, BSCLS | $9,590 | $14,385 | | | | |
| **New York** | | | | | | | | | | |
| Albany College of Pharmacy | 15 | Aug | 12 | Dipl, Cert, CT, BS | $11,600 | $11,600 | | | | |
| Memorial Sloan - Kettering Cancer Ctr (New York) | 6 | Aug | 12 | Cert, CT | $8,000 | $8,000 | | | | |
| SUNY Upstate Medical University (Syracuse) | 11 | Aug | 12 | BS | $6,526 | $15,450 | | | | |
| **North Carolina** | | | | | | | | | | |
| Central Piedmont Community College (Charlotte) | 5 | Aug | 12 | Cert | $1,896 | $10,536 | | | | |
| **North Dakota** | | | | | | | | | | |
| University of North Dakota (Grand Forks) | 8 | Aug | 12 | Cert, BS | $9,493 | $9,493 | | | | |
| **Pennsylvania** | | | | | | | | | | |
| Univ Hlth Ctr of PA/Anisa I Kanbour Sch -Cyto (Pittsburgh) | 8 | Jul | 12 | Cert | $6,495 | $8,495 | | | | |
| **Puerto Rico** | | | | | | | | | | |
| University of Puerto Rico (San Juan) | 6 | Aug | 12 | Cert | $1,400 | $3,000 | | | | |
| **Rhode Island** | | | | | | | | | | |
| Rhode Island School of Cytotechnology (Providence) | 8 | Sep | 11 | Cert, CT, MS | $14,500 | $14,500 | | | | |
| **South Carolina** | | | | | | | | | | |
| Medical University of South Carolina (Charleston) | 12 | Aug | 21 | MS | $14,316 | $29,445 | | | | |
| **Tennessee** | | | | | | | | | | |
| University of Tennessee Health Science Ctr (Memphis) | 6 | Sep | 21 | MS | $7,960 | $19,960 | | | | |
| **Texas** | | | | | | | | | | |
| Brooke Army Medical Center (Ft Sam Houston) | 24 | Aug | 12 | Cert, BS | | | | | | |
| Univ of Texas M D Anderson Cancer Ctr (Houston) | 6 | Sep | 12 | Cert, BS | $3,240 | $15,530 | | | | • |
| **Utah** | | | | | | | | | | |
| University of Utah Health Science Center (Salt Lake City) | 4 | Jun | 12 | Cert, BS | $5,900 | $15,100 | | | | |
| **Vermont** | | | | | | | | | | |
| Fletcher Allen Health Care (Burlington) | 7 | Sep | 12 | Cert, BS | $7,000 | $7,000 | | | | |
| **West Virginia** | | | | | | | | | | |
| Cabell Huntington Hospital | 4 | May | 12 | Cert, BS | $5,000 | $5,000 | | | | |
| CAMC Health Education & Research Institute (Charleston) | 6 | Jun | 12 | Cert | $5,000 | $5,000 | | | | |
| **Wisconsin** | | | | | | | | | | |
| Marshfield Clinic | 4 | Jul | 12 | Cert | $850 | $850 | | | | |
| State Laboratory of Hygiene (Madison) | 12 | Aug | 12 | Cert, BS | $5,500 | $5,500 | | | | |
| University of Wisconsin - Milwaukee | 6 | Sep | 12 | BS | $4,060 | $15,190 | | | | |

*Data are shown only for programs that completed the 2006 AMA Survey of Health Professions Education Programs
‡Key to Offers: 1: Evening or weekend classes; 2: Non-English instruction; 3: Cultural competence instruction

## Occupational Description

Emerging as a distinct profession in the 1940s, dance/movement therapy, a creative arts therapy, is rooted in the expressive nature of dance itself. Dance is the most fundamental of the arts, involving a direct expression and experience of oneself through the body. It is a basic form of authentic communication, and as such it is an especially effective medium for therapy. Based in the belief that the body, the mind, and the spirit are interconnected, dance/movement therapy is defined by the American Dance Therapy Association (ADTA) as "the psychotherapeutic use of movement as a process that furthers the emotional, cognitive, social, and physical integration of the individual."

## Job Description

Dance/movement therapists work with individuals of all ages, groups, and families in a wide variety of settings. They focus on helping their clients improve self-esteem and body image, develop effective communication skills and relationships, expand their movement vocabulary, and gain insight into patterns of behavior, as well as create new options for coping with problems. Dance/movement therapy can be a powerful tool for stress management and the prevention of physical and mental health problems. Movement is the primary medium that dance/movement therapists use for observation, assessment, research, therapeutic interaction, and interventions.

## Employment Characteristics

Dance/movement therapists work in settings that include psychiatric and rehabilitation facilities, schools, nursing homes, drug treatment centers, counseling centers, medical facilities, crisis centers, and wellness and alternative health care centers. Dance/movement therapy is used with people of all ages, races, and ethnic backgrounds in individual, couples, family, and group therapy formats.

There are approximately 1,300 dance/movement therapists in 46 states and 41 foreign countries.

## Educational Programs

**Length.** Professional training for US dance/movement therapists is on the graduate level. Graduates receive a master's degree in dance/movement therapy or related degree title. Graduates from an ADTA-approved dance/movement therapy program are eligible for the DTR (Dance Therapist Registered) credential upon completion of graduate studies. Approved programs have met the basic educational standards of the ADTA, which include that the program's parent institution is accredited by its regional accreditation association.

**Prerequisites.** Extensive dance experience and a liberal arts background with coursework in psychology are required. For specific prerequisites, contact each graduate program.

**Credentials.** The ADTA distinguishes between dance/movement therapists prepared to work in professional settings within a team under supervision and those prepared for the responsibilities of working independently in private practice or providing supervision.

- *DTR: Dance Therapist Registered*—Therapists with this title have a master's degree and are fully qualified to work in a professional treatment system.
- *ADTR: Academy of Dance Therapists Registered*—Therapists with this title have met additional requirements and are fully qualified to teach, provide supervision, and engage in private practice.

## Inquiries

American Dance Therapy Association
2000 Century Plaza, Suite 108
10632 Little Patuxent Parkway
Columbia, MD 21044-3263
412 624-6595
www.adta.org

## Dance/Movement Therapist

## Colorado

**Naropa University**
Dance/Movement Therapy Prgm
2130 Arapahoe Ave
Boulder, CO 80302
www.naropa.edu/somatic
*Prgm Dir:* Z"e Avstreih, MA LPC ADTR
*Tel:* 303 245-4774   *Fax:* 303 245-4827
*E-mail:* zoe@naropa.edu
*Med Dir:* Leah D'Abate, MA DTR

## Illinois

**Columbia College**
Dance/Movement Therapy Prgm
600 S Michigan
Chicago, IL 60605-9988
www.colum.edu/graduate/graddance.html
*Prgm Dir:* Susan Imus, MA ADTR LCPC GLCMA
*Tel:* 312 344-7697   *Fax:* 312 344-8054
*E-mail:* simus@colum.edu

## New Hampshire

**Antioch/New England Graduate School**
Dance/Movement Therapy Prgm
40 Avon St
Keene, NH 03431-3552
www.antiochne.edu/academics/appsy/dmt.html
*Prgm Dir:* Susan Loman, MA ADTR NCC
*Tel:* 603 357-3122, Ext 222   *Fax:* 603 357-0718
*E-mail:* sloman@antiochne.edu

## New York

**Pratt Institute**
Dance/Movement Therapy Prgm
East 3, 200 Willoughby Ave
Brooklyn, NY 11205
*Prgm Dir:* Laurel Thompson, MPS ADTR ATR-BC
*Tel:* 718 636-3428
*E-mail:* lthompso@pratt.edu

## Pennsylvania

**Drexel University**
*Cosponsor:* Hahnemann Creative Arts in Therapy
   Program
Dance/Movement Therapy Prgm
1505 Race St
10th Fl Bellet Building, Mail Stop 905
Philadelphia, PA 19102-1192
www.drexel.edu/cnhp/creative_arts/about.asp
*Prgm Dir:* Ellen Schelly Hill, MMT ADTR NCC LPC
*Tel:* 215 762-7851   *Fax:* 215 762-6933
*E-mail:* es42@drexel.edu

## Dance/Movement Therapist

| Programs* | Class Capacity | Begins | Length (months) | Award | Res Tuition | Non-res Tuition | Stipend | Offers:‡ 1 | 2 | 3 |
|---|---|---|---|---|---|---|---|---|---|---|
| **Colorado** | | | | | | | | | | |
| Naropa University (Boulder) | 30 | Sep | 36 | Dipl, Cert, MA | | | | | | • |
| **Illinois** | | | | | | | | | | |
| Columbia College (Chicago) | 28 | Fall | 30, 14 | Dipl, Cert, MA | $14,000 | $0 | | • | | • |
| **New Hampshire** | | | | | | | | | | |
| Antioch/New England Graduate School (Keene) | 20 | Fall semester | 30 | Cert, MA | $17,500 | $17,500 | | • | | • |
| **Pennsylvania** | | | | | | | | | | |
| Drexel University (Philadelphia) | 40 | Sep | 21 | Cert, MA | $17,960 | $17,960 | | | | • |

*Data are shown only for programs that completed the 2006 AMA Survey of Health Professions Education Programs
‡Key to Offers: 1: Evening or weekend classes; 2: Non-English instruction; 3: Cultural competence instruction

**Includes:**
- Dental assistant
- Dental hygienist
- Dental laboratory technician

# Dental Assistant

### Occupational Description

The dental assistant increases the efficiency of the dental care team by aiding the dentist in the delivery of oral health care. The dental assistant performs a wide range of tasks requiring both interpersonal and technical skills. Duties range from aiding and educating patients to preparing and sterilizing dental instruments and performing administrative work.

### Job Description

Dental assistants are responsible for helping patients feel comfortable before, during, and after treatment; assisting the dentist during treatment; exposing and processing dental radiographs (x-rays); recording the patient's medical history and taking blood pressure and pulse; preparing and sterilizing instruments and equipment for the dentist's use; providing patients with oral care instructions following such procedures as surgery or placement of a restoration (filling); teaching patients proper brushing and flossing techniques; making impressions of patients' teeth for study casts; and performing administrative and scheduling tasks, including using a personal computer, communicating by telephone, and maintaining an inventory supply system.

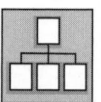

### Employment Characteristics

Most of the more than 247,000 active dental assistants are employed by general dentists. In addition, dental specialists employ dental assistants. Most assistants work chairside, although they may also participate in the business aspects of the practice. Besides dental offices, other employment settings available to dental assistants include schools and clinics (public health dentistry); hospitals (assisting dentists who are treating bedridden patients or in more elaborate dental procedures performed only in hospitals); dental school clinics; insurance companies (processing dental insurance claims); and vocational schools, technical institutes, community colleges, and universities (teaching others to be dental assistants).

Among independent general practitioners, the average number of employees per dentist has remained relatively stable between 1997 and 1998, averaging 4.2 positions. In 1999, this number has increased to 4.3 positions.

Dental assisting also offers both flexibility and stability. Dental assistants have the flexibility to work full or part time. As of 2001, dental assistants had been working in their current practices for an average of 6.6 years.

### Salary

The salary of a dental assistant varies, depending on the responsibilities associated with the specific position, the individual's training, and the geographic location of employment. The average national wage of a full-time dental assistant employed by a general practitioner in 2004 was $15.48 per hour. Refer to Section IV, Table 5 of this *Directory* for more information, or see www.ama-assn.org/go/ salary.

In addition to salary, dental assistants may receive benefit packages from their employers, including health and disability insurance coverage, dues for membership in professional organizations, an allowance for uniforms, profit sharing plans, and paid vacations.

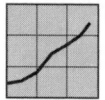

### Employment Outlook

Most areas of the country are currently reporting shortages of dental assistants. Owing to the success of preventive dentistry in reducing the incidence of oral disease, senior citizens—a growing population—will retain their teeth longer and will be even more aware of the importance of regular dental care.

### Educational Programs

**Length.** Nine to 11 months.
**Prerequisites.** High school diploma or equivalent.
**Certification.** Dental assistants who pass an examination administered by the Dental Assisting National Board, Inc, may use the designation of Certified Dental Assistant (CDA). Dental assistants are eligible to take the examination if they have completed a dental assisting program accredited by the Commission on Dental Accreditation or have completed 2 years of full-time work experience as dental assistants. State regulations vary and some states offer registration or licensure in addition to this national certification program.

### Certification

Dental assistants can become certified by passing the Certified Dental Assistant examination, administered by The Dental Assisting National Board (DANB). Passing this examination qualifies a dental assistant to use the designation Certified Dental Assistant (CDA). Dental assistants are eligible to take this examination if they have completed a dental assisting program accredited by the Commission on Dental Accreditation, or have 2 years full-time work experience, or are a graduate from a dental hygiene or dental program accredited by the Commission on Dental Accreditation.

### Miscellaneous Facts

- The 259 dental assisting education programs in the United States accredited by the American Dental Association's Commission on Dental Accreditation enrolled 7,240 students in 2004-2005.
- Women represented 96.4% of students enrolled in dental assisting programs in 2004-2005.
- Minority students represented 23% of enrollees in dental assisting programs in 2004-2005.
- Excellent career opportunities exist for nontraditional dental assisting students, seeking career change or job reentry after a period of unemployment, or from a culturally diverse background. Many dental assisting education programs offer more flexible program designs that meet the needs of nontraditional students by offering a variety of educational options, such as part-time or evening hours.

# Dental Hygienist

## Occupational Description

Dental hygienists provide dental hygiene services as they work with dentists in the delivery of dental care to patients. Hygienists are licensed to use their knowledge and clinical skills to provide dental care to patients and their interpersonal skills to motivate and instruct patients on methods to prevent oral disease and maintain oral health.

## Job Description

Although the range of services performed by dental hygienists varies from state to state, patient services rendered by dental hygienists frequently include

- Performing patient screening procedures, such as assessing oral health conditions, reviewing health and dental history, and taking blood pressure, pulse, and temperature; oral cancer screening; head & neck inspection; and dental charting
- Exposing and developing dental radiographs (x-rays)
- Removing calculus and plaque (hard and soft deposits) from teeth
- Applying preventive materials to teeth (eg, sealants and fluorides)
- Teaching patients appropriate oral hygiene techniques
- Counseling patients regarding proper nutrition and its impact on oral health
- Making impressions of patients' teeth for study casts
- Administration of anesthesia (depending upon state practice act)

## Employment Characteristics

Most of the approximately 158,000 licensed dental hygienists in the United States in 2004 are employed by general dentists. Additionally, dental specialists (such as periodontists or pediatric dentists) employ dental hygienists. Most hygienists work one to one with patients in providing dental hygiene services.

Dental hygienists also may be employed to provide dental hygiene care for patients in hospitals, nursing homes, public health clinics, and schools. Depending on the level of education and experience achieved, dental hygienists also can apply their skills and knowledge to other career activities, such as teaching. Research, public health, and business administration are other options. In addition, employment opportunities may be available with companies that market dental-related materials and equipment.

In some states, dental hygienists may also own their own dental hygiene business or practice on an independent contracting basis. These practitioners are not actually employed by dentists but provide dental hygiene services through contractual agreements.

Among independent general practitioners, the average number of employees per dentist has remained relatively stable between 1994 and 1998, averaging 4.2 positions. In 1999, the average number of employees increased to 4.3. Because 70.2% of solo general practitioners employ a dental hygienist, and 25.1% employ two or more hygienists, employment opportunities in this field are excellent.

As a career, dental hygiene also offers both stability and flexibility. As of 2001, for example, dental hygienists had been working in their current practices for an average of 7.3 years. Many hygienists also have considerable flexibility to undertake a full- or part-time schedule with evening or weekend hours.

## Salary

The salary of a dental hygienist varies, depending on the responsibilities associated with the specific position, the geographic location of employment, and the type of practice or other setting in which the hygienist works. Salary information from the American Dental Association indicates that the median salary for full-time dental hygienists in 2004 was $30.30 per hour. Refer to Section IV, Table 5 of this *Directory* for more information, or see www.ama-assn.org/go/ salary.

In addition, many full-time dental hygienists receive benefit packages from their dentist/employers, which may include health insurance coverage, dues for membership in professional organizations, paid vacations and sick leave, and tuition assistance for continuing education. Most state dental boards require mandatory continuing education for maintenance of the dental hygiene license.

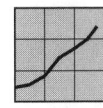

## Employment Outlook

According to the 2006-2007 edition of the *Occupational Outlook Handbook* and the *Monthly Labor Review,* published by the US Department of Labor's Bureau of Labor Statistics, dental hygiene is expected to increase 27% or more between 2004-2014. Some areas of the country are currently reporting shortages of dental hygienists.

Owing to the success of preventive dentistry in reducing the incidence of oral disease, senior citizens—a growing population—will retain their teeth longer and will require regular dental hygiene care.

Excellent career opportunities exist for nontraditional dental hygiene students, who might meet one or more of the following criteria: over 23 years of age, seeking career change or job reentry after a period of unemployment, or from a culturally diverse background. Some dental hygiene education programs offer more flexible program designs that meet the needs of nontraditional students by offering a variety of educational options, such as part-time or evening hours.

## Educational Programs

**Length.** The majority of community college-based dental hygiene programs offer a 2-year associate degree. University-based dental hygiene programs may offer baccalaureate and master's degrees, which generally require at least 2 or more years of further education.

**Prerequisites.** Admission requirements vary, depending on the institution. High school-level courses such as health, biology, psychology, chemistry, mathematics, and speech will be beneficial in a dental hygiene career. Many programs prefer individuals who have completed at least 1 year of college, and some baccalaureate degree programs require applicants to have completed 2 years of college.

**Curriculum.** Dental hygiene education programs provide supervised patient care experiences. Programs also include courses in the liberal arts (English, speech, sociology, and psychology); basic sciences (anatomy, physiology, chemistry, biochemistry, immunology, nutrition, pharmacology, microbiology, and general pathology); and clinical sciences (dental hygiene; tooth morphology; head, neck, and oral anatomy; oral embryology and histology; oral pathology; radiography; periodontology; pain management; radiology; and dental materials). After completing a dental hygiene program, dental hygienists can pursue additional training in such areas as education, health administration, basic sciences, and public health.

### Licensure

Dental hygienists are licensed by each state to provide dental hygiene care and patient education. Eligibility for state licensure usually includes graduation from a Commission-accredited dental hygiene education program. In addition to requiring a passing score on the state-authorized licensure examination, which tests candidates' clinical dental hygiene skills as well as their knowledge of dental hygiene and related subjects, almost all states require candidates for licensure to obtain a passing score on the Dental Hygiene National Board Examination (a comprehensive written examination).

Upon receipt of license, a dental hygienist may use RDH, signifying Registered Dental Hygienist, after his/her name.

### Miscellaneous Facts

- The United States has approximately 286 dental hygiene education programs accredited by the American Dental Association's Commission on Dental Accreditation.
- Approximately 97.4% of the 13,895 students enrolled in dental hygiene programs in 2004-2005 were women.
- Minority students represented 17% of enrollees in dental hygiene programs in 2004-2005.

# Dental Laboratory Technician

### Occupational Description

Dental laboratory technicians make dental prostheses—replacements for natural teeth, including dentures and crowns. The hallmarks of the qualified dental laboratory technician are skill in using small hand instruments, accuracy, artistic ability, and attention to detail to create practical and esthetically pleasing replacements.

### Job Description

Dental laboratory technicians seldom interact directly with patients; rather, they work with dentists by following detailed written instructions to make dental prostheses, which are replacements for natural teeth that enable people who have lost some or all of their teeth to eat, chew, talk, and smile in a manner similar to the way they did before. The dental technician uses impressions (molds) of the patient's teeth or oral soft tissues to create full dentures, removable partial dentures or fixed bridges, crowns, and orthodontic appliances and splints.

Dental technicians use sophisticated instruments and equipment and work with a variety of materials for replacing damaged or missing tooth structure, including waxes, plastics, precious and nonprecious alloys, stainless steel, and porcelain.

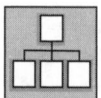

### Employment Characteristics

Most of the more than 43,000 active dental laboratory technicians in the United States today work in commercial dental laboratories, which on average employ between three to five technicians. In addition, some dentists employ dental technicians in their private dental offices. Other employment opportunities for dental technicians include dental schools, hospitals, the military, and companies that manufacture dental prosthetic materials. Dental laboratory technician education programs also offer teaching positions for qualified technicians.

### Salary

The starting salary of a dental technician varies depending on the responsibilities associated with the specific position and the geographic location of employment. Refer to Section IV, Table 5 of this *Directory* for more information, or see www.ama-assn.org/go/ salary. In addition to salary, many dental technicians receive benefit packages from their employers, which may include health and disability insurance coverage, reimbursement for continuing education programs, and paid vacations and holidays. Experienced technicians may become self-employed by opening their own dental laboratories, leading to greater financial rewards.

### Employment Outlook

Since most dentists use laboratory services, employment opportunities in this field are excellent. Owing to the success of preventive dentistry in reducing the incidence of oral disease, senior citizens—a growing population—will retain their teeth longer and will require more sophisticated prostheses for longer periods, thus increasing the demand for dental laboratory services.

Excellent career opportunities exist for nontraditional dental technology students, who might be seeking career change or job reentry after a period of unemployment, or from a culturally diverse background.

### Educational Programs

**Length.** Most dental laboratory technicians receive their education and training through a 2-year program at a community college, vocational school, technical college, or dental school, for which they may receive a certificate or an associate degree.

**Prerequisites.** High school diploma or its equivalent, although the Commission strongly encourages formal college-level education.

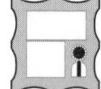

### Certification

Dental laboratory technicians can become certified by passing an examination, administered by the National Board for Certification in Dental Laboratory Technology, that evaluates their technical skills and knowledge. Passing this examination qualifies a dental technician to use the designation Certified Dental Technician (CDT). A CDT specializes in one or more of five areas: complete dentures, partial dentures, crowns and bridges, ceramics, and orthodontics.

Dental technicians are eligible to take the examination if they have completed a dental laboratory technology program accredited by the Commission on Dental Accreditation and have 2 years of professional experience or have completed 5 years of work experience as dental technicians, or have graduated from a nonaccredited program and have 3 years of professional experience and passed a comprehensive examination.

### Miscellaneous Facts

- In 2004-2005, 551 first-year dental laboratory technician students were enrolled in the approximately 23 dental technology education programs in the United States accredited by the American Dental Association's Commission on Dental Accreditation.
- Dental technology presents equal career opportunities for women and men. In 2004-2005, 47% of the students enrolled in dental technology programs were women, 53% were men.

- Minority students represented approximately 47.5% of enrollees in dental technology programs in the 2004-2005 graduating class.

 **Inquiries**

**Careers/Curriculum**
American Dental Association
211 E Chicago Avenue
Chicago, IL 60611-2678
312 440-2390
www.ada.org/prof/ed/careers

American Dental Education Association
1400 K Street NW, Suite 1100
Washington, DC 20005
202 289-7201
www.adea.org

American Dental Assistants Association
35 E Wacker Drive, Suite 1730
Chicago, IL 60601
312 541-1550
www.dentalassistant.org

American Dental Hygienists' Association
444 N Michigan Avenue, Suite 3400
Chicago IL 60611
312 440-8900
www.adha.org

Laboratory Conference Section Board of the American Dental Trade Association
4222 King Street W
Alexandria, VA 22302
703 379-7755
National Association of Dental Laboratories
325 John Knox Rd L103
Tallahassee, FL 32303
800 950-1150
www.nadl.org

**Certification**
Dental Assisting National Board, Inc
676 N St Clair, Suite 1880
Chicago, IL 60611
312 642-3368
800 FOR-DANB (367-3262)
312 642-1475 Fax
www.danb.org

National Board of Certification for Dental Laboratory Technicians
1530 Metropolitan Blvd
Tallahassee, FL 32308
850 224-0711
www.nadl.org

**Program Accreditation**
Commission on Dental Accreditation
American Dental Association
211 E Chicago Avenue
Chicago, IL 60611-2678
312 440-4653
312 440-2915 Fax
www.ada.org

## Dental Assistant*

## Alabama

**James H Faulkner State Community College**
Dental Assisting Prgm
1900 Hwy 31 S
Bay Minette, AL 36507-2619
www.faulknerstate.edu
*Prgm Dir:* Michele Snider, CDA RDH AAS BA
*Tel:* 251 580-2110   *Fax:* 251 580-2228
*E-mail:* msnider@faulkner.edu
*Med Dir:* Graham Jean, MSN

**Calhoun Community College**
Dental Assisting Prgm
PO Box 2216
Decatur, AL 35609-2216
*Prgm Dir:* Patricia Stueck, MA
*Tel:* 256 306-2812   *Fax:* 256 306-2525
*E-mail:* phs@calhoun.edu

**Wallace State Community College**
Dental Assisting Prgm
801 Main St NW
PO Box 2000
Hanceville, AL 35077-2000
www.wallacestate.edu
*Prgm Dir:* Barbara Ebert, RDH MA
*Tel:* 256 352-8380   *Fax:* 256 352-8382
*E-mail:* barbara.ebert@wallacestate.edu

**H Council Trenholm State Technical College**
Dental Assisting Prgm
1225 Air Base Blvd
Montgomery, AL 36108
www.trenholmtech.cc.al.us
*Prgm Dir:* Cecile B Mathews, MS
*Tel:* 334 420-4427   *Fax:* 334 420-4269
*E-mail:* cmathews@trenholmtech.cc.al.us

## Alaska

**University of Alaska Anchorage**
Dental Assisting Prgm
3211 Providence Dr, AHS 124
Anchorage, AK 99508-8371
*Prgm Dir:* Stephanie Olson, CDA BA
*Tel:* 907 786-6929   *Fax:* 907 786-6938
*E-mail:* afsmo1@uaa.alaska.edu

## Arizona

**Mesa Community College**
Dental Assisting Prgm
1833 W Southern Ave
Mesa, AZ 85205
*Prgm Dir:* Monica Williamson Nenad, RDH MEd
*Tel:* 480 472-0806
*E-mail:* monica.nenad@mcmail.maricopa.edu

**Phoenix College**
Dental Assisting Prgm
1202 W Thomas Rd
Phoenix, AZ 85013
www.maricopa.edu
*Prgm Dir:* Rita Perry, CDA RDH MPA
*Tel:* 602 285-7326   *Fax:* 602 285-7330
*E-mail:* rita.perry@pcmail.maricopa.edu

**Pima Community College**
Dental Assisting Prgm
2202 W Anklam Rd
Tucson, AZ 85709
*Prgm Dir:* Pamela Horch
*Tel:* 520 206-3101   *Fax:* 520 206-3027
*E-mail:* pamela.horch@pima.edu

*Additional information about programs that returned the AMA's survey is available in a table beginning on page 118.

# Arkansas

### Arkansas Northeastern College
Dental Assisting Prgm
I-55 and Hwy 148
PO Box 36
Burdette, AR 72321
www.anc.edu
*Prgm Dir:* Kristi Austin, CDA RDA
*Tel:* 870 780-1210  *Fax:* 870 763-1496
*E-mail:* kmaustin@anc.edu

### Pulaski Technical College
Dental Assisting Prgm
3000 W Scenic Dr
North Little Rock, AR 72118-3399
www.pulaskitech.edu
*Prgm Dir:* De Anna Davis, MEd
*Tel:* 501 812-2236  *Fax:* 501 812-2391
*E-mail:* ddavis@pulaskitech.edu

# California

### College of Alameda
Dental Assisting Prgm
555 Atlantic Ave
Alameda, CA 94501
*Prgm Dir:* Yvonne Carter, BA
*Tel:* 510 748-2262  *Fax:* 510 748-2156
*E-mail:* whycarter@aol.com

### Heald College - Concord Campus
Dental Assisting Prgm
5130 Commercial Circle
Concord, CA 94520
*Prgm Dir:* Maria Pardorla
*Tel:* 925 288-5800
*E-mail:* maria_pardorla@heald.edu

### Orange Coast College
Dental Assisting Prgm
2701 Fairview Rd
Costa Mesa, CA 92628-0120
*Prgm Dir:* Judy Rose, MA CDA RDA
*Tel:* 714 432-5565  *Fax:* 714 432-5534
*E-mail:* jrose@mail.occ.cccd.edu

### Cypress College
Dental Assisting Prgm
9200 Valley View St
Cypress, CA 90630
*Prgm Dir:* Mary Kay Davis, BS
*Tel:* 714 484-7293  *Fax:* 714 527-2175
*E-mail:* mdavis@cypress.cc.ca.us

### College of the Redwoods
Dental Assisting Prgm
7351 Tompkins Hill Rd
Eureka, CA 95501
www.redwoods.edu
*Prgm Dir:* Hillary Reed, RDA RDAEF CDA COA CDPMA
*Tel:* 707 476-4253  *Fax:* 707 476-4419
*E-mail:* hillary-reed@redwoods.edu

### Citrus College
Dental Assisting Prgm
1000 W Foothill
Glendora, CA 91741
www.citruscollege.edu
*Prgm Dir:* Claudia Pohl, CDA RDA BVEd
*Tel:* 626 914-8728  *Fax:* 626 914-8724
*E-mail:* cpohl@citruscollege.edu

### Heald College - Hayward Campus
Dental Assisting Prgm
25500 Industrial Blvd
Hayward, CA 94545
*Prgm Dir:* Evangeline Tenchavez
*Tel:* 510 783-2100
*E-mail:* evangeline_tenchavez@heald.edu

### College of Marin
Dental Assisting Prgm
835 College Ave
Kentfield, CA 94904
*Prgm Dir:* Grace Hom, CDA RDAEF MA
*Tel:* 415 485-9327  *Fax:* 415 485-9328
*E-mail:* grace.hom@marin.edu
*Med Dir:* Rosalind Hartman, RN, MS

### Hacienda LaPuente Adult Education
Dental Assisting Prgm
14101 E Nelson Ave
La Puente, CA 91746
*Prgm Dir:* Gretchen Richardson
*Tel:* 626 934-2890  *Fax:* 626 934-5723
*E-mail:* grichardson@hlpusd.k12.ca.us

### Foothill Community College
Dental Assisting Prgm
12345 El Monte Rd
Los Altos Hills, CA 94022
www.foothill.edu/bio/programs/dentala
*Prgm Dir:* Cara Miyasaki-Ching, RDHEF MS
*Tel:* 650 949-7351  *Fax:* 650 949-9788
*E-mail:* miyasakicara@fhda.edu

### Modesto Junior College
Dental Assisting Prgm
435 College Ave
Modesto, CA 95350
*Prgm Dir:* Robert Keach, CDA RDA
*Tel:* 209 575-6914  *Fax:* 209 575-6593
*E-mail:* keachr@mjc.edu

### Monterey Peninsula College
Dental Assisting Prgm
980 Fremont St
Monterey, CA 93940
*Prgm Dir:* Patricia A Lewis, BS
*Tel:* 831 646-4137  *Fax:* 831 645-1353
*E-mail:* plewis@mpc.edu

### Cerritos Community College
Dental Assisting Prgm
11110 E Alondra Blvd
Norwalk, CA 90650
*Prgm Dir:* Donna Wedell
*Tel:* 562 860-2451, Ext 2575  *Fax:* 562 467-5077
*E-mail:* dwedell@cerritos.edu

### Pasadena City College
Dental Assisting Prgm
1570 E Colorado Blvd
Pasadena, CA 91106
www.pasadena.edu
*Prgm Dir:* Lori Gagliardi, MEd
*Tel:* 626 585-7542  *Fax:* 626 585-7966
*E-mail:* ligagliardi@pasadena.edu

### Diablo Valley College
Dental Assisting Prgm
321 Golf Club Rd
Pleasant Hill, CA 94523
*Prgm Dir:* Marylou Pineda, MA CDA RDA
*Tel:* 925 685-1230, Ext 2351  *Fax:* 925 689-6529
*E-mail:* MPineda@dvc.edu

### Chaffey College
Dental Assisting Prgm
5885 Haven Ave
Rancho Cucamonga, CA 91737-3002
www.chaffey.edu
*Prgm Dir:* Beverly Cox, MAEd BS
*Tel:* 909 941-2417, Ext 2417  *Fax:* 909 477-8826
*E-mail:* Beverly.Cox@chaffey.edu

### Sacramento City College
Dental Assisting Prgm
3835 Freeport Blvd
Sacramento, CA 95822
www.scc.losrios.edu
*Prgm Dir:* Michael Dunne, DDS
*Tel:* 916 558-2650  *Fax:* 916 538-2067
*E-mail:* dunnem@scc.losrios.edu

### San Diego Mesa College
Dental Assisting Prgm
7250 Mesa College Dr
San Diego, CA 92111-4999
www.sdccd.edu
*Prgm Dir:* Margaret Fickess, RDA CDA MEd
*Tel:* 619 388-2697  *Fax:* 619 388-2677
*E-mail:* mfickess@sdccd.edu

### City College of San Francisco
Dental Assisting Prgm
50 Phelan Ave
San Francisco, CA 94112
www.ccsf.org
*Prgm Dir:* Anna Nelson, CDA RDA MA
*Tel:* 415 239-3479  *Fax:* 415 239-3719
*E-mail:* anelson@ccsf.edu

### San Jose City College
Dental Assisting Prgm
2100 Moorpark Ave
San Jose, CA 95128-2799
*Prgm Dir:* Laura sanchez, AA
*Tel:* 408 298-2181, Ext 3254  *Fax:* 408 275-9386
*E-mail:* laura.sanchez@sjcc.edu

### Palomar Community College
Dental Assisting Prgm
1140 W Mission Rd
San Marcos, CA 92069
*Prgm Dir:* Denise Rudy
*Tel:* 760 744-1150, Ext 2573  *Fax:* 760 744-1150
*E-mail:* drudy@palomar.edu

### College of San Mateo
Dental Assisting Prgm
1700 W Hillsdale Blvd
San Mateo, CA 94402-3795
*Prgm Dir:* Audrey Behrens, CDA RDA BA
*Tel:* 650 574-6212  *Fax:* 650 574-6485
*E-mail:* audb2@pacbell.net

### Contra Costa College
Dental Assisting Prgm
2600 Mission Bell Dr
San Pablo, CA 94806
*Prgm Dir:* Sandra Everhart, BA
*Tel:* 510 235-7800, Ext 4385  *Fax:* 510 236-6768
*E-mail:* severhart@contracosta.cc.ca.us

### Santa Rosa Junior College
Dental Assisting Prgm
1501 Mendocino Ave
Santa Rosa, CA 95401-4395
www.santarosa.edu
*Prgm Dir:* Doni Bird, MA
*Tel:* 707 522-2828  *Fax:* 707 527-4426
*E-mail:* dbird@santarosa.edu

### Heald College - Stockton Campus
Dental Assisting Prgm
1605 E March Ln
Stockton, CA 95210
*Prgm Dir:* Levy Pineda, CDT DMD
*Tel:* 209 473-5200, Ext 5244  *Fax:* 209 477-2739
*E-mail:* levy_pineda@heald.edu
*Med Dir:* Ruth Brown

# Colorado

### T H Pickens Technical Center
Dental Assisting Prgm
500 Airport Blvd
Aurora, CO 80011
*Prgm Dir:* Karen Einsle, BA CDA EDDA
*Tel:* 303 344-4910, Ext 27787  *Fax:* 303 326-1277
*E-mail:* karene@pickens.aps.k12.co.us

**IntelliTec Medical Institute**
Dental Assisting Prgm
2345 N Academy Blvd
Colorado Springs, CO 80909
*Prgm Dir:* Ed Rito
*Tel:* 719 596-7400

**Pikes Peak Community College**
Dental Assisting Prgm
11195 Highway 83
Colorado Springs, CO 80921-3602
www.ppcc.edu
*Prgm Dir:* Frank Delgesso
*Tel:* 719 538-5412    *Fax:* 719 538-5439
*E-mail:* frank.delgesso@ppcc.edu

**Emily Griffith Opportunity School**
Dental Assisting Prgm
1250 Welton St
Denver, CO 80204
*Prgm Dir:* Brenda D Stevens
*Tel:* 720 423-4736    *Fax:* 303 575-4762
*E-mail:* Brenda_Stevens@dpsk12.org

**Front Range Comm College**
Dental Assisting Prgm
Larimer Campus, 4616 S Shields
Ft Collins, CO 80526-8312
www.frontrange.edu
*Prgm Dir:* Dana Quiller, DDS
*Tel:* 970 204-8220    *Fax:* 970 204-8118
*E-mail:* dana.quiller@frontrange.edu

**Pueblo Community College**
Dental Assisting Prgm
900 W Orman Ave, MT Rm 122B
Pueblo, CO 81004
www.pueblocc.edu
*Prgm Dir:* Janet Trujillo, CDA EFDA BS
*Tel:* 719 549-3263    *Fax:* 719 549-3381
*E-mail:* janet.trujillo@pueblocc.edu
*Med Dir:* Lucinda Mihelich, PT, BS, MEd

# Connecticut

**Porter and Chester Institute - Enfield**
Dental Assisting Prgm
132 Weymouth Rd
Enfield, CT 06082
*Prgm Dir:* Susan M Corriveau, BA CDA
*Tel:* 860 253-6100    *Fax:* 860 741-0234
*E-mail:* pcidentassist@aol.com

**Tunxis Community College**
Dental Assisting Prgm
271 Scott Swamp Rd
Farmington, CT 06032-3187
*Prgm Dir:* Claudia Turcotte, CDA RDH MSDH MSOSH
*Tel:* 860 255-3743    *Fax:* 860 255-3649
*E-mail:* cturcotte@txcc.commnet.edu

**Eli Whitney Vocational Technical School**
Dental Assisting Prgm
71 Jones Rd
Hamden, CT 06514
*Prgm Dir:* Angela Macolino, CDA
*Tel:* 203 397-4031, Ext 344    *Fax:* 203 397-4129
*E-mail:* Angela.Macolino@ct.gov

**A I Prince Technical High School**
Dental Assisting Prgm
500 Bookfield St
Hartford, CT 06106
www.cttech.org/prince
*Prgm Dir:* Janice Ferrara, CDA
*Tel:* 860 951-7112, Ext 323    *Fax:* 860 951-1529
*E-mail:* Janice.Ferrara@po.state.ct.us

**Briarwood College**
Dental Assisting Prgm
2279 Mt Vernon Rd
Southington, CT 06489
*Prgm Dir:* Rosemary Ryan
*Tel:* 800 952-2444, Ext 69    *Fax:* 860 628-6444
*E-mail:* ryanr@briarwood.edu

**Windham Regional Vocational Tech School**
Dental Assisting Prgm
210 Birch St
Willimantic, CT 06226
*Prgm Dir:* Susan Dolliver, BA
*Tel:* 860 456-3879, Ext 322    *Fax:* 860 450-0630
*E-mail:* susan.dolliver@po.state.ct.us

# Florida

**South Florida Community College**
Dental Assisting Prgm
600 W College Dr
Avon Park, FL 33825
www.southflorida.edu
*Prgm Dir:* Rebecca Sroda, RDH MS
*Tel:* 863 784-7021    *Fax:* 863 453-9442
*E-mail:* srodar@southflorida.edu

**Manatee Technical Institute**
Dental Assisting Prgm
5520 Lakewood Ranch Blvd
Bradenton, FL 34211
www.manateetechnicalinstitute.org
*Prgm Dir:* Kimberly Bland, CDA BS
*Tel:* 941 752-8100, Ext 210    *Fax:* 941 727-6254
*E-mail:* blandk@manateeschools.net
*Med Dir:* Ronald Acquaro, DMD

**Brevard Community College**
Dental Assisting Prgm
1519 Clearlake Rd
Cocoa, FL 32922
*Prgm Dir:* Holly Kahler, CDA RDH EdD
*Tel:* 321 433-7571    *Fax:* 321 433-7599
*E-mail:* kahlerh@brevardcc.edu

**Daytona Beach Community College**
Dental Assisting Prgm
1200 W International Speedway Blvd
Daytona Beach, FL 32120
www.dbcc.edu
*Prgm Dir:* Pamela Ridilla, CDA RDH MS
*Tel:* 386 785-2093    *Fax:* 386 785-2090
*E-mail:* ridillp@dbcc.edu

**Americare School of Nursing**
Dental Assisting Prgm
7275 Estapona Circle
Fern Park, FL 32730
*Prgm Dir:* Dana Wickman
*Tel:* 407 262-8101

**Broward Community College**
Dental Assisting Prgm
3501 SW Davie Rd, Bldg 8
Ft Lauderdale, FL 33314
www.broward.edu
*Prgm Dir:* Joyce Abraham, CDA RDH BA
*Tel:* 954 201-6904    *Fax:* 954 201-6397
*E-mail:* jabraham@broward.edu

**Edison College**
Dental Assisting Prgm
PO Box 60210
Ft Myers, FL 33906-6210
www.edison.edu
*Prgm Dir:* Karen Molumby, CDA RDH MBA
*Tel:* 239 985-8322    *Fax:* 239 985-8352
*E-mail:* kmolumby@edison.edu

**Indian River Community College**
Dental Assisting Prgm
3209 Virginia Ave
Ft Pierce, FL 34981-5599
*Prgm Dir:* Karen Allen, MEd
*Tel:* 772 462-7530    *Fax:* 772 462-4900
*E-mail:* kallen@ircc.edu

**Santa Fe Community College**
Dental Assisting Prgm
3000 NW 83rd St, Bldg W
Gainesville, FL 32606
*Prgm Dir:* Karen Autrey, BHS MEd
*Tel:* 352 395-5705    *Fax:* 352 395-5662
*E-mail:* karen.autrey@sfcc.edu

**Palm Beach Community College**
Dental Assisting Prgm
4200 Congress Ave
Lake Worth, FL 33461
www.pbcc.edu
*Prgm Dir:* Colleen Bradshaw, CDA BA
*Tel:* 561 868-3196    *Fax:* 561 868-3753
*E-mail:* bradshac@pbcc.edu

**Traviss Career Center**
Dental Assisting Prgm
3225 Winter Lake Rd
Lakeland, FL 33803
www.pcsb.k12.fl.us/traviss
*Prgm Dir:* Susan Rexroat, CDA CDPMA BA
*Tel:* 863 499-2715, Ext 286    *Fax:* 863 413-2067
*E-mail:* susan.rexroat@polk-fl.net

**Lindsey Hopkins Tech Education Center**
Dental Assisting Prgm
750 NW 20th St
Miami, FL 33127
*Prgm Dir:* Debra Morales
*Tel:* 305 324-7084    *Fax:* 305 324-6249
*E-mail:* dmorales@dadeschools.net

**Robert Morgan Educational Center**
Dental Assisting Prgm
18180 SW 122nd Ave
Miami, FL 33177
http://rmec.dadeschools.net
*Prgm Dir:* Aurea Hurtado, CDA BS
*Tel:* 305 253-9920, Ext 2212    *Fax:* 305 259-3270
*E-mail:* ahurtado@dadeschools.net

**Lorenzo Walker Institute of Technology**
Dental Assisting Prgm
3702 Estey Ave
Naples, FL 34104
www.lwit.edu
*Prgm Dir:* Linda Rae Flood, CDA RDH BS
*Tel:* 239 377-0932, Ext 0932    *Fax:* 239 377-1004
*E-mail:* floodli@collier.k12.fl.us
*Med Dir:* Donald L Lintzenich, DDS

**Okaloosa-Walton College**
Dental Assisting Prgm
100 College Blvd
Niceville, FL 32578
www.owcc.cc.fl.us
*Prgm Dir:* Mary E Thomas
*Tel:* 850 729-6444
*E-mail:* thomasm@owc.edu

**Central Florida Community College**
Dental Assisting Prgm
3001 SW College Rd
PO Box 1388
Ocala, FL 34479
*Prgm Dir:* Deanna Stentiford
*Tel:* 352 854-2322, Ext 1529
*E-mail:* stentifd@cf.edu

**Orlando Tech**
Dental Assisting Prgm
301 W Amelia St
Orlando, FL 32801
www.orlandotech@ocps.net
*Prgm Dir:* Cynthia K Bradley, CDA CDPMA EFDA
*Tel:* 407 246-7060, Ext 4923   *Fax:* 407 317-3372
*E-mail:* bradlec@ocps.k12.fl.us

**Gulf Coast Community College**
Dental Assisting Prgm
5230 W US Hwy 98
Panama City, FL 32401-1041
*Prgm Dir:* Kim Herremans, RDH MS
*Tel:* 850 769-1551, Ext 5832   *Fax:* 850 747-3246
*E-mail:* kherremans@gulfcoast.edu

**Pensacola Junior College**
Dental Assisting Prgm
5555 Hwy 98 W
Pensacola, FL 32507
*Prgm Dir:* Teresa Lucas
*Tel:* 850 484-2348   *Fax:* 850 484-2390
*E-mail:* tlucas@pjc.edu

**Charlotte Technical Center**
Dental Assisting Prgm
18300 Toledo Blade Blvd
Port Charlotte, FL 33948-3399
www.ccps.k12.fl.us
*Prgm Dir:* Lance Petersen, DDS, DDS
*Tel:* 941 255-7500, Ext 140   *Fax:* 941 255-7509
*E-mail:* Lance_Petersen@ccps.k12.fl.us
*Med Dir:* Lance Petersen, DDS, DDS

**Pinellas Tech Educ Ctr - St Petersburg**
Dental Assisting Prgm
901 34th St S
St Petersburg, FL 33711
www.ptec.pinellas.k12.fl.us
*Prgm Dir:* Barbara Thomas, BS CDA MA
*Tel:* 800 893-2500, Ext 1073   *Fax:* 727 552-1613
*E-mail:* bthomas@ptec.pinellas.k12.fl.us

**Tallahassee Community College**
Dental Assisting Prgm
444 Appleyard Dr
Tallahassee, FL 32304
*Prgm Dir:* Cynthia R Biron, RDH MA
*Tel:* 850 201-8577   *Fax:* 850 201-8575
*E-mail:* bironc@tcc.fl.edu

**Erwin Technical Center**
Dental Assisting Prgm
2010 E Hillsborough Ave
Tampa, FL 33610
*Prgm Dir:* Darcy Abel Hunter, MEd
*Tel:* 813 231-1800, Ext 1323   *Fax:* 813 231-1812
*E-mail:* hunter_d@firn.edu
*Med Dir:* Manuel Vilaret, DMD

**Hillsborough Community College**
Dental Assisting Prgm
Dale Mabry Campus
PO Box 30030
Tampa, FL 33614
www.hccfl.edu/dm
*Prgm Dir:* Connie Gore, CDA BS
*Tel:* 813 253-7279   *Fax:* 813 253-7377
*E-mail:* cgore@hccfl.edu

# Georgia

**Albany Technical College**
Dental Assisting Prgm
1704 S Slappey Blvd
Albany, GA 31701-3514
*Prgm Dir:* Priscilla Ryals, RDH CDA EdS
*Tel:* 912 430-3543   *Fax:* 912 430-5115
*E-mail:* pryals@albanytech.edu

**Athens Technical College**
Dental Assisting Prgm
Allied Health and Nursing
800 US Hwy 29 N
Athens, GA 30601
www.athenstech.edu
*Prgm Dir:* Tina A Grile, MHS RDH CDA
*Tel:* 706 355-5142
*E-mail:* tgrile@athenstech.edu

**Atlanta Technical College**
Dental Assisting Prgm
1560 Metropolitan Pkwy SW Rm 2157
Atlanta, GA 30310-4446
www.atlantatech.edu
*Prgm Dir:* Dorothy Smith, CDA BA
*Tel:* 404 756-3723, Ext 3723   *Fax:* 404 758-8522
*E-mail:* dsmith@atlantatech.edu

**Augusta Technical College**
Dental Assisting Prgm
3200 Augusta Tech Dr
Augusta, GA 30906
www.augustatech.edu
*Prgm Dir:* Beverly Dalber, BS CDA
*Tel:* 706 771-4178   *Fax:* 706 771-4181
*E-mail:* bdalber@augustatech.edu
*Med Dir:* Gwendolyn Taylor, MSN

**Columbus Technical College**
Dental Assisting Prgm
928 Manchester Expwy
Columbus, GA 31904
www.columbustech.edu
*Prgm Dir:* Jan J Jones, CDA RDH MPA
*Tel:* 706 649-1977   *Fax:* 706 649-1599
*E-mail:* jjones@columbustech.edu
*Med Dir:* Linn Storey, RN, BSN, MPA

**Gwinnett Technical College**
Dental Assisting Prgm
5150 Sugarloaf Pkwy
Lawrenceville, GA 30043
*Prgm Dir:* Lea Anna Harding, BS Ed
*Tel:* 770 962-7580   *Fax:* 770 962-1338
*E-mail:* lharding@gwinnetttech.edu

**Lanier Technical College**
Dental Assisting Prgm
2990 Landrum Education Dr
Oakwood, GA 30566
*Prgm Dir:* Liza Charlton, BA
*Tel:* 770 531-6370   *Fax:* 770 531-6306
*E-mail:* lcharlto@laniertech.edu

**Coosa Valley Technical College**
Dental Assisting Prgm
466 Brock Rd
Rockmart, GA 30153
www.cvtcollege.org
*Prgm Dir:* Donna Sutton, RDH MS CDA
*Tel:* 770 684-3128   *Fax:* 770 684-8710
*E-mail:* dsutton@coosavalleytech.edu

**Savannah Technical College**
Dental Assisting Prgm
5717 White Bluff Rd
Savannah, GA 31405-5521
www.savtec.org
*Prgm Dir:* Debi White
*Tel:* 912 443-5812
*E-mail:* dwhite@savannahtech.edu

**Medix School**
Dental Assisting Prgm
2108 Cobb Pkwy
Smyrna, GA 30080
www.medixschool.com
*Prgm Dir:* Terroll Farries, CDA
*Tel:* 770 980-0002   *Fax:* 770 980-0811
*E-mail:* email@medixschool.com
*Med Dir:* Phillip Sachs, MD

**Ogeechee Technical College**
Dental Assisting Prgm
1 Joe Kennedy Blvd
Statesboro, GA 30458-8049
*Prgm Dir:* Denise Campopiano, BS RDA CDA
*Tel:* 912 486-7614   *Fax:* 912 486-7604
*E-mail:* dcampopiano@ogeecheetech.edu

**Valdosta Technical College**
Dental Assisting Prgm
PO Box 928-4089 Val Tech Rd
Valdosta, GA 31603-0928
*Prgm Dir:* Sandi Woodward, RDH CDA BSED
*Tel:* 229 259-5533   *Fax:* 229 259-5567
*E-mail:* swoodward@valdostatech.edu
*Med Dir:* Randy Smith, DDS

**Middle Georgia Technical College**
Dental Assisting Prgm
80 Cohen Walker Dr
Warner Robins, GA 31088-2729
*Prgm Dir:* Debbie Wrenn, CDA EFDA
*Tel:* 478 988-6908   *Fax:* 478 988-6875
*E-mail:* dwrenn@middlegatech.edu

# Hawaii

**Maui Community College**
Dental Assisting Prgm
310 Kaahumana Ave
Kahului, HI 96732
*Prgm Dir:* June M Wolken-Vierra
*E-mail:* jwvierra@hawaii.edu

# Idaho

**Apollo College - Boise**
Dental Assisting Prgm
1200 N Liberty
Boise, ID 83704
www.apolloboise.com
*Prgm Dir:* Erlinda Hiles, BA CDA
*Tel:* 208 377-8080, Ext 17   *Fax:* 208 322-7658
*E-mail:* ehiles@apolloboise.com

**Boise State University**
Dental Assisting Prgm
College of Technology
1910 University Dr
Boise, ID 83725
*Prgm Dir:* Bonnie Tollinger, CDA
*Tel:* 208 426-2169   *Fax:* 208 426-5659
*E-mail:* btollinger@boisestate.edu

# Illinois

**John A Logan College**
Dental Assisting Prgm
700 Logan College Rd
Carterville, IL 62918
*Prgm Dir:* Kathy Gibson, MS
*Tel:* 618 985-3741, Ext 8389   *Fax:* 618 985-9181
*E-mail:* kathy.gibson@jal.cc.il.us

**Kaskaskia College**
Dental Assisting Prgm
27210 College Rd
Centralia, IL 62801
www.kaskaskia.edu
*Prgm Dir:* Lori Schmidt, RDA RDH AAS BS
*Tel:* 618 545-3319   *Fax:* 618 532-1990
*E-mail:* lschmidt@kc.cc.il.us

**Elgin Community College**
Dental Assisting Prgm
1700 Spartan Dr
Elgin, IL 60123
www.elgin.edu
*Prgm Dir:* Kim Plate, CDA
*Tel:* 847 214-7351   *Fax:* 847 622-3063
*E-mail:* kplate@elgin.edu
*Med Dir:* Phyllis Thomson, MS RN

**Lewis & Clark Community College**
Dental Assisting Prgm
5800 Godfrey Rd, RiverBend Arena RM 235
Godfrey, IL 62035
www.lc.edu
*Prgm Dir:* Constance Pero-Fox, CDA BS
*Tel:* 618 468-4411   *Fax:* 618 468-2394
*E-mail:* cperofox@lc.edu

**Illinois Valley Community College**
Dental Assisting Prgm
815 N Orlando Smith Ave
Oglesby, IL 61348-9691
www.ivcc.edu
*Prgm Dir:* Patricia Pearson, CDA
*Tel:* 815 224-2720, Ext 359   *Fax:* 815 224-3033
*E-mail:* Pat_Pearson@ivcc.edu

# Indiana

**Ivy Tech Community College - Anderson**
Dental Assisting Prgm
104 W 53rd St
Anderson, IN 46013
www.ivytech.edu/anderson
*Prgm Dir:* Teresa Macauley, CDA EFDA MS
*Tel:* 765 643-7133, Ext 2369
*E-mail:* tmacaule@ivytech.edu

**Columbus Area Career Connection/Ivy Tech St**
Dental Assisting Prgm
230 S Marr Rd
Columbus, IN 47201
*Prgm Dir:* Christy Ross, CDA
*Tel:* 812 376-4202   *Fax:* 812 376-4358
*E-mail:* rossc@bcsc.k12.in.us

**University of Southern Indiana**
Dental Assisting Prgm
8600 University Blvd
Evansville, IN 47712
*Prgm Dir:* Kim Hite, CDA RDH BS
*Tel:* 812 465-1155   *Fax:* 812 465-7092
*E-mail:* kghite@usi.edu

**Indiana Univ - Purdue Univ Ft Wayne**
Dental Assisting Prgm
2101 Coliseum Blvd E
Ft Wayne, IN 46805
*Prgm Dir:* Connie Kracher, MSD CDA BSEd
*Tel:* 260 481-6567   *Fax:* 260 481-4162
*E-mail:* kracher@ipfw.edu

**Indiana University Northwest**
Dental Assisting Prgm
3223 Broadway St
Gary, IN 46409
www.iun.edu
*Prgm Dir:* Juanita Robinson, CDA EFDA LDH MSEd
*Tel:* 219 980-6734   *Fax:* 219 981-4249
*E-mail:* jurobin@iun.edu

**Indiana University**
Dental Assisting Prgm
1121 W Michigan St
Indianapolis, IN 46202-5186
www.iusd.iupui.edu
*Prgm Dir:* Pamela Ford, CDA EFDA RDH MSW
*Tel:* 317 278-2814   *Fax:* 317 274-1363
*E-mail:* ptavenne@iupui.edu
*Med Dir:* Lawrence Goldblatt, DDS, MSD

**Professional Careers Institute**
Dental Assisting Prgm
7302 Woodland Dr
Indianapolis, IN 46278
*Prgm Dir:* Shannon Irvin, BS
*Tel:* 317 299-6001   *Fax:* 317 298-6342
*E-mail:* shannon.irvin@pcicareers.com
*Med Dir:* Lisa Stephens, MA,AAS

**Ivy Tech Community College - Kokomo**
Dental Assisting Prgm
700 E Firmin St
Kokomo, IN 46902
*Prgm Dir:* Darci Post
*Tel:* 765 457-0858, Ext 123

**Ivy Tech Community College - Lafayette**
Dental Assisting Prgm
3101 S Creasy Ln
PO Box 6299
Lafayette, IN 47903-6299
*Prgm Dir:* Judith Buckles
*Tel:* 765 269-5205   *Fax:* 765 269-5248
*E-mail:* jbuckles@ivytech.edu

**Indiana University South Bend**
Dental Assisting Prgm
1700 Mishawaka Ave
South Bend, IN 46634
*Prgm Dir:* Nanci Genich-Yokom, CDA RDH MBA
*Tel:* 574 237-4154   *Fax:* 574 237-4854
*E-mail:* nyokom@iusb.edu

# Iowa

**Des Moines Area Community College**
Dental Assisting Prgm
2006 Ankeny Blvd
Ankeny, IA 50023-3993
*Prgm Dir:* Deborah P Bell, AA
*Tel:* 515 964-6308   *Fax:* 515 964-6640
*E-mail:* dpbell@dmacc.edu

**Scott Community College**
Dental Assisting Prgm
500 Belmont Rd
Bettendorf, IA 52722-5649
*Prgm Dir:* Tina Ford-Ball, CDA RDA EFDA
*Tel:* 563 441-4260   *Fax:* 563 441-4204
*E-mail:* tball@eicc.edu

**Kirkwood Community College**
Dental Assisting Prgm
6301 Kirkwood Blvd SW, PO Box 2068
Cedar Rapids, IA 52406-9973
*Prgm Dir:* Pam Hanson
*Tel:* 319 398-5560   *Fax:* 319 398-1293
*E-mail:* phanson@kirkwood.edu

**Iowa Western Community College**
Dental Assisting Prgm
2700 College Rd, PO Box 4C
Council Bluffs, IA 51503
www.iwcc.edu
*Prgm Dir:* Sue Norman, BA CDA RDA
*Tel:* 712 325-3469   *Fax:* 712 325-3736
*E-mail:* snorman@iwcc.edu

**Vatterott College - Des Moines Campus**
Dental Assisting Prgm
6100 Thornton Ave, Ste 290
Des Moines, IA 50321
www.vatterott-college.edu
*Prgm Dir:* Sarah Bouma
*Tel:* 515 309-9000
*E-mail:* sarah.bouma@vatterott-college.edu

**Marshalltown Community College**
Dental Assisting Prgm
3700 S Center St
Marshalltown, IA 50158
*Prgm Dir:* Elaine B Peterson, BA CDA RDA EFDA
*Tel:* 641 752-7106, Ext 294   *Fax:* 641 752-8149
*E-mail:* elaine.peterson@iavalley.edu

**Northeast Iowa Community College**
Dental Assisting Prgm
10250 Sundown Rd
Peosta, IA 52068
www.nicc.edu
*Prgm Dir:* Gloria Kluesner, CDA BLS RDA EFDA
*Tel:* 800 728-7367, Ext 227   *Fax:* 319 556-5058
*E-mail:* kluesneg@nicc.edu

**Western Iowa Tech Community College**
Dental Assisting Prgm
4647 Stone Ave, PO Box 5199
Sioux City, IA 51102-5199
*Prgm Dir:* Jackie Krueger
*Tel:* 712 274-8733, Ext 1267   *Fax:* 712 274-6412
*E-mail:* kruegej@witcc.edu

**Hawkeye Community College**
Dental Assisting Prgm
1501 E Orange Rd, PO Box 8015
Waterloo, IA 50704
www.hawkeyecollege.edu
*Prgm Dir:* Suzanne Van Syoc, CDA RDA BA
*Tel:* 319 296-2320, Ext 1355   *Fax:* 319 296-2874
*E-mail:* svansyoc@hawkeyecollege.edu

# Kansas

**Flint Hills Technical College**
Dental Assisting Prgm
3301 W 18th Ave
Emporia, KS 66801
*Prgm Dir:* Monica Jones, CDA MBE
*Tel:* 620 341-2300, Ext 214   *Fax:* 620 343-7252
*E-mail:* mjones@fhtc.kansas.net

**Salina Area Technical School**
Dental Assisting Prgm
2562 Centennial Rd
Salina, KS 67401
*Prgm Dir:* Janet Fisher, CDA MS
*Tel:* 785 309-3126   *Fax:* 785 309-3101
*E-mail:* janet.fisher@usd305.com

**Wichita Area Technical College**
Dental Assisting Prgm
324 N Emporia St
Wichita, KS 67202
*Prgm Dir:* Melanie Mitchell, CDA
*Tel:* 316 677-1355, Ext 71368   *Fax:* 316 677-1332
*E-mail:* mmitchell@watc.edu

# Kentucky

**Bluegrass Community and Technical College**
Dental Assisting Prgm
164 Opportunity Way
Lexington, KY 40511-1020
www.bluegrass.kctcs.edu
*Prgm Dir:* Larry Chiswell, DMD
*Tel:* 859 246-6200, Ext 56232   *Fax:* 859 246-4667
*E-mail:* larry.chiswell@kctcs.edu

**West Kentucky Community & Technical College**
Dental Assisting Prgm
4810 Alben Barkley Dr
Paducah, KY 42001
www.westkentucky.kctcs.edu
*Prgm Dir:* Darlene Daniel, BS
*Tel:* 270 534-3358, Ext 43358   *Fax:* 270 554-9754
*E-mail:* Darlene.Daniel@kctcs.edu

# Maine

**University College of Bangor**
Dental Assisting Prgm
Lincoln Hall, 29 Texas Ave
Bangor, ME 04401-4324
*Prgm Dir:* Diana L Olsen, CDA CDPMA RDH MS
*Tel:* 207 262-7878    *Fax:* 207 262-7751
*E-mail:* dolsen@maine.edu

# Maryland

**Medix School**
Dental Assisting Prgm
700 York Rd
Towson, MD 21204
*Prgm Dir:* Linda Einolf
*Tel:* 410 337-5155    *Fax:* 410 337-5104
*E-mail:* info@medixschool.com

# Massachusetts

**Massasoit Community College**
Dental Assisting Prgm
900 Randolph St
Canton, MA 02021
*Prgm Dir:* Mary A Frohn, BS
*Tel:* 781 821-2222, Ext 2754    *Fax:* 781 575-9428
*E-mail:* mfrohn@massasoit.mass.edu
*Med Dir:* Richard Cronin, MD

**Northern Essex Community College**
Dental Assisting Prgm
45 Franklin St
Lawrence, MA 01841
www.necc.mass.edu/healthprofessions
*Prgm Dir:* Kerin Hamidiani, MEd
*Tel:* 978 738-7427    *Fax:* 978 738-7450
*E-mail:* khamidiani@necc.mass.edu

**Middlesex Community College - Lowell**
Dental Assisting Prgm
33 Kearney Square
Lowell, MA 01852
*Prgm Dir:* Margaret Bloy, MS
*Tel:* 978 656-3053, Ext 3053    *Fax:* 978 656-3078
*E-mail:* bloyp@middlesex.mass.edu

**Charles H McCann Technical School**
Dental Assisting Prgm
70 Hodges Cross Rd
North Adams, MA 01247
*Prgm Dir:* Denice Kozak, CDA AS
*Tel:* 413 663-5383, Ext 183    *Fax:* 413 664-9424
*E-mail:* dkozak@mccanntech.org

**Southeastern Technical Institute**
Dental Assisting Prgm
250 Foundry St
South Easton, MA 02375
*Prgm Dir:* Audrey Beaudoin, MEd
*Tel:* 508 238-1860    *Fax:* 508 230-1558
*E-mail:* abeaudoin@sersd.org

**Springfield Technical Community College**
Dental Assisting Prgm
One Armory Square
PO Box 9000
Springfield, MA 01102
www.stcc.edu
*Prgm Dir:* Carol A Giaquinto, CDA RDH MEd
*Tel:* 413 755-4861    *Fax:* 413 755-6312
*E-mail:* giaquinto@stcc.edu

**Quinsigamond Community College**
Dental Assisting Prgm
670 W Boylston St
Worcester, MA 01606
*Prgm Dir:* Adrienne Nichols, MEd
*Tel:* 508 854-2742    *Fax:* 508 854-2704
*E-mail:* anichols@qcc.mass.edu

# Michigan

**Washtenaw Community College**
Dental Assisting Prgm
4800 E Huron River Dr
Ann Arbor, MI 48106
www.wccnet.edu/health/dental.php
*Prgm Dir:* Kathleen Weber, CDA RDA BAS
*Tel:* 734 973-3338    *Fax:* 734 677-5334
*E-mail:* weber@wccnet.edu

**Lake Michigan College**
Dental Assisting Prgm
2755 E Napier Ave
Benton Harbor, MI 49022
www.lakemichigancollege.edu
*Prgm Dir:* Deborah Burch, CDA RDA MS
*Tel:* 616 927-8100, Ext 5100    *Fax:* 616 927-8186
*E-mail:* burch@lakemichigancollege.edu

**Wayne County Community College District**
Dental Assisting Prgm
8551 Greenfield, Rm 308A
Detroit, MI 48228
www.wcccd.edu
*Prgm Dir:* Pamela Zarb, CDA RDA RDH MA LPC
*Tel:* 313 943-4045    *Fax:* 313 943-4025
*E-mail:* pzarb1@wcccd.edu
*Med Dir:* NA NA, NA

**Mott Community College**
Dental Assisting Prgm
1401 E Court St
Flint, MI 48503
*Prgm Dir:* Darlene Boersema, MA
*Tel:* 810 762-0496    *Fax:* 810 232-8874
*E-mail:* csmith@mcc.edu

**Grand Rapids Community College**
Dental Assisting Prgm
143 Bostwick St NE
Grand Rapids, MI 49503
*Prgm Dir:* Bunny Bookwalter, RDA CDA RDH MS
*Tel:* 616 234-4240    *Fax:* 616 234-4317
*E-mail:* bbookwal@grcc.edu
*Med Dir:* Pamela Naujalis

**Northwestern Michigan College**
Dental Assisting Prgm
1701 E Front St
Traverse City, MI 49686
*Prgm Dir:* Sallie Donovan, MS
*Tel:* 231 995-1240    *Fax:* 231 995-1950
*E-mail:* sdonovan@nmc.edu

**Delta College**
Dental Assisting Prgm
1961 Delta Rd
University Center, MI 48710
*Prgm Dir:* Pamela Smith, CDA RDA MAT
*Tel:* 989 686-9565, Ext 9565    *Fax:* 989 686-8736
*E-mail:* pamelasmith@delta.edu

# Minnesota

**Northwest Technical College - Bemidji**
Dental Assisting Prgm
905 Grant Ave SE
Bemidji, MN 56601
www.ntcmn.edu
*Prgm Dir:* Julie Dokken, RDA CDA
*Tel:* 218 333-6657    *Fax:* 218 333-6694
*E-mail:* julie.dokken@ntcmn.edu

**Central Lakes College**
Dental Assisting Prgm
501 W College Dr
Brainerd, MN 56401
*Prgm Dir:* LeAnn Schoenle
*Tel:* 218 855-8106, Ext 8106    *Fax:* 218 855-8220
*E-mail:* lschoenl@clcmn.edu

**Minnesota West Comm & Tech College**
Dental Assisting Prgm
1011 First St W
Canby, MN 56220
*Prgm Dir:* Daniel Prust, DDS MS
*Tel:* 507 223-7252, Ext 139    *Fax:* 507 223-5291
*E-mail:* danp@cb.mnwest.mnscu.edu

**Herzing College - Lakeland Academy Division**
Dental Assisting Prgm
5700 W Broadway Ave
Crystal, MN 55428
www.herzing.edu
*Prgm Dir:* Linda R Boyum, CDA RDA
*Tel:* 763 231-3177    *Fax:* 763 535-9205
*E-mail:* lboyum@mpls.herzing.edu

**Hennepin Technical College**
Dental Assisting Prgm
13100 College View Dr
Eden Prarie, MN 55347
*Prgm Dir:* Susan Thaemert, BS
*Tel:* 763 995-1593
*E-mail:* susan.thaemert@hennepintech.edu
*Med Dir:* Ellen Leger, PhD

**Hibbing Community College**
Dental Assisting Prgm
1515 E 25th St
Hibbing, MN 55747
www.hcc.mnscu.edu
*Prgm Dir:* Anne Badanjak, MEd
*Tel:* 218 262-7242    *Fax:* 218 262-7257
*E-mail:* annebadanjak@hibbing.edu

**Minneapolis Community & Technical Coll**
Dental Assisting Prgm
1501 Hennepin Ave S
Minneapolis, MN 55403
*Prgm Dir:* Kathleen Lapham
*Tel:* 612 659-6071    *Fax:* 612 659-6825
*E-mail:* kathleen.lapham@minneapolis.edu

**Minnesota State Comm & Tech Coll - Moorhead**
Dental Assisting Prgm
1900 28th Ave S
Moorhead, MN 56560-4899
www.minnesota.edu
*Prgm Dir:* Thomas Boe, DDS MBA
*Tel:* 218 299-6819, Ext 6819    *Fax:* 218 299-6532
*E-mail:* thomas.boe@minnesota.edu

**South Central College**
Dental Assisting Prgm
1920 Lee Blvd
North Mankato, MN 56003
*Prgm Dir:* Karon Metz, CDA RDA RF BS
*Tel:* 507 389-5846    *Fax:* 507 389-5850
*E-mail:* karon.metz@southcentral.edu

**Rochester Community & Technical College**
Dental Assisting Prgm
851 30th Ave SE
Rochester, MN 55904
*Prgm Dir:* Bonnie Crawford, CDA RDA
*Tel:* 507 280-3149    *Fax:* 507 280-3180
*E-mail:* Bonnie.Crawford@roch.edu

**Dakota County Technical College**
Dental Assisting Prgm
1300 145th St E
Rosemount, MN 55068
www.dctc.edu
*Prgm Dir:* Diana Sullivan, MS
*Tel:* 651 423-8483    *Fax:* 651 423-8775
*E-mail:* diana.sullivan@dctc.edu

**St Cloud Technical College**
Dental Assisting Prgm
1540 Northway Dr
St Cloud, MN 56303-1240
www.sctc.edu
*Prgm Dir:* Rita Peterson, BS CDA RDA
*Tel:* 320 308-5031    *Fax:* 320 308-5055
*E-mail:* rpeterson@sctc.edu

**Century College**
Dental Assisting Prgm
3300 Century Ave N
White Bear Lake, MN 55110
*Prgm Dir:* Arlene Retzer, CDA RDA
*Tel:* 651 779-5778    *Fax:* 651 799-5779
*E-mail:* a.retzer@century.mnscu.edu

# Mississippi

**Pearl River Community College**
Dental Assisting Prgm
5448 US Hwy 49 S
Hattiesburg, MS 39401
*Prgm Dir:* Emily Addison, CDA BS
*Tel:* 601 544-5483    *Fax:* 601 545-5548
*E-mail:* eaddison@prcc.edu

**Hinds Community College**
Dental Assisting Prgm
1750 Chadwick Dr
Jackson, MS 39204
*Prgm Dir:* Richard Gavant, DMD
*Tel:* 601 371-3526    *Fax:* 601 371-3703
*E-mail:* rgavant@hindscc.edu

# Missouri

**Nichols Career Center**
Dental Assisting Prgm
605 Union St
Jefferson City, MO 65101
www.jcps.k12.mo.us
*Prgm Dir:* Kathy Jeffries, CDA EFDA BA MBA
*Tel:* 573 659-3112    *Fax:* 573 659-3154
*E-mail:* kathy.jeffries@jcps.k12.mo.us

**Concorde Career College - Kansas City**
Dental Assisting Prgm
3239 Broadway Blvd
Kansas City, MO 64111
*Prgm Dir:* Robert Pruitt, CDA
*Tel:* 816 531-5223    *Fax:* 816 756-3231
*E-mail:* rpruitt@concorde.edu

**Metropolitan Community College - Penn Valley**
Dental Assisting Prgm
3201 SW Trafficway
Kansas City, MO 64111-2764
www.mcckc.edu
*Prgm Dir:* Rose Gupta, BDS
*Tel:* 816 759-4237    *Fax:* 816 759-4553
*E-mail:* Rose.Gupta@mcckc.edu

**Ozarks Technical Community College**
Dental Assisting Prgm
1001 E Chestnut Expressway
Springfield, MO 65802
*Prgm Dir:* Kelly Tillery, MEd RDH CDA
*Tel:* 417 895-7124, Ext 7124    *Fax:* 417 895-7057
*E-mail:* tilleryk@otc.edu

**Missouri College**
Dental Assisting Prgm
10121 Manchester Rd
St Louis, MO 63122-1583
*Prgm Dir:* Patricia Frank
*Tel:* 314 909-5568

**St Louis Community College - Forest Park**
Dental Assisting Prgm
5600 Oakland Ave
St Louis, MO 63110-1393
*Prgm Dir:* Mary Ann Taylor, CDA MSN
*Tel:* 314 644-9332    *Fax:* 314 951-9490
*E-mail:* mtaylor@stlcc.edu

# Montana

**Montana State Univ - Great Falls Coll of Tech**
Dental Assisting Prgm
2100 16th Ave S
Great Falls, MT 59405
www.msugf.edu
*Prgm Dir:* Carmen Perry, BS MEd CDA
*Tel:* 406 771-4354    *Fax:* 406 771-4317
*E-mail:* cperry@msugf.edu

**Salish Kootenai College**
Dental Assisting Prgm
Box 70
52000 Hwy 93
Pablo, MT 59855
www.skc.edu
*Prgm Dir:* Donna Kotyk, RDH CDA MA
*Tel:* 406 275-4912    *Fax:* 406 275-4816
*E-mail:* donna_kotyk@skc.edu

# Nebraska

**Central Community College**
Dental Assisting Prgm
PO Box 1024
Hastings, NE 68902-1024
www.cccneb.edu
*Prgm Dir:* Marie Cecil, CDA MAT
*Tel:* 402 461-2467    *Fax:* 402 460-2138
*E-mail:* mcecil@cccneb.edu

**Southeast Community College**
Dental Assisting Prgm
8800 O St
Lincoln, NE 68520-1299
*Prgm Dir:* Susan Asher, CDA BS
*Tel:* 402 437-2740    *Fax:* 402 437-2404
*E-mail:* sasher@southeast.edu

**Mid-Plains Community College**
Dental Assisting Prgm
1101 Halligan Dr
North Platte, NE 69101
www.mpcc.edu
*Prgm Dir:* Rose White, AAS
*Tel:* 308 535-3650    *Fax:* 308 534-5767
*E-mail:* whiter@mpcc.edu
*Med Dir:* Diane Hoffman, Master Degree in Nursing

**Metropolitan Community College**
Dental Assisting Prgm
PO Box 3777
Omaha, NE 68103-0777
*Prgm Dir:* Karen Finley, CDA BGS
*Tel:* 402 738-4676    *Fax:* 402 738-4554
*E-mail:* kfinley@mccneb.edu

**Vatterott College - Omaha Campus**
Dental Assisting Prgm
11818 I St
Omaha, NE 68137
*Prgm Dir:* Wendy Simon
*Tel:* 402 891-9411, Ext 142
*E-mail:* wsimon@vatterott-college.edu

# Nevada

**Community College of Southern Nevada**
Dental Assisting Prgm
6375 W Charleston Blvd
Las Vegas, NV 89146
www.ccsn.nevada.edu
*Prgm Dir:* Carole Brew, BS
*Tel:* 702 651-5851    *Fax:* 702 651-7453
*E-mail:* carole_brew@ccsn.edu

**Truckee Meadows Community College**
Dental Assisting Prgm
7000 Dandini Blvd
Reno, NV 89512-3999
*Prgm Dir:* Julie Muhle, CDA AA AAS
*Tel:* 775 673-7125    *Fax:* 775 673-7034
*E-mail:* jmuhle@tmcc.edu

# New Hampshire

**New Hampshire Technical Institute**
Dental Assisting Prgm
31 College Dr
Concord, NH 03301-7412
www.nhti.edu
*Prgm Dir:* Lynnea Adams, CDA BA
*Tel:* 603 271-7149    *Fax:* 603 271-7182
*E-mail:* ladams@nhctc.edu

# New Jersey

**Camden County College**
Dental Assisting Prgm
PO Box 200
Blackwood, NJ 08012
www.camdencc.edu
*Prgm Dir:* Sandra Rodier, AAS
*Tel:* 856 227-7200, Ext 4471    *Fax:* 856 374-5048
*E-mail:* srodier@camdencc.edu

**Cumberland County Technical Education Center**
Dental Assisting Prgm
601 Bridgeton Ave
Bridgeton, NJ 08302
*Prgm Dir:* Judy Zirkle
*Tel:* 856 451-9000, Ext 346    *Fax:* 856 451-8487
*E-mail:* JZirkle@cumberland.tec.nj.us

**Cape May County Technical Institute**
Dental Assisting Prgm
188 Crest Haven Rd
Cape May Courthouse, NJ 08210
*Prgm Dir:* Candida J Ditzler, CDA RDA
*Tel:* 609 465-2161, Ext 432    *Fax:* 609 465-4962
*E-mail:* cditzler@capemaytech.com

**The Institute for Health Education**
Dental Assisting Prgm
7 Spielman Rd
Fairfield, NJ 07004
*Prgm Dir:* Theresa Lennon, CDA RDA
*Tel:* 973 808-1110    *Fax:* 973 808-1130
*E-mail:* Healthcareer@gmail.com

**Atlantic County Institute of Technology**
Dental Assisting Prgm
5080 Atlantic Ave
Mays Landing, NJ 08330
www.acitech.org
*Prgm Dir:* Stefanie J White, CDA RDA LRT(D) RDH BS
*Tel:* 609 625-2249, Ext 1316    *Fax:* 609 625-8622
*E-mail:* swhite@acitech.org

**Univ of Med & Dent of New Jersey**
Dental Assisting Prgm
1776 Raritan Rd
Scotch Plains, NJ 07076
www.shrp.umdnj.edu
*Prgm Dir:* Carolyn Breen, EdD CDA RDA RDH
*Tel:* 908 889-2419    *Fax:* 908 889-2477
*E-mail:* breen@umdnj.edu

**Technical Institute of Camden County**
*Cosponsor:* Camden County Technical Schools
Dental Assisting Prgm
343 Berlin-Cross Keys Rd
Sicklerville, NJ 08081-9709
www.ccts.tec.nj.us
*Prgm Dir:* Susan Michaleski, BA CDA CDPMA
*Tel:* 609 767-7000, Ext 5553    *Fax:* 609 767-6625
*E-mail:* smichaleski@ccts.tec.nj.us
*Med Dir:* Teri Stallone, Ph D

**Berdan Institute**
Dental Assisting Prgm
201 Willowbrook Blvd 2nd Fl
Wayne, NJ 07470
www.berdaninstitute.com
*Prgm Dir:* Helene Pizzuta, CDA RDA
*Tel:* 973 837-1818, Ext 212    *Fax:* 973 837-1840
*E-mail:* hpizzuta@edaff.com
*Med Dir:* Irene Figliolina, CMA

# New Mexico

**Central New Mexico Community College**
Dental Assisting Prgm
525 Buena Vista SE
Albuquerque, NM 87106
www.tvi.edu
*Prgm Dir:* Melanie Upshaw, BSDH CDA
*Tel:* 505 224-4111, Ext 5247    *Fax:* 505 224-5201
*E-mail:* mupshaw@cnm.edu

**University of New Mexico - Gallup**
Dental Assisting Prgm
200 College Rd
Gallup, NM 87301
www.gallup.unm.edu
*Prgm Dir:* Jean Martinez-Welles, MS CDA
*Tel:* 505 863-7515    *Fax:* 505 863-7513
*E-mail:* jmwelles@unm.edu

**Dona Ana Community College**
Dental Assisting Prgm
3400 S Espina St
Las Cruces, NM 88007
http://dabcc.nmsu.edu
*Prgm Dir:* Martha McCaslin
*Tel:* 505 527-7653
*E-mail:* mmccasli@nmsu.edu

**Santa Fe Community College**
Dental Assisting Prgm
6401 Richards Ave, Rm 421B
Santa Fe, NM 87509
www.sfccnm.edu/dental
*Prgm Dir:* Aamna Nayyar, BSc BDS
*Tel:* 505 428-1258, Ext 258    *Fax:* 505 428-1526
*E-mail:* anayyar@sfccnm.edu

# New York

**SUNY Educational Opportunity Center**
Dental Assisting Prgm
465 Washington St
Buffalo, NY 14203
*Prgm Dir:* Susan Camizzi
*Tel:* 716 849-6727, Ext 181    *Fax:* 716 849-6755

**Columbia University**
Dental Assisting Prgm
630 W 168th St
P & S Box 20
New York, NY 10032
http://dental.columbia.edu/odma
*Prgm Dir:* Dennis A Mitchell, DDS MPH
*Tel:* 212 342-3716    *Fax:* 212 305-1034
*E-mail:* DML48@columbia.edu

**Monroe Community College**
Dental Assisting Prgm
1000 E Henrietta Rd
Rochester, NY 14623-5780
*Prgm Dir:* David B Lawrence, DDS
*Tel:* 585 292-2761    *Fax:* 585 424-5703
*E-mail:* dlawrence@monroecc.edu

# North Carolina

**Asheville-Buncombe Technical Comm College**
Dental Assisting Prgm
340 Victoria Rd
Asheville, NC 28801
www.abtech.edu
*Prgm Dir:* Shaun Tate, RDH MAEd
*Tel:* 828 254-1921, Ext 271    *Fax:* 828 281-9792
*E-mail:* state@abtech.edu

**Univ of North Carolina at Chapel Hill**
Dental Assisting Prgm
CB # 7450
Chapel Hill, NC 27599-7450
www.dent.unc.edu
*Prgm Dir:* Ethel Campbell, RDH MS
*Tel:* 919 966-2777    *Fax:* 919 966-6761
*E-mail:* ethel_campbell@dentistry.unc.edu

**Central Piedmont Community College**
Dental Assisting Prgm
3210 CPCC West Campus Dr
Charlotte, NC 28208
www.cpcc.edu
*Prgm Dir:* Jane Lavin, BSDH RDH
*Tel:* 704 330-4614    *Fax:* 704 330-4641
*E-mail:* jane_lavin@cpcc.cc.nc.us

**Fayetteville Technical Community College**
Dental Assisting Prgm
2201 Hull Rd
PO Box 35236
Fayetteville, NC 28303
www.faytechcc.edu
*Prgm Dir:* Angela Simmons, CDA BS
*Tel:* 910 678-9858    *Fax:* 910 678-8500
*E-mail:* simmonsa@faytechcc.edu

**Wayne Community College**
Dental Assisting Prgm
Caller Box 8002
Goldsboro, NC 27530
www.waynecc.edu
*Prgm Dir:* Connie McCullen, CDA BS
*Tel:* 919 735-5152, Ext 289    *Fax:* 919 736-3204
*E-mail:* conniem@waynecc.edu
*Med Dir:* William Radford, DDS

**Alamance Community College**
Dental Assisting Prgm
PO Box 8000
Graham, NC 27253-8000
www.alamance.cc.nc.us
*Prgm Dir:* Janelle Christopher, BS CDA
*Tel:* 336 506-4162    *Fax:* 336 578-7047
*E-mail:* christopherj@alamance.cc.nc.us

**Coastal Carolina Community College**
Dental Assisting Prgm
444 Western Blvd
Jacksonville, NC 28546
*Prgm Dir:* Elise Beall, BS
*Tel:* 910 938-6276    *Fax:* 910 455-4989
*E-mail:* bealle@coastal.cc.nc.us

**Guilford Technical Community College**
Dental Assisting Prgm
PO Box 309
601 High Point Rd
Jamestown, NC 27282
www.gtcc.edu
*Prgm Dir:* Lynda Snider, CDA BS
*Tel:* 336 334-4822, Ext 2212    *Fax:* 336 819-2041
*E-mail:* lfsnider@gtcc.edu
*Med Dir:* Richard Foster, DMD

**Western Piedmont Community College**
Dental Assisting Prgm
1001 Burkemont Ave
Morganton, NC 28655
*Prgm Dir:* Tammy Glover, CDA RDH
*Tel:* 828 438-6129    *Fax:* 828 430-7183
*E-mail:* tglover@wp.cc.nc.us

**Wake Technical Community College**
Dental Assisting Prgm
9101 Fayetteville Rd
Raleigh, NC 27603-5696
*Prgm Dir:* Trudy Clark, BS
*Tel:* 919 212-9386    *Fax:* 919 250-4329
*E-mail:* tsclark@waketech.edu

**Rowan - Cabarrus Community College**
Dental Assisting Prgm
PO Box 1595
1333 Jake Alexander Blvd
Salisbury, NC 28145-1595
http://rowancabarrus.edu
*Prgm Dir:* Linda Kamp, CDA BS
*Tel:* 704 637-0760, Ext 308    *Fax:* 704 642-0750
*E-mail:* kampl@rowancabarrus.edu

**Wilkes Community College**
Dental Assisting Prgm
PO Box 120
Wilkesboro, NC 28697-0120
www.wilkescc.edu
*Prgm Dir:* Larry Taylor
*Tel:* 336 838-6257    *Fax:* 336 838-6255
*E-mail:* larry.taylor@wilkescc.edu

**Martin Commuity College**
Dental Assisting Prgm
1161 Kehukee Park Rd
Williamston, NC 27892-9988
www.martincc.edu
*Prgm Dir:* Susan Cutler
*Tel:* 252 792-1521    *Fax:* 252 792-4425
*E-mail:* scutler@martincc.edu

**Cape Fear Community College**
Dental Assisting Prgm
411 N Front St
Wilmington, NC 28401-3993
http://cfcc.edu
*Prgm Dir:* Nancy Fetter, CDA RDH BS BEd
*Tel:* 910 362-7416    *Fax:* 910 362-7418
*E-mail:* nfetter@cfcc.edu
*Med Dir:* Regina McBarron, ADN BSN MS

**Forsyth Technical Community College**
Dental Assisting Prgm
2100 Silas Creek Pkwy
Winston-Salem, NC 27103
*Prgm Dir:* Jannette Whisenhunt, PhD
*Tel:* 336 734-7414    *Fax:* 336 734-7444
*E-mail:* jwhisenhunt@forsythtech.edu

# North Dakota

**North Dakota State College of Science**
Dental Assisting Prgm
800 N 6th St
Wahpeton, ND 58076
*Prgm Dir:* Susan Swanson, DDS
*Tel:* 701 671-2333    *Fax:* 701 671-2570
*E-mail:* Susan.Swanson@ndscs.edu

# Ohio

**Polaris Career Center**
Dental Assisting Prgm
7285 Old Oak Blvd
Middleburg Heights, OH 44130-3375
*Prgm Dir:* Kathy Ann Becker, CDA EFDA
*Tel:* 440 891-7600  *Fax:* 440 243-3952
*E-mail:* kbecker@polaris.edu

**Jefferson Community College**
Dental Assisting Prgm
4000 Sunset Blvd
Steubenville, OH 43952
www.jcc.edu
*Prgm Dir:* Donna Robinson, BA CDA EFDA
*Tel:* 740 264-5591, Ext 164  *Fax:* 740 264-9504
*E-mail:* drobinson@jcc.edu

**Choffin Career & Technical Center**
Dental Assisting Prgm
200 E Wood St
Youngstown, OH 44501
*Prgm Dir:* Paula J Oliver, CDA CDPMA CODA
*Tel:* 330 744-8749  *Fax:* 330 744-8749

# Oklahoma

**Rose State College**
Dental Assisting Prgm
6420 SE 15th St
Midwest City, OK 73110
www.rose.edu
*Prgm Dir:* Janet Turley, RDH CDA MEd
*Tel:* 405 733-7337  *Fax:* 405 736-0389
*E-mail:* jturley@rose.edu

**Moore Norman Technology Center**
Dental Assisting Prgm
4701 12th Ave NW
Norman, OK 73609
www.mntechnology.com
*Prgm Dir:* Pat Trent, RDH BS CDA
*Tel:* 405 364-5763, Ext 6523  *Fax:* 405 447-6289
*E-mail:* ptrent@mntechnology.com

**Metro Technology Centers**
Dental Assisting Prgm
1720 Springlake Dr
Oklahoma City, OK 73111
*Prgm Dir:* Erica Hannah
*Tel:* 405 605-4613  *Fax:* 405 424-9403
*E-mail:* ehannah@metrotech.org

# Oregon

**Linn - Benton Community College**
Dental Assisting Prgm
6500 SW Pacific Blvd
Albany, OR 97321
http://linnbenton.edu
*Prgm Dir:* Sheri Billetter, CDA EFDA FADAA
*Tel:* 541 917-4496  *Fax:* 541 917-4508
*E-mail:* billets@linnbenton.edu
*Med Dir:* Ann Malosh, BA, MEd

**Central Oregon Community College**
Dental Assisting Prgm
2600 NW College Way
Bend, OR 97701
*Prgm Dir:* Deb Davies, RDH AAS
*Tel:* 541 330-4368  *Fax:* 541 317-3064
*E-mail:* ddavies@cocc.edu
*Med Dir:* H M Kemple, MD

**Lane Community College**
Dental Assisting Prgm
4000 E 30th Ave
Eugene, OR 97405
*Prgm Dir:* Sandra Stice, COA
*Tel:* 541 463-5638  *Fax:* 541 463-5616
*E-mail:* stices@lanecc.edu

**Blue Mountain Community College**
Dental Assisting Prgm
2411 NW Carden
PO Box 100
Pendleton, OR 97801
www.bluecc.edu
*Prgm Dir:* Crystal D Patton-Doherty, CDA EFDA BS
*Tel:* 541 278-5876  *Fax:* 541 278-5874
*E-mail:* cpatton@bluecc.edu

**Concorde Career Institute - Portland**
Dental Assisting Prgm
1425 NE Irving St, Bldg 300
Portland, OR 97232
www.concorde.edu
*Prgm Dir:* Tracy Cook, BS CDA EFDA EFODA
*Tel:* 503 281-4181, Ext 130  *Fax:* 503 281-6739
*E-mail:* tcook@concorde.edu

**Portland Community College**
Dental Assisting Prgm
PO Box 19000
Portland, OR 97280-0990
www.pcc.edu
*Prgm Dir:* Josette Beach, RDH MS
*Tel:* 503 977-4235  *Fax:* 503 977-8300
*E-mail:* jbeach@pcc.edu

**Chemeketa Community College**
Dental Assisting Prgm
4000 Lancaster Dr NE
PO Box 14007
Salem, OR 97309-7070
www.chemeketa.edu
*Prgm Dir:* Lorene Vollmar, CDA EFDA MS
*Tel:* 503 399-5265  *Fax:* 503 399-5496
*E-mail:* lorene@chemeketa.edu

# Pennsylvania

**Harcum College**
Dental Assisting Prgm
750 Montgomery Ave
Bryn Mawr, PA 19010
www.harcum.edu
*Prgm Dir:* Dorothea Cavallucci, MS
*Tel:* 610 526-6029  *Fax:* 610 526-6182
*E-mail:* dcavallucci@harcum.edu

**Harrisburg Area Community College**
Dental Assisting Prgm
One HACC Dr
Harrisburg, PA 17110-2999
*Prgm Dir:* Debra Nickey, CDA RDH MAEd
*Tel:* 717 221-1731  *Fax:* 717 780-1110
*E-mail:* danickey@hacc.edu

**Manor College**
Dental Assisting Prgm
Expanded Functions Dental Assististing Program
700 Fox Chase Rd
Jenkintown, PA 19046-3399
www.manor.edu
*Prgm Dir:* Diane Meehan, CDA EFDA
*Tel:* 215 885-2360, Ext 288  *Fax:* 215 885-6084
*E-mail:* dmeehan@manor.edu

**Commonwealth Tech Inst at Hiram G Andrews Ctr**
Dental Assisting Prgm
727 Goucher St
Johnstown, PA 15905
*Prgm Dir:* James Howanek, DDS
*Tel:* 814 255-8281  *Fax:* 814 255-3406
*E-mail:* jhowanek@state.pa.us
*Med Dir:* Donald Rullman, BA

**Luzerne County Community College**
Dental Assisting Prgm
1333 S Prospect St
Nanticoke, PA 18634-3899
www.luzerne.edu
*Prgm Dir:* Cathryn M Brown, CDA RDH MEd
*Tel:* 570 740-0447  *Fax:* 570 740-0265
*E-mail:* CBrown@luzerne.edu

**Bradford School - Pittsburgh**
Dental Assisting Prgm
125 West Station Square Dr, Ste 129
Pittsburgh, PA 15219
*Prgm Dir:* Linda DeFalle, MEd
*Tel:* 412 319-6361  *Fax:* 412 471-6714
*E-mail:* ldefalle@bradfordpittsburgh.edu
*Med Dir:* Carol DeAngelis, RDH, EFDA, CDA

**Westmoreland County Community College**
Dental Assisting Prgm
145 Pavilion Ln
Youngwood, PA 15697-1895
www.wccc-pa.edu
*Prgm Dir:* Angela Rinchuse, RDH MEd
*Tel:* 724 925-4163  *Fax:* 724 925-5808
*E-mail:* rinchusea@wccc-pa.edu

# Puerto Rico

**University of Puerto Rico**
Dental Assisting Prgm
Medical Sciences Campus
PO Box 365067
San Juan, PR 00936-5067
*Prgm Dir:* Angel Rafael Aja, DDS
*Tel:* 787 758-2525, Ext 1157  *Fax:* 787 751-3061
*E-mail:* angela@cprs.rcm.upr.edu

# Rhode Island

**Community College of Rhode Island**
Dental Assisting Prgm
1762 Louisquisset Pike
Lincoln, RI 02865
www.ccri.edu
*Prgm Dir:* Donna Medas Patton, CDA RDH MEd
*Tel:* 401 333-7220  *Fax:* 401 333-7146
*E-mail:* dmedaspatton@ccri.edu

# South Carolina

**Aiken Technical College**
Dental Assisting Prgm
PO Drawer 696
Aiken, SC 29802
*Prgm Dir:* Amelia Capers, BS
*Tel:* 803 593-9231  *Fax:* 803 593-6526
*E-mail:* capersa@atc.edu

**Trident Technical College**
Dental Assisting Prgm
PO Box 118067
Charleston, SC 29423-8067
www.tridenttech.edu
*Prgm Dir:* Barbara Jarrett, MS
*Tel:* 843 574-6294  *Fax:* 843 574-6585
*E-mail:* Barbara.Jarrett@tridenttech.edu

**Midlands Technical College**
Dental Assisting Prgm
Expanded Duty Dental Assisting
PO Box 2408
Columbia, SC 29202
*Prgm Dir:* Maria Marchant, BHS
*Tel:* 803 822-3453   *Fax:* 803 822-3079
*E-mail:* marchantm@midlandstech.edu

**Horry - Georgetown Technical College**
Dental Assisting Prgm
2050 Hwy 501 E, Box 261966
Conway, SC 29528-6066
www.hgtc.edu
*Prgm Dir:* Pamela Moyers, CDA AS
*Tel:* 843 349-5331   *Fax:* 843 349-7576
*E-mail:* pamela.moyers@hgtc.edu

**Florence-Darlington Technical College**
Dental Assisting Prgm
PO Box 100548
Florence, SC 29501-0548
www.fdtc.edu
*Prgm Dir:* J Richard Moore, DMD
*Tel:* 843 661-8223   *Fax:* 843 292-0851
*E-mail:* richard.moore@fdtc.edu

**Greenville Technical College**
Dental Assisting Prgm
South Pleasantburg Ave
PO Box 5616 - Station B
Greenville, SC 29606-5616
*Prgm Dir:* Cynthia Baker, DDS
*Tel:* 864 250-8000, Ext 8074   *Fax:* 864 250-8588
*E-mail:* cynthia.baker@gvltec.edu

**Tri-County Technical College**
Dental Assisting Prgm
PO Box 587
Pendleton, SC 29670
*Prgm Dir:* Donna Shannon, CDA RDH BS
*Tel:* 864 646-1347, Ext 1347   *Fax:* 864 646-1892
*E-mail:* dshanno1@tctc.edu

**York Technical College**
Dental Assisting Prgm
Expanded Duty DA Program
452 S Anderson Rd
Rock Hill, SC 29730
*Prgm Dir:* Kathleen Peter, RDH MHA
*Tel:* 803 981-7068   *Fax:* 803 327-8059
*E-mail:* peter@yorktech.com

**Spartanburg Community College**
Dental Assisting Prgm
PO Drawer 4386, Highway I-85
Spartanburg, SC 29305-4386
www.sccsc.edu
*Prgm Dir:* Kim Golightly
*Tel:* 864 592-4872   *Fax:* 864 591-3881
*E-mail:* golightlyk@sccsc.edu

# South Dakota

**Lake Area Technical Institute**
Dental Assisting Prgm
230 11th St NE
Watertown, SD 57201
http://sweb.lakeareatech.edu/dental
*Prgm Dir:* Rhonda Bradberry, CDA
*Tel:* 605 882-5284, Ext 214   *Fax:* 605 882-6299
*E-mail:* bradberr@lati.tec.sd.us

# Tennessee

**Chattanooga State Technical Comm College**
Dental Assisting Prgm
4501 Amnicola Hwy
Chattanooga, TN 37406-1097
*Prgm Dir:* Karen Castleberry, CDA RDA BS
*Tel:* 423 697-4474   *Fax:* 423 697-2629
*E-mail:* karen.castleberry@chattanoogastate.edu
*Med Dir:* William Johnson, DMD

**Tennessee Technology Center - Dickson**
Dental Assisting Prgm
740 Hwy 46 South
Dickson, TN 37055
www.dickson.tec.tn.us
*Prgm Dir:* Vanessa Pilkinton, CDA RDA
*Tel:* 615 441-6220, Ext 121   *Fax:* 615 441-6223
*E-mail:* vanessa.pilkinton@dickson.tec.tn.us

**Northeast State Technical Comm College**
Dental Assisting Prgm
1000 West Jason Witten Way
Elizabethton, TN 37643
www.northeaststate.edu
*Prgm Dir:* Paulette Kehm-Yelton, MPA CDA EFDA
*Tel:* 423 547-4910   *Fax:* 423 543-2266
*E-mail:* pkyelton@NortheastState.edu
*Med Dir:* Harold Lane, DDS

**Volunteer State Community College**
Dental Assisting Prgm
Nashville Pike
Gallatin, TN 37066
www.volstate.edu
*Prgm Dir:* Desiree Sutphen, CDA RDA BA
*Tel:* 615 452-8600, Ext 3439   *Fax:* 615 230-3344
*E-mail:* Desiree.Sutphen@volstate.edu
*Med Dir:* Ron Coleman, EdD

**Tennessee Technology Center - Knoxville**
Dental Assisting Prgm
1100 Liberty St
Knoxville, TN 37919
www.knoxville.tec.tn.us
*Prgm Dir:* Mary Vaughn, BS CDA RDA
*Tel:* 865 546-5567, Ext 131   *Fax:* 865 971-4474
*E-mail:* mvaughn@knoxville.tec.tn.us

**Concorde Career College - Memphis**
Dental Assisting Prgm
5100 Poplar Ave, Ste 132
Memphis, TN 38137
*Prgm Dir:* Shay Ashford
*Tel:* 901 761-9494   *Fax:* 901 761-3293
*E-mail:* sashford@concorde.edu

**Tennessee Technology Center - Memphis**
Dental Assisting Prgm
550 Alabama Ave
Memphis, TN 38105-3604
www.memphis.tec.tn.us
*Prgm Dir:* Bettie J Brooks
*Tel:* 901 543-6143   *Fax:* 901 543-6126
*E-mail:* bettie.brooks@memphis.tec.tn.us

**Tennessee Technology Center - Murfreesboro**
Dental Assisting Prgm
1303 Old Fort Pkwy
Murfreesboro, TN 37129
www.murfreesboro.tec.tn.us
*Prgm Dir:* Suzanne Dowdle, CDA RDA
*Tel:* 615 898-8010, Ext 133   *Fax:* 615 893-4194
*E-mail:* suzanne.dowdle@murfreesboro.tec.tn.us

# Texas

**Del Mar College**
Dental Assisting Prgm
101 Baldwin Blvd
Corpus Christi, TX 78404-3897
www.delmar.edu/da
*Prgm Dir:* David G Arreguin, DDS
*Tel:* 361 698-1358   *Fax:* 361 698-1175
*E-mail:* darreguin@delmar.edu

**Grayson County College**
Dental Assisting Prgm
6101 Grayson Dr
Box 48
Denison, TX 75020-8299
www.grayson.edj
*Prgm Dir:* Gwen Kirk, CDA AS BBM RDA
*Tel:* 903 463-8780
*E-mail:* kirkgw@grayson.edu
*Med Dir:* Robert Steele, DDS

**El Paso Community College**
Dental Assisting Prgm
PO Box 20500
El Paso, TX 79998
*Prgm Dir:* Sharon Dickinson, CDA CDPMA RDA
*Tel:* 915 831-4065   *Fax:* 915 831-4445
*E-mail:* sdickins@epcc.edu

**Texas State Technical College - Harlingen**
Dental Assisting Prgm
1902 N Loop 499
Harlingen, TX 78550-3697
*Prgm Dir:* Robert Bennett, DMD
*Tel:* 956 364-4690   *Fax:* 956 364-5148
*E-mail:* bob.bennett@harlingen.tstc.edu
*Med Dir:* Bob Bennett, DMD

**Houston Community College**
Dental Assisting Prgm
1900 Pressler
Houston, TX 77030
www.hccs.edu
*Prgm Dir:* Rosalva Perez, BS CDA RDA
*Tel:* 713 718-7350   *Fax:* 713 718-7608
*E-mail:* rosie.perez@hccs.edu
*Med Dir:* Douglas Simmons, DDS, MPH

**Lamar State College - Orange**
Dental Assisting Prgm
410 Fronge St
Orange, TX 77630
*Prgm Dir:* Carolyn Flippen
*Tel:* 409 882-3022
*E-mail:* Carolyn.Flippen@lsco.edu

**San Antonio College**
Dental Assisting Prgm
1300 San Pedro Ave, NTC 125
San Antonio, TX 78212-4299
*Prgm Dir:* Stella Lovato, MA Ed MS
*Tel:* 210 785-6184   *Fax:* 210 733-2926
*E-mail:* slovato@accd.edu

**882d Training Group**
Dental Assisting Prgm
917 Missile Rd
Sheppard AFB, TX 76311
*Prgm Dir:* Thomas Grimm, DDS
*Tel:* 940 676-6932   *Fax:* 940 676-8104
*E-mail:* thomas.grimm@sheppard.af.mil

**Texas State Technical College at Waco**
Dental Assisting Prgm
3801 Campus Dr
Waco, TX 76705
http://waco.tstc.edu
*Prgm Dir:* Donna Estes, AAS
*Tel:* 254 867-4864   *Fax:* 254 867-2239
*E-mail:* donna.estes@tstc.edu

# Utah

**Davis Applied Technology College**
Dental Assisting Prgm
550 East 300 South
Kaysville, UT 84037
*Prgm Dir:* Cathleen Turnbow, CDA
*Tel:* 801 593-2349   *Fax:* 801 593-7849
*E-mail:* turnboc@datc.net

**Bridgerland Applied Technology College**
Dental Assisting Prgm
1301 West 600 North
Logan, UT 84321
*Prgm Dir:* Wendi Wilde, ASC
*Tel:* 435 750-3121   *Fax:* 435 750-2016
*E-mail:* wwilde@bridgerlandatc.org

**Ogden - Weber Applied Technology College**
Dental Assisting Prgm
200 N Washington Blvd
Ogden, UT 84404-6704
*Prgm Dir:* Brenda Fell, CDA CDPMA AS
*Tel:* 801 627-8444   *Fax:* 801 395-3703
*E-mail:* fellb@owatc.edu

**Careers Unlimited**
Dental Assisting Prgm
575 E University Pkwy, Ste 163
Orem, UT 84097
*Prgm Dir:* Krista McClure
*Tel:* 801 787-3223

**Ameritech College**
Dental Assisting Prgm
1675 N Freedom Blvd, Ste 3B
Provo, UT 84604
www.ameritech.edu
*Prgm Dir:* Lynn Tyler, BS CDA
*Tel:* 801 377-2900, Ext 109   *Fax:* 801 375-3077
*E-mail:* ltyler@ameritech.edu

**Provo College**
Dental Assisting Prgm
1450 W 820 N
Provo, UT 84604
*Prgm Dir:* Judy Simpson, AAS
*Tel:* 801 818-8915   *Fax:* 801 375-9728
*E-mail:* judys@provocollege.edu

# Vermont

**Center for Technology - Essex**
Dental Assisting Prgm
3 Educational Dr
Essex Junction, VT 05452
*Prgm Dir:* Beth Ladd, CDA EFDA
*Tel:* 802 879-8153   *Fax:* 802 879-5593
*E-mail:* bladd@cte.k12.vt.us

# Virginia

**Tidewater Technical**
Dental Assisting Prgm
7020 N Military Hwy
Norfolk, VA 23518-4202
*Prgm Dir:* Linda Hamilton
*Tel:* 757 853-2121   *Fax:* 757 852-9017

**J Sargeant Reynolds Community College**
Dental Assisting Prgm
PO Box 85622
Richmond, VA 23285-5622
www.jsr.vccs.edu
*Prgm Dir:* Nancy L Daniel, BS
*Tel:* 804 523-5380   *Fax:* 804 786-5298
*E-mail:* ndaniel@jsr.vccs.edu

# Washington

**Bellingham Technical College**
Dental Assisting Prgm
3028 Lindbergh Ave
Bellingham, WA 98225
www.btc.ctc.edu
*Prgm Dir:* Karen McGuinn, CDA
*Tel:* 360 752-8315   *Fax:* 360 752-7115
*E-mail:* kmcguinn@btc.ctc.edu

**Lake Washington Technical College**
Dental Assisting Prgm
11605 132nd Ave NE
Kirkland, WA 98034
www.lwtc.ctc.edu/dental
*Prgm Dir:* Scarlet Kendrick, CDA MA
*Tel:* 425 739-8369   *Fax:* 425 739-8292
*E-mail:* scarlet.kendrick@lwtc.edu

**Clover Park Technical College**
Dental Assisting Prgm
4500 Steilacoom Blvd SW
Lakewood, WA 98498-4098
www.cptc.ctc.edu
*Prgm Dir:* Roberta Wirth
*Tel:* 253 589-6023   *Fax:* 253 589-5866
*E-mail:* roberta.wirth@cptc.edu

**South Puget Sound Community College**
Dental Assisting Prgm
2011 Mottman Rd SW
Olympia, WA 98512-6292
www.spscc.ctc.edu
*Prgm Dir:* Joan Martin, CDA BA MAED
*Tel:* 360 596-5295   *Fax:* 360 596-5707
*E-mail:* jmartin@spscc.ctc.edu
*Med Dir:* Allen Mason, Doctorate

**Renton Technical College**
Dental Assisting Prgm
3000 NE Fourth St
Renton, WA 98056
*Prgm Dir:* Kathy Leviton, BS CDA
*Tel:* 425 235-2352, Ext 5560   *Fax:* 425 235-2436
*E-mail:* kleviton@rtc.ctc.edu

**Seattle Vocational Institute**
Dental Assisting Prgm
2120 S Jackson St
Seattle, WA 98144
*Prgm Dir:* Jeanne Nichols, BA
*Tel:* 206 587-4930   *Fax:* 206 587-4949
*E-mail:* jnichols@sccd.ctc.edu
*Med Dir:* Pollene Speed McIntyre, MS, DDS

**Spokane Community College**
Dental Assisting Prgm
N 1810 Greene St, MS 2090
Spokane, WA 99207
*Prgm Dir:* Donna Phinney, MEd
*Tel:* 509 533-7300   *Fax:* 509 533-8621
*E-mail:* dphinney@scc.spokane.edu

**Bates Technical College**
Dental Assisting Prgm
1101 S Yakima Ave
Tacoma, WA 98405
www.bates.ctc.edu
*Prgm Dir:* Jean Watley
*Tel:* 253 680-7215   *Fax:* 253 680-7201
*E-mail:* jwatley@bates.ctc.edu

# West Virginia

**Mercer County Technical Education Center**
Dental Assisting Prgm
1397 Stafford Dr
Princeton, WV 24740
*Prgm Dir:* Bertha Robertson, CDA RDA
*Tel:* 304 425-9551, Ext 127   *Fax:* 304 425-0833

# Wisconsin

**Fox Valley Technical College**
Dental Assisting Prgm
1825 N Bluemound Dr
Appleton, WI 54913
www.fvtc.edu
*Prgm Dir:* Harold Peaslee, CDA BA MS
*Tel:* 920 735-5666   *Fax:* 920 735-2582
*E-mail:* peaslee@fvtc.edu

**Northeast Wisconsin Technical College**
Dental Assisting Prgm
2740 W Mason St, PO Box 19042
Green Bay, WI 54307-9042
*Prgm Dir:* Carol Johnson, CDA BS
*Tel:* 920 498-5472, Ext 5472   *Fax:* 920 498-5673
*E-mail:* carol.johnson@nwtc.edu

**Blackhawk Technical College**
Dental Assisting Prgm
6004 Prairie Rd
PO Box 5009
Janesville, WI 53547-5009
*Prgm Dir:* Lois D Swanson, CDA RDH MS
*Tel:* 608 757-7732   *Fax:* 608 743-4407
*E-mail:* ldswanson@blackhawk.edu

**Gateway Technical College**
Dental Assisting Prgm
3520 30th Ave
Kenosha, WI 53144-1690
www.gtc.edu
*Prgm Dir:* Heidi Gottfried, BA CDA
*Tel:* 262 564-2544   *Fax:* 262 564-2299
*E-mail:* gottfriedh@gtc.edu
*Med Dir:* Kathleen Russ

**Western Technical College**
Dental Assisting Prgm
304 N Sixth St, PO Box C-0908
La Crosse, WI 54601
www.wwtc.edu
*Prgm Dir:* Sue Harpstreith
*Tel:* 608 785-9137   *Fax:* 608 785-9087
*E-mail:* harpstreiths@wwtc.edu

## Dental Assistant

| Programs* | Class Capacity | Begins | Length (months) | Award | Res Tuition | Non-res Tuition | Stipend | Offers:‡ 1 | 2 | 3 |
|---|---|---|---|---|---|---|---|---|---|---|
| **Alabama** | | | | | | | | | | |
| H Council Trenholm State Technical College (Montgomery) | 20 | Aug | 12, 18 | Cert, AAT | $4,050 | $7,290 | | | | |
| James H Faulkner State Community College (Bay Minette) | 30 | Aug | 12, 24 | Cert, AAS | $5,859 | $10,322 | | | | • |
| Wallace State Community College (Hanceville) | 30 | Aug | 12 | Cert | $2,052 | $4,104 | | | | • |
| **Arizona** | | | | | | | | | | |
| Phoenix College | 24 | Aug Jan | 9, 24 | Cert, AAS | $1,980 | $0 | | | | • |
| **Arkansas** | | | | | | | | | | |
| Arkansas Northeastern College (Burdette) | 20 | Aug | 10 | Cert | $1,950 | $2,150 | $70 | | | |
| Pulaski Technical College (North Little Rock) | 26 | Aug | 9 | Cert | $2,808 | $4,641 | | | | |
| **California** | | | | | | | | | | |
| Chaffey College (Rancho Cucamonga) | 55 | Aug Jan | 9, 18 | Cert, AS | $954 | $5,600 | $4,000 | • | | • |
| Citrus College (Glendora) | 30 | Aug Jan | 9 | Cert | $978 | $4,500 | | | | |
| City College of San Francisco | 36 | Aug | 10, 20 | Cert, AS | $871 | $4,790 | | | | |
| College of Marin (Kentfield) | 30 | Aug | 10, 18 | Cert | $1,068 | $66,500 | | | | |
| College of the Redwoods (Eureka) | 24 | Aug | 9, 24 | Cert, AS | $884 | $6,222 | | | | • |
| Foothill Community College (Los Altos Hills) | 30 | Sep | 10 | Cert, AS | $1,200 | $4,800 | | | | |
| Hacienda LaPuente Adult Education (La Puente) | 75 | Varies | 12 | Cert | $675 | $675 | | • | | |
| Heald College - Stockton Campus | 24 | Jul | 18 | AAS | $24,000 | $0 | | | | |
| Pasadena City College | 30 | Aug | 10 | Cert | $650 | $2,610 | | | | • |
| Sacramento City College | 30 | Aug | 11 | Cert, AS | $477 | $4,796 | | • | | |
| San Diego Mesa College | 24 | Jun | 10 | Cert, AS | $900 | $4,098 | | | | |
| Santa Rosa Junior College | 30 | Aug | 11 | Cert | $975 | $5,317 | | | | • |
| **Colorado** | | | | | | | | | | |
| Front Range Comm College (Ft Collins) | 24 | Aug Jan | 11 | Cert | $3,315 | $9,945 | | • | | |
| Pikes Peak Community College (Colorado Springs) | 30 | Aug | 11, 19 | Cert, AAS | $3,731 | $16,534 | | • | | |
| Pueblo Community College | 18 | Aug | 11, 24 | Cert, AAS | $8,335 | $17,380 | $3,935 | | | • |
| **Connecticut** | | | | | | | | | | |
| A I Prince Technical High School (Hartford) | 18 | Sep | 10 | Cert | $2,250 | $10,060 | | | | |
| Eli Whitney Vocational Technical School (Hamden) | 18 | Aug | 9 | Cert | $2,650 | $2,650 | | | | |
| Tunxis Community College (Farmington) | 24 | Aug | 10 | Cert | $3,000 | $8,100 | | • | | • |
| **Florida** | | | | | | | | | | |
| Broward Community College (Ft Lauderdale) | 36 | Aug | 10 | Cert | $3,500 | $9,500 | | | | |
| Charlotte Technical Center (Port Charlotte) | 40 | Jul Jan | 11 | Cert | $2,300 | $2,300 | | | | • |
| Daytona Beach Community College | 24 | Aug | 10, 12 | Cert | $2,850 | $11,216 | | | | • |
| Edison College (Ft Myers) | 12 | Aug | 10 | Cert | $2,545 | $9,554 | | | | |
| Erwin Technical Center (Tampa) | 14 | Every 18 wks | 12 | Dipl | $2,238 | $7,854 | | | | |
| Hillsborough Community College (Tampa) | 36 | Aug Jan May | 12 | Cert | $3,200 | $6,700 | | | | |
| Lorenzo Walker Institute of Technology (Naples) | 16 | Jan | 16 | Cert | $2,300 | $8,900 | | • | | |
| Manatee Technical Institute (Bradenton) | 20 | Aug | 11, 15 | Cert | $2,350 | $4,700 | | • | | • |
| Okaloosa-Walton College (Niceville) | 24 | Fall term | 12 | Cert | $2,187 | $8,189 | | • | | • |
| Orlando Tech | 25 | Aug Jan | 10 | Cert | $2,450 | $0 | | • | | |
| Palm Beach Community College (Lake Worth) | 25 | Aug | 9 | Cert | $4,000 | $9,000 | | | | |
| Pinellas Tech Educ Ctr - St Petersburg | 15 | Quarterly | 13 | Cert | $2,349 | $9,422 | | | | |
| Robert Morgan Educational Center (Miami) | 15 | Sep Jan Apr | 12 | Cert | $2,188 | $7,660 | | | | • |
| Santa Fe Community College (Gainesville) | 30 | Aug | 12 | Cert | $2,001 | $8,007 | | | | • |
| South Florida Community College (Avon Park) | 12 | Aug | 11 | Cert | $4,500 | $9,000 | | | | • |
| Traviss Career Center (Lakeland) | 16 | Aug | 11 | Cert | $3,793 | $4,793 | | | | • |
| **Georgia** | | | | | | | | | | |
| Athens Technical College | 15 | Summer | 4 | Dipl | $1,812 | $0 | | | | |
| Atlanta Technical College | 20 | Spring Fall quarters | 15 | Dipl, Cert | $3,265 | $6,000 | | | | |
| Augusta Technical College | 30 | Sep Mar | 15 | Dipl | $1,840 | $3,312 | | | | |
| Columbus Technical College | 14 | Jan Jul | 15 | Dipl | $2,821 | $0 | | | | |
| Coosa Valley Technical College (Rockmart) | 20 | Jan | 12 | Dipl | $2,100 | $4,200 | | | | |
| Gwinnett Technical College (Lawrenceville) | 26 | Jul | 12 | Dipl | $2,200 | $3,800 | | | | • |
| Medix School (Smyrna) | 35 | Every 5 wks | 10 | Dipl | $10,567 | $10,567 | | • | | |
| Middle Georgia Technical College (Warner Robins) | 16 | Winter quarter | 12 | Dipl | $1,636 | $1,636 | | | | |
| Savannah Technical College | 25 | Apr | 12 | Dipl | $1,116 | $2,124 | | | | |
| Valdosta Technical College | 14 | Oct | 12 | Dipl | $1,856 | $3,712 | | | | • |
| **Idaho** | | | | | | | | | | |
| Apollo College - Boise | 30 | Every 6 wks | 9 | Cert | $11,210 | $11,210 | | • | | • |
| **Illinois** | | | | | | | | | | |
| Elgin Community College | 24 | Jan Aug | 11 | Cert | $3,042 | $9,988 | | | | |
| Illinois Valley Community College (Oglesby) | 25 | Aug | 11 | Cert | $1,863 | $6,425 | | | | • |
| Kaskaskia College (Centralia) | 24 | Aug | 9 | Cert | $3,139 | $0 | | | | • |
| Lewis & Clark Community College (Godfrey) | 24 | Aug Jan | 9 | Cert | $3,315 | $8,094 | | | | • |

*Data are shown only for programs that completed the 2006 AMA Survey of Health Professions Education Programs

‡Key to Offers: 1: Evening or weekend classes; 2: Non-English instruction; 3: Cultural competence instruction

## Dental Assistant

| Programs* | Class Capacity | Begins | Length (months) | Award | Res Tuition | Non-res Tuition | Stipend | Offers:‡ 1 | 2 | 3 |
|---|---|---|---|---|---|---|---|---|---|---|
| **Indiana** | | | | | | | | | | |
| Indiana University (Indianapolis) | 35 | Aug | 10, 20 | Cert | $6,121 | $17,324 | | | | • |
| Indiana University Northwest (Gary) | 20 | Aug | 12 | Cert | $4,835 | $12,784 | | | | • |
| Ivy Tech Community College - Anderson | 20 | Jan | 12 | Cert | $5,200 | $9,000 | | | | |
| Professional Careers Institute (Indianapolis) | 35 | Nov | 11 | Dipl, DA | $9,050 | $9,050 | | • | | |
| University of Southern Indiana (Evansville) | 24 | Aug | 9, 18 | Cert, AS | $5,085 | $12,442 | | | | |
| **Iowa** | | | | | | | | | | |
| Des Moines Area Community College (Ankeny) | 32 | Aug | 12 | Dipl | $4,606 | $9,212 | | | | • |
| Hawkeye Community College (Waterloo) | 24 | Aug | 11 | Dipl | $4,092 | $8,184 | | | | |
| Iowa Western Community College (Council Bluffs) | 12 | Aug | 10 | Dipl | $5,280 | $7,464 | | | | |
| Marshalltown Community College | 20 | Aug | 11 | Dipl | $5,904 | $7,200 | | | | • |
| Northeast Iowa Community College (Peosta) | 25 | Aug | 11 | Dipl | $4,232 | $4,232 | | | | • |
| Vatterott College - Des Moines Campus | | | 14 | Dipl | $20,005 | $0 | | | | • |
| Western Iowa Tech Community College (Sioux City) | 24 | Aug | 9 | Dipl | $3,808 | $5,032 | | • | | • |
| **Kansas** | | | | | | | | | | |
| Flint Hills Technical College (Emporia) | 25 | Aug | 9 | Cert | $4,076 | $4,076 | | | | |
| Salina Area Technical School | 18 | Aug | 9 | Cert | $3,400 | $3,400 | | | | |
| Wichita Area Technical College | 18 | Aug | 9 | Cert, AAS | $4,500 | $8,800 | | | | • |
| **Kentucky** | | | | | | | | | | |
| Bluegrass Community and Technical College (Lexington) | 50 | Aug | 11 | Dipl | $4,508 | $13,708 | | | | |
| West Kentucky Community & Technical College (Paducah) | 20 | Aug | 11 | Dipl | $5,104 | $15,042 | | | | |
| **Massachusetts** | | | | | | | | | | |
| Middlesex Community College - Lowell | 20 | Sep | 10 | Cert, AS | $4,320 | $4,752 | | | | |
| Northern Essex Community College (Lawrence) | 30 | Sep | 9 | Cert | $2,800 | $3,400 | | | | |
| Quinsigamond Community College (Worcester) | 12 | Sep | 9 | Dipl | $5,910 | $13,103 | | | | • |
| Southeastern Technical Institute (South Easton) | 24 | Sep | 9 | Dipl | $2,500 | $3,600 | | | | |
| Springfield Technical Community College | 25 | Jun | 12 | Cert | $4,404 | $13,084 | | | | |
| **Michigan** | | | | | | | | | | |
| Delta College (University Center) | 20 | Aug | 12 | Cert | $3,420 | $4,905 | | | | |
| Lake Michigan College (Benton Harbor) | | Varies | 12, 24 | Cert, AAS | $5,330 | $6,643 | | • | | |
| Washtenaw Community College (Ann Arbor) | 24 | Sep | 12 | Cert | $4,477 | $6,076 | | | | |
| Wayne County Community College District (Detroit) | 24 | Aug | 11 | Dipl, Cert | $2,053 | $2,447 | | | | • |
| **Minnesota** | | | | | | | | | | |
| Century College (White Bear Lake) | 30 | Aug Jan Jun | 15, 24 | Dipl, AAS | $9,477 | $17,986 | | | | |
| Dakota County Technical College (Rosemount) | 45 | Aug | 12 | Dipl, AAS | $4,173 | $8,346 | | | | • |
| Herzing College - Lakeland Academy Division (Crystal) | 180 | Jan Jun Sep | 12, 21 | Dipl, AAS | $17,500 | $17,500 | | | | |
| Hibbing Community College | 32 | Sep | 9 | Dipl | $5,200 | $6,700 | | | | |
| Minnesota State Comm & Tech Coll - Moorhead | 24 | Aug | 11 | Dipl | $5,192 | $10,384 | | | | |
| Minnesota West Comm & Tech College (Canby) | 40 | Aug | 11 | Dipl, AAS | $3,376 | $6,752 | | | | |
| Northwest Technical College - Bemidji | 30 | Aug | 11 | Dipl | $6,426 | $6,426 | | | | |
| South Central College (North Mankato) | 24 | Jun Aug | 12, 24 | Dipl, AAS | $6,123 | $12,246 | | | | |
| St Cloud Technical College | 30 | Aug | 20 | Dipl, AAS | $8,985 | $8,985 | | | | |
| **Mississippi** | | | | | | | | | | |
| Pearl River Community College (Hattiesburg) | 15 | Aug | 12 | Cert | $2,200 | $2,200 | | | | |
| **Missouri** | | | | | | | | | | |
| Concorde Career College - Kansas City | 15 | | 10 | Dipl | $10,797 | $10,797 | | • | | |
| Metropolitan Community College - Penn Valley (Kansas City) | 25 | Aug | 9, 24 | Cert, AS | $5,000 | $11,000 | | | | |
| Nichols Career Center (Jefferson City) | 18 | Aug | 9 | Cert | $5,600 | $5,600 | | | | |
| **Montana** | | | | | | | | | | |
| Montana State Univ - Great Falls Coll of Tech | 18 | Sep | 11 | Cert | $2,424 | $8,181 | | • | | • |
| Salish Kootenai College (Pablo) | 30 | Varies | 12, 24 | Cert, AAS | $4,209 | $9,933 | | • | | |
| **Nebraska** | | | | | | | | | | |
| Central Community College (Hastings) | 22 | Aug | 10, 18 | Dipl, AAS | $2,604 | $3,654 | | • | | |
| Mid-Plains Community College (North Platte) | 14 | Aug | 11 | Dipl, AAS | $2,898 | $3,645 | | | | |
| **Nevada** | | | | | | | | | | |
| Community College of Southern Nevada (Las Vegas) | 24 | Sep Jan | 9 | Cert | $1,770 | $4,692 | | • | | |
| **New Hampshire** | | | | | | | | | | |
| New Hampshire Technical Institute (Concord) | 30 | Aug | 11 | Cert | $5,160 | $11,868 | | | | |
| **New Jersey** | | | | | | | | | | |
| Atlantic County Institute of Technology (Mays Landing) | 15 | Sep | 10 | Cert | $4,500 | $6,000 | | | | |
| Berdan Institute (Wayne) | 25 | Monthly | 9 | Cert | $10,000 | $0 | | • | | |
| Camden County College (Blackwood) | 22 | Sep | 9, 24 | Cert, AAS | $5,600 | $5,800 | | | | |
| Cape May County Technical Institute (Cape May Courthouse) | 15 | Sep | 10 | Cert | $1,500 | $3,450 | | | | |
| Technical Institute of Camden County (Sicklerville) | 24 | Sep | 10, 15 | Dipl, Cert | $3,300 | $3,516 | | | | |
| The Institute for Health Education (Fairfield) | 25 | Sep Jan May | 8, 12 | Cert | $6,700 | $6,700 | | • | | • |
| Univ of Med & Dent of New Jersey (Scotch Plains) | 24 | Jan | 10 | Cert | $5,520 | $8,280 | | | | • |

*Data are shown only for programs that completed the 2006 AMA Survey of Health Professions Education Programs

‡Key to Offers: 1: Evening or weekend classes; 2: Non-English instruction; 3: Cultural competence instruction

# Dental Assistant

Dental Assistant

| Programs* | Class Capacity | Begins | Length (months) | Award | Res Tuition | Non-res Tuition | Stipend | Offers:‡ 1 | 2 | 3 |
|---|---|---|---|---|---|---|---|---|---|---|
| **New Mexico** | | | | | | | | | | |
| Central New Mexico Community College (Albuquerque) | 24 | Aug | 12 | Cert | $497 | $7,960 | | | | • |
| Dona Ana Community College (Las Cruces) | 18 | Jan | 12 | Dipl, Cert | $2,200 | $0 | | | | • |
| Santa Fe Community College | 12 | Aug | 10, 24 | Cert, AAS | $1,560 | $3,114 | | • | | • |
| University of New Mexico - Gallup | 10 | Aug | 12 | Cert | $1,464 | $3,724 | | • | • | • |
| **New York** | | | | | | | | | | |
| Columbia University (New York) | 25 | Sep Mar | 12 | Cert | | | | | | |
| Monroe Community College (Rochester) | 14 | Sep | 10 | Cert | $2,700 | $5,400 | | | | |
| **North Carolina** | | | | | | | | | | |
| Alamance Community College (Graham) | 30 | Aug | 15 | Dipl | $1,496 | $8,156 | | • | | |
| Asheville-Buncombe Technical Comm College | 30 | Aug | 12 | Dipl | $1,824 | $10,128 | | | | |
| Cape Fear Community College (Wilmington) | 18 | Aug | 12 | Dipl | $1,659 | $9,219 | | | | • |
| Central Piedmont Community College (Charlotte) | 36 | Aug | 12 | Dipl | $1,824 | $10,128 | | | | • |
| Fayetteville Technical Community College | 42 | Aug | 12 | Dipl | $1,670 | $8,870 | | | | • |
| Forsyth Technical Community College (Winston-Salem) | 20 | Fall | 11 | Dipl | $1,920 | $3,520 | | | | |
| Guilford Technical Community College (Jamestown) | 36 | Aug | 11 | Dipl | $1,580 | $8,780 | | | | |
| Martin Commuity College (Williamston) | 15 | August | 12 | Dipl | $1,900 | $1,900 | | | • | |
| Rowan - Cabarrus Community College (Salisbury) | 18 | Aug | 12 | Dipl | $1,890 | $10,128 | | | | |
| Univ of North Carolina at Chapel Hill | 24 | Jul | 10 | Cert | $2,100 | $2,100 | | | | |
| Wayne Community College (Goldsboro) | 24 | Aug | 12 | Dipl | $1,612 | $8,812 | | | | |
| Western Piedmont Community College (Morganton) | 16 | Aug | 12 | Dipl | $1,866 | $7,120 | | | | |
| Wilkes Community College (Wilkesboro) | 15 | Aug | 12 | Dipl | $1,896 | $10,536 | | | | |
| **Ohio** | | | | | | | | | | |
| Jefferson Community College (Steubenville) | 24 | Aug | 10 | Cert | $2,419 | $2,624 | | | | |
| **Oklahoma** | | | | | | | | | | |
| Moore Norman Technology Center | 12 | Aug | 9 | Cert | $1,060 | $1,565 | | | | |
| Rose State College (Midwest City) | 12 | Aug | 10 | Cert, AAS | $2,772 | $6,368 | | | | |
| **Oregon** | | | | | | | | | | |
| Blue Mountain Community College (Pendleton) | 17 | Sep | 9 | Cert | $3,391 | $6,782 | | | | |
| Chemeketa Community College (Salem) | 30 | Sep | 9 | Cert | $3,008 | $9,635 | | | | |
| Concorde Career Institute - Portland | 15 | Nov Jan Feb Mar Apr | 9 | Dipl | $11,300 | $0 | | • | | |
| Linn - Benton Community College (Albany) | 24 | Sep | 11, 24 | Cert, AS | $3,695 | $9,420 | | | | • |
| Portland Community College | 45 | Sep | 9 | Cert | $2,943 | $8,695 | | | | • |
| **Pennsylvania** | | | | | | | | | | |
| Bradford School - Pittsburgh | 45 | Sep | 16 | Dipl | $13,120 | $13,120 | | | | |
| Commonwealth Tech Inst at Hiram G Andrews Ctr (Johnstown) | 10 | Jan May Sep | 12 | Dipl | $11,900 | $0 | | | | |
| Harcum College (Bryn Mawr) | 30 | Sep | 10, 24 | Cert, AS | $15,350 | $15,350 | | | | |
| Harrisburg Area Community College | 26 | Aug | 10 | Cert | $3,900 | $7,320 | | | | |
| Luzerne County Community College (Nanticoke) | 24 | Jul | 10 | Cert | $2,730 | $5,460 | | | | |
| Manor College (Jenkintown) | 18 | Sep | 19, 10 | Cert, AS | $10,130 | $13,200 | | • | | |
| Westmoreland County Community College (Youngwood) | 16 | Aug | 12 | Dipl | $2,856 | $5,712 | | | | • |
| **Rhode Island** | | | | | | | | | | |
| Community College of Rhode Island (Lincoln) | 24 | Sep | 9 | Cert | $2,254 | $6,628 | | | | • |
| **South Carolina** | | | | | | | | | | |
| Florence-Darlington Technical College | 18 | Aug | 12 | Dipl | $3,074 | $5,170 | | | | |
| Greenville Technical College | 40 | Aug | 12 | Dipl, Cert | $4,275 | $8,940 | | • | | • |
| Horry - Georgetown Technical College (Conway) | 18 | Aug | 12 | Dipl | $4,200 | $6,396 | | | | |
| Spartanburg Community College | 22 | Aug | 12 | Dipl | $1,400 | $1,800 | | | | |
| Trident Technical College (Charleston) | 24 | Aug | 12 | Dipl | $4,670 | $8,697 | | | | |
| York Technical College (Rock Hill) | 20 | Aug | 12 | Dipl | $2,900 | $6,528 | | | | |
| **South Dakota** | | | | | | | | | | |
| Lake Area Technical Institute (Watertown) | 42 | Aug | 11, 24 | Dipl, AAS | $2,656 | $2,656 | | | | |
| **Tennessee** | | | | | | | | | | |
| Northeast State Technical Comm College (Elizabethton) | 15 | Aug | 12 | Cert, AAS | $2,892 | $9,396 | | | | |
| Tennessee Technology Center - Dickson | 15 | May | 12 | Dipl | $1,986 | $1,986 | | | | • |
| Tennessee Technology Center - Knoxville | 30 | Sep | 12 | Dipl | $1,986 | $1,986 | | | | |
| Tennessee Technology Center - Memphis | 30 | May | 12 | Dipl | $2,058 | $0 | | • | | |
| Tennessee Technology Center - Murfreesboro | 24 | Sep | 12 | Dipl | $4,300 | $4,300 | | | | |
| Volunteer State Community College (Gallatin) | 24 | Aug | 12 | Cert | $3,550 | $13,195 | | | | |
| **Texas** | | | | | | | | | | |
| 882d Training Group (Sheppard AFB) | 24 | Varies | 20 | Cert | | | | | | |
| Del Mar College (Corpus Christi) | 24 | Sep | 12 | Cert, AAS | $965 | $2,165 | | | | |
| El Paso Community College | 10 | Aug | 12, 24 | Cert, AAS | $2,103 | $2,781 | | | | |
| Grayson County College (Denison) | 25 | Aug | 10 | Cert | $1,548 | $3,708 | | | | • |
| Houston Community College | 30 | Aug | 12 | Cert | $1,835 | $3,969 | | | | • |
| Texas State Technical College at Waco | 220 | Open entry | 12 | Cert | $4,184 | $8,747 | | | | |

*Data are shown only for programs that completed the 2006 AMA Survey of Health Professions Education Programs

‡Key to Offers: 1: Evening or weekend classes; 2: Non-English instruction; 3: Cultural competence instruction

## Dental Assistant

| Programs* | Class Capacity | Begins | Length (months) | Award | Res Tuition | Non-res Tuition | Stipend | Offers:‡ 1 | 2 | 3 |
|---|---|---|---|---|---|---|---|---|---|---|
| **Utah** | | | | | | | | | | |
| Ameritech College (Provo) | 40 | Every 5 wks | 9 | Dipl | $8,477 | $8,477 | | | | |
| **Virginia** | | | | | | | | | | |
| J Sargeant Reynolds Community College (Richmond) | 20 | Aug Jan | 12 | Cert | $3,544 | $10,824 | | | | • |
| **Washington** | | | | | | | | | | |
| Bates Technical College (Tacoma) | 15 | Sep Jan Apr | 11 | Cert, AS | $4,230 | $4,230 | | | | • |
| Bellingham Technical College | 22 | Jan Mar | 11 | Cert | $2,780 | $10,480 | | | | • |
| Clover Park Technical College (Lakewood) | 75 | Sep Jan Mar Jun | 9 | Cert, Tech | $3,618 | $0 | | | | • |
| Lake Washington Technical College (Kirkland) | 30 | Mar Sep | 11, 17 | Cert, AAS | $3,601 | $3,601 | | | | • |
| Renton Technical College | 44 | Sep Jan | 10 | Cert, AAS, AAST | $3,291 | $3,291 | | | | • |
| Seattle Vocational Institute | 15 | Open entry | 12 | Cert | $3,981 | $0 | | | | |
| South Puget Sound Community College (Olympia) | 32 | Sep | 11 | Cert, ATA | $3,478 | $3,991 | | | | • |
| Spokane Community College | 42 | Sep | 9, 18 | Cert, AAS | $2,666 | $4,420 | | | | • |
| **Wisconsin** | | | | | | | | | | |
| Fox Valley Technical College (Appleton) | 25 | Jan Aug | 9 | Dipl | $3,009 | $13,900 | | | | |
| Gateway Technical College (Kenosha) | 18 | Aug | 10 | Dipl | $3,022 | $13,668 | | | | |
| Northeast Wisconsin Technical College (Green Bay) | 16 | Jun Aug | 10 | Dipl | $3,063 | $14,067 | | | • | • |
| Western Technical College (La Crosse) | 24 | Aug | 10 | Dipl | $2,030 | $12,151 | | | | • |

*Data are shown only for programs that completed the 2006 AMA Survey of Health Professions Education Programs
‡Key to Offers: 1: Evening or weekend classes; 2: Non-English instruction; 3: Cultural competence instruction

## Dental Hygienist*

# Alabama

**Wallace State Community College**
Dental Hygiene Prgm
PO Box 2000
Hanceville, AL 35077-2000
www.wallacestate.edu
*Prgm Dir:* Barbara Ebert, BS
*Tel:* 256 352-8380    *Fax:* 256 352-8382
*E-mail:* barbara.ebert@wallacestate.edu

# Alaska

**University of Alaska Anchorage**
Dental Hygiene Prgm
3211 Providence Dr, AHS 124
Anchorage, AK 99508-8371
www.uaa.alaska.edu/dental
*Prgm Dir:* Sandra Pence, BS RDH
*Tel:* 907 786-6925    *Fax:* 907 786-6938
*E-mail:* pence@uaa.alaska.edu

# Arizona

**Mohave Community College**
Dental Hygiene Prgm
3400 Highway 95
Bullhead City, AZ 86442
www.mohave.edu
*Prgm Dir:* Tracy Gift, RDH MS
*Tel:* 928 704-7793    *Fax:* 928 704-7790
*E-mail:* tgift@mohave.edu

**Northern Arizona University**
Dental Hygiene Prgm
Box 15065
Flagstaff, AZ 86011-5065
www.nau.edu/hp/dept/dh
*Prgm Dir:* Denise Helm, MA
*Tel:* 928 523-7425    *Fax:* 928 523-6195
*E-mail:* denise.helm@nau.edu

**Mesa Community College**
Dental Hygiene Prgm
7110 E McKellips Rd
Mesa, AZ 85207
www.maricopa.edu
*Prgm Dir:* Phebe Blitz, RDH CDA MS
*Tel:* 480 654-7772    *Fax:* 480 654-7372
*E-mail:* phebeblitz@mail.mc.maricopa.edu

**Phoenix College**
Dental Hygiene Prgm
1202 W Thomas Rd
Phoenix, AZ 85013
www.maricopa.edu
*Prgm Dir:* Linda Garcia
*Tel:* 602 285-7325    *Fax:* 602 285-7330
*E-mail:* linda.garcia@pcmail.maricopa.edu

**Rio Salado College**
Dental Hygiene Prgm
2323 W 14th St
Temple, AZ 85281-6950
www.riosalado.maricopa.edu
*Prgm Dir:* Mary (Liz) Elizabeth Kaz, RDH MS
*Tel:* 480 517-8020    *Fax:* 480 517-8029
*E-mail:* liz.kaz@riomail.maricopa.edu

**Pima Community College**
Dental Hygiene Prgm
2202 W Anklam Rd
Tucson, AZ 85709
*Prgm Dir:* Joyce Flieger
*Tel:* 520 206-6916    *Fax:* 520 206-3027
*E-mail:* joyce.flieger@pima.edu

# Arkansas

**University of Arkansas - Fort Smith**
Dental Hygiene Prgm
5210 Grand Ave, PO Box 3649
Ft Smith, AR 72913-3649
*Prgm Dir:* Carol Amerine, RDH BS
*Tel:* 479 788-7272    *Fax:* 479 788-7273
*E-mail:* camerine@uafortsmith.edu

**University of Arkansas for Medical Sciences**
Dental Hygiene Prgm
4301 W Markham St Slot 609
Little Rock, AR 72205
www.uams.edu
*Prgm Dir:* Susan Long, RDH EdD
*Tel:* 501 686-5735    *Fax:* 501 686-8519
*E-mail:* longsusanl@uams.edu

# California

**Cabrillo College**
Dental Hygiene Prgm
6500 Soquel Dr
Aptos, CA 95003
www.cabrillo.edu
*Prgm Dir:* Bridgete H Clark, RHD DDS
*Tel:* 831 479-6471    *Fax:* 831 477-5687
*E-mail:* brclark@cabrillo.edu

**Southwestern College**
Dental Hygiene Prgm
900 Otay Lakes Rd
Chula Vista, CA 91910
www.swccd.edu
*Prgm Dir:* Teresa Poulos, MEd
*Tel:* 619 216-6665, Ext 4860    *Fax:* 619 216-6678
*E-mail:* tpoulos@swccd.edu

**West Los Angeles College**
Dental Hygiene Prgm
9000 Overland Ave
Culver City, CA 90230
www.wlac.edu
*Prgm Dir:* Ulla E Lemborn, MS
*Tel:* 310 287-4242    *Fax:* 310 287-4461
*E-mail:* lemboru@wlac.edu

**Cypress College**
Dental Hygiene Prgm
9200 Valley View St
Cypress, CA 90630
www.cypresscollege.edu
*Prgm Dir:* Ina Rydalch, RDH MA
*Tel:* 714 484-7299    *Fax:* 562 421-0988
*E-mail:* irydalch@cypresscollege.edu

*Additional information about programs that
returned the AMA's survey is available in a table
beginning on page 132.

**Fresno City College**
Dental Hygiene Prgm
1101 E University Ave
Fresno, CA 93741
www.fresnocitycollege.com
*Prgm Dir:* Monta Denver, RDH MS Jean Kulbeth, RDH
    MS
*Tel:* 559 244-2602    *Fax:* 559 244-2614
*E-mail:* monta.denver@fresnocitycollege.edu

**Chabot College**
Dental Hygiene Prgm
25555 Hesperian Blvd
Hayward, CA 94545
http://chabotweb.clpccd.cc.ca.us
*Prgm Dir:* JoAnn Galliano, MEd RDH
*Tel:* 510 723-6866    *Fax:* 510 723-7089
*E-mail:* jgalliano@chabotcollege.edu

**Loma Linda University**
Dental Hygiene Prgm
School of Dentistry
Loma Linda, CA 92350
www.llu.edu
*Prgm Dir:* Joni Stephens, EdS RDH
*Tel:* 909 558-4631, Ext 48234    *Fax:* 909 558-0313
*E-mail:* jstephens@llu.edu

**Foothill Community College**
Dental Hygiene Prgm
12345 El Monte Rd
Los Altos Hills, CA 94022
*Prgm Dir:* Phyllis Spragge, MA
*Tel:* 650 949-7467    *Fax:* 650 947-9788
*E-mail:* spraggephyllis@fhda.edu

**University of Southern California**
Dental Hygiene Prgm
University Park MC0641
Los Angeles, CA 90089-6041
*Prgm Dir:* Diane Melrose
*Tel:* 213 740-1089    *Fax:* 213 740-1094
*E-mail:* mmelrose@usc.edu

**Riverside Comm Coll - Moreno Valley Campus**
Dental Hygiene Prgm
16130 Lasselle St
Moreno Valley, CA 92551
*Prgm Dir:* Donna Lesser
*Tel:* 951 571-6425
*E-mail:* donna.lesser@rcc.edu

**Cerritos Community College**
Dental Hygiene Prgm
11110 E Alondra Blvd
Norwalk, CA 90650
www.cerritos.edu
*Prgm Dir:* Patricia J Stewart, RDH PhD
*Tel:* 562 860-2451, Ext 2557    *Fax:* 562 467-5077
*E-mail:* pstewart@cerritos.edu

**Oxnard College**
Dental Hygiene Prgm
4000 S Rose Ave
Oxnard, CA 93033-6699
*Prgm Dir:* Betsy Lindbergh, RDH DDS
*Tel:* 805 986-5823    *Fax:* 805 986-5867
*E-mail:* blindbergh@vcccd.net

**Pasadena City College**
Dental Hygiene Prgm
1570 E Colorado Blvd
Pasadena, CA 91106
*Prgm Dir:* Jeanne Porush, RDH
*Tel:* 626 585-7537    *Fax:* 626 585-7966
*E-mail:* jkporush@pasadena.edu

**Diablo Valley College**
Dental Hygiene Prgm
321 Golf Club Rd
Pleasant Hill, CA 94523
www.dvc.edu
*Prgm Dir:* Gay Teel, RDH MA
*Tel:* 925 685-1230, Ext 2345    *Fax:* 925 689-6529
*E-mail:* gteel@dvc.edu

**Shasta College**
Dental Hygiene Prgm
11555 Old Oregon Trail
PO Box 496006
Redding, CA 96049-6006
*Prgm Dir:* Charles Cort, RDH MA
*Tel:* 530 245-7334, Ext 205    *Fax:* 530 245-7333
*E-mail:* ccort@shastacollege.edu

**Sacramento City College**
Dental Hygiene Prgm
3835 Freeport Blvd
Sacramento, CA 95822
www.scc.losrios.edu
*Prgm Dir:* Michael Dunne
*Tel:* 916 558-2650    *Fax:* 916 558-2067
*E-mail:* dunnem@scc.losrios.edu

**Western Career College - Sacramento**
Dental Hygiene Prgm
8909 Folsom Blvd
Sacramento, CA 95826
www.westerncollege.edu
*Prgm Dir:* Dorothy J Rowe, PhD
*Tel:* 916 361-5163    *Fax:* 916 361-6666
*E-mail:* djrowe@westerncollege.com

**Santa Rosa Junior College**
Dental Hygiene Prgm
1501 Mendocino Ave
Santa Rosa, CA 95401-4395
www.santarosa.edu
*Prgm Dir:* Doni Bird, MA
*Tel:* 707 522-2828    *Fax:* 707 527-4426
*E-mail:* dbird@santarosa.edu

**Taft College**
Dental Hygiene Prgm
29 Emmons Park Dr, Box 1437
Taft, CA 93268
www.taft.cc.ca.us
*Prgm Dir:* Stacy Eastman, RDH DDS
*Tel:* 661 763-7706    *Fax:* 661 763-7808
*E-mail:* seastman@taft.org

**San Joaquin Valley College - Visalia**
Dental Hygiene Prgm
8400 W Mineral King
Visalia, CA 93291
www.sjvc.edu
*Prgm Dir:* Cindy Callaghan, RDH
*Tel:* 559 622-1947    *Fax:* 559 651-3645
*E-mail:* cindyc@sjvc.edu

# Colorado

**Community College of Denver**
Dental Hygiene Prgm
Bldg 753 1062 Akron Way
Denver, CO 80230
www.ccd.edu/dental
*Prgm Dir:* Stephanie Harrison, RDH MA
*Tel:* 303 365-8334    *Fax:* 303 365-8330
*E-mail:* stephanie.harrison@ccd.edu

**U of Colorado at Denver and Health Scis Ctr**
Dental Hygiene Prgm
School of Dentistry
4200 E Ninth Ave, Campus Box C284
Denver, CO 80262
*Prgm Dir:* Terri Tilliss, RDH MS MA
*Tel:* 303 315-8017    *Fax:* 303 315-0328
*E-mail:* Terri.Tilliss@UCHSC.edu
*Med Dir:* Donna Stach, RDH MEd

**Pueblo Community College**
Dental Hygiene Prgm
900 W Orman Ave
Pueblo, CO 81004
www.pueblocc.edu
*Prgm Dir:* Sue Kochevar, RDH MA
*Tel:* 719 549-3286    *Fax:* 719 549-3136
*E-mail:* Sue.Kochevar@pueblocc.edu

**Colorado Northwestern Community College**
Dental Hygiene Prgm
500 Kennedy Dr
Rangley, CO 81648
www.cncc.edu
*Prgm Dir:* Mark Patterson, RDH BS
*Tel:* 800 562-1105, Ext 247    *Fax:* 970 675-3330
*E-mail:* mark.patterson@cncc.edu

# Connecticut

**U of Bridgeport/Fones Sch of Dental Hygiene**
Dental Hygiene Prgm
30 Hazel St
Bridgeport, CT 06601
*Prgm Dir:* Meg Zayan
*Tel:* 203 576-4138    *Fax:* 203 576-4220
*E-mail:* mzayan@bridgeport.edu

**Tunxis Community College**
Dental Hygiene Prgm
271 Scott Swamp Rd
Farmington, CT 06032-3187
www.tunxis.edu
*Prgm Dir:* Mary A Bencivengo, RDH MS
*Tel:* 860 255-3626    *Fax:* 860 255-3649
*E-mail:* mbencivengo@txcc.commnet.edu

**University of New Haven**
Dental Hygiene Prgm
300 Orange Ave
West Haven, CT 06516
*Prgm Dir:* Sandra D'Amato-Palambo, MPS
*Tel:* 203 931-6023    *Fax:* 203 931-6083
*E-mail:* spalumbo@newhaven.edu

# Delaware

**Delaware Tech & Comm Coll - Wilmington**
Dental Hygiene Prgm
333 Shipley St
Wilmington, DE 19801
*Prgm Dir:* Judith A Hall, RDH BS
*Tel:* 302 657-5177    *Fax:* 302 577-6431
*E-mail:* jhall@dtcc.edu

# District of Columbia

**Howard University**
Dental Hygiene Prgm
600 W St NW Rm 401
Washington, DC 20059
www.howard.edu
*Prgm Dir:* Marie Varley Gillis, RDH MS
*Tel:* 202 806-0079    *Fax:* 202 806-0354
*E-mail:* mgillis@howard.edu

# Florida

**South Florida Community College**
Dental Hygiene Prgm
600 W College Dr
Avon Park, FL 33825
www.southflorida.edu
*Prgm Dir:* Rebecca Sroda, MS
*Tel:* 863 784-7021    *Fax:* 863 453-9442
*E-mail:* srodar@southflorida.edu

**Manatee Community College**
Dental Hygiene Prgm
5840 26th St W
Bradenton, FL 34207
www.mccfl.edu
*Prgm Dir:* Anita J Weaver, MS
*Tel:* 941 752-5350    *Fax:* 941 747-6643
*E-mail:* weavera@mccfl.edu

**Brevard Community College**
Dental Hygiene Prgm
1519 Clearlake Rd
Cocoa, FL 32922
*Prgm Dir:* Janice Elkins
*Tel:* 321 632-1111, Ext 7568    *Fax:* 321 433-7599
*E-mail:* elkinsj@brevardcc.edu

**Daytona Beach Community College**
Dental Hygiene Prgm
1155 County Rd 4139
DeLand, FL 32724
www.dbcc.edu
*Prgm Dir:* Pamela S Ridilla
*Tel:* 386 785-2093    *Fax:* 386 785-2090
*E-mail:* ridillp@dbcc.edu

**Broward Community College**
Dental Hygiene Prgm
3501 SW Davie Rd
Ft Lauderdale, FL 33314
www.broward.edu
*Prgm Dir:* Joyce Abraham, CDA RDH BA
*Tel:* 954 201-6904    *Fax:* 954 201-6397
*E-mail:* jabraham@broward.edu

**Edison College**
Dental Hygiene Prgm
8099 College Pkwy SW
Ft Myers, FL 33906-6210
www.edison.edu
*Prgm Dir:* Karen Molumby, CDA RDH MBA
*Tel:* 239 985-8322    *Fax:* 239 985-8352
*E-mail:* kmolumby@edison.edu

**Indian River Community College**
Dental Hygiene Prgm
3209 Virginia Ave
Ft Pierce, FL 34981-5599
*Prgm Dir:* Marta Ferguson, EdD
*Tel:* 772 462-7523    *Fax:* 772 462-4900
*E-mail:* mferguso@ircc.edu

**Santa Fe Community College**
Dental Hygiene Prgm
3000 NW 83rd St, Bldg W81
Gainesville, FL 32606
*Prgm Dir:* Karen Autrey, BHS MEd
*Tel:* 352 395-5705    *Fax:* 352 395-5758
*E-mail:* karen.autrey@sfcc.edu

**Florida Community College - Jacksonville**
Dental Hygiene Prgm
4501 Capper Rd
Jacksonville, FL 32218
www.fccj.org
*Prgm Dir:* Jeffrey Smith, DMD
*Tel:* 904 766-6655    *Fax:* 904 713-4856
*E-mail:* jesmith@fccj.edu

**Palm Beach Community College**
Dental Hygiene Prgm
4200 Congress Ave
Lake Worth, FL 33461
www.pbcc.edu
*Prgm Dir:* Beth Kuzmirek, RDH BS Ed
*Tel:* 561 868-3752    *Fax:* 561 868-3753
*E-mail:* kuzmireb@pbcc.edu

**Miami Dade College**
Dental Hygiene Prgm
Medical Ctr Campus, 950 NW 20th St
Miami, FL 33127
*Prgm Dir:* Susan Kass, EdD
*Tel:* 305 237-4029    *Fax:* 305 237-4278
*E-mail:* skass@mdc.edu

**Pasco-Hernando Community College**
Dental Hygiene Prgm
10230 Ridge Rd
M-144
New Port Richey, FL 34654-5199
*Prgm Dir:* James Hall, DDS
*Tel:* 727 816-3281    *Fax:* 727 816-3478
*E-mail:* hallj@phcc.edu

**Valencia Community College**
Dental Hygiene Prgm
1800 S Kirkman Rd
Orlando, FL 32811
*Prgm Dir:* Pamela Sandy, MA
*Tel:* 407 582-1544    *Fax:* 407 582-1295
*E-mail:* PSandy@valenciacc.edu

**Gulf Coast Community College**
Dental Hygiene Prgm
5230 W US Hwy 98
Panama City, FL 32401-1041
*Prgm Dir:* Mary Benjamin, RDH DDS
*Tel:* 850 769-1551, Ext 5832    *Fax:* 850 747-3246
*E-mail:* mbenjamin@gulfcoast.edu
*Med Dir:* Mary Benjamin, DDS

**Pensacola Junior College**
Dental Hygiene Prgm
5555 Hwy 98 W
Pensacola, FL 32507
*Prgm Dir:* Linda Lambert, RDH MS
*Tel:* 850 484-2244    *Fax:* 850 484-2390
*E-mail:* llambert@pjc.edu

**St Petersburg College**
Dental Hygiene Prgm
PO Box 13489
St Petersburg, FL 33781
*Prgm Dir:* Tami Grzesikowski, RDH
*Tel:* 727 341-3671    *Fax:* 727 341-3744
*E-mail:* grzesikowskit@spjc.edu

**Tallahassee Community College**
Dental Hygiene Prgm
444 Appleyard Dr
Tallahassee, FL 32304-2895
*Prgm Dir:* Cynthia R Biron, MA
*Tel:* 850 201-8577    *Fax:* 850 201-8575
*E-mail:* bironc@tcc.fl.edu

**Hillsborough Community College**
Dental Hygiene Prgm
4001 Tampa Bay Blvd
Tampa, FL 33614-2754
*Prgm Dir:* Donna Solovan-Gleason, RDH PhD
*Tel:* 813 253-7426
*E-mail:* dsolovangleason@hccfl.edu

# Georgia

**Darton College**
Dental Hygiene Prgm
2400 Gillionville Rd
Albany, GA 31707
www.darton.edu
*Prgm Dir:* Stacey Marshall, RDH MSEd DMD
*Tel:* 229 317-6840    *Fax:* 229 317-6620
*E-mail:* stacey.marshall@darton.edu

**Athens Technical College**
Dental Hygiene Prgm
800 US Hwy 29 N
Athens, GA 30601-1500
www.athenstech.edu
*Prgm Dir:* Tina Grile, MHS RDH CDA
*Tel:* 706 355-5142    *Fax:* 706 425-3104
*E-mail:* tgrile@athenstech.edu

**Medical College of Georgia**
Dental Hygiene Prgm
1120 15th St, AD3103
Augusta, GA 30912-0200
www.mcg.edu
*Prgm Dir:* Marie Collins, RDH MS
*Tel:* 706 721-2938    *Fax:* 706 721-8857
*E-mail:* mcollins@mcg.edu
*Med Dir:* Tina Moses, DMD

**Columbus Technical College**
Dental Hygiene Prgm
928 Manchester Expwy
Columbus, GA 31904-6572
www.columbustech.edu
*Prgm Dir:* Jan Jones, CDA RDH MPA
*Tel:* 706 649-1977    *Fax:* 706 649-1599
*E-mail:* jjones@columbustech.edu
*Med Dir:* Linn Storey, BSN, MPA

**West Central Technical College**
Dental Hygiene Prgm
4600 Timber Ridge Dr
Douglasville, GA 30135
www.westcentral.org
*Prgm Dir:* Carol C Johnson, RDH MSHA
*Tel:* 770 947-7225    *Fax:* 770 947-7377
*E-mail:* cjohnson@westcentraltech.edu
*Med Dir:* Ken Smith, Dean of Allied Health

**Georgia Perimeter College**
Dental Hygiene Prgm
2101 Womack Rd
Dunwoody, GA 30338-4497
*Prgm Dir:* Linda D Boyd, RDH RD EdD
*Tel:* 770 551-3096    *Fax:* 770 604-3797
*E-mail:* lboy2@gpc.edu

**Central Georgia Technical College**
Dental Hygiene Prgm
300 Macon Tech Dr
Macon, GA 31206
www.centralgatech.edu
*Prgm Dir:* Marsha R McCrimmon, MPH RDH
*Tel:* 478 757-2487    *Fax:* 478 757-4395
*E-mail:* mccrimmon@centralgatech.edu

**Clayton State University**
Dental Hygiene Prgm
2000 Clayton State Blvd W
Morrow, GA 30260
www.clayton.edu
*Prgm Dir:* Susan Duley, RDH EdD LPC CEDS
*Tel:* 678 466-4911    *Fax:* 678 466-4911
*E-mail:* susanduley@clayton.edu

**Lanier Technical College**
Dental Hygiene Prgm
2990 Landrum Education Dr
Oakwood, GA 30566
*Prgm Dir:* David Byers, DMD
*Tel:* 770 535-6905    *Fax:* 770 531-6306
*E-mail:* dbyers@laniertech.edu

**Georgia Highlands College**
Dental Hygiene Prgm
415 East Third Ave
Rome, GA 30161
www.highlands.edu/dental
*Prgm Dir:* Donna Miller, RDH MS
*Tel:* 706 295-6760    *Fax:* 706 802-5140
*E-mail:* dmiller@highlands.edu

**Armstrong Atlantic State University**
Dental Hygiene Prgm
11935 Abercorn St
Savannah, GA 31419-1997
*Prgm Dir:* Suzanne Edenfield, EdD RDH
*Tel:* 912 921-7440  *Fax:* 912 921-7466
*E-mail:* edenfisu@mail.armstrong.edu

**Valdosta State University**
*Cosponsor:* Valdosta Technical College
Dental Hygiene Prgm
4089 Val Tech Rd, PO Box 928
Valdosta, GA 31603
*Prgm Dir:* Renee' Graham, BSEd RDH
*Tel:* 912 245-3716  *Fax:* 912 259-5567
*E-mail:* rgraham@valdostatech.org
*Med Dir:* Donnie McGajee, EdD

**Valdosta Technical College**
Dental Hygiene Prgm
4089 Val Tech Rd
Valdosta, GA 31603-0928
*Prgm Dir:* Rene Graham, BS
*Tel:* 229 259-5534  *Fax:* 229 259-5567
*E-mail:* rgraham@valdostatech.edu

**Middle Georgia Technical College**
Dental Hygiene Prgm
80 Cohen Walker Dr
Warner Robins, GA 31088-2729
www.middlegatech.edu
*Prgm Dir:* Barbara Jansen, DDS
*Tel:* 478 988-7054  *Fax:* 478 988-6875
*E-mail:* bjansen@middlegatech.edu

# Hawaii

**University of Hawaii at Manoa**
Dental Hygiene Prgm
2445 Campus Rd, Rm 200-B
Honolulu, HI 96822
*Prgm Dir:* Carolyn Kuba, MEd
*Tel:* 808 956-8821  *Fax:* 808 956-5707
*E-mail:* ckuba@hawaii.edu

# Idaho

**Apollo College - Boise**
Dental Hygiene Prgm
1200 N Liberty St
Boise, ID 83704
www.aiht.com
*Prgm Dir:* David Reff, DDS
*Tel:* 208 377-8080, Ext 30  *Fax:* 208 947-6822
*E-mail:* Dreff@apolloboise.com

**Idaho State University**
Dental Hygiene Prgm
Box 8048
Pocatello, ID 83209-8380
www.isu.edu/departments/dentalhy
*Prgm Dir:* Kathleen Hodges, RDH MS
*Tel:* 208 282-2744  *Fax:* 208 282-4071
*E-mail:* hodgkat1@isu.edu

# Illinois

**Southern Illinois Univ at Carbondale**
Dental Hygiene Prgm
School of Allied Health
Mail Code 6615
Carbondale, IL 62901
www.siu.edu
*Prgm Dir:* Dwayne Summers, DMD
*Tel:* 618 453-7213  *Fax:* 618 453-7020
*E-mail:* dsummers@siu.edu

**John A Logan College**
Dental Hygiene Prgm
700 Logan College Rd
Caterville, IL 62918
www.jalc.edu
*Prgm Dir:* Pamela Karns, RDH BSDH
*Tel:* 618 985-2828, Ext 8639  *Fax:* 618 985-4654
*E-mail:* pamkarns@jalc.edu

**Parkland College**
Dental Hygiene Prgm
2400 W Bradley Ave
Champaign, IL 61821
*Prgm Dir:* Mary L Emmons, MEd
*Tel:* 217 373-3717  *Fax:* 217 373-3830
*E-mail:* MEmmons@parkland.edu

**Kennedy-King College**
Dental Hygiene Prgm
UIC College of Dentistry, Rm 201 S, Mail Code 621
801 S Paulina
Chicago, IL 60612
*Prgm Dir:* Shirley M Beaver, RDH PhD
*Tel:* 312 996-8071  *Fax:* 312 996-1023
*E-mail:* sbeaver@uic.edu

**Prairie State College**
Dental Hygiene Prgm
202 S Halsted St
Chicago Heights, IL 60411
*Prgm Dir:* Barbara Gorbitz, RDH BS
*Tel:* 708 709-3714, Ext 3714  *Fax:* 708 709-3777
*E-mail:* bgorbitz@prairiestate.edu

**Carl Sandburg College**
Dental Hygiene Prgm
2400 Tom L Wilson Blvd
209 E Main St
Galesburg, IL 61401
*Prgm Dir:* Lauri Wiechmann, RDH MPA
*Tel:* 309 344-2595, Ext 222  *Fax:* 309 344-2611
*E-mail:* lwiechmann@sandburg.edu

**College of DuPage**
Dental Hygiene Prgm
425 Fawell Blvd
Glen Ellyn, IL 60137
*Prgm Dir:* Patricia S Wellner
*Tel:* 630 942-4237  *Fax:* 630 858-5409
*E-mail:* wellner@cdnet.cod.edu

**Lewis & Clark Community College**
Dental Hygiene Prgm
5800 Godfrey Rd
Godfrey, IL 62035
www.lc.edu
*Prgm Dir:* Michelle Singley, RDH EdM
*Tel:* 618 468-4413  *Fax:* 618 468-2394
*E-mail:* msingley@lc.edu

**College of Lake County**
Dental Hygiene Prgm
19351 W Washington St
Grayslake, IL 60030-1198
*Prgm Dir:* Sue Nierstheimer, RDH
*Tel:* 847 543-2638  *Fax:* 847 223-1357
*E-mail:* snierstheimer@clcillinois.edu

**Lake Land College**
Dental Hygiene Prgm
5001 Lake Land Blvd
Mattoon, IL 61938-9366
www.lakeland.cc.il.us
*Prgm Dir:* Jane Slaughter, RDH BS
*Tel:* 217 234-5202  *Fax:* 217 234-5248
*E-mail:* jslaught@lakeland.cc.il.us

**Harper College**
Dental Hygiene Prgm
1200 W Algonquin Rd
Palatine, IL 60067
*Prgm Dir:* Kathleen Hock, MAD
*Tel:* 847 925-6543  *Fax:* 847 925-6077
*E-mail:* khock@harper.cc.il.us

**Illinois Central College**
Dental Hygiene Prgm
201 SW Adams St
Peoria, IL 61635-0001
*Prgm Dir:* Debra Spears, BS
*Tel:* 309 999-4662  *Fax:* 309 673-9626
*E-mail:* dspears@icc.edu

**Rock Valley College**
Dental Hygiene Prgm
3301 N Mulford Rd
Rockford, IL 61114-5699
*Prgm Dir:* Marie Navickis, BSDH
*Tel:* 815 921-3206  *Fax:* 815 921-3249
*E-mail:* m.navickis@rvc.cc.il.us

# Indiana

**University of Southern Indiana**
Dental Hygiene Prgm
8600 University Blvd
Evansville, IN 47712
*Prgm Dir:* D Carl, RDH MEd
*Tel:* 812 464-1707  *Fax:* 812 465-7092
*E-mail:* dcarl@usi.edu

**Indiana Univ - Purdue Univ Ft Wayne**
Dental Hygiene Prgm
2101 Coliseum Blvd E
Ft Wayne, IN 46805
www.ipfw.edu/dhy
*Prgm Dir:* Elaine S Foley, RDH MSEd
*Tel:* 219 481-6837  *Fax:* 219 481-4162
*E-mail:* foley@ipfw.edu

**Indiana University Northwest**
Dental Hygiene Prgm
3223 Broadway St
Gary, IN 46409
www.iun.edu
*Prgm Dir:* Juanita Robinson, CDA EFDA LDH MSEd
*Tel:* 219 980-6734  *Fax:* 219 981-4249
*E-mail:* jurobin@iun.edu

**Indiana University**
Dental Hygiene Prgm
1121 W Michigan St
Indianapolis, IN 46202-5186
*Prgm Dir:* Nancy Young, MEd
*Tel:* 317 274-7801  *Fax:* 317 274-1363
*E-mail:* nayoung@iupui.edu

**Indiana University South Bend**
Dental Hygiene Prgm
1700 Mishawaka Ave
South Bend, IN 46634
*Prgm Dir:* Nanci Genich-Yokom, CDA RDH MBA
*Tel:* 574 237-4154  *Fax:* 574 237-4854
*E-mail:* nyokom@iusb.edu

# Iowa

**Des Moines Area Community College**
Dental Hygiene Prgm
2006 Ankeny Blvd
Ankeny, IA 50023-3993
*Prgm Dir:* D Penney, DH MS
*Tel:* 515 964-6582  *Fax:* 515 964-6582
*E-mail:* dapenney@dmacc.edu

**Kirkwood Community College**
Dental Hygiene Prgm
6301 Kirkwood Blvd SW, PO Box 2068
Cedar Rapids, IA 52406
www.kirkwood.edu
*Prgm Dir:* Shaunda Clark, CDA RDH BS MEd
*Tel:* 319 398-5514  *Fax:* 319 398-1293
*E-mail:* shaunda.clark@kirkwood.edu

**Iowa Western Community College**
Dental Hygiene Prgm
2700 College Rd, PO Box 4C
Council Bluffs, IA 51502-3004
www.iwcc.edu
*Prgm Dir:* Janet Hillis, RDH MA
*Tel:* 712 325-3738, Ext 3738   *Fax:* 712 325-3736
*E-mail:* jhillis@iwcc.edu

**Hawkeye Community College**
Dental Hygiene Prgm
1501 E Orange Rd
PO Box 8015
Waterloo, IA 50704
www.hawkeyecollege.edu
*Prgm Dir:* Sarah A Turner, RDH MAE
*Tel:* 319 296-4432   *Fax:* 319 296-4450
*E-mail:* sturner@hawkeyecollege.edu

# Kansas

**Johnson County Community College**
Dental Hygiene Prgm
12345 College Blvd
Overland Park, KS 66210-1299
www.jccc.edu
*Prgm Dir:* Margaret LoGiudice, RDH MS
*Tel:* 913 469-2582   *Fax:* 913 469-2378
*E-mail:* mlog@jccc.edu

**Wichita State University**
Dental Hygiene Prgm
1845 N Fairmount
Wichita, KS 67260-0144
*Prgm Dir:* Denise Maseman, RDH MS
*Tel:* 316 978-3614   *Fax:* 316 978-5459
*E-mail:* denise.maseman@wichita.edu

# Kentucky

**Western Kentucky University**
Dental Hygiene Prgm
Academic Complex, Rm 201
Bowling Green, KY 42101
*Prgm Dir:* Lynn Austin, RDH MPH
*Tel:* 270 745-2427   *Fax:* 270 745-6869
*E-mail:* lynn.austin@wku.edu

**Henderson Community College**
Dental Hygiene Prgm
2660 S Green St
Henderson, KY 42420
*Prgm Dir:* Kim Conley, RDH MS
*Tel:* 270 831-9707   *Fax:* 270 831-9745
*E-mail:* kim.conley@kctcs.net

**Bluegrass Community and Technical College**
Dental Hygiene Prgm
Room 250 Oswald Bldg
Cooper Dr
Lexington, KY 40506
www.bluegrass.kctcs.edu/LCC/DHY
*Prgm Dir:* C Lawrence Chiswell, MSEd DMD
*Tel:* 859 246-6232   *Fax:* 859 246-4667
*E-mail:* larry.chiswell@kctcs.edu

**University of Louisville**
Dental Hygiene Prgm
School of Dentistry
501 S Preston St
Louisville, KY 40292
*Prgm Dir:* Susan Crim, PhD
*Tel:* 502 852-1229   *Fax:* 502 852-1317
*E-mail:* sjbail02@louisville.edu

**Big Sandy Community & Technical College**
Dental Hygiene Prgm
1 Bert T Combs Dr
Prestonsburg, KY 41653
www.bigsandy.kctcs.edu
*Prgm Dir:* Eric Dixon, DMD
*Tel:* 606 886-3863, Ext 67179   *Fax:* 606 889-0742
*E-mail:* edixon0001@kctcs.edu

# Louisiana

**Louisiana State University**
Dental Hygiene Prgm
8000 GSRI Ave, Building 3110
Baton Rouge, LA 70820
www.lsuhsc.edu
*Prgm Dir:* Caroline Mason, RDH BS MEd
*Tel:* 225 334-1792   *Fax:* 225 334-1794
*E-mail:* cmason@lsuhsc.edu

**University of Louisiana at Monroe**
Dental Hygiene Prgm
700 University Ave
Monroe, LA 71209
*Prgm Dir:* Beverly B Jarrell, RDH MEd
*Tel:* 318 342-1619   *Fax:* 318 342-1687
*E-mail:* jarrell@ulm.edu

**Southern Univ at Shreveport**
Dental Hygiene Prgm
3050 Martin Luther King Jr Dr
Shreveport, LA 71107
www.susla.edu
*Prgm Dir:* Kheysia Washington, BS
*Tel:* 318 674-3417   *Fax:* 318 676-5495
*E-mail:* kwashington@susla.edu
*Med Dir:* Guy Gerald Gipson, DMD

# Maine

**University College of Bangor**
Dental Hygiene Prgm
29 Texas Ave
Bangor, ME 04401-4324
www.uma.maine.edu
*Prgm Dir:* Ann Curtis, RDH RD MS CAS
*Tel:* 207 262-7870   *Fax:* 207 262-7871
*E-mail:* curtisa@maine.edu

**University of New England**
Dental Hygiene Prgm
Westbrook College Campus
716 Stevens Ave
Portland, ME 04103-2670
www.une.edu
*Prgm Dir:* Bernice Mills, RDH MS
*Tel:* 207 221-4314   *Fax:* 207 221-4889
*E-mail:* bamills@une.edu

# Maryland

**Baltimore City Community College**
Dental Hygiene Prgm
2901 Liberty Heights Ave
Baltimore, MD 21215
*Prgm Dir:* Annette Russell, RDH BS MS
*Tel:* 410 462-7718   *Fax:* 410 462-7682
*E-mail:* arussell@bccc.edu

**University of Maryland**
Dental Hygiene Prgm
Division of Dental Hygiene
666 W Baltimore St, Rm 3-G-31
Baltimore, MD 21201
www.umaryland.edu
*Prgm Dir:* Jacquelyn Fried, RDH MS
*Tel:* 410 706-7773   *Fax:* 410 706-0349
*E-mail:* jfried@umaryland.edu

**Allegany College of Maryland**
Dental Hygiene Prgm
12401 Willowbrook Rd SE
Cumberland, MD 21502-2596
*Prgm Dir:* J Steven Skupas, DDA
*Tel:* 301 784-5580   *Fax:* 301 784-5015
*E-mail:* sskupas@ac.cc.md.us

# Massachusetts

**Mass College of Pharmacy & Health Sciences**
Dental Hygiene Prgm
179 Longwood Ave
Boston, MA 02115-5896
*Prgm Dir:* Gail Barnes
*Tel:* 617 892-8229
*E-mail:* gail.barnes@mcphs.edu

**Bristol Community College**
Dental Hygiene Prgm
777 Elsbree St
Fall River, MA 02720
www.bristol.mass.edu
*Prgm Dir:* Kristine Bishop Chapman, RDH MA
*Tel:* 508 678-2811, Ext 2143   *Fax:* 508 730-3281
*E-mail:* kchapman@bristol.mass.edu

**Middlesex Community College - Lowell**
Dental Hygiene Prgm
33 Kearney Square
Lowell, MA 01852
*Prgm Dir:* Kathleen Sweeney, RDH BS MS
*Tel:* 978 656-3096   *Fax:* 978 656-3078
*E-mail:* sweeneyk@middlesex.mass.edu

**Mt Ida College**
Dental Hygiene Prgm
777 Dedham St
Newton Centre, MA 02459
*Prgm Dir:* Robin Matloff
*Tel:* 617 928-7346   *Fax:* 617 928-7370
*E-mail:* rmatloff@mountida.edu

**Springfield Technical Community College**
Dental Hygiene Prgm
One Armory Square, Ste 1
PO Box 9000
Springfield, MA 01102-9000
www.stcc.edu
*Prgm Dir:* Carol B Szlachetka, CDA RDH MNS
*Tel:* 413 755-4858   *Fax:* 413 755-4926
*E-mail:* szlachetka@stcc.edu

**Cape Cod Community College**
Dental Hygiene Prgm
2240 Iyannough Rd
West Barnstable, MA 02668-1599
www.capecod.edu
*Prgm Dir:* Elaine Madden, RDH BS MEd
*Tel:* 508 362-2131, Ext 4628   *Fax:* 508 375-4008
*E-mail:* emadden@capecod.edu

**Quinsigamond Community College**
Dental Hygiene Prgm
670 W Boylston St
Worcester, MA 01606
*Prgm Dir:* Jane Gauthier, RDH MEd
*Tel:* 508 854-4231   *Fax:* 508 852-2704
*E-mail:* jgauthier@qcc.mass.edu

# Michigan

**University of Michigan**
Dental Hygiene Prgm
1011 N University
Ann Arbor, MI 48109-1078
www.dent.umich.edu/hygiene
*Prgm Dir:* Wendy Kerschbaum, RDH MA MPH
*Tel:* 734 763-3392   *Fax:* 734 763-5503
*E-mail:* wendyek@umich.edu

**Kellogg Community College**
Dental Hygiene Prgm
450 North Ave
Battle Creek, MI 49017
www.kellogg.edu
*Prgm Dir:* Paula Sullivan, RDH MHP
*Tel:* 269 965-3931, Ext 2303   *Fax:* 269 565-2055
*E-mail:* sullivanp@kellogg.edu

**Ferris State University**
Dental Hygiene Prgm
200 Ferris Dr
Big Rapids, MI 49307-2740
www.ferris.edu
*Prgm Dir:* Kimberly Beistle, BA RDH MSA CDA PhD(c)
*Tel:* 231 591-2224   *Fax:* 231 591-2325
*E-mail:* beistlk@ferris.edu

**University of Detroit Mercy**
Dental Hygiene Prgm
8200 W Outer Dr
PO Box 19900
Detroit, MI 48219-0900
*Prgm Dir:* Kathi R Shepherd, RDH MS
*Tel:* 313 494-6693   *Fax:* 313 494-6697
*E-mail:* shephekr@udmercy.edu

**Wayne County Community College District**
Dental Hygiene Prgm
8551 Greenfield, Rm 310
Detroit, MI 48228
*Prgm Dir:* Jo Ann Allen Nyquist, BSDH MA EDS
*Tel:* 919 943-4055   *Fax:* 919 943-4025
*E-mail:* jnyquis1@wcccd.edu

**Mott Community College**
Dental Hygiene Prgm
1401 E Court St
Flint, MI 48503
www.mcc.edu
*Prgm Dir:* Catherine Smith, RDH MA
*Tel:* 810 762-0328   *Fax:* 810 232-8874
*E-mail:* csmith@email.mcc.edu

**Grand Rapids Community College**
Dental Hygiene Prgm
143 Bostwick St NE
Grand Rapids, MI 49503
*Prgm Dir:* Bunny Bookwalter, RDA RDH MS
*Tel:* 616 234-4240   *Fax:* 616 234-4317
*E-mail:* bbookwal@grcc.edu
*Med Dir:* Pamela Naujalis

**Kalamazoo Valley Community College**
Dental Hygiene Prgm
6767 West O Ave, Box 4070
Kalamazoo, MI 49003-4070
*Prgm Dir:* Wanda Scott, MA
*Tel:* 269 488-4267   *Fax:* 269 488-4720
*E-mail:* wscott@kvcc.edu

**Lansing Community College**
Dental Hygiene Prgm
PO Box 40010
Lansing, MI 48901-7210
www.lcc.edu/humanhealth/dental
*Prgm Dir:* Sherry Kohlmann, MA RDH RDA
*Tel:* 517 483-1457   *Fax:* 517 483-9925
*E-mail:* skohlman@lcc.edu

**Baker College of Port Huron**
Dental Hygiene Prgm
3403 Lapeer Rd
Port Huron, MI 48060
www.baker.edu
*Prgm Dir:* Sheree Duff, RDH MS
*Tel:* 810 985-7000, Ext 105   *Fax:* 810 985-7066
*E-mail:* sheree.duff@baker.edu

**Delta College**
Dental Hygiene Prgm
1961 Delta Rd
University Center, MI 48710
www.delta.edu
*Prgm Dir:* Mary Jo Miller, RDH MA
*Tel:* 989 686-9383   *Fax:* 989 667-2210
*E-mail:* maryjomiller@delta.edu
*Med Dir:* Karen Wilson

**Oakland Community College**
Dental Hygiene Prgm
7350 Cooley Lake Rd
Waterford, MI 48327
*Prgm Dir:* Stephanie Markwardt
*Tel:* 248 942-3267   *Fax:* 248 942-3113
*E-mail:* jmmckay@oaklandcc.edu

# Minnesota

**Normandale Community College**
Dental Hygiene Prgm
9700 France Ave S
Bloomington, MN 55431
*Prgm Dir:* Colleen Brickle, RDH EdD
*Tel:* 952 487-8366   *Fax:* 952 487-7022
*E-mail:* colleen.brickle@normandale.edu

**Lake Superior College**
Dental Hygiene Prgm
2101 Trinity Rd
Duluth, MN 55811
www.lsc.edu
*Prgm Dir:* Kathy Griffin, RDH BS MEd RF
*Tel:* 218 733-5938   *Fax:* 218 723-4921
*E-mail:* k.griffin@lsc.edu

**Argosy University/Twin Cities**
Dental Hygiene Prgm
1515 Cental Parkway
Eagan, MN 55121
www.argosyu.edu
*Prgm Dir:* Dinah Bunn, RDH EdD
*Tel:* 651 846-3416   *Fax:* 651 994-0885
*E-mail:* dbunn@argosyu.edu

**Minnesota State University - Mankato**
Dental Hygiene Prgm
3 Morris Hall
Mankato, MN 56001
*Prgm Dir:* Lynnette M Engeswick, RDH MS
*Tel:* 507 389-5848   *Fax:* 507 389-5850
*E-mail:* lynnette.engeswick@mnsu.edu

**Herzing College - Lakeland Academy Division**
Dental Hygiene Prgm
5700 W Broadway Ave
Minneapolis, MN 55428
www.herzing.edu
*Prgm Dir:* Sandra Johnston, RDH BS
*Tel:* 763 535-3000   *Fax:* 763 535-9205
*E-mail:* sjohnston@mpls.herzing.edu

**University of Minnesota - Minneapolis**
Dental Hygiene Prgm
9-372 Moos Tower
515 Delaware St SE
Minneapolis, MN 55455
*Prgm Dir:* Christine Blue, MS
*Tel:* 612 625-5954   *Fax:* 612 625-1605
*E-mail:* bluex005@umn.edu

**Minnesota State Comm & Tech Coll - Moorhead**
Dental Hygiene Prgm
Allied Dental Careers Dept
1900 28th Ave S
Moorhead, MN 56560-4899
*Prgm Dir:* Thomas Boe, DDS MBA
*Tel:* 218 299-6819   *Fax:* 218 299-6532
*E-mail:* thomas.boe@minnesota.edu

**Rochester Community & Technical College**
Dental Hygiene Prgm
851 30th Ave SE
Rochester, MN 55904
*Prgm Dir:* Anne M High, RDH MS
*Tel:* 507 280-3114   *Fax:* 507 280-3180
*E-mail:* anne.high@roch.edu

**St Cloud Technical College**
Dental Hygiene Prgm
1540 Northway Dr
St Cloud, MN 56303-1240
www.sctc.edu
*Prgm Dir:* Barbara L Henkemeyer, RDH MS
*Tel:* 320 308-5906   *Fax:* 320 308-5971
*E-mail:* bhenkemeyer@sctc.edu

**Century College**
Dental Hygiene Prgm
3300 Century Ave N
White Bear Lake, MN 55110
www.century.edu
*Prgm Dir:* Linda Jorgenson, RDH BS
*Tel:* 651 779-3983   *Fax:* 651 779-5779
*E-mail:* l.jorgenson@century.edu

# Mississippi

**Northeast Mississippi Community College**
Dental Hygiene Prgm
Cunningham Blvd
Booneville, MS 38829
*Prgm Dir:* Nick Alexander
*Tel:* 662 720-7283   *Fax:* 662 278-1165
*E-mail:* nalexan@necc.cc.ms.us

**Pearl River Community College**
Dental Hygiene Prgm
5448 US Hwy 49 S
Hattiesburg, MS 39401
*Prgm Dir:* Stanley L Hill, DMD
*Tel:* 601 554-5509   *Fax:* 601 554-5548
*E-mail:* slhill@prcc.edu

**University of Mississippi Medical Center**
Dental Hygiene Prgm
2500 N State St, SHRP
Jackson, MS 39216-4505
*Prgm Dir:* Beckie Barry, MEd
*Tel:* 601 984-6310   *Fax:* 601 815-1717
*E-mail:* bbarry@shrp.umsmed.edu

**Meridian Community College**
Dental Hygiene Prgm
910 Hwy 19 N
Meridian, MS 39307
*Prgm Dir:* William Lindsay, DDS
*Tel:* 601 484-8751, Ext 751   *Fax:* 601 484-8704
*E-mail:* blindsay@mcc.cc.ms.us

**Mississippi Delta Community College**
Dental Hygiene Prgm
PO Box 668
Highway 3 at Cherry St
Moorhead, MS 38761
*Prgm Dir:* Melissa-Marble Warrington, RDH BS
*Tel:* 662 246-6511   *Fax:* 662 246-6299
*E-mail:* mwarrington@msdelta.edu

# Missouri

**Missouri Southern State University**
Dental Hygiene Prgm
3950 E Newman and Duquesne Rds
Joplin, MO 64801-1595
www.mssu.edu
*Prgm Dir:* Sandra Scorse, DDS
*Tel:* 417 625-9709   *Fax:* 417 625-3078
*E-mail:* scorse-s@mssu.edu

**University of Missouri - Kansas City**
Dental Hygiene Prgm
650 E 25th St
Kansas City, MO 64108
www.umkc.edu/dentistry
*Prgm Dir:* Cynthia Amyot, BSDH EdD
*Tel:* 816 235-2050  *Fax:* 816 235-2157
*E-mail:* amyotc@umkc.edu

**State Fair Community College**
Dental Hygiene Prgm
3201 W 16th St
Sedalia, MO 65301
www.sfccmo.edu
*Prgm Dir:* Rene Fiquet-Freeman, RDH BS
*Tel:* 660 530-5800, Ext 234
*E-mail:* rfiquet@sfccmo.edu

**Ozarks Technical Community College**
Dental Hygiene Prgm
1001 E Chestnut Expressway
Springfield, MO 65802
www.otc.edu
*Prgm Dir:* Kelly Tillery, MEd RDH CDA
*Tel:* 417 447-8829  *Fax:* 417 447-8806
*E-mail:* tilleryk@otc.edu

**St Louis Community College - Forest Park**
Dental Hygiene Prgm
5600 Oakland Ave
St Louis, MO 63110-1393
*Prgm Dir:* Pat Heaton, RDH MA
*Tel:* 314 644-9330  *Fax:* 314 951-9490
*E-mail:* pheaton@stlcc.edu

# Montana

**Montana State Univ - Great Falls Coll of Tech**
Dental Hygiene Prgm
2100 16th Ave S
Great Falls, MT 59405
www.msugf.edu
*Prgm Dir:* Kim L Woloszyn, RDH BA
*Tel:* 406 771-4389  *Fax:* 406 771-4317
*E-mail:* kwoloszyn@msugf.edu

# Nebraska

**Central Community College**
Dental Hygiene Prgm
PO Box 1024
Hastings, NE 68902-1024
www.cccneb.edu
*Prgm Dir:* Wanda Cloet, RDH MS
*Tel:* 402 461-2470  *Fax:* 402 460-2128
*E-mail:* wcloet@cccneb.edu

**University of Nebraska - Lincoln**
Dental Hygiene Prgm
40th and Holdrege Sts
Lincoln, NE 68583-0740
*Prgm Dir:* Gwen Hlava, RDH
*Tel:* 402 472-1270  *Fax:* 402 472-5290
*E-mail:* ghlava@unmc.edu

# Nevada

**Community College of Southern Nevada**
Dental Hygiene Prgm
6375 W Charleston Blvd
W1A
Las Vegas, NV 89146
www.ccsn.edu
*Prgm Dir:* Doreen Craig, RDH MA
*Tel:* 702 651-5593  *Fax:* 702 651-7401
*E-mail:* doreen_craig@ccsn.edu

**Truckee Meadows Community College**
Dental Hygiene Prgm
7000 Dandini Blvd, RMDT 417-H
Reno, NV 89512-3999
*Prgm Dir:* Vickie Kimbrough, RDH MBA
*Tel:* 775 674-7554  *Fax:* 775 673-8242
*E-mail:* vkimbrough@tmcc.edu

# New Hampshire

**New Hampshire Technical Institute**
Dental Hygiene Prgm
31 College Dr
Concord, NH 03301-7412
www.nhctc.edu
*Prgm Dir:* Donna Clougherty, RDH MEd
*Tel:* 603 271-7164  *Fax:* 603 271-7182
*E-mail:* dclougherty@nhctc.edu

# New Jersey

**Camden County College**
Dental Hygiene Prgm
PO Box 200
Blackwood, NJ 08012
*Prgm Dir:* Catherine Boos, DMD
*Tel:* 856 227-7200, Ext 4472  *Fax:* 856 374-5061
*E-mail:* cboos@camdencc.edu

**Middlesex County College**
Dental Hygiene Prgm
2600 Woodbridge Ave, PO Box 3050
Edison, NJ 08818-3050
*Prgm Dir:* Hope-Claire Holbeck, RDH MS
*Tel:* 732 906-2580  *Fax:* 732 906-2633
*E-mail:* hope-claire_holbeck@middlesexcc.edu

**Bergen Community College**
Dental Hygiene Prgm
400 Paramus Rd
Paramus, NJ 07652
*Prgm Dir:* Susan Callahan Barnard, DHSc RDH
*Tel:* 201 447-7937  *Fax:* 201 612-3876
*E-mail:* sbarnard@bergen.edu

**Burlington County College**
Dental Hygiene Prgm
601 Pemberton-Browns Mills Rd
Pemberton, NJ 08068
*Prgm Dir:* Linda Hecker
*Tel:* 609 894-9311, Ext 1419
*E-mail:* lhecker@bcc.edu

**Univ of Med & Dent of New Jersey**
Dental Hygiene Prgm
1776 Raritan Rd
Scotch Plains, NJ 07076
www.shrp.umdnj.edu
*Prgm Dir:* Carolyn K Breen, EdD CDA RDA RDH
*Tel:* 908 889-2419  *Fax:* 908 889-2477
*E-mail:* breen@umdnj.edu

# New Mexico

**University of New Mexico**
Dental Hygiene Prgm
Health Science Center
Division of Dental Hygiene
Albuquerque, NM 87131-1391
*Prgm Dir:* Demetra Logothethis, MS
*Tel:* 505 272-4513  *Fax:* 505 272-5584
*E-mail:* dlogothetis@salud.unm.edu

**San Juan College**
Dental Hygiene Prgm
4601 College Blvd
Farmington, NM 87402-4699
*Prgm Dir:* Paula Spaight, PhD
*Tel:* 505 566-3763  *Fax:* 505 566-3790
*E-mail:* spaightp@sanjuancollege.edu

# New York

**Broome Community College**
Dental Hygiene Prgm
PO Box 1017
Binghamton, NY 13902
http://sunybroome.edu
*Prgm Dir:* Maureen Hankin, RDH MPH
*Tel:* 607 778-5393  *Fax:* 607 778-5467
*E-mail:* hankin_m@sunybroome.edu

**Eugenio Maria De Hostos Community College**
Dental Hygiene Prgm
475 Grand Concourse
Bronx, NY 10451
*Prgm Dir:* Mary Errico, DDS
*Tel:* 718 518-4234  *Fax:* 718 518-4194
*E-mail:* merrico@hostos.cuny.edu

**New York City College of Technology**
Dental Hygiene Prgm
Dental Hygiene Department P201
300 Jay St
Brooklyn, NY 11201-2983
*Prgm Dir:* Leonard Fiedman, DMD
*Tel:* 718 260-5074  *Fax:* 718 260-5069
*E-mail:* lfriedman@citytech.cuny.edu

**SUNY at Canton**
Dental Hygiene Prgm
School of Science Health and Professional Studies
34 Cornell Dr
Canton, NY 13617-1096
www.canton.edu/can/can_start.taf?page=study_school_
  healthmed
*Prgm Dir:* Pamela P Quinn, RDH BSE MSEd
*Tel:* 800 388-7123
*E-mail:* quinnp@canton.edu

**Farmingdale State University of New York**
Dental Hygiene Prgm
2350 Broadhollow Rd
Farmingdale, NY 11735
www.farmingdale.edu
*Prgm Dir:* Laura Joseph, RDH MS EdD
*Tel:* 631 420-2060  *Fax:* 631 420-2582
*E-mail:* Josephlm@farmingdale.edu

**Orange County Community College**
Dental Hygiene Prgm
115 South St
Middletown, NY 10940
www.sunyorange.edu
*Prgm Dir:* Roberta Smith, MPS
*Tel:* 845 341-4306  *Fax:* 845 341-4799
*E-mail:* rsmith@sunyorange.edu

**New York University**
Dental Hygiene Prgm
345 E 24th St
New York, NY 10010
www.nyu.edu/dental
*Prgm Dir:* Cheryl Westphal, RDH
*Tel:* 212 998-9390  *Fax:* 212 995-4593
*E-mail:* cmw1@nyu.edu

**Monroe Community College**
Dental Hygiene Prgm
1000 E Henrietta Rd
Rochester, NY 14623-5780
*Prgm Dir:* David B Lawrence, DDS
*Tel:* 585 292-2761  *Fax:* 585 424-5703
*E-mail:* dlawrence@monroecc.edu

**Hudson Valley Community College**
Dental Hygiene Prgm
80 Vandenburgh Ave
Troy, NY 12180
*Prgm Dir:* J Romano, RDH BS MA
*Tel:* 518 629-7442  *Fax:* 518 629-8191
*E-mail:* romanjud@hvcc.edu

**Erie Community College - North Campus**
Dental Hygiene Prgm
6205 Main St
Williamsville, NY 14221-7095
www.ecc.edu
*Prgm Dir:* Joseph Sowinski, MS DDS
*Tel:* 716 851-1390   *Fax:* 716 851-1392
*E-mail:* sowinski@ecc.edu

# North Carolina

**Asheville-Buncombe Technical Comm College**
Dental Hygiene Prgm
340 Victoria Rd
Asheville, NC 28801
www.abtech.edu
*Prgm Dir:* Shaun Tate, RDH MAEd
*Tel:* 828 254-1921, Ext 271   *Fax:* 828 281-9792
*E-mail:* state@abtech.edu

**Univ of North Carolina at Chapel Hill**
Dental Hygiene Prgm
3220 Old Dental Bldg, CB 7450
Chapel Hill, NC 27599-7450
www.dent.unc.edu
*Prgm Dir:* Sally Mauriello, RDH EdD
*Tel:* 919 966-2800   *Fax:* 919 966-6761
*E-mail:* sally_mauriello@dentistry.unc.edu

**Central Piedmont Community College**
Dental Hygiene Prgm
1201 Elizabeth Ave, Kings Dr
Charlotte, NC 28204
*Prgm Dir:* Judith Qualtieri, RDH MS
*Tel:* 704 330-6365   *Fax:* 704 330-6533
*E-mail:* judy.qualtieri@cpcc.edu

**Fayetteville Technical Community College**
Dental Hygiene Prgm
2201 Hull Rd, PO Box 35236
Fayetteville, NC 28303
www.faytechcc.edu
*Prgm Dir:* Susan Ellis, RDH MEd
*Tel:* 910 678-8575   *Fax:* 910 678-8500
*E-mail:* elliss@faytechcc.edu

**Wayne Community College**
Dental Hygiene Prgm
3000 Wayne Memorial Dr
Goldsboro, NC 27533
www.waynecc.edu
*Prgm Dir:* Sue Fowler, BS
*Tel:* 919 735-5152, Ext 206   *Fax:* 919 736-9425
*E-mail:* suef@waynecc.edu
*Med Dir:* William Radford, DDS

**Catawba Valley Community College**
Dental Hygiene Prgm
2550 Hwy 70 SE
Hickory, NC 28602
www.cvcc.edu/prog_study/health/dhygiene
*Prgm Dir:* Luis E Arzola, DMD
*Tel:* 828 327-7000, Ext 4339   *Fax:* 828 624-5205
*E-mail:* larzola@cvcc.edu

**Coastal Carolina Community College**
Dental Hygiene Prgm
444 Western Blvd
Jacksonville, NC 28540
*Prgm Dir:* Joseph Hewitt, DDS
*Tel:* 910 938-6271   *Fax:* 910 455-4849
*E-mail:* hewittj@coastal.cc.nc.us

**Guilford Technical Community College**
Dental Hygiene Prgm
PO Box 309
Jamestown, NC 27282
*Prgm Dir:* Lois Smith, RDH CDA MS
*Tel:* 336 334-4822, Ext 2452   *Fax:* 336 819-2015
*E-mail:* lhsmith@gtcc.edu

**Wake Technical Community College**
Dental Hygiene Prgm
9101 Fayetteville Rd
Raleigh, NC 27603-5696
www.waketech.edu
*Prgm Dir:* Brenda Maddox, RDH MS
*Tel:* 919 212-3263   *Fax:* 919 212-3261
*E-mail:* bpmaddox@waketech.edu

**Halifax Community College**
Dental Hygiene Prgm
PO Drawer 809
Weldon, NC 27890-0809
www.hcc.cc.nc.us
*Prgm Dir:* Doris Markham, RDH MSEd
*Tel:* 252 538-4305
*E-mail:* markhamdj@halifaxcc.edu

**Cape Fear Community College**
Dental Hygiene Prgm
411 N Front St
Wilimington, NC 28401-3993
*Prgm Dir:* Catherine R Cotter, CDA RDH MEd
*Tel:* 910 362-7417   *Fax:* 910 362-7418
*E-mail:* ccotter@cfcc.edu

**Forsyth Technical Community College**
Dental Hygiene Prgm
2100 Silas Creek Pkwy
Winston-Salem, NC 27103-5197
*Prgm Dir:* Janette Whisenhunt, PhD (RDH CDA BS
  MEd)
*Tel:* 336 734-7414   *Fax:* 336 734-7444
*E-mail:* jwhisenhunt@forsythtech.edu

# North Dakota

**North Dakota State College of Science**
Dental Hygiene Prgm
800 N 6th St
Wahpeton, ND 58076
*Prgm Dir:* Susan Swanson, DDS
*Tel:* 701 671-2334   *Fax:* 701 671-2570
*E-mail:* susan.swanson@ndscs.nodak.edu

# Ohio

**University of Cincinnati**
*Cosponsor:* Raymond Walters College
Dental Hygiene Prgm
9555 Plainfield Rd
Cincinnati, OH 45236
*Prgm Dir:* Janelle Schierling, EdD
*Tel:* 513 745-5631   *Fax:* 513 792-8623
*E-mail:* schierjm@ucrwcu.rwc.uc.edu

**Cuyahoga Community College**
Dental Hygiene Prgm
2900 Community College Ave
Cleveland, OH 44115
*Prgm Dir:* Mary Lou Gerosky
*Tel:* 216 987-4494   *Fax:* 216 987-4386
*E-mail:* mary-lou.gerosky@tri-c.cc.oh.us

**Columbus State Community College**
Dental Hygiene Prgm
550 E Spring St, PO Box 1609
Union Hall 410
Columbus, OH 43216
*Prgm Dir:* Connie Grossman, MEd RDH
*Tel:* 614 287-5645   *Fax:* 614 287-5198
*E-mail:* cgrossma@cscc.edu

**Ohio State University**
Dental Hygiene Prgm
305 W 12th Ave
PO Box 182357
Columbus, OH 43218
www.dent.osu.edu/dhy
*Prgm Dir:* Michele Carr, RDH MA
*Tel:* 614 688-4897   *Fax:* 614 292-8013
*E-mail:* carr.3@osu.edu
*Med Dir:* Carole Anderson, PhD

**Sinclair Community College**
Dental Hygiene Prgm
444 W Third St
Dayton, OH 45402
*Prgm Dir:* Rena Shuchat, MS
*Tel:* 937 512-2779   *Fax:* 937 512-4175
*E-mail:* rena.shuchat@sinclair.edu

**Lorain County Community College**
Dental Hygiene Prgm
1005 N Abbe Rd
Elyria, OH 44035-1691
*Prgm Dir:* Denise Price
*Tel:* 440 365-5222, Ext 7196   *Fax:* 440 366-4116
*E-mail:* dprice@lorainccc.edu

**Lakeland Community College**
Dental Hygiene Prgm
7700 Clocktower Dr
Kirtland, OH 44094-5198
*Prgm Dir:* Cathy Patterson
*Tel:* 440 525-7190   *Fax:* 440 525-7433
*E-mail:* cpatterson@lakelandcc.edu

**Rhodes State College**
Dental Hygiene Prgm
4240 Campus Dr
Lima, OH 45804
*Prgm Dir:* Denise Bowers, MSEd
*Tel:* 419 995-8385   *Fax:* 419 995-8162
*E-mail:* bowers.d@rhodesstate.edu

**Stark State College of Technology**
Dental Hygiene Prgm
6200 Frank Ave NW
North Canton, OH 44720-7299
www.starkstate.edu
*Prgm Dir:* Nichole Oocumma, RDH MA CHES
*Tel:* 330 966-5458, Ext 4707   *Fax:* 330 966-6586
*E-mail:* noocumma@starkstate.edu

**Shawnee State University**
Dental Hygiene Prgm
940 Second St
Portsmouth, OH 45662
www.shawnee.edu
*Prgm Dir:* Nancy Murray, RDH MS
*Tel:* 740 351-3236, Ext 3273   *Fax:* 740 351-3354
*E-mail:* nmurray@shawnee.edu
*Med Dir:* James Kadel, DDS

**Owens Community College**
Dental Hygiene Prgm
PO Box 10,000
Toledo, OH 43699
www.owens.edu
*Prgm Dir:* Elizabeth Tronolone, MS
*Tel:* 419 661-7290   *Fax:* 419 661-7304
*E-mail:* elizabeth_tronolone@owens.edu

**Youngstown State University**
Dental Hygiene Prgm
Dept of Health Professions
One University Plaza
Youngstown, OH 44555
*Prgm Dir:* Madeleine Haggerty, RDH
*Tel:* 330 941-1766   *Fax:* 330 941-1767
*E-mail:* mbhaggerty@ysu.edu

# Oklahoma

**Rose State College**
Dental Hygiene Prgm
6420 SE 15th St
Midwest City, OK 73110
www.rose.edu
*Prgm Dir:* Janet Turley, RDH CDA MEd
*Tel:* 405 733-7337   *Fax:* 405 736-0389
*E-mail:* jturley@rose.edu

**Univ of Oklahoma Health Sciences Center**
Dental Hygiene Prgm
PO Box 26901
Oklahoma City, OK 73117
*Prgm Dir:* Jane A Bowers
*Tel:* 405 271-4436   *Fax:* 405 271-4785
*E-mail:* jane-bowers@ouhsc.edu

**Tulsa Community College**
Dental Hygiene Prgm
909 S Boston Ave, Rm MP 458
Tulsa, OK 74119-2094
www.tulsacc.edu
*Prgm Dir:* Denise Lysikowski, RDH MPH
*Tel:* 918 595-7019   *Fax:* 918 595-8300
*E-mail:* dlysikow@tulsacc.edu

# Oregon

**Lane Community College**
Dental Hygiene Prgm
4000 E 30th Ave
Eugene, OR 97405
*Prgm Dir:* Sharon Hagan, RDH MS
*Tel:* 541 463-5616   *Fax:* 541 463-5616
*E-mail:* hagans@lanecc.edu

**Mt Hood Community College**
Dental Hygiene Prgm
26000 SE Stark St
Gresham, OR 97030
*Prgm Dir:* T Tong, RDH MS
*Tel:* 503 491-7691   *Fax:* 503 491-6005
*E-mail:* tongt@mhcc.edu

**Oregon Institute of Technology**
Dental Hygiene Prgm
3201 Campus Dr
Semon Hall #222
Klamath Falls, OR 97601
www.oit.edu
*Prgm Dir:* Terri Armstrong, MS
*Tel:* 541 885-1886   *Fax:* 541 885-1849
*E-mail:* armstrot@oit.edu

**Portland Community College**
Dental Hygiene Prgm
PO Box 19000
Portland, OR 97219-0990
www.pcc.edu
*Prgm Dir:* Josette Beach, RDH MS
*Tel:* 503 977-4235   *Fax:* 503 977-8300
*E-mail:* jbeach@pcc.edu

# Pennsylvania

**Northampton Community College**
Dental Hygiene Prgm
3835 Green Pond Rd
Bethlehem, PA 18020
www.northampton.edu
*Prgm Dir:* Terry Sigal Greene, RDH MEd
*Tel:* 610 861-5440   *Fax:* 610 861-4139
*E-mail:* tsigalgreene@northampton.edu

**Montgomery County Community College**
Dental Hygiene Prgm
340 DeKalb Pike
Blue Bell, PA 19422-0758
*Prgm Dir:* Jenny Sheaffer, RDH MS
*Tel:* 215 641-6623   *Fax:* 215 619-7171
*E-mail:* jsheaffe@mc3.edu

**Harcum College**
Dental Hygiene Prgm
750 Montgomery Ave
Bryn Mawr, PA 19010
*Prgm Dir:* Jean Byrnes-Ziegler, MA
*Tel:* 610 526-6110, Ext 6046   *Fax:* 610 526-6182
*E-mail:* jbyrnes-ziegler@harcum.edu

**Tri-State Business Institute**
Dental Hygiene Prgm
5757 West Ridge Rd
Erie, PA 16506
www.tsbi.edu
*Prgm Dir:* Susan Gorman, RDH BS Ed
*Tel:* 814 838-7673, Ext 266   *Fax:* 814 838-4276
*E-mail:* sgorman@tsbi.edu

**Harrisburg Area Community College**
Dental Hygiene Prgm
One HACC Dr, SM-102
Harrisburg, PA 17110-2999
www.hacc.edu
*Prgm Dir:* Kathleen Schlotthauer, RDH MEd
*Tel:* 717 780-2419   *Fax:* 717 780-1170
*E-mail:* kmschlot@hacc.edu

**Manor College**
Dental Hygiene Prgm
700 Fox Chase Rd
Jenkintown, PA 19046-3399
*Prgm Dir:* Virginia Saunders, RDH BS MEd
*Tel:* 215 885-2360, Ext 284   *Fax:* 215 576-6564
*E-mail:* vsaunders@manor.edu

**Luzerne County Community College**
Dental Hygiene Prgm
1333 S Prospect St
Nanticoke, PA 18634-3899
www.luzerne.edu
*Prgm Dir:* Cathryn Brown, CDA RDH MEd
*Tel:* 570 740-0447   *Fax:* 570 740-0265
*E-mail:* cbrown@luzerne.edu

**Community College of Philadelphia**
Dental Hygiene Prgm
1700 Spring Garden St
Philadelphia, PA 19130
www.ccp.edu
*Prgm Dir:* Theresa M Grady
*Tel:* 215 751-8927   *Fax:* 215 751-8937
*E-mail:* tgrady@ccp.edu

**University of Pittsburgh**
Dental Hygiene Prgm
B-82 Salk Hall
Pittsburgh, PA 15261
www.dental.pitt.edu/students/dental_hygiene.php
*Prgm Dir:* Angelina Riccelli, RDH MS
*Tel:* 412 648-8432   *Fax:* 412 383-8737
*E-mail:* riccelli@pitt.edu

**Pennsylvania College of Technology**
Dental Hygiene Prgm
One College Ave
Williamsport, PA 17701
www.pct.edu/schools/hs/dental
*Prgm Dir:* Shawn Kiser, RDH BS MEd
*Tel:* 570 320-8007   *Fax:* 570 320-2401
*E-mail:* skiser@pct.edu

**Westmoreland County Community College**
Dental Hygiene Prgm
145 Pavilion Ln
Youngwood, PA 15697-1895
www.wccc-pa.edu
*Prgm Dir:* Angela S Rinchuse, RDH MEd
*Tel:* 724 925-4163   *Fax:* 724 925-5808
*E-mail:* rinchusea@wccc-pa.edu

# Rhode Island

**Community College of Rhode Island**
Dental Hygiene Prgm
1762 Louisquisset Pike
Lincoln, RI 02865-4585
www.ccri.edu
*Prgm Dir:* Kathleen Gazzola, RDH EMT CDA BS MA
*Tel:* 401 333-7227   *Fax:* 401 333-7146
*E-mail:* kgazzola@ccri.edu

# South Carolina

**Trident Technical College**
Dental Hygiene Prgm
PO Box 118067
Charleston, SC 29423-8067
www.tridenttech.edu
*Prgm Dir:* Barbara Ankersen, RDH MS
*Tel:* 843 574-6439   *Fax:* 843 574-6585
*E-mail:* barbara.ankersen@tridenttech.edu

**Midlands Technical College**
Dental Hygiene Prgm
PO Box 2408
Columbia, SC 29202
*Prgm Dir:* Martha H Hanks, DDS
*Tel:* 803 822-3451   *Fax:* 803 822-3079
*E-mail:* hanksm@midlandstech.edu

**Horry - Georgetown Technical College**
Dental Hygiene Prgm
PO Box 261966
2050 Highway 501 East
Conway, SC 29528-6066
www.hgtc.edu
*Prgm Dir:* Alice Derouen, RDH MEd
*Tel:* 843 349-5371   *Fax:* 843 349-7576
*E-mail:* alice.derouen@hgtc.edu

**Florence-Darlington Technical College**
Dental Hygiene Prgm
PO Box 100548
Florence, SC 29501-0548
*Prgm Dir:* Richard Moore, DMD
*Tel:* 843 661-8223   *Fax:* 843 292-0851
*E-mail:* richard.moore@fdtc.edu

**Greenville Technical College**
Dental Hygiene Prgm
PO Box 5616
Greenville, SC 29606-5616
www.greenvilletech.com
*Prgm Dir:* Debra Grubbs, RDH BSHSE
*Tel:* 864 250-8588   *Fax:* 864 250-8261
*E-mail:* debra.grubbs@gvltec.edu

**York Technical College**
Dental Hygiene Prgm
452 S Anderson Rd
Rock Hill, SC 29730
*Prgm Dir:* Kathleen Peter, RDH MHA
*Tel:* 803 327-8039   *Fax:* 803 327-8059
*E-mail:* peter@york.tech.com

# South Dakota

**University of South Dakota**
Dental Hygiene Prgm
East Hall 414, E Clark St
Vermillion, SD 57069
*Prgm Dir:* Ann Brunick, RDH MS
*Tel:* 605 677-5379   *Fax:* 605 677-5638
*E-mail:* Ann.Brunick@usd.edu

# Tennessee

**Chattanooga State Technical Comm College**
Dental Hygiene Prgm
4501 Amnicola Hwy
Chattanooga, TN 37406-1097
*Prgm Dir:* William Johnson
*Tel:* 423 697-4768   *Fax:* 423 634-3071
*E-mail:* william.johnson@chattanoogastate.edu

**East Tennessee State University**
Dental Hygiene Prgm
PO Box 70690
Johnson City, TN 37614-0690
www.etsu.edu/cpah/dental
*Prgm Dir:* Charles Faust, RDH EdD
*Tel:* 423 439-4497   *Fax:* 423 439-4030
*E-mail:* faust@etsu.edu

**University of Tennessee Health Science Ctr**
Dental Hygiene Prgm
930 Madison Ave, Ste 600
Memphis, TN 38163
www.utmem.edu/allied/dental_hygiene_home.html
*Prgm Dir:* Margaret Waring, EdD BS RDH
*Tel:* 901 448-6230   *Fax:* 901 448-7545
*E-mail:* mwaring@utmem.edu
*Med Dir:* Elaine Freiden, BS, RDH

**Tennessee State University**
Dental Hygiene Prgm
3500 John A Merritt Blvd
Nashville, TN 37209-1561
*Prgm Dir:* Marian W Patton, RDH EdD
*Tel:* 615 963-5801   *Fax:* 615 963-5836
*E-mail:* mpatton@tnstate.edu

**Roane State Community College**
Dental Hygiene Prgm
701 Briarcliff Ave
Oak Ridge, TN 37830-8795
www.rscc.cc.tn.us
*Prgm Dir:* Michael D Curran, DDS
*Tel:* 865 483-0816, Ext 2127   *Fax:* 865 481-2019
*E-mail:* Curranm@roanestate.edu

# Texas

**Amarillo College**
Dental Hygiene Prgm
PO Box 447
Amarillo, TX 79178
*Prgm Dir:* Donna Cleere, RDH BSOE MEd
*Tel:* 806 354-6064   *Fax:* 806 354-6076
*E-mail:* cleere-dk@actx.edu

**Austin Community College**
Dental Hygiene Prgm
Highland Business Center
5930 Middle Fiskville Rd
Austin, TX 78702
*Prgm Dir:* Renee Cornett
*Tel:* 512 223-5711
*E-mail:* rcornett@austincc.edu

**Lamar Institute of Technology**
Dental Hygiene Prgm
PO Box 10061
Beaumont, TX 77710
www.lit.edu
*Prgm Dir:* Betty Reynard, RDH EdD
*Tel:* 409 880-8846   *Fax:* 409 880-8955
*E-mail:* betty.reynard@lit.edu

**Coastal Bend College**
Dental Hygiene Prgm
3800 Charco Rd
Beeville, TX 78102
http://coastalbend.edu
*Prgm Dir:* Andrea Westmoreland, RDH MEd
*Tel:* 361 354-2556   *Fax:* 361 354-2540
*E-mail:* andygump@coastalbend.edu

**Howard College**
Dental Hygiene Prgm
1001 Birdwell Ln
Big Spring, TX 79720
*Prgm Dir:* Jeri Farmer, RDH MEd
*Tel:* 915 264-5075   *Fax:* 915 264-5630
*E-mail:* jfarmer@howardcollege.edu

**Blinn College**
Dental Hygiene Prgm
PO Box 6030
Bryan, TX 77805-6030
www.blinn.edu
*Prgm Dir:* Lisa Wiese, RDH MS
*Tel:* 979 209-7272   *Fax:* 979 209-7289
*E-mail:* Lwiese@blinn.edu

**Del Mar College**
Dental Hygiene Prgm
101 Baldwin Blvd
Corpus Christi, TX 78404
www.delmar.edu/dh
*Prgm Dir:* David G Arreguin, DDS
*Tel:* 361 698-1315   *Fax:* 361 698-1175
*E-mail:* darreguin@delmar.edu

**Baylor College of Dentistry, Texas A&M HSC**
Dental Hygiene Prgm
The Texas A & M University System Health Science
   Center
PO Box 660677
Dallas, TX 75266-0677
*Prgm Dir:* Janice Dewald, BSDH DDS MS
*Tel:* 214 828-8340   *Fax:* 214 828-8196
*E-mail:* jdewald@bcd.tamhsc.edu

**Texas Woman's University**
Dental Hygiene Prgm
Box 425796, TWU Station
Denton, TX 76204
*Prgm Dir:* Nahid Nikpour, PhD
*Tel:* 940 898-2874   *Fax:* 940 898-2869
*E-mail:* nnikpour@mail.twu.edu

**El Paso Community College**
Dental Hygiene Prgm
PO Box 20500
El Paso, TX 79998
www.epcc.edu
*Prgm Dir:* William Ekvall, DDS
*Tel:* 915 831-4060   *Fax:* 915 831-4445
*E-mail:* williame@epcc.edu

**Texas State Technical College - Harlingen**
Dental Hygiene Prgm
1902 N Loop 499
Harlingen, TX 78550-3697
*Prgm Dir:* Barbara L Bennett, CDA
*Tel:* 956 364-4697   *Fax:* 956 364-5162
*E-mail:* bbennett@harlingen.tstc.edu

**Univ of Texas Hlth Sci Ctr at Houston**
Dental Hygiene Prgm
PO Box 20068
Houston, TX 77225-0068
www.db.uth.tmc.edu
*Prgm Dir:* Nina Bay Infante, RDH MS
*Tel:* 713 500-4085   *Fax:* 713 500-0410
*E-mail:* Nina.B.Infante@uth.tmc.edu

**Tarrant County College**
Dental Hygiene Prgm
828 Harwood Rd
Hurst, TX 76054
*Prgm Dir:* Cindy A O'Neal, RDH BS
*Tel:* 817 515-6640   *Fax:* 817 515-6700
*E-mail:* cindy.oneal@tccd.edu

**Kingwood College - NHMCCD**
Dental Hygiene Prgm
20000 Kingwood Dr
Health & Science Building 118A
Kingwood, TX 77339
www.nhmccd.edu
*Prgm Dir:* Maribeth Stitt, RDH MEd
*Tel:* 281 312-1517   *Fax:* 281 312-1722
*E-mail:* Maribeth.Stitt@nhmccd.edu

**Collin County Community College**
Dental Hygiene Prgm
2200 W University Dr
McKinney, TX 75069-8001
*Prgm Dir:* Joanne Fletcher, RDH MS
*Tel:* 972 548-6738   *Fax:* 972 548-6536
*E-mail:* jfletcher@ccccd.edu

**Univ of Texas Hlth Sci Ctr at San Antonio**
Dental Hygiene Prgm
7703 Floyd Curl Dr MSC 6244
San Antonio, TX 78229-3900
www.uthscsa.edu
*Prgm Dir:* Juanita Wallace, PhD RDH
*Tel:* 210 567-8820   *Fax:* 210 567-8843
*E-mail:* wallacej@uthscsa.edu

**Temple College**
Dental Hygiene Prgm
2600 S First St
Temple, TX 76504-7435
*Prgm Dir:* Norma Maedgen, RDH PhD
*Tel:* 254 298-8677   *Fax:* 254 298-8676
*E-mail:* norma.maedgen@templejc.edu

**Tyler Junior College**
Dental Hygiene Prgm
PO Box 9020
Tyler, TX 75711
*Prgm Dir:* Carrie Hobbs, RDH BS MEd
*Tel:* 903 510-2341   *Fax:* 903 510-2879
*E-mail:* chob@tjc.edu

**Wharton County Junior College**
Dental Hygiene Prgm
911 Boling Hwy
Wharton, TX 77488
www.wcjc.edu
*Prgm Dir:* Leigh Ann Collins, MAI
*Tel:* 979 532-6398   *Fax:* 979 532-6489
*E-mail:* lacollins@wcjc.edu

**Midwestern State University**
Dental Hygiene Prgm
3410 Taft Blvd
Wichita Falls, TX 76308-2099
www.mwsu.edu
*Prgm Dir:* Barbara DeBois, RDH MS
*Tel:* 940 397-4764   *Fax:* 940 397-4973
*E-mail:* dental.hygiene@mwsu.edu

# Utah

**Weber State University**
Dental Hygiene Prgm
3920 University Circle
Ogden, UT 84408-3920
*Prgm Dir:* Stephanie Bossenberger, MS RDH
*Tel:* 801 626-6130   *Fax:* 801 626-7304
*E-mail:* bossenberger@weber.edu

**Utah Valley State College**
Dental Hygiene Prgm
800 W University Pkwy
Orem, UT 84058
*Prgm Dir:* George Veit, DDS
*Tel:* 801 764-7592   *Fax:* 801 863-7592
*E-mail:* veitge@uvsc.edu

**Salt Lake Community College**
Dental Hygiene Prgm
4600 S Redwood Rd, PO Box 30808
Salt Lake City, UT 84130-0808
*Prgm Dir:* Sandra "Bobi" Merritt, RDH BA MA
*Tel:* 801 957-4831   *Fax:* 801 957-2819
*E-mail:* Bobi.Merritt@slcc.edu

**Dixie State College of Utah**
Dental Hygiene Prgm
225 S 700 E
St George, UT 84770
*Prgm Dir:* Gordon Jennings, DDS
*Tel:* 435 652-7869   *Fax:* 435 656-4031
*E-mail:* jennings@dixie.edu

# Vermont

**Vermont Technical College**
Dental Hygiene Prgm
301 Lawrence Place
Williston Campus
Williston, VT 05495
*Prgm Dir:* Ellen Grimes, EdD
*Tel:* 802 879-5632   *Fax:* 802 879-2317
*E-mail:* ellen.grimes@vtc.edu

# Virginia

**Old Dominion University**
Dental Hygiene Prgm
Gene W Hirschfeld School of Dental Hygiene
Technology Bldg, 4608 Hampton Blvd
Norfolk, VA 23529-0499
*Prgm Dir:* Deanne Shuman, BSDH MS PhD
*Tel:* 757 683-3338   *Fax:* 757 683-5239
*E-mail:* dshuman@odu.edu

**Virginia Commonwealth University**
Dental Hygiene Prgm
School of Dentistry
PO Box 980566
Richmond, VA 23298-0566
*Prgm Dir:* Kim Isringhausen, BSDH RDH MPH
*Tel:* 804 828-9096   *Fax:* 804 827-0969
*E-mail:* ktisring@vcu.edu

**Virginia Western Community College**
Dental Hygiene Prgm
3097Colonial Ave SW
Roanoke, VA 24015
*Prgm Dir:* Martha Roberson, RDH MSHA
*Tel:* 540 857-6282   *Fax:* 540 857-6224
*E-mail:* mroberson@vw.vccs.edu

**Northern Virginia Community College**
Dental Hygiene Prgm
6699 Springfield Center Dr
Springfield, VA 22150
www.nvcc.edu
*Prgm Dir:* Edith Tynan, RDH MS MA
*Tel:* 703 822-6570   *Fax:* 703 822-6610
*E-mail:* etynan@nvcc.edu

**Wytheville Community College**
Dental Hygiene Prgm
1000 E Main St
Wytheville, VA 24382
*Prgm Dir:* Patricia M Bradshaw, CDA RDH MS
*Tel:* 276 223-4832   *Fax:* 276 223-4778
*E-mail:* wcbradp@wcc.vccs.edu

# Washington

**Lake Washington Technical College**
Dental Hygiene Prgm
11605 132nd Ave NE
Kirkland, WA 98034
www.lwtc.ctc.edu/dental
*Prgm Dir:* Mary Young, RDH
*Tel:* 425 739-8403   *Fax:* 425 739-8292
*E-mail:* mary.young@lwtc.edu

**Pierce College**
Dental Hygiene Prgm
9401 Farwest Dr SW
Lakewood, WA 98498-1999
www.pierce.ctc.edu
*Prgm Dir:* Sharon S Golightly, RDH EdD
*Tel:* 253 964-6661   *Fax:* 253 964-6313
*E-mail:* sgolight@pierce.ctc.edu

**Columbia Basin College**
Dental Hygiene Prgm
2600 N 20th Ave
Pasco, WA 99301
www.columbiabasin.edu
*Prgm Dir:* Lynn Stedman, RDH BSDH MEd MA
*Tel:* 509 547-0511, Ext 2991   *Fax:* 509 544-2025
*E-mail:* lynn.stedman@columbiabasin.edu

**Seattle Central Community College**
Dental Hygiene Prgm
1701 Broadway BE 3210
Seattle, WA 98122
http://seattlecentral.org
*Prgm Dir:* Ona Canfield, RDH MEd
*Tel:* 206 587-6922   *Fax:* 206 587-6337
*E-mail:* scccdh@sccd.ctc.edu

**Shoreline Community College**
Dental Hygiene Prgm
16101 Greenwood Ave N
Shoreline, WA 98133
http://success.shore.ctc.edu/dental
*Prgm Dir:* Marianne Baker, RDH MEd
*Tel:* 206 546-4709   *Fax:* 206 546-5830
*E-mail:* mbaker@shoreline.edu

**Eastern Washington University**
Dental Hygiene Prgm
310 N Riverpoint Blvd Box E
Spokane, WA 99202
*Prgm Dir:* Rebecca Stolberg, RDH MS
*Tel:* 509 368-6528   *Fax:* 509 368-6514
*E-mail:* rstolberg@mail.ewu.edu

**Clark College**
Dental Hygiene Prgm
1800 E McLoughlin Blvd
Vancouver, WA 98663
*Prgm Dir:* G Liberman
*Tel:* 360 699-0474   *Fax:* 360 699-2880
*E-mail:* gliberman@clark.edu

**Yakima Valley Community College**
Dental Hygiene Prgm
PO Box 22520
Yakima, WA 98907
www.yvcc.edu
*Prgm Dir:* Patricia Hakala, RDH BA
*Tel:* 509 574-4918   *Fax:* 509 574-6875
*E-mail:* phakala@yvcc.edu
*Med Dir:* John Nelson, BS MIE ABD for EdD

# West Virginia

**Community & Technical College at WVU Tech**
Dental Hygiene Prgm
604 Davis Hall
Montgomery, WV 25136
*Prgm Dir:* Kristin Mallory, RDH EdD
*Tel:* 304 442-3254   *Fax:* 304 442-3093
*E-mail:* Kristin.Mallory@mail.wvu.edu

**West Virginia University**
Dental Hygiene Prgm
Health Science Ctr N, PO Box 9425
Morgantown, WV 26506-9425
www.hsc.wvu.edu
*Prgm Dir:* Marcia Gladwin, EdD
*Tel:* 304 293-3417   *Fax:* 304 293-4882
*E-mail:* mgladwin@hsc.wvu.edu
*Med Dir:* James Koelbl, DDS, MS, MJ

**Southern West Virginia Comm & Tech College**
Dental Hygiene Prgm
Dempsey Branch Rd
Mt Gay, WV 25637
*Prgm Dir:* Lisa Haddox-Heston
*Tel:* 304 792-7098, Ext 259
*E-mail:* lisah@southern.wvnet.edu

**West Liberty State College**
Dental Hygiene Prgm
PO Box 295
CSC #121
West Liberty, WV 26074-0295
www.wlsc.edu
*Prgm Dir:* Margaret Six, RDH MSDH
*Tel:* 304 336-8030   *Fax:* 304 336-8905
*E-mail:* sixmj@wlsc.edu

# Wisconsin

**Fox Valley Technical College**
Dental Hygiene Prgm
1825 Bluemound Dr
Appleton, WI 54914
www.fvtc.edu
*Prgm Dir:* Joan Rohrer, RDH BS
*Tel:* 920 735-2452   *Fax:* 920 831-4314
*E-mail:* rohrer@fvtc.edu

**Northeast Wisconsin Technical College**
Dental Hygiene Prgm
2740 W Mason St, PO Box 19042
Green Bay, WI 54307-9042
www.nwtc.edu
*Prgm Dir:* Sheila Gross, RDH MS
*Tel:* 920 498-6839   *Fax:* 920 498-6890
*E-mail:* sheila.gross@nwtc.edu

**Madison Area Technical College**
Dental Hygiene Prgm
211 N Carroll St
Madison, WI 53703
www.matcmadison.edu
*Prgm Dir:* E Lynn Goetsch, RDH MS
*Tel:* 608 258-2470   *Fax:* 608 258-2482
*E-mail:* egoetsch@matcmadison.edu

**Milwaukee Area Technical College**
Dental Hygiene Prgm
700 W State St
Milwaukee, WI 53233
*Prgm Dir:* Laurie Klos, MS
*Tel:* 414 297-7126   *Fax:* 414 297-7205
*E-mail:* klosl@matc.edu

## Waukesha County Technical College
Dental Hygiene Prgm
800 Main St
Pewaukee, WI 53072
www.wctc.edu
*Prgm Dir:* Pamela Brilowski, RDH MS
*Tel:* 262 691-5561, Ext 5561   *Fax:* 262 691-5077
*E-mail:* pbrilowski@wctc.edu

## Northcentral Technical College
Dental Hygiene Prgm
1000 Campus Dr
Wausau, WI 54401
www.ntc.edu
*Prgm Dir:* Michelle Hilts, RDH
*Tel:* 715 675-3331, Ext 1329   *Fax:* 715 675-3772
*E-mail:* hilts@ntc.edu

# Wyoming

## Laramie County Community College
Dental Hygiene Prgm
1400 E College Dr
Cheyenne, WY 82007-3299
*Prgm Dir:* Connie Henry
*Tel:* 307 778-1229   *Fax:* 307 772-7304
*E-mail:* chenry@lccc.wy.edu

## Sheridan College
Dental Hygiene Prgm
3059 Coffeen Ave
Sheridan, WY 82801
www.sheridan.edu
*Prgm Dir:* Jan Dill, RDH BSDH MS
*Tel:* 307 674-6446, Ext 3405   *Fax:* 307 674-3352
*E-mail:* jldill@sheridan.edu

## Dental Hygienist

| Programs* | Class Capacity | Begins | Length (months) | Award | Res Tuition | Non-res Tuition | Stipend | Offers:‡ 1 | 2 | 3 |
|---|---|---|---|---|---|---|---|---|---|---|
| **Alabama** | | | | | | | | | | |
| Wallace State Community College (Hanceville) | 30 | Aug | 21 | AAS | $2,430 | $4,860 | | | | • |
| **Alaska** | | | | | | | | | | |
| University of Alaska Anchorage | 12 | Aug | 18 | AAS | $3,161 | $10,527 | | | | |
| **Arizona** | | | | | | | | | | |
| Mesa Community College | 18 | Aug | 18 | AAS | $60 | $85 | | • | | • |
| Mohave Community College (Bullhead City) | | | 20 | | | | | | | • |
| Northern Arizona University (Flagstaff) | 26 | Aug | 30 | BS | $4,376 | $13,316 | | • | | • |
| Phoenix College | 22 | Aug | 18 | AAS | $1,830 | $6,600 | | | | • |
| Rio Salado College (Temple) | 36 | Feb | 15 | AAS | $1,700 | $0 | | • | | • |
| **Arkansas** | | | | | | | | | | |
| University of Arkansas for Medical Sciences (Little Rock) | 32 | Aug | 19 | AS, BS | $6,000 | $0 | | | | |
| **California** | | | | | | | | | | |
| Cabrillo College (Aptos) | 22 | Jun | 24 | AS | $1,615 | $12,026 | | | | |
| Cerritos Community College (Norwalk) | 24 | Aug | 20 | AA | $689 | $4,001 | | | | • |
| Chabot College (Hayward) | 20 | Aug | 24 | AA | $780 | $7,320 | | | | • |
| Cypress College | 18 | Aug | 18 | Cert, AS | $377 | $2,166 | | | | • |
| Diablo Valley College (Pleasant Hill) | 22 | Aug | 21 | Cert, AS Den | $800 | $11,000 | | | | • |
| Fresno City College | 30 | Aug | 18 | AS | $264 | $3,036 | | | | |
| Loma Linda University | 42 | Sep | 21 | BS | $25,431 | $25,431 | | | | • |
| Sacramento City College | 24 | Aug | 20 | AS | $517 | $4,969 | | | | |
| San Joaquin Valley College - Visalia | 30 | Jun Oct Feb | 16 | AS | $24,200 | $24,200 | | | | |
| Santa Rosa Junior College | 24 | Aug | 24 | AS | $26 | $171 | | | | • |
| Southwestern College (Chula Vista) | 36 | Aug | 24 | AS | $740 | $5,060 | | | | |
| Taft College | 24 | Aug | 18 | AS | $741 | $3,800 | | | | |
| West Los Angeles College (Culver City) | 24 | Sep | 21 | AS | $754 | $4,843 | | | | • |
| Western Career College - Sacramento | 30 | Jan May Aug | 17 | AS | | | | | | |
| **Colorado** | | | | | | | | | | |
| Colorado Northwestern Community College (Rangley) | 27 | Aug | 24 | AAS | $1,104 | $4,190 | | • | | |
| Community College of Denver | 24 | Aug | 22, 34 | Cert, AAS | $4,282 | $10,853 | | | | • |
| Pueblo Community College | 15 | Aug | 18 | AAS | $4,500 | $10,500 | | | | • |
| **Connecticut** | | | | | | | | | | |
| Tunxis Community College (Farmington) | 36 | Sep | 24 | AS | $2,672 | $7,976 | | | | |
| **Delaware** | | | | | | | | | | |
| Delaware Tech & Comm Coll - Wilmington | 24 | Aug | 20 | AAS | $2,579 | $6,448 | | • | | |
| **District of Columbia** | | | | | | | | | | |
| Howard University (Washington) | 20 | Aug | 20 | Cert | $14,918 | $14,918 | | | | |
| **Florida** | | | | | | | | | | |
| Broward Community College (Ft Lauderdale) | 16 | Aug | 12 | AS | $4,000 | $10,000 | | | | • |
| Daytona Beach Community College (DeLand) | 15 | Aug | 20 | AS | $6,110 | $23,186 | | | | • |
| Edison College (Ft Myers) | 18 | Aug | 24 | AS | $3,500 | $12,000 | | | | |
| Florida Community College - Jacksonville | 36 | Aug | 21 | AS, AAS | $3,100 | $11,723 | | | | • |
| Manatee Community College (Bradenton) | 14 | Aug | 24 | AS | $1,260 | $4,708 | | | | • |
| Palm Beach Community College (Lake Worth) | 24 | Aug | 21 | AS | $11,000 | $33,000 | | | | • |
| Pensacola Junior College | 36 | May | 24 | AAS | $3,000 | $14,825 | | | | |
| Santa Fe Community College (Gainesville) | 24 | Aug | 24, 16 | AS | $5,575 | $20,658 | | | | • |
| South Florida Community College (Avon Park) | 12 | Aug | 21 | AAS | $4,100 | $20,000 | | | | |
| Valencia Community College (Orlando) | 24 | Jun | 21 | AS | $2,550 | $0 | | | | • |

*Data are shown only for programs that completed the 2006 AMA Survey of Health Professions Education Programs
‡Key to Offers: 1: Evening or weekend classes; 2: Non-English instruction; 3: Cultural competence instruction

## Dental Hygienist

| Programs* | Class Capacity | Begins | Length (months) | Award | Res Tuition | Non-res Tuition | Stipend | Offers:‡ 1 | 2 | 3 |
|---|---|---|---|---|---|---|---|---|---|---|
| **Georgia** | | | | | | | | | | |
| Armstrong Atlantic State University (Savannah) | 26 | Aug | 21 | AS | $2,924 | $10,240 | | | | |
| Athens Technical College | 16 | Fall | 21 | AS | $1,812 | $0 | | | | |
| Central Georgia Technical College (Macon) | 18 | Aug | 21 | AS | $1,457 | $2,945 | | | | |
| Clayton State University (Morrow) | 28 | Aug | 60 | BSDH | $2,536 | $5,121 | | | | • |
| Columbus Technical College | 14 | Oct | 20 | ADH | $3,751 | $3,751 | | | | • |
| Darton College (Albany) | 24 | Aug | 24 | AS | $2,477 | $6,722 | | | | |
| Georgia Highlands College (Rome) | 14 | Aug | 20 | AS | | | | | | • |
| Medical College of Georgia (Augusta) | 28 | Aug | 21 | BS | $4,547 | $18,190 | | | | • |
| Middle Georgia Technical College (Warner Robins) | 14 | Oct | 21 | AAT | $1,860 | $3,720 | | | | |
| West Central Technical College (Douglasville) | 24 | Jul | 24 | AAS | $1,828 | $1,828 | | | | • |
| **Idaho** | | | | | | | | | | |
| Apollo College - Boise | 30 | | 20 | AAS | $45,000 | $45,000 | | | | • |
| Idaho State University (Pocatello) | 26 | Aug | 24 | BS, MS | $3,880 | $11,320 | | | | • |
| **Illinois** | | | | | | | | | | |
| College of Lake County (Grayslake) | 24 | Fall semester | 24 | | $1,500 | $2,800 | | | | |
| Illinois Central College (Peoria) | 24 | Aug | 22 | AAS | $1,428 | $5,440 | | | | |
| John A Logan College (Caterville) | 24 | Aug | 17 | AAS | $1,881 | $5,584 | | | | |
| Lake Land College (Mattoon) | 30 | Aug | 21 | AAS | $2,600 | $5,350 | | | | • |
| Lewis & Clark Community College (Godfrey) | 24 | Aug | 18 | AAS | $3,825 | $10,230 | | | | |
| Prairie State College (Chicago Heights) | 36 | May | 24 | AAS | $6,000 | $9,000 | | | | |
| Southern Illinois Univ at Carbondale | 36 | Aug | 32 | BS | $4,254 | $7,356 | | | | |
| **Indiana** | | | | | | | | | | |
| Indiana Univ - Purdue Univ Ft Wayne | 30 | Aug | 16 | AS | $6,846 | $15,681 | | | | • |
| Indiana University (Indianapolis) | 50 | Aug | 21 | AS, BS | $7,800 | $16,766 | | | | • |
| Indiana University Northwest (Gary) | 24 | Aug | 24 | AS | $8,610 | $22,722 | | | | • |
| University of Southern Indiana (Evansville) | 24 | Aug | 16 | AS, BS | $5,076 | $11,965 | | | | • |
| **Iowa** | | | | | | | | | | |
| Des Moines Area Community College (Ankeny) | 24 | Aug | 17 | AAS | $3,420 | $6,840 | | | | • |
| Hawkeye Community College (Waterloo) | 22 | Aug | 21 | AAS | $2,886 | $5,772 | | | | • |
| Iowa Western Community College (Council Bluffs) | 18 | Aug | 21 | AAS | $1,800 | $0 | | | | • |
| Kirkwood Community College (Cedar Rapids) | 24 | Fall | 18 | AAS | $4,356 | $4,356 | | | | • |
| **Kansas** | | | | | | | | | | |
| Johnson County Community College (Overland Park) | 26 | Aug | 24 | AAS | $3,969 | $9,072 | | | | • |
| Wichita State University | 36 | Aug | 19 | AS, BSDH | $4,500 | $12,000 | | | | • |
| **Kentucky** | | | | | | | | | | |
| Big Sandy Community & Technical College (Prestonsburg) | 16 | Aug | 24 | AAS | $3,724 | $11,324 | | | | |
| Bluegrass Community and Technical College (Lexington) | 24 | Jun | 24 | AAS | $4,060 | $10,680 | | | | |
| University of Louisville | 30 | Aug | 20 | BS | $4,000 | $6,000 | | | | • |
| Western Kentucky University (Bowling Green) | 24 | Aug | 17, 33 | Dipl, AS, BS | $3,500 | $8,350 | | | | |
| **Louisiana** | | | | | | | | | | |
| Louisiana State University (Baton Rouge) | 38 | Aug | 18 | BS | $2,546 | $0 | | | | • |
| Southern Univ at Shreveport | 12 | Aug | 21 | AAS | $3,452 | $3,982 | | | | • |
| University of Louisiana at Monroe | 26 | Aug | 24, 48 | BS | $4,328 | $12,017 | | | | |
| **Maine** | | | | | | | | | | |
| University College of Bangor | 24 | Sep | 27 | AS, BS | $4,985 | $11,124 | | | | • |
| University of New England (Portland) | 50 | Sep | 27, 36 | AS, BS | $22,940 | $22,940 | | • | | • 3 |
| **Maryland** | | | | | | | | | | |
| University of Maryland (Baltimore) | 30 | Aug | 24 | Dipl, BS, MS | $4,300 | $14,500 | | • | | • |
| **Massachusetts** | | | | | | | | | | |
| Bristol Community College (Fall River) | 22 | Sep | 16 | AS | $4,550 | $8,510 | | | | • |
| Cape Cod Community College (West Barnstable) | 22 | Sep | 18 | AS | $3,660 | $9,840 | | • | | • |
| Middlesex Community College - Lowell | 42 | Sep | 24 | AS | $4,320 | $11,736 | | | | • |
| Quinsigamond Community College (Worcester) | 30 | Sep | 18 | AS | $8,310 | $16,344 | | | | • |
| Springfield Technical Community College | 20 | Sep | 18 | AS | $5,576 | $13,280 | | | | |
| **Michigan** | | | | | | | | | | |
| Baker College of Port Huron | 30 | Sep | 27 | AAS | $9,200 | $9,200 | | | | • |
| Delta College (University Center) | 18 | Aug Sep | 18 | AAS | $2,734 | $5,407 | | | | • |
| Ferris State University (Big Rapids) | 60 | Fall semester | 21 | AS | $6,100 | $12,288 | | | | |
| Kalamazoo Valley Community College | 24 | Aug | 21 | Dipl, AAS | $2,200 | $4,000 | | | | |
| Kellogg Community College (Battle Creek) | 20 | Aug | 24 | AAS | $3,748 | $5,171 | | | | • |
| Lansing Community College | 24 | Aug | 16 | AAS | $5,061 | $7,207 | | | | • |
| Mott Community College (Flint) | 36 | Jul | 24 | AAS | $3,620 | $7,100 | | | | • |
| University of Michigan (Ann Arbor) | 33 | Sep | 36, 24 | BS, MS | $10,370 | $29,522 | | | | • |

*Data are shown only for programs that completed the 2006 AMA Survey of Health Professions Education Programs
‡Key to Offers: 1: Evening or weekend classes; 2: Non-English instruction; 3: Cultural competence instruction

## Dental Hygienist

| Programs* | Class Capacity | Begins | Length (months) | Award | Res Tuition | Non-res Tuition | Stipend | Offers:‡ 1 | 2 | 3 |
|---|---|---|---|---|---|---|---|---|---|---|
| **Minnesota** | | | | | | | | | | |
| Argosy University/Twin Cities (Eagan) | 48 | Sep | 19 | AS | | | | | | |
| Century College (White Bear Lake) | 24 | Aug | 20 | AAS | $2,632 | $7,498 | | • | | • |
| Herzing College - Lakeland Academy Division (Minneapolis) | 36 | Jan | 16 | AAS | | | | | | • |
| Lake Superior College (Duluth) | 18 | Aug | 18 | AAS | $7,375 | $13,275 | | | | • |
| St Cloud Technical College | 14 | Aug | 18 | AAS | $8,500 | $8,500 | | | | |
| University of Minnesota - Minneapolis | 24 | Sep | 27 | BS | $8,000 | $20,000 | | | | • |
| **Mississippi** | | | | | | | | | | |
| Pearl River Community College (Hattiesburg) | 16 | Aug | 20 | AAS | $1,900 | $2,900 | | | | |
| **Missouri** | | | | | | | | | | |
| Missouri Southern State University (Joplin) | 30 | Aug | 22 | AS | $3,510 | $7,020 | | | | |
| Ozarks Technical Community College (Springfield) | 18 | Fall | 21 | Assoc | $75 | $95 | | | | • |
| State Fair Community College (Sedalia) | 10 | Fall | 24 | AAS | $70 | $150 | | | | |
| University of Missouri - Kansas City | 30 | Aug | 20 | BS, MSDH | $10,024 | $14,505 | | | | • |
| **Montana** | | | | | | | | | | |
| Montana State Univ - Great Falls Coll of Tech | 14 | Aug | 22 | AAS | $2,940 | $8,382 | | | | |
| **Nebraska** | | | | | | | | | | |
| Central Community College (Hastings) | 15 | Aug | 24 | AAS | $2,232 | $3,276 | | | | • |
| **Nevada** | | | | | | | | | | |
| Community College of Southern Nevada (Las Vegas) | 30 | Sep | 21 | AAS/AS, BSDH | $1,600 | $5,000 | | | | • |
| Truckee Meadows Community College (Reno) | 12 | Fall | 24 | AS | $1,457 | $3,803 | | | | |
| **New Hampshire** | | | | | | | | | | |
| New Hampshire Technical Institute (Concord) | 48 | Aug | 21, 29 | AS | $7,010 | $14,038 | | | | |
| **New Jersey** | | | | | | | | | | |
| Bergen Community College (Paramus) | 44 | Sep | 18 | AAS | $3,060 | $5,934 | | | | • |
| Camden County College (Blackwood) | 22 | Sep | 24 | AAS | $3,139 | $3,311 | | | | |
| Middlesex County College (Edison) | 30 | Sep | 15 | AAS | $2,965 | $5,993 | | | | |
| Univ of Med & Dent of New Jersey (Scotch Plains) | 44 | Jan | 32 | AAS | $6,480 | $9,720 | | • | | • |
| **New York** | | | | | | | | | | |
| Broome Community College (Binghamton) | 40 | Sep | 30 | AAS | $2,690 | $5,380 | | | | • |
| Erie Community College - North Campus (Williamsville) | 60 | Sep Jan | 24 | AAS | $2,900 | $5,800 | | | | • |
| Farmingdale State University of New York | 52 | Sep | 16 | AS | $4,350 | $10,300 | | | | • |
| Hudson Valley Community College (Troy) | 45 | Aug | 20 | Dipl, AAS | $2,700 | $8,090 | | • | | • |
| Monroe Community College (Rochester) | 48 | Sep | 24 | AAS | $2,700 | $5,400 | | | | |
| New York University | 100 | Sep | 20, 40 | Dipl, AAS, BS | $39,000 | $39,000 | | • | | |
| Orange County Community College (Middletown) | 20 | Sep | 9 | AAS | $2,900 | $5,800 | | | | |
| SUNY at Canton | 25 | Aug | 24 | AAS | $4,350 | $7,210 | | | | • |
| **North Carolina** | | | | | | | | | | |
| Asheville-Buncombe Technical Comm College | 20 | Aug | 21 | AAS | $1,824 | $10,128 | | | | • |
| Catawba Valley Community College (Hickory) | 20 | Aug | 24 | AS | $1,824 | $10,128 | | | | • |
| Central Piedmont Community College (Charlotte) | 30 | May | 24 | AAS | $1,850 | $9,450 | | | | • |
| Coastal Carolina Community College (Jacksonville) | 24 | Aug | 23 | AAS | $1,497 | $8,156 | | | | |
| Fayetteville Technical Community College | 34 | Aug | 22 | AAS | $1,790 | $9,748 | | | | • |
| Forsyth Technical Community College (Winston-Salem) | 12 | Aug | 21 | AAS | $1,920 | $10,560 | | | | |
| Guilford Technical Community College (Jamestown) | 36 | Aug | 22 | AAS | $2,174 | $9,734 | | | | |
| Halifax Community College (Weldon) | 18 | Aug | 21 | AAS | $1,737 | $8,098 | | | | • |
| Univ of North Carolina at Chapel Hill | 36 | Aug | 18 | Cert, BS, MS | $3,455 | $18,103 | | | | • |
| Wake Technical Community College (Raleigh) | 24 | Aug | 22 | AAS | $1,580 | $8,780 | | | | • |
| Wayne Community College (Goldsboro) | 30 | Aug | 20 | AAS | $1,540 | $8,121 | | | • | • |
| **Ohio** | | | | | | | | | | |
| Cuyahoga Community College (Cleveland) | 24 | Aug | 21 | AAS | $2,191 | $2,896 | | | | • |
| Lakeland Community College (Kirtland) | 20 | Sep | 18 | AAS | $3,400 | $8,800 | | | | • |
| Ohio State University (Columbus) | 32 | Sep | 33 | BSDH | $8,662 | $20,562 | | | | • |
| Owens Community College (Toledo) | 30 | Jun | 24 | AAS | $4,182 | $7,837 | | • | | |
| Rhodes State College (Lima) | 24 | Sep | 21 | AAS | $4,000 | $8,000 | | | | • |
| Shawnee State University (Portsmouth) | 24 | Sep | 22 | Dipl, AAS | $5,508 | $9,396 | | | | • |
| Sinclair Community College (Dayton) | 35 | Sep | 21 | AAS | $2,886 | $4,715 | | | | |
| Stark State College of Technology (North Canton) | 20 | Aug | 22 | AAS | $3,556 | $4,676 | | | | • |
| University of Cincinnati | 40 | Sep | 20 | AAS | $5,232 | $13,566 | | | | • |
| Youngstown State University | 27 | Aug | 19 | AAS | $6,861 | $12,502 | | | | • |
| **Oklahoma** | | | | | | | | | | |
| Rose State College (Midwest City) | 12 | Aug | 24 | AAS | $2,272 | $6,368 | | | | |
| Tulsa Community College | 14 | Aug | 16 | AAS | $1,673 | $4,488 | | | | • |
| **Oregon** | | | | | | | | | | |
| Oregon Institute of Technology (Klamath Falls) | 27 | Sep | 36, 18 | BS, AAS | $6,531 | $18,057 | | | | • |
| Portland Community College | 20 | Sep | 18 | AAS | $4,860 | $14,800 | | | | • |

*Data are shown only for programs that completed the 2006 AMA Survey of Health Professions Education Programs
‡Key to Offers: 1: Evening or weekend classes; 2: Non-English instruction; 3: Cultural competence instruction

## Dental Hygienist

| Programs* | Class Capacity | Begins | Length (months) | Award | Res Tuition | Non-res Tuition | Stipend | Offers:‡ 1 | 2 | 3 |
|---|---|---|---|---|---|---|---|---|---|---|
| **Pennsylvania** | | | | | | | | | | |
| Community College of Philadelphia | 32 | Jul | 24 | AAS | $8,000 | $15,000 | | | | |
| Harrisburg Area Community College | 24 | Aug | 20 | AA | $7,298 | $10,701 | | | | • |
| Luzerne County Community College (Nanticoke) | 30 | Jul | 24 | AAS | $3,354 | $6,708 | | | | |
| Northampton Community College (Bethlehem) | 40 | Aug | 17 | AAS | $6,768 | $14,400 | | | | • |
| Pennsylvania College of Technology (Williamsport) | 36 | Aug | 22, 36 | AAS, BS | $10,620 | $13,350 | | • | | • |
| Tri-State Business Institute (Erie) | 24 | Aug | 21 | AST | $14,350 | $14,350 | | | | |
| University of Pittsburgh | 36 | Aug | 22, 21 | Cert, BSDH | $16,104 | $30,126 | | | | • |
| Westmoreland County Community College (Youngwood) | 24 | Aug | 21 | AAS | $5,372 | $10,744 | | • | | • |
| **Rhode Island** | | | | | | | | | | |
| Community College of Rhode Island (Lincoln) | 36 | Sep | 24 | AAS | $2,951 | $6,224 | | • | | • |
| **South Carolina** | | | | | | | | | | |
| Florence-Darlington Technical College | 18 | Aug | 22 | AAS | $3,500 | $6,700 | | | | |
| Greenville Technical College | 40 | Aug | 27 | AS | $4,275 | $8,940 | | • | | • |
| Horry - Georgetown Technical College (Conway) | 16 | Aug | 21 | AS | $4,200 | $6,396 | | | | |
| Trident Technical College (Charleston) | 24 | Jan | 21 | AS | $4,671 | $8,847 | | • | | |
| York Technical College (Rock Hill) | 20 | Aug | 21 | AAS | $2,900 | $6,528 | | | | |
| **Tennessee** | | | | | | | | | | |
| East Tennessee State University (Johnson City) | 24 | Aug | 19 | BSDH | $3,678 | $9,633 | | | | |
| Roane State Community College (Oak Ridge) | 12 | Aug | 19 | AAS | $1,952 | $5,846 | | | | • |
| Tennessee State University (Nashville) | 36 | Aug | 18, 24 | AAS | $2,207 | $6,863 | | | | • |
| University of Tennessee Health Science Ctr (Memphis) | 34 | Sep | 21 | Dipl, BS | $4,466 | $14,564 | | | | • |
| **Texas** | | | | | | | | | | |
| Baylor College of Dentistry, Texas A&M HSC (Dallas) | 30 | Aug | 21 | BS, MS | $3,700 | $13,000 | | | | • |
| Blinn College (Bryan) | 12 | Sep | 24 | AAS | $1,900 | $4,275 | | | | • |
| Coastal Bend College (Beeville) | 30 | Aug | 19 | Dipl, AAS | $5,784 | $6,576 | | | | • |
| Del Mar College (Corpus Christi) | 24 | Sep | 16 | AAS | $1,655 | $1,855 | | | | |
| El Paso Community College | 16 | Aug | 21 | AAS | $1,496 | $2,615 | | | | |
| Kingwood College - NHMCCD | 12 | Fall semester | 21 | AAS | $1,960 | $7,700 | | | | • |
| Lamar Institute of Technology (Beaumont) | 30 | Jul | 24 | AAS | $2,400 | $0 | | | | |
| Midwestern State University (Wichita Falls) | 18 | Aug Sep | 18 | BS | $5,200 | $14,300 | | | | • |
| Tarrant County College (Hurst) | 30 | Aug | 24 | AAS | $1,426 | $4,650 | | | | • |
| Temple College | 12 | May | 24 | AAS | $3,900 | $4,700 | | | | • |
| Texas Woman's University (Denton) | 28 | Aug | 36 | BS | $3,690 | $10,170 | | | | |
| Univ of Texas Hlth Sci Ctr at Houston | 40 | Aug | 20 | Cert, BS | $4,488 | $13,563 | | | | |
| Univ of Texas Hlth Sci Ctr at San Antonio | 48 | Aug | 22, 48 | Cert, BS, MS | $3,926 | $11,792 | | • | | • |
| Wharton County Junior College | 28 | Aug | 24 | AAS | $1,440 | $1,440 | | | | • |
| **Utah** | | | | | | | | | | |
| Salt Lake Community College (Salt Lake City) | 26 | Aug | 18 | Dipl, AAS, AS | $2,312 | $7,232 | | | | • |
| **Virginia** | | | | | | | | | | |
| Northern Virginia Community College (Springfield) | 54 | Aug | 22 | AAS | $2,600 | $7,800 | | | | • |
| Virginia Western Community College (Roanoke) | 18 | Aug | 18 | AAS | $2,781 | $17,444 | | | | • |
| Wytheville Community College | 20 | Aug | 21 | AAS | $2,500 | $7,332 | | | | |
| **Washington** | | | | | | | | | | |
| Columbia Basin College (Pasco) | 18 | Fall | 22 | AAS | $3,072 | $9,808 | | | | • |
| Lake Washington Technical College (Kirkland) | 30 | Sep | 21 | AAS | $5,463 | $5,463 | | | • | • |
| Pierce College (Lakewood) | 26 | Sep | 21 | AAS | $2,866 | $4,813 | | | | • |
| Seattle Central Community College | 18 | Fall | 21 | AAS-T | $3,460 | $10,882 | | | | • |
| Shoreline Community College | 24 | Sep | 20 | AAAS | | | | | | • |
| Yakima Valley Community College | 36 | Sep | 18 | AAS | $3,400 | $8,178 | | | | • |
| **West Virginia** | | | | | | | | | | |
| Community & Technical College at WVU Tech (Montgomery) | 22 | Aug | 24 | AS | $3,168 | $10,438 | | | | |
| West Liberty State College | 40 | Aug | 18 | AS | $3,686 | $9,054 | | | | • |
| West Virginia University (Morgantown) | 24 | Aug | 43 | BSDH, MSDH | $3,668 | $9,922 | | | | • |
| **Wisconsin** | | | | | | | | | | |
| Fox Valley Technical College (Appleton) | 18 | Aug | 12 | AS | $2,100 | $11,100 | | • | | |
| Madison Area Technical College | 36 | Aug | 18 | AAS | $1,802 | $10,200 | | | | |
| Milwaukee Area Technical College | 48 | Aug Jan | 18 | AAS | $2,997 | $17,789 | | • | • | |
| Northcentral Technical College (Wausau) | 24 | Aug | 24, 27 | AS | $5,962 | $20,104 | | | | |
| Northeast Wisconsin Technical College (Green Bay) | 24 | Jun | 26 | AAS | $4,500 | $14,547 | | | • | • |
| Waukesha County Technical College (Pewaukee) | 30 | Aug | 20 | AS | $3,656 | $20,368 | | | | • |
| **Wyoming** | | | | | | | | | | |
| Sheridan College | 24 | Aug | 18 | AAS | $1,368 | $4,104 | | | | |

*Data are shown only for programs that completed the 2006 AMA Survey of Health Professions Education Programs
‡Key to Offers: 1: Evening or weekend classes; 2: Non-English instruction; 3: Cultural competence instruction

## Dental Laboratory Technician

# Alabama

**H Council Trenholm State Technical College**
Dental Lab Technician Prgm
1225 Air Base Blvd
Montgomery, AL 36108
*Prgm Dir:* Roosevelt Daniel, DDS
*Tel:* 334 420-4428
*E-mail:* rdaniel@trenholmtech.cc.al.us

# Arizona

**Pima Community College**
Dental Lab Technician Prgm
2202 W Anklam Rd, HRP 220
Tucson, AZ 85709-0080
*Prgm Dir:* Max Atwell, BS
*Tel:* 520 206-3100    *Fax:* 520 206-3027
*E-mail:* Max.Atwell@pima.edu

# California

**Los Angeles City College**
Dental Lab Technician Prgm
855 N Vermont Ave
Los Angeles, CA 90029
*Prgm Dir:* Arax S Cohen, BSB
*Tel:* 323 953-4000, Ext 2501
*E-mail:* cohenas@lacitycollege.edu

**Pasadena City College**
Dental Lab Technician Prgm
1570 E Colorado Blvd, R505
Pasadena, CA 91106
www.pasadena.edu
*Prgm Dir:* Anita M Bobich, BA AA CDT
*Tel:* 626 585-7884    *Fax:* 626 585-3168
*E-mail:* ambobich@pasadena.edu

# Florida

**McFatter Vocational Technical Center**
Dental Lab Technician Prgm
6500 Nova Dr
Davie, FL 33317
www.mcfattertech.com
*Prgm Dir:* Fred Isaac
*Tel:* 754 321-3023, Ext 3022    *Fax:* 754 321-5820
*E-mail:* fred.isaac@browardschools.com
*Med Dir:* Mary Smith

**Indian River Community College**
Dental Lab Technician Prgm
3209 Virginia Ave
Ft Pierce, FL 34981-5599
*Prgm Dir:* Don Symington, EdS
*Tel:* 772 462-7527    *Fax:* 772 462-4900
*E-mail:* dsymingt@ircc.edu

# Georgia

**Atlanta Technical College**
Dental Lab Technician Prgm
1560 Metropolitan Pkwy SW
Atlanta, GA 30310-4446
*Prgm Dir:* Becky Tolson, CDT
*Tel:* 404 756-3811    *Fax:* 404 758-8522
*E-mail:* btolson@atlantatech.edu

# Idaho

**Idaho State University**
Dental Lab Technician Prgm
Dental Lab Tech Box 8380
Pocatello, ID 83209-8380
*Prgm Dir:* Diane Edmunds
*Tel:* 208 282-3141    *Fax:* 208 282-3975
*E-mail:* edmudian@isu.edu
*Med Dir:* Debbie Thompson

# Indiana

**Indiana Univ - Purdue Univ Ft Wayne**
Dental Lab Technician Prgm
2101 Coliseum Blvd E
Ft Wayne, IN 46805
www.ipfw.edu/dlt
*Prgm Dir:* Charles Champion, CDT MS
*Tel:* 260 481-6837    *Fax:* 260 481-5767
*E-mail:* champion@ipfw.edu

# Iowa

**Kirkwood Community College**
Dental Lab Technician Prgm
6301 Kirkwood Blvd SW, PO Box 2068
Cedar Rapids, IA 52406-9973
www.kirkwood.edu
*Prgm Dir:* Betty Mitchell, BS CDT
*Tel:* 319 398-5400    *Fax:* 319 398-1293
*E-mail:* betty.mitchell@kirkwood.edu
*Med Dir:* Nancy Glab, RN, MSN

# Kentucky

**Bluegrass Community and Technical College**
Dental Lab Technician Prgm
470 Cooper Dr, 330 Oswald Bldg
Lexington, KY 40506-0235
www.bluegrass.kctcs.edu/LCC/DLT
*Prgm Dir:* Robin Gornto, MSEd CDT
*Tel:* 859 246-6244    *Fax:* 859 246-4671
*E-mail:* robin.gornto@kctcs.edu

# Louisiana

**Louisiana State University**
Dental Lab Technician Prgm
1100 Florida Ave
New Orleans, LA 70119
www.lsusd.lsumc.edu/academic/lab_tech.htm
*Prgm Dir:* Allan Rappold, DDS
*Tel:* 504 619-8684    *Fax:* 504 619-8780
*E-mail:* arappo@lsuhsc.edu

# Massachusetts

**Middlesex Community College - Lowell**
Dental Lab Technician Prgm
33 Kearney Square
Lowell, MA 01852
www.middlesex.mass.edu
*Prgm Dir:* Jerry Kessler, BS
*Tel:* 978 656-3056, Ext 3056    *Fax:* 978 656-3078
*E-mail:* kesslerj@middlesex.cc.ma.us

# New York

**New York City College of Technology**
Dental Lab Technician Prgm
300 Jay St
Brooklyn, NY 11201-2983
*Prgm Dir:* Nicholas Manos
*Tel:* 718 260-5137    *Fax:* 718 260-5995
*E-mail:* NManos@citytech.cuny.edu

**Erie Community College - South Campus**
Dental Lab Technician Prgm
4041 Southwestern Blvd
Orchard Park, NY 14127-2199
www.ecc.edu
*Prgm Dir:* Marvin Herman, MD
*Tel:* 716 851-1759    *Fax:* 716 851-1704
*E-mail:* herman@ecc.edu

# North Carolina

**Durham Technical Community College**
Dental Lab Technician Prgm
1637 Lawson St
Durham, NC 27703
www.durhamtech.edu
*Prgm Dir:* Michael Patrick, AA AAS CDT
*Tel:* 919 686-3399    *Fax:* 919 686-3737
*E-mail:* patrickm@durhamtech.edu

# Oregon

**Portland Community College**
Dental Lab Technician Prgm
PO Box 19000
Portland, OR 97280-0990
www.pcc.edu
*Prgm Dir:* Josette Beach, RDH MS
*Tel:* 503 977-4235    *Fax:* 503 977-8300
*E-mail:* jbeach@pcc.edu

# Texas

**Univ of Texas Hlth Sci Ctr at San Antonio**
Dental Lab Technician Prgm
7703 Floyd Curl Dr
San Antonio, TX 78284-7914
*Prgm Dir:* Roosevelt Davis, MS
*Tel:* 210 567-3056    *Fax:* 210 567-3061
*E-mail:* davisrd@uthscsa.edu

**School of Health Care Sciences - Air Force**
Dental Lab Technician Prgm
381 TRS/XWAE
917 Missile Rd
Sheppard AFB, TX 76311-2246
*Prgm Dir:* Lt Col Thomas S Bingham III
*Tel:* 940 676-6967    *Fax:* 940 676-6928

# Virginia

**J Sargeant Reynolds Community College**
Dental Lab Technician Prgm
PO Box 85622
Richmond, VA 23285-5622
www.jsr.vccs.edu
*Prgm Dir:* Ernie Wolfe, CDT AAS BS MEd
*Tel:* 804 523-5931    *Fax:* 804 786-5298
*E-mail:* ewolfe@jsr.vccs.edu

# Washington

**Bates Technical College**
Dental Lab Technician Prgm
1101 S Yakima Ave
Tacoma, WA 98405
*Prgm Dir:* Jean Watley
*Tel:* 253 680-7215    *Fax:* 253 680-7293
*E-mail:* jwatley@bates.ctc.edu

## Dental Laboratory Technician

| Programs* | Class Capacity | Begins | Length (months) | Award | Res Tuition | Non-res Tuition | Stipend | Offers:‡ 1 | 2 | 3 |
|---|---|---|---|---|---|---|---|---|---|---|
| **California** | | | | | | | | | | |
| Pasadena City College | 26 | Sep | 19 | Cert, AA, AS | $360 | $1,700 | | | | |
| **Florida** | | | | | | | | | | |
| McFatter Vocational Technical Center (Davie) | 24 | Aug | 18 | Cert | $4,180 | $8,000 | | | | |
| **Indiana** | | | | | | | | | | |
| Indiana Univ - Purdue Univ Ft Wayne | 20 | Aug | 24 | AS | $6,241 | $14,296 | | | | |
| **Iowa** | | | | | | | | | | |
| Kirkwood Community College (Cedar Rapids) | 15 | Aug | 22 | AAS | $4,183 | $8,366 | | | | • |
| **Kentucky** | | | | | | | | | | |
| Bluegrass Community and Technical College (Lexington) | 15 | Aug | 16 | AAS | $2,352 | $7,056 | | | | |
| **Louisiana** | | | | | | | | | | |
| Louisiana State University (New Orleans) | 12 | Jul | 24, 36 | AS, BS | $4,695 | $5,995 | | | | |
| **Massachusetts** | | | | | | | | | | |
| Middlesex Community College - Lowell | 15 | Sep | 18 | AS | $2,870 | $3,235 | | | | |
| **New York** | | | | | | | | | | |
| Erie Community College - South Campus (Orchard Park) | 30 | Sep | 24 | AAS | $2,500 | $5,000 | | • | | • |
| **North Carolina** | | | | | | | | | | |
| Durham Technical Community College | 24 | Aug | 12, 21 | Cert, AAS | $2,076 | $10,716 | | | | |
| **Oregon** | | | | | | | | | | |
| Portland Community College | 26 | Sep | 18 | Cert | $2,542 | $7,585 | | | | |
| **Virginia** | | | | | | | | | | |
| J Sargeant Reynolds Community College (Richmond) | 20 | Aug | 21 | Cert, AAS | $2,935 | $8,777 | | | | |

*Data are shown only for programs that completed the 2006 AMA Survey of Health Professions Education Programs
‡Key to Offers: 1: Evening or weekend classes; 2: Non-English instruction; 3: Cultural competence instruction

# Diagnostic Medical Sonographer

## Occupational Description

The diagnostic medical sonographer provides patient services using medical ultrasound (high-frequency sound waves that produce images of internal structures). Working under the supervision of a physician responsible for the use and interpretation of ultrasound procedures, the sonographer helps gather sonographic data to diagnose a variety of conditions and diseases, as well as monitor fetal development.

## Job Description

The sonographer provides patient services in a variety of medical settings in which the physician is responsible for the use and interpretation of ultrasound procedures. In assisting physicians in gathering sonographic data, the diagnostic medical sonographer is able to obtain, review, and integrate pertinent patient history and supporting clinical data to facilitate optimum diagnostic results; perform appropriate procedures and record anatomical, pathological, and/or physiological data for interpretation by a physician; record and process sonographic data and other pertinent observations made during the procedure for presentation to the interpreting physician; exercise discretion and judgment in the performance of sonographic services; provide patient education related to medical ultrasound; and promote principles of good health.

## Employment Characteristics

Diagnostic medical sonographers may be employed in hospitals, clinics, private offices, and industry. Most full-time sonographers work about 40 hours a week; they may have evening weekend hours and times when they are on call and must be ready to report to work on short notice.

The demand for sonographers, including suitably qualified educators, researchers, and administrators, continues to exceed the supply., with faster than average job growth anticipated. The supply and demand ratio affects salaries, depending on experience and responsibilities.

## Salary

According to the Society of Diagnostic Medical Sonographers, the hourly salary for diagnostic medical sonographers is $29.00; median income in 2005 was $61,984. Refer to Section IV, Table 5 of this *Directory* for more information, or see www.ama-assn.org/go/ salary.

## Educational Programs

**Length.** Accredited programs are between 1 and 4 years (certificate, associate, and baccalaureate level), depending on program design, objectives, and the degree or certificate awarded.

**Prerequisites.** Applicants to a 1-year program must possess qualifications in a clinically related allied health profession. Applicants to 2-year programs must be high school graduates (or equivalent) with an educational background in basic science, general physics, and algebra. All applicants must demonstrate satisfactory completion of the following courses at college level: general physics, biological science, algebra, and communication skills.

Skills potential and practicing sonographers should exhibit include social perceptiveness, learning strategies, critical thinking skills, instructional skills, active listening, active learning, reading comprehension, and written/oral expression.

**Curriculum.** Curricula of accredited programs include physical sciences, applied biological sciences, patient care, clinical medicine, applications of ultrasound, instrumentation, related diagnostic procedures, and image evaluation. A plan for well-structured, competency-based clinical education is an essential part of the curriculum of all sonography programs.

## Inquiries

**Careers/Curriculum**

Society of Diagnostic Medical Sonography
2745 Dallas Parkway, Suite 350
Plano, TX 75093-4706
214 473-8057
214 473-8563 Fax
E-mail: info@sdms.org
www.sdms.org

Society for Vascular Ultrasound
4601 Presidents Drive, Suite 260
Lanham, MD 20706-4365
301 459-7550 or 800 SVT- VEIN
301 459-5651 Fax
www.svtnet.org

American Society of Echocardiography
1500 Sunday Drive, Suite 102
Raleigh, NC 27607
919 861-5574
919 787-4916 Fax
www.asecho.org

**Certification**

American Registry for Diagnostic Medical Sonography
51 Monroe Street, Plaza East Ore
Rockville, MD 20852
301 738-8401

American Registry of Radiologic Technologists
1255 Northland Drive
St Paul, MN 55120-1155
651 687-0048
www.arrt.org

**Program Accreditation**

Commission on Accreditation of Allied Health Education Programs (CAAHEP) in collaboration with:
Joint Review Committee on Education in Diagnostic Medical Sonography
2025 Woodlane Drive
St Paul, MN 55125
651 731-1582
E-mail: jrc-dms@jcahpo.org

**Diagnostic Medical Sonographer***

# Alabama

**Wallace State Community College**
Diagnostic Med Sonography Prgm
General concentration
PO Box 2000
Hanceville, AL 35077-2000
www.wallacestate.edu
*Prgm Dir:* Janet E Money, RDMS CNMT
*Tel:* 256 352-8318    *Fax:* 256 352-8320
*E-mail:* janet.money@wallacestate.edu
*Med Dir:* Leon Bell, MD

**Insitute of Ultrasound Diagnostics**
Diagnostic Med Sonography Prgm
General concentration
1230 Montlimar Dr, Ste A
Mobile, AL 36609
www.iudmed.com
*Prgm Dir:* Kathryn A Gill
*Tel:* 251 460-2485    *Fax:* 251 460-0672
*E-mail:* kgill@iudmed.com
*Med Dir:* Larry J Arcement, MD

**Baptist Medical Center South**
Diagnostic Med Sonography Prgm
General concentration
2169 Normandie Dr
Montgomery, AL 36111
www.baptistfirst.org/educationopportunities.htm
*Prgm Dir:* Brandi S Merrill, BS RT(R) RDMS (AB
    OB/GYN) RVT
*Tel:* 334 281-3470, Ext 2    *Fax:* 334 281-3671
*E-mail:* bmerrill@baptistfirst.org
*Med Dir:* David C Montiel, MD

# Arizona

**GateWay Community College**
Diagnostic Med Sonography Prgm
General concentration
108 N 40th St
Phoenix, AZ 85034
*Prgm Dir:* Kathleen Murphy
*Tel:* 602 286-8490    *Fax:* 602 288-8003
*E-mail:* Murphy@gatewaycc.edu
*Med Dir:* John Crowe, MD

# Arkansas

**University of Arkansas for Medical Sciences**
Diagnostic Med Sonography Prgm
General, Vascular & Echocardiography concentrations
4301 W Markham St Mail Slot 563-B
Little Rock, AR 72205
www.uams.edu/chrp/dms
*Prgm Dir:* Terry DuBose, MS RDMS
*Tel:* 501 686-6510    *Fax:* 501 686-6513
*E-mail:* duboseterryj@uams.edu
*Med Dir:* Teresita L Angtuaco, MD

**Arkansas State University**
Diagnostic Med Sonography Prgm
General concentration
PO Box 910
State University, AR 72467
www.astate.edu
*Prgm Dir:* Jennifer DeClerk, BSRS RDMS RT(R)
*Tel:* 870 972-2914    *Fax:* 870 972-2004
*E-mail:* jdeclerk@astate.edu
*Med Dir:* Ellen McDaniel, MD

---

*Additional information about programs that
returned the AMA's survey is available in a table
beginning on page 146.

# California

**Orange Coast College**
Diagnostic Med Sonography Prgm
General concentration
2701 Fairview Rd
Costa Mesa, CA 92628-5005
*Prgm Dir:* Joan M Clasby, BVE RDMS
*Tel:* 714 432-5893    *Fax:* 714 432-5534
*E-mail:* jclasby@mail.occ.cccd.edu
*Med Dir:* Gideon Strich, MD

**Cypress College**
Diagnostic Med Sonography Prgm
General concentration
9200 Valley View
Cypress, CA 90630
www.cypresscollege.edu
*Prgm Dir:* Lynn Mitts, MA RT ARDMS
*Tel:* 714 484-7283, Ext 48901    *Fax:* 714 527-2175
*E-mail:* lmitts@cypresscollege.edu
*Med Dir:* Michael Poh, MD

**Loma Linda University**
Diagnostic Med Sonography Prgm
General concentration
Sch of Allied Hlth Profs
Loma Linda, CA 92354
www.llumc.edu
*Prgm Dir:* Marie T DeLange, BS RT RDMS RDCS
*Tel:* 909 558-4000, Ext 43027    *Fax:* 909 558-4166
*E-mail:* mdelange@ahs.llumc.edu
*Med Dir:* Glenn A Rouse, MD

**Foothill Community College**
Diagnostic Med Sonography Prgm
General concentration
12345 El Monte Rd
Los Altos Hills, CA 94022-4599
www.foothill.edu
*Prgm Dir:* Kathleen Austin, BS RDMS RT
*Tel:* 650 949-7304    *Fax:* 650 949-7686
*E-mail:* austinkathleen@fhda.edu
*Med Dir:* Volney Van Dalsem III, MD

**Merced College**
Diagnostic Med Sonography Prgm
General, Cardiac concentrations
3600 M St
Merced, CA 95348-2806
*Prgm Dir:* Joy Guthrie, MS RDMS RDCS
*Tel:* 209 384-6057
*E-mail:* guthriej@hotmail.com
*Med Dir:* J Charles Smith, MD

**Kaiser Permanente Sch of Allied Hlth
    Sciences**
Diagnostic Med Sonography Prgm
General concentration
938 Marina Way S
Richmond, CA 94804
www.kp.org
*Prgm Dir:* Carmelo Fernandez, MD RDMS
*Tel:* 510 231-5055    *Fax:* 510 231-5103
*E-mail:* Carmelo.F.Fernandez@kp.org
*Med Dir:* Darryl Jones, MD

**Univ of California San Diego Med Ctr**
Diagnostic Med Sonography Prgm
General concentration
200 W Arbor Dr
San Diego, CA 92103-8759
www.radtech.ucsd.edu
*Prgm Dir:* Nannette Forsythe, BA RDMS RVT
*Tel:* 619 543-6617    *Fax:* 619 543-7464
*E-mail:* nforsythe@ucsd.edu
*Med Dir:* Mary O'Boyle, MD

# Colorado

**U of Colorado at Denver and Health Scis Ctr**
Diagnostic Med Sonography Prgm
General concentration
Univ of Colorado Hosp - Anhschutz Center of Advance
    Medicine
Aurora, CO 80045
www.uchsc.edu/radiology
*Prgm Dir:* James Cacari, BS RT RVT RDMS
*Tel:* 720 848-1872    *Fax:* 720 848-1882
*E-mail:* James.Cacari@uch.edu
*Med Dir:* Elizabeth R Stamm, MD

# Connecticut

**St Francis Hospital & Medical Center**
*Cosponsor:* The Hofman Heart Institute of Connecticut
Diagnostic Med Sonography Prgm
Cardiac concentration
114 Woodland St
Hartford, CT 06105
*Prgm Dir:* Richard Palma, BS RDCS APS FASE
*Tel:* 860 714-4568
*E-mail:* rpalma@stfranciscare.org
*Med Dir:* Bernard Clark, III, MD

# Delaware

**Delaware Tech & Comm Coll - Wilmington**
Diagnostic Med Sonography Prgm
General, Cardiac & Vascular concentrations
333 Shipley St
Wilmington, DE 19801
www.dtcc.edu/Wilmington/ah
*Prgm Dir:* Lily O Lee, BS RDMS RVT
*Tel:* 302 765-4588    *Fax:* 302 765-4599
*E-mail:* lillee@christianacare.org
*Med Dir:* Howard M Levy, MD

# District of Columbia

**George Washington University**
Diagnostic Med Sonography Prgm
General, Cardiac & Vascular concentrations
900 23rd St NW #6180
Washington, DC 20037
www.gwu.edu/~sonoprog
*Prgm Dir:* Catheeja Ismail, MA RDMS
*Tel:* 202 994-8697    *Fax:* 202 994-1073
*E-mail:* cismail@gwu.edu
*Med Dir:* Michael C Hill, MB

# Florida

**Broward Community College**
Diagnostic Med Sonography Prgm
General & Cardiac concentrations
North Campus Bldg 41
Coconut Creek, FL 33066
*Prgm Dir:* Sharon Calton, MS RDMS RDCS FSDMS
*Tel:* 954 201-2089    *Fax:* 954 201-2348
*E-mail:* scalton@broward.edu
*Med Dir:* Maria Rodriguez, MD

**Keiser College - Daytona Beach**
Diagnostic Med Sonography Prgm
General concentration
1800 Business Park Blvd
Daytona Beach, FL 32114
www.keisercollege.edu
*Prgm Dir:* Marianne Peiffer, RDMS
*Tel:* 386 274-5060, Ext 123    *Fax:* 386 274-2725
*E-mail:* mariannep@keisercollege.edu
*Med Dir:* Tanya Marchand, MD

**Keiser University**
Diagnostic Med Sonography Prgm
General concentration
1500 NW 49th St
Ft Lauderdale, FL 33309
*Prgm Dir:* Rosy Silverman, RDMS
*Tel:* 954 776-4456, Ext 524   *Fax:* 954 776-5157
*E-mail:* rsilverman@keisercollege.edu
*Med Dir:* Sylvia Petterson, MD

**Nova Southeastern University**
Diagnostic Med Sonography Prgm
Vascular concentration
3200 S University Dr
Ft Lauderdale, FL 33328
*Prgm Dir:* Terrence D Case, M Ed RVT
*Tel:* 954 262-1220
*E-mail:* tcase@nova.edu
*Med Dir:* James Benenati, MD

**St Vincent's Medical Center**
Diagnostic Med Sonography Prgm
General & Vascular concentrations
1800 Barrs St
Jacksonville, FL 32204
www.jaxhealth.com
*Prgm Dir:* Chemene Wilson, RDMS RT
*Tel:* 904 308-8272   *Fax:* 904 308-5109
*E-mail:* cwils002@stvincentshealth.com
*Med Dir:* Michael Donohue, MD

**Miami Dade College**
Diagnostic Med Sonography Prgm
General & Cardiac concentrations
Medical Center Campus
Miami, FL 33127-4693
*Prgm Dir:* Dalia Sanchez-Suarez, BS RDMS RDCS
*Tel:* 305 237-4245   *Fax:* 305 237-4278
*E-mail:* dsanche1@mdc.edu
*Med Dir:* Pamela Dickson, MD

**Valencia Community College**
Diagnostic Med Sonography Prgm
General concentration
PO Box 3028
Orlando, FL 32811
*Prgm Dir:* Barbara Ball, BA RT(R) RDMS
*Tel:* 407 582-1191   *Fax:* 407 582-1278
*E-mail:* bball@valencia.cc.fl.us
*Med Dir:* Holly Saunders, MD

**Palm Beach Community College**
Diagnostic Med Sonography Prgm
General concentration
3160 PGA Blvd
Palm Beach Gardens, FL 33410
*Prgm Dir:* Patty Moraino-Braga, BS RDMS RVT RDCS RT(R)
*Tel:* 561 207-5053
*E-mail:* bragap@pbcc.edu

**Hillsborough Community College**
Diagnostic Med Sonography Prgm
General concentration
PO Box 30030
Tampa, FL 33630-3030
*Prgm Dir:* Louis J Gomez, RDMS RVT
*Tel:* 813 253-7412   *Fax:* 813 253-7473
*E-mail:* lgomez@hccfl.edu
*Med Dir:* Claude Guidi, MD

# Georgia

**Grady Health System**
Diagnostic Med Sonography Prgm
General concentration
80 Jesse Hill Jr Dr SE, PO Box 26095
Atlanta, GA 30303-3050
www.gradyhealthsystem.org
*Prgm Dir:* Judy K Billings, BS RT(R) RDMS
*Tel:* 404 616-5032   *Fax:* 404 616-3512
*E-mail:* jbillings@gmh.edu
*Med Dir:* William Small, MD

**Sanford-Brown Institute**
Diagnostic Med Sonography Prgm
General concentration
1140 Hammond Dr, Ste A 1150
Atlanta, GA 30328
*Prgm Dir:* B Dwight Gunter
*Tel:* 770 576-6451
*E-mail:* dgunter@sb-atlanta.com

**Medical College of Georgia**
Diagnostic Med Sonography Prgm
General concentration
1120 15th St, AE1003
Augusta, GA 30912
*Prgm Dir:* Eric L Meaders
*Tel:* 706 721-2759   *Fax:* 706 721-8293
*E-mail:* emeaders@mcg.edu
*Med Dir:* Lloyd Schnuck, MD

**Coosa Valley Technical College**
Diagnostic Med Sonography Prgm
General, Vascular & Echocardiography concentrations
1 Maurice Culberson Dr
Rome, GA 30161
http://test.cvtcollege.org/Ac_Programs/dms_vascular
*Prgm Dir:* Leif Penrose, BA RDMS RDCS RVT
*Tel:* 706 295-6970   *Fax:* 706 295-6894
*E-mail:* lpenrose@coosavalleytech.edu
*Med Dir:* Vick Whitley, MD

**Ogeechee Technical College**
Diagnostic Med Sonography Prgm
General concentration
1 Joe Kennedy Blvd
Statesboro, GA 30458
*Prgm Dir:* Tina W Welch, AS RDMS RVT RT( R )
*Tel:* 912 688-6019   *Fax:* 912 486-7604
*E-mail:* twelch@ogeecheetech.edu
*Med Dir:* John E Martin, MD FACS

# Idaho

**Boise State University**
Diagnostic Med Sonography Prgm
General concentration
Radiologic Sciences Department
Boise, ID 83725-1845
http://radsci.boisestate.edu
*Prgm Dir:* Joie Burns, MS RT(R)(S) RDMS RVT
*Tel:* 208 426-1996   *Fax:* 208 426-4459
*E-mail:* jburns@boisestate.edu
*Med Dir:* Lisa Scales, MD

# Illinois

**Southern Illinois Univ at Carbondale**
Diagnostic Med Sonography Prgm
General concentration
1365 Douglas Dr, Mailcode 6615
Carbondale, IL 62901
www.siu.edu
*Prgm Dir:* Karen Having, MS Ed RT(R) RDMS
*Tel:* 618 453-4980   *Fax:* 618 453-7268
*E-mail:* khaving@siu.edu
*Med Dir:* Jagan Allnani, MD

**John A Logan College**
Diagnostic Med Sonography Prgm
Cardiac concentration
700 Logan College Rd
Carterville, IL 62918
www.jalc.edu
*Prgm Dir:* Valerie Newberry, RVT RDCS
*Tel:* 618 985-2828, Ext 8622   *Fax:* 618 985-4654
*E-mail:* valerienewberry@jalc.edu
*Med Dir:* Mohammad Yusuf Mansuri, MD

**Northwestern Memorial Hospital**
Diagnostic Med Sonography Prgm
General concentration
676 St Clair St, Ste 550
Chicago, IL 60611
www.nmh.org
*Prgm Dir:* Casey Clarke, BSRT RT(R) RDMS RDCS
*Tel:* 312 926-1196   *Fax:* 312 926-1741
*E-mail:* cclarke@nmh.org
*Med Dir:* Helena Gabriel, MD

**Rush University**
Diagnostic Med Sonography Prgm
Vascular concentration
600 S Paulina - Suite 440
Chicago, IL 60612
*Prgm Dir:* Eileen French-Sherry, MA RVT
*Tel:* 312 942-7286
*E-mail:* eileen_french-sherry@rush.edu

**College of DuPage**
Diagnostic Med Sonography Prgm
General & Vascular concentrations
425 Fawell Blvd
Glen Ellyn, IL 60137
www.cod.edu
*Prgm Dir:* Terrie Ciez, MS RDMS RDCS RT CNMT
*Tel:* 630 942-2436   *Fax:* 630 858-5409
*E-mail:* ciezte@cdnet.cod.edu
*Med Dir:* Michael Schwartz, MD FACR

**Triton College**
Diagnostic Med Sonography Prgm
General concentration
2000 N Fifth Ave
River Grove, IL 60171
*Prgm Dir:* Debra L Krukowski, BS RDMS RTR
*Tel:* 708 456-0300, Ext 3979   *Fax:* 708 583-3336
*E-mail:* dkrukows@triton.edu
*Med Dir:* S Asokan, MD

**South Suburban College**
Diagnostic Med Sonography Prgm
General concentration
15800 S State St
South Holland, IL 60473
*Prgm Dir:* Casey Clark, RDMS RDCS RT(R)
*Tel:* 708 596-2000, Ext 2318
*E-mail:* dms@southsuburbancollege.edu

# Iowa

**Mercy College of Health Sciences**
Diagnostic Med Sonography Prgm
General & Cardiovascular concentrations
928 Sixth Ave
Des Moines, IA 50309-1239
*Prgm Dir:* Kathleen Lane, MSE S RDCS RVT
*Tel:* 515 643-6610   *Fax:* 515 643-6698
*E-mail:* klane@mercydesmoines.org
*Med Dir:* Richard Marcus, MD

**University of Iowa Hospitals & Clinics**
Diagnostic Med Sonography Prgm
General, Cardiac & Vascular concentrations
C-723 Radiology
Iowa City, IA 52242-1077
http://radiology.uiowa.edu/RadTech/
  SonographyBrochure.htm
*Prgm Dir:* Stephanie Ellingson, MS RDMS RDCS RVT
  RTR
*Tel:* 319 356-4871   *Fax:* 319 384-9574
*E-mail:* stephanie-ellingson@uiowa.edu
*Med Dir:* Monzer Abu-Yousef, MD

# Kansas

**University of Kansas Medical Center**
Diagnostic Med Sonography Prgm
General & Vascular concentrations
2105 Bell Memorial Hospital
Kansas City, KS 66160
*Prgm Dir:* Candace S Spalding, BA RDMS RVT RT(R)
*Tel:* 913 588-6802
*E-mail:* cpaldin@kumc.edu
*Med Dir:* Stanton Rosenthal, MD

**University of Kansas Medical Center**
Diagnostic Med Sonography Prgm
Cardiac concentration
Mid-American Cardiology
Kansas City, KS 66160
*Prgm Dir:* Mary Chivington, BS RDCS RVT
*Tel:* 913 588-9635   *Fax:* 913 588-1605
*E-mail:* mchivington@mac.md
*Med Dir:* Steven D Owens, MD

**Washburn University**
Diagnostic Med Sonography Prgm
General, Cardiac and Vascular concentrations
1700 SW College Ave
Topeka, KS 66621
*Prgm Dir:* Doug James
*Tel:* 785 231-1010, Ext 2170
*E-mail:* doug.james@washburn.edu
*Med Dir:* Steven Watkins, MD

# Kentucky

**St Catharine College**
Diagnostic Med Sonography Prgm
General, Cardiac & Vascular concentrations
310 Xavier Dr
Bardstown, KY 40004
*Prgm Dir:* Dennis Walter, AAS RRT RVT
*Tel:* 502 348-0475   *Fax:* 502 348-0466
*E-mail:* dwalters@bardstowncable.net
*Med Dir:* Thomas Bergamini, MD

**Bowling Green Technical College**
Diagnostic Med Sonography Prgm
General concentration
1845 Loop Dr
Bowling Green, KY 42101-3601
www.bowlinggreen.kctcs.edu
*Prgm Dir:* Becky Stevens, BS RT(R)(M) RDMS
*Tel:* 270 901-1082   *Fax:* 270 901-1139
*E-mail:* becky.stevens@kctcs.edu
*Med Dir:* Rodney Veitschegger, MD

**West Kentucky Community & Technical
  College**
Diagnostic Med Sonography Prgm
General concentration
4810 Alben Barkley Dr, PO Box 7380
Paducah, KY 42002-7308
www.westkentucky.kctcs.edu
*Prgm Dir:* Alice Robertson, BS RT(R) RDMS
*Tel:* 270 534-3487, Ext 43487   *Fax:* 270 534-3498
*E-mail:* alice.robertson@kctcs.edu
*Med Dir:* Danny R Hatfield, MD

# Louisiana

**Louisiana State University - Eunice**
Diagnostic Med Sonography Prgm
General concentration
PO Box 1129
Eunice, LA 70535
*Prgm Dir:* Drew Thibodeaux
*Tel:* 337 550-1431
*E-mail:* dthibode@lsue.edu
*Med Dir:* John Higgins, MD

**Delgado Community College**
Diagnostic Med Sonography Prgm
General concentration
615 City Park Ave
New Orleans, LA 70119
www.dcc.edu
*Prgm Dir:* John Geshner, BA RDMS RDCS
*Tel:* 504 568-6473   *Fax:* 504 568-5494
*E-mail:* hgaspa@dcc.edu
*Med Dir:* Robert S Perret, MD

# Maryland

**Johns Hopkins Hospital**
Diagnostic Med Sonography Prgm
General concentration
Radiology Admin B-179
Baltimore, MD 21287
http://radiologycareers.rad.jhmi.edu
*Prgm Dir:* Carol Blank, MS RDMS
*Tel:* 410 528-8263   *Fax:* 410 528-8308
*E-mail:* cblank1@jhmi.edu
*Med Dir:* Ulrike Hamper, MD

**University of Maryland Baltimore County**
Diagnostic Med Sonography Prgm
General, Cardiac & Vascular concentrations
South Campus, Technology Center
Baltimore, MD 21227
www.umbctrainingcenters.com
*Prgm Dir:* Monica Guzman, RDMS
*Tel:* 443 543-5400   *Fax:* 443 543-5410
*E-mail:* monicag@umbtrainingcenters.com
*Med Dir:* Joel Fradin, MD

**Montgomery College**
Diagnostic Med Sonography Prgm
General, Cardiac & Vascular concentrations
7977 Georgia Ave
Silver Spring, MD 20910
*Prgm Dir:* Linda Zanin, EdD RDMS
*Tel:* 301 562-5569   *Fax:* 301 562-5569
*E-mail:* linda.zanin@montgomerycollege.edu
*Med Dir:* Michael Smith, MD

# Massachusetts

**Middlesex Community College**
Diagnostic Med Sonography Prgm
General concentration
Springs Rd
Bedford, MA 01730
www.middlesex.mass.edu
*Prgm Dir:* Thomas Walsh, MA RDMS RVS
*Tel:* 781 280-3983   *Fax:* 781 280-3845
*E-mail:* walsht@middlesex.mass.edu
*Med Dir:* David I Rose, MD

**Bunker Hill Community College**
Diagnostic Med Sonography Prgm
General & Cardiac concentrations
250 New Rutherford Ave
Boston, MA 02129-2991
*Prgm Dir:* Michelle Gagnon, BS RDCS RVT
*Tel:* 617 228-2407   *Fax:* 617 228-2052
*E-mail:* mgagnon@bhcc.mass.edu
*Med Dir:* Edgar Schick, MD

**Springfield Technical Community College**
Diagnostic Med Sonography Prgm
General concentration
One Armory Square
Springfield, MA 01105
*Prgm Dir:* David J Sloan, BA RDMS RVT
*Tel:* 413 755-4915   *Fax:* 413 755-6312
*E-mail:* djsloan@stcc.edu
*Med Dir:* Frederick Hampf, Jr, MD

# Michigan

**Henry Ford Hospital**
Diagnostic Med Sonography Prgm
General concentration
2799 W Grand Blvd
Detroit, MI 48202
www.henryfordhealth.org
*Prgm Dir:* Michael Moffatt, MA RDMS
*Tel:* 313 916-3519   *Fax:* 313 916-9480
*E-mail:* mike@rad.hfh.org
*Med Dir:* Denise Collins, MD

**Jackson Community College**
Diagnostic Med Sonography Prgm
General, Cardiac & Vascular concentrations
2111 Emmons Rd
Jackson, MI 49201
*Prgm Dir:* Lynne Schreiber, MS RDMS RT(R)
*Tel:* 517 796-8535   *Fax:* 517 796-8633
*E-mail:* lynne_schreiber@jccmi.edu
*Med Dir:* Libby S Anderson, MD

**Lansing Community College**
Diagnostic Med Sonography Prgm
General concentration
PO Box 40010
Lansing, MI 48901-7210
www.lcc.edu
*Prgm Dir:* Julie A Atkinson, BBA RDMS
*Tel:* 517 483-1410   *Fax:* 517 483-1508
*E-mail:* atkinsoj@lcc.edu
*Med Dir:* John Crockett, MD

**Oakland Community College - Bloomfield Hills**
Diagnostic Med Sonography Prgm
General concentration
22322 Rutland Dr
Southfield, MI 48075
www.oaklandcc.edu
*Prgm Dir:* Carolyn E Nacy, RDMS
*Tel:* 248 233-2918   *Fax:* 248 233-2891
*E-mail:* ceoneill@oaklandcc.edu
*Med Dir:* Michael Edwards, MD

**Providence Hospital**
Diagnostic Med Sonography Prgm
General & Vascular concentrations
16001 W Nine Mile Rd
Southfield, MI 48075
*Prgm Dir:* Janette Jablonski, BAS RDMS RVT RDCS
*Tel:* 248 849-5385   *Fax:* 248 849-5395
*E-mail:* janette.jablonski@providence-stjohnhealth.org
*Med Dir:* James Selis, MD

**Delta College**
Diagnostic Med Sonography Prgm
General concentration
1961 Delta Rd
University Center, MI 48710
*Prgm Dir:* Kim Boldt, BS RMS RVT RDCS RT
*Tel:* 989 686-9361   *Fax:* 989 667-2230
*E-mail:* kboldt@alpha.delta.edu
*Med Dir:* Harold Blumenstein, MD

# Minnesota

### Argosy University/Twin Cities
Diagnostic Med Sonography Prgm
General & Echocardiography concentrations
1515 Central Parkway
Eagan, MN 55121
*Prgm Dir:* Susan D Hummel, BS RT(R) RDMS
*Tel:* 651 846-3405   *Fax:* 651 994-0144
*E-mail:* shummel@argosyu.edu
*Med Dir:* David Lee, MD

### College of St Catherine - Minneapolis
Diagnostic Med Sonography Prgm
General concentration
601 25th Ave S
Minneapolis, MN 55454
*Prgm Dir:* Dick Mabbs, BA RVT RDMS
*Tel:* 612 690-7889   *Fax:* 612 690-7765
*E-mail:* dvmabbs@stkate.edu

### Mayo School of Health Sciences
Diagnostic Med Sonography Prgm
General, Cardiac & Vascular concentrations
200 First St SW
Rochester, MN 55905
www.mayo.edu/mshs
*Prgm Dir:* Kathryn Kuntz, BS RT RDMS RVT
*Tel:* 507 284-6520   *Fax:* 507 284-0656
*E-mail:* dms.admissions@mayo.edu
*Med Dir:* Michael Farrell, MD

### St Cloud Technical College
Diagnostic Med Sonography Prgm
General concentration
1540 Northway Dr
St Cloud, MN 56303
*Prgm Dir:* Jeff Gunderson, BS RT(R) RDMS RVT
*Tel:* 320 308-0971   *Fax:* 320 308-6172
*E-mail:* jgunderson@sctc.edu
*Med Dir:* John Brindley, MD

# Mississippi

### Hinds Community College
Diagnostic Med Sonography Prgm
General concentration
1750 Chadwick Dr
Jackson, MS 39204
*Prgm Dir:* Melissa K Mabry, RT(R) RDMS RVT
*Tel:* 601 371-3536   *Fax:* 601 371-3508
*E-mail:* mmabry@hindscc.edu

### Itawamba Community College
Diagnostic Med Sonography Prgm
General concentration
2176 S Eason Blvd
Tupelo, MS 38804
*Prgm Dir:* Nita Megginson, RT(R) RDMS
*Tel:* 662 620-5145
*E-mail:* nmmegginson@iccms.edu
*Med Dir:* Eric Emig, MD

# Missouri

### CoxHealth
Diagnostic Med Sonography Prgm
General, Cardiac & Vascular concentrations
3801 S National Ave
Springfield, MO 65807
*Prgm Dir:* Tammy J Stearns, BSRT(R) RDMS RVT
*Tel:* 417 269-8669   *Fax:* 417 269-8900
*E-mail:* Tammy.Stearns@CoxHealth.com
*Med Dir:* Joseph Mailloux, MD

### St Louis Community College - Forest Park
Diagnostic Med Sonography Prgm
General, Cardiac & Vascular concentrations
5600 Oakland Ave
St Louis, MO 63110
*Prgm Dir:* Beth Anderhub, MEd RDMS RT
*Tel:* 314 644-9399   *Fax:* 314 644-9752
*E-mail:* banderhub@stlcc.edu
*Med Dir:* William Middleton, MD

# Nebraska

### NE Methodist Coll Nursing & Allied Hlth
Diagnostic Med Sonography Prgm
General, Cardiac & Vascular concentrations
8501 W Dodge Rd
Omaha, NE 68114
*Prgm Dir:* Patricia Sullivan, MA RDMS RT
*Tel:* 402 354-4851   *Fax:* 402 354-8875
*E-mail:* psulliv@methodistcollege.edu
*Med Dir:* Nick Nelson, MD

### University of Nebraska Medical Center
Diagnostic Med Sonography Prgm
General concentration
Division of Radiation Sciences Technology Education
Omaha, NE 68198-4545
www.unmc.edu/alliedhealth/rste
*Prgm Dir:* Kim Michael, MA RT(R) RDMS RVT
*Tel:* 402 559-1189   *Fax:* 402 559-4667
*E-mail:* kkmichael@unmc.edu
*Med Dir:* Joseph C Anderson, MD

# Nevada

### Community College of Southern Nevada
Diagnostic Med Sonography Prgm
General, Cardiac & Vascular concentrations
6375 W Charleston Blvd, W3K
Las Vegas, NV 89146
*Prgm Dir:* Tracy Lopez, MEd RDMS RDCS RVT
*Tel:* 702 651-5925   *Fax:* 702 651-7459
*E-mail:* tracy_lopez@ccsn.edu
*Med Dir:* Bruce Topper, MD

# New Hampshire

### New Hampshire Technical Institute
Diagnostic Med Sonography Prgm
General concentration
31 College Dr
Concord, NH 03301-7412
www.nhti.edu
*Prgm Dir:* Kevin P Barry, MEd RT(R) RDMS RDCS
*Tel:* 603 271-7154   *Fax:* 603 271-7182
*E-mail:* kbarry@nhctc.edu
*Med Dir:* Albert Tu, MD

# New Jersey

### Sanford-Brown Institute
Diagnostic Med Sonography Prgm
General concentration
675 US 1 Plaza Gill Ln, 2nd Fl
Iselin, NJ 08830
www.sbinj.com
*Prgm Dir:* Anthony Rodriguez, RVT RDMS
*Tel:* 732 623-5740   *Fax:* 866 759-5357
*E-mail:* arodriguez@sb-nj.com
*Med Dir:* Peter Mezzacappa, MD

### Bergen Community College
Diagnostic Med Sonography Prgm
General & Cardiac concentrations
400 Paramus Rd
Paramus, NJ 07652-1595
*Prgm Dir:* Katherine Benz-Campbell, MA BS RDMS
*Tel:* 201 447-7944, Ext 7939   *Fax:* 201 612-3876
*E-mail:* kbcampbell@bergen.edu
*Med Dir:* Frederick P Ayers, MD

### Muhlenberg Regional Medical Center
Diagnostic Med Sonography Prgm
General concentration
Park Ave and Randolph Rd
Plainfield, NJ 07061
www.muhlenbergschools.org
*Prgm Dir:* Harry H Holdorf, MPA RT(ARRT) RDMS
*Tel:* 908 668-2884   *Fax:* 908 226-4568
*E-mail:* hholdorf@solarishs.org
*Med Dir:* Eric B Schmell, MD

### Univ of Med & Dent of New Jersey
Diagnostic Med Sonography Prgm
General concentration
School of Health Related Professions
Scotch Plains, NJ 07076
www.shrp.umdnj.edu
*Prgm Dir:* Cynthia Silkowski, MA RDMS RVT
*Tel:* 908 889-2521   *Fax:* 908 889-2527
*E-mail:* silkowcy@umdnj.edu
*Med Dir:* Alan G Dembner, MD

### Gloucester County College
Diagnostic Med Sonography Prgm
General concentration
1400 Tanyard Rd
Sewell, NJ 08080
www.gccnj.edu
*Prgm Dir:* Michael Keith, BSRT RDMS
*Tel:* 856 415-2195   *Fax:* 856 464-8463
*E-mail:* mkeith@gccnj.edu
*Med Dir:* Thomas Neidbala, MD

# New Mexico

### Central New Mexico Community College
Diagnostic Med Sonography Prgm
General concentration
525 Buena Vista SE
Albuquerque, NM 87106
*Prgm Dir:* Darlene Blagg
*Tel:* 505 224-4127   *Fax:* 505 224-4120
*E-mail:* DBlagg@cnm.edu
*Med Dir:* Robin Gaupp

### Dona Ana Community College
Diagnostic Med Sonography Prgm
General & Vascular concentrations
3400 S Espina - MSC 3DA
Las Cruces, NM 88003-8001
*Prgm Dir:* Carolyn T Coffin, MPH RT RDMS RVT RCDS
*Tel:* 505 582-7047
*E-mail:* cacoffin@nmsu.edu
*Med Dir:* Eduardo Martinez, MD

# New York

### Long Island University
Diagnostic Med Sonography Prgm
Vascular concentration
1 University Plaza (M-101)
Brooklyn, NY 11201-5372
*Prgm Dir:* Michael J Hartman
*Tel:* 718 488-1118
*E-mail:* michael.hartman@liu.edu

### SUNY Downstate Medical Center
Diagnostic Med Sonography Prgm
General & Cardiac concentrations
College of Health Related Professions
Brooklyn, NY 11203
www.downstate.edu/chrp/dmi
*Prgm Dir:* Joyce Miller, EdD RDMS
*Tel:* 718 270-7765   *Fax:* 718 270-7746
*E-mail:* joyce.miller@downstate.edu
*Med Dir:* Harris L Cohen, MD

**New York University**
Diagnostic Med Sonography Prgm
General & Cardiac concentrations
726 Broadway Rm 652
New York, NY 10003-6688
http://web.scps.nyu.edu/mcghee/degree.programs
*Prgm Dir:* Kerry Weinberg, MPA RT RDMS RDCS
*Tel:* 212 992-8723   *Fax:* 212 995-4890
*E-mail:* kerry.weinberg@nyu.edu
*Med Dir:* Joseph Yee, MD

**Western Suffolk BOCES**
Diagnostic Med Sonography Prgm
General, Cardiac & Vascular concentrations
152 Laurel Hill Rd
Northport, NY 11768
*Prgm Dir:* John Thomas, MS BED RDMS RVT RDCS RT
*Tel:* 631 261-3730   *Fax:* 631 623-4908
*E-mail:* jthomas@wsboces.org
*Med Dir:* Mark Williams, MD

**Rochester Institute of Technology**
Diagnostic Med Sonography Prgm
General & Echocardiography concentrations
153 Lomb Memorial Dr
Rochester, NY 14623
www.rit.edu
*Prgm Dir:* Hamad Ghazle, MS BS RDMS
*Tel:* 585 475-2241   *Fax:* 585 475-5809
*E-mail:* hhgscl@rit.edu
*Med Dir:* Susan Voci, MD

**Hudson Valley Community College**
Diagnostic Med Sonography Prgm
General & Cardiac concentrations
80 Vandenburgh Ave
Troy, NY 12180
*Prgm Dir:* Sheila Hughes, BPS LRT RDMS RVT
*Tel:* 518 629-7345   *Fax:* 518 629-4871
*E-mail:* hugheshe@hvcc.edu
*Med Dir:* Ronald Karo, MD

## North Carolina

**Asheville-Buncombe Technical Comm College**
Diagnostic Med Sonography Prgm
General & Vascular concentrations
340 Victoria Rd
Asheville, NC 28801
*Prgm Dir:* Chastity Coates Case, RT(R) RDMS RVT
*Tel:* 828 254-1921, Ext 470   *Fax:* 828 251-6355
*E-mail:* ccase@asheville.cc.nc.us
*Med Dir:* Keith Kohatsu, MD

**Pitt Community College**
Diagnostic Med Sonography Prgm
General & Cardiac concentrations
PO Drawer 7007
Greenville, NC 27835-7007
*Prgm Dir:* Ruggie MacKenzie, RDMS RDCS
*Tel:* 252 321-4254   *Fax:* 252 321-4451
*E-mail:* rmackenz@email.pittcc.edu
*Med Dir:* Douglas Shusterman, MD

**Caldwell Comm College & Tech Institute**
Diagnostic Med Sonography Prgm
General & Cardiac concentrations
2855 Hickory Blvd
Hudson, NC 28761
www.cccti.com
*Prgm Dir:* Kimberlee B Watts, BS RDMS
*Tel:* 828 726-2322   *Fax:* 828 726-2489
*E-mail:* kwatts@cccti.edu
*Med Dir:* John Ende, MD

**Johnston Community College**
Diagnostic Med Sonography Prgm
General concentration
PO Box 2350
Smithfield, NC 27577
*Prgm Dir:* Cathy Godwin, BS RTR RDMS RDCS RVT
*Tel:* 919 209-2158   *Fax:* 919 209-2153
*E-mail:* godwinc@johnstoncc.edu
*Med Dir:* John Matzko, MD

**Forsyth Technical Community College**
Diagnostic Med Sonography Prgm
General concentration
2100 Silas Creek Parkway
Winston-Salem, NC 27103
*Prgm Dir:* John Cassell, BS RDMS RDCS RVT RT
*Tel:* 336 734-7430   *Fax:* 336 734-7444
*E-mail:* jcassell@forsythtech.edu
*Med Dir:* Joseph Contento, MD

## Ohio

**Mercy Medical Center**
Diagnostic Med Sonography Prgm
General concentration
1320 Mercy Dr
Canton, OH 44708
www.thequalityhospital.com
*Prgm Dir:* Susan Black, RDMS
*Tel:* 330 489-1000, Ext 6609   *Fax:* 330 580-4726
*E-mail:* susan.black@csauh.com
*Med Dir:* David Brine, MD

**Cincinnati State Tech & Comm College**
Diagnostic Med Sonography Prgm
General & Cardiac concentrations
3250 Central Pkwy
Cincinnati, OH 45247
*Prgm Dir:* Susan Watson
*Tel:* 513 569-1665   *Fax:* 513 569-1659
*E-mail:* susan.watson@cincinnatistate.edu
*Med Dir:* Creighton Wright, MD

**Lorain County Community College**
Diagnostic Med Sonography Prgm
General concentration
1005 N Abbe Rd
Elyria, OH 44035
www.lorainccc.edu/LCCC/Academic/
    Allied_Health_Nursing/default.Normal.421.lccc
*Prgm Dir:* Craig Peneff, BSAS RDMS RVT
*Tel:* 800 995-5222, Ext 7189   *Fax:* 440 366-4116
*E-mail:* cpeneff@lorainccc.edu
*Med Dir:* Cathy Miller, MD

**Kettering College of Medical Arts**
Diagnostic Med Sonography Prgm
General, Cardiac & Vascular concentrations
3737 Southern Blvd
Kettering, OH 45429
www.kcma.edu
*Prgm Dir:* Joyce Grube, MS RDMS
*Tel:* 937 298-3399, Ext 55656   *Fax:* 937 395-8635
*E-mail:* joyce.grube@kcma.edu
*Med Dir:* Judith Greene, MD

**Sanford-Brown Institute**
Diagnostic Med Sonography Prgm
General concentration
17535 Rosbough Dr
Middleburg Heights, OH 44130
www.sbcleveland.com
*Prgm Dir:* Renato M Agustin, MD RDMS RVT
*Tel:* 440 202-3236   *Fax:* 440 239-9648
*E-mail:* ragustin@sbc-cleveland.com
*Med Dir:* Bradley Blackburn, MD

**Central Ohio Technical College**
Diagnostic Med Sonography Prgm
General, Cardiac & Vascular concentrations
1179 University Dr
Newark, OH 43055-1767
*Prgm Dir:* Cathie Scholl, RDMS RVT
*Tel:* 740 366-9274   *Fax:* 740 366-5047
*E-mail:* cscholl@cotc.edu
*Med Dir:* Owen Lee, MD

**Cuyahoga Community College**
Diagnostic Med Sonography Prgm
General, Cardiac & Vascular concentrations
11000 Pleasant Valley Rd
Parma, OH 44130
www.tri-c.edu/dms
*Prgm Dir:* Denise M Kinches, BS RT RDMS
*Tel:* 216 987-5564   *Fax:* 216 987-5066
*E-mail:* denise.kinches@tri-c.edu
*Med Dir:* Scott Kolodny, MD

**Owens Community College**
Diagnostic Med Sonography Prgm
General concentration
PO Box 10,000, Oregon Rd
Toledo, OH 43699-1947
*Prgm Dir:* Susan Perry, RDMS BSRT
*Tel:* 419 661-7560   *Fax:* 419 661-7251
*E-mail:* sperry@owens.edu
*Med Dir:* Malcolm Doyle, MD

## Oklahoma

**Moore Norman Technology Center**
Diagnostic Med Sonography Prgm
General concentration
4701 12th Ave NW
Norman, OK 73069
www.mntechnology.com
*Prgm Dir:* Meleah Meadows, BSRT RDMS RVT
*Tel:* 405 364-5763, Ext 7200
*E-mail:* mmeadows@mntechnology.com
*Med Dir:* Paul Massad, MD

**Univ of Oklahoma Health Sciences Center**
Diagnostic Med Sonography Prgm
General & Cardiac concentrations
PO Box 26901
Oklahoma City, OK 73190
www.ah.ouhsc.edu/main
*Prgm Dir:* Kari Boyce, PhD RDMC RDCS
*Tel:* 405 271-6477   *Fax:* 405 271-1424
*E-mail:* kari-boyce@ouhsc.edu
*Med Dir:* Jay Harolds, MD

## Pennsylvania

**Northampton Community College**
Diagnostic Med Sonography Prgm
General concentration
3835 Green Pond Rd
Bethlehem, PA 18020
www.northampton.edu
*Prgm Dir:* Catherine Rienzo, MS RT(R) RDMS
*Tel:* 610 332-6177   *Fax:* 610 861-4581
*E-mail:* crienzo@northampton.edu
*Med Dir:* Robert Rienzo, MD

**College Misericordia**
Diagnostic Med Sonography Prgm
General concentration
301 Lake St
Dallas, PA 18612
*Prgm Dir:* Sheryl E Goss, MS RT RDMS RDCS RVT
*Tel:* 570 674-6790   *Fax:* 570 674-3052
*E-mail:* sgoss@misericordia.edu
*Med Dir:* Ronald Konecke, MD

### Great Lakes Institute of Technology
Diagnostic Med Sonography Prgm
General concentration
5100 Peach St
Erie, PA 16509
www.glit.edu
*Prgm Dir:* Doug McCraney, RDMS RVT
*Tel:* 814 864-6666, Ext 264    *Fax:* 814 868-1717
*E-mail:* dougm@glit.edu
*Med Dir:* Howard Hudson, MD

### Harrisburg Area Community College
Diagnostic Med Sonography Prgm
General concentration
1 HACC Dr
Harrisburg, PA 17110-2999
www.hacc.edu
*Prgm Dir:* Julia R Imboden
*Tel:* 717 221-1322    *Fax:* 717 909-4010
*E-mail:* jrimbode@hacc.edu
*Med Dir:* Anand Jagannath, MD

### Lancaster Gen Coll of Nursing & Hlth Sciences
Diagnostic Med Sonography Prgm
General concentration
410 N Lime St
Lancaster, PA 17602
www.LancasterGeneral.org
*Prgm Dir:* Robert M Hess, BS RDMS
*Tel:* 717 544-5637    *Fax:* 717 544-5970
*E-mail:* rmhess@LancasterGeneral.org
*Med Dir:* Rebecca G Pennell, MD

### Comm Coll of Allegheny County
Diagnostic Med Sonography Prgm
General & Cardiac concentrations
595 Beatty Rd
Monroeville, PA 15146
http://ccac.edu
*Prgm Dir:* Lynn Gigandet, MSEd RDMS
*Tel:* 724 325-6731    *Fax:* 724 325-6701
*E-mail:* lgigandet@ccac.edu
*Med Dir:* William Katz, MD

### Thomas Jefferson University
Diagnostic Med Sonography Prgm
General, Cardiac & Vascular concentrations
130 S Ninth St, Ste 1011
Philadelphia, PA 19107
*Prgm Dir:* Michael J Hartman, MS RDMS RVT RT(R)
*Tel:* 215 503-8724    *Fax:* 215 503-1031
*E-mail:* michael.hartman@jefferson.edu
*Med Dir:* Barry Goldberg, MD

### Western Sch of Hlth & Business - Pittsburgh
Diagnostic Med Sonography Prgm
General concentration
421 Seventh Ave
Pittsburgh, PA 15219
*Prgm Dir:* Sandra M Gosnell, RDMS
*Tel:* 412 209-0472    *Fax:* 412 281-0319
*E-mail:* sgosnell@western-school.com
*Med Dir:* James Ferris, MD

### Lackawanna College
Diagnostic Med Sonography Prgm
Vascular concentration
501 Vine St
Scranton, PA 18509
*Prgm Dir:* Janine Oliveri, MSEd BSRT RDMS RVT
*Tel:* 570 504-7920    *Fax:* 570 961-7832
*E-mail:* oliverij@lackawanna.edu
*Med Dir:* Edward Batzel, MD

### Crozer - Chester Med Ctr
Diagnostic Med Sonography Prgm
General concentration
One Medical Center Blvd
Upland, PA 19013
*Prgm Dir:* Barbara Annunziato, BS RT RDMS RVT
*Tel:* 610 447-2502, Ext 2    *Fax:* 610 447-2296
*E-mail:* barbara.annunziato@crozer.org
*Med Dir:* Lorna Blum, MD

### Wilkes-Barre General Hospital
*Cosponsor:* Wyoming Valley Health Care
Diagnostic Med Sonography Prgm
General concentration
575 N River St
Wilkes-Barre, PA 18764
www.wvhcs.org
*Prgm Dir:* Lisa Capizzi, BS RT(R)(M) RDMS RVT
*Tel:* 570 552-4654    *Fax:* 570 552-4653
*E-mail:* lcapizzi@wvhcs.org
*Med Dir:* Satish Patel, MD

## Rhode Island

### Rhode Island Hospital
Diagnostic Med Sonography Prgm
General concentration
4 Davol Square, Bldg "A", Box 162
Providence, RI 02903
www.lifespan.org/diagimag
*Prgm Dir:* Paul Maria, RDMS RVT
*Tel:* 401 528-8531, Ext 228    *Fax:* 401 457-0219
*E-mail:* pmaria@lifespan.org
*Med Dir:* Holly Gil, MD

## South Carolina

### Greenville Technical College
Diagnostic Med Sonography Prgm
General concentration
PO Box 5616
Greenville, SC 29606
www.greenvilletech.com
*Prgm Dir:* Ronda Keller, RT(R) RDMS RVT
*Tel:* 864 250-8110    *Fax:* 864 250-8462
*E-mail:* ronda.keller@gvltec.edu
*Med Dir:* Daniel M Thomason, MD

## South Dakota

### Southeast Technical Institute
Diagnostic Med Sonography Prgm
General concentration
2320 N Career Ave
Sioux Falls, SD 57107
*Prgm Dir:* Jo Ellen Hagemeyer, RT(R) RDMS
*Tel:* 605 367-7624
*E-mail:* jo.ellen.hagemeyer@southeasttech.com
*Med Dir:* Brad Paulson, MD

## Tennessee

### Chattanooga State Technical Comm College
Diagnostic Med Sonography Prgm
General concentration
HPF 177
Chattanooga, TN 37406
www.chattanoogastate.edu/Allied_Health/sonography/
    diagnost.asp
*Prgm Dir:* Jody Hancock, MAEd RDMS RVT RT(R)
*Tel:* 423 697-3341    *Fax:* 423 697-3324
*E-mail:* jody.hancock@chattanoogastate.edu
*Med Dir:* Kenneth Rule, MD

### Volunteer State Community College
Diagnostic Med Sonography Prgm
General concentration
1480 Nashville Pike
Gallatin, TN 37066-3188
*Prgm Dir:* Gene Spain, BBA RDMS
*Tel:* 615 452-8600, Ext 3339    *Fax:* 615 230-3574
*E-mail:* gene.spain@volstate.edu
*Med Dir:* William McLendon, MD

### Baptist College of Health Sciences
Diagnostic Med Sonography Prgm
General concentration
1003 Monroe Ave
Memphis, TN 38104
*Prgm Dir:* Mitzi Roberts, MS RT RDMS
*Tel:* 901 572-2653    *Fax:* 901 227-5533
*E-mail:* mitzi.roberts@bchs.edu
*Med Dir:* James E Machin, MD

### Methodist Le Bonheur Healthcare
Diagnostic Med Sonography Prgm
General concentration
Dept of Radiology
Memphis, TN 38104
www.methodisthealth.org
*Prgm Dir:* Dawn L Driver, RDMS RDCS RVT
*Tel:* 901 516-8099    *Fax:* 901 516-2870
*E-mail:* driverd@methodisthealth.org
*Med Dir:* Don Emerson, MD

### Vanderbilt University Medical Center
Diagnostic Med Sonography Prgm
General concentration
1161 21st Ave S
Nashville, TN 37232-2675
www.mc.vanderbilt.edu
*Prgm Dir:* Jill D Herzog, BS RT(R) RDMS RVT
*Tel:* 615 343-0905    *Fax:* 615 322-3764
*E-mail:* jill.herzog@vanderbilt.edu
*Med Dir:* Arthur C Fleischer, MD

## Texas

### Alvin Community College
Diagnostic Med Sonography Prgm
Echocardiography & Vascular concentrations
3110 Mustang Rd
Alvin, TX 77511-4898
www.alvincollege.edu
*Prgm Dir:* Jessica L Murphy, BS RRT RDCS RVT
*Tel:* 281 756-3656
*E-mail:* jmurphy@alvincollege.edu
*Med Dir:* Salim Dabaghi, MD

### Austin Community College
Diagnostic Med Sonography Prgm
General & Cardiac concentrations
3401 Webberville Rd
Austin, TX 78702
www.austincc.edu/health/sono
*Prgm Dir:* Regina Swearengin, BS RDMS
*Tel:* 512 223-5944    *Fax:* 512 223-5895
*E-mail:* ginas@austincc.edu
*Med Dir:* Marcus Lines, MD

### Lamar Institute of Technology
Diagnostic Med Sonography Prgm
General concentration
PO Box 10061
Beaumont, TX 77710
www.lit.edu
*Prgm Dir:* Sheila Trahan, RDMS RT
*Tel:* 409 839-2924    *Fax:* 409 880-8955
*E-mail:* sheila.trahan@lit.edu
*Med Dir:* Ramon Garcia, MD

### Univ Tx at Brownsville/Tx Southmost Coll
Diagnostic Med Sonography Prgm
General concentration
80 Fort Brown
Brownsville, TX 78620
*Prgm Dir:* Marti Flores, MS RDMS RDCS RT(R)
*Tel:* 956 882-5011    *Fax:* 956 882-5012
*E-mail:* marti.flores@utb.edu
*Med Dir:* George Skye, MD

**Del Mar College**
Diagnostic Med Sonography Prgm
General concentration
101 Baldwin and Ayers Sts
Corpus Christi, TX 78404
*Prgm Dir:* Rita S Orchard, MS RT(R) RDMS
*Tel:* 361 698-1101    *Fax:* 361 698-1598
*E-mail:* rorchard@delmar.edu
*Med Dir:* Kenneth R Cook, MD

**Cy-Fair College**
Diagnostic Med Sonography Prgm
General concentration
9191 Barker-Cypress Rd
Cypress, TX 77433
www.cy-faircollege.com
*Prgm Dir:* Christina Hagerty, MEd RT(R) RDMS
*Tel:* 281 290-3971
*E-mail:* christina.hagerty@nhmccd.edu

**El Centro College**
Diagnostic Med Sonography Prgm
General concentration
Main and Lamar Sts
Dallas, TX 75202
www.elcentrocollege.edu/sonography
*Prgm Dir:* Jan Bryant, MS RDMS
*Tel:* 214 860-2300    *Fax:* 214 860-2268
*E-mail:* jdb5529@dcccd.edu
*Med Dir:* Mark Zibilich, MD

**Sanford-Brown Institute - Dallas**
Diagnostic Med Sonography Prgm
General concentration
1250 W Mockingbird Ln
Dallas, TX 75247
http://sbdallas.com
*Prgm Dir:* Amjad Majeed, MB BS DMRD RDMS
*Tel:* 214 459-8490, Ext 8517    *Fax:* 214 638-6006
*E-mail:* amajeed@sbdallas.com
*Med Dir:* JOSEPH CHAN, MD, BOARD CERTIFIED IN RADIOLOGY

**El Paso Community College**
Diagnostic Med Sonography Prgm
General concentration
PO Box 20500
El Paso, TX 79998
www.epcc.edu
*Prgm Dir:* Nora M Balderas, BS RT(R) RDMS
*Tel:* 915 831-4108, Ext 4141    *Fax:* 915 831-4114
*E-mail:* norab@epcc.edu
*Med Dir:* William Sullivan, MD

**Sanford-Brown Institute - Houston**
Diagnostic Med Sonography Prgm
General concentration
10500 Forum Place Dr #200
Houston, TX 77036
*Prgm Dir:* Wendy Fuller
*Tel:* 713 779-1110    *Fax:* 713 779-2408
*E-mail:* wfuller@sbhouston.com

**Midland College**
Diagnostic Med Sonography Prgm
General concentration
3600 N Garfield
Midland, TX 79705
*Prgm Dir:* Elizabeth Brown
*Tel:* 432 685-4600    *Fax:* 432 685-4762
*E-mail:* lbrown@midland.edu
*Med Dir:* Jess J Dalehite, MD

**Tyler Junior College**
Diagnostic Med Sonography Prgm
General concentration
PO Box 9020
Tyler, TX 75711
*Prgm Dir:* Pam Brower, RVT RVS
*Tel:* 903 510-2668    *Fax:* 903 510-2889
*E-mail:* pbro@tjc.edu
*Med Dir:* Ted Willis, MD

# Virginia

**Southwest Virginia Community College**
Diagnostic Med Sonography Prgm
General concentration
PO Box SVCC (A-139)
Richlands, VA 24641-1510
*Prgm Dir:* Linda Brewster, BS RDMS RT(R)
*Tel:* 540 964-7642    *Fax:* 540 964-7715
*E-mail:* linda_brewster@sw.cc.va.us
*Med Dir:* John E Hutchison, Jr, MD

**Tidewater Community College**
Diagnostic Med Sonography Prgm
General concentration
1700 College Crescent
Virginia Beach, VA 23453
www.tcc.edu
*Prgm Dir:* Felicia Toreno, MSEd RDMS RVT
*Tel:* 757 822-7271    *Fax:* 757 822-7556
*E-mail:* ftoreno@tcc.edu
*Med Dir:* Bernard Stephanie, MD

# Washington

**Bellevue Community College**
Diagnostic Med Sonography Prgm
General & Cardiac concentrations
3000 Landerholm Circle SE, Rm B243
Bellevue, WA 98007-6484
*Prgm Dir:* Ann Polin, MSRS RDMS
*Tel:* 425 564-4181    *Fax:* 425 564-3128
*E-mail:* apolin@bcc.ctc.edu
*Med Dir:* Robert Bree, MD

**Seattle University**
Diagnostic Med Sonography Prgm
General, Cardiac & Vascular concentrations
901 12th Ave, PO Box 222000
Seattle, WA 98122-4360
www.seattleu.edu
*Prgm Dir:* Carolyn Coffin, MPH RDMS RDCS RVT
*Tel:* 206 296-5960    *Fax:* 206 296-6429
*E-mail:* coffinc@seattleu.edu
*Med Dir:* Edward Gibbons, MD

# West Virginia

**Mountain State University**
Diagnostic Med Sonography Prgm
General concentration
PO Box 9003
Beckley, WV 25802-9003
www.mountainstate.edu
*Prgm Dir:* Robert Lilly, MS RDCS RDMS
*Tel:* 304 929-1389, Ext 1389    *Fax:* 304 253-3485
*E-mail:* boblilly@mountainstate.edu
*Med Dir:* John Tamminen, MD

**West Virginia University Hospitals**
Diagnostic Med Sonography Prgm
General concentration
PO Box 8062
Morgantown, WV 26506
www.wvuhradtech.com
*Prgm Dir:* Debra L Williams, RDMS
*Tel:* 304 598-4187    *Fax:* 304 598-4072
*E-mail:* williamsd@rcbhsc.wvu.edu
*Med Dir:* Michael Cunningham, MD

# Wisconsin

**Chippewa Valley Technical College**
Diagnostic Med Sonography Prgm
General concentration
620 W Clairemont Ave
Eau Claire, WI 54701
*Prgm Dir:* Tina Salava, RT(R) RDMS RVT
*Tel:* 715 833-6430    *Fax:* 715 833-6430
*E-mail:* tsalava@cvtc.edu
*Med Dir:* Thomas Edwards, MD

**Northeast Wisconsin Technical College**
Diagnostic Med Sonography Prgm
General concentration
2740 W Mason St
Green Bay, WI 543007-904
*Prgm Dir:* Shelly Mondeik
*Tel:* 920 496-8685
*E-mail:* shelly.mondeik@nwtc.edu
*Med Dir:* John R Gassner, MD

**University of Wisconsin Hospital and Clinics**
Diagnostic Med Sonography Prgm
General & Cardiac concentrations
Dept of Radiology
Madison, WI 53792
www.uwhealth.org/healthprof
*Prgm Dir:* Carol Mitchell, PhD RDMS RDCS RVT RT(R)
*Tel:* 608 263-9033    *Fax:* 608 263-9208
*E-mail:* cc.mitchell@hosp.wisc.edu
*Med Dir:* Thomas C Winter, MD

**Aurora St Luke's Medical Center**
Diagnostic Med Sonography Prgm
General & Vascular concentrations
180 W Grange Ave
Milwaukee, WI 53207
www.aurora.org/sonography
*Prgm Dir:* Laura Woodruff, RT(R)(M) RDMS RVT
*Tel:* 414 747-4352    *Fax:* 414 747-4366
*E-mail:* laura.woodruff@aurora.org
*Med Dir:* Peter Cooley, MD

**Columbia St Mary's Hospitals**
Diagnostic Med Sonography Prgm
General & Vascular concentrations
2121 E Newport Ave
Milwaukee, WI 53211
www.columbia-stmarys.org
*Prgm Dir:* Rosemarie Fridrick, RDMS RT(R)
*Tel:* 414 961-3945    *Fax:* 414 961-4205
*E-mail:* rfridric@columbia-stmarys.org
*Med Dir:* Kenneth Clark, MD

**St Francis Hospital**
Diagnostic Med Sonography Prgm
General concentration
3237 S 16th St
Milwaukee, WI 53215
*Prgm Dir:* Marc Wojciechowski, RDMS RDCS RVT MS
*Tel:* 414 647-5711    *Fax:* 414 647-5732
*E-mail:* marc.wojciechowski@WFHC.org
*Med Dir:* Robert Gould, MD

## Diagnostic Medical Sonographer

| Programs* | Class Capacity | Begins | Length (months) | Award | Res Tuition | Non-res Tuition | Stipend | Offers:‡ 1 | 2 | 3 |
|---|---|---|---|---|---|---|---|---|---|---|
| **Alabama** | | | | | | | | | | |
| Baptist Medical Center South (Montgomery) | 8 | Oct | 12 | Cert | $5,000 | $5,000 | | | | |
| Insitute of Ultrasound Diagnostics (Mobile) | 30 | Jan Jul | 12 | Cert | $9,500 | $9,500 | | | | |
| Wallace State Community College (Hanceville) | 25 | Aug | 18 | Dipl, AAS | $3,600 | $7,200 | | | | • |
| **Arkansas** | | | | | | | | | | |
| Arkansas State University | 12 | Jun | 18 | BS | $9,827 | $21,107 | | | | |
| University of Arkansas for Medical Sciences (Little Rock) | 20 | Aug | 24, 18 | Cert, BS DMS, AC DMS | $6,630 | $14,820 | | | | |
| **California** | | | | | | | | | | |
| Cypress College | 20 | Aug | 12 | Cert | $2,000 | $0 | | | | • |
| Foothill Community College (Los Altos Hills) | 30 | Jun | 15 | Cert, AS | $2,500 | $9,744 | | | | • |
| Kaiser Permanente Sch of Allied Hlth Sciences (Richmond) | 15 | Oct | 18 | Cert | $9,000 | $9,000 | | | | • |
| Loma Linda University | 10 | Sep | 24 | Cert | $10,000 | $10,000 | | | | |
| Univ of California San Diego Med Ctr | 6 | Jan | 12 | Cert | $10,000 | $10,000 | | | | |
| **Colorado** | | | | | | | | | | |
| U of Colorado at Denver and Health Scis Ctr (Aurora) | 14 | Sep | 12 | Cert | $3,000 | $3,000 | | | | |
| **Delaware** | | | | | | | | | | |
| Delaware Tech & Comm Coll - Wilmington | 14 | May | 24 | AAS | $2,934 | $7,335 | | | | • |
| **District of Columbia** | | | | | | | | | | |
| George Washington University | 45 | Aug | 48, 16 | Cert, BS | $22,650 | $22,650 | | | | • |
| **Florida** | | | | | | | | | | |
| Broward Community College (Coconut Creek) | 20 | Jun | 14, 24 | Cert, AS | $2,500 | $8,500 | | | | • |
| Keiser College - Daytona Beach | 14 | Jan May Sep | 24 | AS | $18,660 | $18,660 | | | | |
| Keiser University (Ft Lauderdale) | 14 | Jan May Sep | 24 | AS | $12,000 | $0 | | | | |
| Palm Beach Community College (Palm Beach Gardens) | 15 | May | 24, 14 | Cert, AS, CCC | | | | | | |
| St Vincent's Medical Center (Jacksonville) | 6 | Jul | 18 | Cert | $3,500 | $3,500 | | | | |
| Valencia Community College (Orlando) | 12 | Aug | 24 | AS | $2,379 | $8,929 | | | | • |
| **Georgia** | | | | | | | | | | |
| Coosa Valley Technical College (Rome) | 46 | Fall | 18 | Dipl, AAT | $1,088 | $2,176 | | | | |
| Grady Health System (Atlanta) | 9 | Sep | 12 | Cert | $3,500 | $3,500 | | | | • |
| Ogeechee Technical College (Statesboro) | 13 | Sep | 24 | Dipl | $1,384 | $2,768 | | | | • |
| **Idaho** | | | | | | | | | | |
| Boise State University | 7 | Aug | 12 | Cert, BS | $5,590 | $13,360 | | | | |
| **Illinois** | | | | | | | | | | |
| College of DuPage (Glen Ellyn) | 38 | Fall | 15, 6 | Cert | $5,500 | $12,000 | | • | | |
| John A Logan College (Carterville) | 7 | Aug | 18, 30 | Dipl, Cert, AS | $64 | $255 | $60 | | | |
| Northwestern Memorial Hospital (Chicago) | 6 | Jul | 18 | Cert | $7,500 | $7,500 | | | | • |
| Southern Illinois Univ at Carbondale | 20 | Aug | 12 | Cert | $4,500 | $9,000 | | | • | • |
| **Iowa** | | | | | | | | | | |
| University of Iowa Hospitals & Clinics (Iowa City) | 11 | Sep | 18 | Cert, BS | $4,500 | $6,750 | | | | • |
| **Kentucky** | | | | | | | | | | |
| Bowling Green Technical College | 7 | Aug | 13 | Dipl | $4,116 | $4,116 | | | | • |
| West Kentucky Community & Technical College (Paducah) | 12 | Aug | 12 | Dipl, AAS | $3,724 | $11,172 | | • | | |
| **Louisiana** | | | | | | | | | | |
| Delgado Community College (New Orleans) | 12 | Aug | 16 | Cert | $3,200 | $8,600 | | | | |
| **Maryland** | | | | | | | | | | |
| Johns Hopkins Hospital (Baltimore) | 10 | Jun | 14 | Cert | $7,500 | $7,500 | | | | |
| University of Maryland Baltimore County | 40 | Jul | 14 | Cert, BS/BA | $10,990 | $10,990 | | | | • |
| **Massachusetts** | | | | | | | | | | |
| Middlesex Community College (Bedford) | 24 | Sep | 12, 24 | Cert, AS | $3,060 | $3,366 | | • | | |
| Springfield Technical Community College | 10 | Sep | 22 | AS | $1,296 | $7,056 | | | | |
| **Michigan** | | | | | | | | | | |
| Henry Ford Hospital (Detroit) | 7 | Sep | 15 | Cert | $1,000 | $1,000 | | | | • |
| Lansing Community College | 30 | Jun | 15 | Cert, AS | $3,000 | $4,000 | | | | • |
| Oakland Community College - Bloomfield Hills (Southfield) | 22 | May | 15 | AAS | $2,426 | $4,107 | | • | | • |
| **Minnesota** | | | | | | | | | | |
| Mayo School of Health Sciences (Rochester) | 24 | Sep | 18 | Cert | $11,025 | $11,025 | | | | • |
| **Nebraska** | | | | | | | | | | |
| University of Nebraska Medical Center (Omaha) | 8 | Aug | 12 | BS | $6,795 | $20,160 | | | | • |
| **New Hampshire** | | | | | | | | | | |
| New Hampshire Technical Institute (Concord) | 8 | Sep | 16 | Cert | $7,090 | $14,051 | | | | |
| **New Jersey** | | | | | | | | | | |
| Gloucester County College (Sewell) | 15 | Sep | 22 | AAS | $4,924 | $12,746 | | | | |
| Muhlenberg Regional Medical Center (Plainfield) | | Sep | 28 | Dipl, Cert, AS | $15,651 | $15,651 | | | | • |
| Sanford-Brown Institute (Iselin) | 20 | Aug Feb | 18 | Cert | | | | | | |

*Data are shown only for programs that completed the 2006 AMA Survey of Health Professions Education Programs

‡Key to Offers: 1: Evening or weekend classes; 2: Non-English instruction; 3: Cultural competence instruction

## Diagnostic Medical Sonographer

| Programs* | Class Capacity | Begins | Length (months) | Award | Res Tuition | Non-res Tuition | Stipend | Offers:‡ 1 | 2 | 3 |
|---|---|---|---|---|---|---|---|---|---|---|
| Univ of Med & Dent of New Jersey (Scotch Plains) | 19 | Sep | 15 | Cert, BSAHT | $9,786 | $14,559 | | | | • |
| **New York** | | | | | | | | | | |
| Hudson Valley Community College (Troy) | 22 | Aug | 12 | Cert | $4,500 | $9,100 | | | | |
| New York University | 20 | Sep | 30 | AAS | $22,205 | $22,205 | | | | • |
| Rochester Institute of Technology | 25 | Sep | 12, 48 | Cert, BS | $23,949 | $23,949 | | • | | |
| SUNY Downstate Medical Center (Brooklyn) | 28 | Sep | 21 | BS | $6,927 | $16,319 | | | | • |
| **North Carolina** | | | | | | | | | | |
| Caldwell Comm College & Tech Institute (Hudson) | 30 | Aug | 21 | AAS | $1,896 | $10,536 | | | | • |
| Forsyth Technical Community College (Winston-Salem) | 9 | Aug | 21 | AAS | $1,562 | $8,668 | | | | • |
| **Ohio** | | | | | | | | | | |
| Cuyahoga Community College (Parma) | 48 | Aug | 21, 18 | Cert, AAS | $2,900 | $7,850 | | | | • |
| Kettering College of Medical Arts | 20 | Aug | 36, 48 | Dipl, AS, BS | $16,000 | $16,000 | | | | • |
| Lorain County Community College (Elyria) | 11 | Jan | 17 | AAS | $4,061 | $9,878 | | | | • |
| Mercy Medical Center (Canton) | 4 | Jul | 12 | Cert | $12,000 | $12,000 | | | | |
| Sanford-Brown Institute (Middleburg Heights) | 24 | Jul Dec | 18 | Dipl | $12,962 | $12,962 | | • | | |
| **Oklahoma** | | | | | | | | | | |
| Moore Norman Technology Center | 10 | Varies | 15 | Dipl | $225 | $3,036 | | | | • |
| Univ of Oklahoma Health Sciences Center (Oklahoma City) | 22 | Aug | 48 | BS | $6,203 | $9,500 | | | | • |
| **Pennsylvania** | | | | | | | | | | |
| Comm Coll of Allegheny County (Monroeville) | 32 | Aug | 20 | Cert, AS | $2,108 | $6,324 | | | | • |
| Crozer - Chester Med Ctr (Upland) | 8 | Sep | 18 | Cert | $8,000 | $8,000 | | | | |
| Great Lakes Institute of Technology (Erie) | 18 | Sep Jan May | 18 | Dipl | $21,790 | $21,790 | | | | |
| Harrisburg Area Community College | 12 | Jan | 12, 24 | Cert, AAS | $6,080 | $121,260 | | | | |
| Lancaster Gen Coll of Nursing & Hlth Sciences | 16 | Aug | 24 | Dipl, AS | $8,745 | $8,745 | | | | • |
| Northampton Community College (Bethlehem) | 11 | Fall semester | 24, 18 | Dipl, AAS | $3,145 | $9,879 | | | | |
| Wilkes-Barre General Hospital | 4 | Sep 7 | 18 | Cert | $0 | $10,000 | | | | |
| **Rhode Island** | | | | | | | | | | |
| Rhode Island Hospital (Providence) | 6 | Jul | 14 | Cert | $4,000 | $4,000 | | | | |
| **South Carolina** | | | | | | | | | | |
| Greenville Technical College | 15 | Aug | 16 | Dipl | $4,820 | $10,070 | | | | |
| **Tennessee** | | | | | | | | | | |
| Chattanooga State Technical Comm College | 25 | Aug | 12 | | $5,215 | $14,131 | | | | |
| Methodist Le Bonheur Healthcare (Memphis) | 10 | Aug | 18, 15 | Cert | $5,000 | $5,000 | | | | • |
| Vanderbilt University Medical Center (Nashville) | 12 | Jul | 18 | Cert | $8,400 | $8,400 | | | | • |
| **Texas** | | | | | | | | | | |
| Alvin Community College | 10 | Jan May | 18, 24 | Cert, ATC, AAS | $1,500 | $3,500 | | | • | • |
| Austin Community College | 30 | Summer | 22, 28 | AAS | $3,618 | $14,616 | | | • | • |
| Cy-Fair College (Cypress) | 20 | Aug | 21 | Dipl, Cert, AAS | $3,172 | $5,612 | | | | • |
| Del Mar College (Corpus Christi) | 19 | Aug Sep | 18 | AAS | $528 | $1,000 | | | | |
| El Centro College (Dallas) | 10 | Jun | 16, 28 | Cert, AAS | $1,626 | $3,006 | | | | • |
| El Paso Community College | 10 | Jul | 18, 24 | Dipl, Cert, AAS | $3,996 | $6,215 | | • | | • |
| Lamar Institute of Technology (Beaumont) | 15 | Jul | 18 | Cert, AAS | $1,280 | $5,460 | | | | |
| Sanford-Brown Institute - Dallas | 20 | Spring Fall | 17 | Cert | $14,885 | $14,885 | | • | | • |
| Tyler Junior College | 12 | Aug | 16 | AAS | $3,533 | $5,759 | $1,500 | | | |
| **Virginia** | | | | | | | | | | |
| Tidewater Community College (Virginia Beach) | 15 | Aug | 21 | AAS | $5,832 | $17,716 | | | | • |
| **Washington** | | | | | | | | | | |
| Seattle University | 25 | Sep | 39 | BS | $22,905 | $0 | | | | |
| **West Virginia** | | | | | | | | | | |
| Mountain State University (Beckley) | 29 | May | 18, 30 | Cert, AAS, BS | $21,070 | $21,070 | | | | |
| West Virginia University Hospitals (Morgantown) | 3 | Jul | 18 | Cert | $1,600 | $1,600 | | | | • |
| **Wisconsin** | | | | | | | | | | |
| Aurora St Luke's Medical Center (Milwaukee) | 11 | Sep | 18 | Cert | $2,400 | $2,400 | | | | |
| Columbia St Mary's Hospitals (Milwaukee) | 5 | Oct | 18 | Cert | $2,000 | $2,000 | | | | |
| St Francis Hospital (Milwaukee) | 5 | Sep | 18 | Cert | $3,000 | $3,000 | | | | |
| University of Wisconsin Hospital and Clinics (Madison) | 16 | Sep | 24 | Dipl, BS | $5,834 | $18,586 | | | | |

*Data are shown only for programs that completed the 2006 AMA Survey of Health Professions Education Programs

‡Key to Offers: 1: Evening or weekend classes; 2: Non-English instruction; 3: Cultural competence instruction

**Includes:**
- Registered dietitian/nutritionist
- Dietetic technician, registered

# Registered Dietitian/Nutritionist

## Occupational/Job Description

Dietetics is the science of applying food and nutrition to health. Registered Dietitians are nutritionists who integrate and apply the principles derived from the sciences of food, nutrition, biochemistry, physiology, food management, and behavior to achieve and maintain the health status of the public they serve.

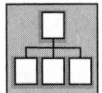

## Employment Characteristics

*Clinical registered dietitians* are a vital part of the medical team in hospitals, nursing homes, health maintenance organizations, and other health care facilities. As a key member of the health care team, the clinical RD provides medical nutrition therapy and the use of specific nutrition services to treat chronic conditions, illnesses, or injuries. Opportunities for advancement are available by choosing a particular area of nutrition practice, such as diabetes, heart disease, or pediatrics, or by expanding into hospital administration.

*Community registered dietitians* work in public and home health agencies, day care centers, health and recreation clubs, and in government-funded programs that feed and counsel families, the elderly, pregnant women, children, and individuals with special needs. Wherever proper nutrition can help improve quality of life, community RDs reach out to the public to teach, monitor, and advise.

*Educator registered dietitians* work in colleges, universities, and medical centers, teaching future physicians, nurses, dietitians, and dietetic technicians the science of foods and nutrition.

*Research registered dietitians* work in government agencies, food and pharmaceutical companies, and major universities and medical centers. They conduct or direct experiments to answer critical nutrition questions, find alternative foods, and form dietary recommendations for the public.

*Consultant registered dietitians* work under contract with health care or food companies or in their own business. In private practice, they perform nutrition screening and assessment of their own clients and those referred to them by physicians. They counsel on weight loss, cholesterol reduction, and a variety of other diet-related concerns. Those under contract with health care facilities consult with food service or restaurant managers, food vendors or distributors, athletes, nursing home residents, or company employees.

*Management registered dietitians* work in health care institutions, schools, cafeterias, and restaurants, playing a key role where food is served. They are responsible for personnel management, menu planning, budgeting, and purchasing.

*Business registered dietitians* work in food- and nutrition-related industries, in such areas as communications, consumer affairs, product development, sales, marketing, advertising, and public relations.

## Salary

According to the American Dietetic Association (ADA) 2005 Dietetics Compensation and Benefits Survey, among registered dietitians employed full time in dietetics in their primary position for less than 5 years, half earn between $35,000 and $46,000 per year. Salary levels may vary with location, scope of responsibility, and supply of job applicants. Salary also increases as experience increases; many RDs, particularly those in management and business earn incomes between $50,000 and $72,000 per year. Refer to Section IV, Table 5 of this *Directory* for more information, or see www.ama-assn.org/go/ salary.

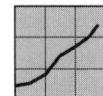

## Employment Outlook

According to the US Bureau of Labor Statistics, employment of dietitians is expected to grow faster than the average profession through the year 2015, especially in the community, consulting, and business areas.

## Educational Programs

**Length.** The professional component is a minimum of 2 years at the baccalaureate or graduate's degree level. Postbaccalaureate dietetic internship programs vary from 6 months to 2 years, depending on study design and integration in a graduate program. Following completion of academic and supervised practice requirements, individuals are eligible to take a national certification examination for registered dietitians. Many states also regulate dietitians and nutritionists.

**Prerequisites.** Variable for programs at the baccalaureate and graduate degree levels, depending on the degree offered and institutional requirements. Applicants to postbaccalaureate dietetic internship programs must have completed a baccalaureate degree from a US regionally accredited college or university or equivalent foreign degree and Commission on Accreditation for Dietetics Education (CADE)-approved didactic coursework.

**Curriculum.** Didactic curriculum requirements focus on food and nutrition sciences and management, supported by the physical and biological as well as behavioral and social sciences, in addition to business, economics, and communication. The supervised practice curriculum provides experiences to develop the skills and competence to practice dietetics.

### Program Types

*Coordinated Program in Dietetics (CP)*
- An academic program in a regionally accredited college or university granting a baccalaureate or graduate degree
- Includes didactic instruction and a minimum of 900 hours of supervised practice experiences, which may be planned concurrently with or following the didactic component
- Graduates are eligible to write the registration examination for dietitians

*Didactic Program in Dietetics (DPD)*
- An academic program in a regionally accredited college or university granting a baccalaureate or graduate degree
- Graduates can apply for a dietetic internship program leading to eligibility to write the registration examination for dietitians

*Dietetic Internship (DI)*
- A supervised practice program sponsored by a health care facility, college or university, federal or state agency, business, or corporation
- Minimum of 900 hours of supervised practice experiences

- Entry requires completion of CADE-approved Didactic Program in Dietetics and at least a baccalaureate degree from a US regionally accredited college or university or foreign equivalent
- May be full-time or part-time and vary from 6 months to 2 years
- Graduates are eligible to write the registration examination for dietitians

# Dietetic Technician, Registered

### Occupational/Job Description

Dietetic technicians assist in shaping the public's food choices and provide nutrition assessment and counseling to persons with illnesses or injuries. Technicians often screen patients to identify nutrition problems, provide patient education and counseling to individuals and groups, develop menus and recipes, supervise food service personnel, purchase food, and monitor inventory and food quality.

### Employment Characteristics

As an integral part of the nutrition care team, dietetic technicians work with registered dietitians in a number of different settings, such as hospitals, public health nutrition programs, and long-term care facilities. Technicians also work in child nutrition and school lunch programs, community wellness centers, health clubs, nutrition programs for the elderly, food companies, restaurants, and food service management.

### Salary

According to the ADA's 2005 Dietetics compensation and benefit survey, half of all registered dietetic technicians employed full-time in their primary position for 4 years or less earn annual incomes of between $26,000 and $37,000. Salary levels may vary based on location, scope of responsibility, and supply of DTRs. Refer to Section IV, Table 5 of this *Directory* for more information, or see www.ama-assn.org/go/salary.

### Educational Programs

**Length.** Two years (associate degree), combining classroom and supervised practical experience, at a US regionally accredited college or university. After completing this program, individuals are eligible to take the registration examination for dietetic technicians. Those who pass the exam become Dietetic Technicians, Registered, and can use the initials "DTR" after their names.

**Prerequisites.** Applicants must have a high school diploma or equivalency and must meet institutional entrance requirements.

**Curriculum.** Didactic instruction and a minimum of 450 hours of supervised practice experiences make up the curriculum. Food, nutrition, and management courses are emphasized, supported by the sciences, especially biology, anatomy, and chemistry. Mathematics, English, sociology, psychology, communications, and business courses are also important.

### Inquiries

#### Careers

Inquiries regarding careers in dietetics and nutrition should be addressed to:

Accreditation and Education Programs
American Dietetic Association
120 S Riverside Plaza, Suite 2000
Chicago, IL 60606-6995
312 899-0040, Ext 5400
E-mail: education@eatright.org

#### Certification/Registration

Inquiries regarding dietitian and dietetic technician registration should be addressed to:

Commission on Dietetic Registration
120 S Riverside Plaza, Suite 2000
Chicago, IL 60606-6995
312 899-0040, Ext 5500
E-mail: cdr@eatright.org

#### Accredited and Approved Programs

A list of accredited and approved programs with selected information is available on the ADA Web site at:
www.eatright.org/cade

## Dietitian/Nutritionist

# Alabama

**Auburn University**
Dietetics-Didactic Prgm
Nutrition and Food Science
328 Spidle Hall
Auburn, AL 36849-5605
www.humsci.auburn.edu
*Prgm Dir:* Robin B Fellers, PhD RD LD
*Tel:* 334 844-4261  *Fax:* 334 844-4261
*E-mail:* rfellers@auburn.edu

**Samford University**
Dietetics-Didactic Prgm
Family and Consumer Educ
800 Lakeshore Dr
Birmingham, AL 35229-2239
www.samford.edu
*Prgm Dir:* Patricia H Terry, PhD RD LD
*Tel:* 205 726-2930  *Fax:* 205 726-2068
*E-mail:* phterry@samford.edu

**University of Alabama at Birmingham**
Dietetic Internship Prgm
Dept of Nutrition Sciences
Webb Bldg Rm 212
Birmingham, AL 35294
www.uab.edu/nutrition
*Prgm Dir:* M Amanda Brown, PhD RD
*Tel:* 205 934-3006  *Fax:* 205 934-3006
*E-mail:* marbrown@uab.edu

**Oakwood College**
Dietetics-Didactic Prgm
Family and Consumer Science
7000 Adventist Blvd
Huntsville, AL 35896
www.oakwood.edu/fcs
*Prgm Dir:* Patricia A Dyette, DrPH MS RD
*Tel:* 256 726-7230  *Fax:* 256 726-7233
*E-mail:* dsmith@oakwood.edu

**Oakwood College**
Dietetics-Preprofessional Practice Prgm
Family and Consumer Science
7000 Adventist Blvd
Huntsville, AL 35896
www.oakwood.edu/fcs/dip2002.html
*Prgm Dir:* Marta L Sovyanhadi, DrPH MPH RD
*Tel:* 256 726-7228  *Fax:* 256 726-7233
*E-mail:* msovyanhadi@oakwood.edu

**Jacksonville State University**
Dietetics-Didactic Prgm
Family and Consumer Sciences
125 E Mason Hall
Jacksonville, AL 36265
www.jsu.edu
*Prgm Dir:* Debra K Goodwin, MA RD
*Tel:* 256 782-5053  *Fax:* 256 782-5916
*E-mail:* dgoodwin@jsu.edu

**University of Montevallo**
Dietetics-Didactic Prgm
Family and Consumer Sciences
Station 6385 Bloch Hall
Montevallo, AL 35115-6000
www.montevallo.edu
*Prgm Dir:* Frances E Andrews, PhD RD LD
*Tel:* 205 665-6385   *Fax:* 205 665-6385
*E-mail:* andrews@montevallo.edu

**Alabama A&M University**
Dietetics-Didactic Prgm
Nutrition & Hospitality Mgmt
Div of Family & Consumer Serv PO Box 639
Normal, AL 35762-0639
www.aamu.edu
*Prgm Dir:* Ann P Warren, MS RD LD
*Tel:* 256 858-4103   *Fax:* 256 372-4103
*E-mail:* ann.warren@email.aamu.edu

**University of Alabama**
Dietetics-Didactic Prgm
Human Nutrition & Hosp Mgmt
PO Box 870158
Tuscaloosa, AL 35487-0158
www.ches.ua.edu
*Prgm Dir:* Alvin Niuh, MS RD
*Tel:* 205 348-4710   *Fax:* 205 348-6157
*E-mail:* aniuh@ches.ua.edu

**University of Alabama**
Dietetics-Coordinated Prgm
Human Nutrition & Hosp Mgmt
PO Box 870158, 206 Doster Hall
Tuscaloosa, AL 35487-0158
www.ches.ua.edu
*Prgm Dir:* Debra W Morrison, MS RD
*Tel:* 205 348-8130   *Fax:* 205 348-3789
*E-mail:* dmorriso@ches.ua.edu

**Tuskegee University**
Dietetics-Didactic Prgm
Food & Nutrition Sciences
204 Campbell Hall
Tuskegee, AL 36088
www.tuskegee.edu
*Prgm Dir:* Beatrice W Phillips, EdD RD LD
*Tel:* 334 727-8326   *Fax:* 334 727-8493
*E-mail:* ghebwp@tusk.edu

# Alaska

**University of Alaska Anchorage**
Dietetic Internship Prgm
3211 Providence Dr
108 Cuddy Center
Anchorage, AK 99508
www.uaa.alaska.edu/di
*Prgm Dir:* Carrie D Benton-King, MS RD
*Tel:* 907 786-1362   *Fax:* 907 786-1402
*E-mail:* afcdk@uaa.alaska.edu

# Arizona

**Arizona State University East**
Dietetics-Didactic Prgm
Department of Nutrition
7001 E Williams Field Rd
Mesa, AZ 85212
www.east.asu.edu/ecollege/nutrition
*Prgm Dir:* Linda A Vaughan, PhD RD
*Tel:* 480 727-1731   *Fax:* 480 727-1064
*E-mail:* linda.vaughan@asu.edu

**Arizona State University East**
Dietetic Internship Prgm
Department of Nutrition
7001 E Williams Field Rd, Bldg 20
Mesa, AZ 85212
www.east.asu.edu/ecollege/nutrition
*Prgm Dir:* Christina Shepard, MS RD
*Tel:* 480 727-1748   *Fax:* 480 727-1064
*E-mail:* tina.shepard@asu.edu

**Focus on Nutrition**
Dietetics-Preprofessional Practice Prgm
6446-40 E Trailridge Circle
Phoenix, AZ 85215-0811
http://azdpac.org
*Prgm Dir:* Marsha R Stieber, MSA RD CNSD
*Tel:* 480 854-1112   *Fax:* 480 396-5438
*E-mail:* marshastieber@att.net

**Paradise Valley Unified School District**
Dietetics-Preprofessional Practice Prgm
20621 N 32nd St
Phoenix, AZ 85050
www.pvusd.k12.az.us
*Prgm Dir:* Kathleen M Glindmeier, MBA RD
*Tel:* 602 493-6330   *Fax:* 602 493-6334
*E-mail:* kglindmeier@pvusd.k12.az.us

**Yavapai County Health Department**
Dietetic Internship Prgm
930 Division St
Prescott, AZ 86301
*Prgm Dir:* Judy S Lee-Norris, MPH RD
*Tel:* 928 442-5488   *Fax:* 928 771-3369
*E-mail:* judy.norris@co.yavapai.az.us

**Maricopa County Dept of Public Health**
Dietetic Internship Prgm
Office of Nutrition Services
1414 W Broadway, Ste 237
Tempe, AZ 85282
*Prgm Dir:* Shirley K Strembel, MS RD
*Tel:* 602 506-9325, Ext 231   *Fax:* 480 966-3233
*E-mail:* shirleystrembel@mail.maricopa.gov

**Carondelet St Mary's Hospital**
Dietetic Internship Prgm
Morrisons Custom Mgmt Co
1601 W St Mary's Rd
Tucson, AZ 85745
www.azdpac.org/programinfo.cfm?PID=100020
*Prgm Dir:* Linda L Kautz Osterkamp, PhD RD FADA
*Tel:* 520 872-6109   *Fax:* 520 740-6108
*E-mail:* lkosterkamp@carondelet.org

**University of Arizona**
Dietetic Internship Prgm
University Medical Center Support Services
1501 N Campbell Ave
Tucson, AZ 85724-5088
www.azdpac.org
*Prgm Dir:* Susan E Bristol, MS RD CNSD
*Tel:* 520 694-6622   *Fax:* 520 694-6276
*E-mail:* sbristol@umcaz.edu

**University of Arizona**
Dietetics-Didactic Prgm
Dept of Nutritional Sciences
PO Box 210038, Shantz 309
Tucson, AZ 85721-0038
www.ag.arizona.edu/NSC/nschome.htm
*Prgm Dir:* Wanda Hain Howell, PhD RD CNSD
*Tel:* 520 621-1619   *Fax:* 520 621-9446
*E-mail:* whhowell@ag.arizona.edu
*Med Dir:* Jennifer Muir Bowers, PhD RD CNSD

# Arkansas

**Henderson State University**
Dietetics-Didactic Prgm
Dept of Family & Consumer Sciences
PO Box 7504
Arkadelphia, AR 71999-0001
www.hsu.edu/dept/fcs
*Prgm Dir:* Vicki L Posey, MS RD LD
*Tel:* 870 230-5542   *Fax:* 870 230-5172
*E-mail:* poseyv@hsu.edu

**Ouachita Baptist University**
Dietetics-Didactic Prgm
Dept of Biology
PO Box 3769
Arkadelphia, AR 71998-0001
www.obu.edu/biology/program_in_dietetics.htm
*Prgm Dir:* Stacy L Freeman, MS RD LD
*Tel:* 870 245-5542   *Fax:* 870 245-5500
*E-mail:* freemans@obu.edu

**University of Central Arkansas**
Dietetics-Didactic Prgm
Family and Consumer Sciences
McAlister Hall 100
Conway, AR 72035
www.uca.edu/divisions/academic/chas/diet.html
*Prgm Dir:* Mary H Harlan, EdD RD LD
*Tel:* 501 450-5950   *Fax:* 501 450-5958
*E-mail:* maryh@uca.edu

**University of Central Arkansas**
Dietetic Internship Prgm
Family and Consumer Sciences
McAlister Hall 100
Conway, AR 72035
www.uca.edu/divisions/academic/chas/diet.html
*Prgm Dir:* Elizabeth L Butler, MSE RD LD
*Tel:* 501 450-5954   *Fax:* 501 450-5958
*E-mail:* lizc@uca.edu

**University of Arkansas**
Dietetics-Didactic Prgm
Human Environmental Sciences
118 HOEC
Fayetteville, AR 72701
www.uark.edu/depts/hesweb
*Prgm Dir:* Marjorie E Fitch-Hilgenberg, PhD RD
*Tel:* 479 575-6815   *Fax:* 479 575-7171
*E-mail:* mfitch@uark.edu

**University of Arkansas for Medical Sciences**
Dietetic Internship Prgm
Veterans Affairs Med Ctr
4301 W Markham St, Slot 627
Little Rock, AR 72205-7199
www.uams.edu/chrp/diet.htm
*Prgm Dir:* Polly A Carroll, MA RD LD
*Tel:* 501 686-5714   *Fax:* 501 686-5716
*E-mail:* carrollpollya@uams.edu

**University of Arkansas at Pine Bluff**
Dietetics-Didactic Prgm
Dept of Home Economics
1200 N University Dr, PO Box 4971
Pine Bluff, AR 71601
*Prgm Dir:* Lucille Meadows, MS RD
*Tel:* 870 575-8817   *Fax:* 870 543-8823
*E-mail:* meadows_l@uapb.edu

**Harding University**
Dietetics-Didactic Prgm
Family and Consumer Sciences
900 E Center Ave, Box 12233
Searcy, AR 72149-0001
*Prgm Dir:* Lisa Ritchie, EdD RD LD
*Tel:* 501 279-4677   *Fax:* 501 279-4098
*E-mail:* lritchie@harding.edu

# California

**Clinica Sierra Vista**
Dietetic Internship Prgm
1430 Truxtun Ave, Ste 120
Bakersfield, CA 93301-3834
*Prgm Dir:* Laurie J Wallace, MS RD
*Tel:* 661 587-5781   *Fax:* 661 322-1418
*E-mail:* wallacel@clinicasierravista.org

**University of California - Berkeley**
Dietetics-Didactic Prgm
Dept of Nutritional Sciences and Toxicology
119 Morgan Hall
Berkeley, CA 94720-3104
http://nutrition.berkeley.edu
*Prgm Dir:* Mary S Mead, Med RD
*Tel:* 510 642-0980   *Fax:* 510 642-0535
*E-mail:* mmead@nature.berkeley.edu

**University of California - Berkeley**
Dietetic Internship Prgm
Public Health Dietetic Internship
140 Warren Hall
Berkeley, CA 94720-7360
*Prgm Dir:* Kathleen A McBurney, DrPH RD
*Tel:* 510 643-0969   *Fax:* 510 642-9891
*E-mail:* kmcburne@uclink.berkeley.edu

**California State University - Chico**
Dietetic Internship Prgm
Nutrition & Food Sciences
Dept of Biological Sciences
Chico, CA 95929-0002
www.csuchico.edu/biol/nfsc
*Prgm Dir:* Michelle R Neyman Morris, PhD RD
*Tel:* 530 898-4757   *Fax:* 530 898-5586
*E-mail:* mrmorris@csuchico.edu

**California State University - Chico**
Dietetics-Didactic Prgm
Dept of Biological Sciences
Tehama Hall 124
Chico, CA 95929-0002
*Prgm Dir:* Faye C Johnson, EdD RD
*Tel:* 530 898-6805   *Fax:* 530 898-4363
*E-mail:* fjohnson@csuchico.edu

**University of California - Davis**
Dietetics-Didactic Prgm
Dept of Nutrition
One Shields Ave
Davis, CA 95616-8669
www.ucdavis.edu
*Prgm Dir:* Francene M Steinberg, PhD RD
*Tel:* 530 752-0160   *Fax:* 530 752-8966
*E-mail:* fmsteinberg@ucdavis.edu

**California State University - Fresno**
Dietetic Internship Prgm
Dept of Ecology Food Science & Nutrition
5300 N Campus Dr MS FF17
Fresno, CA 93740-8019
http://cast.csufresno.edu/fsn
*Prgm Dir:* Mollie K Smith, MS RD
*Tel:* 559 278-8009   *Fax:* 559 278-7623
*E-mail:* mollies@csufresno.edu

**California State University - Fresno**
Dietetics-Didactic Prgm
Dept of Ecology Food Science & Nutrition
5300 N Campus Dr MS FF17
Fresno, CA 93740-8019
http://cast.csufresno.edu/fsn
*Prgm Dir:* Sandra S Witte, PhD RD
*Tel:* 559 278-2164   *Fax:* 559 278-7623
*E-mail:* sandraw@csufresno.edu

**Glendale Memorial Hospital and Health Center**
Dietetic Internship Prgm
Food & Nutrition Services
1420 S Central Ave
Glendale, CA 91204-2594
www.glendalememorial.com/dietary
*Prgm Dir:* Remi Hayashi, MS RD
*Tel:* 818 409-7643   *Fax:* 818 507-4665
*E-mail:* rhayashi@chw.edu

**Public Health Foundation Enterprises**
Dietetic Internship Prgm
WIC Program
12781 Schabarum Ave
Irwindale, CA 91706-6802
www.phfewic.org
*Prgm Dir:* Robin B Evans, RD MPH
*Tel:* 626 856-6618, Ext 292   *Fax:* 626 813-9390
*E-mail:* robin@phfewic.org

**Loma Linda University**
Dietetics-Coordinated Prgm
School of Pub Hlth Nutri Dept
Nichol Hall, Rm 1102
Loma Linda, CA 92350
www.llu.edu/llu/nutrition
*Prgm Dir:* Ella H Haddad, DrPH MS RD
*Tel:* 909 558-4598   *Fax:* 909 558-4095
*E-mail:* ehaddad@sph.llu.edu

**Loma Linda University**
Dietetics-Coordinated Prgm
School of Allied Health Professions
Dept of Nutrition & Dietetics
Loma Linda, CA 92354
www.llu.edu/llu/nutrition
*Prgm Dir:* Bertrum C Connell, PhD RD FADA
*Tel:* 909 558-4593   *Fax:* 909 558-4291
*E-mail:* bconnell@llu.edu

**California State University - Long Beach**
Dietetic Internship Prgm
Family and Consumer Sciences
1250 N Bellflower Blvd
Long Beach, CA 90840-0501
www.csulb.edu/~gcfrank
*Prgm Dir:* Gail C Frank, DrPH RD CHES
*Tel:* 562 985-4494   *Fax:* 562 985-4414
*E-mail:* gcfrank@csulb.edu

**California State University - Long Beach**
Dietetics-Didactic Prgm
Family and Consumer Sciences
1250 Bellflower Blvd
Long Beach, CA 90840-0501
www.csulb.edu/colleges/chhs/dpd
*Prgm Dir:* Jacqueline D Lee, PhD RD
*Tel:* 562 985-4545   *Fax:* 562 985-4414
*E-mail:* jjlee@csulb.edu

**California State University - Los Angeles**
Dietetics-Didactic Prgm
Health & Nutritional Sciences
5151 State University Dr
Los Angeles, CA 90032-8162
www.calstatela.edu/dept/hnut_sci/dept_pro.htm
*Prgm Dir:* Jane Burrell Uzcategui, MA RD
*Tel:* 323 343-4666   *Fax:* 323 343-6482
*E-mail:* juzcate@exchange.calstatela.edu

**California State University - Los Angeles**
Dietetics-Coordinated Prgm
Kinesiology & Nutritional Science
5151 State University Dr
Los Angeles, CA 90032-8172
www.calstatela.edu/dept/hnut_sci/dept_pro.htm
*Prgm Dir:* Laura L Calderon, DrPH RD
*Tel:* 323 343-5439   *Fax:* 323 343-6482
*E-mail:* lcalder@exchange.calstatela.edu

**Center for Child Development and Nutrition**
*Cosponsor:* Children's Hosp of Los Angeles
Dietetic Internship Prgm
Child Development & Dvlpmntl Disorders
Mailstop 53, PO Box 54700
Los Angeles, CA 90054-0700
*Prgm Dir:* Leslie H Wills, MA RD CSP
*Tel:* 323 671-3816   *Fax:* 323 671-3842
*E-mail:* hwills@chla.usc.edu

**Los Angeles County+USC Healthcare Network**
Dietetic Internship Prgm
Morrison Management Specialists
1200 N State St, Rm 1506
Los Angeles, CA 90033-4525
https://www.lacuscdi.org
*Prgm Dir:* Elizabeth H Ma, MS RD
*Tel:* 323 226-6901   *Fax:* 323 222-3422
*E-mail:* ema@lacusc.org

**VA Greater Los Angeles Healthcare System**
Dietetic Internship Prgm
Nutrition & Food Dept 120
11301 Wilshire Blvd
Los Angeles, CA 90073
*Prgm Dir:* Colleen M Ross, MS RD
*Tel:* 310 268-3120   *Fax:* 310 268-4787
*E-mail:* colleen.ross@med.va.gov

**Pepperdine University**
Dietetics-Didactic Prgm
RAC108C, Nutritional Sciences
24255 Pacific Coast Hwy
Malibu, CA 90263-4325
www.pepperdine.edu
*Prgm Dir:* Susan E Helm, PhD RD
*Tel:* 310 506-4325   *Fax:* 310 506-4785
*E-mail:* susan.helm@pepperdine.edu

**Napa State Hospital**
Dietetic Internship Prgm
2100 Napa Vallejo Hwy
Napa, CA 94558-6293
*Prgm Dir:* Wen F Pao, MS RD CDE
*Tel:* 707 253-5428   *Fax:* 707 254-2422
*E-mail:* wpao@dmhnsh.state.ca.us

**California State University - Northridge**
Dietetics-Didactic Prgm
Family Environmental Sciences
18111 Nordhoff St
Northridge, CA 91330-8308
www.csun.edu
*Prgm Dir:* Deirdre M Larkin, MS RD
*Tel:* 818 677-3124   *Fax:* 818 677-4778
*E-mail:* deidre.larkin@csun.edu

**California State University - Northridge**
Dietetic Internship Prgm
Family Environmental Sciences
18111 Nordhoff St
Northridge, CA 91330-8308
www.csundi.com
*Prgm Dir:* Annette A Besnilian, MPH RD
*Tel:* 818 677-3051
*E-mail:* annette.besnilian@csun.edu

**Patton State Hospital**
Dietetic Internship Prgm
3102 E Highland Ave
Patton, CA 92369
*Prgm Dir:* Jeanie J Kim, MA RD
*Tel:* 909 425-7575   *Fax:* 909 425-7069
*E-mail:* jkim@dmhpsh.state.ca.us

**California State Polytechnic University**
Dietetic Internship Prgm
Foods & Nutri/Home Economics
3801 W Temple Ave
Pomona, CA 91768-2557
www.csupomona.edu/~kcaldwellfreeman
*Prgm Dir:* Mark S Meskin, PhD RD FADA
*Tel:* 909 869-4877, Ext 2226   *Fax:* 909 869-5078
*E-mail:* msmeskin@csupomona.edu

**California State Polytechnic University**
Dietetics-Didactic Prgm
Foods and Nutrition Dept
3801 W Temple Ave
Pomona, CA 91768-2557
www.csupomona.edu
*Prgm Dir:* Mark S Meskin, PhD RD FADA
*Tel:* 909 869-4877   *Fax:* 909 869-5078
*E-mail:* msmeskin@csupomona.edu

**Porterville Development Center**
Dietetic Internship Prgm
PO Box 2000
Porterville, CA 93258-2000
*Prgm Dir:* Jeanny C Chang, MS RD
*Tel:* 559 782-2715   *Fax:* 559 782-2756
*E-mail:* jchang@pdc.dds.ca.gov

**Central Valley WIC Dietetic Internship**
Dietetic Internship Prgm
1560 E Manning Ave
Reedley, CA 93654
*Prgm Dir:* Beth Arrindell, MPH RD
*Tel:* 559 646-6611   *Fax:* 559 646-6615
*E-mail:* earrind@aol.com

**California State University - Sacramento**
Dietetic Internship Prgm
6000 J St
Sacramento, CA 95819-6053
www.csus.edu/indiv/c/cunninghamw
*Prgm Dir:* Wendy Mueller-Cunningham, PhD RD ETT
*Tel:* 916 278-6732   *Fax:* 916 278-7520
*E-mail:* wcunningham@csus.edu

**California State University - Sacramento**
Dietetics-Didactic Prgm
Human Environmental Sciences
6000 J St
Sacramento, CA 95819-6053
www.csus.edu/facs
*Prgm Dir:* Dianne A Hyson, PhD RD
*Tel:* 916 278-7320   *Fax:* 916 278-7520
*E-mail:* dhyson@csus.edu

**UC Davis Medical Center**
Dietetic Internship Prgm
2315 Stockton Blvd
Sacramento, CA 95817
*Prgm Dir:* Julie S Rogers, MS RD IBCLC
*Tel:* 916 734-5256
*E-mail:* julie.rogers@ucdmc.ucdavis.edu

**California State University - San Bernardino**
Dietetics-Didactic Prgm
Health Science & Human Ecology
5500 University Pkwy
San Bernardino, CA 92407-2318
http://health.csusb.edu/dchen
*Prgm Dir:* Dorothy C Chen-Maynard, PhD RD
*Tel:* 909 537-5340   *Fax:* 909 880-7037
*E-mail:* dchen@csusb.edu

**San Diego State University**
Dietetic Internship Prgm
Exercise and Nutritional Sciences
5500 Campanile Dr
San Diego, CA 92182-7251
www.rohan.sdsu.edu
*Prgm Dir:* Janet I Nash, MS RD
*Tel:* 619 594-2416   *Fax:* 619 594-6553
*E-mail:* jingnash@mail.sdsu.edu

**San Diego State University**
Dietetics-Didactic Prgm
Exercise and Nutritional Sciences
5500 Campanile Dr
San Diego, CA 92182-7251
www.sdsu.edu
*Prgm Dir:* Joan W Rupp, MS RD
*Tel:* 760 436-6162   *Fax:* 619 594-6553
*E-mail:* rupp@mail.sdsu.edu

**VA San Diego Healthcare System**
Dietetic Internship Prgm
Dietetics & Clinical Nutrition Serv 120
3350 La Jolla Village Dr
San Diego, CA 92161-0002
*Prgm Dir:* Teresa Bush-Zurn, MA RD FADA
*Tel:* 858 552-8585, Ext 2407   *Fax:* 858 552-4340
*E-mail:* teresa.bush-zurn@med.va.gov

**San Francisco State University**
Dietetic Internship Prgm
Cons & Fam Studies
1600 Holloway Ave
San Francisco, CA 94132
www.sfsu.edu/~sda
*Prgm Dir:* Patricia A Booth, MS RD FADA
*Tel:* 415 353-1355   *Fax:* 415 338-0947
*E-mail:* patricia.booth@ucsfmedctr.org

**San Francisco State University**
Dietetics-Didactic Prgm
Cons & Fam Studies
1600 Holloway Ave
San Francisco, CA 94132-1722
www.sfsu.edu
*Prgm Dir:* Joan S Frank, MS RD
*Tel:* 415 338-6988   *Fax:* 415 338-0947
*E-mail:* jfrank@sfsu.edu

**University of California - San Francisco**
Dietetic Internship Prgm
Medical Center, Box 0212 Rm M-294
Dept of Nutrition & Dietetics
San Francisco, CA 94143-0212
http://medweb.his.ucsf.edu/nutrition
*Prgm Dir:* Patricia A Booth, MS RD FADA
*Tel:* 415 353-1355   *Fax:* 415 353-8703
*E-mail:* patricia.booth@ucsfmedctr.org

**San Jose State University**
Dietetics-Didactic Prgm
Nutrition & Food Science
San Jose, CA 95192-0058
*Prgm Dir:* Nancy C Lu, PhD RD
*Tel:* 408 924-3109   *Fax:* 408 924-3114
*E-mail:* nancyclu@email.sjsu.edu

**San Jose State University**
Dietetic Internship Prgm
Nutrition Food & Science
One Washington Square
San Jose, CA 95192-0058
*Prgm Dir:* Kathryn P Sucher, ScD RD
*Tel:* 408 924-3104   *Fax:* 408 924-3114
*E-mail:* ksucher@email.sjsu.edu

**California Polytechnic State University**
Dietetics-Didactic Prgm
Food Science & Nutrition
San Luis Obispo, CA 93407
www.calpoly.edu
*Prgm Dir:* Kristen L Buckshi, MS RD
*Tel:* 805 756-2757   *Fax:* 805 756-1146
*E-mail:* kbuckshi@calpoly.edu

**Olive View/UCLA Medical Center**
Dietetic Internship Prgm
Dept of Food and Nutrition
14445 Olive View Dr, Rm 1C112
Sylmar, CA 91342-1438
*Prgm Dir:* Brook Reitman, MPH RD
*Tel:* 818 364-4220   *Fax:* 818 364-3998
*E-mail:* breitman@ladhs.org

# Colorado

**Centura Health/Penrose-St Francis Hlth Serv**
Dietetic Internship Prgm
Nutrition Services
PO Box 7021
Colorado Springs, CO 80933-7021
*Prgm Dir:* Suzanne M Logan, MS RD
*Tel:* 719 776-5900   *Fax:* 719 776-2500
*E-mail:* suzannelogan@centura.org

**Tri-County Health Department**
Dietetic Internship Prgm
Nutrition Services
4857 S Broadway
Englewood, CO 80110-6894
*Prgm Dir:* Anne E Bennett, MPH RD
*Tel:* 303 220-9200   *Fax:* 303 761-1528
*E-mail:* abennett@tchd.org

**Colorado State University**
Dietetic Internship Prgm
Food Sci & Human Nutrition
Ft Collins, CO 80523-1571
www.colostate.edu
*Prgm Dir:* Mary A Harris, PhD RD
*Tel:* 970 491-7462   *Fax:* 970 491-7252
*E-mail:* harris@cahs.colostate.edu

**Colorado State University**
Dietetics-Didactic Prgm
Food Sci & Human Nutrition
Gifford Bldg 205
Ft Collins, CO 80523-1571
www.colostate.edu/dept/FSHN
*Prgm Dir:* Mary A Harris, PhD RD
*Tel:* 970 491-7462   *Fax:* 970 491-7252
*E-mail:* harris@cahs.colostate.edu

**University of Northern Colorado**
Dietetic Internship Prgm
Community Health & Nutrition
CHN Box 93-501 20th St
Greeley, CO 80639
www.unco.edu/dietetic
*Prgm Dir:* Alana Cline, PhD RD
*Tel:* 970 351-1769   *Fax:* 970 351-1489
*E-mail:* alana.cline@unco.edu

**University of Northern Colorado**
Dietetics-Didactic Prgm
Community Health & Nutrition
CHN, Box 93-501 20th St
Greeley, CO 80639
www.hhs.unco.edu/Diet.htm
*Prgm Dir:* Jamie M Erskine, PhD RD
*Tel:* 970 351-2451   *Fax:* 970 351-1489
*E-mail:* jamie.erskine@unco.edu

# Connecticut

**Danbury Hospital**
Dietetic Internship Prgm
24 Hospital Ave
Danbury, CT 06810-6077
www.danburyhospital.org
*Prgm Dir:* Janice D Hiser, MAA RD CDN
*Tel:* 203 797-7216   *Fax:* 203 797-7619
*E-mail:* janice.hiser@danhosp.org

**Yale-New Haven Hospital**
Dietetic Internship Prgm
Food & Nutrition
20 York St GBB
New Haven, CT 06504
www.ynhh.org/general/training.html#diet
*Prgm Dir:* Deborah F Flanel, MS RD FADA
*Tel:* 203 688-2949   *Fax:* 203 688-2412
*E-mail:* debbie.ford-flannel@ynhh.org

**University of Connecticut**
Dietetics-Coordinated Prgm
School of Allied Health
358 Mansfield Rd Unit 2101
Storrs, CT 06269-2101
www.alliedhealth.uconn.edu
*Prgm Dir:* Robin H Abourizk, MA MS RD
*Tel:* 860 486-0016    *Fax:* 860 486-5375
*E-mail:* robin.abourizk@uconn.edu

**University of Connecticut**
Dietetics-Didactic Prgm
Nutritional Sciences U-4017
3624 Horsebarn Rd Extension
Storrs, CT 06269
www.canr.uconn.edu/nusci
*Prgm Dir:* Ellen Rosa Shanley, MBA RD CDN
*Tel:* 860 486-0119    *Fax:* 860 486-3674
*E-mail:* ellen.shanley@uconn.edu

**University of Connecticut**
Dietetic Internship Prgm
School of Allied Health
358 Mansfield Rd Unit 2101
Storrs, CT 06269-2101
www.alliedhealth.uconn.edu
*Prgm Dir:* Robin H Abourizk, MA MS RD
*Tel:* 860 486-0016    *Fax:* 860 486-5375
*E-mail:* robin.abourizk@uconn.edu

**Saint Joseph College**
Dietetic Internship Prgm
Nutrition and Family Studies
1678 Asylum Ave
West Hartford, CT 06117-2700
*Prgm Dir:* Donna W Corcoran, MS RD
*Tel:* 860 231-5254, Ext 254    *Fax:* 860 231-8396
*E-mail:* dcorcoran@sjc.edu

**Saint Joseph College**
Dietetics-Didactic Prgm
Nutrition and Family Studies
1678 Asylum Ave
West Hartford, CT 06117-2700
www.sjc.edu
*Prgm Dir:* Margery L Lawrence, PhD RD
*Tel:* 860 231-5388    *Fax:* 860 231-8396
*E-mail:* mlawrence@sjc.edu

**Saint Joseph College**
Dietetics-Coordinated Prgm
Nutrition and Family Studies
1678 Asylum Ave
West Hartford, CT 06117-2700
www.sjc.edu
*Prgm Dir:* Margaret E Gaughan, PhD RD
*Tel:* 860 231-5234, Ext 234    *Fax:* 860 231-8396
*E-mail:* mgaughan@sjc.edu

**University of New Haven**
Dietetics-Didactic Prgm
College of Arts and Sciences
300 Orange Ave
West Haven, CT 06516-1916
www.newhaven.edu
*Prgm Dir:* Georgia A Chavent, MS RD
*Tel:* 203 932-7410    *Fax:* 203 932-7083
*E-mail:* gchavent@newhaven.edu

# Delaware

**Delaware State University**
Dietetics-Didactic Prgm
Dept of Family & Consumer Sciences
1200 N Dupont Hwy
Dover, DE 19901-2277
www.dsc.edu
*Prgm Dir:* Carol C Giesecke, PhD RD
*Tel:* 302 857-6439    *Fax:* 302 857-6441
*E-mail:* cgiesecke@desu.edu

**University of Delaware**
Dietetic Internship Prgm
Nutrition & Dietetics
315 Alison Hall
Newark, DE 19716
www.udel.edu
*Prgm Dir:* Antoinette Rucinski, MA RD
*Tel:* 302 831-2079    *Fax:* 302 831-4186
*E-mail:* rucinski@udel.edu

**University of Delaware**
Dietetics-Didactic Prgm
Dept of Nutrition & Dietetics
234A Alison Hall
Newark, DE 19716-3301
www.udel.edu
*Prgm Dir:* Elizabeth M Lieux, PhD RD
*Tel:* 302 831-2732    *Fax:* 302 831-4186
*E-mail:* lieux@udel.edu

# District of Columbia

**Howard University**
Dietetics-Coordinated Prgm
Nutritional Sciences
6th and Bryant Sts NW, Annex I
Washington, DC 20059
www.howard.edu
*Prgm Dir:* Avis P Graham, PhD MS RD LD
*Tel:* 202 806-5648    *Fax:* 202 806-9233
*E-mail:* agraham@howard.edu

**University of the District of Columbia**
Dietetics-Didactic Prgm
Biological & Environmental Sciences
Bldg 44 Rm 200-2, 4200 Connecticut Ave NW
Washington, DC 20008-1173
*Prgm Dir:* Prema Ganganna, PhD RD LD
*Tel:* 202 274-5516    *Fax:* 202 274-5845
*E-mail:* pganganna@udc.edu

# Florida

**University of Florida**
Dietetics-Didactic Prgm
Food Sci & Human Nutrition
PO Box 110370
Gainesville, FL 32611-0370
www.ufl.edu
*Prgm Dir:* Pamela S McMahon, PhD RD
*Tel:* 352 392-1512, Ext 703    *Fax:* 352 392-9467
*E-mail:* PSMcMahon@ifas.ufl.edu

**University of Florida**
Dietetic Internship Prgm
Food Sci & Human Nutrition
359 FSB, PO Box 110370
Gainesville, FL 32611
http://fshn.ifas.ufl.edu
*Prgm Dir:* Gail P Abbott Kauwell, PhD RD LD
*Tel:* 352 392-1991, Ext 227    *Fax:* 352 392-9467
*E-mail:* gpk@ifas.ufl.edu

**St Luke's Hosp/Mayo Clinic Jacksonville**
Dietetic Internship Prgm
4500 San Pablo Rd
Jacksonville, FL 32224
*Prgm Dir:* Sherry M Mahoney, MBA RD CDE
*Tel:* 904 953-2753    *Fax:* 904 953-2954
*E-mail:* mahoney.sherry@mayo.edu
*Med Dir:* James Scolapio, MD

**University of North Florida**
Dietetics-Didactic Prgm
College of Health
4567 St Johns Bluff Rd S
Jacksonville, FL 32224-2645
www.unf.edu
*Prgm Dir:* Sally E Weerts, PhD RD
*Tel:* 904 620-1433    *Fax:* 904 620-2848
*E-mail:* sweerts@unf.edu

**University of North Florida**
Dietetic Internship Prgm
Dept of Health Science
4567 St Johns Bluff Rd S
Jacksonville, FL 32224-2646
www.unf.edu/coh
*Prgm Dir:* Catherine W Christie, PhD RD FADA
*Tel:* 904 620-1423    *Fax:* 904 620-2848
*E-mail:* cchristi@unf.edu

**Florida International University**
Dietetics-Coordinated Prgm
Dept of Dietetics & Nutrition
11200 SW 8th St
Miami, FL 33199
http://w3.fiu.edu/dietetics
*Prgm Dir:* Dona C Greenwood, PhD Rd
*Tel:* 305 348-0194    *Fax:* 305 348-1996
*E-mail:* greenwoo@fiu.edu

**Florida International University**
Dietetic Internship Prgm
Dietetics & Nutrition
University Park HB 206
Miami, FL 33199
*Prgm Dir:* Amy Jaffe, MS RD LD
*Tel:* 305 348-2878    *Fax:* 305 348-1996
*E-mail:* amyjaffe@bellsouth.net

**Florida International University**
Dietetics-Didactic Prgm
Dietetics & Nutrition
11200 SW 8th St
Miami, FL 33199
http://w3.fiu.edu/dietetics
*Prgm Dir:* Dona C Greenwood, PhD RD
*Tel:* 305 348-0194    *Fax:* 305 348-1996
*E-mail:* greenwoo@fiu.edu

**Pasco County Health Department**
Dietetic Internship Prgm
Nutrition Div
10841 Little Rd
New Port Richey, FL 34654-2533
www9.myflorida.com/chpasco/DI_Web/dihome.htm
*Prgm Dir:* Clara R Lawhead, MS RD LD FADA
*Tel:* 727 869-3900, Ext 120    *Fax:* 727 863-9734
*E-mail:* clara_lawhead@doh.state.fl.us

**Sarasota District Schools**
Dietetic Internship Prgm
Food & Nutrition Services
101 Old Venice Rd
Osprey, FL 34229-9023
www.sarasota.k12.fl.us/~fns
*Prgm Dir:* Beverly L Girard, MBA MS RD
*Tel:* 941 486-2199    *Fax:* 941 486-2021
*E-mail:* dietetic_internship@sarasota.k12.fl.us

**Sarasota Memorial Hospital**
Dietetic Internship Prgm
Food & Nutrition Services
1700 S Tamiami Trail
Sarasota, FL 34239-3555
www.smh.com/newdeptpage
*Prgm Dir:* Joan E Disley, MBA RD LD
*Tel:* 941 917-6041    *Fax:* 941 917-6196
*E-mail:* joan-disley@smh.com

**Florida Department of Education**
Dietetic Internship Prgm
325 W Gaines St, Rm 1032
Tallahassee, FL 32399-0400
*Prgm Dir:* Richard E Parks, MS RD
*Tel:* 852 459-9267    *Fax:* 850 921-8824
*E-mail:* richard.parks@fldoe.org

**Florida State University**
Dietetics-Didactic Prgm
Nutrition Food & Exercise Sciences
436 Sandels Bldg
Tallahassee, FL 32306-1493
www.fsu.edu/nfes
*Prgm Dir:* Delores D Bruesdell, MS RD
*Tel:* 850 644-8215   *Fax:* 850 645-5000
*E-mail:* dtruesde@mailer.fsu.edu

**Florida State University**
Dietetic Internship Prgm
Nutrition Food & Exercise Sciences
400 Sandels Bldg
Tallahassee, FL 32306-1493
www.fsu.edu~human
*Prgm Dir:* Laura R Cook, PhD RD LD
*Tel:* 850 644-1828   *Fax:* 850 644-5000
*E-mail:* lcook@mailer.fsu.edu

**James A Haley Veteran's Hospital**
Dietetic Internship Prgm
13000 N Bruce B Downs Blvd
Tampa, FL 33612-4745
*Prgm Dir:* Anne E Brezina, MEd RD
*Tel:* 813 972-2000, Ext 7234   *Fax:* 813 978-5838
*E-mail:* anne.brezina@med.va.gov

# Georgia

**University of Georgia**
Dietetics-Didactic Prgm
Dept of Foods & Nutrition
Dawson Hall
Athens, GA 30602
*Prgm Dir:* Joan G Fischer, PhD RD LD
*Tel:* 706 542-7983   *Fax:* 706 542-5059
*E-mail:* jfischer@fcs.uga.edu

**University of Georgia**
Dietetic Internship Prgm
Dept of Foods & Nutrition
Dawson Hall
Athens, GA 30602
www.fcs.uga.edu
*Prgm Dir:* Barbara M Grossman, PhD RD LD
*Tel:* 706 542-4908   *Fax:* 706 542-5059
*E-mail:* bgrossman@fcs.uga.edu

**Div of Pub Hlth/Georgia Dept of Hum Res**
Dietetic Internship Prgm
Office of Nutrition
Two Peachtree St NW, Ste 11-254
Atlanta, GA 30303-3142
www.ph.dhr.state.ga.us/programs
*Prgm Dir:* Frances Hanks Cook, MA RD LD CPM
*Tel:* 404 657-2884   *Fax:* 404 657-2886
*E-mail:* frcook@dhr.state.ga.us

**Emory University Hospital**
Dietetic Internship Prgm
Food & Nutrition Services
1364 Clifton Rd NE, Rm FG06
Atlanta, GA 30322
http://medshare.emory.org/EHC/FNSWeb.nsf
*Prgm Dir:* Maureen C McAndrews, MPH RD LD
*Tel:* 404 712-4176   *Fax:* 404 712-7452
*E-mail:* dietetic_internship@emory.org

**Georgia State University**
Dietetics-Didactic Prgm
Dept of Nutrition & Dietetics
MSC2A0880, 33 Gilmer St SE, Unit 2
Atlanta, GA 30303-3083
www.gsu.edu/~wwwntr
*Prgm Dir:* Delia H Baxter, PhD RD
*Tel:* 404 651-1108   *Fax:* 404 651-1235
*E-mail:* dbaxter@gsu.edu

**Georgia State University**
Dietetic Internship Prgm
Dept of Nutrition & Dietetics
University Plaza
Atlanta, GA 30303-3083
www.gsu.edu/nutrition
*Prgm Dir:* Barbara L Hopkins, MMSc RD LD
*Tel:* 404 651-1082   *Fax:* 404 651-1235
*E-mail:* bhopkins@gsu.edu

**Augusta Area Dietetic Internship**
*Cosponsor:* Morrison Health Care Foodservice
Dietetic Internship Prgm
University Hospital
1350 Walton Way
Augusta, GA 30901-2629
https://www.universityhealth.org
*Prgm Dir:* Jeanne Bingham-Lee, MS RD LD
*Tel:* 706 774-8897   *Fax:* 706 774-8671
*E-mail:* jlee@uh.org

**Fort Valley State University**
Dietetics-Didactic Prgm
Family and Consumer Sciences
1005 State University Dr, PO Box 4622 FVSU
Ft Valley, GA 31030
*Prgm Dir:* Sharon K Hunt, MS RD
*Tel:* 912 825-6234   *Fax:* 912 825-6078
*E-mail:* hunts@mail.fvsu.edu

**Life University**
Dietetics-Didactic Prgm
Dept of Nutrition
1269 Barclay Circle
Marietta, GA 30060-2903
*Prgm Dir:* Jaleh Dehpahlavan, MMSc RD LD
*Tel:* 706 426-2736   *Fax:* 706 426-2698
*E-mail:* jalehd@life.edu

**Life University**
Dietetic Internship Prgm
1269 Barclay Circle SE
110 Annex B
Marietta, GA 30060-2903
*Prgm Dir:* Jaleh Dehpahlavan, MMSc RD LD
*Tel:* 770 426-2736   *Fax:* 770 426-2697
*E-mail:* jalehd@life.edu

**Southern Regional Medical Center**
Dietetic Internship Prgm
11 Upper Riverdale Rd SW
Riverdale, GA 30274-2600
www.southernregional.org
*Prgm Dir:* Stephanie L Crocker, MMSc RD CNSD
*Tel:* 770 991-8030   *Fax:* 770 991-8690
*E-mail:* stephanie.crocker@southernregional.org

**Georgia Southern University**
Dietetics-Didactic Prgm
Family and Consumer Sciences
Box 8034
Statesboro, GA 30460
www.georgiasouthern.edu
*Prgm Dir:* Padmini Shankar, PhD RD LD
*Tel:* 912 681-5785   *Fax:* 912 681-0276
*E-mail:* pshankar@georgiasouthern.edu

# Hawaii

**University of Hawaii at Manoa**
Dietetics-Didactic Prgm
Food Sci & Human Nutrition
Ag Sciences III, 1955 East West Rd, Rm 3141
Honolulu, HI 96822-2218
www.hawaii.edu/dietetics
*Prgm Dir:* Anne C Shovic, PhD RD
*Tel:* 808 956-3847   *Fax:* 808 956-4024
*E-mail:* shovic@hawaii.edu

# Idaho

**University of Idaho**
Dietetics-Coordinated Prgm
Family and Consumer Science
College of Agriculture
Moscow, ID 83844-3183
www.uidaho.edu/fcs
*Prgm Dir:* Kathleen A Gabel, PhD RD LD
*Tel:* 208 885-6026   *Fax:* 208 885-5751
*E-mail:* kgabel@uidaho.edu

**Idaho State University**
Dietetic Internship Prgm
Health & Nutrition Sciences
Campus Box 8109
Pocatello, ID 83209-8109
www.isu.edu
*Prgm Dir:* Mary L Dundas, PhD RD LD FADA
*Tel:* 208 236-2352   *Fax:* 208 236-4903
*E-mail:* dundmary@isu.edu

**Idaho State University**
Dietetics-Didactic Prgm
Health & Nutrition Science
Campus Box 8109, 1291 East Terry, MLK Way
Pocatello, ID 83209-8109
www.isu.edu
*Prgm Dir:* Laura E G McKnight, MPH RD LD
*Tel:* 208 282-2961   *Fax:* 208 282-4903
*E-mail:* mcknlaur@isu.edu

# Illinois

**Olivet Nazarene University**
Dietetics-Didactic Prgm
Family and Consumer Sciences
One University Dr
Bourbonnais, IL 60914
http://web.olivet.edu/facs
*Prgm Dir:* Catherin N Anstrom, MBA RD LD
*Tel:* 815 928-5398   *Fax:* 815 935-4990
*E-mail:* canstrom@olivet.edu

**Southern Illinois Univ at Carbondale**
Dietetics-Didactic Prgm
Animal Sci Food & Nutrition
Mailcode 4317
Carbondale, IL 62901-4317
*Prgm Dir:* Sara Long, PhD RD
*Tel:* 618 453-7512   *Fax:* 613 453-7517
*E-mail:* saralong@siu.edu

**Southern Illinois Univ at Carbondale**
Dietetic Internship Prgm
Food & Nutrition
Mail Code 4317
Carbondale, IL 62901-4317
www.siu.edu/cwis
*Prgm Dir:* Dawn C Bloyd Null, MS RD
*Tel:* 618 453-5192   *Fax:* 618 453-7517
*E-mail:* dawnnull@siu.edu

**Eastern Illinois University**
Dietetics-Didactic Prgm
Family & Consumer Sci, Klehm Hall 109-B
600 Lincoln Ave, Klehm Hall 1433
Charleston, IL 61920-3099
www.eiu.edu/~famsci
*Prgm Dir:* Melanie T Burns, PhD RD
*Tel:* 217 581-6680   *Fax:* 217 581-6090
*E-mail:* mburns@eiu.edu

**Eastern Illinois University**
Dietetic Internship Prgm
Family and Consumer Sciences
600 Lincoln Ave, Klehm Hall 1433
Charleston, IL 61920-3099
www.eiu.edu/~dietetic
*Prgm Dir:* Karla Kennedy Hagan, PhD RD LD
*Tel:* 217 581-6353   *Fax:* 217 581-6090
*E-mail:* kjkennedyhagan@eiu.edu

**Loyola University of Chicago**
Dietetic Internship Prgm
Dept of Food and Nutrition
6525 N Sheridan Rd
Chicago, IL 60626-5311
www.luc.edu/depts/nutrition/internsh.htm
*Prgm Dir:* Joanne Kouba, MS RD
*Tel:* 773 508-8298    *Fax:* 773 508-8296
*E-mail:* jkouba@luc.edu

**Loyola University of Chicago**
Dietetics-Didactic Prgm
Dept of Food and Nutrition
6525 N Sheridan Rd
Chicago, IL 60626-5311
www.luc.edu/depts/nutrition
*Prgm Dir:* Tracey L Carlyle, MS RD LD
*Tel:* 773 508-8299    *Fax:* 773 508-8296
*E-mail:* tcarlyl@wpo.it.luc.edu

**Rush University Medical Center**
Dietetic Internship Prgm
1653 W Congress Pkwy
425 Triangle Office Bldg
Chicago, IL 60612-3864
www.rushu.rush.edu/nutrition
*Prgm Dir:* Linda J Lafferty, PhD RD FADA
*Tel:* 312 942-7845    *Fax:* 312 942-5203
*E-mail:* Linda_Lafferty@rush.edu

**University of Illinois at Chicago**
Dietetics-Coordinated Prgm
Human Nutrition and Dietetics
1919 W Taylor St, M/C 517
Chicago, IL 60612-7256
*Prgm Dir:* Vijay Ganji, PhD RD LDN
*Tel:* 312 996-1209    *Fax:* 312 413-0319
*E-mail:* vganji@uic.edu

**Northern Illinois University**
Dietetics-Didactic Prgm
Family Consumer and Nutri Sciences
DeKalb, IL 60115-2854
www.niu.edu
*Prgm Dir:* Judith M Lukaszuk, PhD RD LD
*Tel:* 815 753-6352    *Fax:* 815 753-1321
*E-mail:* jmlukaszuk@niu.edu

**Northern Illinois University**
Dietetic Internship Prgm
Family Consumer and Nutri Sciences
DeKalb, IL 60115-2854
www.fcns.niu.edu
*Prgm Dir:* Lucy Robinson, MS RD
*Tel:* 815 753-6386    *Fax:* 815 753-1321
*E-mail:* lrobins@niu.edu

**Ingalls Memorial Hospital**
Dietetic Internship Prgm
One Ingalls Dr
Harvey, IL 60426
www.ingalls.org
*Prgm Dir:* Mary Keith Vaughn, MA RD
*Tel:* 708 915-5723    *Fax:* 708 210-3110
*E-mail:* mvaughn@ingalls.org

**Edward Hines Jr VA Hospital**
Dietetic Internship Prgm
Nutrition and Food Service (120D)
PO Box 5000
Hines, IL 60141
www.hinesvainternship.com
*Prgm Dir:* Sharon Foley, MS RD
*Tel:* 708 216-2343    *Fax:* 708 202-2252
*E-mail:* sharon.foley@med.va.gov

**Benedictine University**
Dietetics-Didactic Prgm
Dept of Nutrition
5700 College Rd
Lisle, IL 60532-0900
www.ben.edu/nutrition
*Prgm Dir:* Catherine LS Arnold, MS EdD RD LDN
*Tel:* 630 829-6534    *Fax:* 630 829-6547
*E-mail:* carnold@ben.edu

**Benedictine University**
Dietetic Internship Prgm
Dept of Nutrition
5700 College Rd Birck-329
Lisle, IL 60532-0900
www.ben.edu
*Prgm Dir:* Julie Moreschi-Mason, MS RD LD
*Tel:* 630 829-6548    *Fax:* 630 829-6547
*E-mail:* jmoreschi@ben.edu

**Western Illinois University**
Dietetics-Didactic Prgm
Family and Consumer Sciences
Macomb, IL 61455
www.wiu.edu/users/mifcs
*Prgm Dir:* Karen R Greathouse, PhD RD
*Tel:* 309 298-1581    *Fax:* 309 298-2688
*E-mail:* KR-Greathouse@wiu.edu

**Illinois State University**
Dietetics-Didactic Prgm
Family and Consumer Sciences
Campus Box 5060
Normal, IL 61790-5060
www.cast.ilstu.edu/fcs
*Prgm Dir:* Robert W Cullen, PhD RD
*Tel:* 309 438-8850    *Fax:* 309 438-5659
*E-mail:* rcullen@ilstu.edu

**Illinois State University**
Dietetic Internship Prgm
Dept of Family & Consumer Sciences
Campus Box 5060
Normal, IL 61790-5060
www.cast.ilstu.edu/isudi
*Prgm Dir:* Julie M R Schumacher, MS RD
*Tel:* 309 438-7031    *Fax:* 309 438-5659
*E-mail:* jmraede@ilstu.edu

**Bradley University**
Dietetics-Didactic Prgm
Family and Consumer Sciences
1501 W Bradley Ave
Peoria, IL 61625-0015
www.bradley.edu
*Prgm Dir:* Jeannette Davidson, PhD RD LD CNS
*Tel:* 309 677-2436    *Fax:* 309 677-3813
*E-mail:* jad@bradley.edu

**OSF St Francis Medical Center**
Dietetic Internship Prgm
530 NE Glen Oak Ave
Peoria, IL 61637-0001
www.osfsaintfrancis.org/Schools/DieteticIntern
*Prgm Dir:* Golda E Ewalt, MS RD LD
*Tel:* 309 655-3707    *Fax:* 309 655-4022
*E-mail:* golda.e. ewalt@osfhealthcare.org

**Dominican University**
Dietetics-Didactic Prgm
Dept of Nutrition Sciences
7900 W Division St
River Forest, IL 60305-1066
www.dom.edu
*Prgm Dir:* Judith A Beto, PhD RD LD FADA
*Tel:* 708 524-6906    *Fax:* 708 366-5360
*E-mail:* judybeto@email.dom.edu

**St John's Hospital**
Dietetic Internship Prgm
800 E Carpenter St
Springfield, IL 62769
www.st-johns.org
*Prgm Dir:* Sara A Lopinski, MS RD LD
*Tel:* 217 544-6464, Ext 44818    *Fax:* 217 757-6871
*E-mail:* slopinsk@st-johns.org

**Univ of Illinois at Urbana - Champaign**
Dietetic Internship Prgm
Dept of Food Science & Human Nutrition
443 Bevier Hall, 905 S Goodwin Ave
Urbana, IL 61801
www.aces.uiuc.edu/~dietetic
*Prgm Dir:* Vijaya Jain, Ms RD CDN
*Tel:* 217 265-6865    *Fax:* 217 265-0925
*E-mail:* vjain@uiuc.edu

**Univ of Illinois at Urbana - Champaign**
Dietetics-Didactic Prgm
Food Sci & Human Nutri, 345 Bevier Hall
905 S Goodwin Ave
Urbana, IL 61801-3852
www.fshn.uiuc.edu/Dietetics
*Prgm Dir:* Karen L Plawecki, MS RD
*Tel:* 217 244-2884    *Fax:* 217 265-0925
*E-mail:* plawecki@uiuc.edu

# Indiana

**Indiana University - Bloomington**
Dietetics-Didactic Prgm
Applied Health Science
HPER 116
Bloomington, IN 47405-7109
*Prgm Dir:* Victoria M Getty, MEd RD
*Tel:* 812 855-1531    *Fax:* 812 855-3936
*E-mail:* vgetty@indiana.edu

**Purdue University - Calumet**
Dietetic Internship Prgm
Dept of Behavioral Sciences
2200 169th St
Hammond, IN 46323-2094
*Prgm Dir:* Rita A Fields, MS RD
*Tel:* 219 989-2940    *Fax:* 219 989-2008
*E-mail:* fields@calumet.purdue.edu

**Indiana University**
Dietetic Internship Prgm
Nutrition/Diet Prgm, Sch of Allied Hlth
Ball Residence, Rm 114
Indianapolis, IN 46202-5180
*Prgm Dir:* Jacquelynn O'Palka, PhD RD
*Tel:* 317 278-0934    *Fax:* 317 278-3940
*E-mail:* jopalka@iupui.edu

**Marian College**
Dietetics-Coordinated Prgm
Dietetics Nursing & Nutritional Sciences
3200 Cold Springs Rd
Indianapolis, IN 46222-1997
www.marian.edu
*Prgm Dir:* Catherine H Strain, MS RD
*Tel:* 317 955-6340    *Fax:* 317 955-6448
*E-mail:* cstrain@marian.edu

**Marian College**
Dietetics-Didactic Prgm
Nursing & Nutrition Sciences
3200 Cold Springs Rd
Indianapolis, IN 46222-1997
www.marian.edu
*Prgm Dir:* Catherine H Strain, MS RD
*Tel:* 317 955-6340    *Fax:* 317 955-6448
*E-mail:* cstrain@marian.edu

**Ball Memorial Hospital**
Dietetic Internship Prgm
Dept of Dietetics
2401 University Ave
Muncie, IN 47303-3499
www.cardinalhealthsystem.org/Diet_Intern_Program/di
eteticinterns.htm
*Prgm Dir:* Karen S Shields, MA RD
*Tel:* 765 747-3273   *Fax:* 765 741-2994
*E-mail:* kshields@chsmail.com

**Ball State University**
Dietetics-Didactic Prgm
Family and Consumer Sciences
Muncie, IN 47306-0250
www.bsu.edu/cast/fcs
*Prgm Dir:* Kimberli L Pik,e, MS RD
*Tel:* 765 285-5931   *Fax:* 765 285-2314
*E-mail:* klpike@bsu.edu

**Ball State University**
Dietetic Internship Prgm
Family and Consumer Sciences
150 Applied Technology Bldg
Muncie, IN 47306
www.bsu.edu/cast/fcs/dieteticinternship.html
*Prgm Dir:* Marla Kurtz, MA RD
*Tel:* 765 285-5940   *Fax:* 765 285-2314
*E-mail:* mskurtz@bsu.edu

**Indiana State University**
Dietetics-Coordinated Prgm
Family and Consumer Sciences
Terre Haute, IN 47809
*Prgm Dir:* Victoria M Getty, MEd RD
*Tel:* 812 855-1531   *Fax:* 812 237-3304
*E-mail:* vgetty@indiana.edu

**Purdue University**
Dietetics-Didactic Prgm
Dept of Foods and Nutrition
1264 Stone Hall
West Lafayette, IN 47907-1264
www.cfs.purdue.edu/fdsnnutr
*Prgm Dir:* Olivia B Wood, MPH RD
*Tel:* 765 494-8238   *Fax:* 765 494-0674
*E-mail:* woodo@purdue.edu

**Purdue University**
Dietetics-Coordinated Prgm
Dept of Foods and Nutrition
700 W State St
West Lafayette, IN 47907-2059
www.cfs.purdue.edu
*Prgm Dir:* Carol J Boushey, PhD MPH RD
*Tel:* 765 496-6569   *Fax:* 765 494-0674
*E-mail:* boushey@purdue.edu

# Iowa

**Iowa State University**
Dietetics-Didactic Prgm
Food Sci & Human Nutrition
220 MacKay Hall
Ames, IA 50011-1120
www.dietetics.iastate.edu
*Prgm Dir:* Anne M Oldham, MS RD LD
*Tel:* 515 294-6414   *Fax:* 515 294-5390
*E-mail:* aoldham@iastate.edu

**Iowa State University**
Dietetic Internship Prgm
Food Sci & Human Nutrition
220 MacKay Hall
Ames, IA 50011
www.dietetics.iastate.edu
*Prgm Dir:* Jean Anderson, MS RD LD
*Tel:* 515 294-7316   *Fax:* 515 294-6193
*E-mail:* janderso@iastate.edu

**University of Northern Iowa**
Dietetics-Didactic Prgm
Design Fam & Cons Sciences
219 Latham Hall, 1227 W 27th St
Cedar Falls, IA 50614-0332
http://csbnt.csbs.uni.edu/dept/dfcs
*Prgm Dir:* Tammie Collum, MS RD LD
*Tel:* 319 273-2418   *Fax:* 319 273-7096
*E-mail:* tammie.collum@uni.edu

**University of Iowa Hospitals & Clinics**
Dietetic Internship Prgm
200 Hawkins Dr, W146GH
Iowa City, IA 52242-1051
www.uihc.uiowa.edu/fns
*Prgm Dir:* Laurie L Parks, MBA RD
*Tel:* 319 356-1240   *Fax:* 319 356-8674
*E-mail:* laurie-kroymann@uiowa.edu

# Kansas

**University of Kansas Medical Center**
Dietetic Internship Prgm
Dept of Dietetics & Nutrition
3901 Rainbow Blvd
Kansas City, KS 66160-7250
www2.kumc.edu/sah/dietetics
*Prgm Dir:* Rachel Barkley, MS RD LD
*Tel:* 913 588-7683   *Fax:* 913 588-8946
*E-mail:* rbarkley@kumc.edu

**Kansas State University**
Dietetics-Didactic Prgm
Hotel Rest Inst Mgmt & Diet
Justin Hall 104
Manhattan, KS 66506-1404
www.ksu.edu/humec/hrimd
*Prgm Dir:* Deborah D Canter, PhD RD LD
*Tel:* 785 532-5507   *Fax:* 785 532-5522
*E-mail:* canter@humec.ksu.edu

**Kansas State University**
Dietetics-Coordinated Prgm
Hotel Rest Inst Mgmt & Diet
Justin Hall 104
Manhattan, KS 66506-1404
www.ksu.edu/humec/hrimd
*Prgm Dir:* Roni M Schwartz, MBA RD LD
*Tel:* 785 532-5576   *Fax:* 785 532-5522
*E-mail:* rmschwar@ksu.edu

# Kentucky

**Berea College**
Dietetics-Didactic Prgm
Child & Family Studies
CPO 2192
Berea, KY 40404
www.berea.edu/cfs/cfs.home.html#faculty
*Prgm Dir:* Janice B Blythe, PhD RD LD
*Tel:* 859 985-3743, Ext 3743   *Fax:* 859 985-3917
*E-mail:* janice_blythe@berea.edu

**Western Kentucky University**
Dietetics-Didactic Prgm
Consumer & Family Sci, Acad Complex 302F
One Big Red Way
Bowling Green, KY 42101-3576
www.wku.edu/dietetics
*Prgm Dir:* Danita M Saxon Kelley, PhD RD
*Tel:* 270 745-4352   *Fax:* 270 745-3999
*E-mail:* Danita.Kelley@wku.edu

**Univ of Kentucky Chandler Med Ctr**
Dietetic Internship Prgm
Nutrition Services
H507 UKCMC, 800 Rose St
Lexington, KY 40536-0293
www.hes.eku.edu
*Prgm Dir:* Beverly S Vivian, MS RD LD
*Tel:* 859 323-5154   *Fax:* 606 257-4588
*E-mail:* bsvivi0@uky.edu

**University of Kentucky**
Dietetics-Didactic Prgm
Nutrition & Food Science
204 Funkhouser Bldg
Lexington, KY 40506-0050
www.uky.edu
*Prgm Dir:* Margaret E Cook Newell, MS RD LD CN
*Tel:* 859 257-1661   *Fax:* 859 257-1275
*E-mail:* mecook0@uky.edu

**University of Kentucky**
Dietetics-Coordinated Prgm
Nutrition and Food Science
210C Erikson Hall
Lexington, KY 40506-0050
www.uky.edu
*Prgm Dir:* Margaret E Cook-Newell, MS RD CDE
*Tel:* 859 257-1661   *Fax:* 859 257-1275
*E-mail:* mecook0@uky.edu

**University of Kentucky**
Dietetic Internship Prgm
Human Environmental Sciences
204 Funkhouser Bldg
Lexington, KY 40506-0054
www.uky.edu
*Prgm Dir:* Hazel W Forsythe, PhD RD LD
*Tel:* 606 257-3800   *Fax:* 606 257-3707
*E-mail:* vbal@uky.edu

**Spalding University**
Dietetic Internship Prgm
851 S Fourth St
Louisville, KY 40203-2115
www.spalding.edu
*Prgm Dir:* Kathy Rapp, MBA RD CN
*Tel:* 502 585-9911, Ext 2357   *Fax:* 502 588-7175
*E-mail:* krapp@spalding.edu

**Morehead State University**
Dietetic Internship Prgm
College of Science and Technology
Reed Hall, UPO 721
Morehead, KY 40351
*Prgm Dir:* Marilyn T Sampley, PhD RD
*Tel:* 606 783-2023   *Fax:* 606 783-5007
*E-mail:* m.sampley@morehead-st.edu

**Morehead State University**
Dietetics-Didactic Prgm
Science & Technology
Reed Hall 246, UPO 721
Morehead, KY 40351-1689
*Prgm Dir:* Marilyn T Sampley, PhD RD LD CN
*Tel:* 606 783-2023   *Fax:* 606 783-5007
*E-mail:* m.sampley@morehead-st.edu

**Murray State University**
Dietetic Internship Prgm
Nutrition Dietetics & Food Management
200 N Oakley, Applied Sciences Bldg
Murray, KY 42071-3345
www.murraystate.edu/cit/fcs/intrn.htm
*Prgm Dir:* C Jeffrey Frame, PhD RD
*Tel:* 270 762-3387   *Fax:* 270 762-6950
*E-mail:* nutrition@murraystate.edu

**Murray State University**
Dietetics-Didactic Prgm
Food Serv Systems Management
200 N Oakley, Applied Sciences Bldg
Murray, KY 42071-3345
www.mursuky.edu/qacd/CIT/FCS
*Prgm Dir:* Kathryn L Timmons, MS RD
*Tel:* 270 762-3387   *Fax:* 270 762-6950
*E-mail:* kathy.timmons@murraystate.edu

**Eastern Kentucky University**
Dietetic Internship Prgm
Human Environmental Sciences
102 Burrier Bldg, 521 Lancaster Ave
Richmond, KY 40475-3107
www.fcs.eku.edu
*Prgm Dir:* Erin K Scarberry, MS RD
*Tel:* 859 622-3445   *Fax:* 859 622-1914
*E-mail:* erin.scarberry@eku.edu

**Eastern Kentucky University**
Dietetics-Didactic Prgm
Human Environmental Sciences
102 Burrier
Richmond, KY 40475
www.fcs.eku.edu
*Prgm Dir:* Sara W Sutton, MS RD
*Tel:* 859 622-3445   *Fax:* 859 622-1163
*E-mail:* sara.sutton@eku.edu

# Louisiana

**Louisiana State University**
Dietetics-Didactic Prgm
School of Human Ecology
Baton Rouge, LA 70803-4300
http://sun.huec.lsu.edu
*Prgm Dir:* Carol E O'Neill, MPH RD
*Tel:* 225 578-1631   *Fax:* 225 578-2697
*E-mail:* coneil1@lsu.edu

**Louisiana State University**
Dietetic Internship Prgm
Tower Dr
Baton Rouge, LA 70803-4301
www.sun.huec.isu.edu/hnf.html
*Prgm Dir:* Prithiva S Chanmugam, PhD RD
*Tel:* 225 578-1729   *Fax:* 225 578-2697
*E-mail:* chanmugam@aol.com

**Southern Univ and A&M College**
Dietetic Internship Prgm
PO Box 11342
Baton Rouge, LA 70813-1342
www.subr.edu
*Prgm Dir:* Bernestine B McGee, PhD RD LDN
*Tel:* 225 771-4660   *Fax:* 225 771-3107
*E-mail:* bernestine_mcgee@cxs.subr.edu

**Southern Univ and A&M College**
Dietetics-Didactic Prgm
Agriculture & Home Econ
PO Box 11342
Baton Rouge, LA 70813-1342
www.subr.edu
*Prgm Dir:* Bernestine B McGee, PhD RD LDN
*Tel:* 225 771-4660   *Fax:* 225 771-3107
*E-mail:* bernestine_mcgee@cxs.subr.edu

**North Oaks Medical Center**
Dietetic Internship Prgm
Nutritional Services
PO Box 2668
Hammond, LA 70404-2668
www.northoaks.org
*Prgm Dir:* Virginia M Pelegrin, MPH RD
*Tel:* 985 230-6548   *Fax:* 985 230-6563
*E-mail:* pelegrinv@northoaks.org

**University of Louisiana at Lafayette**
Dietetic Internship Prgm
School of Human Resources
PO Box 40399, McKinley St
Lafayette, LA 70504-0399
www.louisiana.edu/faculty/rfournet
*Prgm Dir:* Rachel M Fournet, PhD LDN RD
*Tel:* 337 482-5724   *Fax:* 337 482-5395
*E-mail:* rmf0931@louisiana.edu

**University of Louisiana at Lafayette**
Dietetics-Didactic Prgm
Coll of Applied Life Sciences
Sch of Human Res, Box 40399, McKinley St
Lafayette, LA 70504-0399
www.sustainablelouisiana/humr.org
*Prgm Dir:* Rachel M Fournet, PhD RD LDN
*Tel:* 337 482-5724   *Fax:* 337 482-5395
*E-mail:* rmf0931@louisiana.edu

**McNeese State University**
Dietetics-Didactic Prgm
PO Box 92820
Lake Charles, LA 70609
*Prgm Dir:* Beth Fontenot, MS RD LDN
*Tel:* 337 475-5970   *Fax:* 337 475-5681
*E-mail:* bfonteno@mail.mcneese.edu

**McNeese State University**
Dietetic Internship Prgm
PO Box 92820 MSU
Lake Charles, LA 70609-2820
*Prgm Dir:* Debra Hollingsworth, PhD LDN RD
*Tel:* 337 475-5683   *Fax:* 337 475-5681
*E-mail:* dholling@mcneese.com

**Touro Infirmary**
Dietetic Internship Prgm
1401 Foucher St
New Orleans, LA 70115-3515
*Prgm Dir:* Patricia Fitzpatrick, MA LDN RD
*Tel:* 504 897-8034   *Fax:* 504 897-8093
*E-mail:* fitzpatrickp@touro.com

**Tulane University**
Dietetic Internship Prgm
Pub Hlth and Comm Hlth Sciences
1440 Canal St, Ste 2317
New Orleans, LA 70112-2699
*Prgm Dir:* Jodi E Zighelboim, MPH LDN RD
*Tel:* 504 584-2672   *Fax:* 504 584-3540
*E-mail:* jzighel@mailhost.tcs.tulane.edu

**Louisiana Tech University**
Dietetic Internship Prgm
College of Human Ecology
PO Box 3167
Ruston, LA 71272
www.ans.latech.edu/humaneco-intern.html
*Prgm Dir:* Dawn Erickson, MPH LDN RD
*Tel:* 318 257-3043   *Fax:* 318 257-4014
*E-mail:* erickson@ans.latech.edu

**Louisiana Tech University**
Dietetics-Didactic Prgm
College of Human Ecology
PO Box 3167
Ruston, LA 71272
www.ans.latech.edu/humaneco-index.html
*Prgm Dir:* Martha L Dix, PhD RD LDN
*Tel:* 318 257-2713   *Fax:* 318 257-4014
*E-mail:* dix@.latech.edu

**Nicholls State University**
Dietetics-Didactic Prgm
Family and Consumer Sciences
Box 2014
Thibodaux, LA 70310
*Prgm Dir:* Colette G Leistner, PhD RD
*Tel:* 985 448-4732   *Fax:* 985 449-7073
*E-mail:* colette.leistner@nicholls.edu

# Maine

**University of Maine - Orono**
Dietetics-Didactic Prgm
Food Sci & Human Nutrition
5735 Hitchner Hall
Orono, ME 04469-5735
www.fsn.umaine.edu
*Prgm Dir:* Susan S Sullivan, DSc RD
*Tel:* 207 581-3130   *Fax:* 207 581-1636
*E-mail:* susan_sullivan@umenfa.maine.edu

**University of Maine - Orono**
Dietetic Internship Prgm
Food Sci & Human Nutrition
5735 Hitchner Hall, Rm 113
Orono, ME 04469-5749
www.ume.umaine.edu/~nfa/fsn
*Prgm Dir:* Adrienne A White, PhD RD
*Tel:* 207 581-3134   *Fax:* 207 581-1636
*E-mail:* aawhite@umenfa.maine.edu

# Maryland

**Johns Hopkins Bayview Medical Center**
Dietetic Internship Prgm
Clinical Nutrition Dept
4940 Eastern Ave
Baltimore, MD 21224-2735
www.jhbmc.jhu.edu/nutri
*Prgm Dir:* Cheryl R Koch, MS RD CNSD FADA
*Tel:* 410 550-1319   *Fax:* 410 550-0650
*E-mail:* ckoch@jhmi.edu

**Morgan State University**
Dietetics-Didactic Prgm
Dept of Family & Consumer Sci, Jenkins Bldg, Rm 403-A
1700 E Cold Spring Ln
Baltimore, MD 21251
www.morgan.edu
*Prgm Dir:* Ivis T Forrester-Anderson, PhD RD
*Tel:* 443 885-4043   *Fax:* 443 319-3787
*E-mail:* Dr_ivisforrester@hotmail.com

**University of Maryland Medical System**
Dietetic Internship Prgm
Food & Nutrition Services
22 S Greene St
Baltimore, MD 21201-1595
*Prgm Dir:* Laura K Wohlberg, MS RD
*Tel:* 410 328-2561   *Fax:* 410 328-1007
*E-mail:* lwohlberg@umm.edu

**National Institutes of Health**
Dietetic Internship Prgm
Nutrition Dept, Clinical Ctr, Bldg 10 Rm B1S-234
10 Center Dr MSC 1078
Bethesda, MD 20892-1078
www.cc.nih.gov/nutr
*Prgm Dir:* Maureen G Leser, MS RD
*Tel:* 301 496-6349   *Fax:* 301 496-0622
*E-mail:* mleser@nih.gov

**Univ of Maryland at College Park**
Dietetics-Didactic Prgm
Dept of Nutrition and Food Science
College Park, MD 20742-7521
www.agnr.umd.edu/users/nfsc
*Prgm Dir:* Nancy J Brenowitz, MS RD
*Tel:* 301 405-4532   *Fax:* 301 314-9327
*E-mail:* nb81@umail.umd.edu

**Univ of Maryland at College Park**
Dietetic Internship Prgm
Dept of Nutrition and Food Science
0112 Skinner Bldg
College Park, MD 20742
www.agnr.umd.edu/users/nfsc/internship
*Prgm Dir:* Phyllis E Fatzinger-McShane, MS RD LD
*Tel:* 301 405-6201   *Fax:* 301 574-3842
*E-mail:* pmcshane@umail.umd.edu

**Sodexho Health Care Services, Mid Atlantic**
Dietetic Internship Prgm
10500 Little Patuxent Parkway, Ste 620
Columbia, MD 21044
www.sodexhousa.com/car_dietetic.html
*Prgm Dir:* Janet Debelius, MA RD LD
*Tel:* 410 744-2798, Ext 30   *Fax:* 410 715-1694
*E-mail:* janet.debelius@sodexhousa.com

**University of Maryland Eastern Shore**
Dietetic Internship Prgm
Dept of Human Ecology
Princess Anne, MD 21853-1299
www.umes.edu/ecology/ap4.htm
*Prgm Dir:* Malinda D Cecil, MS RD LD
*Tel:* 410 651-7578   *Fax:* 410 651-6207
*E-mail:* mdcecil@umes.edu

**University of Maryland Eastern Shore**
Dietetics-Didactic Prgm
Dept of Human Ecology
Princess Anne, MD 21853-1299
www.umes.edu/ecology/undergraduate.htm
*Prgm Dir:* Malinda D Cecil, MS RD LD
*Tel:* 410 651-7578   *Fax:* 410 651-6207
*E-mail:* mdcecil@umes.edu

# Massachusetts

**University of Massachusetts - Amherst**
Dietetic Internship Prgm
Div of Continuing Education
358 N Pleasant St
Amherst, MA 01003-9296
www.umass.edu/contined
*Prgm Dir:* Nadine S Braunstein, MS RD LDN CDE
*Tel:* 413 545-2484   *Fax:* 413 577-3838
*E-mail:* interndirector@nutrition.umass.edu

**University of Massachusetts - Amherst**
Dietetics-Didactic Prgm
Dept of Nutrition Box 31420
213 Chenoweth Laboratory, 100 Holdsworth Way
Amherst, MA 01003-9282
www.umass.edu
*Prgm Dir:* Sara J Sabelawski, Med RD LDN
*Tel:* 413 577-1077   *Fax:* 413 545-1074
*E-mail:* sara@nutrition.umass.edu

**Beth Israel Deaconess Medical Center**
Dietetic Internship Prgm
330 Brookline Ave
Boston, MA 02215-5491
www.bidmc.harvard.edu/dietetic
*Prgm Dir:* Patricia Queen Samour, M MSC RD
*Tel:* 617 667-2580   *Fax:* 617 667-7180
*E-mail:* psamour@bidmc.harvard.edu

**Boston University/Sargent College**
Dietetic Internship Prgm
Graduate Nutrition Div
635 Commonwealth Ave
Boston, MA 02215-1605
www.bu.edu/sargent
*Prgm Dir:* Joan Salge Blake, MS RD
*Tel:* 617 353-7470   *Fax:* 617 353-7567
*E-mail:* salge@bu.com

**Boston University/Sargent College**
Dietetics-Didactic Prgm
635 Commonwealth Ave
Boston, MA 02215-1605
www.bu.edu/sargent/HS
*Prgm Dir:* Roberta P Durschlag, PhD RD
*Tel:* 617 353-7488   *Fax:* 617 353-7567
*E-mail:* rdurschl@bu.edu

**Brigham & Women's Hospital**
Dietetic Internship Prgm
75 Francis St
Boston, MA 02115-6195
www.brighamandwomens.org/nutrition/
   dieteticinternship.htm
*Prgm Dir:* Alice McCarley, MS RD
*Tel:* 617 732-5577   *Fax:* 617 278-6929
*E-mail:* amccarley@partners.org

**Frances Stern Nutrition Center**
Dietetic Internship Prgm
New England Medical Center
Tufts University
Boston, MA 02111-1533
www.tufts.edu/nutrition/program/dietetic_internship
*Prgm Dir:* Kelly A Kane, MS RD LDN CNSD
*Tel:* 617 636-8309   *Fax:* 617 636-8325
*E-mail:* KKane@tufts-nemc.org

**Massachusetts General Hospital**
Dietetic Internship Prgm
Dept of Dietetics
Boston, MA 02114
www.massgeneral.org/visitor/nutrition_dietetics.htm
*Prgm Dir:* Kathleen E Creedon, MHPE RD
*Tel:* 617 726-2589   *Fax:* 617 726-7649
*E-mail:* kcreedon@partners.org

**Simmons College**
Dietetic Internship Prgm
Dept of Nutrition
300 The Fenway
Boston, MA 02115
www.simmons.edu
*Prgm Dir:* Nancie Harvey Herbold, EdD RD
*Tel:* 617 521-2709   *Fax:* 617 521-3137
*E-mail:* herbold@simmons.edu

**Simmons College**
Dietetics-Didactic Prgm
Dept of Nutrition
300 The Fenway
Boston, MA 02115-5898
www.simmons.edu
*Prgm Dir:* Elizabeth S Metallinos-Katsaras, PhD RD
*Tel:* 617 521-2708   *Fax:* 617 521-3137
*E-mail:* metallin@simmons.edu

**Mount Auburn Hospital**
Dietetic Internship Prgm
330 Mount Auburn St
Cambridge, MA 02138
www.mtauburn.caregroup.org/mededu/mededu.asp
*Prgm Dir:* Rena M Prendergast, MTS LDN RD
*Tel:* 617 492-3500, Ext 3059   *Fax:* 617 499-5673
*E-mail:* rprendergast@post.harvard.edu

**Framingham State College**
Dietetics-Didactic Prgm
Family and Consumer Sciences
100 State St
Framingham, MA 01701-9101
www.framingham.edu
*Prgm Dir:* Marilyn M Abernethy, DrPH RD
*Tel:* 508 626-4757   *Fax:* 508 626-4003
*E-mail:* maberne@frc.mass.edu

**Framingham State College**
Dietetics-Coordinated Prgm
Family and Consumer Sciences
100 State St
Framingham, MA 01701-9101
www.framingham.edu
*Prgm Dir:* Suzanne H Neubauer, PhD RD
*Tel:* 508 626-4754   *Fax:* 508 626-4003
*E-mail:* sneubau@frc.mass.edu

**Sodexho Marriott Services**
Dietetic Internship Prgm
Distance Education Dietetic Internship
45 Hayden Ave
Lexington, MA 02420
www.dieteticintern.com
*Prgm Dir:* Barbara J Woodland, EdD RD
*Tel:* 603 487-5293   *Fax:* 603 487-5293
*E-mail:* dieteticinternship@sodexhousa.com

**Sodexho Marriott Services**
Dietetic Internship Prgm
Southcoast Hospitals Group
101 Page St
New Bedford, MA 02740-3464
www.southcoast.org/jobs/interns-dietary.html
*Prgm Dir:* Elizabeth F Winthrop, MS RD
*Tel:* 508 961-5161   *Fax:* 508 961-5166
*E-mail:* winthrope@southcoast.org

# Michigan

**University of Michigan**
Dietetics-Didactic Prgm
School of Public Health
Human Nutrition Prgm
Ann Arbor, MI 48109-2029
*Prgm Dir:* Theresa L Han-Markey, MS RD
*Tel:* 734 647-0214   *Fax:* 734 764-5233
*E-mail:* hanmark@umich.edu

**University of Michigan**
Dietetic Internship Prgm
School of Public Health
1420 Washington Heights M6150
Ann Arbor, MI 48109-2029
*Prgm Dir:* Andrea J Lasichak, MS RD CDE
*Tel:* 734 764-3277   *Fax:* 734 764-5865
*E-mail:* apfdir@med.umich.edu

**University of Michigan Hospitals & Health Ctr**
Dietetic Internship Prgm
UH2C227/0056
1500 E Medical Ctr Dr
Ann Arbor, MI 48109-0056
www.med.umich.edu/pfans/internship
*Prgm Dir:* Joyce Kerestes-Smith, MS RD
*Tel:* 734 936-5197   *Fax:* 734 936-5195
*E-mail:* joyceks@umich.edu

**Andrews University**
Dietetics-Didactic Prgm
Dept of Nutrition
Berrien Springs, MI 49104-0210
www.andrews.edu.NUFS
*Prgm Dir:* Winston J Craig, PhD RD
*Tel:* 269 471-3370   *Fax:* 269 471-3485
*E-mail:* wcraig@andrews.edu

**Andrews University**
Dietetic Internship Prgm
Dept of Nutrition
Berrien Springs, MI 49104-0210
www.andrews.edu
*Prgm Dir:* Winston J Craig, PhD RD
*Tel:* 269 471-3370   *Fax:* 269 471-3485
*E-mail:* wcraig@andrews.edu

**Detroit Health Department**
Dietetic Internship Prgm
Nutrition Div
1151 Taylor St
Detroit, MI 48202-1732
*Prgm Dir:* Indira D Arya, MS RD
*Tel:* 313 876-4090   *Fax:* 313 876-4645
*E-mail:* aryai@health.ci.detroit.mi.us

**Harper University Hospital**
Dietetic Internship Prgm
Dept of Nutrition & Food Svcs
3901 Beaubien Ave
Detroit, MI 48201-2018
www.harperhospital.org/harper/diet/intro
*Prgm Dir:* Dianne L Trippett, MS RD
*Tel:* 313 745-2037   *Fax:* 313 993-8509
*E-mail:* dtrippet@dmc.org

**Henry Ford Hospital**
Dietetic Internship Prgm
Dept of Food & Nutrition Services
2799 W Grand Blvd
Detroit, MI 48202-2689
*Prgm Dir:* Lynne A Hufnagle, MA RD
*Tel:* 313 916-3473   *Fax:* 313 916-0761
*E-mail:* lhufnag1@hfhs.org

**Marygrove College**
Dietetics-Didactic Prgm
Dept of Human Nutrition & Foods
8425 W McNichols Rd
Detroit, MI 48221
www.marygrove.edu
*Prgm Dir:* Ethel M Nettles, PhD RD
*Tel:* 313 927-1322   *Fax:* 313 927-1345
*E-mail:* mnettles@marygrove.edu

**Wayne State University**
Dietetics-Coordinated Prgm
Nutrition & Food Science
3009 Science Hall
Detroit, MI 48202
www.science.wayne.edu/~nfs/dietetics.htm
*Prgm Dir:* Tonia G Reinhard, MS RD
*Tel:* 313 577-2500   *Fax:* 313 577-8616
*E-mail:* tonia.reinhard@wayne.edu

**Michigan State University**
Dietetics-Didactic Prgm
Food Science & Human Nutrition
210 Trout FSHN Bldg
East Lansing, MI 48824-1224
www.msu.edu/unit/fshn
*Prgm Dir:* Lorraine J Weatherspoon, PhD RD
*Tel:* 517 355-8474, Ext 136   *Fax:* 517 353-8963
*E-mail:* weathe43@msu.edu

**Michigan State University**
Dietetic Internship Prgm
Dept of Food Science and Human Nutrition
2100 Anthony Hall
East Lansing, MI 48824-1225
www.msu.edu/unit/fshn/grad/intern.html
*Prgm Dir:* Gail J Rogers, MS RD CD CLE
*Tel:* 517 355-7713, Ext 183   *Fax:* 517 353-1676
*E-mail:* rogersg3@msu.edu

**Hurley Medical Center**
Dietetic Internship Prgm
Nutrition Service Dept
One Hurley Plaza
Flint, MI 48503-5993
*Prgm Dir:* Dawn R Moszyk, MA RD
*Tel:* 810 257-9772   *Fax:* 810 424-4780
*E-mail:* dmoszyk1@hurleymc.com

**Western Michigan University**
Dietetic Internship Prgm
Family and Consumer Sciences
3025 Kohrman Hall
Kalamazoo, MI 49008-5067
www.wmich.edu
*Prgm Dir:* Caroline B Webber, PhD RD
*Tel:* 269 387-3710   *Fax:* 616 387-3353
*E-mail:* caroline.webber@wmich.edu

**Western Michigan University**
Dietetics-Didactic Prgm
Family and Consumer Sciences
3024 Kohrman Hall
Kalamazoo, MI 49008
www.wmich.edu
*Prgm Dir:* Arezoo Rojhani, PhD RD
*Tel:* 616 387-3729   *Fax:* 616 387-3353
*E-mail:* rojhani@wmich.edu

**Madonna University**
Dietetics-Didactic Prgm
Dept of Biological and Health Sciences
36600 Schoolcraft Rd
Livonia, MI 48150-1173
www.madonna.edu
*Prgm Dir:* Karen J Schmitz, PhD RD
*Tel:* 734 432-5534   *Fax:* 734 432-5393
*E-mail:* kschmitz@madonna.edu

**Northern Michigan University**
Dietetics-Didactic Prgm
HPER
Marquette, MI 49855
*Prgm Dir:* Mohey A Mowafy, PhD RD
*Tel:* 906 227-2366   *Fax:* 906 227-2181
*E-mail:* mmowafy@nmu.edu

**Central Michigan University**
Dietetics-Didactic Prgm
Human Environmental Studies
205 Wightman Hall
Mt Pleasant, MI 48859
http://nutrition.ehs.cmich.edu
*Prgm Dir:* John V Logomarsino, PhD RD
*Tel:* 989 774-2004   *Fax:* 989 774-2435
*E-mail:* henri1cr@cmich.edu

**Central Michigan University**
Dietetic Internship Prgm
Human Environmental Studies
205 Wightman Hall
Mt Pleasant, MI 48859-3652
http://nutrition.cmich.edu
*Prgm Dir:* Ann F Jay, MS RD
*Tel:* 517 879-4357   *Fax:* 517 879-4378
*E-mail:* jay1af@cmich.edu

**Oakland County Health Division**
Dietetic Internship Prgm
Health Education, Nutrition, and Substance Abuse Svcs
27725 Greenfield Rd
Southfield, MI 48076-3625
*Prgm Dir:* Linda H Eaton, MS RD
*Tel:* 248 424-7133   *Fax:* 248 424-7144
*E-mail:* jrmchale@hotmail.com

**Eastern Michigan University**
Dietetics-Coordinated Prgm
206 Roosevelt Hall
Ypsilanti, MI 48197
www.emich.edu
*Prgm Dir:* Lydia A D Kret, MS RD
*Tel:* 734 487-3220   *Fax:* 734 487-7087
*E-mail:* lydia.kret@emich.edu

# Minnesota

**Minnesota State University - Mankato**
Dietetics-Didactic Prgm
Family Consumer Sciences Department
102 Wiecking Center
Mankato, MN 56001
www.mnsu.edu
*Prgm Dir:* Susan B Fredstrom, PhD RD CNSD
*Tel:* 507 389-6016   *Fax:* 507 389-2411
*E-mail:* susan.fredstrom@mnsu.edu

**Fairview University**
Dietetic Internship Prgm
Nutrition Services MMC 84, Univ Campus
420 Delaware St SE
Minneapolis, MN 55455
www.fairview.org/diet
*Prgm Dir:* Rose M Boughton, MA RD
*Tel:* 612 273-7499   *Fax:* 612 273-0066
*E-mail:* rbought1@fairview.org

**Veterans Affairs Medical Center**
Dietetic Internship Prgm
Nutrition & Food Service
One Veterans Dr
Minneapolis, MN 55417
www.va.gov/nfs
*Prgm Dir:* Heidi H Hoover, MS RD
*Tel:* 612 725-2004   *Fax:* 612 727-5997
*E-mail:* heidi.hoover@va.gov

**Concordia College - Moorhead**
Dietetic Internship Prgm
Dept of Family & Nutrition Sciences
901 S 8th St
Moorhead, MN 56562
www.cord.edu/dept/auxservices/ccdiweb.htm
*Prgm Dir:* Debra Lee-Ross, MS RD
*Tel:* 219 299-3706   *Fax:* 219 299-4409
*E-mail:* dlee@cord.edu

**Concordia College - Moorhead**
Dietetics-Didactic Prgm
Family & Nutrition Sciences
Moorhead, MN 56562
www.cord.edu
*Prgm Dir:* Betty J Larson, EdD LRD
*Tel:* 218 299-3748   *Fax:* 218 299-4308
*E-mail:* blarson@gloria.cord.edu

**St Mary's Hospital**
*Cosponsor:* Mayo Medical Center
Dietetic Internship Prgm
1216 2nd St SW
Rochester, MN 55902-1906
www.mayo.edu
*Prgm Dir:* Jeanne E Grant, MS RD LD
*Tel:* 507 255-5617   *Fax:* 507 255-7379
*E-mail:* grant.jeanne@mayo.edu

**College of St Benedict/St John's University**
Dietetics-Didactic Prgm
Nutrition Dept
37 S College Ave
St Joseph, MN 56374-2099
www.csbsju.edu/nutrition
*Prgm Dir:* Diane Veale Jones, MS RD
*Tel:* 320 363-5976   *Fax:* 320 363-5582
*E-mail:* lshepherd@csbsju.edu

**College of St Benedict/St John's University**
Dietetics-Coordinated Prgm
Nutrition Dept ASC 151
37 S College Ave
St Joseph, MN 56374-2099
www.csbsju.edu/nutrition
*Prgm Dir:* Diane Veale Jones, MS RD
*Tel:* 320 363-5976   *Fax:* 320 363-5582
*E-mail:* djones@csbsju.edu

**College of St Catherine - St Paul**
Dietetics-Didactic Prgm
Family Consumer & Nutri Sci, MC 4182
2004 Randolph Ave
St Paul, MN 55105-1750
www.stkate.edu
*Prgm Dir:* Patricia J Ode, MA RD LD
*Tel:* 651 690-6204   *Fax:* 651 690-6958
*E-mail:* peode@stkate.edu

**Regions Hospital**
Dietetic Internship Prgm
640 Jackson St
St Paul, MN 55101-2502
*Prgm Dir:* Elizabeth L Orchard, MA RD
*Tel:* 651 254-2712   *Fax:* 651 254-9927
*E-mail:* betty.l.orchard@healthpartners.com

**University of Minnesota - St Paul**
Dietetics-Coordinated Prgm
268 Food Science & Nutrition
1334 Eckles Ave
St Paul, MN 55108-6099
www.fsci.umn.edu
*Prgm Dir:* Gail J Schlegel, MPH RD LN
*Tel:* 612 624-5362  *Fax:* 612 625-5272
*E-mail:* gschlege@umn.edu

**University of Minnesota - St Paul**
Dietetic Internship Prgm
225 Food Science & Nutrition
1334 Eckles Ave
St Paul, MN 55108-6099
http://fscn.che.umn.edu/digs
*Prgm Dir:* Carolyn M Peterson, MS RD LD
*Tel:* 612 625-5285  *Fax:* 612 624-5272
*E-mail:* cmpeters@umn.edu

**University of Minnesota - St Paul**
Dietetics-Didactic Prgm
225 Food Science & Nutrition
1334 Eckles Ave
St Paul, MN 55108-6099
www.fscn.che.umn.edu
*Prgm Dir:* Teri L Burgess-Champoux, MS RD LD
*Tel:* 612 624-2787  *Fax:* 612 625-5272
*E-mail:* dpd@umn.edu

# Mississippi

**Alcorn State University**
Dietetics-Didactic Prgm
Family and Consumer Sciences
1000 ASU Dr #839
Alcorn State, MS 39096-7500
www.alcorn.edu/HumanSciences
*Prgm Dir:* Ross C Santell, PhD RD
*Tel:* 601 877-6258  *Fax:* 601 877-3960
*E-mail:* rsantell@lorman.alcorn.edu

**Alcorn State University**
Dietetic Internship Prgm
1000 ASU Dr #839
Alcorn State, MS 39096-7500
*Prgm Dir:* Mattie R Rasco, MS RD LD
*Tel:* 601 877-6258  *Fax:* 601 877-3960
*E-mail:* mrasco@lorman.alcorn.edu

**Delta State University**
Dietetics-Coordinated Prgm
Div of Family & Consumer Sciences
PO Box 3273
Cleveland, MS 38733
www.deltastate.edu/academics/educ/fcs
*Prgm Dir:* Katherine A Davis, PhD RD
*Tel:* 662 846-4316  *Fax:* 662 846-4314
*E-mail:* kdavis@deltastate.edu

**University of Southern Mississippi**
Dietetics-Didactic Prgm
Family & Consumer Science
118 College Dr #5172
Hattiesburg, MS 39406
www.usm.edu/nfs
*Prgm Dir:* Elaine F Molaison, PhD RD
*Tel:* 601 266-5377  *Fax:* 601 266-4680
*E-mail:* elaine.molaison@usm.edu

**University of Southern Mississippi**
Dietetic Internship Prgm
Family and Consumer Sciences
PO Box 5172
Hattiesburg, MS 39406-5035
www.usm.edu/nfs/Dintern/Dintern.htm
*Prgm Dir:* Ruth Ann M Broome, MS RD
*Tel:* 601 924-9769  *Fax:* 601 924-1119
*E-mail:* rabroome@netdoor.com

**St Dominic-Jackson Memorial Hospital**
Dietetic Internship Prgm
969 Lakeland Dr
Jackson, MS 39216-4699
www.stdom.com
*Prgm Dir:* Esther A Coats, MS RD LD
*Tel:* 601 200-6935  *Fax:* 601 200-0799
*E-mail:* dietitians@stdom.com

**Mississippi State University**
Dietetic Internship Prgm
School of Human Sciences Box 9745
128 Lloyd Ricks
Mississippi State, MS 39762-9745
www.msstate.edu/dept/fsnhp/index.html
*Prgm Dir:* Michelle D Lee, MS RD
*Tel:* 662 325-7232  *Fax:* 662 325-8188
*E-mail:* mle@fsnhp.msstate.edu

**Mississippi State University**
Dietetics-Didactic Prgm
School of Human Sciences
PO Box 9805, 107 Herzer, Stone Blvd
Mississippi State, MS 39762-9745
www.msstate.edu/dept/fsnhp/index.html
*Prgm Dir:* Michelle D Lee, MS RD
*Tel:* 662 325-7232  *Fax:* 662 325-8188
*E-mail:* mlee@fsnhp.msstate.edu

**University of Mississippi**
Dietetics-Didactic Prgm
Family and Consumer Sciences
110 Meek Hall, PO Box 1848
University, MS 38677-1848
*Prgm Dir:* Laurel G Lambert, PhD RD LD
*Tel:* 662 915-7807  *Fax:* 662 915-7039
*E-mail:* LambertL@olemiss.edu

# Missouri

**Southeast Missouri State University**
Dietetic Internship Prgm
Dept of Human Enviromental Studies
One University Plaza
Cape Girardeau, MO 63701-4799
www5.semo.edu/dietetic_internship
*Prgm Dir:* Anne B Marietta, PhD RD
*Tel:* 573 651-2733  *Fax:* 573 651-2949
*E-mail:* abmarietta@semo.edu

**Southeast Missouri State University**
Dietetics-Didactic Prgm
Dept of Human Environmental Studies
One University Plaza MS 5750
Cape Girardeau, MO 63701-4799
www.semo.edu/study/dietetics/index.htm
*Prgm Dir:* Marcia L Nahikian-Nelms, PhD RD
*Tel:* 573 651-2994  *Fax:* 573 651-2949
*E-mail:* mnelms@semovm.semo.edu

**University of Missouri - Columbia**
Dietetics-Coordinated Prgm
106 McKee
Columbia, MO 65211-0001
www.missouri.edu/~nutsci
*Prgm Dir:* Catherine A Peterson, PhD RD
*Tel:* 573 882-4136  *Fax:* 573 884-4885
*E-mail:* raedekem@missouri.edu

**Missouri Dept of Health & Senior Service**
Dietetic Internship Prgm
PO Box 570
Jefferson City, MO 65102-0570
*Prgm Dir:* Vicki A Strickland, MA RD LD
*Tel:* 573 526-7961  *Fax:* 573 522-3244
*E-mail:* vicki.strickland@dhss.mo.us

**ARAMARK Healthcare Support Services - KC**
Dietetic Internship Prgm
St Joseph Health System
1000 Carondelet Dr
Kansas City, MO 64114-4802
www.aramark.com
*Prgm Dir:* Kathy Linhart, MS RD
*Tel:* 816 943-2146  *Fax:* 816 943-4719
*E-mail:* linhart-kathy@aramark.com

**Northwest Missouri State University**
Dietetics-Didactic Prgm
Coll of Educ & Human Services
Family and Consumer Sciences, Rm 309 Admin Bldg
Maryville, MO 64468-6001
www.nwmissouri.edu
*Prgm Dir:* Jenell D Ciak, PhD RD
*Tel:* 660 562-1167  *Fax:* 660 562-1900
*E-mail:* jciak@mail.nwmissouri.edu

**College of the Ozarks**
Dietetics-Didactic Prgm
Dietetics & Nutrition Educ
PO Box 17
Point Lookout, MO 65726-0017
ww.cofo.edu/dietetics/default.asp
*Prgm Dir:* Elizabeth A Huddleston, MS RD
*Tel:* 417 334-6411, Ext 4416  *Fax:* 417 335-2618
*E-mail:* bhuddleston@cofo.edu

**Missouri State University**
Dietetics-Didactic Prgm
Dept of Biomedical Sciences
901 S National Ave
Springfield, MO 65804
www.missouristate.edu
*Prgm Dir:* Cynthia J Reid, PhD RD
*Tel:* 417 836-5603  *Fax:* 417 836-5588
*E-mail:* cindyheiss@missouristate.edu

**Barnes-Jewish Coll of Nursing/Allied Health**
Dietetic Internship Prgm
306 S Kingshighway Blvd
Mailstop 90-30-625
St Louis, MO 63110
www.barnesjewishcollege.edu
*Prgm Dir:* Kelly Eiden, MS RD LD CNSD
*Tel:* 314 454-5307  *Fax:* 314 454-5239
*E-mail:* kae9581@bjc.org

**DVA Medical Center**
Dietetic Internship Prgm
Jefferson Barracks Div-Nutri & Food Serv 120/JB
One Jefferson Barracks Dr
St Louis, MO 63125
www.va.gov/nfs/StLouisVA
*Prgm Dir:* Amy L Knoblock-Hahn, MS RD LD
*Tel:* 314 894-6631  *Fax:* 314 845-5023
*E-mail:* amy.knoblock-hahn@med.va.gov

**Fontbonne University**
Dietetics-Didactic Prgm
Human Environmental Sciences
6800 Wydown Blvd
St Louis, MO 63105-3098
*Prgm Dir:* Cheryl A Houston, PhD RD
*Tel:* 314 889-1415  *Fax:* 314 719-8015
*E-mail:* chouston@fontbonne.edu

**Saint Louis University**
*Cosponsor:* Doisy College of Health Sciences
Dietetics-Didactic Prgm
Dept of Nutri & Diet, Rm 3076
3437 Caroline St
St Louis, MO 63104-1111
www.slu.edu/doisycollege/ND
*Prgm Dir:* Joy Short, MS RD
*Tel:* 314 577-8523  *Fax:* 314 577-8520
*E-mail:* ebelharj@slu.edu

**Saint Louis University**
*Cosponsor:* Doisy College of Health Sciences
Dietetic Internship Prgm
Dept of Nutri & Diet Rm 3076
3437 Caroline St
St Louis, MO 63104-1111
www.slu.edu/colleges/AH
*Prgm Dir:* Mildred K Mattfeldt-Beman, PhD RD
*Tel:* 314 577-8523  *Fax:* 314 577-8520
*E-mail:* steitzka@slu.edu

**Central Missouri State University**
Dietetics-Didactic Prgm
Dept of Health and Human Performance
Morrow 100
Warrensburg, MO 64093
www.cmsu.edu/dietetics
*Prgm Dir:* Swarna L Mandali, PhD RD LD
*Tel:* 660 543-4361  *Fax:* 660 543-8295
*E-mail:* mandali@cmsu1.cmsu.edu

# Montana

**Montana State University**
Dietetics-Didactic Prgm
Health & Human Development
101 MH H&PE Complex
Bozeman, MT 59717
www.montana.edu/wwwhhd
*Prgm Dir:* Alison H Harmon, PhD RD LN
*Tel:* 406 994-6338  *Fax:* 406 994-6314
*E-mail:* harmon@montana.edu

# Nebraska

**University of Nebraska - Kearney**
Dietetics-Didactic Prgm
Family & Consumer Science
Otto Olsen Bldg Rm 205C 00, 905 W 25th St
Kearney, NE 68849-2130
www.unk.edu
*Prgm Dir:* Sharon L Davis, MS RD
*Tel:* 308 865-8229  *Fax:* 308 865-8040

**University of Nebraska - Lincoln**
Dietetic Internship Prgm
Nutrition Science & Dietetics
120 Ruth Leverton Hall
Lincoln, NE 68583-0806
cehs.unl.edu/nhs/internships/dieteticintern.shtml
*Prgm Dir:* Wanda M Koszewski, PhD RD LMNT
*Tel:* 402 472-7966  *Fax:* 402 472-1587
*E-mail:* wkoszewski1@unl.edu

**University of Nebraska - Lincoln**
Dietetics-Didactic Prgm
Nutrition Science & Dietetics
104H Ruth Leverton Hall
Lincoln, NE 68583-0806
www.unl.edu
*Prgm Dir:* Linda O Young, MS RD
*Tel:* 402 472-2925  *Fax:* 402 472-1587
*E-mail:* lyoung3@unl.edu

**University of Nebraska Medical Center**
Dietetic Internship Prgm
981200 Nebraska Medical Center
Omaha, NE 68198-1200
www.unmc.edu/alliedhealth/mne
*Prgm Dir:* Glenda R Woscyna, MS RD LMNT
*Tel:* 402 559-7365  *Fax:* 402 559-6010
*E-mail:* gwoscyna@nebraskamed.org

# Nevada

**University of Nevada - Las Vegas**
Dietetics-Didactic Prgm
Box 453026
4505 S Maryland Pkwy
Las Vegas, NV 89154-3026
*Prgm Dir:* Laura J Kruskall, PhD RD
*Tel:* 702 895-1169  *Fax:* 702 895-2616
*E-mail:* laura.kruskall@unlv.edu

**University of Nevada - Reno**
Dietetic Internship Prgm
Dept of Nutrition
Mail Stop 142
Reno, NV 89557-0132
www.cabnr.unr.edu/nutrition
*Prgm Dir:* Karon S Felten, MS RD
*Tel:* 775 784-6446  *Fax:* 775 784-6449
*E-mail:* kfelten@cabnr.unr.edu

**University of Nevada - Reno**
Dietetics-Didactic Prgm
Dept of Nutrition
SFB Mail Stop 142
Reno, NV 89557
www.unr.edu/hcs/nutrition
*Prgm Dir:* Judith M Ashley, PhD MSPH RD
*Tel:* 775 784-6442  *Fax:* 775 784-6449
*E-mail:* jashley@unr.edu

# New Hampshire

**University of New Hampshire**
Dietetics-Didactic Prgm
Animal & Nutri Sci Human Nutrition Ctr
Kendall Hall
Durham, NH 03824
www.anscandnutr.unh.edu
*Prgm Dir:* Colette Janson-Sand, PhD RD
*Tel:* 603 862-1723  *Fax:* 603 862-0308
*E-mail:* chjs@cisunix.unh.edu

**University of New Hampshire**
Dietetic Internship Prgm
Nutrition Assessment & Counseling Services
Kendall Hall
Durham, NH 03824
www.dieteticinternship.unh.edu
*Prgm Dir:* Joanne D Burke, PhD RD LD
*Tel:* 603 862-1456  *Fax:* 603 862-0308
*E-mail:* jburke@cisunix.unh.edu

**Keene State College**
Dietetics-Didactic Prgm
Health Sciences/Nutrition
Joslin House MS 2903
Keene, NH 03435-2903
www.keene.edu/programs/hlsc
*Prgm Dir:* Karrie A Kalich, MS RD LD
*Tel:* 603 358-2423  *Fax:* 603 358-2892
*E-mail:* kkalich@keene.edu

**Keene State College**
Dietetic Internship Prgm
229 Main St M2903
Keene, NH 03431
www.keene.edu/academics/dietetics
*Prgm Dir:* Lisa Prospert, MS RD
*Tel:* 603 358-2915  *Fax:* 603 358-2892
*E-mail:* lprosper@keene.edu

# New Jersey

**College of St Elizabeth**
Dietetics-Didactic Prgm
Dept of Foods and Nutrition
2 Convent Rd
Morristown, NJ 07960-6989
www.cse.edu
*Prgm Dir:* Monica W Luby, MS RD
*Tel:* 973 290-4127  *Fax:* 973 290-4138
*E-mail:* mluby@cse.edu

**College of St Elizabeth**
Dietetic Internship Prgm
Dept of Foods and Nutrition
Henderson Hall, 2 Convent Rd
Morristown, NJ 07960-6989
www.cse.edu
*Prgm Dir:* Kathleen M Carozza, MA RD
*Tel:* 973 290-4122  *Fax:* 973 290-4138
*E-mail:* kcarozza@cse.edu

**Rutgers University**
Dietetics-Didactic Prgm
Nutrition Sci, 229B Davison Hall
26 Nichol Ave
New Brunswick, NJ 08901-2882
http://aesop.rutgers.edu/~nutrition
*Prgm Dir:* Barbara L Tangel, MS RD
*Tel:* 732 932-9570  *Fax:* 732 932-6522
*E-mail:* bltangel@rci.rutgers.edu

**Univ of Med & Dent of New Jersey**
Dietetic Internship Prgm
School of Health Related Profs
1776 Raritan Rd
Scotch Plains, NJ 07076
http://shrp.umdnj.edu/programs/dietetic
*Prgm Dir:* Denise D Langevin, MS RD
*Tel:* 908 889-2488  *Fax:* 908 889-2487
*E-mail:* langevdd@umdnj.edu

**Univ of Med & Dent of New Jersey**
Dietetics-Coordinated Prgm
School of Health Related Profs Primary Care Dept
1776 Raritan Rd
Scotch Plains, NJ 07076
http://shrp.umdnj.edu/catalog/Primary_Care
*Prgm Dir:* Joyce A O'Connor, DrPH, RD
*Tel:* 908 889-2487  *Fax:* 908 889-2487
*E-mail:* oconnoja@umdnj.edu

**New Jersey Dept of Health & Senior Services**
Dietetic Internship Prgm
50 E State St
PO Box 364
Trenton, NJ 08625-0364
*Prgm Dir:* Sherry A Valente, MA RD
*Tel:* 609 292-9560  *Fax:* 609 292-3580
*E-mail:* sherry.valente@doh.state.nj.us

**Montclair State University**
Dietetics-Didactic Prgm
Dept of Human Ecology
111 Finley
Upper Montclair, NJ 07043
www.montclair.edu/pages/heco/food-dietetics.html
*Prgm Dir:* Doreen Liou, EdD RD
*Tel:* 973 655-4416  *Fax:* 973 655-7467
*E-mail:* lioud@mail.montclair.edu

**Montclair State University**
Dietetic Internship Prgm
Dept of Home Ecology
University Hall
Upper Montclair, NJ 07043
*Prgm Dir:* Carol A Sokolik, MS RD
*Tel:* 973 655-4375  *Fax:* 973 655-4399
*E-mail:* sokolikc@mail.montclair.edu

# New Mexico

## University of New Mexico
Dietetics-Didactic Prgm
Individual Family & Community Education
Nutrition MSC05 3040
Albuquerque, NM 87131-1231
www.nutrition.unm.edu
*Prgm Dir:* Karen E Heller, PhD RD
*Tel:* 505 277-3160    *Fax:* 505 277-4362
*E-mail:* kheller@unm.edu

## University of New Mexico
Dietetic Internship Prgm
Individual Family & Community Education
Nutrition MSC05 3040
Albuquerque, NM 87131-1231
www.nutrition.unm.edu
*Prgm Dir:* Jean L Cerami, MS RD CDE
*Tel:* 505 277-0937    *Fax:* 505 277-4362
*E-mail:* jcerami@unm.edu

## New Mexico State University
Dietetics-Didactic Prgm
Dept of Family and Consumer Sciences
Box 30003, MSC 3470
Las Cruces, NM 88003-8003
www.nmsu.edu/~famcon
*Prgm Dir:* Margaret Ann Bock, PhD RD LD
*Tel:* 505 646-1178    *Fax:* 505 646-1889
*E-mail:* abock@nmsu.edu

# New York

## CUNY Herbert H Lehman College
Dietetic Internship Prgm
Dept of Health Services
250 Bedford Park Blvd W
Bronx, NY 10468-1589
www.lehman.cuny.edu/deannss/healthsci/di/info.html
*Prgm Dir:* Andrea Boyar, PhD RD CDN
*Tel:* 718 960-8084    *Fax:* 718 960-8908
*E-mail:* andrea.boyer@lehman.cuny.edu

## CUNY Herbert H Lehman College
Dietetics-Didactic Prgm
Dept of Health Serv Dietetics Food & Nutri
Bedford Park Blvd W
Bronx, NY 10468-1589
www.lehman.cuny.edu
*Prgm Dir:* Alice Tobias, EdD RD
*Tel:* 718 960-8775    *Fax:* 718 960-8908
*E-mail:* atobias@lehman.cuny.edu

## Veterans Affairs Medical Center
Dietetic Internship Prgm
Nutrition and Food Program (120)
130 W Kingsbridge Rd
Bronx, NY 10468-3904
www.va.gov/visns/visn03/diethome.asp
*Prgm Dir:* Antoinette C Franklin, MS RD
*Tel:* 718 584-9000, Ext 6845    *Fax:* 718 741-4238
*E-mail:* antoinette.franklin@med.va.gov

## CUNY Brooklyn College
Dietetics-Didactic Prgm
Dept of Health and Nutrition Sciences
2900 Bedford Ave
Brooklyn, NY 11210-2889
http://academic.brooklyn.cuny.edu/hns
*Prgm Dir:* Clifford Rouder, EdD RD
*Tel:* 718 951-5000, Ext 2738    *Fax:* 718 951-4670
*E-mail:* rouderc@brooklyn.cuny.edu

## CUNY Brooklyn College
Dietetic Internship Prgm
Dept of Health and Nutrition Sciences
2900 Bedford Ave
Brooklyn, NY 11210-2889
http://academic.brooklyn.cuny.edu/hns
*Prgm Dir:* Roseanne Schnoll, PhD RD CDN
*Tel:* 718 951-5909    *Fax:* 718 951-4670
*E-mail:* rschnoll@brooklyn.cuny.edu

## Long Island University - C W Post Campus
Dietetics-Didactic Prgm
Dept of Nutrition, Life Science Bldg
720 Northern Blvd
Brookville, NY 11548
www.liu.edu/nutrit
*Prgm Dir:* Barbara J Shorter, EdD CDN RD
*Tel:* 516 299-3046    *Fax:* 516 299-3106
*E-mail:* bshorter@liu.edu

## Long Island University - C W Post Campus
Dietetic Internship Prgm
Dept of Nutrition
720 Northern Blvd
Brookville, NY 11548
www.cwpost.liu.edu/nutrit
*Prgm Dir:* Alessandra R Sarcona, MS RD
*Tel:* 516 299-3224    *Fax:* 516 299-3106
*E-mail:* asarcona@liu.edu

## Buffalo State, SUNY
Dietetics-Didactic Prgm
Dietetics and Nutrition Dept
1300 Elmwood Ave
Buffalo, NY 14222-1095
www.buffalostate.edu
*Prgm Dir:* Tejaswini Rao, PhD RD CDN
*Tel:* 716 878-4333    *Fax:* 716 878-5834
*E-mail:* raot@buffalostate.edu

## Buffalo State, SUNY
Dietetics-Coordinated Prgm
Dietetics and Nutrition Dept, Caudell Hall 207
1300 Elmwood Ave
Buffalo, NY 14222-1095
www.buffalostate.edu
*Prgm Dir:* Donna M Hayes, MS RD CDN
*Tel:* 716 878-5634    *Fax:* 716 878-5834
*E-mail:* hayesdm@buffalostate.edu

## D'Youville College
Dietetics-Coordinated Prgm
320 Porter Ave
Buffalo, NY 14201-1084
www.dyc.edu
*Prgm Dir:* Maria E Haas, MS RD CDN
*Tel:* 716 881-7752, Ext 7017    *Fax:* 716 881-7790
*E-mail:* haasm@dyc.edu

## University at Buffalo - SUNY
Dietetic Internship Prgm
15 Farber Hall
3435 Main St
Buffalo, NY 14214
www.phhp.buffalo.edu/ens/nutrition/ntr-internship.html
*Prgm Dir:* Mary E Platek, MS RD CDN
*Tel:* 716 829-3680, Ext 259    *Fax:* 716 829-3700
*E-mail:* platek@buffalo.edu

## Sodexho Marriott/Metro New York
*Cosponsor:* NY Metropolitan Dietetic Internship
Dietetic Internship Prgm
90 Merrick Ave, Ste 210
East Meadow, NY 11554
www.dieteticintern.com/Nymetro
*Prgm Dir:* Gayanne R Wolset, MS RD CDE LDN
*Tel:* 516 794-9150, Ext 153    *Fax:* 516 794-9154
*E-mail:* dieteticinternship@sodexhousa.com

## CUNY Queens College
Dietetic Internship Prgm
Family Nutri and Exercise Sciences
65-30 Kissena Blvd
Flushing, NY 11367-1597
www.qc.edu/FNES/dietintern.html
*Prgm Dir:* Susan P Braverman, MS RD
*Tel:* 718 997-4155    *Fax:* 718 997-4163
*E-mail:* susan_braverman@qc.edu

## CUNY Queens College
Dietetics-Didactic Prgm
Dept of Family, Nutri and Exercise Sciences
65-30 Kissena Blvd
Flushing, NY 11367-1597
www.qc.edu
*Prgm Dir:* Patricia K Miner, MS RD
*Tel:* 718 997-4152    *Fax:* 718 997-4163
*E-mail:* Patrcia.miner@qc.cuny.edu

## Cornell University
Dietetic Internship Prgm
Division of Nutritional Sciences
225 Savage Hall
Ithaca, NY 14853-4401
www.nutrition.cornell.edu/dns7_dieteticintern.html
*Prgm Dir:* Patsy M Brannon, PhD RD
*Tel:* 607 255-2613    *Fax:* 607 255-0178
*E-mail:* pmb22@cornell.edu

## Cornell University
Dietetics-Didactic Prgm
Div of Nutritional Sciences
373 MVR Hall
Ithaca, NY 14853-4401
www.nutrition.cornell.edu/dns7_dietetic.html
*Prgm Dir:* Emily Wilcox Gier, MBA RD CDN
*Tel:* 607 255-2638    *Fax:* 607 255-0178
*E-mail:* eg47@cornell.edu

## ARAMARK Healthcare Support Services
*Cosponsor:* Metropolitan New York DI Program-Mary
Immaculate Hospital
Dietetic Internship Prgm
90-15 158th Ave
Jamaica, NY 11414
www.aramark.com/careers.asp
*Prgm Dir:* Allison Charny, MS RD CDN CDE
*Tel:* 800 378-5348, Ext 4485    *Fax:* 718 738-5228
*E-mail:* charny-allison@aramark.com

## Columbia University Teachers College
Dietetic Internship Prgm
Dept of Health & Behavior Studies
525 W 120th St Box 137
New York, NY 10027-6625
www.tc.columbia.edu/hbs/nutrition
*Prgm Dir:* Christina Costa, MS RD CNSD
*Tel:* 212 678-3460    *Fax:* 212 678-4048
*E-mail:* CC2005@columbia.edu

## CUNY Hunter College
Dietetic Internship Prgm
School of Health Sciences
425 E 25th St, Box 896
New York, NY 10010-2590
www.hunter.cuny.edu/schoolhp/phn/dietetic_internship
*Prgm Dir:* Allison R Marshall, MS RD CDN
*Tel:* 212 772-3806    *Fax:* 212 481-5260
*E-mail:* allimarshall1234@aol.com

## CUNY Hunter College
Dietetics-Didactic Prgm
Brookdale Hlth Science Ctr
425 E 25th St
New York, NY 10010-2590
*Prgm Dir:* Khursheed P Navder, PhD RD
*Tel:* 212 481-5118    *Fax:* 212 481-5260
*E-mail:* knavder@hunter.cuny.edu

**New York - Presbyterian Hospital**
Dietetic Internship Prgm
525 E 68th St AN-833
Box 92
New York, NY 10021-4873
www.nyp.org/nutrition
*Prgm Dir:* Elaine Rosenthal, MS RD
*Tel:* 212 746-0836  *Fax:* 212 746-8287
*E-mail:* erosenth@nyp.org

**New York University**
Dietetic Internship Prgm
Nutrition and Food Studies
35 W 4th St 10th Fl
New York, NY 10011-1172
www.nyu.edu/education/nutrition
*Prgm Dir:* Lisa M Sasson, MS RD
*Tel:* 212 998-5580  *Fax:* 212 995-4194
*E-mail:* lisa.sasson@nyu.edu

**New York University**
Dietetics-Didactic Prgm
Nutrition & Food Studies
35 W 4th St 10th Fl, Rm 1077
New York, NY 10012-1172
www.nyu.edu/education/nutrition
*Prgm Dir:* Kristie J Lancaster, PhD RD
*Tel:* 212 998-5594  *Fax:* 212 995-4194
*E-mail:* kristie.lancaster@nyu.edu

**New York Institute of Technology**
Dietetics-Didactic Prgm
Clinical Nutrition Dept
NYCOM II, Rm 334
Old Westbury, NY 11568-8000
www.nyit.edu
*Prgm Dir:* Mindy E Haar, MS RD CDN
*Tel:* 516 686-3818  *Fax:* 516 686-3795
*E-mail:* mhaar@nyit.edu

**New York Institute of Technology**
Dietetic Internship Prgm
NYCOM II, Rm 342
PO Box 8000
Old Westbury, NY 11568-8000
www.nyit.edu
*Prgm Dir:* Denise L Donaldson-Kaiser, MS RD CDN
*Tel:* 516 686-3880  *Fax:* 516 686-3795
*E-mail:* ddonalds@nyit.edu

**SUNY College at Oneonta**
Dietetics-Didactic Prgm
Dept of Human Ecology
104C Human Ecology
Oneonta, NY 13820-4015
www.oneonta.edu/academics/huec/dietetics4.asp
*Prgm Dir:* Mary Ann Dowdell, PhD RD
*Tel:* 607 436-2112  *Fax:* 607 436-2051
*E-mail:* dowdelma@oneonta.edu

**SUNY College at Oneonta**
Dietetic Internship Prgm
Dept of Human Ecology
39 Denison Hall
Oneonta, NY 13820
www.oneonta.edu/academics/dieteticinternship
*Prgm Dir:* Jennifer L Bueche, PhD RD CDN
*Tel:* 607 436-2070  *Fax:* 607 436-3084
*E-mail:* buechejl@oneonta.edu

**SUNY College at Plattsburgh**
Dietetics-Didactic Prgm
Nutrition & Food Studies
101 Broad St
Plattsburgh, NY 12901-2681
www.plattsburgh.edu/clg
*Prgm Dir:* Enamuthu Joseph, PhD RD CDN
*Tel:* 518 564-4223  *Fax:* 518 564-3100
*E-mail:* josephe@plattsburgh.edu

**Rochester Institute of Technology**
Dietetics-Coordinated Prgm
Hospitality and Service Management, George Eastman
    Bldg
14 Lomb Memorial Dr
Rochester, NY 14623-5604
www.rit.edu
*Prgm Dir:* Elizabeth A Kmiecinski, MS RD CDN
*Tel:* 585 475-2357  *Fax:* 585 475-5099
*E-mail:* eakism@rit.edu

**Rochester Institute of Technology**
Dietetics-Didactic Prgm
Hospitality and Service Management
14 Lomb Memorial Dr
Rochester, NY 14623-5604
www.rit.edu
*Prgm Dir:* Barbara Cerio, MS RD
*Tel:* 585 475-2352  *Fax:* 585 475-5099
*E-mail:* bxcism@rit.edu

**Stony Brook University**
Dietetic Internship Prgm
Department of Family Medicine
Health Sciences Center Level 4, Rm 050
Stony Brook, NY 11794-8461
www.hsc.stonybrook.edu/SOM/fammed/intern_program
*Prgm Dir:* Josephine Connolly Schoonen, MS RD
*Tel:* 631 444-8246  *Fax:* 631 444-7552
*E-mail:* josephine.connolly-schoonenen@stonybrook.edu

**Syracuse University**
Dietetics-Didactic Prgm
Nutrition & Foodservice Mgmt
034 Slocum Hall
Syracuse, NY 13244-1250
*Prgm Dir:* Tanya M Horacek, PhD RD
*Tel:* 315 443-2386  *Fax:* 315 443-2735
*E-mail:* thoracek@syr.edu

**Syracuse University**
Dietetic Internship Prgm
Nutrition & Foodservice Mgmt
034 Slocum Hall
Syracuse, NY 13244-1250
www.hshp.syr.edu/schools/nhm/academics/grad/mams/
    dietetic
*Prgm Dir:* Debra Connolly, MA RD
*Tel:* 315 443-2386  *Fax:* 315 443-2735
*E-mail:* dzconnol@syr.edu

**Syracuse University**
Dietetics-Coordinated Prgm
Nutrition & Foodservice Mgmt
034 Slocum Hall
Syracuse, NY 13244-1250
www.hshp.syr.edu/schools/nhm
*Prgm Dir:* Kay S Bruening, PhD RD
*Tel:* 315 443-9326  *Fax:* 315 443-2735
*E-mail:* ksbrueni@syr.edu

**Marymount College**
Dietetics-Didactic Prgm
Dept of Human Ecology
100 Marymount Ave
Tarrytown, NY 10591
www.fordham.edu
*Prgm Dir:* Paula A Van Aken, MS RD CDN
*Tel:* 914 332-8278  *Fax:* 914 631-8586
*E-mail:* vanakenr@fordham.edu

**The Sage Colleges**
Dietetic Internship Prgm
45 Ferry St
Troy, NY 12180-4115
www.sage.edu
*Prgm Dir:* Melodie J Bell-Cavallino, MS RD FADA
*Tel:* 518 244-2075  *Fax:* 518 244-4586
*E-mail:* bellcm@sage.edu

**The Sage Colleges**
Dietetics-Didactic Prgm
Div of Health and Rehab Sciences-Ackerman Hall
45 Ferry St
Troy, NY 12180-4115
www.sage.edu/divisions/hrs/nutrition.html
*Prgm Dir:* Rayane Abu-Sabha, PhD RD
*Tel:* 518 244-4573  *Fax:* 518 244-4586
*E-mail:* abusar@sage.edu

**Westchester Medical Center**
Dietetic Internship Prgm
Food Services Div
Grasslands Rd
Valhalla, NY 10595-1689
*Prgm Dir:* Joan C O'Brien, MPA RD CDN
*Tel:* 914 493-8978  *Fax:* 914 493-1376
*E-mail:* obrienj@wcmc.com

# North Carolina

**Appalachian State University**
Dietetic Internship Prgm
Family and Consumer Sciences
PO Box 32056
Boone, NC 28608-2056
www.appstate.edu/foodgrad/nutrition.html
*Prgm Dir:* Susan L Bogardus, PhD MS RD
*Tel:* 828 262-2631  *Fax:* 828 265-8620
*E-mail:* bogardussl@appstate.edu

**Appalachian State University**
Dietetics-Didactic Prgm
Family and Consumer Sciences
Boone, NC 28608-2630
www.appstate.edu
*Prgm Dir:* Lisa McAnulty, PhD RD
*Tel:* 828 262-2631  *Fax:* 828 265-8620
*E-mail:* mcanultyl@appstate.edu

**Univ of North Carolina at Chapel Hill**
Dietetics-Coordinated Prgm
McGavran-Greenberg Hall
Dept of Nutrition CB 7461
Chapel Hill, NC 27599-7461
www.sph.unc.edu/nutr
*Prgm Dir:* Carolyn H Barrett, MS MPH RD
*Tel:* 919 966-7214  *Fax:* 919 966-7216
*E-mail:* carolyn_barrett@unc.edu

**Univ of North Carolina at Chapel Hill**
Dietetics-Didactic Prgm
McGavran-Greenburg Hall
Dept of Nutrition CB 7461
Chapel Hill, NC 27599-7461
www.sph.unc.edu/nutr
*Prgm Dir:* Carolyn H Barrett, MS MPH RD
*Tel:* 919 966-7214  *Fax:* 919 966-7216
*E-mail:* carolyn_barrett@unc.edu

**Western Carolina University**
Dietetic Internship Prgm
Nutrition/Sciences
Dept of Health Sciences
Cullowhee, NC 28723
www.wcu.edu/aps/healths/HS_UG-DI.htm
*Prgm Dir:* Teresa M Breedlove, MS RD LDN
*Tel:* 828 227-7446  *Fax:* 828 227-7446
*E-mail:* tbreedlove@wcu.edu

**Western Carolina University**
Dietetics-Didactic Prgm
Dept of Health Sciences
Cullowhee, NC 28723
www.wcu.edu/aps/health/HS_UG-ND.htm
*Prgm Dir:* Wayne E Billon, PhD RD
*Tel:* 828 227-3528  *Fax:* 828 227-7446
*E-mail:* billon@wcu.edu

**North Carolina Central University**
Dietetics-Didactic Prgm
Dept of Human Sciences
PO Box 19615
Durham, NC 27707-0099
www.nccu.edu/artsci/human/foodsandnut.html
*Prgm Dir:* Esther C Okeiyi, PhD RD LDN
*Tel:* 919 530-7439   *Fax:* 919 530-7983
*E-mail:* eokeiyi@nccu.edu

**North Carolina Central University**
Dietetic Internship Prgm
Dept of Human Sciences
PO Box 19615
Durham, NC 27707-0099
www.nccu.edu/artsci/human/foodsandnut.html
*Prgm Dir:* Esther C Okeiyi, PhD RD LDN
*Tel:* 919 530-7439   *Fax:* 919 560-7983
*E-mail:* eokeiyi@nccu.edu

**North Carolina A&T State University**
Dietetics-Didactic Prgm
Dept of Human Environment and Families
102 Benbow Hall, 1601 E Market St
Greensboro, NC 27411-1064
www.ag.ncat.edu/academics/fcs/dietetics/index.htm
*Prgm Dir:* Lizette Sanchez-Lugo, PhD RD
*Tel:* 336 334-7850   *Fax:* 336 334-7265
*E-mail:* lizette@ncat.edu

**Univ of North Carolina at Greensboro**
Dietetic Internship Prgm
Dept of Nutrition & Food Service Systems
1000 Spring Garden St
Greensboro, NC 27402-6170
www.uncg.edu/nutrition
*Prgm Dir:* Martha L Taylor, PhD RD
*Tel:* 336 256-0326   *Fax:* 336 334-4129
*E-mail:* marth_taylor@uncg.edu

**Univ of North Carolina at Greensboro**
Dietetics-Didactic Prgm
Nutrition & Foodservice Systems
PO Box 26170
Greensboro, NC 27402-6170
www.uncg.edu/nutrition
*Prgm Dir:* Cheryl A Lovelady, PhD RD LDN
*Tel:* 336 334-5313   *Fax:* 336 334-4129
*E-mail:* cheryl_lovelady@uncg.edu

**East Carolina University**
Dietetics-Didactic Prgm
School of Human Environmental Sciences
Nutrition & Hospitality Mgmt
Greenville, NC 27858-4353
www.ecu.edu/che/nuhm/nutrbs.htm
*Prgm Dir:* Sylvia Escott-Stump, MA RD LDN
*Tel:* 252 328-6917   *Fax:* 252 328-4276
*E-mail:* escottstumps@mail.ecu.edu

**East Carolina University**
Dietetic Internship Prgm
School of Human Environmental Sciences
Nutrition & Hospitality Mgmt
Greenville, NC 27858-4353
www.ecu.edu/che/nuhm/di/index.htm
*Prgm Dir:* Sylvia Escott-Stump, MA RD LDN
*Tel:* 252 328-1352   *Fax:* 252 328-4276
*E-mail:* escottstumps@mail.ecu.edu

**Meredith College**
Dietetic Internship Prgm
Human Environmental Sciences
3800 Hillsborough St
Raleigh, NC 27607-5298
www.meredith.edu/hes/foods_dietetic.htm
*Prgm Dir:* Cathleen E Ostrowski, MS RD LDN
*Tel:* 919 760-8419   *Fax:* 919 760-2819
*E-mail:* ostrowsk@meredith.edu

**Meredith College**
Dietetics-Didactic Prgm
Human Environmental Sciences
3800 Hillsborough St
Raleigh, NC 27607-5298
*Prgm Dir:* Susan G Fisher, PhD RD
*Tel:* 919 760-8079   *Fax:* 919 760-2819
*E-mail:* fishers@meredith.edu

# North Dakota

**North Dakota State University**
Dietetic Internship Prgm
EML Hall 351
PO Box 5057
Fargo, ND 58105-5057
www.ndsu.edu/instruct/north/north
*Prgm Dir:* Barbara B North, MS RD LRD
*Tel:* 701 231-7479   *Fax:* 701 231-7174
*E-mail:* barbara.north@ndsu.nodak.edu

**North Dakota State University**
Dietetics-Didactic Prgm
Dept of Health, Nutrition, and Exercise Science
PO Box 5057
Fargo, ND 58105
www.ndsu.nodak.edu
*Prgm Dir:* Lynette S Winters, MS LRD
*Tel:* 701 231-7480   *Fax:* 701 231-7174
*E-mail:* lynette.winters@ndsu.nodak.edu

**North Dakota State University**
Dietetics-Coordinated Prgm
Dept of Health, Nutrition, and Exercise Science
PO Box 5057
Fargo, ND 58105-5057
www.ndsu.nodak.edu
*Prgm Dir:* Lynette S Winters, MS LRD
*Tel:* 701 231-7480   *Fax:* 701 231-7174
*E-mail:* lynette.winters@ndsu.nodak.edu

**University of North Dakota**
Dietetics-Coordinated Prgm
Nutrition and Dietetics
PO Box 8237
Grand Forks, ND 58202-8237
www.und.edu/dept/nursing/nutrition.html
*Prgm Dir:* Judith H Hall, MS RD LRD
*Tel:* 701 777-3752   *Fax:* 701 777-3268
*E-mail:* judy_hall@mail.und.nodak.edu

# Ohio

**University of Akron**
Dietetics-Coordinated Prgm
Home Econ and Family Ecology
215 Schrank Hall S
Akron, OH 44325-6103
www.uakron.edu/fcs
*Prgm Dir:* Sandra L Kudak, MS RD
*Tel:* 330 972-6043   *Fax:* 330 972-4934
*E-mail:* slhudak@uakron.edu

**University of Akron**
Dietetics-Didactic Prgm
School of Family and Consumer Sciences
215 Schrank Hall S
Akron, OH 44325-6103
www3.uakron.edu/fcs
*Prgm Dir:* Sue A Rasor-Greenhalg, MS RD
*Tel:* 330 972-6046   *Fax:* 330 972-4934
*E-mail:* sue@uakron.edu

**Ohio University**
Dietetics-Didactic Prgm
Human and Consumer Sciences
Grover Center W324
Athens, OH 45701-2979
www.ohiou.edu
*Prgm Dir:* David H Holben, PhD RD
*Tel:* 740 593-2875   *Fax:* 740 593-0289
*E-mail:* holben@ohio.edu

**Bluffton College**
Dietetics-Didactic Prgm
Family and Consumer Sciences
1 University Dr, Box 1346
Bluffton, OH 45817-1196
www.bluffton.edu/fcs
*Prgm Dir:* Deborah I Myers, MS RD
*Tel:* 419 358-3229   *Fax:* 419 358-3323
*E-mail:* myersd@bluffton.edu

**Bowling Green State University**
Dietetics-Didactic Prgm
Family and Consumer Sciences
206 Johnston Hall
Bowling Green, OH 43403-0254
www.bgsu.edu/colleges/edhd/FCS
*Prgm Dir:* Christine M Haar, MS RD LD
*Tel:* 419 372-8941   *Fax:* 419 372-7854
*E-mail:* chaar@bgnet.bgsu.edu

**Bowling Green State University**
Dietetic Internship Prgm
Family and Consumer Sciences
Bowling Green, OH 43403-0254
www.bgsu.edu/colleges/edhd/FCS/F&n_internship.htm
*Prgm Dir:* Christine Haar, MS RD
*Tel:* 419 372-8941   *Fax:* 419 372-7854
*E-mail:* chaar@bgnet.bgsu.edu

**Christ Hospital**
Dietetic Internship Prgm
Food & Nutrition Services
2139 Auburn Ave
Cincinnati, OH 45219-2906
www.health-alliance.com
*Prgm Dir:* Susan V Dvorak, MEd RD LD
*Tel:* 513 585-2283   *Fax:* 513 585-3033
*E-mail:* dvoraksv@healthall.com

**Cincinnati Children's Div Dvlpmntl Disorders**
*Cosponsor:* University Center of Excellence on
   Disabilities
Dietetic Internship Prgm
Children's Hospital Med Ctr
3333 Burnett Ave, Pav 3713
Cincinnati, OH 45229-3039
*Prgm Dir:* Shirley M Ekvall, PhD RD LD
*Tel:* 513 636-4614   *Fax:* 513 636-7361
*E-mail:* ekvalls@chmcc.org

**Good Samaritan Hospital**
Dietetic Internship Prgm
Nutrition Dept
375 Dixmyth Ave
Cincinnati, OH 45220-2489
*Prgm Dir:* Jackene M Laverty, MEd RD LD
*Tel:* 513 872-1983   *Fax:* 513 872-4986
*E-mail:* jackene_laverty@trihealth.com

**University of Cincinnati**
Dietetics-Didactic Prgm
Dept of Health Sciences
363C Hastings & William French Bldg, Box 670394
Cincinnati, OH 45267-0394
www.cahs.uc.edu/departments/dietetic.s.cfm
*Prgm Dir:* Sarah Collins Couch, PhD RD LD
*Tel:* 513 558-7506   *Fax:* 513 558-7500
*E-mail:* rcouchsc@email.uc.edu

**Case Western Reserve University**
Dietetic Internship Prgm
Dept of Nutrition-School of Medicine
10900 Euclid Ave
Cleveland, OH 44106-1712
*Prgm Dir:* Isabel M Parraga, PhD RD LD
*Tel:* 216 368-6626   *Fax:* 216 368-6644
*E-mail:* imp@po.cwru.edu

**Case Western Reserve University**
Dietetics-Didactic Prgm
Dept of Nutrition
10900 Euclid Ave
Cleveland, OH 44106-4906
www.cwru.edu/med/nutrition.home.html
*Prgm Dir:* James H Swain, PHD RD LD
*Tel:* 216 368-8554    *Fax:* 216 368-6644
*E-mail:* jhs31@cwru.edu

**Cleveland Clinic Foundation**
Dietetic Internship Prgm
Nutrition Therapy M17
9500 Euclid Ave
Cleveland, OH 44195
www.clevelandclinic.org/education/diet
*Prgm Dir:* Sue Kent, MS RD LD
*Tel:* 216 444-6487    *Fax:* 216 444-9415
*E-mail:* kents@ccf.org
*Med Dir:* Cindy Moore, MS RD LD

**Louis Stokes Cleveland VA Med Ctr**
Dietetic Internship Prgm
10701 East Blvd
Cleveland, OH 44106-1702
www.cwru.edu/med/nutrition/clevamc.html
*Prgm Dir:* Anne Raguso, PhD RD
*Tel:* 216 421-3800    *Fax:* 216 421-3014
*E-mail:* anne.raguso@med.va.gov

**MetroHealth Medical Center**
Dietetic Internship Prgm
2500 MetroHealth Dr
Cleveland, OH 44109-1998
www.metrohealth.org/nutrition/dietetic.asp
*Prgm Dir:* Lisa M Bonacuse, MS RD
*Tel:* 216 778-2718    *Fax:* 216 778-8363
*E-mail:* lbonacuse@metrohealth.org

**University Hospitals of Cleveland**
Dietetic Internship Prgm
11100 Euclid Ave
Lakeside 5021
Cleveland, OH 44106-5000
www.cwru.edu/med/nutrition/uhocle.html
*Prgm Dir:* Felicia Vatakis, MS RD LD
*Tel:* 216 844-1310    *Fax:* 216 844-8188
*E-mail:* felicia.vatakis@uhhs.com

**Mt Carmel College of Nursing**
Dietetic Internship Prgm
127 S Davis Ave
Columbus, OH 43222-1504
www.mccn.edu
*Prgm Dir:* Kathleen M Blanchard, MS RD LD
*Tel:* 614 234-5439    *Fax:* 614 234-2875
*E-mail:* kblancha@mchs.com

**Ohio State University**
Dietetics-Didactic Prgm
Dept of Human Nutrition & Food Management
1787 Neil Ave
Columbus, OH 43210-1220
http://hec.osu.edu/hn
*Prgm Dir:* Anne M Smith, PhD RD
*Tel:* 614 292-0715    *Fax:* 614 292-8880
*E-mail:* smith.23@osu.edu

**Ohio State University**
Dietetics-Coordinated Prgm
Sch of Allied Medical Profs
1583 Perry St
Columbus, OH 43210-1234
www.amp.ohio-state.edu
*Prgm Dir:* M Rosita Schiller, PhD RD LD FADA
*Tel:* 614 292-0635    *Fax:* 614 292-0210
*E-mail:* schiller.1@osu.edu

**Ohio State University**
Dietetic Internship Prgm
School of Allied Medical Profs
Med Dietetics, 1583 Perry St
Columbus, OH 43210-1234
www.amp.ohio-state.edu
*Prgm Dir:* Kay N Wolf, PhD RD LD
*Tel:* 614 292-8131    *Fax:* 614 292-0210
*E-mail:* kwolf@amp.osu.edu

**Ohio State University**
Dietetic Internship Prgm
Dept of Human Nutrition & Food Managemnt
325 Campbell Hall, 1787 Neil Ave
Columbus, OH 43210-1220
http://hec.osu.edu/hn/programs/di
*Prgm Dir:* Gail L Kaye, PhD RD LD
*Tel:* 614 292-8189    *Fax:* 614 292-8880
*E-mail:* kaye3@osu.edu

**Miami Valley Hospital**
Dietetic Internship Prgm
One Wyoming St
Dayton, OH 45409-2793
www.maimivalleyhospital.com/dietetic
*Prgm Dir:* Rebecca M Lee, MS RD LD
*Tel:* 937 208-2448    *Fax:* 937 341-8194
*E-mail:* rmlee@mvh.org

**University of Dayton**
Dietetics-Didactic Prgm
Health & Sports Science
300 College Park Ave
Dayton, OH 45469-1210
http://homepages.udayton.educ
*Prgm Dir:* Patricia Dolan, MS RD
*Tel:* 937 229-4203    *Fax:* 937 229-4244
*E-mail:* patricia.dolan@notes.udayton.edu

**Kent State University**
Dietetics-Didactic Prgm
Family and Consumer Studies
Nixson Hall, Nutrition & Dietetics
Kent, OH 44242
www.dept.kent.edu/f&cs
*Prgm Dir:* Karen L Gordon, PhD RD LD
*Tel:* 330 672-2197    *Fax:* 330 672-2194
*E-mail:* klowry@kent.edu

**Kent State University**
Dietetic Internship Prgm
100 Nixson Hall
Kent, OH 44242
www.kent.edu/f&cs
*Prgm Dir:* Nancy H Burzminski, MS RD LD
*Tel:* 330 672-2064    *Fax:* 330 672-2194
*E-mail:* nburzim@kent.edu

**Miami University**
Dietetics-Didactic Prgm
Phys Ed Hlth & Sports Studies
100A Phillips Hall
Oxford, OH 45056
www.ohiou.edu
*Prgm Dir:* Susan J Rudge, PhD RD LD
*Tel:* 513 529-5036    *Fax:* 513 529-5006
*E-mail:* rudgesj@muohio.edu

**Notre Dame College**
Dietetics-Didactic Prgm
4545 College Rd
South Euclid, OH 44121-4293
www.ndc.edu
*Prgm Dir:* Margaret A Cullis, MSc MSRD
*Tel:* 216 381-1680, Ext 340    *Fax:* 216 381-3227

**Youngstown State University**
Dietetics-Didactic Prgm
Human Ecology Dept
One University Plaza
Youngstown, OH 44555-0001
www.ysu.edu
*Prgm Dir:* Jeanine L Mincher, MS RD LD
*Tel:* 330 941-3346    *Fax:* 330 941-1824
*E-mail:* jlmincher@ysu.edu

**Youngstown State University**
Dietetics-Coordinated Prgm
One University Plaza
Youngstown, OH 44555-0001
www.ysu.edu
*Prgm Dir:* Jean H Hassell, MS RD LD
*Tel:* 330 941-3344    *Fax:* 330 941-1824
*E-mail:* jhassell@cc.ysu.edu

# Oklahoma

**University of Central Oklahoma**
Dietetic Internship Prgm
Human Environmental Sciences
100 N University Dr
Edmond, OK 73034
www.educ.ucok.edu
*Prgm Dir:* Tiffany L Schlinke, MS RD LD
*Tel:* 405 974-5369    *Fax:* 405 974-3850
*E-mail:* tschlinke@ucok.edu

**University of Central Oklahoma**
Dietetics-Didactic Prgm
College of Education
Human Environmental Sciences
Edmond, OK 73034
www.ucok.edu
*Prgm Dir:* Marilyn B Walters, PhD RD
*Tel:* 405 974-5805, Ext 5805    *Fax:* 405 974-3850
*E-mail:* mwaters@ucok.edu

**Langston University**
Dietetics-Didactic Prgm
Dept of Home Ecology
302 Jones Hall
Langston, OK 73050
www.lunet.edu
*Prgm Dir:* Saigeetha Sangiah, PhD RD LD
*Tel:* 405 466-3337    *Fax:* 405 466-3364
*E-mail:* ssangiah@lunet.edu

**Univ of Oklahoma Health Sciences Center**
Dietetic Internship Prgm
College of Allied Hlth-Dept of Nutri Sci
PO Box 26901 CHB
Oklahoma City, OK 73190
www.ah.ouhsc.edu/main/departments/nutrition/ns.htm
*Prgm Dir:* Jani A Van Grevenhof, MS RD CDE
*Tel:* 405 271-2113, Ext 41178    *Fax:* 405 271-1560
*E-mail:* jani-vangrevenhof@ouhsc.edu

**Univ of Oklahoma Health Sciences Center**
Dietetics-Didactic Prgm
College of Allied Health
Nutritional Sci, PO Box 26901
Oklahoma City, OK 73190
www.ah.ouhsc.edu/main/Programs/prog1.asp?prog_id=9
*Prgm Dir:* Stephen R Glore, PhD RD
*Tel:* 405 271-2113, Ext 41176    *Fax:* 405 271-1560
*E-mail:* stephen-glore@ouhsc.edu

**Univ of Oklahoma Health Sciences Center**
Dietetics-Coordinated Prgm
Dept of Nutritional Sciences
801 NE 13th, PO Box 26901
Oklahoma City, OK 73190
www.ouhsc.edu
*Prgm Dir:* Karen M Funderburg, MS RD LD
*Tel:* 405 271-2113    *Fax:* 405 271-1560
*E-mail:* karen-funderburg@ouhsc.edu

**Oklahoma State University**
Dietetic Internship Prgm
Dept of Nutritional Sciences
HES 301
Stillwater, OK 74078-6337
www.okstate.edu/hes/nsci
*Prgm Dir:* Julia A Huber, MS RD LD
*Tel:* 405 744-6954   *Fax:* 405 744-7113
*E-mail:* huber@okstate.edu

**Oklahoma State University**
Dietetics-Didactic Prgm
Nutritional Sciences Dept
301 HES
Stillwater, OK 74078-6141
www.okstate.edu/web/ches/nsci.asp
*Prgm Dir:* Tay S Kennedy, PhD RD
*Tel:* 405 744-5965   *Fax:* 405 744-1357
*E-mail:* ktay@okstate.edu

**Northeastern State University**
Dietetics-Didactic Prgm
Coll of Business & Technology
600 N Grand, 210A PA Bldg
Tahlequah, OK 74464-2399
arapaho.nsuok.edu/~fcs
*Prgm Dir:* Alexandria R Miller, PhD RD LD
*Tel:* 918 456-5511, Ext 2965   *Fax:* 918 458-2337
*E-mail:* millerak@nsuok.edu

# Oregon

**Oregon State University**
Dietetics-Didactic Prgm
Nutrition and Food Mgmt
212 Milam Hall
Corvallis, OR 97331-5103
www.oregonstate.edu
*Prgm Dir:* Mary M Cluskey, PhD RD
*Tel:* 541 737-0960   *Fax:* 541 737-6914
*E-mail:* cluskeym@oregonstate.edu

**Oregon Health & Science University**
Dietetic Internship Prgm
EJH-10-FM 10
3181 SW Sam Jackson Park Rd
Portland, OR 97201-3098
www.ohsu.edu/dietetic
*Prgm Dir:* Dorothy W Hagan, PhD RD LD
*Tel:* 503 494-7596   *Fax:* 503 494-7076
*E-mail:* hagand@ohsu.edu

**Mid Willamette Valley Dietetic Internship**
Dietetic Internship Prgm
1955 Dallas Hwy NW, Ste 1200
Salem, OR 97304
www.capitalmanor.com
*Prgm Dir:* Nancy Dunton, PhD RD
*Tel:* 503 362-4101   *Fax:* 503 371-9021
*E-mail:* MWVDI@capitalmanor.com

# Pennsylvania

**Cedar Crest College**
Dietetics-Didactic Prgm
100 College Dr
Allentown, PA 18104-6196
www2.cedarcrest.edu/academic/ntr
*Prgm Dir:* Jane E Zielger, MS RD CNSD
*Tel:* 610 606-4666   *Fax:* 610 606-4624
*E-mail:* jeziegle@cedarcrest.edu

**Sodexho Marriott Services**
Dietetics-Preprofessional Practice Prgm
6081 Hamilton Blvd
PO Box 3501
Allentown, PA 18106-0501
www.woodco.com
*Prgm Dir:* Alison B Reyes, MS RD
*Tel:* 610 366-5237   *Fax:* 610 366-5454
*E-mail:* areyes@woodco.com

**Geisinger Medical Center**
Dietetic Internship Prgm
100 N Academy Ave
Danville, PA 17822-0115
www.geisinger.org
*Prgm Dir:* Kessey J Kieselhorst, MPA RD
*Tel:* 800 441-6211, Ext 1664   *Fax:* 570 271-7805
*E-mail:* kkieselhorst@geisinger.edu

**Edinboro University of Pennsylvania**
*Cosponsor:* Pennsylvania Consortium
Dietetics-Coordinated Prgm
Biology and Health Services
Edinboro, PA 16444
*Prgm Dir:* Sally J Lanz, MS RD
*Tel:* 814 732-2458   *Fax:* 814 732-2422
*E-mail:* lanz@edinboro.edu

**Gannon University**
*Cosponsor:* Pennsylvania Consortium
Dietetics-Coordinated Prgm
Sciences Engineering and Hlth Science
109 University Square
Erie, PA 16541-0001
www.gannon.edu
*Prgm Dir:* Dawna T Mughal, PhD RD FADA
*Tel:* 814 871-5452   *Fax:* 814 871-5662
*E-mail:* mughal@gannon.edu

**Mercyhurst College**
*Cosponsor:* Pennsylvania Consortium
Dietetics-Coordinated Prgm
Dept of Human Ecology
501 E 38th St
Erie, PA 16546
www.mercyhurst.edu
*Prgm Dir:* Charlene J Glispy, MS RD
*Tel:* 814 824-2462   *Fax:* 814 824-3053
*E-mail:* cglispy@mercyhurst.edu

**Messiah College**
Dietetics-Didactic Prgm
Dept of Natural Sciences
One College Ave
Grantham, PA 17027
www.messiah.edu
*Prgm Dir:* Kathryn A Witt, PhD RD LDN
*Tel:* 717 766-2511, Ext 7140   *Fax:* 717 691-6046
*E-mail:* kwitt@messiah.edu

**Seton Hill University**
Dietetics-Coordinated Prgm
Family and Consumer Sciences
Seton Hill Dr
Greensburg, PA 15601-1599
www.setonhill.edu
*Prgm Dir:* Janice G Sandrick, PhD RD
*Tel:* 724 830-1045   *Fax:* 724 830-1571
*E-mail:* sandrick@setonhill.edu

**Immaculata University**
Dietetics-Didactic Prgm
Fashion Foods & Nutrition
Box 722
Immaculata, PA 19345-0722
www.immaculata.edu/academics/
   Academic_Departments.htm
*Prgm Dir:* Susan W Johnston, MS RD
*Tel:* 610 647-4400, Ext 3444   *Fax:* 610 251-1668
*E-mail:* sjohnston@immaculata.edu

**Immaculata University**
Dietetic Internship Prgm
Nutrition Educ
Box 500, Graduate Div
Immaculata, PA 19345-0901
www.immaculata.edu
*Prgm Dir:* Susan W Johnston, MS RD
*Tel:* 610 647-4400, Ext 3444   *Fax:* 610 993-8550
*E-mail:* sjohnston@immaculata.edu

**Indiana University of Pennsylvania**
Dietetics-Didactic Prgm
Dept of Food & Nutrition
911 South Dr
Indiana, PA 15705-1087
www.hhs.iup.edu/fn
*Prgm Dir:* Rita M Johnson, PhD RD FADA
*Tel:* 724 357-4440   *Fax:* 724 357-7582
*E-mail:* rjohnson@iup.edu

**Indiana University of Pennsylvania**
Dietetic Internship Prgm
Dept of Food & Nutrition
Ackerman Hall 14
Indiana, PA 15705-1087
www.hhs.iup.edu/fn
*Prgm Dir:* Stephanie A Taylor-David, PhD RD
*Tel:* 724 357-4440   *Fax:* 724 357-7582
*E-mail:* stdavis@iup.edu

**Mansfield University**
Dietetics-Didactic Prgm
203C Elliott Hall
Dept of Health Sciences
Mansfield, PA 16933
www.mnsfld.edu/~health/index.htm
*Prgm Dir:* Kathy J Wright, PhD RD
*Tel:* 570 662-4628   *Fax:* 570 662-4137
*E-mail:* kwright@mansfield.edu

**Drexel University**
Dietetics-Didactic Prgm
Nutrition and Food Sciences
3141 Chestnut St
Philadelphia, PA 19104-2875
www.bioscience.drexel.edu
*Prgm Dir:* Karen A Drummond, EdD RD FADA LDN
*Tel:* 215 895-6441   *Fax:* 215 895-1273
*E-mail:* kad65@drexel.edu

**La Salle University**
Dietetics-Coordinated Prgm
1900 W Olney Ave
Philadelphia, PA 19141-1199
www.lasalle.edu/academ/nutrition
*Prgm Dir:* Jule Anne D Henstenburg, MS RD CSP
*Tel:* 215 951-1258   *Fax:* 215 951-1896
*E-mail:* henstenb@lasalle.edu

**La Salle University**
Dietetics-Didactic Prgm
1900 W Olney Ave
Philadelphia, PA 19141-1199
www.lasalle.edu/academ/nursing/nutrition/home.htm
*Prgm Dir:* Jule Anne D Henstenburg, MS RD CSP
*Tel:* 215 951-1258   *Fax:* 215 951-1772
*E-mail:* henstenb@lasalle.edu

**Family Health Council**
Dietetic Internship Prgm
960 Penn Ave, Ste 600
Pittsburgh, PA 15222-1417
www.adagiohealth.org
*Prgm Dir:* Karen A Virostek, MS RD FADA
*Tel:* 412 288-2130, Ext 134   *Fax:* 412 288-9036
*E-mail:* kvirostek@verizon.net

**U of Pitt Med Ctr Presbyterian Shadyside**
Dietetic Internship Prgm
5230 Centre Ave
Pittsburgh, PA 15232-1304
www.upmc.edu/shadyside/Dieteticinternship
*Prgm Dir:* Joyce L Scott-Smith, MS RD
*Tel:* 412 623-2114   *Fax:* 412 623-2429
*E-mail:* scottsmithjl@upmc.edu

**University of Pittsburgh**
Dietetics-Coordinated Prgm
Health and Rehab Sciences
4048 Forbes Tower
Pittsburgh, PA 15260-1802
www.shrs.pitt.edu/cdn
*Prgm Dir:* Deborah A Hutcheson, MS RD LDN CNSD
*Tel:* 412 383-6534   *Fax:* 412 383-6636
*E-mail:* dhutches@pitt.edu

**University of Pittsburgh**
Dietetics-Didactic Prgm
Health and Rehab Science
4048 Forbes Tower
Pittsburgh, PA 15260-1802
www.shrs.pitt.educ/cdn
*Prgm Dir:* Deborah A Hutcheson, MS RD LDN CNSD
*Tel:* 412 383-6534   *Fax:* 412 383-6365
*E-mail:* dhutches@pitt.edu

**Marywood University**
Dietetic Internship Prgm
Dept of Nutrition and Dietetics
2300 Adams Ave
Scranton, PA 18509-1514
www.marywood.edu/departments/nutr_diet/home.html
*Prgm Dir:* Maureen Dunne Touhey, MS RD LDN
*Tel:* 570 348-6211, Ext 2575   *Fax:* 570 340-6029
*E-mail:* dunnetouhey@es.marywood.edu

**Marywood University**
Dietetics-Didactic Prgm
Dept of Nutrition and Dietetics
2300 Adams Ave
Scranton, PA 18509-1514
www.marywood.edu/departments/nutr_diet/home.html
*Prgm Dir:* Kathleen H McKee, PhD RD
*Tel:* 570 348-6211, Ext 2632   *Fax:* 570 348-6029
*E-mail:* khmckee@marywood.edu

**Marywood University**
Dietetics-Coordinated Prgm
Dept of Nutrition and Dietetics
2300 Adams Ave
Scranton, PA 18509-1598
www.marywood.edu/departments/nutr_diet/home.html
*Prgm Dir:* Maureen Dunne-Touhey, MS RD LDN
*Tel:* 570 348-6211, Ext 2575   *Fax:* 570 348-6029
*E-mail:* dunnetouhey@es.marywood.edu

**Penn State University**
Dietetic Internship Prgm
Nutrition Dept Coll of Hlth & Human Dev
5126 Henderson Bldg
University Park, PA 16802
http://nutrition.hhdev.psu.edu/internship/index.html
*Prgm Dir:* Marie Y Kamp, MBA RD
*Tel:* 814 865-9150   *Fax:* 814 863-6103
*E-mail:* myk2@psu.edu

**Penn State University**
Dietetics-Didactic Prgm
Nutrition Dept
Coll of Hlth & Human Devel
University Park, PA 16802-6500
http://nutrition.hhdve.psu.edu/internship/index.html
*Prgm Dir:* Marie Y Kamp, MBA RD
*Tel:* 814 865-9150   *Fax:* 814 863-6103
*E-mail:* myk2@psu.edu

**ARAMARK Corporation Mid - Atlantic Region**
Dietetic Internship Prgm
41 Pebble Ridge Rd
Warrington, PA 18976
www.aramark.com/Careers.asp
*Prgm Dir:* Patricia M Richards, MA RD
*Tel:* 800 666-4272   *Fax:* 215 918-1607
*E-mail:* richards-pat@aramark.com

**Sodexho Marriott Services**
Dietetic Internship Prgm
Fuld Campus
Rte 532 & General Sullivan Rd
Washington Cross, PA 18977
www.dieteticintern.com/NJ-Phila
*Prgm Dir:* Gayanne R Wolset, MS RD CDE
*Tel:* 215 493-4025   *Fax:* 215 612-5302
*E-mail:* dieteticinternship@sodexhousa.com

**West Chester University**
Dietetics-Didactic Prgm
H302 Department of Health
Sturzebecker Health Sciences Center
West Chester, PA 19383
www.wcupa.edu
*Prgm Dir:* Jeffrey E Harris, DrPH MPH RD
*Tel:* 610 436-2655   *Fax:* 610 436-2860
*E-mail:* jharris@wcupa.edu

# Puerto Rico

**Puerto Rico Department of Health**
Dietetic Internship Prgm
PO Box 70184
San Juan, PR 00936
*Prgm Dir:* Ana E Rivera, MS RD
*Tel:* 787 274-6831   *Fax:* 787 274-8345
*E-mail:* aerivera@salud.gov.pr

**University of Puerto Rico**
Dietetic Internship Prgm
College of Health Related Professions
Med Sci Campus, PO Box 365067
San Juan, PR 00936-5067
http://cprsweb.rcm.upr.edu
*Prgm Dir:* Rita L De Delgado, MEd RD
*Tel:* 787 758-2525, Ext 4503   *Fax:* 787 756-8529
*E-mail:* ritalucca@cprs.rcm.upr.edu

**University of Puerto Rico**
Dietetics-Didactic Prgm
Box 23347 UPR Station
Rio Piedras Campus
San Juan, PR 00931-3347
*Prgm Dir:* Nayda I Neris, EdD RD
*Tel:* 787 764-0000, Ext 2630   *Fax:* 787 772-1422
*E-mail:* nineris@uprrp.edu

**Veterans Affairs Medical Center**
Dietetic Internship Prgm
10 Calle Casia
San Juan, PR 00921-3201
*Prgm Dir:* Awilda Ortiz, MPH RD LND
*Tel:* 787 641-2962, Ext 10333   *Fax:* 787 641-8366
*E-mail:* awilda.ortiz@med.va.gov

# Rhode Island

**University of Rhode Island**
Dietetic Internship Prgm
Food Science & Nutrition
106 Ranger Hall
Kingston, RI 02881
www.uri.edu/cels/fsn
*Prgm Dir:* Geoffrey W Greene, PhD RD
*Tel:* 401 874-4028   *Fax:* 401 874-5974
*E-mail:* gwg@uri.edu

**University of Rhode Island**
Dietetics-Didactic Prgm
Food Science & Nutrition
110 Ranger Hall
Kingston, RI 02881-0804
www.uri.edu/cels/fsn
*Prgm Dir:* Catherine English, PhD RD LDN
*Tel:* 401 874-5869   *Fax:* 401 874-5974
*E-mail:* cathy@uri.edu

**Johnson & Wales University**
Dietetics-Didactic Prgm
College of Culinary Arts
One Washington Ave
Providence, RI 02905
www.jwu.edu
*Prgm Dir:* Suzanne P Vieira, MS RD LDN
*Tel:* 401 598-1881   *Fax:* 401 598-1161
*E-mail:* suzanne.vieira@jwu.edu

# South Carolina

**Medical University of South Carolina**
Dietetic Internship Prgm
96 Jonathan Lucas St, Ste 219K
PO Box 250327
Charleston, SC 29425
www.musc.edu/dieteticinternship
*Prgm Dir:* Kelley L Martin, MPH RD
*Tel:* 843 792-1415   *Fax:* 843 792-4184
*E-mail:* martinkl@musc.edu

**Clemson University**
Dietetics-Didactic Prgm
Dept of Food Science
223 Poole, Agricultural Center
Clemson, SC 29634-0371
www.clemson.edu/foodscience
*Prgm Dir:* Mary E Kunkel, PhD RD
*Tel:* 864 656-5690   *Fax:* 864 656-0331
*E-mail:* bkunkel@clemson.edu

**SC Dept of Health & Environmental Control**
Dietetic Internship Prgm
Mills Complex
PO Box 101106
Columbia, SC 29211-0106
www.scdhec.gov
*Prgm Dir:* Sandra H Spann, MS RD
*Tel:* 803 898-0819   *Fax:* 803 898-0557
*E-mail:* spannsh@dhec.sc.gov

**South Carolina State University**
Dietetics-Didactic Prgm
Staley Hall PO Box 7657
300 College Ave
Orangeburg, SC 29117-0001
*Prgm Dir:* Kimberly A McClain, MS RD
*Tel:* 803 516-4590   *Fax:* 803 533-3628
*E-mail:* kmcclain@scsu.edu

**Winthrop University**
Dietetics-Didactic Prgm
Dept of Human Nutrition
302 Life Sciences Building
Rock Hill, SC 29733
www.winthrop.edu/nutrition
*Prgm Dir:* Christine H Goodner, MS RD
*Tel:* 803 323-4022   *Fax:* 803 323-2254
*E-mail:* goodnerc@winthrop.edu

**Winthrop University**
Dietetic Internship Prgm
Dept of Human Nutrition
302 Life Sciences Bldg
Rock Hill, SC 29733
www.winthrop.edu/nutrition/dietetic.htm
*Prgm Dir:* Judy H Thomas, MS RD
*Tel:* 803 323-4521   *Fax:* 803 323-2254
*E-mail:* thomasjh@winthrop.edu

# South Dakota

**South Dakota State University**
Dietetics-Didactic Prgm
Family and Consumer Sciences
PO Box 2275A
Brookings, SD 57007-0497
www3.sdstate.edu
*Prgm Dir:* Kendra K Kattelmann, PhD RD LN
*Tel:* 605 688-5161   *Fax:* 605 688-5603
*E-mail:* kendra_kattelmann@sdstate.edu

**University of South Dakota**
Dietetic Internship Prgm
School of Medicine
1400 W 22nd St
Sioux Falls, SD 57105
www.usd.edu/cd/dieteticinternship
*Prgm Dir:* Marcie C Kemnitz, MS RD LN
*Tel:* 605 357-1581   *Fax:* 605 357-1438
*E-mail:* mkemnitz@usd.edu

**Mt Marty College**
Dietetics-Didactic Prgm
Nutrition & Food Science
1105 W 8th St
Yankton, SD 57078-3724
www.mtmc.edu
*Prgm Dir:* Sr Thecla M Holzbauer, MS RD LN
*Tel:* 605 668-1520   *Fax:* 605 668-1607
*E-mail:* tholzbauer@mtmc.edu

**Mt Marty College**
Dietetics-Coordinated Prgm
Dept of Nutrition & Food Science
1105 W 8th St
Yankton, SD 57078-3724
www.mtmc.edu
*Prgm Dir:* Sr Thecla M Holzbauer, MS RD LN
*Tel:* 605 668-1520   *Fax:* 605 668-1607
*E-mail:* tholzbauer@mtmc.edu

# Tennessee

**University of Tennessee - Chattanooga**
Dietetics-Didactic Prgm
Dept of Human Ecology, #4204
615 McCallie Ave
Chattanooga, TN 37403
http://iwww.utc.edu/~hecodept/academic.html
*Prgm Dir:* Holly A Dieken, PhD RD LDN
*Tel:* 423 425-5379   *Fax:* 423 425-4479
*E-mail:* holly-dieken@utc.edu

**Tennessee Technological University**
Dietetics-Didactic Prgm
School of Home Economics
Box 5035
Cookeville, TN 38505
http://iweb.tntech.edu/abrunt
*Prgm Dir:* Cathy Hix Cunningham, PhD RD
*Tel:* 931 372-3376   *Fax:* 931 372-3150
*E-mail:* ccunningham@tntech.edu

**Carson - Newman College**
Dietetics-Didactic Prgm
PO Box 71881
2130 Branner Ave
Jefferson City, TN 37760-7001
www.cn.edu
*Prgm Dir:* Kitty R Coffey, PhD RD LDN CFCS
*Tel:* 865 471-3295   *Fax:* 865 471-3502
*E-mail:* Kcoffey@cn.edu

**East Tennessee State University**
Dietetic Internship Prgm
Applied Human Sciences-STAT
PO Box 70671
Johnson City, TN 37614-0671
www.etsu.edu/scitech/ahsc/ahsc.htm
*Prgm Dir:* Elizabeth C Lowe, MS RD LDN
*Tel:* 423 439-7537   *Fax:* 423 439-7539
*E-mail:* lowee@etsu.edu

**East Tennessee State University**
Dietetics-Didactic Prgm
Applied Human Sciences
PO Box 70671
Johnson City, TN 37614-0671
www.etsu.edu
*Prgm Dir:* Alison Schaefer, MS RD LDN
*Tel:* 423 439-7502   *Fax:* 423 439-7539
*E-mail:* schaefer@etsu.edu

**University of Tennessee - Knoxville**
Dietetic Internship Prgm
Human Ecology, Dept of Nutrition
1215 Cumberland Ave, Rm 229
Knoxville, TN 37996-1920
http://nutrition.utk.edu
*Prgm Dir:* Karen B Wetherall, MS RD LDN
*Tel:* 865 974-6256   *Fax:* 865 974-3491
*E-mail:* kbalnick@utk.edu

**University of Tennessee - Knoxville**
Dietetics-Didactic Prgm
Dept of Nutrition
1215 W Cumberland Ave
Knoxville, TN 37996-1920
*Prgm Dir:* Melissa B Hansen-Petrik, PhD RD
*Tel:* 865 974-6264   *Fax:* 865 974-3491
*E-mail:* phanse@utk.edu

**University of Tennessee - Martin**
Dietetic Internship Prgm
Human Environmental Sciences
340 Gooch Hall
Martin, TN 38238-5045
www.utm.edu/departments/agr/humenvir/ap4grad.htm
*Prgm Dir:* Lori H Littleton, MS RD
*Tel:* 731 587-7107   *Fax:* 731 587-7106
*E-mail:* loril@utm.edu

**University of Tennessee - Martin**
Dietetics-Didactic Prgm
Dept of Family & Consumer Sci
340 Gooch Hall
Martin, TN 38238-5045
www.utm.edu/departments/agr/humenvir/diet.htm
*Prgm Dir:* Georgina M Awipi, PhD RD
*Tel:* 731 587-7101   *Fax:* 731 587-7106
*E-mail:* gawipi@utm.edu

**University of Memphis**
Dietetics-Didactic Prgm
Consumer Science and Education
Fieldhouse 1611
Memphis, TN 38152
http://hss.memphis.edu
*Prgm Dir:* Robin R Roach, EdD RD
*Tel:* 901 678-3110   *Fax:* 901 678-5324
*E-mail:* rroach@memphis.edu

**University of Memphis**
Dietetic Internship Prgm
Dept of Health and Sport Science
161A Fieldhouse
Memphis, TN 38152
http://coe.memphis.edu/hss/MS-ClinicalNutrition.htm
*Prgm Dir:* Linda H Clemens, EdD RD
*Tel:* 901 678-3108   *Fax:* 901 678-5324
*E-mail:* lhclemns@memphis.edu

**Middle Tennessee State University**
Dietetics-Didactic Prgm
Dept of Human Sciences
PO Box 86
Murfreesboro, TN 37132
www.mtsu.edu
*Prgm Dir:* Lisa M Sheehan-Smith, EdD RD LDN
*Tel:* 615 898-2090   *Fax:* 615 898-5130
*E-mail:* lsheehan@mtsu.edu

**National HealthCare LP**
Dietetic Internship Prgm
PO Box 1398
Murfreesboro, TN 37133-1398
www.nhccare.com/di.htm
*Prgm Dir:* Patty T Poe, MEd RD LDN
*Tel:* 615 890-2020, Ext 1830   *Fax:* 615 890-0123
*E-mail:* ppoe@nhccare.com

**Lipscomb University**
Dietetic Internship Prgm
3901 Granny White Pike
Nashville, TN 37204-3951
www.lipscomb.edu
*Prgm Dir:* Autumn C Marshall, PhD RD
*Tel:* 615 279-6106   *Fax:* 615 269-1808
*E-mail:* autumn.marshall@lipscomb.edu

**Lipscomb University**
Dietetics-Didactic Prgm
Family and Consumer Sciences
3901 Granny White Pike
Nashville, TN 37204-3951
www.lipscomb.edu
*Prgm Dir:* Nancy H Hunt, MS MEd RD
*Tel:* 615 279-5767   *Fax:* 615 269-1808
*E-mail:* nancy.hunt@lipscomb.edu

**Tennessee State University**
Dietetics-Didactic Prgm
Family Consumer Sciences PO Box 9538
3500 John A Merritt Blvd
Nashville, TN 37209-1561
www.tnstate.edu/sacs
*Prgm Dir:* Sandria L Godwin, PhD RD LD
*Tel:* 615 963-5619   *Fax:* 615 963-5033
*E-mail:* sgodwin@tnstate.edu

**Vanderbilt University Medical Center**
Dietetic Internship Prgm
B-802TVC
1301 22nd Ave S
Nashville, TN 37232-5510
www.mc.vanderbilt.edu/alliedhealth
*Prgm Dir:* Cynthia Broadhurst Facemire, MS RD LDN
*Tel:* 615 322-0062   *Fax:* 615 343-8810
*E-mail:* cynthia.facemire@mcmail.vanderbilt.edu

# Texas

**Abilene Christian University**
Dietetics-Didactic Prgm
Family and Consumer Sciences
ACU Box 28155
Abilene, TX 79699
www.acu.edu/exercisescience
*Prgm Dir:* Sheila A Jones, MS RD
*Tel:* 915 674-2089   *Fax:* 915 674-2086
*E-mail:* joness@acu.edu

**Texas Department of Health**
Dietetic Internship Prgm
Bureau of Nutrition Services
1100 W 49th St
Austin, TX 78756
www.tdh.state.tx.us/wichd/nut/intern-intro.htm
*Prgm Dir:* Sherry Clark, MPH RD LD
*Tel:* 512 458-7111, Ext 2142   *Fax:* 512 458-7446
*E-mail:* sherry.clark@dshs.tx.us

**University of Texas at Austin**
Dietetics-Coordinated Prgm
Dept of Human Ecology
1 University Station, A2700
Austin, TX 78712-1097
www.utexas.edu/depts/he/ntr/cpd1.htm
*Prgm Dir:* Katherine Y Southworth, MS RD LD
*Tel:* 512 471-0637   *Fax:* 512 471-5630
*E-mail:* ksworth@mail.utexas.edu

**University of Texas at Austin**
Dietetics-Didactic Prgm
Dept of Human Ecology
1 University Station, A2700
Austin, TX 78712
www.utexas.edu/depts/he
*Prgm Dir:* Jane F Tillman, MS RD
*Tel:* 512 471-7639   *Fax:* 512 471-5630
*E-mail:* mljt@mail.utexas.edu

**Lamar Institute of Technology**
Dietetics-Didactic Prgm
Family and Consumer Sciences
PO Box 10035
Beaumont, TX 77710-0035
www.lamar.edu
*Prgm Dir:* Connie S Ruiz, PhD RD LD
*Tel:* 409 880-8663    *Fax:* 409 880-8666
*E-mail:* ruizcs@hal.lamar.edu

**Lamar Institute of Technology**
Dietetic Internship Prgm
Family and Consumer Sciences
PO Box 10035
Beaumont, TX 77710-0035
www.lamar.edu
*Prgm Dir:* Amy R Shows, PhD RD LD
*Tel:* 409 880-8667    *Fax:* 409 880-8666
*E-mail:* arshows@my.lamar.edu

**Texas A&M University**
Dietetic Internship Prgm
Dept of Animal Science
2253 TAMU
College Station, TX 77843-2253
http://dieteticinternship.tamu.edu
*Prgm Dir:* Linda J Talley, PhD RD RS
*Tel:* 979 458-4642    *Fax:* 979 862-2378
*E-mail:* ljtalley@tamu.edu

**Texas A&M University**
Dietetics-Didactic Prgm
Human Nutrition Section
2253 TAMU
College Station, TX 77843-2471
http://nfs.tamu.edu
*Prgm Dir:* Karen A Beathard, MS RD
*Tel:* 979 845-2142    *Fax:* 979 862-2378
*E-mail:* karen-beathard@tamu.edu

**Baylor University Med Center**
Dietetic Internship Prgm
3500 Gaston Ave
Dallas, TX 75246-2045
www.baylordallas.com
*Prgm Dir:* Julie A Grim, MPH RD
*Tel:* 214 820-4019    *Fax:* 214 820-2263
*E-mail:* juliegr@baylorhealth.edu

**Presbyterian Hospital of Dallas**
Dietetic Internship Prgm
8200 Walnut Hill Ln
Dallas, TX 75231-4402
www.phscare.org/phd/dietetics
*Prgm Dir:* Kristi Wade, MS RD CSP LD
*Tel:* 214 345-7558    *Fax:* 214 345-8473
*E-mail:* kristiwade@texashealth.org

**Univ of Texas Southwestern Med Ctr at Dallas**
Dietetics-Coordinated Prgm
Dept of Clinical Nutrition
5323 Harry Hines Blvd
Dallas, TX 75390-8877
www.utsouthwestern.edu/clinnut
*Prgm Dir:* Bernadette B Latson, MS RD
*Tel:* 214 648-1520    *Fax:* 214 648-1514
*E-mail:* clindiet@utsouthwestern.edu

**Texas Woman's University**
Dietetics-Didactic Prgm
Nutrition & Food Sciences
304 Administration Dr, Clock Tower
Denton, TX 76201
www.twu.edu/hs/nfs
*Prgm Dir:* Victorine L Imrhan, PhD RD
*Tel:* 940 898-2650    *Fax:* 940 898-2634
*E-mail:* vimrhan@mail.twu.edu

**Texas Woman's University**
Dietetic Internship Prgm
Nutrition & Food Sciences
PO Box 425888
Denton, TX 76204-5888
www.twu.edu/hs/nfs/intern.htm
*Prgm Dir:* Martha L Rew, MS RD
*Tel:* 940 898-2657    *Fax:* 940 898-2634
*E-mail:* mrew@twu.edu

**Univ of Texas - Pan American**
Dietetics-Coordinated Prgm
Health & Human Services
1201 W University Dr
Edinburg, TX 78539-2909
www.panam.edu
*Prgm Dir:* Bahram Faraji, DrPH RD LD
*Tel:* 956 316-7083    *Fax:* 956 318-5265
*E-mail:* bfaraji@utpa.edu

**US Military Dietetic Internship Consortium**
Dietetic Internship Prgm
Nutrition Care Div
3851 Roger Brooke Dr
Ft Sam Houston, TX 78234-6200
www.amsc.amedd.army.mil/training.asp
*Prgm Dir:* Debra R Hernandez, MS RD LD
*Tel:* 210 916-4332    *Fax:* 210 916-4152
*E-mail:* debra.r.hernandez@amedd.army.mil

**Texas Christian University**
Dietetics-Didactic Prgm
Dept of Nutrition & Dietetics
TCU Box 298600
Ft Worth, TX 76129
www.tcu.edu
*Prgm Dir:* Marlyn Dart, PhD RD LD
*Tel:* 817 257-7309    *Fax:* 817 257-5849
*E-mail:* l.dart@tcu.edu

**Texas Christian University**
Dietetics-Coordinated Prgm
Dept of Nutrition & Dietetics
TCU Box 298600
Ft Worth, TX 76129
www.tcu.edu
*Prgm Dir:* Anne D VanBeber, PhD RD
*Tel:* 817 257-7309    *Fax:* 817 257-5849
*E-mail:* a.vanbeber@tcu.edu

**Houston Veterans Affairs Medical Center**
Dietetic Internship Prgm
2002 Holcombe Blvd
Houston, TX 77030-4298
www.va.gov/nfs/HoustonVAMC
*Prgm Dir:* Caroline Nelson, MS RD CNSD
*Tel:* 713 791-1414    *Fax:* 713 794-7448
*E-mail:* caroline.nelson@va.gov

**Texas Southern University**
Dietetics-Didactic Prgm
Human Svcs and Consumer Sci
3100 Cleburne Ave
Houston, TX 77004-4575
www.tsu.edu
*Prgm Dir:* Annie M Morgan, MS RD LD
*Tel:* 713 313-7637    *Fax:* 713 313-7228
*E-mail:* morgan_am@tsu.edu

**Texas Woman's University**
Dietetic Internship Prgm
1130 John Freeman Blvd
Houston, TX 77030-2897
www.twu.edu/hs/nfs/hintern.htm
*Prgm Dir:* Rose M Bush, MS RD
*Tel:* 713 794-2371    *Fax:* 713 794-2374
*E-mail:* rbush@mail.twu.edu

**Univ of Texas Hlth Sci Ctr at Houston**
Dietetic Internship Prgm
Sch of Public Hlth, E619 RAS Bldg
1200 Herman Pressler St
Houston, TX 77030
www.sph.uth.tmc.edu
*Prgm Dir:* Jeanne B Martin, PhD RD LD
*Tel:* 713 500-9347    *Fax:* 713 500-9329
*E-mail:* jeanne.b.martin@uth.tmc.edu

**University of Houston**
Dietetic Internship Prgm
Human Dvlpmnt and Consumer Sci
4800 Calhoun Rd
Houston, TX 77204-6020
www.hhp.uh.edu/internship
*Prgm Dir:* Sharon K Bode, PhD RD
*Tel:* 713 743-4112    *Fax:* 713 743-4033
*E-mail:* dietintern@uh.edu

**University of Houston**
Dietetics-Didactic Prgm
Human Dvlpmnt and Consumer Sci
4800 Calhoun Rd
Houston, TX 77204-6861
http://hhp.uh.edu/nutrition
*Prgm Dir:* Sharon K Bode, PhD RD
*Tel:* 713 743-4112    *Fax:* 713 743-4033
*E-mail:* sbode@uh.edu

**Sam Houston State University**
Dietetics-Didactic Prgm
Food Science & Nutrition
SHSU Box 2177
Huntsville, TX 77341-2177
www.shsu.edu/~hec_www
*Prgm Dir:* Zaheer Kirmani, PhD RD LD
*Tel:* 936 294-1245    *Fax:* 936 294-4204
*E-mail:* hec_zak@shsu.edu

**Sam Houston State University**
Dietetic Internship Prgm
PO Box 2177
1700 Sam Houston Ave
Huntsville, TX 77341-2177
www.shsu.edu/~hec_www
*Prgm Dir:* Claudia V Sealey-Potts, PhD RD
*Tel:* 936 294-1250    *Fax:* 936 294-4204
*E-mail:* hec_cvs@shsu.edu

**Texas A&M University - Kingsville**
Dietetics-Didactic Prgm
Dept of Human Sciences
MSC 168, 700 University Blvd
Kingsville, TX 78363
www.tamuk.edu/aghs/hsci
*Prgm Dir:* Farzad Deyhim, PhD RD
*Tel:* 361 593-2230    *Fax:* 361 593-2230
*E-mail:* farzad.deyhim@tamuk.edu

**Texas A&M University - Kingsville**
Dietetic Internship Prgm
Dept of Human Sciences
MSC 168, 700 University Blvd
Kingsville, TX 78363-8202
www.tamuk.edu/aghs/hsci/brochure.html
*Prgm Dir:* Rena K Quinton, PhD RD
*Tel:* 361 593-2205    *Fax:* 361 593-2230
*E-mail:* kagg000@tamuk.edu

**US Military Dietetic Internship Consortium**
Dietetic Internship Prgm
Wilford Hall Medical Center
959 MDTS/MTN, 2200 Bergquist Dr, Ste 1
Lackland AFB, TX 78236-5300
*Prgm Dir:* Mari L Chamberlain, MS RD
*Tel:* 210 292-5737    *Fax:* 210 292-7826
*E-mail:* mari.chamberlain@lackland.af.mil

**Texas Tech University**
Dietetics-Didactic Prgm
Educ Nutri & Rest Hotel Mgmt
PO Box 41162, 15th and Akron St
Lubbock, TX 79409-1162
www.hs.ttu.edu
*Prgm Dir:* Debra B Reed, PhD RD LD
*Tel:* 806 742-3068   *Fax:* 806 742-3042
*E-mail:* debra.reed@ttu.edu

**Texas Tech University**
Dietetic Internship Prgm
Dept of Educ, Nutr, Rest, and Hotel Mgmt
PO Box 41162, 15th and Akron St
Lubbock, TX 79409-1162
www.hs.ttu.edu/intern
*Prgm Dir:* Carmen R Roman-Shriver, PhD RD
*Tel:* 806 742-3068   *Fax:* 806 742-3042
*E-mail:* c.roman-shriver@ttu.edu

**Stephen F Austin State University**
Dietetics-Didactic Prgm
Dept of Human Sciences
SFA Station 13014
Nacogdoches, TX 75962-3014
www.sfasu.edu/sweems
*Prgm Dir:* Darla R Daniel, PhD RD
*Tel:* 936 468-2439   *Fax:* 936 468-2140
*E-mail:* ddaniel@sfasu.edu

**Stephen F Austin State University**
Dietetic Internship Prgm
Dept of Human Sciences
SFA Station PO Box 13014
Nacogdoches, TX 75962-3014
www.sfasu.edu/di
*Prgm Dir:* Brenda G Marques, PhD RD
*Tel:* 936 468-4502   *Fax:* 936 468-2140
*E-mail:* bmarques@sfasu.edu

**Prairie View A&M University**
Dietetic Internship Prgm
Box 4329
Prairie View, TX 77446
www.pvamu.edu
*Prgm Dir:* Sharon McWhinney, PhD RD LD
*Tel:* 936 857-4417   *Fax:* 936 857-2200
*E-mail:* sharon_mcwhinney@pvamu.edu

**Prairie View A&M University**
Dietetics-Didactic Prgm
Dept of Human Sciences
PO Box 4329
Prairie View, TX 77446-4329
www.pvamu.edu
*Prgm Dir:* Sharon McWhinney, PhD RD LD
*Tel:* 936 857-4417   *Fax:* 936 857-4441
*E-mail:* sharon_mcwhinney@pvamu.edu

**University of the Incarnate Word**
Dietetic Internship Prgm
4301 Broadway Box 311
San Antonio, TX 78209
www.uiw.edu
*Prgm Dir:* Joseph C Bonilla, PhD RD
*Tel:* 210 829-3152   *Fax:* 210 829-3153
*E-mail:* josephb@universe.uiwtx.edu

**University of the Incarnate Word**
Dietetics-Didactic Prgm
4301 Broadway St
San Antonio, TX 78209-6318
www.uiw.edu
*Prgm Dir:* Joseph C Bonilla, PhD RD
*Tel:* 210 829-3167   *Fax:* 210 829-3153
*E-mail:* josephb@universe.uiwtx.edu

**Texas State University-San Marcos**
Dietetic Internship Prgm
Family and Consumer Sciences
601 University Dr
San Marcos, TX 78666-4616
www.fcs.txstate.edu/diet_intern.htm
*Prgm Dir:* B J Friedman, PhD RD LD
*Tel:* 512 245-2155   *Fax:* 512 245-3829
*E-mail:* bf04@txstate.edu

**Texas State University-San Marcos**
Dietetics-Didactic Prgm
Family and Consumer Sciences
601 University Dr
San Marcos, TX 78666-4616
www.fcs.txstate.edu/nf.htm
*Prgm Dir:* Sylvia L Crixell, PhD RD LD
*Tel:* 512 245-2482   *Fax:* 512 245-3829
*E-mail:* scrixell@txstate.edu

**Tarleton State University**
Dietetics-Didactic Prgm
Dept of Human Sciences
Box T-0380
Stephenville, TX 76402
www.tarleton.edu
*Prgm Dir:* Linda J Talley, PhD RD
*Tel:* 254 968-9196   *Fax:* 254 968-9728
*E-mail:* ltalley@tarleton.edu

**Baylor University**
Dietetics-Didactic Prgm
Family and Consumer Sciences
BU Box 97346
Waco, TX 76798-7346
*Prgm Dir:* LuAnn Soliah, PhD RD LD
*Tel:* 254 710-6258   *Fax:* 254 710-3629
*E-mail:* luann_soliah@baylor.edu

# Utah

**Utah State University**
Dietetics-Didactic Prgm
8700 Old Main Hill
Logan, UT 84322-8700
www.usu.edu/~dietetic
*Prgm Dir:* Noreen B Schvaneveldt, MS RD CD
*Tel:* 435 797-2105   *Fax:* 435 797-2379
*E-mail:* noreens@cc.usu.edu

**Utah State University**
Dietetics-Coordinated Prgm
Nutrition and Food Science
Logan, UT 84322-8700
www.usu.edu/~dietetic
*Prgm Dir:* Noreen B Schvaneveldt, MS RD CD
*Tel:* 435 797-2105   *Fax:* 435 797-2379
*E-mail:* noreens@cc.usu.edu

**Utah State University - Salt Lake**
Dietetic Internship Prgm
5250 S Commerce Dr Ste 300
Murray, UT 84107
http://extension.usu.edu/intern
*Prgm Dir:* Ann Martin Mildenhall, MS RD
*Tel:* 801 269-9422, Ext 305   *Fax:* 801 266-7907
*E-mail:* annmil@ext.usu.edu

**Brigham Young University**
Dietetics-Didactic Prgm
Food Science and Nutrition
S219 ESC, PO Box 24620
Provo, UT 84602-4620
http://ndfs.byu.edu
*Prgm Dir:* Nora Nyland, PhD RD CD
*Tel:* 801 422-6676   *Fax:* 801 422-8714
*E-mail:* nora_nyland@byu.edu

**Brigham Young University**
Dietetic Internship Prgm
Food Science and Nutrition
S219 ESC, PO Box 24620
Provo, UT 84602-4620
http://ndfs.byu.edu
*Prgm Dir:* Nora Nyland, PhD RD CD
*Tel:* 801 422-6676   *Fax:* 801 422-8714
*E-mail:* nora_nyland@byu.edu

**University of Utah**
Dietetics-Coordinated Prgm
Div of Foods & Nutrition
250 S 1850 E #214
Salt Lake City, UT 84112
www.health.utah.edu/fdnu
*Prgm Dir:* Julie M Metos, MPH RD CD
*Tel:* 801 585-3024   *Fax:* 801 585-3218
*E-mail:* julie.metos@hsc.utah.edu

# Vermont

**University of Vermont**
Dietetics-Didactic Prgm
Dept of Nutritional Sciences
Terrill Hall
Burlington, VT 05405
http://nutrition.uvm.edu
*Prgm Dir:* Jane K Ross, PhD RD
*Tel:* 802 656-0539   *Fax:* 802 656-0407
*E-mail:* jross@uvm.edu

# Virginia

**Virginia Polytechnic Inst & State Univ**
Dietetics-Didactic Prgm
Human Nutrition Foods & Exercise
College of Agricultur and Life Sciences
Blacksburg, VA 24061-0430
www.hnfe.vt.edu
*Prgm Dir:* Susan F Clark, PhD RD
*Tel:* 540 231-8768   *Fax:* 540 231-3916
*E-mail:* afclark@vt.edu

**Virginia Polytechnic Inst & State Univ**
Dietetic Internship Prgm
338 Wallace Hall
Blacksburg, VA 24061-0430
www.hnfe.vt.edu
*Prgm Dir:* Carol Papillon, MPH RD
*Tel:* 540 231-8769   *Fax:* 540 231-3916
*E-mail:* cpapillo@vt.edu

**University of Virginia Health System**
Dietetic Internship Prgm
Box 800673
Charlottesville, VA 22908
http://hsc.virginia.edu/internet/dietetics
*Prgm Dir:* Ana Abad Sinden, MS RD CNSD
*Tel:* 434 924-2286   *Fax:* 434 982-3957
*E-mail:* ara6t@virginia.edu

**James Madison University**
Dietetics-Didactic Prgm
Dept of Health Sciences MSC 4301
HHS 3140
Harrisonburg, VA 22807
www.jmu.edu
*Prgm Dir:* Janet W Gloeckner, PhD RD
*Tel:* 540 568-7084   *Fax:* 540 568-3336
*E-mail:* gloeckjw@jmu.edu

**James Madison University**
Dietetic Internship Prgm
MSC 4301
800 S Main St
Harrisonburg, VA 22807
www.healthsci.jmu.edu/dietetics/graduate.htm
*Prgm Dir:* Cynthia P Cadieux, PhD RD
*Tel:* 540 568-3816   *Fax:* 540 568-3336
*E-mail:* cadieucp@jmu.edu

**Norfolk State University**
Dietetics-Didactic Prgm
Food Sci & Nutri/Chem
700 Park Ave
Norfolk, VA 23504-3992
www.nsu.edu/alliedhealth/foodscience
*Prgm Dir:* Jill E Comess, MS RD
*Tel:* 757 823-8216  *Fax:* 757 823-2909
*E-mail:* jcomess@nsu.edu

**Virginia State University**
Dietetics-Didactic Prgm
Dept of Human Ecology, Box 9211
Petersburg, VA 23806-0001
www.vsu.edu
*Prgm Dir:* Paula F Inserra, PhD RD
*Tel:* 804 524-5729  *Fax:* 804 524-5048
*E-mail:* pinserra@vsu.edu

**Virginia State University**
Dietetic Internship Prgm
PO Box 9211
Petersburg, VA 23806
www.vsu.edu
*Prgm Dir:* Paula F Inserra, PhD RD
*Tel:* 804 524-5729  *Fax:* 804 524-5048
*E-mail:* pinserra@vsu.edu

**Radford University**
Dietetics-Didactic Prgm
Dept of Health Services
PO Box 6962
Radford, VA 24142-5826
www.radford.edu/~fdsn-web
*Prgm Dir:* Mary J Miller, MS RD
*Tel:* 540 831-7680  *Fax:* 540 831-7736
*E-mail:* m-miller@radford.edu

**Radford University**
Dietetic Internship Prgm
Foods and Nutrition Prgm
PO Box 6962
Radford, VA 24142-6962
*Prgm Dir:* Susan F Clark, PhD RD
*Tel:* 540 831-7679  *Fax:* 540 831-7736
*E-mail:* sfclark@radford.edu

**Virginia Commonwealth Univ/Med Coll of VA**
Dietetic Internship Prgm
PO Box 980294
Richmond, VA 23298-0294
http://views.vcu.edu/dietetic
*Prgm Dir:* Ann E Robbins, MS RD
*Tel:* 804 828-9108  *Fax:* 804 628-0921
*E-mail:* aerobbin@hsc.vcu.edu

**Virginia Department of Health**
Dietetic Internship Prgm
Div of WIC and Community Nutrition Services
109 Governor St, 9th Fl
Richmond, VA 23219
*Prgm Dir:* Kathleen M Sergent, MA RD
*Tel:* 804 864-7832  *Fax:* 804 371-6162
*E-mail:* Kathleen.Sergent@vdh.virginia.gov

# Washington

**Central Washington University**
Dietetics-Didactic Prgm
Family and Consumer Sciences
400 East University Way
Ellensburg, WA 98926-7565
www.cwu.edu
*Prgm Dir:* Ethan A Bergman, PhD RD FADA
*Tel:* 509 963-2366  *Fax:* 509 963-2787
*E-mail:* bergmane@cwu.edu

**Central Washington University**
Dietetic Internship Prgm
Family and Consumer Sciences
400 E University Way
Ellensburg, WA 98929-7565
www.cwu.edu/~fandcs/internship.html
*Prgm Dir:* Linda R Cashman, MS RD CD
*Tel:* 509 963-2786  *Fax:* 509 963-2787
*E-mail:* cashmanl@cwu.edu

**Bastyr University**
Dietetic Internship Prgm
Nutrition Program
14500 Juanita Dr NE
Kenmore, WA 98028-4966
*Prgm Dir:* Debra A Boutin, MS RD
*Tel:* 425 602-3124  *Fax:* 425 823-6222
*E-mail:* dboutin@bastyr.edu

**Bastyr University**
Dietetics-Didactic Prgm
14500 Juanita Dr NE
Kenmore, WA 98028-4966
www.bastyr.edu
*Prgm Dir:* Doris M Piccinin, MS RD CDE
*Tel:* 425 602-3099  *Fax:* 425 823-6222
*E-mail:* dpiccinin@bastyr.edu

**Washington State University**
Dietetics-Coordinated Prgm
106 FSHN Bldg
PO Box 646376
Pullman, WA 99164-6376
http://fshn.wsu.edu
*Prgm Dir:* Susan J Scheunemann, MS RD
*Tel:* 253 968-1425  *Fax:* 509 335-4815
*E-mail:* susan.scheunemann@puyallup.wsu.edu

**Washington State University**
Dietetics-Didactic Prgm
Food Sci & Human Nutrition
FSHN 120, PO Box 646376
Pullman, WA 99164-6376
http://fshn.wsu.edu
*Prgm Dir:* Miriam S Edlefsen, PhD RD
*Tel:* 509 335-1395  *Fax:* 509 335-4815
*E-mail:* medlefsen@wsu.edu

**Sea Mar Community Health Center**
Dietetic Internship Prgm
8915 14th Ave S
Seattle, WA 98108-4807
www.seamar.org
*Prgm Dir:* Rana H Nelson Blouch, MS RD
*Tel:* 206 762-0876, Ext 6311  *Fax:* 206 763-1856
*E-mail:* rananelson@seamarchc.org

**Seattle Pacific University**
Dietetics-Didactic Prgm
Family and Consumer Sciences
3307 3rd Ave W
Seattle, WA 98119-1997
www.spu.edu/depts/fcs
*Prgm Dir:* Gaile L Moe, PhD RD CD
*Tel:* 206 281-2238  *Fax:* 206 281-2035
*E-mail:* gmoe@spu.edu

**University of Washington**
Dietetic Internship Prgm
Box 353410
305 Raitt Hall
Seattle, WA 98195
http://depts.washington.edu/nutr
*Prgm Dir:* Louise W Peck, MPH RD
*Tel:* 206 543-2501  *Fax:* 206 685-1698
*E-mail:* lpeck@u.washington.edu

**University of Washington**
Dietetics-Didactic Prgm
305 Raitt Hall, Box 353410
Seattle, WA 98195-3410
http://depts.washington.edu/nutrdpd/DPDHome.html
*Prgm Dir:* Barbara A Bruemmer, PhD RD
*Tel:* 206 616-7362  *Fax:* 206 685-1696
*E-mail:* bbruemme@u.washington.edu

**Washington State University**
Dietetic Internship Prgm
310 N Riverpoint Blvd Box M
Spokane, WA 99202-1675
www.wsu.dietetic-internship.spokane.wsu.edu
*Prgm Dir:* Janet K Beary, PhD RD CHES
*Tel:* 509 358-7562  *Fax:* 509 358-7627
*E-mail:* beary@wsu.edu

# West Virginia

**Marshall University**
Dietetics-Didactic Prgm
One John Marshall Dr
Huntington, WV 25755-9521
www.marshall.edu/dietetics.htm
*Prgm Dir:* Mary K Gould, MS RD LD
*Tel:* 304 696-3364  *Fax:* 304 696-3177
*E-mail:* gouldm@marshall.edu

**Marshall University**
Dietetic Internship Prgm
Corbly Hall 203 Family & Consumer Sci
One John Marshall Dr
Huntington, WV 25755-9521
www.marshall.edu/dietetics/internship.htm
*Prgm Dir:* Kelli Williams, MA RD
*Tel:* 304 696-4336  *Fax:* 304 696-3177
*E-mail:* williamsk@marshall.edu

**West Virginia University**
Dietetics-Didactic Prgm
Coll of Agriculture and Forestry
PO Box 6108
Morgantown, WV 26506-6124
www.caf.wvu.edu
*Prgm Dir:* Betty J Forbes, MA RD LD
*Tel:* 304 293-3402, Ext 1761  *Fax:* 304 293-2750
*E-mail:* betty.forbes@mail.wvu.edu

**West Virginia University**
Dietetic Internship Prgm
Coll of Agriculture and Forestry
PO Box 6108
Morgantown, WV 26506-6124
www.caf.wvu.edu
*Prgm Dir:* Betty J Forbes, MA RD LD
*Tel:* 304 293-3402, Ext 1761  *Fax:* 304 293-2750
*E-mail:* betty.forbes@mail.wvu.edu

**West Virginia University Hospitals**
Dietetic Internship Prgm
Dept of Nutrition & Environmental Serv
Medical Center Dr, PO Box 8016
Morgantown, WV 26506-8016
www.hsc.wvu.edu/di
*Prgm Dir:* Jill L Johnston, MS RD LD
*Tel:* 304 598-4105, Ext 3339  *Fax:* 304 598-4119
*E-mail:* johnstonj@r.wvuh.edu

# Wisconsin

**University of Wisconsin - Green Bay**
Dietetics-Didactic Prgm
Human Biology Dept LS 423
2420 Nicolet Dr
Green Bay, WI 54311-7001
www.uwgb.edu/humbio/program/Majors.htm
*Prgm Dir:* Karen Lacey, MS RD CD
*Tel:* 920 465-2332  *Fax:* 920 465-2769
*E-mail:* laceyk@uwgb.edu

**University of Wisconsin - Green Bay**
Dietetic Internship Prgm
2420 Nicolet Dr
Human Biology Dept ES 301
Green Bay, WI 54311-7001
www.uwgb.edu/humbio/dietetics/Index.htm
*Prgm Dir:* Karen Lacey, MS RD CD
*Tel:* 920 465-2332   *Fax:* 920 465-2769
*E-mail:* laceyk@uwgb.edu

**Viterbo College**
Dietetic Internship Prgm
Nutrition and Dietetics Dept
900 Viterbo Dr
La Crosse, WI 54601-4797
www.viterbo.edu/academic/gr/dietintern
*Prgm Dir:* Karen Gibson, MS RD
*Tel:* 608 796-3662   *Fax:* 608 796-3668
*E-mail:* kmgibson@viterbo.edu

**Viterbo College**
Dietetics-Coordinated Prgm
Nutrition and Dietetics Dept
900 Viterbo Dr
La Crosse, WI 54601-4797
www.viterbo.edu/academic/ug/sls/majors/dietetics
*Prgm Dir:* Karen Gibson, MS RD
*Tel:* 608 796-3662   *Fax:* 608 796-3668
*E-mail:* kmgibson@viterbo.edu

**University of Wisconsin - Madison**
Dietetics-Coordinated Prgm
Dept of Nutritional Sciences
1415 Linden Dr
Madison, WI 53706-1571
http://wiscinfo.doit.wisc.edu/nutrisci
*Prgm Dir:* Lynette M Karls, MS RD
*Tel:* 608 262-5847   *Fax:* 608 262-5860
*E-mail:* karls@nutrisci.wisc.edu

**University of Wisconsin - Madison**
Dietetics-Didactic Prgm
Dept of Nutritional Sciences
1415 Linden Dr
Madison, WI 53706-1571
http://nutrisci.wisc.edu
*Prgm Dir:* Denise M Ney, PhD RD
*Tel:* 608 262-2727   *Fax:* 608 262-5860
*E-mail:* ney@nutrisci.wisc.edu

**University of Wisconsin Hospital and Clinics**
Dietetic Internship Prgm
Food and Nutrition Services F4/120
600 Highland Ave
Madison, WI 53792-1510
www.uwhealth.org
*Prgm Dir:* Marjorie U Morgan, MS RD
*Tel:* 608 263-8237   *Fax:* 608 263-1636
*E-mail:* me.morgan@hosp.wisc.edu

**University of Wisconsin - Stout**
Dietetics-Didactic Prgm
College of Human Development
415 E 10th Ave
Menomonie, WI 54751-0790
www.uwstout.edu/programs/bsd
*Prgm Dir:* Charlene E Schmidt, PhD RD CD
*Tel:* 715 232-1994   *Fax:* 715 232-2317
*E-mail:* schmidtcha@uwstout.edu

**University of Wisconsin - Stout**
Dietetic Internship Prgm
Dept of Food & Nutrition
222 Home Economics Bldg
Menomonie, WI 54751
www.uwstout.edu/programs/msfns/intern
*Prgm Dir:* Karen L Ostenso McDaniel, MS RD
*Tel:* 715 232-2394   *Fax:* 715 232-2317
*E-mail:* ostensomcdanielk@uwstout.edu

**Mount Mary College**
Dietetics-Coordinated Prgm
Dept of Dietetics
2900 N Menomonee River Pkwy
Milwaukee, WI 53222-4597
www.mtmary.edu/dept_dietetics.htm
*Prgm Dir:* Janet A Fischer, MS RD
*Tel:* 414 256-1216, Ext 359   *Fax:* 414 256-1224
*E-mail:* fischej@mtmary.edu

**Mount Mary College**
Dietetic Internship Prgm
Graduate Program in Dietetics
2900 N Menomonee River Pkwy
Milwaukee, WI 53222-4597
www.mtmary.edu/dietetics.htm
*Prgm Dir:* Lisa Stark, MPH MS RD CD
*Tel:* 414 256-1216   *Fax:* 414 256-1224
*E-mail:* starkl@mtmary.edu

**University of Wisconsin - Stevens Point**
Dietetics-Didactic Prgm
Health Promotion & Human Development
202 CPS
Stevens Point, WI 54481
www.uwsp.edu/hphd/academics
*Prgm Dir:* Jayne S Steinmetz, PhD RD CD
*Tel:* 715 346-4087   *Fax:* 715 346-3751
*E-mail:* jsteinme@uwsp.edu

# Wyoming

**University of Wyoming**
Dietetics-Didactic Prgm
Dept of Family & Consumer Sciences
1000 E University Ave
Laramie, WY 82071-3354
www.uwyo.edu/family
*Prgm Dir:* Rhoda M Schantz, PhD RD
*Tel:* 307 766-4145   *Fax:* 307 766-5686
*E-mail:* schantz@uwyo.edu

## Dietetic Technician

# Arizona

**Central Arizona College**
Dietetic Technician-AD Prgm
8470 N Overfield Rd
Coolidge, AZ 85228-9030
www.centralaz.edu
*Prgm Dir:* Glenna McCollum, MPH RD
*Tel:* 800 465-1016   *Fax:* 520 426-4476
*E-mail:* glenna_mccollum@centralaz.edu

# Arkansas

**Black River Technical College**
Dietetic Technician-AD Prgm
PO Box 468
Pocahontas, AR 72455
www.blackrivertech.org
*Prgm Dir:* Angela C Caldwell, MS RD LD
*Tel:* 870 248-4000, Ext 4120   *Fax:* 870 248-4100
*E-mail:* angelac@blackrivertech.org

# California

**Orange Coast College**
Dietetic Technician-AD Prgm
2701 Fairview Rd
Costa Mesa, CA 92628-0120
*Prgm Dir:* Eleanor B Huang, MS RD
*Tel:* 714 432-5835, Ext 5   *Fax:* 714 432-5609
*E-mail:* ehuang@mail.occ.cccd.edu

**Loma Linda University**
Dietetic Technician-AD Prgm
Nutrition and Dietetics
School of Allied Hlth Profs
Loma Linda, CA 92350
www.llu.edu/llu/nutrition
*Prgm Dir:* Georgia Hodgkin, EdD RD
*Tel:* 909 558-4593   *Fax:* 909 558-4291
*E-mail:* ghodgkin@llu.edu

**Long Beach City College**
Dietetic Technician-AD Prgm
Fam and Consumer Studies Div
Liberal Arts Campus
Long Beach, CA 90808-1706
*Prgm Dir:* Linda Allen Huy, EdD RD
*Tel:* 562 938-4550   *Fax:* 562 938-4118
*E-mail:* lhuy@lbcc.edu

**Los Angeles City College**
Dietetic Technician-AD Prgm
Family and Consumer Studies
855 N Vermont Ave
Los Angeles, CA 90029-3590
www.lacc.cc.ca.us
*Prgm Dir:* Janice M Young, MS RD
*Tel:* 323 953-4000, Ext 2291   *Fax:* 323 777-3862
*E-mail:* youngjj@lacitycollege.edu

**Merritt College**
Dietetic Technician-AD Prgm
12500 Campus Dr
Oakland, CA 94619-3107
*Prgm Dir:* Helenka T Livingston, M
*Tel:* 510 436-2521   *Fax:* 510 434-3873
*E-mail:* shouston@peralta.cc.ca.us

**Chaffey College**
Dietetic Technician-AD Prgm
Food Service Management
5885 Haven Ave
Rancho Cucamonga, CA 91737-3002
www.chaffey.edu
*Prgm Dir:* Candice F Hines-Tinsley, MA RD
*Tel:* 909 941-2711   *Fax:* 909 466-2831
*E-mail:* candice.tinsley@chaffey.edu

**Cosumnes River Community College**
Dietetic Technician-AD Prgm
8401 Center Pkwy
Sacramento, CA 95823-5799
http://crc.losrios.edu/~diettech
*Prgm Dir:* Dana Wu Wassmer, MS RD
*Tel:* 916 691-7514    *Fax:* 916 691-7146
*E-mail:* wassmed@crc.losrios.edu

**San Bernardino Valley College**
Dietetic Technician-AD Prgm
Family and Consumer Science
701 S Mount Vernon
San Bernardino, CA 92410-2748
*Prgm Dir:* Denise D Whisler, DrPH MPH RD
*Tel:* 909 384-8542    *Fax:* 909 384-9725

# Colorado

**Front Range Comm College**
Dietetic Technician-AD Prgm
3645 W 112th Ave, #23
Westminster, CO 80031-2199
www.frontrange.edu
*Prgm Dir:* Lou Ann Dixon, MEd RD
*Tel:* 303 404-5260    *Fax:* 303 404-2178
*E-mail:* louann.dixon@wc.frontrange.edu

# Connecticut

**Gateway Community College**
Dietetic Technician-AD Prgm
88 Bassett Rd
New Haven, CT 06473
www.gwctc.commnet.edu
*Prgm Dir:* Marcia Swan Doran, MS RD
*Tel:* 203 234-3309    *Fax:* 203 234-3353
*E-mail:* mdoran@gwcc.commnet.edu

**Briarwood College**
Dietetic Technician-AD Prgm
2279 Mt Vernon Rd
Southington, CT 06489-1007
www.briarwood.edu
*Prgm Dir:* Paula D Kellogg Leibovitz, MS RD CDN
*Tel:* 860 628-4751, Ext 179    *Fax:* 860 628-6444
*E-mail:* leibovitzp@briarwood.edu

# Florida

**Florida Community College - Jacksonville**
Dietetic Technician-AD Prgm
North Campus, 4501 Capper Rd
Jacksonville, FL 32218-4436
www.fccj.org
*Prgm Dir:* Margaret R Wolson, MS RD
*Tel:* 904 766-6743    *Fax:* 904 766-6654
*E-mail:* mwolson@fccj.org

**Palm Beach Community College**
Dietetic Technician-AD Prgm
4200 S Congress Ave
Mail Station 32
Lake Worth, FL 33461-4796
www.pbcc.edu/dietetic
*Prgm Dir:* Patricia Froehlich, MS RD
*Tel:* 561 439-8126    *Fax:* 561 439-8314
*E-mail:* froehlit@pbcc.edu

**Miami Dade College**
Dietetic Technician-AD Prgm
Mitchell Wolfson Campus
300 NE 2nd Ave
Miami, FL 33132-2297
www.mdcc.edu/wolfson/departments/nances
*Prgm Dir:* Susan J Myers, MS RD
*Tel:* 305 237-3162    *Fax:* 305 237-3802
*E-mail:* susan.myers@mdc.edu

**Pensacola Junior College**
Dietetic Technician-AD Prgm
1000 College Blvd
Pensacola, FL 32504-8998
www.pjc.cc.fl.us
*Prgm Dir:* Gloria Gonzalez, MS RD
*Tel:* 850 484-1168    *Fax:* 850 484-1183
*E-mail:* ggonzalez@pjc.edu

# Illinois

**Malcolm X College**
Dietetic Technician-AD Prgm
1900 W Van Buren St
Chicago, IL 60612-3145
www.ccc.edu/malcolmx/new_site001
*Prgm Dir:* Perla M Kushida, EdD RD RN
*Tel:* 312 850-7383    *Fax:* 312 850-7453
*E-mail:* pkushida@ccc.edu

**Harper College**
Dietetic Technician-AD Prgm
1200 W Algonquin Rd
Palatine, IL 60067-7398
www.harpercollege.edu
*Prgm Dir:* Jane Allendorph, MS RD LD
*Tel:* 847 925-6537    *Fax:* 847 925-6047
*E-mail:* jallendo@harpercollege.edu

# Indiana

**Ball State University**
Dietetic Technician-AD Prgm
Dept of Home Economics
150 Applied Technology Bldg
Muncie, IN 47306-0250
www.bsu.edu/cast/fcs/dieteticinternship.html
*Prgm Dir:* Corine M Carr, EdD RD
*Tel:* 765 285-2255    *Fax:* 765 285-2314
*E-mail:* ccarr@bsu.edu

# Louisiana

**Delgado Community College**
Dietetic Technician-AD Prgm
615 City Park Ave
New Orleans, LA 70119
www.dcc.edu
*Prgm Dir:* Donna Pace, MBA RD LDN
*Tel:* 504 483-4330    *Fax:* 504 483-4324
*E-mail:* dmpace@dcc.edu

# Maine

**Washington County Technical College**
Dietetic Technician-AD Prgm
RR 1 Box 22
Calais, ME 04619-9701
www.wctc.org
*Prgm Dir:* Denise E Harris, MS RD
*Tel:* 207 454-1091    *Fax:* 207 454-1026
*E-mail:* dharris@wctc.org

**Southern Maine Community College**
Dietetic Technician-AD Prgm
2 Fort Rd
South Portland, ME 04106
www.smccme.edu
*Prgm Dir:* Marie B Struble, PhD RD
*Tel:* 207 767-4648    *Fax:* 207 767-2731
*E-mail:* mstruble@smccme.edu

# Maryland

**Baltimore City Community College**
Dietetic Technician-AD Prgm
Dept of Allied Health
2901 Liberty Heights Ave
Baltimore, MD 21215-7893
www.bccc.state.md.us
*Prgm Dir:* Jolene R Campbell, MEd RD LD
*Tel:* 410 462-7724    *Fax:* 410 462-7734
*E-mail:* jcampbell@bccc.state.md.us

# Massachusetts

**Caritas Labour, College**
Dietetic Technician-AD Prgm
2120 Dorchester Ave
Boston, MA 02124-5698
www.laboure.edu
*Prgm Dir:* Anne S Manion, MBA RD
*Tel:* 617 296-8300, Ext 4042    *Fax:* 617 296-7947
*E-mail:* anne_manion@laboure.edu

**North Shore Community College**
Dietetic Technician-AD Prgm
One Ferncroft Rd
Danvers, MA 01923-4093
*Prgm Dir:* Bernadette S Lucas, MS RD
*Tel:* 978 762-4000, Ext 1542    *Fax:* 978 762-4022
*E-mail:* blucas@northshore.edu

# Minnesota

**Normandale Community College**
Dietetic Technician-AD Prgm
9700 France Ave S
Bloomington, MN 55431-4309
www.normandale.mnscu.edu
*Prgm Dir:* Krista Jordheim
*Tel:* 952 487-8374    *Fax:* 952 487-8101
*E-mail:* krista.jordheim@normandale.edu

**University of Minnesota - Crookston**
Dietetic Technician-AD Prgm
Center for Health & Human Serv
2900 University Ave
Crookston, MN 56716-5001
www.crk.umn.edu
*Prgm Dir:* Sharon Stewart, PhD RD
*Tel:* 218 281-8202    *Fax:* 218 281-8050
*E-mail:* sstewart@mail.crk.umn.edu

# Missouri

**St Louis Community College - Florissant Valley**
Dietetic Technician-AD Prgm
3400 Pershall Rd
St Louis, MO 63135-1499
www.stlcc.edu
*Prgm Dir:* Jeanne R Pranger-Florini, MS RD
*Tel:* 314 595-4378    *Fax:* 314 595-2047
*E-mail:* jflorini@stlcc.cc.mo.us

# Nebraska

**Southeast Community College**
Dietetic Technician-AD Prgm
8800 O St
Lincoln, NE 68520-1227
www.southeast.edu
*Prgm Dir:* Bernadine J Taylor, MA RD
*Tel:* 402 437-2465    *Fax:* 402 437-2404
*E-mail:* jtaylor@southeast.edu

# Nevada

**Truckee Meadows Community College**
Dietetic Technician-AD Prgm
7000 Dandini Blvd, RDMT 334J
Reno, NV 89512-3999
www.tmcc.edu
*Prgm Dir:* Janice L Grover, MS RD
*Tel:* 775 673-8218    *Fax:* 775 674-7983
*E-mail:* jgrover@tmcc.edu

# New Hampshire

**University of New Hampshire**
Dietetic Technician-AD Prgm
Thompson Schl of Applied Sci
6A Putnam Hall
Durham, NH 03824
www.unh.edu
*Prgm Dir:* Nancy Johnson, MEd RD
*Tel:* 603 862-1050    *Fax:* 603 862-2915
*E-mail:* njohnson@cisunix.unh.edu

# New Jersey

**Camden County College**
Dietetic Technician-AD Prgm
PO Box 200
College Dr
Blackwood, NJ 08012-0200
www.camdencc.edu
*Prgm Dir:* Marsha V Patrick, MS RD
*Tel:* 856 227-7200, Ext 4359    *Fax:* 856 374-4862
*E-mail:* mpatrick@camdencc.edu

**Middlesex County College**
Dietetic Technician-AD Prgm
2600 Woodbridge Ave, PO Box 3050
Edison, NJ 08818-3050
www.middlesex.cc.nj.us
*Prgm Dir:* Mary Pat Maciolek, MBA RD
*Tel:* 732 906-2538    *Fax:* 732 906-7745
*E-mail:* Mary-Pat_Maciolek@middlesexcc.edu

# New York

**LaGuardia Community College**
Dietetic Technician-AD Prgm
City University of New York
31-10 Thomson Ave Rm E300
Long Island City, NY 11101-3071
www.lagcc.cuny.edu
*Prgm Dir:* Rosann T Ippolito, MS RD
*Tel:* 718 482-5758    *Fax:* 718 609-2052
*E-mail:* ippolitoro@lagcc.cuny.edu

**SUNY Agric & Tech College at Morrisville**
Dietetic Technician-AD Prgm
Bailey Annex
Morrisville, NY 13408
*Prgm Dir:* Marie Louise Smith, MEd RD CD
*Tel:* 315 684-6288    *Fax:* 315 684-6592
*E-mail:* smithml@morrisville.edu

**Dutchess Community College**
Dietetic Technician-AD Prgm
53 Pendell Rd
Poughkeepsie, NY 12601-1512
www.sunydutchess.edu
*Prgm Dir:* Felicia Hirning, MS RD
*Tel:* 914 431-8323    *Fax:* 914 431-8329
*E-mail:* hirning@sunydutchess.edu

**Suffolk County Community College**
Dietetic Technician-AD Prgm
Eastern Campus
121 Speonk-Riverhead Rd
Riverhead, NY 11901-3499
www.sunysuffolk.edu
*Prgm Dir:* Jodi E Levine, MS RD CDN
*Tel:* 631 548-2590    *Fax:* 631 548-2617
*E-mail:* levinej@sunysuffolk.edu

**Rockland Community College**
Dietetic Technician-AD Prgm
145 College Rd
Suffern, NY 10901-3699
www.sunyrockland.edu
*Prgm Dir:* Georgette Howell, MS RD
*Tel:* 845 574-4130    *Fax:* 845 574-4498

**Westchester Community College**
Dietetic Technician-AD Prgm
75 Grasslands Rd
Student Center Building, Rm 215
Valhalla, NY 10595-1698
www.sunywcc.edu
*Prgm Dir:* Teresa A Cousins, MS RD CDN
*Tel:* 914 606-6182    *Fax:* 914 785-6423
*E-mail:* teresa.cousins@sunywcc.edu

**Erie Community College - North Campus**
Dietetic Technician-AD Prgm
6205 Main St
Williamsville, NY 14221-7095
www.ecc.edu
*Prgm Dir:* Margaret E Garfoot, MS RD
*Tel:* 716 851-1598    *Fax:* 716 851-1429
*E-mail:* garfoot@ecc.edu

# North Carolina

**Lenoir Community College**
Dietetic Technician-AD Prgm
PO Box 188
Kinston, NC 28502-0188
*Prgm Dir:* Katharine E Ten Pas, MS RD LDN
*Tel:* 252 527-6223    *Fax:* 252 527-2712

# Ohio

**Cincinnati State Tech & Comm College**
Dietetic Technician-AD Prgm
Health Technologies Div
3520 Central Pkwy
Cincinnati, OH 45223-2690
www.cincinnatistate.edu
*Prgm Dir:* Laura P Horn, MEd RD LD
*Tel:* 513 569-1620    *Fax:* 513 569-1659
*E-mail:* laura.horn@cincinnatistate.edu

**Cuyahoga Community College**
Dietetic Technician-AD Prgm
2900 Community College Ave
Cleveland, OH 44115-3196
www.tri-c.edu
*Prgm Dir:* Judith A Kaplan, MS RD LD
*Tel:* 216 987-4316    *Fax:* 216 987-4386
*E-mail:* judith.kaplan@tri-c.edu

**Columbus State Community College**
Dietetic Technician-AD Prgm
550 E Spring St, PO Box 1609
Columbus, OH 43216-1609
www.cscc.edu
*Prgm Dir:* Jan VanHorn, MS RD LD
*Tel:* 614 287-2580    *Fax:* 614 287-5973
*E-mail:* jvanho01@cscc.edu

**Sinclair Community College**
Dietetic Technician-AD Prgm
444 W Third St
Dayton, OH 45402-1460
www.sinclair.edu
*Prgm Dir:* Nora L Schaefer, Med RD
*Tel:* 937 512-5168    *Fax:* 937 512-4592
*E-mail:* nora.schaefer@sinclair.edu

**Rhodes State College**
Dietetic Technician-AD Prgm
4240 Campus Dr
Lima, OH 45804-3576
www.rhodesstate.edu
*Prgm Dir:* Marilyn Gilroy, MS RD LD
*Tel:* 419 995-8328    *Fax:* 419 995-8818
*E-mail:* Gilroy.M@rhodesstate.edu

**Hocking College**
Dietetic Technician-AD Prgm
3301 Hocking Pkwy
Nelsonville, OH 45764-9704
www.hocking.edu
*Prgm Dir:* Martha R Skeeles, MS RD LD
*Tel:* 740 753-3591, Ext 2223    *Fax:* 740 753-5105
*E-mail:* skeeles_m@hocking.edu

**Owens Community College**
Dietetic Technician-AD Prgm
PO Box 10000
Oregon Rd
Toledo, OH 43699-1947
www.owens.edu
*Prgm Dir:* Tekla M Madaras, MEd RD
*Tel:* 419 661-7214    *Fax:* 419 661-7251
*E-mail:* tekla_mandaras@owens.edu

**Youngstown State University**
Dietetic Technician-AD Prgm
Dept of Human Ecology
One University Plaza
Youngstown, OH 44555-3344
www.ysu.edu
*Prgm Dir:* Suzanne M Leson, MS RD LD
*Tel:* 330 941-1823    *Fax:* 330 941-1824
*E-mail:* smleson@ysu.edu

# Oklahoma

**Oklahoma State University - Okmulgee**
Dietetic Technician-AD Prgm
1801 E 4th St
Okmulgee, OK 74447-3901
www.osu-okmulgee.edu/hosp
*Prgm Dir:* Nelda G Downer, MS RD LD
*Tel:* 918 293-5004    *Fax:* 918 293-4618
*E-mail:* downer@osu-okmulgee.edu

# Pennsylvania

**Community College of Philadelphia**
Dietetic Technician-AD Prgm
1700 Spring Garden St
Philadelphia, PA 19130-3991
www.ccp.edu
*Prgm Dir:* Dorothy R Koteski, MS RD
*Tel:* 215 751-8948    *Fax:* 215 751-8937
*E-mail:* dkoteski@ccp.edu

**Comm Coll of Allegheny County**
Dietetic Technician-AD Prgm
808 Ridge Ave
Pittsburgh, PA 15212-6097
www.ccac.edu
*Prgm Dir:* Elizabeth C Vargo, MS RD
*Tel:* 412 237-2640    *Fax:* 412 237-4521
*E-mail:* evargo@ccac.edu

**Penn State University**
Dietetic Technician-AD Prgm
Coll of Health & Human Development
Hotel Rest & Recreation Mgmt
University Park, PA 16802-1307
www.worldcampus.psu.edu
*Prgm Dir:* Beth A Egan, MEd RD LDN
*Tel:* 814 863-7539    *Fax:* 814 863-4257
*E-mail:* bethegan@psu.edu

**Westmoreland County Community College**
Dietetic Technician-AD Prgm
400 Armbrust Rd
Youngwood, PA 15697
www.wccc-pa.edu
*Prgm Dir:* Cheryl B Shipley, MS RD
*Tel:* 724 925-4235    *Fax:* 724 925-4293
*E-mail:* shipleyc@wccc-pa.edu

# Tennessee

**Southwest Tennessee Community College**
Dietetic Technician-AD Prgm
PO Box 780
Union A-106
Memphis, TN 38101-0780
www.southwest.tn.edu
*Prgm Dir:* Linda Lee Pope, MS RD LDN
*Tel:* 901 333-5056    *Fax:* 901 333-5057
*E-mail:* lpope@southwest.tn.edu

# Texas

**Tarrant County Jr College - South**
Dietetic Technician-AD Prgm
2100 Southeast Pkwy
Arlington, TX 76018
www.tccd.net
*Prgm Dir:* Sheryl Harris, MS RD
*Tel:* 817 515-3621    *Fax:* 817 515-3581
*E-mail:* sheryl.harris@tccd.edu

**El Paso Community College**
Dietetic Technician-AD Prgm
100 W Rio Grande Ave
El Paso, TX 79902
www.epcc.edu
*Prgm Dir:* Marsha Cummings, MEd RD LD
*Tel:* 915 831-4470    *Fax:* 915 831-4114
*E-mail:* marshac@epcc.edu

**San Jacinto College Central**
Dietetic Technician-AD Prgm
8060 Spencer Hwy
PO Box 2007
Pasadena, TX 77505-2007
www.sjcd.edu
*Prgm Dir:* Adrianne J Sonnier, MS RD LD
*Tel:* 281 476-1501, Ext 1498    *Fax:* 281 478-2790
*E-mail:* adrianne.sonnier@sjcd.edu

**St Philip's College**
Dietetic Technician-AD Prgm
1801 Martin Luther King Dr
San Antonio, TX 78203-2098
www.accd.edu/spc/tourism
*Prgm Dir:* Mary A Kunz, MS RD LD
*Tel:* 210 531-3315    *Fax:* 210 531-3351
*E-mail:* mkunz@accd.edu

# Virginia

**Northern Virginia Community College**
Dietetic Technician-AD Prgm
HRI/DIT CF208
8333 Little River Trnpk
Annandale, VA 22003-3796
www.nvcc.edu
*Prgm Dir:* Janet Sass, MS RD
*Tel:* 703 323-3458    *Fax:* 703 323-3509
*E-mail:* jsass@nvcc.edu

**Tidewater Community College**
Dietetic Technician-AD Prgm
1700 College Crescent
Virginia Beach, VA 23456-1918
www.tcc.edu
*Prgm Dir:* Christine Medlin, PhD RD
*Tel:* 757 822-7336    *Fax:* 757 427-1338
*E-mail:* cmedlin@tcc.edu

# Washington

**Shoreline Community College**
Dietetic Technician-AD Prgm
16101 Greenwood Ave N, Rm 5334
Seattle, WA 98133-5696
www.shoreline.ctc.edu
*Prgm Dir:* Alison P Lehy, MS RD
*Tel:* 206 546-5891    *Fax:* 206 546-5869
*E-mail:* aleahy@ctc.edu

**Spokane Community College**
Dietetic Technician-AD Prgm
N 1810 Greene St, MS 2090
Spokane, WA 99217-5399
www.scc.spokane.edu
*Prgm Dir:* Erin Clason, MPH RD CDE
*Tel:* 509 533-7314    *Fax:* 509 533-8621

# Wisconsin

**Madison Area Technical College**
Dietetic Technician-AD Prgm
3550 Anderson St
Madison, WI 53704-2599
www.matcmadison.edu
*Prgm Dir:* Michael G Braun, MS RD
*Tel:* 608 246-6313    *Fax:* 608 246-6880
*E-mail:* mbraun@matcmadison.edu

**Milwaukee Area Technical College**
Dietetic Technician-AD Prgm
1200 S 71st St
West Allis, WI 53214-3110
www.matc.edu
*Prgm Dir:* Marian M Benz, MS RD CDE CD
*Tel:* 414 456-5364    *Fax:* 414 456-5425
*E-mail:* benzmj@matc.edu

## Occupational Description

Electroneurodiagnostic (END) technology is the medical diagnostic field devoted to the recording and study of electrical activity in the brain and nervous system. END technologists possess the knowledge, skills, and attributes to obtain interpretable recordings of patients' nervous system function. They work in collaboration with medical researchers, clinicians, physicians, and other health professionals.

## Job Description

The END technologist can be involved in one or more of the following diagnostic procedures: electroencephalography (EEG), evoked potential (EP), polysomnography (PSG), nerve conduction studies (NCS), and intraoperative monitoring (IOM). The technologist takes the medical history; documents the clinical condition of patients; understands and employs the optimal use of EG, EP, PSG, and NCS equipment; and applies adequate recording electrodes. Among other duties, the END technologist also understands the interface between EEG, EP, PSG, and NCS equipment and other electrophysiological devices and procedures; recognizes and understands EEG/EP/NCS/sleep activity displayed; manages medical emergencies in the laboratory; and prepares a descriptive report of recorded activity for the interpreting physician. The responsibilities of the technologist may also include laboratory management and the supervision of END technologists.

## Employment Characteristics

END personnel work primarily in neurology-related departments of hospitals, but many also work in clinics and the private offices of neurologists and neurosurgeons. Growth in employment within the profession is expected to be greater than average, owing to the increased use of EEG and EP techniques in surgery; in diagnosing and monitoring patients with epilepsy; and in diagnosing sleep disorders. Technologists generally work a 40-hour week, but may work 12-hour days for sleep studies and be on-call for emergencies and intraoperative monitoring.

## Salary

According to the American Society of Electroneurodiagnostic Technologists, Inc (ASET), 2003 entry-level salaries average $34,726. Refer to Section IV, Table 5 of this *Directory* for more information, or see www.ama-assn.org/go/ salary.

## Educational Programs

**Length.** Programs may be 12 to 24 months and are typically integrated into a community college-sponsored program leading to an associate degree.

**Prerequisites.** High school diploma or equivalent.

**Curriculum.** The curriculum includes anatomy, physiology, and neuroanatomy (with major emphasis on the brain), as well as instrumentation, personal and patient safety, recording techniques, clinical electroneurodiagnostics, and correlations. Clinical rotations are conducted in medical centers.

## Inquiries

**Careers**

American Society of Electroneurodiagnostic Technologists, Inc
6501 East Commerce Avenue
Suite 120
Kansas City, MO 64120
816 931-1120
816 931-1145 Fax
E-mail: info@aset.org
www.aset.org

**Certification/Registration (Credentials R EEG T®, R EP T®, CNIM®)**

Janice Walbert, R EEG/EP T, Executive Director
American Board of Registration of Electroencephalographic and Evoked Potential Technologists (ABRET)
1904 Croydon Drive
Springfield, IL 62703
217 553-3758
217 585-6663 Fax
E-mail: abreteo@aol.com
www.abret.org

**Certification/Registration (Credentials R EDT)**

Corinne Atkins, R EDT, Executive Director
American Association of Electrodiagnostic Technologists (AAET)
28 Sabins Lane
North Chatham, MA 02650
508 945-2781 Phone/Fax
E-mail: rta1@aol.com
www.aaet.info

**Certification/Registration (Credentials RPSGT™)**

Bobby Stanley, Jr, Executive Director
The Board of Registered Polysomnographic Technologists (BRPT)
8201 Greensboro Drive, Suite 300
McLean, VA 22102
703 610-9020
703 610-9005 Fax
E-mail: brpt@amg-inc.com

**Program Accreditation**

Commission on Accreditation of Allied Health Education Programs (CAAHEP) in collaboration with:
Committee on Accreditation of END Education Programs (CoA-END)
CoA-END
Kristina Port, R EEG/EP T, RPSGT
7600 Hunters Hollow Trail
Novelty, OH 44072
440 338-7541
E-mail: kaport@prodigy.net

## Electroneurodiagnostic Technologist*

# California

**Orange Coast College**
Electroneurodiagnostic Tech Prgm
2701 Fairview Rd
PO Box 5005
Costa Mesa, CA 92628-5005
www.orangecoastcollege.edu
*Prgm Dir:* Walt Banoczi, R EEG/EP T CNIM RPSGT
*Tel:* 714 432-5591   *Fax:* 714 432-5534
*E-mail:* wbanoczi@cccd.edu
*Med Dir:* Hugh McIntyre, MD, PhD

# Florida

**Erwin Technical Center**
Electroneurodiagnostic Tech Prgm
2010 E Hillsborough Ave
Tampa, FL 33610
*Prgm Dir:* Henry Coet III, R EEG T
*Tel:* 813 231-1800, Ext 1341   *Fax:* 813 231-1820
*E-mail:* coet_h@firn.edu
*Med Dir:* Salem Benbadis, MD

# Illinois

**St John's Hospital**
*Cosponsor:* Lincoln Land Community College
Electroneurodiagnostic Tech Prgm
School of ENDT
800 E Carpenter St
Springfield, IL 62769
*Prgm Dir:* Diane Liesen, R EEG T
*Tel:* 217 544-6464, Ext 44707   *Fax:* 217 535-3695
*E-mail:* diane.liesen@st-johns.org
*Med Dir:* Steven Evans, MD

# Indiana

**Clarian Health Partners Inc**
Electroneurodiagnostic Tech Prgm
Wile Hall Rm 604
I-65 at 21st St
Indianapolis, IN 46206-1367
www.clarian.org
*Prgm Dir:* Debby Ferguson, BS REEG/EPT RPSGT
  RNCST MS
*Tel:* 317 962-1291
*E-mail:* dferguson@clarian.org
*Med Dir:* Omkar Markand, MD

# Iowa

**Scott Community College**
Electroneurodiagnostic Tech Prgm
500 Belmont Rd
Bettendorf, IA 52722
www.eicc.edu
*Prgm Dir:* Sherry Kelly, R EEG T RPSGT
*Tel:* 563 441-4268   *Fax:* 563 441-4154
*E-mail:* skelly@eicc.edu
*Med Dir:* Stephen Rasmus, MD

**Kirkwood Community College**
*Cosponsor:* Univ of Iowa Hosp and Clinics
Electroneurodiagnostic Tech Prgm
Dept of Neurology
6301 Kirkwood Blvd SW, PO Box 2068
Cedar Rapids, IA 52406-9973
*Prgm Dir:* Elizabeth Meng, R EEG/EPT
*Tel:* 319 356-8768   *Fax:* 319 351-1209
*E-mail:* elizabeth-meng@uiowa.edu
*Med Dir:* Thoru Yamada, MD

# Massachusetts

**Caritas Labour, College**
Electroneurodiagnostic Tech Prgm
2120 Dorchester Ave
Boston, MA 02124
*Prgm Dir:* Jean Wilkins Farley, MA R EEG T
*Tel:* 617 296-8300, Ext 4043   *Fax:* 617 296-7947
*E-mail:* Jean_Farley@laboure.edu
*Med Dir:* Sanford Auerbach, MD

# Michigan

**Carnegie Institute**
Electroneurodiagnostic Tech Prgm
550 Stephenson Hwy, Stes 100-109
Troy, MI 48083
www.carnegie-institute.com
*Prgm Dir:* Lisa Lovely, BA R EEG/EPT CNIMI
*Tel:* 248 589-1078   *Fax:* 248 589-1631
*E-mail:* lisalovely@umich.edu
*Med Dir:* Jonathan Edwards, MD

# Minnesota

**Mayo School of Health Sciences**
Electroneurodiagnostic Tech Prgm
Mayo Clinic
200 First St SW
Rochester, MN 55905
www.mayo.edu/mshs
*Prgm Dir:* Jan M Buss, R NCS T
*Tel:* 507 284-1255   *Fax:* 507 284-0999
*E-mail:* buss.jan@mayo.edu
*Med Dir:* Michael H Silber, MD

# Pennsylvania

**Crozer - Chester Med Ctr**
Electroneurodiagnostic Tech Prgm
One Medical Ctr Blvd
Upland, PA 19013
*Prgm Dir:* Kellee W Trice, R EEG/EPT RPSGT
*Tel:* 610 447-2691   *Fax:* 610 447-2696
*E-mail:* kellee.trice@crozer.org
*Med Dir:* Lawrence Green, MD

# Texas

**McLennan Community College**
Electroneurodiagnostic Tech Prgm
1400 College Dr
Waco, TX 76708
www.mclennan.edu
*Prgm Dir:* Mary Feltman, BS REEGT
*Tel:* 254 299-8525   *Fax:* 254 299-8747
*E-mail:* mfeltman@mclennan.edu
*Med Dir:* Joel Freitag, Board Certified Neurologist

# Virginia

**Naval School of Health Sciences**
Electroneurodiagnostic Tech Prgm
1001 Holcomb Rd
Portsmouth, VA 23708-5200
*Prgm Dir:* Michael McNamara, R EEGT/ R EPT
*Tel:* 757 953-2144
*E-mail:* End@hsp.med.navy.mil

# Wisconsin

**Western Technical College**
Electroneurodiagnostic Tech Prgm
304 N Sixth St, PO Box 908
La Crosse, WI 54602-0908
www.westerntc.edu
*Prgm Dir:* Stacey Austin, R EEG/EP T RPSGT
*Tel:* 608 789-6141, Ext 96141   *Fax:* 608 785-9879
*E-mail:* austins@westerntc.edu
*Med Dir:* Gregory Fischer, MD

*Additional information about programs that
returned the AMA's survey is available in a table
beginning on page 178.

# Electroneurodiagnostic Technologist

| Programs* | Class Capacity | Begins | Length (months) | Award | Res Tuition | Non-res Tuition | Stipend | Offers:‡ | | |
|---|---|---|---|---|---|---|---|---|---|---|
| | | | | | | | | 1 | 2 | 3 |
| **California** | | | | | | | | | | |
| Orange Coast College (Costa Mesa) | 24 | Aug (even years only) | 22 | Cert, AA | $1,520 | $15,500 | | | | |
| **Illinois** | | | | | | | | | | |
| St John's Hospital (Springfield) | 8 | Aug | 21 | Dipl, Cert, AGE-EN | $2,000 | $2,000 | | | | • |
| **Indiana** | | | | | | | | | | |
| Clarian Health Partners Inc (Indianapolis) | 6 | Varies | 18 | Cert | $3,600 | $3,600 | | | | |
| **Iowa** | | | | | | | | | | |
| Scott Community College (Bettendorf) | 15 | Aug | 22 | AAS | $2,975 | $4,526 | | • | | |
| **Michigan** | | | | | | | | | | |
| Carnegie Institute (Troy) | 12 | Sep | 12 | Dipl | $8,415 | $8,415 | | • | | |
| **Minnesota** | | | | | | | | | | |
| Mayo School of Health Sciences (Rochester) | 8 | Aug | 24 | Cert, AAS | $4,000 | $4,000 | | | | |
| **Texas** | | | | | | | | | | |
| McLennan Community College (Waco) | 20 | Sep | 12, 24 | Cert, AA | $1,960 | $2,380 | | | | • |
| **Wisconsin** | | | | | | | | | | |
| Western Technical College (La Crosse) | 22 | Jun | 24 | AAS | $2,311 | $15,411 | | • | | |

*Data are shown only for programs that completed the 2006 AMA Survey of Health Professions Education Programs
‡Key to Offers: 1: Evening or weekend classes; 2: Non-English instruction; 3: Cultural competence instruction

## Occupational Description

Emergency medical technicians (EMTs) and EMT-paramedics are trained to provide emergency care to people who have suffered from an illness or an injury outside of the hospital setting. EMTs and paramedics work under protocols approved by a physician medical director to recognize, assess, and manage medical emergencies and transport patients to definitive medical care. EMTs provide basic life support, and EMT-paramedics provide advanced life support.

## Profession Description

EMTs and EMT-paramedics may be employed by a private ambulance company, fire department, police department, public EMS agency, private ambulance company, hospital, or combination of the above. EMS responders may be paid or volunteers in the community.

EMTs must be proficient in First Aid, and training is centered on recognizing and treating life-threatening conditions outside the hospital environment. EMTs learn the basics of how to handle cardiac and respiratory arrest, heart attacks, seizures, diabetic emergencies, respiratory problems, and other medical emergencies. They also learn how to manage traumatic injuries such as falls, fractures, lacerations, and burns. EMTs also are introduced into patient assessment, history taking, and vital signs.

EMTs perform CPR, artificial ventilations, oxygen administration, basic airway management, defibrillation using an AED, spinal immobilization, vital signs, bandaging/splinting, and may administer Nitroglycerin, Glucose, Epinephrine, and Albuterol in special circumstances.

EMT-Paramedics perform all of the skills performed by an EMT-Basic. In addition, they perform advance airway management, such as endotracheal intubation, under medical supervision and from a base station, usually in a hospital emergency room. They obtain electrocardiographs (ECGs), introduce intravenous lines, and administer numerous emergency medications. EMT-paramedics assess ECG tracings and defibrillate. They have extensive training in patient assessment and are exposed to a variety of clinical experiences during training.

## Educational Programs

In most locations in the United States, the minimum level of education that most EMS professionals have before entering the workforce is that of a Basic-Level EMT. Individuals who work as firefighters or police officers may perform some emergency medical work when trained as first responders. Some paramedic programs provide an all-inclusive program that includes both EMT and paramedic training in one program. All levels of EMS training are set by the federal government through the National Highway Traffic Safety Administration.

EMT training is offered at community colleges, technical schools, hospitals, and universities as well as EMS, fire, and police academies. Those interested in EMT training should contact their state's EMS Office. Those interested in paramedic training should contact the Committee on Accreditation for EMS Professionals. Both of these agencies can help potential students find local training.

**Length.** EMT training varies from 2 to 6 months, depending on the training site and hours of class scheduled per week. There are training programs that have class every day for several months for those interested in quick completion. Longer programs are available to accommodate students who have family, a full-time job, or other responsibilities that limit their available time for education. Approximate training requirements are:

- First Responder          40 hours of training
- EMT-Basic                110 hours of training
- EMT-Intermediate         200-400 hours of training
- Paramedic                1,000 or more hours of training

**Prerequisites.** An EMT student is expected to be a high school graduate or the equivalent and to meet the physical and mental demands of the occupation. EMT-paramedic students must have completed their EMT training prior to enrollment in most EMT-paramedic courses unless they are enrolled in a joint EMT and paramedic program. Some paramedic programs are part of Bachelor of Science degree programs offered at colleges and universities.

**Curriculum.** EMT and Paramedic training are composed of in-classroom, didactic instruction; in-hospital clinical practice; and a supervised field internship on an ambulance. Courses typically are competency-based and supported by performance assessments. Instruction provides students with knowledge of acute and critical changes in physiology and psychological and clinical symptoms that they might encounter in an emergency medical situation.

## Inquiries

**Careers**

National Association of Emergency Medical Technicians (NAEMT)
PO Box 1400
Clinton, MS 39060-1400
800 34-NAEMT
www.naemt.org

**Certification/Licensure**

National Registry of Emergency Medical Technicians (NREMT)
Rocco V Morando Bldg
Box 29233
6610 Busch Blvd
Columbus, OH 43229-0233
614 888-4484
www.nremt.org

**Program Accreditation**

Commission on Accreditation of Allied Health Education Programs (CAAHEP) in collaboration with:
Committee on Accreditation of Educational Programs for the EMS Professions
1248 Harwood Road
Bedford, TX 76021-4244
817 283-9403
www.caahep.org

## Emergency Medical Technician-Paramedic*

# Alabama

**Lurleen B Wallace Junior College**
Emergency Med Tech-Paramedic Prgm
PO Box 1418
Andalusia, AL 36420
*Prgm Dir:* Wayne Godwin
*Tel:* 334 881-2239   *Fax:* 334 222-0136
*E-mail:* wgodwin@lbwcc.edu
*Med Dir:* Tim Day, MD

**University of Alabama at Birmingham**
Emergency Med Tech-Paramedic Prgm
Department of Emergency Medicine
912 18th St S
Birmingham, AL 35205
*Prgm Dir:* Randal E Gray, BS MA NREMT-P CCEMT-P
*Tel:* 205 934-3611   *Fax:* 205 975-7573
*E-mail:* rgray@uab.edu
*Med Dir:* Guillermo Pierluisi, MD

**Calhoun Community College**
Emergency Med Tech-Paramedic Prgm
PO Box 2216 Hwy 31 N
Decatur, AL 35609
*Prgm Dir:* Jarrod Taylor, BS RN NREMT-P
*Tel:* 256 306-2781   *Fax:* 256 306-2909
*E-mail:* jwt@calhoun.edu
*Med Dir:* Henry Gaillard, MD

**George C Wallace Community College**
Emergency Med Tech-Paramedic Prgm
1141 Wallace Dr
Dothan, AL 36303
www.wallace.edu
*Prgm Dir:* Rebecca Burke, BS Ed NREMT-P
*Tel:* 334 556-2442   *Fax:* 334 556-2405
*E-mail:* rburke@wallace.edu
*Med Dir:* James M Jones, DO

**Gadsden State Community College**
Emergency Med Tech-Paramedic Prgm
PO Box 227
Gadsden, AL 35902-0227
www.gadsdenstate.edu
*Prgm Dir:* Patrick T Brown, BS NREMT-P
*Tel:* 256 549-8654   *Fax:* 256 549-8276
*E-mail:* pbrown@gadsdenstate.edu
*Med Dir:* Niel Christin, MD

**Wallace State Community College**
Emergency Med Tech-Paramedic Prgm
301 Main St NW, PO Box 2000
Hanceville, AL 35077
*Prgm Dir:* Cindy Durham
*Tel:* 256 352-8335   *Fax:* 256 352-8337
*E-mail:* ems@wallacestate.edu
*Med Dir:* Robert Echols, MD

**University of South Alabama**
Emergency Med Tech-Paramedic Prgm
2002 Old Bay Front Dr
Mobile, AL 36615
www.usouthal.edu
*Prgm Dir:* David W Burns, MPH ScD(c) EMT-P
*Tel:* 251 431-6418   *Fax:* 251 431-6525
*E-mail:* dburns@usouthal.edu
*Med Dir:* Frank S Pettyjohn, MD

**H Council Trenholm State Technical College**
Emergency Med Tech-Paramedic Prgm
1225 Air Base Blvd
Montgomery, AL 36108
www.trenholmtech.cc.al.us
*Prgm Dir:* Becky A Morris, BS NREMT-P
*Tel:* 334 420-4432   *Fax:* 334 420-4437
*E-mail:* bmorris@elmore.rr.com,
   bmorris@trenholmtech.cc.al
*Med Dir:* John Campbell, MD

**Northwest-Shoals Community College**
Emergency Med Tech-Paramedic Prgm
PO Box 2545
Muscle Shoals, AL 35662
*Prgm Dir:* J Bret McGill, MS NREMT-P
*Tel:* 256 331-5336   *Fax:* 256 331-5371
*E-mail:* mcgillb@nwscc.edu
*Med Dir:* Steve Wampler, MD

**Southern Union State Community College**
Emergency Med Tech-Paramedic Prgm
1701 Lafayette Pkwy
Opelika, AL 36801
www.suscc.edu
*Prgm Dir:* Steven Simpson, BSBA NREMT-P
*Tel:* 334 745-6437, Ext 5533   *Fax:* 334 741-9795
*E-mail:* ssimpson@suscc.edu
*Med Dir:* John E Campbell, MD

**Northeast Alabama Community College**
Emergency Med Tech-Paramedic Prgm
PO Box 159
Rainsville, AL 35986-0159
www.nacc.edu
*Prgm Dir:* Roger Wootten, BS NREMT-P
*Tel:* 256 228-6001, Ext 355   *Fax:* 256 228-3309
*E-mail:* woottenr@nacc.edu
*Med Dir:* Thomas McFarland, MD

**Bevill State Community College**
Emergency Med Tech-Paramedic Prgm
101 S State St
Sumiton, AL 35148
*Prgm Dir:* Scott Karr, MEd NREMT-P
*Tel:* 205 648-3271, Ext 5570   *Fax:* 205 384-4581
*E-mail:* skarr@mail.bscc.edu
*Med Dir:* Charles Shipman, MD

**Shelton State Community College**
*Cosponsor:* Alabama Fire College
Emergency Med Tech-Paramedic Prgm
2501 Phoenix Dr
Tuscaloosa, AL 35405
*Prgm Dir:* Julie L Coffman, BS AD
*Tel:* 205 391-3743   *Fax:* 205 391-3771
*E-mail:* jcoffman@alabamafirecollege.org
*Med Dir:* Phillip Bobo, MD

# Arkansas

**Univ of Arkansas Comm College-Batesville**
Emergency Med Tech-Paramedic Prgm
2005 White Dr, PO Box 3350
Batesville, AR 72501
*Prgm Dir:* Timothy H Hodges, NREMT-P
*E-mail:* thodges@uaccb.edu
*Med Dir:* Robert Fox, MD

**NorthWest Arkansas Community College**
Emergency Med Tech-Paramedic Prgm
1 College Dr
Bentonville, AR 72712
www.nwacc.edu
*Prgm Dir:* Jamin Snarr, NREMTP AAS
*Tel:* 479 619-4251   *Fax:* 479 619-4254
*E-mail:* jsnarr@nwacc.edu
*Med Dir:* Mark Rucker, MD

**Arkansas Northeastern College**
Emergency Med Tech-Paramedic Prgm
Interstate 55 & Hwy 148
Burdette, AR 72321
www.anc.edu
*Prgm Dir:* Kathy Arnold, NREMT-P
*Tel:* 870 780-1221   *Fax:* 870 763-1496
*E-mail:* karnold@anc.edu
*Med Dir:* John Williams, MD

**U of Arkansas-Monticello Coll of Tech-McGehee**
Emergency Med Tech-Paramedic Prgm
700 W Gaines
PO Box 620
Dermott, AR 71638
*Prgm Dir:* Gursam Singh, BS NREMT-P
*Tel:* 870 538-0248, Ext 5504   *Fax:* 870 538-5666
*E-mail:* singh@seark.net
*Med Dir:* Robert B Scott, MD

**East Arkansas Community College**
Emergency Med Tech-Paramedic Prgm
1700 Newcastle Rd
Forrest City, AR 72335
www.eacc.edu
*Prgm Dir:* Tami Jo Jones, BSEd NREMT-P
*Tel:* 870 633-4480, Ext 269   *Fax:* 870 633-7222
*E-mail:* tjjones@eacc.edu
*Med Dir:* Mohammed Knefati, MD

**North Arkansas College**
Emergency Med Tech-Paramedic Prgm
1515 Pioneer Dr
Harrison, AR 72601
*Prgm Dir:* K C Jones
*Tel:* 870 391-3125   *Fax:* 870 391-3250
*E-mail:* kcjones@northark.edu
*Med Dir:* Barbara Ashe, MD

**National Park Community College**
Emergency Med Tech-Paramedic Prgm
101 College Dr
Hot Springs, AR 71913
www.npcc.edu
*Prgm Dir:* John Dodd, MS NREMT-P/I
*Tel:* 501 760-4158   *Fax:* 501 760-4141
*E-mail:* jdodd@npcc.edu
*Med Dir:* Gene Shelby, MD

**University of Arkansas for Medical Sciences**
Emergency Med Tech-Paramedic Prgm
4301 W Markham St Slot 635
Little Rock, AR 72205
www.uams.edu
*Prgm Dir:* Danny Bercher, MEd
*Tel:* 501 686-5773   *Fax:* 501 686-6513
*E-mail:* dlbercher@uams.edu
*Med Dir:* Gregory Hall, MD

**Arkansas Tech University - Ozark Campus**
Emergency Med Tech-Paramedic Prgm
PO Box 506
Ozark, AR 72956
*Prgm Dir:* Lisa Robles, NREMT-P
*Tel:* 479 667-2117, Ext 326   *Fax:* 479 667-2106
*E-mail:* lisa.robles@mail.atu.edu
*Med Dir:* Todd Carter, MD

**Southeast Arkansas College**
Emergency Med Tech-Paramedic Prgm
1900 Hazel St
Pine Bluff, AR 71603
*Prgm Dir:* Floyd Nutter, EMT-P AAS
*Tel:* 870 543-5917   *Fax:* 870 543-5912
*E-mail:* fnutter@seark.edu
*Med Dir:* Charles Mabry, MD

---

*Additional information about programs that
returned the AMA's survey is available in a table
beginning on page 189.

**Black River Technical College**
Emergency Med Tech-Paramedic Prgm
PO Box 468
Pocahontas, AR 72455
www.blackrivertech.org
*Prgm Dir:* Kimeron J Hubbard
*Tel:* 870 888-5750  *Fax:* 870 886-7481
*E-mail:* captdiver@earthlink.net
*Med Dir:* William S Lewis, MD

# California

**Bakersfield College**
Emergency Med Tech-Paramedic Prgm
1801 Panorama Dr
Bakersfield, CA 93305
*Prgm Dir:* Myron Smith
*Tel:* 661 395-4284
*E-mail:* smithm@hallamb.com

**Southwestern College**
Emergency Med Tech-Paramedic Prgm
School of Technology & Human Services, Rm 570B
900 Otay Lakes Rd
Chula Vista, CA 91910
www.swc.cc.ca.us
*Prgm Dir:* Joanne Stonecipher, RN MSN
*Tel:* 619 421-6700, Ext 5599  *Fax:* 619 216-6616
*E-mail:* jstonecipher@swc.cc.ca.us
*Med Dir:* Mark Handy, MD

**Los Angeles County Paramedic Training Inst**
Emergency Med Tech-Paramedic Prgm
5555 Ferguson Dr, Ste 220
Commerce, CA 90022
www.dhs.co.la.ca.us
*Prgm Dir:* Terry Crammer, RN BSN
*Tel:* 323 890-7506  *Fax:* 323 890-8528
*E-mail:* tcrammer@ladhs.org
*Med Dir:* Scott Youngquist, MD

**Palomar Community College**
Emergency Med Tech-Paramedic Prgm
1951 E Valley Pkwy
Escondido, CA 92056
*Prgm Dir:* Mary Reed
*Tel:* 760 744-1150, Ext 8150
*E-mail:* mreed@palomar.edu
*Med Dir:* Gary Vilke, MD

**California EMS Academy Inc**
Emergency Med Tech-Paramedic Prgm
1098 Foster City Blvd, Ste 106 PMB 708
Foster City, CA 94404
www.caems-academy.com
*Prgm Dir:* Nancy L Black, RN MS
*Tel:* 866 577-9197  *Fax:* 650 701-1968
*E-mail:* Nancy@caems-academy.com
*Med Dir:* Michael Laufer, MD

**Fresno City College**
Emergency Med Tech-Paramedic Prgm
1901 E Shields, Ste 250
Fresno, CA 93726
www.fccti.com
*Prgm Dir:* Mark Allen, EMT-P BA
*Tel:* 559 256-0188  *Fax:* 559 256-0199
*E-mail:* mark.allen@fresnocitycollege.edu
*Med Dir:* Raymond Miranda, MD

**Fresno County Paramedic Program**
Emergency Med Tech-Paramedic Prgm
1221 Fulton Mall
Fresno, CA 93721
*Prgm Dir:* Debra Becker, RN
*Tel:* 559 445-3387  *Fax:* 559 445-3205
*E-mail:* dbecker@co.fresno.ca.us
*Med Dir:* Marc Shalit, MD

**Imperial Valley Community College**
Emergency Med Tech-Paramedic Prgm
380 E Aten Rd
Imperial, CA 92251
*Prgm Dir:* Jackilyn E Cypher, RN CEN MICN NREMT-P
*Tel:* 760 355-6275  *Fax:* 760 355-6346
*E-mail:* jcypher@imperial.cc.ca.us
*Med Dir:* Bruce E Haynes, MD

**UCLA Center for Prehospital Care**
Emergency Med Tech-Paramedic Prgm
Daniel Freeman Memorial Hospital
333 N Prairie Ave
Inglewood, CA 90301
*Prgm Dir:* William Dunne, MS NREMT-P
*Tel:* 310 674-7050, Ext 3580  *Fax:* 310 680-8640
*E-mail:* wdunne@mednet.ucla.edu
*Med Dir:* Steve Rottman, MD

**California Paramedic Institute**
Emergency Med Tech-Paramedic Prgm
23141 Lake Center Dr
Lake Forest, CA 92630
*Prgm Dir:* Robert Nieblas
*E-mail:* rnieblas@cpimedic.com

**Northern California Training Institute - Bay Area Counties**
Emergency Med Tech-Paramedic Prgm
7503 Southfront Rd #A
Livermore, CA 94550
*Prgm Dir:* Kenneth Reed, BS EMT-P
*Tel:* 925 454-6184  *Fax:* 925 454-6297
*E-mail:* kenneth_reed@amr-ems.com
*Med Dir:* Jack Wood, DO

**Saddleback College**
Emergency Med Tech-Paramedic Prgm
28000 Marguerite Pkwy
Mission Viejo, CA 92692
*Prgm Dir:* Barbara Penland, RN MA
*Tel:* 949 582-4385
*E-mail:* bpenland@saddleback.edu
*Med Dir:* Jennifer Mason, MD

**Butte College**
Emergency Med Tech-Paramedic Prgm
3536 Butte Campus Dr
Oroville, CA 95965
www.butte.edu
*Prgm Dir:* Michael Boyd
*Tel:* 530 879-4310  *Fax:* 530 895-2472
*E-mail:* boydmi@butte.edu
*Med Dir:* Peter Russo, MD

**Foothill Community College**
Emergency Med Tech-Paramedic Prgm
4000 Middlefield Rd, Ste 1
Palo Alto, CA 94303
*Prgm Dir:* Mary Jane Arlene Green, NREMT-P AA
*Tel:* 650 949-6972  *Fax:* 650 949-6979
*E-mail:* greenmary@foothill.edu
*Med Dir:* Ram Duraceti, MD

**Riverside Comm Coll - Moreno Valley Campus**
Emergency Med Tech-Paramedic Prgm
3423 Davis Ave
Riverside, CA 92518
www.rcc.edu/academicPrograms/ems
*Prgm Dir:* Chris Nollette, EdD NREMTP LP
*Tel:* 951 571-6100, Ext 4609
*E-mail:* chris.nollette@rcc.edu
*Med Dir:* Reza Vaezazizi, MD

**Antelope Valley College**
Emergency Med Tech-Paramedic Prgm
2997 Desert St
Rosamond, CA 93560
*Prgm Dir:* Lance Hodge, EMT-P
*Tel:* 661 722-6300, Ext 6289
*E-mail:* lhodge@avc.edu

**Northern California Training Institute - Placer County**
Emergency Med Tech-Paramedic Prgm
333 Sunrise Ave, Ste 500
Roseville, CA 95661
*Prgm Dir:* Vickie Wolf, RN
*Tel:* 916 960-6284  *Fax:* 916 960-6296
*E-mail:* vwolf@amr-ems.com
*Med Dir:* Jack Wood, DO FACEP

**American River College**
Emergency Med Tech-Paramedic Prgm
4700 College Oak Dr
Sacramento, CA 95841-4217
www.arc.losrios.edu
*Prgm Dir:* Grant Goold, PhD
*Tel:* 916 484-8843
*E-mail:* GooldG@arc.losrios.edu
*Med Dir:* R Steven Tharratt, MD

**City College of San Francisco**
Emergency Med Tech-Paramedic Prgm
John Adams Campus
1860 Hayes St
San Francisco, CA 94117
*Prgm Dir:* Megan Corry, MA EMTP
*Tel:* 415 561-1938  *Fax:* 415 561-1999
*E-mail:* mcorry@ccsf.edu
*Med Dir:* Marlena Tang, MD

**WestMed College**
Emergency Med Tech-Paramedic Prgm
5300 Stevens Creek Blvd, Ste 200
San Jose, CA 95129
*Prgm Dir:* Jolyn Camacho
*Tel:* 408 977-0723  *Fax:* 408 977-1396
*E-mail:* vershep@aol.com
*Med Dir:* Gregory Gilbert, MD

**Hospital Consortium Education Network**
Emergency Med Tech-Paramedic Prgm
222 W 39th Ave 3A09
San Mateo Medical Center
San Mateo, CA 94403
*Prgm Dir:* Art Hsieh
*Tel:* 650 573-3930
*E-mail:* ahsieh@hospitalconsort.org
*Med Dir:* Chris Fee, MD

**Emergency Training Services Inc**
Emergency Med Tech-Paramedic Prgm
3050 Paul Sweet Rd
Santa Cruz, CA 95065
*Prgm Dir:* Mary Foraker, BSN
*Tel:* 831 476-8813  *Fax:* 831 477-4914
*E-mail:* info@emergencytraining.com
*Med Dir:* Terry Lapid, MD

**Northern California Training Institute - Santa Barbara County**
Emergency Med Tech-Paramedic Prgm
800 S College Ave
Santa Maria, CA 93454
*Prgm Dir:* Vicki Wolf, EMT-P
*Tel:* 805 688-6550
*E-mail:* vwolf@amr-ems.com
*Med Dir:* Jack Wood, DO

**Emergency Medical Sciences Training Institute**
*Cosponsor:* San Joaquin County EMS Agency
Emergency Med Tech-Paramedic Prgm
343 E Main St #906
Stockton, CA 95202
*Prgm Dir:* Craig Stroup, BS NREMT-P
*Tel:* 209 461-5550  *Fax:* 209 461-5553
*E-mail:* Emstiservices@aol.com
*Med Dir:* Richard Buys, MD

## Mendocino Community College

Emergency Med Tech-Paramedic Prgm
1000 Hensley Creek Rd
Ukiah, CA 95482
*Prgm Dir:* Bill Webster, BS
*Tel:* 707 468-3005
*E-mail:* bwebster@mendocino.cc.ca.us
*Med Dir:* Waubli Franklin, MD

## Ventura College

Emergency Med Tech-Paramedic Prgm
4667 Telegraph Rd
Ventura, CA 93003
*Prgm Dir:* Meredith H Mundell, BSN RN
*Tel:* 805 654-6364    *Fax:* 805 654-6328
*E-mail:* MMundell@vcccd.net
*Med Dir:* Elizabeth Patterson, MD

## Victor Valley Community College District

Emergency Med Tech-Paramedic Prgm
18422 Bear Valley Rd
Victorville, CA 92395
www.vvc.edu
*Prgm Dir:* Scott C Jones, MBA EMT-P
*Tel:* 760 245-4271, Ext 2338    *Fax:* 760 951-5861
*E-mail:* joness@vvc.edu
*Med Dir:* R David Kovacik, MD

## Mt San Antonio College

Emergency Med Tech-Paramedic Prgm
1100 N Grand Ave
Walnut, CA 91789
www.mtsac.edu
*Prgm Dir:* Stephen A Williams, RN MEd NREMT
*Tel:* 909 594-5611, Ext 4750    *Fax:* 909 468-3938
*E-mail:* swilliam@mtsac.edu
*Med Dir:* Roger E Toop, MD

## Santa Rosa Junior College

Emergency Med Tech-Paramedic Prgm
5743 Skylane Blvd
Windsor, CA 95492
www.santarosa.edu/ps
*Prgm Dir:* Linda V Anderson, RN
*Tel:* 707 836-2919    *Fax:* 707 836-2948
*E-mail:* landerson@santarosa.edu
*Med Dir:* Ed West, MD

## Crafton Hills College

Emergency Med Tech-Paramedic Prgm
11711 Sand Canyon Rd
Yucaipa, CA 92399
www.craftonhills.edu
*Prgm Dir:* Kathy Crow, BVE NREMT-P
*Tel:* 909 389-3220    *Fax:* 909 389-3253
*E-mail:* kcrow@craftonhills.edu
*Med Dir:* Phong Nguyen, MD

# Colorado

## Pikes Peak Community College

Emergency Med Tech-Paramedic Prgm
5675 S Academy Blvd, CC-13
Colorado Springs, CO 80906
www.ppcc.edu
*Prgm Dir:* Jeff Force, EMT-P
*Tel:* 719 540-7697    *Fax:* 719 540-7699
*E-mail:* jeff.force@ppcc.edu
*Med Dir:* Shawna Langstaff, MD

## Centura Health-St Anthony Hospitals

Emergency Med Tech-Paramedic Prgm
4231 W 16th Ave, Ste 413
Denver, CO 80204
*Prgm Dir:* Scott Phillips, BS NREMT-P
*Tel:* 303 629-4478    *Fax:* 303 629-3622
*E-mail:* scottphillips@centura.org
*Med Dir:* Gerald Estep, MD

## Community College of Aurora

Emergency Med Tech-Paramedic Prgm
9235 E 10th Dr, Rm 154
Denver, CO 80230
www.ccaurora.edu/ems
*Prgm Dir:* Bob Matoba
*Tel:* 303 340-7212    *Fax:* 303 340-7209
*E-mail:* bob.matoba@ccaurora.edu
*Med Dir:* Gilbert Pineda

## Denver Health Medical Center

Emergency Med Tech-Paramedic Prgm
190 W 6th Ave
Mail Code 3652
Denver, CO 80204
http://education.denverems.org
*Prgm Dir:* James Manson, NREMT-P
*Tel:* 303 436-5347    *Fax:* 303 436-8844
*E-mail:* jmanson@dhha.org
*Med Dir:* Kevin McVaney, MD

## HealthONE EMS

*Cosponsor:* Arapahoe Community College
Emergency Med Tech-Paramedic Prgm
333 W Hampden Ave #200
Englewood, CO 80110
www.healthoneems.com
*Prgm Dir:* Lori Burns, MSN RN
*Tel:* 303 788-8820    *Fax:* 303 788-7656
*E-mail:* lori.burns@healthonecares.com
*Med Dir:* Dylan Luyten, MD

## Pueblo Community College

Emergency Med Tech-Paramedic Prgm
900 W Orman Ave
Pueblo, CO 81004
*Prgm Dir:* Dawnelle Mathis, BS NREMT-P
*Tel:* 719 549-3489    *Fax:* 719 549-3147
*E-mail:* dawnelle.mathis@pueblocc.edu
*Med Dir:* Michele K Sweeney, MD FACEP NREMT-B

# Connecticut

## Capital Community College

Emergency Med Tech-Paramedic Prgm
950 Main St
Hartford, CT 06103
www.ccc.commnet.edu
*Prgm Dir:* Terry DeVito, RN MEd CEN EMT-P
*Tel:* 860 906-5153, Ext 1009    *Fax:* 860 906-5148
*E-mail:* tdevito@ccc.commnet.edu
*Med Dir:* Michael Gutman, MD

# Delaware

## Delaware Technical & Community College

Emergency Med Tech-Paramedic Prgm
100 Campus Dr
ETB 706
Dover, DE 19904
*Prgm Dir:* Aaron Z Royston, MS NREMT-P
*Tel:* 302 857-1325    *Fax:* 302 857-1398
*E-mail:* aroyston@college.dtcc.edu
*Med Dir:* Ross Megargel, DO

# District of Columbia

## George Washington University

Emergency Med Tech-Paramedic Prgm
2150 Pennsylvania Ave NW
Washington, DC 20057
*Prgm Dir:* Keith Monosky
*Tel:* 202 741-2945    *Fax:* 202 741-2946
*E-mail:* monosky@gwumc.edu
*Med Dir:* Robert N E French, MD

# Florida

## Manatee Technical Institute

Emergency Med Tech-Paramedic Prgm
5603 34th St W
Bradenton, FL 34210
*Prgm Dir:* Michael Brooks
*Tel:* 941 752-8100, Ext 228    *Fax:* 941 727-6254
*E-mail:* brooks2m@fc.manatee.k12.fl.us
*Med Dir:* Robert Hitchcock, MD

## Brevard Community College

Emergency Med Tech-Paramedic Prgm
1519 Clearlake Rd
Cocoa, FL 32922
*Prgm Dir:* Melissa B Robinson, RN BSH NREMT-P
*Tel:* 321 433-7587, Ext 64175    *Fax:* 321 634-3731
*E-mail:* robinsonm@brevardcc.edu
*Med Dir:* John McPherson, MD

## Broward Community College

Emergency Med Tech-Paramedic Prgm
3501 SW Davie Rd
Davie, FL 33314
*Prgm Dir:* Elizabeth Jordan, MN NREMT-P
*Tel:* 954 201-6776    *Fax:* 954 201-6397
*E-mail:* ejordan@broward.edu
*Med Dir:* Barry R Weiss, MD

## Daytona Beach Community College

Emergency Med Tech-Paramedic Prgm
PO Box 2811
Daytona Beach, FL 32120-2811
*Prgm Dir:* Martha Driscoll, AS NREMT-P
*Tel:* 386 255-3000, Ext 3249    *Fax:* 386 506-3222
*E-mail:* Driscom@dbcc.edu
*Med Dir:* John G Shedd, MD FACEP

## Lake Technical Center

Emergency Med Tech-Paramedic Prgm
2001 Kurt St
Eustis, FL 32726
*Prgm Dir:* L Johnson
*Tel:* 352 589-2250, Ext 220    *Fax:* 352 357-4776
*E-mail:* johnsonl@lake.k12.fl.us
*Med Dir:* John Geeslin, MD

## Edison College

Emergency Med Tech-Paramedic Prgm
8099 College Pkwy SW, PO Box 06210
Ft Myers, FL 33906-6210
*Prgm Dir:* Joe Crutcher
*Tel:* 239 985-8308    *Fax:* 239 489-9331
*E-mail:* jcrutcher@edison.edu
*Med Dir:* Pedro Perez, MD

## Indian River Community College

Emergency Med Tech-Paramedic Prgm
3209 Virginia Ave
Ft Pierce, FL 34981-5599
*Prgm Dir:* Marjorie Bowers, EdD RN EMT-P
*Tel:* 772 462-7533    *Fax:* 772 462-7816
*E-mail:* mbowers@ircc.edu
*Med Dir:* Yvette Witra, DO

## Santa Fe Community College

Emergency Med Tech-Paramedic Prgm
3737 NE 39th Ave
Institute of Public Safety
Gainesville, FL 32609
http://ips.sfcc.edu/ips/EMS
*Prgm Dir:* Louis Mallory, MBA REMT-P
*Tel:* 352 334-0300    *Fax:* 352 334-0329
*E-mail:* louis.mallory@sfcc.edu
*Med Dir:* Pete Gianas, MD

**Florida Community College - Jacksonville**
Emergency Med Tech-Paramedic Prgm
North Campus, 4501 Capper Rd
Jacksonville, FL 32218
www.fccj.edu
*Prgm Dir:* Marjorie Fisher, MA EMT-P
*Tel:* 904 766-6513    *Fax:* 904 766-5573
*E-mail:* mfisher@fccj.edu
*Med Dir:* Tisha Gallanter, MD

**Lake City Community College**
Emergency Med Tech-Paramedic Prgm
149 SE College Place
Lake City, FL 32025
*Prgm Dir:* Alan Espinosa, AS EMT-P
*Tel:* 386 754-4292    *Fax:* 386 754-4792
*E-mail:* espinosaa@lakecitycc.edu
*Med Dir:* Stanley Janasiewicz, MD

**Palm Beach Community College**
Emergency Med Tech-Paramedic Prgm
4200 S Congress Ave, MS60
Lake Worth, FL 33461
*Prgm Dir:* Shari Turner, NREMT-P MEd
*Tel:* 561 868-3873    *Fax:* 561 868-3814
*E-mail:* turners@pbcc.edu
*Med Dir:* Kenneth Scheppke, MD

**Miami Dade College**
Emergency Med Tech-Paramedic Prgm
Medical Center Campus
950 NW 20th St
Miami, FL 33127
*Prgm Dir:* Michael A Yoder, RN EMT-P
*Tel:* 305 237-4337    *Fax:* 305 237-4278
*E-mail:* myoder@mdcc.edu
*Med Dir:* Monica M Manasa, MD

**Pasco-Hernando Community College**
Emergency Med Tech-Paramedic Prgm
10230 Ridge Rd
New Port Richey, FL 34654-5199
www.phcc.edu
*Prgm Dir:* Toni Vineyard, EMT-P
*Tel:* 727 816-3733    *Fax:* 727 816-3417
*E-mail:* vineyat@phcc.edu
*Med Dir:* Charles Boothby, DO

**Central Florida Community College**
Emergency Med Tech-Paramedic Prgm
PO Box 1388
Ocala, FL 34478
*Prgm Dir:* Wayne Ramsey, REMT-P
*Tel:* 352 237-2111, Ext 1252
*E-mail:* ramseyw@cf.edu
*Med Dir:* Joseph Yates, MD

**Valencia Community College**
Emergency Med Tech-Paramedic Prgm
PO Box 3028
Orlando, FL 32802-3028
*Prgm Dir:* Andrea Brody, BS RN NREMT-P
*Tel:* 407 299-5000, Ext 1595    *Fax:* 407 582-1278
*E-mail:* abrody@valenciacc.edu
*Med Dir:* Mark Trach, MD FACEP

**Gulf Coast Community College**
Emergency Med Tech-Paramedic Prgm
5230 W US Hwy 98
Panama City, FL 32401
*Prgm Dir:* Daniel Finley, PhD NREMT-P
*Tel:* 850 913-3315    *Fax:* 850 747-3246
*E-mail:* dfinley@gulfcoast.edu
*Med Dir:* George Tracy, MD

**Pensacola Junior College**
Emergency Med Tech-Paramedic Prgm
Warrington Campus
5555 W Hwy 98
Pensacola, FL 32507-1097
*Prgm Dir:* James Sellers
*Tel:* 850 484-2225    *Fax:* 850 484-2364
*E-mail:* Jsellers@pjc.edu
*Med Dir:* John Hybart, MD

**Seminole Community College**
Emergency Med Tech-Paramedic Prgm
100 Weldon Blvd
Sanford, FL 32773-6199
*Prgm Dir:* Robert D Holborn, MEd EMT-P
*Tel:* 407 328-2200    *Fax:* 407 328-2098
*E-mail:* holbornr@scc-fl.edu
*Med Dir:* Ronald D Brown, MD

**Sarasota County Technical Institute**
Emergency Med Tech-Paramedic Prgm
4748 Beneva Rd
Sarasota, FL 34233
*Prgm Dir:* Linda W Swisher, EdD RN
*Tel:* 941 924-1365, Ext 381    *Fax:* 941 361-6886
*E-mail:* linda_swisher@sarasota.k12.fl.us
*Med Dir:* Steven R Newman, MD

**St Petersburg College**
Emergency Med Tech-Paramedic Prgm
Health Education Ctr
PO Box 13489
St Petersburg, FL 33733
*Prgm Dir:* Nerina Stepanovsky, MSN RN EMT-P
*Tel:* 727 341-3680    *Fax:* 727 341-3784
*E-mail:* stepanovskyn@spcollege.edu
*Med Dir:* Joe A Nelson, DO MS FAOEP

**Tallahassee Community College**
Emergency Med Tech-Paramedic Prgm
444 Appleyard Dr
Tallahassee, FL 32304
www.tcc.fl.edu
*Prgm Dir:* Brian P Dunmyer, BA RP
*Tel:* 850 201-8328    *Fax:* 850 201-8329
*E-mail:* dunmyerb@tcc.fl.edu
*Med Dir:* Sam Ashoo, MD

**Hillsborough Community College**
Emergency Med Tech-Paramedic Prgm
PO Box 30030
Tampa, FL 33630
*Prgm Dir:* William D Corso, RN EMT-P MA
*Tel:* 813 253-7454    *Fax:* 813 253-7464
*E-mail:* bcorso@hccfl.edu
*Med Dir:* I Charles Sand, MD

**Polk Community College**
Emergency Med Tech-Paramedic Prgm
999 Ave H NE
Winter Haven, FL 33881-4299
*Prgm Dir:* Don Guillette, AS EMS AS-Fire Science EMT-P
*Tel:* 863 669-2902    *Fax:* 863 297-1036
*E-mail:* dguillette@polk.edu
*Med Dir:* Manuel Salazar, MD

# Georgia

**Gwinnett Technical College**
Emergency Med Tech-Paramedic Prgm
5150 Sugarloaf Pkwy
Lawrenceville, GA 30043-5702
www.gwinnettTech.edu
*Prgm Dir:* Steven Moyers, NREMT-P MA
*Tel:* 678 226-6732    *Fax:* 770 685-1280
*E-mail:* smoyers@GwinnettTech.edu
*Med Dir:* James J Dugal, MD, FACEP

# Idaho

**Idaho State University**
Emergency Med Tech-Paramedic Prgm
College of Technology
Box 8380
Pocatello, ID 83209-8380
*Prgm Dir:* Jeff Bates, EMT-P
*Tel:* 208 282-4169    *Fax:* 208 282-4641
*E-mail:* batejef2@isu.edu
*Med Dir:* Murry Sturkie, MD

**College of Southern Idaho**
Emergency Med Tech-Paramedic Prgm
315 Falls Ave
Twin Falls, ID 83301
www.csi.edu/paramedic
*Prgm Dir:* Gordon Kokx, BS NREMT-P
*Tel:* 208 732-6710    *Fax:* 208 736-4743
*E-mail:* gkokx@csi.edu
*Med Dir:* Kevin Kraal, MD

# Illinois

**Loyola University Medical Center**
Emergency Med Tech-Paramedic Prgm
2160 S First Ave
Bldg 110LL
Maywood, IL 60153
www.loyolaems.com
*Prgm Dir:* Lauri Beechler, RN, BS
*Tel:* 708 327-2547    *Fax:* 708 327-2548
*E-mail:* lbeechler@lumc.edu
*Med Dir:* Mark Cichon, DO, FACEP, FACOEP

**Trinity College of Nursing & Health Sciences**
Emergency Med Tech-Paramedic Prgm
2122 25th Ave
Rock Island, IL 61201
www.trinitycollegeqc.edu
*Prgm Dir:* Karen Wilson, MSN RN
*Tel:* 309 779-7728    *Fax:* 309 779-7796
*E-mail:* wilsonk@trinityqc.com
*Med Dir:* Walter J Bradley, MD MBA

# Indiana

**St Francis Hospital & Health Centers**
Emergency Med Tech-Paramedic Prgm
1600 Albany St
Beech Grove, IN 46107
www.ssfhs.org
*Prgm Dir:* Brad Sparks, NREMT-P
*Tel:* 317 782-6480    *Fax:* 317 783-6481
*E-mail:* brad.sparks@ssfhs.org
*Med Dir:* Mike Russell, MD

**Elkhart General Hospital**
Emergency Med Tech-Paramedic Prgm
600 East Blvd
Elkhart, IN 46514
*Prgm Dir:* Anthony Hartman
*Tel:* 574 523-3127    *Fax:* 574 523-3458
*E-mail:* thartman@egh.org
*Med Dir:* Eugene Huang

**Ivy Tech Community College SW - Evansville**
*Cosponsor:* St Mary's Medical Center and Deaconess Hospital
Emergency Med Tech-Paramedic Prgm
3501 First Ave
Evansville, IN 47710
*Prgm Dir:* Timothy J Vollmer, MS Ed
*Tel:* 812 429-9870    *Fax:* 812 429-1495
*E-mail:* tvollmer@ivytech.edu
*Med Dir:* Peter Stevenson, MD

**Methodist Hospitals Inc**
Emergency Med Tech-Paramedic Prgm
Midlake Campus
2269 W 2269th Ave
Gary, IN 46404
*Prgm Dir:* Joseph Paunicka, BS EMT-P
*Tel:* 219 944-4160   *Fax:* 219 944-4174
*E-mail:* JPAUNICKA@methodisthospitals.org
*Med Dir:* David E Ross, MD

**Clarian Health Partners Inc**
*Cosponsor:* Methodist Hospital
Emergency Med Tech-Paramedic Prgm
Wile Hall Rm 631
1701 N Senate Blvd
Indianapolis, IN 46206
*Prgm Dir:* Lindi Holt, MS NREMT-P
*Tel:* 317 962-3327   *Fax:* 317 962-2102
*E-mail:* lholt@clarian.org
*Med Dir:* Edward Bartkus, MD

# Iowa

**Mercy Medical Center**
Emergency Med Tech-Paramedic Prgm
207 Crocker Ste 100
Des Moines, IA 50309
*Prgm Dir:* Fritz Nordengern
*Tel:* 515 643-7499   *Fax:* 515 643-7492
*E-mail:* ems@mercydesmoines.org
*Med Dir:* Tim Gerdis, DO

**University of Iowa Hospitals & Clinics**
Emergency Med Tech-Paramedic Prgm
200 Hawkins Dr, S608-1 GH
Iowa City, IA 52242
*Prgm Dir:* Douglas K York, REMT-P
*Tel:* 319 356-2597   *Fax:* 319 353-7508
*E-mail:* douglas-york@uiowa.edu
*Med Dir:* Alfred Hansen, MD PhD

# Kansas

**Coffeyville Community College**
Emergency Med Tech-Paramedic Prgm
400 W 11th St
Coffeyville, KS 67337
*Prgm Dir:* Kacia Adams, MICT I/C AS
*Tel:* 620 251-7700, Ext 2174   *Fax:* 620 252-7059
*E-mail:* kacia@coffeyville.edu
*Med Dir:* J L Christensen, DO

**Flint Hills Technical College**
Emergency Med Tech-Paramedic Prgm
3301 W 18th St
Emporia, KS 66801
www.fhtc.net
*Prgm Dir:* Carman Allen, MICT I/C BA
*Tel:* 620 343-4600, Ext 265   *Fax:* 620 343-4610
*E-mail:* callen@fhtc.net
*Med Dir:* Joel E Homung, MD

**Garden City Community College**
Emergency Med Tech-Paramedic Prgm
801 Campus Dr
Penka Bldg
Garden City, KS 67846
*Prgm Dir:* Lenora Cook
*Tel:* 316 276-9560   *Fax:* 316 276-9569
*E-mail:* lenora.cook@gcccks.edu
*Med Dir:* Harold Perkins, MD

**Barton County Community College**
Emergency Med Tech-Paramedic Prgm
245 NE 30th Rd
Great Bend, KS 67530
*Prgm Dir:* Chy Miller
*Tel:* 620 792-9347
*E-mail:* millerc@bartonccc.edu
*Med Dir:* Steve Herson, MD

**Kansas City Kansas Community College**
Emergency Med Tech-Paramedic Prgm
7250 State Ave
Kansas City, KS 66112
*Prgm Dir:* Donna Olafson
*Tel:* 913 288-7179   *Fax:* 913 288-7677
*E-mail:* dolafson@toto.net
*Med Dir:* Dennis Allen, MD

**Johnson County Community College**
Emergency Med Tech-Paramedic Prgm
12345 College Blvd
Overland Park, KS 66210-1299
*Prgm Dir:* Ray Wright, MICT
*Tel:* 913 469-8500, Ext 3175   *Fax:* 913 469-2315
*E-mail:* rwright@jccc.net
*Med Dir:* Mark Holcomb, MD

**Cowley County Community College**
Emergency Med Tech-Paramedic Prgm
1406 E 8th St
Winfield, KS 67156
www.cowley.edu/departments/allied/ems
*Prgm Dir:* Slade Griffiths, MEd MICT IC
*Tel:* 620 441-6584   *Fax:* 620 441-6589
*E-mail:* griffiths@cowley.edu
*Med Dir:* Chandy Samuel, MD

# Kentucky

**Eastern Kentucky University**
Emergency Med Tech-Paramedic Prgm
Dizney 225
Richmond, KY 40475-3135
www.emc.eku.edu
*Prgm Dir:* Nancye Davis, RN MSN
*Tel:* 859 622-1028   *Fax:* 859 622-6333
*E-mail:* nancye.davis@eku.edu
*Med Dir:* Neville Pohl, MD

# Louisiana

**Our Lady of the Lake College**
Emergency Med Tech-Paramedic Prgm
5345 Brittany Dr
Baton Rouge, LA 70808
*Prgm Dir:* Carl S Cramer
*Tel:* 225 788-1765   *Fax:* 225 768-1775
*E-mail:* ccramer@ololcollege.edu
*Med Dir:* Rohl Adi, MD

**Bossier Parish Community College**
Emergency Med Tech-Paramedic Prgm
6220 East Texas St
Bossier City, LA 71111
www.bpcc.edu
*Prgm Dir:* Austin Beard, BS EMT-P
*Tel:* 318 678-6135   *Fax:* 318 678-6199
*E-mail:* abeard@bpcc.edu
*Med Dir:* Derrel Graham, MD

**Nicholls State University**
Emergency Med Tech-Paramedic Prgm
235 Civic Center Blvd
Houma, LA 70360
www.nicholls.edu/ahs/paramedic
*Prgm Dir:* Richard K Walker, BGS NREMT-P
*Tel:* 985 876-8858   *Fax:* 985 876-8856
*E-mail:* kim.walker@nicholls.edu
*Med Dir:* Mickey VIator, MD

**Delgado Community College**
Emergency Med Tech-Paramedic Prgm
615 City Park Ave
New Orleans, LA 70119
*Prgm Dir:* Eric Randall, RN BS EMT
*Tel:* 504 483-4646   *Fax:* 504 485-2361
*E-mail:* eranda@dcc.edu
*Med Dir:* James Moises, MD

# Maryland

**Anne Arundel Community College**
Emergency Med Tech-Paramedic Prgm
101 College Pkwy
Arnold, MD 21012-1895
*Prgm Dir:* Sally Gresty
*Tel:* 410 777-7084   *Fax:* 410 777-7099
*E-mail:* sjgresty@aacc.edu
*Med Dir:* Roy A M Myers, MD

**Comm Coll of Baltimore County - Essex Campus**
Emergency Med Tech-Paramedic Prgm
7201 Rossville Blvd
Baltimore, MD 21237
www.ccbcmd.edu
*Prgm Dir:* Robert M Henderson Jr, BS NREMT-P
*Tel:* 410 780-6477   *Fax:* 410 780-6405
*E-mail:* rhenderson@ccbcmd.edu
*Med Dir:* Kevin B Gerold, DO JD

**University of Maryland Baltimore County**
Emergency Med Tech-Paramedic Prgm
Emergency Health Services Dept
1000 Hilltop Circle
Baltimore, MD 21250
www.ehs.umbc.edu
*Prgm Dir:* Dwight A Polk, MSW NREMT-P
*Tel:* 410 455-3782   *Fax:* 410 455-3045
*E-mail:* polk@umbc.edu
*Med Dir:* Kevin Seaman, MD

**Howard Community College**
Emergency Med Tech-Paramedic Prgm
10901 Little Patuxent Parkway
Columbia, MD 21044
www.howardcc.edu
*Prgm Dir:* Angel Clark Burba, MS EMT-P
*Tel:* 410 772-4948   *Fax:* 410 772-4494
*E-mail:* aburba@howardcc.edu
*Med Dir:* Kevin Seaman, MD

**Associates in Emergency Care**
Emergency Med Tech-Paramedic Prgm
PO Box 490
Damascus, MD 20872
http://aecare911.org
*Prgm Dir:* Sal E Marini, NREMT-P MA
*Tel:* 301 865-8880   *Fax:* 301 865-8881
*E-mail:* aecare911@aol.com
*Med Dir:* Jeff Joseph, DO

# Michigan

**Huron Valley Ambulance Center**
Emergency Med Tech-Paramedic Prgm
2215 Hogback Rd
Ann Arbor, MI 48105
*Prgm Dir:* Tom Ayers, EMTP I/C
*Tel:* 734 477-6731   *Fax:* 734 477-6927
*E-mail:* tayers@hva.org
*Med Dir:* Robert M Domeier, MD

**Lansing Community College**
Emergency Med Tech-Paramedic Prgm
3100-HHPS Division
500 N Washington, PO Box 40010
Lansing, MI 48901-7210
*Prgm Dir:* Tim Cooper, EMT-P I/C AAS
*Tel:* 517 483-1410   *Fax:* 517 483-1508
*E-mail:* tcooper@lcc.edu
*Med Dir:* Robert K Orr, DO

# Minnesota

### Northland Community & Technical College
Emergency Med Tech-Paramedic Prgm
2022 Central Ave NE
East Grand Forks, MN 56721
www.northlandcollege.edu
*Prgm Dir:* Dan Sponsler, BS NREMT-P
*Tel:* 218 773-4634    *Fax:* 218 773-4502
*E-mail:* dan.sponsler@northlandcollege.edu
*Med Dir:* Steven Weiser, MD

### Inver Hills Community College
Emergency Med Tech-Paramedic Prgm
2500 E 80th St
Inver Grove Heights, MN 55076-3224
*Prgm Dir:* Marti Breiter
*Tel:* 651 450-8675    *Fax:* 651 450-8579
*E-mail:* mbreite@inverhills.edu
*Med Dir:* Koren Kaye, MD

### South Central College
Emergency Med Tech-Paramedic Prgm
1920 Lee Blvd
North Mankato, MN 56003
www.sctc.mnscu.edu
*Prgm Dir:* Laurie Oelslager
*Tel:* 507 389-7306    *Fax:* 507 388-9951
*E-mail:* laurie.oelslager@southcentral.edu
*Med Dir:* Mark Rorem, MD

### Greater Minnesota Paramedic Consortium
Emergency Med Tech-Paramedic Prgm
43356 Long Lake Dr
Ottertail, MN 56571
*Prgm Dir:* Dennis A Ehrichs, NREMT-P
*Tel:* 218 367-3022    *Fax:* 218 367-3022
*E-mail:* eta@eot.com
*Med Dir:* Michael Wilcox, MD

### Rochester Community & Technical College
Emergency Med Tech-Paramedic Prgm
851 30th Ave SE
Rochester, MN 55904
*Prgm Dir:* Richard Peterson
*Tel:* 800 247-1296
*E-mail:* richard.peterson@roch.edu
*Med Dir:* Daniel Hankins, MD

### St Cloud Technical College
Emergency Med Tech-Paramedic Prgm
1540 Northway Dr
St Cloud, MN 56303
*Prgm Dir:* Larry R Starks
*Tel:* 320 308-5405    *Fax:* 320 255-2595
*E-mail:* lstarks@sctc.edu
*Med Dir:* Peter Charvat, MD

### Century College
Emergency Med Tech-Paramedic Prgm
3300 Century Ave N
White Bear Lake, MN 55110
*Prgm Dir:* Denise Howard
*Tel:* 651 779-5794    *Fax:* 651 779-5797
*E-mail:* l.goerisch@century.mnscu.edu
*Med Dir:* Marc Conterato, MD

# Mississippi

### East Central Community College
Emergency Med Tech-Paramedic Prgm
PO Box 129
Decatur, MS 39327
*Prgm Dir:* S Bush
*Tel:* 601 635-2111, Ext 214    *Fax:* 601 635-4031
*E-mail:* sbush@eccc.edu
*Med Dir:* Shawn Anderson, DO

### Jones County Junior College
Emergency Med Tech-Paramedic Prgm
900 S Court St
Ellisville, MS 39437
www.jcjc.edu
*Prgm Dir:* Gregory M Cole, BS NREMT-P CCEMT-P FP-C
*Tel:* 601 477-4074    *Fax:* 601 477-4152
*E-mail:* mike.cole@jcjc.edu
*Med Dir:* Michal Larochelle, DO

### Itawamba Community College
Emergency Med Tech-Paramedic Prgm
602 W Hill St
Fulton, MS 38843
*Prgm Dir:* Deborah Roebuck, RN REMT-P
*Tel:* 601 620-5270    *Fax:* 601 862-8150
*E-mail:* dmroebuck@iccms.edu
*Med Dir:* William Beazley, DO

### Mississippi Gulf Coast Community College
Emergency Med Tech-Paramedic Prgm
2226 Switzer Rd
Gulfport, MS 39507
*Prgm Dir:* Gary Shirley, REMTP BS ThM
*Tel:* 228 896-2554    *Fax:* 228 897-3918
*E-mail:* gary.shirley@mgccc.edu
*Med Dir:* William Bradford, MD

### University of Mississippi Medical Center
Emergency Med Tech-Paramedic Prgm
2500 N State St
Jackson, MS 39216
http://shrp.umc.edu
*Prgm Dir:* Clyde Deschamp, PhD NREMT-P
*Tel:* 601 984-5585    *Fax:* 601 815-1715
*E-mail:* cdeschamp@shrp.umsmed.edu
*Med Dir:* Frederick B Carlton, Jr, MD

### Holmes Community College
Emergency Med Tech-Paramedic Prgm
412 W Ridgeland Ave
Ridgeland, MS 39157
www.holmescc.edu
*Prgm Dir:* Blaine Riggleman, BA BS MPA
*Tel:* 601 856-5400    *Fax:* 601 605-3410
*E-mail:* briggleman@holmescc.edu
*Med Dir:* Stephen Chouteau, MD

### Northwest Mississippi Community College
Emergency Med Tech-Paramedic Prgm
4975 Hwy 51 N
Drawer 7020
Senatobia, MS 38668
www.northwestms.edu
*Prgm Dir:* Brenda Hood, RN NREMT
*Tel:* 662 562-3986    *Fax:* 662 560-1107
*E-mail:* bjhood@northwestms.edu
*Med Dir:* Frank Adcock, MD

# Missouri

### IHM Health Studies Center
Emergency Med Tech-Paramedic Prgm
2500 Abbott Pl
St Louis, MO 63143
*Prgm Dir:* John F Elder, NREMT-P
*Tel:* 314 768-1234, Ext 1128    *Fax:* 314 768-1595
*E-mail:* ihm@abbottems.org
*Med Dir:* Lawrence Lewis, MD

# Montana

### Montana State University - Billings
Emergency Med Tech-Paramedic Prgm
3803 Central Ave
Billings, MT 59102
*Prgm Dir:* David J Gurchiek, MS NREMT-P
*Tel:* 406 247-3076    *Fax:* 406 652-1729
*E-mail:* dgurchiek@msubillings.edu
*Med Dir:* Doug Parker, MD

# Nebraska

### Creighton University
Emergency Med Tech-Paramedic Prgm
Creighton University EMS Education
2514 Cuming St
Omaha, NE 68131
http://ems.creighton.edu
*Prgm Dir:* William Raynovich, NREMTP EdD
*Tel:* 402 280-1280    *Fax:* 402 280-1288
*E-mail:* billr@creighton.edu
*Med Dir:* Richard Walker, MD

# Nevada

### Community College of Southern Nevada
Emergency Med Tech-Paramedic Prgm
6375 W Charleston Blvd
Las Vegas, NV 89146
www.ccsn.nevada.edu
*Prgm Dir:* Rod Hackwith, BS NREMT-P
*Tel:* 702 651-7385    *Fax:* 702 651-5077
*E-mail:* rod_hackwith@ccsn.edu
*Med Dir:* Ross Berkeley, MD

# New Hampshire

### New Hampshire Technical Institute
Emergency Med Tech-Paramedic Prgm
31 College Dr
Concord, NH 03001-7412
www.nhti.edu
*Prgm Dir:* Nancy L Brubaker, MEd EMT-P RN
*Tel:* 603 271-7157    *Fax:* 603 271-7148
*E-mail:* nbrubaker@nhctc.edu
*Med Dir:* Patrick Lanzetta, MD

### New England EMS Institute
Emergency Med Tech-Paramedic Prgm
One Elliot Way
Manchester, NH 03103
www.neemsi.org
*Prgm Dir:* William Thorpe, Jr., BS NREMT-P I/C
*Tel:* 603 663-2641    *Fax:* 603 663-2110
*E-mail:* thorpe@elliot-hs.org
*Med Dir:* Lee Steckowych, MD

# New Mexico

### University of New Mexico
Emergency Med Tech-Paramedic Prgm
School of Medicine
2700 Yale Blvd SE
Albuquerque, NM 87106
*Prgm Dir:* Larry A Cobb, RN CEN EMT-P
*Tel:* 505 272-5757    *Fax:* 505 244-1505
*E-mail:* larryc@unm.edu
*Med Dir:* Darren Braude, MD

### Dona Ana Community College
Emergency Med Tech-Paramedic Prgm
Box 30001, Dept 3DA
Las Cruces, NM 88003
http://dabcc.nmsu.edu
*Prgm Dir:* Joyce S Bradley, NREMT-P AAS BHCS
*Tel:* 505 527-7645    *Fax:* 505 527-7765
*E-mail:* jobradle@nmsu.edu
*Med Dir:* Benjamin Diven, MD

### Eastern New Mexico University - Roswell
Emergency Med Tech-Paramedic Prgm
PO Box 6000
52 University Blvd
Roswell, NM 88202-6000
www.roswell.enmu.edu
*Prgm Dir:* Mike Buldra, MEd NREMT-P
*Tel:* 505 624-7239    *Fax:* 505 624-7100
*E-mail:* mike.buldra@roswell.enmu.edu
*Med Dir:* Matthew Foster, MD

# New York

### New York Methodist Hospital
Emergency Med Tech-Paramedic Prgm
2009 85th St
Brooklyn, NY 11214
*Prgm Dir:* Jerry Rozenberg, PA-C EMT-P
*Tel:* 718 333-2273   *Fax:* 718 654-2092
*E-mail:* ecpmedic@aol.com
*Med Dir:* Joseph Bove, MD

### Bassett Healthcare Center for Rural EMS Educ
*Cosponsor:* State University of New York - Cobleskill
Emergency Med Tech-Paramedic Prgm

Cobleskill, NY 12043
www.cremse.org
*Prgm Dir:* Richard Beebe, MEd RN EMTP
*Tel:* 518 255-5367
*E-mail:* BeebeRW@Cobleskill.Edu
*Med Dir:* Donald Doynow, MD

### CUNY Borough of Manhattan Community Coll
Emergency Med Tech-Paramedic Prgm
199 Chambers St
Dept of Allied Hlth Sciences
New York, NY 10007
*Prgm Dir:* Everett W Flannery, MPS RRT
*Tel:* 212 346-8731   *Fax:* 212 346-8738
*E-mail:* eflannery@bmcc.cuny.edu
*Med Dir:* Diane Sixsmith, MD MPH

### Monroe Community College
Emergency Med Tech-Paramedic Prgm
1190 Scottsville Rd, Ste 216
Rochester, NY 14624
www.mccparamedic.com
*Prgm Dir:* Peter Bonadonna, EMT-P NYS
*Tel:* 585 753-3713   *Fax:* 585 753-3850
*E-mail:* pbonadonna@monroecc.edu
*Med Dir:* Eric Davis, MD

### SUNY Ulster
Emergency Med Tech-Paramedic Prgm
Cottekill Rd
Stone Ridge, NY 12484
*Prgm Dir:* Douglas Sandbrook
*Tel:* 845 687-5276   *Fax:* 845 687-5133
*E-mail:* sandbrod@sunyulster.edu
*Med Dir:* Daniel G Hafner, MD

### SUNY Upstate Medical University
Emergency Med Tech-Paramedic Prgm
750 E Adams St
Syracuse, NY 13210
*Prgm Dir:* Gail J Weinstein
*Tel:* 315 464-6223   *Fax:* 315 464-4854
*E-mail:* weinsteg@upstate.edu
*Med Dir:* Richard Hunt, MD

### Hudson Valley Community College
Emergency Med Tech-Paramedic Prgm
80 Vandenburgh Ave
Troy, NY 12180
*Prgm Dir:* Gregory Chapman, RRT REMT-P
*Tel:* 518 629-4899   *Fax:* 518 629-4881
*E-mail:* chapmgre@hvcc.edu
*Med Dir:* Donald Doynow, MD FACEP

### Dutchess Community College
Emergency Med Tech-Paramedic Prgm
31 Marshall Rd
Wappingers Falls, NY 12590
*Prgm Dir:* Michele Lieberman
*Tel:* 845 298-0717
*E-mail:* lieberman@sunydutchess.edu

# North Carolina

### Western Carolina University
Emergency Med Tech-Paramedic Prgm
122 Moore
Cullowhee, NC 28723
http://emc.wcu.edu
*Prgm Dir:* Michael W Hubble, PhD
*Tel:* 828 227-7113, Ext 3516   *Fax:* 828 227-7446
*E-mail:* mhubble@email.wcu.edu
*Med Dir:* David C Trigg, MD

### Joint Special Operations Medical Training Ctr
Emergency Med Tech-Paramedic Prgm
Uniformed Services University of the Health Sciences
Ft Bragg, NC 28310
*Prgm Dir:* Jeffery Kingsbury, MD MPH
*Tel:* 910 396-7217   *Fax:* 910 396-5395
*E-mail:* kingsbje@soc.mil
*Med Dir:* Alan Davis, MD, FS

### Catawba Valley Community College
Emergency Med Tech-Paramedic Prgm
2550 Hwy 70 SE
Hickory, NC 28602-9699
www.cvcc.edu
*Prgm Dir:* Martha McCrea, RN PA BSN EMT-P
*Tel:* 828 327-7000, Ext 4347   *Fax:* 828 327-7276
*E-mail:* mmccrea@cvcc.edu
*Med Dir:* Jon Giometti, MD

# North Dakota

### University of Mary/St Alexius Medical Ctr
Emergency Med Tech-Paramedic Prgm
PO Box 5510
Bismarck, ND 58506-5510
*Prgm Dir:* Mark Haugen, NREMT-P EMS
*Tel:* 701 530-7700   *Fax:* 701 530-7701
*E-mail:* mhaugen@primecare.org
*Med Dir:* Gordon Leingang, DO

# Ohio

### Akron General Medical Center
Emergency Med Tech-Paramedic Prgm
400 Wabash Ave
Akron, OH 44307
*Prgm Dir:* Scott W Martin, MEd NREMT-P
*Tel:* 330 344-6655   *Fax:* 330 253-8293
*E-mail:* smartin@agmc.org
*Med Dir:* Thomas J Elson, MD

### Summa Health System
Emergency Med Tech-Paramedic Prgm
444 N Main St
Akron, OH 44310
*Prgm Dir:* Brian Tritchler, RN BSN EMT-P
*Tel:* 330 379-5966   *Fax:* 330 379-9543
*E-mail:* tritchlb@summa-health.org
*Med Dir:* Michael Mackan, MD

### University of Cincinnati
Emergency Med Tech-Paramedic Prgm
9555 Plainfield Rd
Cincinnati, OH 45236-0086
*Prgm Dir:* Alan F Mistler, MEd RN REMT-P
*Tel:* 513 936-7131   *Fax:* 513 745-0363
*E-mail:* alan.mistler@uc.edu
*Med Dir:* Hollynn Larrabee, MD

### Columbus State Community College
Emergency Med Tech-Paramedic Prgm
550 E Spring St
Columbus, OH 43215
*Prgm Dir:* Carolyn Steffl, EMT-P AS
*Tel:* 614 227-2510   *Fax:* 614 287-5490
*E-mail:* csteffl@cscc.edu
*Med Dir:* Douglas Rund, MD

### Parma Community General Hospital
Emergency Med Tech-Paramedic Prgm
7300 State Rd
Parma, OH 44134
*Prgm Dir:* Joseph W Toth, REMT-P
*Tel:* 440 743-4970   *Fax:* 440 743-4966
*E-mail:* jtoth@parmahospital.org
*Med Dir:* Carl B Schikowski, MD

### Youngstown State University
Emergency Med Tech-Paramedic Prgm
Department of Health Professions
One University Plaza
Youngstown, OH 44555-3327
*Prgm Dir:* Randall W Benner, MEd NREMT-P
*Tel:* 330 941-3327   *Fax:* 330 941-2921
*E-mail:* rwbenner@cc.ysu.edu
*Med Dir:* Laura Dollison, DO

# Oklahoma

### Oklahoma City Community College
Emergency Med Tech-Paramedic Prgm
7777 S May Ave
Oklahoma City, OK 73159
*Prgm Dir:* Romeo L Opichka, BBA REMT-P
*Tel:* 405 682-1611, Ext 7343   *Fax:* 405 682-7826
*E-mail:* ropichka@okccc.edu
*Med Dir:* Richard Blubaugh, DO

# Oregon

### College of Emergency Services (CESWA)
Emergency Med Tech-Paramedic Prgm
9735 SW Sunshine Ct, Ste 1000
Beaverton, OR 97005
*Prgm Dir:* Carl T Miller, MS EMT-P
*Tel:* 503 644-9999   *Fax:* 503 644-1672
*E-mail:* ces@ces-ems.org
*Med Dir:* Gregory B Lorts, MD

### Chemeketa Community College
Emergency Med Tech-Paramedic Prgm
4000 Lancaster Dr NE
PO Box 14007
Salem, OR 97309-7070
http://newterra.chemeketa.edu/faculty/macj/
   emergencyservices
*Prgm Dir:* Gregg W Lander, BS NREMT-P
*Tel:* 503 399-2664   *Fax:* 503 588-6438
*E-mail:* greggl@chemeketa.edu
*Med Dir:* Linda Johnson, MD

### Oregon Health & Science University
*Cosponsor:* Oregon Institute of Technology
Emergency Med Tech-Paramedic Prgm
12400 SW Tonquin Rd
Sherwood, OR 97140
www.oit.edu/paramedic
*Prgm Dir:* Suzann Schmele, NREMT-P
*Tel:* 503 625-4721   *Fax:* 503 625-2497
*E-mail:* schmelse@tvfr.com
*Med Dir:* James Bryan, MD PhD

# Pennsylvania

### Harrisburg Area Community College
Emergency Med Tech-Paramedic Prgm
One HACC Dr
Harrisburg, PA 17110-2999
*Prgm Dir:* Craig Davis, MEd EMT-P
*Tel:* 717 780-2564   *Fax:* 717 780-2551
*E-mail:* cadavis@hacc.edu
*Med Dir:* Jesse A Weigel, MD

**Ctr for Emer Med of Western Pennsylvania**
Emergency Med Tech-Paramedic Prgm
230 McKee Place
Ste 500
Pittsburgh, PA 15213
www.centerem.com
*Prgm Dir:* Thomas E Platt, MEd NREMT-P
*Tel:* 412 647-4665    *Fax:* 412 647-4670
*E-mail:* plattt@pitt.edu
*Med Dir:* Owen Traynor, MD

**Pennsylvania College of Technology**
Emergency Med Tech-Paramedic Prgm
One College Ave
Williamsport, PA 17701
www.pct.edu/schools/hs/paramedic
*Prgm Dir:* Mark A Trueman, BS NREMT-P
*Tel:* 570 329-4931    *Fax:* 570 320-4432
*E-mail:* mtrueman@pct.edu
*Med Dir:* Gregory Frailey, DO

# South Carolina

**Greenville Technical College**
Emergency Med Tech-Paramedic Prgm
506 S Pleasantburg Dr 106D #702
Greenville, SC 29607
http://greenvilletech.com/academic_avenues/emt.shtml
*Prgm Dir:* Michael Fisher, NREMT-P CCEMT-P AHS
*Tel:* 864 250-8490    *Fax:* 864 250-8218
*E-mail:* Mike.Fisher@gvltec.edu
*Med Dir:* Stephen E Parks, MD

# South Dakota

**Avera McKennan Hospital**
*Cosponsor:* University Health Center
Emergency Med Tech-Paramedic Prgm
800 E 21st St
Sioux Falls, SD 57117
http://averamckennanEMS.org
*Prgm Dir:* Don Jones, BS NREMT-P
*Tel:* 605 322-2086    *Fax:* 605 322-2090
*E-mail:* don.jones@mckennan.org
*Med Dir:* Stephen Karl, MD

# Tennessee

**Northeast State Technical Comm College**
Emergency Med Tech-Paramedic Prgm
PO Box 246
Blountville, TN 37617-0246
*Prgm Dir:* Darren K Ellenburg, BS EMT-P
*Tel:* 423 323-0238    *Fax:* 423 323-0213
*E-mail:* dkellenburg@NortheastState.edu
*Med Dir:* Matthew Riggins, MD

**Chattanooga State Technical Comm College**
Emergency Med Tech-Paramedic Prgm
4501 Amnicola Hwy
Chattanooga, TN 37406
www.chattanoogastate.edu
*Prgm Dir:* Curtis Aukerman, MAEd NREMT-P
*Tel:* 423 697-3332    *Fax:* 423 697-3324
*E-mail:* curtis.aukerman@chattanoogastate.edu
*Med Dir:* Christopher A Wagg, MD

**Columbia State Community College**
Emergency Med Tech-Paramedic Prgm
PO Box 1315
Columbia, TN 38402
*Prgm Dir:* Richard Beck, BS EMT-P
*Tel:* 615 790-5676
*E-mail:* rbeck4@columbiastate.edu
*Med Dir:* Mike Ricardson, MD

**Volunteer State Community College**
Emergency Med Tech-Paramedic Prgm
1480 Nashville Pike
Gallatin, TN 37066
*Prgm Dir:* Richard A Collier, BSN CEN EMT-P
*Tel:* 615 230-3346    *Fax:* 615 230-3344
*E-mail:* ric.collier@volstate.edu
*Med Dir:* John Nixon, MD

**Jackson State Community College**
Emergency Med Tech-Paramedic Prgm
2046 N Parkway St
Jackson, TN 38301-3797
*Prgm Dir:* Thomas H Coley, BEd EMT-P
*Tel:* 901 424-3520, Ext 296    *Fax:* 901 425-9551
*E-mail:* tcoley@jscc.edu
*Med Dir:* Doug Phillips, MD

**Roane State Community College**
Emergency Med Tech-Paramedic Prgm
132 Hayfield Rd
Knoxville, TN 37922-2301
*Prgm Dir:* Maria Smith, BSN CCEMTP EMT-P
*Tel:* 423 539-6905    *Fax:* 423 539-6907
*E-mail:* smith_ma@roanestate.edu
*Med Dir:* Bert Toney, MD

**Southwest Tennessee Community College**
Emergency Med Tech-Paramedic Prgm
PO Box 780
Memphis, TN 38101
*Prgm Dir:* Glenn Faught, MS EMT-P
*Tel:* 901 333-5414, Ext 5414    *Fax:* 901 333-5391
*Med Dir:* Loren Crown, MD

**Walters State Community College**
Emergency Med Tech-Paramedic Prgm
500 S Davy Crockett Pkwy
Morristown, TN 37813-6899
www.ws.edu
*Prgm Dir:* Tim Strange, BS EMT-P I/C
*Tel:* 423 585-2672    *Fax:* 423 318-2738
*E-mail:* tim.strange@ws.edu
*Med Dir:* Mark Harrell, MD

# Texas

**Austin Community College**
Emergency Med Tech-Paramedic Prgm
3401 Webberville Rd
Austin, TX 78702
www.austincc.edu/health/emsp
*Prgm Dir:* Kyle Pierce, LP BA
*Tel:* 512 223-5918    *Fax:* 512 223-5898
*E-mail:* kpierce@austincc.edu
*Med Dir:* B Duke Kimbrough, MD

**Lee College**
Emergency Med Tech-Paramedic Prgm
PO Box 818
Baytown, TX 77522-0818
*Prgm Dir:* Ernest K Whitener, MS EMT-P
*Tel:* 281 425-6484    *Fax:* 281 425-6520
*E-mail:* ewhitener@lee.edu
*Med Dir:* David Hall, MD

**Univ of Texas Southwestern Med Ctr at Dallas**
*Cosponsor:* El Centro College
Emergency Med Tech-Paramedic Prgm
5323 Harry Hines Blvd
Dallas, TX 75390-9134
*Prgm Dir:* Debra Cason, RN MS EMT-P
*Tel:* 214 648-5246    *Fax:* 214 648-5245
*E-mail:* debra.cason@UTSouthwestern.edu
*Med Dir:* James M Atkins, MD

**Galveston College**
Emergency Med Tech-Paramedic Prgm
4015 Ave Q
Galveston, TX 77550
www.gc.edu
*Prgm Dir:* Rory Prue, BS NREMT-P LP
*Tel:* 409 944-4242, Ext 494    *Fax:* 409 944-1511
*E-mail:* rprue@gc.edu
*Med Dir:* Russell Miller, MD

**Houston Community College**
Emergency Med Tech-Paramedic Prgm
1900 Galen Dr, MC 1637/H232
Houston, TX 77030
*Prgm Dir:* George W Hatch, Jr, EdD EMT-P
*Tel:* 713 718-7692, Ext 102    *Fax:* 713 718-7697
*E-mail:* george.hatch@hccs.edu
*Med Dir:* Arlo F Weltge, MD

**North Harris College**
Emergency Med Tech-Paramedic Prgm
2700 W Thorne Dr, Ste WN174
Houston, TX 77073
www.northharriscollege.com
*Prgm Dir:* Steven L Kolar, MBA RN LP EMSC
*Tel:* 281 618-5783    *Fax:* 281 618-1155
*E-mail:* steven.kolar@nhmccd.edu
*Med Dir:* Frederick L Hill, DO

**San Jacinto College North**
Emergency Med Tech-Paramedic Prgm
5800 Uvalde Rd
Houston, TX 77049-4599
*Prgm Dir:* Denise Williams, REMT-P Lic P MEd
*Tel:* 281 459-7151    *Fax:* 281 459-7603
*E-mail:* denise.williams@sjcd.edu
*Med Dir:* Steven Ellerbe, DO

**Tarrant County College**
Emergency Med Tech-Paramedic Prgm
Northeast Campus
828 Harwood Rd
Hurst, TX 76054-3299
*Prgm Dir:* Jeff McDonald, AA LP
*Tel:* 817 515-6448    *Fax:* 817 515-6700
*E-mail:* jeff.mcdonald@tccd.edu
*Med Dir:* Roy Yamada, MD

**Brazosport College**
Emergency Med Tech-Paramedic Prgm
500 College Dr
Lake Jackson, TX 77568
http://brazosport.edu/~ems
*Prgm Dir:* John Creech, AAS LP
*Tel:* 979 230-3426    *Fax:* 979 230-3390
*E-mail:* jcreech@brazosport.edu
*Med Dir:* Alan Barker, DO

**South Plains College**
Emergency Med Tech-Paramedic Prgm
1401 S College Ave
Levelland, TX 79336
www.southplainscollege.edu
*Prgm Dir:* Mike DeLoach, BS NREMT-P
*Tel:* 806 885-3048, Ext 4627    *Fax:* 806 897-5295
*E-mail:* mdeloach@southplainscollege.edu
*Med Dir:* Charles Addington, DO

**San Jacinto College Central**
Emergency Med Tech-Paramedic Prgm
8060 Spencer Hwy
PO Box 2007
Pasadena, TX 77501-2007
*Prgm Dir:* Joseph J Hamilton, MS LP
*Tel:* 713 476-1862    *Fax:* 713 478-2754
*E-mail:* Joe.Hamilton@sjcd.edu
*Med Dir:* Elizabeth Jones, MD

**Univ of Texas Hlth Sci Ctr at San Antonio**
Emergency Med Tech-Paramedic Prgm
7703 Floyd Curl Dr
San Antonio, TX 78229
www.uthscsa.edu/emt
*Prgm Dir:* Lance Villers, PhD
*Tel:* 210 567-8760    *Fax:* 210 567-8749
*E-mail:* villers@uthscsa.edu
*Med Dir:* Donald J Gordon, MD

**College of the Mainland**
Emergency Med Tech-Paramedic Prgm
1200 Amburn Rd
Texas City, TX 77591
www.com.edu
*Prgm Dir:* Cissy Matthews, MBA LP EMSC
*Tel:* 409 938-1211, Ext 461    *Fax:* 409 938-3146
*E-mail:* cmatthews@com.edu
*Med Dir:* Robert Fromm, MD

**Wharton County Junior College**
Emergency Med Tech-Paramedic Prgm
911 Boling Hwy
Wharton, TX 77488
www.wcjc.edu
*Prgm Dir:* Maggie Mejorado, LP EMT-P AAA AAS
*Tel:* 979 532-6540    *Fax:* 979 532-6541
*E-mail:* maggiem@wcjc.edu
*Med Dir:* Larry Lipscomb, MD

# Utah

**Weber State University**
Emergency Med Tech-Paramedic Prgm
3902 University Circle
Ogden, UT 84408-3902
*Prgm Dir:* Vapory Quick, MSN MREMT-P
*Tel:* 801 626-6521    *Fax:* 801 626-6610
*E-mail:* vquick@weber.edu
*Med Dir:* Jon Apfelbaum, MD

**Utah Valley State College**
Emergency Med Tech-Paramedic Prgm
3131 Mike Jense Parkway
Provo, UT 84601
*Prgm Dir:* Barry Stone, PA-C MPAS RN EMT-P
*Tel:* 801 863-7700, Ext 7747    *Fax:* 801 863-7738
*E-mail:* stoneba@uvsc.edu
*Med Dir:* Keith Hooker, MD

**Unified Fire Authority/Utah Valley State Coll**
Emergency Med Tech-Paramedic Prgm
3380 S 900 W
Salt Lake City, UT 84119
*Prgm Dir:* Karla Holmes, RN
*Tel:* 801 743-7200
*E-mail:* kholmes@ufa-slco.org
*Med Dir:* Gerald Kim Rowland, MD

**Dixie State College of Utah**
Emergency Med Tech-Paramedic Prgm
225 S 700 E
St George, UT 84770
*Prgm Dir:* Betty Wallis, RN MN
*Tel:* 435 752-7876    *Fax:* 435 656-4025
*E-mail:* wallis@dixie.edu
*Med Dir:* Gordon Larsen, MD

# Virginia

**Piedmont Virginia Community College**
*Cosponsor:* University of Virginia Dept of Emergency
  Medicine
Emergency Med Tech-Paramedic Prgm
501 College Dr
Charlottesville, VA 22902
www.pvcc.edu
*Prgm Dir:* Rita Krenz, BA BS NREMT-P
*Tel:* 434 961-5227    *Fax:* 434 961-5428
*E-mail:* rkrenz@pvcc.edu
*Med Dir:* Sabina Braithwaite, MD

**Loudoun County Dept of Fire-Rescue**
Emergency Med Tech-Paramedic Prgm
16600 Courage Court
Leesburg, VA 20175
*Prgm Dir:* Jose V Salazar
*Tel:* 703 777-0333    *Fax:* 703 777-0451
*E-mail:* jsalazar@loudoun.gov
*Med Dir:* Donald A Sabella, MD

**Southwest Virginia Community College**
Emergency Med Tech-Paramedic Prgm
PO Box SVCC
US Hwy 19
Richlands, VA 24641
www.sw.edu
*Prgm Dir:* Bill Akers Jr, MS NREMTP
*Tel:* 276 964-7729    *Fax:* 276 964-7531
*E-mail:* Bill.Akers@sw.edu
*Med Dir:* Norman L Rexrode, MD

**J Sargeant Reynolds Community College**
Emergency Med Tech-Paramedic Prgm
PO Box 85622
Richmond, VA 23285
*Prgm Dir:* Gregory S Neiman, BA NREMTP
*Tel:* 804 786-1376    *Fax:* 804 786-5298
*E-mail:* gneiman@jsr.vccs.edu
*Med Dir:* Arthur C Ernest, MD

**Virginia Commonwealth Univ/Med Coll of VA**
Emergency Med Tech-Paramedic Prgm
PO Box 980044
Richmond, VA 23298-0044
*Prgm Dir:* Daniel P Barry, MHA NREMT-P
*Tel:* 804 828-3687
*E-mail:* dbarry@hsc.vcu.edu
*Med Dir:* Rao Ivatury, MD

**Jefferson College of Health Sciences**
Emergency Med Tech-Paramedic Prgm
PO Box 13186
Roanoke, VA 24031-3186
www.jchs.edu
*Prgm Dir:* Glen Mayhew, REMT-P DHS
*Tel:* 540 985-8398    *Fax:* 540 985-9722
*E-mail:* cigrm1@jchs.edu
*Med Dir:* Sydney Vail, MD

**Northern Virginia Community College**
Emergency Med Tech-Paramedic Prgm
6699 Springfield Center Dr, Office 239
Springfield, VA 22150
www.nvcc.edu/medical/health/emt
*Prgm Dir:* Holly C Frost, MS NREMT-P
*Tel:* 703 822-6557    *Fax:* 703 822-6614
*E-mail:* hfrost@nvcc.edu
*Med Dir:* James A Vafier, MD

**Tidewater Community College**
Emergency Med Tech-Paramedic Prgm
1700 College Crescent
Virginia Beach, VA 23453
*Prgm Dir:* Lorna Ramsey, RN MSN NREMT-P
*Tel:* 757 822-7335    *Fax:* 757 822-7460
*E-mail:* lramsey@tcc.edu
*Med Dir:* Mary Kay Ross, MD

# Washington

**Bellingham Technical College**
Emergency Med Tech-Paramedic Prgm
3028 Lindbergh Ave
Bellingham, WA 98225
*Prgm Dir:* Roger Christensen
*Tel:* 360 676-6830    *Fax:* 360 738-7312
*E-mail:* Rchristensen@btc.ctc.edu
*Med Dir:* Marvin Wayne, MD

**Central Washington University**
Emergency Med Tech-Paramedic Prgm
Dept of Health, Human Performance and Nutrition
400 E University Way
Ellensburg, WA 98926
www.cwu.edu/~pehls
*Prgm Dir:* Carolyn E Booth, RN, MPH
*Tel:* 509 963-1451    *Fax:* 509 963-1848
*E-mail:* boothc@cwu.edu
*Med Dir:* Jackson Horsley, MD

**Columbia Basin College**
Emergency Med Tech-Paramedic Prgm
2600 N 20th Ave
Pasco, WA 99301
www.cbc2.org
*Prgm Dir:* Troy Stratford, BS EMT-P
*Tel:* 509 547-0511, Ext 4020    *Fax:* 509 542-5675
*E-mail:* tstratford@columbiabasin.edu
*Med Dir:* Cheryl Snyder, DO

**Harborview Med Ctr - Univ of Washington**
Emergency Med Tech-Paramedic Prgm
325 Ninth Ave, Mailbox 359727
Seattle, WA 98104
*Prgm Dir:* Roy D Waugh, EMT-P
*Tel:* 206 521-1215    *Fax:* 206 521-1914
*E-mail:* rwaugh@u.washington.edu
*Med Dir:* Michael K Copass, MD

**Spokane Community College**
Emergency Med Tech-Paramedic Prgm
N 1810 Greene St, MS 2090
Spokane, WA 99217
*Prgm Dir:* John Huckert, BA EMT-P
*Tel:* 509 533-8129    *Fax:* 509 533-8621
*E-mail:* JHuckert@scc.spokane.edu
*Med Dir:* Tim Chestnut, MD

**Tacoma Community College**
Emergency Med Tech-Paramedic Prgm
6501 S 19th St
Tacoma, WA 98466
*Prgm Dir:* Michael Smith, BS MICP
*Tel:* 253 566-5220    *Fax:* 253 566-5273
*E-mail:* msmith@tcc.ctc.edu
*Med Dir:* Jeffrey Morse, MD

**Tacoma Fire Department**
*Cosponsor:* Pierce College
Emergency Med Tech-Paramedic Prgm
2124 E Marshall Ave
Tacoma, WA 98421
*Prgm Dir:* Lisa Breitinger, RN
*Tel:* 253 591-5149    *Fax:* 253 591-5417
*E-mail:* lbreitinger@ci.tacoma.wa.us
*Med Dir:* James Billingsley, MD

**Northwest Regional Training Center**
Emergency Med Tech-Paramedic Prgm
11606 NE 66th St, Ste 103
Vancouver, WA 98662
*Prgm Dir:* Gregg D Ramirez, BS EMT P
*Tel:* 360 735-8788    *Fax:* 360 892-4350
*E-mail:* gregg.ramirez@ci.vancouver.wa.us
*Med Dir:* Lynn K Wittwer, MD

**Emergency Medical Technician-Paramedic**

| Programs* | Class Capacity | Begins | Length (months) | Award | Res Tuition | Non-res Tuition | Stipend | Offers:‡ 1 | 2 | 3 |
|---|---|---|---|---|---|---|---|---|---|---|
| **Alabama** | | | | | | | | | | |
| Gadsden State Community College | 240 | Aug Jan May | 4, 24 | Dipl, Cert, AAS | $3,780 | $6,840 | | • | | |
| George C Wallace Community College (Dothan) | 15 | Aug May | 24, 15 | Cert, AAS | $2,916 | $5,112 | | • | • | |
| H Council Trenholm State Technical College (Montgomery) | 25 | Jun Sep Dec | 18 | Cert, AAT | $1,438 | $0 | | • | | |
| Northeast Alabama Community College (Rainsville) | 20 | Jan | 17 | Cert, AAS | $3,468 | $6,060 | | | | • |
| Southern Union State Community College (Opelika) | 15 | Sep | 18 | Cert, AS | $3,240 | $6,480 | | • | | |
| University of South Alabama (Mobile) | 20 | Jan May Sep | 18 | Cert | $3,429 | $6,858 | | • | | |
| **Arkansas** | | | | | | | | | | |
| Arkansas Northeastern College (Burdette) | 15 | Jan Aug | 16 | Cert | $2,610 | $3,190 | | | | |
| Black River Technical College (Pocahontas) | 25 | Aug | 24 | Cert, AAS | $2,046 | $5,518 | | | | |
| East Arkansas Community College (Forrest City) | Open | Aug | 16 | Cert, AAS | $1,372 | $1,680 | | | | |
| National Park Community College (Hot Springs) | 20 | Aug | 24 | AAS | $1,350 | $1,560 | | | | |
| NorthWest Arkansas Community College (Bentonville) | 16 | Aug | 11, 24 | Cert, AAS | $1,885 | $3,330 | | | | |
| University of Arkansas for Medical Sciences (Little Rock) | 22 | Aug | 17, 6 | AS | $3,252 | $16,813 | | | | • |
| **California** | | | | | | | | | | |
| American River College (Sacramento) | 36 | Jan | 24, 12 | Cert, AS | $1,800 | $5,580 | | • | • | |
| Butte College (Oroville) | 20 | Aug | 9 | Cert, AS | $1,000 | $6,369 | | | | |
| California EMS Academy Inc (Foster City) | 30 | Jan Jun | 12 | Cert | $9,600 | $9,600 | | | | • |
| Crafton Hills College (Yucaipa) | 48 | Aug Jan | 10, 12 | Cert | $1,000 | $6,000 | | | | |
| Fresno City College | 24 | Nov May | 11 | Cert | $4,800 | $4,800 | | | | |
| Fresno County Paramedic Program | 25 | Apr | 16 | Cert | $5,565 | $5,565 | | | | |
| Los Angeles County Paramedic Training Inst (Commerce) | 40 | Jan Apr Jul Oct | 6 | Cert | $874 | $4,960 | | | | |
| Mt San Antonio College (Walnut) | 35 | Aug Jan May | 6 | Cert, AS EMS | $962 | $6,550 | | | | • |
| Riverside Comm Coll - Moreno Valley Campus | 40 | Aug | 12 | Dipl, Cert, AAS | $2,600 | $0 | | | | |
| Santa Rosa Junior College (Windsor) | 28 | Aug | 12 | Cert, AS | $2,500 | $7,000 | | | | • |
| Southwestern College (Chula Vista) | 48 | Sep | 10, 24 | Cert, AS | $884 | $4,114 | | | | • |
| Victor Valley Community College District (Victorville) | 40 | Jan Jun | 12 | Cert | $925 | $4,615 | | • | | |
| **Colorado** | | | | | | | | | | |
| Community College of Aurora (Denver) | 30 | Jan Aug | 18 | Cert | $6,100 | $16,875 | $86 | | | |
| Denver Health Medical Center | 26 | Sep | 12 | Cert | $7,500 | $7,500 | | • | • | |
| HealthONE EMS (Englewood) | 25 | Jan Jun | 6, 12 | Cert, Assoc | $5,500 | $5,500 | | | | |
| Pikes Peak Community College (Colorado Springs) | 24 | Aug | 12, 24 | Cert, AASEMS | $5,000 | $15,753 | | | | |
| **Connecticut** | | | | | | | | | | |
| Capital Community College (Hartford) | 30 | Aug Jan | 12 | Cert, AS | $3,455 | $7,593 | | • | • | |
| **Florida** | | | | | | | | | | |
| Florida Community College - Jacksonville | 24 | Jan Sep | 16, 24 | Cert, AS/AAS | $2,060 | $8,230 | | • | | |
| Hillsborough Community College (Tampa) | 40 | Aug Jan May | 10 | Cert, AS | $3,400 | $10,200 | | | | |
| Pasco-Hernando Community College (New Port Richey) | 30 | Aug | 11 | Cert, ATD | $2,622 | $10,110 | | • | | |
| Polk Community College (Winter Haven) | 24 | Aug | 12, 24 | Cert, AS | $2,900 | $9,900 | | | | |
| Santa Fe Community College (Gainesville) | 24 | Aug | 12 | Cert, AS | $2,741 | $10,094 | | | | |
| Tallahassee Community College | 15 | Aug | 11, 23 | Cert, AS | $1,550 | $4,650 | | | | |
| **Georgia** | | | | | | | | | | |
| Gwinnett Technical College (Lawrenceville) | 30 | Sep | 24 | Dipl, AAT | $1,500 | $2,748 | | | | |
| **Idaho** | | | | | | | | | | |
| College of Southern Idaho (Twin Falls) | 12 | Jan | 18 | Cert, AAS, TC | $2,000 | $5,600 | | | | • |
| **Illinois** | | | | | | | | | | |
| Loyola University Medical Center (Maywood) | 32 | Jan | 11 | Cert | $3,000 | $3,000 | | | | |
| Trinity College of Nursing & Health Sciences (Rock Island) | 20 | Aug | 10 | Cert, AAS | $8,000 | $8,000 | | | | • |
| **Indiana** | | | | | | | | | | |
| Ivy Tech Community College SW - Evansville | 24 | Aug | 20 | Cert, AAS | $3,014 | $6,092 | | • | | |
| Methodist Hospitals Inc (Gary) | 20 | Aug | 16 | Cert, AS | $3,000 | $3,000 | | | | |
| St Francis Hospital & Health Centers (Beech Grove) | 20 | Sep | 13 | Cert | $3,400 | $3,400 | | | | |
| **Kansas** | | | | | | | | | | |
| Cowley County Community College (Winfield) | 20 | Jan | 12, 18 | AAS | $3,470 | $6,117 | | • | • | |
| Flint Hills Technical College (Emporia) | 20 | Aug | 24 | AS | $3,400 | $3,400 | | | | • |
| **Kentucky** | | | | | | | | | | |
| Eastern Kentucky University (Richmond) | 20 | Aug | 12, 24 | Cert, AS, BS | $2,236 | $6,136 | | • | | |
| **Louisiana** | | | | | | | | | | |
| Bossier Parish Community College (Bossier City) | 40 | Aug | 20 | Cert, AS | $1,394 | $3,534 | | • | • | |
| Nicholls State University (Houma) | 18 | Jun | 24 | AS | $5,200 | $12,600 | | | | |
| **Maryland** | | | | | | | | | | |
| Anne Arundel Community College (Arnold) | 21 | Jan | 18, 24 | Cert, AAS | $5,500 | $5,500 | | • | | • |
| Associates in Emergency Care (Damascus) | 30 | Aug | 10 | Cert | $4,000 | $4,000 | | • | | |
| Comm Coll of Baltimore County - Essex Campus | 25 | Sep | 24 | Dipl, Cert, AAS | $2,600 | $4,900 | | • | | • |
| Howard Community College (Columbia) | 15 | Aug | 11 | Cert, AAS | $4,200 | $7,520 | | | | |
| University of Maryland Baltimore County | 15 | Sep | 48 | BS, MS | $8,622 | $17,354 | | | | • |

*Data are shown only for programs that completed the 2006 AMA Survey of Health Professions Education Programs
‡Key to Offers: 1: Evening or weekend classes; 2: Non-English instruction; 3: Cultural competence instruction

**Emergency Medical Technician-Paramedic**

| Programs* | Class Capacity | Begins | Length (months) | Award | Res Tuition | Non-res Tuition | Stipend | Offers:‡ 1 | 2 | 3 |
|---|---|---|---|---|---|---|---|---|---|---|
| **Minnesota** | | | | | | | | | | |
| Northland Community & Technical College (East Grand Forks) | 20 | Aug | 21 | AAS | $4,700 | $4,700 | | | | |
| South Central College (North Mankato) | 25 | Aug | 16 | Dipl, AAS | $2,216 | $4,432 | | | | • |
| **Mississippi** | | | | | | | | | | |
| Holmes Community College (Ridgeland) | 16 | Aug | 12, 16 | Cert, AAS | $1,524 | $2,124 | | • | | • |
| Jones County Junior College (Ellisville) | 15 | Aug Jan | 18 | AAS | $1,800 | $2,952 | | • | | |
| Mississippi Gulf Coast Community College (Gulfport) | 30 | Aug | 12 | Cert, AAS | $2,235 | $5,004 | | | | |
| Northwest Mississippi Community College (Senatobia) | 25 | Aug | 16 | AS | $1,700 | $2,000 | | | | |
| University of Mississippi Medical Center (Jackson) | 25 | Aug | 12 | Cert | $2,250 | $2,913 | | | | • |
| **Nebraska** | | | | | | | | | | |
| Creighton University (Omaha) | 50 | Aug | 12 | Dipl, Cert, BS EMS, AS | $25,000 | $25,000 | | • | | |
| **Nevada** | | | | | | | | | | |
| Community College of Southern Nevada (Las Vegas) | 24 | Jan Sep | 11, 13 | Cert, AAS | $1,485 | $3,700 | | | | |
| **New Hampshire** | | | | | | | | | | |
| New England EMS Institute (Manchester) | 30 | Varies | 15 | Cert | $8,125 | $8,125 | | • | | |
| New Hampshire Technical Institute (Concord) | 14 | Aug 29 | 18 | AS | $4,200 | $0 | | | | |
| **New Mexico** | | | | | | | | | | |
| Dona Ana Community College (Las Cruces) | 20 | Jul | 12, 36 | Cert, AAS | $2,695 | $6,615 | | • | • | • |
| Eastern New Mexico University - Roswell | 24 | Jun | 14, 24 | Cert, CC, AS | $2,300 | $5,900 | | • | • | • |
| **New York** | | | | | | | | | | |
| Bassett Healthcare Center for Rural EMS Educ (Cobleskill) | 30 | Sep | 12, 24 | Cert, AAS, AS | $4,000 | $5,500 | | • | | |
| Monroe Community College (Rochester) | 30 | Jan | 18 | Dipl, Cert, AAS | $3,000 | $6,000 | | • | | |
| **North Carolina** | | | | | | | | | | |
| Catawba Valley Community College (Hickory) | 25 | Aug | 24 | AAS | $1,461 | $8,121 | | | | • |
| Joint Special Operations Medical Training Ctr (Ft Bragg) | 144 | Every 6 wks | 6 | Cert | | | | | | |
| Western Carolina University (Cullowhee) | 20 | Aug | 36 | BS | $3,945 | $13,528 | | | | |
| **Ohio** | | | | | | | | | | |
| Summa Health System (Akron) | 35 | Jan | 10 | Cert | $3,200 | $3,200 | | • | | |
| University of Cincinnati | 36 | Aug Feb | 10 | Cert, AD | $5,000 | $12,800 | | • | | |
| **Oregon** | | | | | | | | | | |
| Chemeketa Community College (Salem) | 24 | Sep Apr | 18 | AAS | $2,852 | $9,338 | | | | • |
| Oregon Health & Science University (Sherwood) | 24 | Sep | 12 | Dipl, AAS | $14,800 | $18,009 | | | | |
| **Pennsylvania** | | | | | | | | | | |
| Ctr for Emer Med of Western Pennsylvania (Pittsburgh) | 48 | Aug | 10, 24 | Cert, BS | $4,790 | $8,380 | | | | |
| Pennsylvania College of Technology (Williamsport) | 32 | Aug | 24, 13 | Cert, AAS | $10,360 | $13,061 | | | | |
| **South Carolina** | | | | | | | | | | |
| Greenville Technical College | 36 | Aug (day) Jan (eve) | 21 | Cert, AS | $4,125 | $8,625 | | • | | • |
| **South Dakota** | | | | | | | | | | |
| Avera McKennan Hospital (Sioux Falls) | 25 | Feb Sep | 10, 12 | Dipl, Cert, AA, BS | $4,750 | $4,750 | | • | | • |
| **Tennessee** | | | | | | | | | | |
| Chattanooga State Technical Comm College | 75 | Aug | 12 | Dipl, Cert, AAS | $3,000 | $9,000 | | • | | |
| Southwest Tennessee Community College (Memphis) | 75 | Sep | 12 | Cert | $1,536 | $6,144 | | • | | |
| Walters State Community College (Morristown) | 50 | Aug | 12, 24 | Dipl, Cert, AAS | $3,448 | $12,336 | | | | |
| **Texas** | | | | | | | | | | |
| Austin Community College | 25 | Sep May | 9, 24 | Cert, EMT-I, AAS | $2,222 | $3,672 | | | | |
| Brazosport College (Lake Jackson) | 20 | Aug | 12 | Dipl, Cert, AAS | $716 | $1,873 | | • | | |
| College of the Mainland (Texas City) | 20 | Jun | 12 | Cert, AAS | $950 | $1,795 | | • | | |
| Galveston College | 20 | Sep | 8 | Cert, AA | $600 | $770 | | • | | |
| North Harris College (Houston) | 50 | Aug | 21 | Cert, AAS | $852 | $1,692 | | • | | • |
| San Jacinto College Central (Pasadena) | 60 | Jan Sep | 17 | Cert, AAS | $750 | $1,125 | | • | • | • |
| South Plains College (Levelland) | | | | | | | | | | |
| Tarrant County College (Hurst) | 24 | Aug Jan | 16, 13 | Cert, AAS | $987 | $1,239 | | • | | |
| Univ of Texas Hlth Sci Ctr at San Antonio | 30 | Varies | 9, 4 | Cert | $3,168 | $12,243 | | | | |
| Univ of Texas Southwestern Med Ctr at Dallas | 38 | Aug Feb | 6, 13 | Cert | $2,203 | $4,168 | | | | • |
| Wharton County Junior College | 20 | Fall semester | 8 | Cert, AAS | $590 | $980 | | • | • | • |
| **Utah** | | | | | | | | | | |
| Utah Valley State College (Provo) | 24 | Aug Jan | 8 | Cert | $1,400 | $3,600 | | | | |
| **Virginia** | | | | | | | | | | |
| Jefferson College of Health Sciences (Roanoke) | 30 | Aug | 18 | AAS | $13,000 | $0 | | • | | |
| Northern Virginia Community College (Springfield) | 25 | Aug Jan | 16 | Cert, AAS | $1,421 | $6,240 | | • | • | • |
| Piedmont Virginia Community College (Charlottesville) | 20 | Aug | 24 | Cert, AAS | $800 | $1,600 | | • | | |
| Southwest Virginia Community College (Richlands) | 75 | Aug | 22 | AAS | $3,200 | $10,000 | | • | | • |
| **Washington** | | | | | | | | | | |
| Central Washington University (Ellensburg) | 25 | Sep | 12 | Cert, BS | $4,872 | $15,240 | | | | • |
| Columbia Basin College (Pasco) | 24 | Jan | 18 | Dipl, Cert, AAS | $2,800 | $4,000 | | | | • |
| Harborview Med Ctr - Univ of Washington (Seattle) | 20 | Oct | 10 | Cert | $8,000 | $0 | | | | |

*Data are shown only for programs that completed the 2006 AMA Survey of Health Professions Education Programs

‡Key to Offers: 1: Evening or weekend classes; 2: Non-English instruction; 3: Cultural competence instruction

**Includes:**
- Exercise physiology (clinical and applied)
- Exercise science
- Personal fitness training

## Exercise Physiology

### Occupational/Job Description

Exercise physiology is a discipline that includes clinical exercise physiology and applied exercise physiology. Applied exercise physiologists manage programs to assess, design, and implement individual and group exercise and fitness programs for apparently healthy individuals and individuals with controlled disease. Clinical exercise physiologists work under the direction of a physician to apply physical activity and behavioral interventions in clinical situations where they have been scientifically proven to provide therapeutic or functional benefit.

### Employment Characteristics

As a clinical part of the health and wellness team, exercise physiologists can work with personal fitness trainers, exercise science professionals, and physicians in cardiac rehabilitation, typically in a hospital or clinical setting. Exercise physiologists work with clients who have been diagnosed with a chronic metabolic, pulmonary, or cardiac disease.

### Educational Programs

**Length:** Exercise physiologist programs can be completed in a 2-year master's degree level program.
**Prerequisites:** Applicants should have a high school diploma or equivalent, meet the specific institutional entrance requirements, and have a bachelor's degree in exercise science.
**Curriculum:** Exercise physiologist programs will include a comprehensive academic curriculum and at least one culminating internship experience.

## Exercise Science

### Occupational/Job Description

Exercise science encompasses a wide variety of disciplines including, but not limited to, biomechanics, sports nutrition, sport psychology, motor control/development, and exercise physiology. The study of these disciplines is integrated into the academic preparation of exercise science professionals. Exercise science professionals work in the health and fitness industry and are skilled in evaluating health behaviors and risk factors, conducting fitness assessments, writing appropriate exercise prescriptions, and motivating individuals to modify negative health habits and maintain positive lifestyle behaviors for health promotion.

### Employment Characteristics

As an integral part of the health and wellness team, exercise science professionals can work with personal fitness trainers and exercise physiologists in a number of different settings, such as corporate, clinical, community, and commercial fitness and wellness centers. Exercise science professionals work with the apparently healthy population and clients with controlled disease, leading and demonstrating these clients in safe and effective methods of exercise. The exercise science professional can also assess risk factors and identify the health status of clients.

### Educational Programs

**Length:** Exercise science programs can be completed in a 4-year bachelor's degree level program.
**Prerequisites:** Applicants should have a high school diploma or equivalent and meet the specific institutional entrance requirements.
**Curriculum:** Exercise science programs include a comprehensive academic curriculum and at least one culminating internship experience.

## Personal Fitness Training

### Occupational/Job Description

Personal fitness trainers are skilled practitioners who work with a wide variety of client demographics in one-to-one and small group environments. They are familiar with multiple forms of exercise used to improve and maintain health-related components of physical fitness and performance. They are knowledgeable in basic assessment and development of exercise recommendations. In addition, they are proficient in leading and demonstrating safe and effective methods of exercise and motivating individuals to begin and to continue with healthy behaviors. They consult with and refer to other appropriate allied health professionals when client conditions exceed the personal trainer's education, training, and experience.

### Employment Characteristics

As an integral part of the health and wellness team, personal fitness trainers can work with exercise science professionals and exercise physiologists in a number of different settings, such as corporate, clinical, community, and commercial fitness and wellness centers. Personal fitness training involves working with the apparently healthy population, leading and demonstrating these clients in safe and effective methods of exercise.

### Educational Programs

**Length:** Personal fitness training programs can be completed in a 1-year certificate program or in a 2-year associate's degree level program.
**Prerequisites:** Applicants should have a high school diploma or equivalent and meet the specific institutional entrance requirements.

**Curriculum:** Personal fitness training programs will include a comprehensive academic curriculum and at least one culminating internship experience.

 **Inquiries**

**Careers**

American Alliance for Health, Physical Education, Recreation, and Dance (AAHPERD)
1900 Association Drive
Reston, VA 20191-1598
800 213-7193
www.aahperd.org

American Association of Cardiovascular and Pulmonary Rehabilitation (AACVPR)
401 North Michigan Avenue, Suite 2200
Chicago, IL 60611
312 321-5146
www.aacvpr.org

Medical Fitness Association (MFA)
PO Box 73103
Richmond, VA 23235-8026
804 327-0330
www.medicalfitness.org

American Kinesiotherapy Association (AKTA)
CCB-KT
PO Box 1390
Hines, IL 60141-1390
800 296-2582
www.akta.org

**Certification**

American College of Sports Medicine (ACSM)
401 West Michigan Street
Indianapolis, IN 46202
317 637-9200
www.acsm.org

National Strength and Conditioning Association (NSCA)
1885 Bob Johnson Drive
Colorado Springs, CO 80906
800 815-6826
www.nsca.com

National Academy of Sports Medicine (NASM)
26632 Agoura Road
Calabasas, CA 91302
800 460-6276
www.nasm.org

The Cooper Institute
12330 Preston Road
Dallas, TX 75230
972 341-3200
www.cooperinst.org

**Program Accreditation**

Commission on Accreditation of Allied Health Education Programs (CAAHEP) in collaboration with:
Committee on Accreditation of the Exercise Sciences (CoAES)
401 West Michigan Street
Indianapolis, IN 46202
317 352-3826
www.coaes.org

## Exercise Physiology

### Louisiana

**University of Louisiana at Monroe**
Exercise Physiology Prgm
Department of Kinesiology
700 University Ave
Monroe, LA 71209
www.ulm.edu
*Prgm Dir:* Lisa Colvin, PhD
*Tel:* 318 342-1306   *Fax:* 318 342-1308
*E-mail:* lcolvin@ulm.edu
*Med Dir:* Frank Rizzo, MD

### Pennsylvania

**East Stroudsburg University**
Exercise Physiology Prgm
Exercise Physiology
Koehler Fieldhouse
East Stroudsburg, PA 18301
*Prgm Dir:* Donald Cummings, PhD
*Tel:* 570 422-3302
*E-mail:* dcummings@po-box.esu.edu

## Exercise Science

### Louisiana

**University of Louisiana at Monroe**
Exercise Science Prgm
Department of Kinesiology
700 University Ave
Monroe, LA 71209
www.ulm.edu
*Prgm Dir:* Lisa Colvin, PhD
*Tel:* 318 342-1306   *Fax:* 318 342-1308
*E-mail:* lcolvin@ulm.edu
*Med Dir:* Frank Rizzo, MD

### Maryland

**Salisbury University**
Exercise Science Prgm
Maggs Physical Acitivity Center
Salisbury, MD 21801
*Prgm Dir:* Susan Muller, MD
*Tel:* 410 548-5555
*E-mail:* smmuller@salisbury.edu

### Massachusetts

**Westfield State College**
Exercise Science Prgm
Department of Movement Sciences
577 Western Ave
Westfield, MA 01086
*Prgm Dir:* Theresa Fitts, DPE
*Tel:* 413 572-5368
*E-mail:* tfitts@wsc.ma.edu

### North Dakota

**North Dakota State University**
Exercise Science Prgm
Bentson Bunker Fieldhouse
Fargo, ND 58105
*Prgm Dir:* Donna Terbizan
*Tel:* 701 231-7792   *Fax:* 701 231-8872
*E-mail:* d.terbizan@ndsu.edu

### Pennsylvania

**East Stroudsburg University**
Exercise Science Prgm
Koehler Fieldhouse
East Stroudsburg, PA 18301
*Prgm Dir:* Donald Cummings, PhD
*Tel:* 570 422-3302
*E-mail:* dcummings@po-box.esu.edu

**Slippery Rock University**
Exercise Science Prgm
128 E Gym, Stoner Complex
Slippery Rock, PA 16057
*Prgm Dir:* Patricia Pierce, FACSM
*Tel:* 724 738-2830
*E-mail:* patricia.pierce@sru.edu

## Exercise Physiology

| Programs* | Class Capacity | Begins | Length (months) | Award | Res Tuition | Non-res Tuition | Stipend | Offers:‡ 1 | 2 | 3 |
|---|---|---|---|---|---|---|---|---|---|---|
| **Louisiana** | | | | | | | | | | |
| University of Louisiana at Monroe | 20 | Each semester | 12 | MS | | | | • | | |

## Exercise Science

| Programs* | Class Capacity | Begins | Length (months) | Award | Res Tuition | Non-res Tuition | Stipend | Offers:‡ 1 | 2 | 3 |
|---|---|---|---|---|---|---|---|---|---|---|
| **Louisiana** | | | | | | | | | | |
| University of Louisiana at Monroe | | Each semester | | BS | | | | • | | |
| **North Dakota** | | | | | | | | | | |
| North Dakota State University (Fargo) | 50 | Fall semester | 48 | BS | | | | | | |

*Data are shown only for programs that completed the 2006 AMA Survey of Health Professions Education Programs
‡Key to Offers: 1: Evening or weekend classes; 2: Non-English instruction; 3: Cultural competence instruction

# Genetic Counselor

## Occupational Description

A genetic counselor is a health care professional who is academically and clinically prepared to provide genetic counseling services to individuals and families seeking information about the occurrence, or risk of occurrence, of a genetic condition or birth defect.

## Job Description

The genetic counseling process involves the collection and interpretation of family, genetic, medical, and psychosocial history information. Analysis of this information, together with an understanding of genetic principles and the knowledge of current technologies, provides clients and their families with information about risk, prognosis, medical management, and diagnostic and prevention options. Information is discussed in a client-centered manner while respecting the broad spectrum of beliefs and value systems that exist in our society. The genetic counseling process ultimately facilitates informed decision-making and promotes behaviors that reduce the risk of disease.

## Employment Characteristics

Genetic counselors practice as part of a health care team. The settings in which genetic counselors work include hospitals and medical centers, public health agencies, colleges and universities, diagnostic laboratories and biotechnology companies, research institutions, private practice, and governmental agencies.

According to the 2004 Professional Status Survey (PSS), conducted by the National Society of Genetic Counselors (NSGC), 73% of all counselors work in university medical centers, private hospitals, and medical facilities. Others are employed by managed care organizations, diagnostic laboratories, federal and state government offices, public health agencies, private practice, and outreach clinics.

Prenatal, pediatric, adult, and cancer are the most common specialty areas in which genetic counselors work, according the 2004 PSS. Other specialty areas include neurogenetics, psychiatry, infertility, laboratory testing/screening, and disease-specific clinics.

In addition to health care, advances made by the Human Genome Project have relevance in many other areas. As a result, genetic counselors have expanded and adapted their skills into areas such as research, industry, education, policy, pubic health, administration, and advocacy.

## Salary

According to the 2004 PSS, the yearly gross salaries reported by survey respondents range from $33,000 to $97,000, with an average of $53,000. Salaries vary by location and are highest in California, New York, and New Jersey and lowest in the southeastern United States. Refer to Section IV, Table 5 of this *Directory* for more information, or see www.ama-assn.org/go/salary.

## Educational Programs

Genetic counselors have a master's degree from a graduate program specifically accredited to prepare individuals for a career as a genetic counselor. The training is specialized and includes coursework and hands-on supervised clinical experiences.

The coursework includes instruction in the following general content areas:
- human, medical and clinical genetics
- psychosocial theory and techniques
- social, ethical and legal issues
- health care delivery systems and public health principles
- teaching techniques
- research methods

The supervised experiences expose students to the natural history and management of and psychosocial issues associated with common genetic conditions and birth defects. Students develop their genetic counseling skills in a variety of clinical genetics settings. Students also obtain experience in teaching, laboratory methods, and research.

## Inquiries

**Careers**
Kristen Smith, CAE, Executive Director
National Society of Genetic Counselors
401 North Michigan Avenue
Chicago, IL 60611
312 321-6834
312 673-6972 Fax
E-mail: nsgc@nsgc.org

**Credentialing/ Program Accreditation**
Sharon Robinson, MS, Administrator
American Board of Genetic Counseling
9650 Rockville Pike
Bethesda, MD 20814-3998
E-mail: ABGC@genetics.faseb.org
www.abgc.net

## Genetic Counselor*

## Arkansas

**University of Arkansas for Medical Sciences**
*Cosponsor:* College of Health Related Professions
Genetic Counseling Prgm
Department of Genetic Counseling
4301 W Markham St #836
Little Rock, AR 72205
www.uams.edu/chrp/genetics
*Prgm Dir:* Bruce Haas, MS CGC
*Tel:* 501 526-7700   *Fax:* 501 526-7711
*E-mail:* GeneticCounseling@uams.edu
*Med Dir:* Bradley Schaefer, MD

## British Columbia, Canada

**University of British Columbia**
Genetic Counseling Prgm
Department of Medical Genetics
Children's & Women's Health Centre of BC
Vancouver, BC V6H 3NI
www.medgen.ubc.ca/programs/mast-gen.htm
*Prgm Dir:* Anita Dircks, MSc CGC CCGC
*Tel:* 604 875-2832   *Fax:* 604 875-3490
*E-mail:* adircks@cw.bc.ca
*Med Dir:* Dessa Sadovnick, PhD

## California

**California State University - Northridge**
Genetic Counseling Prgm
18111 Nordhoff St
Northridge, CA 91330
www.csun.edu/geneticcounseling
*Prgm Dir:* Aida Metzenberg, PhD
*Tel:* 818 677-3355   *Fax:* 818 677-6692
*E-mail:* aida.metzenberg@csun.edu
*Med Dir:* Linda Cowan, MD

**University of CA-Irvine School of Medicine**
Genetic Counseling Prgm
Div of Human Genetics
Dept of Pediatrics, UCIMC-ZOT 4482
Orange, CA 92868
www.ucihs.uci.edu/pediatrics/gcprogram
*Prgm Dir:* Ann P Walker, MA CGC
*Tel:* 714 456-5789   *Fax:* 714 456-5330
*E-mail:* awalker@uci.edu
*Med Dir:* Maureen Bocian, MD FACMG FAAP

## Colorado

**U of Colorado at Denver and Health Scis Ctr**
Genetic Counseling Prgm
The Children's Hospital
1056 E 19th Ave, Box B300
Denver, CO 80218
www.uchsc.edu/gs/gs
*Prgm Dir:* Carol Walton, MS CGC
*Tel:* 303 861-6395   *Fax:* 303 861-3921
*E-mail:* walton.carol@tchden.org
*Med Dir:* David Manchester, MD

## District of Columbia

**Howard University**
Genetic Counseling Prgm
520 W St NW, Box 75
Washington, DC 20059
*Prgm Dir:* Verle Headings, MD PhD
*Tel:* 202 806-9814   *Fax:* 202 806-7058
*E-mail:* vheadings@fac.howard.edu
*Med Dir:* Verle Headings, MD PhD

## Illinois

**Northwestern University**
Genetic Counseling Prgm
Feinberg School of Medicine
676 N St Clair #1280
Chicago, IL 60611
www.cgm.northwestern.edu/gpgc.htm
*Prgm Dir:* Kelly Ormond, MS
*Tel:* 312 926-7726   *Fax:* 312 926-3553
*E-mail:* geneticcounseling@northwestern.edu
*Med Dir:* Lee Shulman, MD

## Indiana

**Indiana University**
Genetic Counseling Prgm
School of Medicine
Dept of Medical and Molecular Genetics
Indianapolis, IN 46202-5251
www.iupui.edu/~medgen
*Prgm Dir:* Paula Delk, MS
*Tel:* 317 278-8837   *Fax:* 317 274-2387
*E-mail:* pwinter@iupui.edu
*Med Dir:* Wilfredo Torres-Martinez, MD

## Maryland

**University of Maryland**
Genetic Counseling Prgm
660 W Redwood St, Ste 570
Baltimore, MD 21201
*Prgm Dir:* Shannon DeLany Dixon, MS CGC
*Tel:* 410 706-4713   *Fax:* 410 706-1644
*E-mail:* sdelany@som.umaryland.edu
*Med Dir:* Eric Wulfsberg, MD

**Johns Hopkins Univ/Natl Human Genome Research**
*Cosponsor:* National Institutes of Health
Genetic Counseling Prgm
Social and Behavioral Research Branch NHGRI/NIH
2 Center Dr MSC0249, Bldg 2, Rm 4W17
Bethesda, MD 20892
www.nhgri.nih.gov/intramural_research
*Prgm Dir:* Barbara Biesecker, MS
*Tel:* 301 496-3979   *Fax:* 301 480-3108
*E-mail:* barbarab@mail.nih.gov
*Med Dir:* Alan Guttmacher, MD

## Massachusetts

**Boston University**
Genetic Counseling Prgm
Center for Human Genetics
715 Albany St, W408
Boston, MA 02118
www.bumc.bu.edu/hg
*Prgm Dir:* MaryAnn Whalen, MS
*Tel:* 617 638-7170
*E-mail:* maryann@bu.edu
*Med Dir:* Jeff Milunsky, MD

**Brandeis University**
Genetic Counseling Prgm
Biology Dept MS008
Brandeis University
Waltham, MA 02454
www.bio.brandeis.edu/gc01/index.html
*Prgm Dir:* Judith Tsipis, PhD
*Tel:* 781 736-3165   *Fax:* 781 736-3107
*E-mail:* tsipis@brandeis.edu
*Med Dir:* Joan Stoler, MD

## Michigan

**University of Michigan**
Genetic Counseling Prgm
Dept of Human Genetics
4909 Buhl
Ann Arbor, MI 48109-0618
www.hg.med.umich.edu:16080/GCWeb
*Prgm Dir:* Beverly M Yashar, MS PhD
*Tel:* 734 763-2933   *Fax:* 734 763-3784
*E-mail:* yashar@umich.edu
*Med Dir:* Elizabeth Petty, MD

**Wayne State University**
Genetic Counseling Prgm
540 E Canfield, 3216 Scott Hall
Detroit, MI 48201
*Prgm Dir:* Anne Greb, MS
*Tel:* 313 577-6298   *Fax:* 313 577-9137
*E-mail:* agreb@genetics.wayne.edu
*Med Dir:* Laura S Martin, MD

## Minnesota

**University of Minnesota - Minneapolis**
Genetic Counseling Prgm
Department of Genetics, Cell Biology & Development
University of Minnesota Institute of Human Genetics
Minneapolis, MN 55455
www.cbs.umn.edu/gc
*Prgm Dir:* Bonnie S LeRoy, MS CGC
*Tel:* 612 624-7193   *Fax:* 612 625-4490
*E-mail:* leroy001@umn.edu
*Med Dir:* Richard A King, MD PhD

## New York

**Sarah Lawrence College**
Genetic Counseling Prgm
One Mead Way
Bronxville, NY 10708-5999
*Prgm Dir:* Caroline Lieber, MS
*Tel:* 914 395-2371   *Fax:* 914 395-2664
*E-mail:* clieber@sarahlawrence.edu
*Med Dir:* Jessica Davis, MD

**Mt Sinai School of Medicine**
Genetic Counseling Prgm
Dept of Human Genetics, Box 1497
One Gustave Levy Pl
New York, NY 10029
*Prgm Dir:* Randi Zinberg, MS CGC
*Tel:* 212 241-6947   *Fax:* 212 860-3316
*E-mail:* randi.zinberg@mssm.edu
*Med Dir:* Judith P Willner, MD

## North Carolina

**Univ of North Carolina at Greensboro**
Genetic Counseling Prgm
119 McIver St
Greensboro, NC 27402
www.uncg.edu/gen
*Prgm Dir:* Nancy Callanan, MS CGC
*Tel:* 336 256-0175   *Fax:* 336 256-0174
*E-mail:* nancy_callanan@uncg.edu
*Med Dir:* Pamela Reitnauer, MD, PhD

*Additional information about programs that
returned the AMA's survey is available in a table
beginning on page 196.

# Ohio

**Cincinnati Children's Hospital Medical Center**
Genetic Counseling Prgm
Division of Human Genetics
3333 Burnet Ave ML 4006
Cincinnati, OH 45229
*Prgm Dir:* Nancy Steinberg Warren, MS CGC
*Tel:* 513 636-4475   *Fax:* 513 636-0543
*E-mail:* Nancy.Warren@UC.Edu
*Med Dir:* Howard Saal, MD

**Case Western Reserve University**
Genetic Counseling Prgm
10900 Euclid Ave
Cleveland, OH 44106-4955
*Prgm Dir:* Anne Matthews, RN PhD
*Tel:* 216 368-1821   *Fax:* 216 368-3432
*E-mail:* alm14@cwru.edu
*Med Dir:* Georgia Wiesner, MD

# Oklahoma

**Univ of Oklahoma Health Sciences Center**
Genetic Counseling Prgm
940 NE 13th St, Rm 2B2418
Oklahoma City, OK 73104
www.ouhsc.edu
*Prgm Dir:* Susan J Hassed, MS
*Tel:* 405 271-8001, Ext 42182   *Fax:* 405 271-8697
*E-mail:* susan-hassed@ouhsc.edu
*Med Dir:* John Mulvihill, MD

# Ontario, Canada

**University of Toronto**
Genetic Counseling Prgm
555 University Ave
Toronto, ON M5S 1A8
www.utoronto.ca/medicalgenetics/student%20site/
   gencouns.htm
*Prgm Dir:* Cheryl Shuman, MS CGC
*Tel:* 416 813-6374
*E-mail:* cshuman@sickkids.on.ca
*Med Dir:* David Chitayat, MD, FRCPC, FCCMG, FABMG

# Pennsylvania

**Arcadia University**
Genetic Counseling Prgm
450 S Easton Rd
Glenside, PA 19038
*Prgm Dir:* Kathleen Valverde, MS CGC
*Tel:* 215 572-4058   *Fax:* 215 881-8758
*E-mail:* valverde@arcadia.edu
*Med Dir:* E Carolyn Anderson, MD

**University of Pittsburgh**
Genetic Counseling Prgm
Dept of Human Genetics
A300 Crabtree Hall, 130 DeSoto St
Pittsburgh, PA 15261
www.hgen.pitt.edu
*Prgm Dir:* Elizabeth Gettig, MS
*Tel:* 412 624-9951   *Fax:* 412 624-3020
*E-mail:* betsy.gettig@hgen.pitt.edu
*Med Dir:* W Allen Hogge, MD

# Quebec, Canada

**McGill University**
Genetic Counseling Prgm
1205 Dr Penfield Ave, Rm N5/13
Montreal, QC H3A 1Y1
www.mcgill.ca/humgenet
*Prgm Dir:* Jennifer Fitzpatrick, MS
*Tel:* 514 398-3600   *Fax:* 514 398-2430
*E-mail:* jennifer.fitzpatrick@mcgill.ca
*Med Dir:* Laura Russell, MD

# South Carolina

**University of South Carolina**
Genetic Counseling Prgm
Dept of OB/GYN
Two Medical Park
Columbia, SC 29203
*Prgm Dir:* Janice G Edwards, MS
*Tel:* 803 779-4928, Ext 227   *Fax:* 803 434-4596
*E-mail:* jedwards@medpark.sc.edu
*Med Dir:* Anthony Gregg, MD

# Texas

**Univ of Texas Medical School at Houston**
Genetic Counseling Prgm
Dept of Pediatrics
PO Box 20708
Houston, TX 77225
*Prgm Dir:* Claire Singletary, MS CGC
*Tel:* 713 500-5763   *Fax:* 713 500-5689
*Med Dir:* Hope Northup, MD

# Utah

**University of Utah Health Science Center**
Genetic Counseling Prgm
15 N 2030 E #2100
Salt Lake City, UT 84112
*Prgm Dir:* Bonnie J Baty, MS CGC LGC
*Tel:* 801 581-6914   *Fax:* 801 585-7252
*E-mail:* bonnie.baty@hsc.utah.edu
*Med Dir:* David Viskochil, MD PhD

# Virginia

**Virginia Commonwealth Univ/Med Coll of VA**
Genetic Counseling Prgm
Dept of Human Genetics
PO Box 980033
Richmond, VA 23298-0033
www.vipbg.vcu.edu/hg
*Prgm Dir:* Rachel Gannaway, MS
*Tel:* 804 828-9632, Ext 132   *Fax:* 804 828-7094
*E-mail:* rgannaway@mcvh-vcu.edu
*Med Dir:* Joann Bodurtha, MD MPH

# Wisconsin

**University of Wisconsin - Madison**
Genetic Counseling Prgm
445 Henry Mall, Rm 118
Madison, WI 53706
*Prgm Dir:* Catherine A Reiser, MS CGC
*Tel:* 608 262-9722   *Fax:* 608 263-3496
*E-mail:* reiser@waisman.wisc.edu
*Med Dir:* Richard M Pauli, MD

## Genetic Counselor

| Programs* | Class Capacity | Begins | Length (months) | Award | Res Tuition | Non-res Tuition | Stipend | Offers:‡ 1 | 2 | 3 |
|---|---|---|---|---|---|---|---|---|---|---|
| **Arkansas** | | | | | | | | | | |
| University of Arkansas for Medical Sciences (Little Rock) | | Fall semester | 5 | MS | | | | | | • |
| **British Columbia, Canada** | | | | | | | | | | |
| University of British Columbia (Vancouver) | 6 | Sep | 20 | MSc | $15,300 | $15,300 | | | | • |
| **California** | | | | | | | | | | |
| California State University - Northridge | 10 | Sep | 21 | MS | $3,624 | $15,150 | | • | | • |
| University of CA-Irvine School of Medicine (Orange) | 5-6 | Sep | 21 | MS | $9,669 | $26,197 | | | • | • |
| **Colorado** | | | | | | | | | | |
| U of Colorado at Denver and Health Scis Ctr | 8 | Aug | 21 | Dipl, MS | $8,000 | $12,500 | | | | • |
| **District of Columbia** | | | | | | | | | | |
| Howard University (Washington) | 5 | Aug | 24 | MS | $12,400 | $12,400 | | | | • |
| **Illinois** | | | | | | | | | | |
| Northwestern University (Chicago) | 10 | Sep | 18 | Dipl, MS | $44,000 | $44,000 | $6,600 | | | • |
| **Indiana** | | | | | | | | | | |
| Indiana University (Indianapolis) | 6 | Aug | 22 | Dipl, MS | $4,000 | $11,500 | | | | • |
| **Maryland** | | | | | | | | | | |
| Johns Hopkins Univ/Natl Human Genome Research (Bethesda) | 5 | Sep | 22 | ScM | $32,976 | $32,976 | $25,400 | | | • |
| **Massachusetts** | | | | | | | | | | |
| Boston University | 5 | Sep | 21 | MS | $30,000 | $30,000 | $3,500 | | | • |
| Brandeis University (Waltham) | 10 | Aug | 20 | MS | | | | | | • |

*Data are shown only for programs that completed the 2006 AMA Survey of Health Professions Education Programs
‡Key to Offers: 1: Evening or weekend classes; 2: Non-English instruction; 3: Cultural competence instruction

## Genetic Counselor

| Programs* | Class Capacity | Begins | Length (months) | Award | Res Tuition | Non-res Tuition | Stipend | Offers:‡ 1 | 2 | 3 |
|---|---|---|---|---|---|---|---|---|---|---|
| **Michigan** | | | | | | | | | | |
| University of Michigan (Ann Arbor) | 6 | Sep | 20 | MS | $12,149 | $24,415 | $6,007 | | | • |
| **Minnesota** | | | | | | | | | | |
| University of Minnesota - Minneapolis | 8 | Sep | 21 | MS | $8,748 | $15,848 | | | | • |
| **North Carolina** | | | | | | | | | | |
| Univ of North Carolina at Greensboro | 8 | Aug | 21 | MS | $4,000 | $15,000 | | | | • |
| **Ohio** | | | | | | | | | | |
| Cincinnati Children's Hospital Medical Center | 20 | Sep | 21 | MS | $10,773 | $19,878 | $4,500 | • | | • |
| **Oklahoma** | | | | | | | | | | |
| Univ of Oklahoma Health Sciences Center (Oklahoma City) | 4 | Fall | 21 | MS | | | | | | • |
| **Ontario, Canada** | | | | | | | | | | |
| University of Toronto | 4 | Sep | 18 | MSc | $6,147 | $10,560 | | | | • |
| **Pennsylvania** | | | | | | | | | | |
| Arcadia University (Glenside) | 12 | Aug | 18 | MS | $21,100 | $21,100 | $5,000 | | • | |
| University of Pittsburgh | 10 | Aug | 21 | MS | $15,472 | $29,350 | | | | • |
| **Quebec, Canada** | | | | | | | | | | |
| McGill University (Montreal) | 4 | Sep 1 | 20 | MSc | $3,000 | $9,000 | | | • | • |
| **Texas** | | | | | | | | | | |
| Univ of Texas Medical School at Houston | 6 | Aug | 21 | MS | $1,500 | $9,000 | | | | |
| **Virginia** | | | | | | | | | | |
| Virginia Commonwealth Univ/Med Coll of VA (Richmond) | 5 | Aug | 21 | MS | $8,320 | $17,562 | | | | • |

*Data are shown only for programs that completed the 2006 AMA Survey of Health Professions Education Programs
‡Key to Offers: 1: Evening or weekend classes; 2: Non-English instruction; 3: Cultural competence instruction

**Includes:**
- Health information administrator
- Health information technician

## Definition of the Profession

The health information management profession includes managers, technicians, and specialists expert in systems and processes for health information management, including:
- Planning: Formulating strategic, functional, and user requirements for health information
- Engineering: Designing information flow, data models, and definitions
- Administration: Managing data collection and storage, information retrieval, and release
- Application: Analyzing, interpreting, classifying, and coding data and facilitating information use by others
- Policy: Establishing and implementing security, confidentiality, retention, integrity, and access standards

## Health Information Administrator

### Occupational Description

Graduates of baccalaureate degree educational programs in health information management are known as health information administrators and apply their training and expertise in both science and management to develop, implement, and/or provide oversight to health care data collection and reporting systems to assure the integrity and availability of the information resources needed to support authorized users and decision-makers. Health information managers have expertise in developing and managing effective processes and systems to assure the integrity of health care data and to preserve the complete, accurate, and legal source of patient data (patient medical records), as well as expertise in developing and managing effective processes and systems to preserve patient privacy, confidentiality, and the security of health information maintained in paper or computerized systems. Common job titles held by health information administrators in today's job market are related to line, staff, and/or technical positions such as director, assistant director, manager, privacy officer, compliance officer, claims analyst, clinical information specialist, HIM educator, etc. It is anticipated that job titles will change (eg, health information engineer, clinical information coordinator, data administrator, information security officer) as health care enterprises expand their reliance on information systems and technology. Health information administrators have, and will continue to assume, roles that directly contribute to the development of computer-based patient record systems and a national health information infrastructure.

### Job Description

The tasks or functions performed by health information administrators are numerous and are continually changing within the work environment. Although the job title and work setting will dictate the actual tasks performed by the health information administrator, in general this individual performs tasks related to the management of health information and the systems used to collect, store, process, retrieve, analyze, disseminate, and communicate that information, regardless of the physical medium in which information is maintained. In addition, health information administrators assess the uses of information and identify what information is available and where there are inconsistencies, gaps, and duplications in health data sources. They are capable of planning and designing systems and serving as pivotal team members in the development of computer-based patient record systems and other enterprise-wide information systems. Their responsibilities also include serving as brokers of information services. Among the information services provided are a design and requirements definition for clinical and administrative systems development, data administration, data quality management, data security management, decision support design and data analyses, and management of information-intensive areas such as clinical quality/performance assessment and utilization and case management.

### Employment Characteristics

Presently, opportunities for practice are found in numerous settings such as acute care general hospitals, managed care organizations, consulting firms, claims and reimbursement organizations, accounting firms, home health care agencies, long-term care facilities, corrections facilities, drug companies, behavioral health care organizations, insurance companies, state and federal health care agencies, and public health care computing industries. Practice opportunities are unlimited.

### Salary

According to the American Health Information Management Association (AHIMA), entry-level salaries average between $40,000 and $75,000. Refer to Section IV, Table 5 of this *Directory* for more information, or see www.ama-assn.org/go/salary.

### Educational Programs

**Length.** Baccalaureate degree programs are 4 years. Postbaccalaureate and other certificate programs are generally 1 year.

**Prerequisites.** Applicants for the 4-year baccalaureate degree program should have a high school diploma or equivalent. Applicants for the 1-year post-baccalaureate certificate program should have a baccalaureate degree that includes coursework in science and statistics, as specified.

**Curriculum.** The preprofessional curriculum should include appropriate general education credit predicated on the requirements of the academic institution. The professional curriculum requires biomedical sciences (anatomy, physiology, language of medicine, pharmacology, and disease processes); information technology (microcomputer applications, programming, system architectures and operating systems, introduction to database concepts, and data communications); health care delivery systems; legal aspects of health care and ethical issues; organization and management (managerial principles, human resources management and development, financial management for health care, organizational behavior, and interpersonal skills); quantitative methods and research methodologies (introductory and advanced health care statistics/epidemiology, research methods in health care); health care information requirements and standards; health care information systems (computer applications in health care, systems analysis and design); health data content and structures, classification, nomenclature and reimbursement systems, clinical quality assessment, and performance improvement; biomedical and health services research support; health information services management; and a capstone experience/practicum/project.

# Health Information Technician

### Occupational Description

Graduates of associate degree programs are known as health information technicians and conduct health data collection, monitoring, maintenance, and reporting activities in accordance with established data quality principles, legal and regulatory standards, and professional best practice guidelines. These functions encompass, among other areas, monitoring electronic and paper-based documentation and processing and using health data for billing and reporting purposes through use of various electronic systems. Common job titles held by health information technicians in today's job market include reimbursement specialist, information access and disclosure specialist, coder, medical record technician, data quality coordinator, supervisor, etc. It is anticipated that job titles will change as health care enterprises expand their reliance on information systems and technology. Health information technicians have, and will continue to assume, roles that support efforts toward the development of computer-based patient record systems and a national health information infrastructure.

### Job Description

The tasks or functions performed by health information technicians are numerous and continually changing within the work environment. The job title and work setting will dictate the actual tasks performed by the health information technician. However, in general, these individuals perform tasks related to the use, analysis, validation, presentation, data abstracting, analysis, coding, release of information, data privacy and security, retrieval, quality measurement, and control of health care data regardless of the physical medium in which information is maintained. Their task responsibilities may also include supervising personnel.

### Employment Characteristics

Presently, opportunities for practice are found in numerous settings such as acute care general hospitals, managed care organizations, physician office practices, home health care agencies, long-term care facilities, correctional facilities, behavioral health care organizations, insurance companies, ambulatory settings, and state and federal health care agencies, and public health departments. Practice opportunities are unlimited.

### Salary

According to AHIMA, entry-level salaries average between $25,000 and $50,000. Refer to Section IV, Table 5 of this *Directory* for more information, or see www.ama-assn.org/go/salary.

### Educational Programs

**Length.** Programs are generally 2 years, offering an associate degree.

**Prerequisites.** High school diploma or equivalent.

**Curriculum.** In addition to general education courses, the professional component of the technician program requires biomedical sciences (anatomy, physiology, language of medicine, disease processes, and pharmacology); information technology (microcomputer applications and computers in health care); health data content and structure; health care delivery systems, organization and supervision, health care statistics, and data literacy; clinical quality assessment and performance improvement; clinical classification systems; reimbursement methodologies; legal and ethical issues; and supervised professional practice experiences in health information departments of health care facilities and agencies.

### Inquiries

**Careers and Credentialing**
American Health Information Management Association
233 N Michigan Avenue, Suite 2150
Chicago, IL 60601-5800
312 233-1100
www.ahima.org

**Program Accreditation**
Commission on Accreditation for Health Informatics and Information Management Education (CAHIIM)
233 N Michigan Avenue, Suite 2150
Chicago, IL 60601-5800
312 233-1129
312 233-1429 Fax
E-mail: george.payan@cahiim.org
www.cahiim.org

## Health Information Administrator*

# Alabama

**University of Alabama at Birmingham**
Health Information Admin Prgm
1675 University Blvd
Webb 610
Birmingham, AL 35294
www.uab.edu/him
*Prgm Dir:* Sara S Grostick, MA RHIA FAHIMA
*Tel:* 205 934-1678   *Fax:* 205 934-5980
*E-mail:* grostick@uab.edu

**Alabama State University**
Health Information Admin Prgm
PO Box 271
Montgomery, AL 36101-0271
*Prgm Dir:* Nina Vick-Abro, MEd RHIA
*Tel:* 334 229-5058   *Fax:* 334 229-5880
*E-mail:* nabro@alasu.edu

# Arkansas

**Arkansas Tech University**
Health Information Admin Prgm
1311 N El Paso, T5
Russellville, AR 72801
www.atu.edu
*Prgm Dir:* Melinda Wilkins, MEd RHIA
*Tel:* 479 968-0441   *Fax:* 479 964-0504
*E-mail:* melinda.wilkins@atu.edu

# California

**Loma Linda University**
Health Information Admin Prgm
1905 Nichol Hall
Loma Linda, CA 92350
www.llu.edu/llu/sahp/health/#program
*Prgm Dir:* Marilyn R Davidian, MA RHIA
*Tel:* 909 558-4976   *Fax:* 909 558-0404
*E-mail:* mdavidian@llu.edu

# Colorado

**Regis University**
Health Information Admin Prgm
3333 Regis Blvd
Denver, CO 80221-1099
*Prgm Dir:* Sheila A Carlon, PhD RHIA
*Tel:* 303 458-4108   *Fax:* 303 964-5533
*E-mail:* scarlon@regis.edu

# Florida

**Florida International University**
Health Information Admin Prgm
11200 SW 8th St, HLS 146
Miami, FL 33199
*Prgm Dir:* Sandra McDonald, MHSA RHIA
*Tel:* 305 348-0406   *Fax:* 305 348-0437
*E-mail:* mcdons@fiu.edu

**University of Central Florida**
Health Information Admin Prgm
PO Box 162200
Orlando, FL 32816-2205
*Prgm Dir:* Alice Noblin, MBA RHIA
*Tel:* 407 823-2353   *Fax:* 407 823-6138
*E-mail:* anoblin@mail.ucf.edu

**Florida A&M University**
Health Information Admin Prgm
324 Ware-Rhaney East
Tallahassee, FL 32307
www.famusoahs.com
*Prgm Dir:* Marjorie H McNeill, PhD RHIA CCS
*Tel:* 850 599-3818   *Fax:* 850 561-2457
*E-mail:* marjorie.mcneill@famu.edu

# Georgia

**Medical College of Georgia**
Health Information Admin Prgm
Bldg AL-130
1120 15th St
Augusta, GA 30912-0400
*Prgm Dir:* Carol A Campbell, DBA RHIA
*Tel:* 706 721-3436   *Fax:* 706 721-6067
*E-mail:* cacampbe@mail.mcg.edu

**Macon State College**
Health Information Admin Prgm
100 College Station Dr
Macon, GA 31206-5144
http://facultyweb.maconstate.edu/nursing%26HIM/HIM
*Prgm Dir:* Nanette Sayles, EdD RHIA CCS CHP
*Tel:* 478 471-2788   *Fax:* 478 417-2787
*E-mail:* nsayles@mail.maconstate.edu

# Illinois

**Chicago State University**
Health Information Admin Prgm
9501 S King Dr, BHS 610
Chicago, IL 60628-1598
*Prgm Dir:* Leona M Thomas, MHS RHIA
*Tel:* 773 995-2593   *Fax:* 773 995-4484
*E-mail:* lmthomas@msn.com

**University of Illinois at Chicago**
Health Information Admin Prgm
College of Applied Health Sciences
1919 W Taylor St, M/C 250
Chicago, IL 60612
www.bhis.uic.edu
*Prgm Dir:* Karen Patena, MBA RHIA
*Tel:* 312 996-1444   *Fax:* 312 996-8342
*E-mail:* patena@uic.edu

**Illinois State University**
Health Information Admin Prgm
5220 Health Sciences
Normal, IL 61790-5220
*Prgm Dir:* Francis L Waterstraat, Jr, PHD RHIA
*Tel:* 309 438-8809   *Fax:* 309 438-2450
*E-mail:* fwaterst@ilstu.edu

# Indiana

**Indiana University**
Health Information Admin Prgm
School of Informatics
ICTC 475, 535 W Michigan St
Indianapolis, IN 46202
*Prgm Dir:* Danita Forgey, MIS RHIA CCS-P CCS
*Tel:* 317 278-4113   *Fax:* 317 278-7859
*E-mail:* dforgey@iupui.edu

# Kansas

**University of Kansas Medical Center**
Health Information Admin Prgm
Taylor 1012, Mail Stop 2008
3901 Rainbow Blvd
Kansas City, KS 66160
www.him.kumc.edu
*Prgm Dir:* Karl Koob, MMIS RHIA CPEHR
*Tel:* 913 588-2423   *Fax:* 913 588-2428
*E-mail:* kkoob@kumc.edu

# Kentucky

**Eastern Kentucky University**
Health Information Admin Prgm
521 Lancaster Ave
Richmond, KY 40475-3102
www.health.eku.edu/HSA
*Prgm Dir:* Dawn W Jackson, MAEd RHIA
*Tel:* 859 622-1915   *Fax:* 859 622-2013
*E-mail:* dawn.jackson@eku.edu

# Louisiana

**University of Louisiana at Lafayette**
Health Information Admin Prgm
College of Sciences
PO Box 41007
Lafayette, LA 70504
http://him.louisiana.edu
*Prgm Dir:* Carol Venable, MPH RHIA FAHIMA
*Tel:* 337 482-6629   *Fax:* 337 482-5902
*E-mail:* venable@louisiana.edu

**Louisiana Tech University**
Health Information Admin Prgm
Health Information Management Program
PO Box 3171
Ruston, LA 71272
*Prgm Dir:* Angela Kennedy, MBA RHIA CPHQ
*Tel:* 318 257-2854   *Fax:* 318 257-4896
*E-mail:* kennedy@him.latech.edu

# Michigan

**Ferris State University**
Health Information Admin Prgm
Coll of Allied Hlth Sciences
200 Ferris Dr, VFS402
Big Rapids, MI 49307-2740
www.ferris.edu/htmls/colleges/alliedhe/MR
*Prgm Dir:* Ellen J Haneline, MEd RHIA
*Tel:* 231 591-2313   *Fax:* 231 591-2325
*E-mail:* haneline@ferris.edu

# Minnesota

**College of St Scholastica**
Health Information Admin Prgm
1200 Kenwood Ave
Duluth, MN 55811
*Prgm Dir:* Kathleen M LaTour, MA RHIA FAHIMA
*Tel:* 218 723-6011   *Fax:* 218 733-2239
*E-mail:* klatour@css.edu

**College of St Catherine - St Paul**
Health Information Admin Prgm
2004 Randolph Ave
St Paul, MN 55105
*Prgm Dir:* Joanne D Valerius, MPH RHIA
*Tel:* 651 690-6752   *Fax:* 651 690-7849
*E-mail:* jdvalerius@stkate.edu

# Mississippi

**University of Mississippi Medical Center**
Health Information Admin Prgm
Health Information Management Program
2500 N State St
Jackson, MS 39216
*Prgm Dir:* Rebecca J Yates, MEd RHIA
*Tel:* 601 815-1150   *Fax:* 601 815-1717
*E-mail:* byates@shrp.umsmed.edu

*Additional information about programs that returned the AMA's survey is available in a table beginning on page 202.

PROGRAMS

## Missouri

**Stephens College**
Health Information Admin Prgm
Campus Box 2083
1200 East Broadway
Columbia, MO 65215
*Prgm Dir:* Darla D Branda, MA RHIA
*Tel:* 573 876-7125  *Fax:* 573 876-2320
*E-mail:* DBranda@stephens.edu

**Saint Louis University**
*Cosponsor:* Doisy College of Health Sciences
Health Information Admin Prgm
Department of Health Information Management
3437 Caroline St
St Louis, MO 63104
*Prgm Dir:* Jody Smith, PhD RHIA FAHIMA
*Tel:* 314 977-8516  *Fax:* 314 977-8503
*E-mail:* smithjk@slu.edu

## Nebraska

**College of Saint Mary**
Health Information Admin Prgm
7000 Mercy Rd
Omaha, NE 68106
www.csm.edu
*Prgm Dir:* Ellen B Jacobs, MEd RHIA
*Tel:* 402 399-2611  *Fax:* 402 399-2657
*E-mail:* ejacobs@csm.edu

## New Jersey

**Kean University**
Health Information Admin Prgm
Health Information Management Program
1000 Morris Ave
Union, NJ 07083-0411
www.kean.edu/~him
*Prgm Dir:* Barbara Manger, MPA RHIA CCS FAHIMA
*Tel:* 908 737-3450  *Fax:* 908 737-3455
*E-mail:* bmanger@kean.edu

## New York

**Long Island University - C W Post Campus**
Health Information Admin Prgm
720 Northern Blvd
School of Health Professions
Brookville, NY 11548
*Prgm Dir:* Kerry Kimmins-Lawrence, MHA RHIA
*Tel:* 516 299-2485  *Fax:* 516 299-2527
*E-mail:* kkimmins@nshs.edu

**SUNY Institute of Tech - Utica/Rome**
Health Information Admin Prgm
PO Box 3050
Utica, NY 13504-3050
http://web2.sunyit.edu
*Prgm Dir:* Donna L Silsbee, PhD RHIA
*Tel:* 315 792-7391  *Fax:* 315 792-7555
*E-mail:* Donna.Silsbee@sunyit.edu

## North Carolina

**Western Carolina University**
Health Information Admin Prgm
126 Moore Hall
Cullowhee, NC 28723
www.wcu.edu/aps/healths/HS_UG-him.htm
*Prgm Dir:* Irene Mueller, EdD RHIA
*Tel:* 828 227-3510  *Fax:* 828 227-7446
*E-mail:* imueller@email.wcu.edu

**East Carolina University**
Health Information Admin Prgm
Sch Allied Hlth Sciences
LAHN Building 4340-D
Greenville, NC 27858
www.ecu.edu/cs-dhs/hsim/him.cfm
*Prgm Dir:* Elizabeth J Layman, PhD RHIA FAHIMA
*Tel:* 252 744-6177  *Fax:* 252 744-6179
*E-mail:* laymane@ecu.edu

## Ohio

**University of Cincinnati**
Health Information Admin Prgm
College of Allied Health Sciences, French East Bldg
3202 Eden Ave
Cincinnati, OH 45267
*Prgm Dir:* Gail Smith, MA RHIA CCS-P
*Tel:* 513 558-8512
*E-mail:* gail.smith@uc.edu

**Ohio State University**
Health Information Admin Prgm
453 W 10th Ave, Rm 543
Columbus, OH 43210-1234
*Prgm Dir:* Melanie Brodnik, PhD RHIA
*Tel:* 614 292-0567  *Fax:* 614 292-0210
*E-mail:* melanie.brodnik@osumc.edu

**University of Toledo**
Health Information Admin Prgm
2801 W Bancroft St
Mail Stop 119
Toledo, OH 43606-3390
*Prgm Dir:* Maria A Janes, MEd RHIA
*Tel:* 419 530-4523
*E-mail:* marie.janes@utoledo.edu

## Oklahoma

**East Central University**
Health Information Admin Prgm
Department of Health Information Management
1100 E 14th St
Ada, OK 74820
*Prgm Dir:* Sandra A Dixon, MEd MCE RHIA
*Tel:* 580 310-5555  *Fax:* 580 310-5606
*E-mail:* sdixon@mailclerk.ecok.edu

**Southwestern Oklahoma State University**
Health Information Admin Prgm
100 Campus Dr
Weatherford, OK 73096
www.swosu.edu
*Prgm Dir:* Marion Prichard, MEd RHIA
*Tel:* 580 774-3287, Ext 3287  *Fax:* 580 774-7159
*E-mail:* marion.prichard@swosu.edu

## Pennsylvania

**Gwynedd-Mercy College**
Health Information Admin Prgm
1325 Sumneytown Pike
Gwynedd Valley, PA 19437
www.gmc.edu
*Prgm Dir:* Christine Staropoli, MS RHIA CCS
*Tel:* 215 542-4660  *Fax:* 215 641-5559
*E-mail:* staropoli.c@gmc.edu

**Temple University**
Health Information Admin Prgm
College of Health Professions
3307 N Broad St
Philadelphia, PA 19140
*Prgm Dir:* Laurinda B Harman, PhD RHIA
*Tel:* 215 707-4823  *Fax:* 215 707-5852
*E-mail:* Laurinda.Harman@temple.edu

**Duquesne University**
Health Information Admin Prgm
Chair, Dept of Health Mgmt Systems
429 Fisher Hall
Pittsburgh, PA 15282-0001
*Prgm Dir:* Kathleen Begler, MPM RHIA
*Tel:* 412 396-4772  *Fax:* 412 396-5554
*E-mail:* begler@duq.edu

**University of Pittsburgh**
Health Information Admin Prgm
6051 Forbes Tower
Pittsburgh, PA 15260
www.him.pitt.edu
*Prgm Dir:* Mervat Abdelhak, PhD RHIA
*Tel:* 412 383-6650  *Fax:* 412 383-6655
*E-mail:* madelhak@pitt.edu

## Puerto Rico

**University of Puerto Rico**
Health Information Admin Prgm
GPO Box 365067
San Juan, PR 00936-5067
http://cprsweb.rcm.upr.edu/adminfsalud.asp
*Prgm Dir:* Anna Orabona-Ocasio, EdD RHIA CCS
*Tel:* 787 758-2525, Ext 4507  *Fax:* 787 764-3609
*E-mail:* anaorabona@cprs.rcm.upr.edu

## South Dakota

**Dakota State University**
Health Information Admin Prgm
East Hall
Madison, SD 57042-1799
*Prgm Dir:* Dorine Bennett, MBA RHIA
*Tel:* 605 256-5137  *Fax:* 605 256-5060
*E-mail:* dorine.bennett@dsu.edu

## Tennessee

**University of Tennessee Health Science Ctr**
Health Information Admin Prgm
Dept of Health Information Management
920 Madison Ave, Ste 518
Memphis, TN 38163
www.utmem.edu
*Prgm Dir:* Elizabeth Bowman, MPA RHIA
*Tel:* 901 448-6486  *Fax:* 901 448-1629
*E-mail:* HIM@utmem.edu

**Tennessee State University**
Health Information Admin Prgm
3500 John A Merritt Blvd
Nashville, TN 37209-1561
*Prgm Dir:* Elizabeth I Kunnu, MEd RHIA
*Tel:* 615 963-7441  *Fax:* 615 963-7498
*E-mail:* ekunnu@tnstate.edu

## Texas

**Texas Southern University**
Health Information Admin Prgm
3100 Cleburne
Houston, TX 77004
*Prgm Dir:* Fanny Hawkins, EdD RRA RHIA CPHQ
*Tel:* 713 313-7341  *Fax:* 713 313-1094
*E-mail:* Hawkins_FC@tsu.edu

**Texas State University-San Marcos**
Health Information Admin Prgm
601 University Dr
San Marcos, TX 78666
www.health.txstate.edu/health
*Prgm Dir:* Sue E Biedermann, MSHP RHIA
*Tel:* 512 245-8242  *Fax:* 512 245-8258
*E-mail:* sb02@txstate.edu

# Utah

**Weber State University**
Health Information Admin Prgm
3911 University Circle
Ogden, UT 84408-3911
*Prgm Dir:* Patricia L Shaw, MEd RHIA
*Tel:* 801 626-7989   *Fax:* 801 626-6475
*E-mail:* pshaw@weber.edu

# Washington

**University of Washington**
Health Information Admin Prgm
1100 NE 45th, Ste 405
Seattle, WA 98105
http://depts.washington.edu/hia
*Prgm Dir:* Gretchen Murphy, MEd RHIA FAHIMA
*Tel:* 206 543-8810   *Fax:* 206 616-5249
*E-mail:* gcmurphy@u.washington.edu

## Health Information Administrator

| Programs* | Class Capacity | Begins | Length (months) | Award | Res Tuition | Non-res Tuition | Stipend | Offers:‡ 1 | 2 | 3 |
|---|---|---|---|---|---|---|---|---|---|---|
| **Alabama** | | | | | | | | | | |
| University of Alabama at Birmingham | 30 | Fall | 21, 45 | Cert, BS | $5,920 | $11,840 | | • | | |
| **Arkansas** | | | | | | | | | | |
| Arkansas Tech University (Russellville) | 30 | Aug | 37 | BS | $3,432 | $6,864 | | • | | |
| **California** | | | | | | | | | | |
| Loma Linda University | 60 | Sep | 26, 9 | Cert, BS | $21,000 | $21,000 | | | | • |
| **Florida** | | | | | | | | | | |
| Florida A&M University (Tallahassee) | 30 | Aug | 48 | BS | $98 | $498 | | | | |
| University of Central Florida (Orlando) | 32 | Aug | 24 | BS | $2,211 | $12,870 | | | | |
| **Georgia** | | | | | | | | | | |
| Macon State College | 50 | Aug Jan May | 48 | BS | $3,804 | $15,216 | | • | | |
| **Illinois** | | | | | | | | | | |
| Chicago State University | 50 | Aug | 48 | Cert, BS | $5,236 | $9,076 | | | | |
| Illinois State University (Normal) | 30 | Aug | 24 | BS | $6,500 | $13,500 | | | | |
| University of Illinois at Chicago | 45 | Aug | 24 | BS, MS | $10,742 | $23,132 | | | | |
| **Kansas** | | | | | | | | | | |
| University of Kansas Medical Center (Kansas City) | 24 | Aug | 24 | BS | $6,063 | $15,732 | | | | |
| **Kentucky** | | | | | | | | | | |
| Eastern Kentucky University (Richmond) | 20 | Aug Jan | 36 | BS | $5,192 | $14,538 | | | | |
| **Louisiana** | | | | | | | | | | |
| University of Louisiana at Lafayette | 45 | Aug | 48 | BS | $3,264 | $9,444 | | | | |
| **Michigan** | | | | | | | | | | |
| Ferris State University (Big Rapids) | 20 | Aug | 48 | BS | $7,200 | $14,640 | | • | | • |
| **Minnesota** | | | | | | | | | | |
| College of St Catherine - St Paul | 15 | Sep Jan | 30 | BA/BS | | | $392 | • | | • |
| College of St Scholastica (Duluth) | 40 | Sep | 24 | BA, MA | $23,434 | $0 | | | | |
| **Missouri** | | | | | | | | | | |
| Saint Louis University (St Louis) | 25 | Aug Jan | 48 | BS | $24,760 | $24,760 | | | | |
| Stephens College (Columbia) | 16 | Varies | | Cert, BS, MBA | | | | | | |
| **Nebraska** | | | | | | | | | | |
| College of Saint Mary (Omaha) | 30 | Aug Jan May | 24 | BS | $17,750 | $17,750 | | • | | |
| **New Jersey** | | | | | | | | | | |
| Kean University (Union) | 25 | Sep | 48, 60 | BS, BS/MS | $3,753 | $5,069 | | | | |
| **New York** | | | | | | | | | | |
| SUNY Institute of Tech - Utica/Rome | 24 | Sep Jan | 18 | BS, BPS | $4,350 | $10,610 | | | | |
| **North Carolina** | | | | | | | | | | |
| East Carolina University (Greenville) | 30 | Aug | 18 | BS | $4,003 | $14,517 | | | | • |
| Western Carolina University (Cullowhee) | 18 | Aug | 20 | BS | $3,946 | $13,528 | | | | |
| **Ohio** | | | | | | | | | | |
| Ohio State University (Columbus) | 25 | Sep | 18 | Cert, BS | $4,383 | $12,732 | | | | |
| **Oklahoma** | | | | | | | | | | |
| Southwestern Oklahoma State University (Weatherford) | 16 | Aug | 24 | BS | $3,456 | $8,256 | | | • | |
| **Pennsylvania** | | | | | | | | | | |
| Gwynedd-Mercy College (Gwynedd Valley) | 15 | Aug | 24 | BHS | $19,720 | $19,720 | | • | | • |
| University of Pittsburgh | 35 | Sep Jan Apr | 18, 24 | BS, MS/PhD | $14,308 | $26,290 | | | | |
| **Puerto Rico** | | | | | | | | | | |
| University of Puerto Rico (San Juan) | 10 | Aug | 21 | MS | $3,306 | $4,655 | | • | • | |
| **South Dakota** | | | | | | | | | | |
| Dakota State University (Madison) | 30 | Sep | 48 | BS | $2,540 | $3,811 | | | | |

*Data are shown only for programs that completed the 2006 AMA Survey of Health Professions Education Programs
‡Key to Offers: 1: Evening or weekend classes; 2: Non-English instruction; 3: Cultural competence instruction

## Health Information Administrator

| Programs* | Class Capacity | Begins | Length (months) | Award | Res Tuition | Non-res Tuition | Stipend | Offers:‡ 1 | 2 | 3 |
|---|---|---|---|---|---|---|---|---|---|---|
| **Tennessee** | | | | | | | | | | |
| University of Tennessee Health Science Ctr (Memphis) | 15 | Sep | 12 | BS | $7,721 | $20,768 | | | | |
| **Texas** | | | | | | | | | | |
| Texas State University-San Marcos | 50 | Aug | 22 | BS | $6,000 | $15,000 | | | | • |
| **Utah** | | | | | | | | | | |
| Weber State University (Ogden) | 20 | Open | 24 | Dipl, BS | $2,793 | $9,776 | | • | | • |
| **Washington** | | | | | | | | | | |
| University of Washington (Seattle) | 40 | Sep | 12 | Cert, BS | $14,500 | $14,500 | | • | | |

*Data are shown only for programs that completed the 2006 AMA Survey of Health Professions Education Programs
‡Key to Offers: 1: Evening or weekend classes; 2: Non-English instruction; 3: Cultural competence instruction

# Health Information Technician*

## Alabama

**Wallace State Community College**
Health Information Tech Prgm
Bevill Health Education Bldg
Post Office Box 2000
Hanceville, AL 35077-9080
www.wallacestate.edu
*Prgm Dir:* Donna S Stanley, EdS RHIA
*Tel:* 256 352-8327   *Fax:* 256 352-8320
*E-mail:* donna.stanley@wallacestate.edu

**Bishop State Community College**
Health Information Tech Prgm
1365 Martin Luther King Ave
Mobile, AL 36603
*Prgm Dir:* Annalesia Sharp, RHIA
*Tel:* 334 405-4451   *Fax:* 334 405-4505
*E-mail:* asharp@bishop.edu

## Alaska

**University of Alaska Southeast**
Health Information Tech Prgm
UAS Sitka Campus
1332 Seward Ave
Sitka, AK 99835-9498
www.uas.alaska.edu
*Prgm Dir:* Carol Petrie Liberty, RHIA MS
*Tel:* 907 747-7718   *Fax:* 907 747-7720
*E-mail:* carol.liberty@uas.alaska.edu

## Arizona

**Phoenix College**
Health Information Tech Prgm
1202 W Thomas Rd
Phoenix, AZ 85013
www.maricopa.edu
*Prgm Dir:* Bonnie Petterson, PhD RHIA
*Tel:* 602 285-7149   *Fax:* 602 285-7528
*E-mail:* b.petterson@pcmail.maricopa.edu

*Additional information about programs that
returned the AMA's survey is available in a table
beginning on page 210.

## Arkansas

**National Park Community College**
Health Information Tech Prgm
101 College Dr
Hot Springs, AR 71913-9174
*Prgm Dir:* Valerie Bond, MA RHIA
*Tel:* 501 760-4294   *Fax:* 501 760-4141
*E-mail:* vbond@npcc.edu

**University of Arkansas for Medical Sciences**
Health Information Tech Prgm
4301 W Markham St, Slot 733
Little Rock, AR 72205
www.uams.edu
*Prgm Dir:* Kathy C Trawick, EdD RHIA
*Tel:* 501 686-8613   *Fax:* 501 686-8519
*E-mail:* trawickkathyc@uams.edu

## California

**Cypress College**
Health Information Tech Prgm
9200 Valley View St
Cypress, CA 90630
*Prgm Dir:* Rosalie Majid, RHIA
*Tel:* 714 484-7283   *Fax:* 714 527-2175
*E-mail:* rmajid@cypresscollege.edu

**Fresno City College**
Health Information Tech Prgm
1101 E University Ave
Fresno, CA 93741
www.fresnocitycollege.com
*Prgm Dir:* Sarah Edwards, RHIA
*Tel:* 559 244-2640   *Fax:* 559 244-2626
*E-mail:* sarah.edwards@fresnocitycollege.edu

**Chabot College**
Health Information Tech Prgm
25555 Hesperian Blvd
Hayward, CA 94545-5001
*Prgm Dir:* Wanda Ziemba, MFA RHIT BA
*Tel:* 510 723-7496   *Fax:* 510 656-7492
*E-mail:* wziemba@chabotcollege.edu

**Charles R Drew Univ of Med & Science**
Health Information Tech Prgm
1731 E 120th St
Los Angeles, CA 90059
www.cdrewu.edu
*Prgm Dir:* Dorothy Hendrix, MEd RHIT
*Tel:* 323 563-5888   *Fax:* 323 563-5898
*E-mail:* dohendri@cdrewu.edu

**East Los Angeles College**
Health Information Tech Prgm
1301 Avenida Cesar Chavez
Monterey Park, CA 91754-6099
*Prgm Dir:* Elizabeth Garcia
*Tel:* 323 265-8884   *Fax:* 323 265-8876
*E-mail:* garciaea@elac.edu

**DeVry University - Pomona Campus**
Health Information Tech Prgm
901 Corporate Center Dr
Pomona, CA 91768
*Prgm Dir:* Tori Bailey, MS RHIA
*Tel:* 909 868-4064
*E-mail:* tbailey@socal.devry.edu

**Cosumnes River Community College**
Health Information Tech Prgm
8401 Center Pkwy
Sacramento, CA 95823-5799
*Prgm Dir:* George Hines, RHIA RHIT
*Tel:* 916 691-7452   *Fax:* 916 691-7146
*E-mail:* hinesg@crc.losrios.edu

**San Diego Mesa College**
Health Information Tech Prgm
7250 Mesa College Dr
San Diego, CA 92111
www.sdccd.edu
*Prgm Dir:* Teddy L Scribner, MS RHIA
*Tel:* 619 388-2606   *Fax:* 619 279-5668
*E-mail:* tscribne@sdccd.edu

**City College of San Francisco**
Health Information Tech Prgm
John Adams Campus
1860 Hayes St
San Francisco, CA 94117
*Prgm Dir:* Marie T Conde, MPA RHIT CCS
*Tel:* 415 561-1818   *Fax:* 415 561-1861
*E-mail:* mconde@ccsf.edu

**Santa Barbara City College**
Health Information Tech Prgm
721 Cliff Dr
Santa Barbara, CA 93109-2394
www.sbcc.edu
*Prgm Dir:* Sue Willner, RHIA
*Tel:* 805 965-0581, Ext 2851   *Fax:* 805 963-7222
*E-mail:* willner@sbcc.edu

## Colorado

**Arapahoe Community College**
Health Information Tech Prgm
5900 S Santa Fe Dr, PO Box 9002
Littleton, CO 80160-9002
*Prgm Dir:* Annette Bigalk, RHIA
*Tel:* 303 795-5795   *Fax:* 303 797-5842
*E-mail:* annette.bigalk@arapahoe.edu

# Connecticut

**Briarwood College**
Health Information Tech Prgm
2279 Mt Vernon Rd
Southington, CT 06489
www.briarwood.edu
*Prgm Dir:* Phyllis Hilt, MBA RHIA
*Tel:* 860 628-4751, Ext 186    *Fax:* 860 276-8838
*E-mail:* hiltp@briarwood.edu

# Florida

**Broward Community College**
Health Information Tech Prgm
Ctr for Hlth Science Educ
1000 Coconut Creek Blvd
Coconut Creek, FL 33066
*Prgm Dir:* Charline Bumgardner, BA RHIT
*Tel:* 954 201-2086    *Fax:* 954 201-2348
*E-mail:* cbumgard@broward.edu

**Daytona Beach Community College**
Health Information Tech Prgm
PO Box 2811
Daytona Beach, FL 32120-2811
www.dbcc.edu
*Prgm Dir:* Nancy Thomas, EdD RHIA
*Tel:* 386 506-3748    *Fax:* 386 506-3300
*E-mail:* thomasn@dbcc.edu

**International College**
Health Information Tech Prgm
4501 Colonial Blvd
Ft Myers, FL 33912
*Prgm Dir:* Deborah Howard, MA RHIA CCS
*Tel:* 239 482-0019
*E-mail:* dhoward@internationalcollege.edu

**Indian River Community College**
Health Information Tech Prgm
3209 Virginia Ave
Ft Pierce, FL 34981-5599
*Prgm Dir:* Claudia Keating, MEd RHIA
*Tel:* 772 462-4911    *Fax:* 772 462-4900
*E-mail:* ckeating@ircc.edu

**Santa Fe Community College**
Health Information Tech Prgm
3000 NW 83rd St
A07K
Gainesville, FL 32606
www.santafe.edu
*Prgm Dir:* Karen Bakuzonis, MSHA RHIA
*Tel:* 352 381-3835    *Fax:* 352 395-5286
*E-mail:* karen.bakuzonis@sfcc.edu
*Med Dir:* Douglas Robertson

**Florida Community College - Jacksonville**
Health Information Tech Prgm
4501 Capper Rd
Jacksonville, FL 32218
*Prgm Dir:* Eudelia S Thomas, MS RHIA
*Tel:* 904 766-6663    *Fax:* 904 766-5565
*E-mail:* ethomas@fccj.edu

**Lake-Sumter Community College**
Health Information Tech Prgm
9501 US Hwy 441
Leesburg, FL 34788-8751
www.lscc.edu
*Prgm Dir:* Brandy G Ziesemer, MA RHIA CCS
*Tel:* 352 365-3581    *Fax:* 352 365-3501
*E-mail:* ziesemerb@lscc.edu

**Miami Dade College**
Health Information Tech Prgm
Medical Center Campus
950 NW 20th St
Miami, FL 33127
www.mdc.edu/medical
*Prgm Dir:* Mary Worsley, RHIA
*Tel:* 305 237-4156    *Fax:* 305 237-4278
*E-mail:* mworsley@mdc.edu
*Med Dir:* Richard Prentiss, MPH

**Central Florida Community College**
Health Information Tech Prgm
3001 SW College Rd
Ocala, FL 34474
*Prgm Dir:* Suzanne B Garrett, MSA RHIA
*Tel:* 352 854-2322, Ext 1466
*E-mail:* garretts@cf.edu

**St Johns River Community College**
Health Information Tech Prgm
Orange Park Campus
283 College Dr
Orange Park, FL 32065
*Prgm Dir:* Sheila Newberry, MEd RHIT
*Tel:* 904 276-6758    *Fax:* 904 276-6888
*E-mail:* SheilaNewberry@sjrcc.edu

**Pensacola Junior College**
Health Information Tech Prgm
Dept of Allied Health
1000 College Blvd
Pensacola, FL 32504
www.pjc.edu
*Prgm Dir:* Donna M Shumway, MEd RHIA
*Tel:* 904 484-2213    *Fax:* 904 484-2375
*E-mail:* dshumway@pjc.edu

**St Petersburg College**
Health Information Tech Prgm
PO Box 13489
St Petersburg, FL 33733
*Prgm Dir:* Angela Picard, MEd RHIA
*Tel:* 727 341-3623    *Fax:* 727 341-3455
*E-mail:* picard.angela@spcollege.edu

**Polk Community College**
Health Information Tech Prgm
999 Ave H NE
Winter Haven, FL 33881
www.polk.edu
*Prgm Dir:* Hertencia Bowe, MSA RHIA
*Tel:* 863 297-1010, Ext 5370    *Fax:* 863 297-1056
*E-mail:* hbowe@polk.edu

# Georgia

**Darton College**
Health Information Tech Prgm
2400 Gillionville Rd
Albany, GA 31707
www.darton.edu
*Prgm Dir:* Linda Parks, MA RHIT CCS
*Tel:* 229 317-6894    *Fax:* 229 317-6682
*E-mail:* linda.parks@darton.edu

**DeVry University - Atlanta**
Health Information Tech Prgm
250 N Arcadia Ave
Decatur, GA 30030-2198
*Prgm Dir:* Kristyn Murphy-Rodvill, MHA RHIA
*Tel:* 404 292-7900, Ext 2123
*E-mail:* kmurphy-rodvill@faculty.atl.devry.edu

**Macon State College**
Health Information Tech Prgm
100 College Station Dr
Macon, GA 31206-5145
www.maconstate.edu
*Prgm Dir:* Nanette Sayles, EdD RHIA CCS CHP
*Tel:* 478 471-2788    *Fax:* 478 471-2787
*E-mail:* nsayles@mail.maconstate.edu

**Ogeechee Technical College**
Health Information Tech Prgm
1 Joe Kennedy Blvd
Statesboro, GA 30458
www.ogeecheetech.edu
*Prgm Dir:* Lisa Parker Kagay, MBA RHIA
*Tel:* 912 486-7792    *Fax:* 912 486-7604
*E-mail:* lkagay@ogeecheetech.edu
*Med Dir:* John Martin, MD

# Idaho

**Boise State University**
Health Information Tech Prgm
College of Health Science
1910 University Dr
Boise, ID 83725
*Prgm Dir:* Patricia Elison-Bowers, PhD RHIA
*Tel:* 208 426-1130    *Fax:* 208 426-2199
*E-mail:* pelison@boisestate.edu

**Idaho State University**
Health Information Tech Prgm
Campus Box 8380
Pocatello, ID 83209-8380
*Prgm Dir:* Glenna Young, RHIA CCS
*Tel:* 208 282-4524    *Fax:* 208 282-3975
*E-mail:* younglen@isu.edu

# Illinois

**Southwestern Illinois College**
Health Information Tech Prgm
2500 Carlyle Rd
Belleville, IL 62221
www.swic.edu
*Prgm Dir:* Wendy Holder, RHIA
*Tel:* 618 235-2700, Ext 5385    *Fax:* 618 235-2052
*E-mail:* wendy.holder@swic.edu

**DeVry University - Chicago**
Health Information Tech Prgm
3300 N Campbell Ave
Chicago, IL 60618
*Prgm Dir:* Patricia Hertel, MA RHIT
*Tel:* 773 697-2147
*E-mail:* phertel@chi.devry.edu

**Northwestern Business College**
Health Information Tech Prgm
4811 N Milwaukee Ave
Chicago, IL 60630
*Prgm Dir:* Joni Rudd, RHIA
*Tel:* 773 777-4200, Ext 2319    *Fax:* 708 237-5005
*E-mail:* jrudd@nwbc.edu

**Danville Area Community College**
Health Information Tech Prgm
2000 E Main St
Danville, IL 61832
www.dacc.edu
*Prgm Dir:* Janet M Westberg, MS RHIA
*Tel:* 217 443-8574    *Fax:* 217 443-8595
*E-mail:* janwest@dacc.edu

**Oakton Community College**
Health Information Tech Prgm
1600 E Golf Rd
Des Plaines, IL 60016
www.oakton.edu/acad/dept/hit
*Prgm Dir:* Anita Taylor, MAdED RHIA CCS
*Tel:* 847 635-1957    *Fax:* 847 635-1764
*E-mail:* anitat@oakton.edu

**College of DuPage**
Health Information Tech Prgm
425 Fawell Blvd
Glen Ellyn, IL 60137-6599
*Prgm Dir:* Kim D Pack, MS RHIA
*Tel:* 630 942-2532    *Fax:* 630 858-5409
*E-mail:* packki@cdnet.cod.edu

## College of Lake County
Health Information Tech Prgm
19351 W Washington St
Grayslake, IL 60030-1198
www.clcillinois.edu
*Prgm Dir:* Margaret Kyriakos, MBA RHIA
*Tel:* 847 543-2879    *Fax:* 847 223-1357
*E-mail:* mkyriakos@clcillinois.edu

## Southern Illinois Collegiate Common Market
Health Information Tech Prgm
3213 S Park Ave
Herrin, IL 62948
www.siccm.com
*Prgm Dir:* Mary J Sullivan, PhD RHIA
*Tel:* 618 942-6902    *Fax:* 618 942-6658
*E-mail:* sullivan@siccm.com

## DeVry University - Naperville
Health Information Tech Prgm
1200 E Dielhl Rd
Naperville, IL 60563
*Prgm Dir:* Natasha Freeman Cauley, MPH RHIA
*Tel:* 404 494-9417
*E-mail:* ncauley@devry.edu

## Moraine Valley Community College
Health Information Tech Prgm
10900 S 88th Ave
Palos Hills, IL 60465
*Prgm Dir:* Donna Schnepp, RHIA
*Tel:* 708 974-5315    *Fax:* 708 974-0185
*E-mail:* schnepp@morainevalley.edu

# Indiana

## Indiana University Northwest
Health Information Tech Prgm
3400 Broadway
Gary, IN 46408
www.iun.edu/~ahealth/him/hit.shtml
*Prgm Dir:* Margaret A Skurka, MS RHIA CCS
*Tel:* 219 980-6654    *Fax:* 219 980-6649
*E-mail:* mskurk@iun.edu

## Vincennes University
Health Information Tech Prgm
1002 N First St
Vincennes, IN 47591-9986
www.vinu.edu
*Prgm Dir:* Sharon O'Neill, MS RHIA
*Tel:* 812 888-4411    *Fax:* 812 888-4550
*E-mail:* soneill@vinu.edu

# Iowa

## Scott Community College
Health Information Tech Prgm
500 Belmont Rd
Bettendorf, IA 52722
www.eicc.edu/highschool/programs/career/
    health_careers/hit.html
*Prgm Dir:* Barbara A Foster, BA RHIA
*Tel:* 563 441-4157    *Fax:* 563 441-4204
*E-mail:* bfoster@eicc.edu
*Med Dir:* Greg Khoury, MD FACS

## Northeast Iowa Community College
Health Information Tech Prgm
1625 Hwy 150 S
PO Box 400
Calmar, IA 52132
www.nicc.edu
*Prgm Dir:* Rhonda Seibert, MEd RHIA
*Tel:* 563 562-3263, Ext 345    *Fax:* 563 562-4357
*E-mail:* seibertr@nicc.edu

## Kirkwood Community College
Health Information Tech Prgm
6301 Kirkwood Blvd SW, PO Box 2068
Cedar Rapids, IA 52406-9973
www.kirkwood.edu
*Prgm Dir:* Betty Haar, BS RHIA
*Tel:* 319 398-4923    *Fax:* 319 398-1293
*E-mail:* betty.haar@kirkwood.edu

## Indian Hills Community College
Health Information Tech Prgm
Health Occupations Division Arts & Science Center
525 Grandview Ave - Bldg 6
Ottumwa, IA 52501
www.indianhills.edu
*Prgm Dir:* Heidi Jones, BS RHIA
*Tel:* 641 683-5163    *Fax:* 641 683-5254
*E-mail:* hjones@indianhills.edu

## Northwest Iowa Community College
Health Information Tech Prgm
603 W Park St
Sheldon, IA 51201
*Prgm Dir:* Mari Beth Lane, RHIA
*Tel:* 712 324-5061, Ext 223    *Fax:* 712 324-4136
*E-mail:* MLane@nwicc.edu

# Kansas

## Hutchinson Community College
Health Information Tech Prgm
Area Voc School
1300 N Plum
Hutchinson, KS 67501
www.hutchcc.edu
*Prgm Dir:* Loretta A Horton, MEd RHIA
*Tel:* 316 665-4955    *Fax:* 316 665-4988
*E-mail:* hortonl@hutchcc.edu

## Washburn University
Health Information Tech Prgm
1700 College Ave SW
Topeka, KS 66621
*Prgm Dir:* Michelle Shipley, MS RHIA CCS
*Tel:* 785 670-2174
*E-mail:* michelle.shipley@washburn.edu

# Kentucky

## Western Kentucky University
Health Information Tech Prgm
South Campus
Bowling Green, KY 42101
www.wku.edu
*Prgm Dir:* Karen C Sansom, MS RHIA
*Tel:* 270 780-2567    *Fax:* 270 780-2591
*E-mail:* karen.sansom@wku.edu

## Jefferson Community and Technical College
Health Information Tech Prgm
109 E Broadway
Louisville, KY 40202
*Prgm Dir:* Elizabeth Neichter, EdD RHIA
*Tel:* 502 213-2199
*E-mail:* elizabeth.neichter@kctcs.edu

## National College of Business & Technology - Louisville
Health Information Tech Prgm
4205 Dixie Highway
Louisville, KY 40216
*Prgm Dir:* Kristie Couch, RHIA
*Tel:* 502 447-7634    *Fax:* 502 447-7665
*E-mail:* kcouch@ncbt.edu

# Louisiana

## Delgado Community College
Health Information Tech Prgm
615 City Park Ave
New Orleans, LA 70119-4399
*Prgm Dir:* Melissa LaCour, RHIA
*Tel:* 504 483-4435    *Fax:* 504 483-4609
*E-mail:* mlacou@dcc.edu

## Louisiana Tech University
Health Information Tech Prgm
PO Box 3171
Ruston, LA 71272
*Prgm Dir:* Helen Baxter, MA RHIA
*Tel:* 318 257-2854    *Fax:* 318 257-4896
*E-mail:* baxter@him.latech.edu

## Southern Univ at Shreveport
Health Information Tech Prgm
610 Texas St, Ste 328A
Shreveport, LA 71101
*Prgm Dir:* Kimberly Madden, RHIA
*Tel:* 318 674-3487    *Fax:* 318 676-3454
*E-mail:* kmadden@susla.edu

# Maine

## Kennebec Valley Community College
Health Information Tech Prgm
92 Western Ave
Fairfield, ME 04937
www.kvcc.me.edu
*Prgm Dir:* Sharon S Veilleux, MPA RHIT
*Tel:* 207 453-5156    *Fax:* 207 453-5197
*E-mail:* sveilleux@kvcc.me.edu

# Maryland

## Baltimore City Community College
Health Information Tech Prgm
2901 Liberty Heights Ave
Baltimore, MD 21215-7893
*Prgm Dir:* Betty Neely Mitchell, RHIA
*Tel:* 410 462-7735    *Fax:* 410 462-7734
*E-mail:* bmitchell@bccc.edu

## Hagerstown Business College
Health Information Tech Prgm
18618 Crestwood Dr
Hagerstown, MD 21742
*Prgm Dir:* Beth Shanholtzer, RHIA
*Tel:* 301 739-2680, Ext 152    *Fax:* 301 791-7661
*E-mail:* bshanholtzer@hagerstownbusinesscol.edu

## Prince George's Community College
Health Information Tech Prgm
301 Largo Rd
Largo, MD 20774-2199
http://academic.pgcc.edu/alliedhealth
*Prgm Dir:* Muriel Adams, RHIA CCS
*Tel:* 301 322-0735    *Fax:* 301 386-7528
*E-mail:* madams@pgcc.edu

## Montgomery College
Health Information Tech Prgm
7600 Takoma Ave
Takoma Park, MD 20912
*Prgm Dir:* Shirley Suzanne Meiskey, MSA RHIA
*Tel:* 301 562-5519    *Fax:* 301 562-5585
*E-mail:* sue.meiskey@montgomerycollege.edu

# Massachusetts

## Caritas Labour, College
Health Information Tech Prgm
2120 Dorchester Ave
Boston, MA 02124
*Prgm Dir:* Nancy Entwistle, RHIT CCS
*Tel:* 617 296-8300, Ext 4063
*E-mail:* Nancy_Entwistle@laboure.edu

**Bristol Community College**
Health Information Tech Prgm
777 Elsbree St
Fall River, MA 02777
*Prgm Dir:* Edward J Dobbs, BBA RHIA
*Tel:* 508 678-2811, Ext 2329   *Fax:* 508 675-2318
*E-mail:* edobbs@bristol.mass.edu

**Fisher College**
Health Information Tech Prgm
451 Elm St
North Attleboro, MA 02760
*Prgm Dir:* Patricia Parkes, RHIA
*Tel:* 508 699-6200
*E-mail:* pparkes@fisher.edu

# Michigan

**Baker College of Allen Park**
Health Information Tech Prgm
4500 Enterprise Dr
Allen Park, MI 48101
*Prgm Dir:* Charmaine Irvin, MEd RHIA RHIT CCS
*Tel:* 313 425-3732
*E-mail:* charmain.irvin@baker.edu

**Ferris State University**
Health Information Tech Prgm
Coll of Allied Hlth Sciences
200 Ferris Dr, VFS 402
Big Rapids, MI 49307-2740
*Prgm Dir:* Ellen J Haneline, MEd RHIA
*Tel:* 231 591-2269   *Fax:* 231 591-2325
*E-mail:* haneline@ferris.edu

**Baker College of Flint**
Health Information Tech Prgm
1050 W Bristol Rd
Flint, MI 48507
www.baker.edu
*Prgm Dir:* Amy Savage, MAT RHIA CCS
*Tel:* 810 766-4147   *Fax:* 810 766-2055
*E-mail:* amy.savage@baker.edu

**Schoolcraft College**
Health Information Tech Prgm
1751 Radcliff St
Garden City, MI 48135-1197
*Prgm Dir:* Patricia A Rubio, MSA RHIA
*Tel:* 734 462-4770, Ext 6025   *Fax:* 734 462-4775
*E-mail:* prubio@schoolcraft.edu

**Davenport University - Grand Rapids (Fulton St)**
Health Information Tech Prgm
415 Fulton St E
Grand Rapids, MI 49503
www.davenport.edu/tabid/709/default.aspx
*Prgm Dir:* Susan Slajus, BS RHIA
*Tel:* 616 451-3511
*E-mail:* susan.slajus@davenport.edu

# Minnesota

**Anoka Technical College**
Health Information Tech Prgm
1355 W Hwy 10
Anoka, MN 55303
*Prgm Dir:* Mary Ann Jackels, BA RHIA
*Tel:* 763 576-4970
*E-mail:* mjackels@ank.tec.mn.us

**Rasmussen Colleges - Brooklyn Park**
Health Information Tech Prgm
8301 93rd Ave North
Brooklyn Park, MN 55445
*Prgm Dir:* Cindy Glewwe, RHIA
*Tel:* 651 687-9000
*E-mail:* cindyg@rasmussen.edu

**Rasmussen Colleges - Eagan**
Health Information Tech Prgm
3500 Federal Dr
Eagan, MN 55122
www.rasmussen.edu
*Prgm Dir:* Cindy Glewwe, RHIA
*Tel:* 651 687-9000   *Fax:* 651 687-0507
*E-mail:* cindyg@rasmussen.edu

**College of St Catherine - Minneapolis**
Health Information Tech Prgm
601 25th Ave S
Minneapolis, MN 55454
www.stkate.edu
*Prgm Dir:* Marsha Holey, RHIA
*Tel:* 651 690-7795   *Fax:* 651 690-7849
*E-mail:* mkholey@stkate.edu

**Minnesota State Comm & Tech Coll - Moorhead**
Health Information Tech Prgm
1900 28th Ave S
Moorhead, MN 56560
*Prgm Dir:* Susan Hanna, RHIT
*Tel:* 218 299-6558   *Fax:* 218 236-0342
*E-mail:* susan.hanna@minnesota.edu

**Rochester Community & Technical College**
Health Information Tech Prgm
851 30th Ave SE
Rochester, MN 55904
*Prgm Dir:* Judy Gust, MEd BA RHIT
*Tel:* 507 285-7456
*E-mail:* judy.gust@roch.edu

**Ridgewater College - Willmar Campus**
Health Information Tech Prgm
2101 15th Ave NW
Willmar, MN 56201-2098
*Prgm Dir:* Mary Juenemann, RHIA CCS
*Tel:* 320 231-2949   *Fax:* 320 231-7690
*E-mail:* mary.juenemann@ridgewater.edu

# Mississippi

**Hinds Community College**
Health Information Tech Prgm
1750 Chadwick Dr
Jackson, MS 39204
www.hindscc.edu
*Prgm Dir:* Michelle McGuffee, RHIA
*Tel:* 601 371-3514   *Fax:* 601 376-4541
*E-mail:* mlmcguffee@hindscc.edu

**Meridian Community College**
Health Information Tech Prgm
910 Hwy 19 N
Meridian, MS 39307
www.mcc.cc.ms.us/healthinfo
*Prgm Dir:* Robin Allen Jones, RHIA
*Tel:* 601 484-8759   *Fax:* 601 484-8824
*E-mail:* rjones@meridiancc.edu

**Itawamba Community College**
Health Information Tech Prgm
2176 South Eason Blvd
Tupelo, MS 38801
www.iccms.edu
*Prgm Dir:* Nena Scott, MSEd RHIA CCS CCSP
*Tel:* 662 620-5123   *Fax:* 662 620-5077
*E-mail:* npscott@iccms.edu

# Missouri

**Metropolitan Community College - Penn Valley**
Health Information Tech Prgm
3201 SW Trafficway
Kansas City, MO 64111
*Prgm Dir:* Jennifer Scott, RHIA
*Tel:* 816 759-4245   *Fax:* 816 759-4553
*E-mail:* Jennifer.Scott@MCCKC.edu

**Ozarks Technical Community College**
Health Information Tech Prgm
1001 E Chestnut Expressway
Springfield, MO 65802-3625
www.otc.edu
*Prgm Dir:* Sue Kirk, BGS RHIT
*Tel:* 417 447-8821   *Fax:* 417 447-8806
*E-mail:* kirks@otc.edu
*Med Dir:* Steven Bishop, PhD

**Missouri Western State University**
Health Information Tech Prgm
4525 Downs Dr
St Joseph, MO 64507
*Prgm Dir:* Marsha Dolan, MBA RHIA
*Tel:* 816 271-5949   *Fax:* 816 271-5849
*E-mail:* dolan@missouriwestern.edu

**St Charles Community College**
Health Information Tech Prgm
4601 Mid Rivers Mall Dr
St Peters, MO 63376
www.stchas.edu/divisions/msh/hitindex.shtml
*Prgm Dir:* Candace E Neu, MDE RHIA CCS
*Tel:* 636 922-8292   *Fax:* 636 922-8478
*E-mail:* cneu@stchas.edu

# Montana

**Montana State Univ - Great Falls Coll of Tech**
Health Information Tech Prgm
2100 16th Ave S
Great Falls, MT 59405
*Prgm Dir:* Kimberly Baumann, RHIA
*Tel:* 406 771-4358   *Fax:* 406 771-4317
*E-mail:* kbaumann@msugf.edu

# Nebraska

**Central Community College**
Health Information Tech Prgm
PO Box 1024
Hastings, NE 68902-1024
*Prgm Dir:* Shawna Stump, RHIA
*Tel:* 402 461-2514   *Fax:* 402 461-2506
*E-mail:* sstump@cccneb.edu
*Med Dir:* Deborah Brennan, PhD

**Clarkson College**
Health Information Tech Prgm
101 S 42nd St
Omaha, NE 68131-2739
*Prgm Dir:* Mary Miller, RHIA
*Tel:* 402 552-6216   *Fax:* 402 552-6019
*E-mail:* millerm@clarksoncollege.edu

**College of Saint Mary**
Health Information Tech Prgm
7000 Mercy Rd
Omaha, NE 68106
www.csm.edu
*Prgm Dir:* Ellen B Jacobs, MEd RHIA
*Tel:* 402 399-2611   *Fax:* 402 399-2657
*E-mail:* ejacobs@csm.edu

**Western Nebraska Community College**
Health Information Tech Prgm
1601 E 27th St
Scottsbluff, NE 69361-1899
http://wncc.net
*Prgm Dir:* Peg Wolff, MBA RHIA
*Tel:* 308 635-6064   *Fax:* 308 635-6100
*E-mail:* pwolff@wncc.net
*Med Dir:* Anne Hippe, MSN CNS RN

# Nevada

**Community College of Southern Nevada**
Health Information Tech Prgm
6375 W Charleston Blvd
Las Vegas, NV 89146
*Prgm Dir:* Cassie Gentry, MEd RHIA CHP
*Tel:* 702 651-5896 *Fax:* 702 651-5077
*E-mail:* cassie_gentry@ccsn.edu

# New Jersey

**Camden County College**
Health Information Tech Prgm
200 N Broadway
Attn: Health Information Technology College Hill
Camden, NJ 08012
*Prgm Dir:* Lynette M Williamson, MBA RHIA CCS CPC
*Tel:* 856 968-1331
*E-mail:* lwilliamson@camdencc.edu

**Hudson County Community College**
Health Information Tech Prgm
870 Bergen Ave
Cudari Building, Rm 209
Jersey City, NJ 07306
www.hccc.edu
*Prgm Dir:* Judith Hall, RHIA
*Tel:* 201 360-4225
*E-mail:* jhall@hccc.edu
*Med Dir:* Lloyd M Kahn, DPM

**Passaic County Community College**
Health Information Tech Prgm
One College Blvd
Paterson, NJ 07505-1179
*Prgm Dir:* Lisa DeLiberto, BS RRA
*Tel:* 973 684-6297 *Fax:* 973 684-6627
*E-mail:* LDeLiberto@pccc.edu

**Burlington County College**
Health Information Tech Prgm
601 Pemberton Browns Mills Rd
Pemberton, NJ 08068-1599
http://staff.bcc.edu/hit
*Prgm Dir:* Van Nguyen, MHA RHIA
*Tel:* 609 894-9311, Ext 7339 *Fax:* 609 894-4712
*E-mail:* vnguyen@bcc.edu

# New Mexico

**Central New Mexico Community College**
Health Information Tech Prgm
525 Buena Vista SE
BIT
Albuquerque, NM 87106
www.cnm.edu
*Prgm Dir:* Mechel McKinney, BSB RHIA
*Tel:* 505 224-3905 *Fax:* 505 224-3850
*E-mail:* mmckinney@cnm.edu

**San Juan College**
Health Information Tech Prgm
4601 College Blvd
Farmington, NM 87402
*Prgm Dir:* Carroll Schnabel, RHIA
*Tel:* 505 566-3005 *Fax:* 505 566-3820
*E-mail:* schnabelc@sanjuancollege.edu

**University of New Mexico - Gallup**
Health Information Tech Prgm
200 College Dr
Gallup, NM 87301
*Prgm Dir:* Melody Brashear, RHIT
*Tel:* 505 863-7659 *Fax:* 505 863-7513
*E-mail:* mbrash@unm.edu

# New York

**Alfred State College - SUNY**
Health Information Tech Prgm
10 Upper College Dr
Alfred, NY 14802
www.alfredstate.edu
*Prgm Dir:* Michelle Green, MPS RHIA FAHIMA
*Tel:* 607 587-3672 *Fax:* 607 587-3270
*E-mail:* greenma@alfredstate.edu

**Broome Community College**
Health Information Tech Prgm
Decker Health Science Bldg, PO Box 1017
Binghamton, NY 13902
*Prgm Dir:* Mary L Rosato, MA
*Tel:* 607 778-5051 *Fax:* 607 778-5467
*E-mail:* rosato_m@sunybroome.edu

**Suffolk Community College**
Health Information Tech Prgm
Crooked Hill Rd
Brentwood, NY 11717
*Prgm Dir:* Diane P Fabian, MBA RHIA
*Tel:* 631 851-6342 *Fax:* 631 851-6339
*E-mail:* fabiand@sunysuffolk.edu

**Suffolk County Community College**
Health Information Tech Prgm
Crooked Hill Rd
Brentwood, NY 11717
*Prgm Dir:* Diane Fabian, MBA RHIA
*Tel:* 631 851-6342
*E-mail:* fabiand@sunysuffolk.edu

**Trocaire College**
Health Information Tech Prgm
360 Choate Ave
Buffalo, NY 14220-2094
www.trocaire.edu
*Prgm Dir:* Laurie J Rivet, MBA
*Tel:* 716 827-2561 *Fax:* 716 828-6107
*E-mail:* rivetl@trocaire.edu
*Med Dir:* Thomas Mitchell

**CUNY Borough of Manhattan Community Coll**
Health Information Tech Prgm
199 Chambers St, N 749
New York, NY 10007
*Prgm Dir:* Lynda Carlson, MPH MS RHIT
*Tel:* 212 220-8339 *Fax:* 212 748-7465
*E-mail:* lcarlson@bmcc.cuny.edu

**Monroe Community College**
Health Information Tech Prgm
1000 E Henrietta Rd
Rochester, NY 14623
*Prgm Dir:* Sharon L Insero, RHIA
*Tel:* 585 292-2375 *Fax:* 585 292-3834
*E-mail:* sinsero@monroecc.edu

**Molloy College**
Health Information Tech Prgm
1000 Hempstead Ave
PO Box 5002
Rockville Centre, NY 11571-5002
*Prgm Dir:* Peter Micallef, MS RHIA CCS
*Tel:* 516 678-5000, Ext 6463 *Fax:* 516 256-2252
*E-mail:* ppmicallef@molloy.edu

**Onondaga Community College**
Health Information Tech Prgm
Rte 173
Syracuse, NY 13215
*Prgm Dir:* Judith Chrisman, MS RRA
*Tel:* 315 469-2102 *Fax:* 315 469-2180
*E-mail:* chrismaj@sunyocc.edu

**Mohawk Valley Community College**
Health Information Tech Prgm
1101 Sherman Dr
Utica, NY 13501
www.mvcc.edu
*Prgm Dir:* Sue Bice, MS RHIA
*Tel:* 315 792-5513 *Fax:* 315 731-5855
*E-mail:* sbice@mvcc.edu

**Erie Community College - North Campus**
Health Information Tech Prgm
6205 Main St
Williamsville, NY 14221-7095
*Prgm Dir:* Gail M Lauritsen, EdM RHIA
*Tel:* 716 851-5293 *Fax:* 716 851-1429
*E-mail:* lauritsen@ecc.edu

# North Carolina

**Central Piedmont Community College**
Health Information Tech Prgm
PO Box 35009
Charlotte, NC 28235
*Prgm Dir:* Marty Long, MPH RHIA
*Tel:* 704 330-6187 *Fax:* 704 330-6131
*E-mail:* Marty.Long@cpcc.edu

**Pitt Community College**
Health Information Tech Prgm
PO Drawer 7007
Greenville, NC 27835-7007
www.pittcc.edu
*Prgm Dir:* Kay Gooding, MPH MA Ed RHIA
*Tel:* 252 493-7361 *Fax:* 252 321-4451
*E-mail:* kgooding@email.pittcc.edu

**Catawba Valley Community College**
Health Information Tech Prgm
2550 Hwy 70 SE
Hickory, NC 28602
*Prgm Dir:* Debra W Cook, MAEd RHIA
*Tel:* 704 327-7000, Ext 4342 *Fax:* 704 327-7276
*E-mail:* dcook@cvcc.edu

**Davidson County Community College**
Health Information Tech Prgm
PO Box 1287
Lexington, NC 27293-1287
*Prgm Dir:* Heather Kyles-Watson, RHIA
*Tel:* 336 249-8186, Ext 310 *Fax:* 336 249-9060
*E-mail:* hkwatson@davidsonccc.edu

**McDowell Technical Community College**
Health Information Tech Prgm
54 College Dr
Marion, NC 28752
www.mcdowelltech.cc.nc.us
*Prgm Dir:* Valerie Dobson, RHIA
*Tel:* 828 652-0699 *Fax:* 828 652-1014
*E-mail:* valeried@mcdowelltech.edu
*Med Dir:* Penny Cross, MSNed, RN

**Brunswick Community College**
Health Information Tech Prgm
SE Regional Allied Health Consortium
PO Box 30
Supply, NC 28462
*Prgm Dir:* Polly Decker, RHIA
*Tel:* 910 755-7347 *Fax:* 910 755-7469
*E-mail:* deckerp@brunswickcc.edu

**Southwestern Community College**
Health Information Tech Prgm
447 College Dr
Sylva, NC 28779
www.southwesternccc.edu
*Prgm Dir:* Penny Wells, MA Ed RHIA
*Tel:* 828 586-4091, Ext 362 *Fax:* 828 586-3129
*E-mail:* pwells@southwesternccc.edu

**Edgecombe Community College**
Health Information Tech Prgm
2009 W Wilson St
Tarboro, NC 27886
*Prgm Dir:* Kim A Bell, RHIA
*Tel:* 252 823-5166, Ext 186   *Fax:* 252 985-2212
*E-mail:* bellk@edgecombe.edu
*Med Dir:* Chalmers Nunn, MD

# North Dakota

**United Tribes Technical College**
Health Information Tech Prgm
3315 University Dr
Bismarck, ND 58504
www.uttc.edu
*Prgm Dir:* Karla Baxter, RHIA
*Tel:* 701 255-3285, Ext 1245   *Fax:* 701 530-0630
*E-mail:* kbaxter@uttc.edu

**North Dakota State College of Science**
Health Information Tech Prgm
800 N Sixth St
Wahpeton, ND 58076
www.ndscs.edu/hlthinfo
*Prgm Dir:* Geralyn Matejcek, MBA RHIA
*Tel:* 701 671-2269, Ext 2269   *Fax:* 701 671-2570
*E-mail:* geralyn.matejcek@ndscs.edu

# Ohio

**Cincinnati State Tech & Comm College**
Health Information Tech Prgm
3520 Central Pkwy
Cincinnati, OH 45223
*Prgm Dir:* Sherri Mallett, MEd RHIA
*Tel:* 513 569-1674   *Fax:* 513 569-4963
*E-mail:* sherri.mallett@cincinnatistate.edu

**Cuyahoga Community College**
Health Information Tech Prgm
Metro Campus, 2900 Community College Ave
Cleveland, OH 44115
*Prgm Dir:* Kathleen D Loflin, RHIA
*Tel:* 216 987-4456   *Fax:* 216 987-4386
*E-mail:* kathy.loflin@tri-c.edu

**Columbus State Community College**
Health Information Tech Prgm
550 E Spring St
Columbus, OH 43215
*Prgm Dir:* Lisa A Cerrato, MSBS RHIA
*Tel:* 614 287-2541   *Fax:* 614 287-5144
*E-mail:* lcerrato@cscc.edu

**Sinclair Community College**
Health Information Tech Prgm
444 W Third St
Dayton, OH 45402
www.sinclair.edu
*Prgm Dir:* Barbara Wallace, MBA RHIA CCS CCS-P
*Tel:* 937 512-2973   *Fax:* 937 512-2239
*E-mail:* barbara.wallace@sinclair.edu

**Bowling Green State University - Firelands**
Health Information Tech Prgm
One University Dr
Huron, OH 44839
www.firelands.bgsu.edu
*Prgm Dir:* Mona M Burke, MA RHIA
*Tel:* 419 433-5560, Ext 20860   *Fax:* 419 433-9696
*E-mail:* burkem@bgnet.bgsu.edu

**Hocking College**
Health Information Tech Prgm
3301 Hocking Pkwy
Nelsonville, OH 45764
*Prgm Dir:* Karen Wright, MHSA RHIA RHIT
*Tel:* 614 753-6433   *Fax:* 614 753-5105
*E-mail:* wright_k@hocking.edu

**Stark State College of Technology**
Health Information Tech Prgm
6200 Frank Ave NW
North Canton, OH 44720
www.starkstate.edu
*Prgm Dir:* Darlene S Horn, RHIA
*Tel:* 330 966-5458, Ext 4296   *Fax:* 330 966-6586
*E-mail:* dhorn@starkstate.edu

**Mercy College of Northwest Ohio**
Health Information Tech Prgm
2221 Madison Ave
Toledo, OH 43624-1198
*Prgm Dir:* Tammy Mathewson, RHIA CCS-P
*Tel:* 419 251-1452   *Fax:* 419 251-1730
*E-mail:* tammy.mathewson@mercycollege.edu
*Med Dir:* Kimberly Watson, MBA RHIA

**Owens Community College**
Health Information Tech Prgm
Oregon Rd
PO Box 10,000
Toledo, OH 43699
www.owens.edu
*Prgm Dir:* Bonnie Hemp, RHIA CPHQ
*Tel:* 567 661-7286   *Fax:* 567 661-7331
*E-mail:* bonnie_hemp@owens.edu

**Professional Skills Institute**
Health Information Tech Prgm
20 Arco Dr
Toledo, OH 43607
*Prgm Dir:* Cynthia Topel, BS RHIA
*Tel:* 419 531-9610
*E-mail:* ctopel@proskills.com

# Oklahoma

**Rose State College**
Health Information Tech Prgm
6420 SE 15th St
Midwest City, OK 73110
www.rose.edu
*Prgm Dir:* Cecil D Brooks, RHIA CCS
*Tel:* 405 733-7578   *Fax:* 405 736-0338
*E-mail:* cbrooks@rose.edu

**Tulsa Community College**
Health Information Tech Prgm
Allied Health Div Metro Campus
909 S Boston Ave
Tulsa, OK 74119-2095
*Prgm Dir:* Sandra S Smith, MEd BS RHIA
*Tel:* 918 595-7201   *Fax:* 918 595-7091
*E-mail:* ssmith@tulsacc.edu

# Oregon

**Central Oregon Community College**
Health Information Tech Prgm
2600 NW College Way
Bend, OR 97701
*Prgm Dir:* Beverlee Jackson, RHIT
*Tel:* 541 383-7736   *Fax:* 541 318-3785
*E-mail:* bjackson@cocc.edu

**Portland Community College**
Health Information Tech Prgm
12000 SW 49th Ave
Portland, OR 97219
www.pcc.edu
*Prgm Dir:* Susan Williams, RHIA
*Tel:* 503 978-5665   *Fax:* 503 978-5257
*E-mail:* slwillia@pcc.edu

# Pennsylvania

**Gwynedd-Mercy College**
Health Information Tech Prgm
1325 Sumneytown Pike
Gwynedd Valley, PA 19437
www.gmc.edu
*Prgm Dir:* Christine Staropoli, MS RHIA CCS
*Tel:* 215 542-4660   *Fax:* 215 641-5559
*E-mail:* staropoli.c@gmc.edu

**Harrisburg Area Community College**
Health Information Tech Prgm
1641 Old Philadephia Pike
Lancaster, PA 17602
*Prgm Dir:* Matthew Alfonso, RHIA
*Tel:* 717 358-2880   *Fax:* 717 358-2951
*E-mail:* malfonso@hacc.edu

**Community College of Philadelphia**
Health Information Tech Prgm
1700 Spring Garden St
Philadelphia, PA 19130
*Prgm Dir:* Joyce Garozzo, MS RHIA CCS
*Tel:* 215 751-8946   *Fax:* 215 751-8937
*E-mail:* jgarozzo@ccp.edu

**Comm Coll of Allegheny County**
Health Information Tech Prgm
808 Ridge Ave
Pittsburgh, PA 15212-6097
*Prgm Dir:* JoAnn Avoli, MEd RHIA
*Tel:* 412 237-2614   *Fax:* 412 237-6579
*E-mail:* javoli@ccac.edu

**Lehigh Carbon Community College**
Health Information Tech Prgm
4525 Education Park Dr
Schnecksville, PA 18078-2598
*Prgm Dir:* George Peters, MS RHIA
*Tel:* 610 799-1596   *Fax:* 610 799-1527
*E-mail:* gpeters@lccc.edu
*Med Dir:* Paul H Schenck, MD

**South Hills School of Business & Technology**
Health Information Tech Prgm
480 Waupelani Dr
State College, PA 16801
www.southhills.edu
*Prgm Dir:* Kay Strigle, RHIA
*Tel:* 814 234-7755   *Fax:* 814 234-0926
*E-mail:* striglek@hotmail.com

**Pennsylvania College of Technology**
Health Information Tech Prgm
One College Ave
Williamsport, PA 17701
*Prgm Dir:* Daniel K Christopher, MBA RHIA
*Tel:* 570 326-3767, Ext 5774   *Fax:* 570 327-4529
*E-mail:* dchristo@pct.edu

# Puerto Rico

**Huertas Junior College**
Health Information Tech Prgm
PO Box 8429
Caguas, PR 00726
*Prgm Dir:* Norma Rodriguez
*Tel:* 787 743-2156, Ext 266   *Fax:* 787 745-4815
*E-mail:* normadrodriguez@yahoo.com

**Universidad del Este**
Health Information Tech Prgm
PO Box 2010
Carolina, PR 00983
www.suagm.edu/une
*Prgm Dir:* Ada Lily Torres, RHIA MEd
*Tel:* 787 257-7373, Ext 2106   *Fax:* 787 752-0070
*E-mail:* altorres@mail.suagm.edu

**Universidad Adventista de las Antillas**
Health Information Tech Prgm
PO Box 118
Mayaguez, PR 00681-0118
http://www.uaa.edu
*Prgm Dir:* Otoniel Cabrera
*Tel:* 787 834-9595, Ext 2320    *Fax:* 787 834-9597

**Inter American University of Puerto Rico**
Health Information Tech Prgm
Call Box 5100
San German, PR 00683
www.sg.inter.edu
*Prgm Dir:* Magda Lopez, MS RHIA
*Tel:* 787 264-1912, Ext 7368    *Fax:* 787 892-6350
*E-mail:* maglopez@sg.inter.edu

# South Carolina

**Midlands Technical College**
Health Information Tech Prgm
PO Box 2408
Columbia, SC 29202
www.midlandstech.edu
*Prgm Dir:* Natalie Sartori, RHIA
*Tel:* 803 822-3072    *Fax:* 803 822-3247
*E-mail:* sartorin@midlandstech.edu
*Med Dir:* Edward Nicholson

**Florence-Darlington Technical College**
Health Information Tech Prgm
PO Box 100548
Florence, SC 29501-0548
*Prgm Dir:* Malissa Jackson, CCS RHIT
*Tel:* 843 661-8091    *Fax:* 843 292-0851
*E-mail:* Malissa.Jackson@fdtc.edu

**Greenville Technical College**
Health Information Tech Prgm
PO Box 5616
Greenville, SC 29606
*Prgm Dir:* Susan Walther, RHIA CCS
*Tel:* 864 848-2032    *Fax:* 864 848-2038
*E-mail:* susan.walther@gvltec.edu

# South Dakota

**Dakota State University**
Health Information Tech Prgm
East Hall
Madison, SD 57042
*Prgm Dir:* Dorine Bennett, MBA RHIA
*Tel:* 605 256-5137    *Fax:* 605 256-5060
*E-mail:* Dorine.Bennett@dsu.edu

# Tennessee

**Chattanooga State Technical Comm College**
Health Information Tech Prgm
4501 Amnicola Hwy
Chattanooga, TN 37406-1097
*Prgm Dir:* Rose Scalise, RHIA CPHQ
*Tel:* 423 697-3381    *Fax:* 423 697-2586
*E-mail:* rose.scalise@chattanoogastate.edu

**Dyersburg State Community College**
Health Information Tech Prgm
1510 Lake Rd
Dyersburg, TN 38024
www.dscc.edu
*Prgm Dir:* Susan Osborne, BS CCS
*Tel:* 731 286-3294    *Fax:* 731 286-3271
*E-mail:* osborne@dscc.edu

**Volunteer State Community College**
Health Information Tech Prgm
1480 Nashville Pike
Gallatin, TN 37066-3188
www.volstate.edu
*Prgm Dir:* Abby Cooper, BBA RHIA
*Tel:* 615 230-3337    *Fax:* 615 230-3251
*E-mail:* Abby.Cooper@volstate.edu

**Roane State Community College**
Health Information Tech Prgm
276 Patton Ln
Harriman, TN 37748-5011
www.roanestate.edu
*Prgm Dir:* Karen H Feltner, RHIT CCS
*Tel:* 865 354-3000, Ext 4327    *Fax:* 865 882-4535
*E-mail:* feltnerkh@roanestate.edu

**Walters State Community College**
Health Information Tech Prgm
500 S Davy Crockett Pkwy
Morristown, TN 37813-6899
*Prgm Dir:* Anita Gail Winkler, BS RHIA
*Tel:* 423 585-6990    *Fax:* 423 585-6955
*E-mail:* gail.winkler@ws.edu

# Texas

**Texas State Tech College, West Texas**
Health Information Tech Prgm
650 E Hwy 80
Abilene, TX 79601
*Prgm Dir:* Martha Davis, RHIA
*Tel:* 325 734-3637    *Fax:* 325 670-9345
*E-mail:* martha.davis@abilene.tstc.edu

**Lee College**
Health Information Tech Prgm
PO Box 818
Baytown, TX 77522-0818
*Prgm Dir:* Ann Marice Ivey, MS RHIA CMT
*Tel:* 281 425-6569    *Fax:* 281 425-6585
*E-mail:* mivey@lee.edu

**Lamar Institute of Technology**
Health Information Tech Prgm
PO Box 10061
Beaumont, TX 77710
*Prgm Dir:* Debra Long, MS RHIA CCS
*Tel:* 409 839-2918    *Fax:* 409 839-2919
*E-mail:* debra.long@lit.edu

**Panola College**
Health Information Tech Prgm
1109 W Panola St
Carthage, TX 75633
*Prgm Dir:* Sandra Payne, RHIA
*Tel:* 903 693-1116    *Fax:* 903 693-1152
*E-mail:* spayne@panola.edu

**Del Mar College**
Health Information Tech Prgm
101 Baldwin Blvd
Corpus Christi, TX 78404-3897
*Prgm Dir:* Thomas Williams
*Tel:* 361 698-2330    *Fax:* 361 698-1172
*E-mail:* twilliams@delmar.edu

**Dallas County Community College**
Health Information Tech Prgm
4849 W Illinois Ave
Dallas, TX 75211
*Prgm Dir:* Cathy Beaty, RHIA
*Tel:* 214 860-8661
*E-mail:* cbeaty@dcccd.edu

**El Paso Community College**
Health Information Tech Prgm
PO Box 20500
El Paso, TX 79998
*Prgm Dir:* Jean A Garrison, RHIA
*Tel:* 915 831-4074    *Fax:* 915 831-4114
*E-mail:* JeanG@epcc.edu

**Texas State Technical College - Harlingen**
Health Information Tech Prgm
1902 N Loop 499
Harlingen, TX 78550-3697
www.harlingen.tstc.edu/healthinfo
*Prgm Dir:* Deborah Woods, BBA RHIA
*Tel:* 956 364-4768    *Fax:* 956 364-5155
*E-mail:* dwoods@harlingen.tstc.edu

**Houston Community College**
Health Information Tech Prgm
Health Careers Education Div
1900 Pressler Dr
Houston, TX 77030
*Prgm Dir:* Carla Tyson, EdD MHA RHIA
*Tel:* 713 718-7347    *Fax:* 713 718-7521
*E-mail:* carla.tyson@hccs.edu

**North Harris College**
Health Information Tech Prgm
2700 W W Thorne Dr
Houston, TX 77073-3499
www.northharriscollege.com
*Prgm Dir:* Jeanne Qualey, RHIA CCS
*Tel:* 281 765-7829    *Fax:* 281 618-1155
*E-mail:* jeanne.qualey@nhmccd.edu

**San Jacinto College North**
Health Information Tech Prgm
5800 Uvalde Rd
Houston, TX 77049-4599
*Prgm Dir:* James Steen, MBA BS RHIA
*Tel:* 281 459-7608    *Fax:* 281 459-7651
*E-mail:* james.steen@sjcd.edu

**Tarrant County College**
Health Information Tech Prgm
828 Harwood Rd
Hurst, TX 76054
*Prgm Dir:* DeeAnn Carver, RHIA
*Tel:* 817 515-6544    *Fax:* 817 515-6700
*E-mail:* deeann.carver@tccd.edu

**DeVry University - Dallas**
Health Information Tech Prgm
4800 Regent Blvd
Irving, TX 75063
*Prgm Dir:* Barbara Odom-Wesley, PhD RHIA FAHIMA
*Tel:* 972 929-9322
*E-mail:* bodom-wesley@dal.devry.edu

**South Plains College**
Health Information Tech Prgm
Reese Center
528 Gilbert Dr
Lubbock, TX 79416
*Prgm Dir:* Nancy Powell, BS RHIT
*Tel:* 806 885-3048, Ext 4644    *Fax:* 806 885-5608
*E-mail:* npowell@southplainscollege.edu

**South Texas College**
Health Information Tech Prgm
1101 E Vermont
McAllen, TX 78503
*Prgm Dir:* Irma Rodriguez, MEd RHIA CCS
*Tel:* 956 872-3170    *Fax:* 956 872-3180
*E-mail:* irmar@southtexascollege.edu

**Midland College**
Health Information Tech Prgm
3600 N Garfield
Midland, TX 79705
*Prgm Dir:* Melinda Teel, RHIA CCS
*Tel:* 915 685-5573    *Fax:* 915 685-4762
*E-mail:* mteel@midland.edu

**Howard College**
Health Information Tech Prgm
3501 N US Hwy 67
San Angelo, TX 76904
www.howardcollege.edu
*Prgm Dir:* Sonja Davis, RHIA
*Tel:* 325 481-8300, Ext 236    *Fax:* 325 481-8321
*E-mail:* sdavis@howardcollege.edu

### St Philip's College
Health Information Tech Prgm
1801 Martin Luther King Dr
San Antonio, TX 78203
*Prgm Dir:* Cheryl Spears, RHIA
*Tel:* 210 531-3456   *Fax:* 210 531-3459
*E-mail:* cspears@accd.edu

### Tyler Junior College
Health Information Tech Prgm
PO Box 9020
Tyler, TX 75711
www.tjc.edu
*Prgm Dir:* Charlotte Creason, RHIA
*Tel:* 903 510-2669   *Fax:* 903 597-6518
*E-mail:* ccre@tjc.edu

### Vernon College
Health Information Tech Prgm
4105 Maplewood Ave
Vernon, TX 76308
*Prgm Dir:* Roxanne Hill, RHIA
*Tel:* 940 696-8752, Ext 3237   *Fax:* 940 696-3244
*E-mail:* rhill@vernoncollege.edu

### McLennan Community College
Health Information Tech Prgm
1400 College Dr, AS 208
Waco, TX 76708
*Prgm Dir:* Lesley Weber, RHIA ALF
*Tel:* 254 299-8233   *Fax:* 254 299-8435
*E-mail:* lweber@mclennan.edu

### Wharton County Junior College
Health Information Tech Prgm
911 Boling Hwy
Wharton, TX 77488
www.wcjc.edu
*Prgm Dir:* Mary W King, MS RHIA
*Tel:* 979 532-6363   *Fax:* 979 532-6489
*E-mail:* maryk@wcjc.edu

## Utah

### Weber State University
Health Information Tech Prgm
3911 University Circle
Ogden, UT 84408-3911
www.weber.edu
*Prgm Dir:* Patricia L Shaw, MEd RHIA
*Tel:* 801 626-7242   *Fax:* 801 626-6475
*E-mail:* pshaw@weber.edu

## Virginia

### Northern Virginia Community College
Health Information Tech Prgm
6699 Springfield Center Dr
Springfield, VA 22150
www.nvcc.edu
*Prgm Dir:* David Munch, MBA RHIA
*Tel:* 703 822-6641   *Fax:* 703 822-6619
*E-mail:* dmunch@nvcc.edu

### Tidewater Community College
Health Information Tech Prgm
1700 College Crescent
Virginia Beach, VA 23456
*Prgm Dir:* Gussie L Hammond, MS BS RHIA
*Tel:* 757 822-7262   *Fax:* 757 427-1338
*E-mail:* ghammond@tcc.edu

## Washington

### Shoreline Community College
Health Information Tech Prgm
16101 Greenwood Ave N
Seattle, WA 98133
*Prgm Dir:* Donna J Wilde, MPA RHIA
*Tel:* 206 543-4757   *Fax:* 206 533-5103
*E-mail:* dwilde@shoreline.edu

### Spokane Community College
Health Information Tech Prgm
N 1810 Greene St, MS 2090
Spokane, WA 99217
*Prgm Dir:* Sharon Meyer, RHIT
*Tel:* 509 533-8852   *Fax:* 509 533-8621
*E-mail:* Smeyer@scc.spokane.edu

### Tacoma Community College
Health Information Tech Prgm
6501 S 19th St
Tacoma, WA 98466-6100
www.tacomacc.edu
*Prgm Dir:* Marion Miller, MBA RHIA CCS
*Tel:* 253 566-5227   *Fax:* 253 566-5273
*E-mail:* mmiller@tacomacc.edu

## West Virginia

### Fairmont State Univ
*Cosponsor:* Pierpont Comm & Tech College
Health Information Tech Prgm
Locust Ave
Fairmont, WV 26554
*Prgm Dir:* Vickie Findley, MPA RHIA
*Tel:* 304 367-4764   *Fax:* 304 367-4268
*E-mail:* vfindley@fairmontstate.edu

### Marshall Community & Technical College
Health Information Tech Prgm
One John Marshall Dr
Huntington, WV 25755
*Prgm Dir:* Janet Smith, MS RHIA CCS
*Tel:* 304 696-6796
*E-mail:* smithjan@marshall.edu

### Marshall University
Health Information Tech Prgm
One John Marshall Dr
Huntington, WV 25755
*Prgm Dir:* Janet Smith, RHIA
*Tel:* 304 696-6796   *Fax:* 304 696-3013
*E-mail:* smithjan@marshall.edu

### West Virginia Northern Community College
Health Information Tech Prgm
1704 Market St
Wheeling, WV 26003
*Prgm Dir:* Korene Atkins, MA RHIA CCS CPC CPC-H
*Tel:* 304 233-5900, Ext 5509
*E-mail:* katkins@northern.wvnet.edu

## Wisconsin

### Chippewa Valley Technical College
Health Information Tech Prgm
620 W Clairemont Ave
Eau Claire, WI 54701
*Prgm Dir:* Margie Konik, MA RHIA
*Tel:* 715 858-1824   *Fax:* 715 833-6470
*E-mail:* mkonik@cvtc.edu

### Northeast Wisconsin Technical College
Health Information Tech Prgm
2740 W Mason St, PO Box 19042
Green Bay, WI 54307-9042
*Prgm Dir:* Marilyn Toninato, BA RHIA
*Tel:* 920 498-5577   *Fax:* 920 498-5673
*E-mail:* marilyn.toninato@nwtc.edu

### Western Technical College
Health Information Tech Prgm
304 N Sixth St, PO Box C-908
La Crosse, WI 54602-0908
http://learn.westerntc.edu/brownt
*Prgm Dir:* Tamra R Brown, ME RHIA
*Tel:* 608 785-9549   *Fax:* 608 785-9087
*E-mail:* brownt@westerntc.edu

### Gateway Technical College
Health Information Tech Prgm
1001 S Main St
Racine, WI 53403
*Prgm Dir:* Cynthia Fickenscher, MBA RHIA CCA
*Tel:* 262 619-6396   *Fax:* 262 631-6811
*E-mail:* fickenscherc@gtc.edu

### Moraine Park Technical College
Health Information Tech Prgm
2151 N Main St
West Bend, WI 53095
www.morainepark.edu
*Prgm Dir:* Gloria Madison, MS RHIA
*Tel:* 262 335-5730   *Fax:* 262 335-5708
*E-mail:* gmadison@morainepark.edu

## Health Information Technician

| Programs* | Class Capacity | Begins | Length (months) | Award | Res Tuition | Non-res Tuition | Stipend | Offers:‡ 1 | 2 | 3 |
|---|---|---|---|---|---|---|---|---|---|---|
| **Alabama** | | | | | | | | | | |
| Wallace State Community College (Hanceville) | 75 | Aug | 21 | AAS | $2,736 | $5,472 | | | | |
| **Alaska** | | | | | | | | | | |
| University of Alaska Southeast (Sitka) | 20 | Sep Jan | 15, 24 | Cert, AAS | $120 | $0 | | • | | |
| **Arizona** | | | | | | | | | | |
| Phoenix College | 50 | Every semester | 24 | Cert, AAS | $1,950 | $4,410 | | • | | • |
| **Arkansas** | | | | | | | | | | |
| University of Arkansas for Medical Sciences (Little Rock) | 25 | Aug Jan | 22 | AS | $4,466 | $10,788 | | • | | |

*Data are shown only for programs that completed the 2006 AMA Survey of Health Professions Education Programs
‡Key to Offers: 1: Evening or weekend classes; 2: Non-English instruction; 3: Cultural competence instruction

Health Information Technician

| Programs* | Class Capacity | Begins | Length (months) | Award | Res Tuition | Non-res Tuition | Stipend | Offers:‡ 1 | 2 | 3 |
|---|---|---|---|---|---|---|---|---|---|---|
| **California** | | | | | | | | | | |
| Charles R Drew Univ of Med & Science (Los Angeles) | 25 | Aug | 24, 30 | Cert, AS | $14,500 | $0 | | • | • | • |
| Fresno City College | 25 | Aug Jan | 16 | AS | $382 | $0 | | • | | |
| San Diego Mesa College | 25 | Aug | 24 | AS | $26 | $160 | | • | | |
| Santa Barbara City College | 40 | Sep | 24 | Cert, AS | $720 | $5,190 | | • | | |
| **Colorado** | | | | | | | | | | |
| Arapahoe Community College (Littleton) | 60 | Aug Jan | 24 | AAS | $3,600 | $8,400 | | | | |
| **Connecticut** | | | | | | | | | | |
| Briarwood College (Southington) | 15 | Sep | 24 | AAS | $16,520 | $0 | | • | | |
| **Florida** | | | | | | | | | | |
| Broward Community College (Coconut Creek) | 30 | Aug | 24 | AS | $1,500 | $6,000 | | • | | |
| Central Florida Community College (Ocala) | 30 | Fall term | 24 | Dipl, AA | $4,380 | $16,029 | | • | | |
| Daytona Beach Community College | 20 | Aug | 21 | AAS | $2,214 | $8,643 | | | | |
| Lake-Sumter Community College (Leesburg) | 24 | Aug Jan | 24 | AAS | $2,300 | $6,292 | | • | | |
| Miami Dade College | 50 | Aug | 24 | Dipl, AS | $4,154 | $12,988 | | • | | |
| Pensacola Junior College | 20 | Aug | 20, 12 | Cert, AAS | $1,608 | $6,320 | | • | | |
| Polk Community College (Winter Haven) | 30 | Aug | 24 | Dipl, AS, AAS | $66 | $246 | | • | | |
| Santa Fe Community College (Gainesville) | 24 | Aug | 24, 36 | AS | $1,876 | $7,084 | | • | | |
| **Georgia** | | | | | | | | | | |
| Darton College (Albany) | 35 | Aug | 21 | AS | $1,440 | $3,944 | | | | |
| Macon State College | 45 | Aug Jan May | 22 | AS | $3,840 | $15,216 | | | | |
| Ogeechee Technical College (Statesboro) | 25 | Sep | 21 | AS | $1,332 | $2,664 | | | | |
| **Idaho** | | | | | | | | | | |
| Boise State University | 40 | Aug Jan | 24 | AS | $4,778 | $12,186 | | | | |
| **Illinois** | | | | | | | | | | |
| College of DuPage (Glen Ellyn) | 20 | Aug | 21 | AAS | $2,880 | $8,440 | $96 | • | | |
| College of Lake County (Grayslake) | 15 | Aug | 24 | AAS | $2,640 | $8,811 | | • | | |
| Danville Area Community College | 30 | Aug | 20 | Assoc | $2,000 | $4,000 | | | | |
| Northwestern Business College (Chicago) | 24 | Sep Dec Mar Jun | 18 | AAS | $17,410 | $17,410 | | • | • | • |
| Oakton Community College (Des Plaines) | 20 | Aug Jan | 21, 30 | AAS | $4,575 | $17,263 | | • | | |
| Southern Illinois Collegiate Common Market (Herrin) | 24 | Aug | 18 | AS | $3,216 | $11,078 | | | | |
| Southwestern Illinois College (Belleville) | 20 | Aug | 22 | AAS | $2,079 | $5,742 | | | | |
| **Indiana** | | | | | | | | | | |
| Indiana University Northwest (Gary) | 24 | Aug | 12, 21 | AS | $4,100 | $11,000 | | • | | |
| Vincennes University | 24 | Aug | 19 | AS | $4,098 | $10,229 | | | | |
| **Iowa** | | | | | | | | | | |
| Indian Hills Community College (Ottumwa) | 75 | Aug | 21 | AAS | $3,614 | $5,439 | | | | • |
| Kirkwood Community College (Cedar Rapids) | 24 | Aug | 21 | Dipl, AAS | $3,960 | $7,921 | | | | |
| Northeast Iowa Community College (Calmar) | 20 | Aug | 24 | AAS | $3,400 | $3,400 | | | | |
| Scott Community College (Bettendorf) | 20 | Aug | 14, 24 | Dipl, AAS | $97 | $135 | | • | | |
| **Kansas** | | | | | | | | | | |
| Hutchinson Community College | 16 | Aug | 22 | AAS | $2,144 | $3,296 | | • | | |
| Washburn University (Topeka) | 20 | Aug | 12, 24 | Cert, AS | $6,300 | $14,292 | | • | | |
| **Kentucky** | | | | | | | | | | |
| National College of Business & Technology - Louisville | 20 | Dec Mar Jun Sep | 24 | AS | $5,760 | $5,760 | | • | | |
| Western Kentucky University (Bowling Green) | 30 | Aug | 21 | AS | $5,962 | $14,400 | | • | | |
| **Louisiana** | | | | | | | | | | |
| Delgado Community College (New Orleans) | 20 | Aug | 22 | AAS | $1,674 | $4,284 | | | | |
| **Maine** | | | | | | | | | | |
| Kennebec Valley Community College (Fairfield) | 18 | Aug Jan | 18 | AAS | $4,290 | $9,295 | | • | | |
| **Maryland** | | | | | | | | | | |
| Montgomery College (Takoma Park) | 25 | Sep | 21 | AAS | $2,966 | $7,689 | | • | | |
| Prince George's Community College (Largo) | 20 | Sep | 24 | AAS | $3,290 | $8,855 | | • | | |
| **Michigan** | | | | | | | | | | |
| Baker College of Flint | 25 | Sep | 24 | AAS | $8,400 | $8,400 | | • | | |
| Davenport University - Grand Rapids (Fulton St) | | | | AAS | $28,490 | $28,490 | | • | | |
| Ferris State University (Big Rapids) | 30 | Aug | 24 | AAS | $7,200 | $14,640 | | • | | • |
| Schoolcraft College (Garden City) | 30 | Aug | 24 | AAS | $2,924 | $4,300 | | • | | |
| **Minnesota** | | | | | | | | | | |
| College of St Catherine - Minneapolis | 20 | Sep Feb | 24 | AAS | $14,000 | $0 | | • | | |
| Rasmussen Colleges - Eagan | 200 | Oct | 18 | Dipl, AAS, Coding | $12,700 | $0 | | • | | |
| **Mississippi** | | | | | | | | | | |
| Hinds Community College (Jackson) | 16 | Aug | 22 | AAS | $1,660 | $3,866 | | | | |
| Itawamba Community College (Tupelo) | 20 | Aug | 24 | AAS | $950 | $2,340 | | | | |
| Meridian Community College | 15 | Aug | 18 | AAS | $1,450 | $2,740 | | | | |

*Data are shown only for programs that completed the 2006 AMA Survey of Health Professions Education Programs
‡Key to Offers: 1: Evening or weekend classes; 2: Non-English instruction; 3: Cultural competence instruction

# Health Information Technician

| Programs* | Class Capacity | Begins | Length (months) | Award | Res Tuition | Non-res Tuition | Stipend | Offers:‡ 1 | 2 | 3 |
|---|---|---|---|---|---|---|---|---|---|---|
| **Missouri** | | | | | | | | | | |
| Ozarks Technical Community College (Springfield) | | Aug | 22 | Cert, AAS | $2,925 | $3,575 | | • | | |
| St Charles Community College (St Peters) | 50 | Aug | 21 | AS | $2,414 | $3,484 | | • | | |
| **Nebraska** | | | | | | | | | | |
| Central Community College (Hastings) | | Aug | 20 | Dipl, Cert, AAS | $1,600 | $2,336 | | • | | |
| College of Saint Mary (Omaha) | 30 | Aug Jan May | 24 | AS | $17,750 | $17,750 | | • | | |
| Western Nebraska Community College (Scottsbluff) | 25 | Aug Jan | 24, 36 | Dipl, Cert, AAS | $1,360 | $1,552 | | • | | |
| **Nevada** | | | | | | | | | | |
| Community College of Southern Nevada (Las Vegas) | 15 | Sep | 24 | AAS | $1,917 | $6,716 | | • | | • |
| **New Jersey** | | | | | | | | | | |
| Burlington County College (Pemberton) | 20 | Sep | 21 | AAS | $2,669 | $3,104 | | • | | |
| Hudson County Community College (Jersey City) | 20 | Sep Jan | 24 | AAS | $2,788 | $5,576 | | • | | |
| **New Mexico** | | | | | | | | | | |
| Central New Mexico Community College (Albuquerque) | 25 | Jan | 25 | AAS | | | | • | | |
| San Juan College (Farmington) | 20 | Aug Jan | 24 | AAS | $720 | $1,040 | | • | • | • |
| University of New Mexico - Gallup | 15 | Sep | 24 | AAS | $1,170 | $1,170 | | • | | |
| **New York** | | | | | | | | | | |
| Alfred State College - SUNY | 27 | Aug | 20 | AAS | $6,000 | $10,000 | | • | | |
| Mohawk Valley Community College (Utica) | 25 | Aug | 19 | AAS | $3,100 | $6,200 | | • | | • |
| Monroe Community College (Rochester) | 33 | Sep | 18 | AAS | $2,700 | $5,400 | | • | | |
| Suffolk Community College (Brentwood) | 24 | Sep | 24 | AAS | $1,550 | $3,100 | | • | | |
| Trocaire College (Buffalo) | 30 | Aug Jan | 24 | AAS | $11,780 | $0 | | • | | |
| **North Carolina** | | | | | | | | | | |
| Brunswick Community College (Supply) | 20 | Aug | 21 | Dipl, AAS | $1,533 | $8,103 | | | | |
| McDowell Technical Community College (Marion) | 25 | Aug | 21 | AAS | $1,140 | $6,154 | | • | | • |
| Pitt Community College (Greenville) | 35 | Jan May Aug | 21, 12 | Dipl, AAS | $1,937 | $10,241 | | • | | |
| Southwestern Community College (Sylva) | 25 | Aug | 12, 21 | Dipl, Cert, AAS | $664 | $3,544 | | • | | |
| **North Dakota** | | | | | | | | | | |
| North Dakota State College of Science (Wahpeton) | 24 | Aug | 13, 22 | Cert, AAS | $6,257 | $8,697 | | | | |
| United Tribes Technical College (Bismarck) | 50 | May | 24, 12 | Cert, AAS | $3,325 | $3,325 | | | | • |
| **Ohio** | | | | | | | | | | |
| Bowling Green State University - Firelands (Huron) | 18 | Aug | 18 | AAS | $9,060 | $16,368 | | • | | • |
| Columbus State Community College | 40 | Sep | 18 | Cert, AAS | $2,928 | $6,432 | | • | | |
| Owens Community College (Toledo) | 25 | Aug | 24 | Assoc | $2,640 | $5,280 | | • | | |
| Sinclair Community College (Dayton) | 30 | Sep | 22 | AAS | $2,248 | $3,369 | | • | | |
| Stark State College of Technology (North Canton) | 24 | Aug | 18 | AAS | $3,468 | $4,488 | | | | |
| **Oklahoma** | | | | | | | | | | |
| Rose State College (Midwest City) | 15 | Aug | 12, 21 | Cert, AAS | $2,345 | $6,573 | | | | |
| **Oregon** | | | | | | | | | | |
| Portland Community College | 35 | Sep | 18 | AAS | $3,000 | $8,500 | | | | |
| **Pennsylvania** | | | | | | | | | | |
| Comm Coll of Allegheny County (Pittsburgh) | 25 | Aug | 18 | AS | $3,407 | $9,007 | | • | | |
| Gwynedd-Mercy College (Gwynedd Valley) | 15 | Aug | 24 | AS | $19,720 | $19,720 | | • | | • |
| South Hills School of Business & Technology (State College) | 25 | Sep | 20 | AST | $11,050 | $11,050 | | | | |
| **Puerto Rico** | | | | | | | | | | |
| Inter American University of Puerto Rico (San German) | 20 | Aug | 10 | AAS | $3,230 | $0 | | • | | |
| Universidad Adventista de las Antillas (Mayaguez) | 20 | Aug | 26 | AS | $2,050 | $3,122 | | | | |
| Universidad del Este (Carolina) | 40 | Aug | 20 | AS | $4,142 | $4,142 | | | • | |
| **South Carolina** | | | | | | | | | | |
| Greenville Technical College | 20 | Aug | 21 | AS | $4,350 | $8,850 | $450 | • | | • |
| Midlands Technical College (Columbia) | 18 | Aug | 12, 24 | Cert, AS | $3,138 | $9,504 | | | | |
| **South Dakota** | | | | | | | | | | |
| Dakota State University (Madison) | 30 | Sep | 24 | AS | $2,659 | $3,898 | | | | |
| **Tennessee** | | | | | | | | | | |
| Dyersburg State Community College | 30 | Aug | 24 | Cert, AAS | $2,481 | $9,157 | | • | | |
| Roane State Community College (Harriman) | 18 | Aug | 21 | Cert, AAS | $2,230 | $6,676 | | • | | |
| Volunteer State Community College (Gallatin) | 40 | Aug | 24 | Cert, AAS | $2,049 | $7,513 | | • | | |
| **Texas** | | | | | | | | | | |
| Howard College (San Angelo) | 24 | Aug Jan | 24, 21 | AAS | $1,500 | $2,000 | | • | | |
| Lamar Institute of Technology (Beaumont) | 25 | Aug | 24 | AAS | $2,000 | $7,000 | | | | |
| McLennan Community College (Waco) | 25 | Aug | 24 | AAS | $4,464 | $0 | | • | | |
| North Harris College (Houston) | 40 | Aug | 24 | AAS | $1,344 | $2,994 | | • | | |
| South Plains College (Lubbock) | 19 | Aug | 9, 20 | Dipl, Cert, AAS | $1,918 | $2,581 | | • | | • |
| South Texas College (McAllen) | 25 | Sep | 24 | AAS | | | | • | | |
| Texas State Tech College, West Texas (Abilene) | 150 | Sep Jan May | 24 | AAS | $65 | $182 | | | | |

*Data are shown only for programs that completed the 2006 AMA Survey of Health Professions Education Programs
‡Key to Offers: 1: Evening or weekend classes; 2: Non-English instruction; 3: Cultural competence instruction

## Health Information Technician

| Programs* | Class Capacity | Begins | Length (months) | Award | Res Tuition | Non-res Tuition | Stipend | Offers:‡ 1 | 2 | 3 |
|---|---|---|---|---|---|---|---|---|---|---|
| Texas State Technical College - Harlingen | 32 | Aug | 24 | AAS | $3,347 | $8,010 | | | | |
| Tyler Junior College | 15 | Aug | 24 | AAS | $1,704 | $2,792 | $1,500 | | | |
| Wharton County Junior College | 20 | Aug | 20 | AAS | $1,890 | $4,270 | | | | |
| **Utah** | | | | | | | | | | |
| Weber State University (Ogden) | 20 | Aug | 24 | AAS | $2,793 | $9,776 | | • | | |
| **Virginia** | | | | | | | | | | |
| Northern Virginia Community College (Springfield) | 25 | Aug | 18 | AAS | $1,632 | $5,349 | | • | | |
| **Washington** | | | | | | | | | | |
| Shoreline Community College (Seattle) | 20 | Sep | 18 | AAAS | $3,200 | $6,000 | | • | | • |
| Tacoma Community College | 25 | Sep | 21 | Cert, AAS | $3,558 | $10,502 | | • | | |
| **West Virginia** | | | | | | | | | | |
| West Virginia Northern Community College (Wheeling) | 14 | Aug | 24 | AAS | $1,824 | $5,808 | | • | | |
| **Wisconsin** | | | | | | | | | | |
| Chippewa Valley Technical College (Eau Claire) | 24 | Aug | 21 | AS | $3,142 | $17,162 | | | | |
| Moraine Park Technical College (West Bend) | 20 | Aug | 24 | AAS | $3,700 | $10,100 | | • | | |
| Western Technical College (La Crosse) | 16 | Aug | 20 | AAS | $2,827 | $19,054 | | | | |

*Data are shown only for programs that completed the 2006 AMA Survey of Health Professions Education Programs
‡Key to Offers: 1: Evening or weekend classes; 2: Non-English instruction; 3: Cultural competence instruction

PROGRAMS

# Kinesiotherapist

## Occupational Description

The kinesiotherapist is academically and clinically prepared to provide rehabilitation exercise and education under the prescription of a licensed physician in an appropriate setting. Kinesiotherapists are qualified to implement exercise programs designed to reverse or minimize debilitation and enhance the functional capacity of medically stable patients in a wellness, sub-acute, or extended care setting. The role of the kinesiotherapist demands intelligence, judgment, honesty, interpersonal skills, and the capacity to react to emergencies in a calm and reasoned manner. An attitude of respect for self and others, adherence to the concepts of privilege and confidentiality in communicating with patients, and a commitment to the patient's welfare are standard attributes. At a minimum, a kinesiotherapist is educated in areas of basic exercise science and clinical applications of rehabilitation exercise. Training is received in orthopedic, neurological, psychiatric, pediatric, cardiovascular-pulmonary, and geriatric practice settings.

## Job Description

Kinesiotherapy is the application of scientifically based exercise principles adapted to enhance the strength, endurance, and mobility of individuals with functional limitations or those requiring extended physical conditioning.

The kinesiotherapist is a health care professional competent in the administration of musculoskeletal, neurological, ergonomic, biomechanical, psychosocial, and task-specific functional tests and measures. The kinesiotherapist determines the appropriate evaluation tools and interventions necessary to establish, in collaboration with the client, a goal-specific treatment plan.

The intervention process includes the development and implementation of a treatment plan, assessment of progress toward goals, modification as necessary to achieve goals and outcomes, and client education. The foundation of clinician-client rapport is based on education, instruction, demonstration, and mentoring of therapeutic techniques and behaviors to restore, maintain, and improve overall functional abilities.

The *Scope of Practice for Kinesiotherapy* identifies the job tasks that registered kinesiotherapists are qualified to perform. This document reflects the evaluation procedures and treatment interventions for medically stable individuals who need extended physical conditioning. The individual kinesiotherapist may obtain additional training and credentials in areas beyond the scope of practice. The *Standards of Practice for Registered Kinesiotherapists* serves as a guideline for practicing registered kinesiotherapists and provides a basis for assessment of kinesiotherapy practices.

## Employment Characteristics

Registered kinesiotherapists are employed in Department of Veterans Affairs Medical Centers, public and private hospitals, medical fitness facilities, rehabilitation facilities, learning disability centers, schools, colleges and universities, private practice, and as exercise consultants.

The types of treatments carried out by kinesiotherapists focus on but are not limited to
- Therapeutic exercise
- Ambulation training
- Geriatric rehabilitation
- Aquatic therapy
- Adapted fitness and conditioning
- Prosthetic/orthotic rehabilitation
- Psychiatric rehabilitation
- Driver training
- Adapted exercise for the home setting

## Salary

Depending on the particular job setting, the average projected starting salary for registered kinesiotherapists is $32,500 to $38,000 annually. The overall average is $48,000; upper-level salaries are in the range of $60,000. Refer to Section IV, Table 5 of this *Directory* for more information, or see www.ama-assn.org/go/salary.

## Educational Programs

**Length.** The kinesiotherapy program is 4 to 5 years. The total minimum requirements are 128 semester hours. Minimum requirements for years 1 and 2 are 59 semester hours and for years 3 and 4 are 67 semester hours.
**Prerequisites.** Applicants should have a high school diploma or equivalent and meet institutional entrance requirements.
**Curriculum.** The program has a comprehensive academic and clinical curriculum plan that fulfills or exceeds the minimum requirements for kinesiotherapy accreditation.

The curriculum plan includes an organized and sequential series of integrated learning experiences designed to achieve or exceed minimum competencies.

All academic and clinical courses are guided by written measurable behavioral objectives and use case-based, patient-centered, problem-solving activities.

The curriculum plan includes academic learning experiences, which lead to the attainment of all academic competencies listed in the *Minimum Core Competencies of Kinesiotherapists.*

Students must complete the following content areas: human anatomy, human physiology, exercise physiology, kinesiology/biomechanics, therapeutic exercise/adapted physical education, growth and development, motor learning/control/performance, general psychology, organization and administration, tests and measurements, research methods or statistics, and first aid and cardiopulmonary resuscitation, Introduction to Kinesiotherapy, pathophysiology, clinical neurology, rehabilitation procedures, patient assessment and management, and therapeutic activities. Students are strongly encouraged to complete the following academic courses: abnormal psychology or mental health, physiological psychology, exercise testing and prescription, gerontology, medical ethics, medical terminology, pharmacology, health/medical/functional outcomes management, health education, and Kinesiotherapy I and II.

## Inquiries

**Careers**
American Kinesiotherapy Association
118 College Drive, #5142
Hattiesburg, MS 39406
601 266-5371
601 266-4445 Fax
E-mail: helen.fuller@usm.edu

**Credentialing**
Board of Registration for Kinesiotherapy
University of Toledo
2801 W Bancroft
Toledo, OH 43606-2434
419 530-2731
E-mail: dwoods@utnet.utoledo.edu

**Program Accreditation**
Commission on Accreditation of Allied Health Education Programs
   (CAAHEP) in collaboration with:
Committee on Accreditation of Education Programs for
   Kinesiotherapy
University of Southern Mississippi
118 College Drive, #5142
Hattiesburg, MS 39406
601 266-5371
601 266-4445 Fax
E-mail: jerry.purvis@usm.edu

## Kinesiotherapist

## California

**California State University - Long Beach**
Kinesiotherapy Prgm
1250 Bellflower Blvd
Long Beach, CA 90840
*Prgm Dir:* Janet M Fisher, PhD
*Tel:* 562 985-8481
*E-mail:* fisherja@csulb.edu
*Med Dir:* Patricia Nance, MD

**San Diego State University**
Kinesiotherapy Prgm
5500 Campanile Dr
Department of Exercise & Nutritional Sciences
San Diego, CA 92182-7251
*Prgm Dir:* James A Yaggie, PhD RKT
*Tel:* 619 594-2392   *Fax:* 619 594-6553
*E-mail:* jyaggie@mail.sdsu.edu
*Med Dir:* Greg Gastaldo, MD

## Mississippi

**University of Southern Mississippi**
Kinesiotherapy Prgm
118 College Dr #5142
Hattiesburg, MS 39406-5142
*Prgm Dir:* Jerry W Purvis, MS RKT
*Tel:* 601 266-5371   *Fax:* 601 266-4445
*E-mail:* Jerry.Purvis@usm.edu
*Med Dir:* Thomas Sturdavant, MD

## North Carolina

**Shaw University**
Kinesiotherapy Prgm
118 E South St
Raleigh, NC 27601
www.shawu.edu
*Prgm Dir:* Edwards Bennett, PhD
*Tel:* 919 546-8316
*E-mail:* BEdwardsSU@aol.com
*Med Dir:* Fred Long, MD

## Ohio

**University of Toledo**
Kinesiotherapy Prgm
2801 W Bancroft St
Toledo, OH 43606-3390
*Prgm Dir:* Leonard O Greninger, PhD RKT
*Tel:* 419 530-2953   *Fax:* 419 530-2477
*E-mail:* Leonard.Greninger@utoledo.edu

## Virginia

**Norfolk State University**
Kinesiotherapy Prgm
700 Park Ave
Norfolk, VA 23504
*Prgm Dir:* Delano Tucker, PhD
*Tel:* 757 823-8703   *Fax:* 757 823-9412
*E-mail:* dtucker@nsu.edu
*Med Dir:* Cynthia Burwell, EdD CHES

## History

Massage has its roots in the far reaches of human history. Rubbing a sore muscle or stroking another person for comfort are natural responses. The first written records that refer to massage date back more than 4,000 years to China. In ancient Greece, Hippocrates, the father of modern medicine, wrote, "The physician must be experienced in many things, but most assuredly in rubbing."

Massage comes from both Western and Eastern traditions. Western traditions date back to ancient Greece and Rome. Modern Western massage owes a great deal to the work of Peter Henrik Ling, a 19th-century educator and athlete from Sweden. His approach, which combined hands-on techniques with active movements, became known as Swedish Massage, probably the most common therapeutic massage modality in the West.

Eastern traditions can be traced back to the folk medicine of China and the Ayurvedic medicine of India. Shiatsu, acupressure, reflexology, and many other contemporary techniques have their roots in these sources.

The incorporation of massage into health care was fairly well-established in the 19th century, but those connections decreased through most of the 20th century. A growing body of clinical research on the efficacy and value of massage as part of integrated health care, as well as a rapid acceptance and adoption of use of massage in recent years, has fueled a renewed collaboration between massage therapists and other health care professionals.

Surveys of hospitals, conducted through the American Hospital Association, have shown a rapid increase in use of massage in the hospital setting. Consumer surveys conducted by Opinion Research Corporation, International between 1997 and 2005 saw the number of American adults receiving a massage from a massage therapist each year jump from 8% in 1997 to 22% in 2005.

## Occupational Description

The American Massage Therapy Association (AMTA) has defined massage and massage therapy as follows:

Massage is manual soft tissue manipulation, and includes holding, causing movement, and/or applying pressure to the body. Manual means by use of hand or body.

Massage therapy is a profession in which the practitioner applies manual techniques, and may apply adjunctive therapies, with the intention of positively affecting the health and well-being of the client.

An increasing body of research shows massage reduces heart rate, can help lower blood pressure, increases blood circulation and lymph flow, relaxes muscles, improves range of motion, and increases endorphins. Recent studies indicate massage enhances the functioning of the immune system. Although therapeutic massage does not increase muscle strength, it can stimulate weak, inactive muscles and, thus, partially compensate for the lack of exercise and inactivity resulting from illness or injury. It also can hasten and lead to a more complete recovery from exercise or injury.

## Job Description

Some of the most common types of massage are Swedish massage, deep-tissue massage, Shiatsu-acupressure, neuromuscular, trigger point, and sports massage.

Massage therapy is strenuous work. Practitioners must use correct body mechanics to prevent injury and fatigue. If the therapist travels to give massage, they transport either a massage table or massage chair and all supplies necessary to give a massage. The profession requires good listening skills and the ability to make clients comfortable and relaxed. Massage therapists are usually very satisfied with their profession, because they have a positive impact on clients' health and well being. The vast majority of AMTA members say they went into the profession because they want to help people improve their health.

In addition to the actual massage, massage therapists market their practices, keep financial and client records, maintain supplies and equipment, educate their clients about massage and inform them of any physical irregularities they discover, and work with health insurance companies to receive fees. Practitioners take basic medical histories on clients and discuss with the client their current health. During massage, therapists pay close attention to how the client is responding and discuss levels of massage pressure with the client. They also must be aware of medical conditions that might contraindicate massage and advise clients when massage is not appropriate.

## Employment Characteristics

Massage therapists work in many different environments, such as hospitals, physician or chiropractor offices, nursing homes, private practice, pain clinics, resorts, cruise ships, shopping malls, airports, spas, and salons. Some focus exclusively on massage for stress relief and relaxation, while others specialize—pregnancy massage, massage to reduce lymphedema after cancer surgery, massage for pain relief, and sports massage, among other specialties. Approximately 23% of AMTA members report that they work in a medical setting.

Massage therapists may be independent contractors, sole practitioners, or employees. Some travel to clients' homes or to business offices. Onsite chair massage has become a very popular form of massage, because of its convenience of use in a variety of settings, such as corporate offices.

## Salary

Earnings among massage therapists vary widely, depending on where the therapist practices, their level of experience and whether or not they work full-time as massage therapists. Additional time is spent on practice management, billing, marketing, etc.

Based on member surveys, the median full-time massage income of AMTA members is $20,000 to $49,000 per year. However, 43% earn more than $30,000 annually from their work as massage therapists. Approximately 57% of AMTA members describe their massage therapy practice as parttime. Refer to Section IV, Table 5 of this *Directory* for more information, or see www.ama-assn.org/go/salary.

### Educational Programs

Minimum entry-level standards for massage therapy training vary greatly, based on state or local requirements, professional association standards, or insurance requirements. State regulatory requirements for massage practice range from a minimum of 300 in-class hours at a recognized massage school to 1,000 in-class hours of massage training in an accredited massage program.

The Commission on Massage Therapy Accreditation (COMTA) is recognized by the US Department of Education as a specialized accrediting agency for massage therapy and bodywork programs and institutions. It is the only recognized accrediting agency focused solely on the quality of education for massage therapy. Massage schools and programs may voluntarily seek accreditation by COMTA or may choose some other accreditation or none at all. COMTA accredits massage programs and institutions that offer a minimum training of 600 hours of classroom and clinical instruction, conducted or directly supervised by qualified faculty. Six defined competency requirements must be included in the program curriculum and students must be assessed on having met the competencies.

The AMTA requires its members to have minimum training of 500 hours of classroom instruction and recommends training at a school accredited by COMTA. Its Professional-level members must provide evidence of 48 clock hours of continuing education every 4 years.

### Licensure and Certification

**Licensure.** As of November 2005, 36 states and Washington, DC had passed regulations on massage therapy. Some states license massage therapists, while others have basic standards. Regulation of massage therapy by the states has increased dramatically in recent years, with 24 states having passed regulations between 1990 and 2005.

**Certification.** The National Certification Board for Therapeutic Massage & Bodywork (NCBTMB) administers the National Certification Examination in Therapeutic Massage & Bodywork and certifies massage therapists who pass the examination and maintain their status through continuing education. This examination is now required in some states and municipalities for massage therapists to practice. The examination is an entry-level test and requires that a person has completed a minimum of 500 in-class hours of massage training.

National certification, while growing as a requirement in some civil jurisdictions, is by no means a universal requirement.

### Inquiries

**Massage Careers, Organizational Memberships**
American Massage Therapy Association
500 Davis Street, Suite 900
Evanston, IL 60201-4695
847 864-0123
847 864-1178 Fax
E-mail: info@amtamassage.org
www.amtamassage.org

**Certification**
National Certification Board for Therapeutic Massage & Bodywork
8201 Greensboro Drive, Suite 300
McLean, VA 22102
800 296-0664
703 610-9015
703 610-9005 Fax
E-mail: info@ncbtmb.com
www.ncbtmb.com

**Program Accreditation**
Commission on Massage Therapy Accreditation
1007 Church Street, Suite 302
Evanston, IL 60201
847 869-5039
847 869-6739 Fax
E-mail: info@comta.org
www.comta.org

## Massage Therapist*

## Alaska

**University of Alaska Anchorage**
Massage Therapy Prgm
3211 Providence Dr
Adm 216
Anchorage, AK 99508
*Prgm Dir:* Sally Mead
*Tel:* 907 786-6930　*Fax:* 907 272-4742
*E-mail:* afslm1@uaa.alaska.edu

## Arizona

**Desert Institute of the Healing Arts**
Massage Therapy Prgm
140 E 4th St
Tucson, AZ 85705
*Prgm Dir:* Ginger Castle
*Tel:* 800 733-8098　*Fax:* 520 624-2996
*E-mail:* diha@azstarnet.com

## California

**Mueller College of Holistic Massage Therapies**
Massage Therapy Prgm
4607 Park Blvd
San Diego, CA 92116-2630
*Prgm Dir:* Jeffrey Welsh, PhD MA RRT HHP
*Tel:* 619 291-9811　*Fax:* 619 543-1113
*E-mail:* paul@mueller.edu

## Colorado

**Massage Therapy Institute of Colorado**
Massage Therapy Prgm
1441 York St, Ste 301
Denver, CO 80206
*Tel:* 303 329-6345

**Academy of Natural Therapy**
Massage Therapy Prgm
123 Elm Ave
PO Box 237
Eaton, CO 80615
*Tel:* 970 454-2628

## Connecticut

**Connecticut Center for Massage Therapy**
Massage Therapy Prgm
1154 Poquonnock Rd
Groton, CT 06340
*Prgm Dir:* Linda Derick
*Tel:* 860 446-2299　*Fax:* 860 446-9410
*E-mail:* linda@ccmt.com

*Additional information about programs that returned the AMA's survey is available in a table beginning on page 221.

## Connecticut Center for Massage Therapy
Massage Therapy Prgm
75 Kitts Ln
Newington, CT 06111
*Prgm Dir:* Linda Derick
*Tel:* 860 667-1886   *Fax:* 860 667-2175
*E-mail:* linda@ccmt.com

## Connecticut Center for Massage Therapy
Massage Therapy Prgm
25 Sylvan Rd S
Westport, CT 06880
*Prgm Dir:* Linda Derick
*Tel:* 203 221-7325   *Fax:* 203 221-0144
*E-mail:* linda@ccmt.com

# Delaware

## National Massage Therapy Institute
Massage Therapy Prgm
1601 Concord Pike, Ste 82-84
Wilmington, DE 19803
*Prgm Dir:* Lisa Jakober
*Tel:* 302 436-4508
*E-mail:* lisa@studymassage.com

# District of Columbia

## Potomac Massage Training Institute
Massage Therapy Prgm
5028 Wisconsin Ave NW - LL
Washington, DC 20016-4118
www.pmti.org
*Prgm Dir:* Demara Stamler
*Tel:* 202 686-7046, Ext 115   *Fax:* 202 966-4579
*E-mail:* dstamler@pmti.org
*Med Dir:* Philomena Queen, Dean of Students

# Florida

## Florida College of Natural Health
Massage Therapy Prgm
616 67th St Circle E
Bradenton, FL 34208
*Prgm Dir:* Kathryn Knox
*Tel:* 800 966-7117   *Fax:* 941 744-1242
*E-mail:* kknox@fcnh.com

## Florida School of Massage
Massage Therapy Prgm
6421 SW 13th St
Gainesville, FL 32608
www.floridaschoolofmassage.com
*Prgm Dir:* Paul Davenport
*Tel:* 352 378-7891
*E-mail:* info@floridaschoolofmassage.com

## Florida College of Natural Health
Massage Therapy Prgm
2600 Lake Lucien Dr, Ste 140
Maitland, FL 32751
*Prgm Dir:* Diane Owens, PhD
*Tel:* 800 393-7337   *Fax:* 407 261-0342
*E-mail:* dowens@fcnh.com

## Educating Hands School of Massaage
Massage Therapy Prgm
120 SW 8th St
Miami, FL 33130-3513
www.educatinghands.com
*Prgm Dir:* Janet Gonzalez
*Tel:* 800 999-6991   *Fax:* 305 857-0298
*E-mail:* janet@educatinghands.com

## Florida College of Natural Health
Massage Therapy Prgm
7925 NW 12th St, Ste 201
Miami, FL 33126
*Prgm Dir:* Debra Starr Cohen
*Tel:* 800 599-9599   *Fax:* 305 597-9110
*E-mail:* dstarr@fcnh.com

## Florida College of Natural Health
Massage Therapy Prgm
2001 W Sample Rd, Ste 100
Pompano Beach, FL 33064
www.fcnh.com
*Prgm Dir:* Sherry Parker
*Tel:* 800 541-9299   *Fax:* 954 975-9633
*E-mail:* sparker@fcnh.com

## Sarasota School of Massage Therapy
Massage Therapy Prgm
1932 Ringling Blvd
Sarasota, FL 34236
www.sarasotaschoolofmassagetherapy.edu
*Prgm Dir:* Joe Lubow, BS LMT
*Tel:* 877 613-7768   *Fax:* 941 957-1049
*E-mail:* joe@sarasotaschoolofmassagetherapy.edu

## Core Institute
Massage Therapy Prgm
223 W Carolina St
Tallahassee, FL 32301
*Prgm Dir:* Philip R Arcuri
*Tel:* 850 222-8673   *Fax:* 850 561-6160
*E-mail:* phil@coreinstitute.com

# Illinois

## Chicago School of Massage Therapy
Massage Therapy Prgm
17 N State St, 5th Fl
Chicago, IL 60602
www.cortiva.com/locations/csmt
*Prgm Dir:* Jason Scholte
*Tel:* 312 753-7900, Ext 7929   *Fax:* 312 753-7901
*E-mail:* jscholte@cortiva.com

## Morton College
Massage Therapy Prgm
3801 S Central
Cicero, IL 60804
*Prgm Dir:* Linda Moore
*Tel:* 708 656-2000, Ext 396
*E-mail:* Linda.Moore@morton.edu

## Chicago School of Massage Therapy
*Cosponsor:* McHenry County College Satellite
Massage Therapy Prgm
100 S Main St
Crystal Lake, IL 60014
*Prgm Dir:* Connie Meshenie
*Tel:* 773 477-9444   *Fax:* 773 477-7256

## National University of Health Sciences
Massage Therapy Prgm
200 E Roosevelt Rd
Lombard, IL 60148
www.nuhs.edu
*Prgm Dir:* Randy Swenson, BS DC MHPE
*Tel:* 630 889-6545   *Fax:* 630 889-6635
*E-mail:* rswenson@nuhs.edu

## Kishwaukee College
Massage Therapy Prgm
21193 Malta Rd
Malta, IL 60150-9699
*Prgm Dir:* Leslie Ciaccio, BFA CMT
*Tel:* 815 825-2086, Ext 259   *Fax:* 815 825-2072
*E-mail:* lciaccio@kishwaukeecollege.edu

# Indiana

## Alexandria School of Scientific Therapeutics
Massage Therapy Prgm
809 S Harrison
PO Box 287
Alexandria, IN 46001
*Prgm Dir:* Herbert Hobbs
*Tel:* 800 622-8756   *Fax:* 765 724-9156
*E-mail:* alexssin.edudir@insightbb.com

# Iowa

## Carlson College of Massage Therapy
Massage Therapy Prgm
11809 County Rd X-28
Anamosa, IA 52205
www.carlsoncollege.com
*Prgm Dir:* Wayne Pakulis, BA LMT NCTMB
*Tel:* 319 462-3402   *Fax:* 319 462-5990
*E-mail:* waynepakulis@carlsoncollege.com

## Institute of Therapeutic Massage & Wellness
Massage Therapy Prgm
1730 Wilkes Ave
Davenport, IA 52804
*Tel:* 563 445-1055

# Kansas

## BMSI Institute
Massage Therapy Prgm
8665 W 96th St, Ste 300
Overland Park, KS 66212
*Tel:* 913 649-3322

# Kentucky

## Cincinnati Sch of Med Massage-Northern Kentucky
Massage Therapy Prgm
1793 Patrick Rd
Burlington, KY 41005
*Tel:* 859 578-8868

# Louisiana

## Blue Cliff College - Baton Rouge
Massage Therapy Prgm
6160 Perkins Rd, Ste 200
Baton Rouge, LA 70808-4191
*Prgm Dir:* Rose Steptoe
*Tel:* 225 757-3770   *Fax:* 225 757-7422
*E-mail:* roses@bluecliffcollege.com

# Maine

## New Hamp Inst for Therapeutic Arts
Massage Therapy Prgm
27 Sandy Creek Rd
Bridgton, ME 04009
www.nhita.com
*Prgm Dir:* Patrick Ian Cowan, PhD
*Tel:* 207 647-3794   *Fax:* 207 647-2954
*E-mail:* nhita@nhita.com

## Downeast School of Massage
Massage Therapy Prgm
PO Box 24
99 Moose Meadow Ln
Waldoboro, ME 04572-0024
www.downeastschoolofmassage.net
*Prgm Dir:* Nancy Dail, BA LMT NCTMB
*Tel:* 207 832-5531   *Fax:* 207 832-0504
*E-mail:* ndail@aol.com
*Med Dir:* Gordon Paine, MD

# Manitoba, Canada

## Massage Therapy College of Manitoba
Massage Therapy Prgm
691 Wolseley Ave, 2nd Fl
Winnipeg, MB R3G 1C3
*Prgm Dir:* Garth Beddome
*Tel:* 204 772-8999   *Fax:* 204 772-5090
*E-mail:* mcollege@mts.net

# Maryland

**Allegany College of Maryland**
Massage Therapy Prgm
Therapeutic Massage Program
12401 Willowbrook Rd
Cumberland, MD 21502
*Tel:* 301 784-5191

**Baltimore School of Massage**
Massage Therapy Prgm
517 Progress Dr
Ste A-L
Linthicum, MD 21090
www.bsom.com
*Prgm Dir:* Richard Rynders
*Tel:* 410 636-7929   *Fax:* 410 636-7857
*E-mail:* richardr@bsom.com
*Med Dir:* Elizabeth Helper, MA

# Massachusetts

**Stillpoint Program - Greenfield Comm Coll**
Massage Therapy Prgm
270 Main St
Greenfield, MA 01301
*Prgm Dir:* Patricia A Wachter
*Tel:* 413 775-1634   *Fax:* 413 774-2285
*E-mail:* wachter@gcc.mass.edu

**Springfield Technical Community College**
Massage Therapy Prgm
Dept of Myofascial
One Armory Sq
Springfield, MA 01105
*Prgm Dir:* Bernadette Della Bitta Nicholson
*Tel:* 413 755-4885
*E-mail:* Nicholson@stcc.edu

**Cortiva Institute-Muscular Therapy Institute**
Massage Therapy Prgm
103 Morse St
Watertown, MA 02472
*Tel:* 617 668-1000

# Michigan

**Ann Arbor Institute of Massage Therapy**
Massage Therapy Prgm
2835 Carpenter Rd
Ann Arbor, MI 48108
*Prgm Dir:* Jocelyn Granger
*Tel:* 734 677-4430, Ext 24   *Fax:* 734 677-4520
*E-mail:* jgranger@aaimt.edu

**Lakewood School of Therapeutic Massage**
Massage Therapy Prgm
1102 6th St
Port Huron, MI 48060
*Tel:* 810 987-3959

# Minnesota

**Northwestern Health Science University School of Massage Therapy**
Massage Therapy Prgm
School of Massage Therapy
2501 W 84th St
Bloomington, MN 55431
*Tel:* 952 888-4777

# Mississippi

**Mississippi School of Therapeutic Massage**
Massage Therapy Prgm
5140 Galaxie Dr
Jackson, MS 39206-4339
www.mstm.info
*Prgm Dir:* Sara Simpson
*Tel:* 601 362-3624   *Fax:* 601 362-3694
*E-mail:* mstmjxn@aol.com

# Missouri

**St Charles School of Massage Therapy**
Massage Therapy Prgm
2440 Executive Dr
Ste 100
St Charles, MO 63303
www.spastcharles.com
*Prgm Dir:* Kathleen Crawford, LMT
*Tel:* 636 498-0777   *Fax:* 636 498-0708
*E-mail:* oasis@anet-stl.com

# Montana

**Health Works Institute**
Massage Therapy Prgm
111 S Grand Annex 3
Bozeman, MT 59715
www.healthworksinstitute.com
*Prgm Dir:* Ruth Marion
*Tel:* 406 582-1555   *Fax:* 406 522-0493
*E-mail:* director@healthworksinstitute.com

**Big Sky Somatic Institute**
Massage Therapy Prgm
1802 11 Ave
Helena, MT 59601
*Tel:* 406 442-8998

# New Hampshire

**New Hampshire Institute for Therapeutic Arts**
Massage Therapy Prgm
153 Lowell Rd
Hudson, NH 03051
*Prgm Dir:* Janet Alexis
*Tel:* 603 882-3022   *Fax:* 603 598-9101
*E-mail:* janhita@aol.com

# New Jersey

**Institute for Therapeutic Massage - Browns Mills**
Massage Therapy Prgm
200 Trenton Rd
Browns Mills, NJ 08015
*Prgm Dir:* Lisa Helbig
*Tel:* 973 839-6131
*E-mail:* joanne@massageprogram.com

**National Massage Therapy Institute**
Massage Therapy Prgm
6712 Washington Ave, Ste 302
Egg Harbor, NJ 08234
*Prgm Dir:* Lisa Jakober
*Tel:* 609 277-1599
*E-mail:* lisa@studymassage.com

**Academy of Massage Therapy**
Massage Therapy Prgm
321 Main St, 2nd Fl
Hackensack, NJ 07601
www.academyofmassage.com
*Prgm Dir:* Rey Visperas, MBA
*Tel:* 888 268-7898   *Fax:* 201 568-5181
*E-mail:* info@academyofmassage.com
*Med Dir:* Antonella Sena, DC

**Institute for Therapeutic Massage - Morristown**
Massage Therapy Prgm
95 Mount Kemble Ave
Morristown, NJ 07962
*Prgm Dir:* Lisa Helbig
*Tel:* 973 839-6131
*E-mail:* admissions@massageprogram.com

**Inst for Therapeutic Mass Univ of Med & Dent**
Massage Therapy Prgm
150 Bergen St
Newark, NJ 07103
*Prgm Dir:* Lisa Helbig
*Tel:* 973 839-6131
*E-mail:* admissions@massageprogram.com

**Omega Institute**
Massage Therapy Prgm
7050 Rte 38 E
Pennsauken, NJ 08109
*Prgm Dir:* Kathryn Fraser
*Tel:* 856 663-4299, Ext 221   *Fax:* 856 661-9585
*E-mail:* kfraser@omegacareers.com

**Cortiva-Inst Somerset Sch of Massage Therapy - Piscataway**
Massage Therapy Prgm
180 Centennial Ave
Piscataway, NJ 08854
www.ssmt.org
*Prgm Dir:* Tracey Cudo
*Tel:* 732 885-3400, Ext 29   *Fax:* 732 885-0440
*E-mail:* tcudo@cortiva.com

**Institute for Therapeutic Massage - Pompton Lakes**
Massage Therapy Prgm
125 Wanaque Ave
Pompton Lakes, NJ 07442
*Prgm Dir:* Lisa Helbig
*Tel:* 973 839-6131
*E-mail:* admissions@massageprogram.com

**Institute for Therapeutic Massage**
Massage Therapy Prgm
99 Hwy 37 West
Toms River, NJ 08755
*Tel:* 732 936-9111

**National Massage Therapy Institute - Turnersville**
Massage Therapy Prgm
108-L Greentree Rd & Black Horse Pike
Turnersville, NJ 08012
*Prgm Dir:* Lisa Jakober
*Tel:* 856 227-1754
*E-mail:* lisa@studymassage.com

**Cortiva-Inst Somerset Sch of Massage Therapy - Wall Township**
Massage Therapy Prgm
1985 Hwy 34
Wall Township, NJ 07719
*Prgm Dir:* Rhonda Brunelle
*Tel:* 732 282-0100   *Fax:* 732 282-1108
*E-mail:* rbrunelle@cortiva.com

**Healing Hands Institute for Massage Therapy**
Massage Therapy Prgm
41 Bergenline Ave
Westwood, NJ 07675
*Prgm Dir:* Eva Carey, NCMT MBA
*Tel:* 201 722-0099   *Fax:* 201 722-0690
*E-mail:* hhi@aol.com
*Med Dir:* Alice Feuerstein, RN LMT

# New Mexico

**Crystal Mountain Sch of Therapeutic Massage**
Massage Therapy Prgm
4775 Indian School Rd NE
Ste 102
Albuquerque, NM 87110
*Tel:* 505 872-2030

# North Carolina

**Body Therapy Institute**
Massage Therapy Prgm
300 Southwind Rd
Siler City, NC 27344
*Prgm Dir:* Rick Rosen, MA NC LMT
*Tel:* 919 663-3111, Ext 13   *Fax:* 919 663-0369
*E-mail:* rick@bti.edu

# Nova Scotia, Canada

**ICT Northumberland College**
Massage Therapy Prgm
1888 Brunswick St, 5th Fl
Halifax, NS 338
www.ictschools.com
*Prgm Dir:* Florent Villeneuve
*Tel:* 416 762-4857, Ext 241   *Fax:* 416 762-5733
*E-mail:* florentv@ictschools.com

# Ohio

**Cincinnati School of Medical Massage**
Massage Therapy Prgm
11250 Cornell Park Dr, Ste 203
Cincinnati, OH 45242
*Tel:* 513 469-6300

**Dayton School of Medical Message**
Massage Therapy Prgm
4457 Far Hills Ave
Dayton, OH 45429
*Tel:* 937 294-6994

**Dayton School of Medical Message**
Massage Therapy Prgm
Apollo Career Center
3325 Shawnee Rd
Lima, OH 45806
*Tel:* 888 860-4544

**Cleveland Insitute of Medical Massage**
Massage Therapy Prgm
18334-D E Bagley Rd
Middleburg Heights, OH 44130
*Tel:* 440 243-8650

# Ontario, Canada

**ICT Kikkawa College**
Massage Therapy Prgm
2340 Dundas St W, Unit G-04
Toronto, ON M6P 4A9
www.ictschools.com
*Prgm Dir:* Florent Villeneuve
*Tel:* 416 762-4857, Ext 241   *Fax:* 416 762-5733
*E-mail:* florentv@ictschools.com

# Oregon

**East-West Coll of the Healing Arts**
Massage Therapy Prgm
525 NE Oregon St
Portland, OR 97232
www.eastwestcollege.com
*Prgm Dir:* David Slawson
*Tel:* 800 635-9141   *Fax:* 503 232-4392
*E-mail:* dslawson@eastwestcollege.com

# Pennsylvania

**Synergy Healting Arts Center & Massage School**
Massage Therapy Prgm
13593 Monterey Ln
Blue Ridge Summit, PA 17214
*Tel:* 800 286-1931

**Cortiva Inst - Penn School of Muscle Therapy**
Massage Therapy Prgm
1173 Egypt Rd
PO Box 400
Oaks, PA 19456-0400
www.psmt.com
*Prgm Dir:* Joanne Slanga, BA MSLS MTS NCBTMB LMT
*Tel:* 610 666-9060, Ext 18   *Fax:* 610 666-9061
*E-mail:* psmt@psmt.com

**National Massage Therapy Insitute**
Massage Therapy Prgm
Division of PSB
10050 Roosevelt Blvd
Philadelphia, PA 19116
*Prgm Dir:* Lisa Jakober
*Tel:* 800 264-9835   *Fax:* 215 969-1368
*E-mail:* lisa@studymassage.com

**Baltimore School of Massage**
Massage Therapy Prgm
170 Red Rock Rd
York, PA 17402
*Prgm Dir:* Anita Perry-Strong
*Tel:* 717 268-1881   *Fax:* 717 268-1991
*E-mail:* anitas@steinerleisure.com

# Rhode Island

**Community College of Rhode Island**
Massage Therapy Prgm
Therapeutic Massage Program
One John H Chafee Blvd
Newport, RI 02840
*Tel:* 401 851-1681

# Tennessee

**Roane State Community College**
Massage Therapy Prgm
Somatic Massage Therapy Program
701 Briarcliff Ave
Oak Ridge, TN 37830
*Prgm Dir:* Paula Keefe
*Tel:* 865 481-2017
*E-mail:* KeefePK@roanestate.edu

# Virginia

**Virginia School of Massage**
Massage Therapy Prgm
2008 Morton Dr
Charlottesville, VA 22903
*Tel:* 888 599-2001   *Fax:* 804 293-4190
*E-mail:* virginia@vasom.com

**National Massage Therapy Institute**
Massage Therapy Prgm
803 W Broad St, Ste 110
Falls Church, VA 22046
*Prgm Dir:* Lisa Jakober
*Tel:* 703 237-3905
*E-mail:* lisa@studymassage.com

**Cayce/Reilly School of Massotherapy**
Massage Therapy Prgm
215 67th St
Virginia Beach, VA 23451
*Prgm Dir:* Nancy Hallingse-Smith
*Tel:* 757 457-7202   *Fax:* 757 428-0398
*E-mail:* nsmith@edgarcayce.org

# Washington

**Brenneke School of Massage**
Massage Therapy Prgm
425 Pontius Ave N, Ste 100
Seattle, WA 98109
www.brennekeschool.com
*Prgm Dir:* Julie Darrah, LMP
*Tel:* 206 282-1233, Ext 107   *Fax:* 206 282-9183
*E-mail:* info@brennekeschool.com
*Med Dir:* Dawn Schmidt, BS, LMP

**Brian Utting School of Massage**
Massage Therapy Prgm
900 Thomas St, Ste 200
Seattle, WA 98109
*Prgm Dir:* Brian Utting
*Tel:* 206 292-8055   *Fax:* 206 292-0113
*E-mail:* compliance@busm.edu

# West Virginia

**Mountain State School of Massage**
Massage Therapy Prgm
601 50th St
Charleston, WV 25304
*Prgm Dir:* Vanessa Hendley
*Tel:* 304 926-8822   *Fax:* 304 926-8837
*E-mail:* info@mtnstmassage.com
*Med Dir:* Mary Beth Mangus, DC

# Wisconsin

**Blue Sky School of Professional Massage**
Massage Therapy Prgm
220 American Blvd
De Pere, WI 54155
*Tel:* 920 338-9500

**Blue Sky School of Professional Massage**
Massage Therapy Prgm
220 Oak St
Grafton, WI 53024
*Tel:* 262 376-1011

**Blue Sky School of Professional Massage**
Massage Therapy Prgm
2122 Luann Ln
Madison, WI 53713
*Tel:* 608 270-5245

**Lakeside School of Massage Therapy - Madison**
Massage Therapy Prgm
6121 Odana Rd
Madison, WI 53219
*Prgm Dir:* Claude Gagnon
*Tel:* 414 372-4345, Ext 12   *Fax:* 414 372-5350
*E-mail:* claude@lakesideschoolmassage.org

**Lakeside School of Massage Therapy - Milwaukee**
Massage Therapy Prgm
1726 N 1st St, Ste 200
Milwaukee, WI 53212
www.lakeside.edu
*Prgm Dir:* Carole Ostendorf, PhD
*Tel:* 414 372-4345, Ext 14   *Fax:* 414 372-5350
*E-mail:* carole@lakeside.edu
*Med Dir:* Mary Gagnon-Riordan

## Massage Therapist

| Programs* | Class Capacity | Begins | Length (months) | Award | Res Tuition | Non-res Tuition | Stipend | Offers:‡ 1 | 2 | 3 |
|---|---|---|---|---|---|---|---|---|---|---|
| **District of Columbia** | | | | | | | | | | |
| Potomac Massage Training Institute (Washington) | 100 | Feb Aug | 18 | Cert | $7,800 | $7,800 | | • | | • |
| **Florida** | | | | | | | | | | |
| Educating Hands School of Massaage (Miami) | 22 | Quarterly | 6, 11 | Dipl | $6,500 | $6,500 | | • | | |
| Florida College of Natural Health (Pompano Beach) | 30 | Monthly | 6, 9 | Cert, Assoc | | | | • | | |
| Florida School of Massage (Gainesville) | 66 | Jan May Sep | 6 | Dipl | $7,100 | $7,100 | | | | • |
| Sarasota School of Massage Therapy | 22 | Jan Mar Jun Sep | 9, 18 | Dipl | $8,325 | $8,325 | | | | |
| **Illinois** | | | | | | | | | | |
| Chicago School of Massage Therapy | 550 | Jan 15 | 12, 15 | Dipl | $12,950 | $12,950 | | • | | • |
| National University of Health Sciences (Lombard) | 60 | Jan May Sep | 12 | Cert, AAS | $8,742 | $8,742 | | • | | • |
| **Iowa** | | | | | | | | | | |
| Carlson College of Massage Therapy (Anamosa) | 46 | Mar Sep (FT) Sep (PT) | 6, 10 | Dipl | $8,000 | $8,000 | | • | | |
| **Maine** | | | | | | | | | | |
| Downeast School of Massage (Waldoboro) | 80 | Sep Jan | 11, 24 | Dipl | $10,000 | $0 | | • | | |
| New Hamp Inst for Therapeutic Arts (Bridgton) | 24 | Sep Jan | 12 | Cert | | | | • | | |
| **Maryland** | | | | | | | | | | |
| Baltimore School of Massage (Linthicum) | 50 | Monthly | 6, 12 | Dipl, Cert | | | | • | | |
| **Mississippi** | | | | | | | | | | |
| Mississippi School of Therapeutic Massage (Jackson) | | | | Dipl | | | | | | |
| **Missouri** | | | | | | | | | | |
| St Charles School of Massage Therapy | 20 | Sep Jan Apr Jun | 7, 12 | Cert | $10,000 | $0 | | • | | |
| **Montana** | | | | | | | | | | |
| Health Works Institute (Bozeman) | 20 | Sep Jan Mar | 11, 20 | Dipl, Cert | $8,300 | $8,300 | | • | | • |
| **New Jersey** | | | | | | | | | | |
| Academy of Massage Therapy (Hackensack) | 20 | 4x/yr | 8, 13 | Cert | $12,950 | $0 | | • | • | • |
| Cortiva-Inst Somerset Sch of Massage Therapy - Piscataway | 50 | Feb Mar Jun Sep Oct | 12, 15 | Dipl | $9,600 | $9,600 | | • | | |
| **Nova Scotia, Canada** | | | | | | | | | | |
| ICT Northumberland College (Halifax) | 80 | Sep Jan | 20, 18 | Dipl | $7,400 | $7,400 | | | | |
| **Ontario, Canada** | | | | | | | | | | |
| ICT Kikkawa College (Toronto) | 60 | Sep | 37 | Dipl | $7,500 | $7,500 | | | | |
| **Oregon** | | | | | | | | | | |
| East-West Coll of the Healing Arts (Portland) | 18 | Jan Apr Jul Oct | 12, 18 | Cert | | | | • | | • |
| **Pennsylvania** | | | | | | | | | | |
| Cortiva Inst - Penn School of Muscle Therapy (Oaks) | Var | Sep Oct Mar May | 6, 12 | Dipl | $0 | $9,195 | | • | | |
| **Washington** | | | | | | | | | | |
| Brenneke School of Massage (Seattle) | | Oct Apr May Jul | 12, 15 | Cert | | | | • | | • |
| **Wisconsin** | | | | | | | | | | |
| Lakeside School of Massage Therapy - Milwaukee | 36 | 7x/yr | 6, 12 | Dipl, ADMT | $8,250 | $8,250 | | • | | |

*Data are shown only for programs that completed the 2006 AMA Survey of Health Professions Education Programs

‡Key to Offers: 1: Evening or weekend classes; 2: Non-English instruction; 3: Cultural competence instruction

# Medical Assistant

## Occupational Description

Medical assisting is a multi-skilled allied health profession; practitioners work primarily in ambulatory settings such as medical offices and clinics. Medical assistants function as members of the health care delivery team and perform administrative and clinical procedures.

## Job Description

Medical assistants work under the supervision of physicians in their offices or other medical settings. In accordance with respective state laws, they perform a broad range of administrative and clinical duties:

### Administrative duties

- Scheduling and receiving patients
- Preparing and maintaining medical records
- Performing basic secretarial skills and medical transcription
- Handling telephone calls and writing correspondence
- Serving as a liaison between the physician and other individuals
- Managing practice finances

### Clinical duties

- Asepsis and infection control
- Taking patient histories and vital signs
- Performing first aid and CPR
- Preparing patients for procedures
- Assisting the physician with examinations and treatments
- Collecting and processing specimens
- Performing selected diagnostic tests
- Preparing and administering medications as directed by the physician

Both administrative and clinical duties involve maintenance of equipment and supplies for the practice. A medical assistant who is sufficiently qualified by education and/or experience may be responsible for supervising personnel, developing and conducting public outreach programs to market the physician's professional services, and participating in the negotiation of leases and of equipment and supply contracts.

## Employment Characteristics

More medical assistants are employed by practicing physicians than any other type of allied health personnel. Medical assistants are usually employed in physicians' offices and other ambulatory healthcare settings, where they perform a variety of administrative and clinical tasks to facilitate the work of the physician. The responsibilities of medical assistants vary, depending on whether they work in a clinic, hospital, large group practice, or small private office. With demand from more than 200,000 physicians, there are, and will probably continue to be, almost unlimited opportunities for formally educated medical assistants.

## Salary

According to the American Association of Medical Assistants (AAMA), the average entry-level salary for Certified Medical Assistants (CMAs) in 2004 was $22,650. The average annual salary for full-time practitioners with a CMA was $27,951. Refer to Section IV, Table 5 of this *Directory* for more information, or see www.ama-assn.org/go/salary.

## Educational Programs

**Length.** Programs grant an associates degree, certificate, or diploma.

**Prerequisites.** High school diploma or equivalent is usually required.

**Curriculum.** The curricula of accredited programs must ensure achievement of the *Entry-Level Competencies for the Medical Assistant.* The curriculum must include anatomy and physiology, medical terminology, medical law and ethics, psychology, communications (oral and written), medical assisting administrative procedures, and medical assisting clinical procedures. Programs must include an externship that provides practical experience in qualified physicians' offices, accredited hospitals, or other health care facilities.

## Inquiries

### Careers

American Association of Medical Assistants
20 North Wacker Drive, Suite 1575
Chicago, IL 60606-2903
800 228-2262
312 899-1500
www.aama-ntl.org

### Certification/Registration

Director of Certification
American Association of Medical Assistants
20 North Wacker Drive, Suite 1575
Chicago, IL 60606-2903
312 424-3100

### Program Accreditation

Commission on Accreditation of Allied Health Education Programs (CAAHEP) in collaboration with:
American Association of Medical Assistants Endowment
20 North Wacker Drive, Suite 1575
Chicago, IL 60606-2963
312 899-1500
312 899-1259 Fax
E-mail: accreditation@aama-ntl.org

## Medical Assistant*

# Alabama

**George C Wallace Community College**
Medical Assistant Prgm
1141 Wallace Dr
Dothan, AL 36303
www.wallace.edu
*Prgm Dir:* William Arwood, MT MLT(ASCP)CMA
*Tel:* 334 556-2427, Ext 2427   *Fax:* 334 556-2231
*E-mail:* warwood@wallace.edu
*Med Dir:* Michael O'Brien, MD

**Wallace State Community College**
Medical Assistant Prgm
PO Box 2000
Hanceville, AL 35077-2000
*Prgm Dir:* Tracie Fuqua, BS CMA
*Tel:* 205 352-8321, Ext 256   *Fax:* 205 352-8320
*E-mail:* tracie.fuqua@wallacestate.edu
*Med Dir:* James Thomason, MD

**H Council Trenholm State Technical College**
Medical Assistant Prgm
1225 Air Base Blvd, PO Box 9000
Montgomery, AL 36108-3105
www.trenholmtech.cc.al.us
*Prgm Dir:* Annitta Love, RN MSN
*Tel:* 334 420-4425   *Fax:* 334 420-4449
*E-mail:* alove@trenholmtech.cc.al.us
*Med Dir:* Julian McIntyre, MD

**South University**
Medical Assistant Prgm
5355 Vaughn Rd
Montgomery, AL 36116-1120
www.southuniversity.edu
*Prgm Dir:* Linda Reynolds, BA MT(ASCP) CMA
*Tel:* 334 395-8800, Ext 8827   *Fax:* 334 395-8859
*E-mail:* lreynolds@southuniversity.edu
*Med Dir:* Raghu Makkamala, MD

# Alaska

**University of Alaska Anchorage**
Medical Assistant Prgm
3211 Providence Dr
Anchorage, AK 99508-4630
http://alliedhealth.uaa.alaska.edu/ma
*Prgm Dir:* Robin Wahto, BSN RN CMA
*Tel:* 907 786-6932   *Fax:* 907 786-6938
*E-mail:* afrjw@uaa.alaska.edu
*Med Dir:* Judith Whitcomb, MD

**University of Alaska - Fairbanks**
Medical Assistant Prgm
PO Box 758040
Fairbanks, AK 99701-8040
www.uaf.edu
*Prgm Dir:* Christa Bartlett, CMA CPC
*Tel:* 907 455-2887   *Fax:* 907 455-2865
*E-mail:* ffclb2@uaf.edu
*Med Dir:* Natalie Mayer, MD

# Arizona

**Bryman School - Phoenix**
Medical Assistant Prgm
2250 W Peoria Ave, Ste A-200
Phoenix, AZ 85029
*Prgm Dir:* Christine L Martin, CMA AGS
*Tel:* 602 274-4300   *Fax:* 602 248-9087
*E-mail:* phoenixmartins@ad.com
*Med Dir:* Steven McRonels, DO

---

*Additional information about programs that
returned the AMA's survey is available in a table
beginning on page 245.

**Long Technical College**
Medical Assistant Prgm
13450 N Black Canyon Hwy, Ste 104
Phoenix, AZ 85029
*Prgm Dir:* Elizabeth Emma Muniz, CMA RMA AHI
*Tel:* 602 548-1955, Ext 214   *Fax:* 602 548-1956
*E-mail:* lmuniz@longtechnicalcollege.com
*Med Dir:* Wilbur Cole, DO

# Arkansas

**Arkansas Tech University**
Medical Assistant Prgm
1701 N Boulder Ave
McEver Building
Russellville, AR 72801
www.atu.edu
*Prgm Dir:* Phyllis Cox, MAEd MT(ASCP)
*Tel:* 479 498-6073   *Fax:* 479 964-0504
*E-mail:* phyllis.cox@atu.edu
*Med Dir:* Stanley Bradley, MD

# California

**Corinthian College - Bryman Campus**
Medical Assistant Prgm
2215 Mission Rd
Alhambra, CA 91803
www.bryman-college.com
*Prgm Dir:* Blanca Zepeda, BS
*Tel:* 626 979-4940, Ext 121   *Fax:* 626 979-4961
*E-mail:* bzepeda@cci.edu
*Med Dir:* Jayi Shaw, MD

**CSi Bryman College**
Medical Assistant Prgm
511 N Brookhurst St, Ste 300
Anaheim, CA 92801
www.cci.edu
*Prgm Dir:* Judith L Enlow, CMA RT BS
*Tel:* 714 953-6500, Ext 139   *Fax:* 714 953-4163
*E-mail:* jenlow@cci.edu
*Med Dir:* David T Asher, MD

**Western Career College - Antioch**
Medical Assistant Prgm
2157 Country Hills Dr
Antioch, CA 94509
www.westercareercollege.edu
*Prgm Dir:* Rafik Messiha, MDDOMS
*Tel:* 925 522-7777, Ext 47120   *Fax:* 925 755-0079
*E-mail:* rmessiha@ant.westerncollege.com
*Med Dir:* Baldomero DeLeon, Jr, MD

**Cabrillo College**
Medical Assistant Prgm
6500 Soquel Dr
Aptos, CA 95003
www.cabrillo.edu
*Prgm Dir:* Charlotte A Jensen, CMA BS MPA/HSA
*Tel:* 831 479-6438   *Fax:* 831 479-5054
*E-mail:* chjensen@cabrillo.edu
*Med Dir:* Jill McBride, MD

**Heald College - Concord**
Medical Assistant Prgm
5130 Commerical Circle
Concord, CA 94520
www.heald.edu
*Prgm Dir:* Ekbal Fakhoury, MD DTM&H MS
*Tel:* 925 288-5800, Ext 5824   *Fax:* 925 687-2428
*E-mail:* Ekbal_Fakhoury@heald.edu
*Med Dir:* Nabil Athanasious, MD

**Orange Coast College**
Medical Assistant Prgm
2701 Fairview Rd
Costa Mesa, CA 92628-5005
*Prgm Dir:* Margie Willis, CMA-AC
*Tel:* 714 432-5658   *Fax:* 714 432-5534
*E-mail:* mwillis@mail.occ.cccd.edu
*Med Dir:* Brian Coyne, MD

**Western Career College - Emeryville Campus**
Medical Assistant Prgm
1400 65th St, Ste 200
Emeryville, CA 94608
*Prgm Dir:* Ghanzanfar Mahmood, MBBS
*Tel:* 510 601-0133, Ext 27   *Fax:* 510 601-0793
*E-mail:* Gmahmood@svcollege.com
*Med Dir:* Esther Alfonso

**Heald College - Fresno**
Medical Assistant Prgm
255 W Bullard Ave
Fresno, CA 93704
www.heald.edu
*Prgm Dir:* Jo Lynn Dowell
*Tel:* 559 438-4222   *Fax:* 559 438-0948
*E-mail:* Jolynn_Dowell@heald.edu
*Med Dir:* Razia Sheikh, Medical Doctor

**Concorde Career Institute - Garden Grove**
Medical Assistant Prgm
12951 Euclid St, Ste 101
Garden Grove, CA 92804
*Prgm Dir:* Salvador Ruiz, PhD
*Tel:* 714 620-1043   *Fax:* 714 530-8421
*E-mail:* sruiz@concorde.edu
*Med Dir:* Benny Hau, MD

**Bryman College - Gardena**
Medical Assistant Prgm
1045 W Redondo Beach Blvd, Ste 275
Gardena, CA 90247
*Prgm Dir:* Shaun Wright, AS BFA
*Tel:* 310 527-7105   *Fax:* 310 527-7985
*E-mail:* swright@cci.edu
*Med Dir:* Mischelle Turner, PA

**Chabot College**
Medical Assistant Prgm
25555 Hesperian Blvd
Hayward, CA 94545-5001
www.chabotcollege.edu
*Prgm Dir:* Jane Vallely, BS RN CMA
*Tel:* 510 723-7211   *Fax:* 510 782-9315
*E-mail:* medassistvallely@yahoo.com
*Med Dir:* Daniel Martin, MD

**Heald College - Hayward**
Medical Assistant Prgm
25500 Industrial Blvd
Hayward, CA 94545
*Prgm Dir:* Pamela Stoker, MA RMA
*Tel:* 510 259-7265
*E-mail:* pamela_stoker@heald.edu
*Med Dir:* Gloria Carreon, MD

**Bryman College - Los Angeles**
Medical Assistant Prgm
3460 Wilshire Blvd Ste 500
Los Angeles, CA 90010
*Prgm Dir:* Jeanna Andres, CMA
*Tel:* 213 388-9950, Ext 123   *Fax:* 213 388-9907
*E-mail:* jandres@cci.edu
*Med Dir:* Donald Henderson, MD

**Heald College - Milpitas**
Medical Assistant Prgm
341 Great Mall Pkwy
Milpitas, CA 95035
www.heald.edu
*Prgm Dir:* Nelly Mangarova, MD
*Tel:* 408 934-4900, Ext 127   *Fax:* 408 934-1006
*E-mail:* nelly_mangarova@heald.edu
*Med Dir:* Jacqueline Nguyen, MD

**Modesto Junior College**
Medical Assistant Prgm
435 College Ave
Modesto, CA 95350
www.mjc.edu
*Prgm Dir:* Shirley M Buzbee, CMA
*Tel:* 209 575-6377   *Fax:* 209 575-6593
*E-mail:* buzbees@yosemite.cc.ca.us
*Med Dir:* Roland C Nyegaard, MD

**Pasadena City College**
Medical Assistant Prgm
1570 E Colorado Blvd
Pasadena, CA 91106-2003
www.pasadena.edu
*Prgm Dir:* Joyce Y Nakano, CMA-A
*Tel:* 626 585-7431   *Fax:* 626 585-7977
*E-mail:* jynakano@pasadena.edu
*Med Dir:* Deborah J Hileman-Ford, FNP, MN, MS, CCRN

**Western Career College - Pleasant Hill**
Medical Assistant Prgm
380 Civic Dr, Ste 300
Pleasant Hill, CA 94523
*Prgm Dir:* Rebecca Burford
*Tel:* 925 609-6650   *Fax:* 925 609-6666
*E-mail:* rebeccab@westerncollege.com
*Med Dir:* Sue Knight, MD

**Heald College - Rancho Cordova**
Medical Assistant Prgm
2910 Prospect Park Dr
Rancho Cordova, CA 95670
www.heald.edu
*Prgm Dir:* Cynthia Smith, MS
*Tel:* 916 638-1616, Ext 2335   *Fax:* 916 853-7021
*E-mail:* Cynthia_Smith@heald.edu
*Med Dir:* Robert McCarran, MD

**Heald College - Roseville**
Medical Assistant Prgm
7 Sierra Gate Plaza
Roseville, CA 95678
www.heald.edu
*Prgm Dir:* Michael McCann, BS MA
*Tel:* 916 789-8600   *Fax:* 916 789-8606
*E-mail:* michael_mccann@heald.edu
*Med Dir:* Robert McCarran, DO

**Cosumnes River Community College**
Medical Assistant Prgm
8401 Center Pkwy
Sacramento, CA 95823-5704
www.crc.losrios.edu
*Prgm Dir:* Cori Burns, CMA
*Tel:* 916 691-7296   *Fax:* 916 691-7146
*E-mail:* burnsc@crc.losrios.edu
*Med Dir:* Daniel J Fields, MD

**Western Career College - Sacramento**
Medical Assistant Prgm
8909 Folsom Blvd
Sacramento, CA 95826-3203
www.westerncollege.edu
*Prgm Dir:* Danielle Malott, BS
*Tel:* 916 361-5140   *Fax:* 916 361-6666
*E-mail:* dmalott@westerncollege.com
*Med Dir:* Kathy Lauchaire, PA

**Heald College - Salinas**
Medical Assistant Prgm
1450 N Main St
Salinas, CA 93906
www.heald.edu
*Prgm Dir:* Don Stanley, DC DACAN
*Tel:* 831 443-1700, Ext 141   *Fax:* 831 443-1050
*E-mail:* don_stanley@heald.edu
*Med Dir:* Sheilaja Mittal, MD

**Bryman College - San Bernardino**
Medical Assistant Prgm
217 E Club Center Dr, Ste A
San Bernardino, CA 92408
www.cci.edu
*Prgm Dir:* Dora Perez, CMA
*Tel:* 909 777-3300   *Fax:* 909 777-3550
*E-mail:* dorap@cci.edu
*Med Dir:* Cheryl Emoto, MD

**Concorde Career College - San Bernardino**
Medical Assistant Prgm
201 E Airport Dr, Ste A
San Bernardino, CA 92408
*Prgm Dir:* Monireh Karimkhani, BSN CMA CPI NCMA
   NCMOA NCICS
*Tel:* 909 884-8891, Ext 336   *Fax:* 909 384-1768
*E-mail:* mkarimkani@concorde.edu
*Med Dir:* Benny Hau, MD

**San Diego Mesa College**
Medical Assistant Prgm
7250 Mesa College Dr
San Diego, CA 92111
www.sdccd.edu
*Prgm Dir:* Danielle (Acting) Twyman, BS PA-C
*Tel:* 619 388-2267
*E-mail:* dtwyman@sdccd.edu

**Bryman College - San Francisco**
Medical Assistant Prgm
A Corinthian School
814 Mission St, 5th Fl
San Francisco, CA 94103-3038
www.mybryman.com
*Prgm Dir:* Joan Jeong, BA CMA
*Tel:* 415 777-2500, Ext 250   *Fax:* 415 495-3457
*E-mail:* jjeong@cci.edu
*Med Dir:* Donald Dossett, MD

**City College of San Francisco**
Medical Assistant Prgm
1860 Hayes St
San Francisco, CA 94117
www.ccsf.org
*Prgm Dir:* Dory Rincon, MA CMA RHIT CPC
*Tel:* 415 561-1821   *Fax:* 415 561-1999
*E-mail:* drincon@ccsf.edu
*Med Dir:* Sunny Clark, RN

**Heald College - San Francisco**
Medical Assistant Prgm
350 Mission St
San Francisco, CA 94105
www.heald.edu
*Prgm Dir:* George Fakhoury, MD DORCP CMA
*Tel:* 415 808-1400, Ext 1468   *Fax:* 415 808-1594
*E-mail:* george_fakhoury@heald.edu
*Med Dir:* John Copeland, MD

**Western Career College - San Jose**
Medical Assistant Prgm
6201 San Ignacio Ave
San Jose, CA 95119
*Prgm Dir:* Carla Read, CMA
*Tel:* 408 360-0840, Ext 243   *Fax:* 408 360-0848
*E-mail:* cread@sj.westerncollege.com
*Med Dir:* Chris Stetz, MD

**Western Career College - San Leandro**
Medical Assistant Prgm
15555 E 14th St, Ste 500
San Leandro, CA 94578
*Prgm Dir:* Lisa Dianda, BA ThB
*Tel:* 510 276-3888, Ext 117   *Fax:* 510 276-3653
*E-mail:* ldianda@westerncollege.com
*Med Dir:* John Morgan, MD

**West Valley Community College District**
Medical Assistant Prgm
14000 Fruitvale Ave
Saratoga, CA 95070
*Prgm Dir:* Kristina A Kunkle-Gaiero, BA (HSA) CMA
*Tel:* 408 741-4019   *Fax:* 408 741-2145
*E-mail:* jekr@comcast.net
*Med Dir:* Stanford Shoor, MD

**Heald College - Stockton**
Medical Assistant Prgm
1605 E March Ln
Stockton, CA 95210
www.heald.edu
*Prgm Dir:* Ruth Brown, RN MS
*Tel:* 209 473-5200, Ext 5229   *Fax:* 209 477-2739
*E-mail:* ruth_brown@heald.edu
*Med Dir:* Shaukat Shah, MD

**Southern California Regional Occupational Ctr**
Medical Assistant Prgm
2300 Crenshaw Blvd
Torrance, CA 90501
*Prgm Dir:* Rose Hecht, CMA-C
*Tel:* 310 224-4200, Ext 262   *Fax:* 310 782-1589
*E-mail:* rhecht@scroc.k12.ca.us
*Med Dir:* Lynn P Coates, MD

**East San Gabriel Valley Occupational Pgm**
Medical Assistant Prgm
1501 W Del Norte St
West Covina, CA 91790
*Prgm Dir:* Katherine Hill, LVN
*Tel:* 626 962-5080, Ext 136   *Fax:* 626 472-5125
*E-mail:* khill@esgvrop.org
*Med Dir:* Arturo Velazquez, MD

# Colorado

**Parks College - Aurora Campus**
Medical Assistant Prgm
14280 E Jewell Ave, Ste 100
Aurora, CO 80014
*Prgm Dir:* Loren McAllister
*Tel:* 303 745-6244   *Fax:* 303 745-6245
*E-mail:* lmcallister@cci.edu
*Med Dir:* James Arthur, MD

**Everest College**
Medical Assistant Prgm
1815 Jet Wing Dr
Colorado Springs, CO 80916-2300
www.everest-college.com
*Prgm Dir:* Lori Mondragon, RN
*Tel:* 719 638-6580, Ext 139   *Fax:* 719 638-6818
*E-mail:* lmondragon@cci.edu
*Med Dir:* Max Nevarez, MD

**Community College of Denver**
Medical Assistant Prgm
3532 Franklin St
Denver, CO 80205
*Prgm Dir:* Darla A Ruff, CMA
*Tel:* 303 293-8737, Ext 214   *Fax:* 303 292-4315
*E-mail:* darla.ruff@ccd.edu
*Med Dir:* Martin Shure, MD

**Parks College**
Medical Assistant Prgm
9065 Grant St
Denver, CO 80229-4339
*Prgm Dir:* Michael Toth, RN ADRN BS
*Tel:* 303 457-2757, Ext 128   *Fax:* 303 457-4030
*E-mail:* mtoth@cci.edu
*Med Dir:* James W Arthur, MD

**Westwood College**
Medical Assistant Prgm
Institute of Health Careers
7350 N Broadway
Denver, CO 80221
www.westwood.edu
*Prgm Dir:* Christine E Hollander, CMA MA
*Tel:* 800 992-5050, Ext 685   *Fax:* 303 426-1832
*E-mail:* chollander@westwood.edu
*Med Dir:* Marilyn Gandolph, PA-C

**Red Rocks Community College**
Medical Assistant Prgm
13300 Sixth Ave
Lakewood, CO 80228
www.rrcc.edu/health/medass.html
*Prgm Dir:* Rita Stoffel, MBA MT
*Tel:* 303 914-6625   *Fax:* 303 914-6626
*E-mail:* rita.stoffel@rrcc.edu
*Med Dir:* Jim Keller, PA

**Arapahoe Community College**
Medical Assistant Prgm
5900 S Santa Fe Dr
PO Box 9002
Littleton, CO 80160-9002
*Prgm Dir:* Sue Treitz, MA BSN RN
*Tel:* 303 797-5962   *Fax:* 303 797-5842
*E-mail:* sue.treitz@arapahoe.edu
*Med Dir:* Steven Singer, MD

**Front Range Comm College**
Medical Assistant Prgm
2121 Miller Dr
Longmont, CO 80501
www.frontrange.edu
*Prgm Dir:* Kari Williams, BS DC
*Tel:* 303 678-3833   *Fax:* 303 678-3856
*E-mail:* kari.williams@frontrange.edu
*Med Dir:* Gregory Jaramillo, MD

# Connecticut

**Branford Hall Career Institute**
Medical Assistant Prgm
One Summit Pl
Branford, CT 06405
www.branfordhall.com
*Prgm Dir:* Nancy Maisonet, Phy A
*Tel:* 203 488-2525   *Fax:* 203 488-2920
*E-mail:* nmaisonet@branfordhall.com
*Med Dir:* Mark Volpe, MD

**St Vincent's College**
Medical Assistant Prgm
2800 Main St
Bridgeport, CT 06606
*Prgm Dir:* Holly E Mulrenan, MS RN CMA
*Tel:* 203 576-5518   *Fax:* 203 581-6533
*E-mail:* hmulrenan@stvincentscollege.edu
*Med Dir:* Frank Scifo, MD

**Quinebaug Valley Community College**
Medical Assistant Prgm
742 Upper Maple St
Danielson, CT 06239-1436
www.qvctc.commnet.edu
*Prgm Dir:* Cheri Goretti, MA MT(ASCP) CMA
*Tel:* 860 774-1160, Ext 349   *Fax:* 860 774-7768
*E-mail:* cgoretti@qvcc.commnet.edu
*Med Dir:* Joseph Allesandro, MD

**Goodwin College**
Medical Assistant Prgm
745 Burnside Ave
East Hartford, CT 06108
www.goodwincollege.org
*Prgm Dir:* Danielle Wilken, MS MT(ASCP)
*Tel:* 860 528-4111, Ext 307   *Fax:* 860 282-4625
*E-mail:* dwilken@goodwin.edu
*Med Dir:* Deborah Poerio, APRN,MS

**Stone Academy**
Medical Assistant Prgm
1315 Dixwell Ave
Hamden, CT 06514
*Prgm Dir:* Kathryn Piscitello, RN
*Tel:* 203 288-7474   *Fax:* 203 288-8869
*E-mail:* kpiscitello@stoneacademy.com
*Med Dir:* Leonard Fasano, MD

**Capital Community College**
Medical Assistant Prgm
950 Main St
Hartford, CT 06103
www.ccc.commnet.edu
*Prgm Dir:* Susan Perreira, CMA RMA MS
*Tel:* 860 906-5156   *Fax:* 860 906-5148
*E-mail:* sperreira@ccc.commnet.edu
*Med Dir:* Reinbard Kage, MD

**New England Tech Institute**
Medical Assistant Prgm
200 John Downey Dr
New Britain, CT 06051-2904
*Prgm Dir:* Ada M Rossi, CMA BS
*Tel:* 860 225-8641   *Fax:* 860 224-2983
*E-mail:* arossi@netiedu.com
*Med Dir:* Deepa Limaye, MD

**Ridley-Lowell Business & Technical Institute**
Medical Assistant Prgm
470 Bank St
New London, CT 06320
*Prgm Dir:* Dennis Lisee, LPN AS EMT
*Tel:* 860 443-7441   *Fax:* 860 442-3096
*E-mail:* NLmedical@ridley.edu
*Med Dir:* Gregory Azia, MD

**Norwalk Community College**
Medical Assistant Prgm
188 Richards Ave
Norwalk, CT 06854-1655
www.ncc.commnet.edu
*Prgm Dir:* Lauren Perlstein, RN MSN
*Tel:* 203 857-6852   *Fax:* 203 857-3364
*E-mail:* lperlstein@ncc.commnet.edu
*Med Dir:* Maryann Farina, PA

**Briarwood College**
Medical Assistant Prgm
2279 Mt Vernon Rd
Southington, CT 06489
www.briarwood.com
*Prgm Dir:* Eleanor Flores, RN BSN MEd
*Tel:* 860 628-4751, Ext 178   *Fax:* 860 628-6444
*E-mail:* florese@briarwood.edu
*Med Dir:* Pacifico Flores, MD

**Fox Institute of Business**
Medical Assistant Prgm
99 South St
West Hartford, CT 06110-1922
*Prgm Dir:* Michele Howard-Swan, CMA BS
*Tel:* 860 947-2299   *Fax:* 860 947-2290
*E-mail:* foxmedicalhead@yahoo.com
*Med Dir:* Paul Dolinsky, MD

**Porter and Chester Institute - Wethersfield**
Medical Assistant Prgm
125 Silas Deane Hwy
Wethersfield, CT 06109
www.porterchester.com
*Prgm Dir:* Karen Juchniewicz, AS
*Tel:* 860 529-2519   *Fax:* 860 563-2595
*E-mail:* kjuchniewicz@porterchester.com
*Med Dir:* Sandor Nagy, MD

**Northwestern Connecticut Comm**
Medical Assistant Prgm
Park Place E
Winsted, CT 06098
*Prgm Dir:* Barbara C Berger, RN MS CMA
*Tel:* 860 738-6308   *Fax:* 860 738-6439
*E-mail:* bberger@nwcc.commnet.edu
*Med Dir:* Amor Lomibao, MD

# Delaware

**Delaware Tech & Comm Coll - Wilmington**
Medical Assistant Prgm
333 Shipley St
Wilmington, DE 19801-2412
www.dtcc.edu/wilmington/ah
*Prgm Dir:* Valerie J Bergeron, EdD
*Tel:* 302 657-5170   *Fax:* 302 577-6431
*E-mail:* valerie@dtcc.edu
*Med Dir:* Karlo Magat, MD

# Florida

**Florida Metropolitan Univ - Pinellas Campus**
Medical Assistant Prgm
2471 McMullen Booth Rd, Ste 200
Clearwater, FL 33759
www.fmu.edu
*Prgm Dir:* Bonnie Lee Bruns, BS CMAA CBCS
*Tel:* 727 725-2688, Ext 131   *Fax:* 727 796-3722
*E-mail:* bbruns@cci.edu
*Med Dir:* Timothy Carlson, MD

**Brevard Community College**
Medical Assistant Prgm
1519 Clearlake Rd
Cocoa, FL 32922-6503
*Prgm Dir:* Kris A Hardy, AS CDF CMA
*Tel:* 321 433-7545, Ext None   *Fax:* 321 634-3731
*E-mail:* hardyk@brevardcc.edu
*Med Dir:* Robert Paxson, MD

**McFatter Vocational Technical Center**
Medical Assistant Prgm
6500 Nova Dr
Davie, FL 33317
www.mcfattertech.com
*Prgm Dir:* Peter Doolin, MS MT (ASCP) RMA
*Tel:* 754 321-5886   *Fax:* 754 321-5820
*E-mail:* peter.Doolin@browardschools.com
*Med Dir:* Robert Fleigelman, MD

**Daytona Beach Community College**
Medical Assistant Prgm
1200 W International Speedway Blvd
Daytona Beach, FL 32114
www.dbcc.edu
*Prgm Dir:* Suzanne S Fielding, BS NR-CMA CMA
*Tel:* 386 506-3215   *Fax:* 386 506-3300
*E-mail:* fieldis@dbcc.edu
*Med Dir:* Gohar Khan, MD

**Keiser College - Daytona Beach**
Medical Assistant Prgm
1800 Business Park Blvd
Daytona Beach, FL 32114
www.keisercollege.edu
*Prgm Dir:* Barbara McLarnan, NRCMA NRCPT BMO
*Tel:* 386 274-5060   *Fax:* 386 274-2725
*E-mail:* bmclarnan@keisercollege.edu
*Med Dir:* Waikens Jerry, MD

**Broward Community College**
Medical Assistant Prgm
3501 SW Davie Rd
Ft Lauderdale, FL 33314
www.broward.edu
*Prgm Dir:* Adelaida DeLaGuardia-Piz, BSM CMA
*Tel:* 954 201-6906   *Fax:* 954 201-9037
*E-mail:* adelagua@broward.edu
*Med Dir:* Glen Moran, DO

**International College**
Medical Assistant Prgm
4501 Colonial Blvd
Ft Myers, FL 33912
*Prgm Dir:* Carlene Harrison, MPA CMA
*Tel:* 239 482-0019, Ext 6179    *Fax:* 239 513-9108
*E-mail:* charrison@internationalcollege.edu
*Med Dir:* Albert Alessi, DO

**Palm Beach Community College**
Medical Assistant Prgm
4200 Congress Ave
Lake Worth, FL 33461
*Prgm Dir:* Barbara Kalfin, CMA
*Tel:* 561 357-1330    *Fax:* 561 868-3635
*E-mail:* kalfinb@pbcc.edu
*Med Dir:* Michael Chidester, MD

**Florida Career Institute**
Medical Assistant Prgm
4222 S Florida Ave
lakeland, FL 33813
*Prgm Dir:* Segundo Manuel Sion Cansing, MD
*Tel:* 941 646-1400    *Fax:* 941 646-5236
*E-mail:* judys@marcogrp.com
*Med Dir:* Edward L Demmi, MD

**Florida Metropolitan Univ - Lakeland Campus**
Medical Assistant Prgm
995 E Memorial Blvd - Ste 110
Lakeland, FL 33801
*Prgm Dir:* Karla Roush, CMA AA RMA
*Tel:* 863 686-1444, Ext 110    *Fax:* 863 688-9881
*E-mail:* kroush@cci.edu

**Florida Metropolitan Univ - Melbourne Campus**
Medical Assistant Prgm
2401 N Harbor City Blvd
Melbourne, FL 32935-6657
*Prgm Dir:* Jennie Lesser
*Tel:* 321 253-2929    *Fax:* 321 255-2017
*E-mail:* jlesser@cci.edu
*Med Dir:* Hadar Heshmati, MD MPH PhD

**Robert Morgan Educational Center**
Medical Assistant Prgm
18180 SW 122nd Ave
Miami, FL 33177
*Prgm Dir:* Judith W Garcia, RN MS CMA
*Tel:* 305 253-9920, Ext 2207    *Fax:* 305 253-3023
*E-mail:* GarciaJudith@msn.com
*Med Dir:* Marco Vitiello, MD PA

**Lorenzo Walker Institute of Technology**
Medical Assistant Prgm
3702 Estey Ave
Naples, FL 34104
*Prgm Dir:* Ina Kelley, RN BSN
*Tel:* 941 430-6900, Ext 2040    *Fax:* 941 430-6694
*E-mail:* kellyin@collier.k12.fl.us
*Med Dir:* Paul Jones, MD

**Community Technical & Adult Education Center**
Medical Assistant Prgm
1014 SW 7th Rd
Ocala, FL 34474
*Prgm Dir:* Gail McPadden, AS
*Tel:* 352 671-7200
*E-mail:* gail.mcpadden@marion.k12.fl.us
*Med Dir:* Gary M Wright, BA MD

**Florida Metro Univ - North Orlando Campus**
Medical Assistant Prgm
5421 Diplomat Cir
Orlando, FL 32810
*Prgm Dir:* Terri J Baker, BSN
*Tel:* 407 628-5870, Ext 142    *Fax:* 407 628-1344
*E-mail:* tbaker@cci.edu
*Med Dir:* Joan L Wilson, MD

**Florida Metro Univ - Orlando College South**
Medical Assistant Prgm
9200 South Park Center Loop
Orlando, FL 32819-8606
www.fmu.edu
*Prgm Dir:* Elizabeth Henisse, MBA BAS RMA CPC CPC-H REEG/T
*Tel:* 407 851-2525, Ext 174    *Fax:* 407 851-1477
*E-mail:* ehenisse@cci.edu
*Med Dir:* Joel Weinberger, DO

**Pensacola Junior College**
Medical Assistant Prgm
Warrington Campus
5555 W Hwy 98
Pensacola, FL 32507
*Prgm Dir:* Dale Brewer, BS MEd CMA
*Tel:* 850 484-2221    *Fax:* 850 454-2365
*E-mail:* DBrewer@pjc.edu
*Med Dir:* Robert D Flurry, MD

**Indian River Community College**
Medical Assistant Prgm
500 NW California Blvd
Port St Lucie, FL 34986
*Prgm Dir:* Theresa Errante-Parrino, CMA EMT-P BS
*Tel:* 772 336-6237    *Fax:* 772 336-6235
*E-mail:* tparrino@ircc.edu
*Med Dir:* Dwight Dawkins, MD

**Seminole Community College**
Medical Assistant Prgm
100 Weldon Blvd
Sanford, FL 32773-6199
www.scc-fl.edu
*Prgm Dir:* Ann Mautner, RN BSN CMA
*Tel:* 407 328-2016    *Fax:* 407 708-2402
*E-mail:* mautnera@scc-fl.edu
*Med Dir:* Steven Brint, MD

**Sarasota County Technical Institute**
Medical Assistant Prgm
4748 Beneva Rd
Sarasota, FL 34233-1758
*Prgm Dir:* Harlean Satin, MPA RN C
*Tel:* 941 924-1365, Ext 382    *Fax:* 941 361-6886
*E-mail:* harlean_satin@sarasota.k12.fl.us
*Med Dir:* Thomas H Williams, MD

**First Coast Technical Institute**
Medical Assistant Prgm
2980 Collins Ave
St Augustine, FL 32084
*Prgm Dir:* Donna L Oakley, RN BSN
*Tel:* 904 829-1087    *Fax:* 904 829-1077
*E-mail:* oakleyd@fcti.org
*Med Dir:* James A Joyner, III, MD

**Pinellas Tech Educ Ctr - St Petersburg**
Medical Assistant Prgm
901 34th St S
St Petersburg, FL 33711-2209
www.ptec.pinellas.k12.fl.us
*Prgm Dir:* Diane M Klieger, RN MBA CMA
*Tel:* 727 893-2500, Ext 1069    *Fax:* 727 323-6653
*E-mail:* kliegerdmk@yahoo.com
*Med Dir:* PJ Morales, MD PA

**Lively Technical Center**
Medical Assistant Prgm
500 Appleyard Dr
Tallahassee, FL 32304-2810
*Prgm Dir:* Bonnie Strade
*Tel:* 850 487-7489    *Fax:* 850 487-7478
*E-mail:* stradeb@mail.lively.leon.k12.fl.us
*Med Dir:* David Brown, MSN ARNP

**Erwin Technical Center**
Medical Assistant Prgm
2010 E Hillsborough Ave
Tampa, FL 33610-8299
*Prgm Dir:* Elaine Waldbart, RN CMA
*Tel:* 813 231-1800, Ext 1344    *Fax:* 813 231-1820
*E-mail:* waldbart_e@popmail.firn.edu
*Med Dir:* Alan Iezzi, MD

**Florida Metro U - Tampa Coll - Hillsborough**
Medical Assistant Prgm
3319 W Hillsborough Ave
Tampa, FL 33614-5801
www.fmu.edu
*Prgm Dir:* Reuven Cohen, DO MS
*Tel:* 813 879-6000, Ext 122    *Fax:* 813 871-2483
*E-mail:* rcohen@cci.edu
*Med Dir:* Robert Casa¤as, MD, FACI

**Florida Metropolitan Univ - Tampa College**
Medical Assistant Prgm
3924 Coconut Palm Dr
Tampa, FL 33619
*Prgm Dir:* Christina Conklin
*Tel:* 813 621-0041, Ext 68    *Fax:* 813 623-5769
*E-mail:* cconklin@cci.edu
*Med Dir:* Galo Alava, MD MBA

**Southwest Florida College**
Medical Assistant Prgm
3910 Riga Blvd
Tampa, FL 33619
www.swfc.edu
*Prgm Dir:* Susan Caldwell, RN MBA CNOR RMA-pending
*Tel:* 813 630-4401    *Fax:* 813 630-4272
*E-mail:* Scaldwell@swfc.edu
*Med Dir:* Steve Morris, MD

**South University**
Medical Assistant Prgm
1760 N Congress Ave
West Palm Beach, FL 33409
*Prgm Dir:* Carmen Carpenter, BSN MS CMA
*Tel:* 561 697-9200, Ext 3370    *Fax:* 561 697-9944
*E-mail:* ccarpenter@southuniversity.edu
*Med Dir:* Dana Richard, MD

**Central Florida College**
Medical Assistant Prgm
1573 W Fairbanks Ave
Winter Park, FL 32789
*Prgm Dir:* James Harry Phillips, CMA BS
*Tel:* 407 843-3984    *Fax:* 407 843-9828
*E-mail:* jphillips@centralfloridacollege.edu
*Med Dir:* George Ellis, MD

**Winter Park Tech**
Medical Assistant Prgm
901 Webster Ave
Winter Park, FL 32789-3098
*Prgm Dir:* Melinda Gioielli
*Tel:* 407 622-2900, Ext 2228    *Fax:* 407 975-2435
*E-mail:* gioielm@ocps.net
*Med Dir:* Bill Byrd, MD

# Georgia

**Albany Technical College**
Medical Assistant Prgm
1704 S Slappey Blvd
Albany, GA 31701
*Prgm Dir:* Cathy Benson, RN BSN
*Tel:* 229 430-3542    *Fax:* 229 430-2853
*E-mail:* cbenson@albanytech.edu
*Med Dir:* Devell R Young, MD

**South Georgia Technical College**
Medical Assistant Prgm
900 S Ga Tech Pkwy
Americus, GA 31709-8104
*Prgm Dir:* Carolyn Myers, RN BSN MSN
*Tel:* 229 931-2560   *Fax:* 229 931-2732
*E-mail:* cmyers@southgatech.edu
*Med Dir:* Gatewood Dudley, MD

**Atlanta Technical College**
Medical Assistant Prgm
1560 Metropolitan Pkwy SW
Atlanta, GA 30310
www.atlantatech.edu
*Prgm Dir:* Carolyn Helms, BS
*Tel:* 404 225-4570   *Fax:* 404 758-8522
*E-mail:* chelms@atlantatech.edu
*Med Dir:* C Amu, MD

**Augusta Technical College**
Medical Assistant Prgm
3200 Augusta Tech Dr
Augusta, GA 30906
*Prgm Dir:* Lisa Nagle, BSEd CMA
*Tel:* 706 771-4189   *Fax:* 706 771-4181
*E-mail:* lnagle@augustatech.edu
*Med Dir:* Paul Fischer, MD

**Savannah River College**
Medical Assistant Prgm
2528 Centerwest Pkwy, Bldg A
Augusta, GA 30909
www.savannahrivercollege.edu
*Prgm Dir:* Katie Barton, LPN BA
*Tel:* 706 738-5046, Ext 22   *Fax:* 706 736-3599
*E-mail:* kbarton@savannahrivercollege.edu
*Med Dir:* Robert Clark, DO

**North Georgia Technical College - Blairsville Campus**
Medical Assistant Prgm
434 Meeks Ave
Blairsville, GA 30512
*Prgm Dir:* Kathrine Ivester, CMA (AAMA) CLS (NCA) MPA
*Tel:* 706 754-7784   *Fax:* 706 754-7777
*E-mail:* kivester@ngtcollege.org
*Med Dir:* D Bryan Johnson, MD

**North Georgia Technical College**
Medical Assistant Prgm
PO Box 65
1500 Hwy 197 N
Clarkesville, GA 30523
*Prgm Dir:* Kathrine Ivester, CMA (AAMA) CLS (NCA) MPA
*Tel:* 706 754-7784   *Fax:* 706 754-7777
*E-mail:* kivester@northgatech.edu
*Med Dir:* Ed Hendricks, DO

**DeKalb Technical College**
Medical Assistant Prgm
495 N Indian Creek Dr
Clarkston, GA 30021
*Prgm Dir:* Linda Rials, MEd BSN CMA
*Tel:* 404 297-9522, Ext 1164   *Fax:* 404 294-0617
*E-mail:* rialsl@dekalbtech.edu
*Med Dir:* S Catherine Huggins, MD

**Columbus Technical College**
Medical Assistant Prgm
928 Manchester Expwy
Columbus, GA 31904-6572
*Prgm Dir:* Barbara Gaither
*Tel:* 706 649-1499   *Fax:* 706 641-5284
*E-mail:* bgaither@columbustech.org
*Med Dir:* Antonio R Rodriguez, MD

**Dalton State College**
Medical Assistant Prgm
213 N College Dr
Dalton, GA 30720
*Prgm Dir:* Kenneth H Earley, MBA MOA
*Tel:* 706 272-4559   *Fax:* 706 272-4563
*E-mail:* kearley@em.daltonstate.edu
*Med Dir:* Sarah J Polow, DO

**East Central Technical College**
Medical Assistant Prgm
706 West Baker HWY
Douglas, GA 31533
www.eastcentraltech.edu
*Prgm Dir:* Kymberly Anderson, RN MA
*Tel:* 912 383-4300, Ext 2214   *Fax:* 912 389-4308
*E-mail:* kanderson@eastcentraltech.edu
*Med Dir:* Meg Minchew, PA-C

**Heart of Georgia Technical College**
Medical Assistant Prgm
560 Pinehill Rd
Dublin, GA 31021
*Prgm Dir:* Susie Drew, RN
*Tel:* 478 274-7885   *Fax:* 478 275-6642
*E-mail:* sdrew@heartofgatech.edu
*Med Dir:* Mike Lancaster, NP

**Griffin Technical College**
Medical Assistant Prgm
501 Varsity Rd
Griffin, GA 30223
*Prgm Dir:* Faye Johnson, MT(AMT) RN CAHI
*Tel:* 770 233-5498
*E-mail:* fjohnson@griffintech.edu
*Med Dir:* James R Gore Jr, DDS MD

**Appalachian Technical College**
Medical Assistant Prgm
100 Campus Dr
Jasper, GA 30143
www.appalachiantech.edu
*Prgm Dir:* Trish Jackson, BSN RN RMA
*Tel:* 706 692-4500, Ext 4576   *Fax:* 706 692-4433
*E-mail:* tjackson@appalachiantech.edu
*Med Dir:* Cathy Eaton, FNP, RN

**West Georgia Technical College**
Medical Assistant Prgm
303 Fort Dr
LaGrange, GA 30240
*Prgm Dir:* Elaine Gilbert, MSN, RMA
*Tel:* 706 837-4267   *Fax:* 706 845-4339
*E-mail:* egilbert@westgatech.edu
*Med Dir:* Daniel Guy, MD

**Gwinnett Technical College**
Medical Assistant Prgm
5150 Sugarloaf Pkwy
PO Box 1505
Lawrenceville, GA 30043-5702
*Prgm Dir:* Marcie C Jones, BS CMA
*Tel:* 770 962-7580, Ext 346   *Fax:* 770 962-7985
*E-mail:* mjones@gwinnett.tec.ga.us
*Med Dir:* Joel Fine, MD

**Chattahoochee Technical College**
Medical Assistant Prgm
980 S Cobb Dr
Marietta, GA 30060-3300
www.chattcollege.com
*Prgm Dir:* Jacqualine K Adair, RN BSN CMA
*Tel:* 770 528-4590   *Fax:* 770 528-4584
*E-mail:* jadair@chattcollege.com
*Med Dir:* Elizabeth Street, MD

**Clayton State University**
Medical Assistant Prgm
2000 Clayton State University
Morrow, GA 30260
www.tech.clayton.edu
*Prgm Dir:* LaTanya Young, PAC
*Tel:* 678 466-4611   *Fax:* 678 466-4669
*E-mail:* latanyayoung@clayton.edu
*Med Dir:* Al Reynolds, MD

**Moultrie Technical College**
Medical Assistant Prgm
361 Industrial Dr
Moultrie, GA 31768
www.moultrietech.edu
*Prgm Dir:* Robin Kern, RN BSN
*Tel:* 229 891-7000, Ext 4195   *Fax:* 229 217-4211
*E-mail:* rkern@moultrietech.edu
*Med Dir:* Arthur Marsh, MD

**Lanier Technical College**
Medical Assistant Prgm
2990 Landrum Education Dr
Oakwood, GA 30566
www.laniertech.edu
*Prgm Dir:* Celia Celorio, BS CMA NCPT
*Tel:* 770 531-6354   *Fax:* 770 531-6306
*E-mail:* CCELORIO@laniertech.edu
*Med Dir:* Brad Noon, MD

**Northwestern Technical College**
Medical Assistant Prgm
265 Bicentennial Trail
Rock Spring, GA 30739
*Prgm Dir:* Denise M Grant, RN BSN MSN CMA
*Tel:* 706 764-3532   *Fax:* 706 764-3718
*E-mail:* dgrant@northwesterntech.edu

**Coosa Valley Technical College**
Medical Assistant Prgm
One Maurice Culberson Dr
Rome, GA 30161
www.cvtcollege.org
*Prgm Dir:* Jennifer Stephenson, RN MPH RMA
*Tel:* 706 295-6479   *Fax:* 706 295-6894
*E-mail:* jstephenson@coosavalleytech.edu
*Med Dir:* Gene Davidson, MD

**Savannah Technical College**
Medical Assistant Prgm
5717 White Bluff Rd
Savannah, GA 31405-5521
www.savannahtech.org
*Prgm Dir:* Jacqueline Muller, RN BSN
*Tel:* 912 443-5810   *Fax:* 912 443-5826
*E-mail:* jmuller@savannahtech.edu
*Med Dir:* Ester McAlpine, MD

**South University**
*Cosponsor:* Educational Management Corporation
Medical Assistant Prgm
709 Mall Blvd
Savannah, GA 31406
*Prgm Dir:* Hany Eissa, MD
*Tel:* 912 201-8051   *Fax:* 912 201-8070
*E-mail:* heissa@southuniversity.edu
*Med Dir:* Charlene Moon, FNP

**Medix School**
Medical Assistant Prgm
2108 Cobb Pkwy
Smyrna, GA 30080-7630
www.medixschool.edu
*Prgm Dir:* Larry Ritchie, BS
*Tel:* 770 980-0002   *Fax:* 770 980-0811
*E-mail:* lritchie@edaff.com
*Med Dir:* LArry Feldman, MD

**Ogeechee Technical College**
Medical Assistant Prgm
One Joe Kennedy Blvd
Statesboro, GA 30458-3199
www.ogeecheetech.edu
*Prgm Dir:* Marilyn M Turner, RN CMA
*Tel:* 912 486-7616  *Fax:* 912 486-7604
*E-mail:* mturner@ogeecheetech.edu
*Med Dir:* Kristin Updegraff, MD

**Swainsboro Technical College**
Medical Assistant Prgm
346 Kite Rd
Swainsboro, GA 30401-5700
www.swainsborotech.edu
*Prgm Dir:* Kimberly Dismuke Brown, CMA BSHS
*Tel:* 478 289-2243  *Fax:* 478 289-2214
*E-mail:* kbrown@swainsborotech.edu
*Med Dir:* Sanjay Serrao, MD

**Southwest Georgia Technical College**
Medical Assistant Prgm
15689 US Hwy 19 N
Thomasville, GA 31792
www.southwestgatech.edu
*Prgm Dir:* Glenda Hatcher, BSN RN CMA
*Tel:* 229 225-5081  *Fax:* 229 225-5289
*E-mail:* ghatcher@southwestgatech.edu
*Med Dir:* Greg Dodson, PA

**Valdosta Technical College**
Medical Assistant Prgm
4089 Val Tech Rd, PO Box 928
Valdosta, GA 31603-0928
*Prgm Dir:* Cecelia Bruce, RN
*Tel:* 229 333-2100, Ext 2664  *Fax:* 229 259-5567
*E-mail:* cbruce@valdostatech.org
*Med Dir:* Joe S Thomas, MD

**Southeastern Technical College**
Medical Assistant Prgm
3001 East First St
Vidalia, GA 30474
www.southeasterntech.edu
*Prgm Dir:* Dana Roessler, BSN RN
*Tel:* 912 538-3198  *Fax:* 912 538-3106
*E-mail:* droessler@southeasterntech.edu
*Med Dir:* Mike Wiggins, FNP

**West Central Technical College**
Medical Assistant Prgm
176 Murphy Campus Blvd
Waco, GA 30182
www.westcentraltech.edu
*Prgm Dir:* Jamie Ellis Shell, AAT CMA
*Tel:* 770 537-6051  *Fax:* 770 537-7992
*E-mail:* jshell@westcentraltech.edu
*Med Dir:* David Shoenfeld, MD

# Guam

**Guam Community College**
Medical Assistant Prgm
PO Box 23069
GMF Barrigada Guam, GU 96921
*Prgm Dir:* Barbara Mafnas, RN
*Tel:* 671 735-5656  *Fax:* 671 734-2550
*E-mail:* bmafnas@guamcc.net
*Med Dir:* Vincent Akimoto, MD

# Hawaii

**Heald College - Honolulu**
Medical Assistant Prgm
1500 Kapiolani Blvd
Honolulu, HI 96814
www.heald.edu
*Prgm Dir:* Perfecto Y Salvador, CHI RMA
*Tel:* 808 955-1500, Ext 533  *Fax:* 808 955-6964
*E-mail:* Perfecto_Salvador@heald.edu
*Med Dir:* Gregory Caputy, MD

**Kapi'olani Community College**
Medical Assistant Prgm
4303 Diamond Head Rd
Honolulu, HI 96816-4496
*Prgm Dir:* Lynn Hamada, RN CMA MPH
*Tel:* 808 734-9240  *Fax:* 808 734-9126
*E-mail:* lynnh@hawaii.edu
*Med Dir:* Franklin Young, MD

# Idaho

**Eastern Idaho Technical College**
Medical Assistant Prgm
1600 S 25th E
Idaho Falls, ID 83404-5788
*Prgm Dir:* Cindy K Mills, CMA
*Tel:* 208 524-3000, Ext 3340  *Fax:* 208 524-3007
*E-mail:* cmills@eitc.edu
*Med Dir:* Eric G Baird, MD

**Lewis Clark State College**
Medical Assistant Prgm
500 8th Ave
Lewison, ID 83501
*Prgm Dir:* Janet Wyatt, AAS RN
*Tel:* 208 792-2466
*E-mail:* jawyatt@lcsc.edu
*Med Dir:* Vince Fibelstad, MD

**Idaho State University**
Medical Assistant Prgm
Campus Box 8380
Pocatello, ID 83209-8380
*Prgm Dir:* Norma J Bird, MEd BS CMA
*Tel:* 208 282-4317  *Fax:* 208 282-3975
*E-mail:* birdnorm@isu.edu
*Med Dir:* Anthony Joseph, MD

**College of Southern Idaho**
Medical Assistant Prgm
PO Box 1238
Twin Falls, ID 83303-1238
http://hshs.csi.edu/medical_assistant
*Prgm Dir:* Penny Glenn, MEd CMA
*Tel:* 208 732-6728, Ext 6728  *Fax:* 208 736-4743
*E-mail:* pglenn@csi.edu
*Med Dir:* Robert Ward, MD

# Illinois

**Robert Morris College - Aurora**
Medical Assistant Prgm
905 Meridian Lake Dr
Aurora, IL 60504
www.robertmorris.edu
*Prgm Dir:* Janet Haggerty Davis, BSN MS MBA PhD
*Tel:* 312 935-6805  *Fax:* 312 935-6060
*E-mail:* jdavis@robertmorris.edu
*Med Dir:* Ali H Kutom, MD

**Southwestern Illinois College**
Medical Assistant Prgm
2500 Carlyle Rd
Belleville, IL 62221
www.swic.edu
*Prgm Dir:* Cheryl E Hutchison, CMA CPC
*Tel:* 618 235-2700, Ext 5332  *Fax:* 618 235-2052
*E-mail:* cheryl.hutchison@swic.edu
*Med Dir:* Patrick Zimmerman, MD

**Robert Morris College - Bensenville**
Medical Assistant Prgm
1000 Tower Ln
Bensenville, IL 60106
www.robertmorris.edu
*Prgm Dir:* Janet Haggerty Davis, BSN MS MBA PhD
*Tel:* 312 935-6805  *Fax:* 312 935-6060
*E-mail:* jdavis@robertmorris.edu
*Med Dir:* Ali H Kutom, MD

**Northwestern Business College - Southwest Campus**
Medical Assistant Prgm
Southwest Campus
7725 S Harlem Ave
Bridgeview, IL 60455
www.northwesternbc.edu
*Prgm Dir:* Yovanna Caraballo, BS CMA
*Tel:* 773 736-3580, Ext 2504  *Fax:* 773 736-4366
*E-mail:* ycaraballo@northwesternbc.edu
*Med Dir:* Cory Chen, MD

**Northwestern Business College**
Medical Assistant Prgm
4829 N Lipps Ave
Chicago, IL 60630
www.northwesternbc.edu
*Prgm Dir:* Yovanna Caraballo, BS CMA
*Tel:* 773 777-4220, Ext 2344  *Fax:* 773 777-2861
*E-mail:* ycaraballo@nwbc.edu
*Med Dir:* Cory Chen, MD

**Robert Morris College - Chicago**
Medical Assistant Prgm
401 S State St
Chicago, IL 60605
www.robertmorris.edu
*Prgm Dir:* Janet Haggerty Davis, BSN MS MBA PhD
*Tel:* 312 935-6805  *Fax:* 312 935-6060
*E-mail:* jdavis@robertmorris.edu
*Med Dir:* Ali H Kutom, MD

**Spanish Coalition for Jobs Inc**
Medical Assistant Prgm
2011 W Pershing Rd
Chicago, IL 60609
www.scj-usa.org
*Prgm Dir:* Robert Benway, EdD
*Tel:* 773 247-0707, Ext 252  *Fax:* 773 247-4975
*E-mail:* bbenway@scj-usa.org
*Med Dir:* Hugo Alvarez, MD

**Midwest Technical Institute**
Medical Assistant Prgm
405 N Limit St
PO Box 506
Lincoln, IL 62656
www.midwesttechnicalinstitute.edu
*Prgm Dir:* Kathleen J Steinberg, RN BSN
*Tel:* 217 735-3105, Ext 107  *Fax:* 217 735-1055
*E-mail:* kjsteinbergrn@hotmail.com
*Med Dir:* Connie Duda, Nurse Practitioner

**Robert Morris College - Orland Park**
Medical Assistant Prgm
43 Orland Square Dr
Orland Park, IL 60462
www.robertmorris.edu
*Prgm Dir:* Janet Haggerty Davis, BSN MS MBA PhD
*Tel:* 312 935-6805  *Fax:* 312 935-6060
*E-mail:* jdavis@robertmorris.edu
*Med Dir:* Ali H Kutom, MD

**Harper College**
Medical Assistant Prgm
1200 W Algonquin Rd
Health Careers and Public Safety
Palatine, IL 60067-7398
*Prgm Dir:* Geri Kale-Smith, MS CMA
*Tel:* 847 925-6444  *Fax:* 847 925-6047
*E-mail:* gkalesmi@harpercollege.edu
*Med Dir:* John Venson, MD

**Moraine Valley Community College**
Medical Assistant Prgm
10900 S 88th Ave
Palos Hills, IL 60465-0937
*Prgm Dir:* Mary OMalley-Absalon, BSN MBA
*Tel:* 708 974-5708
*Med Dir:* Samuel A Farbstein, MD

**Midstate College**
Medical Assistant Prgm
411 W Northmoor Rd
Peoria, IL 61614-3558
*Prgm Dir:* Joanna Holly, RN CMA MS
*Tel:* 309 692-4092, Ext 2040  *Fax:* 309 692-3893
*E-mail:* jholly@midstate.edu
*Med Dir:* Robert A Lizer, DO

**Robert Morris College - Peoria**
Medical Assistant Prgm
1 Technology Plaza
211 Fulton St
Peoria, IL 61602
www.robertmorris.edu
*Prgm Dir:* L Frank Fegan, MPA MT(ASCP)
*Tel:* 217 726-1631  *Fax:* 217 793-4210
*E-mail:* fegan@warpnet.net
*Med Dir:* Richard M Holloway, MD

**Rockford Business College**
Medical Assistant Prgm
730 N Church
Rockford, IL 61103-6917
www.rbcsuccess.com
*Prgm Dir:* Karla Renee Garcia, BA LPN
*Tel:* 815 965-8616  *Fax:* 815 965-0360
*E-mail:* kgarcia@rbcsuccess.com
*Med Dir:* Amy Elizabeth Semenchuk, RN, NSN

**South Suburban College**
Medical Assistant Prgm
15800 S State St, Rm 4453
South Holland, IL 60473
*Prgm Dir:* Lisa L Campbell
*Tel:* 708 596-2000
*E-mail:* lcampbell@southsuburbancollege.edu

**Robert Morris College - Springfield**
Medical Assistant Prgm
3101 Montvale Dr
Springfield, IL 62704
www.robertmorris.edu
*Prgm Dir:* L Frank Fegan, MPA MT(ASCP)
*Tel:* 217 726-1631  *Fax:* 217 793-4210
*E-mail:* fegan@warpnet.net
*Med Dir:* Richard M Holloway, MD

**Waubonsee Community College**
Medical Assistant Prgm
Rte 47 at Waubonsee Dr
Sugar Grove, IL 60554
*Prgm Dir:* Jess Toussaint
*Tel:* 630 466-7900, Ext 2467  *Fax:* 630 466-4119
*E-mail:* jtoussaint@waubonsee.edu
*Med Dir:* Kathy Thorne, MD

**Robert Morris College - Waukegan**
Medical Assistant Prgm
1507 Waukegan Rd
Waukegan, IL 60085
www.robertmorris.edu
*Prgm Dir:* Janet Haggerty Davis, BSN MS MBA PhD
*Tel:* 312 935-6805  *Fax:* 312 935-6060
*E-mail:* jdavis@robertmorris.edu
*Med Dir:* Ali H Kutom, MD

# Indiana

**Ivy Tech Community College - Anderson**
Medical Assistant Prgm
104 W 53rd St
Anderson, IN 46013
*Prgm Dir:* Neilsen Schulz, CMA MA RCP
*Tel:* 765 643-7133  *Fax:* 765 643-3294
*E-mail:* nschulz@ivytech.edu
*Med Dir:* Mark Jennings, MD

**Indiana Business College**
Medical Assistant Prgm
2222 Poshard Dr
Columbus, IN 47203
*Prgm Dir:* Aimee Anfdermauer, AAS
*Tel:* 812 379-9000
*E-mail:* aimee.aufdermauer@ibcschools.edu
*Med Dir:* Greg Cromstock

**Ivy Tech Community College - Columbus**
Medical Assistant Prgm
4475 Central Ave
Columbus, IN 47203-1868
www.ivytech.edu/columbus
*Prgm Dir:* Katherine Hawkins, BS CMA
*Tel:* 812 372-9925, Ext 5163  *Fax:* 812 372-0311
*E-mail:* khawkins@ivytech.edu
*Med Dir:* William Blaisdell, MD

**Indiana Business College - Evansville**
Medical Assistant Prgm
4601 Theater Dr
Evansville, IN 47715
www.ibcschools.edu
*Prgm Dir:* Brenda G Emge, CMA AAS
*Tel:* 812 476-6000, Ext 109  *Fax:* 812 471-8576
*E-mail:* brenda.emge@ibcschools.edu
*Med Dir:* Melinda Jackson, MD PhD

**Ivy Tech Community College SW - Evansville**
Medical Assistant Prgm
3501 First Ave
Evansville, IN 47710-3319
www.ivytech.edu
*Prgm Dir:* Kitty Lutz, RN CMA MSEd
*Tel:* 812 429-1381  *Fax:* 812 429-9805
*E-mail:* klutz@ivytech.edu
*Med Dir:* Patrick Flamion, MD

**Brown Mackie College - Ft Wayne**
Medical Assistant Prgm
4422 E State Blvd
Ft Wayne, IN 46815
*Prgm Dir:* Sheryl Gerald
*Tel:* 260 484-4000  *Fax:* 260 484-2678
*E-mail:* sgerald@brownmackie.edu
*Med Dir:* Robert Wilkins, MD

**Indiana Business College - Ft Wayne**
Medical Assistant Prgm
6413 N Clinton St
Ft Wayne, IN 46825
*Prgm Dir:* Jason V Amich, MSc CEP
*Tel:* 260 471-7667  *Fax:* 260 471-6918
*E-mail:* jason.amich@ibcschools.edu
*Med Dir:* Carol Riley

**International Business Coll - Ft Wayne**
Medical Assistant Prgm
5699 Coventry Ln
Ft Wayne, IN 46804
*Prgm Dir:* Pamela L Neu, AS CMA
*Tel:* 260 459-4541  *Fax:* 260 436-1896
*E-mail:* pneu@ibcfortwayne.edu
*Med Dir:* Lisa Holtsclaw, DO

**Ivy Tech Community College NE - Ft Wayne**
Medical Assistant Prgm
3800 N Anthony Blvd
Ft Wayne, IN 46805-1430
www.ivytech.edu
*Prgm Dir:* Tova Green, BS CMA
*Tel:* 260 480-4163  *Fax:* 260 480-4149
*E-mail:* tgreen@ivytech.edu
*Med Dir:* Kintinar Thomas, MD

**Clarian Health Partners Inc**
Medical Assistant Prgm
Health Sciences Education Center
2039 N Capitol Ave, #110
Indianapolis, IN 46202
*Prgm Dir:* J Keith Hiatt, RRT CMA CPhT
*Tel:* 317 937-5137  *Fax:* 317 937-5163
*E-mail:* jhiatt@clarian.org
*Med Dir:* Halina Harding, DO

**Indiana Business College**
Medical Assistant Prgm
8150 Brookville Rd
Indianapolis, IN 46239
*Prgm Dir:* Monica Hershberger, BS
*Tel:* 317 375-8000  *Fax:* 317 351-1871
*E-mail:* monica.hershberger@ibcschools.edu
*Med Dir:* Stanley Adkins, MD

**International Business Coll - Indianapolis**
Medical Assistant Prgm
7205 Shadeland Station
Indianapolis, IN 46256
www.ibcindianapolis.edu
*Prgm Dir:* Judy Mackey, CMA AAS
*Tel:* 317 813-2304  *Fax:* 317 841-6419
*E-mail:* jmackey@ibcindianapolis.edu
*Med Dir:* James Dupler, MD

**Ivy Tech Community College of Indiana**
Medical Assistant Prgm
1 W 26th St
PO Box 1763
Indianapolis, IN 46206
*Prgm Dir:* Lori J Andrews, CMA MSEd RN
*Tel:* 317 921-4589  *Fax:* 317 921-4432
*E-mail:* landrews@ivytech.edu

**Professional Careers Institute**
Medical Assistant Prgm
7302 Woodland Dr
Indianapolis, IN 46278-1736
*Prgm Dir:* Terri Burton, CMA AAS
*Tel:* 317 299-6001, Ext 341  *Fax:* 317 298-6342
*E-mail:* tburton714@sbcglobal.net
*Med Dir:* David Gerstein, MD

**Ivy Tech Community College - Kokomo**
Medical Assistant Prgm
700 E Firmin St
Kokomo, IN 46903
www.ivytech.edu
*Prgm Dir:* Patricia L Slusher, BS MT(ASCP) CMA
*Tel:* 765 457-0858, Ext 105  *Fax:* 765 457-1036
*E-mail:* pslusher@ivytech.edu
*Med Dir:* James Cole, Physician Assistant

**Ivy Tech Community College - Lafayette**
Medical Assistant Prgm
3101 S Creasy Ln
PO Box 6299
Lafayette, IN 47903
www.laf.ivytech.edu
*Prgm Dir:* Cindy Ann Abel, BS CMA PBT(ASCP)
*Tel:* 765 269-5206  *Fax:* 765 269-5248
*E-mail:* cabel@ivytech.edu
*Med Dir:* Karla Cheesman, ANP RN

**Ivy Tech Community College - Lawrenceburg**
Medical Assistant Prgm
500 Industrial Dr
Lawrenceburg, IN 47025
www.ivytech.edu
*Prgm Dir:* Theresa Disch, BS BA LPN CMA
*Tel:* 812 537-4010, Ext 234  *Fax:* 812 537-0993
*E-mail:* tdisch@ivytech.edu
*Med Dir:* Jeanne M Thompson, MD

## Ivy Tech Community College - Madison
Medical Assistant Prgm
590 Ivy Tech Dr
Madison, IN 47250
www.ivytech.edu/madison
*Prgm Dir:* Annabet Garner, AAS CMA
*Tel:* 812 273-0105, Ext 27  *Fax:* 812 265-4028
*E-mail:* agarner@ivytech.edu
*Med Dir:* Alice Carlson-Jackson, CNP

## Ivy Tech Community College - Marion
Medical Assistant Prgm
1015 E 3rd St
Marion, IN 46952
*Prgm Dir:* Linda Pruitt, MBA CMA
*Tel:* 765 662-9843, Ext 320  *Fax:* 765 664-4256
*E-mail:* lpruitt@ivytech.edu
*Med Dir:* Mark A Westfall, MD FAAFP

## Ivy Tech Community College - Michigan City
Medical Assistant Prgm
3714 Franklin St
Michigan City, IN 46360-7311
*Prgm Dir:* Viki Pavlakovic, CMA
*Tel:* 219 879-9137, Ext 230  *Fax:* 219 879-9157
*E-mail:* vpavlakovic@ivytech.edu
*Med Dir:* J Timothy Ames, MD

## Indiana Business College
Medical Assistant Prgm
411 W Riggin Rd
Muncie, IN 47303
*Prgm Dir:* Carolyn Edmunds, BA CMA LPN
*Tel:* 765 288-8681, Ext 117
*E-mail:* carolyn.edmunds@ibcschools.edu
*Med Dir:* Max Rudicel, MD

## Ivy Tech Community College EC - Muncie
Medical Assistant Prgm
4301 Cowan Rd
Muncie, IN 47307
*Prgm Dir:* Julayne A Masterman, MA BS CMA
*Tel:* 765 289-2291, Ext 352  *Fax:* 765 289-2291
*E-mail:* jmasterm@ivytech.edu
*Med Dir:* John Dickey, MD

## Ivy Tech Community College - Richmond
Medical Assistant Prgm
2325 Chester Blvd
Richmond, IN 47374
*Prgm Dir:* Kathryn Plankenhorn
*Tel:* 765 966-2656, Ext 375  *Fax:* 765 939-2641
*E-mail:* kplanken@ivytech.edu
*Med Dir:* Snyder Kimberly, NP

## Ivy Tech Community College SC - Sellersburg
Medical Assistant Prgm
8204 Hwy 311
Sellersburg, IN 47172
www.ivytech.edu/sellersburg
*Prgm Dir:* Pamela Burton, CMA LRT CPT
*Tel:* 812 246-3301, Ext 4218  *Fax:* 812 246-9905
*E-mail:* pburton@ivytech.edu
*Med Dir:* Joseph Beaven, MD

## Brown Mackie College - South Bend
Medical Assistant Prgm
1030 E Jefferson Blvd
South Bend, IN 46617
*Prgm Dir:* Sheryl Gerald, RMA BA
*Tel:* 574 237-0774, Ext 6557  *Fax:* 574 237-3585
*E-mail:* sgerald@brownmackie.edu
*Med Dir:* Max Helman, DO

## Ivy Tech Community College - South Bend
Medical Assistant Prgm
220 Dean Johnson Blvd
South Bend, IN 46601
www.ivytech.edu
*Prgm Dir:* Marti C Garrels, CMA MSA MT(ASCP)
*Tel:* 574 289-7001, Ext 5719  *Fax:* 574 236-7166
*E-mail:* garrels4@aol.com
*Med Dir:* John Bulger, MD

## Indiana Business College - Terre Haute
Medical Assistant Prgm
1378 S State Rd 46
Terre Haute, IN 47803
http://ibcschools.edu
*Prgm Dir:* Ursula Cole, MAEd CMA CCS-P RHE
*Tel:* 812 877-2100
*E-mail:* ursula.cole@ibcschools.edu
*Med Dir:* Nedu Gopala, MD

## Ivy Tech Community College - Terre Haute
Medical Assistant Prgm
8000 S Education Dr
Terre Haute, IN 47802-4845
www.goivytech.net
*Prgm Dir:* Becky Schonberger, RN BS CMA
*Tel:* 800 377-4882, Ext 2245  *Fax:* 812 298-2245
*E-mail:* bschonbe@ivytech.edu
*Med Dir:* Harold Loveall, MD

# Iowa

## Des Moines Area Community College
Medical Assistant Prgm
2006 Ankeny Blvd
Building #9
Ankeny, IA 50023
www.dmacc.edu
*Prgm Dir:* Diane M Vander Ploeg, MS CMA
*Tel:* 515 964-6457  *Fax:* 515 964-6440
*E-mail:* dmvanderploeg@dmacc.edu
*Med Dir:* Joellen Heims, DO

## Hamilton College
Medical Assistant Prgm
Cedar Falls Campus
7009 Nordic Dr
Cedar Falls, IA 50613
*Prgm Dir:* Ginger Cameron
*Tel:* 319 277-0220
*E-mail:* gicameron@hamiltoncf.com
*Med Dir:* Annie Kontos, DO

## Hamilton College
Medical Assistant Prgm
Cedar Rapids Campus
3165 Edgewood Pkwy SW
Cedar Rapids, IA 52404
*Prgm Dir:* Erlinda Aponte
*Tel:* 319 363-0481
*Med Dir:* Cindy Squires, MD

## Kirkwood Community College
Medical Assistant Prgm
6301 Kirkwood Blvd SW, PO Box 2068
Cedar Rapids, IA 52406-2068
*Prgm Dir:* Dawn M Eitel, AAS CMA
*Tel:* 319 398-5564  *Fax:* 319 398-1293
*E-mail:* dawn.eitel@kirkwood.edu
*Med Dir:* Brian Lindo, MD

## Iowa Western Community College
Medical Assistant Prgm
2700 College Rd, PO Box 4C
Council Bluffs, IA 51502-3004
www.iwcc.edu
*Prgm Dir:* Karen Nelson, RN AA AS
*Tel:* 712 325-3348  *Fax:* 712 325-3736
*E-mail:* knelson@iwcc.edu
*Med Dir:* Kris Smith, APNP

## Kaplan University, Davenport Campus
Medical Assistant Prgm
1801 E Kimberly Rd (Ste 1)
Davenport, IA 52807
www.kaplancollege.edu
*Prgm Dir:* "Mac" George Henry McNeal, RN MN PhD
*Tel:* 563 441-2453  *Fax:* 563 355-1320
*E-mail:* mmcneal@kucampus.edu
*Med Dir:* Richard Ng, DO

## Mercy College of Health Sciences
Medical Assistant Prgm
928 6th Ave
Des Moines, IA 50309-1239
*Prgm Dir:* Janet Robert-Andersen, MS
*Tel:* 515 643-6705
*E-mail:* janderson@mercydesmoines.org

## Iowa Central Community College
Medical Assistant Prgm
330 Ave M
Ft Dodge, IA 50501
www.iccc.cc.ia.us
*Prgm Dir:* Kelly Kruger, RN
*Tel:* 515 576-7201, Ext 2308  *Fax:* 515 576-5656
*E-mail:* kruger@iowacentral.com
*Med Dir:* Mark Marner, MD

## North Iowa Area Community College
Medical Assistant Prgm
500 College Dr
Mason City, IA 50401
*Prgm Dir:* Deb Stockberger, MSN RN
*Tel:* 641 422-4146  *Fax:* 641 422-4115
*E-mail:* stockdeb@niacc.edu
*Med Dir:* Susan Sieh, MD

## Iowa Lakes Community College
Medical Assistant Prgm
1900 N Grand
Spencer, IA 51301-3881
*Prgm Dir:* Carol Hartig, RN BSN CMA
*Tel:* 712 262-7141  *Fax:* 712 262-4047
*E-mail:* chartig@iowalakes.edu
*Med Dir:* David Robinson, DO

## Hamilton College
Medical Assistant Prgm
4655 121st St
Urbandale, IA 50310-2311
*Prgm Dir:* Tricia Berry, MA OTR/L
*Tel:* 515 727-6802  *Fax:* 515 727-2115
*E-mail:* tberry@hamiltonia.edu
*Med Dir:* Susan Donahue, MD

## Southeastern Community College
Medical Assistant Prgm
1500 West Agency Rd
West Burlington, IA 52655
*Prgm Dir:* Debbie Shaffer, RN
*Tel:* 319 208-5213  *Fax:* 319 752-4957
*E-mail:* dshaffer@scciowa.edu
*Med Dir:* Carl Hays, MD

# Kansas

## Northwest Kansas Technical College
Medical Assistant Prgm
1209 Harrison St
PO Box 668
Goodland, KS 67735-0668
www.nwktc.edu
*Prgm Dir:* Karen Lucas, AAS LPN
*Tel:* 785 890-2072  *Fax:* 785 899-5711
*E-mail:* klucas@mail.nwktc.org
*Med Dir:* Jackie L Jorgensen, ARNP

## Wichita Area Technical College
Medical Assistant Prgm
324 N Emporia St
Wichita, KS 67202-2512
www.watc.edu
*Prgm Dir:* Beth Buchholz, BS CMA
*Tel:* 316 677-1377, Ext 71377  *Fax:* 316 677-1332
*E-mail:* bbuchholz@watc.edu
*Med Dir:* Paul Davis, MD

# Kentucky

**National College of Business & Technology - Danville**
Medical Assistant Prgm
115 E Lexington Ave
Danville, KY 40422
www.ncbt.edu
*Prgm Dir:* Anita G Denson, CMA BA
*Tel:* 606 236-6991  *Fax:* 606 236-1063
*E-mail:* adenson@ncbt.edu
*Med Dir:* Thomas Jackson, MD

**National College of Business & Technology - Florence**
Medical Assistant Prgm
7627 Ewing Blvd
Florence, KY 41042
*Prgm Dir:* Everlee O'Nan, RMA
*Tel:* 859 525-6510, Ext 19  *Fax:* 859 525-8961
*E-mail:* eonan@ncbt.edu
*Med Dir:* Joan Ziegelmeyer, ARNP

**Henderson Community College**
Medical Assistant Prgm
2660 S Green St
Henderson, KY 42420
www.hencc.kctcs.edu
*Prgm Dir:* Randa Hawa, MHSA BS MT(ASCP)
*Tel:* 270 831-9722  *Fax:* 270 831-9718
*E-mail:* Randa.Hawa@kctcs.edu
*Med Dir:* James Stearns, MD

**Bluegrass Community and Technical College**
Medical Assistant Prgm
308 Vo-Tech Rd
Lexington, KY 40511
*Prgm Dir:* Joyce Combs, BS AAS CMA
*Tel:* 859 246-2400, Ext 2321  *Fax:* 859 246-2417
*E-mail:* joyce.combs@kctcs.edu
*Med Dir:* Margaret Terhune, MD

**National College of Business & Technology - Lexington**
Medical Assistant Prgm
2376 Sir Barton Way
Lexington, KY 40508
www.ncbt.edu
*Prgm Dir:* Jessica Hart, AS CMA
*Tel:* 859 253-0621  *Fax:* 859 233-3054
*E-mail:* jhart@ncbt.edu
*Med Dir:* Gretchen Cliburn, PA

**Sullivan University - Lexington**
Medical Assistant Prgm
2355 Harrodsburg Rd
Lexington, KY 40504
*Prgm Dir:* Jill Ferrari
*Tel:* 606 276-4357  *Fax:* 606 276-1153
*E-mail:* jferrari@faculty.sullivan.edu
*Med Dir:* Joseph Gerhardstein, MD

**Jefferson Community and Technical College**
Medical Assistant Prgm
800 W Chestnut St
Louisville, KY 40203-2074
*Prgm Dir:* Jannie Washington, MEd CMA-C
*Tel:* 502 213-4233  *Fax:* 502 213-4502
*E-mail:* jannie.washington@kctcs.edu
*Med Dir:* Robert Hammer, MD

**National College of Business & Technology - Louisville**
Medical Assistant Prgm
4205 Dixie Highway
Louisville, KY 40216-3801
www.nationalbusiness.edu
*Prgm Dir:* Laura Crain, MT
*Tel:* 502 447-7634  *Fax:* 502 447-7665
*E-mail:* cmiles@ncbt.edu
*Med Dir:* Eugene Giles, MD

**Spencerian College**
Medical Assistant Prgm
4627 Dixie Hwy
Louisville, KY 40216
*Prgm Dir:* Lori Warren, MA RN
*Tel:* 502 447-1000, Ext 221  *Fax:* 502 447-4574
*E-mail:* lwarren@spencerian.edu
*Med Dir:* Leo Wine, MD

**Maysville Comm & Tech College - Rowan Campus**
Medical Assistant Prgm
609 Viking Dr
Morehead, KY 40351
*Prgm Dir:* Diana Reeder, CMA AAS
*Tel:* 606 783-1538, Ext 66363  *Fax:* 606 783-1538
*E-mail:* diana.reeder@kctcs.edu
*Med Dir:* Laura Ellis, MD

**West Kentucky Community & Technical College**
Medical Assistant Prgm
Blandville Rd, PO Box 7408
Paducah, KY 42002-7408
*Prgm Dir:* Vicki Kirschner, MS MLT CMA
*Tel:* 270 534-3480  *Fax:* 270 554-2695
*E-mail:* Vicki.Kirschner@kctcs.edu
*Med Dir:* Ronald Wilson, MD

**National College of Business & Technology - Pikeville**
Medical Assistant Prgm
50 National College Blvd
Pikeville, KY 41501
www.ncbt.edu
*Prgm Dir:* Roberta Michele Wolford, CMA AS
*Tel:* 606 478-7200  *Fax:* 606 478-7209
*E-mail:* rmwolford@ncbt.edu
*Med Dir:* Rao J Bhatrajo, MD

**Eastern Kentucky University**
Medical Assistant Prgm
Dizney 225
Richmond, KY 40475-3135
www.eku.edu
*Prgm Dir:* Rebecca Newsome, PhD MT(ASCP) CMA
*Tel:* 859 622-6335  *Fax:* 859 622-6333
*E-mail:* rebecca.newsome@eku.edu
*Med Dir:* Joseph Bark, MD

**National College of Business & Technology - Richmond**
Medical Assistant Prgm
125 S Killarney Ln
Richmond, KY 40475
www.nationalbusiness.edu
*Prgm Dir:* Paula Beth Ciolek, AS CMA
*Tel:* 859 623-8956  *Fax:* 859 624-5544
*E-mail:* pciolek@ncbt.edu
*Med Dir:* Glenn Owen, MD PSC

# Louisiana

**Bossier Parish Community College**
Medical Assistant Prgm
6220 E Texas
Bossier City, LA 71111
www.bpcc.edu
*Prgm Dir:* Constance Marie Winter, MPH RN
*Tel:* 318 678-6382  *Fax:* 318 678-6199
*E-mail:* cwinter@bpcc.edu
*Med Dir:* Vicki Cobb, MD

**Career Technical College**
Medical Assistant Prgm
2319 Louisville Ave
Monroe, LA 71201
*Prgm Dir:* Robin Jackson, MA
*Tel:* 318 323-2889  *Fax:* 318 323-3113
*E-mail:* rjackson@careertc.com
*Med Dir:* Craig S Turner, MD

**Bryman College - New Orleans**
Medical Assistant Prgm
1201 Elmwood Park Blvd, Ste 600
New Orleans, LA 70123
www.bryman-college.com
*Prgm Dir:* Giselle Wilson, CMA BS
*Tel:* 504 733-7117, Ext 112  *Fax:* 504 734-1217
*E-mail:* giwilson@cci.edu
*Med Dir:* Kathy Wilson, MD

**Louisiana Tech College**
Medical Assistant Prgm
59125 Bayou Dr
Plaquemine, LA 70764
*Prgm Dir:* Leah Cullins-Hartford, BSN RN
*Tel:* 225 342-8228  *Fax:* 225 342-8229
*E-mail:* leahscullins@excite.com

# Maine

**Beal College**
Medical Assistant Prgm
99 Farm Rd
Bangor, ME 04401
www.bealcollege.edu
*Prgm Dir:* Barbara Marchelletta, CMA
*Tel:* 207 947-4591  *Fax:* 207 947-0208
*E-mail:* bmarchelletta@bealcollege.edu
*Med Dir:* Christopher Ritter, MD

**Kennebec Valley Community College**
Medical Assistant Prgm
92 Western Ave
Fairfield, ME 04937-1367
www.kvcc.me.edu
*Prgm Dir:* Ann Hickman, CMA BS
*Tel:* 207 453-5005  *Fax:* 207 453-5197
*E-mail:* ahickman@kvcc.me.edu
*Med Dir:* William Alto, MD, MPH

# Maryland

**Anne Arundel Community College**
Medical Assistant Prgm
101 College Pkwy
Arnold, MD 21012
*Prgm Dir:* Heidi McLean, CMA BS
*Tel:* 410 777-7366  *Fax:* 410 777-7099
*E-mail:* hmmclean@aacc.edu
*Med Dir:* Alex Hertman, MD

**Allegany College of Maryland**
Medical Assistant Prgm
12401 Willowbrook Rd SE
Cumberland, MD 21502
www.allegany.edu
*Prgm Dir:* Peggy Hughes, AA BS MEd
*Tel:* 301 784-5319  *Fax:* 301 784-5022
*E-mail:* phughes@allegany.edu
*Med Dir:* Anthony Bolino, MD

**Hagerstown Business College**
Medical Assistant Prgm
18618 Crestwood Dr
Hagerstown, MD 21742
www.hagerstownbusinesscol.org
*Prgm Dir:* Kay Nave, BS CMA MRT AAS
*Tel:* 301 739-2670, Ext 146  *Fax:* 301 791-7661
*E-mail:* KNave@hagerstownbusinesscol.edu
*Med Dir:* Ernest Uzicanin, MD

**Cecil Community College**
Medical Assistant Prgm
1 Seahawk Dr
North East, MD 21901
*Prgm Dir:* Joann E Palaisa, MEd
*Tel:* 410 287-6060, Ext 707
*E-mail:* jpalaisa@cecilcc.edu
*Med Dir:* Elizabeth Lowe, MD

**Medix School**
Medical Assistant Prgm
700 York Rd
Towson, MD 21204
www.medixschool.edu
*Prgm Dir:* Christine Wirtz, BS CMA
*Tel:* 410 337-5155    *Fax:* 410 337-5104
*E-mail:* cwirtz@edaff.com
*Med Dir:* John D Griswold, MD

# Massachusetts

**Massasoit Community College**
Medical Assistant Prgm
900 Randolph St
Canton, MA 02021-1355
*Prgm Dir:* Linda N Dente, BSEd CMA
*Tel:* 781 821-2222, Ext 2601    *Fax:* 781 575-9428
*E-mail:* LDente@massasoit.mass.edu
*Med Dir:* Cecile Hodges, RN,NP

**Porter and Chester Institute - Chicopee**
Medical Assistant Prgm
134 Dulong Circle
Chicopee, MA 01022
*Prgm Dir:* Elizabeth Murphy, CMA
*Tel:* 413 593-3162    *Fax:* 413 593-6439
*E-mail:* medicalassisting@porterchester-ch.com
*Med Dir:* Jay Ungar, MD

**North Shore Community College**
Medical Assistant Prgm
One Ferncroft Rd
Danvers, MA 01923-0840
www.northshore.edu
*Prgm Dir:* Mariann Splaine Henry, AS
*Tel:* 978 762-4179    *Fax:* 978 762-4022
*E-mail:* msplaine@northshore.edu
*Med Dir:* Michael Shrenko, MD

**Bristol Community College**
Medical Assistant Prgm
777 Elsbree St
Fall River, MA 02719
*Prgm Dir:* Lisa Wright, MA MT RMA
*Tel:* 508 678-2811, Ext 2629    *Fax:* 508 730-3281
*E-mail:* lwright@bristol.mass.edu
*Med Dir:* Christopher Joncas, MD

**Mt Wachusett Community College**
Medical Assistant Prgm
444 Green St
Gardner, MA 01440-1000
*Prgm Dir:* Brenda M Tatro, BA
*Tel:* 978 630-9357    *Fax:* 978 630-9555
*E-mail:* b_tatro@mwcc.mass.edu
*Med Dir:* Trisha Vorderstrasse, MD

**Northern Essex Community College**
Medical Assistant Prgm
45 Franklin St
Lawrence, MA 01841-1121
www.necc.mass.edu
*Prgm Dir:* Kathleen Welch-Hudson, MS
*Tel:* 978 738-7512    *Fax:* 978 738-7450
*E-mail:* khudson@necc.mass.edu
*Med Dir:* Cecelia Sederman, RN NP

**Middlesex Community College - Lowell**
Medical Assistant Prgm
33 Kearney Square
Lowell, MA 01852-1901
www.middlesex.mass.edu
*Prgm Dir:* Sue A Hunt, RN MA CMA
*Tel:* 978 656-3024    *Fax:* 978 656-3078
*E-mail:* hunts@middlesex.mass.edu
*Med Dir:* Leslie Schwab, MD

**Charles H McCann Technical School**
Medical Assistant Prgm
70 Hodges Cross Rd
North Adams, MA 01247
http://postsecondary.boxcarexpress.com
*Prgm Dir:* Terry LeClair, MA
*Tel:* 413 663-5383, Ext 182    *Fax:* 413 664-9424
*E-mail:* tleclair@mccanntech.org
*Med Dir:* Daniel Sullivan, MD

**Southeastern Technical Institute**
Medical Assistant Prgm
250 Foundry St
South Easton, MA 02375
www.sersd.org
*Prgm Dir:* Martha Dualsky, MEd BSN
*Tel:* 508 230-1337    *Fax:* 508 230-1558
*E-mail:* mdualsky@sersd.org
*Med Dir:* William Lawrence, MD

**Springfield Technical Community College**
Medical Assistant Prgm
One Armory Sq
PO Box 9000
Springfield, MA 01105-1204
www.stcc.edu
*Prgm Dir:* Cornelia Pettengill, MEd RN
*Tel:* 413 755-4843    *Fax:* 413 755-6312
*E-mail:* pettengill@stcc.edu
*Med Dir:* Paul Farkas, MD

**Cape Cod Community College**
Medical Assistant Prgm
2240 Lyanough Rd
West Barnstable, MA 02668-1599
*Prgm Dir:* Barbara Colocino
*Tel:* 508 362-2131, Ext 4324
*E-mail:* bcolocino@capecod.edu

**Quinsigamond Community College**
Medical Assistant Prgm
670 W Boylston St
Worcester, MA 01606-2031
www.qcc.mass.edu
*Prgm Dir:* Pamela Ann Fleming, CMA MPA RN
*Tel:* 508 854-2738    *Fax:* 508 852-6943
*E-mail:* pfleming@qcc.mass.edu
*Med Dir:* George Abraham, MD

**The Salter School**
Medical Assistant Prgm
155 Ararat St
Worcester, MA 01606-3450
*Prgm Dir:* Claire Maday-Travis, MA MBA CPHQ
*Tel:* 508 853-1074, Ext 20    *Fax:* 508 853-1083
*E-mail:* ctravis@salterschool.com
*Med Dir:* Kimberly Ebb, MD

# Michigan

**Alpena Community College**
Medical Assistant Prgm
666 Johnson St
Alpena, MI 49707-1495
*Prgm Dir:* Carol Putkamer, RHIA
*Tel:* 989 358-7321    *Fax:* 989 358-7561
*E-mail:* putkamec@alpenacc.edu
*Med Dir:* Avery Aten, MD

**Baker College of Auburn Hills**
Medical Assistant Prgm
1500 University Dr
Auburn Hills, MI 48326-2642
www.baker.edu
*Prgm Dir:* Wilsetta McClain, BBA NR-CMA NCICS RMA
  MBA
*Tel:* 248 276-8785    *Fax:* 248 276-8785
*E-mail:* zetta.mcclain@baker.edu
*Med Dir:* David Pinelli, DO

**Baker College of Cadillac**
Medical Assistant Prgm
9600 E 13th St
Cadillac, MI 49601
www.baker.edu
*Prgm Dir:* Becky J Rodenbaugh, CMA BHSA
*Tel:* 231 876-3116    *Fax:* 231 775-8505
*E-mail:* becky.rodenbaugh@baker.edu
*Med Dir:* Gerald Dudek, DO

**Glen Oaks Community College**
Medical Assistant Prgm
62249 Shimmel Rd
Centreville, MI 49032
*Prgm Dir:* Karen K Ganger, MEd MS R
*Tel:* 888 994-7818    *Fax:* 269 467-4114
*E-mail:* kganger@glenoaks.edu
*Med Dir:* Bharat Vakharia, MD

**Baker College of Clinton Township**
Medical Assistant Prgm
34950 Little Mack Ave
Clinton Township, MI 48035-4701
www.baker.edu
*Prgm Dir:* Elizabeth Hoffman, CMA MA Ed CPT (ASPT)
*Tel:* 586 791-6610, Ext 2689    *Fax:* 586 791-6611
*E-mail:* elizabeth.hoffman@baker.edu
*Med Dir:* Kin Tran, DO

**Macomb Community College**
Medical Assistant Prgm
Health & Human Services
44575 Garfield Rd
Clinton Township, MI 48038-1139
*Prgm Dir:* Delena K Austin, BTIS CMA
*Tel:* 586 286-2194    *Fax:* 586 286-2098
*E-mail:* austind@macomb.edu
*Med Dir:* Kevin Lokar, MD MPH

**Henry Ford Community College**
Medical Assistant Prgm
5101 Evergreen Rd
Dearborn, MI 48128-1495
www.hfcc.edu
*Prgm Dir:* Ronald Bodurka, MS RRT
*Tel:* 313 845-9877    *Fax:* 313 317-6569
*E-mail:* rbodurka@hfcc.edu
*Med Dir:* Patricia Barber, MD

**Baker College of Flint**
Medical Assistant Prgm
G 1050 W Bristol Rd
Flint, MI 48507
*Prgm Dir:* Deborah E Nelson, MS CMA RMA
*Tel:* 810 766-4155    *Fax:* 810 766-2055
*E-mail:* deborah.nelson@baker.edu
*Med Dir:* Kenneth Yokosawa, MD

**Schoolcraft College**
Medical Assistant Prgm
1751 Radcliff St
Garden City, MI 48135-1197
www.schoolcraft.edu
*Prgm Dir:* Patricia A Rubio, RHIA MSA
*Tel:* 734 462-4770, Ext 6025    *Fax:* 734 462-4775
*E-mail:* prubio@schoolcraft.edu
*Med Dir:* Gregory Monroe, DO

**Davenport University - Grand Rapids
(Fulton St)**
Medical Assistant Prgm
415 E Fulton St
Grand Rapids, MI 49503
www.davenport.edu
*Prgm Dir:* Suzanne Garman, RN BC BSN MA
*Tel:* 616 451-3511, Ext 1184    *Fax:* 616 732-1145
*E-mail:* suzanne.garman@davenport.edu
*Med Dir:* Timothy Tobolic, MD

**Mid Michigan Community College**
Medical Assistant Prgm
1375 S Clare Ave
Harrison, MI 48625
www.midmich.edu
*Prgm Dir:* Catherine King RNC-E MSN NNP RN MS
*Tel:* 989 386-6643  *Fax:* 989 386-6666
*E-mail:* cking@midmich.edu
*Med Dir:* David L Bremer, MD

**Baker College of Jackson**
Medical Assistant Prgm
2800 Springport Rd
Jackson, MI 49202-1255
*Prgm Dir:* Angela Marin
*Tel:* 517 841-4527  *Fax:* 517 789-7331
*E-mail:* angela.marin@baker.edu
*Med Dir:* Marty Homles, MD

**Jackson Community College**
Medical Assistant Prgm
2111 Emmons Rd
Jackson, MI 49201
www.jccmi.edu
*Prgm Dir:* Jean M Dennerll, BS CMA
*Tel:* 517 796-8557  *Fax:* 517 768-7004
*E-mail:* dennerljeanm@jccmi.edu
*Med Dir:* Brian Adamczyk, MD

**Davenport University - Kalamazoo**
Medical Assistant Prgm
4123 W Main
Kalamazoo, MI 49006-2748
*Prgm Dir:* Marybeth Pieri-Smith, BS CMA CPC
*Tel:* 269 382-2835, Ext 3311  *Fax:* 269 382-3541
*E-mail:* marybeth.pieri@davenport.edu
*Med Dir:* Arthur Bober, MD

**Kalamazoo Valley Community College**
Medical Assistant Prgm
PO Box 4070, 6767 West O Ave
Kalamazoo, MI 49003-4070
www.kvcc.edu
*Prgm Dir:* Mary Dey, CMA-AC
*Tel:* 269 488-4324  *Fax:* 269 488-4458
*E-mail:* mdey@kvcc.edu
*Med Dir:* Michael T Ku, DO

**Davenport University - Lansing**
Medical Assistant Prgm
220 E Kalamazoo St
Lansing, MI 48933
www.davenport.edu
*Prgm Dir:* Linda Spang, BA EMT-P RMA JD
*Tel:* 517 484-2600, Ext 8212  *Fax:* 517 484-9719
*E-mail:* linda.spang@davenport.edu
*Med Dir:* Howard J Burgess II, MD

**Baker College of Muskegon**
Medical Assistant Prgm
1903 Marquette Ave
Muskegon, MI 49442-1453
*Prgm Dir:* Cindy Gordon, MBA CMA
*Tel:* 231 777-5318  *Fax:* 231 777-5265
*E-mail:* cindy.gordon@baker.edu
*Med Dir:* Michael Krohn, DO

**Baker College of Owosso**
Medical Assistant Prgm
1020 S Washington St
Owosso, MI 48867-4400
www.baker.edu
*Prgm Dir:* Susan Gregoricka, RN BSN MPA
*Tel:* 989 729-3466, Ext 3466  *Fax:* 989 729-3411
*E-mail:* susan.gregoricka@baker.edu
*Med Dir:* Kenneth Root, PA

**Baker College of Port Huron**
Medical Assistant Prgm
3403 Lapeer Rd
Port Huron, MI 48060-2597
*Prgm Dir:* Jennifer Kaltz, CMA CPC
*Tel:* 810 985-7000, Ext 125  *Fax:* 810 985-7066
*E-mail:* jennifer.kaltz@baker.edu
*Med Dir:* Jon Lensmeyer, MD

**Montcalm Community College**
Medical Assistant Prgm
2800 College Dr
Sidney, MI 48885
*Prgm Dir:* Beth Ann Mowatt, RN MSN
*Tel:* 989 328-2111  *Fax:* 989 328-2950
*E-mail:* bethm@montcalm.edu

**National Institute of Technology**
Medical Assistant Prgm
26111 Evergreen Rd, Ste 201
Southfield, MI 48076
*Prgm Dir:* Yvette Harris
*Tel:* 248 799-9933  *Fax:* 248 799-2912
*E-mail:* yharris@cci.edu
*Med Dir:* Mark Petrous, MD

**Carnegie Institute**
Medical Assistant Prgm
550 Stephenson Hwy, Ste 100
Troy, MI 48083
www.carnegie-institute.com
*Prgm Dir:* Bonnie Normile, RN CMA CCVT
*Tel:* 248 589-1078  *Fax:* 248 589-1631
*E-mail:* info@carnegie-institute.edu
*Med Dir:* Colleen Kennedy, MD

**Oakland Community College**
Medical Assistant Prgm
7350 Cooley Lake Rd
Waterford, MI 48327
*Prgm Dir:* Karen A Kittle, CMA CPT CHUC
*Tel:* 248 942-3068
*E-mail:* kakittle@oaklandcc.edu
*Med Dir:* C Kohler Champion, MD

# Minnesota

**Anoka Technical College**
Medical Assistant Prgm
1355 W Hwy 10
Anoka, MN 55303-1564
www.anokatech.edu
*Prgm Dir:* Sandra Lehrke, RN MS CMA
*Tel:* 763 576-4700, Ext 4844  *Fax:* 763 576-4715
*E-mail:* slehrke@anokatech.edu
*Med Dir:* Mark Brakke, MD

**Minnesota School of Business - Brooklyn Ctr**
Medical Assistant Prgm
Brooklyn Center
5910 Shingle Creek Pkwy, Ste 200
Brooklyn Center, MN 55430
*Prgm Dir:* Dee Ann Kerr
*Tel:* 763 566-7777  *Fax:* 763 566-7000
*Med Dir:* Terry Skottegaard, BA BS

**Herzing College**
Medical Assistant Prgm
5700 West Broadway
Crystal, MN 55428
www.herzing.edu
*Prgm Dir:* Nicole Caselius
*Tel:* 763 535-3000  *Fax:* 763 535-9205
*E-mail:* ncaselius@mpls.herzing.edu
*Med Dir:* Bryant Beehler, DO

**Duluth Business University**
Medical Assistant Prgm
4724 Mike Colalillo Dr
Duluth, MN 55807
www.dbumn.edu
*Prgm Dir:* Marianne Bovee, AAS
*Tel:* 218 740-4348  *Fax:* 218 628-2127
*E-mail:* maryb@dbumn.edu
*Med Dir:* Nichole Heciminovich, MD

**Lake Superior College**
Medical Assistant Prgm
2101 Trinity Rd
Duluth, MN 55811
*Prgm Dir:* Lorrie Fox, ASN
*Tel:* 218 733-5919  *Fax:* 218 723-4921
*E-mail:* l.fox@lsc.mnscu.edu
*Med Dir:* Kirsten Bich, MD

**Argosy University/Twin Cities**
Medical Assistant Prgm
1515 Central Parkway
Eagan, MN 55121
*Prgm Dir:* Patricia Enocson, RN BSN
*Tel:* 651 846-3560  *Fax:* 651 994-0170
*E-mail:* penocson@argosyu.edu
*Med Dir:* John Vukelich, MD

**Northland Community & Technical College**
Medical Assistant Prgm
2022 Central Ave NE
East Grand Forks, MN 56721-2702
www.northlandcollege.edu
*Prgm Dir:* Elizabeth McMahon, MSEd BSN RN
*Tel:* 218 773-3441, Ext 4632  *Fax:* 218 773-4502
*E-mail:* elizabeth.mcmahon@northlandcollege.edu
*Med Dir:* Kim Konzak-Jones, MD

**Globe College**
Medical Assistant Prgm
Oakdale Center
7166 Tenth St N
Oakdale, MN 55128
*Prgm Dir:* Stephanie Suddendorft, CMA
*Tel:* 651 730-5100  *Fax:* 651 730-5151
*E-mail:* ssuddendorf@globecollege.edu
*Med Dir:* Phil Stoltenberg, MD

**Minnesota School of Business - Richfield**
Medical Assistant Prgm
1401 W 76th St, Ste 500
Richfield, MN 55423-3841
*Prgm Dir:* Susan Ende, CMA
*Tel:* 612 861-2000  *Fax:* 612 861-5548
*E-mail:* sende@msbcollege.edu
*Med Dir:* Rebecca Mitchell, MD

**Dakota County Technical College**
Medical Assistant Prgm
1300 E 145th St
Rosemount, MN 55068-2999
*Prgm Dir:* Patrice Nadeau, MT ASCP
*Tel:* 651 861-8355, Ext 355  *Fax:* 651 423-8210
*E-mail:* patrice.nadeau@dctc.edu
*Med Dir:* William Spinelli, MD

**Minneapolis Business College**
Medical Assistant Prgm
1711 W County Rd B
Roseville, MN 55113
www.minneapolisbusinesscollege.edu
*Prgm Dir:* Caryn Ziegler, CMA
*Tel:* 651 636-7406  *Fax:* 651 636-8185
*Med Dir:* John Berge, MD

**Century College**
Medical Assistant Prgm
3300 Century Ave N
White Bear Lake, MN 55110-1842
*Prgm Dir:* Michelle Blesi, CMA AA
*Tel:* 651 773-1731  *Fax:* 651 779-5779
*E-mail:* michelle.blesi@century.mnscu.edu
*Med Dir:* Stuart Pemberton, MD

**Ridgewater College - Willmar Campus**
Medical Assistant Prgm
2101 15th Ave NW, PO Box 1097
Willmar, MN 56201
www.ridgewater.edu
*Prgm Dir:* Julene D Bredeson, RN BSN CMA
*Tel:* 320 231-2947   *Fax:* 320 231-7677
*E-mail:* julene.bredeson@ridgewater.edu
*Med Dir:* Rachel Green, MD

**Minnesota West Comm & Tech College**
Medical Assistant Prgm
1450 Collegeway
Worthington, MN 56187
www.mnwest.edu
*Prgm Dir:* Lisa M Smith, RN BSN
*Tel:* 507 372-3400, Ext 3489   *Fax:* 507 372-5801
*E-mail:* lisa.smith@mnwest.edu
*Med Dir:* Carol L Lang, DO

# Mississippi

**Northeast Mississippi Community College**
Medical Assistant Prgm
Cunningham Blvd
Booneville, MS 38829
*Prgm Dir:* Kaye C Roberson, BA CMA
*Tel:* 662 720-7393, Ext 393   *Fax:* 662 728-1165
*E-mail:* krobers@nemcc.edu
*Med Dir:* Horton G Taylor, MD

**Hinds Community College**
Medical Assistant Prgm
3805 Hwy 80 E
Pearl, MS 39208-4295
www.hindscc.edu
*Prgm Dir:* Christine M King, CMA
*Tel:* 601 936-5582   *Fax:* 601 936-1828
*E-mail:* cmking@hindscc.edu
*Med Dir:* Marvin H Jeter, Jr, MD

# Missouri

**Everest College - Springfield Campus**
Medical Assistant Prgm
1010 W Sunshine St
Springfield, MO 65807-2446
www.springfield-college.com
*Prgm Dir:* Melissa Teal, BS CT RMA
*Tel:* 417 864-7220, Ext 204   *Fax:* 417 864-5697
*E-mail:* mteal@cci.edu
*Med Dir:* Elizabeth Stobbe, MD

# Montana

**Montana State University - Billings**
Medical Assistant Prgm
3803 Central Ave
Billings, MT 59102
*Prgm Dir:* Susan Floyd, RN BSN
*Tel:* 406 247-3000
*E-mail:* sfloyd@msubillings.edu
*Med Dir:* Robert Giusti, MD

**Montana State Univ - Great Falls Coll of Tech**
Medical Assistant Prgm
2100 16th Ave S
Great Falls, MT 59406-6010
*Prgm Dir:* Cynthia H Myles, BAN RN MA
*Tel:* 406 771-4383   *Fax:* 406 771-4317
*E-mail:* cmyles@msugf.edu

**Flathead Valley Community College**
Medical Assistant Prgm
777 Grandview Dr
Kalispell, MT 59901-2622
*Prgm Dir:* Karla West, MS BA
*Tel:* 406 756-3918   *Fax:* 406 756-3815
*E-mail:* kwest@fvcc.edu
*Med Dir:* Craig Harrison, MD

# Nebraska

**Central Community College**
Medical Assistant Prgm
Hastings Campus, PO Box 1024
Hastings, NE 68902-1024
www.cccneb.edu
*Prgm Dir:* Michel McKinney, CMA
*Tel:* 402 461-2405   *Fax:* 402 460-2138
*E-mail:* mmckinney@cccneb.edu
*Med Dir:* Fred Catlett, MD

**Hamilton College**
Medical Assistant Prgm
1821 K St
Lincoln, NE 68508
*Prgm Dir:* Michele Leah Guthard, CMA
*Tel:* 402 474-5315   *Fax:* 402 474-5302
*E-mail:* mguthard@hamiltonlincoln.edu
*Med Dir:* Mark Heibel, MD

**Southeast Community College**
Medical Assistant Prgm
8800 O St
Lincoln, NE 68520-1227
*Prgm Dir:* Jeanette Goodwin, BS BSN CMA
*Tel:* 402 437-2756   *Fax:* 402 437-2404
*E-mail:* jgoodwin@southeast.edu
*Med Dir:* Richard Jirovec, MD

**Alegent Health**
Medical Assistant Prgm
6901 N 72nd St
Omaha, NE 68122
*Prgm Dir:* Dona Witters, CMA BS HCM
*Tel:* 402 572-2676   *Fax:* 402 572-3486
*E-mail:* dwitters@alegent.org
*Med Dir:* Michael Waltz, MD

**Hamilton College - Omaha**
Medical Assistant Prgm
3350 N 90th St
Omaha, NE 68134-4710
www.hamiltonomaha.edu
*Prgm Dir:* Kristin Gargano, BS RRT
*Tel:* 402 572-8500, Ext 257   *Fax:* 402 573-1341
*E-mail:* kgargano@hcomaha.com
*Med Dir:* David Jasper, MD

**Nebraska Methodist College**
Medical Assistant Prgm
720 N 87th St
Omaha, NE 68114
http://methodistcollege.edu
*Prgm Dir:* Marcia Franklin, RN BSN
*Tel:* 402 354-7076   *Fax:* 402 354-7130
*E-mail:* marcia.franklin@methodistcollege.edu
*Med Dir:* David Filipi, MD

# Nevada

**Community College of Southern Nevada**
Medical Assistant Prgm
6375 W Charleston Blvd, W4K
Las Vegas, NV 89146-1124
*Prgm Dir:* Gail Silva, BSHS CMA
*Tel:* 702 651-5659   *Fax:* 702 651-5501
*E-mail:* gail_silva@ccsn.edu
*Med Dir:* Tina Saddler, PN

# New Hampshire

**New Hampshire Comm Tech Coll - Claremont**
Medical Assistant Prgm
One College Dr
Claremont, NH 03743-9707
*Prgm Dir:* Lisa B Gould, AS CMA
*Tel:* 603 542-7744, Ext 2743   *Fax:* 603 543-1844
*E-mail:* lgould@nhctc.edu
*Med Dir:* Kathryn Porterfield, ARNP

**Hesser College**
Medical Assistant Prgm
3 Sundial Ave
Manchester, NH 03103-7230
www.hesser.edu
*Prgm Dir:* Brenda Lee Salines, BSMT MEd RMA
*Tel:* 603 668-6660, Ext 2259   *Fax:* 603 621-8995
*E-mail:* bsalines@hesser.edu
*Med Dir:* Jerome P Lang, MD

**New Hampshire Comm Tech College - Manchester**
Medical Assistant Prgm
1066 Front St
Manchester, NH 03102-8528
www.manchester.nhctc.edu
*Prgm Dir:* Cindy Feldhousen, MS CMA
*Tel:* 603 668-6706, Ext 231   *Fax:* 603 668-5354
*E-mail:* cfeldhousen@nhctc.edu
*Med Dir:* William Windler, MD

# New Jersey

**Somerset County Technology Institute**
Medical Assistant Prgm
North Bridge & Vogt Dr
PO Box 6350
Bridgewater, NJ 08807-0350
*Prgm Dir:* Joan Purdy, AS RN CMA
*Tel:* 908 526-8900, Ext 7244   *Fax:* 908 526-9494
*E-mail:* jpurdy@scettc.org
*Med Dir:* Alan P Braun, MD

**Dover Business College - Dover**
Medical Assistant Prgm
15 E Blackwell St
Dover, NJ 07801
www.doverbusinesscollege.org
*Prgm Dir:* Henny Kaufman, MS RN CCRN
*Tel:* 973 546-0123   *Fax:* 973 546-0017
*E-mail:* hk@doverbusinesscollege.org
*Med Dir:* Kasing Ho, MD

**Hudson County Community College**
Medical Assistant Prgm
870 Bergen Ave
Jersey City, NJ 07306-4403
www.hccc.edu
*Prgm Dir:* Judith Bender, MA CMA
*Tel:* 201 360-4279   *Fax:* 201 418-7800
*E-mail:* JBender@hccc.edu
*Med Dir:* Raja Rahula, MD

**Sussex County Community College**
Medical Assistant Prgm
1 College Hill Rd
Newton, NJ 07860
http://sussex.edu
*Prgm Dir:* Tammy Miserendino, BS CMA RMA
*Tel:* 973 300-2172   *Fax:* 973 300-2303
*E-mail:* tmiserendino@sussex.edu
*Med Dir:* Denise Autotte, MD

**Bergen Community College**
Medical Assistant Prgm
400 Paramus Rd
Paramus, NJ 07652
*Prgm Dir:* Steven W Toth, MS CMA RMA
*Tel:* 201 612-5490   *Fax:* 201 612-3876
*E-mail:* stoth@bergen.edu
*Med Dir:* Hugh McGee Jr, MD

**Dover Business College - Paramus**
Medical Assistant Prgm
East 81 Route 4 West
Paramus, NJ 07652
www.doverbusinesscollege.org
*Prgm Dir:* Henny Kaufman, MS RN CCRN
*Tel:* 973 546-0123   *Fax:* 973 546-0017
*E-mail:* hk@doverbusinesscollege.org
*Med Dir:* Kasing Ho, MD

**Technical Institute of Camden County**
Medical Assistant Prgm
343 Berlin Cross Keys Rd
Sicklerville, NJ 08081-9707
*Prgm Dir:* Josephine Jackyra, MLT CMA
*Tel:* 856 767-7002, Ext 5539    *Fax:* 856 767-6625
*E-mail:* jjackyra@yahoo.com
*Med Dir:* John LaRatta, DO

**Berdan Institute**
Medical Assistant Prgm
265 Rte 46 W
Totowa, NJ 07512
*Prgm Dir:* Irene Figliolina, CMA
*Tel:* 973 256-3444    *Fax:* 973 256-0816
*E-mail:* ifigliolina@berdaninstitute.com
*Med Dir:* Giovanni Lima, MD

**Mercer County Technical Schools**
Medical Assistant Prgm
1070 Klockner Rd
Trenton, NJ 08619
www.mctec.net
*Prgm Dir:* Gail M Kenney, CMA LPN AHI RMA AAS
*Tel:* 609 587-7640    *Fax:* 609 587-3304
*E-mail:* GKenney@mctec.net
*Med Dir:* V Subramoni, MD

**Warren County Community College**
Medical Assistant Prgm
475 Rte 57 West
Washington, NJ 07882-4343
*Prgm Dir:* Marianne Van Deursen, BS MLT CMA
*Tel:* 908 835-2430    *Fax:* 908 689-8032
*E-mail:* vandeursen@warren.edu
*Med Dir:* James Goodwin, MD

# New Mexico

**Eastern New Mexico University - Roswell**
Medical Assistant Prgm
PO Box 6000
Roswell, NM 88202-6000
*Prgm Dir:* Cheryl A Vineyard, CMA CPC
*Tel:* 505 624-7199    *Fax:* 505 624-7100
*E-mail:* Cheryl.Vineyard@roswell.enmu.edu
*Med Dir:* Phyllis Tulk, CFNP

# New York

**Bryant & Stratton College - Albany**
Medical Assistant Prgm
1259 Central Ave
Albany, NY 12205-5230
*Prgm Dir:* Maqsood A Nathu, MD MT
*Tel:* 518 437-1802    *Fax:* 518 437-1048
*E-mail:* al.nathu.maqsood@mail.bryantstratton.edu

**Broome Community College**
Medical Assistant Prgm
Decker Bldg, PO Box 1017
Binghamton, NY 13902-1017
http://sunybroome.edu
*Prgm Dir:* Andrea Wade, PhD MT(ASCP)
*Tel:* 607 778-5211    *Fax:* 607 778-5467
*E-mail:* wade_a@sunybroome.edu
*Med Dir:* Bruce T Bowling, MD

**Ridley-Lowell Business & Technical Institute**
Medical Assistant Prgm
116 Front St
Binghamton, NY 13905
www.ridley.edu
*Prgm Dir:* Lynne Radley, BS CMA
*Tel:* 607 724-2941    *Fax:* 607 724-0799
*E-mail:* lrad_ridley@hotmail.com
*Med Dir:* Louis P Mateya Jr, MD

**ASA Institute of Business & Computer Tech**
Medical Assistant Prgm
81 Willoughby St
Mezzanine
Brooklyn, NY 11201-5208
www.asa.edu
*Prgm Dir:* Rosalind Collazo, MS CCMA CBCS
*Tel:* 718 694-2929    *Fax:* 718 532-1433
*E-mail:* rcollazo@asa.edu
*Med Dir:* Salem Ashraf, MD

**Bryant & Stratton College - Buffalo**
Medical Assistant Prgm
465 Main St, Ste 400
Buffalo, NY 14203-1795
www.bryantstratton.edu
*Prgm Dir:* LouElla C Cole, BS
*Tel:* 716 884-9120    *Fax:* 716 884-0091
*E-mail:* lcole@bryantstratton.edu
*Med Dir:* Raul Vazquez, MD

**Trocaire College**
Medical Assistant Prgm
360 Choate Ave
Buffalo, NY 14220-2094
*Prgm Dir:* Tasha Pratcher, MS Ed BA
*Tel:* 716 827-2430    *Fax:* 716 828-6107
*E-mail:* pratchert@trocaire.edu
*Med Dir:* Prem Chopra, MD

**Elmira Business Institute**
Medical Assistant Prgm
303 N Main St
Elmira, NY 14901
www.ebi-college.com
*Prgm Dir:* Sherry Mulhollen, CMA BS
*Tel:* 607 733-7177    *Fax:* 607 733-7178
*E-mail:* smulhollen@ebi-college.com
*Med Dir:* Cynthia Terry, MD

**Wood Tobe'-Coburn School**
Medical Assistant Prgm
8 E 40th St
New York, NY 10016-0190
www.woodtobecoburn.edu
*Prgm Dir:* Lynn McManus, LPN CMA
*Tel:* 212 886-9040, Ext 138    *Fax:* 212 686-9171
*E-mail:* lmcwtc@aol.com
*Med Dir:* Brian McNeary, MD

**Ridley-Lowell Business & Technical Institute**
Medical Assistant Prgm
26 S Hamilton St
Poughkeepsie, NY 12601
*Prgm Dir:* Rhonda Hayes, AS CMA
*Tel:* 845 471-0330    *Fax:* 845 471-4990
*E-mail:* pcdirector@ridley.edu
*Med Dir:* Deborah Roraback, RN

**Bryant & Stratton College - Rochester**
Medical Assistant Prgm
Henrietta Campus, 1225 Jefferson Rd
Greece Campus, 150 Bellwood Dr
Rochester, NY 14623-3185
www.bryantstratton.edu
*Prgm Dir:* Karen C Parysek, CMA
*Tel:* 585 292-5627, Ext 148    *Fax:* 585 292-6015
*E-mail:* kcparysek@bryantstratton.edu
*Med Dir:* Michael Mirwald, MD

**Rochester Business Institute**
Medical Assistant Prgm
1630 Portland Ave
Rochester, NY 14621
www.rochester-institute.com
*Prgm Dir:* Mary Hodges, RN BS AAS
*Tel:* 585 266-0430    *Fax:* 585 338-2464
*E-mail:* mhodges@cci.edu
*Med Dir:* Michael Kufka, MD

**Niagara County Community College**
Medical Assistant Prgm
3111 Saunders Settlement Rd
Sanborn, NY 14132
*Prgm Dir:* Salvatore M Passanese, PhD
*Tel:* 716 614-6411    *Fax:* 716 614-6879
*E-mail:* spassane@niagaracc.suny.edu
*Med Dir:* Thomas Gerbasi, MD

**Bryant & Stratton College - Syracuse**
Medical Assistant Prgm
953 James St
Syracuse, NY 13203-2555
www.bryantstratton.edu
*Prgm Dir:* Cheryl Pearsall, RN MSN CAS
*Tel:* 315 472-6603    *Fax:* 315 474-4383
*E-mail:* sy.pearsall.cheryl@mail.bryantstratton.edu
*Med Dir:* Jeffrey Sneider, MD

**Erie Community College - North Campus**
Medical Assistant Prgm
6205 Main St
Williamsville, NY 14221
www.ecc.edu
*Prgm Dir:* Marcia T Bermel, MS MT(ASCP) CMA
*Tel:* 716 851-1553    *Fax:* 716 851-1429
*E-mail:* bermel@ecc.edu
*Med Dir:* William Scheuler, MD

# North Carolina

**Stanly Community College**
Medical Assistant Prgm
141 College Dr
Albemarle, NC 28001
*Prgm Dir:* Lawrencette McSwain, RN CMA MSEd
*Tel:* 704 991-0397    *Fax:* 704 991-0350
*E-mail:* mcswailr@stanly.edu
*Med Dir:* Beatrice Leitch, GENP CFNP

**South College - Asheville**
Medical Assistant Prgm
1567 Patton Ave
Asheville, NC 28806-1748
www.southcollegenc.com
*Prgm Dir:* Annie Butzner, RN MSN CNS ET PNP RMA
*Tel:* 828 277-5521    *Fax:* 828 277-6151
*E-mail:* abutzner@southcollegenc.com
*Med Dir:* Gregory S Motley, MD

**Central Piedmont Community College**
Medical Assistant Prgm
PO Box 35009
Charlotte, NC 28235-5009
www.cpcc.edu
*Prgm Dir:* Leesa Whicker, CMA
*Tel:* 704 330-6797    *Fax:* 704 330-6131
*E-mail:* leesa.whicker@cpcc.edu
*Med Dir:* Bonnie Goodwin, FNP

**King's College**
Medical Assistant Prgm
322 Lamar Ave
Charlotte, NC 28204
www.kingscollegecharlotte.edu
*Prgm Dir:* Marshal Cayton, CPNP MSN RN RMA
*Tel:* 704 688-3620    *Fax:* 704 348-2029
*E-mail:* mcayton@kingscollegecharlotte.edu
*Med Dir:* Maureen L Beurskens, MD

**Haywood Community College**
Medical Assistant Prgm
185 Freelander Dr
Clyde, NC 28721-9454
*Prgm Dir:* Sandra Kelley, BSN RN
*Tel:* 828 627-4658    *Fax:* 828 627-4656
*E-mail:* skelly@haywood.edu
*Med Dir:* Nancy Freeman, MD

## Cabarrus College of Health Sciences
Medical Assistant Prgm
401 Medical Park Dr
Concord, NC 28025
*Prgm Dir:* Stacey Wilson, MT/PBT (ASCP)CMA
*Tel:* 704 783-1639   *Fax:* 704 783-1764
*E-mail:* SWilson@northeastmedical.org
*Med Dir:* Lynn A Hughes, MD

## Gaston College
Medical Assistant Prgm
201 Hwy 321 S
Dallas, NC 28034-1499
www.gaston.edu
*Prgm Dir:* Betty Jones, MA RN CMA
*Tel:* 704 922-6465   *Fax:* 704 922-2333
*E-mail:* bdjones@gaston.edu
*Med Dir:* Lee Barro, MD

## Surry Community College
Medical Assistant Prgm
630 S Main St
Dobson, NC 27017-8432
*Prgm Dir:* Tammy Gant, RHIT CMA CAHI
*Tel:* 336 386-3256, Ext 3256   *Fax:* 336 386-3497
*E-mail:* gantt@surry.edu
*Med Dir:* Dean Culler, PA

## College of The Albemarle
Medical Assistant Prgm
1208 N Road St
Elizabeth City, NC 27906
*Prgm Dir:* Kathleen Carrion, RMA ASAAH-EMS
*Tel:* 252 335-0821, Ext 2307
*E-mail:* kcarrion@albermarie.edu
*Med Dir:* Catherine Brown

## Wayne Community College
Medical Assistant Prgm
Caller Box 8002
Goldsboro, NC 27534-8002
*Prgm Dir:* Louetta M Brown, BS MT-ASCP
*Tel:* 919 735-5151, Ext 753   *Fax:* 919 736-9425
*E-mail:* lbrown@waynecc.edu
*Med Dir:* David T Tayloe, Jr, MD

## Alamance Community College
Medical Assistant Prgm
PO Box 8000
Graham, NC 27253
*Prgm Dir:* Kaye Acton, AAS CMA
*Tel:* 336 506-4213   *Fax:* 336 578-7047
*E-mail:* actonk@alamance.cc.nc.us
*Med Dir:* Edward D Lance, MD

## Pamlico Community College
Medical Assistant Prgm
PO Box 185
Grantsboro, NC 28529-0185
www.pamlico.cc.nc.us
*Prgm Dir:* Dale Holadia, RN RMA
*Tel:* 252 249-1851, Ext 3044   *Fax:* 252 249-2377
*E-mail:* dholadia@pamlicocc.edu
*Med Dir:* Sue Lee, MD

## Pitt Community College
Medical Assistant Prgm
PO Drawer 7007, Hwy 11 S
Greenville, NC 27835-7007
www.pittcc.edu
*Prgm Dir:* Marsha Hemby, BA RN CMA
*Tel:* 252 493-7284   *Fax:* 252 321-4451
*E-mail:* mhemby@email.pittcc.edu
*Med Dir:* Mary Helen Allen-Hutchison, MD

## Richmond Community College
Medical Assistant Prgm
1042 W Hamlet Ave
Hamlet, NC 28345
*Prgm Dir:* Dena A Evans, RN BSN MPH CMA
*Tel:* 910 582-7000, Ext 7174
*E-mail:* denae@richmondcc.edu
*Med Dir:* Kirk Hassenmeuler, MD

## Guilford Technical Community College
Medical Assistant Prgm
601 High Point Rd, PO Box 309
Jamestown, NC 27282-0309
*Prgm Dir:* Kimberly G Cannon, BS CMA
*Tel:* 336 454-1126, Ext 2307   *Fax:* 336 454-2510
*E-mail:* kgcannon@gtcc.edu
*Med Dir:* E B Mabry, MD

## James Sprunt Community College
Medical Assistant Prgm
PO Box 398
Hwy 11 S
Kenansville, NC 28349-0398
*Prgm Dir:* Angelia Carr-Grady, CMA AAS AGE
*Tel:* 910 296-2400   *Fax:* 910 296-1636
*E-mail:* gradyland2@earthlink.net

## Lenoir Community College
Medical Assistant Prgm
PO Box 188
231 Hwy 58S
Kinston, NC 28502-0188
www.lenoircc.edu
*Prgm Dir:* Linda Susan Johnson, CPC CMA
*Tel:* 252 527-6223, Ext 815   *Fax:* 252 233-6893
*E-mail:* ljohnson@lenoircc.edu
*Med Dir:* Joseph Agsten, MD

## Davidson County Community College
Medical Assistant Prgm
PO Box 1287
Lexington, NC 27293-1287
*Prgm Dir:* Suzanne Rohrbaugh, MA MT(ASCP)
*Tel:* 336 249-8186, Ext 6721   *Fax:* 336 249-9060
*E-mail:* srohr@davidsonccc.edu
*Med Dir:* Sundara Rajan, MD

## Central Carolina Community College - Harnett
Medical Assistant Prgm
1075 E Cornelius Harnett Blvd
Lillington, NC 27546
www.cccc.edu
*Prgm Dir:* Melissa Fogarty, CMA RMA CAHI
*Tel:* 919 718-7325   *Fax:* 919 718-7477
*E-mail:* mfogarty@cccc.edu
*Med Dir:* Michael Tyler, MD

## Vance - Granville Community College
Medical Assistant Prgm
PO Box 777
8100 NC 56 Hwy W
Louisburg, NC 27549
*Prgm Dir:* Tammy Care, BSMT(ASCP)
*Tel:* 252 738-3607   *Fax:* 252 496-6604
*E-mail:* care@vgcc.edu
*Med Dir:* Francine Chavis, MD

## Mitchell Community College
Medical Assistant Prgm
219 N Academy St
Mooresville, NC 28115
www.mitchellcc.edu
*Prgm Dir:* Mary M Marks, FNP-C MSN
*Tel:* 704 978-5424   *Fax:* 704 663-5239
*E-mail:* mmarks@mitchellcc.edu
*Med Dir:* Sharon Setzer, FNP-CS

## Carteret Community College
Medical Assistant Prgm
3505 Arendell St
Morehead City, NC 28557
*Prgm Dir:* Vonda Godette, RN BSN
*Tel:* 252 222-6168   *Fax:* 252 222-6074
*E-mail:* godettev@carteret.edu
*Med Dir:* Jeffery Anderson, MD

## Western Piedmont Community College
Medical Assistant Prgm
1001 Burkemont Ave
Morganton, NC 28655
www.wpcc.edu
*Prgm Dir:* Ann Q Giles, MHS CMA
*Tel:* 828 438-6130   *Fax:* 828 430-7183
*E-mail:* agiles@wpcc.edu
*Med Dir:* Alfred Hamer, MD

## Tri-County Community College
Medical Assistant Prgm
4600 Hwy 64 East
Murphy, NC 28906
*Prgm Dir:* Kathleen Hearl, RN CMA
*Tel:* 704 837-6810   *Fax:* 704 837-3266
*E-mail:* khearl@tricountycc.edu
*Med Dir:* Jenkins Clarkson, MD

## Central Carolina Community College - Chatham
Medical Assistant Prgm
764 West St
Pittsboro, NC 27312
*Prgm Dir:* Melissa Fogarty, CMA RMA CAHI
*Tel:* 919 718-7281   *Fax:* 919 718-7477
*E-mail:* mfogarty@cccc.edu
*Med Dir:* Michael Tyler, MD

## South Piedmont Community College
Medical Assistant Prgm
PO Box 126
Polkton, NC 28135-0126
*Prgm Dir:* Connie W Stack, MEd MLT(ASCP)CMA
*Tel:* 704 272-7635, Ext 5420   *Fax:* 704 272-8904
*E-mail:* cstack@spcc.edu
*Med Dir:* Victoria Rommel, MD

## Wake Technical Community College
Medical Assistant Prgm
9101 Fayetteville Rd
Raleigh, NC 27603-5656
www.waketech.edu
*Prgm Dir:* Charmaine Parker, LPN
*Tel:* 919 250-4333   *Fax:* 919 250-4329
*E-mail:* caparker@waketech.edu
*Med Dir:* Thomas Lee Jeffries, MD

## Johnston Community College
Medical Assistant Prgm
PO Box 2350
245 College Rd
Smithfield, NC 27577
www.johnstoncc.edu
*Prgm Dir:* Rhonda Evans, AAS MA
*Tel:* 919 209-2140
*E-mail:* leer@johnstoncc.edu
*Med Dir:* Kimberly Fox, MD

## Mayland Community College
Medical Assistant Prgm
PO Box 547
Spruce Pine, NC 28777
*Prgm Dir:* Dolly Horton, CMA AAS BS
*Tel:* 828 765-7351, Ext 330
*E-mail:* dhorton@mayland.edu
*Med Dir:* Russell Flint, MD

## Edgecombe Community College
Medical Assistant Prgm
2009 W Wilson St
Tarboro, NC 27886
*Prgm Dir:* Dorothy B Tolson, MAEd BSBA CMA
*Tel:* 252 446-0436, Ext 353   *Fax:* 252 985-2212
*E-mail:* tolsond@edgecombe.edu
*Med Dir:* Moses E Wilson, MD

**Montgomery Community College**
Medical Assistant Prgm
1011 Page St
Troy, NC 27371
www.montgomery.edu
*Prgm Dir:* Cyndi Caviness, CRT CMA
*Tel:* 910 576-6222, Ext 208    *Fax:* 910 576-3005
*E-mail:* cavinessc@montgomery.edu
*Med Dir:* Eron Manusov, MD

**Wilkes Community College**
Medical Assistant Prgm
PO Box 120
1328 Collegiate Dr
Wilkesboro, NC 28697-0120
www.wilkescc.edu
*Prgm Dir:* Joyce A Minton, MA BS CMA RMA
*Tel:* 910 838-6251    *Fax:* 910 838-6255
*E-mail:* mintonj@wilkescc.edu
*Med Dir:* Laura Hubbard, MD

**Martin Commuity College**
Medical Assistant Prgm
1161 Kehukee Park Rd
Williamston, NC 27892-8307
www.martincc.edu
*Prgm Dir:* Marty S Flynn, CMA
*Tel:* 252 792-1521, Ext 292    *Fax:* 252 792-4425
*E-mail:* mflynn@martincc.edu
*Med Dir:* Carl T Dover, MD

**Miller-Motte Technical College**
Medical Assistant Prgm
5000 Market St
Wilmington, NC 28405
*Prgm Dir:* Glenn Grady, MEd BSMT(ASCP) CMA
*Tel:* 910 392-4660, Ext 2015    *Fax:* 910 799-6224
*E-mail:* ggrady@miller-motte.com
*Med Dir:* Crane Jonathan, MD

**Forsyth Technical Community College**
Medical Assistant Prgm
2100 Silas Creek Pkwy
Winston-Salem, NC 27103
*Prgm Dir:* Laura S Durham, BS CMA
*Tel:* 336 734-7362    *Fax:* 336 734-7581
*E-mail:* ldurham@bit.forsythtech.edu
*Med Dir:* James S Gibbs, MD

# Ohio

**Akron Institiue**
Medical Assistant Prgm
1600 S Arlington, Ste 100
Akron, OH 44306
www.akroninstitute.com
*Prgm Dir:* Cindy L Garman, CMA ASAM
*Tel:* 330 724-1600, Ext 15    *Fax:* 330 724-9688
*E-mail:* cgarman@akroninstitute.com
*Med Dir:* Marguerite Erme, MD

**Brown Mackie College - Akron**
Medical Assistant Prgm
2791 Mogadore Rd
Akron, OH 44312
*Prgm Dir:* Debra Tymcio, BA RS
*Tel:* 330 733-8766, Ext 19    *Fax:* 330 734-7581
*E-mail:* dtymcio@brownmackie.edu
*Med Dir:* Bonyo Bonyo, DO

**University of Akron**
Medical Assistant Prgm
Polsky 124C
Akron, OH 44325-3702
*Prgm Dir:* Rebecca Gibson-Lee, MSTE CMA ASPT
*Tel:* 330 972-6515    *Fax:* 330 972-2016
*E-mail:* gibsonr@uakron.edu
*Med Dir:* Garimah Jones, MD

**Ashland County - West Holmes Career Center**
Medical Assistant Prgm
1783 St Rte 60
Ashland, OH 44805-9377
*Prgm Dir:* David Kleinschmidt, MEd ABD
*Tel:* 419 289-3313, Ext 207    *Fax:* 419 289-3729
*E-mail:* awhj_Kleinsc@tccsa.net
*Med Dir:* Abbas Shikary, MD

**Mahoning County Career & Technical Center**
Medical Assistant Prgm
7300 N Palmyra Rd
Canfield, OH 44406-9710
www.mahoningctc.com
*Prgm Dir:* Alice Siwiec, CMA
*Tel:* 330 729-4100, Ext 1953    *Fax:* 330 729-4150
*E-mail:* mjvs_ajs@access-k12.org
*Med Dir:* Darrell L Grace, DO

**Canton City Schools**
Medical Assistant Prgm
Adult Community Education Center
116 McKinley Ave NW Rm 206
Canton, OH 44702
*Prgm Dir:* Kathleen Dean, RN
*Tel:* 330 456-1092    *Fax:* 330 454-6767
*E-mail:* kjsdean@yahoo.com
*Med Dir:* Lisa Vaughn, MD

**Stark State College of Technology**
Medical Assistant Prgm
6200 Frank Ave NW
Canton, OH 44720
www.starkstate.edu
*Prgm Dir:* Geraldine Todaro, MSTE CMA CLPlb
*Tel:* 330 966-5458, Ext 4233    *Fax:* 330 966-6586
*E-mail:* gtodaro@starkstate.edu
*Med Dir:* Lawrence Ronning, MD

**Fairfield Career Center (EVSD)**
Medical Assistant Prgm
4000 Columbus-Lancaster Rd
Carroll, OH 43112
www.eastland-fairfield.com
*Prgm Dir:* Tammy Tipton
*Tel:* 740 756-9245    *Fax:* 740 837-9447
*E-mail:* ttipton@efcts.us
*Med Dir:* Assem Houssein, MD

**RETS Tech Center**
Medical Assistant Prgm
555 E Alex-Bell Rd
Centerville, OH 45459
*Prgm Dir:* Sandra Quinn, MT MCS MBS
*Tel:* 937 433-3410, Ext 218    *Fax:* 937 435-6516
*E-mail:* squinn@retstechcenter.com
*Med Dir:* Richard Hoback, MD

**Pickaway Ross Joint Vocational School**
Medical Assistant Prgm
17273 State Rte 104, Bldg 4
Chillicothe, OH 45601
www.pickawayross.com
*Prgm Dir:* Faye Vermillion, RN BSN MEd
*Tel:* 740 642-1440, Ext 450    *Fax:* 740 642-1454
*E-mail:* faye.vermillion@pickawayross.com
*Med Dir:* Anthony C Freeman, MD

**Brown Mackie College - Cincinnati**
Medical Assistant Prgm
1011 Glendale-Milford Rd
Cincinnati, OH 45215
*Prgm Dir:* Loretto D Reddington, BA CMA CPT(IAPS)
*Tel:* 513 771-2424, Ext 7758    *Fax:* 513 771-3413
*E-mail:* lreddington@brownmackie.edu
*Med Dir:* Jan Fu, MD PhD

**Cincinnati State Tech & Comm College**
Medical Assistant Prgm
3520 Central Pkwy
Cincinnati, OH 45223
*Prgm Dir:* Olivia Watts, BSN RN
*Tel:* 513 569-1571    *Fax:* 513 569-1659
*E-mail:* olivia.watts@cincinnatistate.edu
*Med Dir:* Esly Caldwell, MD

**University of Cincinnati**
*Cosponsor:* Raymond Walters College
Medical Assistant Prgm
9555 Plainfield Rd
Cincinnati, OH 45236
*Prgm Dir:* Rachael Allstatter, MEd
*Tel:* 513 745-5664    *Fax:* 513 745-5771
*E-mail:* rachael.allstatter@uc.edu
*Med Dir:* John S Andrews Jr, MD

**Miami Valley Career Technology Center**
Medical Assistant Prgm
6800 Hoke Rd
Clayton, OH 45315
*Prgm Dir:* Frances Childers Shull, RN MS
*Tel:* 937 854-6297    *Fax:* 937 837-5619
*E-mail:* fchilders@mvctc.k12.oh.us
*Med Dir:* Bhadresh Doshi, MD

**Cuyahoga Community College**
Medical Assistant Prgm
Metro Campus
2900 Community College Ave
Cleveland, OH 44115
*Prgm Dir:* Barbara Freeman, MEd MT(ASCP)
*Tel:* 216 987-4441    *Fax:* 216 987-4386
*E-mail:* barbara.freeman@tri-c.edu
*Med Dir:* Carl F Asseff, MD

**American School of Technology**
Medical Assistant Prgm
2100 Morse Rd
Columbus, OH 43229
www.ast.edu
*Prgm Dir:* Barbara Ruble, BS
*Tel:* 614 436-4820    *Fax:* 614 436-4343
*E-mail:* bruble@ast.edu
*Med Dir:* Michelle Chambers, Dermatologist

**Bradford School**
Medical Assistant Prgm
2469 Stelzer Rd
Columbus, OH 43219
*Prgm Dir:* Amanda Rumich, CMA
*Tel:* 614 416-6200, Ext 110    *Fax:* 614 416-6210
*E-mail:* arumich@bradfordschoolcolumbus.edu
*Med Dir:* R D Lausa, MD

**Columbus State Community College**
Medical Assistant Prgm
550 E Spring St
Columbus, OH 43216
www.cscc.edu
*Prgm Dir:* Kay E Biggs, CMA BS AAS
*Tel:* 614 287-3638    *Fax:* 614 287-5198
*E-mail:* kbiggs@cscc.edu
*Med Dir:* Lisa Ward, PA-C

**Ohio Institute of Health Careers - Columbus**
Medical Assistant Prgm
1880 E Dublin-Granville Rd
Ste 100
Columbus, OH 43229
www.ohioinstituteofhealthcareers.edu
*Prgm Dir:* Michelle Heller, CMA RMA
*Tel:* 614 891-5030, Ext 118    *Fax:* 614 891-5130
*E-mail:* mheller@ohioinstitueofhealthcareers.edu
*Med Dir:* Nino DiIullo, MBA MD

## Miami - Jacobs Career College

Medical Assistant Prgm
110 N Patterson Blvd
Dayton, OH 45402
www.miamijacobs.edu
*Prgm Dir:* Debra Southard, RMA
*Tel:* 937 461-5174, Ext 129 *Fax:* 937 461-3384
*E-mail:* debra.southard@miamijacobs.edu
*Med Dir:* Tami Caro, PA

## Ohio Institute of Photography & Technology

Medical Assistant Prgm
Division of Allied Health
2029 Edgefield Rd
Dayton, OH 45439
*Prgm Dir:* Gail Stone
*Tel:* 937 294-6155, Ext 208 *Fax:* 937 294-2259
*E-mail:* gkstone@oipt.com
*Med Dir:* Alexis Myton, MD

## Sinclair Community College

Medical Assistant Prgm
444 W Third St
Dayton, OH 45402-1421
www.sinclair.edu
*Prgm Dir:* Jennifer L Barr, MEd MT CMA
*Tel:* 937 512-2973 *Fax:* 937 512-2239
*E-mail:* jennifer.barr@sinclair.edu
*Med Dir:* Martin Fujimura, MD

## Ohio Valley College of Technology

Medical Assistant Prgm
16808 St Clair Ave, PO Box 7000
East Liverpool, OH 43920
*Prgm Dir:* Linda S Johnston, RN
*Tel:* 330 385-1070 *Fax:* 330 385-4606
*E-mail:* ljohnson@ovct.edu
*Med Dir:* Tae Jung, MD

## Lorain County Community College

Medical Assistant Prgm
1005 N Abbe Rd
Elyria, OH 44035
www.lorainccc.edu
*Prgm Dir:* Cindy Watkins, RN MSN
*Tel:* 440 366-4189 *Fax:* 440 366-4116
*E-mail:* cwatkins@lorainccc.edu
*Med Dir:* Sana Abumari, MD

## Ohio Institute of Health Careers - Elyria

Medical Assistant Prgm
631 Griswold Rd
Elyria, OH 44035
*Prgm Dir:* Jacquelyn Marshall, MS
*Tel:* 440 324-2293
*E-mail:* jmarshall@oiofhc.com

## Portage Lakes Career Center

Medical Assistant Prgm
4401 Shriver Rd
PO Box 248
Green, OH 44232-0248
www.portagelakescareercenter.org
*Prgm Dir:* Janeane L Brodie, CMA EMT-P
*Tel:* 330 896-8200 *Fax:* 330 896-8197
*E-mail:* pl_adult@portagelakescareercenter.org
*Med Dir:* Albert Feltrup, MD

## Southern State Community College

Medical Assistant Prgm
100 Hobart Dr
Hillsboro, OH 45133-9487
www.sscc.edu
*Prgm Dir:* Carry Ann DeAtley, MBA CMA
*Tel:* 937 393-3431, Ext 2639 *Fax:* 937 393-9370
*E-mail:* cdeatley@sscc.edu
*Med Dir:* Robert Moore, MD

## Lakeland Community College

Medical Assistant Prgm
7700 Clocktower Dr
Kirtland, OH 44094
http://lakelandcc.edu/academic/sh/medasst
*Prgm Dir:* Michele G Miller, MEd COMT CMA
*Tel:* 440 525-7428 *Fax:* 440 525-7433
*E-mail:* mmiller@lakelandcc.edu
*Med Dir:* Komoar Mark, MD

## Ohio University - Lancaster

Medical Assistant Prgm
1570 Granville Pike
Lancaster, OH 43130
www.ohio.edu
*Prgm Dir:* Susan H Maxwell, MA CPS CMA
*Tel:* 740 654-6711, Ext 276 *Fax:* 740 687-9497
*E-mail:* maxwell@ohio.edu
*Med Dir:* Brian Higgins, DO

## Apollo Career Center

Medical Assistant Prgm
3325 Shawnee Rd
Lima, OH 45806
www.apollocareercenter.com
*Prgm Dir:* Tara Shepherd
*Tel:* 419 998-2971 *Fax:* 419 998-2994
*E-mail:* ap_shepherd@noacsc.org
*Med Dir:* James Patterson, MD

## Rhodes State College

Medical Assistant Prgm
4240 Campus Dr TL 104F
Lima, OH 45804
www.rhodesstate.edu
*Prgm Dir:* Dorothy Kiel, CMA-C BS
*Tel:* 419 995-8016, Ext 8016 *Fax:* 419 995-8093
*E-mail:* kiel.d@rhodesstate.edu
*Med Dir:* Michael Wieser, MD

## University of Northwestern Ohio

Medical Assistant Prgm
1441 N Cable Rd
Lima, OH 45805
*Prgm Dir:* Julie Shellenbarger, BS RHIA
*Tel:* 419 227-3141 *Fax:* 419 229-6926
*E-mail:* jashelle@unoh.edu
*Med Dir:* Henry Gerad, MD

## Washington County Career Center

Medical Assistant Prgm
1750 Lancaster St
Marietta, OH 45750
*Prgm Dir:* Karen Seagraves, CMA
*Tel:* 740 373-6283 *Fax:* 740 376-2240
*E-mail:* wks@1st.net
*Med Dir:* Kenneth J Leopold, MD

## Marion Technical College

Medical Assistant Prgm
1467 Mt Vernon Ave
Marion, OH 43302-5694
http://mtc.edu
*Prgm Dir:* Peggy Smith, BS
*Tel:* 614 389-4636, Ext 319 *Fax:* 614 389-6136
*E-mail:* smithp@mtc.edu
*Med Dir:* Carol Solie, MD

## Medina County Career Center

Medical Assistant Prgm
1101 W Liberty St
Medina, OH 44256-9969
*Prgm Dir:* Crispina Mayes, BSN MT(ASCP)
*Tel:* 330 725-8461, Ext 314 *Fax:* 330 725-3842
*E-mail:* mayes@mcjvsd.org
*Med Dir:* John Funk, MD

## Polaris Career Center

Medical Assistant Prgm
7285 Old Oak Blvd
Middleburg Heights, OH 44130
www.polaris.edu
*Prgm Dir:* Kathryn Lazroff
*Tel:* 440 891-7635 *Fax:* 440 243-3952
*E-mail:* klazroff@polaris.edu
*Med Dir:* Michael Banks, MD

## EHOVE Ghrist Adult Career Center

Medical Assistant Prgm
316 W Mason Rd
Milan, OH 44846
www.ehove.net
*Prgm Dir:* Janet Ballard, RN BSN MEd
*Tel:* 419 499-4663, Ext 287 *Fax:* 419 499-5391
*E-mail:* jballard@ehove-jvs.k12.oh.us
*Med Dir:* Stephanie Gibson, MD

## Knox County Career Center

Medical Assistant Prgm
308 Martinsburg Rd
Mt Vernon, OH 43050-4298
*Prgm Dir:* Jo Kline, RNC MSN MEd
*Tel:* 740 393-2933 *Fax:* 740 393-1659
*E-mail:* pnmed@hotmail.com
*Med Dir:* Brent Nimeth, MD

## Hocking College

Medical Assistant Prgm
3301 Hocking Pkwy
Nelsonville, OH 45764-9582
www.hocking.edu
*Prgm Dir:* Kathy West, BS MEd
*Tel:* 614 753-3591, Ext 6434 *Fax:* 614 753-6430
*E-mail:* west_k@hocking.edu
*Med Dir:* Daniel Marazon, DO

## Lorain Cnty Joint Voc School - Adult Career

Medical Assistant Prgm
15181 State Route 58
Oberlin, OH 44074
*Prgm Dir:* Nancy Smith
*Tel:* 440 774-1051 *Fax:* 440 776-2070
*E-mail:* nsmith@leeca.org
*Med Dir:* Chuka Onyeneke, MD

## Bryant & Stratton College - Parma

Medical Assistant Prgm
12955 Snow Rd
Parma, OH 44130
*Prgm Dir:* Sheila Batheja, MSMT
*Tel:* 216 265-3151 *Fax:* 216 265-0325
*E-mail:* pr.batheja.sheila@mail.bryantstratton.edu
*Med Dir:* Richard Christie, MD

## Belmont Technical College

Medical Assistant Prgm
120 Fox-Shannon Pl
St Clairsville, OH 43950-9766
http://btc.edu
*Prgm Dir:* Donna Folmar, RN BSN
*Tel:* 740 695-9500, Ext 1171 *Fax:* 740 695-2247
*E-mail:* dfolmar@btc.edu
*Med Dir:* Joseph Gabis, MD

## Jefferson Community College

Medical Assistant Prgm
4000 Sunset Blvd
Steubenville, OH 43952
www.jcc.edu
*Prgm Dir:* Robin Snider Flohr, EdD RN CMA
*Tel:* 740 264-5591, Ext 159 *Fax:* 740 264-1338
*E-mail:* rflohr@jcc.edu
*Med Dir:* Frank Petrola, MD

**Davis College**
Medical Assistant Prgm
4747 Monroe St
Toledo, OH 43623-4389
*Prgm Dir:* Rhonda Lazette, CMA BS
*Tel:* 419 473-2700, Ext 312   *Fax:* 419 473-2472
*E-mail:* rlazette@daviscollege.edu
*Med Dir:* Karen Asher, DO

**Stautzenberger College**
Medical Assistant Prgm
5355 Southwyck Blvd
Toledo, OH 43614
www.sctoday.edu
*Prgm Dir:* Patricia Bell, BA CMA RMA
*Tel:* 800 552-5099, Ext 224   *Fax:* 419 867-9821
*E-mail:* pebell@stautzenberger.com
*Med Dir:* Robert Rae, MD MSBS

**Trumbull Career & Technical Center**
Medical Assistant Prgm
528 Educational Highway
Warren, OH 44484
www.tctcadultraining.org
*Prgm Dir:* Joan Bushey, RN MEd
*Tel:* 330 847-0503, Ext 1620   *Fax:* 330 847-1177
*E-mail:* joan.bushey@neomin.org
*Med Dir:* April Miller, PA

**Youngstown State University**
Medical Assistant Prgm
Dept of Health Professions
One University Plaza
Youngstown, OH 44555-0001
*Prgm Dir:* Kathylynn Feld, MS RN CMA
*Tel:* 330 742-1760   *Fax:* 330 742-2921
*E-mail:* klfeld@ysu.edu
*Med Dir:* Thomas N Detesco, MD

**Zane State College**
Medical Assistant Prgm
1555 Newark Rd
Zanesville, OH 43701
www.zanestate.edu
*Prgm Dir:* Timothy Berger, RN CMA
*Tel:* 740 588-1270   *Fax:* 740 454-0035
*E-mail:* tberger@zanestate.edu
*Med Dir:* Kristofer Sandland, MD

# Oklahoma

**Moore Norman Technology Center**
Medical Assistant Prgm
4701 12th Ave NW
Norman, OK 73069
*Prgm Dir:* Melissa McCoy, MEd
*Tel:* 405 364-5763, Ext 6524   *Fax:* 405 447-6289
*Med Dir:* Mark McKinnon, MD

**Francis Tuttle Technology Center**
Medical Assistant Prgm
12777 N Rockwell Ave
Oklahoma City, OK 73142-2789
www.francistuttle.com
*Prgm Dir:* David E Wiggins, BS RMA CMA
*Tel:* 405 717-4142   *Fax:* 405 717-4789
*E-mail:* dwiggins@francistuttle.com
*Med Dir:* Randy Kakish, MD

**Metro Technology Centers**
Medical Assistant Prgm
1720 Springlake Dr
Oklahoma City, OK 73111-5240
*Prgm Dir:* Pamela Ashley, MEd CMA
*Tel:* 405 605-4644   *Fax:* 405 424-9403
*E-mail:* pam.ashley@metrotech.org
*Med Dir:* Athena Freise, MD

**Tulsa Community College**
Medical Assistant Prgm
909 S Boston Ave
Tulsa, OK 74119-2095
www.tulsacc.edu
*Prgm Dir:* Julie Wood, BS RHIA
*Tel:* 918 595-7321   *Fax:* 918 595-7091
*E-mail:* jwood@tulsacc.edu
*Med Dir:* Timothy Young, MD

# Oregon

**Linn - Benton Community College**
Medical Assistant Prgm
6500 SW Pacific Blvd
Albany, OR 97321-3755
*Prgm Dir:* Rick Durling, BS CPC CMA
*Tel:* 541 917-4288   *Fax:* 541 917-4445
*E-mail:* rick.durling@linnbenton.edu
*Med Dir:* James Gallant, MD

**Central Oregon Community College**
Medical Assistant Prgm
2600 NW College Way
Bend, OR 97701
www.cocc.edu
*Prgm Dir:* Liberty Matthews, RN ADN
*Tel:* 541 383-7292   *Fax:* 541 318-3750
*E-mail:* lmatthews@cocc.edu
*Med Dir:* John Zachem, DO

**Lane Community College**
Medical Assistant Prgm
4000 E 30th Ave
Eugene, OR 97405-0640
*Prgm Dir:* DeLeesa G Meashintubby, RMA AAS
*Tel:* 541 463-5617   *Fax:* 541 463-4151
*E-mail:* meashintubbyd@lanecc.edu
*Med Dir:* Kraig W Jacobson, MD

**Mt Hood Community College**
Medical Assistant Prgm
26000 SE Stark St
Gresham, OR 97030
*Prgm Dir:* Susan A Boulden, RN CMA
*Tel:* 503 491-7136   *Fax:* 503 491-7618
*E-mail:* bouldens@mhcc.edu
*Med Dir:* Keith Klatt, MD

**Clackamas Community College**
Medical Assistant Prgm
19600 S Molalla Ave
Oregon City, OR 97045-8980
www.clackamas.edu
*Prgm Dir:* Maureen Mitchell, RN BSN
*Tel:* 503 657-6958, Ext 2910   *Fax:* 503 655-5153
*E-mail:* maureenm@clackamas.edu
*Med Dir:* Danielle Blackwell, FNP

**Concorde Career Institute - Portland**
Medical Assistant Prgm
1425 NE Irving St
Portland, OR 97232-4203
*Prgm Dir:* Wendy S Campbell, ADN
*Tel:* 503 281-4181, Ext 107   *Fax:* 503 281-6739
*E-mail:* wcampbell@concorde.edu
*Med Dir:* Roddy Tim, MD

**Everest College**
Medical Assistant Prgm
425 SW Washington St
Portland, OR 97204
*Prgm Dir:* Sara L Newman, CMA
*Tel:* 503 222-3225   *Fax:* 503 241-3144
*E-mail:* Snewman@cci.edu
*Med Dir:* Virgil Hamlin, MD

**Heald College - Portland**
Medical Assistant Prgm
625 SW Broadway, Ste 400
Portland, OR 97205
www.heald.edu
*Prgm Dir:* Daniel Kelley McBride, MS PA-C
*Tel:* 503 229-0492, Ext 237   *Fax:* 503 229-0498
*E-mail:* dkelley@Heald.edu
*Med Dir:* Cheri Springer, MS PA-C

**Portland Community College**
Medical Assistant Prgm
12000 SW 49th Ave
Portland, OR 97219-7132
www.pcc.edu
*Prgm Dir:* Denise A Rigsbee, CMA
*Tel:* 503 978-5664   *Fax:* 503 978-5257
*E-mail:* drigsbee@pcc.edu
*Med Dir:* Keith Klatt, MD

# Pennsylvania

**Greater Altoona Career & Technology Center**
Medical Assistant Prgm
1500 Fourth Ave
Altoona, PA 16602
www.gactc.com/cont-ed
*Prgm Dir:* Karen Sybert, MS CMA
*Tel:* 814 946-8469   *Fax:* 814 941-4690
*Med Dir:* J Grant Hormell, MD

**Butler County Community College**
Medical Assistant Prgm
PO Box 1203
Butler, PA 16003-1203
*Prgm Dir:* Deborah B Proctor, EdD RN
*Tel:* 724 287-8711, Ext 8373   *Fax:* 724 285-6047
*E-mail:* debby.proctor@bc3.edu
*Med Dir:* Sherrie Neely, CNP

**Mount Aloysius College**
Medical Assistant Prgm
7373 Admiral Peary Hwy
Cresson, PA 16630-1999
*Prgm Dir:* Cheryl D Kowalczyk, MSN RN CMA
*Tel:* 814 886-6322   *Fax:* 814 886-2978
*E-mail:* ckowalczyk@mtaloy.edu
*Med Dir:* Charles W Stotler, MD

**Tri-State Business Institute**
Medical Assistant Prgm
5757 W 26th St
Erie, PA 16506
*Prgm Dir:* Kelly Lucore
*Tel:* 814 838-7673
*E-mail:* klucore@tsbi.edu
*Med Dir:* Daniel J Barbero, MD

**Harrisburg Area Community College**
Medical Assistant Prgm
One HACC Dr, PC-333
Harrisburg, PA 17110
www.hacc.edu
*Prgm Dir:* Joel L Bacon, BS RRT
*Tel:* 717 221-1326   *Fax:* 717 909-4010
*E-mail:* jlbacon@hacc.edu
*Med Dir:* Wendy Schaenen, MD

**Thompson Institute**
Medical Assistant Prgm
5650 Derry St
Harrisburg, PA 17111
*Prgm Dir:* Barbara Jayne Snyder
*Tel:* 800 272-4632   *Fax:* 717 558-8560
*E-mail:* bsnyder@thompson.edu
*Med Dir:* Emily W Matlin, DO

### Delaware County Community College
Medical Assistant Prgm
Rte 252 and Media Line Rd
Media, PA 19063-1094
www.dccc.edu
*Prgm Dir:* Jennifer E DeCaro, MPH BS CMA
*Tel:* 610 359-5289   *Fax:* 610 359-7350
*E-mail:* jdecaro@dccc.edu
*Med Dir:* John Carl Munshower, MD

### Career Training Academy
Medical Assistant Prgm
950 Fifth Ave
New Kensington, PA 15068
*Prgm Dir:* MaryAgnes Luczak, CMA
*Tel:* 724 337-1000, Ext 12   *Fax:* 724 335-7140
*E-mail:* vp@careerta.com
*Med Dir:* Bernard L Rottschaeffer, MD

### Bucks County Community College
Medical Assistant Prgm
275 Swamp Rd
Newtown, PA 18940
www.bucks.edu
*Prgm Dir:* Marynell T Zieziula, MS BS CMA
*Tel:* 215 968-8341   *Fax:* 215 504-8509
*E-mail:* zieziula@bucks.edu
*Med Dir:* Harvey Lisgar, DO

### Mercyhurst College
Medical Assistant Prgm
North East Campus
16 W Division St
North East, PA 16428
*Prgm Dir:* Mary Ann Lubiejewski, MSN RN
*Tel:* 814 725-6100   *Fax:* 814 725-6112
*E-mail:* mlubiejewski@mercyhurst.edu
*Med Dir:* Jack Yakish, MD

### Community College of Philadelphia
Medical Assistant Prgm
1700 Spring Garden St
Philadelphia, PA 19130-3936
www.ccp.edu
*Prgm Dir:* Deborah Donaldson Rossi, MA CMA
*Tel:* 215 751-8947   *Fax:* 215 751-8937
*E-mail:* drossi@ccp.edu
*Med Dir:* Kimberly Sabadish, MD

### Thompson Institute
Medical Assistant Prgm
3010 Market St
Philadelphia, PA 19104
www.thompson.edu
*Prgm Dir:* Stephanie Suddendorf, CMA AAS
*Tel:* 215 594-4034   *Fax:* 215 594-4093
*E-mail:* ssuddendorf@thompson.edu
*Med Dir:* Randall Drain, MD

### Bradford School - Pittsburgh
Medical Assistant Prgm
125 West Station Square Dr, Ste 129
Pittsburgh, PA 15219
www.bradfordpittsburgh.edu
*Prgm Dir:* Cynthia Boles, CMA MBA MT(ASCP)
*Tel:* 412 391-6366   *Fax:* 412 471-6714
*E-mail:* cboles@mail.com
*Med Dir:* Peter Dickinson, MD

### Everest Institute
Medical Assistant Prgm
100 Forbes Ave, Ste 1200
Pittsburgh, PA 15222
*Prgm Dir:* Peggy Populo, BS
*Tel:* 412 261-4520, Ext 219   *Fax:* 412 261-4546
*E-mail:* mpopulo@cci.edu
*Med Dir:* Supritha Shetty, MD

### ICM School of Business & Medical Careers
Medical Assistant Prgm
10 Wood St
Pittsburgh, PA 15222
*Prgm Dir:* Lynn G Slack, CMA
*Tel:* 412 261-2647, Ext 253   *Fax:* 412 261-6491
*E-mail:* lslack@icmschool.com
*Med Dir:* Peter Gagianas, MD

### Montgomery County Community College
Medical Assistant Prgm
101 College Dr
Pottstown, PA 19464
www.mc3.edu
*Prgm Dir:* Kathleen Schreiner, RN BS MSHA
*Tel:* 610 718-1812   *Fax:* 610 718-1983
*E-mail:* kschreiner@mc3.edu
*Med Dir:* Ruth Benfield, RNP

### Lehigh Carbon Community College
Medical Assistant Prgm
4525 Education Park Dr
Schnecksville, PA 18078-2502
www.lccc.edu
*Prgm Dir:* Judith K Ehninger, BS RN CMA
*Tel:* 610 799-1516   *Fax:* 610 799-1527
*E-mail:* jehninger@lccc.edu
*Med Dir:* Susan Kostenblatt, MD

### Central Pennsylvania College
Medical Assistant Prgm
Campus on College Hill
Summerdale, PA 17093-0309
*Prgm Dir:* Nikki Marhefka, EdM MT(ASCP) CMA
*Tel:* 717 728-2216   *Fax:* 717 732-5254
*E-mail:* nikkimarhefka@centralpenn.edu
*Med Dir:* Charles E Darowish, DO

### Penn Commercial Inc
Medical Assistant Prgm
242 Oak Spring Rd
Washington, PA 15301
*Prgm Dir:* Betty A Shingle, H ASCP BS MT
*Tel:* 724 222-5330, Ext 240   *Fax:* 724 222-4722
*E-mail:* bshingle@penncommerical.net
*Med Dir:* Anthony S Galletta, MD

### Comm Coll of Allegheny County
Medical Assistant Prgm
1750 Clairton Rd
West Mifflin, PA 15122
www.ccac.edu
*Prgm Dir:* Bonnie Gregg, RN MSEd
*Tel:* 412 469-6324   *Fax:* 412 237-4521
*E-mail:* bgregg@ccac.edu
*Med Dir:* Anita Edwards, MD

### Berks Technical Institute
Medical Assistant Prgm
2205 Ridgewood Rd
Wyomissing, PA 19610
*Prgm Dir:* Cheryl Garman, RN
*Tel:* 610 372-1722, Ext 151   *Fax:* 610 376-4684
*E-mail:* cgarman@berkstech.com
*Med Dir:* David Robel, MD

### Westmoreland County Community College
Medical Assistant Prgm
400 Armbrust Rd
Youngwood, PA 15697-1895
*Prgm Dir:* Patricia E Mihalcin, RN PhD
*Tel:* 724 925-4028   *Fax:* 724 925-4293
*E-mail:* mihalcinp@wccc-pa.edu
*Med Dir:* Frank McGrogan, MD

# South Carolina

### Aiken Technical College
Medical Assistant Prgm
PO Drawer 696
Aiken, SC 29802-0696
www.atc.edu
*Prgm Dir:* Carolyn Barnett, BSN
*Tel:* 803 593-9231, Ext 1654   *Fax:* 803 593-3106
*E-mail:* barnett@atc.edu
*Med Dir:* Michael Vasovski, MD

### Forrest Junior College
Medical Assistant Prgm
601 E River St
Anderson, SC 29624-2498
*Prgm Dir:* Emilie A Nelson, RN BS AHI
*Tel:* 864 225-7653, Ext 310   *Fax:* 864 261-7471
*E-mail:* emilienelson@forrestcollege.com
*Med Dir:* Jeanette C Kinsey, MD

### Miller-Motte Technical College
Medical Assistant Prgm
8085 Rivers Ave
Charleston, SC 29406
www.miller-motte.com
*Prgm Dir:* Patti Steele, MA RN BSN
*Tel:* 843 574-0101, Ext 43   *Fax:* 843 266-3424
*E-mail:* psteele@miller-motte.net
*Med Dir:* Byron Williams, MD

### Trident Technical College
Medical Assistant Prgm
7000 Rivers Ave
PO Box 118067
Charleston, SC 29423
www.tridenttech.edu
*Prgm Dir:* Deborah L White, CMA MS HPE
*Tel:* 803 574-6103   *Fax:* 803 574-6585
*E-mail:* deborah.white@tridenttech.edu
*Med Dir:* Sewell Kahn, MD

### South University
Medical Assistant Prgm
3810 Main St
Columbia, SC 29203
*Prgm Dir:* Lorraine E Schoenbeck, BS CMA
*Tel:* 803 376-5163   *Fax:* 803 799-9083
*E-mail:* lschoenbeck@southuniversity.edu
*Med Dir:* Rachel W Vidal, MD

### Greenville Technical College
Medical Assistant Prgm
PO Box 5616
Greenville, SC 29606-5616
*Prgm Dir:* Janis Crowe, ST CMA
*Tel:* 864 250-8431   *Fax:* 864 250-8806
*E-mail:* Janis.Crowe@gvltec.edu
*Med Dir:* John Johnson, MD

### Piedmont Technical College
Medical Assistant Prgm
PO Box 1467
620 N Emerald Rd
Greenwood, SC 29648-1467
www.ptc.edu
*Prgm Dir:* Gail Rumfelt, CMA CPC
*Tel:* 864 941-8526   *Fax:* 864 941-8684
*E-mail:* rumfelt.g@ptc.edu
*Med Dir:* Dan Robinson, MD

### Orangeburg Calhoun Technical College
Medical Assistant Prgm
3250 St Matthews Rd
Orangeburg, SC 29118
*Prgm Dir:* Sharon Cheek, MLT (ASCP)
*Tel:* 803 535-1346, Ext 1346   *Fax:* 803 535-1350
*E-mail:* cheeks@octech.edu
*Med Dir:* Willie B Louis, MD

**Tri-County Technical College**
Medical Assistant Prgm
Hwy 76, PO Box 587
Pendleton, SC 29670
*Prgm Dir:* Kaye Bathe, BSHA CMA
*Tel:* 864 646-8361, Ext 1352 *Fax:* 864 646-1892
*E-mail:* kbathe@tctc.edu
*Med Dir:* Wajdi D Bouk, MD

**Spartanburg Community College**
Medical Assistant Prgm
PO Box 4386
Spartanburg, SC 29305-4386
www.stcsc.edu
*Prgm Dir:* Donna I Buchanan, MLT CMA (ASCP) BHS
*Tel:* 864 592-4791 *Fax:* 864 591-3881
*E-mail:* buchanand@stcsc.edu
*Med Dir:* Darwin Keller, MD

**Central Carolina Technical College**
Medical Assistant Prgm
506 N Guignard Dr
Sumter, SC 29150
www.cctech.edu
*Prgm Dir:* Micheline B Wheeler, RN
*Tel:* 803 778-7809 *Fax:* 803 778-7868
*E-mail:* wheelermb@cctech.edu
*Med Dir:* Phillip Latham, Jr, MD

**Midlands Technical College**
Medical Assistant Prgm
Airport Campus
1260 Lexington Dr
West Columbia, SC 29170
www.midlandstech.edu
*Prgm Dir:* Linda Rollison, MPH
*Tel:* 803 822-3398 *Fax:* 803 822-3417
*E-mail:* rollisonl@midlandstech.edu
*Med Dir:* Michael P Harris, MD

# South Dakota

**Presentation College**
Medical Assistant Prgm
1500 N Main St
Aberdeen, SD 57401
*Prgm Dir:* Mary Gjernes, AS MLT CMA
*Tel:* 605 229-8544
*E-mail:* Mary.Gjernes@presentation.edu
*Med Dir:* Theresa Cameron, PA

**Mitchell Technical Institute**
Medical Assistant Prgm
821 N Capital
Michell, SD 57301
www.mitchelltech.edu
*Prgm Dir:* Cori Hoffman, RN BSN CMA
*Tel:* 605 995-3024, Ext 3002 *Fax:* 605 996-3299
*E-mail:* Cori.Hoffman@mitchelltech.edu
*Med Dir:* Lucio Margallo, MD

**Colorado Technical University**
Medical Assistant Prgm
3901 W 59th St
Sioux Falls, SD 57108
*Prgm Dir:* Brenda K Frerichs, MS BS CMA
*Tel:* 605 361-0200 *Fax:* 605 361-5954
*E-mail:* bfrerichs@sf.coloradotech.edu
*Med Dir:* Mary Wuebben, PA-C

**National American Univ - Sioux Falls Campus**
Medical Assistant Prgm
2801 S Kiwanis Ave, Ste 100
Sioux Falls, SD 57105-0889
www.national.edu/sioux_falls.html
*Prgm Dir:* Cathie Ogdie, MS DLM MT(ASCP)
*Tel:* 605 336-4695 *Fax:* 605 336-4605
*E-mail:* cogdie@national.edu
*Med Dir:* Peter Johnson, MD

**Lake Area Technical Institute**
Medical Assistant Prgm
230 11th St NE
PO Box 730
Watertown, SD 57201-0730
www.lakeareatech.edu
*Prgm Dir:* Kris Lindahl, CMA
*Tel:* 800 657-4344, Ext 217 *Fax:* 605 882-6299
*E-mail:* lindahlk@lakeareatech.edu
*Med Dir:* Aaron B Shives, MD

# Tennessee

**Northeast State Technical Comm College**
Medical Assistant Prgm
PO Box 246, 2425 Hwy 75
Blountville, TN 37617-0246
*Prgm Dir:* Ginger Burleson, RN
*Tel:* 423 354-2526, Ext 3326 *Fax:* 423 323-0213
*E-mail:* gkburleson@northeaststate.edu
*Med Dir:* Robert L Schubert, MD

**National College of Business & Technology - Bristol**
Medical Assistant Prgm
Tri-Cities Campus
1328 Hwy 11 W
Bristol, TN 37620
*Prgm Dir:* Sheri Elayne Jessee, AS LPN RMA
*Tel:* 423 878-4440 *Fax:* 423 793-1060
*E-mail:* shjesse@ncbt.edu

**Chattanooga State Technical Comm College**
Medical Assistant Prgm
4501 Amnicola Hwy
Chattanooga, TN 37406
www.chattanoogastate.edu/Industrial_Technology/
intprog.asp
*Prgm Dir:* Alexis D Jenkins, MEd MT(ASCP) CMA
*Tel:* 423 697-4438 *Fax:* 423 697-2413
*E-mail:* alexis.jenkins@chattanoogastate.edu
*Med Dir:* Robert Younger, MD

**Miller-Motte Technical College - Chattanooga**
Medical Assistant Prgm
6020 Shallowford Rd
Chattanooga, TN 37421
*Prgm Dir:* Melissa Harper
*Tel:* 423 510-9675
*E-mail:* mharper@miller-motte.com

**Miller-Motte Technical College - Clarksville**
Medical Assistant Prgm
1820 Business Park Dr
Clarksville, TN 37040-0415
*Prgm Dir:* Sandra Pigg, RN BSN
*Tel:* 931 553-0071 *Fax:* 931 552-2916
*E-mail:* spigg@mmtcclk.com
*Med Dir:* Jamie Traylor, RN FNP

**Cleveland State Community College**
Medical Assistant Prgm
Office Systems Admin-Medical Ofc Asst
PO Box 3570
Cleveland, TN 37320-3570
www.clevelandstatecc.edu
*Prgm Dir:* Karmon Kingsley, CMA CHI
*Tel:* 423 614-8702, Ext 702 *Fax:* 423 478-6258
*E-mail:* kkingsley@clevelandstatecc.edu
*Med Dir:* John M Powell, MD

**National College of Business & Technology - Knoxville**
Medical Assistant Prgm
8415 Kingston Pike
Knoxville, TN 37919
www.ncbt.edu
*Prgm Dir:* Rhonda Epps, AAS CMA
*Tel:* 865 539-2011
*E-mail:* repps@ncbt.edu

**South College**
Medical Assistant Prgm
3904 Lonas Dr
Knoxville, TN 37909
www.southcollegetn.edu
*Prgm Dir:* Anna Fritz, MPH MT(ASCP)
*Tel:* 865 251-1843 *Fax:* 865 584-9816
*E-mail:* afritz@southcollegetn.edu
*Med Dir:* Charles E Leonard, MD

**Tennessee Technology Center - Knoxville**
Medical Assistant Prgm
1100 Liberty St
Knoxville, TN 37919
www.knoxville.tec.tn.us
*Prgm Dir:* Donna Williams, LPN RMA
*Tel:* 423 546-5567, Ext 145 *Fax:* 423 971-4474
*E-mail:* dwilliams@knoxville.tec.tn.us
*Med Dir:* Randal Hartline, MD

**Tennessee Technology Center - McMinnville**
Medical Assistant Prgm
241 Vo-Tech Dr
McMinnville, TN 37110
*Prgm Dir:* Deborah Womack, AAS LPN RMA
*Tel:* 931 473-5587, Ext 248 *Fax:* 931 473-6380
*E-mail:* dwomack@mcminnville.tec.tn.us
*Med Dir:* Ty Webb, MD

**Concorde Career College - Memphis**
Medical Assistant Prgm
5100 Poplar Ave, Ste 132
Memphis, TN 38137
*Prgm Dir:* Amie Ray, CMA MA MS
*Tel:* 901 761-9494, Ext 243 *Fax:* 901 761-3293
*E-mail:* aray@concordecareercolleges.com
*Med Dir:* Rosalind Cropper, MD

**Tennessee Technology Center - Memphis**
Medical Assistant Prgm
550 Alabama Ave
Memphis, TN 38105
*Prgm Dir:* Cynthia Farley
*Tel:* 901 543-6172 *Fax:* 901 543-6126
*E-mail:* cynthia.farley@memphis.tec.tn.us
*Med Dir:* H S Gibbs, MD

**National College of Business & Technology - Nashville**
Medical Assistant Prgm
3748 Nolensville Pike
Nashville, TN 37211
www.ncbt.edu
*Prgm Dir:* A'Donni Samuels, CMA AAS
*Tel:* 615 333-3344 *Fax:* 615 333-3429
*E-mail:* asamuels@ncbt.edu

# Texas

**Cisco Junior College**
Medical Assistant Prgm
717 E Industrial Blvd
Abilene, TX 79602
www.cisco.cc.tx.us
*Prgm Dir:* Kelly Meyer, BS CPhT
*Tel:* 325 794-4441 *Fax:* 254 442-2546
*E-mail:* kmeyer@cisco.cc.tx.us
*Med Dir:* Leigh Talliaferro, MD

**El Centro College**
Medical Assistant Prgm
801 Main St
Dallas, TX 75202
www.elcentrocollege.edu
*Prgm Dir:* Pat Moeck, PhD MBA CMA
*Tel:* 214 860-2328 *Fax:* 214 860-2268
*E-mail:* pgm5704@dcccd.edu
*Med Dir:* W H Glaze Sr, MD

**Richland College**
Medical Assistant Prgm
12800 Abrams Rd
Dallas, TX 75243-2199
www.richlandcollege.edu/hp
*Prgm Dir:* Shannon Ydoyaga, BBA
*Tel:* 972 238-6117   *Fax:* 972 761-6793
*E-mail:* shannony@dcccd.edu
*Med Dir:* Michael Maxwell, MD

**Career Centers of Texas**
Medical Assistant Prgm
8360 Burnham Rd, Ste 100
El Paso, TX 79907
www.cctep.com
*Prgm Dir:* Linda Villegas, CMA LVN RMA
*Tel:* 915 595-1935   *Fax:* 915 595-6619
*E-mail:* doe@cctep.com
*Med Dir:* Phillippe Petrus, NCCPA BS

**Computer Career Center**
Medical Assistant Prgm
6101 Montana Ave
El Paso, TX 79925-2021
*Prgm Dir:* James Baird, MBA CAHI
*Tel:* 915 779-8031, Ext 11   *Fax:* 915 779-1683
*E-mail:* jbaird@computercareercenter.com
*Med Dir:* Ruben Ramos, BS NCPT

**El Paso Community College**
Medical Assistant Prgm
PO Box 20500
El Paso, TX 79998-0500
*Prgm Dir:* Cathy D Soto, MBA CMA
*Tel:* 915 831-4139   *Fax:* 915 831-4021
*E-mail:* CathyS@epcc.edu
*Med Dir:* M T Lam, MD

**Western Technical College**
Medical Assistant Prgm
9451 Diana Dr
El Paso, TX 79924
www.wtc-ep.edu
*Prgm Dir:* Amy Martin, BS CMA
*Tel:* 915 566-9621, Ext 118   *Fax:* 915 565-9903
*E-mail:* amartin@wtc-ep.edu
*Med Dir:* Kareh Jorge, MD

**Texas School of Business Friendswood**
Medical Assistant Prgm
3208 FM 528
Friendswood, TX 77546
www.tsb.edu
*Prgm Dir:* Bobby Wilmore, BS
*Tel:* 281 648-0880, Ext 3301   *Fax:* 281 648-0821
*E-mail:* bwilmore@tsb.edu
*Med Dir:* Mary Jo Larose, BA, CMA

**Texas State Technical College - Harlingen**
Medical Assistant Prgm
1902 N Loop 499
Harlingen, TX 78550-3697
*Prgm Dir:* Jean A Lashbrook, RN
*Tel:* 956 364-4797   *Fax:* 956 364-5160
*E-mail:* jean.lashbrook@harlingen.tstc.edu
*Med Dir:* Delfina Ortega, RNC FNP

**Bradford School**
Medical Assistant Prgm
4669 SW Freeway, Ste 300
Houston, TX 77027
www.bradfordschoolhouston.com
*Prgm Dir:* Betsy Roxanne James, NR CMA
*Tel:* 713 629-1500   *Fax:* 713 629-0059
*E-mail:* bjames@bradfordschoolhouston.edu
*Med Dir:* Kirk Lee, MD

**Houston Community College**
Medical Assistant Prgm
1900 Pressler, Ste 434
Houston, TX 77030-3717
www.hccs.edu
*Prgm Dir:* Tomye Geringer, BS RN CMA
*Tel:* 713 718-7359   *Fax:* 713 718-7521
*E-mail:* tomye.geringer@hccs.edu
*Med Dir:* C James Chuong, MD

**San Jacinto College North**
Medical Assistant Prgm
5800 Uvalde Rd
Houston, TX 77049-4599
www.sjcd.edu
*Prgm Dir:* Diana Houston, AA
*Tel:* 281 459-5410   *Fax:* 281 459-7640
*E-mail:* diana.houston@sjcd.edu
*Med Dir:* George A Brooks, MD

**Texas School of Business - North Campus**
Medical Assistant Prgm
711 E Airtex
Houston, TX 77073
www.tsb.edu
*Prgm Dir:* Mary Jo LaRose, CMA
*Tel:* 281 443-8900, Ext 224   *Fax:* 281 209-9422
*E-mail:* mlarose@tsb.edu
*Med Dir:* Mary Faye Hewitt, MD

**Texas School of Business Southwest**
Medical Assistant Prgm
6363 Richmond Ave, Ste 300
Houston, TX 77057
www.tsb.edu
*Prgm Dir:* Mary Jo LaRose, BA CMA
*Tel:* 281 443-8900, Ext 224   *Fax:* 281 209-9422
*E-mail:* mlarose@tsb.edu
*Med Dir:* Mary Hewitt, MD

**Kilgore College**
Medical Assistant Prgm
1100 Broadway
Kilgore, TX 75662-3299
www.kilgore.edu
*Prgm Dir:* Millie M White, MS MT(AMT) CMA
*Tel:* 903 983-8147, Ext 147   *Fax:* 903 988-7596
*E-mail:* mwhite@kilgore.edu
*Med Dir:* Michael W McShan, MD PhD

**Hallmark Institute of Technology**
Medical Assistant Prgm
10401 IH - 10 West
San Antonio, TX 78230
*Prgm Dir:* Jane Peek
*Tel:* 210 690-9000
*E-mail:* jpeek@hallmarkinstitute.com

**San Antonio College**
Medical Assistant Prgm
1300 San Pedro Ave
San Antonio, TX 78212
*Prgm Dir:* Cheryl Startzell, MA BS CMA
*Tel:* 210 733-2437   *Fax:* 210 733-2429
*E-mail:* cstartze@accd.edu
*Med Dir:* Michael Centeno, MD

**San Antonio College of Med & Dental Assts - San Antonio**
Medical Assistant Prgm
4205 San Pedro
San Antonio, TX 78212
*Prgm Dir:* Gilbert DeLeon, MA
*Tel:* 210 733-0777   *Fax:* 210 735-2431
*E-mail:* gdeleon@sac-mda.com

# Utah

**Davis Applied Technology College**
Medical Assistant Prgm
550 East 300 South
Kaysville, UT 84037-2699
*Prgm Dir:* Richelle Gaiter, CMA
*Tel:* 801 593-2369   *Fax:* 801 593-2400
*E-mail:* gaiterrm@datc.tec.ut.us
*Med Dir:* Michael Kirkham, MD

**Bridgerland Applied Technology College**
Medical Assistant Prgm
1301 North 600 West
Logan, UT 84321
*Prgm Dir:* Mindi Wright, BS AS RN
*Tel:* 435 750-3047
*E-mail:* mwright@batc.tec.ut.us
*Med Dir:* Bruce A O'very, MD

**Ogden - Weber Applied Technology College**
Medical Assistant Prgm
200 N Washington Blvd
Ogden, UT 84404-4089
www.owatc.edu
*Prgm Dir:* Emma Anderson, CMA LPT
*Tel:* 801 627-8445   *Fax:* 801 395-3703
*E-mail:* andersoe@owatc.edu
*Med Dir:* N Brent Williams, MD

**Stevens Henager College**
Medical Assistant Prgm
PO Box 9428
Ogden, UT 84409
*Prgm Dir:* Mario Merida, MD
*Tel:* 801 394-7791, Ext 1044   *Fax:* 801 621-0853
*E-mail:* drmerida@email.com
*Med Dir:* Kathryn J Swoboda, MD

**Ameritech College**
Medical Assistant Prgm
1675 N Freedom Blvd, Bldg 3
Provo, UT 84604
www.ameritech.edu
*Prgm Dir:* Beverly Roberts, CMA
*Tel:* 801 377-2900, Ext 116   *Fax:* 801 375-3077
*E-mail:* broberts@ameritech.edu
*Med Dir:* Kenneth Crump, MD

**Provo College**
Medical Assistant Prgm
1450 W 820 N
Provo, UT 84601
*Prgm Dir:* Leslie E Ekker, BS CMA
*Tel:* 801 375-1861   *Fax:* 801 375-9728
*E-mail:* lesliee@provocollege.edu

**Latter Day Saints Business College**
Medical Assistant Prgm
411 E South Temple
Salt Lake City, UT 84111
*Prgm Dir:* Edith A Hamelin, BSN RN CMA
*Tel:* 801 524-8131   *Fax:* 801 524-1900
*E-mail:* edith@ldsbc.edu
*Med Dir:* Don Stromquist, MD

**Salt Lake Community College**
Medical Assistant Prgm
4600 S Redwood Rd, PO Box 30808
Salt Lake City, UT 84130
www.slcc.edu
*Prgm Dir:* Jana W Tucker, CMA/LPRT
*Tel:* 801 957-4090, Ext 4596   *Fax:* 801 957-5708
*E-mail:* jana.tucker@slcc.edu
*Med Dir:* Dennis Hamp, DO

**Utah Career College**
Medical Assistant Prgm
1902 W 7800 S
West Jordan, UT 84088
www.utahcollege.edu
*Prgm Dir:* Carrie Hammond, CMA
*Tel:* 801 304-4224, Ext 131 *Fax:* 801 304-4229
*E-mail:* chammond@utahcollege.edu
*Med Dir:* Charles Canfield, MD

**Everest College**
Medical Assistant Prgm
3280 West 3500 South
West Valley City, UT 84119
*Prgm Dir:* Michael Green, MS
*Tel:* 801 840-4800, Ext 133 *Fax:* 801 969-0828
*E-mail:* migreen@cci.edu
*Med Dir:* Chris Hutchison, MD

# Virginia

**National College of Business & Technology - Bluefield**
Medical Assistant Prgm
PO Box 629
100 Logan St
Bluefield, VA 24605
www.ncbt.edu
*Prgm Dir:* Cathy Kelley-Arney, BSHS AS CMA MLT
*Tel:* 276 326-3621, Ext 19 *Fax:* 276 322-5731
*E-mail:* carney@ncbt.edu
*Med Dir:* Melissa Nickels-Stacy, PA-C

**National College of Business & Technology - Charlottesville**
Medical Assistant Prgm
1819 Emmet St
Charlottesville, VA 22903
www.ncbt.edu
*Prgm Dir:* Jacki Rooke, RN RMA
*Tel:* 434 295-0136 *Fax:* 434 979-8061
*E-mail:* jrooke@ncbt.edu
*Med Dir:* Holly Johnson, MD

**National College of Business & Technology - Danville**
Medical Assistant Prgm
336 Old Riverside Dr
Danville, VA 24541
www.ncbt.edu
*Prgm Dir:* Gail Orr, RMA
*Tel:* 434 793-6822 *Fax:* 434 793-3634
*E-mail:* gorr@ncbt.edu
*Med Dir:* Jacob Moll, MD

**National College of Business & Technology - Harrisonburg**
Medical Assistant Prgm
51-B Burgess Rd
Harrisonburg, VA 22801
www.ncbt.edu
*Prgm Dir:* Lurline Breeden, LPN
*Tel:* 540 432-0943 *Fax:* 540 432-1133
*E-mail:* lbreeden@ncbt.edu
*Med Dir:* Stephen Phillips, MD

**Miller-Motte Technical College**
Medical Assistant Prgm
1011 Creekside Ln
Lynchburg, VA 24502
*Prgm Dir:* Dee Zachau, CMA
*Tel:* 434 239-5222, Ext 214 *Fax:* 434 239-1069
*E-mail:* deegzachau@yahoo.com
*Med Dir:* Craig Petry, MD

**National College of Business & Technology - Lynchburg**
Medical Assistant Prgm
104 Candlewood Court
Lynchburg, VA 24502
www.ncbt.edu
*Prgm Dir:* Carolyn Sue Coleman, CMA AAS
*Tel:* 434 239-3500 *Fax:* 434 239-3948
*E-mail:* scoleman@ncbt.edu
*Med Dir:* Peter Houck, MD

**National College of Business & Technology - Martinsville**
Medical Assistant Prgm
10 Church St
PO Box 232
Martinsville, VA 24114
http://ncbt.edu
*Prgm Dir:* Gary D Jenkins, RN
*Tel:* 276 632-5621 *Fax:* 276 632-7915
*E-mail:* gdjenkins@ncbt.edu
*Med Dir:* Jarrett Sell, MD

**Medical Careers Institute**
Medical Assistant Prgm
1001 Omni Blvd, Ste 200
Newport News, VA 23606
*Prgm Dir:* Robyn Gohsman, AAS RMA
*Tel:* 757 873-2423, Ext 318 *Fax:* 757 873-2472
*E-mail:* rgohsman@medical.edu
*Med Dir:* Barbara Smith-Haywood, MSN RN C OGNP PNP

**Bryant & Stratton College - Richmond**
Medical Assistant Prgm
8141 Hull St Rd
Richmond, VA 23235-6411
*Prgm Dir:* Nina Beaman, MS RNC CMA
*Tel:* 804 745-2444, Ext 328 *Fax:* 804 745-6884
*E-mail:* nabeaman@bryantstratton.edu
*Med Dir:* Kevin Harvey, MD

**National College of Business & Technology - Roanoke Valley**
Medical Assistant Prgm
1813 E Main St
Salem, VA 24153
http://ncbt.edu
*Prgm Dir:* M J Williams, RN RMA
*Tel:* 540 986-1800
*E-mail:* mjwilliams@ncbt.edu
*Med Dir:* Sharon Bottomley, NP

**Bryant & Stratton College - Virginia Beach**
Medical Assistant Prgm
301 Centre Pointe Dr
Virginia Beach, VA 23462
*Prgm Dir:* Cornelia C Mutts, BSN RN CMA
*Tel:* 757 499-7900 *Fax:* 757 499-9977
*E-mail:* cwmutts@bryantstratton.edu
*Med Dir:* Kamal LouKa, MD

**Medical Careers Institute**
Medical Assistant Prgm
5501 Greenwich Rd
Virginia Beach, VA 23462
www.medical.edu
*Prgm Dir:* Larry Hill, BS RMA
*Tel:* 757 497-8400, Ext 294 *Fax:* 757 497-8493
*E-mail:* lahill@medical.edu
*Med Dir:* Herbert Knight, MD

**Tidewater Community College**
Medical Assistant Prgm
1700 College Crescent
Virginia Beach, VA 23453-1918
*Prgm Dir:* Kathleen Carol McNamara, RN BA CMA
*Tel:* 757 822-7252 *Fax:* 757 427-1338
*E-mail:* kmcnamara@tcc.edu
*Med Dir:* Joseph McDermott, MD FCAS

# Washington

**Whatcom Community College**
Medical Assistant Prgm
237 W Kellogg Rd
Bellingham, WA 98226-8033
www.whatcom.ctc.edu
*Prgm Dir:* Barbara Dahl, CMA CPC
*Tel:* 360 676-2170, Ext 3306 *Fax:* 360 752-6767
*E-mail:* bdahl@whatcom.ctc.edu
*Med Dir:* Dana Petersen, Medical Doctor

**Olympic College**
Medical Assistant Prgm
1600 Chester Ave
Bremerton, WA 98337-1699
*Prgm Dir:* Connie Lieseke, CMA MLT (ASCP)
*Tel:* 360 475-7741 *Fax:* 360 475-7491
*E-mail:* clieseke@olympic.edu
*Med Dir:* Sally Fleischman, Physician and Surgeon

**Highline Community College**
Medical Assistant Prgm
PO Box 98000
Des Moines, WA 98198-9800
http://flightline.highline.edu/medassist
*Prgm Dir:* Barbara J Cerna, CMA
*Tel:* 206 878-3710, Ext 3493 *Fax:* 206 870-4850
*E-mail:* bcerna@highline.edu
*Med Dir:* Jackie Shuey, PA-C

**Bryman College - Everett**
Medical Assistant Prgm
906 SE Everett Mall Way, Ste 600
Everett, WA 98208
*Prgm Dir:* Gerry L Landes, CMA
*Tel:* 425 789-7960 *Fax:* 425 789-7989
*E-mail:* landes.g@cci.edu
*Med Dir:* Louie Figueroa, PA-C

**Everett Community College**
Medical Assistant Prgm
2000 Tower St
Everett, WA 98201-1352
www.everettcc.edu
*Prgm Dir:* Elizabeth Adolphsen, BA CMA
*Tel:* 425 388-9467 *Fax:* 425 388-9135
*E-mail:* eadolphsen@everettcc.edu
*Med Dir:* Annalea Gunn, MD

**Lake Washington Technical College**
Medical Assistant Prgm
11605 132nd Ave NE
Kirkland, WA 98034-8506
www.lwtc.ctc.edu
*Prgm Dir:* Gail Muller, RN
*Tel:* 425 739-8283 *Fax:* 425 739-8298
*E-mail:* gail.muller@lwtc.edu
*Med Dir:* Eric Stout, PA-C

**Clover Park Technical College**
Medical Assistant Prgm
4500 Steilacoom Blvd SW
Lakewood, WA 98499
*Prgm Dir:* Michele Jones, CMA AS
*Tel:* 253 589-5841 *Fax:* 253 589-5866
*E-mail:* michele.jones@cptc.edu
*Med Dir:* Ron Morris, MD

**Lower Columbia College**
Medical Assistant Prgm
1600 Maple St
Longview, WA 98632
www.lcc.ctc.edu
*Prgm Dir:* Jerri Weyer, MT (ASCP)
*Tel:* 360 442-2889 *Fax:* 360 442-2879
*E-mail:* jweyer@lcc.ctc.edu
*Med Dir:* Rheinhold Ayoub MD, Pediatrics and Family Health

**Bryman College**
Medical Assistant Prgm
10920 33rd Ave West - Suite 250
Lynnwood, Wa 98036
*Prgm Dir:* Andrea Dorn, BA RMA CUA NCMA
*Tel:* 425 778-9894
*E-mail:* ADorn@cci.edu

**Skagit Valley College**
Medical Assistant Prgm
2405 E College Way
Mt Vernon, WA 98273
*Prgm Dir:* Jeanette Hemming, AAS CMAAC RN
*Tel:* 360 416-7720   *Fax:* 360 416-7918
*E-mail:* jeanette.hemming@skagit.edu
*Med Dir:* Paul Williams, MD

**South Puget Sound Community College**
Medical Assistant Prgm
2011 Mottman Rd SW
Olympia, WA 98512-6218
www.spscc.ctc.edu
*Prgm Dir:* Sandra Thunell, CMA
*Tel:* 360 596-5297   *Fax:* 360 664-0780
*E-mail:* sthunell@spscc.ctc.edu
*Med Dir:* Lowell R Dightman, MD

**Bryman College - Port Orchard**
Medical Assistant Prgm
3649 Frontage Rd
Port Orchard, WA 98367
*Prgm Dir:* Lisa Cook, CMA
*Tel:* 360 473-1120   *Fax:* 360 792-2404
*E-mail:* lcook@cci.edu
*Med Dir:* Narinder Duggal, MD

**Bryman College - Renton**
Medical Assistant Prgm
981 Powell Ave SW, Ste 200
Renton, WA 98055-2990
*Prgm Dir:* Amanda Gaugler, BSc
*Tel:* 425 255-3281   *Fax:* 425 255-9327
*E-mail:* agaugler@cci.edu
*Med Dir:* Eric Lui, MD

**Renton Technical College**
Medical Assistant Prgm
3000 NE Fourth St
Renton, WA 98056
*Prgm Dir:* Terri A Calnan
*Tel:* 425 235-2352, Ext 5734   *Fax:* 425 235-7832
*E-mail:* tcalnan@rtc.ctc.edu

**North Seattle Community College**
Medical Assistant Prgm
9600 College Way N
Seattle, WA 98103-3599
www.northseattle.edu/health/medasst
*Prgm Dir:* Deborah J Bedford, AAS CMA
*Tel:* 206 528-4561   *Fax:* 206 527-3715
*E-mail:* dbedford@sccd.ctc.edu
*Med Dir:* Leslie Vietmeier, ARNP

**Seattle Vocational Institute**
Medical Assistant Prgm
2120 S Jackson St
Seattle, WA 98144
www.sviweb.sccd.ctc.edu
*Prgm Dir:* Patricia A Larson, CMA
*Tel:* 206 587-4934   *Fax:* 206 587-4939
*E-mail:* plarson@sccd.ctc.edu
*Med Dir:* Rayburn Lewis, MD

**Spokane Community College**
Medical Assistant Prgm
1810 N Greene St
Spokane, WA 99217-5399
*Prgm Dir:* Carolyn Beth Thomas, CMA BA
*Tel:* 509 533-8138   *Fax:* 509 533-8621
*E-mail:* cthomas@scc.spokane.edu

**Clark College**
Medical Assistant Prgm
1800 E McLoughlin Blvd
Vancouver, WA 98663
www.clark.edu
*Prgm Dir:* John M Clausen, BS RHIA
*Tel:* 360 992-2468   *Fax:* 360 992-2863
*E-mail:* jclausen@clark.edu
*Med Dir:* Keith Klatt, MD

**Everest College**
Medical Assistant Prgm
120 NE 136 Ave, Ste 130
Vancouver, WA 98684
www.cci.edu
*Prgm Dir:* Patricia Stoddard, BA CMA AMT ARRT RT-ARRT
*Tel:* 360 254-3282   *Fax:* 360 254-3035
*E-mail:* pstoddard@cci.edu
*Med Dir:* John Sobeck, MD

**Wenatchee Valley College**
Medical Assistant Prgm
1300 Fifth Ave
Wenatchee, WA 98801
*Prgm Dir:* Jan Kaiser, RN
*Tel:* 509 682-6667   *Fax:* 509 682-6661
*E-mail:* seagle@wvcmail.ctc.edu
*Med Dir:* Andrew Toth, MD

**Yakima Valley Community College**
Medical Assistant Prgm
PO Box 22520
Yakima, WA 98907
*Prgm Dir:* Shirley Mohsenian, BSN
*Tel:* 509 574-4847   *Fax:* 509 574-6877
*E-mail:* smohsenian@yvcc.edu
*Med Dir:* John Place, MD

# West Virginia

**Mountain State University**
Medical Assistant Prgm
PO Box 9003
Beckley, WV 25802-9003
*Prgm Dir:* Mary Frances Hash, MA CMA
*Tel:* 304 929-1423   *Fax:* 304 929-1600
*E-mail:* fhash@mountainstate.edu
*Med Dir:* Husam Nazer, MD

**Benjamin Franklin Career & Technical Ed Ctr**
Medical Assistant Prgm
500 28th St
Dunbar, WV 25064
*Prgm Dir:* Eva Shirley Longsworth, RT(R) CMA AC MS
*Tel:* 304 766-0369   *Fax:* 304 766-0371
*E-mail:* elongsworth@kcs.kana.k12.wv.us

**Huntington Junior College**
Medical Assistant Prgm
900 Fifth Ave
Huntington, WV 25701
www.huntingtonjuniorcollege.edu
*Prgm Dir:* Virginia Jane Thompson
*Tel:* 304 697-7550   *Fax:* 304 697-7554
*E-mail:* lwest@huntingtonjuniorcollege.edu
*Med Dir:* Ronald Brownfield, DO

**Marshall Community & Technical College**
Medical Assistant Prgm
2000 7th Ave
Huntington, WV 25703-1527
*Prgm Dir:* Janet B Smith, RRA CCS CMA
*Tel:* 304 696-3008   *Fax:* 304 696-2396
*E-mail:* smithjan@marshall.edu

# Wisconsin

**Wisconsin Indianhead Technical College - Ashland Campus**
Medical Assistant Prgm
2100 Beaser Ave
Ashland, WI 54806
*Prgm Dir:* Carol Utpadel, RN
*Tel:* 715 682-4591, Ext 3168   *Fax:* 715 682-8040
*E-mail:* cutpadel@witc.edu
*Med Dir:* Jan Penn, FNP

**Lakeshore Technical College**
Medical Assistant Prgm
1290 North Ave
Cleveland, WI 53015
*Prgm Dir:* Linda J Arentsen, RN MS CMA
*Tel:* 920 693-1893, Ext 1893   *Fax:* 920 693-8955
*E-mail:* linda.arentsen@gotoltc.edu
*Med Dir:* Robert A Gahl, MD

**Chippewa Valley Technical College**
Medical Assistant Prgm
Health Education Center
620 W Clairemont Ave
Eau Claire, WI 54701
www.cvtc.edu
*Prgm Dir:* Brigitte Niedzwiecki, RN MSN
*Tel:* 715 833-6398   *Fax:* 715 833-6470
*E-mail:* bniedzwiecki@cvtc.edu
*Med Dir:* Stacie Murray, PA-C

**Gateway Technical College - Elkhorn**
Medical Assistant Prgm
400 County Rd H
Elkhorn, WI 53121-2020
www.gtc.edu
*Prgm Dir:* Robert Formanek, MT (ASCP) CMA
*Tel:* 262 741-8340   *Fax:* 262 741-8201
*E-mail:* Formanekr@gtc.edu
*Med Dir:* Barbara Bock, Nurse Practioner

**Southwest Wisconsin Technical College**
Medical Assistant Prgm
1800 Bronson Blvd
Fennimore, WI 53809-9778
*Prgm Dir:* Kristen M Anderson, RN BSN
*Tel:* 608 822-3262, Ext 2163   *Fax:* 608 822-6019
*E-mail:* kanderson@swtc.edu
*Med Dir:* Ravikant Maski, MD

**Moraine Park Technical College**
Medical Assistant Prgm
235 N National Ave
Fond du Lac, WI 53936
www.morainepark.edu
*Prgm Dir:* Colleen Lace, LPN AA BLS
*Tel:* 920 924-3276
*E-mail:* clace@morainepark.edu
*Med Dir:* Russell Fredrickson, MD

**Northeast Wisconsin Technical College**
Medical Assistant Prgm
2740 W Mason St, PO Box 19042
Green Bay, WI 54307-9042
*Prgm Dir:* Karen M Koenig, RN BSM
*Tel:* 920 498-5523   *Fax:* 920 491-2660
*E-mail:* karen.koenig@nwtc.edu
*Med Dir:* Bertram Milson, MD

**Lac Courte Oreills Ojibwa Community College**
Medical Assistant Prgm
13466 W Trepania Rd
Hayward, WI 54843
www.lco.edu
*Prgm Dir:* Laurie Jensen, MBA MT CMA
*Tel:* 715 634-4790, Ext 172   *Fax:* 715 634-5049
*E-mail:* ljensen@lco.edu
*Med Dir:* John McKichan, MD

**Blackhawk Technical College**
Medical Assistant Prgm
6004 Prairie Rd, PO Box 53547
Janesville, WI 53547
*Prgm Dir:* Stephanie Veronica Richardson, RN BSN CMA
*Tel:* 608 758-6900   *Fax:* 608 743-4407
*E-mail:* srichardson@blackhawk.edu
*Med Dir:* Cindy Lohrentz, RN FNP

**Western Technical College**
Medical Assistant Prgm
304 N Sixth St, PO Box C-908
La Crosse, WI 54602-0908
*Prgm Dir:* Margaret R Lentz, BSN RN CMA
*Tel:* 608 785-9922   *Fax:* 608 785-9087
*E-mail:* lentzm@wwtc.edu
*Med Dir:* Laura Krister, MD

**Madison Area Technical College**
Medical Assistant Prgm
3550 Anderson St
Madison, WI 53791-9674
http://matcmadison.edu
*Prgm Dir:* Sue Buboltz, RN CMA
*Tel:* 608 264-6110   *Fax:* 608 246-6013
*E-mail:* ksellnow@matcmadison.edu
*Med Dir:* Kay Heggestad, MD

**Mid-State Technical College - Marshfield**
Medical Assistant Prgm
2600 W Fifth St
Marshfield, WI 54449
www.mstc.edu
*Prgm Dir:* Pam Alt, RN MSN NP-C
*Tel:* 715 389-7024   *Fax:* 715 389-2864
*E-mail:* pam.alt@mstc.edu
*Med Dir:* John Olson, MD

**Bryant & Stratton College - Milwaukee**
Medical Assistant Prgm
310 W Wisconsin Ave
Ste 500 East
Milwaukee, WI 53203
www.bryantstratton.edu
*Prgm Dir:* Kathleen Olewinski, MS RHIA NHA CHE
*Tel:* 414 276-5200, Ext 244   *Fax:* 414 276-3930
*E-mail:* kmolewinski@bryantstratton.edu
*Med Dir:* Russell Seager, PhD

**Concordia University Wisconsin**
Medical Assistant Prgm
1127 S 35th St
Milwaukee, WI 53215
*Prgm Dir:* Susan E Lowrey, MEd MA
*Tel:* 414 649-0795   *Fax:* 414 647-2545
*E-mail:* susan.lowrey@cuw.edu
*Med Dir:* Amy J, Miller-McCarthey, MD

**Milwaukee Area Technical College**
Medical Assistant Prgm
700 W State St
Milwaukee, WI 53233-1419
*Prgm Dir:* Gale Bradford, BS MBA
*Tel:* 414 297-6934   *Fax:* 414 297-6851
*E-mail:* bradfoga@matc.edu
*Med Dir:* Ken Redlin, MD

**Wisconsin Indianhead Technical College**
Medical Assistant Prgm
1019 S Knowles Ave
New Richmond, WI 54017
www.witc.edu
*Prgm Dir:* Lynette Oakley, RN BSN CMA-C
*Tel:* 715 246-6561, Ext 4351   *Fax:* 715 246-2777
*E-mail:* loakley@witc.edu
*Med Dir:* David Olson, MD

**Fox Valley Technical College**
Medical Assistant Prgm
150 N Campbell Rd
Oshkosh, WI 54903-2217
www.fvtc.edu
*Prgm Dir:* Barbara Tuchscherer, MSN RN
*Tel:* 920 735-5774   *Fax:* 920 831-4314
*E-mail:* tuchschb@fvtc.edu
*Med Dir:* Michael Hiebert, MD

**Waukesha County Technical College**
Medical Assistant Prgm
800 Main St
Pewaukee, WI 53072-4601
*Prgm Dir:* Maribeth Blankenheim, CMA BSN RN
*Tel:* 414 691-5563   *Fax:* 414 691-5451
*E-mail:* mblankenheim@wctc.edu
*Med Dir:* Michael Mather, MD

**Gateway Technical College**
Medical Assistant Prgm
1001 S Main St
Racine, WI 53403
*Prgm Dir:* Lori Andreucci, MEd CMT CMA
*Tel:* 262 619-6604   *Fax:* 262 619-6811
*E-mail:* andreuccil@gtc.edu
*Med Dir:* A Durrani, MD

**Nicolet Area Technical College**
Medical Assistant Prgm
PO Box 518
Rhinelander, WI 54501-0518
*Prgm Dir:* Candace S Dailey, RN CMA
*Tel:* 715 365-4539   *Fax:* 715 365-4603
*E-mail:* csdailey@nicoletcollege.edu
*Med Dir:* Stuart Nicolas Boismenue, MD

**Mid-State Technical College - Stevens Point**
Medical Assistant Prgm
933 Michigan Ave
Stevens Point, WI 54481-3141
www.mstc.edu
*Prgm Dir:* Pam Alt, RN MSN NP
*Tel:* 715 389-7024   *Fax:* 715 342-3134
*E-mail:* pam.alt@mstc.edu
*Med Dir:* John Olson, MD

**Wisconsin Indianhead Technical College - Superior**
Medical Assistant Prgm
600 N 21st St
Superior, WI 54880
*Prgm Dir:* Luci Gunderson, RN MS MA
*Tel:* 715 394-6677, Ext 6324   *Fax:* 715 934-3771
*E-mail:* luci.gunderson@witc.edu
*Med Dir:* Pam Schwartau, FNP

## Medical Assistant

| Programs* | Class Capacity | Begins | Length (months) | Award | Res Tuition | Non-res Tuition | Stipend | Offers:‡ 1 | 2 | 3 |
|---|---|---|---|---|---|---|---|---|---|---|
| **Alabama** | | | | | | | | | | |
| George C Wallace Community College (Dothan) | 24 | Aug Jan May | 24 | AAS | $2,886 | $5,254 | | | | • |
| H Council Trenholm State Technical College (Montgomery) | 30 | Aug Jan May | 18 | Cert, AAT | $4,040 | $5,080 | | | | • |
| South University (Montgomery) | 30 | Jan Apr Jun Oct | 24 | AS | $14,780 | $0 | | • | | |
| **Alaska** | | | | | | | | | | |
| University of Alaska - Fairbanks | 12 | Fall Spring | 14, 24 | Cert, AAS | $2,340 | $6,800 | | • | | |
| University of Alaska Anchorage | 20 | Sep Jan May | 12, 24 | Cert, AAS | $4,360 | $14,530 | | | | • |
| **Arkansas** | | | | | | | | | | |
| Arkansas Tech University (Russellville) | 10 | Aug | 24 | AS | $3,818 | $7,346 | | | • | • |
| **California** | | | | | | | | | | |
| Bryman College - San Bernardino | 24 | Monthly | 8 | Dipl | | | | • | | |
| Bryman College - San Francisco | 24 | Monthly | 8, 10 | Dipl | $13,358 | $13,358 | | • | | |
| Cabrillo College (Aptos) | 80 | Feb Aug | 16, 24 | Cert, AS | $2,000 | $4,256 | | • | | • |
| Chabot College (Hayward) | 20 | Aug | 9 | Cert, AA | $26 | $163 | | | | • |
| City College of San Francisco | 30 | Aug Jan | 12, 18 | Cert, AS | $800 | $3,000 | | • | | • |
| Corinthian College - Bryman Campus (Alhambra) | 30 | Jan Dec | 8 | Dipl | $11,154 | $11,154 | | • | | |
| Cosumnes River Community College (Sacramento) | 25 | Aug Jan | 15 | Cert, AS | $1,100 | $6,200 | | • | | • |
| CSi Bryman College (Anaheim) | 180 | Monthly | 8, 10 | Dipl | $10,935 | $0 | | • | | |
| East San Gabriel Valley Occupational Pgm (West Covina) | 15 | Open | 10 | Cert | $1,500 | $1,500 | | | | |
| Heald College - Concord | 25 | Quarterly | 18 | AAS | $10,275 | $0 | | • | | • |
| Heald College - Fresno | 25 | Quarterly | 18 | AAS | $9,900 | $9,900 | | • | | • |
| Heald College - Hayward | 25 | Quarterly | 18 | AAS | $9,900 | $0 | | • | | • |
| Heald College - Milpitas | 25 | Quarterly | 18 | AAS | $9,900 | $9,900 | | • | | • |

*Data are shown only for programs that completed the 2006 AMA Survey of Health Professions Education Programs
‡Key to Offers: 1: Evening or weekend classes; 2: Non-English instruction; 3: Cultural competence instruction

## Medical Assistant

| Programs* | Class Capacity | Begins | Length (months) | Award | Res Tuition | Non-res Tuition | Stipend | Offers:‡ 1 | 2 | 3 |
|---|---|---|---|---|---|---|---|---|---|---|
| Heald College - Rancho Cordova | 25 | Quarterly | 18 | AAS | $9,900 | $0 | | • | | • |
| Heald College - Roseville | 25 | Quarterly | 18 | AAS | $9,900 | $9,900 | | • | | • |
| Heald College - Salinas | 24 | Quarterly | 18 | AAS | $9,900 | $0 | | • | | • |
| Heald College - San Francisco | 25 | Quarterly | 18 | AAS | $9,900 | $9,900 | | • | | • |
| Heald College - Stockton | 25 | Quarterly | 18 | AAS | $9,900 | $0 | | • | | • |
| Modesto Junior College | 25 | August | 7 | Cert, AS | $512 | $3,977 | | | | • |
| Pasadena City College | 20 | Aug Sep | 11 | Cert | $990 | $5,605 | | | | |
| San Diego Mesa College | 30 | Aug | 9, 24 | Cert, AS | $1,024 | $4,130 | | • | | |
| Southern California Regional Occupational Ctr (Torrance) | 100 | Open | 11 | Cert | $400 | $400 | | | | • |
| Western Career College - Antioch | 25 | Every 3 wks | 9, 17 | Cert, AS | $26,749 | $0 | | • | | |
| Western Career College - Pleasant Hill | 72 | Continuous | 8 | | $14,160 | $0 | | • | | |
| Western Career College - Sacramento | 384 | Semi-monthly | 10, 15 | Cert | $12,917 | $0 | | • | | • |
| Western Career College - San Jose | 30 | | 9, 18 | Dipl, AS | $10,500 | $10,500 | | • | | |
| Western Career College - San Leandro | 30 | Monthly | 9, 15 | Cert, AS | $13,434 | $13,434 | | • | | • |
| **Colorado** | | | | | | | | | | |
| Everest College (Colorado Springs) | 200 | Jan Apr Jul Oct | 24 | AS | $13,104 | $13,104 | | • | | |
| Front Range Comm College (Longmont) | 18 | Aug Jan | 10, 24 | Cert, AAS | $3,545 | $14,850 | | | | |
| Parks College - Aurora Campus | 30 | Monthly | 8 | Dipl | $8,857 | $8,857 | | • | | • |
| Red Rocks Community College (Lakewood) | 25 | Jan Jun Aug | 12 | Cert, AAS | $3,900 | $16,000 | | • | | • |
| Westwood College (Denver) | 20 | Jan Mar May Aug Oct | 17, 12 | Dipl, AAS | $20,276 | $0 | | | | |
| **Connecticut** | | | | | | | | | | |
| Branford Hall Career Institute | 24 | Sep Nov Feb Apr Jun | 11, 22 | Dipl | $12,695 | $12,695 | | | | |
| Briarwood College (Southington) | 15 | Aug | 24, 18 | Dipl, Cert | $19,300 | $19,300 | | • | | • |
| Capital Community College (Hartford) | 100 | Sep | 9, 18 | Cert, AS | $1,608 | $5,232 | | • | | • |
| Goodwin College (East Hartford) | 50 | Every 8 wks | 12, 16 | Cert, AS | $13,600 | $13,600 | | • | | • |
| New England Tech Institute (New Britain) | 20 | Sep Mar | 15, 21 | Dipl | $13,200 | $13,200 | | • | | |
| Northwestern Connecticut Comm (Winsted) | 40 | Sep Jan | 16 | Cert, AS | $2,672 | $7,976 | | • | | |
| Norwalk Community College | 33 | Sep | 10 | Cert | $2,406 | $7,178 | | | | |
| Porter and Chester Institute - Wethersfield | 20 | Jan Apr Jul Oct Nov Aug | 9, 12 | Dipl | $14,460 | $14,460 | $24,000 | | | |
| Quinebaug Valley Community College (Danielson) | 50 | Sep Jan | 24 | AS | $1,520 | $4,440 | | • | | • |
| **Delaware** | | | | | | | | | | |
| Delaware Tech & Comm Coll - Wilmington | 18 | Aug | 12, 24 | Dipl, AAS | $2,070 | $5,174 | | • | | • |
| **Florida** | | | | | | | | | | |
| Brevard Community College (Cocoa) | 32 | Open | 12 | Cert | $2,000 | $5,265 | | • | | |
| Broward Community College (Ft Lauderdale) | 30 | Aug | 10 | Cert | $3,600 | $7,000 | | • | | • |
| Daytona Beach Community College | 48 | Aug | 10 | Cert | $2,436 | $9,731 | | | | • |
| Florida Metro U - Tampa Coll - Hillsborough | 200 | 8x/yr | 24 | AS | $15,000 | $15,000 | | • | | • |
| Florida Metro Univ - Orlando College South | 400 | Jan Apr Jul Oct | 24 | AS | $11,740 | $11,740 | | • | | |
| Florida Metropolitan Univ - Lakeland Campus | 240 | Apr May Jul Aug Oct Nov | 24 | AS | $10,260 | $10,260 | | • | | |
| Florida Metropolitan Univ - Pinellas Campus (Clearwater) | 30 | Quarterly | 24 | AS | $11,160 | $11,160 | | • | | • |
| Keiser College - Daytona Beach | 20 | Varies | 16 | AS | $11,640 | $11,640 | | • | | • |
| Lively Technical Center (Tallahassee) | 25 | Aug | 12 | Cert | $3,191 | $10,510 | | | | • |
| McFatter Vocational Technical Center (Davie) | 34 | Aug | 11, 22 | Cert | $2,435 | $7,305 | | | | |
| Palm Beach Community College (Lake Worth) | 15 | Aug | 18 | Cert | $2,500 | $8,500 | | | | |
| Pensacola Junior College | 25 | Aug | 12 | Cert | $2,455 | $9,628 | | | | |
| Pinellas Tech Educ Ctr - St Petersburg | 50 | 4x/yr | 13 | Cert | $2,340 | $9,282 | | | | • |
| Sarasota County Technical Institute | 15 | Aug Dec Apr | 12 | Cert | $2,618 | $8,655 | | • | | |
| Seminole Community College (Sanford) | 25 | Aug | 12 | Cert | $2,574 | $9,867 | | • | | |
| South University (West Palm Beach) | 100 | Jan Apr Jun Sep | 24 | AS | $9,585 | $9,585 | | • | | • |
| Southwest Florida College (Tampa) | 20 | Jan | 24 | AAS | $9,600 | $9,600 | | • | | • |
| **Georgia** | | | | | | | | | | |
| Appalachian Technical College (Jasper) | 25 | Oct | 12 | Dipl | $2,573 | $4,897 | | • | | • |
| Atlanta Technical College | 25 | Mar Oct Jan Apr | 15 | Dipl | $3,000 | $0 | | • | | • |
| Augusta Technical College | 25 | Mar Sep | 15 | Dipl | $1,870 | $3,740 | | • | • | • |
| Chattahoochee Technical College (Marietta) | 20 | Sep | 12 | Dipl | $1,700 | $3,188 | | | | |
| Clayton State University (Morrow) | 60 | Aug Jan May | 12, 24 | Cert, AAS | $4,443 | $15,417 | | • | | |
| Coosa Valley Technical College (Rome) | 20 | Oct | 15 | Dipl | $1,008 | $2,016 | | | | |
| East Central Technical College (Douglas) | 35 | Sep | 15 | Dipl | $1,600 | $2,600 | | | | • |
| Gwinnett Technical College (Lawrenceville) | 20 | Mar Sep | 15 | Dipl | $1,248 | $2,256 | | | | • |
| Heart of Georgia Technical College (Dublin) | 20 | Oct Apr | 12 | Dipl | $2,232 | $4,464 | | | | • |
| Lanier Technical College (Oakwood) | 22 | Oct Apr | 15 | Dipl | $1,378 | $2,755 | | | | |
| Medix School (Smyrna) | 350 | Monthly | 10 | Dipl | $10,567 | $10,567 | | • | | |
| Moultrie Technical College | 30 | 2x/yr | 15 | Dipl | $1,248 | $2,496 | | | • | • |
| Ogeechee Technical College (Statesboro) | 40 | Sep Apr | 15 | Dipl | $1,384 | $2,768 | | • | | • |
| Savannah River College (Augusta) | 32 | Jul Aug Oct Nov Jan Feb | 12, 18 | Dipl, AS | $9,695 | $9,695 | $9,695 | • | | |

*Data are shown only for programs that completed the 2006 AMA Survey of Health Professions Education Programs

‡Key to Offers: 1: Evening or weekend classes; 2: Non-English instruction; 3: Cultural competence instruction

## Medical Assistant

| Programs* | Class Capacity | Begins | Length (months) | Award | Res Tuition | Non-res Tuition | Stipend | Offers‡ 1 | 2 | 3 |
|---|---|---|---|---|---|---|---|---|---|---|
| Savannah Technical College | 25 | Apr | 15 | Dipl | $1,860 | $3,720 | | • | | • |
| Southeastern Technical College (Vidalia) | 20 | Apr Oct | 15 | Dipl | $2,136 | $4,272 | | • | | |
| Southwest Georgia Technical College (Thomasville) | 24 | Apr | 15, 18 | Dipl, AAS | $1,488 | $2,976 | | • | | • |
| Swainsboro Technical College | 20 | Sep Jan Mar Jul | 15 | Dipl, AS | $1,296 | $2,592 | | • | | • |
| West Central Technical College (Waco) | 20 | Jul | 12 | Dipl | $1,371 | $2,487 | | | | |
| West Georgia Technical College (LaGrange) | 15 | Year-round | 15 | Dipl | $2,430 | $4,860 | | • | | • |
| **Guam** | | | | | | | | | | |
| Guam Community College (GMF Barrigada Guam) | 30 | Aug | 24 | Cert | $1,830 | $2,287 | | | | |
| **Hawaii** | | | | | | | | | | |
| Heald College - Honolulu | 25 | Quarterly | 18 | AAS | $9,900 | $9,900 | | • | | • |
| **Idaho** | | | | | | | | | | |
| College of Southern Idaho (Twin Falls) | 16 | Aug | 10 | Cert | $2,000 | $5,800 | $2,000 | • | | • |
| Idaho State University (Pocatello) | 18 | Aug | 24 | AAS | $4,190 | $8,270 | | • | | |
| **Illinois** | | | | | | | | | | |
| Harper College (Palatine) | 30 | Aug Jan Jun | 12, 24 | Cert | $2,560 | $14,056 | | • | • | • |
| Midstate College (Peoria) | | Mar Jun Aug Nov | 24 | AAS | $10,000 | $10,000 | | • | | • |
| Midwest Technical Institute (Lincoln) | 20 | Jan | 9 | Dipl | $8,950 | $8,950 | | | | |
| Northwestern Business College (Chicago) | 48 | Sep Dec Mar Jun | 18 | AAS | $14,500 | $14,500 | | • | • | • |
| Northwestern Business College - Southwest Campus (Bridgeview) | 48 | Sep Mar | 18 | AAS | $13,920 | $13,920 | | • | | • |
| Robert Morris College - Aurora | 40 | Jul Feb Sep | 10, 15 | Dipl, AAS, BBA | $15,900 | $15,900 | | • | | |
| Robert Morris College - Bensenville | 40 | Sep Feb | 10, 15 | Dipl, AAS, BBA | $15,900 | $15,900 | | | | • |
| Robert Morris College - Chicago | 120 | Jul Sep Feb | 10, 15 | Dipl, AAS, BBA | $15,900 | $15,900 | | • | | • |
| Robert Morris College - Orland Park | 40 | Feb Jul Sep | 10, 15 | Dipl, AAS, BBA | $15,900 | $15,900 | | • | | • |
| Robert Morris College - Peoria | 40 | Sep Feb | 10, 15 | Dipl, AAS, BBA | $15,900 | $15,900 | | | | • |
| Robert Morris College - Springfield | 40 | Sep Feb | 10, 15 | Dipl, AAS, BBA | $15,900 | $15,900 | | • | | • |
| Robert Morris College - Waukegan | 40 | Jul Sep Feb | 10, 15 | Dipl, AAS, BBA | $15,900 | $15,900 | | • | | • |
| Rockford Business College | 62 | Mar Jun Sep Nov | 24 | Dipl, AAS | $6,445 | $0 | | • | | • |
| Southwestern Illinois College (Belleville) | 60 | Aug Jan Jun | 10, 24 | Cert, AAS | $4,300 | $7,500 | | | | |
| Spanish Coalition for Jobs Inc (Chicago) | 20 | Sep Jan Apr | 10 | Cert | $7,500 | $7,500 | | | | • |
| **Indiana** | | | | | | | | | | |
| Brown Mackie College - South Bend | 50 | Monthly | 24 | AAS | $7,488 | $7,488 | | • | | • |
| Indiana Business College - Evansville | 100 | Jan Apr Jun Sep | 12, 18 | AAS | $12,360 | $0 | | • | | |
| Indiana Business College - Terre Haute | | Quarterly | 18 | AAS | $9,660 | $9,660 | | • | | • |
| International Business Coll - Indianapolis | 48 | Sep Mar | 10, 14 | Dipl, Assoc | $9,575 | $9,575 | | | | • |
| Ivy Tech Community College - Columbus | 25 | Aug Jan May | 18, 24 | Cert, TC, AAS | $4,029 | $7,792 | | • | | • |
| Ivy Tech Community College - Kokomo | 150 | Aug Jan May | 18, 24 | Cert, AAS, TC | $3,159 | $6,426 | | • | | • |
| Ivy Tech Community College - Lafayette | 24 | Jan Aug | 12, 24 | Cert, TC | $4,047 | $8,131 | | • | | |
| Ivy Tech Community College - Lawrenceburg | 50 | Aug Jan May | 18, 24 | Cert, TC, AAS | $2,000 | $4,200 | | • | | • |
| Ivy Tech Community College - Madison | 30 | Aug Jan May | 12 | Cert | $4,212 | $8,424 | | | | • |
| Ivy Tech Community College - Michigan City | 25 | Aug | 12, 24 | Cert, AAS, TC | $5,924 | $6,656 | | • | | • |
| Ivy Tech Community College - South Bend | 200 | Aug Jan May | 15, 20 | Cert, TC, AAS | $3,200 | $5,000 | | • | | |
| Ivy Tech Community College - Terre Haute | 30 | Sep | 12, 18 | Cert, AAS | $4,029 | $8,172 | | • | | • |
| Ivy Tech Community College NE - Ft Wayne | 900 | Aug Jan May | 18, 24 | Dipl, Cert, AAS | $4,000 | $5,100 | | • | | • |
| Ivy Tech Community College SC - Sellersburg | | Aug Jan May | 12, 24 | Cert, TC, AAS | $3,158 | $5,400 | | • | | • |
| Ivy Tech Community College SW - Evansville | 40 | Aug | 16, 21 | Cert, AAS | $87 | $178 | $1,500 | • | • | • |
| **Iowa** | | | | | | | | | | |
| Des Moines Area Community College (Ankeny) | 36 | Aug | 11 | Dipl | $4,704 | $9,408 | | • | | • |
| Hamilton College (Urbandale) | | Jan Apr May Oct | 21, 18 | AAS | $13,000 | $0 | | • | | • |
| Iowa Central Community College (Ft Dodge) | 30 | Sep | 11 | Dipl | $4,480 | $6,500 | | | | |
| Iowa Western Community College (Council Bluffs) | 20 | Aug | 10 | Dipl | $4,500 | $6,750 | | | | |
| Kaplan University, Davenport Campus | 25 | Jul Oct Jan Apr | 15, 18 | Dipl, AAS | $12,960 | $12,960 | | • | | |
| Kirkwood Community College (Cedar Rapids) | 39 | Aug Jan May | 12 | Dipl, AAS | $5,705 | $11,411 | | • | | |
| Southeastern Community College (West Burlington) | 25 | Aug | 11 | Dipl | $4,896 | $5,472 | | | | • |
| **Kansas** | | | | | | | | | | |
| Northwest Kansas Technical College (Goodland) | 15 | Aug | 11, 24 | Cert, AAS | $2,668 | $5,336 | | | | |
| Wichita Area Technical College | 30 | Aug Jan | 9 | Dipl, AAS | $3,564 | $16,573 | | • | | • |
| **Kentucky** | | | | | | | | | | |
| Eastern Kentucky University (Richmond) | 16 | Aug Jan | 22 | AS | $4,660 | $13,070 | | • | | • |
| Henderson Community College | 20 | Aug | 22 | AAS | $2,616 | $7,848 | | | | • |
| National College of Business & Technology - Danville | 50 | Dec Mar Jun Sep | 24 | AS | $6,624 | $6,624 | | • | | • |
| National College of Business & Technology - Florence | 50 | Jan Apr Jun Sep | 24 | AS | $8,544 | $8,544 | | • | | |
| National College of Business & Technology - Lexington | 120 | Jan Apr Jul Sep | 24 | AS | $6,624 | $6,624 | | • | | |
| National College of Business & Technology - Louisville | 75 | Dec Jan Mar Apr Jun Jul | 24 | AS | $6,624 | $6,624 | | • | | |
| National College of Business & Technology - Pikeville | 60 | Jan Apr Jun Sep | 24 | AS | $7,104 | $7,104 | | • | | |
| National College of Business & Technology - Richmond | 75 | Dec Mar Jun Sep | 24 | AS | $6,624 | $6,624 | | • | | |

*Data are shown only for programs that completed the 2006 AMA Survey of Health Professions Education Programs
‡Key to Offers: 1: Evening or weekend classes; 2: Non-English instruction; 3: Cultural competence instruction

## Medical Assistant

| Programs* | Class Capacity | Begins | Length (months) | Award | Res Tuition | Non-res Tuition | Stipend | Offers:‡ 1 | 2 | 3 |
|---|---|---|---|---|---|---|---|---|---|---|
| **Louisiana** | | | | | | | | | | |
| Bossier Parish Community College (Bossier City) | 20 | Aug | 16 | AS | $1,720 | $3,860 | | • | | • |
| Bryman College - New Orleans | 230 | | 8 | Dipl | $9,300 | $0 | | • | | |
| **Maine** | | | | | | | | | | |
| Beal College (Bangor) | 30 | Bimonthly | 24 | AS | $5,820 | $5,820 | | • | | |
| Kennebec Valley Community College (Fairfield) | 24 | Aug | 18 | AAS | $74 | $155 | | • | | • |
| **Maryland** | | | | | | | | | | |
| Allegany College of Maryland (Cumberland) | 20 | Sep | 24 | AAS | $3,532 | $7,284 | | | | • |
| Hagerstown Business College | | Oct Jan Apr Jul | 9, 20 | Cert, AAS | $22,264 | $22,264 | | • | | • |
| Medix School (Towson) | 485 | Concurrent | 10 | Cert | $9,950 | $0 | | • | | • |
| **Massachusetts** | | | | | | | | | | |
| Charles H McCann Technical School (North Adams) | 20 | Sep | 9 | Cert | $2,500 | $11,315 | | | | |
| Massasoit Community College (Canton) | 24 | Sep | 9 | Cert | $4,447 | $12,046 | | • | | |
| Middlesex Community College - Lowell | 24 | Sep Jan | 10, 22 | Cert | $2,871 | $8,845 | | | | • |
| North Shore Community College (Danvers) | 20 | Sep | 9 | Cert | $3,596 | $10,324 | | | | • |
| Northern Essex Community College (Lawrence) | 30 | Sep | 10 | Cert | $3,357 | $0 | | • | | • |
| Quinsigamond Community College (Worcester) | 20 | Sep | 9 | Cert | $4,370 | $12,198 | | • | | • |
| Southeastern Technical Institute (South Easton) | 24 | Sep | 10 | Dipl | $1,500 | $2,500 | | | | |
| Springfield Technical Community College | 36 | Sep | 18 | Cert, AS, COC | $4,874 | $11,384 | | • | | • |
| **Michigan** | | | | | | | | | | |
| Alpena Community College | 20 | Aug | 24 | AAS | $2,200 | $3,200 | | • | | |
| Baker College of Auburn Hills | 30 | Sep | 27 | Cert, AAS | $16,275 | $16,275 | | • | | |
| Baker College of Cadillac | 100 | Sep Jan May | 24 | Cert, AAS | $6,300 | $0 | | • | | |
| Baker College of Clinton Township | 30 | Fall | 24 | Cert, AAS | $7,178 | $7,178 | | • | | |
| Baker College of Owosso | 36 | Sep Jan Apr | 24 | Cert, AAS | $8,640 | $8,640 | | • | | |
| Carnegie Institute (Troy) | 60 | Jan Mar Jun Sep | 12, 24 | Dipl | $8,720 | $8,720 | | • | | |
| Davenport University - Grand Rapids (Fulton St) | 90 | Sep Jan Mar | 18 | Dipl, AAS | $7,656 | $0 | | • | | • |
| Davenport University - Lansing | 12 | Jan May Aug | 18 | AS | $14,070 | $14,070 | | • | | • |
| Henry Ford Community College (Dearborn) | 24 | Aug Jan | 10 | Cert | $2,526 | $4,500 | | • | | • |
| Jackson Community College | 30 | Aug Jan May | 18 | Cert | $2,400 | $2,800 | | • | | • |
| Kalamazoo Valley Community College | 40 | Aug | 12 | Cert, AAS | $1,312 | $2,368 | | • | | • |
| Mid Michigan Community College (Harrison) | 25 | Aug | 24 | AAS | $1,785 | $2,970 | | • | | • |
| Schoolcraft College (Garden City) | 30 | Sep | 12 | Cert | $2,448 | $3,600 | | • | | |
| **Minnesota** | | | | | | | | | | |
| Anoka Technical College | 60 | Aug Jan | 18 | Dipl, AAS | $8,042 | $0 | | • | | |
| Argosy University/Twin Cities (Eagan) | 20 | Sep Jan May Jul | 17 | AAS | $7,830 | $7,830 | | • | | |
| Century College (White Bear Lake) | 20 | Aug Jan | 14, 24 | Dipl | $4,233 | $8,043 | | • | | • |
| Dakota County Technical College (Rosemount) | 34 | Aug Jan | 11, 18 | Dipl, AAS | $5,300 | $0 | | | | • |
| Duluth Business University | 45 | Jan Apr Jul Oct | 14, 24 | Dipl, AAS | | | | • | | • |
| Herzing College (Crystal) | 150 | Jan Jul Sep Oct | 16 | Dipl, AAS | $16,395 | $16,395 | | • | | • |
| Minneapolis Business College (Roseville) | 120 | Sep Mar | 10, 14 | Dipl, AAS | $15,300 | $15,300 | | | | |
| Minnesota School of Business - Richfield | 20 | Jan Apr Jul Oct | 14 | Dipl, AAS | $16,000 | $16,000 | | • | | |
| Minnesota West Comm & Tech College (Worthington) | 30 | Aug | 12, 16 | Dipl, AAS | $4,500 | $4,500 | | • | • | • |
| Northland Community & Technical College (East Grand Forks) | 28 | Aug Jan | 16, 24 | Dipl, AAS | $4,486 | $4,486 | | • | | • |
| Ridgewater College - Willmar Campus | 40 | Aug | 14 | Dipl | $7,165 | $15,000 | | | | • |
| **Mississippi** | | | | | | | | | | |
| Hinds Community College (Pearl) | 15 | Aug | 18 | AAS | $1,660 | $3,866 | | • | | • |
| **Missouri** | | | | | | | | | | |
| Everest College - Springfield Campus | 100 | Jan Apr Jul Oct | 24 | AAS | $27,515 | $27,515 | | • | | |
| **Nebraska** | | | | | | | | | | |
| Alegent Health (Omaha) | 20 | Aug | 12 | Dipl | $6,000 | $6,500 | | | | • |
| Central Community College (Hastings) | 30 | Varies | 12, 24 | Dipl, AAS | $1,295 | $1,903 | | • | | • |
| Hamilton College - Omaha | 30 | Jan Mar May Aug Oct | 18, 17 | AMA | $33,000 | $33,000 | | • | | • |
| Nebraska Methodist College (Omaha) | 25 | Aug | 12 | Cert | $8,900 | $8,900 | | • | | • |
| Southeast Community College (Lincoln) | 25 | Apr Oct | 12 | Dipl | $3,031 | $3,800 | | • | | |
| **Nevada** | | | | | | | | | | |
| Community College of Southern Nevada (Las Vegas) | 60 | Sep Jan | 12 | Cert | $2,398 | $6,400 | | • | | • |
| **New Hampshire** | | | | | | | | | | |
| Hesser College (Manchester) | 100 | Sep Nov Jan Mar May Jul | 18, 20 | Dipl, AS | $10 | $10,395 | | • | | |
| New Hampshire Comm Tech Coll - Claremont | 30 | Sep | 12, 4 | Cert | $7,287 | $10,649 | | • | | • |
| New Hampshire Comm Tech College - Manchester | 32 | Sep Jan | 10, 18 | Cert, AS | $5,248 | $7,872 | | • | | • |
| **New Jersey** | | | | | | | | | | |
| Bergen Community College (Paramus) | 35 | Sep | 24 | AAS | $4,250 | $8,500 | | • | | • |
| Dover Business College - Dover | 25 | 4x/yr | 12 | Dipl | $13,605 | $13,605 | | • | | • |
| Dover Business College - Paramus | 25 | 4x/yr | 12 | Dipl | $13,605 | $13,605 | | • | | • |

*Data are shown only for programs that completed the 2006 AMA Survey of Health Professions Education Programs

‡Key to Offers: 1: Evening or weekend classes; 2: Non-English instruction; 3: Cultural competence instruction

**PROGRAMS**

## Medical Assistant

| Programs* | Class Capacity | Begins | Length (months) | Award | Res Tuition | Non-res Tuition | Stipend | Offers:‡ 1 | 2 | 3 |
|---|---|---|---|---|---|---|---|---|---|---|
| Hudson County Community College (Jersey City) | 25 | Sep Jan | 12, 24 | AAS | $2,692 | $5,344 | | • | | • |
| Mercer County Technical Schools (Trenton) | 20 | Aug | 12 | Dipl, Cert | $3,140 | $3,140 | | • | | |
| Sussex County Community College (Newton) | 40 | Fall semester | 10, 18 | Cert, MA | $7,746 | $7,746 | | | | • |
| Warren County Community College (Washington) | 20 | Aug Jan | 7 | Cert | $4,900 | $4,900 | | • | | |
| **New Mexico** | | | | | | | | | | |
| Eastern New Mexico University - Roswell | 30 | Aug | 12, 24 | Cert | $525 | $2,160 | | • | | • |
| **New York** | | | | | | | | | | |
| ASA Institute of Business & Computer Tech (Brooklyn) | 900 | Feb Jun Oct | 18 | AOS | $10,320 | $10,320 | | | | • |
| Broome Community College (Binghamton) | 35 | Aug | 18 | AAS | $2,914 | $5,828 | | | | • |
| Bryant & Stratton College - Buffalo | | Jan May Sep | 16 | AOS | $18,675 | $0 | | • | | • |
| Bryant & Stratton College - Rochester | 200 | Jan May Sep | 16 | AOS | $17,730 | $17,730 | | • | | |
| Bryant & Stratton College - Syracuse | 150 | Sep Jan May | 16 | AOS | $12,744 | $0 | | • | | |
| Elmira Business Institute | 100 | Feb Jun Oct | 16 | AOS | $18,000 | $18,000 | | | | • |
| Erie Community College - North Campus (Williamsville) | 60 | Sep Jan | 12, 21 | Cert, AAS | $2,987 | $5,974 | | • | | • |
| Niagara County Community College (Sanborn) | 24 | Sep Jan | 24 | AAS | $3,096 | $4,644 | | • | | • |
| Ridley-Lowell Business & Technical Institute (Binghamton) | 12 | Sep Jan Apr | 12 | Dipl | $10,400 | $10,400 | | • | | • |
| Rochester Business Institute | 650 | Jan | | AAS | $10,424 | $10,424 | | • | | |
| Wood Tobe'-Coburn School (New York) | 104 | Sep Jul Mar | 10, 14 | Dipl, AD | $14,400 | $14,400 | | | | |
| **North Carolina** | | | | | | | | | | |
| Carteret Community College (Morehead City) | 25 | Aug | 12 | Dipl | $1,580 | $8,500 | | | | • |
| Central Carolina Community College - Chatham (Pittsboro) | 25 | Aug | 12, 18 | Dipl, AAS | $1,896 | $10,536 | | • | | |
| Central Carolina Community College - Harnett (Lillington) | 35 | Aug | 12, 18 | Dipl, AAS | $1,896 | $10,536 | | • | | |
| Central Piedmont Community College (Charlotte) | 120 | Aug Jan | 12 | Dipl | $1,896 | $10,536 | | • | | • |
| Davidson County Community College (Lexington) | 24 | Aug | 12, 21 | Dipl, AAS | $1,800 | $10,500 | | | | • |
| Forsyth Technical Community College (Winston-Salem) | 24 | Sep | 24 | AAS | $2,698 | $13,984 | | | | • |
| Gaston College (Dallas) | 45 | Aug | 21 | AAS | $1,896 | $10,536 | | | | |
| Johnston Community College (Smithfield) | 30 | Aug | 9, 21 | Dipl, AAS | $880 | $5,470 | | | | |
| King's College (Charlotte) | 120 | Mar Jul Sep | 10, 14 | Dipl, AAS | $11,960 | $11,960 | $2,500 | | | • |
| Lenoir Community College (Kinston) | 20 | Aug | 21 | AAS | $1,724 | $9,104 | | • | | • |
| Martin Commuity College (Williamston) | 24 | Aug | 12, 24 | Dipl, AAS | $1,264 | $7,024 | | | | |
| Mitchell Community College (Mooresville) | 20 | Aug | 12 | Dipl, Cert | $1,700 | $7,000 | | | | • |
| Montgomery Community College (Troy) | 25 | Aug | 21 | AAS | $1,750 | $9,099 | | | • | • |
| Pamlico Community College (Grantsboro) | 40 | Aug | 16, 24 | Dipl, AS | $1,914 | $11,883 | | | | |
| Pitt Community College (Greenville) | 30 | Aug Jan May | 24 | Cert, AAS | $1,461 | $8,121 | | | | |
| South College - Asheville | 100 | Jan Apr Jun Oct | 24 | AAS | $9,900 | $9,900 | | • | | |
| South Piedmont Community College (Polkton) | 20 | Aug | 12, 21 | Dipl, AAS | $2,775 | $14,356 | | | | • |
| Vance - Granville Community College (Louisburg) | 25 | Aug | 21, 12 | Dipl, AAS | $1,620 | $8,200 | | | | |
| Wake Technical Community College (Raleigh) | 20 | Aug Jan | 12 | Dipl | $1,768 | $9,726 | | | | • |
| Western Piedmont Community College (Morganton) | 25 | Aug | 12, 21 | Dipl, Cert, AAS | $1,856 | $10,316 | | | | |
| Wilkes Community College (Wilkesboro) | 60 | Aug | 12, 21 | AAS | $1,140 | $6,031 | | • | | |
| **Ohio** | | | | | | | | | | |
| Akron Instititue | 100 | Sep Oct Jan Mar | 12, 18 | Dipl, AAS | $10,955 | $0 | | • | | |
| American School of Technology (Columbus) | 200 | Jan Feb Mar May Jun Aug | 12 | Dipl | $8,600 | $8,600 | | • | | |
| Apollo Career Center (Lima) | 24 | Aug | 10 | Cert | $6,700 | $6,700 | | | | |
| Ashland County - West Holmes Career Center | 25 | Sep | 9 | Cert | $6,188 | $0 | | | | |
| Belmont Technical College (St Clairsville) | 48 | Sep | 18 | AAS | $3,988 | $7,022 | | | | |
| Brown Mackie College - Akron | 300 | Monthly | 24, 12 | Dipl, AAS | $8,592 | $0 | | • | | |
| Brown Mackie College - Cincinnati | 300 | Monthly | 12, 24 | Dipl, AAS | $8,000 | $0 | | • | | • |
| Columbus State Community College | 20 | Sep | 12 | Cert | $4,450 | $11,000 | | | | • |
| EHOVE Ghrist Adult Career Center (Milan) | 25 | Sep | 9 | Cert | $4,600 | $4,600 | | • | | |
| Fairfield Career Center (EVSD) (Carroll) | 25 | Aug | 9 | Cert | $5,750 | $0 | | • | | |
| Hocking College (Nelsonville) | 85 | Sep | 18 | AAS | $3,348 | $6,696 | | • | | • |
| Jefferson Community College (Steubenville) | 25 | Aug | 12, 18 | Cert, AAS | $3,600 | $3,800 | | • | | • |
| Knox County Career Center (Mt Vernon) | 24 | Sep | 10 | Dipl | $5,330 | $0 | | | | • |
| Lakeland Community College (Kirtland) | 24 | Aug | 12 | Cert | $3,054 | $3,169 | | • | | • |
| Lorain County Community College (Elyria) | 20 | Aug | 9, 24 | Cert, AAS | $2,399 | $2,889 | $45 | | | |
| Mahoning County Career & Technical Center (Canfield) | 25 | Sep (Nov if needed) | 9 | Cert | $4,160 | $4,160 | | | | |
| Marion Technical College | 15 | Sep | 12 | Cert | $4,300 | $0 | | • | | |
| Miami - Jacobs Career College (Dayton) | 25 | Every 6 wks | 24 | AAS | $13,248 | $0 | | • | | • |
| Ohio Institute of Health Careers - Columbus | 30 | Quarterly | 11 | Dipl | $10,360 | $0 | | • | | • |
| Ohio University - Lancaster | 45 | Sep Jan Mar Jun | 24 | AAS | $4,581 | $8,904 | | • | | |
| Pickaway Ross Joint Vocational School (Chillicothe) | 18 | Aug | 10 | Cert | $4,900 | $4,900 | | | | |
| Polaris Career Center (Middleburg Heights) | 25 | Aug | 9 | Cert | $5,295 | $0 | | | | • |
| Portage Lakes Career Center (Green) | 25 | Sep Jan | 10 | Cert | $5,100 | $6,200 | | • | | |
| RETS Tech Center (Centerville) | 100 | | 14 | AAS | $7,535 | $7,535 | | • | | |
| Rhodes State College (Lima) | 20 | Fall | 24 | AAS | $4,725 | $8,162 | | • | | • |

*Data are shown only for programs that completed the 2006 AMA Survey of Health Professions Education Programs
‡Key to Offers: 1: Evening or weekend classes; 2: Non-English instruction; 3: Cultural competence instruction

# Medical Assistant

| Programs* | Class Capacity | Begins | Length (months) | Award | Res Tuition | Non-res Tuition | Stipend | Offers:‡ 1 | 2 | 3 |
|---|---|---|---|---|---|---|---|---|---|---|
| Sinclair Community College (Dayton) | 40 | Sep | 21 | AAS | $2,323 | $3,793 | | • | | |
| Southern State Community College (Hillsboro) | 60 | Sep | 18 | AAS | $3,651 | $7,032 | | | | |
| Stark State College of Technology (Canton) | 38 | Aug | 20 | Dipl, AAS | $4,200 | $5,950 | | • | | |
| Stautzenberger College (Toledo) | 200 | 4x/yr | 15, 24 | Dipl, AAS | $10,814 | $10,814 | | • | | |
| Trumbull Career & Technical Center (Warren) | 25 | Aug Jan | 10 | Cert | $6,715 | $6,715 | | • | | |
| University of Akron | 35 | Aug | 24 | AAS | $3,288 | $7,403 | | | | |
| University of Cincinnati | 80 | Sep Jan Mar | 24 | AAS | $1,553 | $4,025 | | • | | • |
| Zane State College (Zanesville) | 50 | Sep | 18, 21 | Cert, AAS | $12,500 | $17,500 | | • | | |
| **Oklahoma** | | | | | | | | | | |
| Francis Tuttle Technology Center (Oklahoma City) | 15 | Controlled Entry | 10 | Cert | $1,520 | $1,520 | | | | • |
| Moore Norman Technology Center | 24 | Aug Jan | 9 | Dipl, Cert | $1,060 | $1,565 | | | | • |
| Tulsa Community College | 20 | Jan Aug | 11, 22 | Cert, MA, MA | $2,273 | $6,005 | | | • | • |
| **Oregon** | | | | | | | | | | |
| Central Oregon Community College (Bend) | 24 | Sep | 12 | Cert | $3,717 | $10,384 | | • | | |
| Clackamas Community College (Oregon City) | 24 | Sep | 9 | Cert | $56 | $194 | | | | • |
| Everest College (Portland) | 30 | Jan Feb Apr May Jul Oct | 15, 18 | Dipl, AAS | | | | • | | • |
| Heald College - Portland | 25 | Quarterly | 12, 18 | Dipl, AAS | $9,900 | $0 | | • | | • |
| Portland Community College | 24 | Sep | 10 | Cert | $2,881 | $8,170 | | | | • |
| **Pennsylvania** | | | | | | | | | | |
| Berks Technical Institute (Wyomissing) | 60 | Jan Apr Jul Oct | 18, 24 | ASB | $17,864 | $17,864 | | • | | • |
| Bradford School - Pittsburgh | 75 | Mar Jul Sep | 10, 14 | Dipl, ASB | $13,120 | $13,120 | | | | • |
| Bucks County Community College (Newtown) | 24 | Aug | 24 | Cert, AA | $2,000 | $4,000 | | • | | • |
| Butler County Community College | 35 | Aug | 10, 24 | Cert, AS | $3,502 | $5,644 | | • | | • |
| Central Pennsylvania College (Summerdale) | 35 | Oct Jan Apr Jul | 21 | AA | $10,380 | $0 | | | | • |
| Comm Coll of Allegheny County (West Mifflin) | 20 | Aug | 9, 15 | Dipl | $2,400 | $4,800 | | • | | • |
| Community College of Philadelphia | 24 | Sep | 18 | AAS | $4,473 | $11,529 | | | | |
| Delaware County Community College (Media) | 24 | Sep | 10, 24 | Cert, AAS | $2,325 | $4,030 | | • | | • |
| Greater Altoona Career & Technology Center | 18 | Aug | 11 | Dipl | $6,800 | $6,800 | | | | |
| Harrisburg Area Community College | 20 | Aug | 12, 24 | Cert, AS | $185 | $235 | | • | | • |
| Lehigh Carbon Community College (Schnecksville) | 24 | Aug Jan | 15 | AAS | $2,280 | $4,560 | | • | | |
| Montgomery County Community College (Pottstown) | 20 | Sep Jan | 11 | Cert | $97 | $191 | | | | • |
| Mount Aloysius College (Cresson) | 30 | Aug | 24 | AS | $14,800 | $14,800 | | • | | • |
| Thompson Institute (Philadelphia) | 360 | Monthly | 8, 10 | Dipl | $12,100 | $12,100 | | • | | |
| **South Carolina** | | | | | | | | | | |
| Aiken Technical College | 30 | Aug | 12 | Cert | $1,500 | $4,800 | | | | |
| Central Carolina Technical College (Sumter) | 24 | May | 12 | Dipl | $4,050 | $4,752 | | | | |
| Midlands Technical College (West Columbia) | 20 | Aug | 12 | Cert | $4,356 | $13,068 | | | | |
| Miller-Motte Technical College (Charleston) | 200 | Jan Apr Jul Oct | 18 | AAS | $11,520 | $11,520 | | | | |
| Piedmont Technical College (Greenwood) | 20 | Aug | 12 | Dipl | $4,110 | $4,758 | | | | |
| Spartanburg Community College | 20 | Aug | 12 | Dipl | $4,299 | $8,175 | | | | • |
| Trident Technical College (Charleston) | 30 | May | 12 | Dipl | $4,425 | $8,379 | | • | | |
| **South Dakota** | | | | | | | | | | |
| Colorado Technical University (Sioux Falls) | | Oct Jan Apr Jul Aug | 18 | AS | | | | • | | |
| Lake Area Technical Institute (Watertown) | 30 | Aug Jan | 14, 18 | Dipl, AAS | $4,385 | $4,385 | | • | | |
| Mitchell Technical Institute (Mitchell) | 20 | Aug | 18 | AAS | $4,368 | $4,368 | | | | |
| National American Univ - Sioux Falls Campus | 30 | Jun Sep Dec Mar | 24 | AAS | | | | • | | |
| Presentation College (Aberdeen) | | Aug Jan | 18 | AS | $12,300 | $12,300 | | • | | |
| **Tennessee** | | | | | | | | | | |
| Chattanooga State Technical Comm College | 20 | Aug | 12 | Dipl | $2,136 | $2,136 | | | | |
| Cleveland State Community College | 15 | Aug | 24 | AAS | $1,600 | $2,396 | | • | | • |
| National College of Business & Technology - Bristol | 65 | Jan MarApr JunJul SepOct | 24 | AS | $9,420 | $9,420 | | • | | |
| National College of Business & Technology - Knoxville | 60 | Dec Jan Mar Apr Jun Sep | 24 | AAS | $9,810 | $9,810 | | | | |
| National College of Business & Technology - Nashville | 25 | Jan Mar Jun Sep | 24 | AS | $7,104 | $7,104 | | • | | |
| South College (Knoxville) | 150 | Jan Apr Jun Oct | 21 | AS | $15,200 | $15,200 | | • | | • |
| Tennessee Technology Center - Knoxville | 20 | Sep | 12 | Dipl | $3,443 | $3,443 | | | | • |
| Tennessee Technology Center - McMinnville | 18 | Sep 5 | 12 | Dipl | $1,986 | $1,986 | | | | |
| **Texas** | | | | | | | | | | |
| Bradford School (Houston) | 30 | Feb Aug Oct | 10 | Dipl | $20,000 | $0 | | | | |
| Career Centers of Texas (El Paso) | 30 | 8x/yr | 9, 14 | Dipl | $11 | $11,250 | | • | | • |
| Cisco Junior College (Abilene) | 20 | Sep Jan Jun Jul | 12, 24 | Cert, AAS | $2,985 | $3,444 | | • | | • |
| El Centro College (Dallas) | 60 | Sep Jan | 12 | Cert | $1,296 | $2,376 | | | • | • |
| Houston Community College | 75 | Aug Jan Jun | 12 | Cert | $3,500 | $5,498 | | | | • |
| Kilgore College | 16 | Aug | 12, 21 | Cert, AAS | $1,910 | $3,327 | | | | |

*Data are shown only for programs that completed the 2006 AMA Survey of Health Professions Education Programs

‡Key to Offers: 1: Evening or weekend classes; 2: Non-English instruction; 3: Cultural competence instruction

## Medical Assistant

| Programs* | Class Capacity | Begins | Length (months) | Award | Res Tuition | Non-res Tuition | Stipend | Offers:‡ 1 | 2 | 3 |
|---|---|---|---|---|---|---|---|---|---|---|
| Richland College (Dallas) | 30 | Jan Sep with flex entry | 8, 15 | Cert | $3,000 | $3,000 | $1,500 | • | • | • |
| San Antonio College | 50 | Jan Aug | 13, 24 | Cert, AAS | $1,929 | $6,609 | | • | | • |
| San Jacinto College North (Houston) | 50 | Aug | 12 | Cert | $1,300 | $2,000 | | | | |
| Texas School of Business - North Campus (Houston) | 30 | 6 wks days/75 nights | 7, 11 | Dipl | $0 | $11,400 | | • | | |
| Texas School of Business Friendswood | 30 | Monthly | 7, 10 | Dipl | $11,570 | $11,570 | | • | | |
| Texas School of Business Southwest (Houston) | 30 | 6 wks days/75 nights | 7, 11 | Dipl | $11,400 | $0 | | • | | |
| Western Technical College (El Paso) | 25 | Monthly | 9 | Cert | $9,090 | $0 | | • | | |
| **Utah** | | | | | | | | | | |
| Ameritech College (Provo) | 160 | Every 5 wks | 9 | Dipl | $10,084 | $10,084 | | | | |
| Ogden - Weber Applied Technology College | 20 | Year-round | 12 | Cert, AAT | $1,248 | $3,744 | | | | • |
| Salt Lake Community College (Salt Lake City) | 30 | Aug Jan May | 12 | Cert | $3,500 | $9,900 | | • | | • |
| Stevens Henager College (Ogden) | 30 | Every 4 wks | 15, 20 | AOS | $14,200 | $14,200 | | • | | |
| Utah Career College (West Jordan) | 16 | Jan Apr Jul Oct | 12, 18 | Dipl, AAS | $16,120 | $16,120 | | • | | • |
| **Virginia** | | | | | | | | | | |
| Bryant & Stratton College - Richmond | 20 | Sep Jan May | 20 | AAS | $18,600 | $18,600 | | • | | • |
| Medical Careers Institute (Virginia Beach) | 30 | Every 5 wks | 14 | AAS | $20,900 | $20,900 | | • | | • |
| Miller-Motte Technical College (Lynchburg) | 120 | 8x/yr | 18 | AOS | $18,144 | $18,144 | | • | | |
| National College of Business & Technology - Bluefield | 75 | Dec Mar Jun Sep | 24 | AS | $6,624 | $6,624 | | • | | |
| National College of Business & Technology - Charlottesville | 60 | Dec Mar Jun Sep | 24 | AS | $6,624 | $6,624 | | • | | |
| National College of Business & Technology - Danville | 12 | Jun Sep Dec Mar | 24 | AS | $7,104 | $7,104 | | • | | |
| National College of Business & Technology - Harrisonburg | 30 | Sep Dec Mar Jun | 24 | AS | $7,776 | $7,776 | | • | | |
| National College of Business & Technology - Lynchburg | 70 | Dec Mar Jun Sep | 24 | AS | $7,104 | $7,104 | | • | | |
| National College of Business & Technology - Martinsville | 30 | Mar Jun Sep Dec | 24 | Assoc | $10,880 | $10,880 | | • | | • |
| National College of Business & Technology - Roanoke Valley (Salem) | 75 | Dec Mar Jun Sep | 24 | AAS | $6,624 | $6,624 | | • | | |
| **Washington** | | | | | | | | | | |
| Bryman College - Everett | 270 | Every 4 wks | 8 | Dipl | $11,655 | $11,655 | | • | | |
| Bryman College - Port Orchard | 32 | Every 4 wks | 8 | Dipl | $12,255 | $0 | | • | | |
| Bryman College - Renton | 364 | Monthly | 8, 10 | Dipl | $11,605 | $11,605 | | • | | |
| Clark College (Vancouver) | 30 | Sep | 20 | Cert, AAS | $3,964 | $4,223 | | | | • |
| Clover Park Technical College (Lakewood) | 40 | Quarterly | 12 | Cert | $5,764 | $0 | | • | | • |
| Everest College (Vancouver) | 30 | Every 6 wks | 15, 18 | Dipl, AAS | $11,808 | $11,808 | | • | | • |
| Everett Community College | Var | Continuous | 15, 24 | Cert, ATA | $2,627 | $9,568 | | • | • | • |
| Highline Community College (Des Moines) | 70 | Quarterly | 18 | AAS | $2,664 | $7,443 | | • | | • |
| Lake Washington Technical College (Kirkland) | 30 | Sep Apr | 12, 15 | Cert, AAS | $3,483 | $3,483 | | | | • |
| Lower Columbia College (Longview) | 24 | Sep | 12 | Cert | $3,526 | $4,535 | | | | • |
| North Seattle Community College | 100 | Sep Jan Mar Jun | 12, 15 | Cert, AAS | $3,240 | $10,160 | | • | | • |
| Olympic College (Bremerton) | 30 | Sep | 12 | Cert | $2,991 | $4,896 | | • | | • |
| Seattle Vocational Institute | 30 | Sep Jan Apr Jun | 12 | Cert | $3,800 | $3,800 | | | | |
| South Puget Sound Community College (Olympia) | 20 | Sep | 12, 24 | Cert, ATA | $2,971 | $3,164 | | | | • |
| Whatcom Community College (Bellingham) | 24 | Sep Jan | 12, 24 | Cert, AAS | $2,725 | $2,725 | | • | | • |
| **West Virginia** | | | | | | | | | | |
| Huntington Junior College | 200 | Sep Jan Apr Jun | 18, 12 | Dipl, AS | $7,800 | $7,800 | | • | | |
| **Wisconsin** | | | | | | | | | | |
| Bryant & Stratton College - Milwaukee | 300 | Varies | 24 | AS | $11,820 | $0 | | • | | • |
| Chippewa Valley Technical College (Eau Claire) | 16 | Jan Aug | 12 | Dipl | $2,432 | $18,984 | | • | | • |
| Concordia University Wisconsin (Milwaukee) | 30 | Aug | 9 | Cert | $6,970 | $6,970 | | | | • |
| Fox Valley Technical College (Oshkosh) | 25 | Jan | 9 | Dipl | $3,073 | $17,659 | | • | | |
| Gateway Technical College - Elkhorn | 15 | Aug | 9 | Dipl | $2,618 | $12,608 | | | | |
| Lac Courte Oreills Ojibwa Community College (Hayward) | 20 | Sep | 18 | AAS | $4,050 | $4,050 | | • | | • |
| Madison Area Technical College | 24 | Aug Jan | 9 | Dipl | $4,418 | $16,561 | | | | |
| Mid-State Technical College - Marshfield | 12 | Aug Jan | 9 | Dipl | $2,500 | $2,500 | | • | | |
| Mid-State Technical College - Stevens Point | 12 | Jan Aug | 9 | Dipl | $2,500 | $2,500 | | • | | |
| Moraine Park Technical College (Fond du Lac) | 30 | Aug | 9 | Dipl | $4,230 | $20,007 | | • | | • |
| Northeast Wisconsin Technical College (Green Bay) | 40 | Aug | 10 | Dipl | $2,576 | $0 | | • | | |
| Western Technical College (La Crosse) | 20 | Jan Aug | 9 | Dipl | $2,656 | $16,840 | | | | • |
| Wisconsin Indianhead Technical College (New Richmond) | 30 | Aug Jan | 9 | Dipl | $3,479 | $0 | | • | | • |
| Wisconsin Indianhead Technical College - Superior | 22 | Jan | 10 | Dipl | $3,200 | $0 | | | | • |

*Data are shown only for programs that completed the 2006 AMA Survey of Health Professions Education Programs

‡Key to Offers: 1: Evening or weekend classes; 2: Non-English instruction; 3: Cultural competence instruction

# Medical Illustrator

## Occupational Description

Medical illustrators specialize in the visual transformation, display, and communication of scientific information. Their graduate level training in biomedical science, art, design, visual technology, education, and communication enables them to understand and visualize scientific data and concepts to teach the general public and professionals in the fields of health care, research, pharmaceuticals, biotechnology, and demonstrative evidence. Medical illustrations appear in medical textbooks, medical advertisements, professional journals, instructional animations, computer-assisted learning programs, scientific exhibits, lecture presentations, general magazines, and courtroom presentations. Depending on the intended use, medical illustrations can be highly realistic and anatomically precise or they can be thematic, interpretive, abstract, or even wildly conceptual. Although medical illustration is commonly seen in print and electronic media, medical illustrators also work in three dimensions, creating anatomical teaching models, models for simulated medical procedures, and prosthetic parts for patients.

## Job Description

Medical illustrators work closely with clients to interpret their needs and create visual solutions for them through effective problem solving. While some medical illustrators specialize in a single art medium or work primarily for one medical specialty, the majority handle an ever-changing variety of assignments from different clients, involving a variety of biomedical content, and requiring a variety of media solutions. In addition to design and production roles, medical illustrators may function as consultants, art directors, supervisors, and administrators within the field of biocommunications.

## Employment Characteristics

Many medical illustrators are employed in medical schools and large medical centers that have teaching and research programs. Other medical artists are employed by hospitals, clinics, dental schools, or schools of veterinary medicine. Some institutional medical illustrators work alone, whereas, others are part of large multimedia departments. Other medical illustrators choose to target specific markets such as medical publishers, pharmaceutical companies, advertising agencies, animation studios, physicians, or attorneys. Some work independently on a freelance basis; others set up small companies designed to provide illustration services to various targeted markets.

The employment outlook for medical illustrators is good. This is in part due to the relatively few medical illustrators who graduate each year, and in part due to the growth in medical research that continually reveals new treatments and technologies that require medical illustrations. A growing demand by patients to better understand their own bodies and medical options has expanded the need for medical illustration aimed at the lay public. In addition, increased need for medical illustrations and models to educate juries during courtroom presentations has expanded the medical-legal subspecialty of medical illustration.

## Salary

Earnings vary significantly according to 1) the experience and ability of the artist, 2) the type of work, and 3) the area of the country where one works. The average starting salary in an institutional setting for a graduate of an accredited program of medical illustration is around $45,000 a year plus fringe benefits. Experienced salaried illustrators usually earn about $53,000 a year. Administrators and faculty members generally earn somewhat more. Salaried illustrators often supplement their income with freelance work. Refer to Section IV, Table 5 of this *Directory* for more information, or see www.ama-assn.org/go/salary.

## Educational Programs

**Length.** Accredited programs generally last 2 full years resulting in a master's degree.

**Prerequisites.** All current medical illustrator programs are at an advanced level and are based on a master's model. Although admission requirements to accredited programs vary, a bachelors degree with an emphasis on art and science is preferred. In addition, a portfolio of artwork and a personal interview are required.

**Curriculum.** While the area of emphasis may vary from program to program, the curriculum includes the following courses: an advanced course in human anatomy with dissection and courses in other biomedical sciences such as embryology, histology, neuroanatomy, cell biology, molecular biology, physiology, pathology, immunology, pharmacology, and genetics. Art and theory courses include anatomical drawing, illustration techniques in line, tone, and color (hand-rendered and computer-generated), surgical illustration, graphic design, computer graphics and multimedia, instructional design, motion media production, three-dimensional models and exhibits, management and business practices, and professional ethics.

## Inquiries

**Careers/Curriculum**

Association of Medical Illustrators
PO Box 1897
Lawrence, KS 66044
866 393-4264
E-mail: hq@ami.org

**Program Accreditation**

Commission on Accreditation of Allied Health Education Programs (CAAHEP) in collaboration with:
Accreditation Review Committee for the Medical Illustrator
c/o CAAHEP
1361 Park Street
Clearwater, FL 33756
727 210-2350
727 210-2354 Fax
www.caahep.org

## Medical Illustrator

### Georgia

**Medical College of Georgia**
Medical Illustrator Prgm
Allied Health and Grad Studies, Ste CJ-1101
1120 15th St
Augusta, GA 30912-0300
*Prgm Dir:* Steven J Harrison, BS MS CMI FAMI FBPA
*Tel:* 706 721-3266   *Fax:* 706 721-7855
*E-mail:* sharriso@mcg.edu

### Illinois

**University of Illinois at Chicago**
Medical Illustrator Prgm
Dept of Biomed Visualization (MC-527)
College of Applied Health Sciences
Chicago, IL 60612
*Prgm Dir:* Scott Barrows, BS FAMI
*Tel:* 312 996-8344   *Fax:* 312 996-8342
*E-mail:* sbarrows@uic.edu

### Maryland

**Johns Hopkins School of Medicine**
Medical Illustrator Prgm
Dept of Art Applied to Med
1830 E Monument St, Ste 7000
Baltimore, MD 21205
www.hopkinsmedicine.org
*Prgm Dir:* Gary P Lees, BS MS CMI FAMI
*Tel:* 410 955-3213   *Fax:* 410 955-1085
*E-mail:* glees@jhmi.edu

### Ontario, Canada

**University of Toronto**
Medical Illustrator Prgm
1 King's College Circle
Medical Sciences Building, Rm 2356
Toronto, ON M5S 3J3
*Prgm Dir:* Linda Wilson-Pauwels
*Tel:* 907 569-4266
*E-mail:* l.wilson.pauwels@utoronto.ca

### Texas

**Univ of Texas Southwestern Med Ctr at Dallas**
Medical Illustrator Prgm
Biomedical Communications Graduate Program
5323 Harry Hines Blvd
Dallas, TX 75235-8881
www.utsouthwestern.edu/biomedcom
*Prgm Dir:* Lewis E Calver, MS
*Tel:* 214 648-4699   *Fax:* 214 648-5353
*E-mail:* biocomm@utsouthwestern.edu

## Medical Illustrator

| Programs* | Class Capacity | Begins | Length (months) | Award | Res. Tuition | Non-res. Tuition | Stipend | Offers:‡ 1 | 2 | 3 |
|---|---|---|---|---|---|---|---|---|---|---|
| **Maryland** | | | | | | | | | | |
| Johns Hopkins School of Medicine (Baltimore) | 6 | Aug | 22 | MA | $32,200 | $32,200 | | | | |
| **Texas** | | | | | | | | | | |
| Univ of Texas Southwestern Med Ctr at Dallas | 7 | May | 24 | Dipl, MA | $1,200 | $7,800 | | | | |

*Data are shown only for programs that completed the 2006 AMA Survey of Health Professions Education Programs
‡Key to Offers: 1: Evening or weekend classes; 2: Non-English instruction; 3: Cultural competence instruction

# Medical Librarian

### Occupational Description

Medical librarians are information professionals who specialize in health resources and provide medical information for physicians, allied health professionals, patients, consumers, students, and corporations. Using materials ranging from traditional print sources to electronic databases, medical librarians devise and use innovative strategies to assess and deliver information to their clients. Physicians often call upon medical librarians to provide life-saving information for patient care.

### Job Description

Medical librarians help improve the quality of patient care by helping health care professionals stay abreast of new developments and treatments. Additionally, they find relevant health information for patients and consumers, serve as educators for students pursuing health care degrees, and provide training in the location and use of medical resources. Increasingly medical librarians use technology to design Web sites and distance education programs and to construct digital libraries. Others work for Internet companies and electronic publishers that index and organize information for the Web. Medical librarians also participate as members of research teams on university campuses and serve health care corporations, such as insurance and pharmaceutical companies, by providing information necessary for developing new products and services.

In administrative roles, medical librarians serve as directors, chief information officers, and deans or associate deans of information technology departments. Librarians ensure their informational mission is accomplished by providing leadership and strategic planning for their institutions, managing multimillion dollar budgets, pursuing grant funding, and developing marketing and public relations plans for their libraries.

Medical librarians work closely with support staff in the library to accomplish day-to-day tasks. Known as medical library assistants, these support staff provide critical operations support in all areas of the library, including circulation, serials management, acquisitions, interlibrary loan, cataloging, billing, and reference services. Some states have associations and special interest groups that support the educational needs of library support staff. Such organizations include the New York State Library Assistants' Association (NYSLAA) and the Metropolitan New York Library Council's Library Assistants, Support Staff and Associates special interest group. NYSLAA sponsors a Certificate of Achievement Program that recognizes library assistants for their contributions to libraries and the library profession.

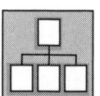

### Employment Characteristics

Medical librarians and medical library assistants are employed anywhere health information is needed, including hospitals, academic medical centers, clinics, colleges, universities, professional schools, consumer health libraries, research centers, foundations, biotechnology centers, insurance companies, medical equipment manufacturers, pharmaceutical companies, publishers, and federal state and local government agencies.

### Salary

Salaries vary according to the type and location of the institution, level of responsibility and technical skill, and length of employment. According to the Medical Library Association (MLA), the average starting salary for entry-level medical librarians was $40,832 in 2005. The overall average salary for experienced medical librarians was $57,982. Library directors can earn up to $158,600. Refer to Section IV, Table 5 of this *Directory* for more information, or see www.ama-assn.org/go/salary.

### Educational Programs

**Length.** Programs are 1 to 2 years and result in a master's degree.

**Prerequisites.** Medical librarians must have a master of library and information science degree from an American Library Association–accredited school. An undergraduate degree in any field is necessary for admission to a master's program. Undergraduate courses in biology, medical sciences, medical terminology, computer science, education, and management are helpful. Medical librarians may also apply for membership in the Academy of Health Information Professionals, a credentialing program for medical librarians sponsored by the Medical Library Association.

**Curriculum.** Programs leading to a master of library and information science degree include a wide variety of courses. All students take core courses in research, information resources, cataloging, and management and choose between tracks for public, school, academic, or special libraries. Upon choosing a track, the student and academic advisor select courses that reflect the student's career goals. Those wishing to focus on systems and technology will take a variety of technology courses in addition to the core and specialty track courses. Medical librarianship falls within the special library curriculum in many schools of library and information science. Medical library curriculum courses include resources for consumer health information, resources and services for health sciences information, medical informatics, and resources and services for special populations.

### Inquiries

**Careers/Credentialing**
Medical Library Association
65 East Wacker Place, Suite 1900
Chicago, IL 60601-7246
312 419-9094
312 419-8950 Fax
E-mail: mlapd2@mlahq.org
www.mlanet.org/career

**Library Assistants**
New York State Library Assistants' Association
www.nyslaa.org

Metropolitan New York Library Council
Library Assistants, Support Staff and Associates
Vergie Savage-Branch
Cornell University Medical College Library
1300 York Avenue
New York, NY 10021
212 746-6091
www.metro.org/SIGs/lassa.html

**Program Accreditation**

American Library Association, Committee on Accreditation/Office
for Accreditation
Karen O'Brien, Director
50 East Huron
Chicago, IL 60611
312 280-2434
E-mail: kobrien@ala.org

## Medical Librarian

## Alabama

**University of Alabama**
Medical Librarian Prgm
School of Library and Information Studies
Box 870252
Tuscaloosa, AL 35487-0252
*Prgm Dir:* Elizabeth Aversa, Dean
*Tel:* 205 348-4610   *Fax:* 205 348-3746
*E-mail:* info@slis.ua.edu

## Alberta, Canada

**University of Alberta**
Medical Librarian Prgm
School for Library & information Studies
3-20 Rutherford South
Edmonton, AB T6G 2J4
*Prgm Dir:* Anna Altmann
*Tel:* 780 492-4578   *Fax:* 780 492-2430
*E-mail:* slis@ualberta.ca

## Arizona

**University of Arizona**
Medical Librarian Prgm
School of Information Resources & Library Science
1515 E First St
Tucson, AZ 85719
*Prgm Dir:* Brooke E Sheldon, Dean
*Tel:* 520 621-3565   *Fax:* 520 621-3279
*E-mail:* sirls@u.arizona.edu

## British Columbia, Canada

**University of British Columbia**
Medical Librarian Prgm
School of Library, Archival & Information Studies
Ste 301, 6190 Agronomy Rd
Vancouver, BC V6T 1Z3
*Prgm Dir:* Edie Rasmussen
*Tel:* 604 822-2404   *Fax:* 604 822-6006
*E-mail:* slais@interchange.ubc.ca

## California

**University of California - Los Angeles**
Medical Librarian Prgm
Graduate School of Education & Information
Studies Bldg
Los Angeles, CA 90095-1520
*Prgm Dir:* Virginia A Walter, Chair
*Tel:* 310 825-8799   *Fax:* 310 206-3076
*E-mail:* vwalter@ucla.edu

**San Jose State University**
Medical Librarian Prgm
School of Library & Information Science
One Washington Square
San Jose, CA 95192-0029
*Prgm Dir:* Blanche Woolls
*Tel:* 408 924-2490   *Fax:* 408 924-2476
*E-mail:* office@slis.sjsu.edu

## Connecticut

**Southern Connecticut State University**
Medical Librarian Prgm
Information and Library Sciences
501 Crescent St
New Haven, CT 06515
*Prgm Dir:* Edward Harris, Dean
*Tel:* 203 392-5781   *Fax:* 203 392-5780
*E-mail:* mckayl1@southernct.edu

## District of Columbia

**Catholic University of America**
Medical Librarian Prgm
School of Library & Information Science
620 Michigan Ave NE
Washington, DC 20064
*Prgm Dir:* Martha Hale, Dean
*Tel:* 202 319-5085   *Fax:* 202 319-5574
*E-mail:* cua-slis@cua.edu

## Florida

**Florida State University**
Medical Librarian Prgm
School of Information Studies
Shores Bldg
Tallahassee, FL 32306-2100
*Prgm Dir:* Jane B Robbins, Dean
*Tel:* 850 644-5775   *Fax:* 850 644-9763
*E-mail:* grad@lis.fsu.edu

**University of South Florida**
Medical Librarian Prgm
School of Library and Information Sciences
4202 E Fowler Ave
Tampa, FL 33620
*Prgm Dir:* Vicki Gregory
*Tel:* 813 974-3520   *Fax:* 813 974-6840
*E-mail:* lis@cas.usf.edu

## Hawaii

**University of Hawaii at Manoa**
Medical Librarian Prgm
Library and Information Science Program
2550 McCarthy Mall
Honolulu, HI 96822
*Prgm Dir:* Diane Nahl, Chair
*Tel:* 808 956-7321   *Fax:* 808 956-5835
*E-mail:* slis@hawaii.edu

## Illinois

**Univ of Illinois at Urbana - Champaign**
Medical Librarian Prgm
Graduate School of Library and Information Science
501 E Daniel St
Champaign, IL 61820-6211
*Prgm Dir:* John Unsworth, Dean
*Tel:* 217 333-3280   *Fax:* 217 244-3302
*E-mail:* apply@alexia.lis.uiuc.edu

**Dominican University**
Medical Librarian Prgm
Graduate School of Library and Information Science
7900 W Division St
River Forest, IL 60305
*Prgm Dir:* Susan Roman, Dean
*Tel:* 708 524-6845   *Fax:* 708 524-6657
*E-mail:* gslis@dom.edu

## Indiana

**Indiana University**
Medical Librarian Prgm
School of Library and Information Science at
Indianapolis
755 W Michigan St
Indianapolis, IN 46202
http://slis.iupui.edu
*Prgm Dir:* Katherine Schilling, MLS EdD AHIP
*Tel:* 317 278-2375   *Fax:* 317 278-1807
*E-mail:* katschil@iupui.edu

## Iowa

**University of Iowa**
Medical Librarian Prgm
School of Library and Information Science
3087 Main Library
Iowa City, IA 52242-1420
*Prgm Dir:* David Eichmann
*Tel:* 319 335-5707   *Fax:* 319 335-5374
*E-mail:* slis@uiowa.edu

## Kansas

**Emporia State University**
Medical Librarian Prgm
School of Library and Information Management
1200 Commerical, Campus Box 4025
Emporia, KS 66801
*Prgm Dir:* Diane Bailiff, Inter Dean
*Tel:* 620 341-5203   *Fax:* 620 641-5233
*E-mail:* kitselmc@emporia.edu

# Kentucky

**University of Kentucky**
Medical Librarian Prgm
School of Library and Information Science
502 King Library
Lexington, KY 40506-0039
*Prgm Dir:* Timothy W Sineath
*Tel:* 859 257-8876   *Fax:* 859 257-4205
*E-mail:* salsman@uky.edu

# Louisiana

**Louisiana State University**
Medical Librarian Prgm
School of Library and Information Science
267 Coates Hall
Baton Rouge, LA 70803
http://slis.lsu.edu
*Prgm Dir:* Beth Paskoff, Dean, PhD
*Tel:* 225 578-3158   *Fax:* 225 578-4581
*E-mail:* slis@lsu.edu
*Med Dir:* Michelynn McKnight, PhD AHIP

# Maryland

**University of Maryland**
Medical Librarian Prgm
College of Information Studies
4105 Hornbake Bldg
College Park, MD 20742
*Prgm Dir:* Bruce Dearstyne, Dean
*Tel:* 301 405-2033   *Fax:* 301 314-9145
*E-mail:* lbscgrad@deans.umd.edu

# Massachusetts

**Simmons College**
Medical Librarian Prgm
Graduate School of Library and Information Science
300 The Fenway
Boston, MA 02115
*Prgm Dir:* Michele Cloonan, Dean
*Tel:* 617 521-2800   *Fax:* 617 521-3192
*E-mail:* gslis@simmons.edu

# Michigan

**University of Michigan**
Medical Librarian Prgm
School of Information
550 E University Ave, 304 W Hall Bldg
Ann Arbor, MI 48109-1092
*Prgm Dir:* John L King, Dean
*Tel:* 734 764-9376   *Fax:* 734 764-2475
*E-mail:* si.admissions@umich.edu

**Wayne State University**
Medical Librarian Prgm
Library and Information Science Program
106 Kresge Library
Detroit, MI 48202
www.lisp.wayne.edu
*Prgm Dir:* Joseph J Mika, PhD MLS
*Tel:* 313 577-1825   *Fax:* 313 577-7563
*E-mail:* asklis@wayne.edu
*Med Dir:* Lynda Baker, PhD MLS

# Mississippi

**University of Southern Mississippi**
Medical Librarian Prgm
School of Library and Information Science
118 College Dr #5146
Hattiesburg, MS 39406-0001
*Prgm Dir:* M J Norton
*Tel:* 601 266-4228   *Fax:* 601 266-5774
*E-mail:* slis@usm.edu

# Missouri

**University of Missouri - Columbia**
Medical Librarian Prgm
Information Science and Learning Technologies
303 Townsend Hall
Columbia, MO 65211
*Prgm Dir:* John Wedman, Dean
*Tel:* 573 882-4546   *Fax:* 573 884-0122
*E-mail:* sislt@missouri.edu

# New Jersey

**Rutgers University**
Medical Librarian Prgm
Dept of Library and Information Science
4 Huntington St
New Brunswick, NJ 08901-1071
*Prgm Dir:* Nicholas J Belkin
*Tel:* 732 932-7500, Ext 1071   *Fax:* 732 932-2644
*E-mail:* scilsmls@scils.rutgers.edu

# New York

**SUNY at Albany**
Medical Librarian Prgm
School of Inform Science and Policy Draper 113
135 Western Ave
Albany, NY 12222
*Prgm Dir:* Peter Bloniarz, Dean
*Tel:* 518 442-5110   *Fax:* 518 442-5367
*E-mail:* infosci@albany.edu

**Long Island University - C W Post Campus**
Medical Librarian Prgm
Palmer School of Library and Information Science
720 Northern Blvd
Brookville, NY 11548
*Prgm Dir:* Michael E D Koenig, Dean
*Tel:* 516 299-2866   *Fax:* 516 299-4168
*E-mail:* palmer@cwpost.liu.edu

**University at Buffalo - SUNY**
Medical Librarian Prgm
Library and Information Studies
534 Baldy Hall
Buffalo, NY 14260
*Prgm Dir:* Judith Robinson, Chair
*Tel:* 716 645-2412   *Fax:* 716 645-3775
*E-mail:* ub-lis@buffalo.edu

**CUNY Queens College**
Medical Librarian Prgm
Graduate School of Library and Information Studies
65-30 Kissena Blvd
Flushing, NY 11367-1597
*Prgm Dir:* Marianne Cooper
*Tel:* 718 997-3790   *Fax:* 718 997-3797
*E-mail:* gslis@qcunix1.qc.edu

**St John's University**
Medical Librarian Prgm
Division of Library and Information Science
8000 Utopia Pkwy
Jamiaca, NY 11439
*Prgm Dir:* Sherry Vellucci
*Tel:* 718 990-6200   *Fax:* 718 990-2071
*E-mail:* libis@stjohns.edu

**Pratt Institute**
Medical Librarian Prgm
School of Information and Library Science
144 W 14th St, 6th Fl
New York, NY 10011
*Prgm Dir:* Marie Radford, Acting Dean
*Tel:* 212 647-7682   *Fax:* 212 367-2492
*E-mail:* infosils@pratt.edu

**Syracuse University**
Medical Librarian Prgm
School of Information Studies
4-206 Center for Science and Technology
Syracuse, NY 13244
*Prgm Dir:* Raymond Von Dran, Dean
*Tel:* 315 443-2911   *Fax:* 315 443-5673
*E-mail:* ist@syr.edu

# North Carolina

**Univ of North Carolina at Chapel Hill**
Medical Librarian Prgm
School of Information and Library Science
100 Manning Hall, CB #3360
Chapel Hill, NC 27599-3360
*Prgm Dir:* Joanne Gard Marshall, Dean
*Tel:* 919 962-8366   *Fax:* 919 962-8071
*E-mail:* info@ils.unc.edu

**North Carolina Central University**
Medical Librarian Prgm
School of Library and Information Sciences
1800 Fayetteville St
Durham, NC 27707
*Prgm Dir:* Robert M Ballard, Dean
*Tel:* 919 530-6485   *Fax:* 919 530-6402
*E-mail:* lsis@nccu.edu

**Univ of North Carolina at Greensboro**
Medical Librarian Prgm
Dept of Library and Information Studies
349 Curry Bldg
Greensboro, NC 27401-6170
*Prgm Dir:* Lee Shiflett
*Tel:* 336 334-3477   *Fax:* 336 334-5060
*E-mail:* lis@uncg.edu

# Nova Scotia, Canada

**Dalhousie University**
Medical Librarian Prgm
School of Library and Information Studies
3rd Fl, Killiam Library
Halifax, NS B3H 3J5
*Prgm Dir:* Fiona Black
*Tel:* 902 494-3656   *Fax:* 902 494-2451
*E-mail:* slis@dal.ca

# Ohio

**Kent State University**
Medical Librarian Prgm
School of Library and Information Sciences
PO Box 5190
Kent, OH 44242-0001
*Prgm Dir:* Richard E Rubin
*Tel:* 330 672-2782   *Fax:* 330 672-7965
*E-mail:* inform@slis.kent.edu

# Oklahoma

**University of Oklahoma**
Medical Librarian Prgm
School of Library and Information Studies
401 W Brooks, Rm 120
Norman, OK 73019-6032
*Prgm Dir:* Danny P Wallace
*Tel:* 405 325-3921   *Fax:* 405 325-7648
*E-mail:* slisinfo@ou.edu

# Ontario, Canada

**University of Western Ontario**
Medical Librarian Prgm
Graduate Programs in Library and Information Science
255 Middlesex College
London, ON N6A 5B7
*Prgm Dir:* Catherine Ross, Dean
*Tel:* 519 661-4017   *Fax:* 519 661-3506
*E-mail:* mlisinfo@uwo.ca

**University of Toronto**
Medical Librarian Prgm
Faculty of Information Studies
140 St George St
Toronto, ON M5S 3G6
www.fis.utoronto.ca
*Prgm Dir:* Brian Cantwell Smith, Dean, PhD
*Tel:* 416 978-3202, Ext 1   *Fax:* 416 978-5762
*E-mail:* dean@fis.utoronto.ca
*Med Dir:* Nadine Wathen, PhD

# Pennsylvania

**Clarion University of Pennsylvania**
Medical Librarian Prgm
Department of Library Science
210 Carlson Library Bldg
Clarion, PA 16214
*Prgm Dir:* Andrea L Miller, Chair
*Tel:* 814 393-2271   *Fax:* 814 393-2150
*E-mail:* reed@clarion.edu

**Drexel University**
Medical Librarian Prgm
College of Information Science and Technology
3141 Chestnut St
Philadelphia, PA 19104-2875
*Prgm Dir:* David E Fenske, Dean
*Tel:* 215 895-2474   *Fax:* 215 895-2494
*E-mail:* info@cis.drexel.edu

**University of Pittsburgh**
Medical Librarian Prgm
Department of Library and Information Science
135 N Bellefield Ave
Pittsburgh, PA 15260
*Prgm Dir:* Ronald Larsen, Dean
*Tel:* 412 624-5142   *Fax:* 412 624-5231
*E-mail:* inquiry@mail.sis.pitt.edu

# Puerto Rico

**University of Puerto Rico**
Medical Librarian Prgm
Information Science and Technologies
PO Box 210906
San Juan, PR 00031-1906
*Prgm Dir:* Consuelo Figueras
*Tel:* 787 763-6199   *Fax:* 787 764-2311
*E-mail:* egcti@rrpac.upr.clu.edu

# Quebec, Canada

**McGill University**
Medical Librarian Prgm
Graduate School of Library and Information Studies
3459 McTavish St, MS 57-F
Montreal, QC H3A 1Y1
*Prgm Dir:* Jamshid Beheshti
*Tel:* 514 398-4204   *Fax:* 514 398-7193
*E-mail:* gslis@mcgill.ca

**University de Montreal**
Medical Librarian Prgm
Ecole de bibiotheconomie et des sciences de
   l'information
CP 6128, succursale centre-ville
Montreal, QC H3C 3J7
*Prgm Dir:* Carol Couture
*Tel:* 514 343-6400   *Fax:* 514 343-5753
*E-mail:* ebsiinfo@ebsi.umontreal.ca

# Rhode Island

**Rhode Island School of Cytotechnology**
Medical Librarian Prgm
Graduate School of Library and Information Studies
Rodman Hall
Kingston, RI 02881
*Prgm Dir:* W Michael Havener
*Tel:* 401 874-2947   *Fax:* 401 874-4964
*E-mail:* gslis@etal.uri.edu

# South Carolina

**University of South Carolina**
Medical Librarian Prgm
School of Library and Information Science
Davis College
Columbia, SC 29208
*Prgm Dir:* Daniel Barron
*Tel:* 803 777-3858   *Fax:* 803 777-7938
*E-mail:* ddbarron@gwm.sc.edu

# Tennessee

**University of Tennessee - Knoxville**
Medical Librarian Prgm
School of Information Sciences
451 Communications Bldg
Knoxville, TN 37996-0341
*Prgm Dir:* Douglas Raber, Interim Dean
*Tel:* 865 974-2148   *Fax:* 865 974-4967
*E-mail:* sis@utk.edu

# Texas

**University of Texas at Austin**
Medical Librarian Prgm
School of Information
1 University Station D7000
Austin, TX 78712-0390
*Prgm Dir:* Andrew Dillon, Dean
*Tel:* 512 471-3821   *Fax:* 512 471-3971
*E-mail:* info@ischool.utexas.edu

**Texas Woman's University**
Medical Librarian Prgm
School of Library and Information Studies
PO Box 425438
Denton, TX 76204-5438
*Prgm Dir:* Lynn Westbrook
*Tel:* 940 898-2602   *Fax:* 940 898-2611
*E-mail:* slis@twu.edu

**University of North Texas**
Medical Librarian Prgm
School of Library and Information Sciences
PO Box 311068
Denton, TX 76203-1068
*Prgm Dir:* Philip Turner, Dean
*Tel:* 940 565-2731   *Fax:* 940 565-3101
*E-mail:* slis@unt.edu

# Washington

**University of Washington**
Medical Librarian Prgm
Information School
370 Mary Gates Hall
Seattle, WA 98195-2840
www.ischool.washington.edu
*Prgm Dir:* Harry Bruce, Dean, PhD
*Tel:* 206 685-9937   *Fax:* 206 616-3152
*E-mail:* mlis@u.washington.edu

# Wisconsin

**University of Wisconsin - Madison**
Medical Librarian Prgm
Library and Information Studies
600 N Park St
Madison, WI 53706
*Prgm Dir:* Catherine Arnott Smith
*Tel:* 608 263-2900   *Fax:* 608 263-4849
*E-mail:* casmith24@wisc.edu

**University of Wisconsin - Milwaukee**
Medical Librarian Prgm
School of Information Studies
PO Box 413
Milwaukee, WI 53201
*Prgm Dir:* Thomas D Walker, Interim Dean
*Tel:* 414 229-4707   *Fax:* 414 229-6699
*E-mail:* info@sois.uwm.edu

## Job Description

Music therapists use music within a therapeutic relationship to address physical, emotional, cognitive, and social needs of individuals of all ages, improving quality of life for persons who are well and meeting the needs of children and adults with disabilities or illnesses. After assessing the strengths and needs of each client, qualified music therapists develop a treatment plan with goals and objectives and then provide the indicated treatment. Music therapists structure the use of both instrumental and vocal music strategies to facilitate changes that are non-musical in nature. They may improvise or compose music with clients, accompany and conduct group music experiences, provide instrument instruction, direct music and movement activities, or structure music listening opportunities. Music therapists provide services for children and adults with psychiatric disorders, developmental disabilities, speech and hearing impairments, physical disabilities, and neurological impairments, among others. Music therapy interventions can be designed to promote wellness, manage stress, alleviate pain, enhance memory, improve communication, and provide unique opportunities for interaction. Depending upon the needs of the clients involved, music therapy sessions are offered on an individual or group basis. Music therapists are usually members of an interdisciplinary team of health care professionals who work collaboratively to address clients' treatment needs.

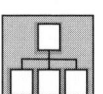

## Personal Qualifications

Music therapists should have a genuine interest in people and a desire to help others empower themselves. The essence of music therapy practice involves establishing caring and professional relationships with people of all ages and abilities. Empathy, patience, tact, a sense of humor, imagination, creativity, and an understanding of oneself are important characteristics for professionals in this field. People thinking about music therapy as a career must be accomplished musicians. They must be versatile and able to adjust to changing circumstances. Music therapists must express themselves well in speech and in writing. In addition, they must be able to work well with other health care providers.

## Employment Characteristics

Music therapists are employed in many different settings including general and psychiatric hospitals, mental health agencies, physical rehabilitation centers, nursing homes, public and private schools, substance abuse programs, forensic facilities, hospice programs, and day care facilities. Typically, full-time therapists work a standard 40-hour workweek. Some therapists prefer part-time work and choose to develop contracts with specific agencies, providing music therapy services for an hourly or contractual fee. In addition, a growing number of clinicians are choosing to start private practices in music therapy to benefit from opportunities provided through self-employment.

## Salary

According to the American Music Therapy Association (AMTA), the overall average salary for full-time music therapists was $42,364 in 2004. Individual salaries can vary by population served, work setting, geographic location, years of experience, and level of graduate education completed. The income range reported in 2004 included salaries up to $200,000. Refer to Section IV, Table 5 of this *Directory* for more information, or see www.ama-assn.org/go/salary.

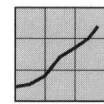

## Future Outlook

As an increasing number of consumers seek non-invasive, alternative and complementary therapies as treatment options, the need for music therapists continues to rise. An increased need for music therapists in early intervention programs, special education settings, geriatric facilities, and community based services offers a variety of employment options. The next 10 years hold positive opportunities for the music therapy profession.

## Educational Programs

**Length.** Those who wish to become music therapists must earn a bachelor's degree or higher in music therapy from one of over 70 AMTA-approved colleges and universities. Entry-level study requires academic coursework and 1,200 hours of clinical training, including a supervised internship.
**Prerequisites.** For entry into undergraduate programs, a high school diploma is required, along with demonstration of musicianship. Candidates for the masters degree must hold a baccalaureate degree or equivalent in music therapy (see "Certification" below) or be working concurrently toward fulfilling degree equivalency requirements.
**Curriculum.** The curriculum is designed to impart entry-level competencies in three main areas: musical foundations, clinical foundations, and music therapy foundations and principles. Graduate programs in music therapy examine, with greater breadth and depth, issues relevant to the clinical, professional, and academic preparation of music therapists, usually in combination with established methods of research inquiry.

## Certification

At the completion of academic and clinical training, students are eligible to take the national examination administered by the Certification Board for Music Therapists (CBMT), an independent, non-profit certifying agency fully accredited by the National Commission for Certifying Agencies. After successful completion of the CBMT examination, graduates are issued the credential necessary for professional practice, Music Therapist-Board Certified (MT-BC). To demonstrate continued competence and to maintain this credential, music therapists are required to complete 100 hours of continuing music therapy education, or to retake and pass the CBMT examination within every 5-year recertification cycle.

## Inquiries

**Education and Careers**
American Music Therapy Association (AMTA)
8455 Colesville Road, Suite 1000
Silver Spring, MD 20910
301 589-3300
301 589-5175 Fax
E-mail: info@musictherapy.org
www.musictherapy.org

**Certification**

Certification Board for Music Therapists (CBMT)
506 East Lancaster Avenue, Suite 102
Downingtown, PA 19335
800 765-2268 or 610 269-8900
610 269-9232 Fax
E-mail: info@cbmt.org
www.cbmt.org

**Program Accreditation**

National Association of Schools of Music (NASM)
11250 Roger Bacon Drive, Suite 21
Reston, VA 20190
703 437-0700
703 437-6312
E-mail: info@arts-accredit.org
www.nasm.arts-accredit.org

## Music Therapist*

## Alabama

**University of Alabama**
Music Therapy Prgm
School of Music
PO Box 870366
Tuscaloosa, AL 35487-0366
www.musictherapy.ua.edu
*Prgm Dir:* Carol A Prickett, PhD MT-BC
*Tel:* 205 348-1432    *Fax:* 205 348-1675
*E-mail:* cpricket@music.ua.edu

## Arizona

**Arizona State University**
Music Therapy Prgm
School of Music
PO Box 870405
Tempe, AZ 85287-0405
www.herbergercollege.com
*Prgm Dir:* Barbara Crowe, MM MT-BC
*Tel:* 480 965-7413    *Fax:* 480 965-2659
*E-mail:* Barbara.J.Crowe@asu.edu

## California

**California State University - Northridge**
Music Therapy Prgm
18111 Nordhoff St
Northridge, CA 91330
*Prgm Dir:* Ronald Borczon, MM RMT-BC
*Tel:* 818 677-3174
*E-mail:* rborczon@csun.edu

**Chapman University**
Music Therapy Prgm
School of Music
One University Dr
Orange, CA 92866
www.chapman.edu/music
*Prgm Dir:* David W Luce, PhD MT-BC
*Tel:* 714 532-6032    *Fax:* 714 744-7671
*E-mail:* luce@chapman.edu

**University of the Pacific**
Music Therapy Prgm
Conservatory of Music
3601 Pacific Ave
Stockton, CA 95211
http://pacific.edu/conservatory/academic_majors/music
    _therapy_college_degree.asp
*Prgm Dir:* Therese M West, PhD MT-BC FAMI
*Tel:* 209 946-2419    *Fax:* 209 946-2770
*E-mail:* twest@pacific.edu

## Colorado

**Colorado State University**
Music Therapy Prgm
School of the Arts
Department of Music, Theatre and Dance
Ft Collins, CO 80523-1178
www.colostate.edu
*Prgm Dir:* William Davis, PhD RMT
*Tel:* 970 491-5888    *Fax:* 970 491-7541
*E-mail:* william.davis@colostate.edu

## District of Columbia

**Howard University**
Music Therapy Prgm
College of Arts & Sciences, Music Dept
6th and Fairmont St NW
Washington, DC 20059
*Prgm Dir:* Donna Washington, MCAT RMT-BC
*Tel:* 202 806-7136    *Fax:* 202 806-9673
*E-mail:* dwashington@howard.edu

## Florida

**University of Miami**
Music Therapy Prgm
Frost School of Music
PO Box 248165
Coral Gables, FL 33124
*Prgm Dir:* Shannon de l'Etoile, PhD MT-BC
*Tel:* 305 284-3943    *Fax:* 305 284-3901
*E-mail:* sdel@miami.edu

**Florida State University**
Music Therapy Prgm
College of Music
Tallahassee, FL 32306-1180
*Prgm Dir:* Jayne M Standley, PhD MT-BC
*Tel:* 850 644-4565    *Fax:* 850 644-6100
*E-mail:* jstandle@mailer.fsu.edu

## Georgia

**University of Georgia**
Music Therapy Prgm
School of Music
Fine Arts Bldg
Athens, GA 30602-3153
*Prgm Dir:* Roy Kennedy
*Tel:* 706 542-2801    *Fax:* 706 542-0276
*E-mail:* rkennedy@uga.edu

**Georgia College & State University**
Music Therapy Prgm
CBX 067
Milledgeville, GA 31061
*Prgm Dir:* Chesley S Mercado, EdD MT-BC
*Tel:* 478 445-2645    *Fax:* 478 445-2645
*E-mail:* chesley.mercado@gcsu.edu

## Illinois

**Western Illinois University**
Music Therapy Prgm
Music Dept
Macomb, IL 61455
*Prgm Dir:* Bruce A Prueter, MS MT-BC
*Tel:* 309 298-1187    *Fax:* 309 298-1968
*E-mail:* BA-Prueter@wiu.edu

**Illinois State University**
Music Therapy Prgm
5660 School of Music
Normal, IL 61790-5660
*Prgm Dir:* Marie Di Giammarino, EdD MT-BC
*Tel:* 309 438-8198    *Fax:* 309 438-8318
*E-mail:* mdigiam@ilstu.edu

## Indiana

**University of Evansville**
Music Therapy Prgm
Dept of Music
1800 Lincoln Ave
Evansville, IN 47722
http://music.evansville.edu
*Prgm Dir:* Mary Ellen Wylie, PhD MT-BC
*Tel:* 812 488-2754    *Fax:* 812 488-2101
*E-mail:* mw26@evansville.edu

**Indiana Univ - Purdue Univ Ft Wayne**
Music Therapy Prgm
2101 Coliseum Blvd E
Ft Wayne, IN 46805-1499
www.ipfw.edu/academics/programs/undergraduate/m/
    music-therapy
*Prgm Dir:* Nancy Jackson, MMT MT-BC
*Tel:* 260 481-6716    *Fax:* 260 481-5422
*E-mail:* jacksonn@ipfw.edu

**Indiana Univ-Purdue U-Indianapolis**
Music Therapy Prgm
IU School of Music Program at IUPUI
535 W Michigan St, IT 379
Indianapolis, IN 46202
www.music.iupui.edu
*Prgm Dir:* Debra Burns, PhD MT-BC
*Tel:* 317 278-2014
*E-mail:* desburns@iupui.edu

**St Mary of the Woods College**
Music Therapy Prgm
Dept of Performing and Visual Arts
St Mary of the Woods, IN 47876
*Prgm Dir:* Tracy Richardson, MS MT BC
*Tel:* 812 535-5154    *Fax:* 812 535-5177
*E-mail:* trichard@smwc.edu

*Additional information about programs that
returned the AMA's survey is available in a table
beginning on page 262.

# Iowa

**University of Iowa**
Music Therapy Prgm
School of Music
Voxman Music Bldg
Iowa City, IA 52242
*Prgm Dir:* Kate Gfeller, PhD RMT
*Tel:* 319 335-1657   *Fax:* 319 335-2637
*E-mail:* kay-gfeller@uiowa.edu

**Wartburg College**
Music Therapy Prgm
Music Department, 100 Wartburg Blvd
Waverly, IA 50677
*Prgm Dir:* Marjorie O'Konski, MME MT-BC
*Tel:* 319 352-8401   *Fax:* 319 352-8501
*E-mail:* marj.okonski@wartburg.edu

# Kansas

**University of Kansas**
Music Therapy Prgm
Art, Music Ed and Music Therapy
311 Bailey Hall
Lawrence, KS 66045
*Prgm Dir:* Cynthia Colwell, PhD MT-BC
*Tel:* 785 864-4784   *Fax:* 785 864-5076
*E-mail:* ccolwell@ku.edu

# Kentucky

**University of Louisville**
Music Therapy Prgm
School of Music
Louisville, KY 40292
www.louisville.edu/music/therapy
*Prgm Dir:* Barbara L Wheeler, PhD MT-BC
*Tel:* 502 852-2316   *Fax:* 502 852-0520
*E-mail:* barbara.wheeler@louisville.edu

# Louisiana

**Loyola University New Orleans**
Music Therapy Prgm
College of Music
6363 St Charles Ave
New Orleans, LA 70118
www.loyno.edu
*Prgm Dir:* Victoria Vega, MMT MT-BC
*Tel:* 504 865-2142   *Fax:* 504 865-2852
*E-mail:* vpvega@loyno.edu

# Massachusetts

**Berklee College of Music**
Music Therapy Prgm
Chair, Music Therapy Dept
1140 Boylston St
Boston, MA 02215-3693
www.berklee.edu
*Prgm Dir:* Suzanne B Hanser, EdD MT-BC
*Tel:* 617 747-2639   *Fax:* 617 747-2605
*E-mail:* shanser@berklee.edu

**Lesley University**
Music Therapy Prgm
Expressive Therapics Division
29 Everett St
Cambridge, MA 02138
www.lesley.edu
*Prgm Dir:* Michele Forinash, DA LMHC MT-BC
*Tel:* 617 349-8166
*E-mail:* mforinas@lesley.edu

**Anna Maria College**
Music Therapy Prgm
Dept of Music, Box 45
Paxton, MA 01612-1198
*Prgm Dir:* Lisa Summer, MCAT MT-BC
*Tel:* 508 849-3454
*E-mail:* lsummer@annamaria.edu

# Michigan

**Michigan State University**
Music Therapy Prgm
School of Music
East Lansing, MI 48824-1043
*Prgm Dir:* Fredrick Tims, PhD MT BC
*Tel:* 517 353-9122   *Fax:* 517 432-2880
*E-mail:* tims@msu.edu

**Western Michigan University**
Music Therapy Prgm
School of Music
1903 W Michigan Ave
Kalamazoo, MI 49008-3834
www.wmich.edu/musictherapy
*Prgm Dir:* Brian Wilson, MM MT-BC
*Tel:* 269 387-4724   *Fax:* 269 387-1113
*E-mail:* brian.wilson@wmich.edu

**Eastern Michigan University**
Music Therapy Prgm
Dept of Music and Dance
N101 Alexander Bulding
Ypsilanti, MI 48197
www.emich.edu/music/html/music_therapy.html
*Prgm Dir:* Michael G McGuire, MM MT-BC
*Tel:* 734 487-0292   *Fax:* 734 487-6939
*E-mail:* michael.mcguire@emich.edu

# Minnesota

**Augsburg College**
Music Therapy Prgm
2211 Riverside Ave
Minneapolis, MN 55454
*Prgm Dir:* Roberta Kagin, MME MT-BC
*Tel:* 612 330-1273   *Fax:* 612 330-1264
*E-mail:* kagin@augsburg.edu

**University of Minnesota - Minneapolis**
Music Therapy Prgm
School of Music
2106 4th St S
Minneapolis, MN 55455
*Prgm Dir:* Annie Heiderscheit, PhD MT-BC FAMI NMT
*Tel:* 612 624-7512
*E-mail:* heide007@umn.edu

# Mississippi

**Mississippi University for Women**
Music Therapy Prgm
Fine & Performing Arts Division Box W-70
Columbus, MS 39701
*Prgm Dir:* Carmen Osburn, MA MT-BC
*Tel:* 662 241-7897
*E-mail:* cosburn@MUW.Edu

**William Carey College**
Music Therapy Prgm
498 Tuscan Ave
Hattiesburg, MS 39401
*Prgm Dir:* Paul D Cotton, PhD RMT
*Tel:* 601 318-6416   *Fax:* 601 582-6414
*E-mail:* paul.cotten@wmcarey.edu

# Missouri

**University of Missouri - Kansas City**
Music Therapy Prgm
Conservatory of Music, 316 Grant Hall
4949 Cherry
Kansas City, MO 64110-2229
*Prgm Dir:* Robert Groene, PhD MT-BC
*Tel:* 816 235-2920   *Fax:* 816 235-5864
*E-mail:* groener@umkc.edu

**Drury University**
Music Therapy Prgm
900 N Benton Ave
Springfield, MO 65802
*Prgm Dir:* Michael Cassity, PhD MT-BC
*Tel:* 417 873-7370
*E-mail:* mcassity@drury.edu

**Maryville University**
Music Therapy Prgm
650 Maryville University Dr
St Louis, MO 63141
www.maryville.edu
*Prgm Dir:* Cynthia A Briggs, MM PsyD MT-BC
*Tel:* 314 589-9441   *Fax:* 314 529-9039
*E-mail:* cbriggs@maryville.edu

# New Jersey

**Montclair State University**
Music Therapy Prgm
Music Dept
Upper Montclair, NJ 07043
www.montclair.edu/music
*Prgm Dir:* Karen Goodman, PhD MT-BC
*Tel:* 973 655-7583
*E-mail:* goodmank@mail.montclair.edu

# New York

**SUNY College at Fredonia**
Music Therapy Prgm
School of Music
Mason Hall
Fredonia, NY 14063
*Prgm Dir:* Joni Milgram-Luterman, PhD MT-BC
*Tel:* 716 673-4648   *Fax:* 716 673-3154
*E-mail:* joni.milgram-luterman@fredonia.edu

**SUNY at New Paltz**
Music Therapy Prgm
Music Dept
75 S Manheim Blvd
New Paltz, NY 12561
www.newpaltz.edu
*Prgm Dir:* Russell Hilliard, PhD LCSW MT_BC
*Tel:* 845 257-2709   *Fax:* 845 257-3121
*E-mail:* rehmt@aol.com

**New York University**
Music Therapy Prgm
35 W 4th St
New York, NY 10012
*Prgm Dir:* Barbara Hesser, MA CMT
*Tel:* 212 998-5452
*E-mail:* barbara.hesser@nyu.edu

**Nazareth College of Rochester**
Music Therapy Prgm
4245 East Ave
Rochester, NY 14618
*Prgm Dir:* Bryan C Hunter, PhD MT-BC
*Tel:* 585 389-2702   *Fax:* 585 586-2452
*E-mail:* bchunter@naz.edu

# North Carolina

### Appalachian State University
Music Therapy Prgm
Hayes School of Music
873 River St
Boone, NC 28608
www.appstate.edu
*Prgm Dir:* Cathy McKinney, PhD MT BC
*Tel:* 828 262-6444   *Fax:* 828 262-6446
*E-mail:* mckinnych@appstate.edu

### Queens University of Charlotte
Music Therapy Prgm
Music Dept
1900 Selwyn Ave
Charlotte, NC 28274-0001
www.queens.edu
*Prgm Dir:* Frances McClain, PhD MT-BC
*Tel:* 704 337-2570   *Fax:* 704 337-2356
*E-mail:* mcclainf@queens.edu

### East Carolina University
Music Therapy Prgm
212 AJ Fletcher Music Center
Greenville, NC 27858
*Prgm Dir:* Barbara Cobb Memory, PhD RMT-BC
*Tel:* 252 328-6343
*E-mail:* memoryb@mail.ecu.edu

# North Dakota

### University of North Dakota
Music Therapy Prgm
3350 Campus Rd
Grand Forks, ND 58202
www.undmusic.org
*Prgm Dir:* Therese Costes, MT-BC MA MSW
*Tel:* 701 777-2828
*E-mail:* therese_costes@und.nodak.edu

# Ohio

### Ohio University
Music Therapy Prgm
School of Music
440 Music Bldg
Athens, OH 45701
*Prgm Dir:* Anita Louise Steele, MT
*Tel:* 740 593-4249   *Fax:* 740 593-1429
*E-mail:* steelea@ohio.edu

### Baldwin-Wallace College
Music Therapy Prgm
Cleveland Consortium
275 Eastland Rd
Berea, OH 44017
*Prgm Dir:* A Louise Steele, M MEd MT-BC
*Tel:* 740 593-4249
*E-mail:* lkay@bw.edu

### University of Dayton
Music Therapy Prgm
Music Dept
300 College Park Ave
Dayton, OH 45469-0290
*Prgm Dir:* Susan Gardstrom, PhD MT-BC
*Tel:* 937 229-3908   *Fax:* 937 229-3916
*E-mail:* SGardstrom@udayton.edu

### College of Wooster
*Cosponsor:* Cleveland Music Therapy Consortium
Music Therapy Prgm
Dept of Music
Wooster, OH 44691
*Prgm Dir:* Lalene DyShere Kay, MM MT-BC
*Tel:* 330 263-2419
*E-mail:* lkay@wooster.edu

# Oklahoma

### Southwestern Oklahoma State University
Music Therapy Prgm
100 Campus Dr
Weatherford, OK 73096
www.swosu.edu/music/therapy
*Prgm Dir:* ChihChen Sophia Lee, PhD MT-BC
*Tel:* 580 774-3218   *Fax:* 580 774-3795
*E-mail:* sophia.lee@swosu.edu

# Ontario, Canada

### University of Windsor
Music Therapy Prgm
School of Music
401 Sunset Ave
Windsor, ON N9B 3P4
www.uwindsor.ca/music
*Prgm Dir:* Sandia Curtis, PhD MT-BC MTA
*Tel:* 519 253-3000, Ext 2796   *Fax:* 519 971-3614
*E-mail:* scurtis@uwindsor.ca

# Oregon

### Marylhurst University
Music Therapy Prgm
17600 Pacific Hwy (Hwy 43)
Marylhurst, OR 97036-0261
www.marylhurst.edu
*Prgm Dir:* Christine Korb, MM MT-BC
*Tel:* 800 634-9982, Ext 3361   *Fax:* 503 636-9526
*E-mail:* ckorb@marylhurst.edu

# Pennsylvania

### Elizabethtown College
Music Therapy Prgm
Dept of Fine and Performing Arts
One Alpha Dr
Elizabethtown, PA 17022-2298
www.etown.edu
*Prgm Dir:* Gene Ann Behrens, PhD MT-BC
*Tel:* 717 361-1991   *Fax:* 717 361-1187
*E-mail:* behrenga@etown.edu

### Immaculata University
Music Therapy Prgm
Dept of Music
Box 703
Immaculata, PA 19345
www.immaculata.edu
*Prgm Dir:* Brian Abrams, PhD MT-BC LPC LCAT FAMI
*Tel:* 610 647-4400, Ext 3490   *Fax:* 610 251-1668
*E-mail:* babrams@immaculata.edu

### Mansfield University
Music Therapy Prgm
Dept of Music
Mansfield, PA 16933
*Prgm Dir:* Elizabeth Grovenstein, MT-BC
*Tel:* 717 662-4717   *Fax:* 717 662-4114
*E-mail:* egrovens@mnsfld.edu

### Drexel University
Music Therapy Prgm
245 N 15th St, MS 905
Philadelphia, PA 19102-1192
*Prgm Dir:* Paul Nolan, MCAT LPC MT-BC
*Tel:* 215 762-6927   *Fax:* 215 762-6933
*E-mail:* paul.nolan@drexel.edu

### Temple University
Music Therapy Prgm
College of Music 012-00
Philadelphia, PA 19122
www.temple.edu/musictherapy
*Prgm Dir:* Kenneth Bruscia, PhD MT-BC
*Tel:* 215 204-8314   *Fax:* 215 204-4957
*E-mail:* kbruscia@temple.edu

### Duquesne University
Music Therapy Prgm
Mary Pappert School of Music
600 Forbes Ave
Pittsburgh, PA 15282
www.duq.edu
*Prgm Dir:* Donna Marie Beck, PhD FAMI MT-BC
*Tel:* 412 396-6080, Ext 6086   *Fax:* 412 396-5479
*E-mail:* beckd@duq.edu

### Marywood University
Music Therapy Prgm
2300 Adams Ave
Scranton, PA 18509
*Prgm Dir:* Mariam Pfeifer, MA MT-BC
*Tel:* 717 348-6211, Ext 2527   *Fax:* 717 961-4768
*E-mail:* pfeifer@marywood.edu

### Slippery Rock University
Music Therapy Prgm
SRU Dept of Music
Slippery Rock, PA 16057
*Prgm Dir:* Sue Shuttleworth, EdD MT-BC
*Tel:* 724 738-2447   *Fax:* 724 738-4469
*E-mail:* sue.shuttleworth@sru.edu

# South Carolina

### Charleston Southern University
Music Therapy Prgm
9200 University Blvd, PO Box 118087
Charleston, SC 29423-8087
www.csuniv.edu
*Prgm Dir:* April Malone, MM MT-BC
*Tel:* 843 863-7969   *Fax:* 843 863-7042
*E-mail:* amalone@csuniv.edu

# Texas

### West Texas A&M University
Music Therapy Prgm
Dept of Music and Dance
WTAMU Box 60879
Canyon, TX 79016-0001
*Prgm Dir:* Edward Kahler, PhD MT-BC
*Tel:* 806 651-2822   *Fax:* 806 651-2958
*E-mail:* ekahler@mail.wtamu.edu

### Southern Methodist University
Music Therapy Prgm
Meadows School of the Arts
Div of Music Therapy/Music
Dallas, TX 75275
*Prgm Dir:* Robert Krout, EdD MT-BC
*Tel:* 214 768-3175   *Fax:* 214 768-4669
*E-mail:* rkrout@smu.edu

### Texas Woman's University
Music Therapy Prgm
PO Box 425768
TWU Station
Denton, TX 76204
*Prgm Dir:* Nancy Hadsell, PhD MT-BC
*Tel:* 940 898-2514   *Fax:* 940 898-2494
*E-mail:* nhadsell@twu.edu

### Sam Houston State University
Music Therapy Prgm
School of Music, Box 2208, SHSU
Huntsville, TX 77341
*Prgm Dir:* Karen Miller, MM MT-BC NMT
*Tel:* 936 294-1376   *Fax:* 936 294-3765
*E-mail:* musictherapy@shsu.edu

### Incarnate Word College
Music Therapy Prgm
4301 Broadway Ave
PO Box 65
San Antonio, TX 78209
*Prgm Dir:* Janice Dvorkin, DPsy ACMT
*Tel:* 210 829-3856   *Fax:* 210 829-3880
*E-mail:* dvorkin@uiwtx.edu

# Utah

**Utah State University**
Music Therapy Prgm
4015 Old Main Hill, Department of Music
Logan, UT 84322
www.usu.edu
*Prgm Dir:* Maureen Hearns, MEd MT-BC
*Tel:* 435 797-3009   *Fax:* 435 797-1862
*E-mail:* mhearns@cc.usu.edu

# Virginia

**Radford University**
Music Therapy Prgm
Dept of Music
Radford, VA 24142
*Prgm Dir:* James E Borling, MM MT-BC
*Tel:* 540 831-5177   *Fax:* 540 831-6133
*E-mail:* jborling@radford.edu

**Shenandoah University**
Music Therapy Prgm
1460 University Dr
Winchester, VA 22601-5195
*Prgm Dir:* Michael J Rohrbacher, PhD RMT-BC
*Tel:* 540 665-4560   *Fax:* 540 665-5402
*E-mail:* Mrohrbac@su.edu

# Wisconsin

**University of Wisconsin - Eau Claire**
Music Therapy Prgm
Dept of Public Health Professions
College of Nursing and Health Sciences
Eau Claire, WI 54702-4004
www.uwec.edu/ph/mt/index.htm
*Prgm Dir:* Lee Anna Rasar, MT-BC WMTR NMT
*Tel:* 715 836-4260   *Fax:* 715 836-3952
*E-mail:* rasarla@uwec.edu

**Alverno College**
Music Therapy Prgm
3401 S 39th St, PO Box 3439222
Milwaukee, WI 53234-3922
*Prgm Dir:* Diane Knight, MS MT-BC
*Tel:* 414 382-6135   *Fax:* 414 382-6954
*E-mail:* diane.knight@alverno.edu

## Music Therapist

| Programs* | Class Capacity | Begins | Length (months) | Award | Res Tuition | Non-res Tuition | Stipend | Offers:‡ 1 | 2 | 3 |
|---|---|---|---|---|---|---|---|---|---|---|
| **Alabama** | | | | | | | | | | |
| University of Alabama (Tuscaloosa) | 25 | Aug | 54 | BM | $1,919 | $5,197 | | | | |
| **Arizona** | | | | | | | | | | |
| Arizona State University (Tempe) | 75 | Aug | 38 | Cert, BM, MT | $1,884 | $6,028 | | • | | • |
| **California** | | | | | | | | | | |
| Chapman University (Orange) | 25 | Sep Feb | 54 | Dipl, BM | $27,300 | $0 | | | | • |
| University of the Pacific (Stockton) | 25 | Aug | 36 | Dipl, Cert, BM, MA | $19,000 | $19,000 | | | | • |
| **Colorado** | | | | | | | | | | |
| Colorado State University (Ft Collins) | 100 | Aug | 38 | BM, MM | $5,000 | $15,000 | | | | |
| **Florida** | | | | | | | | | | |
| Florida State University (Tallahassee) | 40 | Aug | 54, 12 | Cert, BM, MM | $2,860 | $13,888 | | | | • |
| University of Miami (Coral Gables) | 50 | Aug | 24, 54 | BM, MM | $20,034 | $20,034 | | | | • |
| **Georgia** | | | | | | | | | | |
| University of Georgia (Athens) | 50 | Aug | 48 | Dipl, Cert, BMMT | $1,638 | $5,264 | | | | • |
| **Illinois** | | | | | | | | | | |
| Illinois State University (Normal) | 20 | Aug | 54, 24 | BMMT, MMT | $6,966 | $13,816 | | | | |
| **Indiana** | | | | | | | | | | |
| Indiana Univ - Purdue Univ Ft Wayne | 15 | Aug | 54 | BS | $4,300 | $9,800 | | • | | |
| Indiana Univ-Purdue U-Indianapolis | 12 | | 36 | MS | | | | • | | |
| University of Evansville | 40 | Aug | 54 | BMMT | $21,000 | $0 | | | | • |
| **Iowa** | | | | | | | | | | |
| Wartburg College (Waverly) | 50 | Sep | 54 | BME/MT | $20,500 | $20,500 | | • | | • |
| **Kentucky** | | | | | | | | | | |
| University of Louisville | 28 | Aug | 54 | Cert, BM | $5,550 | $15,000 | | | | |
| **Louisiana** | | | | | | | | | | |
| Loyola University New Orleans | | Aug Jan | 22, 42 | Dipl, BMT | | | | | | |
| **Massachusetts** | | | | | | | | | | |
| Berklee College of Music (Boston) | 100 | Sep Jan Jun | 42 | Cert, BMUS | $15,000 | $15,000 | | | | • |
| Lesley University (Cambridge) | 20 | Sep | 24 | MA, PhD | | | | • | | |
| **Michigan** | | | | | | | | | | |
| Eastern Michigan University (Ypsilanti) | 40 | Sep | 46 | BMT | $6,000 | $20,000 | | | | • |
| Western Michigan University (Kalamazoo) | 20 | Sep | 38 | BM | $5,262 | $13,248 | | • | | |
| **Minnesota** | | | | | | | | | | |
| University of Minnesota - Minneapolis | 15 | Sep | 30 | BA | $4,320 | $8,460 | | | | |
| **Missouri** | | | | | | | | | | |
| Maryville University (St Louis) | 20 | Aug | 54, 48 | Dipl, BS, MMT | $17,800 | $0 | | | | |
| University of Missouri - Kansas City | 60 | Aug | 30, 54 | Dipl, BA, MA | $257 | $599 | | | | |
| **New Jersey** | | | | | | | | | | |
| Montclair State University (Upper Montclair) | 20 | Sep Jan | 54, 24 | Cert, BA, MA | $2,790 | $5,014 | | • | | |
| **New York** | | | | | | | | | | |
| SUNY at New Paltz | 30 | Aug 25 | 54 | BS, MS | $4,200 | $8,400 | | • | | • |
| **North Carolina** | | | | | | | | | | |
| Appalachian State University (Boone) | 15 | Aug Jan | 54 | BM | $2,221 | $11,963 | | | | |
| Queens University of Charlotte | 40 | Aug | 60 | BM | $19,450 | $0 | | • | | |

*Data are shown only for programs that completed the 2006 AMA Survey of Health Professions Education Programs
‡Key to Offers: 1: Evening or weekend classes; 2: Non-English instruction; 3: Cultural competence instruction

## Music Therapist

| Programs* | Class Capacity | Begins | Length (months) | Award | Res Tuition | Non-res Tuition | Stipend | Offers:‡ 1 | 2 | 3 |
|---|---|---|---|---|---|---|---|---|---|---|
| **North Dakota** | | | | | | | | | | |
| University of North Dakota (Grand Forks) | 20 | Fall | 48 | BM | $5,792 | $13,786 | | | | • |
| **Ohio** | | | | | | | | | | |
| Ohio University (Athens) | 50 | Sep | 46, 24 | Dipl, BM, MM | $8 | $15,840 | | | | • |
| **Oklahoma** | | | | | | | | | | |
| Southwestern Oklahoma State University (Weatherford) | 20 | Aug | 48 | Cert, BM | $3,240 | $7,740 | $43,896 | | | • |
| **Ontario, Canada** | | | | | | | | | | |
| University of Windsor | | Sep | 6, 48 | BA | $4,825 | $11,389 | | | | • |
| **Oregon** | | | | | | | | | | |
| Marylhurst University | 35 | Sep | 48 | Dipl, BMT | $14,000 | $14,000 | | | • | |
| **Pennsylvania** | | | | | | | | | | |
| Duquesne University (Pittsburgh) | 15 | Aug Jan | 48 | Dipl, Cert, BSMT | $17,223 | $24,223 | | | • | • |
| Elizabethtown College | 40 | Aug | 54 | BM | $23,710 | $23,710 | | | | • |
| Immaculata University | 60 | Ongoing | 52, 24 | Dipl, Cert, BMUS, MA | | | | | • | • |
| Slippery Rock University | 30 | Aug | 37 | BS | $8,032 | $13,388 | | | • | |
| Temple University (Philadelphia) | 60 | Sep Jan | 24, 54 | BMT, MMT | $6,200 | $11,530 | | | • | |
| **South Carolina** | | | | | | | | | | |
| Charleston Southern University | | Aug Jan | 54 | BA | $15,980 | $0 | | | | |
| **Texas** | | | | | | | | | | |
| Sam Houston State University (Huntsville) | 25 | Aug | 42 | Dipl, BMMT | $1,600 | $10,432 | | | | • |
| West Texas A&M University (Canyon) | | Aug Jan May | 54 | Dipl, BMus | | | | | | • |
| **Utah** | | | | | | | | | | |
| Utah State University (Logan) | 30 | Aug | 54 | BS, BA | | | | | | • |
| **Wisconsin** | | | | | | | | | | |
| University of Wisconsin - Eau Claire | 50 | Sep Jan | 54 | BMT | $5,442 | $13,092 | | | | |

*Data are shown only for programs that completed the 2006 AMA Survey of Health Professions Education Programs

‡Key to Offers: 1: Evening or weekend classes; 2: Non-English instruction; 3: Cultural competence instruction

## Occupational Description

Nuclear medicine is the medical specialty that utilizes the nuclear properties of radioactive and stable nuclides to make diagnostic evaluations of the anatomic or physiologic conditions of the body and to provide therapy with unsealed radioactive sources. The skills of the nuclear medicine technologist complement those of the nuclear medicine physician and of other professionals in the field.

## Job Description

Nuclear medicine technologists perform a number of tasks in the areas of patient care, technical skills, and administration. When caring for patients, they acquire adequate knowledge of the patients' medical histories to understand and relate to their illnesses and pending diagnostic procedures for therapy, instruct patients before and during procedures, evaluate the satisfactory preparation of patients before commencing a procedure, and recognize emergency patient conditions and initiate life-saving first aid when appropriate.

Nuclear medicine technologists apply their knowledge of radiation physics and safety regulations to limit radiation exposure, prepare and administer radiopharmaceuticals, use radiation detection devices and other kinds of laboratory equipment that measure the quantity and distribution of radionuclides deposited in the patient or in a patient specimen, perform in vivo and in vitro diagnostic procedures, use quality control techniques as part of a quality assurance program covering all procedures and products in the laboratory, and participate in research activities.

Administrative functions may include supervising other nuclear medicine technologists, students, laboratory assistants, and other personnel; participating in procuring supplies and equipment; documenting laboratory operations; participating in departmental inspections conducted by various licensing, regulatory, and accrediting agencies; and participating in scheduling patient examinations.

## Employment Characteristics

The employment outlook in nuclear medicine technology is good. Opportunities may be found in major medical centers, smaller hospitals, and independent imaging centers. Opportunities also are available for obtaining positions in clinical research, education, and administration.

## Salary

Salaries vary depending on the employer and geographic location. According to a 2004 survey from the American Society of Radiologic Technologists, the overall average salary is $67,429, although salaries vary by geographic location. Refer to Section IV, Table 5 of this *Directory* for more information, or see www.ama-assn.org/go/salary.

## Educational Programs

**Length.** The professional portion of the programs is 1 year. Institutions offering accredited programs may provide an integrated educational sequence leading to an associate or baccalaureate degree over a period of 2 or 4 years.

**Prerequisites.** Applicants for admission must have graduated from high school or the equivalent and have acquired postsecondary competencies in human anatomy and physiology, physics, algebra, medical terminology, computer applications, oral and written communications, and general chemistry.

**Curriculum.** The curriculum includes patient care, statistics, nuclear medicine physics and radiation physics, radiation biology, radiation safety and protection, radionuclide chemistry and radiopharmacy, nuclear instrumentation, positron emission tomography (PET), computer applications for nuclear medicine, diagnostic nuclear medicine imaging and nonimaging in vivo and in vitro procedures, immunology as related to nuclear medicine, radionuclide therapy, and quality control and quality assurance.

## Inquiries

**Careers/Curriculum**

American Society of Radiologic Technologists
15000 Central Avenue SE
Albuquerque, NM 87123
www.asrt.org

Society of Nuclear Medicine—Technologist Section
1850 Samuel Morse Drive
Reston, VA 22090-5316
703 708-9000

**Certification/Registration**

Nuclear Medicine Technology Certification Board
2970 Clairmont Road NE, Suite 610
Atlanta, GA 30329-1634
404 315-1739

American Registry of Radiologic Technologists
1255 Northland Drive
Mendota Heights, MN 55120
www.arrt.org

**Program Accreditation**

Joint Review Committee on Educational Programs in Nuclear
    Medicine Technology (JRCNMT)
716 Black Point Road
PO Box 1149
Polson, MT 59860-1149
406 883-0003
406 883-0022 Fax
E-mail: jrcnmt@centurytel.net
www.jrcnmt.org

## Nuclear Medicine Technologist*

# Alabama

**University of Alabama at Birmingham**
Nuclear Medicine Technology Prgm
RMSB 441
1530 Third Ave S
Birmingham, AL 35294-1212
www.uab.edu/NMTProgram
*Prgm Dir:* Ann M Steves, MS CNMT
*Tel:* 205 934-2004    *Fax:* 205 975-7302
*E-mail:* stevesa@uab.edu
*Med Dir:* Luvenia Bender, MD

# Arizona

**GateWay Community College**
Nuclear Medicine Technology Prgm
108 N 40th St
Phoenix, AZ 85034
*Prgm Dir:* Jeanne Dial, CNMT
*Tel:* 602 286-8512    *Fax:* 602 286-8480
*E-mail:* jeanne.dial@gwmail.maricopa.edu

# Arkansas

**Baptist Health**
Nuclear Medicine Technology Prgm
11900 Colonel Glenn Rd, Ste 1000
Little Rock, AR 72210-2820
www.baptist-health.org
*Prgm Dir:* Sharon S Ward, MA CNMT RT(N) ASCP(N)
*Tel:* 501 202-7447    *Fax:* 501 202-7712
*E-mail:* ssward@baptist-health.org
*Med Dir:* Kevin Forte, MD

**University of Arkansas for Medical Sciences**
Nuclear Medicine Technology Prgm
4301 W Markham St #714
Little Rock, AR 72205
www.uams.edu/chrp/nuclearmedicine
*Prgm Dir:* Martha W Pickett, MHSA CNMT
*Tel:* 501 686-6848    *Fax:* 501 526-7975
*E-mail:* mwpickett@uams.edu
*Med Dir:* Victoria Major, MD

# California

**VA Palo Alto Health Care System**
Nuclear Medicine Technology Prgm
3801 Miranda Ave
Nuclear Medicine (115)
Palo Alto, CA 94304
www.palo-alto.med.va.gov
*Prgm Dir:* Kent Hutchings, MS CNMT
*Tel:* 650 493-5000, Ext 64135    *Fax:* 650 849-1279
*E-mail:* kent.hutchings@med.va.gov
*Med Dir:* George Segall, MD

**Kaiser Permanente Sch of Allied Hlth Sciences**
Nuclear Medicine Technology Prgm
938 Marina Way South
Richmond, CA 94804
www.kpsahs.org
*Prgm Dir:* Linda Bogner, BS CNMT
*Tel:* 510 231-5053    *Fax:* 510 231-5001
*E-mail:* Linda.Bogner@kp.org
*Med Dir:* Juanita Yun, MD

**LA County Harbor UCLA Medical Center**
Nuclear Medicine Technology Prgm
1000 W Carson St, Rm B-108
Torrance, CA 90509
www.csudh.edu/soh/clinicalsciences.htm
*Prgm Dir:* Will Wade, MPA BS CNMT ARRT
*Tel:* 310 222-2842    *Fax:* 310 328-7288
*E-mail:* wwade@humc.edu
*Med Dir:* Frederick S Mishkin, MD

# Connecticut

**Gateway Community College**
Nuclear Medicine Technology Prgm
88 Bassett Rd
North Haven, CT 06473
www.gwctc.commnet.edu/allied-health-new/alliedhealth
*Prgm Dir:* Kathleen Murphy, MS CNMT ARRT(N) NCT
*Tel:* 203 285-2381    *Fax:* 203 285-2400
*E-mail:* kmurphy@gwcc.commnet.edu
*Med Dir:* Vicente Caride, MD

# Delaware

**Delaware Tech & Comm Coll - Wilmington**
Nuclear Medicine Technology Prgm
333 Shipley St
Wilmington, DE 19801
www.dtcc.edu/wilmington/ah
*Prgm Dir:* LaRay A Fox, MEd CNMT
*Tel:* 302 765-4590    *Fax:* 302 765-4599
*E-mail:* lfox@christianacare.org
*Med Dir:* Vidya V Sagar, MD

# Florida

**Santa Fe Community College**
Nuclear Medicine Technology Prgm
3000 NW 83rd St
Gainesville, FL 32606-6200
*Prgm Dir:* Stelio N Marchionno, BS RT(N) CNMT
*Tel:* 352 395-5673    *Fax:* 352 395-5711
*E-mail:* stelio.marchionno@sfcc.edu
*Med Dir:* Suzanne T Mastin, MD

**St Vincent's Medical Center**
Nuclear Medicine Technology Prgm
Medical Sciences Education Department
1800 Barrs St
Jacksonville, FL 32204
*Prgm Dir:* LeRoy Stecker, MBA RT(N) CNMT
*Tel:* 904 308-8484    *Fax:* 904 308-5109
*E-mail:* lstec001@stvincentshealth.com
*Med Dir:* Marc Freeman, MD

**Jackson Mem Medical Center**
Nuclear Medicine Technology Prgm
1611 NW 12th Ave, Central Building #250
Miami, FL 33136
*Prgm Dir:* Sharon S Halula, BS CNMT
*Tel:* 305 585-6345    *Fax:* 305 585-8538
*E-mail:* shalula@um-jmh.org
*Med Dir:* George Sfakianakis, MD

**Florida Hospital College of Health Sciences**
Nuclear Medicine Technology Prgm
711 Lake Estelle Dr
Orlando, FL 32803
www.fhchs.edu
*Prgm Dir:* Joseph Hawkins, BS CNMT
*Tel:* 407 303-9380
*E-mail:* joe_hawkins@fhchs.edu
*Med Dir:* Anton Serafini, MD

**Hillsborough Community College**
Nuclear Medicine Technology Prgm
PO Box 30030
Tampa, FL 33630-3030
www.hccfl.edu
*Prgm Dir:* Larry Gibson, MBA RT(N) CNMT
*Tel:* 813 253-7418    *Fax:* 813 253-7491
*E-mail:* lgibson@hccfl.edu
*Med Dir:* Hemant Chheda, MD

# Georgia

**Medical College of Georgia**
Nuclear Medicine Technology Prgm
Department of Biomedical and Radiological Technologies
AE-1003
Augusta, GA 30912-0600
www.mcg.edu/SAH/brt/NMed.htm
*Prgm Dir:* Mary Anne Owen, MHE RT(N)
*Tel:* 706 721-3691    *Fax:* 706 721-8293
*E-mail:* mowen@mail.mcg.edu
*Med Dir:* Hadyn T Williams, MD

**Coosa Valley Technical College**
Nuclear Medicine Technology Prgm
One Maurice Culberson Dr
Rome, GA 30161
www.cvtcollege.org
*Prgm Dir:* Karen Fluharty, CNMT RT(N) NCT
*Tel:* 706 295-6883    *Fax:* 706 295-6894
*E-mail:* kfluharty@coosavalleytech.edu
*Med Dir:* Russell Roberts, MD

**Armstrong Atlantic State University**
Nuclear Medicine Technology Prgm
11935 Abercorn St
Savannah, GA 31419-1997
*Prgm Dir:* Rochelle Bornett Lee, MBA RT(N)
*Tel:* 912 920-6564    *Fax:* 912 921-5838
*E-mail:* leeroche@mail.armstrong.edu
*Med Dir:* Peter Michael Britt, MD

**Middle Georgia Technical College**
Nuclear Medicine Technology Prgm
80 Cohen Walker Dr
Warner Robins, GA 31088
www.middlegatech.edu
*Prgm Dir:* Patricia Melnick, MEd BSRT(R) (M)
*Tel:* 478 988-6912    *Fax:* 478 988-6813
*E-mail:* pmelnick@middlegatech.edu
*Med Dir:* Michael Zinsmeister, MD

# Illinois

**Northwestern Memorial Hospital**
Nuclear Medicine Technology Prgm
Sch of Nuclear Med Tech, Galter 8th Fl
251 E Huron St
Chicago, IL 60611-2908
www.nmh.org/nmh/forhealthcareprofessionals/sdnmt.htm
*Prgm Dir:* Nancy McDonald, BS CNMT
*Tel:* 312 926-1777    *Fax:* 312 926-8118
*E-mail:* nmcdonal@nmh.org
*Med Dir:* Stewart Spies, MD

**College of DuPage**
Nuclear Medicine Technology Prgm
425 Fawell Blvd
Glen Ellyn, IL 60137-6599
www.cod.edu
*Prgm Dir:* Joanne M Metler, MS CNMT PET
*Tel:* 630 942-3065    *Fax:* 630 858-5409
*E-mail:* metler@cdnet.cod.edu
*Med Dir:* Michael Schwartz, MD

*Additional information about programs that returned the AMA's survey is available in a table beginning on page 269.

PROGRAMS

**Edward Hines Jr VA Hospital**
Nuclear Medicine Technology Prgm
Fifth Ave and Roosevelt Rd
Hines, IL 60141-5000
www.gammaquality.com/snmt
*Prgm Dir:* Gary Eastman, CNMT
*Tel:* 708 202-8387, Ext 21976   *Fax:* 708 202-2390
*E-mail:* gary.eastman@med.va.gov
*Med Dir:* Nicholas Friedman, MD

**Triton College**
Nuclear Medicine Technology Prgm
2000 N Fifth Ave
River Grove, IL 60171-1995
www.triton.edu
*Prgm Dir:* Susan Campos, MEd BA CNMT
*Tel:* 708 456-0300, Ext 3655   *Fax:* 708 583-3336
*E-mail:* scampos2@triton.edu
*Med Dir:* Michael Blend, DO PhD, Jack Laude, MD

# Indiana

**Indiana University**
Nuclear Medicine Technology Prgm
School of Medicine
541 Clinical Dr CL-120
Indianapolis, IN 46202-5111
www.indyrad.iupui.edu/iuradsci
*Prgm Dir:* Judith Kosegi, MS RT(RN) CNMT
*Tel:* 317 274-7431   *Fax:* 317 274-4074
*E-mail:* jkosegi@iupui.edu
*Med Dir:* James Fletcher, MD

# Iowa

**Mercy College of Health Sciences**
Nuclear Medicine Technology Prgm
928 6th Ave
Des Moines, IA 50309
*Prgm Dir:* Robert J Loch, MBA CNMT
*Tel:* 515 643-6679
*E-mail:* rloch@mercydesmoines.org
*Med Dir:* Michael Disbro, MD

**University of Iowa Hospitals & Clinics**
Nuclear Medicine Technology Prgm
Dept of Radiology
200 Hawkins Dr
Iowa City, IA 52242-1009
www.radiology.uiowa.edu/
    RadTech/NucMedTechProgram/NMTechh1.htm
*Prgm Dir:* Anthony W Knight, MBA CNMT
*Tel:* 319 356-2954   *Fax:* 319 384-6389
*E-mail:* anthony-knight@uiowa.edu
*Med Dir:* Michael Graham, MD

# Kansas

**University of Kansas Medical Center**
Nuclear Medicine Technology Prgm
3901 Rainbow Blvd
Kansas City, KS 66160-7234
www.rad.kumc.edu/nucmed
*Prgm Dir:* Tina R Crain, MS CNMT RT(R)(N)(QM)
*Tel:* 913 588-6858   *Fax:* 913 588-8393
*E-mail:* tcrain@kumc.edu
*Med Dir:* Reginald Dusing, MD

# Kentucky

**Jefferson Community and Technical College**
Nuclear Medicine Technology Prgm
Allied Health Div
109 E Broadway
Louisville, KY 40202
www.jctc.kctcs.edu
*Prgm Dir:* Cybil Nielsen, MBA CNMT
*Tel:* 502 629-3683   *Fax:* 502 629-2088
*E-mail:* cnielsen0001@kctcs.edu
*Med Dir:* George Zenger, MD

# Louisiana

**Delgado Community College**
Nuclear Medicine Technology Prgm
615 City Park Ave
New Orleans, LA 70119-4399
*Prgm Dir:* Steve Trichell, BS RT(N)
*Tel:* 504 483-4430   *Fax:* 504 483-4609
*E-mail:* strich@dcc.edu
*Med Dir:* Richard J Campeau, MD

# Maine

**Central Maine Medical Center**
Nuclear Medicine Technology Prgm
300 Main St
Lewiston, ME 04240-0305
www.cmmc.org
*Prgm Dir:* Heather D Poulin, CNMT RT(N) CDT
*Tel:* 207 795-5956   *Fax:* 207 795-2476
*E-mail:* PoulinHe@cmhc.org
*Med Dir:* David M Simms, MD

# Maryland

**Johns Hopkins Hospital**
Nuclear Medicine Technology Prgm
600 N Wolfe St
Radiology Admin B-179
Baltimore, MD 21287-1010
http://radiologycareers.rad.jhmi.edu
*Prgm Dir:* Pamela Vorce, BS CNMT MBA
*Tel:* 410 528-8299   *Fax:* 410 528-8308
*E-mail:* pvorce1@jhmi.edu
*Med Dir:* Richard Wahl, MD

**Prince George's Community College**
Nuclear Medicine Technology Prgm
301 Largo Rd
Lanham Hall
Largo, MD 20774-2199
*Prgm Dir:* Nancy Meman, BS CNMT RT(N)
*Tel:* 301 341-3026   *Fax:* 301 386-7528
*E-mail:* nmeman@pgcc.edu
*Med Dir:* Andrew M Keenan, MD

# Massachusetts

**Mass College of Pharmacy & Health Sciences**
Nuclear Medicine Technology Prgm
179 Longwood Ave
Boston, MA 02115-5896
www.mcphs.edu/MCPHSWeb
*Prgm Dir:* Frances K Keech, RT(N) MBA
*Tel:* 617 732-2928   *Fax:* 617 732-2075
*E-mail:* frances.keech@mcphs.edu
*Med Dir:* Annick Vanden Abbeele, MD

**Salem State College**
*Cosponsor:* North Shore Medical Center, Salem
Nuclear Medicine Technology Prgm
352 Lafayette St
Salem, MA 01970-5353
www.salemstate.edu/biology
*Prgm Dir:* T Ware, PhD
*Tel:* 978 542-6236   *Fax:* 978 542-6863
*E-mail:* debbi.germano@salemstate.edu
*Med Dir:* Robert E Belliveau, MD

**Springfield Technical Community College**
Nuclear Medicine Technology Prgm
One Armory Sq
PO Box 9000
Springfield, MA 01102-9000
www.stcc.edu
*Prgm Dir:* Richard T Serino, MEd CNMT
*Tel:* 413 755-4871   *Fax:* 413 755-6312
*E-mail:* rserino@stcc.mass.edu
*Med Dir:* Laurie Gianturco, MD

**UMass Memorial Medical Center**
*Cosponsor:* Worcester State College
Nuclear Medicine Technology Prgm
55 Lake Ave N
Worcester, MA 01655-0243
www.umassmed.edu/nuclear_med/NMT
*Prgm Dir:* Peter H Simkin, MD
*Tel:* 508 856-4253   *Fax:* 508 856-6867
*E-mail:* simkinp@ummhc.org
*Med Dir:* Peter Simkin, MD

# Michigan

**Ferris State University**
Nuclear Medicine Technology Prgm
200 Ferris Dr, VFS 411
Big Rapids, MI 49307-9989
*Prgm Dir:* Sheila Squicciarini, CNMT
*Tel:* 231 591-2319   *Fax:* 231 591-3788
*E-mail:* squiccis@ferris.edu
*Med Dir:* John Freitas, MD

**William Beaumont Hospital**
Nuclear Medicine Technology Prgm
3601 W 13 Mile Rd
Royal Oak, MI 48073-6769
*Prgm Dir:* Mary Premo, BA CNMT RT(N)
*Tel:* 248 898-4125   *Fax:* 248 898-0487
*E-mail:* mpremo@beaumont.edu
*Med Dir:* Howard Dworkin, MD

# Minnesota

**Mayo School of Health Sciences**
Nuclear Medicine Technology Prgm
Sch of Hlth Sciences
200 First St SW
Rochester, MN 55905
www.mayo.edu/mshs
*Prgm Dir:* Nancy Hockert, BS CNMT
*Tel:* 507 284-3245   *Fax:* 507 284-2818
*E-mail:* hockert.nancy@mayo.edu
*Med Dir:* Brian P Mullan, MD

**St Mary's University of Minnesota**
Nuclear Medicine Technology Prgm
700 Terrace Hts #10
Winona, MN 55987-1399
*Prgm Dir:* Jeanne Minnerath, PhD
*Tel:* 507 457-1770   *Fax:* 507 494-6035
*E-mail:* jminnera@smumn.edu
*Med Dir:* Jeanette V Moulthrop, MD

# Mississippi

**University of Mississippi Medical Center**
Nuclear Medicine Technology Prgm
2500 N State St
Jackson, MS 39216-4505
*Prgm Dir:* John Vanderslice Jr, MS RT(R)(N)CNMT
*Tel:* 601 984-2585   *Fax:* 601 984-4986
*E-mail:* jvanderslice@radiology.umsmed.edu
*Med Dir:* Ramesh B Patel, MD

# Missouri

**University of Missouri - Columbia**
Nuclear Medicine Technology Prgm
605 Lewis Hall
Columbia, MO 65211-4230
www.umshp.org/shpsite/shp_home.htm
*Prgm Dir:* Glen Heggie, RTNM EdD FCAMRT
*Tel:* 573 884-7843   *Fax:* 573 884-1490
*E-mail:* heggieg@health.missouri.edu
*Med Dir:* Amolak Singh, MD

**Research Medical Center**
Nuclear Medicine Technology Prgm
2316 E Meyer Blvd
Kansas City, MO 64132-1199
www.researchmedicalcenter.com
*Prgm Dir:* Charlotte Ament, MA BS CNMT
*Tel:* 816 276-4068   *Fax:* 816 276-3138
*E-mail:* Charlotte.Ament@hcamidwest.com
*Med Dir:* Barry Gubin, MD

**Saint Louis University**
Nuclear Medicine Technology Prgm
Doisy College of Health Sciences
3437 Caroline St
St Louis, MO 63104-1395
www.slu.edu/doisycollege.xml
*Prgm Dir:* William Hubble, MA CNMT RT (R)(N)(CT)
*Tel:* 314 577-8526   *Fax:* 314 577-8503
*E-mail:* hubblewl@slu.edu
*Med Dir:* James Littlefield, MD

# Nebraska

**University of Nebraska Medical Center**
Nuclear Medicine Technology Prgm
984545 Nebraska Medical Center
Omaha, NE 68198-4545
www.unmc.edu/alliedhealth/rste
*Prgm Dir:* Marcia Hess Smith, BS CNMT
*Tel:* 402 559-7224   *Fax:* 402 559-4667
*E-mail:* mhesssmith@unmc.edu
*Med Dir:* Jordan Hankins, MD

# Nevada

**University of Nevada - Las Vegas**
Nuclear Medicine Technology Prgm
4505 Maryland Pkwy, 3037
Las Vegas, NV 89154-3037
www.unlv.edu
*Prgm Dir:* Joanie MacDonald, MS RT(N)
*Tel:* 702 895-3136   *Fax:* 702 895-4819
*E-mail:* joan.macdonald@unlv.edu
*Med Dir:* Joe Parravano, MD

# New Jersey

**Muhlenberg Regional Medical Center**
Nuclear Medicine Technology Prgm
Park Ave & Randolph Rd
Plainfield, NJ 07061
*Prgm Dir:* Maunesh C Soni, MS MHA CNMT
*Tel:* 908 668-2770   *Fax:* 908 226-4568
*E-mail:* msoni@solarishs.org
*Med Dir:* James Chen, MD

**Univ of Med & Dent of New Jersey**
Nuclear Medicine Technology Prgm
School of Health Related Professions
1776 Raritan Rd, Rm 538
Scotch Plains, NJ 07076
*Prgm Dir:* Michael Teters, MS RT(N) DABR
*Tel:* 908 889-2449   *Fax:* 908 889-2487
*E-mail:* tetersms@umdnj.edu
*Med Dir:* Lionel Zuckier, MD

**Gloucester County College**
Nuclear Medicine Technology Prgm
1400 Tanyard Rd
Sewell, NJ 08080
*Prgm Dir:* Laura J Sharkey, BS CNMT
*Tel:* 856 415-2196   *Fax:* 856 464-8463
*E-mail:* lsharkey@gccnj.edu

# New York

**Bronx Community College**
Nuclear Medicine Technology Prgm
University Ave and W 181st St
Bronx, NY 10453
*Prgm Dir:* Sherman Heller, PhD
*Tel:* 718 920-5012   *Fax:* 718 920-2629
*E-mail:* sheller@montefiore.org
*Med Dir:* M Donald Blaufox, MD PhD

**University at Buffalo - SUNY**
Nuclear Medicine Technology Prgm
105 Parker Hall, 3435 Main St
Buffalo, NY 14214-3007
www.buffalo.edu
*Prgm Dir:* Elpida S Crawford, MS CNMT
*Tel:* 716 838-5889, Ext 115   *Fax:* 716 838-4918
*E-mail:* esc@buffalo.edu
*Med Dir:* Hani Nabi, MD

**Institute of Allied Medical Professions**
Nuclear Medicine Technology Prgm
405 Park Ave, Ste 501
New York, NY 10022-4405
*Prgm Dir:* John W Hart, CNMT RT(N)
*Tel:* 212 758-1410   *Fax:* 212 758-1424
*E-mail:* john.hart@msnyuhealth.org
*Med Dir:* Josef Machac, MD

**St Vincent's Hospital Manhattan SVCMC**
Nuclear Medicine Technology Prgm
153 W 11th St
New York, NY 10011
www.nucmedicine.com
*Prgm Dir:* Samy Sadek, PhD
*Tel:* 212 604-8716   *Fax:* 212 604-8714
*E-mail:* samy@nucmedicine.com
*Med Dir:* Hussein Abdel-Dayem, MD

**Molloy College**
Nuclear Medicine Technology Prgm
1000 Hempstead Ave
Rockville Centre, NY 11571-5002
*Prgm Dir:* Marc Fischer, MBA CNMT ARRT
*Tel:* 516 678-5000, Ext 6887   *Fax:* 516 256-2252
*E-mail:* mfischer@molloy.edu
*Med Dir:* William Forman, MD

# North Carolina

**University of North Carolina Hospitals**
Nuclear Medicine Technology Prgm
101 Manning Dr
CB 7510 Radiology-Nuclear Medcine
Chapel Hill, NC 27514
www.unchealthcare.org/site/nuclear_medicine
*Prgm Dir:* Gregory S Beavers, MBA CNMT
*Tel:* 919 843-2963   *Fax:* 919 843-0591
*E-mail:* gbeavers@unch.unc.edu
*Med Dir:* William H McCartney, MD

**Caldwell Comm College & Tech Institute**
Nuclear Medicine Technology Prgm
2855 Hickory Blvd
Hudson, NC 28638-2397
*Prgm Dir:* Jimmy L Council, RT(N) CNMT MBA
*Tel:* 828 726-2370   *Fax:* 828 726-2489
*E-mail:* jcouncil@cccti.edu
*Med Dir:* Richard Curtis, MD

**Forsyth Technical Community College**
Nuclear Medicine Technology Prgm
2100 Silas Creek Pkwy
Winston-Salem, NC 27103
*Prgm Dir:* Donald G Harkness, MEd CNMT
*Tel:* 336 734-7187   *Fax:* 336 734-7444
*E-mail:* dharkness@forsyth.cc.nc.us
*Med Dir:* James D Ball, MD

# Ohio

**Aultman Hospital**
Nuclear Medicine Technology Prgm
2600 Sixth St SW
Canton, OH 44710-1799
*Prgm Dir:* Cheryl Beitzel, BS CNMT RT(N)
*Tel:* 330 363-2335   *Fax:* 330 580-6651
*E-mail:* cbeitzel@aultman.org
*Med Dir:* Marshall L Chalfont, MD

**University of Cincinnati College of Allied Health Sciences**
Nuclear Medicine Technology Prgm
3202 Goodman St
PO Box 670394
Cincinnati, OH 45267
www.cahs.uc.edu/departments/AMIT.cfm
*Prgm Dir:* Alan Vespie, MEd CNMT RT(N)
*Tel:* 513 558-2018   *Fax:* 513 558-6002
*E-mail:* alan.vespie@uc.edu
*Med Dir:* Mariano Fernandez-Ulloa, MD

**Ohio State University Medical Center**
Nuclear Medicine Technology Prgm
212 E Doan Hall
410 W 10th Ave
Columbus, OH 43210-1228
http://radiology.osu.edu/tech/nuc_med/
nuc_tech_home.htm
*Prgm Dir:* Stacey R Copley, BS CNMT
*Tel:* 614 293-3131   *Fax:* 614 293-2529
*E-mail:* stacey.copley@osumc.edu
*Med Dir:* John Olsen, MD

**University of Findlay**
Nuclear Medicine Technology Prgm
Nuclear Medicine Institute
1000 N Main St
Findlay, OH 45840-3695
www.findlay.edu/academics/cohp/nminmed
*Prgm Dir:* Richard States, MBA RT(N) CNMT
*Tel:* 419 434-5328   *Fax:* 419 434-4168
*E-mail:* states@findlay.edu
*Med Dir:* Neil Katz, MD

**Cuyahoga Community College**
Nuclear Medicine Technology Prgm
11000 Pleasant Valley Rd
Parma, OH 44130
www.tri-c.edu/NMED
*Prgm Dir:* David Frazee, MS CRA CNMT
*Tel:* 216 987-5298   *Fax:* 216 987-5066
*E-mail:* david.frazee@tri-c.edu
*Med Dir:* Donald Neumann, MD

**Kent State University - Salem Campus**
Nuclear Medicine Technology Prgm
2491 State Rte 45 South
Salem, OH 44460
www.salem.kent.edu
*Prgm Dir:* Janet Long, RT(RN) CNMT
*Tel:* 330 337-4261   *Fax:* 330 332-9256
*E-mail:* long@salem.kent.edu
*Med Dir:* Peter Apicella, MD

**St Elizabeth Health Center**
Nuclear Medicine Technology Prgm
1044 Belmont Ave
Youngstown, OH 44501-1790
*Prgm Dir:* Richard M Blanco, RT(N) CNMT
*Tel:* 330 480-3266   *Fax:* 330 480-2624
*E-mail:* rick_blanco@hmis.org
*Med Dir:* B David Collier, Jr, MD

# Oklahoma

**Univ of Oklahoma Health Sciences Center**
Nuclear Medicine Technology Prgm
PO Box 26901
CHB-451
Oklahoma City, OK 73190-0901
www.ah.ouhsc.edu
*Prgm Dir:* Vesper Grantham, MEd CNMT RT(N)
*Tel:* 405 271-6477   *Fax:* 405 271-1424
*E-mail:* vesper-grantham@ouhsc.edu
*Med Dir:* Jay Harolds, MD

# Pennsylvania

**Cedar Crest College**
Nuclear Medicine Technology Prgm
100 College Dr
Allentown, PA 18104-6196
www.cedarcrest.edu
*Prgm Dir:* Brian S Misanko, PhD
*Tel:* 610 606-4611   *Fax:* 610 606-4616
*E-mail:* bsmisank@cedarcrest.edu
*Med Dir:* Robert Rienzo, MD

**Lancaster Gen Coll of Nursing & Hlth Sciences**
Nuclear Medicine Technology Prgm
410 N Lime St
Lancaster, PA 17602
www.LancasterGeneralCollege.org
*Prgm Dir:* Penni A Longenecker, MEd CNMT RT(N)
*Tel:* 717 544-5668   *Fax:* 717 544-5970
*E-mail:* pelongen@LancasterGeneral.org
*Med Dir:* Robert Basarab, MD

**Jameson Health System**
Nuclear Medicine Technology Prgm
South Campus
1000 S Mercer St
New Castle, PA 16101
*Prgm Dir:* Michael Czachowski, MBA NCT CNMT RT(N)
   ASCP(N)
*Tel:* 724 656-6134   *Fax:* 724 656-4165
*E-mail:* mczachowski@jamesonhealthsystem.com
*Med Dir:* Albert Cook, MD

**Comm Coll of Allegheny County**
Nuclear Medicine Technology Prgm
808 Ridge Ave
Pittsburgh, PA 15212-6097
*Prgm Dir:* Carl Mazzetti, BS CNMT
*Tel:* 412 237-2751   *Fax:* 412 237-4521
*E-mail:* cmazzett@ccac.edu
*Med Dir:* Gilbert Isaacs, MD

**Wyoming Valley Health Care System**
Nuclear Medicine Technology Prgm
575 N River St
Wilkes-Barre, PA 18764
www.wvhcs.org
*Prgm Dir:* Cindy Turchin, MBA RT(R)(N) CNMT
*Tel:* 570 552-1706   *Fax:* 570 552-1758
*E-mail:* cturchin@wvhcs.org
*Med Dir:* Joan Forgetta, MD

**Abington Memorial Hospital**
Nuclear Medicine Technology Prgm
2500 Maryland Rd, Ste 214
Willow Grove, PA 19090
www.amh.org/education/radiologictech.htm
*Prgm Dir:* Nancy R Butterworth, MS RT(R N) CNMT
*Tel:* 215 481-5421   *Fax:* 215 481-5490
*E-mail:* nbutterworth@amh.org
*Med Dir:* Annette Y Griffith, MD

# Puerto Rico

**University of Puerto Rico**
Nuclear Medicine Technology Prgm
GPO Box 365067
San Juan, PR 00936-5067
http://cprsweb.rcm.upr.edu
*Prgm Dir:* Miriam Espada, MPH CNMT
*Tel:* 787 758-2525, Ext 4607   *Fax:* 787 764-1760
*E-mail:* miriamespada@cprs.rcm.upr.edu
*Med Dir:* Frieda Silva, MD

# Rhode Island

**Rhode Island Hospital**
Nuclear Medicine Technology Prgm
School of Nuclear Medicine Technology
3 Davol Square, Bldg A, Box 162
Providence, RI 02903
www.lifespan.org
*Prgm Dir:* Pamela C Maisano, ARRT (R) (N) CNMT
*Tel:* 401 528-8531, Ext 229   *Fax:* 401 457-0219
*E-mail:* pmaisano@lifespan.org
*Med Dir:* Richard B Noto, MD

# South Carolina

**Midlands Technical College**
Nuclear Medicine Technology Prgm
PO Box 2408
Columbia, SC 29202
*Prgm Dir:* Edward O Nicholson, MEd RRT CPFT
*Tel:* 803 822-3434   *Fax:* 803 822-3417
*E-mail:* nicholsone@midlandstech.edu
*Med Dir:* Samuel Friedman, MD

# South Dakota

**Southeast Technical Institute**
Nuclear Medicine Technology Prgm
2320 Career Ave
Sioux Falls, SD 57107
*Prgm Dir:* Mary Hennings-Frank, BS NMT
*Tel:* 605 367-4632   *Fax:* 605 367-5724
*E-mail:* mary.hennings-frank@southeasttech.com
*Med Dir:* Fred Lovrien, MD

# Tennessee

**Chattanooga State Technical Comm College**
Nuclear Medicine Technology Prgm
4501 Amnicola Hwy
Chattanooga, TN 37406-1097
www.chattanoogastate.edu/Allied_Health/nuclear.asp
*Prgm Dir:* Leesa Ross, MA CNMT ARRT(N)
*Tel:* 423 697-3331   *Fax:* 423 697-3324
*E-mail:* Leesa.Ross@ChattanoogaState.edu
*Med Dir:* Brett Austin, MD

**Univ of Tennessee Medical Ctr at Knoxville**
Nuclear Medicine Technology Prgm
Nuclear Medicine Department, UTMC
1924 Alcoa Highway
Knoxville, TN 37920-6999
www.utmedicalcenter.org/radiology
*Prgm Dir:* Glenn Hathaway, BS CNMT
*Tel:* 865 544-9726   *Fax:* 865 544-9074
*E-mail:* ghathawa@mc.utmck.edu
*Med Dir:* Gary T Smith, MD

**Baptist College of Health Sciences**
Nuclear Medicine Technology Prgm
1003 Monroe Ave
Memphis, TN 38104
www.bchs.edu
*Prgm Dir:* Kathy Thompson, MS CNMT
*Tel:* 901 572-2642   *Fax:* 901 572-2750
*E-mail:* kathy.thompson@bchs.edu
*Med Dir:* Mohammed Moinuddin, MD

**Methodist University Hospital**
Nuclear Medicine Technology Prgm
1265 Union Ave
Memphis, TN 38104
http://methodisthealth.org
*Prgm Dir:* Nancy A Clifton, RT(R)(N)CNMT
*Tel:* 901 516-8521   *Fax:* 901 516-2870
*E-mail:* cliftonn@methodisthealth.org
*Med Dir:* R Michael Fleming, MD

**Vanderbilt University Medical Center**
Nuclear Medicine Technology Prgm
Rad Dept 21st and Garland
CCC-1124 MCN
Nashville, TN 37232-2675
www.mc.vanderbilt.edu/radiology
*Prgm Dir:* James A Patton, PhD
*Tel:* 615 322-0508   *Fax:* 615 322-3764
*E-mail:* jim.patton@vanderbilt.edu
*Med Dir:* William H Martin, MD

# Texas

**Amarillo College**
Nuclear Medicine Technology Prgm
PO Box 447
Amarillo, TX 79178
www.actx.edu
*Prgm Dir:* Mark Rowh, BS RT(R) CNMT
*Tel:* 806 354-6071   *Fax:* 806 354-6076
*E-mail:* rowh-me@actx.edu
*Med Dir:* Bill Byrd, MD

**Baylor University Med Center**
Nuclear Medicine Technology Prgm
3500 Gaston Ave
Dallas, TX 75246
www.baylorhealth.edu/rahs
*Prgm Dir:* Amy J Hatch, BA CNMT RT(N)
*Tel:* 214 820-1420   *Fax:* 214 820-1421
*E-mail:* AmyHa@baylorhealth.edu
*Med Dir:* Stanley Grossman, MD

**Galveston College**
Nuclear Medicine Technology Prgm
4015 Ave Q
Galveston, TX 77550-2782
www.gc.edu
*Prgm Dir:* Bobby R Brown, BS CNMT
*Tel:* 409 944-1491   *Fax:* 409 944-1511
*E-mail:* bbrown@gc.edu
*Med Dir:* Fernando Cesani-Vazquez, MD

**Houston Community College**
Nuclear Medicine Technology Prgm
1900 Pressler
Houston, TX 77030
www.hccs.edu/discipline/nmtt/nmtt.html
*Prgm Dir:* Glenn X Smith, BS CNMT NCT
*Tel:* 713 718-7354   *Fax:* 713 718-7608
*E-mail:* glenn.smith@hccs.edu
*Med Dir:* Donald A Podoloff, MD

**University of the Incarnate Word**
Nuclear Medicine Technology Prgm
4301 Broadway St
San Antonio, TX 78209-6397
*Prgm Dir:* Norma Green, MBA RT(N)
*Tel:* 210 829-3991   *Fax:* 210 829-3174
*E-mail:* green@uiwtx.edu
*Med Dir:* Ray Ware, MD

## Utah

**University of Utah Health Science Center**
Nuclear Medicine Technology Prgm
30 North 1900 East #1A971
Salt Lake City, UT 84132-2140
www.uuhsc.utah.edu/rad/nuctech.html
*Prgm Dir:* Delynn Strate, CNMT
*Tel:* 801 585-2326   *Fax:* 801 581-2403
*E-mail:* dstrate@hsc.utah.edu
*Med Dir:* Kathryn Ann Morton, MD

## Vermont

**University of Vermont**
Nuclear Medicine Technology Prgm
302 Rowell Bldg
Burlington, VT 05405
www.uvm.edu/cnhs
*Prgm Dir:* Louis M Izzo, MS CNMT
*Tel:* 802 656-3811   *Fax:* 802 656-8876
*E-mail:* louis.izzo@uvm.edu
*Med Dir:* Janusz Kikut, MD

## Virginia

**Old Dominion University**
Nuclear Medicine Technology Prgm
2118 Technology Building
Norfolk, VA 23529-0287
http://hs.odu.edu/medlab/academics/nmed
*Prgm Dir:* Scott R Sechrist, EdD CNMT ARRT(N)
*Tel:* 757 683-3589   *Fax:* 757 683-5028
*E-mail:* ssechris@odu.edu
*Med Dir:* Lester Johnson, MD PhD

**Naval School of Health Sciences**
Nuclear Medicine Technology Prgm
1001 Holcomb Rd
Portsmouth, VA 23708-5200
*Prgm Dir:* HM1 Samuel S Castro, RT(N) CNMT
*Tel:* 757 953-5050   *Fax:* 757 953-7380
*E-mail:* sscastro@hsp.med.navy.mil
*Med Dir:* David Turton, MD

**Virginia Commonwealth University**
Nuclear Medicine Technology Prgm
701 W Grace St
Box 980495
Richmond, VA 23284-3057
www.sahp.vcu.edu/radsci
*Prgm Dir:* Mark Crosthwaite, MEd CNMT
*Tel:* 804 828-9104   *Fax:* 804 828-5778
*E-mail:* mhcrosthwaite@vcu.edu
*Med Dir:* Melvin J Fratkin, MD

## Washington

**Bellevue Community College**
Nuclear Medicine Technology Prgm
3000 Landerholm Circle SE
Radiation and Imaging Sciences, B243
Bellevue, WA 98007-6484
www.bcc.ctc.edu/nucmed
*Prgm Dir:* Jennifer L Prekeges, MS CNMT
*Tel:* 425 564-2475   *Fax:* 425 564-4193
*E-mail:* jprekege@bcc.ctc.edu
*Med Dir:* Eugene Lin, MD

## West Virginia

**West Virginia State Comm & Tech College**
Nuclear Medicine Technology Prgm
PO Box 1000
221 Cole Complex
Institute, WV 25112-1000
www.wvstateu.edu
*Prgm Dir:* John Gaskins, RBA CNMT
*Tel:* 304 766-3202   *Fax:* 304 766-5243
*E-mail:* gaskinsj@wvsctc.edu
*Med Dir:* Steven A Artz, MD

**West Virginia University Hospitals**
Nuclear Medicine Technology Prgm
Medical Ctr Dr, PO Box 8062
Morgantown, WV 26506-8062
www.wvuhradtech.com
*Prgm Dir:* Kelsie J Kittle, BS RT(R)(N) CNMT
*Tel:* 304 598-4000, Ext 73179   *Fax:* 304 598-4348
*E-mail:* kittlek@rcbhsc.wvu.edu
*Med Dir:* Gary D Marano, MD

**Wheeling Jesuit University**
Nuclear Medicine Technology Prgm
316 Washington Ave
Wheeling, WV 26003-6295
*Prgm Dir:* Angela Macci Bires, EdD RT(N) CNMT
*Tel:* 304 243-2387   *Fax:* 304 243-2608
*E-mail:* amacci@wju.edu
*Med Dir:* William G Castro, MD

## Wisconsin

**Gundersen Lutheran Medical Foundation**
Nuclear Medicine Technology Prgm
Gundersen Lutheran Medical Center
1900 South Ave C03-003
La Crosse, WI 54601
*Prgm Dir:* Robyn Loeffelholz, CNMT
*Tel:* 608 775-5074   *Fax:* 608 775-5973
*E-mail:* rrloeffe@gundluth.org
*Med Dir:* Sue Beier-Hanratty, MD

**St Joseph's Hospital**
Nuclear Medicine Technology Prgm
611 St Joseph Ave
Marshfield, WI 54449
*Prgm Dir:* Carlyn M Johnson, BS CNMT
*Tel:* 715 387-7787   *Fax:* 715 387-7775
*E-mail:* johnsoca@stjosephs-marshfield.org
*Med Dir:* Michael Spieth, MD

**Aurora St Luke's Medical Center**
Nuclear Medicine Technology Prgm
2900 W Oklahoma Ave
Milwaukee, WI 53201-2901
www.aurorahealthcare.org
*Prgm Dir:* Kerry Michell, BS CNMT
*Tel:* 414 649-6418   *Fax:* 414 649-5118
*E-mail:* kerry.michell@aurora.org
*Med Dir:* Parish Desai, MD

**Froedtert Memorial Lutheran Hospital**
Nuclear Medicine Technology Prgm
9200 W Wisconsin Ave
Milwaukee, WI 53226-3596
*Prgm Dir:* Frank G Steffel, BS CNMT
*Tel:* 414 805-2071   *Fax:* 414 771-3460
*E-mail:* fsteffel@fmlh.edu
*Med Dir:* Arthur Z Krasnow, MD

### Nuclear Medicine Technologist

| Programs* | Class Capacity | Begins | Length (months) | Award | Res Tuition | Non-res Tuition | Stipend | Offers:‡ 1 | 2 | 3 |
|---|---|---|---|---|---|---|---|---|---|---|
| **Alabama** | | | | | | | | | | |
| University of Alabama at Birmingham | 15 | Aug | 16 | Cert, BS | $7,200 | $18,000 | | | | • |
| **Arkansas** | | | | | | | | | | |
| Baptist Health (Little Rock) | 9 | Jul | 12 | Cert | $3,540 | $3,540 | | | | |
| University of Arkansas for Medical Sciences (Little Rock) | 67 | Aug | 12 | BS | $663 | $15,249 | | | | • |
| **California** | | | | | | | | | | |
| Kaiser Permanente Sch of Allied Hlth Sciences (Richmond) | 40 | Oct | 18 | Cert | $6,000 | $6,000 | | | | |
| LA County Harbor UCLA Medical Center (Torrance) | 5 | Jul | 12 | Cert | | | | | | |
| VA Palo Alto Health Care System | 12 | Jul | 12 | Cert | | | | | | • |
| **Connecticut** | | | | | | | | | | |
| Gateway Community College (North Haven) | 31 | Sep | 22 | Cert, AS | $3,535 | $9,343 | | | | • |
| **Delaware** | | | | | | | | | | |
| Delaware Tech & Comm Coll - Wilmington | 10 | May | 24 | AAS | $2,934 | $7,344 | | • | | |
| **Florida** | | | | | | | | | | |
| Florida Hospital College of Health Sciences (Orlando) | 29 | May Aug | 15, 24 | Cert, AS | $9,750 | $9,750 | | | | • |
| Hillsborough Community College (Tampa) | 20 | Aug | 24 | AAS | $2,480 | $9,040 | | | | • |
| Jackson Mem Medical Center (Miami) | 12 | Jul | 15 | Cert | $10,000 | $10,000 | | | | • |
| Santa Fe Community College (Gainesville) | | Aug | 22 | Cert, AS | $6,116 | $18,971 | | | | • |
| St Vincent's Medical Center (Jacksonville) | 8 | Jan May Sep | 12 | Cert | $4,000 | $4,000 | | | | • |
| **Georgia** | | | | | | | | | | |
| Armstrong Atlantic State University (Savannah) | 10 | Jun | 24 | BS | $2,734 | $9,702 | | | | |

*Data are shown only for programs that completed the 2006 AMA Survey of Health Professions Education Programs
‡Key to Offers: 1: Evening or weekend classes; 2: Non-English instruction; 3: Cultural competence instruction

## Nuclear Medicine Technologist

| Programs* | Class Capacity | Begins | Length (months) | Award | Res Tuition | Non-res Tuition | Stipend | Offers:‡ 1 | 2 | 3 |
|---|---|---|---|---|---|---|---|---|---|---|
| Coosa Valley Technical College (Rome) | 14 | Sep | 12 | Cert | $1,888 | $3,200 | | | | |
| Medical College of Georgia (Augusta) | 31 | Aug | 12, 21 | Cert, BS | $6,336 | $22,707 | | • | | • |
| Middle Georgia Technical College (Warner Robins) | 10 | Staggered | 12 | Cert | $2,301 | $4,602 | | • | | • |
| **Illinois** | | | | | | | | | | |
| College of DuPage (Glen Ellyn) | 18 | Jun | 15 | Cert | $3,500 | $9,250 | | | | • |
| Edward Hines Jr VA Hospital | 8 | Jan Jul | 12 | Cert, BS | | | | | | |
| Northwestern Memorial Hospital (Chicago) | 12 | Aug | 12, 48 | Cert, 3+1 | $5,000 | $5,000 | | | | • |
| Triton College (River Grove) | 27 | Aug | 22 | AAS | $2,700 | $6,550 | | | | |
| **Indiana** | | | | | | | | | | |
| Indiana University (Indianapolis) | 7 | Jun | 22 | BS | $6,417 | $18,163 | | | | • |
| **Iowa** | | | | | | | | | | |
| Mercy College of Health Sciences (Des Moines) | 9 | Aug | 12 | Cert | | | | | | |
| University of Iowa Hospitals & Clinics (Iowa City) | 10 | Aug | 12 | Cert, BS | $6,387 | $21,667 | | | | • |
| **Kansas** | | | | | | | | | | |
| University of Kansas Medical Center (Kansas City) | 6 | Sep | 12 | Cert | $3,000 | $3,000 | | | | |
| **Kentucky** | | | | | | | | | | |
| Jefferson Community and Technical College (Louisville) | 15 | Aug | 21 | AS | $2,352 | $7,056 | | | | • |
| **Louisiana** | | | | | | | | | | |
| Delgado Community College (New Orleans) | 9 | Aug | 12 | Cert | $2,230 | $6,150 | | | | • |
| **Maine** | | | | | | | | | | |
| Central Maine Medical Center (Lewiston) | 6 | Aug | 12 | Cert | $6,000 | $6,000 | | | | • |
| **Maryland** | | | | | | | | | | |
| Johns Hopkins Hospital (Baltimore) | 15 | Jun | 14 | Cert | $6,000 | $6,000 | | | | • |
| Prince George's Community College (Largo) | 13 | Jun | 13 | Cert, AAS | $3,600 | $8,400 | | | | |
| **Massachusetts** | | | | | | | | | | |
| Mass College of Pharmacy & Health Sciences (Boston) | 25 | Sep | 32, 20 | Cert, BS, Assoc | $20,000 | $20,000 | | | | • |
| Salem State College | 10 | Sep | 38 | BS | $5,248 | $12,298 | | • | | • |
| Springfield Technical Community College | 18 | Sep | 24 | AS | $3,927 | $10,892 | | • | | |
| UMass Memorial Medical Center (Worcester) | 8 | Jun | 12, 48 | Cert, BS | $4,123 | $7,050 | | | | • |
| **Michigan** | | | | | | | | | | |
| William Beaumont Hospital (Royal Oak) | 10 | Sep | 14 | Cert | $3,000 | $3,000 | | | | • |
| **Minnesota** | | | | | | | | | | |
| Mayo School of Health Sciences (Rochester) | 10 | Sep | 12 | Cert, BA/BS | $6,500 | $6,500 | | | | |
| St Mary's University of Minnesota (Winona) | 8 | Sep | 48 | Dipl, Cert, BA | $20,719 | $20,719 | | | | |
| **Mississippi** | | | | | | | | | | |
| University of Mississippi Medical Center (Jackson) | 10 | Jul | 12 | Cert | | | | | | |
| **Missouri** | | | | | | | | | | |
| Research Medical Center (Kansas City) | 9 | Sep | 12 | Cert | $3,300 | $3,300 | | | | • |
| Saint Louis University (St Louis) | 12 | Sep | 12, 49 | Cert, BS | $26,250 | $26,250 | | | | |
| University of Missouri - Columbia | 12 | Aug | 24 | Cert, BHS | $4,956 | $13,500 | | | | • |
| **Nebraska** | | | | | | | | | | |
| University of Nebraska Medical Center (Omaha) | 12 | Aug | 21, 12 | BS | $6,040 | $17,920 | | | | • |
| **Nevada** | | | | | | | | | | |
| University of Nevada - Las Vegas | 18 | Aug | 48 | BS | $2,121 | $7,117 | | • | | • |
| **New Jersey** | | | | | | | | | | |
| Gloucester County College (Sewell) | 24 | Sep | 21 | AAS | $4,686 | $9,372 | | • | | • |
| Muhlenberg Regional Medical Center (Plainfield) | 17 | Jan | 12 | Dipl, Cert, AS | $20,000 | $20,000 | | | | • |
| Univ of Med & Dent of New Jersey (Scotch Plains) | 15 | Sep | 15 | Dipl, Cert, BS | $10,000 | $14,000 | | | | • |
| **New York** | | | | | | | | | | |
| Bronx Community College | 15 | Jan | 30 | Cert, AAS | $2,800 | $4,560 | | • | | • |
| Institute of Allied Medical Professions (New York) | 27 | Oct Apr | 12 | Cert | $30,000 | $30,000 | | • | | • |
| Molloy College (Rockville Centre) | 46 | Sep | 22 | AAS | $16,000 | $16,000 | | | | |
| St Vincent's Hospital Manhattan SVCMC (New York) | 6 | Oct | 12 | Cert | $12,000 | $12,000 | | | | |
| University at Buffalo - SUNY | 15 | Aug | 18 | Dipl, BS | $6,750 | $13,330 | | | | • |
| **North Carolina** | | | | | | | | | | |
| Caldwell Comm College & Tech Institute (Hudson) | 15 | Aug | 21 | AAS | $1,580 | $8,780 | | | | |
| Forsyth Technical Community College (Winston-Salem) | 12/ | Aug Sep | 20 | AAS | $1,520 | $8,440 | | | | • |
| University of North Carolina Hospitals (Chapel Hill) | 10 | Aug | 12 | Cert | | | | | | |
| **Ohio** | | | | | | | | | | |
| Aultman Hospital (Canton) | 6 | Jul | 12 | Cert | $6,000 | $6,000 | | | | |
| Cuyahoga Community College (Parma) | 15 | Aug | 22 | AAS | $5,890 | $14,250 | | • | | • |
| Kent State University - Salem Campus | 14 | Aug | 12 | Dipl, BS | $7,911 | $19,059 | | | | |
| Ohio State University Medical Center (Columbus) | 10 | Last week of Jun | 12 | Cert | $6,000 | $6,000 | | | | • |
| St Elizabeth Health Center (Youngstown) | 11 | Jul | 24 | Cert | $6,000 | $6,000 | | | | |
| University of Cincinnati College of Allied Health Sciences | 45 | Sep | 48, 12 | Cert, BS | $12,532 | $31,896 | | • | | • |

*Data are shown only for programs that completed the 2006 AMA Survey of Health Professions Education Programs
‡Key to Offers: 1: Evening or weekend classes; 2: Non-English instruction; 3: Cultural competence instruction

## Nuclear Medicine Technologist

| Programs* | Class Capacity | Begins | Length (months) | Award | Res Tuition | Non-res Tuition | Stipend | Offers:‡ 1 | 2 | 3 |
|---|---|---|---|---|---|---|---|---|---|---|
| University of Findlay | 56 | Aug Jan | 12, 48 | Cert, AS, BS | $13,230 | $13,230 | | | | • |
| **Oklahoma** | | | | | | | | | | |
| Univ of Oklahoma Health Sciences Center (Oklahoma City) | 11 | Aug | 48 | BS | $6,203 | $9,500 | | | | • |
| **Pennsylvania** | | | | | | | | | | |
| Abington Memorial Hospital (Willow Grove) | 5 | Sep | 12 | Cert | $8,500 | $8,500 | | | | • |
| Cedar Crest College (Allentown) | 8 | Jun | 12, 48 | Cert, BS | $11,356 | $0 | | • | | |
| Jameson Health System (New Castle) | 2 | Jul | 12 | Cert | $3,600 | $3,600 | | • | | • |
| Lancaster Gen Coll of Nursing & Hlth Sciences | 35 | Aug | 12 | Cert | $12,000 | $12,000 | | | | • |
| Wyoming Valley Health Care System (Wilkes-Barre) | 4 | Sep | 12 | Cert | $8,000 | $8,000 | | | | |
| **Puerto Rico** | | | | | | | | | | |
| University of Puerto Rico (San Juan) | 10 | Aug | 12 | BS | $1,700 | $3,200 | | | • | • |
| **Rhode Island** | | | | | | | | | | |
| Rhode Island Hospital (Providence) | 10 | Jul | 15 | Cert | $5,000 | $5,000 | | | | • |
| **South Dakota** | | | | | | | | | | |
| Southeast Technical Institute (Sioux Falls) | 41 | Aug | 24 | AAS | $16,150 | $16,150 | | • | | • |
| **Tennessee** | | | | | | | | | | |
| Baptist College of Health Sciences (Memphis) | 23 | Aug | 18 | BHS | $6,400 | $6,400 | | • | | • |
| Chattanooga State Technical Comm College | 26 | Aug | 12 | Cert | $3,800 | $12,500 | | | | • |
| Methodist University Hospital (Memphis) | 4-6 | Aug | 15 | Cert | $3,600 | $3,600 | | | | • |
| Univ of Tennessee Medical Ctr at Knoxville | 7 | Aug | 12 | Cert, BS | $4,000 | $4,000 | | | | |
| Vanderbilt University Medical Center (Nashville) | 8 | Aug | 12 | Cert | $1,900 | $0 | | | | |
| **Texas** | | | | | | | | | | |
| Amarillo College | 16 | Aug | 24 | AAS | $1,750 | $3,250 | | | | • |
| Baylor University Med Center (Dallas) | 9 | Aug | 20 | Cert | $4,500 | $4,500 | | | | |
| Galveston College | 16 | Aug | 23 | Dipl, AAS | $1,812 | $2,867 | | | | |
| Houston Community College | 41 | Aug | 18, 24 | Cert, AAS | $1,683 | $4,125 | | | | • |
| **Utah** | | | | | | | | | | |
| University of Utah Health Science Center (Salt Lake City) | 7 | Jul | 12 | Cert | $3,000 | $3,000 | | | | • |
| **Vermont** | | | | | | | | | | |
| University of Vermont (Burlington) | 8 | Sep | 48 | BS | $9,832 | $24,816 | | | | • |
| **Virginia** | | | | | | | | | | |
| Naval School of Health Sciences (Portsmouth) | 28 | Jan Jul | 12 | Cert | | | | | | |
| Old Dominion University (Norfolk) | 12 | Aug | 22 | BSNMT | $5,910 | $16,470 | | | | • |
| Virginia Commonwealth University (Richmond) | 10 | Aug | 33 | BS | $7,405 | $23,226 | | | | • |
| **Washington** | | | | | | | | | | |
| Bellevue Community College | 10 | Sep | 15 | Cert | $6,000 | $6,000 | | | | • |
| **West Virginia** | | | | | | | | | | |
| West Virginia State Comm & Tech College (Institute) | 16 | May | 24 | AAS | $2,642 | $7,048 | | • | | • |
| West Virginia University Hospitals (Morgantown) | 4 | Jul | 12 | Cert | $1,600 | $1,600 | | | | |
| Wheeling Jesuit University | 25 | Aug | 48 | BS | $21,000 | $0 | | • | • | • |
| **Wisconsin** | | | | | | | | | | |
| Aurora St Luke's Medical Center (Milwaukee) | 7 | Jun | 12 | BS | $1,500 | $1,500 | | | | |
| Froedtert Memorial Lutheran Hospital (Milwaukee) | 6 | Sep | 12 | Cert, BS | $2,500 | $2,500 | | | | |
| Gundersen Lutheran Medical Foundation (La Crosse) | 2 | Sep | 12 | Cert, BS | $5,200 | $15,200 | | | | • |
| St Joseph's Hospital (Marshfield) | 6 | Aug | 12 | Cert | $5,000 | $12,000 | | | | |

*Data are shown only for programs that completed the 2006 AMA Survey of Health Professions Education Programs

‡Key to Offers: 1: Evening or weekend classes; 2: Non-English instruction; 3: Cultural competence instruction

PROGRAMS

**Includes:**
- Occupational therapist
- Occupational therapy assistant

# Occupational Therapist

### Occupational Description

The practice of occupational therapy means the therapeutic use of everyday life activities (occupations) with individuals or groups for the purpose of participation in roles and situations in home, school, workplace, community, and other settings. Occupational therapy services are provided for the purpose of promoting health and wellness and to those who have or are at risk for developing an illness, injury, disease, disorder, condition, impairment, disability, activity limitation, or participation restriction. Occupational therapy addresses the physical, cognitive, psychosocial, sensory, and other aspects of performance in a variety of contexts to support engagement in everyday life activities that affect health, well-being, and quality of life.

### Job Description

Occupational therapy services are based on evaluation and assessment methods, including the use of skilled observation or the administration and interpretation of standardized or nonstandardized tests and measurements to identify areas for occupational therapy services.

The practice of occupational therapy includes:

A. Methods or strategies selected to direct the process of interventions, such as:
   1. Establishing, remediating, or restoring a skill or ability that has not yet developed or is impaired.
   2. Compensating, modifying, or adapting activity or environment to enhance performance.
   3. Maintaining and enhancing capabilities without which performance in everyday life activities would decline.
   4. Health promotion and wellness to enable or enhance performance in everyday life activities.
   5. Preventing barriers to performance, including disabilities.

B. Evaluation of factors affecting activities of daily living (ADL), instrumental activities of daily living (IADL), education, work, play, leisure, and social participation, including:
   1. Client factors, including body functions (such as neuromuscular, sensory, visual, perceptual, cognitive) and body structures (such as cardiovascular, digestive, integumentary, and genitourinary systems).
   2. Habits, routines, roles, and behavior patterns.
   3. Cultural, physical, environmental, social, and spiritual contexts and activity demands that affect performance.
   4. Performance skills, including motor, process, and communication/interaction skills.

C. Interventions and procedures to promote or enhance safety and performance in activities of daily living (ADL), instrumental activities of daily living (IADL), education, work, play, leisure, and social participation, including:
   1. Therapeutic use of occupations, exercises, and activities.
   2. Training in self-care, self-management, home management, and community/work reintegration.

3. Development, remediation, or compensation of physical, cognitive, neuromuscular, sensory functions and behavioral skills.
4. Therapeutic use of self, including one's personality, insights, perceptions, and judgments, as part of the therapeutic process.
5. Education and training of individuals, including family members, caregivers, and others.
6. Care coordination, case management, and transition services.
7. Consultative services to groups, programs, organizations, or communities.
8. Modification of environments (home, work, school, or community) and adaptation of processes, including the application of ergonomic principles.
9. Assessment, design, fabrication, application, fitting, and training in assistive technology, adaptive devices, and orthotic devices, and training in the use of prosthetic devices.
10. Assessment, recommendation, and training in techniques to enhance functional mobility, including wheelchair management.
11. Driver rehabilitation and community mobility.
12. Management of feeding, eating, and swallowing to enable eating and feeding performance.
13. Application of physical agent modalities, and use of a range of specific therapeutic procedures (such as wound care management; techniques to enhance sensory, perceptual, and cognitive processing; and manual therapy techniques) to enhance performance skills.

### Employment Characteristics

The wide population served by occupational therapists is located in a variety of settings, such as hospitals, clinics, rehabilitation facilities, long-term care facilities, extended care facilities, private practices, schools, camps, the clients' own homes, and community agencies. Occupational therapists both receive referrals from and make referrals to the appropriate health, educational, or medical specialists.

### Salary

AOTA studies conducted in 2006 indicate that the average entry-level salary for occupational therapists is $46,334. Refer to Section IV, Table 5 of this *Directory* for more information, or see www.ama-assn.org/go/salary.

### Educational Programs

**Length.** Programs at the combined baccalaureate/master's level entail 4 to 5 years of college or university preparation.

Postbaccalaureate programs leading to a master's degree are generally 2 ½ years. Following completion of all educational requirements, individuals take a national certification examination. All states also regulate the practice of occupational therapy.

**Prerequisites.** Prerequisites vary among programs. A baccalaureate degree is a prerequisite for most postbaccalaureate occupational therapy programs. A strong foundation of liberal arts and biological, physical, social, and behavioral sciences may be prerequisite to, or concurrent with, the professional education of the program curriculum.

**Curriculum.** Curricula of accredited occupational therapy programs are required to include a broad foundation in the liberal arts and sciences, basic tenets of occupational therapy, occupational therapy theoretical perspectives, the process of screening and evaluation, the process of formulation and implementation of an intervention plan, context of service delivery, management of occupational therapy services, use of research, professional ethics, values, and responsibilities, and 24 weeks of fieldwork education.

# Occupational Therapy Assistant

## Occupational Description

Under the direction of an occupational therapist, the occupational therapy assistant directs an individual's participation in selected tasks to restore, reinforce, and enhance performance; facilitate learning of those skills and functions essential for adaptation and productivity; diminish or correct pathology; and promote and maintain health. The occupational orientation of the assistant is that of guiding the individual's goal-directed use of time, energy, interest, and attention. A fundamental concern is the development and maintenance of the capacity throughout the lifespan to perform with satisfaction to self and others meaningful tasks and roles essential to social participation and to the mastery of self and the environment. Under the therapist's direction, the assistant participates in the development of adaptive skills and performance capacity and is concerned with factors that promote, influence, or enhance performance, as well as those that serve as barriers or impediments to the individual's occupational performance. The occupational therapy assistant provides service to those individuals whose abilities to perform meaningful tasks of living are threatened or impaired by developmental deficits, the aging process, poverty and cultural differences, physical injury or illness, or psychological and social disability.

## Job Description

A contemporary entry-level occupational therapy assistant must:

• Have acquired an educational foundation in the liberal arts and sciences, including a focus on issues related to diversity

• Be educated as a generalist, with a broad exposure to the delivery models and systems utilized in settings where occupational therapy is currently practiced and where it is emerging as a service

• Have achieved entry-level competence through a combination of academic and fieldwork education

• Be prepared to work under the supervision of and in cooperation with the occupational therapist

• Be prepared to articulate and apply occupational therapy principles, intervention approaches and rationales, and expected outcomes as these relate to occupation

• Be prepared to be a lifelong learner and keep current with best practice

• Uphold the ethical standards, values, and attitudes of the occupational therapy profession

## Employment Characteristics

Occupational therapy assistants assist in the planning and implementation of treatment of a diverse popula-

tion in a variety of settings, such as nursing homes, hospitals and clinics, rehabilitation facilities, long-term care facilities, extended care facilities, sheltered workshops, schools and camps, private homes, and community agencies.

## Salary

AOTA studies conducted in 2006 indicate that the average entry-level salary for occupational therapy assistants is approximately $33,000. Refer to Section IV, Table 5 of this *Directory* for more information, or see www.ama-assn.org/go/salary.

## Educational Programs

**Length.** Education may be acquired in either a 2-year associate degree program or a 1-year certificate program. These technical-level education programs are located in 2-year and 4-year colleges and universities, and postsecondary vocational/technical schools and institutions and include academic and fieldwork components, as do the professional level programs. Following completion of all educational requirements, individuals take a national certification examination. Many states also regulate the practice of occupational therapy assistants.

**Prerequisites.** High school diploma or equivalent. A foundation of liberal arts and biological, physical, social, and behavioral sciences may be prerequisite to, or concurrent with, the technical education of the program curriculum.

**Curriculum.** Curricula of accredited occupational therapy assistant programs are required to include a broad foundation of the liberal arts and sciences, basic tenets of occupational therapy, the process of screening and evaluation, the process of intervention and implementation, context of service delivery, assistance in the management of occupational therapy services, use of professional literature, professional ethics, values, and responsibilities, and 16 weeks of fieldwork education.

## Inquiries

**Careers Education**
American Occupational Therapy Association
4720 Montgomery Lane
PO Box 31220
Bethesda, MD 20824-1220
301 652-2682
www.aota.org

**Certification**
National Board for Certification in Occupational Therapy (NBCOT)
800 S Frederick Avenue, Suite 200
Gaithersburg, MD 20877-4150
301 990-7979
www.nbcot.org

**Program Accreditation**
Accreditation Council for Occupational Therapy Education
4720 Montgomery Lane, PO Box 31220
Bethesda, MD 20824-1220
301 652-2682
301 652-1417 Fax
www.aota.org

## Alabama

**University of Alabama at Birmingham**
Occupational Therapy Prgm
School of Health Professions
1530 3rd Ave S - RMSB 354
Birmingham, AL 35294-1212
www.uab.edu
*Prgm Dir:* Penelope Moyers, EdD OTR/L FAOTA
*Tel:* 205 934-9229   *Fax:* 205 975-7787
*E-mail:* pmoyers@uab.edu
*Med Dir:* Harold Jones, PhD

**University of South Alabama**
Occupational Therapy Prgm
Department of Occupational Therapy
1504 Springhill Ave, Rm 5108
Mobile, AL 36604-3273
www.southalabama.edu
*Prgm Dir:* Marjorie E Scaffa, PhD OTR FAOTA
*Tel:* 251 434-3939   *Fax:* 251 434-3934
*E-mail:* mscaffa@jaguar1.usouthal.edu

**Alabama State University**
Occupational Therapy Prgm
915 S Jackson St
PO Box 271
Montgomery, AL 36101-0271
www.alasu.edu
*Prgm Dir:* Angela T Davis, EdD OTR/L
*Tel:* 334 229-5612   *Fax:* 334 229-5882
*E-mail:* adavis@alasu.edu

**Tuskegee University**
Occupational Therapy Prgm
Basil O'Connor Hall
Tuskegee, AL 36088-1696
www.tuskegee.edu/ot
*Prgm Dir:* Gwendolyn Gray, MA OTR/L
*Tel:* 334 727-8695   *Fax:* 334 727-8259
*E-mail:* grayg@tuskegee.edu

## Arizona

**Midwestern University - Glendale Campus**
Occupational Therapy Prgm
19555 N 59th Ave
Glendale, AZ 85308-6814
www.midwestern.edu
*Prgm Dir:* Christine Merchant, MA OTR/L
*Tel:* 623 572-3630   *Fax:* 623 572-3635
*E-mail:* cmerch@midwestern.edu

**AT Still University of Health Sciences**
Occupational Therapy Prgm
5850 E Still Circle
Mesa, AZ 85206-3618
www.atsu.edu
*Prgm Dir:* Bernadette Mineo, PhD OTR/L
*Tel:* 480 219-6071   *Fax:* 480 219-6100
*E-mail:* bmineo@atsu.edu

*Additional information about programs that
returned the AMA's survey is available in a table
beginning on page 280.

## Arkansas

**University of Central Arkansas**
Occupational Therapy Prgm
201 Donaghey Ave HSC, Ste 300, Box 5001
Conway, AR 72035-0001
www.uca.edu
*Prgm Dir:* Linda D Musselman, PhD OTR FAOTA
*Tel:* 501 450-3192   *Fax:* 501 450-3622
*E-mail:* LindaS@mail.uca.edu
*Med Dir:* Linda Musselman, PhD OTR FAOTA

## California

**California State University - Dominguez Hills**
Occupational Therapy Prgm
1000 E Victoria Blvd
Carson, CA 90747-0005
www.csudh.edu/hhs/dhs/OT
*Prgm Dir:* Claudia G Peyton, PhD OTR/L FAOTA
*Tel:* 310 243-3067, Ext 3067   *Fax:* 310 516-3542
*E-mail:* cpeyton@csudh.edu
*Med Dir:* MItchell Maki, PhD

**Loma Linda University**
Occupational Therapy Prgm
Sch of Allied Hlth Professions
Nichol Hall Rm A902
Loma Linda, CA 92350-0001
www.llu.edu
*Prgm Dir:* Esther Huecker, PhD OTR/L
*Tel:* 909 558-4628, Ext 47588   *Fax:* 909 558-0239
*E-mail:* ehuecker@llu.edu

**University of Southern California**
Occupational Therapy Prgm
1540 Alcazar, CHP 133
Los Angeles, CA 90089
www.usc.edu/ot
*Prgm Dir:* Florence A Clark, PhD OTR/L FAOTA
*Tel:* 323 442-2850   *Fax:* 323 442-1540
*E-mail:* fclark@usc.edu
*Med Dir:* Sarah Kelly, OTR

**Samuel Merritt College**
Occupational Therapy Prgm
450 30th St, 4th Fl
Oakland, CA 94609-3108
*Prgm Dir:* Kate Hayner, EdD OTR/L
*Tel:* 510 869-6511, Ext 4780   *Fax:* 510 869-6951
*E-mail:* khayner@samuelmerritt.edu

**San Jose State University**
Occupational Therapy Prgm
Coll of Applied Sciences and Arts
One Washington Square
San Jose, CA 95192-0059
*Prgm Dir:* Marti Southam, PhD OTR/L FAOTA
*Tel:* 408 924-3072   *Fax:* 408 924-3088
*E-mail:* msoutham@casa.sjsu.edu

**Dominican University of California**
Occupational Therapy Prgm
50 Acacia Ave
San Rafael, CA 94901-2298
www.dominican.edu
*Prgm Dir:* Ruth Ramsey, MS OTR/L
*Tel:* 415 257-1393, Ext 1393   *Fax:* 415 458-3774
*E-mail:* rramsey@dominican.edu

## Colorado

**Colorado State University**
Occupational Therapy Prgm
219 Occupational Therapy Bldg
Ft Collins, CO 80523-1573
www.ot.cahs.colostate.edu
*Prgm Dir:* Jodie Redditi-Hanzlik, PhD OTR FAOTA
*Tel:* 970 491-7304   *Fax:* 970 491-6290
*E-mail:* hanzlik@cahs.colostate.edu

## Connecticut

**Sacred Heart University**
Occupational Therapy Prgm
5151 Park Ave
Fairfield, CT 06432-1000
*Prgm Dir:* Jody Bortone, MA OTR/L
*Tel:* 203 396-8023   *Fax:* 203 396-8206
*E-mail:* bortonej@sacredheart.edu

**Quinnipiac University**
Occupational Therapy Prgm
Sch of Health Sciences
275 Mt Carmel Ave
Hamden, CT 06518-0569
www.quinnipiac.edu
*Prgm Dir:* Kimberly D Hartmann, MHS OTR/L FAOTA
*Tel:* 203 582-8679   *Fax:* 203 582-8706
*E-mail:* hartmann@quinnipiac.edu

**University of Hartford**
Occupational Therapy Prgm
Coll of Ed Nursing & Health Professions
200 Bloomfield Ave, Dana Hall Rm 232
West Hartford, CT 06117-1599
*Prgm Dir:* Betsey C Smith, PhD OTR/L
*Tel:* 860 768-4831   *Fax:* 860 768-5706
*E-mail:* bsmith@hartford.edu

## District of Columbia

**Howard University**
Occupational Therapy Prgm
Division of Allied Health Sciences
Sixth and Bryant Sts NW
Washington, DC 20059-0001
*Prgm Dir:* Felecia M Banks, PhD OTR/L
*Tel:* 202 806-7617   *Fax:* 202 462-5248
*E-mail:* fbanks@howard.edu
*Med Dir:* Floyd J Malveaux, MD

## Florida

**Nova Southeastern University**
Occupational Therapy Prgm
Occupational Therapy Department
Health Profs Div, Coll of Allied Health and Nursing
Ft Lauderdale, FL 33328-2018
www.nova.edu/ot
*Prgm Dir:* Sandra Dunbar, DPA OTR/L
*Tel:* 954 262-1242   *Fax:* 954 262-2290
*E-mail:* sdunbar@nova.edu

**Florida Gulf Coast University**
Occupational Therapy Prgm
10501 FGCU Blvd South
Ft Myers, FL 33965-6565
www.fgcu.edu/chp/ot
*Prgm Dir:* Linda Martin, PhD OTR/L FAOTA
*Tel:* 239 590-7556   *Fax:* 239 590-7460
*E-mail:* lmartin@fgcu.edu

**University of Florida**
Occupational Therapy Prgm
101 S Newell Dr
PO Box 100164 HSC
Gainesville, FL 32610-0164
*Prgm Dir:* Joanne Jackson Foss, PhD OTR
*Tel:* 352 273-6135   *Fax:* 352 846-6042
*E-mail:* jfoss@phhp.ufl.edu

**Florida International University**
Occupational Therapy Prgm
University Park Campus, HLS 237
Miami, FL 33199
www.ot.fiu.edu
*Prgm Dir:* Patricia J Scott, PhD MPH OTR FAOTA
*Tel:* 305 348-2264   *Fax:* 305 348-1240
*E-mail:* scottp@fiu.edu

**Barry University**
Occupational Therapy Prgm
11300 NE 2nd Ave
Miami Shores, FL 33161-6695
www.barry.edu
*Prgm Dir:* Douglas M Mitchell, PhD OTR/L
*Tel:* 305 899-3210   *Fax:* 305 899-2958
*E-mail:* dmitchell@mail.barry.edu

**Univ of St Augustine for Health Sciences**
Occupational Therapy Prgm
1 University Blvd
St Augustine, FL 32086-5783
*Prgm Dir:* Karen S Howell, PhD OTR/L FAOTA
*Tel:* 904 826-0084, Ext 222   *Fax:* 904 827-0069
*E-mail:* khowell@usa.edu

**Florida A&M University**
Occupational Therapy Prgm
Ware-Rhaney East
Rm 318
Tallahassee, FL 32307-3500
*Prgm Dir:* Rosalie J Miller, PhD OTR FAOTA
*Tel:* 850 561-2014   *Fax:* 850 561-2457
*E-mail:* rosalie.miller@famu.edu
*Med Dir:* Cynthia Hughess Harris, PhD OTR FAOTA

# Georgia

**Medical College of Georgia**
Occupational Therapy Prgm
School of Allied Health, EF 102
1120 Fifteenth St
Augusta, GA 30912-0700
www.mcg.edu/sahs/ot
*Prgm Dir:* Kathy P Bradley, EdD OTR/L FAOTA
*Tel:* 706 721-3641   *Fax:* 706 721-9718
*E-mail:* kbradley@mcg.edu

**Medical College of Georgia**
*Cosponsor:* Columbus State University
Occupational Therapy Prgm
Illges Hall, Rm 105
4225 University Ave
Columbus, GA 31907-5645
*Prgm Dir:* Robert McAlister, MHE OTR/L
*Tel:* 706 568-2242   *Fax:* 706 568-2073
*E-mail:* rmcalist@mcg.edu

**Brenau University**
Occupational Therapy Prgm
500 Washington, SE
Gainesville, GA 30501
*Prgm Dir:* Sara Jane Brayman, PhD OTR/L FAOTA
*Tel:* 770 534-6139   *Fax:* 770 534-6186
*E-mail:* sbrayman@brenau.edu

# Idaho

**Idaho State University**
Occupational Therapy Prgm
Campus Box 8045
Pocatello, ID 83209-8045
*Prgm Dir:* Martha Hartgraves, PhD OTR/L CLT
*Tel:* 208 282-3758   *Fax:* 208 282-4962
*E-mail:* hartgrav@isu.edu

# Illinois

**Chicago State University**
Occupational Therapy Prgm
9501 S King Dr, Library, Rm 132
Chicago, IL 60628-1598
*Prgm Dir:* Elizabeth S Wittbrodt, MHS OTR/L
*Tel:* 773 995-2366, Ext 2530   *Fax:* 773 995-2839
*E-mail:* ewittbro@csu.edu

**Rush University**
Occupational Therapy Prgm
600 S Paulina, Ste 1010
Chicago, IL 60612-3833
*Prgm Dir:* Clare Giuffrida, PhD OTR/L FAOTA
*Tel:* 312 942-7138   *Fax:* 312 942-6989
*E-mail:* Clare_Giuffrida@rush.edu

**University of Illinois at Chicago**
Occupational Therapy Prgm
1919 W Taylor St M/C 811
Chicago, IL 60612-7250
www.uic.edu/ahp/ot
*Prgm Dir:* Brent Braveman, PhD OTR/L FAOTA
*Tel:* 312 355-2656   *Fax:* 312 413-0256
*E-mail:* bbravema@uic.edu

**Midwestern University**
Occupational Therapy Prgm
College of Health Sciences
555 31st St
Downers Grove, IL 60515-1235
www.midwestern.edu
*Prgm Dir:* Kimberly A Bryze, PhD OTR/L
*Tel:* 630 515-7226   *Fax:* 630 515-7418
*E-mail:* kbryze@midwestern.edu

**Governors State University**
Occupational Therapy Prgm
College of Health Professions
University Park, IL 60466-0975
*Prgm Dir:* Elizabeth Cada, EdD OTR/L FAOTA
*Tel:* 708 534-7295   *Fax:* 708 534-1647
*E-mail:* b-cada@govst.edu

# Indiana

**University of Southern Indiana**
Occupational Therapy Prgm
8600 University Blvd, HP 2068
Evansville, IN 47712-3534
www.health.usi.edu/acadprog
*Prgm Dir:* Barbara J Williams, OTD OTR
*Tel:* 812 465-1179   *Fax:* 812 465-7092
*E-mail:* bjwilliams4@usi.edu

**Indiana University**
Occupational Therapy Prgm
School of Health & Rehabilitation Sciences
1140 W Michigan St, Coleman Hall 311
Indianapolis, IN 46202-5119
*Prgm Dir:* Thomas F Fisher, PhD OTR FAOTA
*Tel:* 317 274-8006   *Fax:* 317 274-2150
*E-mail:* fishert@iupui.edu

**University of Indianapolis**
Occupational Therapy Prgm
1400 E Hanna Ave
Indianapolis, IN 46227-3697
www.ot.uindy.edu
*Prgm Dir:* Candace Beitman, EdD OTR
*Tel:* 317 788-3266   *Fax:* 317 788-3542
*E-mail:* cbeitman@uindy.edu

# Iowa

**St Ambrose University**
Occupational Therapy Prgm
518 W Locust
Davenport, IA 52803-2898
www.sau.edu
*Prgm Dir:* Phyllis Wenthe, MEd OTR/L
*Tel:* 563 333-6276   *Fax:* 563 333-6410
*E-mail:* WenthePhyllisJ@sau.edu

# Kansas

**University of Kansas Medical Center**
Occupational Therapy Prgm
3033 Robinson, Mail Stop 2003
3901 Rainbow Blvd
Kansas City, KS 66160-7602
www.kumc.edu/SAH/OTEd
*Prgm Dir:* Winnie Dunn, PhD OTR FAOTA
*Tel:* 913 588-7195   *Fax:* 913 588-4568
*E-mail:* wdunn@kumc.edu

# Kentucky

**Spalding University**
Occupational Therapy Prgm
851 S Fourth St
Louisville, KY 40203-2188
*Prgm Dir:* Laura S Strickland, EdD OTR/L
*Tel:* 502 585-7125   *Fax:* 502 585-7104
*E-mail:* lstrickland@spalding.edu

**Eastern Kentucky University**
Occupational Therapy Prgm
Department of Occupational Therapy
Dizney 103
Richmond, KY 40475-3135
www.health.eku.edu/ots
*Prgm Dir:* Colleen M Schneck, ScD OTR/L FAOTA
*Tel:* 859 622-6328   *Fax:* 859 622-1601
*E-mail:* Colleen.Schneck@eku.edu

# Louisiana

**Louisiana State Univ Health Sciences Center**
Occupational Therapy Prgm
Sch of Allied Health Professions
1900 Gravier St
New Orleans, LA 70112
*Prgm Dir:* Eve Taylor, PhD LOTR
*Tel:* 504 568-4302   *Fax:* 504 568-4306
*E-mail:* taylor@lsuhsc.edu

**Louisiana State Univ Hlth Sci Ctr-Shreveport**
Occupational Therapy Prgm
1501 Kings Hwy
Shreveport, LA 71130-3932
www.sh.lsuhsc.edu/ah
*Prgm Dir:* Judith C Vestal, PhD LOTR
*Tel:* 318 675-6828   *Fax:* 318 675-8905
*E-mail:* jvesta@lsuhsc.edu

# Maine

**Husson College**
Occupational Therapy Prgm
1 College Circle
3rd Floor Commons
Bangor, ME 04401
www.husson.edu
*Prgm Dir:* Lynn Gitlow, PhD OTR/L ATP
*Tel:* 207 973-1074   *Fax:* 207 973-1061
*E-mail:* gitlowl@husson.edu
*Med Dir:* Teresa Steele, PhD, Dean, School of Health

**University of New England**
Occupational Therapy Prgm
11 Hills Beach Rd
Biddeford, ME 04005-9599
www.une.edu/chp/ot
*Prgm Dir:* Regula Robnett, MS OTR/L
*Tel:* 207 602-2232   *Fax:* 207 602-5963
*E-mail:* rrobnett@une.edu

**University of Southern Maine**
*Cosponsor:* Lewiston-Auburn College
Occupational Therapy Prgm
51 Westminster St
Lewiston, ME 04240-3534
*Prgm Dir:* Roxie M Black, PhD OTR/L FAOTA
*Tel:* 207 753-6515   *Fax:* 207 753-6555
*E-mail:* rblack@usm.maine.edu

# Maryland

**Towson University**
Occupational Therapy Prgm
8000 York Rd
Towson, MD 21252-0001
www.towson.edu/ot
*Prgm Dir:* S Maggie Reitz, PhD OTR/L FAOTA
*Tel:* 410 704-3499   *Fax:* 410 704-2322
*E-mail:* mreitz@towson.edu

# Massachusetts

**Boston University/Sargent College**
Occupational Therapy Prgm
Dept of Occupational Therapy & Rehab Counseling
635 Commonwealth Ave
Boston, MA 02215-1605
www.bu.edu/sargent
*Prgm Dir:* Wendy Coster, PhD OTR/L FAOTA
*Tel:* 617 353-2000, Ext 2727   *Fax:* 617 353-2926
*E-mail:* wjcoster@bu.edu
*Med Dir:* Richard H Egdahl, MD PhD

**Bay Path College**
Occupational Therapy Prgm
588 Longmeadow St
Longmeadow, MA 01106-2212
*Prgm Dir:* Karen Sladyk, PhD OTR/L FAOTA
*Tel:* 413 565-1450   *Fax:* 413 565-1102
*E-mail:* ksladyk@baypath.edu

**Tufts University**
Occupational Therapy Prgm
Department of Occupational Therapy
26 Winthrop St
Medford, MA 02155-7084
www.ase.tufts.edu/bsot
*Prgm Dir:* Sharan Schwartzberg, EdD OTR FAOTA
*Tel:* 617 627-5920   *Fax:* 617 627-3722
*E-mail:* sharan.schwartzberg@tufts.edu

**Salem State College**
Occupational Therapy Prgm
352 Lafayette St
Salem, MA 01970-5353
www.salemstate.edu
*Prgm Dir:* Jeramie S Silveira, MS OTR/L
*Tel:* 978 542-6694   *Fax:* 978 542-6818
*E-mail:* jeramie.silveira@salemstate.edu

**American International College**
Occupational Therapy Prgm
1000 State St
Springfield, MA 01109-3189
*Prgm Dir:* Cathy A Dow-Royer, MA OTR CHT
*Tel:* 413 205-3262   *Fax:* 413 205-3957
*E-mail:* dowroyer@acad.aic.edu

**Springfield College**
Occupational Therapy Prgm
263 Alden St
Springfield, MA 01109-3797
www.springfieldcollege.edu/ot
*Prgm Dir:* Katherine M Post, PhD OTR/L FAOTA
*Tel:* 413 748-3785   *Fax:* 413 748-3796
*E-mail:* kpost@spfldcol.edu

**Worcester State College**
Occupational Therapy Prgm
486 Chandler St
Worcester, MA 01602-2597
*Prgm Dir:* Joanne Gallagher, EdD OTR/L
*Tel:* 508 929-8783   *Fax:* 508 929-8178
*E-mail:* jgallagher@worcester.edu

# Michigan

**Wayne State University**
Occupational Therapy Prgm
Eugene Applebaum College of Pharmacy and Health
 Sciences
259 Mack Ave
Detroit, MI 48202-3489
www.pellerito.edu
*Prgm Dir:* Joseph M Pellerito Jr, MS OTR
*Tel:* 313 577-1435   *Fax:* 313 577-5822
*E-mail:* pellerito@wayne.edu

**Baker College Center of Graduate Studies**
Occupational Therapy Prgm
1116 W Bristol Rd
Flint, MI 48507-5508
*Prgm Dir:* JoAnne Crain, PhD OTR
*Tel:* 810 766-4298   *Fax:* 810 766-2003
*E-mail:* joanne.crain@baker.edu

**Grand Valley State University**
Occupational Therapy Prgm
School of Health Professions
301 Michigan St NE, Ste 200
Grand Rapids, MI 49503-3314
*Prgm Dir:* Nancy J Powell, PhD OTR FAOTA
*Tel:* 616 331-3356   *Fax:* 616 895-3350
*E-mail:* powellna@gvsu.edu

**Western Michigan University**
Occupational Therapy Prgm
1903 W Michigan Ave
Kalamazoo, MI 49008-5333
www.wmich.edu
*Prgm Dir:* Cindee Quake-Rapp, PhD OTR
*Tel:* 269 387-7263   *Fax:* 269 387-7262
*E-mail:* cindee.quake-rapp@wmich.edu

**Saginaw Valley State University**
Occupational Therapy Prgm
Brown Hall 232
7400 Bay Rd
University Center, MI 48710-0001
www.svsu.edu
*Prgm Dir:* Janet Nagayda, OTD MS OTR
*Tel:* 989 964-7353   *Fax:* 989 964-7325
*E-mail:* jnagayda@svsu.edu

**Eastern Michigan University**
Occupational Therapy Prgm
School of Health Sciences
315 Marshall Hall
Ypsilanti, MI 48197-2239
www.emich.edu
*Prgm Dir:* Judith A Olson, PhD OTR
*Tel:* 734 487-2280   *Fax:* 734 487-4095
*E-mail:* judy.olson@emich.edu

# Minnesota

**College of St Scholastica**
Occupational Therapy Prgm
1200 Kenwood Ave
Duluth, MN 55811-4199
www.css.edu
*Prgm Dir:* Carolyn Dorfman, PhD OTR/L
*Tel:* 218 723-6697   *Fax:* 218 723-6472
*E-mail:* cdorfman@css.edu

**University of Minnesota - Minneapolis**
Occupational Therapy Prgm
MMC 388
420 Delware St SE
Minneapolis, MN 55455-0392
www.ot.umn.edu
*Prgm Dir:* Peggy Martin, MS OTR/L
*Tel:* 612 626-4358   *Fax:* 612 625-7192
*E-mail:* marti370@umn.edu

**College of St Catherine - St Paul**
Occupational Therapy Prgm
2004 Randolph Ave, Mail 4092
St Paul, MN 55105-1794
www.stkate.edu/ot
*Prgm Dir:* Mary Lou Henderson, MS OTR/L
*Tel:* 612 690-6622   *Fax:* 612 690-8804
*E-mail:* mlhenderson@stkate.edu

# Mississippi

**University of Mississippi Medical Center**
Occupational Therapy Prgm
2500 N State St
Jackson, MS 39216-4505
http://shrp.umc.edu
*Prgm Dir:* Bette A Groat, MA OTR/L
*Tel:* 601 984-6350   *Fax:* 601 815-1717
*E-mail:* bgroat@shrp.umsmed.edu

# Missouri

**University of Missouri - Columbia**
Occupational Therapy Prgm
407 Lewis Hall
Columbia, MO 65211-4240
*Prgm Dir:* Guy L McCormack, Phd OTR/L FAOTA
*Tel:* 573 882-4183   *Fax:* 573 884-2610
*E-mail:* mccormackg@health.missouri.edu

**Rockhurst University**
Occupational Therapy Prgm
1100 Rockhurst Rd
Kansas City, MO 64110-2561
www.rockhurst.edu/academic/ot
*Prgm Dir:* Kris M Vacek, OTD OTR/L
*Tel:* 816 501-4819   *Fax:* 816 501-4643
*E-mail:* kris.vacek@rockhurst.edu

**Maryville University**
Occupational Therapy Prgm
650 Maryville University Dr
St Louis, MO 63141
www.maryville.edu/sohp/programs/ot.htm
*Prgm Dir:* Paula C Bohr, PhD OTR/L FAOTA
*Tel:* 314 529-9682   *Fax:* 314 529-9191
*E-mail:* pbohr@maryville.edu

**Saint Louis University**
Occupational Therapy Prgm
Doisy School of Health Sciences
3437 Caroline St, Rm 2020
St Louis, MO 63104-1111
www.slu.edu/doisycollege/osot
*Prgm Dir:* Karen Barney, PhD OTR/L FAOTA
*Tel:* 314 977-8514   *Fax:* 314 977-5414
*E-mail:* barneykf@slu.edu

**Washington University**
Occupational Therapy Prgm
School of Medicine, 4444 Forest Park Ave
Campus Box 8505
St Louis, MO 63108
*Prgm Dir:* M Carolyn Baum, PhD OTR/L FAOTA
*Tel:* 314 286-1600   *Fax:* 314 286-1601
*E-mail:* baumc@msnotes.wustl.edu

# Nebraska

**College of Saint Mary**
Occupational Therapy Prgm
7000 Mercy Rd
Omaha, NE 68106-2606
www.csm.edu
*Prgm Dir:* Kristin Haas, OTD OTR/L
*Tel:* 402 384-5281  *Fax:* 402 399-2654
*E-mail:* khaas@csm.edu

**Creighton University**
Occupational Therapy Prgm
School of Pharmacy and Health Professions
2500 California Plaza
Omaha, NE 68178-0259
*Prgm Dir:* Brenda M Coppard, PhD OTR/L
*Tel:* 402 280-5932  *Fax:* 402 280-5692
*E-mail:* bcoppard@creighton.edu

# Nevada

**Touro University - Nevada**
Occupational Therapy Prgm
874 American Pacific Dr
Henderson, NV 89014
*Prgm Dir:* Karen Picus, EdD OTR/L
*Tel:* 702 777-1811  *Fax:* 702 777-1825
*E-mail:* kpicus@touro.edu

# New Hampshire

**University of New Hampshire**
Occupational Therapy Prgm
Sch of Health and Human Services
Hewitt Hall, 4 Library Way
Durham, NH 03824-3563
www.shhs.unh.edu/ot
*Prgm Dir:* Elizabeth Crepeau, PhD OTR/L FAOTA
*Tel:* 603 862-3420  *Fax:* 603 862-0778
*E-mail:* elizabeth.crepeau@unh.edu

# New Jersey

**Richard Stockton College of New Jersey**
Occupational Therapy Prgm
PO Box 195
Jim Leeds Rd
Pomona, NJ 08240-0195
www.stockton.edu
*Prgm Dir:* Victoria Schindler, PhD OTR/L FAOTA
*Tel:* 609 652-4687  *Fax:* 609 652-4858
*E-mail:* victoria.schindler@stockton.edu

**Seton Hall University**
Occupational Therapy Prgm
School of Graduate Medical Education
McQuaid Hall
South Orange, NJ 07079-2689
http://gradmeded.shu.edu/
   prog_master_occupational_therapy.htm
*Prgm Dir:* Ruth Segal, PhD OTR
*Tel:* 973 275-2443  *Fax:* 973 275-2370
*E-mail:* segalrut@shu.edu

**Kean University**
Occupational Therapy Prgm
1000 Morris Ave
PO Box 411
Union, NJ 07083-9982
*Prgm Dir:* Lynn Richard, MA OT
*Tel:* 908 737-3380  *Fax:* 908 737-3377
*E-mail:* lynricha@kean.edu

# New Mexico

**University of New Mexico**
Occupational Therapy Prgm
School of Medicine
Health Science & Service Bldg Rm 215 MSC09 5240
Albuquerque, NM 87131-0001
http://hsc.unm.edu/som/ot
*Prgm Dir:* Janet Poole, PhD OTR/L FAOTA
*Tel:* 505 272-1753  *Fax:* 505 272-3583
*E-mail:* jpoole@salud.unm.edu
*Med Dir:* Heidi Sanders, OTR/L

# New York

**Touro College - Bay Shore**
Occupational Therapy Prgm
1700 Union Blvd
Bay Shore, NY 11706
*Prgm Dir:* Vera-Jean Clark-Brown, MS OTR/L
*Tel:* 631 665-1600, Ext 231  *Fax:* 631 665-6084
*E-mail:* vjclark-brown@touro.edu

**Long Island University - Brooklyn Campus**
Occupational Therapy Prgm
One University Plaza
Brooklyn, NY 11201-5372
*Prgm Dir:* Ann Burkhardt, OTD OTR/L BCN FAOTA
*Tel:* 718 780-4332  *Fax:* 718 780-4535
*E-mail:* ann.burkhardt@liu.edu

**SUNY Downstate Medical Center**
Occupational Therapy Prgm
Coll of Health Related Professions
450 Clarkson Ave Box 81
Brooklyn, NY 11203-2098
www.downstate.edu/chrp/ot
*Prgm Dir:* Joyce Sabari, PhD OTR BCN FAOTA
*Tel:* 718 270-7731  *Fax:* 718 270-7464
*E-mail:* joyce.sabari@downstate.edu

**D'Youville College**
Occupational Therapy Prgm
One D'Youville Square, 320 Porter Ave
Buffalo, NY 14201-1084
www.dyc.edu
*Prgm Dir:* Merlene C Gingher, EdD OT
*Tel:* 716 829-7830  *Fax:* 716 829-8137
*E-mail:* gingherm@dyc.edu

**University at Buffalo - SUNY**
Occupational Therapy Prgm
515 Stockton Kimball Tower
3435 Main St
Buffalo, NY 14214-3079
http://phhp.buffalo.edu/rs/ot
*Prgm Dir:* Susan M Nochajski, PhD OTR
*Tel:* 716 829-6942  *Fax:* 716 829-3217
*E-mail:* nochajsk@buffalo.edu

**Mercy College**
Occupational Therapy Prgm
555 Broadway
Dobbs Ferry, NY 10522-1134
www.mercy.edu/acadivisions/healthprofessions/grad/
   occupational_therapy.cfm
*Prgm Dir:* Joan Toglia, PhD OTR
*Tel:* 914 674-7815  *Fax:* 914 674-7840
*E-mail:* jtoglia@mercy.edu

**Ithaca College**
Occupational Therapy Prgm
204 F Smiddy Hall
Ithaca, NY 14850-7079
*Prgm Dir:* Diane Long, MS OTR
*Tel:* 607 274-3093  *Fax:* 607 274-3055
*E-mail:* dlong@ithaca.edu

**CUNY York College**
Occupational Therapy Prgm
94-20 Guy R Brewer Blvd
Jamaica, NY 11451-9902
www.york.cuny.edu/~healthsci
*Prgm Dir:* Andrea Krauss, DSW OTR/L BCP
*Tel:* 718 262-2720  *Fax:* 718 262-2767
*E-mail:* akrauss@york.cuny.edu

**Keuka College**
Occupational Therapy Prgm
141 Central Ave
Keuka Park, NY 14478
*Prgm Dir:* Vickie Smith, EdD MBA OTR/L
*Tel:* 315 279-5666  *Fax:* 315 279-5439
*E-mail:* vlsmith@mail.keuka.edu

**Columbia University**
Occupational Therapy Prgm
Neurological Institute 8th Fl
710 W 168th St
New York, NY 10032-2603
www.columbiaot.org
*Prgm Dir:* Janet Falk-Kessler, EdD OTR FAOTA
*Tel:* 212 305-5267  *Fax:* 212 305-4569
*E-mail:* jf6@columbia.edu
*Med Dir:* Donald Kornfeld, MD

**New York University**
Occupational Therapy Prgm
The Steinhardt School of Education
35 W 4th St 11th Fl
New York, NY 10012-1172
*Prgm Dir:* Jim Hinojosa, PhD OT/L FAOTA
*Tel:* 212 998-5845  *Fax:* 212 995-4044
*E-mail:* jh9@nyu.edu

**Touro College - Manhattan**
Occupational Therapy Prgm
27-33 W 23rd St
New York, NY 10010
www.touro.edu
*Prgm Dir:* Vera-Jean Clark-Brown, MS OTR/L
*Tel:* 212 463-0400, Ext 691  *Fax:* 631 665-6084
*E-mail:* vjclark-brown@touro.edu

**New York Institute of Technology**
Occupational Therapy Prgm
Department of Occupational Therapy
PO Box 8000
Old Westbury, NY 11568-8000
www.nyit.edu
*Prgm Dir:* Hermine Plotnick, MA OTR/L
*Tel:* 516 686-3865, Ext -  *Fax:* 516 686-3795
*E-mail:* hplotnic@nyit.edu

**Dominican College**
Occupational Therapy Prgm
470 Western Hwy
Orangeburg, NY 10962-1299
www.dc.edu
*Prgm Dir:* Sandra F Countee, PhD OTR/L
*Tel:* 845 359-7800, Ext 332  *Fax:* 845 359-1879
*E-mail:* sandra.countee@dc.edu

**Stony Brook University**
Occupational Therapy Prgm
Sch of Health Tech and Management
Div of Rehabilitation Sciences
Stony Brook, NY 11794-8201
*Prgm Dir:* Donna Costa, MS OTR/L
*Tel:* 631 444-8126  *Fax:* 631 444-6305
*E-mail:* Donna.Costa@stonybrook.edu

**The Sage Colleges**
Occupational Therapy Prgm
45 Ferry St
Troy, NY 12180-4115
*Prgm Dir:* Wendy Krupnick, PhD OTR/L
*Tel:* 518 244-2267  *Fax:* 518 244-4524
*E-mail:* krupnw@sage.edu

**Utica College**
Occupational Therapy Prgm
1600 Burrstone Rd
Utica, NY 13502-4892
*Prgm Dir:* Saly C Townsend, MA OTR
*Tel:* 315 792-3239    *Fax:* 315 792-3248
*E-mail:* stownsend@utica.edu

# North Carolina

**Univ of North Carolina at Chapel Hill**
Occupational Therapy Prgm
Ste 2050 Bondurant Hall, 301A S Columbia St
CB #7122
Chapel Hill, NC 27599-7120
www.alliedhealth.unc.edu/ocsci
*Prgm Dir:* Cathy Nielson, MPH OTR/L FAOTA
*Tel:* 919 966-2452    *Fax:* 919 966-9007
*E-mail:* cnielson@med.unc.edu
*Med Dir:* William Roper, MD

**East Carolina University**
Occupational Therapy Prgm
School of Allied Hlth Sciences
Greenville, NC 27858-4353
www.ecu.edu/ot
*Prgm Dir:* Anne E Dickerson, PhD OTR/L FAOTA
*Tel:* 252 477-6910    *Fax:* 252 477-6990
*E-mail:* dickersona@ecu.edu

**Lenoir - Rhyne College**
Occupational Therapy Prgm
Box 7547
Hickory, NC 28603-7547
www.lrc.edu/ot
*Prgm Dir:* Toni S Oakes, MS OTR/L
*Tel:* 828 328-7366    *Fax:* 828 328-7364
*E-mail:* oakest@lrc.edu

**Winston-Salem State University**
Occupational Therapy Prgm
School of Health Sciences
601 Martin Luther King Jr Dr
Winston-Salem, NC 27110-0003
www.wssu.edu
*Prgm Dir:* Dorothy Bethea, EdD OTR/L
*Tel:* 336 750-3177, Ext 3172    *Fax:* 336 750-3173
*E-mail:* betheadp@wssu.edu

# North Dakota

**University of Mary**
Occupational Therapy Prgm
7500 University Dr
Bismarck, ND 58504-9652
*Prgm Dir:* Janeene Sibla, MS OTR/L
*Tel:* 701 255-7500, Ext 8212    *Fax:* 701 255-7687
*E-mail:* jsibla@umary.edu

**University of North Dakota**
Occupational Therapy Prgm
Box 7126
Grand Forks, ND 58202-7126
*Prgm Dir:* Janet S Jedlicka, PhD OTR/L
*Tel:* 701 777-2209    *Fax:* 701 777-2212
*E-mail:* jjedlicka@medicine.nodak.edu

# Ohio

**Xavier University**
Occupational Therapy Prgm
3800 Victory Pkwy
Cincinnati, OH 45207-7341
www.xavier.edu/ot_grad
*Prgm Dir:* Carol Scheerer, EdD OTR/L
*Tel:* 513 745-3310    *Fax:* 513 745-3261
*E-mail:* scheerer@xavier.edu

**Cleveland State University**
Occupational Therapy Prgm
2121 Euclid Ave
Health Sciences (HS) 101
Cleveland, OH 44115-2440
www.csuohio.edu/healthsci
*Prgm Dir:* Glenn Goodman, PhD OTR/L
*Tel:* 216 687-2493    *Fax:* 216 687-9316
*E-mail:* g.goodman@csuohio.edu

**Ohio State University**
Occupational Therapy Prgm
453 W 10th St
406 Atwell Hall
Columbus, OH 43210-1234
*Prgm Dir:* Jane Case-Smith, EdD OTR/L FAOTA
*Tel:* 614 292-0357    *Fax:* 614 292-0210
*E-mail:* case-smith.1@osu.edu

**University of Findlay College of Health Professions**
Occupational Therapy Prgm
1000 N Main St
Findlay, OH 45840-3695
www.findlay.edu
*Prgm Dir:* Cynthia Goodwin, MS OTR/L
*Tel:* 419 434-6977    *Fax:* 419 434-4083
*E-mail:* goodwin@findlay.edu

**Shawnee State University**
Occupational Therapy Prgm
940 Second St
Portsmouth, OH 45662-4303
*Prgm Dir:* Debra S Scurlock, MA OTR/L
*Tel:* 740 351-3272    *Fax:* 740 351-3354
*E-mail:* dscurlock@shawnee.edu

**University of Toledo**
Occupational Therapy Prgm
3015 Arlington Ave
Toledo, OH 43614-5803
www.meduohio.edu
*Prgm Dir:* Julie J Thomas, PhD OTR/L FAOTA
*Tel:* 419 383-5068    *Fax:* 419 383-5880
*E-mail:* jthomas@meduohio.edu

# Oklahoma

**Univ of Oklahoma Health Sciences Center**
Occupational Therapy Prgm
801 NE 13th St
Oklahoma City, OK 73104-2512
*Prgm Dir:* Cyndy Robinson, MS OTR/L
*Tel:* 405 271-2411, Ext 47139    *Fax:* 405 271-2432
*E-mail:* cyndy-robinson@ouhsc.edu

**Univ of Oklahoma Health Sciences Center - Tulsa**
*Cosponsor:* Univ of Oklahoma at Schusterman
Occupational Therapy Prgm
4502 E 41st St
Tulsa, OK 74135-5072
*Prgm Dir:* Cyndy Robinson, MS OTR/L
*Tel:* 405 271-2131, Ext 47139    *Fax:* 405 271-2432
*E-mail:* cyndy-robinson@ouhsc.edu

# Oregon

**Pacific University**
Occupational Therapy Prgm
222 SE 8th Ave
Hillsboro, OR 97123
www.ot.pacificu.edu
*Prgm Dir:* John A White Jr, PhD OTR
*Tel:* 503 352-2203, Ext 3141    *Fax:* 503 352-2980
*E-mail:* whiteja@pacificu.edu

# Pennsylvania

**College Misericordia**
Occupational Therapy Prgm
Division of Health Sciences
301 Lake St
Dallas, PA 18612-1098
*Prgm Dir:* Ellen McLaughlin, EdD OTR/L
*Tel:* 570 674-6399    *Fax:* 570 674-3052
*E-mail:* emclaugh@misericordia.edu

**Elizabethtown College**
Occupational Therapy Prgm
One Alpha Dr
Elizabethtown, PA 17022-2298
www.etown.edu
*Prgm Dir:* Nancy Carlson, PhD OTR/L
*Tel:* 717 361-1995    *Fax:* 717 361-1176
*E-mail:* carlsona@etown.edu

**Gannon University**
Occupational Therapy Prgm
109 University Square
Erie, PA 16541-0001
www.gannon.edu
*Prgm Dir:* Jeffrey L Boss, MS OTR/L
*Tel:* 814 871-5670    *Fax:* 814 871-7759
*E-mail:* boss001@gannon.edu

**St Francis University**
Occupational Therapy Prgm
PO Box 600
Loretto, PA 15940-0600
www.francis.edu
*Prgm Dir:* Donald E Walkovich, DHSc MS OTR/L
*Tel:* 814 472-3899    *Fax:* 814 472-3950
*E-mail:* dwalkovich@francis.edu

**Penn State University - Mont Alto**
Occupational Therapy Prgm
1 Campus Dr
Mont Alto, PA 17237-9703
*Prgm Dir:* Angela Hissong, DEd OTR/L
*Tel:* 717 749-6147    *Fax:* 717 749-6166
*E-mail:* anh1@psu.edu

**Philadelphia University**
Occupational Therapy Prgm
School House Ln & Henry Ave
Philadelphia, PA 19144-5497
www.philau.edu/ot
*Prgm Dir:* Catherine Verrier Piersol, MS OTR/L
*Tel:* 215 951-2530    *Fax:* 215 951-2526
*E-mail:* PiersolC@PhilaU.edu
*Med Dir:* Matt Dane Baker, PA-C, MS

**Temple University**
Occupational Therapy Prgm
College of Health Professions
3307 N Broad St
Philadelphia, PA 19140-5101
www.temple.edu/ot
*Prgm Dir:* Moya Kinnealey, PhD OTR/L FAOTA
*Tel:* 215 707-4881    *Fax:* 215 707-7656
*E-mail:* moya.kinnealey@temple.edu

**Thomas Jefferson University**
Occupational Therapy Prgm
Jefferson College of Health Professions
130 S 9th St, Rm 810, Edison Bldg
Philadelphia, PA 19107-5233
*Prgm Dir:* Janice P Burke, PhD OTR/L FAOTA
*Tel:* 215 503-9606    *Fax:* 215 503-3499
*E-mail:* janice.burke@jefferson.edu

**University of the Sciences in Philadelphia**
Occupational Therapy Prgm
600 S 43rd St, Box 24
Philadelphia, PA 19104-4495
*Prgm Dir:* Paula Kramer, PhD OTR/L FAOTA
*Tel:* 215 596-8767    *Fax:* 215 596-8690
*E-mail:* p.kramer@usip.edu

**Chatham College**
Occupational Therapy Prgm
Woodland Rd
Pittsburgh, PA 15232-2826
*Prgm Dir:* Joyce Salls, OTD OTR/L BCP
*Tel:* 412 365-1177  *Fax:* 412 365-1213
*E-mail:* salls@chatham.edu
*Med Dir:* Douglas Kress, MD

**Duquesne University**
Occupational Therapy Prgm
Department of Occupational Therapy
Rangos School of Health Sciences Bldg, Rm 234
Pittsburgh, PA 15282-0020
www.healthsciences.duq.edu/ot/othome.html
*Prgm Dir:* Patricia Crist, PhD OTR/L FAOTA
*Tel:* 412 396-5945  *Fax:* 412 396-4343
*E-mail:* crist@duq.edu

**University of Pittsburgh**
Occupational Therapy Prgm
Sch of Health and Rehab Sciences
5012 Forbes Tower
Pittsburgh, PA 15260
www.shrs.pitt.edu/ot
*Prgm Dir:* Joan C Rogers, PhD OTR/L FAOTA
*Tel:* 412 383-6621  *Fax:* 412 383-6613
*E-mail:* jcr@pitt.edu

**Alvernia College**
Occupational Therapy Prgm
400 St Bernardine St
Reading, PA 19607-1799
*Prgm Dir:* Karen Ann Cameron, PhD OTD OTR/L
*Tel:* 610 796-8386  *Fax:* 610 796-8378
*E-mail:* karen.cameron@alvernia.edu

**University of Scranton**
Occupational Therapy Prgm
Leahy Hall
800 Linden St
Scranton, PA 18510-4501
*Prgm Dir:* Carol Reinson, PhD OTR/L
*Tel:* 570 941-6225  *Fax:* 570 941-4380
*E-mail:* reinsonc2@scranton.edu

# Puerto Rico

**University of Puerto Rico at Humacao**
Occupational Therapy Prgm
100 Carr 908 CUH Station
Humacao, PR 00791-4300
*Prgm Dir:* Milagros Marrero Diaz, MPH OTR/L
*Tel:* 787 850-9456  *Fax:* 787 850-9434

**University of Puerto Rico**
Occupational Therapy Prgm
Medical Sciences Campus-CHRP
PO Box 365067
San Juan, PR 00936-5067
http://cprsweb.rcm.upr.edu
*Prgm Dir:* Dyhalma Irizarry, PhD OTR/L FAOTA
*Tel:* 787 758-2525, Ext 4200  *Fax:* 787 282-8174
*E-mail:* dyhalmaIrizarry@cprs.rcm.upr.edu

# South Carolina

**Medical University of South Carolina**
Occupational Therapy Prgm
151B Rutledge Ave
PO Box 2507965
Charleston, SC 29425
www.musc.edu/chp/ot
*Prgm Dir:* Maralynne D Mitcham, PhD OTR/L FAOTA
*Tel:* 843 792-9734  *Fax:* 843 792-0710
*E-mail:* mitchamm@musc.edu

# South Dakota

**University of South Dakota**
Occupational Therapy Prgm
414 E Clark St
Vermillion, SD 57069-2390
*Prgm Dir:* Barbara Brockevelt, PhD OTR/L
*Tel:* 605 677-5601  *Fax:* 605 677-6581
*E-mail:* barb.brockevelt@usd.edu

# Tennessee

**University of Tennessee Health Science Ctr**
Occupational Therapy Prgm
College of Allied Health Sciences
930 Madison Ave, Ste 601
Memphis, TN 38163-0002
www.utmem.edu/allied/othome.html
*Prgm Dir:* Ann Nolen, PsyD OTR
*Tel:* 901 448-9393  *Fax:* 901 448-7545
*E-mail:* anolen@utmem.edu

**Milligan College**
Occupational Therapy Prgm
PO Box 130
Milligan College, TN 37682
*Prgm Dir:* Jeff Snodgrass, PhD OTR/L CWCE CEES
*Tel:* 423 975-8010  *Fax:* 423 975-8019
*E-mail:* jsnodgrass@milligan.edu

**Belmont University**
Occupational Therapy Prgm
1900 Belmont Blvd
Nashville, TN 37212-3757
www.belmont.edu/ot
*Prgm Dir:* Ruth S Ford, EdD OTR/L BCG
*Tel:* 615 460-6706  *Fax:* 615 460-6475
*E-mail:* fordr@mail.belmont.edu

**Tennessee State University**
Occupational Therapy Prgm
College of Health Sciences
3500 John A Merritt Blvd
Nashville, TN 37209-1561
*Prgm Dir:* Larry R Snyder, PhD OTR/L
*Tel:* 615 963-5891  *Fax:* 615 963-5956
*E-mail:* lsnyder@tnstate.edu

# Texas

**Texas Woman's University - Dallas**
Occupational Therapy Prgm
1810 Inwood Rd
Dallas, TX 76204-5648
*Prgm Dir:* Sally Schultz, PhD OTR LPC
*Tel:* 940 898-2801  *Fax:* 940 898-2806
*E-mail:* sschultz@twu.edu

**Texas Woman's University - Denton**
Occupational Therapy Prgm
Box 425648 TWU Station
Denton, TX 76204-5648
*Prgm Dir:* Sally Schultz, PhD OTR LPC
*Tel:* 940 898-2803  *Fax:* 940 898-2806
*E-mail:* sschultz@twu.edu
*Med Dir:* Martin Grabois, MD

**Univ of Texas - Pan American**
Occupational Therapy Prgm
1201 W University Dr
Edinburg, TX 78541-2999
*Prgm Dir:* Judith Bowen, MPA OTR CHTP
*Tel:* 956 381-2475, Ext 2474  *Fax:* 956 381-2476
*E-mail:* jebowen@panam.edu

**University of Texas at El Paso**
Occupational Therapy Prgm
1101 N Campbell St
El Paso, TX 79902-4299
*Prgm Dir:* Karen P Funk, OTD OTR
*Tel:* 915 747-8226  *Fax:* 915 747-8211
*E-mail:* kfunk@utep.edu

**University of Texas Medical Branch/SAHS**
Occupational Therapy Prgm
301 University Blvd
Galveston, TX 77555-1142
www.sahs.utmb.edu/programs/ot
*Prgm Dir:* Gretchen V M Stone, PhD OTR FAOTA
*Tel:* 409 747-1628  *Fax:* 409 747-1615
*E-mail:* grstone@utmb.edu
*Med Dir:* Gretchen Stone, PhD OTR FAOTA

**Texas Woman's University - Houston**
Occupational Therapy Prgm
1130 John Freeman Blvd
Houston, TX 77030-2897
*Prgm Dir:* Sally Schultz, PhD OTR LPC
*E-mail:* sschultz@twu.edu

**University of Texas Health Science Center**
Occupational Therapy Prgm
Department of Occupational Therapy
Laredo, TX 78040-4395
www.uthscsa.edu
*Prgm Dir:* Karin J Barnes, PhD OTR
*Tel:* 210 567-8881  *Fax:* 210 567-8893
*E-mail:* barnesk@uthscsa.edu

**Texas Tech Univ Health Sciences Center**
Occupational Therapy Prgm
3601 4th St, Stop 6220
Lubbock, TX 79430-6220
www.ttuhsc.edu
*Prgm Dir:* Dawndra Scott, PhD OTR
*Tel:* 806 743-3240  *Fax:* 806 743-6005
*E-mail:* Dawndra.Scott@ttuhsc.edu

**Univ of Texas Hlth Sci Ctr at San Antonio**
Occupational Therapy Prgm
7703 Floyd Curl Dr
San Antonio, TX 78229-3900
*Prgm Dir:* Karen Barnes
*Tel:* 210 567-8890  *Fax:* 210 567-8893
*E-mail:* barnesk@uthscsa.edu
*Med Dir:* Nicolas E Walsh, MD

# Utah

**University of Utah**
Occupational Therapy Prgm
520 Wakara Way
Salt Lake City, UT 84108-1290
https://www.health.utah.edu/octh
*Prgm Dir:* JoAnne Wright, PhD OTR/L CLVT
*Tel:* 801 585-9135  *Fax:* 801 585-1001
*E-mail:* jwright@hsc.utah.edu

# Virginia

**James Madison University**
Occupational Therapy Prgm
Department of Health Sciences
Coll of Integrated Sci & Tech, MSC 4301
Harrisonburg, VA 22807-0001
*Prgm Dir:* Jeffrey D Loveland, OTD OTR
*Tel:* 540 568-8170  *Fax:* 540 568-3336
*E-mail:* lovelajd@jmu.edu
*Med Dir:* David Knitter, MD

## Virginia Commonwealth Univ/Med Coll of VA
Occupational Therapy Prgm
PO Box 980008
1000 E Marshall St
Richmond, VA 23298-0008
www.sahp.vcu.edu/occu
*Prgm Dir:* Shelly J Lane, PhD OTR/L FAOTA
*Tel:* 804 828-2219    *Fax:* 804 828-0782
*E-mail:* sjlane@vcu.edu

## Jefferson College of Health Sciences
Occupational Therapy Prgm
920 S Jefferson St
Roanoke, va 24016
*Prgm Dir:* Ave M Mitta, MS OTR
*Tel:* 540 985-4097    *Fax:* 540 985-9773
*E-mail:* ammitta@jchs.edu

## Shenandoah University
Occupational Therapy Prgm
333 W Cork St 5th Fl
Ste 510
Winchester, VA 22601-5195
www.su.edu/ot
*Prgm Dir:* Deborah Marr, ScD OTR/L
*Tel:* 540 678-4312    *Fax:* 540 665-5564
*E-mail:* dmarr@su.edu

# Washington

## University of Washington
Occupational Therapy Prgm
Department of Rehabilitation Medicine
Box 356490
Seattle, WA 98195-6490
http://depts.washington.edu/rehab/ot
*Prgm Dir:* Elizabeth M Kanny, PhD OTR/L FAOTA
*Tel:* 206 598-5393    *Fax:* 206 685-3244
*E-mail:* ekanny@u.washington.edu

## Eastern Washington University
Occupational Therapy Prgm
310 N Riverpoint Blvd
Box R
Spokane, WA 99004-1675
*Prgm Dir:* Greg Wintz, PhD OTR/L
*Tel:* 509 368-6562    *Fax:* 509 368-6561
*E-mail:* gwintz@mail.ewu.edu

## University of Puget Sound
Occupational Therapy Prgm
School of Occupational Therapy
1500 N Warner St #1070
Tacoma, WA 98416
www.ups.edu/ot
*Prgm Dir:* George Tomlin, PhD OTR/L
*Tel:* 253 879-3522    *Fax:* 253 879-2933
*E-mail:* tomlin@ups.edu

# West Virginia

## West Virginia University
Occupational Therapy Prgm
School of Medicine, Robert C Byrd Health Sciences
    Center
PO Box 9139
Morgantown, WV 26506-9139
www.hsc.wvu.edu/som/ot
*Prgm Dir:* Randy P McCombie, PhD OTR/L
*Tel:* 304 293-8828    *Fax:* 304 293-7105
*E-mail:* rmccombie@hsc.wvu.edu
*Med Dir:* Robert D'Alessandri, MD

# Wisconsin

## University of Wisconsin - La Crosse
Occupational Therapy Prgm
4031 Health Science Center
1725 State St
La Crosse, WI 54601-9959
*Prgm Dir:* Peggy Denton, PhD OTR FAOTA
*Tel:* 608 785-5059    *Fax:* 608 785-6647
*E-mail:* denton.pegg@uwlax.edu

## University of Wisconsin - Madison
Occupational Therapy Prgm
1300 University Ave, 2110 MSC
Madison, WI 53706-1532
www.soemadison.wisc.edu/kinesiology
*Prgm Dir:* Mary Schneider, PhD OTR/L
*Tel:* 608 262-2936    *Fax:* 608 263-6434
*E-mail:* schneider@education.wisc.edu

## Concordia University Wisconsin
Occupational Therapy Prgm
12800 N Lake Shore Dr
Mequon, WI 53092-2402
www.cuw.edu/ot
*Prgm Dir:* Linda Samuel, PhD OTR
*Tel:* 262 243-4469    *Fax:* 262 243-4506
*E-mail:* linda.samuel@cuw.edu

## Mount Mary College
Occupational Therapy Prgm
2900 N Menomonee River Pkwy
Milwaukee, WI 53222-4597
*Prgm Dir:* Jane Olson, PhD OTR
*Tel:* 414 256-1246, Ext 348    *Fax:* 414 256-0194
*E-mail:* olsonj@mtmary.edu

## University of Wisconsin - Milwaukee
Occupational Therapy Prgm
College of Health Sciences
PO Box 413
Milwaukee, WI 53201-0413
www.uwm.edu/CHS
*Prgm Dir:* Phyllis King, PhD OT FAOTA
*Tel:* 414 229-5616    *Fax:* 414 229-5100
*E-mail:* pking@uwm.edu

# Wyoming

## University of North Dakota at Casper College
Occupational Therapy Prgm
125 College Dr
Casper, WY 82601-9958
*Prgm Dir:* Janet S Jedlicka, PhD OTR/L
*Tel:* 701 777-2217    *Fax:* 701 777-2212
*E-mail:* jjedlicka@medicine.nodak.edu

## Occupational Therapist

| Programs* | Class Capacity | Begins | Length (months) | Award | Res Tuition | Non-res Tuition | Stipend | Offers:‡ 1 | 2 | 3 |
|---|---|---|---|---|---|---|---|---|---|---|
| **Alabama** | | | | | | | | | | |
| Alabama State University (Montgomery) | 5 | Jun | 30 | MS | $2,004 | $4,008 | | | | • |
| Tuskegee University | 20+ | Aug | 24 | BS, MS | $19,790 | $19,790 | | | | • |
| University of Alabama at Birmingham | 35 | Aug | 28 | MS | $10,976 | $22,200 | | | | • |
| University of South Alabama (Mobile) | 30 | Aug | 27 | MS | $5,960 | $11,920 | | | | • |
| **Arizona** | | | | | | | | | | |
| AT Still University of Health Sciences (Mesa) | 32 | Aug | 28 | MS | $21,985 | $21,985 | | | | • |
| Midwestern University - Glendale Campus | 35 | Aug Sep | 27 | MOT | $22,900 | $22,900 | | | | • |
| **Arkansas** | | | | | | | | | | |
| University of Central Arkansas (Conway) | 48 | May | 37 | MS | $5,755 | $10,255 | | | | • |
| **California** | | | | | | | | | | |
| California State University - Dominguez Hills (Carson) | 34 | Jan | 27 | MSOT | $3,301 | $12,832 | | • | | • |
| Dominican University of California (San Rafael) | 20 | Jan May Aug | 30 | BSMS, MSOT | $27,176 | $27,176 | $1,500 | | | • |
| Loma Linda University | 30 | Jun | 33 | MOT | $18,753 | $18,753 | | | | • |
| University of Southern California (Los Angeles) | 110 | Jun | 18, 27 | BS, MA | $42,282 | $42,282 | | | | • |
| **Colorado** | | | | | | | | | | |
| Colorado State University (Ft Collins) | 45 | Aug | 24 | Dipl, MS | $4,841 | $16,109 | | | | |
| **Connecticut** | | | | | | | | | | |
| Quinnipiac University (Hamden) | 400 | Sep | 55 | BSHS, MOT | $24,340 | $0 | | | | • |
| **Florida** | | | | | | | | | | |
| Barry University (Miami Shores) | 25 | Aug | 30 | MS | $725 | $0 | | • | | • |
| Florida A&M University (Tallahassee) | 40 | Aug | 33 | MS | $2,383 | $8,591 | | | | • |
| Florida Gulf Coast University (Ft Myers) | 24 | Aug | 23 | MS | $229 | $850 | | | | • |
| Florida International University (Miami) | 70 | Aug | 39, 27 | BS/MS, MS | $3,043 | $16,252 | | | | • |
| Nova Southeastern University (Ft Lauderdale) | 45 | Jun | 28 | MOT, OTD Ph | $16,495 | $18,495 | | | | • |

*Data are shown only for programs that completed the 2006 AMA Survey of Health Professions Education Programs
‡Key to Offers: 1: Evening or weekend classes; 2: Non-English instruction; 3: Cultural competence instruction

## Occupational Therapist

| Programs* | Class Capacity | Begins | Length (months) | Award | Res Tuition | Non-res Tuition | Stipend | Offers:‡ 1 | 2 | 3 |
|---|---|---|---|---|---|---|---|---|---|---|
| **Georgia** | | | | | | | | | | |
| Brenau University (Gainesville) | 32 | Aug | 33 | BS, MS | $24,990 | $24,990 | | • | | • |
| Medical College of Georgia (Augusta) | 40 | Aug | 34, 36 | MHS | $9,990 | $16,122 | | | • | • |
| **Idaho** | | | | | | | | | | |
| Idaho State University (Pocatello) | 14 | Aug | 36 | MOT | $4,318 | $10,558 | | | | • |
| **Illinois** | | | | | | | | | | |
| Chicago State University | 25 | Aug | 30 | BS/MOT, MOT | $6,508 | $11,488 | | • | | • |
| Governors State University (University Park) | 24 | Sep Jan May | 27 | MOT | $256 | $768 | | | | • |
| Midwestern University (Downers Grove) | 35 | Sep | 27 | MOT | $23,000 | $23,000 | | | | |
| University of Illinois at Chicago | 30 | Aug | 24, 48 | Dipl, MS | $18 | $36,213 | | | | |
| **Indiana** | | | | | | | | | | |
| University of Indianapolis | 51 | Aug | 30 | MOT | $21,785 | $21,785 | | | | • |
| University of Southern Indiana (Evansville) | 30 | Jul | 22, 34 | BS, MS | $5,446 | $13,342 | | | | |
| **Iowa** | | | | | | | | | | |
| St Ambrose University (Davenport) | 30 | Sep | 29 | MOT | $19,950 | $0 | | | | |
| **Kansas** | | | | | | | | | | |
| University of Kansas Medical Center (Kansas City) | 32 | Jun | 36, 45 | MOT | $6,256 | $15,183 | | | | |
| **Kentucky** | | | | | | | | | | |
| Eastern Kentucky University (Richmond) | 45 | May Sep | 25, 20 | BS, MS | $2,190 | $6,030 | | • | | • |
| **Louisiana** | | | | | | | | | | |
| Louisiana State Univ Hlth Sci Ctr-Shreveport | 20 | Summer | 27 | MOT | $5,121 | $8,246 | | | | • |
| **Maine** | | | | | | | | | | |
| Husson College (Bangor) | 24 | Sep Jan | 46 | MSOT | $12,288 | $0 | | • | | • |
| University of New England (Biddeford) | 42 | Sep | 60 | MSOT, MSPPOT | $22,940 | $22,940 | | | | • |
| University of Southern Maine (Lewiston) | 24 | Sep | 27 | MOT | $200 | $560 | | | | • |
| **Maryland** | | | | | | | | | | |
| Towson University | 36 | Sep | 37, 30 | BS/MS, MS | $7,808 | $17,326 | | | | • |
| **Massachusetts** | | | | | | | | | | |
| Boston University/Sargent College | 50 | Sep | 66, 30 | BSMSOT, MSOT | $33,330 | $33,330 | | | | • |
| Salem State College | 20 | Jun | 48, 72 | BS, BS/MS | | | | • | | • |
| Springfield College | 42 | Sep | 22, 28 | MEd, MS | $26,142 | $26,142 | | | | • |
| Tufts University (Medford) | 45 | Sep Jan | 21 | MS MA | $25,254 | $25,254 | | | | • |
| **Michigan** | | | | | | | | | | |
| Baker College Center of Graduate Studies (Flint) | 50 | Apr | 46 | MOT | $8,840 | $8,840 | | | | • |
| Eastern Michigan University (Ypsilanti) | 45 | Sep | 30 | MOT, MS | $4,239 | $7,452 | | • | | • |
| Saginaw Valley State University (University Center) | 45 | May | 26 | MS | $5,100 | $9,300 | | | | • |
| Wayne State University (Detroit) | 28 | May | 24, 16 | Dipl, MOT, MS | $3,114 | $6,810 | | • | | • |
| Western Michigan University (Kalamazoo) | 64 | Sep Jan | 28 | BS/MS | $4,657 | $10,109 | | • | | • |
| **Minnesota** | | | | | | | | | | |
| College of St Catherine - St Paul | 54 | Sep | 30, 55 | MA | $19,500 | $0 | | • | | • |
| College of St Scholastica (Duluth) | 32 | Sep | 28 | MA, MS | $19,192 | $19,192 | | | | |
| University of Minnesota - Minneapolis | 25 | Sep | 30 | MS | $13,177 | $23,233 | | | | • |
| **Mississippi** | | | | | | | | | | |
| University of Mississippi Medical Center (Jackson) | 38 | May | 36 | MOT, BS | $5,091 | $10,605 | | | | • |
| **Missouri** | | | | | | | | | | |
| Maryville University (St Louis) | 35 | Aug | 48 | MOT | $17,800 | $17,800 | | | | • |
| Rockhurst University (Kansas City) | 36 | Jul | 24 | Dipl, MOT | $17,500 | $17,500 | | | | • |
| Saint Louis University (St Louis) | 30 | Aug | 48, 24 | Dipl, BS, MS | $26,250 | $26,250 | | | | • |
| **Nebraska** | | | | | | | | | | |
| College of Saint Mary (Omaha) | 25 | Aug | 60 | BS/MOT | $16,770 | $16,770 | | | | • |
| **Nevada** | | | | | | | | | | |
| Touro University - Nevada (Henderson) | | | 24 | MSOT | | | | | | |
| **New Hampshire** | | | | | | | | | | |
| University of New Hampshire (Durham) | 60 | Sep | 58, 21 | Dipl, BS/MS, MS | $9,778 | $21,498 | | | | |
| **New Jersey** | | | | | | | | | | |
| Richard Stockton College of New Jersey (Pomona) | 20 | Sep | 30 | MSOT | $11,280 | $15,720 | | | | |
| Seton Hall University (South Orange) | 30 | Sep | 36 | MSOT | $23,776 | $23,776 | | | | • |
| **New Mexico** | | | | | | | | | | |
| University of New Mexico (Albuquerque) | 24 | May | 27 | MOT | $5,600 | $15,000 | | | | • |
| **New York** | | | | | | | | | | |
| Columbia University (New York) | 45 | Sep | 24, 36 | MS | $30,112 | $0 | | | | • |
| CUNY York College (Jamaica) | 40 | Fall | 72 | BS/MS | $5,200 | $6,800 | | | | • |
| D'Youville College (Buffalo) | 215 | May Aug | 46, 27 | BS/MS, MS | $16,600 | $16,600 | | | | • |
| Dominican College (Orangeburg) | 32 | Sep | 24, 36 | BS/MS | $8,775 | $0 | | • | | |
| Long Island University - Brooklyn Campus | 30 | Sep | 36 | BS/MS | | | | | | • |

*Data are shown only for programs that completed the 2006 AMA Survey of Health Professions Education Programs
‡Key to Offers: 1: Evening or weekend classes; 2: Non-English instruction; 3: Cultural competence instruction

## Occupational Therapist

| Programs* | Class Capacity | Begins | Length (months) | Award | Res Tuition | Non-res Tuition | Stipend | Offers:‡ 1 | 2 | 3 |
|---|---|---|---|---|---|---|---|---|---|---|
| Mercy College (Dobbs Ferry) | 30 | Sep | 28 | MS | $18,900 | $0 | | • | | • |
| New York Institute of Technology (Old Westbury) | 40 | May | 33, 66 | BS/MS, MS | $20,000 | $20,000 | | | | • |
| SUNY Downstate Medical Center (Brooklyn) | 30 | Jun 2 | 30 | MS | $10,350 | $16,380 | | | | • |
| Touro College - Bay Shore | 30 | Sep | 34 | BS HS, MSOT | $18,700 | $18,700 | | | | • |
| Touro College - Manhattan (New York) | 35 | Jan | 34 | BS HS, MSOT | $18,700 | $18,700 | | | | • |
| University at Buffalo - SUNY | 40 | May | 60, 24 | BS/MS, MS | $5,861 | $11,811 | | | | • |
| **North Carolina** | | | | | | | | | | |
| East Carolina University (Greenville) | 20 | Aug | 28 | MS | $1,000 | $4,453 | | | | • |
| Lenoir - Rhyne College (Hickory) | 30 | Aug | 18 | BS HOS, MS OT | $14,200 | $14,200 | | | | • |
| Univ of North Carolina at Chapel Hill | 24 | Aug | 24 | MS | $5,674 | $19,672 | | | | |
| Winston-Salem State University | 30 | Aug | 24 | BSOT, MSOT | $2,334 | $8,814 | | • | | • |
| **North Dakota** | | | | | | | | | | |
| University of Mary (Bismarck) | 30 | Sep | 40 | MS | $8,500 | $8,500 | | | | • |
| University of North Dakota (Grand Forks) | 36 | May | 32 | MOT | $5,672 | $7,942 | | | | • |
| **Ohio** | | | | | | | | | | |
| Cleveland State University | 30 | Aug | 28 | MOT | $16,008 | $30,216 | | | | • |
| Ohio State University (Columbus) | 40 | Jun | 27 | MOT | $8,250 | $20,133 | | | | |
| University of Findlay College of Health Professions | 32 | Sep | 36 | BS/MOT | $19,000 | $0 | | • | | • |
| University of Toledo | 20 | Aug | 32 | OTD | $10,215 | $23,255 | | • | | • |
| Xavier University (Cincinnati) | 22 | Aug | 48 | MOT | $21,850 | $21,850 | | | | • |
| **Oregon** | | | | | | | | | | |
| Pacific University (Hillsboro) | 26 | Aug | 31 | Dipl, MOT | $19,698 | $21,108 | | | | • |
| **Pennsylvania** | | | | | | | | | | |
| Chatham College (Pittsburgh) | 44 | Aug | 22, 16 | MOT, OTD | $21,500 | $21,500 | | | | |
| Duquesne University (Pittsburgh) | 30 | Jan | 25, 53 | BS, MS | $23,132 | $0 | | | | • |
| Elizabethtown College | 40 | Aug | 60 | MS | $34,000 | $0 | | | | • |
| Gannon University (Erie) | 45 | Aug | 60, 36 | MS | $15,000 | $15,000 | | | | |
| Philadelphia University | 30 | Sep | 28 | Dipl, MS | $23,000 | $23,000 | | • | | • |
| St Francis University (Loretto) | 40 | Jun | 60 | MOT | | | | | | |
| Temple University (Philadelphia) | 35 | Jul | 24, 36 | MOT | $17,848 | $24,742 | | | | • |
| University of Pittsburgh | 40 | Jun | 22 | MOT | $18,200 | $25,800 | | | | • |
| **Puerto Rico** | | | | | | | | | | |
| University of Puerto Rico (San Juan) | 16 | Aug | 27 | MS | $4,010 | $4,810 | | | • | • |
| **South Carolina** | | | | | | | | | | |
| Medical University of South Carolina (Charleston) | 30 | May | 27 | MS | $17,277 | $33,423 | | | | • |
| **Tennessee** | | | | | | | | | | |
| Belmont University (Nashville) | 32 | Aug | 33, 23 | OTD, MSOT | $21,150 | $21,150 | | • | | • |
| Milligan College | 30 | Aug | 30 | MSOT | $16,000 | $16,000 | | | | |
| University of Tennessee Health Science Ctr (Memphis) | 30 | Jan | 27 | MOT | $5,169 | $17,106 | | | | |
| **Texas** | | | | | | | | | | |
| Texas Tech Univ Health Sciences Center (Lubbock) | 35 | May | 30 | MOT | $8,500 | $20,000 | | | | • |
| Univ of Texas Hlth Sci Ctr at San Antonio | 35 | Jun | 36 | BHCS, MOT | $1,428 | $10,416 | | | | • |
| University of Texas Health Science Center (Laredo) | 35 | Summer | 33 | BS, MOT | | | | | | • |
| University of Texas Medical Branch/SAHS (Galveston) | 40 | Aug | 30 | MOT | $6,702 | $14,209 | | | • | • |
| **Utah** | | | | | | | | | | |
| University of Utah (Salt Lake City) | 24 | Aug | 33 | MOT | $13,000 | $25,700 | | | | • |
| **Virginia** | | | | | | | | | | |
| James Madison University (Harrisonburg) | 18 | Jun | 30 | Dipl, MOT | $9,504 | $26,748 | | | | • |
| Shenandoah University (Winchester) | 26 | Aug | 30 | MS | $610 | $610 | | | | • |
| Virginia Commonwealth Univ/Med Coll of VA (Richmond) | 40 | Jun | 30, 24 | MSOT, MS | $12,481 | $26,343 | | | | |
| **Washington** | | | | | | | | | | |
| University of Puget Sound (Tacoma) | 40 | Sep | 28 | MSOT, MOT | $28,275 | $28,275 | | | | • |
| University of Washington (Seattle) | 25 | Sep | 27 | Dipl, MOT, MS | $11,344 | $26,076 | | | | • |
| **West Virginia** | | | | | | | | | | |
| West Virginia University (Morgantown) | 30 | Jul | 30 | MOT | $4,712 | $14,640 | | | | • |
| **Wisconsin** | | | | | | | | | | |
| Concordia University Wisconsin (Mequon) | 24 | Aug | 36 | Dipl, MOT | $19,990 | $19,990 | | | | • |
| University of Wisconsin - Madison | 25 | Jun | 30, 48 | MS OT, MS TS | $8,320 | $23,590 | | | | |
| University of Wisconsin - Milwaukee | 40 | Sep | 36 | MS | $5,835 | $18,587 | | • | | • |
| **Wyoming** | | | | | | | | | | |
| University of North Dakota at Casper College | 14 | May | 32 | Masters | $10,680 | $10,680 | | | | • |

*Data are shown only for programs that completed the 2006 AMA Survey of Health Professions Education Programs
‡Key to Offers: 1: Evening or weekend classes; 2: Non-English instruction; 3: Cultural competence instruction

## Occupational Therapy Assistant*

## Alabama

**Wallace State Community College**
Occupational Therapy Asst Prgm
PO Box 2000
Hanceville, AL 35077-2000
*Prgm Dir:* Tammy R Gipson, OTR/L
*Tel:* 256 352-8333   *Fax:* 256 352-8320
*E-mail:* tammy.gipson@wallacestate.edu

## Arkansas

**South Arkansas Community College**
Occupational Therapy Asst Prgm
PO Box 7010
300 S West Ave
El Dorado, AR 71731-7010
*Prgm Dir:* Sandra Pugh, OTD OTR/L
*Tel:* 870 862-8131, Ext 227   *Fax:* 870 864-7104
*E-mail:* spugh@southark.edu

## California

**Grossmont College**
Occupational Therapy Asst Prgm
8800 Grossmont College Dr
El Cajon, CA 92020-1799
*Prgm Dir:* Christine Vicino, BA COTA
*Tel:* 619 644-7305   *Fax:* 619 644-7961
*E-mail:* Christi.Vicino@gcccd.edu

**Loma Linda University**
Occupational Therapy Asst Prgm
Sch of Allied Hlth Professions
SAHP-Nichol Hall, Rm A912
Loma Linda, CA 92350-0001
www.llu.edu
*Prgm Dir:* Liane Hewitt, MPH OTR/L
*Tel:* 909 558-4628, Ext 47327   *Fax:* 909 558-0239
*E-mail:* lhewitt@llu.edu

**Sacramento City College**
Occupational Therapy Asst Prgm
Allied Health Dept
3835 Freeport Blvd
Sacramento, CA 95822-1386
*Prgm Dir:* Lynette Beadles, OTR/L
*Tel:* 916 558-2297   *Fax:* 916 558-2392
*E-mail:* beadlel@scc.losrios.edu

**Santa Ana College**
Occupational Therapy Asst Prgm
1530 W 17th St
Santa Ana, CA 92706-3398
*Prgm Dir:* Michelle Parolise, MBA OTR/L
*Tel:* 714 564-6833   *Fax:* 714 564-6158
*E-mail:* parolise_michelle@sac.edu

## Colorado

**Pueblo Community College**
Occupational Therapy Asst Prgm
900 W Orman Ave
Pueblo, CO 81004-1499
www.pueblocc.edu/Academics/AreasStudy/
   HealthProfessions/OccupationalTherapyAssist
*Prgm Dir:* Becky Robler, MEd OTR
*Tel:* 719 549-3268   *Fax:* 719 549-3381
*E-mail:* becky.robler@pueblocc.edu

## Connecticut

**Housatonic Community College**
Occupational Therapy Asst Prgm
900 Lafayette Blvd
Bridgeport, CT 06604-4704
*Prgm Dir:* Michele M Reed, MS MS Ed OTR/L
*Tel:* 203 332-5214   *Fax:* 203 332-5123
*E-mail:* mreed@hcc.commnet.edu

**Manchester Community College**
Occupational Therapy Asst Prgm
PO Box 1046
Great Path MS 17
Manchester, CT 06045-1046
*Prgm Dir:* Martha Nieman, MS OTR/L
*Tel:* 860 512-2717   *Fax:* 860 512-2621
*E-mail:* MNieman@mcc.commnet.edu

**Briarwood College**
Occupational Therapy Asst Prgm
2279 Mt Vernon Rd
Southington, CT 06489
*Prgm Dir:* Deanne S Anderson, MS OTR/L
*Tel:* 860 628-4751, Ext 175   *Fax:* 860 628-6444
*E-mail:* andersond@briarwood.edu

## Delaware

**Delaware Tech & Comm Coll - Owens Campus**
Occupational Therapy Asst Prgm
Owens Campus, PO Box 610
Georgetown, DE 19947-0610
www.dtcc.edu
*Prgm Dir:* Nancy Broadhurst, BS COTA/L
*Tel:* 302 856-5400, Ext 5533   *Fax:* 302 858-5460
*E-mail:* nbroadhu@dtcc.edu

**Delaware Tech & Comm Coll - Wilmington**
Occupational Therapy Asst Prgm
333 Shipley St
Wilmington, DE 19801-2499
*Prgm Dir:* Jan Gorecki, MS OTR/L
*Tel:* 302 888-5298   *Fax:* 302 577-6431
*E-mail:* jgorecki@college.dtcc.edu

## Florida

**Manatee Community College**
Occupational Therapy Asst Prgm
5840 26th St W
Bldg 28
Bradenton, FL 34207-7046
www.mcc.fl.edu
*Prgm Dir:* Debra Chasanoff, MEd OTR/L
*Tel:* 941 752-5346   *Fax:* 941 727-8304
*E-mail:* chasand@mccfl.edu
*Med Dir:* Bonnie Hesselberg, EdD ARNP

**Daytona Beach Community College**
Occupational Therapy Asst Prgm
1200 International Speedway Blvd
Daytona Beach, FL 32120-2811
*Prgm Dir:* Mary Beth Craig-Oatley, EdM OT/L
*Tel:* 386 506-3624   *Fax:* 286 506-3300
*E-mail:* craigom@dbcc.edu

**Keiser University**
Occupational Therapy Asst Prgm
1500 NW 49th St
Ft Lauderdale, FL 33309
www.keisercollege.edu
*Prgm Dir:* Arlene Kinney, OTR/L
*Tel:* 954 776-4456, Ext 332   *Fax:* 954 776-5157
*E-mail:* arlenek@keisercollege.edu

**Keiser College - Melbourne Campus**
Occupational Therapy Asst Prgm
900 S Babcock St
Melbourne, FL 32901
www.keisercollege.edu
*Prgm Dir:* Jill Cline, OT/L
*Tel:* 321 409-4800, Ext 117   *Fax:* 321 725-3766
*E-mail:* jillc@keisercollege.edu

**Florida Hospital College of Health Sciences**
Occupational Therapy Asst Prgm
711 Lake Estelle Dr
Orlando, FL 32803-1235
www.fhchs.edu
*Prgm Dir:* Tia Hughes, MBA OTR/L
*Tel:* 407 303-7747, Ext 9855   *Fax:* 407 303-5572
*E-mail:* tia_hughes@fhchs.edu

**Polk Community College**
Occupational Therapy Asst Prgm
999 Ave H NE
Winter Haven, FL 33881-4299
www.polk.edu
*Prgm Dir:* Saritza Guzman-Sardina, MEd OTR/L
*Tel:* 863 297-1010, Ext 5753   *Fax:* 863 297-1047
*E-mail:* sguzman-sardina@polk.edu

## Georgia

**Darton College**
Occupational Therapy Asst Prgm
2400 Gillionville Rd
Albany, GA 31707-3098
www.darton.edu
*Prgm Dir:* Stacey R Cowart, MS OTR/L
*Tel:* 229 317-6908   *Fax:* 229 317-6682
*E-mail:* stacey.cowart@darton.edu

**Augusta Technical College**
Occupational Therapy Asst Prgm
3200 Augusta Tech Dr
1400 Building
Augusta, GA 30906-3399
www.augusta.tec.ga.us
*Prgm Dir:* Cindy L Loar, MHS OTR/L PTA
*Tel:* 706 771-4188   *Fax:* 706 771-4181
*E-mail:* cloar@augustatech.edu

**Middle Georgia College**
Occupational Therapy Asst Prgm
1100 Second St SE
Cochran, GA 31014-1599
*Prgm Dir:* Heather Copan, MHE OTR/L
*Tel:* 478 934-3402   *Fax:* 478 934-3461
*E-mail:* hcopan@mgc.edu

**Northwestern Technical College**
Occupational Therapy Asst Prgm
PO Box 569
265 Bicentennial Trail
Rock Spring, GA 30739
www.northwesterntech.edu
*Prgm Dir:* Lisa Carruth, OTR/L
*Tel:* 706 764-3846   *Fax:* 706 764-3718
*E-mail:* lcarruth@northwesterntech.edu

## Hawaii

**Kapi'olani Community College**
Occupational Therapy Asst Prgm
Health Sciences Dept
4303 Diamond Head Rd, Kauila 210C
Honolulu, HI 96816-4421
www.kcc.hawaii.edu
*Prgm Dir:* Shelley Boling, OTR
*Tel:* 808 734-9229   *Fax:* 808 734-9126
*E-mail:* boling@hawaii.edu

*Additional information about programs that
returned the AMA's survey is available in a table
beginning on page 288.

# Illinois

## Parkland College
Occupational Therapy Asst Prgm
2400 W Bradley Ave
Champaign, IL 61821-1899
www.parkland.edu
*Prgm Dir:* Rebecca R Bahnke, MHS OTR/L
*Tel:* 217 351-2394   *Fax:* 217 373-3830
*E-mail:* rbahnke@parkland.edu

## Wright College
Occupational Therapy Asst Prgm
4300 N Narragansett Ave
Chicago, IL 60634-1591
www.ccc.edu/wright
*Prgm Dir:* Joyce Wandel, MS OTR/L
*Tel:* 773 481-8875   *Fax:* 773 481-8892
*E-mail:* jwandel@ccc.edu

## Lewis & Clark Community College
Occupational Therapy Asst Prgm
5800 Godfrey Rd
Godfrey, IL 62035-2466
www.lc.edu
*Prgm Dir:* Linda L Orr, MPA OTR/L
*Tel:* 618 468-4416   *Fax:* 618 468-7811
*E-mail:* lorr@lc.edu

## Southern Illinois Collegiate Common Market
Occupational Therapy Asst Prgm
3213 S Park Ave
Herrin, IL 62948
www.siccm.com
*Prgm Dir:* Verlinda Henshaw, OTD OTR/L
*Tel:* 618 942-6902   *Fax:* 618 942-6658
*E-mail:* teachota@siccm.com

## Illinois Central College
Occupational Therapy Asst Prgm
201 SW Adams St
Peoria, IL 61635-0001
*Prgm Dir:* Donna Augustyn-Sloan, MS OTR/L
*Tel:* 309 999-4674   *Fax:* 309 673-9626
*E-mail:* das@icc.edu

## South Suburban College
Occupational Therapy Asst Prgm
15800 S State St
South Holland, IL 60473-1262
*Prgm Dir:* Cathy M Mistovich, MS OTR/L
*Tel:* 708 596-2000, Ext 2473   *Fax:* 708 210-5792
*E-mail:* cmistovich@southsuburbancollege.edu

## Lincoln Land Community College
Occupational Therapy Asst Prgm
5250 Shepherd Rd
PO Box 19256
Springfield, IL 62794-9256
www.llcc.edu
*Prgm Dir:* Ruth J Bixby, OTR/L
*Tel:* 217 786-2872   *Fax:* 217 786-2824
*E-mail:* ruth.bixby@llcc.edu

# Indiana

## University of Southern Indiana
Occupational Therapy Asst Prgm
College of Nursing & Health Professions
8600 University Blvd, HP 2068
Evansville, IN 47712-3534
*Prgm Dir:* Susan G Ahmad, MS OTR/L
*Tel:* 812 465-1178   *Fax:* 812 465-7092
*E-mail:* sahmad@usi.edu

## Brown Mackie College
Occupational Therapy Asst Prgm
Fort Wayne Campus
4422 E State Blvd
Ft Wayne, IN 46802
*Prgm Dir:* LaNiece Clark, COTA
*Tel:* 260 484-4400   *Fax:* 260 484-2678
*E-mail:* laclark@brownmackie.edu

## Brown Mackie College
Occupational Therapy Asst Prgm
1030 E Jefferson Blvd
South Bend, IN 46617-3123
*Prgm Dir:* Michelle Sheperd, OTR
*Tel:* 574 237-0774   *Fax:* 574 237-3585
*E-mail:* msheperd@brownmackie.edu

# Iowa

## Kirkwood Community College
Occupational Therapy Asst Prgm
6301 Kirkwood Blvd SW, PO Box 2068
Cedar Rapids, IA 52406-9973
*Prgm Dir:* Nichelle Miedema, MPA OTR/L
*Tel:* 319 398-5566, Ext 4941   *Fax:* 319 398-1293
*E-mail:* nicky.miedema@kirkwood.edu

# Kansas

## Newman University
Occupational Therapy Asst Prgm
3100 McCormick Ave
Wichita, KS 67213-2097
*Prgm Dir:* Clint Stucky, MS OTR
*Tel:* 316 942-4291, Ext 2349   *Fax:* 316 942-4483
*E-mail:* stuckyc@newmanu.edu

# Kentucky

## Jefferson Community and Technical College
Occupational Therapy Asst Prgm
109 E Broadway St
Louisville, KY 40202-2005
www.jefferson.kctcs.edu
*Prgm Dir:* Carolyn Thornsberry, OTR/L CLT
*Tel:* 502 213-2342   *Fax:* 502 213-2491
*E-mail:* cthornsberry0004@kctcs.edu
*Med Dir:* Eva Oltman

## Madisonville Community College
Occupational Therapy Asst Prgm
750 N Laffoon St
Madisonville, KY 42431-1636
www.madcc.kctcs.edu
*Prgm Dir:* Helen M Grothem, MS OTR/L
*Tel:* 270 824-1742   *Fax:* 270 824-1879
*E-mail:* helen.grothem@kctcs.edu

# Louisiana

## University of Louisiana at Monroe
Occupational Therapy Asst Prgm
College of Health Sciences
700 University Ave Rm 111 Caldwell Hall
Monroe, LA 71209-0430
www.ulm.edu/healthsciences
*Prgm Dir:* Margaret M Meredith, MA LOTR
*Tel:* 318 342-1617   *Fax:* 318 342-5584
*E-mail:* pmeredith@ulm.edu

## Delgado Community College
Occupational Therapy Asst Prgm
City Park Campus
615 City Park Ave
New Orleans, LA 70119-4399
*Prgm Dir:* Linda Kelly, MA LOTR
*Tel:* 504 485-2372   *Fax:* 504 483-4609
*E-mail:* lkelly@dcc.edu

# Maine

## Kennebec Valley Community College
Occupational Therapy Asst Prgm
92 Western Ave
Fairfield, ME 04937-1367
*Prgm Dir:* Diane Sauter-Davis, MA OTR/L
*Tel:* 207 453-5172   *Fax:* 207 453-5194
*E-mail:* dsauter@kvcc.me.edu

# Maryland

## Comm Coll of Baltimore County - Catonsville
Occupational Therapy Asst Prgm
800 S Rolling Rd
Catonsville, MD 21228-9987
www.ccbcmd.edu
*Prgm Dir:* Judith Blum, MS OTR/L
*Tel:* 410 455-4482   *Fax:* 410 719-6501
*E-mail:* jblum@ccbcmd.edu

## Allegany College of Maryland
Occupational Therapy Asst Prgm
12401 Willowbrook Rd SE
Cumberland, MD 21502-2596
www.allegany.edu
*Prgm Dir:* Rae Ann Smith, MEd OTR/L
*Tel:* 301 784-5536   *Fax:* 301 784-5015
*E-mail:* rasmith@allegany.edu

# Massachusetts

## North Shore Community College
Occupational Therapy Asst Prgm
One Ferncroft Rd, PO Box 3340
Danvers, MA 01923-0840
www.northshore.edu
*Prgm Dir:* Maureen S Nardella, MS OTR/L
*Tel:* 978 762-4176   *Fax:* 978 762-4022
*E-mail:* mnardell@northshore.edu

## Bristol Community College
Occupational Therapy Asst Prgm
777 Elsbree St
C-109
Fall River, MA 02720
www.bristol.mass.edu
*Prgm Dir:* Johanna Duponte, MS OTR/L
*Tel:* 508 678-2811, Ext 2325   *Fax:* 508 730-3281
*E-mail:* jduponte@bristol.mass.edu

## Springfield Technical Community College
Occupational Therapy Asst Prgm
One Armory Sq, Ste 1
PO Box 9000
Springfield, MA 01102-9000
www.stcc.edu
*Prgm Dir:* Marianne Joyce, MA OTR/L
*Tel:* 413 755-4881   *Fax:* 413 755-4764
*E-mail:* mjoyce@stcc.edu
*Med Dir:* Michael Foss, MEd RDMS RVT

## Quinsigamond Community College
Occupational Therapy Asst Prgm
670 W Boylston St
Worcester, MA 01606-2092
www.qcc.mass.edu
*Prgm Dir:* Brenda Marshall, MA OTR/L
*Tel:* 508 854-4546   *Fax:* 508 852-6943
*E-mail:* brendam@qcc.mass.edu

# Michigan

**Macomb Community College**
Occupational Therapy Asst Prgm
44575 Garfield Rd
Bldg E 220-8
Clinton Township, MI 48038-1139
*Prgm Dir:* Phyllis Clements, MA OTR
*Tel:* 586 286-2076   *Fax:* 586 286-2098
*E-mail:* clementsp@macomb.edu

**Wayne County Community College District**
Occupational Therapy Asst Prgm
1001 W Fort St
Detroit, MI 48226-9975
*Prgm Dir:* Lettie Redley, MS OTR CHSA
*Tel:* 313 496-2692   *Fax:* 313 961-9648
*E-mail:* lredley1@wcccd.edu

**Mott Community College**
Occupational Therapy Asst Prgm
Southern Lakes Campus
2100 W Thompson Rd
Fenton, MI 48430-9798
*Prgm Dir:* Wendy Blair Early, MS OTR
*Tel:* 810 762-5018   *Fax:* 810 750-8588
*E-mail:* wearly@mcc.edu

**Grand Rapids Community College**
Occupational Therapy Asst Prgm
143 Bostwick St NE
Grand Rapids, MI 49503-3295
*Prgm Dir:* Karen Walker, MA OTR
*Tel:* 616 234-4236   *Fax:* 616 234-4317
*E-mail:* kwalker@grcc.edu

**Baker College of Muskegon**
Occupational Therapy Asst Prgm
1903 Marquette Ave
Muskegon, MI 49442-1490
*Prgm Dir:* CJ Brocker, BBL COTA
*Tel:* 231 777-5289   *Fax:* 231 777-5291
*E-mail:* cj.brocker@baker.edu

# Minnesota

**Anoka Technical College**
Occupational Therapy Asst Prgm
1355 W Hwy 10
Anoka, MN 55303-1590
*Prgm Dir:* Marietta Cosky Saxon, MEd OTR/L
*Tel:* 763 576-4935   *Fax:* 763 576-4715
*E-mail:* msaxon@anokatech.edu

**Northland Community & Technical College**
Occupational Therapy Asst Prgm
2022 Central Ave NE
East Grand Forks, MN 56721-2702
www.northlandcollege.edu
*Prgm Dir:* Cassie Hilts, OTR/L
*Tel:* 218 773-3441   *Fax:* 218 773-4502
*E-mail:* cassie.hilts@northlandcollege.edu

**College of St Catherine - Minneapolis**
Occupational Therapy Asst Prgm
601 25th Ave S
Minneapolis, MN 55454-1494
www.stkate.edu/ota
*Prgm Dir:* Marianne F Christiansen, MA OTR/L FAOTA
*Tel:* 651 690-7772   *Fax:* 651 690-7849
*E-mail:* mfchristiansen@stkate.edu

# Mississippi

**Itawamba Community College**
Occupational Therapy Asst Prgm
602 West Hill St
Fulton, MS 38843
*Prgm Dir:* Suzanne Chittom, MS OTR/L
*Tel:* 662 862-8344   *Fax:* 662 862-8350
*E-mail:* jschittom@iccms.edu

**Pearl River Community College**
Occupational Therapy Asst Prgm
5448 US Hwy 49 S
Hattiesburg, MS 39401-7806
*Prgm Dir:* Tim Pulver, OTR/L
*Tel:* 601 554-5541   *Fax:* 601 554-5511
*E-mail:* tpulver@prcc.edu

**Holmes Community College**
Occupational Therapy Asst Prgm
Ridgeland Campus
412 W Ridgeland Ave
Ridgeland, MS 39157
www.holmescc.edu
*Prgm Dir:* Sherry Hager, MOTR/L
*Tel:* 601 605-3337   *Fax:* 601 605-3410
*E-mail:* shager@holmescc.edu

# Missouri

**Sanford-Brown College - Hazelwood Campus**
Occupational Therapy Asst Prgm
75 Village Square
Hazelwood, MO 63042
*Prgm Dir:* William C Gielow, Jr, MS OTR/L
*Tel:* 314 687-2991, Ext 2991   *Fax:* 314 731-0550
*E-mail:* wgielow@sbc-Hazelwood.com

**Metropolitan Community College - Penn Valley**
Occupational Therapy Asst Prgm
3201 SW Trafficway
Kansas City, MO 64111-2764
*Prgm Dir:* Sherry Carter, MS OTR/L
*Tel:* 816 759-4462   *Fax:* 816 759-4553
*E-mail:* sherry.carter@mcckc.edu

**Ozarks Technical Community College**
Occupational Therapy Asst Prgm
1001 E Chestnut Expressway
Springfield, MO 65802-3625
www.otc.edu
*Prgm Dir:* Becky Jenkins, OTR/L
*Tel:* 417 447-8855   *Fax:* 417 447-8806
*E-mail:* jenkinsr@otc.edu

**St Louis Community College - Meramec**
Occupational Therapy Asst Prgm
11333 Big Bend Blvd
St Louis, MO 63122-5799
www.stlcc.edu
*Prgm Dir:* Nancy M Klein, MS OTR/L
*Tel:* 314 984-7364   *Fax:* 314 984-7250
*E-mail:* nklein@stlcc.edu

**St Charles Community College**
Occupational Therapy Asst Prgm
4601 Mid Rivers Mall Dr
PO Box 76975
St Peters, MO 63376-0975
www.stchas.edu
*Prgm Dir:* Francesca Woods, MA OTR/L
*Tel:* 636 922-8638   *Fax:* 636 922-8478
*E-mail:* fwoods@stchas.edu

# Nevada

**Community College of Southern Nevada**
Occupational Therapy Asst Prgm
West Charleston Campus
6375 W Charleston Blvd, W1A
Las Vegas, NV 89146-1139
www.ccsn.edu
*Prgm Dir:* Christine R Privott, MA OTR/L
*Tel:* 702 651-5582   *Fax:* 702 651-5506
*E-mail:* christine_privott@ccsn.edu

# New Hampshire

**New Hampshire Comm Tech Coll - Claremont**
Occupational Therapy Asst Prgm
One College Dr
Claremont, NH 03743-9707
*Prgm Dir:* Jennifer J Saylor, MEd OTR/L
*Tel:* 603 542-7744, Ext 2525   *Fax:* 603 542-1844
*E-mail:* jsaylor@nhctc.edu

# New Mexico

**Eastern New Mexico University - Roswell**
Occupational Therapy Asst Prgm
Div of Health
52 University Blvd, PO Box 6000
Roswell, NM 88202-6000
*Prgm Dir:* Pat Herrera, OTR/L
*Tel:* 505 624-7267   *Fax:* 505 624-7257
*E-mail:* herrerpa@rm01.enmuros.cc.nm.us

**Western New Mexico University**
Occupational Therapy Asst Prgm
PO Box 680
Silver City, NM 88062-0680
*Prgm Dir:* Gwen Cassel, MOT OTR/L
*Tel:* 505 574-5171   *Fax:* 505 574-5150
*E-mail:* casselg@silver.wnmu.edu

# New York

**Maria College**
Occupational Therapy Asst Prgm
700 New Scotland Ave
Albany, NY 12208-1798
www.mariacollege.edu
*Prgm Dir:* Sandra C Jung, OTR/L
*Tel:* 518 438-3111, Ext 257   *Fax:* 518 453-1366
*E-mail:* sandyj@mariacollege.edu

**Genesee Community College**
Occupational Therapy Asst Prgm
One College Rd
Batavia, NY 14020-9704
www.genesee.edu
*Prgm Dir:* Mary K Hartman, MS OTR/L
*Tel:* 585 345-6838   *Fax:* 585 343-0433
*E-mail:* mkhartman@genesee.edu

**Suffolk County Community College**
Occupational Therapy Asst Prgm
Michael J Campus
Crooked Hill Rd - MA 308
Brentwood, NY 11717-1092
www.sunysuffolk.edu
*Prgm Dir:* Lisa E Hubbs, MS OTR/L
*Tel:* 631 851-6335   *Fax:* 631 851-6854
*E-mail:* hubbsl@sunysuffolk.edu

**SUNY at Canton**
Occupational Therapy Asst Prgm
School of Science Health and Professional Studies
34 Cornell Dr
Canton, NY 13617-1096
*Prgm Dir:* Cindy Hammecker, OTR/L
*Tel:* 315 386-3863   *Fax:* 315 386-7959
*E-mail:* hammeckc@canton.edu

**Mercy College**
Occupational Therapy Asst Prgm
555 Broadway
Dobbs Ferry, NY 10522-1134
www.mercy.edu
*Prgm Dir:* Christine Sullivan, MS OTR/L
*Tel:* 914 674-7831   *Fax:* 914 674-7840
*E-mail:* csullivan@mercy.edu

### Jamestown Community College
Occupational Therapy Asst Prgm
525 Falconer St
PO Box 20
Jamestown, NY 14701-0020
*Prgm Dir:* Heather L Panczykowski, MS OTR/L
*Tel:* 716 665-5220, Ext 2395  *Fax:* 716 661-3168
*E-mail:* HeatherPanczykowski@mail.sunyjcc.edu

### LaGuardia Community College
Occupational Therapy Asst Prgm
31-10 Thomson Ave
Long Island City, NY 11101-3083
*Prgm Dir:* Naomi S Greenberg, PhD OTR/L
*Tel:* 718 482-5777  *Fax:* 718 609-2052
*E-mail:* ngreenbe@lagcc.cuny.edu

### Orange County Community College
Occupational Therapy Asst Prgm
115 South St
Middletown, NY 10940-6404
*Prgm Dir:* Florence Hannes, MS OTR FAOTA
*Tel:* 845 341-4323  *Fax:* 845 341-4799
*E-mail:* fhannes@sunyorange.edu

### Touro College
Occupational Therapy Asst Prgm
Main Campus
27-33 W 23rd St
New York, NY 10010-4202
www.touro.edu/shs
*Prgm Dir:* Rivka Molinsky, MA OTR/L
*Tel:* 212 463-0400, Ext 518  *Fax:* 212 989-2054
*E-mail:* rmolinsky@touro.edu

### Rockland Community College
Occupational Therapy Asst Prgm
145 College Rd
Suffern, NY 10901-3699
www.sunyrockland.edu
*Prgm Dir:* Ellen Spergel, MS OTR
*Tel:* 845 574-4312  *Fax:* 845 574-4594
*E-mail:* espergel@sunyrockland.edu

### Erie Community College
Occupational Therapy Asst Prgm
6205 Main St
Williamsville, NY 14221-7095
www.ecc.edu/studentlife/
    student_acad_occuptherap.php3
*Prgm Dir:* Betsy Jones, OTR/L MS Ed
*Tel:* 716 851-1320  *Fax:* 716 851-1267
*E-mail:* jones@ecc.edu

## North Carolina

### Cabarrus College of Health Sciences
Occupational Therapy Asst Prgm
401 Medical Park Dr
Concord, NC 28025-2405
*Prgm Dir:* Nancy S Green, MHA OTR/L
*Tel:* 704 783-3599, Ext 3599  *Fax:* 704 783-2077
*E-mail:* ngreen@cabarruscollege.edu

### Durham Technical Community College
Occupational Therapy Asst Prgm
1637 Lawson St
Durham, NC 27703-5023
www.durhamtech.edu
*Prgm Dir:* Sue Cheng, MS OTR/L
*Tel:* 919 686-3717  *Fax:* 919 686-3705
*E-mail:* chengs@durhamtech.edu

### Pitt Community College
Occupational Therapy Asst Prgm
PO Drawer 7007
Greenville, NC 27835-7007
*Prgm Dir:* Tommianne L Haithcock, COTA/L
*Tel:* 252 493-7458  *Fax:* 252 321-4451
*E-mail:* thaithco@email.pittcc.edu

### Cape Fear Community College
Occupational Therapy Asst Prgm
411 N Front St
Wilmington, NC 28401-3993
http://cfcc.edu
*Prgm Dir:* Deborah A Amini, MEd OTR/L CHT
*Tel:* 910 362-7096  *Fax:* 910 362-7087
*E-mail:* damini@cfcc.edu

## North Dakota

### North Dakota State College of Science
Occupational Therapy Asst Prgm
Mayme Green Allied Health Facility
800 6th St N
Wahpeton, ND 58076-0002
*Prgm Dir:* Sr Carolita Mauer, MA OTR/L
*Tel:* 701 671-2982, Ext 2982  *Fax:* 701 671-2570
*E-mail:* carolita.mauer@ndscs.nodak.edu

## Ohio

### Stark State College of Technology
Occupational Therapy Asst Prgm
6200 Frank Ave NW
Canton, OH 44720
www.starkstate.edu
*Prgm Dir:* Doris Huston, MA OTR/L
*Tel:* 330 966-5458, Ext 4200  *Fax:* 330 966-6586
*E-mail:* www.dhuston@starkstate.edu

### Cincinnati State Tech & Comm College
Occupational Therapy Asst Prgm
3520 Central Pkwy
Cincinnati, OH 45223-2690
*Prgm Dir:* Claudia J Miller, MHS OTR/L
*Tel:* 513 569-1598  *Fax:* 513 487-1598
*E-mail:* claudia.miller@cincinnatistate.edu

### Cuyahoga Community College
Occupational Therapy Asst Prgm
2900 Community College Ave
Cleveland, OH 44115-3196
www.tri-c.edu/otat
*Prgm Dir:* Hector L Merced, MS OTR/L
*Tel:* 216 987-4498  *Fax:* 216 987-4386
*E-mail:* hector.merced@tri-c.edu

### Sinclair Community College
Occupational Therapy Asst Prgm
444 W Third St
Dayton, OH 45402-1460
www.sinclair.edu
*Prgm Dir:* S Kay Ashworth, MAT OTR/L
*Tel:* 937 512-5178  *Fax:* 937 512-5170
*E-mail:* kay.ashworth@sinclair.edu

### Kent State University - East Liverpool Campus
Occupational Therapy Asst Prgm
400 E Fourth St
East Liverpool, OH 43920-3497
www.eliv.kent.edu
*Prgm Dir:* Harriet Bynum, MS OTR/L
*Tel:* 330 382-7426  *Fax:* 330 382-7564
*E-mail:* hbynum@kent.edu

### Rhodes State College
Occupational Therapy Asst Prgm
4240 Campus Dr
Lima, OH 45804-3597
*Prgm Dir:* Ann Best, OTR/L CEES
*Tel:* 419 995-8080  *Fax:* 419 995-8093
*E-mail:* Best.A@rhodesstate.edu

### Shawnee State University
Occupational Therapy Asst Prgm
940 Second St
Portsmouth, OH 45662-4303
www.shawnee.edu
*Prgm Dir:* Melinda Sissel, MOT OTR
*Tel:* 740 351-3389  *Fax:* 740 351-2354
*E-mail:* msissel@shawnee.edu

### Owens Community College
Occupational Therapy Asst Prgm
Toledo Campus Oregon Rd
PO Box 10,000
Toledo, OH 43699-1947
www.owens.edu
*Prgm Dir:* Beth Ann Hatkevich, MOT OTR/L
*Tel:* 567 661-7000, Ext 7175  *Fax:* 567 661-2634
*E-mail:* beth_hatkevich@owens.edu
*Med Dir:* Janell Lang, EdS

### Zane State College
Occupational Therapy Asst Prgm
1555 Newark Rd
Zanesville, OH 43701-2694
www.zanestate.edu
*Prgm Dir:* Mary Arnold, MA OTR/L
*Tel:* 740 588-1313  *Fax:* 740 588-1332
*E-mail:* marnold@zanestate.edu

## Oklahoma

### Caddo Kiowa Tech Ctr/SW Oklahoma St Univ
Occupational Therapy Asst Prgm
PO Box 190
Ft Cobb, OK 73038
*Prgm Dir:* Sherri Robertson, OTR/L
*Tel:* 405 643-5511, Ext 310  *Fax:* 405 643-2144
*E-mail:* srobertson@caddokiowa.com

### Oklahoma City Community College
Occupational Therapy Asst Prgm
Health Professions Division
7777 S May Ave
Oklahoma City, OK 73159-4444
*Prgm Dir:* Tom Kraft, MEd OTR/L
*Tel:* 405 682-7506  *Fax:* 405 682-7826
*E-mail:* tkraft@occc.edu

### Tulsa Community College
Occupational Therapy Asst Prgm
Metro Campus, Allied Hlth Services Div
909 S Boston Ave
Tulsa, OK 74119-2095
*Prgm Dir:* Gary Braswell, MS OTR/L
*Tel:* 918 595-7319  *Fax:* 918 535-7091
*E-mail:* gbraswel@tulsacc.edu
*Med Dir:* Jim Pickens, EdD

## Pennsylvania

### Mount Aloysius College
Occupational Therapy Asst Prgm
7373 Admiral Peary Hwy
Cresson, PA 16630-1999
www.mtaloy.edu
*Prgm Dir:* Kristina Knott, MA OTR/L
*Tel:* 814 886-6355  *Fax:* 814 886-4278
*E-mail:* kknott@mtaloy.edu

### Penn State University - DuBois
Occupational Therapy Asst Prgm
College Place
DuBois, PA 15801-3199
*Prgm Dir:* LuAnn Demi, MS OTR/L
*Tel:* 814 375-4748  *Fax:* 814 375-4784
*E-mail:* ldb4@psu.edu

### Comm Coll of Allegheny County
Occupational Therapy Asst Prgm
Boyce Campus
595 Beatty Rd
Monroeville, PA 15146-1395
*Prgm Dir:* Lillian Briola, MOT OTR/L
*Tel:* 724 325-6751  *Fax:* 724 325-6701
*E-mail:* lbriola@ccac.edu

**Penn State University - Mont Alto**
Occupational Therapy Asst Prgm
Mont Alto Campus
Campus Dr
Mont Alto, PA 17237-9703
*Prgm Dir:* Dorothy Rockwell, MS OTR/L BCP
*Tel:* 717 749-6165   *Fax:* 717 749-6166
*E-mail:* dld14@psu.edu

**ICM School of Business & Medical Careers**
Occupational Therapy Asst Prgm
10 Wood St
Pittsburgh, PA 15222-1977
*Prgm Dir:* John Sciulli
*Tel:* 412 261-2647, Ext 254   *Fax:* 412 261-6491

**Penn State University - Berks**
Occupational Therapy Asst Prgm
PO Box 7009
Tulpehoeken Rd
Reading, PA 19610-6009
*Prgm Dir:* Tamera K Humbert, DEd OTR/L
*Tel:* 610 396-6179   *Fax:* 610 396-6024
*E-mail:* tkh110@psu.edu

**Lehigh Carbon Community College**
Occupational Therapy Asst Prgm
4525 Education Park Dr
Schnecksville, PA 18078-2598
www.lccc.edu
*Prgm Dir:* Cindy Rifenburg, MS OTR/L
*Tel:* 610 799-1548   *Fax:* 610 799-1527
*E-mail:* crifenburg@lccc.edu

**Pennsylvania College of Technology**
Occupational Therapy Asst Prgm
One College Ave
Williamsport, PA 17701-5799
www.pct.edu
*Prgm Dir:* Barbara J Natell, MSEd OTR/L
*Tel:* 570 321-5549   *Fax:* 570 321-5556
*E-mail:* bnatell@pct.edu

# Puerto Rico

**University of Puerto Rico at Humacao**
Occupational Therapy Asst Prgm
CUH Postal Station 100, Carr 908
Humacao, PR 00791-4300
*Prgm Dir:* Carlos A Galiano, MC MS OTR/L
*Tel:* 787 850-9392, Ext 9392   *Fax:* 787 850-9434
*E-mail:* ce_alverio@webmail.uprh.edu

# Rhode Island

**Community College of Rhode Island**
Occupational Therapy Asst Prgm
Newport County Campus
One John H Chafee Blvd
Newport, RI 02840
www.ccri.edu/rehabhealth
*Prgm Dir:* Shawn Baxter
*Tel:* 401 851-1668   *Fax:* 401 846-9051
*E-mail:* krouillier@ccri.edu

**New England Institute of Technology**
Occupational Therapy Asst Prgm
2500 Post Rd
Warwick, RI 02886-2251
www.neit.edu
*Prgm Dir:* Nancy R Dooley, PhD OTR/L
*Tel:* 401 739-5000, Ext 3400   *Fax:* 401 732-9792
*E-mail:* ndooley@neit.edu

# South Carolina

**Trident Technical College**
Occupational Therapy Asst Prgm
PO Box 118067
Charleston, SC 29423-8067
*Prgm Dir:* Mary Ellen Hiebert, MHS OTR/L
*Tel:* 843 574-6563   *Fax:* 843 574-6585
*E-mail:* MaryEllen.Hiebert@tridenttech.edu

**Greenville Technical College**
Occupational Therapy Asst Prgm
Greer Campus
506 S Pleasantburg Dr, PO Box 5616
Greenville, SC 29606-5616
*Prgm Dir:* Jennifer Coyne, COTA/L
*Tel:* 864 848-2040   *Fax:* 864 848-2038
*E-mail:* jennifer.coyne@gvltec.edu

# South Dakota

**Lake Area Technical Institute**
Occupational Therapy Asst Prgm
230 11th St NE
Watertown, SD 57201-0730
www.lakeareatech.edu
*Prgm Dir:* Julie Kalahar, MS OTR/L
*Tel:* 605 882-5284, Ext 371   *Fax:* 605 882-6299
*E-mail:* kalaharj@lakeareatech.edu
*Med Dir:* Deb Shephard, BS MEd

# Tennessee

**Roane State Community College**
Occupational Therapy Asst Prgm
276 Patton Ln
Harriman, TN 37748-5011
*Prgm Dir:* Jeremy Keough, MSOT OTR/L
*Tel:* 865 481-2000, Ext 2108   *Fax:* 865 481-2019
*E-mail:* keoughjl@roanestate.edu

**Nashville State Technical Community College**
Occupational Therapy Asst Prgm
120 White Bridge Rd, Ste W-60
Nashville, TN 37209
www.nscc.edu/depart/ot
*Prgm Dir:* Donna Whitehouse, MHA OTR
*Tel:* 615 353-3382   *Fax:* 615 353-3608
*E-mail:* donna.whitehouse@nscc.edu

# Texas

**Amarillo College**
Occupational Therapy Asst Prgm
PO Box 447
Amarillo, TX 79178-0001
*Prgm Dir:* Sheree Hilliard Talkington, MA OTR
*Tel:* 806 354-6079   *Fax:* 806 354-6076
*E-mail:* talkington-sl@actx.edu

**Austin Community College**
Occupational Therapy Asst Prgm
Eastview Campus
3401 Webberville Rd
Austin, TX 78702
www.austincc.edu/health/ota
*Prgm Dir:* Carolyn O Cantu, MS OTR
*Tel:* 512 223-5934   *Fax:* 512 223-5897
*E-mail:* ccantu2@austincc.edu

**Panola College**
Occupational Therapy Asst Prgm
1109 W Panola St
Carthage, TX 75633-2397
*Prgm Dir:* Cheri Lambert, OTR
*Tel:* 903 694-4025   *Fax:* 903 694-4010
*E-mail:* clambert@panola.edu

**Del Mar College**
Occupational Therapy Asst Prgm
Dept of Allied Health
101 Baldwin and Ayers Sts
Corpus Christi, TX 78404-3897
www.delmar.edu
*Prgm Dir:* Abel Villarreal, MOT OTR
*Tel:* 512 698-1845   *Fax:* 512 698-1849
*E-mail:* avillar@delmar.edu

**Navarro College**
Occupational Therapy Asst Prgm
3200 W Seventh Ave
Corsicana, TX 75110-4818
www.navarrocollege.edu
*Prgm Dir:* Anita Lane, MEd OTR
*Tel:* 903 875-7583   *Fax:* 903 875-7577
*E-mail:* anita.lane@navarrocollege.edu

**US Army Medical Dept Center & School**
Occupational Therapy Asst Prgm
Academy of Health Sciences
3151 Scott Rd, Ste 1230
Ft Sam Houston, TX 78234-6138
www.cs.army.mil
*Prgm Dir:* CDR Peggy Westerbeck-Silva, MEd OTR/L
*Tel:* 210 221-5324   *Fax:* 210 221-4447
*E-mail:* peggy.westerbeck-silva@cen.amedd.army.mil
*Med Dir:* Michael Pasquarella, DO USA

**Houston Community College**
Occupational Therapy Asst Prgm
Coleman College for Health Sciences
1900 Pressler
Houston, TX 77030-3717
www.hccs.edu
*Prgm Dir:* Linda Williams, MA OTR/L
*Tel:* 713 718-7392, Ext 87392   *Fax:* 713 718-6495
*E-mail:* linda.williams@hccs.edu

**Kingwood College - NHMCCD**
Occupational Therapy Asst Prgm
20000 Kingwood Dr
Kingwood, TX 77339-3801
www.nhmccd.edu
*Prgm Dir:* Alma Watson, MOT OTR
*Tel:* 281 312-1464   *Fax:* 281 312-1490
*E-mail:* alma.r.watson@nhmccd.edu

**Laredo Community College**
Occupational Therapy Asst Prgm
West End Washington St
Laredo, TX 78040-4395
*Prgm Dir:* Terri Gonzalez, MA OTR RMT
*Tel:* 956 721-5460   *Fax:* 956 721-5431
*E-mail:* trgon@laredo.edu

**South Texas College**
Occupational Therapy Asst Prgm
Nursing/Allied Health Division
1101 S Vermont, PO Box 9701
McAllen, TX 78501
www.southtexascollege.edu/nah/program%20ota.htm
*Prgm Dir:* Espy J Brattin, MEd OTR
*Tel:* 956 872-3149   *Fax:* 956 872-3163
*E-mail:* ebrattin@southtexascollege.edu

**St Philip's College**
Occupational Therapy Asst Prgm
1801 Martin Luther King Dr
San Antonio, TX 78203-2098
www.accd.edu/spc
*Prgm Dir:* Jana Cragg, MA OTR
*Tel:* 210 531-3421   *Fax:* 210 531-3459
*E-mail:* jcragg@accd.edu

**Tomball College - NHMCCD**
Occupational Therapy Asst Prgm
30555 Tomball Pkwy
Tomball, TX 77375-4036
*Prgm Dir:* Terra Ruppert, PhD OTR
*Tel:* 281 357-3733   *Fax:* 281 351-3384
*E-mail:* terra.ruppert@nhmccd.edu

## Utah

**Salt Lake Community College**
Occupational Therapy Asst Prgm
4600 S Redwood Rd, PO Box 30808
Salt Lake City, UT 84123
www.slcc.edu
*Prgm Dir:* Katherine B Bruner, MEd OTR/L
*Tel:* 801 957-4894  *Fax:* 801 957-4704
*E-mail:* kathy.bruner@slcc.edu

## Virginia

**Southwest Virginia Community College**
Occupational Therapy Asst Prgm
PO Box SVCC
Richlands, VA 24641-1101
*Prgm Dir:* Annette Looney, OTR/L
*Tel:* 276 935-7748  *Fax:* 276 935-2019
*E-mail:* annette.looney@sw.edu

**Jefferson College of Health Sciences**
Occupational Therapy Asst Prgm
PO Box 13186
Roanoke, VA 24016-4443
www.jchs.edu
*Prgm Dir:* David A Haynes, MBA OTR/L
*Tel:* 540 985-4020  *Fax:* 540 985-9773
*E-mail:* dhaynes@jchs.edu

**Tidewater Community College**
Occupational Therapy Asst Prgm
Virginia Beach Campus
1700 College Crescent
Virginia Beach, VA 23456-1918
www.tcc.edu
*Prgm Dir:* William M Marcil, PhD OTR/L FAOTA
*Tel:* 757 822-7330  *Fax:* 757 427-1338
*E-mail:* wmarcil@tcc.edu

## Washington

**Green River Community College**
Occupational Therapy Asst Prgm
12401 SE 320th St
Auburn, WA 98092-3622
www.greenriver.edu
*Prgm Dir:* S Noel Hepler, MS OTR/L
*Tel:* 253 833-9111, Ext 4341  *Fax:* 253 288-3413
*E-mail:* nhepler@greenriver.edu

## West Virginia

**Mountain State University**
Occupational Therapy Asst Prgm
PO Box AG
Beckley, WV 25802-2830
www.mountainstate.edu/majors/whystudy/ota
*Prgm Dir:* Kay Blose, MOT OTR/L
*Tel:* 304 929-1362  *Fax:* 304 929-1617
*E-mail:* kblose@mountainstate.edu

## Wisconsin

**Fox Valley Technical College**
Occupational Therapy Asst Prgm
1825 N Bluemound Dr, PO Box 2277
Appleton, WI 54912-2277
www.fvtc.edu
*Prgm Dir:* Patricia Holz, MS OT
*Tel:* 920 735-4843  *Fax:* 920 831-4314
*E-mail:* holz@fvtc.edu

**Wisconsin Indianhead Technical College - Ashland Campus**
Occupational Therapy Asst Prgm
2100 Beaser Ave
Ashland, WI 54806-3699
*Prgm Dir:* Mari Jo Ulrich, MA OTR
*Tel:* 715 682-4591, Ext 3167  *Fax:* 715 682-8040
*E-mail:* mjulrich@witc.edu

**Western Technical College**
Occupational Therapy Asst Prgm
304 N Sixth St, PO Box C-0908
La Crosse, WI 54602-0908
www.westerntc.edu
*Prgm Dir:* Doreen M Olson, MS OTR
*Tel:* 608 789-4757  *Fax:* 608 785-9299
*E-mail:* OlsonD@westerntc.edu

**Madison Area Technical College**
Occupational Therapy Asst Prgm
211 N Carroll St
Madison, WI 53703-2285
www.matcmadison.edu
*Prgm Dir:* Catherine Wilson, OTR
*Tel:* 608 258-2313  *Fax:* 608 258-2480
*E-mail:* cwilson@matcmadison.edu

**Milwaukee Area Technical College**
Occupational Therapy Asst Prgm
700 W State St
Milwaukee, WI 53233-1443
*Prgm Dir:* Susan Heitman, MS OTR
*Tel:* 414 297-7158  *Fax:* 414 297-6851
*E-mail:* heitmans@matc.edu

## Wyoming

**Casper College**
Occupational Therapy Asst Prgm
125 College Dr
Casper, WY 82601-9958
www.caspercollege.edu
*Prgm Dir:* Marla J Wonser, OTR/L
*Tel:* 307 268-2867  *Fax:* 307 268-3034
*E-mail:* mwonser@caspercollege.edu

## Occupational Therapy Assistant

| Programs* | Class Capacity | Begins | Length (months) | Award | Res Tuition | Non-res Tuition | Stipend | Offers:‡ 1 | 2 | 3 |
|---|---|---|---|---|---|---|---|---|---|---|
| **California** | | | | | | | | | | |
| Loma Linda University | 60 | Sep | 15 | Dipl, AA | $20,000 | $20,000 | | | | • |
| Sacramento City College | 30 | Jan | 24 | AS | $432 | $3,696 | | • | | • |
| **Colorado** | | | | | | | | | | |
| Pueblo Community College | 20 | Aug | 22 | AAS | $4,888 | $11,044 | | | | • |
| **Delaware** | | | | | | | | | | |
| Delaware Tech & Comm Coll - Owens Campus (Georgetown) | 18 | Jun | 22 | AAS | $2,190 | $5,475 | | | | • |
| **Florida** | | | | | | | | | | |
| Daytona Beach Community College | 25 | Aug | 22 | AAS | $4,627 | $18,060 | | | | • |
| Florida Hospital College of Health Sciences (Orlando) | 24 | Aug | 24 | Dipl, AS | $10,000 | $10,000 | | | • | • |
| Keiser College - Melbourne Campus | 15 | Jan May Sep | 24 | Dipl, AS | $18,000 | $18,000 | | | | • |
| Keiser University (Ft Lauderdale) | 24 | May Aug Jan | 24 | Dipl, AS | $3,800 | $3,800 | | | | • |
| Manatee Community College (Bradenton) | 24 | Fall semester | 20 | AS/AAS | $69 | $256 | | | | |
| Polk Community College (Winter Haven) | 18 | Jan | 20 | AS, AAS | $4,680 | $17,249 | | | | • |
| **Georgia** | | | | | | | | | | |
| Augusta Technical College | 25 | Sep | 27 | AAT | $1,500 | $5,384 | | | | |
| Darton College (Albany) | 22 | Aug | 24 | AS | $4,037 | $16,148 | | • | | • |
| Northwestern Technical College (Rock Spring) | 20 | Jul | 24 | AAT | $1,812 | $1,812 | | | | |
| **Hawaii** | | | | | | | | | | |
| Kapi'olani Community College (Honolulu) | 16 | Aug | 30 | Dipl, AS | $1,096 | $5,872 | | | | • |
| **Illinois** | | | | | | | | | | |
| Lewis & Clark Community College (Godfrey) | 21 | Jan | 24 | AAS | $71 | $284 | | | | • |
| Lincoln Land Community College (Springfield) | 16 | Jan | 24 | AAS | $2,535 | $2,535 | | | | • |
| Parkland College (Champaign) | 20 | Aug | 16 | AAS | $2,464 | $6,880 | | | | • |
| Southern Illinois Collegiate Common Market (Herrin) | 24 | Aug | 22 | AAS | $2,000 | $0 | | | | • |

*Data are shown only for programs that completed the 2006 AMA Survey of Health Professions Education Programs
‡Key to Offers: 1: Evening or weekend classes; 2: Non-English instruction; 3: Cultural competence instruction

## Occupational Therapy Assistant

| Programs* | Class Capacity | Begins | Length (months) | Award | Res Tuition | Non-res Tuition | Stipend | Offers:‡ 1 | 2 | 3 |
|---|---|---|---|---|---|---|---|---|---|---|
| Wright College (Chicago) | 24 | Aug | 28 | AAS | $2,300 | $4,680 | | | | • |
| **Indiana** | | | | | | | | | | |
| Brown Mackie College (South Bend) | 24 | Feb Jul | 24 | AAS | $7,488 | $7,488 | | | | • |
| University of Southern Indiana (Evansville) | 30 | Sep | 28 | AS | $148 | $354 | | | | • |
| **Iowa** | | | | | | | | | | |
| Kirkwood Community College (Cedar Rapids) | 24 | Aug | 22 | AAS | $1,059 | $2,118 | | | | • |
| **Kansas** | | | | | | | | | | |
| Newman University (Wichita) | | | | AAS | | | | | | |
| **Kentucky** | | | | | | | | | | |
| Jefferson Community and Technical College (Louisville) | 20 | Spring Jan | 24 | AAS | $109 | $327 | | | | • |
| Madisonville Community College | 16 | Aug | 18 | AAS | $2,400 | $2,800 | | | | • |
| **Louisiana** | | | | | | | | | | |
| University of Louisiana at Monroe | 30 | Aug | 25 | AS | $4,050 | $10,002 | | | | • |
| **Maryland** | | | | | | | | | | |
| Allegany College of Maryland (Cumberland) | 16 | Aug | 28 | AAS | $2,160 | $4,848 | | | | • |
| Comm Coll of Baltimore County - Catonsville | 40 | Sep | 24 | Dipl, AAS | $1,690 | $2,740 | | | | • |
| **Massachusetts** | | | | | | | | | | |
| Bristol Community College (Fall River) | 20 | Sep | 24 | AS | $123 | $329 | | | | • |
| North Shore Community College (Danvers) | 36 | Sep | 14, 18 | AS | $4,000 | $12,000 | | | | • |
| Quinsigamond Community College (Worcester) | 15 | Sep | 24 | AS | $125 | $331 | | • | | • |
| Springfield Technical Community College | 25 | Sep | 18 | AS | $812 | $7,865 | | | | |
| **Minnesota** | | | | | | | | | | |
| College of St Catherine - Minneapolis | 21 | Sep | 18 | AAS | $14,000 | $14,000 | | | | • |
| Northland Community & Technical College (East Grand Forks) | 24 | Jan | 21 | AAS | $4,608 | $4,608 | | | | • |
| **Mississippi** | | | | | | | | | | |
| Holmes Community College (Ridgeland) | 20 | Aug | 24 | AAS | $712 | $1,377 | | | | • |
| Itawamba Community College (Fulton) | 12 | Jan | | AAS | | | | | | • |
| **Missouri** | | | | | | | | | | |
| Metropolitan Community College - Penn Valley (Kansas City) | 18 | Aug | 22 | AAS | $5,694 | $13,505 | | | | |
| Ozarks Technical Community College (Springfield) | 20 | Aug | 21 | AAS | $2,340 | $3,840 | | | | • |
| Sanford-Brown College - Hazelwood Campus | 20 | Varies | 18 | AAS | $16,775 | $0 | | | | |
| St Charles Community College (St Peters) | 20 | Aug | 20 | AAS | $65 | $95 | | | | • |
| St Louis Community College - Meramec | 24 | Aug | 21 | AAS | $1,800 | $2,340 | | | | • |
| **Nevada** | | | | | | | | | | |
| Community College of Southern Nevada (Las Vegas) | 12 | Sep | 24 | AAS | $2,555 | $4,155 | | • | | • |
| **New Mexico** | | | | | | | | | | |
| Eastern New Mexico University - Roswell | 25 | Aug | 18 | Dipl, AS | $1,051 | $4,320 | | | | • |
| Western New Mexico University (Silver City) | 20 | Aug | 24 | AS | $1,457 | $5,585 | | | | • |
| **New York** | | | | | | | | | | |
| Erie Community College (Williamsville) | 64 | Sep | 24 | AAS | $2,900 | $5,800 | | • | | |
| Genesee Community College (Batavia) | 32 | Aug Jan | 18 | AAS | $3,200 | $3,600 | | • | | • |
| Maria College (Albany) | 46 | Late Aug | 18, 32 | AAS | $7,800 | $7,800 | | • | | • |
| Mercy College (Dobbs Ferry) | 40 | Sep | 12, 24 | AAS | $12,000 | $0 | | • | | |
| Rockland Community College (Suffern) | 30 | Sep Jan | 24 | AAS | $2,800 | $5,600 | | • | | |
| Suffolk County Community College (Brentwood) | 24 | Aug Sep | 21 | AAS | $3,000 | $5,780 | | | | • |
| Touro College (New York) | 30 | Sep | 24 | AAS | $10,000 | $10,000 | | • | | • |
| **North Carolina** | | | | | | | | | | |
| Cape Fear Community College (Wilmington) | 24 | Aug | 21 | AAS | $400 | $2,500 | | | | • |
| Durham Technical Community College | 24 | May | 24, 36 | AAS | $1,900 | $10,536 | | | | • |
| **Ohio** | | | | | | | | | | |
| Cuyahoga Community College (Cleveland) | 20 | Aug | 24 | AAS | $2,723 | $3,599 | | | | • |
| Kent State University - East Liverpool Campus | 25 | Aug | 21 | AAS | $4,586 | $12,018 | | | | • |
| Owens Community College (Toledo) | 30 | Jun | 21 | AAS | $6,542 | $12,551 | | • | | • |
| Shawnee State University (Portsmouth) | 24 | Sep | 21 | AAS | $4,347 | $7,443 | | | | |
| Sinclair Community College (Dayton) | 30 | Sep | 22 | AAS | $2,334 | $3,814 | | | | • |
| Stark State College of Technology (Canton) | 30 | Aug Jan | 18 | AAS | $3,800 | $4,977 | | | | • |
| Zane State College (Zanesville) | 25 | Sep | 25 | AAS | $4,104 | $8,208 | | • | | • |
| **Pennsylvania** | | | | | | | | | | |
| Lehigh Carbon Community College (Schnecksville) | 36 | Sep | 24 | AAS | $2,340 | $4,680 | | • | | • |
| Mount Aloysius College (Cresson) | 20 | Aug | 28 | AS | $15,610 | $15,610 | | | | • |
| Pennsylvania College of Technology (Williamsport) | 32 | Aug | 22 | AAS | $10,080 | $12,660 | | | | • |
| **Rhode Island** | | | | | | | | | | |
| Community College of Rhode Island (Newport) | 24 | Sep | 15 | AAS | $2,390 | $7,000 | | • | | • |
| New England Institute of Technology (Warwick) | 25 | Jul Jan | 18 | AS | $14,000 | $14,000 | | | | |

*Data are shown only for programs that completed the 2006 AMA Survey of Health Professions Education Programs

‡Key to Offers: 1: Evening or weekend classes; 2: Non-English instruction; 3: Cultural competence instruction

## Occupational Therapy Assistant

| Programs* | Class Capacity | Begins | Length (months) | Award | Res Tuition | Non-res Tuition | Stipend | Offers:‡ 1 | 2 | 3 |
|---|---|---|---|---|---|---|---|---|---|---|
| **South Carolina** | | | | | | | | | | |
| Greenville Technical College | 32 | Aug | 21 | AHS | $1,595 | $1,729 | | | | • |
| **South Dakota** | | | | | | | | | | |
| Lake Area Technical Institute (Watertown) | 22 | Aug | 20 | AAS | $4,247 | $4,247 | | • | | • |
| **Tennessee** | | | | | | | | | | |
| Nashville State Technical Community College | 30 | Aug | 24 | AAS | $1,224 | $0 | | | | • |
| Roane State Community College (Harriman) | 20 | Aug | 20 | AAS | $2,142 | $6,414 | | | | • |
| **Texas** | | | | | | | | | | |
| Austin Community College | 20 | Aug | 24 | AAS | $1,365 | $3,500 | | | | |
| Del Mar College (Corpus Christi) | 20 | Sep | 18 | AAS | $1,000 | $0 | | | | |
| Houston Community College | 23 | Aug | 12 | Cert, AAS | $2,600 | $5,065 | | | | • |
| Kingwood College - NHMCCD | 24 | Jan | 20 | AAS | $1,500 | $3,200 | | | | • |
| Navarro College (Corsicana) | 30 | Aug | 30 | AAS | $1,048 | $2,450 | | | | |
| South Texas College (McAllen) | 15 | Sep | 20 | AAS | $1,639 | $2,139 | | | | • |
| St Philip's College (San Antonio) | 25 | Aug | 24 | AAS | $2,184 | $3,288 | | | | |
| US Army Medical Dept Center & School (Ft Sam Houston) | 30 | Jan Apr Aug | 5, 4 | Cert | | | | | | • |
| **Utah** | | | | | | | | | | |
| Salt Lake Community College (Salt Lake City) | 24 | Fall | 24 | AAS | $1,201 | $3,759 | | | | • |
| **Virginia** | | | | | | | | | | |
| Jefferson College of Health Sciences (Roanoke) | 30 | Aug | 22 | AAS | $20,865 | $20,865 | | | | • |
| Tidewater Community College (Virginia Beach) | 30 | Aug | 12, 24 | AAS | $81 | $246 | | | | • |
| **Washington** | | | | | | | | | | |
| Green River Community College (Auburn) | 32 | Sep | 22 | AAS | $2,574 | $2,979 | | | | • |
| **West Virginia** | | | | | | | | | | |
| Mountain State University (Beckley) | 26 | Aug | 14 | AS | $8,000 | $0 | | | | • |
| **Wisconsin** | | | | | | | | | | |
| Fox Valley Technical College (Appleton) | 32 | Aug | 24 | AAS | $2,700 | $10,721 | | | | |
| Madison Area Technical College | 25 | Aug | 22 | AA | $3,930 | $17,200 | | | | • |
| Western Technical College (La Crosse) | 20 | Aug | 22 | AAS | $3,132 | $17,847 | | | | |
| **Wyoming** | | | | | | | | | | |
| Casper College | 12 | Aug | 26 | AS | $1,536 | $4,300 | | | | • |

*Data are shown only for programs that completed the 2006 AMA Survey of Health Professions Education Programs
‡Key to Offers: 1: Evening or weekend classes; 2: Non-English instruction; 3: Cultural competence instruction

## Occupational Description

Ophthalmic dispensing opticians adapt and fit corrective eyewear, including eyeglasses and contact lenses, as prescribed by an ophthalmologist or optometrist. They help customers select appropriate frames, then prepare work orders for ophthalmic laboratory technicians, who grind and insert lenses into frames. The dispensing optician then adjusts the finished eyewear to fit customer needs.

## Job Description

The ophthalmic dispensing optician combines an understanding of the human eye and vision with customer service skills to order the production of corrective eyewear, aid the patient/customer in selecting appropriate, aesthetically pleasing frames, and adjust the frames to fit the customer's face.

Chief duties of the dispensing optician:

- Analyze and interpret prescriptions
- Communicate effectively with patient/customer
- Determine facial and eye measurements
- Identify the human eye structure, function, and pathology
- Assist the customer in selecting appropriate frames and lenses by assessing individual patient needs
- Use an ophthalmologist's or optometrist's prescription to prepare work orders for the ophthalmic laboratory technician
- Deliver prescription eyewear/vision aids and instruct customers in use and care
- Maintain patient/customer records and address complaints
- Provide follow-up services, including eyewear adjustment, repair, and replacement
- Explain theory of refraction
- Identify procedures associated with dispensing artificial eyes and low vision aids, when appropriate
- Adapt, dispense, and fit contact lenses
- Assist in various business duties, including frame and lens inventory, supply and equipment maintenance, and patient insurance/claim forms submission and record keeping
- Apply rules for equipment safety

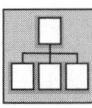

## Employment Characteristics

Dispensing opticians work 40-hour weeks in retail stores, some of which may offer one-stop eye examinations, frames, and on-the-spot lens grinding and fitting, or are self-employed in other optical field areas, such as sales/marketing. Other dispensing opticians provide their eye care services in conjunction with ODs and MDs at eye care centers.

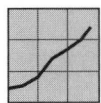

## Employment Outlook

As the baby boomers reach middle age and the percentage of middle-aged and elderly people increases, so will these individuals' need for corrective eyewear. Eyewear as fashion—more and more people today now own two or more pairs of eyeglasses for different occasions—also translates into strong future demand for ophthalmic dispensing opticians, as do the many new vision products, available in plastic and glass, tinted lenses, multifocal extended-wear, and disposable contact lenses.

## Educational Programs

**Length.** Ophthalmic dispensing optician degree programs require 2 years of study.

**Prerequisites.** A high school diploma or its equivalent is generally required for entrance into a program. Ophthalmic dispensing optician students should be familiar with the principles of physics, biology, algebra, and geometry.

**Curriculum.** Ophthalmic dispensing opticianry educational programs include instruction in geometrical optics; ophthalmic optics; anatomy of the eye; and the use of optical instruments, machinery, and tools.

## Inquiries

**Careers**

American Board of Opticians
6506 Loisdale Road, #209
Springfield, VA 22150
703 719-5800

National Academy of Opticianry
8401 Corporate Drive, Ste 605
Landover, MD 20785
301 577-4828

National Federation of Opticianry Schools
9604 Escada Court
Chesterfield, VA 23832
804 790-0026

Opticians Association of America
PO Box 6600
Springfield, VA 22150-6600
703 719-6616

**Program Accreditation**

Commission on Opticianry Accreditation (COA)
PO Box 3073
Merrifield, VA 22116-3073
703 766-1600
703 359-2834 Fax
E-mail: coa@erols.com
www.COAccreditation.com

## Ophthalmic Dispensing Optician

## Connecticut

**Middlesex Community College**
Ophthalmic Dispensing Optician Prgm
Ophthalmic Design and Dispensing
100 Training Hill Rd
Middletown, CT 06457
www.mxcc.commnet.edu
*Prgm Dir:* Raymond P Dennis, MA
*Tel:* 860 343-5845    *Fax:* 860 343-5874
*E-mail:* rdennis@mxcc.commnet.edu

## Florida

**Miami Dade College**
Ophthalmic Dispensing Optician Prgm
Vision Care Technology/Opticianry
950 NW 20th St
Miami, FL 33127
*Prgm Dir:* Jerry Brown
*Tel:* 305 237-4267    *Fax:* 305 237-4278
*E-mail:* jerry.brown@mdc.edu

**Hillsborough Community College**
Ophthalmic Dispensing Optician Prgm
4001 Tampa Bay Blvd
Tampa, FL 33630
*Prgm Dir:* William Underwood
*Tel:* 813 253-7430    *Fax:* 813 253-7379
*E-mail:* bunderwood@hcc.cc.fl.us

## Georgia

**DeKalb Technical College**
Ophthalmic Dispensing Optician Prgm
495 N Indian Creek Dr
Clarkston, GA 30021
*Prgm Dir:* Thomas Schulz
*Tel:* 404 297-9522, Ext 207    *Fax:* 404 294-4234
*E-mail:* schulzt@dekalbtech.edu

**Ogeechee Technical College**
Ophthalmic Dispensing Optician Prgm
One Joe Kennedy Blvd
Statesboro, GA 30458
*Prgm Dir:* Robin Williams
*Tel:* 912 486-7404    *Fax:* 912 486-7604
*E-mail:* rwilliams@ogeecheetech.edu

## Massachusetts

**Holyoke Community College**
Ophthalmic Dispensing Optician Prgm
303 Homestead Ave
Holyoke, MA 01040
*Prgm Dir:* Mary Farrell
*Tel:* 413 552-2288    *Fax:* 413 552-2045
*E-mail:* mfarrell@hcc.mass.edu

## Nevada

**Community College of Southern Nevada**
Ophthalmic Dispensing Optician Prgm
6375 W Charleston Blvd
Las Vegas, NV 89146
*Prgm Dir:* Scott Helkaa
*Tel:* 702 651-5834    *Fax:* 702 651-5762
*E-mail:* scott_helkaa@ccsn.edu

## New Jersey

**Camden County College**
Ophthalmic Dispensing Optician Prgm
PO Box 200, College Dr
Blackwood, NJ 08012
*Prgm Dir:* Nancy Don
*Tel:* 856 227-2649    *Fax:* 856 227-4107
*E-mail:* ndon@camdencc.edu

**Essex County College**
Ophthalmic Dispensing Optician Prgm
303 University Ave
Newark, NJ 07102
*Prgm Dir:* Richard Palumbo
*Tel:* 973 877-3367    *Fax:* 973 877-1920
*E-mail:* palumbo@essex.edu

**Raritan Valley Community College**
Ophthalmic Dispensing Optician Prgm
Ophthalmic Science Program
PO Box 3300
Somerville, NJ 08876
*Prgm Dir:* Brian Thomas, MA ABOM
*Tel:* 908 526-1200, Ext 8277    *Fax:* 908 725-2831
*E-mail:* bthomas@raritanval.edu

## New Mexico

**Southwestern Indian Polytechnic Institute**
Ophthalmic Dispensing Optician Prgm
Optical Technology
9169 Coors Rd NW
Albuquerque, NM 87184
*Prgm Dir:* Samuel Henderson
*Tel:* 505 346-7736    *Fax:* 505 346-2343
*E-mail:* shenders@sipi.bia.edu

## New York

**New York City College of Technology**
Ophthalmic Dispensing Optician Prgm
Department of Vision Care Technology
300 Jay St
Brooklyn, NY 11201
*Prgm Dir:* Robert Russo, MA
*Tel:* 718 260-5298    *Fax:* 718 254-8521
*E-mail:* rrusso@citytech.cuny.edu

**Interboro Institute**
Ophthalmic Dispensing Optician Prgm
450 W 56th St
New York, NY 10019
*Prgm Dir:* Jayne H Weinberger
*Tel:* 212 399-0091    *Fax:* 212 765-5772
*E-mail:* Jweinberger@interboro.edu

**Erie Community College - North Campus**
Ophthalmic Dispensing Optician Prgm
6205 Main St
Williamsville, NY 14221-7095
*Prgm Dir:* John F Godert
*Tel:* 716 851-1570    *Fax:* 716 851-1429
*E-mail:* godert@ecc.edu

## North Carolina

**Durham Technical Community College**
Ophthalmic Dispensing Optician Prgm
1637 Lawson St
Durham, NC 27703
*Prgm Dir:* Michael Szczerbiak
*Tel:* 919 686-3485    *Fax:* 919 686-3737
*E-mail:* szczerbiakm@durhamtech.edu

## Tennessee

**Roane State Community College**
Ophthalmic Dispensing Optician Prgm
Opticianry
276 Patton Ln
Harriman, TN 37748
www.roanestate.edu
*Prgm Dir:* Michael Goggin, ABOC NCLC MS
    Ophthalmic Optics
*Tel:* 865 354-3000, Ext 4319    *Fax:* 865 882-4535
*E-mail:* goggin_mt@roanestate.edu

## Texas

**El Paso Community College**
Ophthalmic Dispensing Optician Prgm
Ophthalmic Technology
PO Box 20500
El Paso, TX 79998
*Prgm Dir:* Jose Baca, BS
*Tel:* 915 831-4075    *Fax:* 915 831-4114
*E-mail:* mannyb@epcc.edu

**Tyler Junior College**
Ophthalmic Dispensing Optician Prgm
Vision Care Technology
PO Box 902
Tyler, TX 75711-9020
*Prgm Dir:* Steve Robbins, ABOC NCLC
*Tel:* 903 510-2961    *Fax:* 909 510-4928
*E-mail:* srob@tjc.edu

## Virginia

**J Sargeant Reynolds Community College**
Ophthalmic Dispensing Optician Prgm
Opticianry Department
PO Box 85622
Richmond, VA 23285-5622
www.reynolds.edu
*Prgm Dir:* Kristina Ostrom, ABO-AC FCLSA
*Tel:* 804 523-5415    *Fax:* 804 786-5298
*E-mail:* kostrom@reynolds.edu

**Tri-Service Optician School (TOPS)**
Ophthalmic Dispensing Optician Prgm
160 Main Rd, Ste 350 Naval Weapons
Yorktown, VA 23691-9984
*Prgm Dir:* HMC O'Guinn
*Tel:* 757 887-7384    *Fax:* 757 887-4511
*Med Dir:* Capt Michael Pattison, USN

## Washington

**Seattle Central Community College**
Ophthalmic Dispensing Optician Prgm
1701 Broadway
Seattle, WA 98122
http://seattlecentral.org
*Prgm Dir:* Gary Clayton
*Tel:* 206 344-4321    *Fax:* 206 344-4316
*E-mail:* KLawler@sccd.ctc.edu

## Wisconsin

**Milwaukee Area Technical College**
Ophthalmic Dispensing Optician Prgm
Opticianry Science
700 W State St
Milwaukee, WI 53233-1443
*Prgm Dir:* Laurie Zielinski, ABOC NCLC
*Tel:* 414 297-7425    *Fax:* 414 297-6851
*E-mail:* zielinsl@matc.edu

## Ophthalmic Dispensing Optician

| Programs* | Class Capacity | Begins | Length (months) | Award | Res Tuition | Non-res Tuition | Stipend | Offers:‡ 1 | 2 | 3 |
|---|---|---|---|---|---|---|---|---|---|---|
| **Connecticut** | | | | | | | | | | |
| Middlesex Community College (Middletown) | 24 | Sep | 20 | AS | $2,672 | $7,976 | | | | |
| **New Jersey** | | | | | | | | | | |
| Raritan Valley Community College (Somerville) | 50 | Aug Jan | 24 | Cert, AAS | $1,040 | $0 | | • | | |
| **New York** | | | | | | | | | | |
| New York City College of Technology (Brooklyn) | 90 | Sep Jan | 21 | AAS | $3,200 | $6,800 | | • | | |
| **Tennessee** | | | | | | | | | | |
| Roane State Community College (Harriman) | 30 | Aug | 22 | AAS | $0 | $3,000 | | | | • |
| **Virginia** | | | | | | | | | | |
| J Sargeant Reynolds Community College (Richmond) | 30 | Every semester | 24 | AAS | $2,916 | $8,856 | | • | | |
| **Washington** | | | | | | | | | | |
| Seattle Central Community College | 25 | Sep | 21 | AAS | $1,900 | $0 | | | | • |
| **Wisconsin** | | | | | | | | | | |
| Milwaukee Area Technical College | 16 | Fall semester | 12 | Dipl | $3,992 | $0 | | | | • |

*Data are shown only for programs that completed the 2006 AMA Survey of Health Professions Education Programs
‡Key to Offers: 1: Evening or weekend classes; 2: Non-English instruction; 3: Cultural competence instruction

**Includes:**
- Ophthalmic assistant
- Ophthalmic medical technician/technologist

## Occupational Description

The ophthalmic assistant, ophthalmic technician, and ophthalmic medical technologist are skilled professionals, qualified by didactic and clinical ophthalmic training, who perform ophthalmic procedures under the direction or supervision of a licensed ophthalmologist who is responsible for the performance of the ophthalmic assistant, ophthalmic technician, and ophthalmic medical technologist.

## Job Description

The functions of the ophthalmic technician and ophthalmic medical technologist are to assist the ophthalmologist by performing delegable tasks, collecting data, administering treatment ordered by an ophthalmologist, and supervising patients.

*The following are duties and tasks that may be delegated by an ophthalmologist, as applicable by state law, to ophthalmic assistants, ophthalmic technicians, and ophthalmic medical technologists.*

### Ophthalmic Assistants, Ophthalmic Technicians, and Ophthalmic Medical Technologists

1. Obtaining a medical history
2. Performing lensometry
3. Obtaining anatomical and functional ocular measurements of the eye, such as axial length
4. Obtaining functional measurements of the eye, such as visual acuity
5. Testing ocular functions, such as visual fields
6. Administering topical ophthalmic and oral medications
7. Instructing the patient (in personal care and the use of contact lenses)
8. Caring for and maintaining ophthalmic instruments
9. Caring for, maintaining, and sterilizing surgical instruments
10. Adjusting and making minor repairs on spectacles
11. Such other tasks as may be delegated consistent with sound medical practice (eg, use of computerized ophthalmic equipment)

### Ophthalmic Technicians and Ophthalmic Medical Technologists

The ophthalmic technician and medical technologist will be expected to perform the duties listed above, at a higher level of expertise. The ophthalmic medical technologist will be expected to exercise considerable clinical skill in the performance of those delegated tasks. The ophthalmic technician and medical technologist may be expected to perform the following additional duties:

12. Performing diagnostic tests
13. Maintaining ophthalmic office equipment
14. Assisting in ophthalmic surgery in the office or hospital
15. Obtaining optical measurements including A-scan
16. Assisting in the fitting of contact lenses
17. Refractometry

### Ophthalmic Medical Technologists

The ophthalmic medical technologist will be expected to perform the duties listed above at a higher level of expertise and exercise considerable clinical skill in the performance of those delegated tasks. They may be expected to perform the following additional duties:

18. Performing ophthalmic clinical photography and fluorescein angiography of the eye
19. Administering advanced ocular motility and binocular function tests
20. Performing ocular electrophysiological procedures
21. Performing advanced microbiological procedures
22. Providing supervision and instruction of other ophthalmic personnel and patients
23. Demonstrating advanced general medical knowledge

## Employment Characteristics

Ophthalmic assistants and ophthalmic medical technicians and technologists are employed primarily by ophthalmologists, medical institutions, clinics, hospitals, ambulatory surgery centers, or physician groups, in which they may be assigned to an ophthalmologist responsible for their supervision and performance. They may be involved with the patients of an ophthalmologist in any setting for which the ophthalmologist is responsible.

## Salary

Salaries vary depending on employer and geographic location. Entry-level salaries range from $21,500 for ophthalmic assistants to $39,000 for ophthalmic technicians and $45,000 for ophthalmic medical technologists. Refer to Section IV, Table 5 of this *Directory* for more information, or see www.ama-assn.org/go/salary.

## Educational Programs

**Length.** Programs are generally less than 1 year for assistants, 1 year for technicians and 2 years for technologists.

**Prerequisites.** High school diploma or equivalent for assistants and technicians and two years of undergraduate study for technologists.

**Curriculum.** Instruction should follow a planned outline, which includes courses in human anatomy and physiology, medical terminology, medical laws and ethics, psychology, ocular anatomy and physiology, ophthalmic optics, microbiology, ophthalmic pharmacology and toxicology, ocular motility, and diseases of the eye. The curriculum also includes diagnostic and treatment procedures, including visual field testing, contact lenses, ophthalmic surgery, and the care and maintenance of ophthalmic instruments and equipment. For technicians and medical technologists, students must also have supervised clinical experience, during which they have opportunities to apply theory to practice through correlated and supervised instruction in clinical practice areas.

## Inquiries

**Careers and Certification**

Inquiries regarding careers, continuing education, and certification criteria should be addressed to:

Joint Commission on Allied Health Personnel in Ophthalmology (JCAHPO)
2025 Woodlane Drive
St Paul, MN 55125-2998
651 731-2944 or 800 284-3937
E-mail: jcahpo@jcahpo.org

Association of Technical Personnel in Ophthalmology (ATPO)
2025 Woodlane Drive
St Paul, MN 55125-2998
651 731-7233 or 800 482-4858
E-mail: ATPOmembership@jcahpo.org

**Program Accreditation**

Commission on Accreditation of Ophthalmic Medical Programs (CoA-OMP)
2025 Woodlane Drive
St Paul, MN 55125-2998
651 731-2944
651 731-0410 Fax
E-mail: CoA-OMP@jcahpo.org

## Ophthalmic Assistant

### California

**Jules Stein Eye Institute, UCLA**
Ophthalmic Assistant Prgm
100 Stein Plaza 3-223
Los Angeles, CA 90095
*Prgm Dir:* Bobbi E Ballenberg, COMT
*Tel:* 310 794-5603   *Fax:* 310 206-8015
*E-mail:* ballenberg@jsei.ucla.edu
*Med Dir:* Kevin Miller, MD

### Louisiana

**Delgado Community College**
Ophthalmic Assistant Prgm
Allied Health Division
615 City Park Ave Bldg #4
New Orleans, LA 70119-4399
*Prgm Dir:* Francesa Langlow, BS COA
*Tel:* 504 483-4003   *Fax:* 504 483-4609
*E-mail:* fmorel@dcc.edu

### Massachusetts

**Holyoke Community College**
Ophthalmic Assistant Prgm
303 Homestead Ave
Holyoke, MA 01040
*Prgm Dir:* Mary Farrell
*Tel:* 413 552-2288   *Fax:* 413 552-2045
*E-mail:* mfarrell@hcc.mass.edu

### North Carolina

**Caldwell Comm College & Tech Institute**
Ophthalmic Assistant Prgm
2855 Hickory Blvd
Hudson, NC 28638
*Prgm Dir:* Barbara T Harris, PA-C MBA COA
*Tel:* 828 726-2356   *Fax:* 828 726-2489
*E-mail:* bharris@cccti.edu
*Med Dir:* John Tye, MD

### Washington

**Renton Technical College**
Ophthalmic Assistant Prgm
3000 NE Fourth St
Renton, WA 98056
*Prgm Dir:* Larry Bovard
*Tel:* 425 235-2352, Ext 7926   *Fax:* 425 235-5836
*E-mail:* lbovard@rtc.edu

## Ophthalmic Medical Technician/Technologist

### Arkansas

**University of Arkansas for Medical Sciences**
Ophthalmic Med Technologist Prgm
4301 W Markham #523
Little Rock, AR 72205-7199
www.uams.edu/chrp/omt
*Prgm Dir:* Suzanne Hansen, BA BS COMT
*Tel:* 501 526-5880   *Fax:* 501 686-6798
*E-mail:* hansensuzannej@uams.edu
*Med Dir:* Christopher T Westfall, MD

### Colorado

**Pima Medical Institute - Denver**
Ophthalmic Med Technician Prgm
1701 W 72nd Ave
Denver, CO 80221
*Prgm Dir:* Wendy Stanton, COT
*Tel:* 303 426-1800   *Fax:* 303 430-4048
*E-mail:* stantonpima@yahoo.com
*Med Dir:* William Hines, MD

### District of Columbia

**Georgetown University Medical Center**
Ophthalmic Med Technician Prgm
Center for Sight
3800 Reservoir Rd NW, LL PHC
Washington, DC 20007
www.georgetown.edu/departments/ophthalmology
*Prgm Dir:* Ella Rosamont-Morgan, BS COMT
*Tel:* 202 444-4862   *Fax:* 202 444-4978
*E-mail:* emm47@georgetown.edu
*Med Dir:* Peter Y Evans, MD

## Ophthalmic Assistant

| Programs* | Class Capacity | Begins | Length (months) | Award | Res Tuition | Non-res Tuition | Stipend | Offers:‡ 1 | 2 | 3 |
|---|---|---|---|---|---|---|---|---|---|---|
| **California** | | | | | | | | | | |
| Jules Stein Eye Institute, UCLA (Los Angeles) | 20 | Sep | 9 | Cert | $1,500 | $1,500 | | | • | • |
| **North Carolina** | | | | | | | | | | |
| Caldwell Comm College & Tech Institute (Hudson) | 15 | May | 12 | Dipl | $1,764 | $3,512 | | | | • |

*Data are shown only for programs that completed the 2006 AMA Survey of Health Professions Education Programs
‡Key to Offers: 1: Evening or weekend classes; 2: Non-English instruction; 3: Cultural competence instruction

# Florida

**University of Florida**
Ophthalmic Med Technologist Prgm
PO Box 100393, JHMHC
Gainesville, FL 32610
www.eye.ufl.edu/optkprog.htm
*Prgm Dir:* Diana J Shamis, MHSE CO COMT
*Tel:* 352 265-7080, Ext 87945    *Fax:* 352 265-7081
*E-mail:* dshamis@eye.ufl.edu
*Med Dir:* Subir Bhatia, MD

# Georgia

**Emory University**
Ophthalmic Med Technologist Prgm
1365-B Clifton Rd NE, Rm B-4628
Atlanta, GA 30322
http://MMScCOMT.emory.edu
*Prgm Dir:* Paul M Larson, MMSc MBA COMT COE
*Tel:* 404 778-4305    *Fax:* 404 778-5128
*E-mail:* plarson@emory.edu
*Med Dir:* Ted Wojno, MD

# Illinois

**Triton College**
Ophthalmic Med Technician Prgm
2000 N Fifth Ave
River Grove, IL 60171
www.triton.edu
*Prgm Dir:* Debra Baker, COMT MA
*Tel:* 708 456-0300, Ext 3442    *Fax:* 708 583-3121
*E-mail:* dbaker1@triton.edu
*Med Dir:* Mark Rosanova, MD

# Louisiana

**Louisiana State Univ Health Sciences Center**
Ophthalmic Med Technologist Prgm
2020 Gravier St, Ste B
New Orleans, LA 70112
*Prgm Dir:* Robin Cooper, COMT ROUB
*Tel:* 504 412-1200, Ext 1213    *Fax:* 504 412-1243
*E-mail:* rcoope@lsuhsc.edu
*Med Dir:* Donald R Bergsma, MD

# Minnesota

**Regions Hospital**
Ophthalmic Med Technologist Prgm
640 Jackson St
St Paul, MN 55101
www.regionshospital.com/Regions
*Prgm Dir:* Kristine A Fey, COMT
*Tel:* 651 254-3000    *Fax:* 651 254-2256
*E-mail:* Kristine.A.Fey@HealthPartners.com
*Med Dir:* Leslie Kopietz, MD

**Regions Hospital**
Ophthalmic Med Technician Prgm
640 Jackson St
St Paul, MN 55101
www.regionshospital.com
*Prgm Dir:* Kristine Fey, COMT
*Tel:* 651 254-3000    *Fax:* 651 254-2256
*E-mail:* Kristine.A.Fey@HealthPartners.com
*Med Dir:* Leslie Kopietz, MD

# New Jersey

**Camden County College**
Ophthalmic Med Technologist Prgm
PO Box 200 College Dr
Blackwood, NJ 08012
*Prgm Dir:* Ray Didonato
*Tel:* 856 374-5058
*E-mail:* rdidonato@camdencc.edu

# North Carolina

**Duke University Medical Center**
Ophthalmic Med Technician Prgm
2351 Erwin Rd, Eye Center Box 3802
Durham, NC 27710
www.dukeeye.org/education
*Prgm Dir:* Ray Fligman, COT ROUB FCLSA
*Tel:* 919 684-6261    *Fax:* 919 684-6844
*E-mail:* ray.fligman@duke.edu
*Med Dir:* Julie Woodward, MD

# Ontario, Canada

**University of Ottawa**
Ophthalmic Med Technologist Prgm
The Ottawa Hospital
General Campus - 501 Smyth Rd
Ottawa, ON K1H 8L6
www.eyeinstitute.net/omtpbrochure.html
*Prgm Dir:* Carla Barbery, BSc COMT
*Tel:* 613 737-8362    *Fax:* 613 737-8836
*E-mail:* cbarbery@ottawahospital.on.ca
*Med Dir:* W Bruce Jackson

# Oregon

**Portland Community College**
Ophthalmic Med Technician Prgm
Cascade Campus, Jackson Hall 208B
705 N Killingsworth St
Portland, OR 97217
www.pcc.edu/omt
*Prgm Dir:* Joanne M Harris, COT
*Tel:* 503 978-5666    *Fax:* 503 978-5257
*E-mail:* jmharris@pcc.edu
*Med Dir:* Andrea Dostroph, MD

# Puerto Rico

**University of Puerto Rico**
Ophthalmic Med Technician Prgm
PO Box 365067
San Juan, PR 00936-5067
http://cprsweb.rcm.upr.edu
*Prgm Dir:* Mercedes Rivera, OMP MAEd
*Tel:* 787 758-2525, Ext 1935    *Fax:* 787 758-3488
*E-mail:* mercedesrivera@cprs.rcm.upr.edu
*Med Dir:* William Townsend, MD

# Tennessee

**Volunteer State Community College**
Ophthalmic Med Technician Prgm
1480 Nashville Pike
Annex Building 400, Rm 104a
Gallatin, TN 37066
www.volstate.edu
*Prgm Dir:* Kathleen Moore, COMT AAS
*Tel:* 615 230-3723    *Fax:* 615 230-3251
*E-mail:* Kathleen.Moore@volstate.edu
*Med Dir:* Gary Jerkins, MD

# Virginia

**Old Dominion University**
Ophthalmic Med Technologist Prgm
Lions Center for Sight
600 Gresham Dr
Norfolk, VA 23507
www.evms.edu/ophthalmology/optech
*Prgm Dir:* Lori Williams, BS COMT
*Tel:* 757 388-3747    *Fax:* 757 388-2109
*E-mail:* optech@evms.edu
*Med Dir:* Ira R Lederman, MD

## Ophthalmic Medical Technician/Technologist

| Programs* | Class Capacity | Begins | Length (months) | Award | Res Tuition | Non-res Tuition | Stipend | Offers:‡ 1 | 2 | 3 |
|---|---|---|---|---|---|---|---|---|---|---|
| **Arkansas** | | | | | | | | | | |
| University of Arkansas for Medical Sciences (Little Rock) | 8 | Aug | 24 | BS | $5,236 | $12,648 | | | | |
| **District of Columbia** | | | | | | | | | | |
| Georgetown University Medical Center (Washington) | 9 | Aug | 23 | Cert | $5,000 | $5,000 | | | | • |
| **Florida** | | | | | | | | | | |
| University of Florida (Gainesville) | 6 | Jul | 24 | Cert | $2,000 | $2,000 | | | | • |
| **Georgia** | | | | | | | | | | |
| Emory University (Atlanta) | 15 | Jun | 24 | MMSc | $15,500 | $15,500 | | | | |
| **Illinois** | | | | | | | | | | |
| Triton College (River Grove) | 30 | Aug | 21 | AAS | $1,548 | $4,393 | | | | • |

*Data are shown only for programs that completed the 2005 AMA Survey of Health Professions Education Programs
‡Key to Offers: 1: Evening or weekend classes; 2: Non-English instruction; 3: Cultural competence instruction

## Ophthalmic Medical Technician/Technologist

| Programs* | Class Capacity | Begins | Length (months) | Award | Res Tuition | Non-res Tuition | Stipend | Offers:‡ 1 | 2 | 3 |
|---|---|---|---|---|---|---|---|---|---|---|
| **Minnesota** | | | | | | | | | | |
| Regions Hospital (St Paul) | 8 | Sep | 21 | Cert | $6,800 | $6,800 | | | | • |
| Regions Hospital (St Paul) | 8 | Sep | 21 | Cert | $6,800 | $6,800 | | | | • |
| **North Carolina** | | | | | | | | | | |
| Duke University Medical Center (Durham) | 15 | Jul | 12 | Cert | $3,500 | $3,500 | | | | • |
| **Ontario, Canada** | | | | | | | | | | |
| University of Ottawa | 8 | Jul | 10 | BSc | $4,850 | $0 | | | | |
| **Oregon** | | | | | | | | | | |
| Portland Community College | 24 | Sep | 21 | AAS | $3,149 | $8,930 | | | | |
| **Puerto Rico** | | | | | | | | | | |
| University of Puerto Rico (San Juan) | 10 | Aug | 12 | Dipl, Assoc | $1,794 | $3,192 | | | • | • |
| **Tennessee** | | | | | | | | | | |
| Volunteer State Community College (Gallatin) | 15 | Aug | 12 | AAS | $3,600 | $13,000 | | | | |
| **Virginia** | | | | | | | | | | |
| Old Dominion University (Norfolk) | 4 | Sep | 22 | Cert, BS | $4,400 | $4,400 | | | | |

*Data are shown only for programs that completed the 2005 AMA Survey of Health Professions Education Programs

‡Key to Offers: 1: Evening or weekend classes; 2: Non-English instruction; 3: Cultural competence instruction

## Job Description

Orthoptics involves the evaluation and treatment of disorders of vision, eye movements, and eye alignment in children and adults. The orthoptist performs a series of diagnostic tests and measurements on patients with visual disorders, including lazy eye, strabismus (misaligned eyes), and double vision. Through interpretation of testing procedures and clinical evaluation, the orthoptist helps the ophthalmologist design a treatment plan, which may involve treatment by the orthoptist, surgical treatment by the ophthalmologist, or some combination of the two.

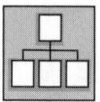

## Employment Characteristics

The orthoptist is the liaison between an ophthalmologist and the patient, assisting in the explanation and execution of the treatment. Orthoptists work in a variety of professional settings:

- As a consultant, the orthoptist may travel to several offices or clinics to see patients or work as a professional advisor to vision-related community agencies.
- An orthoptist may serve as a director of state or local vision screening programs.
- Academic opportunities also exist for individuals who want to offer clinical expertise and instruction to orthoptic students, medical students, and resident physicians. Orthoptists may also participate in clinical research and in the presentation and publication of scientific papers.

Orthoptists possess diagnostic ability, technical understanding, and therapeutic skills. In addition, orthoptists should be able to work well with young children, who make up a large portion of orthoptic patients. It is not uncommon for these young patients to have physical, mental, or emotional disabilities.

## Salary

Orthoptists receive compensation at the high end of that earned by other health professionals, including physical therapists and physician assistants. Refer to Section IV, Table 5 of this *Directory* for more information, or see www.ama-assn.org/go/salary.

## Educational Programs

**Length.** Orthoptist programs require 2 years of postgraduate study.

**Prerequisites.** A baccalaureate degree is required; however, exceptions are considered on an individual basis. The Graduate Record Examination is not required.

**Curriculum.** Lectures, textbooks, journal publications, and proceedings from scientific ophthalmology symposia and conferences form the basis of the didactic teaching. Primary subject areas include anatomy, neuroanatomy, physiology, pharmacology, ophthalmic optics, diagnostic testing and measurement, orthoptic treatment, systemic disease and ocular motor disorders, principles of surgery, and basic ophthalmic examination techniques, as well as genetics, child development, learning disabilities, clinical research methods, and medical writing. In most programs, preparation and presentation of scientific papers are usually required during the second year of education.

Programs are typically structured around an 8-hour day. On average, an orthoptic student will evaluate more than 1,500 patients and observe many more during the course of study. Extensive clinical experience is part of every program.

## Inquiries

Inquiries regarding a career in orthoptics, certification, and program accreditation (brochures about the profession are available upon request):

Leslie France, CO, Executive Director
American Orthoptic Council
3914 Nakoma Rd
Madison, WI 53711
608 233-5383
608 263-4247 Fax (attn: Leslie)
E-mail: lwfranceco@att.net
www.orthoptics.org

American Association of Certified Orthoptists
Ron Biernacki, CO, President
E-mail: Ronald.j.biernacki@vanderbilt.edu
www.orthoptics.org

## Orthoptist*

### British Columbia, Canada

**British Columbia's Children's Hospital**
Orthoptist Prgm
Dept of Ophthamology A135E
4480 Oak St
Vancouver, BC V6H 3N1
www.cw.bc.ca
*Prgm Dir:* Christy Giligson, OC(C)
*Tel:* 604 875-2326   *Fax:* 604 875-3561
*E-mail:* cgiligson@cw.bc.ca
*Med Dir:* Roy A Cline, MD

## California

**Childrens Hospital Los Angeles**
Orthoptist Prgm
4650 Sunset Blvd
Los Angeles, CA 90027
*Prgm Dir:* Paula Edelman, CO
*Tel:* 323 669-5697   *Fax:* 323 662-9080
*E-mail:* pedelman@chla.usc.edu
*Med Dir:* Mark Borchert, MD

*Additional information about programs that returned the AMA's survey is available in a table beginning on page 300.

## Florida

**University of Florida**
Orthoptist Prgm
Box 100393, JHMHC
Gainesville, FL 32610
www.eye.ufl.edu/optkprog.htm
*Prgm Dir:* Diana Shamis, CO COMT MHSE
*Tel:* 352 265-0860   *Fax:* 352 265-7081
*E-mail:* dshamis@eye.ufl.edu
*Med Dir:* Lawrence Levine, MD

# Iowa

**University of Iowa Hospitals & Clinics**
Orthoptist Prgm
Dept of Opthalmology
200 Hawkins Dr
Iowa City, IA 52242
*Prgm Dir:* Pamela Kutschke, BS, CO
*Tel:* 319 356-3863   *Fax:* 319 384-9831
*E-mail:* pamela-kutschke@uiowa.edu
*Med Dir:* Ronald V Keech, MD

# Maryland

**Greater Baltimore Medical Center**
Orthoptist Prgm
Department of Ophthalmology
PPW Suite 505
Baltimore, MD 21204
*Prgm Dir:* Cheryl McCarus, CO COMT
*Tel:* 443 849-8097   *Fax:* 443 849-2648
*E-mail:* cmccarus@gbmc.org
*Med Dir:* Mary Louise Collins, MD

# Michigan

**W K Kellogg Eye Center**
Orthoptist Prgm
1000 Wall St
Ann Arbor, MI 48105
*Prgm Dir:* Bruce A Furr, CO
*Tel:* 734 764-7558   *Fax:* 734 763-7114
*E-mail:* bfurr@umich.edu
*Med Dir:* Monte A Del Monte, MD

# Minnesota

**Park Nicollett Orthoptic Fellowship**
Orthoptist Prgm
3900 Park Nicollet Blvd
Minneapolis, MN 55416
www.parknicollet.com
*Prgm Dir:* Lisa Rovick, CO COMT
*Tel:* 952 993-3737   *Fax:* 952 993-0288
*E-mail:* rovicl@parknicollet.com
*Med Dir:* W Keith Engel, MD

**University of Minnesota - Minneapolis**
Orthoptist Prgm
Dept of Ophthalmology
516 Delaware St SE
Minneapolis, MN 55455
www.tc.umn.edu
*Prgm Dir:* Kimberly Merrill, CO
*Tel:* 612 625-4400   *Fax:* 612 626-3119
*E-mail:* kmerrill@umphysicians.umn.edu
*Med Dir:* Stephen Christiansen, MD

# New York

**University at Buffalo - SUNY**
Orthoptist Prgm
Dept of Ophthalmology
3580 Sheridan Dr
Amherst, NY 14226
www.smbs.buffalo.edu/ophthalmology
*Prgm Dir:* Kyle Arnoldi, COMT CO
*Tel:* 716 834-1445   *Fax:* 716 834-0081
*E-mail:* kylea@buffalo.edu
*Med Dir:* James Reynolds, MD

**New York Eye & Ear Infirmary**
Orthoptist Prgm
Allied Health School in Ophthalmology
310 E 14th St
New York, NY 10003
www.nyee.edu
*Prgm Dir:* Sara Shippman
*Tel:* 212 979-4375   *Fax:* 212 979-4564
*E-mail:* sshippman@nyee.edu
*Med Dir:* Lisabeth s Hall, MD

# Nova Scotia, Canada

**IWK Grace Health Centre**
*Cosponsor:* Dalhousie University
Orthoptist Prgm
Clinical Vision Science Program
5850 University Ave PO Box 9700
Halifax, NS 339
*Prgm Dir:* Karen McMain, OC(C) COMT
*Tel:* 902 470-8959   *Fax:* 902 470-7207
*E-mail:* karen.mcmain@iwk.nshealth.ca
*Med Dir:* Robert LaRoche, MD

# Oklahoma

**Orthoptic Teaching Program of Tulsa**
Orthoptist Prgm
6606 S Yale Ste 110
Tulsa, OK 74136
*Prgm Dir:* Amy G McCarthy, CO COT
*Tel:* 918 481-2781   *Fax:* 918 481-2785
*E-mail:* amccarthy@pediatriceyeassociates.com
*Med Dir:* Gary T Denslow, MD MPH

# Ontario, Canada

**Hospital for Sick Children**
Orthoptist Prgm
555 University Ave Rm M109
Toronto, ON M5G 1X8
www.tcos.ca
*Prgm Dir:* Jennifer Schofield, OC(C)
*Tel:* 416 813-5798   *Fax:* 416 813-6261
*E-mail:* jennifer.schofield@sickkids.ca
*Med Dir:* Stephen Kraft, MD

# Saskatchewan, Canada

**Orthoptic Clinic Eye Care Centre**
Orthoptist Prgm
701 Queen St
Saskatoon, SK S7K 5T6
www.orthoptics.ca
*Prgm Dir:* Ronna Hjertaas, OC(C) COT
*Tel:* 306 655-8058   *Fax:* 306 655-8119
*E-mail:* ronna.hjertaas@saskatoonhealthregion.ca
*Med Dir:* Rob Pekush, MD FRCS(C)

# Wisconsin

**University of Wisconsin Hospital and Clinics**
Orthoptist Prgm
Dept of Ophthalmology & Visual Services
2880 University Ave, Rm 223
Madison, WI 53705-9030
*Prgm Dir:* Jacqueline Shimko, BGS CO
*Tel:* 608 263-7189   *Fax:* 608 263-4247
*E-mail:* jw.shimko@hosp.wisc.edu
*Med Dir:* Thomas D France, MD

# Orthoptist

| Programs* | Class Capacity | Begins | Length (months) | Award | Res Tuition | Non-res Tuition | Stipend | Offers:‡ 1 | 2 | 3 |
|---|---|---|---|---|---|---|---|---|---|---|
| **British Columbia, Canada** | | | | | | | | | | |
| British Columbia's Children's Hospital (Vancouver) | 2 | Jul | 24 | Cert | | | | | | |
| **California** | | | | | | | | | | |
| Childrens Hospital Los Angeles | 2 | Varies | 24 | Cert | $2,000 | $0 | | | | |
| **Florida** | | | | | | | | | | |
| University of Florida (Gainesville) | 2 | Jul | 24 | Cert | $2,000 | $2,000 | | | | • |
| **Iowa** | | | | | | | | | | |
| University of Iowa Hospitals & Clinics (Iowa City) | 4 | Aug | 24 | Cert | | | | | | |
| **Maryland** | | | | | | | | | | |
| Greater Baltimore Medical Center | 2 | Sep | 24 | Cert | $2,500 | $2,500 | | | | |
| **Minnesota** | | | | | | | | | | |
| Park Nicolett Orthoptic Fellowship (Minneapolis) | 1 | Jun 1 | 12, 24 | | $1,000 | $1,000 | | | | |
| University of Minnesota - Minneapolis | 2 | Sep Jul | 24, 12 | Cert | $5,000 | $5,000 | | | | • |
| **New York** | | | | | | | | | | |
| New York Eye & Ear Infirmary | 3 | Sep | 24 | Cert | $5,000 | $5,000 | | | | |
| University at Buffalo - SUNY (Amherst) | 1 | Sep | 24 | Cert | $2,000 | $2,000 | | | | |
| **Oklahoma** | | | | | | | | | | |
| Orthoptic Teaching Program of Tulsa | 1 | First Monday in Jul | 24 | Cert | | | | | | |
| **Ontario, Canada** | | | | | | | | | | |
| Hospital for Sick Children (Toronto) | 2 | Jul | 24 | Cert | $1,100 | $2,500 | | | | |
| **Saskatchewan, Canada** | | | | | | | | | | |
| Orthoptic Clinic Eye Care Centre (Saskatoon) | 3 | Jul | 24 | Cert | $3,000 | $0 | | | | |
| **Wisconsin** | | | | | | | | | | |
| University of Wisconsin Hospital and Clinics (Madison) | 1 | Jul | 24 | Cert | $1,200 | $1,200 | | | | |

*Data are shown only for programs that completed the 2006 AMA Survey of Health Professions Education Programs
‡Key to Offers: 1: Evening or weekend classes; 2: Non-English instruction; 3: Cultural competence instruction

### Occupational Description

Orthotics and prosthetics are applied physical disciplines that address neuromuscular and structural skeletal problems in the human body with a treatment process that includes evaluation and transfer of forces using orthoses and prostheses to achieve optimum function, prevent further disability, and provide cosmesis.

The orthotist and prosthetist work directly with the physician and representatives of other allied health professions in the rehabilitation of the physically challenged. The orthotist designs and fits devices, known as orthoses, to provide care to patients who have disabling conditions of the limbs and spine. The prosthetist designs and fits devices, known as prostheses, for patients who have partial or total absence of a limb.

### Job Description

The role of the orthotist and prosthetist includes, but may not be limited to, five major domains: clinical assessment, patient management, technical implementation, practice management, and professional responsibility.

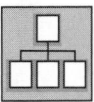

### Employment Characteristics

Orthotists and prosthetists typically provide their services in one or more of the following settings: private facilities, hospitals and clinics, colleges and universities, and medical schools.

### Salary

According to the National Commission on Orthotic and Prosthetic Education, the salary for board-certified orthotists and prosthetists averages between $42,000 and $60,000. Refer to Section IV, Table 5 of this *Directory* for more information, or see www.ama-assn.org/go/salary.

### Educational Programs

**Length.** Orthotic and/or prosthetic education occurs in two forms: baccalaureate degree and certificate programs. Degree programs are based on a standard 4-year curriculum, and certificate courses range from 6 months to 1 year for one discipline to 18 months to 2 years for both disciplines.

**Prerequisites.** Applicants for the 4-year baccalaureate degree programs should have a high school diploma or equivalent and meet institutional entrance requirements. Applicants for postbaccalaureate programs should have a baccalaureate degree that includes appropriate coursework in biology, chemistry, physics, psychology, algebra, human anatomy, and physiology, as well as any other specified by the institution.

**Curriculum.** The professional curriculum includes formal instruction in biomechanics gait analysis/pathomechanics, kinesiology, pathology, materials science, research methods, diagnostic imaging techniques, measurement, impression taking, model rectification, diagnostic fitting, definitive fitting, postoperative management, external power, static and dynamic alignment of sockets related to various amputation levels, and fitting and alignment of orthoses for lower limb, upper limb, and spine with various systems to be included. The curriculum also includes a clinical experience.

### Inquiries

**Careers**

National Commission on Orthotic and Prosthetic Education
330 John Carlyle Street, Suite 200
Alexandria, VA 22314
703 836-7114
703 836-0838 Fax
E-mail: info@ncope.org
www.ncope.org

American Academy of Orthotists and Prosthetists
526 King Street, Suite 200
Alexandria, VA 22314
703 836-0788
703 836-0737 Fax
www.oandp.org

American Orthotic & Prosthetic Association
330 John Carlyle Street, Suite 200
Alexandria, VA 22314
571 431-0876
www.aopanet.org

**Certification**

American Board for Certification in Orthotics & Prosthetics, Inc
330 John Carlyle Street, Suite 200
Alexandria, VA 22314
703 836-7114
www.abcop.org

**Program Accreditation**

Commission on Accreditation of Allied Health Education Programs (CAAHEP) in collaboration with:
National Commission on Orthotic and Prosthetic Education
330 John Carlyle Street, Suite 200
Alexandria, VA 22314
703 836-7114
703 836-0838 Fax
E-mail: info@ncope.org
www.ncope.org

## Orthotist/Prosthetist

# California

**California State University - Dominguez Hills**
Orthotist/Prosthetist Prgm
1000 E Victoria Blvd
Carson, CA 90747
www.csudh.edu/oandp
*Prgm Dir:* Scott Hornbeak, MBA, CPO
*Tel:* 949 643-5374   *Fax:* 949 643-5337
*E-mail:* shornbeak@csudh.edu

# Connecticut

**Hanger Orthopedic Group**
Orthotist/Prosthetist Prgm
dba Newington Certificate Program in Orthotics &
  Prosthetics
181 Patricia M Genova Dr
Newington, CT 06111-1500
www.ncpschool.com
*Prgm Dir:* Robert Lin, CPO FAAOP
*Tel:* 860 667-5361   *Fax:* 860 666-5386
*E-mail:* blin@hanger.com
*Med Dir:* Joseph Smey, EdD

# Georgia

**Georgia Institute of Technology**
Orthotist/Prosthetist Prgm
School of Applied Physiology
281 Ferst Dr
Atlanta, GA 30332-0356
www.ap.gatech.edu/mspo
*Prgm Dir:* Robert J Gregor, PhD
*Tel:* 404 894-3986   *Fax:* 404 894-9982
*E-mail:* robert.gregor@ap.gatech.edu
*Med Dir:* Christopher F Hovorka, MS, CPO

# Illinois

**Northwestern University**
Orthotist/Prosthetist Prgm
345 E Superior St, Rm 1712
Chicago, IL 60611-4496
*Prgm Dir:* Michael Brncick, MEd CPO
*Tel:* 312 238-8006   *Fax:* 312 238-1186
*E-mail:* m-brncick@northwestern.edu

# Minnesota

**Century College**
Orthotist/Prosthetist Prgm
3300 Century Ave N
White Bear Lake, MN 55110
*Prgm Dir:* Edward Haddon, MEd CO
*Tel:* 651 779-5777   *Fax:* 651 779-3343
*E-mail:* e.haddon@century.mnscu.edu

# Texas

**Univ of Texas Southwestern Med Ctr at Dallas**
Orthotist/Prosthetist Prgm
5323 Harry Hines Blvd Ste V5.400
Dallas, TX 75390-9091
www.utsouthwestern.edu/po
*Prgm Dir:* Susan Kapp, BS MEd CPO
*Tel:* 214 648-1580   *Fax:* 214 648-1581
*E-mail:* susan.kapp@utsouthwestern.edu

# Washington

**University of Washington**
Orthotist/Prosthetist Prgm
1959 NE Pacific St
Box 356490
Seattle, WA 98195-6490
http://depts.washington.edu/rehab/po
*Prgm Dir:* Ann Yamane, BS CO LO
*Tel:* 206 616-8586   *Fax:* 206 685-3244
*E-mail:* pando96@u.washington.edu
*Med Dir:* Peter Esselman, MD

## Orthotist/Prosthetist

| Programs* | Class Capacity | Begins | Length (months) | Award | Res Tuition | Non-res Tuition | Stipend | Offers:‡ 1 | 2 | 3 |
|---|---|---|---|---|---|---|---|---|---|---|
| **California** | | | | | | | | | | |
| California State University - Dominguez Hills (Carson) | 16 | Jan and Aug | 7 | Cert, BS | $14,720 | $14,720 | | | | |
| **Connecticut** | | | | | | | | | | |
| Hanger Orthopedic Group (Newington) | 60 | Sep | 9 | Cert | $19,169 | $19,169 | | | | |
| **Georgia** | | | | | | | | | | |
| Georgia Institute of Technology (Atlanta) | 10 | Aug | 24 | MSPO | $6,648 | $24,994 | | | | • |
| **Illinois** | | | | | | | | | | |
| Northwestern University (Chicago) | 45 | Aug Jan | 6 | Cert | $14,388 | $0 | | | | |
| **Texas** | | | | | | | | | | |
| Univ of Texas Southwestern Med Ctr at Dallas | 15 | May | 24 | BS | $3,998 | $13,493 | | | | • |
| **Washington** | | | | | | | | | | |
| University of Washington (Seattle) | 24 | Sep | 20 | BS | $5,985 | $21,285 | | | | |

*Data are shown only for programs that completed the 2006 AMA Survey of Health Professions Education Programs
‡Key to Offers: 1: Evening or weekend classes; 2: Non-English instruction; 3: Cultural competence instruction

# Perfusionist

## Occupational Description

A perfusionist is a skilled person, qualified by academic and clinical education, who operates extracorporeal circulation equipment during any medical situation where it is necessary to support or temporarily replace the patient's circulatory or respiratory function. The perfusionist is knowledgeable concerning the variety of equipment available to perform extracorporeal circulation functions and is responsible, in consultation with the physician, for selecting the appropriate equipment and techniques to be used.

## Job Description

A perfusionist is a skilled allied health professional, trained and educated specifically as a member of an open-heart, surgical team responsible for the selection, setup, and operation of a mechanical device commonly referred to as the heart-lung machine.

During open heart surgery, when the patient's heart is immobilized and cannot function in a normal fashion while the operation is being performed, the patient's blood is diverted and circulated outside the body through the heart-lung machine and returned again to the patient. In effect, the machine assumes the function of both the heart and lungs.

The perfusionist is responsible for operating the machine during surgery, monitoring the altered circulatory process closely, taking appropriate corrective action when abnormal situations arise, and keeping both the surgeon and anesthesiologist fully informed.

In addition to the operation of the heart-lung machine during surgery, perfusionists often function in supportive roles for other medical specialties in operating mechanical devices to assist in the conservation of blood and blood products during surgery, and provide extended, long-term support of patients' circulation outside of the operating room environment.

## Employment Characteristics

Perfusionists primarily work in the operating room during cardiac surgery procedures and may be employed by the hospital, by surgeons, or as employees of a contract independent group practice. The majority of procedures are performed during regular weekly work hours. As critical members of the clinical teams, perfusionists are required to take call and be available for emergency procedures, which can occur at any time. The call schedule is dependant on the number of perfusionists employed by the institution.

## Salary

Perfusionists are well compensated for their services. According to the American Society of Extra-Corporeal Technology (AmSECT), the average base salary range for practicing perfusionists is as follows:

- Recently graduated perfusionist: $60,000-$75,000
- Perfusionist with 2 to 5 years experience: $70,000-$90,000
- Perfusionist with 6 to 10 years experience: $80,000-$100,000
- Perfusionist managers: over $100,000

## Educational Programs

**Length.** Programs are generally 1 to 4 years, depending on the program design, objectives, prerequisites, and student qualifications. Certificate programs require that applicants have a bachelor's degree.

**Prerequisites.** Prerequisites vary depending on the length and design of the program. Most programs require college-level science and mathematics. A background in medical technology, respiratory therapy, or nursing is suggested for some programs.

**Curriculum.** Curricula of accredited programs include courses covering heart-lung bypass for adult, pediatric, and infant patients undergoing heart surgery; long-term supportive extracorporeal circulation; monitoring of the patient undergoing extracorporeal circulation; autotransfusion; and special applications of the technology. Curricula include clinical experience that incorporates and requires performance of an adequate number and variety of circulation procedures.

## Inquiries

**Careers**

American Society of Extra-Corporeal Technology (AmSECT)
National Office
503 Carlisle Drive, Stuie 125
Herndon, VA 20170
703 435-8556
www.amsect.org

American Academy of Cardiovascular Perfusion
PO Box 3596
Allentown, PA 18106-0596
610 395-4853
E-mail: officeAACP@aol.com
http://members.aol.com/OfficeAACP

**Certification**

American Board of Cardiovascular Perfusion
207 N 25th Avenue
Hattiesburg, MS 39401
601 582-2227

**Program Accreditation**

Commission on Accreditation of Allied Health Education Programs (CAAHEP) in collaboration with:
Accreditation Committee - Perfusion Education
6654 S Sycamore Street
Littleton, CO 80120
303 738-0770
303 738-3223 Fax
E-mail: ac-pe@msn.com
www.ac-pe.org

## Perfusionist

# Arizona

**Midwestern University - Glendale Campus**
Perfusion Prgm
19555 N 59th Ave
Glendale, AZ 85308
*Prgm Dir:* Jon W Austin
*Tel:* 623 572-3616  *Fax:* 623 572-3227
*E-mail:* jausti@midwestern.edu
*Med Dir:* James E Meyer, MD

**University of Arizona College of Medicine**
Perfusion Prgm
1501 N Campbell Ave
Room 4402
Tucson, AZ 85724
www.perfusion.arizona.edu
*Prgm Dir:* Douglas F Larson, PhD CCP
*Tel:* 520 626-6339  *Fax:* 520 626-4042
*E-mail:* dflarson@u.arizona.edu
*Med Dir:* Gulshan Sethi, MD

# Connecticut

**Quinnipiac University**
Perfusion Prgm
275 Mt Carmel Ave
Hamden, CT 06518
www.quinnipiac.edu
*Prgm Dir:* Michael J Smith, PhD RRT CCP LP
*Tel:* 203 582-3427  *Fax:* 203 582-8706
*E-mail:* michael.smith@quinnipiac.edu
*Med Dir:* Henry M Rinder, MD

# District of Columbia

**Walter Reed Army Medical Center**
Perfusion Prgm
6900 Georgia Ave NW
Bldg 2, Rm 4655
Washington, DC 20307-5001
*Prgm Dir:* James M Ogletree, MMS PA-C CCP
*Tel:* 202 782-3607  *Fax:* 202 782-8253
*E-mail:* James.Ogletree@NA.AMEDD.ARMY.MIL
*Med Dir:* Robert W Stewart, MD LTC Medical Corps

# Florida

**Barry University**
Perfusion Prgm
11300 NE 2nd Ave
Miami Shores, FL 33161
*Prgm Dir:* Jason L Freed, MS
*Tel:* 800 756-6000, Ext 3214  *Fax:* 305 899-3183
*E-mail:* jfreed@mail.barry.edu
*Med Dir:* Richard A Perryman, MD

# Illinois

**Rush University**
Perfusion Prgm
600 S Paulina Ave
Chicago, IL 60612
www.rush.edu
*Prgm Dir:* Robin Sutton, MS LP (CCP)
*Tel:* 312 942-2305  *Fax:* 312 563-2984
*E-mail:* robin_g_sutton@rush.edu
*Med Dir:* Eric Okum, MD

# Iowa

**University of Iowa Hospitals & Clinics**
Perfusion Prgm
200 Hawkins Dr, 1601 JCP
Iowa City, IA 52242-1086
*Prgm Dir:* Scott Niles, BA CCP
*Tel:* 319 356-8496  *Fax:* 319 353-7174
*E-mail:* scott-niles@uiowa.edu
*Med Dir:* Wayne E Richenbacher, MD

# Massachusetts

**Northeastern University**
Perfusion Prgm
100 Dockser Hall
Boston, MA 02115
*Prgm Dir:* William Gillespie
*Tel:* 617 373-5695  *Fax:* 617 373-2968
*E-mail:* w.gillespie@neu.edu
*Med Dir:* David Morse, MD

# Nebraska

**University of Nebraska Medical Center**
Perfusion Prgm
985155 Nebraska Medical Center
Omaha, NE 68198-5155
*Prgm Dir:* David W Holt, MA CCT NRABT
*Tel:* 402 559-7227  *Fax:* 402 559-6455
*E-mail:* dwholt@unmc.edu
*Med Dir:* John Tinker, MD

# New Jersey

**Cooper University Hospital**
Perfusion Prgm
One Cooper Plaza
310 Sarah Cooper Bldg
Camden, NJ 08103
*Prgm Dir:* Brian Schwartz, BA CCP RN BSN
*Tel:* 856 342-3277  *Fax:* 856 968-8529
*E-mail:* schwartz-brian@cooperhealth.edu
*Med Dir:* Anthony Del Rossi, MD

# New York

**North Shore University Hospital**
*Cosponsor:* LIU CW Post
Perfusion Prgm
School of Cardiovascular Perfusion
225 Community Dr South Entrance
Great Neck, NY 11021
*Prgm Dir:* Richard Chan, BS CCP
*Tel:* 516 918-4356  *Fax:* 516 466-3780
*E-mail:* ehiscvp@aol.com
*Med Dir:* Robert Kalimi, MD

**SUNY Upstate Medical University**
Perfusion Prgm
750 E Adams St
Syracuse, NY 13210
*Prgm Dir:* Bruce Searles, BS CCP
*Tel:* 315 464-6933  *Fax:* 315 464-6914
*E-mail:* Searlesb@upstate.edu
*Med Dir:* Gregory Fink, MD

# Ohio

**Christ Hospital**
Perfusion Prgm
2139 Auburn Ave
Cincinnati, OH 45219
www.health-alliance.com/christ/perfusion_science.html
*Prgm Dir:* Craig Warmuth, MEd CCP
*Tel:* 513 585-1106  *Fax:* 513 585-3241
*E-mail:* warmutcr@healthall.com
*Med Dir:* Donald Mitts, MD

**Cleveland Clinic Foundation**
Perfusion Prgm
9500 Euclid Ave, G-33
Cleveland, OH 44195-5130
www.clevelandclinic.org/heartcenter/perfusion
*Prgm Dir:* Robert Farrow, CCP
*Tel:* 216 444-9215  *Fax:* 216 445-2725
*E-mail:* Farrowr@ccf.org
*Med Dir:* Lee Wallace, MD

**Ohio State University**
Perfusion Prgm
Circulation Technology Division, 152 Atwell Hall
1583 Perry St
Columbus, OH 43210
*Prgm Dir:* Jeffrey B Riley, MHPE CCT
*Tel:* 614 292-7261, Ext 2#  *Fax:* 614 292-0210
*E-mail:* riley.267@osu.edu
*Med Dir:* Glen P Gravlee, MD

# Pennsylvania

**Drexel University**
Perfusion Prgm
College of Nursing & Health Professions
Cardiovascular Perfusion Techology Prgm
Philadelphia, PA 19102-1192
*Prgm Dir:* Robert B Stroud, MS CCP
*Tel:* 215 762-7895  *Fax:* 215 762-1164
*E-mail:* robert.stroud@drexel.edu
*Med Dir:* Rohinton Morris, MD

**U of Pitt Med Ctr Presbyterian Shadyside**
Perfusion Prgm
5230 Centre Ave
Pittsburgh, PA 15232
*Prgm Dir:* Robert Rush, CCP
*Tel:* 412 623-2482  *Fax:* 412 623-1091
*E-mail:* schoolperf@upmc.edu
*Med Dir:* Robert Boretsky, MD

# South Carolina

**Medical University of South Carolina**
Perfusion Prgm
College of Health Professions
151 B Rutledge Ave
Charleston, SC 29425
www.musc.edu/cp
*Prgm Dir:* Joseph J Sistino, CCP
*Tel:* 843 792-2298  *Fax:* 843 792-4417
*E-mail:* sistinoj@musc.edu
*Med Dir:* Arthur J Crumbley, MD

# Tennessee

**Vanderbilt University Medical Center**
Perfusion Prgm
2986 TVC
Nashville, TN 37232-5734
*Prgm Dir:* James J Ramsey, LCP CCP JD
*Tel:* 615 343-9195  *Fax:* 615 333-0742
*E-mail:* james.ramsey@vanderbilt.edu
*Med Dir:* John Byrne, MD

# Texas

**Texas Heart Institute**
Perfusion Prgm
PO Box 20345
Mail Code 1 - 224
Houston, TX 77225
*Prgm Dir:* Terry Crane, BS CCP LP
*Tel:* 832 355-4026  *Fax:* 832 355-8677
*Med Dir:* John R Cooper, MD

# Wisconsin

**Milwaukee School of Engineering**
Perfusion Prgm
1025 N Broadway St
Milwaukee, WI 53202-3109
www.msoe.edu
*Prgm Dir:* Ron Gerrits, PhD
*Tel:* 414 277-7561   *Fax:* 414 277-7465
*E-mail:* gerrits@msoe.edu
*Med Dir:* Alfred J Tector, MD

## Perfusionist

| Programs* | Class Capacity | Begins | Length (months) | Award | Res Tuition | Non-res Tuition | Stipend | Offers:‡ 1 | 2 | 3 |
|---|---|---|---|---|---|---|---|---|---|---|
| **Arizona** | | | | | | | | | | |
| University of Arizona College of Medicine (Tucson) | 4 | Aug | 22 | MS | $2,264 | $9,416 | | | | |
| **Connecticut** | | | | | | | | | | |
| Quinnipiac University (Hamden) | 12 | Aug | 16 | Cert | $18,000 | $18,000 | | | | |
| **District of Columbia** | | | | | | | | | | |
| Walter Reed Army Medical Center (Washington) | 5 | Jul Jan | 18 | Cert | | | | | | |
| **Illinois** | | | | | | | | | | |
| Rush University (Chicago) | 8 | Sep | 21 | BS, MS | $22,515 | $22,515 | | | | |
| **Iowa** | | | | | | | | | | |
| University of Iowa Hospitals & Clinics (Iowa City) | 5 | Aug | 20 | Cert | $7,340 | $18,358 | | | | |
| **New York** | | | | | | | | | | |
| North Shore University Hospital (Great Neck) | 11 | Sep | 24 | Cert, MS | $20,500 | $20,500 | | • | | |
| **Ohio** | | | | | | | | | | |
| Christ Hospital (Cincinnati) | 6 | Sep | 21 | Cert | $14,000 | $14,000 | | | | |
| Cleveland Clinic Foundation | 8 | Jan | 18 | Cert | $0 | $10,000 | | | | • |
| **South Carolina** | | | | | | | | | | |
| Medical University of South Carolina (Charleston) | 12 | Aug | 21 | BS | $10,498 | $28,278 | | | | |
| **Texas** | | | | | | | | | | |
| Texas Heart Institute (Houston) | 8 | Jan Jul | 12 | Cert | $15,000 | $15,000 | | | | |
| **Wisconsin** | | | | | | | | | | |
| Milwaukee School of Engineering | 8 | Sep | 21 | MSP | $20,000 | $20,000 | | | | |

*Data are shown only for programs that completed the 2006 AMA Survey of Health Professions Education Programs
‡Key to Offers: 1: Evening or weekend classes; 2: Non-English instruction; 3: Cultural competence instruction

# Pharmacy Technician

## Occupational Description

Pharmacy technicians assist licensed pharmacists by performing duties that do not require the professional skills and judgment of a licensed pharmacist and assisting in those duties that require the expertise of a pharmacist. Pharmacy technicians are employed in every practice setting where pharmacy is practiced, including institutional, community, home care, long-term care, mail order, and managed care pharmacies. Technicians are also employed in education, research, and the pharmaceutical industry.

Technicians may be trained on the job or by completing a formal program. Some formal training programs meet the program accreditation standards established by the American Society of Health-System Pharmacists. After completing their training, technicians may become a Certified Pharmacy Technician (CPhT) by successfully taking the national certification examination offered by the Pharmacy Technician Certification Board.

## Job Description

According to the 1991-1994 Scope of Pharmacy Practice Project, pharmacy technicians spend their time in the following ways:

- 26%—collect, organize, and evaluate information to assist pharmacists in serving patients
- 21%—develop and manage medication distribution and control systems; about half of this time is spent preparing, dispensing, distributing, and administering medications
- 7%—provide drug information and education

These percentages, however, may vary widely for many reasons, including the wide range of training and qualifications of pharmacists, the use of technicians as directed by a given supervisory pharmacist, and variations in state pharmacy practice laws.

The ASHP Accreditation Standard for Pharmacy Technician Training Programs specifies that graduates of programs should be able to perform the following functions (among others):

- Assist the pharmacist in collecting, organizing, and evaluating information for direct patient care, drug use review, and departmental management
- Receive and screen prescription medication orders for completeness and accuracy
- Use pharmaceutical and medical terms, abbreviations, and symbols appropriately
- Prepare and distribute medications in a variety of health system settings
- Perform arithmetical calculations required for usual dosage determinations and solutions preparation
- Use knowledge of general chemical and physical properties of drugs in manufacturing and packaging operations
- Use knowledge of proper aseptic technique and packaging in the preparation of medications
- Collect payment and/or initiate billing for pharmacy services and goods
- Purchase pharmaceuticals, devices, and supplies according to an established plan in a variety of health systems
- Control medication, equipment, and device inventory according to an established plan in a variety of health systems
- Maintain pharmacy equipment in preparing, storing, and distributing investigational drug products

- Assist the pharmacist in monitoring the practice site and/or service area for compliance with federal, state, and local laws, regulations, and professional standards
- Assist the pharmacist in preparing, storing, and distributing investigational drug products
- Assist the pharmacist in the monitoring of drug therapy
- Assist the pharmacist in identifying patients who desire counseling on the use of medications, equipment, and devices
- Understand the use and side effects of prescription and nonprescription drugs used to treat common disease states
- Appreciate the need to adapt the delivery of pharmacy services for the culturally diverse
- Maintain confidentiality of patient information
- Communicate clearly orally and in writing
- Use computers to perform pharmacy functions
- Demonstrate ethical conduct in all activities related to the delivery of pharmacy services

## Employment Characteristics

Pharmacy technicians typically provide their services in one or more of the following settings: health systems, community pharmacies, chain pharmacies, and home care pharmacies.

## Educational Programs

**Length.** Programs are generally 15 weeks or longer and consist of a minimum of 600 hours of training (contact) time. Graduates generally receive a certificate or AS degree.

**Prerequisites.** Applicants should have a high school diploma or equivalent and meet institutional entrance requirements.

**Curriculum.** The professional curriculum includes formal instruction in didactic, practical, and laboratory areas of pharmacy practice. The curriculum consists of various aspects of pharmacy technician training pertinent to contemporary pharmacy practice. Courses include pharmacy mathematics/calculations, pharmacy for pharmacy technicians, sterile products, pharmaceutical care delivery systems, computer systems for pharmacy, and payment for pharmacy services.

## Inquiries

**Careers**

American Association of Pharmacy Technicians (AAPT)
PO Box 1447
Greensboro, NC 27402
877 368-4771
336 333-9068 Fax
www.pharmacytechnician.com

National Pharmacy Technician Association
3707 FM 1960 RD W, Suite 460
Houston, TX 77068
281 866-7900 or 888 247-8700
281 895-7320 Fax
www.pharmacytechnician.org

**Certification**

Pharmacy Technician Certification Board
1100 15th Street NW, Suite 730
Washington, DC 20005-1707
800 363-8012
202 429-7596 Fax
www.ptcb.org

**Program Accreditation**

American Society of Health System-Pharmacists
Accreditation Services Division
7272 Wisconsin Avenue
Bethesda, MD 20814
301 664-8720
301 664-8872 Fax
E-mail: llifshin@ashp.org

## Pharmacy Technician*

# Alabama

**George C Wallace Community College**
Pharmacy Technician Prgm
801 Main St NW
PO Box 2000
Hanceville, AL 35077
*Prgm Dir:* Brandon Brooks
*Tel:* 256 352-8023   *Fax:* 256 352-8382
*E-mail:* brandon.brooks@wallacestate.edu

# Arizona

**Pima Community College**
Pharmacy Technician Prgm
8181 E Irvington Rd
Tucson, AZ 85709-4000
*Prgm Dir:* Mary Ann Jordan
*Tel:* 520 206-7850   *Fax:* 520 206-7622
*E-mail:* mary.jordan@pima.edu

# California

**North Orange County Community**
Pharmacy Technician Prgm
Continuing Education
1830 W Romneya Dr
Anaheim, CA 92801
www.nocccd.cc.ca.us
*Prgm Dir:* Martha Gutierrez
*Tel:* 714 808-4668   *Fax:* 714 808-4659
*E-mail:* mgutierrez@sce.cc.ca.us

**Western Career College - Antioch**
Pharmacy Technician Prgm
2157 Country Hills Dr
Antioch, CA 94509
www.westerncollege.com
*Prgm Dir:* Sandeep Bansal, RPT
*Tel:* 925 522-7777   *Fax:* 925 522-0097
*E-mail:* sbansal@ant.westerncollege.com

**Western Career College - Citrus**
Pharmacy Technician Prgm
7301 Greenback Ln, Ste A
Citrus Heights, CA 95621
www.westerncollege.com
*Prgm Dir:* Kari Roberts
*Tel:* 916 722-8200   *Fax:* 916 722-6883
*E-mail:* kroberts@westerncollege.com

**Western Career College - Emeryville**
Pharmacy Technician Prgm
1400 65th St
Emeryville, CA 94608
*Prgm Dir:* Monique Oatis
*Tel:* 510 601-0133   *Fax:* 510 601-0793
*E-mail:* moatis@emv.westerncollege.com

*Additional information about programs that
returned the AMA's survey is available in a table
beginning on page 311.

**North-West College - Glendale**
Pharmacy Technician Prgm
124 S Glendale Ave
Glendale, CA 91205
*Prgm Dir:* Marsha Fuerst
*Tel:* 626 960-5046   *Fax:* 626 960-5949
*E-mail:* marshaf@northwestcollege.com

**American University of Health Sciences**
Pharmacy Technician Prgm
3501 Atlantic Ave
Long Beach, CA 90807
www.aihs.edu
*Prgm Dir:* Jeanetta Mastron, CPhT BS
*Tel:* 562 988-2278
*E-mail:* jmastron@auhs.edu

**American Career College**
Pharmacy Technician Prgm
4021 Rosewood Ave 101
Los Angeles, CA 90004
*Prgm Dir:* Lito Cabrera
*Tel:* 323 784-1102   *Fax:* 323 668-7548
*E-mail:* lito@americancareer.com

**Charles R Drew Univ of Med & Science**
Pharmacy Technician Prgm
1731 E 120th St
Los Angeles, CA 90059
www.cdrewu.edu
*Prgm Dir:* Gail Orum Alexander, PharmD
*Tel:* 323 563-4815   *Fax:* 323 563-4827
*E-mail:* gailorum@cdrewu.edu

**Cerritos Community College**
Pharmacy Technician Prgm
11110 Alondra Blvd
Norwalk, CA 90650
www.cerritos.edu/ho
*Prgm Dir:* Hal Malkin, CPhT
*Tel:* 562 860-2451, Ext 3517   *Fax:* 562 947-4228
*E-mail:* hmalkin@Cerritos.edu

**Foothill Community College**
Pharmacy Technician Prgm
Middlefield Campus
4000 Middlefield Rd, Ste I
Palo Alto, CA 94303-4739
www.foothill.edu
*Prgm Dir:* Leonis A Osterdock, RPh
*Tel:* 650 949-6970   *Fax:* 650 949-6979
*E-mail:* mckellarcharlie@foothill.edu

**North-West College - Pasadena**
Pharmacy Technician Prgm
530 East Union
Pasadena, CA 91101
www.northwestcollege.com
*Prgm Dir:* Marsha Fuerst
*Tel:* 626 960-5046
*E-mail:* marshaf@northwestcollege.com

**Western Career College - Pleasant Hill**
Pharmacy Technician Prgm
380 Civic Dr, Ste 300
Pleasant Hill, CA 94523
www.westerncollege.com
*Prgm Dir:* Diana Noack
*Tel:* 925 405-0632   *Fax:* 925 609-6666
*E-mail:* dnoack@westerncollege.com

**North-West College - Pomona**
Pharmacy Technician Prgm
134 W Holt Ave
Pomona, CA 91768
*Prgm Dir:* Marsha Fuerst
*Tel:* 909 623-1552
*E-mail:* marshaf@northwestcollege.com

**Charles A Jones Skills & Business Ed Center**
Pharmacy Technician Prgm
5451 Lemon Hill Ave
Sacramento, CA 95824-1529
www.scusd.edu
*Prgm Dir:* Sandra B Thi Huynh
*Tel:* 916 433-2600, Ext 1400   *Fax:* 916 433-2640
*E-mail:* sandrahu@sac-city.k12.ca.us

**Western Career College - Sacramento**
Pharmacy Technician Prgm
8909 Folsom Blvd
Sacramento, CA 95826
www.westerncollege.edu
*Prgm Dir:* Hieu Nguyen, CPHT BS
*Tel:* 916 361-1660, Ext 41112   *Fax:* 916 361-6666
*E-mail:* hieun@westerncollege.com

**Western Career College - San Jose**
Pharmacy Technician Prgm
6201 San Ignacio Ave
San Jose, CA 95119
*Prgm Dir:* Valerie L Young
*Tel:* 408 360-0840, Ext 334   *Fax:* 408 360-0848
*E-mail:* vyoung@sj.westerncollege.com

**Western Career College - San Leandro**
Pharmacy Technician Prgm
170 Bayfair Mall
San Leandro, CA 94578
*Prgm Dir:* Marlene Lamnin, BS
*Tel:* 510 276-3888, Ext 123
*E-mail:* mlamnin@westerncollege.com

**HealthStaff Training Institute**
Pharmacy Technician Prgm
1505 E 17th St, Ste 122
Santa Ana, CA 92701
*Prgm Dir:* Judee R Tompkins
*Tel:* 714 543-9828   *Fax:* 714 543-9835
*E-mail:* dstan37@msn.com

**Santa Ana College**
Pharmacy Technician Prgm
1530 W 17th St
Santa Ana, CA 92706
*Prgm Dir:* Gail B Askew, PharmD FCSHP
*Tel:* 714 564-6622   *Fax:* 714 564-6158
*E-mail:* askew_gail@sac.edu

**Western Career College - Stockton**
Pharmacy Technician Prgm
1313 West Robinhood Dr, Ste B
Stockton, CA 95207
www.westerncollege.edu
*Prgm Dir:* Cindy Turner, CPhT
*Tel:* 209 956-1240, Ext 44120   *Fax:* 209 956-1244
*E-mail:* cturner@westerncollege.com

**North-West College - West Covina**
Pharmacy Technician Prgm
2121 Garvey Ave N
West Covina, CA 91790
*Prgm Dir:* Marsha Fuerst
*Tel:* 626 870-4113
*E-mail:* marshaf@northwestcollege.com

# Colorado

**Arapahoe Community College**
Pharmacy Technician Prgm
5900 S Santa Fe Dr
Littleton, CO 80160-9002
*Prgm Dir:* Sue Treitz, BSN MA
*Tel:* 303 797-5963    *Fax:* 303 797-5935
*E-mail:* sue.treitz@arapahoe.edu

**Front Range Comm College**
Pharmacy Technician Prgm
Westminster Campus
3645 W 112th Ave
Westminster, CO 80030
www.frontrange.edu
*Prgm Dir:* Linda Calvert, CPhT BS
*Tel:* 303 404-5393    *Fax:* 303 404-2178
*E-mail:* linda.calvert@frontrange.edu

# Connecticut

**Briarwood College**
Pharmacy Technician Prgm
2279 Mount Vernon Rd
Southington, CT 06489
www.briarwood.edu
*Prgm Dir:* Cynthia Enright, RPh
*Tel:* 860 628-4751, Ext 177    *Fax:* 860 628-6444
*E-mail:* enrightc@briarwood.edu

# Florida

**McFatter Vocational Technical Center**
Pharmacy Technician Prgm
6500 Nova Dr
Davie, FL 33317
*Prgm Dir:* William J Shaheen, PharmD RPh
*Tel:* 754 321-5741    *Fax:* 754 321-5820
*E-mail:* william.shaheen@browardschools.com

**Lake City Community College**
Pharmacy Technician Prgm
149 SE College Place
Lake City, FL 32025-8703
www.lakecitycc.edu/departments/pharmtech
*Prgm Dir:* Patty Smith, CPhT
*Tel:* 386 754-4239    *Fax:* 386 754-4739
*E-mail:* smithp@lakecitycc.edu

**Florida Metropolitan Univ - Melbourne Campus**
Pharmacy Technician Prgm
2401 N Harbor City Blvd
Melbourne, FL 32935
*Prgm Dir:* Karen Jenkins, MD MPH
*Tel:* 321 253-2929    *Fax:* 321 255-2017
*E-mail:* kjenkins@cci.edu

**Pinellas Tech Educ Ctr - St Petersburg**
Pharmacy Technician Prgm
901 34th St S
St Petersburg, FL 33711
www.myptec.org
*Prgm Dir:* Jeannie Pappas, CPhT
*Tel:* 813 893-2500, Ext 1131    *Fax:* 813 323-6653
*E-mail:* jpappas@ptec.pinellas.k12.fl.us
*Med Dir:* Richard Witas, Advisor Chair, RPh Assist Dir
HLee Moffitt Cancer Center

**Henry W Brewster Technical Center**
Pharmacy Technician Prgm
2222 N Tampa St
Tampa, FL 33602
*Prgm Dir:* Douglas J Botting, RPh
*Tel:* 813 276-5448    *Fax:* 813 276-5756
*E-mail:* bottingd@mail.brewstertech.org

# Georgia

**Ogeechee Technical College**
Pharmacy Technician Prgm
1 Joe Kennedy Blvd
Statesboro, GA 30458
*Prgm Dir:* Shelly P Jones, CPhT
*Tel:* 912 486-7620    *Fax:* 912 486-7604
*E-mail:* sjones@ogeecheetech.edu

**Southwest Georgia Technical College**
Pharmacy Technician Prgm
15689 US Hwy 19 N
Thomasville, GA 31799
*Prgm Dir:* Kathleen G Allen, BS RPh
*Tel:* 912 225-5093    *Fax:* 912 225-5289
*E-mail:* kallen@swgtc.net

**Valdosta Technical College**
Pharmacy Technician Prgm
4089 Val Tech Rd
Valdosta, GA 31602
*Prgm Dir:* Shelley Swafford
*Tel:* 229 259-5581    *Fax:* 229 259-5567
*E-mail:* swafford@valdostatech.org

**Southeastern Technical College**
Pharmacy Technician Prgm
3001 E First St
Vidalia, GA 30474
*Prgm Dir:* Karen Davis, CPhT
*Tel:* 912 538-3192    *Fax:* 912 538-3106
*E-mail:* kdavis@southeasterntech.edu

# Illinois

**Malcolm X College**
Pharmacy Technician Prgm
1900 W Van Buren St
Ste 3524
Chicago, IL 60612-3145
www.ccc.edu
*Prgm Dir:* Ronald D Grimmette, MAd Ed BS Pharm
*Tel:* 312 850-7385    *Fax:* 312 850-7453
*E-mail:* rgrimmette@ccc.edu

**Blessing Hospital**
Pharmacy Technician Prgm
Broadway and 14th St
Quincy, IL 62301
www.blessinghospital.com
*Prgm Dir:* Patty Loeffler, RN BSN
*Tel:* 217 223-8400, Ext 4832    *Fax:* 217 223-6003
*E-mail:* ploeffler@blessinghospital.com

**South Suburban College**
Pharmacy Technician Prgm
15800 S State St
South Holland, IL 60473
*Prgm Dir:* Jan Keresztes, PharmD
*Tel:* 708 596-2000, Ext 2432    *Fax:* 708 210-5792
*E-mail:* jkeresztes@southsuburbancollege.edu

# Indiana

**Clarian Health Partners Inc**
Pharmacy Technician Prgm
Clarian Health Education Center
Room 506 - Wile Hall - PO Box 1367
Indianapolis, IN 46206-1397
www.clarian.org
*Prgm Dir:* Mary E Mohr, RPh MS
*Tel:* 317 962-0919    *Fax:* 317 962-3483
*E-mail:* mmohr@clarian.org
*Med Dir:* Cheryl Oliver, MS, MT (ASCP)

# Louisiana

**Louisiana State University - Alexandria**
Pharmacy Technician Prgm
8100 Hwy 71 South
Alexandria, LA 71302
www.lsua.edu
*Prgm Dir:* David J Nassif, Pharm D
*Tel:* 318 473-6533    *Fax:* 318 473-6567
*E-mail:* dnassif@lsua.edu

**Bossier Parish Community College**
Pharmacy Technician Prgm
6220 East Texas St
Science and Allied Health Division
Bossier City, LA 71111
*Prgm Dir:* Terri Mundy, RPh
*Tel:* 318 678-6215    *Fax:* 318 678-6199
*E-mail:* tmundy@bpcc.edu

**Delgado Community College**
Pharmacy Technician Prgm
Allied Health Div
615 City Park Ave
New Orleans, LA 70119
*Prgm Dir:* Debbie C Kern, MEd
*Tel:* 504 483-4308    *Fax:* 504 483-4308
*E-mail:* dckern@dcc.edu

# Maryland

**Anne Arundel Community College**
Pharmacy Technician Prgm
101 College Pkwy
Florestano 306B
Arnold, MD 21012
*Prgm Dir:* Stephanie Smith-Baker
*Tel:* 410 777-7497    *Fax:* 410 777-7099
*E-mail:* sesmithbaker@aacc.edu

# Massachusetts

**Holyoke Community College**
Pharmacy Technician Prgm
303 Homestead Ave
Holyoke, MA 01040-1099
*Prgm Dir:* Diane Pacitti
*Tel:* 413 552-2263    *Fax:* 413 552-2045
*E-mail:* dpacitti@hcc.mass.edu

# Michigan

**Washtenaw Community College**
Pharmacy Technician Prgm
4800 E Huron River Dr
PO Box D-1
Ann Arbor, MI 48106
*Prgm Dir:* Dina Cheiman
*Tel:* 734 973-3418    *Fax:* 734 677-5078
*E-mail:* dcheiman@wccnet.edu

**Henry Ford Community College**
Pharmacy Technician Prgm
5101 Evergreen, HCEC Bldg
Dearborn, MI 48128-1495
*Prgm Dir:* Scott McCall, RPh
*Tel:* 313 845-9877
*E-mail:* mccalls@oakwood.org
*Med Dir:* Theresa Mozug, BS CPhT

**Henry Ford Hospital**
Pharmacy Technician Prgm
Wayne County Community College District
1001 W Fort St
Detroit, MI 48226
*Prgm Dir:* Dick Kuschinsky, RPh
*Tel:* 313 496-2686   *Fax:* 313 961-9648
*E-mail:* dkuschi1@wcccd.edu
*Med Dir:* Debraha Watson, PhD

## Minnesota

**Northland Community & Technical College**
Pharmacy Technician Prgm
2022 Central Ave NE
PO Box 111
East Grand Forks, MN 56721
*Prgm Dir:* Joe Farrell, RPh
*Tel:* 218 773-4528   *Fax:* 218 773-4502
*E-mail:* jfarrell@altru.org

**Minnesota State Comm & Tech Coll - Moorhead**
Pharmacy Technician Prgm
405 SW Colfax, PO Box 566
Wadena, MN 56482-0566
*Prgm Dir:* Joe Farrell, RPh
*Tel:* 218 773-3441, Ext 414
*E-mail:* jfarrell@altru.org

**Century College**
Pharmacy Technician Prgm
3300 Century Ave N
White Bear Lake, MN 55110
www.century.edu
*Prgm Dir:* Raymond K Vellenga
*Tel:* 651 779-5780   *Fax:* 651 779-5780
*E-mail:* ray.vellenga@century.edu

## Mississippi

**Jones County Junior College**
Pharmacy Technician Prgm
900 S Court St
Ellisville, MS 39437
*Prgm Dir:* Marsha M Sanders, RPh
*Tel:* 601 477-4230   *Fax:* 601 477-4152
*E-mail:* marsha.sanders@jcjc.edu

## Montana

**College of Tech - Univ of Montana-Missoula**
Pharmacy Technician Prgm
909 S Ave W
Missoula, MT 59801
www.cte.umt.edu
*Prgm Dir:* Mary McHugh, RPh
*Tel:* 406 243-7813   *Fax:* 406 243-7899
*E-mail:* mary.mchugh@mso.umt.edu

## North Carolina

**Durham Technical Community College**
Pharmacy Technician Prgm
1637 Lawson St
Durham, NC 27703
www.durhamtech.edu
*Prgm Dir:* Joe Anne Griffith, BS RPh
*Tel:* 919 686-3686, Ext n/a   *Fax:* 919 686-3705
*E-mail:* griffithj@durhamtech.edu
*Med Dir:* Margaret Skulnik, MS Nursing

## North Dakota

**North Dakota State College of Science**
Pharmacy Technician Prgm
800 6th St N
Wahpeton, ND 58076-0002
www.ndscs.nodak
*Prgm Dir:* Kenneth Strandberg, MBA RPh
*Tel:* 701 671-2114   *Fax:* 701 671-2570
*E-mail:* Kenneth.Strandberg@ndsu.edu

## Ohio

**Collins Career Center**
Pharmacy Technician Prgm
11627 State Route 243
Chesapeake, OH 45619
*Prgm Dir:* Anthony Womack, RPh
*Tel:* 740 867-6641
*E-mail:* tony.womack@omnicare.com

**Cuyahoga Community College**
Pharmacy Technician Prgm
East Campus, Health Careers and Sciences
4250 Richmond Rd
Cleveland, OH 44122
www.tri-c.edu/ptech
*Prgm Dir:* MaryAnn Stuhan, RPh
*Tel:* 216 987-2381
*E-mail:* maryann.stuhan@tri-c.edu

## Pennsylvania

**Great Lakes Institute of Technology**
Pharmacy Technician Prgm
5100 Peach St
Erie, PA 16509
www.glit.edu
*Prgm Dir:* Jodie Corwin, CPhT
*Tel:* 814 864-6666, Ext 236   *Fax:* 814 868-1717
*E-mail:* jodiec@glit.edu

**Western School of Health & Business - Monroeville**
Pharmacy Technician Prgm
1 Monroeville Center, Ste 250
Monroeville, PA 15146
*Prgm Dir:* Peter E DePascale, CPhT MBA
*Tel:* 412 373-9038, Ext 271   *Fax:* 412 373-2544
*E-mail:* pdepascale@western-school.com

**Bidwell Training Center**
Pharmacy Technician Prgm
1650 Metropolitan St
Pittsburgh, PA 15233
*Prgm Dir:* Dolores R Sewchok
*Tel:* 412 323-4000, Ext 215   *Fax:* 412 322-6539
*E-mail:* dsewchok@mcg-btc.org

**Comm Coll of Allegheny Cnty - South Campus**
Pharmacy Technician Prgm
1750 Clairton Rd, Rte 885
West Mifflin, PA 15122
*Prgm Dir:* Jane Coughanour, MT ASCP MEd
*Tel:* 412 469-6280
*E-mail:* jcoughanour@ccac.edu

## South Carolina

**Trident Technical College**
Pharmacy Technician Prgm
7000 Rivers Ave
PO Box 118067
Charleston, SC 29423-8067
*Prgm Dir:* Karen Snipe, CPhT MAEd
*Tel:* 843 574-6481   *Fax:* 843 574-6585
*E-mail:* Karen.Snipe@tridenttech.edu

**Midlands Technical College**
Pharmacy Technician Prgm
PO Box 2408
Columbia, SC 29202
*Prgm Dir:* Don A Ballington, MS
*Tel:* 803 822-3591   *Fax:* 803 822-3417
*E-mail:* ballingtond@midlandstech.edu
*Med Dir:* Deborah Tapley, RPh

**Greenville Technical College**
Pharmacy Technician Prgm
PO Box 5616
Greenville, SC 29606-5616
*Prgm Dir:* Paul Wagner, RPh
*Tel:* 864 250-5073
*E-mail:* paul.wagner@gvltec.edu

**Spartanburg Community College**
Pharmacy Technician Prgm
PO Box 4386
Spartanburg, SC 29305
*Prgm Dir:* Julia B Sherwood, CPht
*Tel:* 864 592-4870   *Fax:* 864 592-4881
*E-mail:* sherwoodj@stcsc.edu

## South Dakota

**Western Dakota Technical Institute**
Pharmacy Technician Prgm
800 Mickelson Dr
Rapid City, SD 57703
*Prgm Dir:* Debborah Cummings, CPhT
*Tel:* 605 394-4034, Ext 308   *Fax:* 605 394-1789
*E-mail:* debborah.cummings@wdti.tec.sd.us

## Tennessee

**Chattanooga State Technical Comm College**
Pharmacy Technician Prgm
4501 Amnicola Hwy
Chattanooga, TN 37406
*Prgm Dir:* Nancy V Watts
*Tel:* 423 697-2568   *Fax:* 423 697-2595
*E-mail:* NANCY.WATTS@chattanoogastate.edu

**Tennessee Technology Center - Jackson**
Pharmacy Technician Prgm
2468 Technology Center Dr
Jackson, TN 38301
www.jackson.tec.tn.us
*Prgm Dir:* Darlene Redd, CPhT
*Tel:* 731 424-0691, Ext 115   *Fax:* 731 424-0807
*E-mail:* dredd@jackson.tec.tn.us

**Concorde Career College - Memphis**
Pharmacy Technician Prgm
5100 Poplar Ave, Ste 132
Memphis, TN 38137
www.concorde.edu
*Prgm Dir:* Sunethra Guy, CPht BA
*Tel:* 901 761-9494, Ext 216   *Fax:* 901 761-3293
*E-mail:* Sguy@concorde.edu
*Med Dir:* Joyce Broyles, Pharm D

**Tennessee Technology Center - Memphis**
Pharmacy Technician Prgm
550 Alabama Ave
Memphis, TN 38105-3604
*Prgm Dir:* Olivia Bowden
*Tel:* 901 543-6100   *Fax:* 901 543-6197
*E-mail:* Olivia.Bowden@memphis.tec.tn.us

**Walters State Community College**
Pharmacy Technician Prgm
500 S Davy Crockett Pkwy
Morristown, TN 37813-6899
*Prgm Dir:* Michelle J Dalton, PharmD
*Tel:* 423 585-6981   *Fax:* 423 585-6955
*E-mail:* michelle.dalton@ws.edu

**Tennessee Technology Center - Murfreesboro**
Pharmacy Technician Prgm
1303 Old Fort Pkwy
Murfreesboro, TN 37129
*Prgm Dir:* Jill Frost, BS CPhT
*Tel:* 615 898-8010, Ext 127    *Fax:* 615 893-4194
*E-mail:* jfrost@murfreesboro.tec.tn.us

**Tennessee Technology Center - Nashville**
Pharmacy Technician Prgm
100 White Bridge Rd
Nashville, TN 37209
*Prgm Dir:* Daniel H Leffler, RPh
*Tel:* 615 284-4076
*E-mail:* dan.leffler@baptisthospital.com

# Texas

**Austin Community College**
Pharmacy Technician Prgm
3401 Webberville Rd
Austin, TX 78702
*Prgm Dir:* Jason Sparks, AAS BA CPhT
*Tel:* 512 223-5949    *Fax:* 512 223-5895
*E-mail:* jsparks3@austincc.edu

**Richland College**
Pharmacy Technician Prgm
12800 Abrams Rd
Dallas, TX 75243-2199
www.richlandcollege.edu/hp
*Prgm Dir:* Jan Parrish, MEd MT(ASCP)
*Tel:* 972 238-6376    *Fax:* 972 761-6793
*E-mail:* janparrish@dcccd.edu

**El Paso Community College**
Pharmacy Technician Prgm
PO Box 20500
El Paso, TX 79998
www.epcc.edu
*Prgm Dir:* Nader Rassaei, MD
*Tel:* 915 831-8836    *Fax:* 915 831-8861
*E-mail:* naderr@epcc.edu
*Med Dir:* Rickardo Hernandez, RPh

**US Army Medical Dept Center & School**
Pharmacy Technician Prgm
MCCS-HCP (Pharmacy Branch)
3151 Scott Rd
Ft Sam Houston, TX 78234-6137
*Prgm Dir:* LTC Jennifer R Styles, MS USA
*Tel:* 210 221-7553
*E-mail:* jennifer.styles1@cen.amedd.army.mil

**University of Texas Medical Branch/SAHS**
Pharmacy Technician Prgm
301 Unversity Blvd, Rte G01
Galveston, TX 77555-0701
*Prgm Dir:* Madeline F Jensen, BSEd
*Tel:* 409 772-6981
*E-mail:* mfjensen@utmb.edu

**Houston Community College**
Pharmacy Technician Prgm
1900 Galen Dr
Houston, TX 77030
*Prgm Dir:* Liz Johnson-Wilroy, BHS
*Tel:* 713 718-7352    *Fax:* 713 718-7608
*E-mail:* liz.wilroy@hccs.edu

**North Harris College**
Pharmacy Technician Prgm
2700 W W Thorne Dr
Houston, TX 77073
www.ashp.org/directories/technicians
*Prgm Dir:* Nancy L Lim, PhD RPh
*Tel:* 281 618-5727    *Fax:* 281 618-5756
*E-mail:* nancy.l.lim@nhmccd.edu
*Med Dir:* Vivian Lilly, PhD, MBA, RN

**San Jacinto College North**
Pharmacy Technician Prgm
5800 Uvalde Rd
Houston, TX 77049
*Prgm Dir:* Donald Becker, CPhT
*Tel:* 281 998-6150, Ext 7453    *Fax:* 281 459-7603
*E-mail:* donald.becker@sjcd.edu

**San Jacinto College South**
Pharmacy Technician Prgm
13735 Beamer Rd
Houston, TX 77089
*Prgm Dir:* Karen Velez, BSM AA CPhT
*Tel:* 281 929-4613    *Fax:* 281 922-3487
*E-mail:* karen.velez@sjcd.edu

**Angelina College**
Pharmacy Technician Prgm
3500 S First (Hwy 595)
PO Box 1768
Lufkin, TX 75902-1768
www.angelina.edu
*Prgm Dir:* Elaine Young, MEd CPhT
*Tel:* 936 633-5433    *Fax:* 936 633-5241
*E-mail:* eyoung@angelina.edu

**South Texas College**
Pharmacy Technician Prgm
PO Box 9701
McAllen, TX 78502-9701
*Prgm Dir:* Theresa M Langlass-Garza, PharmD BCPS
*Tel:* 956 872-3024    *Fax:* 956 872-3138
*E-mail:* tmgarza@southtexascollege.edu

**South Texas Vo-Tech - McAllen**
Pharmacy Technician Prgm
2400 Daffodil St
McAllen, TX 78501
*Prgm Dir:* Lydia R Chavana, MT (ASCP) CLS (NCA)
*Tel:* 956 969-1564    *Fax:* 956 969-1887
*E-mail:* lydia@stvt.edu

**Lamar State College - Orange**
Pharmacy Technician Prgm
410 Front St
Orange, TX 77630
www.lsco.edu
*Prgm Dir:* Randy G Ford
*Tel:* 409 882-3035    *Fax:* 409 882-5000
*E-mail:* Randy.Ford@lsco.edu

**Northwest Vista College**
Pharmacy Technician Prgm
3535 N Ellison Dr
San Antonio, TX 78251
*Prgm Dir:* Jose L Egremy-Hernandez
*Tel:* 210 348-2090    *Fax:* 210 348-2264
*E-mail:* jegremy@accd.edu

**USAF School of Health Care Sciences**
Pharmacy Technician Prgm
382d Training Squadron
917 Missile Rd, Ste 3
Sheppard AFB, TX 76311
*Prgm Dir:* T Shawn Garten, PharmD BCPS MAJ BSC
*Tel:* 940 676-3847    *Fax:* 940 676-3850
*E-mail:* Timothy.Garten@sheppard.af.mil

**Weatherford College**
Pharmacy Technician Prgm
225 College Park Dr
Weatherford, TX 76086
*Prgm Dir:* Gary L Wheat, RPh
*Tel:* 817 594-5471, Ext 229    *Fax:* 817 598-6455
*E-mail:* austin@wc.edu

**South Texas Vo-Tech - Weslaco**
Pharmacy Technician Prgm
2419 E Haggar Ave
Weslaco, TX 78596
www.stvt.edu
*Prgm Dir:* Lydia R Chavana, MT(ASCP) CLS(NCA)
*Tel:* 956 969-1564    *Fax:* 956 969-1887
*E-mail:* lydia@stvt.edu

**Vernon College**
Pharmacy Technician Prgm
4105 Maplewood Dr
Wichita Falls, TX 76308
www.vernoncollege.edu
*Prgm Dir:* Michelle Wood, BS CPhT
*Tel:* 940 696-8752, Ext 3211    *Fax:* 940 689-9129
*E-mail:* mwood@vernoncollege.edu

# Virginia

**Northern Virginia Community College**
Pharmacy Technician Prgm
8333 Little River Turnpike
Annandale, VA 22003
*Prgm Dir:* Debra P Powell
*Tel:* 703 323-3280    *Fax:* 703 323-4576
*E-mail:* dpowell@nvcc.edu

**Naval School of Health Sciences**
Pharmacy Technician Prgm
1001 Holcomb Rd
Portsmouth, VA 23708-5200
*Prgm Dir:* Lt Sharon J Roberts, PharmD
*Tel:* 757 953-6413    *Fax:* 757 953-5033
*E-mail:* SJRoberts@hsp.med.navy.mil

# Washington

**Renton Technical College**
Pharmacy Technician Prgm
3000 NE Fourth St
Renton, WA 98056
*Prgm Dir:* Lauri Masel
*Tel:* 425 235-2352, Ext 5552    *Fax:* 425 235-7832
*E-mail:* lmasel@rtc.ctc.edu

**Spokane Community College**
Pharmacy Technician Prgm
N 1810 Greene St, MS 2090
Spokane, WA 99217-5399
www.scc.spokane.edu/fac/stschritter
*Prgm Dir:* Sandi Tschritter, CPhT BA
*Tel:* 509 533-8199    *Fax:* 509 533-8621
*E-mail:* stschritter@scc.spokane.edu
*Med Dir:* Brigitte Palmer, RPh

**St Joseph Medical Center**
Pharmacy Technician Prgm
1717 South J St
PO Box 2197
Tacoma, WA 98401-2197
*Prgm Dir:* Ronald L Broekemeier, MS RPh
*Tel:* 253 591-6692    *Fax:* 253 207-4949
*E-mail:* ronbroekemeier@fhshealth.org

# West Virginia

**Carver Career Technical Center**
Pharmacy Technician Prgm
4799 Midland Dr
Charleston, WV 25306
http://boe.kana.k12.wv.us/carver
*Prgm Dir:* Janie L March, CPhT BS
*Tel:* 304 348-1965, Ext 130    *Fax:* 304 348-1938
*E-mail:* jmarchcpht@yahoo.com

PROGRAMS

## Pharmacy Technician

| Programs* | Class Capacity | Begins | Length (months) | Award | Res. Tuition | Non-res. Tuition | Stipend | Offers:‡ 1 | 2 | 3 |
|---|---|---|---|---|---|---|---|---|---|---|
| **California** | | | | | | | | | | |
| American University of Health Sciences (Long Beach) | 20 | Open during each module | 6, 8 | Cert, BS CRA | $11,000 | $0 | | • | | |
| Cerritos Community College (Norwalk) | 70 | Aug Jan | 24 | Cert, AA | $1,000 | $2,500 | | • | • | • |
| Charles A Jones Skills & Business Ed Center (Sacramento) | 30 | Every 12 wks | 12 | Cert | $499 | $0 | | | | |
| Charles R Drew Univ of Med & Science (Los Angeles) | 20 | Aug | 24 | AS | $9,600 | $0 | | • | | |
| Foothill Community College (Palo Alto) | 35 | Sep | 9 | Cert, AS | $1,500 | $5,400 | | | | • |
| North Orange County Community (Anaheim) | 40 | Jan Apr Jul Sep | 18 | Cert | | | | • | | • |
| North-West College - Pasadena | 24 | | 8 | Dipl | $9,233 | $0 | | • | | |
| Western Career College - Antioch | 25 | Every 3 wks | 15 | AS | $23,494 | $0 | | • | | |
| Western Career College - Citrus (Citrus Heights) | | | | | | | | | | |
| Western Career College - Pleasant Hill | Var | Every 6 wks | 16 | AS | $24,000 | $0 | | • | | |
| Western Career College - Sacramento | 32 | Every 6 wks | 16 | AAS | $18,224 | $0 | | • | | • |
| Western Career College - San Leandro | 35 | Every 6 wks | | AAS | $16,535 | $0 | | | | |
| Western Career College - Stockton | 25 | Every 3 wks | 15, 12 | Dipl, AS | $22,035 | $0 | | • | | • |
| **Colorado** | | | | | | | | | | |
| Front Range Comm College (Westminster) | 24 | Aug Sep | 7 | Cert | $3,800 | $0 | | • | | • |
| **Connecticut** | | | | | | | | | | |
| Briarwood College (Southington) | 12 | Sep | 9 | Cert | $0 | $13,000 | | • | | • |
| **Florida** | | | | | | | | | | |
| Lake City Community College | 25 | Aug | 12 | Dipl, ATD | $2,578 | $9,635 | | | | |
| McFatter Vocational Technical Center (Davie) | 24 | Aug | 10 | Cert | $2,274 | $6,822 | | | | |
| Pinellas Tech Educ Ctr - St Petersburg | 25 | 5x/yr | 11 | Cert | $2,005 | $7,140 | | | | • |
| **Georgia** | | | | | | | | | | |
| Southeastern Technical College (Vidalia) | 24 | Fall Spring | 15 | Dipl | $0 | $2,265 | | | • | • |
| **Illinois** | | | | | | | | | | |
| Blessing Hospital (Quincy) | 15 | Oct 16 | 5 | Cert | $1,200 | $1,200 | | • | | • |
| Malcolm X College (Chicago) | 25 | Aug | 9 | Cert | $2,376 | $6,388 | | | | • |
| **Indiana** | | | | | | | | | | |
| Clarian Health Partners Inc (Indianapolis) | 18 | Aug | 9 | Cert | $2,300 | $2,300 | | | | |
| **Louisiana** | | | | | | | | | | |
| Bossier Parish Community College (Bossier City) | 20 | Aug Jan | 12 | Dipl, Cert, AAS | $1,800 | $0 | | | | |
| Louisiana State University - Alexandria | 25 | Fall semester | 12 | Cert | $3,000 | $0 | | | | |
| **Michigan** | | | | | | | | | | |
| Henry Ford Hospital (Detroit) | 25 | Jan Sep | 8 | Cert, AS | $2,484 | $3,026 | | | • | • |
| **Minnesota** | | | | | | | | | | |
| Century College (White Bear Lake) | 50 | | 12, 24 | Dipl, Cert, AAS | $2,500 | $0 | | | | |
| Northland Community & Technical College (East Grand Forks) | 20 | Aug | 24, 9 | Dipl, AAS | $2,346 | $0 | | | | |
| **Montana** | | | | | | | | | | |
| College of Tech - Univ of Montana-Missoula | 25 | Sep | 10 | Cert | $3,871 | $7,323 | | | | |
| **North Carolina** | | | | | | | | | | |
| Durham Technical Community College | 24 | Jan May | 12 | Dipl, Cert | $1,896 | $10,536 | | | | • |
| **North Dakota** | | | | | | | | | | |
| North Dakota State College of Science (Wahpeton) | 25 | Aug | 10, 19 | Cert, AAS | $3,516 | $4,116 | | • | | |
| **Ohio** | | | | | | | | | | |
| Cuyahoga Community College (Cleveland) | 48 | Aug | 18, 12 | Cert, AAS | $2,600 | $7,000 | | • | | • |
| **Pennsylvania** | | | | | | | | | | |
| Comm Coll of Allegheny Cnty - South Campus (West Mifflin) | 30 | Aug | 12 | Cert, AS | $96 | $110 | | • | | • |
| Great Lakes Institute of Technology (Erie) | 25 | Feb | 8 | Dipl | $9,610 | $0 | | | | |
| Western School of Health & Business - Monroeville | 20 | Every 10 wks | | AST | $11,500 | $0 | | • | | |
| **Tennessee** | | | | | | | | | | |
| Concorde Career College - Memphis | 24 | Mar May Jun Aug Sep | 8 | Cert | $11,275 | $11,275 | | • | | |
| Tennessee Technology Center - Jackson | 20 | Sep | 12 | Dipl | $2,038 | $0 | | | | |
| **Texas** | | | | | | | | | | |
| Angelina College (Lufkin) | 20 | Fall | 9 | Cert | $1,047 | $1,546 | | | | |
| Austin Community College | 30 | Aug Jan | 9 | Cert | $1,676 | $3,400 | | • | | |
| El Paso Community College | 14 | Aug Jan | 12, 24 | Cert, AAS | $1,800 | $2,600 | | | | |
| Lamar State College - Orange | 30 | Jan Aug | 10 | Cert | $768 | $0 | | | | |
| North Harris College (Houston) | 25 | Fall Spring | 9, 12 | Cert | $1,400 | $0 | | • | | |
| Richland College (Dallas) | 54 | Jan May Aug | 10 | Cert | $3,015 | $3,015 | | • | | • |
| San Jacinto College North (Houston) | 40 | Aug Jan | 9 | Cert | $1,945 | $2,704 | | | | |
| South Texas Vo-Tech - McAllen | 10 | Jan Feb Apr May Oct | 12 | Dipl | | | | | | |
| South Texas Vo-Tech - Weslaco | 10 | Jan Feb Apr May Oct | 12 | Dipl | | | | | | |
| Vernon College (Wichita Falls) | 30 | Fall | 8 | Cert | $1,620 | $1,620 | | • | | |

*Data are shown only for programs that completed the 2006 AMA Survey of Health Professions Education Programs
‡Key to Offers: 1: Evening or weekend classes; 2: Non-English instruction; 3: Cultural competence instruction

## Pharmacy Technician

| Programs* | Class Capacity | Begins | Length (months) | Award | Res. Tuition | Non-res. Tuition | Stipend | Offers:‡ 1 | 2 | 3 |
|---|---|---|---|---|---|---|---|---|---|---|
| **Washington** | | | | | | | | | | |
| Spokane Community College | 25 | Sep | 9 | | $870 | $1,337 | | | | • |
| **West Virginia** | | | | | | | | | | |
| Carver Career Technical Center (Charleston) | 15 | Aug Jan | 6 | Cert | $850 | $0 | | • | | |

*Data are shown only for programs that completed the 2006 AMA Survey of Health Professions Education Programs
‡Key to Offers: 1: Evening or weekend classes; 2: Non-English instruction; 3: Cultural competence instruction

**Includes:**

- Physical therapist
- Physical therapist assistant

# Physical Therapist

## Occupational Description

The physical therapist provides services to many different kinds of patients/clients, from those recovering from accidents or illness and people with disabilities to world-class athletes. Physical therapists help improve patients' strength and mobility, relieve pain, and prevent or limit permanent physical disabilities. Physical therapists take a personal and direct approach to meeting an individual's health goals, working closely with the patient and other health care practitioners. They provide the patient and the patient's family with instruction and home programs to ensure that healing continues after direct patient care has ended.

Physical therapists also work to keep people well and safe from injury, emphasizing the importance of fitness and conditioning and showing people how to avoid injuries at work or play. Physical therapy promotes optimal physical performance and enables health-conscious people to increase their overall fitness level and muscular strength and endurance.

## Job Description

The physical therapist is able to evaluate a patient's

- Aerobic capacity and endurance
- Joint motion
- Muscle strength and endurance
- Posture
- Pain
- Functional ability
- Muscle tone and reflexes
- Appearance and stability of walking
- Need and use of braces and artificial limbs
- Function of the heart and lungs
- Integrity of sensation and perception
- Integrity and health of skin
- Performance of activities required in daily living
- Developmental activities

Physical therapy techniques include:

- Therapeutic exercise
- Mobilization/manipulation and range-of-motion exercises
- Cardiovascular endurance training
- Relaxation exercises
- Therapeutic massage
- Biofeedback
- Training in activities of daily living
- Wound debridement
- Pulmonary physical therapy
- Ambulation training

Modalities, including traction, ultrasound, diathermy, electrotherapy, cryotherapy, hydrotherapy, and laser therapy, also can be applied during the treatment program.

## Employment Characteristics

Physical therapists work in hospitals as well as

- Private physical therapy offices
- Community health centers
- Corporate or industrial health centers
- Sports facilities
- Research institutions
- Rehabilitation centers
- Nursing homes
- Home health agencies
- Schools
- Pediatric centers
- Colleges and universities

## Salary

Average annual income for physical therapists is approximately $56,500, depending on geographic location and practice setting. Physical therapists have the potential to earn more than $100,000 annually. Refer to Section IV, Table 5 of this *Directory* for more information, or see www.ama-assn.org/go/salary.

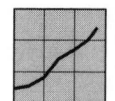

## Employment Outlook

Career opportunities in the field of physical therapy will increase as the "baby boom" generation ages and more patients begin to require treatment for arthritis, stroke, heart disease, and other conditions common to older people. In addition, with the nation's increasing participation in sports and fitness activities, more physical therapists will be needed to treat and help prevent knee, leg, back, shoulder, and other musculoskeletal injuries.

Opportunities also exist for physical therapists from minority groups, who are in great demand but short supply in all aspects of the profession. When physical therapists and their clients share a common language and similar background, the effectiveness of treatment is enhanced.

## Educational Programs

**Length.** All physical therapist education programs culminate in a post-baccalaureate degree. As of July 1, 2006, almost 77% of accredited programs offer the Doctor of Physical Therapy (DPT) degree; all but two of the remaining programs have indicated their intent to convert to offer the DPT in the future.

**Prerequisites.** Candidates should have a high overall grade point average (GPA) and a high GPA in prerequisite coursework. Volunteer experience as a physical therapy aide, letters of recommendation from physical therapists or science teachers, and excellent writing and interpersonal skills are also highly valued.

**Curriculum.** Educational programs include basic and clinical medical science courses and emphasize the theory and practice of physical therapy. The curriculum includes opportunities to apply and integrate theory through extensive clinical education and a variety of practice settings.

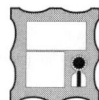

## Licensure, Certification, and Registration

After graduating from an accredited education program, physical therapist candidates must pass a

PROGRAMS

state-administered national exam. Other requirements for physical therapy practice vary from state to state according to physical therapy practice acts or state regulations that govern physical therapy. For more information, contact the state licensing boards.

## Physical Therapist Assistant

### Occupational Description
Physical therapist assistants work under the supervision of a physical therapist. Their duties include assisting the physical therapist by providing selected interventions within the plan of care, training patients in exercises and activities of daily living, conducting treatments, using special equipment, administering modalities and other treatment procedures, and reporting to the physical therapist on the patient's responses.

### Employment Characteristics
Physical therapist assistants work in
- Hospitals
- Private physical therapy offices
- Community health centers
- Corporate or industrial health centers
- Sports facilities
- Research institutions
- Rehabilitation centers
- Nursing homes
- Home health agencies
- Schools
- Pediatric centers
- Colleges and universities

### Salary
The median income for a physical therapist assistant is $26,000; PTAs employed in the southern and western regions of the nation generally earn higher salaries. Refer to Section IV, Table 5 of this *Directory* for more information, or see www.ama-assn.org/go/salary.

### Educational Programs
**Length.** These associate's degree programs—usually offered in a community or junior college—are 2 years long.
**Prerequisites.** Successful completion of high school courses in social sciences, biology, mathematics, physics, English, and chemistry is encouraged but not required.
**Curriculum.** The curriculum includes 1 year of general education and 1 year of technical courses and clinical experience.

### Inquiries

**Careers, Education, and Certification**
American Physical Therapy Association
1111 North Fairfax Street
Alexandria, VA 22314-1488
703 684-2782 or 800 999-2782
703 684-7343 Fax
www.apta.org

**Program Accreditation**
Commission on Accreditation in Physical Therapy Education
Mary Jane Harris, MS PT, Director of Accreditation
1111 North Fairfax Street
Alexandria, VA 22314
703 684-2782
www.apta.org

## Physical Therapist*

## Alabama

**University of Alabama at Birmingham**
Physical Therapy Prgm
Department of Physical Therapy RMSB 360X
1530 3rd Ave South
Birmingham, AL 35294-1212
www.uab.edu/pt
*Prgm Dir:* Sharon Shaw, DrPH
*Tel:* 205 934-3566   *Fax:* 205 975-7787
*E-mail:* sshaw@uab.edu

**University of South Alabama**
Physical Therapy Prgm
Dept of Physical Therapy
1504 Springhill Ave, Rm 1214
Mobile, AL 36604
www.southalabama.edu/alliedhealth/pt
*Prgm Dir:* Dennis Fell, MD PT
*Tel:* 251 434-3575   *Fax:* 251 434-3822
*E-mail:* dfell@jaguar1.usouthal.edu

**Alabama State University**
Physical Therapy Prgm
PO Box 271
915 S Jackson St
Montgomery, AL 36101-0271
*Prgm Dir:* Senobia Crawford, PT PhD
*Tel:* 334 229-4707   *Fax:* 334 229-4945
*E-mail:* asupt@alasu.edu

## Arizona

**Northern Arizona University**
Physical Therapy Prgm
CU Box 15105
Flagstaff, AZ 86011
http://jan.ucc.nau.edu/~hp-p/pt
*Prgm Dir:* Mark Cornwall, PhD PT CPed
*Tel:* 520 523-4092   *Fax:* 520 523-9289
*E-mail:* mark.cornwall@nau.edu

**AT Still University of Health Sciences**
*Cosponsor:* Arizona School of Health Sciences
Physical Therapy Prgm
5850 E Still Circle
Mesa, AZ 85206
www.atsu.edu
*Prgm Dir:* Suzanne Robben Brown, PT MPH
*Tel:* 480 219-6061   *Fax:* 480 219-6100
*E-mail:* sbrown@atsu.edu

## Arkansas

**University of Central Arkansas**
Physical Therapy Prgm
Dept of Physical Therapy
201 Donaghey, PTC 300
Conway, AR 72035-0001
*Prgm Dir:* Nancy Reese, PT PhD
*Tel:* 501 450-3611   *Fax:* 501 450-5822
*E-mail:* nancyr@uca.edu

**Arkansas State University**
Physical Therapy Prgm
PO Box 910
State University, AR 72467-0910
*Prgm Dir:* Jim W Farris, PT PhD
*Tel:* 870 972-3591   *Fax:* 870 972-3652
*E-mail:* jfarris@astate.edu

## California

**Azusa Pacific University**
Physical Therapy Prgm
901 E Alosta Ave
Azusa, CA 91702-7000
www.apu.edu
*Prgm Dir:* Michael Laymon, DPTSc PT OCS
*Tel:* 626 815-5020   *Fax:* 626 815-5017
*E-mail:* mlaymon@apu.edu

*Additional information about programs that returned the AMA's survey is available in a table beginning on page 323.

**California State University - Fresno**
Physical Therapy Prgm
2345 E San Ramon Ave, MS-MH29
Fresno, CA 93740-8031
www.csufresno.edu/physicaltherapy
*Prgm Dir:* Peggy Trueblood, PhD PT
*Tel:* 559 278-3008   *Fax:* 559 278-3635
*E-mail:* peggyt@csufresno.edu

**Loma Linda University**
Physical Therapy Prgm
Dept of Physical Therapy
School of Allied Health Professions
Loma Linda, CA 92350
*Prgm Dir:* Larry Chinnock, EdD PT
*Tel:* 909 558-4632   *Fax:* 909 558-0459
*E-mail:* lchinnock@sahp.llu.edu

**California State University - Long Beach**
Physical Therapy Prgm
College of Health and Human Svcs
1250 Bellflower Blvd
Long Beach, CA 90840
www.csulb.edu/web/colleges/chhs/ada/pt
*Prgm Dir:* Kay Cerny, PhD PT
*Tel:* 562 985-4956, Ext 54956   *Fax:* 562 985-4069
*E-mail:* kcerny@csulb.edu

**Mount St Mary's College**
Physical Therapy Prgm
Dept of Physical Therapy
10 Chester Place
Los Angeles, CA 90007-2598
www.msmc.la.edu/pt
*Prgm Dir:* Deborah Lowe, PT PhD
*Tel:* 213 477-2601   *Fax:* 213 477-2609
*E-mail:* dlowe@msmc.la.edu

**University of Southern California**
Physical Therapy Prgm
Biokinesiology & Physical Therapy
1540 E Alcazar St, CHP 155
Los Angeles, CA 90089
www.usc.edu/pt
*Prgm Dir:* James Gordon, EdD PT FAPTA
*Tel:* 323 442-2900   *Fax:* 323 442-1515
*E-mail:* jamesgor@usc.edu

**California State University - Northridge**
Physical Therapy Prgm
Department of Physical Therapy
18111 Nordhoff St
Northridge, CA 91330-8411
http://hhd.csun.edu/pt
*Prgm Dir:* Janna Beling, PT PhD
*Tel:* 818 677-2203, Ext 7445   *Fax:* 818 677-7411
*E-mail:* janna.beling@csun.edu

**Samuel Merritt College**
Physical Therapy Prgm
450 30th St
Oakland, CA 94609
*Prgm Dir:* Terrence Nordstrom, PT MA
*Tel:* 510 869-6649   *Fax:* 510 869-6282
*E-mail:* tnordstrom@samuelmerritt.edu

**Chapman University**
Physical Therapy Prgm
Department of Physical Therapy
One University Dr
Orange, CA 92866
*Prgm Dir:* Venita Lovelace-Chandler, PT PhD PCS
*Tel:* 714 744-7620   *Fax:* 714 744-7621
*E-mail:* lovelacech@chapman.edu

**Western Univ of Health Sciences**
Physical Therapy Prgm
Department of Physical Therapy Education
309 E Second St
Pomona, CA 91766-1854
www.westernu.edu
*Prgm Dir:* Georgeanne Vlad, PT MA
*Tel:* 909 469-5295   *Fax:* 909 469-5692
*E-mail:* gvlad@westernu.edu

**California State University - Sacramento**
Physical Therapy Prgm
College of Health and Human Services
6000 J St
Sacramento, CA 95819-6020
www.hhs.csus.edu/pt
*Prgm Dir:* Susan M McGinty, EdD MS PT
*Tel:* 916 278-6426, Ext 85056   *Fax:* 916 278-5053
*E-mail:* mcgintys@csus.edu

**University of California - San Francisco**
*Cosponsor:* San Francisco State Univ
Physical Therapy Prgm
1318 7th Ave
San Francisco, CA 94143-0736
*Prgm Dir:* Nancy Byl, PhD PT
*Tel:* 415 476-6650   *Fax:* 415 502-0323
*E-mail:* byl@itsa.ucsf.edu

**University of the Pacific**
Physical Therapy Prgm
Dept of Physical Therapy
3601 Pacific Ave
Stockton, CA 95211
www.pacific.edu
*Prgm Dir:* Cathy Peterson, PT EdD
*Tel:* 209 946-2886   *Fax:* 209 946-2367
*E-mail:* cpeterson@pacific.edu

# Colorado

**Regis University**
Physical Therapy Prgm
3333 Regis Blvd G-9
Denver, CO 80221-1099
www.regis.edu/dpt
*Prgm Dir:* Barbara A Tschoepe, PT PhD
*Tel:* 303 458-4152   *Fax:* 303 964-5474
*E-mail:* btschoep@regis.edu
*Med Dir:* Alison Campbell

**U of Colorado at Denver and Health Scis Ctr**
Physical Therapy Prgm
4200 E 9th Ave, Mailstop C244
Denver, CO 80262
www.uchsc.edu/pt
*Prgm Dir:* Margaret Schenkman, PT PhD
*Tel:* 303 372-9375   *Fax:* 303 372-9016
*E-mail:* margaret.schenkman@uchsc.edu

# Connecticut

**Sacred Heart University**
Physical Therapy Prgm
5151 Park Ave
Fairfield, CT 06825-1000
*Prgm Dir:* Michael J Emery, PT EdD
*Tel:* 203 365-7656   *Fax:* 203 365-4723
*E-mail:* emerym@sacredheart.edu

**Quinnipiac University**
Physical Therapy Prgm
School of Health Sciences
Mt Carmel Ave
Hamden, CT 06518
*Prgm Dir:* Donald Kowalsky, PT EdD
*Tel:* 203 582-8200   *Fax:* 203 582-8706
*E-mail:* donald.kowalsky@quinnipiac.edu

**University of Connecticut**
Physical Therapy Prgm
School of Allied Health Professions
358 Mansfield Rd, U-2101
Storrs, CT 06269-2101
*Prgm Dir:* Scott Hasson, EdD PT
*Tel:* 860 486-0019   *Fax:* 860 486-1588
*E-mail:* scott.hasson@uconn.edu

**University of Hartford**
Physical Therapy Prgm
200 Bloomfield Ave
West Hartford, CT 06117-1599
*Prgm Dir:* Catherine M E Certo, PT ScD FAPTA
*Tel:* 860 768-5367   *Fax:* 860 768-5244
*E-mail:* certo@hartford.edu

# Delaware

**University of Delaware**
Physical Therapy Prgm
Dept of Physical Therapy
301 McKinly Laboratory
Newark, DE 19716
*Prgm Dir:* Stuart A Binder-Macleod, PhD PT FAPTA
*Tel:* 302 831-8046   *Fax:* 302 831-4234
*E-mail:* sbinder@udel.edu

# District of Columbia

**George Washington University**
Physical Therapy Prgm
School of Medicine and Health Sciences
900 23rd St NW - 6th
Washington, DC 20037
*Prgm Dir:* Margaret M Plack, PT
*Tel:* 202 994-8237   *Fax:* 202 994-8400
*E-mail:* hspglr@gwumc.edu

**Howard University**
Physical Therapy Prgm
Coll of Pharmacy, Nursing & Allied Health
6th and Bryant Sts NW
Washington, DC 20059
*Prgm Dir:* Steven Chesbro, PT EdD MHS GCS
*Tel:* 202 806-7613   *Fax:* 202 462-6194
*E-mail:* schesbro@howard.edu

# Florida

**University of Miami**
Physical Therapy Prgm
School of Medicine
5915 Ponce de Leon Blvd, 5th Fl
Coral Gables, FL 33146
*Prgm Dir:* Sherrill H Hayes, PhD PT
*Tel:* 305 284-4535   *Fax:* 305 284-6128
*E-mail:* shayes@miami.edu

**Nova Southeastern University**
Physical Therapy Prgm
Health Professions Division
3200 S University Dr
Ft Lauderdale, FL 33328
*Prgm Dir:* Stanley Wilson, PT EdD CEAS
*Tel:* 954 262-1100, Ext 1662   *Fax:* 954 262-1783
*E-mail:* swilson@nova.edu

**Florida Gulf Coast University**
Physical Therapy Prgm
10501 FGCU Blvd South
Ft Myers, FL 33965-6565
*Prgm Dir:* Ellen Kroog Williamson, PT MS
*Tel:* 941 590-7530   *Fax:* 941 590-7474
*E-mail:* ekwill@fgcu.edu

**University of Florida**
Physical Therapy Prgm
Dept of Physical Therapy
Box 100154 HSC
Gainesville, FL 32610-0154
www.phhp.ufl.edu
*Prgm Dir:* Krista Vandenborne, PhD PT
*Tel:* 352 273-6085    *Fax:* 352 273-6109
*E-mail:* kvandenb@phhp.ufl.edu

**University of North Florida**
Physical Therapy Prgm
College of Health
4567 St Johns Bluff Rd S
Jacksonville, FL 32224
*Prgm Dir:* John P Cummings, PT EdD OCS FAAOMPT
*Tel:* 904 620-2841    *Fax:* 904 620-2848
*E-mail:* jcumming@unf.edu

**Florida International University**
Physical Therapy Prgm
Dept of Physical Therapy/College of Nursing Health
  Sciences
11200 SW 8th St
Miami, FL 33199
www.physicaltherapy.fiu.edu
*Prgm Dir:* Helen Z Cornely, PT EdD
*Tel:* 305 348-1968    *Fax:* 305 348-1979
*E-mail:* cornelyh@fiu.edu

**University of Central Florida**
Physical Therapy Prgm
HPA-1, Ste 256
Orlando, FL 32816-2205
*Prgm Dir:* Gerald Smith, PhD PT
*Tel:* 407 882-0094    *Fax:* 407 823-3464
*E-mail:* ptinfo@mail.ucf.edu

**Univ of St Augustine for Health Sciences**
Physical Therapy Prgm
1 University Blvd
St Augustine, FL 32086-5783
www.usa.edu
*Prgm Dir:* Gerard C Gorniak, PT PhD
*Tel:* 904 826-0084    *Fax:* 904 826-0085
*E-mail:* ggorniak@usa.edu

**Florida A&M University**
Physical Therapy Prgm
School of Allied Health Sciences
Rm 223 Ware-Rhaney Bldg
Tallahassee, FL 32307-3500
*Prgm Dir:* Eric Toran, PhD
*Tel:* 850 599-3820    *Fax:* 850 561-2457
*E-mail:* eric.toran@famu.edu

**University of South Florida**
Physical Therapy Prgm
12901 Bruce B Downs Blvd, MDC 77
Tampa, FL 33612
*Prgm Dir:* William S Quillen, PT PhD SCS FACSM
*Tel:* 813 974-8870    *Fax:* 813 974-8915
*E-mail:* wquillen@hsc.usf.edu

# Georgia

**Emory University**
Physical Therapy Prgm
1441 Clifton Rd NE, Ste 180
Atlanta, GA 30322
*Prgm Dir:* Susan J Herdman, PT PhD FAPTA
*Tel:* 404 712-5660    *Fax:* 404 712-4130
*E-mail:* sherdma@emory.edu

**Georgia State University**
Physical Therapy Prgm
Division of Physical Therapy
PO Box 4019
Atlanta, GA 30302-4019
www.gsu.edu
*Prgm Dir:* Leslie Taylor, PhD PT
*Tel:* 404 651-3092    *Fax:* 404 651-1584
*E-mail:* ltaylorl@gsu.edu

**Medical College of Georgia**
Physical Therapy Prgm
Dept of Physical Therapy
Augusta, GA 30912-0800
*Prgm Dir:* Douglas R Keskula, PT PhD ATC
*Tel:* 706 721-2141    *Fax:* 706 721-3209
*E-mail:* dkeskula@mcg.edu

**North Georgia College & State University**
Physical Therapy Prgm
Barnes Hall Rm A-8
Dahlonega, GA 30597
*Prgm Dir:* Robert J Laird, PhD PT
*Tel:* 706 864-1422    *Fax:* 706 864-1493
*E-mail:* rdtaylor@ngcsu.edu

**Armstrong Atlantic State University**
Physical Therapy Prgm
11935 Abercorn St
Savannah, GA 31419-1997
*Prgm Dir:* David A Lake, PhD PT
*Tel:* 912 921-2327    *Fax:* 912 921-5838
*E-mail:* lakedavi@mail.armstrong.edu

# Idaho

**Idaho State University**
Physical Therapy Prgm
Dept of Physical and Occupational Therapy
Kasiska College of Health Professions, Box 8045
Pocatello, ID 83209
*Prgm Dir:* Alexander G Urfer, PhD PT
*Tel:* 208 282-4095, Ext 4459    *Fax:* 208 282-4962
*E-mail:* urfealex@isu.edu

# Illinois

**Northwestern University**
Physical Therapy Prgm
Feinberg School of Medicine
645 N Michigan Ave, Ste 1100
Chicago, IL 60611-2814
*Prgm Dir:* Julius P A Dewald, PhD PT
*Tel:* 312 908-8160    *Fax:* 312 908-0741
*E-mail:* j-dewald@northwestern.edu

**University of Illinois at Chicago**
Physical Therapy Prgm
College of Applied Health Sciences
1919 W Taylor St, M/C 898
Chicago, IL 60612
*Prgm Dir:* Suzann Campbell, PT PhD FAPTA
*Tel:* 312 996-1502    *Fax:* 312 996-3807
*E-mail:* skc@uic.edu

**Northern Illinois University**
Physical Therapy Prgm
School of Allied Health Professions
DeKalb, IL 60115
*Prgm Dir:* M J Blaschak, PhD PT
*Tel:* 815 753-1383    *Fax:* 815 753-0720
*E-mail:* mblascha@niu.edu

**Midwestern University**
Physical Therapy Prgm
College of Health Sciences
555 31st St
Downers Grove, IL 60515
www.midwestern.edu
*Prgm Dir:* Donna Cech, PT MS PCS
*Tel:* 630 515-7221    *Fax:* 630 515-7224
*E-mail:* dcechx@midwestern.edu

**Rosalind Franklin Univ of Medicine & Science**
Physical Therapy Prgm
College of Health Professions
3333 Green Bay Rd
North Chicago, IL 60064
*Prgm Dir:* Dale Schuit, PhD PT
*Tel:* 847 578-3307    *Fax:* 847 578-8816
*E-mail:* Dale.Schuit@rosalindfranklin.edu

**Bradley University**
Physical Therapy Prgm
Dept of Physical Therapy & Health Science
1501 W Bradley Ave
Peoria, IL 61625
*Prgm Dir:* Mary Jo Mays, PhD PT
*Tel:* 309 677-3489    *Fax:* 309 677-3445
*E-mail:* jun@bumail.bradley.edu

**Governors State University**
Physical Therapy Prgm
1 University Parkway
University Park, IL 60466
*Prgm Dir:* Russell E Carter, PT EdD
*Tel:* 708 534-3147    *Fax:* 708 534-1647
*E-mail:* r-carter@govst.edu

# Indiana

**University of Evansville**
Physical Therapy Prgm
1800 Lincoln Ave
Evansville, IN 47722
http://pt.evansville.edu
*Prgm Dir:* Mary Kessler, MHS PT
*Tel:* 812 488-2345    *Fax:* 812 488-2717
*E-mail:* mk43@evansville.edu

**Indiana University**
Physical Therapy Prgm
1140 W Michigan St, CF 326
Indianapolis, IN 46202-5119
www.dpt.indiana.edu
*Prgm Dir:* Lisa Riolo, PhD PT
*Tel:* 317 278-1875    *Fax:* 317 278-1876
*E-mail:* lriolo@iupui.edu

**University of Indianapolis**
Physical Therapy Prgm
Krannert School of Physical Therapy
1400 E Hanna Ave
Indianapolis, IN 46227-3697
*Prgm Dir:* Christopher Petrosino, PhD PT
*Tel:* 317 788-2182    *Fax:* 317 788-3542
*E-mail:* cpetrosino@uindy.edu

# Iowa

**St Ambrose University**
Physical Therapy Prgm
Physical Therapy Dept
518 W Locust
Davenport, IA 52803
*Prgm Dir:* Sandra L Cassady, PT PhD FAACVPR
*Tel:* 563 333-6403    *Fax:* 563 333-6410
*E-mail:* pt@sau.edu

**Des Moines University**
Physical Therapy Prgm
College of Health Sciences
3200 Grand Ave
Des Moines, IA 50312
*Prgm Dir:* Traci Bush, MSPT OTR/L
*Tel:* 515 271-1634    *Fax:* 515 271-7078
*E-mail:* traci.bush@dmu.edu

**Clarke College**
Physical Therapy Prgm
1550 Clarke Dr
Dubuque, IA 52001-3198
www.clarke.edu
*Prgm Dir:* Andrew Priest, PT EdD
*Tel:* 563 588-6382    *Fax:* 563 584-8684
*E-mail:* andrew.priest@clarke.edu

**University of Iowa**
Physical Therapy Prgm
Carver College of Medicine
1-252 Medical Education Bldg
Iowa City, IA 52242-1190
www.medicine.uiowa.edu/physicaltherapy
*Prgm Dir:* David H Nielsen, PhD PT
*Tel:* 319 335-9791   *Fax:* 319 335-9707
*E-mail:* david-nielsen@uiowa.edu

# Kansas

**University of Kansas Medical Center**
Physical Therapy Prgm
3056 Robinson Hall
3901 Rainbow Blvd
Kansas City, KS 66160-7601
*Prgm Dir:* Lisa Stehno-Bittel, PhD PT
*Tel:* 913 588-6799   *Fax:* 913 588-4568
*E-mail:* ptadmissions@kumc.edu
*Med Dir:* Patricia Pohl, PhD PT

**Wichita State University**
Physical Therapy Prgm
College of Health Professions
Ahlberg Hall, 1845 N Fairmont
Wichita, KS 67260-0043
*Prgm Dir:* Camilla Wilson, PhD
*Tel:* 316 978-3604   *Fax:* 316 978-3025
*E-mail:* Camilla.Wilson@wichita.edu

# Kentucky

**University of Kentucky**
Physical Therapy Prgm
900 S Limestone Ave, CHS Bldg, Rm 204
Lexington, KY 40536-0200
*Prgm Dir:* Anne L Harrison, PT PhD
*Tel:* 859 323-1100, Ext 80596   *Fax:* 859 323-6003
*E-mail:* alharr01@uky.edu

**Bellarmine University**
Physical Therapy Prgm
Lansing School of Nursing & Health Science
2001 Newburg Rd
Louisville, KY 40205
www.bellarmine.edu/pt
*Prgm Dir:* Mark R Wiegand, PhD PT
*Tel:* 502 452-8368   *Fax:* 502 452-8429
*E-mail:* mwiegand@bellarmine.edu

# Louisiana

**Louisiana State Univ Health Sciences Center**
Physical Therapy Prgm
1900 Gravier St
New Orleans, LA 70112
www.alliedhealth.lsuhsc.edu/physicaltherapy
*Prgm Dir:* Elizabeth Weiss, PhD PT
*Tel:* 504 568-4288   *Fax:* 504 568-6552
*E-mail:* eweiss@lsuhsc.edu

**Louisiana State Univ Hlth Sci Ctr-Shreveport**
Physical Therapy Prgm
School of Allied Health Professions
1501 Kings Hwy, PO Box 33932
Shreveport, LA 71130-3932
*Prgm Dir:* Joseph McCulloch, PT PhD
*Tel:* 318 675-6820   *Fax:* 318 675-4208
*E-mail:* jmcclo@lsuhsc.edu
*Med Dir:* John McDonald, MD

# Maine

**Husson College**
Physical Therapy Prgm
1 College Circle
Bangor, ME 04401
*Prgm Dir:* Suzanne P Gordon, PT MA
*Tel:* 207 941-7101   *Fax:* 207 941-7883
*E-mail:* gordons@husson.edu

**University of New England**
Physical Therapy Prgm
Dept of Physical Therapy
716 Stevens Ave
Portland, ME 04103
www.une.edu/chp/pt
*Prgm Dir:* Michael Sheldon, PT MS
*Tel:* 207 221-4591   *Fax:* 207 523-1910
*E-mail:* msheldon@une.edu

# Maryland

**University of Maryland**
Physical Therapy Prgm
School of Medicine
Department of Physical Therapy and Rehabilitation Science
Baltimore, MD 21201
www.pt.umaryland.edu
*Prgm Dir:* Mary Rodgers, PT PhD
*Tel:* 410 706-7720, Ext 5658   *Fax:* 410 706-6387
*E-mail:* mrodgers@umaryland.edu

**University of Maryland Eastern Shore**
Physical Therapy Prgm
Dept of Physical Therapy
Hazel Hall Room 2093
Princess Anne, MD 21853
www.umes.edu/pt
*Prgm Dir:* Raymond L Blakely, PhD PT
*Tel:* 410 651-6360   *Fax:* 410 651-6259
*E-mail:* rlblakely@umes.edu

# Massachusetts

**Boston University**
Physical Therapy Prgm
Sargent Coll of Hlth & Rehab Sciences
635 Commonwealth Ave
Boston, MA 02215
www.bu.edu
*Prgm Dir:* Diane Dalton
*Tel:* 617 353-2720   *Fax:* 617 353-9463
*E-mail:* ddalton@bu.edu

**MGH Institute of Health Professions**
Physical Therapy Prgm
Charlestown Navy Yard
36 1st Ave
Boston, MA 02129
*Prgm Dir:* Leslie G Portney, DPT PhD FAPTA
*Tel:* 617 726-8009   *Fax:* 617 724-6321
*E-mail:* lportney@mghihp.edu

**Northeastern University**
Physical Therapy Prgm
360 Huntington Ave
Rm 6 Robinson Hall
Boston, MA 02115
www.neu.edu
*Prgm Dir:* Meredith H Harris, EdD PT
*Tel:* 617 373-5980   *Fax:* 617 373-3161
*E-mail:* m.harris@neu.edu

**Simmons College**
Physical Therapy Prgm
School for Health Studies
300 The Fenway
Boston, MA 02115
www.simmons.edu
*Prgm Dir:* Shelley Goodgold, PT ScD
*Tel:* 617 521-2635   *Fax:* 617 521-3032
*E-mail:* goodgold@simmons.edu

**University of Massachusetts - Lowell**
Physical Therapy Prgm
Weed Hall, 3 Solomont Way Ste 5
Lowell, MA 01854-5124
www.uml.edu/college/she/PT
*Prgm Dir:* Susan B O'Sullivan, EdD PT
*Tel:* 978 934-4412   *Fax:* 978 934-3006
*E-mail:* Susan_Osullivan@uml.edu

**American International College**
Physical Therapy Prgm
1000 State St
Springfield, MA 01109-983
*Prgm Dir:* Edward Swanson, PT PhD MEd MBA
*Tel:* 413 205-3320   *Fax:* 413 788-9961
*E-mail:* eswanson@acad.aic.edu

**Springfield College**
Physical Therapy Prgm
Dept of Physical Therapy
263 Alden St
Springfield, MA 01109
www.spfldcol.edu/pt
*Prgm Dir:* David Miller, PT PhD
*Tel:* 413 748-3590   *Fax:* 413 748-3371
*E-mail:* dmilller@spfldcol.edu

# Michigan

**Andrews University**
Physical Therapy Prgm
Dept of Physical Therapy
Berrien Springs, MI 49104-0420
www.andrews.edu/PHTH
*Prgm Dir:* Wayne L Perry, PT MBA PhD
*Tel:* 269 471-6033   *Fax:* 269 471-2866
*E-mail:* perryw@andrews.edu
*Med Dir:* Dixie Scott, Physical Therapy Advisor

**Wayne State University**
Physical Therapy Prgm
2248 EACPHS
Detroit, MI 48202
*Prgm Dir:* Thomas Birk, PT PhD
*Tel:* 313 577-1432   *Fax:* 313 577-8685
*E-mail:* ae7647@wayne.edu

**University of Michigan - Flint**
Physical Therapy Prgm
School of Health Professions & Studies
303 E Kearsley St
Flint, MI 48502-2186
www.umflint.edu/pt
*Prgm Dir:* Donna Fry, PhD PT
*Tel:* 810 762-3373   *Fax:* 810 766-6668
*E-mail:* donnafry@umflint.edu

**Grand Valley State University**
Physical Therapy Prgm
301 Michigan St NE, Ste 200
Grand Rapids, MI 49503
www.gvsu.edu/pt
*Prgm Dir:* John Peck, PhD PT
*Tel:* 616 331-2898   *Fax:* 616 331-5999
*E-mail:* peckj@gvsu.edu

**Central Michigan University**
Physical Therapy Prgm
1220 Health Professions Building
Mt Pleasant, MI 48859
*Prgm Dir:* Herman Triezenberg, PhD PT
*Tel:* 989 774-2347   *Fax:* 989 774-2908
*E-mail:* triez1hl@cmich.edu

**Oakland University**
Physical Therapy Prgm
School of Health Sciences
Rochester, MI 48309-4401
*Prgm Dir:* Kristine A Thompson, PT PhD
*Tel:* 248 370-4041   *Fax:* 248 370-4287
*E-mail:* kathomps@oakland.edu

# Minnesota

**College of St Scholastica**
Physical Therapy Prgm
Graduate Studies Adm Asst
1200 Kenwood Ave
Duluth, MN 55811
*Prgm Dir:* Denise Wise, PT PhD
*Tel:* 218 723-6523   *Fax:* 218 723-6472
*E-mail:* dwise@css.edu

**College of St Catherine - Minneapolis**
Physical Therapy Prgm
601 25th Ave S
Minneapolis, MN 55454
*Prgm Dir:* Cort J Cieminski, PT MS ATC CSCS
*Tel:* 651 690-7884   *Fax:* 651 690-7876
*E-mail:* cjcieminski@stkate.edu

**University of Minnesota - Minneapolis**
Physical Therapy Prgm
Box 388
420 Delaware St SE
Minneapolis, MN 55455
www.physther.umn.edu
*Prgm Dir:* James R Carey, PhD PT
*Tel:* 612 626-2746   *Fax:* 612 625-4274
*E-mail:* carey007@umn.edu

**Mayo School of Health Sciences**
Physical Therapy Prgm
200 First St SW
Rochester, MN 55905
www.mayo.edu/mshs/pt-ptmp-rch.html
*Prgm Dir:* John Hollman, PT PhD
*Tel:* 507 284-2054   *Fax:* 507 284-0656
*E-mail:* hollman.john@mayo.edu
*Med Dir:* Robert De Pompolo, MD

# Mississippi

**University of Mississippi Medical Center**
Physical Therapy Prgm
School of Health Related Professions
2500 N State St
Jackson, MS 39216-4505
www.shrp.umc.edu/programs/pt.html
*Prgm Dir:* Neva F Greenwald, PT MSPH
*Tel:* 601 984-6330   *Fax:* 601 984-6344
*E-mail:* ngreenwald@shrp.umsmed.edu

# Missouri

**Southwest Baptist University**
Physical Therapy Prgm
1600 University Ave
Bolivar, MO 65613-2496
www.sbuniv.edu/pt
*Prgm Dir:* Steven G Lesh, PT PhD SCS ATC
*Tel:* 417 328-1672   *Fax:* 417 328-1658
*E-mail:* slesh@sbuniv.edu

**University of Missouri - Columbia**
Physical Therapy Prgm
School of Health Professions
106 Lewis Hall
Columbia, MO 65211
www.umshp.org/pt
*Prgm Dir:* Marian A Minor, PhD PT
*Tel:* 573 882-7103   *Fax:* 573 884-8369
*E-mail:* minorm@health.missouri.edu

**Rockhurst University**
Physical Therapy Prgm
1100 Rockhurst Rd
Kansas City, MO 64110
www.rockhurst.edu/pt
*Prgm Dir:* Brian McKiernan, PT PhD
*Tel:* 816 501-4059   *Fax:* 816 501-4643
*E-mail:* brian.mckiernan@rockhurst.edu

**Missouri State University**
Physical Therapy Prgm
901 S National Ave
Springfield, MO 65804-0089
*Prgm Dir:* Akinniran Oladehin, PhD
*Tel:* 417 836-6179   *Fax:* 417 836-6229
*E-mail:* aoladehin@missouristate.edu

**Maryville University**
Physical Therapy Prgm
650 Maryville University Dr
St Louis, MO 63141
www.maryville.edu/academics/hp/pt
*Prgm Dir:* Judy Woehrle, PT PhD OCS
*Tel:* 314 529-9514   *Fax:* 314 529-9495
*E-mail:* jwoehrle@maryville.edu

**Saint Louis University**
Physical Therapy Prgm
Health Sciences Center
3437 Caroline St
St Louis, MO 63104-1111
*Prgm Dir:* Irma S Ruebling, MA PT
*Tel:* 314 577-8505   *Fax:* 314 577-8513
*E-mail:* ruebling@slu.edu

**Washington University**
Physical Therapy Prgm
School of Medicine, Campus Box 8502
4444 Forest Park Blvd, Ste 1101
St Louis, MO 63108
http://pt.wustl.edu
*Prgm Dir:* Susan S Deusinger, PT PhD FAPTA
*Tel:* 314 286-1407   *Fax:* 314 286-1410
*E-mail:* deusingers@msnotes.wustl.edu

# Montana

**University of Montana**
Physical Therapy Prgm
School of Physical Therapy and Rehabilitation Science
Skaggs Bldg 135
Missoula, MT 59812
www.health.umt.edu
*Prgm Dir:* Reed Humphrey, PT PhD
*Tel:* 406 243-2417   *Fax:* 406 243-2795
*E-mail:* reed.humphrey@umontana.edu

# Nebraska

**Creighton University**
Physical Therapy Prgm
School of Pharmacy and Hlth Profs
2500 California Plaza
Omaha, NE 68178
*Prgm Dir:* Robert Sandstrom, PhD PT
*Tel:* 402 280-4325   *Fax:* 402 280-5692
*E-mail:* rsandstr@creighton.edu

**University of Nebraska Medical Center**
Physical Therapy Prgm
984420 Nebraska Medical Center
Omaha, NE 68198-4420
www.unmc.edu/physicaltherapy
*Prgm Dir:* Patricia A Hageman, PhD PT
*Tel:* 402 559-4259   *Fax:* 402 559-8626
*E-mail:* phageman@unmc.edu

# Nevada

**University of Nevada - Las Vegas**
Physical Therapy Prgm
4505 Maryland Pkwy
Box 453029
Las Vegas, NV 89154-3029
www.unlv.edu
*Prgm Dir:* J Wesley McWhorter, PT PhD
*Tel:* 702 895-3003   *Fax:* 702 895-4883
*E-mail:* james.mcwhorter@unlv.edu

# New Hampshire

**Franklin Pierce College**
Physical Therapy Prgm
5 Chenell Dr
Concord, NH 03301
*Prgm Dir:* Jane Walter Venzke, PT EdD FAPTA
*Tel:* 603 899-4361
*E-mail:* venzkej@fpc.edu

# New Jersey

**Univ of Med & Dent of New Jersey**
Physical Therapy Prgm
65 Bergen St, SSB 319
PO Box 1709
Newark, NJ 07101-1709
www.shrp.umdnj.edu/physicaltherapy
*Prgm Dir:* Alma S Merians, PhD PT
*Tel:* 973 972-7820   *Fax:* 973 972-3717
*E-mail:* merians@umdnj.edu

**Richard Stockton College of New Jersey**
Physical Therapy Prgm
Jim Leeds Rd
Pomona, NJ 08240
*Prgm Dir:* Bess Kathrins, PT MS
*Tel:* 609 652-4638   *Fax:* 609 652-4858
*E-mail:* bkathrins@stockton.edu

**Seton Hall University**
Physical Therapy Prgm
400 S Orange Ave
South Orange, NJ 07079-2689
*Prgm Dir:* Marc Campolo, PT PhD SCS
*Tel:* 973 275-2800   *Fax:* 973 275-2370
*E-mail:* campolma@shu.edu

**SUNJ Rutgers Camden/Univ of Med & Dent NJ**
Physical Therapy Prgm
Primary Care Center, Ste 2105
40 E Laurel Rd
Stratford, NJ 08084
www.umdnj.edu/shrpweb/programs/mpt
*Prgm Dir:* Marie Koval Nardone, PT EdD
*Tel:* 856 566-6456   *Fax:* 856 566-6458
*E-mail:* mptgradm@umdnj.edu

# New Mexico

**University of New Mexico**
*Cosponsor:* Health Sciences Center
Physical Therapy Prgm
1 University of New Mexico
MSC09 5230
Albuquerque, NM 87131-0001
http://hsc.unm.edu/som/physther
*Prgm Dir:* Susan Queen, PT PhD
*Tel:* 505 272-5756   *Fax:* 505 272-8079
*E-mail:* squeen@salud.unm.edu
*Med Dir:* James Dexter, PT, MA

**PROGRAMS**

# New York

### Daemen College
Physical Therapy Prgm
4380 Main St
Amherst, NY 14226-3592
*Prgm Dir:* Sharon L Held, PT DPT MS PCS
*Tel:* 716 839-8344   *Fax:* 716 839-8537
*E-mail:* sheld@daemen.edu

### Long Island University - Brooklyn Campus
Physical Therapy Prgm
Zeckendorf Health Sciences Center
One University Plaza
Brooklyn, NY 11201-5372
*Prgm Dir:* Stacy Jaffee Gropack, PT PhD
*Tel:* 718 488-1682   *Fax:* 718 780-4002
*E-mail:* rgabriel@liu.edu
*Med Dir:* Wing Fu, MA PT PCS

### SUNY Downstate Medical Center
Physical Therapy Prgm
450 Clarkson Ave, Box 16
Brooklyn, NY 11203-2098
*Prgm Dir:* Joanne S Katz, PhD PT
*Tel:* 718 270-7720   *Fax:* 718 270-7439
*E-mail:* jkatz@downstate.edu

### D'Youville College
Physical Therapy Prgm
One D'Youville Square
320 Porter Ave
Buffalo, NY 14201-1084
www.dyc.edu
*Prgm Dir:* Lynn Rivers, PT PhD
*Tel:* 716 829-7702   *Fax:* 716 829-8137
*E-mail:* riversl@dyc.edu

### University at Buffalo - SUNY
Physical Therapy Prgm
Department of Rehabilitation Science
515 Kimball Tower, 3435 Main St
Buffalo, NY 14214-3079
www.sphhp.buffalo.edu/rs
*Prgm Dir:* Louise Gilchrist, PhD PT
*Tel:* 716 829-3141, Ext 191   *Fax:* 716 829-3217
*E-mail:* lag@buffalo.edu

### Mercy College
Physical Therapy Prgm
555 Broadway
Dobbs Ferry, NY 10522
http://grad.mercy.edu/physicaltherapy
*Prgm Dir:* Claudia B Fenderson, EdD PT
*Tel:* 914 674-7823   *Fax:* 914 674-7840
*E-mail:* cfenderson@mercy.edu

### Ithaca College
Physical Therapy Prgm
Dept of Physical Therapy
335 Smiddy Hall
Ithaca, NY 14850-7183
http://departments.ithaca.edu/pt
*Prgm Dir:* Michael A Pagliarulo, PT EdD
*Tel:* 607 274-3716   *Fax:* 607 274-1137
*E-mail:* pags@ithaca.edu

### Columbia University
Physical Therapy Prgm
710 W 168th St, 8th Fl
New York, NY 10032
www.columbiaphysicaltherapy.org
*Prgm Dir:* Risa Granick, EdD MPA PT
*Tel:* 212 305-6907   *Fax:* 212 305-4569
*E-mail:* rg2135@columbia.edu

### CUNY Hunter College
Physical Therapy Prgm
425 E 25th St
New York, NY 10010
www.hunter.cuny.edu/schoolhp/pt
*Prgm Dir:* Gary Krasilovsky, PhD PT
*Tel:* 212 481-7556   *Fax:* 212 481-8618
*E-mail:* gkrasilo@hunter.cuny.edu
*Med Dir:* Mary Cleary, Secretary

### New York University
Physical Therapy Prgm
380 2nd Ave, 4th Fl
New York, NY 10010
http://steinhardt.nyu.edu/pt
*Prgm Dir:* Wen Ling, PT
*Tel:* 212 998-9400, Ext 9415   *Fax:* 212 995-4190
*E-mail:* wen.ling@nyu.edu
*Med Dir:* Marie McLaughlin

### Touro College
Physical Therapy Prgm
27 W 23rd St, 6th Fl
New York, NY 10010-4202
www.touro.edu/shs/pt/pt.asp
*Prgm Dir:* Christopher Kevin Wong, PT MS OCS
*Tel:* 212 463-0400, Ext 606   *Fax:* 631 665-4986
*E-mail:* ckwong@touro.edu

### New York Institute of Technology
Physical Therapy Prgm
Northern Blvd
Box 8000
Old Westbury, NY 11568-8000
*Prgm Dir:* Karen Friel, PT DHS
*Tel:* 516 686-7696   *Fax:* 516 686-7699
*E-mail:* kfriel@nyit.edu

### Dominican College
Physical Therapy Prgm
470 Western Hwy
Orangeburg, NY 10962-1299
www.dc.edu
*Prgm Dir:* Michael Gallucci, PT EdD
*Tel:* 845 398-4800   *Fax:* 845 398-4892
*E-mail:* michael.gallucci@dc.edu

### Clarkson University
Physical Therapy Prgm
PO Box 5880
Potsdam, NY 13699-5880
*Prgm Dir:* Scott D Minor, PT PhD
*Tel:* 315 268-3786   *Fax:* 315 268-1539
*E-mail:* sminor@clarkson.edu

### Nazareth College of Rochester
Physical Therapy Prgm
4245 East Ave
Rochester, NY 14618-3790
www.naz.edu/dept/physical_therapy
*Prgm Dir:* Jennifer Collins, PT MPA EdD
*Tel:* 585 389-2900   *Fax:* 585 389-2908
*E-mail:* jcollin9@naz.edu

### CUNY College of Staten Island
Physical Therapy Prgm
2800 Victory Blvd
Staten Island, NY 10314
www.csi.cuny.edu
*Prgm Dir:* Jeffrey Rothman, EdD PT
*Tel:* 718 982-3153   *Fax:* 718 982-2984
*E-mail:* rothmanj@mail.csi.cuny.edu
*Med Dir:* Dr Jeffrey Rothman, PT, EdD

### Stony Brook University
Physical Therapy Prgm
Sch of Health Tech and Management
Health Sciences Center
Stony Brook, NY 11794-8201
*Prgm Dir:* Richard W Johnson, MA PT
*Tel:* 516 444-3250   *Fax:* 516 444-7621
*E-mail:* richard.johnson@stonybrook.edu

### SUNY Upstate Medical University
Physical Therapy Prgm
College of Health Professions
750 E Adams St
Syracuse, NY 13210
*Prgm Dir:* Susan Miller, DPT OCS
*Tel:* 315 464-6881   *Fax:* 315 464-6887
*E-mail:* millers@upstate.edu

### The Sage Colleges
*Cosponsor:* Sage Graduate School
Physical Therapy Prgm
Dept of Physical Therapy
45 Ferry St
Troy, NY 12180
www.sage.edu
*Prgm Dir:* Marjane Selleck, PT MS PCS
*Tel:* 518 244-2060   *Fax:* 518 244-4524
*E-mail:* sellem@sage.edu

### Utica College
Physical Therapy Prgm
Health & Human Studies Division
1600 Burrstone Rd
Utica, NY 13502-4892
http://utica.edu/academic/gce/pt
*Prgm Dir:* Dale Scalise-Smith, PT PhD
*Tel:* 315 792-3059   *Fax:* 315 792-3248
*E-mail:* dscalise-smith@utica.edu

### New York Medical College
Physical Therapy Prgm
Rm 302, School of Public Health
Valhalla, NY 10595
www.nymc.edu/sph/pt
*Prgm Dir:* Michael Majsak, PT EdD
*Tel:* 914 594-4917   *Fax:* 914 594-4292
*E-mail:* michael_majsak@nymc.edu

# North Carolina

### Univ of North Carolina at Chapel Hill
Physical Therapy Prgm
Div of Physical Therapy
Medical School Wing E, CB 7135
Chapel Hill, NC 27599-7135
*Prgm Dir:* Rick Segal, PT PhD
*Tel:* 919 966-4708   *Fax:* 919 966-3678
*E-mail:* rsegal@med.unc.edu

### Western Carolina University
Physical Therapy Prgm
312 Moore Bldg
Cullowhee, NC 28723-9646
*Prgm Dir:* Karen Y Lunnen, PT EdD
*Tel:* 828 227-2191   *Fax:* 828 227-7071
*E-mail:* klunnen@email.wcu.edu

### Duke University Medical Center
Physical Therapy Prgm
PO Box 3965
Durham, NC 27710
*Prgm Dir:* Jan K Richardson, PhD PT OCS
*Tel:* 919 684-6020   *Fax:* 919 684-1846
*E-mail:* richa052@mc.duke.edu

### Elon University
*Cosponsor:* Alamance Regional Medical Center
Physical Therapy Prgm
Campus Box 2085
Elon, NC 27244-2010
www.elon.edu/graduate
*Prgm Dir:* Elizabeth Rogers, PT EdD
*Tel:* 336 278-6350   *Fax:* 336 278-6414
*E-mail:* rogers@elon.edu

**East Carolina University**
Physical Therapy Prgm
Dept of Physical Therapy
School of Allied Health Sciences
Greenville, NC 27858-4353
*Prgm Dir:* Denis Brunt, EdD PT
*Tel:* 252 328-4450  *Fax:* 252 328-0707
*E-mail:* bruntd@ecu.edu
*Med Dir:* Harold P Jones, PhD

**Winston-Salem State University**
Physical Therapy Prgm
601 Martin Luther King Jr Dr
Winston-Salem, NC 27110
*Prgm Dir:* Robert J Cowie, PhD
*Tel:* 336 750-2190  *Fax:* 336 750-2192
*E-mail:* cowierj@wssu.edu
*Med Dir:* Peggy Valentine, EdD, Dean School of Health
  Sciences

# North Dakota

**University of Mary**
Physical Therapy Prgm
7500 University Dr
Bismarck, ND 58504-9652
www.umary.edu
*Prgm Dir:* Jodi Roller, PT EdD DPT
*Tel:* 701 355-8183  *Fax:* 701 255-7687
*E-mail:* rollerj@umary.edu

**University of North Dakota**
Physical Therapy Prgm
School of Medicine
PO Box 9037, 501 N Columbia Rd
Grand Forks, ND 58202-9037
*Prgm Dir:* Thomas M Mohr, PhD PT
*Tel:* 701 777-2831  *Fax:* 701 777-4199
*E-mail:* tommohr@medicine.nodak.edu
*Med Dir:* H David Wilson, MD

# Ohio

**Ohio University**
Physical Therapy Prgm
School of Physical Therapy
W290 Grover Center
Athens, OH 45701
www.ohiou.edu/phystherapy
*Prgm Dir:* Averell S Overby, DrPH PT
*Tel:* 740 593-2624  *Fax:* 740 593-0293
*E-mail:* overbya@ohio.edu

**College of Mt St Joseph**
Physical Therapy Prgm
5701 Delhi Rd
Cincinnati, OH 45233-1672
www.msj.edu
*Prgm Dir:* Mary Romanello, PT PhD ATC SCS
*Tel:* 513 244-4890  *Fax:* 513 451-2547
*E-mail:* Mary_Romanello@mail.msj.edu

**University of Cincinnati**
Physical Therapy Prgm
College of Allied Health Science
PO Box 670394
Cincinnati, OH 45267-0394
www.cahs.uc.edu/departments/physicalt.cfm
*Prgm Dir:* Lizanne Mulligan, PhD PT PCS
*Tel:* 513 558-7482  *Fax:* 513 587-7474
*E-mail:* mulligea@uc.edu

**Cleveland State University**
Physical Therapy Prgm
Dept of Health Sciences HS 122
2121 Euclid Ave
Cleveland, OH 44115-2407
*Prgm Dir:* Ann Karas Reinthal, PT PhD
*Tel:* 216 687-3554  *Fax:* 216 687-9316
*E-mail:* a.karas@csuohio.edu

**Ohio State University**
Physical Therapy Prgm
516 Atwell Hall
453 West 10th Ave
Columbus, OH 43210
www.amp.osu.edu/pt
*Prgm Dir:* Deborah Givens Heiss, PT PhD DPT OCS
*Tel:* 614 292-5921  *Fax:* 614 292-0210
*E-mail:* heiss.8@osu.edu

**University of Dayton**
Physical Therapy Prgm
300 College Park
Dayton, OH 45469-1210
*Prgm Dir:* Philip Anloague, PT DHSc OCS MTC
*Tel:* 937 229-3250  *Fax:* 937 229-4224
*E-mail:* anloague@udayton.edu

**University of Findlay**
Physical Therapy Prgm
1000 North Main St
Findlay, OH 45840
www.findlay.edu
*Prgm Dir:* Robert M Frampton, DHCE PT
*Tel:* 419 434-4863  *Fax:* 419 434-4336
*E-mail:* frampton@findlay.edu

**Walsh University**
Physical Therapy Prgm
2020 East Maple St
North Canton, OH 44720-3396
*Prgm Dir:* Susan A Bemis, PT EdD
*Tel:* 330 490-7370  *Fax:* 330 490-7371
*E-mail:* sbemis@walsh.edu

**Medical University of Ohio at Toledo**
Physical Therapy Prgm
4416 Collier Bldg
3015 Arlington Ave
Toledo, OH 43614
*Prgm Dir:* Clayton F Holmes, EdD PT
*Tel:* 419 383-3518  *Fax:* 419 383-5880
*E-mail:* cholmes@meduohio.edu

**Youngstown State University**
Physical Therapy Prgm
B080 Cushwa Hall
Youngstown, OH 44555-2558
http://bchhs.ysu.edu/dpt/dpt.html
*Prgm Dir:* Nancy Landgraff, PT PhD
*Tel:* 330 941-2558  *Fax:* 330 941-1898
*E-mail:* nlandgraff@ysu.edu

# Oklahoma

**Langston University**
Physical Therapy Prgm
School of Physical Therapy
PO Box 1500
Langston, OK 73050
*Prgm Dir:* Milagros Jorge, PT EdD
*Tel:* 405 466-3411  *Fax:* 405 466-2915
*E-mail:* mjorge@lunet.edu

**Univ of Oklahoma Health Sciences Center**
Physical Therapy Prgm
College of Allied Health, Rm 235
PO Box 26901
Oklahoma City, OK 73190
*Prgm Dir:* Martha J Ferretti, MPH PT
*Tel:* 405 271-2131  *Fax:* 405 271-2432
*E-mail:* martha-ferretti@ouhsc.edu

# Ontario, Canada

**University of Western Ontario**
Physical Therapy Prgm
Faculty of Health Sciences
Elborn College, Rm 1588
London, ON N6G 1H1
*Prgm Dir:* Tom Overend, PT PhD
*Tel:* 519 661-3360  *Fax:* 519 661-3866
*E-mail:* toverend@uwo.ca

**University of Toronto**
Physical Therapy Prgm
160 - 500 University Ave
8th Fl
Toronto, ON MSG 1V7
www.utoronto.ca/pt
*Prgm Dir:* Katherine Berg, PhD PT
*Tel:* 416 978-0173  *Fax:* 416 946-8562
*E-mail:* pt.chair@utoronto.ca

# Oregon

**Pacific University**
Physical Therapy Prgm
School of Physical Therapy
222 SE 8th Ave, Ste 333
Hillsboro, OR 97123
www.pt.pacificu.edu
*Prgm Dir:* Richard Rutt, PT PhD
*Tel:* 503 352-2846  *Fax:* 503 352-2995
*E-mail:* ruttra@pacificu.edu

# Pennsylvania

**Lebanon Valley College**
Physical Therapy Prgm
101 N College Ave
Annville, PA 17003-0501
*Prgm Dir:* Stan M Dacko, PT PhD
*Tel:* 717 867-6840  *Fax:* 717 867-6365
*E-mail:* dacko@lvc.edu

**Neumann College**
Physical Therapy Prgm
Division of Nursing & Health Sciences
1 Neumann Dr
Aston, PA 19014-1298
*Prgm Dir:* Robert E Post, PT PhD
*Tel:* 610 361-5233  *Fax:* 610 361-5290
*E-mail:* postr@neumann.edu

**Widener University**
Physical Therapy Prgm
One University Pl
Chester, PA 19013
www.widener.edu/ipte
*Prgm Dir:* Robin L Dole, PT EdD PCS
*Tel:* 610 499-1277  *Fax:* 610 499-1231
*E-mail:* rldole@widener.edu

**College Misericordia**
Physical Therapy Prgm
301 Lake St
Dallas, PA 18612-1098
www.misericordia.edu
*Prgm Dir:* Susan Barker, PhD PT
*Tel:* 570 674-6422  *Fax:* 570 674-3052
*E-mail:* sbarker@misericordia.edu

**Gannon University**
Physical Therapy Prgm
Coll of Sciences, Engineering & Health
109 University Square
Erie, PA 16541-0001
www.gannon.edu/programs
*Prgm Dir:* Kristine S Legters, PT DSc NCS
*Tel:* 814 871-5639  *Fax:* 814 871-5662
*E-mail:* legters001@gannon.edu

**Arcadia University**
Physical Therapy Prgm
Dept of Physical Therapy
450 S Easton Rd
Glenside, PA 19038-3295
www.arcadia.edu/pt
*Prgm Dir:* Rebecca L Craik, PhD PT
*Tel:* 215 572-2143   *Fax:* 215 572-2157
*E-mail:* admiss@arcadia.edu
*Med Dir:* Marty Eastlack, PhD, PT

**St Francis University**
Physical Therapy Prgm
PO Box 600
Loretto, PA 15940-0600
*Prgm Dir:* Michael Arnall, PT MS MBA
*Tel:* 814 472-3123   *Fax:* 814 472-3140
*E-mail:* marnall@francis.edu

**Drexel University**
Physical Therapy Prgm
Physical Therapy & Rehabilitation Sciences
Mail Stop 502, 245 N 15th St
Philadelphia, PA 19102
www.drexel.edu/cnhp/depts/rehab/programs/dpt
*Prgm Dir:* Susan S Smith, PT PhD
*Tel:* 215 762-1758   *Fax:* 215 762-3886
*E-mail:* sue.smith@drexel.edu

**Temple University**
Physical Therapy Prgm
College of Health Professions
3307 N Broad St
Philadelphia, PA 19140
*Prgm Dir:* Kim Nixon-Cave, PT PhD PCS
*Tel:* 215 707-4815   *Fax:* 215 707-7500
*E-mail:* deptpt@temple.edu

**Thomas Jefferson University**
Physical Therapy Prgm
College of Health Professions
130 S Ninth St, Ste 830 Edison
Philadelphia, PA 19107-5233
www.tju.edu
*Prgm Dir:* Penny Kroll, PT PhD
*Tel:* 215 503-8061   *Fax:* 215 503-3499
*E-mail:* penny.kroll@jefferson.edu

**University of the Sciences in Philadelphia**
Physical Therapy Prgm
600 S 43rd St
Philadelphia, PA 19104
www.usip.edu
*Prgm Dir:* Marc Campolo, PT PhD
*Tel:* 215 596-8849   *Fax:* 215 895-3121
*E-mail:* m.campol@usip.edu

**Chatham College**
Physical Therapy Prgm
114 Dilworth Hall
Woodland Rd
Pittsburgh, PA 15232-2826
www.chatham.edu
*Prgm Dir:* Patricia Downey, PT PhD OCS
*Tel:* 412 365-1409   *Fax:* 412 365-1213
*E-mail:* downey@chatham.edu

**Duquesne University**
Physical Therapy Prgm
School of Health Sciences
139 Health Sciences Bldg
Pittsburgh, PA 15282
www.healthsciences.duq.edu/phyth
*Prgm Dir:* F Richard Clemente, PT PhD
*Tel:* 412 396-5541   *Fax:* 412 396-4399
*E-mail:* clemente@duq.edu

**University of Pittsburgh**
Physical Therapy Prgm
School of Health and Rehab Sciences
4019 Forbes Tower
Pittsburgh, PA 15260
*Prgm Dir:* Anthony Delitto, PT PhD FAPTA
*Tel:* 412 383-6630, Ext 1   *Fax:* 412 383-6629
*E-mail:* mizakd@upmc.edu

**University of Scranton**
Physical Therapy Prgm
800 Linden St
Scranton, PA 18510-4586
*Prgm Dir:* John P Sanko, PT EdD
*Tel:* 570 941-7934   *Fax:* 570 941-7940
*E-mail:* sankoj1@scranton.edu

**Slippery Rock University**
Physical Therapy Prgm
Graduate School of Physical Therapy
PT Building
Slippery Rock, PA 16057
www.sru.edu/depts/pt
*Prgm Dir:* Carol Martin-Elkins, PT PhD
*Tel:* 724 738-2080, Ext 2916   *Fax:* 724 738-2113
*E-mail:* carol.martin-elkins@sru.edu

# Puerto Rico

**University of Puerto Rico**
Physical Therapy Prgm
Medical Sciences Campus
PO Box 365067
San Juan, PR 00936-5067
http://cprsweb.rcm.upr.edu/terapiafisica.asp
*Prgm Dir:* Annlee Burch, EdD MPH MSPT
*Tel:* 787 758-2525, Ext 4202   *Fax:* 787 753-7262
*E-mail:* annleeburch@cprs.rcm.upr.edu

# Rhode Island

**University of Rhode Island**
Physical Therapy Prgm
Independence Square II
25 West Independence Way
Kingston, RI 02881-0180
www.ptp.uri.edu
*Prgm Dir:* Beth Marcoux, PhD PT
*Tel:* 401 874-5001   *Fax:* 401 874-5630
*E-mail:* bmarcoux@mail.uri.edu

# South Carolina

**Medical University of South Carolina**
Physical Therapy Prgm
Dept of Rehabilitation Sciences
PO Box 250965
Charleston, SC 29425
www.musc.edu/pt
*Prgm Dir:* Kathleen A Cegles, PT DEd
*Tel:* 843 792-2023   *Fax:* 843 792-0710
*E-mail:* cegles@musc.edu
*Med Dir:* Peter Bowman, MS, OTR/L, OTD

**University of South Carolina**
Physical Therapy Prgm
School of Public Health
Columbia, SC 29208
*Prgm Dir:* Bruce A McClenaghan, MPT PED
*Tel:* 803 777-5267   *Fax:* 803 777-8422
*E-mail:* bmcclena@gwm.sc.edu

# South Dakota

**University of South Dakota**
Physical Therapy Prgm
Dept of Physical Therapy
414 E Clark St
Vermillion, SD 57069
www.usd.edu
*Prgm Dir:* Lana R Svien, PT PhD
*Tel:* 605 677-5917   *Fax:* 605 677-6529
*E-mail:* lsvien@usd.edu

# Tennessee

**University of Tennessee - Chattanooga**
Physical Therapy Prgm
615 McCallie Ave
Chattanooga, TN 37403
*Prgm Dir:* Cathie Smith, PT PhD PCS
*Tel:* 423 425-5259   *Fax:* 423 425-2215
*E-mail:* Cathie-Smith@utc.edu

**East Tennessee State University**
Physical Therapy Prgm
Box 70624
Johnson City, TN 37614
www.etsu.edu/cpah/physther
*Prgm Dir:* David A Arnall, PT PhD FACSM ES
*Tel:* 423 439-8275   *Fax:* 423 439-8077
*E-mail:* arnall@etsu.edu
*Med Dir:* Susan Epps, EdD

**University of Tennessee Health Science Ctr**
Physical Therapy Prgm
930 Madison Ave
Ste 640
Memphis, TN 38163
www.utmem.edu/physther
*Prgm Dir:* Barbara H Connolly, PT EdD FAPTA
*Tel:* 901 448-5888   *Fax:* 901 448-7545
*E-mail:* bconnolly@utmem.edu

**Belmont University**
Physical Therapy Prgm
1900 Belmont Blvd
Nashville, TN 37212-3757
www.belmont.edu/pt
*Prgm Dir:* John S Halle, PT PhD ECS
*Tel:* 615 460-6727   *Fax:* 615 460-6729
*E-mail:* hallej@mail.belmont.edu

**Tennessee State University**
Physical Therapy Prgm
3500 John A Merritt Blvd, Box 9564
Nashville, TN 37209-1561
*Prgm Dir:* Rosalyn Pitt, EdD PT
*Tel:* 615 963-5881   *Fax:* 615 963-5935
*E-mail:* rpitt@tnstate.edu

# Texas

**Hardin-Simmons University**
Physical Therapy Prgm
2200 Hickory St
Box 16065 HSU Station
Abilene, TX 79698-6065
www.hsutx.edu/academics/graduate/programs/
    physicaltherapy
*Prgm Dir:* Janelle O'Connell, PhD PT ATC-L
*Tel:* 325 670-5860   *Fax:* 325 670-5868
*E-mail:* joconnel@hsutx.edu

**Univ of Texas Southwestern Med Ctr at Dallas**
Physical Therapy Prgm
Southwestern Allied Health Sciences Sch
5323 Harry Hines Blvd
Dallas, TX 75390-8876
www.utsouthwestern.edu/pt
*Prgm Dir:* Patricia Winchester, PT PhD
*Tel:* 214 648-1551   *Fax:* 214 648-1511
*E-mail:* patricia.winchester@utsouthwestern.edu

**University of Texas at El Paso**
Physical Therapy Prgm
1101 N Campbell
El Paso, TX 79902-0581
*Prgm Dir:* J A Ryberg, PT MA
*Tel:* 915 747-8210   *Fax:* 915 747-8211
*E-mail:* jaryberg@utep.edu

**US Army**
*Cosponsor:* Baylor University
Physical Therapy Prgm
3151 Scott Rd, Ste 1230
Ft Sam Houston, TX 78234-6138
*Prgm Dir:* Lt Col Josef H Moore, PT PhD SCS
*Tel:* 210 221-8410   *Fax:* 210 221-7585
*E-mail:* Cynthia.quiroz@cen.amedd.army.mil

**University of Texas Medical Branch/SAHS**
Physical Therapy Prgm
School of Allied Health Sciences
301 University Blvd
Galveston, TX 77555-1144
*Prgm Dir:* Elizabeth Protas, PT PhD FACSM
*Tel:* 409 772-3068   *Fax:* 409 747-1613
*E-mail:* ejprotas@utmb.edu

**Texas Woman's University**
Physical Therapy Prgm
School of Physical Therapy
1130 John Freeman Blvd
Houston, TX 77030-2897
*Prgm Dir:* Sharon Olson, PhD PT
*Tel:* 713 794-2070   *Fax:* 713 794-2071
*E-mail:* solson@twu.edu

**Texas Tech Univ Health Sciences Center**
Physical Therapy Prgm
3601 Fourth St
Lubbock, TX 79430
*Prgm Dir:* Kerry K Gilbert, PT ScD
*Tel:* 806 743-3223   *Fax:* 806 743-1262
*E-mail:* kerry.gilbert@ttuhsc.edu

**Angelo State University**
Physical Therapy Prgm
2601 West Ave N
ASU Station #10923
San Angelo, TX 76909-0923
*Prgm Dir:* Shelly Weise, PT EdD
*Tel:* 325 942-2545   *Fax:* 325 942-2548
*E-mail:* shelly.weise@angelo.edu
*Med Dir:* Mark Pape, MS PT

**Univ of Texas Hlth Sci Ctr at San Antonio**
Physical Therapy Prgm
7703 Floyd Curl Dr
MSC 6247
San Antonio, TX 78229-3900
www.uthscsa.edu/sah/pt
*Prgm Dir:* Giovanni DeDomenico, PhD
*Tel:* 210 567-8750   *Fax:* 210 567-8774
*E-mail:* dedomenico@uthscsa.edu
*Med Dir:* Denice Trevino

**Texas State University-San Marcos**
Physical Therapy Prgm
Health Science Center
601 University Dr
San Marcos, TX 78666
www.health.txstate.edu/pt
*Prgm Dir:* Barbara Sanders, PhD PT SCS
*Tel:* 512 245-8351   *Fax:* 512 245-8736
*E-mail:* bs04@txstate.edu

# United Kingdom

**Robert Gordon University**
Physical Therapist Prgm
Garthdee Rd
Garthdee Aberdeen, UK AB10 7QG
www.rgu.ac.uk/prospectus/disp_pgProspectusEntry.cfm?
   CourseID=MSHSPH
*Prgm Dir:* Alasdair MacSween, PhD
*E-mail:* a.macsween@rgu.ac.uk

# Utah

**University of Utah**
Physical Therapy Prgm
Div of Physical Therapy
520 Wakara Way
Salt Lake City, UT 84108-1290
www.health.utah.edu/pt
*Prgm Dir:* R Scott Ward, PhD PT
*Tel:* 801 581-4895   *Fax:* 801 585-5629
*E-mail:* scott.ward@hsc.utah.edu
*Med Dir:* Joyce Bawden, Academic Advisor

# Vermont

**University of Vermont**
Physical Therapy Prgm
College of Nursing & Health Sciences
305 Rowell Bldg, 106 Carrigan Dr
Burlington, VT 05405-0068
www.uvm.edu/~cnhs
*Prgm Dir:* Diane U Jette, PT DSc
*Tel:* 802 656-3252   *Fax:* 802 656-6586
*E-mail:* diane.jette@uvm.edu

# Virginia

**Marymount University**
Physical Therapy Prgm
2807 N Glebe Rd
Arlington, VA 22207-4299
*Prgm Dir:* Rita Wong, PT EdD
*Tel:* 703 284-5982   *Fax:* 703 284-5981
*E-mail:* rita.wong@marymount.edu

**Hampton University**
Physical Therapy Prgm
Department of Physical Therapy
Phenix Hall Rm 216
Hampton, VA 23668
*Prgm Dir:* Marilyns G Randolph, PT PhD
*Tel:* 757 727-5260   *Fax:* 757 728-6546
*E-mail:* marilys.randolph@hamptonu.edu

**Old Dominion University**
Physical Therapy Prgm
School of Physical Therapy
129 Wm B Spong Jr Hall
Norfolk, VA 23529-0288
*Prgm Dir:* George Maihafer, PhD PT
*Tel:* 757 683-4519   *Fax:* 757 683-4410
*E-mail:* gmaihafe@odu.edu

**Virginia Commonwealth Univ/Med Coll of VA**
Physical Therapy Prgm
Medical College of Virginia Campus
Box 980224
Richmond, VA 23298-0224
www.vcu.edu/pt
*Prgm Dir:* Thomas P Mayhew, PhD PT
*Tel:* 804 828-0234   *Fax:* 804 828-8111
*E-mail:* tmayhew@vcu.edu

**Shenandoah University**
Physical Therapy Prgm
333 W Cork St
Winchester, VA 22601
*Prgm Dir:* Rose Schmieg, PT DHSc ATC OCS
*Tel:* 540 665-5520   *Fax:* 540 665-5530
*E-mail:* rschmieg@su.edu

# Washington

**University of Washington**
Physical Therapy Prgm
Physical Therapy CC-902, Rehab Medicine
1959 NE Pacific St, Box 356490
Seattle, WA 98195-6490
*Prgm Dir:* Mark R Guthrie, PhD PT
*Tel:* 206 598-5340   *Fax:* 206 685-3244
*E-mail:* mguthrie@u.washington.edu

**Eastern Washington University**
Physical Therapy Prgm
310 N Riverpoint Blvd
Box T, Rm 270
Spokane, WA 99202-0002
www.ewu.edu/pt
*Prgm Dir:* Byron Russell, PhD
*Tel:* 509 368-6608   *Fax:* 509 368-6623
*E-mail:* byron.russell@mail.ewu.edu

**University of Puget Sound**
Physical Therapy Prgm
1500 N Warner
CMB 1070
Tacoma, WA 98416
www.ups.edu/pt
*Prgm Dir:* Kathie Hummel-Berry, PhD PT
*Tel:* 253 879-3281   *Fax:* 253 879-2933
*E-mail:* hummel@ups.edu

# West Virginia

**West Virginia University**
Physical Therapy Prgm
School of Medicine, Robert C Byrd Health Sciences
   Center
PO Box 9226
Morgantown, WV 26506-9226
*Prgm Dir:* MaryBeth Mandich, PhD PT
*Tel:* 304 293-1320   *Fax:* 304 293-7105
*E-mail:* mmandich@hsc.wvu.edu
*Med Dir:* Robert D'Alessandri, MD

**Wheeling Jesuit University**
Physical Therapy Prgm
316 Washington Ave
Wheeling, WV 26003
*Prgm Dir:* Luis G Vargas, PT PhD ATRIC
*Tel:* 304 243-2432, Ext 2291   *Fax:* 304 243-2042
*E-mail:* lvargas@wju.edu

# Wisconsin

**University of Wisconsin - La Crosse**
Physical Therapy Prgm
4033 Health Science Ctr
1725 State St
La Crosse, WI 54601
*Prgm Dir:* Michele Thorman, PT MBA
*Tel:* 608 785-8475   *Fax:* 608 785-8460
*E-mail:* thorman.mich@uwlax.edu

**University of Wisconsin - Madison**
Physical Therapy Prgm
5173 Medical Sciences Center
1300 University Ave
Madison, WI 53706-1532
www.orthorehab.wisc.edu/pt
*Prgm Dir:* Lisa Steinkamp, PT MS MBA
*Tel:* 608 263-7131   *Fax:* 608 262-7809
*E-mail:* steinkam@surgery.wisc.edu
*Med Dir:* Reenie Euhardy, PT MS GCS

**Concordia University Wisconsin**
Physical Therapy Prgm
12800 N Lake Shore Dr
Mequon, WI 53092-7699
*Prgm Dir:* Ruth Gresley, PhD RN
*Tel:* 262 243-4280  *Fax:* 262 243-4506
*E-mail:* ruth.gresley@cuw.edu

**Marquette University**
Physical Therapy Prgm
PO Box 1881
Milwaukee, WI 53201-1881
www.marquette.edu/chs/pt
*Prgm Dir:* Lawrence G Pan, PhD PT
*Tel:* 414 288-7194  *Fax:* 414 288-5987
*E-mail:* lawrence.pan@marquette.edu

**Carroll College**
Physical Therapy Prgm
100 N East Ave
Waukesha, WI 53186
www.cc.edu
*Prgm Dir:* Jane F Hopp, PT PhD
*Tel:* 262 524-7294  *Fax:* 262 524-7690
*E-mail:* jhopp@cc.edu
*Med Dir:* Jennifer Wells-Sperry

## Physical Therapist

| Programs* | Class Capacity | Begins | Length (months) | Award | Res Tuition | Non-res Tuition | Stipend | Offers:‡ 1 | 2 | 3 |
|---|---|---|---|---|---|---|---|---|---|---|
| **Alabama** | | | | | | | | | | |
| University of Alabama at Birmingham | 35 | Jan | 36 | DPT | $11,502 | $28,782 | | | | • |
| University of South Alabama (Mobile) | 36 | Aug | 32 | DPT | $7,849 | $15,698 | | | | |
| **Arizona** | | | | | | | | | | |
| AT Still University of Health Sciences (Mesa) | 54 | Sep | 36 | DPT | $21,985 | $21,985 | | | | • |
| Northern Arizona University (Flagstaff) | 44 | Aug | 33 | DPT | $6,162 | $15,122 | | | | • |
| **California** | | | | | | | | | | |
| Azusa Pacific University | 40 | Feb | 31 | DPT | $23,750 | $23,750 | | • | | |
| California State University - Fresno | 20 | Aug | 24 | MPT | $3,515 | $10,170 | | | | • |
| California State University - Long Beach | 40 | Aug | 33 | MPT | $3,926 | $14,096 | | | | • |
| California State University - Northridge | 40 | Aug | 36 | MPT | $3,042 | $13,890 | | | | • |
| California State University - Sacramento | 32 | Aug | 28 | MPT | $3,702 | $10,770 | | | | • |
| Mount St Mary's College (Los Angeles) | 26 | Aug | 38 | Dipl, DPT | $24,000 | $24,000 | | | | • |
| University of Southern California (Los Angeles) | 95 | Aug | 33 | DPT | $42,082 | $42,082 | | | | • |
| University of the Pacific (Stockton) | 36 | Aug | 25 | DPT | $40,380 | $40,380 | | | | • |
| Western Univ of Health Sciences (Pomona) | 44 | Aug | 36 | DPT | $25,500 | $25,500 | | | | • |
| **Colorado** | | | | | | | | | | |
| Regis University (Denver) | 52 | Aug | 33 | DPT | $20,606 | $20,606 | | | | • |
| U of Colorado at Denver and Health Scis Ctr | 60 | Jun | 36 | DPT | $14,577 | $34,142 | | | | • |
| **Connecticut** | | | | | | | | | | |
| Quinnipiac University (Hamden) | 100 | Sep | 66 | MPT | $23,000 | $23,000 | | | | • |
| University of Connecticut (Storrs) | 30 | May | 36 | MS | $10,620 | $22,745 | | | | • |
| **Florida** | | | | | | | | | | |
| Florida International University (Miami) | 48 | Aug | 30 | MS | $8,415 | $24,747 | | • | | • |
| Univ of St Augustine for Health Sciences | 56 | Jan May Sep | 28 | DPT | $22,800 | $22,800 | | | | |
| University of Florida (Gainesville) | 50 | Aug | 32 | DPT | $13,000 | $21,000 | | | | • |
| **Georgia** | | | | | | | | | | |
| Georgia State University (Atlanta) | 40 | Aug | 36 | DPT | $8,331 | $29,595 | | | | • |
| **Illinois** | | | | | | | | | | |
| Bradley University (Peoria) | 20 | Jun | 36 | DPT | $18,333 | $18,333 | | | | • |
| Midwestern University (Downers Grove) | 45 | Jun | 34 | DPT | $22,934 | $22,934 | | | | • |
| Northwestern University (Chicago) | 60 | Late Aug | 28 | DPT | $30,279 | $30,279 | | | | • |
| **Indiana** | | | | | | | | | | |
| Indiana University (Indianapolis) | 36 | Aug | 35 | DPT | $7,500 | $17,000 | | | | • |
| University of Evansville | 40 | Jun | 42 | MPT | $21,600 | $21,600 | | | | • |
| University of Indianapolis | 46 | Aug | 33 | DPT | $23,990 | $23,990 | | | • | • |
| **Iowa** | | | | | | | | | | |
| Clarke College (Dubuque) | 30 | Aug | 33 | DPT | $19,500 | $19,500 | | | | • |
| University of Iowa (Iowa City) | 36 | Jul | 30 | DPT | $13,698 | $28,781 | | | | • |
| **Kentucky** | | | | | | | | | | |
| Bellarmine University (Louisville) | 48 | Jun | 33 | DPT | $27,000 | $27,000 | | | | • |
| University of Kentucky (Lexington) | 64 | Aug | 39 | DPT | $13,000 | $30,000 | | | | • |
| **Louisiana** | | | | | | | | | | |
| Louisiana State Univ Health Sciences Center (New Orleans) | 40 | May | 36 | DPT | | | | | | • |
| **Maine** | | | | | | | | | | |
| University of New England (Portland) | 30 | Sep | 60 | DPT | $25,000 | $25,000 | | | | |
| **Maryland** | | | | | | | | | | |
| University of Maryland (Baltimore) | 50 | Jun | 36 | DPT | $13,648 | $24,380 | | | | • |
| University of Maryland Eastern Shore (Princess Anne) | 30 | Sep | 36 | DPT | $8,800 | $16,200 | | | | |
| **Massachusetts** | | | | | | | | | | |
| Boston University | 60 | Jun | 36 | DPT | $33,300 | $0 | | • | | |
| Northeastern University (Boston) | 100 | Sep | 72 | DPT | $29,910 | $0 | | | • | • |
| Simmons College (Boston) | 44 | Jul | 36 | DPT | $27,390 | $27,390 | $6,000 | | | • |
| Springfield College | 38 | Sep | 66 | MS | $22,715 | $22,715 | | | | • |
| University of Massachusetts - Lowell | 30 | Sep | 31 | DPT | $9,921 | $18,316 | | | | • |

*Data are shown only for programs that completed the 2006 AMA Survey of Health Professions Education Programs
‡Key to Offers: 1: Evening or weekend classes; 2: Non-English instruction; 3: Cultural competence instruction

# Physical Therapist

## Physical Therapist

| Programs* | Class Capacity | Begins | Length (months) | Award | Res Tuition | Non-res Tuition | Stipend | 1 | 2 | 3 |
|---|---|---|---|---|---|---|---|---|---|---|
| **Michigan** | | | | | | | | | | |
| Andrews University (Berrien Springs) | 36 | Aug | 35, 24 | DPT, DScPT | $24,000 | $24,000 | | | | |
| Grand Valley State University (Grand Rapids) | 40 | Aug | 36 | DPT | | | | | | • |
| University of Michigan - Flint | 40 | Fall | 42 | DPT | $11,544 | $12,734 | | | | |
| **Minnesota** | | | | | | | | | | |
| Mayo School of Health Sciences (Rochester) | 28 | Aug | 33 | DPT | $20,000 | $20,000 | | | | • |
| University of Minnesota - Minneapolis | 50 | Jul | 35 | Dipl, Cert, DPT | $16,419 | $28,674 | | | | • |
| **Mississippi** | | | | | | | | | | |
| University of Mississippi Medical Center (Jackson) | 48 | May | 36 | DPT | $3,603 | $5,250 | | | | • |
| **Missouri** | | | | | | | | | | |
| Maryville University (St Louis) | 40 | Aug | 66 | DPT | $17,000 | $17,000 | | | | • |
| Rockhurst University (Kansas City) | 40 | Jun | 36 | DPT | $19,620 | $19,620 | | | | • |
| Southwest Baptist University (Bolivar) | 32 | Aug | 33 | DPT | $16,500 | $16,500 | | | | • |
| University of Missouri - Columbia | 40 | Jun | 36 | MPT | $10,778 | $16,174 | | | | • |
| Washington University (St Louis) | 81 | Aug | 33 | Dipl, DPT | $27,206 | $27,206 | | | | • |
| **Montana** | | | | | | | | | | |
| University of Montana (Missoula) | 32 | Aug | 33 | DPT | $11,000 | $22,000 | | | | • |
| **Nebraska** | | | | | | | | | | |
| University of Nebraska Medical Center (Omaha) | 40 | Aug | 22 | DPT | $8,003 | $19,756 | | | | • |
| **Nevada** | | | | | | | | | | |
| University of Nevada - Las Vegas | 20 | Jun | 36 | DPT | $9,948 | $19,448 | | | | • |
| **New Jersey** | | | | | | | | | | |
| SUNJ Rutgers Camden/Univ of Med & Dent NJ (Stratford) | 26 | May | 36 | MPT, DPT | $18,712 | $27,815 | | | | • |
| Univ of Med & Dent of New Jersey (Newark) | 44 | Aug | 36 | Dipl, DPT | $15,000 | $23,000 | | | | • |
| **New Mexico** | | | | | | | | | | |
| University of New Mexico (Albuquerque) | 24 | Jun | 36 | MPT | $5,978 | $15,584 | | | • | • |
| **New York** | | | | | | | | | | |
| Columbia University (New York) | 40 | Sep | 33 | DPT | $27,500 | $27,500 | | | | • |
| CUNY College of Staten Island | 25 | Jan Feb-Spring semester | 36 | DPT | $7,670 | $20,720 | | | | • |
| CUNY Hunter College (New York) | 40 | Jun | 36 | DPT | $4,350 | $13,200 | | | | • |
| D'Youville College (Buffalo) | 48 | Sep | 34 | DPT | $19,837 | $19,837 | | | | • |
| Daemen College (Amherst) | 45 | Sep | 33 | DPT | $16,350 | $16,350 | | | | • |
| Dominican College (Orangeburg) | 28 | Jun | 40, 20 | Dipl, DPT, t-DPT | $0 | $15,000 | | • | | • |
| Ithaca College | 85 | Aug | 72 | BS, DPT | $26,832 | $26,832 | | | | • |
| Mercy College (Dobbs Ferry) | 24 | Sep | 38 | BS, MS | $16,000 | $16,000 | | • | | • |
| Nazareth College of Rochester | 40 | Aug | 20, 34 | BS/DPT | $22,000 | $22,000 | | | | • |
| New York Medical College (Valhalla) | 30 | Jun | 36 | Dipl, DPT | $23,500 | $0 | | | | • |
| New York University | 30 | Jun | 39 | DPT | $33,000 | $33,000 | | • | | • |
| SUNY Upstate Medical University (Syracuse) | 32 | Jun | 36 | DPT | $10,920 | $17,540 | | | | • |
| The Sage Colleges (Troy) | 36 | May | 36 | DPT | $670 | $670 | | | | • |
| Touro College (New York) | 80 | Sep | 36 | Dipl, BS/DPT | $19,550 | $19,550 | | | | • |
| University at Buffalo - SUNY | 42 | Aug | 33 | DPT | $10,920 | $17,540 | | | | • |
| Utica College | 36 | Aug | 36 | DPT, DPT | $25,300 | $25,300 | | | | • |
| **North Carolina** | | | | | | | | | | |
| Elon University | 36 | Jan | 36 | DPT | $23,995 | $0 | | | • | • |
| Winston-Salem State University | 26 | May | 30 | MPT | $4,768 | $14,955 | | | | • |
| **North Dakota** | | | | | | | | | | |
| University of Mary (Bismarck) | 30 | Aug | 32 | DPT | $15,000 | $15,000 | | | | • |
| University of North Dakota (Grand Forks) | 48 | Aug | 32 | DPT | $12,500 | $16,500 | | | | • |
| **Ohio** | | | | | | | | | | |
| College of Mt St Joseph (Cincinnati) | 40 | Jun | 36 | DPT | $26,400 | $26,400 | | | | • |
| Ohio State University (Columbus) | 40 | Jun | 30, 39 | MPT, DPT | $12,820 | $28,660 | | | | • |
| Ohio University (Athens) | 36 | Jun | 36 | DPT | $12,264 | $22,920 | $7,200 | | | • |
| University of Cincinnati | 24 | Sep | 27 | MPT | $15,524 | $28,636 | | | | • |
| University of Findlay | 32 | Aug Jan | 33, 29 | MPT, MPT | $30,953 | $30,953 | | • | | • |
| Youngstown State University | 20 | Summer | 36 | MPT | $6,000 | $6,000 | | | | • |
| **Ontario, Canada** | | | | | | | | | | |
| University of Toronto | 83 | Sep | 26 | MScPT | $7,509 | $19,300 | | | | • |
| **Oregon** | | | | | | | | | | |
| Pacific University (Hillsboro) | 38 | Last week of Aug | 27 | DPT | $22,540 | $22,540 | | | | • |
| **Pennsylvania** | | | | | | | | | | |
| Arcadia University (Glenside) | 50 | Sep | 30 | DPT | $21,940 | $0 | 2500 | | | • |
| Chatham College (Pittsburgh) | 30 | Aug | 28 | DPT | $32,000 | $32,000 | | | | • |
| College Misericordia (Dallas) | 50 | Aug | 50 | BS, MSPT | $18,700 | $18,700 | | | | • |

*Data are shown only for programs that completed the 2006 AMA Survey of Health Professions Education Programs
‡Key to Offers: 1: Evening or weekend classes; 2: Non-English instruction; 3: Cultural competence instruction

## Physical Therapist

| Programs* | Class Capacity | Begins | Length (months) | Award | Res Tuition | Non-res Tuition | Stipend | Offers:‡ 1 | 2 | 3 |
|---|---|---|---|---|---|---|---|---|---|---|
| Drexel University (Philadelphia) | 50 | Aug | 32 | DPT | $23,940 | $23,940 | | | | |
| Duquesne University (Pittsburgh) | 40 | Aug | 36 | DPT | $26,272 | $26,272 | | | | • |
| Gannon University (Erie) | 40 | Aug | 36 | DPT | $27,185 | $27,185 | | | | |
| Slippery Rock University | 50 | Aug Sep | 33 | DPT | $16,403 | $25,783 | | | | • |
| Thomas Jefferson University (Philadelphia) | 44 | Sep | 36 | DPT | $24,350 | $0 | | | | |
| University of the Sciences in Philadelphia | 45 | May | 36 | Dipl, DPT | $25,078 | $0 | | | | • |
| Widener University (Chester) | 40 | Sep | 33 | DPT | $22,165 | $22,165 | | | | • |
| **Puerto Rico** | | | | | | | | | | |
| University of Puerto Rico (San Juan) | 48 | Aug | 28 | MS | $2,305 | $0 | | | | |
| **Rhode Island** | | | | | | | | | | |
| University of Rhode Island (Kingston) | 24 | Sep | 28 | DPT | $4,894 | $14,180 | | | | • |
| **South Carolina** | | | | | | | | | | |
| Medical University of South Carolina (Charleston) | 60 | May | 36 | DPT | $16,245 | $32,391 | | | | • |
| **South Dakota** | | | | | | | | | | |
| University of South Dakota (Vermillion) | 26 | Jul | 34 | DPT | $9,687 | $19,345 | | | | • |
| **Tennessee** | | | | | | | | | | |
| Belmont University (Nashville) | 32 | Aug | 36 | DPT | $21,150 | $0 | | | | • |
| East Tennessee State University (Johnson City) | 30 | Jan | 36 | DPT | $7,809 | $20,541 | | • | | • |
| University of Tennessee Health Science Ctr (Memphis) | 60 | Sep | 33 | DPT | $6,884 | $16,232 | | | | • |
| **Texas** | | | | | | | | | | |
| Hardin-Simmons University (Abilene) | 28 | Aug | 28 | DPT | $22,500 | $22,500 | | | | • |
| Texas State University-San Marcos | 40 | Jun | 24 | MSPT | $15,000 | $28,000 | | | | • |
| Univ of Texas Hlth Sci Ctr at San Antonio | 40 | Aug | 30 | MPT | $3,152 | $10,670 | | | | |
| Univ of Texas Southwestern Med Ctr at Dallas | 40 | May | 24 | MPT | $8,743 | $18,118 | | | | |
| University of Texas at El Paso | 24 | Sep | 33 | MPT | $2,500 | $6,600 | | | | |
| **Utah** | | | | | | | | | | |
| University of Utah (Salt Lake City) | 40 | May | 33 | DPT | $12,000 | $24,000 | | | | • |
| **Vermont** | | | | | | | | | | |
| University of Vermont (Burlington) | 24 | Aug | 29 | DPT | $13,520 | $34,122 | | | | • |
| **Virginia** | | | | | | | | | | |
| Virginia Commonwealth Univ/Med Coll of VA (Richmond) | 54 | Jul | 31 | DPT | $8,320 | $17,460 | | | | |
| **Washington** | | | | | | | | | | |
| Eastern Washington University (Spokane) | 38 | Sep | 33 | DPT | $8,500 | $22,824 | | | | • |
| University of Puget Sound (Tacoma) | 30 | Sep | 33 | DPT | $22,620 | $22,620 | | | | • |
| **West Virginia** | | | | | | | | | | |
| Wheeling Jesuit University | 32 | Sep | 24 | DPT | $30,000 | $30,000 | | | | • |
| **Wisconsin** | | | | | | | | | | |
| Carroll College (Waukesha) | 40 | Sep | 28 | MPT | $19,500 | $19,500 | | | | • |
| Concordia University Wisconsin (Mequon) | 26 | Aug Sep | 36 | Dipl, DPT | $20,400 | $20,400 | | | | • |
| Marquette University (Milwaukee) | 62 | Sep | 32 | DPT | $26,570 | $26,570 | | | | |
| University of Wisconsin - Madison | 40 | Jun | 30 | MPT | $8,738 | $24,008 | | | | • |

*Data are shown only for programs that completed the 2006 AMA Survey of Health Professions Education Programs
‡Key to Offers: 1: Evening or weekend classes; 2: Non-English instruction; 3: Cultural competence instruction

## Physical Therapist Assistant*

## Alabama

### Jefferson State Community College
Physical Therapist Assistant Prgm
Center for Health & Biological Sciences
4600 Valleydale Rd
Birmingham, AL 35242
www.jeffstateonline.com
*Prgm Dir:* Glenn Ross, MSPT
*Tel:* 205 520-5995   *Fax:* 205 520-5992
*E-mail:* gross@jeffstateonline.com

*Additional information about programs that
returned the AMA's survey is available in a table
beginning on page 334.

### George C Wallace Community College
Physical Therapist Assistant Prgm
1141 Wallace Dr
Dothan, AL 36303
www.wallace.edu
*Prgm Dir:* Priscilla Tucker, PT
*Tel:* 334 556-2213   *Fax:* 334 983-3600
*E-mail:* ptucker@wallace.edu

### Wallace State Community College
Physical Therapist Assistant Prgm
PO Box 2000
Hanceville, AL 35077-2000
www.wallacestate.edu
*Prgm Dir:* Alina C Adams, PT
*Tel:* 256 352-8332   *Fax:* 256 352-8320
*E-mail:* alina.adams@wallacestate.edu

### Bishop State Community College
Physical Therapist Assistant Prgm
1365 Martin Luther King Ave
Mobile, AL 36603-5362
*Prgm Dir:* Pam Wehner, PT MS
*Tel:* 334 405-4441   *Fax:* 334 405-4442
*E-mail:* pwehner@bishop.edu

### South University
Physical Therapist Assistant Prgm
5355 Vaughn Rd
Montgomery, AL 36116-1120
www.southuniversity.edu
*Prgm Dir:* Joanne Rice, PT DPT
*Tel:* 334 395-8800   *Fax:* 334 834-9559
*E-mail:* jrice@southuniversity.edu

## Arizona

### GateWay Community College
Physical Therapist Assistant Prgm
108 N 40th St
Phoenix, AZ 85034
http://healthcare.gatewaycc.edu/Programs/
PhysicalTherapistAssistant
*Prgm Dir:* Peter Zawicki, MS PT
*Tel:* 602 286-8476   *Fax:* 602 286-8478
*E-mail:* peter.zawicki@gwmail.maricopa.edu

# Arkansas

**NorthWest Arkansas Community College**
Physical Therapist Assistant Prgm
One College Dr
Bentonville, AR 72712-5091
*Prgm Dir:* Deanna Fletcher, MPH PT
*Tel:* 479 619-4153   *Fax:* 479 619-4254
*E-mail:* dfletche@nwacc.edu

**South Arkansas Community College**
Physical Therapist Assistant Prgm
300 West Ave
PO Box 7010
El Dorado, AR 71730
*Prgm Dir:* Jennifer Parks, PT
*Tel:* 870 862-8131, Ext 189   *Fax:* 870 864-7104
*E-mail:* jparks@southark.edu

**Arkansas State University**
Physical Therapist Assistant Prgm
Department of Health Professions
PO Box 910
State University, AR 72467
*Prgm Dir:* Becky Keith, BSPT
*Tel:* 870 972-3591   *Fax:* 870 972-3652
*E-mail:* beckeith@astate.edu

# California

**Loma Linda University**
Physical Therapist Assistant Prgm
School of Allied Health Professions
Nichol Hall Rm 1911
Loma Linda, CA 92350
www.llu.edu/llu/sahp/pt/pta.htm
*Prgm Dir:* Jeannine Mendes, MPT
*Tel:* 909 558-4634   *Fax:* 909 558-0466
*E-mail:* jmendes@llu.edu

**Ohlone College**
Physical Therapist Assistant Prgm
Ohlone Community College Dist
Newark Ohlone Center, 35753 Cedar Blvd
Newark, CA 94560
*Prgm Dir:* Sheryl S Einfalt, MPT
*Tel:* 510 979-7482   *Fax:* 510 742-2315
*E-mail:* cthoel@ohlone.edu

**Cerritos Community College**
Physical Therapist Assistant Prgm
Health Occupations Div
11110 Alondra Blvd
Norwalk, CA 90650
www.cerritos.edu
*Prgm Dir:* Marijean Piorkowski, MS PT DPT
*Tel:* 562 860-2451, Ext 2580   *Fax:* 562 467-5077
*E-mail:* piorkowski@cerritos.edu

**Sonoma College**
Physical Therapist Assistant Prgm
1304 Southpoint Blvd, Ste 280
Petaluma, CA 94954
www.sonomacollege.edu
*Prgm Dir:* I Scott Thompson, MPT
*Tel:* 707 283-0800, Ext 106   *Fax:* 707 283-0808
*E-mail:* sthompson@sonomacollege.edu
*Med Dir:* Tricia Devin, EdD, PhD

**Sacramento City College**
Physical Therapist Assistant Prgm
Science and Allied Health
3835 Freeport Blvd
Sacramento, CA 95822
www.scc.losrios.edu/~sah/physther
*Prgm Dir:* Elizabeth Chape, PT PhD
*Tel:* 916 558-2298   *Fax:* 916 558-2392
*E-mail:* chapee@scc.losrios.edu

**San Diego Mesa College**
Physical Therapist Assistant Prgm
7250 Mesa College Dr
San Diego, CA 92111-4998
*Prgm Dir:* Laura Crandall, PT MS OCS
*Tel:* 619 388-2839   *Fax:* 619 388-2677
*E-mail:* lcrandal@sdccd.edu

# Colorado

**Morgan Community College**
Physical Therapist Assistant Prgm
17800 Rd 20
Ft Morgan, CO 80701
*Prgm Dir:* Carol J Leach, PT MHA PhD CHES
*Tel:* 970 542-3225   *Fax:* 970 867-6608
*E-mail:* carol.leach@morgancc.edu

**Arapahoe Community College**
Physical Therapist Assistant Prgm
2500 W College Dr, PO Box 9002
Littleton, CO 80160-9002
*Prgm Dir:* Paula Provence, MEd PT
*Tel:* 303 797-5897   *Fax:* 303 797-5842
*E-mail:* paula.provence@arapahoe.edu

**Pueblo Community College**
Physical Therapist Assistant Prgm
900 W Orman Ave
Pueblo, CO 81004
*Prgm Dir:* Lucinda (Cindy) Mihelich, PT MEd
*Tel:* 719 549-3433   *Fax:* 719 549-3381
*E-mail:* Cindy.Mihelich@pueblocc.edu

# Connecticut

**Naugatuck Valley Community College**
Physical Therapist Assistant Prgm
750 Chase Pkwy
Waterbury, CT 06708
www.nvcc.commnet.edu/allied_health
*Prgm Dir:* Cynthia M Lacouture, PT MA
*Tel:* 203 596-2157   *Fax:* 203 575-8146
*E-mail:* clacouture@nvcc.commnet.edu

# Delaware

**Delaware Tech & Comm Coll - Owens Campus**
Physical Therapist Assistant Prgm
PO Box 610
Georgetown, DE 19947
*Prgm Dir:* Joanne Howell, PT
*Tel:* 302 855-5933   *Fax:* 302 858-5460
*E-mail:* jhowell@college.dtcc.edu

**Delaware Tech & Comm Coll - Wilmington**
Physical Therapist Assistant Prgm
333 Shipley St
Wilmington, DE 19801
www.wilmington.dtcc.edu/wilmington/ah/pta.html
*Prgm Dir:* Douglas P Huisenga, MPT ATC
*Tel:* 302 888-5292   *Fax:* 302 577-6431
*E-mail:* DHuisenga@Christianacare.org

# Florida

**Manatee Community College**
Physical Therapist Assistant Prgm
5840 26th St W
Bradenton, FL 34207
*Prgm Dir:* Janice L Pollock, PT MHS
*Tel:* 941 752-5340   *Fax:* 941 727-8304
*E-mail:* pollocj@mccfl.edu

**Broward Community College**
Physical Therapist Assistant Prgm
Ctr for Health Science Education ll
1000 Coconut Creek Blvd
Coconut Creek, FL 33066
*Prgm Dir:* Susan Edelstein, PT MEd
*Tel:* 954 201-2086   *Fax:* 954 973-2348
*E-mail:* Sedelste@broward.edu

**Daytona Beach Community College**
Physical Therapist Assistant Prgm
1200 International Speedway Blvd
Daytona Beach, FL 32120-2811
*Prgm Dir:* Ruth A Freeman, PT MEd
*Tel:* 386 506-3752   *Fax:* 386 506-3300
*E-mail:* freemar@dbcc.edu

**Keiser University**
Physical Therapist Assistant Prgm
1500 NW 49th St
Ft Lauderdale, FL 33309
www.keisercollege.edu
*Prgm Dir:* Joseph Vibert, PT MBA
*Tel:* 954 776-4456, Ext 520   *Fax:* 954 351-4046
*E-mail:* jvibert@keisercollege.edu

**Indian River Community College**
Physical Therapist Assistant Prgm
3209 Virginia Ave
Ft Pierce, FL 34981-5596
*Prgm Dir:* Leila J Darress, PT MS
*Tel:* 561 462-4477   *Fax:* 561 462-4900
*E-mail:* ldarress@ircc.edu

**Florida Community College - Jacksonville**
Physical Therapist Assistant Prgm
4501 Capper Rd
Jacksonville, FL 32218-4499
www.fccj.edu
*Prgm Dir:* Gloria J Young, PT EdD
*Tel:* 904 766-5574   *Fax:* 904 766-5573
*E-mail:* gyoung@fccj.edu

**Lake City Community College**
Physical Therapist Assistant Prgm
149 SE College Place
Lake City, FL 32025
*Prgm Dir:* Olga Dreeben, PhD PT
*Tel:* 386 754-4358, Ext 4358   *Fax:* 386 754-4858
*E-mail:* dreebeno@lakecitycc.edu
*Med Dir:* Abraham Pallas, PhD

**Miami Dade College**
Physical Therapist Assistant Prgm
MDCC Medical Campus
950 NW 20th St
Miami, FL 33127
*Prgm Dir:* Kenneth Lee, PT MA
*Tel:* 305 237-4141   *Fax:* 305 237-4116
*E-mail:* kenneth.lee@mdc.edu

**Central Florida Community College**
Physical Therapist Assistant Prgm
PO Box 1388
Ocala, FL 34478-1388
*Prgm Dir:* Jean M McCauley, MHSA PT
*Tel:* 352 854-2322, Ext 1442   *Fax:* 352 237-0510
*E-mail:* mccaulej@cfcc.cc.fl.us

**Gulf Coast Community College**
Physical Therapist Assistant Prgm
5230 W US Hwy 98
Panama City, FL 32401-1041
http://health.gulfcoast.edu
*Prgm Dir:* Laura (Holly) Gunning, PT MHS EdD
*Tel:* 850 913-3312   *Fax:* 850 747-3246
*E-mail:* lgunning@gulfcoast.edu

**Seminole Community College**
Physical Therapist Assistant Prgm
100 Weldon Blvd
Sanford, FL 32773-6199
*Prgm Dir:* Carol A Clayton, PT PhD
*Tel:* 407 708-2234   *Fax:* 407 708-2319
*E-mail:* claytonc@scc-fl.edu

**St Petersburg College**
Physical Therapist Assistant Prgm
PO Box 13489
St Petersburg, FL 33733
www.spcollege.edu/hec/pta
*Prgm Dir:* Rebecca Kramer, MPT
*Tel:* 813 341-3614   *Fax:* 813 341-3744
*E-mail:* kramer.rebecca@spcollege.edu

**Pensacola Junior College**
Physical Therapist Assistant Prgm
Warrington Campus
5555 W Hwy 98
Warrington, FL 32507
*Prgm Dir:* Cena Harmon, PT DPT
*Tel:* 850 484-2373   *Fax:* 850 484-2364
*E-mail:* charmon@pjc.edu
*Med Dir:* T Joseph Dennie, MD PA

**South University**
Physical Therapist Assistant Prgm
West Palm Beach Campus
1760 N Congress Ave
West Palm Beach, FL 33409
*Prgm Dir:* Kenneth Amsler, PhD PT CQAUR
*Tel:* 561 697-9200   *Fax:* 561 697-9944
*E-mail:* kamsler@southuniversity.edu

**Polk Community College**
Physical Therapist Assistant Prgm
999 Ave H NE
Winter Haven, FL 33881
www.polk.edu
*Prgm Dir:* Nelson Marquez, PT EdD
*Tel:* 863 297-1010, Ext 5751   *Fax:* 863 297-1036
*E-mail:* nmarquez@polk.edu

# Georgia

**Darton College**
Physical Therapist Assistant Prgm
2400 Gillionville Rd
Albany, GA 31707
*Prgm Dir:* Kerri Johnson, MPT
*Tel:* 912 430-6904   *Fax:* 912 430-6910
*E-mail:* johnsonk@darton.edu

**Athens Technical College**
Physical Therapist Assistant Prgm
800 US Hwy 29 N
Athens, GA 30601-1500
*Prgm Dir:* Ellen P O'Keefe, PT DPT
*Tel:* 706 355-5055   *Fax:* 706 425-3104
*E-mail:* eokeefe@athenstech.edu

**Gwinnett Technical College**
Physical Therapist Assistant Prgm
5150 Sugarloaf Pkwy
Lawrenceville, GA 30043
*Prgm Dir:* Patricia Ann Balzer, PT MAHCE
*Tel:* 678 226-6635, Ext 6635   *Fax:* 770 962-7985
*E-mail:* pbalzer@GwinnettTech.edu
*Med Dir:* Janice McClure, PhD

**South University**
Physical Therapist Assistant Prgm
709 Mall Blvd
Savannah, GA 31406
*Prgm Dir:* Kenneth R Amsler, PT PhD CQAUR
*Tel:* 912 691-6000   *Fax:* 912 691-6092
*E-mail:* kamsler@southuniversity.edu

# Hawaii

**Kapi'olani Community College**
Physical Therapist Assistant Prgm
4303 Diamond Head Rd
Health Sciences Dept Kauila 122
Honolulu, HI 96816
www.kcc.hawaii.edu
*Prgm Dir:* Jill Wakabayashi, PTA MPH
*Tel:* 808 734-9398   *Fax:* 808 734-9126
*E-mail:* jwakabay@hawaii.edu

# Idaho

**Idaho State University**
Physical Therapist Assistant Prgm
College of Technology
Campus Box 8380
Pocatello, ID 83209-8380
www.isu.edu/departments/PTA
*Prgm Dir:* Jason B Shaw, MPT DA
*Tel:* 208 282-4815   *Fax:* 208 282-3975
*E-mail:* shawjaso@isu.edu

# Illinois

**Southwestern Illinois College**
Physical Therapist Assistant Prgm
2500 Carlyle Rd
Belleville, IL 62221
*Prgm Dir:* Kim Snyder, PTA MEd ACCE
*Tel:* 618 235-2700, Ext 5390   *Fax:* 618 235-2052
*E-mail:* kim.snyder@swic.edu

**Southern Illinois Univ at Carbondale**
Physical Therapist Assistant Prgm
SIU Clinical Ctr, 4602
Carbondale, IL 62901-4602
*Prgm Dir:* Janet L Rogers, PhD PTA
*Tel:* 618 453-6143   *Fax:* 618 453-6126
*E-mail:* jrogers@siu.edu

**Kaskaskia College**
Physical Therapist Assistant Prgm
27210 College Rd
Centralia, IL 62801
*Prgm Dir:* Keith Shaw, PT
*Tel:* 618 532-5931   *Fax:* 618 532-5948
*E-mail:* kshaw@kaskaskia.edu

**Morton College**
Physical Therapist Assistant Prgm
3801 S Central Ave
Cicero, IL 60804
*Prgm Dir:* Kelly Hawthorne
*Tel:* 708 656-8000, Ext 291   *Fax:* 708 656-8031
*E-mail:* Kelly.Hawthorn@morton.edu

**Oakton Community College**
Physical Therapist Assistant Prgm
1600 E Golf Rd
Des Plaines, IL 60016
www.oakton.edu/acad/dept/pta
*Prgm Dir:* Mary DeNotto, MSPT
*Tel:* 847 635-1857   *Fax:* 847 635-1764
*E-mail:* maryd@oakton.edu

**Lake Land College**
Physical Therapist Assistant Prgm
LLC Kluthe Center
1204 Network Center Dr
Effingham, IL 62401
*Prgm Dir:* Martha T Mioux, MHS PT
*Tel:* 217 540-3551   *Fax:* 217 540-3599
*E-mail:* mmioux@lakeland.cc.il.us

**College of DuPage**
Physical Therapist Assistant Prgm
IC 1028, 425 Fawell Blvd
Glen Ellyn, IL 60137-6599
www.cod.edu
*Prgm Dir:* Donald Schmidt, PT MS
*Tel:* 630 942-4076   *Fax:* 630 858-5409
*E-mail:* schmidt@cdnet.cod.edu

**Black Hawk College**
Physical Therapist Assistant Prgm
6600 34th Ave
Moline, IL 61265-5899
www.bhc.edu
*Prgm Dir:* Larry Gillund, MS PT
*Tel:* 309 796-1311, Ext 5393   *Fax:* 309 796-5357
*E-mail:* gillundl@bhc.edu

**Illinois Central College**
Physical Therapist Assistant Prgm
201 SW Adams St
Peoria, IL 61635-0001
www.icc.edu
*Prgm Dir:* Julie A Feeny, PT MS
*Tel:* 309 999-4648   *Fax:* 309 673-9626
*E-mail:* jfeeny@icc.edu

# Indiana

**University of Evansville**
Physical Therapist Assistant Prgm
1800 Lincoln Ave
Evansville, IN 47722
*Prgm Dir:* Barbara Hahn, MA PT
*Tel:* 812 479-2571   *Fax:* 812 479-2717
*E-mail:* bh38@evansville.edu

**University of Saint Francis**
Physical Therapist Assistant Prgm
2701 Spring St
Ft Wayne, IN 46808
*Prgm Dir:* Mary Kay Solon, PT MS
*Tel:* 260 434-7662   *Fax:* 260 434-7585
*E-mail:* mksolon@sf.edu
*Med Dir:* Nancy Gillespie, PhD RN

**Ivy Tech Community College of Indiana**
Physical Therapist Assistant Prgm
1440 E 35th Ave
Gary, IN 46409
*Prgm Dir:* James Dye, DPT
*Tel:* 219 981-4430   *Fax:* 219 981-4821
*E-mail:* jdye@ivytech.edu

**University of Indianapolis**
Physical Therapist Assistant Prgm
Krannert School of Physical Therapy
1400 E Hanna Ave
Indianapolis, IN 46227-3697
http://pt.uindy.edu/pta
*Prgm Dir:* William Staples, PT DPT GSC
*Tel:* 317 788-3500   *Fax:* 317 788-3542
*E-mail:* stapleswh@uindy.edu
*Med Dir:* Anne Hardwick, MS

**Ivy Tech Community College EC - Muncie**
Physical Therapist Assistant Prgm
4301 S Cowan Rd
Muncie, IN 47302
*Prgm Dir:* Mark Wise, PT MA
*Tel:* 765 289-2291, Ext 1404   *Fax:* 765 289-2292
*E-mail:* mwise@ivytech.edu

**Brown Mackie College**
Physical Therapist Assistant Prgm
1030 E Jefferson Blvd
South Bend, IN 46617
www.brownmackie.edu
*Prgm Dir:* Shirley D Mapes, PT MS MBA
*Tel:* 574 237-0774   *Fax:* 574 237-3585
*E-mail:* smapes@amedcts.com

**Vincennes University**
Physical Therapist Assistant Prgm
Health Occupations Dept
Vincennes, IN 47591
*Prgm Dir:* Mark F Goodrich, PT MS ATC/L
*Tel:* 812 888-4416   *Fax:* 812 888-4550
*E-mail:* mgoodrich@vinu.edu

# Iowa

**Kirkwood Community College**
Physical Therapist Assistant Prgm
6301 Kirkwood Blvd SW
Cedar Rapids, IA 52406
www.kirkwood.edu
*Prgm Dir:* Maggie Thomas, MA PT
*Tel:* 319 398-5566   *Fax:* 319 398-1293
*E-mail:* maggie.thomas@kirkwood.edu

**North Iowa Area Community College**
Physical Therapist Assistant Prgm
500 College Dr
Mason, IA 50401
*Prgm Dir:* Carol Patnode, MA PTA
*Tel:* 515 422-4339   *Fax:* 515 422-4115
*E-mail:* patnocar@niacc.edu

**Indian Hills Community College**
Physical Therapist Assistant Prgm
Health Occupations Division
Ottumwa Campus, 525 Grandview
Ottumwa, IA 52501
*Prgm Dir:* Debi Fritz, MS PT
*Tel:* 641 683-5271   *Fax:* 641 683-5254
*E-mail:* dfritz@indianhills.edu

**Western Iowa Tech Community College**
Physical Therapist Assistant Prgm
4647 Stone Ave, PO Box 5199
Sioux City, IA 51102-5199
*Prgm Dir:* Barbara-Anne Huculak, EdD PT
*Tel:* 712 274-6400, Ext 1321   *Fax:* 712 274-6412
*E-mail:* huculab@witcc.edu

# Kansas

**Colby Community College**
Physical Therapist Assistant Prgm
1255 South Range
Colby, KS 67701
www.colbycc.edu
*Prgm Dir:* Patricia A Erickson, PT DPT
*Tel:* 785 460-4797   *Fax:* 785 460-4788
*E-mail:* pate@colbycc.edu

**Kansas City Kansas Community College**
Physical Therapist Assistant Prgm
PO Box 12951
7250 State Ave
Kansas City, KS 66112-9978
*Prgm Dir:* Wanda Gattshall-Peresic, PT DPT
*Tel:* 913 288-7331   *Fax:* 913 288-7649
*E-mail:* wandag@toto.net

**Washburn University**
Physical Therapist Assistant Prgm
School of Applied Studies
1700 SW College Ave
Topeka, KS 66621
*Prgm Dir:* Lori Khan, PT DPT MS
*Tel:* 785 231-1010, Ext 1406   *Fax:* 785 231-1027
*E-mail:* lori.khan@washburn.edu

# Kentucky

**Hazard Community & Technical College**
*Cosponsor:* Southeast Kentucky Community & Technical
   College
Physical Therapist Assistant Prgm
One Community College Dr
Hazard, KY 41701-2402
*Prgm Dir:* Tracy L Bowling, PT
*Tel:* 606 487-3379   *Fax:* 606 439-1600
*E-mail:* tracy.bowling@kctcs.edu

**Jefferson Community and Technical College**
Physical Therapist Assistant Prgm
109 E Broadway St
Louisville, KY 40202-2005
www.jcc.kctcs.net/pta
*Prgm Dir:* Peri Jacobson, PT MBA
*Tel:* 502 213-2193   *Fax:* 502 213-2491
*E-mail:* peri.jacobson@kctcs.edu

**Madisonville Community College**
Physical Therapist Assistant Prgm
750 N Laffoon St
Madisonville, KY 42431-9185
www.madcc.kctcs.edu/~pta
*Prgm Dir:* Angie Moser, PT
*Tel:* 270 824-7552, Ext 164   *Fax:* 270 824-7069
*E-mail:* angie.moser@kctcs.edu

**West Kentucky Community & Technical College**
Physical Therapist Assistant Prgm
PO Box 7380
Paducah, KY 42002-7380
*Prgm Dir:* Peggy R Block, MHS PT
*Tel:* 502 554-6274   *Fax:* 502 554-6227
*E-mail:* peggy.block@kctcs.edu

**Somerset Community College**
Physical Therapist Assistant Prgm
808 Monticello St
Somerset, KY 42501
www.kctcs.edu
*Prgm Dir:* Ronald L Meade, DPT
*Tel:* 606 451-6823   *Fax:* 606 676-3684
*E-mail:* ron.meade@kctcs.edu

# Louisiana

**Our Lady of the Lake College**
Physical Therapist Assistant Prgm
7443 Picardy
Baton Rouge, LA 70808
www.ololcollege.edu
*Prgm Dir:* Kitty Krieg, PT MHS
*Tel:* 225 768-1702   *Fax:* 225 765-5838
*E-mail:* kkrieg@ololcollege.edu

**Bossier Parish Community College**
Physical Therapist Assistant Prgm
6220 E Texas St
Bossier City, LA 71111
*Prgm Dir:* Laura Bryant, PT MEd
*Tel:* 318 678-6000, Ext 6079   *Fax:* 318 678-6199
*E-mail:* lbryant@bpcc.edu

**Delgado Community College**
Physical Therapist Assistant Prgm
615 City Park Ave
New Orleans, LA 70119-4399
*Prgm Dir:* Susan M Welsh, PhD PT
*Tel:* 504 483-4207   *Fax:* 504 483-4609
*E-mail:* swelsh@dcc.edu

# Maine

**Kennebec Valley Community College**
Physical Therapist Assistant Prgm
92 Western Ave
Fairfield, ME 04937-1367
*Prgm Dir:* Nancy Chandler, PT MPH
*Tel:* 207 453-5147   *Fax:* 207 453-5194
*E-mail:* nchandler@kvcc.me.edu

# Maryland

**Chesapeake Area Consortium for Higher Educ**
Physical Therapist Assistant Prgm
101 College Pkwy
Arnold, MD 21012
*Prgm Dir:* David Thomas, PT MGA
*Tel:* 410 777-7039   *Fax:* 410 777-7099
*E-mail:* dcthomas@aacc.edu

**Baltimore City Community College**
Physical Therapist Assistant Prgm
Nursing Bldg Rm 302
2901 Liberty Heights Ave
Baltimore, MD 21215
*Prgm Dir:* Nijole D Kaltreider, PT MEd
*Tel:* 410 462-7727   *Fax:* 410 462-7734
*E-mail:* NKaltreider@bccc.edu

**Allegany College of Maryland**
Physical Therapist Assistant Prgm
12401 Willowbrook Rd SE
Cumberland, MD 21502-2596
*Prgm Dir:* Scott Love, PT DPT
*Tel:* 301 784-5535   *Fax:* 301 784-5015
*E-mail:* slove@allegany.edu

**Montgomery College**
Physical Therapist Assistant Prgm
7600 Takoma Ave
Takoma Park, MD 20912
*Prgm Dir:* LaVerne E Tuckson, MEd PT
*Tel:* 301 562-5520   *Fax:* 301 650-1446
*E-mail:* laverne.tuckson@montgomerycollege.edu

**Carroll Community College**
Physical Therapist Assistant Prgm
1601 Washington Rd
Westminster, MD 21157
www.carrollcc.edu
*Prgm Dir:* Sharon Main, BS PT MBA
*Tel:* 410 386-8259   *Fax:* 410 386-8255
*E-mail:* smain@carrollcc.edu

# Massachusetts

**Bay State College**
Physical Therapist Assistant Prgm
122 Commonwealth Ave
Boston, MA 02116
www.baystate.edu
*Prgm Dir:* George B Coggeshall, MS PT
*Tel:* 617 217-9423   *Fax:* 617 375-0197
*E-mail:* gcoggeshall@baystate.edu

**North Shore Community College**
Physical Therapist Assistant Prgm
One Ferncroft Rd
Danvers, MA 01923-4093
*Prgm Dir:* Mary Meng-Rivas, PT MPH
*Tel:* 978 762-4000, Ext 4165   *Fax:* 978 762-4022
*E-mail:* mmengriv@northshore.edu

**Massachusetts Bay Community College**
Physical Therapist Assistant Prgm
19 Flagg Dr
Framingham, MA 01702
*Prgm Dir:* Brenda Carroll, MS PT
*Tel:* 508 270-4000   *Fax:* 508 270-4131
*E-mail:* bcarroll@massbay.edu

**Mt Wachusett Community College**
Physical Therapist Assistant Prgm
444 Green St
Gardner, MA 01440-1000
www.mwcc.mass.edu
*Prgm Dir:* Jacqueline Shakar, MS PTAT
*Tel:* 978 632-6600, Ext 287    *Fax:* 978 632-6155
*E-mail:* j_shakar@mwcc.mass.edu
*Med Dir:* Debra Orre

**Berkshire Community College**
Physical Therapist Assistant Prgm
1350 West St
Pittsfield, MA 01201-5786
*Prgm Dir:* Michele E Darroch, PT MEd
*Tel:* 413 499-4660, Ext 313    *Fax:* 413 447-7840
*E-mail:* mdarroch@berkshirecc.edu

**Springfield Technical Community College**
Physical Therapist Assistant Prgm
One Armory Sq, Ste 1
PO Box 9000
Springfield, MA 01102
www.stcc.edu/academics/health/descriptions/
    PhysicalTherapy.asp
*Prgm Dir:* Linda C Desmarais, PT DPT MPH
*Tel:* 413 775-4844    *Fax:* 413 755-6312
*E-mail:* ldesmarais@stcc.edu

**Becker College**
Physical Therapist Assistant Prgm
61 Sever St
Worcester, MA 01615-0071
*Prgm Dir:* Elizabeth V Fuller, MS PT EdD
*Tel:* 508 791-9241, Ext 362    *Fax:* 508 831-5213
*E-mail:* efuller@beckercollege.edu

# Michigan

**Kellogg Community College**
Physical Therapist Assistant Prgm
450 North Ave
Battle Creek, MI 49017
www.kellogg.edu/pta
*Prgm Dir:* Kathy A Mann, PT MA
*Tel:* 269 965-3931, Ext 2313    *Fax:* 269 565-2059
*E-mail:* mannk@kellogg.edu

**Macomb Community College**
Physical Therapist Assistant Prgm
44575 Garfield Rd
Clinton Township, MI 48038-1139
*Prgm Dir:* Carol Plisner, PT MA
*Tel:* 586 286-2097    *Fax:* 586 286-2098
*E-mail:* plisnerc@macomb.edu

**Henry Ford Community College**
*Cosponsor:* Oakwood Healthcare System, Inc
Physical Therapist Assistant Prgm
5101 Evergreen Rd
Dearborn, MI 48128
www.hfcc.edu/programs/sheets/
    PhysicalTherapistAssistant.asp
*Prgm Dir:* Cynthia M Scheuer, PT MS
*Tel:* 313 845-9877    *Fax:* 313 317-6569
*E-mail:* cscheuer@hfcc.edu

**Mott Community College - Fenton**
Physical Therapist Assistant Prgm
Southern Lakes Branch Campus
2100 W Thompson Rd
Fenton, MI 48430
www.mcc.edu/slbc_pta.shtm
*Prgm Dir:* Kathleen Vielhaber, PT
*Tel:* 810 762-5021    *Fax:* 810 750-8588
*E-mail:* kvielhab@mcc.edu

**Baker College of Flint**
Physical Therapist Assistant Prgm
1050 W Bristol Rd
Flint, MI 48507-5508
*Prgm Dir:* Elaine Murphy, PhD
*Tel:* 810 766-4193    *Fax:* 810 766-2055
*E-mail:* elaine.murphy@baker.edu

**Finlandia University**
Physical Therapist Assistant Prgm
601 Quincy St
Hancock, MI 49930-1882
www.finlandia.edu
*Prgm Dir:* Cameron T Williams, PT MS
*Tel:* 906 487-7308    *Fax:* 906 487-7552
*E-mail:* cam.williams@finlandia.edu

**Baker College of Muskegon**
Physical Therapist Assistant Prgm
1903 Marquette Ave
Muskegon, MI 49442
*Prgm Dir:* Peter A Schaub, MS PT
*Tel:* 231 777-8800    *Fax:* 231 777-5265
*E-mail:* pete.schaub@baker.edu

**Delta College**
Physical Therapist Assistant Prgm
Rm P-172
University Center, MI 48710
*Prgm Dir:* A Michael Spitz, MSA BS PTA CSCS
*Tel:* 989 686-9478    *Fax:* 989 686-8736
*E-mail:* amspitz@delta.edu

# Minnesota

**Anoka Ramsey Community College**
Physical Therapist Assistant Prgm
11200 Mississippi Blvd NW
Coon Rapids, MN 55433
*Prgm Dir:* Lisa Lentner, MS PT
*Tel:* 763 427-2600    *Fax:* 763 422-3341
*E-mail:* lisa.lentner@anokaramsey.edu

**Lake Superior College**
Physical Therapist Assistant Prgm
2101 Trinity Rd
Duluth, MN 55811
www.lsc.edu/Programs/HealthCareers/PhysicalTherapist
    Assistant
*Prgm Dir:* Jane Worley, MS PT
*Tel:* 218 733-7632    *Fax:* 218 723-4921
*E-mail:* j.worley@lsc.edu

**College of St Catherine - Minneapolis**
Physical Therapist Assistant Prgm
601 25th Ave S
Minneapolis, MN 55454
*Prgm Dir:* Susan R Nelson, MS PT
*Tel:* 612 690-7822    *Fax:* 612 690-7876
*E-mail:* srnelson@stkate.edu

# Mississippi

**Itawamba Community College**
Physical Therapist Assistant Prgm
Dept of Applied Science and Technology
602 W Hill St
Fulton, MS 38843
*Prgm Dir:* Thomas W Hester, MS PT
*Tel:* 601 862-8342    *Fax:* 601 862-8350
*E-mail:* twhester@iccms.edu

**Pearl River Community College - Hattiesburg**
Physical Therapist Assistant Prgm
5448 US Hwy 49 S
Hattiesburg, MS 39401
www.prcc.edu
*Prgm Dir:* Patricia R Crowson, PT MS
*Tel:* 601 554-5486    *Fax:* 601 554-5553
*E-mail:* pcrowson@prcc.edu

**Hinds Community College**
Physical Therapist Assistant Prgm
Nursing Applied Health Center
1750 Chadwick Dr
Jackson, MS 39204-3490
*Prgm Dir:* Lisa G Latham, PT
*Tel:* 601 371-3512    *Fax:* 601 371-3529
*E-mail:* lglatham@hindscc.edu

**Meridian Community College**
Physical Therapist Assistant Prgm
910 Hwy 19 N
Meridian, MS 39307-5890
www.mcc.cc.ms.us
*Prgm Dir:* Kimberly T Ennis, PT MHS
*Tel:* 601 484-8613    *Fax:* 601 484-8743
*E-mail:* kennis@meridiancc.edu

# Missouri

**Linn State Technical College**
Physical Therapist Assistant Prgm
Capital Region Medical Center
Southwest Campus-1432 Southwest Blvd
Jefferson City, MO 65101
www.linnstate.edu
*Prgm Dir:* Marlene Medin, PT MEd
*Tel:* 573 632-5625    *Fax:* 573 632-5623
*E-mail:* marlene.medin@linnstate.edu

**Metropolitan Community College - Penn Valley**
Physical Therapist Assistant Prgm
3201 SW Trafficway
Kansas City, MO 64111-2764
www.mcckc.edu
*Prgm Dir:* Gwen Robertson, MA PT
*Tel:* 816 759-4241    *Fax:* 816 759-4553
*E-mail:* Gwen.Robertson@mcckc.edu

**Ozarks Technical Community College**
Physical Therapist Assistant Prgm
1001 E Chestnut Expressway
Springfield, MO 65802
www.otc.edu
*Prgm Dir:* Rebecca McKnight, PT MS
*Tel:* 417 447-8838    *Fax:* 417 447-8806
*E-mail:* mcknighr@otc.edu

**Missouri Western State University**
Physical Therapist Assistant Prgm
4525 Downs Dr, JGM 304
St Joseph, MO 64507-2294
*Prgm Dir:* Maureen Raffensperger, PT OCS MS
*Tel:* 816 271-4251    *Fax:* 816 271-4168
*E-mail:* raffen@missouriwestern.edu

**St Louis Community College - Meramec**
Physical Therapist Assistant Prgm
11333 Big Bend Blvd
St Louis, MO 63122
*Prgm Dir:* Julie A High, MS PT
*Tel:* 314 984-7385    *Fax:* 314 984-7250
*E-mail:* jhigh@stlcc.edu

# Nebraska

**Northeast Community College**
Physical Therapist Assistant Prgm
801 E Benjamin Ave, PO Box 469
Norfolk, NE 68702-0469
*Prgm Dir:* Jodie Aller, MPT
*Tel:* 402 844-7326, Ext 7326    *Fax:* 402 844-7390
*E-mail:* jodie@northeastcollege.com

**Clarkson College**
Physical Therapist Assistant Prgm
101 S 42nd St
Omaha, NE 68131-2739
*Prgm Dir:* Victoria Trost, DPT
*Tel:* 402 552-6178    *Fax:* 402 552-6019
*E-mail:* trost@clarksoncollege.edu

# Nevada

**Community College of Southern Nevada**
Physical Therapist Assistant Prgm
6375 W Charleston Blvd
Las Vegas, NV 89146
www.ccsn.edu
*Prgm Dir:* Joseph D Cracraft, PhD PT
*Tel:* 702 651-5588    *Fax:* 702 651-5506
*E-mail:* joe_cracraft@ccsn.edu

# New Hampshire

**New Hampshire Comm Tech Coll - Claremont**
Physical Therapist Assistant Prgm
One College Dr
Claremont, NH 03743-9707
www.nhctc.claremont.edu
*Prgm Dir:* Laurel Clute, PT MS
*Tel:* 603 542-7744, Ext 2554    *Fax:* 603 543-1844
*E-mail:* lclute@nhctc.edu

**Hesser College**
Physical Therapist Assistant Prgm
3 Sundial Ave
Manchester, NH 03103
*Prgm Dir:* Paula A Hould, PT
*Tel:* 603 668-6660, Ext 2113    *Fax:* 603 624-2836
*E-mail:* phould@hesser.edu

# New Jersey

**Essex County College**
Physical Therapist Assistant Prgm
303 University Ave
Newark, NJ 07102
*Prgm Dir:* Christine M Stutz-Doyle, MA PT
*Tel:* 973 877-3456    *Fax:* 973 877-1930
*E-mail:* stutz@essex.edu

**Bergen Community College**
Physical Therapist Assistant Prgm
400 Paramus Rd
Rm S 336
Paramus, NJ 07652-1595
*Prgm Dir:* Kenneth H Mailly, PT
*Tel:* 201 612-5319    *Fax:* 201 612-3876
*E-mail:* kmailly@bergen.edu

**Union County College**
Physical Therapist Assistant Prgm
Plainfield Campus
232 E Second St
Plainfield, NJ 07060
www.ucc.edu
*Prgm Dir:* Beth J Rothman, MS PT MEd
*Tel:* 908 412-3582    *Fax:* 908 754-2798
*E-mail:* rothman@ucc.edu

**Mercer County Community College**
Physical Therapist Assistant Prgm
1200 Old Trenton Rd, PO Box B
Trenton, NJ 08690
www.mccc.edu
*Prgm Dir:* Barbara J Behrens, PTA MS
*Tel:* 609 570-3385    *Fax:* 609 570-3831
*E-mail:* behrensb@mccc.edu

# New Mexico

**San Juan College**
Physical Therapist Assistant Prgm
4601 College Blvd
Farmington, NM 87402-4699
www.sanjuancollege.edu/pta
*Prgm Dir:* Wendy Bircher, PT EdD
*Tel:* 505 566-3407    *Fax:* 505 566-3767
*E-mail:* bircherw@sanjuancollege.edu

# New York

**Genesee Community College**
Physical Therapist Assistant Prgm
One College Rd
Batavia, NY 14020-9704
*Prgm Dir:* Peggy Kerr, MS PT
*Tel:* 585 343-0055, Ext 6366    *Fax:* 585 343-0433
*E-mail:* pckerr@genesee.edu

**Broome Community College**
Physical Therapist Assistant Prgm
PO Box 1017
Decker Health Science Bldg 217C
Binghamton, NY 13902
*Prgm Dir:* Denise Abrams, PT MA
*Tel:* 607 778-5211    *Fax:* 607 778-5345
*E-mail:* abrams_d@sunybroome.edu

**Kingsborough Comm Coll/The City Univ of NY**
Physical Therapist Assistant Prgm
2001 Oriental Blvd
Brooklyn, NY 11235-2398
*Prgm Dir:* Steven B Skinner, PT MS EdD
*Tel:* 718 368-4818    *Fax:* 718 368-4873
*E-mail:* sskinner@kbcc.cuny.edu

**Villa Maria College of Buffalo**
Physical Therapist Assistant Prgm
240 Pine Ridge Rd
Buffalo, NY 14225
*Prgm Dir:* James J Kelley, PT EMMDS
*Tel:* 716 961-1835    *Fax:* 716 896-0705
*E-mail:* jkelley@villa.edu

**SUNY at Canton**
Physical Therapist Assistant Prgm
School of Science Health and Professional Studies
34 Cornell Dr
Canton, NY 13617-1096
www.canton.edu
*Prgm Dir:* Deborah S Molnar
*Tel:* 315 386-7394    *Fax:* 315 386-7959
*E-mail:* molnard@canton.edu

**Nassau Community College**
Physical Therapist Assistant Prgm
One Education Dr
Garden City, NY 11530
www.ncc.edu/dptpages/ahs/PT/PTframe1.html
*Prgm Dir:* Laura Gilkes, PT MA
*Tel:* 516 572-9640    *Fax:* 516 572-7565
*E-mail:* gilkesl@ncc.edu

**Herkimer County Community College**
Physical Therapist Assistant Prgm
100 Reservoir Rd
Herkimer, NY 13350
www.herkimer.edu/academics/programs/physicaltherapy
*Prgm Dir:* Catherine E DeLorme, PT MS
*Tel:* 315 866-0300, Ext 8340    *Fax:* 315 866-7523
*E-mail:* delormece@herkimer.edu

**LaGuardia Community College**
Physical Therapist Assistant Prgm
31-10 Thomson Ave E 300 R
Long Island City, NY 11101
www.lagcc.cuny.edu/ptaprogram
*Prgm Dir:* Debra Engel, PT DPT MS
*Tel:* 718 482-5780    *Fax:* 718 609-2052
*E-mail:* dengel@lagcc.cuny.edu

**Orange County Community College**
Physical Therapist Assistant Prgm
115 South St
Middletown, NY 10940
www.sunyorange.edu
*Prgm Dir:* Maria E Masker, PT DPT
*Tel:* 914 341-4290    *Fax:* 914 341-4799
*E-mail:* mmasker@sunyorange.edu

**New York University**
Physical Therapist Assistant Prgm
School of Continuing and Prof Studies
594 Broadway Rm 400
New York, NY 10012
*Prgm Dir:* Maureen Thornby, PhD PT
*Tel:* 212 992-8722    *Fax:* 212 995-4890
*E-mail:* mat2@nyu.edu

**Touro College**
Physical Therapist Assistant Prgm
27 W 23rd St
New York, NY 10010-4202
www.touro.edu/shs/pta.asp
*Prgm Dir:* Christopher Kevin Wong, PT PhD OCS
*Tel:* 212 463-0400, Ext 606    *Fax:* 212 989-2054
*E-mail:* ckwong@touro.edu
*Med Dir:* Kathryn Robshaw-Turnbull, DPT, PT

**Niagara County Community College**
Physical Therapist Assistant Prgm
Div of Life Sciences
3111 Saunders Settlement Rd
Sanborn, NY 14132
*Prgm Dir:* Deborah K Matuch, MS PT
*Tel:* 716 614-6422    *Fax:* 716 614-6798
*E-mail:* matuch@niagaracc.suny.edu

**Suffolk County Community College**
Physical Therapist Assistant Prgm
Dept of Education, Health and Human Services
533 College Rd
Selden, NY 11784
www.sunysuffolk.edu
*Prgm Dir:* Cheryl Gillespie, PT MA DPT
*Tel:* 516 451-4299, Ext 4017    *Fax:* 516 451-4697
*E-mail:* gillesc@sunysuffolk.edu

**Onondaga Community College**
Physical Therapist Assistant Prgm
4941 Onondaga Rd
Syracuse, NY 13215
*Prgm Dir:* Cynthia Warner, PT MEd
*Tel:* 315 498-2388    *Fax:* 315 498-2751
*E-mail:* warnerc@sunyocc.edu

# North Carolina

**Central Piedmont Community College**
Physical Therapist Assistant Prgm
PO Box 35009
Charlotte, NC 28235
www.cpcc.edu/health_services
*Prgm Dir:* Ilene Weiner, MS PT
*Tel:* 704 330-6505    *Fax:* 704 330-6131
*E-mail:* ilene.weiner@cpcc.edu

**Fayetteville Technical Community College**
Physical Therapist Assistant Prgm
PO Box 35236
Fayetteville, NC 28303
*Prgm Dir:* Elaine M Eckel, MA PT
*Tel:* 910 678-8259    *Fax:* 910 678-8500
*E-mail:* eckele@faytechcc.edu

**Caldwell Comm College & Tech Institute**
Physical Therapist Assistant Prgm
2855 Hickory Blvd
Hudson, NC 28638
*Prgm Dir:* Martha Y Zimmerman, PT MA
*Tel:* 828 726-2605    *Fax:* 828 726-2489
*E-mail:* mzimmerman@cccti.edu

**Guilford Technical Community College**
Physical Therapist Assistant Prgm
601 High Point Rd, PO Box 309
Jamestown, NC 27282
*Prgm Dir:* Joey Jeffers, PT MA
*Tel:* 336 334-4822, Ext 2443    *Fax:* 336 454-2510
*E-mail:* jeffersj@gtcc.edu

**Nash Community College**
Physical Therapist Assistant Prgm
522 N Old Carriage Rd
PO Box 7488
Rocky Mount, NC 27804-0488
*Prgm Dir:* Tammie L Clark, PT MS
*Tel:* 252 443-4011, Ext 372    *Fax:* 252 443-0828
*E-mail:* tclark@nashcc.edu

**Southwestern Community College**
Physical Therapist Assistant Prgm
447 College Dr
Sylva, NC 28779
*Prgm Dir:* Debra M Klavohn, PT MA
*Tel:* 828 586-4091, Ext 331    *Fax:* 828 586-3129
*E-mail:* debm@southwesterncc.edu

**Martin Commuity College**
Physical Therapist Assistant Prgm
1161 Kehukee Park Rd
Williamston, NC 27892
*Prgm Dir:* Jean Lambert, PT PhD
*Tel:* 252 792-1521, Ext 237    *Fax:* 252 792-4425
*E-mail:* jlambert@martincc.edu

## North Dakota

**Williston State College**
Physical Therapist Assistant Prgm
PO Box 1326, 1410 University Ave
Williston, ND 58802-1326
*Prgm Dir:* Robert Benson, MA PT LMT NCTMB
*Tel:* 701 774-4291    *Fax:* 701 774-4265
*E-mail:* robert.benson@wsc.nodak.edu

## Ohio

**Kent State University - Ashtabula Campus**
Physical Therapist Assistant Prgm
3325 W 13th St
Ashtabula, OH 44004
www.ashtabula.kent.edu
*Prgm Dir:* Michael Blake, MS PT
*Tel:* 440 964-4333    *Fax:* 440 964-4269
*E-mail:* blake@ashtabula.kent.edu

**Stark State College of Technology**
Physical Therapist Assistant Prgm
Health Technologies Div
6200 Frank Ave NW
Canton, OH 44720
*Prgm Dir:* Wallace H Linville, PT MA
*Tel:* 330 966-5458, Ext 4619    *Fax:* 330 966-6586
*E-mail:* wlinville@starkstate.edu

**University of Cincinnati**
Physical Therapist Assistant Prgm
Department of Rehabilitation Sciences
College of Allied Health Sciences, French East Building
Cincinnati, OH 45267-0394
www.cahs.uc.edu
*Prgm Dir:* Tina Whalen, DPT MPA PT
*Tel:* 513 558-7485    *Fax:* 513 558-7474
*E-mail:* whalentf@uc.edu
*Med Dir:* Teri Slick, M Ed

**Cuyahoga Community College**
Physical Therapist Assistant Prgm
2900 Community College Ave, MHCS 126
Cleveland, OH 44115
www.tri-c.edu/pta
*Prgm Dir:* Toby Sternheimer, MEd PT
*Tel:* 216 987-4502    *Fax:* 216 987-4386
*E-mail:* toby.sternheimer@tri-c.edu

**Sinclair Community College**
Physical Therapist Assistant Prgm
444 W Third St, Rm 3340
Dayton, OH 45402
*Prgm Dir:* Colleen Whittington, PT MHS
*Tel:* 937 512-5355    *Fax:* 937 512-2058
*E-mail:* colleen.whittington@sinclair.edu

**Kent State University - East Liverpool Campus**
Physical Therapist Assistant Prgm
400 E Fourth St
East Liverpool, OH 43920-3497
www.kenteliv.kent.edu
*Prgm Dir:* Thomas Rutledge, MS PT ATC
*Tel:* 330 382-7448    *Fax:* 330 382-7564
*E-mail:* trutledge@eliv.kent.edu

**Lorain County Community College**
Physical Therapist Assistant Prgm
1005 N Abbe Rd
Elyria, OH 44035
*Prgm Dir:* John T Myers, PTA MBA
*Tel:* 800 995-5222, Ext 7881    *Fax:* 216 366-4116
*E-mail:* jmyers@lorainccc.edu

**Rhodes State College**
Physical Therapist Assistant Prgm
4240 Campus Dr
Lima, OH 45804
*Prgm Dir:* Angela M Heaton, PT MSEd
*Tel:* 419 995-8813    *Fax:* 419 995-8818
*E-mail:* heaton.a@rhodesstate.edu

**North Central State College**
Physical Therapist Assistant Prgm
2441 Kenwood Circle, PO Box 698
Mansfield, OH 44901-0698
www.ncstatecollege.edu
*Prgm Dir:* James L Hull, MBA PT
*Tel:* 419 755-4773    *Fax:* 419 755-4790
*E-mail:* jhull@ncstatecollege.edu

**Washington State Community College**
Physical Therapist Assistant Prgm
710 Colegate Dr
Marietta, OH 45750
www.wscc.edu
*Prgm Dir:* Kimberly D Salyers, PTA MA Ed
*Tel:* 740 374-8716, Ext 1692    *Fax:* 740 373-7496
*E-mail:* ksalyers@wscc.edu

**Marion Technical College**
Physical Therapist Assistant Prgm
1467 Mt Vernon Ave
Marion, OH 43302-5694
http://mtc.edu
*Prgm Dir:* Susan P Cotterman, PT MBA
*Tel:* 740 389-4636, Ext 356    *Fax:* 740 389-6136
*E-mail:* cottermans@mtc.edu

**Hocking College**
Physical Therapist Assistant Prgm
3301 Hocking Pkwy SEO 309
Nelsonville, OH 45764
*Prgm Dir:* Kathy Kropf, PT MEd
*Tel:* 740 753-3591, Ext 2867    *Fax:* 740 753-5105
*E-mail:* kropf_k@hocking.edu

**Shawnee State University**
Physical Therapist Assistant Prgm
930 2nd St
Portsmouth, OH 45662
*Prgm Dir:* Sam M Coppoletti, MPT ACCE CSCS
*Tel:* 740 351-3225    *Fax:* 740 351-3354
*E-mail:* scoppoletti@shawnee.edu

**Clark State Community College**
Physical Therapist Assistant Prgm
PO Box 570
Springfield, OH 45501-0570
*Prgm Dir:* Beth M Gustafson, PT MSEd
*Tel:* 937 328-8074    *Fax:* 937 328-6138
*E-mail:* gustafsonb@clarkstate.edu

**Owens Community College**
Physical Therapist Assistant Prgm
PO Box 10,000
30335 Oregon Rd
Toledo, OH 43699
*Prgm Dir:* Nancy Rupp, PT
*Tel:* 419 661-7084    *Fax:* 419 661-7251
*E-mail:* nrupp@owens.edu

**Professional Skills Institute**
Physical Therapist Assistant Prgm
20 Arco Dr
Toledo, OH 43607-1947
*Prgm Dir:* Colleen Macdonald-Mason, MS PT
*Tel:* 419 531-9610    *Fax:* 419 531-4732
*E-mail:* cm-mason@proskills.com

**Zane State College**
Physical Therapist Assistant Prgm
1555 Newark Rd
Zanesville, OH 43701
www.zanestate.edu
*Prgm Dir:* Barbara Shelby, PT MEd
*Tel:* 740 588-1315    *Fax:* 740 454-0035
*E-mail:* bshelby@zanestate.edu

## Oklahoma

**Caddo Kiowa Tech Ctr/SW Oklahoma St Univ**
Physical Therapist Assistant Prgm
PO Box 190
Ft Cobb, OK 73038
www.caddokiowa.com
*Prgm Dir:* LaMae Green, RPT
*Tel:* 405 643-3268    *Fax:* 405 643-2144
*E-mail:* lgreen@caddokiowa.com

**Northeastern Oklahoma A&M College**
Physical Therapist Assistant Prgm
Health Sciences Division
200 I St NE
Miami, OK 74354
www.neoam.edu
*Prgm Dir:* James D Compton, PT
*Tel:* 918 542-8441, Ext 6316    *Fax:* 918 540-6471
*E-mail:* jcompton@neoam.edu

**Oklahoma City Community College**
Physical Therapist Assistant Prgm
7777 S May Ave
Oklahoma City, OK 73159
*Prgm Dir:* Peggy DeCelle Newman, MHR PT
*Tel:* 405 682-1611, Ext 7305    *Fax:* 405 682-7826
*E-mail:* pnewman@okccc.edu

**Carl Albert State College**
Physical Therapist Assistant Prgm
1507 S McKenna
Poteau, OK 74953-5208
*Prgm Dir:* William L Carroll, MPT
*Tel:* 918 647-1358    *Fax:* 918 647-1327
*E-mail:* bcarroll@carlalbert.edu

**Murray State College**
Physical Therapist Assistant Prgm
One Murray Campus, NAH 116
Tishomingo, OK 73460
*Prgm Dir:* Gary Robinson, MS PT PCS
*Tel:* 580 371-2371, Ext 340    *Fax:* 580 371-9844
*E-mail:* grobinson@mscok.edu

**Tulsa Community College**
Physical Therapist Assistant Prgm
909 S Boston Ave
Tulsa, OK 74119
*Prgm Dir:* Suzanne Reese, MS PT
*Tel:* 918 595-7017    *Fax:* 918 595-7091
*E-mail:* sreese@tulsacc.edu

## Oregon

**Mt Hood Community College**
Physical Therapist Assistant Prgm
26000 SE Stark St
Gresham, OR 97030
www.mhcc.edu
*Prgm Dir:* Jane Cedar, PT MS
*Tel:* 503 491-7464    *Fax:* 503 491-6047
*E-mail:* jane.cedar@mhcc.edu

# Pennsylvania

## Harcum College
Physical Therapist Assistant Prgm
750 Montgomery Ave
Bryn Mawr, PA 19010
*Prgm Dir:* Jacqueline Klaczak Kopack, MPT
*Tel:* 610 526-6059   *Fax:* 610 526-6031
*E-mail:* jkopack@harcum.edu

## Butler County Community College
Physical Therapist Assistant Prgm
PO Box 1203
Butler, PA 16003-1203
*Prgm Dir:* Randall J Kruger, MPT
*Tel:* 724 287-8711, Ext 8372   *Fax:* 724 285-6047
*E-mail:* randy.kruger@bc3.edu

## California University of Pennsylvania
Physical Therapist Assistant Prgm
Coll of Ed & Human Serv, Dept of Hlth Sci & Sport
   Studies
250 University Ave
California, PA 15419-1394
*Prgm Dir:* Jodi DeBlassio Dusi, MPT
*Tel:* 724 938-4562   *Fax:* 724 938-4542
*E-mail:* deblassio@cup.edu

## Mount Aloysius College
Physical Therapist Assistant Prgm
7373 Admiral Peary Hwy
Cresson, PA 16630
*Prgm Dir:* Stacy A Sekely, PT DPT
*Tel:* 814 886-6428   *Fax:* 814 886-2978
*E-mail:* ssekely@mtaloy.edu

## Penn State University - DuBois
Physical Therapist Assistant Prgm
College Place
DuBois, PA 15801
www.ds.psu.edu/AcademicAffairs/Programs/PTA
*Prgm Dir:* Barbara Reinard, PT,MA
*Tel:* 814 375-4773
*E-mail:* ber125@psu.edu

## Penn State University - Hazleton
Physical Therapist Assistant Prgm
76 University Dr
Hazleton, PA 18202
*Prgm Dir:* Rosemarie Petrilla
*Tel:* 717 450-3042   *Fax:* 717 450-3182
*E-mail:* rxp21@psu.edu

## Comm Coll of Allegheny County
Physical Therapist Assistant Prgm
595 Beatty Rd
Monroeville, PA 15146
*Prgm Dir:* Norman L Johnson, PT DPT Ded
*Tel:* 724 325-6663   *Fax:* 724 325-6701
*E-mail:* njohnson@ccac.edu

## Penn State University - Mont Alto
Physical Therapist Assistant Prgm
Campus Dr
Mont Alto, PA 17237-9703
*Prgm Dir:* Thomas Glumac, PT
*Tel:* 717 749-6217   *Fax:* 717 749-6039
*E-mail:* txg3@psu.edu

## Mercyhurst College
Physical Therapist Assistant Prgm
North East Campus
16 W Division St
North East, PA 16428
www.northeast.mercyhurst.edu
*Prgm Dir:* Janice Haas, PTA MS
*Tel:* 814 725-6305   *Fax:* 814 725-6339
*E-mail:* jhaas@mercyhurst.edu

## Lehigh Carbon Community College
Physical Therapist Assistant Prgm
4525 Education Park Dr
Schnecksville, PA 18078-2598
www.lccc.edu
*Prgm Dir:* Evelyn M Petrash, PT MS
*Tel:* 610 799-1515   *Fax:* 610 799-1527
*E-mail:* epetrash@lccc.edu

## Penn State University - Shenango Campus
Physical Therapist Assistant Prgm
147 Shenango Ave
Sharon, PA 16146
*Prgm Dir:* Richard L Holzworth, MS PT
*Tel:* 724 983-2866   *Fax:* 724 983-2820
*E-mail:* rlh18@psu.edu

## Central Pennsylvania College
Physical Therapist Assistant Prgm
Campus on College Hill and Valley Rds
Summerdale, PA 17093-0309
www.centralpenn.edu
*Prgm Dir:* Krista M Wolfe, DPT ATC
*Tel:* 717 728-2276   *Fax:* 717 732-5254
*E-mail:* kristawolfe@centralpenn.edu

## University of Pittsburgh - Titusville
Physical Therapist Assistant Prgm
504 E Main St
Titusville, PA 16354-2097
www.upt.pitt.edu/upt_pta
*Prgm Dir:* Malorie Kosht-Fedyshin, PT DPT
*Tel:* 814 827-4445   *Fax:* 814 827-5671
*E-mail:* mkosht@pitt.edu

# Puerto Rico

## University of Puerto Rico at Humacao
Physical Therapist Assistant Prgm
CUH Postal Station
100 CARR 908
Humacao, PR 00791-4300
*Prgm Dir:* Eneida Silva-Collazo, PT MPH
*Tel:* 787 850-9390   *Fax:* 787 850-9423
*E-mail:* e_silva@webmail.uprh.edu

## Ponce Technological University College
Physical Therapist Assistant Prgm
Univ of Puerto Rico Regl Coll Admin
PO Box 7186
Ponce, PR 00732
*Prgm Dir:* Felix A Cuevas Guzman, MPT
*Tel:* 787 844-8181, Ext 2414   *Fax:* 787 844-8679
*E-mail:* fac401@yahoo.com

# Rhode Island

## Community College of Rhode Island
Physical Therapist Assistant Prgm
1 John H Chafee Blvd
Newport, RI 02840
www.ccri.edu
*Prgm Dir:* Kimberly Crealey Rouillier, DPT MS
*Tel:* 401 851-1668   *Fax:* 401 851-1710
*E-mail:* krouillier@ccri.edu

# South Carolina

## Trident Technical College
Physical Therapist Assistant Prgm
PO Box 118067 AH-M
Charleston, SC 29423-8067
www.tridenttech.edu
*Prgm Dir:* Lori Fischer, MS PT
*Tel:* 843 574-6480   *Fax:* 843 574-6585
*E-mail:* lori.fischer@tridenttech.edu

## Midlands Technical College
Physical Therapist Assistant Prgm
PO Box 2408
Columbia, SC 29202
*Prgm Dir:* Lynn van Dijk, PT MBA
*Tel:* 803 822-3590   *Fax:* 803 822-3619
*E-mail:* vandijkl@midlandstech.edu

## Greenville Technical College
Physical Therapist Assistant Prgm
PO Box 5616
Greenville, SC 29606-5616
www.greenvilletech.com
*Prgm Dir:* Alicia Dittmar, MEd PT
*Tel:* 864 848-2036   *Fax:* 864 848-2038
*E-mail:* Alicia.Dittmar@gvltec.edu

# South Dakota

## Lake Area Technical Institute
Physical Therapist Assistant Prgm
230 11th St NE, PO Box 730
Watertown, SD 57201-0730
*Prgm Dir:* Christina Barrett, PT MEd
*Tel:* 605 886-5284, Ext 329   *Fax:* 605 882-6299
*E-mail:* barrettc@lakeareatech.edu

# Tennessee

## Chattanooga State Technical Comm College
Physical Therapist Assistant Prgm
4501 Amnicola Hwy
Chattanooga, TN 37406-1097
*Prgm Dir:* Laura P Warren, MS PT
*Tel:* 423 697-3171, Ext 4450   *Fax:* 423 634-3071
*E-mail:* LAURA.WARREN@chattanoogastate.edu

## Volunteer State Community College
Physical Therapist Assistant Prgm
1480 Nashville Pike
Gallatin, TN 37066
www.volstate.edu
*Prgm Dir:* Dennis Dipert, MS PT
*Tel:* 615 230-3336   *Fax:* 615 230-4816
*E-mail:* dennis.dipert@volstate.edu

## Jackson State Community College
Physical Therapist Assistant Prgm
2046 N Parkway St
Jackson, TN 38301-3797
www.jscc.edu
*Prgm Dir:* Jane David, PT DPT
*Tel:* 731 425-2612, Ext 214   *Fax:* 731 425-9551
*E-mail:* jdavid@jscc.edu

## South College
Physical Therapist Assistant Prgm
3904 Lonas Dr
Knoxville, TN 37909
www.southcollegetn.edu
*Prgm Dir:* Kelly S Nash, MS PT
*Tel:* 865 251-1800, Ext 1839   *Fax:* 865 584-9816
*E-mail:* knash@southcollegetn.edu

## Southwest Tennessee Community College
Physical Therapist Assistant Prgm
Union Ave Campus
737 Union Ave
Memphis, TN 38103-3322
*Prgm Dir:* Edward Zeno, MS PT
*Tel:* 901 333-5394   *Fax:* 901 333-5391
*E-mail:* ezeno@southwest.tn.edu

## Walters State Community College
Physical Therapist Assistant Prgm
500 S Davy Crockett Pkwy
Morristown, TN 37813-6899
www.ws.edu
*Prgm Dir:* Ann Lowdermilk
*Tel:* 423 585-6986   *Fax:* 423 585-6955
*E-mail:* margaret.lowdermilk@ws.edu

**Roane State Community College**
Physical Therapist Assistant Prgm
701 Briarcliff Ave
Oak Ridge, TN 37830
*Prgm Dir:* Kurt Backstrom, MS PT
*Tel:* 865 481-2000, Ext 2117   *Fax:* 865 481-2019
*E-mail:* backstrom_ka@roanestate.edu

# Texas

**Amarillo College**
Physical Therapist Assistant Prgm
PO Box 447
Amarillo, TX 79178
www.actx.edu
*Prgm Dir:* Kelly J Jones, MPT
*Tel:* 806 354-6043   *Fax:* 806 354-6076
*E-mail:* jones-kj@actx.edu

**Austin Community College**
Physical Therapist Assistant Prgm
Eastview Campus
3401 Webberville Rd
Austin, TX 78702
www.austincc.edu/health/ptha
*Prgm Dir:* Jana Israel, PT MEd
*Tel:* 512 223-5938   *Fax:* 512 223-5897
*E-mail:* jisrael@austincc.edu

**Blinn College**
Physical Therapist Assistant Prgm
PO Box 6030
Bryan, TX 77805-6030
www.blinncol.edu/twe/pta
*Prgm Dir:* John K Hubbard, PhD PT
*Tel:* 979 209-7154   *Fax:* 979 209-7524
*E-mail:* John.Hubbard@Blinn.edu

**N Harris Mont Comm Coll Dist/Montgomery Coll**
Physical Therapist Assistant Prgm
3200 College Park Dr
Conroe, TX 77384-4077
*Prgm Dir:* Julie A Pauls, PhD PT
*Tel:* 936 273-7470   *Fax:* 936 273-7050
*E-mail:* Julie.A.Pauls@nhmccd.edu

**Del Mar College**
Physical Therapist Assistant Prgm
Div of Occupational Education & Tech
West Campus
Corpus Christi, TX 78404-3897
*Prgm Dir:* Russell Stowers, PTA MS
*Tel:* 361 698-1847   *Fax:* 361 698-1849
*E-mail:* rstowers@delmar.edu

**El Paso Community College**
Physical Therapist Assistant Prgm
Rio Grande Campus, PO Box 20500
El Paso, TX 79998
*Prgm Dir:* Debra L Tomacelli-Brock, MS PT
*Tel:* 915 831-4172   *Fax:* 915 831-4114
*E-mail:* DebraT@epcc.edu

**Houston Community College**
Physical Therapist Assistant Prgm
Coleman College of Health Science
1900 Pressler Dr
Houston, TX 77030-3799
*Prgm Dir:* Jan H Myers, PT MS
*Tel:* 713 718-7391   *Fax:* 713 718-6495
*E-mail:* jan.myers@hccs.edu

**San Jacinto College South**
Physical Therapist Assistant Prgm
13735 Beamer Rd
Houston, TX 77089-6099
*Prgm Dir:* Carolyn Hoffman, PT MS
*Tel:* 281 929-4697   *Fax:* 281 922-3487
*E-mail:* carolyn.hoffman@sjcd.edu

**Tarrant County College**
Physical Therapist Assistant Prgm
828 Harwood Rd, Northeast Campus
Hurst, TX 76054
*Prgm Dir:* Jill Pool, PTA MEd
*Tel:* 817 515-6555   *Fax:* 817 515-6700
*E-mail:* jill.pool@tccd.edu

**Kilgore College**
Physical Therapist Assistant Prgm
1100 Broadway
Kilgore, TX 75662
www.kilgore.edu/physical_therapist.asp
*Prgm Dir:* Carla Gleaton, MEd PT
*Tel:* 903 983-8148   *Fax:* 903 988-7596
*E-mail:* cgleaton@kilgore.edu

**Laredo Community College**
Physical Therapist Assistant Prgm
West End Washington St
Campus Box 153
Laredo, TX 78040
*Prgm Dir:* Consuelo Moreno, PT MS
*Tel:* 956 764-5724   *Fax:* 956 721-5431
*E-mail:* cmoreno@laredo.edu

**South Texas College**
Physical Therapist Assistant Prgm
PO Box 9701
McAllen, TX 78502-9701
*Prgm Dir:* Diana Salinas Hernandez, PT
*Tel:* 956 683-3152   *Fax:* 956 971-3723
*E-mail:* dianah@stcc.cc.tx.us

**Odessa College**
Physical Therapist Assistant Prgm
201 W University Blvd
Odessa, TX 79764
www.odessa.edu/dept/pta
*Prgm Dir:* Lynn McKelvey, PT MS
*Tel:* 915 335-6842   *Fax:* 915 335-6846
*E-mail:* lmckelvey@odessa.edu

**St Philip's College**
Physical Therapist Assistant Prgm
1801 Martin Luther King Dr
San Antonio, TX 78203-2098
*Prgm Dir:* J Thomas Davis, PhD PT
*Tel:* 210 531-3440   *Fax:* 210 531-3459
*E-mail:* jdavis@accd.edu

**McLennan Community College**
Physical Therapist Assistant Prgm
1400 College Dr
Waco, TX 76708
www.mclennan.edu
*Prgm Dir:* Julie Pickle, PT MS
*Tel:* 254 299-8715   *Fax:* 254 299-8747
*E-mail:* jpickle@mclennan.edu

**Wharton County Junior College**
Physical Therapist Assistant Prgm
911 Boling Hwy
Wharton, TX 77488
*Prgm Dir:* Phil Carter, MEd PT
*Tel:* 979 532-6491, Ext 6373   *Fax:* 979 532-6489
*E-mail:* philc@wcjc.edu

# Utah

**Provo College**
Physical Therapist Assistant Prgm
1450 W 820 N
Provo, UT 84601
*Prgm Dir:* Bradley Cordero, DPT MBA CSCS
*Tel:* 801 818-8910   *Fax:* 801 375-9728
*E-mail:* bradleycordero@provocollege.edu

**Salt Lake Community College**
Physical Therapist Assistant Prgm
PO Box 30808
4600 S Redwood Rd
Salt Lake City, UT 84130-0808
www.slcc.edu/pta
*Prgm Dir:* Diana N Ploeger, PT MEd
*Tel:* 801 957-4054   *Fax:* 801 957-5708
*E-mail:* diana.ploeger@slcc.edu

# Virginia

**Jefferson College of Health Sciences**
Physical Therapist Assistant Prgm
Community Hospital of Roanoke Valley
PO Box 13186, 920 S Jefferson St
Roanoke, VA 24031-3186
*Prgm Dir:* Michael A Krackow, PhD PTA
*Tel:* 540 224-4478   *Fax:* 540 985-9722
*E-mail:* mkrackow@jchs.edu

**Northern Virginia Community College**
Physical Therapist Assistant Prgm
6699 Springfield Center Dr
Springfield, VA 22150
www.nvcc.edu
*Prgm Dir:* Patricia S Ottavio, MPH PT
*Tel:* 703 822-6570   *Fax:* 703 822-6619
*E-mail:* pottavio@nvcc.edu

**Tidewater Community College**
Physical Therapist Assistant Prgm
1700 College Crescent, Bldg E, Rm E101
Virginia Beach, VA 23453
*Prgm Dir:* Melanie C Basinger, PT MS
*Tel:* 757 822-7251   *Fax:* 757 427-1338
*E-mail:* mbasingerl@tcc.edu

**Wytheville Community College**
Physical Therapist Assistant Prgm
1000 E Main St
Wytheville, VA 24382
*Prgm Dir:* Geneva Overton, PTA MA
*Tel:* 276 223-4721   *Fax:* 276 223-4778
*E-mail:* wcoverg@wcc.vccs.edu

# Washington

**Green River Community College**
Physical Therapist Assistant Prgm
Health Sciences Div, Mailstop OE-14
12401 Southeast 320th St
Auburn, WA 98002-3699
www.greenriver.edu/ProgramInformation/
    PhysicalTherapistAssistant.htm
*Prgm Dir:* Barbara Brucker, PT DPT MEd
*Tel:* 253 833-9111, Ext 4343   *Fax:* 253 833-3498
*E-mail:* bbrucker@greenriver.edu

**Whatcom Community College**
Physical Therapist Assistant Prgm
237 W Kellogg Rd
Bellingham, WA 98226
*Prgm Dir:* Rebecca A Graves, PT MS
*Tel:* 360 676-2170, Ext 3311   *Fax:* 360 752-6767
*E-mail:* bgraves@whatcom.ctc.edu

**Spokane Falls Community College**
Physical Therapist Assistant Prgm
3410 W Fort George Wright Dr
MS3190
Spokane, WA 99224-5288
http://spokanefalls.edu/PTA
*Prgm Dir:* Mary Ann Sharkey, PT PhD
*Tel:* 509 533-4144   *Fax:* 509 533-4143
*E-mail:* maryannsh@spokanefalls.edu

# West Virginia

**Mountain State University**
Physical Therapist Assistant Prgm
PO Box 9003
South Kanawha St
Beckley, WV 25801
*Prgm Dir:* Gina Brown, PTA MS ATC
*Tel:* 304 929-1458   *Fax:* 304 929-1600
*E-mail:* gbrown@mountainstate.edu

**Fairmont State Univ**
*Cosponsor:* Pierpont Comm & Tech College
Physical Therapist Assistant Prgm
Caperton Center
501 W Main St
Clarksburg, WV 26301
http://fairmontstate.edu
*Prgm Dir:* Beverly R Born, PT EdD
*Tel:* 304 367-4042   *Fax:* 304 367-4028
*E-mail:* bborn@fairmontstate.edu

**Marshall Community & Technical College**
Physical Therapist Assistant Prgm
2000 7th Ave
Cabell Hall #208
Huntington, WV 25755
*Prgm Dir:* Travis H Carlton, PTA MS
*Tel:* 304 696-3008   *Fax:* 304 696-2396
*E-mail:* carltont@marshall.edu

# Wisconsin

**Northeast Wisconsin Technical College**
Physical Therapist Assistant Prgm
2740 W Mason St
Green Bay, WI 54307
*Prgm Dir:* Julie Siefert, PT MHS LAT ATC
*Tel:* 920 498-5566   *Fax:* 920 491-2660
*E-mail:* julie.siefert@nwtc.edu

**Blackhawk Technical College**
Physical Therapist Assistant Prgm
6004 Prairie Rd, County Trunk G
Janesville, WI 53547-5009
*Prgm Dir:* Ilene Larson, MS PT
*Tel:* 608 757-7698   *Fax:* 608 757-4407
*E-mail:* ilarson@blackhawk.edu

**Gateway Technical College**
Physical Therapist Assistant Prgm
3520 30th Ave
Kenosha, WI 53144
*Prgm Dir:* Jeffrey Kannel, PT MS
*Tel:* 262 564-2482   *Fax:* 262 564-2007
*E-mail:* kannelj@gtc.edu

**Western Technical College**
Physical Therapist Assistant Prgm
304 N Sixth St, PO Box C-0908
La Crosse, WI 54602-0908
*Prgm Dir:* Mary Ann Herlitzke, MS PT
*Tel:* 608 785-9598   *Fax:* 608 785-9299
*E-mail:* herlitzkem@wwtc.edu

**Milwaukee Area Technical College**
Physical Therapist Assistant Prgm
Health Occupations Div
700 W State St
Milwaukee, WI 53233
*Prgm Dir:* Paul J Mansfield, MPT
*Tel:* 414 297-8078   *Fax:* 414 297-8651
*E-mail:* MansfieP@matc.edu

## Physical Therapist Assistant

| Programs* | Class Capacity | Begins | Length (months) | Award | Res Tuition | Non-res Tuition | Stipend | Offers:‡ 1 | 2 | 3 |
|---|---|---|---|---|---|---|---|---|---|---|
| **Alabama** | | | | | | | | | | |
| George C Wallace Community College (Dothan) | 32 | Aug | 21 | AAS | $3,822 | $7,644 | | | | |
| Jefferson State Community College (Birmingham) | 24 | Jan | 24 | AAS | $3,570 | $6,720 | | | | • |
| South University (Montgomery) | 30 | Jan Jun | 18 | AS | $0 | $15,300 | | • | | |
| Wallace State Community College (Hanceville) | 30 | Aug | 12 | AAS | $3,420 | $6,156 | | | | • |
| **Arizona** | | | | | | | | | | |
| GateWay Community College (Phoenix) | 44 | May Aug | 21 | AAS | $2,100 | $6,480 | | • | | • |
| **California** | | | | | | | | | | |
| Cerritos Community College (Norwalk) | 32 | Aug | 24 | AA | $780 | $4,710 | | • | | |
| Loma Linda University | 60 | Jun | 15 | AS | $16,587 | $16,587 | | | | • |
| Sacramento City College | 28 | Aug | 22 | AS | $780 | $5,790 | | • | | • |
| Sonoma College (Petaluma) | 25 | Jun Dec | 21 | AAS | $12,500 | $12,500 | | | | • |
| **Connecticut** | | | | | | | | | | |
| Naugatuck Valley Community College (Waterbury) | 30 | Jan | 24 | Dipl, AS | $1,980 | $5,292 | | | | |
| **Delaware** | | | | | | | | | | |
| Delaware Tech & Comm Coll - Wilmington | 16 | May | 24 | AAS | $1,260 | $0 | | • | | • |
| **Florida** | | | | | | | | | | |
| Florida Community College - Jacksonville | 24 | May | 16 | AS, AAS | $2,150 | $8,029 | | | | • |
| Gulf Coast Community College (Panama City) | 24 | Aug | 24 | AS | $2,300 | $9,000 | | | | • |
| Keiser University (Ft Lauderdale) | 16 | Jan May Aug | 24 | AS | $16,550 | $0 | | | | |
| Lake City Community College | 30 | May | 13 | Dipl, AS | $2,328 | $8,695 | | | | • |
| Polk Community College (Winter Haven) | 18 | Aug | 22 | AS, AAS | $1,699 | $6,374 | | | | • |
| Seminole Community College (Sanford) | 24 | Aug | 24 | AS | $1,768 | $6,331 | | | • | |
| St Petersburg College | 46 | Aug | 18 | AS | $2,627 | $8,919 | | | | • |
| **Hawaii** | | | | | | | | | | |
| Kapi'olani Community College (Honolulu) | 20 | Jun | 15 | AS | $1,680 | $7,479 | | • | | • |
| **Idaho** | | | | | | | | | | |
| Idaho State University (Pocatello) | 20 | Aug | 22 | AAS | $4,000 | $10,048 | | | | |
| **Illinois** | | | | | | | | | | |
| Black Hawk College (Moline) | 24 | Aug | 24 | AAS | $3,100 | $0 | | | | • |
| College of DuPage (Glen Ellyn) | 20 | Aug | 24 | AAS | | | | • | | • |
| Illinois Central College (Peoria) | 24 | Aug | 22 | AAS | $1,980 | $4,290 | | | | |
| Oakton Community College (Des Plaines) | 24 | Aug | 20 | AAS | $2,550 | $7,650 | | | | • |
| **Indiana** | | | | | | | | | | |
| Brown Mackie College (South Bend) | 24 | Apr Dec | 24 | AAS | $13,681 | $13,681 | | | | |
| Ivy Tech Community College EC - Muncie | 25 | Aug | 20 | AS | $2,895 | $5,890 | | | | |
| University of Indianapolis | 25 | Aug | 24 | AS | $19,900 | $19,900 | | • | | • |
| **Iowa** | | | | | | | | | | |
| Kirkwood Community College (Cedar Rapids) | 24 | Aug | 21 | AAS | $3,800 | $7,600 | | • | | • |
| Western Iowa Tech Community College (Sioux City) | 24 | Aug | 24 | AAS | $3,500 | $5,187 | | | | |

*Data are shown only for programs that completed the 2006 AMA Survey of Health Professions Education Programs
‡Key to Offers: 1: Evening or weekend classes; 2: Non-English instruction; 3: Cultural competence instruction

Physical Therapist Assistant

| Programs* | Class Capacity | Begins | Length (months) | Award | Res Tuition | Non-res Tuition | Stipend | Offers:‡ 1 | 2 | 3 |
|---|---|---|---|---|---|---|---|---|---|---|
| **Kansas** | | | | | | | | | | |
| Colby Community College | 24 | Aug | 12 | AAS | $3,283 | $5,194 | | | | • |
| Washburn University (Topeka) | 24 | Aug | 22 | AS | $4,200 | $9,528 | | | • | |
| **Kentucky** | | | | | | | | | | |
| Jefferson Community and Technical College (Louisville) | 22 | Aug | 24 | AAS | $118 | $294 | | | | |
| Madisonville Community College | 16 | Jan | 18 | AAS | $2,352 | $7,128 | | | | • |
| Somerset Community College | 16 | Aug | 24 | AAS | $2,668 | $8,004 | | | | • |
| **Louisiana** | | | | | | | | | | |
| Our Lady of the Lake College (Baton Rouge) | 24 | Jun | 12 | AS | $0 | $226 | | | | • |
| **Maryland** | | | | | | | | | | |
| Carroll Community College (Westminster) | 20 | Sep | 16 | AAS | $2,016 | $4,253 | | | | • |
| **Massachusetts** | | | | | | | | | | |
| Bay State College (Boston) | 40 | Sep | 21 | AS | $15,300 | $15,300 | | | | |
| Mt Wachusett Community College (Gardner) | 24 | Sep | 24 | AS | $5,000 | $0 | | | | • |
| Springfield Technical Community College | 20 | Sep | 24 | Dipl, AS | $3,456 | $9,966 | | | | • |
| **Michigan** | | | | | | | | | | |
| Finlandia University (Hancock) | 24 | Fall | 24 | AAS | $15,734 | $0 | | | | • |
| Henry Ford Community College (Dearborn) | 30 | Late Aug | 24 | AAS | $3,340 | $5,103 | | | | • |
| Kellogg Community College (Battle Creek) | 30 | Aug | 10 | AS | $2,556 | $3,993 | | | | • |
| Mott Community College - Fenton | 24 | Jan | 18 | Dipl, AAS | $4,365 | $6,430 | | | | • |
| **Minnesota** | | | | | | | | | | |
| Lake Superior College (Duluth) | 16 | Aug | 24 | AAS | $3,500 | $3,500 | | | | |
| **Mississippi** | | | | | | | | | | |
| Meridian Community College | 14 | Fall semester | 22 | AAS | $1,450 | $2,740 | | | | |
| Pearl River Community College - Hattiesburg | 20 | Aug | 22 | AAS | $1,286 | $2,500 | | | | • |
| **Missouri** | | | | | | | | | | |
| Linn State Technical College (Jefferson City) | 25 | Aug | 24 | AAS | $136 | $272 | | • | | • |
| Metropolitan Community College - Penn Valley (Kansas City) | 25 | Aug | 22 | Dipl, AAS | $78 | $185 | | | | • |
| Missouri Western State University (St Joseph) | 16 | Aug | 23 | AAS | $188 | $309 | | • | | • |
| Ozarks Technical Community College (Springfield) | 24 | Jan | 17 | Dipl, AAS | $1,092 | $1,792 | | | | • |
| **Nevada** | | | | | | | | | | |
| Community College of Southern Nevada (Las Vegas) | 12 | Sep | 21 | AAS | $4,068 | $6,995 | | | | |
| **New Hampshire** | | | | | | | | | | |
| New Hampshire Comm Tech Coll - Claremont | 15 | Sep | 24 | AS | $5,904 | $13,536 | | | | • |
| **New Jersey** | | | | | | | | | | |
| Mercer County Community College (Trenton) | 24 | Fall | 24 | AAS | $4,300 | $5,000 | | • | | • |
| Union County College (Plainfield) | 32 | Jan | 15 | AAS | $108 | $190 | | | | • |
| **New Mexico** | | | | | | | | | | |
| San Juan College (Farmington) | 12 | Jan | 27 | AAS, AAS | $990 | $1,320 | | • | | • |
| **New York** | | | | | | | | | | |
| Genesee Community College (Batavia) | 30 | Sep | 24 | Dipl, AAS | $2,900 | $3,250 | | • | | • |
| Herkimer County Community College | 28 | Sep | 18 | AAS | $2,900 | $5,000 | | | | • |
| Kingsborough Comm Coll/The City Univ of NY (Brooklyn) | 24 | Mar | 17 | AAS | $3,080 | $4,600 | | | | • |
| LaGuardia Community College (Long Island City) | 20 | Jun Jan | 18, 24 | AAS | $2,800 | $4,560 | | | | • |
| Nassau Community College (Garden City) | 32 | Sep | 21 | AAS | $3,000 | $6,000 | | | | |
| Orange County Community College (Middletown) | 24 | Aug Sep | 26 | AAS | $3,000 | $6,000 | | | | |
| Suffolk County Community College (Selden) | 26 | Sep | 19 | AAS | $2,600 | $5,200 | | | | • |
| SUNY at Canton | 18 | Aug | 21 | Dipl, AAS | $4,350 | $7,000 | | | | • |
| Touro College (New York) | 40 | Fall Spring | 24, 27 | Dipl, AAS | $9,900 | $9,900 | | • | | • |
| **North Carolina** | | | | | | | | | | |
| Central Piedmont Community College (Charlotte) | 30 | May | 24 | AAS | $1,441 | $8,011 | | | | • |
| **North Dakota** | | | | | | | | | | |
| Williston State College | 16 | Aug | 26 | AAS | $3,063 | $4,536 | | | | |
| **Ohio** | | | | | | | | | | |
| Cuyahoga Community College (Cleveland) | 24 | Aug | 24 | AAS | $2,900 | $7,850 | | | | • |
| Kent State University - Ashtabula Campus | 28 | Jan | 24 | Dipl, AAS | $2,293 | $6,009 | | • | | • |
| Kent State University - East Liverpool Campus | 28 | Aug | 16 | AAS | $4,770 | $12,202 | | | | • |
| Marion Technical College | 24 | Sep | 21 | AAS | $4,394 | $6,699 | | • | | |
| North Central State College (Mansfield) | 28 | Jun | 24 | AAS | $4,441 | $8,882 | | | | |
| University of Cincinnati | 24 | Sep | 30 | Dipl, AAS | $9,381 | $23,904 | | | | • |
| Washington State Community College (Marietta) | 20 | Fall | 22 | AAS | $76 | $152 | | • | | |
| Zane State College (Zanesville) | 20 | Fall | 22 | AAS | $5,100 | $10,200 | | | | • |
| **Oklahoma** | | | | | | | | | | |
| Caddo Kiowa Tech Ctr/SW Oklahoma St Univ (Ft Cobb) | 12 | Aug | 10 | AAS | $3,000 | $3,000 | | | | |
| Northeastern Oklahoma A&M College (Miami) | 16 | May | 13 | AAS | | | | | | |

*Data are shown only for programs that completed the 2006 AMA Survey of Health Professions Education Programs
‡Key to Offers: 1: Evening or weekend classes; 2: Non-English instruction; 3: Cultural competence instruction

## Physical Therapist Assistant

| Programs* | Class Capacity | Begins | Length (months) | Award | Res Tuition | Non-res Tuition | Stipend | Offers:‡ 1 | 2 | 3 |
|---|---|---|---|---|---|---|---|---|---|---|
| **Oregon** | | | | | | | | | | |
| Mt Hood Community College (Gresham) | 24 | Sep | 21 | AAS | $3,150 | $9,821 | | | | • |
| **Pennsylvania** | | | | | | | | | | |
| Butler County Community College | 30 | Aug | 21 | Dipl, AAS | $3,139 | $5,658 | | | | • |
| California University of Pennsylvania | 24 | Aug | 21 | Dipl, AAS | $2,519 | $3,779 | | | | • |
| Central Pennsylvania College (Summerdale) | 40 | Jul | 20 | AST | $15,360 | $15,360 | | • | | |
| Lehigh Carbon Community College (Schnecksville) | 36 | Aug | 24 | AAS | $2,280 | $4,560 | | | | |
| Mercyhurst College (North East) | 30 | Sep | 21 | AS | $17,825 | $17,825 | | | | • |
| Mount Aloysius College (Cresson) | 40 | Aug | 20 | AS | $16,890 | $16,890 | | | | • |
| Penn State University - DuBois | 32 | Aug | 24 | Dipl, AS | $10,832 | $16,168 | | | | • |
| University of Pittsburgh - Titusville | 24 | Aug | 20 | AS | $8,218 | $17,098 | | • | | • |
| **Rhode Island** | | | | | | | | | | |
| Community College of Rhode Island (Newport) | 24 | Sep | 24 | AAS | $2,390 | $7,000 | | | | |
| **South Carolina** | | | | | | | | | | |
| Greenville Technical College | 50 | Aug | 11, 15 | AS | $5,100 | $9,390 | | | | • |
| Trident Technical College (Charleston) | 25 | May | 12 | AHS | $4,671 | $5,187 | | | | |
| **Tennessee** | | | | | | | | | | |
| Jackson State Community College | 20 | Aug Sep | 22 | AAS | $1,115 | $3,338 | | | | • |
| South College (Knoxville) | 20 | Oct Apr | 24 | AS | $17,200 | $17,200 | | | | • |
| Volunteer State Community College (Gallatin) | 32 | May | 12 | AAS | $2,587 | $9,774 | | | | |
| Walters State Community College (Morristown) | 20 | Aug | 24 | AAS | $2,381 | $6,553 | | | | |
| **Texas** | | | | | | | | | | |
| Amarillo College | 24 | Jan | 20 | AAS | $837 | $1,829 | | | | • |
| Austin Community College | 20 | Aug | 24 | AAS | $1,300 | $5,200 | | • | | • |
| Blinn College (Bryan) | 12 | Sep | 24 | AAS | $2,035 | $4,657 | | | | |
| Kilgore College | 16 | Aug | 21 | AAS | $1,495 | $3,306 | | | | • |
| McLennan Community College (Waco) | 32 | Aug | 23 | AAS | $1,950 | $3,750 | | | | • |
| Odessa College | 16 | Aug | 21 | AAS | $48 | $58 | | | | • |
| San Jacinto College South (Houston) | 24 | Aug | 24 | Dipl, AAS | $1,420 | $2,675 | | | | • |
| St Philip's College (San Antonio) | 24 | Aug | 22 | AAS | $750 | $1,500 | | | | |
| **Utah** | | | | | | | | | | |
| Salt Lake Community College (Salt Lake City) | 24 | Aug | 16 | AAS | $4,700 | $8,300 | | | | • |
| **Virginia** | | | | | | | | | | |
| Northern Virginia Community College (Springfield) | 34 | Aug | 24 | AAS | $2,500 | $7,000 | | | | • |
| **Washington** | | | | | | | | | | |
| Green River Community College (Auburn) | 32 | Sep | 21 | AAS | $2,895 | $3,318 | | | | • |
| Spokane Falls Community College | 16 | Sep | 21 | AAS | $3,651 | $12,408 | | | | • |
| **West Virginia** | | | | | | | | | | |
| Fairmont State Univ (Clarksburg) | 20 | Aug | 22 | AAS | $3,680 | $8,701 | | | | • |
| **Wisconsin** | | | | | | | | | | |
| Northeast Wisconsin Technical College (Green Bay) | 34 | Aug | 24 | AAS | $3,400 | $169,000 | | | | • |

*Data are shown only for programs that completed the 2006 AMA Survey of Health Professions Education Programs
‡Key to Offers: 1: Evening or weekend classes; 2: Non-English instruction; 3: Cultural competence instruction

# Physician Assistant

## History

The profession of physician assistant (PA) originated in the mid 1960s with leadership from Duke University, the University of Colorado, the University of Washington, and Wake Forest University. The early 1970s brought a rapid growth in the number of such educational programs, which were supported initially with $6.1 million appropriated under the authority of the Health Manpower Act of 1972. The funding also supported some of the initial organization and administration of the national program for the accreditation of educational programs in this field, specifically those designed to prepare individuals as assistants to primary care physicians. Since 1992, the number of accredited PA programs has more than doubled from 55 to 137.

## Occupational Description

The physician assistant is academically and clinically prepared to practice medicine with the direction and responsible supervision of a doctor of medicine or osteopathy. The physician-PA team relationship is fundamental to the PA profession and enhances the delivery of high-quality health care. Within the physician-PA relationship, PAs make clinical decisions and provide a broad range of diagnostic, therapeutic, preventive, and health maintenance services. The clinical role of PAs includes primary and specialty care in medical and surgical practice settings. PA practice is centered on patient care and may include educational, research, and administrative activities.

The role of the physician assistant demands intelligence, sound judgment, intellectual honesty, appropriate interpersonal skills, and the capacity to react to emergencies in a calm and reasoned manner. An attitude of respect for self and others, adherence to the concepts of privilege and confidentiality in communicating with patients, and a commitment to the patient's welfare are essential attributes of the graduate PA.

## Employment Characteristics

The 2005 Physician Assistant Census, published by the American Academy of Physician Assistants, indicates that of the more than 58,600 practicing physician assistants, about 41% are practicing in primary care. Family practice is the most common specialty for physician assistants (28.4%), followed by surgery and surgical subspecialties, emergency medicine, subspecialties of internal medicine, general internal medicine, and dermatology.

The majority of physician assistants practice in ambulatory care settings. Solo and group practices employ 56.4% of all physician assistants. The number of physician assistants employed by hospitals is 22.4%, owing in part to the number of physician assistants working as house staff. The government employs almost 10% of the physician assistant workforce, primarily in the military and the Department of Veterans Affairs. The remaining members of the profession are practicing in community health centers, managed care organizations, freestanding urgent care centers, correctional facilities, and other settings.

Physician assistants work an average of 44.3 hours per week. The number of patient visits for physician assistants in outpatient settings averages 96.5 per week; in inpatient settings the average is 67.3 patient visits per week. Forty percent of physician assistants have on-call responsibilities that average 96 hours per month.

## Salary

Salaries vary depending on the experience of the individual, the practice specialty, job responsibilities, and the regional cost of living. Refer to Section IV, Table 5 of this *Directory* for more information, or see www.ama-assn.org/go/salary.

## Educational Programs

**Length.** Although 25 to 27 months is most common, the length of programs varies, largely owing to a difference in student selection criteria and in the educational objectives of the individual program.

**Prerequisites.** Although requirements differ widely, a majority of programs require 2 years of undergraduate study and some work experience in health care. A balance of study in the applied behavioral sciences and the biological sciences is advised for students who wish to qualify for admission to a physician assistant program.

**Curriculum.** Accreditation standards require competency-based curricula. The professional curriculum for PA education includes basic medical, behavioral, and social sciences; clinical preparatory sciences, patient assessment, and supervised clinical practice; health policy; and professional practice issues. Four-year programs are designed to provide the student with a balance of traditional liberal arts courses and biological and applied behavioral science courses. These courses are prerequisites to clinical didactic and supervised clinical practice instruction common to both 2-year and 4-year programs. Supervised clinical practice rotations in pediatrics, family medicine, internal medicine, prenatal care and gynecology, geriatrics, emergency medicine, psychiatry/behavioral medicine, and surgery offer advanced applied content and supervised clinical work experience in dealing with commonly encountered demands for the primary health care of individuals from infancy through childhood, adolescence, and the various phases of adulthood.

## Licensure

All 50 states, the District of Columbia, and the majority of US territories have enacted laws regulating the practice of physician assistants. In order to practice as a physician assistant, an individual must meet the state's licensing criteria and have a supervising physician. Forty-nine states and the District of Columbia allow physicians to delegate prescriptive authority to the PAs they supervise.

## Inquiries

**Careers**
American Academy of Physician Assistants
950 N Washington Street
Alexandria, VA 22314
703 836-2272
703 684-1924 Fax
E-mail: aapa@aapa.org
www.aapa.org

Physician Assistant Education Association
300 North Washington Street, Suite 505
Alexandria, VA 22314-2544
703 548-5538
703 548-5539 Fax
E-mail: info@PAEAonline.org
www.paeaonline.org

**National Certification**
National Commission on Certification of Physician Assistants
12000 Findley Road, Suite 200
Duluth, GA 30097
678 417-8100
www.nccpa.net

**Program Accreditation**
Accreditation Review Commission on Education for the Physician Assistant (ARC-PA)
John McCarty, Executive Director
12000 Findley Road, Suite 240
Duluth, GA 30097
770 476-1224
770 476-1738 Fax
Email: arc-pa@arc-pa.org
www.arc-pa.org

## Physician Assistant*

## Alabama

### University of Alabama at Birmingham
Surgical Physician Assistant Prgm
Sch of Health Professions
1705 University Blvd, RMSB 481
Birmingham, AL 35294-1212
www.uab.edu
*Prgm Dir:* Herbert Ridings, MA PA-C
*Tel:* 205 934-4605    *Fax:* 205 934-3780
*E-mail:* hridings@uab.edu
*Med Dir:* John J Gleysteen, MD

### University of South Alabama
Physician Assistant Prgm
Dept of Physician Asst Studies
1504 Springhill Ave, Ste 4410
Mobile, AL 36604-3273
www.southalabama.edu/alliedhealth/pa
*Prgm Dir:* Richard Nenstiel, PA-C MBA
*Tel:* 251 434-3641    *Fax:* 251 434-3646
*E-mail:* pastudies@usouthal.edu
*Med Dir:* Richard Esham, MD

## Arizona

### Midwestern University - Glendale Campus
Physician Assistant Prgm
19555 N 59th Ave
Glendale, AZ 85308
www.midwestern.edu
*Prgm Dir:* Kevin Lohenry, MPAS PA-C
*Tel:* 623 572-3311    *Fax:* 623 572-3227
*E-mail:* klohen@midwestern.edu
*Med Dir:* James Meyer, MD

### AT Still University of Health Sciences
Physician Assistant Prgm
5850 E Still Circle
Mesa, AZ 85206
www.ashs.edu
*Prgm Dir:* M Goodwin, PA
*Tel:* 480 219-6040    *Fax:* 480 219-6100
*E-mail:* mgoodwin@atsu.edu
*Med Dir:* Bruce Badagliaqua, DO

## Arkansas

### Harding University
Physician Assistant Prgm
PO Box 12231
Searcy, AR 72149
*Prgm Dir:* Michael Murphy, MD
*Tel:* 501 279-5642    *Fax:* 501 268-2111
*E-mail:* mmurphy@harding.edu

## California

### University of Southern California
Physician Assistant Prgm
1000 S Fremont Ave, Unit 7
Bldg A6, 4th Fl
Alhambra, CA 91803
www.usc.edu/pa
*Prgm Dir:* Rosslynn Byous, PA-C DPA
*Tel:* 626 457-4271    *Fax:* 626 457-4262
*E-mail:* byous@usc.edu
*Med Dir:* Katrina Miller, MD

### Loma Linda University
Physician Assistant Prgm
Nichol Hall, Rm 2033
Loma Linda, CA 92350
*Prgm Dir:* Kenrick Bourne, DrPH PA-C
*Tel:* 909 558-1000, Ext 87289    *Fax:* 909 558-0495
*E-mail:* kbourne@llu.edu
*Med Dir:* Benny Hau, MD

### Charles R Drew Univ of Med & Science
Physician Assistant Prgm
1731 E 120th St
Los Angeles, CA 90059
*Prgm Dir:* Carolyn Spaulding, PhD PA-C
*Tel:* 323 563-5950    *Fax:* 323 563-4833
*E-mail:* caspauld@cdrewu.edu
*Med Dir:* Eugene Hardin, MD

### Riverside County Regional Medical Center
*Cosponsor:* Riverside Community College
Physician Assistant Prgm
Moreno Valley Campus
16130 Lasselle St
Moreno Valley, CA 92551-2045
*Prgm Dir:* Delores Middleton, MSEd PA-C
*Tel:* 951 571-6166    *Fax:* 951 571-6221
*E-mail:* delores.middleton@rcc.edu
*Med Dir:* Paul Aoyagi, MD

### Samuel Merritt College
Physician Assistant Prgm
450 30th St, 4th Fl, Rm 4708
Oakland, CA 94609
www.samuelmerritt.edu
*Prgm Dir:* Lorraine Petti, MS PA-C
*Tel:* 510 869-6623    *Fax:* 510 869-6951
*E-mail:* lpetti@samuelmerritt.edu
*Med Dir:* Monica Rosenthal, MD

### Stanford University School of Medicine
Physician Assistant Prgm
Primary Care Associate Prgm
1215 Welch Rd Ste Modular G
Palo Alto, CA 94305
http://pcap.stanford.edu
*Prgm Dir:* Sherry Stolberg, MGPGP PA-C
*Tel:* 650 725-5340    *Fax:* 650 723-9692
*E-mail:* stolberg@stanford.edu
*Med Dir:* Valerie Berry, MD

### Western Univ of Health Sciences
Physician Assistant Prgm
Primary Care PA Program
309 E Second St
Pomona, CA 91766-1854
www.westernu.edu
*Prgm Dir:* Roy Guizado, MS PA-C
*Tel:* 909 469-5378    *Fax:* 909 469-5407
*E-mail:* roygpac@westernu.edu
*Med Dir:* Alan Cundari, PA DO

### University of California - Davis
Physician Assistant Prgm
2516 Stockton Blvd, Ste 254
Sacramento, CA 95817-2297
http://fnppa.ucdavis.edu
*Prgm Dir:* Betty Ingell, EdD APRN-BC PA-C
*Tel:* 916 734-3551    *Fax:* 916 452-2112
*E-mail:* betty.ingell@ucdmc.ucdavis.edu
*Med Dir:* Joann Seibles, MPH, MD

### Touro University Mare Island
Physician Assistant Prgm
1310 Johnson Ln
Vallejo, CA 94592
*Prgm Dir:* Lauren Padilla
*Tel:* 707 638-5978    *Fax:* 707 638-5955
*E-mail:* lpadilla@touro.edu

### San Joaquin Valley College - Visalia
Physician Assistant Prgm
8400 W Mineral King
Visalia, CA 93291
*Prgm Dir:* Monica Urnson, PA
*Tel:* 559 651-2500    *Fax:* 559 651-3161
*E-mail:* monicau@sjvc.edu
*Med Dir:* David Tenn, MD

*Additional information about programs that returned the AMA's survey is available in a table beginning on page 344.

## Colorado

**U of Colorado at Denver and Health Scis Ctr**
Physician Assistant Prgm
Child Hlth Associate, Physician Assistant Program
Mail Stop F543, PO Box 6508
Aurora, CO 80045-0508
www.uchsc.edu/chapa
*Prgm Dir:* Gerald B Merenstein, MD
*Tel:* 303 315-7963, Ext 3   *Fax:* 303 724-1350
*E-mail:* gerald.merenstein@uchsc.edu
*Med Dir:* Gerald Merenstein, MD

**Red Rocks Community College**
Physician Assistant Prgm
Campus Box 38
13300 W Sixth Ave
Lakewood, CO 80228
www.rrcc.edu/pa
*Prgm Dir:* Jim Keller, MPH PA-C
*Tel:* 303 914-6287   *Fax:* 303 914-6806
*E-mail:* jim.keller@rrcc.edu
*Med Dir:* Robert Beshore, MD

## Connecticut

**Quinnipiac University**
Physician Assistant Prgm
275 Mt Carmel Ave
Hamden, CT 06518-1908
www.quinnipiac.edu
*Prgm Dir:* Cynthia Booth-Lord, MHS PA-C
*Tel:* 203 582-5297   *Fax:* 203 582-5303
*E-mail:* Cynthia.Lord@quinnipiac.edu
*Med Dir:* Ronald Rozett, MD

**Yale University School of Medicine**
Physician Assistant Prgm
47 College St Ste 220
New Haven, CT 06510-3209
www.paprogram.yale.edu
*Prgm Dir:* Mary Warner, MMSc PA-C
*Tel:* 203 785-2860   *Fax:* 203 785-3601
*E-mail:* mary.warner@yale.edu
*Med Dir:* John P Hayslett, MD

## District of Columbia

**George Washington University**
Physician Assistant Prgm
900 23rd St, NW, Ste 6148
Washington, DC 20037
*Prgm Dir:* J Jeffery Heinrich, PA-C EdD
*Tel:* 202 994-6670   *Fax:* 202 994-7647
*E-mail:* heinrich@gwu.edu
*Med Dir:* Brad Moore, MD MPH

**Howard University**
Physician Assistant Prgm
6th & Bryant Sts NW
Washington, DC 20059
*Prgm Dir:* Marvin Barnard, MD
*Tel:* 202 806-7536, Ext 5955   *Fax:* 202 806-4476
*E-mail:* mbarnard@fac.howard.edu
*Med Dir:* Marvin Barnard, MD

## Florida

**Nova Southeastern University - Ft Lauderdale**
Physician Assistant Prgm
3200 S University Dr
Ft Lauderdale, FL 33328
www.nova.edu/pa
*Prgm Dir:* William H Marquardt, MA PA-C
*Tel:* 954 262-1252   *Fax:* 954 262-2285
*E-mail:* marquard@nova.edu
*Med Dir:* Morton A Diamond, MD

**University of Florida**
Physician Assistant Prgm
College of Medicine
PO Box 100176
Gainesville, FL 32610-0176
www.med.ufl.edu/pap/apply
*Prgm Dir:* Wayne D Bottom, PA-C MPH
*Tel:* 352 265-7955   *Fax:* 352 265-7996
*E-mail:* bottow@medicine.ufl.edu
*Med Dir:* Jason Fromm, MD

**Miami Dade College**
Physician Assistant Prgm
950 NW 20th St
Miami, FL 33127-4693
*Prgm Dir:* Pascale GŠhy-AndrŠ, PA-C
*Tel:* 305 237-4381   *Fax:* 305 237-4278
*E-mail:* pgehyand@mdc.edu
*Med Dir:* Fatima Zafar, MD

**Barry University**
Physician Assistant Prgm
11300 NE 2nd Ave Box SGMS-PA
Miami Shores, FL 33161-6695
www.barry.edu
*Prgm Dir:* Doreen C Parkhurst, MD
*Tel:* 305 899-4065, Ext 4065   *Fax:* 305 899-4083
*E-mail:* dparkhurst@mail.barry.edu
*Med Dir:* Doreen C Parkhurst, MD

**Nova Southeastern University - Naples**
Physician Assistant Prgm
2655 Northbrooke Dr
Naples, FL 32610-0176
*Prgm Dir:* Julie Keena
*Tel:* 239 591-4528, Ext 10   *Fax:* 239 495-5925
*E-mail:* jkeena@nsu.nova.edu

## Georgia

**Emory University**
Physician Assistant Prgm
School of Medicine
1462 Clifton Rd, Ste 280
Atlanta, GA 30322
www.emorypa.org
*Prgm Dir:* Virginia Joslin, MPH PA-C
*Tel:* 404 727-7857   *Fax:* 404 727-7836
*E-mail:* vjoslin@learnlink.emory.edu
*Med Dir:* Theresa Berry, MD

**Medical College of Georgia**
Physician Assistant Prgm
Health Sciences Building, Rm EC-3304
987 St Sebastian Way
Augusta, GA 30912
www.mcg.edu/sah/PhyAsst
*Prgm Dir:* Bonnie A Dadig, EdD PA-C
*Tel:* 706 721-3246   *Fax:* 706 721-3990
*E-mail:* bdadig@mcg.edu
*Med Dir:* David Haburchak, MD

**South University**
Physician Assistant Prgm
709 Mall Blvd
Savannah, GA 31406
www.southuniversity.edu/campus/PhysicianAssistant
*Prgm Dir:* John Burns, PA-C MMSc
*Tel:* 912 201-8024   *Fax:* 912 201-8070
*E-mail:* paprogram@southuniversity.edu
*Med Dir:* John Northup, MD

## Idaho

**Idaho State University**
Physician Assistant Prgm
Campus Box 8253
1021 S Red Hill Rd
Pocatello, ID 83209-8253
www.isu.edu/paprog
*Prgm Dir:* John Schroeder, PA-C JD
*Tel:* 208 282-4726   *Fax:* 208 282-4969
*E-mail:* schrjohn@isu.edu
*Med Dir:* Sherwin D'Souza, MD

## Illinois

**Southern Illinois Univ at Carbondale**
Physician Assistant Prgm
School of Allied Health and School of Medicine
Lindegren Hall Rm 129 MC 6516
Carbondale, IL 62901-6516
http://mccoy.lib.siu.edu/~paprogram
*Prgm Dir:* Laurie Dunn, MPAS PA-C
*Tel:* 618 453-8850   *Fax:* 618 453-7216
*E-mail:* ldunn@siumed.edu
*Med Dir:* Penelope K Tippy, MD

**Malcolm X College**
Physician Assistant Prgm
1900 W Van Buren St, Rm 3241
Chicago, IL 60612-3197
*Prgm Dir:* Geri Shangreaux, PA-C
*Tel:* 312 850-3532   *Fax:* 312 850-3536
*E-mail:* gshangreaux@ccc.edu
*Med Dir:* Margaret Dolan, MD

**Midwestern University**
Physician Assistant Prgm
555 31st St
Downers Grove, IL 60515-1235
www.midwestern.edu
*Prgm Dir:* Lisa Wallace, PhD
*Tel:* 630 515-6034   *Fax:* 630 971-6402
*E-mail:* lwalla@midwestern.edu
*Med Dir:* Patrick Towne, MD

**Rosalind Franklin Univ of Medicine & Science**
Physician Assistant Prgm
3333 Green Bay Rd
North Chicago, IL 60064-3095
www.rosalindfranklin.edu
*Prgm Dir:* Patrick Knott, PhD PA-C
*Tel:* 847 578-8689   *Fax:* 847 578-8690
*E-mail:* Patrick.Knott@RosalindFranklin.edu
*Med Dir:* Walid Khayr, MD

## Indiana

**University of Saint Francis**
*Cosponsor:* Department of Physician Assistant Studies
Physician Assistant Prgm
2701 Spring St
Ft Wayne, IN 46808
www.sf.edu/healthscience/pa
*Prgm Dir:* Thomas Hunter
*Tel:* 260 434-7665   *Fax:* 260 434-7585
*E-mail:* thunter@sf.edu
*Med Dir:* Edward Lelonek, MD

**Butler University/Clarian Health**
Physician Assistant Prgm
College of Pharm and Hlth Sci
4600 Sunset Ave
Indianapolis, IN 46208-3485
www.butler.edu/cophs/?pg=135
*Prgm Dir:* John A Lucich, MD
*Tel:* 317 940-6147   *Fax:* 317 940-6172
*E-mail:* jlucich@butler.edu
*Med Dir:* John A Lucich, MD

# Iowa

**Des Moines University**
Physician Assistant Prgm
3200 Grand Ave
Des Moines, IA 50312-4198
www.dmu.edu/pa
*Prgm Dir:* Gregory J Kolbinger, MPAS PA-C
*Tel:* 515 271-1695    *Fax:* 515 271-7115
*E-mail:* gregory.kolbinger@dmu.edu
*Med Dir:* Carolyn Beverly, MD, MPH

**University of Iowa**
Physician Assistant Prgm
Roy J and Lucille A Carver College of Medicine
5167 Westlawn
Iowa City, IA 52242-1100
www.medicine.uiowa.edu/pa/pa.htm
*Prgm Dir:* David P Asprey, PhD PA-C
*Tel:* 319 335-8920    *Fax:* 319 335-8923
*E-mail:* david-asprey@uiowa.edu
*Med Dir:* Daniel S Fick, MD

# Kansas

**Wichita State University**
Physician Assistant Prgm
Campus Box 43
1845 N Fairmount
Wichita, KS 67260-0043
www.chp.wichita.edu/pa
*Prgm Dir:* Richard Muma, PhD MPH
*Tel:* 316 978-3011    *Fax:* 316 978-3025
*E-mail:* richard.muma@wichita.edu
*Med Dir:* Timothy Scanlan, MD

# Kentucky

**University of Kentucky**
Physician Assistant Prgm
Dept of Clinical Sciences
900 S Limestone St, Ste 205
Lexington, KY 40536-0200
*Prgm Dir:* Julie Gurwell, PhD PA-C
*Tel:* 859 323-1100, Ext 80843    *Fax:* 859 257-2454
*E-mail:* jagur@uky.edu
*Med Dir:* Jennifer Joyce, MD

# Louisiana

**Our Lady of the Lake College**
Physician Assistant Prgm
7434 Perkins Rd
Baton Rouge, LA 70808
www.ololcollege.edu/physician_asst.html
*Prgm Dir:* Elaine E Grant, PA-C MPH
*Tel:* 225 214-6988    *Fax:* 225 490-1650
*E-mail:* egrant@ololcollege.edu
*Med Dir:* Susan Nelson, MD, FACP

**Louisiana State Univ Hlth Sci Ctr-Shreveport**
Physician Assistant Prgm
1501 Kings Hwy, PO Box 33932
Shreveport, LA 71130-3932
www.sh.lsuhsc.edu/ah
*Prgm Dir:* Kim Meyer, MPAS PA-C
*Tel:* 318 675-7744    *Fax:* 318 675-6937
*E-mail:* kmeyer1@LSUHSC.edu
*Med Dir:* Richard Turnage, MD

# Maine

**University of New England**
Physician Assistant Prgm
716 Stevens Ave
Portland, ME 04103-2670
*Prgm Dir:* Erich Fogg, PA-C MMSc
*Tel:* 207 221-4529    *Fax:* 207 221-4711
*E-mail:* EFogg@une.edu
*Med Dir:* Handler Jeffery, MD

# Maryland

**Anne Arundel Community College**
Physician Assistant Prgm
101 College Pkwy
Arnold, MD 21012-1895
www.aacc.edu/physassist
*Prgm Dir:* Luis A Ramos, MS PA-C
*Tel:* 410 777-7448    *Fax:* 410 777-7099
*E-mail:* laramos@aacc.edu
*Med Dir:* S David Krimins, MD

**Towson University**
*Cosponsor:* CCBC Essex
Physician Assistant Prgm
7201 Rossville Blvd
Baltimore, MD 21237-3855
*Prgm Dir:* Donna Sewell, MS PA-C
*Tel:* 410 780-6616    *Fax:* 410 780-6405
*E-mail:* dsewell@ccbcmd.edu
*Med Dir:* Edwin W Whiteford, Jr, MD

**University of Maryland Eastern Shore**
Physician Assistant Prgm
Modular 934-5 Backbone Rd
Princess Anne, MD 21853
*Prgm Dir:* Darlene L J Robinson, MSPAS PA-C
*Tel:* 410 651-8932    *Fax:* 410 651-7586
*E-mail:* dljrobinson@umes.edu
*Med Dir:* Christjon Huddleston, MD

# Massachusetts

**Mass College of Pharmacy & Health Sciences**
Physician Assistant Prgm
179 Longwood Ave
Boston, MA 02115
www.mcphs.edu
*Prgm Dir:* Marianne Vail, MS PA-C
*Tel:* 617 732-2918    *Fax:* 617 732-1027
*E-mail:* marianne.vail@bos.mcphs.edu
*Med Dir:* Thomas Patnaude, MD

**Northeastern University**
Physician Assistant Prgm
360 Huntington Ave, 202 Robinson
Boston, MA 02115-5000
www.bouve.neu.edu/Graduate/Health/pap.html
*Prgm Dir:* Rosann M Ippolito, PhD
*Tel:* 617 373-3195    *Fax:* 617 373-3338
*E-mail:* r.ippolito@neu.edu
*Med Dir:* Robin Reed, MD

**Springfield College**
Physician Assistant Prgm
263 Alden St
Springfield, MA 01109-3797
*Prgm Dir:* Jennifer Hixon, MS PA-C
*Tel:* 413 748-3554    *Fax:* 413 748-3595
*E-mail:* jennifer_hixon@spfldcol.edu
*Med Dir:* Mark Kenton, MD

# Michigan

**University of Detroit Mercy**
Physician Assistant Prgm
4001 West McNichols Rd
Detroit, MI 48221
http://healthprofessions.udmercy.edu/paprogram
*Prgm Dir:* Suzanne York, MPH PA-C
*Tel:* 313 993-1930    *Fax:* 313 993-1271
*E-mail:* warnimsk@udmercy.edu
*Med Dir:* Walid A Harb, MD

**Wayne State University**
Physician Assistant Prgm
College of Pharmacy and Health Sciences
259 Mack Ave, Ste 2590
Detroit, MI 48201
*Prgm Dir:* Stephanie Joseph Gilkey, MS PA-C
*Tel:* 313 577-9666    *Fax:* 313 577-5467
*E-mail:* ab4703@wayne.edu
*Med Dir:* Mohamed Siddique, MD

**Grand Valley State University**
Physician Assistant Prgm
Center for Health Sciences
301 Michigan St NE
Grand Rapids, MI 49503
www.gvsu.edu/pa
*Prgm Dir:* Wallace Boeve, MSPA PA-C
*Tel:* 616 331-5988    *Fax:* 616 331-5999
*E-mail:* boevew@gvsu.edu
*Med Dir:* Jeffrey Libra, MD

**Western Michigan University**
Physician Assistant Prgm
3245 Ellsworth Hall
1903 W Michigan Ave
Kalamazoo, MI 49008-5138
*Prgm Dir:* James Van Rhee, PA-C
*Tel:* 269 387-5317    *Fax:* 269 387-3319
*E-mail:* jim.vanrhee@wmich.edu
*Med Dir:* Jeanette Meyer, MD

**Central Michigan University**
Physician Assistant Prgm
HPB 1222
Mt Pleasant, MI 48859
www.cmich.edu
*Prgm Dir:* Ahmad Hakemi, MD
*Tel:* 989 774-1273    *Fax:* 989 774-2433
*E-mail:* ahmad.hakemi@cmich.edu
*Med Dir:* Tammi Moutsatoson, MD

# Minnesota

**Augsburg College**
Physician Assistant Prgm
2211 Riverside Ave, CB 149
Minneapolis, MN 55454
www.augsburg.edu/pa
*Prgm Dir:* Dawn B Ludwig, PhD PA-C
*Tel:* 612 330-1399    *Fax:* 612 330-1757
*E-mail:* ludwig@augsburg.edu
*Med Dir:* Steve Nerheim, MD

# Missouri

**Missouri State University**
Physician Assistant Prgm
901 S National Ave
Springfield, MO 65897
*Prgm Dir:* Steve Dodge, MD
*Tel:* 417 836-6151    *Fax:* 417 836-6406
*E-mail:* roc797f@missouristate.edu
*Med Dir:* William Russell Detten, DO

**Saint Louis University**
Cosponsor: Doisy College of Health Sciences
Physician Assistant Prgm
3437 Caroline St
St Louis, MO 63104-1304
www.slu.edu/doisycollege/pa
Prgm Dir: Dana Sayre-Stanhope, PA-C EdD
Tel: 314 977-8639    Fax: 314 977-8649
E-mail: sayreds@slu.edu
Med Dir: Michael Cox, MD

## Montana

**Rocky Mountain College**
Physician Assistant Prgm
1511 Poly Dr
Billings, MT 59102-1796
Prgm Dir: Scott Murray, MD
Tel: 406 238-7375    Fax: 406 657-1194
E-mail: murrays@rocky.edu
Med Dir: Leonard W Etchart, MD

## Nebraska

**Union College**
Physician Assistant Prgm
3800 S 48th St
Lincoln, NE 68506
www.ucollege.edu/pa
Prgm Dir: Michael J Huckabee, MPAS PA-C
Tel: 402 486-2527    Fax: 402 486-2559
E-mail: mihuckab@ucollege.edu
Med Dir: Dwain Leonhardt, MD

**University of Nebraska Medical Center**
Physician Assistant Prgm
984300 Nebraska Medical Center
Omaha, NE 68198-4300
www.unmc.edu/alliedhealth/pa
Prgm Dir: James E Somers, PhD PA
Tel: 402 559-9495    Fax: 402 559-7996
E-mail: dklandon@unmc.edu
Med Dir: Gerald F Moore, MD

## Nevada

**Touro University - Nevada**
Physician Assistant Prgm
874 American Pacific Dr
Henderson, NV 89014
Prgm Dir: Vicki Chan-Padgett,, PA-C MPAS
Tel: 702 856-3262    Fax: 702 856-3346
E-mail: pastudiesnv@touro.edu

## New Hampshire

**Mass College of Pharmacy & Health Sciences**
Physician Assistant Prgm
1260 Elm St
Manchester, NH 03101
Prgm Dir: Louise Lee, PA-C
Tel: 603 314-0210    Fax: 603 314-0303
E-mail: louise.lee@man.mcphs.edu

## New Jersey

**Univ of Med & Dent of New Jersey**
Physician Assistant Prgm
Robert Wood Johnson Med Sch
675 Hoes Ln
Piscataway, NJ 08854-5635
Prgm Dir: Ruth Fixelle, EdM PA-C
Tel: 732 235-4445    Fax: 732 235-4820
E-mail: r.fixelle@umdnj.edu
Med Dir: Robert O'Connor, MD

**Seton Hall University**
Physician Assistant Prgm
400 S Orange Ave
South Orange, NJ 07079
Prgm Dir: Doreen Stiskal, PT PhD
Tel: 973 275-2027    Fax: 973 275-2370
E-mail: stiskado@shu.edu
Med Dir: John Sensakovic, MD PhD

## New Mexico

**University of New Mexico**
Physician Assistant Prgm
School of Medicine
Dept of Family and Community Medicine MSC 09 5040
Albuquerque, NM 87131-0001
http://hsc.unm.edu/pap
Prgm Dir: Nikki Katalanos, PhD PA-C
Tel: 505 272-9864    Fax: 505 272-9828
E-mail: nkatalanos@salud.unm.edu
Med Dir: Arthur Kaufman, MD

**University of St Francis**
Physician Assistant Prgm
4401 Silver Ave SE, Ste B
Albuquerque, NM 87108
www.stfrancis.edu/pa
Prgm Dir: William Riesterer, PA-C MPAS
Tel: 505 266-5565    Fax: 505 266-5585
E-mail: wriesterer@stfrancis.edu
Med Dir: Alfredo Vigil, MD

## New York

**Albany Medical College**
Physician Assistant Prgm
Mail Code 4
47 New Scotland Ave
Albany, NY 12208-3412
www.amc.edu/pa
Prgm Dir: David Irvine, DHSc RPA-C
Tel: 518 262-5251    Fax: 518 262-6698
E-mail: irvined@mail.amc.edu
Med Dir: Luise Ahlers, MD

**Daemen College**
Physician Assistant Prgm
4380 Main St
Amherst, NY 14226-3592
www.daemen.edu
Prgm Dir: Gregg Shutts, MS RPA-C
Tel: 716 839-8563    Fax: 716 839-8252
E-mail: gshutts@daemen.edu
Med Dir: Nelson P Torre, MD

**Touro College - Bay Shore**
Physician Assistant Prgm
Bay Shore PA Prgm, Winthrop University Hospital Ext Ctr
1700 Union Blvd
Bay Shore, NY 11706
Prgm Dir: Joseph Tommasino, PhD RPAC
Tel: 631 665-1600, Ext 271    Fax: 631 665-6086
E-mail: jtpaphd@aol.com
Med Dir: Scott Ippolito, MD

**Bronx Lebanon Hospital**
Physician Assistant Prgm
1650 Selwyn Ave, Ste 11D
Bronx, NY 10457-7628
Prgm Dir: Paul Foster, RPA-C MPA
Tel: 718 960-1255    Fax: 718 960-1329
E-mail: geminapa@aol.com
Med Dir: Milton Gumbs, MD

**Mercy College**
Physician Assistant Prgm
1200 Waters Place
Bronx, NY 10461
http://grad.mercy.edu/physicianassistant
Prgm Dir: Theresa Horvath, MPH PA-C
Tel: 914 674-7626    Fax: 914 674-7623
E-mail: thorvath@mercy.edu
Med Dir: Paul Gross, MD

**Brooklyn Hosp/Long Island Univ**
Physician Assistant Prgm
121 DeKalb Ave
Brooklyn, NY 11201-5424
Prgm Dir: Elizabeth Salzer, RPA-C MA
Tel: 718 260-2780    Fax: 718 260-2790
E-mail: elizabeth.salzer@liu.edu
Med Dir: Michael Patel, MD

**SUNY Downstate Medical Center**
Physician Assistant Prgm
450 Clarkson Ave/PO Box 1222
Brooklyn, NY 11203
Prgm Dir: Rena Mitchell, MS CHES RPA-C
Tel: 718 270-2324    Fax: 718 270-7459
E-mail: rena.mitchell@downstate.edu
Med Dir: Luther T Clark, MD

**D'Youville College**
Physician Assistant Prgm
320 Porter Ave
Buffalo, NY 14201-1084
www.dyc.edu
Prgm Dir: Maureen F Finney, RPA-C MS
Tel: 716 829-7713    Fax: 716 829-7732
E-mail: finneym@dyc.edu
Med Dir: Ronald P Santasiero, MD

**St Vincent Catholic Med Ctr Brklyn-Queens Reg**
Physician Assistant Prgm
175-05 Horace Harding Expressway
Fresh Meadows, NY 11365
Prgm Dir: Niels Schmidt, PA-C
Tel: 718 357-0500, Ext 106    Fax: 718 357-4588
E-mail: nschmidt@svcmcny.org
Med Dir: Victor Politi, MD

**Hofstra University**
Physician Assistant Prgm
113 Monroe Lecture Center
Hempstead, NY 11549-1270
Prgm Dir: Anne M Bozzarelli, PA-C MHS
Tel: 516 463-4804    Fax: 516 463-5177
E-mail: bioamb@Hofstra.edu
Med Dir: Stanley Shanies, MD FACP

**CUNY York College**
Physician Assistant Prgm
94-20 Guy R Brewer Blvd, Rm 1E12
Jamaica, NY 11451
www.york.cuny.edu/PA
Prgm Dir: Robert Brugna, MBA PA-C
Tel: 718 262-2823    Fax: 718 262-2504
E-mail: paprogram@york.cuny.edu
Med Dir: Manuel St.Martin, MD, JD

**City University of New York, The City College**
Cosponsor: The Sophie Davis School of Biomedical Education
Physician Assistant Prgm
138th St & Convent Ave, Harris Hall, Ste 15
New York, NY 10031
www.med.cuny.edu
Prgm Dir: Adrian Llewellyn, RPA-C MPAS
Tel: 212 650-7745, Ext 7745    Fax: 212 650-6697
E-mail: allewellyn@ccny.cuny.edu
Med Dir: Maurice Wright, MD

**Pace University - Lenox Hill Hospital**
Physician Assistant Prgm
1 Pace Plaza, Rm Y-31
New York, NY 10038-1598
www.pace.edu/dyson/paprogram
*Prgm Dir:* Kathleen T Roche, RPA-C RN FNP MPA
*Tel:* 212 346-1241   *Fax:* 212 346-1503
*E-mail:* kroche@pace.edu
*Med Dir:* Sheila A Cain, MD

**Touro College**
Physician Assistant Prgm
Manhattan PA Program
27-33 W 23rd St
New York, NY 10010
www.touro.edu/shs/pany
*Prgm Dir:* Nadja Graff, PhD
*Tel:* 212 463-0400, Ext 788   *Fax:* 212 741-0195
*E-mail:* ngraff@touro.edu
*Med Dir:* Kyi Win Yu, MD

**Weill Med College/Cornell Univ Med Sch**
Physician Assistant Prgm
575 Lexington Ave
New York, NY 10022
www.med.cornell.edu/pa
*Prgm Dir:* Gerard J Marciano, PA-C MS
*Tel:* 646 962-7277   *Fax:* 646 962-7290
*E-mail:* gjm2001@med.cornell.edu
*Med Dir:* Marie-Lynn Eloi-Stiven, MD

**New York Institute of Technology**
Physician Assistant Prgm
NYCOM II, Rm 352
Northern Blvd, PO Box 8000
Old Westbury, NY 11568-8000
http://iris.nyit.edu/hpbls/pas
*Prgm Dir:* Salvatore Barese, EdD PA-C
*Tel:* 516 686-3881   *Fax:* 516 686-3795
*E-mail:* sudarebe@nyit.edu
*Med Dir:* Timothy T Robinson, DO

**Rochester Institute of Technology**
Physician Assistant Prgm
153 Lomb Memorial Dr
Bldg 75 - CBET
Rochester, NY 14623
www.rit.edu/~676www/main_pa.html
*Prgm Dir:* Heidi B Miller, MPH PA-C
*Tel:* 585 475-5945   *Fax:* 585 475-5809
*E-mail:* hbmscl@rit.edu
*Med Dir:* Paul Levy, MD

**Wagner College/Staten Island University**
Physician Assistant Prgm
Concord Site
1034 Targee St, Spring Bldg
Staten Island, NY 10304
*Prgm Dir:* Nora Lowy, PA-C
*Tel:* 718 390-4615   *Fax:* 718 390-4662
*E-mail:* nlowy@siuh.edu
*Med Dir:* Viola Ortiz, MD

**Stony Brook University**
Physician Assistant Prgm
Sch of Health Tech and Management
HSC, L2-424
Stony Brook, NY 11794-8202
www.stonybrook.edu
*Prgm Dir:* Paul Lombardo, MPS RPAC
*Tel:* 631 444-3190   *Fax:* 631 444-1404
*E-mail:* paul.lombardo@stonybrook.edu
*Med Dir:* Gail Cohan, MD

**Le Moyne College**
Physician Assistant Prgm
1416 Salt Springs Rd
Syracuse, NY 13214-1399
www.lemoyne.edu
*Prgm Dir:* Linda G Allison, MD MPH
*Tel:* 315 445-4745   *Fax:* 315 445-4602
*E-mail:* allisolg@lemoyne.edu
*Med Dir:* James Longo, MD

# North Carolina

**Duke University Medical Center**
Physician Assistant Prgm
Department of Community and Family Medicine
DUMC 3848
Durham, NC 27710-0001
http://pa.mc.duke.edu
*Prgm Dir:* Patricia Dieter, MPA PA-C
*Tel:* 919 681-3259   *Fax:* 919 681-9666
*E-mail:* patricia.dieter@duke.edu
*Med Dir:* Joyce Copeland, MD

**Methodist College**
Physician Assistant Prgm
5105 B College Center Dr
Fayetteville, NC 28311-1498
www.methodist.edu/paprogram
*Prgm Dir:* E Ronald Foster, MA MPAS PAC
*Tel:* 910 630-7495   *Fax:* 910 630-7218
*E-mail:* paprog@methodist.edu
*Med Dir:* Christopher Aul, MD

**East Carolina University**
Physician Assistant Prgm
West Research Campus
1157 VOA Site C Road (NCSR 1212)
Greenville, NC 27834-4353
www.ecu.edu/pa
*Prgm Dir:* Larry Dennis, MPAS PA-C
*Tel:* 252 744-1100   *Fax:* 252 744-1110
*E-mail:* dennisl@ecu.edu
*Med Dir:* Dale Newton, MD

**Wake Forest University School of Medicine**
Physician Assistant Prgm
Medical Center Blvd
Winston-Salem, NC 27157-1006
www.wfubmc.edu/PAprogram
*Prgm Dir:* James Van Rhee, MS PA-C
*Tel:* 336 716-2905   *Fax:* 336 716-4432
*E-mail:* jvanrhee@wfubmc.edu
*Med Dir:* K Patrick Ober, MD

# North Dakota

**University of North Dakota**
Physician Assistant Prgm
School of Medicine & Health Sciences
501 Columbia Rd - Stop 9037
Grand Forks, ND 58202-9037
www.med.und.nodak.edu/depts/pa
*Prgm Dir:* Mary Ann Laxen, MAB PA-C
*Tel:* 701 777-2344   *Fax:* 701 777-2491
*E-mail:* mlaxen@medicine.nodak.edu
*Med Dir:* Elizabeth Burns, MD, MA

# Ohio

**University of Findlay**
Physician Assistant Prgm
1000 N Main St
Findlay, OH 45840
*Prgm Dir:* Paul T Davis, MD
*Tel:* 419 434-4529   *Fax:* 419 434-6557
*E-mail:* davis@findlay.edu
*Med Dir:* Paul T Davis, MD

**Kettering College of Medical Arts**
Physician Assistant Prgm
3737 Southern Blvd
Kettering, OH 45429-1299
*Prgm Dir:* Susan Wulff, MS PA-C
*Tel:* 937 298-3399, Ext 55625   *Fax:* 937 395-8095
*E-mail:* sue.wulff@kcma.edu
*Med Dir:* Mark Clasen, MD

**Marietta College**
Physician Assistant Prgm
215 Fifth St
Marietta, OH 45750
www.marietta.edu/graduate/PA
*Prgm Dir:* Gloria M Stewart, EdD PA-C
*Tel:* 740 376-4950   *Fax:* 740 376-4951
*E-mail:* stewartg@marietta.edu
*Med Dir:* Steve Howe, DO

**Cuyahoga Community College**
Physician Assistant Prgm
11000 Pleasant Valley Rd
Parma, OH 44130
www.tri-c.edu
*Prgm Dir:* Sharon L Luke, PA-C
*Tel:* 216 987-5123   *Fax:* 216 987-5066
*E-mail:* sharon.luke@tri-c.edu
*Med Dir:* Michael Menolasino III, DO

**Medical University of Ohio at Toledo**
Physician Assistant Prgm
3015 Arlington Ave
Toledo, OH 43614
www.meduohio.edu/allh/pa
*Prgm Dir:* Patricia Hogue, MS PA-C
*Tel:* 419 383-5408   *Fax:* 419 383-5880
*E-mail:* patricia.hogue@utoledo.edu
*Med Dir:* Mark Weiner, MD

# Oklahoma

**Univ of Oklahoma Health Sciences Center**
Physician Assistant Prgm
PO Box 26901
Oklahoma City, OK 73190
*Prgm Dir:* Dan McNeill, PhD PA
*Tel:* 405 271-2058   *Fax:* 405 271-3621
*E-mail:* daniel-mcneill@ouhsc.edu
*Med Dir:* James L Brand, MD

# Oregon

**Pacific University**
Physician Assistant Prgm
2043 College Way
Forest Grove, OR 97116
www.pacificu.edu/pa
*Prgm Dir:* Randy Randolph, PA-C MPAS
*Tel:* 503 359-3120   *Fax:* 503 359-2977
*E-mail:* randolph@pacificu.edu
*Med Dir:* Richard Gicking, MD

**Oregon Health & Science University**
Physician Assistant Prgm
3181 SW Sam Jackson Park Rd, GH 219
Portland, OR 97239-3098
www.ohsu.edu/pa
*Prgm Dir:* Ted J Ruback, MS PA-C
*Tel:* 503 494-1484   *Fax:* 503 494-1409
*E-mail:* paprgm@ohsu.edu
*Med Dir:* Colleen Schierholtz, BS, Director of
   Admissions

# Pennsylvania

**DeSales University**
Physician Assistant Prgm
2755 Station Ave
Center Valley, PA 18034-9568
www.desales.edu/physicianassistant
*Prgm Dir:* Christine Bruce, MHSA PA-C
*Tel:* 610 282-1100, Ext 1415   *Fax:* 610 282-1893
*E-mail:* physician.assistant@desales.edu
*Med Dir:* David Brock, MD

**Gannon University**
Physician Assistant Prgm
109 University Square
Erie, PA 16541-0001
www.gannon.edu
*Prgm Dir:* Michele Roth-Kauffman, JD MPAS PA-C
*Tel:* 814 871-5643   *Fax:* 814 871-5502
*E-mail:* roth-kauffman@gannon.edu
*Med Dir:* John C Jageman, MD

**Arcadia University**
Physician Assistant Prgm
450 S Easton Rd
Glenside, PA 19038
*Prgm Dir:* Michael Dryer, MPH PA-C
*Tel:* 215 572-2083   *Fax:* 215 881-8746
*E-mail:* dryer@arcadia.edu
*Med Dir:* Irwin Wolfert, MD

**Seton Hill University**
Physician Assistant Prgm
Seton Hill Dr
Greensburg, PA 15601
*Prgm Dir:* Linda Gabersek
*Tel:* 724 830-1097   *Fax:* 724 838-7846
*E-mail:* lgabersek@setonhill.edu
*Med Dir:* Theodore Stem, MD

**Lock Haven University**
Physician Assistant Prgm
401 N Fairview St
104 Annex Building
Lock Haven, PA 17745
www.lhup.edu
*Prgm Dir:* Walter Eisenhauer, MMSc PA-C
*Tel:* 570 893-2168   *Fax:* 570 893-2540
*E-mail:* weisenha@lhup.edu
*Med Dir:* Michael Greenberg, MD MBA

**St Francis University**
Physician Assistant Prgm
Sullivan Hall Rm 104
PO Box 600
Loretto, PA 15940-0600
www.francis.edu/PhysAsst/PAHome.shtml
*Prgm Dir:* Don Shipman, PA-C EdD
*Tel:* 814 472-3130   *Fax:* 814 472-3137
*E-mail:* pa@francis.edu
*Med Dir:* Lawrence Stem, MD

**Drexel University**
Physician Assistant Prgm
245 N 15th St, MS 504
Philadelphia, PA 19102
*Prgm Dir:* Patrick C Auth, MS PA-C
*Tel:* 215 762-1432   *Fax:* 215 762-1164
*E-mail:* pa27@drexel.edu
*Med Dir:* Ana Nunez, MD

**Philadelphia College of Osteopathic Medicine**
Physician Assistant Prgm
4190 City Ave
Rowland Hall
Philadelphia, PA 19131
www.pcom.edu
*Prgm Dir:* John Cavenagh, PhD PA-C
*Tel:* 215 871-6772   *Fax:* 215 871-6702
*E-mail:* johnca@pcom.edu
*Med Dir:* Kimberly Kaiser, DO

**Philadelphia University**
Physician Assistant Prgm
School of Science and Health
School House Ln and Henry Ave
Philadelphia, PA 19144-5497
www.philau.edu
*Prgm Dir:* Michael Rackover, PA-C MS
*Tel:* 215 951-2908   *Fax:* 215 951-2526
*E-mail:* rackoverm@philau.edu
*Med Dir:* John Krim, DO

**Chatham College**
Physician Assistant Prgm
Woodland Rd
Pittsburgh, PA 15232-2826
*Prgm Dir:* Mark Freeman, MBA MEd PA-C
*Tel:* 412 365-1405   *Fax:* 412 365-1623
*E-mail:* freeman@chatham.edu
*Med Dir:* Douglas Kress, MD

**Duquesne University**
Physician Assistant Prgm
Rangos Sch of Health Sciences
Health Sciences Bldg, Ste 405
Pittsburgh, PA 15282
www.healthsciences.duq.edu/pa/pahome.html
*Prgm Dir:* Bridget Calhoun, MPH PA-C
*Tel:* 412 396-5914   *Fax:* 412 396-4118
*E-mail:* calhoun@duq.edu
*Med Dir:* Michael Essig, MD

**Marywood University**
Physician Assistant Prgm
2300 Adams Ave
Scranton, PA 18509
www.marywood.edu/departments/PA_Program
*Prgm Dir:* Karen Arscott, DO MSc
*Tel:* 570 348-6211, Ext 2175   *Fax:* 570 348-6020
*E-mail:* arscott@marywood.edu
*Med Dir:* Karen Arscott, DO, MSc

**King's College**
Physician Assistant Prgm
133 N River St
Wilkes-Barre, PA 18711-0851
www.kings.edu/paprog
*Prgm Dir:* Frances Feudale, DO FACEP
*Tel:* 570 208-5853, Ext 5768   *Fax:* 570 208-6018
*E-mail:* fafeudal@kings.edu
*Med Dir:* Mark Radziewicz, DO

**Pennsylvania College of Technology**
Physician Assistant Prgm
One College Ave #123
Williamsport, PA 17701
*Prgm Dir:* Joseph Mileto, Jr, MHSc PA-C
*Tel:* 570 327-4779   *Fax:* 570 321-5557
*E-mail:* jmileto@pct.edu
*Med Dir:* Gregory Frailey, DO FACOEP

## South Carolina

**Medical University of South Carolina**
Physician Assistant Prgm
PO Box 250962
151 B Rutledge Ave
Charleston, SC 29425
www.musc.edu/pa
*Prgm Dir:* Reamer L Bushardt, PharmD RPh PA-C
*Tel:* 843 792-3789   *Fax:* 843 792-0506
*E-mail:* busharrl@musc.edu
*Med Dir:* D Glen Askins Jr, MD

## South Dakota

**University of South Dakota**
Physician Assistant Prgm
414 E Clark St
Vermillion, SD 57069-2390
*Prgm Dir:* Wade Nilson, MPAS PA-C
*Tel:* 605 677-5128   *Fax:* 605 677-6569
*E-mail:* wnilson@usd.edu
*Med Dir:* Bruce Vogt, MD

## Tennessee

**Bethel College**
Physician Assistant Prgm
325 Cherry Ave
McKenzie, TN 38201
www.bethel-college.edu/bethelpa
*Prgm Dir:* Thomas Brown, MD
*Tel:* 731 352-4247   *Fax:* 731 352-4589
*E-mail:* brownt@bethel-college.edu
*Med Dir:* Joesph Hames, MD

**Trevecca Nazarene University**
Physician Assistant Prgm
333 Murfreesboro Rd
Nashville, TN 37210-2877
*Prgm Dir:* G Michael Moredock, MD
*Tel:* 615 248-1225   *Fax:* 615 248-1622
*E-mail:* mmoredock@trevecca.edu
*Med Dir:* Wayne Wells, MD

## Texas

**Univ of Texas Southwestern Med Ctr at Dallas**
Physician Assistant Prgm
5323 Harry Hines Blvd, V4.114
Dallas, TX 75390-9090
www.utsouthwestern.edu
*Prgm Dir:* Venetia Orcutt, MBA PA-C
*Tel:* 214 648-1701   *Fax:* 214 648-1003
*E-mail:* venetia.orcutt@utsouthwestern.edu
*Med Dir:* Laurette Dekat, MD, MPH

**Univ of Texas - Pan American**
Physician Assistant Prgm
1201 West University Dr
Edinburg, TX 78541
www.panam.edu/dept/pasp
*Prgm Dir:* Frank Ambriz, BS MSPA PA-C
*Tel:* 956 316-7049   *Fax:* 956 381-2438
*E-mail:* frankambriz@panam.edu
*Med Dir:* Hiram Tavarez, MD

**US Army**
*Cosponsor:* University of Nebraska
Interservice Physician Assistant Prgm
MCCS-HMP Physician Assistant Branch Rm 1202
3151 Scott Rd, Ste 1230
Ft Sam Houston, TX 78234-6138
*Prgm Dir:* Charles H Brakhage, CAPT MSC USN
*Tel:* 210 221-8004   *Fax:* 210 221-8493
*E-mail:* charles.brakhage@amedd.army.mil
*Med Dir:* Alesia C Carrizales, Major MD USAF

**Univ of North Texas Hlth Sci Ctr at Ft Worth**
Physician Assistant Prgm
3500 Camp Bowie Blvd
Ft Worth, TX 76107-2699
*Prgm Dir:* Henry Lemke, MMS PA-C
*Tel:* 817 735-2301   *Fax:* 817 735-2529
*E-mail:* hlemke@hsc.unt.edu
*Med Dir:* Bruce Dubin, JD, DO

**University of Texas Medical Branch/SAHS**
Physician Assistant Prgm
301 University Blvd
Galveston, TX 77555-1145
www.sahs.utmb.edu/pas
*Prgm Dir:* Richard R Rahr, EdD PA-C
*Tel:* 409 772-3047   *Fax:* 409 772-9710
*E-mail:* rrahr@utmb.edu
*Med Dir:* Michael Warren, MD

**Baylor College of Medicine**
Physician Assistant Prgm
One Baylor Plaza-Room BTA 107C
Houston, TX 77030-3411
*Prgm Dir:* Carl E Fasser, PA
*Tel:* 713 798-5405   *Fax:* 713 798-6128
*E-mail:* cfasser@bcm.tmc.edu
*Med Dir:* Stephen Spann, MD

**Texas Tech Univ Health Sciences Center**
Physician Assistant Prgm
3600 N Garfield
Midland, TX 79705
*Prgm Dir:* Elvin E Maxwell, Jr, MA MPAS PA-C
*Tel:* 432 620-9905, Ext 233   *Fax:* 432 620-8605
*E-mail:* ed.maxwell@ttuhsc.edu
*Med Dir:* George Manning, MD

**Univ of Texas Hlth Sci Ctr at San Antonio**
Physician Assistant Prgm
7703 Floyd Curl Dr MC6249
San Antonio, TX 78229-3900
www.uthscsa.edu
*Prgm Dir:* J Dennis Blessing, PhD PA-C
*Tel:* 210 567-8810   *Fax:* 210 567-8846
*E-mail:* blessingd@uthscsa.edu
*Med Dir:* Miguel Ramirez-Colon, MD

## Utah

**University of Utah Health Science Center**
*Cosponsor:* Utah Medical Association
Physician Assistant Prgm
375 Chipeta Way Ste A
Salt Lake City, UT 84108
www.utah.edu/upap
*Prgm Dir:* Don M Pedersen, PhD PA-C
*Tel:* 801 585-7426   *Fax:* 801 581-5807
*E-mail:* dpedersen@upap.utah.edu
*Med Dir:* John Houchins, MD

## Virginia

**James Madison University**
Physician Assistant Prgm
Dept of Health Sciences
MSC 4301
Harrisonburg, VA 22807
www.jmu.edu/healthsci/paweb
*Prgm Dir:* James B Hammond, MA PA-C
*Tel:* 540 568-8171   *Fax:* 540 568-3336
*E-mail:* hammonjb@jmu.edu
*Med Dir:* David Knitter, MD

**Eastern Virginia Medical School**
Physician Assistant Prgm
700 W Olney
Lewis Hall, Ste 1100
Norfolk, VA 23507
www.evms.edu
*Prgm Dir:* Thomas Parish, DHSc PA-C
*Tel:* 757 446-7158, Ext 7126   *Fax:* 757 446-7403
*E-mail:* parishtg@evms.edu
*Med Dir:* Martha Scott, MD

**Jefferson College of Health Sciences**
Physician Assistant Prgm
PO Box 13186
Roanoke, VA 24031-4016
www.jchs.edu
*Prgm Dir:* Wilton C Kennedy, MMSc PA-C
*Tel:* 540 985-4016   *Fax:* 540 224-4551
*E-mail:* wkennedy@jchs.edu
*Med Dir:* Patrick McCarthy, MD

**Shenandoah University**
Physician Assistant Prgm
MOB II, Ste 430
190 Campus Blvd
Winchester, VA 22601
*Prgm Dir:* Anthony Miller, MEd PA-C
*Tel:* 540 545-7257   *Fax:* 540 542-6210
*E-mail:* amiller@su.edu
*Med Dir:* James Laidlaw, MD FACP FACC

## Washington

**University of Washington**
Physician Assistant Prgm
MEDEX Northwest
4311 11th Ave NE, Ste 200
Seattle, WA 98105-4608
www.medex.washington.edu
*Prgm Dir:* Ruth Ballweg, MPA PA-C
*Tel:* 206 616-4001, Ext 6343   *Fax:* 206 616-3889
*E-mail:* medex@u.washington.edu
*Med Dir:* Timothy Evans, MD, PhD

## West Virginia

**Mountain State University**
Physician Assistant Prgm
PO Box 9003
Beckley, WV 25802-2830
www.mountainstate.edu/hs/pa
*Prgm Dir:* Elizabeth Silosky, MSSL
*Tel:* 304 929-1436   *Fax:* 304 256-5571
*E-mail:* bsilosky@mountainstate.edu
*Med Dir:* Bruce Cannon, DO

**Alderson-Broaddus College**
Physician Assistant Prgm
PO Box 2036
500 College Hill Dr
Philippi, WV 26416
*Prgm Dir:* Michael W Holt, MS PA
*Tel:* 304 457-6290   *Fax:* 304 457-6308
*E-mail:* holt_m@ab.edu
*Med Dir:* David Bender, MD

## Wisconsin

**University of Wisconsin - La Crosse**
*Cosponsor:* Gundersen Lutheran & Mayo School of
Health Science
Physician Assistant Prgm
1725 State St, 4031 HSC
La Crosse, WI 54601-3767
*Prgm Dir:* Mark Zellmer, MA PA-C
*Tel:* 608 785-8470   *Fax:* 608 785-6647
*E-mail:* zellmer.mark@uwlax.edu
*Med Dir:* Gregory Thompson, MD

**University of Wisconsin - Madison**
Physician Assistant Prgm
1300 University Ave, 1135 MSC
Madison, WI 53706
*Prgm Dir:* Jeffrey G Nicholson, MEd MPAS PA-C
*Tel:* 608 263-5620   *Fax:* 608 263-6434
*E-mail:* jgnichol@wisc.edu
*Med Dir:* Richard Anstett, MD

**Marquette University**
Physician Assistant Prgm
PO Box 1881
Milwaukee, WI 53201-1881
*Prgm Dir:* Tim Gengembre, PA-C MS MBA PhD
*Tel:* 414 288-5688   *Fax:* 414 288-7951
*E-mail:* tim.gengembre@marquette.edu
*Med Dir:* Paul J Coogan, MD

---

**Physician Assistant**

| Programs* | Class Capacity | Begins | Length (months) | Award | Res Tuition | Non-res Tuition | Stipend | Offers:‡ 1 | 2 | 3 |
|---|---|---|---|---|---|---|---|---|---|---|
| **Alabama** | | | | | | | | | | |
| University of Alabama at Birmingham | 32 | Aug | 27 | MSPAS | $9,400 | $23,500 | | | | • |
| University of South Alabama (Mobile) | 36 | May | 27 | MHS | $8,700 | $17,400 | | | | • |
| **Arizona** | | | | | | | | | | |
| AT Still University of Health Sciences (Mesa) | 70 | Aug | 26 | MS | $20,740 | $0 | | | | • |
| Midwestern University - Glendale Campus | 86 | Jun | 24, 27 | BS, MS | $26,785 | $26,785 | | | | • |
| **California** | | | | | | | | | | |
| Samuel Merritt College (Oakland) | 36 | Sep | 27 | Dipl, MPA | $30,545 | $30,545 | | | | • |
| Stanford University School of Medicine (Palo Alto) | 50 | Sep | 16 | Cert, AS | $17,160 | $24,360 | | | • | • |
| University of California - Davis (Sacramento) | 53 | Jul | 24 | Cert, MSN | $10,281 | $29,007 | | • | | • |
| University of Southern California (Alhambra) | 40 | Aug | 33 | MPAP | $33,500 | $33,500 | | | • | • |
| Western Univ of Health Sciences (Pomona) | 98 | Aug | 24 | MS | $25,340 | $25,340 | | | | |
| **Colorado** | | | | | | | | | | |
| Red Rocks Community College (Lakewood) | 28 | Aug | 24 | Cert, Mas Op | $8,625 | $11,625 | | | • | • |
| U of Colorado at Denver and Health Scis Ctr (Aurora) | 40 | May | 36 | Cert, MPAS | $10,200 | $32,200 | | | | |
| **Connecticut** | | | | | | | | | | |
| Quinnipiac University (Hamden) | 54 | May | 27 | MHS | $21,900 | $21,900 | | | • | • |

*Data are shown only for programs that completed the 2006 AMA Survey of Health Professions Education Programs
‡Key to Offers: 1: Evening or weekend classes; 2: Non-English instruction; 3: Cultural competence instruction

## Physician Assistant

| Programs* | Class Capacity | Begins | Length (months) | Award | Res Tuition | Non-res Tuition | Stipend | Offers:‡ 1 | 2 | 3 |
|---|---|---|---|---|---|---|---|---|---|---|
| Yale University School of Medicine (New Haven) | 32 | Sep | 27 | MMSc | $26,500 | $26,500 | | | | • |
| **Florida** | | | | | | | | | | |
| Barry University (Miami Shores) | 69 | Aug | 28 | Cert, MCMSc | $23,625 | $23,625 | | | • | • |
| Nova Southeastern University - Ft Lauderdale | 90 | Jun | 27 | Dipl, MMS | $20,950 | $21,640 | | | | • |
| University of Florida (Gainesville) | 60 | Jun Jul | 24 | Cert, MPAS | $10,495 | $35,703 | | | | • |
| **Georgia** | | | | | | | | | | |
| Emory University (Atlanta) | 50 | Aug | 28 | MMSC | $19,700 | $19,700 | | | | • |
| Medical College of Georgia (Augusta) | 44 | May | 27 | MPA | $12,000 | $24,000 | | | • | • |
| South University (Savannah) | 60 | Jan | 27 | MSPA, BSPA | $25,000 | $25,000 | | | | • |
| **Idaho** | | | | | | | | | | |
| Idaho State University (Pocatello) | 30 | Aug | 24 | Cert, MPAS | $22,545 | $36,645 | | | | • |
| **Illinois** | | | | | | | | | | |
| Midwestern University (Downers Grove) | 84 | Jun | 27 | Cert, MMS | $22,864 | $25,327 | | | | • |
| Rosalind Franklin Univ of Medicine & Science (North Chicago) | 60 | May | 24 | MS | $20,714 | $20,714 | | | | • |
| Southern Illinois Univ at Carbondale | 24 | Jun | 26 | BS | $11,800 | $23,600 | | | | • |
| **Indiana** | | | | | | | | | | |
| Butler University/Clarian Health (Indianapolis) | 50 | Aug | 30 | MPAS | $26,470 | $26,470 | | | | • |
| University of Saint Francis (Ft Wayne) | 25 | May | 27 | MS | $28,500 | $28,500 | | | | |
| **Iowa** | | | | | | | | | | |
| Des Moines University | 42 | Jun | 25 | MS | $22,300 | $22,300 | | | | • |
| University of Iowa (Iowa City) | 25 | May | 25 | Cert, MPAS | $8,949 | $26,040 | | | | • |
| **Kansas** | | | | | | | | | | |
| Wichita State University | 42 | Jun | 26 | MPA | $8,300 | $23,000 | | | | • |
| **Louisiana** | | | | | | | | | | |
| Louisiana State Univ Hlth Sci Ctr-Shreveport | 36 | May | 27 | Dipl, BS | $6,000 | $8,500 | | | | • |
| Our Lady of the Lake College (Baton Rouge) | 30 | Jan | 28 | Masters | $22,000 | $22,000 | | | | • |
| **Maryland** | | | | | | | | | | |
| Anne Arundel Community College (Arnold) | 40 | May | 25 | Cert | $8,910 | $18,960 | | | | • |
| **Massachusetts** | | | | | | | | | | |
| Mass College of Pharmacy & Health Sciences (Boston) | 40 | Sep | 30 | MPAS | $24,000 | $24,000 | | | | • |
| Northeastern University (Boston) | 34 | Aug | 24 | MS | $22,275 | $22,275 | | | | • |
| **Michigan** | | | | | | | | | | |
| Central Michigan University (Mt Pleasant) | 40 | May | 27 | MS | $19,500 | $35,000 | | | | • |
| Grand Valley State University (Grand Rapids) | 30 | Aug | 32 | MPAS | $10,827 | $21,800 | | | | • |
| University of Detroit Mercy | 45 | Sep | 24, 36 | MS | $28,095 | $28,095 | | • | | • |
| Wayne State University (Detroit) | 50 | May | 24 | MS | $13,360 | $28,044 | | | | • |
| **Minnesota** | | | | | | | | | | |
| Augsburg College (Minneapolis) | 28 | May | 36 | Cert, MS | $24,000 | $24,000 | | | | • |
| **Missouri** | | | | | | | | | | |
| Saint Louis University (St Louis) | 34 | Aug | 27 | MMS | $29,440 | $29,440 | | | | • |
| **Nebraska** | | | | | | | | | | |
| Union College (Lincoln) | 25 | Aug | 32 | Cert, MPAS | $20,000 | $20,000 | | | | • |
| University of Nebraska Medical Center (Omaha) | 40 | Aug | 28 | MPAS | $9,815 | $26,445 | | | | |
| **New Mexico** | | | | | | | | | | |
| University of New Mexico (Albuquerque) | 14 | Jun | 25 | Dipl, Cert, BS | $6,891 | $16,978 | | | | • |
| University of St Francis (Albuquerque) | 30 | Jan | 27 | Cert, MS | $23,000 | $23,000 | | | | |
| **New York** | | | | | | | | | | |
| Albany Medical College | 40 | Jan | 28 | MS | $18,000 | $18,000 | | | | • |
| Brooklyn Hosp/Long Island Univ | 42 | Aug | 24 | Cert, BS | $23,000 | $0 | | | | • |
| City University of New York, The City College | 35 | july | 28 | Dipl, Cert, BS | $5,040 | $12,460 | | | | • |
| CUNY York College (Jamaica) | 30 | Sep | 24 | BS | $10,255 | $17,645 | | | | • |
| D'Youville College (Buffalo) | 40 | Aug | 54 | BS/MS | $16,600 | $16,600 | | • | • | • |
| Daemen College (Amherst) | 225 | Sep | 60, 33 | Dipl, BS/MS, MS | | | | | | • |
| Hofstra University (Hempstead) | 40 | Sep | 27 | Dipl, Cert, BS | $33,000 | $33,000 | | | | |
| Le Moyne College (Syracuse) | 35 | Aug | 24 | Dipl, BS, MS | $27,000 | $27,000 | | | | • |
| Mercy College (Bronx) | 40 | Jun | 27 | Dipl, BS, MS | $24,300 | $24,300 | | | • | • |
| New York Institute of Technology (Old Westbury) | 40 | Sep | 30 | BS, MS | $25,000 | $25,000 | | | | • |
| Pace University - Lenox Hill Hospital (New York) | 36 | Sep | 48 | BS | $26,000 | $26,000 | | | | |
| Rochester Institute of Technology | 25 | Sep | 48 | BS | $24,627 | $24,627 | | • | | • |
| St Vincent Catholic Med Ctr Brklyn-Queens Reg (Fresh Meadows) | 150 | Aug | 22 | Cert | $15,500 | $15,500 | | | | |
| Stony Brook University | 40 | Jul | 25 | MS | $11,218 | $17,031 | | | | • |
| SUNY Downstate Medical Center (Brooklyn) | 33 | May | 27 | BS | $11,509 | $18,200 | | | | • |
| Touro College (New York) | 30 | Aug | 28 | BS | $14,120 | $14,120 | | • | | • |
| Weill Med College/Cornell Univ Med Sch (New York) | 32 | Aug | 26 | Cert | $17,959 | $17,959 | | | | • |

*Data are shown only for programs that completed the 2006 AMA Survey of Health Professions Education Programs
‡Key to Offers: 1: Evening or weekend classes; 2: Non-English instruction; 3: Cultural competence instruction

## Physician Assistant

| Programs* | Class Capacity | Begins | Length (months) | Award | Res Tuition | Non-res Tuition | Stipend | Offers:‡ 1 | 2 | 3 |
|---|---|---|---|---|---|---|---|---|---|---|
| **North Carolina** | | | | | | | | | | |
| Duke University Medical Center (Durham) | 56 | Aug | 25 | Cert, MHS | $26,245 | $26,245 | | | | • |
| East Carolina University (Greenville) | 35 | Aug | 27 | Cert, MS | $6,700 | $24,000 | | | | • |
| Methodist College (Fayetteville) | 30 | Aug | 28 | MMS | $24,000 | $24,000 | | | • | • |
| Wake Forest University School of Medicine (Winston-Salem) | 48 | Jun | 24 | MMS | $20,822 | $20,822 | | | • | • |
| **North Dakota** | | | | | | | | | | |
| University of North Dakota (Grand Forks) | 30 | Aug | 20 | MPAS | $29,000 | $29,000 | | | | • |
| **Ohio** | | | | | | | | | | |
| Cuyahoga Community College (Parma) | 52 | Aug | 22 | AAS | $8,000 | $16,850 | | | | |
| Kettering College of Medical Arts | 40 | May | 27 | MPAS | $22,070 | $22,070 | | | | • |
| Marietta College | 22 | Jun | 27 | MS | $28,088 | $28,088 | | | | • |
| Medical University of Ohio at Toledo | 30 | Aug | 27 | MSBS | $9,280 | $21,135 | | | | • |
| **Oklahoma** | | | | | | | | | | |
| Univ of Oklahoma Health Sciences Center (Oklahoma City) | 50 | Jul | 30 | MHS | $6,594 | $15,592 | | | | |
| **Oregon** | | | | | | | | | | |
| Oregon Health & Science University (Portland) | 34 | Jun | 26 | MPAS | $25,668 | $25,668 | | | • | • |
| Pacific University (Forest Grove) | 40 | May 15 | 28 | MS | $23,305 | $23,305 | | | • | • |
| **Pennsylvania** | | | | | | | | | | |
| DeSales University (Center Valley) | 40 | Aug | 24 | MSPAS | $29,050 | $29,050 | $6,000 | | | • |
| Duquesne University (Pittsburgh) | 35 | May | 27 | MPA | $34,300 | $34,300 | | | | • |
| Gannon University (Erie) | 40 | Aug | 60, 24 | MPAS | $20,680 | $20,680 | | | | • |
| King's College (Wilkes-Barre) | 44 | Aug | 24 | MSPAS | $25,000 | $25,000 | | | | • |
| Lock Haven University | 48 | May | 24 | MHS | $13,925 | $20,888 | | | | • |
| Marywood University (Scranton) | 30 | May | 27 | MS | $25,000 | $25,000 | | | | • |
| Philadelphia College of Osteopathic Medicine | 54 | Jun | 26 | Dipl, Cert, MS | $25,000 | $25,000 | | | | • |
| Philadelphia University | 50 | Aug | 25 | Cert, MS | $27,366 | $27,366 | | | | • |
| St Francis University (Loretto) | 55 | May | 24 | MPAS | $33,972 | $33,972 | | | | |
| **South Carolina** | | | | | | | | | | |
| Medical University of South Carolina (Charleston) | 60 | May | 27 | MS | $15,885 | $31,014 | | | • | • |
| **South Dakota** | | | | | | | | | | |
| University of South Dakota (Vermillion) | 20 | Aug | 28 | MS | $4,095 | $12,073 | | | | • |
| **Tennessee** | | | | | | | | | | |
| Bethel College (McKenzie) | 26 | May | 12 | Cert, MSPAS | $21,900 | $21,900 | | | | • |
| Trevecca Nazarene University (Nashville) | 37 | May | 27 | MS Med | $55,332 | $55,332 | | | • | • |
| **Texas** | | | | | | | | | | |
| Baylor College of Medicine (Houston) | 30 | Jul | 30 | MS | $14,000 | $14,000 | | | • | • |
| Texas Tech Univ Health Sciences Center (Midland) | 48 | Jun | 27 | MPAS | $31,000 | $43,000 | | | • | • |
| Univ of North Texas Hlth Sci Ctr at Ft Worth | 30 | Aug | 34 | MPAS | $6,100 | $19,200 | | | | • |
| Univ of Texas - Pan American (Edinburg) | 33 | Jun | 24 | BS | $4,534 | $15,390 | $1,000 | | | • |
| Univ of Texas Hlth Sci Ctr at San Antonio | 24 | Aug | 33 | MPAS | $7,410 | $13,828 | | | | • |
| Univ of Texas Southwestern Med Ctr at Dallas | 36 | May | 31 | Dipl, MPAS | $4,970 | $17,790 | | | • | • |
| University of Texas Medical Branch/SAHS (Galveston) | 54 | Aug | 24, 48 | MS | $9,900 | $41,000 | | | | • |
| US Army (Ft Sam Houston) | 60 | Jan May Sep | 12 | Cert, BS, MS | | | | | | • |
| **Utah** | | | | | | | | | | |
| University of Utah Health Science Center (Salt Lake City) | 36 | Aug | 27 | Cert, MS | $18,000 | $25,000 | | | | • |
| **Virginia** | | | | | | | | | | |
| Eastern Virginia Medical School (Norfolk) | 50 | Jan | 27 | MPA | $23,057 | $23,057 | | | • | • |
| James Madison University (Harrisonburg) | 25 | May | 24 | MPAS | | | | | | • |
| Jefferson College of Health Sciences (Roanoke) | 64 | Aug | 24 | BS | $22,375 | $22,375 | | | • | • |
| **Washington** | | | | | | | | | | |
| University of Washington (Seattle) | 80 | Jun | 24 | Cert, BCHS | $18,440 | $18,440 | | | | • |
| **West Virginia** | | | | | | | | | | |
| Mountain State University (Beckley) | 35 | Aug | 36 | MSPA | $16,500 | $16,500 | | | | |
| **Wisconsin** | | | | | | | | | | |
| Marquette University (Milwaukee) | 34 | Aug | 34 | MPA | $24,670 | $24,670 | | | | • |

*Data are shown only for programs that completed the 2006 AMA Survey of Health Professions Education Programs
‡Key to Offers: 1: Evening or weekend classes; 2: Non-English instruction; 3: Cultural competence instruction

# Polysomnographic Technologist

## Occupational Description

Polysomnographic technologists perform sleep tests and work with physicians to provide information needed for the diagnosis of sleep disorders. The technologist monitors brain waves, eye movements, muscle activity, multiple breathing variables, and blood oxygen levels during sleep using specialized recording equipment. The technologist interprets the recording as it happens and responds appropriately to emergencies. Technologists provide support services related to the treatment of sleep-related problems, including helping patients use devices for the treatment of breathing problems during sleep and helping individuals develop good sleep habits.

## Job Description

The technologist gathers and analyzes patient information and physician orders to ensure that the appropriate test is performed. Technologists explain the sleep study procedures to the patient. Before a sleep study, technologists prepare and calibrate equipment required for testing and make adjustments if necessary. They apply electrodes and sensors according to accepted published standards; perform appropriate calibrations to ensure proper signals and make adjustments if necessary; and perform positive airway pressure (PAP) mask fitting. PAP consists of a machine (air pump) that is connected to a mask worn over the patient's nose by a hose; the machine gently pushes the air into the patient's nose and down the airway to prevent it from collapsing during sleep, as in the case of sleep apnea.

There are several types of tests, including the Multiple Sleep Latency Test (MSLT, measures daytime sleepiness), Maintenance of Wakefulness Test (MWT, measures ability to stay awake), parasomnia studies (eg, unusual behaviors during sleep), and oxygen and PAP titration (eg, adjusting the air pressure of the PAP machine to find the right pressure that will keep the airway from collapsing). The polysomnographic technologist follows these protocols to ensure appropriate data collection. Technologists follow "lights out" procedures to obtain baseline values. They then perform data collection while keeping track of study quality and making any necessary adjustments. The technologist keeps a log of observations, including sleep stages and clinical events, changes in procedure, and significant events. This helps in the interpretation of polysomnographic results.

During the sleep study the technologist must ensure patient safety, apply PAP at the correct pressure level when appropriate, and administer oxygen as directed. At the end of the study the technologist follows "lights on" procedures to ensure that data have been collected correctly. Polysomnographic technologists must be comfortable working with newborn, child, teenage, adult, and geriatric patients.

After the sleep study the technologist scores sleep/wake stages using professionally accepted guidelines; scores clinical events (such as respiratory events, cardiac events, limb movements, arousals, etc) according to center-specific protocols; and generates accurate reports by tabulating sleep/wake and clinical event data.

All polysomnographic technologists must comply with laws, regulations, guidelines, and standards regarding safety and infection control issues. They perform routine and complex equipment care and maintenance and evaluate sleep study-related equipment and inventory. Current CPR (cardiopulmonary resuscitation) or BCLS (basic cardiac life support) certification is required.

## Employment Characteristics

Most polysomnographic technologists work in sleep disorders centers. Sleep disorders centers may be located within or affiliated with a hospital, or may be "freestanding" (in a physician's office or professional building). Some senior technologists may spend all or part of their time scoring sleep recordings, performing daytime tests, and managing a center, but most of the polysomnographic technologists' work is done at night. Typical shifts are three to four 10- to 12-hour shifts per week. The recommended workload is two patients per night. Salaries and benefits are competitive with other allied health professions.

## Educational Programs

**Length.** A 2-year program leading to an associate's degree is preferred. However, some programs provide a certificate after a year of training.

**Curriculum.** The curriculum of an accredited program focuses on correct performance of polysomnographic procedures and patient safety. Students learn principles of physiological monitoring and the pathophysiology of sleep disorders. Through lecture and observation they gain experience with study protocols.

## Inquiries

### Careers

Christopher Waring, Coordinator
Association of Polysomnographic Technologists
One Westbrook Corporate Center, Suite 920
Westchester, IL 60154
708 492-0796
708 273-9344 Fax
E-mail: cwaring@aptweb.org

### Certification

Bobby Stanley, Jr, Executive Director
Board of Registered Polysomnographic Technologists (BRPT)
8201 Greensboro Drive, Suite 300
McLean, VA 22102
703 610-9020
703 610-9005 Fax
E-mail: brpt@amg-inc.com

### Program Accreditation

Commission on Accreditation of Allied Health Education Programs (CAAHEP) in collaboration with:
Committee on Accreditation of Polysomnographic Technologist Education
Richard S Rosenberg, PhD, Coordinator
One Westbrook Corporate Center, Suite 920
Westchester, IL 60154
708 492-0930
708 492-0943 Fax
E-mail: rrosenberg@aasmnet.org

## Polysomnographic Technologist

# Massachusetts

**Northern Essex Community College**
Polysomnographic Technology Prgm
45 Franklin St
Lawrence, MA 01841
www.necc.mass.edu
*Prgm Dir:* Bonnie McGuire, RPSGT CRT
*Tel:* 978 738-7223, Ext 7223   *Fax:* 978 738-7450
*E-mail:* bmcguire@necc.mass.edu
*Med Dir:* Lawrence Epstein, MD

# Ohio

**Sleep Care Inc**
Polysomnographic Technology Prgm
7634 Rivers Edge Dr
Columbus, OH 43235
*Prgm Dir:* Jennifer Brickner-York, MPH RPSGT CHES
*Tel:* 866 320-8989   *Fax:* 866 291-8990
*E-mail:* jennifer@sleepcareinc.com
*Med Dir:* Susan Borchers, MD

## Polysomnographic Technologist

| Programs* | Class Capacity | Begins | Length (months) | Award | Res. Tuition | Non-res. Tuition | Stipend | Offers:‡ 1 | 2 | 3 |
|---|---|---|---|---|---|---|---|---|---|---|
| **Massachusetts** | | | | | | | | | | |
| Northern Essex Community College (Lawrence) | 20 | | 9 | Cert | $2,520 | $8,304 | | • | | • |
| **Ohio** | | | | | | | | | | |
| Sleep Care Inc (Columbus) | 16 | Varies | 4 | Cert | $10,000 | $10,000 | | | | • |

*Data are shown only for programs that completed the 2006 AMA Survey of Health Professions Education Programs
‡Key to Offers: 1: Evening or weekend classes; 2: Non-English instruction; 3: Cultural competence instruction

# Radiologic Technology

**Includes:**
- Advanced practice specialties in radiologic technology
- Magnetic resonance technologist
- Medical dosimetrist
- Radiation therapist
- Radiographer

## Radiation Therapist

### Occupational Description

Radiation therapists deliver prescribed doses of radiation to patients for therapeutic purposes. In fulfilling this primary responsibility, radiation therapists provide appropriate patient care; apply problem-solving and critical thinking skills in the administration of treatment protocols, tumor localization, and dosimetry; and maintain appropriate patient records. Radiation therapists are particularly concerned with the principles of radiation protection for patients, themselves, and others.

### Job Description

Radiation therapists apply knowledge of anatomy and physiology, oncologic pathology, radiation biology, radiation oncology techniques, treatment planning procedures, and dosimetry in the performance of their duties. They must also communicate effectively with patients, health professionals, and the public.

The radiation therapist accepts responsibility for administering a radiation oncologist (physician)-prescribed course of radiation therapy, providing patient care during treatment, and maintaining treatment records. Radiation therapists also evaluate and assess treatment delivery components, evaluate and assess the daily physiologic and psychologic responsiveness of patients, and ensure quality care for patients undergoing radiation therapy. Additional duties may include tumor localization, dosimetry, patient follow-up, and patient education. Radiation therapists must display competence and compassion in meeting the special needs of the oncology patient.

### Employment Characteristics

Radiation therapists are employed in health care facilities, including hospitals, cancer centers and private offices; they are also employed in settings where their responsibilities focus on education, management, research, and sales.

### Salary

Salaries and benefits vary with experience and employment location but are generally competitive with other health specialties. Some states require licensure as a condition of practice. Refer to Section IV, Table 5 of this Directory for more information, or see www.ama-assn.org/go/salary.

### Educational Programs

**Length.** Programs may be 1, 2, or 4 years, depending on program design, objectives, and the degree or certificate awarded.

**Curriculum.** The curriculum of an accredited program includes an extensive component of technical and professional courses, including an emphasis on structured, competency-based clinical education. Interested individuals should contact a particular program for information on specific courses and prerequisites.

## Medical Dosimetrist

### Occupational Description

Medical dosimetrists, in collaboration with radiation oncologists and medical physicists, generate radiation dose distributions and dose calculations to design radiation treatment plans that will deliver a prescribed dose of radiation to a defined anatomic area.

### Job Description

Medical dosimetrists apply knowledge of anatomy and physiology, oncologic pathology, radiation biology, radiation oncology techniques, treatment planning and dosimetry procedures, and computer computation in the performance of their duties. The medical dosimetrist accepts responsibility for designing a radiation oncologist (physician)-prescribed course of radiation therapy, considering dose-limiting structures and the need for special casts and immobilization devices. They must be able to communicate effectively with other health care professionals.

### Employment Characteristics

Medical dosimetrists are employed in health care facilities, including hospitals and cancer centers.

### Salary

Salary and benefits vary with experience and employment location but are competitive with other health specialties. See Section IV, Table 5 of this *Directory* for more information, or www.ama-assn.org/go/salary.

### Educational Programs

**Length.** Program length varies, depending on program design, objectives, and the degree or certificate awarded.

**Curriculum.** Most programs require prerequisite work in radiation therapy or radiation physics. The curriculum of an accredited program includes an extensive component of technical and professional courses, including an emphasis on structured, competency-based clinical education. Interested individuals should contact a particular program for information on specific courses and prerequisites.

## Radiographer

### Occupational Description

Radiographers use radiation equipment to produce images of the tissues, organs, bones, and vessels of the body, as prescribed by physicians, to assist in the diagnosis of disease or injury. Radiographers continually strive to provide quality patient care and are particularly concerned with limiting radiation exposure to patients, themselves, and others. Radiographers use problem-solving and critical-thinking skills to perform medical imaging procedures by adapting variable technical parameters of the procedure to the condition of the patient.

### Job Description

Radiographers apply knowledge of anatomy, physiology, positioning, radiographic technique, and radiation biology and protection in the performance of their responsibilities. They must be able to communicate effectively with patients, other health professionals, and the public. Additional duties may include evaluating radiologic equipment, conducting a radiographic quality assurance program, providing patient education, and managing a medical imaging department. The radiographer must display competence and compassion in meeting the special needs of the patient.

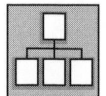

### Employment Characteristics

Radiographers are employed in health care facilities—including hospitals, specialized imaging centers, urgent care clinics, and private physician offices—and as educators or imaging department administrators. Thirty-five states require licensure as a condition of practice.

### Salary

Salaries and benefits are generally competitive with other health professions and vary according to experience and employment location. Refer to Section IV, Table 5 of this Directory for more information, or see www.ama-assn.org/ go/salary.

### Educational Programs

**Length.** Programs are generally 2 to 4 years, depending on program design, objectives, and the degree or certificate awarded.

**Curriculum.** The curriculum of an accredited program includes an extensive component of technical and professional courses, including an emphasis on structured competency-based clinical education. Contact a particular program for information on specific courses and prerequisites.

## Magnetic Resonance Technologist

### Occupational Description

Magnetic resonance technologists use radiowaves, magnetic fields, and computerized equipment to produce images of body tissues. MR technologists strive to provide quality patient care while producing patient images that permit accurate diagnoses. MR technologists use problem-solving and critical-thinking skills to adapt procedural requirements to the patient and specific area of study.

### Job Description

Magnetic resonance technologists apply knowledge of anatomy, physiology, positioning, and MR protocols in the performance of their responsibilities. They must be able to communicate effectively with patients, other health professionals, and the public. The MR technologist must show competence and compassion in meeting the special needs of the patient.

### Employment Characteristics

Magnetic resonance technologists are employed in health care facilities—including hospitals, specialized imaging centers, urgent care clinics, and physician offices—and as educators or imaging department administrators.

### Salary

Salary and benefits are generally competitive with other health professions and vary according to experience and employment location. Refer to Section IV, Table 5 of this *Directory* for more information, or see www.ama-assn.org/go/salary.

### Educational Programs

**Length.** Program length varies, depending on program design, objectives, and the degree or certificate awarded.

**Curriculum.** Many, though not all, programs require certification as a radiographer as a prerequisite. The curriculum of an accredited program includes an extensive component of technical and professional courses, including an emphasis on structured competency-based clinical education. Contact a particular program for information on specific courses and prerequisites.

The JRCERT anticipates accrediting the first magnetic resonance program in 2007.

## Advanced Practice Specialties in Radiologic Technology

Practitioners of the following advanced practice specialties in radiologic technology are eligible for certification by the American Registry of Radiologic Technologists. Candidates for certification must be certified in radiography, radiation therapy, or nuclear medicine and document specific clinical competencies to be eligible for the certification examination.

### Cardiovascular Interventional Technologist

Cardiovascular interventional technologists use radiation to produce images to aid in the diagnosis and/or treatment of vascular disease or other abnormalities, such as coronary artery disease. They may inject the patient with a material that assists in visualizing the images produced.

### Computed Tomography Technologist

Computed tomography technologists use radiation and a computer to produce cross-sectional images of the body. These individuals also may inject the patient with a material that assists in visualizing the images produced.

### Mammographer

Mammographers use radiation to produce images for screening or diagnostic procedures for detection of breast disease. These individuals also provide breast health education.

### Inquiries

**Careers/Curriculum**
American Society of Radiologic Technologists
15000 Central Avenue SE
Albuquerque, NM 87123
505 298-4500

American Association of Medical Dosimetrists
c/o Credentialing Services
One Physics Ellipse
College Park, MD 20740
301 209-3320                              301 209-3343 Fax
E-mail: aamd@aapm.org
www.medicaldosimetry.org

**Certification/Registration**
American Registry of Radiologic Technologists
1255 Northland Drive
Mendota Heights, MN 55120
651 687-0048

**Program Accreditation**
Joint Review Committee on Education in Radiologic Technology
20 N Wacker Drive, Suite 2850
Chicago, IL 60606-3182
312 704-5300                              312 704-5304 Fax
E-mail: mail@jrcert.org
www.jrcert.org

## Radiation Therapist*

## Alabama

**University of Alabama at Birmingham**
Radiation Therapy Prgm
1705 University Blvd
RMSB 437
Birmingham, AL 35294-1212
www.uab.edu/rtt
*Prgm Dir:* Pamela C Cartright, MA Ed RT(R)(T)
*Tel:* 205 934-3443   *Fax:* 205 975-7302
*E-mail:* pamcartr@uab.edu
*Med Dir:* Sharon Spencer, MD

## Arkansas

**Central Arkansas Radiation Therapy Inst**
Radiation Therapy Prgm
PO Box 55050
Little Rock, AR 72215
www.carti.com
*Prgm Dir:* Debra G Tomlinson, MA RT(R)(T)
*Tel:* 501 603-8866   *Fax:* 501 603-9573
*E-mail:* dtomlinson@carti.com
*Med Dir:* Christopher Pope, MD

**Arkansas State University**
Radiation Therapy Prgm
PO Box 910
State University, AR 72467
www.clt.astate.edu/RadSci/RSTherapyhome.htm
*Prgm Dir:* Tracy B White, MS RT(R)(T)
*Tel:* 870 972-2976   *Fax:* 870 972-2004
*E-mail:* twhite@astate.edu

## California

**City of Hope National Medical Center**
Radiation Therapy Prgm
1500 E Duarte Rd
Duarte, CA 91010-0269
www.cityofhope.org
*Prgm Dir:* Christine Forell, BS RT(R)(T)
*Tel:* 626 301-8247   *Fax:* 626 930-5334
*E-mail:* cforell@coh.org
*Med Dir:* Nayana Vora, MD

**Loma Linda University**
Radiation Therapy Prgm
Nichol Hall Rm A829
Loma Linda, CA 92350
www.llumc.edu
*Prgm Dir:* Carol A L Davis, RT(T)
*Tel:* 909 558-7368   *Fax:* 909 558-0264
*E-mail:* cadavis@ahs.llumc.edu
*Med Dir:* Jerry Slater, MD

**California State University - Long Beach**
Radiation Therapy Prgm
1250 Bellflower Blvd
Long Beach, CA 90840-4902
www.csulb.edu
*Prgm Dir:* Stephanie Eatmon, EdD RT(R)(T) FASRT
*Tel:* 562 985-7507   *Fax:* 562 985-2384
*E-mail:* seatmon@csulb.edu
*Med Dir:* Thomas Chung, MD

**Kaiser Permanente Sch of Allied Hlth Sciences**
Radiation Therapy Prgm
938 Marina Way South
Richmond, CA 94804
www.kp.org
*Prgm Dir:* Roma-Dakini Alexander, BS RT(R)(T) CRT
*Tel:* 510 231-5062   *Fax:* 510 231-5001
*E-mail:* Roma.D.Alexander@kp.org
*Med Dir:* Jason Kelly, MD

**City College of San Francisco**
Radiation Therapy Prgm
50 Phelan Ave, Box S91
San Francisco, CA 94112
www.ccsf.edu
*Prgm Dir:* Les K Yim, BS RT(T)(R)
*Tel:* 415 239-3458   *Fax:* 415 239-3930
*E-mail:* lyim@ccsf.edu
*Med Dir:* David A Larson, MD

## Connecticut

**Hartford Hospital**
Radiation Therapy Prgm
80 Seymour St
Hartford, CT 06115-5037
www.harthosp.org
*Prgm Dir:* Nora Uricchio, MEd RT(R)(T)
*Tel:* 860 545-3956   *Fax:* 860 545-6461
*E-mail:* nuricch@harthosp.org
*Med Dir:* Andrew L Salner, MD

**Gateway Community College**
Radiation Therapy Prgm
88 Bassett Rd
North Haven, CT 06473
www.gwcc.commnet.edu
*Prgm Dir:* Gina M Finn, RT(T)
*Tel:* 203 285-2392   *Fax:* 203 285-2400
*E-mail:* gfinn@gwcc.commnet.edu
*Med Dir:* Joseph Colasanto, MD

## District of Columbia

**Howard University**
Radiation Therapy Prgm
515 1/2 W St NW
Washington, DC 20059
*Prgm Dir:* Adrienne D Harrison, MS RT(T)
*Tel:* 202 806-5920   *Fax:* 202 806-4476
*E-mail:* aharrison@howard.edu
*Med Dir:* Ebrahim Ashayeri, MD

## Florida

**21st Cent Oncology Inc Sch-Rad Therapy Tech**
Radiation Therapy Prgm
Cape Coral Office
1419 SE 8th Terrace
Cape Coral, FL 33990
*Prgm Dir:* Claire Marie Skowronski, MS RT(R)(T) CMD
*Tel:* 239 573-5972
*E-mail:* cskowronski@rtsx.com
*Med Dir:* James H Rubenstien, MD

**Halifax Medical Center**
Radiation Therapy Prgm
303 N Clyde Morris Blvd, PO Box 2830
Daytona Beach, FL 32120-2830
www.halifax.org
*Prgm Dir:* Belinda H Phillips, BS RT(R)(T)
*Tel:* 386 254-4075, Ext 3510   *Fax:* 386 254-4231
*E-mail:* belinda.phillips@halifax.org

**Miami Dade College**
Radiation Therapy Prgm
Medical Center Campus
950 NW 20th St
Miami, FL 33127
www.mdc.edu
*Prgm Dir:* Mary Anne Kwon-Carte, MS BSN RN OCN RT(T) CTR
*Tel:* 305 237-4335   *Fax:* 305 237-4278
*E-mail:* maryanne.kwon@mdc.edu
*Med Dir:* Arnold M Markoe, MD

**Hillsborough Community College**
Radiation Therapy Prgm
PO Box 30030
Tampa, FL 33614
*Prgm Dir:* Karen M Nelson, MS RT(R)(T)
*Tel:* 813 253-7372   *Fax:* 813 253-7491
*E-mail:* knelson@hccfl.edu
*Med Dir:* Harvey M Greenberg, MD

*Additional information about programs that returned the AMA's survey is available in a table beginning on page 354.

# Georgia

**Grady Health System**
Radiation Therapy Prgm
80 Jesse Hill Jr Dr, SE, PO Box 26095
Atlanta, GA 30303-3050
www.gradyhealthsystem.org
*Prgm Dir:* Kevin Kindle, BS RT(T)
*Tel:* 404 616-5024    *Fax:* 404 616-3512
*E-mail:* kkindle@gmh.edu
*Med Dir:* Karen Godette, MD

**Medical College of Georgia**
Radiation Therapy Prgm
Ga Radiation Therapy Center
821 St Sebastian Way, Bldg HK
Augusta, GA 30912
www.mcg.edu
*Prgm Dir:* Anne Marie Vann, MEd CMD RT(R)(T)
*Tel:* 706 721-2971    *Fax:* 706 721-7248
*E-mail:* amvann@mcg.edu
*Med Dir:* Byron Dasher, MD

**Coosa Valley Technical College**
Radiation Therapy Prgm
One Maurice Culberson Dr
Rome, GA 30161
*Prgm Dir:* Susan Y Lanham, BS RT(T)
*Tel:* 706 295-6962
*E-mail:* slanham@coosavalleytech.edu

**Armstrong Atlantic State University**
Radiation Therapy Prgm
11935 Abercorn St
Savannah, GA 31419-1997
*Prgm Dir:* Lee Braswell Jr, MPH RT(R)(T)
*Tel:* 912 927-5360, Ext 7429    *Fax:* 912 921-5838
*E-mail:* braswele@mail.armstrong.edu
*Med Dir:* John Duttenhaver, MD

# Illinois

**Northwestern Memorial Hospital**
Radiation Therapy Prgm
251 E Huron St
Galter Pavilion LC-178
Chicago, IL 60611
www.nmh.org
*Prgm Dir:* Alex Zafirovski, BS RT(T)
*Tel:* 312 926-2733    *Fax:* 312 926-6374
*E-mail:* azafirov@nmh.org
*Med Dir:* Bharat Mittal, MD

**Swedish American Hospital**
Radiation Therapy Prgm
1401 E State St
Rockford, IL 61104
*Prgm Dir:* Leia Levy, MAdEd RT(T)
*Tel:* 815 961-2038    *Fax:* 815 966-3966
*E-mail:* llevy@swedishamerican.org

# Indiana

**Indiana University Northwest**
Radiation Therapy Prgm
3400 Broadway
Gary, IN 46408-1197
*Prgm Dir:* Sandy L Piehl, MPA RT(R)(T)
*Tel:* 219 981-4204    *Fax:* 219 980-6649
*E-mail:* spiehl@iun.edu
*Med Dir:* U P Kalolche, MD

**Ball State University**
Radiation Therapy Prgm
1701 N Senate Blvd, PO Box 1367
Wile Hall Rm 645
Indianapolis, IN 46202
www.bsu.edu/physiology-health
*Prgm Dir:* Dan R Strahan, BS RT(T)
*Tel:* 317 962-3377    *Fax:* 317 962-2102
*E-mail:* dstrahan@clarian.org
*Med Dir:* Newell Pugh, MD

**Indiana University**
Radiation Therapy Prgm
School of Medicine-Dept of Radiation Oncology
535 Barnhill Dr, RT 041
Indianapolis, IN 46202
*Prgm Dir:* Donna Kay Dunn, MS RT(T)
*Tel:* 317 274-1302    *Fax:* 317 274-4723
*E-mail:* dodunn@iupui.edu

# Iowa

**University of Iowa Hospitals & Clinics**
Radiation Therapy Prgm
200 Hawkins Dr, LL West Addition PFP
Iowa City, IA 52242-1009
www.uihealthcare.com/depts/med/radiationoncology/
    degreeprograms/rtt_main.html
*Prgm Dir:* Mindi J TenNapel, MBA RT(R)(T)
*Tel:* 319 356-8286    *Fax:* 319 356-1530
*E-mail:* mindi—tennapel@uiowa.edu
*Med Dir:* John Buatti, MD

# Kentucky

**Univ of Kentucky Chandler Med Ctr**
Radiation Therapy Prgm
800 Rose St, Rm C-15
Lexington, KY 40536-0293
www.mc.uky.edu/radiationmedicine
*Prgm Dir:* Carol Scherbak, MSRS
*Tel:* 859 257-7621    *Fax:* 859 257-4931
*E-mail:* cmsche0@email.uky.edu

**James Graham Brown Cancer Center**
Radiation Therapy Prgm
529 S Jackson St
Louisville, KY 40202
*Prgm Dir:* Sharon T Bradshaw, BSH RT(R)(T)
*Tel:* 502 562-4656    *Fax:* 502 562-3209
*E-mail:* sharonbr@ulh.org
*Med Dir:* William Spanos Jr, MD

# Louisiana

**Delgado Community College**
Radiation Therapy Prgm
615 City Park Ave
New Orleans, LA 70119
www.dcc.edu
*Prgm Dir:* Sylvia P Sandberg, BS RT(R)(T)
*Tel:* 504 483-4487
*E-mail:* ssandb@dcc.edu
*Med Dir:* Troy Scroggins, MD

# Maine

**Southern Maine Community College**
Radiation Therapy Prgm
Two Fort Rd
South Portland, ME 04106
www.smccme.edu
*Prgm Dir:* Dennis T Leaver, MS RT(R)(T)
*Tel:* 207 741-5593    *Fax:* 207 741-5593
*E-mail:* dleaver@smccme.edu
*Med Dir:* Celine Goden, MD

# Maryland

**Comm Coll of Baltimore County - Essex
    Campus**
Radiation Therapy Prgm
7201 Rossville Blvd
Baltimore, MD 21237-9987
http://ccbcmd.edu/allied_health/
    radiation_program.html
*Prgm Dir:* Dionne M Johnson, MEd RT(T)
*Tel:* 410 780-6709    *Fax:* 410 780-6946
*E-mail:* djohnson3@ccbcmd.edu
*Med Dir:* Mohan Suntha, MD

# Massachusetts

**Caritas Labour,  College**
Radiation Therapy Prgm
2120 Dorchester Ave
Boston, MA 02124-5698
*Prgm Dir:* Pauline E Clancy, BS RT(T)
*Tel:* 617 296-8300, Ext 4044    *Fax:* 617 296-7947
*E-mail:* pauline_clancy@laboure.edu
*Med Dir:* Hywel Madoc-Jones, MD

**Mass College of Pharmacy & Health Sciences**
Radiation Therapy Prgm
179 Longwood Ave
Boston, MA 02115-5896
www.mcphs.edu
*Prgm Dir:* Susan B Belinsky, EdD RT(R)(T)
*Tel:* 617 732-2261    *Fax:* 617 732-2075
*E-mail:* sbelinsky@mcphs.edu

**Suffolk University**
Radiation Therapy Prgm
Physics Department
41 Temple St
Boston, MA 02114
www.suffolk.edu
*Prgm Dir:* Angela M Lombardo, MS RT(T)
*Tel:* 617 305-1995    *Fax:* 617 367-5063
*E-mail:* alombard@suffolk.edu

**UMass Memorial Medical Center**
Radiation Therapy Prgm
Radiation Oncology Department
55 Lake Ave North
Worcester, MA 01655
*Prgm Dir:* Patricia E Webster, MS RT(T)
*Tel:* 508 856-5551    *Fax:* 508 856-5006
*E-mail:* WebsterP@ummhc.org
*Med Dir:* T J Fitzgerald, MD

# Michigan

**Wayne State University**
Radiation Therapy Prgm
Eugene Applebaum College of Pharmacy and Health
    Sciences
259 Mack Ave
Detroit, MI 48201
www.•ins.wayne.edu/ubk-output/med2.html
*Prgm Dir:* Adam F Kempa, MEd RT(T)
*Tel:* 313 577-1137    *Fax:* 313 577-0908
*E-mail:* aa1156@wayne.edu
*Med Dir:* Kenneth J Levin, MD

**University of Michigan - Flint**
Radiation Therapy Prgm
2102 William S White Bldg
303 E Kearsley St
Flint, MI 48502-1950
www.umflint.edu
*Prgm Dir:* Michele R Fortner, BS RT(R)(T)
*Tel:* 810 237-6502    *Fax:* 810 762-3003
*E-mail:* fortner@umich.edu
*Med Dir:* Howard Sandler, MD

**Grand Valley State University**
Radiation Therapy Prgm
301 Michigan St NE, Ste 200
Grand Rapids, MI 49503
*Prgm Dir:* Bonita Pawloski, BS RT(R)(T)
*Tel:* 616 331-5949
*E-mail:* pawloskb@gvsu.edu

**Baker College of Jackson**
Radiation Therapy Prgm
2800 Springport Rd
Jackson, MI 49202
*Prgm Dir:* Terilynn K Fedchenko, BS RT(R)(T)
*Tel:* 517 780-4562   *Fax:* 517 789-7331
*E-mail:* teri.fedchenko@baker.edu

**William Beaumont Hospital**
Radiation Therapy Prgm
3601 W 13 Mile Rd
Royal Oak, MI 48073-6769
www.beaumonthospitals.com
*Prgm Dir:* Laura L Ochs, MEd BS RT(T)
*Tel:* 248 551-7156   *Fax:* 248 551-7166
*E-mail:* lochs@beaumont.edu
*Med Dir:* Alvaro Martinez, MD

# Minnesota

**Argosy University/Twin Cities**
Radiation Therapy Prgm
1515 Cental Parkway
Eagan, MN 55121
*Prgm Dir:* Patricia Fountinelle, MS RT(R)(T) CMD
*Tel:* 651 846-3404   *Fax:* 651 994-0895
*E-mail:* pfountinelle@argosyu.edu

**Fairview University**
Radiation Therapy Prgm
420 Delaware St SE
MMC Box 494
Minneapolis, MN 55455
*Prgm Dir:* Jodie L Wolla, MS RT(R)(T)
*Tel:* 612 273-6393   *Fax:* 612 273-6411
*E-mail:* jwolla1@fairview.org
*Med Dir:* Kathryn Dusenbery, MD

**Mayo School of Health Sciences**
Radiation Therapy Prgm
Sch of Hlth Sciences
200 First St SW
Rochester, MN 55905
www.mayo.edu/mshs
*Prgm Dir:* Leila A Bussman-Yeakel, BS RT(R)(T)
*Tel:* 507 284-4148   *Fax:* 507 284-0079
*E-mail:* bussmanyeakel.leila@mayo.edu
*Med Dir:* Ivy A Petersen, MD

# Missouri

**CoxHealth**
Radiation Therapy Prgm
3850 S National Ave, Ste 100
Springfield, MO 65807
www.coxhealth.com/schoolseducation/RadiationTherapy
*Prgm Dir:* Benjamin J Morris, BS RT(R)(T)(CT)
*Tel:* 417 269-5363   *Fax:* 417 269-6979
*E-mail:* benjamin.morris@coxhealth.com

**Barnes-Jewish Coll of Nursing/Allied Health**
Radiation Therapy Prgm
306 S Kingshighway Blvd
MS: 90-30-625
St Louis, MO 63110
*Prgm Dir:* Kathleen O Kienstra, MAT RT(R)(T)
*Tel:* 314 454-8457   *Fax:* 314 454-5239
*E-mail:* kkienstra@bjc.org
*Med Dir:* Marie E Taylor, MD

# Nebraska

**University of Nebraska Medical Center**
Radiation Therapy Prgm
984545 Nebraska Medical Center
Omaha, NE 68198-4545
www.unmc.edu/alliedhealth/rste
*Prgm Dir:* Lisa Bartenhagen, MS RT(R)(T)
*Tel:* 402 559-4236   *Fax:* 402 559-2181
*E-mail:* labarten@unmc.edu
*Med Dir:* Charles Enke, MD

# New Hampshire

**New Hampshire Technical Institute**
Radiation Therapy Prgm
31 College Dr
Concord, NH 03301
*Prgm Dir:* Amy C VonKadich, MEd RT(T)
*Tel:* 603 271-0700
*E-mail:* avonkadich@nhctc.edu
*Med Dir:* Charles Leutzinger, MD

# New Jersey

**Cooper University Hospital**
Radiation Therapy Prgm
One Cooper Plaza
Camden, NJ 08103
*Prgm Dir:* Karen D Ljunggren, BS RT(R)(T)(CT)
*Tel:* 856 342-2734
*E-mail:* Ljunggren-Karen@cooperhealth.edu
*Med Dir:* Jeffrey S Wenger, MD

**St Barnabas Medical Center**
Radiation Therapy Prgm
94 Old Short Hills Rd
Livingston, NJ 07039
www.sbhcs.com
*Prgm Dir:* Jennie S Lichtenberger, BS RT(T)(R)(M)
*Tel:* 973 322-5628   *Fax:* 973 322-5648
*E-mail:* jlichtenberger@sbhcs.com
*Med Dir:* Andrew Zablow, MD

**Muhlenberg Regional Medical Center**
Radiation Therapy Prgm
Park Ave & Randolph Rd
Plainfield, NJ 07061
*Prgm Dir:* Beverly S Nias, BS RT(R)(T) ARRT
*Tel:* 908 668-2637   *Fax:* 908 226-4568
*E-mail:* bnias@solarishs.org
*Med Dir:* Louis Schwartz, MD

# New York

**New York Methodist Hospital**
Radiation Therapy Prgm
506 Sixth St, Box 159008
Brooklyn, NY 11215-9008
*Prgm Dir:* Mary Reynolds, BA RT(T)
*Tel:* 718 780-3689   *Fax:* 718 780-3637
*E-mail:* marybreynolds@aol.com
*Med Dir:* M Tchelebi, MD

**Erie Community College**
Radiation Therapy Prgm
121 Ellicott St
Buffalo, NY 14203
*Prgm Dir:* Patricia A Bennewitz, MSEd RT(T)
*Tel:* 716 851-1048, Ext 1048   *Fax:* 716 851-1129
*E-mail:* bennewitz@ecc.edu
*Med Dir:* Vilasini Shanbhag, MD

**Nassau Community College**
Radiation Therapy Prgm
One Education Dr
Garden City, NY 11530
www.ncc.edu
*Prgm Dir:* Catherine Smyth, MA RT(T)
*Tel:* 516 572-7491   *Fax:* 516 572-9750
*E-mail:* cquane@aol.com
*Med Dir:* Jay Bosworth, MD

**Memorial Sloan - Kettering Cancer Ctr**
Radiation Therapy Prgm
1275 York Ave, Box 22
New York, NY 10021
www.mskcc.org/schoolofradiationtherapy
*Prgm Dir:* Wilson H Apollo, BA RT(T)
*Tel:* 212 639-6835   *Fax:* 212 717-3090
*E-mail:* apollow@mskcc.org
*Med Dir:* Joshua Yamada, MD

**SUNY Upstate Medical University**
Radiation Therapy Prgm
750 E Adams St
Syracuse, NY 13210
www.upstate.edu
*Prgm Dir:* Joan E O'Brien, MSEd RT(T)
*Tel:* 315 464-8448   *Fax:* 315 464-6940
*E-mail:* obrienj@upstate.edu

# North Carolina

**Pitt Community College**
Radiation Therapy Prgm
PO Drawer 7007
Greenville, NC 27835-7007
*Prgm Dir:* Elaine Spencer, BS RT(T)
*Tel:* 252 493-7452   *Fax:* 252 321-4451
*E-mail:* espencer@email.pittcc.edu
*Med Dir:* Ron Allison, MD

**Forsyth Technical Community College**
Radiation Therapy Prgm
2100 Silas Creek Pkwy
Winston-Salem, NC 27104
www.forsythtech.edu
*Prgm Dir:* Christina R Gibson, MPH RT(R)(T)
*Tel:* 336 734-7184   *Fax:* 336 734-7444
*E-mail:* cgibson@forsythtech.edu
*Med Dir:* Lisa Evans, MD

# Ohio

**Aultman Hospital**
Radiation Therapy Prgm
2600 Sixth St SW
Canton, OH 44710
*Prgm Dir:* Victoria Migge, MS RT(R)(T)
*Tel:* 330 363-4853   *Fax:* 330 588-2601
*E-mail:* vmigge@aultman.com
*Med Dir:* Douglas Keyser, MD

**University of Cincinnati**
*Cosponsor:* Raymond Walters College
Radiation Therapy Prgm
234 Goodman Ave, ML 757
Cincinnati, OH 45219
*Prgm Dir:* Carolyn Hollan, MS RT(R)(T)
*Tel:* 513 584-9099   *Fax:* 513 584-4007
*E-mail:* carolyn.hollan@uc.edu
*Med Dir:* William Barrett, MD

**Cleveland Clinic Foundation**
Radiation Therapy Prgm
9500 Euclid Ave (T28)
Cleveland, OH 44195
*Prgm Dir:* Patricia A Barrett, MS RT(R)(T) CMD
*Tel:* 216 444-5484   *Fax:* 216 444-5331
*E-mail:* barretp1@ccf.org
*Med Dir:* John Suh, MD

**A G James Cancer Hosp & Research Inst**
Radiation Therapy Prgm
300 W Tenth Ave
Columbus, OH 43210
*Prgm Dir:* Ruth M Hackworth, BS RT(R)(T)
*Tel:* 614 293-6203    *Fax:* 614 293-4044
*E-mail:* ruth.hackworth@osumc.edu
*Med Dir:* Nina Mayr, MD

# Oklahoma

**Univ of Oklahoma Health Sciences Center**
Radiation Therapy Prgm
PO Box 26901 Rm 451
Oklahoma City, OK 73190
www.ah.ouhsc.edu/main
*Prgm Dir:* Stacy L Anderson, MS RT(T) CMD
*Tel:* 405 271-6477    *Fax:* 405 271-1424
*E-mail:* stacy-anderson@ouhsc.edu
*Med Dir:* Carl R Bogardus, MD

# Oregon

**Oregon Health & Science University**
Radiation Therapy Prgm
3181 SW Sam Jackson Park Rd, GH 119
Portland, OR 97239-3098
http://euston.ohsu.edu/radiation_therapy
*Prgm Dir:* Anne M Maddeford, MS MAcOM RT(T) LAc
*Tel:* 503 494-6708    *Fax:* 503 494-2730
*E-mail:* maddefoa@ohsu.edu
*Med Dir:* Carol Marquez, MD

# Pennsylvania

**Gwynedd-Mercy College**
Radiation Therapy Prgm
1325 Sumneytown Pike
PO Box 901
Gwynedd Valley, PA 19437-0901
www.gmc.edu
*Prgm Dir:* Patricia J Giordano, MS RT(R)(T)
*Tel:* 215 542-4658    *Fax:* 215 641-5559
*E-mail:* giordano.p@gmc.edu
*Med Dir:* C Jules Rominger, MD

**Thomas Jefferson University**
Radiation Therapy Prgm
130 S Ninth Street - Suite 1013
Philadelphia, PA 19107
*Prgm Dir:* Jacqueline E Chmiel, RT(R)(T)
*Tel:* 215 503-1434
*E-mail:* jacqueline.chmiel@jefferson.edu

**Comm Coll of Allegheny County**
Radiation Therapy Prgm
808 Ridge Ave, M607
Pittsburgh, PA 15212-6097
www.ccac.edu
*Prgm Dir:* Elizabeth Anne Harkay, BS RT(T)
*Tel:* 412 237-2752    *Fax:* 412 237-6579
*E-mail:* eharkay@ccac.edu
*Med Dir:* Melvin Deutsch, MD

# South Carolina

**Spartanburg Community College**
Radiation Therapy Prgm
PO Box 4386
Spartanburg, SC 29305-4386
www.stcsc.edu
*Prgm Dir:* Bert Wood, BS RT(R)(T)(CT) CMD
*Tel:* 864 591-3886    *Fax:* 864 591-3881
*E-mail:* woodb@stcsc.edu
*Med Dir:* Julian Josey, MD

# Tennessee

**Chattanooga State Technical Comm College**
Radiation Therapy Prgm
4501 Amnicola Hwy
Chattanooga, TN 37406
http://chattanoogastate.edu
*Prgm Dir:* Lisa Legg, BS RT(R)(T)
*Tel:* 423 697-3336    *Fax:* 423 697-3324
*E-mail:* lisa.legg@chattanoogastate.edu
*Med Dir:* Deanna Davidson, MD

**Baptist College of Health Sciences**
Radiation Therapy Prgm
1003 Monroe Ave
Memphis, TN 38104
www.bchs.edu
*Prgm Dir:* Beverly K Coker, MA RT(R)(T)
*Tel:* 901 572-2645    *Fax:* 901 572-2750
*E-mail:* beverly.coker@bchs.edu

**Vanderbilt University Medical Center**
Radiation Therapy Prgm
Center for Radiation Oncology
The Vanderbilt Clinic B 902
Nashville, TN 37232-5671
*Prgm Dir:* April D Tingler, BS RT(R)(T)
*Tel:* 615 343-4304    *Fax:* 615 343-0161
*E-mail:* april.d.tingler@vanderbilt.edu
*Med Dir:* Dennis Hallahan, MD

# Texas

**Amarillo College**
Radiation Therapy Prgm
PO Box 447
Amarillo, TX 79178
www.actx.edu
*Prgm Dir:* Tony Tackitt, MEd RT(T)
*Tel:* 806 354-6063    *Fax:* 806 354-6076
*E-mail:* tackitt-tm@actx.edu
*Med Dir:* Jim Stafford, MD

**Galveston College**
Radiation Therapy Prgm
4015 Ave Q
Galveston, TX 77550-2782
*Prgm Dir:* Hubert Callahan, BS RT(R)(T)
*Tel:* 409 944-1492    *Fax:* 409 772-3014
*E-mail:* hcallaha@gc.edu
*Med Dir:* Martin Colman, MD

**Univ of Texas M D Anderson Cancer Ctr**
Radiation Therapy Prgm
1515 Holcombe Blvd, PO Unit 190
Houston, TX 77030-4009
www.mdanderson.org/healthsciences
*Prgm Dir:* Shaun T Caldwell, MS RT(R)(T)
*Tel:* 713 792-3455    *Fax:* 713 792-0956
*E-mail:* scaldwell@mdanderson.org
*Med Dir:* Eric A Strom, MD

**Texas State University-San Marcos**
Radiation Therapy Prgm
601 University Dr
San Marcos, TX 78666
*Prgm Dir:* Ronnie G Lozano, MSRS RT(T)(ARRT)
*Tel:* 512 245-1345    *Fax:* 512 245-1477
*E-mail:* RL10@txstate.edu

# Virginia

**University of Virginia Health System**
*Cosponsor:* University of Virginia Medical Center
Radiation Therapy Prgm
PO Box 800383
Charlottesville, VA 22908-0383
www.healthsystem.virginia.edu/radonc
*Prgm Dir:* Frances R Taylor, BA RT(R)(T)
*Tel:* 434 243-2760    *Fax:* 434 982-3262
*E-mail:* frt@virginia.edu
*Med Dir:* Bernard Schneider, PhD, MD

**Virginia Commonwealth University**
Radiation Therapy Prgm
701 W Grace St, Ste 2100
Box 843057
Richmond, VA 23298-3057
*Prgm Dir:* Melanie Dempsey, RTT CMD
*Tel:* 804 828-9104    *Fax:* 804 828-5778
*E-mail:* mcdempsey@vcu.edu
*Med Dir:* Monica Morris, MD

**Virginia Western Community College**
Radiation Therapy Prgm
PO Box 14007
Roanoke, VA 24038
*Prgm Dir:* Carole S Graham, MS Ed RT(R)(T)
*Tel:* 540 981-7731
*E-mail:* cgraham@vw.vccs.edu

# Washington

**Bellevue Community College**
Radiation Therapy Prgm
3000 Landerholm Circle SE, Rm B243
Bellevue, WA 98007-6484
www.bcc.ctc.edu/radon
*Prgm Dir:* Julius B Armstrong, MBA RT(T)
*Tel:* 425 564-5079    *Fax:* 425 564-4193
*E-mail:* jarmstro@bcc.ctc.edu
*Med Dir:* Michael Hunter, MD

# West Virginia

**West Virginia University Hospitals**
Radiation Therapy Prgm
PO Box 8150, Medical Ctr Dr
Morgantown, WV 26506-8150
www.wvuhradtech.com
*Prgm Dir:* Christina M Paugh, MA RT(R)(T)
*Tel:* 304 293-8827    *Fax:* 304 293-4717
*E-mail:* paughc@rcbhsc.wvu.edu
*Med Dir:* John C Frich, MD

# Wisconsin

**Moraine Park Technical College**
Radiation Therapy Prgm
235 N National Ave
Fond du Lac, WI 54936
*Prgm Dir:* Kerry J Stehlik, MS RT(R)(T)
*Tel:* 920 924-6395
*E-mail:* kstehlik@excel.net

**University of Wisconsin - La Crosse**
Radiation Therapy Prgm
1725 State St, 4094 HSC
La Crosse, WI 54601
www.uwlax.edu/rt
*Prgm Dir:* Kristine Saeger, MS RT(R)(T)
*Tel:* 608 785-6979    *Fax:* 608 785-8460
*E-mail:* saeger.kris@uwlax.edu
*Med Dir:* Ellen Ziaja, MD

**Radiation Therapist**

| Programs* | Class Capacity | Begins | Length (months) | Award | Res Tuition | Non-res Tuition | Stipend | Offers:‡ 1 | 2 | 3 |
|---|---|---|---|---|---|---|---|---|---|---|
| **Alabama** | | | | | | | | | | |
| University of Alabama at Birmingham | 46 | Aug Jan | 18 | Dipl, Cert, BS | $10,002 | $20,004 | | | | • |
| **Arkansas** | | | | | | | | | | |
| Arkansas State University | 10 | Aug | 48 | BS | $4,879 | $12,546 | | | | |
| Central Arkansas Radiation Therapy Inst (Little Rock) | 15 | Aug | 12 | Cert | $3,000 | $3,000 | | | | • |
| **California** | | | | | | | | | | |
| California State University - Long Beach | 42 | Aug | 48 | BS | $2,572 | $7,626 | | | | • |
| City College of San Francisco | 15 | Aug | 26 | AS | $500 | $4,000 | | | | |
| City of Hope National Medical Center (Duarte) | 16 | Oct | 12 | Cert | $3,000 | $3,000 | | | | • |
| Kaiser Permanente Sch of Allied Hlth Sciences (Richmond) | 22 | Oct | 18 | Cert | $9,000 | $9,000 | | | | |
| Loma Linda University | 18 | Sep | 12 | Cert | $15,500 | $15,500 | | | | • |
| **Connecticut** | | | | | | | | | | |
| Gateway Community College (North Haven) | 28 | Aug | 22 | AS | $3,276 | $7,976 | | • | | • |
| Hartford Hospital | 16 | May Sep | 24, 16 | Cert | $3,000 | $3,000 | | | | • |
| **Florida** | | | | | | | | | | |
| Halifax Medical Center (Daytona Beach) | 9 | Jun | 12 | Cert | $2,800 | $2,800 | | | | • |
| Miami Dade College | 58 | Aug | 24, 12 | AS | $2,654 | $8,807 | | | | • |
| **Georgia** | | | | | | | | | | |
| Armstrong Atlantic State University (Savannah) | 15 | Jun | 24 | BS | $3,483 | $13,935 | | • | | • |
| Coosa Valley Technical College (Rome) | 34 | | 12, 21 | Dipl, Cert, AAS | $1,480 | $0 | | | | |
| Grady Health System (Atlanta) | 16 | Sep | 12 | Cert | $3,500 | $0 | | | | • |
| Medical College of Georgia (Augusta) | 14 | Aug | 21 | Cert, BS | $5,691 | $20,127 | | | | |
| **Illinois** | | | | | | | | | | |
| Northwestern Memorial Hospital (Chicago) | 8 | Aug | 12 | Cert, BS | $5,000 | $5,000 | | | | • |
| Swedish American Hospital (Rockford) | 10 | Aug | 17, 22 | Cert | $3,000 | $3,000 | | | | • |
| **Indiana** | | | | | | | | | | |
| Ball State University (Indianapolis) | 8 | May | 24 | AAS | $6,500 | $6,500 | | | | |
| Indiana University (Indianapolis) | 14 | Jun | 48 | BS | $9,654 | $23,769 | | | | • |
| Indiana University Northwest (Gary) | 10 | Aug | 24 | BS | $4,899 | $12,750 | | | | |
| **Iowa** | | | | | | | | | | |
| University of Iowa Hospitals & Clinics (Iowa City) | 7 | Aug | 12 | Cert, BS | $6,000 | $6,000 | | | | • |
| **Kentucky** | | | | | | | | | | |
| James Graham Brown Cancer Center (Louisville) | 9 | Sep | 12 | Cert | $3,000 | $3,000 | | | | • |
| Univ of Kentucky Chandler Med Ctr (Lexington) | 6 | Sep | 12 | Cert | $1,500 | $1,500 | | | | • |
| **Louisiana** | | | | | | | | | | |
| Delgado Community College (New Orleans) | 8 | Evey 16 mos | 16 | Cert | $1,800 | $7,400 | | | | • |
| **Maine** | | | | | | | | | | |
| Southern Maine Community College (South Portland) | 11 | Sep | 15, 24 | AS | $2,376 | $4,320 | | • | | |
| **Maryland** | | | | | | | | | | |
| Comm Coll of Baltimore County - Essex Campus | 25 | Sep | 12, 24 | Cert, AAS | $2,300 | $4,600 | | | | • |
| **Massachusetts** | | | | | | | | | | |
| Mass College of Pharmacy & Health Sciences (Boston) | 58 | Sep | 33, 21 | BS, BS | $21,200 | $21,200 | | | | • |
| Suffolk University (Boston) | 12 | Rolling admissions | 48, 24 | Cert, BS | $21,140 | $21,140 | | | | • |
| UMass Memorial Medical Center (Worcester) | 7 | Sep | 15 | Cert | $4,500 | $4,500 | | | | • |
| **Michigan** | | | | | | | | | | |
| Baker College of Jackson | 25 | Sep | 36 | BS | $7,948 | $0 | | • | | • |
| University of Michigan - Flint | 15 | Jun | 22 | BS | $19,728 | $39,000 | | | | • |
| Wayne State University (Detroit) | 28 | Sep | 48 | BSRT | $8,100 | $18,300 | | | | • |
| William Beaumont Hospital (Royal Oak) | 6 | Sep | 24 | Cert | $4,250 | $0 | | | | • |
| **Minnesota** | | | | | | | | | | |
| Mayo School of Health Sciences (Rochester) | 8 | Sep | 20 | Cert | $8,000 | $8,000 | | | | • |
| **Missouri** | | | | | | | | | | |
| CoxHealth (Springfield) | 4 | Late Sep | 13 | Cert | $3,000 | $3,000 | | | | • |
| **Nebraska** | | | | | | | | | | |
| University of Nebraska Medical Center (Omaha) | 17 | Aug | 12 | Dipl, BS | $6,621 | $19,264 | | | | • |
| **New Jersey** | | | | | | | | | | |
| St Barnabas Medical Center (Livingston) | 12 | Sep | 12 | Cert | $7,000 | $7,000 | | | | |
| **New York** | | | | | | | | | | |
| Erie Community College (Buffalo) | 17 | Sep | 24 | AAS | $2,500 | $5,000 | | | | |
| Memorial Sloan - Kettering Cancer Ctr (New York) | 10 | Sep | 24 | Cert | $3,000 | $3,000 | | | | • |
| Nassau Community College (Garden City) | 47 | Sep | 24 | AAS | $1,400 | $2,800 | | | | • |
| SUNY Upstate Medical University (Syracuse) | 15 | Aug | 20 | BS | $5,798 | $14,146 | | | | • |
| **North Carolina** | | | | | | | | | | |
| Forsyth Technical Community College (Winston-Salem) | 22 | Aug | 24, 12 | AAS | $1,700 | $9,450 | | | | |
| Pitt Community College (Greenville) | 25 | Aug | 12 | Dipl | $1,500 | $7,800 | | | | |

*Data are shown only for programs that completed the 2006 AMA Survey of Health Professions Education Programs
‡Key to Offers: 1: Evening or weekend classes; 2: Non-English instruction; 3: Cultural competence instruction

# Radiation Therapist

| Programs* | Class Capacity | Begins | Length (months) | Award | Res Tuition | Non-res Tuition | Stipend | Offers:‡ 1 | 2 | 3 |
|---|---|---|---|---|---|---|---|---|---|---|
| **Ohio** | | | | | | | | | | |
| A G James Cancer Hosp & Research Inst (Columbus) | 10 | Sept | 15 | Cert | $3,000 | $3,000 | | | | |
| Cleveland Clinic Foundation | 9 | Jul | 12 | Cert | $2,500 | $2,500 | | | | • |
| University of Cincinnati | 18 | Sep | 14, 24 | Cert, AAS | $5,232 | $13,566 | | • | | • |
| **Oklahoma** | | | | | | | | | | |
| Univ of Oklahoma Health Sciences Center (Oklahoma City) | 12 | Aug | 48 | BS | $3,522 | $7,400 | | | | • |
| **Oregon** | | | | | | | | | | |
| Oregon Health & Science University (Portland) | 8 | Aug | 24 | BS | $11,262 | $20,030 | | | | • |
| **Pennsylvania** | | | | | | | | | | |
| Comm Coll of Allegheny County (Pittsburgh) | 30 | Aug | 12, 24 | Cert, AS | $3,320 | $9,960 | | | | • |
| Gwynedd-Mercy College (Gwynedd Valley) | 15 | Sep | 20, 40 | Cert, BS | $18,580 | $18,580 | | | | • |
| **South Carolina** | | | | | | | | | | |
| Spartanburg Community College | 12 | Sep | 12 | Cert | $5,500 | $5,500 | | | | |
| **Tennessee** | | | | | | | | | | |
| Baptist College of Health Sciences (Memphis) | 8 | Fall | 48 | BS | $4,200 | $0 | | • | | |
| Chattanooga State Technical Comm College | 54 | Aug | 12 | Cert | $3,110 | $11,271 | | | | • |
| **Texas** | | | | | | | | | | |
| Amarillo College | 19 | Aug | 24 | AAS | $966 | $2,194 | | | | • |
| Univ of Texas M D Anderson Cancer Ctr (Houston) | 30 | Sep | 12 | Cert, BS | $2,530 | $15,510 | | | | • |
| **Virginia** | | | | | | | | | | |
| University of Virginia Health System (Charlottesville) | 12 | Aug | 16 | Cert | $4,000 | $4,000 | | | | • |
| Virginia Commonwealth University (Richmond) | 10 | Aug | 33 | BS | $6,221 | $22,221 | | | | • |
| **Washington** | | | | | | | | | | |
| Bellevue Community College | 15 | Sep | 24 | AA | $4,000 | $4,000 | | | | • |
| **West Virginia** | | | | | | | | | | |
| West Virginia University Hospitals (Morgantown) | 4 | Jul | 12 | Cert | $1,600 | $1,600 | | | | • |
| **Wisconsin** | | | | | | | | | | |
| University of Wisconsin - La Crosse | 23 | Sep | 23 | BS | $6,500 | $19,000 | | | | • |

*Data are shown only for programs that completed the 2006 AMA Survey of Health Professions Education Programs
‡Key to Offers: 1: Evening or weekend classes; 2: Non-English instruction; 3: Cultural competence instruction

## Radiographer*

# Alabama

**Carraway Methodist Medical Center**
Radiography Prgm
1600 Carraway Blvd
Birmingham, AL 35242
*Prgm Dir:* Robert N Odom, JD RT(R)
*Tel:* 205 502-6920   *Fax:* 205 502-5365
*E-mail:* odomb@carraway.com
*Med Dir:* Randall W Finley, MD

**Jefferson State Community College**
Radiography Prgm
2601 Carson Rd
GLB Rm 249
Birmingham, AL 35215-3098
www.jeffstateonline.com
*Prgm Dir:* Christie W Bolton, MEd RT(R)
*Tel:* 205 856-6017   *Fax:* 205 856-7725
*E-mail:* cbolton@jeffstateonline.com
*Med Dir:* Jeffery Frye, MD

**University of Alabama at Birmingham**
Radiography Prgm
1705 University Blvd
RMSB 447
Birmingham, AL 35294-1212
http://main.uab.edu/shrp/default.aspx?pid=77881
*Prgm Dir:* Audrey Harris, MA Ed RT(R)(CT)(M)(QM)
*Tel:* 205 934-7539   *Fax:* 205 975-7302
*E-mail:* radprof@uab.edu
*Med Dir:* Robert Koehler, MD

**George C Wallace Community College**
Radiography Prgm
1141 Wallace Dr
Dothan, AL 36303
*Prgm Dir:* G Bates Gilmore, BS RT(R)
*Tel:* 334 556-2299   *Fax:* 334 983-3600
*E-mail:* bgilmore@wallace.edu
*Med Dir:* Sibley N Turner, MD

**Gadsden State Community College**
Radiography Prgm
PO Box 227
Gadsden, AL 35902-0227
www.gadsdenstate.edu
*Prgm Dir:* Deborah Gay Utz, MEd RT(R)
*Tel:* 256 549-8468   *Fax:* 256 549-8458
*E-mail:* gutz@gadsdenstate.edu

**Wallace State Community College**
Radiography Prgm
PO Box 2000
Hanceville, AL 35077-2000
*Prgm Dir:* Terrie Gammon, PhD RT (R)(M)
*Tel:* 256 352-8309   *Fax:* 256 352-8320
*E-mail:* terrie.gammon@wallacestate.edu
*Med Dir:* Fred Moss, MD

**Crestwood Medical Center**
Radiography Prgm
One Hospital Dr SE
Huntsville, AL 35801
www.triadhospitals.com
*Prgm Dir:* Ronald A Murphree, BS RT(R)
*Tel:* 256 880-4587
*E-mail:* ronald.murphree@triadhospitals.com
*Med Dir:* Greg Gum, MD

**Huntsville Hospital**
Radiography Prgm
101 Sivley Rd
Huntsville, AL 35801
http://huntsvillehospital.org
*Prgm Dir:* Cheryl E Dutton Walton, BS RT(R)
*Tel:* 256 265-8928   *Fax:* 256 265-9417
*E-mail:* cdutton@hhsys.org
*Med Dir:* Timothy Baker, MD

**University of South Alabama**
Radiography Prgm
1504 Springhill Ave, Ste 2515
Mobile, AL 36604-3273
www.usouthal.edu
*Prgm Dir:* Charles W Newell, EdD RT(R)(MR)(CV)(CT)
*Tel:* 251 434-3456   *Fax:* 251 434-3458
*E-mail:* cnewell@jaguar1.usouthal.edu
*Med Dir:* Steven K Teplick, MD

**Baptist Medical Center South**
Radiography Prgm
2169 Normandie Dr
Montgomery, AL 36111
www.baptistfirst.org/educationopportunities.htm
*Prgm Dir:* Paul H Littlefield, EdS RT(R)
*Tel:* 334 281-3061   *Fax:* 334 281-3671
*E-mail:* plittlefield@baptistfirst.org
*Med Dir:* David C Montiel, MD

**Southern Union State Community College**
Radiography Prgm
1701 Lafayette Pkwy
Opelika, AL 36801
www.suscc.edu
*Prgm Dir:* Carol Southern, EdD RT(R)(CT)
*Tel:* 334 745-6437, Ext 5517   *Fax:* 334 745-6342
*E-mail:* jcs@suscc.edu
*Med Dir:* Michelle W Agee, MD PhD

**DCH Regional Medical Center**
Radiography Prgm
809 University Blvd E
Tuscaloosa, AL 35401
www.dchsystem.com
*Prgm Dir:* Dena F Ennis, MA Ed RT(R)
*Tel:* 205 759-6009   *Fax:* 205 345-8521
*E-mail:* dennis@dchsystem.com
*Med Dir:* Howard Holley, MD

# Arizona

**Pima Medical Institute - Mesa**
Radiography Prgm
957 S Dobson Rd
Mesa, AZ 85202
*Prgm Dir:* Nancy M Daugherty, BS RT(R)
*Tel:* 480 644-0267, Ext 120   *Fax:* 480 649-5249
*E-mail:* ndaugherty@pmi.edu
*Med Dir:* John Mabry, MD

**Apollo College/US Education Corporation**
Radiography Prgm
2701 W Bethany Home Rd
Phoenix, AZ 85017
*Prgm Dir:* Jamie S Tucker, BAAS RT(R)(CT)
*Tel:* 602 433-1333, Ext 21112   *Fax:* 602 433-1414
*E-mail:* jtucker@apollocollege.edu

**GateWay Community College**
Radiography Prgm
108 N 40th St
Phoenix, AZ 85034
www.gatewaycc.edu
*Prgm Dir:* Mary Carrillo, MBA HCM RT (R)(M)(CDT)
*Tel:* 602 286-8542   *Fax:* 602 286-8480
*E-mail:* carrillo@gatewaycc.edu
*Med Dir:* Wm W Horsley, MD

**Pima Community College**
Radiography Prgm
2202 W Anklam Rd, HRP 220
Tucson, AZ 85709-0080
www.pima.edu
*Prgm Dir:* Randall D Dings, BS RT(R)
*Tel:* 520 206-6916   *Fax:* 520 206-3027
*E-mail:* rdings@pima.edu
*Med Dir:* Jim N Stewart, MD

**Pima Medical Institute - Tucson**
Radiography Prgm
3350 E Grant Rd
Tucson, AZ 85716
*Prgm Dir:* Michael T Hawkes, BSRS RT(R)
*Tel:* 520 881-1284
*E-mail:* mhawkes@pmi.edu
*Med Dir:* Harold Trief, MD

**Arizona Western College**
Radiography Prgm
PO Box 929
Yuma, AZ 85366-0929
*Prgm Dir:* Victoria Holas, M Ed RT(R)
*Tel:* 928 314-9574
*E-mail:* victoria.holas@azwestern.edu

# Arkansas

**South Arkansas Community College**
Radiography Prgm
300 S West Ave, PO Box 7010 W Campus
El Dorado, AR 71731-7010
www.southark.edu
*Prgm Dir:* Deborah M Edney, BS RT(R) ARRT
*Tel:* 870 862-8131, Ext 226   *Fax:* 870 864-7104
*E-mail:* dedney@southark.edu
*Med Dir:* Tie Ong, MD

**University of Arkansas for Medical Sciences**
Radiography Prgm
AHEC-Northwest
2907 E Joyce St
Fayetteville, AR 72703
*Prgm Dir:* Stanley R Olejniczak, MS RT(R)
*Tel:* 479 521-7615   *Fax:* 479 684-6501
*E-mail:* solejniczak@ahecnw.uams.edu
*Med Dir:* Murray Harris, MD

**University of Arkansas - Fort Smith**
Radiography Prgm
5201 Grand Ave
PO Box 3649
Ft Smith, AR 72913-3649
*Prgm Dir:* Nancy G Hawking, EdD RT(R)
*Tel:* 501 788-7852   *Fax:* 501 788-7869
*E-mail:* nhawking@uafortsmith.edu
*Med Dir:* Richard N Brown, MD

**North Arkansas College**
Radiography Prgm
1515 Pioneer Dr
Harrison, AR 72601
www.northark.edu
*Prgm Dir:* Sondra L Richards, MS RT(R)(M)
*Tel:* 870 391-3318   *Fax:* 870 391-3250
*E-mail:* srichard@northark.edu

**National Park Community College**
Radiography Prgm
101 College Dr
Hot Springs, AR 71913-9174
www.npcc.edu
*Prgm Dir:* Timothy J Skaife, MA RT(R)
*Tel:* 501 760-4282   *Fax:* 501 760-4141
*E-mail:* tskaife@npcc.edu
*Med Dir:* Michael Hickman, MD

*Additional information about programs that returned the AMA's survey is available in a table beginning on page 380.

**Baptist Health**
Radiography Prgm
11900 Colonel Glenn Rd, Ste 1000
Little Rock, AR 72210-2820
www.baptist-health.org
*Prgm Dir:* Brenda A Simmons, BS RT(R)
*Tel:* 501 202-7942   *Fax:* 501 202-7712
*E-mail:* bsimmons@baptist-health.org
*Med Dir:* Nokes Steven, MD

**St Vincent Infirmary Medical Center**
Radiography Prgm
Two St Vincent Circle
Little Rock, AR 72205-5499
www.stvincenthealth.com/radtech
*Prgm Dir:* Frank Porter, RT(R) (QM)
*Tel:* 501 552-2994   *Fax:* 501 552-4236
*E-mail:* fporter@stvincenthealth.com
*Med Dir:* George Norton, MD

**University of Arkansas for Medical Sciences**
Radiography Prgm
4301 W Markham St, Slot 563
Little Rock, AR 72205
www.uams.edu/chrp/rad-tech
*Prgm Dir:* Joseph R Bittengle, MEd RT(R)
*Tel:* 501 686-6510   *Fax:* 501 686-6513
*E-mail:* bittenglejosephr@uams.edu
*Med Dir:* C Barry Buckner, MD

**Southeast Arkansas College**
Radiography Prgm
1900 Hazel St
Pine Bluff, AR 71603
*Prgm Dir:* Marilyn H Carter, BS RT(R)
*Tel:* 870 543-5941   *Fax:* 870 543-5912
*E-mail:* mcarter@seark.edu
*Med Dir:* Ira J Gordon, MD

**Arkansas State University**
Radiography Prgm
PO Box 910
State University, AR 72467
www.astate.edu
*Prgm Dir:* Raymond F Winters, MS RT(R)
*Tel:* 870 972-3073, Ext 3329   *Fax:* 870 972-2004
*E-mail:* rwinters@astate.edu
*Med Dir:* Mark White, MD

**University of Arkansas for Medical Sciences**
Radiography Prgm
300 E Sixth St
Texarkana, AR 71854
www.uams.edu/chrp/rad-tech
*Prgm Dir:* William M Pedigo, MPA RT(R)
*Tel:* 870 779-6029   *Fax:* 870 779-6045
*E-mail:* pedigowilliamm@uams.edu
*Med Dir:* Alan Jean, MD

# California

**Cabrillo College**
Radiography Prgm
6500 Soquel Dr
Aptos, CA 95003
www.cabrillo.edu
*Prgm Dir:* Ann S Smeltzer, BS RT(R)(M)
*Tel:* 831 479-5056   *Fax:* 831 479-5748
*E-mail:* ansmeltz@cabrillo.edu
*Med Dir:* Jannelle Rasi, MD

**Bakersfield College**
Radiography Prgm
1801 Panorama Dr
Bakersfield, CA 93305
www.bakersfieldcollege.edu
*Prgm Dir:* Nancy J Perkins, MA RT(R)(M)
*Tel:* 661 395-4284   *Fax:* 661 395-4295
*E-mail:* nperkins@bakersfieldcollege.edu

**Mills-Peninsula Health Services**
Radiography Prgm
1783 El Camino Real
Burlingame, CA 94010
www.mills-peninsula.org
*Prgm Dir:* Patricia Sperry, BS RT(R)(CT)
*Tel:* 650 696-5519   *Fax:* 650 696-5995
*E-mail:* sperryp@sutterhealth.org
*Med Dir:* Brian Johnson, MD

**Pima Medical Institute - Chula Vista**
Radiography Prgm
780 Bay Blvd, Ste 101
Chula Vista, CA 91910
www.pimamedical.com/locations/chulavista.htm
*Prgm Dir:* Lisa F Schmidt, PhD RT(R)(M)
*Tel:* 619 425-3200, Ext 121   *Fax:* 619 425-0182
*E-mail:* lschmidt@pmi.edu
*Med Dir:* Kent Thompson, Radiologist

**Arrowhead Regional Medical Center**
Radiography Prgm
400 N Pepper Ave
Colton, CA 92324-1819
*Prgm Dir:* Morris Hunter, MA RT(R)
*Tel:* 909 580-3540   *Fax:* 909 580-1561
*E-mail:* hunterm@armc.sbcounty.gov
*Med Dir:* Byron Fujimoto, MD

**Orange Coast College**
Radiography Prgm
2701 Fairview Rd, PO Box 5005
Costa Mesa, CA 92628-5005
www.occ.cccd.edu
*Prgm Dir:* Loren A Sachs, MA RT(R)(CV)(CT)(MR)
*Tel:* 714 432-5540   *Fax:* 714 432-5534
*E-mail:* laschs@occ.cccd.edu

**Cypress College**
Radiography Prgm
9200 Valley View St
Cypress, CA 90630-5897
www.cypresscollege.edu
*Prgm Dir:* Robert J Parelli, MA RT(R)
*Tel:* 714 484-7000, Ext 47286   *Fax:* 714 527-2175
*E-mail:* rparelli@cypresscollege.edu
*Med Dir:* Frederick A Birnberg, MD

**Fresno City College**
Radiography Prgm
1101 E University Ave
Fresno, CA 93741
www.fresnocitycollege.com
*Prgm Dir:* Paul N Gonzales, MS RT(R)
*Tel:* 559 244-2652   *Fax:* 559 244-2626
*E-mail:* paul.gonzales@fresnocitycollege.edu

**Loma Linda University**
Radiography Prgm
School of Allied Health Professions
Loma Linda, CA 92350
www.llu.edu/llu/sahp
*Prgm Dir:* Mark J Clements, MA RT(R)
*Tel:* 909 558-4931, Ext 47272   *Fax:* 909 558-7965
*E-mail:* mclements@llu.edu
*Med Dir:* Gregory Watkins, MD

**Foothill Community College**
Radiography Prgm
12345 El Monte Rd
Los Altos Hills, CA 94022-4599
www.foothill.edu
*Prgm Dir:* Eloise J Orrell, MS RT(R)
*Tel:* 650 949-7469   *Fax:* 650 949-7686
*E-mail:* orrelleloise@foothill.edu
*Med Dir:* Volney F VanDalsem III, MD

**Charles R Drew Univ of Med & Science**
Radiography Prgm
1731 E 120th St
Los Angeles, CA 90059
www.cdrewu.edu
*Prgm Dir:* Eugene Hasson, MS RT(R)
*Tel:* 323 563-5835, Ext 5885   *Fax:* 323 357-3419
*E-mail:* euhasson@cdrewu.edu
*Med Dir:* Janis Owens, MD

**East Los Angeles Occupational Center**
Radiography Prgm
2100 Marengo St
Los Angeles, CA 90033
*Prgm Dir:* Alicia Ortiz, MEd RT(R)(M)
*Tel:* 323 223-1283, Ext 161
*E-mail:* aokp2@sbcglobal.net

**Los Angeles City College**
Radiography Prgm
855 N Vermont Ave
Los Angeles, CA 90029
www.lacitycollege.edu
*Prgm Dir:* Edward C Vasquez, BA RT(R)
*Tel:* 323 953-4000, Ext 2942   *Fax:* 323 953-4013
*E-mail:* vasqueec@lacitycollege.edu
*Med Dir:* Keith Terasaki, MD

**Yuba Community College**
Radiography Prgm
2088 N Beale Rd
Marysville, CA 95901
www.yccd.edu/radtech
*Prgm Dir:* Angela Willson, MPA RT(R)(M)(QM)
*Tel:* 530 741-6960   *Fax:* 530 741-3541
*E-mail:* awillson@yccd.edu

**Merced College**
Radiography Prgm
3600 M St
Merced, CA 95348-2898
www.mccd.edu/alliedhealth/radtechhp.htm
*Prgm Dir:* K Judy Rose, MA RT(R)
*Tel:* 209 384-6132   *Fax:* 209 384-6167
*E-mail:* rose.j@mccd.edu
*Med Dir:* Steven Hansen, MD

**Moorpark College**
Radiography Prgm
7075 Campus Rd
Moorpark, CA 93021
www.moorparkcollege.net
*Prgm Dir:* Guadalupe Aldana, MS RT(R)
*Tel:* 805 378-1400, Ext 1632   *Fax:* 805 378-1548
*E-mail:* laldana@vcccd.net
*Med Dir:* Ruth Polan, MD

**Maric College**
Radiography Prgm
6180 Laurel Canyon Blvd, Ste 101
North Hollywood, CA 91606
*Prgm Dir:* Joyce H Cohen, BS RT(R)(M)
*Tel:* 818 763-2563
*E-mail:* jcohen@mariccollege.edu

**California State University - Northridge**
Radiography Prgm
18111 Nordhoff St
Northridge, CA 91330-8285
www.csun.edu/~vchsc02t
*Prgm Dir:* Anita M Slechta, MS RT(R)(M)
*Tel:* 818 677-2475   *Fax:* 818 677-2045
*E-mail:* anita.slechta@csun.edu
*Med Dir:* Vincent Fennell, MD

**Merritt College**
Radiography Prgm
12500 Campus Dr
Oakland, CA 94619
http://merritt.peralta.edu
*Prgm Dir:* Jennifer L Yates, MS RT(R)(M)(BD)
*Tel:* 510 436-2427   *Fax:* 510 434-3870
*E-mail:* jyates@peralta.edu
*Med Dir:* Ronald L Eisenberg, MD

**Pasadena City College**
Radiography Prgm
1570 E Colorado Blvd
Pasadena, CA 91106-2003
*Prgm Dir:* Leavon Spires, MA BS RT(R)(T) CRT
*Tel:* 626 585-7469   *Fax:* 626 585-7977
*E-mail:* lxspires@pasadena.edu
*Med Dir:* Michael Smith, MD

**Chaffey College**
Radiography Prgm
5885 Haven Ave
Rancho Cucamonga, CA 91737
www.chaffey.edu
*Prgm Dir:* Andrea Guillen Dutton, MEd RT(R)(M)
*Tel:* 909 477-8978   *Fax:* 909 477-8984
*E-mail:* andrea.dutton@chaffey.edu
*Med Dir:* Patrick Bryan, MD

**Canada College**
Radiography Prgm
4200 Farm Hill Blvd
Redwood City, CA 94061
www.smccd.net
*Prgm Dir:* Rafael A Rivera, RT(R)
*Tel:* 650 306-3283   *Fax:* 650 306-3281
*E-mail:* riverar@smccd.net
*Med Dir:* Michael David Hollett, MD

**Kaiser Permanente Sch of Allied Hlth Sciences**
*Cosponsor:* Richmond Medical Center
Radiography Prgm
938 Marina Way S
Richmond, CA 94804
www.kpsahs.org
*Prgm Dir:* Gregory J Wheeler, BA RT(R)(CT)
*Tel:* 510 231-5032   *Fax:* 510 231-5019
*E-mail:* Gregory.Wheeler@kp.org
*Med Dir:* Paul Radosevich, MD

**Naval School of Health Sciences, San Diego**
Radiography Prgm
34101 Farenholt Ave
San Diego, CA 92134
http://nshssd.med.navy.mil
*Prgm Dir:* John F Somersall, BS RT(R)
*Tel:* 619 532-5995   *Fax:* 619 532-7796
*E-mail:* jfsomersall@nshs-sd.med.navy.mil
*Med Dir:* CDR John M Maher, MC USN

**San Diego Mesa College**
Radiography Prgm
7250 Mesa College Dr
San Diego, CA 92111
www.sdccd.edu
*Prgm Dir:* Catherine M Bertsch, MEd RT(R)(MR)
*Tel:* 916 388-2666   *Fax:* 916 388-2929
*E-mail:* cbertsch@sdccd.edu
*Med Dir:* Norman Chen, MD

**City College of San Francisco**
Radiography Prgm
50 Phelan Ave, Box S69
San Francisco, CA 94112
www.ccsf.edu
*Prgm Dir:* Kyle R Thornton, MA Ed CRT RTr
*Tel:* 415 239-3694   *Fax:* 415 239-3930
*E-mail:* kthornto@ccsf.edu

**Central California School of Continuing Edu**
Radiography Prgm
3195 McMillan #F
San Luis Obispo, CA 93401
www.ccsce.org
*Prgm Dir:* Robert Desch, MS BA RT(R)
*Tel:* 805 543-9123   *Fax:* 805 543-6330
*E-mail:* ccsce@ccsce.org
*Med Dir:* Richard Sisson, MD

**Santa Barbara City College**
Radiography Prgm
721 Cliff Dr
Santa Barbara, CA 93109-2394
*Prgm Dir:* Debra S McMahan, MS RT(R) PA
*Tel:* 805 965-0581, Ext 2504   *Fax:* 805 963-7222
*E-mail:* mcmahan@sbcc.edu
*Med Dir:* Brian Schnier, MD

**Santa Rosa Junior College**
Radiography Prgm
1501 Mendocino Ave
Santa Rosa, CA 95401-4395
*Prgm Dir:* Xuan M Ho, PhD RT(R)
*Tel:* 707 527-4346   *Fax:* 707 527-4426
*E-mail:* xho@santarosa.edu
*Med Dir:* Gary Shaw, MD

**San Joaquin General Hospital**
Radiography Prgm
PO Box 1020
Stockton, CA 95201
www.deltacollege.edu
*Prgm Dir:* Michael L Walker, BA RT(R)
*Tel:* 209 468-6233   *Fax:* 209 468-6038
*E-mail:* mwalker@sjgh.org
*Med Dir:* Dennis Jacobsen, MD

**El Camino College**
Radiography Prgm
16007 S Crenshaw Blvd
Torrance, CA 90506
*Prgm Dir:* Dawn N Guzman, MEd RT(R)(M)
*Tel:* 310 660-3247   *Fax:* 310 660-3806
*E-mail:* dguzman@elcamino.edu

**LA County Harbor UCLA Medical Center**
Radiography Prgm
1000 W Carson St, Box 27
PO Box 2910
Torrance, CA 90509-2910
*Prgm Dir:* Tuyen T Bui, BS RT(R)
*Tel:* 310 222-2825   *Fax:* 310 618-9500
*E-mail:* tubui@ladhs.org
*Med Dir:* C Mark Mehringer, MD

**Mt San Antonio College**
Radiography Prgm
1100 N Grand Ave
Walnut, CA 91789-1399
*Prgm Dir:* David L McLaughlin, BS RT(R)
*Tel:* 909 594-5611, Ext 4527   *Fax:* 909 468-3938
*E-mail:* dmclaugh@mtsac.edu
*Med Dir:* A Franklin Turner, MD

# Colorado

**Red Rocks Community College**
Radiography Prgm
5420 Miller St
Arvada, CO 80002
www.rrcc.edu/medicalimaging
*Prgm Dir:* John F Trombly, BSRS
*Tel:* 303 914-6034   *Fax:* 303 420-7572
*E-mail:* john.trombly@rrcc.edu
*Med Dir:* Robert Hunter, MD

**Concorde Career Institute - Aurora**
Radiography Prgm
111 N Havana St
Aurora, CO 80010
www.concorde.edu
*Prgm Dir:* Richard M Crabb, MPA RT(R)(MR)
*Tel:* 720 207-0499   *Fax:* 303 839-5478
*E-mail:* rcrabb@concorde.edu

**Memorial Hospital**
Radiography Prgm
1400 E Boulder
175 S Union, Ste 240
Colorado Springs, CO 80909
www.memorialhospital.com
*Prgm Dir:* Elaine R Ivan, BS RT(R)(M)
*Tel:* 719 365-8291   *Fax:* 719 365-5374
*E-mail:* elaine.ivan@memhospcs.org
*Med Dir:* Bert Carollo, MD

**Centura Health-St Anthony Hospitals**
Radiography Prgm
1601 Lowell Blvd
Denver, CO 80204-1597
*Prgm Dir:* Wayne J Stellick, MA Ed RT(R)
*Tel:* 303 899-5265   *Fax:* 303 899-5134
*E-mail:* waynestellick@centura.org
*Med Dir:* Robert Seigel, MD

**Community College of Denver - Lowry Campus**
Radiography Prgm
1070 Alton Way, Bldg 849
Denver, CO 80230
*Prgm Dir:* Donna R Kimrey, RT(R)
*Tel:* 303 365-8379   *Fax:* 303 365-8396
*E-mail:* donna.kimrey@ccd.edu

**Pima Medical Institute - Denver**
Radiography Prgm
1701 W 72nd Ave
Denver, CO 80221
*Prgm Dir:* Ryan Z Minic, MPH RT(R)
*Tel:* 303 426-1800   *Fax:* 303 450-4048
*E-mail:* rminic@pmi.edu
*Med Dir:* Stephen Miller, MD

**Mesa State College**
Radiography Prgm
1100 North Ave
Grand Junction, CO 81501
www.mesastate.edu
*Prgm Dir:* Bette Schans, PhD RT(R)
*Tel:* 970 248-1651   *Fax:* 970 248-1133
*E-mail:* bschans@mesastate.edu

**Aims Community College**
Radiography Prgm
5401 W 20th St, PO Box 69
Greeley, CO 80632
*Prgm Dir:* Hollly A Knaub, BAS RT(R)
*Tel:* 970 339-6324   *Fax:* 970 506-6963
*E-mail:* holly.knaub@aims.edu
*Med Dir:* Mark Howshar, MD

# Connecticut

**St Vincent's College**
Radiography Prgm
2800 Main St
Bridgeport, CT 06606
*Prgm Dir:* Terry K Hine, MAT RT(R)(M)
*Tel:* 203 576-5072   *Fax:* 203 581-6533
*E-mail:* thine@stvincentscollege.edu
*Med Dir:* Robert D Russo, MD

**Danbury Hospital**
Radiography Prgm
24 Hospital Ave
Danbury, CT 06810
*Prgm Dir:* Jay Meyerson, MBA RT(R)
*Tel:* 203 797-7182   *Fax:* 203 739-8944
*E-mail:* jay.meyerson@danhosp.org
*Med Dir:* Stuart A Sherman, MD

**Quinnipiac University**
Radiography Prgm
Mt Carmel Ave
Hamden, CT 06518-0008
www.quinnipiac.edu
*Prgm Dir:* William F Hennessy, MHS RT(R)(M)(QM)
*Tel:* 203 582-5214   *Fax:* 203 582-8706
*E-mail:* Bill.Hennessy@quinnipiac.edu

**Capital Community College**
Radiography Prgm
950 Main St
Hartford, CT 06103-1207
*Prgm Dir:* Paul Creech, MPH JD RT(R)
*Tel:* 860 906-5155   *Fax:* 860 906-5148
*E-mail:* pcreech@ccc.commnet.edu
*Med Dir:* Michael Twohig, MD

**Hartford Hospital**
Radiography Prgm
560 Hudson St
Hartford, CT 06106
www.harthosp.org/Education/ProfEd/AlliedHealth/
   rtech.htm
*Prgm Dir:* Pamela M Cooke, BS RT(R)(M)
*Tel:* 860 545-3955   *Fax:* 860 545-6461
*E-mail:* pcooke@harthosp.org
*Med Dir:* Stuart K Markowitz, MD

**Middlesex Community College**
Radiography Prgm
100 Training Hill Rd
Middletown, CT 06457
*Prgm Dir:* Elaine Lisitano, BA RT(R)(M)
*Tel:* 860 344-6505   *Fax:* 860 358-8887
*E-mail:* elaine_lisitano@midhosp.org
*Med Dir:* Michael Crain, MD

**Gateway Community College**
Radiography Prgm
88 Bassett Rd
North Haven, CT 06473
*Prgm Dir:* Julie A Austin, BS RT(R)(M)
*Tel:* 203 285-2382   *Fax:* 203 285-2400
*E-mail:* jaustin@gwcc.commnet.edu
*Med Dir:* Anne M Curtis, MD

**Stamford Hospital**
Radiography Prgm
Shelburne Rd, W Broad St, Box 9317
Stamford, CT 06904-9317
www.stamhealth.org
*Prgm Dir:* Dorothy A Saia, MA RT(R)(M)
*Tel:* 203 276-7877   *Fax:* 203 276-7352
*E-mail:* dsaia@stamhealth.org
*Med Dir:* Harvey Hecht, MD

**Naugatuck Valley Community College**
Radiography Prgm
750 Chase Pkwy
Waterbury, CT 06708
*Prgm Dir:* James P Pronovost, MS RT(R)
*Tel:* 203 575-8266   *Fax:* 203 575-8146
*E-mail:* jpronovost@nvcc.commnet.edu
*Med Dir:* Gerald Berg, MD

**University of Hartford**
Radiography Prgm
200 Bloomfield Ave
West Hartford, CT 06117
*Prgm Dir:* Peter W Kennedy, PhD
*Tel:* 860 768-4823   *Fax:* 860 768-5706
*E-mail:* pkennedy@hartford.edu
*Med Dir:* Sidney Ulreich, MD

**Windham Community Memorial Hospital**
Radiography Prgm
112 Mansfield Ave
Willimantic, CT 06226
www.wcmh.org
*Prgm Dir:* Mark K Patros, BS RT(R)
*Tel:* 860 456-6713   *Fax:* 860 456-6838
*E-mail:* mpatros@wcmh.org

# Delaware

**Delaware Tech & Comm Coll - Owens Campus**
Radiography Prgm
PO Box 610, Rte 18
Georgetown, DE 19947
*Prgm Dir:* David C Ludema, EdD RT(R)
*Tel:* 302 855-5934   *Fax:* 302 858-5460
*E-mail:* dludema@college.dtcc.edu

**Delaware Tech & Comm Coll - Wilmington**
Radiography Prgm
333 Shipley St
Wilmington, DE 19802
*Prgm Dir:* Theresa A Foy, MEd RT(R)
*Tel:* 302 765-4594   *Fax:* 302 765-4599
*E-mail:* tfoy@christianacare.org
*Med Dir:* Jackie Napoletano, MD

# District of Columbia

**University of the District of Columbia**
Radiography Prgm
4200 Connecticut Ave NW
Washington, DC 20008
*Prgm Dir:* McDonald G Kpadeh, MEd RT(R)
*Tel:* 202 274-5882   *Fax:* 202 274-5952
*E-mail:* kpadeh@udc.edu
*Med Dir:* Tracy M Walton, Jr, MD

**Washington Hospital Center**
Radiography Prgm
110 Irving St NW
Washington, DC 20010
*Prgm Dir:* Mitchell Bieber, BA RT(R)
*Tel:* 202 877-6434, Ext 7.6434   *Fax:* 202 877-8625
*E-mail:* Mitchell.Bieber@medstar.net
*Med Dir:* David C Grant, Jr, MD

# Florida

**West Boca Medical Center**
Radiography Prgm
21644 State Rd 7
Boca Raton, FL 33428
*Prgm Dir:* Raymond Mata, BS RT(R)
*Tel:* 561 488-8173   *Fax:* 561 488-8379
*E-mail:* Ray.Mata@tenethealth.com
*Med Dir:* Carl Rosenkrantz, MD

**Bethesda Memorial Hospital**
Radiography Prgm
2815 S Seacrest Blvd
Boynton Beach, FL 33435
*Prgm Dir:* Charles E Lockett Jr, BS RT(R)(CV)
*Tel:* 561 737-7733, Ext 4792   *Fax:* 561 734-2545
*E-mail:* Eddie.Lockett@BethesdaHealthcare.com
*Med Dir:* David O'Connor, MD

**Manatee Community College**
Radiography Prgm
5840 26th St W
Bradenton, FL 32406-1849
*Prgm Dir:* Patrick W Patterson, MS RT(R)(N)CNMT
*Tel:* 941 752-5245   *Fax:* 941 727-6443
*E-mail:* patterp@mccfl.edu

**Brevard Community College**
Radiography Prgm
1519 Clearlake Rd
Cocoa, FL 32934
www.brevardcc.edu
*Prgm Dir:* Susan A Sheehan, MS RT(R)(M)(QM)
*Tel:* 321 433-7591   *Fax:* 321 433-7599
*E-mail:* sheehans@brevardcc.edu

**Halifax Medical Center**
Radiography Prgm
303 N Clyde Morris Blvd
Daytona Beach, FL 32114
*Prgm Dir:* Darcie J Nethery, PhD RT(R)
*Tel:* 386 254-4075, Ext 3509   *Fax:* 386 254-4231
*E-mail:* darcie.nethery@halifax.org
*Med Dir:* Charles Burkett, MD

**Keiser College - Daytona Beach**
Radiography Prgm
1800 Business Park Blvd
Daytona Beach, FL 32114
www.keisercollege.edu
*Prgm Dir:* Gloria E Wyatt, BS RT(R)
*Tel:* 386 274-5060, Ext 108   *Fax:* 386 274-2725
*E-mail:* gwyatt@keisercollege.edu

**Keiser University**
Radiography Prgm
1500 NW 49th St
Ft Lauderdale, FL 33309
www.keisercollege.edu
*Prgm Dir:* Theresa S Reid-Paul, MBA HCM RT(R)
*Tel:* 954 776-4456, Ext 410   *Fax:* 954 776-5157
*E-mail:* terryr@keisercollege.edu

**MedVance Institute - Ft Lauderdale
   Campus/KIMC Investments Inc**
Radiography Prgm
Ft Lauderdale/KIMC Investments, Inc
4850 West Oakland Park Blvd, Ste 200
Ft Lauderdale, FL 33313
*Prgm Dir:* Cassandra D Forbes, MS RTr
*Tel:* 954 587-7100
*E-mail:* cassandra.forbes@medvance.edu

**Edison College**
Radiography Prgm
PO Box 60210
8099 College Parkway, SW
Ft Myers, FL 33906-6210
www.edison.edu
*Prgm Dir:* James E Mayhew, MS RT(R)(QM)
*Tel:* 239 489-9110   *Fax:* 239 985-8352
*E-mail:* jmayhew@edison.edu
*Med Dir:* Thomas Presbrey, MD

**Indian River Community College**
Radiography Prgm
3209 Virginia Ave
Ft Pierce, FL 34982
*Prgm Dir:* Gary Shaver, EdD RT (R)
*Tel:* 772 462-4368   *Fax:* 772 462-7816
*E-mail:* gshaver@ircc.edu
*Med Dir:* Ronald Beaton, MD

**Santa Fe Community College**
Radiography Prgm
3000 NW 83rd St
Gainesville, FL 32602-6200
*Prgm Dir:* Barbara A Konter, BSRS RT(R)
*Tel:* 352 395-5702   *Fax:* 352 395-5711
*E-mail:* bobbie.konter@sfcc.edu
*Med Dir:* Suzanne T Mastin, MD

**Mayo Clinic Jacksonville**
Radiography Prgm
4500 San Pablo Rd
Jacksonville, FL 32224
*Prgm Dir:* Myke Kudlas, MEd RT(R)(QM)
*Tel:* 904 953-8663   *Fax:* 904 953-2894
*E-mail:* kudlas.myke@mayo.edu
*Med Dir:* Laura Bancroft, MD

**PROGRAMS**

**Shands Jacksonville**
Radiography Prgm
School of Radiologic Technology
655 W Eighth St
Jacksonville, FL 32209
www.jax.ufl.edu
*Prgm Dir:* Thomas A Graham, BS RT(R)
*Tel:* 904 244-3274   *Fax:* 904 244-6070
*E-mail:* thomas.graham@jax.ufl.edu
*Med Dir:* H Martin Northup, MD

**St Vincent's Medical Center**
Radiography Prgm
1800 Barrs St
Jacksonville, FL 32204
www.stvincentshealth.com
*Prgm Dir:* Karen L Frank, MBA RT(R) CRA CHE
*Tel:* 904 308-8552   *Fax:* 904 308-5109
*E-mail:* Kfran002@jaxhealth.com
*Med Dir:* Daniel Donohue, MD

**Lakeland Regional Medical Center**
Radiography Prgm
1324 Lakeland Hills Blvd, PO Box 95448
Lakeland, FL 33805
http://lrmc.com
*Prgm Dir:* Lynn C Sadler, MSRS RT(R)(QM)
*Tel:* 863 687-1100, Ext 3769   *Fax:* 863 687-1471
*E-mail:* lynn.sadler@lrmc.com
*Med Dir:* Howard A Gorell, MD

**Keiser College - Melbourne Campus**
Radiography Prgm
900 S Babcock St
Melbourne, FL 32901
www.keisercollege.edu
*Prgm Dir:* Theresa D Roberts, MHS RT(R)(MR)
*Tel:* 321 409-4800, Ext 136   *Fax:* 321 725-3766
*E-mail:* troberts@keisercollege.edu

**Jackson Mem Medical Center**
Radiography Prgm
1611 NW 12th Ave
Miami, FL 33136-1094
https://www.um-jmh.org
*Prgm Dir:* Courtney P Glenn, MSM RT(R)
*Tel:* 305 585-6811   *Fax:* 305 585-6579
*E-mail:* cglenn@um-jmh.org
*Med Dir:* Rivera Alfonso, MD

**MedVance Institute**
Radiography Prgm
Miami/KIMC Investments, Inc
9035 Sunset Dr, Ste 200
Miami, FL 33173
*Prgm Dir:* Deborah Hughes, MBA RT(R)(M)
*Tel:* 305 596-5553
*E-mail:* deborah.hughes@medvance.edu

**Miami Dade College**
Radiography Prgm
Medical Center Campus
950 NW 20th St
Miami, FL 33127
www.mdc.edu
*Prgm Dir:* Merlinn R Zolfaghari, MS RT(R)
*Tel:* 305 237-4292   *Fax:* 305 237-4116
*E-mail:* Merlin.Zolfaghari@mdc.edu

**Marion County School of Radiologic Tech**
Radiography Prgm
1014 SW Seventh Rd
Ocala, FL 34474
www.mcctae.com
*Prgm Dir:* Timothy Richardson, BS RT(R)
*Tel:* 352 671-7222   *Fax:* 352 671-7216
*E-mail:* Timothy.Richardson@marion.k12.fl.us
*Med Dir:* Kerry B Raduns, MD

**Florida Hospital College of Health Sciences**
Radiography Prgm
671 Winyah Dr
Orlando, FL 32803
*Prgm Dir:* Genese M Gibson, MA RT(R)(M)(QM)
*Tel:* 407 895-7747, Ext 1077   *Fax:* 407 303-7820
*E-mail:* genese_gibson@fhchs.edu
*Med Dir:* Susan Lynne Rebsamen, MD

**University of Central Florida**
Radiography Prgm
HPA2, Rm 210K
4000 Central Florida Blvd
Orlando, FL 32816-2220
www.ucf.edu
*Prgm Dir:* Thomas J Edwards III, EdD RT(R)(QM)
*Tel:* 407 823-2174   *Fax:* 407 823-6138
*E-mail:* tedwards@mail.ucf.edu
*Med Dir:* George A Stanley, MD

**Valencia Community College**
Radiography Prgm
PO Box 3028
Orlando, FL 32802
www.valenciacc.edu/asdegrees/as.asp
*Prgm Dir:* Julie McCaughtry, BS RT(R)(CT)
*Tel:* 407 582-1192   *Fax:* 407 582-1278
*E-mail:* jmccaughtry@valenciacc.edu

**Palm Beach Community College**
Radiography Prgm
3160 PGA Blvd
Palm Beach Gardens, FL 33418-2893
www.pbcc.edu/radiography
*Prgm Dir:* Vicki E Shaver, EdD RT(R)
*Tel:* 561 207-5067   *Fax:* 561 207-5011
*E-mail:* shaverv@pbcc.edu
*Med Dir:* David Mullin, MD

**MedVance Institute - West Palm Beach Campus/KIMC Investments Inc**
Radiography Prgm
1630 S Congress Ave
Palm Springs, FL 33461
www.medvance.org
*Prgm Dir:* Julie M Slusser, BS RT(R)(M)
*Tel:* 561 304-3466, Ext 128   *Fax:* 561 423-9251
*E-mail:* julie.slusser@medvance.edu

**Gulf Coast Community College**
Radiography Prgm
School of Radiography
5230 W US Hwy 98
Panama City, FL 32401-1041
http://health.gulfcoast.edu
*Prgm Dir:* Dee Ann VanDerSchaaf, RT(R)
*Tel:* 850 913-3318   *Fax:* 850 747-3246
*E-mail:* dvanderschaaf@gulfcoast.edu
*Med Dir:* Avery Brinkley, MD

**Pensacola Junior College**
Radiography Prgm
Warrington Campus
5555 W Hwy 98
Pensacola, FL 32507-1097
www.pjc.edu
*Prgm Dir:* Marilyn K Coseo, EdD
*Tel:* 850 484-2305   *Fax:* 850 484-2390
*E-mail:* mcoseo@pjc.edu
*Med Dir:* Harry Cramer, MD

**Keiser University - Sarasota**
Radiography Prgm
6151 Lake Osprey Dr
Sarasota, FL 34240
www.keisercollege.edu
*Prgm Dir:* Kathleen M Drotar, MA Ed RT(R)(N)(T)
*Tel:* 941 907-3900   *Fax:* 941 907-2889
*E-mail:* kdrotar@keisercollege.edu

**Hillsborough Community College**
Radiography Prgm
PO Box 30030
Tampa, FL 33630-3030
www.hccfl.edu
*Prgm Dir:* Beth E Wyckoff, BS RT(R)
*Tel:* 813 253-7371   *Fax:* 813 253-7226
*E-mail:* bwyckoff@hccfl.edu
*Med Dir:* Stephen Stenzler, MD

**Polk Community College**
Radiography Prgm
999 Ave H NE
Winter Haven, FL 33881-4299
www.polk.edu
*Prgm Dir:* Barbara A Koontz, MA RT(R)(M)
*Tel:* 863 297-1010, Ext 5722   *Fax:* 863 297-1036
*E-mail:* bkoontz@polk.edu

# Georgia

**Albany Technical College**
Radiography Prgm
1704 S Slappey Blvd
Albany, GA 31708
*Prgm Dir:* Kelley R Allen, BS RT(R)
*Tel:* 912 430-3619   *Fax:* 912 430-5115
*E-mail:* kallen@albanytech.edu
*Med Dir:* Lorenzo Carson, MD

**Athens Technical College**
Radiography Prgm
800 US Hwy 29 N
Athens, GA 30601-1500
www.athenstech.edu
*Prgm Dir:* Gerald R Cummings, MBA RT(R)(QM)
*Tel:* 706 355-5052   *Fax:* 706 369-5753
*E-mail:* jcummings@athenstech.edu

**Emory University**
Radiography Prgm
Medical Imaging Program
PO Box 25901
Atlanta, GA 30322
http://radtech.radiology.emory.edu
*Prgm Dir:* Dawn Couch Moore, MMSc RT(R)
*Tel:* 404 727-3200   *Fax:* 404 712-7256
*E-mail:* dawn.moore@emoryhealthcare.org
*Med Dir:* Ellen Patrick, MD

**Grady Health System**
Radiography Prgm
80 Jesse Hill Jr Dr, SE
PO Box 26095
Atlanta, GA 30303-3050
www.gradyhealthsystem.org
*Prgm Dir:* Cheryl T Pressly, BMSc RT(R)
*Tel:* 404 616-3611   *Fax:* 404 616-3512
*E-mail:* cpressly@gmh.edu
*Med Dir:* Jack Fountain, MD

**North Metro Technical College**
Radiography Prgm
2000 S Park Pl
Atlanta, GA 30339
*Prgm Dir:* Deanne D Collins, BS RT(R)
*Tel:* 770 956-6606   *Fax:* 770 956-6601
*E-mail:* dcollins@northmetrotech.edu
*Med Dir:* James Tallman, MD

**University Hospital**
Radiography Prgm
1350 Walton Way
Augusta, GA 30901-3599
www.universityhealth.org
*Prgm Dir:* Patricia S Graham, MBA RT(R)
*Tel:* 706 774-8646   *Fax:* 706 774-5079
*E-mail:* pgraham@uh.org
*Med Dir:* Jimpsy Johnson, MD

**Coastal Georgia Community College**
Radiography Prgm
3700 Altama Ave
Brunswick, GA 31520-3644
www.cgcc.peachnet.edu
*Prgm Dir:* Dianne T Castor, BSEd RT(R)
*Tel:* 912 264-7381  *Fax:* 912 262-3283
*E-mail:* dcastor@cgcc.edu
*Med Dir:* Adriana Rodriquez, MD

**Columbus Technical College**
Radiography Prgm
928 Manchester Expwy
Columbus, GA 31904-6572
www.columbustech.edu
*Prgm Dir:* Patricia A Mansell, MS RT(R)(M)
*Tel:* 706 641-5277  *Fax:* 706 641-5284
*E-mail:* pmansell@Columbustech.edu

**Dalton State College**
Radiography Prgm
650 College Dr
Dalton, GA 30720
*Prgm Dir:* Susan West, MEd RT(R)
*Tel:* 706 272-2605  *Fax:* 706 272-2699
*E-mail:* swest@daltonstate.edu
*Med Dir:* Thomas Richey, MD

**DeKalb Medical Center**
Radiography Prgm
2701 N Decatur Rd
Decatur, GA 30033
www.dekalbmedicalcenter.org
*Prgm Dir:* Lashaun D Taylor, BS RT(R)
*Tel:* 404 501-5306  *Fax:* 404 501-1883
*E-mail:* Shaun_Taylor@dkmc.org

**West Central Technical College**
Radiography Prgm
4600 Timber Ridge Dr
Douglasville, GA 30135
www.westcentraltech.edu
*Prgm Dir:* S Paige Saylors, BMSc RT(R)(M)(QM)
*Tel:* 770 947-7222  *Fax:* 770 947-3818
*E-mail:* psaylors@westcentraltech.edu
*Med Dir:* Eugene Maso, MD

**Heart of Georgia Technical College**
Radiography Prgm
560 Pinehill Rd
Dublin, GA 31021
*Prgm Dir:* Roslyn Johnson, BAS RT(R)
*Tel:* 478 274-7882  *Fax:* 478 275-6642
*E-mail:* rjohnson@hgtc.org
*Med Dir:* Barry Parker, MD

**Griffin Technical College**
Radiography Prgm
501 Varsity Rd
Griffin, GA 30223-2042
www.griffintech.edu
*Prgm Dir:* Deborah R Dawson, BMSc RT(R)
*Tel:* 770 229-3225  *Fax:* 770 229-3294
*E-mail:* ddawson@griffintech.edu
*Med Dir:* Ronald Gay, MD

**West Georgia Technical College**
Radiography Prgm
303 Fort Dr
LaGrange, GA 30240
www.westgatech.edu
*Prgm Dir:* Wanda W Barbee, MEd RT(R)
*Tel:* 706 845-4323, Ext 4202  *Fax:* 706 845-4339
*E-mail:* wbarbee@westgatech.org
*Med Dir:* Lloyd C Brewton, MD

**Gwinnett Technical College**
Radiography Prgm
5150 Sugarloaf Pkwy
Lawrenceville, GA 30043
www.gwinnettTech.edu
*Prgm Dir:* James A Sass, MEd RT(R)(M)(QM)
*Tel:* 678 226-6326  *Fax:* 770 685-1338
*E-mail:* jsass@gwinnetttech.edu

**Moultrie Technical College**
Radiography Prgm
800 Veterans Pkwy N
Moultrie, GA 31788
*Prgm Dir:* Alfred L Jones, BS RT(R)
*Tel:* 229 891-7000  *Fax:* 229 891-7010
*E-mail:* ajones@moultrietech.edu
*Med Dir:* C Matthew Paine, MD

**Coosa Valley Technical College**
Radiography Prgm
1 Maurice Culberson Dr
Rome, GA 30161
*Prgm Dir:* Mark H Layne, MEd RT(R)(CT)(MR)
*Tel:* 706 295-6955  *Fax:* 706 295-6984
*E-mail:* mlayne@coosavalleytech.edu
*Med Dir:* John Hulsey, MD

**Armstrong Atlantic State University**
Radiography Prgm
11935 Abercorn St
Savannah, GA 31419
www.armstrong.edu
*Prgm Dir:* Sharyn D Gibson, EdD RT(R)
*Tel:* 912 927-5360  *Fax:* 912 921-5838
*E-mail:* gibsonsh@mail.armstrong.edu
*Med Dir:* Don Starr, MD

**Ogeechee Technical College**
Radiography Prgm
One Joe Kennedy Blvd
Statesboro, GA 30458
www.ogeecheetech.edu
*Prgm Dir:* Lynda Tinker, MA RT(R)(M)(CT)(QM)
*Tel:* 912 681-5500, Ext 1642  *Fax:* 912 871-1162
*E-mail:* ltinker@ogeechee.edu
*Med Dir:* Don R Connell, MD

**Valdosta Technical College**
Radiography Prgm
4089 Val Tech Rd
Valdosta, GA 31603
www.valdostatechnicalcollege.edu
*Prgm Dir:* Linda L Booth, BS RT(R)
*Tel:* 229 245-2461  *Fax:* 229 259-5567
*E-mail:* lbooth@valdostatech.edu
*Med Dir:* Charles F Hobby, MD

**Southeastern Technical College**
Radiography Prgm
3001 E 1st St
Vidalia, GA 30474
www.southeasterntech.edu
*Prgm Dir:* Tara Weldon Carter, RT(R)(M)(CT) RDMS
*Tel:* 912 538-3152  *Fax:* 912 538-3106
*E-mail:* tcarter@southeasterntech.edu
*Med Dir:* James C Madix, American Board of Radiology

**Middle Georgia Technical College**
Radiography Prgm
80 Cohen Walker Dr
Warner Robins, GA 31088
www.middlegatech.edu
*Prgm Dir:* Patricia Melnick, MEd RT(R)(M)
*Tel:* 478 988-6912  *Fax:* 478 988-6875
*E-mail:* pmelnick@middlegatech.edu

**Okefenokee Technical College**
Radiography Prgm
1701 Carswell Ave
Waycross, GA 31503
www.okefenokeetech.edu
*Prgm Dir:* Donna L Yeomans, RT(R)
*Tel:* 912 287-5837  *Fax:* 912 287-4865
*E-mail:* dyeomans@okefenokeetech.edu
*Med Dir:* Howard Griffin, MD

# Hawaii

**Kapi'olani Community College**
Radiography Prgm
4303 Diamond Head Rd
Honolulu, HI 96816
*Prgm Dir:* Harry K Nakayama, BS RT(R)
*Tel:* 808 734-9251  *Fax:* 808 734-9126
*E-mail:* hnakayam@hawaii.edu
*Med Dir:* Robert DiMauro, MD

# Idaho

**Boise State University**
Radiography Prgm
College of Health Sciences
1910 University Dr
Boise, ID 83725
http://radsci.boisestate.edu
*Prgm Dir:* Darlene K Travis, BS RT(R)(T)(CV)
*Tel:* 208 426-3290  *Fax:* 208 426-4459
*E-mail:* dtravis@boisestate.edu
*Med Dir:* John Truska, MD

**College of Southern Idaho**
Radiography Prgm
315 Falls Ave
Twin Falls, ID 83303-1238
www.csi.edu/l4.asp?radiology
*Prgm Dir:* O Gary Lauer, PhD RT(R)
*Tel:* 208 732-6719  *Fax:* 208 736-4743
*E-mail:* glauer@csi.edu

# Illinois

**Southwestern Illinois College**
Radiography Prgm
2500 Carlyle Rd
Belleville, IL 62221-9989
*Prgm Dir:* Rhonda K Kern, MBA RT(R)
*Tel:* 618 235-2700, Ext 5303  *Fax:* 618 235-2052
*E-mail:* rhonda.kern@swic.edu
*Med Dir:* John Mattingly, MD

**Kaskaskia College**
Radiography Prgm
27210 College Rd
Centralia, IL 62801
www.kaskaskia.edu
*Prgm Dir:* Mimi L Polczynski, MS ED RT(R)(M)(CT)
*Tel:* 618 545-3000, Ext 3363  *Fax:* 618 545-3028
*E-mail:* mpolczynski@kc.cc.il.us
*Med Dir:* Richard Rudman, MD

**Parkland College**
Radiography Prgm
2400 W Bradley Ave
Champaign, IL 61821-1899
*Prgm Dir:* Kimberly A Mills, BS RT(R)(CT)
*Tel:* 217 351-2436  *Fax:* 217 373-3830
*E-mail:* kmills@parkland.edu
*Med Dir:* Thomas Deschler, MD

## Advocate Illinois Masonic Medical Center
Radiography Prgm
836 W Wellington Ave
Chicago, IL 60657
www.advocatehealth.com
*Prgm Dir:* Philis George, BS RT(R)(M)
*Tel:* 773 296-8951    *Fax:* 773 296-8960
*E-mail:* philis.george@advocatehealth.com
*Med Dir:* S Nadimpalli, MD

## Advocate Trinity Hospital
Radiography Prgm
2320 E 93rd St
Chicago, IL 60617
*Prgm Dir:* Joann Kern, BS RT(R)
*Tel:* 773 967-5292    *Fax:* 773 967-3954
*E-mail:* joann.kern@advocatehealth.com
*Med Dir:* Ari Mintz, MD

## Malcolm X College
Radiography Prgm
1900 W Van Buren St
Chicago, IL 60612
www.ccc.edu/malcolmx
*Prgm Dir:* Geraldine Williams, MEd RT(R)
*Tel:* 312 850-7373    *Fax:* 312 850-7372
*E-mail:* glwilliams@ccc.edu
*Med Dir:* Henry Wiggins, MD

## Wright College
Radiography Prgm
4300 N Narragansett Ave
Chicago, IL 60634
*Prgm Dir:* Dennis M King, MHS RT(R)
*Tel:* 773 481-8880    *Fax:* 773 481-8892
*E-mail:* dking@ccc.edu

## Danville Area Community College
Radiography Prgm
2000 E Main St
Danville, IL 61832
www.dacc.edu
*Prgm Dir:* Alberto Bello Jr, MEd RT(R)(CV)
*Tel:* 217 443-8552    *Fax:* 217 443-8595
*E-mail:* abello@dacc.edu
*Med Dir:* B J Rao, MD

## Sauk Valley Community College
Radiography Prgm
173 Illinois Rte 2
Dixon, IL 61021-9112
www.svcc.edu
*Prgm Dir:* Stan Shippert, MEd RT(R)(MR)(CT)
*Tel:* 815 288-5511, Ext 342    *Fax:* 815 288-5651
*E-mail:* shippes@svcc.edu
*Med Dir:* Krishna Chadalavada, MD

## St Francis Hospital
*Cosponsor:* Resurrection Health Care
Radiography Prgm
355 Ridge Ave
Evanston, IL 60202
www.reshealthcare.org
*Prgm Dir:* Mary Ellen Newton, MS RT(R)(M)
*Tel:* 847 316-5810    *Fax:* 847 316-5811
*E-mail:* mnewton@reshealthcare.org
*Med Dir:* Thomas G Cronin Jr, MD

## Carl Sandburg College
Radiography Prgm
2400 Tom L Wilson Blvd
Galesburg, IL 61401
www.sandburg.edu
*Prgm Dir:* Jan Jacobs, MA RT(R)
*Tel:* 309 341-5461    *Fax:* 309 341-5429
*E-mail:* janjacobs@sandburg.edu
*Med Dir:* Subbia Jagannathan, MD

## College of DuPage
Radiography Prgm
425 N Fawell Blvd
Glen Ellyn, IL 60137-6599
www.cod.edu
*Prgm Dir:* Gina M Carrier, MBA MS RT(R)
*Tel:* 630 942-2434    *Fax:* 630 858-5409
*E-mail:* carrier@cdnet.cod.edu
*Med Dir:* Daniel E Horan, MD

## College of Lake County
Radiography Prgm
19351 W Washington St
Grayslake, IL 60030-1198
www.clcillinois.edu
*Prgm Dir:* Lynn Platz Wiechert, MS RT(R)(CT)
*Tel:* 847 543-2880    *Fax:* 847 543-3880
*E-mail:* lwiechert@clcillinois.edu
*Med Dir:* Linda Sherbahn, MD

## McDonough District Hospital
Radiography Prgm
525 E Grant St
Macomb, IL 61455
*Prgm Dir:* Richard L Hart, RT(R)
*Tel:* 309 833-4101, Ext 2462    *Fax:* 309 836-1551
*E-mail:* rlhart@mdh.org
*Med Dir:* Ronald E Rigdon, MD

## Kishwaukee College
Radiography Prgm
21193 Malta Rd
Malta, IL 60150-9699
www.kishwaukeecollege.edu
*Prgm Dir:* Carol Guschl, BS RT(R)(QM)
*Tel:* 815 825-2086, Ext 257    *Fax:* 815 825-2072
*E-mail:* guschl@kishwaukeecollege.edu

## Bloomington-Normal School of Radiography
Radiography Prgm
900 Franklin Ave
Normal, IL 61761
*Prgm Dir:* Beth S Kuhfuss, MS RT(R)
*Tel:* 309 452-2834    *Fax:* 309 392-2835
*E-mail:* panacea@trianglenet.net
*Med Dir:* Richard Puckett, MD

## Olney Central College
Radiography Prgm
305 N West St
Olney, IL 62450
www.iecc.edu
*Prgm Dir:* Carol Kocher, BA RT(R)(M)
*Tel:* 618 395-7777, Ext 2239    *Fax:* 618 392-3293
*E-mail:* kocherc@iecc.edu
*Med Dir:* Kevin Wright, MD

## Moraine Valley Community College
Radiography Prgm
10900 S 88th Ave
Palos Hills, IL 60465
*Prgm Dir:* Linda L Metz, MEd RT(R)
*Tel:* 708 974-4300    *Fax:* 708 974-0185
*E-mail:* metz@morainevalley.edu
*Med Dir:* Irving Fuld, MD

## Illinois Central College
Radiography Prgm
Thomas K Thomas Bldg
201 SW Adams St
Peoria, IL 61635-0001
www.icc.edu
*Prgm Dir:* Diane L Schulz, MEd RT(R) FASRT
*Tel:* 309 999-4659    *Fax:* 309 673-9626
*E-mail:* dschulz@icc.edu
*Med Dir:* George A Gentry, MD

## OSF St Francis Medical Center
Radiography Prgm
530 NE Glen Oak Ave
Peoria, IL 61637
www.osfsaintfrancis.org
*Prgm Dir:* Suzanne M Yezek, MS RT(R)(M)
*Tel:* 309 655-2782    *Fax:* 309 655-2172
*E-mail:* Suzanne.M.Yezek@osfhealthcare.org
*Med Dir:* Clinton J Wentz, MD

## Blessing Hospital
Radiography Prgm
Box 7005, Broadway at 11th St
Quincy, IL 62305-7005
www.blessinghospital.org
*Prgm Dir:* Barbara A Mayfield, BS RT(R)
*Tel:* 217 223-8400, Ext 6161    *Fax:* 217 223-6898
*E-mail:* bmayfield@blessinghospital.com
*Med Dir:* Tony Rodriguez, MD

## Triton College
Radiography Prgm
2000 N Fifth Ave
River Grove, IL 60171
*Prgm Dir:* Catherine T Lekostaj, MA RT(R)
*Tel:* 708 456-0300, Ext 3370    *Fax:* 708 583-3336
*E-mail:* clekosta@triton.edu

## Trinity College of Nursing & Health Sciences
Radiography Prgm
2122 25th Ave
Rock Island, IL 61201-5317
www.trinitycollegeqc.edu
*Prgm Dir:* Elaine L Foht, MS Ed RT(R)
*Tel:* 309 779-7754    *Fax:* 309 779-7796
*E-mail:* fohte@trinityqc.com

## Rockford Memorial Hospital
Radiography Prgm
2400 N Rockton Ave
Rockford, IL 61103
www.rhsnet.org
*Prgm Dir:* Patricia Griesman, MS RT(R)
*Tel:* 815 971-5480    *Fax:* 815 968-3407
*E-mail:* pgriesman@rhsnet.org
*Med Dir:* Anthony Murray, MD

## Swedish American Hospital
Radiography Prgm
1401 E State St
Rockford, IL 61104-2315
www.swedishamerican.org
*Prgm Dir:* Debra L Letizio, MA RT(R)(CV)(QM)
*Tel:* 815 489-4966    *Fax:* 815 489-4975
*E-mail:* dletizio@swedishamerican.org
*Med Dir:* Mark Traill, MD

## South Suburban College
Radiography Prgm
15800 S State St
South Holland, IL 60473
*Prgm Dir:* Margaret Radlowski, MA RT(R)(M)
*Tel:* 708 596-2000, Ext 2261    *Fax:* 708 210-5792
*E-mail:* mradlowski@southsuburbancollege.edu
*Med Dir:* Neal A Rosner, MD

## Lincoln Land Community College
Radiography Prgm
Shepherd Rd
PO Box 19256
Springfield, IL 62794-9256
www.llcc.edu
*Prgm Dir:* William J Callaway, BA RT(R)
*Tel:* 217 786-2408, Ext 217    *Fax:* 217 786-2824
*E-mail:* bill.callaway@llcc.edu

# Indiana

**Columbus Regional Hospital**
Radiography Prgm
2400 E 17th St
Columbus, IN 47201
www.crh.org
*Prgm Dir:* Karen A Frazier, MS RT(R)
*Tel:* 812 376-5354   *Fax:* 812 376-5988
*E-mail:* kfrazier@crh.org
*Med Dir:* Richard L Pitman, MD

**University of Southern Indiana**
Radiography Prgm
8600 University Blvd
Evansville, IN 47712
*Prgm Dir:* Martin A Reed, PhD RT(R)
*Tel:* 812 464-1894   *Fax:* 812 465-7092
*E-mail:* mreed@usi.edu
*Med Dir:* Christina Nazenin Shinaver, MD

**Ft Wayne School of Radiography**
Radiography Prgm
700 Broadway
Ft Wayne, IN 46802
*Prgm Dir:* Elizabeth Ann Lewis, BS RT(R)
*Tel:* 260 425-3990   *Fax:* 260 425-3887
*E-mail:* fwsr@fwi.com
*Med Dir:* Michael Kinzer, MD

**University of Saint Francis**
Radiography Prgm
2701 Spring St
Ft Wayne, IN 46808
*Prgm Dir:* Donna J Lyke, MS RT(R)(M)(QM)
*Tel:* 260 434-7671   *Fax:* 260 434-7697
*E-mail:* dlyke@sf.edu
*Med Dir:* John Rock, MD

**Indiana University Northwest**
Radiography Prgm
3400 Broadway
Gary, IN 46408-1197
*Prgm Dir:* Arlene Adler, MEd RT(R) FAERS
*Tel:* 219 980-6540   *Fax:* 219 980-6649
*E-mail:* aadler@iun.edu

**Hancock Memorial Hospital**
Radiography Prgm
801 N State St
Greenfield, IN 46140
www.hancockregional.org
*Prgm Dir:* Vaughn Sutton, BS RT(R)
*Tel:* 317 468-4468   *Fax:* 317 468-4549
*E-mail:* vsutton@hancockregional.org

**Ball State University**
Radiography Prgm
1701 N Senate Blvd
Wile Hall 643
Indianapolis, IN 46202
www.bsu.edu/physiology-health/radiography
*Prgm Dir:* Donna L Long, MSM RT(R)(M)(QM)
*Tel:* 317 962-3284   *Fax:* 317 962-2102
*E-mail:* dlong2@clarian.org

**Community Health Network**
Radiography Prgm
1500 N Ritter Ave
Indianapolis, IN 46219
www.ecommunity.com/radiologyschool
*Prgm Dir:* Meryem Cole, BS RT(R)
*Tel:* 317 355-5867   *Fax:* 317 351-7733
*E-mail:* mcole@ecommunity.com
*Med Dir:* Gordon McLaughlin, MD

**Indiana University**
Radiography Prgm
541 Clinical Dr 120
Indianapolis, IN 46202-5111
www.indyrad.iupui.edu/public/radsci/radhome/
radhome.htm
*Prgm Dir:* Bruce W Long, MS RT(R)(CV) FASRT
*Tel:* 317 274-5254   *Fax:* 317 274-4074
*E-mail:* blong@iupui.edu
*Med Dir:* Valerie Jackson, MD

**Ivy Tech Community College of Indiana**
Radiography Prgm
9301 East 59th St
Indianapolis, IN 46216
www.ivytech.edu/indianapolis
*Prgm Dir:* Ann Sisel, MS RT(R)(M)
*Tel:* 317 921-4438   *Fax:* 317 546-3808
*E-mail:* asisel@ivytech.edu

**St Vincent Health/St Joseph Hospital**
Radiography Prgm
2001 W 86th St
Indianapolis, IN 46260
www.stvincent.org/education/radiography
*Prgm Dir:* Mark E Adkins, MS Ed RT(R)(QM)
*Tel:* 317 338-3879
*E-mail:* meadkins@stvincent.org
*Med Dir:* Timothy Davis, MD, PhD

**Indiana University - Kokomo**
Radiography Prgm
2300 S Washington St
Kokomo, IN 46902
www.iuk.edu/academics/healthsci/Radiography.htm
*Prgm Dir:* John O Hughey, MS RT(R)
*Tel:* 765 455-9329   *Fax:* 765 455-9310
*E-mail:* johughey@iuk.edu
*Med Dir:* William Harvey, MD

**King's Daughters' Hospital & Health Services**
Radiography Prgm
One King's Daughter's Dr
PO Box 447
Madison, IN 47250
*Prgm Dir:* Bobbi J Klein, MS RT(R)
*Tel:* 812 265-5211, Ext 50633   *Fax:* 812 265-0184
*E-mail:* kleinb@kdhhs.org
*Med Dir:* William Skiles, MD

**Ivy Tech Community College - Marion**
Radiography Prgm
1015 E 3rd St
Marion, IN 46952
*Prgm Dir:* Debra Dillman, BS RT(R)
*Tel:* 765 651-3111
*E-mail:* ddillman@ivytech.edu

**Ball Memorial Hospital**
Radiography Prgm
2401 University Ave
Muncie, IN 47303-3499
www.cardina/healthsystem.org/radschool.html
*Prgm Dir:* Susan J Hinds, RT(R)
*Tel:* 765 747-4372   *Fax:* 765 747-4415
*E-mail:* shinds@chsmail.org
*Med Dir:* Charles Liephart, MD

**Reid Hospital & Health Care Services**
Radiography Prgm
1401 Chester Blvd
Richmond, IN 47374
www.reidhosp.com
*Prgm Dir:* Roger A Preston, MSRS RT(R)
*Tel:* 765 983-3167   *Fax:* 765 983-3176
*E-mail:* prestoro@reidhosp.com
*Med Dir:* William Cory Gray, MD

**Indiana University South Bend**
Radiography Prgm
1700 Mishawaka Ave, PO Box 7111
South Bend, IN 46634
*Prgm Dir:* James H Howard, MSEd RT(R)
*Tel:* 574 520-5569   *Fax:* 574 520-5576
*E-mail:* jhhoward@iusb.edu
*Med Dir:* Jerrold Van Dyke, MD

**Ivy Tech Community College - Terre Haute**
Radiography Prgm
7999 US Hwy 41 S
Terre Haute, IN 47802-4898
www.ivytech7.cc.in.us/programs/rad.html
*Prgm Dir:* Lou Ann Wisbey, BS RT(R)(T)
*Tel:* 812 298-2242   *Fax:* 812 299-2441
*E-mail:* lwisbey@ivytech.edu
*Med Dir:* Ranganath Vedala, MD

**Good Samaritan Hospital**
Radiography Prgm
520 S Seventh St
Vincennes, IN 47591
*Prgm Dir:* Roger D Sterling, BS RT(R)
*Tel:* 812 885-8011   *Fax:* 812 885-3445
*E-mail:* rsterling@gshvin.org
*Med Dir:* John Mathis, DO

# Iowa

**Scott Community College**
Radiography Prgm
500 Belmont Rd
Bettendorf, IA 52722-5649
*Prgm Dir:* Donna M Collentine, MA RT(R)
*Tel:* 563 441-4262   *Fax:* 563 441-4204
*E-mail:* dcollentine@eicc.edu

**Mercy/St Luke's Hospital**
Radiography Prgm
PO Box 3026
1026 A Ave NE
Cedar Rapids, IA 52406-3026
www.isrt.org/mstl.htm
*Prgm Dir:* Dana D Schmitz, BS Ed RT(R)
*Tel:* 319 369-7077   *Fax:* 319 368-5721
*E-mail:* schmitdd@crstlukes.com
*Med Dir:* Elwood Stone, MD

**Jennie Edmundson Memorial Hospital**
Radiography Prgm
933 E Pierce St
Council Bluffs, IA 51503
*Prgm Dir:* Kristin E Schnitker, BA RT(R)
*Tel:* 712 396-6746   *Fax:* 712 396-6227
*E-mail:* kris.schnitker@nmhs.org

**Iowa Health - Des Moines**
Radiography Prgm
1200 Pleasant St
Des Moines, IA 50309-1453
www.iowahealth.org
*Prgm Dir:* Suzanne E Crandall, EdD RT(R)
*Tel:* 515 241-6883   *Fax:* 515 241-8015
*E-mail:* crandase@ihs.org

**Mercy College of Health Sciences**
Radiography Prgm
928 Sixth Ave
Des Moines, IA 50309
www.mchs.edu
*Prgm Dir:* Angela Sapp, BS RT(R)
*Tel:* 515 643-6739   *Fax:* 515 643-6698
*E-mail:* asapp@mercydesmoines.org
*Med Dir:* Michael Disbro, MD

### Iowa Central Community College
Radiography Prgm
330 Ave M
Ft Dodge, IA 50501
*Prgm Dir:* Chantel Gruver, BS RT(R) RDMS RVT
*Tel:* 515 576-0099, Ext 2302   *Fax:* 515 576-5656
*E-mail:* Gruver_c@iowacentral.com
*Med Dir:* Raymond D Schamel, MD

### University of Iowa Hospitals & Clinics
Radiography Prgm
Radiology, C-723 GH
200 Hawkins Dr
Iowa City, IA 52242-1077
www.radiology.uiowa.edu/radtech
*Prgm Dir:* Kathy Martensen, MA RT(R)
*Tel:* 319 356-4332   *Fax:* 319 384-9574
*E-mail:* kathy-martensen@uiowa.edu
*Med Dir:* Yutako Sato, MD

### Mercy Medical Center - North Iowa
Radiography Prgm
1000 Fourth St SW
Mason City, IA 50401
*Prgm Dir:* Janet L Moore, BS RT(R)(M)(QM)
*Tel:* 641 422-6079, Ext none   *Fax:* 641 422-5301
*E-mail:* moorejl@mercyhealth.com
*Med Dir:* Kristen White, MD

### Indian Hills Community College
Radiography Prgm
525 Grandview
Ottumwa, IA 52501
www.ihcc.cc.ia.us
*Prgm Dir:* Jeanette DeWitt, BS RT(R)(M)(CT)
*Tel:* 515 683-5164, Ext 5316   *Fax:* 515 683-5184
*E-mail:* jdewitt@ihcc.cc.ia.us
*Med Dir:* Elvin McCarl, MD

### Northeast Iowa Community College
Radiography Prgm
Peosta Campus
10250 Sundown Rd
Peosta, IA 52068
www.nicc.edu
*Prgm Dir:* Angela K Kronlage, MA RT(R)
*Tel:* 563 556-5110, Ext 311   *Fax:* 563 556-5058
*E-mail:* kronlagea@portal.nicc.edu

### St Luke's College
Radiography Prgm
2720 Stone Park Blvd
Sioux City, IA 51104
www.stlukescollege.edu
*Prgm Dir:* Deanna Butcher, MA RT(R)
*Tel:* 712 279-3734   *Fax:* 712 233-8017
*E-mail:* butchedl@stlukes.org

### Allen College
Radiography Prgm
1825 Logan Ave
Waterloo, IA 50703
www.allencollege.edu
*Prgm Dir:* Peggy S Fortsch, MA RT(R)
*Tel:* 319 226-2031   *Fax:* 319 226-2051
*E-mail:* fortscps@ihs.org

### Covenant Medical Center
Radiography Prgm
3421 W 9th St
Waterloo, IA 50702
www.covhealth.com/radiology_school.asp
*Prgm Dir:* Kenneth K Helfrick, BA RT(R)(QM)
*Tel:* 319 272-7296   *Fax:* 319 272-7105
*E-mail:* Helfrick.Ken@wfhc.org

# Kansas

### Ft Hays State University
Radiography Prgm
600 Park St
Hays, KS 67601-4099
*Prgm Dir:* Michael E Madden, PhD RT(R)(CT)(MR)
*Tel:* 785 628-5678   *Fax:* 785 628-4076
*E-mail:* mmadden@fhsu.edu

### Hutchinson Community College
Radiography Prgm
815 N Walnut
Hutchinson, KS 67501
www.hutchcc.edu
*Prgm Dir:* Renee Kautzer, MS RT(R)
*Tel:* 620 665-4954   *Fax:* 620 665-4988
*E-mail:* kautzerr@hutchcc.edu
*Med Dir:* Michael Schekall, MD

### Labette Community College
Radiography Prgm
200 S 14th St
Parsons, KS 67357
www.labette.edu
*Prgm Dir:* V June Downing, EdS RT(R)
*Tel:* 620 820-1158   *Fax:* 620 421-1539
*E-mail:* juned@labette.edu
*Med Dir:* Robert Gibbs, MD

### Washburn University
Radiography Prgm
1700 SW College Ave
Topeka, KS 66621
www.washburn.edu
*Prgm Dir:* Jera J Roberts, EdS RT(R)(M)
*Tel:* 785 670-2173   *Fax:* 785 670-1027
*E-mail:* jera.roberts@washburn.edu
*Med Dir:* Clay Harvey, MD

### Newman University
Radiography Prgm
3100 McCormick Ave
Wichita, KS 67213
www.newmanu.edu
*Prgm Dir:* Ronald Shipley, MS RT(R)
*Tel:* 316 942-4291, Ext 2351   *Fax:* 316 942-4483
*E-mail:* shipleyr@newmanu.edu
*Med Dir:* H David Clifton, MD

# Kentucky

### King's Daughters' Medical Center
Radiography Prgm
2201 Lexington Ave
Ashland, KY 41101
*Prgm Dir:* Krista M Lambert, BS RT(R)(MR)
*Tel:* 606 327-4637   *Fax:* 606 327-7425
*E-mail:* krista.lambert@kdmc.net
*Med Dir:* Chun Hong Kim, MD

### Bowling Green Technical College
Radiography Prgm
1845 Loop Dr
Bowling Green, KY 42101
www.bowlinggreen.kctcs.edu
*Prgm Dir:* Diane Button, MA RT(R)(M)
*Tel:* 270 901-1077   *Fax:* 270 901-1139
*E-mail:* diane.button@kctcs.edu

### Elizabethtown Community & Technical College
Radiography Prgm
620 College St Rd
Elizabethtown, KY 42701
www.elizabethtowncc.com
*Prgm Dir:* Penelope Logsdon, MA RT(R)
*Tel:* 270 706-8649   *Fax:* 270 766-5131
*E-mail:* penelope.logsdon@kctcs.edu
*Med Dir:* Stewart Couch, MD

### Hazard Community & Technical College
*Cosponsor:* Southeast Community College
Radiography Prgm
One Community College Dr
Hazard, KY 41701
www.hazard.kctcs.edu
*Prgm Dir:* Homer Terry, MS RT(R)(M)(QM)
*Tel:* 606 436-5721, Ext 73389   *Fax:* 606 439-1600
*E-mail:* homer.terry@kctcs.edu
*Med Dir:* Ashook Pated, MD

### Northern Kentucky University
Radiography Prgm
227 Albright Health Ctr
Highland Heights, KY 41099-2104
www.nku.edu/~nursing/radtechhome.html
*Prgm Dir:* Trina L Koscielicki, MEd RT(R)
*Tel:* 859 572-5477   *Fax:* 859 572-1314
*E-mail:* koscielicki@nku.edu
*Med Dir:* James L Schmitt, MD

### Bluegrass Community and Technical College
Radiography Prgm
164 Opportunity Way
Lexington, KY 40511-2653
www.central.kctcs.edu/radiology.htm
*Prgm Dir:* Robyn Jean Potter, BS RT(R)(M)
*Tel:* 859 246-2400, Ext 2283   *Fax:* 859 246-2504
*E-mail:* robyn.potter@kctcs.edu

### St Joseph Healthcare
Radiography Prgm
One St Joseph Dr
Lexington, KY 40504
www.saintjosephhealthcare.org
*Prgm Dir:* Robyn A Bradley, BS RT(R)(M)
*Tel:* 859 313-2282   *Fax:* 859 313-3104
*E-mail:* bradleyr@sjhlex.org
*Med Dir:* Christine Riley, MD

### Jefferson Community and Technical College
Radiography Prgm
109 E Broadway St
Louisville, KY 40202
www.jctc.kctcs.net/radtech/Faculty_&_Staff.dwt
*Prgm Dir:* Don Pack, EdD RT(R)
*Tel:* 502 629-3586   *Fax:* 502 629-2088
*E-mail:* don.pack@kctcs.edu
*Med Dir:* Nettie G King, MD

### Spencerian College
Radiography Prgm
4627 Dixie Hwy
Louisville, KY 40216
www.spencerian.edu
*Prgm Dir:* Mary Kaye Griffin, BSH
*Tel:* 502 447-1000, Ext x7846   *Fax:* 502 447-4574
*E-mail:* mgriffin@spencerian.edu
*Med Dir:* Leo Wine, MD

### Madisonville Community College
Radiography Prgm
750 N Laffoon St
Health Campus
Madisonville, KY 42431
*Prgm Dir:* Tonia R Gibson, BS RT(R)
*Tel:* 270 824-7552, Ext 182   *Fax:* 270 824-7069
*E-mail:* tonia.gibson@kctcs.edu
*Med Dir:* Phillip C Trover, MD

### Morehead State University
Radiography Prgm
150 University Blvd
Reed Hall 408
Morehead, KY 40351
www.morehead-st.edu
*Prgm Dir:* Barbara L Dehner, MSRS RT(R)(M)(CT) FAERS
*Tel:* 606 783-2651   *Fax:* 606 783-5051
*E-mail:* b.dehner@moreheadstate.edu

### Owensboro Community & Technical College
Radiography Prgm
4800 New Hartford Rd
Owensboro, KY 42303
www.octc.kctcs.edu
*Prgm Dir:* Nadine Joy Menser, MS RT(R)(T)
*Tel:* 270 686-4633   *Fax:* 270 686-4662
*E-mail:* joy.menser@kctcs.edu
*Med Dir:* Wayne Myers, MD

### West Kentucky Community & Technical College
Radiography Prgm
5200 Alben Barkley Dr, PO Box 7380
Paducah, KY 42002-7408
*Prgm Dir:* Deborah Smith, BS RT(R)(M)
*Tel:* 270 534-3479   *Fax:* 270 534-3498
*E-mail:* deborah.smith@kctcs.edu
*Med Dir:* Sharron D Butler, MD

### Southeast Kentucky Comm & Tech College
Radiography Prgm
3300 US Highway 25E South
Pineville, KY 40977
www.secc.kctcs.edu/AcademicAffairs/AlliedHealth/
   RadiographyP
*Prgm Dir:* Joseph Paul Hutson, MS RT(R)
*Tel:* 606 248-2118   *Fax:* 606 248-2166
*E-mail:* paul.hutson@kctcs.edu

### Somerset Community College
Radiography Prgm
808 Monticello St
Somerset, KY 42501
www.somerset.kctcs.edu/admissions/programs.htm
*Prgm Dir:* Doyle B Decker, BA RT(R)(CT)
*Tel:* 606 451-6774
*E-mail:* doyle.decker@kctcs.edu
*Med Dir:* William Baker, MD

### St Catharine College
Radiography Prgm
2735 Bardstown Rd
St Catharine, KY 40061
*Prgm Dir:* Donna J Crum, MS RT(R)(CT)
*Tel:* 859 336-5082, Ext 1333
*E-mail:* dcrum@sccky.edu

# Louisiana

### Baton Rouge General Medical Center
Radiography Prgm
3600 Florida Blvd, PO Box 2511
Baton Rouge, LA 70806
www.brgeneral.org/sort
*Prgm Dir:* Catherine W Lennier, RT(R)
*Tel:* 225 387-7157   *Fax:* 225 381-6168
*E-mail:* cathy.lennier@brgeneral.org
*Med Dir:* J Sidney Lawton, MD

### MedVance Institute
Radiography Prgm
Baton Rouge/KIMC Investments, Inc
9255 Interline Ave
Baton Rouge, LA 70809
*Prgm Dir:* Kacey A Roberts, BS RT(R)
*Tel:* 225 248-1015, Ext 217   *Fax:* 225 248-9517
*E-mail:* kroberts@medvance.org

### Our Lady of the Lake College
Radiography Prgm
7434 Perkins, Rd
Baton Rouge, LA 70808
www.ololcollege.edu
*Prgm Dir:* Debbie Gallerson, MEd RT(R)
*Tel:* 225 768-1737   *Fax:* 225 768-0819
*E-mail:* dgallers@ololcollege.edu
*Med Dir:* Robert C McReynolds Jr, MD

### Louisiana State University - Eunice
Radiography Prgm
PO Box 1129
Eunice, LA 70535
www.lsue.edu/radtech
*Prgm Dir:* Robert L McLaughlin, Jr, MA RT(R)
*Tel:* 337 550-1340   *Fax:* 337 550-1289
*E-mail:* rmclaugh@lsue.edu
*Med Dir:* John Higgins, MD

### North Oaks Medical Center
Radiography Prgm
PO Box 2668
42161 Veterans Ave
Hammond, LA 70404
www.northoaks.org
*Prgm Dir:* Marsha J Talbert, BA RT(R)
*Tel:* 985 345-9805   *Fax:* 985 345-9894
*E-mail:* talbertm@northoaks.org

### Lafayette General Medical Center
Radiography Prgm
1214 Coolidge Ave
PO Box 52009 OCS
Lafayette, LA 70505
www.lgmc.com
*Prgm Dir:* Charlotte S Powell, MS RS LRT
*Tel:* 337 289-8457   *Fax:* 337 289-8458
*E-mail:* cpowell@lgmc.com

### McNeese State University
Radiography Prgm
Box 92000
Lake Charles, LA 70609
www.faculty.mcneese.edu/gbradley
*Prgm Dir:* Gregory L Bradley, MEd RT(R)
*Tel:* 337 475-5657   *Fax:* 337 475-5664
*E-mail:* bradleyg1@aol.com
*Med Dir:* J R Romero, MD

### Career Technical College
Radiography Prgm
2319 Louisville Ave
Monroe, LA 71201
*Prgm Dir:* Tonya L Krone, MA RT(R)(M)
*Tel:* 318 998-5638
*E-mail:* tkrone@careertc.com

### University of Louisiana at Monroe
Radiography Prgm
700 University Ave
Monroe, LA 71209-0450
www.ulm.edu
*Prgm Dir:* George J Hicks, MS RT(R)
*Tel:* 318 342-3270   *Fax:* 318 342-1635
*E-mail:* hicks@ulm.edu
*Med Dir:* Bruce Golson, MD

### Delgado Community College
Radiography Prgm
615 City Park Ave
New Orleans, LA 70119-4399
www.dcc.edu
*Prgm Dir:* Carleen Boudreaux, MS RT(R)(M)
*Tel:* 504 483-4429   *Fax:* 504 483-4609
*E-mail:* cboudr@dcc.edu

### Our Lady of Holy Cross College
Radiography Prgm
1516 Jefferson Hwy
New Orleans, LA 70121
*Prgm Dir:* Carl J Tholen, BS RT(R)(CT)
*Tel:* 504 842-3705   *Fax:* 504 842-2459
*E-mail:* ctholen@ochsner.org
*Med Dir:* Edward Bluth, MD

### Northwestern State University
Radiography Prgm
College of Nursing
1800 Line Ave
Shreveport, LA 71101-4653
www.nsula.edu
*Prgm Dir:* Laura S Aaron, PhD RT(R)(M)(QM)
*Tel:* 318 677-3100   *Fax:* 318 677-3127
*E-mail:* carwilel@nsula.edu
*Med Dir:* Horacio D'Agostino, MD

### Southern Univ at Shreveport
Radiography Prgm
610 Texas St, #331
Shreveport, LA 71101
*Prgm Dir:* Shelia S Swift, BS RT(R)
*Tel:* 318 678-4646
*E-mail:* sswift@susla.edu

# Maine

### Eastern Maine Community College
Radiography Prgm
354 Hogan Rd
Bangor, ME 04401
www.emcc.edu
*Prgm Dir:* Susan Roeder, MEd RT(R)(N)
*Tel:* 207 974-4659   *Fax:* 207 974-4608
*E-mail:* sroeder@emcc.edu
*Med Dir:* David Warner, MD

### Kennebec Valley Community College
Radiography Prgm
92 Western Ave
Fairfield, ME 04937
*Prgm Dir:* Michelle Luciano, MC RT(R)
*Tel:* 207 453-5043
*E-mail:* mluciano@kvcc.me.edu

### Central Maine Medical Center
Radiography Prgm
300 Main St
Lewiston, ME 04240-0305
www.cmmc.org
*Prgm Dir:* Judith Ripley, BS RT(R)
*Tel:* 207 795-5974   *Fax:* 207 795-2476
*E-mail:* ripleyj@cmhc.org
*Med Dir:* David Simms, MD

### Mercy Hospital
Radiography Prgm
144 State St
Portland, ME 04101
*Prgm Dir:* Brenda M Rice, BS RT(R)
*Tel:* 207 879-3501   *Fax:* 207 879-2452
*E-mail:* riceb@mercyme.com
*Med Dir:* David Langdon, MD

### Southern Maine Community College
Radiography Prgm
Two Fort Rd
South Portland, ME 04106
www.smtc.edu
*Prgm Dir:* Sally A Doe, MS RT(R)
*Tel:* 207 741-5596   *Fax:* 207 741-5590
*E-mail:* sdoe@smccme.edu
*Med Dir:* James Place, MD

# Maryland

### Anne Arundel Community College
Radiography Prgm
101 College Pkwy
Arnold, MD 21012-1895
*Prgm Dir:* Thomas A Luby, Jr., MBA RT(R)(QM)
*Tel:* 410 777-7025   *Fax:* 410 777-7099
*E-mail:* taluby@aacc.edu
*Med Dir:* Mark Radovich, MD

**Greater Baltimore Medical Center**
Radiography Prgm
6701 N Charles St
Baltimore, MD 21204
www.gbmc.org/education/radiology
*Prgm Dir:* Brenda Schuette, BA RT(R)(M)(QM)
*Tel:* 443 849-2463   *Fax:* 443 849-2866
*E-mail:* bschuette@gbmc.org
*Med Dir:* Alexander Munitz, MD

**Johns Hopkins Hospital**
Radiography Prgm
600 N Wolfe St
Blalock B179
Baltimore, MD 21287
http://radiologycareers.rad.jhmi.edu
*Prgm Dir:* Sandra Moore, MA RT(R)(M)
*Tel:* 410 528-8210   *Fax:* 410 528-8308
*E-mail:* semoore@jhmi.edu
*Med Dir:* Jane Benson, MD

**Maryland General Hospital**
Radiography Prgm
827 Linden Ave
Baltimore, MD 21201
*Prgm Dir:* Francis E Potts, BA RT(R)(M)
*Tel:* 410 225-8750   *Fax:* 410 669-8710
*E-mail:* fpotts@marylandgeneral.org
*Med Dir:* Dharmendra Kumar, MD

**Comm Coll of Baltimore County - Essex Campus**
Radiography Prgm
7201 Rossville Blvd
Baltimore County, MD 21237-9987
www.ccbcmd.edu
*Prgm Dir:* Erin V Cordray, BA RT(R)
*Tel:* 410 780-6666   *Fax:* 410 780-6405
*E-mail:* ecordray@ccbcmd.edu

**Allegany College of Maryland**
Radiography Prgm
12401 Willowbrook Rd SE
Cumberland, MD 21502-2596
www.allegany.edu
*Prgm Dir:* Cathy A Kline, BA RT(R)
*Tel:* 301 784-5560   *Fax:* 301 784-5015
*E-mail:* ckline@allegany.edu
*Med Dir:* James K Benjamin, MD

**Hagerstown Community College**
Radiography Prgm
11400 Robinwood Dr
Hagerstown, MD 21742-6590
*Prgm Dir:* Brenda J Hassinger, MS RT(R)(M)
*Tel:* 301 790-2800, Ext 205   *Fax:* 301 739-0737
*E-mail:* hassingerb@hagerstowncc.edu
*Med Dir:* Francis Citro, MD

**Prince George's Community College**
Radiography Prgm
301 Largo Rd
Largo, MD 20774-2199
http://academic.pgcc.edu/alliedhealth
*Prgm Dir:* Angela Dopkowski Anderson, MA RT(R)(CT)(QM)
*Tel:* 301 322-0569   *Fax:* 301 386-7528
*E-mail:* aanderson@pgcc.edu
*Med Dir:* Edward Druy, MD

**Wor-Wic Community College**
Radiography Prgm
32000 Campus Dr
Salisbury, MD 21804
*Prgm Dir:* Junior A Gray, MBA RT(R)
*Tel:* 410 572-8741   *Fax:* 410 572-8730
*E-mail:* jgray@wor-wic.edu
*Med Dir:* Peter Libby, MD

**Holy Cross Hospital**
Radiography Prgm
1500 Forest Glen Rd
Silver Spring, MD 20910
www.holycrosshealth.org
*Prgm Dir:* Staci M Maier, BS RT(R)
*Tel:* 301 754-7889   *Fax:* 301 754-7373
*E-mail:* maiers@holycrosshealth.org
*Med Dir:* Howard DiPiazza, MD

**Montgomery College**
Radiography Prgm
7600 Takoma Ave
Takoma Park, MD 20912
*Prgm Dir:* Rose M Aehle, MS RT(R)(M)
*Tel:* 301 562-5564   *Fax:* 301 562-5558
*E-mail:* rose.aehle@montgomerycollege.edu
*Med Dir:* Ahalya Premkumar, MD

**Washington Adventist Hospital**
Radiography Prgm
7600 Carroll Ave
Takoma Park, MD 20912
*Prgm Dir:* Kristin Mitas, MS RT(R)
*Tel:* 301 891-6556   *Fax:* 301 891-6558
*E-mail:* kmitas@ahm.com
*Med Dir:* Harvey Esrov, MD

**Chesapeake College**
Radiography Prgm
PO Box 8
Wye Mills, MD 21679
www.chesapeake.edu
*Prgm Dir:* Linda Burchett Blythe, BSRS RT(R)
*Tel:* 410 822-5400, Ext 5927   *Fax:* 410 770-3764
*E-mail:* lblythe@chesapeake.edu
*Med Dir:* Stephen C Brigham, MD

# Massachusetts

**Middlesex Community College**
Radiography Prgm
Springs Rd
Bedford, MA 01730
*Prgm Dir:* William Darmody, MEd RT(R)(MR)
*Tel:* 781 280-3942   *Fax:* 781 275-4911
*E-mail:* darmodyw@middlesex.mass.edu
*Med Dir:* Kenneth Peelle, MD

**Bunker Hill Community College**
Radiography Prgm
Medical Imaging Program
250 New Rutherford Ave
Boston, MA 02129-2925
*Prgm Dir:* Donna Misrati, MBA RT(R)(CT)
*Tel:* 617 228-2197   *Fax:* 617 228-2052
*E-mail:* dmisrati@bhcc.mass.edu
*Med Dir:* Miriam Vincent, MD

**Mass College of Pharmacy & Health Sciences**
Radiography Prgm
179 Longwood Ave
Boston, MA 02115
www.mcphs.edu
*Prgm Dir:* Thomas G Sandridge, MS RT(R)
*Tel:* 617 735-1585   *Fax:* 617 732-2075
*E-mail:* tom.sandridge@mcphs.edu

**MGH Institute of Health Professions**
Radiography Prgm
Charlestown Navy Yard
36 1st Ave
Boston, MA 02129-4557
*Prgm Dir:* Richard Terrass, M Ed RT(R)
*Tel:* 617 726-0781
*E-mail:* rterrass@mghihp.edu

**Massasoit Community College**
Radiography Prgm
One Massasoit Blvd
Brockton, MA 02302
www.massasoit.mass.edu
*Prgm Dir:* Marianne E Spatola, MSEd RT(R)(M)
*Tel:* 508 588-9100, Ext 1764   *Fax:* 508 427-1262
*E-mail:* mspatola@massasoit.mass.edu
*Med Dir:* Samuel McFadden, MD

**North Shore Community College**
Radiography Prgm
One Ferncroft Rd, PO Box 3340
Danvers, MA 01923-0840
www.northshore.edu
*Prgm Dir:* Christine E Wiley, MEd RT(R)(M)
*Tel:* 978 762-4163   *Fax:* 978 762-4022
*E-mail:* cwiley@northshore.edu
*Med Dir:* Philip Thomason, MD

**Massachusetts Bay Community College**
Radiography Prgm
19 Flagg Dr
Framingham, MA 01702
www.massbay.edu
*Prgm Dir:* Micheal Glisson, MS RT(R)
*Tel:* 508 270-4031   *Fax:* 508 270-4131
*E-mail:* mglisson@massbay.edu
*Med Dir:* Nicholas Argy, MD, JD

**Holyoke Community College**
Radiography Prgm
303 Homestead Ave
Holyoke, MA 01040
*Prgm Dir:* Kathryn C Root, MEd RT(R)
*Tel:* 413 552-2460
*E-mail:* kroot@hcc.mass.edu
*Med Dir:* Howard Raymond, MD

**Northern Essex Community College**
Radiography Prgm
45 Franklin St
Lawrence, MA 01840
www.necc.mass.edu
*Prgm Dir:* Nancy Garcia, MEd RT(R)
*Tel:* 978 738-7493   *Fax:* 978 738-7146
*E-mail:* ngarcia@necc.mass.edu
*Med Dir:* Arthur Lathrop Zerbey III, MD

**Regis College**
Radiography Prgm
170 Governors Ave
Medford, MA 02155
www.lmregis.org
*Prgm Dir:* James V Lampka, MS RT(R)
*Tel:* 781 306-6658   *Fax:* 781 306-6710
*E-mail:* jlampka@lmh.edu
*Med Dir:* James Coleman, MD

**Springfield Technical Community College**
Radiography Prgm
One Armory Sq PO Box 9000
Springfield, MA 01101-9000
www.stcc.edu
*Prgm Dir:* Anthony J Kapadoukakis, PhD RT(R)
*Tel:* 413 755-4850   *Fax:* 413 733-0688
*E-mail:* akapadoukakis@stcc.edu
*Med Dir:* J Robert Kirkwood, MD

**Quinsigamond Community College**
Radiography Prgm
670 W Boylston St
Worcester, MA 01606-2092
www.qcc.mass.edu/radiography
*Prgm Dir:* Linda LeFave, Med RT(R)(M)(QM)
*Tel:* 508 854-4289   *Fax:* 508 852-6943
*E-mail:* lindal@qcc.mass.edu
*Med Dir:* Michael Popik, MD

# Michigan

**Washtenaw Community College**
Radiography Prgm
4800 E Huron River Dr
Ann Arbor, MI 48106-0978
www.wccnet.edu
*Prgm Dir:* Connie Foster, MA RT(R) RDMS
*Tel:* 734 973-3418   *Fax:* 734 677-5078
*E-mail:* cfoster@wccnet.edu
*Med Dir:* James Shields, MD

**Kellogg Community College**
Radiography Prgm
450 North Ave
Battle Creek, MI 49017-3397
www.kellogg.edu
*Prgm Dir:* Janis M Scholl, MA RT(R)(M)
*Tel:* 269 965-3931, Ext 2315   *Fax:* 269 565-2059
*E-mail:* schollj@kellogg.edu
*Med Dir:* Charles W O Dell Jr, MD

**Lake Michigan College**
Radiography Prgm
2755 E Napier Ave
Benton Harbor, MI 49022-1899
*Prgm Dir:* Kerry T Mohney, RT(R)(M) BS MA
*Tel:* 269 927-8100, Ext 5093   *Fax:* 269 927-8186
*E-mail:* mohney@lakemichigancollege.edu
*Med Dir:* Nathan Jordan, MD

**Ferris State University**
Radiography Prgm
901 S State St
Big Rapids, MI 49307-9989
*Prgm Dir:* Lisa L Wall, MEd RT(R)
*Tel:* 231 591-2326   *Fax:* 231 591-3788
*E-mail:* walll@ferris.edu
*Med Dir:* Craig Karsama, MD

**Henry Ford Community College**
Radiography Prgm
Health Careers Division
5101 Evergreen Rd
Dearborn, MI 48128
*Prgm Dir:* Sharon W Wu, BS RT(R)
*Tel:* 313 317-6595   *Fax:* 313 317-6569
*E-mail:* swu@hfcc.edu
*Med Dir:* David S Yates, MD

**Henry Ford Hospital**
Radiography Prgm
2799 W Grand Blvd
Detroit, MI 48202
*Prgm Dir:* Kathleen Kath, MS RT(R)
*Tel:* 313 916-1348   *Fax:* 313 916-3049
*E-mail:* kathy@rad.hfh.edu
*Med Dir:* William Paul Sanders, MD

**Sinai - Grace Hospital**
Radiography Prgm
6071 W Outer Dr
Detroit, MI 48235
*Prgm Dir:* Mary Elizabeth Beam, BS RT(R)
*Tel:* 313 966-6866   *Fax:* 313 966-1272
*E-mail:* mbeam@dmc.org
*Med Dir:* Burt T Weying III, MD

**St John Health**
Radiography Prgm
22101 Moross Rd
Detroit, MI 48236-2172
*Prgm Dir:* Denise R Allen, MBA RT(R)(M)(QM)
*Tel:* 313 343-4544   *Fax:* 313 343-3359
*E-mail:* denise.allen@stjohn.org
*Med Dir:* Rojanandham Samudrala, MD FACR

**Hurley Medical Center**
Radiography Prgm
One Hurley Plaza
Flint, MI 48503-5993
www.hurleymc.com
*Prgm Dir:* Dawn Sturk, MS RT(R)
*Tel:* 810 257-9835   *Fax:* 810 257-9009
*E-mail:* dsturk1@hurleymc.com
*Med Dir:* Apparao Mukkamala, MD

**Grand Rapids Community College**
Radiography Prgm
143 Bostwick St NE
Grand Rapids, MI 49503
*Prgm Dir:* John F Godisak, MA RT(R)
*Tel:* 616 234-4233   *Fax:* 616 234-4317
*E-mail:* jgodisak@grcc.edu

**Mid Michigan Community College**
Radiography Prgm
1375 S Clare Ave
Harrison, MI 48625-9447
www.midmich.edu
*Prgm Dir:* John Skinner, MEd MSA RT(R)
*Tel:* 989 386-6646   *Fax:* 989 386-9088
*E-mail:* jskinner@midmich.edu

**Lansing Community College**
Radiography Prgm
Dept 3100 Health & Human Service Careers PO Box
40010
Lansing, MI 48901-7210
www.lcc.edu
*Prgm Dir:* Brian W Pickford, MS RT(R)
*Tel:* 517 483-5379   *Fax:* 517 483-1508
*E-mail:* pickforb@lcc.edu
*Med Dir:* Gerald Aben, MD

**Marquette General Health System**
Radiography Prgm
420 W Magnetic Ave
Marquette, MI 49855
www.mgh.org
*Prgm Dir:* JoAnna Perucco, MS RT(R)(M)
*Tel:* 906 225-4916   *Fax:* 906 225-4943
*E-mail:* jlperucco@mgh.org
*Med Dir:* Steve Minn, MD

**Baker College of Muskegon**
Radiography Prgm
1903 Marquette Ave
Muskegon, MI 49442-1490
*Prgm Dir:* Cameron J Vander Stel, BS RT(R)
*Tel:* 231 777-5224
*E-mail:* cameron.vanderstel@baker.edu

**Baker College of Owosso**
Radiography Prgm
1020 S Washington St
Owosso, MI 48867-4400
www.baker.edu
*Prgm Dir:* Kathleen M Wallen, BS RT(R)
*Tel:* 989 729-3416   *Fax:* 989 729-3411
*E-mail:* kathleen.wallen@baker.edu

**Port Huron Hospital**
Radiography Prgm
1221 Pine Grove
PO Box 5011
Port Huron, MI 48061-5011
www.porthuronhospital.org
*Prgm Dir:* Monica S Rowling, BAS RT(R)(M)
*Tel:* 810 989-3163   *Fax:* 810 989-3197
*E-mail:* mrowling@porthuronhosp.org
*Med Dir:* David Tracy, MD

**William Beaumont Hospital**
Radiography Prgm
3601 W 13 Mile Rd
Royal Oak, MI 48073
www.beaumont.edu
*Prgm Dir:* Terese A Trost, MA RT(R)
*Tel:* 248 898-6048   *Fax:* 248 898-5015
*E-mail:* ttrost@beaumont.edu
*Med Dir:* Henrietta Juras, MD

**Oakland Community College - Bloomfield Hills**
Radiography Prgm
22322 Rutland Dr
Southfield, MI 48075-4793
www.oaklandcc.edu
*Prgm Dir:* Carolyn Nacy, BS RT(R) RDMS
*Tel:* 248 233-2918   *Fax:* 248 233-2891
*E-mail:* ceoneill@oaklandcc.edu
*Med Dir:* Edwards Michael, MD

**Providence Hospital**
Radiography Prgm
16001 W Nine Mile Rd, PO Box 2043
Southfield, MI 48037
www.realmedicine.org/Providence
*Prgm Dir:* Mary A Kleven, MAOM RT(R)(M)
*Tel:* 248 849-3293   *Fax:* 248 424-5395
*E-mail:* Mary.Kleven@stjohn.org
*Med Dir:* Thomas Hall, MD

**Delta College**
Radiography Prgm
1961 Delta Rd
University Center, MI 48710
*Prgm Dir:* Kathleen M Gavalas, MEd RT(R)
*Tel:* 517 686-9533   *Fax:* 517 667-2230
*E-mail:* kmgavala@delta.edu
*Med Dir:* Kristin Nelsen, MD

# Minnesota

**Riverland Community College**
Radiography Prgm
1600 8th Ave NW
Austin, MN 55912
*Prgm Dir:* Eugene D Frank, MA RT(R) FASRT
*Tel:* 507 433-0645   *Fax:* 507 433-0515
*E-mail:* gfrank@river.cc.mn.us
*Med Dir:* Richard Bergen, MD

**Lake Superior College**
Radiography Prgm
2101 Trinity Rd
Duluth, MN 55811-2741
www.lsc.edu
*Prgm Dir:* Nancy A Fredrickson, RT(R)
*Tel:* 218 725-7714   *Fax:* 218 723-4921
*E-mail:* n.fredrickson@lsc.edu
*Med Dir:* William Witrak, MD

**Argosy University/Twin Cities**
Radiography Prgm
1515 Central Parkway
Eagan, MN 55121
www.argosyu.edu
*Prgm Dir:* Deborah M Jambor, BS RT(R)
*Tel:* 651 846-3418
*E-mail:* djambor@argosyu.edu
*Med Dir:* Gregory Taylor, MD

**Northland Community & Technical College**
Radiography Prgm
2022 Central Ave NE
East Grand Forks, MN 56721-2702
*Prgm Dir:* Debra King, BS RT(R)
*Tel:* 218 773-3441, Ext 567   *Fax:* 218 773-4502
*E-mail:* deb.king@northlandcollege.edu
*Med Dir:* Mark Schneider, MD

**Fairview University**
Radiography Prgm
School of Radiologic Tech
6545 France Ave Ste 450
Edina, MN 55435
*Prgm Dir:* Linda A Dehrer-Wendt, BS RT(R)
*Tel:* 952 836-3544   *Fax:* 952 836-3925
*E-mail:* ldehrer@fairview.org

**College of St Catherine - Minneapolis**
Radiography Prgm
601 25th Ave S
Minneapolis, MN 55454
www.stkate.edu
*Prgm Dir:* Alan Bode, MA RT(R)(QM)
*Tel:* 651 690-7887   *Fax:* 651 690-7849
*E-mail:* ajbode@stkate.edu
*Med Dir:* Geoffrey Bodeau, MD

**Veterans Affairs Medical Center**
Radiography Prgm
One Veterans Dr
Minneapolis, MN 55417
*Prgm Dir:* Michael C Stori, MS RT(R)(CT)
*Tel:* 612 467-2546   *Fax:* 612 727-5635
*E-mail:* michael.stori@med.va.gov
*Med Dir:* Howard Ansel, MD

**North Memorial Medical Center**
Radiography Prgm
3300 N Oakdale
Robbinsdale, MN 55422
www.northmemorial.com
*Prgm Dir:* Joni E Gosch, MS RT(R)
*Tel:* 763 520-5337   *Fax:* 763 520-8770
*E-mail:* joni.gosch@northmemorial.com
*Med Dir:* Richard Belkin, MD

**Mayo School of Health Sciences**
Radiography Prgm
Siebens Rm 1119
200 First St SW
Rochester, MN 55905
www.mayo.edu/mshs
*Prgm Dir:* Beverly J Tupper, MS RT(R)(CV)
*Tel:* 507 284-3169   *Fax:* 507 284-0656
*E-mail:* tupper.beverly@mayo.edu
*Med Dir:* Eric J Lantz, MD

**St Cloud Hospital**
Radiography Prgm
1406 Sixth Ave N
St Cloud, MN 56303
www.centracare.com
*Prgm Dir:* John Falconer, BA RT(R)
*Tel:* 320 255-5719, Ext 53001   *Fax:* 320 255-5730
*E-mail:* falconerj@centracare.com
*Med Dir:* Ralph Fedor, MD

**Methodist Hospital**
Radiography Prgm
6500 Excelsior Blvd
St Louis Park, MN 55426
*Prgm Dir:* Linda C Olson, RT(R)
*Tel:* 952 993-5410   *Fax:* 952 993-6531
*E-mail:* olsonli@parknicollet.com

**Century College**
Radiography Prgm
3300 Century Ave N
White Bear Lake, MN 55110
www.century.edu
*Prgm Dir:* Diane J Fleury-Evans, MA RT(R)
*Tel:* 651 779-3334   *Fax:* 651 779-5779
*E-mail:* diane.fleury@century.edu

**Rice Memorial Hospital**
Radiography Prgm
301 Becker Ave SW
Willmar, MN 56201-3302
*Prgm Dir:* Luther Linn, RT(R)
*Tel:* 320 231-4553   *Fax:* 320 231-4865
*E-mail:* llinn@rice.willmar.mn.us

# Mississippi

**Northeast Mississippi Community College**
Radiography Prgm
Cunningham Blvd
Booneville, MS 38829
www.nemcc.edu
*Prgm Dir:* Jennifer C Davis, BSRT(R)(T)
*Tel:* 662 720-7364   *Fax:* 662 720-1165
*E-mail:* jcdavis@nemcc.edu
*Med Dir:* Michael Currie, MD

**Jones County Junior College**
Radiography Prgm
900 S Court St
Ellisville, MS 39437
www.jcjc.edu
*Prgm Dir:* Timothy S Cochran, MS RT(R)
*Tel:* 601 477-4159   *Fax:* 601 477-4152
*E-mail:* sandy.cochran@jcjc.edu
*Med Dir:* Clyde R Allen, MD

**Itawamba Community College**
Radiography Prgm
602 W Hill St
Fulton, MS 38843-0999
*Prgm Dir:* Deborah G Shell, MEd RT(R)
*Tel:* 662 862-8345   *Fax:* 662 862-8350
*E-mail:* dgshell@iccms.edu
*Med Dir:* Doug Clark, MD

**Mississippi Gulf Coast Community College**
Radiography Prgm
PO Box 100
Gautier, MS 39553
www.mgccc.edu
*Prgm Dir:* Judy S Lewis, BA RT(R)(M)(CT)(QM)
*Tel:* 228 497-9602, Ext 7710   *Fax:* 228 497-7676
*E-mail:* judy.lewis@mgccc.edu
*Med Dir:* Paul H Moore, MD

**Pearl River Community College**
Radiography Prgm
5448 US Hwy 49 S
Hattiesburg, MS 39401
*Prgm Dir:* C David Armstrong, MEd RT(R)
*Tel:* 601 554-5484   *Fax:* 601 554-5553
*E-mail:* darmstrong@prcc.edu
*Med Dir:* Mark Molpus, MD

**University of Mississippi Medical Center**
Radiography Prgm
2500 N State St
Jackson, MS 39216-4505
http://radtech.umc.edu
*Prgm Dir:* Mark R Gray, BA RT(R)
*Tel:* 601 984-2605   *Fax:* 601 984-4986
*E-mail:* mgray@radiology.umsmed.edu
*Med Dir:* Ramesh Patel, MD

**Meridian Community College**
Radiography Prgm
910 Hwy 19 N
Meridian, MS 39307
www.mcc.cc.ms.us
*Prgm Dir:* Seena Shazowee Edgerton, BSRT(R)(M)
*Tel:* 601 484-8609, Ext 609   *Fax:* 601 484-8704
*E-mail:* sedgerton@meridiancc.edu
*Med Dir:* Mary Ann Cowart, MD

**Mississippi Delta Community College**
Radiography Prgm
PO Box 668
Moorhead, MS 38761
*Prgm Dir:* Alice K Pyles, MS RS RT(R)
*Tel:* 601 246-6504   *Fax:* 601 246-6507
*E-mail:* apyles@msdelta.edu

**Hinds Community College**
Radiography Prgm
PMB 10458
Raymond, MS 39154-9799
www.hindscc.edu
*Prgm Dir:* Stephen C Compton, RT(R)
*Tel:* 601 371-3521   *Fax:* 601 371-3508
*E-mail:* sccompton@hindscc.edu
*Med Dir:* Gary Cirilli, MD

**Copiah-Lincoln Community College**
Radiography Prgm
PO Box 649
Wesson, MS 39191
www.colin.edu
*Prgm Dir:* Billie Faye Sartin, AAS RT(R)
*Tel:* 601 643-8496   *Fax:* 601 643-8214
*E-mail:* bfsartin@colin.edu
*Med Dir:* Prentis L Smith, MD

# Missouri

**Southeast Missouri Hospital**
Radiography Prgm
2001 William St, 2nd Floor
Cape Girardeau, MO 63703
www.sehcollege.org
*Prgm Dir:* Peter J Barger, MSEd RT(R)(CT)
*Tel:* 573 334-6825, Ext 30   *Fax:* 573 339-4628
*E-mail:* pbarger@sehosp.org
*Med Dir:* Mark Gates, MD

**University of Missouri - Columbia**
Radiography Prgm
School of Health Professions
605 Lewis Hall
Columbia, MO 65211
www.hsc.missouri.edu/~shrp/docs/radsci.html
*Prgm Dir:* Patricia A Tew, MS RT(R)(CT)
*Tel:* 573 884-2623   *Fax:* 573 884-1490
*E-mail:* tewp@health.missouri.edu

**Mineral Area Regional Medical Center**
Radiography Prgm
1212 Weber Rd
Farmington, MO 63640-3398
www.marmc.org
*Prgm Dir:* Brandi N Grindel, BS RT(R)
*Tel:* 573 701-7387   *Fax:* 573 756-6109
*E-mail:* Brandi_Grindel@chs.net

**Sanford-Brown College**
Radiography Prgm
South Campus
1203 Smizer Mill Rd
Fenton, MO 63026
*Prgm Dir:* Abby L Freeman, MSA RT(R)
*Tel:* 636 349-4900, Ext 157   *Fax:* 636 349-9170
*E-mail:* afreeman@sbc-fenton.com
*Med Dir:* Armand E Brodeur, MD

**Nichols Career Center**
Radiography Prgm
605 Union St
Jefferson City, MO 65101
www.jcps.k12.mo.us/education/school
*Prgm Dir:* Ronda Wahl, PhD RT(R)
*Tel:* 573 659-3238   *Fax:* 573 659-3154
*E-mail:* ronda.wahl@jcps.k12.mo.us
*Med Dir:* Sid Belshe, MD

**Missouri Southern State University**
Radiography Prgm
3950 E Newman Rd
Joplin, MO 64801-1595
www.mssu.edu
*Prgm Dir:* Alan D Schiska, MS RT(R) RDMS
*Tel:* 417 625-3118
*E-mail:* schiska-A@mssu.edu
*Med Dir:* Curtis Hammerman, MD

**Avila University**
Radiography Prgm
11901 Wornall Rd
Kansas City, MO 64145-9990
*Prgm Dir:* Carole A Hillestad, MSEd RT(R)
*Tel:* 816 501-3624   *Fax:* 816 501-2457
*E-mail:* hillestadca@mail.avila.edu

**Colorado Technical University**
Radiography Prgm
520 E 19th Ave
Kansas City, MO 64116
www.coloradotech.edu
*Prgm Dir:* Jessica L Patton, MS RT(R)(CT)
*Tel:* 816 303-7864   *Fax:* 816 472-0688
*E-mail:* jpatton@kc.coloradotech.edu
*Med Dir:* Lawrence Ricci, MD

**Metropolitan Community College - Penn Valley**
Radiography Prgm
3201 SW Trafficway
Kansas City, MO 64111
www.mcckc.edu
*Prgm Dir:* Judith E Taylor, MEd RT(R)(M)
*Tel:* 816 759-4243   *Fax:* 816 759-4553
*E-mail:* Judy.Taylor@mcckc.edu
*Med Dir:* Gerald Finke, DO

**Research Medical Center**
Radiography Prgm
2316 E Meyer Blvd
Kansas City, MO 64132-1199
*Prgm Dir:* Cheryl S Johnson, MS RT(R)
*Tel:* 816 276-3390   *Fax:* 816 276-3138
*E-mail:* cheryl.johnson2@hcamidwest.com
*Med Dir:* Jay Rozen, MD

**Saint Luke's Hospital**
Radiography Prgm
4401 Wornall Rd
Kansas City, MO 64111
*Prgm Dir:* Valarie J Tolson, MSEd RT(R)
*Tel:* 816 932-3766   *Fax:* 816 932-2322
*E-mail:* vtolson@saint-lukes.org
*Med Dir:* Mark L Redick, MD

**Rolla Technical Center**
Radiography Prgm
500 Forum Dr
Rolla, MO 65401-3699
*Prgm Dir:* Waneta M Odgen, MEd RT(R)(M)
*Tel:* 573 458-0160, Ext 16190   *Fax:* 573 458-0164
*E-mail:* mogden@rolla.k12.mo.us

**State Fair Community College**
Radiography Prgm
3201 W 16th St
Sedalia, MO 65301
www.sfccmo.edu
*Prgm Dir:* Beverly Wilkerson, MSN
*Tel:* 660 530-5800, Ext 403   *Fax:* 660 530-5820
*E-mail:* bwilkerson@sfccmo.edu

**CoxHealth**
Radiography Prgm
1423 N Jefferson Ave
Springfield, MO 65802
www.coxhealth.com/schoolseducation/
   DiagnosticImaging/sdi_rad-tech.htm
*Prgm Dir:* David Frazier, BS RT(R)(QM)
*Tel:* 417 269-4074   *Fax:* 417 269-4250
*E-mail:* david.frazier@coxhealth.com
*Med Dir:* Joseph Mailloux, MD

**St John's Regional Health Center**
Radiography Prgm
1235 E Cherokee St
Springfield, MO 65804-2263
www.stjohns.com
*Prgm Dir:* Joan Hedrick, MEd RT(R)(M)
*Tel:* 417 820-2982   *Fax:* 417 820-3427
*E-mail:* jfhedrick@sprg.mercy.net
*Med Dir:* Douglas Hacker, MD

**Hillyard Technical Center**
Radiography Prgm
3434 Faraon St
St Joseph, MO 64506
*Prgm Dir:* Shirley A Erickson, MBA RT(R)(N)
*Tel:* 816 671-4170
*E-mail:* shirley.erickson@sjsd.k12.mo.us

**Barnes-Jewish Coll of Nursing/Allied Health**
Radiography Prgm
306 S Kingshighway Blvd
St Louis, MO 63110
www.barnesjewishcollege.edu
*Prgm Dir:* Ms Johnnie B Moore, MEd RT(R)
*Tel:* 314 454-7597   *Fax:* 314 454-5239
*E-mail:* jbm0623@bjc.org
*Med Dir:* R Gilbert Jost, MD

**St John's Mercy Medical Center**
Radiography Prgm
615 S New Ballas Rd
St Louis, MO 63141-8277
*Prgm Dir:* James E Ibaviosa, BA RT(R)(CV)
*Tel:* 314 569-6933, Ext 3326   *Fax:* 314 569-6343
*E-mail:* ibavje@stlo.mercy.net
*Med Dir:* James A Nepute, MD

**St Louis Community College - Forest Park**
Radiography Prgm
5600 Oakland Ave
St Louis, MO 63110
*Prgm Dir:* Darrell E McKay, PhD RT(R)(M)(QM) FASRT
*Tel:* 314 644-9325   *Fax:* 314 644-9752
*E-mail:* DMcKay@stlcc.edu

## Montana

**Benefis Healthcare**
Radiography Prgm
500 15th Ave S
Great Falls, MT 59405
*Prgm Dir:* Thomas M Liston, RT(R)
*Tel:* 406 455-2164   *Fax:* 406 455-2162
*E-mail:* listthom@benefis.org
*Med Dir:* John C Hackethorn, MD

## Nebraska

**Mary Lanning Memorial Hospital**
Radiography Prgm
715 N St Joseph Ave
Hastings, NE 68901
www.mlmh.org
*Prgm Dir:* Jean M Korth, MS Ed ARRT(R)
*Tel:* 402 461-5087   *Fax:* 402 461-5059
*E-mail:* jkorth@mlmh.org
*Med Dir:* Eric Rodriguez, MD

**Southeast Community College**
Radiography Prgm
8800 O St
Lincoln, NE 68520
www.southeast.edu
*Prgm Dir:* Bev Harvey, MEd RT(R)
*Tel:* 402 437-2759   *Fax:* 402 437-2404
*E-mail:* bharvey@southeast.edu
*Med Dir:* Micheal DeWald, MD

**Alegent Health**
Radiography Prgm
7500 Mercy Rd
Omaha, NE 68124
*Prgm Dir:* Luann Baylor, BGS RT(R)
*Tel:* 402 398-5527   *Fax:* 402 398-5583
*E-mail:* lbaylor@alegent.org
*Med Dir:* Thomas Forrest, MD

**Clarkson College**
Radiography Prgm
101 S 42nd St
Omaha, NE 68131-2739
*Prgm Dir:* Ellen L Collins, MS RT(R)(M)
*Tel:* 402 552-6140   *Fax:* 402 552-6019
*E-mail:* collins@clarksoncollege.edu

**University of Nebraska Medical Center**
Radiography Prgm
984545 Nebraska Medical Center
Omaha, NE 68198-4545
www.unmc.edu/alliedhealth/rste
*Prgm Dir:* Connie L Mitchell, MA RT(R)(CT)
*Tel:* 402 559-6945   *Fax:* 402 559-4667
*E-mail:* clmitche@unmc.edu
*Med Dir:* Timothy E Moore, MD

**Regional West Medical Center**
Radiography Prgm
4021 Ave B
Scottsbluff, NE 69361
www.rwmc.net
*Prgm Dir:* Daniel R Gilbert, MSEd
   RT(R)(CV)(MR)(CT)(QM)
*Tel:* 308 630-1155   *Fax:* 308 630-1983
*E-mail:* gilberd@rwmc.net
*Med Dir:* Stephen Johnson, MD PhD

## Nevada

**Pima Medical Institute - Las Vegas**
Radiography Prgm
3333 E Flamingo Rd
Las Vegas, NV 89121
*Prgm Dir:* Marsha M Sortor, MHE RT(R)(QM)(M)(N)
*Tel:* 702 458-9650, Ext 235
*E-mail:* photon4me2@aol.com

**University of Nevada - Las Vegas**
Radiography Prgm
4505 Maryland Pkwy
Las Vegas, NV 89154-3017
www.unlv.edu
*Prgm Dir:* George Pales, PhD RT(R)(MR)(T)
*Tel:* 702 895-0821   *Fax:* 702 895-3296
*E-mail:* GPales@ccmail.nevada.edu
*Med Dir:* Alan F Weissman, MD

**Truckee Meadows Community College**
Radiography Prgm
7000 Dandini Blvd (R 417)
Reno, NV 89512-3999
www.tmcc.edu/x-ray
*Prgm Dir:* Deborah K Baker, BS RT(R)
*Tel:* 775 673-7121   *Fax:* 775 673-7034
*E-mail:* dbaker@tmcc.edu
*Med Dir:* Thomas Barcia, MD

## New Hampshire

**New Hampshire Technical Institute**
Radiography Prgm
31 College Dr
Concord, NH 03301-7412
www.nhti.edu
*Prgm Dir:* Kevin P Barry, MEd RT(R) RDMS RDCS
*Tel:* 603 271-7154   *Fax:* 603 271-7148
*E-mail:* kbarry@nhctc.ecu

**Lebanon College**
Radiography Prgm
15 Hanover St
Lebanon, NH 03766
*Prgm Dir:* Karen Amadon Burgess, MEd RT(R)(M)
*Tel:* 603 448-2445, Ext 125
*E-mail:* kburgess@lebanoncollege.edu
*Med Dir:* Douglas Goodwin, MD FACR

# New Jersey

## Cooper University Hospital
Radiography Prgm
One Cooper Plaza
Camden, NJ 08103
www.cooperhealth.org
*Prgm Dir:* Francis Williams, BS RT(R)
*Tel:* 856 342-2397    *Fax:* 856 968-8532
*E-mail:* williams-frank@cooperhealth.edu

## Middlesex County College
Radiography Prgm
2600 Woodbridge Ave, PO Box 3050
Edison, NJ 08818-3050
www.middlesexcc.edu
*Prgm Dir:* Albert M Snopek, BS RT(R)(M)(CV)(QM)
*Tel:* 732 906-2583    *Fax:* 732 906-7784
*E-mail:* asnopek@middlesexcc.edu
*Med Dir:* J K Amorosa, MD

## Englewood Hospital & Medical Center
Radiography Prgm
350 Engle St
Englewood, NJ 07631
*Prgm Dir:* Pamela Woodward, MS
    RT(R)(M)(MR)(QM)(CT)
*Tel:* 201 894-3481    *Fax:* 201 816-0168
*E-mail:* pamela.woodward@ehmc.com
*Med Dir:* Mark Shapiro, MD

## Christ Hospital
Radiography Prgm
176 Palisade Ave
Jersey City, NJ 07306
www.christhospital.org
*Prgm Dir:* Kenneth Lee, DHS RT(R)(M)
*Tel:* 201 795-8520    *Fax:* 201 795-5818
*E-mail:* klee@christhospital.org
*Med Dir:* Eileen Concannon, MD

## Brookdale Community College
Radiography Prgm
765 Newman Springs Rd
Lincroft, NJ 07738
www.brookdalecc.edu/fac/radtech
*Prgm Dir:* Terry M Konn, PhD RT(R)
*Tel:* 732 224-2696    *Fax:* 732 224-2998
*E-mail:* tkonn@brookdalecc.edu

## Essex County College
Radiography Prgm
303 University Ave
Newark, NJ 07102
*Prgm Dir:* Ronald M Kopec, MAEd RT(R)
*Tel:* 973 877-3437    *Fax:* 973 877-1930
*E-mail:* kopec@essex.edu

## Bergen Community College
Radiography Prgm
400 Paramus Rd
Paramus, NJ 07652
www.bergen.edu
*Prgm Dir:* William L Leonard, MA RT(R)
*Tel:* 201 447-7944    *Fax:* 201 612-3876
*E-mail:* wleonard@bergen.edu
*Med Dir:* Robert S Port, MD

## Passaic County Community College
Radiography Prgm
One College Blvd
Paterson, NJ 07505-1179
www.pccc.edu
*Prgm Dir:* Eileen Maloney, MEd RT(R)(M)FASRT
*Tel:* 973 684-5280    *Fax:* 973 684-5843
*E-mail:* emaloney@pccc.edu

## Burlington County College
Radiography Prgm
601 Pemberton Browns Mills Rd
Pemberton, NJ 08068
www.bcc.edu
*Prgm Dir:* Elizabeth Price, MS RT(R)(M)(CT)
*Tel:* 609 894-9311, Ext 1407    *Fax:* 609 726-0628
*E-mail:* eprice@bcc.edu
*Med Dir:* Douglas Moore, MD

## Muhlenberg Regional Medical Center
Radiography Prgm
Park Ave and Randolph Rd
Plainfield, NJ 07061
www.muhlenbergschools.org
*Prgm Dir:* Valerie L Carlisle, MEd RT(R)
*Tel:* 908 668-2966    *Fax:* 908 226-4568
*E-mail:* VCarlisle@solarishs.org

## County College of Morris
Radiography Prgm
214 Center Grove Rd
Randolph, NJ 07869
*Prgm Dir:* Denise M Vill'Neuve, MA RT(R)(M)(CT)
*Tel:* 973 328-5354    *Fax:* 973 328-5379
*E-mail:* dvillneuve@ccm.edu

## The Valley Hospital
Radiography Prgm
223 N Van Dien Ave
Ridgewood, NJ 07450
*Prgm Dir:* Maureen K Wolf, BS RT(R)(N)
*Tel:* 201 447-8221    *Fax:* 201 251-3280
*E-mail:* mwolf@valleyhealth.com
*Med Dir:* Ronald Arams, MD

## Shore Memorial Hospital
Radiography Prgm
1 E New York Ave
Somers Point, NJ 08244-2387
http://shore memorial.org
*Prgm Dir:* Jane Leggieri, MS RT(R)(QM)
*Tel:* 609 653-3924    *Fax:* 609 653-3566
*E-mail:* jleggieri@shorememorial.org
*Med Dir:* David Beglieter, MD

## Mercer County Community College
Radiography Prgm
1200 Old Trenton Rd, PO Box B
Trenton, NJ 08690
www.mccc.edu
*Prgm Dir:* Sandra L Kerr, MA RT(R)(M)
*Tel:* 609 586-4800, Ext 3337    *Fax:* 609 689-0762
*E-mail:* kerr@mccc.edu
*Med Dir:* Gregory Kaufman, MD

## St Francis Medical Center
Radiography Prgm
601 Hamilton Ave
Trenton, NJ 08629-1986
www.stfrancismedical.com
*Prgm Dir:* Theresa M Levitsky, MA RT(R)(CV)(M)(QM)
*Tel:* 609 599-5234    *Fax:* 609 599-5529
*E-mail:* tlevitsky@che-east.org
*Med Dir:* Ethan Tarasov, MD

## Cumberland County College
Radiography Prgm
College Dr, PO Box 1500
Vineland, NJ 08362-1500
*Prgm Dir:* Robert P Champa, MA RT(R)
*Tel:* 609 691-8600, Ext 264    *Fax:* 609 691-9489
*E-mail:* rchampa@cccnj.net
*Med Dir:* Ernesto Go, MD

## Pascack Valley Hospital
Radiography Prgm
Old Hook Rd
Westwood, NJ 07675-3181
www.pvhospital.org
*Prgm Dir:* Vincent J Monte, BS RT(R)
*Tel:* 201 358-3219    *Fax:* 201 358-3216
*E-mail:* vmonte@pvhospital.org
*Med Dir:* Frank Gingerelli, MD MACR

# New Mexico

## Pima Medical Institute - Albuquerque
Radiography Prgm
2201 San Pedro NE, Bldg 3, Ste 100
Albuquerque, NM 87110
*Prgm Dir:* James G Murrell, MSRS RT(R)(M)(CT)(QM)
*Tel:* 505 881-1234    *Fax:* 505 884-8371
*E-mail:* jmurrell@pmi.edu
*Med Dir:* Farooq P Agha, MD

## Clovis Community College
Radiography Prgm
417 Schepps Blvd
Clovis, NM 88101
*Prgm Dir:* Jeannie Kilgore, MEd RT(R)
*Tel:* 505 769-4996    *Fax:* 505 769-4190
*E-mail:* jeannie.kilgore@clovis.edu
*Med Dir:* Michael Rowley, MD

## Northern New Mexico College
Radiography Prgm
921 Paseo de Onate
Espanola, NM 87532
www.nnmc.edu
*Prgm Dir:* Michael P Frain, MA RT(R)
*Tel:* 505 747-2218    *Fax:* 505 747-5415
*E-mail:* frainm@nnmc.edu
*Med Dir:* Robin Gaupp, MD

## Dona Ana Community College
Radiography Prgm
MSC 3DA PO Box 30001
3400 S Espina St
Las Cruces, NM 88003-8001
http://dabcc.nmsu.edu
*Prgm Dir:* Joyce M Ortego, MS RT(R)
*Tel:* 505 527-7581    *Fax:* 505 527-7765
*E-mail:* jortego@nmsu.edu

# New York

## Broome Community College
Radiography Prgm
Upper Front St, PO Box 1017
Binghamton, NY 13902
*Prgm Dir:* Nancy E Button, MS RT(R)
*Tel:* 607 778-5070    *Fax:* 607 778-5467
*E-mail:* button_n@sunybroome.edu
*Med Dir:* David Lisi, MD PhD

## Bronx Community College
Radiography Prgm
University Ave and W 181st St
CP Hall Rm 222
Bronx, NY 10453
*Prgm Dir:* Virginia M Mishkin, MS RT(R) (M) (QM)
*Tel:* 718 289-5396    *Fax:* 718 289-6373
*E-mail:* virginia.mishkin@bcc.cuny.edu
*Med Dir:* Carl Forcade, MD

## Hostos Community College of CUNY
Radiography Prgm
475 Grand Concourse
Bronx, NY 10451
*Prgm Dir:* Allen Solomon, MSEd RT(R)
*Tel:* 718 518-4123    *Fax:* 718 518-4238
*E-mail:* asolomon@hostos.cuny.edu
*Med Dir:* Marie F Gade, MD

## Long Island College Hospital
Radiography Prgm
339 Hicks St
Brooklyn, NY 11201
*Prgm Dir:* Sergeo Guilbaud, BS RT(R)
*Tel:* 718 780-1681   *Fax:* 718 780-1592
*E-mail:* sguilbau@chpnet.org
*Med Dir:* Deborah L Reede, MD

## New York City College of Technology
Radiography Prgm
300 Jay St
Brooklyn, NY 11201-2983
*Prgm Dir:* Mary Alice Browne, MS RT(R)(CV)(CT)(MR)
*Tel:* 718 260-5360, Ext 5360   *Fax:* 718 260-5540
*E-mail:* mabrowne@citytech.cuny.edu

## New York Methodist Hospital
Radiography Prgm
506 Sixth St
Brooklyn, NY 11215
www.nym.org
*Prgm Dir:* Anthony F DeVito, MA RT(R)
*Tel:* 718 780-3887   *Fax:* 718 780-3494
*E-mail:* devito2@juno.com
*Med Dir:* David Garner, MD

## Long Island University - C W Post Campus
Radiography Prgm
720 Northern Blvd
Brookville, NY 11548-1300
www.liu.edu/radtech
*Prgm Dir:* James F Joyce, MS RT(R)(M)(QM)
*Tel:* 516 299-3075   *Fax:* 516 299-3081
*E-mail:* jjoyce@liu.edu
*Med Dir:* Gerald Irwin, MD CM

## Trocaire College
Radiography Prgm
360 Choate Ave
Buffalo, NY 14220
*Prgm Dir:* Nancy L Augustyn, MSEd RT(R)
*Tel:* 716 826-1200, Ext 1243   *Fax:* 716 828-6107
*E-mail:* augustynn@trocaire.edu
*Med Dir:* Noel M Chiantella, MD

## Arnot Ogden Medical Center
Radiography Prgm
600 Roe Ave
Elmira, NY 14905-1676
www.arnothealth.org
*Prgm Dir:* Ellen R Richards, MPS RT(R)
*Tel:* 607 737-4289   *Fax:* 607 737-4116
*E-mail:* erichards@aomc.org
*Med Dir:* Edwin P Hutsal, MD

## Nassau Community College
Radiography Prgm
One Education Dr
Garden City, NY 11530
*Prgm Dir:* Jeffrey T Miller, BS RT(R)
*Tel:* 516 572-7539   *Fax:* 516 572-7565
*E-mail:* millerjt@ncc.edu
*Med Dir:* Howard Gelber, MD

## Glens Falls Hospital
Radiography Prgm
School of Radiologic Technology
126 South St
Glens Falls, NY 12801
*Prgm Dir:* Roger F Weeden, Jr, BA RT(R)
*Tel:* 518 926-7025   *Fax:* 518 926-3747
*E-mail:* rweeden@glensfallshosp.org
*Med Dir:* Daniel Burke, MD

## St James Mercy Health
Radiography Prgm
411 Canisteo St
Hornell, NY 14843
*Prgm Dir:* Lynne M Freeland, MS RT(R)
*Tel:* 607 324-8265   *Fax:* 607 324-8214
*E-mail:* lfreelan@sjmh.org
*Med Dir:* Iddo Netamyshu, MD

## Woman's Christian Association Hospital
Radiography Prgm
207 Foote Ave
Jamestown, NY 14701-0840
*Prgm Dir:* Amanda M Bender, RT(R)
*Tel:* 716 664-8366   *Fax:* 716 664-8312
*Med Dir:* James G Dahlie, MD

## Orange County Community College
Radiography Prgm
115 South St
Middletown, NY 10940
www.sunyorange.edu
*Prgm Dir:* Diedre Costic, MPS RT(R)(M)
*Tel:* 845 341-4148   *Fax:* 845 341-4511
*E-mail:* dcostic@sunyorange.edu
*Med Dir:* J Yacovone, MD

## Winthrop University Hospital
Radiography Prgm
259 First St
Ste 127
Mineola, NY 11501
www.winthrop-radiology.com
*Prgm Dir:* Virginia M Edele, MBA RT(R)(M)
*Tel:* 516 663-2536   *Fax:* 516 663-2587
*E-mail:* vedele@winthrop.org
*Med Dir:* Orlando Ortiz, MBA, MD

## Bellevue Hospital Center
Radiography Prgm
First Ave and 27th St, C&D Bldg, RmD510
New York, NY 10016
*Prgm Dir:* Andrew C Richter, BS RT(R)
*Tel:* 212 562-4895   *Fax:* 212 263-6781
*E-mail:* richtera@bellevue.nychhc.org
*Med Dir:* Nancy B Genieser, MD

## Harlem Hospital Center
Radiography Prgm
506 Lenox Ave
Kountz Pavilion Rm 415
New York, NY 10037
*Prgm Dir:* William Hall, MPH RT(R) (MR)
*Tel:* 212 939-3480   *Fax:* 212 939-3479
*E-mail:* WAH47@columbia.edu
*Med Dir:* Roberta Locko, MD

## Robt J Hochstim Sch Rad/S Nassau Comm Hosp
Radiography Prgm
One Healthy Way
PO Box 9007
Oceanside, NY 11572
http://snch.org
*Prgm Dir:* Gina Collins, MPA RT(R)(M)
*Tel:* 516 632-4678   *Fax:* 516 336-2983
*E-mail:* gcollins@snch.org
*Med Dir:* Stephen Lastig, MD

## Champlain Valley Phys Hospital Med Ctr
Radiography Prgm
75 Beekman St
Plattsburgh, NY 12901
*Prgm Dir:* Fayrene M Ashline, BS RT(R)(M)(N)
*Tel:* 518 562-7510   *Fax:* 518 562-7486
*E-mail:* fashline@cvph.org
*Med Dir:* David Hammack, MD

## Peconic Bay Medical Center
Radiography Prgm
1300 Roanoke Ave
Riverhead, NY 11901
*Prgm Dir:* William L DeCamp, BS RT(R)(CT)
*Tel:* 631 548-6173   *Fax:* 631 548-6751

## Monroe Community College
Radiography Prgm
1000 E Henrietta Rd
Rochester, NY 14623-5780
www.monroecc.edu
*Prgm Dir:* Eileen M Doyle, MPA RT(R)
*Tel:* 585 292-2379   *Fax:* 585 292-3834
*E-mail:* edoyle@monroecc.edu

## Mercy Medical Center
Radiography Prgm
PO Box 9024
Rockville Centre, NY 11571
*Prgm Dir:* Barbara Geiger, MA RT(R)(M)
*Tel:* 516 705-2272   *Fax:* 516 705-1079
*E-mail:* barbara.geiger@chsli.org
*Med Dir:* Kenneth Schwartz, MD

## Niagara County Community College
Radiography Prgm
3111 Saunders Settlement Rd
Sanborn, NY 14132
*Prgm Dir:* Carolyn Cianciosa, MS RT(R)
*Tel:* 716 614-6416   *Fax:* 716 614-6879
*E-mail:* sciera@niagaracc.suny.edu
*Med Dir:* Brian Block, MD

## North Country Community College
Radiography Prgm
23 Santanoni Ave, PO Box 89
Saranac Lake, NY 12983-0089
www.nccc.edu
*Prgm Dir:* Bonnie S Clinebell, BS RT(R)(M)
*Tel:* 518 891-2915, Ext 255   *Fax:* 518 891-2915
*E-mail:* bclinebell@nccc.edu
*Med Dir:* Richard Moccia, MD

## SUNY Upstate Medical University
Radiography Prgm
Medical Imaging Sciences Program
750 E Adams St
Syracuse, NY 13210
*Prgm Dir:* David A Clemente, MSEd RT(R)(MR)
*Tel:* 315 464-6929   *Fax:* 315 464-4561
*E-mail:* clementd@upstate.edu
*Med Dir:* John Cardella, MD

## Faxton - St Luke's Healthcare
Radiography Prgm
Champlain Ave, PO Box 479
Utica, NY 13503-0479
*Prgm Dir:* Rosemary Morin, MS RT(R)
*Tel:* 315 798-6136   *Fax:* 315 798-6295
*E-mail:* xrayed@dreamscape.com
*Med Dir:* Andrew Sternich, MD

## St Elizabeth Medical Center
Radiography Prgm
2209 Genesee St
Utica, NY 13501
www.stemc.org
*Prgm Dir:* Janice M Lutz, MPS RT(R)(M)
*Tel:* 315 798-8258   *Fax:* 315 734-3084
*E-mail:* jlutz@stemc.org
*Med Dir:* Raphael Alcuri, MD

## Westchester Community College
Radiography Prgm
75 Grasslands Rd
Valhalla, NY 10595-1698
*Prgm Dir:* Robert Q Wong, BS RT(R)
*Tel:* 914 606-7881   *Fax:* 914 606-7832
*E-mail:* robert.wong@sunywcc.edu
*Med Dir:* Donna A Brown, MD

## St Joseph's Medical Center
Radiography Prgm
127 S Broadway
Yonkers, NY 10701
*Prgm Dir:* Rose P LaBate, BS RT(R)(M)
*Tel:* 914 378-8151
*E-mail:* labate99@yahoo.com
*Med Dir:* Michael J Schnur, MD

**PROGRAMS**

# North Carolina

**Asheville-Buncombe Technical Comm College**
Radiography Prgm
340 Victoria Rd
Asheville, NC 28801
www.abtech.edu
*Prgm Dir:* Debra J Reese, MPH RT(R)
*Tel:* 828 254-1921, Ext 282    *Fax:* 828 281-9846
*E-mail:* dreese@abtech.edu
*Med Dir:* Keith Kohatsu, MD

**Univ of North Carolina at Chapel Hill**
Radiography Prgm
Medical School
CB 7130 E Wing
Chapel Hill, NC 27599-7130
www.med.unc.edu/ahs/radisci
*Prgm Dir:* Joy J Renner, MA RT(R)
*Tel:* 919 966-5147    *Fax:* 919 966-6951
*E-mail:* jrenner@med.unc.edu
*Med Dir:* Jordan Renner, MD

**Carolinas College of Health Sciences**
Radiography Prgm
PO Box 32861
1200 Blythe Blvd
Charlotte, NC 28232-2861
www.carolinascollege.edu
*Prgm Dir:* Susan P Stricker, BS RT(R)
*Tel:* 704 355-2446    *Fax:* 704 355-5967
*E-mail:* susan.stricker@carolinashealthcare.Org

**Presbyterian Healthcare**
Radiography Prgm
200 Hawthorne Ln, PO Box 33549
Charlotte, NC 28204
www.presbyterian.org
*Prgm Dir:* Elizabeth S Shields, BS RT(R)
*Tel:* 704 384-5104    *Fax:* 704 384-5684
*E-mail:* esshields@novanthealth.org
*Med Dir:* Bennett R Hollenberg, MD

**Fayetteville Technical Community College**
Radiography Prgm
PO Box 35236
Fayetteville, NC 28303-0236
*Prgm Dir:* Anita L McKnight, BHS RT(R)
*Tel:* 910 678-8303    *Fax:* 910 678-8500
*E-mail:* mcknigha@faytechcc.edu

**Moses Cone Health System**
Radiography Prgm
1200 N Elm St
Greensboro, NC 27401-1020
www.mosescone.com/radtech
*Prgm Dir:* Rene Parrish, BS RT(R)
*Tel:* 336 832-7487    *Fax:* 336 832-7465
*E-mail:* rene.parrish@mosescone.com
*Med Dir:* Gary J Fischer, MD

**Pitt Community College**
Radiography Prgm
PO Drawer 7007, Hwy 11 S
Greenville, NC 27835-7007
www.pittcc.edu
*Prgm Dir:* Louise R Cox, BA RT(R)(CV)(M)
*Tel:* 252 493-7464    *Fax:* 252 321-4451
*E-mail:* lcox@email.pittcc.edu
*Med Dir:* Julian R Vainright Jr, MD

**Vance - Granville Community College**
Radiography Prgm
PO Box 917
Henderson, NC 27536
*Prgm Dir:* Lauren Noble, EdD RT(R)
*Tel:* 252 492-2061, Ext 3229    *Fax:* 252 738-3319
*E-mail:* noble@vgcc.edu
*Med Dir:* John Mark Spargo, MD

**Caldwell Comm College & Tech Institute**
Radiography Prgm
2855 Hickory Blvd
Hudson, NC 28638
*Prgm Dir:* Rosanne Y Annas, BS RT(R)
*Tel:* 704 726-2358    *Fax:* 704 726-2489
*E-mail:* rannas@ccti.edu
*Med Dir:* Richard Curtis, MD

**Carteret Community College**
Radiography Prgm
3505 Arendell St
Morehead City, NC 28557-2989
www.carteret.edu
*Prgm Dir:* Elaine M Fuge, MEd RT(R)
*Tel:* 252 222-6165    *Fax:* 252 222-6074
*E-mail:* fugee@carteret.edu
*Med Dir:* Elizabeth D'Angelo, MD

**Wilkes Regional Medical Center**
Radiography Prgm
1370 West D St, PO Box 609
North Wilkesboro, NC 28659
http://wilkesregional.org
*Prgm Dir:* Betty S Winslow, MA RT(R)
*Tel:* 336 651-8431    *Fax:* 336 651-8432
*E-mail:* bwinslow@wilkesregional.com
*Med Dir:* Carl W Hoffman, MD

**Sandhills Community College**
Radiography Prgm
3395 Airport Rd
Pinehurst, NC 28374
www.sandhills.edu
*Prgm Dir:* Kay B Nardo, BA RT(R)(CV)
*Tel:* 910 695-3916    *Fax:* 910 693-2060
*E-mail:* nardok@sandhills.edu
*Med Dir:* Soledad Griffin, MD

**Wake Technical Community College**
Radiography Prgm
9101 Fayetteville Rd
Raleigh, NC 27603-5696
*Prgm Dir:* Deborah J Wood, MEd RT(R)(M)
*Tel:* 919 250-4290    *Fax:* 919 250-4329
*E-mail:* djwood@waketech.edu
*Med Dir:* Donald G Detweiler, MD

**Edgecombe Community College**
Radiography Prgm
225 Tarboro St
Rocky Mount, NC 27801
www.edgecombe.edu
*Prgm Dir:* Donald R Mastman, BS RT(R)(MR)(CT)
*Tel:* 252 446-0436, Ext 375    *Fax:* 252 985-2212
*E-mail:* mastmanr@edgecombe.edu
*Med Dir:* Paul Guay, MD

**Rowan - Cabarrus Community College**
Radiography Prgm
PO Box 1595
Salisbury, NC 28145-1595
*Prgm Dir:* Frankie W Lyons, MHA RT(R)(M)
*Tel:* 704 637-0760, Ext 398    *Fax:* 704 642-0750
*E-mail:* lyonsf@rowancabarrus.edu
*Med Dir:* Douglas Sheafor, MD

**Cleveland Community College**
Radiography Prgm
137 S Post Rd
Shelby, NC 28152
www.clevelandcommunitycollege.edu
*Prgm Dir:* Alease Rousseau, BS RT(R)
*Tel:* 704 484-4091, Ext 4091    *Fax:* 704 484-5304
*E-mail:* rousseau@cleveland.cc.nc.us
*Med Dir:* Michael Douglas Wehmueller, MD

**Johnston Community College**
Radiography Prgm
PO Box 2350
Smithfield, NC 27577-2350
www.johnstoncc.edu
*Prgm Dir:* Sheila W Smith, M Ed RT(R)
*Tel:* 919 209-2033    *Fax:* 919 209-2153
*E-mail:* smiths@johnstoncc.edu

**Southwestern Community College**
Radiography Prgm
447 College Dr
Sylva, NC 28779
www.southwesterncc.edu
*Prgm Dir:* Meg Rollins, BS RT(R)(N)
*Tel:* 828 586-4091, Ext 320    *Fax:* 828 586-3129
*E-mail:* mrollins@southwesterncc.edu
*Med Dir:* George R Dixson, MD

**Cape Fear Community College**
Radiography Prgm
411 N Front St
Wilmington, NC 28401
http://cfcc.edu
*Prgm Dir:* Anita Phillips, RT(R) MEd
*Tel:* 910 362-7298    *Fax:* 910 362-7087
*E-mail:* aphillips@cfcc.edu

**Forsyth Technical Community College**
Radiography Prgm
2100 Silas Creek Pkwy
Winston-Salem, NC 27103
www.forsythtech.edu
*Prgm Dir:* Linda W Yurko, MA RT(R)(M)
*Tel:* 336 734-7180    *Fax:* 336 734-7444
*E-mail:* lyurko@forsythtech.edu
*Med Dir:* David Ott, MD

# North Dakota

**Medcenter One**
Radiography Prgm
300 N 7th St
Bismarck, ND 58506-5525
www.medcenterone.com
*Prgm Dir:* Cindy Hanson, BS RT(R)
*Tel:* 701 323-5470    *Fax:* 701 323-5479
*E-mail:* chanson@mohs.org
*Med Dir:* W H Cain, MD

**MeritCare Medical Center**
Radiography Prgm
801 North Broadway
Fargo, ND 58122
*Prgm Dir:* Mary Jo Bergman, MEd MS RN RT(R)
*Tel:* 701 364-1728
*E-mail:* mary.jo.bergman@meritcare.com
*Med Dir:* Richard Marsden, MD

**Trinity Health**
Radiography Prgm
407 3rd St NE
Minot, ND 58701
www.trinityhealth.org
*Prgm Dir:* Debbie Hornbacher, MS RT(R)
*Tel:* 701 857-5000, Ext 5620    *Fax:* 701 857-5620
*E-mail:* deb.hornbacher@trinityhealth.org
*Med Dir:* Kenneth Keller, MD

# Ohio

**Children's Hospital Medical Ctr of Akron**
Radiography Prgm
One Perkins Square
Akron, OH 44308
www.akronchildrens.org
*Prgm Dir:* David L Whipple, MEd RT(R)
*Tel:* 330 543-8849, Ext 3    *Fax:* 330 543-8282
*E-mail:* dwhipple@chmca.org

**University of Cincinnati**
Cosponsor: Raymond Walters College
Radiography Prgm
9555 Plainfield Rd
Blue Ash, OH 45236
Prgm Dir: Angie Arnold, MEd RT(R)
Tel: 513 745-5659   Fax: 513 936-7113
E-mail: angie.arnold@uc.edu

**Aultman Hospital**
Radiography Prgm
2600 Sixth St SW
Canton, OH 44710
Prgm Dir: Jacquelyn A Hammonds, BA RT(R)
Tel: 330 363-5352
E-mail: Jackie_Hammonds@hotmail.com
Med Dir: Paul Wong, MD

**Mercy Medical Center**
Radiography Prgm
1320 Mercy Dr NW
Canton, OH 44708
Prgm Dir: Gary F Greathouse, MS RT(R)
Tel: 330 489-1273, Ext 1273   Fax: 330 430-2772
E-mail: gary.greathouse@CSAUH.com
Med Dir: David Spriggs, MD

**Collins Career Center**
Radiography Prgm
11627 St Rte 243
Chesapeake, OH 45619
Prgm Dir: Karen G Russell, RT(R)
Tel: 740 867-6641, Ext 362
E-mail: kgrussell@collins-cc.k12.oh.us

**Xavier University**
Radiography Prgm
3800 Victory Pkwy
Cincinnati, OH 45207-4331
www.xavier.edu
Prgm Dir: Donna J Endicott, MEd RT(R)
Tel: 513 745-3358   Fax: 513 745-1079
E-mail: endicott@xavier.edu
Med Dir: Timothy Miller, MD

**Columbus State Community College**
Radiography Prgm
550 E Spring St
GR389
Columbus, OH 43215
www.cscc.edu/ah
Prgm Dir: James J Byrne, MEd RT(R)
Tel: 800 621-6407, Ext 5215   Fax: 614 287-6059
E-mail: jbyrne@cscc.edu
Med Dir: Mary Lee Hess, MD

**Sinclair Community College**
Radiography Prgm
444 W Third St
Dayton, OH 45402-1460
www.sinclair.edu/departments/rat
Prgm Dir: Debra Schwartz, BS RT(R)(M)
Tel: 937 512-2159   Fax: 937 512-2058
E-mail: debra.schwartz@sinclair.edu

**Lorain County Community College**
Radiography Prgm
1005 N Abbe Rd, HS223
Elyria, OH 44035
Prgm Dir: Jeffrey J Walmsley, MEd RT(R)(QM)
Tel: 440 366-7197   Fax: 440 366-4116
E-mail: jwalmsle@lorainccc.edu

**Cleveland Clinic Health System**
Radiography Prgm
18901 Lakeshore Blvd
Euclid, OH 44119
www.cchseast.org/schools
Prgm Dir: Gloria A Albrecht, BS RT(R)
Tel: 216 692-7512   Fax: 216 692-7806
E-mail: galbrech@cchseast.org
Med Dir: Jeffrey S Unger, MD

**Kettering College of Medical Arts**
Radiography Prgm
3737 Southern Blvd
Kettering, OH 45429
www.kcma.edu
Prgm Dir: Lawrence Beneke, MSEd RT(R)(CT)
Tel: 937 298-3399, Ext 55696   Fax: 937 395-8484
E-mail: larry.beneke@kcma.org

**Lakeland Community College**
Radiography Prgm
7700 Clocktower Dr
Kirtland, OH 44094-5198
Prgm Dir: Jack A Thomas, MS RT(R)(QM)
Tel: 440 525-7074   Fax: 440 525-7433
E-mail: jthomas@lakelandcc.edu
Med Dir: Victor J Demarco, MD

**Rhodes State College**
Radiography Prgm
4240 Campus Dr
Lima, OH 45804-3597
Prgm Dir: Dennis F Spragg, MSEd RT(R)(CT)(QM)
Tel: 419 995-8257   Fax: 419 995-8818
E-mail: Spragg.D@rhodesstate.edu
Med Dir: Thomas Church, MD

**North Central State College**
Radiography Prgm
2441 Kenwood Cir, PO Box 698
Mansfield, OH 44901-0698
Prgm Dir: Ellen M Johnson, BS RT(R)
Tel: 419 755-4809   Fax: 419 755-5630
E-mail: ejohnson@ncstatecollege.edu
Med Dir: Paul Buehrer, MD

**Marietta Memorial Hospital**
Radiography Prgm
401 Matthew St
Marietta, OH 45750
Prgm Dir: Paul E Richards Jr, MA RT(R)
Tel: 740 374-8716, Ext 1720   Fax: 740 373-7496
E-mail: prichards@wscc.edu
Med Dir: Paul Prachun, MD

**Marion Technical College**
Radiography Prgm
1467 Mt Vernon Ave
Marion, OH 43302-5694
http://mtc.edu
Prgm Dir: Linda Rizzo, BS RT(R)
Tel: 740 389-4636, Ext 246   Fax: 740 725-4018
E-mail: rizzol@mtc.edu
Med Dir: Edwin G Davy, MD

**Central Ohio Technical College**
Radiography Prgm
1179 University Dr
Newark, OH 43055-1767
Prgm Dir: Linnea A Hopewell, MEd RT(R)(M)(QM)
Tel: 740 366-9387   Fax: 740 366-5047
E-mail: hopewell@cotc.edu
Med Dir: Joseph Fondriest, MD

**Cuyahoga Community College**
Radiography Prgm
Western Campus
11000 Pleasant Valley Rd
Parma, OH 44130-5199
http://www.tri-c.edu
Prgm Dir: Alice H Kreutzberg, BA RT(R)
Tel: 216 987-5261   Fax: 216 987-5066
E-mail: alice.kreutzberg@tri-c.edu
Med Dir: Craig Irish, MD

**Shawnee State University**
Radiography Prgm
940 Second St
Portsmouth, OH 45662-4344
Prgm Dir: William W Sykes, MBA
RT(R)(M)(CT)(MR)(QM)
Tel: 740 351-3253   Fax: 740 351-3354
E-mail: bsykes@shawnee.edu
Med Dir: George Johnson, MD

**Kent State University - Salem Campus**
Radiography Prgm
2491 State Rte 45 South
Salem, OH 44460-9412
www.salem.kent.edu
Prgm Dir: Janice J Gibson, MEd RT(R)
Tel: 330 337-4223   Fax: 330 337-4122
E-mail: gibson@salem.kent.edu
Med Dir: Peter Apicella, MD

**Firelands Regional Medical Center**
Radiography Prgm
1912 Hayes Ave
Sandusky, OH 44870
Prgm Dir: Cynthia S Felske, RT(R)
Tel: 419 621-7124   Fax: 419 621-7209
E-mail: felskec@firelands.com
Med Dir: James D Frank, MD

**Jefferson Community College**
Radiography Prgm
4000 Sunset Blvd
Steubenville, OH 43952-3598
www.jeffersoncc.org
Prgm Dir: Shelly Gaumer, MS Ed RT(R)
Tel: 740 264-5591, Ext 229   Fax: 740 264-1338
E-mail: sgaumer@jcc.edu
Med Dir: W Hunter Vaughan, MD

**Mercy College of Northwest Ohio**
Radiography Prgm
2221 Madison Ave
Toledo, OH 43624-1132
www.mercycollege.edu
Prgm Dir: Linda S Wheatley, PhD RT(R)(MR)
Tel: 419 251-8958   Fax: 419 251-1730
E-mail: linda.wheatley@mercycollege.edu
Med Dir: Terrance Loh, MD

**Owens Community College**
Radiography Prgm
PO Box 10,000
Toledo, OH 43699-1947
www.owens.edu
Prgm Dir: Catherine A Ford, BS RT(R)
Tel: 567 661-7261   Fax: 567 661-7251
E-mail: catherine_ford@owens.edu

**Zane State College**
Radiography Prgm
1555 Newark Rd
Zanesville, OH 43701
www.zanestate.edu
Prgm Dir: Tricia D Leggett, MSEd RT(R)(QM)
Tel: 740 588-1271   Fax: 740 454-0035
E-mail: tleggett@zanestate.edu
Med Dir: Mellon Richard, MD

# Oklahoma

**Western Oklahoma State College**
Radiography Prgm
2801 N Main
Altus, OK 73521-1397
www.wosc.edu
Prgm Dir: Nancy J Estes, MSRS RT(R)(M)
Tel: 580 477-7823   Fax: 580 477-7865
E-mail: nancy.estes@wosc.edu

**Autry Technology Center**
Radiography Prgm
1201 W Willow
Enid, OK 73703
www.autrytech.com
*Prgm Dir:* Sharon Johnson, MS RT(R)(T)
*Tel:* 580 242-2750, Ext 134   *Fax:* 580 233-8262
*E-mail:* sjohnson@autrytech.com

**Great Plains Technology Center**
Radiography Prgm
4500 W Lee Blvd
Lawton, OK 73505
www.gptech.org/rad
*Prgm Dir:* Carrie L Baxter, MEd RT(R)(M)(CT)
*Tel:* 580 355-6371, Ext 5577   *Fax:* 580 250-5583
*E-mail:* cbaxter@gptech.org
*Med Dir:* Randall Behrmann, DO SWMC

**Rose State College**
Radiography Prgm
6420 SE 15th St
Midwest City, OK 73110-2799
*Prgm Dir:* Jonnye Griffin, MA RT(R)(QM)(M)
*Tel:* 405 733-7568   *Fax:* 405 736-0338
*E-mail:* jgriffin@rose.edu
*Med Dir:* Susan Edwards, MD

**Bacone College**
Radiography Prgm
2299 Old Bacone Rd
Muskogee, OK 74403-1568
www.bacone.edu
*Prgm Dir:* Francis Ozor, EdD MEd MPH BS AAS
   RT(R)(ARRT)
*Tel:* 918 781-7326, Ext 7326   *Fax:* 918 781-7214
*E-mail:* ozorf@bacone.edu
*Med Dir:* Gayle Joslin, MD

**Indian Capital Technology Center**
Radiography Prgm
2403 N 41st St E
Muskogee, OK 74403-1799
*Prgm Dir:* Ernest A Briggs, MS RT(R)
*Tel:* 918 687-6383, Ext 243   *Fax:* 866 877-9987
*E-mail:* ernieb@ictctech.com
*Med Dir:* Jim Bolene, MD

**Metro Technology Centers**
Radiography Prgm
1720 Springlake Dr
Oklahoma City, OK 73111
*Prgm Dir:* Barbara J Harper, MEd RT(R)
*Tel:* 405 605-4634   *Fax:* 405 424-9403
*E-mail:* barbara.harper@metrotech.org

**Univ of Oklahoma Health Sciences Center**
Radiography Prgm
801 NE 13th St, CHB-451
PO Box 26901
Oklahoma City, OK 73126
www.ouhsc.edu/radtech
*Prgm Dir:* Jeffrey L Berry, MS RT(R)(CT)
*Tel:* 405 271-6477   *Fax:* 405 271-1424
*E-mail:* jeff-berry@ouhsc.edu
*Med Dir:* Timothy Tytle, MD

**Carl Albert State College**
Radiography Prgm
1507 S McKenna
Poteau, OK 74953
www.carlalbert.edu
*Prgm Dir:* Linda Pearson, PhD RT(R)(M)(QM)
*Tel:* 918 635-3312   *Fax:* 918 635-3524
*E-mail:* lpearson@carlalbert.edu

**Southwestern Oklahoma State University**
Radiography Prgm
409 E Mississippi
Sayre, OK 73662-1236
www.swosu.edu/sayre
*Prgm Dir:* Chris Stufflebean, MBA RT(R)
*Tel:* 580 928-5533, Ext 155   *Fax:* 580 928-5533
*E-mail:* chris.stufflebean@swosu.edu
*Med Dir:* James E Milton, MD

**Meridian Technology Center**
Radiography Prgm
1312 S Sangre Rd
Stillwater, OK 74074-1841
*Prgm Dir:* Tanya Lynn Vasso, RT(R)(M)
*Tel:* 405 377-3333, Ext 336   *Fax:* 405 377-9604
*E-mail:* tanyav@meridian-technology.com
*Med Dir:* Paul Massad, MD

**Tulsa Community College**
Radiography Prgm
909 S Boston Ave
Tulsa, OK 74119-7263
www.tulsacc.edu
*Prgm Dir:* Benedict J Middleton, MS RT(R)
*Tel:* 918 595-7012   *Fax:* 918 595-7091
*E-mail:* rmiddlet@tulsacc.edu
*Med Dir:* Thomas W White, MD

**Tulsa Technology Center**
Radiography Prgm
801 E 91st St
Tulsa, OK 74132
www.tulsatech.org
*Prgm Dir:* Kathleen M Davis, MS RT(R)
*Tel:* 918 828-4230   *Fax:* 918 828-4219
*E-mail:* kathy.davis@tulsatech.org

# Oregon

**Portland Community College**
Radiography Prgm
12000 SW 49th Ave, PO Box 19000
Portland, OR 97280-0990
*Prgm Dir:* Virginia Vanderford, MEd RT(R)(M)
*Tel:* 503 977-4907   *Fax:* 503 977-8240
*E-mail:* vvanderf@pcc.edu
*Med Dir:* Michael Vervarka, MD

# Pennsylvania

**Northampton Community College**
Radiography Prgm
3835 Green Pond Rd
Bethlehem, PA 18020
www.northampton.edu
*Prgm Dir:* Zoland Z Zile III, MS RT(R)(QM)
*Tel:* 610 861-5387   *Fax:* 610 861-4581
*E-mail:* zzile@northampton.edu
*Med Dir:* robert rienzo, MD

**Bradford Regional Medical Center**
Radiography Prgm
116 Interstate Pkwy
Bradford, PA 16701
*Prgm Dir:* Scott Gregoire, BS RT(R)
*Tel:* 814 362-8292   *Fax:* 814 368-7750
*E-mail:* sgregoire@mail.bfdmed.org

**Harcum College**
Radiography Prgm
130 S Bryn Mawr Ave
Bryn Mawr, PA 19010-3158
http://harcum.edu
*Prgm Dir:* Julie A Taddeo, BS RT(R)
*Tel:* 610 526-6127   *Fax:* 610 526-6141
*E-mail:* jtaddeo@harcum.edu
*Med Dir:* J Thomas Murphy, MD

**Holy Spirit Hospital**
Radiography Prgm
503 N 21st St
Camp Hill, PA 17011-2288
www.hsh.org/home.htm
*Prgm Dir:* Kevin L Otto, MS RT(R)
*Tel:* 717 763-2123   *Fax:* 717 763-2963
*E-mail:* kotto@hsh.org
*Med Dir:* Barbara Kunkle, MD

**Clearfield Hospital**
Radiography Prgm
809 Turnpike Ave, PO Box 992
Clearfield, PA 16830
*Prgm Dir:* Sandra L Alsop, RT(R)(M)(CT)
*Tel:* 814 768-2230   *Fax:* 814 768-2258
*E-mail:* salsop@clearfieldhosp.org
*Med Dir:* Richard G Williams, MD

**College Misericordia**
Radiography Prgm
301 Lake St
Dallas, PA 18612-1098
www.misericordia.edu/mi
*Prgm Dir:* Elaine Halesey, EdD RT(R)(QM)
*Tel:* 570 674-6480   *Fax:* 570 674-3052
*E-mail:* ehalesey@misericordia.edu
*Med Dir:* Ron Konecke, MD

**Geisinger Medical Center**
Radiography Prgm
100 N Academy Ave
Danville, PA 17822-2007
www.geisinger.org
*Prgm Dir:* Kenneth A Roszel, MS RT(R)
*Tel:* 570 271-6301   *Fax:* 570 271-5976
*E-mail:* kroszel@geisinger.edu
*Med Dir:* Cathy Woomert, MD

**Gannon University**
Radiography Prgm
109 University Square
Erie, PA 16541-0001
www.gannon.edu
*Prgm Dir:* Cynthia L Liotta, MS RT(R)(CT)
*Tel:* 814 871-5644   *Fax:* 814 871-5662
*E-mail:* liotta@gannon.edu

**Conemaugh Memorial Medical Center**
Radiography Prgm
1086 Franklin St
Johnstown, PA 15905-4398
*Prgm Dir:* Gloria J Mongelluzzo, MEd RT(R)(M)
*Tel:* 814 534-9582   *Fax:* 814 534-9945
*E-mail:* gmongell@conemaugh.org
*Med Dir:* Gary Kramer, MD

**Armstrong County Memorial Hospital**
Radiography Prgm
One Nolte Dr
Kittanning, PA 16201
www.acmh.org
*Prgm Dir:* Paula Keister, BS RT(R)(M)
*Tel:* 724 543-8206   *Fax:* 724 543-8652
*E-mail:* keisterp@acmh.org

**Lancaster Gen Coll of Nursing & Hlth Sciences**
Radiography Prgm
410 N Lime St
Lancaster, PA 17602
www.LancasterGeneral.org
*Prgm Dir:* Robin L Harclerode, MEd RT(R)
*Tel:* 717 544-4865   *Fax:* 717 544-5970
*E-mail:* rlharcle@lancastergeneral.org
*Med Dir:* Jeffrey Kramer, MD

**Mansfield University**
Radiography Prgm
Elliott Hall
Mansfield, PA 16933
*Prgm Dir:* Jo Ann Hanlon, BS RT(R)(M)
*Tel:* 570 882-4007, Ext 4007   *Fax:* 570 882-6509
*E-mail:* jhanlon@ghs.guthrie.org

## Ohio Valley General Hospital
Radiography Prgm
25 Heckel Rd
McKees Rocks, PA 15136
www.ohiovalleyhospital.org
*Prgm Dir:* Susan Sherwin, MBA RT(R)(MR)
*Tel:* 412 777-6210   *Fax:* 412 777-6866
*E-mail:* ssherwin@ohiovalleyhospital.org
*Med Dir:* Theodore J Molnar, MD

## Comm Coll of Allegheny County
Radiography Prgm
595 Beatty Rd
Monroeville, PA 15146-1395
www.ccac.edu
*Prgm Dir:* August B Kellermann III, PhD RT(R)
*Tel:* 724 325-6754   *Fax:* 724 325-6701
*E-mail:* akellermann@ccac.edu
*Med Dir:* Peter Bonadio, MD

## Jameson Hospital
Radiography Prgm
1000 S Mercer St
New Castle, PA 16101
*Prgm Dir:* David W Hyser, BS RT(R)
*Tel:* 724 656-6134   *Fax:* 724 656-6148
*E-mail:* dhyser@jamesonhealthsystem.com

## Penn State University - New Kensington
Radiography Prgm
3550 Seventh St Rd Rte 780
New Kensington, PA 15068
www.nk.psu.edu/Academics/Degrees/radsci.htm
*Prgm Dir:* Debra Majetic
*Tel:* 724 334-6738   *Fax:* 724 334-6111
*E-mail:* dak25@psu.edu

## Bucks County Community College
Radiography Prgm
275 Swamp Rd, Tyler Hall 302A
Newtown, PA 18940
www.bucks.edu/healthcare/radiography
*Prgm Dir:* Gail J Hoffman, MEd RT(R)
*Tel:* 215 504-8644   *Fax:* 215 504-8504
*E-mail:* hoffmang@bucks.edu
*Med Dir:* Ronald Adelman, MD

## Albert Einstein Medical Center
Radiography Prgm
5501 Old York Rd
Philadelphia, PA 19141
http://einsteinxray.org
*Prgm Dir:* Bonnie L Benson, RT(R)
*Tel:* 215 456-6234   *Fax:* 215 456-3232
*E-mail:* bensonb@einstein.edu
*Med Dir:* Adam R Guttentag, MD

## Community College of Philadelphia
Radiography Prgm
1700 Spring Garden St
Philadelphia, PA 19130-3991
www.ccp.edu
*Prgm Dir:* Sally M Rensch, MPP RT(R)
*Tel:* 215 751-8424   *Fax:* 215 751-8937
*E-mail:* srensch@ccp.edu

## Drexel University
Radiography Prgm
245 N 15th St
Mail Stop 206
Philadelphia, PA 19102-1192
www.drexel.edu
*Prgm Dir:* La Vetta N Reliford, MSRS RT(R)
*Tel:* 215 762-3990   *Fax:* 215 762-3990
*E-mail:* lnr25@drexel.edu
*Med Dir:* Robert Siegle, MD

## Holy Family University
Radiography Prgm
Grant and Frankford Aves
9701 Frankford Ave
Philadelphia, PA 19114
*Prgm Dir:* Mark B Ness, BS RT(R)
*Tel:* 215 637-7700, Ext 3275   *Fax:* 215 827-0478
*E-mail:* mness@holyfamily.edu
*Med Dir:* Michael Kates, MD

## St Christopher Hospital School of Rad Tech
Radiography Prgm
Erie Ave at Front
Philadelphia, PA 19134
*Prgm Dir:* Celestine Coleman, BS RT(R)
*Tel:* 215 427-6751   *Fax:* 215 427-6845
*E-mail:* Celestine.Coleman@tenethealth.com

## Thomas Jefferson University
Radiography Prgm
130 S Ninth St, Ste 1010
Philadelphia, PA 19107-5233
www.tju.edu/chp
*Prgm Dir:* Frances H Gilman, MS RT(R)(CT)(MR)(CV)
*Tel:* 215 503-1865   *Fax:* 215 503-1031
*E-mail:* frances.gilman@jefferson.edu
*Med Dir:* Vijay Rao, MD

## Univ of Penn MC/Hosp of the Univ of Penn
Radiography Prgm
3400 Spruce St Basement Donner Building
Philadelphia, PA 19104
www.uphs.upenn.edu/radiology
*Prgm Dir:* Joanne M Niewood, MEd RT(R)(CT)
*Tel:* 215 662-7511   *Fax:* 215 614-0330
*E-mail:* joanne.niewood@uphs.upenn.edu
*Med Dir:* Michael Bleshman, MD

## UPMC Health System
Radiography Prgm
Murdock Building Ste 206
3434 Forbes Ave
Pittsburgh, PA 15213-2582
http://schoolofmedicalimaging.upmc.com
*Prgm Dir:* Denise Csonka Lake, MEd RT(R)(M)
*Tel:* 412 647-3528   *Fax:* 412 647-3713
*E-mail:* laked@upmc.edu
*Med Dir:* Diane C Strollo, MD

## Western Sch of Hlth & Business - Pittsburgh
Radiography Prgm
421 7th Ave
Pittsburgh, PA 15219
*Prgm Dir:* Bernadette Ann Trainer, MEd RT(R)
*Tel:* 412 281-2600, Ext 144   *Fax:* 412 281-0819
*E-mail:* btrainer@western-school.com
*Med Dir:* Joseph Lenkey, MD

## Montgomery County Community College
Radiography Prgm
101 College Dr
Pottstown, PA 19464
www.mc3.edu/aa/career/programs/rt.htm
*Prgm Dir:* Debra J Poelhuis, MS RT(R)(M)
*Tel:* 610 718-1813   *Fax:* 610 718-1983
*E-mail:* dpoelhui@mc3.edu

## Reading Hospital & Medical Center
Radiography Prgm
PO Box 16052
Reading, PA 19612-6052
www.readinghospital.org
*Prgm Dir:* Kathleen R Jackson, BS RT(R)
*Tel:* 610 988-8993   *Fax:* 610 988-8400
*E-mail:* jacksonk@readinghospital.org
*Med Dir:* Robert Guay, MD

## St Joseph Medical Center
Radiography Prgm
12th and Walnut Sts, PO Box 316
Reading, PA 19603
*Prgm Dir:* Cynthia L Keane, RT(R)
*Tel:* 610 378-2237   *Fax:* 610 378-2803
*E-mail:* cynthiakeane@catholichealth.net
*Med Dir:* Irving Ehrlich, MD

## Penn State University - Schuylkill Haven
Radiography Prgm
200 University Dr
Schuylkill Haven, PA 17972-0308
www.sl.psu.edu/programs/2rsca.html
*Prgm Dir:* David K Rill, MEd RT(R)
*Tel:* 570 385-6108   *Fax:* 570 385-6105
*E-mail:* dkr7@psu.edu

## Johnson College
Radiography Prgm
3427 N Main Ave
Scranton, PA 18508-1495
www.johnson.edu
*Prgm Dir:* Jane Maas, BS RT(R)
*Tel:* 570 342-6404, Ext 176
*E-mail:* jmaas@johnson.edu
*Med Dir:* Larry Kestin, MD

## UPMC Northwest
Radiography Prgm
100 Fairfield Dr
Seneca, PA 16346
*Prgm Dir:* Walter G Jones Sr, BS RT(R)
*Tel:* 814 677-1433, Ext na   *Fax:* 814 677-1770
*E-mail:* joneswg@upmc.edu
*Med Dir:* Donald Bittner, MD

## Sewickley Vlly Hosp/Heritage Valley Hlth Sys
Radiography Prgm
720 Blackburn Rd
Sewickley, PA 15143
*Prgm Dir:* Joyce E Cirelli, MS RT(R)
*Tel:* 724 773-4761, Ext 4761   *Fax:* 412 749-4241
*E-mail:* jcirelli@hvhs.org
*Med Dir:* Mark A Schnurer, MD

## Sharon Regional Health System
Radiography Prgm
740 E State St
Sharon, PA 16146-3395
www.srhs-pa.org
*Prgm Dir:* Sherry Masotto Swetz, MS RT(R)
*Tel:* 724 983-5603   *Fax:* 724 983-5614
*E-mail:* smasotto@srhs-pa.org

## Crozer - Chester Med Ctr
Radiography Prgm
One Medical Center Blvd
Upland, PA 19013
*Prgm Dir:* Lisa M Iacovelli, BS RT(R)(QM)
*Tel:* 610 447-2578   *Fax:* 610 447-6137
*E-mail:* lisa.iacovelli@crozer.org
*Med Dir:* Joseph Stock, MD

## Washington Hospital
Radiography Prgm
155 Wilson Ave
Washington, PA 15301
*Prgm Dir:* Karen C Williams, MA RT(R)(QM)
*Tel:* 724 223-3326   *Fax:* 724 250-4417
*E-mail:* kwilliams@washingtonhospital.org
*Med Dir:* Stephen Kelminson, MD

## Wilkes-Barre General Hospital
Radiography Prgm
575 N River St
Wilkes-Barre, PA 18764
www.wvhcs.org
*Prgm Dir:* Kathleen A Smith, BA RT(R)
*Tel:* 570 552-1760   *Fax:* 570 552-2476
*E-mail:* ksmith@wvhcs.org

**Pennsylvania College of Technology**
Radiography Prgm
One College Ave
Williamsport, PA 17701-5799
*Prgm Dir:* Robert J Slothus, MS RT(R)
*Tel:* 570 326-3761, Ext 7409   *Fax:* 570 321-5556
*E-mail:* rslothus@pct.edu
*Med Dir:* John Becker, DO

**Abington Memorial Hospital**
Radiography Prgm
2500 Maryland Rd
Willow Grove, PA 19090-1284
*Prgm Dir:* Deborah A Bazinet, BS RT(R)(M)
*Tel:* 215 481-5421   *Fax:* 215 481-5490
*E-mail:* dbazinet@amh.org
*Med Dir:* Richard Weiss, MD

# Puerto Rico

**Universidad Central del Caribe**
Radiography Prgm
PO Box 60327
Bayamon, PR 00960-6032
www.uccaribe.edu
*Prgm Dir:* Jose Rafael Moscoso-Alvarez, EdD LT
*Tel:* 787 798-3001, Ext 2330   *Fax:* 787 785-3425
*E-mail:* jose.moscoso@uccaribe.edu

**Universidad del Este**
Radiography Prgm
PO Box 2010
Carolina, PR 00984-2010
*Prgm Dir:* Ynes Cordova-Acosta, MPH LT
*Tel:* 787 257-7373, Ext 3252
*E-mail:* ycordova@suagm.edu

**Inter American University of Puerto Rico**
Radiography Prgm
PO Box 5100
San German, PR 00683-9801
www.sg.inter.edu
*Prgm Dir:* Sara L Torres, MBA RT(R)
*Tel:* 787 264-1912, Ext 7438   *Fax:* 787 892-6350
*E-mail:* sara@sg.inter.edu

**University of Puerto Rico**
Radiography Prgm
GPO Box 5067
San Juan, PR 00936-5067
*Prgm Dir:* Juan Melendez-Sostre, MEd MPH RT(R)
*Tel:* 787 751-4434, Ext 4609   *Fax:* 787 751-4434
*E-mail:* juanmelendez@cprs.rcm.upr.edu
*Med Dir:* Edgar Colon, MD

# Rhode Island

**Community College of Rhode Island**
Radiography Prgm
Flanagan Campus
1762 Louisquisset Pike
Lincoln, RI 02865-4585
*Prgm Dir:* Sharon E Perkins, MEd RT(R)(M)
*Tel:* 401 333-7144   *Fax:* 401 333-7441
*E-mail:* sperkins@ccri.edu

**Rhode Island Hospital**
Radiography Prgm
3 Davol Square
Bldg A, 4th Fl
Providence, RI 02903
www.lifespan.org/diagimag
*Prgm Dir:* Ellen E Alexandre, BS RT(R)
*Tel:* 401 528-8531, Ext 223   *Fax:* 401 457-0219
*E-mail:* ealexandre@lifespan.org
*Med Dir:* Holly Gil, MD

# South Carolina

**Aiken Technical College**
Radiography Prgm
PO Drawer 696
Aiken, SC 29802-0696
www.atc.edu
*Prgm Dir:* Jean Archer Fishel, BS RT(R)
*Tel:* 803 593-9231, Ext 1634   *Fax:* 803 593-3106
*E-mail:* fishelj@atc.edu

**AnMed Health Medical Center**
Radiography Prgm
800 N Fant St
Anderson, SC 29621
*Prgm Dir:* Gaye Nichols, RT(R)
*Tel:* 864 260-3705   *Fax:* 864 261-1319
*E-mail:* gnichols@anmed.com
*Med Dir:* Thomas U Tuten, MD

**Technical College of the Lowcountry**
Radiography Prgm
PO Box 1288
Beaufort, SC 29901
www.tcl.edu
*Prgm Dir:* John W Eichinger, MS RS RT(R)(CT)
*Tel:* 843 470-8397   *Fax:* 843 525-8268
*E-mail:* jeichinger@tcl.edu

**Trident Technical College**
Radiography Prgm
7000 Rivers Ave, PO Box 118067
Charleston, SC 29423-8067
www.tridenttech.edu
*Prgm Dir:* Krista R Gentry, BHS RT(R)
*Tel:* 843 574-6077   *Fax:* 843 574-6585
*E-mail:* krista.gentry@tridenttech.edu

**Midlands Technical College**
Radiography Prgm
PO Box 2408
Columbia, SC 29202
www.midlandstech.edu
*Prgm Dir:* C William Mulkey, EdD RT(R)
*Tel:* 803 822-3482   *Fax:* 803 822-3417
*E-mail:* mulkeyb@midlandstech.edu
*Med Dir:* Robert Waldron II, MD

**Horry - Georgetown Technical College**
Radiography Prgm
Hwy 501 E, PO Box 261966
Conway, SC 29528-6066
www.hor.tec.sc.us/default800.htm
*Prgm Dir:* Beckey Miller, MS RT(R)
*Tel:* 843 349-5384   *Fax:* 843 349-7880
*E-mail:* beckey.miller@hgtc.edu
*Med Dir:* Scott Mencken, MD

**Florence-Darlington Technical College**
Radiography Prgm
PO Box 100548
Florence, SC 29501-0548
www.fdtc.edu
*Prgm Dir:* E Yancy Wells, BS RT(R)
*Tel:* 843 676-8529   *Fax:* 843 292-0851
*E-mail:* Yancy.Wells@fdtc.edu
*Med Dir:* Raymond L Thomas, MD

**Greenville Technical College**
Radiography Prgm
PO Box 5616
Greenville, SC 29606
www.greenvilletech.com
*Prgm Dir:* Janet K Hirt, BS RT(R)(MR)
*Tel:* 864 250-8316   *Fax:* 864 250-8462
*E-mail:* jan.hirt@gvltec.edu

**Piedmont Technical College**
Radiography Prgm
PO Box 1467, Emerald Rd
Greenwood, SC 29648-1467
*Prgm Dir:* William Lee Balentine Sr, MSM RT(R)
*Tel:* 864 941-8523   *Fax:* 864 941-8684
*E-mail:* balentine.l@ptc.edu
*Med Dir:* W A Kitchens, Jr, MD

**Orangeburg Calhoun Technical College**
Radiography Prgm
3250 St Matthews Rd
Orangeburg, SC 29118-8299
*Prgm Dir:* Frances W Andrews, BHS RT(R)
*Tel:* 803 535-1356   *Fax:* 803 535-1350
*E-mail:* andrewsf@octech.edu
*Med Dir:* Dallas Lovelace, MD

**York Technical College**
Radiography Prgm
452 S Anderson Rd
Rock Hill, SC 29730
*Prgm Dir:* Michele Rutan Wells, BS RT(R)(M)RDMS
*Tel:* 803 981-7036   *Fax:* 803 327-8059
*E-mail:* mwells@yorktech.com

**Spartanburg Community College**
Radiography Prgm
PO Drawer 4386
Spartanburg, SC 29305-4386
www.sccsc.edu
*Prgm Dir:* Deborah B Jennings, BS RT(R)(M)(QM)
*Tel:* 864 592-4722   *Fax:* 864 591-3881
*E-mail:* jenningsd@sccsc.edu
*Med Dir:* Larry Warren, MD

# South Dakota

**Presentation College**
Radiography Prgm
1500 N Main St
Aberdeen, SD 57401
www.presentation.edu
*Prgm Dir:* Robert V Hagen, BS RT(R)
*Tel:* 605 229-8355   *Fax:* 605 229-8518
*E-mail:* robert.hagen@presentation.edu

**Mitchell Technical Institute**
Radiography Prgm
821 N Capital
Mitchell, SD 57301
www.mitchelltech.com/departments/radtech
*Prgm Dir:* Eric A Schaffer, BS RT(R)(CT)
*Tel:* 605 995-3068
*E-mail:* eric.schaffer@mitchelltech.edu
*Med Dir:* Carey Buhler, MD

**Rapid City Regional Hospital**
Radiography Prgm
353 Fairmont Blvd, PO Box 6000
Rapid City, SD 57709-6000
www.rcrh.org
*Prgm Dir:* Jerilyn Jo Powell, BS RT(R) RDMS RVT
*Tel:* 605 719-8433   *Fax:* 605 719-1436
*E-mail:* jpowell2@rcrh.org
*Med Dir:* Baxter Ron, MD

**Avera McKennan Hospital**
Radiography Prgm
800 E 21st St, PO Box 5045
Sioux Falls, SD 57117-5045
*Prgm Dir:* Stephen Cooper, BS RT(R)
*Tel:* 605 322-1720   *Fax:* 605 322-1701
*E-mail:* stephen.cooper@mckennan.org
*Med Dir:* Patrick Nelson, MD

**Sioux Valley Hospital**
Radiography Prgm
1305 W 18th St, PO Box 5039
Sioux Falls, SD 57117-5039
*Prgm Dir:* Kenneth G Lee, BS RT(R)
*Tel:* 605 333-6466   *Fax:* 605 333-1554
*E-mail:* leek@siouxvalley.org
*Med Dir:* Randal L Welter, MD

**Avera Sacred Heart Hospital**
Radiography Prgm
501 Summit St
Yankton, SD 57078-9967
www.averasacredheart.com
*Prgm Dir:* Anessa M Van Osdel, BS RT(R)(M)
*Tel:* 605 668-8158   *Fax:* 605 668-8153
*E-mail:* avanosdel@shhservices.com
*Med Dir:* Frank Messner, MD

# Tennessee

**Chattanooga State Technical Comm College**
Radiography Prgm
4501 Amnicola Hwy
Chattanooga, TN 37406-1097
www.chattanoogastate.edu/allied_health/rad_tech/rad_
   tech_main.asp
*Prgm Dir:* Margery Sanders, MBA RT(R)(M)(QM)
*Tel:* 423 697-4450, Ext 4211   *Fax:* 423 634-3071
*E-mail:* margery.sanders@chattanoogastate.edu
*Med Dir:* Deloris Rissling, MD

**Columbia State Community College**
Radiography Prgm
PO Box 1315
1665 Hampshire Pike
Columbia, TN 38402-1315
*Prgm Dir:* Brenda M Coleman, MSRS RT(R)
*Tel:* 931 540-2745, Ext 2745   *Fax:* 931 540-2798
*E-mail:* bcoleman@columbiastate.edu
*Med Dir:* Gary Podgorski, MD

**MedVance Institute**
Radiography Prgm
1025 Highway 111
Cookeville, TN 38501
*Prgm Dir:* William H May, MEd RT(R)
*Tel:* 931 526-3660, Ext 233
*E-mail:* bill.may@medvance.edu

**East Tennessee State University**
Radiography Prgm
Nave Center, 1000 West E St
Elizabethton, TN 37643
*Prgm Dir:* Shirley J Cherry, MBA RT(R)
*Tel:* 423 547-4912   *Fax:* 423 547-4921
*E-mail:* cherrys@etsu.edu
*Med Dir:* Kelly Cassedy, MD

**Volunteer State Community College**
Radiography Prgm
1480 Nashville Pike
Gallatin, TN 37066-3188
www.volstate.edu
*Prgm Dir:* Monica M (White) Korpady, MS RT(R)(M)
*Tel:* 615 230-3651   *Fax:* 615 230-3224
*E-mail:* mkorpady@volstate.edu
*Med Dir:* E Paul Nance, MD

**Jackson State Community College**
Radiography Prgm
2046 N Parkway St
Jackson, TN 38301-3797
www.jscc.edu/allied/rad.htm
*Prgm Dir:* Gerald L Graddy, MS RT(R)
*Tel:* 731 424-3520, Ext 299   *Fax:* 731 425-9551
*E-mail:* ggraddy@jscc.edu
*Med Dir:* Thomas R Thompson, MD

**South College**
Radiography Prgm
3904 Lonas Dr
Knoxville, TN 37909
www.southcollegetn.edu
*Prgm Dir:* Donna Shehane, EdD RT(R)
*Tel:* 865 251-1885   *Fax:* 865 584-9816
*E-mail:* dshehane@southcollegetn.edu
*Med Dir:* James LePage, MD

**Univ of Tennessee Medical Ctr at Knoxville**
Radiography Prgm
Radiology Dept
1924 Alcoa Hwy
Knoxville, TN 37920-6999
www.utmedicalcenter.org/radiology
*Prgm Dir:* Clyde R Hembree, MBA RT(R)
*Tel:* 865 544-9005   *Fax:* 865 544-8581
*E-mail:* chembree@mc.utmck.edu
*Med Dir:* Judson Cash, MD

**Baptist College of Health Sciences**
Radiography Prgm
1003 Monroe Ave
Memphis, TN 38104
www.bchs.edu
*Prgm Dir:* Wanda Lillie, MS RT(R)
*Tel:* 901 572-2646   *Fax:* 901 572-2750
*E-mail:* wanda.lillie@bchs.edu
*Med Dir:* James E Machin, MD

**Methodist University Hospital**
Radiography Prgm
1265 Union Ave
Memphis, TN 38104
www.methodisthealth.org
*Prgm Dir:* Peggy D Franklin, BS RT(R)
*Tel:* 901 726-8099   *Fax:* 901 726-2870
*E-mail:* franklip@methodisthealth.org
*Med Dir:* Frank D Parks, MD

**Southwest Tennessee Community College**
Radiography Prgm
PO Box 780
Memphis, TN 38101-0780
www.southwest.tn.edu
*Prgm Dir:* Thomas H Wolfe, MSRS RT(R)
*Tel:* 901 333-5417   *Fax:* 901 333-5102
*E-mail:* thwolfe@southwest.tn.edu

**Nashville General Hospital**
Radiography Prgm
1818 Albion St
Nashville, TN 37208
*Prgm Dir:* Kenneth W Jones, MEd RT(R)
*Tel:* 615 341-4440   *Fax:* 615 341-4906
*E-mail:* ken.w.jones@nashville.gov
*Med Dir:* Anthony Disher, MD

**Roane State Community College**
Radiography Prgm
701 Briarcliff Ave
Oak Ridge, TN 37830
www.roanestate.edu keyword: radiology
*Prgm Dir:* Jean Reif Robinson, BA RT(R)
*Tel:* 865 481-2015   *Fax:* 865 481-2019
*E-mail:* robinsonjr@roanestate.edu

# Texas

**Hendrick Medical Center**
Radiography Prgm
1900 Pine St
Abilene, TX 79601-2316
www.ehendrick.org/radiography
*Prgm Dir:* Richard K Bower, MEd RT(R)
*Tel:* 352 670-2427   *Fax:* 352 670-2990
*E-mail:* rbower@ehendrick.org
*Med Dir:* Michel Duma, MD

**Amarillo College**
Radiography Prgm
PO Box 447
Amarillo, TX 79178
www.actx.edu
*Prgm Dir:* Bill Crawford, BAAS RT(R)
*Tel:* 806 354-6070   *Fax:* 806 354-6076
*E-mail:* crawford-be@actx.edu
*Med Dir:* Gayle Bickers, MD

**Austin Community College**
Radiography Prgm
3401 Webberville Road
Austin, TX 78702
www.austincc.edu
*Prgm Dir:* Rudy L Garza, MS RT(R)
*Tel:* 512 223-5817   *Fax:* 512 223-5901
*E-mail:* rudygarz@austincc.edu
*Med Dir:* Marcus L Lines, MD

**Lamar Institute of Technology**
Radiography Prgm
PO Box 10061
Beaumont, TX 77710
www.lit.edu
*Prgm Dir:* Brenda A Barrow, MEd RT(R)(CT)
*Tel:* 409 880-8845   *Fax:* 409 880-8955
*E-mail:* brenda.barrow@lit.edu
*Med Dir:* Joseph Nightingale, MD

**Memorial Hermann Baptist Hospital Beaumont**
Radiography Prgm
PO Drawer 1591
Beaumont, TX 77704
www.mhbh.org
*Prgm Dir:* Carolyn M Nicholas, MEd RT(R)(M)
*Tel:* 409 212-5726   *Fax:* 409 212-5743
*E-mail:* carolyn.nicholas@mhbh.org

**Univ Tx at Brownsville/Tx Southmost Coll**
Radiography Prgm
80 Ft Brown
Brownsville, TX 78520
*Prgm Dir:* Manuel Gavito, BS RT(R)
*Tel:* 956 554-5010   *Fax:* 956 554-8910
*E-mail:* otivag@utb.edu
*Med Dir:* William McKinney, MD

**Blinn College**
Radiography Prgm
PO Box 6030
Bryan, TX 77805-6030
*Prgm Dir:* M Elia Flores, MEd RT(R)
*Tel:* 979 209-7566   *Fax:* 979 209-7289
*E-mail:* sflores@blinn.edu
*Med Dir:* Ernest Elmendorf, MD

**N Harris Mont Comm Coll Dist/Montgomery
Coll**
Radiography Prgm
3200 College Park Dr
Conroe, TX 77384
www.nhmccd.edu
*Prgm Dir:* Cynthia A Griffith, EdD RT RDMS
*Tel:* 936 273-7328   *Fax:* 936 273-7054
*E-mail:* cgriffith@nhmccd.edu

**Del Mar College**
Radiography Prgm
101 Baldwin and Ayers Sts
Corpus Christi, TX 78404
*Prgm Dir:* Ismael Garcia, BAAS RT(R)
*Tel:* 361 698-2858, Ext 2826   *Fax:* 361 698-2811
*E-mail:* igarcia@delmar.edu
*Med Dir:* Chandra S Katragadda, MD

**Cy-Fair College**
Radiography Prgm
9191 Barker-Cypress Rd
Cypress, TX 77433
www.cy-faircollege.com
*Prgm Dir:* Cynthia B Robertson, MEd RT(R)
*Tel:* 281 290-3966   *Fax:* 281 290-5282
*E-mail:* cynthia.b.robertson@nhmccd.edu

**Baylor University Med Center**
Radiography Prgm
3616 Worth St
Dallas, TX 75246
*Prgm Dir:* Terri Wyly, MS RT(R)(T)
*Tel:* 214 820-3780   *Fax:* 214 820-7773
*E-mail:* terriwy@baylorhealth.edu
*Med Dir:* Karl Glastad, MD

**El Centro College**
Radiography Prgm
Main and Lamar Sts
Dallas, TX 75202-3604
www.elcentrocollege.edu/Programs/HealthLegalStudies/
  RadSci/
*Prgm Dir:* Jolayne Jackson, BA RT(R)
*Tel:* 214 860-2278   *Fax:* 214 860-2268
*E-mail:* jxj5540@dcccd.edu
*Med Dir:* Robert Parkey, MD

**El Paso Community College**
Radiography Prgm
PO Box 20500
El Paso, TX 79998
www.epcc.edu
*Prgm Dir:* Christl E Thompson, MA RT(R)
*Tel:* 915 831-4098   *Fax:* 915 831-4131
*E-mail:* christlt@epcc.edu
*Med Dir:* Chetan Moorthy, MD

**US Army Medical Dept Center & School**
Radiography Prgm
3151 Scott Rd
Ste 1316, MCCS-HCR
Ft Sam Houston, TX 78234-6137
http://radiology.amedd.army.mil
*Prgm Dir:* Brunhilde S Green, MA RT(R)
*Tel:* 210 221-8958   *Fax:* 210 221-7611
*E-mail:* brunhilde.green@amedd.army.mil
*Med Dir:* James A Breitweser, MD

**Career Centers of Texas - Ft Worth**
Radiography Prgm
2001 Beach St, Ste 201
Ft Worth, TX 76103
*Prgm Dir:* Carolyn A Johnson, BS RT(R)
*Tel:* 817 413-2000
*E-mail:* cjohns04@jpshealthnetwork.org

**Galveston College**
Radiography Prgm
4015 Ave Q
Galveston, TX 77550-2782
*Prgm Dir:* Deborah M Scroggins, MS
  RT(R)(CV)(M)(CT)
*Tel:* 409 944-1496   *Fax:* 409 944-3014
*E-mail:* dscroggi@gc.edu
*Med Dir:* Leonard E Swischuk, MD

**Harris County Hosp Dist/Ben Taub Gen Hosp**
Radiography Prgm
c/o Lyndon B Johnson General Hospital
5656 Kelley St
Houston, TX 77026
*Prgm Dir:* Hazel E Bourne, MS RT(R)
*Tel:* 713 873-2276   *Fax:* 713 873-2416
*E-mail:* hazel_bourne@hchd.tmc.edu

**Houston Community College**
Radiography Prgm
Coleman College For Health Sciences
1900 Pressler Dr
Houston, TX 77030
www.hccs.edu
*Prgm Dir:* Lynne Y Davis, EdD RT(R)
*Tel:* 713 718-7367   *Fax:* 713 718-7608
*E-mail:* lynne.davis@hccs.edu
*Med Dir:* Patrick Conoley, MD

**MedVance Institute**
Radiography Prgm
Houston/KIMC Investments Inc
6220 Westpark, Ste 180
Houston, TX 77057
*Prgm Dir:* Roy M Smither, II, MA RT(R)
*Tel:* 713 266-6594
*E-mail:* rsmither@medvance.org

**Memorial Hermann Healthcare System**
Radiography Prgm
Technical Education Center
921 Gessner
Houston, TX 77024
*Prgm Dir:* Rita J Robinson, BS RT(R)
*Tel:* 713 932-4861   *Fax:* 713 932-3701
*E-mail:* rita_robinson@mhhs.org
*Med Dir:* Samuel L Smiley, MD

**Tarrant County College**
Radiography Prgm
828 W Harwood Rd
Hurst, TX 76054-3299
www.tccd.edu
*Prgm Dir:* Mark Holt, MS RT(R)
*Tel:* 817 515-6569   *Fax:* 817 515-6700
*E-mail:* mark.holt@tccd.edu

**Laredo Community College**
Radiography Prgm
West End Washington St
Laredo, TX 78040-4395
www.laredo.edu
*Prgm Dir:* Oscar Gomez, BAAS RT(R)
*Tel:* 956 721-5386   *Fax:* 956 721-5431
*E-mail:* ogomez@laredo.edu
*Med Dir:* Salah Rafati, MD

**Covenant Medical Center**
Radiography Prgm
3706-20th Ste A
Lubbock, TX 79410
www.covenantsor.com
*Prgm Dir:* Lori Jordan, MS RT(R)(MR)
*Tel:* 806 725-0456   *Fax:* 806 797-4350
*E-mail:* ljordan@covhs.org
*Med Dir:* Rob Posteraro, MD

**South Plains College**
Radiography Prgm
Reese Center
819 Gilbert Dr
Lubbock, TX 79416
www.southplainscollege.edu
*Prgm Dir:* Denny Barnes, BS RT(R)
*Tel:* 806 885-3048, Ext 4629   *Fax:* 806 885-5608
*E-mail:* dbarnes@southplainscollege.edu

**Angelina College**
Radiography Prgm
PO Box 1768
3500 S First
Lufkin, TX 75902
www.angelina.edu
*Prgm Dir:* Angie L Wilcox, MEd RT(R)(CT)
*Tel:* 936 633-5413, Ext 5413   *Fax:* 936 633-3207
*E-mail:* awilcox@angelina.edu
*Med Dir:* B G Kistler, MD

**Midland College**
Radiography Prgm
3600 N Garfield
Midland, TX 79705-6399
www.midland.edu
*Prgm Dir:* Quinn B Carroll, MEd RT(R)
*Tel:* 915 685-4600, Ext 4592   *Fax:* 915 685-4762
*E-mail:* eskimo@midland.edu
*Med Dir:* Marlon Hughes, MD

**Odessa College**
Radiography Prgm
201 W University Blvd
Odessa, TX 79764
www.odessa.edu
*Prgm Dir:* Carolyn Sue Leach, BS RT(R)
*Tel:* 432 335-6449   *Fax:* 432 335-6846
*E-mail:* sleach@odessa.edu

**San Jacinto College Central**
Radiography Prgm
8060 Spencer Hwy, PO Box 2007
Pasadena, TX 77501-2007
*Prgm Dir:* Christopher J Gould, MS RT(R)
*Tel:* 281 476-1871   *Fax:* 281 478-2754
*E-mail:* cgould@sjcd.edu
*Med Dir:* Neil Langley, MD

**Baptist Health System**
Radiography Prgm
8400 Datapoint Dt
Ste 226
San Antonio, TX 78229
www.bshp.edu
*Prgm Dir:* Anna Y Flores, BS, R T (R)(M)
*Tel:* 210 297-9160   *Fax:* 210 297-0941
*E-mail:* aflores@baptisthealthsystem.org
*Med Dir:* Polly Hansen, MD

**St Philip's College**
Radiography Prgm
1801 Martin Luther King Dr
San Antonio, TX 78203-2098
*Prgm Dir:* Donna Laird, BS RT(R)(M)
*Tel:* 210 531-3422   *Fax:* 210 531-3503
*E-mail:* dlaird@accd.edu
*Med Dir:* Ewell Clarke, MD

**USAF School of Health Care Sciences**
Radiography Prgm
939 Missile Rd, 382 TRS/XYAF
Sheppard AFB, TX 76311-2246
*Prgm Dir:* TSgt Danyell A Gardner, BS RT(R)
*Tel:* 940 676-3807   *Fax:* 940 676-2210
*E-mail:* danyell.gardner@sheppard.af.mil

**Tyler Junior College**
Radiography Prgm
PO Box 9020
Tyler, TX 75711
www.tjc.edu
*Prgm Dir:* Nancy A Wardlow, MS RT(R)
*Tel:* 903 510-2346   *Fax:* 903 510-2880
*E-mail:* nwar1@tjc.edu

**Citizens Medical Center**
Radiography Prgm
2701 Hospital Dr
Victoria, TX 77901
http://citizensmedicalcenter.org
*Prgm Dir:* Mary Jane Reynolds, BS RT(R)(T)
*Tel:* 361 572-5062   *Fax:* 361 572-5091
*E-mail:* mjreynolds@cmcvtx.org
*Med Dir:* D Bruce Tharp, MD

**McLennan Community College**
Radiography Prgm
1400 College Dr
Waco, TX 76708
*Prgm Dir:* Heather Zanek, BSRS RT(R) CT
*Tel:* 254 299-8342   *Fax:* 254 299-8397
*E-mail:* hzanek@mclennan.edu
*Med Dir:* David Risinger, MD

## Wharton County Junior College

Radiography Prgm
911 Boling Hwy, Ste J230
Wharton, TX 77488
www.wcjc.edu
*Prgm Dir:* Sharla S Walker, BS RT(R)(M)(QM)
*Tel:* 979 532-6379   *Fax:* 979 532-6489
*E-mail:* sharlaw@wcjc.cc.tx.us
*Med Dir:* O Preston Copeland, MD

## Midwestern State University

Radiography Prgm
3410 Taft Blvd
Wichita Falls, TX 76308-2099
*Prgm Dir:* Nadia Bugg, PhD RT(R) FAERS
*Tel:* 940 397-4571   *Fax:* 940 397-4845
*E-mail:* nadia.bugg@mwsu.edu
*Med Dir:* Paul Bice, MD

# Utah

## Salt Lake Community College

Radiography Prgm
South City Campus
1575 S State St
Salt Lake City, UT 84115
*Prgm Dir:* Lisa Wood, MS RT(R)
*Tel:* 801 957-3254   *Fax:* 801 957-3300
*E-mail:* lisa.wood@slcc.edu
*Med Dir:* Edwin Arthur Stevens, MD

# Vermont

## Champlain College

Radiography Prgm
163 S Willard St, PO Box 670
Burlington, VT 05402-0670
www.champlain.edu
*Prgm Dir:* Susyn Dees, MS RT(R)(CT)
*Tel:* 802 865-6469   *Fax:* 802 860-2750
*E-mail:* dees@champlain.edu
*Med Dir:* Curtis Green, MD

# Virginia

## University of Virginia Health System

*Cosponsor:* University of Virginia Medical Center
Radiography Prgm
Jefferson Park Ave, Radiology Box 800377
Charlottesville, VA 22908
www.med.virginia.edu
*Prgm Dir:* Jody Crane, BS RTR-BD
*Tel:* 434 924-9344   *Fax:* 434 982-0626
*E-mail:* jdc5s@virginia.edu
*Med Dir:* Hubert Shafer Jr, MD

## Danville Regional Medical Center

Radiography Prgm
142 South Main St
Danville, VA 24541
www.danvilleregional.org
*Prgm Dir:* Kevin L Murray, MSEd RT(R)(QM)
*Tel:* 434 799-2271
*E-mail:* murrayk@drhsi.org
*Med Dir:* Henry Huson III, DO

## Mary Washington Hospital

Radiography Prgm
1001 Sam Perry Blvd
Fredericksburg, VA 22401
*Prgm Dir:* Damon L Yaughn, RT(R)
*Tel:* 540 741-1802   *Fax:* 540 741-2560
*Med Dir:* Thomas Medsker, MD

## Rockingham Memorial Hospital

Radiography Prgm
235 Cantrell Ave
Harrisonburg, VA 22801-3293
www.rhc.com
*Prgm Dir:* Russell Crank, BA RT(R)
*Tel:* 540 433-4532   *Fax:* 540 433-4423
*E-mail:* rcrank@rhcc.com
*Med Dir:* Dennis G Rohrer, MD

## Central Virginia Community College

Radiography Prgm
3506 Wards Rd
Lynchburg, VA 24502-2498
www.cvcc.vccs.edu
*Prgm Dir:* Eddie W Haynes, MEd RT(R)
*Tel:* 434 832-7683   *Fax:* 434 832-7835
*E-mail:* HaynesE@cvcc.vccs.edu
*Med Dir:* Kenvin O Hicks, MD

## ACT College

Radiography Prgm
8870 Rixlew Ln Ste 201
Manassas, VA 20109
http://actcollege.edu
*Prgm Dir:* Donna Carroll Van Norman, MEd RT(R)(M)
*Tel:* 703 527-6660, Ext x471   *Fax:* 703 365-9288
*E-mail:* dvannorman@actcollege.edu

## Riverside School of Health Careers

Radiography Prgm
Newport News Public Schools
School of Radiologic Technology
Newport News, VA 23601
*Prgm Dir:* Pamela Gebhart-Cline, MSFE RT(R)(M)
*Tel:* 757 240-2202   *Fax:* 757 240-2201
*E-mail:* pam.cline@rivhs.com
*Med Dir:* Jonothan Demeo, MD

## Southside Regional Medical Center

Radiography Prgm
801 S Adams St
Petersburg, VA 23803
www.srmconline.com
*Prgm Dir:* Pamela J G Shelton, MBA RT(R)
*Tel:* 804 862-5883   *Fax:* 804 862-5171
*E-mail:* pshelton@chs.net
*Med Dir:* Cary Straton, MD

## Naval School of Health Sciences

Radiography Prgm
1001 Holcomb Rd
Portsmouth, VA 23708-5200
*Prgm Dir:* Tony L Ellison, BS RT(R)
*Tel:* 757 953-6420   *Fax:* 757 953-5033
*E-mail:* tlellison@hsp.med.navy.mil

## Southwest Virginia Community College

Radiography Prgm
PO Box SVCC
Richlands, VA 24641-1510
www.sw.vccs.edu
*Prgm Dir:* Donald B Lowe, MEd RT(R)
*Tel:* 540 964-7306   *Fax:* 540 964-7715
*E-mail:* don.lowe@sw.edu

## Bon Secours School of Medical Imaging

Radiography Prgm
8550 Magellan Parkway Ste 1100
Richmond, VA 23227
*Prgm Dir:* Kimberly Likens-Perry, MSEd RT(R)
*Tel:* 804 627-5307   *Fax:* 804 627-5304
*E-mail:* kim_likens@bshsi.com

## Virginia Commonwealth University

Radiography Prgm
701 W Grace St, Ste 2100
Box 843057
Richmond, VA 23284-3057
www.sahp.vcu.edu/radsci
*Prgm Dir:* Terri L Fauber, EdD RT(R)(M)
*Tel:* 804 828-9104   *Fax:* 804 828-5778
*E-mail:* tfauber@vcu.edu
*Med Dir:* James Messmer, MD

## Virginia Western Community College

Radiography Prgm
3095 Colonial Ave SW
PO Box 14007
Roanoke, VA 24038
*Prgm Dir:* Shirl Duke Lamanca, MSEd RT(R)
*Tel:* 540 857-6285   *Fax:* 540 857-6224
*E-mail:* slamanca@vw.vccs.edu
*Med Dir:* Abram Patterson, MD

## Tidewater Community College

Radiography Prgm
1700 College Crescent
Virginia Beach, VA 23453
www.tcc.edu
*Prgm Dir:* Kim B Utley, MS RT(R)
*Tel:* 757 822-7253   *Fax:* 757 822-7503
*E-mail:* kutley@tcc.edu

## Winchester Medical Center Inc

Radiography Prgm
1840 Amherst St
PO Box 3340
Winchester, VA 22601
*Prgm Dir:* John D Orndoff, BS RT(R)
*Tel:* 540 536-8136   *Fax:* 540 536-8827
*E-mail:* jorndorf@valleyhealthlink.com
*Med Dir:* Joseph Poe, MD

# Washington

## Pima Medical Institute - Seattle

Radiography Prgm
9709 Third Ave NE, #400
Seattle, WA 98115
www.pmi.edu
*Prgm Dir:* Jacqueline Kralik, BAS RT(R)(CT)(MR)
*Tel:* 206 322-6100, Ext 114   *Fax:* 206 324-1985
*E-mail:* jkralik@pmi.edu
*Med Dir:* Philip Lowe, MD

## Apollo College

Radiography Prgm
10102 E Knox Ste 200
Spokane, WA 99206
www.apollocollege.com
*Prgm Dir:* Susan Johnson
*Tel:* 509 532-8888   *Fax:* 509 533-5983
*E-mail:* sjohnson@apollocollege.edu

## Spokane Community College

Radiography Prgm
1810 N Greene St, MS 1090
Spokane, WA 99217
www.scc.spokane.edu
*Prgm Dir:* Deborah K Miller, MEd RT(R)(CV)(M)
*Tel:* 509 624-5012   *Fax:* 509 624-3546
*E-mail:* Millerdk2@shmc.org
*Med Dir:* Phillip Curtis, MD

## Tacoma Community College

Radiography Prgm
6501 S 19th St, Building 19
Tacoma, WA 98466
www.tacomacc.edu
*Prgm Dir:* Michael Mixdorf, MEd RT(R)(CT)
*Tel:* 253 566-5168   *Fax:* 253 566-5273
*E-mail:* mmixdorf@tcc.ctc.edu
*Med Dir:* Michael Dowd, MD

# West Virginia

**Mountain State University**
Radiography Prgm
Box 9003
Beckley, WV 25802-9003
*Prgm Dir:* Kelli J Bahr Summers, BS RT(R)
*Tel:* 800 766-6067, Ext 1465   *Fax:* 304 929-1617
*E-mail:* kbahr@mountainstate.edu

**Bluefield State College**
Radiography Prgm
219 Rock St
Bluefield, WV 24701
www.bluefieldstate.edu
*Prgm Dir:* Melissa Oxley Haye, MS RT(R)
*Tel:* 304 327-4145   *Fax:* 304 327-4219
*E-mail:* mhaye@bluefieldstate.edu
*Med Dir:* Afzal Ahmed, MD

**University of Charleston**
Radiography Prgm
2300 MacCorkle Ave SE
Charleston, WV 25304
*Prgm Dir:* Joan L Clark, MS RT(R)(M)
*Tel:* 304 357-4839   *Fax:* 304 357-4769
*E-mail:* joanclark@ucwv.edu
*Med Dir:* J L Leef, Jr, MD

**United Hospital Center**
Radiography Prgm
3 Hospital Plaza, PO Box 1680
Clarksburg, WV 26302-1680
www.uhcwv.org
*Prgm Dir:* Rosemary Trupo, MBA RT(R)
*Tel:* 304 624-2895, Ext 1   *Fax:* 304 624-2856
*E-mail:* trupor@uhcwv.org
*Med Dir:* W Parke Thrush, MD

**St Mary's Medical Center**
*Cosponsor:* Marshall Community and Technical College
Radiography Prgm
2900 First Ave
Huntington, WV 25702
*Prgm Dir:* Rita Fisher-Carroll, PhD RT(R)(CV)(CT)
*Tel:* 304 526-1259   *Fax:* 304 526-1487
*E-mail:* rfisher@st-marys.org
*Med Dir:* Hans Dransfeld, MD

**West Virginia University Hospitals**
Radiography Prgm
Box 8062
Morgantown, WV 26506
www.wvuhradtech.com
*Prgm Dir:* Jay S Morris, MA RT(R)(CV)
*Tel:* 304 598-4251   *Fax:* 304 598-4702
*E-mail:* morrisj@wvuh.com
*Med Dir:* Mathis Frick, MD

**Southern West Virginia Comm & Tech College**
Radiography Prgm
PO Box 2900
Mt Gay, WV 25637
www.southern.wvnet.edu
*Prgm Dir:* Eva M Hallis, MS RT(R)
*Tel:* 304 792-7098, Ext 267   *Fax:* 304 792-7028
*E-mail:* evah@southern.wvnet.edu
*Med Dir:* Mahesh Koppikar, MD (Radiologist)

**Ohio Valley Medical Center**
Radiography Prgm
2000 Eoff St
Wheeling, WV 26003
*Prgm Dir:* Catherine M Ball, MS RT(R)(M)
*Tel:* 304 234-8781   *Fax:* 304 234-8410
*E-mail:* cball@ovrh.org
*Med Dir:* Mark Kenamond, MD

**Wheeling Hospital**
Radiography Prgm
One Medical Park
Wheeling, WV 26003-0668
www.wheelinghospital.com
*Prgm Dir:* Misty D Kahl, BS RT(R)
*Tel:* 304 243-3173   *Fax:* 304 243-3130
*E-mail:* mkahl@wheelinghospital.com
*Med Dir:* Mark L Benson, MD

# Wisconsin

**Lakeshore Technical College**
Radiography Prgm
1290 North Ave
Cleveland, WI 53015-1414
*Prgm Dir:* James R Odau, MS RT(R)
*Tel:* 920 693-1840   *Fax:* 920 693-8955
*E-mail:* james.odau@gotoltc.edu

**Chippewa Valley Technical College**
Radiography Prgm
620 W Clairemont Ave
Eau Claire, WI 54701-6162
*Prgm Dir:* Shelly Y Olson, MS RT(R)
*Tel:* 715 833-6675   *Fax:* 715 833-6470
*E-mail:* solson@cvtc.edu
*Med Dir:* James Cupery, MD

**Wisconsin Indianhead Technical College**
Radiography Prgm
235 N National Ave
Fond du Lac, WI 54936
*Prgm Dir:* Dyan Hannam, MS Ed R(R)
*Tel:* 920 924-3243
*E-mail:* dhannam@morainepark.edu

**Bellin Health Hosp/Bellin Health Systems Inc**
Radiography Prgm
744 S Webster, PO Box 23400
Green Bay, WI 54305-3400
www.bellin.org/careers
*Prgm Dir:* Randy Griswold, MPA RT(R)
*Tel:* 920 433-3497   *Fax:* 920 433-5811
*E-mail:* rcgris@bellin.org
*Med Dir:* Robert Monette, MD

**Blackhawk Technical College**
Radiography Prgm
6004 Prairie Rd PO Box 5009
Janesville, WI 53547
www.blackhawk.edu
*Prgm Dir:* Joseph W Ipsen, MEd RT(R)
*Tel:* 608 757-7703   *Fax:* 608 743-4578
*E-mail:* jipsen@blackhawk.edu
*Med Dir:* Jeffery Scherer, MD

**Western Technical College**
Radiography Prgm
304 N Sixth St, PO Box 0908
La Crosse, WI 54602-0908
www.wwtc.edu/rad
*Prgm Dir:* Stephen R Schreiner, MEPD RT(R)(CT)
*Tel:* 608 789-9256   *Fax:* 608 785-9299
*E-mail:* schreiners@wwtc.edu
*Med Dir:* Brian Manske, MD

**Madison Area Technical College**
Radiography Prgm
3550 Anderson St
Madison, WI 53704-2599
*Prgm Dir:* Kay Parish, MA RT(R) RDMS
*Tel:* 608 258-2478   *Fax:* 608 258-2480
*E-mail:* kparish@matcmadison.edu
*Med Dir:* Jeffery Block, MD

**University of Wisconsin Hospital and Clinics**
Radiography Prgm
600 Highland Ave
E3/311 Radiology
Madison, WI 53792-3252
www.uwhealth.org/healthprof
*Prgm Dir:* Karen L Tvedten, MEd RT(R)
*Tel:* 608 263-9029   *Fax:* 608 263-9208
*E-mail:* kl.tvedten@hosp.wisc.edu
*Med Dir:* Jeff Foster, MD

**St Joseph's Hospital**
Radiography Prgm
611 St Joseph Ave
Marshfield, WI 54449
*Prgm Dir:* Susan M Duncan, BS RT(R)
*Tel:* 715 387-7184   *Fax:* 715 389-5431
*E-mail:* duncans@stjosephs-marshfield.org
*Med Dir:* John Sheflin, MD

**Aurora St Luke's Medical Center**
Radiography Prgm
180 W Grange Ave
Milwaukee, WI 53207
*Prgm Dir:* Debra J Biggins, BA RT(R)(CV)(M)(QM)
*Tel:* 414 747-4335   *Fax:* 414 747-4366
*E-mail:* debra.biggins@aurora.org
*Med Dir:* Lynn Mastey, MD

**Columbia St Mary's Hospitals**
Radiography Prgm
2025 E Newport Ave
Milwaukee, WI 53211
www.ccon.edu/SORT.htm
*Prgm Dir:* James J Lemerond, BS RT(R)
*Tel:* 414 961-8158   *Fax:* 414 961-4121
*E-mail:* jlemeron@columbia-stmarys.org
*Med Dir:* James Aceto, MD

**Froedtert Memorial Lutheran Hospital**
Radiography Prgm
9200 W Wisconsin Ave
Milwaukee, WI 53226
www.froedtert.com
*Prgm Dir:* Susan Lura Sanson, MEd RT(R)(QM)
*Tel:* 414 805-4999   *Fax:* 414 805-4990
*E-mail:* ssanson@fmlh.edu
*Med Dir:* Katherine Shaffer, MD

**Milwaukee Area Technical College**
Radiography Prgm
700 W State St
Milwaukee, WI 53233-1443
*Prgm Dir:* Bradley J Rothe, BS RT(R)(CT)
*Tel:* 414 297-6645   *Fax:* 414 297-6851
*E-mail:* RotheB@matc.edu
*Med Dir:* Lynn M Gilles, MD

**St Michael Hospital**
Radiography Prgm
2400 W Villard Ave
Milwaukee, WI 53209
*Prgm Dir:* Diane E Wingenter, MS RT(R)
*Tel:* 414 527-5149   *Fax:* 414 527-5156
*E-mail:* dwingenter@covhealth.org
*Med Dir:* Bruce Cardone, MD

**Theda Clark Regional Medical Center**
Radiography Prgm
130 Second St
PO Box 2021
Neenah, WI 54956
www.thedacare.org
*Prgm Dir:* Troy Albrecht, BS RT(R)(CT)
*Tel:* 920 729-3146   *Fax:* 920 729-2118
*E-mail:* troy.albrecht@thedacare.org
*Med Dir:* Thomas Tolly, MD

**Mercy Medical Center/Affinity Health System**
Radiography Prgm
500 S Oakwood Rd
Oshkosh, WI 54903
*Prgm Dir:* James R Werner, BS RT(R)
*Tel:* 920 223-0135   *Fax:* 920 223-1727
*E-mail:* jwerner@affinityhealth.org
*Med Dir:* Paul Larson, MD

**Wheaton Franscian Healthcare-All Saints**
Radiography Prgm
3801 Spring St
Racine, WI 53405
*Prgm Dir:* Linda Szolwinski, BS RT(R)
*Tel:* 262 687-8962
*E-mail:* Linda.Szolwinski@wfhc.org
*Med Dir:* Malcolm Hatfield, MD

**Northcentral Technical College**
Radiography Prgm
1000 Campus Dr
Wausau, WI 54401-1899
www.ntc.edu
*Prgm Dir:* Steven J Hommerding, PhD ABDRT(R)
*Tel:* 715 675-3331, Ext 1326   *Fax:* 715 675-5621
*E-mail:* hommerdi@ntc.edu

# Wyoming

**Casper College**
Radiography Prgm
125 College Dr
Casper, WY 82601
*Prgm Dir:* Laurie Weaver, BS RT(R)
*Tel:* 307 268-2587   *Fax:* 307 268-3034
*E-mail:* lweaver@caspercollege.edu
*Med Dir:* Steven Horn, MD

**Laramie County Community College**
Radiography Prgm
1400 E College Dr
Cheyenne, WY 82007
www.lccc.wy.edu/radiography
*Prgm Dir:* Starla L Mason, MS RT(R)(QM)
*Tel:* 307 778-1391   *Fax:* 307 778-4386
*E-mail:* smason@lccc.wy.edu
*Med Dir:* James G Hubbard, MD

## Radiographer

| Programs* | Class Capacity | Begins | Length (months) | Award | Res. Tuition | Non-res. Tuition | Stipend | Offers:‡ 1 | 2 | 3 |
|---|---|---|---|---|---|---|---|---|---|---|
| **Alabama** | | | | | | | | | | |
| Baptist Medical Center South (Montgomery) | 16 | Oct | 22 | Cert | $3,000 | $3,000 | | | | • |
| Crestwood Medical Center (Huntsville) | 16 | Jan | 24 | Cert | $2,000 | $2,000 | | | | |
| DCH Regional Medical Center (Tuscaloosa) | 20 | Oct | 24 | Cert | $2,200 | $2,200 | | | | • |
| Gadsden State Community College | 25 | Aug | 21 | AAS | $3,420 | $6,118 | | | | • |
| Huntsville Hospital | 15 | Jul | 24 | Cert | $2,000 | $2,000 | | | | |
| Jefferson State Community College (Birmingham) | 37 | Aug | 22 | AAS | $1,914 | $3,514 | | | | |
| Southern Union State Community College (Opelika) | 30 | Aug | 21 | AAS | $3,600 | $7,200 | | | | • |
| University of Alabama at Birmingham | 30 | Aug | 28 | BS | $6,000 | $12,424 | | | | • |
| University of South Alabama (Mobile) | 45 | Aug | 24 | Cert, BS | $4,254 | $8,509 | | | | |
| **Arizona** | | | | | | | | | | |
| GateWay Community College (Phoenix) | 64 | Aug | 22 | AAS | $2,600 | $11,100 | | | • | • |
| Pima Community College (Tucson) | 40 | Jul | 24 | AAS | $1,702 | $11,384 | | | | |
| **Arkansas** | | | | | | | | | | |
| Arkansas State University | 44 | Jun | 24 | AAS, BS | $4,800 | $10,400 | | | | |
| Baptist Health (Little Rock) | 15 | Jul | 24 | Cert | $4,000 | $4,400 | | | | • |
| National Park Community College (Hot Springs) | 28 | Aug | 23 | AAS | $1,200 | $2,960 | | | | |
| North Arkansas College (Harrison) | 42 | Aug | 22 | AAS | $3,016 | $5,942 | | | | |
| South Arkansas Community College (El Dorado) | 13 | Aug | 24 | AAS | $2,540 | $5,060 | | | | • |
| St Vincent Infirmary Medical Center (Little Rock) | 13 | Jul | 24 | Cert | $8,000 | $8,000 | | | | |
| University of Arkansas for Medical Sciences (Little Rock) | 19 | Aug | 24, 36 | AS, BS | $4,860 | $11,160 | | • | | • |
| University of Arkansas for Medical Sciences (Texarkana) | 48 | Aug | 36, 48 | AS, BS | $4,860 | $11,160 | | • | | |
| **California** | | | | | | | | | | |
| Arrowhead Regional Medical Center (Colton) | 10 | Aug | 24 | Cert | | | | | | • |
| Bakersfield College | 24 | Jun | 24 | AS | $800 | $6,000 | | | | |
| Cabrillo College (Aptos) | 22 | Aug | 22 | AS | $730 | $4,994 | | | | |
| California State University - Northridge | 30 | Sep | 55 | BS | $3,042 | $10,692 | | • | | • |
| Canada College (Redwood City) | 20 | Jun | 25 | Cert, AS | $780 | $6,664 | | | | • |
| Central California School of Continuing Edu (San Luis Obispo) | 38 | Year-round | 24 | Cert | $10,500 | $10,500 | | • | | |
| Chaffey College (Rancho Cucamonga) | 28 | Aug | 24 | Cert, AS | $1,976 | $13,452 | | | | • |
| Charles R Drew Univ of Med & Science (Los Angeles) | 60 | Aug | 24 | Dipl, AS | $8,200 | $8,200 | | | | |
| City College of San Francisco | 20 | Aug Jan | 30 | AS | $672 | $5,434 | | | | |
| Cypress College | 38 | Aug | 30 | Cert, AS | $689 | $4,531 | 3600 | • | | • |
| Foothill Community College (Los Altos Hills) | 35 | Aug | 22 | AS | $1,500 | $3,984 | | | | • |
| Fresno City College | 37 | Aug | 24 | AS | $400 | $3,660 | | | | |
| Kaiser Permanente Sch of Allied Hlth Sciences (Richmond) | 55 | Oct | 24 | Cert | $4,000 | $4,000 | | • | | • |
| LA County Harbor UCLA Medical Center (Torrance) | 14 | Jul | 24 | Cert | | | | | | • |
| Loma Linda University | 31 | Sep | 18 | AS, BS | $11,439 | $11,439 | | | | |
| Los Angeles City College | 26 | Sep | 27 | Cert, AS | $400 | $3,915 | 7200 | • | | • |
| Merced College | 20 | Aug | 29 | Cert, AS | $806 | $4,988 | 6000 | | | • |
| Merritt College (Oakland) | 32 | Aug | 24 | Cert, AS | $1,350 | $7,380 | | | | • |
| Mills-Peninsula Health Services (Burlingame) | 9 | Jul | 24 | Cert | | | 1200 | | | • |
| Moorpark College | 25 | Jun | 24 | AS | $750 | $5,130 | | | | • |
| Naval School of Health Sciences, San Diego | 35 | Feb Aug | 12 | Cert | | | | | | • |
| Orange Coast College (Costa Mesa) | 28 | Aug | 22 | AA | $682 | $6,195 | | | | |
| Pima Medical Institute - Chula Vista | 81 | Every 8 mos | 24 | AOS | $11,414 | $0 | | | | • |
| San Diego Mesa College | 45 | Sep | 24 | Cert | $390 | $3,000 | | | | |

*Data are shown only for programs that completed the 2006 AMA Survey of Health Professions Education Programs
‡Key to Offers: 1: Evening or weekend classes; 2: Non-English instruction; 3: Cultural competence instruction

## Radiographer

| Programs* | Class Capacity | Begins | Length (months) | Award | Res. Tuition | Non-res. Tuition | Stipend | Offers:‡ 1 | 2 | 3 |
|---|---|---|---|---|---|---|---|---|---|---|
| San Joaquin General Hospital (Stockton) | 15 | Jul | 24 | Cert | $1,047 | $7,044 | | | | • |
| Yuba Community College (Marysville) | 100 | Aug | 24 | Cert, AS | $350 | $2,500 | | | | • |
| **Colorado** | | | | | | | | | | |
| Aims Community College (Greeley) | 25 | Aug | 24 | AAS | $3,225 | $11,250 | | | | • |
| Community College of Denver - Lowry Campus | 40 | Aug | 22 | AAS | $3,000 | $9,239 | | | | |
| Concorde Career Institute - Aurora | 24 | Sep Apr | 22 | Dipl, AAS | $14,865 | $0 | | | | • |
| Memorial Hospital (Colorado Springs) | 18 | Aug | 24 | AAS | $3,250 | $3,250 | | | | • |
| Mesa State College (Grand Junction) | 18 | Aug | 21 | AAS | $4,000 | $10,700 | | | | • |
| Red Rocks Community College (Arvada) | 18 | Aug | 21 | Cert, AAS | $3,300 | $7,500 | | | | • |
| **Connecticut** | | | | | | | | | | |
| Capital Community College (Hartford) | 20 | Sep | 22 | AS | $2,600 | $7,600 | | • | | |
| Gateway Community College (North Haven) | 36 | Sep | 20 | AS | $2,672 | $7,976 | | | | |
| Hartford Hospital | 28 | Sep | 24 | Dipl, Cert | $2,500 | $2,500 | | | | • |
| Middlesex Community College (Middletown) | 19 | Jun | 27 | Dipl, Cert, AS | $3,435 | $8,078 | | | | • |
| Naugatuck Valley Community College (Waterbury) | 24 | Sep | 22 | AS | $3,022 | $8,326 | | | | • |
| Quinnipiac University (Hamden) | 70 | Sep | 35 | Cert, BS | $22,000 | $22,000 | | • | | • |
| Stamford Hospital | 14 | Jul | 24 | Cert | $2,800 | $2,800 | | | | |
| Windham Community Memorial Hospital (Willimantic) | 24 | Oct | 24 | Cert | $6,000 | $6,000 | | | | |
| **Delaware** | | | | | | | | | | |
| Delaware Tech & Comm Coll - Wilmington | 20 | May | 24 | AAS | $2,934 | $7,335 | | | | • |
| **District of Columbia** | | | | | | | | | | |
| Washington Hospital Center | 32 | Sep | 22 | Cert | $2,500 | $2,500 | 3000 | | | • |
| **Florida** | | | | | | | | | | |
| Bethesda Memorial Hospital (Boynton Beach) | 9 | Jul | 24 | Cert | $1,100 | $0 | | | | • |
| Brevard Community College (Cocoa) | 32 | May | 24 | AS | $2,300 | $7,500 | | | | |
| Edison College (Ft Myers) | 38 | Aug | 24 | Dipl, AS | $2,726 | $9,138 | | | | |
| Gulf Coast Community College (Panama City) | 20 | Aug | 24 | AAS | $3,159 | $11,774 | | | | • |
| Halifax Medical Center (Daytona Beach) | 10 | Jan | 30 | Cert | $2,800 | $2,800 | | | | • |
| Hillsborough Community College (Tampa) | 25 | Aug | 22 | AS | $2,519 | $9,196 | | | | |
| Jackson Mem Medical Center (Miami) | 21 | Jun | 24 | Dipl | $2,500 | $3,000 | | | | • |
| Keiser College - Daytona Beach | 94 | | 24 | Dipl, AS | $17,748 | $0 | | | | |
| Keiser College - Melbourne Campus | 120 | Jan May Sep | 24 | AS | $17,748 | $0 | | • | | |
| Keiser University (Ft Lauderdale) | 24 | Jan May Sep | 24 | AS | $17,748 | $17,748 | | | | • |
| Keiser University - Sarasota | 96 | Jan May Aug | 24 | AS | $17,748 | $17,748 | | | | • |
| Lakeland Regional Medical Center | 14 | Jul | 24 | Cert | $500 | $500 | | | | • |
| Marion County School of Radiologic Tech (Ocala) | 30 | Aug | 24 | Cert | $3,400 | $6,800 | | | | • |
| MedVance Institute (Miami) | 50 | Varies | 24 | AS | $11,724 | $0 | | | | |
| MedVance Institute - West Palm Beach Campus/KIMC Investments Inc (Palm Springs) | 35 | Jul | 24 | AS | $15,800 | $0 | | | | |
| Miami Dade College | 55 | Jun | 24 | AAS | $1,985 | $6,793 | | | | • |
| Palm Beach Community College (Palm Beach Gardens) | 40 | Jan | 24 | AS | $1,900 | $6,352 | | | | |
| Pensacola Junior College | 35 | Aug | 23 | AAS | $2,047 | $7,630 | | • | | • |
| Polk Community College (Winter Haven) | 20 | Jan | 24 | AAS | $2,440 | $9,040 | | | | |
| Santa Fe Community College (Gainesville) | 33 | Aug | 22 | AS | $5,156 | $16,055 | | | | |
| Shands Jacksonville | 15 | Jul | 24 | Cert | $750 | $750 | | | | |
| St Vincent's Medical Center (Jacksonville) | 11 | Jul | 24 | Dipl, Cert | $2,500 | $2,500 | | | | • |
| University of Central Florida (Orlando) | 16 | Aug | 24 | BS | $5,116 | $24,958 | | • | | • |
| Valencia Community College (Orlando) | 22 | Aug | 24 | AS | $2,683 | $10,085 | | | | • |
| **Georgia** | | | | | | | | | | |
| Armstrong Atlantic State University (Savannah) | 30 | Jun | 48 | BS | $3,903 | $13,860 | | | | • |
| Athens Technical College | 17 | Sep | 21 | AAT | $1,528 | $3,056 | | | | |
| Coastal Georgia Community College (Brunswick) | 15 | Aug | 24 | AS | $2,631 | $9,567 | | | | |
| Columbus Technical College | 15 | Jan | 24 | AS | $1,900 | $0 | | | | |
| Coosa Valley Technical College (Rome) | 32 | Jul | 24 | Dipl, AAS | $1,400 | $2,800 | | | | |
| DeKalb Medical Center (Decatur) | 42 | Jul | 24 | Cert | $1,600 | $1,600 | | | | |
| Emory University (Atlanta) | 20 | Fall semester | 33 | BMSc | $13,200 | $13,200 | | | | • |
| Grady Health System (Atlanta) | 35 | Sep | 24 | Cert | $1,900 | $1,900 | | | | • |
| Griffin Technical College | 28 | Jul | 24 | AAT | $1,764 | $3,528 | | | | • |
| Gwinnett Technical College (Lawrenceville) | 58 | Sep | 21 | AAS | $1,900 | $3,380 | | | | • |
| Middle Georgia Technical College (Warner Robins) | 16 | Oct | 24 | Dipl, AAT | $1,860 | $3,720 | | | | • |
| North Metro Technical College (Atlanta) | 48 | Jan | 21 | Dipl, AAT | $1,755 | $0 | | | | |
| Ogeechee Technical College (Statesboro) | 15 | Oct | 21 | Dipl | $1,196 | $2,392 | | | | • |
| Okefenokee Technical College (Waycross) | 18 | Jul | 24 | Dipl | $1,472 | $0 | | | | |
| Southeastern Technical College (Vidalia) | 24 | Fall | 21 | Dipl | $2,985 | $0 | | | | |
| University Hospital (Augusta) | 14 | Jul | 24 | Cert | $900 | $900 | | | | • |
| Valdosta Technical College | 24 | Sep | 21 | AAS | $1,344 | $2,688 | | | | |
| West Central Technical College (Douglasville) | 18 | Summer | 24 | AAT | $1,828 | $0 | | | | • |

*Data are shown only for programs that completed the 2006 AMA Survey of Health Professions Education Programs
‡Key to Offers: 1: Evening or weekend classes; 2: Non-English instruction; 3: Cultural competence instruction

PROGRAMS

## Radiographer

| Programs* | Class Capacity | Begins | Length (months) | Award | Res. Tuition | Non-res. Tuition | Stipend | Offers:‡ 1 | 2 | 3 |
|---|---|---|---|---|---|---|---|---|---|---|
| West Georgia Technical College (LaGrange) | 20 | Oct | 21 | AAS | $1,812 | $3,300 | | | | • |
| **Idaho** | | | | | | | | | | |
| Boise State University | 26 | Aug | 36, 48 | AS, BS | $5,558 | $12,966 | | • | | • |
| College of Southern Idaho (Twin Falls) | 34 | Fall | 24 | AAS | $1,000 | $2,800 | | | | |
| **Illinois** | | | | | | | | | | |
| Advocate Illinois Masonic Medical Center (Chicago) | 25 | Sep | 24 | Cert | $3,000 | $3,000 | | | | • |
| Advocate Trinity Hospital (Chicago) | 17 | Aug | 24 | Dipl | $3,000 | $3,000 | | | | • |
| Blessing Hospital (Quincy) | 16 | Jul | 24 | Cert | $2,000 | $2,000 | | | | • |
| Bloomington-Normal School of Radiography | 13 | Jul | 24 | Cert | $1,500 | $1,500 | | | | • |
| Carl Sandburg College (Galesburg) | 20 | Jun | 24 | AAS | $3,147 | $5,960 | | | | • |
| College of DuPage (Glen Ellyn) | 72 | May | 24 | Dipl, AAS | $3,225 | $9,000 | | | | • |
| College of Lake County (Grayslake) | 30 | Aug | 22 | AAS | $4,899 | $13,524 | | | | |
| Danville Area Community College | 15 | Aug | 24 | AAS | $2,500 | $6,100 | | | | • |
| Illinois Central College (Peoria) | 20 | Aug | 23 | AAS | $2,520 | $5,580 | | | | • |
| Kaskaskia College (Centralia) | 58 | Aug | 18, 21 | AAS | $2,394 | $4,104 | | | | |
| Kishwaukee College (Malta) | 22 | Aug | 22 | AAS | $3,000 | $8,150 | | | | • |
| Lincoln Land Community College (Springfield) | 16 | Jun | 24 | AAS | $2,448 | $0 | | • | | • |
| Malcolm X College (Chicago) | 35 | Jun | 24 | AAS | $2,077 | $5,237 | | | | • |
| Olney Central College | 28 | Jun | 24 | AAS | $2,800 | $7,417 | | | | |
| OSF St Francis Medical Center (Peoria) | 11 | Jan | 24 | Cert | $1,500 | $1,500 | | | | • |
| Rockford Memorial Hospital | 10 | Jun | 24 | Cert | $1,600 | $2,400 | | | | |
| Sauk Valley Community College (Dixon) | 34 | Aug | 22 | AAS | $2,800 | $10,535 | | | | |
| St Francis Hospital (Evanston) | 25 | Aug | 24 | Dipl, Cert | $2,500 | $0 | | | | |
| Swedish American Hospital (Rockford) | 10 | May | 24 | Cert | $1,500 | $0 | | | | |
| Trinity College of Nursing & Health Sciences (Rock Island) | 16 | Jun | 23 | AAS | $6,900 | $6,900 | | | | • |
| **Indiana** | | | | | | | | | | |
| Ball Memorial Hospital (Muncie) | 14 | Jun | 24 | AAS | $2,000 | $2,000 | | | | • |
| Ball State University (Indianapolis) | 20 | May | 26 | AS | $6,360 | $16,726 | | | | • |
| Columbus Regional Hospital | 6 | Jul | 24 | Cert | $2,500 | $2,500 | | | | |
| Community Health Network (Indianapolis) | 14 | Jun | 24 | Cert | $2,000 | $2,000 | | | | |
| Ft Wayne School of Radiography | 28 | Jun | 24 | Cert | $3,000 | $3,000 | | | | • |
| Good Samaritan Hospital (Vincennes) | 10 | Jun | 24 | Cert | $1,800 | $1,800 | | | | |
| Hancock Memorial Hospital (Greenfield) | 15 | Jul | 24 | Cert | $3,000 | $3,000 | | | | • |
| Indiana University (Indianapolis) | 37 | Jun | 22 | AS | $5,535 | $15,700 | | | | |
| Indiana University - Kokomo | 24 | Aug | 21, 9 | AS, BS | $4,600 | $0 | | | | |
| Indiana University Northwest (Gary) | 45 | Jul | 24 | AS | $5,668 | $14,126 | | | | • |
| Indiana University South Bend | 19 | Jul | 22 | AS | $4,348 | $11,419 | | • | | • |
| Ivy Tech Community College - Terre Haute | 32 | Aug | 21 | AS | $2,780 | $5,182 | | | | • |
| Ivy Tech Community College of Indiana (Indianapolis) | 20 | Aug | 21 | AS | $2,974 | $5,908 | | | | |
| King's Daughters' Hospital & Health Services (Madison) | 6 | Jul 1 | 24 | Cert | $2,500 | $2,500 | | | | |
| Reid Hospital & Health Care Services (Richmond) | 11 | Oct | 24 | Cert | $2,000 | $2,000 | | | | |
| St Vincent Health/St Joseph Hospital (Indianapolis) | 17 | Aug | 22 | Cert | $4,000 | $4,000 | | | | |
| University of Saint Francis (Ft Wayne) | 24 | Aug | 22 | AS | | | | | • | • |
| University of Southern Indiana (Evansville) | 17 | Jan | 24 | AS | $2,880 | $7,029 | | | | |
| **Iowa** | | | | | | | | | | |
| Allen College (Waterloo) | 17 | Jun | 24 | AS | $14,796 | $14,796 | | | | |
| Covenant Medical Center (Waterloo) | 12 | Jun | 24 | Cert | $2,000 | $2,000 | | | | • |
| Indian Hills Community College (Ottumwa) | 40 | Aug | 24 | AAS | $4,515 | $6,794 | | | • | • |
| Iowa Central Community College (Ft Dodge) | 30 | Sep | 22 | AAS | $2,891 | $4,366 | | | | |
| Iowa Health - Des Moines | 14 | Jul | 24 | Cert | $3,000 | $3,000 | | | | • |
| Jennie Edmundson Memorial Hospital (Council Bluffs) | 8 | Aug | 24 | Cert | $1,500 | $1,500 | | | | • |
| Mercy College of Health Sciences (Des Moines) | 38 | Jun | 24 | AS | $15,305 | $15,305 | | | | • |
| Mercy Medical Center - North Iowa (Mason City) | 12 | Mid-Aug | 24 | Cert | $1,250 | $1,250 | | | | • |
| Mercy/St Luke's Hospital (Cedar Rapids) | 23 | Jun | 24 | Cert | $2,000 | $2,000 | | | | • |
| Northeast Iowa Community College (Peosta) | 25 | Aug | 24 | AAS | $4,000 | $4,000 | | | | • |
| St Luke's College (Sioux City) | 18 | Aug | 24 | ASR | $12,920 | $12,920 | | | | |
| University of Iowa Hospitals & Clinics (Iowa City) | 25 | Jul | 24 | Cert | $4,500 | $4,500 | | | | • |
| **Kansas** | | | | | | | | | | |
| Ft Hays State University | 75 | Jun | 24 | AS | $7,326 | $10,155 | | • | | • |
| Hutchinson Community College | 30 | Aug | 24 | AASRT | $2,245 | $3,450 | | | | |
| Labette Community College (Parsons) | 28 | Jun | 23 | AAS | $2,070 | $2,700 | | | | |
| Newman University (Wichita) | 83 | Aug | 24 | AS | $15,822 | $15,822 | | | | • |
| Washburn University (Topeka) | 28 | Aug | 22 | AS | $5,200 | $10,680 | | | | |
| **Kentucky** | | | | | | | | | | |
| Bluegrass Community and Technical College (Lexington) | 22 | Aug | 24 | Dipl | $2,616 | $7,848 | | | | |
| Bowling Green Technical College | 21 | Aug | 22 | AAS | $2,616 | $7,848 | | | | |
| Elizabethtown Community & Technical College | 16 | Aug | 20 | AAS | $2,548 | $4,000 | | | | • |

*Data are shown only for programs that completed the 2006 AMA Survey of Health Professions Education Programs
‡Key to Offers: 1: Evening or weekend classes; 2: Non-English instruction; 3: Cultural competence instruction

## Radiographer

| Programs* | Class Capacity | Begins | Length (months) | Award | Res. Tuition | Non-res. Tuition | Stipend | Offers:‡ 1 | 2 | 3 |
|---|---|---|---|---|---|---|---|---|---|---|
| Hazard Community & Technical College | 22 | Aug | 24 | AAS | $3,234 | $8,820 | | | | |
| Jefferson Community and Technical College (Louisville) | 59 | Aug | 21 | AAS | $109 | $140 | | | | • |
| Morehead State University | 42 | Aug | 24 | AAS | $4,320 | $11,480 | | | | |
| Northern Kentucky University (Highland Heights) | 30 | Jul | 22 | AAS | $4,968 | $9,696 | | | | • |
| Owensboro Community & Technical College | 40 | Aug | 24 | AAS | $1,450 | $4,350 | | | | • |
| Somerset Community College | 36 | Aug | 21 | AAS | $2,352 | $7,056 | | | | • |
| Southeast Kentucky Comm & Tech College (Pineville) | 25 | Aug | 22 | Dipl, AAS | $2,943 | $8,829 | | | | |
| Spencerian College (Louisville) | 15 | Quarterly | 27 | AD | $32,500 | $0 | | | | • |
| St Joseph Healthcare (Lexington) | 12 | Jun | 24 | Cert | $2,000 | $2,000 | | | | |
| West Kentucky Community & Technical College (Paducah) | 20 | Aug | 24 | Dipl, AAS | $2,040 | $6,120 | | | | • |
| **Louisiana** | | | | | | | | | | |
| Baton Rouge General Medical Center | 16 | Jul | 24 | Cert | $1,500 | $1,500 | | | | |
| Delgado Community College (New Orleans) | 56 | Aug | 24 | Dipl, AAS | $3,336 | $10,126 | | | | • |
| Lafayette General Medical Center | 6 | Sep | 24 | Cert | $4,000 | $4,000 | | | | |
| Louisiana State University - Eunice | 20 | Jun | 24 | AS | $3,269 | $7,769 | | | | • |
| McNeese State University (Lake Charles) | 21 | Aug | 21 | BS | $4,039 | $6,930 | | | | |
| North Oaks Medical Center (Hammond) | 20 | Jul | 24 | Cert | $4,000 | $0 | | | | |
| Northwestern State University (Shreveport) | 40 | Aug | 45 | BS | $3,400 | $9,400 | | • | | |
| Our Lady of Holy Cross College (New Orleans) | 12 | Aug | 21 | Dipl, BS-HS, AS-RS | $12,800 | $12,800 | | | | • |
| Our Lady of the Lake College (Baton Rouge) | 28 | Aug | 18 | AS | $8,475 | $8,475 | | | | • |
| University of Louisiana at Monroe | 48 | Aug | 21 | BSRT | $3,076 | $9,028 | | • | | • |
| **Maine** | | | | | | | | | | |
| Central Maine Medical Center (Lewiston) | 13 | Aug | 24 | Cert | $3,000 | $3,000 | | | | • |
| Eastern Maine Community College (Bangor) | 20 | Aug | 23 | AAS | $2,856 | $6,258 | | | | |
| Mercy Hospital (Portland) | 11 | Jul | 24 | Cert | $2,000 | $2,000 | | | | • |
| Southern Maine Community College (South Portland) | 23 | Sep | 23 | AS | $4,000 | $0 | | | | |
| **Maryland** | | | | | | | | | | |
| Allegany College of Maryland (Cumberland) | 23 | Jul | 22 | AAS | $3,400 | $7,300 | | | | |
| Anne Arundel Community College (Arnold) | 30 | Jun | 24 | AAS | $8,105 | $0 | | • | • | • |
| Chesapeake College (Wye Mills) | 14 | Jun | 24 | AAS | $3,990 | $9,415 | | | | • |
| Comm Coll of Baltimore County - Essex Campus | 24 | Jul | 23 | AAS | $3,150 | $5,922 | | | | |
| Greater Baltimore Medical Center | 12 | Aug | 23 | Cert | $2,500 | $2,500 | | | | |
| Hagerstown Community College | 35 | May | 24 | AAS | $3,115 | $6,510 | | | | |
| Holy Cross Hospital (Silver Spring) | 11 | Aug | 22 | Dipl, Cert | $1,000 | $1,000 | | | | |
| Johns Hopkins Hospital (Baltimore) | 22 | Jun | 18 | Cert | $5,200 | $5,200 | | | | • |
| Montgomery College (Takoma Park) | 30 | Sep | 24 | AAS | $3,204 | $8,820 | | | | • |
| Prince George's Community College (Largo) | 41 | Aug | 19 | Dipl, AAS | $2,400 | $4,291 | | | | • |
| **Massachusetts** | | | | | | | | | | |
| Bunker Hill Community College (Boston) | 55 | Sep | 22, 35 | AS | $4,100 | $10,700 | | • | | • |
| Holyoke Community College | 26 | Sep | 21 | AS | $3,855 | $10,623 | | | | • |
| Mass College of Pharmacy & Health Sciences (Boston) | 30 | Sep | 33 | Cert, BS | $21,200 | $21,200 | | | | • |
| Massachusetts Bay Community College (Framingham) | 45 | Sep | 21, 33 | AS | $4,861 | $12,800 | | • | | • |
| Massasoit Community College (Brockton) | 35 | Sep | 21 | AS | $888 | $8,510 | | | | • |
| North Shore Community College (Danvers) | 20 | Sep | 21 | AS | $4,002 | $12,006 | | | | • |
| Northern Essex Community College (Lawrence) | 25 | Sep | 21 | AS | $4,095 | $13,260 | | | | • |
| Quinsigamond Community College (Worcester) | 26 | Sep | 22 | AS | $4,500 | $8,000 | | | | • |
| Regis College (Medford) | 32 | Aug | 22 | AS | $14,000 | $0 | | | • | • |
| Springfield Technical Community College | 22 | Sep | 23 | AS | $3,415 | $9,590 | | | | |
| **Michigan** | | | | | | | | | | |
| Baker College of Owosso | 40 | Sep | 24 | AAS | $9,030 | $9,030 | | | | • |
| Delta College (University Center) | 18 | Aug | 24 | AAS | $2,370 | $3,328 | | | | • |
| Henry Ford Hospital (Detroit) | 20 | Sep | 24 | Cert | $500 | $500 | | | | |
| Hurley Medical Center (Flint) | 7 | Sep | 24 | Cert | $1,500 | $0 | | | | • |
| Kellogg Community College (Battle Creek) | 22 | Jun | 24 | AAS | $2,628 | $4,106 | | | | |
| Lansing Community College | 72 | Aug | 21 | AAS | $3,250 | $5,250 | | | | • |
| Marquette General Health System | 10 | Aug | 24 | Cert | $1,650 | $1,650 | | | | • |
| Mid Michigan Community College (Harrison) | 28 | Aug | 24 | AAS | $2,791 | $4,992 | | | | • |
| Oakland Community College - Bloomfield Hills (Southfield) | 25 | May | 15 | AAS | $2,426 | $4,107 | | • | | • |
| Port Huron Hospital | 7 | Jul | 24 | Dipl | $4,000 | $4,000 | | | | |
| Providence Hospital (Southfield) | 12 | Sep Oct | 24 | Cert | $1,750 | $0 | | | | • |
| Sinai - Grace Hospital (Detroit) | 14 | Jul | 24 | Cert | $500 | $500 | | | | |
| Washtenaw Community College (Ann Arbor) | 37 | May 30 | 24 | AAS | $4,000 | $5,000 | | | | • |
| William Beaumont Hospital (Royal Oak) | 12 | Jan Jul | 24 | Cert | $500 | $500 | | | | • |
| **Minnesota** | | | | | | | | | | |
| Argosy University/Twin Cities (Eagan) | 21 | Sep Jan May | 24, 36 | Dipl, AAS | $13,500 | $0 | | | | |
| Century College (White Bear Lake) | 50 | Aug | 24 | AAS | $141 | $268 | | • | | • |
| College of St Catherine - Minneapolis | 18 | Sep Feb | 23 | AAS | $14,555 | $0 | | | | • |

*Data are shown only for programs that completed the 2006 AMA Survey of Health Professions Education Programs
‡Key to Offers: 1: Evening or weekend classes; 2: Non-English instruction; 3: Cultural competence instruction

| Programs* | Class Capacity | Begins | Length (months) | Award | Res. Tuition | Non-res. Tuition | Stipend | Offers:‡ 1 | 2 | 3 |
|---|---|---|---|---|---|---|---|---|---|---|
| Lake Superior College (Duluth) | 30 | Aug | 24, 18 | Dipl, AAS | $4,085 | $8,170 | | | | |
| Mayo School of Health Sciences (Rochester) | 40 | Aug | 24 | Cert, AS | $6,300 | $7,800 | | | | • |
| Methodist Hospital (St Louis Park) | 5 | Jan Sep | 24 | Dipl | $1,000 | $1,000 | | | | |
| North Memorial Medical Center (Robbinsdale) | 6 | Mar Sep | 24 | Cert | $1,000 | $1,000 | | | | |
| Northland Community & Technical College (East Grand Forks) | 16 | Aug | 24 | AAS | $10,746 | $21,493 | | | | • |
| Rice Memorial Hospital (Willmar) | 8 | Aug | 24 | Cert | $600 | $600 | | | | |
| St Cloud Hospital | 9 | Sep | 24 | Cert | $4,597 | $4,597 | | | | • |
| Veterans Affairs Medical Center (Minneapolis) | 15 | Sep | 24 | Cert | | | | | | |
| **Mississippi** | | | | | | | | | | |
| Copiah-Lincoln Community College (Wesson) | 24 | Aug | 24 | AAS | $2,480 | $4,730 | | | | |
| Hinds Community College (Raymond) | 51 | Jun | 24 | AS | $2,100 | $3,606 | | | | |
| Itawamba Community College (Fulton) | 21 | Aug | 22 | AAS | $3,000 | $5,480 | | | | • |
| Jones County Junior College (Ellisville) | 14 | May | 24 | Dipl, AAS | $2,005 | $4,616 | | | | • |
| Meridian Community College | 26 | Aug | 24 | Dipl, AAS | $1,700 | $2,800 | | | | |
| Mississippi Gulf Coast Community College (Gautier) | 38 | May | 24 | AAS | $2,235 | $5,004 | | | | • |
| Northeast Mississippi Community College (Booneville) | 15 | Jul | 23 | AAS | $1,700 | $3,420 | | | | • |
| University of Mississippi Medical Center (Jackson) | 25 | Jul | 12, 24 | Cert | $212 | $0 | | | | |
| **Missouri** | | | | | | | | | | |
| Barnes-Jewish Coll of Nursing/Allied Health (St Louis) | 25 | Aug | 12, 24 | ASR | $13,300 | $13,300 | | | | |
| Colorado Technical University (Kansas City) | 35 | Jan Jul | 21 | AAS | $30,040 | $0 | | | | • |
| CoxHealth (Springfield) | 23 | Sep | 24 | Cert | $2,500 | $2,500 | | | | • |
| Metropolitan Community College - Penn Valley (Kansas City) | 68 | Jun | 24 | AAS | $2,886 | $6,825 | | • | | |
| Mineral Area Regional Medical Center (Farmington) | 24 | Aug | 22 | Cert | $3,150 | $3,150 | | | | • |
| Missouri Southern State University (Joplin) | 10 | Aug | 24 | AS | $5,070 | $10,140 | | | | • |
| Nichols Career Center (Jefferson City) | 14 | Aug | 24 | Cert | $6,300 | $6,300 | | | | |
| Research Medical Center (Kansas City) | 15 | Jul | 24 | Cert | $2,200 | $2,200 | | | | • |
| Southeast Missouri Hospital (Cape Girardeau) | 17 | Jul 25 | 22 | Dipl, Assoc | $8,750 | $0 | | | | |
| St John's Regional Health Center (Springfield) | 24 | Jul | 23 | Cert | $1,500 | $1,500 | | | | |
| State Fair Community College (Sedalia) | 48 | Aug | 21 | AAS | $3,542 | $4,460 | | | | • |
| University of Missouri - Columbia | 12 | Jun | 24 | BHS | $7 | $11,977 | | | | • |
| **Montana** | | | | | | | | | | |
| Benefis Healthcare (Great Falls) | 6 | Jul | 24 | Cert | | | | | | • |
| **Nebraska** | | | | | | | | | | |
| Alegent Health (Omaha) | 34 | Aug | 23 | Cert | $2,050 | $2,050 | | | | • |
| Clarkson College (Omaha) | 28 | Aug | 24 | AS | $12,000 | $12,000 | | | | • |
| Mary Lanning Memorial Hospital (Hastings) | 20 | Aug | 24 | Dipl | $2,400 | $2,400 | | | | • |
| Regional West Medical Center (Scottsbluff) | 14 | Aug | 24 | Cert | $1,250 | $1,250 | | | | • |
| Southeast Community College (Lincoln) | 70 | Jul Jan | 24 | AAS | $6,040 | $6,500 | | | | |
| University of Nebraska Medical Center (Omaha) | 14 | Aug | 21 | BS | $9,440 | $28,025 | | • | | • |
| **Nevada** | | | | | | | | | | |
| Pima Medical Institute - Las Vegas | 72 | Nov | 24 | AS | $12,500 | $0 | | | | • |
| Truckee Meadows Community College (Reno) | 16 | Aug | 24 | AAS | $1,704 | $6,666 | | | | • |
| University of Nevada - Las Vegas | 45 | Aug | 24 | Cert | $2,370 | $10,155 | | | | • |
| **New Hampshire** | | | | | | | | | | |
| New Hampshire Technical Institute (Concord) | 35 | Jun | 24 | Dipl, AS | $8,985 | $17,116 | | | | |
| **New Jersey** | | | | | | | | | | |
| Bergen Community College (Paramus) | 42 | Sep | 24 | AAS | $2,908 | $5,816 | | | | |
| Brookdale Community College (Lincroft) | 30 | Sep | 24 | AAS | $3,700 | $5,600 | | | | • |
| Burlington County College (Pemberton) | 14 | May | 24 | AAS | $4,600 | $0 | | | | |
| Christ Hospital (Jersey City) | 18 | Sep | 12, 24 | Dipl, Cert | $7,500 | $7,500 | | | | |
| Cooper University Hospital (Camden) | 53 | Sep | 24 | Cert | $3,000 | $3,000 | | | | |
| Mercer County Community College (Trenton) | 82 | Aug | 22 | AAS | $3,367 | $0 | | • | | • |
| Middlesex County College (Edison) | 36 | Sep | 24 | AAS | $3,500 | $7,000 | | | | • |
| Muhlenberg Regional Medical Center (Plainfield) | 42 | Sep | 36 | Dipl, AS | $2,886 | $5,772 | | | | • |
| Pascack Valley Hospital (Westwood) | 7 | Sep | 24 | Cert | $5,000 | $5,000 | | | | |
| Passaic County Community College (Paterson) | 28 | Sep | 24 | AAS | $3,022 | $6,044 | | | | • |
| Shore Memorial Hospital (Somers Point) | 25 | Aug | 24 | Cert | $4,000 | $4,000 | | | | • |
| St Francis Medical Center (Trenton) | 8 | Jul | 24 | Cert | $3,117 | $3,117 | | | | • |
| The Valley Hospital (Ridgewood) | 13 | Sep | 24 | Cert, AS | $13,718 | $13,718 | | | | • |
| **New Mexico** | | | | | | | | | | |
| Clovis Community College | 12 | Aug | 24 | AAS | $785 | $1,386 | | | | |
| Dona Ana Community College (Las Cruces) | 23 | Aug | 21 | AS | $1,216 | $3,154 | | | | |
| Northern New Mexico College (Espanola) | 16 | Aug | 22 | AAS | $1,116 | $2,635 | | | | • |
| **New York** | | | | | | | | | | |
| Arnot Ogden Medical Center (Elmira) | 7 | Aug | 24 | Cert | $3,055 | $3,055 | | | | |
| Bellevue Hospital Center (New York) | 20 | Oct | 24 | Cert | $4,500 | $4,500 | | | | • |
| Bronx Community College | 40 | Aug | 24 | AAS | $2,800 | $3,000 | | | | • |

*Data are shown only for programs that completed the 2006 AMA Survey of Health Professions Education Programs
‡Key to Offers: 1: Evening or weekend classes; 2: Non-English instruction; 3: Cultural competence instruction

## Radiographer

| Programs* | Class Capacity | Begins | Length (months) | Award | Res. Tuition | Non-res. Tuition | Stipend | Offers:‡ 1 | 2 | 3 |
|---|---|---|---|---|---|---|---|---|---|---|
| Champlain Valley Phys Hospital Med Ctr (Plattsburgh) | 14 | Jul | 24 | Cert, AS | $5,000 | $7,160 | | | | |
| Hostos Community College of CUNY (Bronx) | 60 | Sep | 24 | AAS | $4,200 | $8,000 | | | | |
| Long Island University - C W Post Campus (Brookville) | 25 | Sep | 48 | BS | $24,770 | $24,770 | | • | | • |
| Mercy Medical Center (Rockville Centre) | 17 | Sep | 24 | Cert | $4,250 | $0 | | | | |
| Monroe Community College (Rochester) | 46 | Sep | 21 | AAS | $3,036 | $6,072 | | | | • |
| New York City College of Technology (Brooklyn) | 60 | Sep | 24 | AAS | $4,000 | $6,800 | | | | |
| New York Methodist Hospital (Brooklyn) | 14 | Sep | 24 | Cert | $5,000 | $5,000 | | | | • |
| Niagara County Community College (Sanborn) | 80 | Sep | 24 | AAS | $2,976 | $4,464 | | | | |
| North Country Community College (Saranac Lake) | 40 | Sep | 23 | Dipl, AAS | $3,050 | $8,000 | | | | • |
| Orange County Community College (Middletown) | 24 | Sep | 24 | AAS | $3,000 | $6,000 | | | | |
| Peconic Bay Medical Center (Riverhead) | 12 | Sep | 24 | Cert | $4,000 | $4,000 | | | | |
| Robt J Hochstim Sch Rad/S Nassau Comm Hosp (Oceanside) | 10 | Sep | 24 | Cert | $4,100 | $4,100 | | | | • |
| St Elizabeth Medical Center (Utica) | 30 | Sep | 23 | Cert | $3,750 | $3,750 | | | | • |
| St James Mercy Health (Hornell) | 16 | Aug | 24 | Cert | $4,000 | $4,000 | | | | |
| Westchester Community College (Valhalla) | 45 | Sep | 24 | AAS | $3,350 | $8,376 | | | | |
| Winthrop University Hospital (Mineola) | 11 | Sep | 24 | Cert | $4,500 | $4,500 | | | | • |
| **North Carolina** | | | | | | | | | | |
| Asheville-Buncombe Technical Comm College | 24 | Aug | 21 | AAS | $1,580 | $8,780 | | | | • |
| Caldwell Comm College & Tech Institute (Hudson) | 18 | Aug | 21 | AAS | $1,824 | $10,128 | | | | |
| Cape Fear Community College (Wilmington) | 24 | Fall semester | 21 | AAS | $1,189 | $10,536 | | | | |
| Carolinas College of Health Sciences (Charlotte) | 40 | Aug | 21 | AAS | $5,250 | $5,250 | | | | |
| Carteret Community College (Morehead City) | 20 | Aug | 24 | AAS | $1,713 | $9,225 | | | | • |
| Cleveland Community College (Shelby) | 21 | Aug | 19 | AAS | $1,106 | $3,376 | | | | • |
| Edgecombe Community College (Rocky Mount) | 32 | Aug | 21 | AAS | $1,435 | $7,851 | | | | • |
| Forsyth Technical Community College (Winston-Salem) | 25 | Aug May | 21 | AAS | $1,575 | $7,969 | | | | • |
| Johnston Community College (Smithfield) | 48 | Aug | 21 | AAS | $840 | $4,552 | | | | |
| Moses Cone Health System (Greensboro) | 20 | Jul | 24 | Cert | $1,000 | $1,000 | 840 | | | • |
| Pitt Community College (Greenville) | 35 | Aug | 21 | AAS | $1,422 | $7,902 | | | | |
| Presbyterian Healthcare (Charlotte) | 10 | Aug | 23 | Cert | $550 | $550 | | | | • |
| Sandhills Community College (Pinehurst) | 18 | Aug | 21 | AAS | $1,422 | $7,902 | | | | |
| Southwestern Community College (Sylva) | 28 | Aug | 21 | AAS | $1,096 | $6,104 | | | | • |
| Univ of North Carolina at Chapel Hill | 12 | Jul | 48 | BS | $4,600 | $18,400 | | | | |
| Vance - Granville Community College (Henderson) | 31 | Aug | 21 | AAS | $1,500 | $8,200 | | | | |
| Wilkes Regional Medical Center (North Wilkesboro) | 24 | Jun | 24 | Cert | $1,250 | $1,250 | | | | |
| **North Dakota** | | | | | | | | | | |
| Medcenter One (Bismarck) | 6 | Aug | 24 | Cert | | | | | | |
| MeritCare Medical Center (Fargo) | 8 | Aug | 24 | Cert | | | | | | • |
| Trinity Health (Minot) | 12 | Jul | 24 | Cert | $2,500 | $3,000 | | | | • |
| **Ohio** | | | | | | | | | | |
| Central Ohio Technical College (Newark) | 35 | Sep | 21 | AAS | $5,100 | $8,925 | | | | • |
| Children's Hospital Medical Ctr of Akron | 34 | Jul | 24 | Cert, AAS | $20,202 | $20,202 | | | | |
| Cleveland Clinic Health System (Euclid) | 28 | Aug | 24 | Cert | $5,700 | $5,700 | | | | • |
| Columbus State Community College | 32 | Summer Quarter | 21 | AAS | $4,345 | $8,690 | | • | | |
| Cuyahoga Community College (Parma) | 24 | Aug Jan | 24 | AAS | $2,810 | $3,723 | | • | | |
| Jefferson Community College (Steubenville) | 16 | Aug | 24 | AAS | $2,144 | $5,270 | | | | |
| Kent State University - Salem Campus | 38 | Jun | 24 | AAS | $6,289 | $16,364 | | | | • |
| Kettering College of Medical Arts | 27 | Aug | 22 | Cert, AS | $10,560 | $10,560 | | • | | |
| Lorain County Community College (Elyria) | 36 | Aug | 21 | AAS | $3,180 | $3,831 | | | | • |
| Marietta Memorial Hospital | 42 | Sep | 24 | Cert, AAS | $4,028 | $8,056 | | | | • |
| Marion Technical College | 16 | Jun | 23 | AAS | $5,983 | $0 | | • | | • |
| Mercy College of Northwest Ohio (Toledo) | 30 | Aug | 24 | AAS | $8,502 | $0 | | | | |
| Mercy Medical Center (Canton) | 8 | Jul | 24 | Dipl | $4,500 | $0 | | | | |
| Owens Community College (Toledo) | 24 | Jan Jun | 23 | AAS | $3,180 | $5,500 | | | | • |
| Rhodes State College (Lima) | 35 | Sep | 21 | AAS | $5,077 | $10,154 | | | | • |
| Sinclair Community College (Dayton) | 62 | Jan Sep | 24 | Dipl, AAS | $4,000 | $11,000 | | | | • |
| Xavier University (Cincinnati) | 20 | Aug | 23 | AS | $7,000 | $7,000 | | | | • |
| Zane State College (Zanesville) | 24 | Jun | 24 | AAS | $4,560 | $9,120 | | • | | • |
| **Oklahoma** | | | | | | | | | | |
| Autry Technology Center (Enid) | 9 | Aug | 24 | Cert | $1,996 | $6,868 | | | | |
| Bacone College (Muskogee) | 33 | Jun | 24 | AAS | $10,000 | $10,000 | | | | |
| Carl Albert State College (Poteau) | 10 | Fall | 21 | Dipl, AAS | $2,083 | $5,080 | | | | • |
| Great Plains Technology Center (Lawton) | 22 | Jul | 24 | Cert, AAS | $1,875 | $2,875 | | | | • |
| Indian Capital Technology Center (Muskogee) | 30 | Jul | 24 | Dipl | $1,500 | $3,000 | | | | • |
| Rose State College (Midwest City) | 20 | Aug | 24 | AAS | $2,581 | $3,602 | | | | • |
| Southwestern Oklahoma State University (Sayre) | 17 | Aug | 24 | AAS | $2,100 | $5,040 | | | | |
| Tulsa Community College | 40 | Jun | 24 | Dipl, AAS | $2,821 | $7,480 | | | | • |
| Tulsa Technology Center | 20 | Aug | 22 | Cert | $3,320 | $3,320 | | | • | • |

*Data are shown only for programs that completed the 2006 AMA Survey of Health Professions Education Programs
‡Key to Offers: 1: Evening or weekend classes; 2: Non-English instruction; 3: Cultural competence instruction

## Radiographer

| Programs* | Class Capacity | Begins | Length (months) | Award | Res. Tuition | Non-res. Tuition | Stipend | Offers:‡ 1 | 2 | 3 |
|---|---|---|---|---|---|---|---|---|---|---|
| Univ of Oklahoma Health Sciences Center (Oklahoma City) | 21 | Aug | 48 | BS | $5,960 | $8,900 | | | | • |
| Western Oklahoma State College (Altus) | 29 | Fall | 22 | Dipl, AAS | $3,100 | $0 | | | | |
| **Pennsylvania** | | | | | | | | | | |
| Albert Einstein Medical Center (Philadelphia) | 20 | Sep | 24 | Cert | $3,000 | $3,000 | | • | | |
| Armstrong County Memorial Hospital (Kittanning) | 8 | Jul | 24 | Cert | $3,000 | $3,000 | | | | |
| Bradford Regional Medical Center | 10 | Sep | 24 | Dipl | $2,500 | $2,500 | | | | • |
| Bucks County Community College (Newtown) | 28 | May | 24 | Cert | $4,995 | $4,995 | | | | • |
| Clearfield Hospital | 6 | Sep | 24 | Cert | $2,075 | $2,075 | | | | • |
| College Misericordia (Dallas) | 34 | Aug | 42 | BS | $19,800 | $19,800 | | | | |
| Comm Coll of Allegheny County (Monroeville) | 48 | Aug | 24 | AS | $1,092 | $2,020 | | • | | |
| Community College of Philadelphia | 30 | Jul | 24 | AAS | $2,863 | $8,112 | | | | • |
| Conemaugh Memorial Medical Center (Johnstown) | 17 | Sep | 24 | Cert | $4,075 | $4,075 | | | | |
| Crozer - Chester Med Ctr (Upland) | 24 | Jul Jan | 24 | Cert | $4,500 | $4,500 | | | | |
| Drexel University (Philadelphia) | 30 | Sep | 21 | Assoc | $4,900 | $0 | | | | |
| Gannon University (Erie) | 16 | Aug | 24 | AS | $26,000 | $26,000 | | | | • |
| Geisinger Medical Center (Danville) | 20 | Aug | 24 | Cert | $7,500 | $0 | | | | |
| Harcum College (Bryn Mawr) | 66 | Sep | 24 | Dipl, AAS | $15,250 | $15,250 | | | | • |
| Holy Spirit Hospital (Camp Hill) | 8 | Jan | 24 | Cert | $1,700 | $1,700 | | | | • |
| Johnson College (Scranton) | 48 | Aug | 21 | AS | $12,246 | $12,246 | | | | • |
| Lancaster Gen Coll of Nursing & Hlth Sciences | 21 | Aug | 21 | AS | $13,370 | $13,370 | | | | • |
| Montgomery County Community College (Pottstown) | 26 | Aug | 24 | AAS | $2,835 | $8,505 | | | | • |
| Northampton Community College (Bethlehem) | 34 | Aug | 24 | AAS | $3,450 | $10,350 | | | | • |
| Ohio Valley General Hospital (McKees Rocks) | 9 | Aug | 24 | Cert | $11,170 | $11,170 | | | | |
| Penn State University - New Kensington | 30 | Sep | 24 | AS | $10,200 | $0 | | | | |
| Penn State University - Schuylkill Haven | 66 | Aug | 28 | AS | $16,000 | $24,000 | | | | • |
| Pennsylvania College of Technology (Williamsport) | 30 | Aug | 24 | AAS | $11,850 | $0 | | | | • |
| Reading Hospital & Medical Center | 20 | Aug | 24 | Cert | $2,750 | $3,500 | | | | • |
| Sharon Regional Health System | 6 | Aug | 24 | Cert | $2,500 | $2,500 | | | | • |
| St Christopher Hospital School of Rad Tech (Philadelphia) | 65 | Jul | 24 | Cert | $6,500 | $6,500 | | • | | |
| St Joseph Medical Center (Reading) | 10 | Jul | 24 | Cert | $1,000 | $0 | | | | |
| Thomas Jefferson University (Philadelphia) | 25 | Sep | 12 | BS, MS | $22,884 | $22,884 | | | | • |
| Univ of Penn MC/Hosp of the Univ of Penn (Philadelphia) | 52 | Sep | 24 | Cert | $5,000 | $5,000 | | | | • |
| UPMC Health System (Pittsburgh) | 49 | Jul Jan | 24 | Dipl, Cert | $3,000 | $3,000 | | • | | |
| UPMC Northwest (Seneca) | 7 | Jul | 24 | Cert | $3,000 | $3,000 | | | | |
| Wilkes-Barre General Hospital | 12 | Sep | 24 | Cert | $5,000 | $5,000 | | | | • |
| **Puerto Rico** | | | | | | | | | | |
| Inter American University of Puerto Rico (San German) | 75 | Aug | 22 | AS | $4,800 | $4,800 | | | | |
| Universidad Central del Caribe (Bayamon) | 35 | Aug | 22 | AS | $3,200 | $0 | | | • | • |
| **Rhode Island** | | | | | | | | | | |
| Community College of Rhode Island (Lincoln) | 90 | Jun | 24 | AAS | $2,792 | $8,252 | | | | • |
| Rhode Island Hospital (Providence) | 20 | Jun | 24, 30 | Cert | $2,500 | $2,500 | | • | | • |
| **South Carolina** | | | | | | | | | | |
| Aiken Technical College | 20 | Fall semester | 24 | Dipl, AS | $3,036 | $0 | | | | |
| AnMed Health Medical Center (Anderson) | 19 | Jul | 24 | Cert | $1,000 | $1,000 | | | | |
| Florence-Darlington Technical College | 42 | Aug | 21 | AS | $1,513 | $1,644 | | | | |
| Greenville Technical College | 31 | Aug | 28 | AS | $4,530 | $9,480 | | | | • |
| Horry - Georgetown Technical College (Conway) | 27 | Aug | 18 | AHS | $1,600 | $2,300 | | | | • |
| Midlands Technical College (Columbia) | 20/ | Jun | 24 | AS | $5,400 | $7,800 | | | | • |
| Spartanburg Community College | 25 | Aug | 24 | AAS | $4,293 | $5,367 | | | | • |
| Technical College of the Lowcountry (Beaufort) | 32 | Fall semester | 24 | Assoc | $1,450 | $2,891 | | | | |
| Trident Technical College (Charleston) | 55 | May | 24 | AHS | $4,425 | $4,914 | | | | • |
| York Technical College (Rock Hill) | 20 | May | 24 | AHS | $4,350 | $9,792 | | | | • |
| **South Dakota** | | | | | | | | | | |
| Avera McKennan Hospital (Sioux Falls) | 12 | Sep | 24 | Cert | $1,250 | $1,250 | | | | |
| Avera Sacred Heart Hospital (Yankton) | 8 | Sep | 24 | Cert | $900 | $900 | | | | |
| Mitchell Technical Institute | 15 | Aug | 24 | Dipl, AAS | $4,200 | $4,200 | | | | • |
| Presentation College (Aberdeen) | 20 | Jun | 36 | BS | $11,400 | $11,400 | | | | • |
| Rapid City Regional Hospital | 10 | Jun | 24 | Cert | $1,250 | $1,250 | | | | • |
| Sioux Valley Hospital (Sioux Falls) | 14 | Jul | 24 | Cert | $1,000 | $1,000 | | | | |
| **Tennessee** | | | | | | | | | | |
| Baptist College of Health Sciences (Memphis) | 20 | Aug | 20 | BS | $7,900 | $7,900 | | | | |
| Chattanooga State Technical Comm College | 62 | Aug | 24 | AAS | $2,739 | $11,124 | | | | • |
| Jackson State Community College | 31 | Sep | 24 | AAS | $3,329 | $12,317 | | | | • |
| Methodist University Hospital (Memphis) | 22 | Jul | 24 | Cert | $2,500 | $2,500 | | | | |
| Nashville General Hospital | 28 | Oct | 24 | Cert | $3,500 | $3,500 | | | | • |
| Roane State Community College (Oak Ridge) | 42 | Aug | 22 | AAS | $3,213 | $9,621 | | | | • |
| South College (Knoxville) | 106 | Jan | 18, 30 | AS, BS | $17,200 | $17,200 | | • | | • |

*Data are shown only for programs that completed the 2006 AMA Survey of Health Professions Education Programs
‡Key to Offers: 1: Evening or weekend classes; 2: Non-English instruction; 3: Cultural competence instruction

## Radiographer

| Programs* | Class Capacity | Begins | Length (months) | Award | Res. Tuition | Non-res. Tuition | Stipend | Offers:‡ 1 | 2 | 3 |
|---|---|---|---|---|---|---|---|---|---|---|
| Southwest Tennessee Community College (Memphis) | 35 | Jul | 24 | AAS | $1,200 | $4,400 | | | | |
| Univ of Tennessee Medical Ctr at Knoxville | 14 | Jul | 24 | Cert | $2,000 | $2,000 | | | | • |
| Volunteer State Community College (Gallatin) | 35 | Jul | 24 | Dipl, AAS | $2,441 | $8,781 | | | | |
| **Texas** | | | | | | | | | | |
| Amarillo College | 30 | Aug | 24 | AAS | $1,800 | $3,200 | | | | |
| Angelina College (Lufkin) | 32 | Aug | 23 | AAS | $1,637 | $3,357 | | | | |
| Austin Community College | 44 | Aug | 24 | AAS | $1,908 | $7,308 | | | | |
| Baptist Health System (San Antonio) | 38 | Jan | 24 | Dipl | $14,894 | $0 | | | • | • |
| Blinn College (Bryan) | 18 | Sep | 24 | AAS | $5,500 | $0 | | | | |
| Career Centers of Texas - Ft Worth | 36 | Jul | 24 | Dipl | $13,500 | $0 | | | | |
| Citizens Medical Center (Victoria) | 5 | Jul | 24 | Cert | $1,000 | $1,000 | | | | • |
| Covenant Medical Center (Lubbock) | 19 | Aug | 22 | Cert | $3,300 | $3,300 | | | | |
| Cy-Fair College (Cypress) | 40 | | 24 | AAS | $1,596 | $0 | | | | • |
| Del Mar College (Corpus Christi) | 32 | Jul | 24 | AAS | $1,224 | $7,840 | | | | |
| El Centro College (Dallas) | 54 | Aug | 23 | AAS | $1,080 | $1,800 | | | • | • |
| El Paso Community College | 16 | May | 24 | AAS | $2,180 | $2,995 | | | | • |
| Harris County Hosp Dist/Ben Taub Gen Hosp (Houston) | 30 | Jul | 24 | Cert | $1,500 | $1,650 | | | | • |
| Hendrick Medical Center (Abilene) | 15 | Jan Jul | 24 | Cert | $2,800 | $2,800 | | | | |
| Houston Community College | 125 | Jan Aug | 24 | AAS | $2,200 | $4,100 | | | | • |
| Lamar Institute of Technology (Beaumont) | 38 | Jun | 48 | AAS | $2,500 | $8,000 | | | | |
| Laredo Community College | 16 | Sep | 24 | Dipl, AAS | $700 | $0 | | • | | |
| McLennan Community College (Waco) | 42 | Aug | 24 | AAS | $1,224 | $3,384 | | | • | • |
| Memorial Hermann Baptist Hospital Beaumont | 64 | Jun | 24 | Cert | $4,800 | $0 | | | | |
| Midland College | 12 | Aug | 23 | AAS | $1,710 | $3,192 | | | | |
| Midwestern State University (Wichita Falls) | 65 | Sep | 24 | AAS | $5,800 | $15,404 | | | | • |
| N Harris Mont Comm Coll Dist/Montgomery Coll (Conroe) | 50 | Jan | 24 | AAS | $1,648 | $3,248 | | | | |
| Odessa College | 17 | Jul | 24 | AAS | $1,728 | $2,868 | | | | • |
| South Plains College (Lubbock) | 32 | Sep | 24 | AAS | $1,846 | $3,408 | | | | |
| St Philip's College (San Antonio) | 68 | Jun | 24 | AAS | $1,300 | $4,000 | | | | • |
| Tarrant County College (Hurst) | 36 | Jul | 22 | AAS | $1,704 | $2,166 | | | | • |
| Tyler Junior College | 43 | Aug | 24 | AAS | $1,801 | $3,881 | | | | • |
| US Army Medical Dept Center & School (Ft Sam Houston) | 44 | | 11 | Cert | | | | | | |
| Wharton County Junior College | 18 | Aug | 24 | AAS | $32 | $64 | | | | • |
| **Utah** | | | | | | | | | | |
| Salt Lake Community College (Salt Lake City) | 35 | Aug | 21 | AAS | $3,167 | $9,820 | | • | | • |
| **Vermont** | | | | | | | | | | |
| Champlain College (Burlington) | 23 | Aug | 21, 45 | Dipl, AS, BS | $14,660 | $0 | 900 | | | • |
| **Virginia** | | | | | | | | | | |
| ACT College (Manassas) | 21 | Jun | 20 | AAS | $11,750 | $0 | | | | • |
| Central Virginia Community College (Lynchburg) | 20 | Aug | 24 | AAS | $2,615 | $7,950 | | | | • |
| Danville Regional Medical Center | 50 | Jan | 24 | Cert | $3,500 | $3,500 | | | | |
| Naval School of Health Sciences (Portsmouth) | 35 | Apr Sep | 12 | Cert | | | | | | • |
| Rockingham Memorial Hospital (Harrisonburg) | 15 | Jun | 24 | Cert, AAS | $3,180 | $6,570 | | | | |
| Southside Regional Medical Center (Petersburg) | 33 | Aug | 22 | Cert | $3,510 | $3,510 | | | | |
| Southwest Virginia Community College (Richlands) | 36 | Jul | 24 | AAS | $77 | $89 | | | | |
| Tidewater Community College (Virginia Beach) | 45 | May | 24 | AAS | $2,754 | $8,314 | | | | • |
| University of Virginia Health System (Charlottesville) | 20 | Jul | 24 | Cert | $2,000 | $2,000 | | | | • |
| Virginia Commonwealth University (Richmond) | 48 | Aug | 33 | BS | $7,000 | $22,000 | | | | • |
| Virginia Western Community College (Roanoke) | 24 | Aug | 24 | AAS | $1,600 | $5,100 | | | | |
| **Washington** | | | | | | | | | | |
| Apollo College (Spokane) | 30 | Multiple start dates | 24 | AOS | $20,000 | $0 | | | | • |
| Pima Medical Institute - Seattle | 24 | Mar Jul Nov | 24 | AS | $11,732 | $11,732 | | | | • |
| Spokane Community College | 25 | Sep | 22 | Dipl, AAS | $8,500 | $12,000 | | | | • |
| Tacoma Community College | 24 | Sep | 24 | Dipl, Cert, AAS | $3,303 | $3,830 | | | | • |
| **West Virginia** | | | | | | | | | | |
| Bluefield State College | 34 | May | 24 | AS | $4,404 | $9,058 | | • | | |
| Southern West Virginia Comm & Tech College (Mt Gay) | 40 | Aug | 21 | Dipl, AAS | $2,089 | $6,497 | | • | | • |
| United Hospital Center (Clarksburg) | 15 | Jun | 24 | Cert, AAS | $3,000 | $3,000 | | | | • |
| West Virginia University Hospitals (Morgantown) | 18 | Jul | 24 | Cert | $1,600 | $1,600 | | | | • |
| Wheeling Hospital | 10 | Jul | 24 | Cert | $1,500 | $1,500 | | | | • |
| **Wisconsin** | | | | | | | | | | |
| Aurora St Luke's Medical Center (Milwaukee) | 24 | Sep | 24 | Cert | $2,000 | $2,000 | | | | • |
| Bellin Health Hosp/Bellin Health Systems Inc (Green Bay) | 16 | Aug | 24 | Dipl | $3,500 | $3,500 | | | | • |
| Blackhawk Technical College (Janesville) | 18 | Jun | 24 | AS | $2,817 | $15,043 | | | | |
| Columbia St Mary's Hospitals (Milwaukee) | 20 | Sep | 24 | Cert | $2,000 | $2,000 | | | | |
| Froedtert Memorial Lutheran Hospital (Milwaukee) | 45 | Sep | 24 | Cert | $3,000 | $3,000 | | | | • |
| Lakeshore Technical College (Cleveland) | 45 | Jan | 22 | AAS | $2,958 | $15,276 | | | | • |

*Data are shown only for programs that completed the 2006 AMA Survey of Health Professions Education Programs

‡Key to Offers: 1: Evening or weekend classes; 2: Non-English instruction; 3: Cultural competence instruction

## Radiographer

| Programs* | Class Capacity | Begins | Length (months) | Award | Res. Tuition | Non-res. Tuition | Stipend | Offers:‡ 1 | 2 | 3 |
|---|---|---|---|---|---|---|---|---|---|---|
| Mercy Medical Center/Affinity Health System (Oshkosh) | 14 | Sep | 24 | Cert | $10,000 | $10,000 | | | | |
| Milwaukee Area Technical College | 30 | Aug | 24 | AAS | $3,020 | $0 | | | | |
| Northcentral Technical College (Wausau) | 18 | Aug | 24 | AS | $2,720 | $2,720 | | | | • |
| St Michael Hospital (Milwaukee) | 10 | Sep | 24 | Cert | $1,500 | $1,500 | | | | |
| Theda Clark Regional Medical Center (Neenah) | 16 | Sep | 24 | Cert | $2,000 | $2,000 | | | | • |
| University of Wisconsin Hospital and Clinics (Madison) | 16 | Sep | 24 | Dipl | $2,000 | $2,000 | | | | • |
| Western Technical College (La Crosse) | 32 | Aug | 23 | AAS | $3,354 | $16,068 | | | | |
| Wheaton Franscian Healthcare-All Saints (Racine) | 15 | Jul | 24 | Cert | $1,200 | $1,200 | | | | |
| **Wyoming** | | | | | | | | | | |
| Casper College | 18 | Jun | 24 | AS | $1,536 | $4,296 | | | • | • |
| Laramie County Community College (Cheyenne) | 18 | Aug | 24 | AAS | $2,672 | $6,480 | | | | • |

*Data are shown only for programs that completed the 2006 AMA Survey of Health Professions Education Programs
‡Key to Offers: 1: Evening or weekend classes; 2: Non-English instruction; 3: Cultural competence instruction

# Rehabilitation Counselor

## History

Initially, rehabilitation professionals were recruited from a variety of human service disciplines, including public health nursing, social work, and school counseling. Although educational programs began to appear in the 1940s, it was not until the availability of federal funding for rehabilitation counseling programs in 1954 that the profession began to grow and establish its own identity.

Historically, rehabilitation counselors primarily served working-age adults with disabilities. Today, the need for rehabilitation counseling services extends to persons of all age groups who have disabilities. Rehabilitation counselors also may provide general and specialized counseling to people with disabilities in public human service programs and private practice settings.

## Occupational Description

Working directly with an individual with a disability, the rehabilitation counselor determines and coordinates services to assist people with disabilities in moving from psychological and economic dependence to independence.

## Job Description

Rehabilitation counselors assist people with physical, mental, or emotional disabilities to become or remain self-sufficient, productive citizens. Disabilities may result from birth defects, illness and disease, work-related injuries, automobile accidents, the stresses of war, work, daily life, and the aging process. Rehabilitation counselors help individuals with disabilities deal with societal and personal problems, plan careers, and find and keep satisfying jobs. They also may work with individuals, professional organizations, and advocacy groups to address the environmental and social barriers that create obstacles for people with disabilities. The rehabilitation counselor builds bridges between the often isolated world of people with disabilities and their families, communities, and work environments.

Other responsibilities for the rehabilitation counselor include:

- Evaluating an individual's potential for independent living and employment and arranging for medical and psychological services and vocational assessment, training, and job placement
- Evaluating medical and psychological reports and conferring with physicians and psychologists about the types of work individuals can perform
- Working with employers to identify and/or modify job responsibilities to accommodate individuals with disabilities

The rehabilitation counselor draws on knowledge from several fields, including psychology, medicine, psychiatry, sociology, social work, education, and law. Their specialized knowledge of disabilities and environmental factors that interact with disabilities, as well as specific knowledge and skills, differentiate rehabilitation counselors from other types of counselors.

## Employment Characteristics

Many rehabilitation counselors work in state rehabilitation agencies or community rehabilitation programs. Because all state rehabilitation agencies follow the same general procedures, a rehabilitation counselor has geographical mobility and can find employment throughout the United States and its territories. Other potential employers include comprehensive rehabilitation centers, universities and academic settings, insurance companies, substance abuse rehabilitation centers, correctional facilities, halfway houses, and independent living centers. Reflecting this wide range of job opportunities, rehabilitation counselors are often employed in positions with different job titles, such as counselor, job placement specialist, substance abuse counselor, rehabilitation consultant, independent living specialist, or case manager.

## Salary

In 2006, the Council on Rehabilitation Education collected data from 97 of the 102 accredited MS programs in the country. The average salary was $46,992. The range of salaries for an entry-level assistant professor position was $41,310 to $62,000; the salary ranges in the private sector are considerably higher. Refer to Section IV, Table 5 of this *Directory* for more information, or see www.ama-assn.org/go/salary.

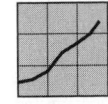

## Employment Outlook

Rehabilitation counselors serve a large portion of the US population. An estimated 43 million Americans have physical, mental, or psychological disabilities that restrict their activities and prevent them from obtaining or maintaining jobs.

Consequently, the employment outlook for the profession is excellent. Based on national employment outlook studies and regional and state surveys, hundreds of rehabilitation counselor positions are expected to be available in the coming years for qualified master's level professionals. Recent studies show that rehabilitation counselor education programs are not graduating sufficient numbers of qualified students to meet current and anticipated marketplace needs.

Recently the roles and responsibilities of rehabilitation counselors have expanded, further increasing the attractiveness of a career in the profession. Rehabilitation counselors, for example, have begun to determine, coordinate, and arrange for rehabilitation and transition services for children within school systems. In addition, rehabilitation counselors are providing geriatric rehabilitation services to older persons with health problems, and workers injured on the job are increasingly receiving rehabilitation services through private rehabilitation counseling companies and employers' disability management and employee assistance programs.

Many former teachers, attorneys, nurses, physical therapists, occupational therapists, clergy, and business people have found second careers as rehabilitation counselors.

## Educational Programs

**Length.** Rehabilitation counselor education programs typically provide between 18 and 24 months of academic and field-based clinical training. Clinical training consists of a practicum and a minimum of 600 hours of supervised internship experience. Clinical field experiences are available in a variety of community, state, federal, and private rehabilitation-related programs.

**Prerequisites.** Although no formal requirements exist, most rehabilitation counseling graduate students have undergraduate degrees in rehabilitation services, psychology, sociology, or other human services-related fields.

**Curriculum.** Rehabilitation counselors are trained in counseling theory, skills, and techniques; individual, group, and environmental assessment; psychosocial and medical aspects of disability, including human growth and development; principles of psychiatric rehabilitation; case management and rehabilitation planning; issues and ethics in rehabilitation service delivery; technological adaptation; vocational evaluation and work adjustment; career counseling; and job development and placement. In addition, students often take required or elective courses in such areas as group counseling, marriage and family counseling, substance abuse rehabilitation, juvenile and adult offender rehabilitation, mental retardation, communication disorders, sign language, stress management, psychological testing, and rehabilitation administration.

## Licensure, Certification, and Registration

Certification and licensure of rehabilitation counselors help protect the public and provide a means of identifying those individuals who possess the minimum training and meet supervised work experience standards established by professional groups and governmental agencies.

**Certification.** The Commission on Rehabilitation Counselor Certification (CRCC), an independent credentialing body incorporated in 1974, certifies rehabilitation counselors throughout the United States and in Canada who meet educational and work experience requirements, pass an examination, and maintain certification by completing 100 hours of acceptable continuing education credit every 5 years, 10 hours of which must be in ethics.

**Licensure.** A counseling license is a credential authorized by a state legislature that regulates the title and/or practice of professional counselors. Rehabilitation counselors are eligible for licensure as professional counselors in nearly all states that regulate counselors; licensure requirements include passing an examination, acquiring needed supervised counseling experience, and, in some states, completing specified coursework.

**Registration.** A number of state workers' compensation laws or regulations specify education, training, and/or credentials requirements for people providing rehabilitation counseling services to workers with disabilities. In these states, rehabilitation counselors pay a fee and provide proof of education and/or certification to register with the state workers' compensation agency. Most of these states also require the certified rehabilitation counselor (CRC)

credential, although the permitted scope of services may vary from one state to the next.

## Inquiries

### Careers
National Rehabilitation Counseling Association
8807 Sudley Road, #102
Manassas, VA 22110-4719
703 361-2077

American Rehabilitation Counseling Association
5999 Stevenson Avenue
Alexandria, VA 22304
703 823-9800

International Association of Rehabilitation Professionals
PO Box 697
Brookline, MA 02146
617 566-4432

National Council on Rehabilitation Education
Charles Arokiasamy, RhD, CRC
Rehabilitation Counseling Program
Kremen School of Education and Human Development
California State University-Fresno
5005 Maple Avenue, M/S ED3
Fresno, CA 90740-8025
559 278-0325

### Certification
Commission on Rehabilitation Counselor Certification
300 North Martingale Road, Suite 460
Schaumburg, IL 60173
847 944-1325

### Program Accreditation
Council on Rehabilitation Education
300 North Martingale Road, Suite 460
Schaumburg, IL 601738
847 944-1345
847 944-1324 Fax
www.core-rehab.org

---

## Rehabilitation Counselor*

# Alabama

**Auburn University**
Rehabilitation Counseling Prgm
Rehabilitation and Special Education
1228 Haley Center
Auburn University, AL 36849
*Prgm Dir:* E Davis Martin, Jr, EdD CRC NCC
*Tel:* 334 844-2083   *Fax:* 334 844-2080
*E-mail:* martiev@auburn.edu

**University of Alabama at Birmingham**
Rehabilitation Counseling Prgm
School of Education, Rm 157
901 S 13th St
Birmingham, AL 35294-1250
*Prgm Dir:* Barry Stephens, PhD
*Tel:* 205 934-3701   *Fax:* 205 934-3702
*E-mail:* barrystephens@uab.edu

**Alabama A&M University**
Rehabilitation Counseling Prgm
PO Box 580
Normal, AL 35762
www.aamu.edu
*Prgm Dir:* William Fennessee, RhD CRC
*Tel:* 256 372-4039   *Fax:* 256 372-5255
*E-mail:* william.fennessee@email.aamu.edu

**Troy University**
Rehabilitation Counseling Prgm
College of Education
#10 McCartha Hall
Troy, AL 36082
*Prgm Dir:* Linda Shumaker-Williams, PhD LPC CRC
*Tel:* 334 670-3350   *Fax:* 334 670-3291
*E-mail:* lindaw@trojan.troyst.edu

**University of Alabama**
Rehabilitation Counseling Prgm
318 Graves Hall, PO Box 870231
Tuscaloosa, AL 35487-0231
*Prgm Dir:* Jamie Satcher, PhD
*Tel:* 205 348-1178   *Fax:* 205 348-7884
*E-mail:* Jsatcher@bamaed.ua.edu

---

*Additional information about programs that returned the AMA's survey is available in a table beginning on page 397.

# Arizona

**University of Arizona**
Rehabilitation Counseling Prgm
Dept of Spec Educ and Rehab
College of Education
Tucson, AZ 85721
www.arizona.edu
*Prgm Dir:* Amos Sales, EdD CRC NCC
*Tel:* 520 621-0941   *Fax:* 520 621-3821
*E-mail:* sales@u.arizona.edu

# Arkansas

**University of Arkansas**
Rehabilitation Counseling Prgm
Rehabilitation Education and Research Program
100 Graduate Education Bldg
Fayetteville, AR 72701
*Prgm Dir:* Lynn Koch, PhD CRC
*Tel:* 479 575-8696   *Fax:* 479 575-3319
*E-mail:* lckoch@uark.edu

**University of Arkansas at Little Rock**
Rehabilitation Counseling Prgm
2801 S University Ave
Little Rock, AR 72204
www.teletrain.com/ualr
*Prgm Dir:* Larry R Dickerson, PhD
*Tel:* 501 569-3428   *Fax:* 501 569-3547
*E-mail:* lrdickerson@ualr.edu

**Arkansas State University**
Rehabilitation Counseling Prgm
Dept of Psychology and Counseling
PO Box 1560
State University, AR 72467
*Prgm Dir:* Lisa Ochs, JD PhD CRC
*Tel:* 870 972-3064   *Fax:* 870 972-3962
*E-mail:* tturner@astate.edu

# California

**California State University - Fresno**
Rehabilitation Counseling Prgm
5005 N Maple Ave, MS ED 3
Fresno, CA 93740
www.csufresno.edu
*Prgm Dir:* Charles Arokiasamy, RhD
*Tel:* 559 278-0325   *Fax:* 559 278-0045
*E-mail:* charlesa@csufresno.edu

**California State University - Los Angeles**
Rehabilitation Counseling Prgm
5151 State University Dr, King Hall C1064
Los Angeles, CA 90032
www.calstatela.edu
*Prgm Dir:* Martin Brodwin, PhD CRC
*Tel:* 323 343-4440   *Fax:* 323 343-5605
*E-mail:* mbrodwi@calstatela.edu

**California State University - Sacramento**
Rehabilitation Counseling Prgm
College of Education
6000 J St
Sacramento, CA 95819-6079
*Prgm Dir:* Guy Deaner, PhD
*Tel:* 916 278-6663   *Fax:* 916 278-3948
*E-mail:* deanerg@csus.edu

**California State University - San Bernardino**
Rehabilitation Counseling Prgm
School of Education
5500 University Pkwy
San Bernardino, CA 92407
*Prgm Dir:* Joseph Turpin, PhD CRC
*Tel:* 909 880-5680   *Fax:* 909 880-5992
*E-mail:* jturpin@csusb.edu

**San Diego State University**
Rehabilitation Counseling Prgm
Dept of Adm Rehab and Postsecondary Ed
3590 Camino del Rio, North
San Diego, CA 92108-1716
www.interwork.sdsu.edu
*Prgm Dir:* Caren Sax, EdD CRC
*Tel:* 619 594-7183   *Fax:* 619 594-4208
*E-mail:* csax@mail.sdsu.edu

**San Francisco State University**
Rehabilitation Counseling Prgm
Dept of Counseling
1600 Holloway Ave
San Francisco, CA 94132
*Prgm Dir:* Leslie Zwillinger, PhD CRC
*Tel:* 415 338-7647   *Fax:* 415 338-0594
*E-mail:* lzwill@sfsu.edu

# Colorado

**University of Northern Colorado**
Rehabilitation Counseling Prgm
School of Human Sciences
Gunter 1250, Box 132
Greeley, CO 80639
www.unco.edu/HHS/hs/hs.htm
*Prgm Dir:* Joseph Ososkie, PhD
*Tel:* 970 351-1579   *Fax:* 970 351-1255
*E-mail:* joe.ososkie@unco.edu

# Connecticut

**Central Connecticut State University**
Rehabilitation Counseling Prgm
165 Stanley St
New Britain, CT 06050
*Prgm Dir:* Judith Rosenberg
*E-mail:* rosenbergj@ccsu.edu

# District of Columbia

**George Washington University**
Rehabilitation Counseling Prgm
Dept of Counseling, Human/Org Studies
2134 G St NW, 3rd Fl
Washington, DC 20052
*Prgm Dir:* Jorge Garcia, RhD CRC LPC
*Tel:* 202 994-7126   *Fax:* 202 994-3436
*E-mail:* garcia@gwu.edu

# Florida

**Florida Atlantic University**
Rehabilitation Counseling Prgm
777 Glades Rd
Boca Raton, FL 33431
*Prgm Dir:* Michael Frain, PhD CRC
*E-mail:* frain@fau.edu

**University of Florida**
Rehabilitation Counseling Prgm
Dept of Rehabilitation Counseling
PO Box 100175
Gainesville, FL 32610-0175
*Prgm Dir:* Linda Shaw, PhD CRC
*Tel:* 352 273-6745   *Fax:* 352 273-6048
*E-mail:* lshaw@phhp.ufl.edu

**University of North Florida**
Rehabilitation Counseling Prgm
Department of Public Health
4567 St Johns Bluff Rd S
Jacksonville, FL 32224
www.unf.edu/brooks/cohrehab.htm
*Prgm Dir:* Jeanne Patterson, EdD CRC
*Tel:* 904 620-1428   *Fax:* 904 620-1035
*E-mail:* jpatters@unf.edu

**Florida State University**
Rehabilitation Counseling Prgm
205 Stone Bldg
Tallahassee, FL 32306
www.coe.fsu.edu/cerds/programs/RehabCounsel.html
*Prgm Dir:* Deborah Ebener, PhD CRC NCC
*Tel:* 850 644-1789   *Fax:* 850 644-8715
*E-mail:* debener@fsu.edu
*Med Dir:* Jane Burkhead, PhD, CRC

**University of South Florida**
Rehabilitation Counseling Prgm
Dept of Rehabilitation & Mental Health Counseling
4202 E Fowler Ave, SOC 107
Tampa, FL 33620-8100
www.cas.usf.edu/rehab_counseling
*Prgm Dir:* Charlotte Dixon, RhD CRC NCC
*Tel:* 813 974-0973   *Fax:* 813 974-8080
*E-mail:* dixon@cas.usf.edu
*Med Dir:* Tennyson Wright, PhD, CRC

# Georgia

**Georgia State University**
Rehabilitation Counseling Prgm
Counseling and Psych Services Dept
PO Box 3980
Atlanta, GA 30302-3980
www.gsu.edu
*Prgm Dir:* Roger Weed, PhD CRC CDMS CLCP LPC CCM
  FNRCA FIALCP
*Tel:* 404 651-2550   *Fax:* 404 651-1160
*E-mail:* rweed@gsu.edu

**Fort Valley State University**
Rehabilitation Counseling Prgm
1005 State College Dr
PO Box 31030-4313
Ft Valley, GA 31030-4313
www.fvsu.edu
*Prgm Dir:* Dothel Edwards, Jr, RhD CRC CLCP
*Tel:* 478 825-6237, Ext 6526   *Fax:* 478 827-3097
*E-mail:* edwardsd@fvsu.edu

**Thomas University**
Rehabilitation Counseling Prgm
Division of Human Services
1501 Millpond Rd
Thomasville, GA 31792
www.thomasu.edu
*Prgm Dir:* Gadson Melvin, PhD
*Tel:* 229 226-1621, Ext 217   *Fax:* 229 226-1653
*E-mail:* mgadson@thomasu.edu

# Hawaii

**University of Hawaii at Manoa**
Rehabilitation Counseling Prgm
Dept of Counselor Education
1776 University Ave, WA2-221
Honolulu, HI 96822
*Prgm Dir:* Brenda Cartwright, PhD CRC
*Tel:* 808 956-4386   *Fax:* 808 956-3814
*E-mail:* bcartwri@hawaii.edu

# Idaho

**University of Idaho**
Rehabilitation Counseling Prgm
Counseling, School Psychology, Special Education, and
  Educat
College of Education Rm 206
Moscow, ID 83844-3083
*Prgm Dir:* Jerry Fischer, RhD CRC LPC
*Tel:* 208 885-5947   *Fax:* 208 885-6869
*E-mail:* jfischer@uidaho.edu

# Illinois

### Southern Illinois Univ at Carbondale
Rehabilitation Counseling Prgm
Rehabilitation Institute
Ste 308
Carbondale, IL 62901-4609
*Prgm Dir:* Thomas Upton, PhD CRC
*Tel:* 618 453-8287   *Fax:* 618 453-8271
*E-mail:* tupton@siu.edu

### Univ of Illinois at Urbana - Champaign
Rehabilitation Counseling Prgm
1206 S 4th, 121 Hoff
Champaign, IL 61820
www.uiuc.edu
*Prgm Dir:* David Strauser, PhD CRC
*Tel:* 217 333-2307   *Fax:* 217 333-2766
*E-mail:* strauser@uiuc.edu

### Illinois Institute of Technology
Rehabilitation Counseling Prgm
Inst of Psych, Rm 252 Life Science Bldg
3101 S Dearborn St
Chicago, IL 60616
www.iit.edu/colleges/psych
*Prgm Dir:* Chow Lam, PhD
*Tel:* 312 567-3515   *Fax:* 312 567-3493
*E-mail:* Lam@iit.edu

### Northeastern Illinois University
Rehabilitation Counseling Prgm
5500 N St Louis Ave
Chicago, IL 60625
www.neiu.edu/~counsedu
*Prgm Dir:* Kenneth F Currier, PhD CRC
*Tel:* 773 442-5550, Ext 5576   *Fax:* 773 442-5559
*E-mail:* k-currier@neiu.edu

### Northern Illinois University
Rehabilitation Counseling Prgm
Dept of Communicative Disorders
Northern Illinois University
DeKalb, IL 60115
www.coms.hhsweb.com/grcgrad.htm
*Prgm Dir:* Deborah Gough, EdD
*Tel:* 815 753-6519   *Fax:* 815 753-9123
*E-mail:* dgough@niu.edu

# Indiana

### Ball State University
Rehabilitation Counseling Prgm
Dept of Counseling Psychology & Guidance Services
TC 622
Muncie, IN 47306-0585
*Prgm Dir:* Molly Tschoop, PhD CRC
*Tel:* 765 285-8044   *Fax:* 765 285-2067
*E-mail:* mktschoop@bsu.edu

# Iowa

### Drake University
Rehabilitation Counseling Prgm
Counseling, Leadership, and Adult Development
3206 University
Des Moines, IA 50311
*Prgm Dir:* Robert Stensrud, EdD CRC
*Tel:* 515 271-3061   *Fax:* 515 271-4140
*E-mail:* robert.stensrud@drake.edu

### University of Iowa
Rehabilitation Counseling Prgm
N362 Lindquist Center
Iowa City, IA 52242-1529
www.coe164.education.uiowa.edu:8180/crsd/rehab
*Prgm Dir:* Vilia Travydas, PhD CRC
*Tel:* 319 335-5285   *Fax:* 319 335-5291
*E-mail:* vilia-tarvydas@uiowa.edu

# Kansas

### Emporia State University
Rehabilitation Counseling Prgm
Campus Box 4036, 1200 Commerical
Emporia, KS 66801
www.emporia.edu/counre
*Prgm Dir:* Marvin D Kuehn, EdD
*Tel:* 620 341-5795   *Fax:* 620 341-6200
*E-mail:* mkuehn@emporia.edu

# Kentucky

### University of Kentucky
Rehabilitation Counseling Prgm
224 Taylor Education Bldg
Lexington, KY 40506
*Prgm Dir:* Ralph M Crystal, PhD CRC
*Tel:* 859 257-3834   *Fax:* 859 257-3835
*E-mail:* crystal@uky.edu

# Louisiana

### Southern Univ and A&M College
Rehabilitation Counseling Prgm
229 Blanks Hall
Baton Rouge, LA 70813
www.subr.edu/science/rehabcounsel
*Prgm Dir:* Madan M Kundu, PhD CRC NCC LRC FNRCA
*Tel:* 225 771-2819   *Fax:* 225 771-2293
*E-mail:* kundusubr@aol.com

### Louisiana State Univ Health Sciences Center
Rehabilitation Counseling Prgm
School of Allied Health Professions
1900 Gravier St, Ste 8A1
New Orleans, LA 70112-2262
*Prgm Dir:* John Dolan, RhD CRC
*Tel:* 504 568-4320   *Fax:* 504 568-4324
*E-mail:* jdolan@lsuhsc.edu

# Maine

### University of Southern Maine
Rehabilitation Counseling Prgm
400 Bailey Hall
Gorham, ME 04038
*Prgm Dir:* Stephen T Murphy, PhD CRC
*Tel:* 207 780-5319   *Fax:* 207 780-5043
*E-mail:* smurphy@usm.maine.edu

# Maryland

### Coppin State University
Rehabilitation Counseling Prgm
2500 W North Ave
Baltimore, MD 21216
*Prgm Dir:* Janet D Spry, EdD
*Tel:* 410 951-3514   *Fax:* 410 951-3511
*E-mail:* jspry@coppin.edu

### University of Maryland
Rehabilitation Counseling Prgm
Counseling and Personnel Services Dept
Coll of Education, Benjamin Bldg Rm 3214
College Park, MD 20742
www.education.umd.edu/edcp/rehab/program/html
*Prgm Dir:* Ellen Fabian, PhD
*Tel:* 301 405-2872   *Fax:* 301 405-9995
*E-mail:* efabian@umd.edu

### University of Maryland Eastern Shore
Rehabilitation Counseling Prgm
Ste 1062, Hazel Hall
Princess Anne, MD 21853
www.umes.edu
*Prgm Dir:* William Talley, RhD CRC
*Tel:* 410 651-6262   *Fax:* 410 651-6736
*E-mail:* wbtalley@umes.edu

# Massachusetts

### Boston University
Rehabilitation Counseling Prgm
Sargent Coll of Hlth and Rehab Sciences
635 Commonwealth Ave
Boston, MA 02215
*Prgm Dir:* Arthur E Dell Orto, PhD CRC
*Tel:* 617 353-7486   *Fax:* 617 353-8914
*E-mail:* ado@bu.edu

### University of Massachusetts - Boston
Rehabilitation Counseling Prgm
Counseling and School Psychology
Graduate College of Education
Boston, MA 02125-3393
*Prgm Dir:* Rick Houser, PhD
*Tel:* 617 287-7668   *Fax:* 617 287-7664
*E-mail:* rick.houser@umb.edu

### Springfield College
Rehabilitation Counseling Prgm
Rehabilitation and Disability Studies Dept
263 Alden St
Springfield, MA 01109
www.spfldcol.edu
*Prgm Dir:* Thomas J Ruscio, EdD
*Tel:* 413 748-3566   *Fax:* 413 748-3787
*E-mail:* truscioj@spfldcol.edu

### Assumption College
Rehabilitation Counseling Prgm
Inst for Social and Rehab Services
500 Salisbury St
Worcester, MA 01609-1296
*Prgm Dir:* Lee Pearson, CRC
*Tel:* 508 767-7063   *Fax:* 508 798-2872
*E-mail:* lpearson@assumption.edu

# Michigan

### Wayne State University
Rehabilitation Counseling Prgm
Theoretical and Behavioral Foundation
323 College of Education, 5425 Gullen Mall
Detroit, MI 48202
*Prgm Dir:* George Parris, PhD LPC CCRC NCP ABDA
*Tel:* 313 577-1619   *Fax:* 313 577-5235
*E-mail:* gparris@wayne.edu

### Michigan State University
Rehabilitation Counseling Prgm
Counseling, Educ Psych and Spec Educ
237 Erickson Hall
East Lansing, MI 48824-1034
*Prgm Dir:* John F Kosciulek, PhD CRC
*Tel:* 517 355-1838   *Fax:* 517 353-6393
*E-mail:* jkosciul@msu.edu

### Western Michigan University
Rehabilitation Counseling Prgm
The Graduate College
3404 Sangren Hall, Mail Stop 5218
Kalamazoo, MI 49008
*Prgm Dir:* William R Wiener, PhD CRC
*Tel:* 616 387-3455   *Fax:* 616 387-3567
*E-mail:* william.wiener@wmich.edu

# Minnesota

### Minnesota State University - Mankato
Rehabilitation Counseling Prgm
Dept of Speech, Hearing, and Rehabilitation Svcs
College of Allied Health and Nursing
Mankato, MN 56001
http://ahn.mnsu.edu/rehabilitation
*Prgm Dir:* Gerald Schneck, PhD CRC-MAC CEA FVE
   NCC
*Tel:* 507 389-5438   *Fax:* 507 389-2821
*E-mail:* gerald.schneck@mnsu.edu

**St Cloud State University**
Rehabilitation Counseling Prgm
Dept of Counselor Education & Ed Psych
720 4th Ave S
St Cloud, MN 56301-4498
www.stcloudstate.edu/ceep/rehab
*Prgm Dir:* John C Hotz, RhD CRC
*Tel:* 320 308-2240    *Fax:* 320 308-4082
*E-mail:* jhotz@stcloudstate.edu

## Mississippi

**Jackson State University**
Rehabilitation Counseling Prgm
PO Box 17501
Jackson, MS 39217-7501
*Prgm Dir:* Frank L Giles, CRC
*Tel:* 601 968-2370    *Fax:* 601 968-2213
*E-mail:* fgiles@jsums.edu

**Mississippi State University**
Rehabilitation Counseling Prgm
Counseling, Ed Psych and Special Education
Mail Stop 9727
Mississippi State, MS 39762
www.educ.msstate.edu/cepse
*Prgm Dir:* Charles Palmer, PhD
*Tel:* 662 325-7917    *Fax:* 662 325-3263
*E-mail:* cpalmer@colled.msstate.edu

## Missouri

**University of Missouri - Columbia**
Rehabilitation Counseling Prgm
Dept of Educational, School, and Counseling Psychology
98 Corporate Lake Dr
Columbia, MO 65211
*Prgm Dir:* C David Roberts, PhD
*Tel:* 573 882-3807    *Fax:* 573 884-5989
*E-mail:* RobertsC@missouri.edu

**Maryville University**
Rehabilitation Counseling Prgm
650 Maryville University Dr
St Louis, MO 63141
*Prgm Dir:* Michael Kiener, PhD CRC
*Tel:* 314 529-9443    *Fax:* 314 529-9139
*E-mail:* mkiener@maryville.edu

## Montana

**Montana State University - Billings**
Rehabilitation Counseling Prgm
Dept of Rehabilitation and Human Services
1500 N 30th St
Billings, MT 59101-0298
*Prgm Dir:* Kyle Colling
*Tel:* 406 657-2056    *Fax:* 406 657-2255
*E-mail:* kcolling@msubillings.edu

## New Jersey

**Univ of Med & Dent of New Jersey**
Rehabilitation Counseling Prgm
Dept of Psychiatric Rehabilitation
1776 Raritan Rd
Scotch Plains, NJ 07076
*Prgm Dir:* Janice Oursler, PhD CRC
*Tel:* 908 889-2462    *Fax:* 908 889-2432
*E-mail:* ourslejd@umdnj.edu

## New York

**SUNY at Albany**
Rehabilitation Counseling Prgm
School of Education, Rm 220
1400 Washington Ave
Albany, NY 12222
*Prgm Dir:* Sheldon A Grand, PhD CRC
*Tel:* 518 442-5041    *Fax:* 518 442-4953
*E-mail:* sgrand@nycap.rr.com

**University at Buffalo - SUNY**
Rehabilitation Counseling Prgm
Dept of Counseling, School, and Educ Psych
409 Baldy Hall
Buffalo, NY 14260-1000
*Prgm Dir:* Timothy Janikowski
*Tel:* 716 645-2484, Ext 1067    *Fax:* 716 645-6166
*E-mail:* tjanikow@buffalo.edu

**Hofstra University**
Rehabilitation Counseling Prgm
119 Hofstra University, 160 Hagedorn
Hempstead, NY 11549-1190
www.hofstra.edu
*Prgm Dir:* Jamie S Mitus, PhD
*Tel:* 516 463-7453    *Fax:* 516 463-6184
*E-mail:* JamieS.Mitus@Hofstra.edu

**St John's University**
Rehabilitation Counseling Prgm
8000 Utopia Pkwy
Jamaica, NY 11439
*Prgm Dir:* Caroline K Wilde, PhD CRC
*Tel:* 718 990-6455    *Fax:* 718 990-7468
*E-mail:* wildec@stjohns.edu

**CUNY Hunter College**
Rehabilitation Counseling Prgm
Dept of Educ Found and Counseling Prgms
695 Park Ave
New York, NY 10021
*Prgm Dir:* John O'Neill, PhD CRC
*Tel:* 212 772-5188    *Fax:* 212 650-3198

**Syracuse University**
Rehabilitation Counseling Prgm
259 Huntington Hall
Syracuse, NY 13244
*Prgm Dir:* Dennis Gilbride, PhD CRC
*Tel:* 315 443-5264    *Fax:* 315 443-5732
*E-mail:* ddgilbri@syr.edu

## North Carolina

**Univ of North Carolina at Chapel Hill**
Rehabilitation Counseling Prgm
108 Medical School Wing E, CB 7205
Chapel Hill, NC 27599-7205
*Prgm Dir:* Eileen Burker, PhD CRC
*Tel:* 919 966-8788    *Fax:* 919 966-9007
*E-mail:* Eileen_burker@med.unc.edu

**North Carolina A&T State University**
Rehabilitation Counseling Prgm
1601 E Market St
Greensboro, NC 27411
*Prgm Dir:* Tyra Turner Whittaker, RhD CRC
*E-mail:* tnwhitta@ncat.edu

**East Carolina University**
Rehabilitation Counseling Prgm
Dept of Rehabilitation Studies
School of Allied Health Sciences
Greenville, NC 27858-4353
www.ecu.edu/rehb
*Prgm Dir:* Paul Alston, PhD CRC
*Tel:* 252 328-4452    *Fax:* 252 328-0725
*E-mail:* alstonp@ecu.edu

**Winston-Salem State University**
Rehabilitation Counseling Prgm
School of Education
Campus Box 19386
Winston-Salem, NC 27110
*Prgm Dir:* Katrina Miller, PhD
*Tel:* 336 750-2941    *Fax:* 336 750-2914
*E-mail:* millerk@wssu.edu

## Ohio

**Ohio University**
Rehabilitation Counseling Prgm
201 McCracken Hall
Athens, OH 45701
*Prgm Dir:* Jerry A Olsheski, PhD CRC LPC
*Tel:* 740 593-0032    *Fax:* 740 593-0799
*E-mail:* olsheski@ohio.edu

**Bowling Green State University**
Rehabilitation Counseling Prgm
School of Intervention Services
Bowling Green, OH 43403-0255
*Prgm Dir:* Jay R Stewart, PhD
*Tel:* 419 372-7301    *Fax:* 419 372-8265
*E-mail:* jstewar@bgnet.bgsu.edu

**Ohio State University**
Rehabilitation Counseling Prgm
356 Arps Hall, 1945 N High St
Columbus, OH 43210-1172
*Prgm Dir:* Michael Klein, PhD CRC
*Tel:* 614 292-8183    *Fax:* 614 292-4255
*E-mail:* klein.3@osu.edu

**Wright State University**
Rehabilitation Counseling Prgm
M052 Creative Arts Center
Dayton, OH 45435-0001
www.cehs.wright.edu/academic/human_services
*Prgm Dir:* Jan La Forge, PhD CRC PC NCC
*Tel:* 937 775-2075    *Fax:* 937 775-2042
*E-mail:* jan.laforge@wright.edu

**Kent State University**
Rehabilitation Counseling Prgm
405 White Hall
Kent, OH 44242
www.kent.edu/rehab
*Prgm Dir:* Connie McReynolds, PhD CRC LP
*Tel:* 330 672-0602    *Fax:* 330 672-2512
*E-mail:* cmcreyno@kent.edu

## Oklahoma

**East Central University**
Rehabilitation Counseling Prgm
Box C-1
Ada, OK 74820
*Prgm Dir:* Randal Elston, EdD CRC CVE
*Tel:* 580 310-5463    *Fax:* 580 436-3329
*E-mail:* relston@mailclerk.ecok.edu

**Langston University**
Rehabilitation Counseling Prgm
4205 N Lincoln Blvd
Oklahoma City, OK 73105
*Prgm Dir:* Corey L Moore, RhD CRC
*Tel:* 405 962-1671    *Fax:* 405 962-1628
*E-mail:* clmoore@lunet.edu

## Oregon

**Western Oregon University**
Rehabilitation Counseling Prgm
Education Bldg 220
Monmouth, OR 97361
www.wou.edu
*Prgm Dir:* Linda Keller, PhD CRC LPC
*Tel:* 503 838-8444, Ext 88746    *Fax:* 503 838-8228
*E-mail:* kellerl@wou.edu

### Portland State University
Rehabilitation Counseling Prgm
PO Box 751
Portland, OR 97207-0751
www.webmail.pdx.edu
*Prgm Dir:* Hanoch Livneh, PhD CRC
*Tel:* 503 725-4719   *Fax:* 503 725-5599
*E-mail:* livnehh@pdx.edu

# Pennsylvania

### Edinboro University of Pennsylvania
Rehabilitation Counseling Prgm
322 Butterfield Hall
Edinboro, PA 16444
www.edinboro.edu/cwis/education/counseling/rehab_co
unseling.html
*Prgm Dir:* Susan H Packard, PhD CRC LPC NCC CAC
*Tel:* 814 732-2430   *Fax:* 814 732-2233
*E-mail:* spackard@edinboro.edu

### University of Pittsburgh
Rehabilitation Counseling Prgm
5044 Forbes Tower
3600 Forbes at Atwwod St
Pittsburgh, PA 15260
*Prgm Dir:* Michael McCue, PhD CRC
*Tel:* 412 383-6589   *Fax:* 412 383-6597
*E-mail:* mmccue@pitt.edu

### University of Scranton
Rehabilitation Counseling Prgm
Dept of Counseling and Human Services
Scranton, PA 18510-4523
*Prgm Dir:* Lori A Bruch, EdD CRC CVE
*Tel:* 717 941-4308   *Fax:* 717 941-5882
*E-mail:* BruchL1@scranton.edu

### Penn State University
Rehabilitation Counseling Prgm
327 Cedar Bldg
University Park, PA 16802-3110
www.ed.psu.edu/cecprs
*Prgm Dir:* Elias Mpofu, PhD CRC
*Tel:* 814 863-2411   *Fax:* 814 863-7750
*E-mail:* exm31@psu.edu

# Puerto Rico

### University of Puerto Rico
Rehabilitation Counseling Prgm
College of Social Science
PO Box 23345
San Juan, PR 00931-3345
*Prgm Dir:* Marilyn Mendoza Lugo
*Tel:* 787 764-0000, Ext 4206   *Fax:* 787 763-4199
*E-mail:* core@rrpac.upr.clu.edu

# Rhode Island

### Salve Regina University
Rehabilitation Counseling Prgm
100 Ochre Point Ave
Newport, RI 02840
www.salve.edu/graduatestudies/programs/rc
*Prgm Dir:* Dimity Peter, PhD CRC
*Tel:* 401 341-3189   *Fax:* 401 341-2973
*E-mail:* dimity.peter@salve.edu

# South Carolina

### University of South Carolina
Rehabilitation Counseling Prgm
School of Medicine
3555 Harden St Ext, Ste B20
Columbia, SC 29208
*Prgm Dir:* Linda L Leech, PhD
*Tel:* 803 434-6170   *Fax:* 803 434-4301
*E-mail:* lleech@gw.mp.sc.edu

### South Carolina State University
Rehabilitation Counseling Prgm
Dept of Human Services
300 College St NE
Orangeburg, SC 29117-0001
*Prgm Dir:* David Staten, PhD
*Tel:* 803 516-4917   *Fax:* 803 533-3636
*E-mail:* dstaten@scsu.edu

# Tennessee

### University of Tennessee - Knoxville
Rehabilitation Counseling Prgm
A522 Claxton Complex
Knoxville, TN 37996-3452
http://web.utk.edu/~edpsych/rehabilitation_counseling
*Prgm Dir:* Amy L Skinner, PhD LPC-MHSP CRC NCC
*Tel:* 865 974-8090   *Fax:* 865 974-0135
*E-mail:* askinner@utk.edu

### University of Memphis
Rehabilitation Counseling Prgm
Dept of Counseling, Educ Psych and Research
Patterson Hall, Rm 119
Memphis, TN 38152-6010
*Prgm Dir:* Erin Martz, PhD CRC
*Tel:* 901 678-4820   *Fax:* 901 678-3215
*E-mail:* emartz@memphis.edu

# Texas

### University of Texas at Austin
Rehabilitation Counseling Prgm
Special Education
1 University Station, D5300
Austin, TX 78712
*Prgm Dir:* Randall Parker, PhD
*Tel:* 512 232-5687   *Fax:* 512 232-5686
*E-mail:* r.parker@mail.utexas.edu

### Univ of Texas Southwestern Med Ctr at Dallas
Rehabilitation Counseling Prgm
5323 Harry Hines Blvd
Dallas, TX 75390-9088
*Prgm Dir:* Claire Korman, PhD
*Tel:* 214 648-1753   *Fax:* 214 648-1771
*E-mail:* claire.korman@utsouthwestern.edu

### University of North Texas
Rehabilitation Counseling Prgm
Dept of Rehab Social Work & Addiction
PO Box 311456
Denton, TX 76203
www.unt.edu
*Prgm Dir:* Rodney Isom, PhD CRC CDMS
*Tel:* 940 565-2234   *Fax:* 940 565-3960
*E-mail:* isom@unt.edu

### Univ of Texas - Pan American
Rehabilitation Counseling Prgm
1201 W University Dr
Edinburg, TX 78539-2999
*Prgm Dir:* Irmo Marini, PhD CRC
*Tel:* 956 316-7036   *Fax:* 956 380-6499
*E-mail:* imarini@utpa.edu

### Texas Tech University
Rehabilitation Counseling Prgm
Rehabilitation Sciences
3601 Fourth St, M/S 6225, Ste 2C-200
Lubbock, TX 79430-6225
*Prgm Dir:* Evans Spears, PhD CRC
*Tel:* 806 743-4208   *Fax:* 806 743-3518
*E-mail:* evans.spears@ttuhsc.edu

### Stephen F Austin State University
Rehabilitation Counseling Prgm
Dept of Human Services
PO Box 13019, SFA Station
Nacogdoches, TX 75962-3019
*Prgm Dir:* Robert O Choate, EdD NCC CRC LPC
*Tel:* 409 468-1147   *Fax:* 409 468-1342
*E-mail:* rchoate@sfasu.edu

# Utah

### Utah State University
Rehabilitation Counseling Prgm
Dept of Special Educ and Rehab
2865 Old Main Hill
Logan, UT 84322-2865
www.rce.usu.edu
*Prgm Dir:* Julie Smart, PhD CRC LPC NCC ABDA CCFC
*Tel:* 435 797-3269   *Fax:* 435 797-3572
*E-mail:* jsmart@cc.usu.edu

# Virginia

### Virginia Commonwealth University
Rehabilitation Counseling Prgm
Dept of Rehab Counseling
1112 E Clay St, Box 980330
Richmond, VA 23298-0330
*Prgm Dir:* Christine Reid, PhD
*Tel:* 804 827-0915   *Fax:* 804 828-1321
*E-mail:* creid@hsc.vcu.edu

# Washington

### Western Washington University
Rehabilitation Counseling Prgm
6912 220th St SW
Ste 105
Mountlake Terrace, WA 98043
www.wwu.edu/rc
*Prgm Dir:* Elizabeth Swett, PhD CRC
*Tel:* 425 771-7435   *Fax:* 425 774-9303
*E-mail:* elizabeth.swett@wwu.edu

# West Virginia

### West Virginia University
Rehabilitation Counseling Prgm
502 Allen Hall, PO Box 6122
Morgantown, WV 26506-6122
*Prgm Dir:* Margaret K Glenn, EdD CRC
*Tel:* 304 293-2276   *Fax:* 304 293-4082
*E-mail:* mkglenn@mail.wvu.edu

# Wisconsin

### University of Wisconsin - Madison
Rehabilitation Counseling Prgm
Rehab Psych and Special Educ
432 N Murray St
Madison, WI 53706
*Prgm Dir:* Norman L Berven, PhD
*Tel:* 608 262-7917   *Fax:* 608 262-8108
*E-mail:* berven@education.wisc.edu

### University of Wisconsin - Stout
Rehabilitation Counseling Prgm
Dept of Rehabilitation & Counseling
250F Vocational Rehabilitation
Menomonie, WI 54751
*Prgm Dir:* Michelle Hamilton, PhD CVE CRC
*Tel:* 715 232-1895   *Fax:* 715 232-2356
*E-mail:* hamiltonmi@uwstout.edu

## Rehabilitation Counselor

| Programs* | Class Capacity | Begins | Length (months) | Award | Res. Tuition | Non-res. Tuition | Stipend | Offers:‡ 1 | 2 | 3 |
|---|---|---|---|---|---|---|---|---|---|---|
| **Alabama** | | | | | | | | | | |
| Alabama A&M University (Normal) | 30 | Fall semester | 24 | Masters | $6,180 | $11,040 | | • | | • |
| **Arizona** | | | | | | | | | | |
| University of Arizona (Tucson) | 25 | Aug | 24 | BA, MA | $1,020 | $4,210 | $3,500 | • | | |
| **Arkansas** | | | | | | | | | | |
| University of Arkansas at Little Rock | 50 | Sep Jan Jun | 18, 24 | Cert, MA | $280 | $280 | | • | | • |
| **California** | | | | | | | | | | |
| California State University - Fresno | 40 | Jan Aug | 24 | MS | $3,190 | $15,000 | $5,670 | • | | • |
| California State University - Los Angeles | 28 | Sep | 24 | BS, MS | $3,300 | $7,166 | | • | | |
| San Diego State University | 30 | Aug | 25, 36 | MS | $3,700 | $10,000 | $3,600 | • | | • |
| **Colorado** | | | | | | | | | | |
| University of Northern Colorado (Greeley) | 25 | Aug | 24, 48 | BS, MA | $2,708 | $11,268 | | • | | • |
| **Florida** | | | | | | | | | | |
| Florida State University (Tallahassee) | 70 | Aug | 24, 22 | BS, MS/PHD | $7,875 | $29,527 | | • | | • |
| University of North Florida (Jacksonville) | 40 | Fall | 18 | MS | $9,029 | $30,318 | | • | | • |
| University of South Florida (Tampa) | 200 | Aug Jan | 28 | MA | $6,050 | $21,530 | | • | | • |
| **Georgia** | | | | | | | | | | |
| Fort Valley State University (Ft Valley) | 25 | Aug | 24 | MS | $3,600 | $12,000 | | • | | • |
| Georgia State University (Atlanta) | 20 | Aug | 24 | MS | $4,368 | $17,464 | $800 | • | | • |
| Thomas University (Thomasville) | 30 | Aug | 20 | MS | $7,800 | $7,800 | | • | | |
| **Idaho** | | | | | | | | | | |
| University of Idaho (Moscow) | 12 | Aug | 24 | MEd, MS | $6,000 | $10,000 | $9,000 | • | | • |
| **Illinois** | | | | | | | | | | |
| Illinois Institute of Technology (Chicago) | 32 | Aug | 24, 36 | Cert, MS | $21,810 | $21,810 | $5,000 | • | | • |
| Northeastern Illinois University (Chicago) | 50 | Fall Spring | 24, 36 | MA | $4,000 | $4,000 | $10,000 | • | | • |
| Northern Illinois University (DeKalb) | 20 | Aug Jan May | 22 | BS, MA | $1,991 | $4,775 | | • | | • |
| Southern Illinois Univ at Carbondale | 60 | May Aug Jan | 24 | Dipl, MS | $2,430 | $5,832 | | • | | |
| Univ of Illinois at Urbana - Champaign | 15 | Sep | 18 | MS | $1,885 | $5,221 | | • | | |
| **Iowa** | | | | | | | | | | |
| University of Iowa (Iowa City) | 22 | Jun Sep | 23, 48 | MA, PhD | $5,689 | $15,723 | | | | • |
| **Kansas** | | | | | | | | | | |
| Emporia State University | 25 | Aug Jan Jun | 24 | MS | $4,162 | $11,122 | | • | | • |
| **Louisiana** | | | | | | | | | | |
| Southern Univ and A&M College (Baton Rouge) | 25 | Aug | 24 | MS | $4,559 | $11,068 | $5,500 | • | | |
| **Maryland** | | | | | | | | | | |
| Coppin State University (Baltimore) | 12 | Sep | 30 | MEd | $0 | $2,196 | | • | | |
| University of Maryland (College Park) | 15 | Sep | 24 | Cert, MA, MEd | $3,705 | $9,355 | | • | | • |
| University of Maryland Eastern Shore (Princess Anne) | 14 | Fall | 24 | MS | $5,480 | $9,872 | | • | | • |
| **Massachusetts** | | | | | | | | | | |
| Springfield College | 25 • | Sep Jan May | 24, 18 | Dipl, Cert, BS, MS/MEd | $20,370 | $20,370 | $16,296 | • | | |
| **Michigan** | | | | | | | | | | |
| Wayne State University (Detroit) | 25 | Sep | 24 | MA, PhD | $2,205 | $2,205 | | • | | • |
| **Minnesota** | | | | | | | | | | |
| Minnesota State University - Mankato | 25 | Late Aug | 16, 20 | MS | $7,673 | $10,464 | | | | • |
| St Cloud State University | 12 | Sep | 21 | MS | $5,600 | $8,800 | | • | | |
| **Mississippi** | | | | | | | | | | |
| Mississippi State University | 15 | Aug | 24 | MS | $3,874 | $8,780 | | • | | • |
| **Missouri** | | | | | | | | | | |
| Maryville University (St Louis) | | | | | | | | | | |
| **New York** | | | | | | | | | | |
| CUNY Hunter College (New York) | 35 | Sep Feb | 24 | MS | $4,350 | $7,600 | | • | | |
| Hofstra University (Hempstead) | 20 | Sep | 21, 24 | Dipl, MS, MS | $16,560 | $16,560 | | • | | |
| University at Buffalo - SUNY | 10 | Aug | 24 | MS | $3,993 | $5,793 | | • | | • |
| **North Carolina** | | | | | | | | | | |
| East Carolina University (Greenville) | 65 • | Aug | 24 | Dipl, Cert, BS, MS/PhD | $4,000 | $15,000 | | • | | |
| **Ohio** | | | | | | | | | | |
| Kent State University | 25 | Sep Jan Jun | 24 | MEd | $6,000 | $10,000 | $7,500 | • | | • |
| Wright State University (Dayton) | 20 | Sep Jan Mar Jun | 48, 24 | BS, MRC | $11,464 | $19,964 | | • | | • |
| **Oregon** | | | | | | | | | | |
| Portland State University | 14 | Sep | 24 | MS | $8,000 | $12,000 | | • | | |
| Western Oregon University (Monmouth) | 15 | Sep | 20 | MS | $14,000 | $28,000 | | | | |
| **Pennsylvania** | | | | | | | | | | |
| Edinboro University of Pennsylvania | 25 | Aug Jan | 30 | MA | $5,888 | $9,422 | | • | | • |

*Data are shown only for programs that completed the 2006 AMA Survey of Health Professions Education Programs

‡Key to Offers: 1: Evening or weekend classes; 2: Non-English instruction; 3: Cultural competence instruction

## Rehabilitation Counselor

| Programs* | Class Capacity | Begins | Length (months) | Award | Res. Tuition | Non-res. Tuition | Stipend | Offers:‡ 1 | 2 | 3 |
|---|---|---|---|---|---|---|---|---|---|---|
| Penn State University (University Park) | 10 | Late Aug | 24 | BS, MEd/Do | $10,048 | $19,724 | | • | | • |
| **Rhode Island** | | | | | | | | | | |
| Salve Regina University (Newport) | 15 | Fall Spring | 24 | MA | $0 | $8,400 | | • | | • |
| **Tennessee** | | | | | | | | | | |
| University of Memphis | 25 | Aug Jan | 18, 22 | MS | $7,101 | $15,837 | $3,600 | • | | • |
| University of Tennessee - Knoxville | 20 | Aug | 16 | MS | $5,100 | $12,000 | | • | | • |
| **Texas** | | | | | | | | | | |
| University of North Texas (Denton) | 60 | Aug | 24, 48 | MS | $7,500 | $12,500 | | • | | • |
| University of Texas at Austin | 40 | Jun Aug | 16, 40 | Dipl, MA/MEd, PhD | $4,000 | $8,000 | $20,000 | • | | • |
| **Utah** | | | | | | | | | | |
| Utah State University (Logan) | 80 | Sep | 18 | MRC | $2,250 | $6,800 | $1,200 | • | | • |
| **Washington** | | | | | | | | | | |
| Western Washington University (Mountlake Terrace) | 30 | Fall Winter Spring | 27 | MA | $5,954 | $5,954 | | • | | • |

*Data are shown only for programs that completed the 2006 AMA Survey of Health Professions Education Programs
‡Key to Offers: 1: Evening or weekend classes; 2: Non-English instruction; 3: Cultural competence instruction

# Respiratory Therapy

**Includes:**
- Respiratory therapist (advanced)
- Respiratory therapist (entry-level)

## Respiratory Therapist (Advanced)

### Occupational Description

Respiratory therapists work in a wide variety of settings to evaluate, treat, and manage patients of all ages with respiratory illnesses and other cardiopulmonary disorders. The advanced respiratory therapist participates in clinical decision-making and patient education, develops and implements respiratory care plans, applies patient-driven protocols, utilizes evidence-based clinical practice guidelines, and participates in health promotion, disease prevention, and disease management. The advanced level respiratory therapist may be required to exercise considerable independent judgment, under the supervision of a physician, in the respiratory care of patients.

### Job Description

In fulfillment of the advanced therapist role, the respiratory therapist may perform the following procedures:
- Acquiring and evaluating clinical data
- Assessing the cardiopulmonary status of patients
- Performing and assisting in the performance of prescribed diagnostic studies, such as obtaining blood samples, blood gas analysis, pulmonary function testing, and polysomnography
- Evaluating data to assess the appropriateness of prescribed respiratory care
- Establishing therapeutic goals for patients with cardiopulmonary disease
- Participating in the development and modification of respiratory care plans
- Performing case management of patients with cardiopulmonary and related diseases
- Initiating prescribed respiratory care treatments, evaluating and monitoring patient responses to such therapy, and modifying the prescribed therapy to achieve the desired therapeutic objectives
- Initiating and conducting prescribed pulmonary rehabilitation
- Providing patient, family, and community education
- Promoting cardiopulmonary wellness, disease prevention, and disease management
- Participating in life support activities as required and promoting evidence-based medicine, research, and clinical practice guidelines

### Employment Characteristics

Respiratory therapists are employed in a variety of settings that include acute care, chronic care, subacute care, extended care, and rehabilitation facilities; educational institutions; clinics; physician's offices; home care; sleep labs; diagnostic and research labs; and pharmaceutical companies.

### Salary

The Human Resources Study from the American Association for Respiratory Care (AARC) indicated that advanced level respiratory therapists with an RRT (Registered Respiratory Therapist) credential earned an average salary of $57,803 in year 2005. Refer to Section IV, Table 5 of this *Directory* for more information, or see www.ama-assn.org/go/salary.

### Educational Programs

**Length.** Advanced level respiratory therapists complete 2 or more years of formal training and education leading to an associate, baccalaureate, or graduate degree.

**Prerequisite.** Varies depending on degree awarded.

**Curriculum.** The knowledge and skills for performing these functions are achieved through formal college- or university-based programs of classroom, laboratory, and clinical preparation. Biological and physical sciences required include anatomy, physiology, chemistry, physics, microbiology, computer science, pharmacology, and pathophysiology. Coursework may also be required in mathematics, communications, psychology, medical ethics, and the social sciences. Professional coursework may include patient assessment, monitoring, and evaluation, diagnostic and therapeutic procedures, airway management and mechanical ventilatory support, infection control, basic and advanced life support, patient and caregiver education, rehabilitation and disease management, and health promotion/disease prevention. Clinical training in all aspects of respiratory care applicable to pediatric, adult, and geriatric patients is also provided.

## Respiratory Therapist (Entry-level)

### Occupational Description

The entry-level respiratory therapist performs general respiratory care procedures. Entry-level therapists may assume clinical responsibility for specified respiratory care modalities involving the application of therapeutic techniques under the supervision of an advanced-level therapist and/or a physician.

### Job Description

In fulfillment of the entry-level role, the respiratory therapist may perform the following procedures:
- Review existing data in the patient records, including patient history and physical, blood gas results, pulmonary function results, imaging studies, and monitoring data
- Collect and evaluate additional pertinent clinical data and recommend procedures to obtain additional data
- Select, assemble, use, and troubleshoot equipment used in the delivery of respiratory care, ensure infection control, and perform quality control procedures
- Maintain records and communicate information regarding patient's clinical status to appropriate members of the health care team
- Maintain a patient's airway, including care of artificial airways
- Remove bronchial secretions, assure provision of adequate respiratory support, and evaluate and monitor patient's objective and subjective responses to respiratory care
- Modify therapeutic procedures and recommend modifications in the respiratory care plan based on patient's response

- Initiate, conduct, or modify respiratory care techniques in an emergency setting
- Act as an assistant to the physician performing special procedures
- Initiate and conduct pulmonary rehabilitation and home care within the prescription

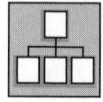

### Employment Characteristics

Entry-level respiratory therapists are employed in hospitals, nursing care facilities, clinics, physicians' offices, home care companies, pulmonary function laboratories, and sleep labs.

### Salary

The Human Resources Study from the American Association for Respiratory Care (AARC) indicated that the average salary for respiratory therapists with a CRT (Certified Respiratory Therapist) credential was $56,035 in the year 2005. This same study indicated that the mean salary of all respiratory therapists at entry-level was $41,537 in year 2005. Refer to Section IV, Table 5 of this *Directory* for more information, or see www.ama-assn.org/go/salary.

### Educational Programs

**Length.** Educational programs for entry-level therapists are usually 2 years in length, leading to an associate degree.

**Prerequisites.** High school degree or equivalent.

**Curriculum.** The knowledge and skills of an entry-level therapist are acquired through formal programs of didactic, laboratory, and clinical preparation. Courses include biological and physical sciences basic to understanding the function of the cardiopulmonary system. Included are chemistry, biology, anatomy and physiology, physics, microbiology, mathematics, computer science, and psychology, as well as courses in basic equipment, therapeutic modalities, clinical medicine, and pathophysiology. Clinical training in routine and special procedures applicable to pediatric, adult, and geriatric patients is also provided.

### Inquiries

**Careers**

American Association for Respiratory Care
11030 Ables Lane
Dallas, TX 75229
972 243-2272
www.aarc.org

**Licensure/Certification/Registration**

National Board for Respiratory Care
8310 Nieman Road
Lenexa, KS 66214
www.nbrc.org

**Program Accreditation**

Commission on Accreditation of Allied Health Education Programs (CAAHEP) in collaboration with:
Committee on Accreditation for Respiratory Care (CoARC)
1248 Harwood Road
Bedford, TX 76021
817 283-2835
817 354-8519 Fax
E-mail: bill@heasc.org

---

## Respiratory Therapist (Advanced)*

## Alabama

**University of Alabama at Birmingham**
Respiratory Therapist (Advanced) Prgm
RMSB 486
1705 University Blvd
Birmingham, AL 35294-1212
www.uab.edu/rt
*Prgm Dir:* Wesley M Granger, PhD RRT
*Tel:* 205 934-3783   *Fax:* 205 975-7302
*E-mail:* grangerw@uab.edu
*Med Dir:* Philip McARdle, MD

**George C Wallace Community College**
Respiratory Therapist (Advanced) Prgm
1141 Wallace Dr
Dothan, AL 36303
*Prgm Dir:* Drayton Odom, BS RRT
*Tel:* 334 556-2291   *Fax:* 334 556-2422
*E-mail:* dodom@wallace.edu
*Med Dir:* Allen Latimer, MD

**Wallace State Community College**
Respiratory Therapist (Advanced) Prgm
PO Box 2000
Hanceville, AL 35077-2000
www.wallacestate.edu
*Prgm Dir:* Paul D Taylor, DHSc RRT
*Tel:* 205 352-8310   *Fax:* 205 352-8320
*E-mail:* paul.taylor@wallacestate.edu
*Med Dir:* Russell Beaty, MD

**University of South Alabama**
Respiratory Therapist (Advanced) Prgm
1504 Springhill Ave, Ste 2545
Mobile, AL 36604
www.usouthal.edu
*Prgm Dir:* William V Wojciechowski, MS RRT
*Tel:* 251 434-3405   *Fax:* 251 434-3941
*E-mail:* wwojciec@usouthal.edu
*Med Dir:* Ronald Allison, MD

**Shelton State Community College**
Respiratory Therapist (Advanced) Prgm
3401 M L King Jr Blvd
Tuscaloosa, AL 35401
*Prgm Dir:* Bruce E Spruell, AS RRT
*Tel:* 205 391-2366   *Fax:* 205 391-2658
*E-mail:* bspruell@sheltonstate.edu
*Med Dir:* Philip Bobo, MD

## Arizona

**Pima Medical Institute - Mesa**
Respiratory Therapist (Advanced) Prgm
957 S Dobson Rd
Mesa, AZ 85202
www.pmi.edu
*Prgm Dir:* Cynthia Smathers, MEd RRT
*Tel:* 480 898-9898   *Fax:* 480 649-5249
*E-mail:* csmathers@pmi.edu
*Med Dir:* David Drachler, MD

**GateWay Community College**
Respiratory Therapist (Advanced) Prgm
108 N 40th St
Phoenix, AZ 85034
www.gatewaycc.edu
*Prgm Dir:* Edward Hoskins, MEd RRT
*Tel:* 602 286-8503   *Fax:* 602 286-8478
*E-mail:* hoskins@gatewaycc.edu
*Med Dir:* Philip J Fracica, MD

**Long Technical College**
Respiratory Therapist (Advanced) Prgm
13450 N Black Canyon Hwy, Ste 104
Phoenix, AZ 85029
www.longtechnicalcollege.com
*Prgm Dir:* Robert Brown, BS RRT
*Tel:* 602 548-1955   *Fax:* 602 926-1456
*E-mail:* bbrown@longtechnicalcollege.com
*Med Dir:* David Rabinowitz, DO

---

*Additional information about programs that returned the AMA's survey is available in a table beginning on page 414.

**Pima Community College**
Respiratory Therapist (Advanced) Prgm
2202 W Anklam Rd, HRP 242
Tucson, AZ 85709-0080
www.pima.edu
*Prgm Dir:* Jody Kosanke, MEd RRT-NPS
*Tel:* 520 206-3107   *Fax:* 520 206-3027
*E-mail:* Jody.Kosanke@pima.edu
*Med Dir:* Linda S Snyder, MD

**Pima Medical Institute - Tucson**
Respiratory Therapist (Advanced) Prgm
3350 E Grant Rd
Tucson, AZ 85716
*Prgm Dir:* Tammy Redasky, BS RRT
*Tel:* 520 881-1483   *Fax:* 520 881-1483
*E-mail:* leanna@qwest.net
*Med Dir:* Robert Aaronson, MD

# Arkansas

**NorthWest Arkansas Community College**
Respiratory Therapist (Advanced) Prgm
One College Dr
Bentonville, AR 72712
www.nwacc.edu
*Prgm Dir:* Alan Clark, BA RRT
*Tel:* 479 619-4250   *Fax:* 479 619-4254
*E-mail:* aclark@nwacc.edu
*Med Dir:* John Sexton, MD

**University of Arkansas for Medical Sciences**
Respiratory Therapist (Advanced) Prgm
Central Arkansas Veterans Healthcare Services
CHRP
Little Rock, AR 72205
*Prgm Dir:* Erna L Boone, MEd RRT
*Tel:* 501 257-2348   *Fax:* 501 257-2349
*E-mail:* booneernal@uams.edu
*Med Dir:* Robert H Warren, MD

**Southeast Arkansas College**
Respiratory Therapist (Advanced) Prgm
1900 Hazel St
Pine Bluff, AR 71603
www.seark.edu
*Prgm Dir:* Kathleen M Boyle, MEd MS RRT-NPS
*Tel:* 870 543-5924   *Fax:* 870 543-5912
*E-mail:* kboyle@seark.edu
*Med Dir:* J Clyde Campbell, MD

**University of Arkansas for Medical Sciences**
Respiratory Therapist (Advanced) Prgm
UAMS AHEC-Southwest
300 E 6th St
Texarkana, AR 71854
www.uams.edu
*Prgm Dir:* Patrick Evans, RRT RN MEd
*Tel:* 870 779-6033   *Fax:* 870 779-6045
*E-mail:* evansdennisp@uams.edu
*Med Dir:* Malcolm Smith, MD

# California

**San Joaquin Valley College - Bakersfield**
Respiratory Therapist (Advanced) Prgm
201 New Stine Rd
Bakersfield, CA 93309
*Prgm Dir:* Albert Ayotte, MBA RRT RCP
*Tel:* 661 834-0126, Ext 140
*E-mail:* albert.ayotte@sjvc.edu
*Med Dir:* Hans E Einstein, MD

**Orange Coast College**
Respiratory Therapist (Advanced) Prgm
2701 Fairview Rd, PO Box 5005
Costa Mesa, CA 92626
www.orangecoastcollege.edu
*Prgm Dir:* Daniel S Adelmann, MS RRT
*Tel:* 714 432-5541   *Fax:* 714 432-5534
*E-mail:* dadelman@occ.cccd.edu
*Med Dir:* Paul A Selecky, MD PhD

**Grossmont College**
Respiratory Therapist (Advanced) Prgm
8800 Grossmont College Dr
El Cajon, CA 92020-1799
www.grossmont.edu/healthprofessions
*Prgm Dir:* Lorenda Seibold-Phalan, MA RRT-NPS RCP
*Tel:* 619 644-7448   *Fax:* 619 644-7961
*E-mail:* lorenda.seibold-phalan@gcccd.net
*Med Dir:* David C Willms, MD

**Ohlone College**
Respiratory Therapist (Advanced) Prgm
43600 Mission Blvd, PO Box 3909
Fremont, CA 94539
*Prgm Dir:* Carol McNamee-Cole, MA RRT
*Tel:* 510 659-6029   *Fax:* 510 659-6070
*E-mail:* hstory@ohlone.edu
*Med Dir:* Francis C Johnson, MD

**Fresno City College**
Respiratory Therapist (Advanced) Prgm
1101 E University Ave
Fresno, CA 93741
*Prgm Dir:* Steven L Boyd, MBA RRT
*Tel:* 559 244-2631   *Fax:* 559 244-2626
*E-mail:* steven.boyd@fresnocitycollege.edu
*Med Dir:* Enok Lohne, MD

**Loma Linda University**
Respiratory Therapist (Advanced) Prgm
Nichol Hall, Rm 1926
Loma Linda, CA 92350
www.llu.edu
*Prgm Dir:* David Stanton, MS RPFT RCP RRT-NPS
*Tel:* 909 558-4932   *Fax:* 909 558-4701
*E-mail:* dstanton@llu.edu
*Med Dir:* N Leonard Specht, MD

**Foothill Community College**
Respiratory Therapist (Advanced) Prgm
12345 El Monte Rd
Los Altos Hills, CA 94022
www.foothill.edu/bio/programs/respther
*Prgm Dir:* Virginia Becchine, MA RRT
*Tel:* 650 949-7466   *Fax:* 650 949-7686
*E-mail:* becchinevirginia@foothill.edu
*Med Dir:* Bernhard Votteri, MD

**Modesto Junior College**
Respiratory Therapist (Advanced) Prgm
435 College Ave
Modesto, CA 95350
*Prgm Dir:* David Nordin, BA RCP RRT
*Tel:* 209 575-6381   *Fax:* 209 575-6593
*E-mail:* nordind@yosemite.cc.ca.us
*Med Dir:* Robert Tanaka, MD

**East Los Angeles College**
Respiratory Therapist (Advanced) Prgm
1301 Avenida Cesar Chavez
Monterey Park, CA 91754
www.elac.edu
*Prgm Dir:* Michael R Carr, BA RRT
*Tel:* 323 265-8612, Ext 8612   *Fax:* 323 265-8684
*E-mail:* michael_carr@sbcglobal.net
*Med Dir:* Gerard W Frank, MD

**Napa Valley College**
Respiratory Therapist (Advanced) Prgm
2277 Napa Vallejo Hwy
Napa, CA 94558
*Prgm Dir:* Kate Benscoter, MEd RRT
*Tel:* 707 253-3145   *Fax:* 707 259-8933
*E-mail:* kbenscoter@napavalley.edu
*Med Dir:* Nazir Habib, MD

**Butte College**
Respiratory Therapist (Advanced) Prgm
3536 Butte Campus Dr
Oroville, CA 95965
www.butte.edu
*Prgm Dir:* Donna Davis, BS RRT
*Tel:* 530 895-2827   *Fax:* 530 895-2472
*E-mail:* davisdo@butte.edu
*Med Dir:* Gerard Valcarenghi, MD

**San Joaquin Valley College**
Respiratory Therapist (Advanced) Prgm
10641 Church St
Rancho Cucamonga, CA 91730
*Prgm Dir:* Richardo Guzman, BA CRT RRT
*Tel:* 909 948-7582
*E-mail:* richardog@sjvc.edu
*Med Dir:* Robert C Jones, MD

**American River College**
Respiratory Therapist (Advanced) Prgm
4700 College Oak Dr
Sacramento, CA 95841
*Prgm Dir:* James L Warman, PhD RRT
*Tel:* 916 484-8876   *Fax:* 916 484-8030
*E-mail:* warmanj@arc.losrios.edu
*Med Dir:* Theodore Bacharach, MD

**Skyline College**
Respiratory Therapist (Advanced) Prgm
3300 College Dr
San Bruno, CA 94066
*Prgm Dir:* Raymond Hernandez
*Tel:* 650 738-4457   *Fax:* 650 738-4299
*E-mail:* hernandezr@smccd.net
*Med Dir:* Gordon Mak, MD

**California College San Diego**
Respiratory Therapist (Advanced) Prgm
2820 Camino Del Rio S
San Diego, CA 92108
*Prgm Dir:* Ed Moser
*Tel:* 619 295-5785   *Fax:* 619 295-5762
*E-mail:* emoser@cc-sd.edu
*Med Dir:* Kenneth Grudko, MD

**Los Angeles Valley College**
Respiratory Therapist (Advanced) Prgm
5800 Fulton Ave
Valley Glen, CA 91401-4096
www.lavc.edu
*Prgm Dir:* Frank Sinsheimer, EdD RRT RCP
*Tel:* 818 947-2562   *Fax:* 818 947-2850
*E-mail:* sinshefa@lavc.edu
*Med Dir:* Alan Rothfeld, MD

**Victor Valley Community College District**
Respiratory Therapist (Advanced) Prgm
18422 Bear Valley Rd
Victorville, CA 92392-5849
*Prgm Dir:* Larry Boutcher, PhD RRT
*Tel:* 760 245-4271, Ext 2222   *Fax:* 760 951-5861
*E-mail:* boutcherl@vvc.edu
*Med Dir:* Nagamani Dandamudi, MD

**San Joaquin Valley College - Visalia**
Respiratory Therapist (Advanced) Prgm
8400 W Mineral King
Visalia, CA 93291
www.sjvc.edu
*Prgm Dir:* Gaylene Mooney, MEd RRT-NPS
*Tel:* 559 622-1943   *Fax:* 559 651-3645
*E-mail:* Gaylene.Mooney@sjvc.edu
*Med Dir:* William R Winn, MD

**Mt San Antonio College**
Respiratory Therapist (Advanced) Prgm
1100 N Grand Ave
Walnut, CA 91789
*Prgm Dir:* Terrance M Krider, BS RRT RCP
*Tel:* 909 594-5611, Ext 3930   *Fax:* 909 468-3938
*E-mail:* thood@mtsac.edu
*Med Dir:* Earl S Young, MD

**Crafton Hills College**
Respiratory Therapist (Advanced) Prgm
11711 Sand Canyon Rd
Yucaipa, CA 92399
www.craftonhills.edu
*Prgm Dir:* Kenneth R Bryson, MEd BVE RCP RRT
*Tel:* 909 389-3284   *Fax:* 909 389-3229
*E-mail:* kbryson@crafton.sbccd.cc.ca.us
*Med Dir:* Richard L Sheldon, MD

# Colorado

**T H Pickens Technical Center**
Respiratory Therapist (Advanced) Prgm
500 Airport Blvd
Aurora, CO 80011-9307
*Prgm Dir:* Leigh Otto, BS RRT
*Tel:* 303 344-4910, Ext 27781   *Fax:* 303 326-1277
*E-mail:* leigho@pickens.aps.k12.co.us
*Med Dir:* James Ellis, Jr, MD

**Pima Medical Institute - Denver**
Respiratory Therapist (Advanced) Prgm
1701 W 72nd Ave
Denver, CO 80221
*Prgm Dir:* Amy Brown, BS RRT
*Tel:* 303 426-1800   *Fax:* 303 430-4048
*E-mail:* abrown@pmi.edu
*Med Dir:* Joseph Heit, MD

**Pueblo Community College**
Respiratory Therapist (Advanced) Prgm
900 W Orman Ave
Pueblo, CO 81004
www.pueblocc.edu
*Prgm Dir:* Delia Lechtenberg, MA RRT
*Tel:* 719 549-3266   *Fax:* 719 549-3381
*E-mail:* delia.lechtenberg@pueblocc.edu
*Med Dir:* Craig Shapiro, MD

# Connecticut

**Quinnipiac University**
Respiratory Therapist (Advanced) Prgm
Mt Carmel Ave
Hamden, CT 06518
*Prgm Dir:* Ronald G Beckett, RRT PhD
*Tel:* 203 582-8682   *Fax:* 203 582-8706
*E-mail:* ronald.beckett@quinnipiac.edu
*Med Dir:* Michael J McNamee, MD

**Manchester Community College**
Respiratory Therapist (Advanced) Prgm
Great Path, PO Box 1046
Manchester, CT 06045-1046
www.mcc.commnet.edu
*Prgm Dir:* Nancy LaRoche-Shovak, MS RRT
*Tel:* 860 512-2714   *Fax:* 860 512-2714
*E-mail:* nlaroche-shovak@mcc.commnet.edu
*Med Dir:* Richard L ZuWallack, MD

**Norwalk Hospital/Norwalk Community
College**
Respiratory Therapist (Advanced) Prgm
24 Maple St
Norwalk, CT 06856
www.norwalkhealth.org
*Prgm Dir:* Geraldine Bernacki, RRT
*Tel:* 203 852-2479   *Fax:* 203 852-2738
*E-mail:* geraldine.bernacki@norwalkhealth.org
*Med Dir:* Stephen M Winter, MD

**Naugatuck Valley Community College**
Respiratory Therapist (Advanced) Prgm
750 Chase Pkwy
Waterbury, CT 06708
www.nvcc.commnet.edu
*Prgm Dir:* Margaret Guerrera, BS RRT
*Tel:* 203 596-8662   *Fax:* 203 575-8146
*E-mail:* mguerrera@nvcc.commnet.edu
*Med Dir:* David Hill, MD

**University of Hartford**
Respiratory Therapist (Advanced) Prgm
200 Bloomfield Ave
West Hartford, CT 06117-1559
*Prgm Dir:* Peter W Kennedy, PhD RRT
*Tel:* 860 768-4823   *Fax:* 860 768-5706
*E-mail:* pkennedy@hartford.edu
*Med Dir:* R Frederic Knauft, MD

# Delaware

**Delaware Tech & Comm Coll - Owens Campus**
Respiratory Therapist (Advanced) Prgm
PO Box 610
Georgetown, DE 19947
www.dtcc.edu/owens/rescare
*Prgm Dir:* James G Little, MEd RRT
*Tel:* 302 856-5400, Ext 5540   *Fax:* 302 858-5460
*E-mail:* jlittle@dtcc.edu
*Med Dir:* Vikas Batra, MD

**Delaware Tech & Comm Coll - Wilmington**
*Cosponsor:* Christiana Care Health Services
Respiratory Therapist (Advanced) Prgm
Ste 101, Medical Arts Complex
700 West Lea Blvd
Wilmington, DE 19802
*Prgm Dir:* Joseph Ciarlo, BS RRT
*Tel:* 302 765-4598   *Fax:* 302 765-4599
*E-mail:* jciarlo@christianacare.org
*Med Dir:* Albert Rizzo, MD

# District of Columbia

**University of the District of Columbia**
Respiratory Therapist (Advanced) Prgm
4200 Connecticut Ave NW
Washington, DC 20008
*Prgm Dir:* Susan D Lockwood, MA RN RRT
*Tel:* 202 274-5925   *Fax:* 202 274-5952
*E-mail:* slockwood@udc.edu
*Med Dir:* Bernard Grand, MD

# Florida

**Broward Community College**
Respiratory Therapist (Advanced) Prgm
1000 Coconut Creek Blvd
Coconut Creek, FL 33066
*Prgm Dir:* John Prince, RRT
*Tel:* 954 969-2082   *Fax:* 954 973-2348
*E-mail:* jprince@broward.edu
*Med Dir:* Milton Braunstein, MD

**Daytona Beach Community College**
Respiratory Therapist (Advanced) Prgm
PO Box 2811
1200 W Internatonal Speedway Blvd
Daytona Beach, FL 32184-2811
www.dbcc.edu
*Prgm Dir:* Michael R McCumber, EdD RRT
*Tel:* 386 506-3759   *Fax:* 386 506-3300
*E-mail:* mccumbm@dbcc.edu
*Med Dir:* Michael A Diamond, MD

**Edison College**
Respiratory Therapist (Advanced) Prgm
8099 College Pkwy SW
PO Box 60210
Ft Myers, FL 33906-6210
www.edison.edu
*Prgm Dir:* Jeffrey B Elsberry, PhD RRT
*Tel:* 239 489-9251   *Fax:* 239 985-8352
*E-mail:* jelsberry@edison.edu
*Med Dir:* George Mestas, MD

**Indian River Community College**
Respiratory Therapist (Advanced) Prgm
3209 Virginia Ave
Ft Pierce, FL 34981-5599
*Prgm Dir:* Georgette Rosenfeld, MEd RRT RN
*Tel:* 561 462-7542   *Fax:* 561 462-4900
*E-mail:* grosenfe@ircc.edu
*Med Dir:* Donald Hoffman, MD

**Florida Community College - Jacksonville**
Respiratory Therapist (Advanced) Prgm
North Campus, 4501 Capper Rd
Jacksonville, FL 32218
*Prgm Dir:* James R Woods, MS RRT RPFT
*Tel:* 904 766-6513   *Fax:* 904 766-5573
*E-mail:* jwoods@fccj.edu
*Med Dir:* Miren Schinco, MD

**ATI Health Education Centers**
Respiratory Therapist (Advanced) Prgm
1395 NW 167th St, Ste 200
Miami, FL 33169
*Prgm Dir:* David Robbins, BS RRT
*Tel:* 305 628-1000, Ext 220   *Fax:* 305 628-1461
*E-mail:* rrobbins@atienterprises.edu
*Med Dir:* Edgar Bolton, Jr, DO

**Miami Dade College**
Respiratory Therapist (Advanced) Prgm
Medical Center Campus
950 NW 20th St
Miami, FL 33127
www.mdc.edu/medical/academic_programs/respiratory/
   respiratory.htm
*Prgm Dir:* Randy De Kler, MS RRT
*Tel:* 305 237-4423   *Fax:* 305 237-4278
*E-mail:* rdekler@mdc.edu
*Med Dir:* Kunjana Mavunda, MD

**University of Central Florida**
Respiratory Therapist (Advanced) Prgm
Dept of Health Professions
HPA II-206, PO Box 162200
Orlando, FL 32816-2205
*Prgm Dir:* Jeffery Ludy, EdD RRT
*Tel:* 407 823-3888   *Fax:* 407 823-6138
*E-mail:* jludy@mail.ucf.edu
*Med Dir:* Lawrence M Gilliard, MD

**Valencia Community College**
Respiratory Therapist (Advanced) Prgm
PO Box 3028
Orlando, FL 32802-9961
www.valenciacc.edu
*Prgm Dir:* Lynn W Capraun, MS RRT
*Tel:* 407 582-1550   *Fax:* 407 582-1278
*E-mail:* lcapraun@valenciacc.edu
*Med Dir:* Daniel T Layish, MD

**Palm Beach Community College**
Respiratory Therapist (Advanced) Prgm
3160 PGA Blvd
Palm Beach Gardens, FL 33410-2893
*Prgm Dir:* Thomas J Reid, BSRT RRT
*Tel:* 561 207-5068   *Fax:* 561 207-5011
*E-mail:* reidt@pbcc.edu
*Med Dir:* Rogelio Choy, MD

**Gulf Coast Community College**
Respiratory Therapist (Advanced) Prgm
5230 W US Hwy 98
Panama City, FL 32401
*Prgm Dir:* Bradley E Killion, BS RRT
*Tel:* 904 872-3837   *Fax:* 904 747-3246
*E-mail:* Bkillion@gulfcoast.edu
*Med Dir:* S A Daffin, III, MD

**St Petersburg College**
Respiratory Therapist (Advanced) Prgm
7200 66th St N
Pinellas Park, FL 33781
*Prgm Dir:* Stephen P Mikles, EdS RRT
*Tel:* 727 341-3627   *Fax:* 727 341-3744
*E-mail:* mikless@spjc.edu
*Med Dir:* Anthony N Ottaviani, DO

**Seminole Community College**
Respiratory Therapist (Advanced) Prgm
100 Weldon Blvd
Sanford, FL 32773
www.scc-fl.edu/respiratory
*Prgm Dir:* Steve Shideler, MS RRT
*Tel:* 407 708-2293   *Fax:* 407 708-2402
*E-mail:* shideles@scc-fl.edu
*Med Dir:* Matthew Lube, MD

**Florida A&M University**
Respiratory Therapist (Advanced) Prgm
Ware-Rhaney Extension
Ste 328
Tallahassee, FL 32307
*Prgm Dir:* Alphonso Baldwin, PhD RRT RPFT
*Tel:* 850 561-2186   *Fax:* 850 561-2457
*E-mail:* alphonso.baldwin@famu.edu
*Med Dir:* David Saint, MD

**Tallahassee Community College**
Respiratory Therapist (Advanced) Prgm
444 Appleyard Dr
Tallahassee, FL 32304
*Prgm Dir:* Dewey Streetman, RRT
*Tel:* 850 201-8327   *Fax:* 850 201-8329
*E-mail:* streetmd@tcc.fl.edu
*Med Dir:* Kenneth Wasson, MD

**Hillsborough Community College**
Respiratory Therapist (Advanced) Prgm
PO Box 30030
4001 Tampa Bay Blvd
Tampa, FL 33630-3030
www.hccfl.edu
*Prgm Dir:* Richard Palma
*Tel:* 813 253-7318   *Fax:* 813 253-7491
*E-mail:* rpalma@hccfl.edu
*Med Dir:* John Downs, MD

# Georgia

**Darton College**
Respiratory Therapist (Advanced) Prgm
2400 Gillionville Rd
Albany, GA 31707
www.darton.edu
*Prgm Dir:* William F Thomas, MS RRT
*Tel:* 229 317-6896   *Fax:* 229 317-6682
*E-mail:* william.thomas@darton.edu
*Med Dir:* Mark M Shoemaker, MD

**Athens Technical College**
Respiratory Therapist (Advanced) Prgm
800 US Hwy 29 N
Athens, GA 30601-1500
www.athenstech.edu
*Prgm Dir:* Bruce A Ott, EdD RRT
*Tel:* 706 355-5104   *Fax:* 706 425-3104
*E-mail:* bott@athenstech.edu
*Med Dir:* Jane Parks, MD

**Georgia State University**
Respiratory Therapist (Advanced) Prgm
PO Box 4019
Atlanta, GA 30302-4019
http://chhs.gsu.edu/cardio
*Prgm Dir:* Lynda T Goodfellow, EdD RRT
*Tel:* 404 651-3037   *Fax:* 404 651-1584
*E-mail:* ltgoodfellow@gsu.edu
*Med Dir:* Cindy D Powell, MD

**Augusta Technical College**
Respiratory Therapist (Advanced) Prgm
3200 Augusta Tech Dr
Augusta, GA 30906
www.augustatech.edu
*Prgm Dir:* Rita Waller, MSN RRT RPSGT
*Tel:* 706 771-4194   *Fax:* 706 771-4181
*E-mail:* rwaller@augustatech.edu
*Med Dir:* Samuel S Harvey, MD

**Medical College of Georgia**
Respiratory Therapist (Advanced) Prgm
815 St Sebastian Way, Rm HM-143
Augusta, GA 30912-0850
*Prgm Dir:* Arthur A Taft, PhD RRT
*Tel:* 706 721-3553   *Fax:* 706 721-0495
*E-mail:* ataft@mail.mcg.edu
*Med Dir:* Bashir Ahmad Chaudhary, MD

**Heart of Georgia Technical College**
Respiratory Therapist (Advanced) Prgm
560 Pinehill Rd
Dublin, GA 31021
*Prgm Dir:* Natalie A Smith, BS RRT RPFT
*Tel:* 478 274-7881   *Fax:* 478 275-6642
*E-mail:* natalies@hgtc.org
*Med Dir:* Carroll Reese, MD

**Griffin Technical College**
Respiratory Therapist (Advanced) Prgm
501 Varsity Rd
Griffin, GA 30223
www.griffintech.edu
*Prgm Dir:* Duane Reed, MA RRT
*Tel:* 770 233-6169   *Fax:* 770 229-3294
*E-mail:* dreed@griffintech.edu
*Med Dir:* John Dupre, MD

**Gwinnett Technical College**
Respiratory Therapist (Advanced) Prgm
5150 Sugarloaf Pkwy
Lawrenceville, GA 30043-5702
*Prgm Dir:* Robert P DeLorme, EdS RRT-NPS
*Tel:* 678 226-6658
*E-mail:* bdelorme@gwinnetttech.edu
*Med Dir:* Gregory Mauldon, MD

**Macon State College**
Respiratory Therapist (Advanced) Prgm
100 College Station Dr
Macon, GA 31297
*Prgm Dir:* Charles R Matson, MEd RRT
*Tel:* 478 471-2789   *Fax:* 478 471-2787
*E-mail:* cmatson@mail.maconstate.edu
*Med Dir:* R Jonathon Dean, MD

**Coosa Valley Technical College**
Respiratory Therapist (Advanced) Prgm
One Maurice Culberson Dr
Rome, GA 30161
www.cvtcollege.org
*Prgm Dir:* Zenia Bratton, BS RRT
*Tel:* 706 295-6910   *Fax:* 706 295-6894
*E-mail:* zbratton@coosavalleytech.edu
*Med Dir:* Gary Voccio, MD

**Armstrong Atlantic State University**
Respiratory Therapist (Advanced) Prgm
11935 Abercorn St
Savannah, GA 31419
*Prgm Dir:* Ross L Bowers, III, MHS RRT
*Tel:* 912 921-7446   *Fax:* 912 921-5585
*E-mail:* bowersro@mail.armstrong.edu
*Med Dir:* Stephen Morris, MD

**Southwest Georgia Technical College**
Respiratory Therapist (Advanced) Prgm
15689 US Hwy 19 N
Thomasville, GA 31792
www.southwestgatech.edu
*Prgm Dir:* Tammy A Miller, MEd CPFT RRT
*Tel:* 912 225-5094   *Fax:* 912 225-5289
*E-mail:* tmiller@southwestgatech.edu
*Med Dir:* Craig Wolff, MD

# Hawaii

**Kapi'olani Community College**
Respiratory Therapist (Advanced) Prgm
4303 Diamond Head Rd
Honolulu, HI 96816
*Prgm Dir:* Stephen F Wehrman, RRT RPFT AE-C
*Tel:* 808 734-9243   *Fax:* 808 734-9126
*E-mail:* wehrman@hawaii.edu
*Med Dir:* Christine Fukui, MD

# Idaho

**Boise State University**
Respiratory Therapist (Advanced) Prgm
College of Health Sciences
1910 University Dr, Rm HSR 116
Boise, ID 83725
http://respther.boisestate.edu
*Prgm Dir:* Lonny Ashworth
*Tel:* 208 426-3383   *Fax:* 208 426-4093
*E-mail:* lashwor@boisestate.edu
*Med Dir:* William Dittrich, MD

# Illinois

**Southwestern Illinois College**
Respiratory Therapist (Advanced) Prgm
St Elizabeth's Hospital
211 South Third St
Belleville, IL 62222
www.swic.edu
*Prgm Dir:* Margaret J McMillin, MEd RRT
*Tel:* 618 234-8911, Ext 1989   *Fax:* 618 222-4658
*E-mail:* mmcmilli@sebh.org
*Med Dir:* Barbara A Sudholt, MD

**Southern Illinois Univ at Carbondale**
Respiratory Therapist (Advanced) Prgm
School of Allied Health, 1365 Douglas Dr
College of Applied Sciences & Arts, MC6615
Carbondale, IL 62901
*Prgm Dir:* Stanley M Pearson, II, MSEd RRT
*Tel:* 618 453-7221   *Fax:* 618 453-7020
*E-mail:* stanman@siu.edu
*Med Dir:* Parviz Sanjabi, MD

**Kaskaskia College**
Respiratory Therapist (Advanced) Prgm
27210 College Rd
Centralia, IL 62801
www.kaskaskia.edu
*Prgm Dir:* Sharon (Beth) Urban, RRT
*Tel:* 618 545-3352   *Fax:* 618 532-2365
*E-mail:* burban@kaskaskia.edu
*Med Dir:* Aziz Rahman, MD

**Parkland College**
Respiratory Therapist (Advanced) Prgm
2400 W Bradley Ave
Champaign, IL 61821
*Prgm Dir:* Terry Des Jardins, MEd RRT
*Tel:* 217 351-2224   *Fax:* 217 351-2581
*E-mail:* TDesJardins@parkland.edu
*Med Dir:* Maury Topolosky, MD

**Malcolm X College**
Respiratory Therapist (Advanced) Prgm
1900 W Van Buren St
Chicago, IL 60612-3197
*Prgm Dir:* George A West, MS RRT
*Tel:* 312 850-7382    *Fax:* 312 850-7453
*E-mail:* gwest@ccc.edu
*Med Dir:* Corey Franklin, MD

**Olive Harvey College**
Respiratory Therapist (Advanced) Prgm
10001 W Woodlawn Ave, Rm 3317
Chicago, IL 60628
*Prgm Dir:* Sandra Barnes, MSHSA RRT NPS
*Tel:* 773 291-6568, Ext NA    *Fax:* 773 291-6304
*E-mail:* sbarnes@ccc.edu
*Med Dir:* Naresh Upadhyay, MD

**Kankakee Community College**
Respiratory Therapist (Advanced) Prgm
100 College Dr
Kankakee, IL 60901-6505
*Prgm Dir:* Nancy Stephens, BA MBA RRT
*Tel:* 815 802-8842    *Fax:* 815 802-8101
*E-mail:* nstephens@kcc.edu
*Med Dir:* Joel P Villegas, MD

**Moraine Valley Community College**
Respiratory Therapist (Advanced) Prgm
10900 S 88th Ave
Palos Hills, IL 60465
www.morainevalley.edu
*Prgm Dir:* Debra Guerrero, MS RRT RPSGT
*Tel:* 708 974-5774    *Fax:* 708 974-0185
*E-mail:* guerrero@morainevalley.edu
*Med Dir:* Muhammad Hamadeh, MD

**Illinois Central College**
Respiratory Therapist (Advanced) Prgm
201 SW Adams St
Peoria, IL 61635-0001
*Prgm Dir:* Carole (Kelly) Crawford-Jones, MS RRT
*Tel:* 309 999-4663    *Fax:* 309 673-9626
*E-mail:* kcrawfordjones@icc.edu
*Med Dir:* Patrick Whitten, MD

**Triton College**
Respiratory Therapist (Advanced) Prgm
2000 N Fifth Ave
River Grove, IL 60171
*Prgm Dir:* Kristine Anderson, MEd RRT-NPS CPFT
*Tel:* 708 456-0300, Ext 3429    *Fax:* 708 583-3121
*E-mail:* kanderso@triton.edu
*Med Dir:* Patrick J Fahey, MD

**Rock Valley College**
Respiratory Therapist (Advanced) Prgm
3301 N Mulford Rd
Rockford, IL 61114-5699
www.rockvalleycollege.edu
*Prgm Dir:* James R Sills, MEd CPFT RRT
*Tel:* 815 921-3220    *Fax:* 815 921-3249
*E-mail:* j.sills@rvc.cc.il.us
*Med Dir:* Ric Kullberg, MD

**St John's Hospital**
*Cosponsor:* Lincoln Land Community College
Respiratory Therapist (Advanced) Prgm
School of Respiratory Care
800 E Carpenter St
Springfield, IL 62769
www.st-johns.org
*Prgm Dir:* Jan Szoke, RRT
*Tel:* 217 544-6464, Ext 44254    *Fax:* 217 535-3881
*E-mail:* jan.szoke@st-johns.org
*Med Dir:* Pradeep S Kulkarni, MD

# Indiana

**University of Southern Indiana**
Respiratory Therapist (Advanced) Prgm
8600 University Blvd
Evansville, IN 47712
*Prgm Dir:* Robert Hooper, MA RRT
*Tel:* 812 464-1702    *Fax:* 812 465-7092
*E-mail:* rhooper@usi.edu
*Med Dir:* Duane H Kuhlenschmidt, MD

**Ivy Tech Community College NE - Ft Wayne**
Respiratory Therapist (Advanced) Prgm
3800 N Anthony Blvd
Ft Wayne, IN 46805
*Prgm Dir:* Jennifer Brink, BS RRT RPFT
*Tel:* 260 482-9171, Ext 4270    *Fax:* 260 480-4149
*E-mail:* jbrink@ivytech.edu
*Med Dir:* Richard C Cardillo, MD

**Indiana University Northwest**
Respiratory Therapist (Advanced) Prgm
3400 Broadway NW Campus
Hawthorne Hall
Gary, IN 46408
*Prgm Dir:* Susan T Pearson, MPA RRT
*Tel:* 219 980-6548    *Fax:* 219 980-6649
*E-mail:* spearson@iun.edu
*Med Dir:* Raja G Devanathan, MD

**Clarian Health Partners Inc**
Respiratory Therapist (Advanced) Prgm
I-65 at 21st St
PO Box 1367
Indianapolis, IN 46206-1367
www.clarian.org
*Prgm Dir:* Linda Van Scoder, EdD RRT
*Tel:* 317 962-8475    *Fax:* 317 962-2102
*E-mail:* lvanscoder@clarian.org
*Med Dir:* Chris C Naum, MD

**Indiana University**
Respiratory Therapist (Advanced) Prgm
1140 W Michigan St, CF 224
Indianapolis, IN 46202
*Prgm Dir:* Deborah L Cullen, EdD RRT
*Tel:* 317 274-7381    *Fax:* 317 278-7383
*E-mail:* dcullen@iupui.edu
*Med Dir:* Simon Hillier, MB ChB

**Ivy Tech Community College of Indiana**
Respiratory Therapist (Advanced) Prgm
Central Indiana Region
One W 26th St, PO Box 1763
Indianapolis, IN 46208-1763
www.ivytech.edu
*Prgm Dir:* Marcus D Stowe, MS RRT
*Tel:* 317 921-4410    *Fax:* 317 921-4753
*E-mail:* mstowe@ivytech.edu
*Med Dir:* James D Pike, DO

**Ivy Tech Community College - Lafayette**
Respiratory Therapist (Advanced) Prgm
3101 S Creasy Ln
Lafayette, IN 47903
*Prgm Dir:* Peggy James, MBA RRT CPFT
*Tel:* 765 772-9207    *Fax:* 765 772-9248
*E-mail:* pjames@ivytech.edu
*Med Dir:* David Emery, MD

**Ivy Tech Community College NW - Gary**
Respiratory Therapist (Advanced) Prgm
3714 Franklin St
Michigan City, IN 46360
*Prgm Dir:* Susan Layhew, MA RRT
*Tel:* 219 879-9137    *Fax:* 219 879-9157
*E-mail:* slayhew@ivytech.edu
*Med Dir:* Charles Rebsco, MD

# Iowa

**Des Moines Area Community College**
Respiratory Therapist (Advanced) Prgm
2006 Ankeny Blvd
Ankeny, IA 50023
www.dmacc.edu/programs/respiratorytherapy
*Prgm Dir:* Kerry E George, MEd RRT FAARC
*Tel:* 515 964-6298    *Fax:* 515 964-6327
*E-mail:* kegeorge@dmacc.edu
*Med Dir:* Ricardo Flores, MD

**Kirkwood Community College**
Respiratory Therapist (Advanced) Prgm
6301 Kirkwood Blvd SW, PO Box 2068
Cedar Rapids, IA 52406-9973
*Prgm Dir:* H Kenneth Bronkhorst, MBA RRT
*Tel:* 319 398-4987    *Fax:* 319 398-1293
*E-mail:* ken.bronkhorst@kirkwood.edu
*Med Dir:* Lois Geist, MD

**Northeast Iowa Community College**
Respiratory Therapist (Advanced) Prgm
10250 Sundown Rd
Peosta, IA 52068
*Prgm Dir:* Amy Rausch, AAS RRT BS
*Tel:* 563 556-5110, Ext 274    *Fax:* 563 556-5058
*E-mail:* rauscha@nicc.edu
*Med Dir:* Mark Janes, MD

**St Luke's College**
Respiratory Therapist (Advanced) Prgm
2720 Stone Park Blvd
Sioux City, IA 51104
*Prgm Dir:* Allen W Barbaro, MS RRT
*Tel:* 712 279-7964
*E-mail:* barbaraw@stlukes.org

**Hawkeye Community College**
Respiratory Therapist (Advanced) Prgm
1501 E Orange Rd
PO Box 8015
Waterloo, IA 50704
www.hawkeyecollege.edu
*Prgm Dir:* Jamie Bute, AAS RRT
*Tel:* 319 296-2320, Ext 1919    *Fax:* 319 296-4400
*E-mail:* jbute@hawkeyecollege.edu
*Med Dir:* James Cafaro, MD

**Southeastern Community College**
Respiratory Therapist (Advanced) Prgm
1500 West Agency PO 180
PO Box 180
West Burlington, IA 52655-0180
*Prgm Dir:* Stacy Sells, BHS RRT-NPS CPFT
*Tel:* 319 208-5204    *Fax:* 319 752-4957
*E-mail:* ssells@scciowa.edu
*Med Dir:* Valerie Velho, MD

# Kansas

**Bethany Med Ctr/Kansas City Kansas Comm Coll**
Respiratory Therapist (Advanced) Prgm
7250 State Ave
Kansas City, KS 66112
www.kckcc.edu
*Prgm Dir:* C Michael Parrett, MBA RPFT RRT-NPS
*Tel:* 913 288-7245    *Fax:* 913 288-7649
*E-mail:* mparrett@kckcc.edu
*Med Dir:* Sabato Sissillo, MD

**University of Kansas Medical Center**
Respiratory Therapist (Advanced) Prgm
3901 Rainbow Blvd, 4006 Delp, Mail Stop 1013
Kansas City, KS 66160
www.kumc.edu/SAH/resp_care
*Prgm Dir:* Barbara A Ludwig, MA RRT
*Tel:* 913 588-4634    *Fax:* 913 588-4631
*E-mail:* bludwig@kumc.edu
*Med Dir:* William Atkinson, MD

**Seward County Community College**
Respiratory Therapist (Advanced) Prgm
PO Box 1137
Liberal, KS 67901-1137
www.sccc.edu
*Prgm Dir:* Edward B Anderson, MS RRT
*Tel:* 316 626-3080    *Fax:* 316 626-3040
*E-mail:* eanderso@sccc.net
*Med Dir:* David J Fitzgerald, DO

**Johnson County Community College**
Respiratory Therapist (Advanced) Prgm
Respiratory Care, 12345 College Blvd
Overland Park, KS 66210
www.jccc.net/home/depts/001256
*Prgm Dir:* Clarissa M Craig, PhD RRT
*Tel:* 913 469-2583    *Fax:* 913 469-2315
*E-mail:* ccraig@jccc.edu
*Med Dir:* Larry D Botts, MD

**Labette Community College**
Respiratory Therapist (Advanced) Prgm
200 S 14th St
Parsons, KS 67357
www.labette.edu
*Prgm Dir:* Connie S Crooks, BS RRT
*Tel:* 620 820-1160    *Fax:* 620 421-1539
*E-mail:* conniec@labette.edu
*Med Dir:* Daniel Pauls, MD

**Washburn University**
Respiratory Therapist (Advanced) Prgm
1700 College Ave SW
Topeka, KS 66621
www.washburn.edu/respiratory
*Prgm Dir:* Pat Munzer, DHSc RRT
*Tel:* 785 670-1404    *Fax:* 785 670-1027
*E-mail:* pat.munzer@washburn.edu
*Med Dir:* Iris Gonzalez, MD

**Newman University**
Respiratory Therapist (Advanced) Prgm
3100 McCormick Ave
Wichita, KS 67213-2097
www.newmanu.edu
*Prgm Dir:* Meg Trumpp, MEd RRT AE-C
*Tel:* 316 942-4291, Ext 2344    *Fax:* 316 942-4483
*E-mail:* trumppm@newmanu.edu
*Med Dir:* Guy M Grabau, MD, FCCP

# Kentucky

**Northern Kentucky University**
Respiratory Therapist (Advanced) Prgm
Nunn Dr, AHC-225
Highland Heights, KY 41099-8002
*Prgm Dir:* Debra Kasel, MEd RRT CPFT AE-C
*Tel:* 859 572-5608    *Fax:* 859 572-1314
*E-mail:* kaseld@nku.edu
*Med Dir:* Roy Moser III, MD

**Bluegrass Community and Technical College**
*Cosponsor:* University of Kentucky
Respiratory Therapist (Advanced) Prgm
Rm 330 Oswald Bldg, Cooper Dr
Lexington, KY 40506-0235
*Prgm Dir:* James K Matchuny, BS RRT
*Tel:* 859 257-4872, Ext 4106    *Fax:* 859 257-9580
*E-mail:* jkmatc1@uky.edu
*Med Dir:* James McCormick, MD

**Jefferson Community and Technical College**
Respiratory Therapist (Advanced) Prgm
109 E Broadway St
Louisville, KY 40202
*Prgm Dir:* Joan Kruse, MEd RRT
*Tel:* 502 213-2197    *Fax:* 502 213-2491
*E-mail:* joan.kruse@kctcs.edu
*Med Dir:* Robert W Powell, MD

**Madisonville Community College**
Respiratory Therapist (Advanced) Prgm
Health Campus
750 N Laffoon St
Madisonville, KY 42431
*Prgm Dir:* David F Pennaman, MS RRT
*Tel:* 270 824-7552
*E-mail:* david.pennaman@kctcs.edu

**West Kentucky Community & Technical College**
Respiratory Therapist (Advanced) Prgm
PO Box 7408
5200 Blandville Rd
Paducah, KY 42002-7408
*Prgm Dir:* Ruth Thompson, MS RRT
*Tel:* 270 534-3486    *Fax:* 270 554-6227
*E-mail:* ruth.thompson@kctcs.edu
*Med Dir:* Bradley Rankin, MD

**Big Sandy Community & Technical College**
Respiratory Therapist (Advanced) Prgm
Mayo Campus
513 Third St
Paintsville, KY 41240
www.bigsandy.kctcs.edu
*Prgm Dir:* Melissa Skeens, RRT NPS RPFT
*Tel:* 606 789-5321, Ext 82822    *Fax:* 606 789-9753
*E-mail:* melissa.skeens@kctcs.edu
*Med Dir:* Loey Kousa, MD

**Southeast Kentucky Comm & Tech College**
Respiratory Therapist (Advanced) Prgm
3300 US Highway 25E South
Pineville, KY 40977
*Prgm Dir:* Michael Good, MS RRT-NPS RPFT
*Tel:* 606 248-2122    *Fax:* 606 248-2166
*E-mail:* mike.good@kctcs.edu
*Med Dir:* Abdi Vazey, MD

# Louisiana

**Bossier Parish Community College**
Respiratory Therapist (Advanced) Prgm
6220 East Texas St
Bossier City, LA 71111
*Prgm Dir:* Sandra Partain, MHS RRT-NPS
*Tel:* 318 675-6813    *Fax:* 318 675-6937
*E-mail:* sparta@lsuhsc.edu
*Med Dir:* Keith Payne, MD

**Nicholls State University**
Respiratory Therapist (Advanced) Prgm
Dept of Allied Health Sciences
235 Civic Center Blvd
Houma, LA 70360
www.nicholls.edu
*Prgm Dir:* Errol Champagne, MEd RRT-NPS
*Tel:* 985 876-8855    *Fax:* 985 876-8856
*E-mail:* errol.champagne@nicholls.edu
*Med Dir:* Ralph Bourgeois, MD

**Delgado Community College**
Respiratory Therapist (Advanced) Prgm
615 City Park Ave
New Orleans, LA 70119
*Prgm Dir:* Diane M Olsen-Rawls, MHS RRT
*Tel:* 504 483-4114, Ext 4007    *Fax:* 504 483-4609
*E-mail:* dolsen@dcc.edu
*Med Dir:* Mack Thomas, MD

**Louisiana State Univ Health Sciences Center**
Respiratory Therapist (Advanced) Prgm
1900 Gravier St
New Orleans, LA 70112
www.lsuhsc.edu/no
*Prgm Dir:* John Zamjahn, PhD RRT
*Tel:* 504 568-4228    *Fax:* 504 599-0410
*E-mail:* jzamja@lsuhsc.edu
*Med Dir:* Carol Mason, MD

**Our Lady of Holy Cross College**
*Cosponsor:* Alton Ochsner School of Allied Health Sciences
Respiratory Therapist (Advanced) Prgm
4123 Woodland Dr
New Orleans, LA 70131
*Prgm Dir:* Mary LaBiche, MEd RRT
*Tel:* 504 842-3736    *Fax:* 504 842-4372
*E-mail:* mlabiche@ochsner.org
*Med Dir:* Clifford B Burns, MD

**Louisiana State Univ Hlth Sci Ctr-Shreveport**
Respiratory Therapist (Advanced) Prgm
1501 Kings Hwy
Shreveport, LA 71130
*Prgm Dir:* Dennis R Payne, MD
*Tel:* 318 675-6814
*E-mail:* kpayne@lsuhsc.edu
*Med Dir:* D Keith Payne, MD

**Southern Univ at Shreveport**
Respiratory Therapist (Advanced) Prgm
3050 Martin Luther King Jr Dr
Shreveport, LA 71107
*Prgm Dir:* JoAnn Warren, BA RRT
*Tel:* 318 674-3452    *Fax:* 318 676-5307
*E-mail:* jwarren@susla.edu
*Med Dir:* Robert Long, MD

# Maine

**Kennebec Valley Community College**
Respiratory Therapist (Advanced) Prgm
92 Western Ave
Fairfield, ME 04937-1367
www.kvcc.me.edu
*Prgm Dir:* Barbara A Larsson, MEd RRT
*Tel:* 207 453-5161    *Fax:* 207 453-5197
*E-mail:* blarsson@kvcc.me.edu
*Med Dir:* Kristin Holm, MD

**Southern Maine Community College**
Respiratory Therapist (Advanced) Prgm
Two Fort Rd
South Portland, ME 04106
www.smccme.edu
*Prgm Dir:* Walter C Chop, MS RRT
*Tel:* 207 741-5592    *Fax:* 207 741-4560
*E-mail:* wchop@smccme.edu
*Med Dir:* Patricia Lerwick, MD

# Maryland

**Baltimore City Community College**
Respiratory Therapist (Advanced) Prgm
2901 Liberty Heights Ave
Baltimore, MD 21215
www.bccc.edu
*Prgm Dir:* William Cooper, RRT
*Tel:* 410 462-7776    *Fax:* 410 462-7401
*E-mail:* wcooper@bccc.edu
*Med Dir:* Clifford Boehm, MD RRT

**Comm Coll of Baltimore County - Essex Campus**
Respiratory Therapist (Advanced) Prgm
7201 Rossville Blvd
Baltimore, MD 21237
www.ccbcmd.edu
*Prgm Dir:* Barbara Schenk, BA RRT
*Tel:* 410 780-6760    *Fax:* 410 682-8044
*E-mail:* bschenk@ccbcmd.edu
*Med Dir:* Martin Sheridan, MD

### Allegany College of Maryland
Respiratory Therapist (Advanced) Prgm
12401 Willowbrook Rd SE
Cumberland, MD 21502-2596
*Prgm Dir:* William R Rocks, EdD RRT
*Tel:* 301 784-5522    *Fax:* 301 784-5015
*E-mail:* brocks@allegany.edu
*Med Dir:* James M Raver, MD

### Frederick Community College
Respiratory Therapist (Advanced) Prgm
7932 Opossumtown Pike
Frederick, MD 21702
*Prgm Dir:* Mark L Paugh, PhD RRT
*Tel:* 301 846-2528    *Fax:* 301 846-2498
*E-mail:* mpaugh@frederick.edu
*Med Dir:* Neil V Waravdekar, MD

### Prince George's Community College
Respiratory Therapist (Advanced) Prgm
301 Largo Rd
Largo, MD 20774
http://academic.pgcc.edu/alliedhealth
*Prgm Dir:* James Courtwright, RRT MSHP
*Tel:* 301 322-0860    *Fax:* 301 386-7528
*E-mail:* jcourtwright@pgcc.edu
*Med Dir:* Joseph J Colella, MD

### Salisbury University
Respiratory Therapist (Advanced) Prgm
1101 Camden Ave
Salisbury, MD 21801
www.salisbury.edu
*Prgm Dir:* Robert Joyner, PhD RRT
*Tel:* 410 543-6365    *Fax:* 410 548-9185
*E-mail:* rljoyner@salisbury.edu
*Med Dir:* Rodney Layton, MD

### Columbia Union College
Respiratory Therapist (Advanced) Prgm
7600 Flower Ave
Takoma Park, MD 20912
www.cuc.edu
*Prgm Dir:* Vicki Rosette, RRT RPFT
*Tel:* 301 891-4187    *Fax:* 301 891-4181
*E-mail:* vrosette@cuc.edu
*Med Dir:* Alfred Munzer, MD

# Massachusetts

### Northeastern University
Respiratory Therapist (Advanced) Prgm
100 Dockser Hall
360 Huntington Ave
Boston, MA 02115
www.ace.neu.edu/bouve/health/rt
*Prgm Dir:* Scott Stanley, EdD RRT
*Tel:* 617 373-5382    *Fax:* 617 373-2968
*E-mail:* s.stanley@neu.edu
*Med Dir:* Alan Lisbon, MD

### Massasoit Community College
Respiratory Therapist (Advanced) Prgm
One Massasoit Blvd
Brockton, MA 02301
www.massasoit.mass.edu
*Prgm Dir:* Martha DeSilva, MEd RRT
*Tel:* 508 588-9100, Ext 1787    *Fax:* 508 427-1262
*E-mail:* mdesilva@massasoit.mass.edu
*Med Dir:* Ronald Coutu, MD

### North Shore Community College
Respiratory Therapist (Advanced) Prgm
One Ferncroft Rd, PO Box 3340
Danvers, MA 01923-0840
www.northshore.edu
*Prgm Dir:* Geraldine Twomey, MEd RRT RN
*Tel:* 978 762-4166    *Fax:* 978 762-4022
*E-mail:* gtwomey@northshore.edu
*Med Dir:* Faysal Hasan, MD

### Northern Essex Community College
Respiratory Therapist (Advanced) Prgm
45 Franklin St
Lawrence, MA 01841
www.necc.mass.edu
*Prgm Dir:* Saleh Daher, BS RRT CPFT NPS
*Tel:* 978 738-7217, Ext 7217    *Fax:* 978 738-7450
*E-mail:* sdaher@necc.mass.edu
*Med Dir:* Daniel Coleman, MD

### Berkshire Community College
Respiratory Therapist (Advanced) Prgm
1350 West St
Pittsfield, MA 01201
www.berkshirecc.edu
*Prgm Dir:* Thomas P Carey Jr, RRT MPH
*Tel:* 413 499-4660, Ext 252    *Fax:* 413 448-2700
*E-mail:* tcarey@berkshirecc.edu
*Med Dir:* Micheal Mortilini, MD

### Springfield Technical Community College
Respiratory Therapist (Advanced) Prgm
One Armory Sq
Springfield, MA 01105
www.stcc.edu
*Prgm Dir:* Lee J Robinson, MEd RRT
*Tel:* 413 755-4829    *Fax:* 413 755-4764
*E-mail:* robinson@stcc.edu
*Med Dir:* Bruce M Meth, MD

### Quinsigamond Community College
Respiratory Therapist (Advanced) Prgm
670 W Boylston St
Worcester, MA 01606-2092
www.qcc.mass.edu
*Prgm Dir:* Lynda A Nesbitt, MA RRT
*Tel:* 508 854-4398    *Fax:* 508 852-6943
*E-mail:* lyndan@qcc.mass.edu
*Med Dir:* Richard Rosiello, MD

# Michigan

### Ferris State University
Respiratory Therapist (Advanced) Prgm
200 Ferris Dr, VFS 210A
Big Rapids, MI 49307-2740
*Prgm Dir:* Brenda Brown, MS RRT
*Tel:* 231 591-2318    *Fax:* 231 591-2325
*E-mail:* brownbk@ferris.edu
*Med Dir:* Maurice J Norman, MD

### Macomb Community College
Respiratory Therapist (Advanced) Prgm
44575 Garfield Rd, E Bldg Rm 219
Clinton Township, MI 48038-1139
www.macomb.edu
*Prgm Dir:* Mary E Alstead, BS RRT
*Tel:* 586 286-2150    *Fax:* 586 286-2098
*E-mail:* alsteadm@macomb.edu
*Med Dir:* Howard M Kaplan, MD

### Henry Ford Community College
Respiratory Therapist (Advanced) Prgm
Health Careers Education Center
5101 Evergreen Rd
Dearborn, MI 48128-1495
*Prgm Dir:* Debra Szymanski, MA RRT
*Tel:* 313 317-6580    *Fax:* 313 317-6569
*E-mail:* dszyman@hfcc.edu
*Med Dir:* Bradford K Grassmick, MD

### Mott Community College
Respiratory Therapist (Advanced) Prgm
1401 E Court St
Flint, MI 48503
*Prgm Dir:* David L Panzlau, MA RRT
*Tel:* 810 232-6563    *Fax:* 810 762-5619
*E-mail:* dpanzlau@mcc.edu
*Med Dir:* Ahmed Hannan, MD

### Kalamazoo Valley Community College
Respiratory Therapist (Advanced) Prgm
Texas Township Campus
6767 West O Ave, PO Box 4070
Kalamazoo, MI 49003-4070
*Prgm Dir:* Albert W Moss, MA RRT
*Tel:* 269 488-4288    *Fax:* 269 488-4458
*E-mail:* amoss@kvcc.edu
*Med Dir:* John W Dircks, MD

### Monroe County Community College
Respiratory Therapist (Advanced) Prgm
1555 S Raisinville Rd
Monroe, MI 48161
*Prgm Dir:* Bonnie Boggs, RRT
*Tel:* 734 384-4268    *Fax:* 734 384-4187
*E-mail:* bboggs@monroecc.edu
*Med Dir:* Milo Engoren, MD

### Muskegon Community College
Respiratory Therapist (Advanced) Prgm
221 S Quarterline Rd
Muskegon, MI 49442
http://muskegon.cc.mi.us/~devriesd/resp-home.htm
*Prgm Dir:* Daniel B Knue, MM RRT-NPS
*Tel:* 231 777-0370    *Fax:* 231 777-0490
*E-mail:* Dan.Knue@muskegoncc.edu
*Med Dir:* Mark Ivey, MD

### Oakland Community College
Respiratory Therapist (Advanced) Prgm
22322 Rutland Dr
Southfield, MI 48075
*Prgm Dir:* Sue J Work, MSA RRT-NPS
*Tel:* 248 233-2919    *Fax:* 248 233-2891
*E-mail:* sjwork@oaklandcc.edu
*Med Dir:* Alan D Betensley, MD

### Delta College
Respiratory Therapist (Advanced) Prgm
1961 Delta Rd
University Center, MI 48710
*Prgm Dir:* Earl B Gregory, MS RRT
*Tel:* 989 686-9489    *Fax:* 989 667-2211
*E-mail:* ebgregor@alpha.delta.edu
*Med Dir:* Kizhaketat P Sukumaran, MD

# Minnesota

### Lake Superior College
Respiratory Therapist (Advanced) Prgm
2101 Trinity Rd
Duluth, MN 55811
www.lsc.edu
*Prgm Dir:* Cynthia Annable, RRT NPS RPFT
*Tel:* 218 733-5925    *Fax:* 218 723-4921
*E-mail:* c.annable@lsc.edu
*Med Dir:* Paul Windberg, MD

### Northland Community & Technical College
Respiratory Therapist (Advanced) Prgm
2022 Central Ave NE
East Grand Forks, MN 56721
www.northlandcollege.edu
*Prgm Dir:* Anthony Sorum, BA RRT
*Tel:* 218 773-4791    *Fax:* 218 773-4502
*E-mail:* tony.sorum@northlandcollege.edu
*Med Dir:* Wayne Bretweiser, MD

### College of St Catherine - Minneapolis
Respiratory Therapist (Advanced) Prgm
601 25th Ave S
Minneapolis, MN 55454
*Prgm Dir:* John E Boatright,, MA RRT
*Tel:* 651 690-7819    *Fax:* 651 690-7849
*E-mail:* jeboatright@stkate.edu
*Med Dir:* Asra Mohiuddin, MD

**University of Minnesota - Minneapolis**
Respiratory Therapist (Advanced) Prgm
Mayo Foundation
1115 Siebens
Rochester, MN 55905
*Prgm Dir:* Vanessa King, MED RRT
*Tel:* 507 284-0174   *Fax:* 507 284-0656
*E-mail:* king.vanessa@mayo.edu
*Med Dir:* James Findlay, MD

**St Paul College**
Respiratory Therapist (Advanced) Prgm
235 Marshall Ave
St Paul, MN 55102
*Prgm Dir:* Joseph Buhain, RRT RCP EMT-B
*Tel:* 651 846-1501   *Fax:* 651 221-1416
*E-mail:* joseph.buhain@saintpaul.edu
*Med Dir:* Avi Nahum, MD

# Mississippi

**Northeast Mississippi Community College**
Respiratory Therapist (Advanced) Prgm
Cunningham Blvd
Booneville, MS 38829
www.nemcc.edu
*Prgm Dir:* Beverly Prince, RRT BS
*Tel:* 662 720-7387   *Fax:* 662 728-1165
*E-mail:* biprince@nemcc.edu
*Med Dir:* B Wayne McAlpin, MD

**Itawamba Community College**
Respiratory Therapist (Advanced) Prgm
602 W Hill St
Fulton, MS 38843
*Prgm Dir:* James Harold Plunkett, BA RRT
*Tel:* 601 862-8348   *Fax:* 601 862-8350
*E-mail:* hjplunkett@iccms.edu
*Med Dir:* Benjamin Moore, MD

**Mississippi Gulf Coast Community College**
Respiratory Therapist (Advanced) Prgm
PO Box 100
Gautier, MS 39553
*Prgm Dir:* Donald Hayes, AS RRT
*Tel:* 228 497-7711, Ext 290   *Fax:* 228 497-7676
*E-mail:* donald.hayes@mgccc.edu
*Med Dir:* Gary Rodberg, MD

**Pearl River Community College**
Respiratory Therapist (Advanced) Prgm
Forrest County Voc-Tech Ctr
5448 US Hwy 49 S
Hattiesburg, MS 39401
*Prgm Dir:* Lori Anderson
*Tel:* 601 554-5521   *Fax:* 601 554-5487
*E-mail:* landerson@prcc.edu
*Med Dir:* Steven W Stogner, MD

**Hinds Community College**
Respiratory Therapist (Advanced) Prgm
1750 Chadwick Dr
Jackson, MS 39204
www.hindscc.edu
*Prgm Dir:* Shirley Miller, RRT BSEd
*Tel:* 601 371-3517   *Fax:* 601 376-4541
*E-mail:* slmiller@hindscc.edu
*Med Dir:* William C Pinkston, MD

**Meridian Community College**
Respiratory Therapist (Advanced) Prgm
910 Hwy 19 N
Meridian, MS 39307
www.meridiancc.edu
*Prgm Dir:* Steve W Arinder, RRT BS MPH
*Tel:* 601 484-8752   *Fax:* 601 482-3936
*E-mail:* sarinder@meridiancc.edu
*Med Dir:* Richmond Alexander III, MD

**Copiah-Lincoln Community College**
Respiratory Therapist (Advanced) Prgm
Natchez Campus Career and Technical
30 Campus Dr
Natchez, MS 39120-5398
www.colin.edu
*Prgm Dir:* Walton B Wilson, BS RRT
*Tel:* 601 446-1161   *Fax:* 601 446-1298
*E-mail:* walt.wilson@colin.edu
*Med Dir:* Barry F Tillman, MD

**Northwest Mississippi Community College**
Respiratory Therapist (Advanced) Prgm
5197 WE Ross Parkway
Southaven, MS 38671
*Prgm Dir:* Regina K Clark, MEd RRT
*Tel:* 601 280-6151   *Fax:* 601 280-6161
*E-mail:* r_clark@northwestms.edu
*Med Dir:* H Edward Garrett, MD

# Missouri

**Cape Girardeau Career & Technology Center**
Respiratory Therapist (Advanced) Prgm
1080 S Silver Springs Rd
Cape Girardeau, MO 63703
www.cape.k12.mo.us
*Prgm Dir:* Kenneth L Pfau, BS RRT
*Tel:* 573 334-0449, Ext 320   *Fax:* 573 334-5930
*E-mail:* pfauk@cape.k12.mo.us
*Med Dir:* Richard E Moore, MD

**University of Missouri - Columbia**
Respiratory Therapist (Advanced) Prgm
605 Lewis Hall
Columbia, MO 65211
www.umshp.org/rt
*Prgm Dir:* Rosemary G Hogan, RRT MEd
*Tel:* 573 882-8034   *Fax:* 573 884-1490
*E-mail:* hoganr@health.missouri.edu
*Med Dir:* Rajiv Dhand, MD

**Hannibal Career & Technical Center**
*Cosponsor:* Hannibal LaGrange College
Respiratory Therapist (Advanced) Prgm
4550 McMasters Ave
Hannibal, MO 63401
*Prgm Dir:* David D Bach, BHS RRT RCP
*Tel:* 573 221-4430, Ext 180   *Fax:* 573 221-1385
*E-mail:* dbach@hannibal.k12.mo.us
*Med Dir:* Manoocher Nassery, MD

**Missouri Southern State University**
Respiratory Therapist (Advanced) Prgm
3950 E Newman Rd
Joplin, MO 64801-1595
*Prgm Dir:* Glenda Pippin, BS RRT CPFT
*Tel:* 417 659-4405   *Fax:* 417 659-4408
*E-mail:* pippin-g@mssu.edu
*Med Dir:* M Kreimid, MD

**Concorde Career College - Kansas City**
Respiratory Therapist (Advanced) Prgm
3239 Broadway Blvd
Kansas City, MO 64111
www.concordecareercolleges.com
*Prgm Dir:* Lana C Conrad, BS RRT
*Tel:* 816 531-5223, Ext 403   *Fax:* 816 756-3231
*E-mail:* lconrad@concorde.edu
*Med Dir:* Lida N Osbern, MD

**Rolla Technical Center**
Respiratory Therapist (Advanced) Prgm
500 Forum Dr
Rolla, MO 65401
*Prgm Dir:* Diane R Oldfather
*Tel:* 573 458-0160
*E-mail:* doldfather@rolla.k12.mo.us

**Ozarks Technical Community College**
Respiratory Therapist (Advanced) Prgm
1001 E Chestnut Expressway
Springfield, MO 65802
www.otc.edu
*Prgm Dir:* Doug Pursley, MEd RRT
*Tel:* 417 447-8823   *Fax:* 417 447-8841
*E-mail:* pursleyd@otc.edu
*Med Dir:* John Wolfe, MD

**St Louis Community College - Forest Park**
Respiratory Therapist (Advanced) Prgm
5600 Oakland Ave
St Louis, MO 63110
*Prgm Dir:* James R Brennan, MEd RRT
*Tel:* 314 644-9079   *Fax:* 314 951-9412
*E-mail:* jbrennan@stlcc.edu
*Med Dir:* Oscar Schwartz, MD

# Montana

**Montana State Univ - Great Falls Coll of Tech**
Respiratory Therapist (Advanced) Prgm
2100 16th Ave S
Great Falls, MT 59405
*Prgm Dir:* Leonard Bates, MEd RRT
*Tel:* 406 771-4360   *Fax:* 406 771-4313
*E-mail:* lbates@msugf.edu
*Med Dir:* Richard Blevins, MD

**University of Montana**
Respiratory Therapist (Advanced) Prgm
909 S Ave W
Missoula, MT 59801
*Prgm Dir:* Robert Wafstet, MS RRT
*Tel:* 406 243-7821   *Fax:* 406 243-7899
*E-mail:* Robert.Wafstet@umontana.edu
*Med Dir:* William Bekemeyer, MD

# Nebraska

**Southeast Community College**
Respiratory Therapist (Advanced) Prgm
8800 O St
Lincoln, NE 68520-1299
*Prgm Dir:* Charlotte Pasco, BA RRT
*Tel:* 402 437-2781   *Fax:* 402 437-2404
*E-mail:* cpasco@southeast.edu
*Med Dir:* John Rudersdorf, MD

**Alegent Health**
*Cosponsor:* Midland Lutheran College
Respiratory Therapist (Advanced) Prgm
6901 N 72nd St
Omaha, NE 68122
*Prgm Dir:* Todd Klopfenstein, Bs RRT
*Tel:* 402 572-2312   *Fax:* 402 572-3157
*E-mail:* tklopfen@alegent.org
*Med Dir:* Louis A D Violi, MD

**Metropolitan Community College**
Respiratory Therapist (Advanced) Prgm
PO Box 3777
Omaha, NE 68103
*Prgm Dir:* Jerald A Moss, RRT MPA
*Tel:* 402 738-4653   *Fax:* 402 738-4005
*E-mail:* jmoss@metropo.mccneb.edu
*Med Dir:* Lon W Keim, MD

**NE Methodist Coll Nursing & Allied Hlth**
Respiratory Therapist (Advanced) Prgm
720 N 87th St
Omaha, NE 68114
www.methodistcollege.edu
*Prgm Dir:* Christine Hamilton, MA RRT AE-C
*Tel:* 402 354-7065   *Fax:* 402 354-7130
*E-mail:* Chris.Hamilton@methodistcollege.edu
*Med Dir:* Patrick G Meyers, MD

**PROGRAMS**

# Nevada

**Community College of Southern Nevada**
Respiratory Therapist (Advanced) Prgm
6375 W Charleston Blvd
Las Vegas, NV 89146-1124
www.ccsn.nevada.edu
*Prgm Dir:* Carlton R Insley III, PhD RRT
*Tel:* 702 651-5665   *Fax:* 702 651-5028
*E-mail:* randy_insley@ccsn.edu
*Med Dir:* Rachakonda D Prabhu, MD

# New Hampshire

**New Hampshire Comm Tech Coll - Claremont**
Respiratory Therapist (Advanced) Prgm
One College Dr
Claremont, NH 03743
*Prgm Dir:* John Marcley, MEd RRT
*Tel:* 603 542-7744, Ext 2530   *Fax:* 603 543-1844
*E-mail:* jmarcley@nhctc.edu
*Med Dir:* H Worth Parker, MD

# New Jersey

**Northwest NJ Consortium Resp Care Educ**
Respiratory Therapist (Advanced) Prgm
Saint Clares Hospital
400 Blackwell St
Dover, NJ 07801
*Prgm Dir:* Dianne Adams, MA RRT
*Tel:* 973 537-3906   *Fax:* 973 537-3996
*E-mail:* dadams@saintclares.org
*Med Dir:* Jack Goldshlack, DO

**Brookdale Community College**
Respiratory Therapist (Advanced) Prgm
765 Newman Springs Rd
Lincroft, NJ 07738
www.brookdalecc.edu
*Prgm Dir:* Carol Schedel, MA RRT
*Tel:* 732 224-2692   *Fax:* 732 224-2998
*E-mail:* cschedel@brookdalecc.edu
*Med Dir:* J DeTullio, MD

**Univ of Med & Dent of New Jersey**
Respiratory Therapist (Advanced) Prgm
School of Health Related Professions
65 Bergen St
Newark, NJ 07107-3006
*Prgm Dir:* Albert J Heuer, PhD RRT
*Tel:* 973 972-2418   *Fax:* 973 972-5258
*E-mail:* heueraj@umdnj.edu
*Med Dir:* Marc H Lavietes, MD

**Bergen Community College**
Respiratory Therapist (Advanced) Prgm
400 Paramus Rd
Paramus, NJ 07652
*Prgm Dir:* Robert A Muller, MA RRT
*Tel:* 201 612-5337   *Fax:* 201 612-3876
*E-mail:* rmuller@bergen.edu
*Med Dir:* Joseph Manno, MD

**Univ of Med & Dent of New Jersey**
Respiratory Therapist (Advanced) Prgm
Sch of Hlth Related Professions
UEC 40 East Laurel Rd
Stratford, NJ 08084
www.shrp.umdnj.edu
*Prgm Dir:* Alan Realey, BS RRT
*Tel:* 856 566-2891   *Fax:* 856 566-2894
*E-mail:* realeyam@umdnj.edu
*Med Dir:* Thomas Morley, DO

# New Mexico

**Central New Mexico Community College**
Respiratory Therapist (Advanced) Prgm
525 Buena Vista SE
Albuquerque, NM 87106
www.tvi.edu
*Prgm Dir:* John Blewett, RRT
*Tel:* 505 224-4138   *Fax:* 505 224-4120
*E-mail:* jblewett@cnm.edu
*Med Dir:* Diane Klepper, MD

**Dona Ana Community College**
Respiratory Therapist (Advanced) Prgm
Box 30001 Dept 3DA
3400 S Espina
Las Cruces, NM 88003-0001
*Prgm Dir:* Virginia Durant, MA RRT
*Tel:* 505 527-7607   *Fax:* 505 527-7765
*E-mail:* vdurant@nmsu.edu
*Med Dir:* James Bradley, MD

**Eastern New Mexico University - Roswell**
Respiratory Therapist (Advanced) Prgm
PO Box 6000
Roswell, NM 88202-6000
www.roswell.enmu.edu
*Prgm Dir:* Gina Buldra, BS RRT RCP
*Tel:* 505 624-7217   *Fax:* 505 624-7700
*E-mail:* gina.buldra@roswell.enmu.edu
*Med Dir:* Dan Raes, MD

# New York

**Genesee Community College**
Respiratory Therapist (Advanced) Prgm
1 College Rd
Batavia, NY 14020-1519
www.genesee.edu
*Prgm Dir:* Ronald M Jacobs, MBA RRT-NPS
*Tel:* 585 343-0055, Ext 6633   *Fax:* 585 343-0433
*E-mail:* rmjacobs@genesee.edu
*Med Dir:* Peter J Papadakos, MD

**Long Island University - Brooklyn Campus**
Respiratory Therapist (Advanced) Prgm
University Plaza
Brooklyn, NY 11201
*Prgm Dir:* Thomas J Johnson, RRT
*Tel:* 718 488-1492   *Fax:* 718 488-1432
*E-mail:* tjohnson@liu.edu
*Med Dir:* Albert Heurich, MD

**Nassau Community College**
Respiratory Therapist (Advanced) Prgm
One Education Dr
Garden City, NY 11530
www.ncc.edu
*Prgm Dir:* Warren Hostetter, RRT
*Tel:* 516 572-7560   *Fax:* 516 572-7565
*E-mail:* hostetw@ncc.edu
*Med Dir:* Martin M Moskowitz, MD

**CUNY Borough of Manhattan Community Coll**
Respiratory Therapist (Advanced) Prgm
199 Chambers St
New York, NY 10007
*Prgm Dir:* Everett W Flannery, MPS RRT
*Tel:* 212 346-8731   *Fax:* 212 346-8738
*E-mail:* eflannery@bmcc.cuny.edu
*Med Dir:* Paul Goldiner, MD

**New York University**
Respiratory Therapist (Advanced) Prgm
11 W 42nd St, Rm 518
New York, NY 10036
www.nyu.edu
*Prgm Dir:* Carole Smith, RRT
*Tel:* 212 998-1212
*Med Dir:* Vincent Donnabella, MD

**Molloy College**
Respiratory Therapist (Advanced) Prgm
1000 Hempstead Ave
PO Box 5002
Rockville Centre, NY 11571-5002
www.molloy.edu
*Prgm Dir:* Robert Tralongo, MBA RRT-NPS
*Tel:* 516 678-5000, Ext 6337   *Fax:* 516 256-2252
*E-mail:* rtralongo@molloy.edu
*Med Dir:* Joseph Genovese, DO

**Stony Brook University**
Respiratory Therapist (Advanced) Prgm
Sch of Health Tech and Management
Stony Brook, NY 11794-8203
*Prgm Dir:* James A Ganetis, MS RRT
*Tel:* 631 444-3180   *Fax:* 631 444-7621
*E-mail:* kaxton@epo.hsc.sunysb.edu
*Med Dir:* Gerald Smaldone, MD PhD

**Onondaga Community College**
Respiratory Therapist (Advanced) Prgm
Rte 173
Syracuse, NY 13215
www.sunyocc.edu
*Prgm Dir:* Daniel Cleveland, RRT BS
*Tel:* 315 498-2458   *Fax:* 315 498-2751
*E-mail:* clevelad@sunyocc.edu
*Med Dir:* Russell Acevedo, MD

**SUNY Upstate Medical University**
Respiratory Therapist (Advanced) Prgm
750 E Adams St
Syracuse, NY 13210
www.upstate.edu/chp/csrc
*Prgm Dir:* Joseph G Sorbello, MSEd RRT RT
*Tel:* 315 464-5580   *Fax:* 315 464-6876
*E-mail:* sorbellj@upstate.edu
*Med Dir:* Edward D Sivak, MD

**Hudson Valley Community College**
Respiratory Therapist (Advanced) Prgm
80 Vandenburgh Ave
Troy, NY 12180
www.hvcc.edu
*Prgm Dir:* Patricia G Hyland, MEd RRT
*Tel:* 518 629-7454   *Fax:* 518 629-7594
*E-mail:* hylanpat@hvcc.edu
*Med Dir:* Tom Smith, MD

**Mohawk Valley Community College**
Respiratory Therapist (Advanced) Prgm
1101 Sherman Dr
Utica, NY 13501
www.mvcc.edu/Respcare
*Prgm Dir:* Lorie Phillips, MS RRT
*Tel:* 315 792-5664   *Fax:* 315 731-5855
*E-mail:* lphillips@mvcc.edu
*Med Dir:* Michael Bauer, MD

**Westchester Community College**
Respiratory Therapist (Advanced) Prgm
75 Grasslands Rd
Valhalla, NY 10595
*Prgm Dir:* Jose Quinones, MS RRT
*Tel:* 914 606-6883   *Fax:* 914 606-7832
*E-mail:* jose.quinones@sunywcc.edu
*Med Dir:* George Maguire, MD

**Erie Community College - North Campus**
Respiratory Therapist (Advanced) Prgm
6205 Main St
Williamsville, NY 14221
*Prgm Dir:* James J Bierl, MS RRT NPS
*Tel:* 716 851-1531   *Fax:* 716 851-1429
*E-mail:* bierl@ecc.edu
*Med Dir:* Eric TenBrock, MD

# North Carolina

### Stanly Community College
Respiratory Therapist (Advanced) Prgm
141 College Dr
Albemarle, NC 28001
*Prgm Dir:* Tammy P Crump, PhD RRT
*Tel:* 704 991-0267, Ext 267    *Fax:* 704 991-0110
*E-mail:* crumptp@stanly.edu
*Med Dir:* William S Miles, MD

### Central Piedmont Community College
Respiratory Therapist (Advanced) Prgm
PO Box 35009
Charlotte, NC 28235
www.cpcc.edu
*Prgm Dir:* Brian Stearns, RRT
*Tel:* 704 330-6274    *Fax:* 704 330-6131
*E-mail:* brian.stearns@cpcc.edu
*Med Dir:* Toan Huynh, MD

### Durham Technical Community College
Respiratory Therapist (Advanced) Prgm
1637 Lawson St, Drawer 11307
Durham, NC 27703
www.durhamtech.edu
*Prgm Dir:* Richard D Miller, PhD RRT RCP
*Tel:* 919 686-3643    *Fax:* 919 686-3693
*E-mail:* millerr@durhamtech.edu
*Med Dir:* James R Yankaskas, MD

### Fayetteville Technical Community College
Respiratory Therapist (Advanced) Prgm
2201 Hull Rd
Fayetteville, NC 28303
*Prgm Dir:* Ruth A Baldwin, MA RRT
*Tel:* 910 678-8316    *Fax:* 910 678-8500
*E-mail:* baldwinr@faytechcc.edu
*Med Dir:* Albert Curseen, MD

### Pitt Community College
Respiratory Therapist (Advanced) Prgm
PO Drawer 7007
Greenville, NC 27834
*Prgm Dir:* Donna V Neal, PhD RRT RCP
*Tel:* 252 493-7378    *Fax:* 252 321-4451
*E-mail:* dneal@email.pittcc.edu
*Med Dir:* Robert Shaw, MD

### Catawba Valley Community College
Respiratory Therapist (Advanced) Prgm
2550 Hwy 70 SE
Hickory, NC 28602
www.cvcc.edu
*Prgm Dir:* Catherine A Bitsche, MA RRT
*Tel:* 828 327-7000, Ext 4391    *Fax:* 828 327-7276
*E-mail:* cbitsche@cvcc.edu
*Med Dir:* John Dew, MD

### Robeson Community College
Respiratory Therapist (Advanced) Prgm
PO Box 1420
Lumberton, NC 28359
*Prgm Dir:* Kelli Heustess, RCP RRT CPFT
*Tel:* 910 272-3400    *Fax:* 910 272-3416
*E-mail:* kheustes@robeson.cc.nc.us
*Med Dir:* Charles R Beasley, MD

### Carteret Community College
Respiratory Therapist (Advanced) Prgm
3505 Arendell St
Morehead City, NC 28557
www.carteret.edu
*Prgm Dir:* Laurie A Freshwater, RRT RPFT
*Tel:* 252 222-6281    *Fax:* 252 222-6074
*E-mail:* lap@carteret.edu
*Med Dir:* Terrence Goodman, MD

### Sandhills Community College
Respiratory Therapist (Advanced) Prgm
3395 Airport Rd
Pinehurst, NC 28374
www.sandhills.edu
*Prgm Dir:* William L Croft, MS RRT-NPS RCP
*Tel:* 910 695-3836    *Fax:* 910 692-6918
*E-mail:* croftb@sandhills.edu
*Med Dir:* Farrell Collins, MD

### Edgecombe Community College
Respiratory Therapist (Advanced) Prgm
225 Tarboro St
Rocky Mount, NC 27801
www.edgecombe.edu
*Prgm Dir:* Ralph D Webb, BAS RRT RCP
*Tel:* 252 446-0436, Ext 330    *Fax:* 252 985-2212
*E-mail:* webbrd@edgecombe.edu
*Med Dir:* Donald Rabil, MD

### Southwestern Community College
Respiratory Therapist (Advanced) Prgm
447 College Dr
Sylva, NC 28779-9578
*Prgm Dir:* Terri McQuiddy
*Tel:* 828 586-4091, Ext 459    *Fax:* 828 586-3381
*E-mail:* tmcquiddy@southwesterncc.edu
*Med Dir:* Harry G Lipham, MD

### Rockingham Community College
Respiratory Therapist (Advanced) Prgm
PO Box 38
Wentworth, NC 27375-0038
*Prgm Dir:* Thomas W Harding, MS RRT RPFTRCP
*Tel:* 336 342-4261, Ext 2339    *Fax:* 336 634-1253
*E-mail:* hardingt@rockinghamcc.edu
*Med Dir:* Edward L Hawkins, MD FCCP

### Forsyth Technical Community College
Respiratory Therapist (Advanced) Prgm
2100 Silas Creek Pkwy
302A Bob Greene Hall
Winston-Salem, NC 27103-5197
www.forsythtech.edu
*Prgm Dir:* Perry W Sheppard, MEd RRT-NPS RPFT RCP
*Tel:* 336 734-7427    *Fax:* 336 734-7444
*E-mail:* psheppard@forsythtech.edu
*Med Dir:* Barry Sigal, MD

# North Dakota

### University of Mary/St Alexius Medical Ctr
Respiratory Therapist (Advanced) Prgm
900 E Broadway, PO Box 5510
Bismarck, ND 58502
http://st.alexius.org/about_stas/allied_schools/
    respiratory/index.html
*Prgm Dir:* Wilmer D Beachey, PhD RRT
*Tel:* 701 530-7757    *Fax:* 701 530-7701
*E-mail:* wbeachey@primecare.org
*Med Dir:* James A Hughes, MD, FAACP

### NDSU/Merit Care Hospital Consortium
Respiratory Therapist (Advanced) Prgm
MeritCare Hospital
PO Box MC
Fargo, ND 58122-0207
www.ndsu.edu/rc
*Prgm Dir:* Gary Brown, BA RRCP RRT
*Tel:* 701 234-6147    *Fax:* 701 234-6942
*E-mail:* gary.brown@meritcare.com
*Med Dir:* Patrick Stoy, MD

# Ohio

### University of Akron
Respiratory Therapist (Advanced) Prgm
302 E Buchtel Mall
Akron, OH 44325-3702
*Prgm Dir:* LaVerne Yousey, MS RRT
*Tel:* 330 972-7906    *Fax:* 330 972-2016
*E-mail:* laverne@uakron.edu
*Med Dir:* Bradley R Martin, MD

### Stark State College of Technology
Respiratory Therapist (Advanced) Prgm
6200 Frank Ave NW
Canton, OH 44720-7299
www.starkstate.edu
*Prgm Dir:* Peter R Castillo, MEd RRT
*Tel:* 330 966-5458, Ext 4311    *Fax:* 330 966-6586
*E-mail:* pcastillo@starkstate.edu
*Med Dir:* Antonio Lazcano, MD

### Collins Career Center
*Cosponsor:* Marshall Community & Technical College
Respiratory Therapist (Advanced) Prgm
11627 State Route 243
Chesapeake, OH 45619
www.collins-cc.k12.oh.us
*Prgm Dir:* Keith Terry, BS RRT RN
*Tel:* 740 867-6641, Ext 411    *Fax:* 740 867-9626
*E-mail:* KATerry@collins-cc.k12.oh.us
*Med Dir:* Rahul D Patil, MD MS MBBS

### Cincinnati State Tech & Comm College
Respiratory Therapist (Advanced) Prgm
3520 Central Pkwy
Cincinnati, OH 45223
*Prgm Dir:* Debra Lierl, MEd RRT
*Tel:* 513 569-1690    *Fax:* 513 569-1559
*E-mail:* debra.lierl@cincinnatistate.edu
*Med Dir:* Peter Enyeart, MD

### Columbus State Community College
Respiratory Therapist (Advanced) Prgm
550 E Spring St
Columbus, OH 43215
*Prgm Dir:* Susan L Donohue, MEd RCP RRT
*Tel:* 614 287-2633    *Fax:* 614 287-6080
*E-mail:* sdonohue@cscc.edu
*Med Dir:* Thomas Boes, MD

### Ohio State University
Respiratory Therapist (Advanced) Prgm
431 Atwell Hall
453 W Tenth Ave
Columbus, OH 43210
http://amp.osu.edu/RT
*Prgm Dir:* F Herbert Douce, MS RRT-NPS RPFT
*Tel:* 614 292-8445    *Fax:* 614 292-0210
*E-mail:* douce.2@osu.edu
*Med Dir:* Jeffrey E Weiland, MD

### Sinclair Community College
Respiratory Therapist (Advanced) Prgm
444 W Third St
Dayton, OH 45402
*Prgm Dir:* Beth Zickefoose, BS RRT-NPS RPFT
*Tel:* 937 512-2550    *Fax:* 937 512-2058
*E-mail:* beth.zickefoose@sinclair.edu
*Med Dir:* James Murphy, MD

### Bowling Green State University
Respiratory Therapist (Advanced) Prgm
One University Dr
Huron, OH 44839-9791
www.firelands.bgsu.edu/programs/rt
*Prgm Dir:* Rod C Roark, MS RRT
*Tel:* 419 433-5560, Ext 20865    *Fax:* 419 433-9696
*E-mail:* rroark@bgnet.bgsu.edu
*Med Dir:* Anthony J Linz, DO

**Kettering College of Medical Arts**
Respiratory Therapist (Advanced) Prgm
3737 Southern Blvd
Kettering, OH 45429
www.kcma.edu
*Prgm Dir:* Nancy E Colletti, MS RRT CPFT RCIS
*Tel:* 937 298-3399, Ext 55644    *Fax:* 937 395-8635
*E-mail:* nancy.colletti@kcma.edu
*Med Dir:* George Burton, MD

**Lakeland Community College**
Respiratory Therapist (Advanced) Prgm
7700 Clocktower Dr
Kirtland, OH 44094-5198
www.lakelandcc.edu
*Prgm Dir:* Catherine J Kenny, MEd RRT
*Tel:* 440 525-7343    *Fax:* 440 525-7433
*E-mail:* ckenny@lakelandcc.edu
*Med Dir:* David Denholm, MD

**Rhodes State College**
Respiratory Therapist (Advanced) Prgm
4240 Campus Dr
Lima, OH 45804
*Prgm Dir:* Richard N Woodfield Jr, MS RRT
*Tel:* 419 995-8366    *Fax:* 419 995-8818
*E-mail:* woodfield.r@RhodesState.edu
*Med Dir:* Rick D Watson, MD

**North Central State College**
Respiratory Therapist (Advanced) Prgm
2441 Kenwood Circle, PO Box 698
Mansfield, OH 44901
www.ncstatecollege.edu
*Prgm Dir:* Robert A Slabodnick, MEd BS RRT/NPS
*Tel:* 419 755-4891, Ext 4891    *Fax:* 419 755-5630
*E-mail:* rslabod@ncstatecollege.edu
*Med Dir:* Henry Heinzmann, MD

**Washington State Community College**
Respiratory Therapist (Advanced) Prgm
710 Colegate Dr
Marietta, OH 45750
www.wscc.edu
*Prgm Dir:* James R Kinker, PhD RRT
*Tel:* 614 374-8716, Ext 1605    *Fax:* 614 373-7496
*E-mail:* rkinker@wscc.edu
*Med Dir:* Jon Paul Tipton, MD

**Cuyahoga Community College**
Respiratory Therapist (Advanced) Prgm
11000 Pleasant Valley Rd
Parma, OH 44130
www.tri-c.edu/respcare
*Prgm Dir:* David A Lucas, MS RRT
*Tel:* 216 987-5267    *Fax:* 216 987-5066
*E-mail:* david.lucas@tri-c.edu
*Med Dir:* Joseph A Sopko, MD

**Shawnee State University**
Respiratory Therapist (Advanced) Prgm
940 Second St
Portsmouth, OH 45662
*Prgm Dir:* Donald L Thomas, MS RRT
*Tel:* 740 351-3235    *Fax:* 740 351-3354
*E-mail:* dthomas@shawnee.edu
*Med Dir:* Elie M Saab, MD

**Jefferson Community College**
Respiratory Therapist (Advanced) Prgm
4000 Sunset Blvd
Steubenville, OH 43952
www.jcc.edu
*Prgm Dir:* Cynthia Carducci, MEd RRT
*Tel:* 740 264-5591, Ext 171    *Fax:* 740 264-9504
*E-mail:* ccarducci@jcc.edu
*Med Dir:* Thomas Walthers, MD

**University of Toledo**
Respiratory Therapist (Advanced) Prgm
College of Health Science & Human Service
2801 W Bancroft St
Toledo, OH 43606
www.utoledo.edu
*Prgm Dir:* Suzanne Spacek, MEd RRT-NPS CPFT
*Tel:* 419 530-4556    *Fax:* 419 530-4780
*E-mail:* suzanne.spacek@utoledo.edu
*Med Dir:* Robert A May, MD

**Youngstown State University**
Respiratory Therapist (Advanced) Prgm
One University Plaza
Youngstown, OH 44555
www.ysu.edu
*Prgm Dir:* Louis N Harris, EdD RRT
*Tel:* 330 941-1764    *Fax:* 330 941-2921
*E-mail:* lnharris01@ysu.edu
*Med Dir:* Tejinder S Bal, MD

# Oklahoma

**Rose State College**
Respiratory Therapist (Advanced) Prgm
6420 SE 15th St
Midwest City, OK 73110
www.rose.edu/cstudent/hsdiv/res_prog
*Prgm Dir:* Kathe Rowe, BSEd RRT-NPS
*Tel:* 405 733-7571    *Fax:* 405 736-0338
*E-mail:* krowe@rose.edu
*Med Dir:* Matthew J Britt, MD

**Francis Tuttle Technology Center**
*Cosponsor:* Oklahoma City Community College
Respiratory Therapist (Advanced) Prgm
12777 N Rockwell Ave
Oklahoma City, OK 73142-2789
*Prgm Dir:* Lezli Heyland, BS RRT
*Tel:* 405 717-4269    *Fax:* 405 717-4789
*E-mail:* lheyland@francistuttle.com
*Med Dir:* Stephen Adler, MD

**Tulsa Community College**
Respiratory Therapist (Advanced) Prgm
909 S Boston Ave
Tulsa, OK 74119
www.tulsacc.edu
*Prgm Dir:* Gary Persing, BS RRT
*Tel:* 918 595-7015    *Fax:* 918 595-7091
*E-mail:* gpersing@tulsacc.edu
*Med Dir:* Fred Garfinkel, MD

# Oregon

**Lane Community College**
Respiratory Therapist (Advanced) Prgm
4000 E 30th Ave
Eugene, OR 97405-0640
www.lanecc.edu
*Prgm Dir:* Roger H Hecht, AS RRT
*Tel:* 541 463-5624    *Fax:* 541 463-4151
*E-mail:* hechtr@lanecc.edu
*Med Dir:* Khuram Ameen, MD

**Mt Hood Community College**
Respiratory Therapist (Advanced) Prgm
26000 SE Stark St
Gresham, OR 97030
*Prgm Dir:* George Hicks, MS RRT
*Tel:* 503 491-7172    *Fax:* 503 491-6047
*E-mail:* hicksg@mhcc.edu
*Med Dir:* Alan Barker, MD

**Oregon Institute of Technology**
Respiratory Therapist (Advanced) Prgm
202 S Riverside
Medford, OR 97501
*Prgm Dir:* James L Hulse, MPH RRT-NPS RPFT
*Tel:* 541 245-7516    *Fax:* 541 774-4203
*E-mail:* james.hulse@oit.edu
*Med Dir:* John Ordal, MD

# Pennsylvania

**West Chester University**
*Cosponsor:* Bryn Mawr Hospital
Respiratory Therapist (Advanced) Prgm
130 S Bryn Mawr Ave
Bryn Mawr, PA 19010
http://health-sciences.wcupa.edu/health
*Prgm Dir:* Brian Kellar, MS RRT-NPS RPFT AE-C
*Tel:* 610 526-3347    *Fax:* 610 526-3821
*E-mail:* kellarb@mlhs.org
*Med Dir:* David Prince, MD

**Gannon University**
Respiratory Therapist (Advanced) Prgm
109 University Square
Erie, PA 16541
www.gannon.edu/resource/dept/respcare
*Prgm Dir:* Charles Cornfield, MS RRT
*Tel:* 814 871-5637    *Fax:* 814 871-5662
*E-mail:* cornfield@gannon.edu
*Med Dir:* Joseph Rowane, DO

**Gwynedd-Mercy College**
Respiratory Therapist (Advanced) Prgm
1325 Sumneytown Pike
PO Box 901
Gwynedd Valley, PA 19437
www.gmc.edu
*Prgm Dir:* William F Galvin, MSEd RRT CPFT AE-C
*Tel:* 215 641-5536    *Fax:* 215 641-5559
*E-mail:* Galvin.w@gmc.edu
*Med Dir:* Donald Peterson, MD

**Harrisburg Area Community College**
Respiratory Therapist (Advanced) Prgm
One HACC Dr
Harrisburg, PA 17110
*Prgm Dir:* Bradley A Leidich, MSEd RRT
*Tel:* 717 780-2315    *Fax:* 717 780-2551
*E-mail:* baleidic@hacc.edu
*Med Dir:* William M Anderson, III, MD

**University of Pittsburgh - Johnstown**
Respiratory Therapist (Advanced) Prgm
227 Krebs Hall
450 Schoolhouse Rd
Johnstown, PA 15904
www.upj.pitt.edu
*Prgm Dir:* Bruce J Colbert, MS RRT
*Tel:* 814 269-2960    *Fax:* 814 269-2044
*E-mail:* bcolbert@pitt.edu
*Med Dir:* Jayesh B Desai, MD

**Millersville University of Pennsylvania**
*Cosponsor:* Lancaster General Hospital
Respiratory Therapist (Advanced) Prgm
PO Box 1002
Millersville, PA 17551-0302
www.muweb.millersville.edu/~rtp
*Prgm Dir:* John M Hughes, RRT MEd AE-C
*Tel:* 717 735-3167    *Fax:* 717 544-5970
*E-mail:* muprt@comcast.net
*Med Dir:* Lee Duke, MD

**Western Sch of Hlth & Business - Pittsburgh**
Respiratory Therapist (Advanced) Prgm
One Monroeville Center, Ste 125
Monroeville, PA 15246
*Prgm Dir:* Michael Mehall, BS RRT
*Tel:* 412 373-9038    *Fax:* 412 281-0319
*E-mail:* mmehall@western-school.com
*Med Dir:* C Vaughn Strimlan, MD

**Luzerne County Community College**
Respiratory Therapist (Advanced) Prgm
1333 S Prospect St
Nanticoke, PA 18634
www.luzerne.edu
*Prgm Dir:* Christopher Tino, BS RRT
*Tel:* 570 740-0467    *Fax:* 570 740-0526
*E-mail:* ctino@luzerne.edu
*Med Dir:* Terrence Fagan, MD

**Community College of Philadelphia**
Respiratory Therapist (Advanced) Prgm
1700 Spring Garden St
Philadelphia, PA 19130
www.ccp.edu
*Prgm Dir:* Frank M Alsis, EdD RRT
*Tel:* 215 751-8423   *Fax:* 215 751-8937
*E-mail:* falsis@ccp.edu
*Med Dir:* Paul S Karlin, DO

**Comm Coll of Allegheny County**
Respiratory Therapist (Advanced) Prgm
808 Ridge Ave
Pittsburgh, PA 15212
www.ccac.edu
*Prgm Dir:* Thomas A Roop, RRT
*Tel:* 412 237-2607   *Fax:* 412 237-6579
*E-mail:* troop@ccac.edu
*Med Dir:* David Laman, MD

**Indiana University of Pennsylvania**
Respiratory Therapist (Advanced) Prgm
4800 Friendship Ave
Pittsburgh, PA 15224
*Prgm Dir:* William J Malley, MS RRT
*Tel:* 412 578-7000   *Fax:* 412 578-4651
*E-mail:* gmeyers@wpahs.org
*Med Dir:* Paul C Fiehler, MD

**Reading Area Community College**
Respiratory Therapist (Advanced) Prgm
Ten S Second St, PO Box 1706
Reading, PA 19603
www.racc.edu
*Prgm Dir:* James H Cicman, BHS RRT
*Tel:* 610 372-4721, Ext 5435   *Fax:* 610 372-4264
*E-mail:* jcicman@racc.edu
*Med Dir:* Joseph Mariglio, MD

**Mansfield University**
*Cosponsor:* Robert Packer Hospital
Respiratory Therapist (Advanced) Prgm
One Guthrie Square
Sayre, PA 18840
www.guthrie.org/RespRx
*Prgm Dir:* Larry Vosburgh, BS RRT
*Tel:* 570 882-4513   *Fax:* 570 882-6509
*E-mail:* vosburgh_larry@guthrie.org
*Med Dir:* James Walsh, MD

**Crozer - Chester Med Ctr**
*Cosponsor:* Delaware County Community College
Respiratory Therapist (Advanced) Prgm
One Medical Center Blvd
Upland, PA 19013
*Prgm Dir:* Patti L Curran, MEd RRT RPFT
*Tel:* 610 447-2440   *Fax:* 610 447-6353
*E-mail:* Patti.Curran@Crozer.org
*Med Dir:* Jerome Rudnitzky, MD

**York College of Pennsylvania**
*Cosponsor:* York Hospital
Respiratory Therapist (Advanced) Prgm
1001 S George St
York, PA 17405
*Prgm Dir:* Mark Simmons, MSEd RPFT RRT-NPS
*Tel:* 717 851-2464   *Fax:* 717 851-2934
*E-mail:* msimmons@wellspan.org
*Med Dir:* Richard J Murray, MD

# Rhode Island

**Community College of Rhode Island**
Respiratory Therapist (Advanced) Prgm
1762 Louisquisset Pike
Lincoln, RI 02865-4585
www.ccri.edu/alliedhealth
*Prgm Dir:* Joanne Jacobs, MA RRT AE-C
*Tel:* 401 333-7024   *Fax:* 401 333-7441
*E-mail:* jjacobs@ccri.edu
*Med Dir:* Thomas J Raimondo, DO

# South Carolina

**Trident Technical College**
Respiratory Therapist (Advanced) Prgm
PO Box 118067
7000 Rivers Ave
Charleston, SC 29423
www.tridenttech.edu
*Prgm Dir:* Ann R Moore, MS RRT CPFT
*Tel:* 843 574-6101   *Fax:* 843 574-6585
*E-mail:* ann.moore@tridenttech.edu
*Med Dir:* Alice Boylan, MD

**Midlands Technical College**
Respiratory Therapist (Advanced) Prgm
PO Box 2408
Columbia, SC 29202
*Prgm Dir:* Linda H Ackerman, MA RRT
*Tel:* 803 822-3433   *Fax:* 803 822-3079
*E-mail:* ackermanl@midlandstech.edu
*Med Dir:* J Daniel Love, MD

**Florence-Darlington Technical College**
Respiratory Therapist (Advanced) Prgm
PO Box 100548
Florence, SC 29501-0548
www.fdtc.edu
*Prgm Dir:* John A Evans, BS RRT
*Tel:* 843 661-8148   *Fax:* 843 661-8306
*E-mail:* john.evans@fdtc.edu
*Med Dir:* William M Hazelwood, MD

**Greenville Technical College**
Respiratory Therapist (Advanced) Prgm
PO Box 5616
Greenville, SC 29606
www.greenvilletech.com
*Prgm Dir:* Jim Woody, MinED RRT
*Tel:* 864 228-5073   *Fax:* 864 228-5079
*E-mail:* Jim.Woody@gvltec.edu
*Med Dir:* L Hayes, MD

**Piedmont Technical College**
Respiratory Therapist (Advanced) Prgm
PO Box 1467, Emerald Rd
Greenwood, SC 29647
*Prgm Dir:* Jane C Walker, BSN RN
*Tel:* 864 941-8526   *Fax:* 864 941-8684
*E-mail:* walker.j@ptc.edu
*Med Dir:* Daniel Robinson, MD

**Tri-County Technical College**
Respiratory Therapist (Advanced) Prgm
PO Box 587
Pendleton, SC 29670
www.tctc.edu
*Prgm Dir:* Tom Baxter, MHRD RRT
*Tel:* 864 646-1354   *Fax:* 864 646-1892
*E-mail:* tbaxter@tctc.edu
*Med Dir:* Ravi Chandran, MD

**Spartanburg Community College**
Respiratory Therapist (Advanced) Prgm
PO Drawer 4386
Spartanburg, SC 29305-4386
www.stcsc.edu
*Prgm Dir:* Randall W Anderson, RRT
*Tel:* 864 592-4938   *Fax:* 864 592-4881
*E-mail:* andersonr@stcsc.edu
*Med Dir:* Charles Fogarty, MD

# South Dakota

**Dakota State University**
Respiratory Therapist (Advanced) Prgm
Science Center
Madison, SD 57042-1799
www.dsu.edu
*Prgm Dir:* Bruce Feistner, MSS RRT
*Tel:* 605 322-8613   *Fax:* 605 322-6666
*E-mail:* bruce.feistner@dsu.edu
*Med Dir:* Ashraf Elshami, MD

# Tennessee

**Chattanooga State Technical Comm College**
Respiratory Therapist (Advanced) Prgm
4501 Amnicola Hwy
Chattanooga, TN 37406
*Prgm Dir:* Sharon Hall
*Tel:* 423 697-4772   *Fax:* 423 634-3071
*E-mail:* sharon.hall@chattanoogastate.edu
*Med Dir:* Suresh Enjeti, MD

**Columbia State Community College**
Respiratory Therapist (Advanced) Prgm
1665 Hampshire Pike
Columbia, TN 38402
www.columbiastate.edu/respiratory
*Prgm Dir:* R David Johnson
*Tel:* 931 540-2663   *Fax:* 931 540-2795
*E-mail:* david.johnson@columbiastate.edu
*Med Dir:* Erik Iversen, MD

**East Tennessee State University**
Respiratory Therapist (Advanced) Prgm
1000 West E St, ETSU Nave Center
Elizabethton, TN 37643
*Prgm Dir:* Douglas E Masini, EdD RPFT RRT FAARC
*Tel:* 423 547-4900   *Fax:* 423 547-4921
*E-mail:* masini@etsu.edu
*Med Dir:* Jeff Farrow, MD

**Volunteer State Community College**
Respiratory Therapist (Advanced) Prgm
1480 Nashville Pike
Gallatin, TN 37066-3188
www.volstate.edu
*Prgm Dir:* Cory E Martin, BS RRT
*Tel:* 888 335-8722, Ext 3349   *Fax:* 615 230-3224
*E-mail:* Cory.Martin@volstate.edu
*Med Dir:* Christopher Harris, MD

**Roane State Community College**
Respiratory Therapist (Advanced) Prgm
276 Patton Ln
Harriman, TN 37748
www.roanestate.edu
*Prgm Dir:* Lesha Hill, BS RRT
*Tel:* 865 539-6904   *Fax:* 865 539-6907
*E-mail:* hillla@roanestate.edu
*Med Dir:* Richard A Obenour, MD

**Jackson State Community College**
Respiratory Therapist (Advanced) Prgm
2046 N Parkway St
Jackson, TN 38301-3797
www.jscc.edu
*Prgm Dir:* Cathy K Garner, BS RRT-NPS
*Tel:* 731 424-3520, Ext 235   *Fax:* 731 425-9551
*E-mail:* cgarner@jscc.edu
*Med Dir:* Robert J Gilroy, Jr., MD

**Baptist College of Health Sciences**
Respiratory Therapist (Advanced) Prgm
1003 Monroe
Memphis, TN 38104
www.bchs.edu
*Prgm Dir:* Brian J Parker, MPH RRT-NPS RPFT AE-C
*Tel:* 901 572-2568   *Fax:* 901 572-2750
*E-mail:* brian.parker@bchs.edu
*Med Dir:* C Michael Smith, Jr., MD

**Concorde Career College - Memphis**
Respiratory Therapist (Advanced) Prgm
5100 Poplar Ave, Ste 132
Memphis, TN 38137
*Prgm Dir:* Michael Camp, BS RRT
*Tel:* 901 761-9494   *Fax:* 901 843-9442
*E-mail:* mcamp@concordecareer.edu
*Med Dir:* Amado Freire, MD

PROGRAMS

### Tennessee State University
Respiratory Therapist (Advanced) Prgm
3500 John A Merritt Blvd
Nashville, TN 37209
*Prgm Dir:* Thomas John, PhD RRT
*Tel:* 615 963-7420   *Fax:* 615 963-7422
*E-mail:* tjohn@tnstate.edu
*Med Dir:* Bijoy John, MD

# Texas

### Alvin Community College
Respiratory Therapist (Advanced) Prgm
3110 Mustang Rd
Alvin, TX 77511
www.alvincollege.edu
*Prgm Dir:* Diane Flatland, MS RRT-NPS CPFT LP
*Tel:* 281 756-3658   *Fax:* 281 756-3860
*E-mail:* dflatland@alvincollege.edu
*Med Dir:* Wayne K Hite, DO

### Amarillo College
Respiratory Therapist (Advanced) Prgm
PO Box 447
Amarillo, TX 79178
*Prgm Dir:* William A Young, MS RRT
*Tel:* 806 354-6058   *Fax:* 806 354-6076
*E-mail:* young-wa@actx.edu
*Med Dir:* Bruce Baker, MD

### Lamar Institute of Technology
Respiratory Therapist (Advanced) Prgm
PO Box 10061
Beaumont, TX 77710
www.lit.edu
*Prgm Dir:* Gwen Walden, BS RRT NPS CPFT RCP
*Tel:* 409 880-8852   *Fax:* 409 880-8955
*E-mail:* gwen.walden@lit.edu
*Med Dir:* N Jeff Alford, MD

### Univ Tx at Brownsville/Tx Southmost Coll
Respiratory Therapist (Advanced) Prgm
83 Ft Brown
Brownsville, TX 78520
*Prgm Dir:* John L McCabe, PhD RRT
*Tel:* 956 554-5010   *Fax:* 956 554-5012
*E-mail:* john.mccabe@utb.edu
*Med Dir:* Lorenzo Pelly, MD

### Del Mar College
Respiratory Therapist (Advanced) Prgm
101 Baldwin and Ayers Sts
Corpus Christi, TX 78404
*Prgm Dir:* Jeffrey T Watson, BA RRT
*Tel:* 361 698-2835   *Fax:* 361 698-1825
*E-mail:* jwatson@delmar.edu
*Med Dir:* William Burgin, Jr, MD

### ATI-Career Training
Respiratory Therapist (Advanced) Prgm
10003 Technology Blvd West
Dallas, TX 75220
*Prgm Dir:* Rudy Schatke, MS RRT RCP
*Tel:* 214 902-8191, Ext 201   *Fax:* 214 366-0881
*E-mail:* rschatke@atienterprises.edu
*Med Dir:* Peter Heidbrink, MD

### El Centro College
Respiratory Therapist (Advanced) Prgm
Main and Lamar Sts
Dallas, TX 75202
*Prgm Dir:* Gary L Peschka, MEd RRT RCP
*Tel:* 214 860-2279   *Fax:* 214 860-2268
*E-mail:* glp5547@dcccd.edu
*Med Dir:* Peter Heidbrink, MD

### University of Texas Medical Branch/SAHS
Respiratory Therapist (Advanced) Prgm
School of Allied Health Sciences
301 University Blvd
Galveston, TX 77555-1146
www.sahs.utmb.edu/programs/rc
*Prgm Dir:* Jon O Nilsestuen, PhD RRT FAARC
*Tel:* 409 772-5693   *Fax:* 409 772-3014
*E-mail:* jnilsest@utmb.edu
*Med Dir:* Aristides Koutrouvelis, MD

### Houston Community College
Respiratory Therapist (Advanced) Prgm
1900 Galen Dr
Houston, TX 77030
*Prgm Dir:* Romar S Reyes, MEd RRT
*Tel:* 713 718-7384   *Fax:* 713 718-7136
*E-mail:* romar.reyes@hccs.edu
*Med Dir:* Kenneth L Toppell, MD

### Texas Southern University
Respiratory Therapist (Advanced) Prgm
3100 Cleburne
Houston, TX 77004
*Prgm Dir:* Reginald G Allen
*Tel:* 713 313-7265, Ext 7377   *Fax:* 713 313-1094
*E-mail:* allen_rg@tsu.edu
*Med Dir:* Edward L Patten, MD

### Tarrant County College
Respiratory Therapist (Advanced) Prgm
828 Harwood Rd
Hurst, TX 76054
*Prgm Dir:* John D Hiser, MEd RRT FAARC
*Tel:* 817 515-6435   *Fax:* 817 515-6700
*E-mail:* john.hiser@tccd.edu
*Med Dir:* Woody V Kageler, MD

### Kingwood College - NHMCCD
Respiratory Therapist (Advanced) Prgm
20,000 Kingwood Dr
Kingwood, TX 77339
www.nhmccd.edu
*Prgm Dir:* Kenny P McCowen, RRT
*Tel:* 281 312-1608   *Fax:* 281 312-1490
*E-mail:* kmccowen@nhmccd.edu
*Med Dir:* Edward Flores, MD

### South Plains College
Respiratory Therapist (Advanced) Prgm
Reese Center
819 Gilbert Dr
Lubbock, TX 79416
www.southplainscollege.edu
*Prgm Dir:* Khris Segrist, BS RRT
*Tel:* 806 885-3048, Ext 4625   *Fax:* 806 885-5608
*E-mail:* ksegrist@southplainscollege.edu
*Med Dir:* Kenneth Nugent, MD

### Angelina College
Respiratory Therapist (Advanced) Prgm
PO Box 1768
Lufkin, TX 75902-1768
*Prgm Dir:* Michael Parks, MEd RRT
*Tel:* 936 633-5419
*E-mail:* mparks@angelina.edu
*Med Dir:* M J Thomas, MD

### Collin County Community College
Respiratory Therapist (Advanced) Prgm
2200 W University Dr
McKinney, TX 75071
www.ccccd.edu/rcp
*Prgm Dir:* David R Gibson, BS RRT
*Tel:* 972 548-6870   *Fax:* 972 548-6722
*E-mail:* dgibson@ccccd.edu
*Med Dir:* Timothy Chappell, MD

### Midland College
Respiratory Therapist (Advanced) Prgm
3600 N Garfield
Midland, TX 79705
www.midland.edu
*Prgm Dir:* Robert Weidmann, BS RPFT RRT-NPS RCP
*Tel:* 432 685-5549   *Fax:* 432 685-5575
*E-mail:* rweidmann@midland.edu
*Med Dir:* Gregory Bartha, MD

### Odessa College
Respiratory Therapist (Advanced) Prgm
201 W University Blvd
Odessa, TX 79764
www.odessa.edu
*Prgm Dir:* Jacquelyn R Sullivan, AAS RRT RPFT RPSGT
*Tel:* 432 335-6456   *Fax:* 432 335-6846
*E-mail:* jsullivan@odessa.edu
*Med Dir:* John D Bray, MD

### San Jacinto College Central
Respiratory Therapist (Advanced) Prgm
8060 Spencer Hwy, PO Box 2007
Pasadena, TX 77505-2007
*Prgm Dir:* Larry P Vandiver, MPH RRT
*Tel:* 281 476-1864   *Fax:* 281 478-2754
*E-mail:* lvandi@sjcd.edu
*Med Dir:* Alfred Maksoud, MD

### Univ of Texas Hlth Sci Ctr at San Antonio
Respiratory Therapist (Advanced) Prgm
7703 Floyd Curl Dr Mail Code 6248
San Antonio, TX 78229-3900
*Prgm Dir:* David Vines
*Tel:* 210 567-8850   *Fax:* 210 567-8852
*E-mail:* vines@uthscsa.edu
*Med Dir:* Jay I Peters, MD

### Texas State University-San Marcos
Respiratory Therapist (Advanced) Prgm
601 University Dr
Health Professions Bldg Rm 350A
San Marcos, TX 78666-4616
www.health.txstate.edu/RC
*Prgm Dir:* S Gregory Marshall, PhD RRT CRT RCP
*Tel:* 512 245-8243   *Fax:* 512 245-7978
*E-mail:* sm10@txstate.edu
*Med Dir:* Peter A Petroff, MD

### Temple College
Respiratory Therapist (Advanced) Prgm
2600 S First St
Temple, TX 76504
www.templejc.edu
*Prgm Dir:* William M Cornelius III, MHSM RRT-NPS
*Tel:* 254 298-8928   *Fax:* 254 298-8676
*E-mail:* bill.cornel@templejc.edu
*Med Dir:* William G Petersen, MD

### Tyler Junior College
Respiratory Therapist (Advanced) Prgm
PO Box 9020
Tyler, TX 75711
*Prgm Dir:* Phyllis Brunner, BS RRT
*Tel:* 903 510-2472   *Fax:* 903 510-2592
*E-mail:* pbru@tjc.edu
*Med Dir:* James Stocks, MD

### Victoria College
Respiratory Therapist (Advanced) Prgm
2200 E Red River
Victoria, TX 77901
*Prgm Dir:* Chris E Kallus, MEd RRT RCP
*Tel:* 361 572-6491   *Fax:* 361 582-2542
*E-mail:* chris.kallus@victoriacollege.edu
*Med Dir:* Bruce E Wheeler, MD

**Weatherford College**
Respiratory Therapist (Advanced) Prgm
225 College Park Dr
Weatherford, TX 76086
*Prgm Dir:* Tonya Edwards, BS RRT RPFT RCP
*Tel:* 817 594-5471, Ext 452   *Fax:* 817 598-6455
*E-mail:* edwardst@wc.edu
*Med Dir:* Roger Gleason, MD

**Midwestern State University**
Respiratory Therapist (Advanced) Prgm
3410 Taft Blvd
Wichita Falls, TX 76308
*Prgm Dir:* Ann Medford, MA RRT
*Tel:* 940 397-4653   *Fax:* 940 397-4513
*E-mail:* ann.medford@mwsu.edu
*Med Dir:* Lowell Harvey, MD

# Utah

**Weber State University**
Respiratory Therapist (Advanced) Prgm
3904 University Circle
Ogden, UT 84408-3904
http://weber.edu/dchp.xml
*Prgm Dir:* Michell Oki, RRT RPFT NPS RPsgT MPAcc
*Tel:* 801 626-6835   *Fax:* 801 626-7075
*E-mail:* moki@weber.edu
*Med Dir:* Christopher Anderson, MD

# Vermont

**Vermont Technical College**
Respiratory Therapist (Advanced) Prgm
201 Lawrence Place
Box 1A
Williston, VT 05495
www.vtc.edu
*Prgm Dir:* Faye Bacon, MEd RRT
*Tel:* 802 879-5972   *Fax:* 802 879-5645
*E-mail:* fbacon@vtc.edu
*Med Dir:* Gerald Davis, MD

# Virginia

**Mountain Empire Community College**
Respiratory Therapist (Advanced) Prgm
3441 Mountain Empire Rd
Big Stone Gap, VA 24219
*Prgm Dir:* Michael W Cook, MA RRT
*Tel:* 540 523-2400, Ext 277   *Fax:* 540 523-5458
*E-mail:* mcook@me.vccs.edu
*Med Dir:* Joseph F Spiddy, MD

**Central Virginia Community College**
Respiratory Therapist (Advanced) Prgm
3506 Wards Rd
Lynchburg, VA 24502
www.cv.cc.va.us
*Prgm Dir:* Martha N Crawley, MEd RRT
*Tel:* 804 832-7685   *Fax:* 804 832-7835
*E-mail:* crawleym@cvcc.vccs.edu
*Med Dir:* Michael G Milam, MD

**Southwest Virginia Community College**
Respiratory Therapist (Advanced) Prgm
Box SVCC
Richlands, VA 24641-1101
*Prgm Dir:* Joseph S DiPietro, PhD RRT
*Tel:* 276 964-7306   *Fax:* 276 964-7715
*E-mail:* Joe.DiPietro@sw.edu
*Med Dir:* Randy Forehand, MD

**J Sargeant Reynolds Community College**
Respiratory Therapist (Advanced) Prgm
PO Box 85622
Richmond, VA 23285
www.reynolds.edu
*Prgm Dir:* Donald K O'Donohue, RRT
*Tel:* 804 523-5375   *Fax:* 804 786-5298
*E-mail:* dodonohue@reynolds.edu
*Med Dir:* Clifton L Parker, MD

**Northern Virginia Community College**
Respiratory Therapist (Advanced) Prgm
6699 Springfield Center Dr
Springfield, VA 22150
www.nvcc.edu/medical/health/respiratory
*Prgm Dir:* Kathy Grilliot, MS Ed RRT-NPS
*Tel:* 703 822-6563   *Fax:* 703 822-6619
*E-mail:* kgrilliot@nvcc.edu
*Med Dir:* James Lamberti, MD

**Tidewater Community College**
Respiratory Therapist (Advanced) Prgm
1700 College Crescent
Virginia Beach, VA 23456
*Prgm Dir:* Gary Cross, BS RRT
*Tel:* 757 822-7263   *Fax:* 757 427-1338
*E-mail:* gcross@tcc.edu
*Med Dir:* Ignacio Ripoll, MD

**Shenandoah University**
Respiratory Therapist (Advanced) Prgm
1775 N Sector Ct
Winchester, VA 22601-5195
www.su.edu
*Prgm Dir:* William A O'Neill, MA RRT
*Tel:* 540 665-5516   *Fax:* 540 665-5519
*E-mail:* woneill@su.edu
*Med Dir:* Thomas M Murphy, MD

# Washington

**Highline Community College**
Respiratory Therapist (Advanced) Prgm
2400 S 240th St
Des Moines, WA 98198-9800
*Prgm Dir:* Robert W Hirnle
*Tel:* 206 878-3710, Ext 3471   *Fax:* 206 870-3780
*E-mail:* bhirnle@highline.edu
*Med Dir:* Stefanie Nunez, MD

**Seattle Central Community College**
Respiratory Therapist (Advanced) Prgm
1701 Broadway, 2BE3210
Seattle, WA 98122
www.sccd.ctc.edu
*Prgm Dir:* Robert IC McCallum, BS RRT
*Tel:* 206 587-4056   *Fax:* 206 587-6337
*E-mail:* rmccallum@sccd.ctc.edu
*Med Dir:* Bill Watts, MD

**Spokane Community College**
Respiratory Therapist (Advanced) Prgm
N 1810 Greene St
Spokane, WA 99217
www.scc.spokane.edu/?resp
*Prgm Dir:* Gary White, MEd RRT RPFT
*Tel:* 509 533-7310   *Fax:* 509 533-8621
*E-mail:* gwhite@scc.spokane.edu
*Med Dir:* James C Bonvallet, MD

**Tacoma Community College**
Respiratory Therapist (Advanced) Prgm
6501 S 19th St
Tacoma, WA 98466
*Prgm Dir:* Ken Lizzi, MPH RRT
*Tel:* 253 566-5231   *Fax:* 253 566-5273
*E-mail:* klizzi@tcc.ctc.edu
*Med Dir:* James R Taylor, MD

# West Virginia

**Carver Career Technical Center**
Respiratory Therapist (Advanced) Prgm
4799 Midland Dr
Charleston, WV 25306
*Prgm Dir:* Donna Peters, BA RRT
*Tel:* 304 348-1965, Ext 115   *Fax:* 304 348-1938
*Med Dir:* Edward Grey, MD

**West Virginia Northern Community College**
Respiratory Therapist (Advanced) Prgm
1704 Market St
Wheeling, WV 26003
*Prgm Dir:* Ralph C Lucki, MA RRT
*Tel:* 304 233-5900, Ext 5511   *Fax:* 304 232-0965
*E-mail:* rlucki@northern.wvnet.edu
*Med Dir:* Richard Ryncarz, MD

**Wheeling Jesuit University**
Respiratory Therapist (Advanced) Prgm
316 Washington Ave
Wheeling, WV 26003
*Prgm Dir:* Marybeth Emmerth, MS RRT CPFT
*Tel:* 304 243-2208   *Fax:* 304 243-6246
*E-mail:* memmerth@wju.edu
*Med Dir:* Richard E Ryncarz, MD

# Wisconsin

**Northeast Wisconsin Technical College**
Respiratory Therapist (Advanced) Prgm
2740 W Mason St, PO Box 19042
Green Bay, WI 54307
*Prgm Dir:* Katherine A Schlitz
*Tel:* 920 498-5533   *Fax:* 920 498-5673
*E-mail:* katherine.schlitz@nwtc.edu
*Med Dir:* John Andrews, MD

**Western Technical College**
Respiratory Therapist (Advanced) Prgm
304 N Sixth St, PO Box 908
La Crosse, WI 54602-0908
www.wwtc.edu/rcp
*Prgm Dir:* Robert A Milisch, MEd RRT
*Tel:* 608 785-9244   *Fax:* 608 785-9087
*E-mail:* milischr@wwtc.edu
*Med Dir:* Edward Winga, MD

**Madison Area Technical College**
Respiratory Therapist (Advanced) Prgm
3550 Anderson St
Madison, WI 53704
www.matcmadison.edu
*Prgm Dir:* Gail Walker, RRT
*Tel:* 608 246-6698   *Fax:* 608 246-6013
*E-mail:* gwalker@matcmadison.edu
*Med Dir:* John P Schilling, MD

**Mid-State Technical College**
Respiratory Therapist (Advanced) Prgm
2600 W 5th St
Marshfield, WI 54449
www.mstc.edu
*Prgm Dir:* Scott S Osborne, RRT MEd
*Tel:* 715 389-7033   *Fax:* 715 389-2864
*E-mail:* Scott.Osborne@mstc.edu
*Med Dir:* John A Campbell, MD, PhD

**Milwaukee Area Technical College**
Respiratory Therapist (Advanced) Prgm
700 W State St
Milwaukee, WI 53233
*Prgm Dir:* Mark J Hoffman, MS RRT
*Tel:* 414 297-7130   *Fax:* 414 297-7990
*E-mail:* HoffmanM@matc.edu
*Med Dir:* Randolph Lipchik, MD

| Programs* | Class Capacity | Begins | Length (months) | Award | Res. Tuition | Non-res. Tuition | Stipend | Offers:‡ 1 | 2 | 3 |
|---|---|---|---|---|---|---|---|---|---|---|
| **Alabama** | | | | | | | | | | |
| University of Alabama at Birmingham | 30 | Aug | 24 | Cert, BS | $5,376 | $10,752 | | | | • |
| University of South Alabama (Mobile) | 30 | Aug | 24 | BS | $6,930 | $10,400 | | | | |
| Wallace State Community College (Hanceville) | 35 | Aug | 21 | AAS | $2,160 | $4,320 | | | | |
| **Arizona** | | | | | | | | | | |
| GateWay Community College (Phoenix) | 30 | Aug Jan | 24 | AAS | $2,400 | $10,320 | | | | |
| Long Technical College (Phoenix) | 20 | 14x/yr | 18, 24 | AOS | $14,315 | $14,315 | | • | | |
| Pima Community College (Tucson) | 32 | Aug | 22 | AAS | $1,750 | $6,752 | | | | • |
| Pima Medical Institute - Mesa | 30 | Feb Jun Oct | 24 | Dipl, AS | $22,900 | $0 | | | | • |
| **Arkansas** | | | | | | | | | | |
| NorthWest Arkansas Community College (Bentonville) | 20 | Aug | 21 | AAS | $5,250 | $7,900 | | • | | • |
| Southeast Arkansas College (Pine Bluff) | 16 | Aug | 16 | AAS | $2,300 | $4,600 | | | | |
| University of Arkansas for Medical Sciences (Texarkana) | 12 | Aug | 21, 33 | BS | $5,508 | $16,524 | | | | |
| **California** | | | | | | | | | | |
| Butte College (Oroville) | 24 | Aug | 22 | Cert, AS | $772 | $6,272 | | | | • |
| Crafton Hills College (Yucaipa) | 30 | Aug | 12, 24 | Cert, AS | $1,846 | $12,567 | | | | |
| East Los Angeles College (Monterey Park) | 45 | Jun Jul | 24 | Cert, AS | $26 | $180 | | • | | • |
| Foothill Community College (Los Altos Hills) | 25 | Sep | 22 | AS | $750 | $6,200 | | | | • |
| Grossmont College (El Cajon) | 45 | Aug | 18 | AS | $663 | $3,850 | | | | • |
| Loma Linda University | 20 | Sep | 21 | Cert, BS | $21,534 | $21,534 | | | | • |
| Los Angeles Valley College (Valley Glen) | 27 | Varies | 24 | Cert, AS | $28 | $3,920 | | • | | |
| Orange Coast College (Costa Mesa) | 30 | Aug | 22 | AS | $754 | $4,466 | | | | |
| San Joaquin Valley College - Bakersfield | 24 | Varies | 18 | AS | $11,685 | $0 | | | | |
| San Joaquin Valley College - Visalia | 24 | Varies | 18 | AS | $11,445 | $0 | | | | |
| **Colorado** | | | | | | | | | | |
| Pima Medical Institute - Denver | 24 | Apr Aug Dec | 22 | AOS | $24,340 | $24,340 | | | | • |
| Pueblo Community College | 20 | Aug | 20 | AAS | $2,118 | $8,586 | | | | |
| **Connecticut** | | | | | | | | | | |
| Manchester Community College | 20 | Sep | 20 | AS | $3,252 | $4,578 | | • | | |
| Naugatuck Valley Community College (Waterbury) | 37 | Sep | 24 | AS | $2,536 | $7,568 | | • | • | • |
| Norwalk Hospital/Norwalk Community College | 15 | Sep | 24 | AS | $2,232 | $6,696 | | | | • |
| **Delaware** | | | | | | | | | | |
| Delaware Tech & Comm Coll - Owens Campus (Georgetown) | 20 | Jun | 24 | AAS | $2,934 | $7,335 | | | | |
| **Florida** | | | | | | | | | | |
| Daytona Beach Community College | 28 | Aug | 21 | AAS | $2,650 | $10,010 | | | | • |
| Edison College (Ft Myers) | 25 | Aug | 22 | AS | $2,925 | $10,252 | | • | • | |
| Hillsborough Community College (Tampa) | 25 | Aug | 21 | AS, AAS | $3,200 | $10,202 | | | | • |
| Miami Dade College | 50 | Aug | 24 | AS | $2,112 | $7,824 | | | | • |
| Palm Beach Community College (Palm Beach Gardens) | 25 | Aug | 24 | AS | $2,527 | $9,006 | | | | • |
| Seminole Community College (Sanford) | 24 | Aug | 20 | AS | $3,000 | $8,000 | | • | | |
| University of Central Florida (Orlando) | 20 | Aug | 20 | BS | $2,982 | $15,488 | | | | • |
| Valencia Community College (Orlando) | 24 | Jan | 20 | AS | $5,244 | $19,912 | | • | | |
| **Georgia** | | | | | | | | | | |
| Athens Technical College | 16 | Sep | 24 | AAS | $1,812 | $3,624 | | | | • |
| Augusta Technical College | 24 | Jul | 24 | AAT | $1,444 | $2,888 | | | | |
| Coosa Valley Technical College (Rome) | 24 | Jul | 24 | AAT | $1,500 | $3,000 | | | | • |
| Darton College (Albany) | 20 | Aug | 21 | AS | $4,000 | $0 | | • | | • |
| Georgia State University (Atlanta) | 50 | Aug | 22 | BS | $5,006 | $20,015 | | | | • |
| Griffin Technical College | 20 | Jul | 14 | AS | $1,528 | $3,056 | | | | |
| Gwinnett Technical College (Lawrenceville) | 20 | Mar | 15 | AAT | $1,900 | $3,388 | | | | |
| Southwest Georgia Technical College (Thomasville) | 20 | Jul | 24 | AAS | $1,208 | $2,416 | | • | | |
| **Hawaii** | | | | | | | | | | |
| Kapi'olani Community College (Honolulu) | 16 | Jul | 23 | AS | $956 | $5,732 | | • | | • |
| **Idaho** | | | | | | | | | | |
| Boise State University | 24 | Aug | 33 | AS, BS | $4,700 | $11,000 | | • | | |
| **Illinois** | | | | | | | | | | |
| Kankakee Community College | 16 | Jan | 24 | AAS | $2,100 | $6,195 | | | | • |
| Kaskaskia College (Centralia) | 24 | Aug | 24 | AAS | $4,028 | $7,676 | | | | |
| Moraine Valley Community College (Palos Hills) | 32 | Aug | 21 | AAS | $2,664 | $7,659 | | | | • |
| Olive Harvey College (Chicago) | 20 | Aug | 24 | Dipl, AAS | $2,448 | $5,530 | | | | • |
| Rock Valley College (Rockford) | 15 | Aug | 22 | AAS | $2,450 | $8,600 | | • | | |
| Southern Illinois Univ at Carbondale | 25 | Aug | 16, 28 | Cert, AAS | $7,205 | $18,273 | | | | • |
| Southwestern Illinois College (Belleville) | 25 | Aug | 22 | Dipl, AAS | $4,473 | $11,147 | | | | • |
| St John's Hospital (Springfield) | 6 | Aug | 21 | AGE | $1,400 | $1,400 | | | | • |

*Data are shown only for programs that completed the 2006 AMA Survey of Health Professions Education Programs
‡Key to Offers: 1: Evening or weekend classes; 2: Non-English instruction; 3: Cultural competence instruction

## Respiratory Therapist (Advanced)

| Programs* | Class Capacity | Begins | Length (months) | Award | Res. Tuition | Non-res. Tuition | Stipend | Offers:‡ 1 | 2 | 3 |
|---|---|---|---|---|---|---|---|---|---|---|
| Triton College (River Grove) | 30 | Aug | 22 | AAS | $1,988 | $4,508 | | | | • |
| **Indiana** | | | | | | | | | | |
| Clarian Health Partners Inc (Indianapolis) | 30 | Aug | 48 | BS | $173 | $345 | | | | • |
| Ivy Tech Community College - Lafayette | 24 | Aug | 21 | AS | $3,172 | $6,413 | | • | | |
| Ivy Tech Community College NE - Ft Wayne | 28 | Aug | 21 | AS | $3,171 | $6,412 | | • | | |
| Ivy Tech Community College NW - Gary (Michigan City) | 20 | Aug | 20 | AS | $6,932 | $14,100 | | | | • |
| Ivy Tech Community College of Indiana (Indianapolis) | 26 | May | 24, 31 | AS | $2,894 | $5,268 | | | | |
| University of Southern Indiana (Evansville) | 14 | Aug | 24 | AS | $5,500 | $13,111 | | | | |
| **Iowa** | | | | | | | | | | |
| Des Moines Area Community College (Ankeny) | 24 | Aug | 23 | AAS | $3,950 | $7,900 | | | | • |
| Hawkeye Community College (Waterloo) | 24 | Aug | 22 | AAS | $8,058 | $16,117 | | | | |
| Southeastern Community College (West Burlington) | 20 | Aug | 22 | AAS | $6,000 | $7,000 | | | | • |
| **Kansas** | | | | | | | | | | |
| Bethany Med Ctr/Kansas City Kansas Comm Coll | 15 | Aug Jan | 24 | AAS | $3,850 | $6,200 | | • | | • |
| Johnson County Community College (Overland Park) | 20 | Jun | 24 | AAS | $2,400 | $5,438 | | • | | • |
| Labette Community College (Parsons) | 25 | Aug | 22 | Dipl, AAS | $2,800 | $5,400 | | | | • |
| Newman University (Wichita) | 20 | Aug | 28 | ASHS | $15,822 | $15,822 | | | | • |
| Seward County Community College (Liberal) | 18 | Aug | 24 | AAS | $4,650 | $6,375 | | | | |
| University of Kansas Medical Center (Kansas City) | 24 | Aug | 22 | BS | $6,890 | $18,103 | | | | |
| Washburn University (Topeka) | 18 | Aug | 22 | AS | $6,460 | $8,160 | | | | |
| **Kentucky** | | | | | | | | | | |
| Big Sandy Community & Technical College (Paintsville) | 25 | Aug | 24 | AAS | $3,270 | $9,810 | | • | | • |
| Jefferson Community and Technical College (Louisville) | 20 | Aug | 19 | AAS | $3,270 | $6,540 | $25 | | | • |
| Northern Kentucky University (Highland Heights) | 18 | Aug | 21 | AAS | $5,400 | $10,800 | | | | |
| Southeast Kentucky Comm & Tech College (Pineville) | 24 | Aug | 22 | AAS | $9,000 | $0 | | • | | |
| **Louisiana** | | | | | | | | | | |
| Bossier Parish Community College (Bossier City) | 15 | May | 12 | Cert | $1,800 | $3,800 | | | | • |
| Louisiana State Univ Health Sciences Center (New Orleans) | 15 | May | 24 | BS, MHS | $5,120 | $8,245 | | | | • |
| Nicholls State University (Houma) | 15 | Aug | 12 | Cert, AS, BS | $4,944 | $13,873 | | | | |
| **Maine** | | | | | | | | | | |
| Kennebec Valley Community College (Fairfield) | 14 | Sep | 18 | AS | $3,034 | $6,355 | | • | | • |
| Southern Maine Community College (South Portland) | 20 | Sep | 21 | AS | $2,900 | $5,800 | | | | |
| **Maryland** | | | | | | | | | | |
| Allegany College of Maryland (Cumberland) | 24 | Aug Sep | 17 | AAS | $3,690 | $8,282 | | | | |
| Baltimore City Community College | 25 | Aug | 20 | AAS | $1,673 | $3,923 | | | | |
| Columbia Union College (Takoma Park) | 20 | Sep | 20 | AAS | $17,340 | $17,340 | | | | |
| Comm Coll of Baltimore County - Essex Campus | 32 | Sep | 21 | AAS | $744 | $1,296 | | | | • |
| Prince George's Community College (Largo) | 24 | Aug | 21 | AAS | $2,967 | $7,349 | | | | • |
| Salisbury University | 20 | Sep | 24 | BS | $4,974 | $10,908 | | | | |
| **Massachusetts** | | | | | | | | | | |
| Berkshire Community College (Pittsfield) | 20 | Sep Jan | 17 | Dipl, AS | $3,900 | $8,350 | | • | | |
| Massasoit Community College (Brockton) | 28 | Sep | 18 | Cert, AS | $4,180 | $12,046 | | | | • |
| North Shore Community College (Danvers) | 20 | Sep | 21 | AS | $2,720 | $8,500 | | • | | • |
| Northeastern University (Boston) | 20 | Sep | 57, 15 | BS, MS | $19,320 | $19,320 | | • | | |
| Northern Essex Community College (Lawrence) | 24 | Sep | 20 | AS | $3,675 | $12,110 | | • | | • |
| Quinsigamond Community College (Worcester) | 20 | Sep | 18 | AS | $120 | $326 | | • | | • |
| Springfield Technical Community College | 20 | Sep | 18 | Cert, AS | $4,500 | $8,300 | | • | | • |
| **Michigan** | | | | | | | | | | |
| Kalamazoo Valley Community College | 24 | Aug | 19 | AAS | $1,877 | $4,212 | $4,460 | • | | • |
| Macomb Community College (Clinton Township) | 40 | Aug | 21 | AAS | $2,380 | $3,640 | | | | • |
| Muskegon Community College | 20 | Sep | 28 | AAS | $5,500 | $6,500 | | • | | |
| Oakland Community College (Southfield) | 22 | May | 16, 28 | AS | $1,725 | $2,870 | | | | • |
| **Minnesota** | | | | | | | | | | |
| Lake Superior College (Duluth) | 28 | Aug | 21 | AAS | $6,500 | $10,800 | | | | • |
| Northland Community & Technical College (East Grand Forks) | 30 | Aug | 27 | AAS | $4,448 | $8,896 | | • | | • |
| **Mississippi** | | | | | | | | | | |
| Copiah-Lincoln Community College (Natchez) | 15 | Aug | 12, 21 | AS | $1,700 | $3,500 | | | | |
| Hinds Community College (Jackson) | 30 | Aug | 22 | Dipl, AAS | $2,000 | $4,546 | | | | • |
| Meridian Community College | 15 | Aug | 21 | AAS | $1,750 | $2,640 | | | | |
| Northeast Mississippi Community College (Booneville) | 15 | Aug | 22 | AAS | $850 | $1,710 | | | | |
| **Missouri** | | | | | | | | | | |
| Cape Girardeau Career & Technology Center | 22 | Aug | 21 | AAS | $5,800 | $5,800 | | | | |
| Concorde Career College - Kansas City | 10 | Fall Spring | 6 | Cert | $6,516 | $6,516 | | | | |
| Hannibal Career & Technical Center | 20 | Aug | 24 | AAS | $180 | $180 | | | | • |
| Ozarks Technical Community College (Springfield) | 20 | Jun | 21 | AAS | $2,964 | $4,864 | | | | • |
| University of Missouri - Columbia | 12 | Aug | 24 | BHS | $8,397 | $10,563 | | | | • |

*Data are shown only for programs that completed the 2006 AMA Survey of Health Professions Education Programs
‡Key to Offers: 1: Evening or weekend classes; 2: Non-English instruction; 3: Cultural competence instruction

**PROGRAMS**

# Respiratory Therapist (Advanced)

Respiratory Therapist (Advanced)

| Programs* | Class Capacity | Begins | Length (months) | Award | Res. Tuition | Non-res. Tuition | Stipend | Offers:‡ 1 | 2 | 3 |
|---|---|---|---|---|---|---|---|---|---|---|
| **Montana** | | | | | | | | | | |
| Montana State Univ - Great Falls Coll of Tech | 15 | Sep | 20 | AAS | $3,232 | $10,921 | | | | |
| University of Montana (Missoula) | 16 | Aug | 24 | AAS | $3,178 | $8,351 | | | | • |
| **Nebraska** | | | | | | | | | | |
| Metropolitan Community College (Omaha) | 21 | Sep | 21 | AAS | $3,500 | $5,000 | | | | |
| NE Methodist Coll Nursing & Allied Hlth (Omaha) | 14 | Aug Jan Jun | 24, 48 | AS, BS | $11,940 | $11,940 | | • | • | • |
| Southeast Community College (Lincoln) | 30 | Jul | 18 | AAS | $6,303 | $6,480 | | | | • |
| **Nevada** | | | | | | | | | | |
| Community College of Southern Nevada (Las Vegas) | 35 | Sep | 21 | AAS | $1,300 | $4,800 | | • | | |
| **New Hampshire** | | | | | | | | | | |
| New Hampshire Comm Tech Coll - Claremont | 15 | Aug | 21 | AS | $6,260 | $17,922 | | | | |
| **New Jersey** | | | | | | | | | | |
| Bergen Community College (Paramus) | 30 | Sep | 22 | Cert, AAS | $3,869 | $7,263 | | | | |
| Brookdale Community College (Lincroft) | 25 | Sep | 18 | AAS | $2,900 | $5,800 | | | | • |
| Univ of Med & Dent of New Jersey (Newark) | 32 | May | 12, 15 | AS, BS | $7,000 | $14,000 | | | | • |
| Univ of Med & Dent of New Jersey (Stratford) | 35 | Jun | 24 | AAS | $8,000 | $12,500 | | | | • |
| **New Mexico** | | | | | | | | | | |
| Central New Mexico Community College (Albuquerque) | 25 | Sep | 20 | AS | $1,000 | $7,476 | | | | |
| Eastern New Mexico University - Roswell | 20 | Aug | 18 | AS | $684 | $1,944 | | | | • |
| **New York** | | | | | | | | | | |
| Erie Community College - North Campus (Williamsville) | 32 | Sep | 21 | AAS | $2,900 | $5,800 | | • | | • |
| Genesee Community College (Batavia) | 34 | Aug | 22 | AAS | $1,450 | $1,625 | | | | |
| Hudson Valley Community College (Troy) | 36 | Aug | 21 | AAS | $2,700 | $8,100 | | | | |
| Mohawk Valley Community College (Utica) | 30 | Aug | 22 | AAS | $3,100 | $6,200 | | • | | • |
| Molloy College (Rockville Centre) | 24 | Sep | 24 | AAS | $15,940 | $15,940 | | | | |
| Nassau Community College (Garden City) | 24 | Sep | 24 | AAS | $2,900 | $5,800 | | • | | • |
| New York University | 45 | Sep | 21 | AAS | $17,000 | $17,000 | | • | | |
| Onondaga Community College (Syracuse) | 25 | Jan | 24 | AAS | $3,742 | $7,485 | | • | | |
| SUNY Upstate Medical University (Syracuse) | 20 | Aug | 22 | BS | $4,350 | $10,300 | | • | | |
| Westchester Community College (Valhalla) | 36 | Sep | 24 | AAS | $3,150 | $7,876 | | • | | |
| **North Carolina** | | | | | | | | | | |
| Carteret Community College (Morehead City) | 20 | Aug | 23 | AAS | $1,750 | $10,373 | | | | • |
| Catawba Valley Community College (Hickory) | 15 | Aug | 21 | AAS | $1,659 | $9,219 | | | | • |
| Central Piedmont Community College (Charlotte) | 25 | Aug | 21 | AAS | $1,422 | $7,110 | | | | |
| Durham Technical Community College | 25 | Aug | 21 | AAS | $1,896 | $10,536 | | | • | • |
| Edgecombe Community College (Rocky Mount) | 16 | Aug | 21 | AAS | $1,843 | $9,763 | | | | |
| Forsyth Technical Community College (Winston-Salem) | 20 | Aug | 21 | AAS | $1,580 | $8,780 | | | | • |
| Pitt Community College (Greenville) | 18 | Aug | 21 | AAS | $1,216 | $6,752 | | | | • |
| Robeson Community College (Lumberton) | 20 | Aug | 21 | AAS | $1,008 | $5,328 | | | | |
| Sandhills Community College (Pinehurst) | 18 | Aug | 21 | AAS | $1,132 | $6,200 | | | | |
| Stanly Community College (Albemarle) | 24 | Aug | 21 | AAS | $1,600 | $8,000 | | • | | |
| **North Dakota** | | | | | | | | | | |
| NDSU/Merit Care Hospital Consortium (Fargo) | 12 | May | 13 | Cert, BS | $4,774 | $12,747 | | | | |
| University of Mary/St Alexius Medical Ctr (Bismarck) | 10 | Aug | 20 | Cert, BS | $13,000 | $13,000 | | | | • |
| **Ohio** | | | | | | | | | | |
| Bowling Green State University (Huron) | 25 | Aug | 28 | AAS | $5,036 | $13,760 | | • | | • |
| Collins Career Center (Chesapeake) | 24 | Aug | 22 | Dipl, AAS | $4,800 | $4,800 | $4,800 | | | • |
| Columbus State Community College | 35 | Sep | 21 | AAS | $4,385 | $9,625 | | | | • |
| Cuyahoga Community College (Parma) | 25 | Aug | 22 | AAS | $2,761 | $7,475 | | | | |
| Jefferson Community College (Steubenville) | 21 | Aug | 21 | AAS | $2,754 | $3,672 | | | | • |
| Kettering College of Medical Arts | 24 | Aug | 22 | AS | $0 | $11,550 | | | | • |
| Lakeland Community College (Kirtland) | 20 | Aug | 21 | AAS | $3,685 | $9,657 | | • | | • |
| North Central State College (Mansfield) | 20 | Sep | 21 | AAS | $4,194 | $8,388 | | • | | |
| Ohio State University (Columbus) | 20 | Sep | 45 | Cert, BS | $7,479 | $18,066 | | | | |
| Rhodes State College (Lima) | 26 | Jun | 24 | AAS | $6,091 | $12,183 | | | | • |
| Sinclair Community College (Dayton) | 50 | Sep | 21 | AAS | $2,271 | $3,710 | | | | |
| Stark State College of Technology (Canton) | 20 | Aug | 21 | AAS | $4,800 | $6,800 | | • | | |
| University of Akron | 25 | Sep | 22, 48 | AAS, BS | $5,125 | $13,278 | | | | |
| University of Toledo | 25 | Aug | 39 | BS | $6,816 | $15,627 | | • | | |
| Washington State Community College (Marietta) | 24 | Sep | 21 | AAS | $4,134 | $8,268 | | • | | • |
| Youngstown State University | 18 | Aug | 46 | Cert, BSRC | $6,509 | $11,717 | | • | | • |
| **Oklahoma** | | | | | | | | | | |
| Rose State College (Midwest City) | 22 | Aug | 12 | AAS | $3,064 | $6,770 | | | | • |
| Tulsa Community College | 25 | Aug | 24 | Dipl, AAS | $3,300 | $5,000 | | | | |
| **Oregon** | | | | | | | | | | |
| Lane Community College (Eugene) | 20 | Sep | 21 | AAS | $6,745 | $22,310 | | • | | • |

*Data are shown only for programs that completed the 2006 AMA Survey of Health Professions Education Programs
‡Key to Offers: 1: Evening or weekend classes; 2: Non-English instruction; 3: Cultural competence instruction

| Programs* | Class Capacity | Begins | Length (months) | Award | Res. Tuition | Non-res. Tuition | Stipend | Offers:‡ 1 | 2 | 3 |
|---|---|---|---|---|---|---|---|---|---|---|
| **Pennsylvania** | | | | | | | | | | |
| Comm Coll of Allegheny County (Pittsburgh) | 25 | Aug | 22 | AS | $3,400 | $6,500 | | | | |
| Community College of Philadelphia | 36 | Sep | 22 | Cert, AAS | $5,000 | $10,000 | | • | | • |
| Crozer - Chester Med Ctr (Upland) | 16 | Sep | 22 | AAS | $4,077 | $7,995 | | • | | |
| Gannon University (Erie) | 16 | Aug | 24, 48 | Cert, AS, BS | $20,680 | $20,680 | | | | |
| Gwynedd-Mercy College (Gwynedd Valley) | 12 | Aug | 19, 27 | Cert, AS, PostAS | $19,720 | $19,720 | | • | | • |
| Luzerne County Community College (Nanticoke) | 24 | Jun | 24 | AAS | $2,920 | $5,840 | | | | |
| Mansfield University (Sayre) | 18 | Aug | 21 | AAS | $4,906 | $12,266 | | • | | • |
| Millersville University of Pennsylvania | 15 | Aug | 16 | Cert, BS | $6,236 | $13,658 | | | | • |
| Reading Area Community College | 15 | Jun Sep Jan Mar | 12, 24 | Cert, AAS | $2,448 | $4,896 | | | | |
| University of Pittsburgh - Johnstown | 30 | Sep | 20 | Dipl, Cert, AS | $10,572 | $21,136 | | | | • |
| West Chester University (Bryn Mawr) | 25 | Aug | 48 | BS | $4,598 | $11,496 | | | | |
| York College of Pennsylvania | 12 | Sep | 36, 48 | AS, BS | $12,000 | $12,000 | | | | |
| **Rhode Island** | | | | | | | | | | |
| Community College of Rhode Island (Lincoln) | 24 | Jun | 24 | AAS | $2,040 | $5,992 | | | | • |
| **South Carolina** | | | | | | | | | | |
| Florence-Darlington Technical College | 20 | Aug | 24 | AS | $5,004 | $7,755 | | | | |
| Greenville Technical College | 25 | Aug | 24 | AS | $4,275 | $8,940 | | | • | • |
| Spartanburg Community College | 24 | Aug | 12, 24 | AHS | $4,293 | $5,367 | | | | |
| Tri-County Technical College (Pendleton) | 20 | Aug | 24 | AS | $1,309 | $2,982 | | | | |
| Trident Technical College (Charleston) | 24 | Aug | 22 | AS | $2,970 | $3,276 | | | | |
| **South Dakota** | | | | | | | | | | |
| Dakota State University (Madison) | 32 | Sep Jun | 21, 39 | AS, BS | $3,495 | $10,675 | | | | • |
| **Tennessee** | | | | | | | | | | |
| Baptist College of Health Sciences (Memphis) | 20 | Aug | 22 | BHS | $6,000 | $6,000 | | • | | • |
| Columbia State Community College | 24 | Aug | 21 | AAS | $2,100 | $8,500 | | | | |
| Jackson State Community College | 20 | Aug | 21 | AAS | $3,725 | $10,014 | | | | • |
| Roane State Community College (Harriman) | 20 | Aug | 21 | AAS | $1,024 | $3,072 | | | | |
| Tennessee State University (Nashville) | 24 | Aug | 37 | BS | $2,207 | $6,863 | | | | |
| Volunteer State Community College (Gallatin) | 14 | Aug | 4, 8 | Cert | $1,183 | $4,390 | | | | |
| **Texas** | | | | | | | | | | |
| Alvin Community College | 22 | Aug | 21 | AAS | $2,733 | $5,466 | | | | • |
| Amarillo College | 21 | Aug | 23 | AAS | $1,440 | $2,977 | | | | |
| Angelina College (Lufkin) | 20 | Aug | 21 | AAS | $546 | $1,096 | | | | |
| Collin County Community College (McKinney) | 25 | Aug | 22 | Cert, AAS | $1,242 | $2,514 | | | | |
| Del Mar College (Corpus Christi) | 18 | Sep | 21 | AAS | $2,500 | $5,000 | | | | |
| Kingwood College - NHMCCD | 24 | Jan | 24 | AAS | $2,976 | $5,616 | | | | • |
| Lamar Institute of Technology (Beaumont) | 45 | Aug | 24 | AAS | $3,298 | $9,796 | | | | • |
| Midland College | 20 | Aug | 20, 23 | AAS | $1,615 | $1,790 | | | | |
| Odessa College | 20 | Aug | 22 | AAS | $2,400 | $2,700 | | | | • |
| South Plains College (Lubbock) | 30 | Aug | 22 | AAS | $1,900 | $2,200 | | | | • |
| Tarrant County College (Hurst) | 30 | Aug | 21 | AAS | $1,829 | $2,297 | | | | |
| Temple College | 20 | Aug | 24 | Cert, AAS | $2,178 | $5,647 | | | | • |
| Texas State University-San Marcos | 40 | Sep | 48 | Cert, BSRC | $5,866 | $13,882 | | | | |
| Tyler Junior College | 30 | Aug | 22 | AAS | $3,200 | $5,800 | | • | | • |
| University of Texas Medical Branch/SAHS (Galveston) | 20 | Aug | 24 | BSRC | $5,681 | $17,276 | | | | • |
| Victoria College | 18 | Aug | 21 | AAS | $1,512 | $2,448 | | | | • |
| **Utah** | | | | | | | | | | |
| Weber State University (Ogden) | 20 | Aug | 33 | AS, BS | $3,500 | $10,500 | | | | • |
| **Vermont** | | | | | | | | | | |
| Vermont Technical College (Williston) | 27 | Aug | 18 | AS | $15,751 | $22,711 | | | | • |
| **Virginia** | | | | | | | | | | |
| Central Virginia Community College (Lynchburg) | 20 | Aug | 21 | AAS | $1,506 | $6,486 | | | | • |
| J Sargeant Reynolds Community College (Richmond) | 36 | Jan | 7 | Cert | $2,202 | $6,659 | | • | | |
| Northern Virginia Community College (Springfield) | 30 | Aug | 22 | AAS | $5,902 | $26,926 | | | | • |
| Shenandoah University (Winchester) | 25 | Aug | 20, 38 | Cert, BS, AS | $19,900 | $19,900 | | | | • |
| Southwest Virginia Community College (Richlands) | 24 | Aug | 22 | AAS | $2,616 | $8,719 | | • | | • |
| **Washington** | | | | | | | | | | |
| Seattle Central Community College | 25 | Sep | 22 | AAS | $2,649 | $7,872 | | | | • |
| Spokane Community College | 24 | Sep | 22 | AAS | $3,000 | $9,000 | | | | • |
| Tacoma Community College | 24 | Jun | 24 | AAS | $2,336 | $9,185 | | | | • |
| **West Virginia** | | | | | | | | | | |
| Carver Career Technical Center (Charleston) | 25 | Jul | 22 | Dipl, AS | $4,150 | $4,150 | | • | | • |
| West Virginia Northern Community College (Wheeling) | 40 | Aug | 21 | AAS | $1,752 | $5,592 | | • | | • |
| Wheeling Jesuit University | 12 | Sep | 48 | BS | $19,000 | $19,000 | | | | • |

*Data are shown only for programs that completed the 2006 AMA Survey of Health Professions Education Programs

‡Key to Offers: 1: Evening or weekend classes; 2: Non-English instruction; 3: Cultural competence instruction

## Respiratory Therapist (Advanced)

| Programs* | Class Capacity | Begins | Length (months) | Award | Res. Tuition | Non-res. Tuition | Stipend | Offers:‡ 1 | 2 | 3 |
|---|---|---|---|---|---|---|---|---|---|---|
| **Wisconsin** | | | | | | | | | | |
| Madison Area Technical College | 26 | Aug | 22 | AAS | $4,995 | $13,076 | | | | |
| Mid-State Technical College (Marshfield) | 20 | Aug | 22 | AS | $3,125 | $15,932 | | | • | • |
| Western Technical College (La Crosse) | 20 | Aug | 21 | AAS | $2,664 | $0 | | | | • |

*Data are shown only for programs that completed the 2006 AMA Survey of Health Professions Education Programs
‡Key to Offers: 1: Evening or weekend classes; 2: Non-English instruction; 3: Cultural competence instruction

## Respiratory Therapist (Entry-Level)*

### Arkansas

**Univ of Arkansas Comm Coll at Hope**
Respiratory Therapist (Entry-Level) Prgm
Hwy 29 S, PO Box 140
Hope, AR 71801
www.uacch.edu
*Prgm Dir:* Ken LeJeune, MS RRT
*Tel:* 870 722-8275   *Fax:* 870 777-5957
*E-mail:* klejeune@uacch.edu
*Med Dir:* Michael Young, MD

**Pulaski Technical College**
Respiratory Therapist (Entry-Level) Prgm
3000 W Scenic Dr
North Little Rock, AR 72118
www.pulaskitech.edu
*Prgm Dir:* Jimmy C Davis, BS RRT
*Tel:* 501 812-2223   *Fax:* 501 812-2316
*E-mail:* jdavis@pulaskitech.edu
*Med Dir:* Anthony R Giglia, MD

**Black River Technical College**
Respiratory Therapist (Entry-Level) Prgm
PO Box 468
Hwy 304 East
Pocahontas, AR 72455
*Prgm Dir:* Eric G Johnson, BS RRT
*Tel:* 870 248-4000, Ext 4153   *Fax:* 870 248-4100
*E-mail:* eric.johnson@blackrivertech.edu
*Med Dir:* William S Hubbard, MD

### California

**Hacienda LaPuente Adult Education**
Respiratory Therapist (Entry-Level) Prgm
Excelsior College
Willow Adult Campus
La Puente, CA 91746
*Prgm Dir:* Kevin Booth, AB RRT RCP
*Tel:* 626 934-2800, Ext 2995   *Fax:* 626 855-3169
*E-mail:* kmbooth@hlpusd.k12.ca.us
*Med Dir:* Brian Tiep, MD

**Concorde Career Institute - North Hollywood**
Respiratory Therapist (Entry-Level) Prgm
12412 Victory Blvd
North Hollywood, CA 91606
*Prgm Dir:* Marlyn H Haberwood, RRT BSHS
*Tel:* 818 766-8151, Ext 253   *Fax:* 818 766-1587
*E-mail:* mhaberwood@concorde.edu
*Med Dir:* Peter M Browne, MD

**California College San Diego**
Respiratory Therapist (Entry-Level) Prgm
2820 Camino Del Rio S
San Diego, CA 92108
*Prgm Dir:* Edward Moser, MBA RRT
*Tel:* 619 295-5785   *Fax:* 619 477-4360
*E-mail:* emoser@cc-sd.edu
*Med Dir:* Jerry E Fein, MD

**Simi Valley Adult School**
Respiratory Therapist (Entry-Level) Prgm
3192 Los Angeles Ave
Simi Valley, CA 93065
*Prgm Dir:* Christine R Kingston, BS RRT
*Tel:* 805 579-6262   *Fax:* 805 522-8902
*E-mail:* ronk@simi.tec.ca.us
*Med Dir:* Michael Littner, MD

**El Camino College**
Respiratory Therapist (Entry-Level) Prgm
16007 S Crenshaw Blvd
Torrance, CA 90506
*Prgm Dir:* Louis M Sinopoli, EdD RRT FAARC AE-C
*Tel:* 310 660-3248   *Fax:* 310 660-3378
*E-mail:* sinopoli@elcamino.edu
*Med Dir:* Mason Greg, MD

**Crafton Hills College**
Respiratory Therapist (Entry-Level) Prgm
11711 Sand Canyon Rd
Yucaipa, CA 92399
www.craftonhills.edu
*Prgm Dir:* Kenneth R Bryson, MEd BE RCP RRT
*Tel:* 909 389-3284   *Fax:* 909 389-3229
*E-mail:* kbryson@crafton.sbccd.cc.ca.us
*Med Dir:* Richard L Sheldon, MD

### Colorado

**T H Pickens Technical Center**
Respiratory Therapist (Entry-Level) Prgm
500 Airport Blvd
Aurora, CO 80011
*Prgm Dir:* Leigh Otto, BS RRT
*Tel:* 303 344-4910, Ext 27781   *Fax:* 303 326-1277
*E-mail:* leigho@pickens.aps.k12.co.us
*Med Dir:* James Ellis, MD

### Illinois

**St Augustine College**
Respiratory Therapist (Entry-Level) Prgm
1333 W Argyle
Chicago, IL 60640
www.staugustine.edu
*Prgm Dir:* Shirley Thomason, AS RRT
*Tel:* 773 878-7182   *Fax:* 773 878-0937
*E-mail:* sthomason@staugustine.edu
*Med Dir:* Raul Wolf, MD

### Kansas

**Bethany Med Ctr/Kansas City Kansas Comm Coll**
Respiratory Therapist (Entry-Level) Prgm
7250 State Ave
Kansas City, KS 66112
www.kckcc.edu
*Prgm Dir:* C Michael Parrett, MBA RPFT RRT-NPS
*Tel:* 913 288-7245   *Fax:* 913 288-7649
*E-mail:* mparrett@kckcc.edu
*Med Dir:* Sabato Sisillo, MD

### Kentucky

**Maysville Comm & Tech College - Rowan Campus**
Respiratory Therapist (Entry-Level) Prgm
609 Viking Dr
Morehead, KY 40351
*Prgm Dir:* Marlene Vice, RRT
*Tel:* 606 783-1538, Ext 442   *Fax:* 606 784-9876
*E-mail:* marlene.vice@kctcs.edu
*Med Dir:* Anthony Weaver, MD

**Laurel Technical College - Rockcastle**
Respiratory Therapist (Entry-Level) Prgm
PO Box 275
Mt Vernon, KY 40456
*Prgm Dir:* Carl Baker, RRT
*Tel:* 606 256-4346   *Fax:* 606 256-4337
*E-mail:* carl.baker@kctcs.edu
*Med Dir:* Abdi Vaezy, MD

**West Kentucky Community & Technical College**
Respiratory Therapist (Entry-Level) Prgm
5200 Blandville Rd
PO Box 7408
Paducah, KY 42002-7408
*Prgm Dir:* Ruth Thompson, MS RRT
*Tel:* 270 534-3486   *Fax:* 270 554-6227
*E-mail:* ruth.thompson@kctcs.edu
*Med Dir:* Bradley T Rankin, MD

**Southeast Kentucky Comm & Tech College**
Respiratory Therapist (Entry-Level) Prgm
3300 US Highway 25E South
Pineville, KY 40977
*Prgm Dir:* Michael S Good, MS RRT RPFT
*Tel:* 606 248-2122   *Fax:* 606 248-2166
*E-mail:* mike.good@kctcs.edu
*Med Dir:* Abdi Vaezy, MD

*Additional information about programs that returned the AMA's survey is available in a table beginning on page 420.

# Louisiana

**Our Lady of the Lake College**
*Cosponsor:* LSU Health Sciences Center
Respiratory Therapist (Entry-Level) Prgm
7434 Perkins Rd
Baton Rouge, LA 70808
www.ololcollege.edu
*Prgm Dir:* Jackie Bush, BS RRT
*Tel:* 225 768-1786    *Fax:* 225 768-0819
*E-mail:* jbush@ololcollege.edu
*Med Dir:* Richard C Kearley, MD

**Bossier Parish Community College**
Respiratory Therapist (Entry-Level) Prgm
6220 E Texas St
Bossier City, LA 71111
*Prgm Dir:* Sandra Partain, MHS RRT-NPS
*Tel:* 318 675-6813    *Fax:* 318 675-6937
*E-mail:* sparta@lsuhsc.edu
*Med Dir:* Keith Payne, MD

**Louisiana State University - Eunice**
Respiratory Therapist (Entry-Level) Prgm
PO Box 1129
Eunice, LA 70535
*Prgm Dir:* Kathleen B Warner, BS RRT-NPS
*Tel:* 337 457-7311, Ext 341    *Fax:* 337 550-1289
*E-mail:* kreynold@lsue.edu
*Med Dir:* Gary Guidry, MD

**Nicholls State University**
Respiratory Therapist (Entry-Level) Prgm
235 Civic Center Blvd
Houma, LA 70360
*Prgm Dir:* Errol Champagne, MEd RRT-NPS
*Tel:* 985 876-8852    *Fax:* 985 876-8856
*E-mail:* errol.champagne@nicholls.edu
*Med Dir:* Ralph Bourgeois, MD

**Louisiana Tech College - Jefferson Campus**
Respiratory Therapist (Entry-Level) Prgm
5200 Blair Dr
Metairie, LA 70001
*Prgm Dir:* Toy B Smoot, MEd RRT
*Tel:* 504 736-7080    *Fax:* 504 736-7120
*E-mail:* tsmoot@theltc.net
*Med Dir:* William Borron, MD

# Michigan

**Monroe County Community College**
Respiratory Therapist (Entry-Level) Prgm
1555 S Raisinville Rd
Monroe, MI 48161
*Prgm Dir:* Bonnie Boggs, BS RRT
*Tel:* 734 384-4268    *Fax:* 734 384-4187
*E-mail:* bboggs@monroeccc.edu
*Med Dir:* Milo Engoren, MD

# Minnesota

**Northland Community & Technical College**
Respiratory Therapist (Entry-Level) Prgm
2022 Central Ave NE
East Grand Forks, MN 56721
www.northlandcollege.edu
*Prgm Dir:* Anthony Sorum, BA RRT
*Tel:* 218 773-3441, Ext 415    *Fax:* 218 773-4502
*E-mail:* tony.sorum@northlandcollege.edu
*Med Dir:* Wayne Bretweiser, MD

# Missouri

**Cape Girardeau Career & Technology Center**
*Cosponsor:* Mineral Area College
Respiratory Therapist (Entry-Level) Prgm
1080 S Silver Spring Rd
Cape Girardeau, MO 63703
*Prgm Dir:* Kenneth L Pfau, BS RRT
*Tel:* 573 334-0449, Ext 320    *Fax:* 573 334-5930
*E-mail:* pfauk@cape.k12.mo.us
*Med Dir:* Richard E Moore, MD

**Sanford-Brown College**
Respiratory Therapist (Entry-Level) Prgm
1203 Smizer Mill Rd
Fenton, MO 63026
*Prgm Dir:* Monica Engh, BHS RRT
*Tel:* 636 349-4900, Ext 125    *Fax:* 636 349-9170
*E-mail:* mengh@sbcstlouis.sanfordbrown.com
*Med Dir:* Anthony Masi, MD

**Missouri Southern State University**
Respiratory Therapist (Entry-Level) Prgm
3950 E Newman Rd
Joplin, MO 64801
*Prgm Dir:* Glenda Pippin
*Tel:* 417 659-4405    *Fax:* 417 659-4408
*E-mail:* pippin-g@mssu.edu
*Med Dir:* S Subramanian, MD

**Concorde Career College - Kansas City**
Respiratory Therapist (Entry-Level) Prgm
3239 Broadway Blvd
Kansas City, MO 64111
www.concordecareercolleges.com
*Prgm Dir:* Lana Conrad, BS RRT
*Tel:* 816 531-5223, Ext 403    *Fax:* 816 756-3231
*E-mail:* lconrad@concorde.edu
*Med Dir:* Lida Osbern, MD

**Rolla Technical Center**
*Cosponsor:* East Central College
Respiratory Therapist (Entry-Level) Prgm
500 Forum Dr
Rolla, MO 65401-4678
*Prgm Dir:* Diane R Oldfather, RRT
*Tel:* 573 458-0160, Ext 16194    *Fax:* 573 458-0164
*E-mail:* doldfather@rolla.k12.mo.us
*Med Dir:* Edward Bruns, DO

# Ohio

**University of Toledo**
Respiratory Therapist (Entry-Level) Prgm
2801 W Bancroft St
Toledo, OH 43606
*Prgm Dir:* Suzanne Spacek, MEd RRT-NPS CPFT
*Tel:* 419 530-4556    *Fax:* 419 530-4780
*E-mail:* suzanne.spacek@utoledo.edu
*Med Dir:* Robert A May, MD

# Oklahoma

**Great Plains Technology Center**
Respiratory Therapist (Entry-Level) Prgm
4500 W Lee Blvd
Lawton, OK 73505
*Prgm Dir:* Jack Powers, RRT
*Tel:* 580 250-5572    *Fax:* 580 250-5583
*E-mail:* jpowers@gptech.org
*Med Dir:* Aaron Trachte, MD

# Pennsylvania

**Gwynedd-Mercy College**
Respiratory Therapist (Entry-Level) Prgm
1325 Sumneytown Pike
PO Box 901
Gwynedd Valley, PA 19437
*Prgm Dir:* William F Galvin, MSEd RRT CPFT AE-C
*Tel:* 215 641-5536    *Fax:* 215 641-5559
*E-mail:* galvin.w@gmc.edu
*Med Dir:* Donald Peterson, MD

**Harrisburg Area Community College**
Respiratory Therapist (Entry-Level) Prgm
One HACC Dr
Harrisburg, PA 17110
*Prgm Dir:* Bradley A Leidich, RRT MSEd
*Tel:* 717 780-2315    *Fax:* 717 780-2551
*E-mail:* baleidic@hacc.edu
*Med Dir:* William Anderson, MD

**Reading Area Community College**
Respiratory Therapist (Entry-Level) Prgm
Ten S Second St, PO Box 1706
Reading, PA 19603
www.racc.edu
*Prgm Dir:* James H Cicman, BHS RRT
*Tel:* 610 372-4721, Ext 5435    *Fax:* 610 607-6254
*E-mail:* jcicman@racc.edu
*Med Dir:* Joseph Mariglio, MD

**York College of Pennsylvania**
*Cosponsor:* York Hospital
Respiratory Therapist (Entry-Level) Prgm
Country Club Rd
York, PA 17405
*Prgm Dir:* Mark Simmons, MSEd RPFT RRT-NPS
*Tel:* 717 851-2464    *Fax:* 717 851-2934
*E-mail:* msimmons@wellspan.org
*Med Dir:* Richard J Murray, MD

# Tennessee

**Volunteer State Community College**
Respiratory Therapist (Entry-Level) Prgm
1480 Nashville Pike
Gallatin, TN 37066-3188
www.volstate.edu
*Prgm Dir:* Cory E Martin, BS RRT
*Tel:* 615 452-8600, Ext 3349    *Fax:* 615 230-3224
*E-mail:* Cory.Martin@volstate.edu
*Med Dir:* Christopher Harris, MD

**Walters State Community College**
Respiratory Therapist (Entry-Level) Prgm
500 S Davy Crockett Pkwy
Morristown, TN 37813-6899
www.ws.edu
*Prgm Dir:* Robert G McGee, RRT MS
*Tel:* 423 798-7941    *Fax:* 423 798-7944
*E-mail:* robert.mcgee@ws.edu
*Med Dir:* Elliot Smith, MD

# Texas

**SWT/HMC Respiratory Care School Consortium**
Respiratory Therapist (Entry-Level) Prgm
1900 Pine St
Abilene, TX 79601
*Prgm Dir:* Jeff Lawrence, MS RRT
*Tel:* 915 670-2368    *Fax:* 915 670-2114
*E-mail:* jlawrenc@hendrickhealth.org
*Med Dir:* Preston Pate, MD

**Univ Tx at Brownsville/Tx Southmost Coll**
Respiratory Therapist (Entry-Level) Prgm
80 Ft Brown
Brownsville, TX 78520
*Prgm Dir:* John L McCabe, PhD RRT
*Tel:* 956 554-5010   *Fax:* 956 554-5012
*E-mail:* john.mccabe@utb.edu
*Med Dir:* Lorenzo Pelly, MD

**US Army Medical Dept Center & School**
Respiratory Therapist (Entry-Level) Prgm
Department of Medical Science
Physician's Extender Branch Bldg 1151, 2651 McIdoe Rd
Ft Sam Houston, TX 78234
www.cs.amedd.army.mil/dms/91v
*Prgm Dir:* Jon P O'Hora, RRT MSHP
*Tel:* 210 295-4411   *Fax:* 210 295-4317
*E-mail:* jon.ohora@cen.amedd.army.mil
*Med Dir:* Michael J Morris, MD

**USAF School of Health Care Sciences**
Respiratory Therapist (Entry-Level) Prgm
USAF Cardiopulmonary Lab Technologist Program
383 TRS/XUFC- 939 Missile Rd
Sheppard AFB, TX 76311
*Prgm Dir:* MSgt Shane A Pearson, BS RRT CPFT CCT
*Tel:* 940 676-3812   *Fax:* 940 676-2210
*E-mail:* shane.pearson@sheppard.af.mil
*Med Dir:* Capt Jeremy Conklin, MD

# Utah

**Weber State University**
Respiratory Therapist (Entry-Level) Prgm
3904 University Circle
Ogden, UT 84408-3904
*Prgm Dir:* Georgine L Bills, MBA RRT
*Tel:* 801 626-7071   *Fax:* 801 626-7075
*E-mail:* gbills@weber.edu
*Med Dir:* Gary K Goucher, MD

# Virginia

**J Sargeant Reynolds Community College**
Respiratory Therapist (Entry-Level) Prgm
PO Box 85622
Richmond, VA 23285
www.reynolds.edu
*Prgm Dir:* Donald K O'Donohue, RRT
*Tel:* 804 523-5375   *Fax:* 804 786-5298
*E-mail:* dodonohue@reynolds.edu
*Med Dir:* Clifton L Parker, MD

## Respiratory Therapist (Entry-Level)

| Programs* | Class Capacity | Begins | Length (months) | Award | Res. Tuition | Non-res. Tuition | Stipend | Offers:‡ 1 | 2 | 3 |
|---|---|---|---|---|---|---|---|---|---|---|
| **Arkansas** | | | | | | | | | | |
| Pulaski Technical College (North Little Rock) | 26 | Aug | 24 | AAS | $3,528 | $3,528 | | | | |
| Univ of Arkansas Comm Coll at Hope | 25 | Jul | 24 | AAS | $2,123 | $4,306 | | | | |
| **California** | | | | | | | | | | |
| Crafton Hills College (Yucaipa) | 35 | Aug | 24 | Cert, AS | $1,612 | $8,928 | | | | |
| **Colorado** | | | | | | | | | | |
| T H Pickens Technical Center (Aurora) | 20 | Aug | 23 | Cert, AAS | $2,627 | $5,254 | | | | |
| **Illinois** | | | | | | | | | | |
| St Augustine College (Chicago) | 60 | Aug Jan | 30 | AAS | $7,200 | $7,200 | | | | |
| **Kansas** | | | | | | | | | | |
| Bethany Med Ctr/Kansas City Kansas Comm Coll | 15 | Aug Jan | 12 | Cert, AAS | $1,950 | $5,100 | | | | • |
| **Louisiana** | | | | | | | | | | |
| Bossier Parish Community College (Bossier City) | 21 | Aug | 9 | AS | $1,800 | $3,860 | | | | • |
| Louisiana Tech College - Jefferson Campus (Metairie) | 15 | Spring | 12, 24 | AAT, AAS | $1,037 | $2,074 | | | | • |
| Nicholls State University (Houma) | 15 | Aug | 24 | Cert, AS | $4,944 | $13,873 | | | | |
| Our Lady of the Lake College (Baton Rouge) | 20 | Jan | 24 | Dipl, AS | $3,280 | $3,280 | | | | • |
| **Minnesota** | | | | | | | | | | |
| Northland Community & Technical College (East Grand Forks) | 30 | Aug | 24 | AAS | $4,448 | $8,896 | | | | • |
| **Missouri** | | | | | | | | | | |
| Concorde Career College - Kansas City | 30 | Every 10-12 wks | 15 | AS | $19,936 | $19,936 | | | | |
| **Ohio** | | | | | | | | | | |
| University of Toledo | | | 21 | | | | | | | |
| **Pennsylvania** | | | | | | | | | | |
| Gwynedd-Mercy College (Gwynedd Valley) | 14 | Aug | 9 | Cert, AS | $19,720 | $19,720 | | • | | |
| Reading Area Community College | 20 | Jun | 24 | Cert, AAS | $2,448 | $4,896 | | | | |
| York College of Pennsylvania | 12 | Sep | 24 | AS | $12,000 | $12,000 | | | | |
| **Tennessee** | | | | | | | | | | |
| Volunteer State Community College (Gallatin) | 20 | Jun | 12, 8 | Dipl, AAS | $1,183 | $4,390 | | | | |
| Walters State Community College (Morristown) | 22 | Jun | 24 | AAS | $2,700 | $0 | | | | |
| **Texas** | | | | | | | | | | |
| US Army Medical Dept Center & School (Ft Sam Houston) | 70 | Feb May Sep | 9 | AAS | | | | | | • |
| USAF School of Health Care Sciences (Sheppard AFB) | 14 | Varies | 11 | Cert, AS | | | | | | |
| **Virginia** | | | | | | | | | | |
| J Sargeant Reynolds Community College (Richmond) | 60 | Aug | 16 | AAS | $4,322 | $13,070 | | • | | |

*Data are shown only for programs that completed the 2006 AMA Survey of Health Professions Education Programs

‡Key to Offers: 1: Evening or weekend classes; 2: Non-English instruction; 3: Cultural competence instruction

# Speech-Language Pathologist

## Job Description

Speech-language pathologists are professionals educated in the study of human communication, its development, and its disorders. Speech-language pathologists work with people who cannot make speech sounds or cannot make them clearly; those with speech rhythm and fluency problems, such as stuttering; people with voice quality problems, such as inappropriate pitch or harsh voice; those with problems understanding and producing language; those who wish to improve their communication skills by modifying an accent; those with cognitive communication impairments, such as attention, memory, and problem-solving disorders; and those with hearing loss who use hearing aids or cochlear implants, in order to develop auditory skills and improve communication. They also work with people who have swallowing difficulties.

Speech and language difficulties can result from a variety of causes, including stroke, brain injury or deterioration, developmental delays, cerebral palsy, cleft palate, voice pathology, mental retardation, hearing impairment, or emotional problems. Speech-language pathologists use written and oral tests, as well as special instruments, to diagnose the nature and extent of impairment and to record and analyze speech, language, and swallowing irregularities. For individuals with little or no speech capability, speech-language pathologists may select augmentative or alternative communication methods, including automated devices and sign language, and teach their use. They help patients develop, or recover, reliable communication skills so patients can fulfill their educational, vocational, and social roles.

Speech-language pathologists often work with education and other health care professionals, such as teachers, physicians, social workers, and psychologists, to evaluate and treat clients. They counsel individuals and their families concerning communication disorders and how to cope with the stress and misunderstanding that often accompany them. They also work with family members to recognize and change behavior patterns that impede communication and treatment and show them communication-enhancing techniques to use at home.

A graduate degree is required to work in most settings as a speech-language pathologist. A doctoral degree (PhD) is preferred in some career paths, such as college teaching, research, and private practice.

Working with an understanding of the full range of human communication and its disorders, speech-language pathologists:
- Evaluate and diagnose speech, language, and swallowing disorders in individuals of all ages, from infants to the elderly
- Treat speech, language, and swallowing disorders
  In addition, speech-language pathologists may:
- Prepare future professionals in colleges and universities
- Engage in research to enhance knowledge about human communication processes and investigate behavioral patterns associated with communication disorders

## Employment Characteristics

Speech-language pathologists may work in a wide range of settings, including schools, universities, hospitals, rehabilitation centers, skilled nursing facilities, community clinics, geriatric facilities, home health care services, and public health departments, or in private practice.

## Salary

Salaries of speech-language pathologists depend on educational background, specialty, and experience, along with the geographical location and type of setting in which they work. According to the ASHA 2006 Schools Survey Salary Report, the median salary for ASHA-certified speech-language pathologists was $52,131 for those employed on an academic year (ie, 9-10 month) basis and $57,000 for those on a calendar year (ie, 11-12 month). The 2006 median starting salary for certified speech-language pathologists in school settings with 1 to 3 years' experience was $40,041 for an academic year appointment. The median calendar year salary for management positions was $80,000. According to the ASHA 2005 Health Care Survey Salary Report, the median salary for ASHA-certified speech-language pathologists was $60,000. The 2005 median starting salary for certified speech-language pathologists in health care settings with 1 to 3 years' experience was $52,694. The median calendar year salary for management positions was $72,985. Good benefits packages, such as insurance programs and leave, are usually available to speech-language pathologists.

Refer to Section IV, Table 5 of this Directory for more information, or see www.ama-assn.org/go/salary.

## Educational Programs

Approximately 245 universities in the United States offer graduate education programs in speech-language pathology that prepare students for entry into practice.

**Length.** Full-time study usually takes at least 2 years, including summers, to complete a master's degree program in speech-language pathology. In addition, most agencies require a 9- to 12-month postgraduate clinical experience to fulfill credentialing requirements.

**Prerequisites.** Course work in the biological sciences, physical sciences, mathematics, and behavioral or social sciences are required for graduate study. Undergraduate programs in communication sciences and disorders will provide a background in linguistics, phonetics, psychology, normal speech, and language development, and introductory course work in speech-language pathology. Excellent oral and written communication skills are expected.

**Curriculum.** Graduate programs should offer a curriculum to allow a student to meet the knowledge and skills necessary to enter practice in speech-language pathology. A typical graduate program of study includes course content in normal and abnormal communication development; diagnostic and treatment procedures, articulation, expressive and receptive language, voice disorders, fluency, swallowing, and ethics. Opportunities to work in a variety of different clinical settings and with a diverse range of clients should be provided during the graduate program of study.

## Licensure and Certification

In most states, speech-language pathologists must comply with state regulatory (licensure) standards and/or have state teacher certification to practice in specific settings. A graduate degree, completion of a 9- to 12-month clinical experience, and passage of a national examination are typically required to achieve the credentials. Individuals should contact the appropriate state licensure board or teacher certification agency for more information about requirements.

ASHA offers the Certificate of Clinical Competence in Speech-Language Pathology (CCC-SLP), a nationally recognized credential that offers certificate holders ease in qualifying for state credentials because those requirements are similar or identical to ASHA's CCC requirements, recognition as a "highest qualified provider" of speech-language services for reimbursement, and increased opportunities for employment or promotion, as certain positions in hospitals, educational programs, or private practices may require ASHA certification.

### Employment Outlook

Employment of speech-language pathologists is expected to grow about as fast as the average for all occupations through the year 2014 (www.bls.gov/oco/ocos099.htm#outlook). More frequent recognition of problems in preschool and school-age children by teachers and parents, combined with the increased numbers of older citizens, and medical advances, has created a growing need for speech and language services. Additionally, opportunities for employment in research and higher education are expected to increase as baby boomers currently in these positions retire. Clinical opportunities will be especially strong for those with bilingual and multicultural expertise. There are shortages of qualified personnel in some areas of the country, especially in inner city, rural, and less populated areas. Job opportunities in medically related areas are expected to grow at an above average rate. Although competition for positions in some areas is keen, the potential for private practice and contract work is increasing rapidly. Many states now require that all newborns be screened for hearing loss and receive appropriate early intervention services. Greater awareness of the importance of early identification and diagnosis of speech, language, swallowing, and hearing disorders will also increase employment opportunities.

### Inquiries

For information about a specific program, write to the director of the speech-language pathology program in care of the institution listed.

For additional information about the professions or academic program accreditation, contact:

American Speech-Language-Hearing Association (ASHA)
10801 Rockville Pike
Rockville, MD 20852
800 498-2071
301 571-0457 Fax
www.asha.org

## Speech-Language Pathologist*

## Alabama

**Auburn University**
Speech-Language Pathology Prgm
Dept of Communication
1199 Haley Center
Auburn University, AL 36849-5232
*Prgm Dir:* Lawrence F Molt, PhD
*Tel:* 334 844-9600 *Fax:* 334 844-4585
*E-mail:* moltlaw@auburn.edu

**University of South Alabama**
Speech-Language Pathology Prgm
Dept of Speech Pathology & Audiology
2000 University Commons
Mobile, AL 36688-0002
*Prgm Dir:* Paul Dagenais, PhD CCC-SLP
*Tel:* 334 380-2600, Ext 4-2608 *Fax:* 334 380-2699
*E-mail:* pdagenais@usouthal.edu

**University of Montevallo**
Speech-Language Pathology Prgm
Comm Science and Disorders
Station 6720
Montevallo, AL 35115-6720
*Prgm Dir:* Mary Beth Armstrong, PhD CCC-SLP
*Tel:* 205 665-6720 *Fax:* 205 665-6721
*E-mail:* armstrom@montevallo.edu

**Alabama A&M University**
Speech-Language Pathology Prgm
Dept of Special Education
PO Box 580
Normal, AL 35762
*Prgm Dir:* Terry D Douglas, PhD CCC-SLP
*Tel:* 256 851-5533 *Fax:* 256 851-5538
*E-mail:* tdouglas@aamu.edu

**University of Alabama**
Speech-Language Pathology Prgm
Dept of Comm Disorders
PO Box 870242
Tuscaloosa, AL 35487-0242
www.as.ua.edu/comdis
*Prgm Dir:* Karen Steckol, PhD CCC-SLP
*Tel:* 205 348-7131 *Fax:* 205 348-1845
*E-mail:* ksteckol@bama.ua.edu

## Arizona

**Northern Arizona University**
Speech-Language Pathology Prgm
Communication Sciences and Disorders
NAU Box 15045
Flagstaff, AZ 86011-5045
*Prgm Dir:* Katherine Mahosky, MS
*Tel:* 928 523-2969, Ext 7444 *Fax:* 928 523-0034
*E-mail:* katherine.mahosky@nau.edu

**Arizona State University**
Speech-Language Pathology Prgm
Speech and Hearing Science
PO Box 870102
Tempe, AZ 85287-0102
www.asu.edu/clas/shs
*Prgm Dir:* Sid Bacon, PhD
*Tel:* 602 965-2905 *Fax:* 602 965-8516
*E-mail:* spb@asu.edu

**University of Arizona**
Speech-Language Pathology Prgm
Dept of Speech & Hearing
PO Box 210071
Tucson, AZ 85721-0071
*Prgm Dir:* Katheryn Bayles
*Tel:* 520 621-1644 *Fax:* 520 621-9901
*E-mail:* bayles@email.arizona.edu

## Arkansas

**University of Central Arkansas**
Speech-Language Pathology Prgm
Box 4985
Conway, AR 72035-0001
www.uca.edu/chas
*Prgm Dir:* John Lowe III, PhD CCC-SLP
*Tel:* 501 450-3176 *Fax:* 501 450-5474
*E-mail:* jlowe@uca.edu

**University of Arkansas**
Speech-Language Pathology Prgm
Program in Communication Disorders
410 Arkansas Ave
Fayetteville, AR 72701
www.uark.edu/depts/coehp/cdis.html
*Prgm Dir:* Barbara Shadden, PhD CCC-SLP BND
*Tel:* 501 575-4509 *Fax:* 501 575-4507
*E-mail:* bshadde@uark.edu

**University of Arkansas for Medical Sciences**
*Cosponsor:* University of Arkansas at Little Rock
Speech-Language Pathology Prgm
Dept of Audiology & Speech Pathology
2801 S University Ave
Little Rock, AR 72204-1099
www.uams.edu/chrp
*Prgm Dir:* Thomas Guyette, PhD
*Tel:* 501 569-3155 *Fax:* 501 569-3157
*E-mail:* guyettethomasw@uams.edu

**Arkansas State University**
Speech-Language Pathology Prgm
PO Box 910
State University, AR 72467-0904
*Prgm Dir:* Richard A Neeley, PhD
*Tel:* 870 972-3106 *Fax:* 870 972-3788
*E-mail:* meeley@astate.edu

*Additional information about programs that returned the AMA's survey is available in a table beginning on page 432.

# California

**California State University - Chico**
Speech-Language Pathology Prgm
Dept of Communication
1st & Normal Sts
Chico, CA 95929-0350
*Prgm Dir:* Suzanne Miller, PhD
*Tel:* 530 898-4379   *Fax:* 530 898-6612
*E-mail:* sbmiller@csuchico.edu

**California State University - Fresno**
Speech-Language Pathology Prgm
Dept of Communicative Sci & Disorder
5048 N Jackson
Fresno, CA 93740-8022
*Prgm Dir:* Donald Freed
*Tel:* 559 278-2423   *Fax:* 559 278-5187
*E-mail:* donfr@csufresno.edu

**California State University - Fullerton**
Speech-Language Pathology Prgm
Human Communication
800 N State College Blvd
Fullerton, CA 92834
*Prgm Dir:* Edith Li, PhD
*Tel:* 714 278-3617   *Fax:* 714 278-3377
*E-mail:* edithli@fullerton.edu

**California State University - East Bay**
Speech-Language Pathology Prgm
Communicative Sciences and Disorders
25800 Carlos Bee Blvd
Hayward, CA 94542-3065
*Prgm Dir:* Janet Patterson, PhD CCC-SLP
*Tel:* 510 885-7557   *Fax:* 510 885-2186
*E-mail:* janet.patterson@csueastbay.edu

**Loma Linda University**
Speech-Language Pathology Prgm
Speech-Language Pathology & Audiology
Nichol Hall, Rm A506
Loma Linda, CA 92350
*Prgm Dir:* Jean Lowry, PhD
*Tel:* 909 558-4998, Ext 42074   *Fax:* 909 558-4291
*E-mail:* jlowry@sahp.llu.edu

**California State University - Long Beach**
Speech-Language Pathology Prgm
Communicative Disorders
1250 Bellflower Blvd
Long Beach, CA 90840-2501
www.csulb.edu/cdweb
*Prgm Dir:* Carolyn Conway Madding, PhD
*Tel:* 562 985-5283   *Fax:* 562 985-4584
*E-mail:* madding@csulb.edu

**California State University - Los Angeles**
Speech-Language Pathology Prgm
Communication Disorders
5151 State University Dr
Los Angeles, CA 90032
www.calstatela.edu
*Prgm Dir:* Miles Peterson, PhD
*Tel:* 323 343-4690   *Fax:* 323 343-4698
*E-mail:* mpeters@calstatela.edu

**California State University - Northridge**
Speech-Language Pathology Prgm
Dept of Communicative Disorders
18111 Nordhoff St
Northridge, CA 91330-8279
*Prgm Dir:* J Stephen Sinclair, PhD
*Tel:* 818 667-2852   *Fax:* 818 677-2632
*E-mail:* steve.sinclair@csun.edu

**University of Redlands**
Speech-Language Pathology Prgm
University of Redlands -Truesdail Center for Comm
   Disorders
PO Box 3080, 1200 E Colton Ave
Redlands, CA 92373-0999
www.redlands.edu
*Prgm Dir:* Christopher Walker, PhD
*Tel:* 909 748-8061   *Fax:* 909 335-5192
*E-mail:* christopher_walker@redlands.edu

**California State University - Sacramento**
Speech-Language Pathology Prgm
Dept of Speech Pathology & Audiology
6000 J St
Sacramento, CA 95819-6071
*Prgm Dir:* Laureen O'Hanlon, PhD
*Tel:* 916 278-7341   *Fax:* 916 278-7730
*E-mail:* ohanlon@csus.edu

**San Diego State University**
Speech-Language Pathology Prgm
School of Speech, Language & Hearing Sciences
5500 Campanile Dr
San Diego, CA 92182-1518
http://chhs.sdsu.edu/SLHS
*Prgm Dir:* Beverly Wulfeck
*Tel:* 619 594-7746   *Fax:* 619 594-7109
*E-mail:* bwulfeck@mail.sdsu.edu

**San Francisco State University**
Speech-Language Pathology Prgm
1600 Holloway Ave
Burk Hall Rm 104
San Francisco, CA 94132-4158
*Prgm Dir:* Marcia Raggio, PhD
*Tel:* 415 338-1001   *Fax:* 415 338-0916
*E-mail:* mraggio@sfsu.edu

**San Jose State University**
Speech-Language Pathology Prgm
Speech and Hearing Ctr
One Washington Square
San Jose, CA 95192-0079
*Prgm Dir:* Gloria Weddington, PhD
*Tel:* 408 924-3688   *Fax:* 408 924-3641
*E-mail:* novakjm@sjsu.edu

**University of the Pacific**
Speech-Language Pathology Prgm
Dept of Speech-Language Pathology
3601 Pacific Ave
Stockton, CA 95211
www.pacific.edu
*Prgm Dir:* Robert E Hanak, AuD CCC-A
*Tel:* 209 946-2381   *Fax:* 209 946-2647
*E-mail:* rhanyak@pacific.edu

# Colorado

**University of Colorado at Boulder**
Speech-Language Pathology Prgm
2501 Kittredge Loop Rd
Campus Box 409
Boulder, CO 80309-0409
www.colorado.edu/slhs
*Prgm Dir:* Susan M Moore, JD MA-CCC-SLP
*Tel:* 303 492-5284   *Fax:* 303 492-3274
*E-mail:* susan.moore@colorado.edu

**University of Northern Colorado**
Speech-Language Pathology Prgm
Audiology & Speech-Language Sciences
Gunter 1400, Box 140
Greeley, CO 80639-0030
*Prgm Dir:* Ellen Meyer Gregg, PhD CCC-SLP
*Tel:* 970 351-1597   *Fax:* 970 351-2974
*E-mail:* ellen.gregg@unco.edu

# Connecticut

**Southern Connecticut State University**
Speech-Language Pathology Prgm
Communication Disorders
501 Crescent St
New Haven, CT 06515
*Prgm Dir:* Frank E Sansone, PhD
*Tel:* 203 392-5954   *Fax:* 203 392-5968
*E-mail:* sansonef1@southernct.edu

**University of Connecticut**
Speech-Language Pathology Prgm
Communication Sciences
850 Bolton Rd, Unit 1085
Storrs, CT 06269-1085
*Prgm Dir:* Harvey R Gilbert
*Tel:* 860 486-2817   *Fax:* 860 486-5422
*E-mail:* harvey.gilbert@uconn.edu

# District of Columbia

**Gallaudet University**
Speech-Language Pathology Prgm
Audiology & Speech-Lang Path
800 Florida Ave NE
Washington, DC 20002-3695
*Prgm Dir:* James J Mahshie, PhD
*Tel:* 202 651-5329   *Fax:* 202 651-5324
*E-mail:* james.mahshie@gallaudet.edu

**George Washington University**
Speech-Language Pathology Prgm
Dept of Speech and Hearing Science
1922 F St NW
Washington, DC 20052
www.gwu.edu/sphr
*Prgm Dir:* Geralyn M Schulz, PhD CCC-SLP
*Tel:* 202 994-7362   *Fax:* 202 994-2589
*E-mail:* schulz@gwu.edu
*Med Dir:* Michael Bamdad, MA, CCC-SLP

**Howard University**
Speech-Language Pathology Prgm
Comm Sciences and Disorders
525 Bryant St NW
Washington, DC 20059
*Prgm Dir:* Ovetta Harris
*Tel:* 202 806-6990   *Fax:* 202 806-4046
*E-mail:* oharris@howard.edu

**University of the District of Columbia**
Speech-Language Pathology Prgm
Communication Sciences
4200 Connecticut Ave NW
Washington, DC 20008
*Prgm Dir:* April Massey
*Tel:* 202 274-7405   *Fax:* 202 274-5230
*E-mail:* amassey@udc.edu

# Florida

**Florida Atlantic University**
Speech-Language Pathology Prgm
Communication Sciences & Disorders
777 Glades Rd, PO Box 3091
Boca Raton, FL 33431-0991
www.coe.fau.edu/csd/spa.htm
*Prgm Dir:* Deena Louise Wener, PhD CCC-SLP
*Tel:* 561 297-6074   *Fax:* 561 297-2268
*E-mail:* wener@fau.edu

**University of Florida**
Speech-Language Pathology Prgm
Comm Processes and Disorders
335 Dauer Hall, PO Box 117420
Gainesville, FL 32611-7420
*Prgm Dir:* Kenneth J Logan, PhD
*Tel:* 352 392-2113   *Fax:* 352 846-0243
*E-mail:* logan@csd.ufl.edu

**Florida International University**
Speech-Language Pathology Prgm
HLS 143, University Park
Miami, FL 33199
*Prgm Dir:* Elaine Ramos, PhD
*Tel:* 305 348-2710   *Fax:* 305 348-2740
*E-mail:* csd@fiu.edu

**Nova Southeastern University**
Speech-Language Pathology Prgm
Comm Sciences and Disorders
1750 NE 167th st
North Miami, FL 33162-3017
*Prgm Dir:* Wren Newman
*Tel:* 954 262-7726   *Fax:* 954 272-3826
*E-mail:* newmanw@nova.edu

**University of Central Florida**
Speech-Language Pathology Prgm
Communicative Disorders
PO Box 162215
Orlando, FL 32826-2215
*Prgm Dir:* R Jane Lieberman
*Tel:* 407 823-2215   *Fax:* 407 823-4816
*E-mail:* jlieberm@mail.ucf.edu

**Florida State University**
Speech-Language Pathology Prgm
Communication Disorders
107 RRC (R-89)
Tallahassee, FL 32306-1200
*Prgm Dir:* Howard Goldstein
*Tel:* 850 644-2253   *Fax:* 850 644-8994
*E-mail:* howard.goldstein@comm.fsu.edu

**University of South Florida**
Speech-Language Pathology Prgm
Comm Sciences and Disorders
4202 E Fowler Ave, PCD 1017
Tampa, FL 33620-8150
www.cas.usf.edu/csd
*Prgm Dir:* Theresa Chisolm, PhD
*Tel:* 813 974-2006   *Fax:* 813 974-0822
*E-mail:* chisolm@cas.usf.edu

# Georgia

**University of Georgia**
Speech-Language Pathology Prgm
Comm Sciences and Disorders
516 Aderhold Hall
Athens, GA 30602
*Prgm Dir:* Albert R DeChicchis, PhD
*Tel:* 706 542-4561   *Fax:* 706 542-5348
*E-mail:* alde@uga.edu

**Georgia State University**
Speech-Language Pathology Prgm
Communication Disorders
33 Gilmer St SE
Atlanta, GA 30303-3086
*Prgm Dir:* Collen M O'Rourke, PhD
*Tel:* 404 651-2310   *Fax:* 404 651-4901
*E-mail:* corourke@gsu.edu

**Armstrong Atlantic State University**
Speech-Language Pathology Prgm
11935 Abercorn St
Savannah, GA 31419-1997
*Prgm Dir:* Donna R Brooks, PhD
*Tel:* 912 921-7319   *Fax:* 912 961-3054
*E-mail:* brooksdo@mail.armstrong.edu

**Valdosta State University**
Speech-Language Pathology Prgm
Dept of Special Education
1500 N Patterson St
Valdosta, GA 31698-0102
*Prgm Dir:* Corine Myers-Jennings
*Tel:* 229 219-1301   *Fax:* 229 219-1995
*E-mail:* cmjennin@valdosta.edu

# Hawaii

**University of Hawaii at Manoa**
Speech-Language Pathology Prgm
Dept of Speech Pathology & Audiology
1410 Lower Campus Dr
Honolulu, HI 96822
*Prgm Dir:* James T Yates, PhD
*Tel:* 808 956-8279   *Fax:* 808 956-5482
*E-mail:* jyates@hawaii.edu

# Idaho

**Idaho State University**
Speech-Language Pathology Prgm
Communication Sciences & Disorders & Education of
   the Deaf
650 Memorial Dr, Bldg 68, Box 8116
Pocatello, ID 83209-8116
*Prgm Dir:* Joni G Loftin
*Tel:* 208 282-4196   *Fax:* 208 282-4571
*E-mail:* loftjoni@isu.edu

# Illinois

**Southern Illinois Univ at Carbondale**
Speech-Language Pathology Prgm
Comm Disorders and Sciences
1025 Lincoln Dr, Rehn Hall #308
Carbondale, IL 62901-4609
*Prgm Dir:* Kenneth O Simpson, PhD
*Tel:* 618 536-8262   *Fax:* 618 453-8271
*E-mail:* ksimpson@siu.edu

**Univ of Illinois at Urbana - Champaign**
Speech-Language Pathology Prgm
220 Speech & Hearing Sci Bldg
901 S 6th St
Champaign, IL 61820
*Prgm Dir:* Ron D Chambers, PhD
*Tel:* 217 333-2230   *Fax:* 217 244-2235
*E-mail:* rdc@uiuc.edu

**Eastern Illinois University**
Speech-Language Pathology Prgm
Comm Disorders and Sciences
600 Lincoln Ave
Charleston, IL 61920-3099
www.eiu.edu/~commdis
*Prgm Dir:* Gail J Richard, PhD
*Tel:* 217 581-2712   *Fax:* 217 581-7105
*E-mail:* gjrichard@eiu.edu

**Rush University Medical Center**
Speech-Language Pathology Prgm
Communication Disorders and Sciences
1653 W Congress Pkwy
Chicago, IL 60612
*Prgm Dir:* Richard K Peach, PhD
*Tel:* 312 942-3289   *Fax:* 312 942-7211
*E-mail:* richard_k_peach@rush.edu

**Saint Xavier University**
Speech-Language Pathology Prgm
Comm Disorders and Sciences
3700 W 103rd St
Chicago, IL 60655
*Prgm Dir:* Michael Flahive, PhD
*Tel:* 773 298-3566   *Fax:* 773 298-3007
*E-mail:* flahive@sxu.edu

**Northern Illinois University**
Speech-Language Pathology Prgm
Communicative Disorders
DeKalb, IL 60115-2899
*Prgm Dir:* Pamela Jackson, PhD
*Tel:* 815 753-6510   *Fax:* 815 753-9123
*E-mail:* plj@niu.edu

**Southern Illinois Univ at Edwardsville**
Speech-Language Pathology Prgm
Founders Hall, Rm 1300
Campus Box 1147
Edwardsville, IL 62026-1147
www.siue.edu/SECD
*Prgm Dir:* Jean M Harrison, EdD
*Tel:* 618 650-3668   *Fax:* 618 650-3307
*E-mail:* jeharri@siue.edu

**Northwestern University**
Speech-Language Pathology Prgm
Comm Sciences and Disorders
2240 Campus Dr
Evanston, IL 60208
www.communication.northwestern.edu/csd
*Prgm Dir:* Dean C Garstecki, PhD
*Tel:* 847 491-2468   *Fax:* 847 467-7141
*E-mail:* d-garstecki@northwestern.edu

**Western Illinois University**
Speech-Language Pathology Prgm
Dept of Communication
121 Memorial Hall
Macomb, IL 61455-1390
*Prgm Dir:* Maureen Marx, PhD
*Tel:* 309 298-1955, Ext 244   *Fax:* 309 298-2049
*E-mail:* m-marx@wiu.edu

**Illinois State University**
Speech-Language Pathology Prgm
Speech Pathology & Audiology
Fairchild Hall #204
Normal, IL 61790
www.ilstu.edu
*Prgm Dir:* Walter Smoski, PhD
*Tel:* 309 438-8643   *Fax:* 309 438-5221
*E-mail:* abowman@ilstu.edu
*Med Dir:* Heidi Verticchio, MS

**Governors State University**
Speech-Language Pathology Prgm
Communication Disorders
University Park, IL 60466
*Prgm Dir:* Jay Lubinsky
*Tel:* 708 534-4590   *Fax:* 702 235-2195
*E-mail:* j-lubinsky@govst.edu

# Indiana

**Indiana University - Bloomington**
Speech-Language Pathology Prgm
Speech and Hearing Sciences
200 S Jordan Ave
Bloomington, IN 47405-7002
*Prgm Dir:* Phil Connell, PhD
*Tel:* 812 855-4156   *Fax:* 812 855-5531
*E-mail:* pconnell@indiana.edu

**Ball State University**
Speech-Language Pathology Prgm
Speech Pathology & Audiology
2000 University Ave
Muncie, IN 47306
*Prgm Dir:* Mary Jo Germani
*Tel:* 765 285-8160   *Fax:* 765 285-5623
*E-mail:* mgermani@bsu.edu

**Indiana State University**
Speech-Language Pathology Prgm
Department of Communication Disorders
8th & Sycamore St
Terre Haute, IN 47809
http://web.indstate.edu/coe/cd
*Prgm Dir:* Mark A Stimley, PhD
*Tel:* 812 237-2800   *Fax:* 812 237-8137
*E-mail:* csstiml@isugw.indstate.edu

**Purdue University**
Speech-Language Pathology Prgm
Audiology and Speech Sciences
500 Oval Dr
West Lafayette, IN 47907-2038
*Prgm Dir:* Robert Novak
*Tel:* 765 494-3789    *Fax:* 765 494-9771
*E-mail:* novakr@purdue.edu

# Iowa

**University of Northern Iowa**
Speech-Language Pathology Prgm
Communicative Disorders
Cedar Falls, IA 50614-0356
*Prgm Dir:* Clifford Highnam, PhD
*Tel:* 319 273-2577    *Fax:* 319 273-6384
*E-mail:* cliford.highnam@uni.edu

**University of Iowa**
Speech-Language Pathology Prgm
Speech Pathology & Audiology
Iowa City, IA 52242
www.shc.uiowa.edu
*Prgm Dir:* Paul Abbas, PhD
*Tel:* 319 335-8718    *Fax:* 319 335-8851
*E-mail:* paul-abbas@uiowa.edu

# Kansas

**Ft Hays State University**
Speech-Language Pathology Prgm
Communication Disorders
600 Park St
Hays, KS 67601
www.fhsu.edu/commdis
*Prgm Dir:* Amy Finch, PhD
*Tel:* 785 628-5366, Ext 4496    *Fax:* 785 628-5271
*E-mail:* afinch@fhsu.edu

**University of Kansas**
Speech-Language Pathology Prgm
Intercampus Program
3901 Rainbow Rd
Kansas City, KS 66160-7605
*Prgm Dir:* Hugh Catts
*Tel:* 913 588-5937    *Fax:* 913 588-5923
*E-mail:* catts@ku.edu

**Kansas State University**
Speech-Language Pathology Prgm
Comm Sciences and Disorders
Justin Hall 303
Manhattan, KS 66506-1403
*Prgm Dir:* Robert Garcia
*Tel:* 785 532-6879    *Fax:* 785 532-5505
*E-mail:* rgarcia@humec.ksu.edu

**Wichita State University**
Speech-Language Pathology Prgm
Communication Sciences and Disorders
1845 N Fairmount
Wichita, KS 67260-0075
www.wichita.edu/csd
*Prgm Dir:* Kathy Coufal, PhD
*Tel:* 316 978-3171    *Fax:* 316 978-3291
*E-mail:* kathy.coufal@wichita.edu

# Kentucky

**Western Kentucky University**
Speech-Language Pathology Prgm
1 Big Red Way
TPH, Rm 113
Bowling Green, KY 42101-3576
*Prgm Dir:* Joseph Etienne, PhD
*Tel:* 502 745-4302    *Fax:* 502 745-6474
*E-mail:* joseph.etienne@wku.edu

**University of Kentucky**
Speech-Language Pathology Prgm
Communication Disorders
900 S Limestone St, Ste 124G
Lexington, KY 40504-0200
*Prgm Dir:* Judith L Page, PhD
*Tel:* 859 323-1100, Ext 80571    *Fax:* 859 323-8957
*E-mail:* jpage01@uky.edu

**University of Louisville**
Speech-Language Pathology Prgm
Surgery/Graduate Program in Communicative Disorders
Health Sciences Center, Myers Hall
Louisville, KY 40292
*Prgm Dir:* Barbara M Baker, PhD
*Tel:* 502 852-5274    *Fax:* 502 852-0865
*E-mail:* barbara.baker@louisville.edu

**Murray State University**
Speech-Language Pathology Prgm
Communication Disorders
125 Alexander Hall
Murray, KY 42071-3340
*Prgm Dir:* Pearl Gordon Payne, PhD
*Tel:* 502 762-2446    *Fax:* 502 762-3963
*E-mail:* pearl.payne@murraystate.edu

**Eastern Kentucky University**
Speech-Language Pathology Prgm
Communication Disorders
245 Wallace Bldg
Richmond, KY 40475-3102
www.specialed.eku.edu/CD
*Prgm Dir:* Charlotte Hubbard, Ph.D., CCC-SLP
*Tel:* 859 622-4442    *Fax:* 859 622-4443
*E-mail:* charlotte.hubbard@eku.edu

# Louisiana

**Louisiana State Univ and A&M College**
Speech-Language Pathology Prgm
Communication Disorders
Music/Dramatic Arts Bldg #163
Baton Rouge, LA 70803-2606
*Prgm Dir:* Paul R Hoffman
*Tel:* 225 578-2545    *Fax:* 225 578-2528
*E-mail:* cdhoff@lsu.edu

**Southern Univ and A&M College**
Speech-Language Pathology Prgm
Speech Pathology & Audiology
PO Box 11295
Baton Rouge, LA 70813
*Prgm Dir:* Regina Enwefa, PhD
*Tel:* 225 771-3950    *Fax:* 225 771-5652
*E-mail:* carolynp@subr.edu

**Southeastern Louisiana University**
Speech-Language Pathology Prgm
Dept of Special Education
PO Box 10879 - SLU
Hammond, LA 70402
*Prgm Dir:* Paula S Currie
*Tel:* 504 549-2214    *Fax:* 504 549-5030
*E-mail:* pcurrie@selu.edu

**University of Louisiana at Lafayette**
Speech-Language Pathology Prgm
Department of Communicative Disorders
PO Box 43170
Lafayette, LA 70504
http://speechandlanguage.louisiana.edu
*Prgm Dir:* Martin Ball, PhD
*Tel:* 337 482-6721    *Fax:* 337 482-6195
*E-mail:* mjball@louisiana.edu

**University of Louisiana at Monroe**
Speech-Language Pathology Prgm
Department of Communicative Disorders
College of Health Sciences
Monroe, LA 71209
www.ulm.edu/codi
*Prgm Dir:* Judy Fellows, PhD
*Tel:* 318 342-1392    *Fax:* 318 342-3199
*E-mail:* fellows@ulm.edu

**Louisiana State Univ Health Sciences Center**
Speech-Language Pathology Prgm
Communication Disorders
1900 Gravier St
New Orleans, LA 70112
www.alliedhealth.lsuhsc.edu
*Prgm Dir:* Sylvia M Davis, PhD
*Tel:* 504 568-4348    *Fax:* 504 568-4337
*E-mail:* sdavis2@lsuhsc.edu

**Louisiana Tech University**
Speech-Language Pathology Prgm
Dept of Speech
PO Box 3165
Ruston, LA 71272
*Prgm Dir:* J Clarice Dans, PhD
*Tel:* 318 257-4764    *Fax:* 318 257-4492
*E-mail:* cdans@ltparts.latech.edu

**Louisiana State Univ Hlth Sci Ctr-Shreveport**
Speech-Language Pathology Prgm
Mollie E Webb Speech and Hearing Center
3735 Blair Dr
Shreveport, LA 71106
*Prgm Dir:* Thomas W Powell, PhD
*Tel:* 318 632-2015    *Fax:* 318 632-2003
*E-mail:* tpowel@lsuhsc.edu

# Maine

**University of Maine - Orono**
Speech-Language Pathology Prgm
Communication Sciences and Disorders
5724 Dunn Hall
Orono, ME 04469-5724
www.umaine.edu/comscidis
*Prgm Dir:* Nancy E Hall, PhD
*Tel:* 207 581-2006    *Fax:* 207 581-2060
*E-mail:* nhall@maine.edu

# Maryland

**Loyola College of Maryland**
Speech-Language Pathology Prgm
Speech-Language Pathology & Audiology
4501 N Charles St
Baltimore, MD 21210
*Prgm Dir:* Marie R Kerins, EdD CCC-SLP
*Tel:* 410 617-2632, Ext 2632    *Fax:* 410 617-7634
*E-mail:* mkerins@loyola.edu

**Univ of Maryland at College Park**
Speech-Language Pathology Prgm
Hearing and Speech Science
Lefrak Hall
College Park, MD 20742
www.bsos.umd.edu/hesp
*Prgm Dir:* Nan Ratner, EdD CCC
*Tel:* 301 405-4214    *Fax:* 301 314-2023
*E-mail:* nratner@hesp.umd.edu
*Med Dir:* Froma Roth, PhD

**Towson University**
Speech-Language Pathology Prgm
Audiology, Speech Language Pathology & Deaf Studies
8000 York Rd
Towson, MD 21252-0001
www.towson.edu/alsd
*Prgm Dir:* Sharon Glennen, PhD
*Tel:* 410 704-4153    *Fax:* 410 704-4131
*E-mail:* sglennen@towson.edu

# Massachusetts

**University of Massachusetts - Amherst**
Speech-Language Pathology Prgm
Communication Disorders
715 N Pleasant St
Amherst, MA 01003-9304
*Prgm Dir:* Jane Baran, PhD
*Tel:* 413 545-0131  *Fax:* 413 545-0803
*E-mail:* baran@comdis.umass.edu

**Boston University**
Speech-Language Pathology Prgm
Sargent Coll of Hlth and Rehab Sciences
635 Commonwealth Ave
Boston, MA 02215
www.bu.edu/sargent
*Prgm Dir:* Kristine Strand, EdD
*Tel:* 617 353-3188  *Fax:* 617 353-5074
*E-mail:* ksushi@bu.edu

**Emerson College**
Speech-Language Pathology Prgm
Communication Disorders
120 Beacon St
Boston, MA 02116-4624
*Prgm Dir:* Cynthia L Bartlett, PhD
*Tel:* 617 824-8730  *Fax:* 617 824-8735
*E-mail:* Cynthia_Bartlett@emerson.edu

**MGH Institute of Health Professions**
Speech-Language Pathology Prgm
Comm Sciences and Disorders
36 1st Ave, Charlestown Navy Yard
Boston, MA 02129
www.mghihp.edu
*Prgm Dir:* Gregory Lof, PhD CCC-SLP
*Tel:* 617 726-8019  *Fax:* 617 726-8022
*E-mail:* glof@mghihp.edu

**Northeastern University**
Speech-Language Pathology Prgm
Speech-Language Pathology & Audiology
106 Forsyth Bldg, 360 Huntington Ave
Boston, MA 02115
*Prgm Dir:* Linda J Ferrier, PhD
*Tel:* 617 373-3698  *Fax:* 617 373-2239
*E-mail:* l.ferrier@neu.edu

**Worcester State College**
Speech-Language Pathology Prgm
Communication Disorders
486 Chandler St
Worcester, MA 01602-2597
*Prgm Dir:* Linda Larrivee, PhD
*Tel:* 508 929-8055  *Fax:* 508 929-8175
*E-mail:* llarrivee@worcester.edu

# Michigan

**Wayne State University**
Speech-Language Pathology Prgm
Communication Sciences and Disorders
207 Rackham Building
Detroit, MI 48202
www.clas.wayne.edu/CSD
*Prgm Dir:* Alex Johnson, PhD
*Tel:* 313 577-3339  *Fax:* 313 577-8885
*E-mail:* ajohnson@wayne.edu

**Michigan State University**
Speech-Language Pathology Prgm
101 Oyer Bldg
East Lansing, MI 48824-1220
*Prgm Dir:* Michael W Casby, PhD
*Tel:* 517 353-8780  *Fax:* 517 353-3176
*E-mail:* casby@msu.edu

**Western Michigan University**
Speech-Language Pathology Prgm
Speech Pathology & Audiology
Kalamazoo, MI 49008-5355
*Prgm Dir:* John M Hanley, PhD
*Tel:* 616 387-8045  *Fax:* 616 381-8044
*E-mail:* john.hanley@wmich.edu

**Central Michigan University**
Speech-Language Pathology Prgm
Communication Disorders
HPB 2186
Mt Pleasant, MI 48859
www.cmich.edu/cdodept.html
*Prgm Dir:* Renny H Tatchell, PhD CCC-SLP
*Tel:* 989 774-1323  *Fax:* 989 774-2799
*E-mail:* tatch1rh@cmich.edu

**Eastern Michigan University**
Speech-Language Pathology Prgm
110 Porter Bldg
Ypsilanti, MI 48197
*Prgm Dir:* Bill P Cupples
*Tel:* 734 487-3300, Ext 2674  *Fax:* 734 487-2473
*E-mail:* willie.cupples@emich.edu

# Minnesota

**University of Minnesota - Duluth**
Speech-Language Pathology Prgm
Communicative Disorders
1207 Ordean Court, 221 Bohannon Hall
Duluth, MN 55812-9989
*Prgm Dir:* Mark I Mizuko, PhD
*Tel:* 218 726-7974  *Fax:* 218 726-8693
*E-mail:* mmizuko@d.umn.edu

**Minnesota State University - Mankato**
Speech-Language Pathology Prgm
Communication Disorders
103 Armstrong Hall
Mankato, MN 56001
http://ahn.mnsu.edu/cd
*Prgm Dir:* Bruce Poburka, PhD
*Tel:* 507 389-1414  *Fax:* 507 389-2821
*E-mail:* bruce.poburka@mnsu.edu

**University of Minnesota - Minneapolis**
Speech-Language Pathology Prgm
115 Shevlin Hall
164 Pillsbury Dr SE
Minneapolis, MN 55455
www.slhs.umn.edu
*Prgm Dir:* Jennifer Windsor, PhD
*Tel:* 612 624-3322  *Fax:* 612 624-7586
*E-mail:* slhs@umn.edu

**Minnesota State University - Moorhead**
Speech-Language Pathology Prgm
Speech-Language-Hearing Sciences
1104 7th Ave S
Moorhead, MN 56563
www.mnstate.edu/slhs
*Prgm Dir:* Bruce R Hanson, MS
*Tel:* 218 477-2286  *Fax:* 218 477-4392
*E-mail:* hansonbr@mnstate.edu

**St Cloud State University**
Speech-Language Pathology Prgm
A216 Education Bldg
720 4th Ave S
St Cloud, MN 56301
www.stcloudstate.edu/commdisorders
*Prgm Dir:* Monica Devers, PhD
*Tel:* 320 308-2092  *Fax:* 320 308-6441
*E-mail:* mcdevers@stcloudstate.edu

# Mississippi

**Mississippi University for Women**
Speech-Language Pathology Prgm
Speech-Language Pathology/Audiology
PO Box W-1340
Columbus, MS 39701
*Prgm Dir:* Robert F Oyler, PhD
*Tel:* 601 329-7270  *Fax:* 601 329-7460
*E-mail:* royler@muw.edu

**University of Southern Mississippi**
Speech-Language Pathology Prgm
Dept of Speech & Hearing Sciences
PO Box 5092
Hattiesburg, MS 39406-5092
www.usm.edu/shs
*Prgm Dir:* Brett Kemker, PhD
*Tel:* 601 266-5216  *Fax:* 601 266-5224
*E-mail:* brett.kemker@usm.edu

**Jackson State University**
Speech-Language Pathology Prgm
Dept of Communication Disorders
University Center, 3825 Ridgewood Rd, Box 23
Jackson, MS 39211-6453
*Prgm Dir:* Zenobia Bagli, PhD
*Tel:* 601 432-6713  *Fax:* 601 432-6844
*E-mail:* zbagli@jsums.edu

**University of Mississippi**
Speech-Language Pathology Prgm
Communicative Disorders
PO Box 1848
University, MS 38677
*Prgm Dir:* Carolyn Wiles Higdon, PhD
*Tel:* 662 915-7652  *Fax:* 662 915-5717
*E-mail:* chigdon@olemiss.edu

# Missouri

**Southeast Missouri State University**
Speech-Language Pathology Prgm
Dept of Communication Disorders
One University Plaza
Cape Girardeau, MO 63701-4799
www.semo.edu/commdisorders
*Prgm Dir:* Sakina S Drummond, PhD
*Tel:* 573 651-2155  *Fax:* 573 651-2155
*E-mail:* ssdrummond@semovm.semo.edu

**University of Missouri - Columbia**
Speech-Language Pathology Prgm
Communicative Disorders
303 Lewis Hall
Columbia, MO 65211
www.umshp.org/csd
*Prgm Dir:* Philip S Dale, PhD
*Tel:* 573 882-3873  *Fax:* 573 884-8686
*E-mail:* dalep@health.missouri.edu

**Rockhurst University**
Speech-Language Pathology Prgm
Communication Sciences & Disorders
1100 Rockhurst Rd
Kansas City, MO 64110
*Prgm Dir:* Dennis Ingrisano, PhD
*Tel:* 816 501-4742  *Fax:* 816 501-4169
*E-mail:* dennis.ingrisano@rockhurst.edu

**Truman State University**
Speech-Language Pathology Prgm
Communication Disorders
Barnett Hall 222
Kirksville, MO 63501
http://comdis.truman.edu
*Prgm Dir:* Janet L Gooch, PhD
*Tel:* 660 785-4669  *Fax:* 660 785-7424
*E-mail:* jquinzer@truman.edu

**Missouri State University**
Speech-Language Pathology Prgm
Communication Disorders
901 S National Ave
Springfield, MO 65804-0095
*Prgm Dir:* Neil J DiSarno, PhD
*Tel:* 417 836-5368, Ext 66511    *Fax:* 417 836-4242
*E-mail:* neildisarno@missouristate.edu

**Fontbonne University**
Speech-Language Pathology Prgm
Communication Disorders & Deaf Education
6800 Wydown Blvd
St Louis, MO 63105
www.fontbonne.edu
*Prgm Dir:* Lynne Shields, PhD CCC-SLP
*Tel:* 314 889-1407    *Fax:* 314 719-8016
*E-mail:* lshields@fontbonne.edu

**Saint Louis University**
Speech-Language Pathology Prgm
3570 Lindell Blvd
McGannon 23
St Louis, MO 63108
www.slu.edu/colleges/cops/cd
*Prgm Dir:* Travis Threats, PhD CCC-SLP
*Tel:* 314 977-2940    *Fax:* 314 977-3360
*E-mail:* threatst@slu.edu
*Med Dir:* Kathy Murphy, MA, CCC-SLP

**Central Missouri State University**
Speech-Language Pathology Prgm
Speech Pathology & Audiology
Martin Bldg 41
Warrensburg, MO 64093
*Prgm Dir:* Carl Harlan
*Tel:* 660 543-4918    *Fax:* 660 543-8234
*E-mail:* harlan@cmsu1.cmsu.edu

# Nebraska

**University of Nebraska - Kearney**
Speech-Language Pathology Prgm
Communication Disorders Department
College of Education
Kearney, NE 68849-4597
www.unk.edu/departments/cdis
*Prgm Dir:* Laurence M Hilton, PhD CCC-SLP
*Tel:* 308 865-8300    *Fax:* 308 865-8397
*E-mail:* hiltonlm@unk.edu

**University of Nebraska - Lincoln**
Speech-Language Pathology Prgm
Dept of Special Education & Communication Disorders
301 Barkley Center
Lincoln, NE 68583-0738
www.unl.edu/barkley
*Prgm Dir:* John Bernthal, PhD
*Tel:* 402 472-5496    *Fax:* 402 472-7697
*E-mail:* jbernthal1@unl.edu

**University of Nebraska - Omaha**
Speech-Language Pathology Prgm
Special Education & Communication Disorders
6001 Dodge St
Omaha, NE 68182-0054
*Prgm Dir:* Mary J Friehe, PhD
*Tel:* 402 554-2211    *Fax:* 402 554-3572
*E-mail:* mfriehe@mail.unomaha.edu

# Nevada

**University of Nevada - Reno**
Speech-Language Pathology Prgm
Dept of Speech Pathology & Audiology
Redfield Bldg, 152
Reno, NV 89557-0046
*Prgm Dir:* Thomas Watterson, PhD
*Tel:* 775 784-4887    *Fax:* 775 784-4095
*E-mail:* tw@unr.edu

# New Hampshire

**University of New Hampshire**
Speech-Language Pathology Prgm
Dept of Communication Sciences and Disorders
4 Library Way, Hewitt Hall
Durham, NH 03824
www.unh.edu/communication-disorders
*Prgm Dir:* Stephen Calculator, PhD CCC-SLP
*Tel:* 603 862-3836    *Fax:* 603 862-4511
*E-mail:* stephen.calculator@unh.edu

# New Jersey

**College of New Jersey**
Speech-Language Pathology Prgm
Special Education, Language and Literacy, Speech
    Pathology
2000 Pennington Rd
Ewing, NJ 08628-0718
*Prgm Dir:* Jasper B Phelps, PhD
*Tel:* 609 771-2308    *Fax:* 609 637-5172
*E-mail:* phelps@tcnj.edu

**Seton Hall University**
Speech-Language Pathology Prgm
400 S Orange Ave
South Orange, NJ 07079-2689
www.shu.edu
*Prgm Dir:* Robert Orlikoff, PhD CCC-SLP
*Tel:* 973 313-6185    *Fax:* 973 275-2171
*E-mail:* orlikoro@shu.edu

**Kean University**
Speech-Language Pathology Prgm
Dept of Communication Disorders and Deafness
1000 Morris Ave
Union, NJ 07083
www.kean.edu/~keangrad/grad_CE_slp.htm
*Prgm Dir:* Barbara Glazewski, EdD
*Tel:* 908 737-5407    *Fax:* 908 527-3232
*E-mail:* bglazews@kean.edu

**Montclair State University**
Speech-Language Pathology Prgm
Dept of Communication Sciences & Disorders
Speech Bldg
Upper Montclair, NJ 07043
*Prgm Dir:* Sarita Eisenberg
*Tel:* 973 655-7363    *Fax:* 973 655-7072
*E-mail:* speechclinic@mail.montclair.edu

**William Paterson Univ of New Jersey**
Speech-Language Pathology Prgm
Communication Disorders
300 Pompton Rd
Wayne, NJ 07470
*Prgm Dir:* Carole E Gelfer
*Tel:* 973 720-2208    *Fax:* 973 720-3357
*E-mail:* gelferc@wpunj.edu

# New Mexico

**University of New Mexico**
Speech-Language Pathology Prgm
Dept of Speech & Hearing Sciences
MSC01 1195, 1 University of New Mexico
1700 Lomas Blvd NE
Albuquerque, NM 87131-0001
www.unm.edu/~sphrsci
*Prgm Dir:* Janet L Patterson, PhD
*Tel:* 505 277-4453    *Fax:* 505 277-0968
*E-mail:* jpatters@unm.edu

**New Mexico State University**
Speech-Language Pathology Prgm
Dept of Special Education/Communication Disorders
PO Box 30001, MSC3SPE
Las Cruces, NM 88003
www.nmsu.edu
*Prgm Dir:* Connie E Stout, PhD
*Tel:* 505 646-2402    *Fax:* 505 646-7712
*E-mail:* cestout@nmsu.edu

**Eastern New Mexico University**
Speech-Language Pathology Prgm
Dept of Communicative Disorders
Station 3
Portales, NM 88130
*Prgm Dir:* Linda J Weems, PhD
*Tel:* 505 562-2156    *Fax:* 505 562-2380
*E-mail:* linda.weems@enmu.edu

# New York

**College of St Rose**
Speech-Language Pathology Prgm
Communication Disorders Dept
432 Western Ave, Box 100
Albany, NY 12203-1490
*Prgm Dir:* David DeBonis, PhD
*Tel:* 518 458-5461    *Fax:* 518 458-5446
*E-mail:* coopermd@rosnet.strose.edu

**CUNY Herbert H Lehman College**
Speech-Language Pathology Prgm
Dept of Speech-Language-Hearing Sciences
250 Bedford Park Blvd W
Bronx, NY 10468-1589
*Prgm Dir:* Joyce West, PhD
*Tel:* 718 960-8134    *Fax:* 718 960-7376
*E-mail:* speech@lehman.cuny.edu

**CUNY Brooklyn College**
Speech-Language Pathology Prgm
Speech-Language Pathology & Audiology
2900 Bedford Ave
Brooklyn, NY 11210
www.brooklyn.cuny.edu
*Prgm Dir:* Gail B Gurland, PhD
*Tel:* 718 951-5186    *Fax:* 718 951-4363
*E-mail:* ggurland@brooklyn.cuny.edu

**Long Island University - Brooklyn Campus**
Speech-Language Pathology Prgm
Dept of Communication Sciences and Disorders
One University Plaza
Brooklyn, NY 11201-8423
*Prgm Dir:* Elaine Geller, PhD
*Tel:* 718 780-4122    *Fax:* 718 780-4007
*E-mail:* elaine.geller@liu.edu

**Touro College**
Speech-Language Pathology Prgm
1610 E 19th St
Brooklyn, NY 11229
*Prgm Dir:* Hindy Lubinsky
*Tel:* 718 787-1602    *Fax:* 718 787-1137
*E-mail:* Hlubintouro@yahoo.com

**Long Island University - C W Post Campus**
Speech-Language Pathology Prgm
Communication Sciences and Disorders
720 Northern Blvd
Brookville, NY 11548-1300
*Prgm Dir:* Dianne Slavin, PhD
*Tel:* 516 299-2436    *Fax:* 516 299-3151
*E-mail:* dslavin@liu.edu

**Buffalo State, SUNY**
Speech-Language Pathology Prgm
208 Ketchum Hall
1300 Elmwood Ave
Buffalo, NY 14222
www.buffalostate.edu/speech
*Prgm Dir:* Constance Dean Qualls, PhD
*Tel:* 716 878-5502   *Fax:* 716 878-5711
*E-mail:* westlenl@buffalostate.edu

**University at Buffalo - SUNY**
Speech-Language Pathology Prgm
Dept of Communicative Disorders & Sciences
3435 Main St, 122 Cary Hall
Buffalo, NY 14214-3005
*Prgm Dir:* Elaine Stathopoulos, PhD
*Tel:* 716 829-2797   *Fax:* 716 829-3979
*E-mail:* stathop@buffalo.edu

**Mercy College**
Speech-Language Pathology Prgm
555 Broadway, Main Hall, Rm G15
Dobbs Ferry, NY 10522
*Prgm Dir:* Helen Buhler, PhD
*Tel:* 914 674-7743   *Fax:* 914 674-7597
*E-mail:* hbuhler@mercy.edu

**SUNY College at Fredonia**
Speech-Language Pathology Prgm
Dept of Speech Pathology & Audiology
W123 Thompson Hall
Fredonia, NY 14063
*Prgm Dir:* Kim L Tillery, PhD
*Tel:* 716 673-3202   *Fax:* 716 673-3235
*E-mail:* tillery@fredonia.edu

**Adelphi University**
Speech-Language Pathology Prgm
Hy Weinberg Center
158 Cambridge Ave
Garden City, NY 11530
www.adelphi.edu
*Prgm Dir:* Susan Lederer, PhD
*Tel:* 516 877-4770   *Fax:* 516 877-4783
*E-mail:* lederer@adelphi.edu

**SUNY College of Geneseo**
Speech-Language Pathology Prgm
Dept of Communicative Disorders & Sciences
1 College Circle, 218 Sturgis
Geneseo, NY 14454
*Prgm Dir:* Linda I House, PhD
*Tel:* 716 245-5328   *Fax:* 716 345-5434
*E-mail:* house@geneseo.edu

**Hofstra University**
Speech-Language Pathology Prgm
Dept of Speech- Language- Hearing Sciences
110 Hofstra University
Hempstead, NY 11550
www.hofstra.edu
*Prgm Dir:* Carole Ferrand, PhD
*Tel:* 516 463-5508   *Fax:* 516 463-5260
*E-mail:* sphctf@hofstra.edu

**Ithaca College**
Speech-Language Pathology Prgm
Speech Pathology & Audiology
953 Danby Rd
Ithaca, NY 14850-7185
http://departments.ithaca.edu/slpa
*Prgm Dir:* E W Testut, PhD
*Tel:* 607 274-3248   *Fax:* 607 274-1137
*E-mail:* testut@ithaca.edu

**SUNY at New Paltz**
Speech-Language Pathology Prgm
Dept of Communication Disorders
600 Hawk Dr
New Paltz, NY 12561-2499
www.newpaltz.edu
*Prgm Dir:* Elizabeth Hester, PhD
*Tel:* 845 257-3600   *Fax:* 845 257-3605
*E-mail:* hestere@newpaltz.edu

**Columbia University Teachers College**
Speech-Language Pathology Prgm
Speech and Language Pathology and Audiology
525 W 120th St, Box 206
New York, NY 10027
*Prgm Dir:* John H Saxman, PhD
*Tel:* 212 678-3895   *Fax:* 212 678-8233
*E-mail:* saxman@tc.columbia.edu

**CUNY Hunter College**
Speech-Language Pathology Prgm
Communication Sciences Program
425 E 25th St
New York, NY 10010-2590
www.hunter.cuny.edu/schoolhp/comsc
*Prgm Dir:* Dava E Waltzman, PhD
*Tel:* 212 481-4467   *Fax:* 212 481-4458
*E-mail:* dwaltzma@hunter.cuny.edu

**New York University**
Speech-Language Pathology Prgm
Dept of Speech-Language Pathology & Audiology
719 Broadway, Ste 200
New York, NY 10003
www.nyu.edu/education/speech
*Prgm Dir:* Celia Stewart, PhD
*Tel:* 212 998-5230   *Fax:* 212 995-4356
*E-mail:* cs8@nyu.edu

**SUNY College at Plattsburgh**
Speech-Language Pathology Prgm
Dept of Communication Disorders and Sciences
224 Sibley Hall, 101 Broad St
Plattsburgh, NY 12901
*Prgm Dir:* Patrick Coppens, PhD
*Tel:* 518 564-2170   *Fax:* 518 564-5110
*E-mail:* patrick.coppens@Plattsburgh.edu

**St John's University**
Speech-Language Pathology Prgm
Speech and Hearing Ctr
8000 Utopia Pkwy
Queens, NY 11439
http://www.stjohns.edu
*Prgm Dir:* Donna Geffner, CCC-SLP CCC-A
*Tel:* 718 990-6480   *Fax:* 718 990-1917

**Nazareth College of Rochester**
Speech-Language Pathology Prgm
Dept of Comm Sci & Disorder Speech-Lang Pathology
4245 East Ave
Rochester, NY 14618-3790
www.naz.edu/dept/speech
*Prgm Dir:* Lisa Durant-Jones, MS
*Tel:* 585 389-2775   *Fax:* 585 389-2791
*E-mail:* ldurant4@naz.edu

**Syracuse University**
Speech-Language Pathology Prgm
Dept of Communication Sciences & Disorders
805 S Crouse Ave
Syracuse, NY 13244-2280
*Prgm Dir:* Raymond H Colton, PhD
*Tel:* 315 443-9637   *Fax:* 315 443-1113
*E-mail:* rhcolton@syr.edu

**New York Medical College**
Speech-Language Pathology Prgm
School of Public Health
Valhalla, NY 10595
*Prgm Dir:* Ben C Watson
*Tel:* 914 594-4239   *Fax:* 914 594-4853
*E-mail:* slp_sph@nymc.edu

# North Carolina

**Appalachian State University**
Speech-Language Pathology Prgm
Dept of Language, Reading & Exceptionalities
124 Edwin Duncan Hall, Box 32085
Boone, NC 28608-2085
*Prgm Dir:* Donna M Brown
*Tel:* 828 262-2182   *Fax:* 828 262-6767
*E-mail:* browndm@appstate.edu

**Univ of North Carolina at Chapel Hill**
Speech-Language Pathology Prgm
Division of Speech & Hearing Sciences
CB 7190 Wing D Medical School
Chapel Hill, NC 27599-7190
*Prgm Dir:* Jackson Roush, PhD
*Tel:* 919 966-1006   *Fax:* 919 966-0100
*E-mail:* jroush@med.unc.edu

**Western Carolina University**
Speech-Language Pathology Prgm
Human Services, Communication Sciences and
   Disorders
Killian 204, G30 McKee Bldg
Cullowhee, NC 28723-9043
*Prgm Dir:* Billy T Ogletree, PhD
*Tel:* 828 227-7310   *Fax:* 828 227-7021
*E-mail:* fischer@email.wcu.edu

**North Carolina Central University**
Speech-Language Pathology Prgm
Department of Communication Disorders
712 Cecil St
Durham, NC 27707
www.nccu.edu/soe/departments/communications/
   communication_index.htm
*Prgm Dir:* Diane M Scott, PhD
*Tel:* 919 530-7473   *Fax:* 919 530-7975
*E-mail:* discott@nccu.edu

**Univ of North Carolina at Greensboro**
Speech-Language Pathology Prgm
Communication Sciences & Disorders
300 Ferguson Bldg, UNCG - Box 26170
Greensboro, NC 27402-6170
www.uncg.edu/csd
*Prgm Dir:* Celia R Hooper, PhD
*Tel:* 336 334-4657   *Fax:* 336 334-4475
*E-mail:* crhooper@uncg.edu

**East Carolina University**
Speech-Language Pathology Prgm
Dept of Communication Sciences & Disorders
Greenville, NC 27858-4353
*Prgm Dir:* Gregg D Givens
*Tel:* 252 328-4405   *Fax:* 252 328-4469
*E-mail:* givensg@ecu.edu

# North Dakota

**University of North Dakota**
Speech-Language Pathology Prgm
Dept of Communication Sciences & Disorders
PO Box 8040
Grand Forks, ND 58202-8040
*Prgm Dir:* Wayne Swisher, PhD
*Tel:* 701 777-3232   *Fax:* 701 777-3650
*E-mail:* wayne_swisher@und.nodak.edu

**Minot State University**
Speech-Language Pathology Prgm
Dept of Communication Disorders and
   Special Education
500 University Ave W
Minot, ND 58707
www.minotstateu.edu
*Prgm Dir:* Thomas A Linares, PhD
*Tel:* 701 858-3031   *Fax:* 701 858-3845
*E-mail:* thomas.linares@minotstateu.edu

# Ohio

**University of Akron**
Speech-Language Pathology Prgm
School of Speech-Language & Audiology
Polsky Building, Rm 188K
Akron, OH 44325-3001
www.uakron.edu/sslpa
*Prgm Dir:* Roberta DePompei, PhD
*Tel:* 330 972-6114   *Fax:* 330 972-7884
*E-mail:* rdepom1@uakron.edu

**Ohio University**
Speech-Language Pathology Prgm
School of Hearing, Speech and Language Sciences
Grover Center W218
Athens, OH 45701-2979
*Prgm Dir:* Brooke Hallowell, PhD
*Tel:* 740 593-0903   *Fax:* 740 593-0287
*E-mail:* hallowel@ohio.edu

**Bowling Green State University**
Speech-Language Pathology Prgm
Dept of Communication Disorders
200 Health Center
Bowling Green, OH 43403-0149
www.bgsu.edu/departments
*Prgm Dir:* Larry Small, PhD
*Tel:* 419 372-6031   *Fax:* 419 372-8089
*E-mail:* lsmall@bgsu.edu

**University of Cincinnati**
Speech-Language Pathology Prgm
Communication Sciences & Disorders
Mail Station 379
Cincinnati, OH 45221-0379
*Prgm Dir:* Nancy Creaghead, PhD
*Tel:* 513 558-8501   *Fax:* 513 558-8500
*E-mail:* nancy.creaghead@uc.edu

**Case Western Reserve University**
Speech-Language Pathology Prgm
Dept of Communication Sciences
11206 Euclid Ave
Cleveland, OH 44106-7154
www.case.edu/artsci/cosi
*Prgm Dir:* Angela Ciccia, PhD
*Tel:* 216 368-2470   *Fax:* 216 368-6078
*E-mail:* cosigrad@case.edu

**Cleveland State University**
Speech-Language Pathology Prgm
Dept of Speech and Hearing
2121 Euclid Ave, MC 430
Cleveland, OH 44115
*Prgm Dir:* Ben Wallace, PhD
*Tel:* 216 687-6986   *Fax:* 216 687-6983
*E-mail:* b.wallace@csuohio.edu

**Ohio State University**
Speech-Language Pathology Prgm
Dept of Speech and Hearing Sciences
110 Pressey Hall, 1070 Carmack Rd
Columbus, OH 43210-1002
*Prgm Dir:* Robert Allen Fox, PhD
*Tel:* 614 292-8207   *Fax:* 614 292-7504
*E-mail:* fox.2@osu.edu

**Kent State University**
Speech-Language Pathology Prgm
School of Speech Pathology & Audiology
A104 Music & Speech Bldg
Kent, OH 44242
http://dept.kent.edu/spa
*Prgm Dir:* Lynn E Rowan, PhD
*Tel:* 330 672-2672   *Fax:* 330 672-2643
*E-mail:* lrowan@kent.edu

**Miami University**
Speech-Language Pathology Prgm
Dept of Speech Pathology and Audiology
2 Bachelor Hall
Oxford, OH 45056-3414
*Prgm Dir:* Kathleen M Hutchinson
*Tel:* 513 529-2500   *Fax:* 513 529-2502
*E-mail:* hutchik@muohio.edu

**University of Toledo**
Speech-Language Pathology Prgm
Public Health and Rehabilitative Services
2801 W Bancroft St
Toledo, OH 43606
www.utoledo.edu
*Prgm Dir:* Lee W Ellis, PhD
*Tel:* 419 530-4065   *Fax:* 419 530-8774
*E-mail:* lee.ellis@utoledo.edu

# Oklahoma

**University of Central Oklahoma**
Speech-Language Pathology Prgm
Special Services, Speech-Language Pathology
100 N University Dr
Edmond, OK 73034
*Prgm Dir:* Scott F McLaughlin, PhD
*Tel:* 405 974-5705   *Fax:* 405 974-3822
*E-mail:* SMcLaughlin@ucok.edu

**Univ of Oklahoma Health Sciences Center**
Speech-Language Pathology Prgm
Communication Disorders
825 NE 14th, PO Box 26901
Oklahoma City, OK 73190
www.ouhsc.edu/ahealth/faccsd.htm
*Prgm Dir:* Stephen Painton, PhD
*Tel:* 405 271-4214, Ext 46054   *Fax:* 405 271-3360
*E-mail:* stephen-painton@ouhsc.edu

**Oklahoma State University**
Speech-Language Pathology Prgm
Dept of Communication Sciences & Disorders
110 Hanner Bldg
Stillwater, OK 74078-5062
www.cas.okstate.edu/cdis
*Prgm Dir:* Randolph Deal, PhD
*Tel:* 405 744-6021   *Fax:* 405 744-8070
*E-mail:* randolph.deal@okstate.edu

**Northeastern State University**
Speech-Language Pathology Prgm
Speech-Language Pathology Program
600 N Vinita, Special Services Bldg
Tahlequah, OK 74464-7051
*Prgm Dir:* Karen Patterson, PhD
*Tel:* 918 456-5511, Ext 3769   *Fax:* 918 458-9605
*E-mail:* pattersk@nsuok.edu

**University of Tulsa**
Speech-Language Pathology Prgm
Department of Communication Disorders
600 S College Ave
Tulsa, OK 74104-3189
*Prgm Dir:* Paula Cadogan, EdD
*Tel:* 918 631-2504   *Fax:* 918 631-3668
*E-mail:* paula-cadogan@utulsa.edu

# Oregon

**University of Oregon**
Speech-Language Pathology Prgm
Communication Disorders & Sciences
5284 University of Oregon
Eugene, OR 97403-5284
*Prgm Dir:* Kathleen Roberts, PhD CCC-A
*Tel:* 541 346-2480   *Fax:* 541 346-2564
*E-mail:* robertsk@uoregon.edu

**Portland State University**
Speech-Language Pathology Prgm
Speech and Hearing Sciences
PO Box 751
Portland, OR 97207-0751
*Prgm Dir:* Thomas Dolan, PhD
*Tel:* 503 725-3533   *Fax:* 503 725-9171
*E-mail:* dolant@pdx.edu

# Pennsylvania

**Bloomsburg University**
Speech-Language Pathology Prgm
Dept of Audiology & Speech Pathology
400 E 2nd St, Centennial Hall
Bloomsburg, PA 17815-1301
*Prgm Dir:* Dianne Angelo, PhD
*Tel:* 570 389-4436   *Fax:* 570 389-5022
*E-mail:* dangelo@bloomu.edu

**California University of Pennsylvania**
Speech-Language Pathology Prgm
Dept of Communication Disorders
250 University Ave
California, PA 15419
*Prgm Dir:* Barbara H Bonfanti, PhD
*Tel:* 724 938-4175   *Fax:* 724 938-1526
*E-mail:* bonfanti@cup.edu

**Clarion University of Pennsylvania**
Speech-Language Pathology Prgm
Dept of Communication Sciences & Disorders
840 Wood St, 118 Keeling Health Center
Clarion, PA 16214-1232
*Prgm Dir:* Colleen A McAleer, PhD
*Tel:* 814 393-2581   *Fax:* 814 393-2206
*E-mail:* cmcaleer@clarion.edu

**East Stroudsburg University**
Speech-Language Pathology Prgm
Speech Pathology & Audiology
LaRue Hall
East Stroudsburg, PA 18301-2999
*Prgm Dir:* Jane Page, PhD
*Tel:* 570 422-3247   *Fax:* 570 422-3850
*E-mail:* jpage@po-box.esu.edu

**Edinboro University of Pennsylvania**
Speech-Language Pathology Prgm
Speech, Language and Hearing Department
115A Compton Hall
Edinboro, PA 16444
www.edinboro.edu
*Prgm Dir:* Charlotte Molrine, PhD
*Tel:* 814 732-2432   *Fax:* 814 732-2612
*E-mail:* cmolrine@edinboro.edu

**Indiana University of Pennsylvania**
Speech-Language Pathology Prgm
Special Education & Clinical Svcs
203 Davis Hall, 570 S 11th St
Indiana, PA 15705
www.iup.edu
*Prgm Dir:* David W Stein, PhD
*Tel:* 724 357-2450   *Fax:* 724 357-7716
*E-mail:* dwstein@iup.edu

**La Salle University**
Speech-Language Pathology Prgm
1900 W Olney Ave
Philadelphia, PA 19141
www.lasalle.edu/speech
*Prgm Dir:* Barbara Amster, PhD CCC-SLP
*Tel:* 215 951-1986   *Fax:* 215 951-5171
*E-mail:* amster@lasalle.edu

**Temple University**
Speech-Language Pathology Prgm
Communication Sciences
1701 N 13th St, 109 Weiss Hall
Philadelphia, PA 19122
*Prgm Dir:* Brian Goldstein, PhD
*Tel:* 215 204-7543   *Fax:* 215 204-5954
*E-mail:* CHP@temple.edu

**Duquesne University**
Speech-Language Pathology Prgm
600 Forbes Ave
Pittsburgh, PA 15282
www.slp.duq.edu
*Prgm Dir:* Mikael Kimelman, PhD
*Tel:* 412 396-4225   *Fax:* 412 396-4196
*E-mail:* speech-lang@duq.edu

**University of Pittsburgh**
Speech-Language Pathology Prgm
Communication Science and Disorders
4033 Fobes Tower - Atwood St
Pittsburgh, PA 15260
www.shrs.pitt.edu/csd
*Prgm Dir:* Malcolm McNeil, PhD
*Tel:* 412 383-6540   *Fax:* 412 383-6555
*E-mail:* mcneil@pitt.edu

**Marywood University**
Speech-Language Pathology Prgm
Dept of Communication Sciences & Disorders
McGowan Center - 2300 Adams Ave
Scranton, PA 18509-1598
www.marywood.edu
*Prgm Dir:* Mona Griffer, EdD CCC-SLP
*Tel:* 570 348-6211, Ext 2363   *Fax:* 570 961-4708
*E-mail:* griffer@marywood.edu

**Penn State University**
Speech-Language Pathology Prgm
Dept of Communication Sciences and Disorders
110 Moore Bldg
University Park, PA 16802-3100
http://csd.hhdev.psu.edu/grad
*Prgm Dir:* Gordon Blood, PhD
*Tel:* 814 865-3177   *Fax:* 814 863-3759
*E-mail:* F2X@psu.edu

**West Chester University**
Speech-Language Pathology Prgm
Dept of Communicative Disorders
201 Carter Dr
West Chester, PA 19383
*Prgm Dir:* Michael S Weiss, PhD
*Tel:* 610 436-3401   *Fax:* 610 436-3388
*E-mail:* mweiss@wcupa.edu

# Rhode Island

**University of Rhode Island**
Speech-Language Pathology Prgm
Dept of Communicative Disorders
Independence Sq, Ste 1
Kingston, RI 02881-0821
*Prgm Dir:* Jay Singer, PhD
*Tel:* 401 874-5969   *Fax:* 401 874-4404
*E-mail:* DrJay@URI.edu

# South Carolina

**Medical University of South Carolina**
Speech-Language Pathology Prgm
Communication Sciences and Disorders Program
77 President St, Ste 117
Charleston, SC 29425
*Prgm Dir:* Jennifer Horner, PhD
*Tel:* 843 792-0365   *Fax:* 843 792-0710
*E-mail:* hornerj@musc.edu

**University of South Carolina**
Speech-Language Pathology Prgm
Dept of Communication Sciences and Disorders
Arnold School of Public Health
800 Sumter Street
Columbia, SC 29208
*Prgm Dir:* Elaine M Frank, PhD
*Tel:* 803 777-5052   *Fax:* 803 777-3081
*E-mail:* efrank@sc.edu

**South Carolina State University**
Speech-Language Pathology Prgm
Speech Pathology & Audiology
300 College St NE, PO Box 7427
Orangeburg, SC 29117
*Prgm Dir:* Gwendolyn Wilson, EdD
*Tel:* 803 536-8074   *Fax:* 803 536-8593
*E-mail:* gdwilson@scsu.edu

# South Dakota

**University of South Dakota**
Speech-Language Pathology Prgm
Dept of Communication Disorders
414 E Clark St
Vermillion, SD 57069-2390
www.usd.edu/dcom
*Prgm Dir:* Teri J Bellis, PhD
*Tel:* 605 677-5474   *Fax:* 605 677-5767
*E-mail:* tbellis@usd.edu

# Tennessee

**East Tennessee State University**
Speech-Language Pathology Prgm
Dept of Communicative Disorders
PO Box 70643
Johnson City, TN 37614-0643
*Prgm Dir:* Nancy J Scherer, PhD
*Tel:* 423 439-4272   *Fax:* 423 439-4607
*E-mail:* scherern@etsu.edu

**University of Tennessee - Knoxville**
Speech-Language Pathology Prgm
Dept of Audiology & Speech
578 S Stadium Hall
Knoxville, TN 37996-0740
*Prgm Dir:* Patricia M Visser
*Tel:* 865 974-5019   *Fax:* 865 974-1539
*E-mail:* pvisser@utk.edu

**University of Memphis**
Speech-Language Pathology Prgm
School of Audiology & Speech-Language Pathology
807 Jefferson Ave
Memphis, TN 38105
www.ausp.memphis.edu
*Prgm Dir:* Maurice Mendel, PhD
*Tel:* 901 678-5800   *Fax:* 901 525-1282
*E-mail:* mmendel@memphis.edu

**Tennessee State University**
Speech-Language Pathology Prgm
330 10th Ave N, Ste A
PO Box 131
Nashville, TN 37203-3401
*Prgm Dir:* Harold R Mitchell, PhD
*Tel:* 615 963-7081   *Fax:* 615 963-7119
*E-mail:* hmitchell@tnstate.edu

**Vanderbilt University Medical Center**
Speech-Language Pathology Prgm
Hearing and Speech Sciences
1114 19th Ave S
Nashville, TN 37212
*Prgm Dir:* Fred H Bess, PhD
*Tel:* 615 936-5000   *Fax:* 615 936-5014
*E-mail:* fred.h.bess@vanderbilt.edu

# Texas

**Abilene Christian University**
Speech-Language Pathology Prgm
ACU Box 28058
Abilene, TX 79699-8058
*Prgm Dir:* MaLesa Breeding
*Tel:* 325 674-2074   *Fax:* 325 674-2552
*E-mail:* breedingm@acu.edu

**University of Texas at Austin**
Speech-Language Pathology Prgm
Communication Sciences & Disorders
1 University Station, A1100
Austin, TX 78712-1089
*Prgm Dir:* Craig Champlin, PhD
*Tel:* 512 471-4119   *Fax:* 512 471-2957
*E-mail:* champlin@mail.utexas.edu

**Lamar Institute of Technology**
Speech-Language Pathology Prgm
Dept of Speech and Hearing Sciences
Box 10076, Lamar Station
Beaumont, TX 77710
*Prgm Dir:* William Harn, PhD
*Tel:* 409 880-7655   *Fax:* 409 880-2265
*E-mail:* william.harn@lamar.edu

**West Texas A&M University**
Speech-Language Pathology Prgm
PO Box 60757
Canyon, TX 79016-0001
*Prgm Dir:* Howard Wilson, PhD
*Tel:* 806 651-2799   *Fax:* 806 651-5105
*E-mail:* hwilson@mail.wtamu.edu

**University of Texas at Dallas**
Speech-Language Pathology Prgm
Program in Communication Disorders
1966 Inwood Rd
Dallas, TX 75235-7298
*Prgm Dir:* Robert D Stillman, PhD
*Tel:* 214 905-3060   *Fax:* 214 905-3006
*E-mail:* stillman@utdallas.edu

**Texas Woman's University**
Speech-Language Pathology Prgm
Comm Sciences and Disorders
Box 425737, TWU Station
Denton, TX 76204-5737
*Prgm Dir:* Kathleen K Millary, PhD
*Tel:* 940 898-2025   *Fax:* 940 898-2070
*E-mail:* F_Millay@twu.edu

**University of North Texas**
Speech-Language Pathology Prgm
Dept of Speech & Hearing Sciences
PO Box 305010
Denton, TX 76203-5010
*Prgm Dir:* Jeffrey Cokely
*Tel:* 940 565-2481   *Fax:* 940 565-4058
*E-mail:* cokely@unt.edu

**Univ of Texas - Pan American**
Speech-Language Pathology Prgm
Health Science & Human Services West 1.264
1201 W University Dr
Edinburg, TX 78541
www.panam.edu/dept/commdisorder
*Prgm Dir:* Teri Mata-Pistokache, PhD
*Tel:* 956 316-7040   *Fax:* 956 318-5238
*E-mail:* tmpistok@utpa.edu

**University of Texas at El Paso**
Speech-Language Pathology Prgm
1101 N Campbell St
El Paso, TX 79902
www.utep.edu
*Prgm Dir:* Anthony P Salvatore, PhD
*Tel:* 915 747-7250   *Fax:* 915 747-7207
*E-mail:* asalvatore@utep.edu

**Texas Christian University**
Speech-Language Pathology Prgm
Dept of Communication Sciences & Disorders
TCU Box 297450
Ft Worth, TX 76129
www.csd.tcu.edu
*Prgm Dir:* William Ryan, PhD
*Tel:* 817 257-7621    *Fax:* 817 257-5692
*E-mail:* w.ryan@tcu.edu

**University of Houston**
Speech-Language Pathology Prgm
Communication Disorders
100 Clinical Research Services
Houston, TX 77204-6018
*Prgm Dir:* Lynn S Bliss
*Tel:* 713 743-2897    *Fax:* 713 743-2926
*E-mail:* lbliss@uh.edu

**Texas A&M University - Kingsville**
Speech-Language Pathology Prgm
Communication Sciences & Disorders
MSC 177A, 700 University Blvd
Kingsville, TX 78363
www.tamuk.edu
*Prgm Dir:* Shari Schlehuser Beams, PhD
*Tel:* 361 593-3401    *Fax:* 361 593-3404
*E-mail:* kfsls01@tamuk.edu

**Texas Tech Univ Health Sciences Center**
Speech-Language Pathology Prgm
Dept of Speech, Language, and Hearing Sciences
STOP 6073, 3601 4th St
Lubbock, TX 79430
www.ttuhsc.edu/SAH
*Prgm Dir:* Sherry Sancibrian
*Tel:* 806 743-5660, Ext 232    *Fax:* 806 743-5670
*E-mail:* sherry.sancibrian@ttuhsc.edu

**Stephen F Austin State University**
Speech-Language Pathology Prgm
Communication Sciences and Disorders Program
PO Box 13019 SFA
Nacogdoches, TX 75962
*Prgm Dir:* Michael McKaig, PhD
*Tel:* 936 468-1252    *Fax:* 936 468-7096
*E-mail:* mmckaig@sfasu.edu

**Our Lady of the Lake University**
Speech-Language Pathology Prgm
Comm and Learning Disorders
411 S 24th St
San Antonio, TX 78207
www.ollusa.edu
*Prgm Dir:* Mary Ann Acevedo, PhD
*Tel:* 210 431-3938    *Fax:* 210 434-9360
*E-mail:* acevm@lake.ollusa.edu

**Texas State University-San Marcos**
Speech-Language Pathology Prgm
Department of Communication Disorders
Health Professions Building 601 University Dr
San Marcos, TX 78666-4616
www.health.txstate.edu/cdis/cdis.html
*Prgm Dir:* Maria Diana Gonzales, PhD CCC-SLP
*Tel:* 512 245-2330, Ext 2035    *Fax:* 512 245-2029
*E-mail:* mg29@txstate.edu

**Baylor University**
Speech-Language Pathology Prgm
Dept of Communication Sciences & Disorders
One Bear Place # 97332
Waco, TX 76798
www.baylor.edu/communication_disorders
*Prgm Dir:* J David Garrett, PhD
*Tel:* 254 710-2567    *Fax:* 254 710-2590
*E-mail:* David_Garrett@baylor.edu

# Utah

**Utah State University**
Speech-Language Pathology Prgm
Comm Disorders & Deaf Educ
1000 Old Main Mill
Logan, UT 84322-1000
*Prgm Dir:* Beth E Foley, PhD
*Tel:* 435 797-1388    *Fax:* 435 797-0221
*E-mail:* bethf@cc.usu.edu

**Brigham Young University**
Speech-Language Pathology Prgm
Comm Sciences and Disorders
136 John Taylor Bldg
Provo, UT 84602-8641
www.byu.edu/mse
*Prgm Dir:* David McPherson, PhD
*Tel:* 801 422-5117    *Fax:* 801 422-0197
*E-mail:* david_mcpherson@byu.edu

**University of Utah Health Science Center**
Speech-Language Pathology Prgm
Communication Sciences and Disorders
1201 Behavioral Science Bldg
Salt Lake City, UT 84112
www.health.utah.edu/cmdis
*Prgm Dir:* Bruce Smith, PhD
*Tel:* 801 581-6725    *Fax:* 801 571-7955
*E-mail:* bruce.smith@hsc.utah.edu

# Vermont

**University of Vermont**
Speech-Language Pathology Prgm
Communication Sciences
Pomeroy Hall, 489 Main St
Burlington, VT 05405
*Prgm Dir:* Patricia Prelock, PhD
*Tel:* 802 656-3861    *Fax:* 802 656-2528
*E-mail:* Patricia.Prelock@uvm.edu

# Virginia

**University of Virginia**
Speech-Language Pathology Prgm
Communication Disorders
2205 Fontaine Ave, Ste 202
Charlottesville, VA 22908-0781
*Prgm Dir:* Randall R Robey, PhD
*Tel:* 434 924-6354    *Fax:* 434 924-4612
*E-mail:* rrr7w@virginia.edu

**Hampton University**
Speech-Language Pathology Prgm
Dept of Communicative Sciences & Disorders
Science Technology Building, Room 201
Hampton, VA 23668
*Prgm Dir:* Robert Martin Screen, PhD
*Tel:* 757 727-5435    *Fax:* 757 727-5765
*E-mail:* robert.screen@hamptonu.edu

**James Madison University**
Speech-Language Pathology Prgm
Communication Sciences & Disorders
Harrisonburg, VA 22807
www.csd.jmu.edu
*Prgm Dir:* Vicki Reed, EdD
*Tel:* 540 568-6440    *Fax:* 540 568-8077
*E-mail:* reedva@jmu.edu

**Old Dominion University**
Speech-Language Pathology Prgm
Speech Pathology & Audiology
Child Study Ctr
Norfolk, VA 23529-0136
*Prgm Dir:* Nicholas G Bountress, PhD
*Tel:* 757 683-4117    *Fax:* 757 683-5593
*E-mail:* nbountre@odu.edu

**Radford University**
Speech-Language Pathology Prgm
Dept of Communication Sciences & Disorders
PO Box 6961
Radford, VA 24142
*Prgm Dir:* Calaire M Waldron, PhD CCC-SLP
*Tel:* 540 831-7666    *Fax:* 540 831-7669
*E-mail:* cwaldron@radford.edu

# Washington

**Western Washington University**
Speech-Language Pathology Prgm
Dept of Communication Sciences and Disorders
Parks Hall 17, MS 9078
Bellingham, WA 98225-9078
www.wwu.edu/~csd
*Prgm Dir:* Michael T Seilo, PhD
*Tel:* 360 650-3199    *Fax:* 360 650-2843
*E-mail:* mseilo@cc.wwu.edu

**Washington State University**
Speech-Language Pathology Prgm
Dept of Speech & Hearing Sciences
201 Daggy Hall, Box 642420
Pullman, WA 99164-2420
*Prgm Dir:* Gail Chermak
*Tel:* 509 335-4525    *Fax:* 509 335-8357
*E-mail:* chermak@wsu.edu

**University of Washington**
Speech-Language Pathology Prgm
Speech and Hearing Sciences
Seattle, WA 98195-6246
*Prgm Dir:* Pamela E Souza, PhD
*Tel:* 206 543-7829    *Fax:* 206 543-1093
*E-mail:* psouza@u.washington.edu

**Eastern Washington University**
Speech-Language Pathology Prgm
Dept of Communication Disorders
310 N Riverpoint Blvd, Box V
Spokane, WA 99202
www.ewu.edu/commdisorders
*Prgm Dir:* Donald R Fuller, PhD, CCC-SLP
*Tel:* 509 368-6889    *Fax:* 509 368-6791
*E-mail:* dfuller@mail.ewu.edu

# West Virginia

**Marshall University**
Speech-Language Pathology Prgm
Communication Disorders
400 Hal Greer Blvd
Huntington, WV 25755-2675
*Prgm Dir:* Kathryn Chezik
*Tel:* 304 696-3640    *Fax:* 304 696-2986
*E-mail:* chezik@marshall.edu

**West Virginia University**
Speech-Language Pathology Prgm
Department of Speech Pathology & Audiology
805 Allen Hall, Box 6122
Morgantown, WV 26506-6122
*Prgm Dir:* Lynn R Cartwright, EdD
*Tel:* 304 293-2377    *Fax:* 304 293-7565
*E-mail:* kjohnso2@wvu.edu

# Wisconsin

**University of Wisconsin - Eau Claire**
Speech-Language Pathology Prgm
Dept of Communication Sciences and Disorders
105 Garfield Ave
Eau Claire, WI 54702-4004
www.uwec.edu/academic/cdis
*Prgm Dir:* Kristine S Retherford, PhD
*Tel:* 715 836-4186    *Fax:* 715 836-4846
*E-mail:* retherk@uwec.edu

## University of Wisconsin - Madison

Speech-Language Pathology Prgm
Communicative Disorders
1975 Willow Dr - Goodnight Hall
Madison, WI 53706
www.comdis.wisc.edu
*Prgm Dir:* Robert Lutfi
*Tel:* 608 262-6485   *Fax:* 608 262-6466
*E-mail:* ralutfi@wisc.edu

## Marquette University

Speech-Language Pathology Prgm
Dept of Speech-Language Pathology & Audiology
PO Box 1881
Milwaukee, WI 53201-1881
www.marquette.edu/chs/speech
*Prgm Dir:* Steven H Long, PhD
*Tel:* 414 288-3428   *Fax:* 414 288-3980
*E-mail:* steven.long@marquette.edu

## University of Wisconsin - Milwaukee

Speech-Language Pathology Prgm
Dept of Communication Sciences and Disorders
PO Box 413
Milwaukee, WI 53201-0413
http://cfprod.imt.uwm.edu/chs/academics/
    undergraduate/communication
*Prgm Dir:* Marylou Gelfer, PhD
*Tel:* 414 229-6465   *Fax:* 414 229-2620
*E-mail:* gelfer@uwm.edu

## University of Wisconsin - River Falls

Speech-Language Pathology Prgm
Communicative Disorders
401 S Third St
River Falls, WI 54022-5001
*Prgm Dir:* Michael Harris, PhD
*Tel:* 715 425-3830   *Fax:* 715 425-0657
*E-mail:* michael.d.harris@uwrf.edu

## University of Wisconsin - Stevens Point

Speech-Language Pathology Prgm
Communicative Disorders
1901 4th Ave
Stevens Point, WI 54481-3897
www.uwsp.edu/commD
*Prgm Dir:* Gary Cumley, PhD
*Tel:* 715 346-2328   *Fax:* 715 346-2157
*E-mail:* gcumley@uwsp.edu
*Med Dir:* Cynthia Foster, MS

## University of Wisconsin - Whitewater

Speech-Language Pathology Prgm
Communicative Disorders
1011 Roseman Bldg, 800 W Main St
Whitewater, WI 53190-1790
*Prgm Dir:* Pat Casey, PhD
*Tel:* 414 472-1301   *Fax:* 414 472-5210
*E-mail:* caseyp@uww.edu

# Wyoming

## University of Wyoming

Speech-Language Pathology Prgm
Division of Communication Disorders
Dept 3311, 1000 E University Ave
Laramie, WY 82071
www.uwyo.edu/comdis
*Prgm Dir:* Mary Hardin-Jones, PhD
*Tel:* 307 766-6427   *Fax:* 307 766-5584
*E-mail:* mhardinj@uwyo.edu

## Speech-Language Pathologist

| Programs* | Class Capacity | Begins | Length (months) | Award | Res Tuition | Non-res Tuition | Stipend | Offers:‡ 1 | 2 | 3 |
|---|---|---|---|---|---|---|---|---|---|---|
| **Alabama** | | | | | | | | | | |
| University of Alabama (Tuscaloosa) | 30 | Summer Fall semesters | 21, 16 | BS, MA | | | | | | • |
| **Arizona** | | | | | | | | | | |
| Arizona State University (Tempe) | 35 | Aug | 24 | Dipl | | | | | | • |
| **Arkansas** | | | | | | | | | | |
| University of Arkansas (Fayetteville) | 20 | Fall semester | 21 | Dipl, MS | | | | | | • |
| University of Arkansas for Medical Sciences (Little Rock) | 18 | Fall semester | 21 | MS | $245 | $529 | | | | • |
| University of Central Arkansas (Conway) | 40 | Aug | 24, 48 | MS, BS | $5,754 | $10,254 | | | | • |
| **California** | | | | | | | | | | |
| California State University - Long Beach | 24 | Fall | 24 | MA | $3,446 | $13,990 | | • | | |
| California State University - Los Angeles | 18 | Sep | 24 | Dipl, MA | $4,800 | $10,170 | | • | | • |
| Loma Linda University | 60 | Sep | 24 | BS, MS | $16,000 | $16,000 | | | | |
| San Diego State University | 40 | Aug | 24 | Cert, MA | $3,704 | $8,136 | | | • | • |
| University of Redlands | 45 | Sep | 21 | MS | $15,000 | $15,000 | $5,000 | | | • |
| University of the Pacific (Stockton) | 30 | Aug | 15, 24 | MS, BS | $26,000 | $26,000 | | | | • |
| **Colorado** | | | | | | | | | | |
| University of Colorado at Boulder | 30+ | Fall | 24 | Dipl, MA | $6,000 | $20,000 | | • | | • |
| University of Northern Colorado (Greeley) | 20 | Aug | 24 | MA | $5 | $14,832 | | • | | • |
| **District of Columbia** | | | | | | | | | | |
| George Washington University | 35 | Sep | 20 | Dipl, MA | $18,000 | $0 | | | | • |
| **Florida** | | | | | | | | | | |
| Florida Atlantic University (Boca Raton) | 30 | Fall | 24, 48 | Dipl, MS | $8,300 | $31,055 | | | | • |
| University of South Florida (Tampa) | 60 | Aug | 24 | MS | $9,000 | $29,000 | $10,000 | • | | • |
| **Illinois** | | | | | | | | | | |
| Eastern Illinois University (Charleston) | 200 | Fall | 36, 21 | BS, MS | $7,068 | $17,481 | $720 | | | • |
| Illinois State University (Normal) | 40 | Aug Jan | 24 | Dipl, MS | $3,000 | $6,000 | | • | | • |
| Northwestern University (Evanston) | 50 | Fall | 21, 27 | MA | $37,872 | $37,872 | | | | • |
| Southern Illinois Univ at Edwardsville | 60 | Fall semester | 6, 11 | Dipl, MS | | | | • | | |
| **Indiana** | | | | | | | | | | |
| Indiana State University (Terre Haute) | 15 | Fall semester | 23 | Dipl, MS, MA | | | | • | | • |
| **Iowa** | | | | | | | | | | |
| University of Iowa (Iowa City) | 25 | Fall semester | 24, 36 | MA, AuD | $6,959 | $18,353 | | | | |
| **Kansas** | | | | | | | | | | |
| Ft Hays State University | 15 | Sep Jan | 24 | BS, MS | $2,108 | $5,566 | | | | • |
| Wichita State University | 25 | Aug | 48, 24 | Dipl, BA, MA | $6,300 | $17,420 | | | | |

*Data are shown only for programs that completed the 2006 AMA Survey of Health Professions Education Programs
‡Key to Offers: 1: Evening or weekend classes; 2: Non-English instruction; 3: Cultural competence instruction

## Speech-Language Pathologist

| Programs* | Class Capacity | Begins | Length (months) | Award | Res Tuition | Non-res Tuition | Stipend | Offers:‡ 1 | 2 | 3 |
|---|---|---|---|---|---|---|---|---|---|---|
| **Kentucky** | | | | | | | | | | |
| Eastern Kentucky University (Richmond) | 16 | Fall semester | 48 | MA | $7,545 | $21,261 | | • | | • |
| **Louisiana** | | | | | | | | | | |
| Louisiana State Univ Health Sciences Center (New Orleans) | 25 | May | 24 | MCD | $3,555 | $5,430 | | | | |
| University of Louisiana at Lafayette | 24 | Aug | 16, 30 | MS, PhD | | | | | • | |
| University of Louisiana at Monroe | 20 | Fall | 48, 22 | Dipl, BS, MS | $1,700 | $4,676 | | | | • |
| **Maine** | | | | | | | | | | |
| University of Maine - Orono | 20 | Sep | 48, 24 | Dipl, BA, MA | $5,962 | $15,844 | | | | |
| **Maryland** | | | | | | | | | | |
| Towson University | 40 | Fall | 24 | Dipl, MS | $4,020 | $6,720 | | | | |
| Univ of Maryland at College Park | 25 | Aug | 20 | MA | $10,250 | $17,500 | $8,000 | | | |
| **Massachusetts** | | | | | | | | | | |
| Boston University | 35 | Sep 1 | 24 | MS | $33,330 | $0 | | | | • |
| MGH Institute of Health Professions (Boston) | 42 | Sep | 24 | MS | $27,600 | $27,600 | | • | | • |
| **Michigan** | | | | | | | | | | |
| Central Michigan University (Mt Pleasant) | 30 | Aug | 21 | MA | $366 | $366 | | | | • |
| Wayne State University (Detroit) | 35 | Fall term | 20, 48 | MA | $14,000 | $29,675 | | | | • |
| **Minnesota** | | | | | | | | | | |
| Minnesota State University - Mankato | 25 | Aug | 20 | MS | $7,092 | $10,464 | | • | | • |
| Minnesota State University - Moorhead | 30 | Aug | 24 | Masters | $6,000 | $12,000 | | | | |
| St Cloud State University | 17 | Fall | 48, 24 | BS, MS | $2,522 | $5,166 | | • | | • |
| University of Minnesota - Minneapolis | 30 | Fall semester | 27 | MA | $8,748 | $15,848 | | | • | • |
| **Mississippi** | | | | | | | | | | |
| University of Southern Mississippi (Hattiesburg) | 35 | Fall | 15 | Dipl, Masters | $2,156 | $2,715 | $5,400 | | | |
| **Missouri** | | | | | | | | | | |
| Fontbonne University (St Louis) | 25 | Summer Fall | 27 | MS | $489 | $0 | | • | | • |
| Saint Louis University (St Louis) | 20 | Summer | 22 | Dipl, MA | $800 | $800 | | | | • |
| Southeast Missouri State University (Cape Girardeau) | 30 | Fall | 25 | MA | | | | | | • |
| Truman State University (Kirksville) | 15 | Rolling enrollment | 24 | MA | $3,052 | $5,205 | $2,000 | | | |
| University of Missouri - Columbia | | | | | | | | | | |
| **Nebraska** | | | | | | | | | | |
| University of Nebraska - Kearney | 20 | Fall semester | 24 | MS | $3,852 | $7,968 | $8,200 | | | • |
| University of Nebraska - Lincoln | 50 | Fall | 21 | MS, PhD | $150 | $435 | | • | | • |
| University of Nebraska - Omaha | 25- | Fall Spring | 21 | Masters | $3,640 | $9,560 | | • | | • |
| **New Hampshire** | | | | | | | | | | |
| University of New Hampshire (Durham) | 24 | Fall | 24 | MS | | | | | | • |
| **New Jersey** | | | | | | | | | | |
| Kean University (Union) | 40 | Fall semester | 57 | MA | $10,385 | $12,617 | | | | • |
| Seton Hall University (South Orange) | 30 | Fall | 24 | MS | $787 | $787 | | | | |
| **New Mexico** | | | | | | | | | | |
| New Mexico State University (Las Cruces) | 20 | Aug | 24, 36 | MA | $2,000 | $6,500 | | | | • |
| University of New Mexico (Albuquerque) | 65 | Aug | 20, 36 | Dipl, MS SLP | $2,838 | $6,887 | | • | | • |
| **New York** | | | | | | | | | | |
| Adelphi University (Garden City) | | | 24 | MS, DA/AuD | $0 | $600,850 | | • | | • |
| Buffalo State, SUNY | 45 | Aug | | MSEd | | | | • | | • |
| Columbia University Teachers College (New York) | 45 | Sep | 24, 30 | MS | $970 | $970 | | • | | • |
| CUNY Brooklyn College | 40 | Sep 1 | 30 | MS | $6,000 | $0 | | | | • |
| CUNY Hunter College (New York) | 30 | Sep | 17 | Dipl, MS | $8,000 | $15,000 | | • | • | • |
| Hofstra University (Hempstead) | 40 | Sep | 24, 48 | MA, AuD | $20,000 | $20,000 | | • | | • |
| Ithaca College | 30 | Late Aug | 21 | MS | $18,096 | $18,096 | $24,000 | • | | • |
| Mercy College (Dobbs Ferry) | 40 | Sep | 24, 60 | Dipl, BS, MS | $17,010 | $17,010 | | • | | • |
| Nazareth College of Rochester | 60 | Fall Spring | 18 | Masters | $14,160 | $14,160 | | • | | • |
| New York University | 55 | Sep | 29 | Dipl, BS, MA | | | | | | • |
| St John's University (Queens) | 130 | Fall Spring | 48 | MA | | | | • | | |
| SUNY at New Paltz | 20 | Late Aug | 21, 27 | Dipl, Cert, MS | $6,900 | $10,920 | | • | | • |
| **North Carolina** | | | | | | | | | | |
| North Carolina Central University (Durham) | 30 | Aug Jan | 21 | MEd | $5,372 | $16,652 | $2,500 | • | • | • |
| Univ of North Carolina at Greensboro | 28 | Aug | 21 | MS | $2,112 | $13,162 | $3,500 | • | | • |
| **North Dakota** | | | | | | | | | | |
| Minot State University | Open | Fall semester | | BS, MS | $6,000 | $0 | | | | |
| **Ohio** | | | | | | | | | | |
| Bowling Green State University | 20 | Aug | 20 | MS | $10,808 | $18,116 | $4,876 | | | |
| Case Western Reserve University (Cleveland) | 15 | Aug | 22 | Dipl, BA, MA | | | | | | • |
| Kent State University | 35 | Sep Apr | 24 | MA | $8,460 | $15,470 | $5,500 | • | | |
| University of Akron | 150 | Fall semester | 48, 24 | Dipl, BA, MA SLP | $10,662 | $18,964 | | | | • |

*Data are shown only for programs that completed the 2006 AMA Survey of Health Professions Education Programs
‡Key to Offers: 1: Evening or weekend classes; 2: Non-English instruction; 3: Cultural competence instruction

# Speech-Language Pathologist

| Programs* | Class Capacity | Begins | Length (months) | Award | Res Tuition | Non-res Tuition | Stipend | Offers:‡ 1 | 2 | 3 |
|---|---|---|---|---|---|---|---|---|---|---|
| University of Toledo | 40 | Aug | 21 | MA | $9,360 | $18,172 | | • | | • |
| **Oklahoma** | | | | | | | | | | |
| Oklahoma State University (Stillwater) | 20 | Fall | 40, 21 | Dipl, BS, MS | $141 | $416 | $4,800 | • | | |
| Univ of Oklahoma Health Sciences Center (Oklahoma City) | 30 | Aug | 24 | Dipl, MA, MS | $4,770 | $12,250 | | | | |
| **Pennsylvania** | | | | | | | | | | |
| Duquesne University (Pittsburgh) | 25 | Aug Sep | 24 | MS | | | | | | • |
| Edinboro University of Pennsylvania | 20 | Fall semester | 22 | Dipl, MA | $5,800 | $9,500 | $3,500 | | | • |
| Indiana University of Pennsylvania | 18 | Fall semester | 36, 21 | BS Ed, MS | $6,800 | $10,400 | $2,400 | • | | • |
| La Salle University (Philadelphia) | 35 | Aug | 24 | MS | $15,840 | $15,840 | | • | | • |
| Marywood University (Scranton) | 40 | Fall semester | 24, 60 | BS/MS, MS | | | | | | • |
| Penn State University (University Park) | 40 | Aug | 22, 36 | MS, PhD | $18,777 | $34,506 | $12,510 | | | |
| University of Pittsburgh | 100 | Aug | 24, 60 | Dipl, MA/MS, PhD | $23,208 | $30,993 | $16,605 | • | | • |
| **South Dakota** | | | | | | | | | | |
| University of South Dakota (Vermillion) | 50 | Aug Sep | 24 | MA | $3,006 | $5,448 | | | | • |
| **Tennessee** | | | | | | | | | | |
| University of Memphis | 45 | Fall semester | 24, 48 | MA, AuD | $9,119 | $23,532 | $4,500 | | | |
| **Texas** | | | | | | | | | | |
| Baylor University (Waco) | | Fall Spring Summer | | Dipl, MS CSD, MA | | | | | | • |
| Lamar Institute of Technology (Beaumont) | 20 | Fall semester | 24 | MS | $5,844 | $14,094 | $1,000 | • | | • |
| Our Lady of the Lake University (San Antonio) | 28 | Summer Fall | 26 | Dipl, MA | $13,000 | $13,000 | | • | | • |
| Texas A&M University - Kingsville | 20 | Aug Jan | 24 | Dipl, MS | | | | | | • |
| Texas Christian University (Ft Worth) | 12 | Aug | 21 | MS | $800 | $0 | $800 | | • | • |
| Texas State University-San Marcos | 20 | Fall semester | 21 | MSCD, MA | | | | | | • |
| Texas Tech Univ Health Sciences Center (Lubbock) | 30 | Fall | 27 | Masters | $8,000 | $18,000 | $3,000 | | | • |
| Univ of Texas - Pan American (Edinburg) | 30 | Fall | 24 | Dipl, MS | $810 | $2,772 | | | | |
| University of Texas at El Paso | 30 | Aug | 24 | MS | | | | • | • | • |
| **Utah** | | | | | | | | | | |
| Brigham Young University (Provo) | 30 | Sep 1 | 24 | MS | $5,000 | $8,200 | $2,400 | | | • |
| University of Utah Health Science Center (Salt Lake City) | 30 | Fall | 21 | MS, AuD | | | | | | • |
| **Virginia** | | | | | | | | | | |
| James Madison University (Harrisonburg) | 28 | Aug | 24 | MS | $5,700 | $17,100 | $7,000 | | | • |
| **Washington** | | | | | | | | | | |
| Eastern Washington University (Spokane) | 70 | Sep | 24 | BA, MS | $4,044 | $13,317 | | | | • |
| Western Washington University (Bellingham) | 20 | Fall | 24 | MA | $2,084 | $5,600 | | | | • |
| **Wisconsin** | | | | | | | | | | |
| Marquette University (Milwaukee) | 25 | Jan Jun Aug | 24 | Cert, MS | $16,100 | $16,100 | | | | • |
| University of Wisconsin - Eau Claire | 16 | Fall semester | 22 | MS | $6,532 | $9,142 | $5,505 | • | | |
| University of Wisconsin - Madison | 30 | Sep 1 | 24, 48 | Dipl, MS, PhD | $4,369 | $12,004 | | | | • |
| University of Wisconsin - Milwaukee | 25 | Jun Sep | 24 | Dipl, MS | $10,500 | $27,500 | | | | • |
| University of Wisconsin - Stevens Point | | | | SLP, AuD | $328/ credit | $991/ credit | | | | • |
| **Wyoming** | | | | | | | | | | |
| University of Wyoming (Laramie) | 20 | Aug | 24 | Dipl, MS | $4,193 | $10,727 | | | | • |

*Data are shown only for programs that completed the 2006 AMA Survey of Health Professions Education Programs
‡Key to Offers: 1: Evening or weekend classes; 2: Non-English instruction; 3: Cultural competence instruction

# Surgical Assistant

## Occupational Description

As defined by the American College of Surgeons, the surgical assistant provides aid in exposure, hemostasis, closure, and other intraoperative technical functions that help the surgeon carry out a safe operation with optimal results for the patient. In addition to intraoperative duties, the surgical assistant also performs preoperative and postoperative duties to better facilitate proper patient care. The surgical assistant to the surgeon during the operation does so under the direction and supervision of that surgeon and in accordance with hospital policy and appropriate laws and regulations.

## Job Description

In general, surgical assistants have the following responsibilities:

- Determine specific equipment needed per procedure
- Review permit to confirm procedure and special needs
- Select and place of x-rays for reference
- Assist in moving and positioning of patient
- Insert and remove Foley urinary bladder catheter
- Place pneumatic tourniquet
- Confirm procedure with surgeon
- Drape patient within surgeon's guidelines
- Provide retraction of tissue and organs for optimal visualization with regard to tissue type and appropriate retraction instrument and/or technique
- Assist in maintaining hemostasis by direct pressure, use and application of appropriate surgical instrument for the task, placement of ties, placement of suture ligatures, application of chemical hemostatic agents, or other measures as directed by the surgeon
- Use electrocautery mono- and bi-polar
- Clamp, ligate, and cut tissue per surgeon's directive
- Harvest saphenous vein, including skin incision, per surgeon's directive
- Dissect common femoral artery and bifurcate per surgeon's directive
- Maintain integrity of sterile field
- Close all wound layers (facia, subcutaneous and skin) as per surgeon's directive
- Insert drainage tubes per surgeon's directive
- Select and apply wound dressings
- Assist with resuscitation of patient during cardiac arrest or other life-threatening events in the operating room
- Perform any other duties or procedures incident to the surgical procedure deemed necessary and as directed by the surgeon

## Employment Characteristics

Certified surgical assistants assist in a variety of surgery specialties:

- General surgery
- Orthopaedic surgery
- Neurosurgery
- Spinal surgery
- Otolaryngology
- Obstetrical surgery
- Gynecological surgery
- Craniofacial surgery
- Radial neck surgery
- Genitourinary surgery
- Cardiac surgery
- Thoracic surgery
- Vascular surgery
- Trauma surgery
- Plastic surgery
- Ophthalmologic surgery

## Educational Programs

**Length.** Current CAAHEP-accredited programs range from 10 months to 22 months. Surgical assisting is a specialty profession that requires specific training over and above a degree in science, nursing, physician assisting, or another health profession.

**Prerequisites.** Recommended eligibility requirements for admission into a surgical assisting program are:

- Bachelor of Science degree (or higher)
- Associate degree in an allied health field, with 3 years of recent experience
- CST, CNOR, or PA-C, with current certification
- Three years of current operating room scrub and/or assisting experience within the last 5 years
- Military medical training with surgical assistant experience
- Proof of liability insurance
- Current CPR/BLS certification
- Acceptable health and immunization records
- Computer literacy
- Students also must be able to show proof of successful completion of basic science (college level) instruction, including:
- Microbiology
- Pathophysiology
- Pharmacology
- Anatomy and physiology
- Medical terminology

**Curriculum.** Course content includes:

- Advanced surgical anatomy
- Surgical microbiology
- Surgical pharmacology
- Anesthesia methods and agents
- Bioscience
- Ethical and legal considerations
- Fundamental technical skills
- Complications during surgery
- Interpersonal skills
- Clinical application of computers

Students must possess a working knowledge of operating room fundamentals, including aseptic principles and techniques, before moving on to the advanced levels of the program.

## Credentialing

The National Board for Surgical Technology and Surgical Assisting (formerly Liaison Council on Certification for the Surgical Technologist) offers the Certified First Assistant (CST/CFA) credential, and the National Surgical Assistant Association (NSAA) offers a Certified Surgical Assistant (CSA) credential. To be eligible for NBSTSA testing, individuals must be graduates of a CAAHEP-accredited surgical assistant program or a CST with current certification who meets a number of other eligibility requirements.

## Inquiries

**Careers/Curriculum**
Association of Surgical Assistants
6 W Dry Creek Circle
Littleton, CO 80120
800 637-7433 or 303 694-9130
303 694-9169 Fax
www.surgicalassistant.org

National Surgical Assistant Association
2615 Amesbury Road
Winston Salem, NC 27103
888 633-0479
www.nsaa.net

**Certification**
National Board for Surgical Technology and Surgical Assisting
(formerly Liaison Council on Certification for the Surgical
    Technologist)
6 West Dry Creek Circle
Littleton, CO 80120
800 707-0057
303-325-2536 Fax
www.nbstsa.org

National Surgical Assistant Association
2615 Amesbury Road
Winston Salem, NC 27103
888 633-0479
www.nsaa.net

**Program Accreditation**
Commission on Accreditation of Allied Health Education Programs
    (CAAHEP) in collaboration with:
Subcommittee on Accreditation for Surgical Assisting
6 West Dry Creek Circle, Suite 210
Littleton, CO 80120
303 694-9262
303 741-3655 Fax
E-mail: ccollinsworth@ast.org

## Surgical Assistant

## Indiana

**Vincennes University**
Surgical Assistant Prgm
1002 N First St
Vincennes, IN 47512
www.vinu.edu
*Prgm Dir:* Chris Keegan, CST MS
*Tel:* 812 888-5893   *Fax:* 812 888-4550
*E-mail:* ckeegan@vinu.edu
*Med Dir:* Santi Vibul, MD

## Kentucky

**Madisonville Community College**
Surgical Assistant Prgm
Health Campus
750 N Laffoon St
Madisonville, KY 42431
*Prgm Dir:* Jeff Bidwell, CST CSA MA
*Tel:* 270 824-1740   *Fax:* 270 824-8642
*E-mail:* jeff.bidwell@kctcs.edu
*Med Dir:* Mohan Rao, MD

## Michigan

**William Beaumont Hospital**
Surgical Assistant Prgm
22322 Rutland Dr
Southfield, MI 48075
www.beaumonthospitals.com
*Prgm Dir:* Rebecca Pieknik, CST CSA MS
*Tel:* 248 898-7685   *Fax:* 248 898-4406
*E-mail:* rpieknik@beaumonthospitals.com
*Med Dir:* James Catto, MD

## Tennessee

**Meridian Institute of Surgical Assisting**
Surgical Assistant Prgm
PO Box 758
Joelton, TN 37080
*Prgm Dir:* Dennis A Stover, CST SA-C
*Tel:* 615 299-1416   *Fax:* 615 299-1418
*E-mail:* meridianinst@aol.com
*Med Dir:* Timothy Schoettle, MD

**Nashville State Technical Community College**
Surgical Assistant Prgm
120 White Bridge Rd
Nashville, TN 37209
*Prgm Dir:* T Van Bates, BA CST
*Tel:* 615 353-3735
*E-mail:* van.bates@nscc.edu
*Med Dir:* Mark Cooper, MD

## Texas

**South Plains College**
Surgical Assistant Prgm
819 Gilbert Dr, Bldg 5
Lubbock, TX 79416
www.southplainscollege.edu
*Prgm Dir:* Stacey May, CST
*Tel:* 806 885-3048, Ext 4642   *Fax:* 806 885-5608
*E-mail:* smay@southplainscollege.edu
*Med Dir:* John A Griswold, MD

## Virginia

**Eastern Virginia Medical School**
Surgical Assistant Prgm
700 W Olney Rd, Ste 1100
Norfolk, VA 23507
www.evms.edu
*Prgm Dir:* R Clinton Crews, MPH
*Tel:* 757 446-6100   *Fax:* 757 446-6179
*E-mail:* SurgAsst@evms.edu
*Med Dir:* L D Britt, MD

## Surgical Assistant

| Programs* | Class Capacity | Begins | Length (months) | Award | Res. Tuition | Non-res. Tuition | Stipend | Offers:‡ 1 | 2 | 3 |
|---|---|---|---|---|---|---|---|---|---|---|
| **Indiana** | | | | | | | | | | |
| Vincennes University | 15 | Aug | 9 | Cert | $3,073 | $7,671 | | | | • |
| **Kentucky** | | | | | | | | | | |
| Madisonville Community College | 20 | Aug | 10 | Cert | $1,840 | $5,520 | | • | | |
| **Michigan** | | | | | | | | | | |
| William Beaumont Hospital (Southfield) | 10 | Jan | 21 | Cert | $2,982 | $5,985 | | • | | |
| **Texas** | | | | | | | | | | |
| South Plains College (Lubbock) | 15 | Aug | 12 | Cert | $1,502 | $1,854 | | | | |
| **Virginia** | | | | | | | | | | |
| Eastern Virginia Medical School (Norfolk) | 24 | July | 22 | Cert | $10,200 | $10,200 | | | | |

*Data are shown only for programs that completed the 2006 AMA Survey of Health Professions Education Programs
‡Key to Offers: 1: Evening or weekend classes; 2: Non-English instruction; 3: Cultural competence instruction

**PROGRAMS**

# Surgical Technologist

## Occupational Description

Surgical technologists are an integral part of the team of medical practitioners providing surgical care to patients in a variety of settings.

## Job Description

Surgical technologists prepare the operating room by selecting and opening sterile supplies. Preoperative duties also include assembling, adjusting, and checking nonsterile equipment to ensure that it is in proper working order. Common duties include operating sterilizers, lights, suction machines, electrosurgical units, and diagnostic equipment.

When patients arrive in the surgical suite, surgical technologists assist in preparing them for surgery by providing physical and emotional support, checking charts, and observing vital signs. They have been educated to properly position the patient on the operating table, assist in connecting and applying surgical equipment and/or monitoring devices, and prepare the incision site. Surgical technologists have primary responsibility for maintaining the sterile field, being constantly vigilant that all members of the team adhere to aseptic technique.

They most often function as the sterile member of the surgical team who passes instruments, sutures, and sponges during surgery. After "scrubbing," they don sterile gown and gloves and prepare the sterile setup for the appropriate procedure. After other members of the sterile team have scrubbed, they assist them with gowning and gloving and with the application of sterile drapes that isolate the operative site.

In order that surgery may proceed smoothly, surgical technologists anticipate the needs of surgeons, passing instruments and providing sterile items in an efficient manner. They share with the circulator the responsibility of accounting for sponges, needles, and instruments before, during, and after surgery.

Surgical technologists may hold retractors or instruments, sponge or suction the operative site, or cut suture materials as directed by the surgeon. They connect drains and tubing and receive and prepare specimens for subsequent pathologic analysis. They are responsible for preparing and applying sterile dressings following the procedure and may assist in the application of nonsterile dressings, including plaster or synthetic casting materials. After surgery, they prepare the operating room for the next patient.

Surgical technologists are most often members of the sterile team but may function in the nonsterile role of circulator. The circulator is not gowned and gloved during the surgical procedure but is available to respond to the needs of the anesthesia provider, keep a written account of the surgical procedure, and participate jointly with the scrubbed person in counting sponges, needles, and instruments before, during, and after surgery. In operating rooms where local anesthetics are administered, they meet the needs of the conscious patient.

Certified surgical technologists with additional specialized education or training also may act in the role of the surgical first assistant. The surgical first assistant provides aid in exposure, hemostasis, and other technical functions under the surgeon's direction that will help the surgeon carry out a safe operation with optimal results for the patient.

Surgical technologists also may provide staffing in postoperative recovery rooms where patients' responses are carefully monitored in the critical phases following general anesthesia.

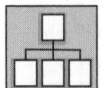

## Employment Characteristics

A majority of surgical technologists work in hospitals, principally in the surgical suite and also in emergency rooms and other settings that call for knowledge of, and ability in, maintaining asepsis, such as materials management and central service. A number work in a wide variety of settings and arrangements, including outpatient surgicenters, private employment by physicians, or as self-employed technologists.

Those who work in hospital and other institutional settings are usually expected to work rotating shifts or to accommodate on-call assignments to ensure adequate staffing for emergency surgical procedures during evening, night, weekend, and holiday hours. Otherwise, surgical technologists follow a standard hospital workday.

## Salary

Salaries vary depending on the experience and education of the individual, the economy of a given region, the responsibilities of the position, and the working hours. According to the US Bureau of Labor Statistics, the median annual salary for surgical technologists was $34,010 in 2004. Refer to Section IV, Table 5 of this *Directory* for more information, or see www.ama-assn.org/go/salary.

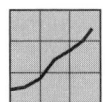

## Job Outlook

According to a recent study conducted by the US Bureau of Labor Statistics, the forecast for employment opportunities for surgical technologists is one of rapid growth. Demand for technologists varies among communities and geographic regions. Prospective students are advised to assess the market for graduates within the region in which they would like to work before matriculating in an educational program. Such information is likely to be available through local employment offices, local accredited programs, and hospital councils or hospitals.

## Educational Programs

**Length.** Programs range from 12 to 24 months.
**Prerequisites.** High school diploma or equivalent.
**Curriculum.** Accreditation standards require didactic instruction and supervised clinical practice. Subject areas include medical terminology, professional ethics, and legal aspects of surgical patient care; anatomy and physiology, microbiology, anesthesia, and pharmacology; sterilization methods and aseptic technique; instruments, supplies, and equipment used in surgery; surgical patient care and safety precautions; and operative procedures and biomedical sciences. Supervised clinical practice in the operating room must include commonly performed procedures in general surgery, obstetrics and gynecology, ophthalmology, otorhinolaryngology, plastic surgery, urology, orthopedics, neurosurgery, thoracic surgery, and cardiovascular and peripheral vascular surgery.

## Inquiries

**Careers/Curriculum**

Association of Surgical Technologists
6 West Dry Creek Circle
Littleton, CO 80120
800 637-7433 or 303 694-9130
303 694-9169 Fax
www.ast.org

**Certification/Registration**

Inquiries regarding certification as a certified surgical technologist (CST) or a CST certified first assistant (CST/CFA) may be addressed to:

National Board for Surgical Technology and Surgical Assisting (formerly Liaison Council on Certification for the Surgical Technologist)
6 West Dry Creek Circle
Littleton, CO 80120
800 707-0057
303 325-2536 Fax
www.nbstsa.org

**Program Accreditation**

Commission on Accreditation of Allied Health Education Programs (CAAHEP) in collaboration with:
Accreditation Review Committee on Education in Surgical Technology
6 West Dry Creek Circle
Littleton, CO 80120
303 694-9262
303 741-3655 Fax
E-mail: ccollinsworth@ast.org

## Surgical Technologist*

# Alabama

**James H Faulkner State Community College**
Surgical Technology Prgm
1900 Hwy 31 S
Bay Minette, AL 36507
www.faulknerstate.edu
*Prgm Dir:* Jean Graham, MSN CNOR
*Tel:* 251 580-2293   *Fax:* 251 580-2199
*E-mail:* jgraham@faulknerstate.edu
*Med Dir:* Judy Brevik, MSN, CNOR

**Virginia College**
Surgical Technology Prgm
PO Box 19249
Birmingham, AL 35209
*Prgm Dir:* Nancy H Wright
*Tel:* 205 271-8250
*E-mail:* nwright@vc.edu

**Calhoun Community College**
Surgical Technology Prgm
PO Box 2216
Decatur, AL 35609-2219
*Prgm Dir:* Grant Wilson, MEd CST
*Tel:* 256 306-2950   *Fax:* 256 306-2507
*E-mail:* sgw@calhoun.edu
*Med Dir:* William Alison, Jr, MD

**Flowers Hospital**
Surgical Technology Prgm
Home Health Buldg
4370 W Main St, Ste 1
Dothan, AL 36302
*Prgm Dir:* Elizabeth Andrews, RN BSN CNOR CST
*Tel:* 334 794-5000, Ext 1129   *Fax:* 334 615-7285
*E-mail:* stprogram@flowershospital.com

**Virginia College**
Surgical Technology Prgm
2970 Cottage Hill Rd
Mobile, AL 36606
*Prgm Dir:* Sharon Jones
*Tel:* 251 343-7227, Ext 2427   *Fax:* 251 343-7287
*E-mail:* sjones@vc.edu
*Med Dir:* Jack Kotlarz, MD

**Southern Union State Community College**
Surgical Technology Prgm
1701 Lafayette Pkwy
Opelika, AL 36801
www.suscc.edu
*Prgm Dir:* Dot Nichols, BBA RN CNOR
*Tel:* 334 745-6437, Ext 5536   *Fax:* 334 745-6342
*E-mail:* dotnichols@suscc.edu
*Med Dir:* William Lazenby, MD

**Bevill State Community College**
Surgical Technology Prgm
PO Box 800
Sumiton, AL 35148
*Prgm Dir:* Linda Neumann, RN CNOR
*Tel:* 205 648-3271, Ext 5569   *Fax:* 205 648-2288
*E-mail:* lneuman@bscc.edu

# Arizona

**Mohave Community College**
Surgical Technology Prgm
1977 West Acoma Blvd
Lake Havasu City, AZ 86403
www.mohave.edu
*Prgm Dir:* Sandra Namio, CST CFA
*Tel:* 928 505-3374   *Fax:* 928 505-3381
*E-mail:* snamio@imail.mohave.edu
*Med Dir:* Linda Riesdorph, RN, MSN Dean/Administrator of Nursing

**GateWay Community College**
Surgical Technology Prgm
108 N 40th St
Phoenix, AZ 85034
*Prgm Dir:* Susan Wallen, RN MS CNOR CRNFA
*Tel:* 602 392-5116   *Fax:* 602 392-5244
*E-mail:* wallen@gatewaycc.edu

**Lamson College**
Surgical Technology Prgm
1126 N Scottsdale Rd #17
Tempe, AZ 85281
www.lamsoncollege.com
*Prgm Dir:* Kelly Harris, CST
*Tel:* 480 898-7000, Ext 215   *Fax:* 480 967-6645
*E-mail:* ccrabtree@lamsoncollege.com
*Med Dir:* Francisco Sudiacal, DC

**Pima Community College**
Surgical Technology Prgm
5901 S Calle Santa Cruz
Tucson, AZ 85709-6350
http://cc.pima.edu/CTD
*Prgm Dir:* Ed Doran
*Tel:* 520 206-5117   *Fax:* 520 506-5196
*E-mail:* Edward.Doran@pima.edu

# Arkansas

**Univ of Arkansas Comm College-Batesville**
Surgical Technology Prgm
PO Box 3350
Batesville, AR 72503
*Prgm Dir:* Paula Russell, RN MSN
*Tel:* 870 612-2064   *Fax:* 870 793-4988
*E-mail:* prussell@uaccb.edu

**University of Arkansas - Fort Smith**
Surgical Technology Prgm
PO Box 3649
Ft Smith, AR 72913
www.uafortsmith.edu
*Prgm Dir:* Sydney Fulbright, MSN RN CNOR
*Tel:* 479 788-7855   *Fax:* 479 788-7153
*E-mail:* sfulbrig@uafortsmith.edu
*Med Dir:* Leon Woods (volunteer), MD

**North Arkansas College**
Surgical Technology Prgm
1515 Pioneer Dr
Harrison, AR 72601
www.northark.edu
*Prgm Dir:* Charlitta Parton, RN CNOR
*Tel:* 870 391-3269   *Fax:* 870 391-3250
*E-mail:* lparton@northark.edu

**Baptist Health**
Surgical Technology Prgm
11900 Colonel Glenn Rd, Ste 1000
Little Rock, AR 72210-2820
www.baptist-health.org
*Prgm Dir:* Gordon Ward, BSN RN CNOR
*Tel:* 501 202-7746   *Fax:* 501 202-7712
*E-mail:* gbward@baptist-health.org

*Additional information about programs that returned the AMA's survey is available in a table beginning on page 455.

**University of Arkansas for Medical Sciences**
Surgical Technology Prgm
2200 Fort Roots Dr, Slot 14B/NLR
North Little Rock, AR 72114-1706
www.uams.chrp.edu
*Prgm Dir:* Gennie Castleberry, BS CST
*Tel:* 501 257-2354   *Fax:* 501 257-2349
*E-mail:* castleberrygennier@uams.edu

**Southeast Arkansas College**
Surgical Technology Prgm
1900 Hazel St
Pine Bluff, AR 71603
*Prgm Dir:* Clemetine Wesley, RN ORT AAS
*Tel:* 870 543-5967   *Fax:* 870 543-5912
*E-mail:* cwesley@seark.edu
*Med Dir:* Charles Mabry, MD

**Northwest Technical Institute**
Surgical Technology Prgm
PO Box 2000
709 S Old Missouri Rd
Springdale, AR 72765
www.nti.tec.ar.us
*Prgm Dir:* Katie Fritz, CST
*Tel:* 479 751-8824, Ext 211   *Fax:* 479 751-7780
*E-mail:* kfritz@nti.tec.ar.us

# California

**San Joaquin Valley College - Bakersfield**
Surgical Technology Prgm
201 New Stine Rd
Bakersfield, CA 93309
www.sjvc.edu
*Prgm Dir:* Patricia Siefkas, ASSc CST
*Tel:* 661 834-0126, Ext 141   *Fax:* 661 834-1021
*E-mail:* PatriciaS@sjvc.edu
*Med Dir:* E Byron O'Neil, MD

**Southwestern College**
Surgical Technology Prgm
900 Otay Lakes Rd
Chula Vista, CA 91910
*Prgm Dir:* Bill Maddox, MHA BS RN
*Tel:* 619 421-6700, Ext 5616   *Fax:* 619 482-6439
*E-mail:* wmaddox@swc.cc.ca.us
*Med Dir:* Jon Grief, DO

**Mt Diablo Adult Education**
Surgical Technology Prgm
1266 San Carlos Ave
Concord, CA 94518
*Prgm Dir:* Susan Garske, BA MS
*Tel:* 925 685-7340, Ext 2734   *Fax:* 925 363-9757

**Fresno City College**
Surgical Technology Prgm
1101 E University Ave
Fresno, CA 93741
*Prgm Dir:* Mary Jane McClain, RN CNOR
*Tel:* 559 244-2643   *Fax:* 559 244-2626
*E-mail:* maryjane.mcclain@fresnocitycollege.edu

**San Joaquin Valley College**
Surgical Technology Prgm
295 East Sierra Ave
Fresno, CA 93710
www.sjcv.edu
*Prgm Dir:* Teri Junge, CST CFA
*Tel:* 559 448-8282   *Fax:* 559 448-8250
*E-mail:* terij@sjvc.edu

**Glendale Career College**
Surgical Technology Prgm
1015 Grandview Ave
Glendale, CA 91201
www.success.edu
*Prgm Dir:* Connie Bell, CST
*Tel:* 818 956-4915, Ext 261   *Fax:* 818 243-6028
*E-mail:* cbell@success.edu

**Bryman College - Hayward**
Surgical Technology Prgm
22336 Main St
Hayward, CA 94541
*Prgm Dir:* Rechelle A Bowen, CST
*Tel:* 510 582-9500   *Fax:* 510 582-9645
*E-mail:* rbowen@cci.edu

**Premiere Career College**
Surgical Technology Prgm
12901 Ramona Blvd
Irwindale, CA 91706
*Prgm Dir:* Antonio Torres, BS MEd CST CRCST
*Tel:* 626 814-2080   *Fax:* 626 814-3242
*E-mail:* antotorres@netzero.net
*Med Dir:* Wilson Morales, MD

**Career Colleges of America - Los Angeles**
Surgical Technology Prgm
1801 S LaCienega Blvd
Los Angeles, CA 90035
*Prgm Dir:* Angela Gomez
*Tel:* 310 287-9901
*E-mail:* angelag@careercolleges.org

**Concorde Career Institute - North Hollywood**
Surgical Technology Prgm
12412 Victory Blvd
North Hollywood, CA 91606
www.concorde.edu
*Prgm Dir:* Ernesto McFarlane, CST
*Tel:* 818 766-8151, Ext 251   *Fax:* 818 766-1587
*E-mail:* emcfarlane@concorde.edu
*Med Dir:* Fazal Aasi, MD

**Glendale Career College - Oceanside Campus**
Surgical Technology Prgm
2204 El Camino Real, Ste 315
Oceanside, CA 92054
www.landmarked.com
*Prgm Dir:* Donna McCasland
*Tel:* 760 450-0340   *Fax:* 760 450-0396
*E-mail:* dmccasland@success.edu
*Med Dir:* Steve Giddings, Campus Director

**American College of Health Professions**
Surgical Technology Prgm
700 E Redland Blvd, #U227
Redlands, CA 92374
*Prgm Dir:* Michele Ray, CST
*Tel:* 909 307-6022   *Fax:* 909 307-6032
*E-mail:* achpfaculty@msn.com
*Med Dir:* Ronald Jaecks, MD

**Bryman College - Reseda**
Surgical Technology Prgm
18040 Sherman Way
Reseda, CA 91335
*Prgm Dir:* David Camarena, MD
*Tel:* 818 774-0550
*E-mail:* dcamarena@cci.edu

**Career Colleges of America - San Bernardino**
Surgical Technology Prgm
184 W Club Center Dr
San Bernardino, CA 92408
*Prgm Dir:* Debbie Bessent
*Tel:* 909 824-0897   *Fax:* 909 872-1144
*E-mail:* debbieb@careercolleges.org
*Med Dir:* Shawn N Molnar

**Skyline College**
Surgical Technology Prgm
3300 College Dr
San Bruno, CA 94066
*Prgm Dir:* Alice Erskine, CST MSN CNOR
*Tel:* 650 738-4470   *Fax:* 650 738-4179
*E-mail:* erskine@smccd.net
*Med Dir:* Lorne Rosenfield, MD

**Concorde Career College**
Surgical Technology Prgm
4393 Imperial Ave
Ste 100
San Diego, CA 92113
www.concorde.edu
*Prgm Dir:* Niall E Davis, CST
*Tel:* 619 688-0800, Ext 307   *Fax:* 619 220-4177
*E-mail:* ndavis@concorde.edu

**Naval School of Health Sciences, San Diego**
Surgical Technology Prgm
34101 Farenholt Ave
San Diego, CA 92134-5291
http://nshssd.med.navy.mil
*Prgm Dir:* CDR Thomas A Sweet, NC, USN, MSN CNOR
*Tel:* 619 532-7831   *Fax:* 619 532-7796
*E-mail:* tsweet@nshs-sd.med.navy.mil
*Med Dir:* CDR Eric J Kuncir, MD, MS, FACS

**UCSF Medical Center**
Surgical Technology Prgm
505 Parnassus Ave
San Francisco, CA 94143
*Prgm Dir:* Jane Kuhn, MSN
*Tel:* 415 353-1309   *Fax:* 415 353-1834
*E-mail:* jane.kuhn@ucsfmedctr.org
*Med Dir:* Jeffrey Katz, MD

**Central County Occupation Center**
Surgical Technology Prgm
760 Hillsdale Ave #MT1
San Jose, CA 95136-1106
*Prgm Dir:* Donna Hewitt, RN
*E-mail:* dhewitt@metroed.net

**Newbridge College**
Surgical Technology Prgm
1840 E 17th St, Ste 140
Santa Ana, CA 92705
*Prgm Dir:* Claro Nunez, CST
*Tel:* 714 550-6735   *Fax:* 714 550-6740
*E-mail:* cnunez@newbridgecollege.edu
*Med Dir:* Robert Beltran, MD

**Simi Valley Adult School**
Surgical Technology Prgm
3192 Los Angeles Ave
Simi Valley, CA 93065
*Prgm Dir:* Ronald O Kruzel, BA CST
*Tel:* 805 579-6200, Ext 249   *Fax:* 805 522-8902
*E-mail:* ronk@simi.tec.ca.us
*Med Dir:* Peter Richman, MD

**Career Colleges of America - South Gate**
Surgical Technology Prgm
5612 E Imperial Hwy
South Gate, CA 90802
*Prgm Dir:* Claro M Nunez, CST
*Tel:* 562 861-8702, Ext 126   *Fax:* 562 869-7013
*E-mail:* claron@careercolleges.org

# Colorado

**Concorde Career Institute - Aurora**
Surgical Technology Prgm
111 N Havana St
Aurora, CO 80010
*Prgm Dir:* Karyn Songer, CST FAST
*Tel:* 303 861-1151, Ext 233   *Fax:* 303 839-5478
*E-mail:* ksonger@concorde.edu

**Everest College**
Surgical Technology Prgm
9065 Grant St
Thornton, CO 80229
*Prgm Dir:* Damen L Sanchez
*Tel:* 303 457-2757   *Fax:* 303 457-4030
*E-mail:* dsanchez@cci.edu

# Connecticut

**Bridgeport Hospital**
Surgical Technology Prgm
200 Mill Hill Ave
Bridgeport, CT 06610
*Prgm Dir:* Janet Serra, RN BSN CNOR
*Tel:* 203 384-3218   *Fax:* 203 384-3046
*E-mail:* njserr@bpthosp.org

**Danbury Hospital**
Surgical Technology Prgm
24 Hospital Ave
Danbury, CT 06810
www.danburyhospital.org
*Prgm Dir:* Mary E Janell, MS RN CNOR
*Tel:* 203 797-7724   *Fax:* 203 797-0706
*E-mail:* mary.janell@danhosp.org
*Med Dir:* Pierre Saldinger, MD

**Eli Whitney Vocational Technical School**
Surgical Technology Prgm
71 Jones Rd
Hamden, CT 06514
*Prgm Dir:* Karen Dempsey, CNOR BSN MBA
*Tel:* 203 397-4031, Ext 386   *Fax:* 203 397-4129
*E-mail:* Karen.Dempsey@po.ct.state.us

**A I Prince Technical High School**
Surgical Technology Prgm
500 Brookfield St
Hartford, CT 06106
*Prgm Dir:* Elia Acosta, CST BA MS
*Tel:* 860 286-9712
*E-mail:* Elia.Acosta@ct.gov

**Manchester Community College**
Surgical Technology Prgm
MS 17, PO Box 1046
Manchester, CT 06045
www.commnet.edu/programs/healthcareers
*Prgm Dir:* Richard Clark, CST MA
*Tel:* 860 512-2715   *Fax:* 860 512-2621
*E-mail:* rclark@mcc.commnet.edu
*Med Dir:* George Perdrizet, MD, PhD

# Florida

**MedVance Institute - Atlantis Campus**
Surgical Technology Prgm
170 JFK Dr
Atlantis, FL 33462
*Prgm Dir:* Yolanda Newton
*Tel:* 561 304-3466   *Fax:* 561 304-3471
*E-mail:* yolanda.newton@medvance.edu
*Med Dir:* John Corbitt, Jr, MD

**Lee County High Tech Center-North**
Surgical Technology Prgm
360 Santa Barbara Blvd N
Cape Coral, FL 33993
*Prgm Dir:* Dona J Hoyt, RN CNOR
*Tel:* 239 574-4440, Ext 227   *Fax:* 239 458-3721
*E-mail:* donah@lee.k12.fl.us
*Med Dir:* Gail Turk, RN CNOR

**Pinellas Tech Ed Ctr - Clearwater Campus**
Surgical Technology Prgm
6100 154th Ave N
Clearwater, FL 33760
www.ptec.pinellas.k12.fl.us
*Prgm Dir:* Candace Gioia, RN BSN CNOR
*Tel:* 727 538-7167, Ext 1038   *Fax:* 727 538-7203
*E-mail:* CGioia@ptec.pinellas.k12.fl.us

**Brevard Community College**
Surgical Technology Prgm
1519 Clearlake Rd
Cocoa, FL 32922
*Prgm Dir:* Judith C Schatte, RN CNOR CRNFA
*Tel:* 321 632-1111, Ext 64123   *Fax:* 321 634-3731
*E-mail:* schattej@brevardcc.edu

**Daytona Beach Community College**
Surgical Technology Prgm
PO Box 2811
Daytona Beach, FL 32120
*Prgm Dir:* Fabrizio Pluchino, CST CFA MBA PhD
*Tel:* 386 506-3747   *Fax:* 386 506-3000
*E-mail:* pluchif@dbcc.edu

**Sanford-Brown Institute - Ft Lauderdale**
Surgical Technology Prgm
1201 W Cypress Creek Rd
Ft Lauderdale, FL 33309
*Prgm Dir:* Pierrot Alexis, CST
*Tel:* 954 308-7396   *Fax:* 954 375-6940
*E-mail:* palexis@sbftlaud.com
*Med Dir:* Roberto Puglisi, MD

**Lee County High Tech Center - Central**
Surgical Technology Prgm
3800 Michigan Ave
Ft Myers, FL 33916
*Prgm Dir:* Vicki Santini, MS RN CNOR CST
*Tel:* 941 334-4544, Ext 384   *Fax:* 941 334-6900
*E-mail:* vickisa@lee.k12.fl.us
*Med Dir:* Jacob Goldberger, MD

**Indian River Community College**
Surgical Technology Prgm
3209 Virginia Ave
Ft Pierce, FL 34981
*Prgm Dir:* Kathy A Gelety, ST RN CNOR
*Tel:* 772 336-6227   *Fax:* 772 336-6235
*E-mail:* kgelety@ircc.net

**National School of Technology**
Surgical Technology Prgm
Hialeah Campus
4410 W 16th Ave, Ste 52
Hialeah, FL 33012
*Prgm Dir:* Ingrid Mendez, ST
*Tel:* 305 558-9500, Ext 146   *Fax:* 305 558-4419
*E-mail:* imendez@cci.edu
*Med Dir:* Juan Jesus Salinas, MD

**Sheridan Technical Center**
Surgical Technology Prgm
5400 Sheridan St
Hollywood, FL 33021
www.sheridantechnical.com
*Prgm Dir:* Maureen Onischuck, RN
*Tel:* 754 321-5482   *Fax:* 754 321-5467
*E-mail:* maureen.onischuck@browardschools.com

**Concorde Career Institute - Jacksonville**
Surgical Technology Prgm
7960 Arlington Expressway
Ste 120
Jacksonville, FL 32211
www.concorde.edu
*Prgm Dir:* Luis Melendez, CST CFA
*Tel:* 904 807-5132   *Fax:* 904 721-9944
*E-mail:* lmelendez@concorde.edu

**Florida Community College - Jacksonville**
Surgical Technology Prgm
4501 Capper Rd
Jacksonville, FL 32218
*Prgm Dir:* Neal Henning
*Tel:* 904 766-6524   *Fax:* 904 713-4859
*E-mail:* nhenning@fccj.edu

**Palm Beach Community College**
Surgical Technology Prgm
4200 S Congress Ave, MS60
Lake Worth, FL 33461
*Prgm Dir:* Toni Crowley, MS RNFA CNOR
*Tel:* 561 868-3561   *Fax:* 561 868-3635
*E-mail:* crowleyt@pbcc.edu

**Traviss Career Center**
Surgical Technology Prgm
3225 Winter Lake Rd
Lakeland, FL 33803
*Prgm Dir:* Pamela Troxell, RN BSN CNOR
*Tel:* 863 499-2700, Ext 299   *Fax:* 863 499-2067
*E-mail:* pam.troxell@polk-fl.net

**Lindsey Hopkins Tech Education Center**
Surgical Technology Prgm
750 NW 20th St
Miami, FL 33127
http://lindsey.dadeschools.net
*Prgm Dir:* Ibia Bustelo, RN BSN CNOR
*Tel:* 305 324-6070, Ext 8015
*E-mail:* IBustelo@dadeschools.net

**National School of Technology**
Surgical Technology Prgm
Kendall Campus
9020 SW 137th Ave
Miami, FL 33186
*Prgm Dir:* Rosie Sirota
*Tel:* 305 386-9900   *Fax:* 305 558-4419
*E-mail:* rsirota@cci.edu
*Med Dir:* Andres Lasserre, MD

**National School of Technology**
Surgical Technology Prgm
Miami Campus
111 NW 183rd St
Miami, FL 33169
*Prgm Dir:* John Mays, CST OSA BSHA
*Tel:* 305 949-9500   *Fax:* 305 558-4419
*E-mail:* jmays@cci.edu

**Lorenzo Walker Institute of Technology**
Surgical Technology Prgm
3702 Estey Ave
Naples, FL 34104
*Prgm Dir:* Nancy Turmelle, RN CNOR
*Tel:* 941 430-6900, Ext 2049   *Fax:* 941 430-6922
*E-mail:* turmelna@collier.k12.fl.us

**Okaloosa-Walton College**
Surgical Technology Prgm
100 College Blvd
Niceville, FL 32578
www.owc.edu
*Prgm Dir:* April Carter, MSN
*Tel:* 850 729-6444   *Fax:* 850 729-6460
*E-mail:* cartera@owc.edu

**Central Florida Community College**
Surgical Technology Prgm
PO Box 1388
3001 SW College Rd
Ocala, FL 34478-1388
*Prgm Dir:* Brenda Frazier, MEd RN CNOR CST BSN
*Tel:* 352 237-2111, Ext 1271   *Fax:* 352 873-5889
*E-mail:* frazierb@cf.edu

**Orlando Tech**
Surgical Technology Prgm
301 W Amelia St
Orlando, FL 32801
*Prgm Dir:* Betty Arnett, RMA CST
*Tel:* 407 246-7060, Ext 4887   *Fax:* 407 317-3372
*E-mail:* arnettb@ocps.net
*Med Dir:* Ferol Lynne Voltaggio, MEd

**Central Florida Institute**
Surgical Technology Prgm
30522 US 19th N, Ste 310
Palm Harbor, FL 34684
www.cfinstitute.com
*Prgm Dir:* Karen Niles, CST MHS MBA
*Tel:* 727 786-4707, Ext 240   *Fax:* 727 781-9421
*E-mail:* kniles@cfinstitute.com

**Gulf Coast Community College**
Surgical Technology Prgm
5230 W US Hwy 98
Panama City, FL 32401
http://health.gulfcoast.edu
*Prgm Dir:* Mary E McNaron, RN CNOR CST BSN MS
*Tel:* 850 769-1551, Ext 3551    *Fax:* 850 747-3246
*E-mail:* lmcnaron@gulfcoast.edu
*Med Dir:* Michael Slavens, MD

**Pensacola Junior College**
Surgical Technology Prgm
5555 Hwy 98 W
Pensacola, FL 32507
*Prgm Dir:* Gayle Griffin, BSN RN
*Tel:* 850 484-2257    *Fax:* 850 484-2326
*E-mail:* ggriffin@pjc.edu

**Virginia College at Penascola**
Surgical Technology Prgm
19 W Garden St
Pensacola, FL 32501
*Prgm Dir:* Barbara Inkel, RN CNOR
*Tel:* 850 436-8444    *Fax:* 850 436-2677
*E-mail:* binkel@vc.edu
*Med Dir:* Jack Kotlarz, MD

**MedVance Institute - Plantation Campus**
Surgical Technology Prgm
4101 NW 3rd Court, Ste 9
Plantation, FL 33317
*Prgm Dir:* Bonnie Merschdorf
*Tel:* 954 587-7100    *Fax:* 954 587-7704
*E-mail:* bmerschdorf@medvance.edu

**Sarasota County Technical Institute**
Surgical Technology Prgm
4748 Beneva Rd
Sarasota, FL 34233
*Prgm Dir:* Harlean Satin, MPA RNCNA BC
*Tel:* 941 924-1365, Ext 372    *Fax:* 941 361-6886
*E-mail:* harlean_satin@srqit.sarasota.k12.fl.us

**Concorde Career Institute - Tampa**
Surgical Technology Prgm
4202 W Spruce St
Tampa, FL 33607
www.concorde.edu
*Prgm Dir:* Keith Orloff, CST FAST
*Tel:* 813 314-2478    *Fax:* 813 872-6884
*E-mail:* korloff@concorde.edu

**Erwin Technical Center**
Surgical Technology Prgm
2010 E Hillsborough Ave
Tampa, FL 33610-8299
*Prgm Dir:* Deborah Pearson, CST FA-CRCST
*Tel:* 813 231-1800, Ext 1331    *Fax:* 813 231-1820
*E-mail:* deborah.pearson@sdhc.k12.fl.us

**Florida Metropolitan Univ - Tampa College**
Surgical Technology Prgm
3924 Coconut Palm Dr
Tampa, FL 33619
*Prgm Dir:* Ken Stanley
*Tel:* 813 621-0041, Ext 68
*E-mail:* kstanley@cci.edu

**Sanford-Brown Institute**
Surgical Technology Prgm
5701 E Hillsborough Ave, Ste 1417
Tampa, FL 33610
www.sbtampa.com
*Prgm Dir:* Joseph Milam, CST CSPDT
*Tel:* 813 496-2777    *Fax:* 813 630-1309
*E-mail:* jmilam@sbtampa.com
*Med Dir:* Jeffrey Neustadt, MD

# Georgia

**Albany Technical College**
Surgical Technology Prgm
1704 S Slappey Blvd
Albany, GA 31701-3514
www.albanytech.edu
*Prgm Dir:* Lori Day, CST
*Tel:* 229 430-3552    *Fax:* 229 430-5115
*E-mail:* lday@albanytech.edu
*Med Dir:* Bennett Cotton, MD

**Athens Technical College**
Surgical Technology Prgm
800 US Hwy 29 N
Athens, GA 30601-1500
*Prgm Dir:* Beth Jackson-Streb, CST RN BSN MEd
*Tel:* 706 355-5072    *Fax:* 706 425-3104
*E-mail:* bjackson-streb@athenstech.edu

**Augusta Technical College**
Surgical Technology Prgm
3200 Augusta Tech Dr, 900 Bldg
Augusta, GA 30906
*Prgm Dir:* Patty Young, RN CNOR
*Tel:* 706 771-4191    *Fax:* 706 771-4181
*E-mail:* pyoung@augusta.tec.ga.us
*Med Dir:* Stephen Gooden, MD

**Coastal Georgia Community College**
Surgical Technology Prgm
3700 Altama Ave
Brunswick, GA 31520-3644
www.cgcc.edu
*Prgm Dir:* Joyce K Tate, RN BSN CNOR
*Tel:* 912 264-7250    *Fax:* 912 262-3283
*E-mail:* jtate@cgcc.edu

**DeKalb Technical College**
Surgical Technology Prgm
495 N Indian Creek Dr
Clarkston, GA 30021
www.dekalbtech.edu
*Prgm Dir:* Frances Offutt, RN ASN
*Tel:* 404 297-9522, Ext 1159    *Fax:* 404 294-6496
*E-mail:* offuttf@dekalbtech.edu
*Med Dir:* Frances Offutt, RN ASN

**Columbus Technical College**
Surgical Technology Prgm
928 Manchester Expwy
Columbus, GA 31904-6572
www.columbustech.edu
*Prgm Dir:* Kellie Byrd, RN BSEd MSN CNOR
*Tel:* 706 649-1498    *Fax:* 706 641-5284
*E-mail:* kbyrd@columbustech.edu
*Med Dir:* Emory Alexander, MD

**Dalton State College**
Surgical Technology Prgm
213 N College Dr
Dalton, GA 30720
*Prgm Dir:* Alice Kirby, RN BSN
*Tel:* 706 272-2658    *Fax:* 706 272-4563
*E-mail:* akirby@em.daltonstate.edu
*Med Dir:* John Turentine, MD

**Griffin Technical College**
Surgical Technology Prgm
501 Varsity Rd
Griffin, GA 30223
*Prgm Dir:* Claudia Holley, RN CST
*Tel:* 770 233-6166    *Fax:* 770 229-3294
*E-mail:* holleyc@griftec.org
*Med Dir:* Joyce Hickey, RN BSN

**Gwinnett Technical College**
Surgical Technology Prgm
5150 Sugarloaf Pkwy
Lawrenceville, GA 30043-5702
www.gwinnettTech.edu
*Prgm Dir:* TC Parker, CST AAS
*Tel:* 770 962-7580, Ext 6348    *Fax:* 770 962-1506
*E-mail:* tcparker@gwinnetttech.edu

**Central Georgia Technical College**
Surgical Technology Prgm
3300 Macon Tech Dr
Macon, GA 31206
*Prgm Dir:* Becky P Darley, RN CNOR
*Tel:* 478 757-3560    *Fax:* 478 757-3575
*E-mail:* bdarley@centralgatech.edu

**Chattahoochee Technical College**
Surgical Technology Prgm
980 S Cobb Ave
Marietta, GA 30060
*Prgm Dir:* Lorraine Wilderman, RN BSN CNOR
*Tel:* 770 528-4488    *Fax:* 770 528-4418
*E-mail:* lwilderman@chattcollege.com

**Georgia Medical Institute**
Surgical Technology Prgm
1600 Terrell Mill Rd
Marietta, GA 30067
*Prgm Dir:* Barbara Armstrong
*Tel:* 770 428-6303    *Fax:* 770 428-8415
*E-mail:* barmstro@cci.edu

**Lanier Technical College**
Surgical Technology Prgm
2990 Landrum Education Dr
Oakwood, GA 30566
*Prgm Dir:* Jamey Watson, CST AAT
*Tel:* 770 531-6421    *Fax:* 770 531-6306
*E-mail:* jwatson@laniertech.edu

**Northwestern Technical College**
Surgical Technology Prgm
265 Bicentennial Trail
Rock Spring, GA 30739
*Prgm Dir:* Mary L Jirsa, RN BSN CNOR
*Tel:* 706 764-3544    *Fax:* 706 764-3718
*E-mail:* mjirsa@northwesterntech.edu
*Med Dir:* Victor A Duncan, MD

**Savannah Technical College**
Surgical Technology Prgm
5717 White Bluff Rd
Savannah, GA 31405
*Prgm Dir:* Brenda L McMilllin, CST MA
*Tel:* 912 303-1816    *Fax:* 912 303-4317
*E-mail:* bmcmillin@savtec.org
*Med Dir:* Clay Burnett, MD

**Ogeechee Technical College**
Surgical Technology Prgm
1 Joe Kennedy Blvd
Statesboro, GA 30458
*Prgm Dir:* Deborah Scott, RNC CNOR
*Tel:* 912 486-7401, Ext 565    *Fax:* 912 486-7604
*E-mail:* dscott@ogeecheetech.edu

**Flint River Technical College**
Surgical Technology Prgm
1533 Hwy 19 S
Thomaston, GA 30286
*Prgm Dir:* Thomas Satterfield, BA CST
*Tel:* 706 646-6185    *Fax:* 706 646-6163
*E-mail:* tsatterfield@flintrivertech.org

**Southwest Georgia Technical College**
Surgical Technology Prgm
15689 US Hwy 19 N
Thomasville, GA 31792
www.southwestgatech.edu
*Prgm Dir:* Sherrie Holliman, RN MSN RNFA
*Tel:* 229 225-5205    *Fax:* 229 225-5289
*E-mail:* sholliman@southwestgatech.edu

**Valdosta Technical College**
Surgical Technology Prgm
4089 Val Tech Rd
Valdosta, GA 31602
www.valdostatech.edu
*Prgm Dir:* Dorothy E Cox-Carter, RN BSN
*Tel:* 229 245-3860   *Fax:* 229 259-5567
*E-mail:* dcox@valdostatech.edu
*Med Dir:* Dallas M Miller, MD

**Southeastern Technical College**
Surgical Technology Prgm
3001 E First St
Vidalia, GA 30474
www.southeasterntech.edu
*Prgm Dir:* Deborah Smith, RN CNOR
*Tel:* 912 538-3182   *Fax:* 912 538-3106
*E-mail:* dsmith@southeasterntech.edu

**West Central Technical College**
Surgical Technology Prgm
176 Murphy Campus Blvd
Waco, GA 30180
www.westcentraltech.edu
*Prgm Dir:* Richard T Bailey, RN BSN
*Tel:* 770 537-6044   *Fax:* 770 537-7992
*E-mail:* rbailey@westcentraltech.edu
*Med Dir:* John Henry Burson III

**Middle Georgia Technical College**
Surgical Technology Prgm
80 Cohen Walker Dr
Warner Robins, GA 31088
www.middlegatech.edu
*Prgm Dir:* Lorna Cox, CST
*Tel:* 912 988-6910   *Fax:* 912 988-6875
*E-mail:* lcox@middlegatech.edu
*Med Dir:* Virgil McEver, III, MD, FACS

**Okefenokee Technical College**
Surgical Technology Prgm
1701 Carswell Ave
Waycross, GA 31503
www.okefenokeetech.edu
*Prgm Dir:* Sally M Smith, RN MEd
*Tel:* 912 287-5839   *Fax:* 912 287-4865
*E-mail:* smsmith@okefenokeetech.edu
*Med Dir:* John Butler, MD

# Hawaii

**Kapi'olani Community College**
Surgical Technology Prgm
4303 Diamond Head Rd
Honolulu, HI 96816
*Prgm Dir:* Christine Nadamoto, CST RN MS
*Tel:* 808 734-9305
*E-mail:* nadamoto@hawaii.edu

# Idaho

**Boise State University**
Surgical Technology Prgm
College of Technology
Boise, ID 83725
*Prgm Dir:* Mona Bourbonnais
*Tel:* 208 426-1519   *Fax:* 208 426-3155
*E-mail:* mbourbonnais@boisestate.edu

**Eastern Idaho Technical College**
Surgical Technology Prgm
1600 S 25th E
Idaho Falls, ID 83404-5788
www.eitc.edu
*Prgm Dir:* Becky Chapman, CST
*Tel:* 208 524-3000, Ext 3427   *Fax:* 208 525-7038
*E-mail:* bchapman@eitc.edu
*Med Dir:* David J Chamberlain, DO

**College of Southern Idaho**
Surgical Technology Prgm
315 Falls Ave, Aspen Bldg #158
PO Box 1238
Twin Falls, ID 83303-1238
*Prgm Dir:* Janet Milligan, RN CNOR
*Tel:* 208 732-6706   *Fax:* 208 733-4743
*E-mail:* jmilligan@csi.edu
*Med Dir:* David McClusky, MD

# Illinois

**Parkland College**
Surgical Technology Prgm
2400 W Bradley Ave
Champaign, IL 61821
*Prgm Dir:* Jody Randolph, BSN RN CNOR
*Tel:* 217 351-2375   *Fax:* 217 373-3830
*E-mail:* jorandolph@parkland.edu

**Malcolm X College**
Surgical Technology Prgm
1900 W Van Buren St
Chicago, IL 60612
*Prgm Dir:* Marietta McDuffy, CST BS MEd
*Tel:* 312 850-7351   *Fax:* 312 850-7453
*E-mail:* mmcduffy@ccc.edu

**Prairie State College**
Surgical Technology Prgm
202 S Halsted St
Chicago Heights, IL 60411
www.prairiestate.edu
*Prgm Dir:* Susan Chap, BS PA RN
*Tel:* 708 709-3780   *Fax:* 708 709-3777
*E-mail:* schap@prairiestate.edu

**Richland Community College**
Surgical Technology Prgm
1 College Park
Decatur, IL 62521-8512
*Prgm Dir:* Katherine Lee, MS CST
*Tel:* 217 875-7200, Ext 763   *Fax:* 217 875-6965
*E-mail:* klee@richland.edu

**Shawnee Community College**
Surgical Technology Prgm
East St Louis Higher Education Ctr
601 James R Thompson Blvd
East St Louis, IL 62201
*Prgm Dir:* Denisa Huff
*Tel:* 618 874-8714   *Fax:* 618 634-3300
*E-mail:* dhuff@eslccc.com

**Elgin Community College**
Surgical Technology Prgm
1700 Spartan Dr
Elgin, IL 60123
www.elgin.edu
*Prgm Dir:* Maureen Lange, RNC CNOR MS
*Tel:* 847 214-7303   *Fax:* 847 214-7527
*E-mail:* mlange@elgin.edu

**College of Lake County**
Surgical Technology Prgm
19351 W Washington St
Grayslake, IL 60030
www.clcillinois.edu
*Prgm Dir:* Soheila Kayoud, CST BA
*Tel:* 847 543-2776   *Fax:* 847 223-1357
*E-mail:* skayoud@clcillinois.edu

**Southern Illinois Collegiate Common Market**
Surgical Technology Prgm
3213 S Park Ave
Herrin, IL 62948
www.siccm.com
*Prgm Dir:* Pamela Appleton, RN BS CNOR
*Tel:* 618 942-6902   *Fax:* 618 942-6658
*E-mail:* appleton@siccm.com

**Illinois Central College**
Surgical Technology Prgm
Peoria Campus Div
201 SW Adams St
Peoria, IL 61635-0001
www.icc.edu
*Prgm Dir:* William Hammer, CST MEd
*Tel:* 309 999-4633   *Fax:* 309 673-9626
*E-mail:* bhammer@icc.edu

**John Wood Community College**
Surgical Technology Prgm
1301 S 48th St
Quincy, IL 62305
*Prgm Dir:* Sandra McKelvie, RN
*Tel:* 217 641-4550, Ext 4550   *Fax:* 217 224-4208
*E-mail:* mckelvie@jwcc.edu

**Triton College**
Surgical Technology Prgm
2000 N Fifth Ave
River Grove, IL 60171
http://triton.edu
*Prgm Dir:* Jessica Brown, CST
*Tel:* 708 456-0300, Ext 3563   *Fax:* 708 583-3336
*E-mail:* jbrown@triton.edu

**Trinity College of Nursing & Health Sciences**
Surgical Technology Prgm
2122 25th Ave
Rock Island, IL 61201-1216
*Prgm Dir:* Farla Champ, MS RN CNOR CST
*Tel:* 309 779-7768   *Fax:* 309 779-7746
*E-mail:* champf@trinityqc.com

**Rock Valley College**
Surgical Technology Prgm
3301 N Mulford Rd
Rockford, IL 61114-5699
*Prgm Dir:* James Shear, RN BS CNOR
*Tel:* 815 921-3205   *Fax:* 815 654-4408
*E-mail:* J.Shear@rvc.cc.il.us

**Capital Area Career Center**
*Cosponsor:* St John's Hospital
Surgical Technology Prgm
2201 Toronto Rd
Springfield, IL 62712
*Prgm Dir:* Paula Malone, RCST SFA LPN
*Tel:* 217 529-5431, Ext 155   *Fax:* 217 585-2165
*E-mail:* Paulam@caspn.org
*Med Dir:* Brian Russell, MD, FACS

**Waubonsee Community College**
Surgical Technology Prgm
Rte 47 at Waubonsee Dr
Sugar Grove, IL 60554
*Prgm Dir:* Jess Toussaint, MS
*Tel:* 630 466-2350   *Fax:* 630 466-4119
*E-mail:* jtoussaint@waubonsee.edu
*Med Dir:* Allen Bloom, MD

**College of DuPage**
Surgical Technology Prgm
550 E Washington St
West Chicago, IL 60185
*Prgm Dir:* Kathy Cabai, CST RN CNOR MSinEd CST
*Tel:* 630 293-4115, Ext 2601   *Fax:* 630 293-4129
*E-mail:* cabaik@cod.edu

# Indiana

**Bloomington Hospital**
Surgical Technology Prgm
PO Box 1149
605 W Second St
Bloomington, IN 47402
www.bloomingtonhospital.org
*Prgm Dir:* Lillian Bartlett, RN BSN CNOR
*Tel:* 812 335-5571   *Fax:* 812 335-5444
*E-mail:* lbartlett@bloomhealth.org

**Ivy Tech Community College - Columbus**
Surgical Technology Prgm
4475 Central Ave
Columbus, IN 47203-1868
*Prgm Dir:* Susan D Sheets, RN BSN CNOR
*Tel:* 812 372-9925, Ext 185    *Fax:* 812 372-0311
*E-mail:* ssheets@ivytech.edu
*Med Dir:* Richard Wiethoff, MD

**Ivy Tech Community College SW - Evansville**
Surgical Technology Prgm
3501 First Ave
Evansville, IN 47710
www.ivytech.edu
*Prgm Dir:* Julia Hinkle, RN MHS CNOR BSN
*Tel:* 812 429-1490    *Fax:* 812 429-9805
*E-mail:* jhinkle@ivytech.edu

**Indiana Business College - Ft Wayne**
Surgical Technology Prgm
6413 N Clinton St
Ft Wayne, IN 46825
*Prgm Dir:* Jill Lobacz, AAS CST
*Tel:* 260 471-7667
*E-mail:* jill.lobacz@ibcschools.edu

**University of Saint Francis**
Surgical Technology Prgm
2701 Spring St
Ft Wayne, IN 46808
www.sf.edu/healthscience/surgtech
*Prgm Dir:* Elizabeth Slagle, MS RN CST
*Tel:* 260 434-7673    *Fax:* 260 434-7697
*E-mail:* eslagle@sf.edu

**Clarian Health Partners Inc**
Surgical Technology Prgm
PO Box 1367
Indianapolis, IN 46206-1367
*Prgm Dir:* Terry Myers, RN
*Tel:* 317 962-1864    *Fax:* 317 962-3483
*E-mail:* tmyers@clarian.org

**Ivy Tech Community College of Indiana**
Surgical Technology Prgm
Central Indiana Region
One W 26th St, PO Box 1763
Indianapolis, IN 46206-1763
*Prgm Dir:* Wanda L Haver, CST BS ED
*Tel:* 317 921-4404    *Fax:* 317 921-4753
*E-mail:* whaver@ivytech.edu
*Med Dir:* Ted Grissell, MD

**Ivy Tech Community College - Kokomo**
Surgical Technology Prgm
700 E Firmin St
Kokomo, IN 46902
www.ivytech.edu
*Prgm Dir:* Judi Townsend, BSN CNOR
*Tel:* 765 457-0858, Ext 107    *Fax:* 765 457-1036
*E-mail:* jtownsen@ivytech.edu

**Ivy Tech Community College - Lafayette**
Surgical Technology Prgm
3101 S Creasy Ln, PO Box 6299
Lafayette, IN 47903
www.laf.ivy.tec.in.us
*Prgm Dir:* Dorothy S Hall, MSN RN CST
*Tel:* 765 269-5208    *Fax:* 765 269-5248
*E-mail:* dhall@ivytech.edu

**Olympia College**
Surgical Technology Prgm
707 E 80th Pl, Ste 46410
Merrillville, IN 46410
*Prgm Dir:* Patricia Rich
*Tel:* 219 756-6811    *Fax:* 219 756-6812
*E-mail:* prich@cci.edu

**Ivy Tech Community College - Michigan City**
Surgical Technology Prgm
3714 Franklin
Michigan City, IN 46350
www.ivytech.edu
*Prgm Dir:* Lora Plank, RN CNOR CST
*Tel:* 219 879-9137, Ext 227    *Fax:* 219 879-9157
*E-mail:* lplank@ivytech.edu
*Med Dir:* Nicholas Retson, MD

**Ivy Tech Community College EC - Muncie**
Surgical Technology Prgm
4301 S Cowan Rd
Muncie, IN 47302
*Prgm Dir:* Caryn Humphrey, BSN RN CNOR
*Tel:* 765 289-2291, Ext 542    *Fax:* 765 289-2292
*E-mail:* chumphre@ivytech.edu
*Med Dir:* Michael Hodkin, MD

**Vincennes University**
Surgical Technology Prgm
1002 N First St, WAB-1
Vincennes, IN 47591
*Prgm Dir:* Chris Keegan, MS CST
*Tel:* 812 888-5893    *Fax:* 812 888-4550
*E-mail:* ckeegan@indian.vinu.edu
*Med Dir:* H Dan Adams, MD, MBA

# Iowa

**Kirkwood Community College**
Surgical Technology Prgm
6301 Kirkwood Blvd SW
Cedar Rapids, IA 52406
*Prgm Dir:* Terri Grell, RN BSN CNOR
*Tel:* 319 398-5566, Ext 5513    *Fax:* 319 398-1293
*E-mail:* tgrell@kirkwood.cc.ia.us

**Mercy College of Health Sciences**
Surgical Technology Prgm
928 Sixth Ave
Des Moines, IA 50309-1239
www.mchs.edu
*Prgm Dir:* Nancy Zylstra, RN CNOR RNFA
*Tel:* 515 643-6718    *Fax:* 515 643-6686
*E-mail:* nzylstra@mercydesmoines.org
*Med Dir:* Deborah Turner, MD FACOG

**Western Iowa Tech Community College**
Surgical Technology Prgm
4647 Stone Ave
PO Box 5199
Sioux City, IA 51106-5199
*Prgm Dir:* Renee Nemitz, RN CST
*Tel:* 712 274-8733, Ext 1391    *Fax:* 712 274-6412
*E-mail:* nemitzra@witcc.com
*Med Dir:* Gloria Stewart

**Iowa Lakes Community College**
Surgical Technology Prgm
1900 N Grand Ave, Ste 8
Spencer, IA 51301-2294
www.iowalakes.edu
*Prgm Dir:* Dana Grafft, CST
*Tel:* 712 262-7141, Ext 6624    *Fax:* 712 262-4047
*E-mail:* dgrafft@iowalakes.edu

# Kansas

**Hutchinson Community College**
Surgical Technology Prgm
815 N Walnut, Davis Hall
Hutchinson, KS 67501
www.hutchcc.edu
*Prgm Dir:* Katherine Gill, RN BSN CNOR
*Tel:* 620 665-4950
*E-mail:* gillk@hutchcc.edu
*Med Dir:* Tyler Hughes, MD

**Seward County Community College**
Surgical Technology Prgm
PO Box 1137
1801 N Kansas Ave
Liberal, KS 67905-1137
http://alliedhealth.sccc.edu
*Prgm Dir:* Carmen Summer, RN BSN CNOR
*Tel:* 620 626-3078    *Fax:* 620 626-3040
*E-mail:* carmen.sumner@sccc.edu
*Med Dir:* James Harrington, DO

**Kaw Area Technical School**
Surgical Technology Prgm
5724 Huntoon
Topeka, KS 66604
*Prgm Dir:* Donna Hess, CST
*Tel:* 785 228-6340    *Fax:* 785 273-7080
*E-mail:* dhess@kats.tec.ks.us
*Med Dir:* Wanda Schumacher, RN BSN

**Wichita Area Technical College**
Surgical Technology Prgm
324 N Emporia St
Wichita, KS 67202-2591
*Prgm Dir:* Cathy Mattingly, RN
*Tel:* 316 677-1355    *Fax:* 316 677-1332
*E-mail:* cmattingly@watc.edu
*Med Dir:* Bradley Bruner, MD

# Kentucky

**Ashland Technical College**
Surgical Technology Prgm
4818 Roberts Dr
Ashland, KY 41102-9046
*Prgm Dir:* Jaqueline Cavins, RN BSN CNOR
*Tel:* 606 326-2006    *Fax:* 606 928-4330
*E-mail:* jacqueline.cavins@kctcs.edu

**Bowling Green Technical College**
Surgical Technology Prgm
1845 Loop Dr
Bowling Green, KY 42101
*Prgm Dir:* Sherry Wells, CST CFA
*Tel:* 270 901-1079    *Fax:* 270 901-1139
*E-mail:* sherry.wells@kctcs.edu
*Med Dir:* Robert Watson, MD

**Bluegrass Community and Technical College**
Surgical Technology Prgm
308 Vo-Tech Rd
Lexington, KY 40511-2626
*Prgm Dir:* Kevin R Craycraft, CST
*Tel:* 859 296-5600    *Fax:* 859 246-2504
*E-mail:* kevinr.craycraft@kctcs.edu

**Jefferson Community and Technical College**
Surgical Technology Prgm
800 W Chestnut St
Louisville, KY 40203
*Prgm Dir:* Richard McClure, CST
*Tel:* 502 595-4275    *Fax:* 502 213-4502
*E-mail:* richard.mcclure@kctcs.edu

**Spencerian College**
Surgical Technology Prgm
4627 Dixie Hwy
Louisville, KY 40216
www.spencerian.edu
*Prgm Dir:* R Matt Matthews, CST CFA BSN BSEd
*Tel:* 502 447-1000, Ext 7875    *Fax:* 502 447-4574
*E-mail:* mmatthews@spencerian.edu
*Med Dir:* Leo J Wine, MD

**Madisonville Community College**
Surgical Technology Prgm
750 Laffoon St
Madisonville, KY 42431
*Prgm Dir:* Jeff Bidwell, CST CSA MA
*Tel:* 270 824-7552, Ext 183    *Fax:* 270 824-7069
*E-mail:* jeff.bidwell@kctcs.edu
*Med Dir:* Jack L Hamman, MD

**Kentucky Community Technical College**
Surgical Technology Prgm
4800 New Hartford Rd
Owensboro, KY 42303
*Prgm Dir:* Peggy Howard, RN CNOR BSN
*Tel:* 270 686-4634   *Fax:* 270 686-4662
*E-mail:* peggy.howard@kctcs.edu

**West Kentucky Community & Tech College**
Surgical Technology Prgm
Blandville Rd, PO Box 7408
Paducah, KY 42002
*Prgm Dir:* Debbie Swain, CST
*Tel:* 270 534-3482   *Fax:* 270 534-3498
*E-mail:* debbie.swain@kctcs.edu
*Med Dir:* W Robyn Howe, MD

**Southeast Kentucky Comm & Tech College**
Surgical Technology Prgm
3300 US Highway 25E South
Pineville, KY 40977
*Prgm Dir:* Roberta Dean, RN BS MSN
*Tel:* 606 248-2117   *Fax:* 606 248-2166
*E-mail:* roberta.dean@kctcs.edu
*Med Dir:* Talmadge V Hays, MD

**Somerset Community College**
Surgical Technology Prgm
808 Monticello St
Somerset, KY 42501
*Prgm Dir:* Dennis Hargis, RN BSN
*Tel:* 606 679-8501, Ext 3766
*E-mail:* dennis.hargis@kctcs.edu

# Louisiana

**MedVance Institute**
Surgical Technology Prgm
9255 Interline Ave
Baton Rouge, LA 70809
*Prgm Dir:* Janice Fairchild, CST
*Tel:* 225 248-1015   *Fax:* 225 248-9517
*E-mail:* jfairchild@medvance.org

**Our Lady of the Lake College**
Surgical Technology Prgm
7434 Perkins Rd
Baton Rouge, LA 70808
www.ololcollege.edu
*Prgm Dir:* Alice Comish, RN BSN CNOR
*Tel:* 225 768-1734   *Fax:* 225 768-0819
*E-mail:* acomish@ololcollege.edu
*Med Dir:* Alec Hirsch, MD

**Bossier Parish Community College**
Surgical Technology Prgm
6220 East Texas
Bossier City, LA 71111
www.bpcc.edu
*Prgm Dir:* Al Smith, RN BSN MEd
*Tel:* 318 678-6330   *Fax:* 318 678-6199
*E-mail:* asmith@bpcc.edu

**Louisiana Tech College - Lafayette Campus**
Surgical Technology Prgm
1101 Bertrand Dr
PO Box 4909
Lafayette, LA 70506
www.theltc.net
*Prgm Dir:* Lorie Lavergne, CST
*Tel:* 337 262-5962, Ext 238   *Fax:* 337 262-5122
*E-mail:* lorie.lavergne@theltc.net
*Med Dir:* Pemella Williams, RN, BSN, MHSA

**Career Technical College**
Surgical Technology Prgm
2319 Louisville Ave
Monroe, LA 71201
*Prgm Dir:* Regena Pardon, CST CFA RN
*Tel:* 318 998-5602   *Fax:* 318 323-3113
*E-mail:* rpardon@careertc.com
*Med Dir:* Kelli Batten

**Delgado Community College**
Surgical Technology Prgm
615 City Park Ave
New Orleans, LA 70119-4399
*Prgm Dir:* Sandra Palmer, MA
*Tel:* 504 568-6559   *Fax:* 504 568-5494
*E-mail:* spalme@dcc.edu
*Med Dir:* M Naraghi, MD

**Southern Univ at Shreveport**
Surgical Technology Prgm
610 Texas St, Ste 333
Shreveport, LA 71101
*Prgm Dir:* Didaciane G Keys, RN BSN
*Tel:* 318 674-3390   *Fax:* 318 676-5308
*E-mail:* dkeys@susla.edu

**Louisiana Tech College - Lafourche Campus**
Surgical Technology Prgm
1425 Tiger Dr
Thibodaux, LA 70301
www.theltc.net
*Prgm Dir:* Delores Gordon, CST
*Tel:* 985 447-0924, Ext 111   *Fax:* 985 447-0927
*E-mail:* delores.gordon@theltc.net

# Maine

**Eastern Maine Community College**
Surgical Technology Prgm
354 Hogan Rd
Bangor, ME 04401
*Prgm Dir:* Suzanne Brunner
*Tel:* 207 941-4600   *Fax:* 207 941-4608
*E-mail:* sbrunner@emcc.edu
*Med Dir:* Eric Steele, DO

**Maine Medical Center**
Surgical Technology Prgm
School of Surgical Technology
Fort Rd
Portland, ME 04106
*Prgm Dir:* Diane Fecteau, RN MSA
*Tel:* 207 662-8030   *Fax:* 207 662-8198
*E-mail:* fected@mmc.org
*Med Dir:* Brad Cushing, MD

**Northern Maine Community College**
Surgical Technology Prgm
33 Edgemont Dr
Presque Isle, ME 04769
*Prgm Dir:* Sonja Fongemie
*Tel:* 207 768-2767   *Fax:* 207 768-2723
*E-mail:* sfongemie@nmcc.edu

# Maryland

**Baltimore City Community College**
Surgical Technology Prgm
2901 Liberty Heights Ave
Baltimore, MD 21215
*Prgm Dir:* Audretta F Smith, BSN
*Tel:* 410 462-7722   *Fax:* 410 462-7734
*E-mail:* asmith@bccc.edu

**Comm Coll of Baltimore County - Essex Campus**
Surgical Technology Prgm
7201 Rossville Blvd
Baltimore, MD 21237-3899
*Prgm Dir:* Susan Fisher, CST CFA
*Tel:* 410 780-6869   *Fax:* 410 780-6856
*E-mail:* sfisher@ccbcmd.edu
*Med Dir:* Anthony Sclama, MD

**Frederick Community College**
Surgical Technology Prgm
7932 Opossumtown Pike
Frederick, MD 21702
*Prgm Dir:* Nancy N Dankanich, RN BSN MA
*Tel:* 301 846-2506   *Fax:* 301 846-2498
*E-mail:* ndankanich@frederick.edu

**Montgomery College**
Surgical Technology Prgm
900 Hungerford Dr
Rockville, MD 20850
*Prgm Dir:* Patrice Upshaw, RN MSN
*Tel:* 301 562-5541   *Fax:* 301 650-1335
*E-mail:* patrice.upshaw@montgomerycollege.edu
*Med Dir:* Ira Brecher, MD

**Chesapeake College**
Surgical Technology Prgm
PO Box 8
Wye Mills, MD 21632
*Prgm Dir:* Lisa T Szewczyk, CST
*Tel:* 410 822-5400, Ext 712   *Fax:* 410 770-3764
*E-mail:* lszewczyk@chesapeake.edu
*Med Dir:* Stanley M Bysshe, MD

# Massachusetts

**Bunker Hill Community College**
Surgical Technology Prgm
175 Hawthorne St
Chelsea Campus
Chelsea, MA 02150-2917
*Prgm Dir:* Jane S Roman, RN MEd CNOR
*Tel:* 617 228-3363   *Fax:* 617 228-2106
*E-mail:* Jroman@bhcc.mass.edu

**North Shore Community College**
Surgical Technology Prgm
1 Ferncroft Rd
PO Box 3340
Danvers, MA 01923-0840
*Prgm Dir:* Anne Baras, BSN CNOR MS
*Tel:* 978 762-4000, Ext 1503   *Fax:* 978 762-4181
*E-mail:* abaras@northshore.edu

**Massachusetts Bay Community College**
Surgical Technology Prgm
19 Flagg Dr
Framingham, MA 01700
www.massbay.edu
*Prgm Dir:* Patricia Caggiano, RN MS CNOR
*Tel:* 508 875-5300
*E-mail:* pcaggiano@massbay.edu

**Charles H McCann Technical School**
Surgical Technology Prgm
70 Hodges Cross Rd
North Adams, MA 01247
www.mccanntech.org
*Prgm Dir:* Tom Lescarbeau, CST CFA AS
*Tel:* 413 663-5383, Ext 180   *Fax:* 413 664-9424
*E-mail:* tlescarbeau@mccanntech.org

**Quincy College**
Surgical Technology Prgm
1495 Hanock St, 4th Fl Heritage Bldg
Quincy, MA 02169
*Prgm Dir:* Catherine DeLorey
*Tel:* 617 984-1718   *Fax:* 617 984-1792
*E-mail:* cdelorey@quincycollege.edu

**Springfield Technical Community College**
Surgical Technology Prgm
One Armory Sq
Springfield, MA 01105
*Prgm Dir:* Paula Hutchinson
*Tel:* 413 755-4887   *Fax:* 413 733-0688
*E-mail:* phutchinson@stcc.edu

**Quinsigamond Community College**
Surgical Technology Prgm
670 W Boylston St
Worcester, MA 01606-2092
*Prgm Dir:* Deborah A Coleman, MSN
*Tel:* 508 854-2734   *Fax:* 508 854-2704
*E-mail:* dcoleman@qcc.mass.edu

# Michigan

**Baker College of Cadillac**
Surgical Technology Prgm
9600 E 13th St
Cadillac, MI 49601
*Prgm Dir:* Mary Gauthier, CST RN
*Tel:* 231 775-8458   *Fax:* 231 779-9157
*E-mail:* mary.gauthier@baker.edu
*Med Dir:* Gerald Dudek, DO

**Baker College of Clinton Township**
Surgical Technology Prgm
34950 Little Mack Ave
Clinton Township, MI 48035
www.baker.edu
*Prgm Dir:* Lynda Custer, CST MA Ed
*Tel:* 586 790-2802   *Fax:* 586 791-6611
*E-mail:* lynda.custer@baker.edu

**Macomb Community College**
Surgical Technology Prgm
44575 Garfield Rd
Clinton Township, MI 48038-1139
*Prgm Dir:* Elizabeth Ness, CST BA
*Tel:* 586 286-2192   *Fax:* 586 286-2098
*E-mail:* nesse@macomb.edu
*Med Dir:* Michael Haynes, MD

**Henry Ford Community College**
Surgical Technology Prgm
5101 Evergreen Rd
Dearborn, MI 48128
*Prgm Dir:* Dorothy C Rothgery, MA CST/CFA
*Tel:* 313 317-6598   *Fax:* 313 317-6569
*E-mail:* drothger@hfcc.net

**Wayne County Community College District**
Surgical Technology Prgm
1001 W Fort St
Detroit, MI 48226
*Prgm Dir:* Mark Shikhman, MD PhD
*Tel:* 313 496-2680, Ext 2680   *Fax:* 313 961-9648
*E-mail:* mshikhm1@wcccd.edu
*Med Dir:* Thomas Siegel, MD

**Baker College of Flint**
Surgical Technology Prgm
1050 W Bristol Rd
Flint, MI 48507-5508
*Prgm Dir:* Julia A Jackson, CST BA
*Tel:* 810 766-4151   *Fax:* 810 766-2055
*E-mail:* Julia.Jackson@baker.edu

**Baker College of Jackson**
Surgical Technology Prgm
2800 Springport Rd
Jackson, MI 49202
www.baker.edu
*Prgm Dir:* Paula Hayes, CST BBA
*Tel:* 517 841-4535   *Fax:* 517 789-6302
*E-mail:* Paula.Hayes@baker.edu

**Kalamazoo Valley Community College**
Surgical Technology Prgm
6767 West O Ave
PO Box 4070
Kalamazoo, MI 49003-4070
www.kvcc.edu
*Prgm Dir:* James Taylor, MA
*Tel:* 269 488-4208   *Fax:* 269 488-4458
*E-mail:* jtaylor@kvcc.edu

**Lansing Community College**
Surgical Technology Prgm
3400 Human Health & Public Service
PO Box 40010
Lansing, MI 48901-7210
www.lcc.edu
*Prgm Dir:* Joseph Long, CST BA MPA EdD
*Tel:* 517 483-1432   *Fax:* 517 483-1508
*E-mail:* longj9@lcc.edu

**Northern Michigan University**
Surgical Technology Prgm
1401 Presque Isle Ave
Marquette, MI 49855
*Prgm Dir:* Sandra Kontio, CST
*Tel:* 906 227-1669   *Fax:* 906 227-1309
*Med Dir:* Craig Coccia, MD

**Baker College of Muskegon**
Surgical Technology Prgm
1903 Marquette Ave
Muskegon, MI 49442
www.baker.edu
*Prgm Dir:* Ruth Deters, RN BAHA CST
*Tel:* 616 777-5277   *Fax:* 616 777-5291
*E-mail:* ruth.deters@baker.edu

**Baker College of Port Huron**
Surgical Technology Prgm
3403 Lapeer Rd
Port Huron, MI 48060
www.baker.edu
*Prgm Dir:* Donna McFadden, RN CNOR
*Tel:* 810 982-9005
*E-mail:* donna.mcfadden@baker.edu
*Med Dir:* Krishna Valjee, MD

**Oakland Community College - Bloomfield Hills**
*Cosponsor:* William Beaumont Hospital
Surgical Technology Prgm
3601 W 13 Mile Rd
Royal Oak, MI 48073
*Prgm Dir:* Rebecca Pieknik, MS CST CSA
*Tel:* 248 898-7685   *Fax:* 248 898-0112
*E-mail:* rpieknik@beaumonthospitals.com
*Med Dir:* James Robbins, MD

**Delta College**
Surgical Technology Prgm
1961 Delta Rd
University Center, MI 48710
www.delta.edu
*Prgm Dir:* Margrethe May, CST MS
*Tel:* 989 686-9505   *Fax:* 989 667-2230
*E-mail:* mmay@delta.edu

# Minnesota

**Anoka Technical College**
Surgical Technology Prgm
1355 W Hwy 10
Anoka, MN 55303
www.anokatech.edu
*Prgm Dir:* Rita M Schutz, RN
*Tel:* 763 576-4974   *Fax:* 763 576-4715
*E-mail:* rschutz@anokatech.edu

**Lake Superior College**
Surgical Technology Prgm
2101 Trinity Rd
Duluth, MN 55811
*Prgm Dir:* Candace S Melde, RN
*Tel:* 218 733-5906   *Fax:* 218 723-4921
*E-mail:* c.melde@lsc.mnscu.edu

**Northland Community & Technical College**
Surgical Technology Prgm
2022 Central Ave NE
East Grand Forks, MN 56721
www.northlandcollege.edu
*Prgm Dir:* Ruth LeTexier, BSN
*Tel:* 218 773-4623   *Fax:* 218 773-4502
*E-mail:* ruth.letexier@northlandcollege.edu
*Med Dir:* Mark Siegel, MD

**St Mary's University of Minnesota**
Surgical Technology Prgm
2500 Park Ave
Minneapolis, MN 55404-4403
*Prgm Dir:* Becky Brodine
*Tel:* 612 728-5162, Ext 162   *Fax:* 612 728-5167
*E-mail:* bbrodin@smumn.edu

**Rochester Community & Technical College**
Surgical Technology Prgm
851 30th Ave SE
Rochester, MN 55904
www.rctc.edu
*Prgm Dir:* Jane Kruger, RN BSN CNOR
*Tel:* 507 280-3118   *Fax:* 507 280-3180
*E-mail:* jane.kruger@roch.edu

**St Cloud Technical College**
Surgical Technology Prgm
1540 Northway Dr
St Cloud, MN 56303
www.sctc.edu
*Prgm Dir:* Terry Wilson, MEd RN
*Tel:* 320 308-5921   *Fax:* 320 308-5971
*E-mail:* twilson@sctc.edu

# Mississippi

**East Central Community College**
Surgical Technology Prgm
PO Box 129
Decatur, MS 39327
www.eccc.edu
*Prgm Dir:* Melanie Gilmore
*Tel:* 601 635-2111, Ext 294   *Fax:* 601 635-4031
*E-mail:* mgilmore@eccc.edu

**Holmes Community College**
Surgical Technology Prgm
1060 Avent Dr
Grenada, MS 38901
www.holmescc.edu
*Prgm Dir:* Jessica Elliott, CST
*Tel:* 662 227-2310   *Fax:* 662 227-2296
*E-mail:* jelliott@holmescc.edu

**Pearl River Community College**
Surgical Technology Prgm
5448 US Hwy 49 S
Hattiesburg, MS 39401
*Prgm Dir:* Debbie Hinton, BS CST
*Tel:* 601 554-5542   *Fax:* 601 554-5548
*E-mail:* dhinton@prcc.edu

**Hinds Community College**
Surgical Technology Prgm
1750 Chadwick Dr
Jackson, MS 39204
*Prgm Dir:* Teresa Jones, RN
*Tel:* 601 371-3513   *Fax:* 601 371-3529
*E-mail:* tajones@hindscc.edu

**Mississippi Gulf Coast Community College**
Surgical Technology Prgm
PO Box 77
11203 Old Hwy 63 S
Lucedale, MS 39452
*Prgm Dir:* Karen Howell, RN MS CNOR
*Tel:* 601 766-6430   *Fax:* 601 947-4899
*E-mail:* karen.howell@mgccc.edu
*Med Dir:* Chris Wiggins, MD

**Itawamba Community College**
Surgical Technology Prgm
2176 S Eason Blvd
Tupelo, MS 38804
*Prgm Dir:* Tonya Tice, RN
*Tel:* 662 620-5121   *Fax:* 662 620-5077
*E-mail:* tltice@iccms.edu

# Missouri

**Southeast Missouri Hospital**
Surgical Technology Prgm
2001 William St
Cape Girardeau, MO 63703
*Prgm Dir:* Susan Jackson, BSN CNOR
*Tel:* 573 334-6825, Ext 28   *Fax:* 573 339-7805
*E-mail:* sjackson@sehosp.org

**Columbia Public Schools**
Surgical Technology Prgm
1818 W Worley St
Columbia, MO 65203-6100
*Prgm Dir:* Jane Klick, RN CNOR
*Tel:* 573 886-2276   *Fax:* 573 886-2080
*E-mail:* jklick@columbia.k12.mo.us

**Sanford-Brown College - Hazelwood Campus**
Surgical Technology Prgm
75 Village Square
Hazelwood, MO 63042
*Prgm Dir:* Amy Gray, MD CST
*Tel:* 314 687-2940   *Fax:* 314 731-0550
*E-mail:* agray@sbc-hazelwood.com

**Franklin Tech Center**
*Cosponsor:* Missouri Southern State Univ
Surgical Technology Prgm
3950 E Newman Rd
Joplin, MO 64801
*Prgm Dir:* Gaylard Roper, CST
*Tel:* 417 625-9723   *Fax:* 417 659-4408
*Med Dir:* Mark S Cotner, MD

**Metropolitan Community College - Penn Valley**
Surgical Technology Prgm
2700 E 18th St
Kansas City, MO 64127
*Prgm Dir:* Carolyn A Parks, RN BAN CNOR
*Tel:* 816 482-5073   *Fax:* 816 482-5110
*E-mail:* Carolyn.Parks@mcckc.edu
*Med Dir:* Kelly James, MD

**Rolla Technical Center**
Surgical Technology Prgm
500 Forum Dr
Rolla, MO 65401
*Prgm Dir:* Tammy Mangold, MEd CST/CFA
*Tel:* 573 458-0160, Ext 16535   *Fax:* 573 458-0164
*E-mail:* tmangold@rolla.k12.mo.us
*Med Dir:* Sherry Phippen, MD

**Ozarks Technical Community College**
Surgical Technology Prgm
1001 E Chestnut Expressway
Springfield, MO 65802
www.otc.edu/degrees/alldhlth/surgtec
*Prgm Dir:* Arlene Chriswell, RN BSN MSBA CNOR
*Tel:* 417 447-8845   *Fax:* 417 447-8806
*E-mail:* chriswea@otc.edu
*Med Dir:* Joel Waxman, MD

**Hillyard Technical Center**
Surgical Technology Prgm
3434 Faraon St
St Jospeh, MO 64506
*Prgm Dir:* Linda Ann VanDyke, CST CFA
*Tel:* 816 671-4170
*E-mail:* linda.vandyke@sjsd.k12.mo.us
*Med Dir:* Steven Long, MD

**St Louis Community College - Forest Park**
Surgical Technology Prgm
5600 Oakland Ave
St Louis, MO 63110
*Prgm Dir:* Diane Gerardot, MA,CST
*Tel:* 314 644-9340   *Fax:* 314 951-9412
*E-mail:* dgerardot@stlcc.edu
*Med Dir:* J Alexander Marchosky, MD

**South Central Career Center**
Surgical Technology Prgm
610 East Olden St
West Plains, MO 65775
*Prgm Dir:* Septembre Lasater
*Tel:* 417 256-8883   *Fax:* 417 256-5786
*E-mail:* slasater@mail.wphs.k12.mo.us

# Montana

**Montana State Univ - Great Falls Coll of Tech**
Surgical Technology Prgm
2100 16th Ave S
Great Falls, MT 59406-6010
*Prgm Dir:* Diona Davis, CST CFA
*Tel:* 406 771-4358   *Fax:* 406 771-4317
*E-mail:* ddavis@msugf.edu

**Flathead Valley Community College**
Surgical Technology Prgm
777 Grandview Dr
Kalispell, MT 59901
www.fvcc.edu
*Prgm Dir:* Erin Howardson, CST
*Tel:* 406 751-6994   *Fax:* 406 756-0317
*E-mail:* eahowardson@yahoo.com

**University of Montana**
Surgical Technology Prgm
College of Technology
909 South Ave W
Missoula, MT 59801
*Prgm Dir:* Debbie Fillmore
*Tel:* 406 243-7860   *Fax:* 406 243-7899
*E-mail:* deborah.fillmore@umontana.edu

# Nebraska

**Southeast Community College**
Surgical Technology Prgm
8800 O St
Lincoln, NE 68520
*Prgm Dir:* Kathleen Uribe, MA CST
*Tel:* 402 437-2785   *Fax:* 402 437-2404
*E-mail:* kuribe@southeast.edu

**Nebraska Methodist College**
Surgical Technology Prgm
515 S 26th St
Omaha, NE 68105
*Prgm Dir:* Patricia Boettger, RN CNOR CRNFA
*Tel:* 402 354-6528   *Fax:* 402 354-6550
*E-mail:* pat.boettger@methodistcollege.edu
*Med Dir:* Gregory Eakins

# Nevada

**Western Nevard Community College**
Surgical Technology Prgm
2201 W College Pkwy
Carson City, NV 89703
www.wncc.edu
*Prgm Dir:* Duane Sorensen, CST
*Tel:* 775 445-3248   *Fax:* 775 887-3021
*E-mail:* dsorense@wncc.edu
*Med Dir:* Susan Buchwald, MD

**Community College of Southern Nevada**
Surgical Technology Prgm
6375 W Charleston Blvd
Las Vegas, NV 89146
*Prgm Dir:* Eugene Lewis, CST
*Tel:* 702 651-5945   *Fax:* 702 651-5501
*E-mail:* eugene_lewis@ccsn.edu

**Nevada Career Institute**
Surgical Technology Prgm
3025 E Desert Inn Rd, Ste A
Las Vegas, NV 89121
www.landmarked.com
*Prgm Dir:* Donna McCasland
*Tel:* 702 893-4725, Ext 231   *Fax:* 702 893-3881
*E-mail:* dmccasland@success.edu
*Med Dir:* Joanne Leming, Regional Director

# New Hampshire

**Concord Hospital**
Surgical Technology Prgm
250 Pleasant St
Concord, NH 03301
*Prgm Dir:* Colleen Rutherford, RN MS CNOR
*Tel:* 603 225-2711, Ext 3891   *Fax:* 603 228-7306
*E-mail:* crutherf@crhc.org

**Dartmouth Hitchcock Medical Center**
Surgical Technology Prgm
One Medical Center Dr
Lebanon, NH 03756-0001
www.dhmc.org
*Prgm Dir:* Carol Majewski
*Tel:* 603 650-8134
*E-mail:* carol.a.majewski@hitchcock.org
*Med Dir:* Carter Dodge, MD

**New Hampshire Comm Tech College - Stratham**
Surgical Technology Prgm
277 Portsmouth Ave
Stratham, NH 03885-2297
*Prgm Dir:* Sandra Carlson, RN BSN CNOR
*Tel:* 603 775-2367   *Fax:* 603 772-1198
*E-mail:* scarlson@nhctc.edu

# New Jersey

**Sanford-Brown Institute**
Surgical Technology Prgm
675 US Hwy 1
Iselin, NJ 08830
*Prgm Dir:* Carol Conover
*Tel:* 732 634-1131   *Fax:* 732 634-1040
*E-mail:* cconover@sb-nj.com
*Med Dir:* Irving Karten, MD

**Sussex County Community College**
Surgical Technology Prgm
1 College Hill Rd
Newton, NJ 07860
*Prgm Dir:* Barbara J Cook, BA CMA
*Tel:* 973 300-2263   *Fax:* 973 300-2303
*E-mail:* bcook@sussex.edu
*Med Dir:* John Semian, MD

**Bergen Community College**
Surgical Technology Prgm
400 Paramus Rd
Paramus, NJ 07652
www.bergen.cc.nj.us/academics/AC
*Prgm Dir:* Joan Ann Verderame, MA RN
*Tel:* 201 447-7944, Ext 7921   *Fax:* 201 612-3876
*E-mail:* jverderame@bergen.edu

# New Mexico

**Central New Mexico Community College**
Surgical Technology Prgm
525 Buena Vista SE
Albuquerque, NM 87106
*Prgm Dir:* Elizabeth L Alongi, RN BSN CNOR
*Tel:* 505 224-4166   *Fax:* 505 224-4120
*E-mail:* ealongi@cnm.edu

# New York

### Long Island University - Brooklyn Campus
*Cosponsor:* Center for Health Science Technologies
Surgical Technology Prgm
School of Health Professions
1 University Plaza (M-315)
Brooklyn, NY 11201
www.liu.edu
*Prgm Dir:* Karen Chambers, CST
*Tel:* 718 488-3438, Ext 3438   *Fax:* 718 780-4575
*E-mail:* karen.chambers@liu.edu
*Med Dir:* Anbalagan George, MBBS

### SUNY Downstate Medical Center
Surgical Technology Prgm
440 Lenox Rd
Brooklyn, NY 11203
*Prgm Dir:* Tomas Wharton, MD
*Tel:* 718 270-2983   *Fax:* 718 270-8114
*E-mail:* twharton@downstate.edu

### Trocaire College
Surgical Technology Prgm
360 Choate Ave
Buffalo, NY 14220-2094
www.trocaire.edu
*Prgm Dir:* Linda Kerwin, RN MSN MA
*Tel:* 716 826-1200, Ext 1254   *Fax:* 716 828-6107
*E-mail:* kerwinl@trocaire.edu

### Nassau Community College
Surgical Technology Prgm
One Education Dr
Dept AHS
Garden City, NY 11530
www.ncc.edu
*Prgm Dir:* Caroline Kaufmann, RN CNOR
*Tel:* 516 572-7918   *Fax:* 516 572-7565
*E-mail:* kaufmac@ncc.edu

### Bronx Lebanon Hospital
Surgical Technology Prgm
275 7th Ave, 16th Fl
New York, NY 10001
www.cwe.org
*Prgm Dir:* Smyrna E Clause, RN MS
*Tel:* 212 647-1900, Ext 275   *Fax:* 212 647-1916
*E-mail:* sclause@cwe.org
*Med Dir:* Chantal Turnier, MD

### New York University Medical Center
Surgical Technology Prgm
660 1st Ave, 2nd Floor
New York, NY 10016
www.surgtech.com
*Prgm Dir:* Zaida Jacoby, RN MA MEd
*Tel:* 212 263-6644   *Fax:* 212 263-0600
*E-mail:* zaida.jacoby@med.nyu.edu

### Western Suffolk BOCES
Surgical Technology Prgm
152 Laurel Hill Rd
Northport, NY 11768
*Prgm Dir:* Kathleen Baker, RN BSN MSEd
*Tel:* 631 261-3727   *Fax:* 631 623-4908
*E-mail:* kbaker@wsboces.org

### Ulster County Board of Cooperative Ed Service
Surgical Technology Prgm
PO Box 601, Rt 9W
Port Ewen, NY 12466
www.mhric.org
*Prgm Dir:* Marita Kitchell, EdMS SAS BSN
*Tel:* 845 331-5050, Ext 2232   *Fax:* 845 339-8797
*E-mail:* mkitchel@mhric.org
*Med Dir:* Rhonda O'Connor, LPN

### Niagara County Community College
Surgical Technology Prgm
3111 Saunders Settlement Rd
Sanborn, NY 14132
*Prgm Dir:* Gemma Fournier, RN CST
*Tel:* 716 614-6417   *Fax:* 716 614-6879
*E-mail:* fournier@niagaracc.suny.edu

### Stony Brook University
Surgical Technology Prgm
School of Nursing
Health Sciences Center Level 2
Stony Brook, NY 11794-8240
*Prgm Dir:* Ora James Bouey, RN
*Tel:* 631 444-3549   *Fax:* 631 444-7763
*E-mail:* ora.bouey@stonybrook.edu
*Med Dir:* Norman Edelman, MD

### Onondaga Community College
Surgical Technology Prgm
4941 Onondaga Rd
Syracuse, NY 13215
*Prgm Dir:* Mary Pat M Annable, RN MSN CNOR
*Tel:* 315 498-2463   *Fax:* 315 498-2751
*E-mail:* annablem@sunyocc.edu
*Med Dir:* Brian Anderson, MD

# North Carolina

### Asheville-Buncombe Technical Comm College
Surgical Technology Prgm
340 Victoria Rd
Asheville, NC 28801
www.abtech.edu
*Prgm Dir:* Robin B Keith, RN BSN CNOR CST
*Tel:* 828 254-1921, Ext 892   *Fax:* 828 251-6355
*E-mail:* rkeith@abtech.edu
*Med Dir:* Daniel Eglinton, MD, Orthopaedic Surgeon

### South College - Asheville
Surgical Technology Prgm
1567 Patton Ave
Asheville, NC 28803
*Prgm Dir:* Tara Luhrs
*Tel:* 828 252-2486
*E-mail:* tluhrs@southcollegenc.com

### Carolinas College of Health Sciences
Surgical Technology Prgm
1200 Blythe Rd, PO Box 32861
Charlotte, NC 28232-2861
*Prgm Dir:* Rebecca M Cuthbertson, BSN
*Tel:* 704 355-1547   *Fax:* 704 355-5967
*E-mail:* rcuthbertson@carolinas.org
*Med Dir:* Michael H Thomason, MD

### Presbyterian Healthcare
Surgical Technology Prgm
1901 E 5th St
PO Box 33549
Charlotte, NC 28233-3549
www.presbyterian.org
*Prgm Dir:* Eugene Pease, RN MSN CNOR
*Tel:* 704 384-3173
*E-mail:* epease@novanthealth.org
*Med Dir:* Thomas Zweng, MD

### Cabarrus College of Health Sciences
Surgical Technology Prgm
401 Medical Park Dr
Concord, NC 28025
http://cabarruscollege.edu
*Prgm Dir:* Karen H Galloway, RN CNOR PhD
*Tel:* 704 783-1555, Ext 3503   *Fax:* 704 783-1764
*E-mail:* kgallowa@northeastmedical.org
*Med Dir:* John J Wassel, MD

### Durham Technical Community College
Surgical Technology Prgm
1637 Lawson St
Durham, NC 27703
*Prgm Dir:* Tammy Holden, CST
*Tel:* 919 686-3690   *Fax:* 919 686-3705
*E-mail:* holdent@durhamtech.edu

### College of The Albemarle
Surgical Technology Prgm
1208 N Road St, PO Box 2327
Elizabeth City, NC 27906-2327
*Prgm Dir:* Irene M Jack, CST
*Tel:* 252 335-0821, Ext 2995   *Fax:* 252 335-2011
*E-mail:* irenejack@albemarle.edu

### Fayetteville Technical Community College
Surgical Technology Prgm
PO Box 35236
Fayetteville, NC 28303
www.faytechcc.edu
*Prgm Dir:* Terry Herring, MS CST/CFA
*Tel:* 910 678-8358
*E-mail:* herringt@faytechcc.edu

### Blue Ridge Community College
Surgical Technology Prgm
100 College Dr
Flat Rock, NC 28731-9624
*Prgm Dir:* Debra Houdek, RN BSN MSN CST FNP
*Tel:* 828 694-1831   *Fax:* 828 692-2441
*E-mail:* debrah@blueridge.edu
*Med Dir:* Charles Albers, MD

### Catawba Valley Community College
Surgical Technology Prgm
2550 Hwy 70 SE
Hickory, NC 28602
*Prgm Dir:* Carol Harrison, RN
*Tel:* 704 327-7000, Ext 4332   *Fax:* 704 327-7301
*E-mail:* charriso@cvcc.cc.nc.us

### Coastal Carolina Community College
Surgical Technology Prgm
444 Western Blvd
Jacksonville, NC 28546-6877
*Prgm Dir:* Karen Gurney, RN
*Tel:* 910 938-6274   *Fax:* 910 938-6806
*E-mail:* gurneyk@coastal.cc.nc.us

### Guilford Technical Community College
Surgical Technology Prgm
601 High Point Rd, PO Box 309
Jamestown, NC 27282-0309
www.gtcc.edu
*Prgm Dir:* Arthur A Makin, BS CST
*Tel:* 336 334-4822, Ext 2764   *Fax:* 336 454-3554
*E-mail:* aamakin@gtcc.edu

### Lenoir Community College
Surgical Technology Prgm
PO Box 188
Kinston, NC 28502-0188
*Prgm Dir:* Patsy Burrus, RN BSN
*Tel:* 252 527-6223, Ext 802   *Fax:* 252 233-6893
*E-mail:* pburrus@lcc.edu

### Sandhills Community College
Surgical Technology Prgm
2200 Airport Rd
Pinehurst, NC 28374
*Prgm Dir:* Brenda Luck
*Tel:* 910 695-3838   *Fax:* 910 692-2756
*E-mail:* luckb@sandhills.edu

### South Piedmont Community College
Surgical Technology Prgm
PO Box 126
Polkton, NC 28135
*Prgm Dir:* Carol Courtney, BSN
*Tel:* 704 272-7635, Ext 280   *Fax:* 704 272-8904
*E-mail:* ccourtney@spcc.edu
*Med Dir:* Thomas Friedrich, MD

**Wake Technical Community College**
Surgical Technology Prgm
9109 Fayetteville Rd
Raleigh, NC 27603
*Prgm Dir:* Martha Shurtleff, RN MSN CNOR
*Tel:* 919 250-4331   *Fax:* 919 250-4329
*E-mail:* mashurtl@waketech.edu
*Med Dir:* James Fogartie Jr., MD, FACS

**Cleveland Community College**
Surgical Technology Prgm
137 S Post Rd
Shelby, NC 28152
www.cleveland.cc.nc.us
*Prgm Dir:* Wanda Leonard, CST
*Tel:* 704 484-6615   *Fax:* 704 484-5304
*E-mail:* leonardw@cleveland.cc.nc.us
*Med Dir:* Micheal Barringer, MD

**Edgecombe Community College**
Surgical Technology Prgm
2009 W Wilson St
Tarboro, NC 27886
*Prgm Dir:* Linda Harrison, CST
*Tel:* 252 823-5166, Ext 232   *Fax:* 252 823-6817
*E-mail:* harrisonl@edgecombe.edu
*Med Dir:* Charles Middleton, MD

**Rockingham Community College**
Surgical Technology Prgm
PO Box 38
Wentworth, NC 27375
www.rockinghamcc.edu
*Prgm Dir:* Mary Margaret Martin, RN BSN CNOR
APRN-BC
*Tel:* 336 342-4261, Ext 2259   *Fax:* 336 634-1253
*E-mail:* martinm@rockinghamcc.edu

**Miller-Motte Technical College**
Surgical Technology Prgm
5000 Market St
Wilmington, NC 28405
www.mmtcwilmington.net
*Prgm Dir:* Glenn Grady, MEd BSMT(ASCP) CMA
*Tel:* 910 392-4660, Ext 2015   *Fax:* 910 799-6224
*E-mail:* ggrady@miller-motte.edu
*Med Dir:* Thomas Clancy, MD

**Wilson Technical Community College**
Surgical Technology Prgm
904 Herring Ave
Wilson, NC 27893
www.wilsontech.edu
*Prgm Dir:* Lynanne Boyette, RN
*Tel:* 252 246-1323   *Fax:* 252 243-7148
*E-mail:* lboyette@wilsontech.edu
*Med Dir:* Glenda J Bondurant, MSN RN

# North Dakota

**Bismarck State College**
Surgical Technology Prgm
PO Box 5587
Bismarck, ND 58506-5587
www.bsc.nodak.edu
*Prgm Dir:* Jean Hinton, RN CNOR CST
*Tel:* 701 224-5722   *Fax:* 701 224-5550
*E-mail:* Jean.Hinton@bsc.nodak.edu

# Ohio

**University of Akron**
Surgical Technology Prgm
Polsky Bldg Rm 124 H
Akron, OH 44325-3702
http://sc.uakron.edu/?/alliedhealth/index.html
*Prgm Dir:* Sherry Gamble, RN MSN CNOR
*Tel:* 330 972-6514   *Fax:* 330 972-2016
*E-mail:* slg@uakron.edu
*Med Dir:* Kirby Sweitzer, MD

**Univ of Cincinnati - Clermont College**
Surgical Technology Prgm
4200 Clermont College Dr
Batavia, OH 45103
*Prgm Dir:* Linda M Baker-Numrich, RN BS CNOR
*Tel:* 513 732-5331   *Fax:* 513 732-5304
*E-mail:* bakerlm@email.uc.edu
*Med Dir:* Jesus Hontanosis, MD

**Collins Career Center**
Surgical Technology Prgm
11627 State Route 243
Chesapeake, OH 45619
www.collins-cc.k12.oh.us
*Prgm Dir:* Katherine Hall, RN
*Tel:* 740 867-6641, Ext 520   *Fax:* 740 867-9626
*E-mail:* kehall@collins-cc.k12.oh.us
*Med Dir:* Dr Patil, MD

**Cincinnati State Tech & Comm College**
Surgical Technology Prgm
3520 Central Pkwy
Cincinnati, OH 45223
www.cincinnatistate.edu
*Prgm Dir:* Wanda Dantzler, RN MEd BSN CNOR CRCST
*Tel:* 513 569-1673   *Fax:* 513 569-1659
*E-mail:* wanda.dantzler@cincinnatistate.edu
*Med Dir:* Marianne Krismer, RD, LD

**Cuyahoga Community College**
Surgical Technology Prgm
2900 Community College Ave
Cleveland, OH 44115-3196
*Prgm Dir:* Beth Stokes, BFA AAS CST
*Tel:* 216 987-6146, Ext 6146   *Fax:* 216 987-4366
*E-mail:* beth.stokes@tri-c.edu

**Columbus State Community College**
Surgical Technology Prgm
550 E Spring St, PO Box 1609
Columbus, OH 43216-1609
*Prgm Dir:* Dennis Murphy, BS CST
*Tel:* 614 287-2514, Ext 2514   *Fax:* 614 287-3854
*E-mail:* dmurphy@cscc.edu

**Mt Carmel College of Nursing**
Surgical Technology Prgm
127 S Davis Ave
Columbus, OH 43222
*Prgm Dir:* Linda Atkinson, RN AD
*Tel:* 614 234-1388   *Fax:* 614 234-2875
*E-mail:* latkinson@mchs.com

**Sinclair Community College**
Surgical Technology Prgm
444 W Third St
Rm 3340 Surgical Technology
Dayton, OH 45402-1460
*Prgm Dir:* Susan Willin-Mulay, RN BSN
*Tel:* 937 512-2850   *Fax:* 937 512-2058
*E-mail:* susan.willin-mulay@sinclair.edu

**Lorain County Community College**
Surgical Technology Prgm
1005 N Abbe Rd
Elyria, OH 44035
*Prgm Dir:* Patricia Sedlak, RN BSN CNOR RNFA
*Tel:* 800 995-5222, Ext 7159   *Fax:* 440 366-4116
*E-mail:* psedlak@lorainccc.edu

**Lakeland Community College**
Surgical Technology Prgm
7700 Clocktower Dr
Kirtland, OH 44094
*Prgm Dir:* Nancymarie Phillips, RN BA BSN MEd
CNOR RNFA
*Tel:* 440 525-7016   *Fax:* 440 525-4733
*E-mail:* nphillips@lakelandcc.edu

**Apollo Career Center**
Surgical Technology Prgm
3325 Shawnee Rd
Lima, OH 45806
www.apollocareercenter.com
*Prgm Dir:* Barbara E Cook, RN BSN CNOR
*Tel:* 419 999-5618   *Fax:* 419 998-2994
*E-mail:* ap_cook@noacsc.org
*Med Dir:* James Patterson, MD

**Scioto County Joint Vocational School**
Surgical Technology Prgm
951 Vern Riffe Dr
Lucasville, OH 45648
www.scjvs.com
*Prgm Dir:* Maxine Mathis, MS
*Tel:* 740 259-5526, Ext 202   *Fax:* 740 259-8312
*E-mail:* emathis_sj@scoca-k12.org
*Med Dir:* Thomas Khoury, MD

**EHOVE Ghrist Adult Career Center**
Surgical Technology Prgm
316 W Mason Rd
Milan, OH 44846
*Prgm Dir:* Beth Lucas, CST AAS
*Tel:* 419 499-8173   *Fax:* 419 499-8173
*E-mail:* blucas@ehove-jvs.k12.oh.us

**Central Ohio Technical College**
Surgical Technology Prgm
1179 University Dr
Newark, OH 43055-1797
*Prgm Dir:* Jacob Benjamin
*Tel:* 740 366-1351, Ext 229   *Fax:* 740 366-9275
*E-mail:* benjamin.17@osu.edu

**Buckeye Hills Career Center**
Surgical Technology Prgm
351 Buckeye Hills Rd
Rio Grande, OH 45674
*Prgm Dir:* Sue Gilliam, RN CNOR
*Tel:* 740 245-5334, Ext 252   *Fax:* 740 245-9465
*E-mail:* bj_sgilliam@seovec.org
*Med Dir:* Alice Gricoski, MD

**Owens Community College**
Surgical Technology Prgm
PO Box 10,000, Oregon Rd
Toledo, OH 43699
www.owens.edu
*Prgm Dir:* Kristine Flickinger, RN BA CNOR
*Tel:* 567 661-7000, Ext 7310   *Fax:* 567 661-7665
*E-mail:* kristine_flickinger@owens.edu

**Choffin Career & Technical Center**
*Cosponsor:* Youngstown City Schools
Surgical Technology Prgm
200 E Wood St
Youngstown, OH 44503
*Prgm Dir:* Carole DuBose, LPN/CST
*Tel:* 330 744-8763   *Fax:* 330 744-8705
*E-mail:* youn_cgd@access-k12.org
*Med Dir:* Denise Vaclave-Danko, MEd

# Oklahoma

**Southern Oklahoma Technology Center**
Surgical Technology Prgm
2610 Sam Noble Pkwy
Ardmore, OK 73401
www.sotc.org
*Prgm Dir:* Tracie Kelch, RN
*Tel:* 580 220-6273   *Fax:* 580 220-6534
*E-mail:* tkelch@sotc.org

**Canadian Valley Technology Center**
Surgical Technology Prgm
1401 Michigan Ave
Chickasha, OK 73018
www.cvtech.org
*Prgm Dir:* Anita Followwill, RN
*Tel:* 405 224-7220   *Fax:* 405 222-7511
*E-mail:* afollowwill@cvtech.org
*Med Dir:* Howell, Cardiovascular surgeon

**Central Technology Center**
Surgical Technology Prgm
3 Court Circle
Drumright, OK 74030
*Prgm Dir:* LaDonna Gear
*Tel:* 918 352-2551, Ext 289   *Fax:* 918 352-2441
*E-mail:* ladonnag@ctechok.org

**Autry Technology Center**
Surgical Technology Prgm
1201 W Willow
Enid, OK 73703
*Prgm Dir:* Kim McFarland
*Tel:* 405 242-2750, Ext 127   *Fax:* 405 233-8262
*E-mail:* kmcfarland@autrytech.com
*Med Dir:* Barry Pollard, MD

**Great Plains Technology Center**
Surgical Technology Prgm
4500 W Lee Blvd
Lawton, OK 73505
*Prgm Dir:* Ann Tahah, LPN
*Tel:* 580 250-5574   *Fax:* 580 250-5583
*E-mail:* atahah@gptech.org

**Oklahoma Health Academy**
Surgical Technology Prgm
1939 N Moore Ave
Moore, OK 73160
www.oklahomahealthacademy.org
*Prgm Dir:* Kirk Webster, MSEd
*Tel:* 405 912-2777   *Fax:* 405 912-2770
*E-mail:* jkwebster@oklahomahealthacademy.org
*Med Dir:* Robert Holybee, CST

**Muskogee Regional Medical Center**
Surgical Technology Prgm
300 Rockefeller Dr
Muskogee, OK 74401
*Prgm Dir:* Sandra Blakeslee, RN BSN CNOR
*Tel:* 918 684-2280   *Fax:* 918 684-2338
*E-mail:* sblakeslee@muskogeehealth.com

**Moore Norman Technology Center**
Surgical Technology Prgm
4701 12th Ave NW
Norman, OK 73069
*Prgm Dir:* Kim Shannon, RN
*Tel:* 405 364-5763, Ext 6525   *Fax:* 405 447-6289
*E-mail:* kshannon@mntechnology.com

**Heritage College**
Surgical Technology Prgm
7100 I-35 Service Rd
Oklahoma City, OK 73149
*Prgm Dir:* Courtney Powers
*Tel:* 405 631-3399   *Fax:* 405 631-6711
*E-mail:* courtneyp@heritage-education.com

**Metro Technology Centers**
Surgical Technology Prgm
Health Careers Center
1720 Springlake Dr
Oklahoma City, OK 73111
*Prgm Dir:* Vicki Bushey, CST
*Tel:* 405 424-8324, Ext 620   *Fax:* 405 424-9403
*E-mail:* vbushey@metrotech.org

**Platt College - Oklahoma City**
Surgical Technology Prgm
309 S Ann Arbor
Oklahoma City, OK 73128
*Prgm Dir:* Terri Taylor, CST
*Tel:* 405 946-7799   *Fax:* 405 943-2150
*E-mail:* terrrit@plattcollege.org

**Platt College - Tulsa**
Surgical Technology Prgm
3801 S Sheridan
Tulsa, OK 74145
*Prgm Dir:* Pam Buff, CST
*Tel:* 918 663-9000   *Fax:* 918 622-1240
*E-mail:* pamb@plattcollege.org

**Tulsa Technology Center**
Surgical Technology Prgm
3420 S Memorial Dr
Tulsa, OK 74145-1390
www.tulsatech.org
*Prgm Dir:* Carol Dollar, RN BSN
*Tel:* 918 828-1036   *Fax:* 918 828-5239
*E-mail:* carol.dollar@tulsatech.org

**Wes Watkins Technology Center Distist 25**
Surgical Technology Prgm
7892 Hwy 9
Wetumka, OK 74883
www.wwtech.org
*Prgm Dir:* Natalie Kennedy, CST
*Tel:* 405 452-5500, Ext 291   *Fax:* 405 452-3561
*E-mail:* nkennedy@wwtech.org
*Med Dir:* Linda Sanford, RN

# Oregon

**Mt Hood Community College**
Surgical Technology Prgm
26000 SE Stark St
Gresham, OR 97030
*Prgm Dir:* Jacqueline Morfitt, RN CNOR
*Tel:* 503 491-7179   *Fax:* 503 491-6047
*E-mail:* morfittj@mhcc.edu

**Concorde Career Institute - Portland**
Surgical Technology Prgm
1425 NE Irving St, Bldg 300
Portland, OR 97232
www.concorde.edu
*Prgm Dir:* Jonathan Lee
*Tel:* 503 248-6139   *Fax:* 503 281-6739
*E-mail:* jlee@concorde.edu
*Med Dir:* Kim Ierien

# Pennsylvania

**Northampton Community College**
Surgical Technology Prgm
3835 Green Pond Rd
Bethlehem, PA 18020
www.northampton.edu
*Prgm Dir:* Judith Rex, RNC MSN
*Tel:* 610 861-5094   *Fax:* 610 332-6556
*E-mail:* jrex@northampton.edu

**St Luke's Hospital**
Surgical Technology Prgm
801 Ostrum St
Bethlehem, PA 18015
*Prgm Dir:* Barbara Benfield, LPN CST
*Tel:* 610 954-4466   *Fax:* 610 954-6097
*Med Dir:* Balshi James, MD

**Mount Aloysius College**
Surgical Technology Prgm
7373 Admiral Peary Hwy
Cresson, PA 16630-1999
*Prgm Dir:* Doris Etienne, AS ST
*Tel:* 814 886-6536   *Fax:* 814 886-2978
*E-mail:* detienne@mtaloy.edu

**Great Lakes Institute of Technology**
Surgical Technology Prgm
5100 Peach St
Erie, PA 16509
*Prgm Dir:* Brian Lock, II, RN PhD
*Tel:* 814 864-6666, Ext 233   *Fax:* 814 868-1717
*E-mail:* brianl@glit.edu

**Harrisburg Area Community College - Harrisburg**
Surgical Technology Prgm
349 Wiconisco St
One HACC Dr
Harrisburg, PA 17110
www.hacc.edu
*Prgm Dir:* Amy Kennedy, RN MSN CNOR
*Tel:* 717 221-1321   *Fax:* 717 909-4010
*E-mail:* alkenned@hacc.edu

**Conemaugh Memorial Medical Center**
Surgical Technology Prgm
1086 Franklin St
Johnstown, PA 15905
www.conemaugh.org
*Prgm Dir:* Patricia Pavlikowski, MA RN CNOR CST
*Tel:* 814 534-9772   *Fax:* 814 534-9945
*E-mail:* Ppavlik@conemaugh.org/education/
*Med Dir:* Thomas Helling, MD

**Harrisburg Area Community College - Lancaster**
Surgical Technology Prgm
Lancaster Campus
1641 Old Philadelphia Pike
Lancaster, PA 17602
www.hacc.edu
*Prgm Dir:* Amy Kennedy, RN MSN CNOR
*Tel:* 717 358-2871   *Fax:* 717 358-2865
*E-mail:* alkenned@hacc.edu

**Lancaster Gen Coll of Nursing & Hlth Sciences**
Surgical Technology Prgm
410 N Lime St
Lancaster, PA 17602
www.LancasterGeneral.org
*Prgm Dir:* Christina Baumer, RN PhD CNOR CHES
*Tel:* 717 544-4912, Ext 76984   *Fax:* 717 290-5970
*E-mail:* clbaumer@lancastergeneral.org

**Delaware County Community College**
Surgical Technology Prgm
901 S Media Line Rd
Media, PA 19063-1094
*Prgm Dir:* Jane Rothrock, DNSc RN CNOR FAAN
*Tel:* 610 359-5286   *Fax:* 610 359-7350
*E-mail:* drjaner2@comcast.net

**Comm Coll of Allegheny County**
Surgical Technology Prgm
595 Beatty Rd
Monroeville, PA 15146
*Prgm Dir:* Linda C Radzvin
*Tel:* 724 325-6779   *Fax:* 724 325-6799
*E-mail:* lradzvin@ccac.edu

**Western School of Health & Business**
Surgical Technology Prgm
1 Monroeville Center, Ste 125
Monroeville, PA 15146
www.western-school.com
*Prgm Dir:* Rachelle Graft-Hall, RN BSN
*Tel:* 412 373-9038, Ext 227   *Fax:* 412 373-2544
*E-mail:* rhall@western-school.com

**Luzerne County Community College**
Surgical Technology Prgm
1333 S Prospect St
Nanticoke, PA 18634
*Prgm Dir:* Mary Ann Owens, RN
*Tel:* 570 740-0506   *Fax:* 570 740-0553
*E-mail:* mowens@luzerne.edu

**Connelley Technical Institute**
Surgical Technology Prgm
1501 Bedford Ave
Pittsburgh, PA 15219
*Prgm Dir:* G A Candie Gagne, CST CRCST MS
*Tel:* 412 338-3720    *Fax:* 412 338-3742
*E-mail:* candiegagne@comcast.net

**Montgomery County Community College**
Surgical Technology Prgm
101 College Dr
Pottstown, PA 19464
www.mc3.edu
*Prgm Dir:* Bessie Lindberg, CST
*Tel:* 610 718-1806
*E-mail:* blindber@mc3.edu

**Reading Hospital & Medical Center**
Surgical Technology Prgm
PO Box 16052
Reading, PA 19612
www.readinghospital.org
*Prgm Dir:* Bonita McCoy, RN BS CNOR
*Tel:* 610 988-8546    *Fax:* 610 988-5004
*E-mail:* mccoyb@readinghospital.org

**McCann School of Business and Technology**
Surgical Technology Prgm
1147 N Fourth St
Sunbury, PA 17601
*Prgm Dir:* Josette Crebs, RN CNOR
*Tel:* 570 286-3058    *Fax:* 570 286-4723
*E-mail:* jcrebs@mccannschool.com
*Med Dir:* Darla Miller, DPM

**Pennsylvania College of Technology**
Surgical Technology Prgm
1 College Ave
Williamsport, PA 17701
*Prgm Dir:* Thomas J Campana, MD FACS
*Tel:* 570 320-2400, Ext 7744
*E-mail:* tcampana@pct.edu

**Westmoreland County Community College**
Surgical Technology Prgm
400 Armbrust Rd
Youngwood, PA 15697-1895
*Prgm Dir:* Patricia E Mihalcin, RN PhD
*Tel:* 724 925-4028    *Fax:* 724 925-4293
*E-mail:* mihalcinp@wccc-pa.edu
*Med Dir:* Adel Armanious, MD

# Rhode Island

**New England Institute of Technology**
Surgical Technology Prgm
2500 Post Rd
Warwick, RI 02886-2266
*Prgm Dir:* Lisa Reed, RN MS CNOR
*Tel:* 401 739-5000, Ext 3485    *Fax:* 401 739-7738
*E-mail:* lreed@neit.edu

# South Carolina

**Technical College of the Lowcountry**
Surgical Technology Prgm
PO Box 1288
Beaufort, SC 29901
*Prgm Dir:* Julio A Acosta
*Tel:* 843 470-8415
*E-mail:* jacosta@tcl.edu

**Miller-Motte Technical College**
Surgical Technology Prgm
8085 Rivers Ave
Charleston, SC 29406
*Prgm Dir:* Rob Sanchez
*Tel:* 843 574-0101, Ext 45
*E-mail:* cdurrand@miller-motte.net

**Midlands Technical College**
Surgical Technology Prgm
PO Box 2408
Columbia, SC 29202
www.midlandstech.edu
*Prgm Dir:* Kathy Patnaude, CST AOT
*Tel:* 803 822-3438    *Fax:* 803 822-3417
*E-mail:* patnaudek@midlandstech.edu

**Florence-Darlington Technical College**
Surgical Technology Prgm
PO Box 100548
Florence, SC 29501-0548
www.fdtc.edu
*Prgm Dir:* Neva B Lawson, CST FA RN
*Tel:* 843 661-8149    *Fax:* 843 292-0851
*E-mail:* neva.lawson@fdtc.edu

**Horry - Georgetown Technical College**
Surgical Technology Prgm
4003 S Fraser St
Georgetown, SC 29440
www.hgtc.edu
*Prgm Dir:* Dolores Aurand
*Tel:* 843 520-1435    *Fax:* 843 546-1437
*E-mail:* dolores.aurand@hgtc.edu

**Aiken Technical College**
Surgical Technology Prgm
2276 Jefferson Davis Highway
PO Box 400
Graniteville, SC 29829
www.atc.edu
*Prgm Dir:* Amy O'Rourke, RN BSN MSA
*Tel:* 803 593-9954, Ext 1648    *Fax:* 803 593-3106
*E-mail:* orourkea@atc.edu
*Med Dir:* Wayne Frei, MD

**Greenville Technical College**
Surgical Technology Prgm
PO Box 5616
Greenville, SC 29606-8616
www.greenvilletech.com
*Prgm Dir:* Orrie Burdine, RN CNOR
*Tel:* 864 250-8298, Ext 8298    *Fax:* 864 250-8549
*E-mail:* Orrie.Burdine@gvltec.edu

**Piedmont Technical College**
Surgical Technology Prgm
PO Box 1467, Emerald Rd
Greenwood, SC 29646
www.ptc.edu
*Prgm Dir:* Susan Boggs, RN BSN CNOR
*Tel:* 864 941-8535    *Fax:* 864 941-8684
*E-mail:* boggs.s@ptc.edu

**Tri-County Technical College**
Surgical Technology Prgm
PO Box 587
Pendleton, SC 29670-0587
www.tctc.edu
*Prgm Dir:* Louise Vandiver, CST
*Tel:* 864 646-8361, Ext 1401    *Fax:* 864 646-1892
*E-mail:* lvandive@tctc.edu
*Med Dir:* Lynn Lewis, MSN

**York Technical College**
Surgical Technology Prgm
452 S Anderson Rd
Rock Hill, SC 29730
*Prgm Dir:* Liz Boatwright, CST
*Tel:* 803 981-7071    *Fax:* 803 981-7216
*E-mail:* boatwright@yorktech.com

**Spartanburg Community College**
Surgical Technology Prgm
PO Drawer 4386
Spartanburg, SC 29305
*Prgm Dir:* Emily W Rogers, BS RN CNOR CST
*Tel:* 864 592-4870    *Fax:* 864 592-4881
*E-mail:* rogerse@stcsc.edu

**Central Carolina Technical College**
Surgical Technology Prgm
506 N Guignard Dr, Bldg 600
Sumter, SC 29150
www.cctech.edu
*Prgm Dir:* Lynn Shorter, CST
*Tel:* 803 778-1961, Ext 214
*E-mail:* shortersl@cctech.edu
*Med Dir:* Michael Frisina, MA

# South Dakota

**Presentation College**
Surgical Technology Prgm
1500 N Main St
Aberdeen, SD 57401
*Prgm Dir:* Christina Rice
*Tel:* 605 229-8415    *Fax:* 605 229-8518
*E-mail:* ricec@presentation.edu

**Western Dakota Technical Institute**
Surgical Technology Prgm
800 Mickelson Dr
Rapid City, SD 57703
*Prgm Dir:* Lynn Seime
*Tel:* 605 394-4034, Ext 307    *Fax:* 605 394-1789
*E-mail:* lseime@wdti.tec.sd.us

**Southeast Technical Institute**
Surgical Technology Prgm
2320 N Career Ave
Sioux Falls, SD 57107
*Prgm Dir:* Judy R Tyler, RN BSN MEd
*Tel:* 605 367-5990    *Fax:* 605 367-6108
*E-mail:* judy.tyler@southeasttech.com

# Tennessee

**Northeast State Technical Comm College**
Surgical Technology Prgm
PO Box 246, 2425 Hwy 75
Blountville, TN 37617-0246
*Prgm Dir:* Laurie M Bollman, CST AS
*Tel:* 423 279-3681    *Fax:* 423 477-4859
*E-mail:* lmbollman@northeaststate.edu
*Med Dir:* Robert Saunders, MD

**Chattanooga State Technical Comm College**
Surgical Technology Prgm
4501 Amnicola Hwy
Chattanooga, TN 37406-1097
*Prgm Dir:* Lucy Hampton, BSN RN CNOR
*Tel:* 423 697-4491    *Fax:* 423 697-2413
*E-mail:* lucy.hampton@chattanoogastate.edu
*Med Dir:* Don Barker, MD

**Miller-Motte Technical College - Chattanooga**
Surgical Technology Prgm
6020 Shallowford Rd
Chattanooga, TN 37421
*Prgm Dir:* Kimberly Hudgins Laster
*Tel:* 423 510-9675    *Fax:* 423 510-1985
*E-mail:* klaster@miller-motte.com

**Miller-Motte Technical College - Clarksville**
Surgical Technology Prgm
1820 Business Park Dr
Clarksville, TN 37040
*Prgm Dir:* Diana Holter, AS
*Tel:* 931 553-0071

**MedVance Institute**
Surgical Technology Prgm
1025 Highway 111
Cookeville, TN 38501
www.medvance.org
*Prgm Dir:* Rita Reagan, CST
*Tel:* 931 526-3660, Ext 244    *Fax:* 931 372-2603
*E-mail:* rreagan@medvance.org

## Tennessee Technology Center - Crossville
Surgical Technology Prgm
910 Miller Ave
PO Box 2959
Crossville, TN 38557
www.crossville.tec.tn.us
*Prgm Dir:* Willie Green, CST
*Tel:* 931 484-7502, Ext 119   *Fax:* 931 484-8911
*E-mail:* willie.green@mail.tec.tn.us

## Tennessee Technology Center - Dickson
Surgical Technology Prgm
740 Hwy 46 South
Dickson, TN 37055
www.dickson.tec.tn.us
*Prgm Dir:* Laura Travis, RN BSN
*Tel:* 615 441-6220, Ext 110   *Fax:* 615 441-6223
*E-mail:* ltravis@dickson.tec.tn.us

## Dyersburg State Community College
Surgical Technology Prgm
1510 Lake Rd
Dyersburg, TN 38024
*Prgm Dir:* Laura Dudley, BSN
*Tel:* 731 286-3390   *Fax:* 731 288-7744
*E-mail:* dudley@dscc.edu
*Med Dir:* Jeffrey Swetnam, MD

## Tennessee Technology Center - Hohenwald
Surgical Technology Prgm
813 W Main St
Hohenwald, TN 38462
www.hohenwald.tec.tn.us
*Prgm Dir:* Rick Brewer
*Tel:* 931 796-5351, Ext 125   *Fax:* 931 796-4892
*E-mail:* Rbrewer@hohenwald.tec.tn.us
*Med Dir:* Ann Mashburn, RN, BSN -
   Coordinator of Surgical Technology

## Tennessee Technology Center - Jackson
Surgical Technology Prgm
2468 Technology Center Dr
Jackson, TN 38301
*Prgm Dir:* Barbara Avent, BSN
*Tel:* 731 424-0691, Ext 120   *Fax:* 731 424-0807
*E-mail:* bavent@jackson.tec.tn.us

## Ft Sanders Regional Medical Center
Surgical Technology Prgm
9352 Park W Blvd
PO Box 22993
Knoxville, TN 37916
*Prgm Dir:* Mary Rippy, RN BSAOM CNOR
*Tel:* 865 541-1821   *Fax:* 865 541-2212
*E-mail:* mrippy@covhlth.com

## Tennessee Technology Center - Knoxville
Surgical Technology Prgm
1100 Liberty St
Knoxville, TN 37919
http://knoxville.tec.tn.us
*Prgm Dir:* Dorothy McGhee, RN ASN
*Tel:* 423 546-5567, Ext 122   *Fax:* 423 971-4474
*E-mail:* dmcghee@knoxville.tec.tn.us

## Tennessee Technology Center - McMinnville
Surgical Technology Prgm
241 Vo-Tech Dr
McMinnville, TN 37110
*Prgm Dir:* Patricia Merlo, BS RN
*Tel:* 931 473-5587, Ext 243   *Fax:* 931 473-6380
*E-mail:* pmerlo@mcminnville.tec.tn.us
*Med Dir:* Ralph Bard, MD

## Concorde Career College - Memphis
Surgical Technology Prgm
5100 Poplar Ave, Ste 132
Memphis, TN 38137
www.concordecareercolleges.com
*Prgm Dir:* Chuck Lane, CST BSOM
*Tel:* 901 761-9494
*E-mail:* clane@concorde.edu

## Tennessee Technology Center - Memphis
Surgical Technology Prgm
550 Alabama Ave
Memphis, TN 38105-3799
www.memphis.tec.tn.us
*Prgm Dir:* Elizabeth Oxner
*Tel:* 901 543-6164   *Fax:* 901 543-6126
*E-mail:* eoxner@memphis.tec.tn.us
*Med Dir:* Robert Smith, MD

## Tennessee Technology Center - Murfreesboro
Surgical Technology Prgm
1303 Old Fort Pkwy
Murfreesboro, TN 37129
*Prgm Dir:* Mike Ford, AD RN
*Tel:* 615 898-8010, Ext 146   *Fax:* 615 893-4194
*E-mail:* mford@murfreesboro.tec.tn.us

## Nashville State Technical Community College
Surgical Technology Prgm
120 White Bridge Rd
Nashville, TN 37209
*Prgm Dir:* Van Bates, BA CST
*Tel:* 615 353-3329   *Fax:* 615 353-3376
*E-mail:* van.bates@nscc.edu
*Med Dir:* Allen Anderson, MD

## Tennessee Technology Center - Paris
Surgical Technology Prgm
312 S Wilson St
Paris, TN 38242
www.paris.tec.tn.us
*Prgm Dir:* Alice McCutcheon, RN
*Tel:* 731 644-7365, Ext 140   *Fax:* 731 644-7368
*E-mail:* amccutcheon@paris.tec.tn.us

# Texas

## Cisco Junior College
Surgical Technology Prgm
717 East Industrial Blvd
Abilene, TX 79602
www.cisco.cc.tx.us
*Prgm Dir:* Kelly Meyer, BS
*Tel:* 325 794-4400, Ext 4441   *Fax:* 254 442-5100
*E-mail:* kmeyer@cisco.cc.tx.us
*Med Dir:* Leigh Taliaferro, MD

## Amarillo College
Surgical Technology Prgm
PO Box 447
Amarillo, TX 79178
*Prgm Dir:* Kelly Ellis, MS
*Tel:* 806 356-3663   *Fax:* 806 354-6076
*E-mail:* ellis-kd@actx.edu

## Concorde Career Institute - Arlington
Surgical Technology Prgm
601 Ryan Plaza Dr
Ste 200
Arlington, TX 76011
www.concorde.edu
*Prgm Dir:* Benmjamin McCrory, Jr, CST
*Tel:* 817 792-2524   *Fax:* 817 461-3443
*E-mail:* bmccrory@concorde.edu

## Austin Community College
Surgical Technology Prgm
3401 Webberville Rd
Austin, TX 78702
www.austincc.edu/health
*Prgm Dir:* Kathleen Baumbach, RN CNOR
*Tel:* 512 223-5801   *Fax:* 512 223-5901
*E-mail:* kbaumbac@austincc.edu
*Med Dir:* Tim Faulkenberry, MD

## Virginia College at Austin
Surgical Technology Prgm
6301 E Hwy 290
Austin, TX 78723
*Prgm Dir:* Javier Espinales, CST MEd
*Tel:* 512 279-2807   *Fax:* 512 317-3502
*E-mail:* jespinales@vc.edu

## St Joseph Regional Health Center
Surgical Technology Prgm
2801 Franciscan Dr
Bryan, TX 77802-2544
*Prgm Dir:* Steve Heacock
*Tel:* 979 776-2433   *Fax:* 979 776-4986
*E-mail:* sheacock@st-joseph.org
*Med Dir:* Mark Montgomery, MD

## North Central Texas College
Surgical Technology Prgm
1500 N Corinth St
Corinth, TX 76208-5408
www.nctc.edu
*Prgm Dir:* Judie Rodgers, RN BS
*Tel:* 940 498-6260   *Fax:* 940 498-6444
*E-mail:* jrodgers@nctc.edu

## Del Mar College
Surgical Technology Prgm
101 Baldwin and Ayers Sts
Corpus Christi, TX 78404
*Prgm Dir:* Elena Mendieta, MS RN
*Tel:* 512 698-1105   *Fax:* 512 886-1598
*E-mail:* emendie@delmar.edu

## El Centro College
Surgical Technology Prgm
801 Main St
Dallas, TX 75202-3604
www.elcentrocollege.edu
*Prgm Dir:* Cindy Calcaterra, MBA RN CNOR
*Tel:* 214 860-2281   *Fax:* 214 860-2268
*E-mail:* CCalcaterra@dcccd.edu

## Career Centers of Texas
Surgical Technology Prgm
8360 Burnham Rd, Ste 100
El Paso, TX 79907
www.careercenters.edu
*Prgm Dir:* Steven Wherrey, CST
*Tel:* 915 595-1935   *Fax:* 915 595-6619
*E-mail:* swherrey@cct-ep.com

## El Paso Community College
Surgical Technology Prgm
PO Box 20500
El Paso, TX 79998
www.epcc.edu
*Prgm Dir:* Cynthia Rivera, RN BSN
*Tel:* 915 831-4086   *Fax:* 915 831-4114
*E-mail:* CynthiaR@epcc.edu

## Galveston College
Surgical Technology Prgm
4015 Ave Q
Galveston, TX 77550
www.gc.edu
*Prgm Dir:* Suzanna Martinez, CST
*Tel:* 409 944-1493, Ext 493   *Fax:* 409 762-9367
*E-mail:* sumartin@gc.edu

## Texas State Technical College - Harlingen
Surgical Technology Prgm
1902 N Loop 499
Harlingen, TX 78550
www.harlingen.tstc.edu/surgtech
*Prgm Dir:* Robert Sanchez, BSN CST RN
*Tel:* 956 364-4805   *Fax:* 956 364-5160
*E-mail:* robert.sanchez@harlingen.tstc.edu
*Med Dir:* Ruben Martinez, MD

## Academy of Health Care Professions
Surgical Technology Prgm
240 Northwest Mall
Houston, TX 77092
www.ahcp.edu
*Prgm Dir:* Terry Rogers, CST AAS
*Tel:* 713 424-3100, Ext 3132   *Fax:* 713 425-3193
*E-mail:* trogers@academyofhealth.com
*Med Dir:* Sarah McDonald, MD

**Houston Community College**
Surgical Technology Prgm
1900 Pressler
Houston, TX 77030
www.hccs.edu
*Prgm Dir:* Christine Castillo-Sainz, RN BA
*Tel:* 713 718-7362   *Fax:* 713 718-7653
*E-mail:* christine.castillo@hccs.edu

**MedVance Institute**
Surgical Technology Prgm
6220 Westpark, Ste 180
Houston, TX 77057
*Prgm Dir:* Shawn Kelly, CST BSBM
*Tel:* 713 266-6594   *Fax:* 713 782-5873
*E-mail:* shawn.kelly@medvance.edu
*Med Dir:* Imran Ayub, MD

**Memorial Hermann Healthcare System**
Surgical Technology Prgm
921 Gessner
Houston, TX 77024
www.memorialhermann.org
*Prgm Dir:* Judith Farmer, RN
*Tel:* 713 242-3797   *Fax:* 713 242-3701
*E-mail:* judith.farmer@memorialhermann.org
*Med Dir:* Carl Giesler, MD

**Sanford-Brown Institute - Houston**
Surgical Technology Prgm
10500 Forum Place Dr #200
Houston, TX 77036
*Prgm Dir:* Linda Hill
*Tel:* 713 779-1110   *Fax:* 713 779-2408
*E-mail:* lhill@sbhouston.com

**Tarrant County College**
Surgical Technology Prgm
828 Harwood Rd
Hurst, TX 76054
*Prgm Dir:* Don Braziel, CST BS
*Tel:* 817 515-6160   *Fax:* 817 515-6700
*E-mail:* donnie.braziel@tccd.edu

**Trinity Valley Community College**
Surgical Technology Prgm
Hlth Science Ctr, 800 Hwy 243 W
Kaufman, TX 75142
www.tvcc.edu/healthscience/surgtech.htm
*Prgm Dir:* Nancy L Couch, RN CNOR
*Tel:* 972 932-4309, Ext 29   *Fax:* 972 932-5010
*E-mail:* couch@tvcc.edu

**Kilgore College**
Surgical Technology Prgm
1100 Broadway
Kilgore, TX 75662
www.kilgore.edu
*Prgm Dir:* Lane J Barnett, MSN RN CNOR
*Tel:* 903 983-8163   *Fax:* 903 988-7596
*E-mail:* lbarnett@kilgore.edu

**South Plains College**
Surgical Technology Prgm
Reese Center
819 Gilbert Dr
Lubbock, TX 79416
*Prgm Dir:* Stacey May, CST
*Tel:* 806 885-3048, Ext 4642   *Fax:* 806 885-5608
*E-mail:* smay@southplainscollege.edu

**Paris Junior College**
Surgical Technology Prgm
2400 Clarksville St
Paris, TX 75460
www.parisjc.edu
*Prgm Dir:* Norman Gilbert, CFA SA
*Tel:* 903 782-0734   *Fax:* 903 782-0733
*E-mail:* ngilbert@parisjc.edu
*Med Dir:* Clarence Temple, MD

**San Jacinto College Central**
Surgical Technology Prgm
8060 Spencer Hwy, PO Box 2007
Pasadena, TX 77505-2007
*Prgm Dir:* Diane DeYoung, MEd RN
*Tel:* 713 478-2759   *Fax:* 713 478-2754
*E-mail:* ddeyou@sjcd.edu
*Med Dir:* Charles Cowles, MD

**Lamar State College - Port Arthur**
Surgical Technology Prgm
PO Box 310
Port Arthur, TX 77641-0310
www.pa.lamar.edu
*Prgm Dir:* Brandon Buckner, CST
*Tel:* 409 984-6367   *Fax:* 409 984-6005
*E-mail:* brandon.buckner@lamarpa.edu

**Howard College**
Surgical Technology Prgm
3501 N US Hwy 67
San Angelo, TX 76905
*Prgm Dir:* Kay Millican, RN MSN FNP
*Tel:* 915 481-8300, Ext 233   *Fax:* 915 481-8321
*E-mail:* kmillican@howardcollege.edu

**Baptist Health System**
Surgical Technology Prgm
730 N Main, Ste 212
San Antonio, TX 78205-1115
*Prgm Dir:* Christallia Starks, RN MSN CRCST
*Tel:* 210 297-9166   *Fax:* 210 297-0940
*E-mail:* cstarks@baptisthealthsystem.com
*Med Dir:* Sabas Abuabara, MD

**St Philip's College**
Surgical Technology Prgm
1801 Martin Luther King Dr
San Antonio, TX 78203-2098
www.accd.edu/spc
*Prgm Dir:* Lana Seay, RN BSOE
*Tel:* 210 531-3418   *Fax:* 210 531-3459
*E-mail:* lseay@accd.edu

**Temple College**
Surgical Technology Prgm
2600 S First St
Temple, TX 76504
www.templejc.edu
*Prgm Dir:* Carol A Reinking, RN MS CNOR
*Tel:* 254 298-8652   *Fax:* 254 298-8278
*E-mail:* carol.reinking@templejc.edu

**Tyler Junior College**
Surgical Technology Prgm
PO Box 9020
1530 SSW Loop 323 (Zip 75701)
Tyler, TX 75711
*Prgm Dir:* Sherry Seaton, RN BSN CNOR
*Tel:* 903 510-2962   *Fax:* 903 510-2931
*E-mail:* ssea@tjc.edu

**Wharton County Junior College**
Surgical Technology Prgm
911 Boling Hwy
Wharton, TX 77488
www.wcjc.edu
*Prgm Dir:* Melissa Wade, LVN
*Tel:* 979 532-6310   *Fax:* 979 532-6489
*E-mail:* melissaw@wcjc.edu

**Vernon College**
Surgical Technology Prgm
4105 Maplewood Ave
Wichita Falls, TX 76308
www.vernoncollege.edu
*Prgm Dir:* Jeff Feix, LVN CST/CFA
*Tel:* 940 696-8752, Ext 3266   *Fax:* 940 696-3244
*E-mail:* jfeix@vernoncollege.edu

# Utah

**Davis Applied Technology College**
Surgical Technology Prgm
550 E 300 So
Kaysville, UT 84037
www.datc.net
*Prgm Dir:* Pamela June Carter, RN MEd CNOR
*Tel:* 801 593-2330   *Fax:* 801 593-2400
*E-mail:* pjcarter@datc.net

**Stevens Henager College**
Surgical Technology Prgm
1350 W 1890 South
Ogden, UT 84401
*Prgm Dir:* Mario Merida, MD
*Tel:* 801 394-7791, Ext 1044   *Fax:* 801 621-0853
*E-mail:* mmerida@stevenshenager.edu
*Med Dir:* Glenn Morrell, MD

**Ameritech College**
Surgical Technology Prgm
1675 N Freedom Blvd, Ste 3B
Provo, UT 84604
www.ameritech.edu
*Prgm Dir:* Adrian Dominguez, CST
*Tel:* 801 377-2900, Ext 207   *Fax:* 801 375-3077
*E-mail:* adominguez@ameritech.edu

**Intermountain Health Care**
Surgical Technology Prgm
LDS Hospital
8th Ave and C St
Salt Lake City, UT 84143
*Prgm Dir:* Carolyn H Scheese, RN BSN CNOR
*Tel:* 801 408-3240   *Fax:* 801 408-5240
*E-mail:* carolyn.scheese@intermountainmail.org
*Med Dir:* William Hamilton, MD

**Dixie State College of Utah**
Surgical Technology Prgm
225 S 700 East St
St George, UT 84770
www.dixie.edu
*Prgm Dir:* Jeanne Mortenson, RN BSN CNOR
*Tel:* 435 652-7567   *Fax:* 435 652-7873
*E-mail:* jmortenson@dixie.edu

**Salt Lake Community College**
Surgical Technology Prgm
9301 S Wights Fort Rd
West Jordan, UT 84088
www.slcc.edu
*Prgm Dir:* Mia Carsey, CST BS
*Tel:* 801 256-5922   *Fax:* 801 256-5930
*E-mail:* mia.carsey@slcc.edu
*Med Dir:* G Remington Brooks, MD, FACS

# Virginia

**Piedmont Virginia Community College**
Surgical Technology Prgm
501 College Dr
Charlottesville, VA 22902
www.pvcc.edu
*Prgm Dir:* Ann Smith, RN MSN CNOR
*Tel:* 434 961-5239   *Fax:* 434 961-5441
*E-mail:* asmith@pvcc.edu

**Riverside School of Health Careers**
Surgical Technology Prgm
316 Main St
Newport News, VA 23601
*Prgm Dir:* Carolyn M Branson, CNOR CST RN BS
*Tel:* 757 240-2214   *Fax:* 757 240-2201
*E-mail:* carol.branson@rivhs.com

**Naval School of Health Sciences**
Surgical Technology Prgm
1001 Holcomb Rd, Bldg 104
Portsmouth, VA 23708-5200
*Prgm Dir:* Helena Ely, CDR NC USN
*Tel:* 757 953-6434    *Fax:* 757 953-7380
*E-mail:* hgely@hsp.med.navy.mil
*Med Dir:* Beth Jaklic, MD

**J Sargeant Reynolds Community College**
Surgical Technology Prgm
700 E Jackson St
Richmond, VA 23285-5622
*Prgm Dir:* Ed DeGennaro
*Tel:* 804 786-1375    *Fax:* 804 768-5298
*E-mail:* edegennaro@jsr.vccs.edu
*Med Dir:* Frances Stanley, RN MSN

# Washington

**Bellingham Technical College**
Surgical Technology Prgm
3028 Lindbergh Ave
Bellingham, WA 98225
*Prgm Dir:* Lorrie Zwiers, RN CNOR
*Tel:* 360 738-3105, Ext 415    *Fax:* 360 676-2798
*E-mail:* lzwiers@btc.ctc.edu

**Clover Park Technical College**
Surgical Technology Prgm
4500 Steilacoom Blvd SW
Lakewood, WA 98499
www.cptc.ctc.edu
*Prgm Dir:* Kezia Clark, ST
*Tel:* 253 589-5530    *Fax:* 253 589-5866
*E-mail:* kezia.clark@cptc.edu

**Renton Technical College**
Surgical Technology Prgm
3000 NE Fourth St
Renton, WA 98056
www.rtc.edu
*Prgm Dir:* Rosemary Thurston, RN
*Tel:* 206 235-7812    *Fax:* 206 235-5836
*E-mail:* rthurston@rtc.edu
*Med Dir:* Melvin Freeman, MD

**Seattle Central Community College**
Surgical Technology Prgm
1701 Broadway
Seattle, WA 98122
www.seattlecentral.org
*Prgm Dir:* Annejeannette Serba, RN CNOR
*Tel:* 206 587-6950    *Fax:* 206 587-6337
*E-mail:* aserba@sccd.ctc.edu

**Spokane Community College**
Surgical Technology Prgm
N 1810 Greene St
Spokane, WA 99207
*Prgm Dir:* Jeannie Hurd, RN BS CNOR
*Tel:* 509 533-7303    *Fax:* 509 533-8621
*E-mail:* jhurd@scc.spokane.edu
*Med Dir:* Courtney Clyde, MD

**Yakima Valley Community College**
Surgical Technology Prgm
PO Box 22520
Yakima, WA 98907-2520
*Prgm Dir:* Libby A McRae, CST
*Tel:* 509 574-4913
*E-mail:* lmcrae@yvcc.edu

# West Virginia

**Carver Career Technical Center**
Surgical Technology Prgm
4799 Midland Dr
Charleston, WV 25306
*Prgm Dir:* Letitia Hodovan, BS MS
*Tel:* 304 348-1965, Ext 110    *Fax:* 304 348-1932
*E-mail:* lhodovan@kcs.kana.k12.wv.us

**Monongalia County Tech Education Center**
*Cosponsor:* Monongalia County Board of Education
Surgical Technology Prgm
1000 Mississippi St
Morgantown, WV 26501
*Prgm Dir:* Lynda Overking, RN BSN MS
*Tel:* 304 291-9240, Ext 22    *Fax:* 304 291-9247
*E-mail:* loverkin@access.k12.wv.us
*Med Dir:* Lynda J Overking, RN BSN MS

**Southern West Virginia Comm & Tech College**
Surgical Technology Prgm
PO Box 2900
Dempsey Branch Rd
Mt Gay, WV 25637
*Prgm Dir:* Judith Curry, RN BSN
*Tel:* 304 792-7098, Ext 113    *Fax:* 304 792-7043
*E-mail:* judyc@southern.wvnet.edu
*Med Dir:* Kenneth Sells, DO

**West Virginia University - Parkersburg**
Surgical Technology Prgm
300 Campus Dr
Parkersburg, WV 26104
www.wvup.edu/nursing
*Prgm Dir:* Margaret Ponce, RN
*Tel:* 304 424-8300    *Fax:* 304 424-8315
*E-mail:* peggy.ponce@mail.wvu.edu

**West Virginia Northern Community College**
Surgical Technology Prgm
15th and Jacob Sts
Wheeling, WV 26003
*Prgm Dir:* Nancy Krupinski, PhD RN-BSN-MS
CNOR-CHES CST
*Tel:* 304 233-5900, Ext 4456    *Fax:* 304 233-5837
*E-mail:* nkrupinski@northern.wvnet.edu

# Wisconsin

**Chippewa Valley Technical College**
Surgical Technology Prgm
620 W Clairemont Ave
Eau Claire, WI 54701-6162
*Prgm Dir:* David Yount
*Tel:* 715 858-1897    *Fax:* 715 833-6431
*E-mail:* dyount@cvtc.edu

**Northeast Wisconsin Technical College**
Surgical Technology Prgm
2740 W Mason St, PO Box 19042
Green Bay, WI 54307-9042
www.nwtc.edu
*Prgm Dir:* Cynthia McDonald, CST BSN MS CNOR
*Tel:* 920 498-5540    *Fax:* 920 491-2660
*E-mail:* cindy.mcdonald@nwtc.edu
*Med Dir:* Per R Anderas, MD

**Gateway Technical College**
Surgical Technology Prgm
3520 30th Ave
Kenosha, WI 53144-1690
http://gtc.edu
*Prgm Dir:* Kristina Vines, CST
*Tel:* 262 564-2748    *Fax:* 262 564-2299
*E-mail:* vinesk@gtc.edu

**Western Technical College**
Surgical Technology Prgm
304 N Sixth St, PO Box 908
La Crosse, WI 54602-0908
*Prgm Dir:* Joan Miksis, CST RN BSN
*Tel:* 608 785-9193    *Fax:* 608 785-9087
*E-mail:* miksisj@wwtc.edu

**Madison Area Technical College**
Surgical Technology Prgm
Health Human & Protective Services
3550 Anderson St
Madison, WI 53704
*Prgm Dir:* Kathy Moninger, RN
*Tel:* 608 246-6280    *Fax:* 608 246-6013
*E-mail:* kmoninger@matcmadison.edu

**Mid-State Technical College**
Surgical Technology Prgm
2600 W Fifth St
Marshfield, WI 54449
www.mstc.edu
*Prgm Dir:* Kelly Altmann, RN BSN CNOR
*Tel:* 715 387-2538, Ext 7030    *Fax:* 715 389-2864
*E-mail:* Kelly.Altmann@mstc.edu
*Med Dir:* Steven Standford, MD

**Milwaukee Area Technical College**
Surgical Technology Prgm
700 W State St
Milwaukee, WI 53233
www.matc.edu
*Prgm Dir:* Patricia Stapleton, RN MSN
*Tel:* 414 297-7151    *Fax:* 414 297-8955
*E-mail:* stapletp@matc.edu
*Med Dir:* Marvin Wagner, MD

**Waukesha County Technical College**
Surgical Technology Prgm
800 Main St
Pewaukee, WI 53072
www.wctc.edu
*Prgm Dir:* Sharon A Corrao, RN BSN MEd CNOR CST
*Tel:* 262 691-5407    *Fax:* 262 691-5241
*E-mail:* scorrao@wctc.edu
*Med Dir:* David Schmitt, MD

**Northcentral Technical College**
Surgical Technology Prgm
1000 Campus Dr
Wausau, WI 54401
*Prgm Dir:* Julie Osness-Thorson, RN MSN
*Tel:* 715 675-3331, Ext 4497    *Fax:* 715 675-9776
*E-mail:* osness@ntc.edu

# Wyoming

**Laramie County Community College**
Surgical Technology Prgm
1400 E College Dr
Cheyenne, WY 82007
www.lccc.wy.edu
*Prgm Dir:* Katherine Snyder, CST FAST BS
*Tel:* 307 778-1155    *Fax:* 307 778-1395
*E-mail:* ksnyder@lccc.wy.edu

## Surgical Technologist

| Programs* | Class Capacity | Begins | Length (months) | Award | Res Tuition | Non-res Tuition | Stipend | Offers:‡ 1 | 2 | 3 |
|---|---|---|---|---|---|---|---|---|---|---|
| **Alabama** | | | | | | | | | | |
| James H Faulkner State Community College (Bay Minette) | 25 | Aug | 12, 24 | Cert, AAS | $93 | $164 | | | | |
| Southern Union State Community College (Opelika) | 20 | Aug | 9, 12 | Cert | $3,100 | $3,400 | | | | • |
| **Arizona** | | | | | | | | | | |
| Lamson College (Tempe) | 48 | Every 4 wks | 16 | Dipl, Assoc | $23,000 | $23,000 | | • | | |
| Mohave Community College (Lake Havasu City) | 15 | Aug | 10 | Cert | $5,150 | $8,350 | | | | • |
| Pima Community College (Tucson) | 12 | Dec | 11 | Cert | $9,241 | $9,241 | | | | |
| **Arkansas** | | | | | | | | | | |
| Baptist Health (Little Rock) | 30 | Jan | 11 | Dipl | $5,500 | $8,445 | | | | |
| North Arkansas College (Harrison) | 9 | Aug | 9, 24 | Cert, AAS | $1,590 | $2,710 | | | | |
| Northwest Technical Institute (Springdale) | 20 | Aug | 11 | Dipl | $3,940 | $3,940 | | | | |
| University of Arkansas - Fort Smith (Ft Smith) | 18 | Aug | 10, 18 | Cert, AAS | $1,830 | $6,840 | | | | • |
| University of Arkansas for Medical Sciences (North Little Rock) | 16 | Aug | 10, 20 | Cert, AS | $6,290 | $13,484 | | | | |
| **California** | | | | | | | | | | |
| Career Colleges of America - South Gate | | Monthly | 16 | Dipl | $17,099 | $17,099 | | | | |
| Concorde Career College (San Diego) | 24 | Jan Jun Oct | 13 | Dipl | $19,549 | $19,549 | | | | |
| Concorde Career Institute - North Hollywood | 24 | Jan Apr Jul Oct | 13 | Dipl | $19,439 | $19,439 | | | | |
| Glendale Career College | 30 | Every 9 wks | 14 | Dipl | $20,667 | $20,667 | | • | | |
| Glendale Career College - Oceanside Campus | 20 | Open | 17 | Dipl | $22,075 | $22,075 | | | | |
| Mt Diablo Adult Education (Concord) | 15 | Sep 6 | 15 | Cert | $5,300 | $5,300 | | | | |
| Naval School of Health Sciences, San Diego | 55 | Jan May Sep | 6 | Cert | | | | | | • |
| San Joaquin Valley College (Fresno) | 16 | Nov | 16 | AS | $22,200 | $22,200 | | | | |
| San Joaquin Valley College - Bakersfield | 24 | Mar Oct | 15 | Dipl, AS | $12,580 | $12,580 | | | | • |
| Skyline College (San Bruno) | 25 | Aug | 10 | Cert, AS | $3,780 | $5,096 | | | | • |
| **Colorado** | | | | | | | | | | |
| Concorde Career Institute - Aurora | 24 | 3x/yr | 13 | Cert | $21,820 | $21,820 | | | | |
| **Connecticut** | | | | | | | | | | |
| A I Prince Technical High School (Hartford) | 18 | Aug | 10 | Cert | $2,600 | $0 | | | | |
| Danbury Hospital | 12 | Sep | 12 | Dipl | $3,000 | $3,000 | | | | |
| Manchester Community College | 24 | Sep | 21 | AS | $1,888 | $5,816 | | | | |
| **Florida** | | | | | | | | | | |
| Central Florida Institute (Palm Harbor) | 16 | Varies | 10, 13 | AAS | $13,979 | $0 | | • | | |
| Concorde Career Institute - Jacksonville | 24 | Feb Jul Oct | 12 | Dipl | $18,800 | $18,800 | | | | |
| Concorde Career Institute - Tampa | 24 | Feb Apr Jun Sep Nov | 12 | Dipl | $18,284 | $18,284 | | | | |
| Florida Community College - Jacksonville | 45 | Aug | 10 | Cert | $2,590 | $9,265 | | | | • |
| Gulf Coast Community College (Panama City) | 16 | Jan | 12 | Cert | $3,471 | $9,966 | | | | • |
| Lindsey Hopkins Tech Education Center (Miami) | 16 | Oct May | 15 | Cert | $2,600 | $5,500 | | | | |
| National School of Technology (Hialeah) | 90 | Monthly | 13 | Dipl | $18,445 | $18,445 | | | | |
| Okaloosa-Walton College (Niceville) | 14 | Aug | 12 | Cert | $4,037 | $6,757 | | | | |
| Orlando Tech | 15 | Every 65 mos | 13 | Cert | $2,487 | $9,961 | | | | |
| Pinellas Tech Ed Ctr - Clearwater Campus | 20 | Every Aug | 10 | Cert | $1,674 | $6,324 | | | | |
| Sanford-Brown Institute (Tampa) | 16 | Varies | 12 | Dipl | $19,850 | $19,850 | | • | | |
| Sanford-Brown Institute - Ft Lauderdale | 16 | Varies | 12 | Cert | $22,000 | $22,000 | | • | | |
| Sarasota County Technical Institute | 10 | Aug Dec Apr | 12 | Cert | $2,425 | $9,315 | | • | | |
| Sheridan Technical Center (Hollywood) | 26 | Aug | 12 | Cert | $2,554 | $10,201 | | | | |
| **Georgia** | | | | | | | | | | |
| Albany Technical College | 25 | Oct Mar | 15 | Dipl | $1,812 | $3,624 | | | | |
| Coastal Georgia Community College (Brunswick) | 12 | May | 15 | Cert | $3,084 | $12,332 | | | | |
| Columbus Technical College | 30 | Apr Oct | 15, 24 | Dipl, AAS | $1,800 | $3,600 | | | | |
| DeKalb Technical College (Clarkston) | 15 | Jul | 18 | Dipl | $1,876 | $6,340 | | | | |
| Gwinnett Technical College (Lawrenceville) | 30 | Jul | 15 | Dipl | $3,600 | $7,000 | | | | • |
| Middle Georgia Technical College (Warner Robins) | 20 | Jan | 12 | Dipl | $1,488 | $2,976 | | | | |
| Okefenokee Technical College (Waycross) | 20 | Fall | 24, 15 | Dipl, AAS | $2,592 | $0 | | | | |
| Southeastern Technical College (Vidalia) | 21 | Jul | 18 | Dipl | $1,296 | $2,592 | | | | • |
| Southwest Georgia Technical College (Thomasville) | 16 | Jul | 17, 24 | Dipl, Assoc | $1,872 | $3,744 | | • | | • |
| Valdosta Technical College | 25 | Jul | 15 | Dipl | $1,845 | $3,405 | | | | |
| West Central Technical College (Waco) | 12 | Sep | 12 | Dipl | $3,400 | $0 | | | | |
| **Idaho** | | | | | | | | | | |
| Eastern Idaho Technical College (Idaho Falls) | 12 | Aug | 18 | AAS | $789 | $2,892 | | | | |
| **Illinois** | | | | | | | | | | |
| Capital Area Career Center (Springfield) | 13 | Jul Feb | 11 | Cert | $9,000 | $9,000 | | | | • |
| College of DuPage (West Chicago) | 30 | Jan | 12 | Cert, AAS | $4,600 | $12,000 | | | | |
| College of Lake County (Grayslake) | 16 | Jun | 15 | Cert, AS | $2,380 | $10,534 | | | | |
| Elgin Community College | 24 | Jan | 12 | Cert | $3,528 | $11,631 | | | | |
| Illinois Central College (Peoria) | 20 | Aug | 17, 22 | Cert, AAS | $2,310 | $4,375 | | | | |
| Prairie State College (Chicago Heights) | 20 | Aug | 11 | Cert | $3,042 | $7,630 | | | | |

*Data are shown only for programs that completed the 2006 AMA Survey of Health Professions Education Programs
‡Key to Offers: 1: Evening or weekend classes; 2: Non-English instruction; 3: Cultural competence instruction

| Programs* | Class Capacity | Begins | Length (months) | Award | Res Tuition | Non-res Tuition | Stipend | Offers:‡ 1 | 2 | 3 |
|---|---|---|---|---|---|---|---|---|---|---|
| Richland Community College (Decatur) | 30 | Aug | 21 | Cert, AAS | $6,618 | $16,462 | | | | |
| Southern Illinois Collegiate Common Market (Herrin) | 24 | Aug | 12 | Cert | $1,803 | $1,916 | | | | |
| Trinity College of Nursing & Health Sciences (Rock Island) | 16 | Jun | 11 | Cert, AAS | $8,395 | $8,395 | | | | • |
| Triton College (River Grove) | 40 | Aug | 11 | Cert | $3,000 | $6,021 | | | | • |
| **Indiana** | | | | | | | | | | |
| Bloomington Hospital | 12 | Feb | 11 | Cert | $4,000 | $4,000 | | | | • |
| Clarian Health Partners Inc (Indianapolis) | 24 | Aug | 11 | Cert | $3,951 | $3,951 | | | | |
| Ivy Tech Community College - Columbus | 16 | Aug | 24 | AS | $3,071 | $6,247 | | | | |
| Ivy Tech Community College - Kokomo | 12 | Aug | 24 | AS | $2,457 | $4,998 | | | | • |
| Ivy Tech Community College - Lafayette | 20 | Aug | 24 | AAS | $2,518 | $5,107 | | • | | |
| Ivy Tech Community College - Michigan City | 32 | Aug | 24 | AAS | $3,247 | $6,605 | | • | | |
| Ivy Tech Community College SW - Evansville | 20 | Aug | 24 | AAS | $1,843 | $4,284 | | • | • | |
| University of Saint Francis (Ft Wayne) | 22 | Aug | 20 | AS | $17,760 | $17,760 | | • | • | • |
| Vincennes University | 20 | Aug | 11, 24 | Cert, AS | $4,552 | $11,365 | | | | • |
| **Iowa** | | | | | | | | | | |
| Iowa Lakes Community College (Spencer) | 24 | Aug | 12 | Dipl, AAS | $2,898 | $2,990 | | • | | • |
| Mercy College of Health Sciences (Des Moines) | 20 | Sep | 11 | Cert, AS | $16,965 | $16,965 | | | | |
| **Kansas** | | | | | | | | | | |
| Hutchinson Community College | 18 | Aug | 10 | Cert | $2,501 | $4,100 | | | | |
| Kaw Area Technical School (Topeka) | 12 | Aug 10 | 10 | Cert | $4,043 | $17,466 | | | | • |
| Seward County Community College (Liberal) | 16 | Aug | 11, 22 | Cert, AAS | $2,604 | $3,570 | | | | • |
| **Kentucky** | | | | | | | | | | |
| Bluegrass Community and Technical College (Lexington) | 20 | Aug | 16, 24 | Dipl, AAS | $960 | $2,175 | | | | |
| Bowling Green Technical College | 18 | Aug | 11 | Dipl, GOTS | $109 | $327 | | | | |
| Madisonville Community College | 16 | Aug | 10 | Dipl, ASGOTS | $92 | $370 | | | | |
| Southeast Kentucky Comm & Tech College (Pineville) | 12 | Aug | 15, 24 | Dipl, AAS | $109 | $131 | | • | | |
| Spencerian College (Louisville) | | Jan | 12 | Dipl | $20,000 | $20,000 | | | | |
| West Kentucky Community & Technical College (Paducah) | 16 | Aug | 11 | Dipl, AAS | $1,185 | $3,595 | | | | |
| **Louisiana** | | | | | | | | | | |
| Bossier Parish Community College (Bossier City) | 16 | Jun | 18 | Dipl | $860 | $1,930 | | | | • |
| Louisiana Tech College - Lafayette Campus | 30 | Aug | 24 | AAS | $1,115 | $1,905 | | | | • |
| Louisiana Tech College - Lafourche Campus (Thibodaux) | 12 | Jan | 24 | AAS | $1,552 | $2,636 | | | | |
| Our Lady of the Lake College (Baton Rouge) | 20 | Jun | 12 | AD | | | | • | | |
| **Maine** | | | | | | | | | | |
| Maine Medical Center (Portland) | 20 | Sep Mar | 12 | Dipl | $3,250 | $3,250 | | | | |
| **Massachusetts** | | | | | | | | | | |
| Charles H McCann Technical School (North Adams) | 10 | Sep | 9 | Cert | $2,500 | $11,288 | | | | |
| Massachusetts Bay Community College (Framingham) | 30 | Sep | 12 | Cert | $2,765 | $9,835 | | • | | |
| North Shore Community College (Danvers) | 20 | Sep May | 9 | Cert | $4,500 | $6,888 | | | | |
| **Michigan** | | | | | | | | | | |
| Baker College of Clinton Township | 20 | Jun | 24 | AAS | $8,680 | $8,680 | | | | • |
| Baker College of Jackson | 20 | Sep | 18 | AAS | $7,285 | $0 | | | | |
| Baker College of Muskegon | 20- | Sep Dec Apr Jun | 24 | AAS | $8,415 | $8,415 | | • | | • |
| Baker College of Port Huron | 12 | Jun | 12 | AAS | $5,400 | $5,400 | | • | | |
| Delta College (University Center) | 15 | Aug | 16 | Cert, AAS | $2,393 | $3,432 | | | | • |
| Kalamazoo Valley Community College | 16 | Aug | 10 | Cert | $2,135 | $4,760 | | | | |
| Lansing Community College | 24 | Aug | 24 | AAS | $3,100 | $3,900 | | | | |
| Oakland Community College - Bloomfield Hills (Royal Oak) | 25 | Sep | 10 | AAS | $3,490 | $6,435 | | | | • |
| **Minnesota** | | | | | | | | | | |
| Anoka Technical College | 30 | Aug Jan | 13 | Dipl, AAS | $7,886 | $15,773 | | • | | |
| Northland Community & Technical College (East Grand Forks) | 28 | Sep | 18 | AAS | $3,780 | $7,560 | | • | | |
| Rochester Community & Technical College | 25 | Aug | 18 | AAS | $9,389 | $16,468 | | • | | • |
| St Cloud Technical College | 24 | Jun Aug | 12, 19 | Dipl, AAS | $4,245 | $8,168 | | • | | • |
| **Mississippi** | | | | | | | | | | |
| East Central Community College (Decatur) | 24 | Jan | 12 | Cert, AAS | $2,000 | $3,000 | | | | |
| Holmes Community College (Grenada) | 15 | Aug | 12, 24 | Cert, AAS | $1,806 | $4,356 | | | | • |
| Itawamba Community College (Tupelo) | 16 | Aug | 12, 24 | Dipl, Cert, AS | $2,190 | $3,760 | | | | |
| **Missouri** | | | | | | | | | | |
| Ozarks Technical Community College (Springfield) | 20 | Aug | 9, 18 | Cert, AAS | $3,870 | $4,730 | | | | • |
| St Louis Community College - Forest Park | 25 | Aug | 12 | Dipl, Cert | $4,298 | $7,106 | | | | • |
| **Montana** | | | | | | | | | | |
| Flathead Valley Community College (Kalispell) | 8 | Sep | 24 | AAS | $1,272 | $4,137 | | | | |
| **Nebraska** | | | | | | | | | | |
| Southeast Community College (Lincoln) | 22 | Every third quarter | 21 | AAS | $5,094 | $9,965 | | • | | • |
| **Nevada** | | | | | | | | | | |
| Community College of Southern Nevada (Las Vegas) | 12 | Sep | 10 | Cert | $1,857 | $5,795 | | | | |

*Data are shown only for programs that completed the 2006 AMA Survey of Health Professions Education Programs

‡Key to Offers: 1: Evening or weekend classes; 2: Non-English instruction; 3: Cultural competence instruction

## Surgical Technologist

| Programs* | Class Capacity | Begins | Length (months) | Award | Res Tuition | Non-res Tuition | Stipend | Offers:‡ 1 | 2 | 3 |
|---|---|---|---|---|---|---|---|---|---|---|
| Nevada Career Institute (Las Vegas) | 30 | Every 12 wks | 17 | Dipl | $22,075 | $22,075 | | | | |
| Western Nevard Community College (Carson City) | 16 | Aug | 10 | Cert | $2,029 | $4,154 | | | | • |
| **New Hampshire** | | | | | | | | | | |
| Concord Hospital | 8 | Sep | 12 | Cert | $5,000 | $5,000 | | | | |
| Dartmouth Hitchcock Medical Center (Lebanon) | 5 | Sep | 11 | Cert | $5,000 | $5,000 | | | | • |
| **New Jersey** | | | | | | | | | | |
| Bergen Community College (Paramus) | 44 | Sep | 10 | Cert | $2,936 | $4,972 | | • | | • |
| **New York** | | | | | | | | | | |
| Bronx Lebanon Hospital (New York) | 15 | Sep | 11 | Cert | | | | • | | |
| Long Island University - Brooklyn Campus | 50 | Jan Jun Sep | 15, 23 | Cert | $11,000 | $0 | | • | | |
| Nassau Community College (Garden City) | 64 | Sep | 24 | AAS | $1,570 | $3,200 | | | | |
| New York University Medical Center | 30 | Sep Mar | 12 | Cert | $10,000 | $0 | | • | | |
| Stony Brook University | 20 | Jan | 11 | Cert | $8,995 | $8,995 | | | | |
| Trocaire College (Buffalo) | 40 | Sep Jan | 24 | AAS | $11,210 | $11,210 | | | | |
| Ulster County Board of Cooperative Ed Service (Port Ewen) | 15 | Sep | 10 | Cert | $6,800 | $0 | | | | |
| Western Suffolk BOCES (Northport) | 12 | Mar Sep | 10 | Cert | $0 | $9,525 | | | | • |
| **North Carolina** | | | | | | | | | | |
| Asheville-Buncombe Technical Comm College | 16 | Aug | 12 | Dipl | $1,896 | $10,536 | | | | • |
| Cabarrus College of Health Sciences (Concord) | 20 | Aug | 9, 18 | Dipl, AS | $7,600 | $7,600 | | | | • |
| Catawba Valley Community College (Hickory) | 20 | Aug | 12 | Dipl | $1,800 | $9,000 | | | | • |
| Cleveland Community College (Shelby) | 15 | Sep | 12 | Dipl | $568 | $3,152 | | | | |
| College of The Albemarle (Elizabeth City) | 20 | Aug | 12 | Dipl | $1,500 | $8,400 | | | | |
| Durham Technical Community College | 24 | Aug | 12 | Dipl | $1,896 | $10,536 | | | | • |
| Edgecombe Community College (Tarboro) | 15 | Aug | 12 | Dipl | $339 | $2,046 | | | | • |
| Fayetteville Technical Community College | 15 | Aug | 12, 24 | Dipl, AAS | $1,986 | $10,626 | | | | |
| Guilford Technical Community College (Jamestown) | 36 | Aug | 12, 24 | Dipl, AS | $1,147 | $6,410 | | | | |
| Miller-Motte Technical Community College (Wilmington) | 30 | Jan Jul | 24 | AAS | $12,800 | $12,800 | | | | |
| Presbyterian Healthcare (Charlotte) | 24 | Aug | 10 | Dipl | $4,220 | $4,220 | | | | • |
| Rockingham Community College (Wentworth) | 20 | Aug | 12 | Dipl | $1,050 | $8,317 | | | | |
| Wake Technical Community College (Raleigh) | 16 | Aug | 12 | Dipl | $1,800 | $8,000 | | | | • |
| Wilson Technical Community College | 20 | Aug | 12 | Dipl | $1,824 | $10,128 | | | | |
| **North Dakota** | | | | | | | | | | |
| Bismarck State College | 24 | Aug | 20 | AAS | $4,035 | $9,465 | | | | |
| **Ohio** | | | | | | | | | | |
| Apollo Career Center (Lima) | 18 | Sep | 10 | Cert | $7,500 | $7,500 | | | | |
| Choffin Career & Technical Center (Youngstown) | 25 | Aug | 10 | Cert | $6,500 | $6,500 | | | | • |
| Cincinnati State Tech & Comm College | 36 | Sep | 20 | AAS | $4,161 | $8,322 | | • | | |
| Collins Career Center (Chesapeake) | 24 | Aug | 11 | Cert | $4,600 | $4,600 | | | | |
| Columbus State Community College | 30 | Autumn (Sep) | 12, 18 | Cert, AS | $4,460 | $9,576 | | • | • | ‡ |
| Cuyahoga Community College (Cleveland) | 24 | Aug | 21, 12 | Cert, AAS | | | | • | • | |
| EHOVE Ghrist Adult Career Center (Milan) | 30 | Aug-FT Dec-PT | 9, 14 | Cert | $7,600 | $7,600 | | | | |
| Lakeland Community College (Kirtland) | 14 | Aug | 22 | AAS | $2,000 | $5,000 | | | | |
| Owens Community College (Toledo) | 24 | Aug | 20 | AAS | $116 | $217 | | | | • |
| Scioto County Joint Vocational School (Lucasville) | 15 | Jan | 11 | Cert | $6,176 | $0 | | | | |
| University of Akron | 20 | Aug | 22 | AAS | $5,558 | $7,506 | | | | • |
| **Oklahoma** | | | | | | | | | | |
| Canadian Valley Technology Center (Chickasha) | 12 | open entry/exit | 18 | Dipl | $1,755 | $22,632 | | | | • |
| Moore Norman Technology Center | 20 | Aug | 9 | Dipl | $2,113 | $2,618 | | | | |
| Oklahoma Health Academy (Moore) | 25 | Jan Aug | 10 | Dipl | $16,445 | $16,445 | | | | |
| Southern Oklahoma Technology Center (Ardmore) | 12 | Aug | 10 | Cert | $2,866 | $8,229 | | | | • |
| Tulsa Technology Center | 18 | Aug Nov Jan | 9, 12 | Cert | $3,000 | $3,000 | | • | | |
| Wes Watkins Technology Center Distist 25 (Wetumka) | 14 | Aug | 10 | Cert | | | | | | |
| **Oregon** | | | | | | | | | | |
| Concorde Career Institute - Portland | 24 | Apr Aug Dec | 12 | Dipl | $19,506 | $19,506 | | | | |
| **Pennsylvania** | | | | | | | | | | |
| Conemaugh Memorial Medical Center (Johnstown) | 14 | Aug 28 | 12 | Cert, AS | $13,688 | $27,376 | | | | • |
| Harrisburg Area Community College - Harrisburg | 10 | Aug | 21 | Cert, Assoc | $1,020 | $1,920 | | | | • |
| Harrisburg Area Community College - Lancaster | 8 | Jan | 21 | Cert, Assoc | $1,920 | $1,920 | | | | • |
| Lancaster Gen Coll of Nursing & Hlth Sciences | 16 | Aug | 12, 24 | Dipl, AD | $19,480 | $19,480 | | • | | • |
| Montgomery County Community College (Pottstown) | 20 | Sep (Fall semester) | 12, 16 | Cert, AAS | $5,056 | $5,056 | | | | • |
| Mount Aloysius College (Cresson) | 30 | Aug | 24 | Dipl, AS | $14,800 | $14,800 | | | | • |
| Northampton Community College (Bethlehem) | 12 | Aug | 24 | AAS | $6,006 | $18,348 | | | | • |
| Reading Hospital & Medical Center | 10 | Aug | 10 | Dipl | $4,500 | $4,500 | | | | • |
| St Luke's Hospital (Bethlehem) | 16 | Aug | 11 | Cert, AAS | $7,300 | $0 | | | | |
| Western School of Health & Business (Monroeville) | 28 | Every 5 mos | 15 | AST | $27,000 | $27,000 | | | | |

*Data are shown only for programs that completed the 2006 AMA Survey of Health Professions Education Programs
‡Key to Offers: 1: Evening or weekend classes; 2: Non-English instruction; 3: Cultural competence instruction

PROGRAMS

# Surgical Technologist

| Programs* | Class Capacity | Begins | Length (months) | Award | Res Tuition | Non-res Tuition | Stipend | Offers:‡ 1 | 2 | 3 |
|---|---|---|---|---|---|---|---|---|---|---|
| **South Carolina** | | | | | | | | | | |
| Aiken Technical College (Graniteville) | 20 | Aug | 12 | Dipl | $3,036 | $3,558 | | | | |
| Central Carolina Technical College (Sumter) | 22 | May | 12 | Dipl | $6,050 | $7,100 | | | | • |
| Florence-Darlington Technical College | 24 | Aug | 12 | Dipl | $6,292 | $6,864 | | | | • |
| Greenville Technical College | 35 | Aug | 12 | Dipl | $1,500 | $1,625 | | • | | • |
| Horry - Georgetown Technical College | 24 | Aug | 12 | Cert | $3,600 | $4,650 | | | | |
| Midlands Technical College (Columbia) | 24 | Aug | 12 | Dipl | $4,356 | $5,436 | | • | | • |
| Miller-Motte Technical College (Charleston) | 18 | Apr | 25 | Dipl | $6,840 | $6,840 | | | | |
| Piedmont Technical College (Greenwood) | 20 | Aug | 12 | Dipl | $1,700 | $0 | | | | |
| Spartanburg Community College | 25 | Aug | 12, 24 | Dipl, AS | $4,293 | $5,367 | | | | |
| Tri-County Technical College (Pendleton) | 23 | Aug | 12 | Dipl | $2,988 | $3,600 | | | | |
| York Technical College (Rock Hill) | 20 | Aug | 12 | Dipl | $4,554 | $8,886 | | | | |
| **South Dakota** | | | | | | | | | | |
| Western Dakota Technical Institute (Rapid City) | 16 | Aug | 11 | Dipl | $6,700 | $6,700 | | | | |
| **Tennessee** | | | | | | | | | | |
| Concorde Career College - Memphis | 24 | Oct Jan May Aug | 12 | Dipl | $18,651 | $18,651 | | • | | |
| MedVance Institute (Cookeville) | 21 | Jan Jul | 13 | Dipl | $16,000 | $16,000 | | | | |
| Tennessee Technology Center - Crossville | 25 | May | 12 | Dipl | $4,018 | $4,018 | | | | |
| Tennessee Technology Center - Dickson | 16 | Sep | 12 | Dipl | $1,986 | $1,986 | | | | • |
| Tennessee Technology Center - Hohenwald | 25 | Jan | 12 | Dipl | $3,620 | $3,620 | | | | |
| Tennessee Technology Center - Knoxville | 20 | Jan Jul | 12 | Dipl | $2,001 | $2,001 | | | | |
| Tennessee Technology Center - McMinnville | 12 | Sep | 12 | Dipl | $1,076 | $1,076 | | | | |
| Tennessee Technology Center - Memphis | 30 | May Sep | 12 | Dipl | $2,058 | $2,058 | | | | |
| Tennessee Technology Center - Paris | 12 | Sep | 12 | Dipl | $2,058 | $2,058 | | | | • |
| **Texas** | | | | | | | | | | |
| Academy of Health Care Professions (Houston) | 20 | Feb Jun Oct | 16 | Cert | $20,000 | $20,000 | | | | |
| Austin Community College | 16 | Aug Jan | 16, 24 | Cert, AAS | $2,618 | $11,944 | | | | • |
| Career Centers of Texas (El Paso) | 18 | | 10, 15 | Dipl | $20,790 | $0 | | • | | |
| Cisco Junior College (Abilene) | 12 | Aug | 12 | Cert | $986 | $1,139 | | | | |
| Concorde Career Institute - Arlington | 24 | Jul Feb Mar | 12 | Dipl | $14,991 | $14,991 | | | | |
| Del Mar College (Corpus Christi) | 20 | May | 12, 24 | Cert, AAS | $1,016 | $1,304 | | • | | • |
| El Centro College (Dallas) | 30 | Aug | 15 | Cert | $1,728 | $3,168 | | • | | • |
| El Paso Community College | 12 | Jun | 12, 24 | Cert, AAS | $3,500 | $5,000 | | | | • |
| Galveston College | 20 | May | 12 | Cert | $1,990 | $3,340 | | | | |
| Houston Community College | 35 | Aug | 12 | Cert | $3,500 | $4,000 | | | | |
| Kilgore College | 12 | Jun | 18, 24 | Cert, AAS | $4,350 | $5,850 | | | | |
| Lamar State College - Port Arthur | 24 | Jul | 12, 24 | Cert, AAS | $2,352 | $11,466 | | | | |
| MedVance Institute (Houston) | 30 | Sep Mar | 13 | Dipl | $19,800 | $19,800 | | | | |
| Memorial Hermann Healthcare System (Houston) | 20 | Jan | 12 | Dipl | $2,400 | $2,400 | | | | • |
| North Central Texas College (Corinth) | 24 | Aug | 12 | Cert | $1,930 | $3,025 | | | | |
| Paris Junior College | 12 | Aug | 21 | Cert | $1,292 | $2,006 | | | | |
| South Plains College (Lubbock) | 40 | Aug | 12 | Cert, AAS | $3,660 | $3,940 | | | | • |
| St Philip's College (San Antonio) | 25 | Aug | 12 | Cert | $2,500 | $6,000 | | | | • |
| Tarrant County College (Hurst) | 30 | Aug | 11 | Cert | $3,800 | $5,700 | | | | • |
| Temple College | 16 | May | 11 | Cert | $2,948 | $6,997 | | | | |
| Texas State Technical College - Harlingen | 30 | Sep | 12 | AAS | $2,940 | $6,788 | | | | |
| Trinity Valley Community College (Kaufman) | 12 | Aug | 12, 24 | Cert, AAS | $1,731 | $2,191 | | | | • |
| Tyler Junior College | 15 | Aug | 9, 21 | Cert, AAS | $3,702 | $5,414 | $1,500 | | 2 | • |
| Vernon College (Wichita Falls) | 15 | Aug | 12 | Cert | $2,376 | $3,278 | | | | • |
| Wharton County Junior College | 20 | Aug | 12 | Cert | $2,052 | $3,420 | | | | |
| **Utah** | | | | | | | | | | |
| Ameritech College (Provo) | 16 | Every 30 wks | 11 | Dipl | $12,138 | $12,138 | | | | |
| Davis Applied Technology College (Kaysville) | 30 | Open enrollment | 15 | Cert | $1,450 | $3,969 | | | | • |
| Dixie State College of Utah (St George) | 10 | Aug | 9 | Cert | $943 | $3,517 | | | | • |
| Intermountain Health Care (Salt Lake City) | 13 | Varies | 6 | Cert | $400 | $400 | | • | | • |
| Salt Lake Community College (West Jordan) | 22 | Aug | 9, 16 | Cert | $2,405 | $7,232 | | | | • |
| **Virginia** | | | | | | | | | | |
| Piedmont Virginia Community College (Charlottesville) | 35 | Aug | 12 | Cert | $3,423 | $10,685 | | | | |
| Riverside School of Health Careers (Newport News) | 15 | Aug | 11 | Dipl | $6,500 | $6,500 | | | | • |
| **Washington** | | | | | | | | | | |
| Clover Park Technical College (Lakewood) | 20 | Sep Apr | 18 | AAT | $9,546 | $0 | $750 | | | |
| Renton Technical College | 36 | Sep Jan | 11, 24 | Cert, AAS-ST, AASTST | $3,550 | $3,550 | | | | • |
| Seattle Central Community College | 25 | Sep | 9 | Cert | $1,832 | $6,769 | | | | |
| Spokane Community College | 22 | Sep | 20 | AAS | $2,148 | $5,556 | | | | • |
| **West Virginia** | | | | | | | | | | |
| Carver Career Technical Center (Charleston) | 15 | Aug | 10 | Cert | $1,438 | $1,438 | | | | • |
| Monongalia County Tech Education Center (Morgantown) | 20 | Aug | 9 | Cert | $4,000 | $0 | | | | |

*Data are shown only for programs that completed the 2006 AMA Survey of Health Professions Education Programs
‡Key to Offers: 1: Evening or weekend classes; 2: Non-English instruction; 3: Cultural competence instruction

## Surgical Technologist

| Programs* | Class Capacity | Begins | Length (months) | Award | Res Tuition | Non-res Tuition | Stipend | Offers:‡ 1 | 2 | 3 |
|---|---|---|---|---|---|---|---|---|---|---|
| Southern West Virginia Comm & Tech College (Mt Gay) | 20 | Aug | 18 | Dipl, AS | $1,634 | $2,732 | | | | • |
| West Virginia University - Parkersburg | 20 | Aug | 10 | Cert | $2,448 | $8,281 | | | | • |
| **Wisconsin** | | | | | | | | | | |
| Gateway Technical College (Kenosha) | 18 | Aug | 18 | AS | $2,300 | $12,328 | | | | |
| Mid-State Technical College (Marshfield) | 20 | Jun | 12 | Dipl | $3,314 | $17,350 | | • | | |
| Milwaukee Area Technical College | 16 | Aug Jan | 18 | AAS | $2,982 | $17,408 | | | | |
| Northeast Wisconsin Technical College (Green Bay) | 26 | Jun Aug | 12 | Dipl | $3,863 | $19,106 | | • | | |
| Waukesha County Technical College (Pewaukee) | 14 | Aug | 24 | AAS | $5,673 | $15,400 | | • | | |
| **Wyoming** | | | | | | | | | | |
| Laramie County Community College (Cheyenne) | 12 | Aug | 20 | AAS | $1,416 | $4,272 | | | | |

*Data are shown only for programs that completed the 2006 AMA Survey of Health Professions Education Programs
‡Key to Offers: 1: Evening or weekend classes; 2: Non-English instruction; 3: Cultural competence instruction

**PROGRAMS**

## Occupational Description

Therapeutic recreation uses treatment, education, and recreation services to help people with illnesses, disabilities, and other conditions develop and use their leisure in ways that enhance their health, functional abilities, independence, and quality of life.

Therapeutic recreation services contribute to the broad spectrum of health care through treatment (recreational therapy), education, and providing recreational opportunities, all of which are instrumental to improving and maintaining physical, cognitive, emotional, and social functioning, preventing secondary health conditions, and enhancing independent living skills and the overall quality of life.

Recreational therapy services use various interventions to treat physical, cognitive, emotional, and social conditions associated with illness, injury, or chronic disabilities. Recreational therapy includes an education component, which enables individuals to become more informed and active partners in their health care by using activity to cope with the stress of illness and disability. Furthermore, these services assist individuals with managing their disabilities so they may achieve and maintain optimal levels of independence, productivity, and well-being and enter/re-enter the mainstream of community life.

Therapeutic recreation services also include the provision of recreational opportunities (eg, wheelchair sports, exercise and fitness programs, social activities) that can minimize health care costs by allowing individuals with disabilities mechanisms to prevent declines in their physical, cognitive, social, and emotional health, thereby reducing the need for medical services.

## Job Description

The day-to-day work experience of therapeutic recreation specialists can vary dramatically, depending on the setting and clients they serve. All therapeutic recreation specialists, however, conduct assessments of physical, mental, emotional, and social functioning to determine the client's needs, interests, and abilities. The therapeutic recreation specialist works with the client, family, and others to design and implement an individualized treatment, education, or program plan, depending on the setting.

Professional therapeutic recreation services are divided into three specific service areas, which represent a comprehensive continuum approach based on individual needs:

*Treatment* is intended to improve functional skills for individuals with disabilities who require treatment or remediation of functional skills as a prerequisite to their involvement in meaningful leisure experiences.

*Leisure education* provides persons in clinical, residential, and community settings—including individuals with disabilities—opportunities to attain skills, knowledge, and attitudes of leisure involvement.

*Recreation participation* provides opportunities for voluntary involvement in recreation interests and activities. Specialized recreation participation programs are provided when assistance and/or adapted recreation equipment are needed or when appropriate community recreation opportunities are not available.

During a typical day, a therapeutic recreation specialist will be responsible for one or more group activities. These might include a stress management group, a high or low ropes course activity, a community outing, a family activity, an exercise group, or a leisure education group. The therapeutic recreation specialist might also meet with individual clients to conduct an assessment, develop a leisure discharge plan, or plan evening and weekend activities. Charting client progress and communicating with professionals in other disciplines and clients' family members are also part of a typical day.

A therapeutic recreation specialist working in a community recreation agency also conducts assessments to determine client needs and interests and is responsible for adapting activities as needed and for providing adaptive equipment to enable individuals with disabilities or limitations to participate. In addition, the therapeutic recreation specialist provides in-service training for recreation staff who have individuals with disabilities in their programs to orient them to the needs of these individuals and to promote general sensitivity. The therapeutic recreation specialist will generally seek to include clients in existing recreation programs, activities, and classes when possible.

An important responsibility for a therapeutic recreation specialist in both community and clinical settings is to serve as an advocate on behalf of individuals with disabilities. This includes addressing such issues as limited transportation resources, inaccessible facilities, and legislation that affects people with disabilities or limitations. A therapeutic recreation specialist frequently serves on advisory committees and consults with outside agencies to ensure that resources and services are provided for people with disabilities.

One of the most attractive qualities of the therapeutic recreation profession is the opportunity for variety and diversity. The many changes in the health care delivery system have provided—and will continue to offer—an array of challenges and opportunities for continued growth in therapeutic recreation. In addition, the opportunity to positively affect the quality of life of an individual with a disability or limitation is extremely rewarding.

## Employment Characteristics

In clinical settings, such as hospitals and rehabilitation centers, therapeutic recreation specialists treat and rehabilitate individuals with specific medical problems, usually in cooperation with physicians, nurses, psychologists, social workers, and physical and occupational therapists. In long-term care facilities, residential facilities, and community recreation departments, they use leisure activities, individual as well as group-oriented, to improve general health and well-being, but also may treat medical problems. A bachelor's degree in therapeutic recreation (or in recreation with an option in therapeutic recreation) is the usual requirement for an entry-level position in a hospital and in other clinical positions.

Therapeutic recreation specialists assess patients, based on information from medical records, medical staff, family, and patients themselves. They then develop and implement therapeutic recreation programs consistent with patients' needs and interests. For instance, a patient having trouble socializing may be helped to play games with others, or a client with right-side paralysis may be helped to use the left arm to throw a ball or swing a racket.

Therapeutic recreation specialists observe and document patients' participation, reactions, and progress. These records are used by the interdisciplinary team and others to monitor progress, to justify changes or end therapeutic recreation services, and for billing, if applicable.

Community-based therapeutic recreation specialists work in park and recreation departments, special education programs, or programs for older adults or people with disabilities. In these programs, therapeutic recreation specialists help clients become involved in leisure activities and provide them with opportunities for exercise, mental stimulation, creativity, and fun.

Therapeutic recreation specialists often lift and carry equipment as well as implement activities. They generally work a 40-hour week, which may include some evenings, weekends, and holidays.

Therapeutic recreation specialists should be comfortable working with people with disabilities and be patient, tactful, and persuasive. Ingenuity and imagination are helpful in adapting activities to individual needs.

Therapeutic recreation specialists held about 39,000 jobs in 1998. About 38% were in hospitals and 26% were in nursing and personal care facilities. Others were in community mental health centers, adult day care programs, correctional facilities, residential facilities, community programs for people with disabilities, and substance abuse centers. About one out of three therapeutic recreation specialists was self-employed, generally contracting with long-term care facilities or community agencies to develop and oversee programs.

## Salary

According to a 1999 study of members of the National Therapeutic Recreation Society, the average salary of therapeutic recreation specialists was $35,349. In long-term care facilities, the average annual salary was $34,362 in 1999. Average annual earnings for therapeutic recreation specialists in the federal government in nonsupervisory, supervisory, and managerial positions were approximately $36,000 in 1995.

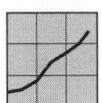

## Employment Outlook

Employment of therapeutic recreation specialists is expected to grow as fast as the average for all occupations through the year 2008, because of anticipated expansion in long-term care, physical and psychiatric rehabilitation, and services for people with disabilities. The US Bureau of Labor Statistics projects that there are approximately 39,000 positions in therapeutic recreation. Nonetheless, the total number of job openings will be relatively low because the occupation is small.

Health care facilities will provide a growing number of jobs in hospital-based adult day care and outpatient programs and units offering short-term mental health and alcohol or drug abuse services; rehabilitation, home health care, transitional programs, and psychiatric facilities will provide additional jobs.

The rapidly growing number of older people is expected to spur job growth for therapeutic recreation specialists and paraprofessionals in long-term care facilities, retirement communities, assisted living facilities, adult day care programs, and social service agencies. Continued growth is also expected in community residential facilities, as well as adult day care programs for people with disabilities.

## Educational Programs

**Length.** A major in therapeutic recreation or recreation with an option in therapeutic recreation entails completion of a bachelor's degree, including a minimum of 18 semester or 24 quarter units in therapeutic recreation and general recreation content coursework; completion of support-ive courses to include a minimum of 18 semester units or 27 quarter units; and completion of a minimum 480-hour, 12-consecutive-week internship/field placement experience in a clinical, residential, or community-based therapeutic recreation program.

**Curriculum.** In addition to therapeutic recreation courses in clinical practice, program design, management, and professional issues, students study human anatomy, physiology, abnormal psychology, medical and psychiatric terminology, human development, characteristics of illness and disabilities, and the concepts of inclusion and normalization. Additional courses cover professional ethics, assessment and referral procedures, and the use of adaptive and medical equipment. In addition, 360 hours of internship under the supervision of a certified therapeutic recreation specialist are required.

## Licensure, Certification, and Registration

A few states regulate the therapeutic recreation profession through licensure, certification, or registration of titles. Applicants for licensure must pass a state examination. Licensure is required in Utah. For more information, contact

Division of Occupational and Professional Licensure
160 East 300 South
Salt Lake City, UT 84145-0801
801 530-6628

National certification is available through the National Council for Therapeutic Recreation Certification (NCTRC), which awards the title of Certified Therapeutic Recreation Specialist (CTRS).

Through registration, qualified individuals are listed on an official roster maintained by a governmental or nongovernmental agency. Information regarding registration requirements may be obtained from state recreation and park associations.

## Career Planning Publications

The *2002 SPRE Curriculum Catalog* (published by the Society for Park and Recreation Educators) provides valuable information on curricula and faculty in the parks, recreation, and leisure studies profession. Degree levels offered and accreditation status are indicated for each program. Each listing includes the location and mailing address for the program, enrollment data, a description of the character of the campus and community, and a detailed listing of the faculty and specialties.

The NRPA Career Center can be accessed at www.nrpa.org.

*Preparing for a Career in Therapeutic Recreation* describes the continuum of services within therapeutic recreation and includes a listing of colleges and universities that offer therapeutic recreation programs, including those accredited by the NRPA/ AALR Council on Accreditation.

## Inquiries

**Careers**
National Therapeutic Recreation Society
22377 Belmont Ridge Road
Ashburn, VA 20148-4501
703 858-2151
800 626-NRPA—membership information and other services
703 858-0794 Fax
E-mail: ntrs@nrpa.org

American Therapeutic Recreation Association
1414 Prince Street, Suite 204
Alexandria, VA 22314
703 683-9420
703 683-9431 Fax

## Certification
National Council for Therapeutic Recreation Certification
7 Elmwood Drive
New City, NY 10956
845 639-1439
845 639-1471 Fax
E-mail: nctrc@nctrc.org
www.nctrc.org

## Program Accreditation
NRPA/AAPAR Council on Accreditation
Professional Services Division
22377 Belmont Ridge Road
Ashburn, VA 20148-4501
703 858-2150
703 858-0794 Fax
www.councilonaccreditation.org

# Therapeutic Recreation Specialist*

## California

### California State University - Chico
Therapeutic Recreation Specialist Prgm
Dept of Recreation and Parks Mgmt
Room 101, Tehama Hall
Chico, CA 95929-0560
*Prgm Dir:* Emilyn Sheffield, PhD
*Tel:* 530 898-6408   *Fax:* 530 898-6557
*E-mail:* esheffield@csuchico.edu

### California State University - Long Beach
Therapeutic Recreation Specialist Prgm
Dept of Recreation and Leisure Studies
1250 Bellflower Blvd
Long Beach, CA 90840-4903
www.csulb.edu/colleges/chhs/departments/rec
*Prgm Dir:* Maridith Janssen, EdD RTC CTRS
*Tel:* 562 985-4071, Ext 4079   *Fax:* 562 985-8154
*E-mail:* mjanssen@csulb.edu

## District of Columbia

### Gallaudet University
Therapeutic Recreation Specialist Prgm
Dept of Physical Educ and Recreation
800 Florida Ave NE
Washington, DC 20002
*Prgm Dir:* Anne Simonsen, PhD CTRS CPRP
*Tel:* 202 651-5591   *Fax:* 202 651-5861
*E-mail:* anne.simonsen@gallaudet.edu

## Florida

### University of Florida
Therapeutic Recreation Specialist Prgm
Dept of Recreation, Parks and Tourism
Room 300 Florida Gym, PO Box 118208
Gainesville, FL 32611-8208
*Prgm Dir:* Stephen Holland
*Tel:* 352 392-4042   *Fax:* 352 392-7588
*E-mail:* sholland@hhp.ufl.edu

## Georgia

### Georgia Southern University
Therapeutic Recreation Specialist Prgm
Dept of Hospitality, Tourism and Family & Consumer
 Sciences
PO Box 8034
Statesboro, GA 30460-8077
www.georgiasouthern.edu
*Prgm Dir:* Henry Eisenhart, PhD
*Tel:* 912 681-5345   *Fax:* 912 681-0276
*E-mail:* henry_e@georgiasouthern.edu

## Illinois

### Southern Illinois Univ at Carbondale
Therapeutic Recreation Specialist Prgm
Dept of Health Educ and Recreation
Pulliam Hall 307, Postal Code 4632
Carbondale, IL 62901
www.siu.edu
*Prgm Dir:* David Birch, PhD
*Tel:* 618 453-4331   *Fax:* 618 453-1829
*E-mail:* dabirch@siu.edu
*Med Dir:* Marjorie Malkin, EdD, CTRS

### Eastern Illinois University
Therapeutic Recreation Specialist Prgm
Recreation Administration
600 Lincoln Ave
Charleston, IL 61920
*Prgm Dir:* William Higelmire, EdD CTRS
*Tel:* 217 581-6344   *Fax:* 217 581-7804
*E-mail:* cfwfh@eiu.edu

### Western Illinois University
Therapeutic Recreation Specialist Prgm
Recreation, Park and Tourism Admin
400 Currens Hall
Macomb, IL 61455
*Prgm Dir:* Dale Adkins, PhD
*Tel:* 309 298-1967   *Fax:* 309 298-2967
*E-mail:* digrino@ccmail.wiu.edu
*Med Dir:* Dean Zoerink, PhD CTRS

### Illinois State University
Therapeutic Recreation Specialist Prgm
Recreation and Park Admin Prgm
McCormick Hall Rm 101, CB 5121
Normal, IL 61790-5121
*Prgm Dir:* Barbara E Schlatter, PhD
*Tel:* 309 438-5608   *Fax:* 309 438-5561
*E-mail:* beschla@ilstu.edu

## Indiana

### Indiana University - Bloomington
Therapeutic Recreation Specialist Prgm
Department of Recreation, Park, and Tourism Studies
HPER Building 133
Bloomington, IN 47405
www.iub.edu/~recpark
*Prgm Dir:* Lynn Jamieson, ReD
*Tel:* 812 855-4711   *Fax:* 812 855-3998
*E-mail:* lyjamies@indiana.edu

## Iowa

### University of Northern Iowa
Therapeutic Recreation Specialist Prgm
Leisure, Youth and Human Services Division
School of HPELS/217 WRC
Cedar Falls, IA 50614-0161
*Prgm Dir:* Joe Wilson, EdD
*Tel:* 319 273-2313   *Fax:* 319 273-5958
*E-mail:* Joe.Wilson@uni.edu

### University of Iowa
Therapeutic Recreation Specialist Prgm
Dept of Sport, Health, Leisure and Physical Studies
E102A Field House
Iowa City, IA 52242-1111
*Prgm Dir:* Bonnie Slatton, PhD
*Tel:* 319 335-9335   *Fax:* 319 335-6669
*E-mail:* bonnie-slatton@uiowa.edu

## Kansas

### Pittsburg State University
Therapeutic Recreation Specialist Prgm
1701 S Broadway
Pittsburg, KS 66762-7557
*Prgm Dir:* Charles Killingsworth, EdD
*Tel:* 620 235-4665   *Fax:* 620 235-4520

## Kentucky

### Eastern Kentucky University
Therapeutic Recreation Specialist Prgm
Begley 405
Richmond, KY 40475
www.recreation.eku.edu
*Prgm Dir:* Charlie Everett, EdD
*Tel:* 859 622-1833   *Fax:* 859 622-2971
*E-mail:* charlie.everett@eku.edu

*Additional information about programs that
returned the AMA's survey is available in a table
beginning on page 464.

# Louisiana

**Grambling State University**
Therapeutic Recreation Specialist Prgm
Leisure Studies, Recreation Career Prgm
PO Box 1193
Grambling, LA 71245
www.gram.edu
*Prgm Dir:* Willie F Daniel
*Tel:* 318 274-2294   *Fax:* 318 274-6053
*E-mail:* danielw@gram.edu
*Med Dir:* Renisha Sampract

# Massachusetts

**Springfield College**
Therapeutic Recreation Specialist Prgm
Dept of Recreation and Leisure Services
263 Alden St, Locklin Hall
Springfield, MA 01109-3797
*Prgm Dir:* Matthew J Pantera, EdD
*Tel:* 413 748-3693   *Fax:* 413 748-3685
*E-mail:* mpantera@spfldcol.edu

# Michigan

**Central Michigan University**
Therapeutic Recreation Specialist Prgm
Dept of Recreation and Park Admin
Mt Pleasant, MI 48859
www.cmich.edu
*Prgm Dir:* Roger Coles, EdD CPRP
*Tel:* 517 774-3858   *Fax:* 517 774-2161
*E-mail:* coles1rl@cmich.edu

# Minnesota

**University of Minnesota - Minneapolis**
Therapeutic Recreation Specialist Prgm
Recreation and Sport Studies
224 Cooke Hall, 1900 University Ave SE
Minneapolis, MN 55455
*Prgm Dir:* Leo McAvoy, PhD
*Tel:* 612 625-5300   *Fax:* 612 626-7700
*E-mail:* mcavo001@tc.umn.edu

# Mississippi

**University of Southern Mississippi**
Therapeutic Recreation Specialist Prgm
School of Human Performance and Recreation
Box 5142
Hattiesburg, MS 39406-5142
*Prgm Dir:* Terry Kinney, PhD
*Tel:* 601 266-5386   *Fax:* 601 266-4445
*E-mail:* terry.kinney@usm.edu

# Missouri

**University of Missouri - Columbia**
Therapeutic Recreation Specialist Prgm
Dept of Parks, Recreation and Tourism
105 Anheuser Busch Nat Res Bldg
Columbia, MO 65211-7230
*Prgm Dir:* C Randal Vessell, PhD
*Tel:* 573 882-7088   *Fax:* 573 882-9526
*E-mail:* randy_vessell@muccmail.missouri.edu

# New Hampshire

**University of New Hampshire**
Therapeutic Recreation Specialist Prgm
Dept of Recreation Management & Policy
108 Hewitt Hall
Durham, NH 03824
www.shhs.unh.edu/rmp
*Prgm Dir:* Janet Sable, EdD CTRS
*Tel:* 603 862-3401   *Fax:* 603 862-2722
*E-mail:* jrsable@unh.edu

# New York

**SUNY College at Cortland**
Therapeutic Recreation Specialist Prgm
Recreation and Leisure Studies Dept
PO Box 2000
Cortland, NY 13045
*Prgm Dir:* Lynn S Anderson, CTRS
*Tel:* 607 753-4941   *Fax:* 607 753-5982
*E-mail:* RLS@cortland.edu

**Ithaca College**
Therapeutic Recreation Specialist Prgm
Dept of Therapeutic Recreation & Leisure
36 Hill Center
Ithaca, NY 14850
*Prgm Dir:* Linda Heyne, PhD
*Tel:* 607 274-3050   *Fax:* 607 274-1943
*E-mail:* lheyne@ithaca.edu

# North Carolina

**Univ of North Carolina at Greensboro**
Therapeutic Recreation Specialist Prgm
Dept of Recreation, Tourism, and Hospitality
    Management
420 HHP Building
Greensboro, NC 27402-6170
www.uncg.edu/rth
*Prgm Dir:* Stuart J Schleien, PhD CTRS CPRP
*Tel:* 336 334-5327   *Fax:* 336 334-3238
*E-mail:* sjs@uncg.edu
*Med Dir:* Charlsena Stone, PhD, CTRS, LRT

**Univ of North Carolina - Wilmington**
Therapeutic Recreation Specialist Prgm
Parks and Recreation Mgmt Curriculum
PO Box 1425, 601 S College Rd
Wilmington, NC 28403-3297
*Prgm Dir:* Dan Johnson, TRS/CTRS
*Tel:* 910 962-3251   *Fax:* 910 962-7073
*E-mail:* johnsonde@uncwil.edu

**Winston-Salem State University**
Therapeutic Recreation Specialist Prgm
Dept of Human Performance and Sport Science
Old Nursing Building
Winston-Salem, NC 27110
www.wssu.edu
*Prgm Dir:* Cynthia Stanley, PhD LRT CTRS
*Tel:* 336 750-2588   *Fax:* 336 750-2591
*E-mail:* stanleyc@wssu.edu

# Ohio

**University of Toledo**
Therapeutic Recreation Specialist Prgm
Dept of Pub Hlth & Rehab Serv-Hlth Ed Ctr
2801 W Bancroft St, 252 Health Ed Ctr
Toledo, OH 43606
*Prgm Dir:* Bruce Groves, PhD
*Tel:* 419 530-2757   *Fax:* 419 530-4759
*E-mail:* bruce.groves2@utoledo.edu

# Oklahoma

**Oklahoma State University**
Therapeutic Recreation Specialist Prgm
Dept of Leisure Studies
103 Colvin Center
Stillwater, OK 74078
*Prgm Dir:* Christine M Cashel, EdD
*Tel:* 405 744-5503   *Fax:* 405 744-6507
*E-mail:* ccashel@okstate.edu

# Pennsylvania

**Slippery Rock University**
Therapeutic Recreation Specialist Prgm
Dept of Parks & Recreation/Environmental Education
101 Eisenberg
Slippery Rock, PA 16057
*Prgm Dir:* Bruce G Boliver
*Tel:* 724 738-2068   *Fax:* 724 738-2938
*E-mail:* bruce.boliver@sru.edu

# South Carolina

**Clemson University**
Therapeutic Recreation Specialist Prgm
Parks, Recreation and Tourism Mgmt
276 Lehotsky Hall
Clemson, SC 29634-0735
*Prgm Dir:* Francis McGuire, PhD
*Tel:* 864 656-3400   *Fax:* 864 656-2226
*E-mail:* rwmcl@clemson.edu

# Tennessee

**University of Tennessee - Knoxville**
Therapeutic Recreation Specialist Prgm
1914 Andy Holt Ave
Knoxville, TN 37996-2710
*Prgm Dir:* Gene Hayes, PhD
*Tel:* 865 974-1288   *Fax:* 865 974-8981
*E-mail:* ghayes1@utk.edu

# Texas

**Texas State University-San Marcos**
Therapeutic Recreation Specialist Prgm
601 University Dr
Jowers Center
San Marcos, TX 78666
*Prgm Dir:* Tom Gustafson, PhD
*Tel:* 512 245-2561   *Fax:* 512 245-8678
*E-mail:* tg08@swt.edu

# Utah

**Brigham Young University**
Therapeutic Recreation Specialist Prgm
Recreation Mgmt and Youth Leadership
273 Richards Bldg
Provo, UT 84602
*Prgm Dir:* Brian Hill, PhD
*Tel:* 801 378-4369   *Fax:* 801 378-7461
*E-mail:* brian.hill@byu.edu

**University of Utah**
Therapeutic Recreation Specialist Prgm
Dept of Parks, Recreation, and Tourism
250 S 1850 E - HPER N-226
Salt Lake City, UT 84112-0920
*Prgm Dir:* Gary D Ellis, PhD
*Tel:* 801 581-8542   *Fax:* 801 581-4930
*E-mail:* gary.ellis@health.utah.edu

## Virginia

**Longwood University**
Therapeutic Recreation Specialist Prgm
Dept of Health, Recreation and Kinesiology
201 High St
Farmville, VA 23909-1899
*Prgm Dir:* Glenda Taylor, PhD
*Tel:* 434 395-2545   *Fax:* 434 395-2530
*E-mail:* taylorgp@longwood.edu

**George Mason University**
*Cosponsor:* School of Recreation, Health, and Tourism
Therapeutic Recreation Specialist Prgm
Parks, Recreation and Leisure Studies Program
10900 University Blvd MS 4E5
Manassas, VA 20110
http://rht.gmu.edu
*Prgm Dir:* Ellen Drogin Rodgers
*Tel:* 703 993-2085   *Fax:* 703 993-2025
*E-mail:* erodger1@gmu.edu
*Med Dir:* Brenda Wiggins, PhD

**Radford University**
Therapeutic Recreation Specialist Prgm
Dept of Recreation Parks and Tourism
Box 6963
Radford, VA 24142
www.radford.edu/~recparks
*Prgm Dir:* James Newman, PhD CTRS
*Tel:* 540 831-7717   *Fax:* 540 831-7719
*E-mail:* jnewman19@radford.edu

**Virginia Commonwealth Univ/Med Coll of VA**
Therapeutic Recreation Specialist Prgm
Dept of Rec, Parks and Sport Mgmt Prgm
817 W Franklin St, Box 2015
Richmond, VA 23284-2015
*Prgm Dir:* Michael S Wise, EdD CPRP
*Tel:* 804 828-1130   *Fax:* 804 828-1946
*E-mail:* mwise@saturn.vcu.edu
*Med Dir:* Allison Wilder, MS CTRS

## Washington

**Eastern Washington University**
Therapeutic Recreation Specialist Prgm
Recreation and Leisure Studies
Dept of PEHR
Cheney, WA 99004
*Prgm Dir:* Matthew R Chase, EdD
*Tel:* 509 359-2341   *Fax:* 509 359-4833
*E-mail:* Matthew.Chase@mail.ewu.edu

## Wisconsin

**University of Wisconsin - La Crosse**
Therapeutic Recreation Specialist Prgm
Recreation Mgmt and Therapeutic Recreation
128 Wittich Hall
La Crosse, WI 54601
www.uwlax.edu/sah/rmtr/ug_tr
*Prgm Dir:* George Arimond, PhD
*Tel:* 608 785-8205   *Fax:* 608 785-8206
*E-mail:* arimond.geor@uwlax.edu

### Therapeutic Recreation Specialist

| Programs* | Class Capacity | Begins | Length (months) | Award | Res. Tuition | Non-res. Tuition | Stipend | Offers:‡ 1 | 2 | 3 |
|---|---|---|---|---|---|---|---|---|---|---|
| **California** | | | | | | | | | | |
| California State University - Long Beach | 142 | Sep Jan | 30 | Cert, BA, MA | $3,200 | $7,200 | | • | | • |
| **Georgia** | | | | | | | | | | |
| Georgia Southern University (Statesboro) | 30 | Aug Jan May | 48 | BS | $2,350 | $9,290 | | | | • |
| **Illinois** | | | | | | | | | | |
| Eastern Illinois University (Charleston) | 65 | Aug Jan | 9 | BS | $154 | $462 | | | | |
| Illinois State University (Normal) | | | | BS | | | | • | | • |
| Southern Illinois Univ at Carbondale | 60 | Aug | 30, 24 | BS, MS | $5,200 | $12,000 | $10,800 | • | | |
| **Indiana** | | | | | | | | | | |
| Indiana University - Bloomington | | Aug | | BS, MS | $4,196 | $13,930 | | • | | |
| **Iowa** | | | | | | | | | | |
| University of Northern Iowa (Cedar Falls) | | Aug Jan | 12, 24 | Cert, BA | $5,387 | $12,705 | | | | • |
| **Kentucky** | | | | | | | | | | |
| Eastern Kentucky University (Richmond) | | | 48 | BS | $1,896 | $5,232 | | | | |
| **Louisiana** | | | | | | | | | | |
| Grambling State University | 300 | Aug Jan | 45 | BS | $3,506 | $8,856 | | • | | • |
| **Michigan** | | | | | | | | | | |
| Central Michigan University (Mt Pleasant) | 25 | Aug Jan | 12, 30 | BS, MA | $4,500 | $8,500 | | | | |
| **New Hampshire** | | | | | | | | | | |
| University of New Hampshire (Durham) | 45 | Sep Jan | | BS, MS | $5,770 | $14,840 | | • | | • |
| **New York** | | | | | | | | | | |
| SUNY College at Cortland | 40 | Aug Jan | 48, 18 | BS, MS/MSE | $5,300 | $11,200 | | • | | • |
| **North Carolina** | | | | | | | | | | |
| Univ of North Carolina at Greensboro | 60 | Aug Jan | 48 | BS, MS | $2,028 | $13,296 | | • | | • |
| Winston-Salem State University | 100 | Aug Jan | 20 | BS | $2,294 | $10,659 | | • | | • |
| **Tennessee** | | | | | | | | | | |
| University of Tennessee - Knoxville | | Aug | 48, 24 | BS, MS | $2,604 | $0 | | • | | |
| **Virginia** | | | | | | | | | | |
| George Mason University (Manassas) | | | | | | | | | | |
| Longwood University (Farmville) | 40 | Aug | 48 | BS | $6,233 | $9,600 | | • | | • |
| Radford University | 25 | Aug Jan | 36 | BS | $5,130 | $18,402 | | | | |
| **Wisconsin** | | | | | | | | | | |
| University of Wisconsin - La Crosse | 35 | Sep | 36 | BS, MS | $3,299 | $9,946 | | | | • |

*Data are shown only for programs that completed the 2006 AMA Survey of Health Professions Education Programs
‡Key to Offers: 1: Evening or weekend classes; 2: Non-English instruction; 3: Cultural competence instruction

# Section II

## Institutions Sponsoring Accredited Programs

# Alabama

## Andalusia

**Lurleen B Wallace Junior College**
Seth Hammett, MBA, President
PO Box 1418
Andalusia, AL 36420
205 222-6591
- Emergency Med Tech-Paramedic Prgm

## Auburn University

**Auburn University**
Edward Richardson, President
107 Sanford Hall
Auburn University, AL 36849
334 844-4650
- Audiologist Prgm
- Counseling Prgm
- Dietetics-Didactic Prgm
- Rehabilitation Counseling Prgm
- Speech-Language Pathology Prgm

## Bay Minette

**James H Faulkner State Community College**
Gary L Branch, President
1900 Hwy 31 S
Bay Minette, AL 36507-2619
334 580-2100
- Dental Assisting Prgm
- Surgical Technology Prgm

## Birmingham

**Carraway Methodist Medical Center**
Cathy Fickes, Administrator
1600 N 26th St
Birmingham, AL 35234
205 226-6000
- Radiography Prgm

**Jefferson State Community College**
Judy M Merritt, PhD, President
2601 Carson Rd
Birmingham, AL 35215
205 856-7777
- Clin Lab Technician/Med Lab Technician
- Physical Therapist Assistant Prgm
- Radiography Prgm

**Samford University**
Thomas E Corts, PhD, President
800 Lakeshore Dr
Birmingham, AL 35229
205 870-5727
- Athletic Training Prgm
- Dietetics-Didactic Prgm

**University of Alabama at Birmingham**
Carol Garrison, PhD, President
AB 1070
Birmingham, AL 35294-0110
205 934-4636
- Clin Lab Scientist/Med Technologist Prgm
- Cytotechnology Prgm
- Dietetic Internship Prgm
- Emergency Med Tech-Paramedic Prgm
- Health Information Admin Prgm
- Nuclear Medicine Technology Prgm
- Occupational Therapy Prgm
- Physical Therapy Prgm
- Radiation Therapy Prgm
- Radiography Prgm
- Rehabilitation Counseling Prgm
- Respiratory Therapist (Advanced) Prgm
- Surgical Physician Assistant Prgm

## Decatur

**Calhoun Community College**
Marilyn Beck, EdD, President
PO Box 2216 Hwy 31 N
Decatur, AL 35609
256 306-2500
- Dental Assisting Prgm
- Emergency Med Tech-Paramedic Prgm
- Surgical Technology Prgm

## Dothan

**Flowers Hospital**
L Keith Granger, BS, President
PO Box 6907
Dothan, AL 36302
334 793-5000
- Surgical Technology Prgm

**George C Wallace Community College**
Linda Young, EdD, President
1141 Wallace Dr
Dothan, AL 36303
800 543-2426
- Emergency Med Tech-Paramedic Prgm
- Medical Assistant Prgm
- Pharmacy Technician Prgm
- Physical Therapist Assistant Prgm
- Radiography Prgm
- Respiratory Therapist (Advanced) Prgm

## Gadsden

**Gadsden State Community College**
Renee D Culverhouse, BA JD, President
PO Box 227
1001 George Wallace Dr
Gadsden, AL 35902-0227
256 549-8221
- Clin Lab Technician/Med Lab Technician
- Emergency Med Tech-Paramedic Prgm
- Radiography Prgm

## Hanceville

**Wallace State Community College**
Vicki Hawsey, EdD, President
Commerce Bldg
PO Box 2000
Hanceville, AL 35077-2000
256 352-8130
- Clin Lab Technician/Med Lab Technician
- Dental Assisting Prgm
- Dental Hygiene Prgm
- Diagnostic Med Sonography Prgm
- Emergency Med Tech-Paramedic Prgm
- Health Information Tech Prgm
- Medical Assistant Prgm
- Occupational Therapy Asst Prgm
- Physical Therapist Assistant Prgm
- Radiography Prgm
- Respiratory Therapist (Advanced) Prgm

## Huntsville

**Crestwood Medical Center**
One Hospital Dr SE
Huntsville, AL 35801
- Radiography Prgm

**Huntsville Hospital**
David Spillers, MHA, CEO
101 Sivley Rd
Huntsville, AL 35801
256 265-8123
- Radiography Prgm

**Oakwood College**
Benjamin F Reaves, President
Huntsville, AL 35896
205 726-7000
- Dietetics-Didactic Prgm
- Dietetics-Preprofessional Practice Prgm

## Jacksonville

**Jacksonville State University**
Harold J McGee, President
Jacksonville, AL 36265
205 782-5781
• Dietetics-Didactic Prgm

## Livingston

**University of West Alabama**
Richard Holland, PhD, President
Station 1
Livingston, AL 35470
205 652-3527
• Athletic Training Prgm

## Mobile

**Bishop State Community College**
Yvonne Kennedy, PhD, President
351 N Broad St
Mobile, AL 36603-5898
205 690-6416
• Health Information Tech Prgm
• Physical Therapist Assistant Prgm

**Insitute of Ultrasound Diagnostics**
Smyth R Gill, BA, CFO
1230 Montlimar Dr, Ste A
Ste 200
Mobile, AL 36609
251 460-2485
• Diagnostic Med Sonography Prgm

**University of Mobile**
Mark R Foley, PhD, President
5735 College Pkwy
Mobile, AL 36663-0220
251 442-2201
• Athletic Training Prgm

**University of South Alabama**
V Gordon Moulton, MBA, President
307 N University Blvd
Mobile, AL 36688-0002
251 460-6111
• Audiologist Prgm
• Clin Lab Scientist/Med Technologist Prgm
• Emergency Med Tech-Paramedic Prgm
• Occupational Therapy Prgm
• Physical Therapy Prgm
• Physician Assistant Prgm
• Radiography Prgm
• Respiratory Therapist (Advanced) Prgm
• Speech-Language Pathology Prgm

**Virginia College**
Madeline Little, President, South Region
5901 Airport Blvd
Mobile, AL 36608
251 343-7227
• Surgical Technology Prgm (2)

## Montevallo

**University of Montevallo**
Robert M McChesney, President
Station 6001
Montevallo, AL 35115
205 665-6000
• Counseling Prgm
• Dietetics-Didactic Prgm
• Speech-Language Pathology Prgm

## Montgomery

**Alabama State University**
Joe A Lee, PhD, President
915 S Jackson St
Montgomery, AL 36101-0271
334 229-4202
• Health Information Admin Prgm
• Occupational Therapy Prgm
• Physical Therapy Prgm

**Auburn University Montgomery**
Guin Nance, PhD, Chancellor
PO Box 244023
Montgomery, AL 36124
334 244-3602
• Clin Lab Scientist/Med Technologist Prgm
• Cytotechnology Prgm

**Baptist Medical Center South**
Russ Tyner, BA MHA, CEO
301 Brown Springs Rd
Montgomery, AL 36117
334 273-4400
• Clin Lab Scientist/Med Technologist Prgm
• Diagnostic Med Sonography Prgm
• Radiography Prgm

**H Council Trenholm State Technical College**
Anthony Molina, PhD, President
1225 Air Base Blvd
Montgomery, AL 36108
334 420-4220
• Dental Assisting Prgm
• Dental Lab Technician Prgm
• Emergency Med Tech-Paramedic Prgm
• Medical Assistant Prgm

**Huntingdon College**
J Cameron West, Th M M Div, President
1500 E Fairview Ave
Montgomery, AL 36106
334 833-4409
• Athletic Training Prgm

**South University**
Victor Biebighauser, BA, President
5355 Vaughn Rd
Montgomery, AL 36116-1120
334 395-8800
• Medical Assistant Prgm
• Physical Therapist Assistant Prgm

## Muscle Shoals

**Northwest-Shoals Community College**
Humphrey Lee, EdD, President
PO Box 2545
Muscle Shoals, AL 35662
256 331-5200
• Emergency Med Tech-Paramedic Prgm

## Normal

**Alabama A&M University**
David B Henson, President
PO Box 1357
Normal, AL 35762
205 851-5000
• Dietetics-Didactic Prgm
• Rehabilitation Counseling Prgm
• Speech-Language Pathology Prgm

## Opelika

**Southern Union State Community College**
Susan E Salatto, MACT, President
1701 LaFayette Pkwy
Opelika, AL 36801
334 745-6437
• Emergency Med Tech-Paramedic Prgm
• Radiography Prgm
• Surgical Technology Prgm

## Rainsville

**Northeast Alabama Community College**
David Campbell, PhD, President
PO Box 159
Rainsville, AL 35986-0159
256 228-6001
• Emergency Med Tech-Paramedic Prgm

## Sumiton

**Bevill State Community College**
Harold Wade, EdD, President
PO Box 800
Sumiton, AL 35148-0800
205 648-3271
• Emergency Med Tech-Paramedic Prgm
• Surgical Technology Prgm

## Troy

**Troy University**
Jack Hawkins, Jr, PhD, Chancellor
Troy State Univ System, University Ave
Troy, AL 36082
334 670-3200
• Athletic Training Prgm
• Counseling Prgm (2)
• Rehabilitation Counseling Prgm

## Tuscaloosa

**DCH Regional Medical Center**
Bryan N Kindred, MBA, President/CEO
809 University Blvd E
Tuscaloosa, AL 35401
205 759-7177
• Radiography Prgm

**Shelton State Community College**
Rick Rogers, EdD, President
9500 Old Greensboro Rd
Tuscaloosa, AL 35405-8522
205 391-2472
• Emergency Med Tech-Paramedic Prgm
• Respiratory Therapist (Advanced) Prgm

**University of Alabama**
Robert Witt, PhD, President
PO Box 870100
Tuscaloosa, AL 35487
205 348-5100
• Athletic Training Prgm
• Counseling Prgm
• Dietetics-Coordinated Prgm
• Dietetics-Didactic Prgm
• Medical Librarian Prgm
• Music Therapy Prgm
• Rehabilitation Counseling Prgm
• Speech-Language Pathology Prgm

## Tuskegee

**Tuskegee University**
Benjamin F Payton, PhD, President
308 Kresge Ctr
Tuskegee, AL 36088-1677
334 727-8501
- Clin Lab Scientist/Med Technologist Prgm
- Dietetics-Didactic Prgm
- Occupational Therapy Prgm

# Alaska

## Anchorage

**University of Alaska Anchorage**
Elaine P Maimon, PhD, Chancellor
3211 Providence Dr
ADM 217
Anchorage, AK 99508
907 786-1437
- Clin Lab Scientist/Med Technologist Prgm
- Clin Lab Technician/Med Lab Technician
- Dental Assisting Prgm
- Dental Hygiene Prgm
- Dietetic Internship Prgm
- Massage Therapy Prgm
- Medical Assistant Prgm

## Fairbanks

**University of Alaska - Fairbanks**
Steve Jones, PhD, Chancellor
320 Signers Hall
Fairbanks, AK 99775-7500
907 474-7112
- Medical Assistant Prgm

## Juneau

**University of Alaska Southeast**
John Pugh, Chancellor
11120 Glacier Hwy
Juneau, AK 99801-8691
907 796-6457
- Health Information Tech Prgm

# Alberta, Canada

## Edmonton

**University of Alberta**
Indira Samarasekera, PhD, President
Edmonton, AB T6G 2M7
- Medical Librarian Prgm

# Arizona

## Coolidge

**Central Arizona College**
Kathleen Arns, PhD, President
Woodruff at Overfield Rd
Coolidge, AZ 85228
602 426-4444
- Dietetic Technician-AD Prgm

## Flagstaff

**Northern Arizona University**
Haeger John, PhD, President
PO Box 4092
Flagstaff, AZ 86011
928 523-3232
- Athletic Training Prgm
- Counseling Prgm
- Dental Hygiene Prgm
- Physical Therapy Prgm
- Speech-Language Pathology Prgm

## Glendale

**Midwestern University - Glendale Campus**
Kathleen H Goeppinger, PhD, President and CEO
19555 N 59th Ave
Glendale, AZ 85308
623 572-3400
- Occupational Therapy Prgm
- Perfusion Prgm
- Physician Assistant Prgm

## Kingman

**Mohave Community College**
Thomas C Henry, PhD, Chancellor
1971 Jagerson Ave
Kingman, AZ 86401
928 757-0800
- Dental Hygiene Prgm
- Surgical Technology Prgm

## Mesa

**Arizona State University East**
Charles E Backus, Provost
7001 E Williams Field Rd
Bldg 20
Mesa, AZ 85212
480 727-3278
- Dietetic Internship Prgm
- Dietetics-Didactic Prgm

**AT Still University of Health Sciences**
James J McGovern, PhD, President
5850 E Still Circle
Mesa, AZ 85206
480 219-6000
- Occupational Therapy Prgm
- Physical Therapy Prgm
- Physician Assistant Prgm

**Mesa Community College**
Larry K Christiansen, EdD, President
1833 W Southern Ave
Mesa, AZ 85205
480 461-7299
- Dental Assisting Prgm
- Dental Hygiene Prgm

## Phoenix

**Apollo College/US Education Corporation**
Steven B Nestor, MSEE, President
2701 W Bethany Home Rd
Phoenix, AZ 85017
602 433-1333
- Radiography Prgm

**Bryman School - Phoenix**
Dennis Pobiak, CEO
2250 W Peoria Ave
Phoenix, AZ 85029
- Medical Assistant Prgm

**Focus on Nutrition**
3923 E Thunderbird Rd, Ste 26-113
Phoenix, AZ 85032
602 788-7096
- Dietetics-Preprofessional Practice Prgm

**GateWay Community College**
Eugene Giovannini, EdD, President
108 N 40th St
Phoenix, AZ 85034
602 286-8000
- Diagnostic Med Sonography Prgm
- Nuclear Medicine Technology Prgm
- Physical Therapist Assistant Prgm
- Radiography Prgm
- Respiratory Therapist (Advanced) Prgm
- Surgical Technology Prgm

**Grand Canyon University**
Gil Stafford, PhD, President
PO Box 111097
Phoenix, AZ 85061-1097
602 249-3300
- Athletic Training Prgm

**Long Technical College**
Jeff Conlon, MBA, CEO
13450 N Black Canyon Hwy #104
Phoenix, AZ 85029
602 548-1955
- Medical Assistant Prgm
- Respiratory Therapist (Advanced) Prgm

**Paradise Valley Unified School District**
20621 N 32nd St
Phoenix, AZ 85032
602 493-2600
- Dietetics-Preprofessional Practice Prgm

**Phoenix College**
Anna Solley, EdD, President
1202 W Thomas Rd
Phoenix, AZ 85013
602 285-7364
- Dental Assisting Prgm
- Dental Hygiene Prgm
- Health Information Tech Prgm

**University of Phoenix**
Laura Palmer Noone, PhD JD, President
4615 E Elwood St
PO Box 52076
Phoenix, AZ 85040
480 966-9577
- Counseling Prgm

## Prescott

**Yavapai County Health Department**
930 Division St
Prescott, AZ 86301
602 771-3122
- Dietetic Internship Prgm

## Tempe

**Arizona State University**
Michael M Crow, PhD, President
Box 872203
Tempe, AZ 85287-2203
480 965-5606
- Audiologist Prgm
- Clin Lab Scientist/Med Technologist Prgm
- Counseling Prgm
- Music Therapy Prgm
- Speech-Language Pathology Prgm

**Lamson College**
Ralph Bilbao, EdD, Executive Director
1126 N Scottsdale Rd #17
Tempe, AZ 85281
480 898-7000
- Surgical Technology Prgm

**Maricopa County Dept of Public Health**
Office of Nutrition Services
1414 W Broadway Ste 237
Tempe, AZ 85282
602 966-3090
- Dietetic Internship Prgm

**Rio Salado College**
Linda M Thor, President
2323 W 14th St
Tempe, AZ 85281-6950
480 517-8000
- Dental Hygiene Prgm

## Tucson

**Carondelet St Mary's Hospital**
Morrisons Custom Management Company
1601 W St Mary's Rd
Tucson, AZ 85745
520 622-5833
- Dietetic Internship Prgm

**Desert Institute of the Healing Arts**
Tucson, AZ 85705
- Massage Therapy Prgm

**Pima Community College**
Roy Flores, PhD, Chancellor
4905 E Broadway
Tucson, AZ 85709-1005
520 206-4747
- Dental Assisting Prgm
- Dental Hygiene Prgm
- Dental Lab Technician Prgm
- Histotechnician Prgm
- Pharmacy Technician Prgm
- Radiography Prgm
- Respiratory Therapist (Advanced) Prgm
- Surgical Technology Prgm

**Pima Medical Institute**
Richard L Luebke, Jr, BS, President
3350 E Grant Rd #200
Tucson, AZ 85716
520 326-1600
- Radiography Prgm (2)
- Respiratory Therapist (Advanced) Prgm (2)

**Pima Medical Institute - Denver**
Richard L Luebke, Sr, BS, President
3350 E Grant Rd #200
Tucson, AZ 85716
520 326-1600
- Ophthalmic Med Technician Prgm
- Radiography Prgm
- Respiratory Therapist (Advanced) Prgm

**University of Arizona**
Peter Likins, PhD, President
Adminstration Bldg Rm 712
Tucson, AZ 85721-0066
520 621-5511
- Audiologist Prgm
- Dietetic Internship Prgm
- Dietetics-Didactic Prgm
- Medical Librarian Prgm
- Orientation and Mobility Specialist Prgm
- Perfusion Prgm
- Rehabilitation Counseling Prgm
- Speech-Language Pathology Prgm

## Yuma

**Arizona Western College**
Don Schoening, PhD, President
PO Box 929
Yuma, AZ 85366-0929
- Radiography Prgm

# Arkansas

## Arkadelphia

**Henderson State University**
Charles Dunn, PhD, President
HSU Box 7532
Arkadelphia, AR 71999-0001
870 230-5091
- Athletic Training Prgm
- Dietetics-Didactic Prgm

**Ouachita Baptist University**
Andrew Westmoreland, EdD, President
410 Ouachita Ct
OBU Box 3753
Arkadelphia, AR 71998
870 245-5000
- Athletic Training Prgm
- Dietetics-Didactic Prgm

## Batesville

**Univ of Arkansas Comm College-Batesville**
Anthony G Kinkel, EdD, Chancellor
PO Box 3350
Batesville, AR 72501
870 793-7581
- Emergency Med Tech-Paramedic Prgm
- Surgical Technology Prgm

## Beebe

**Arkansas State University - Beebe**
Eugene McKay, PhD, Chancellor
PO Box 1000
Beebe, AR 72012
501 882-8356
- Clin Lab Technician/Med Lab Technician

## Bentonville

**NorthWest Arkansas Community College**
Rebecca Paneitz, PhD, President
One College Dr
Bentonville, AR 72712-5091
479 619-4251
- Emergency Med Tech-Paramedic Prgm
- Physical Therapist Assistant Prgm
- Respiratory Therapist (Advanced) Prgm

## Blytheville

**Arkansas Northeastern College**
Robin Myers, EdD, President
PO Box 1109
Blytheville, AR 72316
870 762-1020
- Dental Assisting Prgm
- Emergency Med Tech-Paramedic Prgm

## Conway

**University of Central Arkansas**
Lu Hardin, JD, President
332 Torreyson Library
201 Donaghey Ave
Conway, AR 72035
501 450-3170
- Athletic Training Prgm
- Dietetic Internship Prgm
- Dietetics-Didactic Prgm
- Occupational Therapy Prgm
- Physical Therapy Prgm
- Speech-Language Pathology Prgm

## El Dorado

**South Arkansas Community College**
Thomas Alan Rasco, EdD, President and CEO
PO Box 7010
300 S West Ave
El Dorado, AR 71731-7010
870 862-8131
- Clin Lab Technician/Med Lab Technician
- Occupational Therapy Asst Prgm
- Physical Therapist Assistant Prgm
- Radiography Prgm

## Fayetteville

**University of Arkansas**
John White, PhD, Chancellor
Adminstration Bldg
Fayetteville, AR 72701
479 575-4148
- Athletic Training Prgm
- Counseling Prgm
- Dietetics-Didactic Prgm
- Rehabilitation Counseling Prgm
- Speech-Language Pathology Prgm

## Forrest City

**East Arkansas Community College**
Coy Grace, President
1700 Newcastle Rd
Forrest City, AR 72335
870 633-4480
- Emergency Med Tech-Paramedic Prgm

## Ft Smith

**University of Arkansas - Fort Smith**
Paul Beran, PhD, Chancellor
5201 Grand Ave
PO Box 3649
Ft Smith, AR 72913
479 788-7004
• Dental Hygiene Prgm
• Radiography Prgm
• Surgical Technology Prgm

## Harrison

**North Arkansas College**
Jeff Olson, EdD, President
1515 Pioneer Dr
Harrison, AR 72601-5599
870 743-3000
• Clin Lab Technician/Med Lab Technician
• Emergency Med Tech-Paramedic Prgm
• Radiography Prgm
• Surgical Technology Prgm

## Helena

**Phillips Community College/U of Arkansas**
Steven Murray, EdD, Chancellor
PO Box 785
1000 Campus Dr
Helena, AR 72342
870 338-6474
• Clin Lab Technician/Med Lab Technician
• Phlebotomist Prgm

## Hope

**Univ of Arkansas Comm Coll at Hope**
Welch Charles, EdD, Chancellor
PO Box 140
2500 S Main
Hope, AR 71801
870 777-5722
• Respiratory Therapist (Entry-Level) Prgm

## Hot Springs

**National Park Community College**
Sally Carder, PhD, President
101 College Dr
Hot Springs, AR 71913-9174
501 760-4222
• Clin Lab Technician/Med Lab Technician
• Emergency Med Tech-Paramedic Prgm
• Health Information Tech Prgm
• Radiography Prgm

## Little Rock

**Baptist Health**
Russell D Harrington, Jr, FACHE, President
11900 Colonel Glenn Rd #1000
Little Rock, AR 72210
501 202-2274
• Clin Lab Scientist/Med Technologist Prgm
• Histotechnician Prgm
• Nuclear Medicine Technology Prgm
• Radiography Prgm
• Surgical Technology Prgm

**Central Arkansas Radiation Therapy Inst**
Janice Burford, CEO/President
PO Box 55050
Little Rock, AR 72215
501 664-8573
• Radiation Therapy Prgm

**St Vincent Infirmary Medical Center**
Ken Haynes, Interim CEO
Two St Vincent Circle
Little Rock, AR 72205-5499
501 552-3910
• Radiography Prgm

**University of Arkansas at Little Rock**
Joel Anderson, Chancellor
2801 S University Ave
Little Rock, AR 72204
501 569-3000
• Orientation and Mobility Specialist Prgm
• Rehabilitation Counseling Prgm
• Rehabilitation Teacher Prgm

**University of Arkansas for Medical Sciences**
I Dodd Wilson, MD, Chancellor
4301 W Markham Slot 541
Little Rock, AR 72205
501 686-5680
• Audiologist Prgm
• Clin Lab Scientist/Med Technologist Prgm
• Cytotechnology Prgm
• Dental Hygiene Prgm
• Diagnostic Med Sonography Prgm
• Dietetic Internship Prgm
• Emergency Med Tech-Paramedic Prgm
• Genetic Counseling Prgm
• Health Information Tech Prgm
• Nuclear Medicine Technology Prgm
• Ophthalmic Med Technologist Prgm
• Radiography Prgm (3)
• Respiratory Therapist (Advanced) Prgm (2)
• Speech-Language Pathology Prgm
• Surgical Technology Prgm

## Magnolia

**Southern Arkansas University**
David F Rankin, PhD CFA, President
100 E University
Magnolia, AR 71753-5000
870 235-4001
• Athletic Training Prgm

## Monticello

**U of Arkansas-Monticello Coll of Tech-McGehee**
H Jack Lassiter, PhD, Chancellor
346 University Dr
PO Box 3596
Monticello, AR 71656
870 460-1020
• Emergency Med Tech-Paramedic Prgm

## North Little Rock

**Pulaski Technical College**
Dan Bakke, EdD, President
3000 W Scenic Rd
North Little Rock, AR 72118
501 812-2216
• Dental Assisting Prgm
• Respiratory Therapist (Entry-Level) Prgm

## Ozark

**Arkansas Tech University - Ozark Campus**
Jo Alice Blondin, PhD, Chancellor
1701 Helberg Lane
Ozark, AR 72949
479 667-2117
• Emergency Med Tech-Paramedic Prgm

## Pine Bluff

**Southeast Arkansas College**
Phil E Shirley, PhD, President
1900 Hazel St
Pine Bluff, AR 71603
870 543-5907
• Emergency Med Tech-Paramedic Prgm
• Radiography Prgm
• Respiratory Therapist (Advanced) Prgm
• Surgical Technology Prgm

**University of Arkansas at Pine Bluff**
Lawrence A Davis, Chancellor
1200 N University Dr
Pine Bluff, AR 71601
501 575-8470
• Dietetics-Didactic Prgm

## Pocahontas

**Black River Technical College**
Richard Gaines, MS, Director
PO Box 468
Pocahontas, AR 72455
870 248-4000
• Dietetic Technician-AD Prgm
• Emergency Med Tech-Paramedic Prgm
• Respiratory Therapist (Entry-Level) Prgm

## Russellville

**Arkansas Tech University**
Robert Charles Brown, PhD, President
Administration 210
1605 Coliseum Dr
Russellville, AR 72801
479 968-0237
• Health Information Admin Prgm
• Medical Assistant Prgm

## Searcy

**Harding University**
David B Burks, President
Searcy, AR 72149
501 279-4000
• Dietetics-Didactic Prgm
• Physician Assistant Prgm

## Springdale

**Northwest Technical Institute**
Charles Kelley, EdD, President
709 S Old Missouri Rd
Springdale, AR 72764
479 751-8824
• Surgical Technology Prgm

INSTITUTIONS

## State University

**Arkansas State University**
Leslie Wyatt, PhD, President
Administration PO Box 10
State University, AR 72467-0010
870 972-3030
- Athletic Training Prgm
- Clin Lab Scientist/Med Technologist Prgm
- Clin Lab Technician/Med Lab Technician
- Counseling Prgm
- Diagnostic Med Sonography Prgm
- Physical Therapist Assistant Prgm
- Physical Therapy Prgm
- Radiation Therapy Prgm
- Radiography Prgm
- Rehabilitation Counseling Prgm
- Speech-Language Pathology Prgm

# British Columbia, Canada

## Langley

**Trinity Western University**
7600 Glover Rd
Langley, BC V2Y 1Y1
604 888-7511
- Counseling Prgm

## Vancouver

**British Columbia's Children's Hospital**
4480 Oak St
Vancouver, BC V6N 3V4
604 875-2111
- Orthoptist Prgm

**University of British Columbia**
Stephen J Toope, PhD, President
T325-Third Floor-Koerner Pavillion
2211 Westbrook Mall
Vancouver, BC V6T 1Z1
- Counseling Prgm
- Genetic Counseling Prgm
- Medical Librarian Prgm

# California

## Alameda

**College of Alameda**
George Herring, President
555 Atlantic Ave
Alameda, CA 94501
510 522-7221
- Dental Assisting Prgm

## Alhambra

**Corinthian College - Bryman Campus**
Melody Rider, BA, School President
2215 Mission Rd
Alhambra, CA 91803
626 979-4940
- Medical Assistant Prgm

## Anaheim

**CSi Bryman College**
Cheryl K Smith, BS, School President
511 N Brookhurst St, Ste 300
Anaheim, CA 92801
714 953-6500
- Medical Assistant Prgm

**North Orange County Community**
Jerome Hunter, PhD, Chancellor
1830 W Romneya Dr
Anaheim, CA 92801
714 808-4797
- Pharmacy Technician Prgm

## Aptos

**Cabrillo College**
Brian King, PhD, President/Superintendent
6500 Soquel Dr
Aptos, CA 95003
831 479-6306
- Dental Hygiene Prgm
- Medical Assistant Prgm
- Radiography Prgm

## Arcata

**Humboldt State University**
Rollin C Richmond, PhD, President
1 Harpst St
Arcata, CA 95521
707 826-3311
- Athletic Training Prgm

## Azusa

**Azusa Pacific University**
Jon R Wallace, DBA, President
901 E Alosta Ave
Azusa, CA 91702
626 812-3031
- Athletic Training Prgm
- Physical Therapy Prgm

## Bakersfield

**Bakersfield College**
William Andrews, EdD, President
1801 Panorama Dr
Bakersfield, CA 93305
661 395-4211
- Emergency Med Tech-Paramedic Prgm
- Radiography Prgm

**Clinica Sierra Vista**
1430 Truxtun Ave, Ste 120
Bakersfield, CA 93301-3834
- Dietetic Internship Prgm

**San Joaquin Valley College - Bakersfield**
Mark A Perry, President
201 New Stine Rd
Bakersfield, CA 93309
805 834-0126
- Respiratory Therapist (Advanced) Prgm
- Surgical Technology Prgm

## Belmont

**Notre Dame de Namur University**
Jack Oblack, President
Ralston Ave
Belmont, CA 94002
415 593-1601
- Art Therapy Prgm

## Berkeley

**University of California - Berkeley**
Robert Berdahl, Chancellor
Berkeley, CA 94720
510 642-6000
- Dietetic Internship Prgm
- Dietetics-Didactic Prgm

## Burlingame

**Mills-Peninsula Health Services**
Robert W Merwin, MS, CEO
1783 El Camino Real
Burlingame, CA 94010
650 696-5678
- Radiography Prgm

## Carson

**California State University - Dominguez Hills**
James E Lyons, Sr, PhD, President
1000 E Victoria St
Carson, CA 90747
310 243-3301
- Clin Lab Scientist/Med Technologist Prgm
- Occupational Therapy Prgm
- Orthotist/Prosthetist Prgm

## Chico

**California State University - Chico**
Manuel A Esteban, President
Chico, CA 95929-0222
916 898-5871
- Dietetic Internship Prgm
- Dietetics-Didactic Prgm
- Speech-Language Pathology Prgm
- Therapeutic Recreation Specialist Prgm

## Chula Vista

**Pima Medical Institute - Chula Vista**
Richard L Luebke, Jr, BS, CEO
780 Bay Blvd, Ste 101
Chula Vista, CA 91910
619 425-3200
- Radiography Prgm

**Southwestern College**
Neil Yoneji, MS, Interim Supt/President
900 Otay Lakes Rd
Chula Vista, CA 91910
619 482-6301
- Dental Hygiene Prgm
- Emergency Med Tech-Paramedic Prgm
- Surgical Technology Prgm

## Citrus Heights

**Western Career College - Citrus**
7301 Greenback Lane, Ste A
Citrus Heights, CA 95621
- Pharmacy Technician Prgm

## Colton

**Arrowhead Regional Medical Center**
June Griffith-Collison, Director
400 N Pepper Ave
Colton, CA 92324-1819
909 580-6160
• Radiography Prgm

## Commerce

**Los Angeles County Paramedic Training Inst**
Cathy Chidester, MSN, Assistant Director
5555 Ferguson Dr, Ste 220
Commerce, CA 90022
323 890-7543
• Emergency Med Tech-Paramedic Prgm

## Concord

**Mt Diablo Adult Education**
Joanne Durkee, Director of Adult Educ
1266 San Carlos Ave
Concord, CA 94518
925 685-7340
• Surgical Technology Prgm

## Costa Mesa

**Orange Coast College**
Robert Dees, MA, President
2701 Fairview Rd
Costa Mesa, CA 92628-5005
714 432-5712
• Cardiovascular Technology Prgm
• Dental Assisting Prgm
• Diagnostic Med Sonography Prgm
• Dietetic Technician-AD Prgm
• Electroneurodiagnostic Tech Prgm
• Medical Assistant Prgm
• Radiography Prgm
• Respiratory Therapist (Advanced) Prgm

**Vanguard University**
Murray Dempster, PhD, President
55 Fair Dr
Costa Mesa, CA 92626
714 556-3610
• Athletic Training Prgm

## Culver City

**West Los Angeles College**
Frank Quiambao, President
9000 Overland Ave
Culver City, CA 90230
310 287-4200
• Dental Hygiene Prgm

## Cypress

**Cypress College**
Marjorie Lewis, PhD, President
9200 Valley View St
Cypress, CA 90630
714 484-7000
• Dental Assisting Prgm
• Dental Hygiene Prgm
• Diagnostic Med Sonography Prgm
• Health Information Tech Prgm
• Radiography Prgm

## Duarte

**City of Hope National Medical Center**
James Miser, MD, CEO
1500 E Duarte Rd
Duarte, CA 91010
626 359-8111
• Radiation Therapy Prgm

## El Cajon

**Grossmont College**
Dean Colli, EdD, Interim President
8800 Grossmont College Dr
El Cajon, CA 92020
619 644-7100
• Cardiovascular Technology Prgm
• Occupational Therapy Asst Prgm
• Respiratory Therapist (Advanced) Prgm

## Encino

**Phillips Graduate Institute**
Lisa Porche-Burke, PhD, President
5445 Balboa Blvd
Encino, CA 91316-1509
818 386-5600
• Art Therapy Prgm

## Eureka

**College of the Redwoods**
Cedric A Sampson, President
Tompkins Hill Rd
Eureka, CA 95501
707 445-6700
• Dental Assisting Prgm

## Foster City

**California EMS Academy Inc**
Nancy L Black, RN MS, President
1098 Foster City Blvd
Ste 106 PMB 708
Foster City, CA 94404
866 577-9197
• Emergency Med Tech-Paramedic Prgm

## Fremont

**Ohlone College**
Douglas Treadway, PhD,
    Superintendent/President
43600 Mission Blvd
PO Box 3909
Fremont, CA 94539
510 659-6200
• Physical Therapist Assistant Prgm
• Respiratory Therapist (Advanced) Prgm

## Fresno

**California State University - Fresno**
John D Welty, EdD, President
5241 N Maple Ave
Fresno, CA 93740-0080
559 278-2423
• Athletic Training Prgm
• Counseling Prgm
• Dietetic Internship Prgm
• Dietetics-Didactic Prgm
• Physical Therapy Prgm
• Rehabilitation Counseling Prgm
• Speech-Language Pathology Prgm

**Fresno City College**
Ned Doffoney, PhD, President
1101 E University Ave
Fresno, CA 93741
559 442-8244
• Dental Hygiene Prgm
• Emergency Med Tech-Paramedic Prgm
• Health Information Tech Prgm
• Radiography Prgm
• Respiratory Therapist (Advanced) Prgm
• Surgical Technology Prgm

**Fresno County Paramedic Program**
Brad Maggy, MPA, Interim Director
1221 Fulton Mall
Fresno, CA 93721
559 445-3200
• Emergency Med Tech-Paramedic Prgm

## Fullerton

**California State University - Fullerton**
Milton A Gordon, PhD, President
PO Box 6810
Fullerton, CA 92834-6810
714 278-3456
• Athletic Training Prgm
• Speech-Language Pathology Prgm

## Garden Grove

**Concorde Career Institute - Garden Grove**
Harry Strong III, MD, Executive Director
12951 Euclid St, Ste 101
Garden Grove, CA 92840
714 620-1005
• Medical Assistant Prgm

## Gardena

**Bryman College - Gardena**
Bill Wherritt, President
1045 W Redondo Beach Blvd, Ste 275
Gardena, CA 90247
310 527-7105
• Medical Assistant Prgm

## Glendale

**Glendale Career College**
Seth Webber, MBA, Campus Director
1015 Grandview Ave
Glendale, CA 91201
818 956-4915
• Surgical Technology Prgm (2)

**Glendale Memorial Hospital and Health Center**
1420 S Central Ave
Glendale, CA 91204
818 502-2334
• Dietetic Internship Prgm

## Glendora

**Citrus College**
Michael J Viera, PhD, President/Superintendent
1000 W Foothill
Glendora, CA 91740
818 963-0323
• Dental Assisting Prgm

## Hayward

**Bryman College - Hayward**
Hector Albizo, President
22336 Main St
Hayward, CA 94541
510 582-9500
• Surgical Technology Prgm

**California State University - East Bay**
Norma S Rees, President
25800 Carlos Bee Blvd
Hayward, CA 94542
510 881-3086
• Speech-Language Pathology Prgm

**Chabot College**
Robert E Carlson, EdD, President
25555 Hesperian Blvd
PO Box 5001
Hayward, CA 94545-5001
510 786-6640
• Dental Hygiene Prgm
• Health Information Tech Prgm
• Medical Assistant Prgm

## Imperial

**Imperial Valley Community College**
Paul Pai, EdD, President
380 E Aten Rd
Imperial, CA 92251
760 355-6219
• Emergency Med Tech-Paramedic Prgm

## Irvine

**Concordia Univesity - Irvine**
Jacob Preus, STM ThD, President
1530 Concordia West
Irvine, CA 92612
949 854-8002
• Athletic Training Prgm

**University of CA-Irvine School of Medicine**
Michael V Drake, MD, Chancellor
510 Administration
Campus Dr
Irvine, CA 92697
949 824-5111
• Genetic Counseling Prgm

## Irwindale

**Premiere Career College**
Fe Ludovico-Aragon, MD, Executive Director
12901 Ramona Blvd
Irwindale, CA 91706
626 814-2080
• Surgical Technology Prgm

**Public Health Foundation Enterprises**
WIC Program
12781 Schabarum Ave
Irwindale, CA 91706-6802
818 856-6376
• Dietetic Internship Prgm

## Kentfield

**College of Marin**
Frances White, PhD, Superintendent/President
835 College Ave
Kentfield, CA 94904
415 457-8811
• Dental Assisting Prgm

## La Puente

**Hacienda LaPuente Adult Education**
Alan Kern, MA, Adult School Director
Willow Campus
14101 E Nelson Ave
La Puente, CA 91746-2640
626 933-3915
• Dental Assisting Prgm
• Respiratory Therapist (Entry-Level) Prgm

## La Verne

**University of La Verne**
Stephen Morgan, EdD, President
1950 Third St
La Verne, CA 91750
909 593-4900
• Athletic Training Prgm

## Lake Forest

**California Paramedic Institute**
Rich Wiederhold, Executive Director
23141 Lake Center Dr
Lake Forest, CA 92630
• Emergency Med Tech-Paramedic Prgm

## Lancaster

**Antelope Valley College**
Steve Buffalo, PhD, President
3041 W Ave K, Admin Bldg
Lancaster, CA 93534
805 943-3241
• Emergency Med Tech-Paramedic Prgm

## Livermore

**Northern California Training Institute**
Floyd Graves, VP Admin & Support Serv
7575 Southfront Rd
Livermore, CA 94550
800 827-0111
• Emergency Med Tech-Paramedic Prgm (3)

## Loma Linda

**Loma Linda University**
Richard H Hart, MD Dr PH, Chancellor
Office of the President
Loma Linda, CA 92350-0001
909 558-4540
• Clin Lab Scientist/Med Technologist Prgm
• Cytotechnology Prgm
• Dental Hygiene Prgm
• Diagnostic Med Sonography Prgm
• Dietetic Technician-AD Prgm
• Dietetics-Coordinated Prgm (2)
• Health Information Admin Prgm
• Occupational Therapy Asst Prgm
• Occupational Therapy Prgm
• Phlebotomist Prgm
• Physical Therapist Assistant Prgm
• Physical Therapy Prgm
• Physician Assistant Prgm
• Radiation Therapy Prgm
• Radiography Prgm
• Respiratory Therapist (Advanced) Prgm
• Speech-Language Pathology Prgm

## Long Beach

**American University of Health Sciences**
Kim Dang, PhD, President/Owner
3501 Atlantic Ave
Long Beach, CA 90807
562 988-2278
• Pharmacy Technician Prgm

**California State University - Long Beach**
F King Alexander, PhD, President
1250 Bellflower Blvd SS/AD 300
Long Beach, CA 90840-0115
562 985-4121
• Athletic Training Prgm
• Dietetic Internship Prgm
• Dietetics-Didactic Prgm
• Kinesiotherapy Prgm
• Phlebotomist Prgm
• Physical Therapy Prgm
• Radiation Therapy Prgm
• Speech-Language Pathology Prgm
• Therapeutic Recreation Specialist Prgm

**Long Beach City College**
E Jan Kehoe, PhD, Superintendent/President
4901 E Carson St
Long Beach, CA 90808
562 938-4121
• Dietetic Technician-AD Prgm

## Los Altos Hills

**Foothill Community College**
Penny Patz, PhD, Interim President
12345 El Monte Rd
Los Altos Hills, CA 94022
650 949-7200
• Dental Assisting Prgm
• Dental Hygiene Prgm
• Diagnostic Med Sonography Prgm
• Emergency Med Tech-Paramedic Prgm
• Pharmacy Technician Prgm
• Radiography Prgm
• Respiratory Therapist (Advanced) Prgm

## Los Angeles

**American Career College**
David A Pyle, President
4021 Rosewood Ave 101
Los Angeles, CA 90004
• Pharmacy Technician Prgm

**Bryman College - Los Angeles**
Mariam Mohammadi, MET, President
3640 Wilshire Blvd, Ste 500
Los Angeles, CA 90010
213 388-9950
• Medical Assistant Prgm

**California State University - Los Angeles**
James M Rosser, President
5151 State University Dr
Los Angeles, CA 90032
323 343-4690
• Audiologist Prgm
• Counseling Prgm
• Dietetics-Coordinated Prgm
• Dietetics-Didactic Prgm
• Orientation and Mobility Specialist Prgm
• Rehabilitation Counseling Prgm
• Speech-Language Pathology Prgm

**Center for Child Development and Nutrition**
Walter W Noce, Jr, MPH, President, CEO
4650 Sunset Blvd
Los Angeles, CA 90027
213 669-2301
• Dietetic Internship Prgm

**Charles R Drew Univ of Med & Science**
Susan Kelly, MD, President/COO
1731 E 120th St
Los Angeles, CA 90059
323 563-4800
• Health Information Tech Prgm
• Pharmacy Technician Prgm
• Physician Assistant Prgm
• Radiography Prgm

**Childrens Hospital Los Angeles**
4650 Sunset Blvd
Los Angeles, CA 90027
• Orthoptist Prgm

**East Los Angeles Occupational Center**
2100 Marengo St
Los Angeles, CA 90033
213 223-1283
• Radiography Prgm

**Jules Stein Eye Institute, UCLA**
Bartly J Mondino, MD, Director
100 Stein Plaza
Los Angeles, CA 90095
• Ophthalmic Assistant Prgm

**Los Angeles City College**
Steve Maradian, PhD, President
855 N Vermont Ave
Los Angeles, CA 90029
323 953-4000
• Dental Lab Technician Prgm
• Dietetic Technician-AD Prgm
• Radiography Prgm

**Los Angeles County+USC Healthcare Network**
Douglas D Bagley, MS, Interim Exec Dir
1200 N State St, Rm 112
Los Angeles, CA 90033
213 226-6501
• Dietetic Internship Prgm

**Loyola Marymount University**
Robert B Lawton, SJ, President
1 LMU Dr
Los Angeles, CA 90045
310 338-2700
• Art Therapy Prgm

**Mount St Mary's College**
Jacqueline Powers Doud, PhD, President
Chalon Campus
12001 Chalon Rd
Los Angeles, CA 90049-1599
310 954-4000
• Physical Therapy Prgm

**UCLA Center for Health Sciences**
Sheldon King, MD, Interim Med Ctr Dir
Administration 17-165 CHS
10833 Le Conte Ave
Los Angeles, CA 90024-1730
310 825-5041
• Cytotechnology Prgm

**UCLA Center for Prehospital Care**
Michael Karpf, MD, Vice Provost
17-165 CHS 10833 LeConte Ave
Los Angeles, CA 90095
310 824-5041
• Emergency Med Tech-Paramedic Prgm

**University of California - Los Angeles**
Albert Carnesale, Chancellor
Box 951405, Murphy Hall 2147
Los Angeles, CA 90095-1405
• Medical Librarian Prgm

**University of Southern California**
Steven B Sample, PhD, President
University Park Campus
110 Bovard ADM-110
Los Angeles, CA 90089
213 740-2111
• Dental Hygiene Prgm
• Occupational Therapy Prgm
• Physical Therapy Prgm
• Physician Assistant Prgm

**VA Greater Los Angeles Healthcare System**
Kenneth J Clark, MA JD, Director
11301 Wilshire Blvd
Los Angeles, CA 90073
310 268-3132
• Dietetic Internship Prgm

## Malibu

**Pepperdine University**
David Davenport, President
24255 Pacific Coast Hwy
Malibu, CA 90263
310 456-4000
• Dietetics-Didactic Prgm

## Marysville

**Yuba Community College**
Nicky Harrington, Superintendent/President
2088 N Beale Rd
Marysville, CA 95901
916 741-6716
• Radiography Prgm

## Merced

**Merced College**
Benjamin Duran, PhD, President/Superintendent
3600 M St
Merced, CA 95348-2898
209 384-6101
• Diagnostic Med Sonography Prgm
• Radiography Prgm

## Mission Viejo

**Saddleback College**
Dixie Bullock, RN MN, President
28000 Marguerite Pkwy
Mission Viejo, CA 92692
949 582-4500
• Emergency Med Tech-Paramedic Prgm

## Modesto

**Modesto Junior College**
Richard Rose, PhD, President
435 College Ave
Modesto, CA 95350-9977
209 575-6067
• Dental Assisting Prgm
• Medical Assistant Prgm
• Respiratory Therapist (Advanced) Prgm

## Monterey

**Monterey Peninsula College**
Kirk Avery, Superintendant/Pres
980 Fremont Ave
Monterey, CA 93940
408 646-4010
• Dental Assisting Prgm

## Monterey Park

**East Los Angeles College**
Ernest H Moreno, MA, President
1301 Avenida Cesar Chavez
Monterey Park, CA 91754
323 265-8662
• Health Information Tech Prgm
• Respiratory Therapist (Advanced) Prgm

## Moorpark

**Moorpark College**
James Walker, PhD, President
7075 Campus Rd
Moorpark, CA 93021
805 378-1400
• Radiography Prgm

**INSTITUTIONS**

## Moreno Valley

**Riverside County Regional Medical Center**
Salvatore Rotella, PhD, President
Moreno Valley Campus
16130 Lasselle St
Moreno Valley, CA 92551-2045
951 571-6166
• Physician Assistant Prgm

## Napa

**Napa State Hospital**
2100 Napa Vallejo Hwy
Napa, CA 94558
707 253-5428
• Dietetic Internship Prgm

**Napa Valley College**
Diane Carey, PhD, President/Superintendent
2277 Napa Vallejo Hwy
Napa, CA 94558
707 253-3360
• Respiratory Therapist (Advanced) Prgm

## National City

**California College San Diego**
Carl Barney, CEO
2423 Hoover Ave
National City, CA 91950
619 477-4800
• Respiratory Therapist (Advanced) Prgm
• Respiratory Therapist (Entry-Level) Prgm

## North Hollywood

**Concorde Career Institute - North Hollywood**
Sam Alahmad, PhD, Executive Campus Director
12412 Victory Blvd
North Hollywood, CA 91606
818 766-8151
• Respiratory Therapist (Entry-Level) Prgm
• Surgical Technology Prgm

## Northridge

**California State University - Northridge**
Jolene Koester, PhD, President
18111 Nordhoff St
Northridge, CA 91330
818 677-2121
• Athletic Training Prgm
• Audiologist Prgm
• Counseling Prgm
• Dietetic Internship Prgm
• Dietetics-Didactic Prgm
• Genetic Counseling Prgm
• Music Therapy Prgm
• Physical Therapy Prgm
• Radiography Prgm
• Speech-Language Pathology Prgm

## Norwalk

**Cerritos Community College**
Noelia Vela, MA, Superintendent/Pres-Intrm
11110 E Alondra Blvd
Norwalk, CA 90650
562 860-2451
• Dental Assisting Prgm
• Dental Hygiene Prgm
• Pharmacy Technician Prgm
• Physical Therapist Assistant Prgm

## Oakland

**Merritt College**
Evelyn Wesley, PhD, President
12500 Campus Dr
Oakland, CA 94619
510 436-2414
• Dietetic Technician-AD Prgm
• Radiography Prgm

**Samuel Merritt College**
Sharon Diaz, MS RN, President
450 30th St
Ste 2840
Oakland, CA 94609-3108
510 869-6512
• Occupational Therapy Prgm
• Physical Therapy Prgm
• Physician Assistant Prgm

## Orange

**Chapman University**
James L Doti, PhD, President
One University Dr
Orange, CA 92866
714 997-6611
• Athletic Training Prgm
• Music Therapy Prgm
• Physical Therapy Prgm

**Univ of California Irvine Med Ctr**
Mark Laret, Executive Director
UCI Medical Ctr
101 City Dr S
Orange, CA 92868-3298
714 456-5678
• Clin Lab Scientist/Med Technologist Prgm

## Oroville

**Butte College**
Diana Van Der Ploeg, PhD, President
3536 Butte Campus Dr
Oroville, CA 95965
530 895-2484
• Emergency Med Tech-Paramedic Prgm
• Respiratory Therapist (Advanced) Prgm

## Oxnard

**Oxnard College**
Steven F Arvizu, PhD, President
4000 S Rose Ave
Oxnard, CA 93033-6699
805 986-5800
• Dental Hygiene Prgm

## Palo Alto

**VA Palo Alto Health Care System**
Elizabeth Joyce Freeman, Director
3801 Miranda Ave
Palo Alto, CA 94304
650 493-5000
• Nuclear Medicine Technology Prgm

## Pasadena

**Pasadena City College**
James Kossler, PhD, President
1570 E Colorado Blvd
Pasadena, CA 91106
626 585-7123
• Dental Assisting Prgm
• Dental Hygiene Prgm
• Dental Lab Technician Prgm
• Medical Assistant Prgm
• Radiography Prgm

## Patton

**Patton State Hospital**
3102 E Highland Ave
Patton, CA 92369
909 425-7297
• Dietetic Internship Prgm

## Petaluma

**Sonoma College**
Joe Keats, Chief Executive Officer
1304 Southpoint Blvd, Suite 280
Petaluma, CA 94954
707 283-0800
• Physical Therapist Assistant Prgm

## Pleasant Hill

**Diablo Valley College**
Mark Edelstein, President
321 Golf Club Rd
Pleasant Hill, CA 94523
510 685-1230
• Dental Assisting Prgm
• Dental Hygiene Prgm

## Pomona

**California State Polytechnic University**
Bob H Suzuki, President
3801 W Temple Ave
Pomona, CA 91768
909 869-2000
• Dietetic Internship Prgm
• Dietetics-Didactic Prgm

**DeVry University - Pomona Campus**
Rose Dishman, Regional Vice President
901 Corporate Center Dr
Pomona, CA 91768
909 868-4180
• Health Information Tech Prgm

**Western Univ of Health Sciences**
Philip Pumerantz, PhD, President
309 E Second St
Pomona, CA 91766-1854
909 469-5200
• Physical Therapy Prgm
• Physician Assistant Prgm

## Porterville

**Porterville Development Center**
Porterville, CA 93258
209 782-2753
• Dietetic Internship Prgm

## Rancho Cucamonga

**Chaffey College**
Marie Kane, PhD, Superintendent/President
5885 Haven Ave
Rancho Cucamonga, CA 91737
909 941-2100
• Dental Assisting Prgm
• Dietetic Technician-AD Prgm
• Radiography Prgm

## Rancho Mirage

**Eisenhower Memorial Hospital**
Andrew Deems, President/CEO
39000 Bob Hope Dr
Rancho Mirage, CA 92270
619 340-3911
• Clin Lab Scientist/Med Technologist Prgm

## Redding

**Shasta College**
Douglas M Treadway, Superintendent/President
PO Box 496006
Redding, CA 96049-6006
530 225-4600
• Dental Hygiene Prgm

## Redlands

**American College of Health Professions**
Michele Brooks, President/CEO
700 E Redlands Blvd, #U227
Redlands, CA 92373
909 307-6022
• Surgical Technology Prgm

**University of Redlands**
Stuart Dorsey, President
PO Box 3080
1200 E Colton Ave
Redlands, CA 92373-0999
909 793-2121
• Speech-Language Pathology Prgm

## Redwood City

**Canada College**
Tom Mohr, President
4200 Farm Hill Blvd
Redwood City, CA 94061
650 306-3283
• Radiography Prgm

## Reedley

**Central Valley WIC Dietetic Internship**
1560 E Manning Ave
Reedley, CA 93654
• Dietetic Internship Prgm

## Reseda

**Bryman College - Reseda**
Steven R Schilling, BA MA, President
18040 Sherman Way
Reseda, CA 91335
878 774-0550
• Surgical Technology Prgm

## Richmond

**Kaiser Permanente Sch of Allied Hlth Sciences**
Gwenette S Jackson, CRT, School Administrator
938 Marina Way South
Richmond, CA 94804
510 307-2412
• Diagnostic Med Sonography Prgm
• Nuclear Medicine Technology Prgm
• Radiation Therapy Prgm
• Radiography Prgm

## Riverside

**Riverside Comm Coll - Moreno Valley Campus**
Salvatore G Rotella, PhD, President
4800 Magnolia Ave
Riverside, CA 92506-1299
909 222-8000
• Dental Hygiene Prgm
• Emergency Med Tech-Paramedic Prgm

## Rohnert Park

**Sonoma State University**
1801 E Cotati Ave
Rohnert Park, CA 94928-3609
707 664-2880
• Counseling Prgm

## Sacramento

**American River College**
David Viar, EdD, President
4700 College Oak Dr
Sacramento, CA 95841
916 484-8211
• Emergency Med Tech-Paramedic Prgm
• Respiratory Therapist (Advanced) Prgm

**California State University - Sacramento**
Alexander Gonzalez, PhD, President
CSUS 6000 J St
Sacramento, CA 95819
916 278-7737
• Athletic Training Prgm
• Audiologist Prgm
• Dietetic Internship Prgm
• Dietetics-Didactic Prgm
• Physical Therapy Prgm
• Rehabilitation Counseling Prgm
• Speech-Language Pathology Prgm

**Charles A Jones Skills & Business Ed Center**
Kirk Williams, CEO/Principal
5451 Lemon Hill Ave
Sacramento, CA 95824-1529
• Pharmacy Technician Prgm

**Cosumnes River Community College**
Francisco Rodriguez, PhD, President
8401 Center Pkwy
Sacramento, CA 95823
916 688-7321
• Dietetic Technician-AD Prgm
• Health Information Tech Prgm
• Medical Assistant Prgm

**Sacramento City College**
Art Tylor, PhD, President
3835 Freeport Blvd
Sacramento, CA 95822-1386
916 558-2100
• Dental Assisting Prgm
• Dental Hygiene Prgm
• Occupational Therapy Asst Prgm
• Physical Therapist Assistant Prgm

**UC Davis Medical Center**
2315 Stockton Blvd
Sacramento, CA 95817
• Dietetic Internship Prgm

**Univ of California Davis Health System**
William McGowan, Acting COO
2315 Stockton Blvd
Sacramento, CA 95817
916 453-0750
• Clin Lab Scientist/Med Technologist Prgm

**University of California - Davis**
Claire Pomeroy, MD, Dean
School of Medicine
2315 Stockton Blvd
Sacramento, CA 95817
916 734-7131
• Dietetics-Didactic Prgm
• Physician Assistant Prgm

**Western Career College - Sacramento**
Greg Nathanson, BA, President
7801 Folsom Blvd, Ste 210
Sacramento, CA 95826-9823
916 388-2800
• Dental Hygiene Prgm
• Medical Assistant Prgm
• Pharmacy Technician Prgm

## Salinas

**Hartnell College**
Edward J Valeau, EdD, President
156 Homestead Ave
Salinas, CA 93901
831 755-6700
• Clin Lab Technician/Med Lab Technician

## San Bernardino

**Bryman College - San Bernardino**
217 E Club Center Dr
San Bernardino, CA 92408
• Medical Assistant Prgm

**California State University - San Bernardino**
Anthony H Evans, President
5500 University Pkwy
San Bernardino, CA 92407
909 880-5000
• Dietetics-Didactic Prgm
• Rehabilitation Counseling Prgm

INSTITUTIONS

**Concorde Career College - San Bernardino**
Robert Bosic, JD, Executive Campus Director
201 E Airport Dr, Ste A
San Bernardino, CA 92408
909 884-8891
• Medical Assistant Prgm

**San Bernardino Valley College**
Donald L Singer, President
701 S Mount Vernon Ave
San Bernardino, CA 92410
909 888-6511
• Dietetic Technician-AD Prgm

## San Bruno

**Skyline College**
Victoria Morrow, PhD, President
3300 College Dr
San Bruno, CA 94066
650 738-4111
• Respiratory Therapist (Advanced) Prgm
• Surgical Technology Prgm

## San Diego

**Concorde Career College**
Timothy Vogeley, MBA, Campus President
4393 Imperial Ave
Ste 100
San Diego, CA 92113
619 688-0800
• Surgical Technology Prgm

**Maric College**
Craig Wood, Director of Operations
16835 W Bernardo Dr #205
San Diego, CA 92127
858 654-3602
• Radiography Prgm

**Mueller College of Holistic Massage
  Therapies**
Jeff Welsh, PhD MA HHP, President and CEO
4607 Park Blvd
San Diego, CA 92116
619 291-9811
• Massage Therapy Prgm

**Naval School of Health Sciences, San Diego**
FC Crosby, CAPT, MSC USN, Commanding Officer
34101 Farenholt Ave
San Diego, CA 92134—529
619 532-7700
• Cardiovascular Technology Prgm
• Clin Lab Technician/Med Lab Technician
• Radiography Prgm
• Surgical Technology Prgm

**Point Loma Nazarene University**
Bob Brower, PhD, President
3900 Lomaland Dr
San Diego, CA 92106
619 849-2216
• Athletic Training Prgm

**San Diego Mesa College**
Rita Cepeda, EdD, President
7250 Mesa College Dr
San Diego, CA 92111
619 388-2755
• Dental Assisting Prgm
• Health Information Tech Prgm
• Medical Assistant Prgm
• Physical Therapist Assistant Prgm
• Radiography Prgm

**San Diego State University**
Stephen L Weber, PhD, President
5500 Campanile Dr
San Diego, CA 92182-1518
619 594-7746
• Athletic Training Prgm
• Audiologist Prgm
• Dietetic Internship Prgm
• Dietetics-Didactic Prgm
• Kinesiotherapy Prgm
• Rehabilitation Counseling Prgm
• Speech-Language Pathology Prgm

**Univ of California San Diego Med Ctr**
Richard Liekweg, Director
200 W Arbor Dr, H-910C
San Diego, CA 92103-8970
619 543-6654
• Diagnostic Med Sonography Prgm

**VA San Diego Healthcare System**
3350 La Jolla Village Dr
San Diego, CA 92161
619 552-8585
• Dietetic Internship Prgm

## San Francisco

**Bryman College - San Francisco**
Thomas Horstman, MA, Executive Director
814 Mission St
San Francisco, CA 94103
415 777-2500
• Medical Assistant Prgm

**City College of San Francisco**
Philip R Day, Jr, PhD, Chancellor
50 Phelan Ave
San Francisco, CA 94112
415 239-3000
• Dental Assisting Prgm
• Emergency Med Tech-Paramedic Prgm
• Health Information Tech Prgm
• Medical Assistant Prgm
• Radiation Therapy Prgm
• Radiography Prgm

**Heald College**
Amy McCombs, BA MA, Interim President/CEO
670 Howard St
San Francisco, CA 94105
415 808-1400
• Dental Assisting Prgm (3)
• Medical Assistant Prgm (9)

**San Francisco State University**
Robert A Corrigan, PhD, President
1600 Holloway Ave
San Francisco, CA 94132
415 338-1381
• Audiologist Prgm
• Clin Lab Scientist/Med Technologist Prgm
• Counseling Prgm
• Dietetic Internship Prgm
• Dietetics-Didactic Prgm
• Orientation and Mobility Specialist Prgm
• Rehabilitation Counseling Prgm
• Speech-Language Pathology Prgm

**UCSF Medical Center**
Mark Laret, MA, CEO
500 Parnassus Ave
San Francisco, CA 94143
415 353-2733
• Surgical Technology Prgm

**University of California - San Francisco**
Hailet Debas, MD, Chancellor
Box 0402
San Francisco, CA 94143
415 476-2401
• Dietetic Internship Prgm
• Physical Therapy Prgm

## San Jose

**Central County Occupation Center**
Richard Friberg, Superintendent
760 Hillsdale Ave
San Jose, CA 95136-1190
408 997-2360
• Surgical Technology Prgm

**San Jose City College**
Chui Tsang, President
2100 Moorpark Ave
San Jose, CA 95128
408 298-2181
• Dental Assisting Prgm

**San Jose State University**
Don W Kassing, PhD, President
Tower Hall 206
One Washington Square
San Jose, CA 95192-0002
408 924-1177
• Athletic Training Prgm
• Clin Lab Scientist/Med Technologist Prgm
• Dietetic Internship Prgm
• Dietetics-Didactic Prgm
• Medical Librarian Prgm
• Occupational Therapy Prgm
• Speech-Language Pathology Prgm

**Western Career College - San Jose**
Steve Shishani, MA, Executive Director
6201 San Ignacio Ave
San Jose, CA 95119
408 360-0840
• Medical Assistant Prgm
• Pharmacy Technician Prgm

**WestMed College**
Veronica Shepardson, RN, School Director
1330 S Bascom Ave, Ste G
San Jose, CA 95128
408 977-0723
• Emergency Med Tech-Paramedic Prgm

## San Leandro

**Western Career College - San Leandro**
Greg Nathanson, President
15555 E 14th St, Ste 500
San Leandro, CA 94578-9930
510 276-3888
• Medical Assistant Prgm
• Pharmacy Technician Prgm

## San Luis Obispo

**California Polytechnic State University**
Warren J Baker, President
San Luis Obispo, CA 93407
805 756-1111
• Counseling Prgm
• Dietetics-Didactic Prgm

**Central California School of Continuing Edu**
Gene R Appleby, Administrator
3195 McMillan #F
San Luis Obispo, CA 93401
805 543-9123
• Radiography Prgm

## San Marcos

**Palomar Community College**
Robert Deegan, EdD, Superintendent/President
1140 W Mission Rd
San Marcos, CA 92069
760 744-1150
• Dental Assisting Prgm
• Emergency Med Tech-Paramedic Prgm

## San Mateo

**College of San Mateo**
Peter J Landsberger, President
1700 W Hillsdale Blvd
San Mateo, CA 94402
415 574-6161
• Dental Assisting Prgm

**Hospital Consortium Education Network**
Francine Serafin-Dickson, MBA RN,
 Executive Director
222 W 39th Ave
Ste 3A09, San Mateo Medical Center
San Mateo, CA 94403
650 573-3930
• Emergency Med Tech-Paramedic Prgm

## San Pablo

**Contra Costa College**
D Candy Rose, President
2600 Mission Bell Dr
San Pablo, CA 94806
510 235-7800
• Dental Assisting Prgm

## San Rafael

**Dominican University of California**
Joseph R Fink, PhD, President
50 Acacia Ave
San Rafael, CA 94901-2298
415 485-3200
• Occupational Therapy Prgm

## Santa Ana

**HealthStaff Training Institute**
Diana Standoff, Vice President
1505 E 17th St, Ste 122
Santa Ana, CA 92701
• Pharmacy Technician Prgm

**Newbridge College**
J Ramon Villanueva, School Director
1840 E 17th St, Ste 140
Santa Ana, CA 92705
714 550-8000
• Surgical Technology Prgm

**Santa Ana College**
Erlinda Martinez, EdD, President
1530 W 17th St
Santa Ana, CA 92706-3398
714 564-6975
• Occupational Therapy Asst Prgm
• Pharmacy Technician Prgm

## Santa Barbara

**Santa Barbara City College**
John Romo, Superintendent/President
721 Cliff Dr
Santa Barbara, CA 93109-2394
805 965-0581
• Health Information Tech Prgm
• Radiography Prgm

**Santa Barbara Cottage Hospital**
James L Ash, President/CEO
PO Box 689 Pueblo at Bath Sts
Santa Barbara, CA 93102
805 569-7290
• Clin Lab Scientist/Med Technologist Prgm

## Santa Cruz

**Emergency Training Services Inc**
David Barbin, BS, CEO
3050 Paul Sweet Rd
Santa Cruz, CA 95065
831 476-8813
• Emergency Med Tech-Paramedic Prgm

## Santa Rosa

**Santa Rosa Junior College**
Robert F Agrella, EdD, Superintendent/President
1501 Mendocino Ave
Santa Rosa, CA 95401
707 527-4431
• Dental Assisting Prgm
• Dental Hygiene Prgm
• Emergency Med Tech-Paramedic Prgm
• Radiography Prgm

## Saratoga

**West Valley Community College District**
Phil Hartley, PhD, President
14000 Fruitvale Ave
Saratoga, CA 95070
408 867-2200
• Medical Assistant Prgm

## Simi Valley

**Simi Valley Adult School**
Sondra Jones, MA, Director
3192 Los Angeles Ave
Simi Valley, CA 93065
805 579-6200
• Respiratory Therapist (Entry-Level) Prgm
• Surgical Technology Prgm

## South Gate

**Career Colleges of America**
Jeff Meisel, CEO
5612 E Imperial Hwy
South Gate, CA 90280
562 861-8702
• Surgical Technology Prgm (3)

## Stanford

**Stanford University School of Medicine**
Phillip Pizzo, MD, Dean
School of Medicine Rm M121
Stanford, CA 94305-5302
415 723-6436
• Physician Assistant Prgm

## Stockton

**Emergency Medical Sciences Training
 Institute**
Craig Stroup, BS EMT-P, Program Director
343 E Main St #906
Stockton, CA 95202
209 461-5550
• Emergency Med Tech-Paramedic Prgm

**San Joaquin General Hospital**
Richard Aldred, Hospital Director
PO Box 1020
Stockton, CA 95201
209 468-6600
• Radiography Prgm

**University of the Pacific**
Donald De Rosa, President
3601 Pacific Ave
Stockton, CA 95211
209 946-2222
• Athletic Training Prgm
• Music Therapy Prgm
• Physical Therapy Prgm
• Speech-Language Pathology Prgm

**Western Career College**
Dave Semrau, BS, Executive Director
1313 W Robinhood Dr, Ste B
Stockton, CA 95207
209 356-1240
• Medical Assistant Prgm (3)
• Pharmacy Technician Prgm (4)

## Sylmar

**Olive View/UCLA Medical Center**
Melinda D Anderson, CEO
14445 Olive View Dr, Rm 2C155
Sylmar, CA 91342-1495
818 364-4224
• Dietetic Internship Prgm

## Taft

**Taft College**
Darnell Roe, EdD, President/Superintendent
29 Emmons Park Dr
Box 1437
Taft, CA 93268
661 763-7700
• Dental Hygiene Prgm

## Thousand Oaks

**California Lutheran University**
Howard Wennes, PhD, Interim President
60 W Olsen Rd, MC-1300
Thousand Oaks, CA 91360
805 493-3100
• Athletic Training Prgm

## Torrance

**El Camino College**
Thomas M Fallo, Superintendent/President
16007 Crenshaw Blvd
Torrance, CA 90506
310 532-3670
• Radiography Prgm
• Respiratory Therapist (Entry-Level) Prgm

**LA County Harbor UCLA Medical Center**
Tecla Mickoseff, MBA, Acting Hosp Admin
1000 W Carson St Box 1
Torrance, CA 90509-2910
310 222-2101
• Nuclear Medicine Technology Prgm
• Radiography Prgm

**Southern California Regional Occupational Ctr**
Christine Hoffman, EdD, Superintendent
2300 Crenshaw Blvd
Torrance, CA 90501
310 224-4220
• Medical Assistant Prgm

## Ukiah

**Mendocino Community College**
Marylyn G Brock, EdD, President
1000 Hensley Creek Rd
Ukiah, CA 95482
707 468-3071
• Emergency Med Tech-Paramedic Prgm

## Vallejo

**Touro University Mare Island**
Bernard Laner, President
1310 Johnson Lane
Vallejo, CA 94592
707 638-5442
• Physician Assistant Prgm

## Valley Glen

**Los Angeles Valley College**
Tyree Weider, EdD, President
5800 Fulton Ave
Valley Glen, CA 91401-4096
818 947-2321
• Respiratory Therapist (Advanced) Prgm

## Ventura

**Ventura College**
Larry Claderon, EdD, President
School of Prehospital & Emergency Med
4667 Telegraph Rd
Ventura, CA 93003
805 654-6460
• Emergency Med Tech-Paramedic Prgm

## Victorville

**Victor Valley Community College District**
Patricia Spencer, PhD, President/Superintendent
18422 Bear Valley Rd
Victorville, CA 92392-5849
760 245-4271
• Emergency Med Tech-Paramedic Prgm
• Respiratory Therapist (Advanced) Prgm

## Visalia

**San Joaquin Valley College**
Mark Perry, President
3828 W Caldwell Ave
Visalia, CA 93277
559 734-9000
• Dental Hygiene Prgm
• Physician Assistant Prgm
• Respiratory Therapist (Advanced) Prgm (2)
• Surgical Technology Prgm

## Walnut

**Mt San Antonio College**
Christopher C O'Hearn, MA MEd, President/CEO
1100 N Grand Ave
Walnut, CA 91789
909 594-5611
• Emergency Med Tech-Paramedic Prgm
• Histotechnician Prgm
• Radiography Prgm
• Respiratory Therapist (Advanced) Prgm

## West Covina

**East San Gabriel Valley Occupational Pgm**
Laurel Adler, EdD, Superintendent
1501 W Del Norte Ave
West Covina, CA 91790
626 472-5121
• Medical Assistant Prgm

**North-West College**
Marsha Fuerst, President/CEO
2121 W Garvey Ave N
West Covina, CA 91790
626 960-5046
• Pharmacy Technician Prgm (4)

## Yucaipa

**Crafton Hills College**
Gloria Macias Harrison, MA, President
11711 Sand Canyon Rd
Yucaipa, CA 92399
909 389-3200
• Emergency Med Tech-Paramedic Prgm
• Respiratory Therapist (Advanced) Prgm
• Respiratory Therapist (Entry-Level) Prgm

# Colorado

## Alamosa

**Adams State College**
Richard Wueste, JD, President
208 Edgemont Blvd
Richardson Hall 210
Alamosa, CO 81102-0001
719 587-7341
• Counseling Prgm

## Aurora

**Concorde Career Institute - Aurora**
Don Roberts, BA, Interim Campus Director
111 N Havana St
Aurora, CO 80010
303 861-1151
• Radiography Prgm
• Surgical Technology Prgm

**Parks College - Aurora Campus**
Pat Hardy, President
14280 E Jewell Ave
Aurora, CO 80014
303 745-6244
• Medical Assistant Prgm

**T H Pickens Technical Center**
Art Bogardus, PhD, Executive Director
Career and Technical Education
500 Airport Blvd
Aurora, CO 80011
303 344-4910
• Dental Assisting Prgm
• Respiratory Therapist (Advanced) Prgm
• Respiratory Therapist (Entry-Level) Prgm

## Boulder

**Naropa University**
Thomas Coburn, PhD, President
2130 Arapahoe Ave
Boulder, CO 80302—669
303 444-0202
• Art Therapy Prgm
• Dance/Movement Therapy Prgm

**University of Colorado at Boulder**
G P "Bud" Peterson, Chancellor
Campus Box 409
Boulder, CO 80309-0409
303 492-6445
• Audiologist Prgm
• Speech-Language Pathology Prgm

## Colorado Springs

**Centura Health/Penrose-St Francis Hlth Serv**
Rick O'Connell, MBA, CEO
Penrose Hospital, Member Centura Health
2222 N Nevada Ave, PO Box 7021
Colorado Springs, CO 80907-7021
719 776-5111
• Clin Lab Scientist/Med Technologist Prgm
• Dietetic Internship Prgm

**Everest College**
Larry Jackson, MA BA, President
1815 Jet Wing Dr
Colorado Springs, CO 80916
719 638-6580
- Medical Assistant Prgm
- Surgical Technology Prgm

**IntelliTec Medical Institute**
2345 N Academy Blvd
Colorado Springs, CO 80909
- Dental Assisting Prgm

**Memorial Hospital**
Richard Eitel, MBA, Exec Director
1400 East Boulder
175 S Union, Ste 240
Colorado Springs, CO 80909
719 365-5000
- Radiography Prgm

**Pikes Peak Community College**
Edwin Ray, PhD, VP of Educational Service
5765 S Academy Blvd
Colorado Springs, CO 80906
719 540-7218
- Dental Assisting Prgm
- Emergency Med Tech-Paramedic Prgm

**University of Colorado at Colorado Springs**
Pam Shockley Zalabat, PhD, Chancellor
1420 Austin Bluffs Parkway
Colorado Springs, CO 80933-7150
719 262-3000
- Counseling Prgm

# Denver

**Centura Health-St Anthony Hospitals**
George Zara, CEO, St Anthony Hospitals
4231 W 16th Ave
Denver, CO 80204
303 629-4350
- Emergency Med Tech-Paramedic Prgm
- Radiography Prgm

**Colorado Health Foundation**
Anne Warhover, MD, President/CEO
600 S Cherry St #217
Denver, CO 80246
303 322-3515
- Clin Lab Scientist/Med Technologist Prgm

**Community College of Aurora**
Les Moroye, MA, Assoc VP
9235 E 10th Dr, Room 118
Denver, CO 80230
303 340-7119
- Emergency Med Tech-Paramedic Prgm

**Community College of Denver**
Christine Johnson, PhD, President
PO Box 173363
Denver, CO 80217
303 556-2600
- Dental Hygiene Prgm
- Medical Assistant Prgm
- Radiography Prgm

**Denver Health Medical Center**
Patricia Gabow, MD, CEO
660 Bannock St Mail Code 0278
Denver, CO 80204
303 436-6611
- Emergency Med Tech-Paramedic Prgm

**Emily Griffith Opportunity School**
Sharon Johnson
1250 Welton St
Denver, CO 80204
303 575-4721
- Dental Assisting Prgm

**Heritage College**
Richard Shepard, BS, President
4704 Harlan St #100
Denver, CO 80212
720 855-6014
- Surgical Technology Prgm

**Massage Therapy Institute of Colorado**
Mark Manton, Director
1441 York St, Ste 301
Denver, CO 80206
- Massage Therapy Prgm

**Metropolitan State College of Denver**
Raymond Kieft, EdD, Interim President
Campus Box 1
PO Box 173362
Denver, CO 80217-3362
303 556-3022
- Athletic Training Prgm

**Regis University**
Michael J Sheeran, SJ, President
3333 Regis Blvd
Denver, CO 80221-1099
303 458-4190
- Health Information Admin Prgm
- Physical Therapy Prgm

**U of Colorado at Denver and Health Scis Ctr**
Hank Brown, MS, President
4200 E. Ninth Ave
Denver, CO 80262
303 372-2000
- Counseling Prgm
- Dental Hygiene Prgm
- Diagnostic Med Sonography Prgm
- Genetic Counseling Prgm
- Physical Therapy Prgm
- Physician Assistant Prgm

**Westwood College**
Tony Caggiano, Campus President
7350 N Broadway
Denver, CO 80221-3653
800 992-5050
- Medical Assistant Prgm

# Durango

**Ft Lewis College**
Brad Bartel, PhD, President
1000 Rim Dr
Durango, CO 81301-3999
970 247-7100
- Athletic Training Prgm

# Eaton

**Academy of Natural Therapy**
123 Elm Ave
PO Box 237
Eaton, CO 80615
- Massage Therapy Prgm

# Englewood

**HealthONE EMS**
Mary White, President
501 E Hampden
Englewood, CO 80110
303 788-6484
- Emergency Med Tech-Paramedic Prgm

**Tri-County Health Department**
7000 E Bellview Ave, Ste 301
Englewood, CO 80111-1628
303 220-9200
- Dietetic Internship Prgm

# Ft Collins

**Colorado State University**
Larry Penley, PhD, President
102 Administration Bldg
Ft Collins, CO 80523
970 491-6211
- Counseling Prgm
- Dietetic Internship Prgm
- Dietetics-Didactic Prgm
- Music Therapy Prgm
- Occupational Therapy Prgm

# Ft Morgan

**Morgan Community College**
John R McKay, PhD, President
17800 Rd 20
Ft Morgan, CO 80701-4399
970 867-3081
- Physical Therapist Assistant Prgm

# Grand Junction

**Mesa State College**
Timothy Foster, JD, President
1100 North Avenue
Grand Junction, CO 81501
970 248-1498
- Athletic Training Prgm
- Radiography Prgm

# Greeley

**Aims Community College**
Marilynn Liddell, PhD, President
PO Box 69
Greeley, CO 80632
970 330-8008
- Radiography Prgm

**University of Northern Colorado**
Kay Norton, JD, President
Carter Hall 4000
Greeley, CO 80639
970 351-2121
- Athletic Training Prgm
- Audiologist Prgm
- Counseling Prgm
- Dietetic Internship Prgm
- Dietetics-Didactic Prgm
- Orientation and Mobility Specialist Prgm
- Rehabilitation Counseling Prgm
- Speech-Language Pathology Prgm

INSTITUTIONS

## Lakewood

**Red Rocks Community College**
Cliff Richardson, PhD, President
13300 W Sixth Ave
Lakewood, CO 80228-1255
303 914-6600
- Medical Assistant Prgm
- Physician Assistant Prgm
- Radiography Prgm

## Littleton

**Arapahoe Community College**
Berton Glandon, PhD, President
5900 S Santa Fe Dr
PO Box 9002
Littleton, CO 80160-9002
303 797-5701
- Clin Lab Technician/Med Lab Technician
- Health Information Tech Prgm
- Medical Assistant Prgm
- Pharmacy Technician Prgm
- Physical Therapist Assistant Prgm

**Denver Seminary**
Craig Williford, PhD
6399 S Santa Fe Dr
Littleton, CO 80120
303 761-2482
- Counseling Prgm

## Pueblo

**Colorado State University - Pueblo**
Joe Garcia, PhD, President
Adm 301 - 2200 Bonforte Blvd
Pueblo, CO 81001
719 549-2306
- Athletic Training Prgm

**Parkview Medical Center**
C W Smith, MBA, President/CEO
400 W 16th St
Pueblo, CO 81003
719 584-4573
- Clin Lab Scientist/Med Technologist Prgm

**Pueblo Community College**
Marjorie Villani, MA, Interim President
900 W Orman Ave
Pueblo, CO 81004-1499
719 549-3213
- Dental Assisting Prgm
- Dental Hygiene Prgm
- Emergency Med Tech-Paramedic Prgm
- Occupational Therapy Asst Prgm
- Physical Therapist Assistant Prgm
- Respiratory Therapist (Advanced) Prgm

## Rangley

**Colorado Northwestern Community College**
John Boyd, President
500 Kennedy Dr
Rangley, CO 81648
970 675-2261
- Dental Hygiene Prgm

## Thornton

**Parks College**
Allan Short, PhD, Director
9065 Grant St
Thornton, CO 80229
303 457-2757
- Medical Assistant Prgm

## Westminster

**Front Range Comm College**
Karen Reinertson, PhD, President
3645 W 112th Ave
Westminster, CO 80030
303 404-5422
- Dental Assisting Prgm
- Dietetic Technician-AD Prgm
- Medical Assistant Prgm
- Pharmacy Technician Prgm

# Connecticut

## Branford

**Branford Hall Career Institute**
Gary Camp, CEO
One Summit Pl
Branford, CT 06405
203 488-2525
- Medical Assistant Prgm

## Bridgeport

**Bridgeport Hospital**
Robert Trefry, President
267 Grant St
Bridgeport, CT 06610
203 384-3464
- Surgical Technology Prgm

**Housatonic Community College**
Janis M Wertz-Hadley, EdD, President
900 Lafayette Blvd
Bridgeport, CT 06604-4704
203 332-5224
- Clin Lab Technician/Med Lab Technician
- Occupational Therapy Asst Prgm

**St Vincent's College**
John K Fisher, EdD, President
2800 Main St
Bridgeport, CT 06606
203 576-5578
- Medical Assistant Prgm
- Radiography Prgm

**U of Bridgeport/Fones Sch of Dental Hygiene**
Neil Albert Salonen, President
30 Hazel St
Bridgeport, CT 06601
203 576-4000
- Dental Hygiene Prgm

## Danbury

**Danbury Hospital**
Frank Kelly, President/CEO
24 Hospital Ave
Danbury, CT 06810
203 797-7210
- Clin Lab Scientist/Med Technologist Prgm
- Dietetic Internship Prgm
- Radiography Prgm
- Surgical Technology Prgm

**Western Connecticut State University**
181 White St
Danbury, CT 06810-6885
203 837-8200
- Counseling Prgm

## Danielson

**Quinebaug Valley Community College**
Dianne E Williams, BS MS, President
742 Upper Maple St
Danielson, CT 06239
860 774-1160
- Medical Assistant Prgm

## East Hartford

**Goodwin College**
Mark E Scheinberg, BA, President
745 Burnside Ave
East Hartford, CT 06108
860 528-4111
- Histotechnician Prgm
- Medical Assistant Prgm

## Enfield

**Porter and Chester Institute - Enfield**
Henry Kamerzel, Executive Director
138 Weymouth Rd
Enfield, CT 06082
203 741-2561
- Dental Assisting Prgm

## Fairfield

**Fairfield University**
1073 N Benson Rd
Fairfield, CT 06430-5195
203 254-4000
- Counseling Prgm

**Sacred Heart University**
Anthony J Cernera, PhD, President
5151 Park Ave
Fairfield, CT 06825
203 371-7999
- Athletic Training Prgm
- Occupational Therapy Prgm
- Physical Therapy Prgm

## Farmington

**Tunxis Community College**
Cathryn L Addy, PhD, President
271 Scott Swamp Rd
Farmington, CT 06032-3187
860 255-3500
- Dental Assisting Prgm
- Dental Hygiene Prgm

**University of Connecticut Health Center**
Peter Deckers, MD, EVP Health Affairs
263 Farmington Ave
Farmington, CT 06030
860 679-2594
• Cytotechnology Prgm

## Hamden

**Eli Whitney Vocational Technical School**
E Paulett Moore, Director
71 Jones Rd
Hamden, CT 06514
203 397-4031
• Dental Assisting Prgm
• Surgical Technology Prgm

**Quinnipiac University**
John L Lahey, PhD, President
275 Mt Carmel Ave
Hamden, CT 06518-1908
203 582-3344
• Athletic Training Prgm
• Occupational Therapy Prgm
• Pathologists' Assistant Prgm
• Perfusion Prgm
• Physical Therapy Prgm
• Physician Assistant Prgm
• Radiography Prgm
• Respiratory Therapist (Advanced) Prgm

**Stone Academy**
William Mangini, School Director
1315 Dixwell Ave
Hamden, CT 06514
203 288-7474
• Medical Assistant Prgm

## Hartford

**A I Prince Technical High School**
Laura Vega, PhD, Director
500 Bookfield St
Hartford, CT 06106
860 951-7112
• Dental Assisting Prgm
• Surgical Technology Prgm

**Capital Community College**
Calvin Woodland, EdD PsyD, President
950 Main St
Hartford, CT 06103
860 906-5000
• Emergency Med Tech-Paramedic Prgm
• Medical Assistant Prgm
• Radiography Prgm

**Hartford Hospital**
John J Meehan, MHA, President/CEO
80 Seymour St
PO Box 5037
Hartford, CT 06102-5037
860 545-2100
• Clin Lab Scientist/Med Technologist Prgm
• Radiation Therapy Prgm
• Radiography Prgm

**St Francis Hospital & Medical Center**
David D'Eramo, PhD, President
114 Woodland St
Hartford, CT 06105
860 714-4900
• Diagnostic Med Sonography Prgm

## Manchester

**Manchester Community College**
Jonathan Daube, EdD, President
Great Path Mail Station 1
PO Box 1046
Manchester, CT 06045-1046
860 512-3100
• Occupational Therapy Asst Prgm
• Respiratory Therapist (Advanced) Prgm
• Surgical Technology Prgm

## Middletown

**Middlesex Community College**
Wilfredo Nieves, EdD, President
100 Training Hill Rd
Middletown, CT 06457
860 343-5701
• Ophthalmic Dispensing Optician Prgm
• Radiography Prgm

## New Britain

**Central Connecticut State University**
Jack Miller, EdD, President
1615 Stanley St
New Britain, CT 06050-4010
860 832-3003
• Athletic Training Prgm
• Rehabilitation Counseling Prgm

**New England Tech Institute**
Craig Avery, BS, Regional Exec Director
200 John Downey Dr
New Britain, CT 06051
860 225-8641
• Medical Assistant Prgm

## New Haven

**Albertus Magnus College**
Julia M McNamara, PhD
700 Prospect St
New Haven, CT 06511
203 773-8529
• Art Therapy Prgm

**Gateway Community College**
Dorsey Kendrick, PhD, President
60 Sargent Dr
New Haven, CT 06511
203 285-2060
• Dietetic Technician-AD Prgm
• Nuclear Medicine Technology Prgm
• Radiation Therapy Prgm
• Radiography Prgm

**Southern Connecticut State University**
Cheryl Norton, PhD, President
501 Crescent St
New Haven, CT 06515
203 397-4234
• Athletic Training Prgm
• Audiologist Prgm
• Counseling Prgm
• Medical Librarian Prgm
• Speech-Language Pathology Prgm

**Yale University School of Medicine**
Robert Alpern, MD, Dean
333 Cedar St
New Haven, CT 06510
203 785-4672
• Physician Assistant Prgm

**Yale-New Haven Hospital**
20 York St GBB
New Haven, CT 06504
203 785-5074
• Dietetic Internship Prgm

## New London

**Ridley-Lowell Business & Technical Institute**
Wilfred T Weymouth, MS, President
470 Bank St
New London, CT 06320
860 443-7441
• Medical Assistant Prgm

## Newington

**Connecticut Center for Massage Therapy**
Stephen Kitts, Executive Director
75 Kitts Lane
Newington, CT 06111
860 667-1886
• Massage Therapy Prgm (3)

**Hanger Orthopedic Group**
Ivan Sabel, President/CEO
181 Patricia M Genova Dr
Newington, CT 06111
860 667-5304
• Orthotist/Prosthetist Prgm

## Norwalk

**Norwalk Community College**
David Levinson, PhD, President
188 Richards Ave
Norwalk, CT 06854-1655
203 857-7003
• Medical Assistant Prgm

**Norwalk Hospital/Norwalk Community College**
David W Osborne, MS, CEO
Maple St
Norwalk, CT 06856
203 852-2211
• Respiratory Therapist (Advanced) Prgm

## Southington

**Briarwood College**
Lynn Brooks, JD, President
2279 Mt Vernon Rd
Southington, CT 06489
860 628-4751
• Dental Assisting Prgm
• Dietetic Technician-AD Prgm
• Health Information Tech Prgm
• Medical Assistant Prgm
• Occupational Therapy Asst Prgm
• Pharmacy Technician Prgm

**INSTITUTIONS**

## Stamford

**Stamford Hospital**
Brian G Grissler, MA, President/CEO
Shelburne Rd PO Box 9317
Stamford, CT 06904-9317
203 276-7000
• Radiography Prgm

## Storrs

**University of Connecticut**
Philip H Austin, President
Storrs, CT 06269
860 486-2000
• Athletic Training Prgm
• Audiologist Prgm
• Cytogenetic Technology Prgm
• Dietetic Internship Prgm
• Dietetics-Coordinated Prgm
• Dietetics-Didactic Prgm
• Physical Therapy Prgm
• Speech-Language Pathology Prgm

## Waterbury

**Naugatuck Valley Community College**
Richard L Sanders, EdD, President
750 Chase Pkwy
Waterbury, CT 06708
203 575-8044
• Physical Therapist Assistant Prgm
• Radiography Prgm
• Respiratory Therapist (Advanced) Prgm

## West Hartford

**Fox Institute of Business**
Christopher Coutts, President
99 South St
West Hartford, CT 06110
860 947-2299
• Medical Assistant Prgm

**Saint Joseph College**
Winifred E Coleman, President
1678 Asylum Ave
West Hartford, CT 06117
203 232-4571
• Dietetic Internship Prgm
• Dietetics-Coordinated Prgm
• Dietetics-Didactic Prgm

**University of Hartford**
Walter Harrison, PhD, President
200 Bloomfield Ave
West Hartford, CT 06117-1599
860 768-4417
• Clin Lab Scientist/Med Technologist Prgm
• Occupational Therapy Prgm
• Physical Therapy Prgm
• Radiography Prgm
• Respiratory Therapist (Advanced) Prgm

## West Haven

**University of New Haven**
Lawrence J Denardis, President
300 Orange Ave
West Haven, CT 06516
203 932-7000
• Dental Hygiene Prgm
• Dietetics-Didactic Prgm

## Wethersfield

**Porter and Chester Institute - Wethersfield**
Raymond Clark, Executive Director
125 Silas Deane Hwy
Wethersfield, CT 06109
203 529-2519
• Medical Assistant Prgm

## Willimantic

**Windham Community Memorial Hospital**
Richard Brvenik, MA FACHE, President/CEO
112 Mansfield Ave
Willimantic, CT 06226
860 456-6800
• Radiography Prgm

**Windham Regional Vocational Tech School**
210 Birch St
Willimantic, CT 06226
203 456-3789
• Dental Assisting Prgm

## Winsted

**Northwestern Connecticut Comm**
Barbara Douglass, PhD, President
Park Place E
Winsted, CT 06098
860 738-6410
• Medical Assistant Prgm

# Delaware

## Dagsboro

**National Massage Therapy Institute**
Dagsboro, DE 19939
• Massage Therapy Prgm

## Dover

**Delaware State University**
William B Delauder, President
Dover, DE 19901
302 739-4924
• Dietetics-Didactic Prgm

**Delaware Technical & Community College**
Daniel Simpson, MS, VP, Asst Campus Director
100 Campus Dr
Dover, DE 19904
302 857-1126
• Emergency Med Tech-Paramedic Prgm

## Georgetown

**Delaware Tech & Comm Coll - Owens Campus**
Ileana Smith, EdD, VP/Campus Dir
Owens Campus
PO Box 610
Georgetown, DE 19947
302 855-1674
• Clin Lab Technician/Med Lab Technician
• Occupational Therapy Asst Prgm
• Physical Therapist Assistant Prgm
• Radiography Prgm
• Respiratory Therapist (Advanced) Prgm

## New Castle

**Wilmington College**
320 Dupont Hwy
New Castle, DE 19720-6491
302 328-9401
• Counseling Prgm

## Newark

**University of Delaware**
David P Roselle, PhD, President
104A Hullihen Hall
Newark, DE 19716
302 831-2111
• Athletic Training Prgm
• Clin Lab Scientist/Med Technologist Prgm
• Dietetic Internship Prgm
• Dietetics-Didactic Prgm
• Physical Therapy Prgm

## Wilmington

**Delaware Tech & Comm Coll - Wilmington**
Lawrence Miller, MS, Vice President
333 Shipley St
Wilmington, DE 19801-2499
302 573-5470
• Dental Hygiene Prgm
• Diagnostic Med Sonography Prgm
• Histotechnician Prgm
• Medical Assistant Prgm
• Nuclear Medicine Technology Prgm
• Occupational Therapy Asst Prgm
• Physical Therapist Assistant Prgm
• Radiography Prgm
• Respiratory Therapist (Advanced) Prgm

# District of Columbia

## Washington

**Armed Forces Institute of Pathology**
Washington, DC 20306
• Histotechnician Prgm

**Catholic University of America**
Br Patrick F Ellis, FSC, President
103 Executive Bldg
Washington, DC 20064
202 319-5100
• Medical Librarian Prgm

**Gallaudet University**
I King Jordan, PhD, President
800 Florida Ave NE
Washington, DC 20002
202 651-5329
• Audiologist Prgm
• Counseling Prgm
• Speech-Language Pathology Prgm
• Therapeutic Recreation Specialist Prgm

**George Washington University**
John F Williams, MDEdD, VP, Provost Hlth Affairs
2121 I St NW, Ste 701
Washington, DC 20052
202 994-1000
• Art Therapy Prgm
• Athletic Training Prgm
• Clin Lab Scientist/Med Technologist Prgm
• Counseling Prgm
• Diagnostic Med Sonography Prgm
• Emergency Med Tech-Paramedic Prgm
• Physical Therapy Prgm
• Physician Assistant Prgm
• Rehabilitation Counseling Prgm
• Speech-Language Pathology Prgm

**Georgetown University Medical Center**
Sam Wiesel, MD, Exec VP Hlth Sciences
120 Bldg D, 4000 Reservoir Dr NW
Washington, DC 20007
202 687-4601
• Ophthalmic Med Technician Prgm

**Howard University**
H Patrick Swygert, JD, President
Mordecai Wyatt Johnson Admin Bldg
2400 Sixth St NW
Washington, DC 20059
202 806-2500
• Audiologist Prgm
• Clin Lab Scientist/Med Technologist Prgm
• Dental Hygiene Prgm
• Dietetics-Coordinated Prgm
• Genetic Counseling Prgm
• Music Therapy Prgm
• Occupational Therapy Prgm
• Physical Therapy Prgm
• Physician Assistant Prgm
• Radiation Therapy Prgm
• Speech-Language Pathology Prgm

**Potomac Massage Training Institute**
Demara Stamler, Executive Director
5028 Wisconsin Ave NW, LL
Washington, DC 20016-4118
202 686-7046
• Massage Therapy Prgm

**University of the District of Columbia**
Timothy L Jenkins, Jr, President
4200 Connecticut Ave NW
Washington, DC 20008
202 274-5100
• Dietetics-Didactic Prgm
• Radiography Prgm
• Respiratory Therapist (Advanced) Prgm
• Speech-Language Pathology Prgm

**Walter Reed Army Medical Center**
Virgil T Deal, MD COL MC, Healthcare Sys
   Commander
6900 Georgia Ave, NW
Bulding 2, Command Suite
Washington, DC 20307-5001
202 782-6104
• Clin Lab Scientist/Med Technologist Prgm
• Perfusion Prgm
• Specialist in BB Tech Prgm

**Washington Hospital Center**
James Caldas, MHSA, President, CEO
110 Irving St NW
Washington, DC 20010
202 877-6101
• Clin Lab Scientist/Med Technologist Prgm
• Radiography Prgm

# Florida

## Avon Park

**South Florida Community College**
Norman L Stephens, Jr, EdD, President
600 W College Dr
Avon Park, FL 33825-9399
863 453-6661
• Dental Assisting Prgm
• Dental Hygiene Prgm

## Boca Raton

**Florida Atlantic University**
Frank T Brogan, MEd, President
777 Glades Rd
Boca Raton, FL 33431
561 297-3450
• Counseling Prgm
• Rehabilitation Counseling Prgm
• Speech-Language Pathology Prgm

**West Boca Medical Center**
Richard Gold, CEO
21644 State Rd 7
Boca Raton, FL 33428
561 488-8000
• Radiography Prgm

## Boynton Beach

**Bethesda Memorial Hospital**
Robert Hill, MHA, President
2815 S Seacrest Blvd
Boynton Beach, FL 33435
561 737-7733
• Radiography Prgm

## Bradenton

**Manatee Community College**
Sarah H Pappas, EdD, President
5840 26th St W
PO Box 1849
Bradenton, FL 34207-7046
941 752-5201
• Dental Hygiene Prgm
• Occupational Therapy Asst Prgm
• Physical Therapist Assistant Prgm
• Radiography Prgm

**Manatee Technical Institute**
Mary Cantrell, PhD, Director
5603 34th St W
Bradenton, FL 34210
941 751-7900
• Dental Assisting Prgm
• Emergency Med Tech-Paramedic Prgm

## Cape Coral

**21st Cent Oncology Inc Sch-Rad Therapy Tech**
Daniel E Dosoretz, MD ABR, CEO
1419 SE 8th Terrace
Cape Coral, FL 33990
941 489-3420
• Radiation Therapy Prgm

**Lee County High Tech Center-North**
Michael Schiffer, MEd, Director
360 Santa Barbara Blvd N
Cape Coral, FL 33993-2479
239 574-4440
• Surgical Technology Prgm

## Clearwater

**Florida Metropolitan Univ - Pinellas Campus**
John Buck, BME, President
2471 McMullen Booth Rd, Ste 200
Clearwater, FL 34619
727 725-2688
• Medical Assistant Prgm

**Pinellas Tech Ed Ctr - Clearwater Campus**
Warren Laux, EdD, Director
6100 - 154th Ave N
Clearwater, FL 33760
727 538-7161
• Surgical Technology Prgm

## Cocoa

**Brevard Community College**
Thomas E Gamble, PhD, District President
1519 Clearlake Rd
Cocoa, FL 32922
321 433-7000
• Clin Lab Technician/Med Lab Technician
• Dental Assisting Prgm
• Dental Hygiene Prgm
• Emergency Med Tech-Paramedic Prgm
• Medical Assistant Prgm
• Radiography Prgm
• Surgical Technology Prgm

## Coral Gables

**University of Miami**
Donna Shalala, President
University Station
Coral Gables, FL 33124
305 284-2211
• Athletic Training Prgm
• Music Therapy Prgm
• Physical Therapy Prgm

## Davie

**McFatter Vocational Technical Center**
Mark Thomas, MS, Director
6500 Nova Dr
Davie, FL 33317
754 321-5700
• Dental Lab Technician Prgm
• Medical Assistant Prgm
• Pharmacy Technician Prgm

## Daytona Beach

**Daytona Beach Community College**
Kent D Sharples, EdD, President
PO Box 2811
Daytona Beach, FL 32120-2811
386 506-3200
- Dental Assisting Prgm
- Dental Hygiene Prgm
- Emergency Med Tech-Paramedic Prgm
- Health Information Tech Prgm
- Medical Assistant Prgm
- Occupational Therapy Asst Prgm
- Physical Therapist Assistant Prgm
- Respiratory Therapist (Advanced) Prgm
- Surgical Technology Prgm

**Halifax Medical Center**
Dan Lang, MHA, Chief Operating Officer
303 N Clyde Morris Blvd
PO Box 2830
Daytona Beach, FL 32114
386 254-4065
- Radiation Therapy Prgm
- Radiography Prgm

**Keiser College - Daytona Beach**
Arthur Keiser, PhD, Chief Executive Officer
1800 Business Park Blvd
Bldg 2
Daytona Beach, FL 32114
386 274-5060
- Diagnostic Med Sonography Prgm
- Medical Assistant Prgm
- Radiography Prgm

## Deland

**Stetson University**
H Douglas Lee, BD ThM PhD, President
421 N Woodland Blvd, Unit 8258
Deland, FL 32720-3701
904 822-7250
- Counseling Prgm

## Eustis

**Lake Technical Center**
Steve Hand, MEd, Director
2001 Kurt St
Eustis, FL 32726
352 742-6486
- Emergency Med Tech-Paramedic Prgm

## Fern Park

**Americare School of Nursing**
Gerald Newman, ABA, President
7275 Estapano Circle
Fern Park, FL 32730
407 673-7406
- Dental Assisting Prgm

## Ft Lauderdale

**Broward Community College**
Larry Calderon, PhD, President
111 E Las Olas Blvd
Ft Lauderdale, FL 33301
954 201-7401
- Dental Assisting Prgm
- Dental Hygiene Prgm
- Diagnostic Med Sonography Prgm
- Emergency Med Tech-Paramedic Prgm
- Health Information Tech Prgm
- Medical Assistant Prgm
- Physical Therapist Assistant Prgm
- Respiratory Therapist (Advanced) Prgm

**Keiser University**
Peter Crocitto, MBA BS, Executive Vice
     Chancellor
1500 NW 49th St
Ft Lauderdale, FL 33309
954 776-4456
- Clin Lab Technician/Med Lab Technician
- Diagnostic Med Sonography Prgm
- Occupational Therapy Asst Prgm
- Physical Therapist Assistant Prgm
- Radiography Prgm (2)

**Nova Southeastern University**
Ray Ferrero, JSD, President
3301 College Ave
Ft Lauderdale, FL 33314-7796
954 262-7575
- Anesthesiologist Asst Prgm
- Audiologist Prgm
- Diagnostic Med Sonography Prgm
- Occupational Therapy Prgm
- Physical Therapy Prgm
- Physician Assistant Prgm (2)
- Speech-Language Pathology Prgm

## Ft Myers

**Edison College**
Kenneth P Walker, PhD, President
8099 College Pkwy SW
PO Box 60210
Ft Myers, FL 33906-6210
239 941-9211
- Cardiovascular Technology Prgm
- Dental Assisting Prgm
- Dental Hygiene Prgm
- Emergency Med Tech-Paramedic Prgm
- Radiography Prgm
- Respiratory Therapist (Advanced) Prgm

**Florida Gulf Coast University**
William C Merwin, PhD, President
10501 FGCU Blvd S
Ft Myers, FL 33965-6565
941 590-1055
- Athletic Training Prgm
- Counseling Prgm
- Occupational Therapy Prgm
- Physical Therapy Prgm

**International College**
Terry P McMahan, JD, President
8695 College Pkwy, Ste 217
Ft Myers, FL 33919
941 482-0019
- Health Information Tech Prgm
- Medical Assistant Prgm

**Lee County High Tech Center - Central**
Ronald E Pentiuk, EdS, Director
3800 Michigan Ave
Ft Myers, FL 33916
941 334-4544
- Surgical Technology Prgm

**Southwest Florida College**
Don Jones, BS MBA, CEO
1685 Medical Lane
Ft Myers, FL 33907
239 939-4766
- Medical Assistant Prgm

## Ft Pierce

**Indian River Community College**
Edwin R Massey, PhD, President
3209 Virginia Ave
Ft Pierce, FL 34981-5599
772 462-7201
- Clin Lab Technician/Med Lab Technician
- Dental Assisting Prgm
- Dental Hygiene Prgm
- Dental Lab Technician Prgm
- Emergency Med Tech-Paramedic Prgm
- Health Information Tech Prgm
- Medical Assistant Prgm
- Physical Therapist Assistant Prgm
- Radiography Prgm
- Respiratory Therapist (Advanced) Prgm
- Surgical Technology Prgm

## Gainesville

**Florida School of Massage**
Gainesville, FL 32608
- Massage Therapy Prgm

**Santa Fe Community College**
Jackson Sasser, PhD, President
3000 NW 83rd St
Gainesville, FL 32606-6200
352 395-5164
- Cardiovascular Technology Prgm
- Dental Assisting Prgm
- Dental Hygiene Prgm
- Emergency Med Tech-Paramedic Prgm
- Health Information Tech Prgm
- Nuclear Medicine Technology Prgm
- Radiography Prgm

**University of Florida**
Bernard Machen, DDS MS PhD, President
PO Box 113150
Gainesville, FL 32611-3150
352 392-1311
- Athletic Training Prgm
- Audiologist Prgm
- Counseling Prgm (3)
- Dietetic Internship Prgm
- Dietetics-Didactic Prgm
- Occupational Therapy Prgm
- Ophthalmic Med Technologist Prgm
- Orthoptist Prgm
- Physical Therapy Prgm
- Physician Assistant Prgm
- Rehabilitation Counseling Prgm
- Speech-Language Pathology Prgm
- Therapeutic Recreation Specialist Prgm

## Hollywood

**Sheridan Technical Center**
D Robert Boegli, Director
5400 Sheridan St
Hollywood, FL 33021
754 321-2007
• Surgical Technology Prgm

## Jacksonville

**Concorde Career Institute**
Michael Beaty, MA, Campus President
7960 Arlington Expressway
Ste 120
Jacksonville, FL 32211
904 725-0525
• Surgical Technology Prgm (2)

**Florida Community College - Jacksonville**
Steven R Wallace, PhD, President
501 W State St
Jacksonville, FL 32202
904 632-3000
• Clin Lab Technician/Med Lab Technician
• Dental Hygiene Prgm
• Dietetic Technician-AD Prgm
• Emergency Med Tech-Paramedic Prgm
• Health Information Tech Prgm
• Histotechnician Prgm
• Physical Therapist Assistant Prgm
• Respiratory Therapist (Advanced) Prgm
• Surgical Technology Prgm

**Shands Jacksonville**
Jim Burkhart, Interim COO
655 W Eighth St
Jacksonville, FL 32209
904 244-0000
• Clin Lab Scientist/Med Technologist Prgm
• Radiography Prgm

**St Luke's Hosp/Mayo Clinic Jacksonville**
4201 Belfort Rd
Jacksonville, FL 32216-1431
904 296-3733
• Dietetic Internship Prgm

**St Vincent's Medical Center**
John J Maher, MBA MHA, President
1800 Barrs St
Jacksonville, FL 32204
904 308-4001
• Clin Lab Scientist/Med Technologist Prgm
• Diagnostic Med Sonography Prgm
• Nuclear Medicine Technology Prgm
• Radiography Prgm

**University of North Florida**
John Delaney, JD, President
Office of the President
4567 St Johns Bluff Rd S
Jacksonville, FL 32224-2645
904 620-2500
• Athletic Training Prgm
• Counseling Prgm
• Dietetic Internship Prgm
• Dietetics-Didactic Prgm
• Physical Therapy Prgm
• Rehabilitation Counseling Prgm

## Lake City

**Lake City Community College**
Charles Hall, EdD, President
149 SE College Place
Lake City, FL 32025
386 754-4200
• Emergency Med Tech-Paramedic Prgm
• Pharmacy Technician Prgm
• Physical Therapist Assistant Prgm

## Lake Worth

**Palm Beach Community College**
Dennis P Gallon, PhD, President
4200 Congress Ave
Lake Worth, FL 33461-4796
561 868-3350
• Dental Assisting Prgm
• Dental Hygiene Prgm
• Diagnostic Med Sonography Prgm
• Dietetic Technician-AD Prgm
• Emergency Med Tech-Paramedic Prgm
• Medical Assistant Prgm
• Radiography Prgm
• Respiratory Therapist (Advanced) Prgm
• Surgical Technology Prgm

## Lakeland

**Florida Career Institute**
Pamela J Corrigan, RN MEd, CEO
4222 S Florida Ave
Lakeland, FL 33813
863 646-1400
• Medical Assistant Prgm

**Florida Metropolitan Univ - Lakeland Campus**
Edmund K Gross, EdD, President
995 E Memorial Blvd, Ste 110
Lakeland, FL 33801
941 686-1444
• Medical Assistant Prgm

**Florida Southern College**
Anne Kerr, PhD, President
111 Lake Hollingsworth Dr
Lakeland, FL 33801-5698
863 680-4100
• Athletic Training Prgm

**Lakeland Regional Medical Center**
Jack T Stephens, President/CEO
1324 Lakeland Hills Blvd
PO Box 95448
Lakeland, FL 33804
863 687-1100
• Radiography Prgm

**Traviss Career Center**
Kenneth Lloyd, PhD
3225 Winter Lake Rd
Lakeland, FL 33803
863 449-2700
• Dental Assisting Prgm
• Surgical Technology Prgm

## Lauderdale Lakes

**Sanford-Brown Institute - Ft Lauderdale**
1201 W Cypress Creek Rd
Lauderdale Lakes, FL 33309
954 733-8900
• Surgical Technology Prgm

## Leesburg

**Lake-Sumter Community College**
Charles Mojock, PhD, President
9501 US Hwy 441
Leesburg, FL 34788-8751
352 365-3523
• Health Information Tech Prgm

## Maitland

**Florida College of Natural Health**
Dale Weiberg, Campus Director
2600 Lake Lucian Dr
Maitland, FL 33126
407 261-0319
• Massage Therapy Prgm (4)

## Melbourne

**Florida Metropolitan Univ - Melbourne Campus**
Mark Judge, President
2401 N Harbor City Blvd
Melbourne, FL 32935
321 253-2929
• Medical Assistant Prgm
• Pharmacy Technician Prgm

**Keiser College - Melbourne Campus**
Colleen Rupp, BS, Vice President
900 S Babcock St
Melbourne, FL 32901
321 409-4800
• Occupational Therapy Asst Prgm
• Radiography Prgm

## Miami

**Educating Hands School of Massaage**
Iris Burman
120 SW 8th St
Miami, FL 33130
305 285-6991
• Massage Therapy Prgm

**Florida International University**
Modesto Maidique, PhD, President
Office of Academic Affairs
University Park Campus
Miami, FL 33199
305 348-2111
• Counseling Prgm
• Dietetic Internship Prgm
• Dietetics-Coordinated Prgm
• Dietetics-Didactic Prgm
• Health Information Admin Prgm
• Occupational Therapy Prgm
• Physical Therapy Prgm
• Speech-Language Pathology Prgm

**Jackson Mem Medical Center**
Marvin O'Quinn, MHA, President and CEO
Jackson Health System
1611 NW 12th Ave
Miami, FL 33136
305 585-6754
• Nuclear Medicine Technology Prgm
• Radiography Prgm

**Lindsey Hopkins Tech Education Center**
Rosa Borgen, MS, Principal
750 NW 20th St
Miami, FL 33127
305 324-6070
- Dental Assisting Prgm
- Surgical Technology Prgm

**Miami Dade College**
Eduardo J Padron, PhD, College President
Wolfson Campus
300 NE 2nd Ave
Miami, FL 33132-2296
305 237-3316
- Clin Lab Technician/Med Lab Technician
- Dental Hygiene Prgm
- Diagnostic Med Sonography Prgm
- Dietetic Technician-AD Prgm
- Emergency Med Tech-Paramedic Prgm
- Health Information Tech Prgm
- Histotechnician Prgm
- Ophthalmic Dispensing Optician Prgm
- Physical Therapist Assistant Prgm
- Physician Assistant Prgm
- Radiation Therapy Prgm
- Radiography Prgm
- Respiratory Therapist (Advanced) Prgm

**Robert Morgan Educational Center**
Greg Zawyer, MS, Principal
18180 SW 122nd Ave
Miami, FL 33177
305 253-9920
- Dental Assisting Prgm
- Medical Assistant Prgm

**Sanford-Brown Institute**
David T Ruggieri, President
4400 Biscayne Blvd
Miami, FL 33137
800 445-6108
- Surgical Technology Prgm

## Miami Shores

**Barry University**
Linda M Bevilacqua, OP, President
11300 NE 2nd Ave
Miami Shores, FL 33161-6695
305 899-3010
- Athletic Training Prgm
- Counseling Prgm
- Histotechnology Prgm
- Occupational Therapy Prgm
- Perfusion Prgm
- Physician Assistant Prgm

## Naples

**Lorenzo Walker Institute of Technology**
Denise Duzick, MS, Administrator
3702 Estey Ave
Naples, FL 34104
239 377-0926
- Dental Assisting Prgm
- Medical Assistant Prgm
- Surgical Technology Prgm

## New Port Richey

**Pasco County Health Department**
10841 Little Rd
New Port Richey, FL 34654-2533
813 869-3900
- Dietetic Internship Prgm

**Pasco-Hernando Community College**
Katherine M Johnson, EdD, President
10230 Ridge Rd
New Port Richey, FL 34654-5199
727 847-2727
- Dental Hygiene Prgm
- Emergency Med Tech-Paramedic Prgm

## Niceville

**Okaloosa-Walton College**
James R Richburg, EdD, President
100 College Blvd
Niceville, FL 32578-1295
850 729-5360
- Dental Assisting Prgm
- Surgical Technology Prgm

## North Miami Beach

**National School of Technology**
Martin Knobel, B Ed MS, CEO
12000 Biscayne Blvd #302
North Miami Beach, FL 33181
305 893-0005
- Surgical Technology Prgm (3)

## Ocala

**Central Florida Community College**
Charles Dassance, PhD, President
3001 SW College Rd
PO Box 1388
Ocala, FL 34474
352 854-2322
- Dental Assisting Prgm
- Emergency Med Tech-Paramedic Prgm
- Health Information Tech Prgm
- Physical Therapist Assistant Prgm
- Surgical Technology Prgm

**Community Technical & Adult Education Center**
Samuel Lauff Jr, MA, Administrator
1014 SW 7th Rd
Ocala, FL 34474
352 671-7200
- Medical Assistant Prgm

**Marion County School of Radiologic Tech**
Deborah Jenkins, MEd, Administrator
1014 SW 7th Rd
Ocala, FL 34474-3172
352 671-7200
- Radiography Prgm

## Orlando

**Florida Hospital**
Lars Houmann, PhD, CEO
601 E Rollins St
Orlando, FL 32803
407 303-1531
- Phlebotomist Prgm

**Florida Hospital College of Health Sciences**
David E Greenlaw, DMiN, President
800 Lake Estelle Dr
Orlando, FL 32803
407 303-7747
- Nuclear Medicine Technology Prgm
- Occupational Therapy Asst Prgm
- Radiography Prgm

**Florida Metro Univ - North Orlando Campus**
Ouida B Kirby, BS, President
5421 Diplomat Circle
Orlando, FL 32810
407 628-5870
- Medical Assistant Prgm

**Florida Metro Univ - Orlando College South**
J Gary Adcox, PhD, President
9200 SouthPark Center Loop
Orlando, FL 32819
407 851-2525
- Medical Assistant Prgm

**Orlando Tech**
Ferol Lynne Voltaggio, MEd, Senior Director
301 W Amelia St
Orlando, FL 32801
407 246-7060
- Dental Assisting Prgm
- Surgical Technology Prgm

**University of Central Florida**
John C Hitt, PhD, President
308 Millican Hall
4000 Central Florida Blvd
Orlando, FL 32816-0002
407 823-1823
- Athletic Training Prgm
- Clin Lab Scientist/Med Technologist Prgm
- Counseling Prgm
- Health Information Admin Prgm
- Physical Therapy Prgm
- Radiography Prgm
- Respiratory Therapist (Advanced) Prgm
- Speech-Language Pathology Prgm

**Valencia Community College**
Sanford C Shugart, PhD, President
PO Box 3028
Orlando, FL 32802-3028
407 299-5000
- Dental Hygiene Prgm
- Diagnostic Med Sonography Prgm
- Emergency Med Tech-Paramedic Prgm
- Radiography Prgm
- Respiratory Therapist (Advanced) Prgm

## Osprey

**Sarasota District Schools**
101 Old Venice Rd
Osprey, FL 34229-9023
- Dietetic Internship Prgm

## Palatka

**St Johns River Community College**
R L McLendon, Jr, PhD, President
5001 St Johns Ave
Palatka, FL 32177
386 312-4113
- Health Information Tech Prgm

## Palm Harbor

### Central Florida Institute
Alfred A McCloy, MBA, President
30522 US 19 N, Ste 200
Palm Harbor, FL 34684
727 786-4707
- Cardiovascular Technology Prgm
- Surgical Technology Prgm

## Panama City

### Gulf Coast Community College
Robert L McSpadden, EdD, President
5230 W US Hwy 98
Panama City, FL 32401-1041
850 769-1551
- Dental Assisting Prgm
- Dental Hygiene Prgm
- Emergency Med Tech-Paramedic Prgm
- Physical Therapist Assistant Prgm
- Radiography Prgm
- Respiratory Therapist (Advanced) Prgm
- Surgical Technology Prgm

## Pensacola

### Pensacola Junior College
G Thomas Delaino, PhD, President
1000 College Blvd
Pensacola, FL 32501
850 484-1700
- Dental Assisting Prgm
- Dental Hygiene Prgm
- Dietetic Technician-AD Prgm
- Emergency Med Tech-Paramedic Prgm
- Health Information Tech Prgm
- Medical Assistant Prgm
- Physical Therapist Assistant Prgm
- Radiography Prgm
- Surgical Technology Prgm

### University of West Florida
John Cavanaugh, PhD, President
11000 University Pkwy
Pensacola, FL 32514-5750
904 474-2202
- Athletic Training Prgm
- Clin Lab Scientist/Med Technologist Prgm

### Virginia College at Penascola
Bruce G Capps, BS, President
19 W Garden St
Pensacola, FL 32501
850 916-9868
- Surgical Technology Prgm

## Port Charlotte

### Charlotte Technical Center
Judith Willis, Director
18300 Toledo Blade Blvd
Port Charlotte, FL 33948-3399
941 255-7500
- Dental Assisting Prgm

## Sanford

### Seminole Community College
E Ann McGee, EdD, President
100 Weldon Blvd
Sanford, FL 32773
423 708-4722
- Emergency Med Tech-Paramedic Prgm
- Medical Assistant Prgm
- Physical Therapist Assistant Prgm
- Respiratory Therapist (Advanced) Prgm

## Sarasota

### Sarasota County Technical Institute
Bruce Andersen, MA, Director
4748 Beneva Rd
Sarasota, FL 34233-1798
813 924-1365
- Emergency Med Tech-Paramedic Prgm
- Medical Assistant Prgm
- Surgical Technology Prgm

### Sarasota Memorial Hospital
1700 S Tamiami Trail
Sarasota, FL 34239-3555
813 917-1080
- Dietetic Internship Prgm

### Sarasota School of Massage Therapy
Joe Lubow, Director
1932 Ringling Blvd
Sarasota, FL 34236
941 957-0577
- Massage Therapy Prgm

## St Augustine

### First Coast Technical Institute
Christine Cothron, President
2980 Collins Ave
St Augustine, FL 32095-1919
904 824-4401
- Medical Assistant Prgm

### Univ of St Augustine for Health Sciences
Stanley V Paris, PhD PT FAPTA, President
1 University Blvd
St Augustine, FL 32086
904 826-0084
- Occupational Therapy Prgm
- Physical Therapy Prgm

## St Petersburg

### Bayfront Medical Center
Sue S Brody, MHA, President/CEO
701 Sixth St S
St Petersburg, FL 33701
813 893-6085
- Clin Lab Scientist/Med Technologist Prgm

### Pinellas Tech Educ Ctr - St Petersburg
Dorothy Bailey, EdD, Director
901 34th St S
St Petersburg, FL 33711-2298
727 893-2500
- Dental Assisting Prgm
- Medical Assistant Prgm
- Pharmacy Technician Prgm

### St Petersburg College
Sandra W Pepicello, PhD RN, Provost, Health
  Educ Ctr
PO Box 13489
St Petersburg, FL 33733
727 341-3664
- Clin Lab Technician/Med Lab Technician
- Dental Hygiene Prgm
- Emergency Med Tech-Paramedic Prgm
- Health Information Tech Prgm
- Physical Therapist Assistant Prgm
- Respiratory Therapist (Advanced) Prgm

### Transfusion Med Acad Ctr FL Blood Svcs
German F Leparc, MD, Chief Medical Officer
10100 Dr Martin Luther King Jr St N
St Petersburg, FL 33716-3806
727 568-5433
- Specialist in BB Tech Prgm

## Tallahassee

### Core Institute
George Kousaleos, MassT, Executive Director
223 W Carolina St
Tallahassee, FL 32301
850 222-8673
- Massage Therapy Prgm

### Florida A&M University
Castell V Bryant, EdD, Interim President
401 Lee Hall
Tallahassee, FL 32307-3100
850 599-3225
- Health Information Admin Prgm
- Occupational Therapy Prgm
- Physical Therapy Prgm
- Respiratory Therapist (Advanced) Prgm

### Florida Department of Education
325 W Gaines St, Rm 1032
Tallahassee, FL 32399-0400
- Dietetic Internship Prgm

### Florida State University
TK Wetherell, PhD, President
211 WES
Tallahassee, FL 32306
850 644-2525
- Art Therapy Prgm
- Athletic Training Prgm
- Counseling Prgm
- Dietetic Internship Prgm
- Dietetics-Didactic Prgm
- Medical Librarian Prgm
- Music Therapy Prgm
- Orientation and Mobility Specialist Prgm
- Rehabilitation Counseling Prgm
- Rehabilitation Teacher Prgm
- Speech-Language Pathology Prgm

### Lively Technical Center
Jean Ferguson, MS, Principal
500 N Appleyard Dr
Tallahassee, FL 32304
904 487-7401
- Medical Assistant Prgm

**INSTITUTIONS**

**Tallahassee Community College**
William Law, PhD, President
444 Appleyard Dr
Tallahassee, FL 32304
850 201-8660
- Dental Assisting Prgm
- Dental Hygiene Prgm
- Emergency Med Tech-Paramedic Prgm
- Respiratory Therapist (Advanced) Prgm

## Tampa

**Erwin Technical Center**
Michael D Donohue, MEd, Principal
2010 E Hillsborough Ave
Tampa, FL 33610-8299
813 231-1800
- Clin Lab Technician/Med Lab Technician
- Dental Assisting Prgm
- Electroneurodiagnostic Tech Prgm
- Medical Assistant Prgm
- Surgical Technology Prgm

**Florida Metro U - Tampa Coll - Hillsborough**
Thomas M Barlow, President
3319 W Hillsborough Ave
Tampa, FL 33614
813 879-6000
- Medical Assistant Prgm

**Florida Metropolitan Univ - Tampa College**
Stan Banks, II, BS, President
3924 Coconut Palm Dr
Tampa, FL 33619
813 621-0041
- Medical Assistant Prgm
- Surgical Technology Prgm

**Henry W Brewster Technical Center**
Janice Carter Collier, Prinicipal
2222 N Tampa St
Tampa, FL 33602
- Pharmacy Technician Prgm

**Hillsborough Community College**
Gwendolyn Stephenson, PhD, President
PO Box 31127
39 Columbia Dr (Davis Island)
Tampa, FL 33631-3127
813 253-7050
- Dental Assisting Prgm
- Dental Hygiene Prgm
- Diagnostic Med Sonography Prgm
- Emergency Med Tech-Paramedic Prgm
- Nuclear Medicine Technology Prgm
- Ophthalmic Dispensing Optician Prgm
- Radiation Therapy Prgm
- Radiography Prgm
- Respiratory Therapist (Advanced) Prgm

**James A Haley Veteran's Hospital**
13000 N Bruce B Downs Blvd
Tampa, FL 33612-4745
813 972-2000
- Dietetic Internship Prgm

**Tampa General Hospital**
Ron Hytoff, MD, President, CEO
PO Box 1289
Tampa, FL 33601
813 251-7383
- Clin Lab Scientist/Med Technologist Prgm

**University of South Florida**
Judith Genshaft, PhD, President
ADM 241
Tampa, FL 33620
813 974-2791
- Athletic Training Prgm
- Audiologist Prgm
- Counseling Prgm
- Medical Librarian Prgm
- Physical Therapy Prgm
- Rehabilitation Counseling Prgm
- Speech-Language Pathology Prgm

**University of Tampa**
Ronald Vaughn, PhD, President
401 W Kennedy Blvd
Tampa, FL 33606
813 253-6201
- Athletic Training Prgm

## West Palm Beach

**MedVance Institute**
Deborah K Schwarzberg, PhD, President
KIMC Investments LP - Esperante Ste 200
222 Lakeview Ave
West Palm Beach, FL 33401
561 832-3535
- Radiography Prgm (3)
- Surgical Technology Prgm (2)

**Palm Beach Atlantic University**
David W Clark, PhD, President
PO Box 24708
West Palm Beach, FL 33416-4708
561 803-2302
- Athletic Training Prgm

**South University**
Tracey Schoonmaker, President
1760 N Congress Ave
West Palm Beach, FL 33409
561 697-9200
- Medical Assistant Prgm
- Physical Therapist Assistant Prgm

## Winter Haven

**Polk Community College**
Eileen Holden, PhD, President
999 Ave H NE
Winter Haven, FL 33881-4299
863 297-1098
- Emergency Med Tech-Paramedic Prgm
- Health Information Tech Prgm
- Occupational Therapy Asst Prgm
- Physical Therapist Assistant Prgm
- Radiography Prgm

## Winter Park

**Central Florida College**
Roger Bradley, President
1573 W Fairbanks Ave
Winter Park, FL 32789
407 843-3984
- Medical Assistant Prgm

**Rollins College**
1000 Holt Ave
Winter Park, FL 32789-4499
407 646-2000
- Counseling Prgm

**Winter Park Tech**
Eleanor Cain, Director
901 Webster Ave
Winter Park, FL 32789
407 647-6366
- Medical Assistant Prgm

# Georgia

## Acworth

**North Metro Technical College**
Steve Dougherty, MA, President
5198 Ross Rd
Acworth, GA 30102
770 975-4126
- Radiography Prgm

## Albany

**Albany Technical College**
Anthony Parker, PhD, President
1704 S Slappey Blvd
Albany, GA 31701-3514
229 430-3500
- Dental Assisting Prgm
- Medical Assistant Prgm
- Radiography Prgm
- Surgical Technology Prgm

**Darton College**
Peter J Sireno, EdD, President
2400 Gillionville Rd
Albany, GA 31707
229 317-6905
- Clin Lab Technician/Med Lab Technician
- Dental Hygiene Prgm
- Health Information Tech Prgm
- Histotechnician Prgm
- Occupational Therapy Asst Prgm
- Physical Therapist Assistant Prgm
- Respiratory Therapist (Advanced) Prgm

## Americus

**South Georgia Technical College**
Sparky Reeves, MS, President
900 S Georgia Tech Pkwy
Americus, GA 31709
912 931-2004
- Medical Assistant Prgm

## Athens

**Athens Technical College**
Flora Tydings, EdD, President
800 US Hwy 29 N
Athens, GA 30601
706 355-5000
- Dental Assisting Prgm
- Dental Hygiene Prgm
- Physical Therapist Assistant Prgm
- Radiography Prgm
- Respiratory Therapist (Advanced) Prgm
- Surgical Technology Prgm

**University of Georgia**
Michael F Adams, PhD, President
Administration Bldg
Athens, GA 30602
706 542-1214
- Athletic Training Prgm
- Counseling Prgm
- Dietetic Internship Prgm
- Dietetics-Didactic Prgm
- Music Therapy Prgm
- Speech-Language Pathology Prgm

## Atlanta

**Atlanta Technical College**
Brenda W Jones, PhD, President
1560 Metropolitan Parkway SW
Atlanta, GA 30310
404 225-4400
- Dental Assisting Prgm
- Dental Lab Technician Prgm
- Medical Assistant Prgm

**Div of Pub Hlth/Georgia Dept of Hum Res**
2 Peachtree St NE
Atlanta, GA 30303-3141
404 657-2884
- Dietetic Internship Prgm

**Emory University**
James W Wagner, PhD, President
408 Administration Bldg
Atlanta, GA 30322
404 727-6012
- Anesthesiologist Asst Prgm
- Ophthalmic Med Technologist Prgm
- Physical Therapy Prgm
- Physician Assistant Prgm
- Radiography Prgm

**Emory University Hospital**
John D Henry, Sr, FACHE, CEO
1364 Clifton Rd NE, Rm B216
Atlanta, GA 30322
404 712-4881
- Dietetic Internship Prgm

**Georgia Institute of Technology**
Wayne G Clough, PhD, President
225 North Ave NW
Atlanta, GA 30332-0325
404 894-5051
- Orthotist/Prosthetist Prgm

**Georgia Medical Institute**
Anthony Galang, BS, President
1706 Northeast Expwy
Atlanta, GA 30329
404 327-8787
- Surgical Technology Prgm

**Georgia State University**
Carl V Patton, PhD, President
PO Box 3999
300 Alumni Hall
Atlanta, GA 30302-3999
404 651-2560
- Counseling Prgm
- Dietetic Internship Prgm
- Dietetics-Didactic Prgm
- Physical Therapy Prgm
- Rehabilitation Counseling Prgm
- Respiratory Therapist (Advanced) Prgm
- Speech-Language Pathology Prgm

**Grady Health System**
John Henry, FACHE, President/CEO
80 Jesse Hill Jr Dr, SE
PO Box 26189
Atlanta, GA 30303-3050
404 616-4252
- Cytotechnology Prgm
- Diagnostic Med Sonography Prgm
- Radiation Therapy Prgm
- Radiography Prgm

**Sanford-Brown Institute**
Glenn W Alderson, BA, Executive Director
1140 Hammond Dr, Ste 1150
Atlanta, GA 30328
770 350-0009
- Diagnostic Med Sonography Prgm

## Augusta

**Augusta Area Dietetic Internship**
Larry Read, CEO, Univ Hlth Care Syst
University Hospital
1350 Walton Way (10)
Augusta, GA 30901-2629
706 774-8045
- Dietetic Internship Prgm

**Augusta Technical College**
Terry D Elam, MEd, President
3200 Augusta Tech Dr
Augusta, GA 30906
706 771-4005
- Dental Assisting Prgm
- Medical Assistant Prgm
- Occupational Therapy Asst Prgm
- Respiratory Therapist (Advanced) Prgm
- Surgical Technology Prgm

**Harry T Harper Jr Sch of Cardiac & Vas Tech**
Larry Read, MA, President/CEO
1350 Walton Way
Augusta, GA 30901
706 774-8045
- Cardiovascular Technology Prgm

**Medical College of Georgia**
Daniel Rahn, MD, President
1120 15th St, Rm AA-312
Augusta, GA 30912-0079
706 721-2301
- Clin Lab Scientist/Med Technologist Prgm
- Dental Hygiene Prgm
- Diagnostic Med Sonography Prgm
- Health Information Admin Prgm
- Medical Illustrator Prgm
- Nuclear Medicine Technology Prgm
- Occupational Therapy Prgm (2)
- Physical Therapy Prgm
- Physician Assistant Prgm
- Radiation Therapy Prgm
- Respiratory Therapist (Advanced) Prgm

**Savannah River College**
Darryl H Kerr, BS, President
2528 Centerwest Pkwy Bldg A
Augusta, GA 30909
706 738-5046
- Medical Assistant Prgm

**University Hospital**
Donald C Bray, MS, President/CEO
1350 Walton Way
Augusta, GA 30901
706 722-9011
- Radiography Prgm

## Brunswick

**Coastal Georgia Community College**
Dorothy L Lord, PhD, President
3700 Altama Ave
Brunswick, GA 31520-3644
912 264-7201
- Clin Lab Technician/Med Lab Technician
- Radiography Prgm
- Surgical Technology Prgm

## Carrollton

**State University of West Georgia**
Beheruz N Sethna, President
Carrollton, GA 30118-0001
770 836-6500
- Counseling Prgm

## Clarkesville

**North Georgia Technical College**
Ruth Nichols, EdD, President
PO Box 65
1500 Hwy 197 N
Clarkesville, GA 30523
706 754-7701
- Clin Lab Technician/Med Lab Technician
- Medical Assistant Prgm (2)

## Clarkston

**DeKalb Technical College**
Robin Hoffman, PhD, President
495 N Indian Creek Dr
Clarkston, GA 30021
404 297-9522
- Clin Lab Technician/Med Lab Technician
- Medical Assistant Prgm
- Ophthalmic Dispensing Optician Prgm
- Surgical Technology Prgm

## Cochran

**Middle Georgia College**
Richard Federinko, PhD, President
1100 Second St SE
Cochran, GA 31014-1599
478 934-3011
- Occupational Therapy Asst Prgm

## Columbus

**Columbus State University**
Frank D Brown, PhD, President
4225 University Ave
Columbus, GA 31907-5645
706 568-2211
- Counseling Prgm

**Columbus Technical College**
J Robert Jones, EdS, President
928 Manchester Expwy
Columbus, GA 31904-6572
706 649-1837
- Dental Assisting Prgm
- Dental Hygiene Prgm
- Medical Assistant Prgm
- Radiography Prgm
- Surgical Technology Prgm

## Dahlonega

**North Georgia College & State University**
David Potter, PhD, President
Prince Memorial Hall
Dahlonega, GA 30597
706 864-1993
- Athletic Training Prgm
- Physical Therapy Prgm

## Dalton

**Dalton State College**
James A Burran, PhD, President
650 College Dr
Dalton, GA 30720-3778
706 272-4438
- Clin Lab Technician/Med Lab Technician
- Medical Assistant Prgm
- Phlebotomist Prgm
- Radiography Prgm
- Surgical Technology Prgm

## Decatur

**DeKalb Medical Center**
Eric Norwood, MHA, CEO
2701 N Decatur Rd
Decatur, GA 30033
404 501-5206
- Radiography Prgm

**DeVry University - Atlanta**
250 N Arcadia Ave
Decatur, GA 30030-2198
- Health Information Tech Prgm

**Georgia Perimeter College**
Jacquelyn M Belcher, President
3251 Panthersville Rd
Decatur, GA 30034
404 244-2365
- Dental Hygiene Prgm

## Dublin

**Heart of Georgia Technical College**
Randall L Peters, MA, President
560 Pinehill Rd
Dublin, GA 31021
478 275-6590
- Medical Assistant Prgm
- Radiography Prgm
- Respiratory Therapist (Advanced) Prgm

## Fitzgerald

**East Central Technical College**
Diane W Harper, EdD, President
PO Box 1069
Fitzgerald, GA 31750
229 468-2078
- Medical Assistant Prgm

## Ft Valley

**Fort Valley State University**
Larry E Rivers, PhD, President
1005 State University Dr
Ft Valley, GA 31030-4313
478 825-6315
- Dietetics-Didactic Prgm
- Rehabilitation Counseling Prgm

## Gainesville

**Brenau University**
Ed L Schrader, PhD, President
500 Washington, SE
Gainesville, GA 30501
770 534-6110
- Occupational Therapy Prgm

## Griffin

**Griffin Technical College**
Robert Arnold, EdD, CEO
501 Varsity Rd
Griffin, GA 30223
770 228-7365
- Medical Assistant Prgm
- Radiography Prgm
- Respiratory Therapist (Advanced) Prgm
- Surgical Technology Prgm

## Jasper

**Appalachian Technical College**
Sanford Chandler, PhD, President
100 Campus Dr
Jasper, GA 30143
706 253-4500
- Medical Assistant Prgm

## Kennesaw

**Kennesaw State University**
Betty L Siegel, President
1000 Chastain Rd
Kennesaw, GA 30144-5591
770 423-6000
- Cytogenetic Technology Prgm

## LaGrange

**West Georgia Technical College**
Daryl Gilley, PhD, President
303 Fort Dr
LaGrange, GA 30240
706 845-4323
- Medical Assistant Prgm
- Radiography Prgm

## Lawrenceville

**Gwinnett Technical College**
Sharon Rigsby, ABJ MEd, President
5150 Sugarloaf Parkway
Lawrenceville, GA 30043
678 226-6602
- Dental Assisting Prgm
- Emergency Med Tech-Paramedic Prgm
- Medical Assistant Prgm
- Physical Therapist Assistant Prgm
- Radiography Prgm
- Respiratory Therapist (Advanced) Prgm
- Surgical Technology Prgm

## Macon

**Central Georgia Technical College**
Melton Palmer, Jr, PhD, President
3300 Macon Tech Dr
Macon, GA 31206
478 757-3400
- Clin Lab Technician/Med Lab Technician
- Dental Hygiene Prgm
- Surgical Technology Prgm

**Macon State College**
David Bell, PhD, President
100 College Station Dr
Macon, GA 31206-5145
478 471-2700
- Health Information Admin Prgm
- Health Information Tech Prgm
- Respiratory Therapist (Advanced) Prgm

## Marietta

**Chattahoochee Technical College**
Harlon D Crimm, EdD, President
980 S Cobb Dr
Marietta, GA 30060
770 528-4500
- Medical Assistant Prgm
- Surgical Technology Prgm

**Life University**
Sid E Williams, President
1269 Barclay Cir
Marietta, GA 30060
404 424-0554
- Dietetic Internship Prgm
- Dietetics-Didactic Prgm

## Midgeville

**Georgia College & State University**
Dorothy Leland, President
Midgeville, GA 31061
912 445-5350
- Athletic Training Prgm
- Music Therapy Prgm

## Morrow

**Clayton State University**
Thomas K Harden, EdD, President
2000 Clayton State Blvd
Morrow, GA 30260
678 466-4600
- Dental Hygiene Prgm
- Medical Assistant Prgm

## Moultrie

**Moultrie Technical College**
Tina Anderson, EdD, President
800 Veterans Parkway N
Moultrie, GA 31788
229 891-7000
• Medical Assistant Prgm
• Radiography Prgm

## Oakwood

**Lanier Technical College**
Michael Moye, EdD, President
2990 Landrum Education Dr
Oakwood, GA 30566-0058
770 531-6304
• Clin Lab Technician/Med Lab Technician
• Dental Assisting Prgm
• Dental Hygiene Prgm
• Medical Assistant Prgm
• Surgical Technology Prgm

## Riverdale

**Southern Regional Medical Center**
11 Upper Riverdale Rd SW
Riverdale, GA 30274-2600
770 991-8053
• Dietetic Internship Prgm

## Rock Spring

**Northwestern Technical College**
Ray Brooks, EdD, President
PO Box 569
Rock Spring, GA 30739
706 764-3530
• Medical Assistant Prgm
• Occupational Therapy Asst Prgm
• Surgical Technology Prgm

## Rome

**Coosa Valley Technical College**
Craig McDaniel, EdD, President
One Maurice Culberson Dr
Rome, GA 30161
706 295-6927
• Dental Assisting Prgm
• Diagnostic Med Sonography Prgm
• Medical Assistant Prgm
• Nuclear Medicine Technology Prgm
• Radiation Therapy Prgm
• Radiography Prgm
• Respiratory Therapist (Advanced) Prgm

**Georgia Highlands College**
Randy Pierce, President
3175 Cedartown Hwy SE
Rome, GA 30161
706 802-5000
• Dental Hygiene Prgm

## Savannah

**Armstrong Atlantic State University**
Thomas Jones, PhD, President
11935 Abercorn St
Savannah, GA 31419-1997
912 927-5258
• Clin Lab Scientist/Med Technologist Prgm
• Dental Hygiene Prgm
• Nuclear Medicine Technology Prgm
• Physical Therapy Prgm
• Radiation Therapy Prgm
• Radiography Prgm
• Respiratory Therapist (Advanced) Prgm
• Speech-Language Pathology Prgm

**Savannah Technical College**
C B Rathburn, PhD, President
5717 White Bluff Rd
Savannah, GA 31405-5521
912 351-4404
• Dental Assisting Prgm
• Medical Assistant Prgm
• Surgical Technology Prgm

**South University**
John South, Chancellor
709 Mall Blvd
Savannah, GA 31406
912 201-8000
• Anesthesiologist Asst Prgm
• Medical Assistant Prgm
• Physical Therapist Assistant Prgm
• Physician Assistant Prgm

## Smyrna

**Medix School**
Duncan Anderson, President
2108 Cobb Pkwy
Smyrna, GA 30080
770 980-0002
• Dental Assisting Prgm
• Medical Assistant Prgm

## Statesboro

**Georgia Southern University**
Bruce F Grube, PhD, President
PO Box 8033
Marvin Pittman Admin Bldg
Statesboro, GA 30460-8033
912 681-5211
• Athletic Training Prgm
• Dietetics-Didactic Prgm
• Therapeutic Recreation Specialist Prgm

**Ogeechee Technical College**
Randy Peters, PhD, Interim President
One Joe Kennedy Blvd
Statesboro, GA 30458
912 681-5500
• Dental Assisting Prgm
• Diagnostic Med Sonography Prgm
• Health Information Tech Prgm
• Medical Assistant Prgm
• Ophthalmic Dispensing Optician Prgm
• Pharmacy Technician Prgm
• Radiography Prgm
• Surgical Technology Prgm

## Swainsboro

**Swainsboro Technical College**
Glenn Deibert, EdD, President
346 Kite Rd
Swainsboro, GA 30401
478 289-2250
• Medical Assistant Prgm

## Thomaston

**Flint River Technical College**
Kathy Love, PhD, President
1533 Hwy 19 S
Thomaston, GA 30286
706 646-6148
• Surgical Technology Prgm

## Thomasville

**Southwest Georgia Technical College**
Glenn Deibert, MS, President
15689 US Hwy 19 North
Thomasville, GA 31792
229 225-5069
• Clin Lab Technician/Med Lab Technician
• Medical Assistant Prgm
• Pharmacy Technician Prgm
• Respiratory Therapist (Advanced) Prgm
• Surgical Technology Prgm

**Thomas University**
Gary Bonvillian, PhD, President
Forbes Bldg
1501 Millpond Rd
Thomasville, GA 31792
229 226-1621
• Clin Lab Scientist/Med Technologist Prgm
• Rehabilitation Counseling Prgm

## Valdosta

**Valdosta State University**
Ronald M Zaccari, PhD, President
West Hall
Valdosta, GA 31698
912 333-5952
• Athletic Training Prgm
• Dental Hygiene Prgm
• Speech-Language Pathology Prgm

**Valdosta Technical College**
Tina Anderson, PhD-Ed, Interim President
4089 Valtech Rd
PO Box 928
Valdosta, GA 31603
229 333-2100
• Clin Lab Technician/Med Lab Technician
• Dental Assisting Prgm
• Dental Hygiene Prgm
• Medical Assistant Prgm
• Pharmacy Technician Prgm
• Radiography Prgm
• Surgical Technology Prgm

**INSTITUTIONS**

## Vidalia

**Southeastern Technical College**
Cathryn T Meehan, EdD, President
3001 E First St
Vidalia, GA 30474
912 538-3100
• Clin Lab Technician/Med Lab Technician
• Medical Assistant Prgm
• Pharmacy Technician Prgm
• Radiography Prgm
• Surgical Technology Prgm

## Waco

**West Central Technical College**
Skip Sullivan, PhD, President
176 Murphy Campus Blvd
Waco, GA 30182
770 537-7940
• Clin Lab Technician/Med Lab Technician
• Dental Hygiene Prgm
• Medical Assistant Prgm
• Radiography Prgm
• Surgical Technology Prgm

## Warner Robins

**Middle Georgia Technical College**
Ivan Allen, EdD, President
80 Cohen Walker Dr
Warner Robins, GA 31088
478 988-6912
• Dental Assisting Prgm
• Dental Hygiene Prgm
• Nuclear Medicine Technology Prgm
• Radiography Prgm
• Surgical Technology Prgm

## Waycross

**Okefenokee Technical College**
Gail Thaxton, EdD, President
1701 Carswell Ave
Waycross, GA 31503
912 287-5828
• Clin Lab Technician/Med Lab Technician
• Radiography Prgm
• Surgical Technology Prgm

# Guam

## GMF Barrigada

**Guam Community College**
Herominiano Delos Santos, EdD, President
PO Box 23069
GMF Barrigada, GU 96921
671 735-5636
• Medical Assistant Prgm

# Hawaii

## Honolulu

**Heald College - Honolulu**
Evelyn A Schemmel, BS, VP/Executive Director
1500 Kapiolani Blvd
Honolulu, HI 96814
808 955-1500
• Medical Assistant Prgm

**Kapi'olani Community College**
Leon Richards, PhD, Acting Chancellor
4303 Diamond Head Rd
Ilima Bldg, Rm 213
Honolulu, HI 96816-4421
808 734-9565
• Clin Lab Technician/Med Lab Technician
• Medical Assistant Prgm
• Occupational Therapy Asst Prgm
• Phlebotomist Prgm
• Physical Therapist Assistant Prgm
• Radiography Prgm
• Respiratory Therapist (Advanced) Prgm
• Surgical Technology Prgm

**University of Hawaii at Manoa**
Denise Konan, PhD, Interim Chancellor
Hawaii 202C, 2500 Campus Rd
Honolulu, HI 96822
808 956-7651
• Athletic Training Prgm
• Clin Lab Scientist/Med Technologist Prgm
• Counseling Prgm
• Dental Hygiene Prgm
• Dietetics-Didactic Prgm
• Medical Librarian Prgm
• Rehabilitation Counseling Prgm
• Speech-Language Pathology Prgm

## Kahului

**Maui Community College**
310 Kaahumana Ave
Kahului, HI 96732
808 984-3500
• Dental Assisting Prgm

# Idaho

## Boise

**Apollo College - Boise**
Chuck Ericson, Executive Director
1200 N Liberty St
Boise, ID 83704
208 377-8080
• Dental Assisting Prgm
• Dental Hygiene Prgm

**Boise State University**
Robert Kustra, PhD, President
1910 University Dr
Boise, ID 83725-1000
208 426-1491
• Athletic Training Prgm
• Counseling Prgm
• Dental Assisting Prgm
• Diagnostic Med Sonography Prgm
• Health Information Tech Prgm
• Radiography Prgm
• Respiratory Therapist (Advanced) Prgm
• Surgical Technology Prgm

## Idaho Falls

**Eastern Idaho Technical College**
William Robertson, MEd, President
1600 South 2500 East
Idaho Falls, ID 83404-5788
208 524-3000
• Medical Assistant Prgm
• Surgical Technology Prgm

## Lewiston

**Lewis Clark State College**
Dene Thomas, PhD, President
500 8th Ave
Lewiston, ID 83501
208 792-2216
• Medical Assistant Prgm

## Moscow

**University of Idaho**
Timothy White, PhD, President
PO Box 443151
Moscow, ID 83844-3151
208 885-6365
• Athletic Training Prgm
• Counseling Prgm
• Dietetics-Coordinated Prgm
• Rehabilitation Counseling Prgm

## Nampa

**Northwest Nazarene University**
Richard A Hagood, President
623 Holly St
Nampa, ID 83686-5897
208 467-8011
• Counseling Prgm

## Pocatello

**Idaho State University**
Arthur Vailas, PhD, President
Campus Box 8310
Pocatello, ID 83209-8063
208 282-2171
• Audiologist Prgm
• Clin Lab Scientist/Med Technologist Prgm
• Counseling Prgm
• Dental Hygiene Prgm
• Dental Lab Technician Prgm
• Dietetic Internship Prgm
• Dietetics-Didactic Prgm
• Emergency Med Tech-Paramedic Prgm
• Health Information Tech Prgm
• Medical Assistant Prgm
• Occupational Therapy Prgm
• Physical Therapist Assistant Prgm
• Physical Therapy Prgm
• Physician Assistant Prgm
• Speech-Language Pathology Prgm

## Twin Falls

**College of Southern Idaho**
Jerry Beck, EdD, President
315 Falls Ave
PO Box 1238
Twin Falls, ID 83303-1238
208 732-6728
• Emergency Med Tech-Paramedic Prgm
• Medical Assistant Prgm
• Radiography Prgm
• Surgical Technology Prgm

# Illinois

## Aurora

**Aurora University**
Rebecca L Sherrick, PhD, President
347 S Gladstone Ave
Aurora, IL 60506-4892
630 844-5476
- Athletic Training Prgm

## Belleville

**Southwestern Illinois College**
Elmer Kirchoff, PhD, President
2500 Carlyle Ave
Belleville, IL 62221-5899
800 222-5131
- Clin Lab Technician/Med Lab Technician
- Health Information Tech Prgm
- Medical Assistant Prgm
- Physical Therapist Assistant Prgm
- Radiography Prgm
- Respiratory Therapist (Advanced) Prgm

**St Elizabeth Hospital**
Timothy Brady, FACHE, Exec Vice President
211 S Third St
Belleville, IL 62222
618 234-2120
- Clin Lab Scientist/Med Technologist Prgm

## Bourbonnais

**Olivet Nazarene University**
John C Bowling, EdD, President
One University Ave
Bourbonnais, IL 60914
815 939-5011
- Athletic Training Prgm
- Dietetics-Didactic Prgm

## Carbondale

**Southern Illinois Univ at Carbondale**
Walter Wendler, PhD, Chancellor
116 Anthony Hall, MC 6801
Carbondale, IL 62901
618 453-2341
- Athletic Training Prgm
- Counseling Prgm
- Dental Hygiene Prgm
- Diagnostic Med Sonography Prgm
- Dietetic Internship Prgm
- Dietetics-Didactic Prgm
- Physical Therapist Assistant Prgm
- Physician Assistant Prgm
- Rehabilitation Counseling Prgm
- Respiratory Therapist (Advanced) Prgm
- Speech-Language Pathology Prgm
- Therapeutic Recreation Specialist Prgm

## Carterville

**John A Logan College**
Robert Mees, PhD, President
700 Logan College Rd
Carterville, IL 62918
618 985-2828
- Dental Assisting Prgm
- Dental Hygiene Prgm
- Diagnostic Med Sonography Prgm

## Centralia

**Kaskaskia College**
James Underwood, EdD, President
27210 College Rd
Centralia, IL 62801
618 545-3000
- Dental Assisting Prgm
- Physical Therapist Assistant Prgm
- Radiography Prgm
- Respiratory Therapist (Advanced) Prgm

## Champaign

**Parkland College**
Robert Exley, PhD, President
2400 W Bradley Ave
Champaign, IL 61821-1899
217 351-2231
- Dental Hygiene Prgm
- Occupational Therapy Asst Prgm
- Radiography Prgm
- Respiratory Therapist (Advanced) Prgm
- Surgical Technology Prgm

**Univ of Illinois at Urbana - Champaign**
Richard Herman, PhD, Chancellor
317 Swanlund Admin
MC 304
Champaign, IL 61801
217 333-6290
- Athletic Training Prgm
- Audiologist Prgm
- Dietetic Internship Prgm
- Dietetics-Didactic Prgm
- Medical Librarian Prgm
- Rehabilitation Counseling Prgm
- Speech-Language Pathology Prgm

## Charleston

**Eastern Illinois University**
Louis V Hencken, MSEd, President
600 Lincoln Ave
Charleston, IL 61920
217 581-2011
- Athletic Training Prgm
- Counseling Prgm
- Dietetic Internship Prgm
- Dietetics-Didactic Prgm
- Speech-Language Pathology Prgm
- Therapeutic Recreation Specialist Prgm

## Chicago

**Adler School of Professional Psychology**
Ray Crossman, President
65 E Wacker Pl, Ste 2100
Chicago, IL 60601-7203
312 201-5900
- Art Therapy Prgm

**Advocate Illinois Masonic Medical Center**
Susan Nordstrom-Lopez, Chief Executive
836 W Wellington Ave
Chicago, IL 60657
773 296-7081
- Radiography Prgm

**Advocate Trinity Hospital**
Kenneth Rojek, MBA, Chief Executive Officer
2320 E 93rd St
Chicago, IL 60617
773 967-5000
- Radiography Prgm

**Chicago School of Massage Therapy**
Jason Scholte, President
17 N State, 5th Fl
Chicago, IL 60602
312 753-7900
- Massage Therapy Prgm (2)

**Chicago State University**
Elnora D Daniel, PhD, President
9501 S King Dr
Chicago, IL 60628-1598
773 995-2400
- Counseling Prgm
- Health Information Admin Prgm
- Occupational Therapy Prgm

**Columbia College**
Warrick L Carter, PhD, President
600 S Michigan
Chicago, IL 60605-9988
- Dance/Movement Therapy Prgm

**DeVry University**
Jim Treleaven, President
3300 N Campbell Ave
Chicago, IL 60618
773 697-2080
- Health Information Tech Prgm (2)

**Illinois Institute of Technology**
Lewis M Collens, President
3300 S Federal St
Chicago, IL 60616
312 567-3000
- Rehabilitation Counseling Prgm

**Kennedy-King College**
Wayne Watson, President
6800 S Wentworth Ave
Chicago, IL 60621
773 602-5229
- Dental Hygiene Prgm

**Loyola University of Chicago**
John J Piderit, President
820 N Michigan Ave
Chicago, IL 60611
312 915-6000
- Dietetic Internship Prgm
- Dietetics-Didactic Prgm

**Malcolm X College**
Zerrie D Campbell, MS MA, President
1900 W Van Buren St
Chicago, IL 60612
312 850-7037
- Dietetic Technician-AD Prgm
- Pharmacy Technician Prgm
- Physician Assistant Prgm
- Radiography Prgm
- Respiratory Therapist (Advanced) Prgm
- Surgical Technology Prgm

**North Park University**
David L Parkyn, President
3225 W Foster Ave
Chicago, IL 60625
773 244-5710
- Athletic Training Prgm

**Northeastern Illinois University**
5500 N St Louis Ave
Chicago, IL 60625-4699
773 583-4050
• Counseling Prgm
• Rehabilitation Counseling Prgm

**Northwestern Memorial Hospital**
Dean Harrison, MBA, President
251 E Huron St
Feinberg 3-708F
Chicago, IL 60611
312 926-3007
• Diagnostic Med Sonography Prgm
• Nuclear Medicine Technology Prgm
• Radiation Therapy Prgm

**Olive Harvey College**
Valerie R Roberson, PhD, Interim President
10001 W Woodlawn Ave
Chicago, IL 60628
773 291-6313
• Respiratory Therapist (Advanced) Prgm

**Robert Morris College**
Michael Viollt, President
401 S State St
Chicago, IL 60605
312 935-6600
• Medical Assistant Prgm (7)

**Roosevelt University**
Charles R Middleton, PhD, President
430 S Michigan Ave
Chicago, IL 60605-1394
312 341-3500
• Counseling Prgm

**Rush University**
Larry J Goodman, MD, President
1725 W Harrison St, Ste 364
Chicago, IL 60612
312 942-7073
• Clin Lab Scientist/Med Technologist Prgm
• Diagnostic Med Sonography Prgm
• Occupational Therapy Prgm
• Perfusion Prgm

**Rush University Medical Center**
Larry Goodman, MD, President and CEO
1725 W Harrison St, Ste 364
Chicago, IL 60612
312 942-7073
• Audiologist Prgm
• Dietetic Internship Prgm
• Speech-Language Pathology Prgm

**Saint Xavier University**
Richard A Yanikoski, President
3700 W 103rd St
Chicago, IL 60655
312 298-3561
• Speech-Language Pathology Prgm

**School of the Art Institute of Chicago**
Carol Becker, PhD, Dean of Faculty
37 S Wabash Ave
Chicago, IL 60603
312 899-1236
• Art Therapy Prgm

**Spanish Coalition for Jobs Inc**
Mary Gonzalez-Koening, President
2011 W Pershing Rd
Chicago, IL 60609
773 247-0707
• Medical Assistant Prgm

**St Augustine College**
Z Clara Brennan, PhD, President
1333-45 W Argyle St
Chicago, IL 60640
773 878-8756
• Respiratory Therapist (Entry-Level) Prgm

**University of Illinois at Chicago**
R Michael Tanner, PhD, Provost
University Hall
601 S Morgan St M/C 105
Chicago, IL 60612-7128
312 413-3450
• Dietetics-Coordinated Prgm
• Health Information Admin Prgm
• Medical Illustrator Prgm
• Occupational Therapy Prgm
• Physical Therapy Prgm
• Specialist in BB Tech Prgm

**Wright College**
Charles P Guengerich, PhD, President
4300 N Narragansett Ave
Chicago, IL 60634
773 777-7900
• Occupational Therapy Asst Prgm
• Radiography Prgm

## Chicago Heights

**Prairie State College**
Paul J McCarthy, PhD, President
202 S Halsted
Chicago Heights, IL 60411
708 709-3500
• Dental Hygiene Prgm
• Surgical Technology Prgm

## Cicero

**Morton College**
Brent Knight, PhD, President
3801 S Central Ave
Cicero, IL 60804
708 656-8000
• Massage Therapy Prgm
• Physical Therapist Assistant Prgm

## Danville

**Danville Area Community College**
Alice M Jacobs, PhD, President
2000 E Main St
Danville, IL 61832
217 443-8848
• Health Information Tech Prgm
• Radiography Prgm

## Decatur

**Millikin University**
Douglas E Zemke, President
1184 W Main
Decatur, IL 62522
800 373-7733
• Athletic Training Prgm

**Richland Community College**
Gayle Saunders, PhD, President
1 College Park
Decatur, IL 62521-8512
217 875-7200
• Surgical Technology Prgm

## Deerfield

**Trinity International University**
Gregory L Waybright, PhD, President
2065 Half Day Rd
Deerfield, IL 60015
• Athletic Training Prgm

## DeKalb

**Northern Illinois University**
John Peters, PhD, President
Altgeld Hall
DeKalb, IL 60115
815 753-9501
• Athletic Training Prgm
• Audiologist Prgm
• Clin Lab Scientist/Med Technologist Prgm
• Counseling Prgm
• Dietetic Internship Prgm
• Dietetics-Didactic Prgm
• Orientation and Mobility Specialist Prgm
• Physical Therapy Prgm
• Rehabilitation Counseling Prgm
• Rehabilitation Teacher Prgm
• Speech-Language Pathology Prgm

## Des Plaines

**Oakton Community College**
Margaret B Lee, PhD, President
1600 E Golf Rd
Des Plaines, IL 60016
847 635-1732
• Clin Lab Technician/Med Lab Technician
• Health Information Tech Prgm
• Physical Therapist Assistant Prgm

## Dixon

**Sauk Valley Community College**
George J Mihel, EdD, President
173 IL Rte 2
Dixon, IL 61021-9110
815 288-5511
• Radiography Prgm

## Downers Grove

**Midwestern University**
Kathleen H Goepplinger, PhD, President
555 31st St
Downers Grove, IL 60515-1235
630 515-7300
• Occupational Therapy Prgm
• Physical Therapy Prgm
• Physician Assistant Prgm

## East Peoria

**Illinois Central College**
John S Erwin, President
One College Dr
East Peoria, IL 61635-0001
309 694-5431
- Clin Lab Technician/Med Lab Technician
- Dental Hygiene Prgm
- Occupational Therapy Asst Prgm
- Physical Therapist Assistant Prgm
- Radiography Prgm
- Respiratory Therapist (Advanced) Prgm
- Surgical Technology Prgm

## Edwardsville

**Southern Illinois Univ at Edwardsville**
Vaughn Vandegrift, PhD, Chancellor
Box 1151
Edwardsville, IL 62026
618 650-2475
- Art Therapy Prgm
- Speech-Language Pathology Prgm

## Elgin

**Elgin Community College**
Michael Shirley, PhD, President
1700 Spartan Dr
Elgin, IL 60123-7193
847 697-1000
- Clin Lab Technician/Med Lab Technician
- Dental Assisting Prgm
- Surgical Technology Prgm

## Evanston

**Evanston Northwestern Healthcare - Evanston**
Raymond Grady, President/CEO
Hospitals and Clinics, Evanston Hospital
2650 Ridge Ave
Evanston, IL 60201
847 570-2005
- Clin Lab Scientist/Med Technologist Prgm

**Northwestern University**
Henry S Bienen, PhD, President
2299 Sheridan Rd
Evanston, IL 60201
708 491-7456
- Audiologist Prgm
- Genetic Counseling Prgm
- Orthotist/Prosthetist Prgm
- Physical Therapy Prgm
- Speech-Language Pathology Prgm

**St Francis Hospital**
Jeffrey Murphy, CEO
355 Ridge Ave
Evanston, IL 60202
847 492-4000
- Radiography Prgm

## Galesburg

**Carl Sandburg College**
Thomas A Schmidt, MBA, President
2400 Tom L Wilson Blvd
Galesburg, IL 61401
309 344-2518
- Dental Hygiene Prgm
- Radiography Prgm

## Glen Ellyn

**College of DuPage**
Sunil Chand, PhD, President
425 Fawell Blvd
Glen Ellyn, IL 60137-6599
630 942-2200
- Dental Hygiene Prgm
- Diagnostic Med Sonography Prgm
- Health Information Tech Prgm
- Nuclear Medicine Technology Prgm
- Physical Therapist Assistant Prgm
- Radiography Prgm
- Surgical Technology Prgm

## Godfrey

**Lewis & Clark Community College**
Dale T Chapman, EdD, President
5800 Godfrey Rd
Godfrey, IL 62035-2466
618 468-2000
- Dental Assisting Prgm
- Dental Hygiene Prgm
- Occupational Therapy Asst Prgm

## Grayslake

**College of Lake County**
Richard Fonte, PhD, President
19351 W Washington St
Grayslake, IL 60030
847 543-2201
- Dental Hygiene Prgm
- Health Information Tech Prgm
- Phlebotomist Prgm
- Radiography Prgm
- Surgical Technology Prgm

## Harvey

**Ingalls Memorial Hospital**
One Ingalls Dr
Harvey, IL 60426
708 333-2300
- Dietetic Internship Prgm

## Herrin

**Southern Illinois Collegiate Common Market**
Jack D Hill, PhD, Executive Director
3213 S Park Ave
Herrin, IL 62948
618 942-6902
- Clin Lab Technician/Med Lab Technician
- Health Information Tech Prgm
- Occupational Therapy Asst Prgm
- Surgical Technology Prgm

## Hines

**Edward Hines Jr VA Hospital**
Jack Hetrick, Director
Fifth Ave and Roosevelt Rd
PO Box 5000
Hines, IL 60141
708 216-2153
- Clin Lab Scientist/Med Technologist Prgm
- Dietetic Internship Prgm
- Nuclear Medicine Technology Prgm

## Kankakee

**Kankakee Community College**
Jerry Weber, PhD, President
PO Box 888
Kankakee, IL 60901-0888
815 802-8112
- Clin Lab Technician/Med Lab Technician
- Respiratory Therapist (Advanced) Prgm

## Lebanon

**McKendree College**
James M Dennis, PhD, President
701 College Rd
Lebanon, IL 62254-1299
618 537-6936
- Athletic Training Prgm

## Lincoln

**Midwest Technical Institute**
Brian Huff, President
405 N Limit St
PO Box 506
Lincoln, IL 62656
217 735-3105
- Medical Assistant Prgm

## Lisle

**Benedictine University**
William J Carroll, President
5700 College Rd
Lisle, IL 60532
708 960-1500
- Dietetic Internship Prgm
- Dietetics-Didactic Prgm

## Lombard

**National University of Health Sciences**
James F Winterstein, DC, President
200 E Roosevelt Rd
Lombard, IL 60148
630 889-6604
- Massage Therapy Prgm

## Macomb

**McDonough District Hospital**
Stephen R Hopper, MS, President
525 E Grant St
Macomb, IL 61455
309 833-4101
- Radiography Prgm

**Western Illinois University**
Al Goldfarb, PhD, President
209 Sherman Hall
1 University Circle
Macomb, IL 61455
309 298-1824
- Athletic Training Prgm
- Counseling Prgm
- Dietetics-Didactic Prgm
- Music Therapy Prgm
- Speech-Language Pathology Prgm
- Therapeutic Recreation Specialist Prgm

## Malta

**Kishwaukee College**
Dave Louis, PhD, President
21193 Malta Rd
Malta, IL 60150
815 825-2086
• Massage Therapy Prgm
• Radiography Prgm

## Mattoon

**Lake Land College**
Robert K Luther, President
5001 Lake Land Blvd
Mattoon, IL 61938-9366
217 234-5253
• Dental Hygiene Prgm
• Physical Therapist Assistant Prgm

## Maywood

**Loyola University Medical Center**
Anthony Barbato, MD, Executive Vice President
2160 S First Ave
Maywood, IL 60153
708 216-9000
• Emergency Med Tech-Paramedic Prgm

## Moline

**Black Hawk College**
Keith Miller, PhD, President
6600 34th Ave
Moline, IL 61265-5899
309 796-1311
• Physical Therapist Assistant Prgm

## Naperville

**North Central College**
Harold R Wilde, PhD, President
30 N Brainard
Naperville, IL 60540
630 637-5454
• Athletic Training Prgm

## Normal

**Bloomington-Normal School of Radiography**
Mike Johnson, MBA RT(R), Chair, Board of
    Directors
900 Franklin Ave
Normal, IL 61761
309 452-2834
• Radiography Prgm

**Illinois State University**
C Alvin Bowman, PhD, President
1000 President's Office
418 Hovey Hall
Normal, IL 61790-1000
309 438-5677
• Athletic Training Prgm
• Audiologist Prgm
• Clin Lab Scientist/Med Technologist Prgm
• Dietetic Internship Prgm
• Dietetics-Didactic Prgm
• Health Information Admin Prgm
• Music Therapy Prgm
• Speech-Language Pathology Prgm
• Therapeutic Recreation Specialist Prgm

## North Chicago

**Rosalind Franklin Univ of Medicine & Science**
K Michael Welch, MBChB FRCP, President/CEO
3333 Green Bay Rd
North Chicago, IL 60064
847 578-3000
• Clin Lab Scientist/Med Technologist Prgm
• Pathologists' Assistant Prgm
• Physical Therapy Prgm
• Physician Assistant Prgm

## Oglesby

**Illinois Valley Community College**
Jean Goodnow, President
815 N Orlando Smith Ave
Oglesby, IL 61348-9691
815 224-2720
• Dental Assisting Prgm

## Olney

**Olney Central College**
Jackie Davis, EdD, President
305 N West St
Olney, IL 62450
618 395-7777
• Radiography Prgm

## Palatine

**Harper College**
Robert Breuder, PhD, President
1200 W Algonquin Rd
Palatine, IL 60067
847 925-6000
• Dental Hygiene Prgm
• Dietetic Technician-AD Prgm
• Medical Assistant Prgm

## Palos Hills

**Moraine Valley Community College**
Vernon Crawley, PhD, President/CEO
9000 W College Pkwy
Palos Hills, IL 60465
708 974-5201
• Health Information Tech Prgm
• Medical Assistant Prgm
• Phlebotomist Prgm
• Radiography Prgm
• Respiratory Therapist (Advanced) Prgm

## Peoria

**Bradley University**
David Broski, President
1501 W Bradley Ave
Peoria, IL 61625
309 676-7611
• Counseling Prgm
• Dietetics-Didactic Prgm
• Physical Therapy Prgm

**Midstate College**
R Dale Bunch, President
411 W Northmoor Rd
Peoria, IL 61614
309 692-4092
• Medical Assistant Prgm

**OSF St Francis Medical Center**
Keith Steffen, Administrator
530 NE Glen Oak Ave
Peoria, IL 61637
309 655-2020
• Clin Lab Scientist/Med Technologist Prgm
• Dietetic Internship Prgm
• Histotechnician Prgm
• Radiography Prgm

## Quincy

**Blessing Hospital**
Maureen Kahn, RN MHA, President/CEO
Broadway at 11th St
PO Box 7005
Quincy, IL 62305
217 223-8400
• Clin Lab Technician/Med Lab Technician
• Pharmacy Technician Prgm
• Radiography Prgm

**John Wood Community College**
William M Simpson, EdD, President
1301 S 48th St
Quincy, IL 62305
217 224-6500
• Surgical Technology Prgm

## River Forest

**Concordia University**
Manfred Booa, PhD, Acting President
7400 Augusta
River Forest, IL 60305-1499
708 771-8300
• Counseling Prgm

**Dominican University**
Donna M Carroll, President
7900 W Division
River Forest, IL 60305
708 366-2490
• Dietetics-Didactic Prgm
• Medical Librarian Prgm

## River Grove

**Triton College**
Patricia Granados, EdD, President
2000 N Fifth Ave
River Grove, IL 60171
708 456-0300
• Diagnostic Med Sonography Prgm
• Nuclear Medicine Technology Prgm
• Ophthalmic Med Technician Prgm
• Radiography Prgm
• Respiratory Therapist (Advanced) Prgm
• Surgical Technology Prgm

## Rock Island

**Trinity College of Nursing & Health Sciences**
Carol Dwyer, RN MSN MM, President
2122 25th Ave
Rock Island, IL 61201
309 779-2256
• Emergency Med Tech-Paramedic Prgm
• Radiography Prgm
• Surgical Technology Prgm

## Rockford

**OSF St Anthony Medical Center**
David Schertz, Administrator
5666 E State St
Rockford, IL 61108
815 226-2000
• Clin Lab Scientist/Med Technologist Prgm

**Rock Valley College**
Roland J Chapdelaine, MS EdD, President
3301 N Mulford Rd
Rockford, IL 61114-5699
815 654-4260
• Dental Hygiene Prgm
• Respiratory Therapist (Advanced) Prgm
• Surgical Technology Prgm

**Rockford Business College**
James Devaney, PhD, CEO
730 N Church
Rockford, IL 61103
815 965-8616
• Medical Assistant Prgm

**Rockford Memorial Hospital**
Gary Kaatz, MS, President
2400 N Rockton Ave
Rockford, IL 61103
815 971-5000
• Radiography Prgm

**Swedish American Hospital**
William Gorski, MD, President/CEO
1401 E State St
Rockford, IL 61104-2298
815 968-4400
• Radiation Therapy Prgm
• Radiography Prgm

## Rosemont

**Northwestern Business College**
Lawrence W Schumacher, BA, President
9700 W Higgins Rd, Ste 750
Rosemont, IL 60018
847 318-8550
• Health Information Tech Prgm
• Medical Assistant Prgm (2)

## South Holland

**South Suburban College**
George Dammer, MS, President
15800 S State St
South Holland, IL 60473-1262
708 596-2000
• Diagnostic Med Sonography Prgm
• Medical Assistant Prgm
• Occupational Therapy Asst Prgm
• Pharmacy Technician Prgm
• Phlebotomist Prgm
• Radiography Prgm

## Springfield

**Capital Area Career Center**
John Bailey, EdS, Director
2201 Toronto Rd
Springfield, IL 62712
217 585-2160
• Surgical Technology Prgm

**Lincoln Land Community College**
Charlotte Warren, PhD, President
5250 Shepherd Rd
PO Box 19256
Springfield, IL 62794-9256
217 786-2273
• Occupational Therapy Asst Prgm
• Radiography Prgm

**St John's Hospital**
Richard Carlson, FACHE, Exec Vice President
800 E Carpenter
Springfield, IL 62769
217 544-6464
• Clin Lab Scientist/Med Technologist Prgm
• Dietetic Internship Prgm
• Electroneurodiagnostic Tech Prgm
• Respiratory Therapist (Advanced) Prgm

**University of Illinois at Springfield**
Richard Ringeisen, PhD, Chancellor
One University Plaza
Springfield, IL 62703
217 786-6634
• Clin Lab Scientist/Med Technologist Prgm
• Counseling Prgm

## Sugar Grove

**Waubonsee Community College**
Christine J Sobek, EdD, President
Rte 47 at Waubonsee Dr
Sugar Grove, IL 60554
630 466-7900
• Medical Assistant Prgm
• Surgical Technology Prgm

## Ullin

**Shawnee Community College**
Terry G Ludwig, PhD, President
8364 Shawnee College Rd
Ullin, IL 62992
618 634-3221
• Surgical Technology Prgm

## University Park

**Governors State University**
Stuart Fagen, PhD, President
1 University Pkwy
University Park, IL 60466
708 534-4130
• Counseling Prgm
• Occupational Therapy Prgm
• Physical Therapy Prgm
• Speech-Language Pathology Prgm

# Indiana

## Alexandria

**Alexandria School of Scientific Therapeutics**
Herbert Hobbs, CEO
809 S Harrison
PO Box 287
Alexandria, IN 46001
765 724-9152
• Massage Therapy Prgm

## Anderson

**Anderson University**
James L Edwards, PhD, President
1100 E 5th St
Anderson, IN 46012
765 641-4011
• Athletic Training Prgm

**Ivy Tech Community College - Anderson**
Jack Voelz, MA, Exec Dean
104 W 53rd St
Anderson, IN 46013
317 643-7133
• Dental Assisting Prgm
• Medical Assistant Prgm

## Beech Grove

**St Francis Hospital & Health Centers**
Robert J Brody, MHA, President/CEO
1600 Albany St
Beech Grove, IN 46107
317 783-8220
• Clin Lab Scientist/Med Technologist Prgm
• Emergency Med Tech-Paramedic Prgm

## Bloomington

**Bloomington Hospital**
Nancy Carlstedt, MS, President
PO Box 1149
Bloomington, IN 47402
812 336-6821
• Surgical Technology Prgm

**Indiana University - Bloomington**
Adam Herbert, PhD, President
President's Office, Bryan Hall
Bloomington, IN 47405
812 855-4613
• Athletic Training Prgm
• Audiologist Prgm
• Counseling Prgm
• Dietetics-Didactic Prgm
• Speech-Language Pathology Prgm
• Therapeutic Recreation Specialist Prgm

## Columbus

**Columbus Area Career Connection/Ivy Tech St**
Christy Ross, CDA, Instructor, Director
230 S Marr Rd
Columbus, IN 47201
812 376-4202
• Dental Assisting Prgm

**Columbus Regional Hospital**
Douglas J Leonard, Vice President
2400 E 17th St
Columbus, IN 47201
812 376-5439
• Radiography Prgm

**Ivy Tech Community College - Columbus**
John Hogan, PhD, Chancellor
4475 Central Ave
Columbus, IN 47203-1868
812 372-9925
• Medical Assistant Prgm
• Surgical Technology Prgm

INSTITUTIONS

## Elkhart

**Elkhart General Hospital**
Greg Lintjer, President
600 E Blvd
Elkhart, IN 46514
574 293-8961
• Emergency Med Tech-Paramedic Prgm

## Evansville

**Indiana Business College - Evansville**
Ken Konesco, BS, President
4601 Theater Dr
Evansville, IN 47715
• Medical Assistant Prgm

**Ivy Tech Community College SW - Evansville**
Daniel Schenk, PhD, Chancellor
3501 First Ave
Evansville, IN 47710
812 426-2865
• Emergency Med Tech-Paramedic Prgm
• Medical Assistant Prgm
• Surgical Technology Prgm

**University of Evansville**
Stephen G Jennings, President
1800 Lincoln Ave
Evansville, IN 47722
812 488-2151
• Athletic Training Prgm
• Music Therapy Prgm
• Physical Therapist Assistant Prgm
• Physical Therapy Prgm

**University of Southern Indiana**
H Ray Hoops, PhD, President
8600 University Blvd
Evansville, IN 47712-3534
812 464-1756
• Dental Assisting Prgm
• Dental Hygiene Prgm
• Occupational Therapy Asst Prgm
• Occupational Therapy Prgm
• Radiography Prgm
• Respiratory Therapist (Advanced) Prgm

## Franklin

**Franklin College**
Jay Moseley, President
101 Brainigin Blvd
Franklin, IN 46131
317 738-8010
• Athletic Training Prgm

## Ft Wayne

**Ft Wayne School of Radiography**
Tom Miller, MD, CEO
St Joseph Hospital
700 Broadway
Ft Wayne, IN 46802
219 484-6636
• Radiography Prgm

**Indiana Business College - Ft Wayne**
Janet S Hein, BS, Executive Director
6413 N Clinton St
Ft Wayne, IN 46825
260 471-7667
• Medical Assistant Prgm
• Surgical Technology Prgm

**Indiana Univ - Purdue Univ Ft Wayne**
Michael A Wartell, PhD, Chancellor
2101 Coliseum Blvd E
Ft Wayne, IN 46805
260 481-6103
• Dental Assisting Prgm
• Dental Hygiene Prgm
• Dental Lab Technician Prgm
• Music Therapy Prgm

**International Business Coll - Ft Wayne**
Jim C Zillman, President
5699 Coventry Ln
Ft Wayne, IN 46804
260 459-4500
• Medical Assistant Prgm

**Ivy Tech Community College NE - Ft Wayne**
Mark Keen, PhD, Chancellor
3800 N Anthony Blvd
Ft Wayne, IN 46805
260 482-9171
• Medical Assistant Prgm
• Respiratory Therapist (Advanced) Prgm

**Parkview Hospital**
Dale Wilcox, MD, COO
2200 Randallia Dr
Ft Wayne, IN 46805
260 373-3602
• Clin Lab Scientist/Med Technologist Prgm

**University of Saint Francis**
Sr M Elise Kriss, OSF PhD, President
2701 Spring St
Ft Wayne, IN 46808
260 434-3101
• Physical Therapist Assistant Prgm
• Physician Assistant Prgm
• Radiography Prgm
• Surgical Technology Prgm

## Gary

**Indiana University Northwest**
Bruce Bergland, PhD, Chancellor
3400 Broadway
Gary, IN 46408
219 980-6700
• Clin Lab Technician/Med Lab Technician
• Dental Assisting Prgm
• Dental Hygiene Prgm
• Health Information Tech Prgm
• Phlebotomist Prgm
• Radiation Therapy Prgm
• Radiography Prgm
• Respiratory Therapist (Advanced) Prgm

**Ivy Tech Community College - Michigan City**
G Lupe Valtierra, Chancellor
1440 E 35th Ave
Gary, IN 46409
219 981-1111
• Medical Assistant Prgm
• Surgical Technology Prgm

**Ivy Tech Community College NW - Gary**
Lupe Valtierra, JD, Chancellor
1440 East 35th Ave
Gary, IN 46409
219 981-1111
• Respiratory Therapist (Advanced) Prgm

**Methodist Hospitals Inc**
Edward Charbonneau, President
Northlake Campus
600 Grant St
Gary, IN 46402
219 886-4602
• Emergency Med Tech-Paramedic Prgm

## Greencastle

**DePauw University**
Robert G Bottoms, President
313 S Locust
Administration Bldg
Greencastle, IN 46135
765 658-4055
• Athletic Training Prgm

## Greenfield

**Hancock Memorial Hospital**
Bobby Keen, President
801 N State St
Greenfield, IN 46140
317 462-0457
• Radiography Prgm

## Hammond

**St Margaret Mercy Healthcare Centers**
Tom Gryzbek, President/CEO
Mercy Healthcare Ctrs
5454 Hohman Ave
Hammond, IN 46320
219 932-2300
• Clin Lab Scientist/Med Technologist Prgm

## Indianapolis

**Butler University/Clarian Health**
Bobby Fong, PhD, President
4600 Sunset Ave
Indianapolis, IN 46208-3485
317 940-9900
• Counseling Prgm
• Physician Assistant Prgm

**Clarian Health Partners Inc**
Daniel F Evans, JD, President/CEO
PO Box 1367
Indianapolis, IN 46206-1367
317 962-5900
• Clin Lab Scientist/Med Technologist Prgm
• Electroneurodiagnostic Tech Prgm
• Emergency Med Tech-Paramedic Prgm
• Medical Assistant Prgm
• Pharmacy Technician Prgm
• Respiratory Therapist (Advanced) Prgm
• Surgical Technology Prgm

**Community Health Network**
William E Corley, President
1500 N Ritter Ave
Indianapolis, IN 46219
317 355-5529
• Radiography Prgm

**Indiana Blood Center**
Byron Buhner, AB MS, President/CEO
3450 N Meridian St
Indianapolis, IN 46208
317 916-5001
• Specialist in BB Tech Prgm

**Indiana Business College**
Pamela J Soladine, Director
8150 Brookville Rd
Indianapolis, IN 46239
317 375-8000
• Medical Assistant Prgm (3)

**Indiana Univ-Purdue U-Indianapolis**
Charles R Bantz, Chancellor
425 University Blvd
Cavanaugh Hall 441
Indianapolis, IN 46202
• Music Therapy Prgm

**Indiana University**
Charles Bantz, PhD, Chancellor
Administrative Building, Rm 104
355 N Lansing St
Indianapolis, IN 46202-2896
317 274-4417
• Clin Lab Scientist/Med Technologist Prgm
• Cytotechnology Prgm
• Dental Assisting Prgm
• Dental Hygiene Prgm
• Dietetic Internship Prgm
• Genetic Counseling Prgm
• Health Information Admin Prgm
• Histotechnician Prgm
• Medical Librarian Prgm
• Nuclear Medicine Technology Prgm
• Occupational Therapy Prgm
• Pathologists' Assistant Prgm
• Physical Therapy Prgm
• Radiation Therapy Prgm
• Radiography Prgm
• Respiratory Therapist (Advanced) Prgm

**International Business Coll - Indianapolis**
Eric Stovall, President
7205 Shadeland Station
Indianapolis, IN 46256
317 841-6400
• Medical Assistant Prgm

**Ivy Tech Community College of Indiana**
Hank Dunn, EdD, Chancellor
50 W Fall Creek Pkwy North Dr
Indianapolis, IN 46206-1763
317 921-5935
• Medical Assistant Prgm
• Physical Therapist Assistant Prgm
• Radiography Prgm
• Respiratory Therapist (Advanced) Prgm
• Surgical Technology Prgm

**Marian College**
Daniel A Felicetti, PhD, President
3200 Cold Spring Rd
Indianapolis, IN 46222
317 929-0237
• Dietetics-Coordinated Prgm
• Dietetics-Didactic Prgm

**Professional Careers Institute**
Barri Shirk, BA MS, Executive Directror
7302 Woodland Dr
Indianapolis, IN 46278-1736
317 299-6001
• Dental Assisting Prgm
• Medical Assistant Prgm

**University of Indianapolis**
Beverly Pitts, PhD, President
1400 E Hanna Ave
Indianapolis, IN 46227
317 788-3211
• Athletic Training Prgm
• Occupational Therapy Prgm
• Physical Therapist Assistant Prgm
• Physical Therapy Prgm

## Kokomo

**Indiana University - Kokomo**
Ruth J Person, PhD, Chancellor
2300 S Washington St
Kokomo, IN 46902
765 453-2000
• Radiography Prgm

**Ivy Tech Community College - Kokomo**
Steve Daily, MS, Chancellor
1815 E Morgan St
Kokomo, IN 46901
765 459-0561
• Dental Assisting Prgm
• Medical Assistant Prgm
• Surgical Technology Prgm

**St Vincent Health/St Joseph Hospital**
Darcy Burthay, MSN RN, President
1907 W Sycamore St
Kokomo, IN 46901
317 452-5611
• Radiography Prgm

## Lafayette

**Ivy Tech Community College - Lafayette**
David Bathe, DA, Chancellor
3101 S Creasy Ln
PO Box 6299
Lafayette, IN 47903
765 269-5600
• Dental Assisting Prgm
• Medical Assistant Prgm
• Respiratory Therapist (Advanced) Prgm
• Surgical Technology Prgm

## Lawrenceburg

**Ivy Tech Community College - Lawrenceburg**
James Helms, EdS, Chancellor
500 Industrial Dr
Lawrenceburg, IN 47025
812 537-4010
• Medical Assistant Prgm

## Madison

**Ivy Tech Community College - Madison**
Donald Heiderman, Executive Dean
590 Ivy Tech Dr
Madison, IN 47250
812 265-2580
• Medical Assistant Prgm

**King's Daughters' Hospital & Health Services**
Roger J Allman, MHA, President/CEO
One King's Daughters Dr, PO Box 447
Madison, IN 47250
812 265-5211
• Radiography Prgm

## Marion

**Indiana Wesleyan University**
James Barnes, EdD, President
4201 S Washington St
Marion, IN 46953
765 677-2100
• Athletic Training Prgm
• Counseling Prgm

**Ivy Tech Community College - Marion**
John Lightle, PhD, Campus Dean
1015 E 3rd St
Marion, IN 46952
765 662-9843
• Medical Assistant Prgm
• Radiography Prgm

## Merrillville

**Olympia College**
Mary Klinefelter, President
707 E 80th Pl, Ste 200
8315 Virginia St Ste A
Merrillville, IN 46410
219 756-6811
• Surgical Technology Prgm

## Muncie

**Ball Memorial Hospital**
Brent Batman, MHA, President
2401 University Ave
Muncie, IN 47303-3499
765 747-3393
• Clin Lab Scientist/Med Technologist Prgm
• Dietetic Internship Prgm
• Radiography Prgm

**Ball State University**
Jo Ann Gora, PhD, President
Administration Bldg 101
Muncie, IN 47306
317 285-5555
• Athletic Training Prgm
• Audiologist Prgm
• Counseling Prgm
• Dietetic Internship Prgm
• Dietetic Technician-AD Prgm
• Dietetics-Didactic Prgm
• Radiation Therapy Prgm
• Radiography Prgm
• Rehabilitation Counseling Prgm
• Speech-Language Pathology Prgm

**Ivy Tech Community College EC - Muncie**
Gail Chesterfield, MA, Acting Chancellor
4301 S Cowan Rd
Muncie, IN 47302
765 289-2291
• Medical Assistant Prgm
• Physical Therapist Assistant Prgm
• Surgical Technology Prgm

## North Manchester

**Manchester College**
Parker C Marden, PhD, President
Box 176
North Manchester, IN 46962
219 982-5050
• Athletic Training Prgm

## Richmond

**Ivy Tech Community College - Richmond**
James L Steck, BS MS, Chancellor
2325 Chester Blvd
Richmond, IN 47374
765 966-2656
• Medical Assistant Prgm

**Reid Hospital & Health Care Services**
Barry S MacDowell, MHA, President
1401 Chester Blvd
Richmond, IN 47374
765 983-3122
• Radiography Prgm

## Sellersburg

**Ivy Tech Community College SC - Sellersburg**
Rita Shourds, MS, Chancellor
8204 Hwy 311
Sellersburg, IN 47172
812 246-3301
• Medical Assistant Prgm

## South Bend

**Brown Mackie College**
Connie S Adelman, BGS, Campus President
1030 E Jefferson Blvd
South Bend, IN 46617-3123
574 237-0774
• Medical Assistant Prgm (2)
• Occupational Therapy Asst Prgm (2)
• Physical Therapist Assistant Prgm

**Indiana University South Bend**
Una Mae Reck, PhD, Chancellor
1700 Mishawaka Ave, Box 7111
South Bend, IN 46634
574 237-4220
• Counseling Prgm
• Dental Assisting Prgm
• Dental Hygiene Prgm
• Radiography Prgm

**Ivy Tech Community College - South Bend**
Virginia Calvin, EdD, Chancellor
220 Dean Johnson Blvd
South Bend, IN 46601
219 289-7001
• Clin Lab Technician/Med Lab Technician
• Medical Assistant Prgm
• Phlebotomist Prgm

## St Mary of the Woods

**St Mary of the Woods College**
Joan Lescinski, President
St Mary of the Woods, IN 47876
812 535-5154
• Music Therapy Prgm

## Terre Haute

**Indiana Business College - Terre Haute**
Laura Hale, AAS BBA, Executive Director
1378 South State Road 46
Terre Haute, IN 47803
812 877-2100
• Medical Assistant Prgm

**Indiana State University**
Lloyd Benjamin, PhD, President
Condit House
Terre Haute, IN 47809
812 237-4000
• Athletic Training Prgm
• Counseling Prgm
• Dietetics-Coordinated Prgm
• Speech-Language Pathology Prgm

**Ivy Tech Community College - Terre Haute**
Jeff Pittman, PhD, Chancellor
7999 US Hwy 41 S
Terre Haute, IN 47802-4894
812 299-1121
• Clin Lab Technician/Med Lab Technician
• Medical Assistant Prgm
• Radiography Prgm

## Vincennes

**Good Samaritan Hospital**
Matt Bailey, MHA, President and CEO
520 S Seventh St
Vincennes, IN 47591
812 885-3195
• Clin Lab Scientist/Med Technologist Prgm
• Radiography Prgm

**Vincennes University**
Richard E Helton, PhD, President
1002 First St N
Vincennes, IN 47591
812 888-4201
• Health Information Tech Prgm
• Physical Therapist Assistant Prgm
• Surgical Assistant Prgm
• Surgical Technology Prgm

## West Lafayette

**Purdue University**
Martin Jischke, PhD, President
Hovde Hall
West Lafayette, IN 47907
765 494-9708
• Athletic Training Prgm
• Audiologist Prgm
• Counseling Prgm
• Dietetic Internship Prgm
• Dietetics-Coordinated Prgm
• Dietetics-Didactic Prgm
• Speech-Language Pathology Prgm

# Iowa

## Ames

**Iowa State University**
Gregory Geoffroy, PhD, President
2035 President
117 Beardshear
Ames, IA 50011
515 294-2042
• Athletic Training Prgm
• Dietetic Internship Prgm
• Dietetics-Didactic Prgm

## Anamosa

**Carlson College of Massage Therapy**
Ruth Carlson
11809 County Rd
Box 28
Anamosa, IA 52205
319 462-3402
• Massage Therapy Prgm

## Ankeny

**Des Moines Area Community College**
Robert Denson, JD, President
Bldg 22
2006 S Ankeny Blvd
Ankeny, IA 50023
515 964-6638
• Clin Lab Technician/Med Lab Technician
• Dental Assisting Prgm
• Dental Hygiene Prgm
• Medical Assistant Prgm
• Respiratory Therapist (Advanced) Prgm

## Bettendorf

**Scott Community College**
Thomas Coley, PhD, President
500 Belmont Rd
Bettendorf, IA 52722
563 441-4061
• Dental Assisting Prgm
• Electroneurodiagnostic Tech Prgm
• Health Information Tech Prgm
• Radiography Prgm

## Calmar

**Northeast Iowa Community College**
Penelope Wills, PhD, President
1625 Hwy 150 South
PO Box 400
Calmar, IA 52132
563 562-3263
• Dental Assisting Prgm
• Health Information Tech Prgm
• Radiography Prgm
• Respiratory Therapist (Advanced) Prgm

## Cedar Falls

**University of Northern Iowa**
Benjamin Allen, PhD, President
Seerley Hall
Cedar Falls, IA 50614-0002
319 273-2566
• Athletic Training Prgm
• Counseling Prgm
• Dietetics-Didactic Prgm
• Speech-Language Pathology Prgm
• Therapeutic Recreation Specialist Prgm

## Cedar Rapids

**Coe College**
James Phifer, PhD, President
1220 First Ave NE
Cedar Rapids, IA 52402
319 399-8686
• Athletic Training Prgm

**Kirkwood Community College**
Mick Starcevich, EdD, President
6301 Kirkwood Blvd SW
PO Box 2068
Cedar Rapids, IA 52406-9973
319 398-5501
- Dental Assisting Prgm
- Dental Hygiene Prgm
- Dental Lab Technician Prgm
- Electroneurodiagnostic Tech Prgm
- Health Information Tech Prgm
- Medical Assistant Prgm
- Occupational Therapy Asst Prgm
- Physical Therapist Assistant Prgm
- Respiratory Therapist (Advanced) Prgm
- Surgical Technology Prgm

**Mercy/St Luke's Hospital**
Ted Townsand, MD, President/CEO
PO Box 3026
1026 A Ave NE
Cedar Rapids, IA 52406-3026
319 369-7204
- Clin Lab Scientist/Med Technologist Prgm
- Radiography Prgm

## Council Bluffs

**Iowa Western Community College**
Dan Kinney, PhD, President
2700 College Rd Box 4-C
Council Bluffs, IA 51502-3004
712 325-3200
- Dental Assisting Prgm
- Dental Hygiene Prgm
- Medical Assistant Prgm

**Jennie Edmundson Memorial Hospital**
David M Holcomb, MHA, CEO
933 E Pierce St
Council Bluffs, IA 51501
712 328-6239
- Radiography Prgm

## Davenport

**Institute of Therapeutic Massage & Wellness**
1730 Wilkes Ave
Davenport, IA 52804
- Massage Therapy Prgm

**Kaplan University, Davenport Campus**
Ed DeJaegher, MBA, Exec Dir/Campus President
1801 E Kimberly Rd, Ste 1
Davenport, IA 52807
563 355-3500
- Medical Assistant Prgm

**St Ambrose University**
Edward Rogalski, PhD, President
518 W Locust St
Davenport, IA 52803
563 333-6212
- Occupational Therapy Prgm
- Physical Therapy Prgm

## Decorah

**Luther College**
Richard L Torgerson, PhD, President
700 College Dr
Decorah, IA 52101
563 387-1001
- Athletic Training Prgm

## Des Moines

**Des Moines University**
Terry Branstad, DSC, President
3440 Grand Ave
Des Moines, IA 50312
515 271-1500
- Physical Therapy Prgm
- Physician Assistant Prgm

**Drake University**
David Maxwell, President
25th St and University Ave
Des Moines, IA 50311
515 271-2011
- Rehabilitation Counseling Prgm

**Iowa Health - Des Moines**
Eric Crowell, MS MHA, President and CEO
1200 Pleasant St
Des Moines, IA 50309-1453
515 241-6201
- Radiography Prgm

**Mercy College of Health Sciences**
Barbara Quijano Decker, JD, President
928 Sixth Ave
Des Moines, IA 50309
515 643-6601
- Diagnostic Med Sonography Prgm
- Medical Assistant Prgm
- Nuclear Medicine Technology Prgm
- Radiography Prgm
- Surgical Technology Prgm

**Mercy Medical Center**
David Vellinga, CEO
1111 6th Ave
Des Moines, IA 50314
515 247-4278
- Clin Lab Scientist/Med Technologist Prgm
- Cytotechnology Prgm
- Emergency Med Tech-Paramedic Prgm

**Vatterott College - Des Moines Campus**
Peter Mitchell, President
6100 Thornton Ave
Ste 290
Des Moines, IA 50321
515 309-9000
- Dental Assisting Prgm

## Dubuque

**Clarke College**
Joanne Burrows, PhD, President
1550 Clarke Dr
Dubuque, IA 52001-3198
319 588-6300
- Athletic Training Prgm
- Physical Therapy Prgm

**Loras College**
James E Collins, President
1450 Alta Vista
Dubuque, IA 52004-0178
563 588-7647
- Athletic Training Prgm

## Estherville

**Iowa Lakes Community College**
Mike Hupfer, BSE JD, President
19 S 7th St
Estherville, IA 51334-2295
712 362-0438
- Medical Assistant Prgm
- Surgical Technology Prgm

## Fayette

**Upper Iowa University**
Suzanne James, PhD, Interim President
605 Washington St
PO Box 1857
Fayette, IA 52142
563 425-5354
- Athletic Training Prgm

## Ft Dodge

**Iowa Central Community College**
Robert A Paxton, PhD, President
330 Ave M
Ft Dodge, IA 50501
515 576-7201
- Clin Lab Technician/Med Lab Technician
- Medical Assistant Prgm
- Radiography Prgm

## Indianola

**Simpson College**
John Byrd, President
701 North C St
Hillman Hall
Indianola, IA 50125
515 961-1611
- Athletic Training Prgm

## Iowa City

**University of Iowa**
Gary Fethke, PhD, Interim President
101 Jessup Hall
Iowa City, IA 52242
319 335-3549
- Athletic Training Prgm
- Audiologist Prgm
- Counseling Prgm
- Medical Librarian Prgm
- Music Therapy Prgm
- Physical Therapy Prgm
- Physician Assistant Prgm
- Rehabilitation Counseling Prgm
- Speech-Language Pathology Prgm
- Therapeutic Recreation Specialist Prgm

**University of Iowa Hospitals & Clinics**
Donna Katen-Bahensky, CEO
200 Hawkins Dr 1353 JCP
Iowa City, IA 52242-1059
319 356-3155
- Clin Lab Scientist/Med Technologist Prgm
- Diagnostic Med Sonography Prgm
- Dietetic Internship Prgm
- Emergency Med Tech-Paramedic Prgm
- Nuclear Medicine Technology Prgm
- Orthoptist Prgm
- Perfusion Prgm
- Radiation Therapy Prgm
- Radiography Prgm

**INSTITUTIONS**

## Lamoni

**Graceland University**
John Menzies, PhD, President
1 University Place
Lamoni, IA 50140
641 784-5111
• Athletic Training Prgm

## Marshalltown

**Marshalltown Community College**
Tim A Wynes, JD, President
3702 S Center St
Marshalltown, IA 50158
515 752-4643
• Dental Assisting Prgm

## Mason City

**Mercy Medical Center - North Iowa**
James Fitzpatrick, CHE, President/CEO
1000 Fourth St SW
Mason City, IA 50401
800 637-2994
• Radiography Prgm

**North Iowa Area Community College**
Michael Morrison, PhD, President
500 College Dr
Mason City, IA 50401
515 421-4200
• Medical Assistant Prgm
• Physical Therapist Assistant Prgm

## Orange City

**Northwestern College**
Bruce Murphy, PhD, President
101 7th St SW
Orange City, IA 51041-1996
712 707-7100
• Athletic Training Prgm

## Ottumwa

**Indian Hills Community College**
Jim Lindenmayer, PhD, President
525 Grandview Ave, Building #1
Ottumwa, IA 52501
641 683-5185
• Health Information Tech Prgm
• Physical Therapist Assistant Prgm
• Radiography Prgm

## Pella

**Central College**
David Roe, PhD, President
812 University Box 6300
Pella, IA 50219
641 628-5269
• Athletic Training Prgm

## Sheldon

**Northwest Iowa Community College**
William Giddings, PhD, President
603 W Park St
Sheldon, IA 51201
712 324-5061
• Health Information Tech Prgm

## Sioux City

**Mercy Medical Center-Sioux City**
Paul Dougherty, President/CEO
801 Fifth St
Sioux City, IA 51101
712 279-2018
• Clin Lab Scientist/Med Technologist Prgm

**St Luke's College**
Peter Thoreen, MBA, President, CEO
2720 Stone Park Blvd
Sioux City, IA 51104
712 279-3500
• Clin Lab Scientist/Med Technologist Prgm
• Radiography Prgm
• Respiratory Therapist (Advanced) Prgm

**Western Iowa Tech Community College**
Robert Dunker, PhD, President
4647 Stone Ave
PO Box 5199
Sioux City, IA 51102-5199
712 274-6400
• Dental Assisting Prgm
• Physical Therapist Assistant Prgm
• Surgical Technology Prgm

## Storm Lake

**Buena Vista University**
Fredrick V Moore, President
610 W Fourth St
Storm Lake, IA 50588
712 749-2103
• Athletic Training Prgm

## Urbandale

**Hamilton College**
Colleen McDermott, MBA, Campus President
4655 121st Street
Urbandale, IA 50323
515 727-2100
• Medical Assistant Prgm (3)

## Waterloo

**Allen College**
Jerry Durham, PhD, Chancellor
1825 Logan Ave
Waterloo, IA 50703
319 226-2015
• Radiography Prgm

**Covenant Medical Center**
Jack Dusenbery, CEO/Administrator
3421 W Ninth
Waterloo, IA 50702
319 272-7302
• Radiography Prgm

**Hawkeye Community College**
Greg Schmitz, MA, President
1501 E Orange Rd
PO Box 8015
Waterloo, IA 50704-8015
319 296-4200
• Clin Lab Technician/Med Lab Technician
• Dental Assisting Prgm
• Dental Hygiene Prgm
• Respiratory Therapist (Advanced) Prgm

## Waverly

**Wartburg College**
Jack Ohle, President
902 12th Street NW
Waverly, IA 50677
319 352-0979
• Music Therapy Prgm

## West Burlington

**Southeastern Community College**
Beverly Simone, President
1500 W Agency Rd
PO Box 180
West Burlington, IA 52655-0180
319 752-2731
• Medical Assistant Prgm
• Respiratory Therapist (Advanced) Prgm

# Kansas

## Arkansas City

**Cowley County Community College**
Patrick J McAtee, PhD, President
PO Box 1147
Arkansas City, KS 67005
800 593-2222
• Emergency Med Tech-Paramedic Prgm

## Atchison

**Benedictine College**
Daniel Carey, PhD, President
1020 N Second St
Atchison, KS 66002-1499
913 367-5340
• Athletic Training Prgm

## Coffeyville

**Coffeyville Community College**
Howard G Bass, PhD, President
400 W 11th St
Coffeyville, KS 67337
620 251-7700
• Emergency Med Tech-Paramedic Prgm

## Colby

**Colby Community College**
Lynn Krider, PhD, President
1255 S Range
Colby, KS 67701
785 462-3984
• Physical Therapist Assistant Prgm

## Emporia

**Emporia State University**
Kay Schallencamp, PhD, President
1200 Commercial St
Emporia, KS 66801
620 341-5551
• Art Therapy Prgm
• Athletic Training Prgm
• Counseling Prgm
• Medical Librarian Prgm
• Rehabilitation Counseling Prgm

### Flint Hills Technical College
Dean Hollenbeck, PhD, Interim President
3301 W 18th Ave
Emporia, KS 66801
620 343-4600
- Dental Assisting Prgm
- Emergency Med Tech-Paramedic Prgm

## Garden City

### Garden City Community College
Carol Ballantyne, PhD, President
801 Campus Dr
Garden City, KS 67846
620 276-9602
- Emergency Med Tech-Paramedic Prgm

## Goodland

### Northwest Kansas Technical College
Kenneth Clouse, MS, President
1209 Harrison St
PO Box 668
Goodland, KS 67735-0668
785 890-3641
- Medical Assistant Prgm

## Great Bend

### Barton County Community College
Carl Heilman, PhD, President
245 NE 30th Rd
Great Bend, KS 67530-9283
620 792-2701
- Clin Lab Technician/Med Lab Technician
- Emergency Med Tech-Paramedic Prgm

## Hays

### Ft Hays State University
Lawrence Gould, PhD, Provost
600 W Park St
Hays, KS 67601
913 628-4231
- Athletic Training Prgm
- Radiography Prgm
- Speech-Language Pathology Prgm

## Hillsboro

### Tabor College
Larry W Nikkel, MPH, President
400 S Jefferson
Hillsboro, KS 67063
620 947-3121
- Athletic Training Prgm

## Hutchinson

### Hutchinson Community College
Edward E Berger, EdD, President
1300 N Plum St
Hutchinson, KS 67501
620 665-3505
- Health Information Tech Prgm
- Radiography Prgm
- Surgical Technology Prgm

## Kansas City

### Bethany Med Ctr/Kansas City Kansas Comm Coll
Shirley Wendel, PhD, Consortial Comm Chair
7250 State Ave
Kansas City, KS 66112
913 288-7126
- Respiratory Therapist (Advanced) Prgm
- Respiratory Therapist (Entry-Level) Prgm

### Kansas City Kansas Community College
Thomas R Burke, PhD, President
7250 State Ave
PO Box 12951
Kansas City, KS 66112-9978
913 288-7123
- Emergency Med Tech-Paramedic Prgm
- Physical Therapist Assistant Prgm

### University of Kansas Medical Center
Barbara Atkinson, MD, Exec Vice Chancellor
3901 Rainbow Blvd
3015 Murphy Administration Bldg MS 1049
Kansas City, KS 66160
913 588-1433
- Clin Lab Scientist/Med Technologist Prgm
- Cytotechnology Prgm
- Diagnostic Med Sonography Prgm (2)
- Diagnostic Molecular Scientist Prgm
- Dietetic Internship Prgm
- Health Information Admin Prgm
- Nuclear Medicine Technology Prgm
- Occupational Therapy Prgm
- Physical Therapy Prgm
- Respiratory Therapist (Advanced) Prgm

## Lawrence

### University of Kansas
Robert E Hemenway, Chancellor
3031 Dole Ctr
Lawrence, KS 66045
785 864-0630
- Athletic Training Prgm
- Audiologist Prgm
- Music Therapy Prgm
- Speech-Language Pathology Prgm

## Liberal

### Seward County Community College
Duane Dunn, EdD, President
PO Box 1137
Liberal, KS 67901
316 624-1951
- Clin Lab Technician/Med Lab Technician
- Respiratory Therapist (Advanced) Prgm
- Surgical Technology Prgm

## Manhattan

### Kansas State University
Jon Wefald, President
Anderson Hall
Manhattan, KS 66506-5301
913 532-6221
- Athletic Training Prgm
- Dietetics-Coordinated Prgm
- Dietetics-Didactic Prgm
- Speech-Language Pathology Prgm

## North Newton

### Bethel College
E LaVerne Epp, JD, President
300 E 27th St
North Newton, KS 67117
316 283-2500
- Athletic Training Prgm

## Olathe

### Mid-America Nazarene University
Edwin H Robinson, President
2030 E College Way
Olathe, KS 66062
- Athletic Training Prgm

## Overland Park

### BMSI Institute
C Michael Pizzuto, Director
8665 W 96th, Ste 300
Overland Park, KS 66212
913 649-3322
- Massage Therapy Prgm

### Johnson County Community College
Larry Tyree, EdD, Interim President
12345 College Blvd
Overland Park, KS 66210-1299
913 469-8500
- Dental Hygiene Prgm
- Emergency Med Tech-Paramedic Prgm
- Respiratory Therapist (Advanced) Prgm

## Parsons

### Labette Community College
George Knox, EdD, President
200 S 14th St
Parsons, KS 67357
620 421-6700
- Radiography Prgm
- Respiratory Therapist (Advanced) Prgm

## Pittsburg

### Pittsburg State University
Thomas Bryant
1701 S Broadway
Pittsburg, KS 66762-7500
316 231-7000
- Counseling Prgm
- Therapeutic Recreation Specialist Prgm

## Salina

### Kansas Wesleyan University
Phillip P Kerstetter, PhD, President
100 E Claflin Ave
Salina, KS 67401-6198
785 827-5541
- Athletic Training Prgm

### Salina Area Technical School
Duane Custer, Director
2562 Centennial Rd
Salina, KS 67401
785 309-3108
- Dental Assisting Prgm

## Sterling

**Sterling College**
Bruce Douglas, PhD, President
125 W Cooper
Sterling, KS 67579
620 278-4213
• Athletic Training Prgm

## Topeka

**Kaw Area Technical School**
Richard Hoffman, BS MS, General Director
5724 Huntoon
Topeka, KS 66604
785 228-6300
• Surgical Technology Prgm

**Washburn University**
Jerry Farley, PhD, President
1700 SW College
Topeka, KS 66621
785 670-1556
• Athletic Training Prgm
• Diagnostic Med Sonography Prgm
• Health Information Tech Prgm
• Physical Therapist Assistant Prgm
• Radiography Prgm
• Respiratory Therapist (Advanced) Prgm

## Wichita

**Newman University**
Lee Cooper, PhD, Provost, Interim Pres
3100 McCormick Ave
Wichita, KS 67213-2097
316 942-4291
• Occupational Therapy Asst Prgm
• Radiography Prgm
• Respiratory Therapist (Advanced) Prgm

**Wichita Area Technical College**
Camille E Kluge, BS MS, President
324 N Emporia
Wichita, KS 67202
316 677-9500
• Clin Lab Technician/Med Lab Technician
• Dental Assisting Prgm
• Medical Assistant Prgm
• Surgical Technology Prgm

**Wichita State University**
Donald Beggs, PhD, President
1845 N Fairmont
Wichita, KS 67260-0001
316 978-3001
• Audiologist Prgm
• Clin Lab Scientist/Med Technologist Prgm
• Dental Hygiene Prgm
• Physical Therapy Prgm
• Physician Assistant Prgm
• Speech-Language Pathology Prgm

## Winfield

**Southwestern College**
W R Merriman, Jr, PhD, President
100 College St
Winfield, KS 67156
620 229-6223
• Athletic Training Prgm

# Kentucky

## Ashland

**Ashland Technical College**
Stu Taylor, School Director
4818 Roberts Dr
Ashland, KY 41102-9046
606 929-2055
• Surgical Technology Prgm

**King's Daughters' Medical Center**
Fred Jackson, President
2201 Lexington Ave
Ashland, KY 41101
606 327-4000
• Radiography Prgm

## Berea

**Berea College**
Larry D Shinn, President
Berea, KY 40404
606 986-9341
• Dietetics-Didactic Prgm

## Bowling Green

**Bowling Green Technical College**
Nathan Hodges, EdD, President
1845 Loop Dr
Bowling Green, KY 42101
270 901-1162
• Diagnostic Med Sonography Prgm
• Radiography Prgm
• Surgical Technology Prgm

**Western Kentucky University**
Gary A Ransdell, MD, President
Wetherby Administration Bldg 135
Bowling Green, KY 42101-3576
270 745-4346
• Counseling Prgm
• Dental Hygiene Prgm
• Dietetics-Didactic Prgm
• Health Information Tech Prgm
• Speech-Language Pathology Prgm

## Burlington

**Cincinnati Sch of Med Massage-Northn Kentucky**
1793 Patrick Rd
Burlington, KY 41005
• Massage Therapy Prgm

## Columbia

**Lindsey Wilson College**
210 Lindsey Wilson St
Columbia, KY 42728-1298
502 384-2126
• Counseling Prgm

## Edgewood

**St Elizabeth Medical Center**
Joseph Gross, FACHE, President/CEO
One Medical Village Dr
Edgewood, KY 41017
606 344-2111
• Clin Lab Scientist/Med Technologist Prgm

## Elizabethtown

**Elizabethtown Community & Technical College**
Thelma White, PhD, President
620 College St Rd
Elizabethtown, KY 42701
270 769-2371
• Radiography Prgm

## Hazard

**Hazard Community & Technical College**
Jay K Box, PhD, President
One Community College Dr, Hwy 15 S
Hazard, KY 41701
606 436-5721
• Physical Therapist Assistant Prgm
• Radiography Prgm

## Henderson

**Henderson Community College**
Patrick Lake, EdD, President
2660 S Green St
Henderson, KY 42420
270 827-1867
• Dental Hygiene Prgm
• Medical Assistant Prgm

## Highland Heights

**Northern Kentucky University**
James Votruba, PhD, President
Office of the President, Admin Ctr 800B
Highland Heights, KY 41099-2104
606 572-5123
• Athletic Training Prgm
• Radiography Prgm
• Respiratory Therapist (Advanced) Prgm

## Lexington

**Bluegrass Community and Technical College**
James Kerley, PhD, President
470 Cooper Dr, Oswald Bldg
Rm 209
Lexington, KY 40506-0235
859 246-6501
• Dental Assisting Prgm
• Dental Hygiene Prgm
• Dental Lab Technician Prgm
• Medical Assistant Prgm
• Radiography Prgm
• Respiratory Therapist (Advanced) Prgm
• Surgical Technology Prgm

**Pathology & Cytology Laboratories**
Anne Marshall, MD, President
290 Big Run Rd
Lexington, KY 40503
859 278-9513
• Cytotechnology Prgm

**St Joseph Healthcare**
Gene Woods, President/CEO
One St Joseph Dr
Lexington, KY 40504
859 313-1000
• Radiography Prgm

**Sullivan University - Lexington**
James Ploskonka, PhD, Executive Director
2355 Harrodsburg Rd
Lexington, KY 40504
606 276-4357
• Medical Assistant Prgm

**Univ of Kentucky Chandler Med Ctr**
Murray Clark, PhD, Assoc VP, Med Ctr
N106 Chandler Medical Center
Lexington, KY 40536-0293
859 323-2044
• Dietetic Internship Prgm
• Radiation Therapy Prgm

**University of Kentucky**
Lee Todd, PhD, President
Administration Bldg
Lexington, KY 40506
859 257-9000
• Clin Lab Scientist/Med Technologist Prgm
• Dietetic Internship Prgm
• Dietetics-Coordinated Prgm
• Dietetics-Didactic Prgm
• Medical Librarian Prgm
• Physical Therapy Prgm
• Physician Assistant Prgm
• Rehabilitation Counseling Prgm
• Speech-Language Pathology Prgm

## Louisville

**Bellarmine University**
Peter Cimbolic, PhD, University Provost
2001 Newburg Rd
Louisville, KY 40205
502 452-8000
• Clin Lab Scientist/Med Technologist Prgm
• Cytotechnology Prgm
• Physical Therapy Prgm

**James Graham Brown Cancer Center**
Patty Melvin, Vice President-Planning
529 S Jackson St
Louisville, KY 40202
502 562-4585
• Radiation Therapy Prgm

**Jefferson Community and Technical College**
Anthony Newberry, PhD, President
109 E Broadway
Louisville, KY 40202
502 213-5333
• Health Information Tech Prgm
• Medical Assistant Prgm
• Nuclear Medicine Technology Prgm
• Occupational Therapy Asst Prgm
• Physical Therapist Assistant Prgm
• Radiography Prgm
• Respiratory Therapist (Advanced) Prgm
• Surgical Technology Prgm

**Spalding University**
Joann Ronney, JD, President
851 S Fourth St
Louisville, KY 40203-2188
502 585-9911
• Dietetic Internship Prgm
• Occupational Therapy Prgm

**Spencerian College**
Jan M Gordon, MEd, Executive Director
4627 Dixie Hwy
Louisville, KY 40216
502 447-1000
• Cardiovascular Technology Prgm
• Medical Assistant Prgm
• Radiography Prgm
• Surgical Technology Prgm

**University of Louisville**
James Ramsey, PhD, President
Grawmeyer Hall
Health Sciences Center
Louisville, KY 40292
502 852-5417
• Art Therapy Prgm
• Audiologist Prgm
• Dental Hygiene Prgm
• Music Therapy Prgm
• Speech-Language Pathology Prgm

## Madisonville

**HCC/MCC Consortium**
Judith Rhoads, EdD, President
2000 College Dr
Madisonville, KY 42431
270 824-8562
• Clin Lab Technician/Med Lab Technician

**Madisonville Community College**
Judith Rhoads, EdD, President
2000 College Dr
Madisonville, KY 42431
270 824-8562
• Occupational Therapy Asst Prgm
• Physical Therapist Assistant Prgm
• Radiography Prgm
• Respiratory Therapist (Advanced) Prgm
• Surgical Assistant Prgm
• Surgical Technology Prgm

## Morehead

**Maysville Comm & Tech College - Rowan Campus**
Kenneth J Brown, Director
609 Viking Dr
Morehead, KY 40351
606 783-1538
• Medical Assistant Prgm
• Respiratory Therapist (Entry-Level) Prgm

**Morehead State University**
Wayne Andrews, PhD, President
201 Howell McDowell Admin Bldg
Morehead, KY 40351
606 783-2022
• Dietetic Internship Prgm
• Dietetics-Didactic Prgm
• Radiography Prgm

## Mt Vernon

**Laurel Technical College - Rockcastle**
Donna Hopkins, MA, Principal
PO Box 275
Mt Vernon, KY 40456
606 676-9065
• Respiratory Therapist (Entry-Level) Prgm

## Murray

**Murray State University**
Alexander King, PhD, President
President's Office
218 Wells Hall
Murray, KY 42071
270 762-3763
• Athletic Training Prgm
• Counseling Prgm
• Dietetic Internship Prgm
• Dietetics-Didactic Prgm
• Speech-Language Pathology Prgm

## Owensboro

**Kentucky Community Technical College**
Jackie Addington, PhD, President/CEO
4800 New Hartford Rd
Owensboro, KY 42303
270 686-4400
• Surgical Technology Prgm

**Owensboro Community & Technical College**
Jacqueline Addington, PhD, President
4800 New Hartford Rd
Owensboro, KY 42303-1899
270 686-4400
• Radiography Prgm

**Owensboro Medical Health System**
Jeffery Barber, CEO
811 E Parrish Ave
PO Box 20007
Owensboro, KY 42303
502 688-2100
• Clin Lab Scientist/Med Technologist Prgm

## Paducah

**West Kentucky Community & Technical College**
Barbara Veazey, PhD RN, President/CEO
PO Box 7380
4810 Alben Barkley Dr
Paducah, KY 42002-7380
270 534-3082
• Dental Assisting Prgm
• Diagnostic Med Sonography Prgm
• Medical Assistant Prgm
• Physical Therapist Assistant Prgm
• Radiography Prgm
• Respiratory Therapist (Advanced) Prgm
• Respiratory Therapist (Entry-Level) Prgm
• Surgical Technology Prgm

## Pineville

**Southeast Kentucky Comm & Tech College**
W Bruce Ayers, President
3300 US 25E S
Pineville, KY 40977
606 337-3106
• Clin Lab Technician/Med Lab Technician
• Radiography Prgm
• Respiratory Therapist (Advanced) Prgm
• Respiratory Therapist (Entry-Level) Prgm
• Surgical Technology Prgm

INSTITUTIONS

## Prestonsburg

**Big Sandy Community & Technical College**
George D Edwards, PhD, President/CEO
1 Bert T Combs Dr
Prestonsburg, KY 41653
888 641-4132
• Dental Hygiene Prgm
• Respiratory Therapist (Advanced) Prgm

## Richmond

**Eastern Kentucky University**
Joann K Glasser, JD, President
Coates Admin Bldg Rm 107
Richmond, KY 40475
859 622-2194
• Athletic Training Prgm
• Clin Lab Scientist/Med Technologist Prgm
• Clin Lab Technician/Med Lab Technician
• Counseling Prgm
• Dietetic Internship Prgm
• Dietetics-Didactic Prgm
• Emergency Med Tech-Paramedic Prgm
• Health Information Admin Prgm
• Medical Assistant Prgm
• Occupational Therapy Prgm
• Speech-Language Pathology Prgm
• Therapeutic Recreation Specialist Prgm

## Somerset

**Somerset Community College**
Jo Marshall, PhD, President
808 Monticello St
Somerset, KY 42501
606 679-8501
• Clin Lab Technician/Med Lab Technician
• Physical Therapist Assistant Prgm
• Radiography Prgm
• Surgical Technology Prgm

## St Catharine

**St Catharine College**
William D Huston, MA, President
2735 Bardstown Rd
St Catharine, KY 40061
859 336-5082
• Diagnostic Med Sonography Prgm
• Radiography Prgm

# Louisiana

## Alexandria

**Louisiana State University - Alexandria**
J Robert Cavanaugh, Chancellor
8100 Hwy 71 S
Alexandria, LA 71302-9121
318 445-3672
• Clin Lab Technician/Med Lab Technician
• Pharmacy Technician Prgm

**Rapides Regional Medical Center**
A C Buchanan, BS MS, President
211 Fourth St
PO Box 30101
Alexandria, LA 71301
318 473-3150
• Clin Lab Scientist/Med Technologist Prgm
• Phlebotomist Prgm

## Baton Rouge

**Baton Rouge General Medical Center**
William Holman, President/CEO
3600 Florida St
PO Box 2511-70821
Baton Rouge, LA 70806
225 387-7767
• Radiography Prgm

**Blue Cliff College - Baton Rouge**
Kathie Lea Love
6160 Perkins Rd, Ste 200
Baton Rouge, LA 70808-4191
225 757-3770
• Massage Therapy Prgm

**Louisiana State Univ and A&M College**
William E Davis, Chancellor
Baton Rouge, LA 70803
504 388-3202
• Audiologist Prgm
• Speech-Language Pathology Prgm

**Louisiana State University**
Eric Hovland, Dean
8000 GSRI Rd
Bldg 3110
Baton Rouge, LA 70820
225 334-1816
• Athletic Training Prgm
• Counseling Prgm
• Dental Hygiene Prgm
• Dental Lab Technician Prgm
• Dietetic Internship Prgm
• Dietetics-Didactic Prgm
• Medical Librarian Prgm

**Louisiana Tech College**
Margaret Montgomery-Richard, Chancellor
150 3rd St
Baton Rouge, LA 70801
800 351-7611
• Clin Lab Technician/Med Lab Technician
• Medical Assistant Prgm
• Respiratory Therapist (Entry-Level) Prgm
• Surgical Technology Prgm (2)

**MedVance Institute**
Deborah Schwarzberg, President
9255 Interline Ave
Baton Rouge, LA 70809
225 248-1015
• Clin Lab Technician/Med Lab Technician
• Radiography Prgm
• Surgical Technology Prgm

**Our Lady of the Lake College**
Sandra Harper, PhD, President
7434 Perkins Rd
Baton Rouge, LA 70808
225 768-1710
• Clin Lab Scientist/Med Technologist Prgm
• Clin Lab Technician/Med Lab Technician
• Emergency Med Tech-Paramedic Prgm
• Physical Therapist Assistant Prgm
• Physician Assistant Prgm
• Radiography Prgm
• Respiratory Therapist (Entry-Level) Prgm
• Surgical Technology Prgm

**Southern Univ and A&M College**
Edward Jackson, PhD, Chancellor
Baton Rouge, LA 70813
225 771-5020
• Dietetic Internship Prgm
• Dietetics-Didactic Prgm
• Rehabilitation Counseling Prgm
• Speech-Language Pathology Prgm

## Bossier City

**Bossier Parish Community College**
Tom Carleton, MA, Chancellor
6220 E Texas St
Bossier City, LA 71111
318 678-6014
• Emergency Med Tech-Paramedic Prgm
• Medical Assistant Prgm
• Pharmacy Technician Prgm
• Phlebotomist Prgm
• Physical Therapist Assistant Prgm
• Respiratory Therapist (Advanced) Prgm
• Respiratory Therapist (Entry-Level) Prgm
• Surgical Technology Prgm

## Eunice

**Louisiana State University - Eunice**
William J Nunez III, PhD, Interim Chancellor
PO Box 1129
Eunice, LA 70535
318 457-7311
• Diagnostic Med Sonography Prgm
• Radiography Prgm
• Respiratory Therapist (Entry-Level) Prgm

## Grambling

**Grambling State University**
Horace Judson, President
Grambling, LA 71245
318 247-3811
• Therapeutic Recreation Specialist Prgm

## Hammond

**North Oaks Medical Center**
James E Cathey, Jr, CEO
15790 Paul Vega MD Dr
Hammond, LA 70403
985 345-2700
• Dietetic Internship Prgm
• Radiography Prgm

**Southeastern Louisiana University**
Randay Moffett, EdD, President
SLU 10784
Hammond, LA 70402
985 549-2280
• Athletic Training Prgm
• Counseling Prgm
• Speech-Language Pathology Prgm

## Houma

**L E Fletcher Technical Community College**
F Travis Lavigne Jr, BS MS, Chancellor
310 Saint Charles St
Houma, LA 70360-2863
• Phlebotomist Prgm

## Lafayette

**Lafayette General Medical Center**
John J Burdin, Jr, MHA, President
1214 Coolidge Ave
PO Box 52009 OCS
Lafayette, LA 70505
318 289-7381
• Radiography Prgm

**University of Louisiana at Lafayette**
Ray Authement, PhD, President
UL Drawer 41008
Lafayette, LA 70504
337 482-6203
• Athletic Training Prgm
• Dietetic Internship Prgm
• Dietetics-Didactic Prgm
• Health Information Admin Prgm
• Speech-Language Pathology Prgm

## Lake Charles

**Lake Charles Mem Hosp Sch of Med Tech**
Elton Williams, CPA, President
1701 Oak Park Blvd
Lake Charles, LA 70601
337 494-3200
• Clin Lab Scientist/Med Technologist Prgm

**McNeese State University**
Robert D Hebert, PhD, President
Box 93300
Lake Charles, LA 70609
318 475-5556
• Clin Lab Scientist/Med Technologist Prgm
• Dietetic Internship Prgm
• Dietetics-Didactic Prgm
• Radiography Prgm

## Monroe

**Career Technical College**
Rick Nail, College Director
2319 Louisville Ave
Monroe, LA 71201
318 323-2889
• Medical Assistant Prgm
• Radiography Prgm
• Surgical Technology Prgm

**St Francis Medical Center**
Wester Scott, MHA, President/CEO
309 Jackson St
Monroe, LA 71201
318 327-4141
• Clin Lab Scientist/Med Technologist Prgm

**University of Louisiana at Monroe**
James E Cofer, Sr, EdD, President
700 University Ave, NE Station
Monroe, LA 71209-0430
318 342-1010
• Counseling Prgm
• Dental Hygiene Prgm
• Exercise Physiology Prgm
• Exercise Science Prgm
• Occupational Therapy Asst Prgm
• Radiography Prgm
• Speech-Language Pathology Prgm

## Natchitoches

**Northwestern State University**
Randall J Webb, EdD, President
Natchitoches, LA 71497
318 357-5701
• Counseling Prgm
• Radiography Prgm

## New Orleans

**Bryman College - New Orleans**
Nicky Good, President
1201 Elmwood Park Blvd, Ste 600
New Orleans, LA 70123
504 822-4500
• Medical Assistant Prgm

**Delgado Community College**
Alex Johnson, PhD, Chancellor
615 City Park Ave
New Orleans, LA 70119
504 361-6609
• Clin Lab Technician/Med Lab Technician
• Diagnostic Med Sonography Prgm
• Dietetic Technician-AD Prgm
• Emergency Med Tech-Paramedic Prgm
• Health Information Tech Prgm
• Nuclear Medicine Technology Prgm
• Occupational Therapy Asst Prgm
• Ophthalmic Assistant Prgm
• Pharmacy Technician Prgm
• Phlebotomist Prgm
• Physical Therapist Assistant Prgm
• Radiation Therapy Prgm
• Radiography Prgm
• Respiratory Therapist (Advanced) Prgm
• Surgical Technology Prgm

**Louisiana State Univ Health Sciences Center**
Larry H Hollier, MD, Chancellor
433 Bolivar St
New Orleans, LA 70112-2223
504 568-4800
• Audiologist Prgm
• Cardiovascular Technology Prgm
• Clin Lab Scientist/Med Technologist Prgm
• Occupational Therapy Prgm
• Ophthalmic Med Technologist Prgm
• Physical Therapy Prgm
• Rehabilitation Counseling Prgm
• Respiratory Therapist (Advanced) Prgm
• Speech-Language Pathology Prgm

**Loyola University New Orleans**
Kevin Wildes, SJ, President
6363 St Charles Ave
New Orleans, LA 70118
504 865-2011
• Counseling Prgm
• Music Therapy Prgm

**Medical Center of Louisiana**
Dwayne Thomas, PhD
1532 Tulane Ave
New Orleans, LA 70112-2860
504 903-2311
• Specialist in BB Tech Prgm

**Our Lady of Holy Cross College**
Rev Anthony DeConcilliis, PhD, President
4123 Woodland Dr
New Orleans, LA 70131-7399
504 394-7744
• Counseling Prgm
• Radiography Prgm
• Respiratory Therapist (Advanced) Prgm

**Touro Infirmary**
Gary M Stein, MA, President
1401 Foucher St
New Orleans, LA 70115
504 897-8244
• Dietetic Internship Prgm

**Tulane University**
Eamon M Kelly, President
New Orleans, LA 70118
504 865-5000
• Dietetic Internship Prgm

**University of New Orleans**
Timothy P Ryan, PhD, Chancellor
Lake Front
New Orleans, LA 70148-0001
504 280-6000
• Counseling Prgm

## Pineville

**Louisiana College**
Joe Aguillard, President
Alexandria Hall, Rm 125
Pineville, LA 71359
318 487-7401
• Athletic Training Prgm

## Ruston

**Louisiana Tech University**
Daniel D Reneau, PhD, President
PO Box 3168
Ruston, LA 71272
318 257-3785
• Audiologist Prgm
• Dietetic Internship Prgm
• Dietetics-Didactic Prgm
• Health Information Admin Prgm
• Health Information Tech Prgm
• Speech-Language Pathology Prgm

## Shreveport

**Louisiana State Univ Hlth Sci Ctr-Shreveport**
John McDonald, MD, Chancellor and Dean
1501 Kings Highway
Shreveport, LA 71130
318 675-6141
• Clin Lab Scientist/Med Technologist Prgm
• Occupational Therapy Prgm
• Physical Therapy Prgm
• Physician Assistant Prgm
• Respiratory Therapist (Advanced) Prgm
• Speech-Language Pathology Prgm

**Overton Brooks VA Medical Center**
George Moore, MHA, Director
510 E Stoner Ave
Shreveport, LA 71101-4295
318 424-6037
• Clin Lab Scientist/Med Technologist Prgm

**INSTITUTIONS**

**Southern Univ at Shreveport**
Ray Belton, PhD, Chancellor
3050 Martin Luther King Jr Dr
Shreveport, LA 71107-8032
318 674-3300
- Clin Lab Technician/Med Lab Technician
- Dental Hygiene Prgm
- Health Information Tech Prgm
- Radiography Prgm
- Respiratory Therapist (Advanced) Prgm
- Surgical Technology Prgm

## Thibodaux

**Nicholls State University**
Stephen Hulbert, EdD, President
PO Box 2001
Thibodaux, LA 70310
504 448-4003
- Athletic Training Prgm
- Cytotechnology Prgm
- Dietetics-Didactic Prgm
- Emergency Med Tech-Paramedic Prgm
- Respiratory Therapist (Advanced) Prgm
- Respiratory Therapist (Entry-Level) Prgm

# Maine

## Auburn

**Central Maine Community College**
Scott E Knapp, PhD, President
1250 Turner St
Auburn, ME 04210-6498
207 755-5100
- Clin Lab Technician/Med Lab Technician

## Augusta

**University College of Bangor**
Richard Randall, MA, President
University of Maine-Augusta
46 University Dr
Augusta, ME 04330
207 621-3000
- Dental Assisting Prgm
- Dental Hygiene Prgm

## Bangor

**Beal College**
Allen Stehle, BS, President
99 Farn Road
Bangor, ME 04401
207 947-4591
- Medical Assistant Prgm

**Eastern Maine Community College**
Joyce B Hedlund, EdD, President
354 Hogan Rd
Bangor, ME 04401
207 974-4691
- Radiography Prgm
- Surgical Technology Prgm

**Eastern Maine Medical Center**
Debbie Johnson, President, CEO
489 State St
Bangor, ME 04401
207 973-7051
- Clin Lab Scientist/Med Technologist Prgm

**Husson College**
William H Beardsly, PhD, President
One College Circle
206 Peabody Hall
Bangor, ME 04401-2999
207 973-7138
- Occupational Therapy Prgm
- Physical Therapy Prgm

## Biddeford

**University of New England**
Danielle Ripich, PhD, President
Hills Beach Rd
Biddeford, ME 04005-9599
207 283-0171
- Athletic Training Prgm
- Dental Hygiene Prgm
- Occupational Therapy Prgm
- Physical Therapy Prgm
- Physician Assistant Prgm

## Calais

**Washington County Technical College**
Calais, ME 04619-9701
- Dietetic Technician-AD Prgm

## Fairfield

**Kennebec Valley Community College**
Barbara W Woodlee, EdD, President
92 Western Ave
Fairfield, ME 04937-1367
207 453-5129
- Health Information Tech Prgm
- Medical Assistant Prgm
- Occupational Therapy Asst Prgm
- Physical Therapist Assistant Prgm
- Radiography Prgm
- Respiratory Therapist (Advanced) Prgm

## Lewiston

**Central Maine Medical Center**
Peter Chalke, MHA, President
300 Main St
Lewiston, ME 04240
207 795-2700
- Nuclear Medicine Technology Prgm
- Radiography Prgm

## Orono

**University of Maine - Orono**
Robert Kennedy, PhD, President
5703 Alumni Hall, Rm 200
Orono, ME 04469
207 581-1512
- Athletic Training Prgm
- Dietetic Internship Prgm
- Dietetics-Didactic Prgm
- Speech-Language Pathology Prgm

## Portland

**Maine Medical Center**
Vincent Conti, President
22 Bramhall St
Portland, ME 04102
207 871-2491
- Surgical Technology Prgm

**Mercy Hospital**
Eileen F Skinner, MHACHE, President
144 State St
Portland, ME 04029
207 879-3000
- Radiography Prgm

**University of Southern Maine**
Richard L Pattenaude, PhD, President
96 Falmouth St, PO Box 9300
Portland, ME 04104-9300
207 780-4480
- Athletic Training Prgm
- Counseling Prgm
- Occupational Therapy Prgm
- Rehabilitation Counseling Prgm

## Presque Isle

**Northern Maine Community College**
Timothy Crowley, MEd, President
33 Edgemont Dr
Presque Isle, ME 04769
207 768-2811
- Surgical Technology Prgm

**University of Maine - Presque Isle**
Karl Burgher, PhD, President
181 Main St
22 Preble Hall
Presque Isle, ME 04769
207 768-9525
- Athletic Training Prgm
- Clin Lab Technician/Med Lab Technician

## South Portland

**Southern Maine Community College**
James Ortiz, Eed, President
Fort Rd
South Portland, ME 04106
207 741-5500
- Dietetic Technician-AD Prgm
- Radiation Therapy Prgm
- Radiography Prgm
- Respiratory Therapist (Advanced) Prgm

## Waldoboro

**Downeast School of Massage**
Nancy W Dail, BA LMT NCTMB, Director
99 Moose Meadow Lane
Waldoboro, ME 04572
207 832-5531
- Massage Therapy Prgm

# Manitoba, Canada

## Winnipeg

**Massage Therapy College of Manitoba**
Garth Beddome, BA RMT, Director
Winnipeg, MB R3G 1C3
204 772-8999
- Massage Therapy Prgm

# Maryland

## Arnold

### Anne Arundel Community College
Martha R Smith, PhD, President
101 College Pkwy
Arnold, MD 21012-1875
410 777-2222
- Emergency Med Tech-Paramedic Prgm
- Medical Assistant Prgm
- Pharmacy Technician Prgm
- Physician Assistant Prgm
- Radiography Prgm

### Chesapeake Area Consortium for Higher Educ
101 College Pkwy
Arnold, MD 21012
- Physical Therapist Assistant Prgm

## Baltimore

### Baltimore City Community College
Richard Turner, DME, President
2901 Liberty Heights Ave
Baltimore, MD 21215-7893
410 462-7799
- Dental Hygiene Prgm
- Dietetic Technician-AD Prgm
- Health Information Tech Prgm
- Physical Therapist Assistant Prgm
- Respiratory Therapist (Advanced) Prgm
- Surgical Technology Prgm

### Comm Coll of Baltimore County - Catonsville
Sandra Kurtinitis, PhD, President
7200 Soellers Point Rd
Baltimore, MD 21222
410 285-9993
- Occupational Therapy Asst Prgm

### Coppin State University
Stanley Battle, President
2500 W North Ave
Baltimore, MD 21216
410 951-3838
- Rehabilitation Counseling Prgm

### Greater Baltimore Medical Center
Lawerence Merlis, MBA, President
6701 N Charles St
Baltimore, MD 21204
443 849-2121
- Orthoptist Prgm
- Radiography Prgm

### Johns Hopkins Bayview Medical Center
4940 Eastern Ave
Baltimore, MD 21224-2735
- Dietetic Internship Prgm

### Johns Hopkins Hospital
Ronald R Peterson, MHA, President
733 N. Broadway, BRB 104
Baltimore, MD 21287
410 955-9540
- Cytotechnology Prgm
- Diagnostic Med Sonography Prgm
- Nuclear Medicine Technology Prgm
- Radiography Prgm
- Specialist in BB Tech Prgm

### Johns Hopkins School of Medicine
Edward D Miller, Jr, MD, President, CEO
100 Medical Admin Bldg
720 N Rutland Ave
Baltimore, MD 21205
410 955-3180
- Medical Illustrator Prgm

### Johns Hopkins Univ/Natl Human Genome Research
William R Brody, President
242 Garland
Baltimore, MD 21218
410 516-8068
- Genetic Counseling Prgm

### Loyola College of Maryland
Brian Linnane, SJPhD, President
4501 N Charles St
Baltimore, MD 21210
410 617-2000
- Counseling Prgm (2)
- Speech-Language Pathology Prgm

### Maryland General Hospital
Colene Y Daniel, FACHE, President and CEO
827 Linden Ave
Baltimore, MD 21201
410 995-8600
- Radiography Prgm

### Morgan State University
Earl S Richardson, EdD, President
Coldspring Ln and Hillen Rd
Baltimore, MD 21239
410 319-3200
- Clin Lab Scientist/Med Technologist Prgm
- Dietetics-Didactic Prgm

### University of Maryland
David J Ramsay, DM DPhil, President
515 Lombard Bldg
Second Floor
Baltimore, MD 21201
410 706-7131
- Clin Lab Scientist/Med Technologist Prgm
- Dental Hygiene Prgm
- Genetic Counseling Prgm
- Medical Librarian Prgm
- Pathologists' Assistant Prgm
- Physical Therapy Prgm
- Rehabilitation Counseling Prgm

### University of Maryland Baltimore County
Freeman Hrabowski, PhD, President
1000 Hilltop Circle
Baltimore, MD 21250
410 455-2274
- Diagnostic Med Sonography Prgm
- Emergency Med Tech-Paramedic Prgm

### University of Maryland Medical System
22 S Greene St
Baltimore, MD 21201-1544
410 328-2561
- Dietetic Internship Prgm

## Baltimore County

### Comm Coll of Baltimore County - Essex Campus
Sandra Kurtinitis, PhD, President
7201 Rossville Blvd
Baltimore County, MD 21237-3898
410 780-6322
- Emergency Med Tech-Paramedic Prgm
- Radiation Therapy Prgm
- Radiography Prgm
- Respiratory Therapist (Advanced) Prgm
- Surgical Technology Prgm

## Bel Air

### Harford Community College
James F LaCalle, EdD, President
401 Thomas Run Rd
Bel Air, MD 21015
410 836-4200
- Histotechnician Prgm

## Bethesda

### National Institutes of Health
10 Center Dr
Bethesda, MD 20892
301 496-3311
- Dietetic Internship Prgm

### NIH Clinical Center Blood Bank
Walter L Jones, Deputy Director
NIH/CC/DTM Bldg 10 Rm IC-711
10 Ctr Dr MSC 1184
Bethesda, MD 20892-1184
301 496-4506
- Specialist in BB Tech Prgm

## College Park

### Univ of Maryland at College Park
William E Kirwan, President
College Park, MD 20742
301 405-1000
- Audiologist Prgm
- Counseling Prgm
- Dietetic Internship Prgm
- Dietetics-Didactic Prgm
- Speech-Language Pathology Prgm

## Columbia

### Howard Community College
Mary Ellen Duncan, PhD, President
10901 Little Patuxent Parkway
Columbia, MD 21044
410 772-4820
- Cardiovascular Technology Prgm
- Emergency Med Tech-Paramedic Prgm

### Sodexho Health Care Services, Mid Atlantic
Columbia, MD 21044
- Dietetic Internship Prgm

INSTITUTIONS

## Cumberland

**Allegany College of Maryland**
Donald L Alexander, EdD, President
12401 Willowbrook Rd SE
Cumberland, MD 21502-2596
301 784-5270
- Clin Lab Technician/Med Lab Technician
- Dental Hygiene Prgm
- Massage Therapy Prgm
- Medical Assistant Prgm
- Occupational Therapy Asst Prgm
- Physical Therapist Assistant Prgm
- Radiography Prgm
- Respiratory Therapist (Advanced) Prgm

## Frederick

**Frederick Community College**
Patricia Stanley, EdD, President
7932 Opossumtown Pike
Frederick, MD 21702
301 846-2400
- Respiratory Therapist (Advanced) Prgm
- Surgical Technology Prgm

## Frostburg

**Frostburg State University**
Catherine R Gira, PhD, President
101 Braddock Rd
Frostburg, MD 21532
301 687-4111
- Athletic Training Prgm

## Hagerstown

**Hagerstown Business College**
W Christopher Motz, MEd, President
18618 Crestwood Dr
Hagerstown, MD 21742
301 739-2670
- Health Information Tech Prgm
- Medical Assistant Prgm
- Phlebotomist Prgm

**Hagerstown Community College**
Guy Altieri, EdD, President
11400 Robinwood Dr
Hagerstown, MD 21742-6590
301 790-2800
- Radiography Prgm

## Largo

**Prince George's Community College**
Ronald Williams, PhD, President
301 Largo Rd
Largo, MD 20774
301 322-0400
- Health Information Tech Prgm
- Nuclear Medicine Technology Prgm
- Radiography Prgm
- Respiratory Therapist (Advanced) Prgm

## Linthicum

**Baltimore School of Massage**
Richard Rynders, Campus Director
517 Progress Dr, Ste A - L
Linthicum, MD 21090
410 636-7929
- Massage Therapy Prgm

## New Market

**Associates in Emergency Care**
Renee Joyce, RN BSN, Administrator
5628 Wellspring Court
New Market, MD 20872
301 856-6548
- Emergency Med Tech-Paramedic Prgm

## North East

**Cecil Community College**
W Stephen Pannill, EdD, President
One Seahawk Dr
North East, MD 21901
410 287-6060
- Medical Assistant Prgm

## Princess Anne

**University of Maryland Eastern Shore**
Thelma B Thompson, PhD, President
Princess Anne, MD 21853
410 651-2200
- Dietetic Internship Prgm
- Dietetics-Didactic Prgm
- Physical Therapy Prgm
- Physician Assistant Prgm
- Rehabilitation Counseling Prgm

## Rockville

**Montgomery College**
Charlene Nunley, PhD, President
Central Admin Office
900 Hungerford
Rockville, MD 20850
301 279-5264
- Diagnostic Med Sonography Prgm
- Health Information Tech Prgm
- Physical Therapist Assistant Prgm
- Radiography Prgm
- Surgical Technology Prgm

## Salisbury

**Salisbury University**
Janet Dudley-Eshbach, PhD, President
1101 Camden Ave
Holloway Hall
Salisbury, MD 21801
410 543-6012
- Athletic Training Prgm
- Clin Lab Scientist/Med Technologist Prgm
- Exercise Science Prgm
- Respiratory Therapist (Advanced) Prgm

**Wor-Wic Community College**
Murray K Hoy, PhD, President
32000 Campus Dr
Salisbury, MD 21801
410 334-2800
- Radiography Prgm

## Silver Spring

**Holy Cross Hospital**
Kevin Sexton, MHA, President and CEO
1500 Forest Glen Rd
Silver Spring, MD 20910
301 754-7000
- Radiography Prgm

## Takoma Park

**Columbia Union College**
Randall Wisbey, PhD, President
7600 Flower Ave
Takoma Park, MD 20912
301 891-4128
- Respiratory Therapist (Advanced) Prgm

**Washington Adventist Hospital**
Brian Breckenridge, President
7600 Carroll Ave
Takoma Park, MD 20912
301 891-7600
- Radiography Prgm

## Towson

**Medix School**
Sean London, Director
700 York Rd
Towson, MD 21204
410 337-5155
- Dental Assisting Prgm
- Medical Assistant Prgm

**Towson University**
Robert Caret, President
8000 York Rd
Towson, MD 21252-0001
410 704-2356
- Athletic Training Prgm
- Audiologist Prgm
- Occupational Therapy Prgm
- Physician Assistant Prgm
- Speech-Language Pathology Prgm

## Westminster

**Carroll Community College**
Faye Pappallardo
1601 Washington Rd
Westminster, MD 21157
410 386-8255
- Physical Therapist Assistant Prgm

## Wye Mills

**Chesapeake College**
Stuart M Bounds, EdD, President
PO Box 8
Wye Mills, MD 21679
410 822-5400
- Radiography Prgm
- Surgical Technology Prgm

# Massachusetts

## Amherst

**University of Massachusetts - Amherst**
John Lombardi, Chancellor
Amherst, MA 01003
413 545-0111
- Audiologist Prgm
- Dietetic Internship Prgm
- Dietetics-Didactic Prgm
- Speech-Language Pathology Prgm

## Auburndale

**Lasell College**
Thomas E J De Witt, PhD, President
1844 Commonwealth Ave
Auburndale, MA 02466
617 243-2000
• Athletic Training Prgm

## Bedford

**Middlesex Community College**
Carole A Cowan, EdD, President
Springs Rd
Bedford, MA 01730
781 280-3100
• Diagnostic Med Sonography Prgm
• Radiography Prgm

## Beverly

**Endicott College**
Richard Wylie, EdD, President
376 Hale St
Beverly, MA 01915
978 232-2000
• Athletic Training Prgm

## Boston

**Bay State College**
Howard E Horton, Esq, President
122 Commonwealth Ave
Boston, MA 02116
617 217-9000
• Physical Therapist Assistant Prgm

**Berklee College of Music**
Roger Brown, President
1140 Boylston St
Boston, MA 02215
617 266-1400
• Music Therapy Prgm

**Beth Israel Deaconess Medical Center**
David Dolins, President
Meissner Bldg - Room G-21
1 Deaconess Rd
Boston, MA 02215-5399
617 667-2203
• Dietetic Internship Prgm

**Boston University**
Robert Brown, President
147 Bay State Rd
Boston, MA 02215
617 353-2000
• Athletic Training Prgm
• Genetic Counseling Prgm
• Physical Therapy Prgm
• Rehabilitation Counseling Prgm
• Speech-Language Pathology Prgm

**Boston University/Sargent College**
Robert A Brown, PhD, President
One Sherborn St
Boston, MA 02215-1605
617 353-2000
• Dietetic Internship Prgm
• Dietetics-Didactic Prgm
• Occupational Therapy Prgm

**Brigham & Women's Hospital**
Gary Gottleib, MD, CEO
75 Francis St
Boston, MA 02115-6195
617 732-5595
• Dietetic Internship Prgm

**Bunker Hill Community College**
Mary L Fifield, PhD, President
250 New Rutherford Ave
Boston, MA 02129-2925
617 228-2400
• Diagnostic Med Sonography Prgm
• Radiography Prgm
• Surgical Technology Prgm

**Caritas Labour, College**
Joseph McNabb, PhD, President
2120 Dorchester Ave
Boston, MA 02124-5698
617 296-8300
• Dietetic Technician-AD Prgm
• Electroneurodiagnostic Tech Prgm
• Health Information Tech Prgm
• Radiation Therapy Prgm

**Emerson College**
Jacqueline W Liebergott, President
100 Beacon St
Boston, MA 02116
617 578-8500
• Speech-Language Pathology Prgm

**Fisher College**
Charles Perkins, JD, President
118 Beacon St
Boston, MA 02116
617 236-8800
• Health Information Tech Prgm

**Frances Stern Nutrition Center**
Thomas O'Donnell, MD, President
750 Washington St
PO Box 451
Boston, MA 02111
617 636-7655
• Dietetic Internship Prgm

**Mass College of Pharmacy & Health Sciences**
Charles F Monahan, Jr, President
179 Longwood Ave
Boston, MA 02115
617 732-2880
• Dental Hygiene Prgm
• Nuclear Medicine Technology Prgm
• Physician Assistant Prgm
• Radiation Therapy Prgm
• Radiography Prgm

**Massachusetts General Hospital**
J Robert Buchanan, MD, General Director
32 Fruit St
Boston, MA 02114
617 726-2101
• Dietetic Internship Prgm

**MGH Institute of Health Professions**
Ann Caldwell, MS, President
Charlestown Navy Yard
36 First Ave
Boston, MA 02129
617 726-2947
• Physical Therapy Prgm
• Radiography Prgm
• Speech-Language Pathology Prgm

**Northeastern University**
Joseph E Aoun, PhD, President
360 Huntington Ave
110 Churchill Hall
Boston, MA 02115
617 373-2101
• Athletic Training Prgm
• Audiologist Prgm
• Clin Lab Scientist/Med Technologist Prgm
• Perfusion Prgm
• Physical Therapy Prgm
• Physician Assistant Prgm
• Respiratory Therapist (Advanced) Prgm
• Speech-Language Pathology Prgm

**Simmons College**
Daniel Cheever, Jr, EdD, President
300 The Fenway
Boston, MA 02115
617 521-2000
• Dietetic Internship Prgm
• Dietetics-Didactic Prgm
• Medical Librarian Prgm
• Physical Therapy Prgm

**Suffolk University**
David J Sargent, PhD, President
Beacon Hill
Boston, MA 02108-2770
617 573-8000
• Radiation Therapy Prgm

**University of Massachusetts - Boston**
Keith Motley, Chancellor
100 Morrissey Blvd
Boston, MA 02125
617 287-5000
• Rehabilitation Counseling Prgm

## Bridgewater

**Bridgewater State College**
Adrian Tinsley, PhD, President
Bridgewater, MA 02325
508 697-1201
• Athletic Training Prgm

## Brockton

**Massasoit Community College**
Charles Wall, PhD, President
1 Massasoit Blvd
Brockton, MA 02301
508 588-9100
• Dental Assisting Prgm
• Medical Assistant Prgm
• Radiography Prgm
• Respiratory Therapist (Advanced) Prgm

## Cambridge

**Lesley University**
Margarert A McKenna, President
29 Everett St
Cambridge, MA 02138
617 868-9600
• Art Therapy Prgm
• Music Therapy Prgm

INSTITUTIONS

**Mount Auburn Hospital**
Francis P Lynch, President
330 Mt Auburn St
Cambridge, MA 02238
617 492-3500
• Dietetic Internship Prgm

## Chicopee

**Porter and Chester Institute - Chicopee**
Henry Kamerzel, VP and Executive Director
134 Dulong Circle
Chicopee, MA 01022
413 593-3339
• Medical Assistant Prgm

## Danvers

**North Shore Community College**
Wayne M Burton, EdD, President
One Ferncroft Rd, PO Box 3340
Danvers, MA 01923-0840
978 762-4000
• Dietetic Technician-AD Prgm
• Medical Assistant Prgm
• Occupational Therapy Asst Prgm
• Physical Therapist Assistant Prgm
• Radiography Prgm
• Respiratory Therapist (Advanced) Prgm
• Surgical Technology Prgm

## Fall River

**Bristol Community College**
John J Sbrega, PhD, President
777 Elsbree St
Fall River, MA 02720
508 678-2811
• Clin Lab Technician/Med Lab Technician
• Dental Hygiene Prgm
• Health Information Tech Prgm
• Medical Assistant Prgm
• Occupational Therapy Asst Prgm

## Framingham

**Framingham State College**
Paul F Weller, President
100 State St
Framingham, MA 01701
508 620-1220
• Dietetics-Coordinated Prgm
• Dietetics-Didactic Prgm

## Gardner

**Mt Wachusett Community College**
Daniel M Asquino, PhD, President
444 Green St
Gardner, MA 01440
978 632-6600
• Medical Assistant Prgm
• Physical Therapist Assistant Prgm

## Greenfield

**Stillpoint Program - Greenfield Comm Coll**
270 Main St
Greenfield, MA 01301
• Massage Therapy Prgm

## Haverhill

**Northern Essex Community College**
David F Hartleb, JD, President
100 Elliott Way
Haverhill, MA 01830
978 556-3855
• Dental Assisting Prgm
• Medical Assistant Prgm
• Polysomnographic Technology Prgm
• Radiography Prgm
• Respiratory Therapist (Advanced) Prgm

## Holyoke

**Holyoke Community College**
William Messner, EdD, President
303 Homestead Ave
Holyoke, MA 01040
413 538-7000
• Ophthalmic Assistant Prgm
• Ophthalmic Dispensing Optician Prgm
• Pharmacy Technician Prgm
• Radiography Prgm

## Longmeadow

**Bay Path College**
Carol A Leary, PhD, President
588 Longmeadow St
Longmeadow, MA 01106
413 565-1241
• Occupational Therapy Prgm

## Lowell

**Middlesex Community College - Lowell**
Carole Cowan, PhD, President
33 Kearney Square
Lowell, MA 01852
978 656-3200
• Dental Assisting Prgm
• Dental Hygiene Prgm
• Dental Lab Technician Prgm
• Medical Assistant Prgm

**University of Massachusetts - Lowell**
David Mackenzie, JD, Interim Chancellor
One University Ave
Lowell, MA 01854
978 934-2201
• Clin Lab Scientist/Med Technologist Prgm
• Physical Therapy Prgm

## Medford

**Tufts University**
Lawrence Bacow, PhD, President
Ballou Hall
Medford, MA 02155-7084
617 627-3300
• Occupational Therapy Prgm

## Newton Centre

**Mt Ida College**
Carol Matteson, PhD, President
Sch of Science and Allied Health
777 Dedham St
Newton Centre, MA 02159-3310
617 928-4500
• Dental Hygiene Prgm

## North Adams

**Charles H McCann Technical School**
James J Brosnan, CAGS, Superintendent
70 Hodges Cross Rd
North Adams, MA 01247
413 663-5383
• Dental Assisting Prgm
• Medical Assistant Prgm
• Surgical Technology Prgm

## North Andover

**Merrimack College**
Richard Santagati, MS DCS, President
315 Turnpike St
North Andover, MA 01845
978 837-5111
• Athletic Training Prgm

## North Dartmouth

**University of Massachusetts - Dartmouth**
Jean MacCormack, EdD, Chancellor
Office of the Chancellor
North Dartmouth, MA 02747-2300
508 999-8004
• Clin Lab Scientist/Med Technologist Prgm

## Paxton

**Anna Maria College**
Bernard Parker, President
Sunset Ln
Paxton, MA 01612-1198
508 849-3335
• Music Therapy Prgm

## Pittsfield

**Berkshire Community College**
Paul Raverta, President
1350 West St
Pittsfield, MA 01201
413 499-4660
• Physical Therapist Assistant Prgm
• Respiratory Therapist (Advanced) Prgm

**Berkshire Medical Center**
Helen Downey, COO
725 North St
Pittsfield, MA 01201
413 447-2144
• Clin Lab Scientist/Med Technologist Prgm
• Cytotechnology Prgm

## Quincy

**Quincy College**
G Jeremiah Ryan, President
34 Coddington St
Quincy, MA 02169
617 984-1776
• Surgical Technology Prgm

## Salem

**Salem State College**
Nancy D Harrington, EdD, President
352 Lafayette St
Salem, MA 01970-5353
978 542-6134
- Athletic Training Prgm
- Nuclear Medicine Technology Prgm
- Occupational Therapy Prgm

## South Easton

**Southeastern Technical Institute**
James Hager, MEd, Superintendent
250 Foundry St
South Easton, MA 02375
508 238-1860
- Dental Assisting Prgm
- Medical Assistant Prgm

## Springfield

**American International College**
Vincent M Maniaci, EdD, President
1000 State St
Springfield, MA 01109-3189
413 205-3202
- Occupational Therapy Prgm
- Physical Therapy Prgm

**Springfield College**
Richard B Flynn, EdD, President
Marsh Memorial
263 Alden St
Springfield, MA 01109
413 748-3241
- Art Therapy Prgm
- Athletic Training Prgm
- Communication Disorders Prgm
- EMS Management Prgm
- Occupational Therapy Prgm
- Physical Therapy Prgm
- Physician Assistant Prgm
- Rehabilitation Counseling Prgm
- Therapeutic Recreation Specialist Prgm

**Springfield Technical Community College**
Ira Rubenzahl, PhD, President
One Armory Square, Ste 1
PO Box 9000
Springfield, MA 01102-9000
413 755-4906
- Clin Lab Technician/Med Lab Technician
- Clinical Assisting Prgm
- Dental Assisting Prgm
- Dental Hygiene Prgm
- Diagnostic Med Sonography Prgm
- Massage Therapy Prgm
- Medical Assistant Prgm
- Nuclear Medicine Technology Prgm
- Occupational Therapy Asst Prgm
- Physical Therapist Assistant Prgm
- Radiography Prgm
- Respiratory Therapist (Advanced) Prgm
- Surgical Technology Prgm

## Waltham

**Brandeis University**
Jehuda Reinharz, President
South St
Waltham, MA 02454-9110
781 736-2000
- Genetic Counseling Prgm

**Sodexho Marriott Services**
153 Second Ave
Waltham, MA 02254-3730
800 926-7429
- Dietetic Internship Prgm (2)

## Watertown

**Cortiva Institute-Muscular Therapy Institute**
Mary Ann DiRoberts, President
103 Morse St
Watertown, MA 02472
617 668-1000
- Massage Therapy Prgm

## Wellesley Hills

**Massachusetts Bay Community College**
Carole Berotte Joseph, PhD, President
Wellesley Hills Campus
50 Oakland St
Wellesley Hills, MA 02481
781 230-3100
- Physical Therapist Assistant Prgm
- Radiography Prgm
- Surgical Technology Prgm

## West Barnstable

**Cape Cod Community College**
Kathleen Schatzberg, EdD, President
2240 Iyanough Rd
West Barnstable, MA 02668-1599
508 362-2131
- Dental Hygiene Prgm
- Medical Assistant Prgm

## Westfield

**Westfield State College**
Vicky Carwein, DNS, President
Office of the President
333 Western Ave
Westfield, MA 01086-1630
413 572-5200
- Athletic Training Prgm
- Exercise Science Prgm

## Weston

**Regis College**
Mary Jane England, PhD, President
235 Wellsley St
Weston, MA 02493
781 768-7000
- Radiography Prgm

## Worcester

**Assumption College**
Joseph H Hagan, President
500 Salisbury St
Worcester, MA 01615
508 767-7000
- Rehabilitation Counseling Prgm

**Becker College**
Franklin M Loew, PhD, President
61 Sever St Box 15071
Worcester, MA 01615-0071
508 791-9241
- Physical Therapist Assistant Prgm

**Quinsigamond Community College**
Sheila Sykes, MEd, Interim President
670 W Boylston St
Worcester, MA 01606
508 854-4203
- Dental Assisting Prgm
- Dental Hygiene Prgm
- Medical Assistant Prgm
- Occupational Therapy Asst Prgm
- Radiography Prgm
- Respiratory Therapist (Advanced) Prgm
- Surgical Technology Prgm

**The Salter School**
Charlene Keefe, BS, President
155 Ararat St
Worcester, MA 01606-3450
508 853-1074
- Medical Assistant Prgm

**UMass Memorial Medical Center**
John O'Brien, MD, CEO, Clinical Systems
One Biotech Park, 365 Plantation St
Worcester, MA 01655
508 856-4114
- Nuclear Medicine Technology Prgm
- Radiation Therapy Prgm

**Worcester State College**
Janelle Ashley, PhD, President
486 Chandler St
Worcester, MA 01602-2597
508 929-8020
- Occupational Therapy Prgm
- Speech-Language Pathology Prgm

# Michigan

## Albion

**Albion College**
Peter T Mitchell, PhD, President
203 Ferguson Bldg
Albion, MI 49224
517 629-0210
- Athletic Training Prgm

## Allen Park

**Baker College of Allen Park**
4500 Enterprise Dr
Allen Park, MI 48101
313 425-3700
- Health Information Tech Prgm

INSTITUTIONS

## Allendale

**Grand Valley State University**
Thomas J Haas, PhD, President
One Campus Dr
Allendale, MI 49401-9403
616 331-2182
- Athletic Training Prgm
- Clin Lab Scientist/Med Technologist Prgm
- Occupational Therapy Prgm
- Physical Therapy Prgm
- Physician Assistant Prgm
- Radiation Therapy Prgm

## Alma

**Alma College**
Saundra Tracy, PhD, President
614 W Superior St
Alma, MI 48801
989 463-7146
- Athletic Training Prgm

## Alpena

**Alpena Community College**
Donald L Newport, PhD, President
666 Johnson St
Alpena, MI 49707
989 358-7246
- Medical Assistant Prgm

## Ann Arbor

**Ann Arbor Institute of Massage Therapy**
Douglas Buhlman, Administrator
2835 Carpenter Rd
Ann Arbor, MI 48108
- Massage Therapy Prgm

**Huron Valley Ambulance Center**
Dale J Berry, EMT-P, President, CEO
2215 Hogback Rd
Ann Arbor, MI 48105
734 477-6262
- Emergency Med Tech-Paramedic Prgm

**University of Michigan**
Mary Sue Coleman, PhD, President
2074 Fleming Adm Bldg
503 Thompson St
Ann Arbor, MI 48109-1340
734 764-6270
- Athletic Training Prgm
- Dental Hygiene Prgm
- Dietetic Internship Prgm
- Dietetics-Didactic Prgm
- Genetic Counseling Prgm
- Medical Librarian Prgm

**University of Michigan Hospitals & Health Ctr**
1500 E Medical Ctr Dr
Ann Arbor, MI 48109-0056
313 936-5199
- Dietetic Internship Prgm

**W K Kellogg Eye Center**
1000 Wall St
Ann Arbor, MI 48105
- Orthoptist Prgm

**Washtenaw Community College**
Larry Whitworth, EdD, President
PO Box D-1
Ann Arbor, MI 48106-0610
734 973-3491
- Dental Assisting Prgm
- Pharmacy Technician Prgm
- Radiography Prgm

## Auburn Hills

**Baker College of Auburn Hills**
Jeffrey M Love, President
1500 University Dr
Auburn Hills, MI 48326-2642
248 340-0600
- Medical Assistant Prgm

## Battle Creek

**Kellogg Community College**
G Edward Haring, PhD, President
450 North Ave
Battle Creek, MI 49017
269 965-3931
- Clin Lab Technician/Med Lab Technician
- Dental Hygiene Prgm
- Physical Therapist Assistant Prgm
- Radiography Prgm

## Benton Harbor

**Lake Michigan College**
Randall Miller, EdD, President
2755 E Napier Ave
Benton Harbor, MI 49022-1899
269 927-8100
- Dental Assisting Prgm
- Radiography Prgm

## Berrien Springs

**Andrews University**
Niels-Erik Andreasen, PhD, President
Berrien Springs, MI 49104
269 471-3100
- Clin Lab Scientist/Med Technologist Prgm
- Counseling Prgm
- Dietetic Internship Prgm
- Dietetics-Didactic Prgm
- Physical Therapy Prgm

## Big Rapids

**Ferris State University**
David Eisler, PhD, President
1201 S. State St.
CSS 301
Big Rapids, MI 49307-2737
616 592-2500
- Clin Lab Scientist/Med Technologist Prgm
- Clin Lab Technician/Med Lab Technician
- Dental Hygiene Prgm
- Health Information Admin Prgm
- Health Information Tech Prgm
- Nuclear Medicine Technology Prgm
- Radiography Prgm
- Respiratory Therapist (Advanced) Prgm

## Bloomfield Hills

**Oakland Community College - Bloomfield Hills**
Mary S Spangler, PhD, Chancellor
George A Bee Administrative Ctr
2480 Opdyke Rd
Bloomfield Hills, MI 48304-2266
248 341-2115
- Diagnostic Med Sonography Prgm
- Radiography Prgm
- Surgical Technology Prgm

## Cadillac

**Baker College of Cadillac**
Robert VanDellen, PhD, President
9600 E 13th St
Cadillac, MI 49601
231 775-8458
- Medical Assistant Prgm
- Surgical Technology Prgm

## Centreville

**Glen Oaks Community College**
Glenn S Oxender, BS MA, President
62249 Shimmel Rd
Centreville, MI 49032
269 467-9945
- Medical Assistant Prgm

## Clinton Township

**Baker College of Clinton Township**
Donald Torline, MBA, President
34950 Little Mack Ave
Clinton Township, MI 48035
586 791-6610
- Medical Assistant Prgm
- Surgical Technology Prgm

## Dearborn

**Henry Ford Community College**
Gail Mee, EdD, President
5101 Evergreen Rd
Dearborn, MI 48128-1495
313 845-9218
- Medical Assistant Prgm
- Pharmacy Technician Prgm
- Physical Therapist Assistant Prgm
- Radiography Prgm
- Respiratory Therapist (Advanced) Prgm
- Surgical Technology Prgm

## Detroit

**Detroit Health Department**
1151 Taylor St
Detroit, MI 48202-1732
313 876-4090
- Dietetic Internship Prgm

**DMC University Laboratories**
Verdell Tolbert, MT, VP, Laboratory Services
4201 St Antoine Blvd
Detroit, MI 48201
313 745-4539
- Clin Lab Scientist/Med Technologist Prgm
- Cytotechnology Prgm

**Harper University Hospital**
Paul Broughton, President
3990 John R St
Detroit, MI 48201-2097
313 745-9375
• Dietetic Internship Prgm

**Henry Ford Hospital**
Nancy Schlicting, MBA, CEO
2799 W Grand Blvd
c/o One Ford Place 5A
Detroit, MI 48202
313 876-1257
• Diagnostic Med Sonography Prgm
• Dietetic Internship Prgm
• Pharmacy Technician Prgm
• Radiography Prgm

**Marygrove College**
Glenda Price, PhD, President
8425 W McNichols Rd
Detroit, MI 48221-2599
313 927-1208
• Dietetics-Didactic Prgm

**Sinai - Grace Hospital**
Conrad Mallett, JD, President
6071 W Outer Dr
Detroit, MI 48235
313 966-3525
• Radiography Prgm

**University of Detroit Mercy**
Rev Gerard L Stockhausen, SJ PhD, President
4001 W McNichols Rd
PO Box 19900
Detroit, MI 48221
313 993-1455
• Counseling Prgm
• Dental Hygiene Prgm
• Physician Assistant Prgm

**Wayne County Community College District**
Curtis L Ivery, PhD, Chancellor
801 W Fort St
Detroit, MI 48226-9975
313 496-2510
• Dental Assisting Prgm
• Dental Hygiene Prgm
• Occupational Therapy Asst Prgm
• Surgical Technology Prgm

**Wayne State University**
Irvin D Reid, PhD, President
4200 Faculty Admin Bldg
Detroit, MI 48202-3489
313 577-2230
• Art Therapy Prgm
• Audiologist Prgm
• Clin Lab Scientist/Med Technologist Prgm
• Counseling Prgm
• Dietetics-Coordinated Prgm
• Genetic Counseling Prgm
• Medical Librarian Prgm
• Occupational Therapy Prgm
• Pathologists' Assistant Prgm
• Physical Therapy Prgm
• Physician Assistant Prgm
• Radiation Therapy Prgm
• Rehabilitation Counseling Prgm
• Speech-Language Pathology Prgm

## East Lansing

**Michigan State University**
Kim Wilcox, PhD, Provost
438 Admin Bldg
East Lansing, MI 48824
517 355-1524
• Athletic Training Prgm
• Audiologist Prgm
• Clin Lab Scientist/Med Technologist Prgm
• Dietetic Internship Prgm
• Dietetics-Didactic Prgm
• Music Therapy Prgm
• Orientation and Mobility Specialist Prgm
• Rehabilitation Counseling Prgm
• Speech-Language Pathology Prgm

## Fenton

**Mott Community College - Fenton**
Richard Shaink, PhD, President
Southern Lakes Branch Campus
2100 W Thompson Rd
Fenton, MI 48430
810 762-0200
• Physical Therapist Assistant Prgm

## Ferndale

**Medright, Inc**
Gail Lucas, Program Director
427 Allen
Ferndale, MI 48220
248 547-0834
• Phlebotomist Prgm

## Flint

**Baker College Center of Graduate Studies**
Michael Heberling, PhD, President
1116 W Bristol Rd
Flint, MI 48507-5508
810 766-2033
• Occupational Therapy Prgm

**Baker College of Flint**
Julianne T Princinsky, EdD, President
1050 W Bristol Rd
Flint, MI 48507-5508
810 766-4036
• Health Information Tech Prgm
• Medical Assistant Prgm
• Physical Therapist Assistant Prgm
• Surgical Technology Prgm

**Hurley Medical Center**
Patrick Wardell, President/CEO
One Hurley Plaza
Flint, MI 48503
810 257-9237
• Clin Lab Scientist/Med Technologist Prgm
• Dietetic Internship Prgm
• Radiography Prgm

**Mott Community College**
M Richard Shaink, President
1401 E Court St
Flint, MI 48503
810 762-0200
• Dental Assisting Prgm
• Dental Hygiene Prgm
• Occupational Therapy Asst Prgm
• Respiratory Therapist (Advanced) Prgm

**University of Michigan - Flint**
Juan Mestas, MA PhD, Chancellor
303 E Kearsley St
Flint, MI 48502-2186
810 762-3000
• Physical Therapy Prgm
• Radiation Therapy Prgm

## Grand Rapids

**Aquinas College**
Harry J Knopke, PhD, President
1607 Robinson Rd SE
Grand Rapids, MI 49506-1799
616 632-2880
• Athletic Training Prgm

**Davenport University**
Randolph Flechsig, PhD, President
415 E Fulton St
Grand Rapids, MI 49503
616 451-3511
• Health Information Tech Prgm
• Medical Assistant Prgm (3)

**Grand Rapids Community College**
Juan Olivarez, PhD, President
143 Bostwick Ave NE
Grand Rapids, MI 49503-3295
616 234-4040
• Dental Assisting Prgm
• Dental Hygiene Prgm
• Occupational Therapy Asst Prgm
• Radiography Prgm

## Hancock

**Finlandia University**
Robert Ubbelohde, President
Quincy 601
Hancock, MI 49930-1882
906 482-6300
• Physical Therapist Assistant Prgm

## Harrison

**Mid Michigan Community College**
Ronald Verch, MS, President
1375 S Clare Ave
Harrison, MI 48625
517 386-6642
• Medical Assistant Prgm
• Radiography Prgm

## Holland

**Hope College**
James Bultman, EdD, President
PO Box 9000
Holland, MI 49422-9000
616 395-7780
• Athletic Training Prgm

## Jackson

**Baker College of Jackson**
Patricia Kaufman, PHD, President
2800 Springport Rd
Jackson, MI 49202
517 789-6123
• Medical Assistant Prgm
• Radiation Therapy Prgm
• Surgical Technology Prgm

**Jackson Community College**
Daniel J Phelan, PhD, President
2111 Emmons Rd
Jackson, MI 49201-8399
517 787-0800
- Diagnostic Med Sonography Prgm
- Medical Assistant Prgm

## Kalamazoo

**Kalamazoo Valley Community College**
Marilyn J Schlack, EdD, President
Texas Township Campus
6767 West O Ave PO Box 4070
Kalamazoo, MI 49003-4070
269 488-4434
- Dental Hygiene Prgm
- Medical Assistant Prgm
- Respiratory Therapist (Advanced) Prgm
- Surgical Technology Prgm

**Western Michigan University**
Judith Bailey, PhD, President
3600 Siebert Admin Bldg
Kalamazoo, MI 49008-5130
269 387-2351
- Athletic Training Prgm
- Audiologist Prgm
- Counseling Prgm
- Dietetic Internship Prgm
- Dietetics-Didactic Prgm
- Music Therapy Prgm
- Occupational Therapy Prgm
- Orientation and Mobility Specialist Prgm
- Physician Assistant Prgm
- Rehabilitation Counseling Prgm
- Rehabilitation Teacher Prgm
- Speech-Language Pathology Prgm

## Lansing

**Lansing Community College**
Judith Cardenas, PhD, Interim President
521 N Washington Sq
PO Box 40010
Lansing, MI 48901
517 483-1852
- Dental Hygiene Prgm
- Diagnostic Med Sonography Prgm
- Emergency Med Tech-Paramedic Prgm
- Histotechnician Prgm
- Radiography Prgm
- Surgical Technology Prgm

## Livonia

**Madonna University**
Mary Francilene, President
36600 Schoolcraft Rd
Livonia, MI 48150
313 591-5000
- Dietetics-Didactic Prgm

**Schoolcraft College**
Conway A Jeffress, PhD, President
18600 Haggerty Rd
Livonia, MI 48152-2696
734 462-4400
- Health Information Tech Prgm
- Medical Assistant Prgm

## Marquette

**Marquette General Health System**
William Nemacheck, CEO
420 W Magnetic St
Marquette, MI 49855
906 225-4774
- Radiography Prgm

**Northern Michigan University**
Wong Leslie, PhD, President
1401 Presque Isle Ave
Marquette, MI 49855
906 227-2242
- Athletic Training Prgm
- Clin Lab Scientist/Med Technologist Prgm
- Clin Lab Technician/Med Lab Technician
- Clinical Assisting Prgm
- Cytogenetic Technology Prgm
- Diagnostic Molecular Scientist Prgm
- Dietetics-Didactic Prgm
- Surgical Technology Prgm

## Monroe

**Monroe County Community College**
David Nixon, MA, President
1555 S Raisinville Rd
Monroe, MI 48161
734 384-4166
- Respiratory Therapist (Advanced) Prgm
- Respiratory Therapist (Entry-Level) Prgm

## Mt Pleasant

**Central Michigan University**
Michael Rao, PhD, President
106 Warriner Hall
Mt Pleasant, MI 48859
989 774-3131
- Athletic Training Prgm
- Audiologist Prgm
- Dietetic Internship Prgm
- Dietetics-Didactic Prgm
- Physical Therapy Prgm
- Physician Assistant Prgm
- Speech-Language Pathology Prgm
- Therapeutic Recreation Specialist Prgm

## Muskegon

**Baker College of Muskegon**
Rick E Amidon, PhD, President
1903 Marquette Ave
Muskegon, MI 49442-9982
231 777-5247
- Medical Assistant Prgm
- Occupational Therapy Asst Prgm
- Physical Therapist Assistant Prgm
- Radiography Prgm
- Surgical Technology Prgm

**Muskegon Community College**
Frank Marczak, PhD, President
221 S Quarterline Rd
Muskegon, MI 49442
616 777-0303
- Respiratory Therapist (Advanced) Prgm

## Owosso

**Baker College of Owosso**
Denise Bannan, PhD, President, CEO
1020 S Washington St
Owosso, MI 48867
989 729-3350
- Clin Lab Technician/Med Lab Technician
- Medical Assistant Prgm
- Phlebotomist Prgm
- Radiography Prgm

## Pontiac

**Oakland County Health Division**
1200 N Telegraph Rd
Pontiac, MI 48341-0432
810 858-1832
- Dietetic Internship Prgm

## Port Huron

**Baker College of Port Huron**
Connie Harrison, PhD, President
3403 Lapeer Rd
Port Huron, MI 48060
810 989-2120
- Dental Hygiene Prgm
- Medical Assistant Prgm
- Surgical Technology Prgm

**Lakewood School of Therapeutic Massage**
Nancy Levitt, Director
1102 6th St
Port Huron, MI 48060
810 987-3959
- Massage Therapy Prgm

**Port Huron Hospital**
Brian Connolly, MBA, President
1221 Pine Grove
Port Huron, MI 48061-5011
313 987-5000
- Radiography Prgm

## Rochester

**Oakland University**
Gary D Russi, PhD, President
North Foundation Hall
Rochester, MI 48063
313 370-3500
- Counseling Prgm
- Physical Therapy Prgm

## Royal Oak

**William Beaumont Hospital**
Kenneth Matzick, President, CEO
3601 W 13 Mile Rd
Royal Oak, MI 48073
248 551-0681
- Clin Lab Scientist/Med Technologist Prgm
- Histotechnician Prgm
- Histotechnology Prgm
- Nuclear Medicine Technology Prgm
- Radiation Therapy Prgm
- Radiography Prgm
- Surgical Assistant Prgm

## Sault Ste Marie

**Lake Superior State University**
Betty Youngblood, PhD, President
650 W Easterday Ave
Sault Ste Marie, MI 49783
906 635-2202
• Athletic Training Prgm

## Sidney

**Montcalm Community College**
Donald Burns, PhD, President
2800 College Dr
Sidney, MI 48885
517 328-1221
• Medical Assistant Prgm

## Southfield

**National Institute of Technology**
Marchelle (Mickey) Weaver, BA, President
26111 Evergreen Rd
Ste 201
Southfield, MI 48076
248 799-9933
• Medical Assistant Prgm

**Providence Hospital**
Robert Casalou, CEO, Providence Oper Unit
16001 W Nine Mile Rd
Southfield, MI 48075
248 849-3000
• Diagnostic Med Sonography Prgm
• Radiography Prgm

## Traverse City

**Northwestern Michigan College**
Timothy Nelson, MA, President
1701 E Front St
Traverse City, MI 49686
616 995-1010
• Dental Assisting Prgm

## Troy

**Carnegie Institute**
Gloria J McEachern-Wiggins, CMA,
    President/CEO
550 Stephenson Hwy, Ste 100-110
Troy, MI 48083
248 589-1078
• Cardiovascular Technology Prgm
• Electroneurodiagnostic Tech Prgm
• Medical Assistant Prgm

## University Center

**Delta College**
Jean Goodnow, PhD, President
1961 Delta Rd
University Center, MI 48710
989 686-9201
• Dental Assisting Prgm
• Dental Hygiene Prgm
• Diagnostic Med Sonography Prgm
• Physical Therapist Assistant Prgm
• Radiography Prgm
• Respiratory Therapist (Advanced) Prgm
• Surgical Technology Prgm

**Saginaw Valley State University**
Eric R Gilbertson, JD, President
7400 Bay Rd
University Center, MI 48710-0001
989 964-4145
• Athletic Training Prgm
• Occupational Therapy Prgm

## Warren

**Macomb Community College**
Albert L Lorenzo, PhD, President
14500 E Twelve Mile Rd
Building D-300
Warren, MI 48088-3896
586 445-7241
• Medical Assistant Prgm
• Occupational Therapy Asst Prgm
• Physical Therapist Assistant Prgm
• Respiratory Therapist (Advanced) Prgm
• Surgical Technology Prgm

**St John Health**
Paul VanTiem, MHSA, EVP, COO
28000 Dequindre Road
Warren, MI 48092
586 753-0718
• Clin Lab Scientist/Med Technologist Prgm
• Radiography Prgm

## Waterford

**Oakland Community College**
Gordan May, President
7350 Cooley Lake Rd
Waterford, MI 48327
248 942-3300
• Dental Hygiene Prgm
• Medical Assistant Prgm
• Respiratory Therapist (Advanced) Prgm

## Ypsilanti

**Eastern Michigan University**
John Fallon III, PhD, President
202 Welch Hall
Ypsilanti, MI 48197-2239
734 487-2211
• Athletic Training Prgm
• Clin Lab Scientist/Med Technologist Prgm
• Counseling Prgm
• Dietetics-Coordinated Prgm
• Music Therapy Prgm
• Occupational Therapy Prgm
• Speech-Language Pathology Prgm

# Minnesota

## Alexandria

**Alexandria Technical College**
Kevin Kopischke, PhD, President
1601 Jefferson St
Alexandria, MN 56308
320 762-0221
• Clin Lab Technician/Med Lab Technician

## Anoka

**Anoka Technical College**
Anne Weyandt, BA JD, President
1355 W Hwy 10
Anoka, MN 55303-1590
763 576-4709
• Health Information Tech Prgm
• Medical Assistant Prgm
• Occupational Therapy Asst Prgm
• Surgical Technology Prgm

## Austin

**Riverland Community College**
Gary Rhodes, PhD, President
1600 8th Ave NW
Austin, MN 55912
507 433-0508
• Radiography Prgm

## Bemidji

**Northwest Technical College - Bemidji**
Jon E Quistgaard, PhD, President
905 Grant Ave SE
Bemidji, MN 56601
218 333-6600
• Dental Assisting Prgm

## Bloomington

**Normandale Community College**
Kathi Hiyani-Brown, President
9700 France Ave S
Bloomington, MN 55431
612 832-6301
• Dental Hygiene Prgm
• Dietetic Technician-AD Prgm

**Northwestern Health Science University**
Alfred Traina, DC
2501 W 84th ST
Bloomington, MN 55431
• Massage Therapy Prgm

## Brainerd

**Central Lakes College**
Sally J Ihne, President
300 Quince St
Brainerd, MN 56401
218 828-2525
• Dental Assisting Prgm

## Brooklyn Center

**Minnesota School of Business - Brooklyn Ctr**
Terry L Myhre, President
5910 Shingle Creek Pkwy #200
Brooklyn Center, MN 55430-2319
763 566-7777
• Medical Assistant Prgm

## Brooklyn Park

**Hennepin Technical College**
Ronald Kraft, Interim President
9000 Brooklyn Blvd
Brooklyn Park, MN 55445
763 488-2414
• Dental Assisting Prgm

INSTITUTIONS

**North Hennepin Community College**
Yvette Jackson, PhD, Interim President
7411 85th Ave N
Brooklyn Park, MN 55445
763 424-0820
• Clin Lab Technician/Med Lab Technician

## Coon Rapids

**Anoka Ramsey Community College**
Patrick M Johns, PhD, President
11200 Mississippi Blvd NW
Coon Rapids, MN 55433
763 422-3435
• Physical Therapist Assistant Prgm

## Crookston

**University of Minnesota - Crookston**
Donald G Sargeant, Chancellor
Crookston, MN 56716
218 281-6510
• Dietetic Technician-AD Prgm

## Crystal

**Herzing College**
Thomas Kosel, MA (Admin), President
5700 West Broadway
Crystal, MN 55428
763 535-3000
• Dental Assisting Prgm
• Dental Hygiene Prgm
• Medical Assistant Prgm

## Duluth

**College of St Scholastica**
Larry Goodwin, PhD, President
1200 Kenwood Ave
Duluth, MN 55811
218 723-6033
• Health Information Admin Prgm
• Occupational Therapy Prgm
• Physical Therapy Prgm

**Duluth Business University**
James R Gessner, President
4724 Mike Colalillo Dr
Duluth, MN 55807
218 722-4000
• Medical Assistant Prgm

**Lake Superior College**
Kathleen Nelson, EdD, President
2101 Trinity Rd
Duluth, MN 55811-3399
218 723-7667
• Clin Lab Technician/Med Lab Technician
• Dental Hygiene Prgm
• Medical Assistant Prgm
• Physical Therapist Assistant Prgm
• Radiography Prgm
• Respiratory Therapist (Advanced) Prgm
• Surgical Technology Prgm

**University of Minnesota - Duluth**
Kathryn A Martin, Chancellor
Duluth, MN 55812
218 726-8000
• Athletic Training Prgm
• Counseling Prgm
• Speech-Language Pathology Prgm

## Eagan

**Argosy University/Twin Cities**
William Cowan, DVM, Campus President
1515 Central Pkwy
Eagan, MN 55121
651 846-3407
• Clin Lab Technician/Med Lab Technician
• Dental Hygiene Prgm
• Diagnostic Med Sonography Prgm
• Histotechnician Prgm
• Medical Assistant Prgm
• Radiation Therapy Prgm
• Radiography Prgm

## Edina

**Fairview University**
Gordon Alexander, MD, President
6545 France Ave S
Edina, MN 55435
952 836-3537
• Dietetic Internship Prgm
• Radiation Therapy Prgm
• Radiography Prgm

## Fergus Falls

**Minnesota State Community and Technical Coll**
Ken Peeders, PhD, President
1414 College Way
Fergus Falls, MN 56537
218 736-7500
• Clin Lab Technician/Med Lab Technician

## Hibbing

**Hibbing Community College**
Ken Simberg, MS, President
1515 E 25th St
Hibbing, MN 55746
218 262-7200
• Clin Lab Technician/Med Lab Technician
• Dental Assisting Prgm

## Inver Grove Heights

**Inver Hills Community College**
Cheryl Frank, PhD, President
2500 E 80th St
Inver Grove Heights, MN 55076
651 460-8641
• Emergency Med Tech-Paramedic Prgm

## Mankato

**Mankato State University**
Richard Davenport
South Rd and Ellis Ave
Mankato, MN 56002-8400
507 389-2423
• Counseling Prgm

**Minnesota State University - Mankato**
Richard Davenport, PhD, President
301 Wigley Adminstration Building
Mankato, MN 56001-8400
507 389-2463
• Athletic Training Prgm
• Counseling Prgm
• Dental Hygiene Prgm
• Dietetics-Didactic Prgm
• Rehabilitation Counseling Prgm
• Speech-Language Pathology Prgm

## Minneapolis

**Augsburg College**
William V Frame, PhD, President
2211 Riverside Ave, CB 131
Minneapolis, MN 55454
612 330-1212
• Music Therapy Prgm
• Physician Assistant Prgm

**Fairview Health Services**
David Page, President
2450 Riverside Ave
Minneapolis, MN 55455
612 672-6000
• Clin Lab Scientist/Med Technologist Prgm

**Hennepin County Medical Center**
Lynn Abrahamsen, Administrator
701 Park Ave S
Minneapolis, MN 55415
612 873-2340
• Clin Lab Scientist/Med Technologist Prgm

**Minneapolis Community & Technical Coll**
Phil Davis, President
1501 Hennepin Ave S
Minneapolis, MN 55403
612 659-6300
• Dental Assisting Prgm

**Park Nicolett Orthoptic Fellowship**
2001 Blasdell Ave
Minneapolis, MN 55406
952 933-8000
• Orthoptist Prgm

**University of Minnesota - Minneapolis**
Robert Bruininks, PhD, President
202 Morrill Hall
100 Church St SE
Minneapolis, MN 55455
612 625-1616
• Audiologist Prgm
• Clin Lab Scientist/Med Technologist Prgm
• Dental Hygiene Prgm
• Genetic Counseling Prgm
• Music Therapy Prgm
• Occupational Therapy Prgm
• Orthoptist Prgm
• Physical Therapy Prgm
• Respiratory Therapist (Advanced) Prgm
• Speech-Language Pathology Prgm
• Therapeutic Recreation Specialist Prgm

**Veterans Affairs Medical Center**
Charles A Milbrandt, FACHE, Director
One Veteran's Dr, Mail Rte 114
Minneapolis, MN 55417
612 725-2000
• Dietetic Internship Prgm
• Radiography Prgm

## Moorhead

**Concordia College - Moorhead**
Paul J Dovre, President
Moorhead, MN 56562
218 299-3947
• Dietetic Internship Prgm
• Dietetics-Didactic Prgm

**Minnesota State Comm & Tech Coll - Moorhead**
Ann Valentine, PhD, President
1900 28th Ave S
Moorhead, MN 56560
888 696-7282
• Dental Assisting Prgm
• Dental Hygiene Prgm
• Health Information Tech Prgm
• Pharmacy Technician Prgm

**Minnesota State University - Moorhead**
211C Lommen Hall
Moorhead, MN 56563
• Athletic Training Prgm
• Counseling Prgm
• Speech-Language Pathology Prgm

## North Mankato

**South Central College**
Keith Stover, MA, President
1920 Lee Blvd
PO Box 1920
North Mankato, MN 56002-1920
507 389-7207
• Clin Lab Technician/Med Lab Technician
• Dental Assisting Prgm
• Emergency Med Tech-Paramedic Prgm

## Oakdale

**Globe College**
Terry Myhre, School Owner
Oakdale Center
7166 10th St N
Oakdale, MN 55128
651 730-5100
• Medical Assistant Prgm

## Ottertail

**Greater Minnesota Paramedic Consortium**
Susan Tarnowski, VP of Student Services
43356 Long Lake Dr
Ottertail, MN 56571
• Emergency Med Tech-Paramedic Prgm

## Richfield

**Minnesota School of Business - Richfield**
Terry L Myhre, President
1401 W 76th St
Ste 500
Richfield, MN 55423-3846
612 861-2000
• Medical Assistant Prgm

## Robbinsdale

**North Memorial Medical Center**
Scott R Anderson, MHA, CEO, President
3300 N Oakdale
Robbinsdale, MN 55422
612 520-5000
• Radiography Prgm

## Rochester

**Mayo Clinic Jacksonville**
Michael B Wood, MD, President
Mayo Foundation
200 First St SW
Rochester, MN 55905
• Radiography Prgm

**Mayo School of Health Sciences**
Denis A Cortese, MD, President/CEO
200 First St SW
Rochester, MN 55905
507 284-2663
• Cytotechnology Prgm
• Diagnostic Med Sonography Prgm
• Electroneurodiagnostic Tech Prgm
• Nuclear Medicine Technology Prgm
• Physical Therapy Prgm
• Radiation Therapy Prgm
• Radiography Prgm

**Rochester Community & Technical College**
Donald Supalla, MS, President
851 30th Ave SE
Rochester, MN 55904-4999
507 285-7215
• Dental Assisting Prgm
• Dental Hygiene Prgm
• Emergency Med Tech-Paramedic Prgm
• Health Information Tech Prgm
• Surgical Technology Prgm

**St Mary's Hospital**
1216 2nd St SW
Rochester, MN 55902-1906
507 255-5221
• Dietetic Internship Prgm

## Rosemount

**Dakota County Technical College**
Ron Thomas, PhD, College President
1300 E 145th St
Rosemount, MN 55068-2999
612 423-2200
• Dental Assisting Prgm
• Medical Assistant Prgm

## Roseville

**Minneapolis Business College**
David Whitman, MA, President
1711 W County Rd B
Ste 100N
Roseville, MN 55113
651 636-7406
• Medical Assistant Prgm

**Rasmussen Colleges**
Kristi A Waite, President
1700 W Hwy 36, Ste 830
Roseville, MN 55133
651 636-3305
• Health Information Tech Prgm (2)

## St Cloud

**St Cloud Hospital**
Craig Broman, President
1406 Sixth Ave N
St Cloud, MN 56303
320 255-5666
• Radiography Prgm

**St Cloud State University**
Roy Saigo, President
720 Fourth Ave S
St Cloud, MN 56301
320 255-2240
• Athletic Training Prgm
• Counseling Prgm
• Rehabilitation Counseling Prgm
• Speech-Language Pathology Prgm

**St Cloud Technical College**
Joyce Helens, MS, President
1540 Northway Dr
St Cloud, MN 56303
320 308-5017
• Cardiovascular Technology Prgm
• Dental Assisting Prgm
• Dental Hygiene Prgm
• Diagnostic Med Sonography Prgm
• Emergency Med Tech-Paramedic Prgm
• Surgical Technology Prgm

## St Joseph

**College of St Benedict/St John's University**
Colman O'Connell, President
37 S College Ave
St Joseph, MN 56374
612 363-5011
• Dietetics-Coordinated Prgm
• Dietetics-Didactic Prgm

## St Louis Park

**Methodist Hospital**
David Wessner, CEO
6500 Excelsior Blvd
St Louis Park, MN 55426
612 993-3601
• Radiography Prgm

## St Paul

**Bethel University**
George K Brushaber, PhD, President
3900 Bethel Dr
St Paul, MN 55112
651 638-6230
• Athletic Training Prgm

INSTITUTIONS

**College of St Catherine**
Andrea J Lee, IHM, President
2004 Randolph Ave
St Paul, MN 55105-1794
651 690-6525
- Diagnostic Med Sonography Prgm
- Dietetics-Didactic Prgm
- Health Information Admin Prgm
- Health Information Tech Prgm
- Occupational Therapy Asst Prgm
- Occupational Therapy Prgm
- Phlebotomist Prgm
- Physical Therapist Assistant Prgm
- Physical Therapy Prgm
- Radiography Prgm
- Respiratory Therapist (Advanced) Prgm

**Regions Hospital**
Brock D Nelson, President/CEO
640 Jackson St
St Paul, MN 55101
651 254-2189
- Dietetic Internship Prgm
- Ophthalmic Med Technician Prgm
- Ophthalmic Med Technologist Prgm

**St Paul College**
Donovan Schwichtenberg, PhD, President
235 Marshall Ave
St Paul, MN 55102
651 846-1364
- Clin Lab Technician/Med Lab Technician
- Respiratory Therapist (Advanced) Prgm

**University of Minnesota - St Paul**
1334 Eckles Ave
St Paul, MN 55108
612 624-9278
- Dietetic Internship Prgm
- Dietetics-Coordinated Prgm
- Dietetics-Didactic Prgm

## St Peter

**Gustavus Adolphus College**
James Peterson, PhD, President
800 W College Ave
St Peter, MN 56082
507 933-7537
- Athletic Training Prgm

## Thief River Falls

**Northland Community & Technical College**
Anne Temte, PhD, President
1101 Highway 1 East
Thief River Falls, MN 56721-2702
218 681-0701
- Cardiovascular Technology Prgm
- Clin Lab Technician/Med Lab Technician
- Emergency Med Tech-Paramedic Prgm
- Medical Assistant Prgm
- Occupational Therapy Asst Prgm
- Pharmacy Technician Prgm
- Radiography Prgm
- Respiratory Therapist (Advanced) Prgm
- Respiratory Therapist (Entry-Level) Prgm
- Surgical Technology Prgm

## White Bear Lake

**Century College**
Larry Litecky, PhD, President
3300 Century Ave
White Bear Lake, MN 55110
651 779-3342
- Dental Assisting Prgm
- Dental Hygiene Prgm
- Emergency Med Tech-Paramedic Prgm
- Medical Assistant Prgm
- Orthotist/Prosthetist Prgm
- Pharmacy Technician Prgm
- Radiography Prgm

## Willmar

**Rice Memorial Hospital**
Lawrence Massa, CEO
301 Becker Ave SW
Willmar, MN 56201
320 235-4543
- Radiography Prgm

**Ridgewater College - Willmar Campus**
Douglas Allen, President
PO Box 1097
Willmar, MN 56201-1097
320 235-5114
- Health Information Tech Prgm
- Medical Assistant Prgm

## Winona

**St Mary's University of Minnesota**
Craig Franz, PhD, President
700 Terrace Heights #30
Winona, MN 55987-1399
507 457-1503
- Nuclear Medicine Technology Prgm
- Surgical Technology Prgm

**Winona State University**
Darrell W Krueger, PhD, President
PO Box 5838
Winona, MN 55987-5838
507 457-5003
- Athletic Training Prgm
- Counseling Prgm

## Worthington

**Minnesota West Comm & Tech College**
Ron Wood, PhD, President
1450 Collegeway
Worthington, MN 56187
507 372-3400
- Clin Lab Technician/Med Lab Technician
- Dental Assisting Prgm
- Medical Assistant Prgm

# Mississippi

## Booneville

**Northeast Mississippi Community College**
Johnny Allen, PhD, President
Cunningham Blvd
Booneville, MS 38829
601 728-7751
- Clin Lab Technician/Med Lab Technician
- Dental Hygiene Prgm
- Medical Assistant Prgm
- Radiography Prgm
- Respiratory Therapist (Advanced) Prgm

## Cleveland

**Delta State University**
John Hilpert, PhD, President
Box A-1
Cleveland, MS 38733-0002
601 846-3000
- Athletic Training Prgm
- Counseling Prgm
- Dietetics-Coordinated Prgm

## Clinton

**Mississippi College**
Royce Lee, PhD, President
200 W College St
Clinton, MS 39058-0001
601 925-3000
- Counseling Prgm

## Columbus

**Mississippi University for Women**
Clyda S Rent, President
PO Box W-1340
Columbus, MS 39701
601 329-4750
- Music Therapy Prgm
- Speech-Language Pathology Prgm

## Decatur

**East Central Community College**
Phil A Suthpin, EdD, President
PO Box 129
Decatur, MS 39327
601 635-2111
- Emergency Med Tech-Paramedic Prgm
- Surgical Technology Prgm

## Ellisville

**Jones County Junior College**
Jesse Smith, EdD, President
900 S Court St
Ellisville, MS 39437
601 477-4100
- Emergency Med Tech-Paramedic Prgm
- Pharmacy Technician Prgm
- Radiography Prgm

## Fulton

**Itawamba Community College**
David Cole, PhD, President
602 W Hill St
Fulton, MS 38843
662 862-8001
- Diagnostic Med Sonography Prgm
- Emergency Med Tech-Paramedic Prgm
- Health Information Tech Prgm
- Occupational Therapy Asst Prgm
- Physical Therapist Assistant Prgm
- Radiography Prgm
- Respiratory Therapist (Advanced) Prgm
- Surgical Technology Prgm

## Goodman

**Holmes Community College**
Glenn Boyce, EdD, President
Goodman Campus
PO Box 369
Goodman, MS 39079
601 472-2312
- Emergency Med Tech-Paramedic Prgm
- Occupational Therapy Asst Prgm
- Surgical Technology Prgm

## Hattiesburg

**University of Southern Mississippi**
Shelby F Thames, PhD, President
118 College Dr, #5001
Hattiesburg, MS 39406-5001
601 266-5001
- Athletic Training Prgm
- Audiologist Prgm
- Clin Lab Scientist/Med Technologist Prgm
- Counseling Prgm
- Dietetic Internship Prgm
- Dietetics-Didactic Prgm
- Kinesiotherapy Prgm
- Medical Librarian Prgm
- Speech-Language Pathology Prgm
- Therapeutic Recreation Specialist Prgm

**William Carey College**
James W Edwards, PhD, President
Tuscan Ave
Hattiesburg, MS 39401-9913
601 582-6223
- Music Therapy Prgm

## Jackson

**Jackson State University**
James E Lyons, Sr, President
1440 JR Lynch St
Jackson, MS 39217
601 968-2100
- Rehabilitation Counseling Prgm
- Speech-Language Pathology Prgm

**Mississippi Baptist Medical Center**
Gerald Cotton, Exec Director
1225 N State St
Jackson, MS 39202
601 968-5130
- Clin Lab Scientist/Med Technologist Prgm

**Mississippi School of Therapeutic Massage**
5120 Galaxie Dr
Jackson, MS 39206
- Massage Therapy Prgm

**St Dominic-Jackson Memorial Hospital**
969 Lakeland Dr
Jackson, MS 39216-4699
601 364-6935
- Dietetic Internship Prgm

**University of Mississippi Medical Center**
Daniel W Jones, MD, Vice Chancellor
2500 N State St
Jackson, MS 39216-4505
601 984-1010
- Clin Lab Scientist/Med Technologist Prgm
- Cytotechnology Prgm
- Dental Hygiene Prgm
- Emergency Med Tech-Paramedic Prgm
- Health Information Admin Prgm
- Nuclear Medicine Technology Prgm
- Occupational Therapy Prgm
- Physical Therapy Prgm
- Radiography Prgm

## Lorman

**Alcorn State University**
Rudolph E Waters, Interim President
Lorman, MS 39096
601 877-6100
- Dietetic Internship Prgm
- Dietetics-Didactic Prgm

## Meridian

**Meridian Community College**
Scott Elliott, PhD, President
910 Hwy 19 N
Meridian, MS 39307
601 484-8618
- Clin Lab Technician/Med Lab Technician
- Dental Hygiene Prgm
- Health Information Tech Prgm
- Physical Therapist Assistant Prgm
- Radiography Prgm
- Respiratory Therapist (Advanced) Prgm

## Mississippi State

**Mississippi State University**
Donald W Zacharias, President
Mississippi State, MS 39762
601 325-2131
- Counseling Prgm
- Dietetic Internship Prgm
- Dietetics-Didactic Prgm
- Rehabilitation Counseling Prgm

## Moorhead

**Mississippi Delta Community College**
Larry G Bailey, PhD, President
PO Box 668
Moorhead, MS 38761
601 246-6300
- Clin Lab Technician/Med Lab Technician
- Dental Hygiene Prgm
- Radiography Prgm

## Perkinston

**Mississippi Gulf Coast Community College**
Willis H Lott, EdD, President
PO Box 609
Perkinston, MS 39573
601 928-5211
- Clin Lab Technician/Med Lab Technician
- Emergency Med Tech-Paramedic Prgm
- Radiography Prgm
- Respiratory Therapist (Advanced) Prgm
- Surgical Technology Prgm

## Poplarville

**Pearl River Community College**
William Lewis, EdD, President
101 Hwy 11 N
Poplarville, MS 39470
601 403-1201
- Clin Lab Technician/Med Lab Technician
- Dental Assisting Prgm
- Dental Hygiene Prgm
- Occupational Therapy Asst Prgm
- Physical Therapist Assistant Prgm
- Radiography Prgm
- Respiratory Therapist (Advanced) Prgm
- Surgical Technology Prgm

## Raymond

**Hinds Community College**
Vernon Clyde Muse, EdD, President
PO Box 1100
501 E Main St
Raymond, MS 39154-1100
601 857-3230
- Clin Lab Technician/Med Lab Technician
- Dental Assisting Prgm
- Diagnostic Med Sonography Prgm
- Health Information Tech Prgm
- Medical Assistant Prgm
- Physical Therapist Assistant Prgm
- Radiography Prgm
- Respiratory Therapist (Advanced) Prgm
- Surgical Technology Prgm

## Senatobia

**Northwest Mississippi Community College**
Gary L Spears, EdD, President
4975 Highway 51 N
Senatobia, MS 38668
601 562-3227
- Emergency Med Tech-Paramedic Prgm
- Respiratory Therapist (Advanced) Prgm

## Tupelo

**North Mississippi Medical Center**
John Heer, Jr, MS, CEO
830 S Gloster
Tupelo, MS 38801
662 377-3136
- Clin Lab Scientist/Med Technologist Prgm

## University

**University of Mississippi**
Robert C Khayat, Chancellor
University, MS 38677
601 232-7211
• Counseling Prgm
• Dietetics-Didactic Prgm
• Speech-Language Pathology Prgm

## Wesson

**Copiah-Lincoln Community College**
Howell C Garner, EdD, President
PO Box 649
Wesson, MS 39191-0457
601 643-5101
• Clin Lab Technician/Med Lab Technician
• Radiography Prgm
• Respiratory Therapist (Advanced) Prgm

# Missouri

## Bolivar

**Southwest Baptist University**
Pat Taylor, President
1600 University Ave
Bolivar, MO 65613-2597
417 328-1500
• Athletic Training Prgm
• Physical Therapy Prgm

## Canton

**Culver-Stockton College**
William L Fox, PhD, President
1 College Hill
Canton, MO 63435
217 231-6510
• Athletic Training Prgm

## Cape Girardeau

**Cape Girardeau Career & Technology Center**
Richard Payne, Director CTC
Special School Administrator
1080 S Silver Springs Rd
Cape Girardeau, MO 63703
573 334-0449
• Respiratory Therapist (Advanced) Prgm
• Respiratory Therapist (Entry-Level) Prgm

**Southeast Missouri Hospital**
Tonya Buttry, MSN RNC, President
College of Nursing & Health Sciences
2001 William St
Cape Girardeau, MO 63703
573 334-6825
• Clin Lab Scientist/Med Technologist Prgm
• Radiography Prgm
• Surgical Technology Prgm

**Southeast Missouri State University**
Kenneth Dobbins, PhD, President
One University Plaza, Mail Stop 3300
Cape Girardeau, MO 63701
573 651-2222
• Athletic Training Prgm
• Counseling Prgm
• Dietetic Internship Prgm
• Dietetics-Didactic Prgm
• Speech-Language Pathology Prgm

## Columbia

**Columbia Public Schools**
James R Ritter, EdD, Superintendent of Schools
1818 W Worley
Columbia, MO 65203
573 886-2149
• Surgical Technology Prgm

**Stephens College**
Wendy B Libby, PhD, President
Campus Box 2001
1200 E Broadway
Columbia, MO 65215
573 876-7210
• Health Information Admin Prgm

**University of Missouri - Columbia**
Deaton Brady, PhD, Chancellor
105 Jesse Hall
Columbia, MO 65211
573 882-3387
• Dietetics-Coordinated Prgm
• Medical Librarian Prgm
• Nuclear Medicine Technology Prgm
• Occupational Therapy Prgm
• Physical Therapy Prgm
• Radiography Prgm
• Rehabilitation Counseling Prgm
• Respiratory Therapist (Advanced) Prgm
• Speech-Language Pathology Prgm
• Therapeutic Recreation Specialist Prgm

## Farmington

**Mineral Area Regional Medical Center**
Bob Moore, FACHE, CEO
1212 Weber Rd
Farmington, MO 63640
573 756-4581
• Radiography Prgm

## Fayette

**Central Methodist University**
Marianne E Inman, PhD, President
411 CMC Square
Fayette, MO 65248
660 248-6221
• Athletic Training Prgm

## Fulton

**William Woods University**
Jahnae Barnett, PhD, President
One University Ave
Fulton, MO 65251
573 592-4216
• Athletic Training Prgm

## Hannibal

**Hannibal Career & Technical Center**
Roger McGregor, Director
4550 McMasters Ave
Hannibal, MO 63401
573 221-4430
• Respiratory Therapist (Advanced) Prgm

## Hazelwood

**Sanford-Brown College - Hazelwood Campus**
Chad Freeman, MA, Campus President
75 Village Square
Hazelwood, MO 63042
314 731-5200
• Occupational Therapy Asst Prgm
• Surgical Technology Prgm

## Jefferson City

**Missouri Dept of Health & Senior Service**
Office of Administration
1706 E Elm St, PO Box 687
Jefferson City, MO 65102
314 751-8145
• Dietetic Internship Prgm

**Nichols Career Center**
Bert Kimble, EdD, Superintendent
609 Union
Jefferson City, MO 65101
573 659-3012
• Dental Assisting Prgm
• Radiography Prgm

## Joplin

**Franklin Tech Center**
James Simpson, School Superintendent
1717 E 15th St
Joplin, MO 64802
417 625-5200
• Surgical Technology Prgm

**Missouri Southern State University**
Julio S Leon, President
3950 E Newman Rd
Joplin, MO 64801-1595
417 625-9328
• Dental Hygiene Prgm
• Radiography Prgm
• Respiratory Therapist (Advanced) Prgm
• Respiratory Therapist (Entry-Level) Prgm

**St John's Regional Medical Center**
Gary Rowe, MS, President
2727 McClelland Blvd
Joplin, MO 64804
417 781-2727
• Clin Lab Scientist/Med Technologist Prgm

## Kansas City

**ARAMARK Healthcare Support Services - KC**
St Joseph Health System
1000 Carondelet Dr
Kansas City, MO 64114-4802
816 943-2146
• Dietetic Internship Prgm

**Avila University**
Thomas Gordon, JD LLM, President
11901 Wornall Rd
Kansas City, MO 64145
816 942-8400
• Radiography Prgm

**Colorado Technical University**
Paul Goddard, President
520 E 19th St
Kansas City, MO 64116
816 472-7400
• Radiography Prgm

**Concorde Career College - Kansas City**
Deborah Crow, MS, Campus President
3239 Broadway
Kansas City, MO 64111
816 531-5223
• Dental Assisting Prgm
• Respiratory Therapist (Advanced) Prgm
• Respiratory Therapist (Entry-Level) Prgm

**Metropolitan Community College - Penn Valley**
Bernard Franklin, PhD, President
3201 SW Trafficway
Kansas City, MO 64111-2764
816 759-4201
• Dental Assisting Prgm
• Health Information Tech Prgm
• Occupational Therapy Asst Prgm
• Physical Therapist Assistant Prgm
• Radiography Prgm
• Surgical Technology Prgm

**Research Medical Center**
Niels Vernegaard, CEO
2316 E Meyer Blvd
Kansas City, MO 64132-1199
816 276-4101
• Nuclear Medicine Technology Prgm
• Radiography Prgm

**Rockhurst University**
Thomas Curran, OSFS, President
1100 Rockhurst Rd
Kansas City, MO 64110-2561
816 501-4250
• Occupational Therapy Prgm
• Physical Therapy Prgm
• Speech-Language Pathology Prgm

**Saint Luke's Hospital**
G Richard Hastings, CEO
4401 Wornall Rd
Kansas City, MO 64111
816 932-2101
• Clin Lab Scientist/Med Technologist Prgm
• Phlebotomist Prgm
• Radiography Prgm

**University of Missouri - Kansas City**
Guy Bailey, PhD, Chancellor
5115 Oak
Kansas City, MO 64110-2499
816 235-1000
• Dental Hygiene Prgm
• Music Therapy Prgm

## Kirksville

**Truman State University**
Barbara Dixon, DMA, President
200 McClain Hall
Kirksville, MO 63501
660 785-4000
• Athletic Training Prgm
• Counseling Prgm
• Speech-Language Pathology Prgm

## Linn

**Linn State Technical College**
Donald Claycomb, PhD, President
One Technology Dr
Linn, MO 65051
573 897-5000
• Physical Therapist Assistant Prgm

## Marshall

**Missouri Valley College**
Bonnie Humphrey, PhD, President
500 E College St
Marshall, MO 65340
660 831-4108
• Athletic Training Prgm

## Maryville

**Northwest Missouri State University**
Dean L Hubbard, President
Maryville, MO 64468
816 562-1212
• Dietetics-Didactic Prgm

## North Kansas City

**North Kansas City Hospital**
David Carpenter, MS JD, President and CEO
2800 Clay Edwards Dr
North Kansas City, MO 64116
816 691-2000
• Clin Lab Scientist/Med Technologist Prgm

**Sanford-Brown College**
Dennis Townsend, MEd, Campus President
520 E 19th Ave
North Kansas City, MO 64116
816 472-7400
• Radiography Prgm
• Respiratory Therapist (Entry-Level) Prgm

## Parkville

**Park University**
Beverley Byers-Pevitts, PhD, President
8700 NW Riverpark Dr
Parkville, MO 64152
816 741-2000
• Athletic Training Prgm

## Point Lookout

**College of the Ozarks**
Jerry C Davis, President
Point Lookout, MO 65726
417 334-6411
• Dietetics-Didactic Prgm

## Poplar Bluff

**Three Rivers Community College**
John Cooper, EdD, President
2080 Three Rivers Blvd
Poplar Bluff, MO 63901-2393
573 840-9698
• Clin Lab Technician/Med Lab Technician

## Rolla

**Rolla Technical Center**
Janece Martin, PhD, Director Vocational Prgms
500 Forum Dr
Rolla, MO 65401
573 458-0160
• Radiography Prgm
• Respiratory Therapist (Advanced) Prgm
• Respiratory Therapist (Entry-Level) Prgm
• Surgical Technology Prgm

## Sedalia

**State Fair Community College**
Marsha Drennon, EdD, President
3201 W 16th St
Sedalia, MO 65301
660 530-5800
• Dental Hygiene Prgm
• Radiography Prgm

## Springfield

**CoxHealth**
Robert H Bezanson, MHA, Administrator
1423 N Jefferson St
Springfield, MO 65802
417 269-3108
• Clin Lab Scientist/Med Technologist Prgm
• Diagnostic Med Sonography Prgm
• Radiation Therapy Prgm
• Radiography Prgm

**Drury University**
John Sellars, PhD, President
Burnham Hall 102
Springfield, MO 65802
417 873-7201
• Music Therapy Prgm

**Everest College - Springfield Campus**
Gary Myers, MS, President
1010 W Sunshine
Springfield, MO 65807
417 864-7220
• Medical Assistant Prgm

**Missouri State University**
Michael Nietzel, PhD, President
901 S National Ave
Springfield, MO 65897
417 836-8500
• Athletic Training Prgm
• Audiologist Prgm
• Dietetics-Didactic Prgm
• Physical Therapy Prgm
• Physician Assistant Prgm
• Speech-Language Pathology Prgm

**Ozarks Technical Community College**
Hal Higdon, PhD, President
1001 E Chestnut Expressway
Springfield, MO 65802
417 447-2602
- Dental Assisting Prgm
- Dental Hygiene Prgm
- Health Information Tech Prgm
- Occupational Therapy Asst Prgm
- Physical Therapist Assistant Prgm
- Respiratory Therapist (Advanced) Prgm
- Surgical Technology Prgm

**St John's Regional Health Center**
Robert Brodhead, President
1235 E Cherokee
Springfield, MO 65804
417 820-2709
- Radiography Prgm

## St Charles

**Lindenwood University**
Dennis Spellmann, President
209 S Kings Hwy
St Charles, MO 63301
636 949-4949
- Athletic Training Prgm

**St Charles School of Massage Therapy**
Kathleen Crawford, LMT, Director
2440 Executive Dr
St Charles, MO 63303
636 498-0777
- Massage Therapy Prgm

## St Joseph

**Hillyard Technical Center**
Robert D Stewart, MS, Director
3434 Faragon St
St Joseph, MO 64506
816 671-4170
- Radiography Prgm
- Surgical Technology Prgm

**Missouri Western State University**
James Scanlon, President
4525 Downs Dr
St Joseph, MO 64507
816 271-4237
- Health Information Tech Prgm
- Physical Therapist Assistant Prgm

## St Louis

**Barnes-Jewish Coll of Nursing/Allied Health**
Ronald Evens, MD, President
MS: 90-30-625
306 S Kingshighway Blvd
St Louis, MO 63110
314 454-7054
- Cytotechnology Prgm
- Dietetic Internship Prgm
- Radiation Therapy Prgm
- Radiography Prgm

**DVA Medical Center**
Jefferson Barracks Div
One Jefferson Barracks Dr
St Louis, MO 63125
314 894-6631
- Dietetic Internship Prgm

**Fontbonne University**
Dennis Golden, President
6800 Wydown Blvd
St Louis, MO 63105
314 862-3456
- Dietetics-Didactic Prgm
- Speech-Language Pathology Prgm

**IHM Health Studies Center**
Richard S Kurz, PhD, Chairman
3663 Lindell Blvd
St Louis, MO 63108-3342
314 768-1000
- Emergency Med Tech-Paramedic Prgm

**Maryville University**
Brian Nedwek, PhD, President
13550 Conway Rd
St Louis, MO 63141-7299
314 529-9300
- Music Therapy Prgm
- Occupational Therapy Prgm
- Physical Therapy Prgm
- Rehabilitation Counseling Prgm

**Missouri College**
Michael Vander Velde, MA, President
10121 Manchester Rd
St Louis, MO 63122
314 821-7700
- Dental Assisting Prgm

**Saint Louis University**
Joseph Weixlmann, PhD, Provost
221 N Grand Blvd #106
DuBourg Hall Room 106
St Louis, MO 63103
314 977-3718
- Clin Lab Scientist/Med Technologist Prgm
- Dietetic Internship Prgm
- Dietetics-Didactic Prgm
- Health Information Admin Prgm
- Nuclear Medicine Technology Prgm
- Occupational Therapy Prgm
- Physical Therapy Prgm
- Physician Assistant Prgm
- Speech-Language Pathology Prgm

**St John's Mercy Medical Center**
Denny DeNarvaez, President/CEO
615 S New Ballas Rd
St Louis, MO 63141
314 251-1952
- Clin Lab Scientist/Med Technologist Prgm
- Radiography Prgm

**St Louis Community College**
Henry D Shannon, PhD, Chancellor
300 S Broadway
St Louis, MO 63102-2800
314 539-5000
- Clin Lab Technician/Med Lab Technician
- Dental Assisting Prgm
- Dental Hygiene Prgm
- Diagnostic Med Sonography Prgm
- Dietetic Technician-AD Prgm
- Occupational Therapy Asst Prgm
- Physical Therapist Assistant Prgm
- Radiography Prgm
- Respiratory Therapist (Advanced) Prgm
- Surgical Technology Prgm

**University of Missouri - St Louis**
Blanche M Touhill, President
8001 Natural Bridge Rd
St Louis, MO 63121-4499
314 516-5000
- Counseling Prgm

**Washington University**
Mark S Wrighton, PhD, Chancellor
Brookings Dr Campus Box 1192
St Louis, MO 63130-4899
314 935-5100
- Audiologist Prgm
- Occupational Therapy Prgm
- Physical Therapy Prgm

## St Peters

**St Charles Community College**
John McGuire, PhD, President
4601 Mid Rivers Mall Dr
PO Box 76975
St Peters, MO 63376-0975
636 922-8380
- Health Information Tech Prgm
- Occupational Therapy Asst Prgm

## Warrensburg

**Central Missouri State University**
Aaron Podolefsky, President
Warrensburg, MO 64093
816 543-4111
- Audiologist Prgm
- Dietetics-Didactic Prgm
- Speech-Language Pathology Prgm

## West Plains

**South Central Career Center**
Steve Bryant, MS EdS, Director
610 Olden
West Plains, MO 65775
417 256-6152
- Surgical Technology Prgm

# Montana

## Billings

**Montana State University - Billings**
Ronald P Sexton, PhD, Chancellor
1500 University Dr
Billings, MT 59101
406 657-2300
- Athletic Training Prgm
- Emergency Med Tech-Paramedic Prgm
- Medical Assistant Prgm
- Rehabilitation Counseling Prgm

**Rocky Mountain College**
Thomas Oates, PhD, President
1511 Poly Dr
Billings, MT 59102
406 657-1015
- Physician Assistant Prgm

## Bozeman

**Health Works Institute**
Ruth Marion, Owner-Director
111 S Grand, Annex 3
Bozeman, MT 59715
406 582-1555
• Massage Therapy Prgm

**Montana State University**
Geoffrey Gamble, President
Bozeman, MT 59717
406 994-0211
• Counseling Prgm
• Dietetics-Didactic Prgm

## Great Falls

**Benefis Healthcare**
John Goodnow, CEO
1101 26th St S
Great Falls, MT 59405
406 455-5000
• Clin Lab Scientist/Med Technologist Prgm
• Radiography Prgm

**Montana State Univ - Great Falls Coll of Tech**
Mary Sheehy Moe, EdD, Dean
2100 16th Ave S
Great Falls, MT 59405
406 771-4310
• Dental Assisting Prgm
• Dental Hygiene Prgm
• Health Information Tech Prgm
• Medical Assistant Prgm
• Respiratory Therapist (Advanced) Prgm
• Surgical Technology Prgm

## Helena

**Big Sky Somatic Institute**
1802 11th Ave
Helena, MT 59601
• Massage Therapy Prgm

## Kalispell

**Flathead Valley Community College**
Jane A Karas, PhD, President
777 Grandview Dr
Kalispell, MT 59901
406 756-3822
• Medical Assistant Prgm
• Surgical Technology Prgm

## Missoula

**College of Tech - Univ of Montana-Missoula**
Dennis Lerum, Dean
909 S Ave W
Missoula, MT 59801
• Pharmacy Technician Prgm

**University of Montana**
George Dennison, PhD, President
University Hall 109
Missoula, MT 59812
406 243-2311
• Athletic Training Prgm
• Counseling Prgm
• Physical Therapy Prgm
• Respiratory Therapist (Advanced) Prgm
• Surgical Technology Prgm

## Pablo

**Salish Kootenai College**
Joseph F McDonald, PhD, President
PO Box 117
52000 Hwy 93
Pablo, MT 59855
406 675-4800
• Dental Assisting Prgm

# Nebraska

## Grand Island

**Central Community College**
LaVern Franzen, PhD, President
PO Box 4903
Grand Island, NE 68802-4903
308 389-6300
• Clin Lab Technician/Med Lab Technician
• Dental Assisting Prgm
• Dental Hygiene Prgm
• Health Information Tech Prgm
• Medical Assistant Prgm

## Hastings

**Mary Lanning Memorial Hospital**
W Michael Kearney, MHA, Administrator
715 N St Joseph
Hastings, NE 68901
402 463-4521
• Radiography Prgm

## Kearney

**University of Nebraska - Kearney**
Douglas Kristensen, JD, Chancellor
905 W 25th St
Founders Hall
Kearney, NE 68849
308 865-8208
• Athletic Training Prgm
• Counseling Prgm
• Dietetics-Didactic Prgm
• Speech-Language Pathology Prgm

## Lincoln

**BryanLGH College of Health Sciences**
Phylis Hollamon, MSN, President
5035 Everett St.
Lincoln, NE 68506
402 481-3867
• Cardiovascular Technology Prgm

**Hamilton College**
Thomas Lardenoit, Director
1821 K St
Lincoln, NE 68508
402 474-5315
• Medical Assistant Prgm

**Nebraska Wesleyan University**
Jeanie Watson, PhD, President
5000 St Paul Ave
Lincoln, NE 68504
402 466-2371
• Athletic Training Prgm

## Southeast Community College

Jack Huck, PhD, President
301 S 68th St Place
Lincoln, NE 68510
402 323-3415
• Clin Lab Technician/Med Lab Technician
• Dental Assisting Prgm
• Dietetic Technician-AD Prgm
• Medical Assistant Prgm
• Radiography Prgm
• Respiratory Therapist (Advanced) Prgm
• Surgical Technology Prgm

**Union College**
David Smith, EdD, President
3800 S 48th St
Lincoln, NE 68506
402 486-2500
• Physician Assistant Prgm

**University of Nebraska - Lincoln**
Harvey Perlman, JD, Chancellor
201 ADM
Lincoln, NE 68588-0419
402 472-2116
• Athletic Training Prgm
• Audiologist Prgm
• Dental Hygiene Prgm
• Dietetic Internship Prgm
• Dietetics-Didactic Prgm
• Speech-Language Pathology Prgm

## Norfolk

**Northeast Community College**
Bill R Path
801 E Benjamin Ave
PO Box 469
Norfolk, NE 68702-0469
• Physical Therapist Assistant Prgm

## North Platte

**Mid-Plains Community College**
Michael Chipps, PhD, President
601 W State Farm Rd
North Platte, NE 69101
308 535-3719
• Clin Lab Technician/Med Lab Technician
• Dental Assisting Prgm

## Omaha

**Alegent Health**
Wayne A Sensor, FACHE, President/CEO
1010 N 96th St
Omaha, NE 68122
402 343-4410
• Medical Assistant Prgm
• Radiography Prgm
• Respiratory Therapist (Advanced) Prgm

**Clarkson College**
John W Upright, EdD, President
101 S 42nd St
Omaha, NE 68131-2739
402 552-3394
• Health Information Tech Prgm
• Physical Therapist Assistant Prgm
• Radiography Prgm

**INSTITUTIONS**

**College of Saint Mary**
Maryanne Stevens, RSM PhD, President
7000 Mercy Rd
Omaha, NE 68106
402 399-2435
• Health Information Admin Prgm
• Health Information Tech Prgm
• Occupational Therapy Prgm

**Creighton University**
Rev John P Schlegel, SJ PhD, President
2500 California Plaza
Omaha, NE 68178-0001
402 280-2974
• Athletic Training Prgm
• Emergency Med Tech-Paramedic Prgm
• Occupational Therapy Prgm
• Physical Therapy Prgm

**Hamilton College - Omaha**
Michael Ziaisky, BA, President
3350 N 90th St
Omaha, NE 68134
402 572-8500
• Medical Assistant Prgm

**Metropolitan Community College**
JoAnn McDowell, EdD, President
PO Box 3777
Omaha, NE 68103
402 457-2415
• Dental Assisting Prgm
• Respiratory Therapist (Advanced) Prgm

**NE Methodist Coll Nursing & Allied Hlth**
Dennis Joslin, PhD, President
720 N 87th St
Omaha, NE 68114
402 354-7257
• Diagnostic Med Sonography Prgm
• Respiratory Therapist (Advanced) Prgm

**Nebraska Methodist College**
Dennis Joslin, PhD, President
720 N 87th St
Omaha, NE 68114-3426
402 354-7257
• Medical Assistant Prgm
• Surgical Technology Prgm

**Nebraska Methodist Hospital**
John M Fraser, President/CEO
8303 Dodge St
Omaha, NE 68114
402 354-4531
• Clin Lab Scientist/Med Technologist Prgm

**University of Nebraska - Omaha**
Nancy Belck, PhD, Chancellor
EAB 201
Omaha, NE 68182-0108
402 554-3211
• Athletic Training Prgm
• Counseling Prgm
• Speech-Language Pathology Prgm

**University of Nebraska Medical Center**
Harold M Maurer, MD, Chancellor
986605 Nebraska Medical Center
Omaha, NE 68198-6605
402 559-4201
• Clin Lab Scientist/Med Technologist Prgm
• Cytotechnology Prgm
• Diagnostic Med Sonography Prgm
• Dietetic Internship Prgm
• Nuclear Medicine Technology Prgm
• Perfusion Prgm
• Physical Therapy Prgm
• Physician Assistant Prgm
• Radiation Therapy Prgm
• Radiography Prgm

**Vatterott College - Omaha Campus**
William J Stuckey, MS, President
225 N 80th St
Omaha, NE 68114-3617
402 392-1300
• Dental Assisting Prgm

## Scottsbluff

**Regional West Medical Center**
Todd Sorensen, MD, CEO
4021 Ave B
Scottsbluff, NE 69361
308 635-3711
• Radiography Prgm

**Western Nebraska Community College**
Eileen Ely, PhD, President
1601 E 27th St
Scottsbluff, NE 69361-1899
308 635-6101
• Health Information Tech Prgm

# Nevada

## Carson City

**Western Nevard Community College**
Carol A Lucey, PhD, President
2201 W College Pkwy
Carson City, NV 89703
775 445-4450
• Surgical Technology Prgm

## Henderson

**Touro University - Nevada**
Jay Sexter, CEO
874 American Pacific Dr
Henderson, NV 89014
702 777-8687
• Occupational Therapy Prgm
• Physician Assistant Prgm

## Las Vegas

**Community College of Southern Nevada**
Richard G Carpenter, PhD, President
6375 W Charleston
Las Vegas, NV 89146-1139
702 651-5600
• Clin Lab Technician/Med Lab Technician
• Dental Assisting Prgm
• Dental Hygiene Prgm
• Diagnostic Med Sonography Prgm
• Emergency Med Tech-Paramedic Prgm
• Health Information Tech Prgm
• Medical Assistant Prgm
• Occupational Therapy Asst Prgm
• Ophthalmic Dispensing Optician Prgm
• Physical Therapist Assistant Prgm
• Respiratory Therapist (Advanced) Prgm
• Surgical Technology Prgm

**Nevada Career Institute**
Joanne Leming, LPN, Campus Director
3025 E Desert Inn Rd, Ste A
Las Vegas, NV 89121
702 893-3300
• Surgical Technology Prgm

**Pima Medical Institute - Las Vegas**
Richard Luebke, Sr, BS, President
3333 E Flamingo Rd
Las Vegas, NV 89121
• Radiography Prgm

**Quest Diagnostics - Las Vegas**
John P Schwartz, MBA, President
4230 Burnham Ave
Las Vegas, NV 89119
702 733-7866
• Cytotechnology Prgm

**University of Nevada - Las Vegas**
David B Ashley, PhD, President
4505 Maryland Pkwy
PO Box 451001
Las Vegas, NV 89154
702 895-3201
• Athletic Training Prgm
• Clin Lab Scientist/Med Technologist Prgm
• Counseling Prgm
• Dietetics-Didactic Prgm
• Nuclear Medicine Technology Prgm
• Physical Therapy Prgm
• Radiography Prgm

## Reno

**Truckee Meadows Community College**
Philip Ringle, PhD, President
7000 Dandini Blvd (R-200)
Reno, NV 89512-3999
702 673-7025
• Dental Assisting Prgm
• Dental Hygiene Prgm
• Dietetic Technician-AD Prgm
• Radiography Prgm

**University of Nevada - Reno**
Joseph N Crowley, PhD, President
Reno, NV 89557-0046
702 784-4805
• Counseling Prgm
• Dietetic Internship Prgm
• Dietetics-Didactic Prgm
• Speech-Language Pathology Prgm

# New Hampshire

## Claremont

**New Hampshire Comm Tech Coll - Claremont**
Harvey Hill, BSN MEd EdD, Interim President
1 College Dr
Claremont, NH 03743-9707
603 542-7744
• Clin Lab Technician/Med Lab Technician
• Medical Assistant Prgm
• Occupational Therapy Asst Prgm
• Physical Therapist Assistant Prgm
• Respiratory Therapist (Advanced) Prgm

## Concord

**Concord Hospital**
Michael Green, BA MS, CEO
250 Pleasant St
Concord, NH 03301
603 225-2711
• Surgical Technology Prgm

**New Hampshire Technical Institute**
Lynn Kilchenstein, MA, President
31 College Dr
Concord, NH 03301-7412
603 271-7737
• Dental Assisting Prgm
• Dental Hygiene Prgm
• Diagnostic Med Sonography Prgm
• Emergency Med Tech-Paramedic Prgm
• Radiation Therapy Prgm
• Radiography Prgm

## Durham

**University of New Hampshire**
Bonnie Newman, PhD, Interim President
Thompson Hall
Durham, NH 03824-3547
603 862-2450
• Athletic Training Prgm
• Clin Lab Scientist/Med Technologist Prgm
• Dietetic Internship Prgm
• Dietetic Technician-AD Prgm
• Dietetics-Didactic Prgm
• Occupational Therapy Prgm
• Speech-Language Pathology Prgm
• Therapeutic Recreation Specialist Prgm

## Hudson

**New Hamp Inst for Therapeutic Arts**
Patrick Ian Cowan, PhD, Executive Director
153 Lowell Rd
Hudson, NH 03051
603 882-3022
• Massage Therapy Prgm

**New Hampshire Institute for Therapeutic Arts**
153 Lowell Rd
Hudson, NH 03051
• Massage Therapy Prgm

## Keene

**Antioch/New England Graduate School**
Peter Temes, PhD, President
40 Avon St
Keene, NH 03431-3516
603 357-3122
• Dance/Movement Therapy Prgm

**Keene State College**
Stanley J Yarosewick, PhD, President
229 Main
Keene, NH 03435
603 358-2000
• Athletic Training Prgm
• Dietetic Internship Prgm
• Dietetics-Didactic Prgm

## Lebanon

**Dartmouth Hitchcock Medical Center**
James W Varnum, MBA CHE, CEO
One Medical Center Dr
Lebanon, NH 03756-0001
603 650-5000
• Surgical Technology Prgm

**Lebanon College**
Donald Wenz, PhD, President, CEO
15 Hanover St
Lebanon, NH 03766
603 448-2445
• Radiography Prgm

## Manchester

**Hesser College**
Mary Jo Greco, MAT, President
3 Sundial Ave
Manchester, NH 03103
603 668-6660
• Medical Assistant Prgm
• Physical Therapist Assistant Prgm

**Mass College of Pharmacy & Health Sciences**
1260 Elm St
Manchester, NH 03101
• Physician Assistant Prgm

**New England EMS Institute**
Doug Dean, CEO
One Elliot Way
Manchester, NH 03103
603 628-2220
• Emergency Med Tech-Paramedic Prgm

**New Hampshire Comm Tech College - Manchester**
Darlene Miller, EdD, President
1066 Front St
Manchester, NH 03102
603 668-6706
• Medical Assistant Prgm

## New London

**Colby-Sawyer College**
Anne Ponder, PhD, President
541 Main St
New London, NH 03257
603 526-3737
• Athletic Training Prgm

## Plymouth

**Plymouth State University**
Donald P Wharton, PhD, President
MSC #1
Plymouth, NH 03264
603 535-2211
• Athletic Training Prgm

## Rindge

**Franklin Pierce College**
George J Hagerty
20 College Rd
Rindge, NH 03461-0060
800 437-0048
• Physical Therapy Prgm

## Stratham

**New Hampshire Comm Tech College - Stratham**
Catherine Smith, EdD, President
277 Portsmouth Ave
Stratham, NH 03885
603 772-1194
• Surgical Technology Prgm

# New Jersey

## Blackwood

**Camden County College**
Raymond A Yannuzzi, PhD, President
College Dr
Blackwood, NJ 08012
856 227-7200
• Clin Lab Technician/Med Lab Technician
• Dental Assisting Prgm
• Dental Hygiene Prgm
• Dietetic Technician-AD Prgm
• Health Information Tech Prgm
• Ophthalmic Dispensing Optician Prgm
• Ophthalmic Med Technologist Prgm

## Bridgeton

**Cumberland County Technical Education Center**
601 Bridgeton Ave
Bridgeton, NJ 08302
609 451-9000
• Dental Assisting Prgm

## Bridgewater

**Somerset County Technology Institute**
David D'Alonzo, EdM BS, Superintendent
North Bridge St & Vogt Dr
PO Box 6350
Bridgewater, NJ 08807-0350
908 526-8900
• Medical Assistant Prgm

## Camden

**Cooper University Hospital**
Christopher Olivia, MD, President
One Cooper Plaza
Camden, NJ 08103
856 342-2000
• Perfusion Prgm
• Radiation Therapy Prgm
• Radiography Prgm

**SUNJ Rutgers Camden/Univ of Med & Dent NJ**
Rutgers University, UM & D
UMDNJ - SHRP/401 Haddon Ave
Camden, NJ 08103-1506
• Physical Therapy Prgm

## Cape May Courthouse

**Cape May County Technical Institute**
William Desmond, Superintendant
188 Crest Haven Rd
Cape May Courthouse, NJ 08210
609 465-2161
• Dental Assisting Prgm

## Cranford

**Union County College**
Thomas H Brown, PhD, President
1033 Springfield Ave
Cranford, NJ 07016-1599
908 709-7100
• Physical Therapist Assistant Prgm

## Dover

**Dover Business College**
Timothy Luing, MBA, Executive Director
15 E Blackwell St
Dover, NJ 07801
201 843-8500
• Medical Assistant Prgm (2)

## East Rutherford

**Sodexho Marriott/Metro New York**
405 Murray Hill Pkwy, Ste 2010
East Rutherford, NJ 07073
201 507-5600
• Dietetic Internship Prgm

## Edison

**Middlesex County College**
Joann LaPerla-Morales, EdD, President
2600 Woodbridge Ave
PO Box 3050
Edison, NJ 08818-3050
908 906-2517
• Clin Lab Technician/Med Lab Technician
• Dental Hygiene Prgm
• Dietetic Technician-AD Prgm
• Radiography Prgm

## Egg Harbor

**National Massage Therapy Institute**
6712 Washington Ave, Ste 302
Egg Harbor, NJ 08234
• Massage Therapy Prgm (2)

## Englewood

**Englewood Hospital & Medical Center**
Douglas Duchak, President/CEO
350 Engle St
Englewood, NJ 07631
201 894-3002
• Radiography Prgm

## Ewing

**College of New Jersey**
Barbara Gittenstein, President
2000 Pennington Rd
Ewing, NJ 08628
609 771-1855
• Counseling Prgm
• Speech-Language Pathology Prgm

## Fairfield

**The Institute for Health Education**
7 Spielman Rd
Fairfield, NJ 07004
973 808-1110
• Dental Assisting Prgm

## Glassboro

**Rowan University**
Donald J Farish, PhD JD, President
Bole Hall
201 Mullica Hill Rd
Glassboro, NJ 08028
856 256-4100
• Athletic Training Prgm

## Hackensack

**Academy of Massage Therapy**
Joanna Sechuck-Tringali, NCBTMB BS, Executive
    Director
321 Main Street, 2nd Floor
Hackensack, NJ 07601
888 268-7898
• Massage Therapy Prgm

## Iselin

**Sanford-Brown Institute**
Dennis Mascali, MEd, President
675 US Hwy 1
Iselin, NJ 08830
732 623-5740
• Diagnostic Med Sonography Prgm
• Surgical Technology Prgm

## Jersey City

**Christ Hospital**
Peter A Kelly, MBA, President/CEO
176 Palisade Ave
Jersey City, NJ 07306
201 795-8200
• Radiography Prgm

**Hudson County Community College**
Glen Gabert, PhD, President
70 Sip Ave
Jersey City, NJ 07306
201 360-4001
• Health Information Tech Prgm
• Medical Assistant Prgm

## Lawrenceville

**Rider University**
Mordechai Rozanski, President
2083 Lawrenceville Rd
Lawrenceville, NJ 08648-3099
609 896-5000
• Counseling Prgm

## Lincroft

**Brookdale Community College**
Peter F Burnham, PhD, President
765 Newman Springs Rd
Lincroft, NJ 07738-1522
732 224-2696
• Radiography Prgm
• Respiratory Therapist (Advanced) Prgm

## Livingston

**St Barnabas Medical Center**
John F Bonamo, MD, Executive Director
Old Short Hill Rd
Livingston, NJ 07039
973 332-5628
• Radiation Therapy Prgm

## Long Branch

**Monmouth Medical Center**
Frank Vozos, MD, Executive Director
300 Second Ave
Long Branch, NJ 07740
732 923-7367
• Clin Lab Scientist/Med Technologist Prgm

## Mays Landing

**Atlantic County Institute of Technology**
Philip Guenther, EdD, Superintendent
5080 Atlantic Ave
Mays Landing, NJ 08330
609 625-2249
• Dental Assisting Prgm

## Morristown

**College of St Elizabeth**
Jacqueline Burns, President
2 Convent Rd
Morristown, NJ 07960
201 292-6300
• Dietetic Internship Prgm
• Dietetics-Didactic Prgm

**Morristown Memorial Hospital**
Joanne Conroy, MD, Exec VP, COO
100 Madison Ave
Morristown, NJ 07962-1956
973 971-5450
• Cardiovascular Technology Prgm
• Clin Lab Scientist/Med Technologist Prgm

## Neptune

**Jersey Shore University Medical Center**
Steve Littleson, FACHE, President
1945 Corlies Ave
Neptune, NJ 07753
732 775-5500
• Clin Lab Scientist/Med Technologist Prgm

## New Brunswick

**Rutgers University**
Francis L Lawrence, PhD, President
Old Queens
New Brunswick, NJ 08901
732 932-7454
• Dietetics-Didactic Prgm
• Medical Librarian Prgm

## Newark

**Essex County College**
A Zachery Yamba, President
303 University Ave
Newark, NJ 07102
201 877-3021
• Ophthalmic Dispensing Optician Prgm
• Physical Therapist Assistant Prgm
• Radiography Prgm

**Inst for Therapeutic Mass Univ of Med & Dent**
150 Bergen St
Newark, NJ 07103
• Massage Therapy Prgm

**Univ of Med & Dent of New Jersey**
Bruce C Vladeck, PhD, Interim President
65 Bergen St Rm 1535
Newark, NJ 07107
973 972-4444
• Cardiovascular Technology Prgm
• Clin Lab Scientist/Med Technologist Prgm
• Cytotechnology Prgm
• Dental Assisting Prgm
• Dental Hygiene Prgm
• Diagnostic Med Sonography Prgm
• Dietetic Internship Prgm
• Dietetics-Coordinated Prgm
• Nuclear Medicine Technology Prgm
• Phlebotomist Prgm
• Physical Therapy Prgm
• Physician Assistant Prgm
• Rehabilitation Counseling Prgm
• Respiratory Therapist (Advanced) Prgm (2)

## Newton

**Sussex County Community College**
Harry Damato, MA BS, Interim President
One College Hill
Newton, NJ 07860
973 300-2122
• Medical Assistant Prgm
• Surgical Technology Prgm

## Paramus

**Bergen Community College**
Judith K Winn, PhD, President/CEO
400 Paramus Rd
Paramus, NJ 07652-1595
201 447-7100
• Dental Hygiene Prgm
• Diagnostic Med Sonography Prgm
• Medical Assistant Prgm
• Physical Therapist Assistant Prgm
• Radiography Prgm
• Respiratory Therapist (Advanced) Prgm
• Surgical Technology Prgm

## Paterson

**Passaic County Community College**
Steven M Rose, EdD, President
One College Blvd
Paterson, NJ 07505-1179
201 684-6300
• Health Information Tech Prgm
• Radiography Prgm

## Pemberton

**Burlington County College**
Robert C Messina, Jr, PhD, President
601 Pemberton Browns Mills Rd
Pemberton, NJ 08068-1599
609 894-9311
• Dental Hygiene Prgm
• Health Information Tech Prgm
• Radiography Prgm

## Pennsauken

**Omega Institute**
F Burke, Owner
7050 Rte 38 East
Pennsauken, NJ 08109
856 663-4299
• Massage Therapy Prgm

## Piscataway

**Cortiva-Inst Somerset Sch of Massage Therapy**
Chris Froelich, President
180 Centennial Ave
Piscataway, NJ 08854
212 277-7729
• Massage Therapy Prgm (2)

## Plainfield

**Muhlenberg Regional Medical Center**
John P McGee, President
Park Ave and Randolph Rd
Plainfield, NJ 07061
732 632-1501
• Diagnostic Med Sonography Prgm
• Nuclear Medicine Technology Prgm
• Radiation Therapy Prgm
• Radiography Prgm

## Pomona

**Richard Stockton College of New Jersey**
Herman J Saatkamp, Jr, PhD, President
PO Box 195
Jim Leeds Rd
Pomona, NJ 08240-0195
609 652-4521
• Occupational Therapy Prgm
• Physical Therapy Prgm

## Pompton Lakes

**Institute for Therapeutic Massage**
Lisa Helbig, LMT, Director
125 Wanaque Ave
Pompton Lakes, NJ 07442
973 839-6131
• Massage Therapy Prgm (4)

## Randolph

**County College of Morris**
Edward J Yaw, EdD, President
214 Center Grove Rd
Randolph, NJ 07869
973 328-5370
• Radiography Prgm

## Ridgewood

**The Valley Hospital**
Audrey Meyers, MBA, President
223 N Van Dien Ave
Ridgewood, NJ 07450
201 447-8021
• Clin Lab Scientist/Med Technologist Prgm
• Radiography Prgm

## Sewell

**Gloucester County College**
William F Anderson, MS, President
1400 Tanyard Rd
Sewell, NJ 08080
856 415-2196
• Diagnostic Med Sonography Prgm
• Nuclear Medicine Technology Prgm

## Sicklerville

**Technical Institute of Camden County**
Gary Bennett, PhD, Superintendent
343 Berlin Cross Keys Rd
Berlin 343
Sicklerville, NJ 08081-9709
856 767-7000
• Dental Assisting Prgm
• Medical Assistant Prgm

## Somers Point

**Shore Memorial Hospital**
Al Gutierrez, CHE MBA RT(R), President
1 E New York Ave
Somers Point, NJ 08244-2387
609 653-3924
• Radiography Prgm

## Somerville

**Raritan Valley Community College**
G Jeremiah Ryan, President
PO Box 3300
Somerville, NJ 08876
908 526-1200
• Ophthalmic Dispensing Optician Prgm

## South Orange

**Seton Hall University**
Msgr Robert T Sheeran, STD, President
400 S Orange Ave
South Orange, NJ 07079-2687
973 761-9620
• Athletic Training Prgm
• Occupational Therapy Prgm
• Physical Therapy Prgm
• Physician Assistant Prgm
• Speech-Language Pathology Prgm

## Trenton

**Mercer County Community College**
Thomas Wilfrid, PhD, Acting President
PO Box B
Trenton, NJ 08690
609 586-4800
• Clin Lab Technician/Med Lab Technician
• Physical Therapist Assistant Prgm
• Radiography Prgm

**Mercer County Technical Schools**
Kimberly Schneider, EdD, Superintendent
1085 Old Trenton Rd
Trenton, NJ 08690
609 586-2123
• Medical Assistant Prgm

**New Jersey Dept of Health & Senior Services**
50 E State St
Trenton, NJ 08625-0364
• Dietetic Internship Prgm

**St Francis Medical Center**
Judith Persichilli, President/CEO
601 Hamilton Ave
Trenton, NJ 08629-1986
609 599-5000
• Radiography Prgm

## Union

**Kean University**
Dawood Farahi, PhD, President
1000 Morris Ave
Union, NJ 07083-0411
908 737-7000
• Athletic Training Prgm
• Counseling Prgm
• Health Information Admin Prgm
• Occupational Therapy Prgm
• Speech-Language Pathology Prgm

## Upper Montclair

**Montclair State University**
Susan Cole, PhD, President
Upper Montclair, NJ 07043
973 655-4000
• Athletic Training Prgm
• Dietetic Internship Prgm
• Dietetics-Didactic Prgm
• Music Therapy Prgm
• Speech-Language Pathology Prgm

## Vineland

**Cumberland County College**
Roland J Chapdelaine, EdD, President
College Dr PO Box 517
Vineland, NJ 08362-0517
609 691-8600
• Radiography Prgm

## Washington

**Northwest NJ Consortium Resp Care Educ**
Stephen Serman, Chairman Regl Oper Cncl
c/o Warren County Community College
475 Rte 57 W
Washington, NJ 07882
908 835-2314
• Respiratory Therapist (Advanced) Prgm

**Warren County Community College**
William Austin, EdD, President
475 Route 57 W
Washington, NJ 07882-4343
908 835-9222
• Medical Assistant Prgm

## Wayne

**Berdan Institute**
Duncan Anderson, CEO
201 Willowbrook Blvd, 2nd Fl
Wayne, NJ 07470
973 837-1818
• Dental Assisting Prgm
• Medical Assistant Prgm

**William Paterson Univ of New Jersey**
Arnold Speert, PhD, President
300 Pompton Rd
Wayne, NJ 07470
973 720-2222
• Athletic Training Prgm
• Counseling Prgm
• Speech-Language Pathology Prgm

## Westwood

**Healing Hands Institute for Massage Therapy**
Alice Feuerstein, RN LMT, President
41 Bergenline Ave, 2nd Fl
Westwood, NJ 07675
201 722-0099
• Massage Therapy Prgm

**Pascack Valley Hospital**
Sidney Mitchell, President
Old Hook Rd
Westwood, NJ 07675
201 358-3010
• Radiography Prgm

# New Mexico

## Alamogordo

**New Mexico State U at Alamogordo**
Rodger Bates, PhD, Campus Executive Officer
2400 N Scenic Dr
PO Box 477
Alamogordo, NM 88310
505 439-3696
• Clin Lab Technician/Med Lab Technician

## Albuquerque

**Central New Mexico Community College**
Michael J Glennon, MBA, President
525 Buena Vista SE
Albuquerque, NM 87106-4096
505 224-4411
• Clin Lab Technician/Med Lab Technician
• Dental Assisting Prgm
• Diagnostic Med Sonography Prgm
• Health Information Tech Prgm
• Respiratory Therapist (Advanced) Prgm
• Surgical Technology Prgm

**Crystal Mountain Sch of Therapeutic Massage**
Linda Delker, Director
4775 Indian School Rd NE, Ste 102
Albuquerque, NM 87110
505 872-2030
• Massage Therapy Prgm

**Pima Medical Institute - Albuquerque**
Richard L Luebke, Jr, BS, Executive Director
201 San Pedro NE, Bldg 3
Albuquerque, NM 87110
505 881-1234
• Radiography Prgm

**Southwestern Indian Polytechnic Institute**
Carolyn Elgin, President
9169 Coors NW, Box 10146
Albuquerque, NM 87184
505 897-5347
• Ophthalmic Dispensing Optician Prgm

**University of New Mexico**
David Harris, BBA, Acting President
President's Office
1 University of New Mexico MSC 05 3300
Albuquerque, NM 87131-0001
505 277-2626
• Athletic Training Prgm
• Clin Lab Scientist/Med Technologist Prgm
• Counseling Prgm
• Dental Hygiene Prgm
• Dietetic Internship Prgm
• Dietetics-Didactic Prgm
• Emergency Med Tech-Paramedic Prgm
• Occupational Therapy Prgm
• Physical Therapy Prgm
• Physician Assistant Prgm
• Speech-Language Pathology Prgm

**University of St Francis**
4401 Silver Ave SE, Ste B
Albuquerque, NM 87108
505 266-5565
• Physician Assistant Prgm

## Clovis

**Clovis Community College**
Jim Turner, Interim President
417 Schepps Blvd
Clovis, NM 88101
505 769-4000
• Radiography Prgm

## Espanola

**Northern New Mexico College**
Jose Griego, PhD, President
921 Paseo de Onate
Espanola, NM 87532
505 747-2100
• Radiography Prgm

## Farmington

**San Juan College**
Carol Spencer, PhD, President
4601 College Blvd
Farmington, NM 87402
505 566-3209
• Dental Hygiene Prgm
• Health Information Tech Prgm
• Physical Therapist Assistant Prgm

## Gallup

**University of New Mexico - Gallup**
Beth Miller, EdD, Executive Director
200 College Rd
Gallup, NM 87301
505 863-7501
• Clin Lab Technician/Med Lab Technician
• Dental Assisting Prgm
• Health Information Tech Prgm

## Las Cruces

**Dona Ana Community College**
Margie Huerta, PhD, Campus Executive Officer
Box 30001 Dept 3DA
Las Cruces, NM 88003-0001
505 527-7521
• Dental Assisting Prgm
• Diagnostic Med Sonography Prgm
• Emergency Med Tech-Paramedic Prgm
• Radiography Prgm
• Respiratory Therapist (Advanced) Prgm

**New Mexico State University**
William Conroy, Interim President
Box 30001, Dept 3z
Las Cruces, NM 88003-0001
505 646-2035
• Athletic Training Prgm
• Counseling Prgm
• Dietetics-Didactic Prgm
• Speech-Language Pathology Prgm

## Portales

**Eastern New Mexico University**
Everett L Frost, President
Portales, NM 88130
505 562-1011
• Emergency Med Tech-Paramedic Prgm
• Medical Assistant Prgm
• Occupational Therapy Asst Prgm
• Respiratory Therapist (Advanced) Prgm
• Speech-Language Pathology Prgm

## Santa Fe

**Santa Fe Community College**
John M Pacheco, President
6401 Richards Ave
Santa Fe, NM 87505-4488
505 428-1000
• Dental Assisting Prgm

**Southwestern College**
James Nolan, President
PO Box 4788
Santa Fe, NM 87502-4788
505 471-5756
• Art Therapy Prgm

## Silver City

**Western New Mexico University**
John Counts, PhD, President
PO Box 680
Silver City, NM 88062-0680
505 538-6238
• Occupational Therapy Asst Prgm

# New York

## Albany

**Albany College of Pharmacy**
James Gozzo, President
106 New Scotland Ave
Albany, NY 12208-3492
518 445-7200
• Cytotechnology Prgm

**Albany Medical College**
John L Buono, MD, Dean
47 New Scotland Ave, MC139
Albany, NY 12158
518 262-6008
• Physician Assistant Prgm

**Bryant & Stratton College - Albany**
Michael Gutierrez, Campus Director
1259 Central Ave
Albany, NY 12205
518 437-1802
• Medical Assistant Prgm

**College of St Rose**
Louis C Vaccaro, PhD, President
432 Western Ave
Albany, NY 12203
518 454-5111
• Speech-Language Pathology Prgm

**Maria College**
Sr Laureen Fitzgerald, RSM MA MS, President
700 New Scotland Ave
Albany, NY 12208-1798
518 438-3111
• Occupational Therapy Asst Prgm

**SUNY at Albany**
Karen R Hitchcock, President
1400 Washington Ave
Albany, NY 12222
518 442-3300
• Medical Librarian Prgm
• Rehabilitation Counseling Prgm

## Alfred

**Alfred State College - SUNY**
John Clark, PhD, President
10 Upper College Dr
Alfred, NY 14802
607 587-4211
• Health Information Tech Prgm

**Alfred University**
David Szczerbacki, PhD, Provost
1 Saxon Dr
Alfred, NY 14802
607 871-2137
• Athletic Training Prgm

## Amherst

**Daemen College**
Martin J Anisman, PhD, President
4380 Main St
Amherst, NY 14226-3592
716 829-8210
• Physical Therapy Prgm
• Physician Assistant Prgm

## Batavia

**Genesee Community College**
Stuart Steiner, JD EdD, President
One College Rd
Batavia, NY 14020-9704
585 343-0055
• Occupational Therapy Asst Prgm
• Physical Therapist Assistant Prgm
• Respiratory Therapist (Advanced) Prgm

## Binghamton

**Broome Community College**
Laurence D Spraggs, DA, President
Wales Bldg PO Box 1017
Binghamton, NY 13902
607 778-5000
• Clin Lab Technician/Med Lab Technician
• Dental Hygiene Prgm
• Health Information Tech Prgm
• Medical Assistant Prgm
• Physical Therapist Assistant Prgm
• Radiography Prgm

**Ridley-Lowell Business & Technical Institute**
Wilfred T Weymouth, MS, President
116 Front St
Binghamton, NY 13905
607 724-2941
• Medical Assistant Prgm (2)

## Brentwood

**Suffolk Community College**
Shirley Robinson Pippins, PhD, President
Western Campus
Crooked Hill Rd
Brentwood, NY 11717
516 851-6789
• Health Information Tech Prgm

## Brockport

**SUNY at Brockport**
Paul Yu, PhD, President
350 New Campus Dr
Brockport, NY 14420-2914
716 395-2211
• Athletic Training Prgm
• Counseling Prgm

## Bronx

**Bronx Community College**
Carolyn Williams, President
University Ave and W 181st St
Bronx, NY 10453
718 289-5151
• Nuclear Medicine Technology Prgm
• Radiography Prgm

**Bronx Lebanon Hospital**
Steven Anderman, MBA MS, CEO
1650 Grand Concourse
Bronx, NY 10457
718 518-5586
• Physician Assistant Prgm
• Surgical Technology Prgm

**CUNY Herbert H Lehman College**
Ricardo R Fernandez, President
Bedford Park Blvd W
Bronx, NY 10468
718 960-8000
• Dietetic Internship Prgm
• Dietetics-Didactic Prgm
• Speech-Language Pathology Prgm

**Eugenio Maria De Hostos Community College**
475 Grand Concourse
Bronx, NY 10451
• Dental Hygiene Prgm

**Hostos Community College of CUNY**
Dolores Fernandez, PhD, President
475 Grand Concourse
Bronx, NY 10451
718 518-4300
• Radiography Prgm

**Veterans Affairs Medical Center**
130 W Kingsbridge Rd
Bronx, NY 10468-3904
718 579-1640
• Dietetic Internship Prgm

## Bronxville

**Sarah Lawrence College**
Michele Myers, President
One Mead Way
Bronxville, NY 10708
914 337-0700
• Genetic Counseling Prgm

## Brooklyn

**ARAMARK Healthcare Support Services**
Kingsbrook Jewish Medical Ctr
585 Schenectady Ave
Brooklyn, NY 11203
718 604-5757
• Dietetic Internship Prgm

**ASA Institute of Business & Computer Tech**
Alex Shchegol, MS, President
81 Willoughby St
81 Willoughby St
Brooklyn, NY 11201
718 522-9073
• Medical Assistant Prgm

**Brooklyn Hosp/Long Island Univ**
Frederick Alley, MS FACHE, President
121 DeKalb Ave
Brooklyn, NY 11201
718 250-8005
• Physician Assistant Prgm

**CUNY Brooklyn College**
Vernon E Lattin, President
2900 Bedford Ave
Brooklyn, NY 11210
718 951-5000
• Dietetic Internship Prgm
• Dietetics-Didactic Prgm
• Speech-Language Pathology Prgm

**Kingsborough Comm Coll/The City Univ of NY**
Byron McClenney, President
2001 Oriental Blvd
Brooklyn, NY 11235-2398
718 368-5109
• Physical Therapist Assistant Prgm

**Long Island College Hospital**
Rita Battle, EVP/CEO
339 Hicks St
Brooklyn, NY 11201
718 780-1137
• Radiography Prgm

**Long Island University**
Gale Steven Hynes, JD, Provost
1 University Plaza
Rm M-315
Brooklyn, NY 11201
718 488-3438
• Art Therapy Prgm
• Athletic Training Prgm
• Clin Lab Scientist/Med Technologist Prgm
• Counseling Prgm
• Diagnostic Med Sonography Prgm
• Dietetic Internship Prgm
• Dietetics-Didactic Prgm
• Health Information Admin Prgm
• Medical Librarian Prgm
• Occupational Therapy Prgm
• Physical Therapy Prgm
• Radiography Prgm
• Respiratory Therapist (Advanced) Prgm
• Speech-Language Pathology Prgm (2)
• Surgical Technology Prgm

## New York City College of Technology

Russ Hotzler, PhD, President
300 Jay St
Brooklyn, NY 11201-2983
718 260-5400
• Dental Hygiene Prgm
• Dental Lab Technician Prgm
• Ophthalmic Dispensing Optician Prgm
• Radiography Prgm

**New York Methodist Hospital**
Mark J Mundy, MSHA, President
506 Sixth St
Brooklyn, NY 11215
718 780-3101
• Clin Lab Scientist/Med Technologist Prgm
• Emergency Med Tech-Paramedic Prgm
• Radiation Therapy Prgm
• Radiography Prgm

**Pratt Institute**
Thomas F Schutte, President
200 Willoughby Ave
Brooklyn, NY 11205
718 636-3600
• Art Therapy Prgm
• Dance/Movement Therapy Prgm
• Medical Librarian Prgm

**SUNY Downstate Medical Center**
John LaRosa, MD, President
450 Clarkson Ave Box 1
Brooklyn, NY 11203-2098
718 270-2611
• Diagnostic Med Sonography Prgm
• Occupational Therapy Prgm
• Physical Therapy Prgm
• Physician Assistant Prgm
• Surgical Technology Prgm

## Buffalo

**Bryant & Stratton College - Buffalo**
Jeffrey P Tredo, Campus Director
465 Main St, Ste 400
Buffalo, NY 14203
716 884-9120
• Medical Assistant Prgm

**Buffalo State, SUNY**
Muriel Howard, PhD, President
1300 Elmwood Ave
Buffalo, NY 14222
716 878-4000
• Dietetics-Coordinated Prgm
• Dietetics-Didactic Prgm
• Speech-Language Pathology Prgm

**Canisius College**
Vincent M Cooke, SJ PhD, President
2001 Main St
Buffalo, NY 14208-1098
716 888-2100
• Athletic Training Prgm

**D'Youville College**
Sr Denise A Roche, GNSH PhD, President
320 Porter Avenue
Buffalo, NY 14201-1084
716 829-7673
• Dietetics-Coordinated Prgm
• Occupational Therapy Prgm
• Physical Therapy Prgm
• Physician Assistant Prgm

**Erie Community College**
Louis M Ricci, PhD, President
121 Ellicott St
Buffalo, NY 14203
716 851-1202
• Occupational Therapy Asst Prgm
• Radiation Therapy Prgm

**Erie Community College - North Campus**
William Mariani, MS, President
121 Ellicott St
Buffalo, NY 14203
716 851-1200
• Clin Lab Technician/Med Lab Technician
• Dental Hygiene Prgm
• Dietetic Technician-AD Prgm
• Health Information Tech Prgm
• Medical Assistant Prgm
• Ophthalmic Dispensing Optician Prgm
• Respiratory Therapist (Advanced) Prgm

**SUNY Educational Opportunity Center**
Sherryl Weems, PhD
465 Washington St
Buffalo, NY 14203
716 849-6725
• Dental Assisting Prgm

**Trocaire College**
Paul B Hurley, Jr, PhD, President
360 Choate Ave
Buffalo, NY 14220-2094
716 826-1200
• Health Information Tech Prgm
• Medical Assistant Prgm
• Phlebotomist Prgm
• Radiography Prgm
• Surgical Technology Prgm

**University at Buffalo - SUNY**
John B Simpson, PhD, President
501 Capen Hall
Buffalo, NY 14260-0001
716 645-2901
• Athletic Training Prgm
• Audiologist Prgm
• Clin Lab Scientist/Med Technologist Prgm
• Dietetic Internship Prgm
• Medical Librarian Prgm
• Nuclear Medicine Technology Prgm
• Occupational Therapy Prgm
• Orthoptist Prgm
• Physical Therapy Prgm
• Rehabilitation Counseling Prgm
• Speech-Language Pathology Prgm

**Villa Maria College of Buffalo**
Sr Marcella Marie Garus, President
240 Pine Ridge Rd
Buffalo, NY 14225-3999
716 896-0700
• Physical Therapist Assistant Prgm

## Canton

**SUNY at Canton**
Joseph L Kennedy, PhD, President
FOB 616, 34 Cornell Dr
Canton, NY 13617-1096
315 386-7204
• Dental Hygiene Prgm
• Occupational Therapy Asst Prgm
• Physical Therapist Assistant Prgm

## Cobleskill

**SUNY Agric & Tech College at Cobleskill**
Thomas Haas, PhD, President
Knapp Hall 202, SUNY Cobleskill
Cobleskill, NY 12043
518 255-5111
• Histotechnician Prgm

## Cooperstown

**Bassett Healthcare Center for Rural EMS Educ**
William F Streck, MD, CEO
One Atwell Dr
Cooperstown, NY 13326
607 547-3100
• Emergency Med Tech-Paramedic Prgm

## Cortland

**SUNY College at Cortland**
Erik Bitterbaum, PhD, President
PO Box 2000
Cortland, NY 13045
607 753-2201
• Athletic Training Prgm
• Therapeutic Recreation Specialist Prgm

## Dansville

**Nicholas H Noyes Memorial Hospital**
James Wissler, CEO
111 Clara Barton St
Dansville, NY 14437
• Phlebotomist Prgm

## Dix Hills

**Western Suffolk BOCES**
Michael Mensch, PhD, COO
507 Deer Park Rd
Dix Hills, NY 11746
631 549-4900
• Diagnostic Med Sonography Prgm
• Surgical Technology Prgm

## Dobbs Ferry

**Mercy College**
Louise Feroe, PhD, President
555 Broadway
Dobbs Ferry, NY 10522
914 674-7307
• Occupational Therapy Asst Prgm
• Occupational Therapy Prgm
• Physical Therapy Prgm
• Physician Assistant Prgm
• Speech-Language Pathology Prgm

## Elmira

**Arnot Ogden Medical Center**
Anthony J Cooper, CHE, President/CEO
600 Roe Ave
Elmira, NY 14905-1676
607 737-4231
• Radiography Prgm

**Elmira Business Institute**
Brad C Phillips, BS, President
303 N Main St
Elmira, NY 14901
607 733-7177
• Medical Assistant Prgm

## Farmingdale

**Farmingdale State University of New York**
Jonathan C Gibralter, PhD, President
2350 Broadhollow Rd
Horton Hall
Farmingdale, NY 11735
631 420-2145
• Clin Lab Technician/Med Lab Technician
• Dental Hygiene Prgm

## Flushing

**CUNY Queens College**
James Muyskens, PhD, President
65-30 Kissena Blvd
Flushing, NY 11367
718 997-5000
• Dietetic Internship Prgm
• Dietetics-Didactic Prgm
• Medical Librarian Prgm

## Fredonia

**SUNY College at Fredonia**
Donald A MacPhee, President
Fredonia, NY 14063
716 673-3111
• Music Therapy Prgm
• Speech-Language Pathology Prgm

## Fresh Meadows

**St Vincent Catholic Med Ctr
Brklyn-Queens Reg**
Niels Schmidt, Director of Allied Health
175-05 Horace Harding Expwy
Fresh Meadows, NY 11365
718 357-0500
• Clin Lab Scientist/Med Technologist Prgm
• Physician Assistant Prgm

## Garden City

**Adelphi University**
Robert Scott, President
1 South Ave
Garden City, NY 11530
516 877-3000
• Audiologist Prgm
• Speech-Language Pathology Prgm

**Nassau Community College**
Sean A Fanelli, PhD, President
One Education Dr
Garden City, NY 11530
516 572-7205
• Physical Therapist Assistant Prgm
• Radiation Therapy Prgm
• Radiography Prgm
• Respiratory Therapist (Advanced) Prgm
• Surgical Technology Prgm

## Geneseo

**SUNY College of Geneseo**
Christopher C Dahl, Interim President
One College Circle
Geneseo, NY 14454
716 245-5211
• Speech-Language Pathology Prgm

## Glens Falls

**Glens Falls Hospital**
David Kruczlnicki, CEO
100 Park St
Glens Falls, NY 12801
518 792-3151
• Radiography Prgm

## Hempstead

**Hofstra University**
Stuart Rabinowitz, PhD, President
101 Hofstra University
West Library Wing
Hempstead, NY 11550
516 463-6800
• Art Therapy Prgm
• Athletic Training Prgm
• Audiologist Prgm
• Physician Assistant Prgm
• Rehabilitation Counseling Prgm
• Speech-Language Pathology Prgm

## Herkimer

**Herkimer County Community College**
Ronald F Williams, EdD, President
100 Reservoir Rd
Herkimer, NY 13350
315 866-0300
• Physical Therapist Assistant Prgm

## Hornell

**St James Mercy Health**
Pamela Urban, MBA, CEO, Interim President
411 Canisteo St
Hornell, NY 14843
607 324-8110
• Radiography Prgm

## Ithaca

**Cornell University**
Hunter R Rawlings, President
Ithaca, NY 14853
607 255-2000
• Dietetic Internship Prgm
• Dietetics-Didactic Prgm

**Ithaca College**
Peggy Ryan Williams, EdD, President
953 Danby Rd
Ithaca, NY 14850-7001
607 274-3111
• Athletic Training Prgm
• Occupational Therapy Prgm
• Physical Therapy Prgm
• Speech-Language Pathology Prgm
• Therapeutic Recreation Specialist Prgm

## Jamaica

**CUNY York College**
Marcia Keizs, EdD, President
94-20 Guy R Brewer Blvd
Jamaica, NY 11451-9902
718 262-2350
• Occupational Therapy Prgm
• Physician Assistant Prgm

## Jamestown

**Jamestown Community College**
Gregory T DeCinque, PhD, President
525 Falconer St
PO Box 20
Jamestown, NY 14702-0020
716 665-5220
• Occupational Therapy Asst Prgm

**Woman's Christian Association Hospital**
Betsy Wright, President/CEO
PO Box 840
Jamestown, NY 14702
716 664-8110
• Clin Lab Scientist/Med Technologist Prgm
• Radiography Prgm

## Keuka Park

**Keuka College**
Joseph Burke, PhD, President
Hegeman Hall
Keuka Park, NY 14478-0098
315 536-5201
• Occupational Therapy Prgm

## Long Island City

**LaGuardia Community College**
Gail Mellow, PhD, President
31-10 Thomson Ave
Long Island City, NY 11101-3083
718 482-5050
• Dietetic Technician-AD Prgm
• Occupational Therapy Asst Prgm
• Physical Therapist Assistant Prgm

## Manhasset

**North Shore University Hospital**
Dennis Dowling, CEO
145 Community Dr
Manhasset, NY 11030
516 465-8130
• Perfusion Prgm

## Middletown

**Orange County Community College**
William Richards, PhD, President
115 South St
Middletown, NY 10940-6404
845 341-4700
• Clin Lab Technician/Med Lab Technician
• Dental Hygiene Prgm
• Occupational Therapy Asst Prgm
• Phlebotomist Prgm
• Physical Therapist Assistant Prgm
• Radiography Prgm

## Mineola

**Winthrop University Hospital**
Daniel Walsh, President, CEO
259 First St
Mineola, NY 11501
516 663-2201
• Radiography Prgm

## Morrisville

**SUNY Agric & Tech College at Morrisville**
Frederick W Woodward, President
Morrisville, NY 13408
315 684-6000
• Dietetic Technician-AD Prgm

## New Paltz

**SUNY at New Paltz**
Steven Poskanzer, JD, President
75 South Manheim Blvd
New Paltz, NY 12561
845 257-2121
• Music Therapy Prgm
• Speech-Language Pathology Prgm

## New Rochelle

**College of New Rochelle**
Steve Sweeny, PhD, President
29 Castle Pl
New Rochelle, NY 10805
914 632-5300
• Art Therapy Prgm

## New York

**Bellevue Hospital Center**
Lynda Curtis, MS, Exec Director
First Ave and 27th St
New York, NY 10016
212 562-4132
• Radiography Prgm

**City University of New York, The City College**
Gregory Williams, PhD, President
Administration Bldg A-300
138th St and Convent Ave
New York, NY 10031
212 650-7285
• Physician Assistant Prgm

**Columbia University**
Lee C Bollinger, JD, President
Morningside Campus
411 Low Library
New York, NY 10027
212 854-9970
• Dental Assisting Prgm
• Occupational Therapy Prgm
• Physical Therapy Prgm

**Columbia University Teachers College**
Susan Fuhrman, PhD, President
525 W 120th Street
124 Zankel Bldg
New York, NY 10027
212 678-3000
• Dietetic Internship Prgm
• Speech-Language Pathology Prgm

**CUNY Borough of Manhattan Community Coll**
Antonio Perez, PhD, President
199 Chambers St
New York, NY 10007
212 346-8800
• Emergency Med Tech-Paramedic Prgm
• Health Information Tech Prgm
• Respiratory Therapist (Advanced) Prgm

**CUNY Hunter College**
William Kelly, President
365 Fifth Ave
New York, NY 10016
212 817-7100
• Audiologist Prgm
• Dietetic Internship Prgm
• Dietetics-Didactic Prgm
• Orientation and Mobility Specialist Prgm
• Physical Therapy Prgm
• Rehabilitation Counseling Prgm
• Rehabilitation Teacher Prgm
• Speech-Language Pathology Prgm

**Harlem Hospital Center**
John Palmer, PhD, Exec Director
506 Lenox Ave
New York, NY 10037
212 939-1340
• Radiography Prgm

**Institute of Allied Medical Professions**
Thomas Haggerty, MS, President
405 Park Ave, Ste 501
New York, NY 10022-4405
212 758-1410
• Nuclear Medicine Technology Prgm

**Interboro Institute**
Bruce R Kalisch, President
450 W 56th St
New York, NY 10019
212 399-0091
• Ophthalmic Dispensing Optician Prgm

**Memorial Sloan - Kettering Cancer Ctr**
Harold Varmus, MD, President
1275 York Ave
New York, NY 10021
212 639-6561
• Cytotechnology Prgm
• Radiation Therapy Prgm

**Mt Sinai School of Medicine**
K Davis, MD, President
One Gustave Levy Pl
New York, NY 10029
• Genetic Counseling Prgm

**New York - Presbyterian Hospital**
David B Skinner, MD, President/CEO
525 E 68th St
New York, NY 10021
212 746-4000
• Dietetic Internship Prgm

**New York Eye & Ear Infirmary**
Joseph P Corcoran, President
310 E 14th St
New York, NY 10003
212 979-4300
• Orthoptist Prgm

**New York University**
John Sexton, PhD, President
70 Washington Square S
Rm 216
New York, NY 10012-1172
212 998-2345
• Art Therapy Prgm
• Dental Hygiene Prgm
• Diagnostic Med Sonography Prgm
• Dietetic Internship Prgm
• Dietetics-Didactic Prgm
• Music Therapy Prgm
• Occupational Therapy Prgm
• Physical Therapist Assistant Prgm
• Physical Therapy Prgm
• Respiratory Therapist (Advanced) Prgm
• Speech-Language Pathology Prgm

**New York University Medical Center**
Theresa Bischoff, EVP Deputy Provost
560 First Ave
New York, NY 10016
212 263-7300
• Surgical Technology Prgm

**Pace University - Lenox Hill Hospital**
David A Caputo, PhD, President
1 Pace Plaza
New York, NY 10038
212 346-1098
• Physician Assistant Prgm

**St Vincent's Hospital Manhattan SVCMC**
Len Walsh, FACHE, Executive Director
153 W 11th St
New York, NY 10011
212 604-7515
• Clin Lab Scientist/Med Technologist Prgm
• Nuclear Medicine Technology Prgm

**Touro College**
Bernard Lander, PhD, President
Executive Offices
27-33 W 23rd St
New York, NY 10010-4202
212 463-0400
• Occupational Therapy Asst Prgm
• Occupational Therapy Prgm (2)
• Physical Therapist Assistant Prgm
• Physical Therapy Prgm
• Physician Assistant Prgm (2)
• Speech-Language Pathology Prgm

**Weill Med College/Cornell Univ Med Sch**
Antonio Gotto, Jr, MD, Dean
1300 York Ave, Rm F-105
New York, NY 10021
212 746-6005
• Physician Assistant Prgm

**Wood Tobe'-Coburn School**
Sandi Gruninger, President
8 E 40th St
New York, NY 10016
212 686-9040
• Medical Assistant Prgm

## Oceanside

**Robt J Hochstim Sch Rad/S Nassau Comm Hosp**
Joseph Quagliata, MA MSHA MSHyg, President and CEO
One Healthy Way
PO Box 9007
Oceanside, NY 11572
516 632-3939
• Radiography Prgm

## Old Westbury

**New York Institute of Technology**
Edward Guiliano, PhD, President
Northern Blvd PO Box 8000
Old Westbury, NY 11568-8000
516 686-7650
• Dietetic Internship Prgm
• Dietetics-Didactic Prgm
• Occupational Therapy Prgm
• Physical Therapy Prgm
• Physician Assistant Prgm

## Oneonta

**SUNY College at Oneonta**
Alan B Donovan, President
Oneonta, NY 13820
607 436-3500
• Dietetic Internship Prgm
• Dietetics-Didactic Prgm

## Orangeburg

**Dominican College**
Mary Eileen O'Brien, OP PhD, President
470 Western Hwy
Orangeburg, NY 10962-1299
845 848-7801
• Athletic Training Prgm
• Occupational Therapy Prgm
• Physical Therapy Prgm

## Orchard Park

**Erie Community College - South Campus**
William Mariani, President
4041 S Western Blvd
Orchard Park, NY 14127
716 851-1200
• Dental Lab Technician Prgm

## Plattsburgh

**Champlain Valley Phys Hospital Med Ctr**
Mundy Stephens, JD, President
75 Beekman St
Plattsburgh, NY 12901
518 561-2000
• Radiography Prgm

**Clinton Community College**
Maurice Hickey, PhD, President
Lake Shore Rd Rte 9 S
136 Clinton Point Dr
Plattsburgh, NY 12901
518 562-4100
• Clin Lab Technician/Med Lab Technician

**SUNY College at Plattsburgh**
Horace A Judson, President
Plattsburgh, NY 12901
518 564-2000
- Counseling Prgm
- Dietetics-Didactic Prgm
- Speech-Language Pathology Prgm

## Port Chester

**St Joseph's Medical Center**
Kevin Dahill, BS, VP Operations
Port Chester, NY 10701
914 934-3000
- Radiography Prgm

## Port Ewen

**Ulster County Board of Cooperative Ed Service**
Martin Ruglis, SDA, District Superintendent
PO Box 601
RT GW
Port Ewen, NY 12466
845 331-6680
- Surgical Technology Prgm

## Potsdam

**Clarkson University**
Dennis G Brown, President
Potsdam, NY 13699-5557
315 268-6400
- Physical Therapy Prgm

## Poughkeepsie

**Dutchess Community College**
David Conklin, PhD, President
53 Pendell Rd
Poughkeepsie, NY 12601
914 431-8980
- Clin Lab Technician/Med Lab Technician
- Dietetic Technician-AD Prgm
- Emergency Med Tech-Paramedic Prgm

**Marist College**
Dennis J Murray, PhD, President
3399 North Rd
Graystone
Poughkeepsie, NY 12601
845 575-3000
- Athletic Training Prgm
- Clin Lab Scientist/Med Technologist Prgm

## Queens

**St John's University**
Donald J Harrington, CM, President
8000 Utopia Pkwy
Queens, NY 11439
718 990-6161
- Audiologist Prgm
- Counseling Prgm
- Medical Librarian Prgm
- Rehabilitation Counseling Prgm
- Speech-Language Pathology Prgm

## Riverhead

**Peconic Bay Medical Center**
Andrew J Mitchell, MBA FACHE, President
1300 Roanoke Ave
Riverhead, NY 11901
631 548-6000
- Radiography Prgm

## Rochester

**Bryant & Stratton College - Rochester**
Beth Tarquino, MS Ed, Market Director
Henrietta Campus
1225 Jefferson Rd
Rochester, NY 14623-3136
716 292-5627
- Medical Assistant Prgm

**Monroe Community College**
Thomas Flynn, MS, President
1000 E Henrietta Rd
Rochester, NY 14623
585 292-2100
- Dental Assisting Prgm
- Dental Hygiene Prgm
- Emergency Med Tech-Paramedic Prgm
- Health Information Tech Prgm
- Radiography Prgm

**Nazareth College of Rochester**
Daan Braveman, PhD, President
4245 East Ave
Rochester, NY 14618-2790
585 389-2525
- Art Therapy Prgm
- Music Therapy Prgm
- Physical Therapy Prgm
- Speech-Language Pathology Prgm

**Rochester Business Institute**
Carl A Silvio, BA, President
1630 Portland Ave
Rochester, NY 14621
585 266-0430
- Medical Assistant Prgm

**Rochester General Hospital**
Mark Clements, MPHA, President and CEO
1425 Portland Ave
Rochester, NY 14621
585 922-4000
- Clin Lab Scientist/Med Technologist Prgm
- Phlebotomist Prgm

**Rochester Institute of Technology**
Albert J Simone, PhD, President
One Lomb Memorial Dr
Rochester, NY 14623
585 475-2394
- Diagnostic Med Sonography Prgm
- Dietetics-Coordinated Prgm
- Dietetics-Didactic Prgm
- Physician Assistant Prgm

**University of Rochester**
Joel Seligman, PhD, President
Rochester, NY 14627
- Counseling Prgm

## Rockville Centre

**Mercy Medical Center**
Martin Bieber, MBA CPA, President/CEO
PO Box 9024
Rockville Centre, NY 11571
516 705-2525
- Radiography Prgm

**Molloy College**
Drew Bogner, PhD, President
1000 Hempstead Ave
PO Box 5002
Rockville Centre, NY 11571-5002
516 678-5000
- Cardiovascular Technology Prgm
- Health Information Tech Prgm
- Nuclear Medicine Technology Prgm
- Respiratory Therapist (Advanced) Prgm

## Sanborn

**Niagara County Community College**
James Klyczek, PhD, President
3111 Saunders Settlement Rd
Sanborn, NY 14132
716 614-6222
- Medical Assistant Prgm
- Physical Therapist Assistant Prgm
- Radiography Prgm
- Surgical Technology Prgm

## Saranac Lake

**North Country Community College**
Gail Rogers Rice, EdD, President
20 Winona Ave
PO Box 89
Saranac Lake, NY 12983
518 891-2915
- Radiography Prgm

## Selden

**Suffolk County Community College**
Shirley Robinson Pippins, EdD, President
533 College Rd
Bldg NFL 37
Selden, NY 11784-2899
516 451-4112
- Dietetic Technician-AD Prgm
- Health Information Tech Prgm
- Occupational Therapy Asst Prgm
- Physical Therapist Assistant Prgm

## Staten Island

**CUNY College of Staten Island**
Marlene Springer, PhD, President
2800 Victory Blvd
Staten Island, NY 10314
718 982-2400
- Physical Therapy Prgm

**Wagner College/Staten Island University**
Norman R Smith, EdD, President
631 Howard Ave
Staten Island, NY 10301
718 390-3131
- Physician Assistant Prgm

## Stone Ridge

**SUNY Ulster**
Donald C Katt, EdD, President
Cottekill Rd
Stone Ridge, NY 12484
845 687-5279
• Emergency Med Tech-Paramedic Prgm

## Stony Brook

**Stony Brook University**
Shirley Strum Kenny, PhD, President
Nichols Rd
Admin Bldg Third Fl
Stony Brook, NY 11794-0701
631 632-6255
• Athletic Training Prgm
• Clin Lab Scientist/Med Technologist Prgm
• Cytotechnology Prgm
• Dietetic Internship Prgm
• Occupational Therapy Prgm
• Physical Therapy Prgm
• Physician Assistant Prgm
• Respiratory Therapist (Advanced) Prgm
• Surgical Technology Prgm

## Suffern

**Rockland Community College**
Cliff Wood, PhD, President
145 College Rd
Suffern, NY 10901-3699
845 574-4575
• Dietetic Technician-AD Prgm
• Occupational Therapy Asst Prgm

## Syracuse

**Bryant & Stratton College - Syracuse**
Michael Sattler, Campus Director
953 James St
Syracuse, NY 13203
315 472-6603
• Medical Assistant Prgm

**Le Moyne College**
Charles Beirne, SJ PhD, President
Le Moyne Heights
Syracuse, NY 13214-1399
315 445-4120
• Physician Assistant Prgm

**Onondaga Community College**
Deborah Sydow, PhD, President
4941 Onondaga Rd
Syracuse, NY 13215
315 498-2211
• Health Information Tech Prgm
• Physical Therapist Assistant Prgm
• Respiratory Therapist (Advanced) Prgm
• Surgical Technology Prgm

**SUNY Upstate Medical University**
David Smith, MD, President
750 E Adams St
Syracuse, NY 13210
315 464-4513
• Clin Lab Scientist/Med Technologist Prgm
• Cytotechnology Prgm
• Emergency Med Tech-Paramedic Prgm
• Perfusion Prgm
• Physical Therapy Prgm
• Radiation Therapy Prgm
• Radiography Prgm
• Respiratory Therapist (Advanced) Prgm

**Syracuse University**
Kenneth A Shaw, Chancellor & President
Syracuse, NY 13244
315 443-1870
• Audiologist Prgm
• Counseling Prgm
• Dietetic Internship Prgm
• Dietetics-Coordinated Prgm
• Dietetics-Didactic Prgm
• Medical Librarian Prgm
• Rehabilitation Counseling Prgm
• Speech-Language Pathology Prgm

## Tarrytown

**Marymount College**
Brigid Driscoll, President
Tarrytown, NY 10591
914 631-3200
• Dietetics-Didactic Prgm

## Troy

**Hudson Valley Community College**
Drew Matonak, PhD, President
80 Vandenburgh Ave
Troy, NY 12180-6096
518 629-4822
• Dental Hygiene Prgm
• Diagnostic Med Sonography Prgm
• Emergency Med Tech-Paramedic Prgm
• Respiratory Therapist (Advanced) Prgm

**The Sage Colleges**
Jeanne H Neff, PhD, President
45 Ferry St
Troy, NY 12180-4115
518 244-2214
• Dietetic Internship Prgm
• Dietetics-Didactic Prgm
• Occupational Therapy Prgm
• Physical Therapy Prgm

## Utica

**Faxton - St Luke's Healthcare**
Andrew E Peterson, MS, Exec Director
Champlin Ave PO Box 479
Utica, NY 13503-0479
315 798-6001
• Radiography Prgm

**Mohawk Valley Community College**
Michael I Schafer, EdD, President
Payne Hall 305
1101 Sherman Dr
Utica, NY 13501
315 792-5333
• Health Information Tech Prgm
• Respiratory Therapist (Advanced) Prgm

**St Elizabeth Medical Center**
Sr M Johanna Deleys, OSF, President, CEO
2209 Genesee St
Utica, NY 13501
315 798-8123
• Radiography Prgm

**SUNY Institute of Tech - Utica/Rome**
Peter Spina, PhD, President
PO Box 3050
Utica, NY 13504-3050
315 792-7400
• Health Information Admin Prgm

**Utica College**
Todd S Hutton, PhD, President
1600 Burrstone Rd
Utica, NY 13502-4892
315 792-3222
• Occupational Therapy Prgm
• Physical Therapy Prgm

## Valhalla

**New York Medical College**
Harry C Barrett, DMin MPH, President, CEO
Sunshine Administration Building
Valhalla, NY 10595
914 993-4000
• Physical Therapy Prgm
• Speech-Language Pathology Prgm

**Westchester Community College**
Joseph N Hankin, EdD, President
75 Grasslands Rd
Valhalla, NY 10595
914 606-6706
• Dietetic Technician-AD Prgm
• Radiography Prgm
• Respiratory Therapist (Advanced) Prgm

**Westchester Medical Center**
Grasslands Rd
Valhalla, NY 10595
914 285-7276
• Dietetic Internship Prgm

## Yonkers

**St John's Riverside Hospital**
James Foy, President
967 N Broadway
Yonkers, NY 10701
914 964-4444
• Cytotechnology Prgm

# New Zealand

## Palmerston North

**Massey University**
Steven La Grow, MA EdD COMS, Professor,
   Head of School
School of Health Sciences
Private Bag 11222
Palmerston North, NZ
646 350-5799
- Orientation and Mobility Specialist Prgm
- Rehabilitation Teacher Prgm

# North Carolina

## Albemarle

**Stanly Community College**
Michael R Taylor, EdD, President
141 College Dr
Albemarle, NC 28001-9402
704 982-0121
- Medical Assistant Prgm
- Respiratory Therapist (Advanced) Prgm

## Asheville

**Asheville-Buncombe Technical Comm College**
K Ray Bailey, MEd, President
340 Victoria Rd
Asheville, NC 28801
828 254-1921
- Clin Lab Technician/Med Lab Technician
- Dental Assisting Prgm
- Dental Hygiene Prgm
- Diagnostic Med Sonography Prgm
- Phlebotomist Prgm
- Radiography Prgm
- Surgical Technology Prgm

**South College - Asheville**
Robert Davis, MSL, Executive Director
1567 Patton Ave
Asheville, NC 28806
828 252-2486
- Medical Assistant Prgm
- Surgical Technology Prgm

## Banner Elk

**Lees - McRae College**
David Bushman, PhD, President
PO Box 128
191 Main St
Banner Elk, NC 28604
828 898-8785
- Athletic Training Prgm

## Boiling Springs

**Gardner - Webb University**
Frank R Campbell, PhD, President
Boiling Springs, NC 28017
- Athletic Training Prgm

## Boone

**Appalachian State University**
Kenneth Peacock, PhD, Chancellor
Boone, NC 28608
828 262-2040
- Athletic Training Prgm
- Counseling Prgm
- Dietetic Internship Prgm
- Dietetics-Didactic Prgm
- Music Therapy Prgm
- Speech-Language Pathology Prgm

## Buies Creek

**Campbell University**
Jerry M Wallace, PhD, President
PO Box 127
143 Main Street
Buies Creek, NC 27506
910 893-1205
- Athletic Training Prgm

## Chapel Hill

**Univ of North Carolina at Chapel Hill**
James Moeser, PhD, Chancellor
130 South Bldg CB 9100
Chapel Hill, NC 27599-9100
919 962-1365
- Athletic Training Prgm
- Audiologist Prgm
- Clin Lab Scientist/Med Technologist Prgm
- Counseling Prgm
- Cytotechnology Prgm
- Dental Assisting Prgm
- Dental Hygiene Prgm
- Dietetics-Coordinated Prgm
- Dietetics-Didactic Prgm
- Medical Librarian Prgm
- Occupational Therapy Prgm
- Physical Therapy Prgm
- Radiography Prgm
- Rehabilitation Counseling Prgm
- Speech-Language Pathology Prgm

**University of North Carolina Hospitals**
Eric B Munson, MBA, Director
101 Manning Dr
Chapel Hill, NC 27514
919 966-5111
- Nuclear Medicine Technology Prgm

## Charlotte

**Carolinas College of Health Sciences**
Ellen Sheppard, MA, President
1200 Blythe Rd
PO Box 32861
Charlotte, NC 28232-2861
704 355-5316
- Clin Lab Scientist/Med Technologist Prgm
- Phlebotomist Prgm
- Radiography Prgm
- Surgical Technology Prgm

**Central Piedmont Community College**
P Anthony Zeiss, EdD, President
PO Box 35009
Charlotte, NC 28235
704 330-6566
- Clin Lab Technician/Med Lab Technician
- Cytotechnology Prgm
- Dental Assisting Prgm
- Dental Hygiene Prgm
- Health Information Tech Prgm
- Medical Assistant Prgm
- Physical Therapist Assistant Prgm
- Respiratory Therapist (Advanced) Prgm

**King's College**
Barbara Rockecharlie, MBA, Director
322 Lamar Ave
Charlotte, NC 28204
704 688-3613
- Medical Assistant Prgm

**Presbyterian Healthcare**
Carl Armato, President/CEO
200 Hawthorne Ln
PO Box 33549
Charlotte, NC 28204
704 384-4942
- Radiography Prgm
- Surgical Technology Prgm

**Queens University of Charlotte**
Pamela Lewis Davies, EdD, President
1900 Selwyn Ave
Charlotte, NC 28274-0001
704 337-2200
- Music Therapy Prgm

**Univ of North Carolina at Charlotte**
Phil Dubois, PhD, Chancellor
9201 University Blvd
Charlotte, NC 28223-0001
704 687-2000
- Athletic Training Prgm
- Counseling Prgm

## Clyde

**Haywood Community College**
Nathan Hodges, EdD, President
Freedlander Dr
Clyde, NC 28721-9454
704 627-4515
- Medical Assistant Prgm

## Concord

**Cabarrus College of Health Sciences**
Anita A Brown, RN MEd, Chancellor
401 Medical Park Dr
Concord, NC 28025
704 783-1558
- Medical Assistant Prgm
- Occupational Therapy Asst Prgm
- Surgical Technology Prgm

## Cullowhee

**Western Carolina University**
John W Bardo, PhD, Chancellor
HF Robinson Admin Bldg
Cullowhee, NC 28723
704 227-7100
- Clin Lab Scientist/Med Technologist Prgm
- Counseling Prgm
- Dietetic Internship Prgm
- Dietetics-Didactic Prgm
- Emergency Med Tech-Paramedic Prgm
- Health Information Admin Prgm
- Physical Therapy Prgm
- Speech-Language Pathology Prgm

## Dallas

**Gaston College**
Patricia A Skinner, PhD, President
201 Hwy 321 S
Dallas, NC 28034-1499
704 922-6475
- Medical Assistant Prgm

## Dobson

**Surry Community College**
G Frank Sells, EdD, President
630 S Main St
Dobson, NC 27017-8432
336 386-3213
- Medical Assistant Prgm

## Durham

**Duke University Medical Center**
Victor Dzau, MD, Chancellor
Health Affairs
PO Box 3701 M106A Davison Bldg
Durham, NC 27710
919 684-2255
- Ophthalmic Med Technician Prgm
- Pathologists' Assistant Prgm
- Physical Therapy Prgm
- Physician Assistant Prgm

**Durham Technical Community College**
Phail Wynn, Jr, EdD MBA, President
1637 Lawson St
Durham, NC 27703-5023
919 686-3374
- Dental Lab Technician Prgm
- Occupational Therapy Asst Prgm
- Ophthalmic Dispensing Optician Prgm
- Pharmacy Technician Prgm
- Respiratory Therapist (Advanced) Prgm
- Surgical Technology Prgm

**North Carolina Central University**
James H Ammons, PhD, Chancellor
1801 Fayetteville St
113 Hoey Administration Bldg
Durham, NC 27707
919 530-6104
- Athletic Training Prgm
- Dietetic Internship Prgm
- Dietetics-Didactic Prgm
- Medical Librarian Prgm
- Orientation and Mobility Specialist Prgm
- Speech-Language Pathology Prgm

## Elizabeth City

**College of The Albemarle**
Lynne M Bunch, MA, President
PO Box 2327
Elizabeth City, NC 27906-2327
252 335-0821
- Medical Assistant Prgm
- Surgical Technology Prgm

## Elon

**Elon University**
Leo Lambert, PhD, President
Campus Box 2185
Elon, NC 27244
336 278-7900
- Athletic Training Prgm
- Physical Therapy Prgm

## Fayetteville

**Fayetteville Technical Community College**
Larry Norris, EdD, President
PO Box 35236
Fayetteville, NC 28303-0236
910 678-8321
- Dental Assisting Prgm
- Dental Hygiene Prgm
- Phlebotomist Prgm
- Physical Therapist Assistant Prgm
- Radiography Prgm
- Respiratory Therapist (Advanced) Prgm
- Surgical Technology Prgm

**Methodist College**
Elton Hendricks, PhD, President
5400 Ramsey St
Fayetteville, NC 28311
910 630-7005
- Athletic Training Prgm
- Physician Assistant Prgm

## Flat Rock

**Blue Ridge Community College**
David W Sink, Jr, EdD, President
College Dr
Flat Rock, NC 28737-9624
828 692-3572
- Surgical Technology Prgm

## Ft Bragg

**Joint Special Operations Medical Training Ctr**
Kevin Keenan, MD, Dean
CDR, USAJFKSWCS
AOJK-MED (attn: Keith Cox)
Ft Bragg, NC 28310-5200
910 396-7775
- Emergency Med Tech-Paramedic Prgm

## Goldsboro

**Wayne Community College**
Edward H Wilson, EdD, President
Caller Box 8002
Goldsboro, NC 27530
919 735-5151
- Dental Assisting Prgm
- Dental Hygiene Prgm
- Medical Assistant Prgm

## Graham

**Alamance Community College**
Martin H Nadelman, EdD, President
PO Box 8000
Graham, NC 27253-8000
336 578-2002
- Clin Lab Technician/Med Lab Technician
- Dental Assisting Prgm
- Medical Assistant Prgm

## Grantsboro

**Pamlico Community College**
Marion Altman, EdD, President
PO Box 185
Grantsboro, NC 28529-0185
252 249-1851
- Medical Assistant Prgm

## Greensboro

**Greensboro College**
Craven E Williams, President
815 W Market St
Greensboro, NC 27401
- Athletic Training Prgm

**Moses Cone Health System**
Timothy Rice, MBA, President/CEO
1200 N Elm St
Greensboro, NC 27401-1020
336 832-9500
- Radiography Prgm

**North Carolina A&T State University**
Edward B Fort, Chancellor
1601 E Market St
Greensboro, NC 27411
910 334-7500
- Counseling Prgm
- Dietetics-Didactic Prgm
- Rehabilitation Counseling Prgm

**Univ of North Carolina at Greensboro**
Patricia A Sullivan, PhD, Chancellor
1000 Spring Garden St
Greensboro, NC 27412
910 334-5000
- Athletic Training Prgm
- Counseling Prgm
- Dietetic Internship Prgm
- Dietetics-Didactic Prgm
- Genetic Counseling Prgm
- Medical Librarian Prgm
- Speech-Language Pathology Prgm
- Therapeutic Recreation Specialist Prgm

**INSTITUTIONS**

## Greenville

**East Carolina University**
Steven Ballard, PhD, Chancellor
Spillman Building
Greenville, NC 27858-4353
252 328-6212
- Athletic Training Prgm
- Audiologist Prgm
- Clin Lab Scientist/Med Technologist Prgm
- Dietetic Internship Prgm
- Dietetics-Didactic Prgm
- Health Information Admin Prgm
- Music Therapy Prgm
- Occupational Therapy Prgm
- Physical Therapy Prgm
- Physician Assistant Prgm
- Rehabilitation Counseling Prgm
- Speech-Language Pathology Prgm

**Pitt Community College**
Dennis Massey, PhD, President
PO Drawer 7007
Greenville, NC 27835-7007
252 493-7220
- Diagnostic Med Sonography Prgm
- Health Information Tech Prgm
- Medical Assistant Prgm
- Occupational Therapy Asst Prgm
- Radiation Therapy Prgm
- Radiography Prgm
- Respiratory Therapist (Advanced) Prgm

## Hamlet

**Richmond Community College**
Diane Honeycutt, EdD, President
PO Box 1189
Hamlet, NC 28345
910 582-7000
- Medical Assistant Prgm

## Henderson

**Vance - Granville Community College**
Randy Parker, MEd, President
PO Box 917
Popular Creek Rd
Henderson, NC 27536
252 492-2061
- Medical Assistant Prgm
- Radiography Prgm

## Hickory

**Catawba Valley Community College**
Garrett Hinshaw, EdD, President
2550 Hwy 70 SE
Hickory, NC 28602
828 327-7000
- Dental Hygiene Prgm
- Emergency Med Tech-Paramedic Prgm
- Health Information Tech Prgm
- Respiratory Therapist (Advanced) Prgm
- Surgical Technology Prgm

**Lenoir - Rhyne College**
Wayne Powell, PhD, President
Box 7163
Hickory, NC 28603
828 328-7334
- Athletic Training Prgm
- Occupational Therapy Prgm

## High Point

**High Point University**
Nido Qubein, LLD, President
833 Montlieu Ave
High Point, NC 27262
336 841-9201
- Athletic Training Prgm

## Hudson

**Caldwell Comm College & Tech Institute**
Kenneth A Boham, EdD, President
2855 Hickory Blvd
Hudson, NC 28638-1399
828 726-2200
- Diagnostic Med Sonography Prgm
- Nuclear Medicine Technology Prgm
- Ophthalmic Assistant Prgm
- Physical Therapist Assistant Prgm
- Radiography Prgm

## Jacksonville

**Coastal Carolina Community College**
Ronald K Lingle, PhD, President
444 Western Blvd
Jacksonville, NC 28546-6877
910 938-6210
- Clin Lab Technician/Med Lab Technician
- Dental Assisting Prgm
- Dental Hygiene Prgm
- Surgical Technology Prgm

## Jamestown

**Guilford Technical Community College**
Donald C Cameron, EdD, President
PO Box 309
601 High Point Rd
Jamestown, NC 27282
336 334-4822
- Dental Assisting Prgm
- Dental Hygiene Prgm
- Medical Assistant Prgm
- Physical Therapist Assistant Prgm
- Surgical Technology Prgm

## Kenansville

**James Sprunt Community College**
Mary T Wood, EdD, President
PO Box 398, James Sprunt Dr
Kenansville, NC 28349
910 296-2450
- Medical Assistant Prgm
- Phlebotomist Prgm

## Kinston

**Lenoir Community College**
Brantley Briley, EdD, President
PO Box 188
231 Hwy 58S
Kinston, NC 28502
252 527-6223
- Dietetic Technician-AD Prgm
- Medical Assistant Prgm
- Surgical Technology Prgm

## Lexington

**Davidson County Community College**
Mary Rittling, EdD, President
PO Box 1287
Lexington, NC 27293-1287
336 249-8186
- Clin Lab Technician/Med Lab Technician
- Health Information Tech Prgm
- Medical Assistant Prgm

## Lumberton

**Robeson Community College**
Charles Chrestman, President
PO Box 1420
Lumberton, NC 28359
910 738-7101
- Respiratory Therapist (Advanced) Prgm

## Marion

**McDowell Technical Community College**
Bryan Wilson, EdD, President
54 College Dr
Marion, NC 28752
828 652-0635
- Health Information Tech Prgm

## Mars Hill

**Mars Hill College**
Dan G Lunsford, EdD, President
PO Box 6748
Mars Hill, NC 28754
828 689-1141
- Athletic Training Prgm

## Morehead City

**Carteret Community College**
Joseph T Barwick, PhD, President
3505 Arendell St
Morehead City, NC 28557
252 222-6140
- Medical Assistant Prgm
- Radiography Prgm
- Respiratory Therapist (Advanced) Prgm

## Morganton

**Western Piedmont Community College**
Jim Burnett, EdD, President
1001 Burkemont Ave
Morganton, NC 28655
704 438-6010
- Clin Lab Technician/Med Lab Technician
- Dental Assisting Prgm
- Medical Assistant Prgm

## Murphy

**Tri-County Community College**
Norman Oglesby, PhD, President
2300 Hwy 64 E
Murphy, NC 28906
704 837-6810
- Medical Assistant Prgm

## North Wilkesboro

**Wilkes Regional Medical Center**
Ted Chapin, MPH, CEO
PO Box 609
North Wilkesboro, NC 28659
336 651-8100
• Radiography Prgm

## Pinehurst

**Sandhills Community College**
John R Dempsey, PhD, President
3395 Airport Rd
Pinehurst, NC 28374
910 695-3700
• Clin Lab Technician/Med Lab Technician
• Radiography Prgm
• Respiratory Therapist (Advanced) Prgm
• Surgical Technology Prgm

## Polkton

**South Piedmont Community College**
John McKay, EdD, President
PO Box 126
Polkton, NC 28135
704 272-7635
• Medical Assistant Prgm
• Surgical Technology Prgm

## Raleigh

**Meredith College**
John E Weems, President
Raleigh, NC 27607
919 829-8600
• Dietetic Internship Prgm
• Dietetics-Didactic Prgm

**North Carolina State University**
Raleigh, NC 27695-7801
919 515-2011
• Counseling Prgm

**Shaw University**
Clarence G Newsome, PhD, President
118 E South St
Raleigh, NC 27601
919 546-8300
• Kinesiotherapy Prgm

**Wake Technical Community College**
Stephen Scott, EdD, President
9101 Fayetteville Rd
Raleigh, NC 27603
919 662-3400
• Clin Lab Technician/Med Lab Technician
• Dental Assisting Prgm
• Dental Hygiene Prgm
• Medical Assistant Prgm
• Phlebotomist Prgm
• Radiography Prgm
• Surgical Technology Prgm

## Rocky Mount

**Nash Community College**
William Carver II, MBA, President
522 N Old Carriage Rd
PO Box 7488
Rocky Mount, NC 27804-0488
919 443-4011
• Phlebotomist Prgm
• Physical Therapist Assistant Prgm

## Salisbury

**Catawba College**
Robert Knott, PhD, President
2300 W Innes St
Salisbury, NC 28144
704 637-4414
• Athletic Training Prgm

**Rowan - Cabarrus Community College**
Richard L Brownell, EdD, President
PO Box 1595
1333 Jake Alexander Blvd
Salisbury, NC 28145-1595
704 637-0760
• Dental Assisting Prgm
• Radiography Prgm

## Sanford

**Central Carolina Community College**
Mathew Garrett, BS MA EdD, President
1105 Kelly Dr
Sanford, NC 27330
919 775-5401
• Medical Assistant Prgm (2)

## Shelby

**Cleveland Community College**
L Steve Thornburg, EdD, President
137 S Post Rd
Shelby, NC 28150
704 484-4089
• Radiography Prgm
• Surgical Technology Prgm

## Siler City

**Body Therapy Institute**
Rick Rosen, MA LMT, Codirector
300 Southwind Rd
Siler City, NC 27344
919 663-3111
• Massage Therapy Prgm

## Smithfield

**Johnston Community College**
Donald Reichard, President
PO Box 2350
Smithfield, NC 27577
919 209-2050
• Diagnostic Med Sonography Prgm
• Medical Assistant Prgm
• Radiography Prgm

## Spruce Pine

**Mayland Community College**
Thomas Williams, EdD, President
PO Box 547
Spruce Pine, NC 28777
828 765-7351
• Medical Assistant Prgm

## Statesville

**Mitchell Community College**
Douglas O Eason, PhD, President
West Broad St
Statesville, NC 28677
704 878-3200
• Medical Assistant Prgm

## Supply

**Brunswick Community College**
Steven Greiner, EdD, President
PO Box 30
50 College Road
Supply, NC 28462
910 754-6900
• Health Information Tech Prgm
• Phlebotomist Prgm

## Sylva

**Southwestern Community College**
Cecil Groves, PhD, President
447 College Dr
Sylva, NC 28779
800 586-4091
• Clin Lab Technician/Med Lab Technician
• Health Information Tech Prgm
• Phlebotomist Prgm
• Physical Therapist Assistant Prgm
• Radiography Prgm
• Respiratory Therapist (Advanced) Prgm

## Tarboro

**Edgecombe Community College**
Deborah L Lamm, EdD, President
2009 W Wilson St
Tarboro, NC 27886
252 823-5166
• Health Information Tech Prgm
• Medical Assistant Prgm
• Radiography Prgm
• Respiratory Therapist (Advanced) Prgm
• Surgical Technology Prgm

## Troy

**Montgomery Community College**
Mary Kirk, EdD, President
1011 Page St
Troy, NC 27371
910 576-6222
• Medical Assistant Prgm

## Washington

**Beaufort County Community College**
David McLawhorn, EdD, President
PO Box 1069
Highway 264 East
Washington, NC 27889
252 946-6194
• Clin Lab Technician/Med Lab Technician

## Weldon

**Halifax Community College**
Harold Mitchell, EdD, Interim President
PO Box 809
Weldon, NC 27890
252 536-4221
• Clin Lab Technician/Med Lab Technician
• Dental Hygiene Prgm
• Phlebotomist Prgm

## Wentworth

**Rockingham Community College**
Robert C Keys, PhD, President
PO Box 38
Wentworth, NC 27375-0038
910 342-4261
• Phlebotomist Prgm
• Respiratory Therapist (Advanced) Prgm
• Surgical Technology Prgm

## Whiteville

**Southeastern Community College**
Kathy Matlock, PhD, President
PO Box 151
Whiteville, NC 28472-0151
910 642-7141
• Clin Lab Technician/Med Lab Technician
• Phlebotomist Prgm

## Wilkesboro

**Wilkes Community College**
Gordon Burns, PhD, President
1328 Collegiate Dr
PO Box 120
Wilkesboro, NC 28697-0120
336 838-6100
• Dental Assisting Prgm
• Medical Assistant Prgm

## Williamston

**Martin Commuity College**
Ann R Britt, EdD, President
1161 Kehukee Park Rd
Williamston, NC 27892
919 792-1521
• Dental Assisting Prgm
• Medical Assistant Prgm
• Physical Therapist Assistant Prgm

## Wilmington

**Cape Fear Community College**
Eric B McKeithan, EdD, President
411 N Front St
Wilmington, NC 28401-3993
910 362-7555
• Dental Assisting Prgm
• Dental Hygiene Prgm
• Occupational Therapy Asst Prgm
• Phlebotomist Prgm
• Radiography Prgm

**Miller-Motte Technical College**
Ruth Hodge, BS, Executive Director
5000 Market St
Wilmington, NC 28405
910 392-4660
• Medical Assistant Prgm
• Surgical Technology Prgm

**Univ of North Carolina - Wilmington**
Rosemary DePaolo, PhD, Chancellor
601 S College Rd
Wilmington, NC 28403
910 962-3030
• Athletic Training Prgm
• Therapeutic Recreation Specialist Prgm

## Wilson

**Barton College**
Norval C Kneten, PhD, President
PO Box 5000
Wilson, NC 27893
252 399-6309
• Athletic Training Prgm

**Wilson Technical Community College**
Frank L Eagles, EdD, President
904 Herring Ave
Wilson, NC 27893
252 291-1195
• Surgical Technology Prgm

## Wingate

**Wingate University**
Jerry McGee, EdD, President
PO Box 3055
Wingate, NC 28714
704 233-8111
• Athletic Training Prgm

## Winston-Salem

**Forsyth Technical Community College**
Gary M Green, EdD, President
2100 Silas Creek Pky
Winston-Salem, NC 27103
336 734-7414
• Dental Assisting Prgm
• Dental Hygiene Prgm
• Diagnostic Med Sonography Prgm
• Medical Assistant Prgm
• Nuclear Medicine Technology Prgm
• Radiation Therapy Prgm
• Radiography Prgm
• Respiratory Therapist (Advanced) Prgm

**Wake Forest University**
Reynolda Station
Winston-Salem, NC 27109
910 759-5000
• Counseling Prgm

**Wake Forest University Baptist Medical Center**
Medical Center Blvd
Winston-Salem, NC 27157-1072
• Clin Lab Scientist/Med Technologist Prgm

**Wake Forest University School of Medicine**
Richard H Dean, MD, President
Medical Ctr Blvd
Winston-Salem, NC 27157-1006
336 716-4424
• Physician Assistant Prgm

**Winston-Salem State University**
Michelle Howard-Vital, PhD, Interim Chancellor
601 Martin Luther King Jr Dr
Winston-Salem, NC 27110
336 750-2141
• Clin Lab Scientist/Med Technologist Prgm
• Occupational Therapy Prgm
• Physical Therapy Prgm
• Rehabilitation Counseling Prgm
• Therapeutic Recreation Specialist Prgm

# North Dakota

## Bismarck

**Bismarck State College**
Donna Thigpen, EdD, President
PO Box 5587
Bismarck, ND 58506-5587
701 224-5400
• Clin Lab Technician/Med Lab Technician
• Phlebotomist Prgm
• Surgical Technology Prgm

**Medcenter One**
James Cooper, President/CEO
300 N Seventh St
Bismarck, ND 58506-5525
701 323-6104
• Radiography Prgm

**United Tribes Technical College**
David M Gipp, PhD, President
3315 University Dr
Bismarck, ND 58504
701 255-3285
• Health Information Tech Prgm

**University of Mary**
Sr Thomas Welder, OSB MM, President
7500 University Dr
Bismarck, ND 58504-9652
701 355-8100
• Athletic Training Prgm
• Occupational Therapy Prgm
• Physical Therapy Prgm

**University of Mary/St Alexius Medical Ctr**
Richard A Tschider, MA, Administrator
PO Box 5510
Bismarck, ND 58506-5510
701 224-7600
• Emergency Med Tech-Paramedic Prgm
• Respiratory Therapist (Advanced) Prgm

## Fargo

**MeritCare Medical Center**
Roger Gilbertson, MD, President
801 N Broadway
Fargo, ND 58122
701 234-6954
- Clin Lab Scientist/Med Technologist Prgm
- Radiography Prgm

**NDSU/Merit Care Hospital Consortium**
Gary Lee, BARRT, CEO
MeritCare Hospital
PO Box MC
Fargo, ND 58122-0118
701 234-5190
- Respiratory Therapist (Advanced) Prgm

**North Dakota State University**
Joseph Chapman, PhD, President
PO Box 5167
Fargo, ND 58105-5167
701 231-8011
- Athletic Training Prgm
- Counseling Prgm
- Dietetic Internship Prgm
- Dietetics-Coordinated Prgm
- Dietetics-Didactic Prgm
- Exercise Science Prgm

## Grand Forks

**University of North Dakota**
Charles Kupchella, PhD, President
Box 8193
Grand Forks, ND 58202-8193
701 777-2121
- Athletic Training Prgm
- Clin Lab Scientist/Med Technologist Prgm
- Cytotechnology Prgm
- Dietetics-Coordinated Prgm
- Music Therapy Prgm
- Occupational Therapy Prgm
- Physical Therapy Prgm
- Physician Assistant Prgm
- Speech-Language Pathology Prgm

## Minot

**Minot State University**
David Fuller, President
500 University Ave West
Minot, ND 58707
701 858-3031
- Speech-Language Pathology Prgm

**Trinity Health**
Terry G Hoff, FCHA, President
1 Burdick Expwy West
Minot, ND 58701
701 857-5000
- Radiography Prgm

## Wahpeton

**North Dakota State College of Science**
John Richman, PHD, Interim President
800 N Sixth St
Wahpeton, ND 58076-0002
701 671-2222
- Dental Assisting Prgm
- Dental Hygiene Prgm
- Health Information Tech Prgm
- Occupational Therapy Asst Prgm
- Pharmacy Technician Prgm

## Williston

**Williston State College**
Joseph McCann, PhD, President
1410 University Ave
PO Box 1326
Williston, ND 58802-1326
701 774-4200
- Physical Therapist Assistant Prgm

# Nova Scotia, Canada

## Halifax

**Dalhousie University**
Tom Traves, President
Halifax, NS B3H 3J5
902 424-2511
- Medical Librarian Prgm

**IWK Grace Health Centre**
5850 University Ave
PO Box 3070
Halifax, NS B3J 3G9
- Orthoptist Prgm

# Ohio

## Ada

**Ohio Northern University**
Kendall Baker, PhD, President
525 S Main St
Ada, OH 45810
419 772-2031
- Athletic Training Prgm
- Clin Lab Scientist/Med Technologist Prgm

## Akron

**Akron General Medical Center**
Alan Bleyer, MHA, President
400 Wabash Ave
Akron, OH 44307
330 846-6548
- Cytotechnology Prgm
- Emergency Med Tech-Paramedic Prgm

**Akron Institiue**
David LaRue, Campus President
1600 South Arlington Suite 100
Akron, OH 44306
330 724-1600
- Medical Assistant Prgm

**Brown Mackie College**
Robin Krout, BA, President
2791 Mogadore Rd
Akron, OH 44312
330 771-2424
- Medical Assistant Prgm (2)

**Children's Hospital Medical Ctr of Akron**
William H Considine, MHA, President
One Perkins Square
Akron, OH 44308-1062
330 543-8293
- Clin Lab Scientist/Med Technologist Prgm
- Radiography Prgm

**Summa Health System**
Thomas Strauss, PhD, President and CEO
525 E Market St
PO Box 2090
Akron, OH 44309-2090
330 375-3000
- Emergency Med Tech-Paramedic Prgm

**University of Akron**
Luis M Proenza, PhD, President
Buchtel Hall 114
Akron, OH 44325-4702
330 972-7074
- Athletic Training Prgm
- Audiologist Prgm
- Counseling Prgm
- Dietetics-Coordinated Prgm
- Dietetics-Didactic Prgm
- Medical Assistant Prgm
- Respiratory Therapist (Advanced) Prgm
- Speech-Language Pathology Prgm
- Surgical Technology Prgm

## Alliance

**Mount Union College**
Richard F Giese, PhD, President
1972 Clark Ave
Alliance, OH 44601
330 823-6050
- Athletic Training Prgm

## Ashland

**Ashland County - West Holmes Career Center**
Michael McDaniel, BS MS, Superintendent
1783 State Rte 60
Ashland, OH 44805
419 289-3313
- Medical Assistant Prgm

**Ashland University**
G William Benz, PhD, President
205 Founders Hall
Ashland, OH 44805
419 289-5050
- Athletic Training Prgm

**INSTITUTIONS**

## Athens

**Ohio University**
Roderick McDavis, PhD, President
Cutler Hall 108
Athens, OH 45701
740 593-1804
- Athletic Training Prgm
- Audiologist Prgm
- Counseling Prgm
- Dietetics-Didactic Prgm
- Music Therapy Prgm
- Physical Therapy Prgm
- Rehabilitation Counseling Prgm
- Speech-Language Pathology Prgm

## Batavia

**Univ of Cincinnati - Clermont College**
David H Devier, PhD, Dean
4200 Clermont College Dr
Batavia, OH 45103
513 732-5209
- Surgical Technology Prgm

## Berea

**Baldwin-Wallace College**
Mark Collier, PhD, President
275 Eastland Rd
Berea, OH 44017
440 826-2424
- Athletic Training Prgm
- Music Therapy Prgm

## Bluffton

**Bluffton College**
Elmer Neufeld, President
280 W College Ave
Bluffton, OH 45817
419 358-3000
- Dietetics-Didactic Prgm

## Bowling Green

**Bowling Green State University**
Sidney A Ribeau, PhD, President
220 McFall Ctr
Bowling Green, OH 43403
419 372-2211
- Athletic Training Prgm
- Clin Lab Scientist/Med Technologist Prgm
- Dietetic Internship Prgm
- Dietetics-Didactic Prgm
- Health Information Tech Prgm
- Rehabilitation Counseling Prgm
- Respiratory Therapist (Advanced) Prgm
- Speech-Language Pathology Prgm

## Canfield

**Mahoning County Career & Technical Center**
Roan Craig, PhD, Superintendent
7300 N Palmyra Rd
Canfield, OH 44406-9710
330 729-4100
- Medical Assistant Prgm

## Canton

**Aultman Hospital**
Edward Roth, MBA, President
2600 Sixth St SW
Canton, OH 44710
330 438-6241
- Nuclear Medicine Technology Prgm
- Radiation Therapy Prgm
- Radiography Prgm

**Canton City Schools**
Dianne Talarico, Superintendent
617 McKinley Ave SW
Canton, OH 44707
330 438-2530
- Medical Assistant Prgm

**Mercy Medical Center**
Thomas E Cecconi, President/CEO
1320 Mercy Dr NW
Canton, OH 44708
330 489-1001
- Diagnostic Med Sonography Prgm
- Radiography Prgm

## Cedarville

**Cedarville University**
William Brown, PhD, President
251 N Main St
Cedarville, OH 45314
937 766-7900
- Athletic Training Prgm

## Centerville

**RETS Tech Center**
Duncan Anderson, President
555 East Alex-Bell Rd
Centerville, OH 45459
937 433-3410
- Medical Assistant Prgm

## Chesapeake

**Collins Career Center**
Steve Dodgion, BA MSEd, Superintendent
11627 State Rte 243
Chesapeake, OH 45619
740 867-6641
- Pharmacy Technician Prgm
- Phlebotomist Prgm
- Radiography Prgm
- Respiratory Therapist (Advanced) Prgm
- Surgical Technology Prgm

## Chillicothe

**Pickaway Ross Joint Vocational School**
Brett Smith, Superintendent
895 Crouse Chapel Rd
Chillicothe, OH 45601-9010
740 642-2111
- Medical Assistant Prgm

## Cincinnati

**Christ Hospital**
Susan Croushore, Senior Executive Officer
2139 Auburn Ave
Cincinnati, OH 45219
513 585-1399
- Dietetic Internship Prgm
- Perfusion Prgm

**Cincinnati Children's Div Dvlpmntl Disorders**
Children's Hospital Medical Ctr
Elland and Bethesda Aves
Cincinnati, OH 45229-2899
513 559-4614
- Dietetic Internship Prgm

**Cincinnati Children's Hospital Medical Center**
Nancy L Zimpher, President
3333 Burnet Ave
Cincinnati, OH 45229
513 556-2201
- Genetic Counseling Prgm

**Cincinnati School of Medical Massage**
111250 Cornell Park Dr, Ste 203
Cincinnati, OH 45242
- Massage Therapy Prgm

**Cincinnati State Tech & Comm College**
Ronald D Wright, PhD, President
3520 Central Pkwy
Cincinnati, OH 45223
513 569-1511
- Clin Lab Technician/Med Lab Technician
- Diagnostic Med Sonography Prgm
- Dietetic Technician-AD Prgm
- Health Information Tech Prgm
- Medical Assistant Prgm
- Occupational Therapy Asst Prgm
- Respiratory Therapist (Advanced) Prgm
- Surgical Technology Prgm

**Good Samaritan Hospital**
375 Dixmyth Ave
Cincinnati, OH 45220-2489
513 872-1983
- Dietetic Internship Prgm

**University of Cincinnati**
Nancy L Zimpher, PhD, President
PO Box 210063
Cincinnati, OH 45221-0063
513 556-2201
- Athletic Training Prgm
- Audiologist Prgm
- Counseling Prgm
- Dental Hygiene Prgm
- Dietetics-Didactic Prgm
- Emergency Med Tech-Paramedic Prgm
- Health Information Admin Prgm
- Medical Assistant Prgm
- Nuclear Medicine Technology Prgm
- Physical Therapist Assistant Prgm
- Physical Therapy Prgm
- Radiation Therapy Prgm
- Radiography Prgm
- Speech-Language Pathology Prgm

**University of Cincinnati Medical Center**
Nancy Zimpher, PhD, President
Admin Bldg
Cincinnati, OH 45221-0063
513 558-2201
• Clin Lab Scientist/Med Technologist Prgm
• Specialist in BB Tech Prgm

**Xavier University**
Michael J Graham, SJ PhD, President
3800 Victory Pkwy
Cincinnati, OH 45207-4511
513 745-3501
• Athletic Training Prgm
• Occupational Therapy Prgm
• Radiography Prgm

## Clayton

**Miami Valley Career Technology Center**
John Boggess, PhD, Superintendent
6800 Hoke Rd
Clayton, OH 45315
937 854-6272
• Medical Assistant Prgm

## Cleveland

**Case Western Reserve University**
Edward Hundert, President
University Circle
Cleveland, OH 44106
216 368-2000
• Anesthesiologist Asst Prgm
• Dietetic Internship Prgm
• Dietetics-Didactic Prgm
• Genetic Counseling Prgm
• Speech-Language Pathology Prgm

**Cleveland Clinic Foundation**
Delos M Cosgrove, MD, Chairman
9500 Euclid Ave/H18
Cleveland, OH 44195-5108
216 444-2300
• Dietetic Internship Prgm
• Perfusion Prgm
• Radiation Therapy Prgm

**Cleveland State University**
Michael Schwartz, PhD, President
2121 Euclid Ave
Rhodes Tower 1201
Cleveland, OH 44115-2440
216 687-3544
• Counseling Prgm
• Occupational Therapy Prgm
• Physical Therapy Prgm
• Speech-Language Pathology Prgm

**Cuyahoga Community College**
Jerry Sue Thornton, PhD, President, District Admin
700 Carnegie Ave
Cleveland, OH 44115
216 987-4850
• Clin Lab Technician/Med Lab Technician
• Dental Hygiene Prgm
• Diagnostic Med Sonography Prgm
• Dietetic Technician-AD Prgm
• Health Information Tech Prgm
• Medical Assistant Prgm
• Nuclear Medicine Technology Prgm
• Occupational Therapy Asst Prgm
• Pharmacy Technician Prgm
• Phlebotomist Prgm
• Physical Therapist Assistant Prgm
• Physician Assistant Prgm
• Radiography Prgm
• Respiratory Therapist (Advanced) Prgm
• Surgical Technology Prgm

**John Carroll University**
20700 N Park Blvd
Cleveland, OH 44118-4581
216 397-1886
• Counseling Prgm

**Louis Stokes Cleveland VA Med Ctr**
10701 E Blvd
Cleveland, OH 44106-1702
216 421-3028
• Dietetic Internship Prgm

**MetroHealth Medical Center**
Terry R White, MBA, President/CEO
2500 Metro Health Dr
Cleveland, OH 44109-1998
216 459-5700
• Dietetic Internship Prgm

**University Hospitals of Cleveland**
Farah M Walters, MS, President/CEO
11100 Euclid Ave
Cleveland, OH 44106
216 844-7565
• Dietetic Internship Prgm

## Columbus

**A G James Cancer Hosp & Research Inst**
Dennis Smith, BS, Assoc Exec Director
300 W 10th Ave 519 CHRI
Columbus, OH 43210-1228
614 293-3121
• Radiation Therapy Prgm

**American School of Technology**
Susan Stella, BS BA, Director
2100 Morse Rd
Columbus, OH 43229
614 436-4820
• Medical Assistant Prgm

**ARC Blood Svcs - Central Ohio Region**
Ambrose Ng, MD, Chief Executive Officer
995 E Broad St
Columbus, OH 43205
614 253-2740
• Specialist in BB Tech Prgm

**Bradford School**
Dennis Bartels, President
2469 Stelzer Rd
Columbus, OH 43219
614 416-6200
• Medical Assistant Prgm

**Capital University**
Theodore Fredrickson, PhD, President
Yochum Hall
1 College & Main
Columbus, OH 43209
614 236-6908
• Athletic Training Prgm

**Columbus State Community College**
M Valeriana Moeller, PhD, President
550 E Spring St
PO Box 1609
Columbus, OH 43216-1609
614 287-2402
• Clin Lab Technician/Med Lab Technician
• Dental Hygiene Prgm
• Dietetic Technician-AD Prgm
• Emergency Med Tech-Paramedic Prgm
• Health Information Tech Prgm
• Histotechnician Prgm
• Medical Assistant Prgm
• Phlebotomist Prgm
• Radiography Prgm
• Respiratory Therapist (Advanced) Prgm
• Surgical Technology Prgm

**Mt Carmel College of Nursing**
Ann E Schiele, President
127 S Davis Ave
Columbus, OH 43222
614 234-5032
• Dietetic Internship Prgm
• Surgical Technology Prgm

**Ohio Institute of Health Careers**
Goldean Gibbs, BS, School Director
1880 E Dublin-Granville Rd
Ste 100
Columbus, OH 43229
614 891-5030
• Medical Assistant Prgm (2)

**Ohio State University**
Karen Holbrook, PhD, President
190 N Oval Mall
Columbus, OH 43210-1234
614 292-2424
• Athletic Training Prgm
• Audiologist Prgm
• Clin Lab Scientist/Med Technologist Prgm
• Dental Hygiene Prgm
• Dietetic Internship Prgm (2)
• Dietetics-Coordinated Prgm
• Dietetics-Didactic Prgm
• Health Information Admin Prgm
• Occupational Therapy Prgm
• Pathologists' Assistant Prgm
• Perfusion Prgm
• Physical Therapy Prgm
• Rehabilitation Counseling Prgm
• Respiratory Therapist (Advanced) Prgm
• Speech-Language Pathology Prgm

**Ohio State University Medical Center**
Fred Sanfilippo, MD PhD, CEO
200 Meiling Hall
370 West 9th Ave
Columbus, OH 43210
614 292-1200
• Nuclear Medicine Technology Prgm

**Sleep Care Inc**
7634 Rivers Edge Dr
Columbus, OH 43235
• Polysomnographic Technology Prgm

## Dayton

**Dayton School of Medical Message**
4457 Far Hills Ave
Dayton, OH 45429
• Massage Therapy Prgm (2)

**Miami - Jacobs Career College**
Darlene Waite, MBA, President
110 N. Patterson Ave
PO Box 1433
Dayton, OH 45401
937 461-5174
• Medical Assistant Prgm

**Miami Valley Hospital**
Karl R Tague, President/CEO
One Wyoming St
Dayton, OH 45409
513 223-6192
• Dietetic Internship Prgm

**Ohio Institute of Photography & Technology**
Robert A Martin, BS, Executive Director
2029 Edgefield Rd
Dayton, OH 45439
937 294-6155
• Medical Assistant Prgm

**Sinclair Community College**
Steven Lee Johnson, PhD, President
444 W Third St
Dayton, OH 45402-1460
937 512-2525
• Dental Hygiene Prgm
• Dietetic Technician-AD Prgm
• Health Information Tech Prgm
• Medical Assistant Prgm
• Occupational Therapy Asst Prgm
• Physical Therapist Assistant Prgm
• Radiography Prgm
• Respiratory Therapist (Advanced) Prgm
• Surgical Technology Prgm

**University of Dayton**
Raymond Fitz, PhD SM, President
300 College Pk
Dayton, OH 45469-1624
937 229-1000
• Dietetics-Didactic Prgm
• Music Therapy Prgm
• Physical Therapy Prgm

**Wright State University**
Kim Goldenberg, MD, President
260 University Hall
Dayton, OH 45435
937 775-2312
• Athletic Training Prgm
• Clin Lab Scientist/Med Technologist Prgm
• Counseling Prgm
• Rehabilitation Counseling Prgm

## Defiance

**Defiance College**
Gerald E Wood, EdD, President
701 N Clinton St
Defiance, OH 43512
419 783-2300
• Athletic Training Prgm

## East Liverpool

**Ohio Valley College of Technology**
Debra A Sanford, BS, President
16808 St Clair Ave
PO Box 7000
East Liverpool, OH 43920
330 385-1070
• Medical Assistant Prgm

## Elyria

**Lorain County Community College**
Roy A Church, EdD, President
1005 N Abbe Rd
Elyria, OH 44035
800 995-5222
• Clin Lab Technician/Med Lab Technician
• Dental Hygiene Prgm
• Diagnostic Med Sonography Prgm
• Medical Assistant Prgm
• Phlebotomist Prgm
• Physical Therapist Assistant Prgm
• Radiography Prgm
• Surgical Technology Prgm

## Findlay

**University of Findlay**
Debow Freed, PhD, President
1000 North Main St
Findlay, OH 45840
419 434-4510
• Athletic Training Prgm
• Nuclear Medicine Technology Prgm
• Occupational Therapy Prgm
• Physical Therapy Prgm
• Physician Assistant Prgm

## Granville

**Denison University**
Dale T Knobel, PhD, President
Doane Hall
Granville, OH 43023
740 587-6281
• Athletic Training Prgm

## Green

**Portage Lakes Career Center**
Mark Lukens, MA, Superintendent
4401 Shriver Rd
PO Box 248
Green, OH 44232-0248
330 896-8200
• Medical Assistant Prgm

## Groveport

**Fairfield Career Center (EVSD)**
Mark Weedy, MEd, Superintendent
4465 S Hamilton Rd
Groveport, OH 43125
614 836-5725
• Medical Assistant Prgm

## Hillsboro

**Southern State Community College**
Lawrence N Dukes, EdD, President
100 Hobart Dr
Hillsboro, OH 45133
937 393-3431
• Medical Assistant Prgm

## Kent

**Kent State University**
Lester Lefton, PhD, President
Office of the President
Kent, OH 44242-0001
330 672-2210
• Athletic Training Prgm
• Counseling Prgm
• Dietetic Internship Prgm
• Dietetics-Didactic Prgm
• Medical Librarian Prgm
• Nuclear Medicine Technology Prgm
• Occupational Therapy Asst Prgm
• Physical Therapist Assistant Prgm (2)
• Radiography Prgm
• Rehabilitation Counseling Prgm
• Speech-Language Pathology Prgm

## Kettering

**Kettering College of Medical Arts**
Charles Scriven, PhD, President
3737 Southern Blvd
Kettering, OH 45429
937 298-3399
• Diagnostic Med Sonography Prgm
• Physician Assistant Prgm
• Radiography Prgm
• Respiratory Therapist (Advanced) Prgm

## Kirtland

**Lakeland Community College**
Morris Beverage, Jr, EDM, President
7700 Clocktower Dr
Kirtland, OH 44094
440 525-7177
• Clin Lab Technician/Med Lab Technician
• Dental Hygiene Prgm
• Medical Assistant Prgm
• Radiography Prgm
• Respiratory Therapist (Advanced) Prgm
• Surgical Technology Prgm

## Lancaster

**Ohio University - Lancaster**
MaryAnn Janosik, PhD, Dean
1570 Granville Pike
Lancaster, OH 43130
740 654-6711
• Medical Assistant Prgm

## Lima

**Apollo Career Center**
J Chris Pfister, Superintendent
3325 Shawnee Rd
Lima, OH 45806
479 998-2911
• Medical Assistant Prgm
• Surgical Technology Prgm

**Rhodes State College**
Debra McCurdy, PhD, President
4240 Campus Dr
Lima, OH 45804
419 995-8200
• Dental Hygiene Prgm
• Dietetic Technician-AD Prgm
• Medical Assistant Prgm
• Occupational Therapy Asst Prgm
• Physical Therapist Assistant Prgm
• Radiography Prgm
• Respiratory Therapist (Advanced) Prgm

**University of Northwestern Ohio**
Loren R Jarvis, LLD, President
1441 N Cable Rd
Lima, OH 45805
419 227-3141
• Medical Assistant Prgm

## Lucasville

**Scioto County Joint Vocational School**
Stan Jennings, MS, Superintendent
951 Vern Riffe Dr
Lucasville, OH 45648
740 259-5526
• Surgical Technology Prgm

## Mansfield

**North Central State College**
Ronald E Abrams, PhD, President
2441 Kenwood Circle
PO Box 698
Mansfield, OH 44901
419 755-4800
• Physical Therapist Assistant Prgm
• Radiography Prgm
• Respiratory Therapist (Advanced) Prgm

## Marietta

**Marietta College**
Jean Scott, PhD, President
215 Fifth St
Marietta, OH 45750
740 376-4699
• Athletic Training Prgm
• Physician Assistant Prgm

**Marietta Memorial Hospital**
Larry J Unroe, MHA, President
401 Matthew St
Marietta, OH 45750
740 374-1411
• Radiography Prgm

**Washington County Career Center**
DeWayne O Poling, MA, Director
1750 Lancaster St
Marietta, OH 45750
740 373-6283
• Medical Assistant Prgm

**Washington State Community College**
Charlotte Hatfield, PhD, President
710 Colegate Dr
Marietta, OH 45750
740 374-8716
• Clin Lab Technician/Med Lab Technician
• Physical Therapist Assistant Prgm
• Respiratory Therapist (Advanced) Prgm

## Marion

**Marion Technical College**
J Richard Bryson, PhD, President
1467 Mt Vernon Ave
Marion, OH 43302-5694
740 389-4636
• Clin Lab Technician/Med Lab Technician
• Medical Assistant Prgm
• Physical Therapist Assistant Prgm
• Radiography Prgm

## Mayfield Heights

**Cleveland Clinic Health System**
Thomas Selden, President and CEO
6803 Mayfield Rd
Ste 500, Bldg 1
Mayfield Heights, OH 44124
440 312-8722
• Radiography Prgm

## Medina

**Medina County Career Center**
Thomas Horwedel, MEd, Superintendent
1101 W Liberty St
Medina, OH 44256-9969
216 725-8461
• Medical Assistant Prgm

## Middleburg Heights

**Cleveland Insitute of Medical Massage**
18334-D E Bagley Rd
Middleburg Heights, OH 44130
• Massage Therapy Prgm

**Polaris Career Center**
Linda G Schwarzbach, PhD, Superintendent
7285 Old Oak Blvd
Middleburg Heights, OH 44130
440 891-7619
• Dental Assisting Prgm
• Medical Assistant Prgm

**Sanford-Brown Institute**
Ken Gibson, Executive Director
17535 Rosbough Dr, Ste 100
Middleburg Heights, OH 44130
440 239-9640
• Diagnostic Med Sonography Prgm

**Southwest General Health Center**
L Ken Taylor, MA, President, CEO
18697 Bagley Rd
Middleburg Heights, OH 44130
216 816-8051
• Clin Lab Scientist/Med Technologist Prgm

## Milan

**EHOVE Ghrist Adult Career Center**
Joseph DeRose, Superintendent
316 W Mason Rd
Milan, OH 44846
419 499-4663
• Medical Assistant Prgm
• Surgical Technology Prgm

## Mt St Joseph

**College of Mt St Joseph**
Sr Francis M Thrailkill, PhD OSH, President
5701 Delhi Pike
Mt St Joseph, OH 45051
513 244-4232
• Athletic Training Prgm
• Physical Therapy Prgm

## Mt Vernon

**Knox County Career Center**
Ray Richardson, Superintendent
306 Martinsburg Rd
Mt Vernon, OH 43050
740 397-5820
• Medical Assistant Prgm

## Nelsonville

**Hocking College**
John J Light, PhD, President
3301 Hocking Pkwy
Nelsonville, OH 45764
740 753-3591
• Dietetic Technician-AD Prgm
• Health Information Tech Prgm
• Medical Assistant Prgm
• Physical Therapist Assistant Prgm

## Newark

**Central Ohio Technical College**
Bonnie Coe, PhD, President
1179 University Dr
Newark, OH 43055-1767
740 364-9509
• Diagnostic Med Sonography Prgm
• Radiography Prgm
• Surgical Technology Prgm

## North Canton

**Stark State College of Technology**
John O'Donnell, EdD, President
6200 Frank Ave NW
North Canton, OH 44720-7299
330 494-6170
• Clin Lab Technician/Med Lab Technician
• Dental Hygiene Prgm
• Health Information Tech Prgm
• Medical Assistant Prgm
• Occupational Therapy Asst Prgm
• Physical Therapist Assistant Prgm
• Respiratory Therapist (Advanced) Prgm

**Walsh University**
Richard Jusseaume, President
2020 East Maple St
North Canton, OH 44720-3396
330 490-7102
• Physical Therapy Prgm

## Oberlin

**Lorain Cnty Joint Voc School - Adult Career**
William B Randall, MS, Superintendent
15181 State Route 58
Oberlin, OH 44074
440 774-1056
• Medical Assistant Prgm

## Oxford

**Miami University**
David Hodge, PhD, President
201 Roudebush
Oxford, OH 45056
513 529-2345
• Athletic Training Prgm
• Dietetics-Didactic Prgm
• Speech-Language Pathology Prgm

## Parma

**Bryant & Stratton College - Parma**
Lisa Mason, Campus Director
12955 Snow Rd
Parma, OH 44130
216 265-3151
• Medical Assistant Prgm

**Parma Community General Hospital**
Thomas A Selden, Jr, President, CEO
7007 Powers Blvd
Parma, OH 44129-5495
440 888-1800
• Emergency Med Tech-Paramedic Prgm

## Pepper Pike

**Ursuline College**
Anne Marie Diederich, President
2550 Lander Rd
Pepper Pike, OH 44124
216 449-4200
• Art Therapy Prgm

## Portsmouth

**Shawnee State University**
Rita Rice Morris, PhD, President
940 Second St
Portsmouth, OH 45662-4303
740 351-3208
• Athletic Training Prgm
• Clin Lab Technician/Med Lab Technician
• Dental Hygiene Prgm
• Occupational Therapy Asst Prgm
• Occupational Therapy Prgm
• Physical Therapist Assistant Prgm
• Radiography Prgm
• Respiratory Therapist (Advanced) Prgm

## Rio Grande

**Buckeye Hills Career Center**
D Kent Lewis, BS MS, Superintendent
PO Box 157
Rio Grande, OH 45674
740 245-5334
• Surgical Technology Prgm

**University of Rio Grande**
Barry M Dorsey, PhD, President
218 N College Ave
PO Box 500
Rio Grande, OH 45674-0500
740 245-7204
• Clin Lab Technician/Med Lab Technician

## Sandusky

**Firelands Regional Medical Center**
1101 Decatur St
Sandusky, OH 44870
419 626-7400
• Radiography Prgm

## South Euclid

**Notre Dame College**
Robert E Karsten, Interim President
4545 College Rd
South Euclid, OH 44121
216 381-1680
• Dietetics-Didactic Prgm

## Springfield

**Clark State Community College**
Karen Rafinski, PhD, President
570 E Leffel Ln
Springfield, OH 45505
937 328-6001
• Clin Lab Technician/Med Lab Technician
• Physical Therapist Assistant Prgm

## St Clairsville

**Belmont Technical College**
Joseph Bukowski, EdD, President
120 Fox-Shannon Pl
St Clairsville, OH 43950
740 695-9500
• Medical Assistant Prgm

## Steubenville

**Jefferson Community College**
Laura M Meeks, PhD, President
4000 Sunset Blvd
Steubenville, OH 43952
740 264-5591
• Clin Lab Technician/Med Lab Technician
• Dental Assisting Prgm
• Medical Assistant Prgm
• Radiography Prgm
• Respiratory Therapist (Advanced) Prgm

## Tiffin

**Heidelberg College**
Dominic Dottavio, PhD, President
310 E Market St
Tiffin, OH 44883
419 448-2202
• Athletic Training Prgm

## Toledo

**Davis College**
Diane Brunner, MEd, President/CEO
4747 Monroe St
Toledo, OH 43623
419 473-2700
• Medical Assistant Prgm

**Medical University of Ohio at Toledo**
Lloyd A Jacobs, MD, President
3045 Arlington Ave
Toledo, OH 43614-5805
419 383-3421
• Physical Therapy Prgm
• Physician Assistant Prgm

**Mercy College of Northwest Ohio**
Paul Kessler, EdD, President
2221 Madison Ave
Toledo, OH 43624-1133
419 259-1279
• Health Information Tech Prgm
• Radiography Prgm

**Owens Community College**
Christa Adams, PhD, President
Oregon Rd , PO Box 10000
Toledo, OH 43699
567 661-7210
• Dental Hygiene Prgm
• Diagnostic Med Sonography Prgm
• Dietetic Technician-AD Prgm
• Health Information Tech Prgm
• Occupational Therapy Asst Prgm
• Physical Therapist Assistant Prgm
• Radiography Prgm
• Surgical Technology Prgm

**Professional Skills Institute**
Patricia A Finch, RN, President
20 Arco Dr
Toledo, OH 43607
419 531-9610
• Health Information Tech Prgm
• Physical Therapist Assistant Prgm

**St Vincent Mercy Medical Center**
Steven C Mickus, President
2213 Cherry St
Toledo, OH 43608
419 251-4152
• Clin Lab Scientist/Med Technologist Prgm

**Stautzenberger College**
George Simon, President
5355 Southwyck
Toledo, OH 43614
800 552-5099
• Medical Assistant Prgm

**University of Toledo**
Lloyd Jacobs, MD, President
2801 W Bancroft St
Toledo, OH 43606-3390
419 530-2211
• Athletic Training Prgm
• Cardiovascular Technology Prgm
• Counseling Prgm
• Health Information Admin Prgm
• Kinesiotherapy Prgm
• Occupational Therapy Prgm
• Respiratory Therapist (Advanced) Prgm
• Respiratory Therapist (Entry-Level) Prgm
• Speech-Language Pathology Prgm
• Therapeutic Recreation Specialist Prgm

## Urbana

**Urbana University**
Robert L Head, PhD, President
579 College Way
Urbana, OH 43078
937 484-1300
• Athletic Training Prgm

## Warren

**Trumbull Career & Technical Center**
Wayne McClain, PhD, Superintendent
528 Educational Hwy
Warren, OH 44483
330 847-0503
• Medical Assistant Prgm

**Trumbull Memorial Hospital**
Kris Hoce, MBA, President and CEO
1350 E Market St
Warren, OH 44482
216 841-9117
• Phlebotomist Prgm

## Westerville

**Otterbein College**
C Brent DeVore, PhD, President
Roush Hall 302
Westerville, OH 43081
614 823-1410
• Athletic Training Prgm

## Wilmington

**Wilmington College**
Daniel DiBiasio, PhD, President
251 Ludovic St
Pyle Center Box 1185
Wilmington, OH 45177
937 382-6661
• Athletic Training Prgm

## Wooster

**College of Wooster**
R Stanton Hales, Director
1189 Beall Ave
Wooster, OH 44691
330 263-2311
• Music Therapy Prgm

## Youngstown

**Choffin Career & Technical Center**
Wendy Webb, PhD, Superintendent
20 W Wood St
Youngstown, OH 44501
330 744-6915
• Dental Assisting Prgm
• Surgical Technology Prgm

**St Elizabeth Health Center**
Robert Shroder, MHA, President/CEO
1044 Belmont Ave
Youngstown, OH 44501-1790
330 746-7211
• Nuclear Medicine Technology Prgm

**Youngstown State University**
David C Sweet, PhD, President
One University Plaza
Youngstown, OH 44555
330 941-3101
• Clin Lab Technician/Med Lab Technician
• Counseling Prgm
• Dental Hygiene Prgm
• Dietetic Technician-AD Prgm
• Dietetics-Coordinated Prgm
• Dietetics-Didactic Prgm
• Emergency Med Tech-Paramedic Prgm
• Histotechnician Prgm
• Medical Assistant Prgm
• Physical Therapy Prgm
• Respiratory Therapist (Advanced) Prgm

## Zanesville

**Zane State College**
Paul Brown, EdD, President
1555 Newark Rd
Zanesville, OH 43701-2694
614 454-2501
• Clin Lab Technician/Med Lab Technician
• Medical Assistant Prgm
• Occupational Therapy Asst Prgm
• Phlebotomist Prgm
• Physical Therapist Assistant Prgm
• Radiography Prgm

# Oklahoma

## Ada

**East Central University**
Bill S Cole, EdD, President
Ada, OK 74820
405 332-8000
• Athletic Training Prgm
• Health Information Admin Prgm
• Rehabilitation Counseling Prgm

**Valley View Regional Hospital**
Ron Webb, MHA, President
Health Administration
430 N Monta Vista
Ada, OK 74820
405 332-2323
• Clin Lab Scientist/Med Technologist Prgm

## Altus

**Western Oklahoma State College**
Randy Cumby, MEd, President
2801 N Main
Altus, OK 73521-1397
580 477-2000
• Radiography Prgm

## Ardmore

**Southern Oklahoma Technology Center**
David Powell, PhD, Superintendent
2610 Sam Noble Pkwy
Ardmore, OK 73401
580 223-2070
• Surgical Technology Prgm

## Bethany

**Southern Nazarene University**
Loren P Gresham, PhD, President
6729 NW 39th Expwy
Bethany, OK 73008
405 491-6300
• Athletic Training Prgm

## Drumright

**Central Technology Center**
Phil Waul, MA, Superintendent
3 CT Circle
Drumright, OK 74030
918 352-2551
• Surgical Technology Prgm

## Edmond

**University of Central Oklahoma**
W Roger Webb, President
100 N University Dr
Edmond, OK 73034
405 974-2000
• Dietetic Internship Prgm
• Dietetics-Didactic Prgm
• Speech-Language Pathology Prgm

## El Reno

**Canadian Valley Technology Center**
Earl Cowan, EdD, Superintendent
6505 E Hwy 66
El Reno, OK 73036
405 262-2629
• Surgical Technology Prgm

## Enid

**Autry Technology Center**
James Strate, EdD, Superintendent
1201 W Willow
Enid, OK 73703
405 242-2750
• Radiography Prgm
• Surgical Technology Prgm

## Ft Cobb

**Caddo Kiowa Tech Ctr/SW Oklahoma St Univ**
Jerry L Martin, MEd, Superintendent
PO Box 190
Ft Cobb, OK 73038
405 643-5511
• Occupational Therapy Asst Prgm
• Physical Therapist Assistant Prgm

## Langston

**Langston University**
Ernest L Holloway, President
Langston, OK 73050
405 466-2231
• Dietetics-Didactic Prgm
• Physical Therapy Prgm
• Rehabilitation Counseling Prgm

## Lawton

**Comanche County Memorial Hospital**
Randy Segler, MHA, Chief Executive Officer
3401 W Gore Blvd Box 129
Lawton, OK 73502
405 355-8620
• Clin Lab Scientist/Med Technologist Prgm

**Great Plains Technology Center**
James Nisbett, MSEd, Superintendent
4500 W Lee Blvd
Lawton, OK 73505
580 355-6371
• Radiography Prgm
• Respiratory Therapist (Entry-Level) Prgm
• Surgical Technology Prgm

## Miami

**Northeastern Oklahoma A&M College**
Glenn E Mayle, EdD, President
200 I St
PO Box 3841
Miami, OK 74354
918 542-8441
• Clin Lab Technician/Med Lab Technician
• Physical Therapist Assistant Prgm

## Midwest City

**Rose State College**
Terry Britton, PhD, President
6420 SE 15th
Midwest City, OK 73110-2797
405 733-7300
• Clin Lab Technician/Med Lab Technician
• Dental Assisting Prgm
• Dental Hygiene Prgm
• Health Information Tech Prgm
• Radiography Prgm
• Respiratory Therapist (Advanced) Prgm

## Moore

**Oklahoma Health Academy**
Mike Pugliese, President
1939 N Moore Ave
Moore, OK 73160
405 912-2777
• Surgical Technology Prgm

## Muskogee

**Bacone College**
Robert Duncan, DMin, President
Office of the President
Muskogee, OK 74403-1568
888 682-5514
• Radiography Prgm

**Indian Capital Technology Center**
Tom Stiles, Superintendent
2403 N 41st St East
Muskogee, OK 74403
918 687-7565
• Radiography Prgm

**Muskogee Regional Medical Center**
Anthony Armstrong, CEO
300 Rockefeller Dr
Muskogee, OK 74401
918 682-5501
• Clin Lab Scientist/Med Technologist Prgm
• Surgical Technology Prgm

## Norman

**Moore Norman Technology Center**
John Hunter, MEd, Superintendant
4701 12th Ave NW
Norman, OK 73069
405 364-5763
• Dental Assisting Prgm
• Diagnostic Med Sonography Prgm
• Medical Assistant Prgm
• Surgical Technology Prgm

**University of Oklahoma**
David L Boren, President
660 Parrington Oval
Norman, OK 73019
405 325-3916
• Medical Librarian Prgm

## Oklahoma City

**Francis Tuttle Technology Center**
Kay Martin, PhD, Superintendent/CEO
12777 N Rockwell Ave
Oklahoma City, OK 73142
405 717-4266
• Medical Assistant Prgm
• Respiratory Therapist (Advanced) Prgm

**Metro Technology Centers**
James Branscum, EdD, Superintendent
1900 Springlake Dr
Oklahoma City, OK 73111
405 605-4400
• Dental Assisting Prgm
• Medical Assistant Prgm
• Radiography Prgm
• Surgical Technology Prgm

**Oklahoma City Community College**
Paul W Sechrist, PhD, President
7777 S May Ave
Oklahoma City, OK 73159-4444
405 682-7502
• Emergency Med Tech-Paramedic Prgm
• Occupational Therapy Asst Prgm
• Physical Therapist Assistant Prgm

## Platt College

Micheal A Pugliese, President
309 S Ann Arbor
Oklahoma City, OK 73128
405 946-7799
• Surgical Technology Prgm (2)

**Univ of Oklahoma Health Sciences Center**
Joseph J Ferretti, PhD, SVP/Provost
1000 Stanton L Young, Ste 221
Oklahoma City, OK 73117-1213
405 271-2332
• Audiologist Prgm
• Dental Hygiene Prgm
• Diagnostic Med Sonography Prgm
• Dietetic Internship Prgm
• Dietetics-Coordinated Prgm
• Dietetics-Didactic Prgm
• Genetic Counseling Prgm
• Nuclear Medicine Technology Prgm
• Occupational Therapy Prgm (2)
• Physical Therapy Prgm
• Physician Assistant Prgm
• Radiation Therapy Prgm
• Radiography Prgm
• Speech-Language Pathology Prgm

## Okmulgee

**Oklahoma State University - Okmulgee**
Robert E Klabenes, Provost & VP
Okmulgee, OK 74447
918 756-6211
• Dietetic Technician-AD Prgm

## Poteau

**Carl Albert State College**
Joe E White, PhD, President
1507 S McKenna
Poteau, OK 74953-5208
918 647-1210
• Physical Therapist Assistant Prgm
• Radiography Prgm

## Seminole

**Seminole State College**
James W Utterback, PhD, President
2701 Boren Blvd
PO Box 351
Seminole, OK 74818-0351
405 382-9950
• Clin Lab Technician/Med Lab Technician

## Stillwater

**Meridian Technology Center**
Andrea Kelly, EdD, Superintendent
1312 S Sangre St
Stillwater, OK 74074
405 377-3333
• Radiography Prgm

**Oklahoma State University**
David Schmidely, President, CEO
Stillwater, OK 74078
405 744-5000
- Athletic Training Prgm
- Counseling Prgm
- Dietetic Internship Prgm
- Dietetics-Didactic Prgm
- Speech-Language Pathology Prgm
- Therapeutic Recreation Specialist Prgm

## Tahlequah

**Northeastern State University**
W Roger Webb, President
Tahlequah, OK 74464
918 456-5511
- Dietetics-Didactic Prgm
- Speech-Language Pathology Prgm

## Tishomingo

**Murray State College**
William Pennington, PhD, President
One Murray Campus
NAH #116
Tishomingo, OK 73460
580 371-2371
- Physical Therapist Assistant Prgm

## Tulsa

**Orthoptic Teaching Program of Tulsa**
6606 S Yale Ste 110
Tulsa, OK 74136
- Orthoptist Prgm

**Saint Francis Hospital**
Jake Henry, MD, CEO
6161 S Yale Ave
Tulsa, OK 74136
918 494-6342
- Clin Lab Scientist/Med Technologist Prgm

**Tulsa Community College**
Tom McKeon, EdD, President, CEO
Central Office
6111 E Skelly Dr
Tulsa, OK 74135-6198
918 595-7868
- Clin Lab Technician/Med Lab Technician
- Dental Hygiene Prgm
- Health Information Tech Prgm
- Medical Assistant Prgm
- Occupational Therapy Asst Prgm
- Phlebotomist Prgm
- Physical Therapist Assistant Prgm
- Radiography Prgm
- Respiratory Therapist (Advanced) Prgm

**Tulsa Technology Center**
Gene Callahan, EdD, Superintendent
6111 E Skelly Dr
Tulsa, OK 74135-6100
918 828-5007
- Radiography Prgm
- Surgical Technology Prgm

**University of Tulsa**
Steadman Upham, PhD, President
600 S College Ave
Tulsa, OK 74104
918 631-2000
- Athletic Training Prgm
- Speech-Language Pathology Prgm

## Weatherford

**Southwestern Oklahoma State University**
John Hays, EdD, President
100 Campus Dr
Weatherford, OK 73096
580 774-3766
- Athletic Training Prgm
- Health Information Admin Prgm
- Music Therapy Prgm
- Radiography Prgm

## Wetumka

**Wes Watkins Technology Center Distist 25**
James R Moore, MS, Superintendent
7892 Hwy 9
Wetumka, OK 74883
405 452-5500
- Surgical Technology Prgm

# Ontario, Canada

## Brantford

**Mohawk College**
411 Elgin St
Brantford, ON N3T 5V2
- Orientation and Mobility Specialist Prgm
- Rehabilitation Teacher Prgm

## London

**University of Western Ontario**
London, ON N6G 1H1
- Medical Librarian Prgm
- Physical Therapy Prgm

## Ottawa

**University of Ottawa**
451 Smyth Rd, Rm 2047
Ottawa, ON K1H 8M5
- Ophthalmic Med Technologist Prgm

## Toronto

**Hospital for Sick Children**
555 University Ave
Toronto, ON M5G 1X8
416 813-5798
- Orthoptist Prgm

**ICT Kikkawa College**
Shirley Desborough, Chief Executive Officer
2340 Dundas St W
Unit G-04
Toronto, ON M6P 4A9
416 762-4857
- Massage Therapy Prgm

**ICT Northumberland College**
Shirley Desborough, RMT CST AT, Chief
   Executive Officer
2340 Dundas St W
Toronto, ON M6P 4A9
416 762-4857
- Massage Therapy Prgm

**University of Toronto**
David Naylor, MD DPhil, President
Simcoe Hall, Room 206
27 King's College Circle
Toronto, ON M5S 1A1
416 978-2121
- Genetic Counseling Prgm
- Medical Illustrator Prgm
- Medical Librarian Prgm
- Physical Therapy Prgm

## Windsor

**University of Windsor**
Windsor, ON N9B 3P4
519 253-3000
- Music Therapy Prgm

# Oregon

## Albany

**Linn - Benton Community College**
Rita Cavin, PhD, President
6500 SW Pacific Blvd
Albany, OR 97321
541 917-4999
- Dental Assisting Prgm
- Medical Assistant Prgm

## Beaverton

**College of Emergency Services (CESWA)**
Carl T Miller, MS EMT-P, CEO/Program Director
9735 SW Sunshine Ct #1000
Beaverton, OR 97005
503 644-9999
- Emergency Med Tech-Paramedic Prgm

## Bend

**Central Oregon Community College**
James Middleton, EdD, President
2600 NW College Way
Bend, OR 97701
541 383-7201
- Dental Assisting Prgm
- Health Information Tech Prgm
- Medical Assistant Prgm

## Corvallis

**Oregon State University**
Edward Ray, PhD, President
646 Kerr Administration Bldg
Corvallis, OR 97331
541 737-4133
- Athletic Training Prgm
- Counseling Prgm
- Dietetics-Didactic Prgm

## Eugene

**Lane Community College**
Mary Spilde, PhD, President
4000 E 30th Ave
Eugene, OR 97405-0640
541 463-3000
- Dental Assisting Prgm
- Dental Hygiene Prgm
- Medical Assistant Prgm
- Respiratory Therapist (Advanced) Prgm

**University of Oregon**
David B Frohnmayer, President
Eugene, OR 97403
503 346-3111
- Speech-Language Pathology Prgm

## Forest Grove

**Pacific University**
Philip Creighton, PhD, President
2043 College Way
Forest Grove, OR 97116-1797
503 352-2214
- Occupational Therapy Prgm
- Physical Therapy Prgm
- Physician Assistant Prgm

## Gresham

**Mt Hood Community College**
Robert M Silverman, PhD, President
26000 SE Stark St
Gresham, OR 97030-3300
503 491-7212
- Dental Hygiene Prgm
- Medical Assistant Prgm
- Physical Therapist Assistant Prgm
- Respiratory Therapist (Advanced) Prgm
- Surgical Technology Prgm

## Klamath Falls

**Oregon Institute of Technology**
Martha Anne Dow, PhD, President
3201 Campus Dr
Klamath Falls, OR 97601-8801
503 885-1101
- Dental Hygiene Prgm
- Respiratory Therapist (Advanced) Prgm

## Marylhurst

**Marylhurst University**
Nancy Wilgenbusch, PhD, President
17600 Pacific Hwy
Marylhurst, OR 97036-0261
503 636-8141
- Art Therapy Prgm
- Music Therapy Prgm

## McMinnville

**Linfield College**
Thomas Hellie, PhD, President
900 SE Baker St
McMinnville, OR 97128
503 434-2234
- Athletic Training Prgm

## Monmouth

**Western Oregon University**
Philip Conn, PhD, President
345 N Monmouth Ave
Monmouth, OR 97361
503 838-8888
- Rehabilitation Counseling Prgm

## Newberg

**George Fox University**
David Brandt, PhD, President
414 N Meridian St #6305
Newberg, OR 97132
503 554-2102
- Athletic Training Prgm

## Oregon City

**Clackamas Community College**
Earl P Johnson, PhD, President
19600 S Molalla Ave
Oregon City, OR 97045
503 657-6958
- Clinical Assisting Prgm
- Medical Assistant Prgm

## Pendleton

**Blue Mountain Community College**
John Turner, President
2411 NW Carden
PO Box 100
Pendleton, OR 97801
541 278-5950
- Dental Assisting Prgm

## Portland

**Concorde Career Institute - Portland**
Kevin Lambert, School Director
1425 NE Irving, Bldg #300
Portland, OR 97232
503 281-4181
- Dental Assisting Prgm
- Medical Assistant Prgm
- Surgical Technology Prgm

**East-West Coll of the Healing Arts**
David Slawson
525 NE Oregon St
Portland, OR 97232
503 233-6500
- Massage Therapy Prgm

**Everest College**
Mickey Sieracki, MA MT(ASCP), President
425 SW Washington St
Portland, OR 97204
503 222-3225
- Medical Assistant Prgm

**Heald College - Portland**
Amy McCombs, BA MA, President
625 SW Broadway
Ste 200
Portland, OR 97205
415 808-1400
- Medical Assistant Prgm

**Oregon Health & Science University**
Joseph E Robertson, MD MBA, President
3181 SW Sam Jackson Pk Rd, L101
Portland, OR 97239
503 494-8252
- Clin Lab Scientist/Med Technologist Prgm
- Dietetic Internship Prgm
- Emergency Med Tech-Paramedic Prgm
- Physician Assistant Prgm
- Radiation Therapy Prgm

**Portland Community College**
Preston Pulliams, EdD, District President
PO Box 19000
Portland, OR 97280-0990
503 977-4365
- Clin Lab Technician/Med Lab Technician
- Dental Assisting Prgm
- Dental Hygiene Prgm
- Dental Lab Technician Prgm
- Health Information Tech Prgm
- Medical Assistant Prgm
- Ophthalmic Med Technician Prgm
- Radiography Prgm

**Portland State University**
Daniel Bernstine, President
Portland, OR 97207-0751
503 725-3000
- Audiologist Prgm
- Counseling Prgm
- Rehabilitation Counseling Prgm
- Speech-Language Pathology Prgm

## Salem

**Chemeketa Community College**
Gretchen Schuette, PhD, President
4000 Lancaster Dr NE
PO Box 14007
Salem, OR 97309-7070
503 399-6591
- Dental Assisting Prgm
- Emergency Med Tech-Paramedic Prgm

**Mid Willamette Valley Dietetic Internship**
Capital Manor Retirement Community
1955 Dallas Hwy NW, Ste 1200
Salem, OR 97304
503 362-4101
- Dietetic Internship Prgm

# Pennsylvania

## Abington

**Abington Memorial Hospital**
Richard Jones, MS, President and CEO
1200 Old York Rd
Abington, PA 19001
215 576-2000
- Nuclear Medicine Technology Prgm
- Radiography Prgm

## Allentown

**Cedar Crest College**
Dorothy G Blaney, PhD, President
100 College Dr
Allentown, PA 18104-6196
610 437-4471
- Dietetics-Didactic Prgm
- Nuclear Medicine Technology Prgm

## Altoona

**Altoona Regional Health System**
James W Barner, MBA, President/CEO
620 Howard Ave
Altoona, PA 16601-4899
814 946-2223
• Clin Lab Scientist/Med Technologist Prgm

**Greater Altoona Career & Technology Center**
Lanny F Ross, DEd, Executive Director
1500 Fourth Ave
Altoona, PA 16602
814 946-8450
• Medical Assistant Prgm

## Annville

**Lebanon Valley College**
Stephen C MacDonald, President
Humanities 102
Annville, PA 17003-0501
717 867-6211
• Physical Therapy Prgm

## Aston

**Neumann College**
Rosalie Mirenda, DNSC, President
One Neumann Dr
Aston, PA 19014
610 558-5501
• Athletic Training Prgm
• Clin Lab Scientist/Med Technologist Prgm
• Physical Therapy Prgm

## Bethlehem

**Northampton Community College**
Arthur Scott, EdD, President
3835 Green Pond Rd
Bethlehem, PA 18020
610 861-5458
• Dental Hygiene Prgm
• Diagnostic Med Sonography Prgm
• Radiography Prgm
• Surgical Technology Prgm

**St Luke's Hospital**
Richard Anderson, President
801 Ostrum St
Bethlehem, PA 18015
610 954-4900
• Surgical Technology Prgm

## Bloomsburg

**Bloomsburg University**
Jessica S Kozloff, President
Bloomsburg, PA 17815
717 389-4000
• Audiologist Prgm
• Speech-Language Pathology Prgm

## Blue Bell

**Montgomery County Community College**
Karen Stout, EdD, President
340 DeKalb Pike
East House
Blue Bell, PA 19422
215 641-6506
• Clin Lab Technician/Med Lab Technician
• Dental Hygiene Prgm
• Medical Assistant Prgm
• Phlebotomist Prgm
• Radiography Prgm
• Surgical Technology Prgm

## Blue Ridge Summit

**Synergy Healting Arts Center & Massage School**
13593 Monterey Ln
Blue Ridge Summit, PA 17214
• Massage Therapy Prgm

## Bradford

**Bradford Regional Medical Center**
George E Leonhardt, President/CEO
116 Interstate Pkwy
PO Box 0218
Bradford, PA 16701-0218
814 368-4143
• Radiography Prgm

## Bryn Mawr

**Harcum College**
Edward D'Alessio, PhD, Acting President
750 Montgomery Ave
Bryn Mawr, PA 19010
610 526-6024
• Clin Lab Technician/Med Lab Technician
• Dental Assisting Prgm
• Dental Hygiene Prgm
• Physical Therapist Assistant Prgm
• Radiography Prgm

## Butler

**Butler County Community College**
Cynthia Azari, EdD, President
PO Box 1203
Butler, PA 16003-1203
412 287-8711
• Medical Assistant Prgm
• Physical Therapist Assistant Prgm

## California

**California University of Pennsylvania**
Angelo Armenti, Jr, PhD, President
250 University Ave
California, PA 15419
412 938-4400
• Athletic Training Prgm
• Physical Therapist Assistant Prgm
• Speech-Language Pathology Prgm

## Camp Hill

**Holy Spirit Hospital**
Romaine Niemeyer, MHA, President
503 N 21st St
Camp Hill, PA 17011-2288
717 763-2106
• Radiography Prgm

## Center Valley

**DeSales University**
Bernard O'Connor, OSFS PhD, President
2755 Station Ave
Center Valley, PA 18034-9568
610 282-1100
• Physician Assistant Prgm

## Chester

**Widener University**
James T Harris, EdD, President
One University Pl
Chester, PA 19013
610 499-4102
• Physical Therapy Prgm

## Clarion

**Clarion University of Pennsylvania**
Joseph Grunenwald, President
Clarion, PA 16214
814 393-2000
• Medical Librarian Prgm
• Speech-Language Pathology Prgm

## Clearfield

**Clearfield Hospital**
Robert Murray, III, CHE NHA, President/CEO
PO Box 992
Clearfield, PA 16830
814 768-2497
• Radiography Prgm

## Cresson

**Mount Aloysius College**
Mary Ann Dillon, MD RSM
7373 Admiral Peary Hwy
Cresson, PA 16630
814 886-4131
• Medical Assistant Prgm
• Occupational Therapy Asst Prgm
• Physical Therapist Assistant Prgm
• Surgical Technology Prgm

## Dallas

**College Misericordia**
Michael MacDowell, EdD, President
301 Lake St
Dallas, PA 18612-1098
570 674-6265
• Diagnostic Med Sonography Prgm
• Occupational Therapy Prgm
• Physical Therapy Prgm
• Radiography Prgm

## Danville

**Geisinger Medical Center**
Glenn Steele, MD, President
100 N Academy Ave
Danville, PA 17822-2201
570 271-5200
- Cardiovascular Technology Prgm
- Dietetic Internship Prgm
- Radiography Prgm

## East Stroudsburg

**East Stroudsburg University**
Robert J Dillman, PhD, President
200 Prospect St
East Stroudsburg, PA 18301
717 422-3546
- Athletic Training Prgm
- Exercise Physiology Prgm
- Exercise Science Prgm
- Speech-Language Pathology Prgm

## Edinboro

**Edinboro University of Pennsylvania**
Frank G Pogue, Jr, President
219 Meadville St
Edinboro, PA 16444
814 732-2711
- Counseling Prgm
- Dietetics-Coordinated Prgm
- Rehabilitation Counseling Prgm
- Speech-Language Pathology Prgm

## Elizabethtown

**Elizabethtown College**
Theodore E Long, PhD, President
One Alpha Dr
Elizabethtown, PA 17022-2298
717 361-1193
- Music Therapy Prgm
- Occupational Therapy Prgm

## Elkins Park

**Pennsylvania College of Optometry**
8360 Old York Rd
Elkins Park, PA 19027-1598
- Low Vision Therapy Prgm
- Orientation and Mobility Specialist Prgm
- Rehabilitation Teacher Prgm

## Erie

**Gannon University**
Antione M Garibaldi, PhD, President
109 University Square
Erie, PA 16541-0001
814 871-5800
- Dietetics-Coordinated Prgm
- Occupational Therapy Prgm
- Physical Therapy Prgm
- Physician Assistant Prgm
- Radiography Prgm
- Respiratory Therapist (Advanced) Prgm

**Great Lakes Institute of Technology**
Tony Piccirillo, President/CEO
5100 Peach St
Erie, PA 16509
814 864-6666
- Diagnostic Med Sonography Prgm
- Pharmacy Technician Prgm
- Surgical Technology Prgm

**Mercyhurst College**
Thomas Gamble, PhD, President
501 E 38th St
Glenwood Hills
Erie, PA 16546
814 824-2000
- Athletic Training Prgm
- Dietetics-Coordinated Prgm
- Medical Assistant Prgm
- Physical Therapist Assistant Prgm

**Saint Vincent Health Center**
Sr Catherine Manning, MBA, President
232 W 25th St
Erie, PA 16544
814 452-5111
- Clin Lab Scientist/Med Technologist Prgm

**Tri-State Business Institute**
Guy Euliano, President
5757 W 26th St
Erie, PA 16506
814 838-7673
- Dental Hygiene Prgm
- Medical Assistant Prgm

## Glenside

**Arcadia University**
Jerry M Greiner, PhD, President
450 S Easton Rd
Glenside, PA 19038
215 572-2908
- Genetic Counseling Prgm
- Physical Therapy Prgm
- Physician Assistant Prgm

## Grantham

**Messiah College**
Kim Phipps, PhD, President
Grantham, PA 17027
717 766-2511
- Athletic Training Prgm
- Dietetics-Didactic Prgm

## Greensburg

**Seton Hill University**
JoAnne W Boyle, PhD, President
Box 231K
Greensburg, PA 15601
724 838-4211
- Art Therapy Prgm
- Dietetics-Coordinated Prgm
- Physician Assistant Prgm

## Gwynedd Valley

**Gwynedd-Mercy College**
Kathleen Cieplak Owens, PhD, President
1325 Sumneytown Pike
PO Box 901
Gwynedd Valley, PA 19437-0901
610 641-5560
- Cardiovascular Technology Prgm
- Health Information Admin Prgm
- Health Information Tech Prgm
- Radiation Therapy Prgm
- Respiratory Therapist (Advanced) Prgm
- Respiratory Therapist (Entry-Level) Prgm

## Harrisburg

**Harrisburg Area Community College**
Edna Baehre, PhD, President
One HACC Dr
Harrisburg, PA 17110-2999
717 780-2340
- Clin Lab Technician/Med Lab Technician
- Dental Assisting Prgm
- Dental Hygiene Prgm
- Diagnostic Med Sonography Prgm
- Emergency Med Tech-Paramedic Prgm
- Health Information Tech Prgm
- Medical Assistant Prgm
- Respiratory Therapist (Advanced) Prgm
- Respiratory Therapist (Entry-Level) Prgm
- Surgical Technology Prgm (2)

## Immaculata

**Immaculata University**
Sr R Patricia Fadden, IHM EdD, President
1145 King Rd
Immaculata, PA 19345
610 647-4400
- Dietetic Internship Prgm
- Dietetics-Didactic Prgm
- Music Therapy Prgm

## Indiana

**Indiana University of Pennsylvania**
Tony Atwater, PhD, President
201 Sutton Hall
Indiana, PA 15705
724 357-2200
- Athletic Training Prgm
- Dietetic Internship Prgm
- Dietetics-Didactic Prgm
- Respiratory Therapist (Advanced) Prgm
- Speech-Language Pathology Prgm

## Jenkintown

**Manor College**
Sr Mary Cecilia Jurasinski, OSBM, President
700 Fox Chase Rd
Jenkintown, PA 19046
215 885-2360
- Dental Assisting Prgm
- Dental Hygiene Prgm

## Johnstown

**Commonwealth Tech Inst at Hiram G Andrews Ctr**
Donald Rullman, BA, Director
727 Goucher St
Johnstown, PA 15905
814 255-8231
• Dental Assisting Prgm

**Conemaugh Memorial Medical Center**
Scott Becker, MBA MPH, CEO
1086 Franklin St
Johnstown, PA 15905
814 534-9000
• Clin Lab Scientist/Med Technologist Prgm
• Histotechnician Prgm
• Radiography Prgm
• Surgical Technology Prgm

## Kittanning

**Armstrong County Memorial Hospital**
Jack Hoard, President/CEO
One Nolte Dr
Kittanning, PA 16201
412 543-8404
• Radiography Prgm

## Lancaster

**Lancaster Gen Coll of Nursing & Hlth Sciences**
Mary Grace Simcox, EdD, President
410 N Lime St
Lancaster, PA 17602
717 544-4787
• Cardiovascular Technology Prgm
• Clin Lab Scientist/Med Technologist Prgm
• Diagnostic Med Sonography Prgm
• Nuclear Medicine Technology Prgm
• Radiography Prgm
• Surgical Technology Prgm

## Lock Haven

**Lock Haven University**
Keith Miller, PhD, President
Sullivan 202
Lock Haven, PA 17745
717 893-2000
• Athletic Training Prgm
• Physician Assistant Prgm

## Loretto

**St Francis University**
Rev Gabriel Zeis, TOR, President
PO Box 600
Loretto, PA 15940-0600
814 472-3001
• Occupational Therapy Prgm
• Physical Therapy Prgm
• Physician Assistant Prgm

## Mansfield

**Mansfield University**
John Halstead, PhD, President
North Hall 500
Mansfield, PA 16933
570 662-4046
• Dietetics-Didactic Prgm
• Music Therapy Prgm
• Radiography Prgm
• Respiratory Therapist (Advanced) Prgm

## McKees Rock

**Ohio Valley General Hospital**
William F Provenzano, FACHE, President
25 Heckel Rd
McKees Rock, PA 15136
• Radiography Prgm

## Media

**Delaware County Community College**
Richard D DeCosmo, EdD, President
901 S Media Line Rd
Media, PA 19603-1094
610 359-5100
• Medical Assistant Prgm
• Surgical Technology Prgm

## Millersville

**Millersville University of Pennsylvania**
Francine McNairey, PhD, President
Biemesderfer Executive Ctr
Millersville, PA 17551
717 872-3592
• Respiratory Therapist (Advanced) Prgm

## Monaca

**Community College of Beaver County**
Joe D Forrester, EdD, President
One Campus Dr
Monaca, PA 15061
724 775-8561
• Phlebotomist Prgm

## Monroeville

**Western School of Health & Business**
Kenneth Richards, BS, President
1 Monroeville Center, Ste 250
Monroeville, PA 15146
412 373-6400
• Pharmacy Technician Prgm
• Surgical Technology Prgm

## Nanticoke

**Luzerne County Community College**
Patricia Donohue, PhD, President
1333 S Prospect St
Nanticoke, PA 18634-3899
570 740-0384
• Dental Assisting Prgm
• Dental Hygiene Prgm
• Respiratory Therapist (Advanced) Prgm
• Surgical Technology Prgm

## New Castle

**Jameson Health System**
Thomas White, FACHE, President and CEO
1000 S Mercer St
South Campus
New Castle, PA 16101
724 658-9001
• Nuclear Medicine Technology Prgm

**Jameson Hospital**
Thomas White, FACHE, President/CEO
1211 Wilmington Ave
New Castle, PA 16105
412 658-3511
• Radiography Prgm

## New Kensington

**Career Training Academy**
John M Reddy, BS, President
950 Fifth Ave
New Kensington, PA 15068
724 337-1000
• Medical Assistant Prgm

## Newton

**Bucks County Community College**
James J Linksz, EdD, President
275 Swamp Rd
Tyler Hall 220
Newton, PA 18940
215 968-8222
• Medical Assistant Prgm
• Radiography Prgm

## Oaks

**Cortiva Inst - Penn School of Muscle Therapy**
Jeff Mann, PDMT NCBTMB, President
1173 Egypt Rd
PO Box 400
Oaks, PA 19456-0400
610 666-9060
• Massage Therapy Prgm

## Philadelphia

**Albert Einstein Medical Center**
Barry Freedman, MBA, President
5501 Old York Rd
Philadelphia, PA 19141-3098
215 456-7010
• Radiography Prgm

**Community College of Philadelphia**
Stephen M Curtis, PhD, President
1700 Spring Garden St
Philadelphia, PA 19130
215 751-8028
• Clin Lab Technician/Med Lab Technician
• Dental Hygiene Prgm
• Dietetic Technician-AD Prgm
• Health Information Tech Prgm
• Medical Assistant Prgm
• Phlebotomist Prgm
• Radiography Prgm
• Respiratory Therapist (Advanced) Prgm

**INSTITUTIONS**

**Drexel University**
Constantine N Papadakis, President
32nd and Chestnut Sts
Philadelphia, PA 19104
215 895-2000
- Art Therapy Prgm
- Dance/Movement Therapy Prgm
- Dietetics-Didactic Prgm
- Medical Librarian Prgm
- Music Therapy Prgm
- Perfusion Prgm
- Physical Therapy Prgm
- Physician Assistant Prgm
- Radiography Prgm

**Graduate Hospital**
Brian Finestein
1800 Lombard St
Philadelphia, PA 19146
- Clin Lab Scientist/Med Technologist Prgm

**Holy Family University**
Sr M Francesca Onley, PhD CSFN, President
Grant & Frankford Aves
Philadelphia, PA 19114
215 637-7700
- Radiography Prgm

**La Salle University**
Michael McGinniss, President
1900 W Olney Ave
Philadelphia, PA 19141-1108
- Dietetics-Coordinated Prgm
- Dietetics-Didactic Prgm
- Speech-Language Pathology Prgm

**National Massage Therapy Insitute**
Division of PSB
10050 Roosevelt Blvd
Philadelphia, PA 19116
- Massage Therapy Prgm

**Pennsylvania Hospital**
John R Ball, MD JD, President
800 Spruce St
Philadelphia, PA 19107
215 829-3312
- Clin Lab Scientist/Med Technologist Prgm

**Philadelphia College of Osteopathic Medicine**
Matthew Schure, PhD, President, CEO
4170 City Ave
Evans Hall - President's Office
Philadelphia, PA 19131-1694
215 871-6800
- Physician Assistant Prgm

**Philadelphia University**
James P Gallagher, PhD, President
School House Ln and Henry Ave
Philadelphia, PA 19144-5497
215 951-2970
- Occupational Therapy Prgm
- Physician Assistant Prgm

**St Christopher Hospital School of Rad Tech**
Jill Tillman, MBA, Interim CEO
Erie Ave at Front
Philadelphia, PA 19134
215 427-5480
- Radiography Prgm

**Temple University**
David D W Adamany, PhD, President
President's Office
2nd Floor - Suillivan Hall
Philadelphia, PA 19122
215 204-7405
- Athletic Training Prgm
- Health Information Admin Prgm
- Music Therapy Prgm
- Occupational Therapy Prgm
- Physical Therapy Prgm
- Speech-Language Pathology Prgm

**Thomas Jefferson University**
Robert L Barchi, MD PhD, President
1020 Walnut St, 641 Scott Bldg
Philadelphia, PA 19107-5587
215 955-6617
- Clin Lab Scientist/Med Technologist Prgm
- Cytotechnology Prgm
- Diagnostic Med Sonography Prgm
- Occupational Therapy Prgm
- Physical Therapy Prgm
- Radiation Therapy Prgm
- Radiography Prgm

**Thompson Institute**
Shawn Bartley, EdD, President
3010 Market St
Philadelphia, PA 19104
215 594-4000
- Medical Assistant Prgm (2)

**Univ of Penn MC/Hosp of the Univ of Penn**
Ralph Muller
3400 Spruce St Silverstein One
21 Penn Tower
Philadelphia, PA 19104
215 662-2203
- Radiography Prgm

**University of the Sciences in Philadelphia**
Philip P Gerbino, PhD, President
600 S 43rd St
Philadelphia, PA 19104-4495
215 596-8970
- Occupational Therapy Prgm
- Physical Therapy Prgm

# Pittsburgh

**Bidwell Training Center**
William E Strickland, Jr, President and CEO
1815 Metropolitan St
Pittsburgh, PA 15233-2200
412 323-4000
- Pharmacy Technician Prgm

**Bradford School - Pittsburgh**
Vincent Graziano, MBA, President
125 West Station Square Dr, Ste 129
Pittsburgh, PA 15219
412 391-6710
- Dental Assisting Prgm
- Medical Assistant Prgm

**Chatham College**
Esther L Barazzone, PhD, President
Woodland Rd
Pittsburgh, PA 15232-2826
412 365-1160
- Occupational Therapy Prgm
- Physical Therapy Prgm
- Physician Assistant Prgm

**Comm Coll of Allegheny County**
Wendy Weiner, PhD, President
800 Allegheny Ave
Pittsburgh, PA 15233
412 323-2323
- Clin Lab Technician/Med Lab Technician
- Diagnostic Med Sonography Prgm
- Dietetic Technician-AD Prgm
- Health Information Tech Prgm
- Medical Assistant Prgm
- Nuclear Medicine Technology Prgm
- Occupational Therapy Asst Prgm
- Physical Therapist Assistant Prgm
- Radiation Therapy Prgm
- Radiography Prgm
- Respiratory Therapist (Advanced) Prgm
- Surgical Technology Prgm

**Connelley Technical Institute**
Joseph Poerio, MS, Director of Adult Educ
1501 Bedford Ave
Pittsburgh, PA 15219
412 338-3703
- Surgical Technology Prgm

**Ctr for Emer Med of Western Pennsylvania**
Douglas Garretson, BA NREMT-P, VP, COO
230 McKee Pl, Ste 500
Pittsburgh, PA 15213
412 647-5300
- Emergency Med Tech-Paramedic Prgm

**Duquesne University**
Charles Dougherty, PhD, President
600 Forbes Ave
Administration Bldg Rm 510
Pittsburgh, PA 15282
412 396-6060
- Athletic Training Prgm
- Counseling Prgm
- Health Information Admin Prgm
- Music Therapy Prgm
- Occupational Therapy Prgm
- Physical Therapy Prgm
- Physician Assistant Prgm
- Speech-Language Pathology Prgm

**Everest Institute**
James Callahan, President
100 Forbes Ave, Ste 1200
Pittsburgh, PA 15222
412 261-4520
- Medical Assistant Prgm

**Family Health Council**
625 Stanwix St
Pittsburgh, PA 15222-1417
412 288-9039
- Dietetic Internship Prgm

**ICM School of Business & Medical Careers**
Jackie Flynn, President
10 Wood St
Pittsburgh, PA 15222-1977
412 261-2647
- Medical Assistant Prgm
- Occupational Therapy Asst Prgm

**U of Pitt Med Ctr Presbyterian Shadyside**
Liz Concordia, CEO
5230 Centre Ave
Pittsburgh, PA 15232
412 622-2010
- Dietetic Internship Prgm
- Perfusion Prgm

**Univ Hlth Ctr of PA/Anisa I Kanbour Sch -Cyto**
Leslie Davis, MS, CEO
300 Halket St
Pittsburgh, PA 15213
412 641-4664
• Cytotechnology Prgm

**University of Pittsburgh**
Mark A Nordenberg, JD, Chancellor
Rm 107 Cathedral of Learning
Pittsburgh, PA 15260
412 624-4200
• Athletic Training Prgm (2)
• Audiologist Prgm
• Dental Hygiene Prgm
• Dietetics-Coordinated Prgm
• Dietetics-Didactic Prgm
• Genetic Counseling Prgm
• Health Information Admin Prgm
• Medical Librarian Prgm
• Occupational Therapy Prgm
• Orientation and Mobility Specialist Prgm
• Physical Therapist Assistant Prgm
• Physical Therapy Prgm
• Rehabilitation Counseling Prgm
• Respiratory Therapist (Advanced) Prgm
• Speech-Language Pathology Prgm

**UPMC Health System**
Elizabeth Concordia, MAS, President and CEO
200 Lothrop St
Ste N739MUH
Pittsburgh, PA 15213-2582
412 647-8788
• Radiography Prgm

**Western Sch of Hlth & Business - Pittsburgh**
Kenneth Richards, MBA, President
Pittsburgh Campus
421 Seventh Ave
Pittsburgh, PA 15219-1907
412 373-6400
• Diagnostic Med Sonography Prgm
• Radiography Prgm
• Respiratory Therapist (Advanced) Prgm

## Pottsville

**McCann School of Business and Technology**
Linda Walinsky, MPA, Regional Executive Direct
2638 Woodglen Rd
Pottsville, PA 17901
570 622-3293
• Surgical Technology Prgm

## Reading

**Alvernia College**
Thomas F Flynn, PhD, President
400 Saint Bernardine St
Reading, PA 19607-1799
610 796-8324
• Athletic Training Prgm
• Occupational Therapy Prgm

**Reading Area Community College**
Richard Kratz, EdD, President
10 S Second St
PO Box 1706
Reading, PA 19603
610 372-4721
• Clin Lab Technician/Med Lab Technician
• Respiratory Therapist (Advanced) Prgm
• Respiratory Therapist (Entry-Level) Prgm

**Reading Hospital & Medical Center**
Charles Sullivan, MS, CEO
PO Box 16052
Reading, PA 19612-6052
610 988-8258
• Clin Lab Scientist/Med Technologist Prgm
• Radiography Prgm
• Surgical Technology Prgm

**St Joseph Medical Center**
John Morahan, President/CEO
12th and Walnut Sts
PO Box 316
Reading, PA 19603-0316
610 378-2000
• Radiography Prgm

## Sayre

**Robert Packer Hospital**
Mary N Mannix, MSHA, President
Guthrie Square
Sayre, PA 18840
717 888-6666
• Clin Lab Scientist/Med Technologist Prgm

## Schnecksville

**Lehigh Carbon Community College**
Donald W Snyder, MBA JD LLM, President
4525 Education Pk Dr
Schnecksville, PA 18078-2598
610 799-2121
• Health Information Tech Prgm
• Medical Assistant Prgm
• Occupational Therapy Asst Prgm
• Physical Therapist Assistant Prgm

## Scranton

**Johnson College**
Ann L Pipinski, EdD, President and CEO
3427 N Main Ave
Scranton, PA 18508-1495
570 342-6404
• Radiography Prgm

**Lackawanna College**
Raymond S Angeli, MS, President
501 Vine St
Scranton, PA 18509
570 961-7850
• Diagnostic Med Sonography Prgm

**Marywood University**
Sr Mary Reap, PhD, President
2300 Adams Ave
Scranton, PA 18509
570 348-6231
• Art Therapy Prgm
• Athletic Training Prgm
• Counseling Prgm
• Dietetic Internship Prgm
• Dietetics-Coordinated Prgm
• Dietetics-Didactic Prgm
• Music Therapy Prgm
• Physician Assistant Prgm
• Speech-Language Pathology Prgm

**University of Scranton**
Rev Scott J Pilarz, SJ PhD, President
Scranton, PA 18510-4622
570 941-7500
• Counseling Prgm
• Occupational Therapy Prgm
• Physical Therapy Prgm
• Rehabilitation Counseling Prgm

## Seneca

**UPMC Northwest**
Neil E Todhunter, CHE NHA, President/CEO
100 Fairfield Dr
Seneca, PA 16346
814 676-7140
• Radiography Prgm

## Sewickley

**Sewickley Vlly Hosp/Heritage Valley Hlth Sys**
Norman Mitry, President and CEO
720 Blackburn Rd
Sewickley, PA 15143
724 773-2024
• Radiography Prgm

## Sharon

**Sharon Regional Health System**
John A Zidansek, MHA, President/CEO
740 E State St
Sharon, PA 16146
412 983-3911
• Radiography Prgm

## Shippensburg

**Shippensburg University**
Shippensburg, PA 17257-2210
717 532-9121
• Counseling Prgm

## Slippery Rock

**Slippery Rock University**
G Warren Smith, PhD, President
300 Old Main
Slippery Rock, PA 16057
412 738-2000
• Athletic Training Prgm
• Counseling Prgm
• Exercise Science Prgm
• Music Therapy Prgm
• Physical Therapy Prgm
• Therapeutic Recreation Specialist Prgm

## St David

**Eastern University**
David Black, President
1300 Eagle Rd
St David, PA 19087
• Athletic Training Prgm

## State College

**Mount Nittany Medical Center**
1800 E Park Ave
State College, PA 16803
814 231-7000
• Clin Lab Scientist/Med Technologist Prgm

**South Hills School of Business & Technology**
S Paul Mazza, JD, President
480 Waupelani Dr
State College, PA 16801
814 234-7755
• Health Information Tech Prgm

## Summerdale

**Central Pennsylvania College**
Todd A Milano, BS, President
College Hill and Valley Rds
Summerdale, PA 17093-0309
800 759-2727
• Medical Assistant Prgm
• Physical Therapist Assistant Prgm

## University Park

**Penn State University**
Graham Spanier, President
201 Old Main
University Park, PA 16802
814 865-4700
• Athletic Training Prgm
• Audiologist Prgm
• Clin Lab Technician/Med Lab Technician
• Counseling Prgm
• Dietetic Internship Prgm
• Dietetic Technician-AD Prgm
• Dietetics-Didactic Prgm
• Occupational Therapy Asst Prgm (3)
• Occupational Therapy Prgm
• Physical Therapist Assistant Prgm (4)
• Radiography Prgm (2)
• Rehabilitation Counseling Prgm
• Speech-Language Pathology Prgm

## Upland

**Crozer - Chester Med Ctr**
Joan K Richards, President
One Medical Center Blvd
Upland, PA 19013
610 447-2785
• Diagnostic Med Sonography Prgm
• Electroneurodiagnostic Tech Prgm
• Radiography Prgm
• Respiratory Therapist (Advanced) Prgm

## Warrington

**ARAMARK Corporation Mid - Atlantic Region**
Warrington, PA 18976
• Dietetic Internship Prgm

## Washington

**Penn Commercial Inc**
Robert S Bazant, BS, Director
242 Oak Spring Rd
Washington, PA 15301
724 222-5330
• Medical Assistant Prgm

**Washington Hospital**
Telford W Thomas, MHA, President/CEO
155 Wilson Ave
Washington, PA 15301
724 223-3007
• Radiography Prgm

## Washington Cross

**Sodexho Marriott Services**
Washington Cross, PA 18977
• Dietetic Internship Prgm
• Dietetics-Preprofessional Practice Prgm

## Waynesburg

**Waynesburg College**
Timothy R Thyreen, LHD, President
51 W College St
Waynesburg, PA 15370
412 852-3212
• Athletic Training Prgm

## West Chester

**West Chester University**
Madeleine Wing Adler, PhD, President
102 Phillips Hall
West Chester, PA 19383
610 436-2471
• Athletic Training Prgm
• Dietetics-Didactic Prgm
• Respiratory Therapist (Advanced) Prgm
• Speech-Language Pathology Prgm

## West Mifflin

**Comm Coll of Allegheny Cnty - South Campus**
James Holmberg, PhD, President, Exec Dean
1750 Clairton Rd
West Mifflin, PA 15122-3097
412 469-1100
• Pharmacy Technician Prgm

## Wilkes-Barre

**King's College**
Rev Thomas J O'Hara, CSC PhD, President
133 N River St
Wilkes-Barre, PA 18711
570 208-5900
• Athletic Training Prgm
• Physician Assistant Prgm

**Wilkes-Barre General Hospital**
William Host, MD, CEO
575 N River St
Wilkes-Barre, PA 18764
570 552-3006
• Diagnostic Med Sonography Prgm
• Radiography Prgm

**Wyoming Valley Health Care System**
William Host, MD, President/CEO
575 N River St
Wilkes-Barre, PA 18764
570 552-3000
• Nuclear Medicine Technology Prgm

## Williamsport

**Pennsylvania College of Technology**
Davie Jane Gilmour, PhD, President
One College Ave
Williamsport, PA 17701-5799
570 326-3761
• Dental Hygiene Prgm
• Emergency Med Tech-Paramedic Prgm
• Health Information Tech Prgm
• Occupational Therapy Asst Prgm
• Physician Assistant Prgm
• Radiography Prgm
• Surgical Technology Prgm

**Williamsport Hosp & Medical Center**
Kirby Smith, President
777 Rural Ave
Williamsport, PA 17701
717 326-8101
• Clin Lab Scientist/Med Technologist Prgm

## Wyomissing

**Berks Technical Institute**
Kenneth S Snyder, President
Four Park Plaza
Wyomissing, PA 19610
610 372-1722
• Medical Assistant Prgm

## York

**Baltimore School of Massage**
170 Red Rock Rd
York, PA 17402
• Massage Therapy Prgm

**York College of Pennsylvania**
George W Waldner, PhD, President
Country Club Rd
York, PA 17405-7199
717 846-7788
• Respiratory Therapist (Advanced) Prgm
• Respiratory Therapist (Entry-Level) Prgm

**York Hospital**
Richard Seim, MBA, President
1001 S George St
York, PA 17405
717 851-2650
• Clin Lab Scientist/Med Technologist Prgm

## Youngwood

**Westmoreland County Community College**
Steven C Ender, EdD, President
145 Pavilion Lane
Youngwood, PA 15697-1895
724 925-4000
• Dental Assisting Prgm
• Dental Hygiene Prgm
• Dietetic Technician-AD Prgm
• Medical Assistant Prgm
• Surgical Technology Prgm

# Puerto Rico

## Bayamon

**Universidad Central del Caribe**
Nilda Candelario, MD, President
Call Box 60-327
Bayamon, PR 00960-6032
787 798-3001
• Radiography Prgm

## Caguas

**Huertas Junior College**
Felix Rodriguez Matos, PhD, President
PO Box 8429
Caguas, PR 00726
809 743-1242
• Health Information Tech Prgm

## Carolina

**Universidad del Este**
Alberto Maldonado-Ruiz, Esq, Chancellor
PO Box 2010
Carolina, PR 00984-2010
787 257-7373
• Health Information Tech Prgm
• Radiography Prgm

## Humacao

**University of Puerto Rico at Humacao**
Hilda Colon Plumey, EdD, Chancellor
CUH Postal Station 100, Carr 908
Humacao, PR 00791-4300
787 850-9374
• Occupational Therapy Asst Prgm
• Occupational Therapy Prgm
• Physical Therapist Assistant Prgm

## Mayaguez

**Universidad Adventista de las Antillas**
Myrna Colon-Contreras, PhD, President
PO Box 118
Mayaguez, PR 00681
787 834-9595
• Health Information Tech Prgm

## Ponce

**Ponce Technological University College**
University of Puerto Rico
PO Box 7186
Ponce, PR 00732
• Physical Therapist Assistant Prgm

**Pontifical Catholic University**
Marcelina Velez de Santiago, MS CHEM,
   President
2250 Las Americas Ave, Ste 564
Ponce, PR 00717-9997
787 841-2000
• Clin Lab Scientist/Med Technologist Prgm

## San German

**Inter American University of Puerto Rico**
Prof Agnes Mojica, MA, Chancellor
Call Box 5100
San German, PR 00683
787 892-5634
• Clin Lab Scientist/Med Technologist Prgm
• Health Information Tech Prgm
• Radiography Prgm

## San Juan

**Inter - American University - Metro Campus**
Manuel Fernos, Esq, President
GPO Box 3255
San Juan, PR 00936
787 766-1912
• Clin Lab Scientist/Med Technologist Prgm

**Puerto Rico Department of Health**
PO Box 70184
San Juan, PR 00936
787 274-6831
• Dietetic Internship Prgm

**University of Puerto Rico**
Jose Carlo, MD, Chancellor
Medical Sciences Campus
PO Box 365067
San Juan, PR 00936-5067
787 758-2525
• Clin Lab Scientist/Med Technologist Prgm
• Cytotechnology Prgm
• Dental Assisting Prgm
• Dietetic Internship Prgm
• Dietetics-Didactic Prgm
• Health Information Admin Prgm
• Medical Librarian Prgm
• Nuclear Medicine Technology Prgm
• Occupational Therapy Prgm
• Ophthalmic Med Technician Prgm
• Physical Therapy Prgm
• Radiography Prgm
• Rehabilitation Counseling Prgm

**Veterans Affairs Medical Center**
One Veterans Plaza
San Juan, PR 00927-5800
787 758-7575
• Dietetic Internship Prgm

# Quebec, Canada

## Montreal

**McGill University**
Jennifer Fitzpatrick, MS, Director, GC Program
Human Genetics, Stewart Biology N5/13
1205 Dr Penfield Ave
Montreal, QC H3G 1Y5
514 398-3600
• Genetic Counseling Prgm
• Medical Librarian Prgm

**University de Montreal**
Pavillon Marguerite-D'Youville
CP 6128-Succ Centreville
Montreal, QC H3C 3J7
• Medical Librarian Prgm

## West Montreal

**Concordia University**
Frederick H Lowy, MD
1455 de Maisonneuve Blvd
West Montreal, QC H3G 1M8
514 848-4850
• Art Therapy Prgm

# Rhode Island

## Kingston

**University of Rhode Island**
Robert L Carothers, President
8 Washburn Hall
Kingston, RI 02881-0817
401 792-1000
• Audiologist Prgm
• Dietetic Internship Prgm
• Dietetics-Didactic Prgm
• Physical Therapy Prgm
• Speech-Language Pathology Prgm

## Newport

**Salve Regina University**
M Therese Antone, EdD RSM, President
100 Ochre Point Ave
Newport, RI 02840
• Rehabilitation Counseling Prgm

## North Providence

**Our Lady of Fatima Hospital**
H John Keimig, MHA, President
200 High Service Ave
North Providence, RI 02904
401 456-3000
• Clin Lab Scientist/Med Technologist Prgm

## Providence

**Johnson & Wales University**
Morris J Gaebe, Chancellor
8 Abbott Park Palace
Providence, RI 02903-3703
401 598-1000
• Dietetics-Didactic Prgm

**Rhode Island Hospital**
Joseph Amaral, MD, President/CEO
593 Eddy St
Providence, RI 02905
401 444-5123
• Clin Lab Scientist/Med Technologist Prgm
• Diagnostic Med Sonography Prgm
• Nuclear Medicine Technology Prgm
• Radiography Prgm

**Rhode Island School of Cytotechnology**
Thomas G Parris, Jr, President/CEO
101 Dudley St
Providence, RI 02905
401 274-1100
• Cytotechnology Prgm
• Medical Librarian Prgm

INSTITUTIONS

## Warwick

### Community College of Rhode Island
Ray M Di Pasquale, PhD, President
400 East Ave
Warwick, RI 02886
401 825-2188
- Clin Lab Technician/Med Lab Technician
- Dental Assisting Prgm
- Dental Hygiene Prgm
- Massage Therapy Prgm
- Occupational Therapy Asst Prgm
- Physical Therapist Assistant Prgm
- Radiography Prgm
- Respiratory Therapist (Advanced) Prgm

### New England Institute of Technology
Richard I Gouse, BA, President
2500 Post Rd
Warwick, RI 02886-2251
401 467-7744
- Occupational Therapy Asst Prgm
- Surgical Technology Prgm

# Saskatchewan, Canada

## Saskatoon

### Orthoptic Clinic Eye Care Centre
Saskatoon, SK S7K 0M7
306 655-8058
- Orthoptist Prgm

# South Carolina

## Anderson

### AnMed Health Medical Center
John Miller, MHA, President
800 N Fant St
Anderson, SC 29621
864 261-1109
- Radiography Prgm

### Forrest Junior College
John Re, PhD, President
601 E River St
Anderson, SC 29624
864 225-7653
- Medical Assistant Prgm

## Beaufort

### Technical College of the Lowcountry
Anne McNutt, PhD, President
PO Box 1288
Beaufort, SC 29901
843 525-8211
- Radiography Prgm
- Surgical Technology Prgm

## Charleston

### Charleston Southern University
Jairy C Hunter, PhD, President
PO Box 118087
Charleston, SC 29423-8087
803 863-7000
- Athletic Training Prgm
- Music Therapy Prgm

### College of Charleston
Leo I Higdon, Jr, President
66 George St
Charleston, SC 29424-0001
843 953-5500
- Athletic Training Prgm

### Medical University of South Carolina
Ray Greenberg, MD, President
135 Cannon Street, PO Box d50001
Charleston, SC 29425-2701
803 792-2211
- Cytotechnology Prgm
- Dietetic Internship Prgm
- Histotechnology Prgm
- Occupational Therapy Prgm
- Perfusion Prgm
- Physical Therapy Prgm
- Physician Assistant Prgm
- Speech-Language Pathology Prgm

### Miller-Motte Technical College
Julie S Corner, MEd, Campus Director
8085 Rivers Ave
Charleston, SC 29406
843 574-0101
- Medical Assistant Prgm
- Surgical Technology Prgm

### Trident Technical College
Mary D Thornley, EdD, President
PO Box 118067
Charleston, SC 29423-8067
803 574-6241
- Clin Lab Technician/Med Lab Technician
- Dental Assisting Prgm
- Dental Hygiene Prgm
- Medical Assistant Prgm
- Occupational Therapy Asst Prgm
- Pharmacy Technician Prgm
- Physical Therapist Assistant Prgm
- Radiography Prgm
- Respiratory Therapist (Advanced) Prgm

## Clemson

### Clemson University
Constantine W Curris, President
201 Sikes Hall
Clemson, SC 29634
803 656-3311
- Counseling Prgm
- Dietetics-Didactic Prgm
- Therapeutic Recreation Specialist Prgm

## Columbia

### Midlands Technical College
Marshall (Sonny) White, PhD, President
PO Box 2408
Columbia, SC 29202-2408
803 738-7600
- Clin Lab Technician/Med Lab Technician
- Dental Assisting Prgm
- Dental Hygiene Prgm
- Health Information Tech Prgm
- Medical Assistant Prgm
- Nuclear Medicine Technology Prgm
- Pharmacy Technician Prgm
- Physical Therapist Assistant Prgm
- Radiography Prgm
- Respiratory Therapist (Advanced) Prgm
- Surgical Technology Prgm

### Palmetto Health Baptist
Charles D Beaman, Jr, President
Taylor at Marion St
Columbia, SC 29220
803 771-5042
- Clin Lab Scientist/Med Technologist Prgm

### SC Dept of Health & Environmental Control
Mills Complex, PO Box 101106
Columbia, SC 29211-0106
803 737-3954
- Dietetic Internship Prgm

### Sister of Charity Providence Hospital
Stephen Purves, MHA, President
2435 Forest Dr
Columbia, SC 29204
803 256-5313
- Cardiovascular Technology Prgm

### South University
Anne F Patton, BBA, President
3810 Main St
Columbia, SC 29203
803 799-9082
- Medical Assistant Prgm

### University of South Carolina
Andrew Sorenson, PhD, President
Osborne 203
Columbia, SC 29208
803 777-2001
- Athletic Training Prgm
- Counseling Prgm
- Genetic Counseling Prgm
- Medical Librarian Prgm
- Physical Therapy Prgm
- Rehabilitation Counseling Prgm
- Speech-Language Pathology Prgm

## Conway

### Horry - Georgetown Technical College
H Neyle Wilson, BS MEd, President
PO Box 261966
Conway, SC 29528
843 349-5201
- Dental Assisting Prgm
- Dental Hygiene Prgm
- Radiography Prgm
- Surgical Technology Prgm

## Due West

### Erskine College
Randy T Ruble, PhD, President
2 Washington St
Due West, SC 29639
864 379-8833
- Athletic Training Prgm

## Florence

**Florence-Darlington Technical College**
Charles W Gould, PhD, President
PO Box 100548
Florence, SC 29501-0548
843 661-8000
- Clin Lab Technician/Med Lab Technician
- Dental Assisting Prgm
- Dental Hygiene Prgm
- Health Information Tech Prgm
- Radiography Prgm
- Respiratory Therapist (Advanced) Prgm
- Surgical Technology Prgm

**McLeod Regional Medical Center**
Robert Colones, MA, President and CEO
555 E Cheves St
PO Box 100551
Florence, SC 29506
843 777-2297
- Clin Lab Scientist/Med Technologist Prgm

## Gaffney

**Limestone College**
Walt Griffin, PhD, President
1115 College Dr
Gaffney, SC 29340
864 488-4616
- Athletic Training Prgm

## Graniteville

**Aiken Technical College**
Susan Winsor, PhD, President
2276 Jefferson Davis Hwy
PO 400
Graniteville, SC 29829
803 593-9954
- Dental Assisting Prgm
- Medical Assistant Prgm
- Radiography Prgm
- Surgical Technology Prgm

## Greenville

**Greenville Technical College**
Thomas E Barton, Jr, EdD, President
PO Box 5616
Greenville, SC 29606-5616
864 250-8175
- Clin Lab Technician/Med Lab Technician
- Dental Assisting Prgm
- Dental Hygiene Prgm
- Diagnostic Med Sonography Prgm
- Emergency Med Tech-Paramedic Prgm
- Health Information Tech Prgm
- Medical Assistant Prgm
- Occupational Therapy Asst Prgm
- Pharmacy Technician Prgm
- Physical Therapist Assistant Prgm
- Radiography Prgm
- Respiratory Therapist (Advanced) Prgm
- Surgical Technology Prgm

## Greenwood

**Piedmont Technical College**
Lex D Walters, PhD, President
PO Box 1467
Emerald Rd
Greenwood, SC 29647-1467
864 941-8324
- Medical Assistant Prgm
- Radiography Prgm
- Respiratory Therapist (Advanced) Prgm
- Surgical Technology Prgm

## Greenwood,

**Lander University**
Daniel W Ball, President
320 Stanley Ave
Greenwood,, SC 296499
864 388-8300
- Athletic Training Prgm

## Orangeburg

**Orangeburg Calhoun Technical College**
Anne Crook, PhD, President
3250 St Matthews Rd
Orangeburg, SC 29118
803 535-1200
- Clin Lab Technician/Med Lab Technician
- Medical Assistant Prgm
- Radiography Prgm

**South Carolina State University**
Andrew Hugine, PhD, President
300 College St NE
Orangeburg, SC 29117
803 536-7014
- Counseling Prgm
- Dietetics-Didactic Prgm
- Orientation and Mobility Specialist Prgm
- Rehabilitation Counseling Prgm
- Speech-Language Pathology Prgm

## Pendleton

**Tri-County Technical College**
Ronnie L Booth, PhD, President
PO Box 587
Pendleton, SC 29670
864 646-8361
- Clin Lab Technician/Med Lab Technician
- Dental Assisting Prgm
- Medical Assistant Prgm
- Respiratory Therapist (Advanced) Prgm
- Surgical Technology Prgm

## Rock Hill

**Winthrop University**
Anthony J Digiorgio, PhD, President
701 Oakland Ave
Rock Hill, SC 29733
803 323-2225
- Athletic Training Prgm
- Counseling Prgm
- Dietetic Internship Prgm
- Dietetics-Didactic Prgm

**York Technical College**
Dennis Merrell, EdD, President
452 S Anderson Rd
Rock Hill, SC 29730
803 327-8050
- Clin Lab Technician/Med Lab Technician
- Dental Assisting Prgm
- Dental Hygiene Prgm
- Radiography Prgm
- Surgical Technology Prgm

## Spartanburg

**Spartanburg Community College**
Dan L Terhune, EdD, President
PO Drawer 4386
Spartanburg, SC 29305-4386
864 592-4610
- Clin Lab Technician/Med Lab Technician
- Dental Assisting Prgm
- Medical Assistant Prgm
- Pharmacy Technician Prgm
- Radiation Therapy Prgm
- Radiography Prgm
- Respiratory Therapist (Advanced) Prgm
- Surgical Technology Prgm

## Sumter

**Central Carolina Technical College**
Kay Raffield, President
506 N Guignard Dr
Sumter, SC 29150
803 778-1961
- Medical Assistant Prgm
- Surgical Technology Prgm

## West Columbia

**Lexington Medical Center**
Mike Biedeger, President
2720 Sunset Blvd
West Columbia, SC 29169
803 791-2000
- Clin Lab Scientist/Med Technologist Prgm

# South Dakota

## Aberdeen

**Presentation College**
Lorraine Hale, PhD, President
1500 N Main St
Aberdeen, SD 57401-1299
605 229-8404
- Clin Lab Technician/Med Lab Technician
- Medical Assistant Prgm
- Radiography Prgm
- Surgical Technology Prgm

## Brookings

**South Dakota State University**
Peggy Gordon Miller, EdD, President
Office of the President - AD 222
Brookings, SD 57007
605 688-4151
- Athletic Training Prgm
- Counseling Prgm
- Dietetics-Didactic Prgm

## Madison

**Dakota State University**
Douglas Knowlton, PhD, President
314 Heston Hall
Madison, SD 57042-1799
605 256-5112
• Health Information Admin Prgm
• Health Information Tech Prgm
• Respiratory Therapist (Advanced) Prgm

## Mitchell

**Dakota Wesleyan University**
Robert Duffet, PhD, President
1200 W University
Box 911
Mitchell, SD 57301
605 995-2601
• Athletic Training Prgm

**Mitchell Technical Institute**
Chris Paustian, MA, Director
821 N Capital St
Mitchell, SD 57301
605 995-3022
• Clin Lab Technician/Med Lab Technician
• Medical Assistant Prgm
• Radiography Prgm

## Rapid City

**National American University**
Jerry L Gallentine, PhD, President
14 St Joseph St
Rapid City, SD 57709
605 394-4900
• Athletic Training Prgm

**Rapid City Regional Hospital**
Tim Sughrue, MBA, President/CEO
353 Fairmont Blvd
Rapid City, SD 57701
605 341-8100
• Clin Lab Scientist/Med Technologist Prgm
• Radiography Prgm

**Western Dakota Technical Institute**
Rich Gross, MEd PhD, Director
800 Mickelson Dr
Rapid City, SD 57703
605 394-4034
• Pharmacy Technician Prgm
• Phlebotomist Prgm
• Surgical Technology Prgm

## Sioux Falls

**Augustana College**
Ralph H Wagoner, PhD, President
2001 S Summit Ave
Sioux Falls, SD 57197
605 336-4111
• Athletic Training Prgm

**Avera McKennan Hospital**
Fred Slunecka, MHA, President/CEO
800 E 21st St
PO Box 5045
Sioux Falls, SD 57117-5045
605 322-7808
• Emergency Med Tech-Paramedic Prgm
• Radiography Prgm

**Colorado Technical University**
Vicki Strunk, President
3901 W 59th St
Sioux Falls, SD 57108
605 361-0200
• Medical Assistant Prgm

**National American Univ - Sioux Falls Campus**
Patricia Torpey, MA, Vice President
2801 S Kiwanis Ave, Ste 100
Sioux Falls, SD 57105
605 336-4602
• Medical Assistant Prgm

**Sioux Valley Hospital**
Kelby K Krabbenhoft, MBA, President
1100 S Euclid Ave
PO Box 5039
Sioux Falls, SD 57117-5039
605 333-6424
• Clin Lab Scientist/Med Technologist Prgm
• Radiography Prgm

**Southeast Technical Institute**
Jeff Holcomb, MBA, Director and CEO
2320 N Career Ave
Sioux Falls, SD 57107
605 367-7485
• Cardiovascular Technology Prgm
• Diagnostic Med Sonography Prgm
• Nuclear Medicine Technology Prgm
• Surgical Technology Prgm

## Vermillion

**University of South Dakota**
Jim Abbott, JD, President
414 E Clark St
Vermillion, SD 57069-2390
605 677-5641
• Audiologist Prgm
• Counseling Prgm
• Dental Hygiene Prgm
• Dietetic Internship Prgm
• Occupational Therapy Prgm
• Physical Therapy Prgm
• Physician Assistant Prgm
• Speech-Language Pathology Prgm

## Watertown

**Lake Area Technical Institute**
Deb Shehard, President
230 11th St NE
PO Box 730
Watertown, SD 57201-0730
605 882-5284
• Clin Lab Technician/Med Lab Technician
• Dental Assisting Prgm
• Medical Assistant Prgm
• Occupational Therapy Asst Prgm
• Physical Therapist Assistant Prgm

## Yankton

**Avera Sacred Heart Hospital**
Pamela J Rezac, PhD, CEO
501 Summit St
Yankton, SD 57078
605 668-8000
• Radiography Prgm

**Mt Marty College**
Sr Jacquelyn Ernster, PhD, President
Yankton, SD 57078
605 668-1514
• Dietetics-Coordinated Prgm
• Dietetics-Didactic Prgm

# Tennessee

## Blountville

**Northeast State Technical Comm College**
William W Locke, PhD, President
PO Box 246
2425 Hwy 75
Blountville, TN 37617-0246
423 323-3191
• Cardiovascular Technology Prgm
• Clin Lab Technician/Med Lab Technician
• Dental Assisting Prgm
• Emergency Med Tech-Paramedic Prgm
• Medical Assistant Prgm
• Surgical Technology Prgm

## Chattanooga

**Chattanooga State Technical Comm College**
James L Catanzaro, PhD, President
4501 Amnicola Hwy
Chattanooga, TN 37406-1097
423 697-4455
• Dental Assisting Prgm
• Dental Hygiene Prgm
• Diagnostic Med Sonography Prgm
• Emergency Med Tech-Paramedic Prgm
• Health Information Tech Prgm
• Medical Assistant Prgm
• Nuclear Medicine Technology Prgm
• Pharmacy Technician Prgm
• Physical Therapist Assistant Prgm
• Radiation Therapy Prgm
• Radiography Prgm
• Respiratory Therapist (Advanced) Prgm
• Surgical Technology Prgm

**Miller-Motte Technical College - Chattanooga**
June Kearns, AAS, Campus Director
6020 Shallowford Rd
Chattanooga, TN 37421
423 510-9675
• Medical Assistant Prgm
• Surgical Technology Prgm

**University of Tennessee - Chattanooga**
Roger G Brown, PhD, Chancellor
615 McCallie Ave
101 Founders Hall, Dept 5605
Chattanooga, TN 37403-2598
423 425-5559
• Athletic Training Prgm
• Counseling Prgm
• Dietetics-Didactic Prgm
• Physical Therapy Prgm

## Clarksville

**Austin Peay State University**
Sherry Hoppe, EdD, President
Office of the President
Clarksville, TN 37044
615 648-7566
• Clin Lab Scientist/Med Technologist Prgm

**Miller-Motte Technical College - Clarksville**
Raymond Green, BS, President
1820 Business Park Dr
Clarksville, TN 37040
615 553-0071
• Medical Assistant Prgm
• Surgical Technology Prgm

## Cleveland

**Cleveland State Community College**
Carl Hite, PhD, President
3535 Adkisson Dr
PO Box 3570
Cleveland, TN 37320-3570
423 472-7141
• Medical Assistant Prgm

**Lee University**
Carolyn Dirksen, PhD, VP Academic Affairs
1120 N Ocoee St
Cleveland, TN 37320-3450
423 614-8118
• Athletic Training Prgm

## Columbia

**Columbia State Community College**
O Rebecca Hawkins, PhD, President
Hwy 412 PO Box 1315
Columbia, TN 38401
615 540-2722
• Emergency Med Tech-Paramedic Prgm
• Radiography Prgm
• Respiratory Therapist (Advanced) Prgm

## Cookeville

**MedVance Institute**
Deborah Schwarzberg, MA MT(ASCP), President
1025 Highway 111
Cookeville, TN 38501
615 586-3660
• Clin Lab Technician/Med Lab Technician
• Radiography Prgm
• Surgical Technology Prgm

**Tennessee Technological University**
Robert R Bell, President
1000 N Dixie Ave
Cookeville, TN 38505-0001
931 372-3101
• Dietetics-Didactic Prgm

## Crossville

**Tennessee Technology Center - Crossville**
James Purcell, Director
910 Miller Ave
PO Box 2959
Crossville, TN 38557
931 484-7502
• Surgical Technology Prgm

## Dayton

**Bryan College**
Stephen D Livesay, PhD, President
721 Bryan Dr
Box 7792
Dayton, TN 37321
423 775-7201
• Athletic Training Prgm

## Dickson

**Tennessee Technology Center - Dickson**
Bobby F Sullivan, MS, Director
740 Hwy 46 S
Dickson, TN 37055
615 441-6220
• Dental Assisting Prgm
• Surgical Technology Prgm

## Dyersburg

**Dyersburg State Community College**
Karen A Bowyer, PhD, President
1510 Lake Rd
Dyersburg, TN 38024
731 286-3301
• Health Information Tech Prgm
• Surgical Technology Prgm

## Gallatin

**Volunteer State Community College**
Warren Nichols, EdD, President
1480 Nashville
Gallatin, TN 37066-3188
615 230-3500
• Clin Lab Technician/Med Lab Technician
• Dental Assisting Prgm
• Diagnostic Med Sonography Prgm
• Emergency Med Tech-Paramedic Prgm
• Health Information Tech Prgm
• Ophthalmic Med Technician Prgm
• Physical Therapist Assistant Prgm
• Radiography Prgm
• Respiratory Therapist (Advanced) Prgm
• Respiratory Therapist (Entry-Level) Prgm

## Greenville

**Tusculum College**
Dolphus E Henry, PhD, President
106 McCormick
60 Old Shiloh Rd
Greenville, TN 37743
800 729-0256
• Athletic Training Prgm

## Harriman

**Roane State Community College**
Gary Goff, EdD, President
276 Patton Ln
Harriman, TN 37748-5011
865 882-4501
• Dental Hygiene Prgm
• Emergency Med Tech-Paramedic Prgm
• Health Information Tech Prgm
• Massage Therapy Prgm
• Occupational Therapy Asst Prgm
• Ophthalmic Dispensing Optician Prgm
• Physical Therapist Assistant Prgm
• Radiography Prgm
• Respiratory Therapist (Advanced) Prgm

## Harrogate

**Lincoln Memorial University**
Nancy Moody, DSN, President
PO Box 2028
Cumberland Gap Parkway
Harrogate, TN 37752
423 869-6391
• Athletic Training Prgm
• Clin Lab Scientist/Med Technologist Prgm

## Hohenwald

**Tennessee Technology Center - Hohenwald**
Rick C Brewer, MEd ED SPEC, Director
813 W Main St
Hohenwald, TN 38462
931 796-5351
• Surgical Technology Prgm

## Jackson

**Jackson State Community College**
Bruce Blanding, PhD, President
2046 N Parkway St
Jackson, TN 38301-3797
731 424-3520
• Clin Lab Technician/Med Lab Technician
• Emergency Med Tech-Paramedic Prgm
• Physical Therapist Assistant Prgm
• Radiography Prgm
• Respiratory Therapist (Advanced) Prgm

**Tennessee Technology Center - Jackson**
Don Williams, PhD, Director
2468 Technology Center Dr
Jackson, TN 38301
731 424-0691
• Pharmacy Technician Prgm
• Surgical Technology Prgm

**Union University**
David S Dockery, PhD, President
1050 Union University Dr
Jackson, TN 38305
731 661-5180
• Athletic Training Prgm

## Jefferson City

**Carson - Newman College**
J Cordell Maddox, President
1646 Russell Ave
Jefferson City, TN 37760
615 471-4000
• Athletic Training Prgm
• Dietetics-Didactic Prgm

## Joelton

**Meridian Institute of Surgical Assisting**
Dennis A Stover, CST SA-C, President
PO Box 758
Joelton, TN 37080
615 298-1416
• Surgical Assistant Prgm

INSTITUTIONS

## Johnson City

**East Tennessee State University**
Paul E Stanton, Jr, MD, President
College of Public and Allied Health
Box 70734
Johnson City, TN 37614
423 439-4211
- Audiologist Prgm
- Counseling Prgm
- Dental Hygiene Prgm
- Dietetic Internship Prgm
- Dietetics-Didactic Prgm
- Physical Therapy Prgm
- Radiography Prgm
- Respiratory Therapist (Advanced) Prgm
- Speech-Language Pathology Prgm

## Knoxville

**Ft Sanders Regional Medical Center**
Jim Burkhart, MHA FACHE, Sr Vice President
9352 Park W Blvd
PO Box 22993
Knoxville, TN 37933-0993
865 541-1100
- Surgical Technology Prgm

**South College**
Stephen A South, BS, President
3904 Lonas Dr
Knoxville, TN 37909
865 251-1800
- Medical Assistant Prgm
- Physical Therapist Assistant Prgm
- Radiography Prgm

**Tennessee Technology Center - Knoxville**
David Esa, MSEd, Director
1100 Liberty St
Knoxville, TN 37919
615 546-5567
- Dental Assisting Prgm
- Medical Assistant Prgm
- Surgical Technology Prgm

**Univ of Tennessee Medical Ctr at Knoxville**
Joseph Landsman, President and CEO
1924 Alcoa Highway
Knoxville, TN 37920-6999
865 544-9430
- Clin Lab Scientist/Med Technologist Prgm
- Nuclear Medicine Technology Prgm
- Radiography Prgm

**University of Tennessee - Knoxville**
Loren Crabtree, PhD, Chancellor
810 Andy Holt Tower
Knoxville, TN 37996
865 974-3265
- Audiologist Prgm
- Counseling Prgm
- Dietetic Internship Prgm
- Dietetics-Didactic Prgm
- Medical Librarian Prgm
- Rehabilitation Counseling Prgm
- Speech-Language Pathology Prgm
- Therapeutic Recreation Specialist Prgm

## Lebanon

**Cumberland University**
Harvill C Eaton, PhD, President
One Cumberland Square
Lebanon, TN 37087-3408
615 444-2562
- Athletic Training Prgm

## Martin

**University of Tennessee - Martin**
Nick Dunagan, EdD, Chancellor
325 Adminstration Bldg
Martin, TN 38238
731 881-7500
- Athletic Training Prgm
- Dietetic Internship Prgm
- Dietetics-Didactic Prgm

## McKenzie

**Bethel College**
Rev Robert D Prosser, President
325 Cherry St
McKenzie, TN 38201-1705
901 352-4000
- Physician Assistant Prgm

## McMinnville

**Tennessee Technology Center - McMinnville**
Andy Forrester, MS, Director
241 Vo-Tech Dr
McMinnville, TN 37110
931 473-5587
- Medical Assistant Prgm
- Surgical Technology Prgm

## Memphis

**Baptist College of Health Sciences**
Betty Sue McGarvey, DSN, President
1003 Monroe Ave
Memphis, TN 38104-3199
901 572-2585
- Diagnostic Med Sonography Prgm
- Nuclear Medicine Technology Prgm
- Radiation Therapy Prgm
- Radiography Prgm
- Respiratory Therapist (Advanced) Prgm

**Concorde Career College - Memphis**
Tommy Stewart, AAS, Director
5100 Poplar Ave, Ste 132
Memphis, TN 38137
901 761-9494
- Dental Assisting Prgm
- Medical Assistant Prgm
- Pharmacy Technician Prgm
- Respiratory Therapist (Advanced) Prgm
- Surgical Technology Prgm

**Methodist Le Bonheur Healthcare**
Peggy Troy, RN MSN, CEO
Methodist Professional Bldg
1211 Union Ave - Ste 700
Memphis, TN 38104
901 516-0546
- Diagnostic Med Sonography Prgm

**Methodist University Hospital**
Cecelia Sawyer, CEO
1265 Union Ave
Memphis, TN 38104
901 516-0543
- Nuclear Medicine Technology Prgm
- Radiography Prgm

**Southwest Tennessee Community College**
Nathan L Essex, President
5983 Macon Cove
Memphis, TN 38134-7693
901 333-7822
- Clin Lab Technician/Med Lab Technician
- Dietetic Technician-AD Prgm
- Emergency Med Tech-Paramedic Prgm
- Phlebotomist Prgm
- Physical Therapist Assistant Prgm
- Radiography Prgm

**Tennessee Technology Center - Memphis**
Russell Shelton, Director
550 Alabama Ave
Memphis, TN 38105-3799
901 543-6156
- Dental Assisting Prgm
- Medical Assistant Prgm
- Pharmacy Technician Prgm
- Surgical Technology Prgm

**University of Memphis**
Shirley C Raines, President
Memphis, TN 38152
901 678-2000
- Audiologist Prgm
- Counseling Prgm
- Dietetic Internship Prgm
- Dietetics-Didactic Prgm
- Rehabilitation Counseling Prgm
- Speech-Language Pathology Prgm

**University of Tennessee Health Science Ctr**
William Owen, MD, Chancellor
219 Hyman Admin Bldg
Memphis, TN 38163
901 448-4796
- Clin Lab Scientist/Med Technologist Prgm
- Cytotechnology Prgm
- Dental Hygiene Prgm
- Health Information Admin Prgm
- Occupational Therapy Prgm
- Physical Therapy Prgm

## Milligan College

**Milligan College**
Donald R Jeanes, DD, President
PO Box 1
Milligan College, TN 37682
423 461-8710
- Occupational Therapy Prgm

## Morristown

**Walters State Community College**
Wade B McCamey, EdD, President
500 S Davy Crockett Pkwy
Morristown, TN 37813-6899
423 585-2600
- Emergency Med Tech-Paramedic Prgm
- Health Information Tech Prgm
- Pharmacy Technician Prgm
- Physical Therapist Assistant Prgm
- Respiratory Therapist (Entry-Level) Prgm

## Murfreesboro

### Middle Tennessee State University
Sydney McPhee, PhD, President
Cope Administration Bldg 110
Murfreesboro, TN 37132
615 898-2622
- Athletic Training Prgm
- Counseling Prgm
- Dietetics-Didactic Prgm

### National HealthCare LP
PO Box 1398
Murfreesboro, TN 37133-1398
615 890-2020
- Dietetic Internship Prgm

### Tennessee Technology Center - Murfreesboro
Monty Thomas, BS MS, Director
1303 Old Fort Parkway
Murfreesboro, TN 37130
615 898-8010
- Dental Assisting Prgm
- Pharmacy Technician Prgm
- Surgical Technology Prgm

## Nashville

### Belmont University
Robert C Fisher, PhD, President
1900 Belmont Blvd
Nashville, TN 37212
615 460-6793
- Occupational Therapy Prgm
- Physical Therapy Prgm

### Lipscomb University
L Randolph Lowry III, BA MPA JD, President
3901 Granny White Pike
Nashville, TN 37204
615 279-6194
- Athletic Training Prgm
- Dietetic Internship Prgm
- Dietetics-Didactic Prgm

### Nashville General Hospital
Reginald Coopwood, MD, CEO
1818 Albion Street
Nashville, TN 37208
615 341-4490
- Radiography Prgm

### Nashville State Technical Community College
George H Van Allen, EdD, President
120 White Bridge Rd
PO Box 90285
Nashville, TN 37209-4515
615 353-3236
- Occupational Therapy Asst Prgm
- Surgical Assistant Prgm
- Surgical Technology Prgm

### Tennessee State University
Melvin Johnson, DBA, President
3500 John A Merritt Blvd
Nashville, TN 37209-1561
615 963-7406
- Clin Lab Scientist/Med Technologist Prgm
- Dental Hygiene Prgm
- Dietetics-Didactic Prgm
- Health Information Admin Prgm
- Occupational Therapy Prgm
- Physical Therapy Prgm
- Respiratory Therapist (Advanced) Prgm
- Speech-Language Pathology Prgm

### Tennessee Technology Center - Nashville
Charles F Malin, Director
100 White Bridge Rd
Nashville, TN 37209
- Pharmacy Technician Prgm

### Trevecca Nazarene University
Dan Boone, DMin, President
333 Murfreesboro Rd
Nashville, TN 37210
615 248-1251
- Physician Assistant Prgm

### Vanderbilt University
Nashville, TN 37240-0001
615 322-7311
- Counseling Prgm

### Vanderbilt University Medical Center
Harry R Jacobson, MD, Vice Chancellor, Hlth Aff
1161 21st Ave S
D-3300 MCN
Nashville, TN 37232
615 322-2151
- Audiologist Prgm
- Clin Lab Scientist/Med Technologist Prgm
- Diagnostic Med Sonography Prgm
- Dietetic Internship Prgm
- Nuclear Medicine Technology Prgm
- Perfusion Prgm
- Radiation Therapy Prgm
- Speech-Language Pathology Prgm

## Paris

### Tennessee Technology Center - Paris
Bradley White, PHd, Interim Director
312 S Wilson St
Paris, TN 38242
731 644-7365
- Surgical Technology Prgm

# Texas

## Abilene

### Abilene Christian University
Royce L Money, President
Box 8363
Abilene, TX 79699
915 674-2000
- Dietetics-Didactic Prgm
- Speech-Language Pathology Prgm

### Hardin-Simmons University
Craig Turner, PhD, President
2200 Hickory
HSU Box 16000
Abilene, TX 79698
325 670-1227
- Athletic Training Prgm
- Physical Therapy Prgm

### Hendrick Medical Center
Tim Lancaster, FACHE, President
1900 Pine
Abilene, TX 79601-2316
325 670-2364
- Radiography Prgm

### SWT/HMC Respiratory Care School Consortium
Jeff Lawrence, Program Director
1900 N Pine St
Abilene, TX 79601
915 670-2368
- Respiratory Therapist (Entry-Level) Prgm

## Alvin

### Alvin Community College
A Rodney Allbright, JD, President
3110 Mustang Rd
Alvin, TX 77511
281 756-3598
- Diagnostic Med Sonography Prgm
- Respiratory Therapist (Advanced) Prgm

## Amarillo

### Amarillo College
Steven Jones, EdD, President
PO Box 447
Amarillo, TX 79178-0001
806 371-5123
- Clin Lab Technician/Med Lab Technician
- Dental Hygiene Prgm
- Nuclear Medicine Technology Prgm
- Occupational Therapy Asst Prgm
- Physical Therapist Assistant Prgm
- Radiation Therapy Prgm
- Radiography Prgm
- Respiratory Therapist (Advanced) Prgm
- Surgical Technology Prgm

## Arlington

### Concorde Career Institute - Arlington
Rebecca Zielinski, RN, Campus President
601 Ryan Plaza Dr, Ste 200
Arlington, TX 76011
817 267-1594
- Surgical Technology Prgm

### University of Texas at Arlington
James D Spaniolo, President
321 Davis Hall - Box 19125
Arlington, TX 76019-0125
817 272-2101
- Athletic Training Prgm

## Athens

### Trinity Valley Community College
Ronald C Baugh, MS, President
100 Cardinal Dr
Athens, TX 75751
903 675-6211
- Surgical Technology Prgm

INSTITUTIONS

## Austin

### Austin Community College
Stephen Kinslow, PhD, President
5930 Middle Fiskville Rd
Austin, TX 78752-4390
512 223-7598
- Clin Lab Technician/Med Lab Technician
- Dental Hygiene Prgm
- Diagnostic Med Sonography Prgm
- Emergency Med Tech-Paramedic Prgm
- Occupational Therapy Asst Prgm
- Pharmacy Technician Prgm
- Phlebotomist Prgm
- Physical Therapist Assistant Prgm
- Radiography Prgm
- Surgical Technology Prgm

### Austin State Hospital
Carl Schock, Superintendent
4110 Guadalupe St
Austin, TX 78751
512 419-2041
- Clin Lab Scientist/Med Technologist Prgm

### Texas Department of Health
Austin, TX 78756
- Dietetic Internship Prgm

### University of Texas at Austin
William Powers, PhD, President
1 University Station
Austin, TX 78712
512 471-3434
- Athletic Training Prgm
- Audiologist Prgm
- Dietetics-Coordinated Prgm
- Dietetics-Didactic Prgm
- Medical Librarian Prgm
- Rehabilitation Counseling Prgm
- Speech-Language Pathology Prgm

### Virginia College at Austin
David B Champlin, BA JD, Campus President
6301 E Hwy 290
Austin, TX 78723
512 279-2802
- Surgical Technology Prgm

## Baytown

### Lee College
Martha Ellis, PhD, President
PO Box 818
Baytown, TX 77520
281 425-6300
- Emergency Med Tech-Paramedic Prgm
- Health Information Tech Prgm

## Beaumont

### Christus Hospital - St Elizabeth
Joel Fagerstrom, MBA, Administrator
2830 Calder St
PO Box 5405
Beaumont, TX 77726-5405
409 892-7171
- Clin Lab Scientist/Med Technologist Prgm
- Phlebotomist Prgm

### Lamar Institute of Technology
Paul Szuch, EdD, President
PO Box 10043
Beaumont, TX 77710
409 880-8405
- Audiologist Prgm
- Dental Hygiene Prgm
- Diagnostic Med Sonography Prgm
- Dietetic Internship Prgm
- Dietetics-Didactic Prgm
- Health Information Tech Prgm
- Radiography Prgm
- Respiratory Therapist (Advanced) Prgm
- Speech-Language Pathology Prgm

### Memorial Hermann Baptist Hospital Beaumont
David Parmer, MS, President
Hospital Admin and Accounting
PO Drawer 1591
Beaumont, TX 77704
409 212-5012
- Radiography Prgm

## Beeville

### Coastal Bend College
John Borckman, PhD, President
3800 Charco Rd
Beeville, TX 78102
361 354-2200
- Dental Hygiene Prgm

## Belton

### University of Mary Hardin-Baylor
Jerry Bawcom, PhD, President
900 Colllege St
Belton, TX 76502
254 295-8642
- Athletic Training Prgm

## Big Spring

### Howard College
Cheryl T Sparks, EdD, President
1001 Birdwell Ln
Big Spring, TX 79720
432 264-5000
- Dental Hygiene Prgm
- Health Information Tech Prgm
- Surgical Technology Prgm

## Brenham

### Blinn College
Don E Voelter, PhD, President
902 College Ave
Brenham, TX 77833
979 830-4112
- Dental Hygiene Prgm
- Physical Therapist Assistant Prgm
- Radiography Prgm

## Brownsville

### Univ Tx at Brownsville/Tx Southmost Coll
Juliet Garcia, PhD, President
83 Ft Brown
Brownsville, TX 78520
956 544-8200
- Clin Lab Technician/Med Lab Technician
- Diagnostic Med Sonography Prgm
- Radiography Prgm
- Respiratory Therapist (Advanced) Prgm
- Respiratory Therapist (Entry-Level) Prgm

## Bryan

### St Joseph Regional Health Center
John J Buckley, Jr, FACHE
2801 Franciscan Dr
Bryan, TX 77802-2544
979 776-3777
- Surgical Technology Prgm

## Canyon

### West Texas A&M University
J Patrick O'Brien, President/CEO
Canyon, TX 79016
806 656-2000
- Athletic Training Prgm
- Music Therapy Prgm
- Speech-Language Pathology Prgm

## Carthage

### Panola College
Gregory S Powell, EdD, President
1109 W Panola
Carthage, TX 75633-2397
903 693-2022
- Health Information Tech Prgm
- Occupational Therapy Asst Prgm

## Cisco

### Cisco Junior College
Colleen Smith, PhD, President
Rte 3 - Box 3
Cisco, TX 76437
254 442-2567
- Medical Assistant Prgm
- Surgical Technology Prgm

## College Station

### Baylor College of Dentistry, Texas A&M HSC
Nancy Dickey, MD, President
TAMU 1364
College Station, TX 77843-1364
979 458-7200
- Dental Hygiene Prgm

### Texas A&M University
Robert Gates, President
College Station, TX 77843
409 845-3211
- Athletic Training Prgm
- Clin Lab Scientist/Med Technologist Prgm
- Counseling Prgm (2)
- Dietetic Internship Prgm (2)
- Dietetics-Didactic Prgm (2)
- Speech-Language Pathology Prgm

## Conroe

**N Harris Mont Comm Coll Dist/ Montgomery Coll**
Thomas E Butler, EdD, President
3200 College Park Dr
Conroe, TX 77384
936 273-7222
• Physical Therapist Assistant Prgm
• Radiography Prgm

## Corpus Christi

**Del Mar College**
Carlos A Garcia, PhD, President
101 Baldwin Blvd
Corpus Christi, TX 78404
361 698-1203
• Clin Lab Technician/Med Lab Technician
• Dental Assisting Prgm
• Dental Hygiene Prgm
• Diagnostic Med Sonography Prgm
• Health Information Tech Prgm
• Occupational Therapy Asst Prgm
• Physical Therapist Assistant Prgm
• Radiography Prgm
• Respiratory Therapist (Advanced) Prgm
• Surgical Technology Prgm

## Corsicana

**Navarro College**
Richard Sanchez, EdD, President
3200 W 7th Ave
Corsicana, TX 75110-4818
903 875-7308
• Clin Lab Technician/Med Lab Technician
• Occupational Therapy Asst Prgm

## Cypress

**Cy-Fair College**
Diane K Troyer, PhD, President
9191 Barker-Cypress Rd
Cypress, TX 77433
281 290-3940
• Diagnostic Med Sonography Prgm
• Radiography Prgm

## Dallas

**ATI Health Education Centers**
Joe Mehlmann, President
2777 Stemmons Freeway
Dallas, TX 75207
214 630-5651
• Respiratory Therapist (Advanced) Prgm

**ATI-Career Training**
Gerald Parr, BS, Executive Director
10003 Technology Blvd W
Dallas, TX 75220
214 902-8191
• Respiratory Therapist (Advanced) Prgm

**Baylor University Med Center**
Merrick H Reese, MD, President/CEO
Sammons Cancer Ctr
3535 Worth St
Dallas, TX 75204
214 828-0377
• Dietetic Internship Prgm
• Nuclear Medicine Technology Prgm
• Radiography Prgm

**Dallas County Community College**
Sharon Blackman, EdD, Dean of Technical Educ
12800 Abrams Rd
Dallas, TX 75243-2199
972 238-6954
• Health Information Tech Prgm

**El Centro College**
Michael Jackson, MS, Acting President
801 Main St
Dallas, TX 75202-3604
214 860-2011
• Cardiovascular Technology Prgm
• Clin Lab Technician/Med Lab Technician
• Diagnostic Med Sonography Prgm
• Medical Assistant Prgm
• Radiography Prgm
• Respiratory Therapist (Advanced) Prgm
• Surgical Technology Prgm

**Presbyterian Hospital of Dallas**
8200 Walnut Hill Ln
Dallas, TX 75231-4402
214 345-7558
• Dietetic Internship Prgm

**Richland College**
Stephen Mittelstet, PhD, President
12800 Abrams Rd
Dallas, TX 75243
214 238-6364
• Medical Assistant Prgm
• Pharmacy Technician Prgm

**Sanford-Brown Institute - Dallas**
Gary C Jack, BS MS, Executive Director
2998 N Stemmons Frwy
Dallas, TX 75247
214 638-6400
• Diagnostic Med Sonography Prgm

**Southern Methodist University**
R Gerald Turner, President
6425 Boaz St
PO Box 296
Dallas, TX 75275
214 768-2000
• Music Therapy Prgm

**Univ of Texas Southwestern Med Ctr at Dallas**
Kern Wildenthal, MD PhD, President
5323 Harry Hines Blvd, B12.100
Dallas, TX 75390-9082
214 648-2508
• Clin Lab Scientist/Med Technologist Prgm
• Dietetics-Coordinated Prgm
• Emergency Med Tech-Paramedic Prgm
• Medical Illustrator Prgm
• Orthotist/Prosthetist Prgm
• Physical Therapy Prgm
• Physician Assistant Prgm
• Rehabilitation Counseling Prgm
• Specialist in BB Tech Prgm

## Denison

**Grayson County College**
Alan Scheibmeir, PhD, President
6101 Grayson Dr
Denison, TX 75020
903 465-6030
• Clin Lab Technician/Med Lab Technician
• Dental Assisting Prgm

## Denton

**Texas Woman's University**
Ann Stuart, PhD, President and Chancellor
Box 425587
TWU Station (ACT-15)
Denton, TX 76204-3587
940 898-3201
• Counseling Prgm
• Dental Hygiene Prgm
• Dietetic Internship Prgm (2)
• Dietetics-Didactic Prgm
• Medical Librarian Prgm
• Music Therapy Prgm
• Occupational Therapy Prgm (3)
• Physical Therapy Prgm
• Speech-Language Pathology Prgm

**University of North Texas**
Lee Jackson, Chancellor
Denton, TX 76203-5008
940 565-2000
• Audiologist Prgm
• Counseling Prgm
• Medical Librarian Prgm
• Rehabilitation Counseling Prgm
• Speech-Language Pathology Prgm

## Edinburg

**Univ of Texas - Pan American**
Blandina Cardenas, PhD, President
1201 W University Dr
Edinburg, TX 78541-2999
956 381-2100
• Clin Lab Scientist/Med Technologist Prgm
• Dietetics-Coordinated Prgm
• Occupational Therapy Prgm
• Physician Assistant Prgm
• Rehabilitation Counseling Prgm
• Speech-Language Pathology Prgm

## El Paso

**Career Centers of Texas**
Sally Crickard, Executive Director
8360 Burnham Rd, Ste 100
El Paso, TX 79907
915 595-1935
• Medical Assistant Prgm
• Surgical Technology Prgm

**Computer Career Center**
Lee Chayes, MA, President
6101 Montana
El Paso, TX 79925
915 779-8031
• Medical Assistant Prgm

INSTITUTIONS

**El Paso Community College**
Richard M Rhodes, PhD, President
PO Box 20500
El Paso, TX 79998-0500
915 831-6511
• Clin Lab Technician/Med Lab Technician
• Dental Assisting Prgm
• Dental Hygiene Prgm
• Diagnostic Med Sonography Prgm
• Dietetic Technician-AD Prgm
• Health Information Tech Prgm
• Medical Assistant Prgm
• Ophthalmic Dispensing Optician Prgm
• Pharmacy Technician Prgm
• Physical Therapist Assistant Prgm
• Radiography Prgm
• Surgical Technology Prgm

**University of Texas at El Paso**
Diana Natalicio, PhD, President
500 W University Ave
El Paso, TX 79968
915 747-5555
• Clin Lab Scientist/Med Technologist Prgm
• Occupational Therapy Prgm
• Physical Therapy Prgm
• Speech-Language Pathology Prgm

**Western Technical College**
Randy L Kuykendall, Board Chair, CEO
9624 Plaza Circle
El Paso, TX 79927
915 532-3737
• Medical Assistant Prgm

## Friendswood

**Texas School of Business Friendswood**
Bobby Wilmore, Campus Executive Officer
3208 FM 528
Friendswood, TX 77546
281 648-0880
• Medical Assistant Prgm

## Ft Sam Houston

**Brooke Army Medical Center**
William Fox, Jr, BG MC MD, Commanding
    General
Ft Sam Houston, TX 78234-6200
210 916-0499
• Cytotechnology Prgm

**US Army**
MCCS-HMT, Physical Therapy Branch
3151 Scott Rd
Ft Sam Houston, TX 78234-6138
• Interservice Physician Assistant Prgm
• Physical Therapy Prgm

**US Army Medical Dept Center & School**
Russell Czerw, DC, US Army, Major General
2250 Stanley Rd, Ste 301
Ft Sam Houston, TX 78234-6100
210 221-6325
• Cardiovascular Technology Prgm
• Clin Lab Technician/Med Lab Technician
• Occupational Therapy Asst Prgm
• Pharmacy Technician Prgm
• Radiography Prgm
• Respiratory Therapist (Entry-Level) Prgm

## Ft Worth

**Career Centers of Texas - Ft Worth**
Nancy Tedros, MBA, Executive Director
2001 Beach St, Ste 201
Ft Worth, TX 76103
• Radiography Prgm

**Tarrant County Jr College - South**
Jerry Mullen, Dean, Instruction
5301 Campus Dr
Ft Worth, TX 76119-5998
817 534-4861
• Dietetic Technician-AD Prgm

**Texas Christian University**
Victor Boschini, EdD, Chancellor
TCU 297080
Ft Worth, TX 76129
817 257-7783
• Athletic Training Prgm
• Dietetics-Coordinated Prgm
• Dietetics-Didactic Prgm
• Speech-Language Pathology Prgm

**Univ of North Texas Hlth Sci Ctr at Ft Worth**
Ronald Blanck, DO, President
3500 Camp Bowie Blvd
Ft Worth, TX 76107
817 735-2555
• Physician Assistant Prgm

## Gainesville

**North Central Texas College**
Eddie Hadlock, EdD, President
1525 W California St
Gainesville, TX 76240
940 668-7731
• Surgical Technology Prgm

## Galveston

**Galveston College**
Elva Concha LeBlanc, PhD, President
4015 Ave Q
Galveston, TX 77550-2782
409 944-1200
• Emergency Med Tech-Paramedic Prgm
• Nuclear Medicine Technology Prgm
• Radiation Therapy Prgm
• Radiography Prgm
• Surgical Technology Prgm

**University of Texas Medical Branch/SAHS**
Vicki Freeman, MD, Chair, Department of CLS
301 University Blvd
Galveston, TX 77555-1140
409 772-3055
• Clin Lab Scientist/Med Technologist Prgm
• Occupational Therapy Prgm
• Pharmacy Technician Prgm
• Physical Therapy Prgm
• Physician Assistant Prgm
• Respiratory Therapist (Advanced) Prgm
• Specialist in BB Tech Prgm

## Georgetown

**Southwestern University**
Jake B Schrum, BA MDiv, President
1001 East University Ave.
Georgetown, TX 78627-0770
512 863-1454
• Athletic Training Prgm

## Harlingen

**Texas State Technical College - Harlingen**
J Gilbert Leal, PhD, President
1902 N Loop 499
Harlingen, TX 78550-3697
956 364-4020
• Dental Assisting Prgm
• Dental Hygiene Prgm
• Health Information Tech Prgm
• Medical Assistant Prgm
• Surgical Technology Prgm

## Houston

**Academy of Health Care Professions**
A John Emerald, BA, CEO
1900 N Loop W, Ste 100
Houston, TX 77092
713 425-3100
• Surgical Technology Prgm

**Baylor College of Medicine**
Peter Traber, MD, President/CEO
BCMC 143A, One Baylor Plaza
Houston, TX 77030
713 798-4433
• Physician Assistant Prgm

**Bradford School**
Robert Puig, MA, Dir/Chief Acad Officer
4669 SW Freeway, Ste 300
Houston, TX 77027
713 629-1500
• Medical Assistant Prgm

**Gulf Coast School of Blood Bank Technology**
Bill T Teague, BS MT(ASCP)SBB, President/CEO
1400 La Concha Ln
Houston, TX 77054-1802
713 790-1200
• Specialist in BB Tech Prgm

**Harris County Hosp Dist/Ben Taub Gen Hosp**
David Lopez, MS, CEO/President
2525 Holly Hall
Houston, TX 77266
713 566-6400
• Radiography Prgm

## Houston Community College
Norm Nielsen, PhD, Interim Chancellor
3100 Main, Ste 12D
PO Box 667517
Houston, TX 77266
713 718-5059
- Clin Lab Technician/Med Lab Technician
- Dental Assisting Prgm
- Emergency Med Tech-Paramedic Prgm
- Health Information Tech Prgm
- Histotechnician Prgm
- Medical Assistant Prgm
- Nuclear Medicine Technology Prgm
- Occupational Therapy Asst Prgm
- Pharmacy Technician Prgm
- Physical Therapist Assistant Prgm
- Radiography Prgm
- Respiratory Therapist (Advanced) Prgm
- Surgical Technology Prgm

## Houston Veterans Affairs Medical Center
Robert F Stott, MPH, Hospital Director
2002 Holcombe Blvd 00/580
Houston, TX 77030
713 794-7100
- Dietetic Internship Prgm

## MedVance Institute
Deborah Schwarzberg, PhD, President
6220 Westpark
Houston, TX 77057
- Radiography Prgm
- Surgical Technology Prgm

## Memorial Hermann Healthcare System
Dan Wolterman, CEO/President
7737 SW Freeway PB II
Houston, TX 77074
713 776-5484
- Radiography Prgm
- Surgical Technology Prgm

## Methodist Hospital
Ron Girotto, CEO/President
6565 Fannin, D 200
Houston, TX 77030
713 790-2481
- Clin Lab Scientist/Med Technologist Prgm

## North Harris College
David Sam, PhD, President
2700 W W Thorne Dr
Houston, TX 77073-3499
281 618-5444
- Emergency Med Tech-Paramedic Prgm
- Health Information Tech Prgm
- Pharmacy Technician Prgm

## San Jacinto College North
Charles Grant, PhD, President
5800 Uvalde Rd
Houston, TX 77049-4599
281 458-4050
- Emergency Med Tech-Paramedic Prgm
- Health Information Tech Prgm
- Medical Assistant Prgm
- Pharmacy Technician Prgm

## San Jacinto College South
Linda Watkins, PhD, President
13735 Beamer Rd
Houston, TX 77089
281 922-3400
- Pharmacy Technician Prgm
- Physical Therapist Assistant Prgm

## Sanford-Brown Institute - Houston
James Garrett, BS, Executive Director
10500 Forum Pl, Ste 200
Houston, TX 77036
713 779-1110
- Diagnostic Med Sonography Prgm
- Surgical Technology Prgm

## Texas Heart Institute
Denton A Cooley, MD, President
PO Box 20345, MC 1-224
Houston, TX 77225
832 355-4026
- Perfusion Prgm

## Texas School of Business - North Campus
Ray Green, JD, Executive Director
711 E Airtex
Houston, TX 77026
281 443-8900
- Medical Assistant Prgm

## Texas School of Business Southwest
Ken Kniesel, BA MS, Interim Executive Dir
6363 Richmond Ave, Ste 300
Houston, TX 77057
713 974-2593
- Medical Assistant Prgm

## Texas Southern University
Priscilla Slade, PhD, President
3100 Cleburne St
Houston, TX 77004
713 313-7044
- Clin Lab Scientist/Med Technologist Prgm
- Dietetics-Didactic Prgm
- Health Information Admin Prgm
- Respiratory Therapist (Advanced) Prgm

## Univ of Texas Hlth Sci Ctr at Houston
James Willerson, MD, President
PO Box 20036
Houston, TX 77225-0708
713 500-3000
- Dental Hygiene Prgm
- Dietetic Internship Prgm

## Univ of Texas M D Anderson Cancer Ctr
John Mendelsohn, MD, President
1515 Holcombe Blvd, Box 213
Houston, TX 77030
713 792-6000
- Clin Lab Scientist/Med Technologist Prgm
- Cytogenetic Technology Prgm
- Cytotechnology Prgm
- Diagnostic Molecular Scientist Prgm
- Histotechnician Prgm
- Radiation Therapy Prgm

## Univ of Texas Medical School at Houston
PO Box 20708
Houston, TX 77225
- Genetic Counseling Prgm

## University of Houston
Thomas M Stauffer, PhD, President
2700 Bay Area Blvd
Houston, TX 77058-1098
713 488-9336
- Dietetic Internship Prgm
- Dietetics-Didactic Prgm
- Speech-Language Pathology Prgm

# Huntsville

## Sam Houston State University
James Gaertner, PhD, President
Box 2026
Huntsville, TX 77341
936 294-1013
- Dietetic Internship Prgm
- Dietetics-Didactic Prgm
- Music Therapy Prgm

# Hurst

## Tarrant County College
Larry Darlage, PhD, President, NE Campus
828 W Harwood Rd
Hurst, TX 76054-3299
817 515-6200
- Dental Hygiene Prgm
- Emergency Med Tech-Paramedic Prgm
- Health Information Tech Prgm
- Physical Therapist Assistant Prgm
- Radiography Prgm
- Respiratory Therapist (Advanced) Prgm
- Surgical Technology Prgm

# Irving

## DeVry University - Dallas
4800 Regent Blvd
Irving, TX 75063
- Health Information Tech Prgm

# Kilgore

## Kilgore College
William M Holda, EdD, President
1100 Broadway
Kilgore, TX 75662
903 983-8100
- Medical Assistant Prgm
- Physical Therapist Assistant Prgm
- Surgical Technology Prgm

# Killeen

## Central Texas College
James R Anderson, PhD, Chancellor
US Hwy 190 W
Killeen, TX 76541-9990
817 526-1211
- Clin Lab Technician/Med Lab Technician

# Kingwood

## Kingwood College - NHMCCD
Linda Stegall, EdD, President
20000 Kingwood Dr
Kingwood, TX 77339
281 312-1640
- Dental Hygiene Prgm
- Occupational Therapy Asst Prgm
- Respiratory Therapist (Advanced) Prgm

# Lackland AFB

## US Military Dietetic Internship Consortium
Lackland AFB, TX 78236-5300
- Dietetic Internship Prgm (2)

INSTITUTIONS

## Lake Jackson

**Brazosport College**
Millicent Valek, PhD, President
500 College Dr
Lake Jackson, TX 77568
979 230-3200
- Emergency Med Tech-Paramedic Prgm

## Laredo

**Laredo Community College**
Ramon H Dovalina, PhD, President
West End Washington St
Laredo, TX 78040-4395
956 721-5101
- Clin Lab Technician/Med Lab Technician
- Occupational Therapy Asst Prgm
- Physical Therapist Assistant Prgm
- Radiography Prgm

## Levelland

**South Plains College**
Kelvin Sharp, EdD, President
1401 S College Ave
Levelland, TX 79336
806 894-9611
- Emergency Med Tech-Paramedic Prgm
- Health Information Tech Prgm
- Radiography Prgm
- Respiratory Therapist (Advanced) Prgm
- Surgical Assistant Prgm
- Surgical Technology Prgm

## Lubbock

**Covenant Medical Center**
Charley Trimble, President, CEO
3615 19th St
PO Box 1201
Lubbock, TX 79408
806 725-0579
- Radiography Prgm

**Texas Tech Univ Health Sciences Center**
Bernhard Mittemeyer, MD, Interim President
3601 4th St
Lubbock, TX 79409-2013
806 743-2900
- Athletic Training Prgm
- Audiologist Prgm
- Clin Lab Scientist/Med Technologist Prgm
- Diagnostic Molecular Scientist Prgm
- Occupational Therapy Prgm
- Physical Therapy Prgm
- Physician Assistant Prgm
- Speech-Language Pathology Prgm

**Texas Tech University**
David Schmidly, PhD, President
PO Box 2005
Lubbock, TX 79409
806 742-2121
- Counseling Prgm
- Dietetic Internship Prgm
- Dietetics-Didactic Prgm
- Orientation and Mobility Specialist Prgm
- Rehabilitation Counseling Prgm

## Lufkin

**Angelina College**
Larry M Phillips, EdD, President
PO Box 1768
3500 South First
Lufkin, TX 75902-1768
936 639-1301
- Pharmacy Technician Prgm
- Radiography Prgm
- Respiratory Therapist (Advanced) Prgm

## Marshall

**East Texas Baptist University**
Bob Riley, EdD, President
1209 N Grove
Marshall, TX 75670
903 923-2222
- Athletic Training Prgm

## McAllen

**South Texas College**
Shirley Reed, EdD, President
3201 W Pecan Blvd
PO Box 9701
McAllen, TX 78501-9701
956 872-8366
- Health Information Tech Prgm
- Occupational Therapy Asst Prgm
- Pharmacy Technician Prgm
- Physical Therapist Assistant Prgm

**South Texas Vo-Tech - McAllen**
Elma G Rodriquez, CEO/Owner
2901 N 23rd St, Ste B
McAllen, TX 78501
- Pharmacy Technician Prgm

## McKinney

**Collin County Community College**
Cary A Israel, JD, President
2200 W University Dr
McKinney, TX 75071
972 758-3801
- Dental Hygiene Prgm
- Respiratory Therapist (Advanced) Prgm

## Midland

**Midland College**
David E Daniel, EdD, President
3600 N Garfield
Midland, TX 79705
432 685-4520
- Diagnostic Med Sonography Prgm
- Health Information Tech Prgm
- Radiography Prgm
- Respiratory Therapist (Advanced) Prgm

## Nacogdoches

**Stephen F Austin State University**
Tito Guerrerro, President
1936 North St
Nacogdoches, TX 75962
936 468-2011
- Athletic Training Prgm
- Counseling Prgm
- Dietetic Internship Prgm
- Dietetics-Didactic Prgm
- Orientation and Mobility Specialist Prgm
- Rehabilitation Counseling Prgm
- Speech-Language Pathology Prgm

## Odessa

**Odessa College**
Vance W Gipson, EdD, President
201 W University
Odessa, TX 79764-8299
915 335-6410
- Physical Therapist Assistant Prgm
- Radiography Prgm
- Respiratory Therapist (Advanced) Prgm

## Orange

**Lamar State College - Orange**
J Michael Shahan, PhD, President
410 Front St
Orange, TX 77630-5802
409 883-7750
- Clin Lab Technician/Med Lab Technician
- Dental Assisting Prgm
- Pharmacy Technician Prgm

## Paris

**Paris Junior College**
Pamela Anglin, EdD, President
2400 Clarksville St
Paris, TX 75460
903 782-0330
- Surgical Technology Prgm

## Pasadena

**San Jacinto College Central**
Monte Blue, EdD, President
8060 Spencer Hwy
PO Box 2007
Pasadena, TX 77501-2007
281 476-1501
- Clin Lab Technician/Med Lab Technician
- Dietetic Technician-AD Prgm
- Emergency Med Tech-Paramedic Prgm
- Radiography Prgm
- Respiratory Therapist (Advanced) Prgm
- Surgical Technology Prgm

## Port Arthur

**Lamar State College - Port Arthur**
W Sam Monroe, LLD, President
PO Box 310
Port Arthur, TX 77641-0310
409 984-7101
- Surgical Technology Prgm

## Prairie View

**Prairie View A&M University**
Charles A Hines, President
Prairie View, TX 77446
409 857-3311
• Dietetic Internship Prgm
• Dietetics-Didactic Prgm

## Richardson

**University of Texas at Dallas**
Franklyn G Jenifer, President
PO Box 8630688
Richardson, TX 75083
214 883-2111
• Audiologist Prgm
• Speech-Language Pathology Prgm

## San Angelo

**Angelo State University**
E James Hindman, President
2601 West Ave N
San Angelo, TX 76909-0001
325 942-2555
• Athletic Training Prgm
• Physical Therapy Prgm

## San Antonio

**Baptist Health System**
Trip Pilgrim, MBA, President/CEO
215 E Quincy
San Antonio, TX 78215
210 297-1140
• Radiography Prgm
• Surgical Technology Prgm

**Hallmark Institute of Technology**
Derick Thomas, BS, Campus Director
10401 IH - 10 West
San Antonio, TX 78230
210 690-9000
• Medical Assistant Prgm

**Incarnate Word College**
4301 Broadway Ave
San Antonio, TX 78209
• Music Therapy Prgm

**Northwest Vista College**
Jacqueline Claunch, President
3535 N Ellison Dr
San Antonio, TX 78251-4217
210 348-2001
• Pharmacy Technician Prgm

**Our Lady of the Lake University**
Elizabeth A Sueltenfuss, President
411 S W 24th St
San Antonio, TX 78207
210 434-6711
• Speech-Language Pathology Prgm

**San Antonio College**
Robert E Zeigler, PhD, President
1300 San Pedro Ave
San Antonio, TX 78212-4299
210 733-2190
• Dental Assisting Prgm
• Medical Assistant Prgm

**San Antonio College of Med & Dental Assts**
Esther P Jones, Executive Director
4205 San Pedro
San Antonio, TX 78212
210 733-0777
• Medical Assistant Prgm

**St Mary's University**
One Camino Santa Maria
San Antonio, TX 78228
210 436-3011
• Counseling Prgm

**St Philip's College**
Angie S Runnels, PhD, President
1801 Martin Luther King Dr
San Antonio, TX 78203-2098
210 531-3591
• Clin Lab Technician/Med Lab Technician
• Dietetic Technician-AD Prgm
• Health Information Tech Prgm
• Occupational Therapy Asst Prgm
• Physical Therapist Assistant Prgm
• Radiography Prgm
• Surgical Technology Prgm

**Univ of Texas Hlth Sci Ctr at San Antonio**
Francisco Cigarroa, MD, President
7703 Floyd Curl Dr
San Antonio, TX 78229-3900
210 567-2050
• Clin Lab Scientist/Med Technologist Prgm
• Cytogenetic Technology Prgm
• Dental Hygiene Prgm
• Dental Lab Technician Prgm
• Emergency Med Tech-Paramedic Prgm
• Histotechnician Prgm
• Occupational Therapy Prgm
• Physical Therapy Prgm
• Physician Assistant Prgm
• Respiratory Therapist (Advanced) Prgm
• Specialist in BB Tech Prgm

**University of Texas Health Science Center**
Francisco Cigarroa, MD, President
7703 Floyd Curl Dr
San Antonio, TX 78229-3900
210 616-2000
• Occupational Therapy Prgm

**University of the Incarnate Word**
Louis J Agnese, Jr, PhD, President
4301 Broadway
San Antonio, TX 78209
210 829-3900
• Athletic Training Prgm
• Dietetic Internship Prgm
• Dietetics-Didactic Prgm
• Nuclear Medicine Technology Prgm

## San Marcos

**Texas State University-San Marcos**
Denise M Trauth, PhD, President
601 University Dr
San Marcos, TX 78666
512 245-2121
• Athletic Training Prgm
• Clin Lab Scientist/Med Technologist Prgm
• Counseling Prgm
• Dietetic Internship Prgm
• Dietetics-Didactic Prgm
• Health Information Admin Prgm
• Physical Therapy Prgm
• Radiation Therapy Prgm
• Respiratory Therapist (Advanced) Prgm
• Speech-Language Pathology Prgm
• Therapeutic Recreation Specialist Prgm

## Seguin

**Texas Lutheran University**
Jon Moline, PhD, President
1000 W Court St
Seguin, TX 78155
830 372-8000
• Athletic Training Prgm

## Sheppard AFB

**882d Training Group**
Nancy Dezell, Col, USAF, CEO
939 Missile Rd
Sheppard AFB, TX 76311-2245
940 676-2700
• Clin Lab Technician/Med Lab Technician
• Dental Assisting Prgm

**School of Health Care Sciences - Air Force**
Sheppard AFB, TX 76311
• Dental Lab Technician Prgm

**USAF School of Health Care Sciences**
Col Nancy Dezell, RN BAN MSN, Group
   Commander
882TRG/CC
939 Missile Rd
Sheppard AFB, TX 76311
940 676-2700
• Pharmacy Technician Prgm
• Radiography Prgm
• Respiratory Therapist (Entry-Level) Prgm

## Stephenville

**Tarleton State University**
Dennis P McCabe, PhD, President
Tarleton Station
Stephenville, TX 76402
817 968-9100
• Clin Lab Scientist/Med Technologist Prgm
• Dietetics-Didactic Prgm
• Histotechnology Prgm

## Sweetwater

**Texas State Tech College, West Texas**
Mike Reeser, MEd, President
Homer K Taylor
Sweetwater, TX 79556
325 235-7336
• Health Information Tech Prgm

## Temple

**Scott & White Memorial Hospital**
John Roberts, MD, President
2401 S 31st St
Temple, TX 76508
254 774-2111
• Clin Lab Scientist/Med Technologist Prgm

**Temple College**
Marc A Nigliazzo, PhD, President
2600 S First St
Temple, TX 76504
254 298-8600
• Dental Hygiene Prgm
• Respiratory Therapist (Advanced) Prgm
• Surgical Technology Prgm

## Texas City

**College of the Mainland**
Homer Hayes, PhD, President
1200 Amburn Rd
Texas City, TX 77591
409 938-1211
• Emergency Med Tech-Paramedic Prgm

## Tomball

**Tomball College - NHMCCD**
Raymond M Hawkins, PhD, President
30555 Tomball Pkwy
Tomball, TX 77375-4036
281 351-3333
• Occupational Therapy Asst Prgm

## Tyler

**Tyler Junior College**
William R Crowe, PhD, President
1400 E Fifth St
Tyler, TX 75798
903 510-2380
• Clin Lab Technician/Med Lab Technician
• Dental Hygiene Prgm
• Diagnostic Med Sonography Prgm
• Health Information Tech Prgm
• Ophthalmic Dispensing Optician Prgm
• Radiography Prgm
• Respiratory Therapist (Advanced) Prgm
• Surgical Technology Prgm

## Vernon

**Vernon College**
Steve Thomas, PhD, President
4400 College Dr
Vernon, TX 76384
940 552-6291
• Health Information Tech Prgm
• Pharmacy Technician Prgm
• Surgical Technology Prgm

## Victoria

**Citizens Medical Center**
David Brown, FACHE, Administrator
2701 Hospital Dr
Victoria, TX 77901
512 573-9181
• Radiography Prgm

**Victoria College**
Jimmy Goodson, EdD, President
2200 E Red River
Victoria, TX 77901
361 573-3291
• Clin Lab Technician/Med Lab Technician
• Respiratory Therapist (Advanced) Prgm

## Waco

**Baylor University**
John Lilley, PhD, President/CEO
1 Bear Pl #97096
Waco, TX 76798
254 710-3555
• Athletic Training Prgm
• Dietetics-Didactic Prgm
• Speech-Language Pathology Prgm

**Hillcrest Baptist Medical Center**
Art Hohenberger, FACHE, President
3000 Herring Ave
Waco, TX 76708
254 202-9400
• Clin Lab Scientist/Med Technologist Prgm

**McLennan Community College**
Dennis F Michaelis, PhD, President
1400 College Dr
Waco, TX 76708
254 299-8601
• Clin Lab Technician/Med Lab Technician
• Electroneurodiagnostic Tech Prgm
• Health Information Tech Prgm
• Physical Therapist Assistant Prgm
• Radiography Prgm

**Texas State Technical College at Waco**
Elton Stuckly, MS, President
3801 Campus Dr
Waco, TX 76705
254 867-4836
• Dental Assisting Prgm

## Weatherford

**Weatherford College**
Joe Birmingham, EdD, President
225 College Park
Weatherford, TX 76086-5699
817 594-5471
• Pharmacy Technician Prgm
• Respiratory Therapist (Advanced) Prgm

## Weslaco

**South Texas Vo-Tech - Weslaco**
Elma G Rodriquez, CEO/Owner
2419 E Haggar Ave
Weslaco, TX 78596
956 969-1564
• Pharmacy Technician Prgm

## Wharton

**Wharton County Junior College**
Betty McCrohan, MS, President
911 Boling Hwy
Wharton, TX 77488
979 532-6400
• Dental Hygiene Prgm
• Emergency Med Tech-Paramedic Prgm
• Health Information Tech Prgm
• Physical Therapist Assistant Prgm
• Radiography Prgm
• Surgical Technology Prgm

## Wichita Falls

**Midwestern State University**
Jesse Rogers, PhD, President
3410 Taft Blvd
Wichita Falls, TX 76308-2099
940 397-4211
• Athletic Training Prgm
• Dental Hygiene Prgm
• Radiography Prgm
• Respiratory Therapist (Advanced) Prgm

**United Regional Health Care Systems**
Jeff Hausler, CEO
1600 Eighth St
Wichita Falls, TX 76301
817 723-1461
• Clin Lab Scientist/Med Technologist Prgm

# United Kingdom

## Garthdee Aberdeen

**Robert Gordon University**
William Stevely, Vice Chancellor/President
Garthdee Rd
Garthdee Aberdeen, UK AB107QG
• Physical Therapy Prgm

# Utah

## Cedar City

**Southern Utah University**
Steven D Bennion, PhD, President
351 W University Blvd
Cedar City, UT 84720
435 586-7702
• Athletic Training Prgm

## Kaysville

**Davis Applied Technology College**
Michael J Bouwhuis, MEd, Campus President
550 East 300 South
Kaysville, UT 84037
801 593-2500
• Dental Assisting Prgm
• Medical Assistant Prgm
• Surgical Technology Prgm

## Logan

**Bridgerland Applied Technology College**
Richard L Maughan, MS PhD, Campus President
1301 West 600 North
Logan, UT 84321
435 753-6780
- Dental Assisting Prgm
- Medical Assistant Prgm

**Utah State University**
Stan Albrecht, President
Logan, UT 84322
435 797-1000
- Audiologist Prgm
- Dietetics-Coordinated Prgm
- Dietetics-Didactic Prgm
- Music Therapy Prgm
- Rehabilitation Counseling Prgm
- Speech-Language Pathology Prgm

## Murray

**Utah State University - Salt Lake**
Murray, UT 84107
- Dietetic Internship Prgm

## Ogden

**Ogden - Weber Applied Technology College**
C Brent Wallis, MA, President
200 N Washington Blvd
Ogden, UT 84404
801 627-8304
- Dental Assisting Prgm
- Medical Assistant Prgm

**Stevens Henager College**
Vicky Dewsnup, President, Regional Dir
1890 South 1350 West
PO Box 9428
Ogden, UT 84409-0428
801 394-7791
- Medical Assistant Prgm
- Surgical Technology Prgm

**Weber State University**
F Ann Milner, EdD, President
1001 University Circle
Ogden, UT 84408-1001
801 626-7071
- Athletic Training Prgm
- Clin Lab Scientist/Med Technologist Prgm
- Clin Lab Technician/Med Lab Technician
- Dental Hygiene Prgm
- Emergency Med Tech-Paramedic Prgm
- Health Information Admin Prgm
- Health Information Tech Prgm
- Respiratory Therapist (Advanced) Prgm
- Respiratory Therapist (Entry-Level) Prgm

## Orem

**Careers Unlimited**
575 E University Pkwy, Ste 1-163
Orem, UT 84097
- Dental Assisting Prgm

**Utah Valley State College**
William A Sederburg, PhD, President
800 W University Pkwy
Orem, UT 84058-5999
801 863-8550
- Dental Hygiene Prgm
- Emergency Med Tech-Paramedic Prgm

## Provo

**Ameritech College**
Connie A Garland, BS CDA, CEO Administrator
1675 N Freedom Blvd, Ste 3B
Provo, UT 84604—692
801 377-2900
- Dental Assisting Prgm
- Medical Assistant Prgm
- Surgical Technology Prgm

**Brigham Young University**
Cecil O Samuelson, MD, President
D 346 ASB
Provo, UT 84602
801 422-2521
- Athletic Training Prgm
- Clin Lab Scientist/Med Technologist Prgm
- Dietetic Internship Prgm
- Dietetics-Didactic Prgm
- Speech-Language Pathology Prgm
- Therapeutic Recreation Specialist Prgm

**Provo College**
Gordon C Peters, MBA, Campus President
1450 W North
Provo, UT 84601-1305
801 818-8912
- Dental Assisting Prgm
- Medical Assistant Prgm
- Physical Therapist Assistant Prgm

## Salt Lake City

**Intermountain Health Care**
Vicki Kershaw, CPA MBA, UCR Operations
    Officer
8th Ave and C St
Salt Lake City, UT 84143
801 408-1371
- Surgical Technology Prgm

**Latter Day Saints Business College**
Steven K Woodhouse, MBA, President
411 E South Temple
Salt Lake City, UT 84111
801 524-8101
- Medical Assistant Prgm

**Salt Lake Community College**
Cynthia Bioteau, PhD, President
4600 S Redwood Rd
PO Box 30808
Salt Lake City, UT 84130-0808
801 957-4226
- Clin Lab Technician/Med Lab Technician
- Dental Hygiene Prgm
- Medical Assistant Prgm
- Occupational Therapy Asst Prgm
- Physical Therapist Assistant Prgm
- Radiography Prgm
- Surgical Technology Prgm

**Unified Fire Authority/Utah Valley State Coll**
Don Berry, Fire Chief
3380 S 900 W
Salt Lake City, UT 84119
801 743-7200
- Emergency Med Tech-Paramedic Prgm

**University of Phoenix - Utah**
Brian Mueller, CEO
Salt Lake City, UT 84123
- Counseling Prgm

**University of Utah**
Michael K Young, JD, President
201 S President Circle, Rm 203
Salt Lake City, UT 84112
801 581-5701
- Athletic Training Prgm
- Audiologist Prgm
- Dietetics-Coordinated Prgm
- Occupational Therapy Prgm
- Physical Therapy Prgm
- Therapeutic Recreation Specialist Prgm

**University of Utah Health Science Center**
Michael K Young, JD, President
203 Park Bldg
Salt Lake City, UT 84112
801 581-5701
- Audiologist Prgm
- Clin Lab Scientist/Med Technologist Prgm
- Cytotechnology Prgm
- Genetic Counseling Prgm
- Nuclear Medicine Technology Prgm
- Physician Assistant Prgm
- Speech-Language Pathology Prgm

## St George

**Dixie State College of Utah**
Lee G Caldwell, PhD, President
225 S 700 E
St George, UT 84770-3876
435 652-7501
- Dental Hygiene Prgm
- Emergency Med Tech-Paramedic Prgm
- Surgical Technology Prgm

## West Jordan

**Utah Career College**
Nate Herrmann, Director
1902 West 7800 South
West Jordan, UT 84088
801 304-4224
- Medical Assistant Prgm

## West Valley City

**Everest College**
Larry Banks, President
3280 West 3500 South
West Valley City, UT 84119
801 840-4800
- Medical Assistant Prgm

**INSTITUTIONS**

# Vermont

## Burlington

**Champlain College**
David F Finney, PhD, President
163 S Willard St
PO Box 670
Burlington, VT 05402
802 865-2700
• Radiography Prgm

**Fletcher Allen Health Care**
Melinda Estes, MD, CEO and President
111 Colchester Ave
Burlington, VT 05401-1429
802 847-5959
• Cytotechnology Prgm

**University of Vermont**
Daniel Fogel, PhD, President
349 Waterman Bldg
Burlington, VT 05405
802 656-3186
• Athletic Training Prgm
• Clin Lab Scientist/Med Technologist Prgm
• Counseling Prgm
• Dietetics-Didactic Prgm
• Nuclear Medicine Technology Prgm
• Physical Therapy Prgm
• Speech-Language Pathology Prgm

## Castleton

**Castleton State College**
David Wolk, MS, President
Woodruff Hall
Castleton, VT 05735
802 468-1201
• Athletic Training Prgm

## Essex Junction

**Center for Technology - Essex**
Three Educational Dr
Essex Junction, VT 05452
• Dental Assisting Prgm

## Northfield

**Norwich University**
Richard W Schneider, PhD, President
158 Harmon Dr
Northfield, VT 05663
802 485-2065
• Athletic Training Prgm

## Randolph Center

**Vermont Technical College**
Ty Handy, MBA, President
PO Box 500
Randolph Center, VT 05061
802 728-1258
• Dental Hygiene Prgm
• Respiratory Therapist (Advanced) Prgm

# Virginia

## Annandale

**Northern Virginia Community College**
Robert Templin, PhD, President
4001 Wakefield Chapel Rd
Annandale, VA 22003-3796
703 323-3101
• Clin Lab Technician/Med Lab Technician
• Dental Hygiene Prgm
• Dietetic Technician-AD Prgm
• Emergency Med Tech-Paramedic Prgm
• Health Information Tech Prgm
• Pharmacy Technician Prgm
• Physical Therapist Assistant Prgm
• Respiratory Therapist (Advanced) Prgm

## Arlington

**ACT College**
Jeffrey Moore, MBA, President and CEO
1100 Wilson Blvd
Arlington, VA 22209
• Radiography Prgm

**Marymount University**
James Bundschuh, President
2807 N Glebe Rd
Arlington, VA 22207-4299
703 522-5600
• Counseling Prgm
• Physical Therapy Prgm

## Big Stone Gap

**Mountain Empire Community College**
Terrance Suarez, PhD, President
3441 Mountain Empire Road
Big Stone Gap, VA 24219
540 523-2400
• Respiratory Therapist (Advanced) Prgm

## Blacksburg

**Virginia Polytechnic Inst & State Univ**
Charles W Steger, President
Blacksburg, VA 24061
540 231-6000
• Counseling Prgm
• Dietetic Internship Prgm
• Dietetics-Didactic Prgm

## Charlottesville

**Piedmont Virginia Community College**
Frank Friedman, PhD, President
501 College Dr
Charlottesville, VA 22902
434 961-5200
• Emergency Med Tech-Paramedic Prgm
• Surgical Technology Prgm

**University of Virginia**
John Casteen, President
PO Box 400224
Charlottesville, VA 22904-4224
434 924-3337
• Counseling Prgm
• Speech-Language Pathology Prgm

**University of Virginia Health System**
Edward R Howell, MS, VP and CEO, Hlth Systems
PO Box 800809
Charlottesville, VA 22908
434 243-9308
• Dietetic Internship Prgm
• Radiation Therapy Prgm
• Radiography Prgm

**Virginia School of Massage**
Charlottesville, VA 22903
• Massage Therapy Prgm

## Danville

**Averett University**
Richard A Pfau, PhD, President
420 W Main St
Danville, VA 24541
434 791-5670
• Athletic Training Prgm

**Bridgewater College**
Philip C Stone, President
Box 34, Flory Hall, Rm 100
Danville, VA 24521
540 828-5605
• Athletic Training Prgm

**Danville Regional Medical Center**
Warren E Callaway, FACHE, President
142 S Main St
Danville, VA 24541
434 799-2100
• Radiography Prgm

## Emory

**Emory & Henry College**
Rosalind Reichard, PhD, President
PO Box 947
Emory, VA 24327-0947
276 944-6107
• Athletic Training Prgm

## Fairfax

**George Mason University**
Alan G Merten, PhD, President
4400 University Dr
Fairfax, VA 22030-4444
703 993-8700
• Athletic Training Prgm
• Therapeutic Recreation Specialist Prgm

## Falls Church

**Inova Fairfax Hospital**
Jolene Tournabeni, Administrator
3300 Gallows Rd
Falls Church, VA 22046
703 698-3371
• Clin Lab Scientist/Med Technologist Prgm

**National Massage Therapy Institute**
803 W Board St
Falls Church, VA 22046
• Massage Therapy Prgm

## Farmville

**Longwood University**
Patricia Cormier, PhD, President
201 High St
Farmville, VA 23909-1899
804 395-2000
- Athletic Training Prgm
- Therapeutic Recreation Specialist Prgm

## Fishersville

**Augusta Medical Center**
David Deering, MBA, COO
PO Box 1000
78 Medical Center Dr
Fishersville, VA 22939
540 332-4818
- Clin Lab Scientist/Med Technologist Prgm

## Fredericksburg

**Mary Washington Hospital**
Fred M Rankin III, MPH, President/CEO
1001 Sam Perry Blvd
Fredericksburg, VA 22401
540 741-1414
- Radiography Prgm

## Hampton

**Hampton University**
William R Harvey, President
Hampton, VA 23668
804 727-5000
- Physical Therapy Prgm
- Speech-Language Pathology Prgm

**Thomas Nelson Community College**
Charles Taylor, EdD, President
PO Box 9407
Hampton, VA 23670
757 825-2711
- Clin Lab Technician/Med Lab Technician

## Harrisonburg

**James Madison University**
Linwood Rose, EdD, President
Office of the President, MSC 7608
Harrisonburg, VA 22807
540 568-6868
- Athletic Training Prgm
- Audiologist Prgm
- Counseling Prgm
- Dietetic Internship Prgm
- Dietetics-Didactic Prgm
- Occupational Therapy Prgm
- Physician Assistant Prgm
- Speech-Language Pathology Prgm

**Rockingham Memorial Hospital**
T Carter Melton, Jr, MHA, President
235 Cantrell Ave
Harrisonburg, VA 22801
540 433-4101
- Clin Lab Scientist/Med Technologist Prgm
- Radiography Prgm

## Leesburg

**Loudoun County Dept of Fire-Rescue**
Joseph Pozzo, MA, Chief
16600 Courage Court
Leesburg, VA 20175
703 777-0333
- Emergency Med Tech-Paramedic Prgm

## Lynchburg

**Centra Health Systems of Lynchburg**
George W Dawson, MHA, President/CEO
1920 Atherholt Rd
Lynchburg, VA 24501
804 947-4705
- Clin Lab Technician/Med Lab Technician

**Central Virginia Community College**
Darrel W Staat, DA, President
3506 Wards Rd
Lynchburg, VA 24502
804 832-7601
- Radiography Prgm
- Respiratory Therapist (Advanced) Prgm

**Liberty University**
Jerry Falwell, TheoD, Chancellor
1971 University Blvd
Lynchburg, VA 24502
434 582-2300
- Athletic Training Prgm

**Lynchburg College**
Kenneth R Garren, President
1501 Lakeside Dr
Lynchburg, VA 24501-3199
804 544-8100
- Athletic Training Prgm
- Counseling Prgm

**Miller-Motte Technical College**
Richard Craig, BS ED MEd, President
1912 Memorial Ave
Lynchburg, VA 24501
804 847-7701
- Medical Assistant Prgm

## Newport News

**Medical Careers Institute**
Barbara Larar, Director
1001 Omni Blvd, Ste 200
Newport News, VA 23606
757 873-2423
- Medical Assistant Prgm

**Riverside School of Health Careers**
Tracee Carmean, VP Education
316 Main St
Newport News, VA 23601
757 240-2202
- Radiography Prgm
- Surgical Technology Prgm

## Norfolk

**Eastern Virginia Medical School**
Harry T Lester, MD, President
PO Box 1980
Norfolk, VA 23507
757 446-5600
- Art Therapy Prgm
- Physician Assistant Prgm
- Surgical Assistant Prgm

**Norfolk State University**
Carolyn Meyers, PhD, President
700 Park Ave
Echols Hall Rm 165
Norfolk, VA 23504
757 823-8670
- Clin Lab Scientist/Med Technologist Prgm
- Dietetics-Didactic Prgm
- Kinesiotherapy Prgm

**Old Dominion University**
Roseann Runte, PhD, President
Koch Hall
Norfolk, VA 23529-0001
757 683-3159
- Clin Lab Scientist/Med Technologist Prgm
- Counseling Prgm
- Cytotechnology Prgm
- Dental Hygiene Prgm
- Nuclear Medicine Technology Prgm
- Ophthalmic Med Technologist Prgm
- Physical Therapy Prgm
- Speech-Language Pathology Prgm

**Sentara Norfolk General Hospital**
Rodney Hochman, MD, CEO
600 Gresham Dr
Norfolk, VA 23507
757 668-3361
- Cardiovascular Technology Prgm

**Tidewater Community College**
Deborah M DiCroce, EdD, President
121 College Place
Norfolk, VA 23510
757 822-1050
- Diagnostic Med Sonography Prgm
- Dietetic Technician-AD Prgm
- Emergency Med Tech-Paramedic Prgm
- Health Information Tech Prgm
- Medical Assistant Prgm
- Occupational Therapy Asst Prgm
- Physical Therapist Assistant Prgm
- Radiography Prgm
- Respiratory Therapist (Advanced) Prgm

**Tidewater Technical**
Jerry Yagen
1760 E Little Creek Rd
Norfolk, VA 23518
757 588-2121
- Dental Assisting Prgm

## Petersburg

**Southside Regional Medical Center**
David S Dunham, MHA, Exec Director
801 S Adams St
Petersburg, VA 23803
804 862-5903
- Radiography Prgm

**Virginia State University**
Eddie N Moore, Jr, President
Petersburg, VA 23806
804 524-5000
- Dietetic Internship Prgm
- Dietetics-Didactic Prgm

## Portsmouth

**Naval School of Health Sciences**
Brad L Bennett, CAPT, MSC USN, Commanding
  Officer
1001 Holcomb Rd
Portsmouth, VA 23708-5200
757 953-5040
- Electroneurodiagnostic Tech Prgm
- Nuclear Medicine Technology Prgm
- Pharmacy Technician Prgm
- Radiography Prgm
- Surgical Technology Prgm

## Radford

**Radford University**
Penelope Kyle, MBA JD, President
PO Box 6890
Martin Hall
Radford, VA 24142
540 831-5401
- Athletic Training Prgm
- Counseling Prgm
- Dietetic Internship Prgm
- Dietetics-Didactic Prgm
- Music Therapy Prgm
- Speech-Language Pathology Prgm
- Therapeutic Recreation Specialist Prgm

## Richlands

**Southwest Virginia Community College**
Charles R King, EdD, President
PO Box SVCC
Richlands, VA 24641-1101
276 964-7315
- Diagnostic Med Sonography Prgm
- Emergency Med Tech-Paramedic Prgm
- Occupational Therapy Asst Prgm
- Radiography Prgm
- Respiratory Therapist (Advanced) Prgm

## Richmond

**Bon Secours School of Medical Imaging**
Micheal Kerner, Exec VP, Administrator
5801 Bremo Rd
Richmond, VA 23226
804 285-2011
- Radiography Prgm

**Bryant & Stratton College - Richmond**
John Staschak, President
8141 Hull St Road
Richmond, VA 23235
804 745-2444
- Medical Assistant Prgm

**J Sargeant Reynolds Community College**
Gary Rhodes, PhD, President
PO Box 85622
Richmond, VA 23285-5622
804 523-5200
- Clin Lab Technician/Med Lab Technician
- Dental Assisting Prgm
- Dental Lab Technician Prgm
- Emergency Med Tech-Paramedic Prgm
- Ophthalmic Dispensing Optician Prgm
- Respiratory Therapist (Advanced) Prgm
- Respiratory Therapist (Entry-Level) Prgm
- Surgical Technology Prgm

**Virginia Commonwealth Univ/Med Coll of VA**
Euguene P Trani, PhD, President
PO Box 842512
Richmond, VA 23284-2512
804 828-1200
- Dietetic Internship Prgm
- Emergency Med Tech-Paramedic Prgm
- Genetic Counseling Prgm
- Occupational Therapy Prgm
- Physical Therapy Prgm
- Therapeutic Recreation Specialist Prgm

**Virginia Commonwealth University**
Eugene P Trani, PhD, President
910 W Franklin St
Richmond, VA 23284-2512
804 828-1200
- Athletic Training Prgm
- Clin Lab Scientist/Med Technologist Prgm
- Dental Hygiene Prgm
- Nuclear Medicine Technology Prgm
- Radiation Therapy Prgm
- Radiography Prgm
- Rehabilitation Counseling Prgm

**Virginia Department of Health**
Div of Public Health Nutrition
1500 E Main St Rm 132
Richmond, VA 23219
804 786-5420
- Dietetic Internship Prgm

## Roanoke

**Carilion Medical Center**
Nancy H Agee, RN MN
PO Box 13367
Belleview & Jefferson Sts
Roanoke, VA 24033-3367
540 981-8385
- Clin Lab Scientist/Med Technologist Prgm

**Jefferson College of Health Sciences**
Carol M Seavor, EdD, President
920 S Jefferson St
Roanoke, VA 24016
888 985-8483
- Emergency Med Tech-Paramedic Prgm
- Occupational Therapy Asst Prgm
- Occupational Therapy Prgm
- Physical Therapist Assistant Prgm
- Physician Assistant Prgm

**National College of Business & Technology**
Frank E Longaker, MBA, President
PO Box 6400
Roanoke, VA 24017
540 986-1800
- Health Information Tech Prgm
- Medical Assistant Prgm (7)

**Virginia Western Community College**
Robert Sandel, PhD, President
3095 Colonial Ave SW
PO Box 14045
Roanoke, VA 24015
540 857-7311
- Dental Hygiene Prgm
- Radiation Therapy Prgm
- Radiography Prgm

## Salem

**Roanoke College**
Sabine O'Hara, PhD, President
221 College Lane
Salem, VA 24153
540 375-2200
- Athletic Training Prgm

## Virginia Beach

**Bryant & Stratton College - Virginia Beach**
Lee E Hicklin, Director
301 Centre Pointe Dr
Virginia Beach, VA 23462
757 499-7900
- Medical Assistant Prgm

**Cayce/Reilly School of Massotherapy**
215 67th St
Virginia Beach, VA 23451
- Massage Therapy Prgm

**Medical Careers Institute**
Kevin Beaver, Director
5501 Greenwich Rd
Virginia Beach, VA 23462
757 497-8400
- Medical Assistant Prgm

**Regent University**
Rosemarie Hughes, PhD, Dean
School of Psychology and Counseling
1000 Regent University Dr, CRB 174
Virginia Beach, VA 23464-9800
757 226-4000
- Counseling Prgm

## Williamsburg

**College of William & Mary**
Timothy J Sullivan, President
PO Box 8795
Williamsburg, VA 23187-8794
757 221-4000
- Counseling Prgm

## Winchester

**Shenandoah University**
James Davis, PhD, President
1460 University Dr
Winchester, VA 22601
540 665-4505
- Athletic Training Prgm
- Music Therapy Prgm
- Occupational Therapy Prgm
- Physical Therapy Prgm
- Physician Assistant Prgm
- Respiratory Therapist (Advanced) Prgm

**Winchester Medical Center Inc**
1840 Amherst St
PO Box 3340
Winchester, VA 22601
• Radiography Prgm

## Wytheville

**Wytheville Community College**
Charlie White, PhD, President
1000 E Main St
Wytheville, VA 24382
276 223-4700
• Clin Lab Technician/Med Lab Technician
• Dental Hygiene Prgm
• Physical Therapist Assistant Prgm

## Yorktown

**Tri-Service Optician School (TOPS)**
CDR Lee Cornforth, USN OD MBA, Training
Officer
Naval Ophthalmic Support & Trng Activity
160 Main Rd, Ste 360
Yorktown, VA 23691-9984
757 887-7329
• Ophthalmic Dispensing Optician Prgm

# Washington

## Auburn

**Green River Community College**
Richard Rutkowski, MBA, President
12401 SE 320th St
Auburn, WA 98002-3622
253 833-9111
• Occupational Therapy Asst Prgm
• Physical Therapist Assistant Prgm

## Bellevue

**Bellevue Community College**
B Jean Floten, MS, President
3000 Landerholm Circle SE
Bellevue, WA 98007-6484
425 564-2301
• Diagnostic Med Sonography Prgm
• Nuclear Medicine Technology Prgm
• Radiation Therapy Prgm

## Bellingham

**Bellingham Technical College**
Gerald Pumphrey, EdD, President
3028 Lindbergh Ave
Bellingham, WA 98225
360 738-0221
• Dental Assisting Prgm
• Emergency Med Tech-Paramedic Prgm
• Surgical Technology Prgm

**Western Washington University**
Karen W Morse, PhD, President
516 High Street
Old Main 450, MS 9000
Bellingham, WA 98225
360 650-3480
• Audiologist Prgm
• Counseling Prgm
• Rehabilitation Counseling Prgm
• Speech-Language Pathology Prgm

**Whatcom Community College**
Harold G Heiner, PhD, President
237 W Kellogg Rd
Bellingham, WA 98226
360 676-2170
• Medical Assistant Prgm
• Physical Therapist Assistant Prgm

## Bremerton

**Olympic College**
David Mitchell, PhD, President
1600 Chester Ave
Bremerton, WA 98337-1699
360 792-6050
• Medical Assistant Prgm

## Cheney

**Eastern Washington University**
Rodolfo Ar,valo, PhD, President
214 Showalter Hall
526 5th St
Cheney, WA 99004-2431
509 359-2371
• Athletic Training Prgm
• Counseling Prgm
• Dental Hygiene Prgm
• Occupational Therapy Prgm
• Physical Therapy Prgm
• Speech-Language Pathology Prgm
• Therapeutic Recreation Specialist Prgm

## Des Moines

**Highline Community College**
Priscilla Bell, President
PO Box 98000
Des Moines, WA 98198-9800
206 878-3710
• Medical Assistant Prgm
• Respiratory Therapist (Advanced) Prgm

## Ellensburg

**Central Washington University**
Jerilyn McIntyre, PhD, President
400 E University Way
Ellensburg, WA 98926-7501
509 963-2111
• Dietetic Internship Prgm
• Dietetics-Didactic Prgm
• Emergency Med Tech-Paramedic Prgm

## Everett

**Bryman College**
Kimberly Lothyan, BS MA EdC, President
906 SE Everett Mall Way
Ste 600
Everett, WA 98251
425 789-7960
• Medical Assistant Prgm (4)

**Everett Community College**
Charles Earl, President
2000 Tower St
Everett, WA 98201
425 388-9572
• Medical Assistant Prgm

## Kirkland

**Lake Washington Technical College**
L Michael Metke, EdD, President
11605 132nd Ave NE
Kirkland, WA 98034
425 739-8100
• Dental Assisting Prgm
• Dental Hygiene Prgm
• Medical Assistant Prgm

## Lakewood

**Clover Park Technical College**
Sharon McGavick, EdD, President
4500 Steilacoom Blvd SW
Lakewood, WA 98499-4098
253 589-5678
• Clin Lab Technician/Med Lab Technician
• Dental Assisting Prgm
• Medical Assistant Prgm
• Surgical Technology Prgm

**Pierce College**
Michele Johnson, PhD, President
9401 Farwest Dr SW
Lakewood, WA 98498-1999
253 964-6500
• Dental Hygiene Prgm

## Longview

**Lower Columbia College**
Vernon R Pickett, PhD, President
1600 Maple St
Longview, WA 98632
360 577-2320
• Medical Assistant Prgm

## Mt Vernon

**Skagit Valley College**
James M Ford, PhD, President
2405 E College Way
Mt Vernon, WA 98273
360 416-7997
• Medical Assistant Prgm

## Olympia

**South Puget Sound Community College**
Kenneth J Minnaert, PhD, President
2011 Mottman Rd SW
Olympia, WA 98512-6292
360 596-5206
• Dental Assisting Prgm
• Medical Assistant Prgm

## Pasco

**Columbia Basin College**
Lee R Thornton, PhD, President
2600 N 20th Ave
Pasco, WA 99301
509 547-0511
• Dental Hygiene Prgm
• Emergency Med Tech-Paramedic Prgm

INSTITUTIONS

## Pullman

**Washington State University**
V Lane Rawlins, PhD, President
422 French Administration
PO Box 641048
Pullman, WA 99164-1048
509 335-6666
- Athletic Training Prgm
- Audiologist Prgm
- Dietetic Internship Prgm
- Dietetics-Coordinated Prgm
- Dietetics-Didactic Prgm
- Speech-Language Pathology Prgm

## Renton

**Renton Technical College**
Donald E Bressler, PhD, President
4000 NE Fourth St
Renton, WA 98056
206 235-2235
- Dental Assisting Prgm
- Medical Assistant Prgm
- Ophthalmic Assistant Prgm
- Pharmacy Technician Prgm
- Surgical Technology Prgm

## Seattle

**Antioch University - Seattle**
Toni Murdock, President
2326 Sixth Ave
Seattle, WA 98121-1814
- Art Therapy Prgm

**Bastyr University**
Joseph E Pizzorno, Jr, President
144 NE 54th St
Seattle, WA 98105
206 523-9585
- Dietetic Internship Prgm
- Dietetics-Didactic Prgm

**Brenneke School of Massage**
Julie Darrah, LMP, Director
425 Pontius Ave N, Ste 100
Seattle, WA 98109
206 282-1233
- Massage Therapy Prgm

**Brian Utting School of Massage**
900 Thomas St
Seattle, WA 98109
- Massage Therapy Prgm

**Harborview Med Ctr - Univ of Washington**
David Jaffe, MPA, Admin/Executive Director
325 Ninth Ave
Seattle, WA 98104
206 731-3036
- Emergency Med Tech-Paramedic Prgm

**North Seattle Community College**
Ron LaFayette, PhD, President
9600 College Way N
Seattle, WA 98103
206 527-3602
- Medical Assistant Prgm

**Pima Medical Institute - Seattle**
Richard Luebke, Sr, BS, President
9709 Third Ave NE, #400
Seattle, WA 98115
206 322-6100
- Radiography Prgm

**Sea Mar Community Health Center**
8720 14th Ave S
Seattle, WA 98108-4807
206 726-3730
- Dietetic Internship Prgm

**Seattle Central Community College**
Mildred Ollee, EdD, President
1701 Broadway, 2BE4180
Seattle, WA 98122
206 587-4144
- Dental Hygiene Prgm
- Ophthalmic Dispensing Optician Prgm
- Respiratory Therapist (Advanced) Prgm
- Surgical Technology Prgm

**Seattle Pacific University**
3307 Third Ave W
Seattle, WA 98119
206 281-2050
- Dietetics-Didactic Prgm

**Seattle University**
Stephen Sundborg, SJ STD, President
901 12th Avenue
PO Box 222000
Seattle, WA 98122-4460
206 296-5960
- Diagnostic Med Sonography Prgm

**Seattle Vocational Institute**
Norward J Brooks, PhD, Director
2120 S Jackson
Seattle, WA 98144
206 587-4940
- Dental Assisting Prgm
- Medical Assistant Prgm

**University of Washington**
Harry Bruce, PhD, President
306 Gerberding Hall
Box 351230
Seattle, WA 98195-1230
206 543-5010
- Audiologist Prgm
- Clin Lab Scientist/Med Technologist Prgm
- Dietetic Internship Prgm
- Dietetics-Didactic Prgm
- Health Information Admin Prgm
- Medical Librarian Prgm
- Occupational Therapy Prgm
- Orthotist/Prosthetist Prgm
- Physical Therapy Prgm
- Physician Assistant Prgm
- Speech-Language Pathology Prgm

## Shoreline

**Shoreline Community College**
Lee Lambert, JD, Interim President
16101 Greenwood Ave N
Shoreline, WA 98133
206 546-4551
- Clin Lab Technician/Med Lab Technician
- Dental Hygiene Prgm
- Dietetic Technician-AD Prgm
- Health Information Tech Prgm

## Spokane

**Apollo College**
George Montgomery, BA, President and CEO
10102 E Knox Ste 200
Spokane, WA 99206
509 532-8888
- Radiography Prgm

**Sacred Heart Medical Center**
Michael Wilson, MA, President/CEO
W 101 Eighth Ave
PO Box 2555
Spokane, WA 99220-2555
509 455-3040
- Clin Lab Scientist/Med Technologist Prgm

**Spokane Community College**
Steven Hanson, MS MA, President
N 1810 Greene St MS 1050
Spokane, WA 99217
509 533-3841
- Cardiovascular Technology Prgm
- Dental Assisting Prgm
- Dietetic Technician-AD Prgm
- Emergency Med Tech-Paramedic Prgm
- Health Information Tech Prgm
- Medical Assistant Prgm
- Pharmacy Technician Prgm
- Radiography Prgm
- Respiratory Therapist (Advanced) Prgm
- Surgical Technology Prgm

**Spokane Falls Community College**
Mark Palek, President
3410 W Fort George Wright Dr, MS3160
Spokane, Wa 99224-5288
509 533-3535
- Physical Therapist Assistant Prgm

**Whitworth College**
William P Robinson, PhD, President
300 W Hawthorne Rd
Spokane, WA 99251
509 777-1000
- Athletic Training Prgm

## Tacoma

**Bates Technical College**
David Borofsky, EdD, President
1101 S Yakima Ave
Tacoma, WA 98405
253 680-7100
- Dental Assisting Prgm
- Dental Lab Technician Prgm

**St Joseph Medical Center**
Joseph W Wilezek, President and CEO
1717 South J St
Tacoma, WA 98401-2197
- Pharmacy Technician Prgm

**Tacoma Community College**
Pamela Transue, PhD, President
6501 S 19th St
Tacoma, WA 98466
253 566-5100
- Emergency Med Tech-Paramedic Prgm
- Health Information Tech Prgm
- Radiography Prgm
- Respiratory Therapist (Advanced) Prgm

**Tacoma Fire Department**
Ron Stephens, Fire Chief
901 Fawcett Ave
Tacoma, WA 98402
253 591-5737
• Emergency Med Tech-Paramedic Prgm

**University of Puget Sound**
Ronald R Thomas, PhD, President
1500 N Warner St, CMB 1094
Tacoma, WA 98416-0510
253 879-3201
• Occupational Therapy Prgm
• Physical Therapy Prgm

## Vancouver

**Clark College**
David N Beyer, PhD, President
1800 E McLoughlin Blvd
Vancouver, WA 98663-3598
360 992-2000
• Dental Hygiene Prgm
• Medical Assistant Prgm

**Everest College**
Edward Yakimchick, BA, President
120 NE 136th Ave, Ste 130
Vancouver, WA 98684
360 254-3282
• Medical Assistant Prgm

**Northwest Regional Training Center**
Marty James, Administrator
11606 NE 66th St, Ste 103
Vancouver, WA 98662
360 759-4404
• Emergency Med Tech-Paramedic Prgm

## Wenatchee

**Wenatchee Valley College**
James Richardson, MSEd, President
1300 Fifth St
Wenatchee, WA 98801
509 682-6668
• Clin Lab Technician/Med Lab Technician
• Medical Assistant Prgm

## Yakima

**Yakima Regional CLS Program**
Rick Garnier, CEO
1120 W Spruce St
Yakima, WA 98902
• Clin Lab Scientist/Med Technologist Prgm

**Yakima Valley Community College**
Linda J Kaminski, EdD, President
PO Box 22520
Yakima, WA 98907
509 574-4635
• Dental Hygiene Prgm
• Medical Assistant Prgm
• Surgical Technology Prgm

# West Virginia

## Athens

**Concord University**
Jerry Beasley, PhD, President
PO Box 1000
Campus Box Wall
Athens, WV 24712
304 384-9188
• Athletic Training Prgm

## Beckley

**Mountain State University**
Charles H Polk, EdD, President
PO Box 9003
Beckley, WV 25802
304 253-7351
• Diagnostic Med Sonography Prgm
• Medical Assistant Prgm
• Occupational Therapy Asst Prgm
• Physical Therapist Assistant Prgm
• Physician Assistant Prgm
• Radiography Prgm

## Bluefield

**Bluefield Regional Medical Center**
Lynn Whitaker, MBA, CEO
500 Cherry St
Bluefield, WV 24701
304 327-1701
• Clin Lab Technician/Med Lab Technician

**Bluefield State College**
Albert Walker, EdD, President
219 Rock St
Bluefield, WV 24701
304 327-4030
• Radiography Prgm

## Buckhannon

**West Virginia Wesleyan College**
William R Haden, MA, President
59 College Ave
Buckhannon, WV 26201-2995
304 473-8181
• Athletic Training Prgm

## Charleston

**CAMC Health Education & Research Institute**
Sharon A Hall, President
3200 MacCorkle Ave SE
Charleston, WV 25304
304 388-9900
• Cytotechnology Prgm

**Carver Career Technical Center**
Jim Casdorph, MA, Principal
4799 Midland Dr
Charleston, WV 25306
304 348-1965
• Pharmacy Technician Prgm
• Respiratory Therapist (Advanced) Prgm
• Surgical Technology Prgm

**Mountain State School of Massage**
Robert Rogers, LMTNCTMB, Executive Director
601 50th St
Charleston, WV 25304
304 926-8822
• Massage Therapy Prgm

**University of Charleston**
Edwin H Welch, PhD, President
2300 McCorkle Ave SE
Charleston, WV 25304
304 357-4713
• Athletic Training Prgm
• Radiography Prgm

## Clarksburg

**United Hospital Center**
Bruce C Carter, MS, President
Three Hospital Plaza
Clarksburg, WV 26301
304 624-2332
• Radiography Prgm

## Dunbar

**Benjamin Franklin Career & Technical Ed Ctr**
Alvin L Brown, BA MA MS, Principal
500 28th St
Dunbar, WV 25064
304 766-0369
• Medical Assistant Prgm

## Fairmont

**Fairmont State Univ**
Blair Montgomery, MS, President
1201 Locust Ave
Fairmont, WV 26554
304 367-4692
• Clin Lab Technician/Med Lab Technician
• Health Information Tech Prgm
• Physical Therapist Assistant Prgm

## Huntington

**Cabell Huntington Hospital**
Brent Marstellar, BS, President
1340 Hal Greer Blvd
Huntington, WV 25701
304 526-2111
• Cytotechnology Prgm

**Huntington Junior College**
Carolyn Smith, BA, President
900 Fifth Ave
Huntington, WV 25701
304 697-7550
• Medical Assistant Prgm

**Marshall Community & Technical College**
Dan D Angel, President
One John Marshall Dr
Huntington, WV 25755
304 696-2300
• Health Information Tech Prgm
• Medical Assistant Prgm
• Physical Therapist Assistant Prgm

INSTITUTIONS

**Marshall University**
Stephen Kopp, PhD, President
One John Marshall Dr
Huntington, WV 25755
304 696-2300
- Athletic Training Prgm
- Clin Lab Scientist/Med Technologist Prgm
- Clin Lab Technician/Med Lab Technician
- Dietetic Internship Prgm
- Dietetics-Didactic Prgm
- Health Information Tech Prgm
- Speech-Language Pathology Prgm

**St Mary's Medical Center**
Michael Sellards, BS MHA, Exec Director/CEO
2900 First Ave
Huntington, WV 25702
304 526-1270
- Radiography Prgm

## Institute

**West Virginia State Comm & Tech College**
Ervin V Griffin, PhD, President
PO Box 1000
105 Cole Complex
Institute, WV 25112
304 766-3111
- Nuclear Medicine Technology Prgm

## Montgomery

**Community & Technical College at WVU Tech**
Beverly Jo Harris, EdD, President
208 Davis Hall
Montgomery, WV 25136
304 442-3149
- Dental Hygiene Prgm

## Morgantown

**Monongalia County Tech Education Center**
John George, MA, Director/Principal
1000 Mississippi St
Morgantown, WV 26505
304 291-9240
- Surgical Technology Prgm

**West Virginia University**
David C Hardesty Jr, JD, President
102A Stewart Hall/PO Box 6201
Morgantown, WV 26506-6201
304 293-5531
- Athletic Training Prgm
- Audiologist Prgm
- Clin Lab Scientist/Med Technologist Prgm
- Counseling Prgm
- Dental Hygiene Prgm
- Dietetic Internship Prgm
- Dietetics-Didactic Prgm
- Occupational Therapy Prgm
- Physical Therapy Prgm
- Rehabilitation Counseling Prgm
- Speech-Language Pathology Prgm

**West Virginia University Hospitals**
Bruce McClymonds, President
Rudy Memorial Hospital
PO Box 8136
Morgantown, WV 26505
304 598-4355
- Diagnostic Med Sonography Prgm
- Dietetic Internship Prgm
- Nuclear Medicine Technology Prgm
- Radiation Therapy Prgm
- Radiography Prgm

## Mt Gay

**Southern West Virginia Comm & Tech College**
Joanne C Tomblin, MA, President
PO Box 2900
Mt Gay, WV 25637
304 792-7160
- Clin Lab Technician/Med Lab Technician
- Dental Hygiene Prgm
- Radiography Prgm
- Surgical Technology Prgm

## Parkersburg

**West Virginia University - Parkersburg**
Marie Foster Gnage, PhD, President
300 Campus Dr
Parkersburg, WV 26104
304 424-8200
- Surgical Technology Prgm

## Philippi

**Alderson-Broaddus College**
Stephen E Markwood, PhD, President
PO Box 307
Philippi, WV 26416
304 457-6201
- Athletic Training Prgm
- Physician Assistant Prgm

## Princeton

**Mercer County Technical Education Center**
1397 Stafford Dr
Princeton, WV 24720
- Dental Assisting Prgm

## West Liberty

**West Liberty State College**
John McCollough, PhD, Interim President
Main Hall
West Liberty, WV 26074
304 336-8000
- Clin Lab Scientist/Med Technologist Prgm
- Dental Hygiene Prgm

## Wheeling

**Ohio Valley Medical Center**
Thomas Galinski, MS, President
2000 Eoff St
Wheeling, WV 26003
304 234-8294
- Radiography Prgm

**West Virginia Northern Community College**
Martin Oshinsky, PhD, President
1704 Market St
Wheeling, WV 26003
304 233-5900
- Health Information Tech Prgm
- Respiratory Therapist (Advanced) Prgm
- Surgical Technology Prgm

**Wheeling Hospital**
Donald H Hofreuter, MD, Administrator
Medical Park
Wheeling, WV 26003
304 243-3000
- Radiography Prgm

**Wheeling Jesuit University**
Rev Joseph R Hacala, SJ, President
316 Washington Ave
Wheeling, WV 26003
304 243-2233
- Nuclear Medicine Technology Prgm
- Physical Therapy Prgm
- Respiratory Therapist (Advanced) Prgm

# Wisconsin

## Appleton

**Affinity Health System**
Robert Turner, MS, COO
St Elizabeth Hospital
1506 S Oneida St
Appleton, WI 54915
920 831-8912
- Clin Lab Scientist/Med Technologist Prgm

**Fox Valley Technical College**
David Buettner, PhD, President
1825 N Bluemound Dr
PO Box 2277
Appleton, WI 54913-2277
920 735-5731
- Dental Assisting Prgm
- Dental Hygiene Prgm
- Medical Assistant Prgm
- Occupational Therapy Asst Prgm

## Cleveland

**Lakeshore Technical College**
Michael Lanser, EdD, Director
1290 North Ave
Cleveland, WI 53015
920 693-1123
- Medical Assistant Prgm
- Radiography Prgm

## Eau Claire

**Chippewa Valley Technical College**
William A Ihlenfeldt, PhD, President
620 W Clairemont Ave
Eau Claire, WI 54701
715 833-6211
- Clin Lab Technician/Med Lab Technician
- Diagnostic Med Sonography Prgm
- Health Information Tech Prgm
- Medical Assistant Prgm
- Radiography Prgm
- Surgical Technology Prgm

**Sacred Heart Hospital**
Steve Ronstrom, MHA, Administrator
900 W Clairemont Ave
Eau Claire, WI 54701
715 839-4131
• Clin Lab Scientist/Med Technologist Prgm

**University of Wisconsin - Eau Claire**
Brian Levin-Stankevich, PhD, Chancellor
Schofield 204
105 Garfield Ave
Eau Claire, WI 54702-4004
715 836-2327
• Athletic Training Prgm
• Music Therapy Prgm
• Speech-Language Pathology Prgm

## Fennimore

**Southwest Wisconsin Technical College**
Karen Knox, EdD, President
1800 Bronson Blvd
Fennimore, WI 53809
602 288-3262
• Medical Assistant Prgm

## Fond du Lac

**Moraine Park Technical College**
Gayle Hytrek, PhD, President
235 N National Ave
PO Box 1940
Fond du Lac, WI 54936-1940
414 922-8611
• Clin Lab Technician/Med Lab Technician
• Health Information Tech Prgm
• Medical Assistant Prgm
• Radiation Therapy Prgm

## Grafton

**Blue Sky School of Professional Massage**
220 Oak St
Grafton, WI 53024
• Massage Therapy Prgm (3)

## Green Bay

**Bellin Health Hosp/Bellin Health Systems Inc**
George Kerwin, MBA, President and CEO
744 S Webster Ave
PO Box 23400
Green Bay, WI 54305-3400
414 433-7898
• Radiography Prgm

**Northeast Wisconsin Technical College**
H Jeffrey Rafn, PhD, President
2740 W Mason St PO Box 19042
Green Bay, WI 54307-9042
920 498-5411
• Clin Lab Technician/Med Lab Technician
• Dental Assisting Prgm
• Dental Hygiene Prgm
• Diagnostic Med Sonography Prgm
• Health Information Tech Prgm
• Medical Assistant Prgm
• Physical Therapist Assistant Prgm
• Respiratory Therapist (Advanced) Prgm
• Surgical Technology Prgm

**University of Wisconsin - Green Bay**
Mark L Perkins, Chancellor
Green Bay, WI 54311
414 465-2000
• Dietetic Internship Prgm
• Dietetics-Didactic Prgm

## Hayward

**Lac Courte Oreills Ojibwa Community College**
Danielle Hornett, PhD, President
13466 Trepania Rd
Hayward, WI 54843
715 634-4790
• Medical Assistant Prgm

## Janesville

**Blackhawk Technical College**
Eric Larson, EdD, President
6004 Prairie Rd
PO Box 5009
Janesville, WI 53547-5009
608 756-4121
• Dental Assisting Prgm
• Medical Assistant Prgm
• Physical Therapist Assistant Prgm
• Radiography Prgm

## Kenosha

**Carthage College**
F Gregory Campbell, President
2001 Alford Park Dr
Kenosha, WI 53140-1994
262 551-5858
• Athletic Training Prgm

**Gateway Technical College**
Bryan Albrecht, PhD, President
3520 - 30th Ave
Kenosha, WI 53144
262 564-2200
• Dental Assisting Prgm
• Health Information Tech Prgm
• Medical Assistant Prgm (2)
• Physical Therapist Assistant Prgm
• Surgical Technology Prgm

## La Crosse

**Gundersen Lutheran Medical Foundation**
Jeffrey Thompson, PhD, CEO
1900 South Ave
La Crosse, WI 54601
608 782-7300
• Nuclear Medicine Technology Prgm

**University of Wisconsin - La Crosse**
Elizabeth Hitch, PhD, Chancellor
1725 State St
135 Main Hall
La Crosse, WI 54601
608 785-8004
• Athletic Training Prgm
• Occupational Therapy Prgm
• Physical Therapy Prgm
• Physician Assistant Prgm
• Radiation Therapy Prgm
• Therapeutic Recreation Specialist Prgm

**Viterbo College**
Robert E Gibbons, PhD, President
818 S 9th St
La Crosse, WI 54601
608 784-0040
• Dietetic Internship Prgm
• Dietetics-Coordinated Prgm

**Western Technical College**
J Lee Rasch, EdD, President/District Dir
304 N Sixth St
PO Box C-908
La Crosse, WI 54602-0908
608 785-9210
• Clin Lab Technician/Med Lab Technician
• Dental Assisting Prgm
• Electroneurodiagnostic Tech Prgm
• Health Information Tech Prgm
• Medical Assistant Prgm
• Occupational Therapy Asst Prgm
• Physical Therapist Assistant Prgm
• Radiography Prgm
• Respiratory Therapist (Advanced) Prgm
• Surgical Technology Prgm

## Madison

**Madison Area Technical College**
Bettsey Barhorst, PhD, President
3550 Anderson St
Madison, WI 53704-2599
608 246-6676
• Clin Lab Technician/Med Lab Technician
• Dental Hygiene Prgm
• Dietetic Technician-AD Prgm
• Medical Assistant Prgm
• Occupational Therapy Asst Prgm
• Radiography Prgm
• Respiratory Therapist (Advanced) Prgm
• Surgical Technology Prgm

**State Laboratory of Hygiene**
R H Laessig, PhD, Director
Center for Health Studies - UW Madison
465 Henry Hall
Madison, WI 53706
608 262-1293
• Cytotechnology Prgm

**University of Wisconsin - Madison**
John Wiley, PhD, Chancellor
500 Lincoln Dr
161 Bascom Hall
Madison, WI 53706-1380
608 262-9946
• Athletic Training Prgm
• Audiologist Prgm
• Clin Lab Scientist/Med Technologist Prgm
• Dietetics-Coordinated Prgm
• Dietetics-Didactic Prgm
• Genetic Counseling Prgm
• Medical Librarian Prgm
• Occupational Therapy Prgm
• Physical Therapy Prgm
• Physician Assistant Prgm
• Rehabilitation Counseling Prgm
• Speech-Language Pathology Prgm

**University of Wisconsin Hospital and Clinics**
Donna K Sollenberger, MHA, President and CEO
600 Highland Ave
Dept of Administration
Madison, WI 53792-8350
608 263-8025
- Diagnostic Med Sonography Prgm
- Dietetic Internship Prgm
- Orthoptist Prgm
- Radiography Prgm

## Marshfield

**Marshfield Clinic**
Reed Hall, MS JD, Executive Director
1000 N Oak Ave
Marshfield, WI 54449
715 387-5123
- Cytotechnology Prgm

**St Joseph's Hospital**
Michael A Schmidt, CPA, President/CEO
611 St Joseph Ave
Marshfield, WI 54449-1898
715 387-1713
- Clin Lab Scientist/Med Technologist Prgm
- Histotechnician Prgm
- Nuclear Medicine Technology Prgm
- Radiography Prgm

## Menasha

**Mercy Medical Center/Affinity Health System**
Daniel Neufelder, MS, President
1570 Midway Place
Menasha, WI 54952
920 720-1713
- Radiography Prgm

## Menomonie

**University of Wisconsin - Stout**
Charles W Sorensen, Chancellor
Menomonie, WI 54751
715 232-1123
- Dietetic Internship Prgm
- Dietetics-Didactic Prgm
- Rehabilitation Counseling Prgm

## Mequon

**Concordia University Wisconsin**
Patrick Ferry, PhD, President
12800 N Lake Shore Dr
Mequon, WI 53097
262 243-4368
- Athletic Training Prgm
- Medical Assistant Prgm
- Occupational Therapy Prgm
- Physical Therapy Prgm

## Milwaukee

**Alverno College**
Mary Meehan, President
3401 S 39th St, Box 343922
Milwaukee, WI 53234-3922
414 382-6000
- Music Therapy Prgm

**Aurora St Luke's Medical Center**
Nick Turkal, MD, President
2900 W Oklahoma Ave
PO Box 2901
Milwaukee, WI 53201-2901
414 649-7500
- Diagnostic Med Sonography Prgm
- Nuclear Medicine Technology Prgm
- Radiography Prgm

**Blood Center of Southeast Wisconsin**
William V Miller, MD, President
PO Box 2178
Milwaukee, WI 53201-2178
414 937-6338
- Specialist in BB Tech Prgm

**Bryant & Stratton College - Milwaukee**
Peter Pavone, MS, Campus Director
310 W Wisconsin Ave, Ste 500 E
Milwaukee, WI 53203
414 276-5200
- Medical Assistant Prgm

**Clement J Zablocki VA Medical Center**
Glen Grippen, Director
5000 W National Ave
Milwaukee, WI 53295
414 384-2000
- Clin Lab Scientist/Med Technologist Prgm

**Columbia St Mary's Hospitals**
Leo Brideau, President, CEO
2025 E Newport Ave
Milwaukee, WI 53211
414 961-3638
- Diagnostic Med Sonography Prgm
- Radiography Prgm

**Froedtert Memorial Lutheran Hospital**
William D Petasnick, MHA, President
9200 W Wisconsin Ave
PO Box 26099
Milwaukee, WI 53226
414 805-3000
- Nuclear Medicine Technology Prgm
- Radiography Prgm

**Lakeside School of Massage Therapy**
Carole Ostendorf, PhD PT, Chief Executive
Officer
1726 N 1st St, Ste 200
Milwaukee, WI 53212
414 372-4345
- Massage Therapy Prgm (2)

**Marquette University**
Robert A Wild, SJ, President
O'Hara Hall
Milwaukee, WI 53201-1881
414 288-7223
- Athletic Training Prgm
- Clin Lab Scientist/Med Technologist Prgm
- Physical Therapy Prgm
- Physician Assistant Prgm
- Speech-Language Pathology Prgm

**Milwaukee Area Technical College**
Darnell Cole, PhD, President
700 W State St
Milwaukee, WI 53233-1443
414 297-6320
- Cardiovascular Technology Prgm
- Clin Lab Technician/Med Lab Technician
- Dental Hygiene Prgm
- Dietetic Technician-AD Prgm
- Medical Assistant Prgm
- Occupational Therapy Asst Prgm
- Ophthalmic Dispensing Optician Prgm
- Phlebotomist Prgm
- Physical Therapist Assistant Prgm
- Radiography Prgm
- Respiratory Therapist (Advanced) Prgm
- Surgical Technology Prgm

**Milwaukee School of Engineering**
Hermann Viets, PhD, President
1025 N Broadway
Milwaukee, WI 53202-3109
414 277-7100
- Perfusion Prgm

**Mount Mary College**
Patricia D O'Donoghue, President
2900 N Menomonee River Pkwy
Milwaukee, WI 53222-4597
414 258-4810
- Art Therapy Prgm
- Dietetic Internship Prgm
- Dietetics-Coordinated Prgm
- Occupational Therapy Prgm

**St Francis Hospital**
Debra Standridge, MBA, President
3237 S 16th St
Milwaukee, WI 53215
414 647-5106
- Diagnostic Med Sonography Prgm

**St Michael Hospital**
Jeffery K Jenkins, MHA, President/CEO
2400 W Villard Ave
Milwaukee, WI 53209
414 527-8123
- Radiography Prgm

**University of Wisconsin - Milwaukee**
Carlos Santiago, PhD, Chancellor
PO Box 413
Milwaukee, WI 53201
414 229-4331
- Athletic Training Prgm
- Clin Lab Scientist/Med Technologist Prgm
- Cytotechnology Prgm
- Medical Librarian Prgm
- Occupational Therapy Prgm
- Speech-Language Pathology Prgm

## Neenah

**Theda Clark Regional Medical Center**
John Toiusant, MD, President
130 Second St
PO Box 2021
Neenah, WI 54957-2021
920 729-3100
- Radiography Prgm

## Oshkosh

**University of Wisconsin - Oshkosh**
Richard Wells, Chancellor
800 Algoma Blvd
Oshkosh, WI 54901
920 424-0200
• Athletic Training Prgm
• Counseling Prgm

## Pewaukee

**Waukesha County Technical College**
Barbara Prindiville, PhD, President
800 Main St
Pewaukee, WI 53072
414 691-5435
• Dental Hygiene Prgm
• Medical Assistant Prgm
• Surgical Technology Prgm

## Racine

**Wheaton Franscian Healthcare-All Saints**
Kenneth R Buser, FACHE, President and CEO
3801 Spring St
Racine, WI 53405
262 687-4285
• Radiography Prgm

## Rhinelander

**Nicolet Area Technical College**
Adrian Lorbetske, JD, President
PO Box 518
Rhinelander, WI 54501
715 365-4410
• Medical Assistant Prgm

## River Falls

**University of Wisconsin - River Falls**
Gary A Thibodeau, Chancellor
River Falls, WI 54022
715 425-3911
• Speech-Language Pathology Prgm

## Shell Lake

**Wisconsin Indianhead Technical College**
Charles Levine, PhD, Interim President
505 Pine Ridge Dr
Shell Lake, WI 54871-9300
715 468-2815
• Medical Assistant Prgm (3)
• Occupational Therapy Asst Prgm
• Radiography Prgm

## Stevens Point

**University of Wisconsin - Stevens Point**
Linda Bunnell, PhD, Chancellor
213 Main
Stevens Point, WI 54481
715 346-2123
• Athletic Training Prgm
• Audiologist Prgm
• Clin Lab Scientist/Med Technologist Prgm
• Dietetics-Didactic Prgm
• Speech-Language Pathology Prgm

## Superior

**University of Wisconsin - Superior**
Julius E Erlenbach, Chancellor
1800 Grand Ave
Superior, WI 54880-2898
715 394-8101
• Counseling Prgm

## Waukesha

**Carroll College**
Douglas N Hastad, DEd, President
100 N East Ave
Waukesha, WI 53186-5593
414 547-1211
• Athletic Training Prgm
• Physical Therapy Prgm

## Wausau

**Aspirus Wausau Hospital**
Duane Erwin, JD, President and CEO
333 Pine Ridge Blvd
Wausau, WI 54401
715 847-2259
• Clin Lab Scientist/Med Technologist Prgm

**Northcentral Technical College**
Lori A Weyers, PhD, President
1000 Campus Dr
Wausau, WI 54401-1899
715 675-3331
• Dental Hygiene Prgm
• Phlebotomist Prgm
• Radiography Prgm
• Surgical Technology Prgm

## Whitewater

**University of Wisconsin - Whitewater**
Jack Miller, Chancellor
800 W Main
Whitewater, WI 53190-1791
414 472-1918
• Counseling Prgm
• Speech-Language Pathology Prgm

## Wisconsin Rapids

**Mid-State Technical College**
John Clark, PhD, President
500 32nd St N
Wisconsin Rapids, WI 54494
715 423-5650
• Medical Assistant Prgm (2)
• Phlebotomist Prgm
• Respiratory Therapist (Advanced) Prgm
• Surgical Technology Prgm

# Wyoming

## Casper

**Casper College**
Walter Nolte, PhD, President
125 College Dr
Casper, WY 82601
307 268-2547
• Occupational Therapy Asst Prgm
• Radiography Prgm

**University of North Dakota at Casper College**
Walter Nolte, President
125 College Dr
Casper, WY 82601-9958
• Occupational Therapy Prgm

## Cheyenne

**Laramie County Community College**
Darrel L Hammon, PhD, President
1400 E College Dr
Cheyenne, WY 82007
307 778-5222
• Dental Hygiene Prgm
• Radiography Prgm
• Surgical Technology Prgm

## Laramie

**University of Wyoming**
Thomas Buchanan, PhD, President
Dept. 3434
1000 E University Ave
Laramie, WY 82071
307 766-4121
• Athletic Training Prgm
• Audiologist Prgm
• Counseling Prgm
• Dietetics-Didactic Prgm
• Speech-Language Pathology Prgm

## Sheridan

**Sheridan College**
Kevin Drumm, EdD, President
3059 Coffeen Ave
Box 1500
Sheridan, WY 82801
307 674-6446
• Dental Hygiene Prgm

INSTITUTIONS

# Section III

# Accrediting Agencies

Section III, Accrediting Agencies, provides a summary description of the agencies that accredit/approve education programs listed in the *Directory*. Refer to Table 1 on the following page for a complete list of the occupations for which these agencies accredit/approve programs.

## Participating in the Directory

Accrediting agencies meeting the criteria described below may participate in the process that results in the publication of educational programs in the *Directory*.

Accrediting agencies, including those that use the term *approve* instead of *accredit*, may do either of the following:

- Participate in the annual Survey of Health Professions Education Programs and agree to subsequent publication of program information in the *Directory*
- Provide program information for publication in the *Directory*

To be included in the *Directory*, professions are generally required to meet the following criteria.

### Educational Programs and Credentials

The profession requires formal postsecondary education through a program in an art, science, or therapy related to health care that meets national standards and that culminates in a certificate, diploma, associate degree, baccalaureate degree, master's degree, doctorate, or post-degree certificate. (Not included are degrees of doctor of medicine, osteopathy, dentistry, veterinary medicine, optometry, podiatry, pharmacy, chiropractic, or clinical psychology; any level of nursing education; or graduate degrees in public health or health administration.)

Educational programs for the professions in this *Directory* are sponsored by a variety of institutions, including 4-year colleges, universities, academic health centers, and health care institutions. Programs are also sponsored by junior and community colleges, vocational and technical schools, proprietary schools, medical schools, blood banks, consortia, and US government institutions.

### Professional Responsibility

Members of the profession participate in the delivery of health care, diagnostic and rehabilitation services, therapeutic treatments, or related services, such as

- Identifying, evaluating, treating, or preventing diseases or disorders
- Providing dietetic and nutrition counseling and services
- Promoting healthy behaviors
- Providing health maintenance, rehabilitative, or therapeutic services
- Managing health or information services
- Addressing psychosocial and cognitive needs (eg, music and art therapy)

### Recognition of Accreditation or Approval Process

The agency evaluating the educational programs in accordance with national standards developed by the profession must be recognized for that purpose by the US Department of Education, the Council on Higher Education Accreditation, or an equivalent agency.

### Educational Standards

To become accredited, programs are assessed for their compliance with national educational standards, which generally state the responsibilities and qualifications of a program director, a medical director, and other program officials and faculty, along with requirements related to curriculum, physical resources, financing, and records. Standards are usually flexible enough to allow educational programs to respond to continuous technological changes.

### Definition and Benefits of Accreditation

Accreditation is a process of external peer review in which a private, nongovernmental agency or association grants public recognition to an institution or specialized program of study that meets established qualifications and educational standards, as determined through initial and subsequent periodic evaluations. The process encourages educational institutions and programs to continuously evaluate and improve their processes and outcomes. Accreditation helps prospective students identify institutions with programs that meet standards established by and for the field(s) in which they are interested and assists those who wish to transfer from one institution to another. For institutions, accreditation protects against pressures to modify programs for reasons that are not educationally sound, involves faculty and staff in comprehensive program and institutional evaluation and planning, and stimulates self-improvement by providing national standards against which the institution can evaluate the program(s) it sponsors.

Accreditation also benefits society by providing reasonable assurance of quality educational preparation for professional certification, registration, or licensure. It may be used along with other considerations as a basis for determining eligibility for some types of federal assistance and for identifying institutions and programs for the investment of public and private funds.

### AMA Involvement in the Health Professions

This *Directory* continues to reflect more than 60 years of cooperation in the complex processes related to health professions education and accreditation and in efforts to disseminate information about accredited programs to as wide an audience as possible. The current edition comprises 25 agencies accrediting more than 6,600 educational programs in 71 professions.

For nearly 20 years (1976 to 1994), the American Medical Association (AMA), in collaboration with more than 50 national health professions organizations and medical specialty societies, cooperated with as many as 22 review committees to accredit more than 3,000 programs in 27 professions. That process today is conducted by the accrediting agencies described below, including CAAHEP, which retains 18 of the professions and 16 of the review committees formerly in CAAHEP's predecessor, the CAHEA system (Committee on Allied Health Education and Accreditation). ACOTE, JRCERT, JRCNMT, and NAACLS were formerly a part of that system.

All former members of the CAHEA system participate in the annual survey of accredited programs conducted by the AMA, which requests correct program and institution names, addresses, and personnel; availability of evening/weekend classes; changes in tuition, class capacity, starting date, and credential awarded; and

## Table 1. Twenty-six accrediting agencies and the occupations for which they accredit/approve programs

**Accrediting Agency**
    **Occupations**

Accreditation Council for Occupational Therapy Education (ACOTE)
    Occupational Therapist
    Occupational Therapy Assistant

Accreditation Review Commission on Education for the Physician Assistant (ARC-PA)
    Physician Assistant

American Art Therapy Association (AATA)
    Art Therapist

American Board of Genetic Counseling (ABGC)
    Genetic Counselor

American Dance Therapy Association (ADTA)
    Dance Therapist

American Orthoptic Council (AOC)
    Orthoptist

American Society of Health-System Pharmacists (ASHP)
    Pharmacy Technician

Association for Education and Rehabilitation of the Blind and Visually Impaired (AER)
    Low Vision Therapist
    Orientation and Mobility Specialist
    Rehabilitation Teacher
    Teacher of the Visually Impaired

Commission on Accreditation for Dietetics Education (CADE) of the American Dietetic Association
    Dietetic Technician
    Dietitian/Nutritionist

Commission on Accreditation for Health Informatics and Information Management Education (CAHIIM)
    Health Information Administrator
    Health Information Technician

Commission on Accreditation in Physical Therapy (CAPTE)
    Physical Therapist
    Physical Therapist Assistant

Commission on Accreditation of Allied Health Education Programs (CAAHEP)
    Anesthesiologist Assistant
    Cardiovascular Technologist
    Cytotechnologist
    Diagnostic Medical Sonographer
    Electroneurodiagnostic Technologist
    Emergency Medical Technician-Paramedic
    Exercise Physiologist (Applied and Clinical)
    Exercise Science Professional
    Kinesiotherapist
    Medical Assistant
    Medical Illustrator
    Orthotist/Prosthetist
    Perfusionist
    Polysomnographic Technologist
    Respiratory Therapist (Advanced)
    Respiratory Therapist (Entry-Level)
    Specialist in Blood Bank Technology
    Surgical Assistant
    Surgical Technologist

Commission on Accreditation of Athletic Training Education (CAATE)
    Athletic Trainer

**Accrediting Agency**
    **Occupations**

Commission on Accreditation of Ophthalmic Medical Programs (CoA-OMP)
    Ophthalmic Assistant
    Ophthalmic Medical Technician/Technologist

Commission on Dental Accreditation (CODA) of the American Dental Association
    Dental Assistant
    Dental Hygienist
    Dental Laboratory Technician

Commission on Massage Therapy Accreditation (COMTA)
    Massage Therapist

Commission on Opticianry Accreditation (COA)
    Ophthalmic Dispensing Optician
    Ophthalmic Laboratory Technician

Council for Accreditation of Counseling and Related Educational Programs (CACREP)
    Career Counselor
    College Counselor
    Community Counselor
    Gerontological Counselor
    Marital, Couple and Family Counselor/Therapist
    Mental Health Counselor
    School Counselor
    Student Affairs
    Student Affairs-College Counselor
    Student Affairs-Professional Practice
    Counselor Education and Supervision

Council on Academic Accreditation in Audiology and Speech-Language Pathology (CAA)
    Audiologist
    Speech-language Pathologist

Council on Rehabilitation Education (CORE)
    Rehabilitation Counselor

Joint Review Committee on Education in Radiologic Technology (JRCERT)
    Radiation Therapist
    Radiographer

Joint Review Committee on Educational Programs in Nuclear Medicine Technology (JRCNMT)
    Nuclear Medicine Technologist

Medical Library Association/American Library Association
    Medical Librarian

National Accrediting Agency for Clinical Laboratory Sciences (NAACLS)
    Clinical Assistant
    Clinical Laboratory Technician/Medical Laboratory Technician
    Clinical Laboratory Scientist/Medical Technologist
    Cytogenetic Technologist
    Diagnostic Molecular Scientist
    Histologic Technician/Technologist
    Pathologists' Assistant
    Phlebotomist

National Association of Schools of Music (NASM)
    Music Therapist

National Recreation and Park Association/American Association for Physical Activity and Recreation Council on Accreditation
    Therapeutic Recreation Specialist

statistics on enrollments, graduates, and attrition by gender. Accrediting agency and AMA staff cooperate in updating the program and institution population in the database, verifying the accuracy of the data, and designing the survey instrument.

# Accreditation Council for Occupational Therapy Education

Neil Harvison, PhD, OTR, Director of Accreditation
American Occupational Therapy Association
4720 Montgomery Lane, PO Box 31220
Bethesda, MD 20824-1220
301 652-2682
301 652-1417 Fax
www.aota.org

## Programs Accredited
- Occupational therapist
- Occupational therapy assistant

## Structure and Functions
The Accreditation Council for Occupational Therapy Education (ACOTE) of the American Occupational Therapy Association (AOTA), recognized by the US Department of Education (ED) and Council for Higher Education Accreditation (CHEA), is an associated body of the AOTA Board of Directors. ACOTE evaluates professional and technical curricula, reviews and revises the *Standards for an Accredited Educational Program for the Occupational Therapist* and *Standards for an Accredited Educational Program for the Occupational Therapy Assistant*, and accredits occupational therapist and occupational therapy assistant programs.

## History
The National Society for the Promotion of Occupational Therapy was founded in 1917 and incorporated under the laws of the District of Columbia.

The objective of the association as set forth in its Constitution "shall be to study and advance curative occupations for invalids and convalescents; to gather news of progress in occupational therapy and to use such knowledge to the common good; to encourage original research; [and] to promote cooperation among occupational therapy societies and with other agencies of rehabilitation."

About 3 years after its incorporation, the association was urged by several leading physicians and authorities on hospital administration to establish a national register or directory of occupational therapists "for the protection of hospitals and institutions from unqualified persons posing as occupational therapists."

After careful consideration and on the advice of other national organizations in the field of medicine, the association decided that the first step toward the establishment of a national register or directory was the establishment of minimum standards of training for occupational therapists.

In 1921, the name of the association was changed to the American Occupational Therapy Association. In 1923, accreditation of educational programs became a stated function of the AOTA, and basic educational standards were developed.

The AOTA approached the AMA Council on Medical Education in 1933 to request cooperation in the development and improvement of educational programs for occupational therapists.

The *Essentials of an Acceptable School of Occupational Therapy* were adopted by the AMA House of Delegates in 1935. This action represented the first cooperative accreditation activity by the AMA.

In 1958, the AOTA assumed responsibility for approval of educational programs for the occupational therapy assistant. The

**ACCREDITING AGENCIES**

standards on which accreditation was based were modeled after the *Essentials* established for baccalaureate programs.

In 1964, the AOTA/AMA collaborative relationship in accreditation was officially recognized by the National Commission on Accrediting (NCA). The NCA was a private agency serving as a coordinating agency for accrediting activities in higher education. Although it had no legal authority, it had great influence on educational accreditation through the listing of accrediting agencies it recommended to its members. The NCA continued its activities in merger with the Federation of Regional Accrediting Commissions of Higher Education in January 1975. The new organization was the Council on Postsecondary Accreditation (COPA).

In 1990, the AOTA petitioned the Committee on Allied Health Education and Accreditation (CAHEA) to include the accreditation of the occupational therapy assistant programs in the CAHEA system. Following approval of the change by the AMA Council on Medical Education, CAHEA petitioned both the COPA and the ED for recognition as the accrediting body for occupational therapy assistant education.

In 1991, occupational therapy assistant programs with approval status from the AOTA Accreditation Committee became accredited by CAHEA/AMA in collaboration with the AOTA Accreditation Committee.

On January 1, 1994, the AOTA Accreditation Committee changed its name to the Accreditation Council for Occupational Therapy Education (ACOTE) and became operational as an accrediting agency independent of CAHEA/AMA.

During 1994, ACOTE became listed by the ED as a nationally recognized accrediting agency for professional programs in the field of occupational therapy. ACOTE was also granted initial recognition by the successor of COPA, the Commission on Recognition of Postsecondary Accreditation (CORPA) (now Council for Higher Education Accreditation [CHEA]), the nongovernmental recognition agency for accrediting bodies.

In 1994, 197 previously accredited/approved and developing occupational therapy and occupational therapy assistant educational programs were transferred into the ACOTE accreditation system. In that year, the AOTA membership also approved the creation of AOTA's new accrediting body and establishment of ACOTE as a body of the AOTA Executive Board.

## Purpose and Responsibilities

The purposes of the ACOTE accreditation process are to

1. Encourage continuous self-analysis and improvement of the occupational therapy program by representatives of the institution's administrative staff, teaching faculty, students, governing body, and other appropriate constituencies, with the ultimate aim of assuring students of quality education and assuring recipients of services appropriate occupational therapy care.

2. Accredit entry-level occupational therapy educational programs, evaluate other occupational therapy programs as appropriate, and review and revise the *Standards for an Accredited Educational Program for the Occupational Therapist* and *Standards for an Accredited Educational Program for the Occupational Therapy Assistant.*

3. Determine whether the occupational therapy educational program meets the appropriate approved educational standards.

4. Provide counsel and assistance during the accreditation process to the faculty and administrative personnel of the occupational therapy program and the institution in which it is located and to aid them in recognizing the program's strengths and limitations.

5. Encourage faculty to anticipate and accommodate new trends and developments in the practice of occupational therapy that should be incorporated into the educational process.

6. Assure the educational community, general public, and other agencies or organizations that the program has both clearly defined and appropriate objectives, maintains conditions under which these objectives can reasonably be expected to be achieved, appears to be accomplishing them substantially, and can be expected to continue to do so.

## Standards

*Standards*, revised in 1998, are the minimum educational requirements adopted by AOTA's ACOTE. Each new program is assessed in accordance with the *Standards*, and accredited programs are periodically reviewed to determine whether they remain in substantial compliance. The *Standards* are available at AOTA's home page at www.aota.org and on written request from the AOTA Accreditation Program.

## Composition

ACOTE consists of 18 members, 15 who are selected from the Roster of Accreditation Evaluators, which includes the chair of ACOTE, and three who are appointed public members. The AOTA staff liaison serves as member without voting privileges. The members represent professional and technical levels of education (fieldwork experience) and practice. To provide for continuity within the agency, not more than one third of the membership may rotate off the ACOTE in a single year.

The ACOTE Roster of Accreditation Evaluators consists of up to 105 trained evaluators who do not serve on ACOTE. The Roster includes occupational therapists and occupational therapy assistants with expertise in either occupational therapy or occupational therapy assistant education, fieldwork, practice, administration, or other areas of special expertise. The Roster is maintained on a 3-year staggered rotational basis. New members are recommended for the Roster by ACOTE. Individuals may self-nominate directly to the AOTA Staff Liaison.

### ACOTE Members

Margaret Newsham Beckley, PhD, OTR/L
Skip Capone III, JD, Public Member
Constance Daby, MHS, OTR
Carol A Doehler, MS, OTR/L, Vice Chairperson
Diane Gaffney, MBA, OTR
Jamie Geraci, MS, OTR/L
Beth Ann Hatkevich, MOT, OTR/L
Cindy Kief, MS, COTA/L
Sheama Krishnagiri, PhD, OTR/L
Patricia J Martin, BS, COTA/L
Jane Olson, PhD, OTR, FAOTA, Chairperson
Laurel Cargill Radley, MS, OTR/L
Perri Stern, EdD, OTR/L, FAOTA
Joyce A Wandel, MS, OTR/L
Keith Williams, MA, Public Member
Eunice Zee-Chen, MS, OTR/L, FAOTA

# Accreditation Review Commission on Education for the Physician Assistant

John McCarty, Executive Director
12000 Findley Road, Suite 240
Duluth, GA 30097
770 476-1224
770 476-1738 Fax
E-mail: arc-pa@arc-pa.org
www.arc-pa.org

## Programs Accredited
- Physician assistant

## Structure and Functions
The Accreditation Review Commission on Education for the Physician Assistant (ARC-PA) is the designated authority for evaluating PA education and the sole recognized accrediting agency. In January 2004, the ARC-PA was awarded recognition by the Council for Higher Education Accreditation (CHEA). The ARC-PA collaborating organizations include the American Academy of Family Physicians, the American Academy of Pediatrics, the American Academy of Physician Assistants, the American College of Physicians, the American College of Surgeons, the American Medical Association, and the Physician Assistant Education Association (formerly the Association of Physician Assistant Programs). The ARC-PA also has a public member.

## History
The profession of physician assistant (PA) originated in the mid-1960s. The early 1970s brought a rapid growth in the number of such educational programs, which were supported initially with $6.1 million appropriated under the authority of the Health Manpower Act of 1972. This funding also supported some of the initial organization and administration of the national program for the accreditation of PA educational programs, specifically those designed to prepare individuals as assistants to primary care practitioners.

Interest in the development of national accreditation standards for the education of assistants to primary care physicians was first expressed by the American Society of Internal Medicine (ASIM).

By 1971, standards had been developed collaboratively by a committee composed of representatives from the American Academy of Family Physicians (AAFP), the American Academy of Pediatrics (AAP), the American College of Physicians (ACP), the AMA, the Association of American Medical Colleges (AAMC), the American College of Obstetrics and Gynecology (ACOG), the ASIM, the nursing profession, and educators of the physician assistant. These standards were adopted in that year by the AMA, AAFP, AAP, ACP, and ASIM. (The ASIM withdrew its sponsorship of accreditation in September 1981.)

Early in 1972, the medical specialty organizations that had adopted the new educational standards established the Joint Review Committee on Educational Programs for the Assistant to the Primary Care Physician. A principal function of the committee was to assess the extent to which applicant programs were in compliance with the accreditation standards, then called the *Essentials and Guidelines for the Assistant to the Primary Care Physician,* and to formulate recommendations for accreditation to the AMA Council on Medical Education (CME). This committee was composed of three representatives from each of the four sponsoring organizations. In April 1973, the committee appointed three graduate PAs to serve as members-at-large for 1-year terms. By March 1974, the sponsors of the committee and the AMA had recognized the American Academy of Physician Assistants as the fifth sponsor of the review committee.

Standards for the surgeon assistant were adopted by the American College of Surgeons in 1973 and by the AMA in 1974. Originally, the American College of Surgeons Committee on Allied Health Personnel reviewed applicant programs' compliance with these standards.

As a result of discussions initiated in 1975, the review committees for the assistant to the primary care physician and surgeon assistant were brought together in 1976 into a unified accreditation review committee. On petition from the Association of Physician Assistant Programs, the collaborating sponsoring organizations of the accreditation review committee and the AMA recognized it as the seventh sponsor of the committee in 1978. The AMA became a sponsor of the review committee in 1994 following the dissolution of the Committee on Allied Health Education and Accreditation and the establishment of CAAHEP. The committee was renamed the Accreditation Review Committee on Education for the Physician Assistant in 1988 and became the Accreditation Review Commission on Education for the Physician Assistant in January 2001 when it left the CAAHEP system to become an independent organization.

Following a 2-year consultation with accredited educational programs, sponsors of the accreditation service, and other interested parties, revised *Standards* were adopted for the education of assistants to primary care physicians in 1978. Following a similar consultation, the revised *Standards* were adopted in 1985 as standards for the education of PAs. In 1990, the accreditation standards were consolidated for PA and surgeon assistant education to ensure that both received a comparable base of knowledge and skill in primary care medicine. The ARC-PA has a regular 5-year schedule for revision of the Standards, with the most recent changes going into effect in 2006.

## Standards
The *Accreditation Standards for Physician Assistant Education* defines the requirements to achieve and maintain programmatic accreditation. Each program is assessed and periodically reviewed to determine whether it remains in compliance with the *Standards*. The *Standards* are available on the ARC-PA Web site (www.arc-pa.org).

The *Standards* include criteria for administration, physical and instructional resources, faculty, curriculum, fair practices, educational effectiveness, self-study, outcomes, and other areas. Accreditation assures that a program operates in compliance with these *Standards.*

## Purpose and Responsibilities
The ARC-PA encourages excellence in PA education through its accreditation process. It establishes and maintains minimum standards of quality for educational programs. It awards accreditation to programs through a peer review process that includes documentation and periodic site visit evaluation to substantiate compliance with the *Accreditation Standards for Physician Assistant Education*. The accreditation process is designed to encourage sound educational experimentation and innovation and to stimulate continuous self-study and improvement.

Provisional accreditation is a time-limited accreditation status available to new PA programs that can demonstrate their ability to comply with the *Standards*. It is awarded prior to enrollment of the charter class of students.

**ACCREDITING AGENCIES**

Accreditation provides assurance that a PA educational program accepts and fulfills its commitment to educational quality. Graduation from an accredited program is an eligibility requirement for the Physician Assistant National Certifying Examination and for state licensure.

In addition to establishing educational standards and fostering excellence in PA programs, the ARC-PA provides information and guidance to individuals and organizations regarding PA program accreditation and makes public its accreditation actions.

## Composition
The ARC-PA is composed of 17 commissioners. Each collaborating organization nominates two commissioners, with the exception of the American Academy of Physician Assistants and the Physician Assistant Education Association, which nominate three commissioners. There is one public commissioner. Each commissioner serves a 3-year term; commissioners may serve for no more than two consecutive 3-year terms.

Patrick C Auth, PhD, PA-C (American Academy of Physician Assistants)

James L Brand, MD, FAAFP (American Academy of Family Physicians)

Elizabeth A Burns, MD, MA (American Medical Association)

Richard E Davis, EdD, PA-C, Treasurer (American Academy of Physician Assistants)

Patricia M Dieter, MPA, PA-C (Physician Assistant Education Association)

Timothy C Evans, MD, PhD, FACP, Vice Chair (American College of Physicians)

Paul B Halverson, MD (American College of Physicians)

Catherine R Judd, MS, PA-C (American Academy of Physician Assistants)

Lawrence Thomas Kim, MD, FACS (American College of Surgeons)

James V Lustig, MD (American Academy of Pediatrics)

Michael E Ruhlen, MD, MHCM, FAAP (American Academy of Pediatrics)

Michael R Greenberg, MD (American Medical Association)

Quincy O Scott, DO, FAAFP (American Academy of Family Physicians)

Gloria M Stewart, EdD, PA-C Chair (Physician Assistant Education Association)

Richard H Turnage, MD, FACS (American College of Surgeons)

David Weigle, PhD, Public Member

Suzanne York, MPH, PA-C, Secretary (Physician Assistant Education Association)

# American Art Therapy Association

Paula Howie, ATR-BC, LPC, President
5999 Stevenson Avenue
Alexandria, VA 22304
888 290-0878 or 703 212-2238
E-mail: info@arttherapy.org
www.arttherapy.org

## Programs Accredited
- Art therapist

## Structure and Functions
The American Art Therapy Association, (AATA) Inc, was founded in 1969 and develops and promotes educational, professional, and ethical standards for art therapy education and practice. It is a national organization of over 4,600 members and is directed by an 11-member Board elected by the membership. The AATA sponsors annual conferences, approves educational programs, and publishes *Art Therapy: Journal of the American Art Therapy Association* (first published in 1983), the quarterly *AATA Newsletter*, the *AATA E-Newsletter*, books, and monographs.

## Purpose and Responsibilities
The AATA is an organization of professionals and students dedicated to the belief that the creative process involved in the making of art is healing and life enhancing. Its mission is to serve its members and the public by providing standards of professional competence and developing and promoting knowledge in, and of, art therapy. To that end, it establishes and develops requirements for art therapy education and practice and promotes excellence in clinical, professional, educational, and research activities. The AATA provides information to its members and the public regarding the field of art therapy through publications, a scholarly journal, conferences, and a Web site.

## Standards
AATA *Education Standards* published by the AATA describe the requirements for approval of graduate level art therapy programs by the AATA's Educational Program Approval Board (EPAB). Each new art therapy education program is evaluated in accordance with the AATA *Education Standards*, and approved programs are periodically reviewed to ensure that they remain in substantial compliance. Mental health and healthcare facilities, state boards, and other regulatory agencies use the educational standards established by the AATA to define professional requirements for art therapists and the practice of art therapy.

## Composition
President, Paula Howie, ATR-BC, LPC
President Elect, Peggy Dunn Snow, PhD, ATR-BC
Secretary, Gussie Klorer, PhD, ATR-BC, LCSW, LCPC
Treasurer, Kay Stovall, ATR-BC, MFT
Director, Barbara Ball, PhD, ATR-BC, LPC
Director, Ellie Ehrlich, MA, ATR-BC
Director, Irene Rosner David, PhD, ATR-BC
Director, Shannon Scott, ATR-BC
Director, Savneet Talwar, MA, ATR-BC
Director, Terry Tibbetts, PhD, ATR-BC
Speaker, Assembly of Chapters, Yasmine Awais, ATR-BC, MAAT
Speaker-Elect Pro Tem, Assembly of Chapters, Erica Maxion-Curtis, LMFT, ATR

# American Board of Genetic Counseling

Sharon Robinson, MS, Administrator
Elizabeth Balkite, MS, Executive Director
9650 Rockville Pike
Bethesda, MD 20814-3998
E-mail: info@abgc.net
www.abgc.net

## Programs Accredited

- Genetic counseling

## Sponsorship

The American Board of Genetic Counseling (ABGC), Inc, was incorporated in 1993. Prior to the establishment of ABGC, certification of genetic counselors was available from the American Board of Medical Genetics.

## Purpose and Responsibilities

ABGC serves as the official credentialing organization for the genetic counseling profession in the United States and Canada. ABGC establishes the standards of competence for clinical practice through accreditation of graduate programs in genetic counseling. ABGC advances the role of genetic counselors in healthcare through the certification and recertification or qualified professionals. In this way, the work of ABGC protects the public and promotes the ongoing growth and development of the genetic counseling profession.

## Standards

ABGC is dedicated to maintaining the highest level of professionalism in the field of genetic counseling through a rigorous accreditation and credentialing process. This mission involves setting the standards for both graduate training programs and individual practitioners.

## Composition

The ABGC Board of Directors is composed of 10 certified genetic counselors:
Lisa A Amacker-North, MS, 2006-2010
Troy A Becker, MS, 2003-2007
Leslie Cohen, MS, 2004-2008
Brenda M Finucane, MS, 2002-2006
Anne E Greb, MS, 2004-2008
Robin E Grubs, MS, 2005-2009
Barbara J Pettersen, MS, 2006-2010
Daniel L Riconda, MS, 2002-2006
Carol S Walton, MS, 2005-2009
LuAnn Weik, MS, 2003-2007

# American Dance Therapy Association

Robyn Flaum Cruz, PhD, ADTR, President
2000 Century Plaza, Suite 108
10632 Little Patuxent Parkway
Columbia, MD 21044-3263
412 624-6595
E-mail: robyncruz@stargate.net
www.adta.org

## Programs Approved

- Dance/movement therapy

## Structure and Functions

Since its founding in 1966, the American Dance Therapy Association (ADTA) has worked to maintain the highest standards of professional education and competence in the field. ADTA maintains a registry of dance/movement therapists who have met stringent standards of education and experience, sets and monitors standards for the master's level programs that prepare people to become dance/movement therapists, publishes the *American Journal of Dance Therapy*, holds a professional conference every year, and publishes timely monographs for its members and for allied professionals.

## Standards

ADTA bylaws (Article XV, Section I) state: "The Committee on Approval shall be established for the purpose of evaluating Graduate Dance/Movement Therapy Programs. The necessary qualifications and standards to be met by any program that desires to be approved shall be set and stated according to the *ADTA Standards for Graduate Dance/Movement Therapy Programs*. The committee shall review and evaluate the applications for approval at their annual meeting. The committee shall be responsible to the Board of Directors through its chairperson who serves on the board as a non-voting member."

Further requirements of the committee are that a program shall be eligible to apply for approval after two classes have graduated, and any institution applying for the first time may be granted approval for a 3-year period. Following the 3-year approval, application may be made for 6-year approval at the end of the second year. An annual report is required from each approved program to inform the committee of changes in curriculum or faculty or other changes affecting programs. Annual reports are reviewed for conformance with the standards of approval. Graduate programs that change institutions are considered new programs and must apply as such.

## Composition

### 2007-2008 ADTA Board of Directors

Robyn Flaum Cruz, PhD, ADTR, President
Elissa Q White, ADTR, Immediate Past President
Sharon Goodill, PhD, ADTR Vice President
Patricia P Capello, ADTR, Secretary
Meg Chang, EdD, ADTR, Treasurer
Susan Kierr , ADTR, Standards and Ethics Committee
Beth Lucci, PsyD, ADTR, Credentials Committee
Christina Devereaux, ADTR, Public Relations Committee
Lenore Hervey, PhD, ADTR, Education Research and Practice Committee
Ellen Gold Yacoe, ADTR, Government Affairs Committee
Julie Miller, ADTR, Eastern Region MAL

**ACCREDITING AGENCIES**

# American Orthoptic Council

Leslie France, CO, Executive Director
3914 Nakoma Road
Madison, WI 53711
608 233-5383
608 263-4247 Fax (attn: Leslie)
E-mail: lwfranceco@att.net
www.orthoptics.org

## Programs Accredited

- Orthoptist

## Composition

The American Orthoptic Council (AOC) consists of 20 members, who are representatives of the American Academy of Ophthalmology, American Association of Certified Orthoptists, American Association for Pediatric Ophthalmology and Strabismus, American Ophthalmological Society, and Canadian Orthoptic Council:

James Reynolds, MD, President
Richard Freeman, MD, Vice President
David Weakley, Jr, MD, Treasurer
Lisa Rovick, CO Secretary
George S Ellis, Jr, MD, Immediate Past President
Kyle Arnoldi, CO
Ron Biernacki, CO
Edward G Buckley, MD
Monte Del Monte, MD
George Ellis, Jr, MD
Thomas D France, MD
Katherine Fray, CO
Richard Freeman, MD
Bruce Furr, CO
David B Granet, MD
Natalie Kerr, MD
G Robert LaRoche, MD
Cheryl McCarus, CO
Christie Morse, MD
Edward L Raab, MD
James Reynolds, MD
Lisa Rovick, CO
R Michael Siatkowski, MD
David Weakley, Jr, MD

The AOC sends a representative to the Canadian Orthoptic Council, two representatives to the Joint Commission on Allied Health Professions, and one representative to the National Association of Advisors for Health Professions.

### The 11 AOC standing committees
Accreditation
Bylaws
Continuing Education
Editorial
Ethics
Examination
Instruction
Long-Range Planning
Nominating
Program (Sunday Night Symposium) and other symposia
Public Relations/Recruitment

The AOC also has several ad hoc committees:
Ad hoc Program Development
Ad hoc International Relations

Ad hoc Feasibility of Distance Learning Opportunities/
   Master's-level Program
Ad hoc Syllabus Revision

The American Orthoptic Journal is a peer-reviewed, scientific journal available online at www.aoj.org.

## Standards

Each new orthoptic educational program is assessed in accordance with the standards of the AOC. Each accredited program registers with the council annually, and programs are periodically reviewed and site-visited to ensure that they remain in compliance with all requirements.

# American Society of Health-System Pharmacists

Lisa S Lifshin, RPh
Director, Technician Program Development
Accreditation Services Division
7272 Wisconsin Avenue
Bethesda, MD 20814
301 664-8720
301 664-8872 Fax
E-mail: llifshin@ashp.org
www.ashp.org

## Programs Accredited
- Pharmacy technician

## Structure and Functions

The American Society of Health-System Pharmacists (ASHP) is the accrediting body for pharmacy technician educational programs, in addition to pharmacy residency training programs, and is recognized in this capacity by the Centers for Medicare and Medicaid Services (formerly the Health Care Financing Administration). ASHP reviews and revises the *ASHP Accreditation Standard for Pharmacy Technician Training Programs*. The first ASHP accreditation standard was approved in April 1982. There are over 90 pharmacy technician training programs in the United States accredited by ASHP; the first program was accredited in September 1983. Programs, which are reviewed every 6 years, must adhere to the established standards and regulations to be considered for accreditation.

## Purpose and Responsibilities

The purpose of the ASHP accreditation program is to identify and grant public recognition to sites with pharmacy technician training programs that have been evaluated and found to meet the qualifications of the ASHP's technician accreditation standards. Thus, accreditation of a pharmacy technician program by the ASHP offers assurance to applicants that a program meets certain basic requirements and is therefore an acceptable site for training in pharmacy technician practice in organized health care.

## Standards

The *ASHP Accreditation Standard for Pharmacy Technician Training Programs*, revised in 2002, designates the minimum educational requirements adopted by ASHP. Each new program is assessed in accordance with the *Standards*, and accredited programs are periodically reviewed to determine whether they remain in substantial compliance. This document is available at www.ashp.org and on written request from the Accreditation Services Division.

## Composition

The Commission on Credentialing (COC) consists of 18 members—15 of whom are selected from the pharmacists used as accreditation evaluators—including the chair of COC, two public members, and the past president of ASHP. The ASHP staff liaison (the director of the Accreditation Services Division) serves as a member without voting privileges. The members represent professional levels of education (fieldwork experience) and practice. To provide for continuity within the agency, members of the COC, except for the past president of ASHP, are appointed for membership for a 3-year term.

**ACCREDITING AGENCIES**

# Association for Education and Rehabilitation of the Blind and Visually Impaired

Jim Gandorf, Executive Director
1703 North Beauregard Street, Suite 440
Alexandria, VA 22311-1744
703 671-4500
703 671-6391 Fax
E-mail: jgandorf@aerbvi.org
www.aerbvi.org

## Programs Accredited
- Low vision therapist
- Orientation and mobility specialist
- Rehabilitation teacher
- Teacher of the visually impaired

## Structure and Functions
The Association for Education and Rehabilitation of the Blind and Visually Impaired (AER) is an international organization of professionals in the fields of visual impairment and blindness education and rehabilitation. It was founded in 1984 as the result of a merger between the Association for Education of the Visually Handicapped and the American Association of Workers for the Blind. Review and approval of professional preparation programs is conducted through AER's professional specialty divisions. The review process and the standards were established to help ensure adequate administration, faculty, facilities, and curricula to prepare qualified low vision therapists, orientation and mobility specialists, rehabilitation teachers, and teachers of students and adults with visual impairments.

College and university programs that have completed the review process are regularly reviewed at 2- or 5-year intervals, depending on which university program is being reviewed. Standards for initial approval and subsequent reviews are performed by the appropriate AER division through committees established by the division membership:
- Low vision therapist program approval and review are administered by AER Low Vision Division (7).
- Orientation and mobility specialist programs approval and review are administered by AER Orientation and Mobility Division (9).
- Rehabilitation teaching program approval and review are administered by AER Rehabilitation Teaching Division (11).

Professionals who have completed an approved program are eligible to apply for certification from the Academy for Certification of Vision Rehabilitation and Education Professionals.

## Composition
The AER Board of Directors is composed of professionals from across the United States and Canada and from various areas in the field of education and rehabilitation. Elected into office every 2 years, these individuals provide leadership in moving AER forward to constantly improve benefits and services to the professionals who work with blind and visually impaired people.

### 2006-2008 Officers
Sandra Ruconich, President
Marybeth Dean, President-elect
Gregory Goodrich, Immediate Past President
Marie J Amerson, Secretary

Virginia (Ginny) Backscheider, Treasurer

**Directors**
Jane I Parsard, Canadian Representative
Michelle Clyne, Council of Chapter Presidents Representative

**Board Members At Large**
Patricia Leader, District 1 Representative
Olivia Chavez, District 2 Representative
Jennifer Ottowitz, District 3 Representative
Cammy Holway-Moraros, District 4 Representative
Sandra Lewis, District 5 Representative
Jay Stiteley, District 6 Representative

**Division Representatives**
Tony Candela, CDC & Adult Services (22+): Divisions 2, 6, 11, 15
Wendy K Sapp, Related Services: Divisions 1, 4, 5, 7, 17
Douglas A McJannet, Instructional Services (0-21): Divisions 3, 8, 10, 16
Nora Griffin-Shirley, Large Divisions: Division 9

# Commission on Accreditation for Dietetics Education of the American Dietetic Association

Beverly E Mitchell, MBA, RD
Senior Director, Accreditation and Education Programs
120 South Riverside Plaza, Suite 2000
Chicago, IL 60606-6995
312 899-0040, ext 4872
312 899-4817 Fax
E-mail: bmitchell@eatright.org
www.eatright.org/cade

## Programs Accredited
- Dietetic technician
- Dietitian/nutritionist

## Structure and Functions
The Commission on Accreditation for Dietetics Education (CADE) is the American Dietetic Association (ADA) accrediting body for post-bachelor's dietetic internships, coordinated and didactic programs at the undergraduate and graduate levels, and associate degree dietetic technician programs. The purpose of CADE is to serve the public by establishing and enforcing standards for the educational preparation of dietetics professionals and by recognizing dietetics education programs that meet those standards.

## History
Founded in 1917, the ADA is the largest organization of food and nutrition professionals promoting optimal nutrition to improve public health and well-being. The early leaders laid a strong foundation for dietetics education. By 1928, a list of hospitals with the approved course for student dietitians in hospitals was published. The approved course required that students have a baccalaureate degree with a major in foods and nutrition and receive at least 6 months of training in a hospital under the supervision of a dietitian.

In 1974, *Essentials* were published for dietetic technician programs for food service management and nutrition care support personnel.

In an effort to maintain appropriate standards for program review, the ADA became involved in program accreditation. In 1974, the ADA was first recognized by the US Department of Health, Education, and Welfare, now the US Department of Education (ED), as the accrediting agency for dietetic internships and coordinated undergraduate programs. The ED now recognizes the CADE as an accrediting agency for coordinated and didactic graduate and undergraduate programs, post-baccalaureate dietetic internships, and associate degree technician programs.

In 1994, the ADA bylaws were amended to demonstrate the administrative autonomy of the body charged with accreditation, the CADE.

## Accreditation of Dietetics Education Programs
The CADE accredits programs by evaluating their compliance with the *Eligibility Requirements and Accreditation Standards*. Professional peer review is used throughout the process.

The purpose of accreditation is to evaluate how effectively an educational program meets stated minimum criteria for educational quality and to encourage program improvement. Programs

**ACCREDITING AGENCIES**

voluntarily seek accreditation by demonstrating compliance with the stated criteria. Accreditation criteria ensure that the institution or program fulfills responsibilities to students and the profession within the framework of traditions basic to higher education, such as protecting academic and intellectual freedom and encouraging flexibility in program development.

Members of the public are included on the accreditation decision-making body; persons not associated with the profession of dietetics or a dietetics program play an active role in the accreditation decision-making process. Public disclosure of accreditation decisions is also required.

The CADE has chosen to be recognized as an accrediting body by the Council for Higher Education Accreditation (CHEA) and ED. The CHEA, the nongovernmental body, and ED, the governmental body, recognize the authority of CADE as the accrediting body for dietetics education programs. Recognition by both agencies provides opportunity for interaction with other accrediting bodies and review of CADE accrediting procedures. Recognition by ED allows CADE-accredited dietetic internships to apply for eligibility for federal funding.

## Composition

The CADE board consists of the chair, chair-elect, immediate past chair, six program representatives, a dietetics student, a program administrator, and two public members. In addition, a cadre of 120 educators and practitioners are responsible for reviewing programs based on policies and procedures developed by the CADE board.

The CADE board functions as the governing unit and grants final accreditation awards. Its 2006-2007 members are listed below.

### Chair
Louise W Peck, PhD, RD, Director, Dietetic Internship, University of Washington, Seattle

### Chair-Elect
Nora K Nyland, PhD, RD, CD, Director, Dietetics Program, Brigham Young University, Provo, UT

### Public Members
Holly Mattson, Deputy Executive Director, Foundation for Interior Design Education and Research, Grand Rapids, MI
Mary Ann Tuft, CAE, President, Tuft & Associates, Chicago, IL

### Program Representatives - Dietetic Internship Programs
Clare Costello, MEd, RD, Training Coordinator, Food and Nutrition Services, Medical College of Virginia, Richmond
Kessey Kieselhorst, MPA, RD, CDE, LDN, Director, Regulatory Performance Improvement, Geisinger Health System, Danville, PA
### Program Representative - Coordinated Programs in Dietetics
Kay Stearns Bruening, PhD, RD, Associate Professor/CP Director, Syracuse University, Syracuse, NY

### Program Representative - Dietetic Technician Programs
Jane Allendorph, MS, RD, Director, Dietetic Technician Program, Harper College, Palatine, IL

### Program Representatives - Didactic Programs in Dietetics
Debra Hollingsworth, PhD, RD, Associate Professor, McNeese State University, Lake Charles, LA
Sandra Witte, PhD, RD, Associate Professor and Chair, California State University, Fresno, CA

### Program Administrator
David D Gale, PhD, Dean, College of Health Sciences at Eastern Kentucky University, Richmond, KY

### Dietetics Student
Christine R Clarahan, Didactic Program in Dietetics, Iowa State University, Ames, IA

### Immediate Past Chair/Representative to ADA House of Delegates
Linda W Gabrielson, MS, RD, Dean, Academic Affairs, Allan Hancock College, Santa Maria, California

# Commission on Accreditation for Health Informatics and Information Management Education

Claire Dixon-Lee, PhD, RHIA, FAHIMA
Executive Director, CAHIIM
Vice President, Education and Accreditation, AHIMA
233 North Michigan Avenue, Suite 2150
Chicago, IL 60601-5800
312 233-1100
312 233-1090 Fax
E-mail: claire.dixon-lee@cahiim.org
www.cahiim.org

## Programs Accredited
- Health information management/health informatics

## Purpose and Responsibilities
The Commission on Accreditation for Health Informatics and Information Management Education (CAHIIM) is the accrediting organization for degree-granting programs in health informatics and information management. CAHIIM and its sponsoring organizations cooperate to establish, maintain, and promote appropriate standards of quality for postsecondary educational programs in health informatics and information management to provide a competent, skilled professional workforce for the healthcare industry.

The CAHIIM Mission:
- Establish and enforce accreditation Standards for education in health informatics and information management programs;
- Recognize through accreditation programs that meet the Standards;
- Encourage educational innovation and diversity; and
- Advance the value of health informatics and information management practice through quality education.

CAHIIM strives to carry out its mission by promoting, evaluating, and improving the quality of undergraduate and graduate health informatics and information management education in the United States. CAHIIM serves the public interest by operating in a consistent manner with all applicable ethical, business and accreditation best practices.

The monitoring of degree instruction is tied closely to national professional association competencies for practice in the profession. CAHIIM's sponsoring organization, the American Health Information Management Association (AHIMA), is the national association of HIM professionals founded in 1928 and dedicated to the effective management of personal health information needed to deliver quality healthcare to the public. AHIMA is committed to advancing the HIM profession in an increasingly electronic and global environment through leadership in advocacy, education, certification, and lifelong learning.

## Standards
The *Standards for Health Information Management Associate Degree* and *Baccalaureate Degree* are the minimum standards of quality used in accrediting programs that prepare individuals to enter the health information management profession. Graduates of accredited programs may apply to AHIMA for eligibility to write the national certification examination for Registered Health Information Technician (RHIT) or Registered Health Information Administrator (RHIA).

## Composition

### Board of Commissioners
C Jeanne Solberg, RHIA, Chair
Elizabeth Bowman, MPA, RHIA, Chair-Elect
Dorine Bennett, MBA, RHIA
Diann Brown, MS, RHIA, CHP
Marie T Conde, MPA, RHIA, CCS
Nancy Korn-Smith, RHIT
Ann Peden, MBA, RHIA, CCS
Angela Picard, MEd, RHIA
Ann M Waters, RHIT, CCS
Barbara W Mosley, PhD, RHIA, Past Chair/Academic Administrator
Keith L Olenik, MA, RHIA, CHP, AHIMA Board Liaison to the CAHIIM Board of Commissioners

**ACCREDITING AGENCIES**

# Commission on Accreditation in Physical Therapy Education

Mary Jane Harris, PT, MS
Director, Department of Accreditation
1111 North Fairfax Street
Alexandria, VA 22314
703 706-3245
703 838-8910 Fax
E-mail: accreditation@apta.org
www.capteonline.org

## Programs Accredited

- Physical therapist (PT)
- Physical therapist assistant (PTA)

## Structure and Functions

The Commission on Accreditation in Physical Therapy Education (CAPTE) is the only agency recognized by the US Department of Education (USDE) and the Council for Higher Education Accreditation to accredit entry-level programs for physical therapists and physical therapist assistants. Although CAPTE is a standing committee of the American Physical Therapy Association (APTA), it is independent in its responsibility for management and implementation of its accreditation activities.

## History

Recognition and accreditation of programs in physical therapy has existed since 1928. In the early years, the process was overseen by the APTA; then the American Medical Association (AMA) provided these services; and later the APTA and AMA collaborated to accredit physical therapy programs. In 1979, CAPTE was recognized by USDE as an independent agency and in 1982 AMA ceased accrediting physical therapy programs. Since 1982, then, CAPTE has been the only recognized accrediting agency in physical therapy.

## Purpose and Responsibilities

Primary aspects of CAPTE's accreditation process include the self-study and site visit processes, evaluation by a review committee, and evaluation by the Commission. Evaluation is based upon published *Evaluative Criteria*, which are the minimum criteria CAPTE uses when awarding programmatic accreditation.

CAPTE also conducts various functions of programmatic accreditation, which include:

- Drafting and reviewing its criteria for the operation of PT and PTA programs
- Selecting and training knowledgeable volunteers to review Self-Study Reports and serve as site visitors
- Selecting representatives to serve on the review committees and the Commission
- Granting accreditation awards based on a program's self-study and site visit processes

## Standards

The *Evaluative Criteria* are minimum educational standards adopted by CAPTE for PT and PTA programs. Each new program is assessed in accordance with the *Evaluative Criteria*, and accredited programs are periodically reviewed to determine whether they remain in substantial compliance. The Evaluative Criteria are reviewed periodically (typically at least every 5 years) and updated when feedback from the education and practice community indicates that revision is necessary.

## Composition

CAPTE is composed of 26 members representing various constituencies: PT and PTA academic educators, PT and PTA clinical educators, PT and PTA clinicians, basic scientists, higher education administrators, and the public. Except for the public members and the consumer member, all Commissioners come from the pool of on-site reviewers and are appointed by the APTA Board of Directors to 4-year terms. CAPTE elects its own leadership annually.

In order to do its work, CAPTE is divided into three sub-groups: The 10-member PT Panel reviews PT programs, the 10-member PTA Panel reviews PTA programs, and the 6-member Central Panel is charged with hearing reconsideration of adverse decisions and formal complaints; developing, reviewing, and revising policies, procedures, and position papers; overseeing the review and revision of Evaluative Criteria; and assisting the PT and PTA Panels with their work. All recommendations for action on accreditation status, for change in policies and procedures, etc, that are made by these sub-groups are considered and acted upon by the full Commission.

### Commissioners (2006-2007)

Leslie G Portney, PT, DPT, PhD, FAPTA, Chair
Martha R Hinman, PT, EdD, Vice-Chair
Joyce Maring, PT, EdD, Vice-Chair
Augustine Agho, PhD
Cindy Calmese
Douglas S Christie, PhD
Kathleen S Cornett, PT, MS
Susan Crabtree, PT, MEd
Ann Roberts Divine, PhD
Daryl Dixon
David G Greathouse, PT, PhD, ECS, FAPTA
Barbara Gresham, PT, MS
Karen Grube, PT, MS
Abby M Heydman, PhD, RN
Rebecca Storey Hooper, PT, PhD
Carol C Likens, PT, PhD
Jackie L Long-Goding, MEd, RRT
Terry R Malone, PT, EdD, FAPTA
Lucinda Mihelich, PT, MEd
Joy E Nobles, PTA
Claire Peel, PT, PhD
Gita W Pitter, PhD
Sandra Ann Radtka, PT, PhD
Suzanne Reese, PT, MS
Jeanne K Smith, PT, DPT, MPA, OCS
Kathleen M Vielhaber, PT, MS

# Commission on Accreditation of Allied Health Education Programs

1361 Park Street
Clearwater, FL 33756
727 210-2350
727 210-2354 Fax
E-mail: mail@caahep.org
www.caahep.org

## Programs Accredited

- Anesthesiologist assistant
- Cardiovascular technologist
- Cytotechnologist
- Diagnostic medical sonographer
- Electroneurodiagnostic technologist
- Emergency medical technician-paramedic
- Exercise physiologist (applied and clinical)
- Exercise science professional
- Kinesiotherapist
- Medical assistant
- Medical illustrator
- Perfusionist
- Personal fitness trainer
- Polysomnographic technologist
- Respiratory therapist (advanced)
- Respiratory therapist (entry-level)
- Specialist in blood bank technology
- Surgical assistant
- Surgical technologist

## Structure and Functions

The Commission on Accreditation of Allied Health Education Programs (CAAHEP) was established as a nonprofit agency on July 1, 1994. Today, CAAHEP is one of the largest specialized accreditors in the country, recognizing some 2,000 educational programs in more approximately 1,100 institutions.

The Commission is composed of professionals (commissioners) from allied health and education organizations and medical specialty groups, including 16 Committees on Accreditation (CoA). Commissioners elect 13 members to CAAHEP's Board of Directors (BoD); two additional members are appointed by the BoD to serve as public representatives. Commissioners also establish CAAHEP's bylaws, mission, and vision statements; approve sponsor organizations for membership; vote on the eligibility of new allied health disciplines to join CAAHEP; and monitor the BoD to ensure quality and equity within the CAAHEP system.

## Purposes and Responsibilities

The purposes of CAAHEP are to:

- Promote and support the education of competent and caring allied health professionals and the continued improvement of allied health education programs
- Inform the public of the status of the Commission's educational programs
- Establish standards of accreditation based on input from the profession and other communities of interest
- Maintain the integrity and assure the credibility of the process of accrediting allied health education programs

- Enhance and promote dialogue among all parties and accrediting agencies in the allied health professions on the issues affecting the accreditation of allied health education programs and take a leadership role in coordinating a collective approach to addressing changes in the education of allied health professionals
- Provide recognition and related coordination services for allied health educational programs
- Compile, analyze, and disseminate information and data descriptive of allied health education and accreditation within the CAAHEP system
- Promote the study of critical issues in allied health education and accreditation and respond to the changing health care needs of society by assisting bodies offering allied health education that seek to respond creatively and appropriately to public policy initiatives

CAAHEP discontinued its recognition from the US Department of Education in April 1998 because few of its accredited programs benefited from CAAHEP's participation in this costly and time-consuming process. CAAHEP is, however, nationally recognized by the Council for Higher Education Accreditation. A list of accredited programs and other vital information about the agency is available at www.caahep.org.

## Composition

CAAHEP has five categories of membership.

Categories 1, 2, and 5 are organizational memberships, including CAAHEP's collaborating Committees on Accreditation (Category 2) and Sponsoring (Category 1) and Associate (Category 5) Member Organizations. Each of these organizational members is entitled to appoint one Commissioner to represent the interests and concerns of that organization.

Category 3 members are the educational program sponsors—the colleges, universities, hospitals, military bases, and other institutions that sponsor CAAHEP-accredited educational programs.

Category 4 represents the general public and recent graduates of CAAHEP-accredited programs. The Board of Directors appoints two Commissioners to represent the general public and one to represent recent graduates.

### CAAHEP Board of Directors 2006-2007

Gregory P Paulauskis, PhD, RRT, RCP, President
William J Horgan, CCP, Vice President
Susan M Fuchs, MD, Secretary
Dan Points, Treasurer
Hugh Bonner
George Burton, MD
Gregory Frazer, PhD
Calvin Harris, EdD
Kathleen Jung
Joseph Long, EdD, CST, MPA
M LaCheeta McPherson
Stephen J Rodgers
David (Jeff) Smith
Nancy J Smith, MS, SCT(ASCP)
Steve Valand

**Commissioners and the Groups They Represent
(as of September 2006)**

Accreditation Committee-Perfusion Education
William J Horgan, CCP

Accreditation Review Committee for Anesthesiologist Assistant
Brad Maxwell

Accreditation Review Committee for the Medical Illustrator
Kathleen Jung

Accreditation Review Committee on Education in Surgical
  Technology
Joseph Long

American Academy of Anesthesiologists Assistants
Jeff Smith, MMSc

American Academy of Cardiovascular Perfusion
David A Ogella
Robert McCoach (Alternate)

American Academy of Orthotists and Prosthetists
Michael Oros

American Academy of Pediatrics
Susan M Fuchs, MD

American Academy of Sleep Medicine
Richard Rosenberg

American Alliance for Health, Physical Education, Recreation and
  Dance
Judith Young

American Association for Respiratory Care
Linda Van Scoder
Carolyn O'Daniel (alternate)

American Association for Thoracic Surgery
Clifford H Van Meter, Jr, MD

American Association of Blood Banks
Lee Ann Prihoda

American Association of Cardiovascular and Pulmonary
  Rehabilitation
John Porcari

American Association of Medical Assistants Endowment
Judy A Jondahl

American Board for Certification in Orthotics and Prosthetics
Steven Whiteside

American Board of Cardiovascular Perfusion
Deborah L Adams (Alternate)
Linda G Cantu

American Clinical Neurophysiology Society
Fumisuke Matsuo

American College of Cardiology
James A Atkins

American College of Chest Physicians
George Burton, MD

American College of Emergency Physicians
Paul Hinchey

American College of Obstetricians and Gynecologists
J Martin Tucker

American College of Radiology
Deborah Levine

American College of Sports Medicine
Mike Niederpruem

American College of Surgeons
Lawrence Kim

American Institute of Ultrasound in Medicine
Marie DeLange

American Kinesiotherapy Association
Melissa Fuller

American Medical Association - Medical Education Group
Richard J D Pan, MD

American Orthopedic Society for Sports Medicine
Mitchell D Seemann

American Society of Anesthesiologists
Michael Lasecki

American Society of Cytopathology
Nancy Smith

American Society of Echocardiography
Carol Mitchell

American Society of Electroneurodiagnostic Technologists
Faye McNall

American Society of Extra-Corporeal Technology
Bruce Bartel, CCP

American Society of Neurophysiological Monitoring
Bernard Cohen
Michael Isley (alternate)

American Society of Radiologic Technologists
Sal Martino, EdD

American Thoracic Society
Vacant

Association of Medical Illustrators
William Andrea

Association of Polysomnographic Technologists
Rose Ann Zumstein

Association of Schools of Allied Health Professions
Hugh Bonner
Gregory B Frazer
David Gibson
Marilyn Harrington
Kathleen McEnerney

Association of Surgical Technologists
Kevin Frey

At-Large Representative (Two-Year Institutions)
Angela Kiernan

Board of Registered Polysomnographic Technologists
Cameron Harris

Committee on Accreditation for Education in
  Electroneurodiagnostic Technology
Kristina Port

Committee on Accreditation for Education in the Exercise Sciences
Walter Thompson

Committee on Accreditation for Respiratory Care

Gregory P Paulauskis, PhD, RRT

Committee on Accreditation of Education for Polysomnographic
Technologists
David Young

Committee on Accreditation of Education Programs for
Kinesiotherapy
Jerry Purvis

Committee on Accreditation of Educational Programs for the
Emergency Medical Services Professions
James Atkins, MD

Committee on Accreditation of Specialist in Blood Bank Technology
Schools
William Turcan

(The) Cooper Institute
vacant

Curriculum Review Board of the American Association of Medical
Assistants Endowment
Rebecca Gibson-Lee

Cytotechnology Programs Review Committee
Kalyani Naik

Hospital-based Programs
Cheryl Oliver

Joint Review Committee on Education in Cardiovascular
Technology
Marie Buckley

Joint Review Committee on Education in Diagnostic Medical
Sonography
Kerry Weinberg

Medical Fitness Association
Cary Wing

National Association of Emergency Medical Technicians
Daniel R Gerard

National Association of EMS Educators
Debra Cason

National Association of State EMS Directors
D Randy Kuykendall

National Commission on Orthotic and Prosthetic Education
Bryan Malas

National Network of Health Career Programs in Two-Year Colleges
Wendy Blume
Anne Loochtan
M LaCheeta McPherson
Dan Points
Steve Valand

National Registry of Emergency Medical Technicians
Philip Dickison

National Surgical Assistant Association
Clint Crews

Perfusion Program Director's Council
James Ramsey

Proprietary Institutions
John Padgett

Public Members
Calvin Harris

Stephen Rodgers

Recent Graduate
Molly Marko

Society for Vascular Surgery
R Eugene Zierler

Society of Cardiovascular Anesthesiologists
Martin Allard, MD

Society of Diagnostic Medical Sonography
Lynne Schreiber

Society of Invasive Cardiovascular Professionals
Jeff Davis

Society of Thoracic Surgeons
Gregory Trachiotis

Society of Vascular Ultrasound
LeAnn Maupin

US Department of Defense
Marcia Waldgeir

Vocational Technical Education
Richard Hernandez

CAAHEP Staff
Kathleen Megivern, JD, CAE
Executive Director

Lori Schroeder, MA
Accreditation Services Director

Cynthia Powell
Executive Assistant

History of CAAHEP-Accredited Professions/Review Committee
Contact Addresses

## Anesthesiologist Assistant
Accreditation Review Committee for the Anesthesiologists Assistant
Brad Maxwell
c/o CAAHEP
1361 Park Street
Clearwater, FL 33756
727 210-2350

Anesthesiologist assistant (AA) educational programs began in 1969 at Case Western University in Cleveland, Ohio, and at Emory University in Atlanta, Georgia. Beginning with the National Academy of Sciences original description of the physician assistant role in both general and specialty areas of medicine, these programs were created in response to task analysis studies showing the increasing need for anesthetists with technical backgrounds. Although the Case Western curriculum was originally designed toward a baccalaureate degree, both programs currently award a master's degree for successful completion of their program. The functional relationship of graduates to the anesthesiologist is similar to that of physician assistants to the general medical as well as surgical specialties.

In 1975, the American Society of Anesthesiologists (ASA) took action in support of this new emerging profession on the basis of its review of the educational and clinical objectives. In 1976, the ASA petitioned the AMA Council on Medical Education (CME) for recognition of the anesthesiologist assistant as an emerging health profession. The CME's recognition followed in 1978. This authorized the initiation of a collaborative activity between the AMA's Division of Allied Health Education and Accreditation and the ASA on the development of a body of educational standards.

In 1981, ASA elected not to be a primary sponsor of the accreditation process. This action prompted members of the ASA who strongly supported the AA programs to establish new physician-sponsoring groups. The Association for Anesthesiologist Assistants Education (AAAE) provides national support for the accreditation process.

In 1983, the AAAE and the American Academy of Anesthesiologist Assistants (AAAA) petitioned the CME to recognize them as collaborative sponsors for the programs designed to educate the anesthesiologist assistant. Recognition followed in 1984, and in 1987, standards for the education of the AA were adopted by the AMA, the AAAE, and the AAAA. The Emory and Case Western programs were evaluated and initially accredited by the Committee on Allied Health Education and Accreditation (CAHEA) in 1988 and continue to be fully accredited by the Commission on Accreditation of Allied Health Education Programs (CAAHEP), CAHEA's successor organization. The ASA recognizes AAs as members of the anesthesia care team and has returned as a sponsoring organization of the Committee.

### Blood Bank Technology, Specialist in

Committee on Accreditation of SBB Schools
American Association of Blood Banks
8101 Glenbrook Road
Bethesda, MD 20814-2749
301 215-6589
301 951-3729 Fax
E-mail: ctretter@aabb.org
www.aabb.org

It is significant that both the examination and the process for approving blood bank schools were the result of a cooperative effort between the Board of Registry and the American Association of Blood Banks (AABB).

By 1969, the number of programs had increased to the extent that the (SBB) Specialist in Blood Banking Committee on Accreditation created an official set of *Essentials and Guidelines* (presently known as *Standards and Guidelines*) to generally outline the structure of SBB programs. These were eventually adopted by the AMA House of Delegates in 1971. Subsequently, the *Essentials and Guidelines* underwent revisions in 1977, 1983, and 1991 and were adopted by the AMA Council on Medical Education and the AABB.

### Cardiovascular Technologist

Joint Review Committee on Education in Cardiovascular
  Technology
1248 Harwood Road
Bedford, TX 76021-4244
817 283-2835
817 354-8519 Fax
E-mail: richwalker@coarc.com

In December 1981, the AMA Council on Medical Education (CME) officially recognized cardiovascular technology as an allied health profession. Subsequently, organizations that had indicated an interest in sponsoring accreditation activities for the cardiovascular technologist were invited to appoint a representative to an ad hoc committee to develop *Standards (Essentials)*. Interested individuals were also invited to join the committee.

The ad hoc committee on development of *Standards* for the cardiovascular technologist held its first meeting on April 29, 1982, in Atlanta, Georgia. Twenty-one individuals attended the first meeting, representing the following organizations: American College of Cardiology; AMA; American Society of Echocardiography; American College of Radiology; American Registry of Diagnostic Medical Sonographers; Grossmont College, El Cajon, California; American

Society of Radiologic Technologists; Society of Diagnostic Medical Sonographers; National Alliance of Cardiovascular Technologists; Society of Non-Invasive Vascular Technology; American College of Chest Physicians; American Cardiology Technologists Association; Santa Fe Community College, Gainesville, Florida; and National Society for Cardiopulmonary Technology.

An initial draft of the proposed *Standards and Guidelines of an Accredited Educational Program in Cardiovascular Technology* was developed as a result of this meeting. Subsequent meetings were held to refine and polish the *Standards*. In September 1983, the committee members reached agreement on the *Standards*. The Joint Review Committee on Education in Cardiovascular Technology (JRC-CVT) held its first meeting in November 1985 in preparation for its ongoing review of programs seeking accreditation in cardiovascular technology.

The *Standards* are adopted by CAAHEP and the following sponsor organizations of the JRC-CVT: the American College of Cardiology, American College of Chest Physicians, American College of Radiology, American Society of Echocardiography, American Society of Cardiovascular Professionals, and Society of Vascular Technology (formerly the Society of Non-Invasive Vascular Technology).

### Cytotechnologist

Cytotechnology Programs Review Committee
American Society of Cytopathology
400 West 9th Street, Suite 201
Wilmington, DE 19801-1555
302 429-8802
302 429-8807 Fax
E-mail: dmacintyre@cytopathology.org

In the pioneer days of clinical pathology, it was the rare pathologist who did not have an assistant. These first technical "assistants," some of whom were trained by George N Papanicolaou, MD, famed American anatomist and cytologist, were always the product of an apprentice-type training. As their number and the number of apprentice programs grew, there was a need to certify that the apprentices had learned their tasks well. The Board of Registry of the American Society of Clinical Pathologists (ASCP) offered the examination for the cytology technician for the first time in 1957.

Five years later, in 1962, the *Essentials of an Acceptable School for the Cytotechnologist* were developed by the Cytology Committee of the ASCP and the ASCP Board of Schools and were adopted by the AMA House of Delegates. Until 1975, representatives of the ASCP served on the Cytotechnology Review Committee of the National Accrediting Agency for Clinical Laboratory Sciences (NAACLS), which replaced the ASCP Board of Schools in 1974. In 1975, the American Society of Cytology (ASC) was recognized as the organization that would collaborate with the AMA Council on Medical Education (CME), and the ASC formed the Cytotechnology Programs Review Committee, which assumed the responsibilities formerly handled by NAACLS. In 1977, 1983, and 1992, the cytotechnology *Essentials and Guidelines* were revised and adopted by the AMA CME and the ASC. In 1996, the *Essentials* were renamed *Standards* and adopted by CAAHEP in conjunction with the American Society of Cytopathology and sponsoring organizations.

### Diagnostic Medical Sonographer

Joint Review Committee on Education in Diagnostic Medical
  Sonography
2025 Woodlane Drive
St Paul, MN 55125
651 731-1582
651 731-0410 Fax
E-mail: aglassing@jcahpo.org

In 1972, the American Society of Ultrasound Technical Specialists (ASUTS) appointed a committee to explore the mechanism of accreditation of educational programs for the ultrasound technical specialist through the AMA Council on Medical Education (CME). In October 1973, members of ASUTS (now known as the Society of Diagnostic Medical Sonographers) met with a representative from the AMA Department of Health Manpower and initiated activities to receive formal recognition as an occupation. One year later the occupation of diagnostic medical sonography received recognition by the AMA.

From 1974 to 1979, the *Standards (Essentials) of an Accredited Educational Program for the Diagnostic Medical Sonographer* was developed. Because of the multidisciplinary nature of diagnostic ultrasound, many interested medical and allied health organizations collaborated in drafting the *Standards*, which were formally adopted by the following organizations: the American College of Cardiology (withdrew as a sponsoring organization in 1983; resumed sponsorship in 1986), American College of Radiology, American Institute of Ultrasound in Medicine, AMA, American Society of Echocardiography, American Society of Radiologic Technologists, Society of Diagnostic Medical Sonographers, and Society of Nuclear Medicine (withdrew as a sponsoring organization in 1981). These organizations, with the exception of the AMA and the Society of Nuclear Medicine, and the addition of the Society of Vascular Technology in 1993, and the American College of Obstetricians & Gynecologists in 2001, currently sponsor the Joint Review Committee on Education in Diagnostic Medical Sonography. New Standards were adopted in 1996 and were fully implemented January 1, 1998. Educational programs were first accredited in January 1982.

### Electroneurodiagnostic Technologist

Committee on Accreditation for Education in
    Electroneurodiagnostic Technology
7600 Hunters Hollow Trail
Novelty, OH 44072-9541
440 338-5845 Phone/Fax
E-mail: kaport@prodigy.net

The AMA's involvement in the evaluation and accreditation of educational programs in electroencephalographic (EEG) technology began in 1972 with the recognition of EEG technology as an allied health profession by the AMA Council on Health Manpower. Subsequently, AMA staff worked with representatives of the professional organizations representing this clinical discipline to develop a draft of the *Standards (Essentials) of an Accredited Educational Program for the Electroencephalographic Technologist.*

In 1973, representatives of the American EEG Society, the American Medical EEG Association, and the American Society of Electroneurodiagnostic Technologists (then the American Society of EEG Technologists) presented statements supporting the *Standards*. These organizations and the AMA House of Delegates then considered and adopted the *Standards* for entry-level educational programs for the electroencephalographic technologist.

The Joint Review Committee on Education in Electroencephalographic Technology was established and held its initial meeting in September 1973. In 1988, the name of the committee was changed to the Joint Review Committee on Education in Electroneurodiagnostic Technology.

This review body is composed of six members: four members appointed by the American Society of Electroneurodiagnostic Technologists and two members appointed by the American Clinical Neurophysiology Society. Meetings are held annually. The committee develops recommendations on accreditation status of programs, which are subsequently forwarded to CAAHEP for final action. The *Standards* were revised in 1980, 1987, and 1995. In 1987, evoked

potential (EP) techniques were included in Standards for programs desiring recognition in both EEG and EP techniques. In 1995, polysomnography (PSG) techniques were included for programs desiring recognition in EEG, EP, and PSG techniques. The standards are currently in committee for additional revisions and additions.

### Emergency Medical Technician-Paramedic

Committee on Accreditation of Educational Programs for the EMS
    Professions
1248 Harwood Road
Bedford, TX 76021-4244
817 283-9403
817 354-8519 Fax

The emergency medical technician-paramedic (EMT-paramedic) was first recognized as an allied health occupation in 1975 by the AMA for the purpose of accrediting entry-level educational programs in the profession. Beginning in 1976, a concerted effort by many organizations was begun to develop the educational *Standards (Essentials)* that would be used to evaluate EMT-paramedic programs seeking accreditation. Following several drafts of the proposed *Standards*, with wide distribution to the appropriate communities of interest, the *Standards* were adopted in 1978. Adoption was by the AMA Council on Medical Education (CME) on behalf of the AMA and by the following organizations collaborating with the AMA in this accreditation process and sponsoring the newly formed Joint Review Committee on Educational Programs for the EMT-Paramedic (JRC/EMT-P): the American College of Emergency Physicians, American College of Surgeons, American Psychiatric Association, American Society of Anesthesiologists, National Association of Emergency Medical Technicians, and National Registry of Emergency Medical Technicians.

The JRC/EMT-P is currently sponsored by the American Academy of Pediatrics, American College of Cardiology, American College of Emergency Physicians, American College of Surgeons, American Society of Anesthesiologists, National Association of Emergency Medical Technicians, National Association of EMS Educators, and National Registry of Emergency Medical Technicians.

### Exercise Sciences

Committee on Accreditation for the Exercise Sciences
401 West Michigan Street
Indianapolis, IN 46202
317 637-9200, ext 123
317 634-7817 Fax
E-mail: mniederpruem@acsm.org

In February 2003, the American College of Sports Medicine (ACSM) petitioned CAAHEP for the inclusion of two new professions in the CAAHEP system. The petition envisioned one profession at the baccalaureate level, tentatively called "health and fitness specialist" and one at the master's degree level, designated "clinical exercise specialist." Both professions, as described in the ACSM petition, were voted unanimously to be eligible for participation in CAAHEP at the April 2003 Annual Meeting.

In April 2004, a Committee on Accreditation for the Exercise Sciences was voted admission into the CAAHEP system as a collaborating Committee on Accreditation with the American College of Sports Medicine as their sponsoring organization. Two sets of draft *Standards and Guidelines*, on which that Committee had been working for over a year, were brought to an open hearing in July 2004 and were thereafter formally approved by the CAAHEP Board of Directors. The *Standards* used new terminology for the two professions that had been voted into the system in 2003, although in all other respects, the disciplines were identical to what had been proposed. The "health and fitness specialist" was now designated as the "exercise science professional" and the "clinical exercise specialist" was designated as the "exercise physiologist." For exercise physiology, the *Standards* laid out two tracks: applied and clinical. As of September 2005 there are not yet any accredited programs in either profession.

## Kinesiotherapist

Committee on Accreditation of Education Programs for
  Kinesiotherapy
University of Southern Mississippi
Box 5142
Hattiesburg, MS 39406
601 266-5371
601 266-4445 Fax
E-mail: jerry.purvis@usm.edu

Kinesiotherapy (formally corrective therapy) emerged as an allied health profession in 1946. In 1953, the American Corrective Therapy Association (formally changed to the American Kinesiotherapy Association in 1987), realizing the need for a credentialing process, formally adopted a certification examination to establish a consistent level of competency. The process of credentialing and establishing academic programs continued to evolve as baccalaureate programs were established, and in 1980 the clinical training requirements increased from 400 to 1,000 clock hours.

In 1982, the Council on Professional Standards was established to oversee accreditation and credentialing activities. In that same year standards and guidelines were set for accrediting academic programs preparing students to enter the kinesiotherapy profession. In 1985, minimum core competencies for kinesiotherapists were established.

In 1986, mandatory continuing education requirements were set to maintain registration, and, in 1987, a contract was established with the Professional Evaluation Service to standardize and administer the national certification examination.

CAAHEP voted in April 1995 to recognize kinesiotherapy as an allied health profession. In that same year the American Kinesiotherapy Association and the American Academy of Physical Medicine and Rehabilitation joined CAAHEP as cosponsors of kinesiotherapy. In April 1997, during the third annual CAAHEP conference, the commissioners voted to formally recognize the Committee on Accreditation of Education Programs for Kinesiotherapy. On April 23, 1998, the CAAHEP executive board approved the *Standards and Guidelines for Accredited Educational Programs for Kinesiotherapy*.

## Medical Assistant

American Association of Medical Assistants' Endowment
20 North Wacker Drive, Suite 1575
Chicago, IL 60606-2963
312 899-1500
312 899-1259 Fax
E-mail: accreditation@aama-ntl.org

Since its founding in 1956, the American Association of Medical Assistants (AAMA) has been the only professional association devoted exclusively to the profession of medical assisting. The AAMA holds an important role in improving the educational preparation and continuing education opportunities for the medical assistant.

The first certification examination was administered in 1963, preceding the establishment of the accreditation program. In 1966, the AAMA began work on formal curriculum standards in collaboration with the AMA. A task force of physicians and medical assistants surveyed existing medical assistant programs and drew up tentative

*Standards*, which were adopted in 1969. Approximately 3 years were spent in laying a solid groundwork for a 2-year associate degree program. In 1971, after 2 years of actual accreditation activity, the initial *Standards* were revised to allow for the accreditation of 1-year educational programs. The most recent version of the *Standards* was completed in 1999.

## Medical Illustrator

Accreditation Review Committee for the Medical Illustrator
CAAHEP
1361 Park Street
Clearwater, FL 33756
727-210-2350
727-210-2354
E-mail: schroeder@caahep.org

Formal educational programs for the medical illustrator date back to the early 1900s, with Max Broedel's school at Johns Hopkins University. The Association of Medical Illustrators (AMI) was established in 1945. Under the auspices of the AMI, standards were developed by which the organization has accredited medical illustration programs in this country since 1967.

In 1986, the AMI expressed a desire to have educational programs for the medical illustrator accredited by the Committee on Allied Health Education and Accreditation (CAHEA) of the AMA. This desire stemmed from the recognition that professional medical illustrator programs were more closely related to allied health than to the visual arts.

An ad hoc committee on outside accreditation of the AMI worked with staff of the AMA Division of Allied Health Education and Accreditation to modify the existing *Standards* to comply with the format recommended by the CAHEA. The resulting *Essentials and Guidelines of an Accredited Educational Program for the Medical Illustrator* was adopted by the AMI and the AMA Council on Medical Education (CME) in 1987. Revised *Standards* were adopted by the AMI and the AMA in 1992. Today, the *Standards* are adopted by CAAHEP in collaboration with the AMI.

## Orthotist and Prosthetist

National Commission on Orthotic and Prosthetic Education
330 John Carlyle Street, Suite 200
Alexandria, VA 22314
703 836-7114
703 836-0838 Fax
E-mail: rseabrook@ncope.org

The Educational Accreditation Commission (EAC) was created in August 1972 by the American Board of Certification in Orthotics and Prosthetics, Inc, to meet the orthotics and prosthetics profession's need for an institutional accreditation program. That same year the EAC set out to establish criteria to assess and compare orthotics and prosthetics curricula. From these criteria, *Standards (Essentials)* were developed and revised to meet the profession's needs in training orthotists and prosthetists.

In 1991, the EAC was reorganized and renamed the National Commission on Orthotic and Prosthetic Education (NCOPE). NCOPE's primary mission and obligation is to ensure that educational programs meet the minimum standards of quality to prepare individuals to enter the orthotics and prosthetics profession.

During 1992, the NCOPE and its collaborating organizations, the American Orthotic and Prosthetic Association (AOPA) and the American Academy of Orthotists and Prosthetists (AAOP), applied to the AMA Council on Medical Education (CME) for recognition of orthotics and prosthetics as an allied health profession. Recognition was granted in August 1992.

In February 1993, the NCOPE reformatted its existing *Standards* to meet the requirements of the Committee on Allied Health Education and Accreditation (CAHEA) Recommended Format for *Standards and Guidelines*. The *Standards* were adopted by the AOPA in April 1993. The AMA CME adopted the *Standards* in August 1993, and the AAOP adopted the *Standards* in September 1993. The Standards were revised in 2001 and 2005

## Perfusionist

Accreditation Committee - Perfusion Education
6654 South Sycamore Street
Littleton, CO 80120
303 738-0770
303 738-3223 Fax
E-mail: ac-pe@msn.com
www.ac-pe.org

The field of cardiovascular perfusion emerged in the mid-1960s, with most of its practitioners trained on the job until the mid-1970s. Trainees often come from other disciplines: nursing, respiratory therapy, biomedical engineering, surgical technology, monitoring technicians, and the laboratory sciences.

In 1972, the American Society of Extra-Corporeal Technologists (AmSECT) began a program of certification for perfusionists. In 1975, this program was turned over to a new agency established to conduct certification as an independent activity: the American Board of Cardiovascular Perfusion (ABCP). The ABCP also adopted minimum standards for training programs as developed by AmSECT and began evaluation and accreditation activities. The AmSECT, with the cosponsorship of the American Association of Thoracic Surgeons (AATS) and the Society of Thoracic Surgeons (STS), petitioned the AMA for recognition of the occupation in January 1975. The petition was amended on several occasions, and in December 1976 the Committee on Emerging Health Manpower recommended approval; the AMA Council on Medical Education (CME) granted recognition in that same month.

In 1977, four collaborating organizations sponsored the formation of the Joint Review Committee for Perfusion Education (JRC-PE)—the AATS, ABCP, AmSECT, and STS. From 1978 to 1979, the JRC-PE and others developed the *Standards (Essentials) and Guidelines for an Accredited Educational Program for the Perfusionist*. The *Standards* were adopted in 1980 and accreditation of programs began in 1981. The *Standards* were revised in 1984, 1989, and 1994. In 1984, the Perfusion Program Directors Council became an additional sponsor, as did the Society of Cardiovascular Anesthesiologists in 1989. In 1991, the review committee became known as the Accreditation Committee - Perfusion Education (AC-PE). In 1996, the American Academy of Cardiovascular Perfusion (AACP) became an additional sponsor of the review committee.

## Polysomnography

Committee on Accreditation of Education for Polysomnographic Technologists
One Westbrook Corporate Center, Suite 920
Westchester, IL 60154
708 492-0796
708 273-9344 Fax
E-mailrrosenberg@aasmnet.org

In April 2002, the Association of Polysomnographic Technologists (APT) was voted into CAAHEP as an associate member organization. Over the course of the next year, APT worked with CAAHEP on the development of a Committee on Accreditation. At the April 2004 meeting of CAAHEP, the new Committee on Accreditation for Polysomnographic Technology was approved, along with three sponsoring organizations: the American Academy of Sleep Medicine, the Association of Polysomnographic Technologists, and the Board of Registered Polysomnographic Technologists. At that same meeting, an open hearing was held on proposed Standards, which were formally approved by the CAAHEP Board of Directors on April 24, 2004. As of September 2005 there were not yet any accredited programs in this profession.

## Respiratory Care

Committee on Accreditation for Respiratory Care
1248 Harwood Road
Bedford, TX 76021
800 874-5615
817 354-8519 Fax

In 1957, a resolution to develop schools of inhalation therapy was introduced to the AMA House of Delegates by the Medical Society of New York. Following approval, the resolution was referred to the Council on Medical Education (CME) and subsequently resulted in the proposed report titled *Standards (Essentials) for an Approved School of Inhalation Therapy Technicians*. The proposed *Standards* were tested during the next few years, after which they were recommended for adoption by the CME and formally approved by the AMA House of Delegates in December 1962. The *Standards* were revised in 1967 and included the requirements of an 18-month program.

In 1970, the Board of Schools was reorganized and incorporated as the Joint Review Committee for Inhalation Therapy Education. In 1972, the American Thoracic Society became another sponsor of the Joint Review Committee. In 1972, the *Standards* underwent a third revision, and additional standards were developed for a shorter educational program for training individuals to function as technicians. The *Standards* were approved by the sponsors of the Board of Schools for Inhalation Therapy, the American Society of Anesthesiologists, the American College of Chest Physicians, the American Association of Inhalation Therapy, and the AMA CME and were adopted by the AMA House of Delegates in June 1972. In 1997, the review committee's name was changed to the Committee on Accreditation for Respiratory Care (CoARC).

In 1986, *Standards* for both the respiratory therapy technician and the respiratory therapist were consolidated and the revision was adopted by the sponsors of the review committee and by the AMA. Revised *Standards* were approved in 2000.

## Surgical Technologist

Accreditation Review Committee on Education in Surgical Technology
6 West Dry Creek Circle, Suite 210
Littleton, CO 80120
303 694-9262
303 741-3655 Fax
E-mail: ccollinsworth@ast.org

The profession of surgical technology was developed during World War II when there was a critical need for assistance in performing surgical procedures and a shortage of qualified personnel to meet that need. Individuals were educated specifically to assist in surgical procedures and to function in the operative theater.

The Association of Surgical Technologists (AST) was organized in July 1969, with an advisory board of representatives from the American College of Surgeons (ACS), the Association of Operating Room Nurses (AORN), the American Hospital Association (AHA), and the AMA.

In December 1972, the AMA's Council on Medical Education (CME) adopted the recommended educational standards for this field, and the Accreditation Review Committee on Education in Surgical Technology (ARC-ST) was formed. The ARC-ST is jointly

**ACCREDITING AGENCIES**

sponsored by the AST and ACS. Revised, outcomes-based standards were approved in 2000.

A subcommittee of ARC-ST has been formed to review surgical assistant programs.

# Commission on Accreditation of Athletic Training Education

Lynn Caruthers, Administrative Director
2201 Double Creek, Suite 5006
Round Rock, TX 78664
512 733-9700
512 733-9701 Fax
E-mail: caate@sbcglobal.net
www.jrc-at.org

## Programs Accredited
- Athletic trainer

## History
Work on establishing standards for athletic training educational programs was initiated in 1959 by the National Athletic Trainers' Association (NATA). In 1969, the NATA Committee on Curriculum Development approved the first two programs. By 1979, there were 23 undergraduate programs and two graduate programs approved by NATA. In 1982, 1990, and 1994, the NATA completed role delineation studies that defined entry-level competencies for athletic trainers. By 1997, NATA had approved 87 entry-level and 13 graduate athletic training educational programs.

During 1989, NATA, through its Professional Education Committee, applied to the AMA Council on Medical Education (CME) for recognition of athletic training as an allied health occupation. Recognition was granted on June 22, 1990.

In October 1990, an initial meeting was conducted for the development of the *Standards* (Essentials) for accreditation of educational programs for athletic trainers. Individuals attending that meeting represented the AMA's Division of Allied Health Education and Accreditation (DAHEA), American Academy of Pediatrics (AAP), American Academy of Family Physicians (AAFP), American Orthopaedic Society for Sports Medicine (AOSSM), and NATA. In late 1991, the *Standards* were adopted by the AMA CME and the sponsors of the Joint Review Committee, including AAFP, AAP, and NATA. In January 1995, the AOSSM also became a cosponsor of the JRC-AT. Following the separation of the AMA from the Committee on Allied Health Education and Accreditation (CAHEA), the independent agency, CAAHEP (Commission on Accreditation of Allied Health Education Programs) was formed, with the JRC-AT functioning as a Commission on Accreditation of that group. On July 1, 2006, the JRC-AT separated from CAAHEP and became the independent accreditor CAATE (Commission on Accreditation of Athletic Training Education). At the time of separation, all CAAHEP-accredited athletic training education programs became CAATE-accredited; by the end of 2006, there were 352 CAATE-accredited athletic training education programs in the United States. Currently, there are no CAATE-recognized programs outside of the United States.

## Structure and Functions
The CAATE is composed of professionals (commissioners) representing the profession of athletic training, the stakeholders, and the sponsoring organizations. The Commission establishes the CAATE By-Laws, Mission, and Vision Statements; awards accreditation; and monitors the CAATE Committees to ensure quality, fairness, equity, and due process within the CAATE system.

## Purposes and Responsibilities
The purpose of the CAATE is to develop, maintain, and assure that the quality and content of all accredited athletic training education

programs are consistent with the *Standards*. The mission of the CAATE is to provide accreditation services to institutions that offer entry-level athletic training education programs and to maintain and assure that the quality and content of all such programs are consistent with the Standards.

The goals of the Commission are to ensure:

- Comprehensive review, accreditation decisions, and annual processes that are defined, consistent, and free of personal biases, conflicts of interest, and non-sanctioned interpretations with respect for institutional autonomy
- Regular and consistent quality assurance processes
- Collegial relationships and regular communication with the sponsoring organizations, institutions, and other stakeholders
- Availability of educational opportunities related to program development and accreditation

## Composition

The Commission is composed of representatives from each of the sponsoring organizations and includes one representative from each of the sponsoring organizations (AAFP, AAP, AOSSM, and NATA), five Certified Athletic Trainers who are elected to the Commission by the institutions that sponsor athletic training education programs, and one public and one administrator representing the administration of the sponsoring institutions. The CAATE *Standards* are established by the athletic training profession and endorsed by the sponsoring organizations.

### CAATE Commissioners

Paula S Turocy, EdD, ATC
Patrick Sexton, EdD, ATC
Robert Moss, PhD, ATC
Katie Walsh, EdD, ATC
Greg Gardner, EdD, ATC
David Kaiser, EdD, ATC
Sean Willeford, MS, ATC
Douglas Gregory, MD, FAAP
Mia Griggs, MD
Mark Pinto, MD
Kaye Herth, PhD, RN, FAAN

# Commission on Accreditation of Ophthalmic Medical Programs

Lynn D Anderson, PhD, Executive Director
2025 Woodlane Drive
St Paul, MN 55125
651 731-7237
651 731-0410 Fax
E-mail: coa-omp@jcahpo.org
www.jcahpo.org/newsite/coaomp.html

## Programs and Institutions Accredited

- Ophthalmic Assistant
- Ophthalmic Technician
- Ophthalmic Medical Technologist

## Structure and Functions

The Commission on Accreditation of Ophthalmic Medical Programs (CoA-OMP) accredits postsecondary education ophthalmic medical personnel programs by evaluating program compliance using the *Standards and Guidelines for Accrediting Educational Ophthalmic Medical Personnel Programs,* last revised in 2005. CoA-OMP publishes its *Standards and Guidelines* and *Policies and Procedures*, which outlines the accreditation processes and the expectations of accredited schools and programs. This information is available on the CoA-OMP Web site.

## Purpose and Responsibilities

CoA-OMP is vested with the responsibility and authority to evaluate programs that have requested accreditation services. CoA-OMP's principal means of program evaluation consists of analyzing Self-Study Reports, sending representative teams to conduct site visits of programs, and deliberating at committee meetings.

CoA-OMP reviewers formulate accreditation recommendations, and the reasons for them, to CoA-OMP's Board of Directors for final accreditation action.

## Composition

CoA-OMP consists of five members, who are appointed by CoA-OMP's three sponsoring organizations: two by the Joint Commission on Allied Health Personnel in Ophthalmology (JCAHPO), two by the Association of Technical Personnel in Ophthalmology (ATPO), and one ex-officio member appointed by the Consortium of Ophthalmic Training Programs. They meet twice annually to review accredited institutions and programs, revise and update accreditation standards and procedures, and conduct other CoA-OMP business.

### CoA-OMP Roster, 2005-2006

Lisa Rovick, COMT, CO, President
John D Fisher, MD, PhD, Vice President
Robin Cooper, COMT, ROUB, Secretary
Christopher Westfall, MD, Treasurer
Paul Larson, MMSc, MBA, COMT, COE, Member

# Commission on Dental Accreditation of the American Dental Association

211 East Chicago Avenue
Chicago, IL 60611-2678
312 440-4653
312 440-2915 Fax
www.ada.org

## Programs Accredited

- Dental assistant
- Dental hygienist
- Dental laboratory technician

## History

From the early 1940s until 1975, the American Dental Association's (ADA's) Council on Dental Education was the agency recognized as the national accrediting organization for dentistry and dental-related educational programs. On January 1, 1975, the Council on Dental Education's accreditation authority was transferred to the Commission on Accreditation of Dental and Dental Auxiliary Educational Programs, an expanded agency established to provide representation of all groups affected by its accrediting activities. In 1979, the name of the Commission was changed to the Commission on Dental Accreditation (CODA).

The accreditation standards for educational programs in dental assisting, dental hygiene, and dental laboratory technology have been revised five times—in 1969, 1973, 1979, 1991, and 1998—to reflect the dental profession's changing needs and educational trends. The communities of interest provided input into the latest revision of the standards through an ad hoc committee, open hearings, and review of and comment on drafts of the proposed revised standards. Prior to approving the revised standards in July 1998, the Commission carefully considered comments received from all sources. The revised accreditation standards were implemented in January 2000.

### Dental Assisting

In 1957, the Council on Dental Education sponsored the first national workshop on dental assisting. Practicing dentists, dental educators, and dental assistants made recommendations for the education and certification of dental assistants. These recommendations were considered in developing the first *Requirements for an Accredited Program in Dental Assisting Education*, which were approved by the ADA House of Delegates in 1960.

Prior to 1960, the American Dental Assistants Association (ADAA) approved courses of training for dental assistants, varying in length from 104 clock hours to 2 academic years. Subsequent to the adoption in 1960 of the first accreditation standards, the Council on Dental Education granted provisional approval to those programs approved by the ADAA that were at least 1 academic year in length until site visits could be conducted. Thus, 26 programs appeared on the first list of accredited dental assisting programs published in 1961.

### Dental Hygiene

The first dental hygiene accreditation standards were developed by three groups: the American Dental Hygienists' Association, the National Association of Dental Examiners, and the ADA's Council on Dental Education. The standards were submitted to and approved by the ADA House of Delegates in 1947, 5 years prior to

the launching of the dental hygiene accreditation program in 1952. The first list of accredited dental hygiene programs was published in 1953, with 21 programs.

### Dental Laboratory Technology

The first educational standards for the education of dental laboratory technicians, adopted by the ADA House of Delegates in 1946, were rescinded and revised in 1957. Between 1946 and 1957, four programs for training dental laboratory technicians were developed and the establishment of new programs remained static through 1965. From 1966 to 1979, the number of accredited dental laboratory technology programs increased from 4 to 59. Since that time, the number has decreased to 20.

## Structure and Functions

Maintaining and improving the quality of dental-related educational programs is a primary aim of the Commission. In meeting its responsibilities as a specialized accrediting agency accepted by the dental profession and the US Department of Education, the Commission on Dental Accreditation

- Evaluates dental assisting, dental hygiene, and dental laboratory technology education programs based on the extent to which program goals, institutional objectives, and approved accreditation standards are met
- Supports continuing evaluation of and improvements in dental-related educational programs through institutional self-evaluation
- Encourages innovations in program design based on sound educational principles
- Provides consultation in initial and ongoing program development

The programs listed in this *Directory* for the occupations of dental assistant, dental hygienist, and dental laboratory technician are conducted in accordance with published accreditation standards. A student who enrolls in and completes any of these programs will be considered a graduate of an accredited program.

If an institution in this *Directory* offers more than one version of its dental assisting or dental laboratory technology program, only the program identified in the *Directory* is accredited by the Commission. All listed programs are conducted at the post-secondary level.

Although other programs may be either under development or in operation, this *Directory* lists only those programs accredited by the Commission as of October 2004.

## Standards

The *Accreditation Standards for Dental Assisting Education Programs, Accreditation Standards for Dental Hygiene Education Programs* and *Accreditation Standards for Dental Laboratory Technology Education Programs* have been developed to

- Protect the public
- Serve as a guide for program development
- Serve as a stimulus for the improvement of established programs
- Provide criteria for the evaluation of new and established programs

Accreditation by the Commission on Dental Accreditation requires that a program meet the national standards set forth by the Commission, which represent the minimum requirements for accreditation.

## Composition

The Commission on Dental Accreditation has 30 members:

- Four from the American Dental Association (ADA)
- Four from the American Dental Education Association (ADEA)

- Four from the American Association of Dental Examiners (AADE)
- Four public members
- One dental student member

and one from each of the following organizations:

- American Association of Endodontists (AAE)
- American Association of Hospital Dentists (AAHD)
- American Academy of Oral and Maxillofacial Pathology (AAOMP)
- American Academy of Oral and Maxillofacial Radiology (AAOMR)
- American Association of Oral and Maxillofacial Surgeons (AAOMS)
- American Association of Orthodontists (AAO)
- American Academy of Pediatric Dentistry (AAPD)
- American Academy of Periodontology (AAP)
- American Association of Public Health Dentistry (AAPHD)
- American College of Prosthodontists (ACP)
- American Dental Assistants Association (ADAA)
- American Dental Hygienists' Association (ADHA)
- National Association of Dental Laboratories (NADL)

### Commission Members, 2005-2006

The members of the Commission on Dental Accreditation:
Morris L Robbins, DDS (ADA), Chairman
James R Cole, II, DDS (AADE), Vice Chairman
Steve Adair, DDS (AAPD)
Bruce J Barrette, DDS (AADE)
Anne Boyle, DDS (ADEA)
Lanier Byrd, PhD (public)
Jack G Caton, Jr, DDS (AAP)
Heidi C Crow, DMD (AAHD/ADEA)
Teresa Dolan, DDS (AAPHD)
Cecile A Feldman, DDS (ADEA)
Jennifer Fong (student) (ADEA/ASDA)
Gary Gann, CDT (NADL)
M Joan Gillespie, DDS (ADA)
Jeffery Hutter, DDS (AAE)
Ronald Johnson, DDS (ADEA)
James J Koelbl, DDS (ADEA)
Patrick J Louis, DDS (AAOMS)
Kay J McKay (public)
Sharon M McPherron (public)
Raymond J Melrose, DDS (AAOMP)
Larry Nissen, DDS (ADA)
Samuel E Pick, DDS (AADE)
Brad J Potter, DDS (AAOMR)
Matthew B Roberts, DDS (ADA)
Thomas H Robinson, PhD (public)
Roger B Simonian, DDS (ADA)
Richard D Smith, DDS (AADE)
Diane Macalus Sullivan, CDA, MS (ADAA)
James Vaden, DDS (AAO)
Ronald D Woody, DDS (ACP)
Nancy Zinser, RDH, MS (ADHA)

# Commission on Massage Therapy Accreditation

Stephen Fridley, Executive Director
1007 Church Street, Suite 302
Evanston, IL 60201
847 869-5039
847 869-6739 Fax
E-mail: info@comta.org
www.comta.org

## Programs and Institutions Accredited
- Therapeutic massage and bodywork

## History

In 1982, the American Massage Therapy Association (AMTA) Council of Schools (COS) was established, in recognition of a shared concern among educators and school executive directors for the quality of massage therapy education. Early Council work focused on the need to develop and maintain educational standards.

In 1989, the Commission on Massage Training Approval/Accreditation (COMTAA) was established. In the following 2 years, with the assistance of AMTA's Program Approval Review Committee (PARC), COS, and additional AMTA volunteers and staff, COMTAA created and implemented standards, policies, and procedures that would meet the rigorous standards of the USDE for accrediting agencies. Recognition by the Council on Post-Secondary Accreditation (COPA) was pursued as a developmental stage toward the ultimate goal of US Department of Education recognition.

In 1992, in an effort to combat widespread fraud, waste, and abuse in the federal Title IV financial aid programs, Congress passed the Higher Education Amendments. This law requires USDE-recognized accrediting agencies to act as "gatekeeper" of federal funds. For many years, the responsibility for oversight of student loan programs had been shared by states, accrediting agencies, and the USDE. Congress determined in 1992 that this triad was not able to guarantee program integrity and financial accountability in higher education institutions. The Amendments created a new triad in which the states (which have a new enforcement capability) and the accrediting agencies (which have new requirements to meet) each have responsibility for monitoring and reporting to each other and to the USDE.

While waiting for the USDE regulations to be issued, COMTAA continued to accredit and approve programs and continued to refine its policies and procedures to be ready to come into compliance with those regulations.

In October 1996, an elected COMTAA Commission was seated. The members were elected by the then current COMTAA approved and accredited programs. The initial representation on the Commission included two massage school administrators, two massage school educators, and two public members, one each of professional academic, massage therapist employer, and massage therapist practitioner.

In 1997 the decision was made to end the approval status on March 31, 1999, and change the name to the Commission on Massage Therapy Accreditation (COMTA). In 2004 COMTA became a completely independent organization.

USDE recognition was granted to COMTA on July 10, 2002. On November 8, 2004, COMTA's USDE recognition was continued for 5 years, and the scope of recognition was expanded to include accreditation of associate degree programs.

## Structure and Functions

COMTA is authorized to accredit programs and institutions offering postsecondary education in the areas of therapeutic massage and bodywork through its recognition by the US Department of Education. COMTA publishes a *Policy, Procedure and Standards* manual that outlines the accreditation processes and the expectations of accredited schools and programs. This information is also available on the Web site, www.comta.org.

## Purpose and Responsibilities

COMTA, a nonprofit independent body, seeks to improve the quality of education for students seeking education in the fields of massage therapy and bodywork through an accreditation process that reflects current and emerging professional practice standards. COMTA extends its services to those institutions and programs which award post-secondary certificates, diplomas, or degrees in the practice of massage therapy and bodywork. COMTA provides institutional accreditation for freestanding institutions of massage therapy and bodywork. COMTA also accredits educational programs in massage therapy and bodywork offered at other institutions.

Accredited programs have undergone a rigorous process of self-assessment and peer review and have met standards established by the profession itself as essential to preparing graduates for competence in the profession. The Commission accredits programs and institutions and performs ongoing monitoring of their compliance with COMTA Standards and Policies. Review of annual reports and of institutions'/programs' responses to the Commission's expressed concerns is part of this monitoring process.

## Composition

The Commission consists of 10 members, who are elected by the currently accredited schools and programs: two COMTA school or program educators, two COMTA school administrators, two practitioners, two public members, one massage therapy employer, and one academician. Each member serves a 6-year term. They meet twice annually to review accredited institutions and programs, revise and update accreditation standards and procedures, and conduct business of the agency.

### Commission Roster

John Goss, PhD, Chair
Melissa Wade, Vice Chair
Dawn Schmidt, Executive Committee
Erika Baern
Wayne Blankenship, MA
Pat Hughes
Alan Kite
Judith Mazza
Joanna Sechuck-Tringali, NCTMB
Irv Freeman, PhD

# Commission on Opticianry Accreditation

Tamara A L Halstead, Director of Accreditation
8665 Sudley Road, #341
Manassas, VA 20110
703 940-9134
703 940-9135
E-mail: coa@coaccreditation.com
www.COAccreditation.com

## Programs Accredited

- Ophthalmic dispensing optician
- Ophthalmic laboratory technician

## Structure and Functions

The Commission on Opticianry Accreditation (COA) was formed in 1979 for the sole purpose of accrediting ophthalmic dispensing optician and ophthalmic laboratory technician educational programs, formerly a function of the National Academy of Opticianry. In 1985, the US Department of Education recognized the commission as the accrediting body for 2-year ophthalmic dispensing degree programs and 1-year ophthalmic laboratory technology certificate programs.

## Purpose and Responsibilities

The COA develops and maintains standards for ophthalmic dispensing and laboratory technician programs that meet the changing needs of programs, students, and the public. The Commission strives to ensure the competence of opticianry professionals and the provision of quality professional care to the public.

The COA's accreditation process certifies that a program has voluntarily undergone a comprehensive study indicating that the program has set appropriate educational objectives for students; that the program performs the functions it claims; and that the program furnishes materials and services to enable students to meet those stated objectives.

Accreditation assures the public that program graduates are skilled practitioners and makes programs eligible for many grants and other funding. In addition, many of the states that license opticians recognize graduation from an accredited program as a licensure requirement.

## Standards

Each new ophthalmic dispensing optician and ophthalmic laboratory technician educational program is assessed in accordance with standards published by the COA, and accredited programs are periodically reviewed to ensure that they remain in substantial compliance.

## Composition

The Commission consists of 12 voting commissioners, appointed by the following:
- Four—National Academy of Opticianry Board of Directors
- Four—Opticians Association of America Board of Directors
- Two (from COA-accredited schools)—National Federation of Opticianry Schools
- Two (public members)—appointed by the other members of the commission

The COA's chief staff executive is an ex-officio member of the commission, and the commission may appoint other ex-officio members.

**Commissioners**
Danne Ventura
Jayne Weinberger
Scott Helkaa
Thomas Blair, Jr
Sam Henderson, Treasurer
Edmund Wnuczek, Jr
Wilfredo Nieves
Howard Steed, PhD
Michael Szczerbiak, Chair
Sharon Breivogel-Leonard, Vice Chair

# Council for Accreditation of Counseling and Related Educational Programs

5999 Stevenson Avenue
Alexandria, VA 22304
800 347-6647, ext 301
E-mail: cacrep@cacrep.org
www.cacrep.org

## Programs Accredited

Master's Degree:
- Career Counseling
- College Counseling
- Community Counseling
- Gerontological Counseling
- Marital, Couple, and Family Counseling/Therapy
- Mental Health Counseling
- School Counseling
- Student Affairs

Doctoral Degree:
- Counselor Education and Supervision

## Structure and Functions

The Council for Accreditation of Counseling and Related Educational Programs (CACREP) was incorporated in 1981. This independent council was created by the American Counseling Association and its divisions to develop, implement, and maintain standards of preparation for the counseling profession's graduate-level degree programs. CACREP's purpose is to work with institutions offering graduate-level programs in counseling and related fields to improve their programs and obtain accredited status. As a committed member of the international accreditation community, CACREP frequently consults with other accrediting agencies such as the Council on Rehabilitation Education (CORE), the American Psychological Association (APA), and the National Council for Accreditation of Teacher Education (NCATE). In 1987, CACREP was granted recognition status by the Council on Postsecondary Education (now the Council for Higher Education Accreditation [CHEA]).

## History

In 1968, Robert Stripling, often called "the father of accreditation in counselor accreditation," commented that counseling had begun to realize its "obligation to protect society, insofar as possible, from poorly prepared counselors...through accrediting of counselor education programs." At the time, many institutions of higher learning had already developed counseling courses and programs in counseling at the undergraduate and graduate levels. With the adoption of *Standards* both during the 1960s and in a broadened, refined form in the early 1970s, voluntary and developmental self-evaluation of counselor education programs began to gain the interest of a small but growing number of programs. This movement resulted in the creation of CACREP in 1981.

## Mission

The mission of CACREP is to promote the professional competence of counseling and related practitioners through
- developing preparation standards
- encouraging excellence in program development
- accrediting professional preparation programs

ACCREDITING AGENCIES

## Composition

### Board Membership
Membership on the CACREP Board of Directors consists of up to 15 voting individuals who are elected by the Board. Each representative may serve one 5-year term.

Chair
John R Culbreth, PhD
University of North Carolina at Charlotte

Vice-chair
Rebecca Stanard, PhD
University of West Georgia

Treasurer
Eli Zambrano
San Antonio, TX

Board Members

Carla Adkison Bradley
Western Michigan University

Louis A Busacca, PhD
Mayfield Heights, OH

Craig Cashwell EdD
University of North Carolina at Greensboro

Joseph Dear, PhD
Sacramento, CA

Gary Grand
Alexandria, LA

W Bryce Hagedorn, PhD
Florida International University

Suzan Nolan, EdD
Rapid City, SD

Donna Gooden Payne, J.D.
University of North Carolina at Pembroke

Sue Strong, PhD
Eastern Kentucky University

James Wigtil, PhD
Valley City State University, ND

# Council on Academic Accreditation in Audiology and Speech-Language Pathology

Susan Flesher, Manager, Accreditation Administration
American Speech-Language-Hearing Association (ASHA)
10801 Rockville Pike
Rockville, MD 20852
800 498-2071
301 571-0481 Fax
E-mail: accreditation@asha.org
www.asha.org

## Programs Accredited
- Audiologist
- Speech-language pathologist

## Structure and Functions
As described in its statement of purpose, the American Speech-Language-Hearing Association (ASHA) promotes appropriate academic and clinical preparation of individuals entering the discipline of human communication sciences and disorders. ASHA established and has maintained a program of academic accreditation for entry-level graduate education since the early 1960s.

The Council on Academic Accreditation in Audiology and Speech-Language Pathology (CAA) of ASHA accredits graduate programs that prepare individuals to enter professional practice in audiology and/or speech-language pathology. The CAA was established by ASHA and is authorized to function autonomously in setting and implementing standards and awarding accreditation. ASHA's accreditation program has been continuously recognized by the Council for Higher Education Accreditation since 1964 and by the US Secretary of Education since 1967. The current scope of recognition is as the accrediting agency for the accreditation and preaccreditation (accreditation candidate) of education programs leading to the first professional or clinical degree at the master's or doctoral level and for the accreditation of these programs offered via distance education, throughout the United States.

Accreditation is awarded initially for up to 5 years and then up to 8 years for reaccreditation. Programs seeking reaccreditation are reviewed by the CAA during the year prior to expiration of their accreditation period. The CAA also offers a pre-accreditation status, or Candidacy, to new graduate education programs in the professions. As of 2004, the CAA no longer awards Candidacy to new master's level programs in audiology. As of January 1, 2007, the CAA no longer accredits master's level programs in audiology.

Accreditation by the CAA offers a graduate the assurance that the academic coursework and clinical practicum experiences in the accredited program meet nationally established standards. Accreditation means that a program has
- Engaged in extensive self-study over a period of years
- Prepared and submitted a complex application
- Undergone an on-site visit by a team of specially trained peers
- Received, and responded to, a digest of the report submitted by the site visitors
- Had its application, the site visit report, and its response studied by the CAA
- Undergone final evaluation and approval by the CAA

- Submitted and had approved annual reports during the period of its accreditation

## Composition

The CAA consists of 14 voting members and two nonvoting ex officio representatives, as follows:

- Eight academic members from the faculty/staff of accredited educational programs
  - At least one of these shall have clinical teaching (supervision) as his or her primary role
  - At least five shall have served as academic program accreditation site visitors
- Four clinical practitioners from nonacademic settings
  - At least one of these shall have experience in supervising students and/or clinical fellows
  - At least one shall be an audiologist
  - At least one shall be a speech-language pathologist
- One public member
- Nonvoting ex officio members include:
  - The ASHA Executive Director or his or her designee
  - The chair of the body that establishes ASHA's certification standards

The Council shall consist of members with the following qualifications:

- Six of the voting members shall represent the area of hearing, five of whom must hold the ASHA CCC in audiology
- Seven of the voting members shall represent the area of speech-language pathology, six of whom must hold the ASHA CCC in speech-language pathology

Current CAA members include the following:

Amy B Wohlert, Chair
Academic - SLP
University of New Mexico, Albuquerque

Frederick C Britten
Academic - Audiology
Fort Hays State University, Fort Hays, KS

Arlene E Carney
Academic - Audiology
University of Minnesota, Minneapolis

Kristine Fawson
Public Member
Mapleton, UT

Ellayne S Ganzfried
Practitioner - SLP
Great Neck, NY

Lee Ann C Golper
Academic - SLP
Vanderbilt-Bill Wilkerson Center, Nashville, TN

Mary Anne Hanner, Chair-Elect
Academic - SLP
Eastern Illinois University, Charleston

Stacey Matson
Practitioner - Audiology
Carl T Hayden VA Medical Center, Phoenix, AZ

Colleen O'Rourke, Vice-Chair/Audiology
Academic - Audiology
Georgia State University, Atlanta

George O Purvis
Practitioner - Audiology
VA Medical Center, Louisville, KY

Susan R Snover
Practitioner - SLP
Central Florida Speech and Hearing Center, Lakeland

Wayne E Swisher
Academic - SLP
University of North Dakota, Grand Forks

Richard Talbott
Academic - Audiology
University of South Alabama, Mobile

Jennifer B Watson, Vice-Chair/Speech-Language Pathology
Academic - SLP
Texas Christian University, Fort Worth

Robert E Novak
Ex Officio (Chair, CFCC)
Purdue University, West Lafayette, IN

Judith A Brassaeur
Observer (CAPCSD Representative)
Chico, CA

Stephanie A Davidson
Vice President for Academic Affairs (Monitoring Officer)
Ohio State University, Columbus

# Council on Rehabilitation Education

Marvin D Kuehn, EdD, Executive Director
300 North Martingale Road, Suite 460
Schaumburg, IL 60173
847 944-1345
847 944-1324 Fax
E-mail: mkuehn@emporia.edu
http://core-rehab.org

## Programs Accredited
- Rehabilitation counseling

## History
In 1969, a group of rehabilitation professionals met to discuss the need for accreditation of rehabilitation counselor education (RCE) programs. After 2 years of planning, the Council on Rehabilitation Education (CORE) was formed in 1971 and incorporated in 1972. Five professional rehabilitation organizations were represented on CORE:
- American Rehabilitation Association (ARA), formerly the International Association of Rehabilitation Facilities
- American Rehabilitation Counseling Association (ARCA)
- Council of State Administrators of Vocational Rehabilitation (CSAVR)
- National Council on Rehabilitation Education (NCRE), formerly the Council on Rehabilitation Educators
- National Rehabilitation Counseling Association (NRCA)
   Today, these five organizations—except ARA, which has been replaced by the National Council of State Agencies of the Blind (NCSAB) and two public members—comprise CORE and as such represent the professional and organizational constituencies concerned with the training, evaluation, and employment of rehabilitation counselors.

## Structure and Functions
The CORE accredits 102 university- and college-based rehabilitation counselor educational programs at the master's degree level. Accreditation serves to promote the effective delivery of rehabilitation services to people with disabilities by stimulating and fostering continual review and improvement of master's degree rehabilitation counselor educational programs.

## Purpose and Responsibilities
Each new rehabilitation counseling education program is assessed in accordance with the *Standards for Rehabilitation Counselor Education Programs*, published by CORE, and accredited programs are periodically reviewed to ensure that they remain in substantial compliance. The *Standards* are not intended to limit program creativity or limit variability; programs may adopt innovative procedures or experiences that meet the *Standards* in a different manner.
   As stated in its bylaws, the mission of CORE is the accreditation of rehabilitation counselor education (RCE) programs to promote the effective delivery of rehabilitation services to individuals with disabilities by promoting and fostering continuing review and improvement of master's degree-level RCE programs. The accreditation process serves to
- Promote a high standard of professional education in rehabilitation counseling and to foster program development based on a vocationally oriented, service-to-people attitude

- Encourage sound educational experimentation and innovations and to stimulate continuous self-study and improvement
- Reassess, redefine, and reevaluate program criteria as the needs of the profession and the public change
- Evolve a consultative model for developing programs
- Review admissions and other requirements of RCE programs to ensure that all qualified applicants may participate
- Foster mutual respect and cooperation between RCE programs and the programs of other helping professions
- Emphasize the vocational aspect of services in the broader context of human development and thereby help reduce dependency among all vulnerable consumer groups, especially individuals with the most severe and multiple disabilities
- Meet the personnel needs of public and private rehabilitation agencies
- Publish periodically a roster of recognized programs for members of the profession, the public, government agencies, and prospective students
- Enhance the position of mutual respect and acceptance of RCE programs in the academic community and on campus
- Develop an accreditation system based on the objective assessment of outcomes of the educational program

## Composition
CORE is composed of representatives, appointed to 4-year terms, from each of the five national professional organizations concerned with rehabilitation counseling: ARCA, CSAVR, NCRE, NRCA, and NCSAB. CORE also has two public members, one who represents the consumer public and one, the public at large.

### CORE Members, 2006-2007 Academic Year
**American Rehabilitation Counseling Association (ARCA)**
Linda Shaw, PhD, CRC
University of Florida, Gainesville
Dennis R Maki, PhD, CRC
University of Iowa, Iowa City
**Council of State Administrators of Vocational Rehabilitation (CSAVR)**
Charlene Dwyer, Administrator
Division of Vocational Rehabilitation, Department of Workforce
   Development, Madison, WI
**National Council on Rehabilitation Education (NCRE)**
Tyra Turner Whittaker, RhD, CRC
North Carolina A&T University, Greensboro
Tom Evenson, PhD, CRC
University of North Texas, Denton
**National Rehabilitation Counseling Association (NRCA)**
Joseph S Lechowicz, PhD
Hofstra University, Hempstead, NY
Christine Reid, PhD, CRC
Virginia Commonwealth University, Richmond
**National Council of State Agencies for the Blind, Inc (NCSAB)**
To be determined
**Public Member**
William Courtney, Esquire
Flemington, NJ
Patricia Nunez, CRC, CDMS, CCM
Brea, CA
**Chair of the Commission**
Lance Carluccio, PhD, CRC
Mount Ida College, Newton, MA

### Commission on Standards and Accreditation, Commissioners 2006-2007
**American Rehabilitation Counseling Association (ARCA)**

David B Peterson, PhD, CRC, NCC
Illinois Institute of Technology, Chicago
Irmo Marini, PhD, CRC
University of Texas Pan American, Edinboro
**Commission on Rehabilitation Counselor Certification (CRCC)**
Mary Barros-Bailey, MA, CRC, CDMS, NCC, CLCP, ABVE-D
Intermountain Vocational Services, Boise, ID
**Committee on Undergraduate Education**
David Perry
University of North Dakota, Grand Forks
**Department of Veterans Affairs**
Dorothy T Williams, PhD
VR&E Service, Washington, DC
**International Association of Rehabilitation Professionals (IARP)**
Chrisann Schiro-Geist, PhD, CRC
University of Memphis, Memphis, TN
Cherie King, MEd, CRC, CDMS, ABVE-D
Central Connecticut State University, Coventry, CT
**National Association of Multi-Cultural Rehabilitation Concerns (NAMCRC)**
Brenda Cartwright, EdD, CRC
University of Hawaii at Manoa, Honolulu
**National Association of Non-White Rehabilitation Workers (NANWRW)**
To be determined
**National Council on Rehabilitation Education (NCRE)**
Robinson A Vazquez Ramos, PhD, CRC
University of Puerto Rico, San Juan, PR
Charles Degeneffe
San Diego State University, CA
**National Rehabilitation Counseling Association (NRCA)**
Charles Palmer
Mississippi State University
Lance Carluccio, PhD, CRC
Mount Ida College, Newton, MA
**Vocational Evaluation and Career Assessment Professionals (VECAP)**
Juliet Fried, EdD, CRC, CVE
University of Northern Colorado, Greeley

# Joint Review Committee on Education in Radiologic Technology

Joanne S Greathouse, EdS, RT(R), FASRT, FAERS, Chief Executive Officer
20 North Wacker Drive, Suite 2850
Chicago, IL 60606-3182
312 704-5300
312 704-5304 Fax
E-mail: mail@jrcert.org
www.jrcert.org

## Programs Accredited
- Radiographer
- Radiation therapist
- Magnetic resonance
- Medical dosimetry

## Structure and Functions
The Joint Review Committee on Education in Radiologic Technology (JRCERT) is an independent agency recognized by the US Department of Education (ED) to accredit educational programs in radiography, radiation therapy, magnetic resonance, and medical dosimetry. It is composed of eight directors and is supported by six professional and eight support staff.

The JRCERT develops and maintains standards for accreditation of programs, reviews and evaluates educational programs in relation to those standards, and awards accreditation to programs in substantial compliance with them.

## History
The JRCERT traces its roots to 1944, when the American Society of X-ray Technicians (now the American Society of Radiologic Technologists [ASRT]) and the American Medical Association (AMA) developed the first standards for education in the field of radiologic sciences. From 1944 until 1969, program evaluation was carried out by the American College of Radiology's Commission on Technologists Affairs' Committee on Technologist Training. In 1969, the ASRT and the ACR established the JRCERT within the structure of allied health education accreditation provided by the AMA's Council on Medical Education and later by the Committee on Allied Health Education and Accreditation (CAHEA). The JRCERT was incorporated in 1971 and assumed the duties of review and evaluation of educational programs in the radiologic sciences.

When CAHEA was dissolved in 1994, the JRCERT sought and received recognition from the ED as the national accrediting agency for radiography and radiation therapy educational programs.

## Purposes and Responsibilities
The JRCERT promotes excellence in education and enhances quality and safety of patient care through the accreditation of educational programs.

The JRCERT awards accreditation to educational programs through a peer review process that includes programmatic self-study and an on-site evaluation to substantiate compliance with the *Standards for an Accredited Program in Radiologic Sciences,* the *Standards for an Accredited Program in Magnetic Resonance,* or the *Standards for an Educational Program in Medical Dosimetry.* Peer review evaluators are an integral part of the accreditation process. The JRCERT identifies these individuals and provides education and ongoing support to them. The JRCERT Board of Directors meets periodically to review program materials, including reports of site visit teams and other documentation provided by the program. Accreditation is awarded to those programs that demonstrate substantial compliance with accreditation standards. To maintain accreditation, each program must periodically provide reports, pay fees, and submit to the continuing accreditation process.

## The Value of JRCERT Accreditation

JRCERT accreditation demonstrates that a program adheres to national standards established by the profession to assure that graduates have the appropriate knowledge and skills to perform quality diagnostic and therapeutic procedures. The pursuit of programmatic accreditation encourages self-evaluation of the program, leading to ongoing program improvement.

Accreditation offers assurance to the public that a program accepts and fulfills its commitment to educational quality. Students can be assured that the program meets educational standards established by the profession. Graduates of accredited programs are more marketable, as employers can be assured that they possess the necessary knowledge, skills, attributes, and values to competently perform as entry-level practitioners as established by the profession. Graduates of accredited programs are also assured that they will be eligible for licensure in each of the 50 states. Student eligibility for financial assistance from state and federal government and the private sector is facilitated by participating in an accredited program. National and state certification/licensing agencies are also assured that accredited programs meet minimum standards through the accreditation process.

## Standards

The *Standards for an Accredited Program in Radiologic Sciences*, the *Standards for an Accredited Program in Magnetic Resonance*, and the *Standards for an Educational Program in Medical Dosimetry* serve as the basis of evaluation for accreditation. The standards identify criteria related to mission and goals, outcomes and effectiveness, program integrity, organization and administration, curriculum and academic practices, resources and student services, human resources, students, radiation safety, and fiscal responsibility.

## Composition

The JRCERT is composed of eight directors. Seven directors are nominated by four cooperating organizations: American College of Radiology (ACR), American Healthcare Radiology Administrators (AHRA), American Society of Radiologic Technologists (ASRT), and Association of Educators in Radiologic Sciences, Inc (AERS). The eighth director is a member of the public and is elected by the JRCERT Board of Directors.

### JRCERT Board of Directors

Shaun T Caldwell, MS, RT(R)(T)
The University of Texas MD Anderson Cancer Center

Barbara L Dehner, MSRS, RT(R)(M)(CT), FAERS
Morehead State University

Gregory J Ferenchak, MS, RT(R)(QM)
Broward Community College

Michael J Krajsa
Kutztown University Innovation Center

Denise E Moore, MS RT(R)
Sinclair Community College

H Martin Northup, MD, FACR
University of Florida Health Science Center

Penny K Sneed, MD, FACR
University of California, San Francisco

# Joint Review Committee on Educational Programs in Nuclear Medicine Technology

Elaine Cuklanz, MT(ASCP)NM, Executive Director
716 Black Point Rd
PO Box 1149
Polson, MT 59860-1149
406 883-0003
406 883-0022 Fax
E-mail: jrcnmt@centurytel.net
www.jrcnmt.org

## Programs Accredited

- Nuclear medicine technologist

## Structure and Functions

The Joint Review Committee on Educational Programs in Nuclear Medicine Technology (JRCNMT) was established in 1970 to provide accreditation recognition for educational programs in nuclear medicine technology in collaboration with the American Medical Association. In 1994, the JRCNMT was recognized by the US Department of Education (ED) and the Council for Higher Education Accreditation (CHEA) as an independent accrediting agency responsible for accrediting nuclear medicine technology programs. The JRCNMT is sponsored by four organizations:

- American College of Radiology
- American Society of Radiologic Technologists
- Society of Nuclear Medicine
- Society of Nuclear Medicine-Technologist Section

The Board of Directors is composed of three representatives from each professional organization and two public members. The JRCNMT reviews and revises the *Essentials and Guidelines for an Accredited Educational Program for the Nuclear Medicine Technologist*, evaluates nuclear medicine technology programs, grants accreditation status to qualified programs, maintains a directory of accredited nuclear medicine technology programs, and publishes accreditation actions to provide information to prospective students, employers, and the public.

## History

The JRCNMT was formed by the Society of Nuclear Medicine, Society of Nuclear Medical Technologists, American College of Radiology, American Society of Clinical Pathologists, American Society for Medical Technology, and American Society of Radiologic Technologists. The first meeting of the Joint Review Committee was in 1970.

The Society of Nuclear Medical Technologists, one of the original sponsors, terminated its corporate status as a professional organization in 1975. The American Society for Medical Technology and the American Society of Clinical Pathologists relinquished sponsorship in 1994.

The first *Essentials of an Accredited Educational Program for the Nuclear Medicine Technologist* was adopted by the collaborating organizations in 1969. The *Essentials* were substantially revised in 1976, 1984, 1991, 1997, and 2003.

## Purpose and Responsibilities

The JRCNMT establishes, maintains, and promotes standards of quality for educational programs in nuclear medicine technology to ensure the preparation of skilled professionals, and provides recognition for educational programs that meet or exceed the minimum standards outlined in the *Essentials*. The JRCNMT does the following:

- Establishes appropriate standards through comprehensive review and revision of the *Essentials and Guidelines for an Accredited Educational Program for the Nuclear Medicine Technologist* every 5 years. The JRCNMT conducts formal workshops, invites written comments, and disseminates proposed revisions of the *Essentials* to the community of interest and the public. At the conclusion of the review process, the *Essentials* are submitted to each of the sponsoring organizations for adoption.

- Provides accreditation recognition for nuclear medicine technology programs that meet or exceed the minimum educational standards. The JRCNMT and its staff provide consultative services to programs seeking initial or continuing reaccreditation. The accreditation process is enhanced by the use of program self-analysis and volunteer peer on-site evaluation. Accreditation actions are reached after consideration of the self-analysis, the onsite evaluation report, and follow-up communication. The JRCNMT disseminates information on accreditation actions to the USDE, CHEA, and the public. Information on approved educational programs in nuclear medicine technology is also provided to the AMA for publication in this *Directory*.

- Monitors the outcomes of the educational process on a regular basis. These outcomes are measured by a variety of means including graduates' success in career placement, job satisfaction, employer satisfaction, and certification results. The final product of quality educational programs is a graduate qualified to provide skilled professional services to patients.

## Standards

The *Essentials* are minimum educational standards adopted by the collaborating organizations. Each new program is assessed in accordance with the *Essentials*, and accredited programs are periodically reviewed to determine whether they remain in substantial compliance. The *Essentials and Guidelines* are available on written request from the JRCNMT and are available at www.jrcnmt.org.

## Composition

The JRCNMT is composed of 14 members: 12 professional members representing nuclear medicine and two public members. The selection of members includes practitioners, educators, and public interests.

### JRCNMT Members

Jan M Winn, MEd, RT(N), CNMT, Chair
Richard S Rome, MD, Vice-Chair
Frances K Keech, RT(N), MBA, Secretary-Treasurer
Christine Z Dickinson, MD
Gary L Dillehay, MD
Peter T Kirchner, MD
John R Leahy, MD
Karen Martin, AS, RT(N), CNMT
Mary Anne Owen, MHE, RT(N)
Richard States, MBA, CNMT, RT(N)
Lisa R Stocks-Brush RT(R), CNMT
Hadyn T Williams, MD

### Public Members

Rev Robert J Schrank (Em) BA, Th Dip
Thomas G Richmond, PhD

### Administrative Staff

Elaine Cuklanz, MT(ASCP)NM, Executive Director
John S Cowan, CPA, Accountant/Financial Advisor

**ACCREDITING AGENCIES**

# Medical Library Association/ American Library Association

Medical Library Association
Professional Development Department
65 East Wacker Place, Suite 1900
Chicago, IL 60601-7298
312 419-9094, ext 28
312 419-8950 Fax
E-mail: mlapd2@mlahq.org
www.mlanet.org

American Library Association, Committee on Accreditation/Office
   for Accreditation
Karen O'Brien, Director
50 East Huron
Chicago, IL 60611
312 280-2434
800 545-2433 toll free
E-mail: kobrien@ala.org
www.ala.org

## Programs Accredited

• Librarian, including medical librarian

## American Library Association

A master's degree in library and information sciences is offered
through the American Library Association (ALA)-accredited programs at graduate schools of library and information sciences. The
titles of degrees offered include Master of Arts, Master of Librarian
Science, Master of Library and Information Sciences, and Master of
Science. The ALA Office for Human Resource Development and
Recruitment (hrdr@ala.org or 800 545-2433, ext. 4282) offers information on selecting an ALA-accredited program.

ALA accreditation indicates that the program has undergone a
self-evaluation process, been reviewed by peers, and meets the
*Standards for Accreditation of Master's Programs in Library and
Information Studies* that were established by the Committee on
Accreditation and adopted by the ALA Council in 1992. The Committee on Accreditation evaluates each program for conformity to
the Standards, which address mission, goals, and objectives, curriculum, faculty, students, administration and financial support, and
physical resources and facilities.

## Composition

The members of the ALA Committee on Accreditation are listed
below.

Thomas W Leonhardt, Chair
Director, Scarborough-Phillips Library
St Edward's University

Karen Adams
Director, University Libraries
University of Alberta

Gail W Avery
Executive Assistant to the Director
District of Columbia Public Library

Lynn Silpigni Connaway
Consulting Research Scientist
OCLC

Ruth E Faklis
Director, Prairie Trails Public Library District

Paula J Fenza
Grants Manager
Mather Institute on Aging

Katy E Marre
Professor of English
University of Dayton

Nancy K Roderer
Director, Welch Medical Library
Johns Hopkins University

Richard E Rubin
Director, School of Library and Information Studies
Kent State University

Donna M Shannon
Associate Professor, School of Library and Information Science
University of South Carolina

Danny P Wallace
Professor, School of Library and Information Studies
University of Oklahoma

Virginia A Walter
Professor, Department of Information Science
University of California, Los Angeles

## Medical Library Association

The Medical Library Association (MLA) is a nonprofit educational
organization with more than 4,500 health sciences information professional members worldwide. Founded in 1898, MLA provides lifelong educational opportunities, supports a knowledge base of
health information research, and works with a global network of
partners to promote the importance of quality information for
improved health to the health care community and the public.

Library schools throughout North America offer courses in health
sciences information. MLA maintains a list of these courses by sur-
veying ALA-accredited programs; the list is available at
www.mlanet.org/education/libschools/index.html.

# National Accrediting Agency for Clinical Laboratory Sciences

Dianne M Cearlock, PhD, MT(ASCP), CLS(NCA), Chief Executive Officer (CEO)
8410 West Bryn Mawr Avenue, Suite 670
Chicago, IL 60631-3415
773 714-8880
773 714-8886 Fax
E-mail: dcearlock@naacls.org
www.naacls.org

## Programs Accredited

- Clinical laboratory scientist/medical technologist
- Clinical laboratory technician/medical laboratory technician
- Cytogenetic technologist
- Diagnostic molecular scientist
- Histologic technician/histotechnologist
- Pathologists' assistant

## Programs Approved

- Clinical assistant
- Phlebotomist

## Structure and Functions

The National Accrediting Agency for Clinical Laboratory Sciences (NAACLS), recognized by the Council for Higher Education Accreditation, is a nonprofit organization that independently accredits clinical laboratory science/medical technology, clinical laboratory technician/medical laboratory technician, histotechnician, histotechnologist, diagnostic molecular scientist, cytogenetic technologist, and pathologists' assistant programs. NAACLS approves phlebotomy and clinical assistant educational programs. NAACLS was established in 1973 as the successor to the American Society of Clinical Pathologists' (ASCP) Board of Schools. ASCP and the American Society for Clinical Laboratory Science (ASCLS) are sponsoring organizations of NAACLS. The National Society for Histotechnology and the Association of Genetic Technologists are participating organizations. The American Association of Pathologists' Assistants is an affiliating organization.

## History

The ASCP began program accreditation review as one of the functions of its Board of Registry of Medical Technologists in 1933, working with the American Medical Association (AMA) Council on Medical Education (CME). The first set of *Essentials of an Acceptable School for Clinical Laboratory Technicians* was prepared by the ASCP Board of Registry and subsequently adopted by the AMA House of Delegates in 1937. The first list of 211 accredited programs was issued by the ASCP Board of Registry in 1933. The program title was changed to medical technology in 1947 with that year's revised *Essentials*.

The ASCLS was organized in 1932, joining ASCP in periodically revising the *Essentials*, and was represented on the ASCP Board of Registry and the ASCP Board of Schools (established in 1949). The ASCLS was recognized by the CME as one of the organizations collaborating with the AMA in accrediting educational programs.

In 1972, representatives of ASCLS and ASCP began talks that culminated in the incorporation of NAACLS, in October 1973, as an organization independent of the professional organizations to which the authority to approve *Essentials* was delegated. In 2001, the term *Essentials* was replaced with *Standards* for all program levels.

## Purpose and Responsibilities

### Program Accreditation

Primary aspects of NAACLS's programmatic accreditation process include the self-study and site visit processes, evaluation by a review committee, and evaluation by the Board of Directors. Evaluation is based upon the *Standards*, which are the minimum criteria NAACLS uses when awarding programmatic accreditation.

NAACLS conducts various functions of programmatic accreditation, which include:

- Drafting and reviewing *Standards* for the operation of specialized programs
- Selecting and training knowledgeable volunteers to review Self-Study Reports and serve as site visitors
- Selecting representatives to serve on the review committees and the Board of Directors
- Granting accreditation awards based on a program's self-study and site visit processes

### Program Approval

NAACLS's program approval process identifies phlebotomy and clinical assistant educational programs that are structured to ensure that graduates possess stated career entry-level competencies. Program approval provides this measure of assurance to potential students, prospective employers, and the public. Primary aspects of the NAACLS program approval process are the self-study process, evaluation by the Programs Approval Review Committee (PARC), and evaluation by the Board of Directors. Evaluation is based upon the *Standards*, which are the minimum criteria NAACLS uses when awarding program approval.

The NAACLS program approval functions include:

- Drafting and reviewing *Standards and Competencies* for the operation of specialized programs
- Developing competencies of the entry-level laboratorian
- Selecting representatives to serve on the Programs Approval Review Committee and the Board of Directors
- Granting approval awards based on a program's self-study process

## Standards

The *Standards* are minimum educational standards adopted by NAACLS. Each new program is assessed in accordance with the *Standards*, and accredited programs are periodically reviewed to determine whether they remain in substantial compliance.

## Composition

The composition of NAACLS includes three review committees, the Board of Directors, and the executive office staff. The Clinical Laboratory Sciences Programs Review Committee (CLSPRC) reviews clinical laboratory science/medical technology, clinical laboratory technician/medical laboratory technician, diagnostic molecular scientist, cytogenetic technology, and histotechnician/histotechnologist programs for accreditation. The Affiliated Professions Review Committee (APRC) reviews pathologists' assistant programs for accreditation. The PARC reviews phlebotomy and clinical assistant programs for approval. The Board of Directors functions as the governing unit and grants final accreditation and approval awards. The executive office staff facilitates both the accreditation and approval processes.

The CLSPRC is composed of educators and practitioners elected from the clinical laboratory science disciplines. The APRC is

**ACCREDITING AGENCIES**

composed of educators elected from the profession and two members selected by the affiliated profession. The PARC is composed of elected educators and practitioners representing the disciplines of phlebotomy and clinical assisting. Each of the three review committees is represented by a Board of Directors liaison. Members are elected by the Board of Directors for staggered terms to assure continuity on each committee. The chairman and vice chairman are selected annually by committee members at the summer meeting. Each review committee meets twice annually, in the winter and summer.

The Board of Directors has 13 members, each serving a term of 4 years or until a successor is chosen. Officers are elected every year at the Board of Directors' fall meeting. The Board of Directors, which meets annually in the spring and fall, includes three clinical laboratory scientists who are active members of and nominated for election by the ASCLS (two clinical laboratory science/medical technology educators and one laboratory administrator); three clinical pathologists, nominated for election by the ASCP, who are fellows or associate members of that organization (two clinical pathologist educators and one laboratory director); two educators, who are neither clinical laboratory scientists nor physicians, one of whom is employed by a community or junior college and one by a senior college, elected by the board; two public representatives, whose primary livelihoods are not derived from the laboratory industry, elected by the board; one representative practitioner, who is a clinical laboratory science technician, elected by the board; and one member elected after nomination by each participating organization from among its active members.

## Board of Directors

Shauna Anderson, PhD, MT(ASCP)C, CLS(NCA), President
Lucille Contois, MA, MT(ASCP), Vice President
Cheryl Caskey, MA, CLSp(NCA), Secretary
Terrance E Suarez, PhD, Treasurer
Peggy Simpson, MS, MT(ASCP), Member at Large
David Gale, PhD, Past President
Brian Andrew, Public Member
Larry Bernstein, MD
Betty Dunn, MS, CLSp(CG)
Paula Garrott, EdM, MT(ASCP), CLS(NCA)
Robert Lott, HTL(ASCP)
Joel M Schilling, MD(FASCP)

### Clinical Laboratory Sciences Programs Review Committee (CLSPRC)

Duncan Samo, Med, MT(ASCP), CLS(NCA), Chair
Martha Lake, EdD, CLS(NCA), MT(ASCP), Vice Chair
Maria E Delost, MS, MT(ASCP), CLS(NCA)
Zoe Ann Durkin, MEd, HT(ASCP)
Floyd Grimm III, MEd
M Jane Hudson, PhD, MT(ASCP)SM, CLS(NCA)
Helen M Jenks, MT(ASCP), CLSp(CG)
Marcia Armstrong, MS, MT(ASCP), CLS(NCA)
Cecilia Landin, MS, MT(ASCP)
John Landis, MS, MT(ASCP)
Karen McClure, MS, MT(ASCP)
Carol McCoy, PhD, MT(ASCP), CLS(NCA)
Lydia McMorrow, PhD, ABMG
Lela M Morgan, MA, MT(ASCP), CLS(NCA)
Phyllis Pacifico, EdD, MT(ASCP)
Margie Scott, MD
Yasmen Simonian, PhD, MT(ASCP), CLS(NCA)
Teresa Taff, MA, MT(ASCP)SM

### Programs Approval Review Committee (PARC)

Karen Myers, MA, MT(ASCP)SC, CLS(NCA), Chair
Suzanne Campbell, MS, MT(ASCP)
Susan A K Higgins, MS, MT(ASCP)SC
Terry Kotrla, MS, MT(ASCP)BB, Vice Chair
Katharen LeSavoy, PBT(ASCP)
Wendy Miller, MS, MT(ASCP), CLS(NCA)
Mary Jean Rutherford, Med, MT(ASCP)SC
Nisi Zell, EdS, CLS(NCA), MT(ASCP)SH

### Affiliated Professions Review Committee (APRC)

Jerry Phipps, BS, BHS, Chair
Mousa Al-Abbadi, MD
Colleen O Galvis
Kenneth V Kaloustian, PhD
Mark Kellogg, PhD, MT(ASCP)

# National Association of Schools of Music

Samuel Hope, Executive Director
11250 Roger Bacon Drive, Suite 21
Reston, VA 20190-5248
703 437-0700
703 437-6312 Fax
E-mail: info@arts-accredit.org
http://nasm.arts-accredit.org

## Programs Accredited
• Music therapy

## Structure and Functions
The National Association of Schools of Music (NASM) was founded in 1924 to secure a better understanding among institutions of higher education engaged in work in music (including music therapy); to establish a more uniform method of granting credit; and to set minimum standards for granting degrees and other credentials. In 1975, representatives of member institutions agreed to create a category of membership for nondegree-granting institutions. The services of NASM are available to all types of degree-granting institutions in higher education and to nondegree-granting institutions offering preprofessional programs or general music training programs. Membership in NASM is voluntary.

NASM is recognized by the US Department of Education (ED) as the agency responsible for accrediting all music curricula.

## Purpose and Responsibilities
NASM works to:
• Provide a national forum for the discussion and consideration of concerns relevant to the preservation and advancement of standards in the field of music in higher education
• Develop a national unity and strength to maintain the position of music study in the family of fine arts and humanities in universities, colleges, and schools of music
• Maintain professional leadership in music training and develop a national context for the artist's professional growth
• Establish minimum achievement standards in music curricula without restricting an administration or school in its freedom to develop new ideas, to experiment, or to expand its program
• Recognize that inspired teaching may rightly reject a "status quo" philosophy
• Establish that the prime objective of all educational programs in music is to provide the opportunity for every music student to develop individual potential to the utmost

## Standards
The NASM publishes a biannual handbook that includes specific standards and guidelines required for accreditation of educational programs in music therapy. Each new program applying for accreditation is assessed in accordance with these standards, and accredited programs are periodically reviewed to ensure that they remain in substantial compliance. The NASM publishes an annual directory of accredited music schools and programs.

## Composition
NASM is governed by three commissions, the board of directors, and the executive office staff. The NASM board of directors is composed of the officers (president, vice president, treasurer, secretary, executive director [ex officio], and nine regional chairs) as well as the immediate past president, the chair and the associate chair of the Commission on Accreditation, the chairs of the Commissions on Community/Junior College Accreditation and Non–Degree-Granting Accreditation, and three public members.

## NASM Board of Directors
**President**
Daniel P Sher, University of Colorado, Boulder

**Vice President**
Jo Ann Domb, University of Indianapolis

**Treasurer**
Mellansenah Y Morris, Ohio State University

**Secretary**
Mark Wait, Vanderbilt University

**Past President**
William Hipp, University of Miami

**Chair, Commission on Non–Degree-Granting Accreditation**
Margaret Quackenbush, David Hochstein Memorial Music School

**Chair, Commission on Community/Junior College Accreditation**
Eric W Unruh, Casper College

**Chair, Commission on Accreditation**
James C Scott, University of North Texas

**Associate Chair, Commission on Accreditation**
Charlotte A Collins, Shenandoah University

**Regional Chairs**
Dale E Monson, Brigham Young University
Ramona Holmes, Seattle Pacific University
Michael D Wilder, Friends University
Robert W Kase, University of Wisconsin, Stevens Point
Donald R Grant, Northern Michigan University
Terry B Ewell, Towson University
Dennis J Zeisler, Old Dominion University
Jimmie James, Jr, Jackson State University
Arthur L Shearin, Harding University

**Public Members**
Melinda A Campbell, Duxbury, Massachusetts
Mary E Farley, Mount Kisco, New York

**Ex Officio**
Samuel Hope, NASM Executive Director

**ACCREDITING AGENCIES**

# National Recreation and Park Association/American Association for Physical Activity and Recreation Council on Accreditation

Danielle Timmerman, MS, Staff Liaison
Professional Services Division
22377 Belmont Ridge Road
Ashburn, VA 20148-4501
703 858-2150
703 858-0794 Fax
E-mail: dtimmerman@nrpa.org
www.councilonaccreditation.org

## Programs Accredited

- Therapeutic recreation specialist

## Structure and Functions

The Council on Accreditation of the National Recreation and Park Association (NRPA), in cooperation with the American Association for Physical Activity and Recreation (AAPAR), is authorized by the NRPA Board of Trustees to accredit programs at the baccalaureate level in recreation, therapeutic recreation, park resources, and leisure services. Established in 1974, the Council is recognized by the Council for Higher Education Accreditation (CHEA). The Council on Accreditation publishes the *Handbook of the NRPA/AAPAR Accreditation Process*, *Procedural Guidelines for the Accreditation Process*, and *Standards and Evaluative Criteria for Baccalaureate Programs in Recreation, Park Resources and Leisure Services*.

## Purpose and Responsibilities

The Council believes that accreditation serves the public by promoting and maintaining standards of professional preparation; assists the academic unit and the institution's administration in establishing and attaining appropriate goals for therapeutic recreation programs; ensures continual self-study for development and improvement of professional preparation programs; and encourages innovation and experimentation within acceptable academic and professional standards.

Accredited programs have undergone a rigorous process of self-assessment and peer review and have met standards established by the profession itself as essential to prepare graduates for competence in entry-level positions in the profession. Programs are reviewed every 5 years to ensure continued compliance with the standards.

## Composition

The Council on Accreditation meets twice annually to review accredited programs, revise and update accreditation standards and procedures, and conduct business. The Council consists of 10 members, who are appointed by the two sponsoring organizations: five educators, three practitioners, one college/university administrator, and one public representative. Each member serves a 3-year term.

## Council Roster, 2005–2006

Carlton F Yoshioka, PhD, Council Chair
Roger Coles, EdD, CPRP, Vice Chair
William L Clevenger, EDS, CPRP
Cheryl Beeler, ReD, CPRP
Marcia J Carter, PhD, CTRS, CPRP
Leigh Wion, CTRS, CBIS
John B Johnston, III
Jamie Sabbach, CPRP
Gary Ellis, PhD, FALS
J Robert Rossman, PhD
Danielle Timmerman, MS, Staff Liaison

# Section IV

# Health Professions Education Data

---

## Health Professions Education Data

Every year, the American Medical Association (AMA) surveys health professions' educational programs accredited by the following 24 organizations:

- Accreditation Council for Occupational Therapy Education
- Accreditation Review Commission on Education for the Physician Assistant
- American Art Therapy Association
- American Board of Genetic Counseling
- American Dance Therapy Association
- American Orthoptic Council
- American Society of Health-System Pharmacists
- Association for Education and Rehabilitation of the Blind and Visually Impaired
- Commission on Accreditation for Health Informatics and Information Management Education
- Commission on Accreditation in Physical Therapy Education
- Commission on Accreditation of Allied Health Education Programs
- Commission on Accreditation of Athletic Training Education
- Commission on Accreditation of Ophthalmic Medical Programs
- Commission on Dental Accreditation of the American Dental Association
- Commission on Massage Therapy Accreditation
- Commission on Opticianry Accreditation
- Council for Accreditation of Counseling and Related Educational Programs
- Council on Academic Accreditation in Audiology and Speech-Language Pathology
- Council on Rehabilitation Education
- Joint Review Committee on Education in Radiologic Technology
- Joint Review Committee on Educational Programs in Nuclear Medicine Technology
- National Accrediting Agency for Clinical Laboratory Sciences
- National Association of Schools of Music
- National Recreation and Park Association/American Association for Physical Activity and Recreation Council on Accreditation

The program population is established from periodic accreditation data and updates provided by each agency throughout the year. Accrediting agencies also forwarded to the AMA administrative changes received from educational programs and sponsoring institutions.

The AMA recognizes and thanks the accrediting agencies for their work in ensuring an accurate population and an up-to-date *Directory*. Specifically, the agencies helped AMA staff develop and test the survey instrument and ensured a high survey response rate by contacting programs that had not yet responded by the initial survey deadline of September 29, 2006.

## Annual Survey of Health Professions Education Programs

In 2002, for the first time, the AMA used an online survey instrument to collect data on health professions education programs. This process was also used for the 2006 survey cycle, beginning in May and extending through November 2006. The online survey, available at www.ama-assn.org/go/hpsurvey, helps reduce survey mailing costs and allows for the collection of racial/ethnic program data, which had been discontinued in 1996 due to the financial burden and the difficulty of obtaining accurate data. Program directors received the survey via e-mail, which directs them to the survey site and provides institution ID, program ID, and login password. Those program directors without a valid e-mail address on file received this information via US mail.

To ensure a high response rate and data quality, nonrespondents received reminder e-mails/mailings/phone calls/faxes after the original surveys were sent, both from the AMA and from the accrediting agencies. Telephone, fax, and e-mail inquiries of program personnel helped clarify questionable responses.

Data collected were for the 2005-2006 academic year, generally September 1, 2005, through August 31, 2006.

### Data Collected

The survey collected the following data:

*Student data*
- data on program enrollments, attrition, and graduates by gender and race/ethnicity

*Program data*
- Tuition cost (in-state and out-of-state)
- Class capacity per start date or session
- Availability of evening/weekend classes
- Program length(s)
- Program start date(s)
- Degree/credential(s) awarded
- Program stipend offered
- Data on employment/additional education of recent program graduates
- Name, address, telephone/fax numbers, and e-mail address of program official(s)
- Program Web site
- Availability of education/courses in medical/health care terms in non-English languages
- Availability of education in cultural competence or patient communication

### Health Professions Education Data Book

Data collected on the survey, which had in the past been published in this section of the Directory, are now available in the *Health Professions Education Data Book*.

The 2006-2007 edition of the Data Book, published in May 2006, included the following tables, for academic year 2004-2005:

DATA

Section A: Data on Enrollments, Attrition, and Graduates, Programs, and Sponsoring Institutions
- Accredited or Approved Programs by Occupation, 1985, 1990, 1995, and 2005
- Number of Programs and Enrollments, Attrition, and Graduates by Occupation
- Enrollments by Occupation and Degree/Award
- Graduates by Occupation and Degree/Award
- Accredited Programs by Type of Sponsoring Institution
- Institutions Sponsoring Accredited Programs by Ownership Type
- Accredited Programs by Occupation and Type of Sponsoring Institution
- Enrollments, Graduates, and Number of Programs by State/Province and Occupation

Section B: Data on Enrollments, Attrition, and Graduates by Occupation, Gender, and Race-Ethnic Origin
- Enrollments, Attrition, and Graduates by Occupation and Gender
- Enrollments by Occupation and Gender/Race-Ethnic Origin
- Attrition by Occupation and Gender/Race-Ethnic Origin
- Graduates by Occupation and Gender/Race-Ethnic Origin
- Enrollments, Attrition, and Graduates by Race/Ethnic Origin and Gender
- Enrollments, Attrition, and Graduates by Gender and Type of Sponsoring Institution
- Enrollments, Attrition, and Graduates by Gender and Type of Sponsoring Institution for Whites
- Enrollments, Attrition, and Graduates by Gender and Type of Sponsoring Institution for Blacks
- Enrollments, Attrition, and Graduates by Gender and Type of Sponsoring Institution for Hispanics
- Enrollments, Attrition, and Graduates by Gender and Type of Sponsoring Institution for Native Americans/Alaskan Natives
- Enrollments, Attrition, and Graduates by Gender and Type of Sponsoring Institution for Asians/Pacific Islanders
- Enrollments, Attrition, and Graduates by Gender and Institutional Control Type
- Enrollments, Attrition, and Graduates by Gender and Institutional Control Type for Whites
- Enrollments, Attrition, and Graduates by Gender and Institutional Control Type for Blacks
- Enrollments, Attrition, and Graduates by Gender and Institutional Control Type for Hispanics
- Enrollments, Attrition, and Graduates by Gender and Institutional Control Type for Native Americans/Alaskan Natives
- Enrollments, Attrition, and Graduates by Gender and Institutional Control Type for Asians/Pacific Islanders

Section C: Data on Employment, Cultural Competence/Patient Communication Curricula, Tuition, Program Length, and Salary
- Percentage of Graduates Finding Employment Within 6 Months and/or Seeking Additional Education, by Occupation
- Number of Programs Offering Cultural Competence/Patient Communication Curricula, by Occupation
- Average First-Year Tuition, by Occupation
- Average Program Length, by Occupation
- Health Professions Salary Ranges

For more information or to order, call 312 464-5333 or e-mail enza.perrone@ama-assn.org, or visit www.ama-assn.org/go/hpdatabook.

## The AMA's Role in Health Professions Education Data Collection

Since the early seventies, the AMA has annually collected program enrollment and graduate data for the allied health professions. Since 1989, enrollment, attrition, and graduate data have been compiled by gender. Collection of data on race/ethnic origin was discontinued after the 1995 survey; starting in 2002, these data are now available in the *Health Professions Education Data Book* (see above).

In addition, a wide variety of data collected on the survey are available through the AMA's Health Professions Data Service. For example, data on such variables as enrollment, attrition, and graduates, as well as program size, length, and tuition, can be broken out by state, profession, and/or accrediting agency. For more information, contact

American Medical Association
Medical Education Products
515 N State St
Chicago, IL 60610
312 464-5333
312 464-5830 Fax
E-mail: fred.lenhoff@ama-assn.org

## Table 1. Accredited or Approved Programs by Occupation, 1985, 1990, 1995, 2000, and 2006

| Occupation | 1985 | 1990 | 1995 | 2000 | 2006 |
|---|---|---|---|---|---|
| Anesthesiologist Assistant (AA) | 0 | 2 | 2 | 2 | 4 |
| Art Therapist (ATR) | – | – | – | 28 | 31 |
| Athletic Trainer (AT) | – | – | 29 | 122 | 350 |
| Audiologist (AUD) | 149 | 111 | 120 | 109 | 88 |
| Cardiovascular Technologist (CVT) | 0 | 2 | 14 | 24 | 29 |
| Clinical Assistant (CA) | – | – | – | – | 3 |
| Clinical Laboratory Scientist/Medical Technologist (CLS/MT) | 584 | 420 | 357 | 255 | 226 |
| Clin Lab Tech/Med Lab Tech (CLT/MLT) | 225 | 215 | 223 | 242 | 201 |
| Counselor (C) | 43 | 146 | 239 | 334 | 193 |
| Cytogenetic Technologist (CG) | – | – | 6 | 5 | 5 |
| Cytotechnologist (CYTO) | 58 | 46 | 67 | 48 | 47 |
| Dance Therapist (DANCE) | – | – | – | 5 | 5 |
| Dental Assistant (DA) | 290 | 244 | 229 | 256 | 266 |
| Dental Hygienist (DH) | 198 | 202 | 212 | 255 | 279 |
| Dental Laboratory Technician (DLT) | 58 | 49 | 37 | 30 | 21 |
| Diagnostic Medical Sonographer (DMS) | 24 | 43 | 77 | 76 | 144 |
| Diagnostic Molecular Scientist (DMOS) | – | – | – | 3 | 4 |
| Dietetic Technician (DT) | 80 | 62 | 70 | 67 | 66 |
| Dietitian/Nutritionist (DN) | 444 | 439 | 529 | 544 | 545 |
| Electroneurodiagnostic Technologist (ET) | 20 | 14 | 14 | 11 | 13 |
| Emergency Medical Technician-Paramedic (EMTP) | 20 | 72 | 96 | 122 | 209 |
| Exercise Physiologist (EXPHY) | – | – | – | – | 2 |
| Exercise Science Professional (EXSCI) | – | – | – | – | 6 |
| Genetic Counselor (GC) | – | – | – | 23 | 30 |
| Health Information Administrator (HIA) | 54 | 55 | 53 | 51 | 47 |
| Health Information Technician (HIT) | 85 | 108 | 142 | 175 | 194 |
| Histotechnician (HT) | ++ | ++ | ++ | 24 | 23 |
| Histotechnologist (HTL) | 43++ | 37++ | 31++ | 2 | 4 |
| Kinesiotherapist (K) | – | – | – | 4 | 6 |
| Low Vision Therapist (LVT) | – | – | – | – | 1 |
| Massage Therapist (MASST) | – | – | – | 63 | 84 |
| Medical Assistant (MA) | 168 | 185 | 221 | 475 | 545 |
| Medical Illustrator (MI) | 5 | 6 | 5 | 5 | 5 |
| Medical Librarian (ML) | – | – | – | 56 | 55 |
| Music Therapist (MT) | – | – | – | 61 | 70 |
| Nuclear Medicine Technologist (NMT) | 141 | 107 | 120 | 88 | 97 |
| Occupational Therapist (OT) | 61 | 69 | 98 | 142 | 155 |
| Occupational Therapy Assistant (OTA) | 60 | 69 | 108 | 185 | 134 |
| Ophthalmic Assistant | – | – | – | – | 5 |
| Ophthalmic Dispensing Optician (OD) | – | – | – | 24 | 22 |
| Ophthalmic Medical Technician/Technologist (OMT) | 9 | 10 | 10 | 16 | 16 |
| Orientation and Mobility Specialist (OMS) | – | – | – | – | 18 |
| Orthoptist (OR) | – | – | – | 16 | 15 |
| Orthotist/Prosthetist (OP) | – | – | 1 | 7 | 7 |
| Pathologists' Assistant (PTHA) | 0 | 0 | 0 | 5 | 7 |
| Perfusionist (PERF) | 19 | 26 | 33 | 23 | 21 |
| Pharmacy Technician (PHARMT) | 8 | 29 | 56 | 96 | 96 |
| Phlebotomist (PHLEB) | – | – | 69 | 53 | 55 |
| Physical Therapist (PT) | – | – | – | 194 | 202 |
| Physical Therapist Assistant (PTA) | – | – | – | 261 | 222 |
| Physician Assistant (PA) | 52 | 48 | 64 | 126 | 136 |
| Polysomnographic Technologist (PSOM) | – | – | – | – | 2 |
| Radiation Therapist (RADT) | 101 | 104 | 120 | 72 | 80 |
| Radiographer (RAD) | 744 | 672 | 677 | 583 | 612 |
| Rehabilitation Counselor (RC) | – | – | – | 83 | 100 |
| Rehabilitation Teacher (RTEACH) | – | – | – | 9 | 8 |
| Respiratory Therapist (Advanced) (REST) | 232 | 259 | 286 | 310 | 315 |
| Respiratory Therapist (Entry—Level) (RESTT) | 182 | 159 | 174 | 105 | 42 |
| Specialist in Blood Bank Technology (SBBT) | 59 | 29 | 24 | 15 | 14 |
| Speech-Language Pathologist (SLP) | ** | 194 | 222 | 224 | 237 |
| Surgical Assistant (SA) | 3 | 3 | + | + | 7 |
| Surgical Technologist (ST) | 102 | 113 | 143 | 257 | 407 |
| Therapeutic Recreation Specialist (TRS) | – | – | – | 43 | 40 |
| **Total** | **4,377** | **4,390** | **5,015** | **6,414** | **6,873** |

—These occupations were not part of the accreditation systems reflected in this Directory.

**The number of speech-language pathologist programs for 1985 is included in the number of audiologist programs.

+ Surgeon assistant programs were listed with physician assistant programs.

++ Before 1996, numbers for HT and HTL were not broken out.

DATA

## Table 2. Percentage of Graduates Finding Employment Within 6 Months and/or Seeking Additional Education, by Occupation, Academic Year 2005-2006

| Occupation | # of responding prgms | Found Job (%) | Additional Education (%) | Both (%) |
|---|---|---|---|---|
| Anesthesiologist Assistant | 3 | 100.0 | 1.7 | 0.0 |
| Art Therapist | 15 | 83.4 | 5.5 | 6.3 |
| Athletic Trainer | 214 | 44.7 | 39.6 | 17.1 |
| Audiologist | 48 | 74.4 | 0.6 | 3.4 |
| Cardiovascular Technologist | 18 | 83.1 | 2.2 | 15.6 |
| Clinical Assistant | 2 | 92.5 | 30.0 | 55.0 |
| Clinical Laboratory Scientist/Medical Technologist | 209 | 96.5 | 3.6 | 5.8 |
| Clinical Laboratory Technician/Medical Laboratory Technician | 179 | 91.5 | 11.9 | 15.2 |
| Counselor | 61 | 80.1 | 6.8 | 7.0 |
| Cytogenetic Technologist | 3 | 87.1 | 2.9 | 4.7 |
| Cytotechnologist | 38 | 87.5 | 6.3 | 6.4 |
| Dance/Movement Therapist | 4 | 82.3 | 3.8 | 4.8 |
| Dental Assistant | 177 | 83.8 | 8.9 | 8.3 |
| Dental Hygienist | 195 | 90.9 | 2.5 | 6.2 |
| Dental Laboratory Technician | 11 | 84.0 | 12.6 | 3.4 |
| Diagnostic Medical Sonographer | 87 | 94.8 | 4.6 | 5.1 |
| Diagnostic Molecular Scientist | 2 | 71.5 | 28.5 | 14.0 |
| Electroneurodiagnostic Technologist | 8 | 97.5 | 1.3 | 0.0 |
| Emergency Medical Technician-Paramedic | 103 | 89.7 | 14.0 | 16.0 |
| Exercise Physiologist | 1 | 100.0 | 0.0 | 0.0 |
| Exercise Science Professional | 2 | 50.0 | 0.0 | 0.0 |
| Genetic Counselor | 22 | 89.7 | 1.9 | 1.1 |
| Health Information Administrator | 32 | 86.6 | 8.3 | 5.8 |
| Health Information Technician | 112 | 83.5 | 6.5 | 7.0 |
| Histotechnician | 18 | 74.5 | 15.4 | 9.4 |
| Histotechnologist | 2 | 100.0 | 5.0 | 5.0 |
| Kinesiotherapist | 3 | 65.0 | 20.3 | 4.0 |
| Low Vision Therapist | 1 | 100.0 | 0.0 | 0.0 |
| Massage Therapist | 22 | 63.4 | 9.7 | 15.5 |
| Medical Assistant | 360 | 83.9 | 9.7 | 8.9 |
| Medical Illustrator | 2 | 100.0 | 0.0 | 0.0 |
| Medical Librarian | 7 | 37.1 | 0.9 | 1.3 |
| Music Therapist | 41 | 86.2 | 8.7 | 10.4 |
| Nuclear Medicine Technologist | 92 | 95.3 | 1.4 | 2.9 |
| Occupational Therapist | 110 | 93.5 | 1.9 | 3.5 |
| Occupational Therapy Assistant | 87 | 88.3 | 3.7 | 4.2 |
| Ophthalmic Assistant | 2 | 100.0 | 0.0 | 0.0 |
| Ophthalmic Dispensing Optician | 7 | 95.9 | 7.9 | 5.0 |
| Ophthalmic Medical Technician/Technologist | 13 | 93.2 | 6.5 | 0.1 |
| Orientation and Mobility Specialist | 7 | 99.3 | 5.7 | 6.4 |
| Orthoptist | 13 | 84.6 | 0.0 | 0.0 |
| Orthotist/Prosthetist | 6 | 99.5 | 3.3 | 18.3 |
| Pathologists' Assistant | 6 | 100.0 | 0.0 | 0.0 |
| Perfusionist | 11 | 97.7 | 0.0 | 0.9 |
| Pharmacy Technician | 48 | 71.5 | 12.1 | 11.9 |
| Phlebotomist | 46 | 61.7 | 21.1 | 18.0 |
| Physical Therapist | 119 | 99.5 | 2.2 | 3.5 |
| Physical Therapist Assistant | 107 | 94.4 | 4.0 | 6.3 |
| Physician Assistant | 100 | 91.0 | 1.9 | 1.8 |
| Polysomnographic Technologist | 2 | 93.0 | 3.5 | 0.0 |
| Radiation Therapist | 61 | 91.6 | 3.6 | 7.6 |
| Radiographer | 478 | 94.0 | 9.4 | 15.4 |
| Rehabilitation Counselor | 48 | 89.0 | 4.4 | 4.7 |
| Rehabilitation Teacher | 3 | 100.0 | 0.0 | 0.0 |
| Respiratory Therapist (Advanced) | 218 | 96.5 | 7.0 | 10.3 |
| Respiratory Therapist (Entry-Level) | 21 | 89.4 | 32.8 | 35.6 |
| Specialist in Blood Bank Technology | 10 | 80.0 | 2.5 | 10.0 |
| Speech-Language Pathologist | 104 | 96.0 | 1.5 | 3.3 |
| Surgical Assistant | 5 | 86.4 | 1.6 | 2.0 |
| Surgical Technologist | 233 | 83.1 | 6.3 | 10.5 |
| Therapeutic Recreation Specialist | 19 | 77.4 | 8.0 | 3.7 |

## Table 3. Average First-Year Tuition, by Occupation, Academic Year 2005-2006

| Occupation | Tuition |
|---|---|
| Anesthesiologist Assistant | $27,927 |
| Art Therapist | $12,701 |
| Athletic Trainer | $11,066 |
| Audiologist | $8,549 |
| Cardiovascular Technologist | $7,628 |
| Clinical Assistant | $2,961 |
| Clinical Laboratory Scientist/Medical Technologist | $5,721 |
| Clinical Laboratory Technician/Medical Laboratory | $3,322 |
| Counselor | $5,436 |
| Cytogenetic Technologist | $3,824 |
| Cytotechnologist | $7,347 |
| Dance/Movement Therapist | $16,487 |
| Dental Assistant | $3,673 |
| Dental Hygienist | $4,474 |
| Dental Laboratory Technician | $2,872 |
| Diagnostic Medical Sonographer | $6,124 |
| Diagnostic Molecular Scientist | $6,261 |
| Electroneurodiagnostic Technologist | $4,077 |
| Emergency Medical Technician-Paramedic | $3,492 |
| Exercise Physiologist | $11,430 |
| Exercise Science Professional | $4,250 |
| Genetic Counselor | $12,222 |
| Health Information Administrator | $7,352 |
| Health Information Technician | $3,486 |
| Histotechnician | $2,952 |
| Histotechnologist | $20,000 |
| Kinesiotherapist | $4,802 |
| Massage Therapist | $7,723 |
| Medical Assistant | $6,179 |
| Medical Illustrator | $12,005 |
| Medical Librarian | $7,648 |
| Music Therapist | $9,782 |
| Nuclear Medicine Technologist | $6,485 |
| Occupational Therapist | $12,434 |
| Occupational Therapy Assistant | $3,867 |
| Ophthalmic Assistant | $1,632 |
| Ophthalmic Dispensing Optician | $2,324 |
| Ophthalmic Medical Technician/Technologist | $5,637 |
| Orientation and Mobility Specialist | $6,683 |
| Orthoptist | $2,816 |
| Orthotist/Prosthetist | $9,944 |
| Pathologists' Assistant | $18,183 |
| Perfusionist | $13,435 |
| Pharmacy Technician | $4,435 |
| Phlebotomist | $1,374 |
| Physical Therapist | $14,974 |
| Physical Therapist Assistant | $3,964 |
| Physician Assistant | $19,033 |
| Polysomnographic Technologist | $6,260 |
| Radiation Therapist | $5,830 |
| Radiographer | $4,426 |
| Rehabilitation Counselor | $5,831 |
| Rehabilitation Teacher | $5,245 |
| Respiratory Therapist (Advanced) | $4,482 |
| Respiratory Therapist (Entry-Level) | $4,992 |
| Specialist in Blood Bank Technology | $3,890 |
| Speech-Language Pathologist | $7,808 |
| Surgical Assistant | $3,753 |
| Surgical Technologist | $5,483 |
| Therapeutic Recreation Specialist | $3,628 |

## Table 4. Average Program Length, by Occupation, Academic Year 2005-2006

| Occupation | | # of Months |
|---|---|---|
| Anesthesiologist Assistant | AA | 19.8 |
| Art Therapist | ART | 24.6 |
| Athletic Trainer | ATH | 30.4 |
| Audiologist | AUD | 44.6 |
| Cardiovascular Technologist | CVT | 19.8 |
| Clinical Assistant | CA | 9.0 |
| Clinical Laboratory Scientist/Medical Technologist | CLS/MT | 16.2 |
| Clinical Laboratory Technician/Medical Laboratory Tech | CLT/MLT | 21.0 |
| Counselor | C | 27.1 |
| Cytogenetic Technologist | CG | 13.6 |
| Cytotechnologist | CYTO | 13.2 |
| Dance/Movement Therapist | DANCE | 27.3 |
| Dental Assistant | DA | 11.5 |
| Dental Hygienist | DH | 21.9 |
| Dental Laboratory Technician | DLT | 19.6 |
| Diagnostic Medical Sonographer | DMS | 18.8 |
| Diagnostic Molecular Scientist | DMOS | 9.0 |
| Electroneurodiagnostic Technologist | ET | 18.5 |
| Emergency Medical Technician-Paramedic | EMTP | 15.0 |
| Exercise Physiologist | EXPHY | 9.3 |
| Exercise Science Professional | EXSCI | 22.8 |
| Genetic Counselor | GC | 20.1 |
| Health Information Administrator | HIA | 28.2 |
| Health Information Technician | HIT | 19.7 |
| Histotechnician | HT | 16.0 |
| Histotechnologist | HTL | 11.0 |
| Kinesiotherapist | K | 40.5 |
| Massage Therapist | MASST | 7.5 |
| Medical Assistant | MA | 15.2 |
| Medical Illustrator | MI | 17.6 |
| Medical Librarian | ML | 2.6 |
| Music Therapist | MT | 40.6 |
| Nuclear Medicine Technologist | NMT | 19.3 |
| Occupational Therapist | OT | 33.0 |
| Occupational Therapy Assistant | OTA | 21.8 |
| Ophthalmic Assistant | OA | 4.2 |
| Ophthalmic Dispensing Optician | OD | 18.0 |
| Ophthalmic Medical Technician/Technologist | OMT | 19.6 |
| Orientation and Mobility Specialist | OMS | 16.2 |
| Orthoptist | OR | 24.8 |
| Orthotist/Prosthetist | OP | 14.3 |
| Pathologists' Assistant | PTHA | 20.0 |
| Perfusionist | PERF | 20.0 |
| Pharmacy Technician | PHARMT | 11.0 |
| Phlebotomist | PHLEB | 4.4 |
| Physical Therapist | PT | 33.5 |
| Physical Therapist Assistant | PTA | 21.1 |
| Physician Assistant | PA | 26.7 |
| Polysomnographic Technologist | PSOM | 6.5 |
| Radiation Therapist | RADT | 21.2 |
| Radiographer | RAD | 23.7 |
| Rehabilitation Counselor | RC | 23.3 |
| Rehabilitation Teacher | RTEACH | 16.8 |
| Respiratory Therapist (Advanced) | REST | 22.5 |
| Respiratory Therapist (Entry-Level) | RESTT | 19.0 |
| Specialist in Blood Bank Technology | SBBT | 15.4 |
| Speech-Language Pathologist | SLP | 24.9 |
| Surgical Assistant | SA | 12.3 |
| Surgical Technologist | ST | 14.1 |
| Therapeutic Recreation Specialist | TRS | 31.9 |

## Table 5. Health Professions Salary Ranges*

| Occupation | Starting Salary | Overall Average | Upper Ranges | Year of Data |
|---|---|---|---|---|
| Anesthesiologist assistant | $95,000-$120,000 | | $160,000-$180,000 | 2006 |
| Art therapist | $32,000 | $38,000-$48,000 | $50,000-$80,000 | 2006 |
| Athletic trainer | $35,000 | $45,000 | $55,000-$75,000 | 2005 |
| Audiologist | $45,000 | $62,000 | $78,000 (management) | 2004 |
| Cardiovascular technologist | $36,000-$45,000 | $50,000-$65,000 | $75,000 plus | 2006 |
| Clinical lab scientist/med technologist | | $44,500-$52,000 | | 2005 |
| CLS/MT manager | | $69,500-$72,000 | | 2005 |
| Clinical lab technician/medical lab technician | | $37,100-$41,000 | | 2005 |
| Counselor | $23,560 | $42,110 | $67,170 | 2004 |
| Cytotechnologist | $46,000 | $68,645 | $105,600 | 2005 |
| Cytotech supervisor | $48,000 | $70,646 | $100,000 | 2002 |
| Dental assistant | | $15.48/hr | | 2004 |
| Dental hygienist | | $30.30/hr | | 2004 |
| Dental lab technician | | $15.40/hr | | 2003 |
| Diagnostic medical sonographer | | $61,984 | $67,253 | 2005 |
| Diagnostic molecular scientist | $35,000-$47,000 | $45,000-$82,000 (management) | | 2002 |
| Dietetic technician | $26,000-$37,000* | | $36,300-$41,100 | 2005 |
| Dietitian/nutritionist | $35,300-$46,000* | | $50,000-$72,000 | 2005 |
| Electroneurodiagnostic technologist | $34,726 | $44,621 | $49,192-$52,187 | 2003 |
| Emergency medical technician-paramedic | | $30,400 | $40,000-$46,000 | 1997 |
| Genetic counselor | $40,900 | $53,800 | $108,000 | 2004 |
| Health information administrator | $40,000 | $54,700 | $85,000 | 2003 |
| Health information technician | $30,000 | $39,100 | $50,000 | 2003 |
| Histologic technician | | $40,500 | | 2005 |
| Histotechnologist | | $44,970-$49,360 | | 2005 |
| Kinesiotherapist | $32,500-$38,000 | $48,000 | $60,000 | 2006 |
| Low vision therapist | $44,777 | | | 2002 |
| Magnetic resonance technologist | $44,410 | $66,861 | $78,210 | 2004 |
| Marriage and family counselor/therapist | | $57,119 | | 1997 |
| Massage therapist | | $20,000- $49,000 | | 2005 |
| Medical assistant | $22,650 | $27,951 | $30,000 | 2004 |
| Medical illustrator | $40,000-$45,000 | $45,000-$75,000 | $90,000-$130,000 | 2002 |
| Medical librarian | $41,000 | $58,000 | $158,000 | 2005 |
| Music therapist | | $42,364 | | 2004 |
| Nuclear medicine technologist | | $67,429 | $71,102 | 2004 |
| Occupational therapist | $46,334 | $58,080 | $80,000 | 2006 |
| Occupational therapy assistant | $33,000 | $39,728 | $55,100 | 2006 |
| Ophthalmic assistant | $21,500 | $40,040 | $44,833 | 2005 |
| Ophthalmic dispensing optician | $27,000 | | | 1997 |
| Ophthalmic laboratory technician | $15,100 | | $25,000 | 2000 |
| Ophthalmic medical technologist | $45,000 | $58,481 | $70,400- $125,000 | 2005 |
| Ophthalmic technician | $39,000 | $47,438 | $56,643 | 2005 |
| Orientation and mobility specialist | $46,564 | | | 2002 |
| Orthoptist | $35,000-$42,000 | $45,000-$50,000 | $80,000 | 2000 |
| Orthotist and prosthetist | $22,000-$35,000 | $42,000-$60,000 | | 2003 |
| Pathologists' assistant | $55,000-$75,000 | | $75,000-$100,000 | 2005 |
| Perfusionist | $60,000-$75,000 | $70,000-$90,000 | $100,000 (manager) | 2006 |
| Pharmacy technician | $19,000 | $27,000 | $35,000 | 2002 |
| Phlebotomist | | $24,315-$29,120 | | 2005 |
| Physical therapist | $54,000 | $70,000 | $100,000 | 2006 |
| Physical therapist assistant | $30,000 | $37,000 | | 2005 |
| Physician assistant | $68,116 | $81,129 | $110,000 | 2005 |
| Radiographer | $36,918 | $59,735 | $64,580 | 2004 |
| Radiation therapist | $65,381 | $72,300 | $80,067 | 2004 |
| Rehabilitation counselor | $41,000 | $47,000 | | 2006 |
| Rehabilitation teacher | $37,055 | | | 2002 |
| Respiratory therapist | $41,537 | $56,222 | $74,880 | 2005 |
| Specialist in blood bank technology | $45,000 (bench techs) | $54,000 (supervisors) | $66,000 (managers) | 2006 |
| Speech-language pathologist (schools) | $40,041 | $52,131-$57,000 | $80,000 (management) | 2006 |
| Speech-language pathologist (health care) | $52,694 | $60,000 | $72,985 (management) | 2005 |
| Surgical technologist | | $34,010 | | 2004 |
| Teacher of the visually impaired | $47,086 | | | 2002 |
| Therapeutic recreation specialist (CTRS) | $30,000 | $39,000 | $60,000-$70,000 | 2004 |

*Data valid for those employed full time in their current primary position for 5 years or less.

# Appendix

# Additional Health Professions Information

## Resources for Information on the Health Professions

### Web Sites

**Career Voyages**
US Departments of Labor and Education
www.careervoyages.com/healthcare-alliedhealth.cfm

**Diagnostic Detectives**
Michigan Department of Community Health
www.medlabcareers.msu.edu

**ExploreHealthCareers.org**
The Association of Academic Health Centers, in collaboration with the Federation of Associations of Schools of the Health Professions (FASHP); funded by the Josiah Macy Jr. Foundation
www.explorehealthcareers.org

**Health Occupations for Today and Tomorrow**
Michigan Health Council
www.mihott.com

**Health Occupations for Today and Tomorrow**
South Dakota Department of Health/Office of Rural Health
www.sdjobs.org/sdhott

**H.O.T. Jobs: Health Opportunities in Texas**
East Texas Area Health Education Center
West Texas Area Health Education Center
Center for South Texas Programs
www.texashotjobs.org

**Iowa Health Careers Information Center**
Iowa Department of Public Health
www.idph.state.ia.us/healthcareers

**Labs are Vital**
Abbott Diagnostics Division
www.labsarevital.com

**LifeWorks**
Office of Science Education, National Institutes of Health
http://science.education.nih.gov/lifeworks.nsf

### Scholarship Information

**Health Career Opportunity Program**
Health Resources and Services Administration, Bureau of Health Professions
301 443-2100
http://bhpr.hrsa.gov/diversity/hcop/default.htm

**Federal Student Aid**
US Department of Education
800 443-3243
http://tinyurl.com/4tfej

**College is Possible**
American Council on Education
202 939-9300
www.collegeispossible.org

### National Organizations

**Association of Academic Health Centers**
1400 16th Street NW, Suite 720
Washington, DC 20036-2401
202 265-9600
202 265-7514 Fax
www.ahcnet.org

Founded in 1958, the Association of Academic Health Centers (AHC) is a national, nonprofit organization that consists of over 100 institutional members throughout the United States that are the health complexes of the major universities. Generally, academic health centers consist of an allopathic or osteopathic school of medicine, at least one other health professions school or program, and one or more teaching hospitals. These institutions are the nation's primary resources for education in the health professions, biomedical and health services research, and many aspects of patient care.

The AHC is dedicated to improving the health of the people by advancing the leadership of academic health centers in health professions' education, biomedical and health services research, and health care delivery.

**Association of Schools of Allied Health Professions**
1730 M Street NW, Suite 500
Washington, DC 20036
202 293-4848
202 293-4852 Fax
www.asahp.org

The Association of Schools of Allied Health Professions (ASAHP) works to promote collaboration and partnerships across allied health education and practice, influence health care policy, promote and strengthen research and scholarship, promote and support academic leadership, and promote high quality and innovation in education. ASAHP publishes the *Journal of Allied Health* and holds an annual conference and other meetings.

**Health Occupations Students of America**
6021 Morriss Road, Suite 111
Flower Mound, TX 75028
800 321-HOSA
972 874-0063 Fax
www.hosa.org

Health Occupations Students of America (HOSA) is a national student organization in secondary and postsecondary schools, with members in 2,086 chapters and 40 state associations. As part of the public school system, HOSA offers health career information on a daily basis to over 2,000,000 high school students. Its Career Awareness Campaign is designed to promote health career opportunities in grades K-12 and to mobilize thousands of HOSA members in spreading the word about the value of a career in health care.

The HOSA Web site features a "Career Center" to help its nearly 70,000 members obtain information about potential career choices in health care. It includes information about successful HOSA members who are now practicing health professionals, workforce links, and career information with links to professional association's Web sites.

## Health Professions Network

1850 Samuel Morse Drive
Reston, VA 20190-5316
703 708-9000
703 708-9015 Fax
www.healthpronet.org

The Health Professions Network is a group of volunteers representing allied health professional associations interested in interdisciplinary communication, discussion, and collaboration. Participants meet at least annually to engage in discussion of issues relating to health care and to serve as a conduit for interdisciplinary problem solving and preparation for future health care delivery. The Network also sponsors Allied Health Professions Week, an annual event held in November that honors health care providers working in the more than 80 allied health professions and promotes these professions to students considering a career in health care.

## National Association of Advisors for the Health Professions

PO Box 1518
Champaign, IL 61824-1518
217 355-0063
217 355-1287 Fax
E-mail: NAAHPja@aol.com
www.naahp.org

Established in 1974, the National Association of Advisors for the Health Professions (NAAHP) is an organization of over 800 pre-health professions advisors at colleges and universities throughout the US, including schools of allied health, allopathic medicine, chiropractic, dentistry, nursing, optometry, osteopathic medicine, pharmacy, podiatric medicine, and veterinary medicine. The NAAHP coordinates the activities and efforts of four independent regional associations—Central (CAAHP), Northeast (NEAAHP), Southeast (SAAHP) and West (WAAHP)—to help health professions' advisors across the nation function together and speak with one voice.

## National Commission for Certifying Agencies

2025 M Street NW, Suite 800
Washington, DC 20036
202 857-1165
202 367-2165 Fax
www.noca.org/ncca/ncca.htm

The National Commission for Certifying Agencies (NCCA), the accreditation body of the National Organization for Competency Assurance (NOCA), is the only national accreditation body for private certification organizations in all disciplines. The NCCA works to ensure the health, welfare, and safety of the public through the accreditation of a variety of certification programs/organizations that assess professional competency. The NCCA uses a peer review process to establish accreditation standards, evaluate compliance with the standards, recognize organizations/programs that demonstrate compliance, and serve as a resource on quality certification.

## National Network of Health Career Programs in Two-Year Colleges

Cullen Johnson, Executive Director
714 Harsh Road
Marblehead, OH 43440
419 798-5490
419 798-5490 Fax
E-mail: texascj@bright.net
www.nn2.org

The National Network of Health Career Programs in Two-Year Colleges is an organization of health education leaders from 2-year colleges across the nation dedicated to:

- promoting and encouraging innovation, collaboration, cooperation, and communication with 2-year colleges sponsoring health career programs;
- developing new leaders in health career education; and
- expressing and advocating the interests of health career programs in 2-year colleges (ie, accreditation issues, practice issues, federal policy issues, etc).

## National Society of Allied Health

2139 Georgia Avenue NW, Suite 1C
Washington, DC 20001
202 806-4960
www.nsah.org

Formed in 1978, this nonprofit professional organization draws its members from historically and predominantly black colleges and universities that offer allied health programs. Included are individuals who work as educators, clinicians, and researchers; allied health students become members through induction into the National Society of Allied Health Honor Society.

The organization's major goal is to improve the health care status of African Americans and other at-risk populations through education, employment, community service, and research.

## Publications

### Occupational Outlook Handbook

US Department of Labor, Bureau of Labor Statistics
www.bls.gov/oco

### Diversity: Allied Health Careers

Quarterly magazine
www.diversityalliedhealth.com

## Health Workforce Research Organizations

The Bureau of Health Professions, Health Resources and Services Administration, funds six federally designated regional workforce analysis centers throughout the US:

- California Center for Health Workforce Studies, University of California at San Francisco http://futurehealth.ucsf.edu/cchws.html
- Center for Health Workforce Studies, State University of New York at Albany http://chws.albany.edu
- Illinois Regional Health Workforce Center, University of Illinois at Chicago http://www.uic.edu/sph/ichws/
- Regional Center for Health Workforce Studies, Center for Health Economics and Policy, The University of Texas Health Science Center at San Antonio http://www.uthscsa.edu/rchws/index.asp
- Southeast Regional Center for Health Workforce Studies http://www.healthworkforce.unc.edu/data.html

- WWAMI Center for Health Workforce Studies,
  University of Washington
  http://www.fammed.washington.edu/CHWS/index.html

In addition to these federally designated centers, some states have centers that focus on state-specific health workforce concerns:
- Michigan Center for Health Professions,
  Michigan Health Council
  www.mhc.org
- Center for Health Workforce Planning,
  Iowa Department of Public Health
  www.idph.state.ia.us/hpcdp/workforce_planning.asp

## Other Sources of Health Professions' Data

### Enrollment Snapshot of Radiography, Radiation Therapy and Nuclear Medicine Programs, Fall 2005
http://www.asrt.org/media/pdf/research/enrollmentsurvey05.pdf
American Society of Radiologic Technologists
www.asrt.org
800 444-2778

### 2000-01 Demographic Survey of Undergraduate and Graduate Programs in Communication Sciences and Disorders
www.capcsd.org/survey/2002/2000-01DemographicsSurvey.pdf

These data cover audiology and speech-language pathology educational programs.

Council of Academic Programs in Communication Sciences
and Disorders
952 920-0966
925 920-6098 Fax
E-mail: cap@incnet.com
www.capcsd.org

### 21st Annual Report on Physician Assistant Educational Programs in the United States, 2004-2005
www.paeaonline.org/publications.html

Physician Assistant Education Association
Attn: Geraldene Darden
703 548-5538, ext 306
703 548-5539 Fax
E-mail: gdarden@paeaonline.org
www.paeaonline.org

### Trends in Dietetics Education: 2005-2006
www.eatright.org/ada/files/2006_Annual_Report.pdf

Data collected by the Commission on Accreditation for Dietetics Education for dietitian and dietetic technician programs are published annually in the *CADE Annual Report*.

Commission on Accreditation for Dietetics Education
American Dietetic Association
800 877-1600, ext 4872
312 899-0040
312 899-4817 Fax
E-mail: bmitchell@eatright.org

## Descriptions of Other Health Professions

The American Medical Association (AMA) receives frequent requests for information on health professions that are not listed in the *Directory*. The following organizations and resources may be able to provide more information and/or listings of educational programs in these professions. Please note that these organizations are not affiliated with the AMA and that the organizations and professions listed are not in any sense "recognized" or "approved" by the AMA.

### Anesthesia Technologist/Technician

Anesthesia technicians and technologists are integral members of the anesthesia patient care team. Their role is to assist anesthesiologist and anesthetists in acquiring, preparing, and using the equipment and supplies required to administer anesthesia. In this role, they contribute to safe, efficient, and cost-effective anesthesia care.

Depending on individual expertise and training, the tasks of the anesthesia technician and technologist may include equipment maintenance and servicing such as cleaning and sterilizing, assembling, calibrating and testing, troubleshooting, requisitioning and recording of inspections and maintenance, and operating a variety of mechanical, pneumatic, and electronic equipment used to monitor the patient undergoing anesthesia.

Anesthesia technologists with the appropriate training may perform inspection, maintenance, and on-site repairs.

The anesthesia technician or technologist may also be responsible for purchasing and distributing supplies and equipment and maintaining inventories and service records. Individuals functioning as anesthesia technicians or technologists should be capable of a level of self-direction and supervision commensurate with their training.

American Society of Anesthesia Technologists and Technicians
55 Harristown Road, 2nd Floor
Glen Rock, NJ 07452
201 652-6622
201 447-3831 Fax
E-mail: asattinfo@aol.com
www.asatt.org

### Cancer Registrar

Cancer registrars are data management experts who report cancer statistics for various health care agencies. Registrars work closely with physicians, administrators, researchers, and health care planners to provide support for cancer program development, ensure compliance of reporting standards, and serve as a valuable resource for cancer information with the ultimate goal of preventing and controlling cancer. The cancer registrar is involved in managing and analyzing clinical cancer information for the purpose of education, research, and outcome measurement.

The cancer registrar's primary responsibility is to ensure that timely, accurate, and complete data are incorporated and maintained on all types of cancer diagnosed and/or treated within an institution or other defined population. Information is entered into the database manually and through database linkage and computer interfaces.

Cancer registrars bridge the information gap by capturing a complete summary of the patient's disease from diagnosis through their lifetime. The information is not limited to the episodic information contained in the health care facility record. The summary or abstract is an ongoing account of the cancer patient's history, diagnosis, treatment, and current status.

In addition to managing and reporting cancer data, registrars serve in multiple other professional activities. Cancer registrars participate in cancer program, institution, and community benefit activities as part of the active leadership structure. Registrars provide benchmarking services, monitor quality of care and clinical practice guidelines, assess patterns of care and referrals, and monitor adverse outcomes including mortality and co-morbidity. Cancer registrars can provide consultative services on many issues including registry management and program standards.

National Cancer Registrars Association
1340 Braddock Place, Suite 203
Alexandria, VA 22314
703 299-6640
703 299-6620 Fax
E-mail: info@ncra-usa.org
www.ncra-usa.org

## Chaplain

Chaplains are specialized ministers with advanced training who serve in a variety of institutional settings, such as hospitals, correctional institutions, long term care facilities, rehabilitation centers, hospice programs, the military, and other specialized settings. They are considered members of the treatment team and provide pastoral/spiritual care to patients, family members, and staff.

They respond to crisis situations related to codes, traumas, deaths, grieving, and end of life care. They also attend rounds, accept referrals, are available for consults, and visit patients in their assigned areas. Chaplains work as religious specialists in the institution, with training in spiritual screening and spiritual assessments, and utilize those skills in working with patients on spiritual issues and needs, while being careful to keep professional boundaries clear. Chaplains may also have other skills in medical ethics, clinical quality improvement, patient rights, and hospital accreditation.

Four units of Clinical Pastoral Education (approved by the Association for Clinical Pastoral Education, or ACPE), a graduate theological degree, endorsement by a faith group, and appearance before a Certification Committee are standard requirements for certification as a chaplain by the Association of Professional Chaplains (APC).

The APC serves chaplains in all types of health and human service settings. Almost 4,000 of its members are chaplains involved in pastoral care and represent more than 150 faith groups. As a national, not-for-profit professional association, the APC advocates for quality spiritual care of all persons in institutional settings.

The ACPE accredits institutions, sets standards, certifies supervisors, and trains ministers for specialized work as institutional chaplains.

Association for Professional Chaplains
1701 E Woodfield Road, Suite 400
Schaumburg, IL 60173
847 240-1014
847 240-1015 Fax
E-mail: info@professionalchaplains.org
www.professionalchaplains.org

Association for Clinical Pastoral Education
1549 Clairmont Road, Suite 103
Decatur, GA 30033-4611
404 320-1472
404 320-0849 Fax
www.acpe.edu

## Health Advocate

The health advocate promotes patients' rights in the increasingly complex health care system of the US. Currently, Sarah Lawrence University in New York offers the nation's only master's degree program in health advocacy. Graduates of this program work in diverse professional capacities in health care: as patient representatives and administrators, on ethics committees and medical policy boards, and at hospitals and consumer health agencies. Salaries range from $35,000 to $45,000 for entry level to $65,000 to $75,000 for managers.

The health advocacy program at Sarah Lawrence is a 52-credit program. The curriculum includes core courses, discipline-based courses (eg, health law, economics, history, bioethics, evaluation and assessment, psychology, health care delivery and policy, and topics in physiology), and fieldwork.

Sarah Lawrence College
Health Advocacy Program
One Mead Way
Bronxville, NY 10708
914 395-2371
914 395-2664 Fax
E-mail: grad@slc.edu
www.slc.edu/grad_healthadvocacy.php

## Horticultural Therapist

Horticultural therapy is a complementary therapy profession using gardening and horticultural activities to improve the social, educational, psychological, and physical adjustment of persons. The professional horticultural therapist uses a treatment plan in working with persons who are mentally and physically disabled and others with special needs, including people who are developmentally disabled, the elderly, substance abusers, public offenders, and socially disadvantaged. Horticultural therapists practice in diverse locations, including hospitals and institutions, vocational training facilities, nursing homes and halfway houses, rehabilitation facilities, older adult centers, correctional facilities, schools, arboreta and botanical gardens, parks and recreational settings, farms and horticultural businesses, and community gardens.

Nearly 20 schools in the US offer educational programs or curricula in horticultural therapy.

American Horticultural Therapy Association
3570 E 12th Avenue, Suite 206
Denver, CO 80206
800 634-1603
303 322-2485 Fax
E-mail: ahta@ahta.org
www.ahta.org

## Medical Coder

A medical coder, a member of the health information service team, uses a classification system to assign code numbers and letters to each symptom, diagnosis, disease, procedure, and operation that appears in the patient's chart. These codes are used for insurance reimbursement, research, health planning analysis, and to make clinical decisions. A high degree of accuracy and a working knowledge of medical terminology, anatomy, and physiology are important skills for these professionals.

Career opportunities include:
• Inpatient Hospital Coder
• Reimbursement Specialist
• Outpatient Coder
• Coding Abstracting Analyst
• Insurance Claim Analyst

- Managed Care Organization Coder
- Procedural Coder
- Physician's Office/Clinic Coder

American Health Information Management Association (AHIMA)
233 N Michigan Avenue, Suite 2150
Chicago, IL 60601-5800
312 233-1100
312 233-1500 Fax
E-mail: info@ahima.org
www.ahima.org/careers

## Medical Transcriptionist

Medical transcriptionists are specialists in medical language and healthcare documentation who interpret and transcribe dictation by physicians and other health professionals regarding patient assessment, workup, therapeutic procedures, clinical course, diagnosis, prognosis, and so on, editing dictated material for grammar and clarity as necessary and appropriate.

American Association for Medical Transcription
100 Sycamore Ave
Modesto, CA 95354
800 982-2182
209 527-9620
209 527-9633 Fax
E-mail: aamt@aamt.org
www.aamt.org

## Nurse

Nursing is the largest health care occupation, with 2.4 million jobs. With a continued shortage of qualified nurses throughout health care, employment opportunities are expected to be excellent.

Individuals training in the profession of nursing can hold various positions, including:

- Registered nurse
- Office nurse
- Nursing home nurse
- Home health nurse
- Public health nurse
- Occupational health or industrial nurse
- Head nurses or nurse supervisor

At the advanced level are nurse practitioners, clinical nurse specialists, certified registered nurse anesthetists, and certified nurse-midwives. Advanced practice nurses must meet higher educational and clinical practice requirements beyond the basic nursing education and licensing required of all RNs.

The three major educational paths to registered nursing are:

- associate degree in nursing (ADN), about 2-3 years' length
- bachelor of science degree in nursing (BSN), 4 years
- diploma (offered in hospitals), 2 to 3 years

In management, nurses can advance to assistant head nurse or head nurse, and from there to assistant director, director, and vice president. Increasingly, management-level nursing positions require a graduate degree in nursing or health services administration. Graduate programs preparing executive-level nurses usually last 1 to 2 years. Some nurses move into the business side of health care, with health care corporations or in academia.

Median annual earnings of registered nurses were $52,330 in 2004. Median annual earnings in the industries employing the largest numbers of registered nurses in 2004 were as follows:

- Employment services $63,170
- General medical and surgical hospitals $53,450
- Home health care services $48,990
- Offices of physicians $48,250
- Nursing care facilities $48,220

### Additional Information
American Association of Colleges of Nursing
1 Dupont Circle NW, Suite 530
Washington, DC 20036
202 463-6930
202 785-8320 Fax
www.aacn.nche.edu

(Source: *Occupational Outlook Handbook*, US Department of Labor, Bureau of Labor Statistics)

## Pharmacist

### Role of Pharmacists in Health Care

Pharmacists play a vital role in the health care system through the medicine and information they provide. Although responsibilities vary among the different areas of pharmacy practice, the bottom line is that pharmacists help patients get well. Pharmacist responsibilities include a range of care for patients, from dispensing medications to monitoring patient health and progress to maximize their response to the medication. Pharmacists also educate consumers and patients on the use of prescriptions and over-the-counter medications and advise physicians, nurses, and other health professionals on drug decisions. Pharmacists also provide expertise about the composition of drugs, including their chemical, biological, and physical properties and their manufacture and use. They ensure drug purity and strength and make sure that drugs do not interact in a harmful way. Pharmacists are drug experts ultimately concerned about their patients' health and wellness.

### Job Outlook

The demand for trained pharmacy professionals has increased dramatically in recent years due the rapid growth of the health care and pharmaceutical industries, especially for the growing elderly population. Pharmacists are more actively involved in drug therapy decision-making for patients of all ages. The Department of Health and Human Services (HHS) recently released a report, *The Pharmacy Workforce: A Study of the Supply and Demand for Pharmacists*, which concluded that there is an increasing demand for pharmacists' service that is outpacing the current and possibly future pharmacist supply. The report also states that factors causing the shortage are not likely to abate in the near future. As a result, pharmacy graduates can expect to receive multiple job offers at the time of graduation, with tremendous potential for advancement and competitive salaries.

### Pharmacy Education

Students who wish to pursue pharmacy must earn a Doctor of Pharmacy (PharmD) degree. This professional degree program requires at least 2 years of specific preprofessional (undergraduate) course work followed by 4 academic years (or 3 calendar years) of professional study. Pharmacy colleges and schools may accept students directly from high school for both the prepharmacy and pharmacy curriculum, or after completion of the college course prerequisites. The majority of students enter a pharmacy program with 3 or more years of college experience.

### Additional Information

Contact the American Association of Colleges of Pharmacy (AACP) for specific information on all US colleges and schools of pharmacy.

American Association of Colleges of Pharmacy
1426 Prince Street
Alexandria, VA 22314
703 739-2330, x1024
www.aacp.org

## Physician

Physicians, often referred to as doctors, work in neighborhood clinics, hospitals, offices, even homeless shelters and schools to care for people in need. Others work in research, academic settings, or with health maintenance organizations, pharmaceutical companies, medical device manufacturers, health insurance companies, or in corporations directing health and safety programs.

About one third of the nation's physicians are generalists—"primary care" doctors who provide lifelong medical services. These include internists, family physicians, and pediatricians. Generalists provide a wide range of services children and adults need. When patients' specific health needs require further treatment, generalist physicians send them to see a specialist physician.

Specialist physicians, such as neurologists, cardiologists, and ophthalmologists differ from generalists in that they focus on treating a particular system or part of the body. They collaborate with generalist physicians to ensure that patients receive treatment for specific medical problems as well as complete and comprehensive care throughout life.

Both generalist and specialist physicians complete 4 years of medical school in the US, graduating with a doctor of medicine degree (MDs). Some physicians receive a doctor of osteopathic medicine (DO) degree from a college of osteopathic medicine. Next, physicians complete a residency program of 3 to 5 years in their chosen specialty (eg, neurology, surgery, internal medicine); some also complete an additional fellowship of 1 to 3 years in a subspecialty, such as cardiology, pain management, or neonatal-perinatal medicine.

American Medical Association
515 N State St
Chicago, IL 60610
www.ama-assn.org/go/becominganmd
E-mail: becominganmd@ama-assn.org